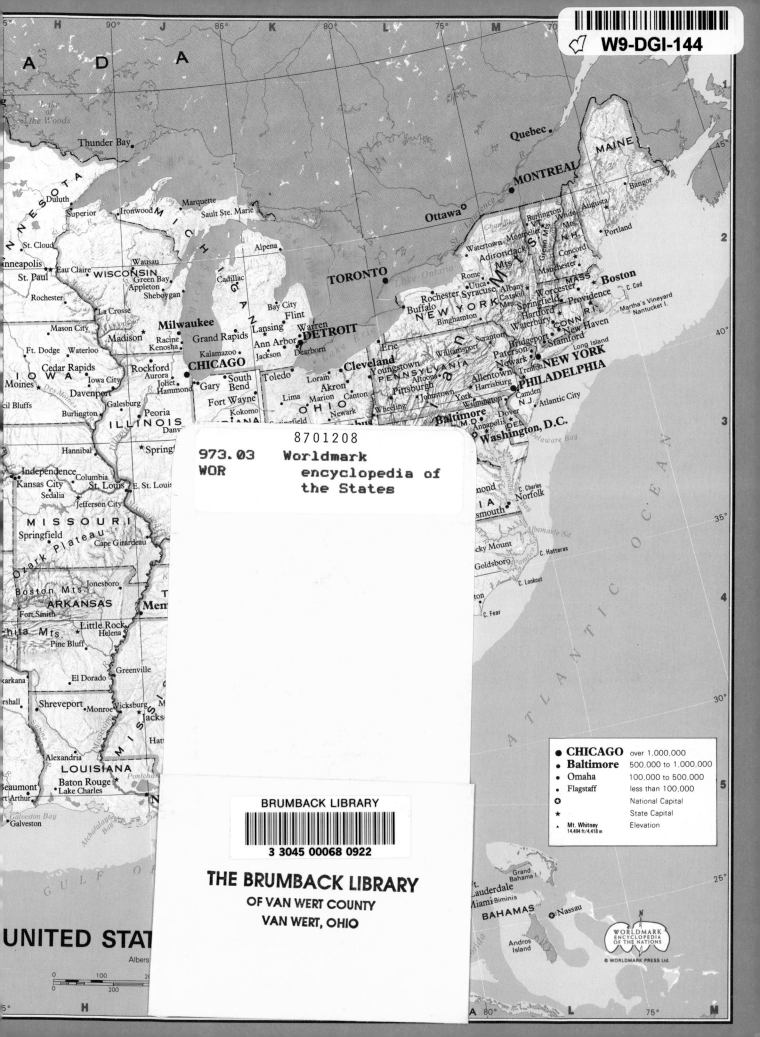

- **CHICAGO** over 1,000,000
- **Baltimore** 500,000 to 1,000,000
- Omaha 100,000 to 500,000
- Flagstaff less than 100,000
- ✪ National Capital
- ★ State Capital
- ▲ Mt. Whitney 14,494 ft/4,418 m Elevation

WORLDMARK
ENCYCLOPEDIA
OF THE NATIONS
© WORLDMARK PRESS Ltd.

UNITED STAT

WORLDMARK ENCYCLOPEDIA OF THE STATES

WORLDMARK PRESS, LTD.
PUBLISHER

JOHN WILEY & SONS, INC.
EXCLUSIVE WORLD DISTRIBUTOR
New York Chichester
Brisbane Toronto Singapore

Typography by Donnelley/Rocappi, Inc., Cherry Hill, N.J.

Production by R. R. Donnelley & Sons, Willard, Ohio.

Preparation of lithographic negatives for black-and-white maps
by John Dreyer & Co., Inc., New York, N.Y.

Printed in the United States of America.

EXCLUSIVE WORLD DISTRIBUTOR
JOHN WILEY & SONS, INC.
NEW YORK   CHICHESTER
BRISBANE   TORONTO   SINGAPORE

LIBRARY OF CONGRESS CATALOGING-IN-PUBLICATION DATA

WORLDMARK ENCYCLOPEDIA OF THE STATES

1. United States—Dictionaries and encyclopedias.
E156.W67   1986   973′.03′21   85-26455
ISBN 0-471-83213-8 (Wiley)

Staff

Editor and Publisher	MOSHE Y. SACHS
Executive Editor	WILLIAM E. SHAPIRO
Senior Editor	GEOFFREY M. HORN
Managing Editor	LAUREN I. SILVERMAN
Staff Editor	AMY E. GOETZ
Consulting Editors	MARY JANE ALEXANDER PAULINE PIEKARZ
Editorial Assistant	PAIGE STOCKLEY
Contributing Editors	DANIEL J. DOMOFF WALTER R. FOX JEFFREY H. HACKER ROBERT HALASZ WILLIAM A. McGEVERAN, JR. KATHRYN PAULSEN PATRICIA A. RODRIGUEZ JOHN A. SHANKS JOHN J. SMITH DONALD YOUNG
Cartographic Editor	MIKLOS PINTHER
Cartographic Staff	SUDSAI POOLSUK VERA TOSTANOSKI
Copy Editors	MARTIN MITCHELL SHARON L. WIRT
Proofreaders	MARION GELLER THERESA LIBERATORE WILLIAM O'HERN VIRGINIA READ PETER ELIOT STONE ELIZABETH SZABLA
Typists	ROSEMARIE T. KRIST ADELA ZYLBER-PESKORZ

Contributors

to the First Edition

ALLEN, HAROLD B. Emeritus Professor of English and Linguistics, University of Minnesota (Minneapolis–St. Paul). LANGUAGES.***

BASSETT, T. D. SEYMOUR. Former University Archivist, University of Vermont (Burlington). VERMONT**

BENSON, MAXINE. Curator of Document Resources, Colorado Historical Society. COLORADO.**

BROWN, RICHARD D. Professor of History, University of Connecticut (Storrs). MASSACHUSETTS.**

CASHIN, EDWARD J. Professor of History, Augusta College. GEORGIA.*

CHANNING, STEVEN A. Professor of History, University of Kentucky (Lexington). KENTUCKY.**

CLARK, CHARLES E. Professor of History, University of New Hampshire (Durham). MAINE.*

COGSWELL, PHILIP, JR. Forum Editor, *The Oregonian.* OREGON.*

CONLEY, PATRICK T. Professor of History and Law, Providence College. RHODE ISLAND.**

CORLEW, ROBERT E. Dean, School of Liberal Arts, Middle Tennessee State University (Murfreesboro). TENNESSEE.*

CREIGH, DOROTHY WEYER. Author and historian; member, Nebraska State Board of Education. NEBRASKA.**

CUNNINGHAM, JOHN T. Author and historian. NEW JERSEY.**

FISHER, PERRY. Director, Columbia Historical Society. DISTRICT OF COLUMBIA.**

FRANTZ, JOE B. Professor of History, University of Texas (Austin). TEXAS.**

GOODELL, LELE. Member, Editorial Board, *Hawaiian Journal of History.* HAWAII (in part).**

GOODRICH, JAMES W. Associate Director, State Historical Society of Missouri. MISSOURI.**

HAMILTON, VIRGINIA. Professor of History, University of Alabama (Birmingham). ALABAMA.**

HAVIGHURST, WALTER. Research Professor of English Emeritus, Miami University (Oxford). OHIO.**

HINTON, HARWOOD P. Editor, *Arizona and the West,* University of Arizona (Tucson). ARIZONA.**

HOOGENBOOM, ARI. Professor of History, Brooklyn College of the City University of New York. PENNSYLVANIA.**

HOOVER, HERBERT T. Professor of History, University of South Dakota (Vermillion). SOUTH DAKOTA.**

HUNT, WILLIAM R. Historian; former Professor of History, University of Alaska. ALASKA.*

JENSEN, DWIGHT. Author and historian. IDAHO.*

JENSEN, RICHARD J. Professor of History, University of Illinois (Chicago). ILLINOIS.*

LARSON, ROBERT W. Professor of History, University of Northern Colorado (Greeley). NEW MEXICO.**

MAPP, ALF J., JR. Author, lecturer, and historian; Professor of English, Creative Writing, and Journalism, Old Dominion University (Norfolk). VIRGINIA.**

MAY, GEORGE S. Professor of History, Eastern Michigan University (Ypsilanti). MICHIGAN.*

MEYER, GLADYS. Professor emeritus, Columbia University. ETHNIC GROUPS.***

MOODY, ERIC N. Historian, Nevada Historical Society. NEVADA.**

MUNROE, JOHN A. H. Rodney Sharp Professor of History, University of Delaware. DELAWARE.**

MURPHY, MIRIAM. Associate Editor, *Utah Historical Quarterly*. UTAH.**

O'BRIEN, KATHLEEN ANN. Project Director, Upper Midwest Women's History Center for Teachers. MINNESOTA.**

PADOVER, SAUL K. Distinguished Service Professor Emeritus, Graduate Faculty, New School (New York City). UNITED STATES OF AMERICA.**

PECKHAM, HOWARD H. Professor emeritus, University of Michigan. INDIANA.**

PRYOR, NANCY. Research consultant and librarian, Washington State Library. WASHINGTON.**

RAWLS, JAMES J. Instructor of History, Diablo Valley College (Pleasant Hill). CALIFORNIA.**

RICE, OTIS K. Professor of History, West Virginia Institute of Technology (Montgomery). WEST VIRGINIA.*

RICHMOND, ROBERT W. Assistant Executive Director, Kansas State Historical Society. KANSAS.**

RIGHTER, ROBERT W. Assistant Professor of History, University of Wyoming (Laramie). WYOMING.**

ROTH, DAVID M. Director, Center for Connecticut Studies, Eastern Connecticut State College (Willimantic). CONNECTICUT.*

SCHEFFER, BARBARA MOORE. Feature Writer. OKLAHOMA (in part).*

SCHEFFER, WALTER F. Regents' Professor of Political Science and Director, Graduate Program in Public Administration, University of Oklahoma (Norman). OKLAHOMA (in part).*

SCHMITT, ROBERT C. Hawaii State Statistician. HAWAII (in part).**

SCUDIERE, PAUL J. Senior Historian, New York State Education Department. NEW YORK.**

SKATES, JOHN RAY. Professor of History, University of Southern Mississippi (Hattiesburg). MISSISSIPPI.**

SMITH, DOUG. Writer, *Arkansas Gazette* (Little Rock). ARKANSAS.**

STOUDEMIRE, ROBERT H. Professor of State and Local Government and Senior Research Associate, Bureau of Governmental Research, University of South Carolina (Columbia). SOUTH CAROLINA.*

SULLIVAN, LARRY E. Librarian, New-York Historical Society. MARYLAND.**

TAYLOR, JOE GRAY. Professor of History, McNeese State University (Lake Charles). LOUISIANA.**

TEBEAU, CHARLTON W. Emeritus Professor of History, University of Miami. FLORIDA.**

THOMPSON, WILLIAM FLETCHER. Director of Research, State Historical Society of Wisconsin. WISCONSIN.**

VIVO, PAQUITA. Author and consultant. PUERTO RICO.**

WALL, JOSEPH FRAZIER. Professor of History, Grinnell College. IOWA.**

WALLACE, R. STUART. Assistant Director/Editor, New Hampshire Historical Society. NEW HAMPSHIRE.**

WATSON, HARRY L. Assistant Professor of History, University of North Carolina (Chapel Hill). NORTH CAROLINA.*

WEAVER, KENNETH L. Associate Professor of Political Science, Montana State University (Bozeman). MONTANA.**

WILKINS, ROBERT P. Professor of History, University of North Dakota (Grand Forks). NORTH DAKOTA.**

WOODS, BOB. Editor, *Sierra Club Wildlife Involvement News.* FLORA AND FAUNA.***

*Full contributor.

**Consultant contributor.

***Special contributor.

Contents

Preface

More than a quarter century ago, the editors of Worldmark Press set out to create a new kind of reference work—one that would view every nation of the world as if through a "world mirror" and not from the perspective of any one country or group of countries. To that work, the *Worldmark Encyclopedia of the Nations,* we offered a companion volume in 1981, the *Worldmark Encyclopedia of the States,* which was selected as an "Outstanding Reference Source" by the Reference Sources Committee of the American Library Association, Reference and Adult Services Division. We now offer a completely revised and updated second edition of the *Worldmark Encyclopedia of the States.*

The fitness of the United States of America as a subject for encyclopedic study is plain. No discussion of world politics, economics, culture, technology, or military affairs would be complete without an intensive examination of the American achievement. What is not so obvious is why we chose to present this work as an encyclopedia of the *states* rather than of the United States. In so doing, we emphasize the fact that the United States is a federal union of separate states with divergent histories, traditions, resources, laws, and economic interests.

Every state, large or small, is treated in an individual chapter, within a framework of 50 standard subject headings; generally, the more populous the state, the longer the article. The District of Columbia and the Commonwealth of Puerto Rico each has its own chapter, and two additional articles describe in summary form the other Caribbean and Pacific dependencies. The concluding chapter is a 50-page overview of the nation as a whole. Supplementing this textual material are tables of conversions and abbreviations, a glossary, and 58 black-and-white maps prepared especially for this encyclopedia.

The editors invited more than 50 distinguished historians, political scientists, scholars, and journalists—all of them intimately familiar with their own states—to examine the text of the first edition for general accuracy, fairness, and completeness. We recognized three categories of contributors. Consultant contributors (the majority) reviewed the chapters on their respective states and in most cases each also wrote part of the article; full contributors wrote entire articles on their respective states and then reviewed and corrected the edited manuscript; and three special contributors prepared sections for every state on the complex and rapidly changing fields of *flora and fauna, ethnic groups,* and *languages.* We encouraged our contributors to amend and amplify the text, but responsibility for the completed articles remained that of the editors, as did responsibility for the preparation of the second edition.

Publication of this encyclopedia was a collective effort that enlisted the talents of scholars, editor-writers, artists, cartographers, typesetters, proofreaders, and many others. Perhaps only those involved in the production of reference books fully appreciate how complex that endeavor can be. We ourselves relied on many reference books in the course of this project, assuming their accuracy in every case though painfully aware of the possibilities of error. (Some of these are discussed in the notes that follow.) Readers customarily expect that a reference book will be correct in every particular; and yet, by the time it has been on the shelves for a few months, a conscientious editor may already have a long list of improvements and corrections to be made in a subsequent edition. Although the editors of Worldmark Press cannot acknowledge all comments and questions personally, we do invite you, the reader, to add your suggestions to our list.

THE EDITORS

Notes

GENERAL NOTE: Although the following comments refer repeatedly to "the states" and "state sources" in describing the process by which official information was gathered for the *Worldmark Encyclopedia of the States,* the editors also wish to acknowledge the kind assistance provided by the governments of the District of Columbia, the Commonwealth of Puerto Rico, the other US Caribbean and Pacific dependencies, and the United States as a whole. Space does not permit a complete listing of all federal, national, and state sources used in preparation of this encyclopedia. We have therefore confined ourselves to source notes that clarify a statistical problem or give due credit to a nongovernmental organization.

MAPS: Each state map shows principal topographical features, including drainage and elevations; county names, county lines, and county seats; the state capital; major cities and metropolitan areas; and, wherever possible, other topographical features, points of interest, place-names, and regional designations mentioned in the text. The key to city populations in both state and endsheet maps is based on US Bureau of the Census mid-1984 estimates, the latest such estimates available in late 1985.

FLAGS, SEALS, AND SYMBOLS: Depictions of the flag and official seal at the head of each state article are based on reproductions solicited from that state and current as of 1985.

LEGAL HOLIDAYS: Lists of legal holidays in each state were compiled from official state sources. A legal holiday is defined as a day (other than a Saturday or Sunday) on which the transaction of government business is limited by law. The Birthday of Martin Luther King, Jr., made a legal public holiday by the 98th Congress (Public Law 98-144) on 2 November 1983 and effective in 1986, will be celebrated on the 3d Monday in January; as of the end of 1985, many states observed this holiday on days other than the 3d Monday in January, and some states did not observe it at all.

WEIGHTS AND MEASURES: Recognizing the trend toward use of the metric system throughout the United States, the text provides metric equivalents for customary measures of length and area, and both Fahrenheit and Centigrade expressions for temperature. Production figures are expressed exclusively in the prevailing customary units.

LOCATION, SIZE, AND EXTENT: The lengths of interstate boundary segments and the total lengths of state boundaries appear in roman type when derived from official government sources; italic type indicates data derived from other sources. Discrepancies in the boundary lengths of neighboring states as specified by official sources arise from divergent methodologies of measurement.

FLORA AND FAUNA: Discussions of endangered species are based on the *List of Endangered and Threatened Wildlife and Plants* maintained by the Fish and Wildlife Service of the US Department of the Interior, and on data supplied by the states.

POPULATION: The January 1985 state population estimates were prepared by Donnelley Marketing Information Services of Dun & Bradstreet. Tables of counties, county seats, county areas, and county populations accompany the articles on the 14 most populous states; the editors regret that space limitations prevented the publication of such a table for each state. Because of rounding, county areas in these tables may not add to the total; because the 1984 population estimates are computer-generated extrapolations from 1980 census data, county populations may not add to the total.

LANGUAGES: Examples of lexical and pronunciation patterns cited in the text are meant to suggest the historic development of principal linguistic features, and should not be taken as a comprehensive statement of current usage. Data on languages spoken in the home were obtained from the "Preliminary Listing of Selected Languages Spoken by Persons 3 Years Old and Over by State: 1980 Census," issued by the US Bureau of the Census.

RELIGIONS: For data on the members and adherents of Christian denominations, the editors are indebted to *Churches and Church Membership in the United States, 1980,* published in 1982 by the Glenmary Research Center, and to the *Official Catholic Directory* (1984 edition), an annual publication of P. J. Kenedy & Sons; for Jewish population estimates, the editors relied on the *American Jewish Year Book* (vol. 85), issued annually by the Jewish Publication Society of America and the American Jewish Committee.

JUDICIAL SYSTEM: *Uniform Crime Reports for the United States,* published annually by the Federal Bureau of Investigation and embodying the FBI Crime Index (tabulations of offenses known to the police), was the principal source for the crime statistics cited in the text.

ARMED FORCES: Estimates published by the Veterans Administration of the number of veterans of US military service in each state represent extrapolations from 1980 census data.

LABOR: The table derived from the 1982 or 1983 federal census of workers includes only those workers covered by unemployment insurance.

ENERGY AND POWER: Data for proved reserves and production of fossil fuels were derived from publications of the American Gas Association, American Petroleum Institute, National Coal Association, and US Department of Energy. Data on nuclear power facilities were obtained from the Nuclear Information and Resource Service and from state sources.

HEALTH: The principal statistical sources for hospitals and medical personnel were annual publications of the American Dental Association, American Hospital Association, and American Medical Association.

LIBRARIES AND MUSEUMS: In most cases, library and museum names are listed in the *American Library Directory* (37th ed.), published in 1984 by R. R. Bowker, and the *Official Museum Directory, 1985,* compiled by the National Register Publishing Co. in cooperation with the American Association of Museums.

PRESS: Circulation data follow the 1985 *Editor & Publisher International Yearbook.*

FAMOUS PERSONS: Entries are current through November 1985. Where a person described in one state is known to have been born in another, the state of birth follows the personal name, in parentheses.

BIBLIOGRAPHY: Bibliographies are intended as a guide to further reading and not as a listing of sources in preparing the articles. Such listings would have far exceeded space limitations, inasmuch as each article draws upon at least 75 documents.

Acknowledgments

The editors wish to express their deepest thanks to the many hundreds of federal, state, and local officials, as well as to the numerous national associations, without whose cooperation this project would not have been possible. The editors are especially grateful to the staff of the US Bureau of the Census, to the Eastern Mapping Center of the US Geological Survey, and to cartographer Richard Edes Harrison.

The publisher wishes to express his esteem for Louis Barron for his unique contribution in setting the scholarly standards of Worldmark Press. Grateful acknowledgment is offered to Richard Zeldin, formerly of John Wiley & Sons, for impressing upon us the need for this encyclopedia, and to Robert B. Polhemus, former vice president of John Wiley & Sons, for his constant encouragement. The editors are profoundly indebted to M. S. Wyeth, Jr., Vice President of Harper & Row, for making the first edition possible.

Key to Subject Headings

All information contained within a state article is uniformly keyed by means of small superior numerals to the left of the subject headings. A heading such as "Population," for example, carries the same key numeral (6) in every article. Thus, to find information about the population of Oklahoma, consult the table of contents for the page number where the Oklahoma article begins and look for section 6 thereunder.

Introductory matter for each state includes: Origin of state name
Nickname
Capital
Date and order of statehood
Song
Motto
Flag
Official seal
Symbols (animal, tree, flower, etc.)
Legal holidays
Time

FLAG COLOR SYMBOLS

 gold/yellow   red   green   blue   orange   brown   white   black

SUBJECT HEADINGS IN NUMERICAL ORDER

1	Location, size, and extent	26	Forestry
2	Topography	27	Mining
3	Climate	28	Energy and power
4	Flora and fauna	29	Industry
5	Environmental protection	30	Commerce
6	Population	31	Consumer protection
7	Ethnic groups	32	Banking
8	Languages	33	Insurance
9	Religions	34	Securities
10	Transportation	35	Public finance
11	History	36	Taxation
12	State government	37	Economic policy
13	Political parties	38	Health
14	Local government	39	Social welfare
15	State services	40	Housing
16	Judicial system	41	Education
17	Armed forces	42	Arts
18	Migration	43	Libraries and museums
19	Intergovernmental cooperation	44	Communications
20	Economy	45	Press
21	Income	46	Organizations
22	Labor	47	Tourism, travel, and recreation
23	Agriculture	48	Sports
24	Animal husbandry	49	Famous persons
25	Fishing	50	Bibliography

SUBJECT HEADINGS IN ALPHABETICAL ORDER

Agriculture	23	Insurance	33
Animal husbandry	24	Intergovernmental cooperation	19
Armed forces	17	Judicial system	16
Arts	42	Labor	22
Banking	32	Languages	8
Bibliography	50	Libraries and museums	43
Climate	3	Local government	14
Commerce	30	Location, size, and extent	1
Communications	44	Migration	18
Consumer protection	31	Mining	27
Economic policy	37	Organizations	46
Economy	20	Political parties	13
Education	41	Population	6
Energy and power	28	Press	45
Environmental protection	5	Public finance	35
Ethnic groups	7	Religions	9
Famous persons	49	Securities	34
Fishing	25	Social welfare	39
Flora and fauna	4	Sports	48
Forestry	26	State government	12
Health	38	State services	15
History	11	Taxation	36
Housing	40	Topography	2
Income	21	Tourism, travel, and recreation	47
Industry	29	Transportation	10

EXPLANATION OF SYMBOLS

Data not available: NA
Nil (or negligible) —
Subtotals on tables are usually given in parentheses ( ).
A fiscal or split year is indicated by a stroke (e.g., 1985/86).
The use of a small dash (e.g., 1985–86) normally signifies the full period of calendar years covered (including the end year indicated).

Conversion Tables*

LENGTH

1 inch	2.540005 centimeters
1 foot (12 inches)	30.4801 centimeters
1 US yard (3 feet)	0.914402 meter
1 statute mile (5,280 feet; 1,760 yards)	1.609347 kilometers
1 British mile	1.609344 kilometers
1 nautical mile (1.1508 statute miles or 6,076.10333 feet)	1.852 kilometers
1 British nautical mile (6,080 feet)	1.85319 kilometers
1 centimeter	0.3937 inch
1 centimeter	0.03280833 foot
1 meter (100 centimeters)	3.280833 feet
1 meter	1.093611 US yards
1 kilometer (1,000 meters)	0.62137 statute mile
1 kilometer	0.539957 nautical mile

AREA

1 sq inch	6.451626 sq centimeters
1 sq foot (144 sq inches)	0.092903 sq meter
1 sq yard (9 sq feet)	0.836131 sq meter
1 acre (4,840 sq yards)	0.404687 hectare
1 sq mile (640 acres)	2.589998 sq kilometers
1 sq centimeter	0.154999 sq inch
1 sq meter (10,000 sq centimeters)	10.76387 sq feet
1 sq meter	1.1959585 sq yards
1 hectare (10,000 sq meters)	2.47104 acres
1 sq kilometer (100 hectares)	0.386101 sq mile

VOLUME

1 cubic inch	16.387162 cubic centimeters
1 cubic foot (1,728 cubic inches)	0.028317 cubic meter
1 cubic yard (27 cubic feet)	0.764559 cubic meter
1 cord (128 cubic feet)	3.62458 cubic meters
1 cubic centimeter	0.061023 cubic inch
1 cubic meter (1,000,000 cubic centimeters)	35.31445 cubic feet
1 cubic meter	1.307943 cubic yards

LIQUID MEASURE

1 US quart	0.946333 liter
1 imperial quart	1.136491 liters
1 US gallon	0.037853 hectoliter
1 imperial gallon	0.04546 hectoliter
1 liter	1.05671 US quarts
1 liter	0.8799 imperial quart
1 hectoliter	26.4178 US gallons
1 hectoliter	21.9975 imperial gallons

TEMPERATURE

Fahrenheit (F)	9/5 Centigrade + 32
Centigrade (C)	Fahrenheit − 32 × 5/9

WEIGHT

1 avoirdupois ounce	0.0283495 kilogram
1 troy ounce	0.0311035 kilogram
1 avoirdupois pound	0.453592 kilogram
1 avoirdupois pound	0.00453592 quintal
1 hundredweight (cwt, 112 lb)	0.50802 quintal
1 ton (short ton, 2,000 lb)	0.907185 metric ton
1 long ton (2,240 lb)	1.016047 metric tons
1 kilogram (1,000 grams)	35.27396 avoirdupois ounces
1 kilogram	32.15074 troy ounces
1 kilogram	2.204622 avoirdupois pounds
1 quintal (100 kg)	220.4622 avoirdupois pounds
1 quintal	1.9684125 hundredweights
1 metric ton (1,000 kg)	1.102311 (short) tons
1 metric ton	0.984206 long ton

BUSHELS

	LB	% OF SHORT TON	BUSHELS PER SHORT TON
Barley (US)	48	2.4	41.667
(UK)	50	2.5	40.0
Corn (UK, US)	56	2.8	35.714
Linseed (UK)	52	2.6	38.462
(Australia, US)	56	2.8	35.714
Oats (US)	32	1.6	62.5
(Canada)	34	1.7	58.824
Potatoes (UK, US)	60	3.0	33.333
Rice (Australia)	42	2.1	47.619
(US)	45	2.25	44.444
Rye (UK, US)	56	2.8	35.714
(Australia)	60	3.0	33.333
Soybeans (US)	60	3.0	33.333
Wheat (UK, US)	60	3.0	33.333

BALES OF COTTON

	LB	% OF SHORT TON	BALES PER SHORT TON
India	392	19.6	5.102
US (net)	480	24.0	4.167

ELECTRIC ENERGY

1 horsepower (hp)	0.7457 kilowatt
1 kilowatt (kw)	1.34102 horsepower

PETROLEUM

One barrel = 42 US gallons = 34.97 imperial gallons = 158.99 liters = 0.15899 cubic meter (or 1 cubic meter = 6.2898 barrels).

*Includes units of measure cited in the text, as well as certain other units employed in parts of the English-speaking world.

ALABAMA

State of Alabama

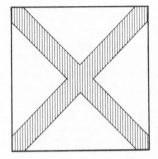

ORIGIN OF STATE NAME: Probably after the Alabama Indian tribe. **NICKNAME:** The Heart of Dixie. **CAPITAL:** Montgomery. **ENTERED UNION:** 14 December 1819 (22d). **SONG:** "Alabama." **MOTTO:** *Audemus jura nostra defendere* (We dare defend our rights). **COAT OF ARMS:** Two eagles, symbolizing courage, support a shield bearing the emblems of the five governments (France, England, Spain, Confederacy, US) that have held sovereignty over Alabama. Above the shield is a sailing vessel modeled upon the ships of the first French settlers of Alabama; beneath the shield is the state motto. **FLAG:** Crimson cross of St. Andrew on a square white field. **OFFICIAL SEAL:** Map of Alabama, including names of major rivers and neighboring states, surrounded by the words "Alabama Great Seal." **BIRD:** Yellowhammer. **FISH:** Tarpon. **FLOWER:** Camellia. **TREE:** Southern (longleaf) pine. **STONE:** Marble. **MINERAL:** Hematite. **LEGAL HOLIDAYS:** New Year's Day, 1 January; Birthdays of Robert E. Lee and Martin Luther King, Jr., 3d Monday in January; Washington's Birthday, 3d Monday in February; Mardi Gras, February or March; Thomas Jefferson's Birthday, 14 April; Confederate Memorial Day, 4th Monday in April; Jefferson Davis's Birthday, 1st Monday in June; Independence Day, 4 July; Labor Day, 1st Monday in September; Columbus Day, 2d Monday in October; Veterans Day, 11 November; Thanksgiving Day, 4th Thursday in November; Christmas Day, 25 December. **TIME:** 6 AM CST = noon GMT.

¹LOCATION, SIZE, AND EXTENT

Located in the eastern south-central US, Alabama ranks 29th in size among the 50 states.

The total area of Alabama is 51,705 sq mi (133,915 sq km), of which land constitutes 50,767 sq mi (131,486 sq km) and inland water 938 sq mi (2,429 sq km). Alabama extends roughly 200 mi (320 km) E–W; the maximum N–S extension is 300 mi (480 km). Alabama is bordered on the N by Tennessee; on the E by Georgia (with part of the line formed by the Chattahoochee River); on the S by Florida (with part of the line defined by the Perdido River) and the Gulf of Mexico; and on the W by Mississippi (with the northernmost part of the line passing through the Tennessee River).

Dauphin Island, in the Gulf of Mexico, is the largest offshore island. The total boundary length of Alabama is 1,044 mi (1,680 km). The state's geographic center is in Chilton County, 12 mi (19 km) SW of Clanton.

²TOPOGRAPHY

Alabama is divided into four major physiographic regions: the Gulf Coastal Plain, Piedmont Plateau, Ridge and Valley section, and Appalachian (or Cumberland) Plateau. The physical characteristics of each province have significantly affected settlement and industrial development patterns within the state.

The coastal plain, comprising the southern half of Alabama, consists primarily of lowlands and low ridges. Included within the coastal plain is the Black Belt—historically, the center of cotton production and plantation slavery in Alabama—an area of rich, chalky soil that stretches across the entire width of central Alabama. Just to the north, the piedmont of east-central Alabama contains rolling hills and valleys. Alabama's highest elevation, Cheaha Mountain, 2,407 feet (734 meters) above sea level, is located at the northern edge of this region. North and west of the piedmont is a series of parallel ridges and valleys running in a northeast-southwest direction. Mountain ranges in this area include the Red, Shades, Oak, Lookout, and other noteworthy southern extensions of the Appalachian chain; elevations of 1,200 feet (366 meters) are found as far south as Birmingham. The Appalachian Plateau covers most of northwestern Alabama, with a portion of the Highland Rim in the extreme north near the

Tennessee border. The floodplain of the Tennessee River cuts a wide swath across both these northern regions.

The largest lake wholly within Alabama is Guntersville Lake, covering about 108 sq mi (280 sq km) and formed during the development of the Tennessee River region by the Tennessee Valley Authority. The TVA lakes—also including Wheeler, Pickwick, and Wilson—are all long and narrow, fanning outward along a line that runs from the northeast corner of the state westward to Florence. The longest rivers are the Alabama, extending from the mid-central region to the Mobile River for a distance of about 160 mi (260 km); the Tennessee, which flows across northern Alabama for about the same distance; and the Tombigbee, which flows south from north-central Alabama for some 150 mi (240 km). The Alabama and Tombigbee rivers, which come together to form the Mobile River, and the Tensaw River flow into Mobile Bay, an arm of the Gulf of Mexico.

About 450 million years ago, Alabama was covered by a warm, shallow sea. Over millions of years, heavy rains washed gravel, sand, and clay from higher elevations onto the rock floor of the sea to help form the foundation of modern Alabama. The skeletons and shells of sea animals, composed of limy material from rocks that had been worn away by water, settled into great thicknesses of limestone and dolomite. Numerous caves and sinkholes formed as water slowly eroded the limestone subsurface of northern Alabama. Archaeologists believe that Russell Cave, in northeastern Alabama, was the earliest site of human habitation in the southeastern US. Other major caves in northern Alabama are Manitou and Sequoyah; near Childersburg is DeSoto Caverns, a huge onyx cave once considered a sacred place by Creek Indians.

Wheeler Dam on the Tennessee River is now a national historic monument. Other major dams include Guntersville, Martin, Millers Ferry, Jordan, Mitchell, and Holt.

³CLIMATE

Alabama's three climatic divisions are the lower coastal plain, largely subtropical and strongly influenced by the Gulf of Mexico; the northern plateau, marked by occasional snowfall in winter; and the Black Belt and upper coastal plain, lying between the two extremes. Among the major population centers, Birmingham has an annual mean temperature of 62°F (17°C), with a normal July daily maximum of 90°F (32°C) and a normal January daily minimum of 34°F (1°C). Montgomery has an annual mean of 65°F (180°C), with a normal July daily maximum of 91°F (33°C) and a normal January daily minimum of 37°F (3°C). The mean in Mobile is 67°F (19°C), the normal July daily maximum 91°F (33°C) and the normal January daily minimum 41°F (51°C). The record low temperature for the state is −27°F (−33°C), registered at New Market, in the northeastern corner, on 30 January 1966; the all-time high is 112°F (44°C), registered at Centreville, in the state's midsection, on 5 September 1925. Mobile, one of the rainiest cities in the US, recorded an average precipitation of 65 in (165 cm) a year between 1951 and 1980.

Two of the most destructive hurricanes to hit Alabama were Camille in August 1969 and Frederic in September 1979, the latter causing extensive property damage in the Mobile Bay area. In 1983, the state was hit by 45 tornadoes.

⁴FLORA AND FAUNA

Alabama was once covered by vast forests of pine, which still form the largest proportion of the state's forest growth. Alabama also has an abundance of poplar, cypress, hickory, oak, and various gum trees. Red cedar grows throughout the state; southern white cedar is found in the southwest, hemlock in the north. Other native trees include hackberry, ash, and holly, with species of palmetto and palm in the Gulf Coast region. There are more than 150 shrubs, mountain laurel and rhododendron among them. Cultivated plants include wisteria and camellia, the state flower.

In a state where large herds of bison, elk, bear, and deer once roamed, only the white-tailed deer remains abundant. Other mammals still found are the Florida panther, bobcat, beaver, muskrat, and most species of weasel. The fairly common raccoon, opossum, rabbit, squirrel, and red and gray foxes are also native, while nutria and armadillo have been introduced to the state. Alabama's birds include golden and bald eagles, osprey and various other hawks, yellowhammer or flicker (the state bird), and black and white warblers; game birds include quail, duck, wild turkey, and goose. Freshwater fish such as bream, shad, bass, and sucker are common. Along the Gulf Coast there are seasonal runs of tarpon (the state fish), pompano, redfish, and bonito.

Endangered or threatened animals include the fine-rayed and shiny pigtoe, 8 kinds of pearly mussels, 3 kinds of darters, chub spotfin, Alabama cavefish, red hills salamander, eastern indigo snake, American alligator, ivory-billed and red-cockaded woodpecker, bald eagle, brown pelican, wood stork, red wolf, and Florida panther.

⁵ENVIRONMENTAL PROTECTION

State agencies concerned with environmental protection include the Alabama Surface Mining Reclamation Commission, the Department of Conservation and Natural Resources, the Environmental Health Administration, and the Department of Environmental Management, which has responsibility for managing the state's air, land, and water resources. The most active environmental groups in the state are the Alabama Conservancy, Safe Energy Alliance, Sierra Club, and League of Women Voters.

Nuclear power has been a source of conflict in Alabama, and in 1983 one Browns Ferry reactor, near Athens, was ordered shut down temporarily for inspection; in 1975, a fire had forced the shutdown of a Browns Ferry reactor. Major concerns of environmentalists in the state are the improvement of land-use planning and the protection of groundwater. Another issue is the transportation, storage, and disposal of hazardous wastes. One of the nation's five largest landfills for toxic wastes is in Emelle, in Sumter County. Air quality is generally satisfactory, but Birmingham has higher levels of air pollution than most US cities.

⁶POPULATION

Alabama ranked 22d in population among the 50 states in 1980, with a census total of 3,893,888, a 13% increase since 1970. By early 1985, Alabama's population had grown to an estimated 4,004,435, reflecting a growth of 2.8% since 1980.

Alabama experienced its greatest population growth between 1810 and 1820, following the defeat of the Creek Nation by General Andrew Jackson and his troops. Population in what is now Alabama boomed from 9,046 in 1810 to 127,901 in 1820, as migrants from older states on the eastern seaboard poured into the territory formerly occupied by the Creek Indians and taken over by the US. Thousands of farmers, hoping to find fertile land or to become wealthy cotton planters, brought their families and often their slaves into the young state, more than doubling Alabama's population between 1820 and 1830. By 1860, Alabama had almost 1,000,000 residents, nearly one-half of whom were black slaves. The Civil War brought Alabama's population growth almost to a standstill, largely because of heavy losses on the battlefield; the total population gain between 1860 and 1870 was only about 30,000. Between 1870 and 1970, Alabama's population rose 150,000–300,000 every decade.

In 1985, Alabama had an estimated population density of 79 persons per sq mi (30 per sq km). About 3 out of every 5 Alabamians lived in urban areas in 1980. First in size among Alabama's metropolitan areas comes Birmingham (43d in the US), which had 895,200 residents in mid-1984. Other major metropolitan areas were Mobile, 465,700, and Montgomery, 284,800. Birmingham proper had a population of 279,813 (54th in the US) in 1984. The next largest cities in 1984 were Mobile (73d), 204,923; Montgomery (78th), 184,963; and Huntsville (109th), 149,527.

⁷ETHNIC GROUPS

Alabama's population is largely divided between whites of English and Scotch-Irish descent and blacks descended from African slaves. The 1980 census counted only 7,483 Indians, mostly of Creek or Cherokee descent. Creek Indians are centered around the small community of Poarch in southern Alabama; most of the Cherokee live in the northeastern part of the state.

The black population of Alabama in 1980 was estimated at 995,623, about one-fourth of the total population. As of 1980, Birmingham was 56% nonwhite, Mobile 37%, and Montgomery 40%. As before the Civil War, rural blacks are most heavily represented in the Black Belt of central Alabama.

In 1980, Alabama had 1,992 Asian Indians, 1,782 Koreans, and 1,503 Chinese; the Hispanic population was about 33,000. All told, the foreign-born numbered about 39,000 (1% of the state's population) in 1980. Among persons reporting a single ancestry group, the leaders were English, 857,864, and Irish, 224,453.

Alabama's Cajuns, of uncertain racial origin (Anglo-Saxon, French, Spanish, Choctaw, Apache, and African elements may all be represented), are ethnically unrelated to the Cajuns of Louisiana. Numbering perhaps 3,000, they live primarily in the pine woods area of upper Mobile and lower Washington counties. Many Alabama Cajuns suffer from poverty, poor health, and malnutrition.

⁸LANGUAGES

Four Indian tribes—the Creek, Chickasaw, Choctaw, and Cherokee—occupied the four quarters of Alabama as white settlement began, but by treaty agreement they were moved westward between 1814 and 1835, leaving behind such place-names as Alabama, Talladega, Mobile, and Tuscaloosa.

Alabama English is predominantly Southern, with a transition zone between it and a smaller area into which South Midland speech was taken across the border from Tennessee. Some features common to both dialects occur throughout the state, such

LOCATION: 30°10′ to 35°N; 84°51′ to 88°28′w. **BOUNDARIES:** Tennessee line, 148 mi (238 km); Georgia line, 298 mi (480 km); Florida line, 219 mi (352 km); Gulf of Mexico coastline, 53 mi (85 km); Mississippi line, 326 mi (525 km).

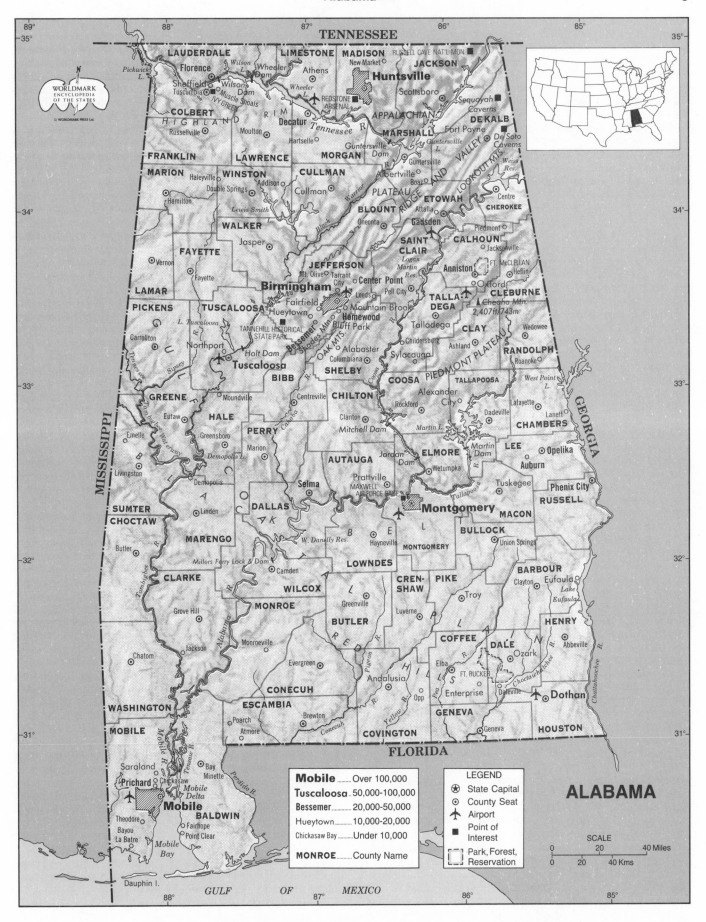

ALABAMA

Mobile ———— Over 100,000
Tuscaloosa ——50,000-100,000
Bessemer ————20,000-50,000
Hueytown————10,000-20,000
Chickasaw Bay ———— Under 10,000
MONROE ———— County Name

LEGEND
⊛ State Capital
⊙ County Seat
✈ Airport
■ Point of Interest
⬚ Park, Forest, Reservation

SCALE
0   20   40 Miles
0   20   40 Kms

See endsheet maps: J4.

as *crocus sack* (burlap bag), *batter cakes* (made of corn meal), *harp* (harmonica), and *snap beans*. In the major Southern speech region are found the decreasing loss of final /r/, the /boyd/ pronunciation of *bird, soft peach* (freestone), *press peach* (clingstone), *mosquito hawk* (dragonfly), *fire dogs* (andirons), and *gopher* (burrowing turtle). In the northern third of the state are found South Midland *arm* and *barb* rhyming with *form* and *orb, redworm* (earthworm), *peckerwood* (woodpecker), *snake doctor* and *snake feeder* (dragonfly), *tow sack* (burlap bag), *plum peach* (clingstone), *French harp* (harmonica), and *dog irons* (andirons).

Alabama has experienced only minor foreign immigration, and 98% of all residents 3 years old or older spoke only English at home in 1980. Principal languages other than English spoken at home were as follows:

Spanish	23,063	Korean	1,631
French	10,622	Greek	1,434
German	10,599	Chinese	1,369
Italian	2,002	Japanese	1,364

⁹RELIGIONS

Although predominantly Baptist today, Alabama was officially Roman Catholic throughout most of the 18th century, under French and Spanish rule. A century passed between the building of the first Catholic church in 1702 and the earliest sustained efforts by Protestant evangelists. The first Baptist church in the state, the Flint River Church in Madison County, was organized in 1808; the following year, the Old Zion Methodist Church was founded in the Tombigbee area. During the second decade of the 19th century, settlers from the southeastern states brought the influence of the Great Revival to Alabama, along with the various Methodist, Presbyterian, and Baptist sects that had developed in its wake. The first black church in Alabama probably dates from 1820. As in other southern states, black slaves who had previously attended the churches of their masters formed their own churches after the Civil War. One of the earliest of these, the Little Zion Methodist Church, was established in 1867 in Mobile; most freed blacks became Baptists, however.

As of 1980, the major Protestant denominations were the Southern Baptist Convention, with 1,182,018 adherents; the United Methodist Church, with 344,790; and Churches of Christ, 113,919. In 1984, Roman Catholics in Alabama numbered 106,123, and there were an estimated 9,560 Jews.

¹⁰TRANSPORTATION

The first rail line in the state—the Tuscumbia Railroad, chartered in 1830—made its first run, 44 mi (71 km) around the Muscle Shoals from Tuscumbia to Decatur, on 15 December 1834. By 1852, however, Alabama had only 165 mi (266 km) of track, less than most other southern states. Further development awaited the end of the Civil War. Birmingham, as planned by John T. Milner, chief engineer of the South and North Railroad, was founded in 1871 as a railroad intersection in the midst of Alabama's booming mining country; it subsequently became the state's main rail center, followed by Mobile. As of the end of 1983, Alabama had 4,185 mi (6,735 km) of Class I track. An Amtrak passenger train connected Birmingham, Anniston, and Tuscaloosa with Washington and New Orleans; total Alabama ridership was 90,366 in 1983/84.

In settlement days the principal roads into Alabama were the Federal Road, formerly a Creek horse path, from Georgia and South Carolina; and the Natchez Trace, bought by the federal government (1801) from the Choctaw and Chickasaw, leading from Kentucky and Tennessee. Throughout most of the 19th century, road building was in the hands of private companies. Only after the establishment of a state highway department in 1911 and the securing of federal aid for rural road building in 1916 did Alabama begin to develop modern road systems.

As of 1983 there were 88,427 mi (142,310 km) of streets, roads, and highways. In the same year, the state had 2,271,950 registered automobiles, 864,316 trucks, and 2,393,878 licensed drivers. Most of the major interstate highways in Alabama intersect at Birmingham: I-65, running from the north to Montgomery and Mobile, and I-59 from the northeast and I-20 from the east, which, after merging at Birmingham, run southwestward to Tuscaloosa and into Mississippi. I-85 connects Montgomery with Atlanta; I-10, Mobile with New Orleans and Tallahassee, Fl.

The coming of the steamboat to Alabama waters, beginning in 1818, stimulated settlement in the Black Belt; however, the high price of shipping cotton by water contributed to the eventual displacement of the steamboat by the railroad. Thanks to the Tennessee Valley Authority, the Tennessee River has been transformed since the 1930s into a year-round navigable waterway, with three locks and dams in Alabama. The 234-mi (377-km), $2-billion Tennessee-Tombigbee project, which opened in 1985, provided a new barge route, partly through Alabama, from the Midwest to the Gulf of Mexico, for which the US Army Corps of Engineers cut a 39-mi (63-km) canal and built 10 locks and dams. This was not only the largest civilian engineering project in the US during the early 1980s but also by far the largest earth-moving project in US history, displacing more earth than was moved to build the Panama Canal. The Alabama-Coosa and Black Warrior–Tombigbee systems also have been made navigable by locks and dams; river barges carry bulk cargoes. Mobile, on the Gulf of Mexico, is Alabama's only international port; in 1983, it handled 9.5 million lb of exports and 6.3 million lb of imports, with values respectively, of $1.4 billion and $668 million.

At the end of 1983, Alabama had 138 airports and 31 heliports. The largest and busiest facility is Birmingham Municipal Airport, where 612,126 passengers enplaned during 1983.

¹¹HISTORY

The region now known as Alabama has been inhabited for some 9,000–10,000 years. The earliest evidence of human habitation, charcoal from an ancient campfire at Russell Cave in northeastern Alabama, is about 9,000 years old. These early peoples, probably descended from humans who crossed from Asia to North America via the Bering Strait, moved from caves and open campsites to permanent villages about AD 1000. Some of their descendants, popularly called Mound Builders, erected huge earthern temple mounds and simple huts along Alabama's rivers, beginning around 1100. Moundville (near Tuscaloosa), one of the most important Mound Builder sites in the southeastern US, includes 20 "platform mounds" for Indian buildings, dating from 1200 to 1500. When the first Europeans arrived, Alabama was inhabited by Indians, half of them either Creek or members of smaller groups living within the Creek confederacy. Cherokee Indians inhabited northeastern Alabama, Chickasaw lived in the northwest, and Choctaw settled in the southwest.

During the 16th century, five Spanish expeditions entered Mobile Bay or explored the region now called Alabama. The most extensive was that of Hernando de Soto, whose army marched from the Tennessee Valley to the Mobile Delta in 1540. In 1702, two French naval officers—Pierre Le Moyne, Sieur d'Iberville, and Jean Baptiste Le Moyne, Sieur de Bienville—established Ft. Louis de la Mobile, the first permanent European settlement in present-day Alabama. Mobile remained in French hands until 1763, when it was turned over to the British under the terms of the Treaty of Paris. Because a British garrison held Mobile during the American Revolution, that city was captured in 1780 by the forces of Spain, an ally of the rebellious American colonists. In 1803, the United States claimed the city as part of the Louisiana Purchase, but in vain. Spanish control of Mobile lasted until the city was again seized during the War of 1812, this time by American troops in 1813. West Florida, including Mobile, was the only territory added to the US as a result of that war.

At the start of the 19th century, Indians still held most of present-day Alabama. War broke out in 1813 between American settlers and a Creek faction known as the Red Sticks, who were determined to resist white encroachment. After General Andrew Jackson and his Tennessee militia crushed the Red Sticks in 1814 at the Battle of Horseshoe Bend, in central Alabama, he forced the Creek to sign a treaty ceding some 40,000 sq mi (103,600 sq km) of land to the US, thereby opening about three-fourths of the present state to white settlement. By 1839, nearly all Alabama Indians had been removed to Indian Territory.

From 1814 onward, pioneers, caught up by what was called "Alabama fever," poured out of the Carolinas, Virginia, Georgia, Tennessee, and Kentucky into what Andrew Jackson called "the best unsettled country in America." Wealthy migrants came in covered wagons, bringing their slaves, cattle, and hogs. But the great majority of pioneers were ambitious farmers hoping to acquire fertile land in the newly opened area. In 1817, Alabama became a territory; on 2 August 1819, a state constitution was adopted; and the following 14 December, Alabama was admitted to statehood.

During the antebellum era, 95% of white Alabamians lived and worked in rural areas, primarily as farmers, planting and harvesting cotton, corn, sorghum, oats, and vegetables and raising razorback hogs and cattle. By 1860, 80% of Alabama farmers owned the land they tilled. Only about one-third of all white Alabamians were slaveowners. Large planters (owners of 50 slaves or more) made up less than 1% of Alabama's white population in 1860. However, they owned 28% of the state's total wealth and occupied one-fourth of the seats in the legislature. In 1820 there were only 41,879 slaves, but in 1860 there were 435,080, 45% of the state population. Dominant in Alabama's economic, political, and social life, the planters led in the secession movement. Most white farm folk, fearing the consequences of an end to slavery, eventually embraced the Confederate cause, although 2,500 white Alabamians did serve in the Union Army and an estimated 8,000–10,000 others acted as Union scouts, deserted Confederate units, or hid from conscription agents.

Alabama seceded from the Union in January 1861 and shortly thereafter joined the Confederate States of America. The Confederacy was organized in Alabama's senate chamber in Montgomery, and Jefferson Davis was inaugurated president on the steps of the capitol. Montgomery served as capital of the Confederacy until May, when the seat of government was moved to Richmond, Va.

Remote from major theaters of war, Alabama experienced only occasional Union raids during the first three years of the conflict. In the summer of 1864, however, Confederate and Union ships fought a major naval engagement in Mobile Bay which ended in surrender by the outnumbered southern forces. During the Confederacy's dying days in the spring of 1865, federal troops swept through Tuscaloosa, Selma, and Montgomery. Their major goal, Selma, one of the Confederacy's main industrial centers, was left almost as heavily devastated as Richmond or Atlanta. Estimates of the number of Alabamians killed in the Civil War range from 25,000 upward.

During Reconstruction, Alabama was under military rule until readmitted to the Union in 1868. For the next six years, Republicans held most top political positions in the state. With the help of the Ku Klux Klan, Democrats regained political control of the state in November 1874.

Like many other southerners, Alabamians sought to create a "New South" in which agriculture would be balanced by industry. In the 1880s and 1890s, at least 20 Alabama towns were touted as ironworking centers. Birmingham, founded in 1871, became the New South's leading industrial center. Its promoters invested in pig iron furnaces, coal mines, steel plants, and real estate. Small companies merged with bigger ones, which were taken over, in turn, by giant corporations. In 1907, Birmingham's Tennessee Coal, Iron, and Railroad Co. was purchased by the nation's largest steelmaker, US Steel.

Another major Alabama enterprise was cotton milling. By 1900, 9,000 men, women, and children were employed in Alabama mills; most of these white workers were farm folk who had lost their land after the Civil War because of mounting debts and low cotton prices. Wages in mills were so low that entire families had to work hours as long as those they had endured as farmers.

As thousands of other Alabama farmers lost their land and became sharecroppers and tenants, unrest mounted. Discontented farmers and factory workers allied during the 1890s in the Populist Party, in an attempt to overthrow the Bourbon Democrats who had dominated Alabama politics for two decades. Although a number of Populists were elected to the Alabama legislature, no Populist candidate succeeded in winning the governorship, primarily because Democrats manipulated the black vote to their own advantage. In 1901, Alabama adopted a new state constitution containing numerous restrictions on voting—supposedly to end vote manipulation and restore honest elections. The tangible result of these new rules was to disfranchise almost all Alabama black voters and thousands of poor whites. For example, the total of blacks registered in 14 counties fell from 78,311 in 1900 to 1,081 in 1903. As recently as 1941, fewer than one-fourth of Alabama adults were registered voters. In 1960, no blacks voted in Lowndes or Wilcox counties, 80% and 78% black, respectively.

During the 1950s and 1960s, national attention focused on civil rights demonstrations in Alabama, including the Montgomery bus boycott of 1955, the Birmingham and University of Alabama demonstrations of 1963, and the voting rights march from Selma to Montgomery in 1965. The primary antagonists were Dr. Martin Luther King, Jr., head of the Southern Christian Leadership Conference, and Governor George C. Wallace, an opponent of integration. These black protests and the sometimes violent reactions to them—such as the 1963 bombing of a church in Birmingham in which four young black girls were killed—helped influence the US Congress to pass the Civil Rights Act of 1964 and the Voting Rights Act of 1965.

Once the most tightly segregated city in the nation, Birmingham has become thoroughly integrated in public facilities, and in 1979 the city elected its first black mayor, Richard Arrington. The civil rights era brought other momentous changes to Alabama. Hundreds of thousands of black voters are now an important force in state politics. Blacks attend schools, colleges, and universities of their choice and enjoy equal access to all public facilities. New racial attitudes among most whites have contributed to a vast improvement in the climate of race relations since 1960. Indeed, a significant amount of black support contributed to Wallace's election to a fourth term as governor in 1982. In 1984 there were 314 black elected officials, including 25 mayors and an associate justice of the state supreme court.

[12]STATE GOVERNMENT

Alabama has had six constitutions, the most recent one dating from 1901. That document was 172,000 words long by the end of 1983—longer than that of any other state—and had been amended 443 times.

Alabama's bicameral legislature consists of a 35-seat senate and a 105-seat house of representatives, all of whose members are elected at the same time for four-year terms. Senators must be at least 25 years of age, representatives 21. Under federal pressure, the legislature in 1983 approved a reapportionment plan, effective in 1986, that was expected to increase black representation; in 1984 there were 24 black legislators.

Elected executive officials are the governor and lieutenant-governor (separately elected), secretary of state, attorney general, treasurer, auditor, eight members of the Board of Education, and three members of the Public Service Commission. The governor, who serves for four years, must be at least 30 years of age and must have been a US citizen for 10 years and a citizen of the state for 7. The governor is limited to a maximum of two consecutive terms.

A bill becomes a law when it is passed by at least a majority of a quorum of both houses and is signed by the governor, left unsigned for 6 days while the legislature is in session, or passed over the governor's veto by a majority of the elected members of each house. The governor may pocket veto a measure submitted fewer than 5 days before adjournment by not signing it within 10 days after adjournment. The submission of a constitutional amendment to the electorate requires the approval of three-fifths of the membership of each house, but such amendments can also be adopted by initiative and referendum, or by a constitutional convention.

Voters in Alabama must be US citizens, at least 18 years of age, and must have resided in the state at least one day prior to the election.

¹³POLITICAL PARTIES

The major political parties in Alabama are the Democratic and Republican parties, each affiliated with the national party organization. The Republicans are weak below the federal-office level, and in early 1985, only 16 of the 140 state legislators were Republicans.

Pre–Civil War political divisions in the state reflected those elsewhere in the South. Small and subsistence farmers, especially in the northern hill country and pine forest areas, tended to be Jacksonian Democrats, while the planters of the Black Belt and the river valleys often voted Whig. After a period of Radical Republican rule during Reconstruction, the Bourbon Democrats, whose party then served largely the interests of wealthy property owners, businessmen, and white supremacists, ran the state for the rest of the century, despite a challenge in the 1890s by the Populist Party.

During the 20th century, the Democratic Party has continued to command virtually every statewide office, major and minor. The continued strength of the party during the 1960s and 1970s depended, in part, on the allegiance of black voters; the number of registered black voters increased from 111,000 in 1960 to about 482,000 in 1984. In the 1980 elections, Republican Ronald Reagan won a slim majority of Alabama's popular vote, while Jeremiah Denton, a former prisoner of war in Viet-Nam, became the first Alabama Republican since Reconstruction to win election to the US Senate. Reagan carried the state handily in 1984, and Senator Howell Heflin, a Democrat, was reelected; Alabama returned five Democrats and two Republicans to the House of Representatives.

On two occasions, 1948 and 1964, the Alabama Democratic Party bolted the national Democratic ticket, each time because of disagreement over civil rights. Barry Goldwater in 1964 was the first Republican presidential candidate in the 20th century to carry Alabama. In 1968, George Wallace carried Alabama overwhelmingly on the American Independent Party slate.

¹⁴LOCAL GOVERNMENT

Alabama had 67 counties, 434 municipalities, and at least 390 special districts in 1982. Counties are governed by county

Alabama Presidential Vote by Political Parties, 1948–84

YEAR	ELECTORAL VOTE	ALABAMA WINNER	DEMOCRAT	REPUBLICAN	STATES' RIGHTS DEMOCRAT	PROHIBITION	PROGRESSIVE
1948	11	Thurmond (SRD)	—	40,930	171,443	1,026	1,522
1952	11	Stevenson (D)	275,075	149,231	—	1,814	—
					UNPLEDGED		
1956	11	Stevenson (D)	279,542	195,694	20,323	—	—
					NAT'L STATES' RIGHTS		
1960	11	*Kennedy (D)	318,303	236,110	4,367	—	—
					UNPLEDGED DEMOCRAT		
1964	10	Goldwater (R)	—	479,085	210,782		—
					AMERICAN IND.		AM. IND. DEMOCRAT
1968	10	Wallace (AI)	195,918	146,591	687,664	3,814	10,518
					AMERICAN		
1972	9	*Nixon (R)	256,923	728,701	11,928	8,559	—
					AMERICAN IND.		COMMUNIST
1976	9	*Carter (D)	659,170	504,070	9,198	6,669	1,954
1980	9	*Reagan (R)	636,730	654,192	—	—	—
					LIBERTARIAN		
1984	9	*Reagan (R)	551,899	872,849	9,504	—	—

* Won US presidential election.

commissions, usually consisting of three to seven commissioners, elected by district. Other county officials include a clerk, assessor, tax collector, sheriff, and superintendent of education. Elections for municipal officers are held every four years.

Until the late 1970s, the predominant form of municipal government, especially in the larger cities, was the commission, whose members are elected either at-large or by district. Partly in response to court orders requiring district elections in order to permit the election of more black officials, there has since been a trend toward the mayor-council form—though the US Supreme Court ruled in May 1980 that Mobile may elect its public officials at-large.

An alteration in local government had a significant effect on the racial climate in Birmingham during the 1960s, when the Young Men's Business Club led a movement to change to the mayor-council system, in order to oust a commission (including Eugene "Bull" Connor as public safety commissioner) that for nearly a decade had reacted negatively to every black demand. After a narrow vote in favor of the change, a moderate was elected mayor in April 1963, but the former commissioners then contested the initial vote that had changed the system. At the height of Birmingham's racial troubles, both the former commissioners and the newly elected council claimed to govern Birmingham, but neither did so effectively. When peace came, it was as the result of an unofficial meeting held between local black leaders and 77 of the city's most influential whites, with federal officials serving as mediators. Although the council, like the commissioners, publicly opposed these negotiations, once they were over and the council's election confirmed, the new moderate leadership permitted peaceful racial accommodation to go forward.

[15] STATE SERVICES

Alabama's Ethics Commission administers the state's ethics law, makes financial disclosure records available to the public, and receives monthly reports from lobbyists. Educational services are administered primarily by the Department of Education and the Alabama Commission on Higher Education. The Department of Aeronautics, Highway Department, and Public Service Commission (PSC) administer transportation services; the PSC supervises, regulates, and controls all transportation companies doing business in the state. Drivers' licenses are issued by the Department of Public Safety.

Health and welfare services are offered primarily through the Department of Public Health, Department of Mental Health, Department of Veterans Affairs, the Commission on Aging, Department of Youth Services, and Department of Pensions and Security. Planning for the state's future health-care needs is carried out by the Health Planning and Development Agency.

Public protection services are administered by the Military Department, Board of Corrections, Alabama Law Enforcement Planning Agency, and Department of Public Safety, among other agencies. Numerous government bodies offer resource protection services: the Department of Conservation and Natural Resources, Department of Environmental Management, Alabama Forestry Commission, Oil and Gas Board, Surface Mining Reclamation Commission, and Alabama State Soil and Water Conservation Committee.

[16] JUDICIAL SYSTEM

The high court of Alabama is the supreme court, consisting of a chief justice and eight associate justices, all elected for staggered six-year terms. It issues opinions on constitutional issues, and hears cases appealed from the lower courts. The court of civil appeals has exclusive appellate jurisdiction in all suits involving sums up to $10,000; its three judges are elected for six-year terms, and the one who has served the longest is the presiding judge. The five judges of the court of criminal appeals are also elected for six-year terms; they choose the presiding judge by majority vote.

Circuit courts, of which there were 39 systems with 113

judgeships in 1982, have exclusive original jurisdiction over civil actions involving sums of more than $5,000, and over criminal prosecutions involving felony offenses. They also have original jurisdiction, concurrent with the district courts, in all civil matters exceeding $500. They have appellate jurisdiction over most cases from district and municipal courts.

A new system of district courts replaced county and juvenile courts as of January 1977. In 1982 there were 65 district courts. At least one judge was elected per county (except that Clay and Coosa counties shared a single judge), and 11 counties elected two or more judges, who, like circuit court judges, serve six-year terms. Municipal court judges are appointed by the municipality. At the end of 1983, 9,856 prisoners were held in state and federal prisons in Alabama, including 1,001 inmates held in local jails because of overcrowding. Alabama's incarceration rate of 243 per 100,000 population was the 7th highest among the states. In 1976, US District Court Judge Frank M. Johnson, Jr., ruled that conditions in Alabama prisons inflicted "cruel and unusual punishment" upon inmates, spurring the process of prison reform; but at the end of 1983, even though many prisoners were being held in local jails, the prison population was 114% of capacity.

Alabama had an FBI Crime Index rate in 1983 of 4,101 crimes per 100,000 population, including 416 violent crimes and 3,685 property crimes; all these rates were below the national averages. However, Alabama had the 14th-highest murder and nonnegligent manslaughter rate in the nation—9.2 per 100,000 population—and Montgomery's murder rate, 13.8, was among the top 20 for US metropolitan areas.

An Alabama case that became internationally notorious was that of the nine "Scottsboro boys," eight of whom were sentenced to death and one to life imprisonment in 1931 for the alleged rape of two white girls, one of whom later recanted her charges. After multiple appeals and reversals, five indictments were subsequently dropped; of the four remaining defendants, all sentenced to lengthy jail terms, three were paroled and one escaped to Michigan, which refused extradition.

[17] ARMED FORCES

The US Department of Defense had 25,178 active military personnel in Alabama during 1983/84. The major installation was the US Army's Redstone Arsenal at Huntsville, with 3,668 military personnel. Redstone is the center of the Army's missile and rocket programs and contains the George C. Marshall Space Flight Center of the National Aeronautics and Space Administration, which directs all private contractors for the space program. Among the spacecraft developed there were the Redstone rocket, which launched the first US astronaut; Explorer I, the first US earth-orbiting satellite; and the Saturn rocket, which boosted the Apollo missions to the moon. Other installations include Ft. Rucker (near Enterprise); Ft. McClellan (Anniston), site of recruit training; the Anniston Army Depot; and Maxwell Air Force Base (Montgomery), site of the US Air University, Air War Colleges, and national headquarters for the Civil Air Patrol. During 1983/84, Alabama firms received defense contract awards totaling over $1.1 billion.

There were 436,000 veterans of US military service in Alabama as of 30 September 1983, 4,000 of whom served in World War I, 163,000 in World War II, 90,000 in the Korean conflict, and 129,000 during the Viet-Nam era. During 1982/83, benefits paid to Alabama veterans amounted to $511 million. As of early 1985, Alabama Army and Air National Guard units had a strength of 23,955. State and local police forces numbered 12,575 during 1983.

[18] MIGRATION

After 1814, Alabama was the mecca of a great migratory wave, mainly of whites of English and Scotch-Irish descent (some with their black slaves) from Virginia, Georgia, and the Carolinas. Since the Civil War, migration to Alabama has been slight. Many blacks left Alabama from World War I through the 1960s to seek employment in the East and Midwest, and the proportion of blacks fell from 35% in 1940 to 26% in 1980. Following the civil rights revolution, the trend began to reverse; more blacks chose to remain in the state, and some who had gone elsewhere returned. Overall, Alabama may have lost as many as 944,000 residents through migration between 1940 and 1970, but enjoyed a net gain from migration of 78,000 between 1970 and 1983.

[19] INTERGOVERNMENTAL COOPERATION

Among the interstate compacts and commissions in which Alabama participates are the Gulf States Marine Fisheries Compact, Interstate Mining Compact, Interstate Oil and Gas Compact, Southeastern Forest Fire Protection Compact, Southern Growth Policies Compact, Southern Interstate Energy Compact, Southern Regional Education Compact, and Tennessee-Tombigbee Waterway Development Compact. The Office of State Planning and Federal Programs coordinates planning efforts by all levels of government.

During 1983/84, Alabama received federal grants amounting to over $1.5 billion.

[20] ECONOMY

Cotton dominated Alabama's economy from the mid-19th century to the 1870s, when large-scale industrialization began. The coal, iron, and steel industries were the first to develop, followed by other resource industries such as textiles, clothing, paper, and wood products. Although Alabama's prosperity has increased, particularly in recent decades, the state still lags in wage rates and per capita income. One factor that has hindered the growth of the state's economy is declining investment in resource industries owned by large corporations outside the state. Between 1974 and 1983, manufacturing grew at little more than half the rate of all state goods and services. Industries such as primary metals, once the backbone of Alabama's economy, were clearly losing importance. The 1980–82 recession hit the state economy harder than the nation as a whole; 39,000 jobs were lost in manufacturing alone, and real output in manufacturing fell by 10.5%.

The gross state product of nearly $37.5 billion in 1982 consisted of manufacturing, 27%; trade, 16%; government, 16%; finance, insurance, and real estate, 13%; services, 10%; transportation, communications, and public utilities, 10%; farming, mining, and construction, 7%; and others, 1%.

Projections show that between 1984 and 1990, employment in food manufacturing, textiles, lumber, transport equipment, and agriculture will decline, while that in nonelectrical machinery, construction, trade, finance, utilities, and government will grow.

[21] INCOME

Alabama's per capita personal income in 1983 was $9,235, for a rank of 44th among the 50 states. Between 1960 and 1978, per capita income rose from 69% of the US average to 80%, but in 1983 it was only 79%. Median money income of four-person families was $22,443 in 1981 (43d in the US), and in 1980, 14.8% of the population was living below the federal poverty level. There is an apparent correlation between poverty and race, especially in rural Alabama: of the 8 counties with the highest percentage of nonwhites in 1980—Bullock, Greene, Hale, Lowndes, Macon, Perry, Sumter, and Wilcox—all but Bullock and Wilcox were among the 10 poorest counties per capita in 1982, their highest per capita average being $6,643, as against the state average of $8,647.

[22] LABOR

Alabama's civilian labor force in 1984 numbered 1,794,000. Alabama's total employment was 1,594,000, yielding an unemployment rate of 11.1%, 3d highest in the nation. In 1983, when the overall unemployment rate was 13.7%, the rate was 9.7% for whites and 26.7% for blacks.

A federal survey in March 1982 revealed the following nonfarm employment pattern in Alabama:

	ESTABLISH-MENTS	EMPLOYEES	ANNUAL PAYROLL ('000)
Agricultural services, forestry, fishing	637	4,399	$ 40,616
Mining, of which:	293	16,590	378,644
Bituminous coal, lignite	(117)	(13,184)	(303,298)
Contract construction	4,909	66,484	1,067,064
Manufacturing, of which:	5,005	337,919	5,235,705
Primary metals	(156)	(30,245)	(608,274)
Transportation, public utilities	2,626	68,783	1,411,087
Wholesale trade	5,715	69,682	1,117,237
Retail trade	19,793	199,283	1,688,817
Finance, insurance, real estate	5,473	60,859	908,241
Services	18,609	203,471	2,446,380
Other	1,469	1,306	20,633
TOTALS	64,529	1,028,776	$14,314,424

State and local government employees (222,000 in 1983) and federal government employees (60,000 in 1984) were not included in this survey.

In 1871, James Thomas Rapier, a black Alabamian who would later serve a term as a US representative from the state, organized the first black labor union in the South, the short-lived Labor Union of Alabama. The Knights of Labor began organizing in the state in 1882. A serious obstacle to unionization and collective bargaining was the convict leasing system, which was not ended officially until 1923 and in practice not until five years later. In 1888, the Tennessee Coal, Iron, and Railroad Co. (later taken over by US Steel) was granted an exclusive ten-year contract to use the labor of all state convicts, paying the state $9–18 a man per month. Child labor was also exploited. Alabama had limited a child's working day to 8 hours in 1887, but a Massachusetts company that was building a large mill in the state secured the repeal of that law in 1895. A weaker measure passed 12 years later limited the child's workweek to 60 hours and set the minimum working age at 12.

As of 1980, 296,000 Alabamians belonged to labor unions, of which there were 1,004 in 1983; unions were especially strong in the northern industrial cities and in Mobile. About 22% of all nonagricultural workers were unionized—one of the highest rates among states with right-to-work laws.

[23] AGRICULTURE

Alabama ranked 26th among the 50 states in agricultural income in 1983, with $2.1 billion.

There was considerable diversity in Alabama's earliest agriculture. By the mid-19th century, however, cotton had taken over, and production of other crops dropped so much that corn and other staples, even work animals, were often imported. In 1860, cotton was grown in every county, and one-crop agriculture had already worn out much of Alabama's farmland.

Diversification began early in the 20th century, a trend accelerated by the destructive effects of the boll weevil on cotton growing. By 1980, only 325,000 acres (131,500 hectares) were planted in cotton, compared to 3,500,000 acres (1,400,000 hectares) in 1930. As of 1984 there were some 54,000 farms in Alabama, occupying approximately 12 million acres (5 million hectares), or roughly 36% of the state's land area. Soybeans and livestock are raised in the Black Belt; peanuts in the southeast; vegetables, livestock, and timber in the southwest; and cotton and soybeans in the Tennessee River Valley. In 1983, Alabama ranked 2d in the US in production of peanuts for nuts, with 454,500,000 lb, worth about $110,000,000. Other crops included soybeans, 27,550,000 bushels, $220,400,000; corn, 18,325,000 bushels, $63,882,000; wheat, 15,180,000 bushels, $49,335,000; tomatoes for fresh market, 244,000 hundredweight, $16,832,000; sweet potatoes, 515,000 hundredweight, $6,183,000; and pecans, 24,000,000 hundredweight $11,496,000. The 1983 cotton crop of 183,000 bales, 7th highest in the nation, was valued at $57,711,000.

24ANIMAL HUSBANDRY

The principal livestock-raising regions of Alabama are the far north, the southwest, and the Black Belt, where the lime soil provides excellent pasturage. During 1983, Alabama produced 653 million lb of cattle and calves, valued at $317.3 million, and 162.6 million lb of hog and pigs, valued at $78.1 million. At the close of 1983 there were 1,870,000 cattle and 440,000 hogs on Alabama farms and ranches. In addition, 53,000 milk cows yielded 573 million lb of milk in 1983.

Alabama is a leading producer of chickens, broilers, and eggs. In broiler production, the state was surpassed only by Arkansas and Georgia in 1983, with 2 billion lb, valued at $553 million. That year, Alabama ranked 6th in chicken production, with 69.8 million lb, worth $9.4 million; and 10th in egg production, with 2.8 billion, worth $180 million.

25FISHING

Alabama's commercial fish catch was 26,405,000 lb, worth $43,778,000, in 1984. The principal fishing port is Bayou La Batre, which brought in about 13,600,000 lb, worth $28,500,000, 12th highest in the nation. Catfish farming is of growing importance. Sixty-nine processing and wholesaling plants had a combined total of about 1,600 employees in 1982.

26FORESTRY

Forestland in Alabama, predominantly pine, covering 21,361,000 acres (8,645,000 hectares), was nearly 3% of the nation's total in 1983, and 66% of the state's land area. Nearly all of that was classified as commercial timberland, 95% of it privately owned. Four national forests covered a gross area of 1,262,876 acres (511,069 hectares) in 1984. Shipments of lumber and wood products were valued at $1.6 billion in 1982; of paper and allied products, $3.5 billion. Production of softwoods and hardwoods totaled 1,507,000 board feet in 1983, 7th among the states.

27MINING

The total value of Alabama's nonfuel mineral production in 1984 was $411,055,000, 22d among the states. Output in 1984 (excluding fossil fuels) was as follows: stone, 23,008,000 tons; sand and gravel, 9,605,000 tons; cement, 3,840,000 tons; clays, 1,900,000 tons; and lime, 1,195,000 tons.

The mining of iron, the foundation of the state's large steel industry, had ceased by the early 1980s, as steelmakers substituted cheaper foreign iron.

28ENERGY AND POWER

Electrical generating plants in Alabama had an installed capacity of 19.2 million kw in 1983, when production totaled 76.2 billion kwh. About half of the capacity and production came from private sources (the Alabama Power Authority and Southern Electric Generating), with most of the remainder attributable to the Tennessee Valley Authority, which also owned three of the state's five nuclear reactors.

Significant petroleum finds in southern Alabama date from the early 1950s. The 1983 output was 18,746,000 barrels; proved reserves as of 31 December 1983 totaled 51,000,000 barrels. During the same year, 125.9 billion cu feet of natural gas were extracted, from 461 wells, leaving reserves of 785 billion cu feet (excluding offshore reserves in Mobile Bay, believed to be extensive). Value added by mining for oil and gas in 1983 was $858.3 million. Coal production, which began in the 19th century, reached 23,625,000 tons in 1978, 8th among the states, of which all was bituminous and 54% was surface mined. Coal reserves in 1983 totaled 5.2 billion tons, four-fifths bituminous and one-fifth lignite.

29INDUSTRY

Alabama's industrial boom, which began in the 1870s with the exploitation of the coal and iron fields in the north, quickly transformed Birmingham into the leading industrial city in the South, producing pig iron more cheaply than its American and English competitors. An important stimulus to manufacturing in the north was the development of ports and power plants along the Tennessee River. Although Birmingham remains highly dependent on steel, the state's industry has diversified considerably since World War II. By the late 1970s, the older smokestack industries were clearly in decline, but Birmingham received a boost in 1984 when US Steel announced it would spend $1.3 billion to make its Fairfield plant the newest fully integrated steel mill in the nation.

As of 1982, the principal employers among industry groups were apparel and textiles, primary metal industries, and fabricated metal products, together accounting for 43% of the state's manufacturing jobs. The total value of shipments of manufactured goods exceeded $29.7 billion, of which the largest contributors were apparel and textiles, 13%; paper and allied products, 12%; primary metals, 11%; food and kindred products, 11%; and chemicals and allied products, 9%. The following table shows the value of shipments in 1982 by selected industries.

Paper mills, except building paper	$1,663,300,000
Tires and inner tubes	1,283,500,000
Petroleum refining	1,180,900,000
Nonferrous rolling and drawing	1,151,900,000
Meat products	1,012,900,000
Blast furnace and basic steel products	927,900,000
Fabricated structural metal products	900,600,000
Men's and boys' apparel	811,000,000

30COMMERCE

With sales of $19.6 billion, Alabama ranked 27th among the 50 states in wholesale trade in 1982. That year, Alabama ranked 26th in retail trade, with sales of $14.5 billion. The leading types of retail businesses were automotive dealers, 20%; food stores, 26%; and general merchandise stores, 12%. Among counties, Jefferson had the largest share of retail sales, 23%. Alcoholic beverages, except for beer, are sold in ABC (Alcoholic Beverage Control) stores, run by the state. Prohibition is by local option; 26 of the 67 counties were dry in 1984, but some dry counties had wet cities.

Alabama exported $1.9 billion worth of manufactured goods in 1981 (24th in the US). Foreign exports of agricultural products totaled $481 million in 1981/82 (28th).

31CONSUMER PROTECTION

The Office of Consumer Protection, established in 1972, was transferred to the Office of the Attorney General in 1979. The major duties of the office are to enforce the Deceptive Trade Practices Act and to offer programs in consumer education.

32BANKING

Alabama's 282 insured commercial banks had assets of $22.1 billion in 1983; outstanding loans totaled $10.4 billion, and deposits were nearly $17.7 billion. There were 35 savings and loan associations at the end of 1983, with combined assets of $5.7 billion. At the end of 1982, the state's savings and loan associations had savings capital of $4.6 billion, and outstanding mortgage loans of $3.7 billion. In 1983, Alabama also had 266 credit unions, with deposits at midyear of $904 million.

33INSURANCE

During 1983, life insurance in force per family in Alabama averaged $35,100, slightly below the national average. There were 33 life insurance companies in the state at midyear. A total of 11,609,000 policies were in force, amounting to $78.5 billion. Benefits paid by life insurance companies to Alabamians totaled $646.3 million, of which $245.7 million consisted of death benefits paid to 105,000 beneficiaries.

There were 25 property/casualty insurance companies in Alabama in 1983. Property and liability insurers wrote premiums amounting to more than $1.3 billion in 1983. Of that total, $258.9 million was automobile liability insurance.

³⁴SECURITIES

Alabama has no securities exchanges. New York Stock Exchange member firms had 51 sales offices and 557 registered representatives in the state during 1983. Some 424,000 Alabamians held shares of public corporations in 1983.

³⁵PUBLIC FINANCE

The Division of the Budget within the Department of Finance prepares and administers the state budget, which the governor submits to the legislature for amendment and approval. The fiscal year runs from 1 October through 30 September.

The following table summarizes consolidated revenues for 1983/84 (actual) and 1984/85 (estimated), in millions.

REVENUES	1983/84	1984/85
General fund	$ 507.7	$ 594.3
Federal and local funds	2,266.0	2,178.4
Special educational trust fund, of which:	1,719.0	1,823.0
Income tax	(775.0)	(831.3)
Sales tax	(580.0)	(620.0)
Utility tax	(148.1)	(160.5)
Other receipts	(215.9)	(211.3)
Special mental health fund	80.0	83.7
TOTALS	$4,572.7	$4,679.4
EXPENDITURES		
Special educational trust fund	$1,477.6	$1,825.3
Highway Department	505.0	562.5
Medicaid Agency	381.7	503.1
Department of Pensions and Security	234.2	239.6
Department of Mental Health	176.7	198.0
Department of Economic and Community Affairs	114.5	150.0
Department of Public Health	94.4	104.9
Department of Corrections	81.0	94.7
Department of Industrial Relations	43.6	48.3
Other outlays	1,200.9	934.5
TOTALS	$4,309.6	$4,660.9

As of mid-1982, the total debt of Alabama state and local governments was $5.3 billion, or $1,356 per capita.

³⁶TAXATION

Alabama ranked 24th in the US in state tax collections in 1983, with $2.3 billion. Per capita tax revenues of all state and local governments, $764 in 1982, were less than those of every other state except Arkansas and Mississippi, and receipts from property taxes ($89 per capita) were the lowest in the nation. The total tax burden ($1,823 per capita) in 1982/83 was 47th among the states.

As of the end of 1983, the personal income tax, which is designated for education, ranged from 2 to 5%, depending on income and marital status. The tax on corporate net income was 5% for most enterprises, but 6% for financial institutions. The state also imposes a sales tax of 4%; localities may charge up to an additional 3%. Other state levies include a value-added tax, utility tax, use tax, and taxes on oil production, oil and gas, cigarettes, beer, and whiskey.

Alabama residents and businesses paid $7.2 billion in federal taxes in 1983/84, while the state received federal expenditures totaling $11.4 billion—a ratio of $0.63 in taxes for every $1 received, one of the most favorable state ratios. State residents filed 1.4 million federal income tax returns in 1983, paying $3.2 billion in tax.

³⁷ECONOMIC POLICY

Alabama seeks to attract out-of-state business by means of tax incentives and plant-building assistance. The Alabama Development Office plans for economic growth through industrial development. The Alabama Industrial Development Training Institute, within the Department of Education, provides job training especially designed to suit the needs of new or expanding industries in the state. The state Foreign Trade Relations Commission seeks to promote international markets for Alabama products.

³⁸HEALTH

Alabama's infant death rate in 1981, 13 per 1,000 live births, was 10th highest in the US; the rate for whites was 10.2, for blacks 18.2. Rates for marriages and divorces all exceeded the national averages for 1982, but the abortion and birth rates in Alabama were below the national norm.

The states's overall death rate in 1981, 9.4 deaths per 1,000 population, included the nation's 5th-highest death rate from cerebrovascular disease, 87.9 per 100,000 population. Alabama also ranked above the national average in death rates from diabetes mellitus, murder, prenatal conditions, and motor vehicle accidents, but below it for heart disease, cancer, cirrhosis and other liver diseases, and arteriosclerosis. The state had the 8th-lowest rate of death by suicide (10.7 per 100,000 population).

Alabama had 146 hospitals in 1983; there were 26,001 beds and 796,963 admissions, for a 76.5% average occupancy rate in 1982. The rate of 657 beds per 100,000 population was 12th-highest among the states. Hospital personnel included 11,646 registered nurses. The average cost of hospital care was $276 per day and $1,985 per stay, both figures being close to 20% below the US average. Alabama had 5,677 physicians, 2,985 registered pharmacists, 21,946 registered nurses, and 14,960 licensed practical nurses in 1984, and 1,489 professionally active dentists in 1982. The 1983 rate of 127 physicians per 100,000 population was 43d among the states.

³⁹SOCIAL WELFARE

Public welfare expenditures in Alabama are still low by national standards; but in 1983 public aid recipients numbered 7.1% of the population, 7th among the states.

Payments of approximately $72 million were made in 1982 to 155,100 recipients of aid to families with dependent children; the average monthly payment, $109 per family, was 3d lowest in the US. During 1982, $369 million was paid to 311,000 recipients under Medicaid, the state's fastest-growing welfare program. In 1982/83, food stamps were issued to 626,000 persons, at a total cost of $334 million. Alabama's participation in the national school lunch program in 1982/83 was 559,000 students, at a federal cost of $65 million.

During 1983, Social Security benefits exceeding $2.6 billion were paid to 643,000 Alabamians: $1.6 billion to 400,000 retired workers, $364 million to 86,000 disabled workers, and $672 million in survivors' benefits to 157,000 recipients. The average monthly payment to retired workers (excluding persons with special benefits) was $397, tied for 3d lowest among the 50 states. Federal Supplemental Security Income payments in 1983 totaled $241.5 million for 130,988 persons, including 95.8 million to the aged, with an average federal monthly payment of $132; $4.5 million to the blind, with an average federal monthly payment of $213; and $141.2 million to the disabled, for an average of $214 a month.

In 1982, workers' compensation payments totaled $132.4 million. Unemployment insurance benefits reached $280 million, with an average of 59,000 beneficiaries per week.

⁴⁰HOUSING

At the time of the 1980 census there were 1,467,374 housing units in Alabama, of which 1,450,011 were occupied. Seventy percent of these were owner-occupied, and 96% had full plumbing.

A total of 35,900 new privately owned units valued at more than $992 million were authorized between 1981 and 1983. During 1982/83, Alabama received $102 million in aid from the US Department of Housing and Urban Development for low-income housing and $57 million in HUD community development block grants.

The Fairhope Single Tax Corp., near Point Clear, was founded in 1893 by Iowans seeking to put into practice the economic theories of Henry George. Incorporated under Alabama law in 1904, this oldest and largest of US single-tax experiments continues to lease land in return for the payment of a rent (the "single tax") based on the land's valuation; the combined rents are used to pay taxes and to provide and improve community services.

[41]EDUCATION

In 1980, 56.5% of Alabamians 25 and older were high school graduates, the 9th-lowest rate in the nation. A quarter of adult Alabamians had no more than 8 years of grade school.

The total enrollment in Alabama's public schools in 1983/84 was 721,901. In 1982/83, 510,000 attended schools from prekindergarten through grade 8 and 214,000 attended high school. In fall 1980, estimated enrollment in nonpublic schools was 62,669. As of 1980, almost 40% of all minority public-school students were in schools with less than 50% minority-group enrollment; 24% were in schools with 99–100% minority enrollment. Despite George Wallace's opposition, court-ordered integration began early in Wallace's first term as governor (1963–67) at all state universities and large public-school systems.

As of 1983/84 there were 60 institutions of higher education in Alabama, 37 public and 23 private; 21 of the public ones were two-year institutions, many of them community colleges founded under Governor George Wallace. The major state universities are Auburn (founded in 1856), with a 1983/84 enrollment of 18,426, and the three campuses of the University of Alabama: the main campus near Tuscaloosa (1831), 15,497; Birmingham (1966), 14,679; and Huntsville (1950), 6,116. Tuskegee Institute, founded as a normal and industrial school in 1881 under the leadership of Booker T. Washington, became one of the nation's most famous black colleges; its 1983/84 enrollment was 3,400. Total enrollment in institutions of higher education was 167,753 in 1982/83.

[42]ARTS

The Alabama Council on the Arts and Humanities, established by the legislature in 1967, provides aid to local nonprofit arts organizations; there were 70 local arts councils in 1980.

A community arts development and residency program is financed by a state income tax check-off and private contributions. The Alabama Shakespeare Festival State Theater performs in Montgomery and also tours. The Birmingham Festival of Arts was founded in 1951, and the city's Alabama School of Fine Arts has been state-supported since 1971. Birmingham also has a professional symphony orchestra and an active chamber music society. Huntsville, Montgomery, and Tuscaloosa also have symphony orchestras.

Sacred Harp a cappella "sings" of old hymn tunes are held regularly. The Tennessee Valley Old Time Fiddlers Convention takes place in October at Athens State College. Every June, the annual Hank Williams Memorial Celebration is held near the country singer's birthplace at the Olive West Community.

[43]LIBRARIES AND MUSEUMS

As of 1982/83, Alabama had 31 county and multicounty regional libraries. Alabama public libraries had a combined total of 5,973,623 volumes in 1982/83, when the total circulation was 12,196,263. The Amelia Gayle Gorgas Library of the University of Alabama had 1,511,340 volumes; the Birmingham Public and Jefferson County Free Library had 19 branches and 1,028,406 volumes. The Alabama Department of History and Archives Library, at Montgomery, has about 250,000 volumes, with special collections on Alabama history and the Civil War. Collections on aviation and space exploration in Alabama's libraries, particularly its military libraries, may be the most extensive in the US outside of Washington, D.C. Memorabilia of Wernher von Braun are in the library at the Alabama Space and Rocket Center at Huntsville; the Redstone Arsenal's Scientific Information Center holds some 185,000 volumes and 1,400,000 technical reports.

Alabama had 59 museums in 1983. The most important art museum is the Birmingham Museum of Art. Other museums include the George Washington Carver Museum at Tuskegee Institute, the Women's Army Corps Museum and Military Police Corps Museum at Ft. McClellan, the US Army Aviation Museum at Ft. Rucker, the Pike Pioneer Museum at Troy, and the Museum of the City of Mobile. Russell Cave National Monument has an archaeological exhibit. In Florence is the W.C. Handy Home; at Tuscumbia, Helen Keller's birthplace, Ivy Green.

[44]COMMUNICATIONS

In 1980, 87% of Alabama's 1,341,856 occupied housing units had telephones. There were 607 post offices in 1985.

During 1983, Alabama had 241 operating radio stations (154 AM, 87 FM) and 33 television stations, of which the state Educational Television Commission operated 9. In 1984, 131 cable television systems served 518,481 subscribers.

[45]PRESS

The earliest newspaper in Alabama, the short-lived *Mobile Centinel* (sic), made its first appearance on 23 May 1811. The oldest newspaper still in existence in the state is the *Mobile Register,* founded in 1813.

As of 1984, Alabama had 13 morning dailies, with a combined circulation of 256,331; 14 evening dailies, with 480,548; and 19 Sunday papers, with 716,528. The following table shows the leading dailies with their 1984 circulations:

AREA	NAME	DAILY	SUNDAY
Birmingham	News (e,S)	165,212	207,891
	Post-Herald (m)	60,172	
Huntsville	Times (e,S)	55,028	66,232
Mobile	Register (m)	49,596	
	Press (e)	48,503	
	Press Register (S)		101,292
Montgomery	Advertiser (m)	49,529	
	Alabama Journal (e)	22,878	
	Alabama Journal and Advertiser (S)		83,858

[46]ORGANIZATIONS

The 1982 US Census of Service Industries counted 690 organizations in Alabama, including 159 business associations; 369 civic, social, and fraternal associations; and 13 educational, scientific, and research associations. National organizations with headquarters in Alabama include Civitan International (Birmingham); the National Speleological Society (Huntsville); and Klanwatch and the Southern Poverty Law Center, both located in Montgomery. The last-named was one of the major civil rights organizations active in Alabama in 1984, along with the Southern Christian Leadership Conference (SCLC) and the National Association for the Advancement of Colored People (NAACP). Two branches of the Ku Klux Klan were also active in Alabama.

[47]TOURISM, TRAVEL, AND RECREATION

A top tourist attraction is the Alabama Space and Rocket Center at Huntsville, home of the US Space Camp. Among the many antebellum houses and plantations to be seen in the state are Magnolia Grove (a state shrine) at Greensboro, Gaineswood and Bluff Hall at Demopolis, Arlington in Birmingham, Oakleigh at Mobile, Sturdivant Hall at Selma, and Shorter Mansion at Eufaula.

The celebration of Mardi Gras in Mobile, which began in 1704, predates that in New Orleans and now occupies several days before Ash Wednesday. Gulf beaches are a popular attraction, and Point Clear, across the bay from Mobile, has been a fashionable resort, especially for southerners, since the 1840s. The state fair is held at Birmingham every October.

During 1984, there were 1,123,849 visits to Alabama's four national park sites, which include Tuskegee Institute National Historic Site and Russell Cave National Monument, an almost continuous archaeological record of human habitation from at least 7000 BC to about AD 1650. During 1983/84, an estimated 6,200,000 tourists visited Alabama's 22 state parks covering a total of 48,027 acres (19,436 hectares). Tannehill Historical State Park features ante- and postbellum dwellings, a restored iron furnace over a century old, and a museum on iron and steel. Overall, travel and tourism generated some 72,000 jobs in 1984.

The Alabama Deep Sea Fishing Rodeo at Dauphin Island attracts thousands of visitors. In 1982/83, 288,427 hunting licenses and 589,053 fishing licenses were issued.

⁴⁸SPORTS

Although Alabama has no major league professional baseball teams, there is a minor league baseball club at Birmingham. Two major professional stock car races, the Winston 500 and Talladega 500, in May and July respectively, are held at Alabama International Motor Speedway in Talladega. Dog racing was legalized in Mobile in 1971. Four of the major hunting-dog competitions in the US are held annually in the state.

Football reigns supreme among collegiate sports, especially at the University of Alabama, a perennial top-10 entry. Competing in the Southeastern Conference, Alabama's Crimson Tide won the Sugar Bowl in 1962, 1964, 1967, 1978, 1979, and 1980, the Orange Bowl in 1943, 1953, 1963, and 1966, and the Cotton Bowl in 1942 and 1981. Alabama coach Paul "Bear" Bryant, who died in 1983, was named College Football Coach of the Year for 1961, 1971, and 1973. Auburn, which also competes in the Southeastern Conference, won the Sugar Bowl in 1984. The Blue-Gray game, an all-star contest, is held at Montgomery on Christmas Day, the Senior-South game is played in Mobile, and the All-American Bowl takes place at Birmingham.

Boat races include the Lake Eufaula Summer Spectacular Boat Race in August, and the Dixie Cup Regatta in Guntersville in July. The Alabama Sports Hall of Fame is located at Birmingham.

⁴⁹FAMOUS ALABAMIANS

William Rufus De Vane King (b.North Carolina, 1786–1853) served as a US senator from Alabama and as minister to France before being elected US vice president in 1852 on the Democratic ticket with Franklin Pierce; he died six weeks after taking the oath of office. Three Alabamians who served as associate justices of the US Supreme Court were John McKinley (b.Virginia, 1780–1852), John A. Campbell (b.Georgia, 1811–89), and Hugo L. Black (1886–1971). Campbell resigned from the court in 1861, later becoming assistant secretary of war for the Confederacy; Black, a US senator from 1927 to 1937, served one of the longest terms (1937–71) in the history of the court and is regarded as one of its most eminent justices.

Among the most colorful figures in antebellum Alabama was William Lowndes Yancey (b.Georgia, 1814–63), a fiery orator who was a militant proponent of slavery, states' rights, and eventually secession. During the early 20th century, a number of Alabamians became influential in national politics. Among them were US senators John Hollis Bankhead (1842–1920) and John Hollis Bankhead, Jr. (1872–1946); the latter's brother, William B. Bankhead (1874–1940), who became speaker of the US House of Representatives in 1936; and US Senator Oscar W. Underwood (b.Kentucky, 1862–1929), a leading contender for the Democratic presidential nomination in 1912 and 1924. Other prominent US senators from Alabama have included (Joseph) Lister Hill (1894–1984) and John Sparkman (1899–1985), who was the Democratic vice-presidential nominee in 1952. Alabama's most widely known political figure is George Corley Wallace (b.1919), who served as governor 1963–67 and 1971–79, and was elected to a fourth term in 1982. Wallace, an outspoken opponent of racial desegregation in the 1960s, was a candidate for the Democratic presidential nomination in 1964; four years later, as the presidential nominee of the American Independent Party, he carried five states. While campaigning in Maryland's Democratic presidential primary on 15 May 1972, Wallace was shot and paralyzed from the waist down by a would-be assassin.

Civil rights leader Martin Luther King, Jr. (b.Georgia, 1929–68), winner of the Nobel Peace Prize in 1964, first came to national prominence as leader of the Montgomery bus boycott of 1955; he also led demonstrations at Birmingham in 1963 and at Selma in 1965. His widow, Coretta Scott King (b.1927) is a native Alabamian. Federal judge Frank M. Johnson, Jr. (b.1918), has made several landmark rulings in civil rights cases.

Helen Keller (1880–1968), deaf and blind as the result of a childhood illness, was the first such multihandicapped person to earn a college degree; she later became a world famous author and lecturer. Another world figure, black educator Booker T. Washington (b.Virginia, 1856–1915), built Alabama's Tuskegee Institute from a school where young blacks were taught building, farming, cooking, brickmaking, dressmaking, and other trades into an internationally known agricultural research center. Tuskegee's most famous faculty member was George Washington Carver (b.Missouri, 1864–1943), who discovered some 300 different peanut products, 118 new ways to use sweet potatoes, and numerous other crop varieties and applications. Among Alabama's leaders in medicine was Dr. William Crawford Gorgas (1854–1920), head of sanitation in Panama during the construction of the Panama Canal; he later served as US surgeon general. Brought to the US after World War II, the internationally known scientist Wernher von Braun (b.Germany, 1912–77) came to Alabama in 1950 to direct the US missile program.

Two Alabama writers, (Nelle) Harper Lee (b.1926) and Edward Osborne Wilson (b.1929), have won Pulitzer Prizes. Famous musicians from Alabama include blues composer and performer W(illiam) C(hristopher) Handy (1873–1958), singer Nat "King" Cole (1917–65), and singer-songwriter Hank Williams (1923–53). Alabama's most widely known actress was Tallulah Bankhead (1903–68), the daughter of William B. Bankhead.

Among Alabama's sports figures are track and field star Jesse Owens (James Cleveland Owens, 1913–80), winner of four gold medals at the 1936 Olympic Games in Berlin; boxer Joe Louis (Joseph Louis Barrow, 1914–81), world heavyweight champion from 1937 to 1949; and baseball stars Leroy Robert "Satchel" Paige (1906? –82), Willie Mays (b.1931), and (Louis) Henry Aaron (b.1934), all-time US home-run leader.

⁵⁰BIBLIOGRAPHY

Agee, James, and Walker Evans. *Let us Now Praise Famous Men.* New York: Ballantine, 1966.

Barnard, William D. *Dixiecrats and Democrats: Alabama Politics, 1942-1950.* University, Ala.: University of Alabama Press, 1974.

Carter, Dan T. *Scottsboro: A Tragedy of the American South.* Baton Rouge: Louisiana State University Press, 1969.

Federal Writers' Project. *Alabama: A Guide to the Deep South.* New York: Somerset, n.d. (orig. 1941).

Harlan, Louis R. *Booker T. Washington: The Making of a Black Leader, 1856-1901.* New York: Oxford University Press, 1972.

Marks, Henry S. and Marsha. *Alabama Past Leaders.* Huntsville, Ala.: Strode, 1981.

Martin, David L. *Alabama's State and Local Governments.* University, Ala.: University of Alabama Press, 1985.

Rogers, William Warren, and Robert David Ward. *August Reckoning: Jack Turner and Racism in Post–Civil War Alabama.* Baton Rouge: Louisiana State University Press, 1973.

Rosengarten, Theodore. *All God's Dangers: The Life of Nate Shaw.* New York: Knopf, 1974.

Van de Veer Hamilton, Virginia. *Hugo Black: The Alabama Years.* Baton Rouge: Louisiana State University Press, 1972.

ALASKA

State of Alaska

ORIGIN OF STATE NAME: From the Aleut word *alakshak,* meaning "peninsula" or "mainland." **CAPITAL:** Juneau. **ENTERED UNION:** 3 January 1959 (49th). **SONG:** "Alaska's Flag." **MOTTO:** North to the Future. **FLAG:** On a blue field, eight gold stars form the Big Dipper and the North Star. **OFFICIAL SEAL:** In the inner circle, symbols of mining, agriculture, and commerce are depicted against a background of mountains and the northern lights. In the outer circle are a fur seal, a salmon, and the words "The Seal of the State of Alaska." **BIRD:** Willow ptarmigan. **FISH:** King salmon. **FLOWER:** Wild forget-me-not. **TREE:** Sitka spruce. **GEM:** Jade. **MINERAL:** Gold. **SPORT:** Dogteam racing (mushing). **LEGAL HOLIDAYS:** New Year's Day, 1 January; Birthday of Martin Luther King, Jr., 3d Monday in January; Lincoln's Birthday, 12 February; Washington's Birthday, 3d Monday in February; Seward's Day, last Monday in March; Memorial Day, last Monday in May; Independence Day, 4 July; Labor Day, 1st Monday in September; Alaska Day, 18 October; Veterans Day, 11 November; Thanksgiving Day, 4th Thursday in November; Christmas Day, 25 December. **TIME:** noon GMT = 3 AM Alaska Standard Time, 2 AM Hawaii-Aleutian Standard Time.

¹LOCATION, SIZE, AND EXTENT

Situated at the northwest corner of the North American continent, Alaska is separated by Canadian territory from the coterminous 48 states. Alaska is the largest of the 50 states, with a total area of 591,004 sq mi (1,530,699 sq km). Land takes up 570,833 sq mi (1,478,456 sq km) and inland water 20,171 sq mi (52,243 sq km). Alaska is more than twice the size of Texas, the next-largest state, and occupies 16% of the total US land area; The E–W extension is 2,261 mi (3,639 km); the maximum N–S extension is 1,420 mi (2,285 km).

Alaska is bounded on the N by the Arctic Ocean and Beaufort Sea; on the E by Canada's Yukon Territory and province of British Columbia; on the S by the Gulf of Alaska, Pacific Ocean, and Bering Sea; and on the W by the Bering Sea, Bering Strait, Chukchi Sea, and Arctic Ocean.

Alaska's many offshore islands include St. Lawrence, St. Matthew, Nunivak, and the Pribilof group in the Bering Sea; Kodiak Island in the Gulf of Alaska; the Aleutian Islands in the Pacific; and some 1,100 islands constituting the Alexander Archipelago, extending SE along the Alaska panhandle.

The total boundary length of Alaska is 8,187 mi (13,176 km), including a general coastline of 6,640 mi (10,686 km); the tidal shoreline extends 33,904 mi (54,563 km). Alaska's geographic center is about 60 mi (97 km) NW of Mt. McKinley. The northernmost point in the US—Point Barrow, at 71°23′30″N, 156°28′30″ W—lies within the state of Alaska, as does the westernmost point—Cape Wrangell on Attu Island in the Aleutians, at 52°55′30″N, 172°28′E. Little Diomede Island, belonging to Alaska, is less than 2 mi (3 km) from Big Diomede Island, belonging to the Soviet Union.

²TOPOGRAPHY

Topography varies sharply among the six distinct regions of Alaska. In the southeast is a narrow coastal panhandle cut off from the main Alaskan landmass by the St. Elias Range. This region, featuring numerous mountain peaks of 10,000 feet (3,000 meters) in elevation, is paralleled by the Alexander Archipelago. South-central Alaska, which covers a 700-mi (1,100-km) arc along the Gulf of Alaska, includes the Kenai Peninsula and Cook Inlet, a great arm of the Pacific penetrating some 200 mi (320 km) to Anchorage. The southwestern region includes the Alaska Peninsula, filled with lightly wooded, rugged peaks, and the 1,700-mi (2,700-km) sweep of the Aleutian islands, barren masses of volcanic origin. Western Alaska extends from Bristol Bay to the Seward Peninsula, an immense tundra dotted with lakes and containing the deltas of the Yukon and Kuskokwim rivers, the longest in the state at 1,900 mi (3,058 km) and 680 mi (1,094 km), respectively. Interior Alaska extends north of the Alaska Range and south of the Brooks Range, including most of the drainage of the Yukon and its major tributaries, the Tanana and Porcupine rivers. The Arctic region extends from Kotzebue, north of the Seward Peninsula, east to Canada. From the northern slopes of the Brooks Range, the elevation falls to the Arctic Ocean.

The 11 highest mountains in the US—including the highest in North America, Mt. McKinley (20,320 feet—6,194 meters), located in the Alaska Range—are in the state, which also contains half the world's glaciers; the largest, Malaspina, covers more area than the entire state of Rhode Island. Ice fields cover 4% of the state. Alaska has more than 3 million lakes larger than 20 acres (8 hectares), and more than one-fourth of all the inland water wholly within the US lies inside the state's borders. The largest lake is Iliamna, occupying about 1,000 sq mi (2,600 sq km).

The most powerful earthquake in US recorded history, measuring 8.5 on the Richter scale, struck the Anchorage region on 27 March 1964, resulting in 114 deaths and $500 million in property damage in Alaska and along the US west coast.

³CLIMATE

Americans who called Alaska "Seward's icebox" when it was first purchased from the Russians were unaware of the variety of climatic conditions within its six topographic regions. Although minimum daily winter temperatures in the Arctic region and in the Brooks Range average −20°F (−29°C) and the ground at Point Barrow is frozen permanently to 1,330 feet (405 meters), summer maximum daily temperatures in the Alaskan lowlands average above 60°F (16°C) and have been known to exceed 90°F (32°C). The southeastern region is moderate, ranging from a daily average of 30°F (−1°C) in January to 56°F (13°C) in July; the south-central zone has a similar summer range, but winters are somewhat harsher, especially in the interior. The Aleutian Islands have chilly, damp winters and rainy, foggy weather for most of the year; western Alaska is also rainy and cool. The all-time high for the state was 100°F (38°C), recorded at Ft. Yukon on 27 June 1915; the reading of −79.8°F (−62°C) registered at Prospect Creek Camp, in the northwestern part of the state, on 23 January 1971 is the lowest temperature ever officially recorded in the US.

Juneau receives an average of 53 in (135 cm) of precipitation each year. The average annual snowfall there is 104 in (264 cm).

⁴FLORA AND FAUNA

Life zones in Alaska range from grasslands, mountains, and tundra to thick forests, in which Sitka spruce (the state tree), western hemlock, tamarack, white birch, and western red cedar predominate. Various hardy plants and wild flowers spring up during the short growing season on the semiarid tundra plains. Species of poppy and gentian are endangered.

Mammals abound amid the wilderness. Great herds of caribou, reindeer, elk, and moose migrate across the state. Kodiak, polar, black, and grizzly bears, Dall sheep, and an abundance of small mammals are also found. The sea otter and musk ox have been successfully reintroduced. Round Island, along the north shore of Bristol Bay, has the world's largest walrus rookery. North America's largest population of bald eagles nests in Alaska, and whales migrate annually to the icy bays. Pristine lakes and streams are famous for trout and salmon fishing. In all, 386 species of birds, 430 fishes, 105 mammals, 7 amphibians, and 3 reptiles have been found in the state. Endangered species include the Eskimo curlew, American and Arctic peregrine falcons, Aleutian Canada goose, and short-tailed albatross; numerous species considered endangered in the conterminous US remain common in Alaska.

⁵ENVIRONMENTAL PROTECTION

Land-use planning is the liveliest controversy in Alaska today, pitting developers against those who consider the state the last unspoiled US region. When Alaska became a state, the federal government owned 99.8% of the land. Under the terms of the Statehood Act of 1959, the state was authorized to select 103 million acres (42 million hectares), or 28% of the total, for its own use. By 1968, the state had chosen 26 million acres (11 million hectares), including the oil-rich Prudhoe Bay area, when all claims were frozen pending action on land claims of Alaskan natives. The resulting federal Native Claims Settlement Act of 1971 not only gave Alaska's Native Americans the right to choose 44 million acres (18 million hectares) but also required the US secretary of the interior to withdraw up to 80 million acres (32 million hectares) of "National Interest Lands" from public use and gave Congress until 1978 to decide how much of this land would be protected from development. When Congress could not reach a decision, Interior Secretary Cecil Andrus, citing a different law, withdrew 114 million acres (46 million hectares) from public use. The Alaska Lands Act of 1980 imposed development restrictions on 159 million acres (64 million hectares) and superseded Andrus's decision. In 1982, the federal government owned almost 77% of Alaskan land, the state government 17%, and native groups 6%; less than 1% of the land was privately held.

Both the state and federal governments have opened the shorelines of the Gulf of Alaska and the Bering and Beaufort seas to oil leasing and development, under strict environmental standards.

⁶POPULATION

Alaska, with a land area one-fifth the size of the coterminous US, ranked 50th in population in 1980, with a census figure of 401,851. Regions of settlement and development constitute less than 0.001% of Alaska's total land area. The population density was less than 1 person per square mile in 1985, when Alaska had an estimated 514,819 people and was closing in on Wyoming's and Vermont's population totals. The increase of 28% from mid-1980 to January 1985 made Alaska the nation's fastest growing state during the period, and as a result, the Census Bureau's estimate of 522,100 for 1990, made before 1985, seems certain to be surpassed.

Historically, population shifts in Alaska have directly reflected economic and political changes. The Alaska gold rush of the 1890s resulted in a population boom from 32,052 in 1890 to 63,592 a decade later; by the 1920s, however, when mining had declined, Alaska's population had decreased to 55,036. The

region's importance to US national defense during the 1940s led to a rise in population from 72,524 to 128,643 during that decade. Oil development, especially the construction of the Alaska pipeline, brought a 78% population increase between 1960 and 1980. Almost all of this gain was from migration; as of 1980, 26% of all state residents had moved there within the past five years. The state's population is much younger than that of the nation as a whole (the median age was 26.1 in 1980, lower than that in any other state but Utah), and only 2.8% of all Alaskans were 65 years of age or older in 1980—by far the lowest such percentage in any state. Alaska is also one of the few states where men outnumber women, 52.8% to 47.2% in 1980.

Alaska's population was 64% urban in 1980. About half of all state residents live in and around Anchorage, whose population was estimated at 226,700 in mid-1984. At the time of the 1980 census Anchorage had a population of 174,431. Other leading cities in 1980 were Fairbanks, 22,645, and Juneau, 19,528. Only about 10% of the population lives in Western Alaska.

⁷ETHNIC GROUPS

Indians—primarily Athapaskan, Tlingit, Haida, and Tsimshian living along the southern coast—numbered 22,631 in 1980. Eskimos (33,817) and Aleuts (7,909), the other native peoples, live mostly in scattered villages to the north and northwest. Taken together, Alaskan natives numbered 64,357 in 1980, 16% of the population. The Native Claims Settlement Act of 1971 gave 13 native corporations nearly $1 billion in compensation for exploration, mining, and drilling rights, and awarded them royalties on oil and the rights to nearly 12% of Alaska's land area.

In 1980, blacks numbered 13,643, or 3.4% of the population. Among those of Asian and Pacific Islands origin were 3,092 Filipinos, 1,595 Japanese, and 1,536 Koreans. Hispanics totaled 5,048, including 2,253 of Mexican origin. Foreign-born persons numbered 16,216 in 1980, 4% of the population.

⁸LANGUAGES

From the Tlingit, Haida, and Tsimshian groups of lower Alaska almost no language influence has been felt, save for *hooch* (from Tlingit *hoochino*); but of the native words known to the sourdough, some have escaped into general usage, notably Eskimo *mukluk* and Aleut *parka*. Native place-names abound: Skagway and Ketchikan (Tlingit), Kodiak and Katmai (Eskimo), and Alaska and Akutan (Aleut).

In 1980, 87.5% of the population 3 years old and older was reported to speak only English in the home. Other major languages spoken in the home, and the number of people speaking them, included Spanish, 5,454, and German, 2,828.

⁹RELIGIONS

The largest religious organization in the state is the Roman Catholic Church, which had 46,230 members in 1984. Southern Baptists constituted the largest Protestant denomination, with 17,088 adherents in 1980. Other major groups were the Latter-day Saints (Mormons), 8,802; Assembly of God, 6,506; United Presbyterians, 5,711; and Episcopalians, 5,178. The very small Jewish population numbered 960 in 1984.

Many Aleuts were converted to the Russian Orthodox religion during the 18th century, and small Russian Orthodox congregations are still active on the Aleutian Islands, in Kodiak and southeastern Alaska, and along the Yukon River.

¹⁰TRANSPORTATION

Alaska had no rail service until 1923, when the 495-mi (797-km) Alaska Railroad linked Seward, Anchorage, and Fairbanks. This railroad, subsequently extended to 538 mi (866 km), is still the only one in the state and is not connected to any other North American rail line; it was federally operated until 1985, when it was bought by the state government for $22.3 million. The line, which makes one northbound and one southbound run a day—when weather permits—carried 161,068 passengers in 1980/81.

The Alaska Highway, which extends 1,523 mi (2,451 km) from

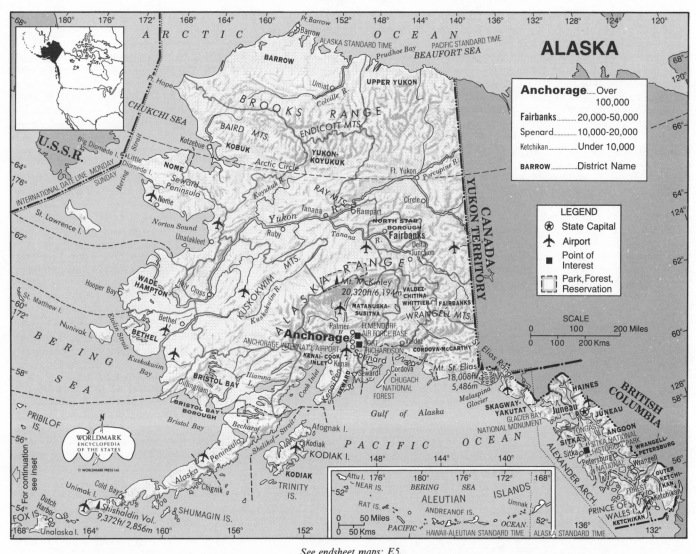

ALASKA

Anchorage	Over 100,000
Fairbanks	20,000-50,000
Spenard	10,000-20,000
Ketchikan	Under 10,000
BARROW	District Name

LEGEND
⊛ State Capital
✈ Airport
■ Point of Interest
⸬ Park, Forest, Reservation

See endsheet maps: E5.
LOCATION: 51°13′05″ to 71°23′30″N; 129°59′W to 172°28′E. **BOUNDARIES:** Arctic Ocean and Pacific Ocean (Gulf of Alaska/
Bering Sea) coastlines, 6,640 mi (10,686 km); Canadian line, 1,538 mi (2,475 km).

Dawson Creek, British Columbia, to Fairbanks, is the only road link with the rest of the US. In-state roads are few and far between; although Fairbanks, Anchorage, and Seward are linked, Juneau is isolated. Only 9,864 mi (15,875 km) of roads were in use as of 1983, including 2,223 mi (3,578 km) of roads in national parks and forests. During the same year, the state had 350,488 registered vehicles and 288,739 licensed drivers. The largest public transit system, that of Anchorage, accommodated 3.5 million unlinked passenger trips in 1984. A state-operated ferry system connects the Alaskan ports to Seattle. The state's major ports, Anchorage and Ketchikan, handled 1,998,185 and 1,473,009 tons of cargo, respectively, in 1982. Valdez, the terminus of the Trans-Alaska Pipeline, handled 90.1 million tons, almost all crude oil. It was the 4th-busiest US port.

Air travel is the primary means of trans-state transportation, with several bush carriers serving the remote communities. Anchorage International Airport, the state's largest, enplaned 1,041,790 passengers and 224,585 tons of freight in 1983, when Alaska had a total of 615 airfields.

11HISTORY

At some time between 10,000 and 40,000 years ago, the ancestors of all of America's aboriginal peoples trekked over a land bridge that connected northeastern Siberia with northwestern America.

These early hunter-gatherers dispersed, eventually becoming three distinct groups: Aleut, Eskimo, and Indian.

Ages passed before overseas voyagers rediscovered Alaska. Separate Russian parties led by Aleksei Chirikov and Vitus Bering (who had sailed in 1728 through the strait that now bears his name) landed in Alaska in 1741. Within a few years, the discoverers were followed by the exploiters, who hunted the region's furbearing animals. In 1784, the first permanent Russian settlement was established on Kodiak Island; 15 years later, the Russian American Company was granted a monopoly over the region. Its manager, Aleksandr Baranov, established Sitka as the company's headquarters. In 1802, the Tlingit Indians captured Sitka, but two years later lost the town and the war with the Russian colonizers. Fluctuations in the fur trade, depletion of the sea otter, and the Russians' inability to make their settlements self-sustaining limited their development of the region. Increasingly, the czarist government viewed the colonies as a drain on the treasury. In 1867, as a result of the persistence of Secretary of State William H. Seward, a devoted American expansionist, Russia agreed to sell its American territories to the US for $7,200,000. From 1867 until the first Organic Act of 1884, which provided for a federally appointed governor, Alaska was administered first by the US Army, then by the US Customs Service.

The pace of economic development quickened after the discovery of gold in 1880 at Juneau. Prospectors began moving into the eastern interior after this success, leading to gold strikes on Forty Mile River in 1886 and at Circle in 1893. But it was the major strike in Canada's Klondike region in 1896 that sparked a mass stampede to the Yukon Valley and other regions of Alaska, including the Arctic. The gold rush led to the establishment of permanent towns in the interior for the first time.

Subsequent development of the fishing and timber industries increased Alaska's prosperity and prospects, although the region suffered from a lack of transportation facilities. A significant achievement came in 1914, when construction started on the Alaska Railroad connecting Seward, a new town with an ice-free port, with Anchorage and Fairbanks. Politically there were advances as well. In 1906, Alaskans were allowed to elect a nonvoting delegate to Congress for the first time. Congress granted territorial status to the region in 1912, and the first statehood bill was introduced in Congress four years later.

Mineral production declined sharply after 1914. Population declined too, and conditions remained depressed through the 1920s, although gold mining was helped by a rise in gold prices in 1934. World War II provided the next great economic impetus for Alaska: the Aleutian campaign following the Japanese invasion of the islands, though not as pivotal as the combat in other areas of the Pacific, did show American policymakers that Alaska's geography was in itself an important resource. Thus the spurt of federal construction and movement of military personnel continued even after the war ended, this time directed at the Soviet Union—only 40 mi (64 km) across the Bering Strait—rather than Japan. The US government built the Alaska Highway and many other facilities, including docks, airfields, and an extension of the Alaska Railroad. Population soared as thousands of civilian workers and military personnel moved to the territory. The newcomers added impetus to a new movement for statehood, and the Alaska Statehood Act was adopted by Congress in June 1958 and ratified by Alaska voters that August. On 3 January 1959, President Dwight Eisenhower signed the proclamation that made Alaska the 49th state.

In 1971, the Native Claims Settlement Act provided an extensive grant to the state's natives but also precipitated a long federal-state controversy over land allocations. A major oil field was discovered in 1968, and in 1974, over the opposition of many environmentalists, construction began on the 789-mi (1,270-km) Trans-Alaska Pipeline from Prudhoe Bay to Valdez. The oil that began flowing through the pipeline in 1977 made Alaska almost immediately one of the nation's leading energy producers.

Alaska's extraordinary oil wealth enabled it to embark on a heavy program of state services and to abolish the state income tax. However, state spending failed to stimulate the private sector to the degree expected. With oil prices declining and oil production forecast to decline, Alaskans faced hard choices in the mid-1980s—cutbacks in spending, increased taxes, or both.

[12]STATE GOVERNMENT

Under Alaska's first and only constitution—adopted in 1956, effective since the time of statehood, and amended 19 times by the end of 1983—the house of representatives consists of 40 members elected for two-year terms; the senate has 20 members elected for staggered four-year terms. The minimum age is 21 for a representative, 25 for a senator; legislators must have resided in the state for at least three years before election and in the district at least one year.

Alaska's executive branch, modeled after New Jersey's, features a strong governor who appoints all cabinet officers (except the comissioner of education) and judges subject to legislative confirmation. The lieutenant governor is the only other elected executive. The governor must be at least 30 years of age, and must have been a US citizen for seven years and an Alaska resident for

seven years. The term of office is four years, and the governor is limited to two consecutive terms. The qualifications for the lieutenant governor are the same as for the governor.

After a bill has been passed by the legislature, it becomes law if signed by the governor; if left unsigned for 15 days (Sundays excluded) while the legislature is in session or for 20 days after it has adjourned; or if passed by a two-thirds vote of the combined houses over a gubernatorial veto (to override a veto of an appropriations bill requires a three-fourths vote). Constitutional amendments require a two-thirds vote of the legislature and ratification by the electorate.

Any US citizen at least 18 years of age who has been a resident of a voting district for 30 days may register to vote in that district.

[13]POLITICAL PARTIES

When Congress debated the statehood question in the 1950's, it was assumed that Alaska would be solidly Democratic, but this expectation has not been borne out: as of 1982, of 251,536 registered voters, only 26% were Democrats, while 17% were Republican, 3% belonged to other parties, and 54% were nonpartisan. In 1985, while the Democrats held the offices of governor and lieutenant governor and 21 of the 40 seats in the state house of representatives, the Republicans held both US Senate seats, Alaska's lone House seat, and 11 out of 20 seats in the state senate. In presidential elections since 1968, Alaskans have voted Republican five consecutive times; they gave Ronald Reagan 62% of the vote in 1980 and 67% in 1984.

William Sheffield, a Democrat, was elected governor in 1982, ending eight years of Republican control of the statehouse. In 1984, US Senator Ted Stevens, a Republican, was reelected, as was US Representative Donald E. Young, also a Republican.

In 1985 there were 11 female state legislators and a black member of the state house of representatives. Three members of the legislature were at least partly of Native American extraction in 1983.

Alaska Presidential Vote by Major Political Parties, 1960–84

YEAR	ELECTORAL VOTE	ALASKA WINNER	DEMOCRAT	REPUBLICAN
1960	3	Nixon (R)	29,809	30,953
1964	3	*Johnson (D)	44,329	22,930
1968	3	*Nixon (R)	35,411	37,600
1972	3	*Nixon (R)	32,967	55,349
1976	3	Ford (R)	44,058	71,555
1980	3	*Reagan (R)	41,842	86,112
1984	3	*Reagan (R)	62,007	138,377

*Won US presidential election.

[14]LOCAL GOVERNMENT

Unlike most other states, Alaska has no counties. Instead, the needs of its small, scattered population were met in 1983 by 11 boroughs (covering 40% of the state) governed by elected assemblies; the rest of the state was considered an unorganized borough. As of 1983 there were 44 cities, most of them governed by elected mayors and councils. Juneau, Sitka, and Anchorage, known as Alaska's three unified municipalities, have consolidated city and borough functions.

[15]STATE SERVICES

By law, Alaska's government may contain no more than 20 administrative departments. As of early 1983 there were 14: Administration, Commerce and Economic Development, Community and Regional Affairs, Education, Environmental Conservation, Fish and Game, Health and Social Services, Labor, Law, Military Affairs, Natural Resources, Public Safety, Revenue, and

Transportation and Public Facilities. In addition, the state has an ombudsman with limited powers to investigate citizen complaints against state agencies.

[16]JUDICIAL SYSTEM

The supreme court, consisting of a chief justice and 4 associate justices, hears appeals for civil matters from the superior court, whose 26 judges in 1983 were organized among the four state judicial districts, and for criminal matters from the 3-member court of appeals. The superior court has original jurisdiction in all civil and criminal matters, and it hears appeals from the district court. The lowest court is the district court, composed in 1983 of 17 judges in four districts. All judges are appointed by the governor from nominations made by the Judicial Council, but are thereafter subject to voter approval; supreme court justices serve terms of 10 years, court of appeals and superior court judges 6, and district judges 4.

In 1983, according to the FBI Crime Index, Alaska had the nation's highest rate of forcible rape and the 3d-highest rate for murder. Overall, the rate was 6,019 crimes per 100,000 population; murder and nonnegligent manslaughter, 13.8; forcible rape, 101.5. Alaska has no capital punishment statute. There were 1,195 inmates of state and federal prisons at the end of 1984.

[17]ARMED FORCES

A huge buildup of military personnel occurred after Wold War II, as the cold war with the Soviet Union led the US to establish the Distant Early Warning (DEW) System, Ballistic Missile Early Warning System, and Joint Surveillance System in the area. Later years saw a cutback in personnel, however, from a high of 40,214 in 1962 to 20,982 in 1984, 11,032 of them in the Air Force. Anchorage is the home of both the largest Army base, Ft. Richardson, and the largest Air Force base, Elmendorf. In the Aleutians are several Navy facilities and the Shemya Air Force Base. Alaska firms received defense contracts worth $437 million in 1983/84.

About 50,000 veterans were living in Alaska as of 30 September 1983, of whom fewer than 500 served in World War I, 11,000 in World War II, 9,000 during the Korean conflict, and 23,000 during the Viet-Nam era. Expenditures on veterans amounted to $44 million in 1982/83.

The Alaska State Troopers provide police protection throughout the state, except in the larger cities, where municipal police forces have jurisdiction. There were 857 full-time state police employees in 1983, of whom 424 were officers. In all, police protection employees numbered 1,289 in October 1983. Army National Guard personnel numbered 2,520 at the end of 1984, and Air National Guard strength was 802 in February 1985.

[18]MIGRATION

The earliest immigrants to North America, more than 10,000 years ago, likely came to Alaska via a land bridge across what is now the Bering Strait. The Russian fur traders who arrived during the 1700s found Aleuts, Eskimos, and Indians already established there. Despite more than a century of Russian sovereignty over the area, however, few Russians came, and those that did returned to the mother country with the purchase of Alaska by the US in 1867.

Virtually all other migration to Alaska has been from the continental US—first during the gold rush of the late 19th century, most recently during the oil boom of the 1970s. Between 1970 and 1983, Alaska's net gain from migration was 78,000.

Mobility is a way of life in Alaska. In 1980, only 32% of those 5 years or older were living in the same house as in 1975, and 29% had been living in another state in 1975. Only 32% of the population was born in Alaska, a lower percentage than any other state but Florida and Nevada.

[19]INTERGOVERNMENTAL COOPERATION

Alaska participates with Washington, Oregon, and Idaho in the Pacific Marine Fisheries Compact. Alaska also belongs to other western regional agreements covering energy, corrections, and education. The most important federal-state effort, the Joint Federal-State Land Use Planning Commission, was involved with the Alaska lands controversy throughout the 1970s.

Federal aid to Alaska was $615.7 million in 1983/84.

[20]ECONOMY

When Alaska gained statehood in 1959, its economy was almost totally dependent on the US government. Fisheries, limited mining (mostly gold and gravel), and some lumber production made up the balance. That all changed with development of the petroleum industry during the 1970s. Construction of the Trans-Alaska Pipeline brought a massive infusion of money and people into the state. Construction, trade, and services boomed—only to decline when the pipeline was completed.

In the mid-1980s, the economy was heavily dependent on government spending, especially by the state, and on the oil industry, which by 1984 supplied 85% of state revenues. Of crucial importance to Alaska's future was the ongoing effort to create a more viable private sector.

[21]INCOME

Alaska boasts the highest per capita income in the US: $16,820 in 1983. Measured in constant dollars, Alaska's total personal income increased 137% between 1970 and 1983, compared to 47% for the nation as a whole; per capita income increased 54%, compared to 28% for the nation. Government accounted for over 42% of earnings; services 12%; trade 11%; contract construction 11%; transportation, communications, and public utilities 10%; manufacturing 7%; finance, insurance and real estate 3%; mining 3%; and agriculture, forestry, and fisheries 1%.

Living costs are high: in 1981, Anchorage's typical family living costs were 25% above the US urban average, and costs in some other Alaskan cities were much higher. A total of 10.7% of all Alaskans were living below the federal poverty level in 1979. In 1982, 3,700 Alaskans were among the top wealth-holders in the nation, with gross assets greater than $500,000.

[22]LABOR

Following completion of the Trans-Alaska Pipeline in 1977, the state entered a period of high unemployment that lasted through the end of the decade. Of the 180,000 in the civilian labor force in 1978, 20,000 were unemployed. The rate of 11.1% was the highest in the US, and the rate for Alaskan natives was much higher. In 1984, the unemployment rate for the civilian labor force of 245,000 was 10%.

A federal survey in March 1982 revealed the following nonfarm employment pattern in Alaska:

	ESTABLISH-MENTS	EMPLOYEES	ANNUAL PAYROLL ('000)
Agricultural services, forestry, fishing	103	603	$ 16,533
Mining, of which:	159	11,541	478,114
Oil and gas extraction	(101)	(9,634)	(380,477)
Contract construction	1,445	12,105	548,017
Manufacturing, of which:	383	8,800	268,537
Food and kindred products	(104)	(3,494)	(86,357)
Lumber and wood products	(65)	(1,221)	(53,644)
Transportation, public utilities	704	16,528	560,939
Wholesale trade	645	7,336	206,740
Retail trade	2,595	28,565	441,336
Finance, insurance, real estate	787	8,373	187,394
Services	3,040	29,857	622,889
Other	518	585	11,395
TOTALS	10,379	124,293	$3,341,894

Government employees, not included in this survey, numbered about 14,000 federal workers in 1982 and the equivalent of 39,000 full-time state and local government workers in 1983; the 815 per

10,000 population full-time state and local government employment far exceeded that of any other state.

Almost 34% of all nonagricultural employees belonged to labor unions in 1980, the 6th-highest percentage in the US. The International Brotherhood of Teamsters is especially strong in the state, covering a range of workers from truck drivers to school administrators. In 1983 there were 118 labor unions in Alaska, fewer than in any other state. Wage rates—averaging $10.54 per hour for production workers in August 1984—were the 3d highest in the US, as they must be to compensate for Alaska's isolation and high living costs.

23AGRICULTURE

Hampered by a short growing season and frequent frosts, Alaska has very limited commercial agriculture. Farm income in 1983 was only $19.8 million, the lowest of all states. Hay, barley, greenhouse and nursery items, and potatoes are the main commodities produced. In 1984 there were about 650 farms and 1,560,000 acres (630,000 hectares) in farms; the leading farming region is the Matanuska Valley, northeast of Anchorage.

24ANIMAL HUSBANDRY

Dairy and livestock products account for about 40% of Alaska's agricultural income. In 1983, an estimated 13.5 million lb of milk, valued at $3 million, were produced by 1,000 milk cows. Meat and poultry production is negligible by national standards.

25FISHING

Alaska was the number-one fishing state in terms of earnings in 1984, and was 2d to Louisiana in the total weight of catch. The salmon catch, the staple of the industry, amounted to 132.5 million fish and over $300 million in 1984, the largest catch in history. Crab, formerly a major export item, has declined in availability, as has shrimp; the shellfish harvest was under 100 million lb in 1984, compared to a peak of nearly 370 million lb in 1980. In all, Alaska's commercial catch in 1984 totaled 1 billion lb, valued at $509.3 million. In that year, Kodiak ranked 3d among US fishing ports in value of catch, with $69.9 million.

Sports fishermen are attracted by Alaska's abundant stocks of salmon and trout.

26FORESTRY

Alaska's timber resources are vast, but full-scale development of the industry awaits fundamental land-use decisions at the federal and state levels, and higher demand for logs, lumber, and wood products, especially in Japan and South Korea, Alaska's two major markets. In 1979, Alaska had 119.1 million acres (48.2 million hectares) of forestland, 16% of the US total. Only 11.2 million acres (4.5 million hectares), or 2% of the US total, were classified as commercial timberland, and three-fourths of that was federally controlled. Alaska contains the nation's two largest national forests, Tongass in the southeast (17.4 million acres—7 million hectares) and Chugach along the Gulf Coast (6.6 million acres—2.7 million hectares).

Lumbering and related industries employed about 1,600 workers in 1982. In 1973, the fledgling industry reached a peak production of 678.8 million board feet, but the closing of some areas to production, the vagaries of the foreign market, and increasing competition from Canada led to a decline in the timber harvest to less than 250 million board feet in 1984. Shipments of lumber and wood products in 1982 were valued at $250.1 million.

27MINING

Alaska's nonfuel mining industry is very limited; 1984 production was valued at only $98.7 million, 41st among the states. Production of sand and gravel reached 42,400 tons in 1984, valued at $76.6 million, and crushed stone production was 2,300,000 tons, valued at $10.8 million. Gold output of 34,702 troy oz had a value of $14.7 million in 1983. Antimony, cement, gemstones, silver, and tin were also mined. The world's largest zinc/lead deposit is near Kotzebue, and an enormous molybdenum deposit is near Ketchikan.

28ENERGY AND POWER

As of 1983, Alaskan production of crude oil was 20% of the nation's total, and 2d only to that of Texas. Of the 625.5 million barrels produced, 96% came from the vast North Shore fields and 4% from the Cook Inlet area.

The Trans-Alaska Pipeline, which runs 789 mi (1,270 km) from the North Slope oil fields to the port of Valdez on the southern coast, carried 1,700,000 barrels of crude oil a day in 1985. Because of a shortage of refinery capacity, much of Alaska's energy products must still be imported.

Natural gas production in 1983 was 276.7 billion cu feet, 8th among states. In 1983, proved reserves were 34.3 trillion cu feet, 3d among states. Electric power production totaled over 3.7 billion kwh in 1983; installed capacity was nearly 3 million kw, and almost all generating facilities were government-owned. Alaska also had proved coal reserves totaling 6.2 billion tons in 1983. Production of coal in 1983 was 786,000 tons, from a single mine, at Healy.

29INDUSTRY

Alaska's small but growing manufacturing sector is centered on petroleum refining and the processing of lumber and food products, especially seafood. As of 1982, Anchorage led all regions with 22% of industrial employment, followed by Kenai, 16%; Kodiak, 11%; and Ketchikan Gateway, 9%. Between 1977 and 1982, value added by manufacture rose by 53%, from $504.2 million to $769.2 million. The value of shipments by manufacturers in 1982 was nearly $2.6 billion. The following table shows value of shipments for selected industries in 1982:

Petroleum and coal products	$1,083,100,000
Food and kindred products	662,100,000
Lumber and wood products	250,100,000

30COMMERCE

Wholesale trade in 1982 amounted to $27.6 million, 49th among the states, and retail sales exceeded $3.2 billion (46th). More than half of all retail sales were in the Anchorage metropolitan area. Grocery stores accounted for 21% of all sales, followed by eating and drinking places, 13%; automotive dealers, 13%; department stores, 7%; gasoline service stations, 7%; and others, 39%.

During 1983, 1.1 billion lb and $97 million worth of imports and 8.4 billion lb and $1 billion worth of exports passed through the Anchorage customs district. Many of Alaska's resource products, including the salmon and crab catch, pass through the Seattle customs district. By federal law, Alaskan petroleum cannot be exported, a provision many Alaskans would like to see repealed. Exports of manufactured goods came to $653 million in 1981. One-third of Alaska's manufactured goods were exported to other countries, the highest ratio of all the states, with paper and food products the leading items. Alaska was the leading fish-exporting state and the largest exporter of salmon.

31CONSUMER PROTECTION

The Consumer Protection Section of the Department of Law maintains offices in Anchorage, Fairbanks, Juneau, and Ketchikan and is charged with enforcement of the state's Unfair Trade Practices and Consumer Protection Act.

32BANKING

As of the end of 1984, Alaska had 15 insured commercial banks; at the close of 1983, the state's insured commercial banks had assets exceeding $3.9 billion. There were 5 savings and loan associations (all of them federally chartered) in 1983; in 1982, the state's savings and loan associations had assets of $490 million and mortgage loans amounting to $216 million. There were 24 credit unions in 1984. Alaska's banks and savings and loan associations are under the regulatory authority of the Department of Commerce and Economic Development's Division of Banking, Securities, and Corporations.

³³INSURANCE

In 1983, one life insurance company and nine property/casualty insurance companies were domiciled in Alaska. Premiums written in 1982 totaled $648.7 million, including $478.9 million for property, casualty, and disability coverage; $104.1 million for life insurance; and $12.8 million for title insurance. Some 903,000 life insurance policies worth $12 billion were in force in 1983, when the average coverage per family was $61,500. The insurance industry is regulated by the Department of Commerce and Economic Development's Division of Insurance.

³⁴SECURITIES

There are no securities exchanges in Alaska. At the end of 1983, New York Stock Exchange firms had 12 sales offices and 119 full-time registered representatives in the state. In 1983, 114,000 Alaskans were shareholders of public corporations.

³⁵PUBLIC FINANCE

Alaska's annual budget is prepared by the Division of Budget and Management, within the Office of the Governor, and submitted by the governor to the legislature for amendment and approval. The fiscal year runs from 1 July through 30 June. The following table summarizes budgeted revenues and appropriations for 1983/84, in millions, for the general fund, including capital-projects funding:

REVENUES	
Taxes	$1,905.1
Rents and royalties	1,419.3
Federal grants	848.6
Other receipts	617.6
TOTAL	$4,790.6

EXPENDITURES	
Department of Transportation and Public Facilities	$1,267.7
Department of Education	830.7
University of Alaska	385.5
Department of Health and Social Services	352.2
Other outlays	2,868.6
TOTAL	$5,704.7

There are a number of special funds, all small except the Permanent Fund, into which 25% of mineral royalties and related income are deposited. In mid-1985, the Permanent Fund held $5.4 billion in principal and $1.1 billion in reserves. Every person in Alaska with at least six months' residency receives an annual dividend; in 1985, this dividend was $400, with the total cost $217 million.

As of 30 June 1983, the outstanding debt of Alaska state and local governments was nearly $8.5 billion, or about $17,700 per capita, by far the highest per capita debt of all states and more than four times that of West Virginia, which ranked 2d.

³⁶TAXATION

The huge sums generated by the sale of oil leases and by oil and gas royalties make Alaska's tax structure highly atypical. In 1985, the state government was deriving 85% of its revenue from oil, and local governments, 65%. There is no state sales tax, but some localities impose a sales tax, as well as a property tax. The corporate tax rate in 1984 ranged from 1% on the first $10,000 of taxable income to 9.4% on amounts over $90,000. Other taxes include ones on alcoholic beverages, motor fuels and vehicles, estates, cigarettes, insurance companies, and fisheries. In April 1980, the state legislature abolished the personal income tax; the business tax was ended in 1979. Nevertheless, Alaska by far led all states in fiscal 1981/82 in per capita state and local tax burden with $6,998, almost triple that of Wyoming.

Alaska's share of the federal tax burden in 1982/83 was nearly $1.9 billion, while its share of federal tax benefits was nearly $2.1 billion. Alaskans filed 235,486 federal income tax returns for 1983, paying nearly $1.1 billion in tax.

³⁷ECONOMIC POLICY

The Alaska Industrial Development Authority, a public corporation of the state, provides various means of financing industry, manufacturing, small businesses, tourism, mining, commercial fishing and other enterprises. The Alaska Resources Corporation, another state public corporation, provides developmental capital for the establishment and expansion of small enterprises in resource industries.

³⁸HEALTH

Alaska's birthrate of 25.4 per 1,000 population in 1982 was 2d only to Utah's. The infant mortality rate of 12.7 per 1,000 live births in 1981 was above the national average. The abortion rate of 172 per 1,000 births in 1982 was 7th lowest.

Alaska's overall death rate of 414.2 per 100,000 population in 1982 was less than half the US average, but the death rate from accidents (90.1 per 100,000) was the highest in the US and over twice the national rate, and the suicide rate of 14.7 was above the national average. The death rate of 15.4 for conditions originating in the prenatal period was the highest of all states in 1982. Alaska's major public health problem is alcoholism; the alcohol-related death rate of 353 per 100,000 in 1979 was the highest of all 50 states.

Alaska's 26 hospitals (9 federal) in 1983 had 1,754 beds and 62,971 admissions; hospital personnel included 1,339 registered nurses in 1981. The average daily cost of a hospital room in 1983 was $234. Alaska had 721 licensed physicians and 285 dentists at the end of 1982.

³⁹SOCIAL WELFARE

Public aid recipients constituted only 3.1% of the population in 1983. The $288 per capita spent on welfare by state and local governments in 1980 was 5th among states. In 1982, a monthly average of 11,804 Alaskans received $31 million in aid to families with dependent children, with the average monthly payment being $220. The school lunch program fed 38,000 students at a federal cost of $5 million, and 19,000 Alaskans used food stamps with a federal subsidy of $20 million in 1982/83. A total of $63.9 million was paid for workers' compensation in 1982. Unemployment benefits for 76,000 claimants totaled $88 million in that year. Medicaid enrollment was 17,000 in 1983, and payments came to $39 million.

Social Security benefits were received by 22,382 Alaskans in December 1982; payments totaled $102 million in 1983. The average benefit per retired worker was $424.76 in December 1982. In December 1983, 3,015 Alaskans received federal Supplemental Security Income; $7.1 million was paid in 1983. In addition, 4,099 persons received state payments, which came to $9.3 million. In 1979, Alaska became the first state to withdraw its government workers from the Social Security system.

⁴⁰HOUSING

Despite the severe winters, housing designs in Alaska do not differ notably from those in other states. Builders do usually provide thicker insulation in walls and ceilings, but the high costs of construction have not encouraged more energy-efficient adaptation to the environment. In 1980, the state legislature passed several measures to encourage energy conservation in housing and in public buildings.

In native villages, traditional dwellings like the half-buried huts of the Aleuts and others have long since given way to conventional, low-standard housing. In point of fact, Alaska's Eskimos never built snow houses as did those of Canada; in the Eskimo language, the word *igloo* refers to any dwelling.

The 1980 census counted 131,463 occupied housing units; 87% of the year-round units had full plumbing. From 1970 to 1978, 43,009 building permits were issued, as construction boomed during the years of pipeline building. Some 25,900 new private housing units, valued at $1.6 billion, were built from 1981 through 1983.

The Alaska State Housing Authority acts as an agent for federal and local governments in securing financial aid for construction and management of low-rent and moderate-cost homes.

41 EDUCATION

As of 1980, 82.5% of the population 25 years or older had completed high school, the highest such percentage among states; 22.4% were college graduates, 2d only to Colorado. No other state came close to Alaska's 1983/84 expenditure of $6,613 per public school pupil, or to its high teachers' salaries.

Enrollment in public schools was 89,517 in the fall of 1983. Private school enrollment was 3,490 in 1982/83. The University of Alaska is the state's leading higher educational institution. The main campus at Fairbanks, established in 1917, had 4,622 students in fall 1983, while the Anchorage campus had 4,088, and the Juneau campus had 2,361. There were 11 state-supported community colleges in 1983, all of them part of the University of Alaska system. Private institutions included two colleges with four-year programs, a theological seminary, and Alaska Pacific University. The University of Alaska's Rural Education Division has a network of education centers and offers 90 correspondence courses in 22 fields of study.

42 ARTS

The Council on the Arts sponsors tours by performing artists, supports artists' residences in the schools, aids local arts projects, and purchases the works of living Alaskans for display in state buildings. Its 1985 per capita appropriation of $1,083 was the highest of all state arts agencies. Fairbanks, Juneau, and Anchorage have symphony orchestras, and Anchorage has a civic opera. The Alaska Repertory Theater tours the state.

43 LIBRARIES AND MUSEUMS

Alaska public libraries had an estimated combined book stock of 1,432,127 and a circulation of 2,021,395 in 1983/84; facilities are located in seven boroughs and in most larger towns. Anchorage had the largest public library system, with eight branches and 295,902 volumes in 1983/84. Also notable are the State Library in Juneau and the library of the University of Alaska at Fairbanks.

Alaska had 37 museums in 1985. The Alaska State Museum in Juneau offers an impressive collection of native crafts and Alaskan artifacts. Sitka National Historical Park features Indian and Russian items, and the nearby Museum of Sheldon Jackson College holds important native collections. Noteworthy historical and archaeological sites include the Totem Heritage Center in Ketchikan. Anchorage has the Alaska Zoo.

44 COMMUNICATIONS

Considering the vast distances traveled and the number of small, scattered communities, the US mail is a bargain for Alaskans; 187 post offices and 1,876 postal service workers moved the state's mail as of March 1985. In 1980, 90% of the state's 957,032 occupied housing units had telephones.

There were 64 radio stations (36 AM, 28 FM) in 1984, along with 13 television stations and 131 cable television systems, which served 518,481 subscribers in 294 communities.

45 PRESS

Two of Alaska's eight daily newspapers and two of its three Sunday newspapers are in Anchorage. Total newspaper circulation as of 1984 was 51,026 mornings, 80,684 evenings, and 130,921 Sundays. The most widely read papers were the *Anchorage Daily News* (mornings) and the *Anchorage Times* (evenings); daily circulation as of 30 September 1984 was 48,077 and 43,121, respectively, with Sunday circulation 56,456 and 51,965, respectively. The *Tundra Times,* also published in Anchorage, is a statewide weekly devoted to native concerns. The University of Alaska Press is the state's only publisher.

46 ORGANIZATIONS

The 1982 Census of Service Industries counted 212 organizations in Alaska, including 45 business associations; 128 civic, social, and fraternal associations; and 7 educational, scientific, and research associations. There are no major national organizations based in Alaska. The largest statewide organization, the Alaska Federation of Natives, with headquarters in Anchorage, represents the state's Eskimos, Aleuts, and Indians.

47 TOURISM, TRAVEL, AND RECREATION

With thousands of miles of unspoiled scenery and hundreds of mountains and lakes, Alaska has vast tourist potential. An estimated 690,000 travelers visited Alaska in 1984. Travel-related industries employed 13,300 full-time workers in 1982, had a payroll of $210 million, and earned revenues of $821 million.

Cruise travel along the Gulf of Alaska is one of the fastest-growing sectors in the tourist trade. One of the most popular tourist destinations is Glacier Bay National Monument. In 1984, 780,887 visits were paid to 13 of the state's 16 national parks, preserves, historical parks, and monuments, which totaled 46.3 million acres (18.7 million hectares). Public recreation sites in 1982 included 22 alpine ski areas, 1,876 mi (3,019 km) of hiking trails, and 3,314 camp units. Licenses were issued to 256,412 fishermen and 90,094 hunters in 1982/83.

48 SPORTS

The only professional sports team in Alaska is a minor league basketball franchise in Anchorage. Fans follow wintertime college basketball in Fairbanks and Anchorage and summertime semipro baseball. Skiing and fishing are favorite participant sports, and dogteam racing, or mushing, is the official state sport.

49 FAMOUS ALASKANS

Alaskan's best-known federal officeholder was Ernest Gruening (b.New York, 1887–1974), a territorial governor from 1939 to 1953 and US senator from 1959 to 1969. Alaska's other original US senator was E.L. "Bob" Bartlett (1904–68). Walter Hickel (b.Kansas, 1919), the first Alaskan to serve in the US cabinet, left the governorship in 1969 to become secretary of the interior. Among historical figures, Vitus Bering (b.Denmark, 1680–1741), a seaman in Russian service who commanded the discovery expedition in 1741, and Aleksandr Baranov (b.Russia, 1746–1819), the first governor of Russian America, are outstanding. Secretary of State William H. Seward (b.New York, 1801–72), who was instrumental in the 1867 purchase of Alaska, ranks as the state's "founding father," although he never visited the region.

Sheldon Jackson (b.New York, 1834–1909), a Presbyterian missionary, introduced the reindeer to the region and founded Alaska's first college in Sitka. Carl Ben Eielson (1897–1929), a famed bush pilot, is a folk hero. Benny Benson (1913–72), born at Chignik, designed the state flag at the age of 13.

50 BIBLIOGRAPHY

Alaska, State of. Department of Commerce and Economic Development. Division of Economic Enterprise. *The Alaska Economy.* Vol. 7, Juneau, 1979.

Alaska, State of. Department of Education. Division of State Libraries and Museums. *Alaska Blue Book 1983.* Juneau, 1983.

Brooks, Alfred Hulse. *Blazing Alaska's Trails.* Fairbanks: University of Alaska Press, 1972.

Gruening, Ernest. *State of Alaska.* New York: Random House, 1968.

Hunt, William R. *Alaska: A Bicentennial History.* New York: Norton, 1976.

Hunt, William R. *North of 53°: The Wild Days of the Alaska-Yukon Mining Frontier.* New York: Macmillan, 1974.

McPhee, John. *Coming into the Country.* New York: Farrar, Straus, & Giroux, 1977.

Naske, Claus M. *A History of Alaska Statehood.* Lanham, Md.: University Press of America, 1985.

Naske, Claus M., and Herman E. Slotnick. *Alaska: A History of the 49th State.* Grand Rapids, Mich.: Eerdmans, 1979.

Oswalt, Wendell H. *Alaskan Eskimos.* San Francisco: Chandler, 1967.

ARIZONA

State of Arizona

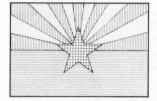

ORIGIN OF STATE NAME: Probably from the Pima or Papago Indian word *arizonac,* meaning "place of small springs." **NICKNAME:** The Grand Canyon State. **CAPITAL:** Phoenix. **ENTERED UNION:** 14 February 1912 (48th). **SONG:** "Arizona." **MOTTO:** *Ditat Deus* (God enriches). **FLAG:** A copper-colored five-pointed star symbolic of the state's copper resources rises from a blue field; six yellow and seven red segments radiating from the star cover the upper half. **OFFICIAL SEAL:** Depicted on a shield are symbols of the state's economy and natural resources, including mountains, a rising sun, and a dam and reservoir in the background, irrigated farms and orchards in the middle distance, a quartz mill, a miner, and cattle in the foreground, as well as the state motto. The words "Great Seal of the State of Arizona 1912" surround the shield. **BIRD:** Cactus wren. **FLOWER:** Blossom of the saguaro cactus. **TREE:** Paloverde. **OFFICIAL NECKWEAR:** Bola tie. **LEGAL HOLIDAYS:** New Year's Day, 1 January; Birthday of Martin Luther King, Jr., 3d Monday in January; Lincoln's Birthday, 1st Monday in February; Washington's Birthday, 3d Monday in February; Memorial Day, last Monday in May; Independence Day, 4 July; Labor Day, 1st Monday in September; Columbus Day, 2d Monday in October; Veterans Day, 11 November; Thanksgiving Day, 4th Thursday in November; Christmas Day, 25 December. **TIME:** 5 AM MST = noon GMT. Arizona is the only state that does not observe daylight savings time.

¹LOCATION, SIZE, AND EXTENT

Located in the Rocky Mountains region of the southwestern US, Arizona ranks 6th in size among the 50 states.

The total area of Arizona is 114,000 sq mi (295,260 sq km), of which land takes up 113,508 sq mi (293,986 sq km) and inland water 492 sq mi (1,274 sq km). Arizona extends about 340 mi (547 km) E–W; the state's maximum N–S extension is 395 mi (636 km).

Arizona is bordered on the N by Utah and on the E by New Mexico (with the two borders joined at Four Corners, the only point in the US common to four states); on the S by the Mexican state of Sonora; and on the W by the Mexican state of Baja California Norte, California, and Nevada (with most of the line formed by the Colorado River). The total boundary length of Arizona is 1,478 mi (2,379 km). The state's geographic center is in Yavapai County, 55 mi (89 km) ESE of Prescott.

²TOPOGRAPHY

Arizona is a state of extraordinary topographic diversity and beauty. The Colorado Plateau, which covers two-fifths of the state in the north, is an arid upland region characterized by deep canyons, notably the Grand Canyon, a vast gorge more than 200 mi (320 km) long, up to 18 mi (29 km) wide, and more than 1 mi (1.6 km) deep. Also within this region are the Painted Desert and Petrified Forest, as well as Humphreys Peak, the highest point in the state, at 12,633 feet (3,851 meters).

The Mogollon Rim separates the northern plateau from a central region of alternating basins and ranges with a general northwest–southeast direction. Ranges in the Mexican Highlands in the southeast include the Chiricahua, Dos Cabezas, and Pinaleño mountains. The Sonora Desert, in the southwest, contains the lowest point in the state, 70 feet (21 meters) above sea level, on the Colorado River near Yuma.

The Colorado is the state's major river, flowing southwest from Glen Canyon Dam on the Utah border through the Grand Canyon and westward to Hoover Dam, then turning south to form the border with Nevada and California. Tributaries of the Colorado include the Little Colorado and Gila rivers. Arizona has few natural lakes, but there are several large artificial lakes formed by dams for flood control, irrigation, and power development. These include Lake Mead (shared with Nevada), formed by Hoover Dam; Lake Powell (shared with Utah); Lake Mohave and

Lake Havasu (shared with California), formed by Davis Dam and Parker Dam, respectively; Roosevelt Lake, formed by Theodore Roosevelt Dam; and the San Carlos Lake, created by Coolidge Dam.

³CLIMATE

Arizona has a dry climate, with little rainfall. Temperatures vary greatly from place to place, season to season, and day to night. Average daily temperatures at Yuma, in the southwestern desert, range from 43° to 67°F (6–19°C) in January and from 81° to 106°F (27–41°C) in July. At Flagstaff, in the interior uplands, average daily January temperatures range from 14° to 41°F (−10–5°C), and average daily July temperatures range from 50° to 81°F (10–27°C). The maximum recorded temperature was 127°F (53°C), registered at Parker on 7 July 1905; the minimum, −40°F (−40°C), was set at Hawley Lake on 7 January 1971.

The higher elevations of the state, running diagonally from the southeast to the northwest, receive between 25 and 30 in (63–76 cm) of precipitation a year, and the rest, for the most part, between 7 and 20 in (18–51 cm). The driest area is the extreme southwest, which receives less than 3 in (8 cm) a year. Snow, sometimes as much as 100 in (254 cm) of it, falls on the highest peaks each winter but is rare in the southern and western lowlands.

The greatest amount of sunshine is registered in the southwest, with the proportion decreasing progressively toward the northeast; overall, the state receives more than 80% of possible sunshine, among the highest in the US, and Phoenix's 86% is higher than that of any other major US city.

⁴FLORA AND FAUNA

Generally categorized as desert, Arizona's terrain also includes mesa and mountains; consequently, the state has a wide diversity of vegetation. The desert is known for many varieties of cacti, from the saguaro, whose blossom is the state flower, to the cholla and widely utilized yucca. Desert flowers include the night-blooming cereus; among medicinal desert flora is the jojoba, also harvested for its oil-bearing seeds. Below the tree line (about 12,000 feet, or 3,658 meters) the mountains are well timbered with varieties of spruce, fir, juniper, ponderosa pine, oak, and piñon. Rare plants, some of them endangered or threatened, include various cacti of commercial or souvenir value.

Arizona's fauna range from desert species of lizards and snakes to the deer, elk, and antelope of the northern highlands. Mountain lion, jaguar, coyote, and black and brown bears are found in the state, along with the badger, black-tailed jackrabbit, and gray fox. Small mammals include various cottontails, mice, and squirrels; prairie dog towns dot the northern regions. Rattlesnakes are abundant, and the desert is rife with reptiles such as the collared lizard and chuckwalla. Native birds include the thick-billed parrot, white pelican, and cactus wren (the state bird). Arizona counts the osprey, desert bighorn, desert tortoise, spotted bat, and Gila monster among its threatened wildlife. Officially listed as endangered or threatened are the southern bald eagle, masked bobwhite (quail), peregrine falcon, Yuma clapper rail, Sonoran pronghorn, ocelot, jaguarundi, black-footed ferret, bonytail chub, humpback chub, Colorado River squawfish, woundfin, Apache trout, Gila topminnow, and fat pocketbook clam.

5ENVIRONMENTAL PROTECTION

Aside from Phoenix, whose air quality is poorer than that of most other US cities, Arizona has long been noted for its clear air, open lands, and beautiful forests. The main environmental concern of the state is to protect these resources in the face of growing population, tourism, and industry.

State agencies with responsibility for the environment include the State Land Department, which oversees natural resource conservation and land mangement of 9.6 million acres (3.9 million hectares) of trust land; the Game and Fish Commission, which administers state wildlife laws; the Department of Health Services, which supervises sewage disposal, water treatment, hazardous and solid waste treatment, and air pollution prevention programs; and the Department of Water Resources, formed in 1980, which is concerned with the development, management, use, and conservation of water.

Legislation enacted in 1980 attempts to apportion water use among cities, mining, and agriculture, the last of which, through irrigation, accounts for about 90% of Arizona's annual water consumption. In 1985, water began to flow from the Colorado River to Phoenix through the canals of the Central Arizona Project (CAP); authorized in 1968, this $3-billion project will bring 1.2 million acre-feet of water to Arizona annually.

6POPULATION

Arizona ranked 29th in the US, with a 1980 census population of 2,718,425, 53% more than in 1970. The state estimate for January 1985 was 3,086,827, 13.6% more than in 1980. Arizona's population growth rate is one of the highest in the nation.

In 1981, 82% of the population was white, 16% Hispanic, 6% American Indian, and 3% black. In 1980, 49.2% of the population was male and 50.8% female. Arizonans 65 years of age or older increased from 9.1% of the population in 1970 to 12.7% in 1984, reflecting to some extent the state's increasing popularity among retirees. From 1973 through 1983 there were 496,152 births and 211,856 deaths. Between 1975 and 1980 there was a net increase from migration of 245,688. Despite its rapid population growth, Arizona still had an estimated population density of only 27 persons per sq mi (10 per sq km) in 1985.

Three out of four Arizonans live in metropolitan areas. The largest cities are Phoenix, with a 1980 population of 789,704 and an estimated 1984 population of 854,990; Tucson, 330,537 and 367,280, respectively; Mesa, 152,453 and 184,310; Tempe, 106,743 and 138,600; Scottsdale, 88,622 and 106,000; and Glendale, 96,988 and 111,590. More than half the state's population resides in Maricopa County, which includes every leading city except Tucson. Phoenix was the nation's 9th-largest city in 1984.

7ETHNIC GROUPS

Arizona has by far the nation's greatest expanse of Indian lands: the state's 22 reservations have a combined area of 19.1 million acres (7.7 million hectares)—26% of the total state area.

The largest single American Indian nation, the Navaho, with a 1980 population of 76,042 in Arizona, is located primarily in the northeastern part of the state. The Navaho reservation, covering 14,221 sq mi (36,832 sq km) within Arizona, extends into Utah and New Mexico and comprises desert, mesa, and mountain terrain. Herders by tradition, the people are also famous for their crafts. Especially since 1965, the Navaho have been active in economic development; reservation resources in uranium and coal have been leased to outside corporations, and loans from the US Department of Commerce have made possible roads, telephones, and other improvements. There are at least 12 and perhaps 17 other tribes (depending on definition). After the Navaho, the leading tribes are the Papago in the south, Apache in the east, and Hopi in the northeast. All together, the Indian population was estimated at 166,330 in January 1985.

The southern part of Arizona has most of the state's largest ethnic majority, a Hispanic population estimated at 453,307 (20% of the population) in mid-1981. There are some old, long-settled Spanish villages, but the bulk of Hispanics are of Mexican origin. These make up most of the unskilled and semiskilled urban and rural labor force. Nevertheless, in 1984 Arizona had 241 Hispanic elected officials, ranking 4th among all states in this category, and 12 were in the 90-member legislature. Raul Castro, a Mexican-American, served as governor, 1975–77.

There were 77,101 black Americans in Arizona in mid-1981. Filipinos, Chinese, Japanese, and other Asian and Pacific Island peoples made up less than 1% of the population.

8LANGUAGES

With the possible exception of the Navaho word *hogan* (earth-and-timber dwelling), the linguistic influence of Arizona's Papago, Pima, Apache, Navaho, and Hopi tribes is almost totally limited to some place-names: Arizona itself, Yuma, Havasu, Tucson, and Oraibi. Indian loanwords spreading from Arizona derive from the Nahuatl speech of the Mexican Aztecs—for example, *coyote, chili, mesquite,* and *tamale.* Spanish, dominant in some sections, has given English *mustang, ranch, stampede, rodeo, marijuana, bonanza, canyon, mesa, patio,* and *fiesta.*

English in the state represents a blend of North Midland and South Midland dialects without clear regional differences, although new meanings developed in the north and east for *meadow* and in the southern strip for *swale* as terms for flat mountain valleys. The recent population surge from eastern states has produced an urban blend with a strong Northern flavor. In 1980, 2,063,503 Arizonans—80% of all residents 3 years old or older—spoke only English at home. Other languages spoken at home, and the number of people speaking them, included Spanish, 342,859; German, 16,497; French, 9,380; and Italian, 7,899.

9RELIGIONS

The first religions of Arizona were the sacred beliefs and practices of the Indians. Catholic missionaries began converting Arizona Indians (Franciscans among the Hopi and Jesuits among the Pima) to the Christian faith in the late 17th century. By the late 18th century, the Franciscans were the main missionary force, and the Roman Catholic Church was firmly established. In 1984, the state had 522,001 Catholics.

The Church of Jesus Christ of Latter-day Saints (Mormons) constitutes the 2d-largest Christian denomination. Mormons were among the state's earliest Anglo settlers; in 1980 there were 139,178 known Mormon adherents. Other major denominations include the Southern Baptist Convention, 110,849, and the United Methodist Church, 52,271.

In 1984, Arizona's estimated Jewish population was 53,285, virtually all of whom lived in the Phoenix or Tucson metropolitan areas.

10TRANSPORTATION

Until the last decade of the 19th century, the principal reason for the development of transportation in Arizona was to open routes to California. The most famous early road was El Camino del

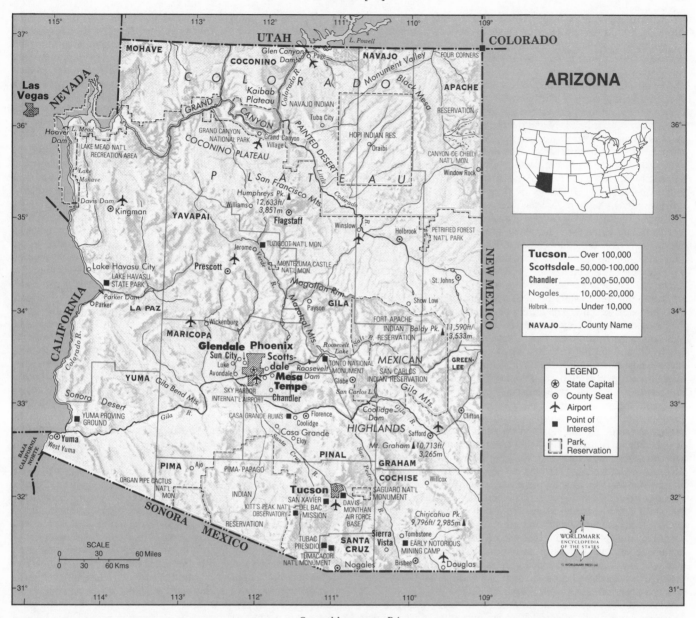

See endsheet maps: D4

LOCATION: 31°20' to 36°59'56"N; 109°2'40" to 114°50'W. **BOUNDARIES:** Utah line, *277 mi (446 km)*; New Mexico line, *389 mi (626 km)*; Mexico line, *373 mi (600 km)*; California line, *234 mi (377 km)*; Nevada line, *205 mi (330 km)*.

Diablo (The Devil's Highway), opened by the missionary Eusebio Kino in 1699. The first wagon road across Arizona was the Gila Trail (Cooke's Wagon Road), opened in 1846 as a southern route to California; Beale's Road was inaugurated in 1857. Also in 1857, the first stagecoach began operations. Until the coming of the railroads in the 1880s, however, the bulk of territorial commerce was by water transport on the Colorado River. Railroad construction reached its peak in the 1920s and declined rapidly thereafter. Railroad trackage totaled 1,785 mi (2,873 km) in 1983, on the Southern Pacific and the Atchison, Topeka, and Santa Fe. Amtrak provides limited passenger service through Flagstaff, Kingman, and other cities in the north and through Tucson, Phoenix, and Yuma on the southern route.

In 1983, the state had 76,334 mi (122,848 km) of streets and roads, of which 67,434 mi (108,525 km) were classified as rural and 8,900 mi (14,323 km) as urban. In addition, interstate highways in Arizona totaled 524 mi (843 km). Of the 2,262,782

motor vehicles registered in that year, 1,423,573 were passenger vehicles, 552,308 trucks, 207,756 trailers, 77,334 motorcycles, and 1,811 buses and taxis. There were an estimated 2,178,815 licensed drivers in 1983, of whom 54.1% were male. Total state and local highway expenditures in 1982/83 were $452.6 million; Arizona received $157.3 million in federal highway trust funds in 1983/84.

Arizona had 168 airports and 69 heliports in 1983. The leading air terminal was Phoenix Sky Harbor International Airport, which handled 8,605,400 arriving and departing passengers; Tucson International Airport ranked second with 2,018,300. In the same year, there were 14,457 active pilots in Arizona.

[11]HISTORY

Evidence of a human presence in Arizona dates back more than 12,000 years. The first Arizonans—the offshoot of migrations across the Bering Strait—were large-game hunters; their remains have been found in the San Pedro Valley in the southeastern part of the state. By AD 500, their descendants had acquired a rudimentary agriculture from what is now Mexico and divided into several cultures. The Basket Makers (Anasazi) flourished in the northeastern part of the state; the Mogollon hunted and foraged in the eastern mountains; the Hohokam, highly sophisticated irrigators, built canals and villages in the central and southern valleys; and the Hakataya, a less-advanced river people, lived south and west of the Grand Canyon. For reasons unknown—a devastating drought is the most likely explanation—these cultures were in decay and the population much reduced by the 14th century. Two centuries later, when the first Europeans arrived, most the natives were living in simple shelters in fertile river valleys, dependent on hunting, gathering, and small-scale farming for subsistence. These Arizona Indians belonged to three linguistic families: Uto-Aztecan (Hopi, Paiute, Chemehuevi, Pima-Papago), Yuman (Yuma, Mohave, Cocopa, Maricopa, Yavapai, Walapai, Havasupai), and Athapaskan (Navaho-Apache). The Hopi were the oldest group, their roots reaching back to the Anasazi; the youngest were the Navaho-Apache, migrants from the Plains, who were not considered separate tribes until the early 18th century.

The Spanish presence in Arizona involved exploration, missionary work, and settlement. Between 1539 and 1605, four expeditions crossed the land, penetrating both the upland plateau and the lower desert in ill-fated attempts to find great riches. In their footsteps came Franciscans from the Rio Grande to work among the Hopi, and Jesuits from the south, led by Eusebio Kino in 1692, to missionize among the Pima. Within a few years, Kino had established a major mission station at San Xavier del Bac, near present-day Tucson. In 1736, a rich silver discovery near the Pima village of Arizonac, about 20 mi (32 km) southwest of present-day Nogales, drew Spanish prospectors and settlers northward. To control the restless Pima, Spain in 1752 placed a military outpost, or presidio, at Tubac on the Santa Cruz River north of Nogales. This was the first major European settlement in Arizona. The garrison was moved north to the new fort at Tucson, also on the Santa Cruz, in 1776. During these years, the Spaniards gave little attention to the Santa Cruz settlements, administered as part of the Mexican province of Sonora, regarding them merely as way stations for colonizing expeditions traveling overland to the highly desirable lands of California. The end of the 18th century and the beginning of the 19th were periods of relative peace on the frontier; mines were developed and ranches begun. Spaniards removed hostile Apache bands onto reservations and made an effort to open a road to Santa Fe.

When Mexico revolted against Spain in 1810, the Arizona settlements were little affected. Mexican authorities did not take control at Arizpe, the Sonoran capital, until 1823. Troubled times followed, characterized by economic stagnation, political chaos, and renewed war with the Apache. Sonora was divided into *partidos* (counties), and the towns on the Santa Cruz were designated as a separate *partido,* with the county seat at Tubac. The area north of the Gila River, inhabited only by Indians, was vaguely claimed by New Mexico. With the outbreak of the Mexican War in 1846, two US armies marched across the region: Col. Stephen W. Kearny followed the Gila across Arizona from New Mexico to California, and Lt. Col. Philip Cooke led a Mormon battalion westward through Tucson to California. The California gold rush of 1849 saw thousands of Americans pass along the Gila toward the new El Dorado. In 1850, most of present-day Arizona became part of the new US Territory of New Mexico; the southern strip was added by the Gadsden Purchase in 1853.

Three years later, the Sonora Exploring and Mining Co. organized a large party, led by Charles D. Poston, to open silver mines around Tubac. A boom followed, with Tubac becoming the largest settlement in the valley; the first newspaper, the *Weekly Arizonian,* was launched there in 1859. The great desire of California for transportation links with the rest of the Union prompted the federal government to chart roads and railroad routes across Arizona, erect forts there to protect Anglo travelers from the Indians, and open overland mail service. Dissatisfied with their representation at Santa Fe, the territorial capital, Arizona settlers joined those in southern New Mexico in 1860 in an abortive effort to create a new territorial entity. The outbreak of the Civil War in 1861 saw the declaration of Arizona as Confederate territory and abandonment of the region by the Union troops. A small Confederate force entered Arizona in 1862 but was driven out by a volunteer Union army from California. On 24 February 1863, President Abraham Lincoln signed into law a measure creating the new Territory of Arizona. Prescott became the capital in 1864, Tucson in 1867, Prescott again in 1877, and finally Phoenix in 1889.

During the early years of territorial status, the development of rich gold mines along the lower Colorado River and in the interior mountains attracted both people and capital to Arizona, as did the discovery of silver bonanzas in Tombstone and other districts in the late 1870s. Additional military posts were constructed to protect mines, towns, and travelers. This activity, in turn, provided the basis for a fledgling cattle industry and irrigated farming. Phoenix, established in 1868, grew steadily as an agricultural center. The Southern Pacific Railroad, laying track eastward from California, reached Tucson in 1880, and the Atlantic and Pacific (later acquired by the Santa Fe), stretching west from Albuquerque through Flagstaff, opened service to California in 1883. By 1890, copper had replaced silver as the principal mineral extracted in Arizona. In the Phoenix area, large canal companies began wrestling with the problem of supplying water for commercial agriculture. This problem was resolved in 1917 with the opening of the Salt River Valley Project, a federal reclamation program that provided enormous agricultural potential.

As a creature of the Congress, Arizona Territory was presided over by a succession of governors, principally Republicans, appointed in Washington. In reaction, the populace was predominantly Democratic. Within the territory, a merchant-capitalist class, with strong ties to California, dominated local and territorial politics until it was replaced with a mining-railroad group whose influence continued well into the 20th century. A move for separate statehood began in the 1880s but did not receive serious attention in Congress for another two decades. In 1910, after Congress passed an enabling act that allowed Arizona to apply for statehood, a convention met at Phoenix and drafted a state constitution. On 14 February 1912, Arizona entered the Union as the 48th state.

During the first half of the 20th century, Arizona shook off its frontier past. World War I spurred the expansion of the copper industry, intensive agriculture, and livestock production. Goodyear Tire and Rubber established large farms in the Salt River Valley to raise pima cotton. The war boom also generated high prices, land speculation, and labor unrest; at Bisbee and Jerome, local authorities forcibly deported more than 1,000 striking miners during the summer of 1917. The 1920s brought depression: banks closed, mines shut down, and agricultural production declined. To revive the economy, local boosters pushed highway construction, tourism, and the resort business. Arizona also shared in the general distress caused by the Great Depression of the 1930s and received large amounts of federal aid for relief and recovery. A copper tariff encouraged the mining industry, additional irrigation projects were started, and public works were begun on Indian reservations, in parks and forests, and at education institutions. Prosperity returned during World War II as camps for military training, prisoners of war, and displaced Japanese-Americans were built throughout the state. Meat, cotton, and copper markets flourished, and the construction of processing and assembly plants suggested a new direction for the state's economy.

Arizona emerged from World War II a modern state. War industries spawned an expanding peacetime manufacturing boom that soon provided the principal source of income, followed by tourism, agriculture, and mining. During the 1950s, the political scene changed. Arizona Republicans captured the governorship, gained votes in the legislature, won congressional seats, and brought a viable two-party system to the state. The rise of Barry Goldwater of Phoenix to national prominence further encouraged Republican influence. Meanwhile, air conditioning changed lifestyles, prompting a significant migration to the state.

¹²STATE GOVERNMENT

The current constitution of Arizona, drafted in 1910 at the height of the Progressive era, contained reform provisions that were very advanced for the time: initiative, referendum, workers' compensation, short terms for elected officials, suffrage for women, and the barring of trusts and monopolies from the state.

Legislative authority is vested in a 30-member senate and a 60-member house of representatives. All senators and representatives serve 2-year terms and are chosen at the general election in November of each even-numbered year. A legislator must be a US citizen, at least 25 years of age, and must have been an Arizona resident for at least 3 years.

Chief executive officials elected statewide include the governor, secretary of state, treasurer, attorney general, and superintendent of public instruction, all of whom serve 4-year terms. The three members of the Corporation Commission, which regulates public services and utilities, are elected for staggered 6-year terms, and the state mine inspector is elected for 2 years. Candidates for executive office must have been US citizens for at least 10 years, must be at least 25 years of age, and must have been residents or citizens of Arizona for at least 5 years.

Bills may originate in either house of the legislature and must be passed by both houses and approved by the governor in order to become law. A two-thirds vote in each house is necessary to override the governor's veto. Under the initiative procedure, legislation and proposed constitutional amendments can be placed on the ballot by petition.

In order to vote in Arizona, a person must be 18 years of age, must be a US citizen, and must have been a resident of the state for 50 days.

¹³POLITICAL PARTIES

Of Arizona's 17 territorial governors, all federally appointed, 14 were Republicans and 3 Democrats. Statehood meant a prolonged period of Democratic dominance. From 1912 through 1950, the state had 9 Democratic and 3 Republican governors; during that 49-year period, Republicans held the statehouse for only 6 years.

Republican Party fortunes improved dramatically after 1950, largely because of the rise to state and national prominence of a conservative Republican, Barry Goldwater, first elected to the US Senate in 1952. From 1951 to 1980, 3 Republican governors occupied the state house for 18 years, and 5 Democratic governors for 12. Arizonans gave the nod to Republican presidential candidates in every election from 1952 through 1976. Several Arizona Republicans were appointed to high office during the Nixon years, and in 1973, another Republican, John J. Rhodes, became minority leader in the US House of Representatives.

Arizonans, who had favored Republican presidential nominees in the seven previous races, made it eight in a row in 1980 by giving more than 60% of the popular vote to Ronald Reagan. Goldwater barely won reelection to the Senate, defeating his Democratic opponent by only 1% of the total votes cast, despite the generally Republican tide. The election left the Republicans with one of the state's two Senate seats, two of Arizona's four US House seats, a narrow majority in the state senate, and a sizeable majority in the state house of representatives. Independent presidential candidate John Anderson captured 76,952 votes, or not quite 9% of the total votes cast. A state-operated lottery was approved. Governor Bruce Babbitt, a Democrat, was reelected in 1982, although both houses of the legislature remained Republican.

President Reagan carried Arizona by a 2-to-1 margin in 1984. Four of the state's five House seats were won by Republicans, and the GOP retained control of both legislative houses.

In 1984, Arizona had 241 Hispanic and 3 Asian elected officials. There were 11 black elected officials in 1985, the same year that there were 18 women in the state legislature.

¹⁴LOCAL GOVERNMENT

Each of Arizona's 15 counties has a sheriff, county attorney, county recorder, treasurer, assessor, superintendent of schools, and three or five supervisors, each elected to a four-year term.

Local governmental units include towns, cities, and charter cities. Towns generally follow the council-mayor form of govern-

Arizona Presidential Vote by Political Party, 1948–84

YEAR	ELECTORAL VOTE	ARIZONA WINNER	DEMOCRAT	REPUBLICAN	PROGRESSIVE
1948	4	*Truman (D)	95,251	77,597	3,310
1952	4	*Eisenhower (R)	108,528	152,042	—
1956	4	*Eisenhower (R)	112,880	176,990	—
1960	4	Nixon (R)	176,781	221,241	—
1964	5	Goldwater (R)	237,753	242,535	—
					AMERICAN IND.
1968	5	*Nixon (R)	170,514	266,721	46,573
					AMERICAN
1972	6	*Nixon (R)	198,540	402,812	21,208
					LIBERTARIAN
1976	6	Ford (R)	295,602	418,642	7,647
1980	6	*Reagan (R)	246,843	529,688	18,784
1984	7	*Reagan (R)	333,854	681,416	10,585

* Won US presidential election.

ment. All of Arizona's largest cities are charter cities. In all, there were 452 local government units in 1982, of which 76 were municipal governments.

Each of the 22 Indian reservations in Arizona has a tribal council or board with members elected by the people.

15 STATE SERVICES

The Arizona Department of Education regulates the school system. The Arizona Board of Regents governs the state's three public universities. The Department of Transportation administers the state's highway and air-transport systems, among other functions.

The Department of Health Services operates programs for environmental health, behavioral health (including alcohol abuse, drug abuse, and mental-illness treatment facilities), and family health services. The National Guard falls under the jurisdiction of the Department of Emergency and Military Affairs, while prisons and rehabilitation programs are administered by the State Department of Corrections and the Board of Pardons and Paroles. The Department of Public Safety oversees the state highway patrol.

Natural resources are the responsibility of several agencies, including the Game and Fish Commission, Department of Mines and Mineral Resources, Oil and Gas Conservation Commission, Parks Board, and Department of Water Resources. The Department of Economic Security handles employment services and public-assistance programs.

16 JUDICIAL SYSTEM

The supreme court is the highest court in Arizona and has administrative responsibility over all other courts in the state. The five supreme court justices, appointed by the governor for staggered six-year terms, choose a chief justice and vice chief justice to preside over the court.

The court of appeals, established in 1964, is organized in two geographical divisions which together have 18 judges. Appeals court judges are appointed for terms of six years.

The superior court is the general trial court of the state, and there must be at least one superior court judge in every Arizona county; in 1984 there were 94 superior court judges in the state's 15 counties. In counties with populations of over 150,000, superior court judges are appointed by the governor; they hold office for terms ending 60 days following the next regular general election after expiration of a two-year term. Those seeking retention run at the next general election on a nonpartisan ballot. In counties with populations of under 150,000, superior court judges are elected by nonpartisan ballot to four-year terms.

Counties are divided into precincts, each of which has a justice court. Every incorporated city and town has a police court. The jurisdiction of justice courts and police courts is limited to minor civil and criminal cases. Local judges are elected for terms of four years.

In 1983, 449,862 civil, criminal, and juvenile cases were filed in Arizona courts. There were 6,612 practicing attorneys in the state in 1984.

According to the FBI Crime Index of 1983, Arizona's crime rate of 6,391.6 per 100,000 population ranked 5th among the 50 states. Rates for murder and nonnegligent manslaughter, forcible rape, aggravated assault, breaking and entering, larceny, and motor-vehicle theft were all above the national average. Both Phoenix (6,841.1) and Tucson (7,655.2) ranked among the most crime-ridden metropolitan areas in 1983. In mid-1983 there were 81 local jails in Arizona, with 2,940 inmates; federal and state institutions held 6,889 prisoners at year-end. Fifty-one prisoners were under sentence of death at the end of 1983.

17 ARMED FORCES

In 1984, 22,618 federal military personnel were stationed at 20 military installations in Arizona. Major military installations include the Army's Fort Huachuca at Sierra Vista; the Air Force's

Williams base near Phoenix and the Luke and Davis-Monthan bases, near Phoenix and Tucson, respectively; and the Marine Corps' Yuma Air Station. Defense Department expenditures in Arizona were nearly $2.9 billion in 1983/84.

As of 1983, about 384,000 veterans lived in the state; 4,000 saw service in World War I; 155,000 in World War II; 73,000 in the Korean conflict; and 120,000 during the Viet-Nam era. Veterans' benefits totaled $324 million in 1982/83.

At the start of 1985 there were 7,184 Army and Air National Guard personnel in Arizona. There were 8,568 state and local police and 4,448 state and local corrections personnel in 1983.

18 MIGRATION

Arizona's first migrants were the ancient peoples who came from Asia across the Bering Strait more than 12,000 years ago. Hispanic settlers began arriving in the late 17th century. The Anglo migration, especially from the South, became significant as the US developed westward to California, and increased at an even faster rate with the building of the railroads during the 1880s. Migration has accelerated since World War II, and Arizona showed a net gain of 803,000 immigrants (27% of the estimated 1983 population) from 1970 to 1983. Increasingly, the new arrivals are northerners seeking favorable weather and a good business climate. Mexico is the main source of foreign immigrants.

19 INTERGOVERNMENTAL COOPERATION

Arizona is a signatory to a boundary agreement with California (1963) and to such interstate accords as the Colorado River Compact, Interstate Oil and Gas Compact, Upper Colorado River Basin Compact, and Western Interstate Energy Compact.

The most important federal project in the state is the Central Arizona Project, approved by Congress in 1968 and designed to divert water from the Colorado River to the Phoenix and Tucson areas for agriculture, energy, and other purposes. Federal aid totaled $989.9 million in 1983/84.

20 ECONOMY

Mining and cattle raising were the principal economic activities in Arizona during the territorial period. With the introduction of irrigation in the early 1900s, farming assumed a greater importance. Improvements in transportation later in the 20th century led to the development of manufacturing and tourism. State and federal sources estimated the value of manufacturing in 1983 at $6.95 billion; tourism and travel, $4.8 billion; mining, $1.5 billion; and agriculture (including livestock), $1.6 billion. Leading products include electronic components and nonelectrical machinery from the manufacturing sector, copper from the mining sector, and cattle and cotton from the farming sector.

Arizona's economy compiled an impressive growth record during the 1970s and early 1980s. Between 1973 and 1983, the state population increased by 39% (4th in the US); nonfarm wage and salary employment grew by 49% (5th in the US); and total personal income by 218% (6th in the US). From 1976 through 1983, value added by manufacturing increased 141%, bank deposits 135%, and retail sales 117%. Only five states had a lower unemployment rate in 1984 than Arizona's 5%.

21 INCOME

In 1983, Arizona ranked 32d among the 50 states with a per capita income of $10,719. Total personal income was $31.8 billion in 1983. Of the 1982 personal income of $29.1 billion, $17.8 billion was concentrated in Maricopa County and $5.7 billion in Pima County (the Tucson metropolitan area). The other 13 counties, all rural and including most of the state's Indian lands, accounted for the rest. Except in Yavapai County, per capita income in these counties was far below that in Maricopa and Pima. In 1979, about 13% of the population was below the federal poverty level.

22 LABOR

In mid-1984, the civilian labor force totaled 1,393,100, 63% of it in Maricopa County and 18% in Pima. A federal survey in 1983 revealed the following nonfarm employment pattern for Arizona:

	ESTABLISH-MENTS	EMPLOYEES	ANNUAL PAYROLL ('000)
Agricultural services, forestry, fishing	992	6,824	$ 74,546
Mining, of which:	222	15,854	404,863
Metals	(84)	(12,682)	(311,924)
Contract construction	6,427	67,224	1,305,737
Manufacturing	3,280	149,065	3,302,578
Transportation, public utilities	1,716	48,864	1,125,785
Wholesale trade	4,777	50,396	963,945
Retail trade	16,936	206,304	2,020,758
Finance, insurance, real estate	6,330	63,727	1,140,881
Services	20,881	221,803	3,088,863
Other	6,336	6,032	151,345
TOTALS	67,897	836,093	$13,579,301

This survey excluded government workers, of whom there were 185,300 in August 1984. Unemployment in 1984 was 5%.

Organized labor has a long history in Arizona. A local of the Western Federation of Miners was founded in 1896, and labor was a powerful force at the constitutional convention in 1910. Nevertheless, the state's work force is much less organized than that of the nation as a whole. In 1980, 119,000 Arizonans belonged to labor unions; there were 330 labor unions in 1983. The state has a right-to-work law.

23 AGRICULTURE
Arizona's agricultural output (including livestock products) was valued at $1.6 billion in 1983 (31st in the US). Cash receipts from farming alone amounted to $928.1 million.

In 1984 there were about 8,000 farms covering 38 million acres (15 million hectares), or about 21% of the state's total area, but only 877,570 acres (355,141 hectares), or 1.2% of the state, were actually farmed for crops in 1983. The average size of 4,518 acres (1,828 hectares) per farm was by far the highest in the US. Arizona's farmed cropland is intensely cultivated and highly productive. In 1980, Arizona was first among all states in cotton yield per acre (1,085 lb). About 95% of all farmland is dependent on irrigation provided by dams and water projects.

Cotton is the leading cash crop in Arizona. In 1983, the state produced 772,000 bales of cotton lint on 313,300 acres (126,800 hectares), with a total value of $260,175,000. (In 1982, 1,161,000 bales were produced, with a value of $348,608,000.) Vegetables, especially lettuce, accounted for a value of $229,214,000 in 1983. Alfalfa is also an important item; total hay production was 1,241,000 tons in 1983, for a value of $114,172,000. Other crops are wheat, sorghum, barley, grapes, and citrus fruits.

24 ANIMAL HUSBANDRY
Livestock production in 1983 was valued at $715.5 million. The total inventory of cattle and calves was 980,000 in 1983, with an estimated value of $377 million. At the close of 1983, the state had 306,000 sheep and lambs and 150,000 hogs and pigs.

In 1983, 486,357 head of cattle were shipped out of Arizona, and 361,100 head were slaughtered within the state; the livestock slaughter also included 196,300 hogs and pigs. A total of 1,225,000 lb of whole milk was sold.

25 FISHING
Arizona has no commercial fishing. The state's lakes and mountain streams lure sport fishermen and are an increasingly important tourist attraction. In 1983, Arizona had 19 commercial fish farms.

26 FORESTRY
The lumber industry in Arizona began during the 19th century, when the building of the transcontinental railroad created a demand for railroad ties. During the 20th century, Arizona's forests have been more valuable for conservation and recreation than for lumber.

The main forest regions stretch from the northwest to the southeast, through the center of the state. All together, in 1979 there were 18,494,000 acres (7,484,000 hectares) of forestland in Arizona, 25% of the state's area and 2.5% of the US total. Commercial timberland accounted for only 3,896,000 acres (1,577,000 hectares). National forests covered nearly 12 million acres (5 million hectares) as of 30 September 1984.

27 MINING
Arizona ranked 5th in the US in nonfuel mineral production in 1984, with output valued at $1.4 billion.

Arizona produced nearly two-thirds of the nation's copper in 1983. Morenci, in Greenlee County, is the nation's 2d-largest copper mine, and San Manuel, in Pinal County, is third. The other major copper mines are in Pima and Pinal counties. In 1984, the state produced 747,757 tons of copper, with a value of $1 billion.

Molybdenum is the next most valuable mineral, with 1984 production of 23,184 lb and a value of $78.8 million, followed by sand and gravel, silver, gold, and crushed stone. In 1984, mineral output included 25,200,000 short tons of sand and gravel, 5,500,000 short tons of crushed stone, and 4,068,000 troy oz of silver. In 1983, gold production was 61,991 troy oz of gold.

28 ENERGY AND POWER
In 1983, Arizona produced 41.3 billion kwh of electric power; installed capacity was 10.9 million kw. The state had 30 power plants, of which 12 were hydroelectric, accounting for 19% of power output. Electric energy sales in the state were 30.7 billion kwh; the surplus production was exported to other states, primarily California.

Arizona's fossil-fuel potential remains largely undeveloped, though oil and natural-gas exploration was under way as the 1980s began. Coal production in 1983 was 11,404,000 short tons, all of it from two surface mines.

Energy resource development in the state is encouraged by the Department of Mines and Mineral Resources, Oil and Gas Conservation Commission, Department of Water Resources, and Solar Energy Commission.

29 INDUSTRY
Manufacturing, which has grown rapidly since World War II, became the state's leading economic activity in the 1970s. Factors contributing to this growth included favorable tax structure, available labor, plentiful electric power, and low land costs. The major manufacturing centers are the Phoenix and Tucson areas. Principal industries include nonelectrical machinery, electrical and electronic equipment, aircraft equipment, food products, and printing and publishing. Military equipment accounts for much of the output.

The total value added by manufacture was $6.2 billion in 1982, and the value of industry shipments was $12.9 billion. Value of industry shipments for selected industries in 1982 was as follows:

Office and computing machines	$1,915,500,000
Semiconductors, related devices	922,500,000
Communication equipment	630,500,000
Instruments and related products	374,400,000

30 COMMERCE
In 1982, wholesale establishments in Arizona had sales of $13.9 billion. Most wholesale establishments were concentrated in Maricopa and Pima counties. Retail sales in 1982 reached $13.9 billion, of which 61% took place in Maricopa County and 19% in Pima. Of the 1982 retail sales total for establishments with payrolls ($13.6 billion), food stores accounted for 25%; automotive dealers, 19%; general merchandise stores, 11%; eating and drinking places, 10%; gasoline service stations, 10%; and other establishments, 25%.

Manufactured goods worth $1.75 billion were exported in 1982, and agricultural products valued at $404 million were sent abroad in 1981/82. Total exports for 1982 were $2.45 billion, including $298 million in minerals. In 1981, Arizona led the nation in exports of copper ore and concentrates, valued at $109 million.

[31] CONSUMER PROTECTION

The Department of Law, headed by the attorney general, has primary responsibility for consumer protection.

[32] BANKING

In 1983 there were 46 commercial banks in Arizona. Total domestic assets of commercial banks came to $22 billion. There were 8 savings and loan associations, with total assets of $10.2 billion. Deposits in all Arizona banks totaled almost $15.9 billion at the close of 1983, and outstanding loans exceeded $11.5 billion.

The banking industry in Arizona is regulated by the Department of Banking.

[33] INSURANCE

In 1983, Arizona policyholders paid $2.4 billion in premiums for life, health, automobile, homeowners, and all other types of insurance. Premiums for life insurance totaled $508 million; accident and health, $575 million; automobile liability and physical damage, $655 million; and homeowners, $159 million.

Purchases of ordinary life insurance in Arizona totaled $10.3 billion in 1983, and the total amount of life insurance in force was $55.2 billion, covering 4.2 million policies. The average amount of life insurance per family increased from $19,600 in 1970 to $47,400 in 1983. Life insurance companies paid $598.7 million in benefits and annuities in 1983, including $170.6 million in death payments.

In 1983 there were 567 life insurance companies and 19 property and casualty insurance companies with home offices in Arizona. The Department of Insurance regulates the state's insurance industry and examines and licenses agents and brokers.

[34] SECURITIES

In 1983, the New York Stock Exchange had 79 sales offices and 994 full-time registered representatives in Arizona. State residents owning shares of public corporations numbered 513,000 in 1983. The state has no securities exchanges.

[35] PUBLIC FINANCE

The state budget is prepared annually and submitted by the governor to the legislature for amendment and approval. The fiscal year runs from 1 July to 30 June.

The following table summarizes consolidated revenues and expenditures for 1983/84 (actual) and 1984/85 (estimated):

REVENUES	1983/84	1984/85
Taxes	$1,872,397,900	$2,202,910,600
Dividends and interest	23,267,400	22,000,000
Licenses, fees, and permits	100,498,100	114,669,900
Grants and aids	767,811,000	827,373,900
Other receipts	207,642,200	225,270,100
TOTALS	$2,971,616,600	$3,392,224,500
EXPENDITURES		
Education	$1,703,263,800	$1,910,114,600
Health and welfare	810,058,200	844,367,100
Transportation	425,968,700	356,904,900
Protection and safety	201,830,900	236,498,600
General government	205,114,600	134,763,800
Other expenses	111,845,300	161,905,500
TOTALS	$3,458,081,500	$3,644,554,500

The total debt of state and local government as of mid-1983 was nearly $7.7 billion, or about $2,600 per capita.

[36] TAXATION

State tax receipts for the general fund in 1983/84 came to almost $1.9 billion, of which income taxes accounted for $633.2 million. The personal income tax ranged from 2% on the first $1,017 to 8% on income over $6,102. For corporations the maximum rate was 10.5% on net income over $6,000. The state sales and use taxes raised $854.4 million in 1983/84, with the retail sales tax rate at 5%. An estate tax, luxury tax, pari-mutuel tax, insurance premium tax, and transport fuel tax are also levied. The state also receives tax revenue from collections of the general property tax and the motor-vehicle license tax, and it has a lottery.

In 1983, Arizona's share of the federal tax burden was $6.4 billion. Arizonans filed 1,204,624 federal income tax returns, paying almost $3 billion in federal income tax.

[37] ECONOMIC POLICY

The Department of Commerce has primary responsibility for attracting business and industry to Arizona, aiding existing business and industry, and assisting companies engaged in international trade. Its programs emphasize job opportunities, energy conservation, support of small businesses, and development of the motion-picture industry.

[38] HEALTH

In 1983 there were 52,919 live births and 22,482 deaths in Arizona. Infant mortality in 1981 was 11 per 1,000 live births among whites and 21.1 among blacks. There were 15,800 legal abortions in 1982; the rate of 298 per 1,000 live births was far below the national average. Arizona's overall death rate was below the US norm in 1981 and was especially low for heart disease and cardiovascular diseases. Deaths from accidents and suicide were above average, however. Serious public-health problems include tuberculosis and San Joaquin Valley fever (coccidioidomycosis), especially among the Indians. Although the state's cancer death rate was lower than the national average, some studies claimed a link between cancer incidence and fallout from nuclear tests held in Nevada during the 1950s and 1960s.

In 1982 there were 80 hospitals, with 12,100 beds. Hospital personnel included 10,114 registered nurses in 1981. The state had 4,911 active nonfederal physicians (175 per 100,000 people) in 1981 and 1,384 dentists in 1982. The average daily semiprivate hospital room charge in January 1984 was $186. In 1981, 343,000 people received $476 million in Medicare benefits.

[39] SOCIAL WELFARE

Arizona was last among states in state and local public welfare expenditures per capita ($68) in 1980. In 1982, aid to families with dependent children totaled $51 million. About 224,000 persons participated in the federal food stamp program in 1983, at a federal cost of $142 million. Some 231,000 pupils (47% of the eligible enrollment) took part in the school lunch program, at a cost to the federal government of $26 million.

In 1983, $21.6 billion in Social Security benefits was paid to 462,000 beneficiaries; the average monthly benefit for retirees was $448. About 29,000 Arizonans received $68.3 million in federal Supplemental Security Income benefits during 1983. Arizona was the only state not participating in the Medicaid program (except for pilot projects) as of 1983. In 1982, unemployment insurance benefits totaled $191 million. Workers' compensation payments amounted to $144.8 million in 1982.

[40] HOUSING

The 1980 census counted 1,066,437 units of year-round housing in Arizona, of which 957,032 were occupied. Of the latter, 68% were owner-occupied and 98% had complete plumbing facilities. Overcrowding was a serious problem: more than 7% of occupied units averaged more than one person per room.

The early 1980s saw the boom in privately owned housing construction continuing, with 133,900 units authorized from 1981 through 1983. In 1983 alone, 64,000 units were authorized. Building permits valued at $2.3 billion were authorized in that year for privately owned housing units.

[41] EDUCATION

In 1980, 72.4% of Arizonans 25 years old and over were high school graduates, and 17.4% were college graduates. Some 28,000 Arizonans graduated from high school in 1982.

The first public school in the state opened in 1871 at Tucson, with 1 teacher and 138 students. In 1983/84, enrollment at public schools was 545,760. There were 25,700 public-school teachers in 1984, 71% of them in the elementary grades. Private schools enrolled 40,261 students in 1980.

The leading public higher educational institutions, the University of Arizona at Tucson and Arizona State University (originally named the Arizona Territorial Normal School) at Tempe, were both established in 1885. As of 1982, the state had 12 colleges and universities and 17 community colleges, with a total enrollment of 210,683. The American Graduate School of International Management, a private institution, is located in Glendale.

42ARTS

Arizona has traditionally been a center for Indian folk arts and crafts. The Arizona State Museum (Tucson), Colorado River Indian Tribes Museum (Parker), Heard Museum of Anthropology and Primitive Art (Phoenix), Mohave Museum of History and Arts (Kingman), Navaho Tribal Museum (Window Rock), and Pueblo Grande Museum (Phoenix) all display Indian creations, both historic and contemporary. Modern Arizona artists are featured at the Tucson Museum of Art and the Yuma Art Center.

Musical and dramatic performances are presented in Phoenix, Tucson, Scottsdale, and other major cities. Phoenix and Tucson have symphony orchestras, and the Arizona Opera Company performs in both cities.

43LIBRARIES AND MUSEUMS

In 1982/83, Arizona's public libraries had a combined book stock of 4.5 million volumes, and total circulation was 16 million. Spending on libraries per capita was $8.70 in 1981. Principal public libraries include the Phoenix Public Library and the State Library and Department of Archives in Phoenix, and the Arizona Historical Society Library in Tucson. The largest university libraries are located at the University of Arizona and Arizona State University.

Arizona has more than 90 museums and historic sites. Attractions in Tucson include the Arizona State Museum, University of Arizona Museum of Art, Arizona Historical Society, Arizona-Sonora Desert Museum, Flandreau Planetarium, and Gene C. Reid Zoological Park. Phoenix has the Heard Museum (anthropology and primitive art), Arizona Mineral Resources Museum, Phoenix Art Museum, Phoenix Zoo, Pueblo Grande Museum, and Desert Botanical Garden. The Museum of Northern Arizona and Lowell Observatory are in Flagstaff. Kitt Peak National Observatory is in Tucson.

Archaeological and historical sites include the cliff dwellings at the Canyon de Chelly, Casa Grande Ruins, Montezuma Castle, Tonto, and Tuzigoot national monuments; the town of Tombstone, the site of the famous O. K. Corral gunfight in the early 1880s; the restored mission church at Tumacacori National Monument; and San Xavier del Bac Church near Tucson.

44COMMUNICATIONS

In 1985 there were 209 post offices and 8,379 postal employees in Arizona. Of the 1,721,594 telephones in service in 1983, 1,107,238 were in Phoenix and 342,626 in Tucson. Ninety percent of all housing units had telephones in 1980.

There were 123 radio stations broadcasting in Arizona in 1983, 70 AM and 53 FM. The state also had 14 commercial television stations in 1984—6 in Phoenix, 3 in Tucson, and 1 each in Flagstaff, Mesa, Nogales, Prescott, and Yuma—as well as 2 educational television stations, one at Arizona State University and the other at the University of Arizona. Sixty-nine cable television systems were serving 346,242 subscribers in 136 communities.

45PRESS

The *Weekly Arizonian,* started in Tubac in 1859, was the first newspaper in the state. The *Daily Arizona Miner,* the state's first daily, was founded at Prescott in 1866. As of 1985 there were 4 morning dailies with a combined 1984 circulation of 402,559 and 14 evening dailies with 276,475 circulation. Eleven dailies had Sunday editions. There were 45 weekly newspapers. The following table shows 1984 circulations for leading dailies:

AREA	NAME	DAILY	SUNDAY
Phoenix	Arizona Republic (m,S)	283,550	430,200
	Gazette (e)	106,142	
Tucson	Citizen (e)	60,515	
	Arizona Daily Star (m,S)	73,392	137,838

Sixty-six magazines and periodicals were published in Arizona in 1984. Among the most notable were *Phoenix Magazine, Phoenix Living,* and *Arizona Living,* devoted to the local and regional life-style; *American West,* dedicated to the Western heritage; *Arizona and the West,* published quarterly by the University of Arizona Library in Tucson; and *Arizona Highways,* a beautifully illustrated monthly published by the Department of Transportation in Phoenix.

46ORGANIZATIONS

The 1982 Census of Service Industries counted 687 organizations in Arizona, including 125 business associations; 394 civic, social, and fraternal associations; and 16 educational, scientific, and research organizations. Among the organizations headquartered in Arizona in 1984 were the National Foundation for Asthma (Tucson), the Pianists Foundation of America (Tucson), the National Indian Athletic Association (Mesa), the Kampground Owners Association (Phoenix), the American Science Fiction Association (Scottsdale), the Southwest Parks and Monuments Association (Globe), and Up With People (Tucson).

47TOURISM, TRAVEL, AND RECREATION

Tourism and travel is the 2d-leading industry in Arizona. In 1983, tourism and travel accounted for $4.8 billion in income according to one estimate; for nearly $2.8 billion in 1982, excluding foreign visitors, according to another. Most tourists were Americans, but travelers from Mexico spent an estimated $622 million in the state in 1981. Total employment in tourism and travel in 1982 was 65,600.

There are 19 national parks and monuments located entirely within Arizona. By far the most popular is Grand Canyon National Park, which had 2,173,584 visitors in 1984. Petrified Forest National Park had 714,761 visitors, and Saguaro National Monument, 1,699,242. In all, the national park service areas entirely in Arizona had 7,674,764 visitors; in addition, 8,103,527 persons visited the Glen Canyon and Lake Mead National Recreation Areas, partly in Arizona. There are also 19 state parks, attracting 2,192,535 visitors in 1983; Lake Havasu, the overwhelming favorite, had 1,000,423 visitors.

Arizona offers excellent camping on both public and private land, and there are many farm vacation sites and dude ranches, particularly in the Tucson and Wickenburg areas. Popular for sightseeing and shopping are the state's Indian reservations, particularly those of the Navaho and Hopi. Boating and fishing on Lake Mead, Lake Powell, Lake Mohave, Lake Havasu, the Colorado River, and the Salt River lakes are also attractions.

Licenses were issued to 186,977 hunters and 454,045 fishermen in 1982/83. Boat registrations in 1983 totaled 107,333.

48SPORTS

The Phoenix Suns, the state's lone major league professional team, play in the National Basketball Association. Several major league baseball teams hold spring training in Arizona, and Phoenix and Tucson have entries in the Pacific Coast League. There is horse racing at Turf Paradise in Phoenix, and dog racing at Phoenix, Tucson, and Yuma. Auto racing is held at Manzanita Raceway and International Raceway, in Phoenix. Both Phoenix and Tucson have hosted tournaments on the Professional Golfers Association's nationwide tour.

Rodeos are held throughout the state. Phoenix hosts the Rodeo of Rodeos, while Tucson has La Fiesta de los Vaqueros (The Festival of the Cowboys).

The Sun Devils of Arizona State won Western Athletic Conference football titles five times between 1969 and 1977 and shared the title twice (once with the University of Arizona Wildcats). Both Arizona State and the University of Arizona joined the Pacific 10 Conference in 1978. The Wildcats captured NCAA Division I baseball championships in 1975 and 1980. The Sun Devils won the championship in 1981. College football's Fiesta Bowl is held annually at Arizona State in Tempe.

⁴⁹FAMOUS ARIZONANS

Although Arizona entered the Union relatively late, many of its citizens have achieved national prominence, especially since World War II. William H. Rehnquist (b.Wisconsin, 1924) was appointed associate justice of the US Supreme Court in 1971; in 1981, Sandra Day O'Connor (b.Texas, 1930) became the first woman to serve on the Supreme Court. Arizona natives who became federal officeholders include Lewis Douglas (1894-1974), a representative who served as director of the budget in 1933-34 and ambassador to the Court of St. James's from 1947 to 1950; Stewart L. Udall (b.1920), secretary of the interior, 1961-69; and Richard B. Kleindienst (b.1923), attorney general, 1972-73, who resigned during the Watergate scandal. Another native son was Carl T. Hayden (1877-1972), who served in the US House of Representatives from statehood in 1912 until 1927 and in the US Senate from 1927 to 1969, thereby setting a record for congressional tenure. Barry Goldwater (b.1909), son of a pioneer family, was elected to the US Senate in 1952, won the Republican presidential nomination in 1964, and returned to the Senate in 1968. His Republican colleague, John J. Rhodes (b.Kansas, 1916), served in the US House of Representatives for 30 years and was House minority leader from 1973 to 1980. Raul H. Castro (b.Mexico, 1916), a native of Sonora, came to the US in 1926, was naturalized, served as Arizona governor from 1975 to 1977, and has held several ambassadorships to Latin America. Morris K. Udall (b.1922), first elected to the US House of Representatives in 1960, contended for the Democratic presidential nomination in 1976.

Prominent state officeholders included General John C. Frémont (b.Georgia, 1813-90), who was territorial governor of Arizona from 1878 to 1883, and George W. P. Hunt (1859-1934), who presided over the state constitutional convention in 1910 and was elected governor seven times during the early decades of statehood. Eusebio Kino (b.Italy, 1645?-1711) was a pioneer Jesuit who introduced missions and European civilization to Arizona. Also important to the state's history and development were Charles D. Poston (1825-1902), who in the late 1850s promoted settlement and separate territorial status for Arizona; Chiricahua Apache leaders Cochise (1812?-74) and Geronimo (1829-1909), who, resisting the forced resettlement of their people by the US government, launched a series of raids that occupied the Army in the Southwest for over two decades; Wyatt Earp (b.Illinois, 1848-1929), legendary lawman of Tombstone during the early 1880s; John C. Greenway (1872-1926), copper magnate and town builder who was a nominee on the Democratic ticket in 1924 for US vice president; and Frank Luke, Jr. (1897-1918), a World War I flying ace who was the first American airman to receive the Medal of Honor.

Distinguished professional people associated with Arizona have included James Douglas (b.Canada, 1837-1918), metallurgist and developer of the Bisbee copper district; Percival Lowell (b.Massachusetts, 1855-1916), who built the Lowell Observatory in Flagstaff; and Andrew Ellicott Douglass (b.Vermont, 1867-1962), astronomer, university president, and inventor of dendrochronology, the science of dating events and environmental variations through the study of tree rings and aged wood. Cesar Chavez (b.1927) is president of the United Farm Workers of America.

Writers whose names have been associated with Arizona include novelist Harold Bell Wright (b.New York, 1872-1944), who lived for an extended period in Tucson; Zane Grey (b.Ohio, 1875-1939), who wrote many of his western adventure stories in his summer home near Payson; and Joseph Wood Krutch (b.Tennessee, 1893-1970), an essayist and naturalist who spent his last two decades in Arizona. Well-known performing artists from Arizona include singers Marty Robbins (1925-82) and Linda Ronstadt (b.1946). Joan Ganz Cooney (b.1929), president of the Children's Television Workshop, was one of the creators of the award-winning children's program *Sesame Street*.

⁵⁰BIBLIOGRAPHY

Arizona: Its People and Resources. 2d ed., rev. Tucson: University of Arizona Press, 1972.

Arizona Yearbook, 1979-80: A Guide to Government in Arizona. Yuma: Arizona Information Press, 1979.

Barnes, Will C. *Arizona Place Names.* Revised by Byrd H. Granger. Tucson: University of Arizona Press, 1960.

Comeaux, Malcolm. *Arizona: A Geography.* Boulder: Westview Press, 1981.

Faulk, Odie B. *Arizona: A Short History.* Norman: University of Oklahoma Press, 1970.

Federal Writers' Project. *Arizona: The Grand Canyon State.* New York: Hastings House, 1968 (orig. 1940).

Fireman, Bert M. *Arizona: Historic Land.* New York: Knopf, 1982.

Goldwater, Barry. *Arizona.* New York: Random House, 1978.

Krutch, Joseph Wood. *The Desert Year.* New York: Viking, 1963 (orig. 1951).

Lamar, Howard R. *The Far Southwest, 1846-1912: A Territorial History.* New Haven: Yale University Press, 1966.

Powell, Lawrence C. *Arizona: A Bicentennial History.* New York: Norton, 1976.

University of Arizona Library. *The Arizona Index: A Subject Index to Periodical Articles About the State.* Boston: Hall Library, 1978.

Valley National Bank of Arizona. *Arizona Statistical Review.* 40th ed. Phoenix, 1984.

Wagoner, Jay J. *Arizona's Heritage.* Layton, Utah: Peregrine Smith, 1977.

Wagoner, Jay J. *Arizona Territory 1863-1912.* Tucson: University of Arizona Press, 1970.

Wagoner, Jay J. *Early Arizona: Prehistory to Civil War.* Tucson: University of Arizona Press, 1975.

Walker, Henry P., and Don Bufkin. *Historical Atlas of Arizona.* Norman: University of Oklahoma Press, 1980.

Wallace, Andrew, ed. *Sources and Readings in Arizona History.* Tucson: Arizona Pioneers' Historical Society, 1965.

Wyllys, Rufus K. *Arizona: The History of a Frontier State.* Phoenix: Hobson and Herr, 1950.

ARKANSAS

State of Arkansas

ORIGIN OF STATE NAME: French derivation of *Akansas* or *Arkansas,* a name given to the Quapaw Indians by other tribes. **NICKNAME:** The Land of Opportunity. **CAPITAL:** Little Rock. **ENTERED THE UNION:** 15 June 1836 (25th). **SONG:** "Arkansas." **MOTTO:** *Regnat populus* (The people rule). **COAT OF ARMS:** In front of an American eagle is a shield displaying a steamboat, plow, beehive, and sheaf of wheat, symbols of Arkansas's industrial and agricultural wealth. The angel of mercy, the goddess of liberty encircled by 13 stars, and the sword of justice surround the eagle, which holds in its talons an olive branch and three arrows, and in its beak a banner bearing the state motto. **FLAG:** On a red field, 25 stars on a blue band border a white diamond containing the word "Arkansas" and four blue stars. **OFFICIAL SEAL:** Coat of arms surrounded by the words "Great Seal of Arkansas" and four blue stars. **BIRD:** Mockingbird. **FLOWER:** Apple blossom. **TREE:** Pine. **GEM:** Diamond. **INSECT:** Honeybee. **LEGAL HOLIDAYS:** New Year's Day, 1 January; Robert E. Lee's birthday, 19 January; Birthday of Martin Luther King, Jr., 3d Monday in January; George Washington's Birthday, 3d Monday in February; Memorial Day, last Monday in May, Independence Day, 4 July; Labor Day, 1st Monday in September; Veterans Day, 11 November; Thanksgiving Day, 4th Thursday in November; Christmas Eve, 24 December; Christmas Day, 25 December. **TIME:** 6 AM CST = noon GMT.

¹LOCATION, SIZE, AND EXTENT

Located in the western south-central US, Arkansas ranks 27th in size among the 50 states.

The total area of Arkansas is 53,187 sq mi (137,754 sq km), of which land takes up 52,078 sq mi (134,882 sq km) and inland water 1,109 sq mi (2,872 sq km). Arkansas extends about 275 mi (443 km) E–W and 240 mi (386 km) N–S.

Arkansas is bordered on the N by Missouri, on the E by Missouri, Tennessee, and Mississippi (with part of the line passing through the St. Francis and Mississippi rivers), on the S by Louisiana, on the SW by Texas (with part of the line formed by the Red River), and on the W by Oklahoma. The total boundary length of Arkansas is *1,168 mi (1,880 km).* The state's geographic center is in Pulaski County, 12 mi (19 km) NW of Little Rock.

²TOPOGRAPHY

The Boston Mountains (an extension of the Ozark Plateau, sometimes called the Ozark Mountains) in the northwest and the Ouachita Mountains in the west-central region not only constitute Arkansas's major uplands but also are the only mountain chains between the Appalachians and the Rockies. Aside from the wide valley of the Arkansas River, which separates the two chains, the Arkansas lowlands belong to two physiographic regions: the Mississippi Alluvial Plain and the Gulf Coastal Plain. The highest elevation in Arkansas, at 2,753 feet (839 meters), is Magazine Mountain, standing north of the Ouachitas in the Arkansas River Valley. The state's lowest point, at 55 feet (17 meters), is on the Oauchita River in south-central Arkansas.

Arkansas's largest lake is the artificial Lake Ouachita, covering 63 sq mi (163 sq km); Lake Chicot, in southeastern Arkansas, an oxbow of the Mississippi River, is the state's largest natural lake, with a length of 18 mi (29 km). Bull Shoals Lake, occupying 71 sq mi (184 sq km), is shared with Missouri. Principal rivers include the Mississippi, forming most of the eastern boundary; the Arkansas, beginning in Colorado and flowing 1,450 mi (2,334 km) through Kansas and Oklahoma and across central Arkansas to the Mississippi; and the Red, White, Ouachita, and St. Francis rivers, all of which likewise drain south and southeast into the Mississippi. Numerous springs are found in Arkansas, of which the best known are Mammoth Springs, near the Missouri border, one of

the largest in the world, with a flow rate averaging 9 million gallons an hour, and Hot Springs in the Ouachitas.

Crowley's Ridge, a unique strip of hills formed by sedimentary deposits and windblown sand, lies west of and parallel to the St. Francis River for about 180 mi (290 km). The ridge is rich in fossils and has an unusual diversity of plant life.

³CLIMATE

Arkansas has a temperate climate, warmer and more humid in the southern lowlands than in the mountainous regions. At Little Rock, the normal daily temperature ranges from 40°F (4°C) in January to 81°F (27°C) in July. A record low of −29°F (−34°C) was set on 13 February 1906 at the Pond weather station, and a record high of 120°F (49°C) on 10 August 1936 at the Ozark station.

Average yearly precipitation is approximately 45 in (114 cm) in the mountainous areas and greater in the lowlands; Little Rock receives an annual average of 49 in (124 cm) and has an average relative humidity ranging from 84% at 7 AM to 57% at 1 PM. Snowfall in the capital averages 5.4 in (13.7 cm) a year.

⁴FLORA AND FAUNA

Arkansas has at least 2,600 native plants, and there are many naturalized exotic species. Cypresses, water oak, hickory, and ash grow in the Mississippi Valley, while the St. Francis Valley is home to the rare cork tree. Crowley's Ridge is thick with tulip trees and beeches. A forest belt of oak, hickory, and pine stretches across south-central and southwestern Arkansas, including the Ozark and Quachita mountains. The Mexican juniper is common along the White River's banks. The state has at least 26 native varieties of orchid; the passion flower is so abundant that it was once considered for designation as the state flower, but the apple blossom was finally named.

Arkansas's native animals include 15 varieties of bat and 3 each of rabbit and squirrel. Common throughout the state are mink, armadillo, white-tailed deer, and eastern chipmunk. Black bear roam the swamp and mountain regions. Among 300 native birds are such game birds as the eastern wild turkey, mourning dove, and bobwhite quail. Among local fish are catfish, gar, and the unusual paddle fish. Arkansas counts 20 frog and toad species, 23 varieties of salamander, and 36 kinds of snake.

The Arkansas Game and Fish Commission lists the leopard darter and fat pocketbook pearly mussel as threatened species. The peregrine falcon and American alligator are listed as endangered, along with the Indiana and gray bats and eastern puma. In 1983, Arkansas established the Non-Game Preservation Committee to promote sound management, conservation, and public awareness of the state's nongame animals and native plants.

⁵ENVIRONMENTAL PROTECTION
The Department of Pollution Control and Ecology, the state's principal environmental protection agency, was established in 1971. A survey by the department in 1980 indicated that the two environmental issues Arkansans were most concerned about were water pollution control and solid waste disposal. As of the early 1980s, the state's water quality standards were among the strictest in the US. The state's solid waste code, promulgated in 1973, was rewritten in 1984 to establish four classes of landfill and to provide for disposal methods tailored to hazardous wastes such as asbestos and laboratory wastes. Citizens' groups actively involved with environmental issues include the Arkansas Ecology Center, Association of Community Organizations for Reform Now (ACORN), the League of Women Voters, the Ozark Society, Ducks Unlimited, and some local chapters of the US Wildlife Federation. The Federation of Water and Air Users presents industry's viewpoints on environmental issues.

The Buffalo, designated as a national river, flows through northern Arkansas. One of the wildest areas in the state is the 113,000-acre (46,000-hectare) White River Wildlife Refuge, which contains more than 100 small lakes. The Department of Arkansas Natural and Cultural Heritage was established in 1975 for, among other purposes, the preservation of rivers and other natural areas in an unspoiled condition.

⁶POPULATION
At the time of the 1980 census, Arkansas had a population of 2,285,500 (33d in the US), an increase of 19% from the 1970 population of 1,923,322. Estimates for January 1985 showed Arkansas with 2,345,431 residents; the average population density was 45 per sq mi (17 per sq km). The Census Bureau estimates that the population will be 2,579,800 in 1990.

As of 1983, Arkansas and Rhode Island were tied for 2d behind Florida in percentage of population aged 65 or over—14.1%—partially reflecting the large numbers of retirees who settled in the state during the 1970s and early 1980s. At the same time, an above-average percentage of Arkansans were 17 or under.

Almost 39% of all state residents lived in metropolitan areas in 1983. The largest city in Arkansas is Little Rock, which had a 1984 population of 170,388. Other major cities include Ft. Smith, 72,607; North Little Rock, 65,025; Pine Bluff, 55,860; Hot Springs, 36,838; and Fayetteville, 35,709. The Little Rock–North Little Rock metropolitan area had 492,700 residents in 1984.

⁷ETHNIC GROUPS
Arkansas's population is predominantly white, composed mainly of descendants of immigrants from the British Isles. The only significant minority group consists of black Americans, numbering 373,768, or 16.4% of the population, in 1980. The 1980 census listed 18,000 Hispanics, 9,346 American Indians, 2,042 Vietnamese, 1,275 Chinese, 921 Filipinos, 832 Asian Indians, and 754 Japanese. The foreign-born population numbered 22,000, or 1% of all Arkansas residents, in 1980.

⁸LANGUAGES
A few place-names—such as Arkansas itself, Choctaw, Caddo, and Ouachita—attest the onetime presence of American Indians, mostly members of the Caddoan tribes, in the Territory of Arkansas.

Arkansas English is essentially a blend of Southern and South Midland speech, with South Midland dominating the mountainous northwest and Southern the southeastern agricultural areas. Common in the east and south are *redworm* (earthworm) and

mosquito hawk (dragonfly). In the northwest appear South Midland *whirlygig* (merry-go-round) and *sallet* (garden greens).

In 1980, 2,139,245 Arkansans—94% of the residents 3 years old or older—spoke only English at home. Other languages spoken at home included:

Spanish	13,568	Vietnamese	1,694
German	5,615	Chinese	1,116
French	5,050	Polish	1,085

⁹RELIGIONS
Although French Roman Catholic priests had worked as missionaries among the Indians since the early 18th century, the state's first mission was founded among the Cherokee by a Congregationalist, Cephas Washburn, in 1820. When the Cherokee were removed to Indian Territory (present-day Oklahoma), the mission moved there as well, remaining active through the Civil War. William Patterson may have been the first Methodist to preach in Arkansas, around 1800, in the area of Little Prairie; the first Methodist circuit, that of Spring River, was organized in 1815. The first Baptist church was likely that of the Salem congregation, begun in 1818 near what is now Pocahontas.

The largest denomination in Arkansas is the Southern Baptist Convention, which had 522,985 known adherents in 1980. Other leading Protestant groups are the United Methodist Church, with 214,526 adherents in 1980, and the Baptist Missionary Association of America, with 70,645. As of 1984, the Roman Catholic population of Arkansas was 52,137 and the estimated Jewish population was 3,175.

¹⁰TRANSPORTATION
Although railroad construction began in the 1850s, not until after the Civil War were any lines completed. The most important railroad—the St. Louis, Iron Mountain, and Southern line—reached Little Rock in 1872 and was subsequently acquired by financier Jay Gould, who added the Little Rock and Ft. Smith line to it in 1882. By 1890, the state had about 2,200 mi (3,500 km) of track; in 1974, trackage totaled 3,559 mi (5,728 km). As of 1984, Arkansas was served by four major railroads with 2,724 mi (4,384 km) of track. Amtrak passenger trains serviced Little Rock, Newport, Walnut Ridge, Malvern, Arkadelphia, and Texarkana en route from St. Louis to Dallas.

Intensive road building began in the 1920s, following the establishment of the State Highway Commission and the inauguration of a gasoline tax. By 1983, Arkansas had 90,796 mi (146,122 km) of roads, streets, and highways. During the same year, 941,393 automobiles and 500,358 trucks were registered in Arkansas, and there were 1,649,567 licensed drivers.

Beginning in the 1820s, steamboats replaced keelboats and flatboats on Arkansas rivers. Steamboat transportation reached its peak during 1870–90, until supplanted by the railroads that were opened during the same two decades. Development of the Arkansas River, completed during the early 1970s, made the waterway commercially navigable all the way to Tulsa.

In 1983, Arkansas had 153 airports and 7 heliports. The principal airport in the state, Adams Field at Little Rock, enplaned 472,351 passengers and handled 8,497 departures.

¹¹HISTORY
Evidence of human occupation of Arkansas reaches back to about 10,000 BC. The bluff dwellers of the Ozark Plateau were among the first human beings to live in what is now Arkansas, making their homes in caves and beneath overhanging rock cliffs along the banks of the upper White River. Farther south are the remains of another primitive people, the Mound Builders. The most significant of the Stone Age monuments they left are those of the Toltec group in Lonoke County, some 25 mi (40 km) southeast of Little Rock. Eventually, both ancient peoples vanished, for reasons that remain unclear.

Foremost among the Indian tribes in Arkansas were the

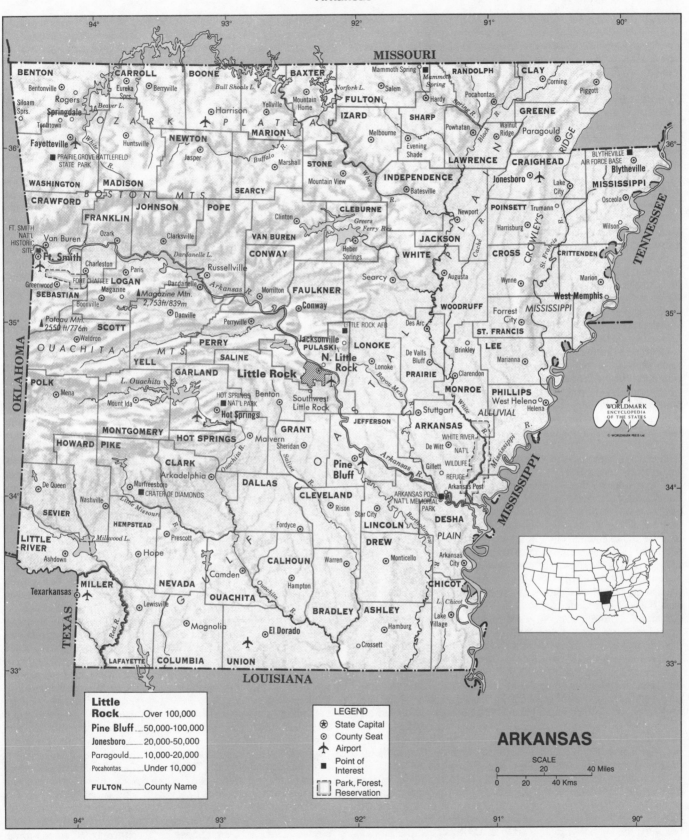

Little Rock — Over 100,000
Pine Bluff — 50,000-100,000
Jonesboro — 20,000-50,000
Paragould — 10,000-20,000
Pocahontas — Under 10,000

FULTON — County Name

LEGEND
⊛ State Capital
⊙ County Seat
✈ Airport
■ Point of Interest
⬚ Park, Forest, Reservation

ARKANSAS

SCALE
0   20   40 Miles
0   20   40 Kms

See endsheet maps: H3.

LOCATION: 33° to 36° 31′N; 89°41′ to 94°42′ W. **BOUNDARIES:** Missouri line, *331 mi (533 km)*; Tennessee line, *163 mi (262 km)*;
Mississippi line, *208 mi (335 km)*; Louisiana line, *166 mi (267 km)*; Texas line, *102 mi (164 km)*;
Oklahoma line, *198 mi (319 km)*.

Quapaw (meaning "downstream people"), agriculturists who had migrated to southern Arkansas in the early 16th century; the Caddo, fighters from Texas, who claimed the western region between the Red and Arkansas rivers; the warlike Osage, who hunted north of the Arkansas River and in present-day Missouri; and the Choctaw and Chickasaw of the northeast. Another prominent tribe, the Cherokee, arrived in the early 19th century, after federal and state authorities had taken their land east of the Mississippi and driven them westward. Nearly all these Indians had been expelled to what is now Oklahoma by the time Arkansas became a state.

The first Europeans to set foot in Arkansas were Spaniards, led by Hernando de Soto. They crossed the Mississippi, probably near present-day Helena, in the spring of 1541, roamed the land for a year or so, and then returned to the mighty river, where De Soto was buried in 1542. More than 100 years later, in 1673, a small band of Frenchmen led by Jacques Marquette, a Jesuit missionary, and Louis Jolliet, a fur trader and explorer, ended their voyage down the Mississippi at the mouth of the Arkansas River and returned north after being advised by friendly Indians that hostile tribes lay to the south. Nine years later, Robert Cavelier, Sieur de la Salle, led an expedition from Canada down the Mississippi to the Gulf of Mexico, stopping at Indian villages in Arkansas along the way and, on 9 April 1682, claiming all the Mississippi Valley for his king, Louis XIV.

Henri de Tonti, who had been second in command to La Salle, came back to Arkansas in 1686 to claim a land grant at the confluence of the Arkansas and White rivers, a few miles inland from the Mississippi. He left six men there; the log house they built was the beginning of Arkansas Post, the first permanent white settlement in the lower Mississippi Valley. Though tiny and isolated, Arkansas Post upheld the French claim to the Mississippi Valley until 1762, when France ceded the territory to Spain. Restored to France in 1800, the territory was sold to the US in the Louisiana Purchase of 1803. White settlers soon began arriving in Arkansas, and in 1806, the Louisiana territorial legislature created the District of Arkansas as a separate entity. When the Louisiana Purchase was further subdivided, Arkansas became part of the Missouri Territory. In 1819, Arkansas gained territorial status in its own right, and its boundaries were fixed by Congress. The territorial capital was moved from Arkansas Post to Little Rock in 1821. By 1835, Arkansas Territory had a population of 52,240, including 9,838 slaves. It was admitted to the Union in 1836 as a slave state, paired with the free state of Michigan in accordance with the Missouri Compromise.

Increasing numbers of slaves were brought into the largely agricultural state as the cultivation of cotton spread. Arkansas, like the rest of the South, was headed for secession, although it waited to commit itself until the Civil War had begun. There was considerable Union sentiment in the state, especially in the hilly northern and western counties, which lacked the large plantations—and the slaves—of southern and eastern Arkansas. But the pro-Union sympathies crumbled after Confederate guns fired on Ft. Sumter, S.C., and a secession convention was held at Little Rock on 6 May 1861. The final vote to leave the Union was 69-1; the lone holdout was Isaac Murphy of Madison County, who became the first Republican governor at the end of the war.

The largest Civil War battle fought in Arkansas, and one of the most significant battles of the war west of the Mississippi, was at Pea Ridge, in the northwest corner of the state. After three days of fighting, the Confederate forces relinquished the field. By September 1863, the Union Army had taken Little Rock, and the capital was moved to Washington, in Hempstead County, until the conclusion of hostilities in 1865.

Like virtually all white southerners, Arkansas's white majority hated the postwar Reconstruction government and repudiated it thoroughly at the first opportunity. Reconstruction officially

ended in 1874, when the reenfranchised white Democratic majority adopted a new state constitution, throwing out the carpetbagger constitution of 1868. The most colorful figure in postwar Arkansas was federal judge Isaac C. Parker, known as the Hanging Judge. From his court at Ft. Smith, he had sole jurisdiction over Indian Territory, which had become a gathering place for the nation's worst cutthroats. Parker and his deputy marshals fought them relentlessly. From 1875 through 1896, the judge hanged 79 men on his Ft. Smith gallows. The struggle was not one-sided: 65 of Parker's deputy marshals were killed.

Industrialization, urbanization, and modernization did not come to Arkansas until after the depression of the 1930s. Following World War II, the state became the first in the South to integrate its public colleges and universities. Little Rock's school board decided in 1954 to comply with the US Supreme Court's desegregation decision. Nevertheless, in September 1957, Governor Orval E. Faubus called out the National Guard to block the integration of Central High School at Little Rock. US President Dwight D. Eisenhower enforced a federal court order to integrate the school by sending in federal troops. The 1957 crisis brought years of notoriety to Arkansas, as Faubus, then in his second term, was elected to a third term and then to three more.

By the end of the Faubus administration, the public mood had changed, and the contrast between Faubus and his successor could not have been greater. Winthrop Rockefeller, millionaire scion of a famous family, moved to Arkansas from New York in the early 1950s, established himself as a gentleman rancher, and devoted himself to luring industry into his adopted state and building a Republican Party organization in one of the most staunchly Democratic states in the Union. Elected governor in 1966, Rockefeller thus became the first Republican to capture the Arkansas statehouse since Reconstruction. The specific accomplishments of his two terms were relatively few—he and the Democratic-controlled legislature warred incessantly—but he helped immeasurably in bringing a new image and a new spirit to the state. Rockefeller's successors have continued the progressive approach he took. New industry has moved into the state, along with new people. But the state remains among the poorest in the nation, with a per capita income in 1983 of only $9,040 (46th among the states).

12 STATE GOVERNMENT

Arkansas's fifth constitution, enacted in 1874 and amended 69 times through 1983, has survived several efforts to replace it with a more modern charter. In November 1980, voters turned down yet another proposed new constitution.

Arkansas's bicameral legislature, the general assembly, consists of a 35-member senate and a 100-member house of representatives. Senators serve four-year terms and must be at least 25 years of age; representatives serve for two years and must be at least 21. Each legislator must be a US citizen and have resided for at least two years in the state and one year in the county or district prior to election.

Under the 1874 constitution, the executive officers elected statewide are the governor, lieutenant governor, secretary of state, treasurer, auditor, and state land commissioner, all of whom serve two-year terms. (In 1986, Arkansas will initiate its first four-year term for elected officials.) The governor and lieutenant governor, who run separately, must be US citizens, must be at least 30 years old, and must have resided in Arkansas for seven years.

A bill passed by both houses of the legislature becomes law if signed by the governor, if passed over his veto by a majority of all elected members of each house, or if neither signed nor returned by the governor within five days when the legislature is in session. Under an initiative procedure, 8% of those who voted for governor in the last election may propose a law, and 10% may initiate a constitutional amendment; initiative petitions must be filed at least four months before the general election in order to be voted upon at that time. A referendum on any measure passed by the general assembly or any item of an appropriations bill or other measure may be petitioned by 6% of the voters; referendum petitions must be filed within 90 days of the session in which the act in question was passed. A successful referendum measure may be repealed by a two-thirds vote of all elected members of the general assembly. Constitutional amendments may also be proposed by the general assembly or by constitutional convention, subject to ratification by the electorate.

To vote in Arkansas, one must be a US citizen at least 18 years of age; registration closes 20 days before an election.

13 POLITICAL PARTIES

The principal political groups in Arkansas are the Democratic Party and the Republican Party, each affiliated with the national party organizations.

Before the Civil War, politics in Arkansas were fraught with violence. As late as 1858, one Arkansas newspaper, the *True Democrat,* asserted that "Arkansas has never to this day had a Senator or Representative in the councils of the nation who has not once, if no more, periled his life on the so-called 'field of honor.' "

Republicans ruled during Reconstruction, which ended in Arkansas after the election of 1872. During that campaign, the Liberal Republicans, nicknamed Brindletails, opposed the Radical Republicans, or Minstrels. After the Minstrel candidate, Elisha Baxter, was elected, he proved so independent a governor that some of the party leaders who had supported him attempted to oust him through a court order in April 1874, declaring his defeated opponent, Joseph Brooks, the winner. Supported by a militia of about 300 blacks under white command, Brooks took over the statehouse; Baxter, bolstered by his own 300-man black army, set up his headquarters three blocks away. The so-called Brooks-Baxter War finally ended with President Ulysses S. Grant's proclamation of Baxter as the lawful governor, followed by Baxter's abdication in favor of Augustus H. Garland, the first of a long series of Bourbon Democrats who were to rule the state well into the 20th century.

After Reconstruction, blacks in Arkansas continued to vote and to be elected to public office; under what became known as the fusion principle, black Republicans and white Democratic leaders in the plantation belt often agreed not to oppose each other's candidates. Segregation in public places was still outlawed, and Little Rock was perhaps the most integrated city in the South. During the 1890s, however, as in the rest of the South, Democrats succeeded in passing laws imposing segregation and disfranchising blacks as well as poor whites. In 1898, the Democrats instituted a nominating primary for whites only.

Not until 1948 and the election of Sidney McMath as governor did Arkansas enjoy relatively progressive government. Although elected to the governorship as a progressive in 1954, McMath's protégé Orval Faubus took a segregationist stand in 1957. In subsequent years, poor whites tended to support Faubus, and blacks and more affluent whites to oppose him. Faubus's successor, progressive Republican Winthrop Rockefeller, was strongly supported by blacks. Rockefeller was followed by three more progressives, all Democrats: Dale Bumpers, David Pryor, and—after Bumpers and Pryor had graduated to the US Senate—Bill Clinton. In a major upset, Clinton was defeated in 1980 by Republican Frank White, but he recaptured the statehouse in 1982 and won reelection in 1984.

14 LOCAL GOVERNMENT

There are 75 counties in Arkansas, some of them with two county seats. Each county is governed by a quorum court, consisting of 9–15 justices of the peace, elected for two-year terms; the county judge, who presides, does not vote but has veto power, which may be overridden by a three-fifths vote of the total membership. Elected county executives, who serve two-year terms, include the sheriff, assessor, coroner, treasurer, and county supervisor.

Arkansas Presidential Vote by Political Parties, 1948–1984

YEAR	ELECTORAL VOTE	ARKANSAS WINNER	DEMOCRAT	REPUBLICAN	STATES' RIGHTS DEMOCRAT
1948	9	*Truman (D)	149,659	50,959	40,068
1952	8	Stevenson (D)	226,300	177,155	—
					CONSTITUTION
1956	8	Stevenson (D)	213,277	186,287	7,008
					NAT'L STATES' RIGHTS
1960	8	*Kennedy (D)	215,049	184,508	28,952
1964	6	*Johnson (D)	314,197	243,264	2,965
					AMERICAN IND.
1968	6	Wallace (AI)	188,228	190,759	240,982
					AMERICAN
1972	6	*Nixon (R)	199,892	448,541	2,887
1976	6	*Carter (D)	498,604	267,903	—
					LIBERTARIAN
1980	6	*Reagan (R)	398,041	403,164	8,970
1984	6	*Reagan (R)	338,646	534,774	2,221

* Won US presidential election.

15STATE SERVICES

Educational services in Arkansas are administered primarily by the Department of Education and the Department of Higher Education. The State Highway and Transportation Department has primary responsibility for roads, rails, and public transit; the offices of motor vehicle registration and driver services are in the Department of Finance and Administration.

Health and welfare services are under the jurisdiction of the Department of Health and the Department of Human Services. Public protection is provided primarily through the Department of Public Safety—which includes the Office of Emergency Services, State Police, National Guard, and Civil Air Patrol—and the Department of Correction, which operates three prisons and three work-release centers. The Public Service Commission, within the Department of Commerce, regulates utilities in the state. Housing services are provided through the Housing Development Agency and the Department of Local Services, whose Division of Manpower offers employment and training programs.

16JUDICIAL SYSTEM

Arkansas's highest court is the supreme court, consisting of a chief justice and six associate justices, elected for staggered eight-year terms. An appeals court of six judges, also elected for eight-year terms, was established in 1978.

Arkansas's courts of original jurisdiction are the circuit courts (law) and the chancery courts (equity), of which there are 86 each; their judges are elected for four-year terms. Courts of limited jurisdiction include justice of the peace, county, municipal, and police courts and courts of common pleas.

Arkansas had an FBI Crime Index rate of 3,501 per 100,000 population in 1983, placing well below the national averages for both violent and nonviolent crimes. In 1982, per capita expenditure on police protection and correction officers was $58, the 2d lowest in the US. The state's once notorious prisons, which had been under federal jurisdiction for more than a decade, were returned in 1978 to state authority, subject to monitoring by an independent ombudsman. By then, the decaying system of the 1960s had been almost entirely replaced by modern facilities. In December 1984 there were 4,454 prisoners in state and federal correctional institutions. There were 3,686 practicing attorneys in the same month.

17ARMED FORCES

As of 1985 there were six military installations in Arkansas, the principal ones being Little Rock Air Force Base, Blytheville Air Force Base, and the Army's Pine Bluff Arsenal. Military personnel in the state numbered 10,120 in 1984. Firms in the state received $692 million in defense contract awards in 1983/84.

As of 1983, some 271,000 Arkansans were veterans of US military service; 4,000 saw service in World War I, 110,000 in World War II, 50,000 during the Korean conflict, and 78,000 during the Viet-Nam era. Arkansas veterans received $451 million in benefits in 1982/83. There were 9,527 Army National Guard personnel at the end of 1984 and 1,982 Air National Guard personnel in February 1985. Arkansas had 6,475 police and correction officers in 1983.

18MIGRATION

Near the end of the 18th century, Indians from east of the Mississippi, displaced by white settlement, entered the area now known as Arkansas. However, as the availability of cheap land in Louisiana Territory drew more and more white settlers—in particular, veterans of the War of 1812, who had been promised 160-acre (65-hectare) tracts—the Indians were pressured to cross the border from Arkansas to present-day Oklahoma.

After the end of the Mexican War, thousands of Arkansans emigrated to Texas, and others were attracted to California in 1849 by the gold rush. Because of a law passed in 1859 requiring free blacks to leave the state by the end of the year or risk being enslaved, Arkansas's population of freedmen dropped from 682 in

1858 to 144 in 1860. During Reconstruction, the state government encouraged immigration by both blacks and whites. Literature sent out by the Office of State Lands and Migration, under the tenure of William H. Grey, a black leader, described the state as a new Africa. Railroads, seeking buyers for the lands they had acquired through government grants, were especially active in encouraging immigration after Reconstruction. Later immigrants included Italians and, in the early 1900s, Germans.

During the depression years and thereafter, Arkansas lost a substantial proportion of its farm population, and many blacks left the state for the industrial cities of the Midwest and the east and west coasts. The net loss from migration totaled 919,000 between 1940 and 1970. Between 1970 and 1980, however, the state gained 180,000 residents through migration, as the Ozarks became one of the fastest-growing rural areas in the US. The state experienced a small net decline of 2,000 in migration between 1980 and 1983.

19INTERGOVERNMENTAL COOPERATION

Among the many interstate agreements in which Arkansas participates are the Arkansas River Basin Compact of 1970 (with Oklahoma), Interstate Oil and Gas Compact, Red River Compact, South Central Forest Fire Protection Compact, Southern Growth Policies Compact, Southern Interstate Energy Compact, and Southern Regional Education Compact. There are boundary agreements with Mississippi, Missouri, and Tennessee.

In 1983/84, Arkansas received federal aid totaling $946.2 million, or $406 per capita.

20ECONOMY

During the 19th century, Arkansas's economic growth was hindered by credit problems. When the state's two central banks, the Arkansas State Bank and the Real Estate Bank, failed during the 1840s, the government defaulted on bonds issued by the latter and amended the constitution to prohibit all banking in Arkansas. Although banking was restored after the Civil War, the state defaulted on its obligations once more in 1877, this time following a decision by the Arkansas supreme court that $10 million worth of railroad bonds issued during Reconstruction were unconstitutional. Not until 1917 did New York banks again accept Arkansas securities.

Cotton dominated Arkansas's agricultural economy until well into the 20th century, when rice, soybean, poultry, and fish farming diversified the output. Coal mining began in the 1870s, bauxite mining near the turn of the century, and oil extraction in the 1920s; lumbering developed in the last quarter of the century, reached its peak about 1909, and then declined until the 1920s, when reforestation started. Industrialization was limited, however, and resources were generally shipped out of state for processing. Not until the 1950s did Arkansas enjoy significant success in attracting industry, thanks in large part to the efforts of Winthrop Rockefeller. Although the Little Rock integration crisis of 1957 was a severe setback to the state's industrial growth, development resumed during the following decades. By the mid-1980s, Arkansas, though still one of the poorest states, was narrowing the gap between its own standard of living and that of the nation as a whole.

21INCOME

In 1983, Arkansas ranked 46th among the 50 states in per capita personal income, with $9,040; only Utah, South Carolina, West Virginia, and Mississippi ranked lower. On the other hand, the 1983 figure represented 77% of the US average per capita income for that year, a large improvement over the 62% recorded in 1960 and the 43% registered in 1929. Total personal income was $21 billion in 1983; measured in constant 1972 dollars, total income increased more than 70% during 1960–70 and almost 59% for 1970–80, both far above the respective national percentages. An additional 6.5% gain in personal income was posted in 1980–83. An estimated 424,000 Arkansans, or 19% of the population, were living below the federal poverty level in 1979. In 1981, median money income of four-person families was $20,583 (50th in the US), $5,691 below the US average.

22LABOR

Arkansas's civilian labor force totaled 1,026,000 in 1983—53.8% of the civilian population. About 922,000 Arkansans were employed and 104,000 unemployed, for an overall unemployment rate of 10.1% in 1983. This rate fell to 8.9% in 1984.

A federal survey in 1982 revealed the following nonfarm employment pattern for Arkansas:

	ESTABLISH-MENTS	EMPLOYEES	ANNUAL PAYROLL ('000)
Agricultural services, forestry, fishing	512	5,315	$ 56,525
Mining, of which:	348	5,303	105,491
Oil, gas extraction	(262)	(3,677)	(74,509)
Contract construction	3,319	27,414	409,690
Manufacturing	2,985	193,262	2,841,188
Transportation, public utilities	1,980	33,991	660,104
Wholesale trade	3,672	39,126	591,169
Retail trade	13,329	120,113	1,041,402
Finance, insurance, real estate	3,056	29,969	459,107
Services	11,812	117,663	1,299,836
Other	1,083	1,065	12,011
TOTALS	42,096	573,221	$7,476,523

Government employees were not covered by this survey; there were 123,000 state and local government employees in 1983 and 18,000 federal government employees in 1982.

Chartered in 1865, the Little Rock Typographical Union, consisting of *Arkansas Gazette* employees, was the first labor union in the state. The United Mine Workers was established in the Ft. Smith area by 1898; six years later, the UMW led in the founding of the Arkansas Federation of Labor. Between 1904 and World War I, a series of progressive labor laws was enacted, including a minimum wage, restrictions on child labor, and prohibitions against blacklisting and payment of wages in scrip. Union strength waned after the war, however, and the labor movement is not a powerful force in the state today. There were 119,000 members of labor unions in Arkansas in 1980 and 583 labor unions in 1983. Arkansas has a right-to-work law.

23AGRICULTURE

Agricultural income in Arkansas was nearly $3 billion in 1983 (17th in the US), with crops and livestock accounting for about 50% each. The state is the nation's leading producer of rice and is among the leaders in cotton, soybeans, and grain sorghum. In 1981/82, Arkansas ranked 11th in agricultural exports.

Cotton was first grown in the state about 1800, along the river valleys. Confined mainly to slaveholding plantations before the Civil War, cotton farming became more widespread in the postwar period, expanding into the hill country of the northwest and eventually into the deforested areas of the northeast, which proved to be some of the most fertile farmland in the nation. As elsewhere in the postbellum South, sharecropping by tenant farmers predominated well into the 20th century, until mechanization and diversification gradually brought an end to the system. Rice was first grown commercially in the early 1900s; by 1920, Arkansas had emerged as a poultry and soybean producer.

During 1983, Arkansas produced 66,500,000 bushels of soybeans, valued at $545,300,000; 58,500,000 bushels of wheat, worth $193,050,000; 1,411,000 tons of hay, worth $79,016,000; and 17,600,000 bushels of sorghum for grain, valued at $53,152,000. The rice harvest in 1982 was 57,037,000 hundredweight, worth $491,089,000. The estimated cotton crop in 1983, 323,000 bales, was worth $102,481,000.

24ANIMAL HUSBANDRY

Poultry farms are found throughout Arkansas, but especially in the northern and western regions. Arkansas is the top-ranked broiler-producing state in the US: in 1983, 673,136,000 broilers brought a gross income of $729,006,000. During the same year, the state ranked 5th in the value of egg production, with $200,018,000 for 3.8 billion eggs; 5th in turkeys, with $83,654,000; and 1st in chickens, with $16,787,000.

In 1983, Arkansas ranchers earned $495,648,000 from the sale of 583,840,000 lb of cattle, calves, beef, and veal, and $84,756,000 from the sale of 174,795,000 lb of hogs, pork, and lard. The livestock inventory included 2,000,000 cattle and calves in 1984 and 395,000 hogs and pigs in 1983. The dairy yield of the state's 84,000 milk cows in 1983 was 873,000,000 lb of milk.

25FISHING

Commercial fish landings in 1984 were estimated at 16,632,000 lb, worth $7,332,000. Of far greater economic importance to Arkansas is fish farming. As of 1985, the state ranked 1st in the US in minnow farming and 2d only to Mississippi in catfish farming. During 1982, the 115 catfish farms in Arkansas had sales totaling $6,420,000. In 1983, all fish farms occupied a total of 42,120 acres (17,045 hectares) and produced 36,780,492 lb of fish, valued at $45,045,783. Some producers rotate fish crops with row crops, periodically draining their fish ponds and planting grains in the rich and well-fertilized soil.

26FORESTRY

Forestland comprised 18,282,000 acres (7,398,000 hectares), or 55% of the state's total land area, in 1983. Of that total, 18,207,-000 acres (7,368,000 hectares) were commercial timberland, 84% of it in private hands. The southwest and central plains, the state's timber belt, constitute one of the most concentrated sources of yellow pine in the US. Shipments of lumber and wood products were valued at nearly $1.25 billion in 1982. Three national forests in Arkansas covered a total of 2,478,454 acres (1,002,998 hectares) of National Forest System lands in 1984.

27MINING

Arkansas ranked 26th among the 50 states in nonfuel mineral production in 1984, with an output valued at $279 million. The only diamond fields in the US are located in Pike County in southwestern Arkansas.

Among the leading minerals in 1984 (excluding fossil fuels) were stone, 15 million tons; sand and gravel, 8.4 million tons; and clays, 1.1 million tons. Other minerals were abrasives, bauxite, bromine, cement, gypsum, lime, and tripoli.

28ENERGY AND POWER

Although Arkansas possesses substantial and varied energy resources—petroleum, natural gas, coal, and water—the state was slow to develop them. As late as 1935, only 1% of Arkansas farms had electric power. The struggle that began during the 1930s over whether Arkansas's rivers would be publicly developed for the production of electricity (in the manner of the Tennessee Valley Authority) was won by the advocates of private power development. As of 1983, Arkansas power plants had a combined capacity of 8.8 million kw, of which about three-fourths was privately owned; production totaled 30.1 billion kwh.

During the same year, an estimated 18,849,000 barrels of crude petroleum were produced, leaving proved reserves of 120,000,000 barrels. Production of natural gas was 206 billion cu feet, with 2.1 trillion cu feet of reserves remaining. As of 1983, Arkansas's known coal reserves were 418,000,000 tons: 288,200,000 bituminous, 104,100,000 anthracite, and 25,700,000 lignite. About 47,000 tons of bituminous coal were mined in 1983.

29INDUSTRY

Manufacturing in Arkansas is diverse, ranging from blue jeans to bicycles, though resource industries such as rice processing and woodworking still play a major role. The total value of shipments of manufactured goods in 1982 was nearly $20 billion, of which food and food products contributed 24%; electric and electronic equipment, 10%; paper and allied products, 10%; petroleum and coal products, 8%; and lumber and wood products, 6%. The following table shows value of shipments for selected industries in 1982:

Meat products	$1,547,800,000
Grain-mill products	1,233,700,000
Preserved fruits and vegetables	449,200,000
Household appliances	437,800,000
Motor vehicles and equipment	312,200,000
Household furniture	230,300,000

30 COMMERCE

Arkansas ranked 34th in the US in wholesale trade during 1982, with sales of $10.7 billion. Retail establishments recorded $9.1 billion in sales in 1982 (32d in the US). The leading retail categories were food stores, 23%; automotive dealers, 22%; department stores, 13%; gasoline service stations, 9%; and eating and drinking places, 7%. Of the counties, Pulaski (including Little Rock and North Little Rock) had the highest percentage of sales, 22%; Little Rock had 13%, highest among the cities.

During 1981/82, Arkansas ranked 11th in the US in agricultural exports, valued at $1.28 billion, or almost 40% of all farm sales. Arkansas is the principal US exporter of rice. In 1981, Arkansas's manufactured exports, worth $1.6 billion, ranked 28th among the 50 states.

31 CONSUMER PROTECTION

Under the mandate of the Consumer Protection Act of 1971, the Consumer Protection Division of the Office of the Attorney General has principal responsibility for consumer affairs.

32 BANKING

In 1836, the first year of statehood, the legislature created the Arkansas State Bank and the Real Estate Bank, which was intended to promote the plantation system. Fraud, mismanagement, and the consequences of the financial panic of 1837 ruined both banks and led to the passage in 1846 of a constitutional amendment prohibiting the incorporation of any lending institution in Arkansas. Money grew scarce, with credit being rendered largely by suppliers and brokers to farmers and planters until after the Civil War, when the prohibition was removed.

Arkansas had 258 insured commercial banks in 1984. At the end of 1983, the state's insured commercial banks had $15.1 billion in assets and $12.7 billion in deposits. The combined assets of 48 savings and loan associations amounted to $4.5 billion at the end of 1982; they held $3.9 billion in savings accounts. As of the early 1980s, Arkansas's usury law, imposing a 10% ceiling on interest rates, was among the most rigid in the US; the US Supreme Court upheld the 10% limit in 1981. The rise of the federal discount rate above that limit, beginning in mid-1979, caused a considerable outflow of capital from Arkansas. The largest bank in the state is the First Commercial Bank, formed by the 1983 merger of the Commercial National and First National banks of Little Rock.

33 INSURANCE

Arkansans held 2.7 million life insurance policies worth $35 billion at the close of 1983. The average of $39,900 in life insurance per family was the lowest of all the states. Benefits paid in 1983 amounted to $324.4 million, including $112.4 million in death payments. During the same year, premiums were written for $179.2 million in automobile liability insurance, $169.3 million for automobile physical-damage coverage, and $124.4 million for homeowners insurance.

34 SECURITIES

There are no securities exchanges in Arkansas. New York Stock Exchange member firms had 48 sales offices and 326 registered representatives in the state at the beginning of 1984. About 246,000 Arkansans owned shares of public corporations in 1983.

35 PUBLIC FINANCE

Under the 1874 constitution, state expenditures may not exceed revenues. The mechanism adopted each biennium to prevent deficit spending is a Revenue Stabilization Act, allocating a proportion of actual state revenues for each program and agency up to the amount appropriated and categorizing programs according to priority as A, B, or C. All A programs must be funded first; then, revenue permitting, B and C programs are funded in that order. In the event of a shortfall, agencies share according to their established proportions.

The Arkansas state budget covers two years, each fiscal year running from 1 July through 30 June. The following table summarizes estimated general fund revenues and appropriations for 1982/83 (actual) and 1983/84 (estimated), in millions:

REVENUES	1982/83	1983/84
Sales and use taxes	$ 437.2	$ 562.2
Individual income tax	460.6	507.2
Corporate income tax	86.9	90.5
Other net receipts	39.0	29.3
TOTALS	$1,023.7	$1,189.2
EXPENDITURES		
Public school fund	$ 477.1	$ 512.1
Institutions of higher education	181.7	193.8
Department of Human Services	178.3	193.0
State general government	71.8	82.6
General education fund	32.7	35.3
Other outlays	82.0	90.0
TOTALS	$1,023.6	$1,106.8

The total outstanding debt of Arkansas state and local governments was $2.45 billion, or about $1,074 per capita, in 1982.

36 TAXATION

As of 1982, Arkansas's state tax revenue per capita was $729, the lowest in the US. In 1984, the state income tax ranged from 1% on amounts under $3,000 to 7% on amounts over $25,000; the corporate income tax ranged from 1% on the first $3,000 to 6% on amounts over $25,000. The state sales tax was 4%. The state also imposes severance taxes on oil, gas, and other natural resources, along with levies on liquor, gasoline, and cigarettes. City and county property taxes in Arkansas are among the lowest in the nation.

Arkansans filed 818,000 federal income tax returns for 1982, paying $1,715,000,000 in tax.

37 ECONOMIC POLICY

First as chairman of the Arkansas Industrial Development Commission and later as governor of the state, Winthrop Rockefeller succeeded in attracting substantial and diverse new industries to Arkansas. In 1979, Governor Bill Clinton established the Department of Economic Development for the purpose of stimulating the growth of small business and finding new export markets. Between 1977 and 1981, Arkansas's exports increased three times as fast as industrial production.

38 HEALTH

The infant death rate in 1981 was 9.8 per 1,000 live births for whites and 18.3 for nonwhites. About 6,700 legal abortions were performed during 1981/82, when the state's abortion rates were among the lowest in the US. During the same year, however, Arkansas's death rate, 9.8 per 1,000 population, was well above the national average of 8.6, and the incidence of cerebrovascular disease—98.8 per 100,000 population—led the US. Death rates from heart disease, cancer, accidents, diabetes, pneumonia, and influenza also exceeded the national average.

Arkansas's 97 hospitals had 13,772 beds and recorded 463,422 admissions in 1983, for an occupancy rate of 68%; personnel included 5,214 registered nurses and 4,280 licensed practical nurses. Hospital costs in 1982—$253 per day and $1,668 per stay—were among the lowest in the US. In 1982, the state had 3,328 licensed physicians and 838 professionally active dentists.

³⁹SOCIAL WELFARE

Social welfare payments in Arkansas generally fall well below national norms. During 1982, payments in aid to families with dependent children totaled $34 million; the average payment per family was $124 monthly. The food stamp program had 296,000 participants at a federal cost of $151 million; and the school lunch program served 298,000 students with a federal outlay of $31 million.

Some 437,000 Arkansans received Social Security benefits totaling $1.8 billion in 1983; the average monthly benefit was $385 for retired workers, $425 for disabled workers, and $335 for widows and widowers. Supplemental Security Income payments of $125.8 million were made to 71,500 Arkansans in 1983; $159 million in unemployment insurance benefits was paid in 1982. Workers' compensation payments totaled $94.6 million in the same year.

⁴⁰HOUSING

The 1980 census counted some 898,593 housing units in Arkansas, 888,740 of them year-round units. From 1981 through 1983, 20,900 new units worth $686 million were authorized.

⁴¹EDUCATION

In 1980, 55.5% of all Arkansans 25 years of age and older were high school graduates (5th lowest in the US). Only 10.8% had completed four or more years of college, the 2d-lowest rate in the US. In 1983, in an effort to raise the quality of education in Arkansas, the state legislature approved a comprehensive program that included smaller classes, more high school level courses, and competency tests for teachers.

Public school enrollment during 1982/83 totaled 432,000: prekindergarten through grade 8, 304,000; and grades 9–12, 128,000. Nonpublic school enrollment was 18,423 in 1980/81.

In some ways, Little Rock was an unlikely site for the major confrontation over school integration that occurred in 1957. The school board had already announced its voluntary compliance with the Supreme Court's desegregation decision, and during Governor Faubus's first term (1955–56), several public schools in the state had been peaceably integrated. Nevertheless, on 5 September 1957, Faubus, claiming that violence was likely, ordered the National Guard to seize Central High School to prevent the entry of nine black students. When a mob did appear following the withdrawal of the National Guardsmen in response to a federal court order later that month, President Dwight Eisenhower dispatched federal troops to Little Rock, and they patrolled the school grounds until the end of the 1958 spring semester. Although Faubus's stand encouraged politicians in other southern states to resist desegregation, in Arkansas integration proceeded at a moderate pace. By 1980, Central High School had a nearly equal balance of black and white students, and the state's school system was one of the most integrated in the South.

In 1984, Arkansas had 35 institutions of higher education, 19 public and 16 private, of which the largest, the University of Arkansas at Fayetteville (established in 1871), had a fall 1983 enrollment of 13,483. Student aid is provided by the State Scholarship Program within the Department of Higher Education, by the Arkansas Student Loan Guarantee Foundation, and by the Arkansas Rural Endowment Fund, Inc.

⁴²ARTS

Little Rock is the home of the Arkansas Symphony and the Arkansas Arts Center, which holds art exhibits and classes and community theater performances. The best-known center for traditional arts and crafts is the Ozark Folk Center at Mountain View; every evening from late spring through October, folk music of the Ozarks may be heard. The Arkansas Folk Festival is held there during two weekends in April, and the Family Harvest Festival for three weeks in October. Arkansas College at Batesville sponsors two-week summer workshops in Ozark crafts, music, and folklore in association with the center. The Grand Prairie Festival of Arts is held at Stuttgart in September.

⁴³LIBRARIES AND MUSEUMS

During 1983, Arkansas had 32 county or regional libraries and 7 municipal libraries. That year, public libraries held a total of 3,732,526 volumes and circulation amounted to 7,643,936. Important collections include those of the University of Arkansas at Fayetteville (1,046,486 volumes), Arkansas State University at Jonesboro (393,012), the Central Arkansas Library System of Little Rock (481,860), and the News Library of the *Arkansas Gazette,* also in Little Rock.

There were 78 museums in 1984 and a number of historic sites. Principal museums include the Arkansas Arts Center and the Museum of Science and History, both at Little Rock; the Arkansas State University Museum at Jonesboro; and the University of Arkansas Museum at Fayetteville, specializing in archaeology, anthropology, and the sciences. Also of interest are the Stuttgart Agricultural Museum; the Arkansas Post County Museum at Gillett, whose artifacts are housed in recreated plantation buildings; Hampson Museum State Park, near Wilson, which has one of the largest collections of Mound Builder artifacts in the US; the Mid-American Museum at Hot Springs, which has visitor-participation exhibits; and the Saunders Memorial Museum at Berryville, with an extensive collection of firearms.

Civil War battle sites include the Pea Ridge National Military Park, the Prairie Grove Battlefield State Park, and the Arkansas Post National Memorial. The Ft. Smith National Historic Site includes buildings and museums from the days when the town was a military outpost on the border of Indian Territory.

⁴⁴COMMUNICATIONS

In 1980, 87% of the state's 816,065 occupied housing units had telephones, among the lowest rates in the nation. As of 1984, the state had 618 post offices. There were 204 commercial radio stations (101 AM, 103 FM) and 20 television stations. Cable television systems served 347,397 subscribers in 258 communities during that year.

⁴⁵PRESS

The first newspaper in Arkansas, the *Arkansas Gazette,* established at Arkansas Post in 1819 by William E. Woodruff, is the state's most widely read and influential journal. In 1984 there were 7 morning dailies with a combined circulation of 286,137, and 26 evening papers with a combined circulation of 204,660 (these totals include 1 all-day newspaper); 16 Sunday papers had a total circulation of 508,757. The following table shows the 1984 circulations of the leading dailies:

AREA	NAME	DAILY	SUNDAY
Ft. Smith	Southwest Times Record		
	(all day,S)	41,258	46,449
Little Rock	Arkansas Democrat (m,S)	75,150	141,625
	Arkansas Gazette (m,S)	126,922	161,246

⁴⁶ORGANIZATIONS

The 1982 Census of Service Industries counted 571 organizations in Arkansas, including 123 business associations; 281 civic, social, and fraternal associations; and 9 educational, scientific, and research organizations. Among the national organizations with headquarters in Arkansas are the American Crossbow Association in Huntsville; the American Fish Farmers Federation in Lonoke; and the Ozark Society, the American Parquet Association, the Federation of American Hospitals, and the Civil War Round Table Associates, all located in Little Rock.

ACORN, the Association of Community Organizations for Reform Now, was founded in Little Rock in 1970 and has since spread to some 20 other states and has become one of the most influential citizens' lobbies in the U.S.

47 TOURISM, TRAVEL, AND RECREATION

During 1983, 13,779,000 persons took trips to and through Arkansas, spending over $1.5 billion. In early 1983/84, national parks in Arkansas attracted 2,040,900 visitors, and the 44 state parks had 6,329,232 visitors.

Leading attractions are the mineral waters and recreational facilities at Hot Springs, Eureka Springs, Mammoth Spring, and Heber Springs. The Crater of Diamonds, near Murfreesboro, is the only known public source of natural diamonds in North America. For a fee, visitors may hunt for diamonds and keep any they find; more than 100,000 diamonds have been found in the area since 1906, of which the two largest are the 40.42-carat Uncle Sam and the 34.25-carat Star of Murfreesboro. In 1984, 1,339 diamonds were found, totaling 202.26 carats.

During 1982/83, licenses were issued to 883,589 hunters and 723,439 fishermen. The World's Championship Duck Calling Contest is held at the beginning of the winter duck season in Stuttgart.

48 SPORTS

Arkansas has no major league professional sports teams, but does claim a class AA minor league baseball entry in the Texas League, the Travelers. Hot Springs has a 62-day Thoroughbred racing season each spring at the Oaklawn Jockey Club, and dog races are held in West Memphis from April through November. Several major rodeos take place in summer and fall, including the Rodeo of the Ozarks in Springdale in early July and the Southeast Arkansas Rodeo and Livestock Show in Pine Bluff each September.

The Razorback football team of the University of Arkansas won the Cotton Bowl in 1947, 1965, and 1976, the Orange Bowl in 1978, the Sugar Bowl in 1969, and the Bluebonnet Bowl in 1982. The men's basketball team won or shared the Southwest Conference championship in 1977, 1978, 1979, 1981, and 1982.

49 FAMOUS ARKANSANS

Arkansas has yet to produce a US president, vice president, or Supreme Court justice, but one Arkansan came close to the latter two offices. US Senator Joseph T. Robinson (1872–1937) was the Democratic nominee for vice president in 1928, on the ticket with Al Smith; later, he was Senate majority leader under President Franklin D. Roosevelt. At the time of his death, Robinson was leading the fight for Roosevelt's bill to expand the Supreme Court's membership and had reportedly been promised a seat on the court if the bill passed. Robinson's colleague, Hattie W. Caraway (b.Tennessee, 1878–1950), was the first woman elected to the US Senate, serving from 1931 to 1945. After World War II, Arkansas's congressional delegation included three men of considerable power and fame: Senator John L. McClellan (1896–1977), investigator of organized labor and organized crime and champion of the Arkansas River navigation project; Senator J. William Fulbright (b.Missouri, 1905), chairman of the Senate Foreign Relations Committee; and Representative Wilbur D. Mills (b.1909), chairman of the House Ways and Means Committee until scandal ended his political career in the mid-1970s. Other federal officeholders include Brooks Hays (1898–1981), former congressman and special assistant to Presidents John F. Kennedy and Lyndon B. Johnson, as well as president of the Southern Baptist Convention, the nation's largest Protestant denomination; and Frank Pace, Jr. (b.1912), secretary of the Army during the Truman administration. General Douglas MacArthur (1880–1964), supreme commander of Allied forces in the Pacific during World War II, supervised the occupation of Japan and was supreme commander of UN troops in Korea until relieved of his command in April 1951 by President Truman.

Orval E. Faubus (b.1910) served six terms as governor (a record), drew international attention during the 1957 integration crisis at Little Rock Central High School, and headed the most powerful political machine in Arkansas history. Winthrop Rockefeller (b.New York, 1912–73) was Faubus's most prominent successor. At the time of his election in 1978, Bill Clinton (b.1946) was the nation's youngest governor.

Prominent business leaders include the Stephens brothers, W. R. "Witt" (b.1907) and Jackson T. (b.1923), whose Stephens, Inc., investment firm in Little Rock is the largest off Wall Street, and Kemmons Wilson (b.1913), founder of Holiday Inns. Other distinguished Arkansans are Edward Durrell Stone (1902–78), renowned architect; C. Vann Woodward (b.1908), Sterling Professor of History at Yale University; and the Right Reverend John M. Allin (b.1921), presiding bishop of the Episcopal Church of the United States. John H. Johnson (b.1918), publisher of the nation's leading black-oriented magazines—*Ebony, Jet,* and others—is an Arkansan, as is Helen Gurley Brown (b.1922), editor of *Cosmopolitan.* Harry S. Ashmore (b.South Carolina, 1916) won a Pulitzer Prize for his *Arkansas Gazette* editorials calling for peaceful integration of the schools during the 1957 crisis; the *Gazette* itself won a Pulitzer for meritorious public service that year. Paul Greenberg (b.Louisiana, 1937), of the *Pine Bluff Commercial,* is another Pulitzer Prize-winning journalist. John Gould Fletcher (1886–1950) was a Pulitzer Prize-winning poet. Other Arkansas writers include Dee Brown (b.Louisiana, 1908), Maya Angelou (b.Missouri, 1928), Charles Portis (b.1933), and Eldridge Cleaver (b.1935).

Arkansas planter Colonel Sanford C. Faulkner (1803–74) is credited with having written the well-known fiddle tune "The Arkansas Traveler" and its accompanying dialogue. Perhaps the best-known country music performers are Johnny Cash (b.1932) and Glen Campbell (b.1938). Film stars Dick Powell (1904–63) and Alan Ladd (1913–64) were also Arkansans.

Notable Arkansas sports personalities include Jerome Herman "Dizzy" Dean (1911–74) and Bill Dickey (b.1907), both members of the Baseball Hall of Fame; Brooks Robinson (b.1937), considered by some the best-fielding third baseman in baseball history; and star pass-catcher Lance Alworth (b.Mississippi, 1940), a University of Arkansas All-American and member of the Professional Football Hall of Fame.

50 BIBLIOGRAPHY

Angelou, Maya. *I Know Why the Caged Bird Sings.* New York: Bantam, 1971.

Arkansas, University of. Industrial Research and Extension Center. *Arkansas State and County Economic Data.* Little Rock, 1984.

Ashmore, Harry S. *Arkansas: A Bicentennial History.* New York: Norton, 1978.

Du Vall, Leland. *Arkansas: Colony and State.* Little Rock: Rose, 1973.

Federal Writers' Project. *Arkansas: A Guide to the State.* New York: Somerset, n.d. (orig. 1941).

Fletcher, John Gould. *Arkansas.* Chapel Hill: University of North Carolina Press, 1947.

Moore, Waddy William (ed.). *Arkansas in the Gilded Age.* Little Rock: Rose, 1976.

Ross, Margaret. *Arkansas Gazette: The Early Years, 1819–66.* Little Rock: Arkansas Gazette Foundation, 1969.

Taylor, Orville W. *Negro Slavery in Arkansas.* Durham, N.C.: Duke University Press, 1958.

Williams, C. Fred. *A Documentary History of Arkansas.* Fayetteville: University of Arkansas Press, 1983.

CALIFORNIA

State of California

CALIFORNIA REPUBLIC

ORIGIN OF STATE NAME: Probably from the mythical island California in a 16th-century romance by Garcí Ordóñez de Montalvo. **NICKNAME:** The Golden State. **CAPITAL:** Sacramento. **ENTERED UNION:** 9 September 1850 (31st). **SONG:** "I Love You, California." **MOTTO:** Eureka (I have found it). **FLAG:** The flag consists of a white field with a red star at upper left and a red stripe and the words "California Republic" across the bottom; in the center, a brown grizzly bear stands on a patch of green grass. **OFFICIAL SEAL:** In the foreground is the goddess Minerva; a grizzly bear stands in front of her shield. The scene also shows the Sierra Nevada, San Francisco Bay, a miner, a sheaf of wheat, and a cluster of grapes, all representing California's resources. The state motto and 31 stars are displayed at the top. The words "The Great Seal of the State of California" surround the whole. **COLORS:** Yale blue and golden yellow. **ANIMAL:** California grizzly bear (extinct). **BIRD:** California valley quail. **FISH:** South Fork golden trout. **FLOWER:** Golden poppy. **TREE:** California redwood. **ROCK:** Serpentine. **MINERAL:** Native gold. **REPTILE:** California desert tortoise. **INSECT:** California dog-face butterfly (flying pansy). **MARINE MAMMAL:** California gray whale. **FOSSIL:** California saber-toothed cat. **LEGAL HOLIDAYS:** New Year's Day, 1 January; Birthday of Martin Luther King, Jr., 3d Monday in January; Lincoln's Birthday, 12 February; Washington's Birthday, 3d Monday in February; Memorial Day, last Monday in May; Independence Day, 4 July; Labor Day, 1st Monday in September; Admission Day, 9 September; Columbus Day, 2d Monday in October; Veterans Day, 11 November; Thanksgiving Day, 4th Thursday in November; Christmas Day, 25 December. **TIME:** 4 AM PST = noon GMT.

¹LOCATION, SIZE, AND EXTENT

Situated on the Pacific coast of the southwestern US, California is the nation's 3d-largest state (after Alaska and Texas).

The total area of California is 158,706 sq mi (411,048 sq km), of which land takes up 156,299 sq mi (404,814 sq km) and inland water 2,407 sq mi (6,234 sq km). California extends about 350 mi (560 km) E–W; its maximum N–S extension is 780 mi (1,260 km).

California is bordered on the N by Oregon; on the E by Nevada; on the SE by Arizona (separated by the Colorado River); on the S by the Mexican state of Baja California Norte; and on the W by the Pacific Ocean.

The eight Santa Barbara islands lie from 20 to 60 mi (32–97 km) off California's southwestern coast; the small islands and islets of the Farallon group are about 30 mi (48 km) W of San Francisco Bay. The total boundary length of the state is 2,050 mi (3,299 km), including a general coastline of 840 mi (1,352 km); the tidal shoreline totals 3,427 mi (5,515 km). California's geographic center is in Madera County, 38 mi (61 km) E of the city of Madera.

²TOPOGRAPHY

California is the only state in the US with an extensive seacoast, high mountains, and deserts. The extreme diversity of the state's landforms is best illustrated by the fact that Mt. Whitney (14,495 feet—4,418 meters), the highest point in the contiguous US, is situated no more than 80 mi (129 km) from the lowest point in the entire country, Death Valley (282 feet, or 86 meters, below sea level). The mean elevation of the state is about 2,900 feet (900 meters).

California's principal geographic regions are the Sierra Nevada in the east, the Coast Ranges in the west, the Central Valley between them, and the Mojave and Colorado deserts in the southeast. The mountain-walled Central Valley, more than 400 mi (640 km) long and about 50 mi (80 km) wide, is probably the state's most unusual topographic feature. It is drained in the north by the Sacramento River, about 320 mi (515 km) long, and in the south by the San Joaquin River, about 350 mi (560 km). The main channels of the two rivers meet at and empty into the northern

arm of San Francisco Bay, flowing through the only significant break in the Coast Ranges, a mountain system that extends more than 1,200 mi (1,900 km) alongside the Pacific. Lesser ranges, including the Siskiyou Mountains in the north and the Tehachapi Mountains in the south, link the two major ranges and constitute the Central Valley's upper and lower limits.

California has 41 mountains exceeding 10,000 feet (3,050 meters). After Mt. Whitney, the highest peaks in the state are Mt. Williamson, in the Sierra Nevada, at 14,375 feet (4,382 meters), and Mt. Shasta (14,162 feet—4,317 meters), an extinct volcano in the Cascades, the northern extension of the Sierra Nevada. Lassen Peak (10,457 feet—3,187 meters), also in the Cascades, is a dormant volcano.

Beautiful Yosemite Valley, a narrow gorge in the middle of the High Sierra, is the activities center of Yosemite National Park. The Coast Ranges, with numerous forested spurs and ridges enclosing dozens of longitudinal valleys, vary in height from about 2,000 to 7,000 feet (600–2,100 meters).

Melted snow from the Sierra Nevada feeds the state's principal rivers, the Sacramento and San Joaquin. The Coast Ranges are drained by the Klamath, Eel, Russian, Salinas, and other rivers. In the south, most rivers are dry creek beds except during the spring flood season; they either dry up from evaporation in the hot summer sun or disappear beneath the surface, like Death Valley's Amargosa River. The Salton Sea, in the Imperial Valley of the southeast, is the state's largest lake, occupying 374 sq mi (969 sq km). This saline sink was created accidentally in the early 1900s when Colorado River water, via an irrigation canal, flooded a natural depression 235 feet (72 meters) below sea level in the Imperial Valley. Lake Tahoe, in the Sierra Nevada at the angle of the California-Nevada border, covers 192 sq mi (497 sq km).

The California coast is indented by two magnificent natural harbors, San Francisco Bay and San Diego Bay, and two smaller bays, Monterey and Humboldt. Two groups of islands lie off the California shore: the Santa Barbara Islands, situated west of Los Angeles and San Diego; and the rocky Farallon Islands, off San Francisco.

The Sierra Nevada and Coast Ranges were formed more than 100 million years ago by the uplifting of the earth's crust. The Central Valley and the Great Basin, including the Mojave Desert and Death Valley, were created by sinkage of the earth's crust; inland seas once filled these depressions but evaporated over eons of time. Subsequent volcanic activity, erosion of land, and movement of glaciers until the last Ice Age subsided some 10,000 years ago gradually shaped the present topography of California. The San Andreas Fault, extending from north of San Francisco Bay for more than 600 mi (970 km) southeast to the Mojave Desert, is a major active earthquake zone and was responsible for the great San Francisco earthquake of 1906.

Because water is scarce in the southern part of the state and because an adequate water supply is essential both for agriculture and for industry, more than 1,000 dams and reservoirs have been built in California. In 1983, the state's reservoirs had an aggregate capacity of 43,664,100 acre-feet of water. In that year, the major dams and their reservoir capacities were (in acre-feet) Shasta, on the Sacramento River, 4,552,000; Oroville, Feather River, 3,537,600; Trinity, Trinity River, 2,448,000; and New Melones, Stanislaus River, 2,400,000.

³CLIMATE

Like its topography, California's climate is varied and tends toward extremes. Generally there are two seasons—a long, dry summer, with low humidity and cool evenings, and a mild, rainy winter—except in the high mountains, where four seasons prevail and snow lasts from November to April. The one climatic constant for the state is summer drought.

California has four main climatic regions. Mild summers and winters prevail in central coastal areas, where temperatures are more equable than virtually anywhere else in the US; in the area between San Francisco and Monterey, for example, the difference between average summer and winter temperatures is seldom more than 10 Fahrenheit (6 Centigrade) degrees. During the summer there are heavy fogs in San Francisco and all along the coast. Mountainous regions are characterized by milder summers and colder winters, with markedly low temperatures at high elevations. The Central Valley has hot summers and cool winters, while the Imperial Valley is marked by very hot, dry summers, with temperatures frequently exceeding 100°F (38°C).

Average annual temperatures for the state range from 47°F (8°C) in the Sierra Nevada to 73°F (23°C) in the Imperial Valley. The highest temperature ever recorded in the US was 134°F (57°C), registered in Death Valley on 10 July 1913. The state's lowest temperature was −45°F (−43°C), recorded on 20 January 1937 at Boca, near the Nevada border.

Among the major population centers, Los Angeles has an average annual temperature of 65°F (18°C), with an average January minimum of 47°F (8°C) and an average July maximum of 83°F (28°C). San Francisco has an annual average of 57°F (14°C), with a January average minimum of 46°F (8°C) and a July average maximum of 64°F (18°C). The annual average in San Diego is 63°F (17°C), the January average minimum 46°F (8°C), and the July average maximum 75°F (24°C). Sacramento's annual average temperature is 60°F (16°C), with January minimums averaging 37°F (3°C) and July maximums of 93°F (34°C).

Annual precipitation varies from only 2 in (5 cm) in the Imperial Valley to 68 in (173 cm) at Blue Canyon, near Lake Tahoe. San Francisco has a normal annual precipitation of 20 in (51 cm), Sacramento 17 in (43 cm), Los Angeles 12 in (30 cm), and San Diego 9 in (23 cm). The largest one-month snowfall ever recorded in the US—390 in (991 cm)—fell in Alpine County in January 1911. Snow averages between 300 and 400 in (760–1,020 cm) annually in the high elevations of the Sierra Nevada, but is rare in the coastal lowlands.

In 1983, Los Angeles, San Francisco, and San Diego were free of frost throughout the year, and Sacramento had a frost-free period of 344 days; the Alturas Ranger station, in the northeast, had only 115 frost-free days. Sacramento has the greatest percentage (79%) of possible annual sunshine among the state's largest cities; Los Angeles has 73%, and San Francisco 67%. San Francisco is the windiest, with an average annual wind speed of 11 mph (18 km/hr).

Severe tropical rainstorms occur often in California during the winter. In late February 1980, six storms struck the southern coast within nine days, killing 24 persons and causing an estimated $425 million in property damage, chiefly in the residential suburbs of Los Angeles, where the downpours produced extensive mudslides. During the winter of 1983/84, four storms killed 11 persons, forced the evacuation of 2,000 more, and closed coastal Highway 1 for 13 months.

⁴FLORA AND FAUNA

Of the 48 conterminous states, California embraces the greatest diversity of climate and terrain. The state's six life zones are the lower Sonoran (desert); upper Sonoran (foothill regions and some coastal lands); transition (coastal areas and moist northeastern counties); and the Canadian, Hudsonian, and Arctic zones, comprising California's highest elevations.

Plant life in the arid climate of the lower Sonoran zone features a diversity of native cactus, mesquite, and paloverde. The Joshua tree (*Yucca brevifolia*) is found in the Mojave Desert. Flowering plants include the dwarf desert poppy and a variety of asters. Fremont cottonwood and valley oak grow in the Central Valley. The upper Sonoran zone includes the unique chapparal belt, characterized by forests of small shrubs, stunted trees, and herbaceous plants. Nemophila, mint, phacelia, viola, and the golden poppy (*Eschscholtzia californica*)—the state flower—also flourish in this zone, along with the lupine, more species of which occur here than anywhere else in the world.

The transition zone includes most of the state's forests, with such magnificent specimens as the redwood (*Sequoia sempervirens*) and "big tree" or giant sequoia (*Sequoia gigantea*), among the oldest living things on earth (some are said to have lived at least 4,000 years). Tanbark oak, California laurel, sugar pine, madrona, broad-leaved maple, and Douglas fir are also common. Forest floors are carpeted with swordfern, alumroot, barrenwort, and trillium, and there are thickets of huckleberry, azalea, elder, and wild currant. Characteristic wild flowers include varieties of mariposa, tulip, and tiger and leopard lilies.

The high elevations of the Canadian zone are abundant with Jeffrey pine, red fir, and lodgepole pine. Brushy areas are covered with dwarf manzanita and ceanothus; the unique Sierra puffball is also found here. Just below timberline, in the Hudsonian zone, grow the whitebark, foxtail, and silver pines. At approximately 10,500 feet (3,200 meters) begins the Arctic zone, a treeless region whose flora includes a number of wild flowers, including Sierra primrose, yellow columbine, alpine buttercup, and alpine shootingstar.

Common plants introduced into California include the eucalyptus, acacia, pepper tree, geranium, and Scotch broom. Among the numerous species found in California that are federally classified as endangered are the Contra Costa wallflower, Antioch Dunes evening primrose, Solano Grass, San Clemente Island larkspar, salt marsh bird's beak, McDonald's rock-cress, and Santa Barbara Island liveforever.

Mammals found in the deserts of the lower Sonoran zone include the jackrabbit, kangaroo rat, squirrel, and opossum. The Texas nightowl, roadrunner, cactus wren, and various species of hawk are common birds, and the sidewinder, desert tortoise, and

LOCATION: 32°32' to 42°N; 114°08' to 124°25'W. **BOUNDARIES:** Oregon line, 220 mi (354 km); Nevada line, 612 mi (985 km); Arizona line, 234 mi (376 km); Mexico line, 144 mi (232 km); Pacific Ocean coastline, 840 mi (1,352 km).

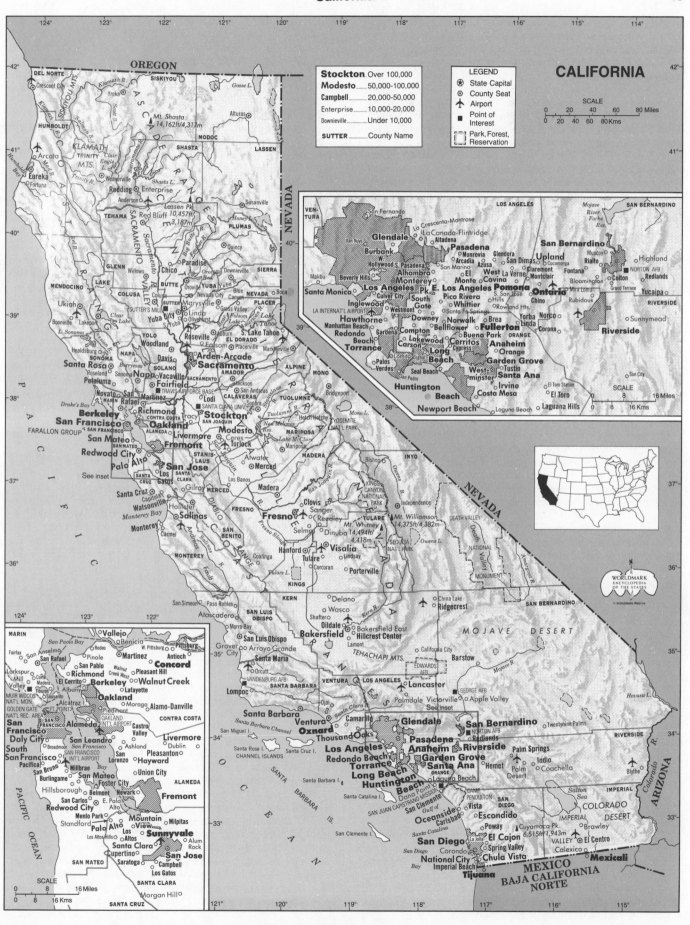

horned toad represent the area's reptilian life. The upper Sonoran zone is home to such mammals as the antelope, brown-footed woodrat, and ring-tailed cat. Birds distinctive to this zone are the California thrasher, bush tit, and California condor.

Animal life is abundant amid the forests of the transition zone. Colombian black-tailed deer, black bear, gray fox, cougar, bobcat, and Roosevelt elk are found. Garter snakes and rattlesnakes are common, as are such amphibians as the water-puppy and redwood salamander. The kingfisher, chickadee, towhee, and hummingbird represent the bird life of this region.

Mammals of the Canadian zone include the mountain weasel, snowshoe hare, Sierra chickaree, and several species of chipmunk. Conspicuous birds include the blue-fronted jay, Sierra hermit thrush, water ouzel, and Townsend solitaire. Birds become scarcer as one ascends to the Hudsonian zone, and the wolverine is now regarded as rare. Only one bird is native to the high Arctic region—the Sierra rosy finch—but others often visit, including the hummingbird and Clark nutcracker. Principal mammals of this region are also visitors from other zones, though the Sierra coney and white-tailed jackrabbit make their homes here. The bighorn sheep also lives in this mountainous terrain. Among fauna found throughout several zones are the mule deer, coyote, mountain lion, red-shafted flicker, and several species of hawk and sparrow.

Aquatic life in California is abundant, from the state's mountain lakes and streams to the rocky Pacific coastline. Many trout species are found, among them rainbow, golden, and Tahoe; migratory species of salmon are also common. Deepsea life-forms include sea bass, yellowfin tuna, barracuda, and several types of whale. Native to the cliffs of northern California are seals, sea lions, and many types of shorebirds, including several migratory species.

The Resources Agency of California's Department of Fish and Game is especially active in listing and providing protection for rare, threatened, and endangered fauna. Joint efforts by state and federal wildlife agencies have established an ambitious—if somewhat controversial—recovery program to revitalize the dwindling population of the majestic condor, the largest bird native to the US.

As of 20 July 1984, some 30 California animals were on the federal endangered list. These included the San Joaquin kit fox, California gray whale (the official state marine mammal), salt marsh harvest mouse, Morro Bay kangaroo rat, bald eagle, California least tern, San Francisco garter snake, desert slender salamander, Santa Cruz long-toed salamander, Mohave tui chub, and Owens River pupfish. All seven butterflies listed as endangered on the federal list are California species. Among threatened animals are the San Clemente sage sparrow, Paiute cutthroat trout, and Southern sea otter. California has a total of 290,821 acres (117,691 hectares) of National Wildlife Refuges.

⁵ENVIRONMENTAL PROTECTION

Efforts to preserve natural wilderness areas in California go back at least to 1890, when the US Congress created three national parks in the Sierra Nevada: Sequoia, Grant (now part of Kings Canyon), and Yosemite. Three years later, some 4 million acres (1.6 million hectares) of the Sierra Nevada were set aside in national forests. In 1892, naturalist John Muir and other wilderness lovers founded the Sierra Club, which with other private groups of conservationists has been influential in saving the Muir Woods and other stands of redwoods from the lumberman's ax.

California's primary resource problem is water: the southern two-thirds of the state accounts for about 75% of annual water consumption but only 30% of the supply. Water has been diverted from the Sierra Nevada snow runoff and from the Colorado River to the cities and dry areas largely by means of aqueducts, some 700 mi (1,100 km) of which have been constructed in federal and state undertakings. In 1960, California embarked on one of the largest public works programs ever undertaken in the US, when

voters approved a bond issue to construct the California Water Project, designed to deliver annually 1.4 trillion gallons of water to central and southern California for residential, industrial, and agricultural use. Other purposes of the project were to provide flood control, generate electric power, and create recreation areas.

Maintaining adequate water resources continued to be a problem in the 1980s. As the result of a US Supreme Court decision, southern California stands to lose close to 20% of its water supply as early as December 1985, when a portion of the water it has been permitted to draw from the Colorado River is diverted to Arizona. In 1982, California voters turned down a proposal to build a canal that would have delivered water that flows into San Francisco Bay to southern California; by late 1984, no other plans to cope with the impending shortage had been approved.

Air pollution has been a serious problem since July 1943, when heavy smog enveloped Los Angeles for the first time; smog conditions in October 1954 forced the closing of the city's airport and harbor. Smog is caused by an atmospheric inversion of cold air that traps unburned hydrocarbons at ground level; perhaps two-thirds of the smog particles are created by automobile exhaust emissions. In 1960, the state legislature passed the first automobile antismog law in the nation, requiring that all cars be equipped with antismog exhaust devices within three years. (Federal laws controlling exhaust emissions on new cars came into effect in the 1970s.) The city's smog problem has since been reduced to manageable proportions, but pollution problems from atmospheric inversions still persist there and in other California cities. In areas with the worst air pollution, antipollution devices are now checked every two years when a vehicle's registration is renewed. Antipollution checks are required statewide for cars changing ownership or being registered in California for the first time.

State land-reclamation programs have been important in providing new agricultural land and controlling flood damage. One of the earliest such programs, begun shortly before 1900, reclaimed 500,000 acres (200,000 hectares) by means of a network of dams, dikes, and canals in the swampy delta lying within the fork of the Sacramento and San Joaquin rivers. In 1887, a state law created irrigation districts in the southeastern region; the Imperial Valley was thus transformed from a waterless, sandy basin into some of the most productive agricultural land in the US.

Flood control was one of the main purposes of the $2.6-billion Feather River Project in the Central Valley, completed during the 1970s. Ironically, in the western portion of the Central Valley, farmland is now threatened by irrigation water tainted by concentrated salts and other soil minerals, for which current drainage systems are inadequate. One drainage system, the San Luis Drain, originally intended to carry the water to San Francisco Bay, was stopped short of completion and goes only as far as the Kesterson National Wildlife Refuge, where, according to the US Fish and Wildlife Service, the tainted water has caused birth defects in birds.

In 1972, the state legislature enacted stringent controls on toxic waste disposal. California has also been a leader in recycling waste products—for example, using acid waste from metal-processing plants as a soil additive in citrus orchards.

The California Department of Water Resources is responsible for maintaining adequate groundwater levels, enforcing water-quality standards, and controlling floodwaters. The state Department of Conservation has overall responsibility for conservation and protection of the state's soil, mineral, petroleum, geothermal, and marine resources. The California Coastal Commission, created in 1972, is designated by federal law to review projects that effect California's coastline, including offshore oil leasing, which has become a source of concern in recent years.

⁶POPULATION

At least 1 out of every 10 Americans lives in California, which ranks 1st in population among the 50 states. California replaced New York as the decennial census leader in 1970, with a total of 19,971,069 residents, and has lengthened its lead ever since. Los Angeles is the 2d most populous city in the US, and Los Angeles County ranks 1st in population among all US counties.

When Europeans first arrived in California, at least 300,000 Indians lived in the area. By 1845, the Indian population had been reduced to about 150,000. Although Spanish missions and settlements were well established in California by the late 18th century, the white population numbered only about 7,000 until the late 1840s. The gold rush brought at least 85,000 adventurers to the San Francisco Bay area by 1850, however, and the state's population increased rapidly thereafter. California's population grew to 379,994 by 1860, and had passed the 1 million mark within 30 years. Since 1890, the number of state residents just about doubled every two decades until the 1970s, when the population continued to increase, though at a slower pace. Although the state's annual growth rate fell from 4% during the 1950s to 2.3% during the 1960s and 1.9% during the 1970s, and was 1.7% between April 1980 and January 1985, when the estimated population reached 25,816,590, projections indicate a steady increase to 27,525,600 by 1990 and to 30,613,100 by the year 2000.

In 1983, California was 2d only to New Jersey in the proportion of residents living in metropolitan areas, more than 95%. At the 1980 census, the population of 23,668,000 was 49.3% male and 50.7% female. The population density in 1970 was 128 persons per sq mi (49 per sq km), or more than double the US average; by 1985, the estimated density was 165 per sq mi (64 per sq km), 12th in the US but the highest of any western state. Densities in urban areas were much higher—6,491 per sq mi (2,506 per sq km) in Los Angeles in 1982, and 18,906 per sq mi (7,300 per sq km) in San Francisco. Population estimates for 1983 indicated that Californians were not conspicuously younger than the US as a whole; the population curve showed a slight bulge toward those between 18 and 64, while the proportions of those older or younger were slightly below the US norms. Californians are highly mobile, however. According to the 1980 census, only 44.6% of state residents 5 years of age or older lived in the same house they had lived in five years earlier. Only 45% of all Californians in 1980 had been born in the state. California gained 1,573,000 people through migration during the 1970s, and about 1,629,000 through net natural increase. The birthrate per 1,000 residents dropped from 23.7 in 1960 to 14.7 in 1978, but then rose to 17.3 in 1983. The death rate declined from 8.6 in 1960 to 7.6 in 1981.

Three out of 4 Californians live in urban areas located within 20 mi (32 km) of the ocean; 3 out of 5 reside in metropolitan San Francisco and Los Angeles. In 1984, the state had 28 cities with more than 100,000 population. Los Angeles had an estimated population of 3,096,721, San Diego 960,452, San Francisco 712,753, San Jose 686,178, Long Beach 378,752, Oakland 351,898, Sacramento 304,131, Fresno 267,377, Anaheim 233,516, and Santa Ana 225,405. The first 4 all placed among the nation's 15 most populous cities in 1984.

Los Angeles, which expanded irregularly and lacks a central business district, nearly quadrupled its population from 319,000 in 1910 to 1,240,000 in 1930, and then doubled it to 2,479,000 by 1960. A major component of the city's population growth was the upsurge in the number of blacks after World War II, especially between 1960 and 1970, when the number of blacks increased from 335,000 to 504,000, many of them crowded into the deteriorating Watts section. During the same period, San Francisco's black population increased from 74,000 to 96,000, while the white population declined from 666,000 to 620,000. Between 1970 and 1980, however, the percentage of blacks fell from 17.9% to 17% in Los Angeles, and from 13.4% to 12.7% in San Francisco.

California Counties, County Seats, and County Areas and Populations

COUNTY	COUNTY SEAT	LAND AREA (SQ MI)	POPULATION (1984 EST.)
Alameda	Oakland	736	1,173,395
Alpine	Markleeville	739	1,136
Amador	Jackson	589	22,121
Butte	Oroville	1,646	158,530
Calaveras	San Andreas	1,021	25,652
Colusa	Colusa	1,153	14,197
Contra Costa	Martinez	730	698,015
Del Norte	Crescent City	1,007	18,351
El Dorado	Placerville	1,715	98,884
Fresno	Fresno	5,978	564,915
Glenn	Willows	1,319	22,678
Humboldt	Eureka	3,579	111,580
Imperial	El Centro	4,173	102,351
Inyo	Independence	10,223	18,291
Kern	Bakersfield	8,130	462,371
Kings	Hanford	1,392	81,726
Lake	Lakeport	1,262	45,643
Lassen	Susanville	4,553	23,892
Los Angeles	Los Angeles	4,070	7,901,220
Madera	Madera	2,145	73,911
Marin	San Rafael	523	224,167
Mariposa	Mariposa	1,456	12,815
Mendocino	Ukiah	3,512	71,682
Merced	Merced	1,944	154,912
Modoc	Alturas	4,064	9,580
Mono	Bridgeport	3,019	9,123
Monterey	Salinas	3,303	319,223
Napa	Napa	744	101,991
Nevada	Nevada City	960	64,964
Orange	Santa Ana	798	2,075,758
Placer	Auburn	1,416	133,166
Plumas	Quincy	2,573	18,470
Riverside	Riverside	7,214	777,173
Sacramento	Sacramento	971	866,831
San Benito	Hollister	1,388	29,134
San Bernardino	San Bernardino	20,064	1,033,704
San Diego	San Diego	4,212	2,063,902
San Francisco	San Francisco¹	46	712,753
San Joaquin	Stockton	1,415	398,630
San Luis Obispo	San Luis Obispo	3,308	179,013
San Mateo	Redwood City	447	604,942
Santa Barbara	Santa Barbara	2,748	322,781
Santa Clara	San Jose	1,293	1,371,522
Santa Cruz	Santa Cruz	446	205,816
Shasta	Redding	3,786	126,495
Sierra	Downieville	959	3,340
Siskiyou	Yreka	6,281	41,835
Solano	Fairfield	834	265,652
Sonoma	Santa Rosa	1,604	326,319
Stanislaus	Modesto	1,506	295,700
Sutter	Yuba City	602	57,006
Tehama	Red Bluff	2,953	43,108
Trinity	Weaverville	3,190	13,169
Tulare	Visalia	4,808	272,430
Tuolumne	Sonora	2,234	38,490
Ventura	Ventura	1,862	584,731
Yolo	Woodland	1,014	120,722
Yuba	Marysville	640	52,568
TOTALS		156,299	25,622,497

¹The city and county of San Francisco are coterminous.

In 1984, the Los Angeles–Anaheim–Riverside urban complex, with a total population of 12,372,600, was the 2d most populous metropolitan area in the US (after New York City). In that year, the San Francisco–Oakland–San Jose area ranked 5th with 5,684,600, metropolitan San Diego 19th with 2,063,900, and metropolitan Sacramento 30th with 1,219,600.

7ETHNIC GROUPS

At least 25% of all foreign-born persons in the US live in California. The state has the nation's largest populations of Mexican-Americans, Chinese, Filipinos, Japanese, Koreans, Vietnamese, Samoans, Guamanians, and American Indians; it has more blacks and Asian Indians than any state except New York, more Eskimos and Aleuts than any state except Alaska, and more native Hawaiians than any state except Hawaii.

The westward movement of American settlers in the third quarter of the 19th century, followed by Germans, Irish, North Italians, and Italian Swiss immigrants, overshadowed but did not obliterate California's Spanish heritage. In 1980, Hispanics were the largest ethnic minority; California had 4,544,331 residents of Spanish origin, more than any other state and 31% of the US total. The majority—3,637,466—were Mexican-Americans; in 1980 there were also 93,038 Puerto Ricans, 61,004 Cubans, and 57,753 Salvadorans. After World War II, the Hispanic communities of Los Angeles, San Diego, and other southern California cities developed strong political organizations. Increasing numbers of Mexican-Americans have won local, state, and federal elective office, though their potential remains unrealized.

Nearly 50% of all Mexican-Americans in the US are farm laborers. In California, with its large corporate growers, fewer than 10% of the growers employ three-fourths of the laborers; it was this concentration of labor that made possible the successful unionization of field workers.

California had 1,253,887 Asians and Pacific Islanders in 1980, nearly 36% of the US total and more than four times the number in New York State.

Chinese workers were imported into the state between 1849 and 1882, when the Chinese Exclusion Act was passed by Congress. In 1980, the state's Chinese population was 322,240, more than double that of New York State and by far the highest in the US. The nation's oldest and largest Chinatown is in San Francisco. Although Chinese-Americans, as they prospered, moved to suburban areas, the seats of the powerful nationwide and worldwide merchant and clan associations are in that city. Los Angeles also has a Chinese district.

The Japanese, spread throughout the western seaboard states, were engaged mainly in agriculture, along with fishing and small business, until their removal and internment during World War II. After the war, some continued in market gardening and other family agriculture, but most, deprived of their landholdings, entered urban occupations, including the professions; many dispersed to other regions of the country. In 1980 there were 261,817 Japanese in California.

While the Chinese and Japanese communities in California are the oldest in the state, they were not the most populous in 1980; this distinction was held by the Filipino community, which numbered 357,514 that year. There were also 103,891 Koreans, 89,587 Vietnamese, 57,989 Asian Indians, 23,091 native Hawaiians, 20,096 Samoans, and 17,662 Guamanians.

Native Americans numbered 201,311 in 1980—198,095 American Indians, 1,734 Eskimos, and 1,482 Aleuts. The figure for American Indians, more than double that for 1970, includes Indians native to California and many others coaxed to resettle there under a policy that sought to terminate tribal status. Along with the remaining indigenous tribes in California, there is also a large urban Indian population, especially in Los Angeles, which has more Indians than any other US city. Many of the urban Indians were unprepared for the new kind of life and unable to earn an adequate living; militant Indians have made dramatic, but on the whole unsuccessful, protests aimed at bettering their condition.

Black Americans constitute a smaller proportion of California's population than of the nation's as a whole: less than 8% in 1980. Considerable migration of blacks took place during World War II, when defense industries on the West Coast offered new opportunities. Many California blacks are middle-class business people and professionals. The Watts district of Los Angeles is a ghetto of low-income blacks; many low-income blacks also live in Oakland.

8LANGUAGES

The speakers of Russian, Spanish, and English who first came to what is now California found an amazing diversity of Indian cultures, ranging from the Wiyot in the north to the Yokuts in the central valley and the Diegueño in the south, and of Indian languages, representing four great language families—Athapaskan, Penutian, Hokan-Siouan, and Aztec. Yet, except for place-names such as Shasta, Napa, and Yuba, they have not lent any of their words to California speech.

As in much of the West, California English is a composite of the eastern dialects and subdialects brought by the continuing westward migration from the eastern states, first for gold and timber, then for farming, for diversified manufacture, for Hollywood, and for retirement. The interior valley is Midland-oriented with such retained terms as *piece* (a between-meals lunch), *quarter till*, *barn lot* (barnyard), *dog irons* (andirons), and *snake feeder* and *snake doctor* (dragonfly), but generally, in both northern and southern California, Northern dominates the mixture of North Midland and South Midland speech in the same communities. Northern *sick to the stomach,* for example, dominates Midland *sick at* and *sick in,* with a 46% frequency; Northern *angleworm* has 53% frequency, as compared with 21% for Midland *fishworm;* and Northern *string beans* has 80% frequency, as compared with 17% North Midland *green beans* and South Midland and Southern *snap beans.* Northern *comforter* was used by 94% of the informants interviewed in a state survey; Midland *comfort* by only 21%. Dominant is Northern /krik/ as the pronunciation of *creek,* but Midland *bucket* has a greater frequency than Northern *pail,* and Midland /greezy/ for *greasy* is scattered throughout the state. Similarly, the distinction between the /hw/ in *wheel* and the /w/ of *weal* is lost in the use of simple /w/ in both words, and *cot* and *caught* sound alike, as do *caller* and *collar.*

There are some regional differences. San Francisco, for instance, has *sody* or *soda water* for a soft drink; there the large sandwich is a *grinder,* while in Sacramento it is either a *poor Joe* or a *submarine.* Notable is the appearance of *chesterfield* (meaning sofa or davenport), found in the Bay region and from San Jose to Sacramento; this sense is common in Canada but now found nowhere else in the US. Boonville, a village about 100 mi (160 km) north of San Francisco, is notorious for "Boontling," a local dialect contrived in the mid-19th century by Scotch-Irish settlers who wanted privacy and freedom from obscenities in their conversation. Now declining in use, Boontling has about 1,000 vocabulary replacements of usual English words, together with some unusual pronunciations and euphemisms.

As the nation's major motion picture, radio, and television entertainment center, Los Angeles has influenced English throughout the nation—even the world—by making English speakers of many dialects audible and visible and by making known new terms and new meanings. It has thus been instrumental in reducing dialectal extremes and in developing increased language awareness.

California's large foreign-language populations have posed major educational problems. In 1974, a landmark San Francisco case, *Lau* v. *Nichols,* brought a decision from the US Supreme Court that children who do not know English should not thereby be handicapped in school, but should receive instruction in their native tongue while learning English.

In 1980, 17,476,947 Californians—or 77% of the population 3 years old or over—reported speaking only English at home. Other languages spoken at home included the following:

Spanish	3,270,267	Vietnamese	68,298
Chinese	264,229	Armenian	51,845
Pilipino or other Filipino language	233,822	Hindi or other Asian Indian language	44,092
German	171,907	Arabic	42,369
Italian	132,838	Farsi (Persian)	37,632
Japanese	130,110	Dutch	37,630
French	114,262	Greek	35,626
Korean	83,422	Russian	32,439
Portuguese	76,933	Yiddish	23,762

⁹RELIGIONS

The first Roman Catholics in California were Spanish friars, who established 21 Franciscan missions from San Diego to Sonoma between 1769 and 1823. After an independent Mexican government began to secularize the missions in 1833, the Indian population at the missions declined from about 25,000 to only about 7,000 in 1840. With the American acquisition of California in 1848, the Catholic Church was reorganized to include the archdiocese of San Francisco. Protestant ministers accompanied migrant miners during the gold rush, founding 32 churches in San Francisco by 1855. These early Protestants included Baptists, Congregationalists, Methodists, Presbyterians, Episcopalians, and Unitarians; a group of Mormons had arrived by ship via Cape Horn in 1846. Small Jewish communities were established throughout California by 1861, and in 1880, the Jewish population was estimated at 18,580. The midwesterners who began arriving in large numbers in the 1880s were mostly Protestants who settled in southern California. By 1900, the number of known Christians in the state totaled 674,000, out of a population of nearly 1,500,000.

The mainstream religions did not satisfy everybody's needs, however, and in the early 20th century, many dissident sects sprang up, including such organizations as Firebrands for Jesus, the Psychosomatic Institute, the Mystical Order of Melchizedek, the Infinite Science Church, and Nothing Impossible, among many others. Perhaps the best-known founder of a new religion was Canadian-born Aimee Semple McPherson, who preached her Foursquare Gospel during the 1920s at the Angelus Temple in Los Angeles, won a large radio audience and thousands of converts, and established 240 branches of her church throughout the state before her death in 1944. She was typical of the many charismatic preachers of new doctrines who gave—and still give—California its exotic religious flavor. Since World War II, religions such as Zen Buddhism and Scientology have won enthusiastic followings, along with various cults devoted to self-discovery and self-actualization.

Nevertheless, the large majority of religious adherents in California continue to follow traditional faiths. In 1984 there were 5,469,475 Roman Catholics (2d only to New York); the archdiocese of Los Angeles had 2,374,021, the San Francisco archdiocese had 368,937, and the remainder were divided among 10 dioceses. In 1980, the state had 3,322,117 known Protestant adherents. The largest non-Catholic Christian denominations were the Church of Jesus Christ of Latter-day Saints, with 405,441 adherents; Southern Baptist, 392,451; United Methodist, 290,720; United Presbyterian, 261,279; American Baptist, 194,594; Assembly of God, 189,577; Episcopal, 184,330; Seventh-Day Adventist, 148,561; Lutheran Church-Missouri Synod, 148,292; and American Lutheran Church, 122,795. In 1984, the Jewish population was estimated at 792,515 (2d in the US), nearly two-thirds of whom lived in the Los Angeles metropolitan area.

¹⁰TRANSPORTATION

California has—and for decades has had—more motor vehicles than any other state, and ranks 2d only to Texas in interstate highway mileage. An intricate 5,400-mi (8,700-km) network of urban freeways is one of the engineering wonders of the modern world—but the traffic congestion in the state's major cities during rush hours may well be the worst in the country.

In pioneer days, the chief modes of transportation were sailing ships and horse-drawn wagons; passage by sea from New York took three months, and the overland route from Missouri was a six-week journey. The gold rush spurred development of more rapid transport. The state's first railroad, completed in 1856, was a 25-mi (40-km) line from Sacramento northeast to Folsom, in the mining country. The Central Pacific–Union Pacific transcontinental railroad, finished 13 years later, was financed in part by several Sacramento business leaders, including Leland Stanford, who became governor of the state in 1861, the same year he assumed the presidency of the Central Pacific. Railroad construction crews, mostly imported Chinese laborers, started from Sacramento and dug and blasted the route through the solid granite of the Sierra Nevada and then across the Nevada desert, linking up with the Union Pacific at Promontory, Utah, on 10 May 1869; Stanford himself helped drive the golden spike that marked the historic occasion. The Southern Pacific completed a line from Sacramento to Los Angeles in 1876, and another to Texas the following year. Other railroads took much longer to build; the coastal railroad from San Francisco to Los Angeles was not completed until 1901, and another line to Eureka was not finished until 1914. The railroads dominated transportation in the state until motor vehicles came into widespread use in the 1920s.

As of 1983, California had 6,464 mi (10,403 km) of Class I track; railways with the greatest track mileage were the Southern Pacific, Santa Fe, Western Pacific, and Union Pacific. Amtrak passenger trains connect the state's major population centers. Amtrak's Los Angeles–San Diego commuter line carries more than 1.2 million passengers annually. On the San Francisco–San Jose peninsula, the Peninsula Commuter Rail Service carries more than 5 million passengers annually.

Urban transit began in San Francisco in 1861 with horse-drawn streetcars. Cable-car service was introduced in 1873; a few cable cars are still in use, mainly for the tourist trade. The 71-mi (114-km) Bay Area Rapid Transit System, or BART, connects San Francisco with Oakland by high-speed, computerized subway trains via a 3.6-mi (5.8-km) tunnel under San Francisco Bay and runs north-south along the San Francisco peninsula. Completed in the 1970s despite many mechanical problems and costly delays, BART carried an average of 219,000 weekday passengers as of December 1984.

Public transit in the Los Angeles metropolitan area was provided by electric trolleys beginning in 1887. By the early 1930s, the Los Angeles Railway carried 70% of the city's transit passengers, and in 1945, its trolleys transported 109 million passengers. Competition from buses—which provided greater mobility, but aggravated the city's smog and congestion problems—forced the trolleys to end service in 1961. Today, Los Angeles lacks an adequate urban transit system, although interurban railroads do connect the city with surrounding suburbs.

California's extensive highway system had its beginnings in the mid-19th century, when stagecoaches began hauling freight to the mining camps from San Francisco, Sacramento, and San Jose. In the early 1850s, 2 stagecoach lines, Adams and Wells Fargo, expanded their routes and began to carry passengers; by 1860, some 250 stagecoach companies were operating in the state. The decline of stagecoach service corresponded with the rise of the railroads. In 1910, at a time when only 36,000 motor vehicles were registered in the state, the California Highway Commission was established. Among its first acts was the issuance of $18 million in bonds for road construction, and the state's first paved highway was constructed in 1912. The number of automobiles surged to 604,000 by 1920; by 1929, about 1 of every 11 cars in the US belonged to a Californian. Ironically, in view of the state's subsequent traffic problems, the initial effect of the automobile was to disperse the population to outlying areas, thus reducing traffic congestion in the cities.

The Pasadena Freeway, the first modern expressway in California, opened in 1941. During the 1960s and 1970s, the state built a complex toll-free highway network linking most cities of more than 5,000 population, tying in with the federal highway system, and costing more than $10 billion. In 1982, total expenditures on roads by all levels of government amounted to nearly $3.1 billion. In 1983/84, federal aid to California from the Highway Trust Fund totaled about $686 million; and from the Urban Mass Transportation Administration, $495 million.

The new freeways, by providing easy access to beach and mountain recreation areas, and in combination with the favorable climate and low price of gasoline, further encouraged the use of the automobile—and led to massive traffic tie-ups, contributed to the decline of public transit, and worsened the coastal cities' air-pollution problems. Los Angeles County claims more automobiles, more miles of streets, and more intersections than any other city in the US. The greatest inducement to automobile travel in and out of San Francisco was the completion in 1936 of the 8-mi (13-km) San Francisco–Oakland Bay Bridge. The following year saw the opening of the magnificent Golden Gate Bridge, which at 4,200 feet (1,280 meters) was the world's longest suspension bridge until New York's Verrazano-Narrows Bridge opened to traffic in 1964.

In 1983, California had 174,033 mi (280,079 km) of roads. Included in this total were 32,126 mi (51,702 km) of federal highways, 15,185 mi (24,438 km) of state highways, 70,044 mi (112,725 km) of county roads, 53,891 mi (86,729 km) of city streets, and 2,787 mi (4,485 km) of other state roads. In that year, the state registered 18,471,865 motor vehicles—about 73 for every 100 state residents—including 13,935,390 automobiles, 3,797,926 trucks, 705,106 motorcycles, and 33,443 buses. There were 16,648,518 California drivers' licenses in force in 1983. In that year, California led the nation in number of traffic injuries, 292,538, and fatalities, 4,571, but the state's death rate per 100,000,000 vehicle miles—2.8—was slightly less than the US average.

The large natural harbors of San Francisco and San Diego monopolized the state's maritime trade until 1912, when Los Angeles began developing port facilities at San Pedro by building a breakwater that eventually totaled 8 mi (13 km) in length. In 1924, Los Angeles surpassed San Francisco in shipping tonnage handled, and became the busiest port on the Pacific coast. In 1983, 59.8 million lb of goods were handled by California's ports, of which the Los Angeles Customs District accounted for 61%, San Francisco 37%, and San Diego 2%.

In 1983, California had 862 aircraft facilities, including 568 airports, 277 heliports, 13 seaplane bases, and 4 stolports. More than 29,000 general aviation aircraft, greater than the number in any other state and nearly 14% of the US total, were registered in California in 1983. California's most active air terminal—and the nation's 3d most active—is Los Angeles International Airport, which handled 139,036 departing passenger aircraft and enplaned 14,265,641 passengers in 1983. Also among the nation's 20 busiest airports in 1983 were those at Van Nuys, Santa Ana, Long Beach, San Francisco, and Oakland.

¹¹HISTORY

The region now known as California has been populated for at least 10,000 years, and possibly far longer. Estimates of the prehistoric Indian population have varied widely, but it is clear that California was one of the most densely populated areas north of Mexico. On the eve of European discovery, at least 300,000 Indians lived there. This large population was divided into no fewer than 105 separate tribes or nations speaking at least 100 different languages and dialects, about 70% of which were as mutually unintelligible as English and Chinese. No area of comparable size in North America, and perhaps the world, contained a greater variety of native languages and cultures than did aboriginal California.

In general, the California tribes depended for their subsistence on hunting, fishing, and gathering the abundant natural food resources. Only in a few instances, notably along the Colorado River, did the Indians engage in agriculture. Reflecting the mild climate of the area, their housing and dress were often minimal. The basic unit of political organization was the village community, consisting of several small villages, or the family unit. For the most part, these Indians were sedentary peoples; they occupied village sites for generations, and only rarely warred with their neighbors.

European contact with California began early in the Age of Discovery, and was a product of the two great overseas enterprises of 16th-century Europe: the search for a western passage to the East and the drive to control the riches of the New World. In 1533, Hernán Cortés, Spanish conqueror of the Aztecs, sent a naval expedition northward along the western coast of Mexico in search of new wealth. The expedition led to the discovery of Baja California (now part of Mexico), mistakenly described by the pilot of the voyage, Fortún Jiménez, as an island. Two years later, Cortés established a settlement on the peninsula at present-day La Paz, but because Baja California seemed barren of any wealth, the project was soon abandoned. The only remaining interest in California was the search for the western mouth of the transcontinental canal—a mythical waterway the Spanish called the Strait of Anian. In 1542, Juan Rodríquez Cabrillo led a voyage of exploration up the western coast in a futile search for the strait. On 28 September, Cabrillo landed at the bay now known as San Diego, thus becoming the first European discoverer of Alta (or Upper) California.

European interest in the Californias waned in the succeeding decades. There was a brief revival of interest following the landing of Francis Drake in 1579; the British navigator put in for repairs probably at what is now Drake's Bay, north of San Francisco, and called the region New Albion, claiming it for England. At about this time, the Spanish made preliminary plans to settle the California coast as a way station for the Manila galleons plying the trade route between Mexico and the Philippines. This project too was abandoned, however, and California remained for generations beyond the periphery of European activity in the New World. Subsequent contact was limited to occasional landfalls by Manila galleons, such as those of Pedro de Unamuno (1587) and Sebastián Cermeño (1595), and the tentative explorations of Sebastián Vizcaíno in 1602–3.

Spanish interest in California revived during the late 18th century, largely because Spain's imperial rivals were becoming increasingly aggressive. For strategic and defensive reasons, Spain decided to establish permanent settlements in the north. In 1769, José de Gálvez, visitor-general in New Spain, selected the president of the Franciscan missions in Baja California, Father Junípero Serra, to lead a group of missionaries on an expedition to Alta California. Accompanying Serra was a Spanish military force under Gaspar de Portolá. This Portolá-Serra expedition marks the beginning of permanent European settlement in California. Over the next half-century, the 21 missions established by the Franciscans along the Pacific coast from San Diego to San Francisco formed the core of Hispanic California. Among the prominent missions were San Diego de Alcalá (founded in 1769), San Francisco de Asís (1776), Santa Barbara (1786), and San José (1797). During most of the Spanish period, Mission San Carlos

Borromeo (1770), at Carmel, was the ecclesiastical headquarters of the province, serving as the residence of the president-general of the Alta California missions.

These missions were more than just religious institutions. The principal concern of the missionaries was to convert the Indians to Christianity—a successful enterprise, if the nearly 88,000 baptisms performed during the mission period are any measure. The Franciscans also sought to bring about a rapid and thorough cultural transformation. The Indians were taught to perform a wide variety of new tasks: making bricks, tiles, pottery, shoes, saddles, wine, candles, and soap; herding horses, cattle, sheep, and goats; and planting, irrigating, and harvesting.

In addition to transforming the way of life of the California Indians, the missions also reduced their number by at least 35,000. About 60% of this decline was due to the introduction of new diseases, especially those of the nonepidemic and venereal type. Alfred Kroeber, founder of the American Anthropological Society, remarked that it "must have caused many of the Fathers a severe pang to realize, as they could not but do daily, that they were saving souls only at the inevitable cost of lives. And yet such was the overwhelming fact.The brute upshot of missionization, in spite of its kindly flavor and humanitarian root, was only one thing, death."

Spain also established several military and civilian settlements in California. The four military outposts, or presidios, at San Diego (1769), Monterey (1770), San Francisco (1776), and Santa Barbara (1782) served to discourage foreign influence in the region and to contain Indian resistance. The presidio at Monterey also served as the political capital, headquarters for the provincial governors appointed in Mexico City. The first civilian settlement, or pueblo, was established at San José de Guadalupe in 1777, with 14 families from the Monterey and San Francisco presidios. The pueblo settlers, granted supplies and land by the government, were expected to provide the nearby presidios with their surplus agricultural products. The second pueblo was founded at Los Angeles (1781), and a third, Branciforte, was established near present-day Santa Cruz in 1797.

During the 40 years following the establishment of the Los Angeles pueblo, Spain did little to strengthen its outposts in Alta California. The province remained sparsely populated and isolated from other centers of Hispanic civilization. During these years, the Spanish-speaking population of 600 grew nearly five-fold, but this expansion was almost entirely due to natural increase rather than immigration.

Spanish control of California ended with the successful conclusion of the Mexican revolution in 1821. For the next quarter-century, California was a province of the independent nation of Mexico. Although California gained a measure of self-rule with the establishment of a provincial legislature, the real authority still remained with the governor appointed in Mexico City. The most important issue in Mexican California was the secularization of the missions, the replacement of the Franciscans with parish or "secular" clergy, and the redistribution of the vast lands and herds the missions controlled. Following the secularization proclamation of Governor José Figueroa in 1834, the Mexican government authorized more than 600 rancho grants in California to Mexican citizens. The legal limit of an individual grant was 11 square leagues (about 76 sq mi—197 sq km), but many large landholding families managed to obtain multiple grants.

The rancho economy, like that of the missions, was based on the cultivation of grain and the raising of huge herds of cattle. The rancheros traded hides and tallow for manufactured goods from foreign traders along the coast. As at the missions, herding, slaughtering, hide tanning, tallow rendering, and all the manual tasks were performed by Indian laborers. By 1845, on the eve of American acquisition, the non-Indian population of the region stood at about 7,000.

During the Mexican period, California attracted a considerable minority of immigrants from the US. Americans first came to California in the late 18th century in pursuit of the sea otter, a marine mammal whose luxurious pelts were gathered in California waters and shipped to China for sale. Later, the hide and tallow trade attracted Yankee entrepreneurs, many of whom became resident agents for American commercial firms. Beginning in 1826, with the arrival overland of Jedediah Strong Smith's party of beaver trappers, the interior of California also began to attract a growing number of Americans. The first organized group to cross the continent for the purpose of settlement in California was the Bidwell-Bartleson party of 1841. Subsequent groups of overland pioneers included the ill-fated Donner party of 1846, whose members, stranded by a snowstorm near the Sierra Nevada summit, resorted to cannibalism so that 47 of the 87 travelers could survive.

Official American efforts to acquire California began during the presidency of Andrew Jackson in the 1830s, but it was not until the administration of James K. Polk that such efforts were successful. Following the American declaration of war against Mexico on 13 May 1846, US naval forces, under command of Commodores John D. Sloat and Robert F. Stockton, launched an assault along the Pacific coast, while a troop of soldiers under Stephen W. Kearny crossed overland. On 13 January 1847, the Mexican forces in California surrendered. More than a year later, after protracted fighting in central Mexico, a treaty of peace was signed at Guadalupe-Hidalgo on 2 February 1848. Under the terms of the treaty, Mexico ceded California and other territories to the US in exchange for $15 million and the assumption by the US of some $3 million in claims by Mexican citizens.

Just nine days before the treaty was signed, James Wilson Marshall discovered gold along the American River in California. The news of the gold discovery, on 24 January 1848, soon spread around the globe, and a massive rush of people poured into the region. By the end of 1848, about 6,000 miners had obtained $10 million worth of gold. During 1849, production was two or three times as large, but the proceeds were spread among more than 40,000 miners. In 1852, the peak year of production, about $80 million in gold was mined in the state, and during the century following its discovery, the total output of California gold amounted to nearly $2 billion.

California's census population quadrupled during the 1850s, reaching nearly 380,000 by 1860, and continued to grow at a rate twice that of the nation as a whole in the 1860s and 1870s. The new population of California was remarkably diverse. The 1850 census found that nearly a quarter of all Californians were foreign-born, while only a tenth of the national population had been born abroad. In succeeding decades, the percentage of foreign-born Californians increased, rising to just under 40% during the 1860s.

One of the most serious problems facing California in the early years of the gold rush was the absence of adequate government. Miners organized more than 500 "mining districts" to regulate their affairs; in San Francisco and other cities, "vigilance committees" were formed to combat widespread robbery and arson. The US Congress, deadlocked over the slavery controversy, failed to provide any form of legal government for California from the end of the Mexican War until its admission as a state in the fall of 1850. Taking matters into their own hands, 48 delegates gathered at a constitutional convention in Monterey in September 1849 to draft a fundamental law for the state. The completed constitution contained several unique features, but most of its provisions were based on the constitutions of Iowa and New York. To the surprise of many, the convention decided by unanimous vote to exclude slavery from the state. After considerable debate, the delegates also established the present boundaries of California. Adopted on 10 October, the constitution was ratified by the voters on 13

November 1849; at the same time, Californians elected their first state officials. California soon petitioned Congress for admission as a state, having bypassed the preliminary territorial stage, and was admitted after southern objections to the creation of another free state were overcome by adoption of the stringent new Fugitive Slave Law. On 9 September 1850, President Millard Fillmore signed the admission bill, and California became the 31st state to enter the union.

The early years of statehood were marked by racial discrimination and considerable ethnic conflict. Indian and white hostilities were intense; the Indian population declined from an estimated 150,000 in 1845 to less than 30,000 by 1870. In 1850, the state legislature enacted a foreign miners' license tax, aimed at eliminating competition from Mexican and other Latin American miners. The Chinese, who replaced the Mexicans as the state's largest foreign minority, soon became the target of a new round of discrimination. By 1852, 25,000 Chinese were in California, representing about a tenth of the state's population. The legislature enacted new taxes aimed at Chinese miners, and passed an immigration tax (soon declared unconstitutional) on Chinese immigrants.

Controversy also centered on the status of the Mexican ranchos, those vast estates created by the Mexican government that totaled more than 13 million acres (5 million hectares) by 1850. The Treaty of Guadalupe-Hidalgo had promised that property belonging to Mexicans in the ceded territories would be "inviolably protected." Nevertheless, in the early years of statehood, thousands of squatters took up residence on the rancho lands. Ultimately, about three-fourths of the original Mexican grants were confirmed by federal commissions and courts; however, the average length of time required for confirmation was 17 years. During the lengthy legal process, many of the grantees either sold parts of their grants to speculators or assigned portions to their attorneys for legal fees. By the time title was confirmed, the original grantees were often bankrupt and benefited little from the decision.

Despite the population boom during the gold rush, California remained isolated from the rest of the country until completion of the transcontinental railroad in 1869. Under terms of the Pacific Railroad Act of 1862, the Central Pacific was authorized by Congress to receive long-term federal loans and grants of land, about 12,500 acres per mi (3,100 hectares per km) of track, to build the western link of the road. The directors of the California corporation—Leland Stanford, Collis P. Huntington, Charles Crocker, and Mark Hopkins—who became known as the Big Four, exercised enormous power in the affairs of the state. Following completion of the Central Pacific, the Big Four constructed additional lines within California, as well as a second transcontinental line, the Southern Pacific, providing service from southern California to New Orleans.

To a degree unmatched anywhere in the nation, the Big Four established a monopoly of transportation in California and the Far West. Eventually the Southern Pacific, as the entire system came to be known after 1884, received from the federal government a total of 11,588,000 acres (4,690,000 hectares), making it the largest private landowner in the state. Opponents of the railroad charged that it had established not only a transportation monopoly but also a corrupt political machine and a "land monopoly" in California. Farmers in the San Joaquin Valley became involved in a protracted land dispute with the Southern Pacific, a controversy that culminated in a bloody episode in 1880, known as the Battle of Mussel Slough, in which seven men were killed. This incident, later dramatized by novelist Frank Norris in *The Octopus* (1901), threw into sharp relief the hostility between many Californians and the state's largest corporation.

In the late 19th century, California's economy became more diversified. The early dependence on gold and silver mining was overcome through the development of large-scale irrigation projects and the expansion of commercial agriculture. Southern California soon was producing more than two-thirds of the nation's orange crop, and more than 90% of its lemons. The population of southern California boomed in the 1880s, fueled by the success of the new citrus industry, an influx of invalids seeking a warmer climate, and a railroad rate war between the Southern Pacific and the newly completed Santa Fe. For a time, the tariff from Kansas City to Los Angeles fell to a dollar a ticket. Real estate sales in Los Angeles County alone exceeded $200 million in 1887.

During the early 20th century, California's population growth became increasingly urban. Between 1900 and 1920, the population of the San Francisco Bay area doubled, while residents of metropolitan Los Angeles increased fivefold. On 18 April 1906, San Francisco's progress was interrupted by the most devastating earthquake ever to strike California. The quake and the fires that raged for three days killed at least 452 people, razed the city's business section, and destroyed some 28,000 buildings. The survivors immediately set to work to rebuild the city, and completed about 20,000 new buildings within three years.

By 1920, the populations of the two urban areas were roughly equal, about 1 million each. As their populations grew, the need for additional water supplies became critical, and both cities became involved in bitter "water fights" with other state interests. Around the turn of the century, San Francisco proposed the damming of the Tuolumne River at the Hetch Hetchy Valley to form a reservoir for the city's water system. Conservationist John Muir and the Sierra Club objected strongly to the proposal, arguing that the Hetch Hetchy was as important a natural landmark as neighboring Yosemite Valley. The conservationists lost the battle, and the valley was flooded. (The dam there is named for Michael O'Shaughnessy, San Francisco's city engineer from 1912 to 1932 and the builder of many of California's water systems.) When Los Angeles began its search for new water supplies, it soon became embroiled in a long controversy over access to the waters of the Owens River. The city constructed a 250-mi (400-km) aqueduct that eventually siphoned off nearly the entire flow of the river, thus jeopardizing the agricultural development of Owens Valley. Residents of the valley dramatized their objection to the project by dynamiting sections of the completed aqueduct.

Important movements for political reform began simultaneously in San Francisco and Los Angeles in the early 20th century. Corruption in the administration of San Francisco Mayor Eugene Schmitz led to a wide-ranging public investigation and to a series of trials of political and business leaders. Meanwhile, in Los Angeles, a coalition of reformers persuaded the city to adopt a new charter with progressive features such as initiative, referendum, and recall. In 1907, reformers throughout the state combined forces and organized the Lincoln-Roosevelt League, dedicated to the election of Progressive candidates to public office. Hiram Johnson, the Progressive Republican candidate for governor in 1910, toured the state with his promise to "kick the Southern Pacific Railroad out of politics." Johnson won the election, and reformers gained control of both houses of the state legislature in 1911. Subsequent reform legislation established effective regulation of the railroads and other public utilities, greater governmental efficiency, female suffrage, closer regulation of public morality, and workers' compensation.

During the first half of the 20th century, California's population growth far outpaced that of the nation as a whole. The state's climate, natural beauty, and romantic reputation continued to attract many, but new economic opportunities were probably most important. In the early 1920s, major discoveries of oil were made in the Los Angeles Basin, and for several years during the decade, California ranked 1st among the states in production of crude oil. The population of Los Angeles County more than doubled during the decade, rising to 2,208,492 by 1930. Spurred by the availability and low price of petroleum products and by an ever expanding system of public roadways, Los Angeles also became the most thoroughly motorized and automobile-conscious city in the world. By 1925, Los Angeles had one automobile for every three persons—more than twice the national average.

Even during the 1930s, when California shared in the nation-wide economic depression, hundreds of thousands of refugees streamed into the state from the dust bowl of the southern Great Plains. The film industry, which offered at least the illusion of prosperity to millions of Americans, continued to prosper during the depression. By 1940 there were more movie theaters in the US than banks, and the films they showed were almost all California products.

Politics in the Golden State in the 1930s spawned splinter movements like the Townsend Plan and the "Ham 'n' Eggs" Plan, both of which advocated cash payments for the elderly. In 1934, Socialist author Upton Sinclair won the Democratic gubernatorial nomination with a plan called End Poverty In California (EPIC), but he lost the general election to the Republican incumbent, Frank Merriam.

During World War II, the enormous expansion of military installations, shipyards, and aircraft plants attracted millions of new residents to California. The war years also saw an increase in the size and importance of ethnic minorities. By 1942, only Mexico City had a larger urban Mexican population than Los Angeles. During the war, more than 93,000 Japanese-Americans in California—most of whom were US citizens and American-born—were interned in "relocation centers" throughout the Far West.

California continued to grow rapidly during the postwar period, as agricultural, aerospace, and service industries provided new economic opportunities. Politics in the state were influenced by international tensions, and the California legislature expanded the activities of its Fact-Finding Committee on Un-American Activities. The University of California became embroiled in a loyalty-oath controversy, culminating in the dismissal in 1950 of 32 professors who refused to sign an anti-Communist pledge. Black-listing became common in the film industry. The early 1950s saw the rise to the US vice-presidency of Richard Nixon, whose early campaigns capitalized on fears of Communist subversion.

At the beginning of 1963, California (according to census estimates) became the nation's most populous state; its population continued to increase at a rate of 1,000 net migrants a day through the middle of the decade. By 1970, however, California's growth rate had slowed considerably. During the 1960s, the state was beset by a number of serious problems that apparently discouraged would-be immigrants. Economic opportunity gave way to recessions and high unemployment. Such rapid-growth industries as aerospace experienced a rapid decline in the late 1960s and early 1970s. Pollution of air and water called into question the quality of the California environment. The tradi-tional romantic image of California was overshadowed by reports of mass murders, bizarre religious cults, extremist social and political movements, and racial and campus unrest.

By the mid-1970s, migration to California had resumed at a more modest annual rate. The political importance of California's preeminence in population can be measured in the size of its congressional delegation and electoral votes. Defeated in his quest for the presidency in 1960, former vice president Nixon in 1968 became the first native Californian to win election to the nation's highest office. Both Ronald Reagan, governor of the state from 1967 to 1975, and Edmund G. Brown, Jr., elected governor in 1974 and reelected in 1978, were active candidates for the US presidency in 1980. Reagan was the Republican presidential winner that year and in 1984.

[12]STATE GOVERNMENT

The first state constitution, adopted in 1849, outlawed slavery and was unique in granting property rights to married women in their own name. A new constitution, drafted in 1878 and ratified the following year, sought to curb legislative abuses—even going so far as to make lobbying a felony—and provided for a more equitable system of taxation, stricter regulation of the railroads, and an eight-hour workday. Of the 152 delegates to the 1878 constitutional convention, only 2 were natives of California, and 35 were foreign-born; no Spanish-speaking persons or Indians were included. This second constitution, as amended, is the basic document of state government today.

The California legislature consists of a 40-member senate and an 80-member assembly. Senators are elected to four-year terms, half of them every two years, and assembly members are elected to two-year terms. As a result of a 1972 constitutional amend-ment, the legislature meets in a continuous two-year session, thus eliminating the need to reintroduce or reprint bills proposed in the first year of the biennium. Each session begins with an organizational meeting on the first Monday in December of even-numbered years; then, following a brief recess, the legislature reconvenes on the first Monday in January. Special sessions may be called by the governor to consider certain specific matters. Members of the senate and assembly must be over 18 years of age, and have been US citizens and residents of the state for at least three years and residents of the districts they represent for at least one year prior to election. Legislative salaries in 1984 were $28,110 annually, plus a $62 per diem and a transportation allowance during the legislative session.

Bills, which may be introduced by either house, are referred to committees, and must be read before each house three times. Legislation must be approved by an absolute majority vote of each house, except for appropriations bills, certain urgent mea-sures, and proposed constitutional amendments, which require a two-thirds vote for passage. Gubernatorial vetoes may be overrid-den by two-thirds majority votes in both houses. In the 1973/74 session, the legislature overrode a veto for the first time since 1946, but overrides have since become more common.

Constitutional amendments and proposed legislation may also be placed on the ballot through the initiative procedure. For a constitutional amendment, petitions must be signed by at least 8% of the number of voters who took part in the last gubernatorial election; for statutory measures, 5%. In each case, a simple majority vote at the next general election is required for passage.

Officials elected statewide include the governor and lieutenant governor (who run separately), secretary of state, attorney general, controller, treasurer, and superintendent of public instruction. Each serves a four-year term, without limitation. As chief execu-tive officer of the state, the governor is responsible for the state's policies and programs, appoints department heads and members of state boards and commissions, serves as commander in chief of the California National Guard, may declare states of emergency, and may grant executive clemency to convicted criminals. The governor's annual salary in 1984 was $49,100.

The lieutenant governor acts as president of the senate and may assume the duties of the governor in case of the latter's death, resignation, impeachment, inability to discharge the duties of the office, or absence from the state. The annual salary of the lieutenant governor was $42,500 in 1984. Salaries of other elected officials also were $42,500, except for the attorney general, who received $47,500. In order to vote in California, one must be a US citizen, at least 18 years old, and have been a resident of the state for at least 29 days prior to the election.

[13]POLITICAL PARTIES

As the state with the largest number of US representatives, 45 in 1984, and electoral votes, 47, California plays a key role in national and presidential politics.

In 1851, the year after California entered the Union, the state Democratic Party was organized. But the party soon split into a pro-South faction, led by US Senator William Gwin, and a pro-North wing, headed by David Broderick. A political leader in San Francisco, Broderick became a US senator in 1857 but was killed in a duel by a Gwin stalwart two years later. This violent factionalism helped switch Democratic votes to the new Republican Party in the election of 1860, giving California's four electoral votes to Abraham Lincoln. This defeat, followed by the Civil War, demolished Senator Gwin's Democratic faction, and he fled to exile in Mexico.

The Republican party itself split into liberal and conservative wings in the early 1900s. Progressive Republicans formed the Lincoln-Roosevelt League to espouse political reforms, and succeeded in nominating and electing Hiram Johnson as governor on the Republican ticket in 1910. The following year, the legislature approved 23 constitutional amendments, including the initiative, referendum, recall, and other reform measures. Johnson won reelection on a Progessive Party line in 1915. After Johnson's election to the US Senate in 1916, Republicans (both liberal and conservative) controlled the state house uninterruptedly for 22 years, from 1917 to 1939. Democratic fortunes sank so low that in 1924 the party's presidential candidate, John W. Davis, got only 8% of the state's votes, leading humorist Will Rogers to quip, "I don't belong to any organized political party—I am a California Democrat." An important factor in the Progressive Republicans' success was the cross-filing system, in effect from 1913 to 1959, which blurred party lines by permitting candidates to appear on the primary ballots of several parties. This favored such Republican moderates as Earl Warren, who won an unprecedented three terms as governor—in 1946, he won both Republican and Democratic party primaries— before being elevated to US chief justice in 1953.

Political third parties have had remarkable success in California since the secretive antiforeign, anti-Catholic Native American Party—called the Know-Nothings because party members were instructed to say they "knew nothing" when asked what they stood for—elected one of their leaders, J. Neely Johnson, as governor in 1855. The Workingmen's Party of California, as much anti-Chinese as it was antimonopolist and prolabor, managed to

elect about one-third of the delegates to the 1878 constitutional convention. The most impressive third-party triumph came in 1912, when the Progressive Party's presidential candidate, Theodore Roosevelt, and vice-presidential nominee, Governor Hiram Johnson, defeated both the Republican and Democratic candidates among state voters. The Socialist Party also attracted support in the early 20th century. In 1910, more than 12% of the vote went to the Socialist candidate for governor, J. Stitt Wilson. Two years later, Socialist congressional nominees in the state won 18% of the vote, and a Socialist assemblyman was elected from Los Angeles. In 1914, two Socialist assemblymen and one state senator were elected. During the depression year of 1934, the Socialist Party leader and author Upton Sinclair won the Democratic nomination for governor on his End Poverty In California program and received nearly a million votes while losing to Republican Frank Merriam. Nonparty political movements have also won followings: several southern California congressmen were members of the ultraconservative John Birch Society during the 1960s, and in 1980 the Grand Dragon of the Ku Klux Klan won the Democratic Party nomination for a US House seat. Even when they lost decisively, third parties have won enough votes to affect the outcome of elections. In 1968, for example, George Wallace's American Independent Party received 487,270 votes, while Republican presidential candidate Richard Nixon topped Democrat Hubert Humphrey by only 223,346.

As of 7 May 1984, California had 11,530,956 registered voters, including 6,143,830 Democrats, 4,048,413 Republicans, 143,133 American Independent Party members, 63,032 Libertarians, 43,960 Peace and Freedom Party members, 27,656 supporters of other parties, and 1,060,932 independents. Even with a 3-2 advantage in voter registration, however, the Democrats managed to carry California in presidential elections only twice between 1948 and 1984, and to elect only two governors—Edmund G. "Pat" Brown (in 1958 and 1962) and his son, Edmund G. "Jerry" Brown, Jr. (in 1974 and 1978)—during the same period. Three times Californians gave their electoral votes to a California Republican, Richard Nixon, though they turned down his bid for governor in 1962. They elected one former film actor, Republican George Murphy, as US senator in 1964, and another, Republican Ronald Reagan, as governor in 1966 and 1970.

California Presidential Vote by Political Parties, 1948–84

YEAR	ELECTORAL VOTE	CALIFORNIA WINNER	DEMOCRAT	REPUBLICAN	STATES' RIGHTS	PROGRESSIVE	SOCIALIST	PROHIBITION
1948	25	*Truman (D)	1,913,134	1,895,269	1,228	190,381	3,459	16,926
					CONSTITUTION		SOC. LABOR	
1952	32	*Eisenhower (R)	2,197,548	2,897,310	3,504	24,692	273	16,117
1956	32	*Eisenhower (R)	2,420,135	3,027,668	6,087	—	300	11,119
1960	32	Nixon (R)	3,224,099	3,259,722	—	—	1,051	21,706
1964	40	*Johnson (D)	4,171,877	2,879,108	—	—	489	
					AMERICAN IND.		PEACE & FREEDOM	
1968	40	*Nixon (R)	3,244,318	3,467,664	487,270	—	27,707	—
						AMERICAN	PEOPLE'S	LIBERTARIAN
1972	45	*Nixon (R)	3,475,847	4,602,096	—	232,554	55,167	980
						COMMUNIST		
1976	45	Ford (R)	3,742,284	3,882,244	51,096	12,766	41,731	56,388
						CITIZENS	PEACE & FREEDOM	
1980	45	*Reagan (R)	3,039,532	4,444,044	—	9,687	60,059	17,797
1984	47	*Reagan (R)	3,922,519	5,467,009	39,265	—	26,297	49,951

* Won US presidential election.

Democratic Senator Alan Cranston was reelected in 1980; Republican Senator Pete Wilson won a first term in 1982. In November 1984, Reagan carried California's 47 electoral votes with about 58% of the popular vote. However, the Republicans picked up only one seat in the US House, leaving the Democrats with a 27-18 advantage. The Democrats kept control over both houses of the state legislature.

The state's direct primary law had a salutary effect on local politics by helping to end the power of political machines in the large cities. In 1910, Los Angeles voters adopted the nonpartisan primary and overthrew the corrupt rule of Mayor A. C. Harper in favor of reformer George Alexander. At the same time, voters were revolting against bossism and corruption in San Francisco, Sacramento, Oakland, and other cities.

Minority groups of all types are represented in California politics. As of January 1985 there were 296 black elected officials, the most prominent of whom was Los Angeles Mayor Thomas Bradley, first elected in 1973 at a time when the city's population was only 18% black. In 1974, Californians elected a black state senator, Mervyn Dymally, as lieutenant governor, and a woman of Asian ancestry, March Fong Eu, as secretary of state. A former college president of Japanese extraction, S. I. Hayakawa, was elected US senator in 1976. In 1984, 460 Hispanics held elective office in California, 3 of them as US representatives. Organized groups of avowed homosexuals began to play an important political role in San Francisco during the 1970s. A woman, Dianne Feinstein, won reelection as mayor of San Francisco in 1983; 3 women served as US representatives, and 15 as state legislators.

[14] LOCAL GOVERNMENT

As of 1984, California had 58 counties, about 1,200 school districts, and more than 3,000 special districts. There were 428 municipal governments in 1982.

County government is administered by an elected board of supervisors, which also exercises jurisdiction over unincorporated towns within the county. Government operations are administered by several elected officials, the number varying according to the population of the county. Most counties have a district attorney, assessor, treasurer–tax collector, superintendent of schools, sheriff, and coroner. Larger counties may also have an elected planning director, public defender, public works director, purchasing agent, and social welfare services director.

Municipalities are governed under the mayor-council, council-manager, or commission system. Most large cities are run by councils of from 5 to 15 members, elected to four-year terms, the councils being responsible for taxes, public improvements, and the budget. An elected mayor supervises city departments and appoints most city officials. Other elected officials usually include the city attorney, treasurer, and assessor. Los Angeles and San Francisco have the mayor-council form of government, but in San Francisco the city and county governments are consolidated under an elected board of supervisors, and the mayor appoints a manager who has substantial authority. San Diego and San Jose have both an elected mayor and city manager chosen by an elected city council.

[15] STATE SERVICES

In accordance with the Political Reform Act of 1974, the Fair Political Practices Commission investigates political campaign irregularities, regulates lobbyists, and enforces full disclosure of political contributions and public officials' assets and income.

Educational services are provided by the Department of Education, which administers the public school system. The department, which is headed by the superintendent of public instruction, also regulates special schools for blind, deaf, and disabled children. The University of California system is governed by a board of regents headed by the governor.

Transportation services are under the direction of the Depart-ment of Transportation (CALTRANS), which overseas mass transit lines, highways, and airports. Intrastate rate regulation of pipelines, railroads, buses, trucks, airlines, and waterborne transportation is the responsibility of the Public Utilities Commission, which also regulates gas, electric, telephone, water, sewer, and steam-heat utilities. The Department of Motor Vehicles licenses drivers, road vehicles, automotive dealers, and boats.

Health and welfare services are provided by many state departments, most of which are part of the Health and Welfare Agency. The Department of Health Services provides health care for nearly 3 million persons through the state's Medi-Cal program. The department's public health services include controlling infectious disease, conducting cancer research, safeguarding water quality, and protecting the public from unsafe food and drugs. The department also has licensing responsibility for hospitals, clinics, and nursing homes. Care for the mentally ill is provided through the Department of Mental Health by means of state hospitals and community outpatient clinics. Disabled people receive counseling, vocational training, and other aid through the Department of Rehabilitation. Needy families receive income maintenance aid and food stamps from the Department of Social Services. Senior citizens can get help from the Department of Aging, which allocates federal funds for the elderly. The Commission on the Status of Women reports to the legislature on women's educational and employment needs, and on statutes or practices that infringe on their rights. The Youth Authority, charged with the rehabilitation of juvenile offenders, operates training schools and conservation camps. The Department of Alcohol and Drug Programs coordinates prevention and treatment activities.

Public protection services are provided by the Military Department, which includes the Army and Air National Guard and the California Cadet Corps, and by the Youth and Adult Correctional Agency, which maintains institutions and programs to control and treat convicted felons and narcotics addicts. The California Highway Patrol has its own separate department, within the Business, Transportation and Housing Agency. This agency also includes the Department of Housing and Community Development. The State and Consumer Services Agency has jurisdiction over the Department of Consumer Affairs, the Department of Veterans Affairs, and several other state departments. A state innovation was the establishment in 1974 of the Seismic Safety Commission to plan public safety programs in connection with California's continuing earthquake problem.

Programs for the preservation and development of natural resources are centralized in the Resources Agency. State parks and recreation areas are administered by the Department of Parks and Recreation. California's vital water needs are the responsibility of the Department of Water Resources. In 1975, as a result of a national oil shortage, the state established the Energy Resources Conservation and Development Commission to develop contingency plans for dealing with fuel shortages, to forecast the state's energy needs, and to coordinate programs for energy conservation. The California Conservation Corps provides employment opportunities for young people in conservation work.

The Department of Industrial Relations has divisions dealing with fair employment practices, occupational safety and health standards, and workers' compensation. The Employment Development Department provides unemployment and disability benefits and operates job-training and work-incentive programs.

[16] JUDICIAL SYSTEM

California has a complex judicial system and a very large correctional system, which at the end of 1984 ranked 1st in the US in number of prisoners.

The state's highest court is the supreme court, which may review appellate court decisions and superior court cases involving the death penalty. The high court has a chief justice and six associate justices, all of whom serve 12-year terms; justices are appointed by the governor, confirmed or disapproved by the Commission on Judicial Appointments (headed by the chief justice), and then submitted to the voters for ratification. The chief justice also chairs the Judicial Council, which seeks to expedite judicial business and to equalize judges' caseloads.

Courts of appeal, organized in six appellate districts, review decisions of superior courts and, in certain cases, of municipal and justice courts. As of 1984 there were 77 district appeals court judgeships. All district court judges are appointed by the governor, reviewed by the Commission on Judicial Appointments, and subject to popular election for 12-year terms.

Superior courts in each of the 58 county seats have original jurisdiction in felony, juvenile, probate, and domestic relations cases, as well as in civil cases involving more than $15,000. They also handle some tax and misdemeanor cases and appeals from lower courts. Municipal courts, located in judicial districts with populations of more than 40,000, hear misdemeanors (except those involving juveniles) and civil cases involving $15,000 or less. In districts with less than 40,000 population, justice courts have jurisdiction similar to that of municipal courts. All trial court judges are elected to six-year terms.

As of 31 December 1984 there were 43,314 prisoners in state and federal prisons in California. The State Department of Corrections maintains 13 correctional institutions, 3 reception centers, 33 conservation camps, and more than 60 parole offices. At the end of 1983 there were 36,653 men and 1,811 women in state prisons; 19,135 former prisoners (17,924 men, 1,211 women) were on parole. As of the same date, 5,870 juvenile offenders were in youth correction facilities, and 6,871 former inmates were on parole. California had 34,888 state and local corrections personnel in 1983, or 13.9 per 10,000 population, about 10% higher than the US average.

According to the FBI, California's crime rate in 1983 was 6,677 crimes per 100,000 population, 2d only to that of Florida. In that year, 1,680,978 crimes were reported to the police, including 194,491 violent crimes and 1,486,487 crimes against property. The 1983 rate per 100,000 population included murder and manslaughter, 11; forcible rape, 48; robbery, 341; assault, 373; burglary, 1,829; larceny-theft, 3,444; and motor vehicle theft, 631. In 1983, Los Angeles reported 307,511 crimes (including 820 murders), San Francisco 62,646 (83 murders), and San Diego 63,737 (77 murders). In 1965, California became the first state to institute a victim compensation program.

California's death penalty statute received its most serious challenge after the 1948 conviction of Caryl Chessman on a charge of forcible rape. Chessman served 12 years on death row at San Quentin, got eight stays of execution, and wrote a best-seller about his ordeal. Despite highly publicized attempts to overturn capital punishment and save Chessman's life, the legislature refused to act, and he was executed in 1960. The death penalty was carried out 30 times in California from 1960 to 1967—more than in any other state—but, as of early 1985, it had not been used since 1967.

[17] ARMED FORCES

California leads the 50 states in defense contracts received, numbers of National Guardsmen and military veterans, veterans' benefit payments, and funding for police forces.

In 1983/84, the US Department of Defense had 337,594 personnel in California, including 203,791 active-duty military and 133,803 civilians. Army military personnel totaled 27,788, the Navy (including Marines) 121,584, and the Air Force 54,419. The Army's principal base, Fort Ord at Monterey, had 19,575 personnel in 1983/84; other bases are located at Oakland and San Francisco. Naval facilities in the San Diego area had more than 60,000 personnel in 1983/84; there are weapons stations at Concord and Seal Beach, and supply depots at Oakland and San Pedro. The Marine Corps training base, Camp Pendleton at

Oceanside, had 32,415 personnel. The Air Force operates three main bases—McClellan AFB at Sacramento, Travis AFB at Fairfield, and Norton AFB at San Bernardino—and numerous smaller installations. In 1983/84, California companies were awarded $26.4 billion in defense contracts, more than two and a half times the total for New York (which ranked 2d), and more than 22% of the US total.

As of 30 September 1984, 2,952,000 veterans of US military service, nearly 11% of the national total, were living in California. Of those, 25,000 served in World War I, 1,114,000 in World War II, 598,000 in the Korean conflict, and 917,000 during the VietNam era. Veterans' benefits paid to Californians in 1982/83 exceeded $2.3 billion, or more than 9% of the US total.

California's military forces consist of the Army and Air National Guard, the naval and state military reserve (militia), and the California Cadet Corps. National Guard strength was about 25,580 in early 1985.

As of October 1983, state and local police forces totaled 68,322, highest in the nation, but the rate of police protection per 10,000 residents—27.1—was 12th among the states. Expenditures for police protection by state and local governments in 1981/82 were nearly $2.4 billion.

[18] MIGRATION

A majority of Californians today are migrants from other states.

The first great wave of migration, beginning in 1848, brought at least 85,000 prospectors by 1850. Perhaps 20,000 of them were foreign-born, mostly from Europe, Canada, Mexico, and South America, as well as a few from the Hawaiian Islands and China. Many thousands of Chinese were brought in during the latter half of the 19th century to work on farms and railroads. When Chinese immigration was banned by the US Congress in 1882, Japanese migration provided farm labor. These ambitious workers soon opened shops in the cities and bought land for small farms. By 1940, about 94,000 Japanese lived in California. During the depression of the 1930s, approximately 350,000 migrants came to California, most of them looking for work. Many thousands of people came there during World War II to take jobs in the burgeoning war industries; after the war, some 300,000 discharged servicemen settled in the state. All told, between 1940 and 1980, California registered a net gain from migration of 9,486,000, representing 56% of its population growth during that period.

Although the 1970s brought an influx of refugees from Indochina, and, somewhat later, from Central America, the bulk of postwar foreign immigration has come from neighboring Mexico. At first, Mexicans—as many as 750,000 a year—were imported legally to supply seasonal labor for California growers. Later, hundreds of thousands—perhaps even millions—of illegal Mexican immigrants crossed the border in search of jobs and then, unless they were caught and forcibly repatriated, stayed on. Counting these state residents for census purposes is extremely difficult, since many of them are unwilling to declare themselves for fear of being identified and deported.

Intrastate migration has followed two general patterns: rural to urban until the mid-20th century, and urban to suburban thereafter. In particular, the percentage of blacks increased in Los Angeles, San Francisco, and San Diego between 1960 and 1970 as black people settled or remained in the cities while whites moved into the surrounding suburbs. In the 1970s, the percentage of blacks in Los Angeles and San Francisco decreased slightly; in San Diego, the percentage of blacks increased from 7.6% to 8.9%. During that decade, intrastate migration spilled over from the suburbs into outlying rural areas. California's net gain from migration during 1970–80 amounted to about 1,573,000. From 1980 to 1983, the net gain was an estimated 750,000.

[19] INTERGOVERNMENTAL COOPERATION

The Colorado River Board of California represents the state's interests in negotiations with the federal government and other states over utilization of Colorado River water and power resources. California also is a member of the Western States Water Council, the Klamath River Compact Commission (with Oregon), and the Tahoe Regional Planning Compact (with Nevada). Regional agreements signed by the state include the Pacific Marine Fisheries Compact, Western Corrections Compact, Western Interstate Energy Compact, and Western Regional Education Compact. The Arizona-California boundary accord dates from 1963. California also is a member of the Commission of the Californias, along with the State of Baja California Norte and the territory of Baja California Sur, both in Mexico.

During the 1983/84 fiscal year, federal aid to California exceeded $9.8 billion.

20 ECONOMY

California leads the 50 states in economic output and total personal income. In the 1960s, when it became the nation's most populous state, California also surpassed Iowa in agricultural production and New York in value added by manufacturing. The state ranks 1st in the US in such important industries as food products, machinery, electric and electronic equipment, aerospace, dairy production, and beef cattle. California also leads the nation in retail sales, foreign trade, and corporate profits.

California had an estimated gross state product of more than $400 billion in 1984; if it were an independent country, it would be the world's 8th leading economic power.

The gold rush of the mid-19th century made mining (which employed more people than any other industry in the state until 1870) the principal economic activity and gave impetus to agriculture and manufacturing. Many unsuccessful miners took up farming or went to work for the big cattle ranches and wheat growers. In the 1870s, California became the most important cattle-raising state and the 2d-leading wheat producer. Agriculture soon expanded into truck farming and citrus production, while new manufacturing industries began to produce ships, metal products, lumber, leather, cloth, refined sugar, flour, and other processed foods. Manufacturing outstripped both mining and agriculture to produce goods valued at $258 million by 1900, and 10 times that by 1925. Thanks to a rapidly growing work force, industrial output continued to expand during and after both world wars, while massive irrigation projects enabled farmers to make full use of the state's rich soil and favorable climate.

By the late 1970s, one of every four California workers was employed in high-technology industry. California has long ranked 1st among the states in defense procurement, and in 1984/85, southern California alone was to receive 68% of the state's total weapons procurement, surpassing the combined totals of New York and Texas.

From its beginnings in the late 18th century, California's wine industry has grown to encompass some 500 wineries. In 1981, they accounted for about 90% of total US production. By 1985, California had surpassed Chicago to rank 2d in advertising among the states.

A highly diversified economy makes California less vulnerable to national recession than most other states. During the first half of the 1980s, the state generally outperformed the national economy. In 1984, California enjoyed an estimated increase of 12.1% in personal income and a 6.1% increase in nonagricultural employment, and reduced the unemployment rate from 9.7% to an estimated 7.8%.

Inflation in California for 1985 was forecast at 4.9%—lower than the national average. California's estimated $31.7 billion in pretax corporate profits in 1984 amounted to 13.5% of the national total.

21 INCOME

With a per capita personal income of $13,329 in 1983, California ranked 4th among the 50 states and more than 13% above the national average. Total personal income exceeded $333 billion— more than 12% of the US total, two-fifths more than in New York State, and nearly double the total for Texas. Measured in constant 1972 dollars, total personal income increased 48% during the 1970s and 7.6% between 1980 and 1983. Median money income of a four-person family was $27,763 in 1981, for a rank of 14th among the states.

Despite California's relatively high average personal income, 2,627,000 state residents in 521,000 families were below the federal poverty level in 1979. On a percentage basis, however, these totals were slightly below the national averages.

California is justly noted for its large number of wealthy residents, particularly in the Los Angeles, Sacramento, and San Francisco metropolitan areas. In 1982, approximately 301,500 California individuals, more than in any other state, were among the nation's top wealth-holders, with gross assets greater than $500,000 each. According to federal income tax returns for 1982, 113,681 Californians had adjusted gross incomes of more than $100,000; this includes 1,553 individuals with incomes exceeding $1,000,000.

In 1982, total disposable income, after state and federal income taxes, amounted to $261.9 billion, or $10,597 per capita, 5th in the nation. The major sources of personal income in 1983 were wages and salaries, 61%; property, 19%; transfer payments, 14%; and proprietor's income, 4%.

The following table shows the sources of Californians' earned income (without property income and transfer payments) for 1980 and 1983 (in millions):

	1980	1983
Manufacturing	$ 36,602	$ 45,204
Services	30,504	42,958
Government	32,366	39,292
Wholesale and retail trade	27,882	34,272
Finance, insurance, and real estate	10,247	14,208
Transportation, utilities	11,459	14,040
Construction	9,154	9,295
Agriculture	2,452	2,743
Mining	1,094	1,560
Other	1,148	1,405
TOTALS	$162,908	$204,977

22 LABOR

California has the largest work force in the nation and the greatest number of employed workers. During the 1970s, California's work force also grew at a higher annual rate than that of any other state.

In 1984, the state's civilian labor force totaled 12,503,000, of whom 11,532,000 (92.2%) were employed and 972,000 (7.8%) unemployed. The labor force increased by almost 2% in 1984, slightly higher than the US rate, and the number of employed workers increased by a total of 4%. Of the total 1983 labor force, 57% was male and 43% female. That year, the rate of participation in the labor force by females was 59.4%, and by males 77.9%, both rates somewhat higher than the national averages.

A large majority of the work force is employed in metropolitan areas, including, as of September 1984, 3,677,100 (35% of all nonagricultural employees) in the Los Angeles–Long Beach area, 1,592,800 (15%) in the San Francisco–Oakland area, 905,200 (8%) in the Anaheim–Santa Ana–Garden Grove area, 750,000 (7%) in metropolitan San Jose, and 718,900 (7%) in the San Diego metropolitan area.

All together, 10,456,000 workers covered by unemployment insurance were employed in nonagricultural industries as of September 1984. A federal survey in 1982 showed the following nonfarm employment pattern in California:

	ESTABLISH-MENTS	EMPLOYEES	ANNUAL PAYROLL ('000)
Agricultural services, forestry, fishing	6,701	56,618	$ 734,529
Mining, of which:	1,425	51,772	1,476,711
Oil, gas extraction	(949)	(33,126)	(888,908)
Contract construction	37,168	418,966	10,115,054
Manufacturing, of which:	42,800	2,057,863	43,736,972
Food products	(2,359)	(159,494)	(3,322,492)
Apparel and textiles	(3,587)	(111,672)	(1,109,438)
Printing and publishing	(5,837)	(129,386)	(2,367,944)
Fabricated metals	(4,237)	(144,353)	(2,844,945)
Nonelectrical machinery	(6,903)	(255,358)	(5,631,707)
Electric, electronic equipment	(3,239)	(352,612)	(7,682,939)
Transportation equipment	(1,509)	(273,868)	(7,610,831)
Transportation, public utilities, of which:	17,955	513,658	12,629,259
Trucking and warehousing	(7,602)	(112,286)	(2,289,073)
Wholesale trade	41,451	572,854	12,239,365
Retail trade	133,538	1,739,410	18,096,742
Finance, insurance, real estate, of which:	51,894	680,525	12,420,771
Banking	(4,957)	(186,814)	(3,016,822)
Real estate	(23,211)	(139,800)	(2,126,083)
Services, of which:	180,567	2,223,663	35,268,016
Health services	(48,743)	(609,145)	(11,212,895)
Other	12,669	14,675	220,155
TOTALS	526,168	8,330,004	$146,937,574

Among the workers not covered by this federal survey were government employees, of whom California had 1,725,100 in 1983: 327,000 were federal employees, and 1,398,100 were employed by state and local governments.

The unemployment rate during the 1970s and early 1980s ranged from the 1973 low of 7% to a high of 9.9% in 1975 and 1982. From 1967 to 1976, an average of 226,000 Californians entered the labor market each year, but the economy generated only about 175,000 jobs annually, so unemployment rose steadily. In 1983, the unemployment rate was 9.7%, 9% for females and 10.2% for males. In 1984, however, the rate was estimated to have fallen to 7.8%.

The labor movement in California was discredited by acts of violence during its early years. On 1 October 1910, a bomb explosion at a *Los Angeles Times* plant killed 21 workers, resulting in the conviction and imprisonment of two labor organizers a year later. Another bomb explosion, this one killing 10 persons in San Francisco on 22 July 1916, led to the conviction of two radical union leaders, Thomas Mooney and Warren Billings; the death penalty for Mooney was later commuted to life imprisonment (the same sentence Billings had received), and after evidence had been developed attesting to his innocence, he was pardoned in 1939. These violent incidents led to the state's Criminal Syndicalism Law of 1919, which forbade "labor violence" and curtailed militant labor activity for more than a decade.

Unionism revived during the depression of the 1930s. In 1934, the killing of two union picketers by San Francisco police during a strike by the International Longshoremen's Association led to a three-day general strike that paralyzed the city, and the union eventually won the demand for its own hiring halls. In Los Angeles, unions in such industries as automobiles, aircraft, rubber, and oil refining obtained bargaining rights, higher wages, and fringe benefits during and after World War II. In 1958, the California Labor Federation was organized, and labor unions have since increased both their membership and their benefits. As of July 1983 there were 2,877 union locals with 2,076,700 members, of whom 22% worked in manufacturing, 15% in construction, 15% in government, 15% in commerce, 8% in transportation and warehousing, 7% in public utilities, and 5% in motion-picture

production. Of the total union membership, 37% worked in Los Angeles County and 28% in the nine-county San Francisco Bay area. Women, who constituted 29% of union members, were most heavily represented in services, government, textile manufacturing, and food processing.

Of all working groups, migrant farm workers have been the most difficult to organize because their work is seasonal and because they are largely members of minority groups, mostly Mexicans, with few skills and limited job opportunities. During the 1960s, a Mexican-American "stoop" laborer named Cesar Chavez established the National Farm Workers Association (later the United Farm Workers Organizing Committee, now the United Farm Workers of America), which, after a long struggle, won bargaining rights from grape, lettuce, and berry growers in the San Joaquin Valley. Chavez's group was helped by a secondary boycott against these California farm products at some grocery stores throughout the US. When his union was threatened by the rival Teamsters Union in the early 1970s, Chavez got help from the AFL-CIO and from Governor Brown, who in 1975 pushed through the state legislature a law mandating free elections for agricultural workers to determine which union they wanted to represent them. The United Farm Workers and Teamsters formally settled their jurisdictional dispute in 1977.

In 1983, production workers in manufacturing industries worked an average of 40 hours per week and earned average weekly wages of $380.80, ranging from a low of $204.79 in the clothing industry to a high of $608.38 in the oil and coal industry. Average weekly earnings in industries other than manufacturing ranged from $268.47 in retail trade to $807.17 for electrical work in contract construction. In 1981, California had 231 strikes and other work stoppages, involving 60,000 workers and 1,539,100 workdays lost.

[23] AGRICULTURE

California leads the 50 states in agricultural production. With only 3% of the nation's farms and farm acreage, the state accounts for almost 10% of US cash farm receipts. Unique in having no single crop that dominates the agricultural economy, California ranked 1st among the 50 states in production of no fewer than 40 farm commodities in 1983. Famous for its specialty crops, California produces virtually all the almonds, apricots, avocados, broccoli, dates, figs, nectarines, olives, pomegranates, safflower, and walnuts grown commercially in the US. California's total farm marketings amounted to $13.5 billion in 1983, down 6% from the previous year.

Agriculture has always thrived in California. The Spanish missions and Mexican ranchos were farming centers until the mid-19th century, when large ranches and farms began to produce cattle, grain, and cotton for the national market. Wheat was a major commodity by the 1870s, when the citrus industry was established and single-family farms in the fertile Central Valley and smaller valleys started to grow large quantities of fruits and vegetables. European settlers planted vineyards on the slopes of the Sonoma and Napa valleys, thus beginning the important California wine industry, which now provides some 90% of US domestic wines. Around 1900, intensive irrigation transformed the dry, sandy Imperial Valley in southeastern California into a garden of abundance for specialty crops. Since World War II, large corporate farms, or agribusinesses, have largely replaced the small one-family farm. Today, the state grows a large percentage of all fruits marketed in the US, as well as much of the nation's grains and vegetables; farm commodities account for about 13% of the state's foreign exports.

Less than one-third of California's total land area was devoted to farming in 1984, when some 78,000 farms comprised about 33 million acres (13 million hectares). According to US Department of Agriculture estimates, the average size of a farm in 1982 was 390 acres (158 hectares). More than one-fourth of all farmland is used to grow crops, and some 97% of all cropland is under irrigation.

The leading cash crops in 1983 were grapes (used for the table, for raisins, and especially for wine), cotton, hay, and lettuce. The following table shows harvested acreage, production, value, and US rank of the 15 leading cash crops in 1983:

	ACRES (1,000)	PRODUCTION (1,000 TONS)	VALUE (1,000)	US RANK
Grapes	644.5	4,917.0	$946,995	1
Cotton	NA	NA	831,705	2
Hay	1,480.0	7,352.0	731,524	6
Nursery products	NA	NA	547,438	1
Lettuce	148.0	2,214.4	541,680	1
Tomatoes	263.0	6,325.4	525,625	1
Flowers and foliage	NA	NA	470,021	1
Oranges	177.5	2,854.0	465,134	2
Strawberries	12.0	312.0	279,283	1
Almonds (shelled)	356.0	120.0	253,600	1
Rice	328.0	1,154.5	205,492	2
Wheat	720.0	1,437.0	193,175	18
Potatoes	56.2	997.5	176,220	6
Broccoli	78.9	349.1	170,132	1
Celery	19.2	598.3	151,933	1

In recent years, increased capitalization and the use of advanced technology have enabled agribusiness to control the entire agricultural process from planting through marketing. Almost 80% of all farm labor is hired.

Irrigation is essential for farming in California, and agriculture consumes 85% of the state's water supply. The major irrigation systems include the Colorado River Project, which had irrigated 500,000 acres (200,000 hectares) in the Imperial Valley by 1913; the Central Valley Project, completed by 1960, which harnessed the runoff of the Sacramento River; and the Feather River Project, also in the Central Valley and finished during the 1970s. Largest of all is the California Water Project, begun in 1960 and completed in its essentials by 1973; during 1983, this project delivered 1.3 million acre-feet of water. On 16 June 1980, the US Supreme Court ended 13 years of litigation by ruling that federally subsidized irrigation water in the Imperial Valley could not be limited to family farms of fewer than 160 acres (65 hectares) but must be made available to all farms regardless of size; the ruling represented a major victory for agribusiness interests.

24 ANIMAL HUSBANDRY

California is a leading producer of livestock and dairy products, which together accounted for 31% of the state's farm income in 1983. In that year, cash receipts for meat animals totaled $1.3 billion; dairy products, $1.9 billion; and poultry and eggs, $776 million.

Beef cattle, raised principally in the Central Valley, were California's 2d most important farm product in 1983, when sales of cattle and calves totaled nearly $1.3 billion (6th in the US). At the end of 1983 there were an estimated 5,000,000 cattle and calves, 900,000 sheep and lambs, and 155,000 hogs and pigs on California farms and ranches. In 1983, cash sales of sheep and lambs for slaughter totaled $42.1 million (3d in the US); hogs and pigs, $28.7 million.

California is the 2d-leading milk producer (behind Wisconsin) among the 50 states. In 1983, milk and cream products ranked 1st among the state's farm commodities by value, with sales of $1.9 billion. Milk cows, raised mainly in the southern interior, totaled 940,000 head in 1984.

California ranked 1st among the 50 states in egg production in 1983, with cash receipts of $371.9 million, based on an output of 8.2 billion eggs. The state was 3d in production of turkeys; the 20.2 million birds sold weighed 416.1 million lb and were valued at $145.6 million. In 1983, 171.6 million broilers were produced,

valued at $252.6 million (9th in the US). Laying hens and turkey breeder hens, raised chiefly in the central part of the state, numbered 33.4 million and 420,000, respectively, in 1983.

California is the nation's leading producer of honey; in 1982, output totaled 25,136,570 lb, valued at $13.8 million. Wool production, 3d highest in the US in 1983, amounted to 9,476,000 lb, valued at $5.9 million.

25 FISHING

The Pacific whaling industry, with its chief port at San Francisco, was important to the California economy in the 19th century, and commercial fishing is still central to the food-processing industry. In 1984, California ranked 5th in the US in commercial fishing, with a catch of 459.2 million lb; the value of the catch, $176.6 million, ranked 6th.

In 1983/84, the California fishing fleet numbered 8,613 vessels; there were 15,377 licensed commercial fishermen. In 1983, principal species caught and total landings included yellowfin tuna, 121,593,353 lb; skipjack tuna, 98,606,953 lb; albacore tuna, 16,342,875 lb; mackerel, 112,509,338 lb; bocaccio, chilipepper, and unspecified rockfish, 26,532,131 lb; Dover sole, 18,500,107 lb; Pacific herring, 17,658,047 lb; sea urchin, 15,807,971 lb; sablefish, 14,340,783 lb; and Dungeness crab, 5,180,975 lb.

Deep-sea fishing is a popular sport. As of June 1984, world records for giant sea bass, California halibut, white catfish, and sturgeon had been set in California.

26 FORESTRY

California has more forests than any other state except Alaska. Forested lands in 1979 covered 40,152,000 acres (16,249,000 hectares), representing 40% of the state's land area and 5.4% of the US total. Nearly 41% of the state's forested area is used to produce commercial timber, and in 1983, California ranked 3d among the 50 states in lumber production and volume of standing timber.

Forests are concentrated in the northwestern part of the state and in the eastern Sierra Nevada. Commercial forestland in private hands was estimated at 7,632,000 acres (3,089,000 hectares) in 1980; an additional 8,573,000 acres (3,469,000 hectares) were US Forest Service lands, and 499,000 acres (202,000 hectares) were public lands other than national forests. Production of timber totaled 3.8 billion board feet in 1983. The volume of sawtimber totaled 258.7 million board feet, mostly of such softwoods as fir, pine, cedar, and redwood.

According to the US Census of Manufactures, the lumber and wood products industry in 1982 employed 39,200 production workers, paid them a total of $622.6 million, and shipped finished wood products valued at more than $3.7 billion.

About half of the state's forests are protected as national forests and state parks or recreational areas. Although stands of giant redwood trees have been preserved in national and state parks since the late 19th century, only about 15% of the original 2,000,000 acres (800,000 hectares) of redwoods between Monterey Bay and southern Oregon remain.

Reforestation of public lands is supervised by the National Forest Service and the California Department of Forestry. In 1924–25, more than 1.5 million redwood and Douglas fir seedlings were planted in the northwestern corner of the state. During the 1930s, the Civilian Conservation Corps replanted trees along many mountain trails, and the California Conservation Corps performed reforestation work in the 1970s.

As of 30 September 1984 there were 22 national forests in California. Total area within their boundaries amounted to 24,112,033 acres (9,757,826 hectares), of which 85% was National Forest System land.

27 MINING

California ranks 4th among the 50 states in value of total mineral production and in 1984 was the nation's leading producer of nonfuel minerals. The state ranks 1st in the production of sand

and gravel, portland cement, boron ores, diatomite, tungsten concentrate, and rare-earth metal concentrates. In 1982, California's mineral output was estimated at $12.6 billion, of which fossil fuels accounted for $11 billion and all other minerals $1.6 billion.

Some 40 minerals are mined in commercial quantities in California, chiefly in the Sierra Nevada range in the eastern part of the state. Gold and silver are still extracted, although not in the quantities of the gold rush days; in 1983, 38,443 troy oz of gold (valued at $16,300,000) and 27,000 troy oz of silver ($308,000) were produced. (This was about half the amount of silver produced in 1978, but more than eight times the amount of gold.) Other metals are copper, iron, and lead. Industrial boron ores are mined in Death Valley and other desert areas.

Most mining and quarrying centers around such basic construction minerals as cement, gypsum, lime, clays, stone, and sand and gravel. In 1982 there were 1,425 mining establishments, of which 84 mined metals, 308 produced nonmetallic minerals (except fuels), 949 extracted oil and gas, and 9 mined coal.

In March 1982, 51,772 employees covered by unemployment insurance were engaged in mining, including 8,031 workers in nonmetallic minerals and 33,126 in oil and gas extraction.

The following table shows preliminary 1983 production and value figures for the state's principal nonmetallic minerals:

	VOLUME (SHORT TONS)	VALUE
Boron ores	1,303,000	$439,181,000
Cement, portland	7,567,000	420,949,000
Sand and gravel	91,000,000	308,700,000
Stone	35,582,000	146,289,000
Lime	358,000	22,994,000
Clays	1,816,000	18,255,000
Gypsum	1,213,000	10,668,000
Pumice	65,000	1,582,000

[28] ENERGY AND POWER

In 1982, petroleum supplied an estimated 48% of the state's energy needs, natural gas 29%, hydroelectric power 9%, coal 1%, nuclear power less than 1%, and other sources about 12%. California ranks 3d among the 50 states in production of electric power, 4th in crude oil, and 7th in natural gas. Despite its ample energy resources, California is a net importer of electric power because of its extensive industrial, residential, and commercial requirements.

Installed electric power capacity in 1983 was 39.8 million kw. In that year, electrical output totaled 137.6 billion kwh. About 45% was generated from hydroelectric plants, 30% from natural gas, 12% from coal, over 5% from oil, 3% from nuclear power plants, 4% from geothermal sources, and less than 1% from other sources.

Originally ordered by the Pacific Gas and Electric Co. in 1966, the Diablo Canyon nuclear power plant near San Luis Obispo has been a source of controversy ever since 1971, when an earthquake fault line was discovered offshore, 2.5 mi (4 km) from the plant. After years of delays and demonstrations by environmental groups, the first unit of the plant began operation in November 1984. A second reactor at Diablo Canyon was scheduled to begin operation in 1986; there would then be a total of six reactors at three plants operating in the state, with a net capacity of nearly 5.7 Mw.

In 1983, sales of electric power in the state totaled 165.2 billion kwh, of which 35% went to commercial businesses, 33% to home consumers, 29% to industries, and 3% to other users. Largely because of the mild California climate, utility bills are lower than in many other states. In 1982, per capita energy consumption in California was 239 million Btu, or 42d among the states; energy expenditures per capita were $1,757, or 28th.

Although less dominant in recent years, the petroleum and natural gas industries are still of great importance in supplying California's energy requirements. Crude oil was discovered in Humboldt and Ventura counties as early as the 1860s, but it was not until the 1920s that large oil strikes were made at Huntington Beach, near Los Angeles, and at Santa Fe Springs and Signal Hill, near Long Beach. These fields added vast pools of crude oil to the state's reserves, which were further augmented in the 1930s by the discovery of large offshore oil deposits in the Long Beach area. The state's attempts to retain rights to tideland oil reserves as far as 30 mi (48 km) offshore were denied by the US Supreme Court in 1965; state claims were thus restricted to Monterey Bay and other submerged deposits within a 3-mi (5-km) offshore limit. In late 1979, oil from the US Navy's petroleum reserves at Elk Hills, in Kern County, was sold by the US Department of Energy to private oil companies at a cost of $41 a barrel—a move that helped to alleviate the domestic oil shortage but also to drive up crude oil prices throughout the US.

The state's largest oil company in 1984 was Chevron, with headquarters in San Francisco; its sales totaled $26.8 billion (5th among US oil companies and 11th among all US industrial corporations) and assets $36.5 billion. Los Angeles–based Atlantic Richfield (6th among US oil companies) had sales of $24.7 billion; Occidental Petroleum (9th), also in Los Angeles, had sales of $15.4 billion.

California's proved oil reserves as of 31 December 1983 were estimated at more than 5.3 billion barrels, nearly 20% of the US total and 3d behind Alaska and Texas. Petroleum production totaled 400 million barrels, representing more than 13% of the domestic output. Production of natural gas totaled 509 billion cu feet, 2.7% of the US total, and proved reserves were nearly 5.8 trillion cu feet (2.9%). In 1983 there were 51,437 producing oil wells and 1,405 active natural gas wells in the state. Virtually all the coal consumed for electric power generation is shipped in from other states.

California has been a leader in developing solar and geothermal power as alternatives to fossil fuels. State tax credits encourage the installation of solar energy devices in commercial and residential property. Near the end of 1984, the world's largest solar plant began generating electricity near California City; the plant has a maximum output of 6.4 Mw.

[29] INDUSTRY

California is the nation's leading industrial state, ranking 1st in almost every general manufacturing category—number of establishments, number of employees, total payroll, value added by manufacture, value of shipments, and new capital spending. Specifically, California ranks among the leaders in machinery, fabricated metals, food processing, computers, aerospace technology, and many other industries.

With its shipyards, foundries, flour mills, and workshops, San Francisco was the state's first manufacturing center. The number of manufacturing establishments in California nearly doubled between 1870 and 1900, and the value of manufactures increased almost tenfold from 1900 to 1925. New factories for transportation equipment, primary metal products, chemicals, and food products sprang up in the state during and after World War II. Second to New York State in industrial output for many years, California finally surpassed that state in most major manufacturing categories in the 1972 Census of Manufactures.

According to the 1982 census, value added by manufacture totaled $94.4 billion, the value of shipments by manufacturers was $199.7 billion, and new capital spending amounted to more than $8.8 billion. Of the total value added, food and food products accounted for 16%; transportation equipment, 13%; electric and electronic equipment, 13%; nonelectrical machinery, 10%; fabricated metal products, 6%; chemicals and chemical products, 4%; printing and publishing, 4%; clothing and textiles, 3%; and other sectors, 31%. The following table shows value of shipments by manufacturers for selected industries in 1982:

Petroleum refining	$24,343,800,000
Communication equipment	10,695,500,000
Office and computing machines	10,557,500,000
Aircraft and parts	9,929,600,000
Electronic components and accessories	9,797,800,000
Guided missiles and space vehicles	8,705,700,000
Preserved fruits and vegetables	6,311,700,000
Beverages	5,896,700,000
Motor vehicles and equipment	4,691,600,000
Miscellaneous plastics products	4,044,200,000
Meat products	3,747,900,000
Dairy products	3,512,900,000
Women's and misses' outerwear	3,317,600,000
Fabricated structural metal products	3,090,900,000
Commercial printing	2,961,800,000
Measuring and controlling devices	2,798,300,000
Construction and related machinery	2,366,100,000
Grain mill products	2,345,300,000
Paperboard containers	2,202,300,000

More than half of all industrial employees were engaged in manufacturing transportation equipment, electrical and electronic equipment, nonelectrical machinery, and food products; more than two-thirds worked in the Los Angeles–Long Beach–Anaheim and San Francisco–Oakland–San Jose metropolitan areas. Los Angeles is among the nation's leading cities in aircraft and automobile production. Among the many new industries established in San Diego, Oakland, Fresno, and Stockton, as well as Los Angeles, after World War II were the western plants of such corporate giants as Aluminum Co. of America, Goodyear Tire and Rubber, Heinz, International Harvester, Johns-Manville (now Manville), Kraft (now Dart and Kraft), Lever Brothers, Procter and Gamble, Swift (now Esmark), and US Steel.

The important aerospace and electronics industries in southern California produce highly sophisticated missiles and space vehicles. Major commercial aircraft plants located in the region include those of McDonnell Douglas and Lockheed, which has corporate headquarters in Burbank; other principal aerospace contractors are Rockwell International, Hughes, General Dynamics, and TRW. In 1984, 42.3% of Californians employed in aerospace lived in Los Angeles County, and 68.2% in six southern California counties. Another center of the state's electronics industry is in Santa Clara County, known as "Silicon Valley." Memorex, National Semiconductor, Intel, Apple Computer, ROLM, and Tandem Computers are among the firms based in the region. Of the $26 billion in prime defense contracts received by all California firms in 1982/83, 25% went for missile and space systems, 24% for military aircraft, and 17% for electronic and communications equipment.

In 1982, California's motion-picture production and distribution industry, based primarily in Los Angeles, had receipts of at least $5.7 billion—more than 86% of the US total.

[30] COMMERCE

Reflecting its rank in population, California led the nation in retail trade in 1982, but was 2d to New York State in wholesale trade. California's wholesale trade in 1982 was $217.7 billion, about 11% of the US total. Durable goods accounted for 51% of wholesale sales, and nondurable goods for the remaining 49%. Of the total sales of durable goods in that year, motor vehicles and automotive parts made up 29%; electrical goods, 17%; metals and minerals (excluding petroleum), 8%; furniture and home furnishings, 4%; and other categories, 42%. Of the nondurable goods, groceries and related products accounted for more than 35%; petroleum products, 22%; paper and paper products, 6%; chemicals and allied products, 6%; and other items, 31%.

The state's 1982 retail sales amounted to $123.9 billion, 12% of the US total. Of total 1982 sales, in establishments with payrolls, food stores accounted for 23%; automobile dealers, 18%; general merchandise stores, 12%; eating and drinking places, 10%;

gasoline service stations, 8%; and other establishments, 29%. Taxable 1983 sales at all outlets totaled $167.6 billion, of which Los Angeles County accounted for 30%; Santa Clara County, 7%; San Diego, 7%; and San Francisco, 4%.

Foreign trade is important to the California economy. About 400,000 workers are employed directly and indirectly in foreign trade. In 1984, goods exported from California were valued at an estimated $34.5 billion and goods imported at more than $51 billion. By region, this trade included the following (in billions):

	EXPORTS	IMPORTS
Asia	$20.1	$39.0
Western Europe	5.6	5.9
Canada	3.1	3.1
Latin America	2.0	2.7
Australia, Oceania	3.4	0.7
Africa	0.2	0.2
Eastern Europe	0.1	0.1

Leading exports in 1984 included data-processing equipment, $3.8 billion; electrical tubes and transistors, $3.8 billion; scientific equipment, measuring instruments, and optical equipment, $2.4 billion; and aircraft parts and spacecraft, $2.1 billion. California's leading agricultural export in 1984 was cotton ($1.5 billion).

California's customs districts are the ports of Los Angeles, San Francisco, and San Diego. San Francisco and San Jose have been designated as federal foreign-trade zones, where imported goods may be stored duty-free for reshipment abroad, or customs duties avoided until the goods are actually marketed in the US.

[31] CONSUMER PROTECTION

The State and Consumer Services Agency, headed by the secretary of state and consumer services, embraces the Department of Consumer Affairs, Department of Fair Employment and Housing, and certain other state agencies. The Department of Consumer Affairs administers the California Consumer Affairs Act and oversees the administrative and financial affairs of the various state agencies that regulate automotive, radio, and television repair shops, collection agencies, employment services, tax preparers, funeral establishments, cemeteries, and boxing, wrestling, and karate matches. The department also supervises the licensing boards for physicians, nurses, dentists, pharmacists, optometrists, veterinarians, civil engineers, construction contractors, social workers, accountants, barbers, cosmetologists, and other professionals who serve the public. The Consumer Advisory Council recommends to the director of the department those laws it considers necessary to protect and promote the interests of consumers.

The assistant attorney general for consumer law, within the Department of Justice, pursues criminal cases affecting consumers, while the assistant attorney general for environment and consumer protection has responsibility in civil matters.

[32] BANKING

In 1848, California's first financial institution—the Miners' Bank—was founded in San Francisco. Especially since 1904, when A. P. Giannini founded the Bank of Italy, now known as the Bank of America, California banks have pioneered in branch banking for families and small businesses. Today, California is among the leading states in branch banking, savings and loan associations, and credit union operations.

In 1984 there were 477 commercial banks in California, of which 453 were insured. Of these, 158 were national banks and 295 state banks. In 1983, banking offices for insured banks in the state totaled 5,019.

As of 31 December 1983, insured commercial banks in California had total assets of $228 billion, 2d only to New York State, and held more than 11% of all US commercial bank assets; state banks had about 30% of the total, and national banks 70%. Loans in 1983 totaled $148.8 billion, 37% of which were mortgage loans and 33% commercial and industrial loans. In the same year, commercial bank deposits were nearly $178.7 billion, 75% time deposits and 25% demand deposits.

The largest bank in California—and the world's 2d largest in 1984—was the Bank of America in San Francisco, with deposits of $94 billion. BankAmerica Corp., the bank's holding company, held $117.7 billion in assets as of 31 December 1984, 2d behind Citicorp (the parent company of Citibank) in New York City. Security Pacific National Bank and First Interstate Bank of California, both of Los Angeles, and Wells Fargo Bank and Crocker National Bank, both of San Francisco, were also among the nation's top 15 commercial banks in 1984.

California's savings and loans associations, whose assets consist largely of home mortgage loans, benefited during the 1970s and early 1980s from the booming real estate market. In 1982, these institutions accounted for more than 22% of outstanding mortgage loans and 19% of total savings deposits of all US savings and loan associations; these percentages far exceeded those of any other state. In 1983 there were 811 such institutions in California, with total assets of $193.6 billion. Their outstanding mortgage loans totaled $132.2 billion, and their savings deposits amounted to $147.9 billion. There were 1,227 credit unions in 1984.

The State Banking Department administers laws and regulations governing state-chartered banks, foreign banking corporations, and trust companies.

³³INSURANCE

Insurance companies provide a major source of California's investment capital by means of premium payments collected from policyholders. Life insurance companies also invest heavily in real estate; in 1983, life insurance firms held an estimated $21.4 billion in mortgage debt on California properties.

In 1983 there were 55 life insurance companies domiciled in California; these and out-of-state companies doing business in the state paid benefits of $5.3 billion (1st in the US), including $1.5 billion in death benefits. Nearly 29.8 million policies were in force, with a total value of $516.8 billion. The average family held $49,900 worth of life insurance, 8% below the US norm.

Property and casualty companies, 138 of which were domiciled in California in 1983, offer automobile, homeowners', and other types of insurance. That year, $14.5 billion in direct premiums were written in the state, including $3.6 billion in automobile liability insurance, $2.3 billion in automobile physical damage insurance, $1.5 billion in homeowners' insurance, and $77 million in earthquake insurance (76% of the US total).

³⁴SECURITIES

California's Pacific Stock Exchange (PSE), the largest securities market in the US outside New York City, is an association of some 500 member brokers who provide an auction market for the stocks, options, and bonds of national and local corporations. The exchange, which operates trading floors in Los Angeles and San Francisco, was established in 1957 to combine the separate exchanges that had been established in those cities at the end of the 19th century. Between 1957 and 1983, the volume of shares traded increased 30-fold, to more than 1 billion.

As of April 1983, the PSE listed some 1,100 stocks, warrants, and bonds for trading. The total included about 95 issues traded exclusively on the PSE; virtually all of the remainder were also traded on New York stock exchanges. Because of the time difference between the east and west coasts, the PSE begins trading at 6:30 AM PST and stays open until 1:30 PM PST, a half hour after the New York exchanges have closed. In January 1985, a seat on the Pacific Stock Exchange sold for $30,000 to $40,000; at the end of 1984, membership of the exchange stood at 516.

In 1983, New York Stock Exchange member firms had 554 sales offices and 8,434 full-time registered representatives in California. That year, 5,367,000 Californians were shareowners of public corporations.

³⁵PUBLIC FINANCE

California's general budget is the largest of all the states in both expenditures and revenues. The state's public finances became the focus of national attention when, on 6 June 1978, California voters approved Proposition 13, a constitutional amendment that reduced local property taxes by more than 50%. This new measure—which slashed the average tax rate per $100 of assessed value from $10.68 in 1977/78 to $4.78 in 1978/79—threatened to impair the public services of county and municipal governments, whose funds came largely from property taxes, by cutting their total revenues by $6.8 billion. The state legislature acted promptly to save essential public services by funding $4.2 billion from the state government's budget surplus to the counties, cities, towns, and school districts, and by increasing user fees and other public charges. Two byproducts of Proposition 13 were the reduction of government employment by 30,000 as of November 1978 and drastic cuts in the state's budget surplus for 1979/80 and future years. Proposition 9, a constitutional amendment that would have cut the state's personal income tax by half, was defeated by the voters in June 1980.

The state budget is prepared by the Department of Finance and presented by the governor to the state legislature for approval. The fiscal year lasts from 1 July through 30 June. The governor's budget request more than doubled from $11.3 billion in 1975 to about $24 billion in 1980, largely because of inflation and because of the state's increased aid to local governments as a result of Proposition 13. Consolidated state revenues for the 1983/84 fiscal year totaled $27.6 billion, state expenditures $26.8 billion.

The following table summarizes consolidated revenues and expenditures for 1984/85 (estimated) and 1985/86 (recommended), in millions:

REVENUES	1984/85	1985/86
Personal income tax	$10,485.0	$11,165.0
Sales and use taxes	9,830.0	10,618.0
Bank and corporation tax	3,525.0	3,950.0
Motor vehicle license and registration fees	2,124.5	2,279.0
Motor vehicle fuel tax	1,145.0	1,149.0
Insurance tax	635.0	675.0
Cigarette tax	260.7	257.0
Inheritance and gift taxes	275.0	193.0
Horse racing fees	140.4	142.2
Alcoholic beverage tax	136.7	139.8
Other receipts	2,445.7	2,353.0
TOTALS	$31,003.0	$32,921.0

EXPENDITURES		
Education, of which:	$13,888.8	$15,250.4
Elementary, secondary	(9,945.0)	(10,905.1)
Higher education	(3,943.8)	(4,345.3)
Health and welfare	7,978.0	8,628.0
Business, transportation, housing	1,986.5	2,019.6
Local government revenue sharing	1,786.5	1,906.0
Corrections	1,088.6	1,206.2
Tax relief	930.0	977.9
Resources	693.7	699.7
State and consumer services	357.5	313.0
Other expenditures	1,824.8	2,129.9
TOTALS	$30,534.4	$33,130.7

A constitutional amendment and initiative statute passed in 1984 provided for the establishment of a statewide lottery to raise additional funds for education.

A comparison of municipal budgets for 1981/82 shows that Los Angeles ranked 2d (behind New York City) among large US cities in expenditures with $2.97 billion, San Francisco 7th with $1.4 billion, and San Diego 25th with $445 million. The following table shows consolidated revenues and expenditures for Los Angeles in 1984/85 (estimated) and 1985/86 (proposed), in millions:

REVENUES	1984/85	1985/86
Property taxes	$ 320.9	$ 357.5
Utilities tax	259.1	278.6
Sales tax	230.9	244.6
Business tax	173.0	186.2
Sewer construction and maintenance	182.5	175.1
Licenses, permits, fees, fines	145.1	135.9
Other receipts and transfers	633.2	748.7
TOTALS	**$1,944.7**	**$2,126.6**
EXPENDITURES		
Police	$ 629.0	$ 642.0
Public works	297.8	337.4
Fire	259.7	282.6
Capital improvements	190.2	186.3
Recreation and parks	72.1	78.6
Transportation	35.9	65.0
Libraries	38.1	41.1
Other expenditures	421.9	493.6
TOTALS	**$1,944.7**	**$2,126.6**

California's total public debt exceeded $28.8 billion as of 30 June 1982, 2d only to New York State's. The per capita debt of $1,216.89 ranked only 43d among the 50 states, however. At the end of 1984, general obligation bonds outstanding from the California state government totaled $7.1 billion; in addition, 10 state agencies and authorities had $7.4 billion of revenue bonds and notes outstanding.

36 TAXATION

In the mid-1970s, Californians were paying more in taxes than residents of any other state. On a per capita basis, California ranked 3d among the 50 states in state and local taxation in 1977, but this heavy tax burden was reduced by the passage in 1978 of Proposition 13. By 1982, California ranked 5th in per capita state and local taxes. The state ranked 12th in federal tax burden per capita in 1982/83.

In 1984/85, California's revenues from state taxes and fees totaled more than $28.5 billion. Of this total, personal income tax provided 37%, state sales tax 34%, bank and corporation tax 12%, and other taxes 17%.

The state's progressive income tax rates as of January 1985 ranged from 1% to 11% on net taxable income. Low-income tax credits exempted single people with earnings of less than $5,000 and married couples earning less than $10,000 from paying state income taxes. An income tax credit is available for the cost of purchasing and installing solar energy systems in the home. State personal income tax has been indexed since 1978. The state corporate income tax on general corporations was 9.6% of net income from California sources; a minimum franchise tax of $200 applied to all firms except banks and financial corporations, whose net income was taxed at rates ranging up to 13%.

The state sales tax as of 1 January 1985 was 6% on retail sales (excepting food for home consumption, prescription medicines, gas, water, electricity, and certain other exempt products); of that 6%, 4.75% represented the basic state rate, 1% was designated for localities, and 0.25% was a county tax for the support of county transit systems. As of June 1985, a surtax of 0.5% was levied in Alameda, Contra Costa, San Francisco, Santa Clara, Santa Cruz, San Mateo, and Los Angeles counties, used principally for metropolitan transit systems. Santa Clara levied an additional 0.5% transit tax. Other state taxation includes inheritance and gift taxes, insurance tax, motor vehicle fees, cigarette tax, alcohol beverage tax, and pari-mutuel betting fees.

Localities derive most of their revenue from property taxes, which were limited in 1978 by Proposition 13 to 1% of market value, with annual increases in the tax not to exceed 2%. The drastic revision reduced property tax collections by about 57% to an estimated $4.9 billion in the 1978/79 fiscal year. The tax rate per $100 of assessed property value for Los Angeles County fell from $12.40 in 1977/78 to $4.78 in 1978/79; for San Francisco County, from $11.82 to $5.06; and for Alameda County, from $12.59 to $5.19. In 1984/85, the state provided $12.5 billion in fiscal assistance for local governments, partly to make up for local revenues lost by property tax reductions that were mandated by the adoption of Proposition 13.

In 1982/83, California bore a heavier share of the federal tax burden than any other state, contributing $68.8 billion in federal taxation, or more than 11% of the US total. But California also received more federal expenditures than any other state—$86.3 billion, for a net benefit of $17.5 billion. In 1983, California ranked 1st among the states in federal income tax payments, which exceeded $32.6 billion, based on 10.8 million returns.

37 ECONOMIC POLICY

The Department of Commerce (formerly the Department of Economic and Business Development) seeks to stimulate the economy and to serve as an ombudsman between business and government. Responsible both for encouraging new businesses to locate in the state and for assisting established industries to expand, the department supplies information to interested companies concerning site locations, labor conditions, wage rates, and other relevant factors. It provides loan guarantees for small businesses and long-term loans to industries that create permanent jobs. The department also prepares studies on the potential economic impact of a new business, and helps manufacturers and agribusinesses to expand their foreign trade. Several other state agencies administer business aid programs. The Office of Small and Minority Business assists businesses in obtaining state contracts.

The Economic Development Commission makes long-term studies of the state's economy and offers policy recommendations. The purpose of the privately funded California Economic Development Corporation, established in 1984 and consisting of some 55 chief executives of major California corporations appointed by the governor, is to attract business to California.

Perhaps the most important tax incentive offered to business during the 1970s was the large reduction in local property taxes as a result of Proposition 13; nearly 60% of the total tax reduction affected industrial and commercial properties. In addition, a tax credit applies to the cost of installing solar energy equipment in commercial and industrial buildings. A 1984 statute provides for the designation of economically distressed areas as Enterprise Zones, within which tax and other incentives may be used to attract private investment. The government no longer takes the state's film industry for granted: in 1984, the California Film Office was created to encourage the making of films in California.

38 HEALTH

Despite California's reputation for unconventional life-styles, the vital statistics for state residents have grown closer to national norms in recent years. Between 1980 and 1982, California's marriage rate was estimated to have risen from 8.9 to 9.3 per 1,000 population, while the national rate rose less than half as much, from 10.6 to 10.8. While the national divorce rate rose from 4 per 1,000 population to 5 between 1972 and 1982, California's remained the same—5.4—and that rate was only the 19th highest in the nation in 1982. California's abortion rate of 44.5 per 1,000 women 15-44 years of age was 2d only to Nevada's and tied New York's in 1982. Its ratio of 617 abortions to 1,000 live births was lower than both Nevada's and New York's.

Because of its large population, California led the nation in live births in 1982 with 429,897. The California birthrate grew faster between 1975 and 1982—from 14.7 to 17.4 per 1,000 population—than did the national rate—from 14.6 to 15.9. The infant death rate in 1981 was 15.5 per 1,000 live births for blacks and 9.9 for whites; both rates were well below the US norm.

California registered 189,075 deaths in 1982, including 4,218 infant deaths, 2,686 neonatal deaths, and 43 maternal deaths. California ranked below the national average in death rates in 1981 for 8 of the 11 leading causes of death (the exceptions were cirrhosis of the liver, pulmonary diseases, and suicide). Principal causes of death and their rates per 100,000 population during 1982 and percentages of all deaths included diseases of the heart, 67,255 (272; 36%); malignant neoplasms, 42,774 (173; 23%); cerebrovascular diseases, 15,605 (63; 8%); accidents 9,836 (40; 5%); chronic pulmonary disease, 6,660 (27; 3.5%); pneumonia and influenza, 5,469 (22; 3%); liver disease, 4,143 (17; 2%); and suicide, 3,752 (15; 2%). California's rate for murder and nonnegligent manslaughter—10.5 per 100,000 population—was 8th highest in the nation in 1983.

In 1980, the California Department of Health recorded 135,885 cases of gonorrhea and 13,270 of syphilis, with 18 deaths. Other reported diseases included 25,500 cases of streptococcal infections, with 15 deaths; 12,819 viral hepatitis, 192 deaths; 4,856 shigellosis, 5 deaths; 4,273 tuberculosis, 230 deaths; 2,951 salmonellosis, 13 deaths; 1,847 amebiasis, 7 deaths; 1,598 chickenpox and shingles, 29 deaths (of which 3 were from chickenpox); 1,162 viral meningitis, 14 deaths; and 1,051 measles, 1 death. In 1984/85, some 107,000 persons were treated for alcoholism and 43,000 for drug problems in California programs that receive state funds. As of 31 December 1983, 909 narcotics addicts were undergoing treatment at California rehabilitation centers. The state declared as rehabilitated 1,423 alcoholics (or 12% of clients treated) and 495 drug addicts (4%) in 1982/83.

As of 30 June 1983, six state hospitals for the mentally disabled had 5,142 patients. During the 1982/83 fiscal year, these hospitals admitted 10,384 patients and discharged an estimated 10,174. In recent years, an increasing number of patients have been treated through community mental health programs rather than in state hospitals; about 472,000 clients were served by such programs in 1984/85.

In 1984, California's state-licensed health-care facilities included 534 general-care hospitals, with 105,287 beds; 53 acute psychiatric hospitals, 7,165 beds; 1,152 skilled nursing facilities, 107,315 beds; and 44 intermediate-care institutions, 2,412 beds. In 1983, the state's 581 hospitals admitted nearly 3,400,000 patients. In 1981, hospital personnel included 83,681 full-time and 53,234 part-time registered nurses. Health insurance payments to California hospitals in 1982 exceeded $3.6 billion. The average cost per patient was $507 a day, the highest in the US and far above the national average cost of $327. The average cost per stay, $3,295, was more than in any other state but less than in the District of Columbia. Medical personnel licensed to practice in California in 1982 included 65,478 physicians and 15,221 dentists. In that year, at least 1,945 chiropractors—nearly 16% of the national total—were located in California.

Under Medi-Cal, a statewide program that pays for the medical care of persons who otherwise could not afford it, 2,855,600 Californians received medical benefits in 1984/85. California has also been a leader in developing new forms of health care. One increasingly popular system, called the health maintenance organization (HMO), provides preventive care, diagnosis, and treatment for which the patient pays a fixed annual premium.

39 SOCIAL WELFARE

In 1983, 8.9% of Californians received public assistance—the 3d highest proportion among the states.

In 1982/83, payments in aid to families with dependent children (AFDC) went to an average of 581,186 families a month with 1,011,510 children. Total AFDC payments in 1982/83 were nearly $2.8 billion. In January 1983, the average monthly AFDC payment was $463.57 per family and $157.66 per recipient—2d

only to Alaska. In 1983, California county welfare departments granted basic subsistence payments for food, shelter, clothing, and other general relief to an average of 63,059 persons and 5,706 families per month; average monthly aid was $180.64 per person and $309.40 per family, for a total of $137,270,542. A monthly average of 28,273 dependent children lived in boarding houses and institutions; during 1982/83, aid to their families totaled $212,612,711, for an average monthy grant of $626.67 per child.

During 1982/83, food stamps were issued to an average of 621,484 households a month comprising 1,766,870 persons, of whom 67% were receiving other forms of public assistance. During the same year, the value of food stamps issued totaled $710,380,292, all of it provided by federal funds. The federal government also paid $240,000,000 that year for the state's school lunch program, in which 1,773,000 pupils participated.

Federal Social Security benefits were paid in 1982 to 3,217,913 persons (9% of the US total), of whom 61% were retired workers, 31% received survivors' benefits, and 8% were disabled persons. In 1983, payments totaled $15.4 billion—the highest in the nation. In 1982, the average monthly benefit to retired workers was $424.98—16th highest in the nation.

As of December 1983, 653,383 Californians were enrolled in federal Supplemental Security Income (SSI) and state supplemental income programs. Of that number, 41% were aged, 56% disabled, and 3% blind. Federal SSI monthly payments averaged $176.36 for the aged, $207.53 for the blind, and $223.65 for the disabled; the state supplements were $137.40, $196.12, and $156.59, respectively.

During 1982/83, 108,009 disabled persons received vocational training, of whom 11,998 were rehabilitated; the total cost of vocational training was nearly $85 million. California led the 50 states in workers' compensation payments in 1982 with a total of $2 billion, more than 12% of the national total and more than two and a half times as much as the 2d-ranking state, New York. In that year, an average of 467,000 workers per week received unemployment insurance benefits. Payments totaled $2.7 billion, and the average weekly benefit was $100, 16% below the national average; the average duration of benefits was 18.1 weeks, 14% longer than the national average.

40 HOUSING

California had long led the US both in the number of housing units built annually and in the value of their construction, but by 1982 it had been overtaken by Texas in both categories and by Florida in the former. However, California still ranks 1st in the number of housing units (9,279,036) and in the average sales price of an existing house: in Anaheim, a single-family house averaged $132,100 in the first quarter of 1985, nearly 80% above the national norm. Housing construction boomed at record rates during the 1970s but slowed down at the beginning of the 1980s because rising building costs and high mortgage interest rates made it difficult for people of moderate means to enter the housing market. From 1981/82 to 1982/83, private housing starts increased by nearly 95%, and in 1984, housing starts were estimated to be up 30% to 218,000; however, a decline was forecast for 1985.

The earliest homes in southern California were Spanish colonial structures renowned for their simplicity and harmony with the landscape. These houses were one story high and rectangular in plan, with outside verandas supported by wooden posts; their thick adobe walls were covered with whitewashed mud plaster. In the north, the early homes were usually two stories high, with thick adobe walls on the ground floor, balconies at the front and back, and tile roofing. Some adobe houses dating from the 1830s still stand in coastal cities and towns, particularly Monterey.

During the 1850s, jerry-built houses of wood, brick, and stone sprang up in the mining towns, and it was not until the 1870s that more substantial homes, in the Spanish-mission style, were built in large numbers in the cities. About 1900, the California bungalow, with overhanging eaves and low windows, began to sweep the state and then the nation. The fusion of Spanish adobe structures and traditional American wooden construction appeared in the 1930s, and "California style" houses gained great popularity throughout the West. Adapted from the functional international style of Frank Lloyd Wright and other innovative architects, modern domestic designs, emphasizing split-level surfaces and open interiors, won enthusiastic acceptance in California. Wright's finest California homes include the Freeman house in Los Angeles and the Millard house in Pasadena. One of Wright's disciples, Viennese-born Richard Neutra, was especially influential in adapting modern design principles to California's economy and climate.

Between 1960 and 1980, some 3.8 million houses and apartments were built in the state, comprising more than 40% of California housing stock. The total number of housing units in the state increased by 53% during 1940–50; 52%, 1950–60; 28%, 1960–70; and 33%, 1970–80. California's total increase in housing stock between 1940 and 1980—296.5%—was 5th among the states. In 1983, 168,358 new units were authorized, of which 59% were single-family dwellings and 41% were in multi-unit structures. The value of housing construction totaled $11.5 billion.

Of the state's 8,629,866 occupied housing units in 1980, nearly 56% were occupied by the owner. Of year-round housing units, 98.2% had at least one full bathroom and 36.2% had at least two full bathrooms. By 1980, the state's year-round housing stock had risen to an estimated 9,206,826 units, of which 63% were single-family dwellings, 33% were in multifamily dwellings, and 4% were mobile homes. The median monthly cost for a mortgaged, owner-occupied housing unit was $408; the median monthly rent for a housing unit was $284.

California housing policies have claimed national attention on several occasions. In 1964, state voters approved Proposition 14, a measure repealing the Fair Housing Act and forbidding any future restrictions on the individual's right to sell, lease, or rent to anyone of his own choosing. The measure was later declared unconstitutional by state and federal courts. In March 1980, a Los Angeles city ordinance banned rental discrimination on the basis of age. A municipal court judge had previously ruled it was illegal for a landlord to refuse to rent an apartment to a couple simply because they had children. Ordinances banning age discrimination had previously been enacted in the cities of San Francisco, Berkeley, and Davis, and in Santa Monica and Santa Clara counties.

⁴¹EDUCATION

California ranks 1st among the states in enrollment in public schools and in institutions of higher education. However, in fall 1983, California had the 2d-highest public school pupil/teacher ratio—23.99—according to the National Education Association. California's current expenditure on public schools in1983/84 was estimated at $2,912 per pupil (31st in the US) based on average daily attendance.

The history of public education in California goes back at least to the 1790s, when the governor of the Spanish colony assigned retired soldiers to open one-room schools at the Franciscan mission settlements of San Jose, Santa Barbara, San Francisco, San Diego, and Monterey. Most of these schools, and others opened during the next three decades, were short-lived, however. During the 1830s, a few more schools were established for Spanish children, including girls, who were taught needlework.

Easterners and midwesterners who came to California in the 1840s laid the foundation for the state's present school system. The first American school was opened in an old stable at the Santa Clara mission in 1846, and the following year a schoolroom was established in the Monterey customhouse. San Francisco's first school was founded in April 1848 by a Yale graduate,

Thomas Douglass, but six weeks later, caught up in the gold rush fever, he dropped his books and headed for the mines. Two years after this inauspicious episode, the San Francisco city council passed an ordinance providing for the first free public school system in California. Although the first public high school was opened in San Francisco in 1856, the California legislature did not provide for state financial support of secondary schools until 1903.

The state's first colleges, Santa Clara College (now the University of Santa Clara), founded by Jesuits, and California Wesleyan (now the University of the Pacific), located in Stockton, both opened in 1851. A year later, the Young Ladies' Seminary (now Mills College) was founded at Benicia. The nucleus of what later became the University of California was established at Oakland in 1853 and moved to nearby Berkeley in 1873. Subsequent landmarks in education were the founding of the University of Southern California (USC) at Los Angeles in 1880 and of Stanford University in 1885, the opening of the first state junior colleges in 1917, and the establishment in 1927 of the Department of Education, which supervised the vast expansion of the California school system in the years following.

Perhaps the outstanding characteristics of public education in the state have been the emphasis placed on practical knowledge, through extensive vocational training programs, and the establishment of coordinating councils representing various community groups to work with educators in shaping school policies. More recent innovations include a state program to ensure educational continuity for children of migrant Mexican farm workers by enabling them to carry their school records back and forth between Mexico and California.

Almost 74% of the state's adult population had completed four years of high school by 1980; in addition, about 2,753,000 Californians had completed four or more years of college, representing nearly 20% of the adult population.

During the 1982/83 school year, California's 7,388 public schools enrolled 4,065,486 pupils. In fall 1983, enrollment was up to 4,089,017; elementary schools (K–8) had 2,764,036 pupils, high schools 1,249,967, and ungraded or special schools 75,014. There were 173,546 teachers and 28,322 other professionals in the public school system that year. In 1982/83, the number of public high school graduates was 241,343; private, 25,097. That year, 4,497 private schools enrolled 532,074 pupils and had 27,783 full-time teachers. In recent years, the number of public schools and their enrollment have gradually fallen, while those of private schools have risen. As of 1 January 1984 there were 606 Roman Catholic elementary schools with 181,515 pupils, and 124 parochial and diocesan high schools with 75,466 pupils. Some 370,000 students per year participate in special education programs, at a cost of about $1.4 billion.

As of fall 1983, the University of California, a state university, enrolled 141,289 students, 130,913 of them full-time. The California state college and university system—which should not be confused with the University of California—had 313,900 students (199,800 full-time). In addition, private colleges and universities that reported enrollments to the state had 166,031 students (113,103 full-time); public community colleges, 1,199,269 (208,187 full-time); private two-year colleges, 5,053 (4,419 full-time); and other public institutions, 3,622 (3,604 full-time).

The University of California has its main campus at Berkeley and branches at Davis, Irvine, Los Angeles (UCLA), Riverside, San Diego, San Francisco, Santa Barbara, and Santa Cruz. California's 19 state universities include those at Los Angeles, Sacramento, San Diego, San Francisco, and San Jose; state colleges are located at Bakersfield, San Bernardino, and Stanislaus.

Privately endowed institutions with the largest student enrollments are the University of Southern California (USC), with

29,411 students in 1983/84, and Stanford University (12,341). Other independent institutions are Occidental College in Los Angeles, Mills College at Oakland, Whittier College, Claremont University Center (including Harvey Mudd College, Pomona College, and Claremont Men's College), and the California Institute of Technology at Pasadena. Sixteen Roman Catholic colleges and universities, including Loyola Marymount University of Los Angeles, enrolled 32,924 students in 1984.

California's public school system is directed by the Department of Education, which is headed by the state superintendent of public instruction, elected on a nonpartisan ballot every four years. The University of California is governed separately by a board of regents which includes, ex officio, the state governor, lieutenant governor, speaker of the assembly, and superintendent of public instruction, along with the university president and the president and vice president of the alumni association. The four state officials also serve on the board of trustees which administers the state colleges and universities. The California Student Aid Commission supervises three financial assistance programs; funding for these programs in 1984/85 was budgeted at more than $96.8 million; there were more than 63,000 grants. In 1983/84, the average scholarship was $1,418 per student per year. The commission also administers the California Guaranteed Student Loan Program and the State Graduate Fellowship Program. All recipients must have been California residents for at least 12 months.

Revenues for the public schools in 1983/84 totaled $12.8 billion, of which state aid furnished about 67%, local property taxes 26%, and federal assistance 7%; between 1977/78 and 1983/84, local contributions fell by 50% and state aid increased by 68% as a result of Proposition 13. Public school teachers were paid an average annual salary of $26,043 (5th highest in the nation) in 1983/84, and teachers at public institutions of higher education averaged $32,017 (2d only to Alaska) in 1982/83. The 1983 Educational Reform Act established requirements for graduation from high school, lengthened the school year, and increased teachers' power to discipline students.

[42] ARTS

The arts have always thrived in California—at first in the Franciscan chapels with their religious paintings and church music, later in the art galleries, gas-lit theaters, and opera houses of San Francisco and Los Angeles, and today in seaside artists' colonies, regional theaters, numerous concert halls, and, not least, in the motion-picture studios of Hollywood.

In the mid-19th century, many artists came from the East to paint western landscapes, and some stayed on in California. The San Francisco Institute of Arts was founded in 1874; the E. B. Crocker Art Gallery was established in Sacramento in 1884; and the Monterey-Carmel artists' colony sprang up in the early years of the 20th century. Other art colonies developed later in Los Angeles, Santa Barbara, Laguna Beach, San Diego, and La Jolla. Notable art museums and galleries include the Los Angeles County Museum of Art (founded in 1910), Huntington Library, Art Gallery and Botanical Gardens at San Marino (1919), San Francisco Museum of Modern Art (1921), Norton Simon Museum of Art at Pasadena (1924), and San Diego Museum of Art (1925).

The theater arrived in California as early as 1846 in the form of stage shows at a Monterey amusement hall. The first theater building was opened in 1849 in Sacramento by the Eagle Theater Co. Driven out of Sacramento by floods, the company soon found refuge in San Francisco; by 1853, that city had seven theaters. During the late 19th century, many famous performers, including dancer Isadora Duncan and actress Maude Adams, began their stage careers in California. Today, California theater groups with national reputations include the American Conservatory Theater of San Francisco, Berkeley Repertory Theater, Mark Taper Forum in Los Angeles, and Old Globe Theater of San Diego.

The motion-picture industry did not begin in Hollywood—the first commercial films were made in New York City and New Jersey in the 1890s—but within a few decades this Los Angeles suburb had become synonymous with the new art form. California became a haven for independent producers escaping an East-Coast monopoly on patents related to filmmaking. (If patent infringements were discovered, the producer could avoid a lawsuit by crossing the border into Mexico.) In 1908, an independent producer, William Selig, completed in Los Angeles a film he had begun in Chicago, *The Count of Monte Cristo,* which is now recognized as the first commercial film produced in California. He and other moviemakers opened studios in Los Angeles, Santa Monica, Glendale, and—finally—Hollywood, where the sunshine was abundant, land was cheap, and the work force plentiful. These independent producers developed the full-length motion picture and the star system, utilizing the talents of popular actors like Mary Pickford, Douglas Fairbanks, and Charlie Chaplin again and again. In 1915, D. W. Griffith produced the classic "silent" *The Birth of a Nation,* which was both a popular and an artistic success. Motion-picture theaters sprang up all over the country, and an avalanche of motion pictures was produced in Hollywood by such increasingly powerful studios as Warner Brothers, Fox, and Metro-Goldwyn-Mayer. Hollywood became the motion-picture capital of the world. By 1923, film production accounted for one-fifth of the state's annual manufacturing value; in 1930, the film industry was one of the 10 largest in the US.

Hollywood flourished by using the latest technical innovations and by adapting itself to the times. Sound motion pictures achieved a breakthrough in 1927 with *The Jazz Singer,* starring Al Jolson; color films appeared within a few years; and Walt Disney originated the feature-length animated cartoon with *Snow White and the Seven Dwarfs* (1937). Whereas most industries suffered drastically from the depression of the 1930s, Hollywood prospered by providing, for the most part, escapist entertainment on a lavish scale. The 1930s saw the baroque spectacles of Busby Berkeley, the inspired lunacy of the Marx Brothers, and the romantic historical drama *Gone with the Wind* (1939). During World War II, Hollywood offered its vast audience patriotic themes and pro-Allied propaganda.

In the postwar period, the motion-picture industry fell on hard times because of competition from television, but it recovered fairly quickly by selling its old films to television and producing new ones specifically for home viewing. In the 1960s, Hollywood replaced New York City as the main center for the production of television programs. Fewer motion pictures were made, and those that were produced were longer and more expensive, including such top box-office attractions as *The Sound of Music* (1965), *The Godfather* (1972), *Jaws* (1975), *Star Wars* (1977), *The Empire Strikes Back,* (1980), *E.T. the Extra-Terrestrial* (1982), and *Ghostbusters* (1984). No longer are stars held under exclusive contracts, and the power of the major studios has waned as the role of independent filmmakers like Francis Ford Coppola, Steven Spielberg, and George Lucas has assumed increased importance.

Among the many composers who came to Hollywood to write film music were Irving Berlin, George Gershwin, Kurt Weill, George Antheil, Ferde Grofé, and Erich Korngold; such musical luminaries as Igor Stravinsky and Arnold Schoenberg were longtime residents of the state. Symphonic music is well established. In addition to the renowned Los Angeles Philharmonic, whose permanent conductors have included Zubin Mehta and Carlo Maria Giulini, there are the San Francisco Symphony and other professional symphonic orchestras in Oakland and San Jose. Some 180 semiprofessional or amateur orchestras have been organized in other communities. Resident opera companies perform regularly in San Francisco and San Diego. Annual musical events include the Monterey Jazz Festival and summer concerts at the Hollywood Bowl.

California has also played a major role in the evolution of popular music since the 1960s. The "surf sound" of the Beach Boys dominated California pop music in the mid-1960s. By 1967, the "acid rock" of bands like the Grateful Dead, the Jefferson Airplane (later the Jefferson Starship), and the Doors had started to gain national recognition—and that year the heralded "summer of love" in San Francisco attracted young people from throughout the country. It was at the Monterey International Pop Festival, also in 1967, that Jimi Hendrix began his rise to stardom. During the 1970s, California was strongly identified with a group of resident singer-songwriters, including Neil Young, Joni Mitchell, Randy Newman, Jackson Browne, and Warren Zevon, who brought a new sophistication to rock lyrics. Los Angeles is a main center of the popular-music industry, with numerous recording studios and branch offices of the leading record companies. Los Angeles–based Motown Industries, the largest black-owned company in the US, is a major force in popular music.

California has nurtured generations of writers, many of whom moved there from other states. In 1864, Mark Twain, a Missourian, came to California as a newspaperman. Four years later, New York-born Bret Harte published his earliest short stories, many set in mining camps, in San Francisco's *Overland Monthly.* The writer perhaps most strongly associated with California is Nobel Prize-winner John Steinbeck, a Salinas native. Hollywood's film industry has long been a magnet for writers, and San Francisco in the 1950s was the gathering place for a group, later known as the Beats (or "Beat Generation"), that included Jack Kerouac and Allen Ginsberg. The City Lights Bookshop, owned by poet Lawrence Ferlinghetti, was the site of readings by Beat poets during this period.

A California law, effective 1 January 1977, is the first in the nation to provide living artists with royalties on the profitable resale of their work.

43 LIBRARIES AND MUSEUMS

As of September 1983, California had 163 main public libraries with 584 branches, 372 smaller library stations, and 80 bookmobiles, making 2,822 stops. In 1982/83, the book stock was nearly 50 million volumes. The state's 164 academic libraries contained 45.7 million volumes; in addition, 426 special libraries had nearly 10.1 million volumes. Circulation of all library materials in 1982/83 totaled 155.6 million.

California has three of the largest public library systems in the nation, along with some of the country's finest private collections. The Los Angeles Public Library System had 5,388,480 volumes in 1982/83; the San Francisco Public Library, 1,850,148; and the San Diego Public Library, 1,775,900. Outstanding among academic libraries is the University of California's library at Berkeley, with its Bancroft collection of western Americana. Stanford's Hoover Institution has a notable collection of research materials on the Russian Revolution, World War I, and worldwide relief efforts thereafter. Numerous rare books, manuscripts, and documents are held in the Huntington Library in San Marino.

California has nearly 400 museums and 50 public gardens. Outstanding museums include the California Museum of Science and Industry, Los Angeles County Museum of Art, and Natural History Museum, all in Los Angeles; the San Francisco Museum of Modern Art, Fine Arts Museums of San Francisco, and Asian Art Museum of San Francisco; the San Diego Museum of Man; the California State Indian Museum in Sacramento; the Norton Simon Museum in Pasadena; and the J. Paul Getty Museum at Malibu. Among historic sites are Sutter's Mill, northeast of Sacramento, where gold was discovered in 1848, and a restoration of the Mission of San Diego de Alcalá, where in 1769 the first of California's Franciscan missions was established. San Diego has an excellent zoo, and San Francisco's Strybing Arboretum and Botanical Gardens has beautiful displays of Asian, Mediterranean, and California flora.

44 COMMUNICATIONS

Mail service in California, begun in 1851 by means of mule-drawn wagons, was soon taken over by stagecoach companies. The need for speedier delivery led to the founding in April 1860 of the Pony Express, which operated between San Francisco and Missouri. On the western end, relays of couriers picked up mail in San Francisco, carried it by boat to Sacramento, and then conveyed it on horseback to St. Joseph, Mo., a hazardous journey of nearly 2,000 mi (3,200 km) within 10 days. The Pony Express functioned for only 16 months, however, before competition from the first transcontinental telegraph line (between San Francisco and New York) put it out of business; telegraph service between San Francisco and Los Angeles had begun a year earlier. Today, mail is conveyed by the US Postal Service, which had 1,163 post offices and 81,362 employees in the state in March 1985.

California has more telephones than any other state. In 1980, 95% of the state's 8,629,866 occupied housing units had telephones.

The state's first radio broadcasting station, KQW in San Jose, began broadcasting speech and music on an experimental basis in 1912. California stations pioneered in program development with the earliest audience-participation show (1922) and the first "soap opera," *One Man's Family* (1932). When motion-picture stars began doubling as radio performers in the 1930s, Hollywood emerged as a center of radio network broadcasting. Similarly, Hollywood's abundant acting talent, experienced film crews, and superior production facilities enabled it to become the principal production center for television programs from the 1950s onward.

California ranks 1st in the US in the number of commercial television stations, and 2d only to Texas in radio stations. In 1984 there were 241 AM and 329 FM radio stations, and 60 commercial and 13 educational television stations; Los Angeles alone had 32 radio and 15 television stations. Los Angeles is also the home of the Pacifica Foundation, which operated 6 listener-sponsored FM radio stations (2 in Berkeley, one in Los Angeles, and 3 outside the state). Affiliates of the Public Broadcasting System serve Los Angeles, San Francisco, and Sacramento. In 1984, 302 systems provided cable television service to 3,301,579 subscribers in 781 communities.

45 PRESS

California's newspapers rank 1st in number and 2d in circulation among the 50 states. Los Angeles publishes one of the nation's most influential dailies, the *Los Angeles Times,* and San Francisco has long been the heart of the influential Hearst newspaper chain.

In August 1846, the state's first newspaper, the *Californian,* first published in Monterey, printed (on cigarette paper—the only paper available) the news of the US declaration of war on Mexico. The *Californian* moved to San Francisco in 1847 to compete with a new weekly, the *California Star.* When gold was discovered, both papers failed to mention the fact, and both soon went out of business as their readers headed for the hills. On the whole, however, the influx of gold seekers was good for the newspaper business. In 1848, the *Californian* and the *Star* were resurrected and merged into the *Alta Californian,* which two years later became the state's first daily newspaper; among subsequent contributors were Mark Twain and Bret Harte. Four years later there were 57 newspapers and periodicals in the state.

The oldest continuously published newspapers in California are the *Sacramento Bee* (founded in 1857), San Francisco's *Examiner* (1865) and *Chronicle* (1868), and the *Los Angeles Times* (1881). *Times* owner and editor Harrison Gray Otis quickly made his newspaper preeminent in Los Angeles—a tradition continued by his son-in-law, Henry Chandler, and by the Otis-Chandler family today. Of all California's dailies, the *Times* is the only one with a depth of international and national coverage to rival the major east coast papers. In 1887, young William Randolph Hearst took over his father's *San Francisco Daily Examiner* and introduced human interest items and sensational news stories to attract readers. The *Examiner* became the nucleus of the Hearst national newspaper chain, which later included the *News–Call Bulletin* and *Herald Examiner* in Los Angeles. The *Bulletin,* like many other newspapers in the state, ceased publication in the decades following World War II because of rising costs and increased competition for readers and advertisers.

In 1984 there were 40 morning dailies with a combined circulation of 4,062,157 and 80 evening dailies with 1,950,263 circulation (these figures include all-day newspapers), plus 58 Sunday newspapers with 5,858,314 average circulation. The *Los Angeles Times* is the only California paper whose daily circulation exceeds 1 million. The following table shows California's leading newspapers, with their 1984 circulations:

AREA	NAME	DAILY	SUNDAY
Fresno	Bee (m,S)	135,832	161,018
Long Beach– Huntington Beach	Press Telegram (all day,S)	128,724	141,028
Los Angeles	Herald Examiner (m,S)	233,193	215,871
	Times (m,S)	1,046,965	1,298,487
Oakland	Tribune (m,S)	148,088	137,739
Orange County– Santa Ana	Register (all day,S)	284,951	316,865
Sacramento	Bee (m,S)	221,407	225,691
	Union (m,S)	108,094	106,531
San Diego	Tribune (e)	121,008	
	Union (m,S)	221,650	348,926
San Francisco	Chronicle (m,S)	535,796 ⎫	
	Examiner (e,S)	138,700 ⎬	699,256
San Jose	Mercury-News (all day,S)	246,226	302,201

California has more book publishers—about 225—than any state except New York. Among the many magazines published in the state are *Architectural Digest, Bon Appetit, Motor Trend, PC World, Runner's World,* and *Sierra.*

46ORGANIZATIONS

Californians belong to thousands of nonprofit societies and organizations, many of which have their national headquarters in the state.

National service organizations operating out of California include the National Assistance League and Braille Institute of America, both in Los Angeles, and Knights of the Round Table International, Pasadena. Cultural and educational groups headquartered in the state are the American Aviation Historical Society, Garden Grove; American Battleships Association, San Diego; and American Society of Zoologists, Thousand Oaks.

Environmental and scientific organizations include the Sierra Club, Friends of the Earth, and Save-the-Redwoods League, all with headquarters in San Francisco; Animal Protection Institute of America, Sacramento; Geothermal Resources Council, Davis; and Seismological Society of America, Berkeley.

Among entertainment-oriented organizations centered in the state are the Academy of Motion Picture Arts and Sciences and the Academy of Television Arts and Sciences, both in Beverly Hills; Directors Guild of America and Writers Guild of America (West), both in Los Angeles; Screen Actors Guild and American Society of Cinematographers, both in Hollywood; and the National Academy of Recording Arts and Sciences, Burbank. Other commercial and professional groups are the Institute of Mathematical Statistics, San Carlos; Manufacturers' Agents National Association, Irvine; National Association of Civil Service Employees, San Diego; and Pacific Area Travel Association, San Francisco.

The many national sports groups with California headquarters include the Association of Professional Ball Players of America (baseball), Garden Grove; US Hang Gliding Association, Los Angeles; National Hot Rod Association, North Hollywood; Professional Karate Association, Beverly Hills; United States

Youth Soccer Association, Castro Valley; Soaring Society of America, Santa Monica; International Softball Congress, Anaheim Hills; American Surfing Association, Huntington Beach; and US Swimming Association, Fresno.

California also has Gamblers Anonymous, Los Angeles; Overeaters Anonymous, Torrance; and the National Investigations Committee on UFOs, Van Nuys.

47TOURISM, TRAVEL, AND RECREATION

California's cornucopia of scenic wonders attracts millions of state residents, out-of-state visitors, and foreign tourists each year. In 1983, California had 4,005,000 foreign visitors, nearly one out of every five who visited the US. Mexico provided nearly 18% of the state's international tourists, Canada 17%, Japan 12%, United Kingdom 5%, and West Germany 5%. California had $2.63 billion in receipts from international tourism in 1983, 2d only to Florida's $2.65 billion. Domestic tourists spent more than $25.9 billion in California that year. Tourism in 1983 generated almost 524,000 jobs in California, 5.3% of the state's nonagricultural employment.

While the state's mild, sunny climate and varied scenery of seacoast, mountains, and desert lure many visitors, the San Francisco and Los Angeles metropolitan areas offer the most popular tourist attractions. San Francisco's Fisherman's Wharf, Chinatown, and Ghirardelli Square are popular for shopping and dining; tourists also frequent the city's unique cable cars, splendid museums, Opera House, and Golden Gate Bridge. The Golden Gate National Recreation Area, comprising 68 sq mi (176 sq km) on both sides of the entrance to San Francisco Bay, includes Fort Point in the Presidio park, Alcatraz Island (formerly a federal prison) in the bay, the National Maritime Museum with seven historic ships, and the Muir Woods, located 17 mi (27 km) north of the city. This popular region had about 16.7 million visitors in 1984. South of the city, the rugged coastal scenery of the Monterey peninsula attracts many visitors; to the northeast, the wineries of the Sonoma and Napa valleys offer their wares for sampling and sale.

The Los Angeles area has the state's principal tourist attractions: the Disneyland amusement center at Anaheim, and Hollywood, which features visits to motion-picture and television studios and sight-seeing tours of film stars' homes in Beverly Hills. One of Hollywood's most popular spots is Mann's (formerly Grauman's) Chinese Theater, where the impressions of famous movie stars' hands and feet (and sometimes paws or hooves) are embedded in concrete. The New Year's Day Tournament of Roses at Pasadena is an annual tradition. Southwest of Hollywood, the Santa Monica Mountain National Recreation Area was created by Congress in 1978 as the country's largest urban park, covering 150,000 acres (61,000 hectares). The Queen Mary ocean liner, docked at Long Beach, is now a marine-oceanographic exposition center and hotel-convention complex.

The rest of the state offers numerous tourist attractions, including some of the largest and most beautiful national parks in the US. In the north are Redwood National Park and Lassen Volcanic National Park. In east-central California, situated in the Sierra Nevada, are Yosemite National Park, which drew 2.7 million visitors in 1984; towering Mt. Whitney in Sequoia National Park, which had 980,000 visitors; and Lake Tahoe, on the Nevada border. About 80 mi (129 km) east of Mt. Whitney is Death Valley. Among the popular tourist destinations in southern California are the zoo and Museum of Man in San Diego's Balboa Park and the Mission San Juan Capistrano, to which, according to tradition, the swallows return each spring. The San Simeon mansion and estate of the late William Randolph Hearst are now a state historical monument.

In 1983 there were 274 state parks, historic sites, reserves, and recreation areas of all types, covering more than 1,116,300 acres (452,000 hectares). In 1982/83, the state issued 2,529,593 angling licenses and 531,944 hunting licenses. Special permits to hunt wild fowl on public lands numbered 79,859, and dear and bear tags 334,412.

[48] SPORTS

California is more heavily represented in professional sports than any other state, with five major league teams in baseball, four in football, four in basketball, and one in ice hockey. Los Angeles is represented by clubs in all these sports, and leads US cities in total sports attendance.

Los Angeles' baseball Dodgers (who moved from Brooklyn, N.Y., at the end of the 1957 season) play in Dodger Stadium. The National Football League (NFL) Raiders (who moved from Oakland in 1982) play their home games in the Los Angeles Coliseum, while the Rams perform at Anaheim Stadium. The basketball Lakers and ice hockey Kings play at the Los Angeles Forum; the Clippers, Los Angeles' other National Basketball Association (NBA) franchise, use the Sports Arena. The California Angels baseball team uses nearby Anaheim Stadium. San Francisco's baseball Giants (formerly of New York City) and football 49ers play in Candlestick Park. Oakland's baseball A's (Athletics) occupy the Oakland-Alameda County Coliseum; the Golden State Warriors basketball team uses the Oakland Coliseum Arena. The baseball Padres and football Chargers play at San Diego Stadium. The state's newest professional franchise is the basketball Kings, which moved to Sacramento from Kansas City in 1985.

From 1972 to 1974, the Oakland A's, led by Jim "Catfish" Hunter, Vida Blue, and Reggie Jackson, won three World Series; between 1959 and 1985, the Los Angeles Dodgers captured eight National League pennants and four world championships with teams whose stars included Sandy Koufax, Don Drysdale, and Steve Garvey. The Giants, led by Hall of Famer Willie Mays, won a National League title in 1962, as did the Padres in 1984. The Lakers won four NBA championships, in 1972, 1980, 1982, and 1985; the Warriors took the NBA crown in 1975. The Raiders appeared in the Super Bowl four times, winning three of them, in 1977, 1981, and 1984. The 49ers won both of their Super Bowl appearances, in 1982 and 1985. The Rams reached the Super Bowl in 1980 but lost to the Pittsburgh Steelers.

Another popular professional sport is horse racing at such well-known tracks as Santa Anita and Hollywood Park. Because of the equable climate, there is racing in California virtually the whole year round.

California's teams have fielded powerhouses in collegiate sports. USC's baseball team won six national championships between 1970 and 1979, and the UCLA basketball team, coached by John Wooden, won seven consecutive NCAA Division I titles from 1967 through 1973; in all, UCLA captured an unprecedented 10 NCAA titles in 12 years (1964-75). From 1970 through 1985, three California collegiate football teams played in 13 of the 16 Rose Bowl games, winning 12: USC took 7, UCLA won 3, and Stanford took 2. In track and field, California colleges have done exceptionally well.

Californians also excel at individual sports, especially golf, tennis, and swimming. Other popular sports are surfing, sports-car racing, archery, badminton, and hang gliding. Los Angeles hosted the summer Olympic Games in 1932 and 1984. The winter games took place in Squaw Valley in 1960.

[49] FAMOUS CALIFORNIANS

Richard Milhous Nixon (b.1913) is the only native-born Californian ever elected to the presidency. Following naval service in World War II, he was elected to the US House of Representatives in 1946, then to the US Senate in 1950. He served as vice president during the Eisenhower administration (1953-61) but failed, by a narrow margin, to be elected president as the Republican candidate in 1960. Returning to his home state, Nixon ran for the California governorship in 1962 but was defeated. The next year he moved his home and political base to New York, from which he launched his successful campaign for the presidency in 1968. As the nation's 37th president, Nixon withdrew US forces from Viet-Nam while intensifying the US bombing of Indochina, established diplomatic relations with China, and followed a policy of détente with the Soviet Union. In 1972, he scored a resounding reelection victory, but within a year his administration was beset by the Watergate scandal. On 9 August 1974, after the House Judiciary Committee had voted articles of impeachment, Nixon became the first president ever to resign the office.

The nation's 31st president, Herbert Hoover (b.Iowa, 1874-1964), moved to California as a young man. There he studied engineering at Stanford University and graduated with its first class (1895) before beginning the public career that culminated in his election to the presidency on the Republican ticket in 1928. Former film actor Ronald Reagan (b.Illinois, 1911) served two terms as state governor (1967-75) before becoming president in 1981.

In 1953, Earl Warren (1891-1974) became the first Californian to serve as US chief justice (1953-69). Warren, a native of Los Angeles, was elected three times to the California governorship and served in that office (1943-53) longer than any other person. Following his appointment to the US Supreme Court by President Eisenhower, Warren was instrumental in securing the unanimous decision in *Brown* v. *Board of Education of Topeka* (1954) that racial segregation was unconstitutional under the 14th Amendment. Other cases decided by the Warren court dealt with defendants' rights, legislative reapportionment, and First Amendment freedoms.

Before the appointment of Earl Warren, California had been represented on the Supreme Court continuously from 1863 to 1926. Stephen J. Field (b.Connecticut, 1816-99) came to California during the gold rush, practiced law, and served as chief justice of the state supreme court from 1859 to 1863. Following his appointment to the highest court by President Lincoln, Field served what was at that time the longest term in the court's history (1863-97). Joseph McKenna (b.Pennsylvania, 1843-1926) was appointed to the Supreme Court to replace Field upon his retirement. McKenna, who moved with his family to California in 1855, became US attorney general in 1897 and was then elevated by President McKinley to associate justice (1898-1925).

Californians have also held important positions in the executive branch of the federal government. Longtime California resident Victor H. Metcalf (b.New York, 1853-1936) served as Theodore Roosevelt's secretary of commerce and labor. Franklin K. Lane (b.Canada, 1864-1921) was Woodrow Wilson's secretary of the interior, and Ray Lyman Wilbur (b.Iowa, 1875-1949) occupied the same post in the Hoover administration. Californians were especially numerous in the cabinet of Richard Nixon. Los Angeles executive James D. Hodgson (b.Minnesota, 1915) was secretary of labor; former state lieutenant governor Robert H. Finch (b.Arizona, 1925) and San Francisco native Caspar W. Weinberger (b.1917) both served terms as secretary of health, education, and welfare; and Claude S. Brinegar (b.1926) was secretary of transportation. Weinberger and Brinegar stayed on at their respective posts in the Ford administration; Weinberger later served as secretary of defense under Ronald Reagan. An important figure in several national administrations, San Francisco–born John A. McCone (b.1902) was chairman of the Atomic Energy Commission (1958-60) and director of the Central Intelligence Agency (1961-65).

John Charles Frémont (b.Georgia, 1813-90) led several expeditions to the West, briefly served as civil governor of California before statehood, became one of California's first two US senators (serving only until 1851), and ran unsuccessfully as the Republican Party's first presidential candidate in 1856. Other prominent

US senators from the state have included Hiram Johnson (1866-1945), who also served as governor from 1911 to 1917; William F. Knowland (1908-74); and, more recently, former college president and semanticist Samuel Ichiye Hayakawa (b.Canada, 1906) and former state controller Alan Cranston (b.1914). Governors of the state since World War II include Reagan, Edmund G. "Pat" Brown (b.1905), 4th-generation Californian Edmund G. "Jerry" Brown, Jr. (b.1938), and George Deukmejian (b.New York, 1928). Other prominent state office-holders are Rose Elizabeth Bird (b.Arizona, 1936), the first woman to be appointed chief justice of the state supreme court, and Wilson Riles (b.Louisiana, 1917), superintendent of public instruction, the first black Californian elected to a state constitutional office. Prominent among mayors are Thomas Bradley (b.Texas, 1917) of Los Angeles, Pete Wilson (b.Illinois, 1933) of San Diego, Dianne Feinstein (b.1933) of San Francisco, and Janet Gray Hayes (b.Indiana, 1926) of San Jose.

Californians have won Nobel Prizes in five separate categories. Linus Pauling (b.Oregon, 1901), professor at the California Institute of Technology (1927-64) and at Stanford (1969-74), won the prize for chemistry in 1954 and the Nobel Peace Prize in 1962. Other winners of the Nobel Prize in chemistry are University of California (Berkeley) professors William Francis Giauque (b.Canada, 1895-1982), in 1949; Edwin M. McMillan (b.1907) and Glenn T. Seaborg (b.Michigan, 1912), who shared the prize in 1951; and Stanford Professor Henry Taube (b.Canada, 1915), in 1983. Members of the Berkeley faculty who have won the Nobel Prize for physics include Ernest Orlando Lawrence (b.South Dakota, 1901-58), in 1939; Emilio Segrè (b.Italy, 1905) and Owen Chamberlain (b.1920), who shared the prize in 1959; and Luis W. Alvarez (b.1911), in 1968. Stanford professor William Shockley (b.England, 1910) shared the physics prize with two others in 1956; William A. Fowler (b.Pennsylvania, 1911), professor at the California Institute of Technology, won the prize in 1983. The only native-born Californian to win the Nobel Prize for literature was novelist John Steinbeck (1902-68), in 1962. Gerard Debreu (b.France, 1921), professor at the University of California at Berkeley, won the 1983 prize for economics.

Other prominent California scientists are world-famed horticulturist Luther Burbank (b.Massachusetts, 1849-1926) and nuclear physicist Edward Teller (b.Hungary, 1908). Naturalist John Muir (b.Scotland, 1838-1914) fought for the establishment of Yosemite National Park. Influential California educators include college presidents David Starr Jordan (b.New York, 1851-1931) of Stanford, and Robert Gordon Sproul (1891-1975) and Clark Kerr (b.Pennsylvania, 1911) of the University of California.

Major figures in the California labor movement were anti-Chinese agitator Denis Kearney (b.Ireland, 1847-1907); radical organizer Thomas Mooney (b.Illinois, 1882-1942); and Harry Bridges (b.Australia, 1901), leader of the San Francisco general strike of 1934. The best-known contemporary labor leader in California is Cesar Chavez (b.Arizona, 1927).

The variety of California's economic opportunities is reflected in the diversity of its business leadership. Prominent in the development of California railroads were the men known as the Big Four: Charles Crocker (b.New York, 1822-88), Mark Hopkins (b.New York, 1813-78), Collis P. Huntington (b.Connecticut, 1821-1900), and Leland Stanford (b.New York, 1824-93). California's long-standing dominance in the aerospace industry is a product of the efforts of such native Californians as John Northrop (1895-1981) and self-taught aviator Allen Lockheed (1889-1969), along with Glenn L. Martin (b.Iowa, 1886-1955); the San Diego firm headed by Claude T. Ryan (b.Kansas, 1898-1982) built the monoplane, *Spirit of St. Louis*, flown by Charles Lindbergh across the Atlantic in 1927. Among the state's banking and financial leaders was San Jose native Amadeo Peter Giannini (1870-1949), founder of the Bank of America. Impor-

tant figures in the development of California agriculture include Edwin T. Earl (1856-1919), developer of the first ventilator-refrigerator railroad car, and Mark J. Fontana (b.Italy, 1849-1922), whose California Packing Corp., under the brand name of Del Monte, became the largest seller of canned fruit in the US. Leaders of the state's world-famous wine and grape-growing industry include immigrants Ágostan Haraszthy de Mokcsa (b.Hungary, 1812?-69), Charles Krug (b.Prussia, 1830-94), and Paul Masson (b.France, 1859-1940), as well as two Modesto natives, Ernest (b.1910) and Julio (b.1911) Gallo. It was at the mill of John Sutter (b.Baden, 1803-80) that gold was discovered in 1848.

Leading figures among the state's newspaper editors and publishers were William Randolph Hearst (1863-1951), whose publishing empire began with the *San Francisco Examiner*, and Harrison Gray Otis (b.Ohio, 1837-1917), longtime owner and publisher of the *Los Angeles Times*. Pioneers of the state's electronics industry include David Packard (b.Colorado, 1912) and William R. Hewlett (b.Michigan, 1913). Stephen Wozniak (b.1950) and Steven Jobs (b.1955) were cofounders of Apple Computer. Other prominent business leaders include clothier Levi Strauss (b.Germany, 1830-1902), paper producer Anthony Zellerbach (b.Germany, 1832-1911), cosmetics manufacturer Max Factor (b.Poland, 1877-1938), and construction and manufacturing magnate Henry J. Kaiser (b.New York, 1882-1967).

California has been home to a great many creative artists. Native California writers include John Steinbeck, adventure writer Jack London (1876-1916), novelist and dramatist William Saroyan (1908-81), and novelist-essayist Joan Didion (b.1934). One California-born writer whose life and works were divorced from his place of birth was Robert Frost (1874-1963), a native of San Francisco. Many other writers who were residents but not natives of the state have made important contributions to literature. Included in this category are Mark Twain (Samuel Langhorne Clemens, b.Missouri, 1835-1910); local colorist Bret Harte (b.New York, 1836-1902); author-journalist Ambrose Bierce (b.Ohio, 1842-1914?); novelists Frank Norris (b.Illinois, 1870-1902), Mary Austin (b.Illinois, 1868-1934), and Aldous Huxley (b.England, 1894-1963); novelist-playwright Christopher Isherwood (b.England, 1904); and poets Robinson Jeffers (b.Pennsylvania, 1887-1962) and Lawrence Ferlinghetti (b.New York, 1920). California has been the home of several masters of detective fiction, including Raymond Chandler (b.Illinois, 1888-1959), Dashiell Hammett (b.Connecticut, 1894-1961), Erle Stanley Gardner (b.Massachusetts, 1889-1970), creator of Perry Mason, and Ross Macdonald (1915-83). Producer-playwright David Belasco (1853-1931) was born in San Francisco.

Important composers who have lived and worked in California include natives Henry Cowell (1897-1965) and John Cage (b.1912), and immigrants Arnold Schoenberg (b.Austria, 1874-1951), Ernest Bloch (b.Switzerland, 1880-1959), and Igor Stravinsky (b.Russia, 1882-1971). Immigrant painters include landscape artists Albert Bierstadt (b.Germany, 1830-1902) and William Keith (b.Scotland, 1839-1911), as well as abstract painter Hans Hofmann (b.Germany, 1880-1966). Contemporary artists working in California include Berkeley-born Elmer Bischoff (b.1916), Wayne Thiebaud (b.Arizona, 1920), and Richard Diebenkorn (b.Oregon, 1922). San Francisco native Ansel Adams (1902-84) is the best known of a long line of California photographers that includes Edward Curtis (b.Wisconsin, 1868-1952), famed for his portraits of American Indians, and Dorothea Lange (b.New Jersey, 1895-1965), chronicler of the 1930s migration to California.

Many of the world's finest performing artists have also been Californians. Violinist Ruggiero Ricci (b.1920) was born in San Francisco, while fellow virtuosos Yehudi Menuhin (b.New York, 1916) and Isaac Stern (b.Russia, 1920) were both reared in the state. Another master violinist, Jascha Heifetz (b.Russia, 1901), makes his home in Beverly Hills. California jazz musicians include Dave Brubeck (b.1920) and Los Angeles–reared Stan Kenton (b.Kansas, 1912).

Among the many popular musicians who live and record in the state are California natives David Crosby (b.1941), Randy Newman (b.1943), and Beach Boys Brian (b.1942) and Carl (b.1946) Wilson.

The list of talented and beloved film actors associated with Hollywood is enormous. Native Californians on the screen include child actress Shirley Temple (Mrs. Charles A. Black, b.1928) and such greats as Gregory Peck (b.1916) and Marilyn Monroe (Norma Jean Baker, 1926–62). Other longtime residents of the state include Douglas Fairbanks (b.Colorado, 1883–1939), Mary Pickford (Gladys Marie Smith, b.Canada, 1894–1979), Harry Lillis "Bing" Crosby (b.Washington, 1904–77), Cary Grant (Archibald Leach, b.England, 1904), John Wayne (Marion Michael Morrison, b.Iowa, 1907–79), Bette Davis (b.Massachusetts, 1908), and Clark Gable (b.Ohio, 1901–60). Other actors born in California include Clint Eastwood (b.1930), Robert Duvall (b.1931), and Robert Redford (b.1937).

Hollywood has also been the center for such pioneer film producers and directors as D. W. Griffith (David Lewelyn Wark Griffith, b.Kentucky, 1875–1948), Cecil B. De Mille (b.Massachusetts, 1881–1959), Samuel Goldwyn (b.Poland, 1882–1974), Frank Capra (b.Italy, 1897), and master animator Walt Disney (b.Illinois, 1901–66).

California-born athletes have excelled in every professional sport. A representative sampling includes Baseball Hall of Famers Joe Cronin (1906–1984), Vernon "Lefty" Gomez (b.1908), and Joe DiMaggio (b.1914), along with John Donald "Don" Budge (b.1915), Richard A. "Pancho" Gonzales (b.1928), Maureen "Little Mo" Connelly (1934–69), and Billie Jean (Moffitt) King (b.1943) in tennis, Gene Littler (b.1930) in golf, Frank Gifford (b.1930) and Orenthal James "O. J." Simpson (b.1947) in football, Mark Spitz (b.1950) in swimming, and Bill Walton (b.1952) in basketball. Robert B. "Bob" Mathias (b.1930) won the gold medal in the decathlon at the 1948 and 1952 Olympic Games.

50 BIBLIOGRAPHY

Bean, Walton, and James J. Rawls. *California: An Interpretive History.* 4th ed. New York: McGraw-Hill, 1982.

Bowman, Lynn. *Los Angeles: Epic of a City.* Berkeley: Howell-North, 1974.

Bronson, William, *How to Kill a Golden State.* Garden City, N.Y.: Doubleday, 1968.

California, State of. Department of Finance. *California Statistical Abstract 1984.* Sacramento, 1984.

Caughey, John W. *California: A Remarkable State's Life History.* 4th ed. Englewood Cliffs, N.J.: Prentice-Hall, 1982.

Cleland, Robert Glass. *From Wilderness to Empire: A History of California.* Edited by Glenn S. Dumke. New York: Knopf, 1959.

Conot, Robert. *Rivers of Blood, Years of Darkness.* New York: Morrow, 1967.

Cook, Sherburne Friend. *The Indian Versus the Spanish Mission: The Conflict Between the California Indian and White Civilization.* Berkeley: University of California Press, 1943.

Cook, Sherburne Friend. *The Population of the California Indians, 1769–1970.* Berkeley: University of California Press, 1976.

Dana, Richard Henry. *Two Years Before the Mast.* Edited by John H. Kemble. Los Angeles: Ward Ritchie Press, 1964.

Dasmann, Raymond F. *The Destruction of California.* New York: Macmillan, 1969.

Davie, Michael. *California: The Vanishing Dream.* New York: Dodd Mead, 1972.

DeVoto, Bernard. *Year of Decision: 1846.* Boston: Houghton Mifflin, 1943.

Federal Writers' Project. *California: A Guide to the Golden State.* Reprint. New York: Somerset, n.d. (orig. 1939).

Finson, Bruce, ed. *Discovering California.* San Francisco: California Academy of Sciences, 1983.

Griswold, Wesley S. *A Work of Giants: Building the First Transcontinental Railroad.* New York: McGraw-Hill, 1962.

Hart, James D. *A Companion to California.* New York: Oxford University Press, 1978.

Harte, Bret. *The Letters of Bret Harte.* New York: AMS Press, 1975 (orig. 1926).

Hesse, Georgia I. *California and the West 1985.* Robert C. Fisher, ed. New York: NAL, 1984.

Hoeber, Thomas R., and Charles Price, eds. *California Government and Politics Annual 1984-85.* Sacramento: California Journal, 1984.

Houston, James D. *Californians: Searching for the Golden State.* New York: Knopf, 1982.

Hyink, Bernard L., et al. *Politics and Government in California.* 11th ed. New York: Harper & Row, 1985.

Jackson, Donald Dale. *Gold Dust.* New York: Knopf, 1980.

Kahrl, William L. *Water and Power: The Conflict over Los Angeles' Water Supply in the Owens Valley.* Berkeley: University of California Press, 1982.

Lavender, David. *California: A Bicentennial History.* New York: Norton, 1976.

Lavender, David. *California: Land of New Beginnings.* New York: Harper & Row, 1972.

Lee, William Storss. *The Great California Deserts.* Illustrations by Edward Sanborn. New York: Putnam, 1963.

Lewis, Oscar. *The Big Four.* New York: Knopf, 1938.

Lillard, Richard. *Eden in Jeopardy, Man's Prodigal Meddling with His Environment: The Southern California Experience.* New York: Knopf, 1966.

Lotchin, Roger A. *San Francisco, 1846–56.* New York: Oxford University Press, 1974.

MacCann, Richard Dyer. *Hollywood in Transition.* New York: Houghton Mifflin, 1962.

McWilliams, Carey. *California: The Great Exception.* Layton, Utah: Gibbs M. Smith, 1976.

Martin, Stoddard. *California Writers: Jack London, John Steinbeck, the Tough Guys.* New York: St. Martin's, 1984.

Miller, Crane S., and Richard S. Hysolp. *California: The Geography of Diversity.* Palo Alto, Calif.: Mayfield, 1983.

Mowry, George. *The California Progressives.* Berkeley and Los Angeles: University of California Press, 1951.

Muir, John. *Mountains of California.* New York: Penguin, 1985.

Nadeau, Remi. *California: The New Society.* Westport, Conn.: Greenwood, 1974 (orig. 1963).

Nash, Gerald D. *The American West in the Twentieth Century.* Englewood Cliffs, N.J.: Prentice-Hall, 1973.

Nash, Gerald D. *State Government and Economic Development: A History of Administrative Policies in California, 1849–1933.* Reprint. Salem, N.H.: Ayer, 1979 (orig. 1964).

Paul, Rodman W. *California Gold.* Lincoln: University of Nebraska Press, 1965.

Pitt, Leonard. *The Decline of the Californios: A Social History of the Spanish-Speaking Californians, 1846–1900.* Berkeley and Los Angeles: University of California Press, 1966.

Rawls, James J. *Indians of California: The Changing Image.* Norman: University of Oklahoma Press, 1984.

Robinson, W. W. *Land in California.* Berkeley and Los Angeles: University of California Press, 1979.

Robinson, W. W. *Los Angeles: From the Days of the Pueblo.* San Francisco: California Historical Society, 1959.

Rolle, Andrew. *California: A History.* 3d ed. Arlington Heights, Ill.: AHM, 1978.

Roske, Ralph J. *Everyman's Eden: California*. New York: Macmillan, 1968.

Ross, Michael J. *California: Its Government and Politics*. 2d ed. Monterey, Calif.: Brooks/Cole, 1983.

Seidenbaum, Art. *This Is California: Please Keep Out!* New York: Wyden, 1975.

Starr, Kevin. *Americans and the California Dream*. New York: Oxford University Press, 1973.

Starr, Kevin. *Inventing the Dream: California Through the Progressive Era*. New York: Oxford University Press, 1985.

Stewart, George R. *The California Trail*. New York: McGraw-Hill, 1962.

Thompson, Warren S. *Growth and Changes in California's Population*. New York: Haynes Foundation, 1955.

Turner, Henry A., and John A. Vieg. *The Government and Policies of California*. New York: McGraw-Hill, 1964.

Watkins, T. H. *California: An Illustrated History*. New York: Outlet, 1983.

Weston, Charis Wilson and Edward. *California and the West*. New York: Duell, Sloan & Pearce, 1940.

COLORADO

State of Colorado

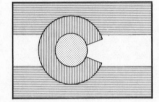

ORIGIN OF STATE NAME: From the Spanish word *colorado,* meaning red or reddish brown. **NICKNAME:** The Centennial State. **CAPITAL:** Denver. **ENTERED UNION:** 1 August 1876 (38th). **SONG:** "Where the Columbines Grow." **MOTTO:** *Nil sine numine* (Nothing without providence). **COAT OF ARMS:** The upper portion of a heraldic shield shows three snow-capped mountains surrounded by clouds; the lower portion has a miner's pick and shovel, crossed. Above the shield are an eye of God and a Roman fasces, symbolizing the republican form of government; the state motto is below. **FLAG:** Superimposed on three equal horizontal bands of blue, white, and blue is a large red "C" encircling a golden disk. **OFFICIAL SEAL:** The coat of arms surrounded by the words "State of Colorado 1876." **ANIMAL:** Rocky Mountain bighorn sheep. **BIRD:** Lark bunting. **FLOWER:** Rocky Mountain columbine. **TREE:** Colorado blue spruce. **GEM:** Aquamarine. **LEGAL HOLIDAYS:** New Year's Day, 1 January; Birthday of Martin Luther King, Jr., 3d Monday in January; Lincoln's Birthday, 12 February; Washington's Birthday, 3d Monday in February; Memorial Day, last Monday in May; Independence Day, 4 July; Colorado Day, 1st Monday in August; Labor Day, 1st Monday in September; Columbus Day, 2d Monday in October; Election Day, 1st Tuesday after 1st Monday in November in even-numbered years; Thanksgiving Day, 4th Thursday in November; Christmas Day, 25 December. **TIME:** 5 AM MST = noon GMT.

¹LOCATION, SIZE, AND EXTENT

Located in the Rocky Mountain region of the US, Colorado ranks 8th in size among the 50 states.

The state's total area is 104,091 sq mi (269,596 sq km), of which 103,595 sq mi (268,311 sq km) consist of land and 496 sq mi (1,285 sq km) comprise inland water. Shaped in an almost perfect rectangle, Colorado extends 387 mi (623 km) E–W and 276 mi (444 km) N–S.

Colorado is bordered on the N by Wyoming and Nebraska; on the E by Nebraska and Kansas; on the S by Oklahoma and New Mexico; and on the W by Utah (with the New Mexico and Utah borders meeting at Four Corners). The total length of Colorado's boundaries is *1,307 mi (2,103 km).* The state's geographic center lies in Park County, 30 mi (48 km) NW of Pikes Peak.

²TOPOGRAPHY

With a mean average elevation of 6,800 feet (2,100 meters), Colorado is the nation's highest state. Dominating the state are the Rocky Mountains. Colorado has 54 peaks 14,000 feet (4,300 meters) or higher, including Elbert, the highest in the Rockies at 14,433 feet (4,399 meters), and Pikes Peak, at 14,110 feet (4,301 meters), one of the state's leading tourist attractions.

The entire eastern third of the state is part of the western Great Plains, a high plateau that rises gradually to the foothills of the Rockies. Colorado's lowest point, 3,350 feet (1,021 meters), on the Arkansas River, is located in this plateau region. Running in a ragged north–south line, slightly west of the state's geographic center, is the Continental Divide, which separates the Rockies into the Eastern and Western slopes. The Eastern Slope Front (Rampart) Range runs south from the Wyoming border and just west of Colorado Springs. Also on the Eastern Slope are the Park, Mosquito, Medicine Bow, and Laramie mountains. Western Slope ranges include the Sawatch, Gore, Elk, Elkhead, and William Fork mountains. South of the Front Range, crossing into New Mexico, is the Sangre de Cristo Range, separated from the San Juan Mountains to its west by the broad San Luis Valley. Several glaciers, including Arapahoe, St. Mary's, Andrews, and Taylor, are located on peaks at or near the Continental Divide.

Colorado's western region is mostly mesa country—broad, flat plateaus accented by deep ravines and gorges, with many subter-

ranean caves. Running northwest from the San Juans are the Uncompahgre Plateau, Grand Mesa, Roan Plateau, the Flat Tops, and Danforth Hills. The Yampa and Green gorges are located in the northwestern corner of the state.

Blue Mesa Reservoir in Gunnison County is Colorado's largest lake. Six major river systems originate in Colorado: the Colorado River, which runs southwest from the Rockies to Utah; the South Platte, northeast to Nebraska; the North Platte, north to Wyoming; the Rio Grande, south to New Mexico; and the Arkansas and Republican, east to Kansas. Dams on these rivers provide irrigation for the state's farmland and water supplies for cities and towns. Eighteen hot springs are still active in Colorado; the largest is at Pagosa Springs.

³CLIMATE

Abundant sunshine and low humidity typify Colorado's highland continental climate. Winters are generally cold and snowy, especially in the higher elevations of the Rocky Mountains. Summers are characterized by warm, dry days and cool nights.

The average annual temperature statewide ranges from 54°F (12°C) at Lamar and at John Martin Dam to about 32°F (0°C) at the top of the Continental Divide; differences in elevation account for significant local variations on any given day. Denver's annual average is 50°F (10°C); normal temperatures range from 16° to 43°F (−9-6°C) in January and from 59° to 88°F (15-31°C) in July. Bennett recorded the highest temperature in Colorado, 118°F (48°C), on 11 July 1888; the record low was −60°F (−51°C), on 1 February 1951 at Taylor Park.

Annual precipitation ranges from a low of 7 in (18 cm) in Alamosa to a high of 25 in (64 cm) in Crested Butte, with Denver receiving about 15 in (38 cm). Denver's snowfall averages 59 in (150 cm) yearly. The average snowfall at Cubres in the southern mountains is nearly 300 in (762 cm); less than 30 mi (48 km) away at Manassa, snowfall is less than 25 in (64 cm).

⁴FLORA AND FAUNA

Colorado's great range in elevation and temperature contributes to a variety of vegetation, distributed among five zones: plains, foothills, montane, subalpine, and alpine. The plains teem with grasses and as many as 500 types of wild flowers. Arid regions contain two dozen varieties of cacti. Foothills are matted with berry shrubs, lichens, lilies, and orchids, while fragile wild flowers, shrubs, and conifers thrive in the montane zone. Aspen and Engelmann spruce are found up to the timberline.

In 1983, Colorado counted 747 nongame wildlife species and 113 sport-game species. Principal big-game species are the elk, mountain lion, Rocky Mountain bighorn sheep (the state animal), antelope, black bear, and white-tailed and mule deers; the mountain goat and the moose—introduced in 1948 and 1975, respectively—are the only nonnative big-game quarry. The lark bunting is the state bird; blue grouse and mourning doves are numerous, and 28 duck species have been sighted. Colorado has about 100 sport-fish species. Scores of lakes and rivers contain bullhead, kokanee salmon, and a diversity of trout. Rare Colorado fauna include the golden trout, white pelican, and wood frog. The lesser prairie chicken and razorback sucker are listed among threatened species. The greater prairie chicken, Canada lynx, wolverine, river otter, and bonytail are among endangered species.

5ENVIRONMENTAL PROTECTION

The Department of Natural Resources and the Department of Health share responsibility for state environmental programs. The first efforts to protect Colorado's natural resources were the result of federal initiatives. On 16 October 1891, US President Benjamin Harrison set aside the White River Plateau as the first forest reserve in the state. Eleven years later, President Theodore Roosevelt incorporated six areas in the Rockies as national forests. By 1906, 11 national forests covering about one-fourth of the state had been created. Mesa Verde National Park, founded in 1906, and Rocky Mountain National Park (1915) were placed under the direct control of the National Park Service. In 1978, Colorado became the first state in the US to encourage taxpayers to allocate part of their state income tax refunds to wildlife conservation.

Air pollution, water supply problems, and hazardous wastes head the list of Colorado's current environmental concerns. The Air Quality Control Commission, within the Department of Health, has primary responsibility for air pollution control. Because of high levels of carbon monoxide, nitrogen dioxide, and particulates in metropolitan Denver, Colorado Springs, Pueblo, and other cities, a motor vehicle emissions inspection system was inaugurated in January 1982 for gasoline-powered vehicles and in January 1985 for diesel-powered vehicles.

Formal efforts to ensure the state's water supply date from the Newlands Reclamation Act of 1902, a federal program designed to promote irrigation projects in the semiarid plains areas; its first effort, the Uncompahgre Valley Project, reclaimed 146,000 acres (59,000 hectares) in Montrose and Delta counties. One of the largest undertakings, the Colorado–Big Thompson Project, started in the 1930s, diverts a huge amount of water from the Western to the Eastern Slope. Colorado's efforts to obtain water rights to the Vermejo River in the Rockies were halted in 1984 by the US Supreme Court, which ruled that New Mexico would retain these rights.

Colorado's rapid population growth during the 1970s and early 1980s taxed an already low water table, especially in the Denver metropolitan area. The Department of Natural Resources' Water Conservation Board and Division of Water Resources are responsible for addressing this and other water-related problems.

The Department of Health has primary responsibility for hazardous waste management. In the spring of 1984, the department, along with federal agencies, began cleaning up nearly 7,000 contaminated sites in Grand Junction and other parts of Mesa County; these sites—homes and properties—were contaminated during the 1950s and 1960s by radioactive mill tailings that had been used as building material and that were not considered hazardous at the time. (It is now known that the low-level radiation emitted by the mill tailings can cause cancer and genetic damage.) In the fall of 1984, Aspen was placed on the federal

Environmental Protection Agency's list of dangerous waste sites because potentially hazardous levels of cadmium, lead, and zinc were found in Aspen's streets, buildings, and water. Cadmium, lead, and zinc mill tailings had been used as filling material during the construction of the popular resort.

6POPULATION

The 9th-fastest-growing state in the US, Colorado rose from 30th in population in 1970 to 28th in 1980, and to 26th in 1985. The 1980 census population was 2,889,964; the 1985 estimate, 3,253,-425. In 1983, 80% lived in metropolitan areas. The population density in 1985 was 31 per sq mi (12 per sq km). The estimated median age in 1983 was 29 years; nearly 27% of the population was under 18 years of age, and less than 9% was over 65. The 1990 population is projected to be 3,755,000.

Denver is the state's largest city and was, in 1984, the 24th-largest US city. Its 1984 population was estimated at 504,588, but its metropolitan area (including Boulder and Longmont) encompassed 1,791,400, or more than half the state's population. Other major cities, with their 1984 population estimates, are Colorado Springs, 247,739; Aurora, 194,772; Lakewood, 121,114; and Pueblo, 101,686.

7ETHNIC GROUPS

Once the sole inhabitants of the state, American Indians in 1980 numbered 18,059, including 235 Eskimos and 98 Aleuts. The black population is also small, less than 4% in 1980; the figure for Denver, however, was 12%. Of far greater importance to the state's history, culture, and economy are its Hispanic residents, of whom there were 340,000 in 1980, comprising nearly 12% of the population. Among residents of Denver, 19% were Hispanics.

The 1980 census counted 29,897 Asians and Pacific Islanders, of whom 9,858 were Japanese, 5,316 Korean, 4,026 Vietnamese, 3,897 Chinese, and 2,901 Filipino. In all, 114,000 persons, or 3.9% of the state population, were foreign-born in 1980.

8LANGUAGES

The first whites to visit Colorado found Arapaho, Kiowa, Comanche, and Cheyenne Indians roaming the plains and often fighting the Ute in the mountains. Despite this diverse heritage, Indian place-names are not numerous: Pagosa Springs, Uncompahgre, Kiowa, and Arapahoe.

Colorado English is a mixture of the Northern and Midland dialects, in proportions varying according to settlement patterns. Homesteading New Englanders in the northeast spread *sick to the stomach, pail,* and *comforter* (tied and filled bedcover), which in the northwest and the southern half are Midland *sick at the stomach, bucket,* and *comfort.* South Midland *butter beans* and *snap beans* appear in the eastern agricultural strip. Denver has *slat fence,* and *Heinz dog* (mongrel). In the southern half of the state, the large Spanish population has bred many loanwords such as *arroyo* and *penco* (pet lamb).

In 1980, 2,466,577 Coloradans—89% of the residents 3 years old and older—spoke only English at home. Other languages spoken at home, and the number of speakers, included Spanish, 183,692; German, 32,862; French, 11,470; and Italian, 7,378.

9RELIGIONS

The Spanish explorers who laid claim to (but did not settle in) Colorado were Roman Catholic, but the first American settlers were mostly Methodists, Lutherans, and Episcopalians. Some evangelical groups sought to proselytize the early mining camps during the mid-19th century.

Roman Catholics comprise the single largest religious group in the state, with 422,696 in 1984. The largest Protestant denomination in 1980 was the United Methodist Church, with 99,445 adherents. Other major denominations included Southern Baptist, 66,560; United Presbyterian, 56,612; Church of Jesus Christ of Latter-day Saints, 51,884; Lutheran Church—Missouri Synod, 46,694; and Protestant Episcopal, 39,167. The World Evangelical Fellowship has its headquarters in Colorado Springs.

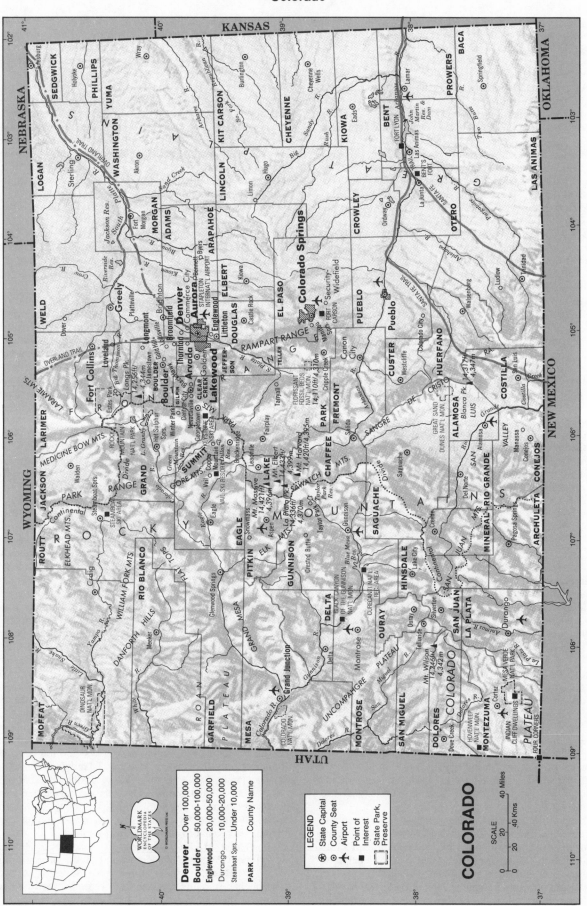

COLORADO

LEGEND
⊛ State Capital
⊙ County Seat
✈ Airport
■ Point of Interest
State Park, Preserve

Denver .... Over 100,000
Boulder 50,000–100,000
Englewood 20,000–50,000
Durango 10,000–20,000
Steamboat Sprs. .... Under 10,000

PARK .... County Name

SCALE
0   20   40 Miles
0   20   40 Kms

© WORLDMARK PRESS Ltd.

See endsheet maps: E3.

LOCATION: 36°59'35" to 41°N; 102°02'31" to 109°02'58"w. **BOUNDARIES:** Wyoming line, 260 mi (419 km); Nebraska line, 173 mi (278 km); Kansas line, 207 mi (333 km); Oklahoma line, 58 mi (93 km); New Mexico line, 333 mi (536 km); Utah line, 276 mi (444 km).

According to the 1984 estimates there were 44,365 Jews in Colorado, nearly all of them in the Denver area.

10 TRANSPORTATION

As the hub of the Rocky Mountain states, Colorado maintains extensive road and rail systems.

Because of its difficult mountain terrain, Colorado was by-passed by the first transcontinental railroads. In 1870, however, the Denver Pacific built a line from Denver to the Union Pacific's cross-country route at Cheyenne, Wyo. Several intrastate lines were built during the 1870s, connecting Denver with the mining towns. In particular, the Denver and Rio Grande built many narrow-gauge lines through the mountains. Denver finally became part of a main transcontinental line in 1934. As of 1984 there were 3,458 mi (5,565 km) of track in the state operated by Class I railroads. Amtrak trains in Colorado had a ridership of 253,714 persons in 1984.

Colorado has an extensive network of roads, including 29 mountain passes. As of 1984 there were 10,920 mi (17,574 km) of municipal roads and 63,346 mi (101,946 km) of rural roads. The major state roads are I-70, US 40, and US 50, all crossing the state from east to west, and I-25, running north–south through the Front Range cities; I-70 and I-25 intersect at Denver. I-76 connects Denver with Nebraska's I-80 to Omaha. Of the 2,767,362 motor vehicles registered in 1983, 1,939,178 were automobiles, 704,024 trucks, and 5,666 buses. There were an estimated 2,229,-042 licensed drivers that year.

A total of 220 public and private airfields served the state in 1983; Stapleton International Airport in Denver is the state's busiest. In 1983, it handled 11,401,005 departing passengers and 102,559 tons of cargo.

11 HISTORY

A hunting people lived in eastern Colorado at least 20,000 years ago, but little is known about them. The Basket Makers, who came to southwestern Colorado after 100 BC, grew corn and squash and lived in pit houses. By AD 800 there were Pueblo tribes who practiced advanced forms of agriculture and pottery making. From the 11th through the 13th centuries (when they migrated southward), the Pueblo Indians constructed elaborate apartment-like dwellings in the cliffs of the Colorado canyons and planted their crops both on the mesa tops and in the surrounding valleys.

In the 1500s, when Spanish conquistadores arrived in the Southwest, northeastern Colorado was dominated by the Cheyenne and Arapaho, allied against the Comanche and Kiowa to the south. These plains-dwellers also warred with the mountain-dwelling Ute Indians, who were divided into the Capote, Moache, and Wiminuche in the southwest; the Yampa, Grand River, and Uintah in the northwest; and the Tabeguache and Uncompahgre along the Gunnison River.

The exact date of the first Spanish entry into the region now called Colorado is undetermined; the explorer Juan de Oñate is believed to have traveled into the southeastern area in 1601. More than a century later, in 1706, Juan de Uribarri claimed southeast-ern Colorado for Spain, joining it with New Mexico. Meanwhile, French traders did little to stake out their claim to the Colorado region, which included most of the area east of the Rocky Mountains. In 1763, France formally ceded the Louisiana Terri-tory to Spain, which returned it to the French in 1801. Two years later, as part of the Louisiana Purchase, Colorado east of the Rockies became US land; the rest of Colorado still belonged to Spain.

Formal boundaries had never been demarcated between the lands of French Louisiana and Spanish New Mexico. In 1806, the US government sent out a group led by Lt. Zebulon M. Pike to explore this southwestern border. Pike's group reached Pueblo on 23 November 1806 and then attempted—without success—to scale the peak that now bears his name. Not until 1819 did the US and Spain agree to establish the boundary along the Arkansas River and then northward along the Continental Divide. The following year, Maj. Stephen Long explored this new border, and Dr. Edwin James made the first known ascent of Pikes Peak.

Eastern Colorado remained a wilderness for the next few decades, although traders and scouts like Charles and William Bent, Kit Carson, and Jim Bridger did venture into the largely uncharted and inhospitable land, establishing friendly relations with the Indians. It was in 1840 at Bent's Fort, the area's major trading center, that the four major eastern tribes ended their warfare and struck an alliance, a bond that lasted through their later struggle against the white settlers and US government. Between 1842 and 1853, John C. Frémont led five expeditions into the region, the first three for the US government. In 1842, he traveled along the South Platte River; on the next two trips, he crossed the Rockies. In his fourth expedition, he and a few of his party barely survived severe winter conditions. Finally, in 1853, Frémont led an expedition over a route traveled by Capt. John Gunnison earlier that year—through the San Luis Valley over Cochetopa Pass and along the Gunnison River. The 1853 trips were made five years after western and southern Colorado had come into US possession through the Mexican War.

The magnet that drew many Americans to Colorado was the greatly exaggerated report of a gold strike in Cherry Creek (present-day Denver) in July 1858. Within a year, thousands of prospectors had crossed the plains to seek their fortune. Many were disappointed and headed back east, but those who stayed benefited from a second strike at North Clear Creek, some 40 mi (64 km) to the west. The subsequent boom led to the founding of such mining towns as Central City, Tarryall, Golden, Blackhawk, Boulder, Nevadaville, Colorado City, and Gold Hill. By 1860, the population exceeded 30,000. A bill to organize a territory called Colorado, along the lines of the state's present-day boundaries, was passed by the US Congress on 28 February 1861. Colorado City, Golden, and Denver served at various times as the territorial capital until 1867, when Denver was selected as the permanent site. Colorado sided with the Union during the Civil War, though some settlers fought for the Confederacy. Union troops from Colorado helped defeat a contingent led by Confederate Gen. Henry H. Sibley at La Glorieta Pass in New Mexico in 1862.

The 1860s also saw the most serious conflict between Indians and white settlers in Colorado history. Cheyenne and Arapaho chiefs had ceded most of their tribal holdings to the US govern-ment in 1861. Sent to a reservation in the Arkansas Valley, these nomadic tribes were expected to farm the land. Unsuccessful at farming, the Indians rebelled against the poor rations supplied them by the US government, and sought to resume a nomadic life-style, hunting buffalo, raiding towns, and attacking travelers along the Overland and Sante Fe trails. Col. John Chivington was placed in charge of controlling the Indian unrest in the summer of 1864, as Territorial Governor John Evans departed for Washing-ton, D.C., leaving the situation in the hands of the military. On 29 November of that year, Chivington led his forces to Sand Creek, on the reservation's northeastern border, where they brutally massacred perhaps 200 Indian men, women, and children who thought they were under the protection of US military forces at nearby Ft. Lyon. Five more years of warfare followed, with the Indians finally defeated at Beecher Island (1868) and Summit Spring (1869). By 1874, most Plains Indians were removed to reservations in what is now Oklahoma. After gold and silver were discovered in areas belonging to the Ute in 1873, they too were forced off the land. By 1880, a series of treaties limited the Ute to a small reservation in the barren mesa country.

The first bill to admit Colorado to statehood was vetoed in 1866 by President Andrew Johnson, who at that time was in the midst of an impeachment fight and feared the entry of two more Republicans into the US Senate. Colorado finally entered the Union as the 38th state on 1 August 1876—less than a month after the nation's 100th birthday—during the presidency of Ulysses S. Grant.

In the early years of statehood, silver strikes at Leadville and Aspen brought settlers and money into Colorado. Rail lines, smelters, and refineries were built, and large coalfields were opened up. The High Plains attracted new farmers, and another new industry—tourism—emerged. As early as the 1860s, resorts had opened near some of the state's mineral springs. By the mid-1870s, scenic canyons and towns became accessible by train. One of the first major spas, Colorado Springs, recorded 25,000 tourists in 1878, and by the mid-1880s, Denver was accommodating up to 200,000 visitors a year.

Colorado's boom years ended with a depression during the early 1890s. Overproduction of silver—coupled with the US government's decision to adopt a gold standard in 1893—wiped out the silver market, causing the closing of mines, banks, and some businesses. Coinciding with this economic disaster was a drought that led to the abandonment of many farms. A more positive development was a gold find at Cripple Creek in 1891.

By the dawn of the 20th century, farmers were returning to the land and making better use of it. Immigrants from Germany and Russia began to grow sugar beets in the Colorado, Arkansas, and South Platte river valleys. Huge reclamation projects brought water to semiarid cropland, and dry-land farming techniques also helped increase yields. The development of the automobile and good roads opened up more of the mountain areas, bringing a big boom in tourism by the 1920s.

Following World War I, the agricultural and mining sectors fell into depression. From 1920 to 1940, statewide employment declined, and the population growth rate lagged behind that of the US as a whole. World War II brought military training camps, airfields, and jobs to the state. After the war, the expansion of federal facilities in Colorado led to new employment opportunities. The siting of both the North American Air Defense Com-

mand and the US Air Force Academy in Colorado Springs helped stimulate the growth of defense, federal research, and aerospace-related industries in the state. As these and other industries grew, so too did Colorado's population and income: between 1960 and 1983, the state's population growth rate was more than twice that of the nation as a whole; and between 1970 and 1983, Colorado moved from 18th to 9th rank among the states in personal income per capita. It was expected that the construction of the Air Force's $1.2 billion Space Operations Center at Colorado Springs, announced in 1983, would accelerate Colorado's economic and population growth.

[12]STATE GOVERNMENT

Colorado's state constitution, ratified on 1 July 1876, is a complex and extremely detailed document specifying the duties and structure of state and local government. Despite numerous amendments and revisions, some anachronistic legislation—people who have been caught dueling, for example, may not hold office—still remains on the books.

The general assembly, which meets annually, consists of a 35-member senate and 65-member house of representatives. There is no constitutional limit to the length of a session, and the legislature may call special sessions by request of two-thirds of the members of each house. Members of the legislature must be US citizens, at least 25 years old, and have been Colorado residents for at least one year.

The executive branch is headed by the governor, who submits the budget and legislative programs to the general assembly, and appoints judges, department heads, boards, and commissions. The governor must be a US citizen, at least 30 years old, and have been a resident of the state for two years or more. Elected with the governor is the lieutenant governor, who assumes the governor's duties in the governor's absence. Other elective officers include the secretary of state, attorney general, and state treasurer, all of whom serve four-year terms.

Colorado Presidential Vote by Political Parties, 1948–84

YEAR	ELECTORAL VOTE	COLORADO WINNER	DEMOCRAT	REPUBLICAN	PROGRESSIVE	SOCIALIST	SOC. LABOR
1948	6	*Truman (D)	267,288	239,714	6,115	1,678	—
						CONSTITUTION	
1952	6	*Eisenhower (R)	245,504	379,782	1,919	2,181	—
1956	6	*Eisenhower (R)	263,997	394,479	—	759	3,308
						SOC. WORKERS	
1960	6	Nixon (R)	330,629	402,242		563	2,803
1964	6	*Johnson (D)	476,024	296,767	—	2,537	—
					AMERICAN IND.		
1968	6	*Nixon (R)	335,174	409,345	60,813	235	3,016
					AMERICAN		
1972	7	*Nixon (R)	329,980	597,189	17,269	666	4,361
							LIBERTARIAN
1976	7	Ford (R)	460,801	584,278	397	1,122	5,338
					STATESMAN	CITIZENS	
1980	7	*Reagan (R)	368,009	652,264	1,180	5,614	25,744
1984	8	*Reagan (R)	454,975	821,817	—	—	11,257

*Won US presidential election.

Bills may originate in either house of the general assembly and become law when passed by majority vote of each house and signed by the governor; a bill may also become law if the governor fails to act on it within 10 days after receiving it. A two-thirds vote in each house is needed to override a gubernatorial veto.

The state constitution may be amended in several ways. An amendment may be introduced in the legislature, passed by a two-thirds majority in both houses, and submitted to the voters for approval. Alternatively, an initiative amendment, signed by a number of eligible voters equaling at least 5% of the number of votes cast for secretary of state in the previous election and then published in every county, may be filed no later than four months before the general election. If approved by the voters, it then becomes law.

Any US citizen 18 or older who is a resident of a Colorado county 32 days prior to an election may register to vote.

¹³POLITICAL PARTIES

The Democratic and Republican parties are the major political organizations in Colorado. Although both parties were in existence when Colorado achieved statehood, the Republicans controlled most statewide offices prior to 1900. Since then, the parties have been more evenly balanced. Of the 1,464,549 registered voters in 1982, 465,566 were Democrats, 462,837 were Republicans, and 536,146 were unaffiliated. Following the 1984 elections, the state had one Democratic and one Republican US senator, and four Republican and two Democratic US representatives; the Democrats controlled the statehouse, the Republicans the state legislature. In 1984, more than 63% of all Coloradan voters cast their ballots for Ronald Reagan; Democrat Walter Mondale won only 36% of the vote. US Senator Gary Hart—one of the few liberal Democratic senators in the West to survive fierce conservative opposition in recent years—was reelected to a second term in 1980, and was runner-up for the 1984 Democratic presidential nomination.

In 1984, Colorado had 175 Hispanic American elected officials, and 1 Asian Pacific American elected official. There were 15 black elected officials in 1985, the same year that 24% of Colorado's legislators were women—the 4th highest percentage in the US. Women also served as lieutenant governor and secretary of state, and as one of the state's six US representatives.

¹⁴LOCAL GOVERNMENT

As of 1985 there were 63 counties, 267 cities and towns, 25 census-designated places, and 181 school districts in Colorado.

The administrative and policymaking body in each county is the board of county commissioners, whose three to five members (dependent on population) are elected to staggered four-year terms. Other county officials include the county clerk, treasurer, assessor, sheriff, coroner, superintendent of schools, surveyor, and attorney.

Statutory cities are those whose structure is defined by the state constitution. Power is delegated by the general assembly to either a council-manager or mayor-council form of government. Colorado municipalities have increasingly opted for home rule, taking control of local functions from the state government. Of the 67 cities operating under their own charters as of 1980, the vast majority had council-manager systems of government; a small number had mayor-council systems. Towns, which generally have fewer than 2,000 residents, are governed by a mayor and a six-member board of trustees. The major source of revenue for both cities and towns is the property tax.

Denver, the only city in Colorado that is also a county, exercises the powers of both levels of government. It is run by a mayor and city council; a city auditor, independently elected, serves as a check on the mayor.

¹⁵STATE SERVICES

The Department of Education supervises and makes policy decisions for all public elementary and secondary schools. The Board of Regents of the University of Colorado governs the operations of that institution as well as its affiliates, the Colorado University Hospital, Children's Diagnostic Center, Psychiatric Hospital, and schools of medicine, nursing, and dentistry. All other state-run colleges, as well as the Colorado Historical Society, Council on the Arts and Humanities, and Advanced Technology Institute, are under the jurisdiction of the Department of Higher Education.

The Department of Highways builds, operates, and maintains state roads. The Department of Social Services administers welfare, medical assistance, rehabilitation, and senior-citizens programs. Human resource planning and development are under the Department of Labor and Employment, and health conditions are monitored by the Department of Health. The Department of Institutions oversees mental health, youth services, and developmental disabilities programs. The state's correctional facilities are administered by the Department of Corrections.

All programs concerned with the protection and control of Colorado's natural resources are the responsibility of the Department of Natural Resources. Other state agencies include the Department of Agriculture, Department of Military Affairs, Department of Regulatory Agencies, Department of Public Safety, and Department of Law.

¹⁶JUDICIAL SYSTEM

The supreme court, the highest court in Colorado, consists of 7 justices elected on a nonpartisan ballot. The number of justices may be increased to 9 upon request of the court and concurrence of two-thirds of the members of the General Assembly. The justices select a chief justice, who also serves as the supervisor of all Colorado courts. The next highest court, the court of appeals, consists of 10 judges, and is confined to civil matters. District courts have original jurisdiction in civil, criminal, juvenile, mental health, domestic relations, and probate cases, except in Denver, where probate and mental health matters are heard by the probate court and all juvenile matters by the juvenile court.

All judges in state courts are appointed to 2-year terms by the governor from a list of names recommended by a judicial nominating commission. The appointees must then be elected by the voters: supreme court justices for 10-year terms, appeals court judges for 8 years, and district court judges for 6.

County courts hear minor civil disputes and misdemeanors. Appeals from the Denver county courts are heard in Denver's superior court. Municipal courts throughout the state handle violations of municipal ordinances.

Colorado's FBI Crime Index crime rate in 1983 was 6,627 per 100,000 people; the violent crime rate was 476 (slightly below the US average), the property crime rate 6,151 (well above the US average). Denver's metropolitan area crime rate (including Boulder) was 7,053 per 100,000: the violent crime rate was 530, the property crime rate 6,523.

¹⁷ARMED FORCES

As of September 1984, 54,548 personnel, of whom 14,489 were civilians, were stationed at the 12 military facilities in the state. The largest Army base is Ft. Carson in Colorado Springs, headquarters of the 4th Infantry Division, with 22,001 personnel. There were 333 personnel at the Army's Rocky Mountain Arsenal near Denver, where chemical weapons have been produced and stored. Colorado Springs is the site of the US Air Force Academy; of the 4,103 cadets in 1985, 499 were women. Peterson Air Force Base is also located in Colorado Springs, as is the North American Air Defense Command (NORAD). Lowry Air Force Base, an important training center, is in the Denver area. Defense contracts awarded in 1983/84 totaled nearly $1.2 billion.

About 399,000 veterans lived in Colorado as of 30 September 1984. Of those who served in wartime, 3,000 were veterans of World War I, 126,000 of World War II, 78,000 of the Korean conflict, and 149,000 of the Viet-Nam era.

As of February 1985, 3,324 Coloradans served in the state's Army National Guard; 1,391 served in the Air National Guard at the end of 1984. State and local police totaled 11,703 in 1983.

[18] MIGRATION

The discovery of gold in 1858 brought an avalanche of prospectors. Some of these migrants later moved westward into the Rockies and Colorado River canyons. In 1873, another gold strike brought settlers into the Ute territory, eventually driving the Indians into a small reservation in the southwestern corner of the state. During the late 19th and early 20th centuries, the sparsely populated eastern plains were settled by farmers from Kansas and Nebraska and by immigrants from Scandinavia, Germany, and Russia. Five years of drought, from 1933 to 1938, helped drive many rural Coloradans off the land into the cities or westward to California.

Since the end of World War II, net migration into the state has been substantial, totaling 905,000 between 1950 and 1983. Growth has been evident in both urban and rural areas, but the largest increase has been in the Denver metropolitan area. A number of migrant workers, mostly Mexican Americans, work seasonally in the western orchards and fields.

[19] INTERGOVERNMENTAL COOPERATION

Among the most important interstate agreements for Coloradans are those governing water resources. Colorado participates with New Mexico in the Animas–La Plata Project Compact, Costilla Creek, and La Plata River compacts; with Kansas in the Arkansas River Compact of 1949; and with Nebraska in the South Platte River Compact. Multistate compacts allocate water from the Colorado and Republican rivers and the Rio Grande. Colorado also is a signatory to such regional agreements as the Interstate Oil and Gas Compact. The Western Interstate Commission for Higher Education has its headquarters in Boulder, as does the National Conference of State Legislatures. Federal grants to Colorado totaled nearly $1.2 billion in 1983/84.

[20] ECONOMY

During the late 1880s, Colorado was the nation's leading silver producer, and an important source of gold. With its abundant reserves of coal, natural gas, and other minerals—and the economic potential of its vast oil-shale deposits—Colorado remains a major mining state, although the mineral industry's share of the state economy has declined throughout this century. Agriculture, primarily livestock, retains its historic importance.

Trade, services, government, and manufacturing were responsible for more than 75% of new jobs created between 1975 and 1985. Trade is the leading source of employment, while manufacturing (4th in employment) is the principal contributor to the gross state product. Nearly one-third of manufacturing jobs in Colorado, compared to 14% nationally, are in high-technology areas.

A major driving force behind Colorado's economy is the US government, which in 1984 employed 93,000 civilian and military personnel in the state. Tourism has expanded rapidly in all areas of the state, and the finance, insurance, and real estate industries have also grown.

[21] INCOME

In 1983, Colorado ranked 9th among the 50 states in per capita income, with $12,580. Personal income that year reached $37.4 billion, or 1.4% of the US total. In 1981, Denver ranked 7th among the 50 largest US cities in per capita income.

Slightly more than 10% of all Coloradans (and 7.4% of Colorado families) were below the federal poverty level in 1980. Median family income was $21,279, 12th in the US. As of 1982, 27,400 Coloradans were among the nation's top wealth-holders, with gross assets greater than $500,000 each.

[22] LABOR

In 1983, an average of 1,557,000 Coloradans were employed in the civilian labor force. Of that number, 875,000 were male, 682,000 were female. Of nonagricultural wage and salary earners, 25% were employed in trade; 22% in services; 18% in government; 14% in manufacturing; 6% in finance, insurance, and real estate; 6% in construction; 6% in transportation and public utilities; and 3% in mining. The state's unemployment rate, estimated to be 4.9% in 1985, has been consistently below the US average.

A federal census in March 1982 revealed the following nonfarm employment pattern in Colorado:

	ESTABLISH-MENTS	EMPLOYEES	ANNUAL PAYROLL ('000)
Agricultural services, forestry, fishing	846	5,454	$ 70,599
Mining, of which:	1,502	50,385	1,385,315
Metals	(135)	(7,244)	(178,090)
Oil, gas extraction	(1,075)	(28,986)	(774,797)
Contract construction	7,540	84,481	1,685,974
Manufacturing	3,904	190,731	3,944,921
Transportation, public utilities	2,372	77,931	1,872,595
Wholesale trade	6,141	81,567	1,602,222
Retail trade	18,904	247,725	2,329,354
Finance, insurance, real estate	7,785	86,629	1,519,748
Services	23,408	280,837	3,729,768
Other	1,704	2,118	31,162
TOTALS	74,106	1,107,858	$18,171,658

The census did not include federal, state, and local government employees, of whom Colorado had 244,400 in 1984.

Colorado's labor history has been marked by major disturbances in the mining industry. From 1881 to 1886, the Knights of Labor led at least 35 strikes in the mines; during the 1890s, the Western Federation of Miners struck hard-rock mines in Telluride and Cripple Creek. The United Mine Workers, who came into the state in 1899, shut down operations at numerous mines in 1900 and 1903. Violence was common in these disputes. In one well-known episode, after striking miners and their families set up a tent colony at Ludlow, near Trinidad, the governor called out the militia; in the ensuing conflict, on 20 April 1914, the miners' tents were burned, killing 2 women and 11 children, an event that touched off a rebellion in the whole area. Federal troops restored order in June, and the strike ended with promises of improved labor conditions. In 1917, the state legislature created the Colorado Industrial Commission, whose purpose is to investigate all labor disputes.

As of 1983 there were 515 unions in Colorado, and in 1980 there were 227,000 union members, constituting more than 18% of nonagricultural employment.

[23] AGRICULTURE

Colorado ranked 19th among the 50 states in agricultural income in 1983, with $3.1 billion, of which more than $926 million came from crops.

As of 1984 there were 27,000 farms and ranches covering about 35 million acres (14 million hectares); the average farm was 1,281 acres (518 hectares). Of the 53,000 farm workers in 1983, 32,000 were family workers and 21,000 were hired hands. The major crop-growing areas are the east and east-central plains for sugar beets, beans, potatoes, and grains; the Arkansas Valley for grains and peaches; and the Western Slope for grains and fruits.

Colorado ranked 6th in the US in production of dry edible beans in 1982, with 2,128,000 hundredweight; 7th in sugar beets, with 920,000 tons; 8th in barley, with 15,910,000 bushels; 9th in potatoes, with 16,000,000 hundredweight; and 11th in wheat, with 84,984,000 bushels. Other field crops include corn, hay, and sorghum. Also in 1982, Colorado produced 220,950 tons of fresh market vegetables, 15,750 tons of vegetables for processing, 40,000,000 lb of commercial apples, and 11,000,000 lb of peaches.

About 1,600,000 lb of tart cherries were harvested in 1983. Colorado is also a major grower of carnations and roses.

24ANIMAL HUSBANDRY

A leading sheep-producing state, Colorado is also a major area for cattle and other livestock. More than two-thirds of all farm income in 1983 came from livestock and livestock products.

From 1858 to about 1890, cattle drives were a common sight in Colorado, as a few cattle barons had their Texas longhorns graze on public-domain lands along the eastern plains and Western Slope. This era came to an end when farmers in these regions fenced in their lands, and the better-quality shorthorns and Herefords took over the market. Today, huge tracts of pastureland are leased from the federal government by both cattle and sheep ranchers, with cattle mostly confined to the eastern plains and sheep to the western part of the state.

Preliminary estimates of livestock output for 1983 included cattle and calves, 1,529,990,000 lb, valued at $941,389,000 (7th in the US); hogs, 109,800,000 lb, $52,496,000 (22d); turkeys, 116,641,000 lb, $46,656,000 (11th); sheep and lambs, 60,083,000 lb, $31,238,000 (3d); and wool, 7,765,000 lb, $4,426,000 (4th).

Other livestock products in 1983 (except where indicated) included chickens, 4,808,000 lb; eggs, 618,000,000; milk, 987,000,000 lb; and butter (1982), 3,785,000 lb.

25FISHING

There is virtually no commercial fishing in Colorado. The state's many warm-water lakes lure sport anglers with perch, black bass, and trout, while walleyes are abundant in mountain streams.

26FORESTRY

About 22,271,000 acres (9,013,000 hectares) of forestland, 3% of the US total, are located in Colorado, most of it in the Rocky Mountains. Of this amount, 11,315,000 acres (4,579,000 hectares) are commercial timberland. Despite the vast supply of wood, commercial forestry is not a major element of the state's economy. Shipments by the lumber and wood products industry totaled $236.4 million in 1982, only 1.3% of all manufactured goods.

27MINING

Colorado's declining nonfuel mineral output was valued at $422 million in 1984, 21st in the US. Colorado's mining industry, particularly molybdenum mining, suffered during the early 1980s from the sagging market for US steel products. In 1984, AMAX Corp. reopened molybdenum mines in Henderson and Climax, but employment in the mining of metallic minerals that year—6,500—remained far below the 1981 peak of 11,800.

Although silver output is far below the boom years of the late 1800s, Colorado still ranked 4th in the US in 1983, producing 2,100,000 troy oz of silver; gold output was 63,063 troy oz. Other important minerals (excluding fossil fuels) were zinc, lead, copper, vanadium, cement, and sand and gravel.

28ENERGY AND POWER

An abundant supply of coal, oil, and natural gas makes Colorado a major energy-producing state.

During 1983, 25.2 billion kwh of electricity were generated in Colorado, about 75% of that in coal-fired plants; installed capacity was 6.8 million kw. There is one nuclear power plant, at Platteville.

Petroleum production in 1983 was 29,050,000 barrels; proved reserves were 186,000,000 barrels. Natural gas production in 1983 was 194.9 billion cu feet; reserves were nearly 2 trillion cu feet.

Colorado's coal output, which reached a peak of some 19 million short tons in 1981, was expected to continue to decline throughout the decade. Output for 1984 was 16.5 million short tons; output for 1985 was estimated at 15.5 million short tons.

Colorado holds the major portion of the nation's proved oil-shale reserves. In 1984, Unocal Corp. completed construction on a large mine and retort facility backed by a federal guarantee to purchase the oil; as of 1985, however, commercial production had not started. Because of its ample sunshine, Colorado is also well-suited to solar energy development. Among the many energy-related facilities in the state is the Solar Energy Research Institute in Golden.

29INDUSTRY

Accounting for an estimated 14% of all nonagricultural employment in 1984, manufacturing is a major segment of the economy. Colorado is the main manufacturing center of the Rocky Mountain states; value of shipments by manufacturers in 1982 was nearly $18 billion. The major sectors were food and food products, 23%; nonelectric machinery, 13%; instruments and related products, 13%; electric and electronic equipment, 8%; fabricated metal products, 7%; transportation equipment, 7%; printing and publishing, 6%; petroleum and coal products, 6%; and other sectors, 17%.

The following table shows value of shipments by manufacturers for selected industries in 1982:

Meat products	$1,724,000,000
Office and computing machines	1,551,200,000
Beverages	990,400,000
Communication equipment	926,600,000
Construction and related machinery	283,400,000

High-technology research and manufacturing grew substantially in Colorado during the 1970s and early 1980s. Storage Technology in Louisville is the largest high-tech company with headquarters in the state, but many large out-of-state companies—including Western Electric, IBM, Hewlett-Packard, Eastman Kodak, TRW, Digital Equipment, Ampex, Ball Aerospace Systems Division, and Cobe Laboratories—have divisions there.

In addition to Storage Technology, three other Colorado companies are on the *Fortune* 500 list of the largest industrial corporations in the US: Manville in Denver, insulation and roofing; Monfort of Colorado in Greeley, food products; and Adolph Coors in Golden, beer.

30COMMERCE

Colorado is the leading wholesale and retail distribution center for the Rocky Mountain states. Wholesale trade was $26.8 billion in 1982; retail sales were $16.6 billion. Major retail sectors included food stores, 22%; automotive dealers, 19%; eating and drinking places, 10%; department stores, 9%; and gasoline service stations, 9%. Metropolitan Denver (including Boulder) accounted for 60% of all retail sales.

Colorado's foreign exports included nearly $1.3 billion in manufactured goods in 1981, and $612 million in agricultural products in 1981/82.

31CONSUMER PROTECTION

The Consumer Protection Unit of the Department of Law, under the jurisdiction of the attorney general, administers the state's Uniform Consumer Credit Code and Consumers Protection Act (1965). The Office of Consumer Counsel, created in 1984, represents consumer interests before the Public Utilities Commission in matters involving electric, gas, and telephone utility service.

32BANKING

In 1984, Colorado had 547 commercial banks, with assets of $22 billion, and 38 savings and loan associations, with assets of $9.6 billion. Most of these institutions were concentrated in metropolitan Denver, the leading banking center between Kansas City and the Pacific.

33INSURANCE

As of 1983, 46 property and casualty insurance companies and 25 life insurance companies were domiciled in Colorado. Coloradans purchased $13.8 billion in ordinary life insurance during 1983, when 4,936,000 policies worth $80 billion were in force. The average Colorado family had $60,600 in life coverage.

34SECURITIES

There are no stock or commodity exchanges in Colorado. In 1983, 96 New York Stock Exchange sales offices and 1,254 registered representatives were in the state. There were 203 registered brokers and dealers and 598,000 shareowners of public corporations in Colorado that year.

35PUBLIC FINANCE

The governor's annual budget is presented to the general assembly in early January. The fiscal year runs from 1 July to 30 June. Since 1977/78 there has been a 7% limitation on state budget increases; excess revenues must be placed in a special reserve fund for tax relief.

The following table summarizes consolidated revenues and expenditures for 1984/85 (estimated) and 1985/86 (recommended):

REVENUES	1984/85	1985/86
Income taxes	$1,039,000,000	$1,110,000,000
Excise taxes	854,200,000	891,000,000
Other revenues	132,500,000	133,000,000
Federal funds	501,358,000	561,950,000
Rebates[1]	(159,500,000)	(168,700,000)
Balance from previous year	764,068,000	842,457,000
TOTALS	$3,131,626,000	$3,369,707,000

EXPENDITURES		
Higher education	$ 710,547,000	$ 753,513,000
Education	677,946,000	729,104,000
Social services	629,634,000	685,816,000
Property tax relief	221,843,000	231,974,000
Institutions	212,220,000	230,023,000
Capital construction	42,621,000	102,551,000
Health	71,896,000	82,851,000
Corrections	56,126,000	61,240,000
Other expenses	521,731,000	589,317,000
TOTALS	$3,144,564,000	$3,466,389,000

[1]For budgeting purposes, rebates are treated as negative revenues and should be subtracted from the totals.

Total public debt for all levels of Colorado government was $6.7 billion, or $2,134 per capita, during 1982/83.

36TAXATION

As of 1985, Colorado's state income tax ranged from 2.5% of the first $1,415 to 8% on income over $14,152. The corporate income tax was 5% of net income, based on a formula taking into consideration both total profits and those derived solely from state sources. The state also imposed a 3% sales and use tax, along with taxes on cigarettes, alcoholic beverages, pari-mutuel racing, fossil fuel production, motor fuel sales, and insurance premiums. Property taxes are the major source of revenue for local governments. Colorado municipalities are also allowed to levy sales and use taxes.

In 1982, Colorado ranked 23d in federal taxes paid, with almost $8.3 billion. Coloradans filed 1,367,273 federal income tax returns in 1983, paying a total of almost $4.3 billion.

37ECONOMIC POLICY

Colorado provides funds for development-related public works and recreational projects, training programs, and programs to encourage new industry. Tax incentives include a reduction of taxes on machinery used in manufacturing.

38HEALTH

Colorado's birthrate was slightly above the US average in 1981, and the infant mortality rate (10.1 for whites and 11.4 for nonwhites) was well below it. There were 25,200 legal abortions in 1982.

Colorado's death rate from all causes was 25% below the national average; specific death rates for heart disease, cancer, cerebrovascular diseases, diabetes, and liver diseases were far below the US norm, while those for accidents and suicide were considerably above it.

In 1983, Colorado had 98 hospitals, with 15,267 beds; hospital personnel included 7,232 full-time and 4,003 part-time registered nurses. In 1982, the average cost per day in Colorado hospitals, $336, was above the US average. The state had 6,803 licensed physicians and 1,929 practicing dentists in 1982. The state's only medical school is the University of Colorado Medical Center in Denver.

39SOCIAL WELFARE

During 1983, 3.6% of all Coloradans received public aid. In 1982, 76,735 individuals were recipients of aid to families with dependent children, at a federal cost of $87.9 million. Some 182,000 Coloradans received food stamps in 1983, at a federal cost of $101 million. The school lunch program served 274,000 pupils that year.

Social Security benefits exceeding $1.5 billion were paid to 347,000 Coloradans in 1983, including retirement benefits of $1.1 billion. In 1982, $203 million was spent on unemployment insurance payments. Workers' compensation benefits totaled $180.5 million that year.

40HOUSING

The 1980 census counted 1,176,309 year-round housing units, an increase of nearly 58% from 1970. Of the total, 1,061,249 units, or 90%, were occupied—684,408 by owners and 376,841 by renters. More than 98% of the occupied units had full plumbing. From 1981 through 1983, 112,500 new privately owned housing units worth more than $4.7 billion were authorized.

41EDUCATION

Colorado residents are better educated than the average American. According to the 1980 census, 23% of the adult population of Colorado had completed four years of college, ranking 1st among the 50 states; almost 79% of all adult Coloradans were high school graduates (3d in the US).

In fall 1984, Colorado's 1,287 public elementary and secondary schools had 545,427 pupils. There were 28,824 elementary and secondary school teachers that year, with an average salary of $22,900. Nonpublic school enrollment was about 35,000 for all grades.

More than 172,000 students were enrolled in 47 colleges and universities in the fall of 1982. The oldest state school is the Colorado School of Mines, founded in Golden in 1869; it had a 1983 enrollment of 2,931. Although chartered in 1861, the University of Colorado did not open until 1876; its Boulder campus is now the largest in the state, with a fall 1983 enrollment of 22,180. Colorado State University, founded at Ft. Collins in 1870, had 18,295 students in 1983. The University of Denver, chartered in 1864 as the Colorado Seminary of the Methodist Episcopal Church, had 8,472 students.

42ARTS

From its earliest days of statehood, Colorado has been receptive to the arts. Such showplaces as the Tabor Opera House in Leadville and the Tabor Grand Opera House in Denver were among the most elaborate buildings in the Old West. Newer centers are Denver's Boettcher Concert Hall, which opened in 1978 as the home of the Denver Symphony, and the adjacent Helen G. Bonfils Theater Complex, which opened in 1980 and houses a repertory theater company.

Other artistic organizations include the Colorado Springs Symphony and Colorado Opera Festival of Colorado Springs; the Central City Opera House Association, which sponsors a summer opera season in this old mining town; and the Four Corners Opera Association in Durango. Aspen is an important summer music center. The amphitheater in Red Rocks Park near Denver, formed by red sandstone rocks, provides a natural and acoustically excellent concert area.

43LIBRARIES AND MUSEUMS

In 1982, public libraries in the state held nearly 6 million volumes and circulated more than 14 million. The largest system was the

Denver Public Library, with 1.8 million volumes in 30 branches. The leading academic library is at the University of Colorado at Boulder, with nearly 2 million volumes.

Colorado has more than 130 museums and historic sites. One of the most prominent museums in the West is the Denver Art Museum, with its large collection of American Indian, South Seas, and Oriental art. Another major art museum is the Colorado Springs Fine Arts Center, specializing in southwestern and western American art. Other notable museums include the Denver Museum of Natural History, University of Colorado Museum in Boulder, Western Museum of Mining and Industry in Colorado Springs, and the Colorado Ski Museum–Ski Hall of Fame in Vail. Museums specializing in state history include the Colorado Heritage Center of the Colorado Historical Society in Denver, Ute Indian Museum in Montrose, Ft. Carson Museum of the Army in the West, Bent's Old Fort National Historic Site in La Junta, Georgetown–Silver Plume Historic District, Healy House–Dexter Cabin and Tabor Opera House Museum in Leadville, and Ft. Vasquez in Platteville.

44 COMMUNICATIONS

Colorado's first mail and freight service was provided in 1859 by the Leavenworth and Pikes Peak Express. As of 1985 there were 10,673 postal workers in the state, and 413 post offices. Over 94% of the state's 1,061,249 occupied housing units had telephones in 1980. Of the 166 radio stations in operation in 1984, 80 were AM and 86 FM. There were 15 commercial and 3 educational television stations; 98 cable television systems served 392,542 subscribers in 199 communities.

45 PRESS

As of 1984 there were 8 morning dailies with a circulation of 738,556, 19 afternoon dailies with a circulation of 210,151, and 10 Sunday papers with a circulation of 1,047,535. The leading newspapers were the *Rocky Mountain News,* 317,638 mornings and 366,003 Sundays; and the *Denver Post,* 243,303 mornings and 354,400 Sundays.

46 ORGANIZATIONS

The 1982 Census of Service Industries counted 1,012 organizations in Colorado, including 208 business associations; 540 civic, social, and fraternal associations; and 34 educational, scientific, and research associations.

Among the environmental associations located in the state are the National Environmental Health Association and the American Humane Association, both in Denver. The International Association of Meteorology and Atmospheric Physics and the American Solar Energy Society are in Boulder. Among the many professional and trade groups in the state are the Geological Society of America in Boulder; National Cattlemen's Association in Englewood; and the American School Food Service Association, American Sheep Producers Council, College Press Service, and National Livestock Producers Association, all in Denver. Colorado Springs is the home of several important sports organizations, including the US Olympic Committee, Amateur Basketball Association of the USA, Amateur Hockey Association of the US, Professional Rodeo Cowboys Association, and the US Ski Association. The Sports Car Club of America is in Englewood.

47 TOURISM, TRAVEL, AND RECREATION

Scenery, history, and skiing combine to make Colorado a prime tourist mecca. In 1984, travel and tourism generated over $4 billion in expenditures in the state.

Ski-related revenue for 1983/84 was $1.2 billion. Vail is the most popular ski resort center, followed by Keystone and Steamboat. Skiing aside, the state's most popular attraction is the US Air Force Academy near Colorado Springs. Nearby are Pikes Peak, the Garden of the Gods (featuring unusual red sandstone formations), and Manitou Springs, a resort center. Besides its many museums, parks, and rebuilt Larimer Square district, Denver's main attraction is the US Mint.

All nine national forests in Colorado are open for camping, as are the state's two national parks: Rocky Mountain, encompassing 263,791 acres (106,753 hectares) in the Front Range; and Mesa Verde, 52,085 acres (21,078 hectares) of mesas and canyons in the southwest. Other attractions include the fossil beds at Dinosaur National Monument, Indian cliff dwellings at Mesa Verde, Black Canyon of the Gunnison, Colorado National Monument at Fruita, Curecanti National Recreation Area, Florissant Fossil Beds, Great Sand Dunes, Hovenweep National Monument, Durango-Silverton steam train, and white-water rafting on the Colorado, Green, and Yampa rivers. In 1982/83, licenses were issued to 358,823 hunters and 715,550 anglers.

48 SPORTS

There are two major league professional sports teams in Colorado, both in Denver: the Broncos of the National Football League and the Nuggets of the National Basketball Association. Denver's minor league baseball team competes in the American Association.

49 FAMOUS COLORADANS

Ft. Collins was the birthplace of Byron R. White (b.1917), who, as an associate justice of the US Supreme Court since 1962, has been the state's most prominent federal officeholder. Colorado's first US senator, Henry M. Teller (b.New York, 1830-1914), also served as secretary of the interior. Gary Hart (b.Kansas, 1937) has been a senator since 1975.

Charles Bent (b.Virginia, 1799-1847), a fur trapper and an early settler in Colorado, built a famous fort and trading post near present-day La Junta. Early explorers of the Colorado region include Zebulon Pike (b.New Jersey, 1779-1813) and Stephen Long (b.New Hampshire, 1784-1864). John Evans (1814-97) was Colorado's second territorial governor and founder of the present-day University of Denver. Ouray (1820-83) was a Ute chief who ruled at the time when mining districts were being opened. Silver magnate Horace Austin Warner Tabor (b.Vermont, 1830-99) served as mayor of Leadville and lieutenant governor of the state, spent money on lavish buildings in Leadville and Denver, but lost most of his fortune before his death. The story of Tabor and his second wife, Elizabeth McCourt Doe Tabor (1862-1935), is portrayed in Douglas Moore's opera *The Ballad of Baby Doe* (1956).

Willard F. Libby (1909-80), winner of the Nobel Prize for chemistry in 1960, and Edward L. Tatum (1909-75), co-winner of the 1958 Nobel Prize for physiology or medicine, were born in Colorado.

Among the performers born in the state were actors Lon Chaney (1883-1930) and Douglas Fairbanks (1883-1939), and band leader Paul Whiteman (1891-1967). Singer John Denver (Henry John Deutschendorf, Jr., b.New Mexico, 1943) is closely associated with Colorado and lives in Aspen.

Colorado's most famous sports personality is Jack Dempsey (1895-1983), born in Manassa and nicknamed the "Manassa Mauler," who held the world heavyweight boxing crown from 1919 to 1926.

50 BIBLIOGRAPHY

Abbott, Carl. *Colorado: A History of the Centennial State.* Boulder: Colorado Associated University Press, 1976.

Athearn, Robert G. *The Coloradans.* Albuquerque: University of New Mexico Press, 1982.

Casewit, Curtis W. *Colorado.* New York: Viking, 1973.

Federal Writers' Project. *Colorado: A Guide to the Highest State.* Reprint. New York: Somerset, 1980 (orig. 1941).

Sprague, Marshall. *Colorado: A Bicentennial History.* New York: Norton, 1976.

Ubbelohde, Carl, Maxine Benson, and Duane A. Smith. *A Colorado History.* 5th ed. Boulder: Pruett, 1982.

Walton, Roger A. *Colorado: A Practical Guide to Its Government and People.* Ft. Collins: Publishers Consultants, 1976.

CONNECTICUT

State of Connecticut

ORIGIN OF STATE NAME: From the Mahican word *quinnehtukqut,* meaning "beside the long tidal river." **NICKNAME:** The Constitution State. (Also: the Nutmeg State.) **CAPITAL:** Hartford. **ENTERED UNION:** 9 January 1788 (5th). **SONG:** "Yankee Doodle." **MOTTO:** *Qui transtulit sustinet* (He who transplanted still sustains). **COAT OF ARMS:** On a rococo shield, three grape vines, supported and bearing fruit, stand against a white field. Beneath the shield is a streamer bearing the state motto. **FLAG:** The coat of arms appears on a blue field. **OFFICIAL SEAL:** The three grape vines and motto of the arms surrounded by the words *Sigillum reipublicae Connecticutensis* (Seal of the State of Connecticut). **ANIMAL:** Sperm whale. **BIRD:** American robin. **FLOWER:** Mountain laurel. **TREE:** White oak. **MINERAL:** Garnet. **INSECT:** European praying mantis. **SHIP:** USS *Nautilus.* **LEGAL HOLIDAYS:** New Year's Day, 1 January; Birthday of Martin Luther King, Jr., 3d Monday in January; Lincoln's Birthday, 12 February; Washington's Birthday, 3d Monday in February; Good Friday, March or April; Memorial Day, last Monday in May; Independence Day, 4 July; Labor Day, 1st Monday in September; Columbus Day, 2d Monday in October; Veterans Day, 11 November; Thanksgiving Day, 4th Thursday in November; Christmas Day, 25 December. **TIME:** 7 AM EST = noon GMT.

¹LOCATION, SIZE, AND EXTENT

Located in New England in the northeastern US, Connecticut ranks 48th in size among the 50 states.

The state's area, 5,018 sq mi (12,997 sq km), consists of 4,872 sq mi (12,619 sq km) of land and 146 sq mi (378 sq km) of inland water. Connecticut has an average length of 90 mi (145 km) E–W, and an average width of 55 mi (89 km) N–S.

Connecticut is bordered on the N by Massachusetts; on the E by Massachusetts and Rhode Island (with part of the line formed by the Pawcatuck River); on the s by New York (with the line passing through Long Island Sound); and on the w by New York. On the sw border, a short panhandle of Connecticut territory juts toward New York City. The state's geographic center is East Berlin in Hartford County. Connecticut has a boundary length of 328 mi (528 km) and a shoreline of 253 mi (407 km).

²TOPOGRAPHY

Connecticut is divided into four main geographic regions. The Connecticut and Quinnipiac river valleys form the Central Lowlands, which bisect the state in a north–south direction. The Eastern Highlands range from 500 feet (150 meters) to 1,100 feet (335 meters) near the Massachusetts border and from 200 feet (60 meters) to 500 feet (150 meters) in the southeast. Elevations in the Western Highlands, an extension of the Green Mountains, range from 200 feet (60 meters) in the south to more than 2,000 feet (600 meters) in the northwest; within this region, near the Massachusetts border, stands Mt. Frissell, the highest point in the state at 2,380 feet (725 meters). The Coastal Lowlands, about 100 mi (160 km) long and generally 2–3 mi (3–5 km) wide, consist of rocky peninsulas, shallow bays, sand and gravel beaches, salt meadows, and good harbors at Bridgeport, New Haven, New London, Mystic, and Stonington.

Connecticut has more than 6,000 lakes and ponds. The two largest bodies of water—both artificial—are Lake Candlewood, covering about 5,000 acres (2,000 hectares), and Barkhamsted Reservoir, a major source of water for the Hartford area. The main river is the Connecticut, New England's longest river at 407 mi (655 km), of which 69 mi (111 km) lie within Connecticut; this waterway, which is navigable as far north as Hartford by means of a 15-foot (5-meter) channel, divides the state roughly in half before emptying into Long Island Sound. Other principal rivers include the Thames, Housatonic, and Naugatuck.

Connecticut's bedrock geology and topography are the product of a number of forces: uplift and depression, erosion and deposit, faulting and buckling, lava flows, and glaciation. About 180 million years ago, the lowlands along the eastern border sank more than 10,000 feet (3,000 meters); the resultant trough or fault extends from northern Massachusetts to New Haven Harbor and varies in width from about 20 mi (32 km) to approximately 4 mi (6 km). During the Ice Ages, the melting Wisconsin glacier created lakes, waterfalls, and sand plains, leaving thin glaciated topsoil and land strewn with rocks and boulders.

³CLIMATE

Connecticut has a generally temperate climate, with mild winters and warm summers. The January mean temperature is 27°F (−3°C) and the July mean is 70°F (21°C). Coastal areas have warmer winters and cooler summers than the interior. Norfolk, in the northwest, has a January mean temperature of 22°F (−6°C) and a July mean of 66°F (19°C), while Bridgeport, on the shore, has a mean of 30°F (−1°C) in January and of 71°F (22°C) in July. The highest recorded temperature in Connecticut was 105°F (41°C) in Waterbury on 22 July 1926; the lowest, −32°F (−36°C) in Falls Village on 16 February 1943. The annual rainfall is about 44–48 in (112–122 cm) and is evenly distributed throughout the year. The state receives some 25–60 in (64–150 cm) of snow each year, with heaviest snowfall in the northwest.

Weather annals reveal a remarkable range and variety of climatic phenomena. Severe droughts were experienced in 1749, 1762, 1929–33, the early 1940s, 1948–50, and 1956–57. The worst recent drought, which occurred in 1963–66, resulted in a severe forest-fire hazard, damage to crops, and rationing of water. Downtown Hartford was inundated by a flood in March 1936. On 21 September 1938, a hurricane struck west of New Haven and followed the Connecticut Valley northward, causing 85 deaths and property losses of more than $125 million. Severe flooding occurred in 1955 and again in 1982. In the latter year, property damage exceeded $266 million.

⁴FLORA AND FAUNA

Connecticut has an impressive diversity of vegetation zones. Along the shore of Long Island Sound are tidal marshes with salt grasses, glasswort, purple gerardia, and sea lavender. On slopes fringing the marshes are black grass, switch grass, marsh elder, and sea myrtle.

The swamp areas contain various ferns, abundant cattails, cranberry, tussock sedge, skunk cabbage, sweet pepperbush, spicebush, and false hellebore. The state's hillsides and uplands support a variety of flowers and plants, including mountain laurel (the state flower), pink azalea, trailing arbutus, Solomon's seal, and Queen Anne's lace. Endangered species in the state include showy lady's slipper, ginseng, showy aster, sickle-leaved golden aster, nodding pogonia, goldenseal, climbing fern, and chaffseed.

The first Englishmen arriving in Connecticut in the 1930s found a land teeming with wildlife. Roaming the forests and meadows were black bear, white-tailed deer, red and gray foxes, timber wolf, cougar, panther, raccoon, and enough rattlesnakes to pose a serious danger. The impact of human settlement on Connecticut wildlife has been profound, however. Only the smaller mammals—the woodchuck, gray squirrel, cottontail, eastern chipmunk, porcupine, raccoon, and striped skunk—remain common. Snakes remain plentiful but are mostly harmless, except for the northern copperhead and timber rattlesnake. Freshwater fish are abundant, and aquatic life in Long Island Sound even more so. Common birds include the robin (the state bird), blue jay, song sparrow, wood thrush, and many species of waterfowl; visible in winter are the junco, pine grosbeak, snowy owl, and winter wren.

Threatened or endangered wildlife listed by the US Department of the Interior include Atlantic and shortnose sturgeon, bog or Muhlenberg's turtle, American peregrine falcon, Indiana bat, bald eagle, and eastern cougar.

5ENVIRONMENTAL PROTECTION

The Department of Environmental Protection, established in 1971, has primary responsibility for state environmental programs. Its 1983/84 expenditures for all programs totaled $21.9 million, including federal grants, but the state continued to experience water and air pollution problems in the 1980s. By mid-1983, only 68% of the state's major rivers met the federal government's "swimmable-fishable" standards, and millions of gallons of sewage continued to be poured into Long Island Sound. Pollution of groundwater forced the state to adopt statewide water-quality standards in September 1980.

In the 1970s, Connecticut's air quality was poor, with high levels of carbon monoxide, sulfur dioxide, and ozone. The state was able to reduce sulfur dioxide levels beginning in 1971 by aggressive regulation of sulfur fuels burned in the state. Less progress was made in controlling other pollutants, however. In 1977, levels of carbon monoxide exceeded federal health standards at eight of nine sampling sites, and ozone levels in 1983, at two times the federal standard, were among the highest in the US. Automobiles produced 90% of the carbon monoxide and 50% of the hydrocarbons in the Connecticut environment. The vehicles emission program established by the state in January 1983, resulted in a decline in carbon monoxide and hydrocarbon pollution. Connecticut's other major air pollution problem is acid rain, much of which originates in New York and New Jersey.

In 1982/83, the Department of Environmental Protection inventoried 700 hazardous waste sites and began issuing permits to hazardous waste facilities; 1,320 inspections related to hazardous waste were carried out in 1983/84. Some 2.5 million tons of solid waste are generated in Connecticut each year, and the state's active 143 landfill sites were expected to be exhausted by 1986. Even though disposal of solid waste is a municipal problem, the Department of Environmental Protection has undertaken studies, including the investigation of mandatory recycling programs, to alleviate the problem.

Legislation encouraging the recycling of beverage containers and banning the use of "fliptop" cans with disposable tabs took effect in 1980.

6POPULATION

The 1980 census total for Connecticut was 3,107,564, representing 1.37% of the total US population. The state had a population gain of only 2.4% (75,359 residents) for the entire decade of the 1970s, compared with a US population growth of 11.4%. One sign of the population lag is that in 1980, Connecticut had the lowest birthrate in the US, 12.5 live births per 1,000 population.

In 1985, Connecticut had an estimated population of 3,160,280; the 1990 population is projected at 3,135,600. Connecticut ranked 5th among the 50 states in population density in 1985 with 649 persons per sq mi (251 per sq km). About 79% of all Connecticut residents lived in urban areas and 21% in rural areas.

Major cities with 1984 populations are Bridgeport, 142,140; Hartford, 135,720; New Haven, 124,188; Waterbury, 102,861; and Stamford, 101,917. Despite urban renewal projects, the three largest cities have lost population since 1960, largely because of the exodus of middle-class whites to the suburbs, which have increased rapidly in population. For example, Bloomfield, to the north of Hartford, gained in population from 5,700 in 1950 to 19,023 in 1984; and Trumbull, near Bridgeport, increased from 8,641 in 1950 to 33,285 in 1984. In the mid 1980s, Bridgeport, Hartford, and New Haven all had growing percentages of minority populations (with relatively low per capita incomes) and shrinking tax bases.

7ETHNIC GROUPS

Connecticut has large populations of second-generation European descent. The biggest groups came from Italy, Ireland, Poland, and Quebec, Canada. Most of these immigrants clustered in the cities of New Haven, Hartford, Bridgeport, and New London. The number of Roman Catholic newcomers drew the hostility of many native-born Connecticuters, particularly during the decade 1910–20, when state officials deported 59 "dangerous aliens" on scant evidence of radicalism and Ku Klux Klan chapters enrolled some 20,000 members.

Since 1950, ethnic groups of non-Yankee ancestry have exercised leadership roles in all facets of Connecticut life, especially in politics. Connecticut elected a Jewish governor in 1954, and its four subsequent governors were of Irish or Italian ancestry. A wave of newcomers to the state during and after World War II consisted chiefly of blacks and Hispanics seeking employment opportunities. In 1980, the black population numbered 217,433, about 7% of the state total. According to the 1980 federal census, there were also about 125,000 Hispanic residents, of whom 88,000 were Puerto Ricans. In addition, Connecticut had 4,822 American Indians and 19,000 Asians and Pacific Islanders. About 268,000 Connecticut residents, or 8.6% of the population, were foreign-born in 1980.

8LANGUAGES

Connecticut English is basically that of the Northern dialect, but features of the eastern New England subdialect occur east of the Connecticut River. In the east, *half* and *calf* have the vowel of *father;* *box* is /bawks/ and *cart* is /kaht/; *yolk* is /yelk/; *care* and *chair* have the vowel of *cat;* and many speakers have the intrusive /r/, as in *swaller it* (swallow it). In the western half, *creek* is /krik/; *cherry* may be /chirry/; *on* has the vowel of *father;* and /r/ is heard after a vowel, as in *cart.* Along the Connecticut River, *butcher* is /boocher/, and *tomorrow* is pronounced /tomawro/. Along the coast, the wind may be *breezing on,* and a *creek* is a saltwater inlet. The sycamore is *buttonball,* one is *sick to his stomach,* gutters are *eavestroughs,* a lunch between meals is a *bite,* and in the northwest, an earthworm is an *angledog.*

In 1980 about 2,560,000 Connecticuters (85% of the resident population 3 years old and older) spoke only English at home. Other languages spoken at home, and the number of people who spoke them, included:

Spanish	107,745	German	20,433
Italian	92,224	Greek	10,359
French	60,339	Hungarian	7,346
Polish	42,655	Ukrainian	4,939
Portuguese	21,485	Yiddish	4,632

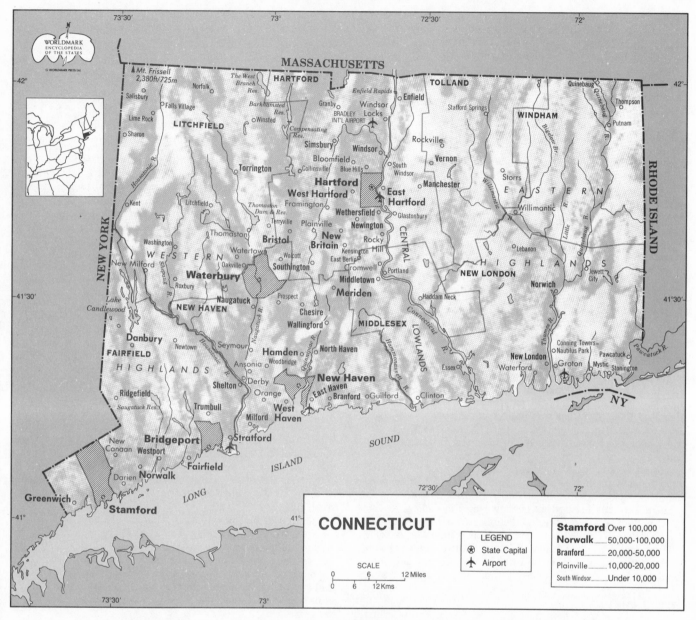

CONNECTICUT

LEGEND
⊗ State Capital
✈ Airport

SCALE
0 ⊢ 6 ⊢ 12 Miles
0 ⊢ 6 ⊢ 12 Kms

Stamford	Over 100,000
Norwalk	50,000-100,000
Branford	20,000-50,000
Plainville	10,000-20,000
South Windsor	Under 10,000

See endsheet maps: M2.

LOCATION: 40°57′ to 42°03′N; 71°47′05″ to 73°43′05″W. **BOUNDARIES:** Massachusetts line, 94 mi (151 km); Rhode Island line, 54 mi (87 km); New York line, 180 mi (290 km).

⁹RELIGIONS

Connecticut's religious development began in the 1630s with the designation of the Congregational Church as the colony's "established church." The Puritan fathers enacted laws decreeing church attendance on Sundays and other appointed days, and requiring all residents to contribute to the financial maintenance of local Congregational ministers. Educational patterns, business practices, social conduct, and sexual activities were all comprehensively controlled in accordance with Puritan principles. "Blue laws" provided penalties for offenses against God's word, such as profanation of the Sabbath and swearing, and capital punishment was mandated for adultery, sodomy, bestiality, lesbianism, harlotry, rape, and incest.

Connecticut authorities harassed and often persecuted such non-Congregationalists as Quakers, Baptists, and Anglicans. However, the church was weakened during the 18th century by increasing numbers of dissenters from the Congregational order. A coalition of dissenters disestablished the church by the Connecticut constitution of 1818. The final blow to Congregational domination came in the late 19th and early 20th centuries, with the arrival of many Roman Catholic immigrants.

Since World War I, Roman Catholics have been the most numerous religious group in the state. As of 1984 there were 1,336,000 Roman Catholics, representing about 42% of the total population. In 1980 there were 493,155 known adherents of Protestant denominations. Leading groups included Congregationalists (United Church of Christ), 130,993; Episcopalians, 107,324; and United Methodists, 61,196. In 1984, the state's estimated Jewish population was 107,575.

¹⁰TRANSPORTATION

Because of both the state's traditional conservatism and the opposition of turnpike and steamboat companies, rail service did not fully develop until the 1840s. Hartford and New Haven were connected in 1839, and the New Haven line was completed to Northampton, Mass., in 1850. In the late 1840s and the 1850s, a network of lines connected Hartford with eastern Connecticut communities. Railroad expansion peaked during the 1890s, when total trackage reached 1,636 mi (2,633 km). The giant in Connecticut railroading from the 1870s until its collapse in 1969 was the New York, New Haven, and Hartford Railroad. As of 1985, the state's rail passenger service was provided by Metro-North Commuter Railroad and Amtrak, which together encompassed 607 route mi (977 km) of track and direct service to 33 of the state's 169 cities and towns. Metro-North carried 23.3 million passengers in 1982/83.

Local bus systems provide intracity transportation. These services are generally subsidized by the state and, in some instances, by the federal Urban Mass Transportation Administration. Intercity bus service (not subsidized by the state or the federal government) is provided in 31 municipalities by 23 companies. In 1983/84, local and commuter bus services carried 42 million passengers.

Connecticut has an extensive system of expressways, state highways, and local roads, aggregating 19,543 mi (31,451 km) in 1983; 99% of the roads are either paved or hard-surfaced. Major highways include I-95, the Connecticut Turnpike, which crosses the entire length of the state near the shore; I-91, linking New Haven and Springfield, Mass.; I-84 from Hartford southwestward through Waterbury and Danbury to New York State; and I-86, connecting I-84 and the Massachusetts Turnpike. In 1983, following the collapse of a Connecticut Turnpike bridge in Greenwich, the Transportation Department initiated an inspection and repair program of the 3,737 highway bridges in the state. As of 1983 there were 2,149,351 automobiles, 146,757 trucks, 118,000 motorcycles, and 8,994 buses registered in the state. Connecticut had 2,249,525 licensed drivers during the same year.

Most of Connecticut's waterborne traffic is handled through the three major ports of New Haven, New London, and Bridgeport, and at the shallow-draft terminals in Norwalk and Stamford.

There were 105 licensed air landing places in Connecticut in 1983, including 54 airports, 43 heliports, 2 stolports, and 6 seaplane bases. Connecticut's principal air terminal is Bradley International Airport at Windsor Locks, located 14 mi (23 km) north of Hartford. Served by seven major airlines and nine commuter lines, Bradley handled 3.2 million passengers during 1983/84.

¹¹HISTORY

The first people known to have lived in the area now called Connecticut were Indians, whose forebears may have come to New England as many as 10,000 years ago. By the early 17th century, Connecticut had between 6,000 and 7,000 Indians organized into 16 tribes, all members of the loose Algonquian Confederation. The most warlike of these tribes were the Pequot, who apparently had migrated not long before from the Hudson River region to escape the Mohawk and had settled along the Connecticut coast. There was also a heavy concentration of Indian groups in the Connecticut River Valley, but fear of Mohawk hunting parties kept them from occupying most of western and northwestern Connecticut.

Because of their fear of the Pequot along the shore and of the Mohawk to the west, most of Connecticut's Indians sought the friendship of English newcomers in the 1630s. The Indians sold land to the English and provided instruction in New World agricultural, hunting, and fishing techniques. The impact of English settlers on Connecticut's friendly Indians was devasting, however. The Indians lost their land, were made dependents in their own territory, and were decimated by such European imports as smallpox and measles. The Pequot, who sought to expel the English from Connecticut by a series of attacks in

1636–37, were defeated during the Pequot War by a Connecticut-Massachusetts force, aided by a renegade Pequot named Uncas. By the 1770s, Connecticut's Indian population was less than 1,500.

The first recorded European penetration of Connecticut was in 1614 by the Dutch mariner Adriaen Block, who sailed from Long Island Sound up the Connecticut River, probably as far as the Enfield Rapids. The Dutch established two forts on the Connecticut River, but they were completely dislodged by the English in 1654.

The early English settlers were part of a great migration of some 20,000 English Puritans who crossed the treacherous Atlantic to New England between 1630 and 1642. The Puritans declared that salvation could be achieved only by returning to the simplicity of the early Christian Church and the truth of God as revealed in the Bible. They sailed to America in order to establish a new society that could serve as a model for the rest of Christendom. Attracted by the lushness of the Connecticut River Valley, the Puritans established settlements at Windsor (1633), Wethersfield (1634), and Hartford (1636). In 1639, these three communities joined together to form the Connecticut Colony, choosing to be governed by the Fundamental Orders, a relatively democratic framework for which the Reverend Thomas Hooker was largely responsible. (According to some historians, the Fundamental Orders comprised the world's first written constitution—hence the state nickname, adopted in 1959.) A separate Puritan colony was planted at New Haven in 1638 under the leadership of John Davenport, a Puritan minister, and Theophilus Eaton, a successful merchant.

In 1662, the Colony of Connecticut secured legal recognition by England. Governor John Winthrop, Jr., persuaded King Charles II to grant a charter that recognized Connecticut's existing framework of government and established its north and south boundaries as Massachusetts and Long Island Sound and its east and west borders as Narragansett Bay and the Pacific Ocean. In 1665, New Haven reluctantly became part of the colony because of economic difficulties and fear of incorporation into Anglican New York.

Connecticut had acrimonious boundary disputes with Massachusetts, Rhode Island, New York, and Pennsylvania. The most serious disagreement was with New York, which claimed the entire area from Delaware Bay to the Connecticut River. The issue was resolved in 1683 when the boundary was set 20 mi (32 km) east of and parallel to the Hudson River, although it was not until 1881 that Connecticut, New York, and Congress established the exact line.

Connecticut functioned throughout the colonial period much like an independent republic. It was the only American colony that generally did not follow English practice in its legislative proceedings, nor did it adopt a substantial amount of English common and statute law for its legal code. Connecticut's autonomy was threatened in 1687 when Sir Edmund Andros, appointed by King James II as the governor of the Dominion of New England, arrived in Hartford to demand surrender of the 1662 charter. Connecticut leaders protected the colony's autonomy by hiding the charter in an oak tree, which subsequently became a landmark known as the Charter Oak.

With its Puritan roots and historic autonomy, Connecticut was a Patriot stronghold during the American Revolution. Tories numbered no more than 7% of the adult male population—2,000 to 2,500 out of a total of 38,000 males. Connecticut sent some 3,600 men to Massachusetts at the outbreak of fighting at Lexington and Concord in April 1775. Jonathan Trumbull, who served as governor from 1769 to 1784, was the only colonial governor in office in 1775 who supported the Patriots. He served throughout the Revolutionary War, during which Connecticut troops participated in most of the significant battles. Connecticut's privateers captured more than 500 British merchant vessels, and its small but potent fleet captured at least 40 enemy ships. Connecticut also produced arms and gunpowder for state and Continental forces, thus beginning an arms-making tradition that would lead to the state's unofficial designation as the "arsenal of the nation." It was also called the Provisions State, in large part because of the crucial supplies of foodstuffs it sent to General George Washington throughout the war. The state's most famous Revolutionary War figure was Nathan Hale, executed as a spy by the British in New York City in 1776.

On 9 January 1788, Connecticut became the 5th state to ratify the Constitution. Strongly Federalist during the 1790s, Connecticut ardently disagreed with the foreign policy of Presidents Thomas Jefferson and James Madison, opposed the War of 1812, and even refused to allow its militia to leave the state. Connecticut's ire over the war was exacerbated by the failure of the government to offer significant help when the British attacked Essex and Stonington in the spring and summer of 1814. The politically vulnerable Federalists were defeated in 1817 by the Toleration Party. This coalition of Republicans and non-Congregationalists headed the drive for the new state constitution (1818) that disestablished the Congregational Church, a Federalist stronghold.

Long before the Civil War, Connecticut was stoutly antislavery. In the early years of independence, the general assembly enacted legislation providing that every black born after 1 March 1784 would be free at age 25. Connecticut had a number of antislavery and abolition societies whose members routed escaped slaves to Canada via the Underground Railroad. The state's pro-Union sentiment was reflected in the enormous support given to the Union war effort; some 55,000 Connecticut men served in the Civil War, suffering more than 20,000 casualties. Arms manufacturers such as Colt and Winchester produced desperately needed rifles and revolvers, and the state's textile, brass, and rubber firms turned out uniforms, buttons, ponchos, blankets, and boots for Union troops.

The contribution by Connecticut industries to the war effort signaled the state's emergence as a manufacturing giant. Its industrial development was facilitated by abundant waterpower, the growth of capital held by banks and insurance companies, a sophisticated transportation network, and, most important, the technological and marketing expertise of the people. The first American hat factory was established in Danbury in 1780, and the nation's brass industry had its roots in Naugatuck Valley between 1806 and 1809. Connecticut clocks became known throughout the world. Micah Rugg organized the first nut and bolt factory in Marion in 1840; Elias Howe invented the first practical sewing machine in Hartford in 1843; and the International Silver Co. of Meriden developed from the efforts of the Rogers family, which had devised a method of silver plating in 1847. Perhaps the most important figure in the development of Connecticut manufacturing was Eli Whitney, best known for inventing the cotton gin (1793). Whitney also developed a system of interchangeable parts while operating a firearms factory near New Haven from the late 1790s until he died in 1825.

Seventy-five years after Whitney's death, Connecticut was a leader in the production of hats, typewriters, electrical fixtures, machine tools, and hardware. The state's textile industry ranked 6th in the nation in 1900, with an annual output of $50 million. By 1904, Connecticut's firearms industry was producing four-fifths of the ammunition and more than one-fourth of the total value of all firearms manufactured by nongovernment factories in the US. These great strides in manufacturing transformed Connecticut from a rural, agrarian society in the early 1800s to an increasingly urban state.

The state's contribution to the Allied forces in World War I more than equaled its Civil War effort. Four Liberty Loan drives raised $437 million, more than from any other state. About 66,000 Connecticuters served in the armed forces, and the state's manufacturers produced 450,000 Enfield rifles, 45,000 Browning automatic rifles, 2 million bayonets, and much other war materiél. By 1917–18, four-fifths of Connecticut's industry was involved in defense production.

The prosperity sparked by World War I continued, for the most part, until 1929. During the 1920s, Connecticuters enjoyed a rising standard of living, as the state became a national leader in the production of specialty parts for the aviation, automotive, and electric power industries. However, from 1919 to 1929, Connecticut lost 14 of its 47 cotton mills to southern states.

The stock market crash of 1929 and the subsequent depression of the 1930s hit highly industrialized Connecticut hard. By the spring of 1932, the state's unemployed totaled 150,000, and cities such as Bridgeport fell deeply in debt. The economic reversal led to significant political change: the ousting of a business-oriented Republican administration, which had long dominated the state, by a revitalized Democratic Party under the leadership of Governor Wilbur L. Cross (1931–39). During his tenure, Connecticut reorganized its state government, improved facilities in state hospitals and penal institutions, and tightened state regulation of business.

Connecticut was pulled out of the unemployment doldrums in 1939, when the state's factories were once again stimulated by defense contracts. The value of war contracts placed in Connecticut was $8 billion by May 1945, and industrial employment increased from 350,000 in 1939 to 550,000 by late 1944. Connecticut's factories turned out submarines, Navy Corsair fighter aircraft, helicopters, 80% of all ball bearings manufactured in the US, and many thousands of small arms. Approximately 220,000 Connecticut men and women served in the US armed forces.

Since 1945, Connecticut has seen substantial population growth, economic diversification with a greater proportion of service industries, the expansion of middle-class suburbs, and an influx of black and Hispanic migrants to the major cities. Urban renewal projects in Hartford and New Haven have resulted in expanded office and recreational facilities, but not much desperately needed new housing. A major challenge facing Connecticut in the 1980s is once again how to effect the social and economic integration of this incoming wave of people and industries.

[12]STATE GOVERNMENT

Connecticut has been governed by four basic documents: the Fundamental Orders of 1639; the Charter of the Colony of Connecticut of 1662; the constitution of 1818 (which remained in effect until 1964, when a federal district court, acting on the basis of the US Supreme Court's "one person, one vote" ruling, ordered Connecticut to reapportion and redistrict its legislature); and the constitution of 1965. This last document adjusted representation to conform with population, and provided for mandatory reapportionment every 10 years.

The state legislature is the general assembly, consisting of a 36-member senate and 151-member house of representatives. Legislators, who must be 18 years old and qualified voters in Connecticut, are elected to both houses for two-year terms from single-member districts of substantially equal populations.

Elected members of the executive branch are the governor and lieutenant governor (who run jointly and must each be at least 30 years of age), secretary of state, treasurer, comptroller, and attorney general. All are elected for four-year terms and may be reelected. The governor, generally with the advice and consent of the general assembly, selects the heads of state departments, commissions, and offices.

A bill becomes law when approved by both houses of the general assembly and signed by the governor. If the governor fails to sign it within 5 days when the legislature is in session, or within 15 days when it has adjourned, the measure also becomes law. A bill vetoed by the governor may be overriden by a two-thirds vote of the members of each house.

A constitutional amendment may be passed in a single legislative session if approved by three-fourths of the total membership of each house. If approved in one session by a majority but by less than three-fourths, the proposed amendment requires approval by majority vote in the next legislative session following a general election. After passage by the legislature, the amendment must be ratified by the voters in the next even-year general election in order to become part of the state constitution.

To vote in state elections, residents must be US citizens, at least 18 years of age, and must satisfy a 21-day registration requirement.

13 POLITICAL PARTIES

Connecticut's major political groups during the first half of the 19th century were successively the Federalist Party, the Democratic-Republican coalition, the Democrats, and the Whigs. The political scene also included a number of minor political parties—the Anti-Masonic, Free Soil, Temperance, and Native American (Know-Nothing) parties—of which the Know-Nothings were the most successful, holding the governorship from 1855 to 1857. The Whig Party collapsed during the controversy over slavery in the 1850s, when the Republican Party emerged as the principal opposition to the Democrats.

From the 1850s to the present, the Democratic and Republican parties have dominated Connecticut politics. The Republicans held power in most of the years between the Civil War and the 1920s. Republican hegemony ended in 1930, when the Democrats elected Wilbur L. Cross as governor. Cross greatly strengthened the Connecticut Democratic Party by supporting organized labor and providing social legislation for the aged and the needy. The success of the increasingly liberal Democrats in the 1930s prodded Connecticut Republicans to become more forward-looking, and the two parties were fairly evenly matched between 1938 and 1954. Connecticut's Democrats have held power in most years since the mid-1950s. As of October 1984, Democratic Party registration in Connecticut was 718,772; Republican, 477,749. There were more than 500,000 unaffiliated voters registered in the state.

In the November 1984 elections, Republican Ronald Reagan carried the state with 61% of the popular vote; Walter Mondale won almost 39%. As a result of Reagan's landslide, Connecticut Republicans won control of both houses of the general assembly for only the second time in 28 years.

14 LOCAL GOVERNMENT

As of 1984, Connecticut had 8 counties, 21 cities, 169 towns, and 11 boroughs. Counties in Connecticut have been geographical subdivisions without governmental functions since county government was abolished in 1960.

Connecticut's cities generally use the council-manager or mayor-council forms of government. The council-manager system provides for an elected council that determines policy, enacts local legislation, and appoints the city manager. The mayor-council system employs an elected chief executive with extensive appointment power and control over administrative agencies.

In most towns, an elected, three-member board of selectmen heads the administrative branch; the town meeting, in which all registered voters may participate, is the legislative body. Boroughs are generally governed by an elected warden, and borough meetings exercise major legislative functions.

15 STATE SERVICES

The Department of Education administers special programs for the educationally disadvantaged, the emotionally and physically disabled, and non–English-speaking students. The Department of Transportation operates state-owned airports, oversees bus system operations, and provides for snow removal from state highways and roads. The Department of Human Resources has a variety of social programs for state residents, including special services for the physically disabled. The Department of Children and Youth Services investigates cases of child abuse and adminis-

Connecticut Presidential Vote by Political Parties, 1948–84

YEAR	ELECTORAL VOTE	CONNECTICUT WINNER	DEMOCRAT	REPUBLICAN	PROGRESSIVE	SOCIALIST
1948	8	Dewey (R)	423,297	437,754	13,713	6,964
1952	8	*Eisenhower (R)	481,649	611,012	1,466	2,244
1956	8	*Eisenhower (R)	405,079	711,837	—	—
1960	8	*Kennedy (D)	657,055	565,813	—	—
1964	8	*Johnson (D)	826,269	390,996	—	—
					AMERICAN IND.	
1968	8	Humphrey (D)	621,561	556,721	76,660	—
						AMERICAN
1972	8	*Nixon (R)	555,498	810,763	—	17,239
						US LABOR
1976	8	Ford (R)	647,895	719,261	7,101	1,789
					LIBERTARIAN	CITIZENS
1980	8	*Reagan (R)	541,732	677,210	8,570	6,130
					CONN–ALLIANCE	COMMUNIST
1984	8	*Reagan (R)	569,597	890,877	1,374	4,826

* Won US presidential election.

ters programs dealing with child protection, adoption, juvenile corrections and rehabilitation, and prevention of delinquency. The Department on Aging has state and regional ombudsmen to handle problems involving nursing homes. Among programs sponsored by the Health Services Department are ones that help people to stop smoking, increase their nutritional awareness, and improve their dental health. The Labor Department provides a full range of services to the unemployed, to jobseekers, and to disadvantaged workers. Other departments deal with consumer protection, economic development, environmental protection, housing, mental retardation, and public safety.

[16] JUDICIAL SYSTEM

Connecticut's judicial system has undergone significant streamlining in recent years, with the abolition of municipal courts (1961), the circuit court (1974), the court of common pleas (1978), and the juvenile court (1978), and the creation of an appellate court (1983). Currently, the Connecticut judicial system consists of the supreme court, appellate court, superior court, and probate courts.

The supreme court comprises the chief justice, five associate justices, and two senior associate justices. The high court hears cases on appeal, primarily from the appellate court but also from the superior court in certain special instances, including the review of a death sentence, reapportionment, election disputes, invalidation of a state statute, or censure of a probate judge. Justices of the supreme court, as well as appellate and superior court judges, are nominated by the governor and appointed by the general assembly for eight-year terms.

The superior court, the sole general trial court, has the authority to hear all legal controversies except those over which the probate courts have exclusive jurisdiction. The superior court sits in 11 state judicial districts and is divided into trial divisions for civil, criminal, and family cases. As of 1985 there were 131 superior court trial judges.

Connecticut has 131 probate courts. These operate on a fee basis, with judges receiving their compensation from fees paid for services rendered by the court. Each probate district has one probate judge, elected for a four-year term.

Connecticut had 10 correctional institutions with an inmate population of 5,718 at the end of 1984; funds for a new state prison were appropriated in 1985/86. State law provides for the death penalty (by electrocution), but as of 1985 there had been no executions in the state for nearly 25 years. Crime rates in 1983 were below the national averages in every category of violent crime, and for most nonviolent crimes except motor vehicle theft.

[17] ARMED FORCES

The principal military installation in Connecticut is the US Navy Submarine Base at Groton, with 6,531 military and civilian personnel in 1983/84. Across the Thames River in New London is one of the nation's four service academies—the US Coast Guard Academy. Founded in 1876 and located at its present site since 1932, this institution offers a four-year curriculum leading to a B.S. degree and a commission as ensign in the Coast Guard. Women were first admitted to the academy in 1976 as members of the class of 1980.

Connecticut ranks first among the 50 states in value of defense contracts per capita. In 1983, the total value of defense contracts was $5.1 billion (7th in the US), and direct defense-oriented business accounted for 10% of the gross state product.

As of 30 September 1983, some 411,000 veterans were living in Connecticut, of whom 4,000 were veterans of World War I, 172,000 of World War II, 78,000 of the Korean conflict, and 103,000 of the Viet-Nam era. Veterans' benefits in 1982/83 totaled $245 million.

In 1985, the Connecticut National Guard had a total personnel of 6,225. In 1983 there were 6,585 police officers in Connecticut, including all state police.

[18] MIGRATION

Connecticut has experienced four principal migrations: the arrival of European immigrants in the 17th century, the out-migration of many settlers to other states beginning in the 18th century, renewed European immigration in the late 19th century, and the intrastate migration of city dwellers to the suburbs since 1945.

Although the first English settlers found an abundance of fertile farmland in the Connecticut Valley, later newcomers were not so fortunate. It is estimated that in 1800, when Connecticut's population was 250,000, nearly three times that many people had moved away from the state, principally to Vermont, western New York, Ohio, and other midwestern states.

The influx of European immigrants increased the number of foreign-born in the state from 38,518 in 1850 to about 800,000 by World War I. After World War II, the rush of middle-class whites (many from neighboring states) to Connecticut suburbs, propelled in part by the "baby boom" that followed the war, was accompanied by the flow of minority groups to the cities. All told, Connecticut had a net increase from migration of 561,000 between 1940 and 1970, followed by a net loss of 132,000 from 1970 to 1983.

[19] INTERGOVERNMENTAL COOPERATION

Among the regional interstate agreements to which Connecticut belongs are the Atlantic States Marine Fisheries Compact, the New England Higher Education Compact, and the New England Corrections Compact. Boundary agreements are in effect with Massachusetts, New York, and Rhode Island. Connecticut participates with New York in the Railroad Passenger Transportation Compact.

In 1983/84, federal aid to Connecticut totaled $1.2 billion.

[20] ECONOMY

Connecticut has had a strong economy since the early 19th century, when the state, unable to support its population by farming, turned to a variety of nonagricultural pursuits. Shipbuilding and whaling were major industries: in the 1840s and 1850s, New London ranked behind only New Bedford and Nantucket, Mass., among US whaling ports. Connecticut has been a leader of the insurance industry since the 1790s.

Connecticut's most important economic pursuit is manufacturing. In 1984, Connecticut was a leader in the manufacture of aircraft engines and parts, bearings, hardware, submarines, helicopters, typewriters, electronic instrumentation, electrical equipment, guns and ammunition, and optical instruments.

Because defense production has traditionally been important to the state, the economy has fluctuated with the rise and fall of international tensions. Connecticut's unemployment rate stood at 8.7% in 1949, dropped to 3.5% in 1951 during the Korean conflict, and rose sharply after the war to 8.3% in 1958. From 1966 to 1968, during the Viet-Nam war, unemployment averaged between 3.1% and 3.7%, but the rate subsequently rose to 9.5% in 1976. In 1984, Connecticut's unemployment rate dropped below 5%, becoming the lowest in the country. Connecticut has lessened its dependence on the defense sector somewhat by attracting nonmilitary domestic and international firms to the state. As of 1984, more than 250 international companies employed more than 30,000 Connecticut workers.

[21] INCOME

Connecticut is one of the wealthiest states, ranking 2d (behind Alaska) in per capita personal income. On a per capita basis, the average Connecticuter received $14,826 in 1983, nearly 27% above the US average. From 1980 to 1983, total personal income in the state increased by nearly 30%, from $35.9 billion to $46.5 billion.

Only 8% of all state residents and 6.2% of Connecticut families were below the federal poverty level in 1979. In 1982, about 46,600 Connecticuters were among the nation's top wealthholders (gross assets greater than $500,000), with combined assets of $49.3 billion.

22LABOR

The state's civilian labor force in 1984 totaled 1,672,000, of whom 95.4% were employed and 4.6% unemployed. The unemployment rate for all males in 1983 was 6.4%, and for females 5.5%.

A federal census in March 1983 revealed the following nonfarm employment pattern in Connecticut:

	ESTABLISH-MENTS	EMPLOYEES	ANNUAL PAYROLL ('000)
Agricultural services, forestry, fishing	1,029	4,085	$ 58,829
Mining	100	1,980	66,867
Contract construction	7,074	52,717	1,274,057
Manufacturing, of which:	8,533	415,433	9,712,546
Fabricated metal products	(948)	(41,738)	(824,357)
Nonelectrical machinery	(1,420)	(49,738)	(1,128,255)
Transportation equipment	(191)	(83,852)	(2,215,171)
Transportation, public utilities	2,348	60,558	1,454,002
Wholesale trade	5,458	72,541	1,719,108
Retail trade	20,181	223,952	2,262,733
Finance, insurance, real estate, of which:	6,442	117,568	2,518,843
Insurance	(1,693)	(65,874)	(1,526,293)
Services	24,587	300,039	4,531,314
Other	2,256	4,413	104,885
TOTALS	78,008	1,253,286	$23,703,184

Among categories of workers excluded by this survey is government workers, of whom Connecticut had 21,000 federal in 1982 and 159,000 state and local in 1983.

During the early 20th century, Connecticut was consistently antiunion and was one of the leading open-shop states in the northeastern US. But great strides were made by organized labor in the 1930s with the support of New Deal legislation recognizing union bargaining rights. In 1980, labor union membership totaled 327,000, and in 1983 there were 698 unions.

23AGRICULTURE

Agriculture is no longer of much economic importance in Connecticut. The number of farms declined from 22,241 in 1945 to 4,300 in 1984, when the average farm covered 116 acres (47 hectares). Cash receipts from all agricultural sales totaled $320.6 million in 1983.

Cash receipts from crop sales in 1983 were $118.3 million. Connecticut ranked 15th in value of tobacco production—$15.6 million. Other principal crops are hay, silage, greenhouse and nursery products, potatoes, sweet corn, tomatoes, apples, pears, and peaches.

24ANIMAL HUSBANDRY

Cash farm income from the sale of livestock and livestock products totaled $202.3 million in 1983, or 63% of the state's total farm production. In 1984 there were 106,000 cattle on Connecticut farms; marketings of cattle and dairy products totaled $108.2 million in 1983. Also during 1983, poultry farmers received $81.3 million from the sale of eggs, $1.6 million for chickens, and $553,000 for turkeys. About 654 million lb of milk were produced on 594 dairy farms.

25FISHING

Commercial fishing does not play a major role in the economy. In 1984, the value of commercial landings was $13.5 million for a catch of 7,771,000 lb of edible finfish and lobster.

During 1983/84, the Conservation and Preservation Division of the Department of Environmental Protection released 692,952 catchable trout, 273,000 kokanee fry, and 82,858 Atlantic salmon smelts into Connecticut waters.

26FORESTRY

By the early 20th century, the forests that covered 95% of Connecticut in the 1630s were generally destroyed. Woodland recovery has been stimulated since the 1930s by an energetic reforestation program. More than half of the state's 1,861,000 acres (753,000 hectares) of forestland in 1984 was wooded with new growth. In 1984, about 1.7 million tree and shrub seedlings were sold by the state nursery. Shipments of lumber and wood products were valued at $118.4 million in 1982; shipments of paper and paper products were valued at $940.1 million.

State woodlands include 87 state parks and 30 state forests covering some 167,343 acres (67,722 hectares).

27MINING

In 1984, Connecticut ranked only 43d among the 50 states in value of nonfuel mineral production, which totaled $75.3 million.

Formerly, there were working iron mines at Salisbury, Sharon, and Kent (iron from these mines was used for cannon and shot during the Revolutionary and Civil wars); copper mines at Granby, Bristol, and Cheshire; nickel mines at Litchfield; and granite quarries and garnet mines at Roxbury. Between the Civil War and World War I, a superior grade of red sandstone was quarried at Portland and barged via the Connecticut River and Long Island Sound to become the "brownstone fronts" of New York City.

At present, the state's commercial mining is limited to clay, lime, sand and gravel, stone, feldspar, and mica. Traprock is crushed for use in building and highway construction.

28ENERGY AND POWER

In 1980, Connecticut's fuel bill was $4.7 billion, of which 63% was for petroleum products, 8% for natural gas, and 29% for electricity. In 1982, prices were 62% higher than the national average for natural gas, 35% higher for electricity, and about 3% higher for petroleum products.

Production of electricity increased from 20 billion kwh in 1970 to 23.9 billion kwh in 1983; installed capacity increased from 4.6 million kw to 6.1 million kw during the same years. The use of coal to generate electric power declined from 85% of the total fuel used in 1965 to zero in 1983 because of the increased utilization of nuclear energy and oil, each of which accounted for about half of all electrical production in 1983. As of 1984, Connecticut had three nuclear reactors, one at Haddam Neck (575,000 kw) and two at Waterford (1,490,000 kw). Electricity consumption increased from 36.2 trillion Btu in 1965 to 72.3 trillion Btu in 1982.

Having no petroleum or gas resources of its own, Connecticut must rely primarily on imported oil from Saudi Arabia, Venezuela, Nigeria, and other countries. About 90% of the natural gas used in Connecticut is piped in from Texas and Louisiana.

29INDUSTRY

Connecticut, one of the most highly industrialized states, is a leading producer of aircraft engines and submarine parts, with the largest employers being United Technologies (its Pratt and Whitney Aircraft Division has headquarters in East Hartford) and the General Dynamics Corp. Electric Boat Division in Groton. The state is also a leading producer of military and civilian helicopters.

The state's value of shipments of manufactured goods totaled $30.1 billion in 1982. The following table shows the value of shipments for Connecticut's major industries in that year:

Transport equipment	$7,417,300,000
Nonelectrical machinery	3,476,400,000
Electric and electronic equipment	3,282,900,000
Fabricated metal products	2,965,400,000
Chemicals and allied products	2,163,500,000
Instruments and related products	2,017,300,000
Primary metals	1,667,700,000
Printing and publishing	1,497,200,000
Food and food products	1,218,400,000
Paper and allied products	940,100,000
Rubber and plastic products	863,100,000
Apparel and other textile products	456,300,000

Leading industrial corporations with headquarters in Connecticut include, in order of assets in 1984, General Electric, Union Carbide, United Technologies, Xerox, Avco, Champion International, AMAX, American Can, Combustion Engineering, Stauffer Chemical, Great Northern Nekoosa, Olin, Pitney Bowes, Singer, Uniroyal, and Chesebrough-Pond's. Connecticut has been particularly successful in attracting big corporations, especially from New York City. More than 250 firms from abroad, mostly from Europe, located in the state or expanded their plants there between 1975 and 1984.

30 COMMERCE

Considering its small size, Connecticut is a busy commercial state. In 1982, its wholesale trade totaled $32.7 billion, and retail trade amounted to $15.8 billion. Of total retail sales in that year, food stores accounted for 22%, automotive dealers 17%, department stores 9%, restaurants and taverns 9%, gasoline service stations 9%, and other establishments 34%.

The estimated value of Connecticut's manufactures exported abroad was over $5 billion in 1983/84, or 11% of the gross state product. In 1981, shipments of transport equipment, nonelectrical machinery, electric and electronic equipment, and instruments accounted for more than 76% of the state's foreign sales. Tobacco was the major agricultural export.

31 CONSUMER PROTECTION

In 1959, Connecticut established a Department of Consumer Protection charged with protecting consumers from injury by product use or merchandising deceit. The department conducts regular inspections of wholesale and retail food establishments, drug-related establishments, bedding and upholstery dealers and manufacturers, and commercial establishments that use weighing and measuring devices. The department also issues and reviews licenses, conducts investigations into alleged fraudulent activities, provides information and referral services to consumers, and responds to their complaints. In 1983/84, the department's 166 employees conducted more than 60,000 inspections and investigations.

32 BANKING

The first banks in Connecticut were established in Hartford, New Haven, Middletown, Bridgeport, Norwich, and New London between 1792 and 1805. By 1850, the state had 45 commercial and 15 savings banks. As of 1984, Connecticut's banking institutions included 14 national banks, 37 state banks and trust companies, and 57 savings banks. Assets of the state's insured commercial banks amounted to $16.1 billion at the end of 1982; their outstanding loans totaled $9.5 billion. The leading commercial bank was the Connecticut National Bank, Hartford, with total deposits of $4.1 billion as of 30 June 1984.

At the start of 1984 there were 32 state and federal savings and loan associations, with combined assets of $7.3 billion. Connecticut also had 145 state credit unions and 255 federal credit unions in 1984.

Banking operations are regulated by the state Department of Banking.

33 INSURANCE

Connecticut's preeminence in the insurance field and Hartford's title as "insurance capital" of the nation date from the late 18th century, when state businessmen agreed to bear a portion of a shipowner's financial risks in return for a share of the profits. Marine insurance companies were established in Hartford and major port cities between 1797 and 1805. The state's first insurance company had been formed in Norwich in 1795 to provide fire insurance. The nation's oldest fire insurance firm is Hartford Fire Insurance, active since 1810. Subsequently, Connecticut companies have been leaders in life, accident, casualty, automobile, and multiple-line insurance.

In 1984 there were 72 insurance companies with headquarters in Connecticut. Life insurance in force in 1983 totaled 5,525,000

policies, valued at $84.3 billion, divided about evenly between individual and group policies. Connecticut families averaged $70,000 in life insurance coverage in 1983 (3d in the US). Property and casualty insurers wrote premiums totaling $2 billion, of which automobile insurance accounted for $845.6 million, and homeowners' insurance $245 million. No-fault automobile insurance has been in effect since 1 January 1973.

The insurance industry is regulated by the state Department of Insurance.

34 SECURITIES

There are no securities or commodities exchanges in Connecticut. New York Stock Exchange member firms had 97 sales offices and 1,194 registered representatives in the state in 1983. In that same year, 835,000 Connecticuters were shareowners of public corporations.

35 PUBLIC FINANCE

Although Connecticut ranks high in per capita income, its unwillingness to impose a personal income tax has kept state expenditures—especially on education—at a relatively low level.

The state budget is prepared annually by the Budget and Financial Management Division of the Office of Policy and Management and submitted by the governor to the general assembly for consideration. The fiscal year runs from 1 July to 30 June. The following is a summary of projected revenues and expenditures for 1984/85 and 1985/86:

REVENUES	1984/85	1985/86
Taxes	$2,981,800,000	$3,066,400,000
Other revenues	525,400,000	835,400,000
Federal grants	446,900,000	487,100,000
Transfers	—	39,900,000
TOTALS	$3,954,100,000	$4,428,800,000

EXPENDITURES		
Education	$1,094,442,200	$1,243,900,000
Public welfare	969,713,923	1,020,000,000
Health and hospitals	358,723,679	432,400,000
Transportation	337,515,300	425,300,000
General government	250,191,723	303,200,000
Corrections	161,948,177	190,900,000
Regulation and protection of persons and property	102,327,395	116,100,000
Judiciary	84,683,388	84,500,000
Conservation and development	33,659,768	39,100,000
Legislature	18,079,450	19,900,000
Debt service, grants, miscellaneous	593,578,349	610,900,000
TOTALS	$4,004,863,352	$4,486,200,000

The state's public debt totaled $4.6 billion as of 30 June 1982. The consolidated debt of Connecticut state and local governments surpassed $6.6 billion in mid-1982; per capita debt averaged $2,125 (19th in the US).

36 TAXATION

Connecticut ranked 30th in state and local taxation in 1982/83, with receipts of $4.5 billion, or $1,431 per capita. Principal taxes are a sales and use tax levied at a rate of 7.5% in 1985, a corporation business tax levied at the basic rate of 11.5%, a tax levied on dividends and interest at a rate of 1–13%, and a motor fuels tax of 16 cents per gallon (rising to 23 cents per gallon by 1 July 1991). Other state taxes are levied on cigarettes, alcoholic beverages, and theater admissions. Property taxes are the main source of local revenue.

In 1983, Connecticuters filed over 1,486,000 federal income tax returns and paid more than $5.7 billion in tax.

37 ECONOMIC POLICY

Connecticut's principal economic goal is to strengthen employment in the manufacturing sector and urban areas. Toward that end, the Department of Economic Development in 1983/84 approved $318 million in low-cost industrial loans to enable companies to construct or expand facilities in the state. Business recruitment missions have been sent to industrialized Europe and Japan to stimulate the state's export program. Officials of the department cooperate with the state's Office of Employment and Training and with the labor and education departments to develop job-training programs.

³⁸HEALTH

Connecticut is one of the healthiest states in the US. The infant mortality rates in 1980 were 10.2 per 1,000 live births for whites and 19.1 for blacks; both were just below the national average. There were 23,200 legal abortions in 1982, a rate of about one abortion for every two live births. The two leading causes of death in 1981 were heart disease and cancer. Death rates for stroke, accidents, diabetes, suicide, and early childhood disease were below the respective national norms, while rates for heart disease, cancer, cirrhosis of the liver, pneumonia and influenza, and atherosclerosis were above them.

Connecticut had five public mental hospitals in 1984—in Hartford, Newington, Middletown, Newtown, and Norwich. Other state medical facilities in 1984 included 2 state training schools and 12 regional centers for the mentally retarded, and 4 other institutions. In 1980, Connecticut had 213 nursing homes with 25 or more beds, 72 rest homes with nursing supervision, and 148 homes for the aged.

In 1982, Connecticut had 65 hospitals, with 18,200 beds and 56,000 personnel. In 1984 there were 52,920 registered nurses and 14,836 licensed practical nurses in the state. Hospital costs in 1981 averaged $379 per day and $2,868 per stay, well above the US average. According to state sources, Connecticut had 13,560 physicians and 4,030 dentists in 1984. Outstanding medical schools are those of Yale University and the University of Connecticut.

³⁹SOCIAL WELFARE

In 1982, aid to families with dependent children was paid to 126,441 state residents, including 85,173 children, and totaled $210.5 million. The national school lunch program provided nutritional assistance during 1982/83 to 200,000 pupils. In that year, 161,000 persons participated in the food stamp program, at a federal cost of $72 million. In 1983, 491,000 state residents received federal Social Security benefits totaling $2.54 billion. Medicare programs provided health care and helped subsidize nutritional programs, senior-citizen centers, and home-care services. The Department of Human Resources provided a wide range of social services to individuals of all ages in the course of administering $72 million in state and federal funds during 1983/84.

As of 1984, unemployment insurance in Connecticut provided recipients with an average weekly benefit of $125.50 for a maximum duration of 26 weeks; payments totaled $179 million.

⁴⁰HOUSING

The 1980 US Census of Population and Housing reported that there were 1,158,884 housing units in Connecticut, an increase of 177,281 units, or 18%, over the 1970 census total. Virtually every town increased its housing stock during that period; the smallest increases were in the big cities, however. In Hartford, the number of units declined from 58,495 in 1970 to 55,254 in 1980.

As of 1980, year-round housing units in Connecticut had a median number of 5.4 rooms and 2.4 people per unit; the median value of owner-occupied dwellings was $65,600, as compared with $25,500 in 1970. The proportion of all housing units having full plumbing facilities was 98.7%.

⁴¹EDUCATION

Believing that the Bible was the only true source of God's truths, Connecticut's Puritan founders viewed literacy as a theological necessity. A law code in 1650 required a town of 50 families to hire a schoolmaster to teach reading and writing, and a town of 100 families to operate a school to prepare students for college. Despite such legislation, many communities in colonial Connecticut did not provide sufficient funding to operate first-rate schools. Public education was greatly strengthened in the 19th century by the work of Henry Barnard, who advocated free public schools, state supervision of common schools, and the establishment of schools for teacher training. By the late 1860s and early 1870s, all of Connecticut's public elementary and high schools were tuition free. In 1865, the Board of Education was established.

As of 1980, 83.7% of adult state residents were high school graduates, and 20.7% had completed four or more years of college. As of fall 1984, Connecticut had 939 public schools, with 31,233 teachers and 467,206 students. During 1980, nonpublic schools had 88,404 students. In 1984, Connecticut had 201 Roman Catholic parochial schools (including 6 colleges and 30 high schools) with 71,984 pupils. The state's private preparatory schools include Choate Rosemary Hall (Wallingford), Taft (Watertown), Westminster (Simsbury), Loomis Chaffee (Windsor), and Miss Porter's (Farmington).

Public institutions of higher education include the University of Connecticut at Storrs, with 23,428 students in 1984; four divisions of the Connecticut State University, at New Britain, New Haven, Danbury, and Willimantic, with a total of 33,661 students; 12 regional community colleges with 35,625 students; and 5 state technical colleges with 8,853 students.

Connecticut's 21 private four-year colleges and universities had 60,044 students in 1984. Among the oldest institutions are Yale, founded in 1701 and settled in New Haven between 1717 and 1719; Trinity College (1823) in Hartford; and Wesleyan University (1831) in Middletown. Other private institutions include the University of Hartford, University of Bridgeport, Fairfield University, and Connecticut College in New London.

A characteristic of public-school financing in Connecticut has been high reliance on local support for education. Differences among towns in their wealth bases and taxation were compounded by the mechanism used to distribute a majority of state funds for public education—the flat-grant-per-pupil formula. After the Connecticut supreme court in *Horton* v. *Meskill* (1978) declared this funding mechanism to be unconstitutional, the general assembly in 1979 replaced it with an equity-based model in order to reduce the disparity among towns in expenses per pupil.

⁴²ARTS

Art museums in Connecticut include the Wadsworth Atheneum in Hartford, the oldest (1842) free public art museum in the US; the Yale University Art Gallery and the Yale Center for British Art in New Haven; the New Britain Museum of American Art; and the Lyman Allyn Museum of Connecticut College in New London. The visual arts are easily accessible through numerous other art museums, galleries, and more than 150 annual arts shows and festivals. The Connecticut Commission on the Arts, established in 1965, has 20 members appointed by the general assembly and 5 by the governor. It administers a state art collection and establishes policies for an art bank program.

The theater is vibrant in contemporary Connecticut, which has numerous dinner theaters, at least 100 community theater groups, and many college and university theater groups. Professional theaters include the American Shakespeare Festival Theater in Stratford, the Long Wharf Theater and the Yale Repertory Theater in New Haven, the Hartford Stage Company, and the Eugene O'Neill Memorial Theater Center in Waterford.

The state's foremost metropolitan orchestras are the Hartford and New Haven symphonies. Professional opera is presented by the Stamford State Opera and by the Connecticut Opera in Hartford. Prominent dance groups include the Connecticut Dance Company in New Haven, the Hartford Ballet Company, and the Pilobolus Dance Theater in the town of Washington.

[43] LIBRARIES AND MUSEUMS

As of 1984, Connecticut had more than 200 public libraries. The leading public library is the Connecticut State Library (Hartford), which houses about 700,000 bound volumes and 2,100 periodicals as well as the official state historical museum. Connecticut's most distinguished academic collection is the Yale University library system (7.9 million volumes), headed by the Sterling Memorial Library and the Beinecke Rare Book and Manuscript Library. Special depositories include the Hartford Seminary Foundation's impressive material on Christian-Muslim relations, the Connecticut Historical Society's especially strong collection of materials pertaining to state history and New England genealogy, the Trinity College Library's collection of church documents, the Indian Museum in Old Mystic, the maritime history collections in the Submarine Library at the US Navy submarine base in Groton, and the G. W. Blunt White Library at Mystic Seaport. In all, Connecticut libraries held 11.3 million volumes and had a combined circulation of 18.6 million in 1983/84.

Connecticut has more than 130 museums, in addition to its historic sites. The Peabody Museum of Natural History at Yale includes an impressive dinosaur hall. Botanical gardens include Harkness Memorial State Park in Waterford, Elizabeth Park in West Hartford, and Hamilton Park Rose Garden in Waterbury. Connecticut's historical sites include the Henry Whitfield House in Guilford (1639), said to be the oldest stone house in the US; the Webb House in Wethersfield, where George Washington met with the Comte de Rochambeau in 1781 to plan military strategy against the British; Noah Webster's birthplace in West Hartford; and the Jonathan Trumbull House in Lebanon.

[44] COMMUNICATIONS

There were 245 post offices and 10,689 postal employees in Connecticut in 1985. As of the 1980 census, 97% of the state's 1,093,678 occupied housing units had telephones.

In 1984, Connecticut had 40 AM and 37 FM radio stations, and 9 television stations. There were educational television stations in Bridgeport, Hartford, New Haven, and Norwich. In addition, in 1984, 27 cable television companies served 556,764 subscribers in 138 communities.

[45] PRESS

The *Hartford Courant,* founded in 1764, is generally considered to be the oldest US newspaper in continuous publication. The leading Connecticut dailies in 1984 were the *Courant,* with an average morning circulation of 218,830 (Sundays, 300,767), and the *New Haven Register,* with an average evening circulation of 92,284 (Sundays, 142,342). Statewide, in 1985 there were 6 morning newspapers (381,151 circulation), 18 evening newspapers (521,678), and 11 Sunday newspapers (801,135). In 1985, Connecticut also had 55 newspapers that appeared weekly or up to three times a week.

Leading periodicals are *American Scientist, Connecticut Magazine, Fine Woodworking, Golf Digest,* and *Tennis.*

[46] ORGANIZATIONS

The 1982 Census of Service Industries counted 952 organizations in Connecticut, including 127 business associations; 605 civic, social, and fraternal associations; and 32 educational, scientific, and research associations. National organizations with headquarters in Connecticut included the Knights of Columbus (New Haven), the American Institute for Foreign Study (Greenwich), Junior Achievement (Stamford), the International Association of Approved Basketball Officials (West Hartford), and Save the Children Federation (Westport).

[47] TOURISM, TRAVEL, AND RECREATION

Tourism has become an increasingly important part of the state economy in recent decades, with tourist expenditures climbing from $77 million in 1956 to more than $2.2 billion in 1983/84. The average number of tourist visits per year exceeds 22 million. Tourism provides about 45,000 jobs for state residents and brings in over $111 million in sales tax revenues. Popular tourist attractions include the Mystic Seaport restoration and its aquarium, the Mark Twain House and state capitol in Hartford, the American Clock and Watch Museum in Bristol, the Lock Museum of America in Terryville, and the Yale campus in New Haven. Outstanding events are the Harvard-Yale regatta held each June on the Thames River in New London, and about 50 fairs held in Guilford and other towns between June and October.

Connecticut abounds in outdoor recreational facilities. As of 1984 there were 84 golf courses open to the public, 117 state parks and forests, and more than 100 freshwater and saltwater boat-launching sites, along with ski areas and snowmobile trails for winter recreation. The state issued licenses to 222,463 hunters and fishermen in 1984.

[48] SPORTS

Connecticut's only major league professional team is the Hartford Whalers of the National Hockey League. New Haven has a minor league hockey franchise. New Britain and Waterbury compete in baseball's Class AA Eastern League. Auto racing takes place at Lime Rock Race Track, Salisbury.

The state runs daily and weekly lotteries, and licenses off-track betting facilities for horse racing (not actually held in the state) and pari-mutuel operations for greyhound racing and jai alai. In 1982/83, the state treasury received $122 million from these activities.

Connecticut schools, colleges, and universities provide amateur athletic competitions, highlighted by Ivy League football games on autumn Saturdays at the Yale Bowl in New Haven.

[49] FAMOUS CONNECTICUTERS

Although Connecticut cannot claim any US president or vice president as a native son, John Moran Bailey (1904-75), chairman of the state Democratic Party (1946-75) and of the national party (1961-68), played a key role in presidential politics as a supporter of John F. Kennedy's successful 1960 campaign.

Two Connecticut natives have served as chief justice of the US: Oliver Ellsworth (1745-1807) and Morrison R. Waite (1816-88). Associate justices include Henry Baldwin (1780-1844), William Strong (1808-95), and Stephen J. Field (1816-99). Other prominent federal officeholders were Oliver Wolcott (1760-1833), secretary of the treasury; Gideon Welles (1802-78), secretary of the Navy; Dean Acheson (1893-1971), secretary of state; and Abraham A. Ribicoff (b.1910), secretary of health, education, and welfare. An influential US senator was Orville H. Platt (1827-1905), known for his authorship of the Platt Amendment (1901), making Cuba a virtual protectorate of the United States. Also well known are Connecticut senators Ribicoff (served 1963-81) and Lowell P. Weicker, Jr. (b.France, 1931), the latter brought to national attention by his work during the Watergate hearings in 1973.

Notable colonial and state governors include John Winthrop, Jr. (b.England, 1606-76), Jonathan Trumbull (1710-85), William A. Buckingham (1804-75), Simeon Eben Baldwin (1840-1927), Marcus Holcomb (1844-1932), Wilbur L. Cross (1862-1948), Chester Bowles (b.1901), and Ribicoff. Ella Tambussi Grasso (1919-81), elected in 1974 and reelected in 1978 but forced to resign for health reasons at the end of 1980, was the first woman governor in the US who did not succeed her husband in the post.

In addition to Winthrop, the founding fathers of Connecticut were Thomas Hooker (b.England, 1586-1647), who was deeply involved in establishing and developing Connecticut Colony, and Theophilus Eaton (b.England, 1590-1658) and John Davenport (b.England, 1597-1670), cofounders and leaders of the strict Puritan colony of New Haven. Other famous historical figures are Israel Putnam (b.Massachusetts, 1718-90), Continental Army major general at the Battle of Bunker Hill, who supposedly admonished his troops not to fire "until you see the whites of their eyes"; diplomat Silas Deane (1737-89); and Benedict Arnold

(1741-1801), known for his treasonous activity in the Revolutionary War but also remembered for his courage and skill at Ft. Ticonderoga and Saratoga. Roger Sherman (b.Massachusetts, 1721-93), a signatory to the Articles of Association, Declaration of Independence (1776), Articles of Confederation (1777), Peace of Paris (1783), and the US Constitution (1787), was the only person to sign all these documents; at the Constitutional Convention, he proposed the "Connecticut Compromise," calling for a dual system of congressional representation. Connecticut's most revered Revolutionary War figure was Nathan Hale (1755-76), the Yale graduate who was executed for spying behind British lines. Radical abolitionist John Brown (1800-1859) was born in Torrington.

Connecticuters prominent in US cultural development include painter John Trumbull (1756-1843), son of Governor Trumbull, known for his canvases commemorating the American Revolution. Joel Barlow (1754-1812) was a poet and diplomat in the early national period. Lexicographer Noah Webster (1758-1843) compiled the *American Dictionary of the English Language* (1828). Frederick Law Olmsted (1822-1903), the first American landscape architect, planned New York City's Central Park. Harriet Beecher Stowe (1811-96) wrote one of the most widely read books in history, *Uncle Tom's Cabin* (1852). Mark Twain (Samuel L. Clemens, b.Missouri, 1835-1910) was living in Hartford when he wrote *The Adventures of Tom Sawyer* (1876), *The Adventures of Huckleberry Finn* (1885), and *A Connecticut Yankee in King Arthur's Court* (1889). Charles Ives (1874-1954), one of the nation's most distinguished composers, used his successful insurance business to finance his musical career and to help other musicians. Eugene O'Neill (b.New York, 1888-1953), the playwright who won the Nobel Prize for literature in 1936, spent summers in New London during his early years. A seminal voice in modern poetry, Wallace Stevens (b.Pennsylvania, 1879-1955), wrote the great body of his work while employed as a Hartford insurance executive. James Merrill (b.New York, 1926), a poet whose works have won the National Book Award (1967), Bollingen Prize (1973), and numerous other honors, lived in Stonington.

Native Connecticuters important in the field of education include Eleazar Wheelock (1711-79), William Samuel Johnson (1727-1819), Emma Willard (1787-1870), and Henry Barnard (1811-1900). Shapers of US history include Jonathan Edwards (1703-58), a Congregationalist minister who sparked the 18th-century religious revival known as the Great Awakening; Samuel Seabury (1729-96), the first Episcopal bishop in the US; Horace Bushnell (1802-76), said to be the father of the Sunday school; Lyman Beecher (1775-1863), a controversial figure in 19th-century American Protestantism who condemned slavery, intemperance, Roman Catholicism, and religious tolerance with equal fervor; and his son Henry Ward Beecher (1813-87), also a religious leader and abolitionist.

Among the premier inventors born in Connecticut were Abel Buel (1742-1822), who cast the first American foundry type in 1769; David Bushnell (1742-1824), who designed the first American submarine; Eli Whitney (1765-1825), inventor of the cotton gin and a pioneer in manufacturing; Charles Goodyear (1800-60), who devised a process for the vulcanization of rubber; Samuel Colt (1814-62), inventor of the six-shooter; Frank Sprague (1857-1934), who designed the first major electric trolley system in the US; and Edwin H. Land (b.1909), inventor of the Polaroid Land Camera. The Nobel Prize in physiology or medicine was won by three Connecticuters: Edward Kendall (1886-1972) in

1949, John Enders (b.1897) in 1954, and Barbara McClintock (b.1902) in 1983.

Other prominent Americans born in Connecticut include clock manufacturer Seth Thomas (1785-1859), circus impresario Phineas Taylor "P. T." Barnum (1810-91), jeweler Charles Lewis Tiffany (1812-1902), financier John Pierpont Morgan (1837-1913), pediatrician Benjamin Spock (b.1903), cartoonist Al Capp (1909-79), soprano Eileen Farrell (b.1920), and consumer advocate Ralph Nader (b.1934). Leading actors and actresses are Ed Begley (1901-70), Katharine Hepburn (b.1909), Rosalind Russell (1911-76), and Robert Mitchum (b. 1917).

Walter Camp (1859-1925), athletic director of Yale University who contributed to the formulation of the rules of US football, was a native of Connecticut.

50 BIBLIOGRAPHY

Anderson, Ruth O. M. *From Yankee to American: Connecticut, 1865-1914.* Chester, Conn.: Pequot Press, 1975.

Bachman, Ben. *Upstream: A Voyage on the Connecticut River.* Boston: Houghton Mifflin, 1985.

Bingham, Harold J. *History of Connecticut.* 4 vols. New York: Lewis, 1962.

Bixby, William. *Connecticut: A New Guide.* New York: Scribner, 1974.

Bushman, Richard L. *From Puritan to Yankee: Character and the Social Order in Connecticut, 1690-1765.* Cambridge: Harvard University Press, 1967.

Connecticut, State of. Secretary of State. *Register and Manual 1984.* Hartford, 1984.

Janick, Herbert F. *A Diverse People: Connecticut, 1914 to the Present.* Chester, Conn.: Pequot Press, 1975.

Lee, W. Storrs. *The Yankees of Connecticut.* New York: Holt, 1957.

Morse, Jarvis Means. *A Neglected Period of Connecticut's History: 1818-1850.* New Haven: Yale University Press, 1933.

Niven, John. *Connecticut for the Union: The Role of the State in the Civil War.* New Haven: Yale University Press, 1965.

Peirce, Neal R. *The New England States: People, Politics, and Power in the Six New England States.* New York: Norton, 1976.

Roth, David M. *Connecticut: A Bicentennial History.* New York: Norton, 1979.

Roth, David M., and Freeman Meyer. *From Revolution to Constitution: Connecticut, 1763-1818.* Chester, Conn.: Pequot Press, 1975.

Stuart, Patricia. *Units of Local Government in Connecticut.* Storrs: Institute of Public Service, University of Connecticut, 1979.

Taylor, Robert J. *Colonial Connecticut.* Millwood, N.Y.: KTO Press, 1979.

Trecker, Janice Law. *Preachers, Rebels, and Traders: Connecticut, 1818-1865.* Chester, Conn.: Pequot Press, 1975.

Van Dusen, Albert E. *Connecticut.* New York: Random House, 1961.

Van Dusen, Albert E. *Puritans Against the Wilderness: Connecticut to 1763.* Chester, Conn.: Pequot Press, 1975.

Warren, William L. *Connecticut Art and Architecture: Looking Backwards Two Hundred Years.* Hartford: American Revolution Bicentennial Commission of Connecticut, 1976.

Whipple, Chandler. *The Indian in Connecticut.* Stockbridge, Mass.: Berkshire Traveller Press, 1972.

Zeichner, Oscar. *Connecticut's Years of Controversy, 1750-1776.* Chapel Hill: University of North Carolina Press, 1949.

DELAWARE

State of Delaware

DECEMBER 7, 1787

ORIGIN OF STATE NAME: Named for Thomas West, Baron De La Warr, colonial governor of Virginia: the name was first applied to the bay. **NICKNAMES:** The First State; the Diamond State. **CAPITAL:** Dover. **ENTERED UNION:** 7 December 1787 (1st). **SONG:** "Our Delaware." **COLORS:** Colonial blue and buff. **MOTTO:** Liberty and Independence. **COAT OF ARMS:** A farmer and a rifleman flank a shield that bears symbols of the state's agricultural resources—a sheaf of wheat, an ear of corn, and a cow. Above is a ship in full sail; below, a banner with the state motto. **FLAG:** Colonial blue with the coat of arms on a buff-colored diamond; below the diamond is the date of statehood. **OFFICIAL SEAL:** The coat of arms surrounded by the words "Great Seal of the State of Delaware 1793, 1847, 1907." The three dates represent the years in which the seal was revised. **BIRD:** Blue hen's chicken. **FLOWER:** Peach blossom. **TREE:** American holly. **ROCK:** Sillimanite. **INSECT:** Ladybug. **LEGAL HOLIDAYS:** New Year's Day, 1 January; Birthday of Martin Luther King, Jr., 3d Monday in January; Lincoln's Birthday, 1st Monday in February; Washington's Birthday, 3d Monday in February; Good Friday, March or April; Memorial Day, last Monday in May; Independence Day, 4 July; Labor Day, 1st Monday in September; Columbus Day, 2d Monday in October; Veterans Day, 11 November; General Election Day, 1st Tuesday after the 1st Monday in November in even-numbered years; Thanksgiving Day, 4th Thursday in November; Christmas Day, 25 December. **TIME:** 7 AM EST = noon GMT.

[1] LOCATION, SIZE, AND EXTENT

Located on the eastern seaboard of the US, Delaware ranks 49th in size among the 50 states. The States' total area is 2,044 sq mi (5,295 sq km), of which land takes up 1,932 sq mi (5,005 sq km) and inland water 112 sq mi (290 sq km). Delaware extends 35 mi (56 km) E–W at its widest; its maximum N–S extension is 96 mi (154 km).

Delaware is bordered on the N by Pennsylvania; on the E by New Jersey (with the line passing through the Delaware River into Delaware Bay) and the Atlantic Ocean; and on the S and W by Maryland.

The boundary length of Delaware, including a general coastline of 28 mi (45 km), totals 200 mi (322 km). The tidal shoreline is 381 mi (613 km). The state's geographic center is in Kent County, 11 mi (18 km) S of Dover.

[2] TOPOGRAPHY

Delaware lies entirely within the Atlantic Coastal Plain except for its northern tip, above the Christina River, which is part of the Piedmont Plateau. The state's highest elevation is 442 feet (135 meters) on Ebright Road, near Centerville, New Castle County. The rolling hills and pastures of the north give way to marshy regions in the south (notably Cypress Swamp), with sandy beaches along the coast. Delaware's mean elevation, 60 feet (18 meters), is the lowest in the US.

Of all Delaware's rivers, only the Nanticoke, Choptank, and Pocomoke flow westward into Chesapeake Bay. The remainder—including the Christina, Appoquinimink, Leipsic, St. Jones, Murderkill, Mispillion, Broadkill, and Indian—flow into Delaware Bay. There are dozens of inland freshwater lakes and ponds.

[3] CLIMATE

Delaware's climate is temperate and humid. The normal daily mean temperature in Wilmington is 54°F (12°C), ranging from 31°F (−1°C) in January to 76°F (24°C) in July. Both the record low and the record high temperatures for the state were established at Millsboro: −17°F (−27°C) on 17 January 1893 and 110°F (43°C) on 21 July 1930. The average annual precipitation is 41 in (104 cm), and about 21 in (53 cm) of snow falls each year. Wilmington's average share of sunshine is 57%—one of the lowest percentages among leading US cities.

[4] FLORA AND FAUNA

Delaware's mixture of northern and southern flora reflects its geographical position. Common trees include black walnut, hickory, sweetgum, and tulip poplar. Shadbush and sassafras are found chiefly in southern Delaware. Two subspecies of orchid are threatened, and one member of the sedge family is endangered.

Mammals native to the state include the white-tailed deer, red and gray foxes, eastern gray squirrel, muskrat, raccoon, woodcock, and common cottontail. The quail, robin, wood thrush, cardinal, and eastern meadowlark are representative birds, while various waterfowl, especially Canada geese, are common. The southern bald eagle and the Delmarva Peninsula fox squirrel are endangered species.

[5] ENVIRONMENTAL PROTECTION

The Coastal Zone Act of 1971 outlaws new industry "incompatible with the protection of the natural environment" of shore areas, but in 1979 the act was amended to permit offshore oil drilling and the construction of coastal oil facilities. The traffic of oil tankers into Delaware Bay represents an environmental hazard.

In 1982, Delaware enacted a bottle law requiring deposits on most soda and beer bottles; deposits for aluminum cans were made mandatory in 1984. In that year, Delaware became the first state to administer the national hazardous waste program at the state level; in 1985, the state was developing legislation to deal with leaking underground storage tanks.

State environmental protection agencies include the Department of Natural Resources and Environmental Control, Coastal Zone Industrial Control Board, and Council on Soil and Water Conservation.

[6] POPULATION

Delaware ranked 48th among the 50 states in January 1985, with an estimated population of 605,711; the population density was 314 people per square mile (121 per sq km). The projected population for 1990 is 629,800, a 5.5% increase over the 1980 census figure of 595,200.

About 71% of all Delawareans live in metropolitan areas. The largest cities in 1984 were Wilmington, with 69,694; Newark, 24,103; and Dover, the capital, with 22,492.

⁷ETHNIC GROUPS

Black Americans constitute Delaware's largest racial minority, numbering 95,971 in 1980 and comprising about 16% of the population. There is a small Hispanic American element (9,670 in 1980).

The 19,000 foreign-born made up 3.2% of the state's population in 1980. The United Kingdom, Ireland, Germany, Italy, and Poland were the leading places of origin.

⁸LANGUAGES

English in Delaware is basically North Midland, with Philadelphia features in Wilmington and the northern portion. In the north, one *wants off* a bus, lowers *curtains* rather than blinds, pronounces *wharf* without /h/, and says /noo/ and /doo/ for *new* and *due* and /krik/ for *creek*. In 1980, 537,228 Delawareans—94% of the resident population 3 years of age or older—spoke only English at home. Other languages spoken at home, and the number of speakers, included Spanish, 8,288, and Italian, 4,222.

⁹RELIGIONS

The earliest permanent European settlers in Delaware were Swedish and Finnish Lutherans and Dutch Calvinists. English Quakers, Scotch-Irish Presbyterians, and Welsh Baptists arrived in the 18th century, though Anglicanization was the predominant trend. The Great Awakening, America's first religious revival, began at Lewes with the arrival of George Whitefield, an Anglican preacher, on 30 October 1739, and the colony soon became a center of early Methodist activity. The Methodist Church was the largest denomination in Delaware by the early 19th century. Subsequent immigration brought Lutherans from Germany, Roman Catholics from Ireland, Germany, Italy, and Poland, and Jews from Germany, Poland, and Russia; most of the Catholic and Jewish immigrants settled in cities, Wilmington in particular.

There were 125,572 Catholics and an estimated 9,500 Jews in Delaware in 1984. Protestantism has the largest number of adherents; in 1980, the leading groups were the United Methodist Church, 60,489; the Episcopal Church, 18,696; and the United Presbyterian Church, 16,957.

¹⁰TRANSPORTATION

The New Castle and Frenchtown Railroad, a portage route, was built in 1832; the state's first passenger line—the Philadelphia, Wilmington, and Baltimore Railroad—opened six years later. As of 1983 there were 220 mi (354 km) of Class I track. In 1984, Amtrak operated approximately 30 trains through Delaware and served both Newark (ridership: 5,031) and Wilmington (ridership: 521,286). Conrail and the Chesapeake and Ohio Railroad are Delaware's main freight carriers. The Delaware Authority for Regional Transit (DART) provides state-subsidized bus service.

In 1983, the state had 5,307 mi (8, 541 km) of highways, roads, and streets. In the same year, there were 426,803 registered vehicles and 432,029 licensed drivers. Delaware's first modern highway—and the first dual highway in the US—running about 100 mi (160 km) from Wilmington to the southern border, was financed by industrialist T. Coleman du Pont between 1911 and 1924. The twin spans of the Delaware Memorial Bridge connect Delaware highways to those in New Jersey; the Delaware Turnpike section of the John F. Kennedy Memorial Highway links the bridge system with Maryland. The Lewes–Cape May Ferry provides auto and passenger service between southern Delaware and New Jersey.

In 1983, the Port of Wilmington, Delaware's chief port, handled 16.6 billion lb of exports and imports valued at more than $2 billion. The Delaware River is the conduit for much of the oil brought by tanker to the US east coast.

Delaware had 37 airfields (22 airports, 15 heliports) in 1983, of which Greater Wilmington Airport was the largest and busiest.

¹¹HISTORY

Delaware was inhabited nearly 10,000 years ago, and a succession of various cultures occupied the area until the first European contact. At that time, the Leni-Lenape (Delaware) Indians occupied northern Delaware, while several tribes, including the Nanticoke and Assateague, inhabited southern Delaware. The Dutch in 1631 were the first Europeans to settle in what is now Delaware, but their little colony (at Lewes) was destroyed by Indians. Permanent settlements were made by the Swedes in 1638 (at Wilmington, under the leadership of a Dutchman, Peter Minuit) and by the Dutch in 1651 (at New Castle). The Dutch conquered the Swedes in 1655, and the English conquered the Dutch in 1664. Eighteen years later, the area was ceded by the duke of York (later King James II), its first English proprietor, to William Penn. Penn allowed Delaware an elected assembly in 1704, but the colony was still subject to him and to his deputy governor in Philadelphia; ties to the Penn family and Pennsylvania were not severed until 1776. Boundary quarrels disturbed relations with Maryland until Charles Mason and Jeremiah Dixon surveyed the western boundary of Delaware (and the Maryland-Pennsylvania boundary) during the period 1763–68. By this time, virtually all the Indians had been driven out of the territory.

In September 1977, during the War for Independence, British soldiers marched through northern Delaware, skirmishing with some of Washington's troops at Cooch's Bridge, near Newark, and seizing Wilmington, which they occupied for a month. In later campaigns, Delaware troops with the Continental Army fought so well that they gained the nickname "Blue Hen's Chickens," after a famous breed of fighting gamecocks. On 7 December 1787, Delaware became the first state to ratify the federal Constitution.

Although Delaware had not abolished slavery, it remained loyal to the Union during the Civil War. By that time, it was the one slave state in which a clear majority of blacks (about 92%) were already free. However, white Delawareans generally resented the Reconstruction policies adopted by Congress after the Civil War, and by manipulation of registration laws denied blacks the franchise until 1890.

The key event in the state's economic history was the completion of a railroad between Philadelphia and Baltimore through Wilmington in 1838, encouraging the industrialization of northern Delaware. Wilmington grew so rapidly that by 1900 it encompassed 41% of the state's population. Considerable foreign immigration contributed to this growth, largely from the British Isles (especially Ireland) and Germany in the mid-19th century and from Italy, Poland, and Russia in the early 20th century.

Flour and textile mills, shipyards, carriage factories, iron foundries, and morocco leather plants were Wilmington's leading enterprises for much of the 19th century. By the early 1900's, however, E. I. du Pont de Nemours and Co., founded near Wilmington in 1802 as a gunpowder manufacturer, made the city famous as a center for the chemical industry.

During the same period, Delaware's agricultural income rose. Peaches and truck crops flourished in the 19th century, along with corn and wheat; poultry and soybeans became major sources of agricultural income in the 20th century. Another economic stimulus was improved transportation.

During the 1950s, Delaware's population grew by an unprecedented 40%. The growth was greatest around Dover, site of a large air base, and on the outskirts of Wilmington. Wilmington itself lost population after 1945 because of the proliferation of suburban housing developments, offices, and factories, including two automobile assembly plants and an oil refinery. Although many neighborhood schools became racially integrated during the 1950s, massive busing was instituted by court order in 1978 to achieve a racial balance in schools throughout northern Delaware.

The 1980s ushered in a period of dramatic economic improvement. According to state sources, Delaware was one of only two states to improve its financial strength during the recession that plagued the early part of the decade.

12 STATE GOVERNMENT

Delaware has had four state constitutions, adopted in 1776, 1792, 1831, and 1897. Under the 1897 document, as amended, the legislative branch is the general assembly, consisting of a 21-member senate and a 41-member house of representatives. Senators are elected for four years, representatives for two.

Delaware's major elected executives include the governor and lieutenant governor (separately elected), treasurer, attorney general, auditor of accounts, and insurance commissioner. All serve four-year terms, except the treasurer and auditor, who are elected for two years. The governor, who may be reelected only once, must be at least 30 years of age and must have been a US citizen for 12 years and a state resident for 6 years before taking office. The legislature may override a gubernatorial veto by a three-fifths vote of the elected members of each house. An amendment to the state constitution must be approved by a two-thirds vote in each house of the general assembly in two successive sessions with an election intervening; Delaware is the only state in which amendments need not be ratified by the voters.

Voters must be US citizens at least 18 years of age. There is no minimum residency requirement, and mail registration is allowed.

13 POLITICAL PARTIES

The Democrats were firmly entrenched in Delaware for three decades after the Civil War; a subsequent period of Republican dominance lasted until the depression of the 1930s. Since then, the two parties have been relatively evenly matched.

As of 1982, the Democrats held an edge of 127,385 to 93,873 in voter registration; there were 64,299 other voters registered as independents or with minor parties. In November 1984, Delaware gave the incumbent, Republican Ronald Reagan, a clear victory with 59.8% of the ballots cast. The 1984 election also produced a large majority for Republican Governor Michael N. Castle and for Democratic Senator Joseph R. Biden, Jr. S. B. Woo, who had immigrated from Shanghai 28 years earlier, became Delaware's new lieutenant governor and the highest-ranking Chinese-American elected official. There were 10 female and 2 Hispanic elected officials in 1984, and 22 elected black officials in 1985.

Delaware Presidential Vote by Major Political Parties, 1948–84

YEAR	ELECTORAL VOTE	DELAWARE WINNER	DEMOCRAT	REPUBLICAN
1948	3	Dewey (R)	67,813	69,588
1952	3	*Eisenhower (R)	83,315	90,059
1956	3	*Eisenhower (R)	79,421	98,057
1960	3	*Kennedy (D)	99,590	96,373
1964	3	*Johnson (D)	122,704	78,078
1968	3	*Nixon (R)	89,194	96,714
1972	3	*Nixon (R)	92,283	140,357
1976	3	*Carter (D)	122,596	109,831
1980	3	*Reagan (R)	105,700	111,185
1984	3	*Reagan (R)	101,656	152,190

* Won US presidential election.

14 LOCAL GOVERNMENT

Delaware is divided into three counties. In New Castle, voters elect a county executive and a county council; in Sussex, the members of the elective county council choose a county administrator, who supervises the executive departments of the county government. Kent operates under an elected levy court, which sets

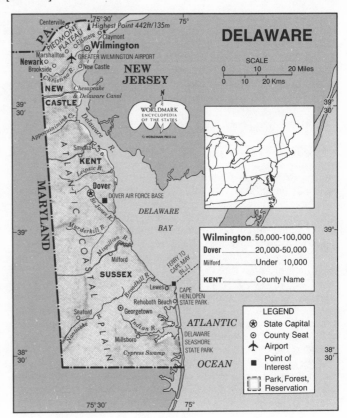

See endsheet maps: L3.

LOCATION: 38°27′ to 39°50′N; 75°02′ to 75°47′W. **BOUNDARIES:** Pennsylvania line, 23 mi (37 km); New Jersey line, 27 mi (44 km); Atlantic coastline, 28 mi (45 km); Maryland line, 122 mi (196 km).

tax rates and runs the county according to regulations spelled out by the assembly. Most of Delaware's 56 municipalities elect a mayor and council.

15 STATE SERVICES

Public education is supervised by state boards of education and vocational education. Highways are the responsibility of the Department of Transportation, while medical care, mental health facilities, drug- and alcohol-abuse programs, and help for the aging fall within the jurisdiction of the Department of Health and Social Services. Public protection services are provided primarily through the Department of Public Safety and Department of Correction. The Department of Labor has divisions covering employment services, vocational rehabilitation, unemployment insurance, and equal employment opportunity. Other services include those of the Department of Services to Children, Youth and Their Families and the Department of Community Affairs.

16 JUDICAL SYSTEM

Delaware's highest court is the supreme court, composed of a chief justice and 4 associate justices, all appointed by the governor and confirmed by the senate for 12-year terms, as are all state judges. Other state courts include the court of chancery, comprising a chancellor and two vice-chancellors, and the superior court, which has a president judge and 10 associate judges. There are 7 judges on the Court of Common Pleas in Wilmington.

Delaware was the last state to abolish the whipping post. During the 1900–1942 period, 1,604 prisoners (22% of the state's prison population) were beaten with a cat-o'-nine-tails. The whipping post, nicknamed "Red Hannah," was used for the last time in 1952 but was not formally abolished until 1972. The death penalty is authorized in Delaware, with hanging as the method of execution. There were 1,070 practicing lawyers in 1984.

[17]ARMED FORCES

Delaware's main defense facility is the military airlift wing at Dover Air Force Base, with 4,830 military personnel in 1984. There were some 76,000 veterans living in Delaware as of 30 September 1983; about 1,000 were veterans of World War I, 31,000 of World War II, 14,000 of the Korean conflict, and 22,000 of the Viet-Nam era. Veterans' benefits totaled $66 million in 1983.

The Delaware National Guard had 3,229 members in early 1985. There were 2,860 police and corrections officers in 1983.

[18]MIGRATION

Delaware has attracted immigrants from a variety of foreign countries: Sweden, Finland, and the Netherlands in the early days; England, Scotland, and Ireland during the later colonial period; and Italy, Poland, and Russia, among other countries, during the first 130 years of statehood. The 1960s and 1970s saw the migration of Puerto Ricans to Wilmington. Delaware enjoyed a net gain from migration of 122,000 persons between 1940 and 1970. Between 1970 and 1983, however, there was a net out-migration of about 8,200.

[19]INTERGOVERNMENTAL COOPERATION

Among the interstate agreements to which Delaware subscribes are the Delaware River and Bay Authority Compact, Delaware River Basin Compact, Atlantic States Marine Fisheries Compact, and Southern Interstate Energy Compact.

Federal aid to Delaware was $298.5 million in 1983/84.

[20]ECONOMY

Since the 1930s, and most especially since the mid-1970s, Delaware has been one of the nation's most prosperous states. Manufacturing—preeminently the chemical industry—is the major contributor to the state's economy. In 1984, Delaware had an unemployment rate of 6.2%, a significant improvement over the 1983 rate of 8.1%.

[21]INCOME

Average personal income per capita in Delaware was $12,442 in 1983, 11th highest among the 50 states. Median money income of four-person families increased from $23,627 in 1979 to $27,174 in 1981 (17th). In 1979, some 68,000 Delawareans, or 11.9% of the state's residents, were living below the federal poverty level. In 1982, 3,000 Delawareans were among the nation's top wealth-holders, with gross assets of more than $500,000.

[22]LABOR

Of the 298,000 members of the civilian labor force in 1983, 164,000 were male and 134,000 female. Some 24,000 Delawareans were unemployed, for an unemployment rate of 8.1%, 34th highest in the US for that year. A federal survey in 1982 revealed the following nonfarm employment pattern for Delaware:

	ESTABLISH-MENTS	EMPLOYEES	ANNUAL PAYROLL ('000)
Agricultural services, forestry, fishing	136	730	$ 9,985
Mining	12	35	922
Contract construction	1,206	13,973	294,477
Manufacturing	589	69,603	1,766,827
Transportation, public utilities	483	12,320	291,962
Wholesale trade	780	12,892	284,489
Retail trade	3,681	43,322	384,059
Finance, insurance, real estate	1,219	13,217	225,894
Services	3,845	47,544	579,986
Other	273	224	3,198
TOTALS	12,224	213,860	$3,841,799

Some 65,000 Delawareans—about 25% of all nonagricultural employees—belonged to labor unions in 1980. In 1983 there were 149 labor unions.

[23]AGRICULTURE

Though small by national standards, Delaware's agriculture is efficient and productive. In 1983, Delaware's total farm market-ings were $455 million (41st in the US), and its net farm income was $148.1 million (33rd in the US).

Tobacco was a leading crop in the early colonial era, but was soon succeeded by corn and wheat. Peaches were a mainstay during the mid-19th century, until the orchards were devastated by "the yellows," a tree disease. Today, the major field crops are corn, soybeans, barley, wheat, melons, potatoes, mushrooms, lima beans, and green peas. Production in 1983 included corn for grain, 10,875,000 bushels, valued at $43,500,000; soybeans, 7,105,000 bushels, $59,327,100; barley, 2,915,000 bushels, $5,393,000; and wheat, 2,106,000 bushels, $7,160,000.

[24]ANIMAL HUSBANDRY

Livestock and livestock products account for about 70% of Delaware's farm income, with poultry farming by far the leading sector. In 1982, Delaware ranked 14th among the states in the value of poultry and poultry products sold, with sales of $221 million. Sales of chickens and broilers in 1983 totaled 879,688,000 lb, valued at $246,373,000; egg production was 142,000,000, valued at $11,301,000. There were an estimated 10,300 milk cows in 1983, when 137,000,000 lb of milk were produced.

[25]FISHING

Fishing, once an important industry in Delaware, has declined in recent decades. The total commercial landings in 1984 were 3,098,000 lb, worth $2,034,000. Clams, plentiful until the mid-1970s, are in short supply because of overharvesting.

[26]FORESTRY

Delaware had 392,000 acres (159,000 hectares) of forestland in 1979, of which 384,000 acres (155,000 hectares) were classified as commercial forest. Sussex County has large stands of southern yellow pine, nearly all of it privately owned.

[27]MINING

In colonial times, iron was taken from bogs in Sussex County and also mined in the Newark area. Today, however, mining is of scant importance to Delaware's economy. In 1984, the state's mineral output was only $3 million, last among the 50 states. Production in that year included 1,200,000 tons of sand and gravel.

[28]ENERGY AND POWER

Installed electric capacity totaled 2 million kw in 1983, when production of electric power reached 9.5 billion kwh. Most of the power is supplied by coal- and oil-fired plants. Delaware has no nuclear reactors, nor does it have any fossil fuel resources.

[29]INDUSTRY

From its agricultural beginnings, Delaware has developed into an important industrial state. Today, Wilmington is called the "Chemical Capital of the World," largely because of E. I. du Pont de Nemours and Co., a chemical industry giant originally founded as a powder mill in 1802. As of 1984, the company was the 7th-largest US industrial corporation, with sales of $36 billion.

The Chrysler Corp. is another leading employer. During the 1983 model year, 416,170 cars were assembled in Delaware. Notable Delaware manufactures, in addition to chemicals and transportation equipment, include cash registers, luggage, apparel, processed meats and vegetables, and railroad and aircraft equipment.

The total value of shipments by manufacturers in 1982 was $8.4 billion. Major sectors and their value of shipments in 1982 were:

Chemicals, chemical products	$1,386,300,000
Food, food products	1,297,500,000
Paper, allied products	456,900,000
Scientific and medical instruments and related products	402,500,000
Rubber, miscellaneous plastic products	269,100,000

³⁰COMMERCE

In 1982, wholesale trade in Delaware totaled over $7.3 billion, and retail establishments had sales exceeding $3.1 billion. The leading retail sectors were food stores, 22%; automotive dealers, 17%; and department stores, 13%.

In 1981, Delaware exported $375 million worth of transportation equipment, chemicals, and other manufactured products to foreign markets—2.5 times the 1977 value.

³¹CONSUMER PROTECTION

The Division of Consumer Affairs within the Department of Community Affairs is responsible for enforcing state and federal consumer protection laws, representing consumer interests before local, state, and federal agencies, and maintaining consumer information and education programs. Other executive agencies with consumer protection responsibilities are the Consumer Affairs Board and the Council on Consumer Affairs.

³²BANKING

Delaware had 37 banks as of 31 December 1984: 14 insured national banks and 4 insured state banks that were members of the federal reserve system; 14 insured state banks that were not members of the federal reserve system; 3 noninsured, nondeposit trust companies; and 2 insured mutual savings banks. In addition, there were 77 credit unions. Total assets of Delaware's commercial banks were $12.8 billion as of December 1983.

³³INSURANCE

As of 1983, a total of 1,526,000 life insurance policies valued at almost $19.7 billion were in force. The average of $83,800 in life insurance per family ranked 1st among the 50 states and was surpassed only by the District of Columbia. Life insurance benefits totaling $167.4 million were paid in 1983; death payments constituted 37% of the total.

³⁴SECURITIES

Delaware has no securities exchanges. As of 30 September 1983, the state had 11 registered brokers and dealers.

³⁵PUBLIC FINANCE

Delaware's annual state budget is prepared by the Office of the Budget and submitted by the governor to the general assembly for amendment and approval. The fiscal year runs from 1 July through 30 June. The following table summarizes state general revenues and expenditures for 1984/85 (budgeted) and 1985/86 (recommended), in millions:

REVENUES	1984/85	1985/86
Individual income taxes	$408.0	$426.0
Franchise taxes	115.0	119.0
Business and occupational gross receipts	70.5	77.6
Corporation income taxes	54.0	50.0
Other receipts	253.1	261.5
TOTALS	$900.6	$934.1
EXPENDITURES		
Public education	$265.6	$291.5
Higher education	93.7	98.1
Health, social services	150.9	157.4
Transportation	60.3	68.3
Other outlays	255.1	268.0
TOTALS	$825.6	$883.3*

* Excluding grants-in-aid.

At the close of fiscal 1982, the outstanding debt of Delaware state and local governments was more than $1.3 billion, or $2,288 per capita (3d among the 50 states).

³⁶TAXATION

Delaware's state tax revenues come primarily from levies on personal and corporate income, inheritance and estates, motor fuels, cigarettes, pari-mutuel betting, and alcoholic beverages.

Delaware's corporate tax rate was 8.7% as of 1 October 1984, and its individual income tax rate, as of 1 January 1985, ranged from 1.3 to 10.7%. The state allows federal income tax as a limited deduction. There is no state sales tax.

Delaware paid $1.7 billion in federal taxes in 1982 and received $1.5 billion in federal expenditures, for a spending/tax ratio of 0.88. State residents filed 263,758 federal income tax returns in 1983, paying an estimated $791,069,000 in tax.

³⁷ECONOMIC POLICY

Legislation passed in 1899 permits companies to be incorporated and chartered in Delaware even if they do no business in the state and hold their stockholders' meetings elsewhere. Another incentive to chartering in Delaware is the state's court of chancery, which has extensive experience in dealing with corporate problems.

The Delaware Development Office, established in 1981, seeks to create jobs by encouraging businesses to relocate to Delaware. The state currently offers state and local tax incentives.

³⁸HEALTH

Infant mortality in 1981 was 10 per 1,000 live births for whites and 24.4 for blacks. With an overall death rate of 8.5 per 1,000 residents, Delaware had lower death rates than the nation as a whole for stroke, atherosclerosis, cirrhosis, accidents, pneumonia, and influenza, but higher death rates for diabetes, perinatal disorders, suicide, diseases of the heart, and malignant neoplasms.

Delaware's 14 hospitals had 97,798 admissions and 3,906 beds in 1983. In 1981, hospital personnel included 2,371 registered nurses. The average cost of hospital care was $302 per day and $2,477 per stay in 1982, both below the US average. Delaware had 1,160 physicians and 279 dentists in 1982.

³⁹SOCIAL WELFARE

Delaware does not have a history of expansive social programs. A survey made in 1938 found that the state, which then ranked 4th nationally in per capita income, spent little more than the poorest southern states on public assistance. By 1983, however, the state ranked 24th in terms of public aid recipients as a percentage of population.

Expenditures on aid to families with dependent children reached $27.9 million in 1982, and over $138 million was expended under Medicare in 1983/84. In 1983, some 47,000 Delawareans participated in the food stamp program, at a federal cost of $28 million; the school lunch program served 54,000 pupils, at a cost of $5 million.

Social Security benefits totaling $441 million went to 90,000 state residents in 1983; the average monthly payment for retired workers was $461. Federal Supplementary Security Income payments totaled over $13.8 million in 1983. Delaware, the only state in 1938 not to have a vocational rehabilitation program, spent $11.9 million for that purpose in 1984. In 1983, a total of $49 million was paid to disabled workers and their spouses and children.

⁴⁰HOUSING

The 1980 census counted 238,611 housing units in 1980, of which 207,081 were occupied. Only 2.4% of the housing units lacked full plumbing. From 1981 through 1983, 8,900 new units were authorized. In 1980, the median rent for a housing unit was $245 per month. The median monthly owner costs were $351 with a mortgage and $146 without one.

⁴¹EDUCATION

The development of public support and financing for an adequate public educational system was the handiwork of industrialist-Progressive Pierre S. du Pont, who undertook the project in 1919. Today's schools compare favorably with those of neighboring states. Nearly 69% of adult Delawareans were high school graduates in 1980, and in 1983 the state's high-school seniors ranked 9th nationally on the Scholastic Aptitude Test.

In 1982/83, 79,252 students were enrolled in public elementary and secondary schools, and 22,095 students were enrolled in public institutes of higher education. In 1980, 23,374 children were enrolled in private and parochial schools. Delaware ranked 3d among the states in per capita state and local government expenditures for higher education in 1981/82. Delaware has two public four-year institutions: the University of Delaware (Newark), with a total enrollment of 15,260 in 1983/84, and Delaware State College (Dover), with an enrollment of 2,128. Alternatives to these institutions include Widener University and the Delaware Technical and Community College, which has four campuses. There are three independent colleges: Goldey Beacon (Wilmington), Wesley (Dover), and Wilmington.

42ARTS

Wilmington has a local symphony orchestra, opera society, and drama league. The Playhouse, located in the Du Pont Building in Wilmington, shows first-run Broadway plays. The restored Grand Opera House in Wilmington, Delaware's Center for the Performing Arts, is the home of the Delaware Symphony and the Delaware Opera Guild as well as host to performances of popular music and ballet.

43LIBRARIES AND MUSEUMS

Delaware had 30 public libraries and branches in 1982/83, with 960,044 books and other materials and a circulation of 2,164,999. The University of Delaware's Hugh M. Morris Library, with 1,554,082 volumes, is the largest academic library in the state. Other distinguished libraries include the Eleutherian Mills Historical Library, the Winterthur Library, and the Historical Society of Delaware Library (Wilmington).

Notable among the state's 21 museums and numerous historical sites are the Hagley Museum and Delaware Art Museum, both in Wilmington, where the Historical Society of Delaware maintains a museum in the Old Town Hall. The Henry Francis du Pont Winterthur Museum features a collection of American antiques and decorative arts. The Brandywine Zoo, adjacent to Rockford Park, is popular with Wilmington's children. The Delaware State Museum is in Dover.

44COMMUNICATIONS

In 1980, 197,049 (95%) of Delaware's housing units had telephones. The state had 10 AM and 11 FM radio stations and 2 educational television stations in 1984. Philadelphia and Baltimore commercial television stations are within range. The state also had 7 cable systems serving 55 communities with 95,757 subscribers in 1984.

45PRESS

Delaware's three leading newspapers, all published in Wilmington by the News-Journal Co. (an affiliate of the Gannett chain), are the *Morning News,* with an average daily circulation of 64,373 in 1984; the *Evening Journal,* 60,489; and the *News-Journal,* 114,963 on Saturdays, 123,820 on Sundays.

46ORGANIZATIONS

The 1982 Census of Service Industries counted 188 organizations in Delaware, including 31 business associations; 114 civic, social, and fraternal associations; and 8 educational, scientific, and research associations. Among national organizations headquartered in Delaware are the International Reading Association, the Jean Piaget Society, and the American Philosophical Association, all located in Newark.

47TOURISM, TRAVEL, AND RECREATION

Delaware's travel and recreation industry is second only to manufacturing in economic importance. In 1982, travel and tourism generated 16,700 jobs (6.5% of all jobs in Delaware), $119 million in payroll, nearly $467 million in expenditures, and $57 million in tax revenues.

Rehoboth Beach on the Atlantic Coast bills itself as the "Nation's Summer Capital" because of the many federal officials and foreign diplomats who summer there; annual festivities include an Easter sunrise service. Among other events are the

Delaware Kite Festival at Cape Henlopen State Park (east of Lewes) every Good Friday, Old Dover Days during the first weekend in May, and Delaware Day ceremonies (7 December, commemorating the day in 1787 when Delaware was the first of the original 13 states to ratify the Constitution) throughout the state.

Fishing, clamming, crabbing, boating, and swimming are the main recreational attractions. In 1982/83, Delaware had 27,247 licensed hunters and 17,552 licensed fishermen.

48SPORTS

Delaware has three major race tracks: Brandywine and Harrington, which have harness racing, and Dover Downs, which has a double track, one for horse racing and one for auto racing. Thoroughbred races are held at Delaware Park in Wilmington.

49FAMOUS DELAWAREANS

Three Delawareans have served as US secretary of state: Louis McLane (1786-1857), John M. Clayton (1796-1856), and Thomas F. Bayard (1828-98). Two Delawareans have been judges on the Permanent Court of International Justice at The Hague: George Gray (1840-1925) and John Bassett Moore (1860-1947). James A. Bayard (b.Pennsylvania, 1767-1815), a US senator from Delaware from 1805 to 1813, was chosen to negotiate peace terms for ending the War of 1812 with the British.

John Dickinson (b.Maryland, 1732-1808), the "penman of the Revolution," and Caesar Rodney (1728-84), wartime chief executive of Delaware, were notable figures of the Revolutionary era. George Read (b.Maryland, 1733-98) and Thomas McKean (b.Pennsylvania, 1734-1817) were, with Rodney, signers for Delaware of the Declaration of Independence. Naval officers of note include Thomas Macdonough (1783-1825) in the War of 1812 and Samuel F. du Pont (b.New Jersey, 1803-65) in the Civil War.

Morgan Edwards (b.England, 1722-95), Baptist minister and historian, was a founder of Brown University. Richard Allen (b.Pennsylvania, 1760-1831) and Peter Spencer (1779-1843) established separate denominations of African Methodists. Welfare worker Emily P. Bissell (1861-1948) popularized the Christmas seal in the US, and Florence Bayard Hilles (1865-1954) was president of the National Woman's Party.

Among scientists and engineers were Oliver Evans (1755-1819), inventor of a high-pressure steam engine; Edward Robinson Squibb (1819-1900), physician and pharmaceuticals manufacturer; Wallace H. Carothers (b.Iowa, 1896-1937), developer of nylon at Du Pont; and Daniel Nathans (b.1928), who shared the Nobel Prize in medicine in 1978 for his research on molecular genetics. Eleuthère I. du Pont (b.France, 1771-1834) founded the company that bears his name; Pierre S. du Pont (1870-1954) was architect of its modern growth.

Delaware authors include Robert Montgomery Bird (1806-54), playwright; Hezekiah Niles (b.Pennsylvania, 1777-1839), journalist; Christopher Ward (1868-1944), historian; Henry Seidel Canby (1878-1961), critic; and novelist Anne Parrish (b.Colorado, 1888-1957). Howard Pyle (1853-1911) was known as a writer, teacher, and artist-illustrator.

50BIBLIOGRAPHY

Federal Writers' Project. *Delaware: A Guide to the First State.* Reprint. New York: Somerset, n.d. (orig. 1938).

Hoffecker, Carol E. *Delaware: A Bicentennial History.* New York: Norton, 1977.

Mosley, Leonard. *Blood Relations: The Rise and Fall of the Du Ponts of Delaware.* New York: Atheneum, 1980.

Munroe, John A. *Colonial Delaware: A History.* Millwood, N.Y.: KTO Press, 1978.

Munroe, John A. *History of Delaware.* Newark: University of Delaware Press, 1979.

Vessels, Jane. *Delaware: Small Wonder.* New York: Abrams, 1984.

FLORIDA

State of Florida

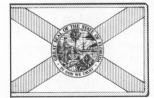

ORIGIN OF STATE NAME: Named in 1513 by Juan Ponce de León, who landed during *Pascua Florida*, the Easter festival of flowers. **NICKNAME:** The Sunshine State. **CAPITAL:** Tallahassee. **ENTERED UNION:** 3 March 1845 (27th). **SONG:** "Old Folks at Home" (also known as "Swanee River"). **POET LAUREATE:** Dr. Edmund Skellings. **MOTTO:** In God We Trust. **FLAG:** The state seal appears in the center of a white field, with four red bars extending from the seal to each corner; the flag is fringed on three sides. **OFFICIAL SEAL:** In the background, the sun's rays shine over a distant highland; in the foreground are a sabal palmetto palm, a steamboat, and an Indian woman scattering flowers on the ground. The words "Great Seal of the State of Florida" and the state motto surround the whole. **ANIMAL:** Florida Panther. **MARINE MAMMALS:** Manatee, dolphin. **BIRD:** Mockingbird. **FISH:** Largemouth bass (freshwater), Atlantic sailfish (saltwater). **FLOWER:** Orange blossom. **TREE:** Sabal palmetto palm. **GEM:** Moonstone. **STONE:** Agatized coral. **SHELL:** Horse conch. **BEVERAGE:** Orange juice. **LEGAL HOLIDAYS:** New Year's Day, 1 January; Robert E. Lee's Birthday, 19 January; Birthday of Martin Luther King, Jr., 3d Monday in January; Lincoln's Birthday, 12 February; Susan B. Anthony's Birthday, 15 February; Washington's Birthday, 3d Monday in February; Shrove Tuesday, February or March; Good Friday, March or April; Pascua Florida Day, 2 April; Confederate Memorial Day, 26 April; Memorial Day, last Monday in May; Jefferson Davis's Birthday, 3 June; Independence Day, 4 July; Labor Day, 1st Monday in September; Columbus Day and Farmers' Day, 2d Monday in October; General Election Day, 1st Tuesday after the 1st Monday in November in even-numbered years; Veterans Day, 11 November; Thanksgiving Day, 4th Thurday in November; Christmas Day, 25 December. **TIME:** 7 AM EST = noon GMT; 6 AM CST = noon GMT.

¹LOCATION, SIZE, AND EXTENT

Located in the extreme southeastern US, Florida is the 2d-largest state (after Georgia) east of the Mississipi River, and ranks 22d in size among the 50 states.

The total area of Florida is 58,664 sq mi (151,939 sq km), of which land takes up 54,153 sq mi (140,256 sq km) and inland water 4,511 sq mi (11,683 sq km). Florida extends 361 mi (581 km) E–W; its maximum N–S extension is 447 mi (719 km). The state comprises a peninsula surrounded by ocean on three sides, with a panhandle of land in the NW.

Florida is bordered on the N by Alabama and Georgia (with the line in the NE formed by the St. Marys River); on the E by the Atlantic Ocean; on the S by the Straits of Florida; and on the W by the Gulf of Mexico and Alabama (separated by the Perdido River).

Offshore islands include the Florida Keys, extending from the state's southern tip into the Gulf of Mexico. The total boundary length of Florida is 1,799 mi (2,895 km). The state's geographic center is in Hernando County, 12 mi (19 km) NNW of Brooksville.

²TOPOGRAPHY

Florida is a huge plateau, much of it barely above sea level. The highest point in the state is believed to be a hilltop in the panhandle, 345 feet (105 meters) above sea level, near the city of Lakewood, in Walton County. No point in the state is more than 70 mi (113 km) from salt water.

Most of the panhandle region is gently rolling country, much like that of southern Georgia and Alabama, except that large swampy areas cut in from the Gulf coast. Peninsular Florida, which contains extensive swampland, has a relatively elevated central spine of rolling country, dotted with lakes and springs. Its east coast is shielded from the Atlantic by a string of sandbars. The west coast is cut by numerous bays and inlets, and near its southern tip are the Ten Thousand Islands, a mass of mostly tiny mangrove-covered islets. Southwest of the peninsula lies Key West, which, at 24°33′N, is the southernmost point of the US mainland.

Almost all the southeastern peninsula and the entire southern end are covered by the Everglades, the world's largest sawgrass swamp, with an area of approximately 5,000 sq mi (13,000 sq km). The Everglades is, in a sense, a huge river, in which water flows south-southwest from Lake Okeechobee to Florida Bay. No point in the Everglades is more than 7 feet (2 meters) above sea level. Its surface is largely submerged during the rainy season—April to November—and becomes a muddy expanse in the dry months. Slight elevations, known as hammocks, support clumps of cypress and the only remaining stand of mahogany in the continental US. To the west and north of the Everglades is Big Cypress Swamp, covering about 2,400 sq mi (6,200 sq km), which contains far less surface water.

Lake Okeechobee, in south-central Florida, is the largest of the state's approximately 30,000 lakes, ponds, and sinks. With a surface area of about 700 sq mi (1,800 sq km), it is the 4th-largest natural lake located entirely within the US. Like all of Florida's lakes, it is extremely shallow, having a maximum depth of 15 feet (5 meters), and was formed through the action of groundwater and rainfall in dissolving portions of the thick limestone layer that underlies Florida's sandy soil. The state's numerous underground streams and caverns were created in a similar manner. Because of the high water table, most of the caverns are filled, but some spectacular examples thick with stalactites can be seen in Florida Caverns State Park, near Marianna. More than 200 natural springs send up some 7 billion gallons of groundwater a day through cracks in the limestone. Silver Springs, near Ocala in north-central Florida, has the largest average flow of all inland springs, 823 cu feet (23 cu meters) per second.

Florida has more than 1,700 rivers, streams, and creeks. The longest river is the St. Johns, which empties into the Atlantic 19 mi (42 km) east of Jacksonville; estimates of its length range from 273 to 318 mi (439–512 km), a clear-cut figure being elusive because of the swampy nature of the headwaters. Other major rivers are the Suwannee, which flows south from Georgia for 177 mi (285 km) through Florida and empties into the Gulf of

Mexico; and the Apalachicola, formed by the Flint and Chattahoochee rivers at the Florida-Georgia border, and flowing southward across the panhandle for 94 mi (151 km) to the Gulf. Jim Woodruff Lock and Dam is located on the Apalachicola about 1,000 feet (300 meters) below the confluence of the two feeder rivers. Completed in 1957, the dam created Lake Seminole, most of which is in Georgia.

More than 4,500 islands ring the mainland. Best known are the Florida Keys, of which Key Largo—about 29 mi (47 km) long and less than 2 mi (3 km) wide—is the largest. Key West—less than 4 mi (6 km) long and 2 mi (3 km) wide—a popular resort, is the westernmost.

For much of the geological history of the US, Florida was under water. During this time, the shells of countless millions of sea animals decayed to form the thick layers of limestone that now blanket the state. The peninsula rose above sea level perhaps 20 million years ago. Even then, the southern portion remained largely submerged, until the buildup of coral and sand around its rim blocked out the sea, leaving dense marine vegetation to decay and form the peaty soil of the present-day Everglades.

³CLIMATE

A mild, sunny climate is one of Florida's most important natural resources, making it a major tourist center and a retirement home for millions of transplanted northerners. Average annual temperatures range from 65° to 70°F (18–21°C) in the north, and from 74° to 77°F (23–25°C) in the southern peninsula and on the Keys. At Jacksonville, the average annual temperature is 68°F (20°C); the average low is 57°F (14°C), the average high 79°F (26°C). At Miami, the annual average is 76°F (24°C), with a low of 69°F (21°C) and a high of 83°F (28°C). The record high temperature, 109°F (43°C), was registered at Monticello on 29 June 1931; the record low, −2°F (−19°C), at Tallahassee on 13 February 1899.

Florida's proximity to the Atlantic and the Gulf of Mexico, and the state's many inland lakes and ponds, together account for the high humidity and generally abundant rainfall, although precipitation can vary greatly from year to year and serious droughts have occurred. At Jacksonville, the average annual precipitation is 53 in (135 cm), with an average of 116 days of precipitation a year. At Miami, precipitation averages 58 in (147 cm), with 130 rainy days a year. Rainfall is unevenly distributed throughout the year, more than half generally occurring from June through September; periods of extremely heavy rainfall are common. The highest 24-hour total ever recorded in the US, 38.7 in (98.3 cm), fell at Yankeetown, west of Ocala on the Gulf coast, on 5–6 September 1950. Despite the high annual precipitation rate, the state also receives abundant sunshine—62% of the maximum possible at Jacksonville, and 73% at Miami. Snow is virtually unheard of in southern Florida but does fall on rare occasions in the panhandle and the northern peninsula.

Winds are generally from the east and southeast in the southern peninsula; in northern Florida, winds blow from the north in winter, bringing cold snaps, and from the south in summer. Average wind velocities are 8.2 mph (13.2 km/hr) at Jacksonville and 9.2 mph (14.8 km/hr) at Miami. Florida's long coastline makes it highly vulnerable to hurricanes and tropical storms, which may approach from either the Atlantic or the Gulf coast, bringing winds of up to 150 mph (240 km/hr). Hurricane Donna, which struck the state 9–10 September 1960 and is considered the most destructive in Florida's history, caused an estimated $300 million in damage. Tornadoes and waterspouts are not uncommon; in 1983, 85 tornadoes caused 51 injuries and 4 deaths.

⁴FLORA AND FAUNA

Generally, Florida has seven floral zones: flatwoods, scrublands, grassy swamps, savannas, salt marshes, hardwood forests (hammocks), and pinelands. Flatwoods consist of open forests and an abundance of flowers, including more than 60 varieties of orchid. Small sand pines are common in the scrublands; other trees here

are the saw palmetto, blackjack, and water oak. The savannas of central Florida support water lettuce, American lotus, and water hyacinth. North Florida's flora includes longleaf and other pines, oaks, and cypresses; one giant Seminole cypress is thought to be 3,500 years old. The state is known for its wide variety of palms, but only 15 are native and more than 100 have been introduced; common types include royal and coconut. Although pine has the most commerical importance, dense mangrove thickets grow along the lower coastal regions, and northern hardwood forests include varieties of rattan, magnolia, and oak. Numerous rare plants have been introduced, among them bougainvillea and oleander. All species of cacti and orchids are regarded as threatened, as are most types of ferns and palms. Endangered species include the prickly apple, key tree cactus, cowhorn orchid, Chapman rhododendron, Harper's beauty, and Florida torreya.

Florida once claimed more than 80 land mammals. Today the white-tailed deer, wild hog, and gray fox can still be found in the wild; such small mammals as the raccoon, eastern gray and fox squirrels, and cottontail and swamp rabbits remain common. Florida's bird population includes many resident and migratory species. The mockingbird was named the state bird in 1927; among game birds are the bobwhite quail, wild turkey, and at least 30 duck species. Several varieties of heron are found, as well as coastal birds such as gulls, pelicans, and frigates. The Arctic tern stops in Florida during its remarkable annual migration between the North and South poles.

Common Florida reptiles are the diamondback rattler and various water snakes. Turtle species include mud, green, and loggerhead, and various lizards abound. More than 300 native butterflies have been identified. The peninsula is famous for its marine life: scores of freshwater and saltwater fish, rays, shrimps, live coral reefs, and marine worms.

All of Florida's lands have been declared sanctuaries for the bald eagle, of which Florida has about 350 pair (2d only to Alaska among the 50 states). The state's unusually long list of threatened and endangered wildlife includes the American crocodile, American alligator, shortnose sturgeon, Atlantic ridley turtle, brown pelican, dusky seaside sparrow, red-cockaded and ivory-billed woodpeckers, Florida panther, Key deer, West Indian (Florida) manatee, Key Largo cotton mouse, Key Largo woodrat, Everglade kite, Cape Sable seaside sparrow, Bachman's warbler, Atlantic salt marsh snake, eastern indigo snake, Okaloosa darter, Stock Island snail, Bahama swallowtail butterfly, and Schaus swallowtail butterfly.

⁵ENVIRONMENTAL PROTECTION

Throughout the 20th century, a rapidly growing population, the expansion of agriculture, and the exploitation of such resources as timber and minerals have put severe pressure on Florida's natural environment.

The state agency principally responsible for safeguarding the environment is the Department of Environmental Regulation (DER), created by the Environmental Reorganization Act of 1975. Its duties include implementing state pollution control laws and improving water-resource management. The department's Division of Environmental Programs oversees and coordinates the activities of the state's five water-management districts, which have planning and regulatory responsibilities. The Department of Natural Resources protects the state's coastal and marine resources. Its Division of Recreation and Parks is authorized by the Conservation Act of 1972 to acquire environmentally endangered tracts of land; the division also administers state parks and wilderness lands. The Department of Agriculture and Consumer

LOCATION: 24°33' to 31°N; 80°02' to 87°28'W. **BOUNDARIES:** Alabama line, 219 mi (353 km); Georgia line, 230 mi (370 km); Atlantic Ocean coastline, 580 mi (933 km); Gulf of Mexico coastline, 770 mi (1,239 km).

Services' Division of Forestry manages four state forests plus the Talquin State Lands. The Game and Fresh Water Fish Commission manages nature preserves and regulates hunting and fishing.

State spending for environmental protection in 1983/84 was $48.9 million. Federal aid in 1983/84 included $90.8 million from the Environmental Protection Agency (EPA) and $13.1 million from the Department of the Interior for land and water conservation, fish and wildlife restoration and management, and the construction of urban parks.

According to a 1983 report by the House Speaker's Task Force on Water Issues, the contamination of groundwater is the state's most serious environmental problem. Groundwater supplies 90% of the drinking water in the state, as well as 82% of industry's needs and 53% of agricultural uses. In 1982/83, groundwater contamination was found at 100 sites, and surface water or soil contamination at 15 sites. Among the major contaminants were the pesticides ethylene dibromide (EDB) and Temik. Of the 1983/84 federal aid from the EPA, $84.9 million was for the construction of sewage treatment plants. A similar amount was granted in 1984/85, and in 1983/84, the state government budgeted $100 million for wastewater treatment works. In 1984, a Groundwater Protection Task Force was established and a three-year program was begun to collect and properly dispose of hazardous wastes in Dade County.

Contamination of groundwater is not the state's only water problem. The steadily increasing demand for water for both residential and farm use has reduced the subterranean runoff of fresh water into the Atlantic and the Gulf of Mexico. As a result, salt water from these bodies has begun seeping into the layers of porous limestone that hold Florida's reserves of fresh water. This problem has been aggravated in some areas by the cutting of numerous inlets by developers of coastal property.

The conversion of large areas of the northern Everglades into farmland has reduced the flow of water to the southern part of the swamp, threatening its delicate ecological balance. To protect crops and create new agricultural land, networks of dikes, levees, spillways, and canals were constructed after World War II to prevent the seasonal flooding by Lake Okeechobee of the region to its south. In an effort to protect a portion of the Everglades from encroachment, Everglades National Park, covering 1,400,533 acres (566,777 hectares), was created in 1947 at the swamp's southern tip.

In 1960, the only undersea park in the US, the John Pennekamp Coral Reef State Park, was established in a 75-sq-mi (194-sq-km) sector off the Atlantic coast of Key Largo, in an effort to protect a portion of the beautiful reefs, rich in tropical fish and other marine life, that adjoin the Keys. Untreated sewage from the Miami area, runoff water polluted by pesticides and other chemicals, dredging associated with coastal development, and the removal of countless pieces of live coral by growing numbers of tourists and souvenir dealers have severely damaged large areas of the reefs.

[6]POPULATION

Florida, the most populous state in the southeastern US, is also one of the fastest growing of the 50 states. In 1960, it was the 10th most populous state; by 1980, it ranked 7th with a population of 9,746,324; and by early 1985, it ranked 6th, with an estimated population of 11,071,358. US Census Bureau projections indicate

Florida Counties, County Seats, and County Areas and Populations

COUNTY	COUNTY SEAT	LAND AREA (SQ MI)	POPULATION (1984 EST.)	COUNTY	COUNTY SEAT	LAND AREA (SQ MI)	POPULATION (1984 EST.)
Alachua	Gainesville	902	166,588	Lee	Ft. Myers	803	252,887
Baker	Macclenny	585	17,136	Leon	Tallahassee	676	164,338
Bay	Panama City	758	108,508	Levy	Bronson	1,100	23,681
Bradford	Starke	293	23,198	Liberty	Bristol	837	4,519
Brevard	Titusville	995	329,497	Madison	Madison	710	15,340
Broward	Ft. Lauderdale	1,211	1,093,340	Manatee	Bradenton	747	169,525
Calhoun	Blountstown	568	9,509	Marion	Ocala	1,610	155,581
Charlotte	Punta Gorda	690	75,235	Martin	Stuart	555	79,101
Citrus	Inverness	629	73,796	Monroe	Key West	1,034	69,451
Clay	Green Cove Springs	592	81,260	Nassau	Fernandina Beach	649	37,448
Collier	East Naples	1,994	110,798	Okaloosa	Crestview	936	127,523
Columbia	Lake City	797	39,048	Okeechobee	Okeechobee	771	24,708
Dade	Miami	1,955	1,705,983	Orange	Orlando	910	532,558
De Soto	Arcadia	636	20,862	Osceola	Kissimmee	1,350	69,531
Dixie	Cross City	701	9,134	Palm Beach	West Palm Beach	1,993	692,217
Duval	Jacksonville	776	612,668	Pasco	Dade City	738	225,109
Escambia	Pensacola	661	255,369	Pinellas	Clearwater	280	792,533
Flagler	Bunnell	491	15,640	Polk	Bartow	1,823	354,572
Franklin	Apalachicola	545	8,214	Putnam	Palatka	733	57,404
Gadsden	Quincy	518	43,301	St. Johns	St. Augustine	617	63,947
Gilchrist	Trenton	354	6,971	St. Lucie	Ft. Pierce	581	111,140
Glades	Moore Haven	763	6,617	Santa Rosa	Milton	1,024	63,300
Gulf	Port St. Joe	559	11,214	Sarasota	Sarasota	573	237,588
Hamilton	Jasper	517	9,232	Seminole	Sanford	298	222,038
Hardee	Wauchula	637	20,732	Sumter	Bushnell	561	28,192
Hendry	La Belle	1,163	21,778	Suwannee	Live Oak	690	24,979
Hernando	Brooksville	477	68,829	Taylor	Perry	1,058	17,847
Highlands	Sebring	1,029	56,341	Union	Lake Butler	246	11,288
Hillsborough	Tampa	1,053	724,454	Volusia	DeLand	1,113	300,342
Holmes	Bonifay	488	15,699	Wakulla	Crawfordville	601	12,494
Indian River	Vero Beach	497	74,094	Walton	De Funiak Springs	1,066	24,675
Jackson	Marianna	942	40,458	Washington	Chipley	590	15,858
Jefferson	Monticello	609	11,472				
Lafayette	Mayo	545	4,275		TOTALS	54,153	10,975,748
Lake	Tavares	954	122,789				

that Florida will have 13,316,000 residents in 1990, for a rank of 4th, and that it will move ahead of New York, taking over 3d place, by the year 2000.

The first US census to include Florida, in 1830, recorded a total population of only 34,730. By 1860, on the eve of the Civil War, the population had more than quadrupled, to 140,424 people; about 80% of them lived in the state's northern rim, where cotton and sugarcane plantations flourished. The population migrated southward in the late 19th and early 20th centuries, following the railroads. A land boom in the early 1920s sharply increased the state's population; the 1930 census was the first in which the state passed the million mark. Migration from other states, especially of retirees, caused a population explosion in the post–World War II period, with much of the increase occurring along the south Atlantic coast. From 1950 to 1960, Florida's population increased 79%—the fastest rate of all the states. From 1960 to 1970, the growth rate was 37%; from 1970 to 1980, 44%; and from 1980 to 1985, 14%.

Of the 1983 population, 91% lived in metropolitan areas; the average population density in 1985 was 204 per sq mi (79 per sq km). Females constituted 52% of the population in 1980, males 48%. Nearly 18% of the 1983 population was 65 years of age or over—the highest such percentage of all the states and 50% above the US average.

The most populous city in Florida is Jacksonville, the 19th-largest city in the US in 1984. Its population, as measured by the 1980 census, was 540,920; the 1984 estimate was 577,971, an increase of nearly 7%. Miami is Florida's 2d-largest city, with a 1984 population of 372,634. The Miami–Ft. Lauderdale metropolitan area, the state's largest metropolitan region, had an estimated 2,799,300 residents (11th in the US) in mid-1984; Jacksonville's metropolitan area population was 795,300, for a rank of 50th in the US. Florida's 2d-largest metropolitan area was Tampa–St. Petersburg–Clearwater, with 1,810,900 residents (20th in the US); the city of Tampa had 275,479 people in 1984, and St. Petersburg had 241,294. Ft. Lauderdale had a city population of 149,872 in 1984. Tallahassee, the state capital, had an estimated population of 112,258.

7ETHNIC GROUPS

Florida's population consists mainly of whites of northern European stock, blacks, and Hispanics. European immigrants came primarily from Germany and the United Kingdom. Germans were particularly important in the development of the citrus fruit industry. Since World War II, the development of southern Florida as a haven for retired northerners has added new population elements to the state, a trend augmented by the presence of numerous military bases.

The largest group of first- and second-generation residents are Cubans, who represented nearly 5% of Florida's population in 1980. Puerto Ricans, Central Americans, and Filipinos have also settled in Florida. The total Hispanic population was 858,158 in 1980, including 470,250 Cubans (more than 100,000 of whom arrived on Florida shores as refugees in 1980), 94,775 Puerto Ricans, and 79,392 Mexicans.

The nonwhite population, as reported in 1980, was 1,426,876, or almost 15%. Black-white relations in the 20th century have been tense. There were race riots following World War I, and the Ku Klux Klan was openly active until World War II. One of the worst race riots in US history devastated black areas of Miami in the spring of 1980.

Florida Indians resisted white encroachment longer and more militantly than Indians in other seaboard states. The leaders in resistance were the Seminole, most of whom by the 1850s had been killed or removed to other states, had fled to the Florida swamplands, or had been assimilated as small farmers. No peace treaty was signed with the Seminole until 1934, following the Indian Reorganization Act that attempted to establish tribal

integrity and self-government for Indian nations. In 1939, the Indian population was reported as only 600, but the 1980 census reported a figure of 24,714 American Indians from 34 tribes. The difference is too large to be explained by natural increase, and there is no evidence of marked in-migration; presumably, then, it reflects a growing consciousness of Indian identity. There are seven Indian reservations: five for the Seminole—Big Cypress, Hollywood, Brighton, Immokallee, and Tampa—and two for the Miccosuckee—one on the Tamiami Trail and one north of Alligator Alley near Big Cypress.

Florida has the 11th-largest population of Asian and Pacific Islanders of the 50 states. In 1980 there were 14,212 Filipinos, 13,471 Chinese, 9,138 Asian Indians, 7,592 Vietnamese, 5,565 Japanese, 4,673 Koreans, and 1,377 Hawaiians.

8LANGUAGES

Spanish and English settlers found what is now Florida inhabited by Indians recently separated from the Muskogean Creeks, who, with the addition of escaped black slaves and remnants of the Apalachee Indians of the panhandle, later became known as the Seminole Indians. Although the bulk of the Seminole were removed to Indian Territory in the 1840s, enough remained to provide the basis of the present population. Florida has such Indian place-names as Okeechobee, Apalachicola, Kissimmee, Sarasota, Pensacola, and Hialeah.

The rapid population change that has occurred in Florida since World War II makes accurate statements about the language difficult. Massive migration from the North Central and North Atlantic areas, including a large number of speakers of Yiddish, has materially affected the previously rather uniform Southern speech of much of the state. Borrowing from the Spanish of the expanding number of Cubans and Puerto Ricans in the Miami area has had a further effect.

Representative words in the Southern speech of most native-born Floridians are *light bread* (white bread), *pallet* (temporary bed on the floor), *fairing off* (clearing up), *serenade* (shivaree), *tote* (carry), *snap beans* (green beans), *mosquito hawk* (dragonfly), *crocus sack* (burlap bag), *pullybone* (wishbone), and *comforter* (tied and filled bedcover), especially in south Florida. Largely limited to the northern half of the state are *pinder* (peanut), *croker sack* instead of crocus sack, *fire dogs* (andirons); also, in the Tampa Bay area, *comfort* (tied and filled bedcover), and, in the panhandle, *whirlygig* (merry-go-round). Some north Florida terms are clearly imported from Georgia: *mutton corn* (green corn), *lightwood* (kindling), and *co-wench!* (a call to cows).

In 1980, 8,161,863 Floridians—87% of the resident population 3 years old and older—spoke only English at home. Other languages spoken at home included:

Spanish	806,744	Yiddish	37,390
French	73,285	Polish	20,425
German	68,604	Greek	15,542
Italian	59,780	Hungarian	10,783

9RELIGIONS

Protestant denominations claim the majority of church members in Florida. The state also has sizable Roman Catholic and Jewish populations.

Dominican and Franciscan friars, intent on converting the Indians, arrived with the Spanish conquistadores and settlers in the 1500s, and for some 200 years Florida's white population was overwhelmingly Catholic. Protestant colonists from Britain arrived in the late 1700s, and a significant influx of Protestant settlers from the southern US followed in the early 1800s. Sephardic Jews from the Carolinas also moved into Florida around this time, although the largest influx of Jews has occurred during the 20th century.

The largest Protestant denominations in 1980 were Southern Baptist Convention, with 786,276 members; United Methodist

Church, 361,447; Episcopal Church, 120,332; Presbyterian Church in the US, 88,608; Assembly of God, 52,595; Lutheran Church–Missouri Synod, 45,627; Churches of Christ, 44,829; United Presbyterian Church in the USA, 44,196; and Lutheran Church in America, 43,208. The Roman Catholic population of Florida in 1984 was 1,584,202. The state has one archdiocese (Miami) and six dioceses (Orlando, Palm Beach, Pensacola-Tallahassee, St. Augustine, St. Petersburg, and Venice).

The estimated number of Jews in Florida in 1984 was 558,820, about 5% of the state's total population and almost 10% of all Jews in the US. The majority of Florida's Jewish population was concentrated along the south Atlantic coast, from Palm Beach to Miami, including 253,340 in the Miami area and 100,000 in Ft. Lauderdale. Florida's Jewish population increased by 28% between 1979 and 1984.

¹⁰TRANSPORTATION

Railroad building in the 19th century opened southern Florida to tourism and commerce. During the 20th century, long-distance passenger trains and, more recently, planes and automobiles have brought millions of visitors to the state each year.

The first operating railway in Florida was the St. Joseph Railroad, which inaugurated service on an 8-mi (13-km) track between St. Joseph Bay and Lake Wimico on 14 April 1836—using mules to pull the train. The railroad soon put into operation the state's first steam locomotive, on 5 September 1836. By the time the Civil War broke out, railroads connected most of northern Florida's major towns, but the rapid expansion of the state's railroad system—and with it the development of southern Florida—awaited two late-19th-century entrepreneurs, Henry B. Plant and Henry M. Flagler. Plant's South Florida Railroad extended service to Tampa in 1884. Flagler consolidated a number of small lines in the 1880s into the Florida East Coast Railway, with service as far south as Daytona. He then extended service down the Atlantic coast, reaching Palm Beach in 1894, Miami in 1896, and, after construction of an extensive series of bridges, Key West in 1912. The "overseas" railway down the Keys was abandoned in 1935 after a hurricane severely damaged the line.

Florida's most extensive railroad is the Seaboard System Railroad, with 2,804 mi (4,513 km) of track within the state as of 31 December 1983; the total in-state trackage in use on that date was 3,335 mi (5,367 km). As of 1984, Amtrak provided passenger service to 31 Florida cities; 788,255 Florida passengers rode the Amtrak system that year.

On 7 June 1979, construction began on a surface rail system for Miami and surrounding areas of Dade County. The first stage of this $1.1-billion mass transit system (known as Metrorail), a 20.5-mi (33-km) line serving Hialeah, Miami International Airport, downtown Miami, and areas to the south, was opened on 20 May 1984.

As of 1984, Florida had 97,396 mi (156,744 km) of public roads. Of this total, 11,460 mi (18,443 km) constituted the state highway system, including 1,302 mi (2,095 km) of interstate highways and the 327-mi (526-km) Florida Turnpike, a toll road. The turnpike's 265-mi (426-km) main section extends from Wildwood in north-central Florida to Ft. Pierce on the Atlantic coast and then south to Miami; a 50-mi (80-km) extension runs between Miramar and Homestead. The Overseas Highway down the Keys, including the famous Seven Mile Bridge (which is actually 35,716 feet, or 10,886 meters—6.8 mi—in length), is part of the state highway system. In 1983, 37 of the 44 bridges connecting the Florida Keys were replaced at a cost of $189 million. Florida also had 59,476 mi (95,718 km) of county roads as of 1984, of which 33,350 mi (53,672 km) were paved. There were some 26,782 mi (43,102 km) of city streets and local roads, of which 22,324 mi (35,927 km) were paved.

Florida had 8,808,486 registered motor vehicles in 1983:

7,113,942 automobiles (24% more than in 1978), 33,276 buses, and 1,661,268 trucks. In addition, there were 232,488 registered motorcycles. As of 1983, 8,347,269 people held active Florida drivers' licenses, 4,333,557 men and 4,013,712 women.

Inland waterways in Florida include the southernmost section of the Atlantic Intracoastal Waterway and the easternmost section of the Gulf Intracoastal Waterway; both are navigable, federally maintained coastal channels for commercial vessels and pleasure craft. Construction began on 27 February 1964 on a barge canal across northern Florida to connect the two intracoastal systems; however, work was ordered stopped by President Richard Nixon on 19 January 1971 because of the threat the canal posed to flora and fauna in the surrounding area.

Florida has several commercially important ports. By far the largest port is Tampa, which handled 43,597,171 tons of cargo in 1982/83. Other major ports and their 1982/83 tonnage handled include Everglades, in Ft. Lauderdale, 12,341,955; Manatee, south of Tampa, 5,744,331; Palm Beach, 2,698,796; Miami, 2,305,645; Jacksonville, 2,213,825; Canaveral, 2,024,529; and Pensacola, 1,185,749.

Florida is the 3d-ranking state in terms of aircraft, pilots, airline passengers, and airports. In addition to civil aviation activity, Florida has 22 military airfields, of which 14 are major air bases. There are approximately 16,000 aircraft based in the state, as well as 41,000 licensed pilots. During 1983, more than 25,000,000 passengers took off from Florida's 23 commercial service airports. There are 564 licensed Florida airfields, including 140 public airports, heliports, seaplane bases, and a blimp facility. Florida's busiest airport is Miami International, which enplaned 7,337,567 passengers in 1983. The same year, Orlando International enplaned 3,721,059 passengers; Tampa International, 3,688,571; and Ft. Lauderdale–Hollywood International, 2,605,343.

¹¹HISTORY

Indians entered Florida from the north 10,000–12,000 years ago, and had reached the end of the peninsula by 1400 BC. They subsisted mainly on large prehistoric animals until around 5000 BC, when they began to hunt smaller animals, as the huge ones became extinct, and to live in villages and, in the north, to practice some agriculture. As they grew in number, the Indians developed more complex economic and social organization. In northeastern Florida and nearby Georgia, they apparently invented pottery independently about 2000 BC, some 800 years earlier than any other Indian group in North America.

In north Florida, an agricultural and hunting economy organized around village life was typical by this time. South of Tampa Bay and Cape Canaveral, Indians lived mostly along the coast and relied heavily on wild plants and on a large variety of aquatic and land animals for meat. The southern groups did not practice agriculture until about 450 BC, when they began to plant corn in villages around Lake Okeechobee. Huge mounds of shells along the southwest coast attest to the importance of clams, conch, and oysters in their diet.

As they spread over Florida and adjusted to widely different local conditions, the Indians fell into six main divisions, with numerous subgroups and distinctive cultural traits. When Europeans arrived in the early 16th century, they found nearly 100,000 Indians: 25,000 Apalachee around Tallahassee; 40,000 Timucua in the northeast; on Tampa Bay, 7,000 Tocobaga; on the southwest coast and around Lake Okeechobee, 20,000 Calusa; on the lower southeast coast, 5,000 Tequesta; and in the Jupiter area, 2,000 Ais and Jeaga.

The Spanish who began arriving in the 16th century found the Indians in upper Florida to be relatively tractable, but those in the lower peninsula remained uniformly hostile and resisted to the last. The Spaniards sought to Christianize the Indians and settle them around missions—to grow food, to supply labor, and to help defend the province. By 1674, 70 Franciscan friars were working in dozens of missions and stations in a line running west from St. Augustine and north along the sea island coast to Carolina.

The impact of the Europeans on the Indian population was, on the whole, disastrous. Indians died of European-introduced diseases, were killed in wars with whites or with other Indians, or moved away. Raids from South Carolina by the Creeks, abetted by the British, between 1702 and 1708 completely destroyed the missions. When the Spanish departed Florida in 1763, the remaining 300 of the original 100,000 Indians left with them.

As early as 1750, however, small groups of Creek tribes from Georgia and Alabama had begun to move into the north Florida area vacated by the first Indian groups. They came in small bands, from widely separated places, at wide intervals, to settle in scattered parts of Florida. Called Seminole—the Creek word for runaway or refugee—these Indians did not then constitute a tribe and had no common government or leadership until resistance to white plans to resettle them brought them together. They numbered only 5,000 when Florida became part of the US.

Pressures on the US president and Congress to remove the Seminole intensified after runaway black slaves began seeking refuge with the Indians. In 1823, the Seminole accepted a reservation north of Lake Okeechobee. Nine years later, an Indian delegation signed a document pledging the Seminole to move within three years to lands in present-day Oklahoma. The Indians' subsequent resistance to removal resulted in the longest and most costly of Indian wars, the Seminole War of 1835-42. The warfare and the Indians' subsequent forced migration left fewer than 300 Seminole in Florida.

The history of the twice-repeated annihilation of Florida Indians is, at the same time, the history of white settlers' rise to power. After Christopher Columbus reached the New World at Hispaniola in 1492, the Caribbean islands became the base for wider searches, one of which brought Juan Ponce de León to Florida. Sailing from Puerto Rico in search of the fabled island of Bimini, he sighted Florida on 27 March 1513 and reached the coast a week later. Ponce de León claimed the land for Spain and named it La Florida, for *Pascua Florida,* the Easter festival of flowers; sailing southward around Florida, he may have traveled as far as Apalachicola, on the shore of the panhandle. In 1521, he returned to found a colony at Charlotte Harbor, on the lower Gulf coast, but the Indians fought the settlers. After Ponce was seriously wounded, the expedition sailed for Cuba, where he died the same year.

Other Spaniards seeking treasure and lands to govern came after Ponce. Pánfilo de Narváez arrived in 1528. Landing near Tampa Bay, he marched inland and northward to Tallahassee, finding only difficult terrain and hostile Indians with little surplus food. Missing a rendezvous with ships at St. Marks, on Apalachee Bay south of Tallahassee, he built five crude boats and took to the Gulf, but the party ran into a storm on the Texas coast that left no boats and only 80 survivors, 4 of whom, marching overland, reached Mexico City in 1536.

Hernando de Soto, a rich and famous associate of Francisco Pizarro in the conquest of Peru, found many men eager to try the same with him in Florida. Appointed governor of Cuba and *adelantado* (loosely, conqueror) of Florida, he followed the route of Narváez to Tallahassee in 1539, finding some food but no promise of wealth.

In 1559, Spain sought to establish a settlement on Pensacola Bay, as well as one at Santa Elena on Port Royal Sound, in South Carolina, to protect the trade route along Florida's east coast. Tristan de Luna led a large expedition to Pensacola Bay, landing on 14 August, but he lost most of his ships and supplies in a storm shortly after he arrived. A smaller expedition thence to Santa Elena was wrecked en route. Unable to find or produce food, Tristan de Luna abandoned the Pensacola settlement at the end of two years.

In 1562, Jean Ribault, with a small expedition of French Huguenots, arrived at the St. Johns River, east of present-day Jacksonville, and claimed Florida for France. Another group of French Huguenot settlers built Ft. Caroline, 5 mi (8 km) upriver, two years later. In the summer of 1565, Ribault brought in naval reinforcements, prepared to defend the French claim against the Spaniards, who had sent the redoubtable Pedro Menéndez de Avilés to find and oust the intruders. Menéndez selected St. Augustine as a base, landing on 28 August, and with the aid of a storm withstood the French effort to destroy him. He then marched overland to take Ft. Caroline by surprise, killing most of the occupants, and later captured Ribault and his shipwrecked men, most of whom he slaughtered. St. Augustine—the first permanent European settlement in the US—served primarily, under Spanish rule, as a military outpost, maintained to protect the wealth of New Spain. Attesting to its importance was the construction of Castillo de San Marcos, the stone fort that still stands guard there. The Spanish established a settlement at Pensacola in 1698, but it too remained only a small frontier garrison town. In 1763, when Spain ceded Florida to England in exchange for Cuba, about 3,000 Spaniards departed from St. Augustine and 800 from Pensacola, leaving Florida to the Seminole.

British Florida reached from the Atlantic to the Mississippi River and became two colonies, East and West Florida. Settlers established farms and plantations, traded with the Indians, and moved steadily toward economic and political self-sufficiency. These settlers did not join the American Revolution, but Florida was affected by the war nonetheless, as thousands of Loyalists poured into East Florida. In 1781, Spain attacked and captured Pensacola. Two years later, Britain ceded both Floridas back to Spain, whereupon most of the Loyalists left for the West Indies.

The second Spanish era was only nominally Spanish. English influence remained strong, and US penetration increased. Florida west of the Perdido River was taken over by the US in 1810, as part of the Louisiana Purchase (1803). Meanwhile, renegade whites, runaway slaves, pirates, and political adventurers operated almost at will.

Present-day Florida was ceded to the US in 1821, in settlement of $5 million in claims by US citizens against the Spanish government. At this time, General Andrew Jackson—who three years earlier had led a punitive expedition against the Seminole and their British allies—came back to Florida as military governor. His main tasks were to receive the territory for the US and to set up a civilian administration, which took office in 1822. William P. DuVal of Kentucky was named territorial governor, and a legislative council was subsequently elected. The new council met first in Pensacola and in St. Augustine, and then, in 1824, in the newly selected capital of Tallahassee, located in the wilderness of north-central Florida, from which the Indians had just been removed. Middle Florida, as it was called, rapidly became an area of slave-owning cotton plantations, and was for several decades the fastest-growing part of the territory. Of the 70,000 people in the state by 1845, Middle Florida had 47% of the total, and four Middle Florida counties with Negro majorities—Jackson, Gadsden, Jefferson, and Leon—were producing 80% of the cotton. The war to remove the Seminole halted the advance of frontier settlement, however, and the Panic of 1837 bankrupted the territorial government and the three banks whose notes it had guaranteed. Floridians drew up a state constitution at St. Joseph in 1838-39, but, being proslavery, had to wait until 1845 to enter the Union, paired with the free state of Iowa.

In 1861, Florida—which had only 140,000 people, about 40% of them blacks (mostly slaves), only 400 mi (644 km) of railroad, and no manufacturing—seceded from the Union and joined the Confederacy. Some 15,000 whites (one-third of whom died) served in the Confederate army, and 1,200 whites and almost as many blacks joined the Union army. Bitterness and some violence accompanied Republican Reconstruction government in 1868–76. The conservative Bourbon Democrats then governed for the rest of the century. They encouraged railroad building and other forms of business, and they kept taxes low by limiting government services. Cotton production never recovered prewar levels, but cattle raising, citrus and vegetable cultivation, forestry, phosphate mining, and, by late in the century, a growing tourist industry took up the slack.

The Spanish-American War in 1898, during which Tampa became the port of embarkation for an expedition to Cuba, stimulated the economy and advertised the state nationwide, not always favorably. Naval activity at Key West and Pensacola became feverish. Lakeland, Miami, Jacksonville, and Fernandina were briefly the sites of training camps.

In 1904, Napoleon Bonaparte Broward was elected governor on a moderately populist platform, which included a program to drain the Everglades lands which the state had received under the Swamp and Overflowed Lands Act of 1850. Drainage did lower water levels, and settlements grew around Lake Okeechobee—developments whose full environmental impact was recognized only much later. By the time Broward took office, Jacksonville had become the state's largest city, with Pensacola and Tampa not far behind, and Key West had dropped from 1st to 4th. During World War I, more than 42,030 Floridians were in uniform.

Boom, bust, blow, and depression characterized the 1920s. Feverish land speculation brought hundreds of thousands of people to Florida in the first half of the decade. Cresting in 1925, the boom was already over in 1926, when a devastating hurricane struck Miami, burying all hope of recovery. Yet population jumped by more than 50% during the decade, and Miami rose from 4th to 2d place among Florida cities. Florida's choice of Republican Herbert Hoover over Al Smith in the 1928 presidential election reflected the Protestant and prohibitionist attitudes of most of the state voters at that time.

The 1930s were marked first by economic depression, then by recovery, new enterprise, and rapidly growing government activity. The Florida depression that began in 1926 was compounded by the national depression that hit late in 1929. Bank and business failures, as well as defaults on city and county bond issues and on mortgage payments, produced growing economic distress. The state joined the federal government in assuming responsibility for relief and recovery. The legalization of pari-mutuel betting in 1931 created a new industry and a new tax source. The state's first paper mill opened in the same year, revolutionizing the forest industry. Private universities in Miami, Tampa, and Jacksonville were started during the depression years.

The 1940s opened with recovery and optimism, arising from the stimulus of production for World War II, production that began well before the actual entry of the US into the war. New Army and Navy installations and training programs brought business growth. After 1941, Florida seemed to become a vast military training school. The number of Army and Navy airfield flying schools increased from 5 to 45. Tourist facilities in all major cities became barracks, mess halls, and classrooms, with 70,000 rooms in Miami Beach alone being used to house troops in 1942. Families of thousands of trainees visited the state. Florida was on the eve of another boom.

First discovered but near last to be developed, the 27th state reached a rank of 27th in population only in 1940. But it was 6th in 1985, with more than 11 million people. Most of the increase was the result of migration, with enough of the migrants being 65 years of age or over to make the proportion of senior citizens in Florida 50% above the national average in the mid-1980s.

¹²STATE GOVERNMENT

Florida's first constitutional convention, which met from December 1838 to January 1839, drew up the document under which the state entered the Union in 1845. A second constitutional convention, meeting in 1861, adopted the ordinance of secession that joined Florida to the Confederacy. After the war, a new constitution was promulgated in 1865, but not until still another document was drawn up and ratified by the state—the fourteenth amendment to the US Constitution—was Florida readmitted to statehood in 1868. A fifth constitution was framed in 1885; extensively revised in 1968, this is the document under which the state is now governed.

The 1968 constitutional revision instituted annual (rather than biennial) regular sessions of the legislature, which consists of a 40-member senate and a 120-member house of representatives. Senators serve four-year terms, with half the senate being elected every two years; representatives serve two-year terms. All legislators must be US citizens, must be at least 21 years of age, must have been residents of Florida for at least two years, and must be registered voters and residents of the district. A legislator's salary in 1985 was $12,000 a year. The maximum length of a regular session is 60 calendar days, unless it is extended by a three-fifths vote of each house. Special sessions may be called by the governor or by joint action of the presiding officers of the two houses (the president of the senate and speaker of the house of representatives). In addition, a special session may be convened by a three-fifths vote of all legislators, the poll being conducted by mail by the secretary of state upon a written request from at least 20% of the members.

The governor is elected for a four-year term; a two-term limit is in effect. The lieutenant governor is elected on the same ticket as the governor. A six-member cabinet—consisting of the secretary of state, attorney general, comptroller, insurance commissioner and treasurer, commissioner of agriculture, and commissioner of education—is independently elected. Each of its officials must be at least 30 years old, a US citizen, and a registered voter, and must have been a resident of Florida for at least seven years; in addition, the attorney general must have been a member of the Florida bar for at least five years.

Each cabinet member heads an executive department. The governor appoints the heads of 10 departments and shares supervision of 7 additional departments with the cabinet. The governor and cabinet also share management of or membership in several other state agencies. These provisions make Florida's elected cabinet one of the strongest such bodies in any of the 50 states.

The Public Service Commission, an arm of the legislative branch with quasi-judicial powers, sets rates for and otherwise regulates (consistent with Interstate Commerce Commission rulings) railroads, telephone companies, and privately owned electric, gas, water, and sewer utilities. The PSC's five members are appointed by the governor from lists prepared by the Florida Public Service Nominating Council, a nine-member body selected, for the most part, by the legislative leadership.

Passage of legislation requires a majority vote of those present and voting in both houses. A bill passed by the legislature becomes law if it is signed by the governor; should the governor take no action on it, it becomes law 7 days (including Sundays) after receipt if the legislature is still in session, or 15 days (including Sundays) after presentation to the governor if the legislature has adjourned. The governor may veto legislation and, in general appropriations bills, may veto individual items. Gubernatorial vetoes may be overridden by a two-thirds vote of the legislators present in each house.

Amendments to the constitution may originate in three ways: by a joint resolution of the legislature passed by a three-fifths majority of the membership of each house; by action of a constitutional revision commission which, under the constitution, must be periodically convened; or by initiative petition, which may call for a constitutional convention. A proposed amendment becomes part of the constitution if it receives a majority vote in a statewide election.

To be eligible to vote in state elections, a person must be at least 18 years of age, a US citizen, and a resident in the county of registration; the registration books close 30 days before a general election.

13 POLITICAL PARTIES

The Democratic and Republican parties are Florida's two principal political organizations. The former is the descendant of one of the state's first two political parties, the Jeffersonian Republican Democrats; this party, along with the Florida Whig Party, was organized shortly before statehood.

Florida's Republican Party was organized after the Civil War and dominated state politics until 1876, when the Democrats won control of the statehouse. Aided from 1889 to 1937 by a poll tax, which effectively disfranchised most of the state's then predominantly Republican black voters, the Democrats won every gubernatorial election but one from 1876 through 1962; the Prohibition Party candidate was victorious in 1916.

By the time Republican Claude R. Kirk, Jr., won the governorship in 1966, Florida had already become, for national elections, a two-party state, although Democrats retained a sizable advantage in party registration. Beginning in the 1950s, many registered Democrats became "presidential Republicans," crossing party lines to give the state's electoral votes to Dwight D. Eisenhower in 1952 and 1956 and to Richard M. Nixon in 1960 (making Nixon the only presidential candidate between 1928 and 1984 to carry Florida but lose the presidency).

A presidential preference primary, in which crossover voting is not permitted, is held on the 2d Tuesday in March of presidential election years. Because it occurs so early in the campaign season, this primary is closely watched as an indicator of candidates' strength. Primaries to select state and local candidates are held in early September, with crossover voting again prohibited; runoff elections are held on the Tuesday five weeks before the general election.

As of 1982, the state had 3,066,351 registered Democrats and 1,500,031 registered Republicans. There were about 464,000 registered black voters. In addition to the Democratic and

Republican parties, organized groups include the Citizens and Libertarian parties. Minor parties running candidates for statewide office can qualify by obtaining petition signatures from 3% of the state's voters.

Democratic candidate Reubin Askew was elected governor in 1970, and four years later became the first Florida governor to succeed himself for a second full term (as permitted by the 1968 constitutional revision). In the 1978 general election, Democrat Bob Graham won the governorship. Democrat Wayne Mixson was elected lieutenant governor on the same ticket, and Democrats captured all six cabinet posts. Graham and Mixson were reelected in 1982. In 1985, Florida had one Democratic US senator, Lawton Chiles, and one Republican US senator, Paula Hawkins.

Florida's US House delegation had 12 Democrats and 7 Republicans. The state senate for 1984–86 contained 31 Democrats and 9 Republicans, and the state house of representatives had 76 Democrats and 44 Republicans. In the 1980 and 1984 elections, Floridians favored Ronald Reagan for president.

Florida had 167 black elected officials in 1985 and 44 Hispanic elected officials in 1984. In 1985, Miami elected its first Cuban-born mayor, Xavier Louis Suarez.

14 LOCAL GOVERNMENT

In 1982, Florida had 67 counties, 391 municipalities, 95 school districts, and 417 special districts.

Generally, legislative authority within each county is vested in a five-member elected board of county commissioners, which also has administrative authority over county departments, except those headed by independently elected officials. In counties without charters, these elected officials usually include a sheriff, tax collector, property appraiser, supervisor of elections, and clerk of the circuit court. County charters may provide for a greater or lesser number of elected officials, and for a professional county administrator (analogous to a city manager). Much state legislation restricting county government operations has been repealed since 1968. Counties may generally enact any law not inconsistent with state law. However, the taxing power of county and other local governments is severely limited.

Municipalities are normally incorporated and chartered by an

Florida Presidential Vote by Political Parties, 1948–84

YEAR	ELECTORAL VOTE	FLORIDA WINNER	DEMOCRAT	REPUBLICAN	STATES' RIGHTS DEMOCRAT	PROGRESSIVE
1948	8	*Truman (D)	281,988	194,280	89,755	11,620
1952	10	*Eisenhower (R)	444,950	544,036	—	—
1956	10	*Eisenhower (R)	480,371	643,849	—	—
1960	10	Nixon (R)	748,700	795,476	—	—
1964	14	*Johnson (D)	948,540	905,941	—	—
					AMERICAN IND.	
1968	14	*Nixon (R)	676,794	886,804	624,207	—
1972	17	*Nixon (R)	718,117	1,857,759	—	—
					AMERICAN	
1976	17	*Carter (D)	1,636,000	1,469,531	21,325	—
						LIBERTARIAN
1980	17	*Reagan (R)	1,417,637	2,043,006	—	30,457
1984	21	*Reagan (R)	1,448,816	2,730,350	—	744

*Won US presidential election.

act of the state legislature. Except where a county charter specifies otherwise, municipal ordinances override county laws. Municipal governments may provide a full range of local services, but as populations rapidly expand beyond municipal boundaries, many of these governments have found that they lack the jurisdiction to deal adequately with area problems. Annexations of surrounding territory are permissible but difficult under state law. Some municipal governments have reached agreement with county or other local governments for consolidation of overlapping, or redundant services or for provision of service by one local government to another on a contract basis. Complete consolidation of a municipal and a county government is authorized by the state constitution, requiring state legislation and voter approval in the area affected. As of 1985, one such consolidation effort had been successful, involving Jacksonville and Duval County.

The problem of overlapping and uncoordinated service is most serious in the case of special districts. These districts, established by state law and by approval of the voters affected, provide a specified service in a specified geographic area. An urban area may have dozens of special districts. State legislation in the 1970s attempted to deal with this problem by permitting counties to set up their own special-purpose districts, whose operations could be coordinated by the county government.

Regional planning councils resulted from the need to cope with problems of greater than local concern. In 1984 there were 11 of these councils, dealing with such issues as land management, resource management, and economic development.

15 STATE SERVICES

A "sunshine" amendment to the constitution and a statutory code of ethics require financial disclosure by elected officials and top-level public employees; the code prohibits actions by officials and employees that would constitute a conflict of interest. The Commission on Ethics, established in 1974, is empowered to investigate complaints of breach of public trust or violation of the code of ethics. In addition, an auditor general appointed by the legislature conducts financial and performance audits of state agencies.

Educational services are provided by the State Department of Education, which sets overall policy and adopts comprehensive objectives for public education, operates the state university and community college systems, and issues bonds (as authorized by the state constitution) to finance capital projects. The Department of Transportation is responsible for developing long-range transportation plans and for construction and maintenance of the state highway system. The Department of Highway Safety and Motor Vehicles licenses drivers, regulates the registration and sale of motor vehicles, and administers the Florida Highway Patrol.

Health and welfare services are the responsibility primarily of the Department of Health and Rehabilitative Services. In 1984, this department operated 16 major institutions and 71 other facilities, including mental hospitals, institutions for the mentally retarded, alcoholic treatment centers, and training schools for juvenile delinquents. In addition, the department administers such social welfare programs as Medicaid, aid to families with dependent children, food stamps, home health care, and foster care and adoption. It is also responsible for disease prevention and for assisting localities in performing health services.

The Department of Corrections maintained 30 major correctional institutions and 55 smaller facilities in 1984. The Department of Law Enforcement is responsible for maintaining public order and enforcing the state criminal code; enforcement activities emphasize combating organized crime, vice, and racketeering. The state's Army and Air National Guard are under the jurisdiction of the Department of Military Affairs. The Florida Highway Patrol, within the Department of Highway Safety and Motor Vehicles, is the only statewide uniformed police force.

The Housing Finance Agency, created in 1980, encourages the investment of private capital in residential housing through the use of public financing. The Northwest Florida Regional Housing Authority provides housing for low-income and other residents of that area. The Florida Housing Advisory Council assists the Department of Community Affairs in carrying out its duties related to housing. The Division of Local Resource Management is designated as the state's housing and urban development agency.

The Department of Labor and Employment Security enforces legislation protecting the state's workers (including child labor and industrial safety laws) and administers the federally funded workers' compensation and unemployment insurance programs and the State Employment Service, whose offices provide employment counseling and assist in job placement. Included in the department are the Division of Employment and Training and the Division of Employment Security. The Department of Administration includes the Division of Personnel, Office of Career Service and State Retirement Commissions, Florida Commission on Human Relations, and Division of Veterans Affairs.

The Department of State manages state historic sites, archives, museums, libraries, and fine arts centers.

16 JUDICIAL SYSTEM

The state's highest court is the supreme court, a panel of seven justices that sits in Tallahassee; every two years, the presiding justices elect one of their number as chief justice. All justices are appointed to six-year terms by the governor upon the recommendation of a judicial nominating commission. They may seek further six-year terms in a yes-no vote in a general election; if the incumbent justice does not receive a majority of "yes" votes, the governor appoints another person to fill the vacancy from the recommended list of qualified candidates. Supreme court justices, who before nomination must have been members of the Florida bar for at least 10 years, received an annual salary in 1984 of $67,588.

The supreme court has appellate jurisdiction only. The state constitution, as amended, prescribes certain types of cases in which an appeal must be heard, including those in which the death penalty has been ordered and those in which a lower appellate court has invalidated a state law or a provision of the state constitution. The court also hears appeals of state agency decisions on utility rates and may, at its discretion, hear appeals in many other types of cases.

Below the supreme court are five district courts of appeal, which sit in Tallahassee, Lakeland, Miami, West Palm Beach, and Daytona Beach. There are 46 district court judges; the method of their selection and retention in office is the same as for supreme court justices. District courts hear appeals of lower court decisions and may review the actions of executive agencies. District court decisions are usually final, since most requests for supreme court review are denied.

The state's principal trial courts are its 20 circuit courts, which have original jurisdiction in many types of cases, including civil suits involving more than $5,000, felony cases, and all cases involving juveniles. Circuit courts may also hear appeals from county courts if no constitutional question is involved. Circuit court judges are elected for six-year terms and must have been members of the Florida bar for at least five years before election. In 1985 there were 348 such judges, the number in each district varying from 57 (Greater Miami) to 4 (the Keys), depending on district population and case load.

Each of Florida's 67 counties has a county court with original jurisdiction in misdemeanor cases, civil disputes involving $5,000 or less, and traffic-violation cases. County court judges are elected for four-year terms and must be members of the bar only in counties with populations of 40,000 or more. In 1985 there were 214 county court judges, the number in each county again varying according to population and case load.

Florida has one of the highest crime rates in the US. In 1983, the number of FBI Index crimes reported in the state was 724,226. Of this total, 88,292 were violent crimes (murder, rape, robbery, and aggravated assault), and 635,934 were nonviolent (breaking and entering, larceny, and auto theft). According to the FBI, Florida's crime rate per 100,000 population was 6,781 in 1983, when only the District of Columbia had a higher crime rate. In 1983, Miami had 49,799 index offenses; Jacksonville had 42,330, and Tampa had 33,159. As of 1984, a total of 27,106 persons were serving prison sentences in institutions run by Florida's correctional authorities. The state has a capital punishment statute, which was upheld by the US Supreme Court in 1976. The first execution under the statute took place in the state on 25 May 1979. Between 1979 and 1984, nine other people were executed.

[17]ARMED FORCES

In 1983 there were 35 US Navy and 20 US Air Force installations in Florida. Naval and Marine Corps personnel stationed in the state (military and civilian) numbered 56,397 in 1984. Military and civilian personnel numbered 17,898 at facilities in Pensacola, 14,447 in Orlando, 12,905 in Jacksonville, and 3,476 at Cecil Field. In October 1979, the Key West Naval Air Station was made the headquarters of a new Caribbean Joint Task Force, established to coordinate US military activities in the Caribbean. The state had 38,961 Air Force personnel in 1984; the largest Air Force bases were Eglin, in Valparaiso; MacDill, near Tampa; and Tyndall, west of Tallahassee. The US Air Force Missile Test Center at Cape Canaveral (called Cape Kennedy from 1963 to 1973) has been the launching site for most US space flights, including all manned flights. US Department of Defense procurement contracts in Florida in 1983/84 totaled $4 billion (nearly 3% of the US total).

Some 1,413,000 US military veterans lived in Florida as of 30 September 1984; of those who saw wartime service, 19,000 served during World War I, 649,000 during World War II, 274,000 during the Korean conflict, and 384,000 during the Viet-Nam era. US Veterans Administration spending in Florida in 1982/83 totaled almost $1.5 billion, including $896 million for compensations and pensions, $371 million for medical services, and $82 million for education and training.

The state's Army and Air National Guard contained approximately 12,054 officers and enlisted personnel in early 1985. Of the Army National Guard's 10,910 members, 9,852 were enlisted and 1,058 were officers.

Florida's police forces had 49,568 full-time-equivalent employees in October 1983, 1,884 of whom were members of the Florida Highway Patrol. The two largest municipal police forces were those of Jacksonville and Miami, with 1,598 and 1,442 employees, respectively. Spending for police protection in 1982 totaled $862 million.

[18]MIGRATION

Florida is populated mostly by migrants. In 1980, only 35% of all state residents were Florida-born, compared with 68% for the US as a whole. Only Nevada had a lower proportion of native residents. Migration from other states accounted for more than 85% of Florida's population increase in the 1970s and for 89% in 1980–83.

The early European immigrants to Florida—first the Spanish, then the English—never populated the state in significant numbers. Immigration from southern states began even before the US acquisition of Florida, and accelerated thereafter. In the 20th century, US immigrants to Florida have come, for the most part, from the Northeast and Midwest. Their motivation has often been to escape harsh northern winters, and a large proportion of the migrants have been retirees and other senior citizens. Between 1970 and 1980, the number of Floridians 65 or over increased by 70%, compared with a 44% increase for the US population as a whole.

Since the 1960s, Florida has also experienced large-scale migration from the Caribbean and parts of Latin America. Although the state has had a significant Cuban population since the second half of the 19th century, the number of immigrants surged after the Cuban revolution of 1959. From December 1965 to April 1973, an airlift agreed to by the Cuban and US governments landed a quarter of a million Cubans in Miami. By 1980, nearly 500,000 Cubans were living in southern Florida, mostly in and around Miami, where the Cuban section had become known as "Little Havana." Another period of large-scale immigration from Cuba, beginning in April 1980, brought more than 100,000 Cubans into Florida harbors. At the same time, Haitian "boat people" were arriving in Florida in significant numbers—often reaching the southern peninsula in packed, barely seaworthy small craft. The number of Haitians in Florida was estimated at 17,280 in 1980. The US government classified some of them as illegal aliens, fleeing extreme poverty in their native country, but the immigrants claimed to be political refugees and sued to halt deportation proceedings against them.

In 1982/83, 27,625 aliens were admitted to Florida, including 5,764 Cubans, 2,670 Jamaicans, 1,399 Colombians, 1,324 Canadians, and 1,293 Haitians. Aliens naturalized in Florida in 1982/83 numbered 12,617 (7% of the US total), including 6,202 Cubans.

[19]INTERGOVERNMENTAL COOPERATION

In 1953, Florida became a signatory to the Alabama-Florida Boundary Compact. Among the interstate regional compacts in which Florida participates are the Southern Interstate Energy Compact, Southeastern Forest Fire Protection Compact, Atlantic States Marine Fisheries Compact, and Gulf States Marine Fisheries Compact. The Central and South American and Caribbean Trade and Development Commission, within Florida's Department of Commerce, was created in 1978 to foster cooperation between Florida and Latin America.

Federal aid to Florida in 1983/84 totaled almost $2.8 billion, including general revenue-sharing funds of $164.1 million. Florida ranked 50th among the US states in per capita federal aid in 1983/84.

[20]ECONOMY

Farming and the lumbering and naval stores industries, all concentrated in northern Florida, were early mainstays of the economy. In the late 19th century, the extension of the railroads down the peninsula opened up an area previously populated only by Indians; given the favorable climate, central and southern Florida soon became major agricultural areas. Tourism, aggressively promoted by the early railroad builders, became a major industry after World War I and remains so today.

Tourists and winter residents with second homes in Florida contribute billions of dollars annually to the state economy and make retailing and construction particularly important economic sectors. However, this dependence on discretionary spending by visitors and part-time dwellers also makes the economy—and especially the housing industry—highly vulnerable to recession. The economic downturn of the early 1980s hit Florida harder than the US generally. New housing starts, for example, which fell by about 2% in the US from 1981 to 1982, dropped by more than 20% in Florida during the same period.

An extremely low level of unionization among Florida workers encouraged growth in manufacturing in the 1970s and early 1980s—but may also help explain Floridians' below-average income levels. The aerospace and electronics industries were expanding rapidly in the early 1980s. Also at that time, the state's economy—particularly that of the Miami area—was benefiting from an influx of Latin American investment funds. Miami was said to have one of the largest "underground economies" in the US, a reference both to the sizable inflow of cash from illicit drug trafficking and to the large numbers of Latin American immigrants working for low, unreported cash wages.

21 INCOME

In 1982, Florida's per capita income was $9,433, 20th among the 50 states but 1st in the Southeast. Personal income in 1982 was $69.8 billion, of which wages and salaries accounted for 85%; other labor income, 8%; and proprietors' income, 7%. Of proprietors' income, 13% was for farm income and 87% for nonfarm income. Florida's median family income in 1981 was $23,504, 39th in the US.

In 1979, 1,287,056 Floridians (13.5%) lived below the federal poverty level. For all persons 65 or over, the proportion was 12.7%.

Nearly 152,000 Floridians were among the nation's top wealthholders in 1982, with gross assets greater than $500,000. The state's most affluent county that year was Palm Beach, on the so-called Gold Coast of southeastern Florida, with a per capita income of $14,150. Next came the Gulf coast county of Sarasota, with $14,098, followed by Broward, with $13,091, and Pinellas, with $12,169. Glades was the poorest county, with a per capita income of $5,019. Union County, in northern Florida, was the 2d poorest ($5,119), even though it had registered a 144% increase in per capita income since 1972.

22 LABOR

Florida's civilian labor force averaged 5,099,000 in 1984. Of that total, 4,777,000 were employed and 322,000 unemployed, for an unemployment rate of 6.3%, down from 8.6% in 1983. As in the US generally, unemployment was higher among nonwhites and among teenagers than among white adults—and highest among black teenagers.

Florida's labor force participation rate is well below the US average. In 1983, about 54% of the state's civilian noninstitutionalized population 16 years of age or older was in the labor force, compared with 58% for the US as a whole. The major reason for the difference is the extremely low participation rate for Floridians 65 or older, reflecting the fact that many people migrate to the state for their retirement years.

Reflecting the importance of tourism to Florida's economy, a higher proportion of the state's workers are employed in the trade and service industries than for the US as a whole; the proportion of workers in manufacturing is a little over half the US average. A federal survey in March 1982 revealed the following nonfarm employment pattern for Florida:

	ESTABLISH-MENTS	EMPLOYEES	ANNUAL PAYROLL ('000)
Agricultural services, forestry, fishing	3,153	28,293	$ 287,133
Mining	307	11,476	226,742
Contract construction	21,890	255,707	3,748,911
Manufacturing, of which:	12,234	464,933	7,827,393
Food and kindred products	(649)	(47,608)	(714,796)
Electric, electronic equipment	(565)	(67,936)	(1,325,405)
Transportation, equipment	(621)	(48,416)	(1,045,485)
Transportation, public utilities	7,403	220,585	4,598,423
Wholesale trade	18,527	210,318	3,644,681
Retail trade	64,296	797,107	7,195,041
Finance, insurance, real estate	23,983	274,508	4,288,590
Services, of which:	74,465	883,625	11,641,598
Health	(17,814)	(246,639)	(4,324,659)
Hotels, lodging	(3,141)	(101,480)	(749,345)
Other	6,812	5,524	93,522
TOTALS	233,070	3,152,076	$43,552,034

The federal survey excluded the self-employed, government workers, and several other employment categories. In March 1983, 538,237 workers were employed in government: 313,222 by localities, 133,131 by the state, and the remaining 91,884 by the federal government.

In 1982, the average annual pay of Florida nonagricultural workers was $14,787. The average weekly earnings for manufacturing in 1983 were $298. Some 420,000 Florida workers belonged to labor unions in 1980, about 12% of the total nonagricultural work force and far below the national average; there were 1,021 labor unions in 1983. The state has a right-to-work law. In 1981 there were 19 work stoppages, involving 4,100 workers.

23 AGRICULTURE

Florida's most important agricultural products, and the ones for which it is most famous, are its citrus fruits. Florida continues to supply the vast majority of orange juice consumed in the US; however, in recent years, a large portion of the product packaged in the state has been imported from Brazil. Florida produced 64.1% of the nation's oranges and 65% of its grapefruits in 1982–83. It is also an important producer of other fruits, vegetables, sugarcane, and soybeans.

The total value of Florida's crops in 1983 exceeded $3.3 billion, 5th highest among the 50 states. Total farm marketings, including livestock marketings and products, exceeded $4.3 billion in 1983 (9th in the US). There were about 36,000 farms covering some 13 million acres (5 million hectares) in 1982; the total represented more than 38% of the state's entire land area.

The orange was introduced to Florida by Spanish settlers around 1570. Oranges had become an important commercial crop by the early 1800s, when the grapefruit was introduced. In 1886, orange production for the first time exceeded 1 million boxes (one box equals 90 lb). Much of this production came from groves along the northern Atlantic coast and the St. Johns River, which offered easy access to maritime shipping routes north. The expansion of the railroads and severe freezes in the 1890s encouraged the citrus industry to move farther south. In the 1980s, Polk, Lake, St. Lucie, Indian River, and Hardee counties in central Florida were the largest producers of citrus fruits. As of January 1982 there were about 71.6 million citrus trees in Florida, covering a total of 847,385 acres (342,926 hectares).

A devastating freeze hit the Florida citrus industry on 24–25 December 1983, destroying over 100,000 acres (40,500 hectares) of citrus. A second disaster occurred in August 1984, when an outbreak of citrus canker destroyed many citrus groves. Severe freezes damaged citrus production in January of 1981, 1982, and 1985.

The value of Florida's citrus crop was $901,372,000 in 1982/83, down from over $1 billion in 1979/80 because of the freezes. The orange crop totaled 139,500,000 90-lb boxes in the 1982/83 season. The grapefruit crop was 39,400,000 85-lb boxes; tangerines, 2,250,000 95-lb boxes; and tangelos and temple oranges, 8,500,000 90-lb boxes. There are 50 plants in Florida where citrus fruits are processed into canned or chilled juice, frozen or pasteurized concentrate, or canned fruit sections. Production of frozen concentrate orange juice totaled 169,577,000 gallons in the 1982/83 season. Stock feed made from peel, pulp, and seeds is an important by-product of the citrus-processing industry; annual production is about 1 million tons. Other citrus by-products are citrus molasses, D-limonene, alcohol, wines, preserves, and citrus seed oil.

Florida is one of the country's leading producers of vegetables. Vegetable farming is concentrated in central and southern Florida, especially in the area south of Lake Okeechobee, where drainage of the Everglades left exceptionally rich soil. In 1982/83, the combined value of the state's vegetable, watermelon, potato, and strawberry crops was over $1 billion, with tomatoes accounting for nearly 36% of that total. In 1982/83, Florida farmers harvested 13,520,000 hundredweight of tomatoes and 8,085,000 hundredweight of watermelons; they sold 5,840,000 hundredweight of potatoes. Florida's tomato and vegetable growers, who had at one time enjoyed a near monopoly of the US winter vegetable market, began in the late 1970s to face increasing competition from Mexican growers, whose lower-priced produce had captured about half the market by 1980.

Florida's major field crop is sugarcane (mostly grown near Lake Okeechobee), which enjoyed a sizable production increase in the 1960s and 1970s, following the cutoff of imports from Cuba. In 1982, Florida's sugarcane production was 12,070,000 tons. Florida's 2d-largest field crop is soybeans (15,606,000 bushels in 1982), followed by hay, peanuts, cotton, and tobacco.

24 ANIMAL HUSBANDRY

Florida is an important cattle-raising state. The Kissimmee Plain, north of Lake Okeechobee, is the largest grazing area. Florida ranked 9th in the US in number of beef cattle in 1984, and 16th in number of dairy cows. Most of the beef cattle are sold to out-of-state feedlots; cash receipts from marketings totaled $359,015,000 in 1983 for 597,730,000 lb of marketed product. Other types of livestock raised commercially are hogs and pigs ($36,466,000 in 1983), sheep, rabbits, and poultry.

The raising of Thoroughbred horses and the production of eggs and honey are also major industries. Florida ranks 3d in the US in number of Thoroughbreds, surpassed only by Kentucky and California. Over $100 million worth of Thoroughbreds were sold at auction in 1983. Florida ranked 8th in egg production in 1983, with an output of nearly 3 billion, worth $123,083,000.

25 FISHING

Florida's extensive shoreline and numerous inland waterways make sport fishing a major recreational activity. Commercial fishing is also economically important.

In 1984, Florida's commercial fish catch was 206,679,000 lb, worth $178,121,000. Florida provided 14% of the nation's fish fillets and steaks and 7% of its industrial fishery products in 1983. The most important commercial species are shrimp, black mullet, grouper, calico scallops, Spanish mackerel, and blue crabs. In 1984, Florida had 39 plants that produced fish fillets and steaks, and 7 that produced canned and industrial fishery products.

Both freshwater and saltwater fishing are important sports. Tarpon, sailfish, and redfish are some of the major saltwater sport species; largemouth bass, panfish, sunfish, catfish, and perch are leading freshwater sport fish.

26 FORESTRY

Over half of Florida's land area—17,040,000 acres (6,896,000 hectares)—was forested in 1979, when the state had 2.4% of all forested land in the US. A total of 15,330,000 acres (6,204,000 hectares) was commercial timberland, of which 86% was privately owned. The most common tree is the pine, which occurs throughout the state but is most abundant in the north.

Florida's forestry industry, concentrated in the northern part of the state, shipped $1.3 billion worth of lumber and wood products in 1982. The most important forestry product is pulpwood for paper manufacturing; production in 1983 was 3.6 million cords. Lumber production that year was 557 million board feet, 540 million board feet of softwoods and 17 million board feet of hardwoods. As of 1984, 6,081,734 acres (2,461,199 hectares) of Florida's forested land consisted of tree farms. In 1983, employment in the forestry and major wood-using industries, including paper and furniture manufacturing, exceeded 50,000.

Four national forests—Apalachicola, Ocala, Osceola, and Choctawatchee—covering 1,224,611 acres (495,584 hectares) are located in Florida. The Division of Forestry operates four state forests covering 306,000 acres (124,000 hectares), and 13,000 acres (5,000 hectares) of state land near Tallahassee. Some timbering is permitted in the state forests, which are also used for recreational activities and wildlife protection.

Virtually all of Florida's natural forest had been cleared by the mid-20th century; the forests existing today are thus almost entirely the result of reforestation. Since 1928, more than 3 billion seedlings have been planted in the state.

27 MINING

Florida ranked 4th among the 50 states in nonfuel mineral output in 1984, when production totaled nearly $1.5 billion.

Florida's principal minerals (excluding fossil fuels) are phosphate rock, limestone, dolomite, clays, and sand and gravel. The state is the leading US producer of phosphate rock (used largely for fertilizer), accounting for 83% of the total domestic and 33% of the world market in 1982. Production is concentrated in Hamilton, Hillsborough, Manatee, Hardee, Polk, and Marion counties. In the year ending 30 June 1983, Florida and North Carolina produced 97,028,000 metric tons of phosphate rock, of which 31,170,000 metric tons were marketable. Florida is also the only US producer of staurolite (used in portland cement manufacturing) and the leading producer of titanium minerals (rutile and ilmenite).

In 1983, Florida ranked 2d in production of fuller's earth (a type of clay), with 27% of US output, and 2d in stone production (7%). Total estimated production of clays in 1984 was 733,000 tons. Portland cement production was 3,600,000 tons; sand and gravel, 18,750,000 tons; and stone, 64,000,000 tons. Limestone, found throughout the state, is used largely in road building and cement manufacture. A variety of limestone known as coquina, consisting partly of mollusk shells (which are easily visible in the stone), was used as a building material by the first Spanish settlers at St. Augustine; a number of structures made from coquina still survive in the city.

28 ENERGY AND POWER

In 1982, a total of 2,342 trillion Btu of energy was consumed in Florida. About 40% of that total was used in transportation, 24% in residences, 18% by commercial establishments, and 18% by industry. The sources of the energy consumed were petroleum, 62%; natural gas, 14%; coal, 10%; nuclear power, 9%; and other sources, 5%.

Per capita energy use in 1982 was 224 million Btu; per capita expenditure in 1981 was $1,744. Although Florida produces some oil and natural gas, it is a net importer of energy resources. Its mild climate and abundant sunshine offer great potential for solar energy development, but this potential had not been extensively exploited by the mid-1980s.

In 1983, Florida had an installed electric energy generating capacity of 31.9 million kw; production that year was 91.5 billion kwh. In 1982, 36% of electricity produced came from residual fuel oil, 25% from coal, 21% from nuclear power, and 18% from natural gas; hydroelectricity and distillate fuels totaled less than 1%. Nuclear generating capacity increased by 27% in August 1983, when the state's fifth nuclear plant, the 830,000-kw St. Lucie 2 facility, operated by the Florida Power and Light Co., south of Ft. Pierce, began commercial operation. Three other nuclear plants owned by Florida Power and Light are St. Lucie 1, with a capacity of 830,000 kw, and Turkey Point Units 3 and 4, each with a capacity of 693,000 kw, located in Dade County. Florida Power Corp. operates the state's other nuclear power plant, the 830,000-kw Crystal River 3 facility, on the northern Gulf coast. Residential customers used 49% of all electricity sold by utilities in 1983, commercial customers 32%, and industrial customers 16% (the remainder went for other purposes).

Florida ranked 10th in oil production in 1981. In 1983, the state produced 19.5 million barrels of crude oil; proved reserves as of 31 December 1983 were 78 million barrels. Natural gas production in 1982 was 22.5 billion cu feet, or 7% of consumption for that year; proved reserves were 49 billion cu feet in 1983.

29 INDUSTRY

Florida ranked 17th among the 50 states in the value of manufacturing shipments in 1982, and 15th in number of manufacturing employees. The state is not a center of heavy industry, and many of its manufacturing activities are related to agriculture and exploitation of natural resources. Leading industries include food processing, electric and electronic equipment, transportation equipment, and chemicals. The value of manufacturing shipments in 1982 totaled $38.7 billion, up from $21 billion in 1977. The state's major industrial categories, in terms of value of shipments, were food and food products, 22%; electric and electronic equipment, 11%; chemicals and allied products, 11%; transportation equipment, 9%; nonelectrical machinery, 8%; printing and publishing, 7%; fabricated metal products, 6%; paper and allied products, 5%; stone, clay, and glass products, 4%; apparel and other textile products, 3%; lumber and wood products, 3%; and other industries, 11%.

Value of shipments by manufacturers for selected industries in 1982 was as follows:

Communication equipment	$2,683,700,000
Preserved fruits and vegetables	2,538,300,000
Agricultural chemicals	2,244,900,000
Office and computing machines	1,901,700,000
Newspapers	1,321,000,000
Fabricated structural metal products	1,247,500,000
Aircraft and parts	1,131,800,000

The cigar-making industry, traditionally important in Florida, has declined considerably with changes in taste and the cutoff of tobacco imports from Cuba. In the late 1930s, the Tampa area alone had well over 100 cigar factories, employing some 10,000 people. The 1982 Census of Manufactures found just 25 plants statewide, with total employment of about 1,600. Value of shipments by the cigar industry was $72.6 million in 1982.

Manufacturing is currently concentrated in and around Florida's largest cities, such as Miami, Tampa–St. Petersburg, Ft. Lauderdale–Hollywood, and Orlando. Dade and Broward counties (Greater Miami), Hillsborough County (Tampa), Pinellas County (St. Petersburg), Duval County (Jacksonville), Orange County (Orlando), and Palm Beach County accounted for almost two-thirds of all manufacturing employment in 1982.

[30] COMMERCE

In 1982, wholesale and retail trade together accounted for about one-fourth of all nonagricultural employment. According to the 1982 US Census of Retail Trade, the state ranked 4th in the US in retail sales volume and 5th in number of retail establishments. The fashionable shops lining Palm Beach's Worth Avenue make it one of the nation's most famous shopping streets.

Sales by Florida's wholesale establishments in 1982 totaled nearly $66 billion, or more than 3% of the US total. Reflecting the importance of agriculture in the state, sales of groceries and produce accounted for 21%, compared with 14% for the US as a whole.

Retail sales in 1982 were $54.5 billion, more than 5% of the US total. At least 33% of retail trade employees worked in the 13,933 restaurants, cafeterias, bars, and similar establishments—a reflection, in part, of the importance of the travel business in Florida's economy. Sales at eating and drinking places represented 10% of total retail sales. Food stores accounted for 23%; automotive dealers, 21%; department, variety, and other general merchandise stores, 11%; gasoline service stations, 9%; furniture, home furnishing, and equipment stores, 5%; building material, hardware, garden supply, and mobile home dealers, 5%; apparel and accessories stores, 5%; drug and proprietary stores, 4%; and other establishments, 7%. The Miami metropolitan area had retail sales of $16.1 billion in 1982. Retail sales in the Tampa–St. Petersburg area were about $8.5 billion; in the Ft. Lauderdale–Hollywood area, $6.7 billion; in the Orlando area, $4.3 billion; in the West Palm Beach–Boca Raton area, $3.8 billion, and in the Jacksonville area, $3.8 billion.

The value of all imports to Florida was $5.8 billion in 1982, 2.4% of the US total. Florida's exports of manufactured goods totaled $3.4 billion in 1981, more than double the 1977 value. Florida's share of US agricultural exports totaled $586 million in fiscal 1982, one and a half times the 1977 value. Duty-free goods for reshipment abroad pass through Port Everglades, Miami, Orlando, Jacksonville, Tampa, and Panama City—free-trade zones established to bring international commerce to the state. In the 1970s and early 1980s, Florida was believed to be the principal entry point for marijuana, cocaine, and other illicit drugs being smuggled into the US from Latin America.

[31] CONSUMER PROTECTION

The Division of Consumer Services, within the Department of Agriculture and Consumer Services, disseminates consumer information, conducts educational programs for consumers, and acts as a clearinghouse for consumer complaints, referring them to the appropriate agency. The Florida Consumers' Council advises the commissioner of agriculture on consumer legislation.

The public counsel to the Public Service Commission (PSC), appointed by a joint committee of the legislature, represents the public interest in commission hearings on utility rates and other regulations. The public counsel can also seek judicial review of PSC rulings, and may appear before other state and federal bodies on the public's behalf in utility and transportation matters.

The Department of Business Regulation oversees pari-mutuel betting; land sales; the operations of condominiums, cooperative apartments, hotels, and restaurants; and the regulation and licensing of alcoholic beverage and tobacco sales.

In 1983, the state legislature enacted the Motor Vehicle Warranty Enforcement Act, which forces automobile dealers to replace new cars or refund the purchase price if the cars are in constant need of repairs.

[32] BANKING

Florida's banking center is Miami. The state's largest commercial banking companies, as of December 1984, were Barnett Banks of Florida, in Jacksonville, with assets of $12.5 billion; Miami's Southeast Banking Corp., $9.9 billion; and Orlando's Sun Banks, $9.4 billion.

At the end of 1984 there were 442 commercial banks (including nondeposit trust companies) in the state. All but 12 were federally insured, including 247 that were state chartered and 183 that were federally chartered. A total of 232 banks were members of the Federal Reserve System.

Assets of Florida's commercial banks as of 31 December 1982 exceeded $59.6 billion. Deposits totaled $48.5 billion, up more than 50% from 1978, and liabilities totaled $55.5 billion. Banks in Dade County accounted for more than twice the shares of deposits and loans of any other county.

The state had 119 federally insured savings and loan associations in 1983. Their total assets on 30 September exceeded $55.5 billion. Institutions in Dade and Broward counties accounted for almost 39% of each of those figures. Florida savings and loan associations made a total of $7.1 billion in loans, almost 10% of the US total; 49% of these loans went for home purchase, 32% for home construction, and 19% for other purposes. Miami area institutions accounted for almost 36% of Florida's total loans.

International banking grew in Florida during the late 1970s and early 1980s with the establishment of Edge Act banks in Miami. These banks have headquarters outside Florida and must engage exclusively in international banking. As of 31 December 1982, there were 28 international banks in Florida, with assets of over $1.9 billion.

The state had 243 active credit unions in 1982, with gross income of $112,198,000, total expenses of $68,452,000, and net income of $43,746,000.

The Department of Banking and Finance charters state banks and also oversees state savings and loan associations, credit unions, consumer finance companies, trust companies, mortgage brokerage firms, and trading stamp companies.

33 INSURANCE

In 1983, 44 life insurance companies and 70 property and casualty insurance companies were domiciled in Florida. These and out-of-state companies doing business in Florida employed 41,356 people as of March 1983, including some 24,533 agents, brokers, and service employees. Almost 28% of insurance-carrier employment was concentrated in Duval County (which includes Jacksonville), and another 14% in Dade County. Nearly 19% of the state's insurance agents and brokers were based in Dade County. Insurance premiums written in 1983 totaled $9.1 billion.

More than 16.5 million life insurance policies were in force in 1983, carrying a total face value of $193.5 billion. Benefits paid during the year were about $2.4 billion, consisting of the following: death payments, $733.3 million; annuity payments, $587.4 million; matured endowments, $40.8 million; disability payments, $26.9 million; surrender values, $579.6 million; and policy and contract dividends, $422.6 million. Average life insurance in force in 1983 was $42,700 per Florida family. Premiums written by property, casualty, and title insurance companies, interinsurance exchanges, and other departments of life insurance totaled $6.9 billion in 1982; direct losses paid during the year came to nearly $5 billion. Of the total premiums written, accident and health policies accounted for $2.5 billion; automotive policies, $2.1 billion; homeowner multiple-peril policies, $509.2 million; and commercial multiple-peril policies, $359.1 million. Florida ranked 1st in the US in flood insurance in 1983, with 543,709 policies in force worth $36.4 billion, or 32% of the US total.

Not surprisingly for a state with so many inland waterways and such a lengthy coastline, marine insurance accounted for $190.2 million in premiums written during 1983. Direct losses paid during 1982 were $108.4 million. The insurance industry is regulated by the state's Department of Insurance.

34 SECURITIES

No securities exchanges are located in Florida, but there were 311 registered brokers and dealers doing business in the state as of 30 September 1983. Employment in securities and commodities brokerage houses and related businesses totaled 11,547 in March 1983. A large portion of the industry's total employment was concentrated in the three Gold Coast counties—Dade, Broward, and Palm Beach. New York Stock Exchange member firms had 420 sales offices and 4,917 registered representatives in Florida as of December 1983. That month, 2,056,000 Floridians were shareowners of public corporations.

The Department of Banking and Finance's Division of Securities, headed by the comptroller, oversees the securities industry.

35 PUBLIC FINANCE

The Office of Planning and Budget of the Governor's Office prepares and submits to the legislature the budget for each fiscal year, which runs from 1 July to 30 June. The largest expenditure items are education, health and social concerns, general government, and transportation. By prohibiting borrowing to finance operating expenses, Florida's constitution requires a balanced budget.

The following table shows the governor's recommended general revenues and allocations for the 1985/86 and 1986/87 fiscal years:

REVENUES	1985/86	1986/87
Trust funds	$ 7,496,200,000	$ 7,701,000,000
Sales and use tax	4,476,500,000	4,917,700,000
Other general revenue	810,900,000	876,400,000
Corporate tax	568,400,000	614,200,000
Beverage tax	466,400,000	486,200,000
Documentary stamps	243,100,000	274,900,000
Intangible tax	117,600,000	130,000,000
Interest	75,500,000	81,000,000
TOTALS	$14,254,600,000	$15,081,400,000

ALLOCATIONS	1985/86	1986/87
Education	$ 5,210,000,000	$ 5,636,100,000
Health and social concerns	3,226,700,000	3,651,100,000
General government	2,468,100,000	2,363,400,000
Transportation	1,114,600,000	1,207,900,000
Economic development	1,011,700,000	1,024,700,000
Public safety	825,000,000	836,400,000
Natural resources	303,800,000	270,700,000
Community development	94,700,000	91,100,000
TOTALS	$14,254,600,000	$15,081,400,000

For county governments, total revenues in the year ending 30 September 1982 were $4,645,190,000; total expenditures came to $4,464,767,000. The county with by far the largest budget was Dade: revenues, $1,561,077,000; expenditures, $1,515,985,000. Jacksonville city government revenues were $359,333,000; expenditures were $356,037,000. Other major city budgets were those of Miami, with revenues of $267,728,000 and expenditures of $290,352,000; Tampa, with revenues of $222,913,000 and expenditures of $236,063,000; and Tallahassee, with revenues of $195,335,000 and expenditures of $204,260,000.

The issuance of state bonds is overseen by the State Board of Administration, which consists of the governor, the state treasurer, and the comptroller. Three principal types of bonds are issued. The first consists of bonds backed by the "full faith and credit" of the state and payable from general revenue. Issuance of such bonds generally requires voter approval. The second type consists of revenue bonds, payable from income derived from the capital project financed—for example, from bridge or highway tolls. The third type consists of bonds payable from a constitutionally specified source—for example, higher education bonds backed by the state gross receipts tax, or elementary and secondary education bonds backed by the motor vehicle license tax. The face value of all state bonds outstanding as of 30 June 1984 was $3.3 billion, of which $2.1 billion comprised education bonds.

The total indebtedness of all Florida state and local governments in 1982 exceeded $13 billion. The state debt outstanding at the end of 1982 was almost $3 billion, or $307.14 per capita.

36 TAXATION

Florida ranked 38th among the 50 states in per capita taxation in 1982 with a tax burden of $946 per person. The 5% sales and use tax is the largest single source of state revenue; property taxes make up the bulk of local receipts. The state constitution prohibits a personal income tax. The 1977 session of the legislature, after narrowly defeating a state sales tax hike, increased taxes on mining and on cigarettes, beer, and wine. However, by the end of the decade, legislative emphasis had shifted to property tax relief. A constitutional amendment approved by voters in March 1980 raised the homestead exemption from school taxes from the first $5,000 of assessed valuation ($10,000 for senior citizens) to the first $25,000. For 1982/83, school districts levied $826.3 million in property taxes, and received state funds of $1,995 million in grants and another $352.3 million in categorical grants for specific programs. Property taxes thus represented nearly one-third of Florida public school moneys.

The state sales tax applies to most retail items (but excludes groceries, medicines, and certain other items), as well as to car and hotel room rentals and theater admissions. The use tax is levied on wholesale items brought into Florida for sale. A 5% tax is levied on corporations' net income over $5,000. Other taxes include those on gasoline and other motor fuels, cigarettes, alcoholic beverages, drivers' licenses and motor vehicles, and pari-mutuel betting. In 1983, the tax on gasoline was changed from a levy based on gallons used to a sales tax. An estate tax is also levied. All told, taxes accounted for more than two-thirds of general revenues in 1982/83.

In 1983, Floridians filed more than 4.6 million federal income tax returns and paid nearly $13.4 billion in federal taxes.

37ECONOMIC POLICY

In the late 1970s and early 1980s, the state government intensified its efforts to attract manufacturing industries to Florida, aiming to decrease the economy's vulnerability to fluctuations in tourism and other discretionary spending. State financial assistance programs for industrial development include state and local tax incentives, state "enterprise zone" legislation, and a statewide development credit corporation. The Division of Economic Development within the Department of Commerce maintains offices in Europe and Japan to promote Florida as a business relocation site and to attract foreign investment funds. This department, in conjunction with the Department of Education, provides occupational training for new employees of firms relocating in Florida and requiring workers with particular skills. At the same time, the state continues to promote tourism, primarily through the Department of Commerce's Division of Tourism. The division conducts national advertising campaigns, publishes promotional materials, and otherwise seeks to publicize the state's attractions. In addition, the Division of Tourism assists municipalities, chambers of commerce, and other tourist-oriented entities in developing cooperative promotional programs.

A special state Department of Citrus assists the important citrus industry with both research and product promotion. Its activities include establishing state grades and standards, approving containers, promulgating rules and regulations, preventing the sale of substandard products, investigating transportation problems, and determining new uses and products.

38HEALTH

Reflecting the age distribution of the state's population, Florida has a relatively low birthrate and a high death rate. Florida's birthrate was 13.9 per 1,000 population in 1983, 44th among the 50 states and well below the overall US figure of 15.5. The state's 1983 provisional infant mortality rate among whites was 9.8 per 1,000 live births; among nonwhites, 20.4 (above the US average). Some 76,900 legal abortions were performed in Florida in 1982; there were 524 abortions per 1,000 live births, well above the US average.

Florida's 1982 death rate, 10.4 per 1,000 population, was the 2d highest of all the states and 21% above the national norm. In 1981, Florida exceeded the national averages in deaths from all major causes except early childhood diseases and pneumonia and flu; the death rate from cerebrovascular diseases was 28% higher than in the nation as a whole. The leading causes of death in 1983 were cardiovascular disease and cancer. The former accounted for 48% all deaths in the state, 54,580 out of 112,969. Cancer claimed 26,351 lives, just over 23% of the total.

The most common types of infectious diseases in 1982 (except where indicated) were gonorrhea, with 62,702 reported cases; streptococcal infections (1980), 24,403; chicken pox (1980), 8,707; syphilis, 4,149; hepatitis, 2,979; meningitis, 1,599; and tuberculosis, 1,467.

Four state mental hospitals admitted 5,399 patients in 1981/82 and discharged 5,959. The number of resident patients as of 30 June 1982 was 4,320, of whom 29% were age 65 or older.

In 1983 there were 254 hospitals in Florida. The total number of beds available was 59,704; admissions totaled 1,906,362. Florida's hospitals had 185,200 full-time equivalent employees in 1983; there were 34,406 full and part-time registered nurses in 1981. In 1980 there were 338 nursing homes, with 35,800 beds and 32,100 resident patients.

As of 31 December 1982, the total number of licensed physicians in the state was 23,481. There were 179 active physicians per 100,000 population in 1980. About 22% of all physicians were practicing in Dade County. One Florida county—Glades—had no practicing physicians in 1983. There were 7,954 licensed dentists in the state in 1984.

Private health insurance benefits paid to hospitals in 1982 exceeded $1.6 billion. The average cost of a hospital stay was $2,512; per day, $335. Florida ranked 10th among the 50 states in number of Medicaid recipients in 1981, with 539,200; Medicaid payments came to $48.9 million. In 1981, the state ranked 3d in number of residents enrolled in the Medicare program. As of 1 July 1982, 1,648,000 Floridians were eligible for Medicare hospital benefits, and 1,651,000 for medical benefits. For 1982, Medicare hospital benefits paid were $1.9 billion; medical benefits, $1 billion.

39SOCIAL WELFARE

More than one-fifth of all Floridians receive Social Security payments. In 1983, the state ranked 3d in the US in both number of beneficiaries and total benefits received.

Direct public assistance payments to Floridians in 1982/83 totaled $576,933,351, including $212,518,993 in aid to families with dependent children (AFDC). During that fiscal year, AFDC cases averaged 94,549 per month; 42% of the cases were in Dade, Broward, Orange, Palm Beach, Escambia, Polk, and Brevard counties.

In 1983, 749,000 people were participating in the federal food stamp program, at a federal cost of $453 million. Some 981,000 students were taking part in the national school lunch program in 1983; federal spending for this program came to $110,000,000.

Recipients of old age, survivors, and disability insurance (OASDI) payments from the Social Security Administration as of December 1982 numbered 2,156,111, consisting of some 1,597,141 retired workers and their dependents, 349,221 survivors of covered workers, 206,839 disabled workers and their dependents, and 2,910 special age-72 beneficiaries. OASDI payments in 1983 totaled more than $10.2 billion, including $7.5 billion to retirees and their dependents, $1.7 billion to survivors, and $964 million to the disabled and their dependents. Retirees' average monthly benefits in December 1982 were $418.80. As of 1983, 170,904 Floridians were receiving Supplemental Security Income. Payments totaled $383 million: $153 million to the aged, $222.7 million to the disabled, and $7.3 million to the blind.

In 1982/83, a total of 11,994 students in grades 7–12 took part in vocational education programs; 52,169 adults also participated. About 49,900 Floridians took part in vocational rehabilitation programs in March 1983, including 6,372 people in job training programs, 12,025 in child day care programs, and 14,561 in residential care programs. In 1984, the federal government's Office of Special Education granted $22,449,000 toward education for the disabled and $30,316,000 toward rehabilitation services and research for the disabled. In 1982, workers' compensation benefits totaled $458.4 million. Total disbursements of unemployment insurance in 1983 came to $406 million; the average number of weekly beneficiaries was about 50,000.

40HOUSING

Florida's housing market fluctuated widely in the 1970s and early 1980s. During the mid-1970s recession, home buying dropped off markedly, and much newly completed housing could not be sold. By late in the decade, however, the unused housing stock had been depleted, and a new building boom was under way. The number of housing units in Florida increased 73.2% between 1970 and 1980; the building of 439,000 new privately owned housing units was authorized between 1981 and 1983.

The 1980 census counted 4,378,691 housing units in Florida. Of that total, 58% were owner-occupied. As of 1 July 1983 there were 24,629 apartment buildings with 526,830 units, and 1,846 rooming houses with 19,442 units. In addition there were 1,709 rental condominiums with 23,345 units and 4,080 transient apartment buildings with 34,357 units. Multifamily housing ranges from beachfront luxury high rises along the Gold Coast to dilapidated residential hotels in the South Beach section of Miami Beach. In 1980, the median monthly cost for an owner-occupied housing unit with a mortgage was $337, and for a unit without a mortgage it was $99. That year, the median rent for a housing unit was $256.

In 1980, Florida had 483,611 year-round condominium housing units (more than any other state), of which 281,462 were owner-occupied and 60,400 were renter-occupied. Large retirement communities, often containing thousands of condominium units, are commonplace in Dade, Broward, and Palm Beach counties.

The Division of Florida Land Sales and Condominiums, within the Department of Business Regulation, registers all sellers of subdivided land and oversees the advertising and selling of land, condominiums, and cooperatives. A major controversy involving condominiums in the early 1970s centered on "rec leases." Until the practice was outlawed in mid-decade, condominium developers often retained ownership of such recreational facilities as the swimming pool, clubhouse, and tennis courts, requiring apartment purchasers to pay rent for their use. The rents were generally set quite low at the time of sale, but raised sharply soon after.

In 1980, 91% of housing owners and 80% of renters were white; 8% of owners and 17% of renters were black; and 5% of owners and 11% of renters were of Hispanic origin, both black and white. The number of mobile homes licensed in 1982/83 was 463,174.

An estimated 189,000 new housing units were authorized in 1983; residential construction contracts that year were valued at over $8.6 billion, 3d to California and Texas and more than 9% of the US total.

41 EDUCATION

In the 1970s, Florida was an innovator in several areas of education, including competency testing, expansion of community colleges, and school finance reform. Further advances were made in 1983 and 1984, when the state increased taxes to help fund education, raised teachers' salaries, initiated the nation's strictest high school graduation requirements, and reformed the curriculum.

Student achievement in reading, writing, and mathematics is measured by the State Student Assessment Test (SSAT), given annually to all 3d, 5th, 8th, and 11th graders. Schools must make test results available to parents and the general public. Under state law, 11th graders must pass the SSAT Part II, which tests ability to apply reading and math skills in practical situations, in order to obtain a regular high school diploma. In 1983, passage of the test became a criterion for graduation. In that year, 87% of all 10th- and 11th-grade students who took the SSAT passed it. A total of 90,700 high school diplomas were awarded in 1982. In 1980, 66.7% of Floridians 25 years of age or older were high school graduates; 14.9% had four or more years of college.

There were 2,280 public schools with 88,537 teachers in 1983/84. The full-time equivalent enrollment was 1,624,094 in fall 1982, including 1,324,601 students in basic education, 164,163 in vocational education, 42,424 in adult education, 70,483 in exceptional education, and 22,423 in alternative educational programs. The basic education total consisted of 418,450 students in grades K–3, 666,179 in grades 4–9, and 239,972 in grades 10–12. In fall 1983, nearly 67% of public school students were white non-Hispanic, 24% black non-Hispanic, and over 8% Hispanic; there were small percentages of Asian and American Indian students. In Dade County, well over one-third of the student body was Hispanic.

There were 206,456 students enrolled in nonpublic preprimary, elementary, and secondary schools in 1983/84. The full-time instructional staff of these schools totaled 13,732. Nonpublic school enrollment, especially in northern Florida, increased markedly in the early 1970s, apparently in response to public school desegregation.

Florida has nine state universities, with a total fall 1984 enrollment of 146,173. The largest, the University of Florida (Gainesville), had a fall 1984 enrollment of 36,500 and a faculty of 2,744. Also part of the state university system are special univer-

sity centers, such as the University of Florida's Institute of Food and Agricultural Science, which provide advanced and graduate courses. The State University System also offers instruction at strategic sites away from the regular campuses. In 1972, Florida completed a community college system that put a public two-year college within commuting distance of virtually every resident. The state's 28 community colleges had a fall 1982 enrollment of more than 219,170. Some 56% of these students were in advanced and professional programs; 28% were in vocational programs. Almost two-thirds of the community college students attended on a part-time basis.

Of Florida's 48 private institutions of higher education, by far the largest is the University of Miami (Coral Gables), which had a fall 1982 enrollment of 14,656.

The Board of Education, consisting of the governor and the cabinet, is responsible for the state's educational system. The commissioner of education (a cabinet member) is also the administrative head of the Department of Education. The policy-making body for the state university system is the Board of Regents; the chancellor is the system's chief administrative officer.

In 1984, the state government provided about 54% of the funding for public education; local governments, 38%; and the federal government, 8%. Estimated operating expenditures in 1984 were $4.9 billion. Spending per pupil in average daily attendance was $3,201 (22d in the US); spending per capita was $455 (39th). Florida's school finance law, the Florida Education Finance Act of 1973, establishes a funding formula aimed at equalizing both per-pupil spending statewide and the property tax burdens of residents of different school districts.

42 ARTS

Key West has long been a gathering place of creative artists, ranging from John James Audubon and Winslow Homer to Ernest Hemingway and Tennessee Williams. On 24 January 1980, the Tennessee Williams Fine Arts Center at Florida Keys Community College, located in Key West, opened with the world premiere of a Williams play, *Will Mr. Merriwether Return from Memphis?* In addition to the playhouse, the center contains gallery space and a cinema.

The Asolo Theater, which is located in Sarasota and is the site of an annual theater festival, was designated the state theater of Florida by the 1965 legislature.

Regional and metropolitan symphony orchestras include the Florida Philharmonic (Miami), Florida Symphony (Orlando), Jacksonville Symphony, Florida Gulf Coast Symphony (St. Petersburg), Florida West Coast Symphony (Sarasota), Ft. Lauderdale Symphony, Miami Beach Symphony, and Greater Palm Beach Symphony. Opera companies include the Asolo Opera Company, Florida Opera West, Orlando Opera Company, Sarasota Opera Association, Palm Beach Opera, and Greater Miami Opera Association.

In 1984, Florida received $418,000 from the National Endowment for the Arts. State art programs and initiatives include a state art school, an art indemnification program that provides state funds to insure touring works of art, an art bank that places artworks in existing state buildings, the 3d-largest budget for public radio in the nation ($495,793), and an art acquisition program sponsored by the house of representatives.

43 LIBRARIES AND MUSEUMS

Florida had 7 multi-county library systems and 58 county systems in 1981/82. The book stock was 14.9 million, and circulation totaled 36 million.

The largest public library systems are those of Miami–Dade County (2,290,256 volumes in 1981/82) and Jacksonville (1,427,-719 volumes). The State Library in Tallahassee housed 382,198 volumes and 137,660 microfiche state and federal documents in 1984; the State Library also distributes federal aid to local libraries and provides other assistance. In 1984, federal aid under the Library Services and Construction Act to Florida public libraries totaled $3.2 million; state aid to public libraries was $5.3 million.

The largest university library in the state is that of the University of Florida, with holdings of more than 2.3 million volumes in 1984. Other major university libraries are those of the University of Miami (1.4 million volumes) and Florida State University (1.4 million).

Florida has about 140 museums, galleries, and historical sites, as well as 24 public gardens. One of the best-known museums is the John and Mabel Ringling Museum of Art (Sarasota), a state-owned facility which houses the collection of the late circus entrepreneur, featuring Italian and North European Renaissance paintings. Also in Sarasota are the Ringling Museum of the Circus and the Circus Hall of Fame, and Ca'd'Zan, the Ringling mansion. The estates and homes of a number of prominent former Florida residents are now open as museums. The Villa Vizcaya Museum and Gardens in Miami, originally the estate of International Harvester founder James R. Deering, displays his collection of 15th–18th-century antiques. Railroad developer Henry Morrison Flagler's home in Palm Beach is now a museum in his name. The Society of the Four Arts is also in Palm Beach. On Key West, Ernest Hemingway's home is also a museum. The John James Audubon house in Key West and Thomas Edison's house in Ft. Myers are two of Florida's other great homes.

The Metrozoo-Miami, with an average annual attendance of 840,000, and the Jacksonville Zoological park, 260,000, are among the state's leading zoos. Both Busch Gardens (Tampa) and Sea World of Florida (Orlando) report average annual attendances of 3,000,000.

The largest historic restoration in Florida is in St. Augustine, where several blocks of the downtown area have been restored to their 18th-century likeness under the auspices of the Historic St. Augustine Preservation Board, a state agency. Castillo de San Marcos, the 17th-century Spanish fort at St. Augustine, is now a national monument under the jurisdiction of the National Park Service and is open to the public. Other Florida cities having historic preservation boards are Pensacola, Tallahassee, and Tampa.

44 COMMUNICATIONS

In 1985, Florida had 473 post offices with 30,922 employees. Postal receipts totaled $847,448,422 in 1982/83, up 8.1% from 1981/82. As of 1980, 91% of the state's occupied housing units had telephones.

Florida's first radio station was WFAW (later WQAM) in Miami, which went on the air in 1920. In 1984, the state had 204 commercial AM stations and 169 FM. Miami was also the site of the state's first television station, WTVJ, which began broadcasting on 27 January 1949.

There were 43 commercial TV stations in Florida in 1984; Miami, with 6 stations, had more than any other Florida city. Eleven educational television stations were operating in 1984, serving virtually the entire state. In 1984, Florida had 180 cable television systems, serving 1,898,447 residents in 516 communities.

45 PRESS

The *East Florida Gazette,* published in St. Augustine in 1783–84, was Florida's earliest newspaper. The oldest paper still publishing is the *Jacksonville Times-Union* (now *Florida Times-Union,*) which first appeared in February 1883.

In 1984, the state had 51 daily newspapers. There were 23 morning papers, with a combined circulation of 2,073,029; 26 evening papers, with a circulation of 751,916; 2 all-day papers (whose circulation is included in the morning and evening totals); and 34 Sunday papers, with a circulation of 3,105,953. The leading English-language dailies and their circulations in 1984 were:

AREA	NAME	DAILY	SUNDAY
Ft. Lauderdale	Sun-Sentinel (m)	108,879	247,133
	News (e)	84,579	
Jacksonville	Florida Times-Union (m, S)	160,989	214,827
	Journal (e)	179,087	
Miami	Herald (m, S)	435,418	527,538
	News (e)	62,175	
Orlando	Sentinel (all day, S)	227,932	282,000
St. Petersburg	Times (m, S)	267,266	340,251
Sarasota	Herald-Tribune (m, S)	100,198	116,592
Tampa	Tribune (all day)	208,798	
	Tribune and Times (S)		273,868
West Palm Beach	Palm Beach Post (m, S)	101,098	164,930

There were 135 weekly newspapers in 1984. Fifteen black newspapers and 4 Spanish-language newspapers were also published in Florida. The most widely read periodical published in Florida is the sensationalist *National Enquirer.* There were 32 book publishers in Florida in 1984, including Academic Press and University Presses of Florida.

46 ORGANIZATIONS

The 1982 census of Service Industries counted 2,683 organizations in Florida, including 549 business associations; 1,511 civic, social, and fraternal associations; and 78 educational, scientific, and research associations.

Commercial, trade, and professional organizations based in Florida include the American Accounting Association (Sarasota), American Welding Society (Miami), American Electroplaters' Society (Winter Park), Florida Citrus Mutual (Lakeland), and Florida Fruit and Vegetable Association (Orlando).

Sports groups include the National Association for Stock Car Auto Racing, better known as NASCAR (Daytona Beach), American Surfing Association (Lake Worth), American Water Ski Association (Winter Haven), and International Game Fish Association (Ft. Lauderdale).

Among other organizations with headquarters in Florida are the American Sunbathing Association, formerly the International Nudist Conference (Kissimmee), and the National Head Start Association (Bradenton).

47 TOURISM, TRAVEL, AND RECREATION

Tourism is a mainstay of the state's economy. In 1983, Florida had almost 39 million visitors—who collectively spent nearly $23 billion there. Almost 22 million visitors entered the state by car in 1983; another 13.2 million visitors arrived by air. Most of Florida's tourists are from elsewhere in the US, although, in the late 1970s and early 1980s, Miami was attracting large numbers of affluent Latin American travelers, lured at least in part by the Latin flavor the large Cuban community has given the city.

Some 338,900 Floridians worked in tourist and recreation-related businesses in 1982, and the state ranked 2d in the nation in the number of travel and tourism employees. In July 1984, the state had 796 licensed hotels, with a total of 91,121 units. In 1983, more than half of all hotels were located in Dade County, where hotels and other tourist accommodations stretch for miles along Collins Avenue in Miami Beach and communities to the north, in the heart of the state's tourist industry. In addition, Florida had 4,431 licensed motels with 190,214 units in 1984.

Florida's biggest tourist attractions are its sun, sand, and surf. According to the state's Department of Commerce, leisure-time activity is the principal reason why more than four-fifths of auto travelers enter the state. A major tourist attraction is Walt Disney World, a huge amusement park near Orlando; in 1982, EPCOT Center, a futuristic exhibition and amusement park, was opened at Disney World. Other major attractions are the Kennedy Space Center at Cape Canaveral and the St. Augustine historic district.

Nine parks and other facilities in Florida operated by the National Park Service drew some 6,565,700 visitors in 1983. The most popular destination was the Gulf Islands National Seashore, located near Pensacola (4,060,400 visitors), followed by the

Canaveral National Seashore (1,075,000). Almost 15 million people visited 92 facilities operated by the Division of Recreation and Parks of the state's Department of Natural Resources. These facilities included 28 state parks, 28 state recreation areas, and 18 state historical sites.

Fishing and boating are major recreational activities. In 1983/84, 499,364 pleasure craft were registered in the state. In 1983, licenses were sold to 729,191 anglers and 257,739 hunters.

In the 1970s and early 1980s, the Miami Beach tourist hotels faced increasing competition from Caribbean and Latin American resorts. The city's business community, seeking to boost tourism, strongly backed a 1978 statewide referendum to authorize casino gambling along part of Collins Avenue in Miami Beach and Hollywood; however, the proposal was defeated by a wide margin. In a local advisory referendum in March 1980, Miami Beach voters approved development in South Beach of an $850-million, 250-acre (100-hectare) complex that would include hotels and a convention center; the complex was not complete as of 1985. As of June 1983, off-track betting, horse racing, dog racing, jai alai and bingo were all legalized and operative forms of gaming.

48 SPORTS

Florida has two National Football League teams, the Miami Dolphins and the Tampa Bay Buccaneers. The Dolphins have been by far the more successful, winning Super Bowls in 1973 (climaxing an undefeated season) and 1974 and appearing in two others, in 1972 and 1983. Although no major league baseball teams play their home games in Florida during the regular season, many have their training camps in the state and play exhibition games (the "grapefruit league") there in the spring.

Interest in collegiate sports focuses on the Florida Gators of the Southeastern Conference, the Florida State Seminoles, and the Miami Hurricanes; the latter two are 1-A Southern independents. Miami won the NCAA national football championship in 1983 and is a perennial power in baseball. Thoroughbred, harness, quarter-horse, and greyhound racing are popular spectator and gambling sports, as is jai alai. The Hialeah Park Race Track, site of the annual Flamingo Stakes, is considered one of the most beautiful in the US.

Several tournaments on both the men's and the women's professional golf tours are played on Florida courses. In stock-car racing, the Daytona 500 is a top race on the NASCAR circuit. Three major collegiate football bowl games are played in the state: the Orange Bowl (Miami), Gator Bowl (Jacksonville), and Florida Citrus Bowl (Orlando).

49 FAMOUS FLORIDIANS

The first Floridian to serve in a presidential cabinet was Alan S. Boyd (b.1922), named the first secretary of transportation (1967–69) by President Lyndon Johnson. Florida also produced one of the major US military figures of World War II, General Joseph Warren Stilwell (1883–1946), dubbed "Vinegar Joe" for his strongly stated opinions. Graduated from West Point in 1904, he served in France during World War I. First posted to China in the 1920s, he became chief of staff to General Chiang Kai-shek and commander of US forces in the China-Burma-India theater during World War II. He was promoted to full general in 1944 but forced to leave China because of his criticism of the Chiang Kai-shek regime.

David Levy Yulee (b.St. Thomas, 1810–86) came to Florida in 1824 and, after serving in the US House of Representatives, was appointed one of the state's first two US senators in 1845, thereby becoming the first Jew to sit in the Senate. He resigned in 1861 to serve in the Confederate Congress. Yulee built the first cross-state railroad, from Fernandina to Cedar Key, in the late 1860s. Ruth Bryan Owen Rohde (b.Illinois, 1885–1954), a longtime Miami resident and member of the US House of Representatives (1929–33), in 1933 became the first woman to head a US diplo-

matic office abroad when she was named minister to Denmark.

Prominent governors of Florida include Richard Keith Call (b.Virginia, 1792–1862), who came to Florida with General Andrew Jackson in 1821 and remained to become governor of the territory in 1826–39 and 1841–44. In the summer of 1836, Call commanded the US campaign against the Seminole. Although a southerner and a slaveholder, he steadfastly opposed secession. Napoleon Bonaparte Broward (1857–1910) was, before becoming governor, a ship's pilot and owner of St. Johns River boats. He used one of these, *The Three Friends,* a powerful seagoing tug, to run guns and ammunition to Cuban rebels in 1896. As governor (1905–9), he was noted for a populist program that included railroad regulation, direct elections, state college reorganization and coordination, and drainage of the Everglades under state auspices. As governor in 1955–61, Thomas LeRoy Collins (b.1909), met the desegregation issue by advocating moderation and respect for the law, helping the state avoid violent confrontations. He served as chairman of both the southern and national governors' conferences, and he was named by President Johnson as the first director of the Community Relations Service under the 1964 Civil Rights Act.

Military figures who have played a major role in Florida's history include the Spanish conquistadores Juan Ponce de León (c.1460–1521), the European discoverer of Florida, and Pedro Menéndez de Avilés (1519–74), founder of the first permanent settlement, St. Augustine. Andrew Jackson (b.South Carolina, 1767–1845), a consistent advocate of US seizure of Florida, led military expeditions into the territory in 1814 and 1818 and, after US acquisition, served briefly in 1821 as Florida's military governor before leaving for Tennessee. During the Seminole War of 1835–42, one of the leading military tacticians was Osceola (c.1800–1838), who, although neither born a chief nor elected to that position, rose to the leadership of the badly divided Seminole by force of character and personality. He rallied them to fierce resistance to removal, making skillful use of guerrilla tactics. Captured under a flag of truce in 1837, he was imprisoned; already broken in health, he died in Fort Moultrie in Charleston harbor. During the Civil War, General Edmund Kirby Smith (1824–93), a native of St. Augustine who was graduated from West Point in 1845, served as commander (1863–65) of Confederate forces west of the Mississippi River. He surrendered the last of the southern forces at Galveston, Texas, on 26 May 1865.

Among the late-19th-century entrepreneurs who played significant roles in Florida's development, perhaps the most important was Henry Morrison Flagler (b.New York, 1830–1913). Flagler made a fortune in Ohio as an associate of John D. Rockefeller in the Standard Oil Co. and did not even visit Florida until he was in his 50s. However, in the 1880s, he began to acquire and build railroads down the length of Florida's east coast and to develop tourist hotels at various points, including St. Augustine, Palm Beach, and Miami, helping to create one of the state's major present-day industries. Henry Bradley Plant (b.Connecticut, 1819–99) did for Florida's west coast what Flagler did for the east. Plant extended railroad service to Tampa in 1884, built a huge tourist hotel there, developed the port facilities, and established steamship lines.

Among Floridians prominent in science was Dr. John F. Gorrie (b.South Carolina, 1802–55), who migrated to Apalachicola in 1833 and became a socially and politically prominent physician, specializing in the treatment of fevers. He blew air over ice brought in by ship from the north to cool the air in sickrooms, and he independently developed a machine to manufacture ice, only to have two others beat him to the patent office by days.

The noted labor and civil rights leader A. Philip Randolph (1889–1979) was a native of Crescent City. Mary McLeod Bethune (b.South Carolina, 1875–1955) was an adviser to President Franklin D. Roosevelt on minority affairs, became the first president (1935) of the National Council of Negro Women, and was a consultant at the 1945 San Francisco Conference that founded the UN. A prominent black educator, she opened a school for girls at Daytona Beach in 1904. The school merged with Cookman Institute in 1923 to become Bethune-Cookman College, which she headed until 1942 and again in 1946-47.

Prominent Florida authors include James Weldon Johnson (1871-1938), perhaps best known for his 1912 novel *Autobiography of an Ex-Colored Man*. He was also the first black to be admitted to the Florida bar (1897) and was a founder and secretary of the NAACP. Majory Stoneman Douglas (b.Minnesota, 1890), who came to Miami in 1915, is the author of several works reflecting her concern for the environment, including *The Everglades: River of Grass* (first published in 1947), *Hurricane* (1958), and *Florida: The Long Frontier* (1967). Marjorie Kinnan Rawlings (b.Washington, D.C., 1895-1953) came to Florida in 1928 to do creative writing. After her first novel, *South Moon Under* (1933), came the Pulitzer Prize–winning *The Yearling* (1938), the poignant story of a 12-year-old boy on the Florida frontier in the 1870s. Zora Neale Hurston (1901-60), born in poverty in the all-Negro town of Eatonville and a graduate of Barnard College, spent four years collecting folklore, which she published in *Mules and Men* (1935) and *Tell My Horse* (1938).

Among the entertainers born in Florida are Sidney Poitier (b.1927), Charles Eugene "Pat" Boone (b.1934), Faye Dunaway (b.1941), and Ben Vereen (b.1946).

Florida's most famous sports figure is Chris Evert Lloyd (Christine Marie Evert, b.1953), who became a dominant force in women's tennis in the mid-1970s. After turning pro in 1973, she won the Wimbledon singles title in 1974, 1976, and 1981, and the US Open from 1975 to 1978 and in 1980 and 1982.

⁵⁰BIBLIOGRAPHY

Bennett, Charles E. *Florida's French Revolution, 1793-1795*. Gainesville: University Presses of Florida, 1981.

Daver, Manning J. (ed.). *Florida's Politics and Government*. 2d ed. Gainesville: University Presses of Florida, 1984.

Douglas, Marjory Stoneman. *The Everglades: River of Grass*. Rev. ed. Miami: Banyan, 1978.

Douglas, Marjory Stoneman. *Florida: The Long Frontier*. New York: Harper & Row, 1967.

Federal Writers' Project. *Florida: A Guide to the Southernmost State*. Reprint. New York: Somerset, 1981 (orig. 1939).

Florida, University of. College of Business Administration. Bureau of Business and Economic Research. *1984 Florida Abstract*. Gainesville: University Presses of Florida, 1984.

Gannon, Michael V. *The Cross in the Sand*. Gainesville: University of Florida Press, 1965.

Hanna, Alfred Jackson, and James Branch Cabell. *The St. Johns: A Parade of Diversities*. New York: Farrar & Rinehart, 1943.

Hanna, Kathryn T. Abbey. *Florida: Land of Change*. Chapel Hill: University of North Carolina Press, 1948.

Harris, Michael H. *Florida History: A Bibliography*. Metuchen, N.J.: Scarecrow Press, 1972.

Jahoda, Gloria. *Florida: A Bicentennial History*. New York: Norton, 1976.

Johns, John E. *Florida During the Civil War*. Gainesville: University of Florida Press, 1963.

Lyon, Eugene. *The Enterprise of Florida: Pedro Menéndez de Avilés and the Spanish Conquest of 1565-68*. Gainesville: University Presses of Florida, 1976.

MacDonald, John D. *Condominium*. Philadelphia: Lippincott, 1977.

Mahon, John K. *The Second Seminole War*. Gainesville: University of Florida Press, 1967.

Morris, Allen (comp.). *The Florida Handbook 1985–86*. Tallahassee: Peninsular Publishing, 1985.

Patrick, Rembert W. *Florida Under Five Flags*. Gainesville: University of Florida Press, 1960.

Rawlings, Marjorie Kinnan. *The Yearling*. New York: Scribner, 1938.

Shofner, Jerrell M. *Nor Is It Over Yet: Florida in the Era of Reconstruction, 1863-77*. Gainesville: University Presses of Florida, 1974.

Smiley, Nixon. *Knights of the Fourth Estate: The Story of the Miami Herald*. Miami: Seemann, 1974.

Smiley, Nixon. *Yesterday's Florida*. Miami: Seemann, 1974.

Tebeau, Charlton. *A History of Florida*. Coral Gables: University of Miami Press, 1981 (orig. 1971).

Thompson, Ralph B., ed. *Florida Statistical Abstract 1984*. 18th ed. Gainesville: University Presses of Florida, 1984.

Wood, Roland, and Edward A. Fernald. *The New Florida Atlas: Patterns of the Sunshine State*. Tampa: Trend House, 1974.

Wright, J. Leitch, Jr. *Florida in the American Revolution*. Gainesville: University Presses of Florida, 1975.

GEORGIA

State of Georgia

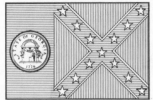

ORIGIN OF STATE NAME: Named for King George II of England in 1732. **NICKNAME:** The Empire State of the South. (Also: Peach State.) **CAPITAL:** Atlanta. **ENTERED UNION:** 2 January 1788 (4th). **SONG:** "Georgia." **MOTTO:** Wisdom, Justice, Moderation. **COAT OF ARMS:** Three columns support an arch inscribed with the word "Constitution"; intertwined among the columns is a banner bearing the state motto. Right of center stands a soldier with a drawn sword, representing the aid of the military in defending the Constitution. Surrounding the whole are the words "State of Georgia 1776." **FLAG:** At the hoist, on a blue bar, is the coat of arms. The remainder comprises the battle flag of the Confederacy. **OFFICIAL SEAL:** Obverse: same as the coat of arms. Reverse: a sailing vessel and a smaller boat are offshore; on land, a man and horse plow a field, and sheep graze in the background. The scene is surrounded by the words "Agriculture and Commerce 1776." **BIRD:** Brown thrasher. **FISH:** Largemouth bass. **FLOWER:** Cherokee rose. **WILDFLOWER:** Azalea. **TREE:** Live oak. **GEM:** Quartz. **INSECT:** Honeybee. **FOSSIL:** Shark tooth. **LEGAL HOLIDAYS:** New Year's Day, 1 January; Robert E. Lee's Birthday, 19 January; Birthday of Martin Luther King, Jr., 3d Monday in January; Washington's Birthday, 3d Monday in February; Confederate Memorial Day, 26 April; National Memorial Day, last Monday in May; Jefferson Davis's Birthday, 3 June; Independence Day, 4 July; Labor Day, 1st Monday in September; Columbus Day, 2d Monday in October; Veterans Day, 11 November; Thanksgiving Day, 4th Thursday in November; Christmas Day, 25 December. **TIME:** 7 AM EST = noon GMT.

¹LOCATION, SIZE, AND EXTENT

Located in the southeastern US, Georgia is the largest state east of the Mississippi River, and ranks 21st in size among the 50 states.

The total area of Georgia is 58,910 sq mi (152,576 sq km), of which land comprises 58,056 sq mi (150,365 sq km) and inland water 854 sq mi (2,211 sq km). Georgia extends 254 mi (409 km) E–W; the maximum N–S extension is 320 mi (515 km).

Georgia is bordered on the N by Tennessee and North Carolina; on the E by South Carolina (with the line formed by the Chattooga, Tugaloo, and Savannah rivers) and by the Atlantic Ocean; on the S by Florida (with the line in the SE defined by the St. Marys River); and on the W by Alabama (separated in the SW by the Chattahoochee River). The state's geographic center is located in Twiggs County, 18 mi (29 km) SW of Macon.

The Sea Islands extend the length of the Georgia coast. The state's total boundary length is 1,039 mi (1,672 km).

²TOPOGRAPHY

Northern Georgia is mountainous, the central region is characterized by the rolling hills of the Piedmont Plateau, and southern Georgia is a nearly flat coastal plain.

The Blue Ridge Mountains tumble to an end in northern Georgia, where Brasstown Bald, at 4,784 feet (1,458 meters), is the highest point in the state. The piedmont slopes slowly to the fall line, descending from about 2,000 feet (610 meters) to 300 feet (90 meters) above sea level. Stone Mountain, where a Confederate memorial is carved into a mass of solid granite 1,686 feet (514 meters) high, is the region's most famous landmark.

The piedmont region ends in a ridge of sand hills running across the state from Augusta to Columbus. The residue of an ancient ocean was caught in the vast shallow basin on the Florida border, known as the Okefenokee Swamp, which filled with fresh water over the centuries. The coastal plain, thinly populated except for towns at the mouths of inland rivers, ends in marshlands along the Atlantic Ocean. Lying offshore are the Sea Islands, called the Golden Isles of Georgia, the most important of which are, from north to south, Tybee, Ossabaw, St. Catherines, Sapelo, St. Simons, Sea Island, Jekyll, and Cumberland.

Two great rivers rise in the northeast: the Savannah, which

forms part of the border with South Carolina, and the Chattahoochee, which flows across the state to become the western boundary. The Flint joins the Chattahoochee at the southwestern corner of Georgia to form the Apalachicola, which flows through Florida into the Gulf of Mexico. The two largest rivers of central Georgia, the Ocmulgee and Oconee, flow together to form the Altamaha, which then flows eastward to the Atlantic. Perhaps the best-known Georgia river, though smaller than any of the above, is the Suwannee, flowing southwest through the Okefenokee Swamp, across Florida, and into the Gulf of Mexico, and famous for its evocation by Stephen Foster in the song "Old Folks at Home." Huge lakes created by dams on the Savannah River are Clark Hill Reservoir and Hartwell Lake; artificial lakes on the Chattahoochee River include Lake Seminole, Walter F. George Reservoir, Lake Harding, West Point Reservoir, and Lake Sidney Lanier.

³CLIMATE

The Chattahoochee River divides Georgia into separate climatic regions. The mountain region to the northwest is colder than the rest of Georgia, averaging 39°F (4°C) in January and 78°F (26°C) in July. The remainder of the state experiences mild winters, ranging from a January average of 44°F (7°C) in the piedmont to 54°F (12°C) on the coast. Summers are hot in the piedmont and on the coast, with July temperatures averaging 80°F (27°C) or above. The record high is 112°F (44°C), set at Louisville on 24 July 1952; the record low is −17°F (−27°C), registered in Floyd County on 27 January 1940.

Humidity is high, ranging from 83% in the morning to 57% in the afternoon in Atlanta. Rainfall varies considerably from year to year, but averages 50 in (127 cm) annually in the lowlands, increasing to 75 in (191 cm) in the mountains; snow falls occasionally in the interior. Tornadoes—there were 26 in 1983—are an annual threat in mountain areas, and Georgia beaches are exposed to hurricane tides.

The growing season is approximately 185 days in the mountains and a generous 300 days in southern Georgia.

⁴FLORA AND FAUNA

Georgia has some 250 species of trees, 90% of which are of commercial importance. White and scrub pines, chestnut, northern red oak, and buckeye cover the mountain zone, while loblolly and shortleaf (yellow) pines and whiteback maple are found throughout the piedmont. Pecan trees grow densely in southern Georgia, and white oak and cypress are plentiful in the eastern part of the state. Trees found throughout the state include red cedar, scaly-bark and white hickories, red maple, sycamore, yellow poplar, sassafras, sweet and black gums, and various dogwoods and magnolias. Common flowering shrubs include yellow jasmine, flowering quince, and mountain laurel. Spanish moss grows abundantly on the coast and around the streams and swamps of the entire coastal plain. The state lists 58 protected plants, of which 23—including buckthorn, golden seal, spiderlily, fringed campion, and starflower—are endangered.

Prominent among Georgia fauna is the white-tailed (Virginia) deer, found in some 50 counties. Other common mammals are the black bear, muskrat, raccoon, opossum, mink, common cottontail, and three species of squirrel—fox, gray, and flying. No fewer than 160 bird species breed in Georgia, among them the mockingbird, brown thrasher (the state bird), and numerous sparrows; the Okefenokee Swamp is home to the sandhill piper, snowy egret, and white ibis. The bobwhite quail is the most popular game bird. There are 79 species of reptile, including such poisonous snakes as the rattler, copperhead, and cottonmouth moccasin. The state's 63 amphibian species consist mainly of various salamanders, frogs, and toads. The most popular freshwater game fish are trout, bream, bass, and catfish, all but the last of which are produced in state hatcheries for restocking. Dolphins, porpoises, shrimp, oysters, and blue crabs are found off the Georgia coast.

Rare or threatened animals include the indigo snake and Georgia's blind cave salamander. The state protects 23 species of wildlife, among them the colonial and Sherman's pocket gophers, right and humpback whales, manatee, brown pelican, American alligator, three species of sea turtle, shortnose sturgeon, and southern cave fish. As of July 1984, 18 of these were on the federal endangered list. Georgia has 468,146 acres (189,453 hectares) of wildlife refuges.

5ENVIRONMENTAL PROTECTION

Erosion of Georgia's farmlands left red clay exposed all across the once-fertile piedmont by the 1930s. The Soil Conservation Acts of 1936 and 1956 paid farmers to grow restorative crops and even to put their lands into tree production. The result was that Georgia's countryside had become green again by the 1980s.

In the early 1970s, environmentalists pointed to the fact that the Savannah River had been polluted by industrial waste and that an estimated 58% of Georgia's citizens lived in districts lacking adequate sewage treatment facilities. In 1972, at the prodding of Governor Jimmy Carter, the general assembly created the Environmental Protection Division (EPD) within the Department of Natural Resources (DNR). This agency administers a dozen environmental laws, most of them passed during the 1970s: the Water Quality Control Act, the Safe Drinking Water Act, the Groundwater Use Act, the Surface Water Allocation Act, the Air Quality Act, the Safe Dams Act, the Radiation Control Act, the Solid Waste Management Act, the Hazardous Waste Management Act, the Sedimentation and Erosion Control Act, the Surface Mining Act, and the Oil and Gas and the Deep Drilling Act. The EPD issues all the environmental permits, with the exception of those required by the Marshlands Protection and Shore Assistance Acts, which are enforced by the Coastal Resources Division of the DNR.

Georgia's greatest environmental problem continues to be water pollution from wastewater discharge. In 1983/84 the EPD made 711 solid-waste management inspections and closed 29 solid-waste dumps; it also conducted 98 environmental radiation surveys. In 1982/83, the EPD decontaminated the Luminous Processes radioactive waste site near Atlanta—the first such decontamination in the nation to be carried out under the Superfund program. Based in Atlanta, the EPD's Emergency Response Team is on call 24 hours a day to assist in environmental emergencies. In 1982/83, the team responded to 193 emergencies: 58% related to water protection, 30% to land, 8% to air, and 4% to radiological problems. In 1983/84, Georgia achieved 98% compliance with its air-quality standards.

6POPULATION

Georgia ranked 13th among the 50 states in 1980 with a population of 5,463,105. The estimated population in 1985 was 5,878,-225, and the population density was 101 per sq mi (39 per sq km).

During the first half of the 18th century, restrictive government policies discouraged settlement. In 1752, when Georgia became a royal colony, the population numbered only 3,500 of whom 500 were blacks. Growth was rapid thereafter, and by 1773 there were 33,000 people, almost half of them black. The American Revolution brought free land and an influx of settlers, so that by 1800 the population had swelled to 162,686. Georgia passed the million mark by 1860, the 2 million mark by 1900, and by 1960, the population had doubled again. Georgia's population increased 19% between 1970 and 1980, well above the US average.

According to the 1980 census, the population was 51.7% female and 48.3% male. With a median age of 28.6, Georgia residents were somewhat younger than the national average. They were also somewhat more mobile: 11.5% of Georgians five years of age or older counted in the 1980 census had lived in another state in 1975.

About 62% of all Georgians lived in urban areas in 1980, and 38% in rural areas. There has always been a strained relationship between rural and urban Georgians, and the state's political system long favored the rural population. Since before the American Revolution, the city people have called the countryfolk "crackers," a term that implies a lack of good manners and which may derive from the fact that these pioneers drove their cattle before them with whips.

Census counts for the state's four largest cities in 1984 were Atlanta, 426,090; Columbus, 174,824; Savannah, 145,014; and Macon, 120,226. The Atlanta metropolitan area had an estimated population of 2,380,000 in 1984.

7ETHNIC GROUPS

Georgia has been fundamentally a white/black state, with minimal ethnic diversity. Most Georgians are of English or Scotch-Irish descent. The number of Georgians who were foreign-born in 1980 was low—91,480 (or 1.7% of the population). However, this was a considerable increase over the 33,000 foreign-born Georgians in 1970. Between 1970 and 1980, the number of Georgians from Asia or the Pacific islands increased from 8,838 to 24,461. By 1980, 2,294 Vietnamese refugees had resettled in Georgia.

Georgia's black population declined from a high of 47% of the total population in 1880 to about 26% in 1970, when there were 1,187,149 blacks. Black citizens composed 27% of the total population and numbered 1,465,000 in 1980. Atlanta, which had 282,912 black residents (66.6%) in 1980, has been a significant center for the development of black leadership, especially at Atlanta University. With its long-established black elite, Atlanta has also been the locus of some large black-owned business enterprises. There are elected and appointed blacks in the state government, and in 1973, Atlanta elected its first black mayor, Maynard Jackson. By 1984 there were 13 black mayors, including Andrew J. Young of Atlanta.

There were only 7,444 American Indians in Georgia in 1980. The great Cherokee nation and its collateral linguistic relatives had been effectively removed from the state 150 years earlier.

LOCATION: 30°21′ to 35°N; 80°51′ to 85°36′W. **BOUNDARIES:** Tennessee line, 73 mi (118 km); North Carolina line, 70 mi (113 km); South Carolina line, 266 mi (428 km); Atlantic Ocean coastline, 100 mi (161 km); Florida line, 242 mi (389 km); Alabama line, 288 mi (463 km).

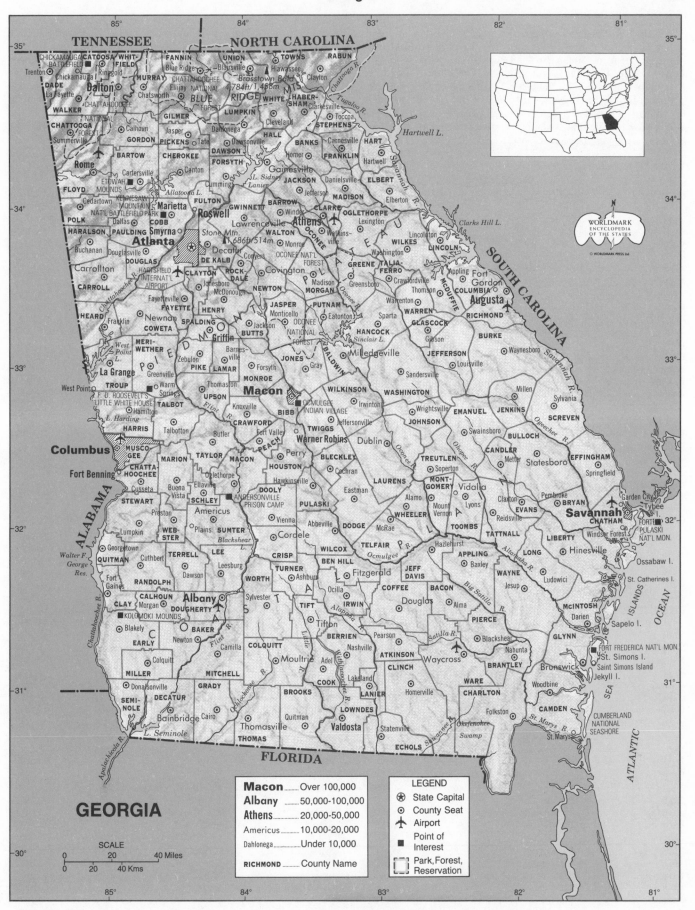

GEORGIA

8 LANGUAGES

The first Europeans entering what is now Georgia found it occupied almost entirely by Creek Indians of the Muskogean branch of Hokan-Siouan stock. Removed by treaty to Indian Territory after their uprising in 1813, the Creek left behind only such place-names as Chattahoochee, Chattooga, and Okefenokee.

Except for the South Midland speech of the extreme northern up-country, Georgia English is typically Southern. Loss of /r/ after a vowel in the same syllable is common. The diphthong /ai/ as in *right* is so simplified that Northern speakers hear the word as *rat*. *Can't* rhymes with *paint*, and *borrow*, *forest*, *foreign*, and *orange* all have the /ah/ vowel as in *father*. However, a highly unusual variety of regional differences, most of them in long vowels and diphthongs, makes a strong contrast between northern up-country and southern low-country speech. In such words as *care* and *stairs*, for example, many up-country speakers have a vowel like that in *cat*, while many low-country speakers have a vowel like in *pane*.

In general, northern Georgia *snake doctor* contrasts with southern Georgia *mosquito hawk* (dragonfly), *goobers* with *pinders*

Georgia Counties, County Seats, and County Areas and Populations

COUNTY	COUNTY SEAT	LAND AREA (SQ MI)	POPULATION (1984 EST.)	COUNTY	COUNTY SEAT	LAND AREA (SQ MI)	POPULATION (1984 EST.)
Appling	Baxley	510	16,274	Fayette	Fayetteville	199	39,486
Atkinson	Pearson	344	6,235	Floyd	Rome	519	78,786
Bacon	Alma	286	9,538	Forsyth	Cumming	226	33,144
Baker	Newton	347	3,595	Franklin	Carnesville	264	15,497
Baldwin	Milledgeville	258	37,873	Fulton	Atlanta[1]	534	614,838
Banks	Homer	234	9,697	Gilmer	Ellijay	427	11,683
Barrow	Winder	163	24,091	Glascock	Gibson	144	2,343
Bartow	Cartersville	456	43,412	Glynn	Brunswick	412	59,285
Ben Hill	Fitzgerald	254	16,858	Gordon	Calhoun	355	32,117
Berrien	Nashville	456	13,788	Grady	Cairo	459	21,028
Bibb	Macon	253	156,843	Greene	Greensboro	390	11,868
Bleckley	Cochran	219	10,674	Gwinnett	Lawrenceville	435	230,277
Brantley	Nahunta	444	9,286	Habersham	Clarkesville	278	26,622
Brooks	Quitman	491	15,296	Hall	Gainesville	379	81,713
Bryan	Pembroke	441	11,769	Hancock	Sparta	469	9,338
Bulloch	Statesboro	678	36,982	Haralson	Buchanan	283	19,361
Burke	Waynesboro	833	20,466	Harris	Hamilton	464	16,205
Butts	Jackson	187	14,872	Hart	Hartwell	230	19,100
Calhoun	Morgan	284	5,570	Heard	Franklin	292	6,707
Camden	Woodbine	649	16,823	Henry	McDonough	321	42,209
Candler	Metter	248	7,772	Houston	Perry	380	83,677
Carroll	Carrollton	502	61,763	Irwin	Ocilla	362	8,851
Catoosa	Ringgold	163	38,402	Jackson	Jefferson	342	26,803
Charlton	Folkston	780	7,664	Jasper	Monticello	371	7,709
Chatham	Savannah	444	212,388	Jeff Davis	Hazlehurst	335	11,775
Chattahoochee	Cusseta	250	20,459	Jefferson	Louisville	529	18,669
Chattooga	Summerville	314	21,454	Jenkins	Millen	353	8,508
Cherokee	Canton	424	62,778	Johnson	Wrightsville	307	8,617
Clarke	Athens	122	76,420	Jones	Gray	394	18,186
Clay	Ft. Gaines	197	3,335	Lamar	Barnesville	186	12,282
Clayton	Jonesboro	148	162,634	Lanier	Lakeland	194	5,648
Clinch	Homerville	821	6,826	Laurens	Dublin	816	38,012
Cobb	Marietta	343	354,289	Lee	Leesburg	358	13,605
Coffee	Douglas	602	28,442	Liberty	Hinesville	517	40,993
Colquitt	Moultrie	556	36,274	Lincoln	Lincolnton	196	6,859
Columbia	Appling	290	49,466	Long	Ludowici	402	5,446
Cook	Adel	232	13,814	Lowndes	Valdosta	507	71,799
Coweta	Newnan	444	42,972	Lumpkin	Dahlonega	287	11,784
Crawford	Knoxville	328	7,210	Macon	Oglethorpe	404	14,156
Crisp	Cordele	275	20,110	Madison	Danielsville	285	19,163
Dade	Trenton	176	11,578	Marion	Buena Vista	366	5,370
Dawson	Dawsonville	210	5,656	McDuffie	Thomson	256	19,423
Decatur	Bainbridge	586	26,416	McIntosh	Darien	425	8,032
DeKalb	Decatur	270	501,871	Meriwether	Greenville	506	20,787
Dodge	Eastman	504	16,795	Miller	Colquitt	284	6,849
Dooly	Vienna	397	10,666	Mitchell	Camilla	512	21,707
Dougherty	Albany	330	103,059	Monroe	Forsyth	397	14,886
Douglas	Douglasville	203	63,430	Montgomery	Mt. Vernon	244	6,955
Early	Blakely	516	13,158	Morgan	Madison	349	12,103
Echols	Statenville	420	2,248	Murray	Chatsworth	345	21,096
Effingham	Springfield	482	20,547	Muscogee	Columbus	218	175,695
Elbert	Elberton	367	18,990	Newton	Covington	277	38,074
Emanuel	Swainsboro	688	21,263	Oconee	Watkinsville	186	14,210
Evans	Claxton	186	8,494	Oglethorpe	Lexington	442	9,424
Fannin	Blue Ridge	384	15,167	Paulding	Dallas	312	29,197

(peanuts), *French harp* with *harmonica*, *plum peach* with *press peach* (both clingstone peaches), *nicker* with *whicker* for a horse's neigh, and *sallet* with *salad*. In Atlanta a big sandwich is a *poorboy*; in Savannah, a peach pit is a *kernel*.

A distinctive variety of black English, called Gullah, is spoken in the islands off the Georgia and South Carolina coast, to which Creole-speaking slaves escaped from the mainland during the 17th and 18th centuries. Characteristic grammatical features include lack of inflection in the personal pronoun, the invariant form of the *be* verb, and the absense of −*s* in the third person singular of the present tense. Many of the private personal names stem directly from West African languages.

In 1980, 5,070,508 Georgians—97% of the population 3 years old and older—spoke only English at home. Other languages spoken at home, and the number of people who spoke them, were as follows:

Spanish	48,561	Chinese	3,813
German	19,631	Italian	2,960
French	19,069	Asian Indian	2,891
Korean	5,152	Greek	2,737

[9] RELIGIONS

The Church of England was the established church in colonial Georgia. During this period, European Protestants were encouraged to immigrate, and German Lutherans and Moravians took advantage of the opportunity. Roman Catholics were barred and Jews were not welcomed, but persons of both denominations came anyway. In the mid-18th century, George Whitefield, called the Great Itinerant, helped touch off the Great Awakening, the religious revival out of which came the Methodist and Baptist denominations. Daniel Marshall, the first "separate" Baptist in Georgia, established a church near Kiokee Creek in 1772. Some 16 years later, James Asbury formed the first Methodist Conference in Georgia.

The American Revolution resulted in the lessening of the authority of Anglicanism and a great increase in the numbers of Baptists, Methodists, and Presbyterians. During the 19th century, fundamentalist sects were especially strong among blacks. Roman Catholics from Maryland, Ireland, and Hispaniola formed a numerically small but important element in the cities, and Jewish citizens were active in the leadership of Savannah and Augusta. Catholics and Jews enjoyed general acceptance from the early 1800s until the first two decades of the 20th century, when they became the targets of political demagogues, notably Thomas E. Watson.

In 1980 there were 2,419,249 known Protestant adherents. The leading denominations were Southern Baptist Convention, 1,374,-905; United Methodist, 488,909; Presbyterian, 95,173; Christian Methodist Episcopal, 84,209; Church of God (Cleveland, Tenn.), 58,349; and Episcopal, 54,231. In 1984, Georgia had 176,260 Roman Catholics and an estimated 42,155 Jews.

[10] TRANSPORTATION

Georgia's location between the Appalachian Mountains and the Atlantic Ocean makes it the link between the eastern seaboard and the Gulf states. In the 18th century, Carolina fur traders crossed the Savannah River at the site of Augusta and followed trails to the Mississippi River. Pioneer farmers soon followed the same trails and used the many river tributaries to send their produce to Savannah, Georgia's first great depot. Beginning in 1816, steamboats plied the inland rivers, but they never replaced the older shallow-drafted Petersburg boats, propelled by poles.

From the 1830s onward, businessmen in the eastern cities of Savannah, Augusta, and Brunswick built railroads west to maintain their commerce. The two principal lines, the Georgia and the Central of Georgia, were required by law to make connection with a state-owned line, the Western and Atlantic, at the new town of Atlanta, which thus became in 1847 the link between Georgia and the Ohio Valley. By the Civil War, Georgia, with more miles of rail than any other Deep South state, was a vital link between the eastern and western sectors of the Confederacy. After the war, the railroads contributed to urban growth as towns sprang up along their routes. Trackage increased from 4,532 mi (7,294 km) in 1890 to 7,591 mi (12,217 km) in 1920. But with competition from motor carriers, the total Class I trackage declined to 4,928 mi (7,931 km) in 1983. In 1979, Atlanta inaugurated the first mass-transit system in the state, including the South's first subway. Public transit

COUNTY	COUNTY SEAT	LAND AREA (SQ MI)	POPULATION (1984 EST.)
Peach	Ft. Valley	151	19,195
Pickens	Jasper	232	12,981
Pierce	Blackshear	344	12,568
Pike	Zebulon	219	8,696
Polk	Cedartown	312	33,018
Pulaski	Hawkinsville	249	8,960
Putman	Eatonton	344	11,424
Quitman	Georgetown	146	2,198
Rabun	Clayton	370	10,771
Randolph	Cuthbert	431	9,430
Richmond	Augusta	326	187,569
Rockdale	Conyers	132	42,957
Schley	Ellaville	169	3,371
Screven	Sylvania	655	14,331
Seminole	Donalsonville	225	8,835
Spalding	Griffin	199	50,752
Stephens	Toccoa	177	22,149
Stewart	Lumpkin	452	5,873
Sumter	Americus	488	30,179
Talbot	Talbotton	395	6,509
Taliaferro	Crawfordville	196	1,997
Tattnall	Reidsville	484	17,865
Taylor	Butler	382	8,051
Telfair	MacRae	444	11,126
Terrell	Dawson	337	11,964
Thomas	Thomasville	551	38,180
Tift	Tifton	268	33,557
Toombs	Lyons	371	23,431
Towns	Hiawassee	165	6,155
Treutlen	Soperton	202	6,016
Troup	La Grange	415	52,612
Turner	Ashburn	289	9,622
Twiggs	Jeffersonville	362	9,670
Union	Blairsville	320	10,349
Upson	Thomaston	326	26,479
Walker	La Fayette	446	55,434
Walton	Monroe	330	32,166
Ware	Waycross	907	37,141
Warren	Warrenton	286	6,472
Washington	Sandersville	683	19,220
Wayne	Jesup	647	21,862
Webster	Preston	210	2,134
Wheeler	Alamo	299	5,078
White	Cleveland	242	10,991
Whitefield	Dalton	291	67,680
Wilcox	Abbeville	382	7,592
Wilkes	Washington	470	11,278
Wilkinson	Irwinton	451	10,701
Worth	Sylvester	575	18,229
	TOTALS	58,056	5,836,548

[1] Atlanta city limits extend into DeKalb County.

ridership amounted to 149,359,899 in 1983, of which 135,900,000 was in Atlanta. Subway ridership in 1984 averaged 140,000 daily. In December 1984, 5 new stations were added to the 14 existing subway stations, extending service into Atlanta's northern suburbs. The Metropolitan Atlanta Rapid Transit Authority announced plans to extend the subway system to Atlanta's Hartsfield International Airport, 9 mi (15 km) south of the downtown business area, by 1988.

Georgia's old intracoastal waterway carries about 1 million tons of shipping annually and is also used by pleasure craft and fishing vessels. Savannah's modern port facilities handled about 16,520 million lb (7,493 million kg) of cargo in 1983; the coastal cities of Brunswick and St. Marys also have deepwater docks.

In the 1920s, Georgia became the gateway to Florida for motorists. Today, I-75 is the main route from Atlanta to Florida, and I-20 is the major east–west highway; both cross at Atlanta with I-85, which proceeds southeast from South Carolina to Alabama. I-95 stretches along the coast from South Carolina through Savannah to Jacksonville, Fla. In 1983, Georgia had 104,955 mi (168,909 km) of roads, 4,207,708 registered motor vehicles, and 3,725,127 licensed drivers. Hartsfield International Airport is the hub of air traffic in the Southeast; it enplaned 18,648,189 passengers and handled 146,482 tons of freight in 1983. In 1984 there were 120 airports open to the public in Georgia.

11 HISTORY

The history of what is now Georgia was influenced by two great prehistoric events: first, the upheaval that produced the mountains of the north, and second, the overflow of an ancient ocean that covered and flattened much of the rest of the state. Human beings have inhabited Georgia for at least 12,000 years. The first nomadic hunters were replaced by shellfish eaters who lived along the rivers. Farming communities later grew up at these sites, reaching their height in the Master Farmer culture about AD 800. These Native Americans left impressive mounds at Ocmulgee, near Macon, and at Etowah, north of Atlanta.

During the colonial period, the most important Indian tribes were the Creek, who lived along the central and western rivers, and the Cherokee, who lived in the mountains. By clever diplomacy, the Creek were able to maintain their position as the balance of power between the English on the one hand and the French and Spanish on the other. With the ascendancy of the English and the achievement of statehood, however, the Creek lost their leverage and were expelled from Georgia in 1826. The Cherokee sought to adopt the white man's ways in their effort to avoid expulsion or annihilation. Thanks to their remarkable linguist Sequoyah, they learned to write their own language, later running their own newspaper, the *Cherokee Phoenix*, and their own schools. Some even owned slaves. Unfortunately for the Cherokee, gold was discovered on their lands; the Georgia state legislature confiscated their territory and outlawed the system of self-government the Cherokee had developed during the 1820s. Despite a ruling by the US Supreme Court, handed down by Chief Justice John Marshall, that Georgia had acted illegally, federal and state authorities expelled the Cherokee between 1832 and 1838. Thousands died on the march to Indian Territory (Oklahoma), known ever since as the Trail of Tears.

Georgia's first European explorer was Hernando de Soto of Spain, who in 1540 crossed the region looking for the fabled Seven Cities of Gold. French Huguenots under Jean Ribault claimed the Georgia coast in 1562 but were driven out by the Spanish captain Pedro Menéndez Avilés in 1565, who by 1586 had established the mission of Santa Catalina de Gaule on St. Catherines Island. (The ruins of this mission—the oldest European settlement in Georgia—were discovered by archaeologists in 1982.) By 1700, Jesuit and Franciscan missionaries had established an entire chain of missions along the Sea Islands and on the lower Chattahoochee.

From Charles Town, in Carolina colony, the English challenged Spain for control of the region, and by 1702 they had forced the Spaniards back to St. Augustine, Fla. In 1732, after the English had become convinced of the desirability of locating a buffer between the valuable rice-growing colony of Carolina and Indian-held lands to the south and west, King George II granted a charter to a group called the Trustees for Establishing the Colony of Georgia in America. The best known of the trustees was the soldier-politician and philanthropist James Edward Oglethorpe. His original intention was to send debtors from English prisons to Georgia, but Parliament refused to support the idea. Instead, Georgia was to be a place where the industrious poor would produce those things England needed, such as silk and wine, and would guard the frontier. Rum and slavery were expressly prohibited.

Oglethorpe and the first settlers landed at Yamacraw Bluff on 12 February 1733 and were given a friendly reception by a small band of Yamacraw Indians and their chief, Tomochichi. Oglethorpe is best remembered for laying out the town of Savannah in a unique design, featuring numerous plazas, that still delights tourists today; however, as a military man, his main interest was defending the colony against the Spanish. After war was declared in 1739, Oglethorpe conducted an unsuccessful siege of St. Augustine. The Spaniards counterattacked at Oglethorpe's fortified town of Frederica on St. Simons Island in July 1742 but were repulsed in a confused encounter known as the Battle of Bloody Marsh, which ended Spanish threats to the British colonies. Soon afterward, Oglethorpe returned permanently to England.

The trustees' restrictions on rum and slavery were gradually removed, and in 1752, control over Georgia reverted to Parliament. Georgia thus became a royal colony, its society, like that of Carolina, shaped by the planting of rice, indigo, and cotton. After the French and Indian War, settlers began to pour into the Georgia backcountry above Augusta. Because these backcountry pioneers depended on the royal government for protection against the Indians, they were reluctant to join the protests by Savannah merchants against new British mercantile regulations. When war came, however, the backcountry seized the opportunity to wrest political control of the new state away from Savannah.

Georgians spent the first three years of the Revolutionary War in annual attempts to invade Florida, each of them unsuccessful. The British turned their attention to Georgia late in 1778, reestablishing control of the state as far as Briar Creek, midway between Savannah and Augusta. After a combined French and American force failed to retake Savannah in October 1779, the city was used by the British as a base from which to recapture Charleston, in present-day South Carolina, and to extend their control further inland. For a year, most of Georgia was under British rule, and there was talk of making the restoration permanent in the peace settlement. However, Augusta was retaken in June 1781, and independent government was restored. A year later, the British were forced out of Savannah.

With Augusta as the new capital of Georgia, a period of rapid expansion began. Georgia ratified the US Constitution on 2 January 1788, the 4th state to do so. The invention of the cotton gin by Eli Whitney in 1793 made cotton cultivation profitable in the lands east of the Oconee River, relinquished by the Creek Indians under the Treaty of New York three years earlier. A mania for land speculation was climaxed by the mid-1790s Yazoo Fraud, in which the state legislature sold 50 million acres (20 million hectares), later the states of Alabama and Mississippi, to land companies of which many of the legislators were members.

Georgia surrendered its lands west of the Chattahoochee River to the federal government in 1802. As the Indians were removed to the west, the lands they had occupied were disposed of by successive lotteries. The settlement of the cotton lands brought prosperity to Georgia, a fact that influenced Georgians to prefer the Union rather than secession during the constitutional crises of 1833 and 1850, when South Carolina was prepared to secede.

After South Carolina did secede in 1860, Georgia also withdrew from the Union and joined the Confederate States of America. Union troops occupied the Sea Islands during 1862. Confederate forces defeated the Union Army's advance into northern Georgia at Chickamauga in 1863, but in 1864, troops under General William Tecumseh Sherman moved relentlessly upon Atlanta, capturing it in September. In November, Sherman began his famous "march to the sea," in which his 60,000 troops cut a swath of destruction 60 mi (97 km) wide. Sherman presented Savannah as a Christmas present to President Abraham Lincoln.

Georgia did not fit the Reconstruction stereotype of a crushed state languishing under blacks and carpetbaggers. The leaders of the weak and short-lived Republican regime (1868–71), which moved the capital to Atlanta, were either Georgia-born or state residents before the war. There were only 29 Negroes among the 172-member house, and but 3 among the 44-member senate. After ratifying the 14th and 15th amendments, Georgia was readmitted to the Union on 15 July 1870.

Commercial interests were strong in antebellum Georgia, but their political power was balanced by that of the great planters. After the Democrats recovered control of the state in 1871, business interests dominated politics. Discontented farmers supported an Independent Party in the 1870s and 1880s, and then the Populist Party in the 1890s. Democratic Representative Thomas E. Watson, who declared himself a Populist during the early 1890s, was defeated three times in congressional races by the party he had deserted. Watson subsequently fomented anti-Negro, anti-Jewish, and anti-Catholic sentiment in order to control a bloc of rural votes with which he dominated state politics for 10 years. In 1920, Watson finally was elected to the US Senate, but he died in 1922. Rebecca L. Felton was appointed to succeed him, thus becoming the first woman to serve in the US Senate, although she was replaced after one day.

Franklin D. Roosevelt learned the problems of Georgia farmers firsthand when he made Warm Springs his second home in 1942. However, his efforts to introduce the New Deal to Georgia after he became president in 1933 were blocked by Governor Eugene Talmadge, who advertised himself as a "real dirt farmer." It was not until the administration of Eurith D. Rivers (1937–41) that progressive social legislation was enacted. Governor Ellis Arnall gained national attention for his forward-looking administration (1943–47), which revised the outdated 1877 state constitution and gave the vote to 18-year-olds. Georgia treated the nation to the spectacle of three governors at once when Eugene Talmadge was elected for a fourth time in 1946 but died before assuming office. His son Herman was then elected by the legislature, but the new lieutenant governor, M. E. Thompson, also claimed the office, and Arnall refused to step aside until the issue was resolved. The courts finally decided in favor of Thompson.

The Supreme Court order to desegregate public schools in 1954 provided Georgia politicians with an emotional issue they exploited to the hilt. Governors Herman Talmadge, Marvin Griffin, and Ernest Vandiver were all elected on pledges to resist integration. A blow was dealt to old-style politics in 1962, however, when the US Supreme Court declared the county-unit system unconstitutional. Under this system, state officers and members of Congress had been selected by county units instead of by popular vote since 1911; the new ruling made city voters as important as those in rural areas. During the 1960s, Atlanta was the home base for the civil-rights efforts of Martin Luther King, Jr., though his campaign to end racial discrimination in Georgia focused most notably on the town of Albany. Federal civil-rights legislation in 1964 and 1965 changed the state's political climate by guaranteeing the vote to black citizens. A black man, Julian Bond, was elected to the state legislature in 1965; in 1973, Maynard Jackson

was elected mayor of Atlanta, thus becoming the first black mayor of a large southern city. For decades the belief that defense of segregation was a prerequisite for state elective office cost white southerners any chance they might have had for national leadership. Governor Jimmy Carter's unequivocal renunciation of racism in his inaugural speech in 1971 thus marked a turning point in Georgia politics and was a key factor in his election to the presidency in 1976.

Another black, former US Ambassador to the United Nations Andrew Young, succeeded Jackson as mayor of Atlanta in 1981, when that city—and the state—was experiencing an economic boom. By 1983, Atlanta's economic growth rate of 2.7% far outpaced the US rate of 0.7% and in 1984 metropolitan Atlanta ranked 3d in population growth rate among major US metropolitan areas.

[12] STATE GOVERNMENT

Georgia's first constitution, adopted in 1777, was considered one of the most democratic in the new nation. Power was concentrated in a unicameral legislature; a senate was added in 1789. The Civil War period brought a flurry of constitution making in 1861, 1865, and 1868. When the Democrats displaced the Republicans after Reconstruction, they felt obliged to replace the constitution of 1868 with a rigidly restrictive one. This document, adopted in 1877, modified by numerous amendments, and revised in 1945 and 1976, continued to govern the state until July 1983, when a new constitution, ratified in 1982, took effect.

The legislature, called the general assembly, consists of a 56-seat senate and a 180-seat house of representatives; all the legislators serve two-year terms. The legislature convenes on the 2d Monday in January and stays in session for 40 legislative days. Recesses called during a session may considerably extend its calendar length. During the 1960s and 1970s, the legislature engaged in a series of attempts to redistrict itself to provide equal representation based on population; it was finally redistricted in 1981 on the basis of 1980 census results.

Elected executives include the governor, lieutenant governor, secretary of state, attorney general, state school superintendent, comptroller general, commissioner of agriculture, and commissioner of labor. Each serves a four-year term. To be eligible for office, the governor and lieutenant governor, who are elected separately, must have been US citizens for 15 years, Georgia citizens for 6 years preceding the election, and at least 30 years of age.

To become law, a bill must be passed by both houses of the legislature and approved by the governor or passed over the executive veto by a two-thirds vote in both houses. All revenue measures originate in the house, but the senate can propose, or concur in, amendments to these bills. Amendments to the constitution may be proposed by two-thirds votes of the elected members of each chamber and must then be ratified by popular vote.

To vote, a person must be a citizen of the US, at least 18 years old, and must have resided in the state for at least 30 days immediately preceding the election.

[13] POLITICAL PARTIES

The first political group to emerge in the state was the Federalist Party, but it was tainted by association with the Yazoo Fraud of the 1790s. The reform party at this time was the Democratic-Republican Party, headed in Georgia by James Jackson (whose followers included many former Federalists), William Crawford, and George Troup. During the presidency of Andrew Jackson (1829–37), one wing, headed by John Clark, supported the president and called itself the Union Party. The other faction, led by Troup, defended South Carolina's right to nullify laws and called itself the States' Rights Party. Subsequently the Union Party affiliated with the Democrats, and the States' Rights Party merged with the Whigs. When the national Whig Party collapsed, many Georgia Whigs joined the Native American (Know-Nothing) Party. During Reconstruction, the Republican Party captured the governor's office, but Republican hopes died when federal troops were withdrawn from the state in 1870.

Georgia voted solidly Democratic between 1870 and 1960, despite challenges from the Independent Party in the 1880s and the Populists in the 1890s. Georgia cast its electoral votes for the Democratic presidential candidate in every election until 1964, when Republican Barry Goldwater won the state. Four years later, George C. Wallace of the American Independent Party received Georgia's 12 electoral votes. Republican Richard Nixon carried the state in 1972, as the Republicans also became a viable party at the local level. In 1976, Georgia's native son Jimmy Carter returned the state to the Democratic camp in presidential balloting. Another native Georgian and former Georgia governor, Lester Maddox, was the American Independent candidate in 1976.

Georgians for the most part remained loyal to President Carter in the 1980 general election, although only 41.8% voted for president—a rank of 49th in the nation. Voters turned out US Senator Herman Talmadge, the Democratic incumbent (who had been involved in a financial scandal), in favor of Republican Mack Matingly. In 1984, some 60% of voting Georgians helped reelect President Ronald Reagan; the state's congressional delegation and legislature remained overwhelmingly Democratic, however. As of January 1985, 340 of Georgia's elected officials were black; 23 women served in the state legislature.

14LOCAL GOVERNMENT

In 1758, colonial Georgia was divided into eight parishes, the earliest political districts represented in the royal assembly. By the constitution of 1777, the parishes were transformed into counties, and as settlement gradually expanded, the number of counties grew to 159. The Georgia constitution of 1877 granted counties from one to three seats in the house of representatives, depending on population. This county-unit system was used in counting votes for elected state and congressional offices until ruled unconstitutional by the US Supreme Court in 1962. Originally administered by judges of county courts, all Georgia counties but one, Towns County, have since adopted a commission system, although some are administered by a single commissioner.

Georgia has 556 incorporated cities, towns, and villages. In 1965, the legislature passed a home-rule law permitting these local governments to amend their own charters. The traditional and most common form of municipal government is the mayor-council form. But city managers are employed by some communities, and a few make use of the commission system.

During the 1970s there were efforts to merge some of the larger cities with their counties. However, most county voters showed an unwillingness to be burdened with city problems, and only Columbus and Muscogee County had achieved consolidated government by 1984.

The Department of Community Affairs serves as the governor's representative to local governments.

15STATE SERVICES

In 1977, the State Ethics Commission was renamed the Campaign and Financial Disclosure Commission. As the title indicates, this commission is charged with providing procedures for public disclosure of all state and local campaign contributions and expenditures.

Educational services are provided by the Board of Education, which exercises jurisdiction over all public schools, including teacher certification and curriculum approval. The superintendent of schools is the board's executive officer. The public colleges are operated by the Board of Regents of the University System of Georgia, whose chief administrator is the chancellor. Air, water, road, and rail services are administered by the Department of Transportation.

The Reorganization Act of 1972 made the Department of Human Resources a catch-all agency for health, rehabilitation, and social-welfare programs. The department, which employs some 17,000 people in 2,000 locations, offers special services to the mentally ill, drug abusers and alcoholics, neglected and abused children and adults, juvenile offenders, the handicapped, the aged, and the poor.

Public protection services are rendered through the Department of Public Safety. Responsibility for natural-resource protection is lodged with the Department of Natural Resources, into which 33 separate agencies were consolidated in 1972. The Environmental Protection Division is charged with maintaining air, land, and water quality standards; the Game and Fish Division manages

Georgia Presidential Vote by Political Parties, 1948–84

YEAR	ELECTORAL VOTE	GEORGIA WINNER	DEMOCRAT	REPUBLICAN	STATES' RIGHTS DEMOCRAT	PROGRESSIVE
1948	12	*Truman (D)	254,646	76,691	85,136	1,636
1952	12	Stevenson (D)	456,823	198,961	—	—
1956	12	Stevenson (D)	444,688	222,778	—	—
1960	12	*Kennedy (D)	458,638	274,472	—	—
1964	12	Goldwater (R)	522,163	616,584	—	—
					AMERICAN IND.	
1968	12	Wallace (AI)	334,440	380,111	535,550	—
1972	12	*Nixon (R)	289,529	881,490	—	—
						AMERICAN
1976	12	*Carter (D)	979,409	483,743	1,071[1]	1,168[1]
					LIBERTARIAN	
1980	12	Carter (D)	890,955	654,168	15,627	—
1984	12	*Reagan (R)	706,628	1,068,722	152[1]	—

*Won US presidential election.

[1] Write-in votes.

wildlife resources; and the Parks and Historic Sites Division administers state parks, recreational areas, and historic sites. Labor services are provided by the Department of Labor, which oversees workers' compensation programs.

[16] JUDICIAL SYSTEM

Georgia's highest court is the supreme court, created in 1845 and consisting of a chief justice, presiding justice (who exercises the duties of chief justice in his absence), and five associate justices. They are elected by the people to staggered six-year terms in nonpartisan elections.

As of 1985 there were 159 superior courts, grouped into 45 circuits. Each circuit has from 1 to 11 judges, depending on its population; there are 124 superior court judges, all of them elected for four-year terms in nonpartisan elections. Superior courts have exclusive jurisdiction in cases of divorce and land title, and in major criminal cases. Cases from local courts can be carried to the court of appeals, consisting of 9 judges elected for staggered six-year terms in nonpartisan elections.

Among Georgia's 524 courts are a probate court in each county and separate juvenile courts in 62 counties. Most judges of the county and city courts are appointed by the governor with the consent of the senate. There were 13,041 practicing attorneys in the state in 1984.

Overcrowding has been a chronic problem in Georgia prisons, especially in the state prison at Reidsville, and the state has been under court order to remedy conditions since 1978. However, despite the early release of 1,500 prisoners during 1983, the prison population numbered 15,347 by the end of the year, and Georgia had the 6th-highest proportion of prisoners in the nation: 259 per 100,000 population.

Since 1930, Georgia has executed more persons that any other state: 366 between 1930 and 1967, and 7 more from 1983 through March 1985. Executions are carried out by electrocution.

According to the FBI Crime Index, the crime rate per 100,000 inhabitants for 1983 was 4,505, down 14% from 1982. Rates for murder and nonnegligent manslaughter and forcible rape were above the US average, but property crime rates were generally below it.

[17] ARMED FORCES

There were 15 military installations in the state in 1983. Major facilities include Dobbins Air Force Base, Ft. Gillem, Ft. McPherson, and the Atlanta Naval Air Station, all located in the Atlanta area; Ft. Stewart and Hunter Army Airfield near Savannah; Ft. Gordon at Augusta; Moody Air Force Base at Macon; Ft. Benning, a major Army training installation at Columbus; Robins Air Force Base, between Columbus and Macon; and a Navy Supply School in Athens. Of the 97,203 Department of Defense personnel in Georgia as of September 1984, 61% were military, and of these, 88% were in the Army. In 1983/84, Georgia firms received defense contracts worth $3,037,594.

There were 632,000 veterans of the US armed forces living in Georgia as of 30 September 1983. Of these, 4,000 served in World War I, 217,000 in World War II, 125,000 in the Korean conflict, and 218,000 during the Viet-Nam era. In all, 77,000 Georgians fought and 1,503 died in World War I, and 320,000 served and 6,754 were killed in World War II. In 1982/83, Georgia veterans received benefits amounting to $676 million, of which $407 million was spent for compensation and pensions.

The National Guard consisted of 14,265 active members at the end of 1984. In 1983, state and local police numbered 14,183 persons, or 24.7 per 10,000 population. The Georgia Bureau of Investigation, part of the Department of Public Safety, operates the Georgia Crime Laboratory, one of the oldest and largest in the US.

[18] MIGRATION

During the colonial period, the chief source of immigrants to Georgia was England; other important national groups were Germans, Scots, and Scotch-Irish. The number of African slaves increased from 1,000 in 1752 to nearly 20,000 in 1776. After the Revolution, a large number of Virginians came to Georgia, as well as lesser numbers of French refugees from Hispaniola and immigrants from Ireland and Germany. Following the Civil War, there was some immigration from Italy, Russia, and Greece.

The greatest population shifts during the 20th century have been from country to town and, after World War I, of black Georgians to northern cities. Georgia suffered a net loss through migration of 502,000 from 1940 to 1960, but enjoyed a net gain of 329,000 during 1970–80 and 120,000 during 1980–83.

[19] INTERGOVERNMENTAL COOPERATION

Multistate agreements in which Georgia participates include the Southern Regional Education Compact, Southeastern Forest Fire Protection Compact, Southern Interstate Energy Compact, and Southern Growth Policies Compact.

In 1982/83, federal aid to Georgia totaled $2.1 billion, of which $112 million was general revenue sharing.

[20] ECONOMY

According to the original plans of Georgia's founders, its people were to be sober spinners of silk. The reality was far different, however: during the period of royally appointed governors, Georgia became a replica of Carolina, a plantation province producing rice, indigo, and cotton. After the Revolution, the invention of the cotton gin established the plantation system even more firmly by making cotton planting profitable in the piedmont. Meanwhile, deerskins and other furs and lumber were produced in the backcountry; rice remained an important staple along the coast; turnpikes, canals, and railroads were built; and textile manufacturing became increasingly important, especially in Athens and Augusta.

At the end of the Civil War, the state's economy was in ruins, and tenancy and sharecropping were common. Manufacturing, especially of textiles, was promoted by "New South" spokesmen like Henry Grady of Atlanta and Patrick Walsh of Augusta. Atlanta, whose nascent industries included production of a thick sweet syrup called Coca-Cola, symbolized the New South idea—then as now. Farmers did not experience the benefits of progress, however. Many of them flocked to the mills while others joined the Populist Party in an effort to air their grievances. To the planters' relief, cotton prices rose from the turn of the century through World War I. Meanwhile, Georgians lost control of their railroads and industries to northern corporations. During the 1920s, the boll weevil wrecked the cotton crops, and farmers resumed their flight to the cities. Not until the late 1930s did Georgia accept Social Security, unemployment compensation, and other relief measures.

Georgia's economy underwent drastic changes as a result of World War II. The raising of poultry and livestock became more important than crop cultivation, and manufacturing replaced agriculture as the chief source of income. In 1982, only about 2.4% of the employed labor force was working in agriculture; 33.4% were blue-collar workers, 17.5% clerical workers, 15.5% professional or technical workers, and 12.6% service workers. Georgia is a leader in the making of paper products, tufted textiles products, processed chickens, naval stores, lumber, and transportation equipment.

After World War II, many northern industries moved to Georgia to take advantage of low wages and low taxes, conditions that meant low benefits for Georgians. A chronic problem for workers in the textile industry has been how to organize labor unions for the purpose of collective bargaining. As of 1985, national unions had not been very successful in their attempts to organize mill workers. A recent barrier to the employment of economically disadvantaged Georgians has been the gradual movement of manufacturing plants from the inner cities to the suburbs.

Georgia's oldest industry, textile manufacturing, remains its most important but has grown slowly in recent years; while most durable-goods industries, such as electrical machinery and appliances, have grown rapidly. Durable-goods industries were expected to provide two out of every three new jobs created in manufacturing in 1985. Exports have become increasingly important to Georgia's economy. From 1977 to 1981, the value of Georgia's exported manufactures increased 76%—considerably faster than the 44% rate of increase in the state's manufacturing.

21 INCOME

The per capita income of Georgians has been low historically, at least since the Civil War. In 1940, the average income was below $350 a year; the average doubled during World War II and stood at $1,034 in 1950. In relative terms, Georgians in 1940 received only 57% as much as other Americans, but by 1950 they were earning nearly 70% of the national average. During the next two decades, Georgia continued to close the gap until, by 1970, per capita income averaged $3,318, or 84% of the national level.

Georgia's per capita income rose to $10,283 in 1983, boosting the state's national rank to 35th. Per capita disposable income in 1982 was $8,067, a 6.7% increase over 1981. The median income for four-person families increased from $9,650 to $25,131 in 1981. Moreover, the number of persons below the federal poverty level declined from 924,000 in 1969 to 884,000 in 1979. In 1982, 35,500 Georgians—1.8% of the US total—were among the top wealth holders in the country, with combined assets of $34.6 billion.

22 LABOR

Georgia's civilian labor force was estimated at 2,760,000 in 1984. Of this total, 166,000 workers, or 6%, were unemployed. In 1982, white males constituted 43.3% of the labor force and were unemployed at a rate of 5.7%, white females 33.4% (unemployed 6.7%), black males 11.2% (unemployed 13.5%), and black females 11.3% (unemployed 13.1%). The most remarkable change in the labor force since World War II has been the rising proportion of women, whose share increased from less than 28% in 1940 to an estimated 44.7% in 1983.

The trend during the 1970s and early 1980s was toward increased employment in service industries and toward multiple-job holding. Employment in agriculture, the leading industry prior to World War II, continued its long-term decline; one indication of declining employment was the decrease in farm population, which went from 515,000 in 1960 to 228,000 in 1970 and 121,000 in 1980. The mining, construction, and manufacturing industries registered employment increases but declined in importance relative to such sectors as trade and government.

A 5.1% growth in construction employment was anticipated for 1985. Manufacturing employment growth was expected to be 2.4%, primarily from durable goods, such as electrical and nonelectrical machinery. For 1970–1985, durable goods employment increased almost four times faster than that of Georgia's traditional nondurable industries, such as textiles and food. The weakest growth was predicted to be in government services—only 0.9%.

A federal census in 1982 revealed the following nonfarm employment pattern for Georgia:

	ESTABLISHMENTS	EMPLOYEES	ANNUAL PAYROLL ('000)
Agricultural services, forestry, fishing	1,087	7,500	$ 89,009
Mining	187	7,104	131,308
Contract construction	8,657	103,738	1,554,295
Manufacturing, of which:	7,831	511,449	7,933,255
Textile mill products	(587)	(96,762)	(1,218,833)
Apparel, other textiles	(577)	(71,684)	(663,214)
Transportation, public utilities	3,735	133,486	3,098,589
Wholesale trade	11,055	147,088	2,827,416

	ESTABLISHMENTS	EMPLOYEES	ANNUAL PAYROLL ('000)
Retail trade	31,323	359,779	$ 3,236,394
Finance, insurance, real estate	9,379	117,799	1,991,971
Services	31,543	347,784	4,461,817
Other	2,203	2,368	30,602
TOTALS	107,000	1,738,095	$25,354,656

Among the workers not covered by this survey were government employees, of whom Georgia had 420,256 (77,887 federal, 342,369 state and local) in October 1982.

Georgia has not been hospitable to union organizers, partly because of a mistrust of "outside agitators," a suspicion reinforced by conservative politicians like Eugene Talmadge who convinced many people that union organizers were promoters of racial equality. Among state laws strictly regulating union activity is a right-to-work law enacted in 1947. Although, union membership in Georgia grew from 273,000 in 1970 to 323,000 in 1980, the proportion of union members in the employed nonagricultural labor force fell from 17.5% to 15%. As of 30 June 1983 there were 946 unions in Georgia. In 1962, the legislature denied state employees the right to strike.

The average earnings for Georgia workers in manufacturing industries climbed from $2.01 an hour in 1965 to $7.13 an hour in 1983. The latter figure was about 19% below the national average. Strikes in Georgia tend to be fewer than in most heavily industrialized states. In 1981, 10,000 workers were involved in 23 work stoppages.

One of the earliest state labor laws was an 1889 act requiring employers to provide seats for females to use when resting. A child-labor law adopted in 1906 prohibited the employment of children under 10 years of age in manufacturing. A general workers' compensation law was enacted in 1920.

23 AGRICULTURE

In 1983, Georgia's farm income reached an estimated $3.3 billion (14th in the US). Georgia ranked 1st in the production of peanuts and pecans, harvesting 51% of all the pecans grown in the US and 48% of the peanuts.

Cotton was the mainstay of Georgia's economy through the early 20th century, and the state's plantations also grew corn, rice, tobacco, wheat, and sweet potatoes. World War I stimulated the cultivation of peanuts along with other crops. By the 1930s, tobacco and peanuts were challenging cotton for agricultural supremacy, and Georgia had also become an important producer of peaches, a product for which the "peach state" is still widely known.

After 1940, farm mechanization and consolidation were rapid. The number of tractors increased from 10,000 in 1940 to 85,000 by 1955. In 1940, 6 out of 10 farms were tenant-operated; by the mid-1960s, less than 1 in 6 was. The number of farms declined from 226,000 in 1945 to 51,000 in 1984, when the average farm size was 265 acres (107 hectares). Georgia's farmland area of 13.5 million acres (5.4 million hectares) represents 36% of its land area. Farm real estate debt in Georgia was estimated at $2.47 billion in 1984.

The following table shows production and value for leading crops in 1982:

	PRODUCTION	VALUE
Soybeans	68,850,000 bu	$375,233,000
Peanuts	1,517,480,000 lb	367,230,000
Tobacco	105,500,000 lb	189,478,000
Corn	69,275,000 bu	183,579,000
Wheat	48,840,000 bu	148,962,000
Pecans	125,000 million lb	79,625,000
Cotton	235,000 bales	61,927,000
Peaches	120,000 million lb	22,460,000

In 1981, 21 cooperatives were headquartered in the state. These included agricultural cooperatives and those that provided services to farmers, such as land bank associations and rural electric cooperatives. Membership totaled 71,226 in 1981, when the co-ops did a net business totaling $1,032,026,000.

²⁴ANIMAL HUSBANDRY

Georgia's cash receipts from livestock and livestock products totaled an estimated $1.7 billion in 1983, or more than half of the total farm income. Georgia ranks 2d only to Arkansas in total cash receipts from chickens and broilers, and 2d to California in receipts from eggs.

At the close of 1983, Georgia farms had 1,725,000 cattle and calves, and 1,310,000 hogs and pigs. Some 2,711,525,000 lb of cattle valued at an estimated $1,565,518,000 were produced in 1983; the totals for hogs were 473,586,000 lb and $214,848,000.

Cows kept for milk production numbered 130,000 in 1982, when Georgia dairies produced 1.4 billion lb of milk. Poultry farmers sold an estimated $676,675,000 worth of broilers and $279,093,000 of eggs in 1983; the total egg production was nearly 4.7 billion. The 2,266,000 turkeys raised in 1983 were valued at $24,110,000.

²⁵FISHING

Georgia's total commercial catch of fish and shellfish in 1984 was 15,844,000 lb, valued at $12,240,000. Although Georgia ranked 22d among the states in value of commercial fishing, the volume represented only 0.5% of the national catch. Commercial fishing in Georgia involves more shellfish—mainly shrimp and crabs—than finfish, the most important of which are caught in the nets of shrimp trawlers. Leading finfish are flounder, king whiting, grouper, porgy, red snapper, and shad. In 1982/83 the commercial shrimp harvest was 4.6 million lb. Commercial fishermen along Georgia's coast numbered about 4,000 in 1983/84, and sports fishermen 110,000.

In brisk mountain streams and sluggish swamps, sports fishermen catch bass, catfish, jackfish, bluegill, crappie, perch, and trout. In 1983/84, 650,000 trout were produced in state hatcheries, and the Game and Fish Division stocked some 2,500 ponds.

²⁶FORESTRY

In 1979, Georgia, which occupied 1.6% of the total US land area, had nearly 3.4% of the nation's forestland and 5.1% of the nation's commercial forests. Georgia's total forest area, 25,256,000 acres (10,221,000 hectares), ranked 4th in the US, and its 23,733,684 acres (9,604,713 hectares) of commercial forests ranked 1st.

Forests cover more than two-thirds of the state's land area. The most densely wooded counties are in the piedmont hills and northern mountains. Ware and Charlton counties in southeastern Georgia, containing the Okefenokee Swamp, are almost entirely forested. In 1982, 38.8% of Georgia's commercial forestland was owned by individuals other than farmers, 25.8% by farmers, 20.9% by the forest industry, 7.9% by other corporations, 5.9% by the federal government, 0.5% by the state government, and 0.2% by local governments.

In 1982 there were 1,036 logging camps and contractors in Georgia, and 258 sawmills and planing mills. The chief products of Georgia's timber industry are pine lumber and pine plywood for the building industry, hardwood lumber for the furniture industry, and pulp for the paper and box industry. In 1983, Georgia produced approximately 2 billion board feet of lumber, of which 88.4% was softwood (pine). In 1981, the state produced 6.1 million cords of pulpwood, of which 87% was softwood.

The two chief recreational forest areas are Chattahoochee National Forest, in the northern part of the state, and Oconee National Forest, in the central region. Georgia has 872,479 acres (353,080 hectares) of National Forest System lands, 99% of which are within the boundaries of the two national forests.

The Georgia Forest Commission employs about 850 persons, including more than 100 professional foresters. In 1983, there were 6,922 fires in Georgia forests. The commission produced 88 million genetically superior seedlings in state seed orchards in 1983/84. The commission's budget for 1984/85 provided for the reforestation of 59,206 acres (23,960 hectares) of marginal agricultural land.

²⁷MINING

The estimated value of nonfuel minerals produced in Georgia in 1984 was $952.6 million, 7th in the US and a new state record. Georgia ranks 1st in the production of marble, kaolin, and crushed granite.

About three-fourths of the nation's supply of kaolin, or "white clay," is produced in Georgia; used mostly for filler and coating in white paper, kaolin is also employed in pottery making. The most important minerals commercially are the clays found along the sand hills at the southern edge of the piedmont. Marble is found chiefly in northeastern Georgia; the marble quarried at Tate is noted for its texture and durability. Small quantities of iron ore are mined in the coastal plain, and coal is found in northwestern Georgia. Other minerals include mica, barite, bauxite, and talc. Estimated output of principal minerals (excluding fossil fuels) in 1984 included stone, 45,500,000 tons; clays, 8,570,000 tons; and sand and gravel, 3,800,000 tons.

Gold was discovered in 1827 in the hills around Dahlonega, and a US mint was established there in 1838. The dome of the state capitol at Atlanta is painted with Dahlonega gold leaf. Gold is no longer mined commercially, but tourists still come to Dahlonega to try their luck.

²⁸ENERGY AND POWER

Georgia is an energy-dependent state which produces only a small proportion of its energy needs, most of it through hydroelectric power. There are no commercially recoverable petroleum or natural-gas reserves, and the state's coal deposits are of no more than marginal importance. Georgia does have large amounts of timberland, however, and it has been estimated that 20–40% of the state's energy demands could be met by using wood that is currently wasted. In 1983/84, 778,944 Georgia residences had wood-burning fireplaces or stoves. The state's southern location and favorable weather conditions also make solar power an increasingly attractive energy alternative. Georgia's extensive river system also offers the potential for further hydroelectric development.

In 1982, Georgia's energy consumption per capita was 284 million Btu, somewhat less than the national average; also less than the national average were the state's 1981 energy expenditures per capita—$1,777.

In 1983, Georgia produced 66.6 billion kwh of electricity and had an installed capacity of 17.5 million kw. As of 1 January 1985 the Georgia Power Co. had two atomic reactors near Baxley with a combined capacity of 1,576,000 kw; two more reactors were under construction at Waynesboro. All utilities are regulated by the Georgia Public Service Commission, which must approve their rates.

Petroleum accounted for 42% of all fuel used in Georgia in 1982, coal for 33%, and natural gas for 19%. Exploration for oil off the coast goes on, but the state's offshore oil resources are expected to be slight. Georgia's demonstrated coal reserves were approximately 3.6 million tons in 1983, and production was 185,000 tons.

²⁹INDUSTRY

Georgia's manufactured goods in 1982 had a total value of more than $48 billion. Important products include textiles, clothing, aircraft, soft drinks, paper, paints and varnishes, bricks and tiles, glassware, and ceramics.

Georgia was primarily an agrarian state before the Civil War, but afterward the cities developed a strong industrial base by taking advantage of abundant waterpower to operate factories. Textiles have long been dominant, but new industries have also been developed. Charles H. Herty, a chemist at the University of Georgia, discovered a new method of extracting turpentine which worked so well that Georgia led the nation in producing turpentine, tar, rosin, and pitch by 1982. Herty also perfected an economical way of making newsprint from southern pines, which was adopted by Georgia's paper mills. With the onset of World War II, meat-processing plants were built at rail centers, and fertilizer plants and cottonseed mills were expanded.

The state's—and Atlanta's—most famous product was created in 1886, when druggist John S. Pemberton developed a formula which he sold to Asa Griggs Candler, who in 1892 formed the Coca-Cola Co. In 1919, the Candlers sold the company to a syndicate headed by Ernest Woodruff, whose son Robert made "Coke" into the world's most widely known commercial product. The transport equipment, chemical, food-processing, apparel, and forest-products industries today rival textiles in economic importance.

The total value of shipments by manufacturers in 1982 was $48.1 billion. The following table shows value of shipments for major industries:

Textile mill products	$8,636,700,000
Food and food products	7,385,600,000
Transportation equipment	5,061,000,000
Paper and paper products	4,409,900,000
Chemicals and allied products	3,880,600,000
Apparel and other textile products	3,174,300,000
Electric and electronic equipment	2,161,100,000
Lumber and wood products	1,969,100,000
Fabricated metal products	1,723,300,000
Nonelectrical machinery	1,624,700,000
Stone, clay, glass products	1,213,700,000

Georgia's heavily forested northern region is dominated by carpet mills, especially around Dalton. In the piedmont plateau, manufacturing is highly diversified, with textiles and transportation equipment the most significant.

In 1984, 12 of the nation's 500 largest industrial corporations listed by *Fortune* magazine had headquarters in Georgia: Coca-Cola, Georgia-Pacific, Gold Kist, National Service Industries, Fuqua Industries, and Oxford Industries, all in Atlanta; the West Point–Pepperell Co. in West Point; Gulfstream Aerospace and Savannah Foods & Industries in Savannah; Flowers Industries in Thomasville; Georgia Kraft in Rome; and Shaw Industries in Dalton.

³⁰COMMERCE

Georgia ranked 10th among the 50 states in wholesale trade in 1982, with total sales of $56.1 billion. The state ranked 13th in retail trade in 1982, with sales totaling $24.4 billion. The most important categories (and their sales shares) were food stores, 22%; automotive dealers, 19%; and general merchandise stores, 11%. In 1984, the sale of distilled spirits was legal countywide in fewer than one-third of Georgia's counties.

Georgia exported manufactured goods worth $2.5 billion in 1981 and farm products worth $706 million in 1981/82, ranking 22d and 19th respectively among the states. Georgia ranked 1st in the nation in the export of peanuts and 2d in poultry products and textile products. Savannah is Georgia's most important export-import center; in 1983 it handled commercial goods valued at $2,253 million.

³¹CONSUMER PROTECTION

Georgia's basic consumer protection law is the Fair Business Practices Act of 1975, which forbids representing products as having official approval when they do not, outlaws advertising without the intention of supplying a reasonable number of the items advertised, and empowers the administrator of the law to investigate and resolve complaints and seek penalties for unfair practices. The administrator heads the Office of Consumer

Affairs, which now also administers five other laws that regulate charitable solicitation, offers to sell or buy business opportunities, buying services or clubs, transient merchants, and promotional contests. In 1983/84, the Office of Consumer Affairs resolved 2,418 cases and mediated 1,117 others.

³²BANKING

The state's first bank was the branch of the Bank of the United States established at Savannah in 1802. Eight years later, the Georgia legislature chartered the Bank of Augusta and the Planters' Bank of Savannah, with the state holding one-sixth of the stock of each bank. The state also subscribed two-thirds of the stock of the Bank of the State of Georgia, which opened branches throughout the region. To furnish small, long-term agricultural loans, the state in 1828 established the Central Bank of Georgia, but this institution collapsed in 1856 because the state kept dipping into its reserves. After the Civil War, the lack of capital and the high cost of credit forced farmers to borrow from merchants under the lien system. By 1900 there were 200 banks in Georgia; with an improvement in cotton prices, their number increased to nearly 800 by World War I. During the agricultural depression of the 1920s, about half these banks failed, and the number has remained relatively stable since 1940. Georgia banking practices came under national scrutiny in 1979, when Bert Lance, President Carter's former budget director and the former president of the National Bank of Georgia, was indicted on 33 counts of bank fraud. The federal government dropped its case after Lance was acquitted on 9 of the charges, and most of the rest were dismissed. In 1985, however, Lance's banking practices were again under investigation.

At the end of 1984 there were 384 insured commercial banks in Georgia, of which 64, including the 55 national banks, were members of the Federal Reserve System. At the end of 1983, the state's 65 savings and loan associations had total assets of almost $9.2 billion. In 1982 there were 442 active credit unions in Georgia.

³³INSURANCE

In mid-1983 there were 33 life insurance companies domiciled in Georgia. Life insurance in force in 1983 totaled more than $127.6 billion, and premiums paid amounted to $1,192 million. Death benefit payments were $377.8 million in 1983, and total benefit payments, including death, disability, and annuity as well as policy dividends and matured endowments, amounted to $927.1 million. The average family had $59,500 in life insurance coverage.

Premiums written by other types of insurance firms in 1983 included $623.4 million for automobile liability coverage, $444.6 million for automobile physical damage coverage, $279.8 million for homeowners insurance, and $144.8 million for commercial multiple-peril insurance. Georgians held $744.75 million in flood insurance as of 31 December 1983. There were 34 property/casualty insurance companies in Georgia in 1984.

³⁴SECURITIES

There are no stock or commodity exchanges in Georgia. In 1983, member firms of the New York Stock Exchange had 119 sales offices and 1,488 registered representatives in the state. In that same year, 804,000 Georgians were shareowners of public corporations.

³⁵PUBLIC FINANCE

Since the Georgia constitution forbids the state to spend more than it takes in from all sources, the governor attempts to reconcile the budget requests of the state department heads with the revenue predicted by economists for the coming fiscal year. The governor's Office of Planning and Budget prepares the budget, which is then presented to the general assembly at the beginning of each year's session. The assembly may decide to change the revenue estimate, but it usually goes along with the governor's forecast. The fiscal year begins on 1 July, and the first question for the assembly when it convenes the following January is whether to raise or lower the current year's budget estimate. If the revenues are better than expected, the legislators enact a supplemental budget; if the income is below expectations, cuts can be made.

The following table summarizes state revenues and expenditures—excluding federal funds—for 1983/84 (actual) and 1984/85 (estimated):

REVENUES	1983/84	1984/85
Taxes, of which:	$3,885,949,063	$4,177,255,000
Income taxes, corporate and individual	(1,782,086,038)	(1,955,000,000)
Sales tax, general	(1,357,160,764)	(1,471,000,000)
Motor fuel tax	(372,751,158)	(383,200,000)
Cigar and cigarette tax	(85,340,116)	(85,200,000)
Regulatory fees and sales	124,653,110	124,745,000
TOTAL	$4,010,602,173	$4,302,000,000
EXPENDITURES		
State Board of Education	$1,491,793,747	$1,656,136,009
Regents, Univ. System of Georgia	575,501,471	628,357,297
Department of Human Resources	536,013,735	592,344,802
Department of Transportation	272,103,107	412,217,504
Department of Medical Assistance	192,331,885	230,300,935
Department of Offender Rehabilitation	171,995,573	196,593,359
Department of Public Safety	53,813,279	58,698,034
Department of Natural Resources	47,174,971	50,701,141
Department of Agriculture	25,040,529	27,047,493
Forestry Commission	22,428,505	25,122,474
Other	394,734,823	424,480,952
TOTAL	$3,782,931,625	$4,302,000,000

In 1982/83, of $11.1 billion in state and local government revenue, the federal government originated 21.9%, the state government 37.4%, and local governments 40.7%. The state government actually received 39.2% of total revenues, and local governments 60.8%.

Georgia's state debt totaled more than $1.8 billion in mid-1983, and local government debt more than $6.7 billion. Total state and local indebtedness in 1981/82 amounted to $1,382 per capita.

36 TAXATION

Georgia was the last of the 13 original colonies to tax its citizens, but today its state tax structure is among the broadest in the US. The first comprehensive state tax was provided by the Property Tax Act of 1852, which allocated 50% of the tax to the counties; as of 1983, less than 1% of property taxes went to the state. Motor vehicle license fees began in 1910; motor fuel has been taxed since 1921, tobacco since 1923. In 1929, Georgia began taxing incomes; a withholding tax on incomes has been required since 1960. In 1951, Georgia enacted what at that time was the most all-inclusive sales tax in the US; this 3% tax is now the state's second-largest source of revenue. State law allows counties to charge an additional 1% local-option sales tax and to use the money to roll back property taxes. In 1983, only 29 counties did not impose an additional sales tax.

As of September 1984, the personal income tax ranged from 1% on the first $1,000 to 6% on taxable income over $10,000; the basic corporate tax rate was 6%.

In 1982/83, Georgians paid 85 cents for every dollar received of federal grants-in-aid. Georgians filed 2.25 million federal income tax returns in 1983, paying $5.7 billion in tax. In 1982/83, Georgia ranked 37th in per capita federal tax burden.

37 ECONOMIC POLICY

Since the time of Henry Grady, spokesman for the New South, Georgia had courted industry. Corporation taxes have been traditionally low, wages also low, and unions weak. Georgia's main attractions for new businesses, as summarized in a 1974 study, are a favorable location for air, highway, and rail transport, a mild climate, a rapidly expanding economy, tax incentives and competitive wage scales, and an abundance of recreational facilities. During the 1970s and early 1980s, Georgia governors aggressively sought out domestic and foreign investors, and German, Japanese, and South American corporations were lured to the state.

The Department of Industry and Trade promotes international trade, cultivates new industry, and is responsible for developing tourism. The state funds city and county development plans, aids recreational projects, promotes research and development, and supports industrial training programs.

38 HEALTH

Georgia's public health facilities developed only after the turn of the century. The Ellis Health Law of 1914 placed the responsibility for public health with the counties, but by 1936 only 36 of the state's 159 counties had full-time health departments. Not until after federal funds became available during the 1940s was malaria, one of the oldest afflictions in Georgia, brought under control through the eradication of mosquito-breeding places. Federal funds also enabled every county to receive public health nursing services and X-ray clinics.

Georgia's birthrate declined to a record low of 15.9 per 1,000 population in 1976; the 1982 birthrate was 16. Fetal deaths have increased dramatically since 1973, when abortion laws were liberalized in Georgia. An estimated 38,500 legal abortions were performed in Georgia in 1982, for a rate of 27.7 per 1,000 women 15-44 years old and a ratio of 415 per 1,000 live births. Georgia's infant mortality rates in 1981—10.3 per 1,000 live births for whites, 19.9 for nonwhites—were slightly better than the national averages. Heart disease, cancer, cerebrovascular disease, accidents, and respiratory diseases were the leading causes of death in 1981. Although the death rate for blacks in Georgia continues to exceed that of whites, the differential decreased considerably between 1966 and 1981. According to state statistics, the rate for whites in 1981 was 8.1 per 1,000 population—the same as in 1966—while the rate for blacks fell from 10 to 8.6. In 1981, 88,469 cases of sexually transmitted diseases were reported, or 1,644 per 100,000 population.

In 1983 there were 191 hospitals in Georgia, which had a total of 1,063,948 admissions and 6,976,893 outpatient visits. In 1981, 16,001 registered nurses were employed in Georgia hospitals. The number of hospital beds increased from 9,673 in 1950 to 33,468 in 1983. The average cost of hospital care in 1982 was $284 per day and $1,900 per stay, in each case below the US average. There were 324 nursing care facilities in Georgia, with 33,181 beds in 1982. In 1984, Georgia had 2,620 dentists, and in 1983, 8,307 doctors, 154 osteopaths, and 1,054 public health nurses.

The Medical College of Georgia, established at Augusta in 1828, is one of the oldest medical schools in the US and the center of medical research in the state. The federal Center for Disease Control was established in Atlanta in 1973.

39 SOCIAL WELFARE

As a responsibility of state government, social welfare came late to Georgia. The state waited two years before agreeing to participate in the federal Social Security system in 1937. Eighteen years later, the state was distributing only $62 million to the aged, blind, and disabled, and to families with dependent children. By 1970, the amount had risen to $150 million, but Georgia still lagged far behind the national average. In 1982/83, state and local government public-welfare payments amounted to $178.47 per capita, well under the US average of $251.02.

In 1983/84, an average of 89,440 Georgia families received an average of $183.02 per month in aid to families with dependent children. In 1983, 605,000 Georgians participated in the federal food stamp program, at a total cost of $318 million. Some 862,000 pupils took part in the federal school lunch program, which cost about $84 million.

Social Security benefits paid to Georgians totaled more than $3.2 billion in 1983. Of that amount, 56% went to retired workers, 23% to survivors of deceased workers, and 13% to the disabled. In December 1982, the average monthly Social Security benefit for Georgia's 402,894 retired workers was $376.89. Supplemental Security Income benefits, distributed to 147,945 Georgians, totaled $286.1 million in 1983. In 1982, workers' compensation payments totaled $240 million. In 1983, unemployment insurance benefits averaged $94.41 weekly, and $299.8 million in benefits was paid.

40 HOUSING

Post World War II housing developments provided Georgia families with modern, affordable dwellings. The home-loan guarantee programs of the Federal Housing Administration and the Veterans Administration made modest down payments, low interest rates, and long-term financing the norm in Georgia. The result was a vast increase in both the number of houses constructed and the percentage of families owning their own homes.

Between 1940 and 1970, the number of housing units in the state doubled to 1,470,754. In 1940, only 3 in 10 Georgia homes were owner-occupied; by 1980, 6 in 10 were. In 1970, 13% of all Georgians were still living in units that lacked full plumbing; in 1980, 3.8% were.

In 1980 there were 2,028,664 housing units in Georgia, of which about 7.6% were mobile homes. About 15% of Georgia's housing was built before 1939. The median value of owner-occupied housing in 1980 was $36,900; the median monthly rent of renter-occupied housing was $153. In 1982, 39,437 building permits were issued for residential construction. By 1985, the median sales price for an existing single-family home in Atlanta was $73,700. This figure was slightly below the national average but 8.5% higher than a year earlier, whereas the national average was only 3.4% higher.

41 EDUCATION

During the colonial period, education was in the hands of private schoolmasters. Georgia's first constitution called for the establishment of a school in each county. The oldest school in the state is Richmond Academy (Augusta), founded in 1788. The nation's oldest chartered public university, the University of Georgia, dates from 1784. Public education was inadequately funded, however, until the inauguration of the 3% sales tax in 1951. By 1960, rural one-teacher schools had disappeared, and children were riding buses to consolidated schools. In the 1953/54 school year, Georgia spent $190 per white student and $132 per black student; by 1983/84, 30 years after the US Supreme Court outlawed public school segregation, the overall cost per pupil had increased to $2,248. Reported receipts for the operation of Georgia's public schools in 1983/84 totaled $2.5 billion, funding for which was 9.3% federal, 58.6% state, and 32.1% local.

Despite increased expenditures, Georgia continues to lag behind the nation in academic achievement. In 1980, only 56.4% of Georgians aged 25 or over had completed high school—for a rank of 43d in the nation. Among steps taken to improve the quality of education in Georgia have been the testing of teacher competence and the requirement, beginning in 1985, that students pass a test of minimum skills in order to graduate from high school. The Board of Regents of the state university system also increased its requirements for students starting college after 1988.

In the fall of 1982, Georgia public schools enrolled an estimated 1,100,634 students, of whom 592,772 were in elementary schools, 406,197 in secondary schools, and 101,665 in kindergarten or special schools. Private schools had 82,505 students in all grades in 1980. In 1984, the average salary for a public school teacher in Georgia was $18,631—37th in the nation.

Georgia had 80 institutions of higher learning, 34 public and 46

private, with a total of 193,367 students in 1982/83. Thirty-two public colleges are components of the University System of Georgia; the largest of these is the University of Georgia (Athens), with a 1983/84 enrollment of 25,042. The largest private university is Emory (Atlanta), with 8,052 students in 1983/84. A scholarship program was established in 1978 for minority students seeking graduate and professional degrees.

42 ARTS

During the 20th century, Atlanta has replaced Savannah as the major art center of Georgia, while Athens, the seat of the University of Georgia, has continued to share in the cultural life of the university. The state has eight major art museums, as well as numerous private galleries; especially notable is the High Museum of Art in Atlanta, dedicated in 1983. The Atlanta Memorial Arts Center was dedicated in 1968 to the 100 members of the association who lost their lives in a plane crash. The Atlanta Art Association exhibits the work of contemporary Georgia artists; Georgia's Art Bus Program delivers art exhibits to Georgia communities, mostly in rural areas, for three-week periods.

The theater has enjoyed popular support since the first professional resident theater troupe began performing in Augusta in 1790. Atlanta has a resident theater, and there are community theaters in some 30 cities and counties. Georgia has actively cultivated the filmmaking industry, and an increasing number of films for cinema and television are being produced in the state— 15 in 1983/84.

Georgia has at least 11 symphony orchestras, ranging from the Atlanta Symphony to community and college ensembles throughout the state. Atlanta and Augusta have professional ballet touring companies, Augusta has a professional opera company, and choral groups and opera societies perform in all major cities.

Macon has become a major recording center, especially for popular music. The north Georgia mountain communities retain their traditional folk music.

43 LIBRARIES AND MUSEUMS

In 1983, the Georgia public library system included 36 regional and 14 county systems, each operating under its own board. Regional libraries and branches totaled 318 in 1984.

The holdings of all public libraries totaled 10,367,634 volumes in 1983, and the combined circulation was 21,527,677 volumes, or 3.63 per capita. The University of Georgia had by far the largest academic collection, including 2,226,090 books in addition to government documents, microforms, and periodicals. Emory University, in Atlanta, has the largest private academic library, with about 1,789,395 bound volumes.

Georgia has at least 135 museums, including the Telfair Academy of Arts and Sciences in Savannah, the Georgia State Museum of Science and Industry in Atlanta, the Columbus Museum of Arts and Sciences, and Augusta-Richmond County Museum in Augusta. Atlanta's Cyclorama depicts the 1864 Battle of Atlanta. The Crawford W. Long Medical Museum in Jefferson is a memorial to Dr. Long, a pioneer in the use of anesthetics. A museum devoted to gold mining is located at Dahlonega.

Georgia abounds in historic sites, 100 of which were selected for acquisition in 1972 by the Georgia Heritage Trust Commission. Sites administered by the National Park Service include the Chickamauga battlefield, Kennesaw Mountain battlefield, Ft. Pulaski National Monument, and Andersonville prison camp near Americus, all associated with the Civil War, as well as the Ft. Frederica National Monument, an 18th-century English barracks on St. Simons Island. Also of historic interest are Factors Wharf in Savannah, the Hay House in Macon, and Franklin D. Roosevelt's "Little White House" at Warm Springs. The Martin Luther King, Jr., National Historic Site was established in Atlanta in 1980. Also in Atlanta, work continued in 1985 on former President Jimmy Carter's library, museum, and conference center complex, scheduled for completion in 1986. The state's most important archaeological sites are the Etowah Mounds at Carterville, the Kolomoki Mounds at Blakely, and the Ocmulgee Indian village near Macon.

⁴⁴COMMUNICATIONS

Airmail service was introduced to Georgia about 1930, and since then the quantity of mail has increased enormously. The US Postal Service had 13,911 employees in Georgia's 639 post offices as of March 1985.

There were 4,339,000 telephones in Georgia, 3,214,000 residential and 1,125,000 business at the end of 1984. At the time of the 1980 census, 88% of Georgia households had telephone service.

In 1983, Georgia had 318 commercial radio stations, 190 AM and 128 FM. There were 23 commercial and 10 educational television stations in 1984.

Cable television systems served 792,607 subscribers in 284 communities in 1984. On 1 June 1980, Atlanta businessman Ted Turner inaugurated the independent Cable News Network, which made round-the-clock news coverage available to 4,100 cable television systems throughout the US. By 1985, Cable News was available to 32.3 million households in the US through 7,731 cable television systems and was broadcast to 22 other countries. In addition, Turner broadcasts CNN Headline News and WTBS.

⁴⁵PRESS

Georgia's first newspaper was the *Georgia Gazette,* published by James Johnston from 1763 until 1776. When royal rule was temporarily restored in Savannah, Johnston published the *Royal Georgia Gazette;* when peace came, he changed the name again, this time to the *Gazette of the State of Georgia.* After the state capital was moved to Augusta in 1785, Greensburg Hughes, a Charleston printer, began publishing the *Augusta Gazette.* Today's *Augusta Chronicle* traces its origin to this paper and claims the honor of being the oldest newspaper in the state. In 1817, the *Savannah Gazette* became the state's first daily. After the Indian linguist Sequoyah gave the Cherokee a written language, Elias Boudinot gave them a newspaper, the *Cherokee Phoenix,* in 1828. Georgia authorities suppressed the paper in 1835, and Boudinot joined his tribe's tragic migration westward.

After the Civil War, Henry Grady made the *Atlanta Constitution* the most famous newspaper in the state, with his "New South" campaign. Joel Chandler Harris's stories of Uncle Remus appeared in the *Constitution,* as did the weekly letters of humorist Charles Henry Smith, writing under the pseudonym of Bill Arp. In 1958, Ralph E. McGill, editor and later publisher of the *Constitution,* won a Pulitzer Prize for his editorial opposition to racial intolerance.

As of 1984, Georgia had 10 morning dailies with 741,635 combined circulation, 26 evening dailies with 893,043, and 15 Sunday newspapers with 1,149,547. The following table shows leading daily newspapers with their 1984 circulations:

AREA	NAME	DAILY	SUNDAY
Atlanta	Constitution (m,S)	227,755	571,646
	Journal (e,S)	183,287	
Augusta	Chronicle (m,S)	60,624	83,770
	Herald (e,S)	18,144	
Columbus	Enquirer (m,S)	33,310	68,055
	Ledger (e,S)	25,591	
Macon	Telegraph and News (m,S)	68,910	90,095
Savannah	News (m,S)	56,337	75,170
	Press (e)	20,641	

Periodicals published in Georgia in 1984 included *Golf World, Atlanta Weekly, Industrial Engineering, Robotics World,* and *Southern Accents.* Among the nation's better-known scholarly presses is the University of Georgia Press, which publishes the *Georgia Review.*

⁴⁶ORGANIZATIONS

The 1982 Census of Service Industries counted 1,213 organizations in Georgia, including 293 business associations; 540 civic, social, and fraternal associations; and 23 educational, scientific, and research associations. National organizations headquartered in Georgia include the National Association of College Deans, Registrars, and Admissions Officers, located in Albany, and the Association of Information and Dissemination Centers, the American Risk and Insurance Association, and the American Business Law Association, located in Athens. The many organizations headquartered in Atlanta include the Industrial Development Research Council, the Southern Association of Colleges and Schools, the Southern Education Foundation, the Southern Regional Council, the Southern Christian Leadership Conference, the American Rheumatism Association, the Arthritis Foundation, the American Academy of Psychotherapists, the Federation of Southern Cooperatives, the International Association of Financial Planning, the National Association of Market Developers, and the Textile Quality Control Association.

⁴⁷TOURISM, TRAVEL, AND RECREATION

Georgia's travel industry earned $8.2 billion in 1984, when more than 288,000 people were employed in tourist-related industries and services. In 1982, travel and tourism was the 2nd-largest employer in the state, generating $877 million in payroll; tourists spent nearly $12.4 million per day.

Major tourist attractions include 2 national forests, 6 national parks, and about 60 state parks and historic areas. Other places of interest include the impressive hotels and convention facilities of downtown Atlanta, the Okefenokee Swamp in southern Georgia, Stone Mountain near Atlanta, former President Jimmy Carter's home in Plains, the birthplace, church, and gravesite of Martin Luther King, Jr., in Atlanta, and the historic squares and riverfront of Savannah. The varied attractions of the Golden Isles include fashionable Sea Island; primitive Cumberland Island, now a national seashore; and Jekyll Island, owned by the state and leased to motel operators and to private citizens for beach homes. Since 1978, the state, under its Heritage Trust Program, has acquired Ossabaw and Sapelo islands, and strictly regulates public access to these wildlife sanctuaries.

Georgia has long been a hunters' paradise. Waynesboro calls itself the "bird dog capital of the world," and Thomasville in South Georgia is a mecca for quail hunters. In 1982/83, Georgia issued licenses to 674,641 fishermen and 370,655 hunters.

⁴⁸SPORTS

Most major professional sports are represented in Georgia. Atlanta–Fulton County Stadium, completed in 1965 at a cost of over $18 million, serves as the home field for two professional teams: baseball's Atlanta Braves, for whom Henry Aaron hit many of his record 755 home runs, and the Atlanta Falcons of the National Football League. The Omni International Sports Complex houses the Atlanta Hawks of the National Basketball Association.

The Atlanta 500 is one of the Winston Cup Grand National auto races. The Masters Golf Tournament, brainchild of golfing's Bobby Jones, has been played at the Augusta National Golf Club since 1934. The Atlanta Golf Classic is also listed on the professional golfers' tour.

Football is king of the college sports. The University of Georgia Bulldogs play in the Southeastern Conference, and Georgia Tech's "Rambling Wrecks" compete in the Atlantic Coast Conference. The Peach Bowl has been an annual postseason football game in Atlanta since 1968.

Professional fishing, sponsored by the Bass Anglers Sportsman's Society, is one of the fastest-growing sports in the state. A popular summer pastime is rafting. Massive raft races on the Chattahoochee at Atlanta and Columbus, and on the Savannah River at Augusta, draw many spectators and participants.

[49] FAMOUS GEORGIANS

James Earl "Jimmy" Carter (b.1924), born in Plains, was the first Georgian to serve as president of the US. He was governor of the state (1971-75) before being elected to the White House in 1976. Georgia has not contributed any US vice presidents; Alexander H. Stephens (1812-83) was vice president of the Confederacy during the Civil War.

Georgians who served on the US Supreme Court include James M. Wayne (1790-1867), John A. Campbell (1811-89), and Joseph R. Lamar (1857-1916). Several Georgians have served with distinction at the cabinet level: William H. Crawford (b.Virginia, 1772-1834), Howell Cobb (1815-68), and William G. McAdoo (1863-1941) as secretaries of the treasury; John M. Berrien (b.New Jersey, 1781-1856) as attorney general; John Forsyth (1781-1841) and Dean Rusk (b.1909) as secretaries of state; George Crawford (1798-1872) as secretary of war; and Hoke Smith (b.North Carolina, 1855-1931) as secretary of the interior.

A leader in the US Senate before the Civil War was Robert Toombs (1810-85). Notable US senators in recent years were Walter F. George (1878-1957), Richard B. Russell (1897-1971), Herman Talmadge (b.1913), and Sam Nunn (b.1938). Carl Vinson (1883-1981) was chairman of the House Armed Services Committee.

Many Georgians found fame in the ranks of the military. Confederate General Joseph Wheeler (1836-1906) became a major general in the US Army during the Spanish-American War. Other Civil War generals included W. H. T. Walker (1816-64), Thomas R. R. Cobb (1823-62), who also codified Georgia's laws, and John B. Gordon (1832-1904), later a US senator and governor of the state. Gordon, Alfred Colquitt (1824-94), and wartime governor Joseph E. Brown (b.South Carolina, 1821-94) were known as the "Bourbon triumvirate" for their domination of the state's Democratic Party from 1870 to 1890. Generals Courtney H. Hodge (1887-1966) and Lucius D. Clay (1897-1978) played important roles in Europe during and after World War II.

Sir James Wright (b.South Carolina, 1714-85) was Georgia's most important colonial governor. Signers of the Declaration of Independence for Georgia were George Walton (b.Virginia, 1741-1804), Button Gwinnett (b.England, 1735-77), and Lyman Hall (b.Connecticut, 1724-90). Signers of the US Constitution were William Few (b.Maryland, 1748-1828) and Abraham Baldwin (b.Connecticut, 1754-1807). Revolutionary War hero James Jackson (b.England, 1757-1806) organized the Democratic-Republican Party (today's Democratic Party) in Georgia.

The first Georgians, the Indians, produced many heroes. Tomochichi (c.1664-1739) was the Yamacraw chief who welcomed Oglethorpe and the first Georgians. Alexander McGillivray (c.1759-93), a Creek chief who was the son of a Scottish fur trader, signed a treaty with George Washington in a further attempt to protect the Creek lands. Osceola (1800-1838) led his Seminole into the Florida swamps rather than move west. Sequoyah (b.Tennessee, 1773-1843) framed an alphabet for the Cherokee, and John Ross (Coowescoowe, b.Tennessee, 1790-1866) was the first president of the Cherokee republic.

Among influential Georgian educators were Josiah Meigs (b.Connecticut, 1757-1822), the first president of the University of Georgia, and Milton Antony (1784-1839), who established the Medical College of Georgia in Augusta in 1828. Crawford W. Long (1815-78) was one of the first doctors to use ether successfully in surgical operations. Paul F. Eve (1806-77) was a leading teacher of surgery in the South, and Joseph Jones (1833-96) pioneered in the study of the causes of malaria.

Distinguished black Georgians include churchmen Henry M. Turner (b.South Carolina, 1834-1915) and Charles T. Walker (1858-1921), educators Lucy Laney (1854-1933) and John Hope (1868-1936), and civil-rights activists William Edward Burghardt DuBois (b.Massachusetts, 1868-1963) and Walter F. White (1893-1955). One of the best-known Georgians was Martin Luther King, Jr. (1929-68), born in Atlanta, leader of the March on Washington in 1963, and winner of the Nobel Peace Prize in 1964 for his leadership in the campaign for civil rights; he was assassinated in Memphis, Tenn., while organizing support for striking sanitation workers. Black Muslim leader Elijah Muhammad (Elijah Poole, 1897-1975) was also a Georgian. Other prominent black leaders include Atlanta Mayor and former UN Ambassador Andrew Young (b.Louisiana, 1932), former Atlanta Mayor Maynard Jackson (b.Texas, 1938), and Georgia Senator Julian Bond (b.Tennessee, 1940).

Famous Georgia authors include Sidney Lanier (1842-81), Joel Chandler Harris (1848-1908), Lillian Smith (1857-1966), Conrad Aiken (1889-1973), Erskine Caldwell (b.1902), Caroline Miller (b.1903), Frank Yerby (b.1916), Carson McCullers (1917-67), James Dickey (b. 1923), and Flannery O'Connor (1925-1964). Also notable is Margaret Mitchell (1900-1949), whose Pulitzer Prize-winning *Gone With the Wind* (1936) typifies Georgia to many readers.

Entertainment celebrities include songwriter Johnny Mercer (1909-76); actors Charles Coburn (1877-1961) and Oliver Hardy (1877-1961); singers and musicians Harry James (1916-83), Ray Charles (Ray Charles Robinson, b.1930), James Brown (b.1933), Little Richard (Richard Penniman, b.1935), Jerry Reed (b.1937), Gladys Knight (b.1944), and Brenda Lee (b.1944); and actors Melvyn Douglas (1901-81), Sterling Holloway (b.1905), Ossie Davis (b.1917), Barbara Cook (b.1927), Jane Withers (b.1927), Joanne Woodward (b.1930), and Burt Reynolds (b.1936).

Major sports figures include baseball's "Georgia peach," Tyrus Raymond "Ty" Cobb (1886-1961); Jack Roosevelt "Jackie" Robinson (1919-72), the first black to be inducted into the Baseball Hall of Fame; and Robert Tyre "Bobby" Jones (1902-71), winner of the "grand slam" of four major golf tournaments in 1930.

Robert E. "Ted" Turner (b.Ohio, 1939), an Atlanta business-man-broadcaster, owns the Atlanta Hawks and the Atlanta Braves and skippered the *Courageous* to victory in the America's Cup yacht races in 1977. Architect John C. Portman, Jr. (b.South Carolina, 1924), was the developer of Atlanta's Peachtree Center.

[50] BIBLIOGRAPHY

Bartley, Numan V. *From Thurmond to Wallace: Political Tendencies in Georgia, 1948-68.* Baltimore: Johns Hopkins Press, 1970.

Brook, Diane L. *Georgia: A Geography.* Boulder, Colo.: Westview, 1985.

Coleman, Kenneth, et al. *A History of Georgia.* Athens: University of Georgia Press, 1977.

Georgia, University of. College of Business Administration. Division of Research. *Georgia Statistical Abstract 1984-85.* Edited by Lorena M. Akioka. Athens, 1984.

Grady, Henry W. *The New South.* Savannah: Beehive Press, 1971.

Hepburn, Lawrence R. *The Georgia History Book.* Athens: University of Georgia Institute of Government, 1982.

King, Coretta Scott. *My Life with Martin Luther King.* New York: Holt, Rinehart & Winston, 1970.

Lane, Mills. *The People's Georgia: An Illustrated Social History.* Savannah: Beehive Press, 1975.

Maguire, Jane. *On Shares: Ed Brown's Story.* New York: Norton, 1976.

Malone, Henry. *Cherokees of the Old South.* Athens: University of Georgia Press, 1956.

Martin, Harold H. *Georgia: A Bicentennial History.* New York: Norton, 1977.

Saye, Albert B. *Georgia History and Government.* Austin: Steck Vaughn, 1973.

Woodward, C. Vann. *Tom Watson: Agrarian Rebel.* New York: Oxford University Press, 1970 (orig. 1938).

HAWAII

State of Hawaii

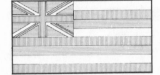

ORIGIN OF STATE NAME: Unknown. The name may stem from Hawaii Loa, traditional discoverer of the islands, or from Hawaiki, the traditional Polynesian homeland. **NICKNAME:** The Aloha State. **CAPITAL:** Honolulu. **ENTERED UNION:** 21 August 1959 (50th). **SONG:** "Hawaii Ponoi." **MOTTO:** *Ua mau ke ea o ka aina i ka pono* (The life of the land is perpetuated in righteousness). **COAT OF ARMS:** The heraldic shield of the Hawaiian kingdom is flanked by the figures of Kamehameha I, who united the islands, and Liberty, holding the Hawaiian flag. Below the shield is a phoenix surrounded by taro leaves, banana foliage, and sprays of maidenhair fern. **FLAG:** Eight horizontal stripes, alternately white, red, and blue, represent the major islands, with the British Union Jack (reflecting the years that the islands were under British protection) in the upper lefthand corner. **OFFICIAL SEAL:** Same as coat of arms, with the words "State of Hawaii 1959" above and the state motto below. **BIRD:** Nene (Hawaiian goose). **FLOWER:** Pua aloalo (hibiscus). **TREE:** Kukui (candlenut). **ISLAND EMBLEMS:** Each of the eight major islands has its own color and emblem. Hawaii: red; lehua (ohia blossom). Kahoolawe: gray; hinahina (beach heliotrope). Kauai: purple; mokihana (fruit capsule of the *Pelea anisata*). Lanai: yellow; kaunaoa *(Cuscuta sandwichiana)*. Maui: pink; lokelani (pink cottage rose). Molokai: green; kukui (candlenut) blossom. Niihau: white; white pupa shell. Oahu: yellow; ilima *(Sida fallax)*. **LEGAL HOLIDAYS:** New Year's Day, 1 January; Birthday of Martin Luther King, Jr., 3d Monday in January; Presidents' Day, 3d Monday in February; Kuhio Day, 26 March; Good Friday, March or April; Memorial Day, last Monday in May; Kamehameha Day, 11 June; Independence Day, 4 July; Admission Day, 3d Friday in August; Labor Day, 1st Monday in September; Discoverers Day, 2d Monday in October; Election Day, 1st Tuesday after 1st Monday in November; Veterans Day, 11 November; Thanksgiving Day, 4th Thursday in November; Christmas Day, 25 December. **TIME:** 2 AM Hawaii-Aleutian Standard Time = noon GMT.

¹LOCATION, SIZE, AND EXTENT

The State of Hawaii is an island group situated in the northern Pacific Ocean, about 2,400 mi (3,900 km) WSW of San Francisco. The smallest of the 5 Pacific states, Hawaii ranks 47th in size among the 50 states.

The 132 Hawaiian Islands have a total area of 6,470 sq mi (16,758 sq km), including 6,425 sq mi (16,641 sq km) of land and only 45 sq mi (117 sq km) of inland water. The island chain extends over 1,576 mi (2,536 km) N–S and 1,425 mi (2,293 km) E–W. The largest island, Hawaii (known locally as the "Big Island"), extends 76 mi (122 km) E–W and 93 mi (150 km) N–S; Oahu, the most populous island, extends 44 mi (71 km) E–W and 30 mi (48 km) N–S.

The eight largest islands of the Hawaiian group are Hawaii (4,035 sq mi—10,451 sq km), Maui (734 sq mi—1,901 sq km), Oahu (617 sq mi—1,598 sq km), Kauai (558 sq mi—1,445 sq km), Molokai (264 sq mi—684 sq km), Lanai (141 sq mi—365 sq km), Niihau (73 sq mi—189 sq km), and Kahoolawe (45 sq mi—117 sq km). The general coastline of the island chain is 750 mi (1,207 km); the tidal shoreline totals 1,052 mi (1,693 km). The state's geographic center is off Maui, at 20°15′N, 156°20′W.

²TOPOGRAPHY

The 8 major and 124 minor islands that make up the State of Hawaii were formed by volcanic eruptions. Mauna Loa, on the island of Hawaii, is the world's largest active volcano, at a height of 13,675 feet (4,168 meters). Kilauea, on the eastern slope of Mauna Loa, is the world's largest active volcanic crater: beginning on 24 May 1969, it spewed forth 242 million cu yards (185 million cu meters) of lava spreading over an area of 19.3 sq mi (50 sq km). The longest volcanic eruption in Hawaii lasted 867 days. Further indications of Hawaii's continuing geological activity are the 14 earthquakes, each with a magnitude of 5 or more on the Richter scale, that shook the islands from 1969 to 1979; one quake, at Puna, on Hawaii, in 1975, reached a magnitude of 7.2.

Hawaii, Maui, Kauai, and Molokai are the most mountainous islands. The highest peak in the state is Mauna Kea (13,796 feet—4,205 meters), on Hawaii; the largest natural lake, Halulu (182 acres—74 hectares), Niihau; the largest artificial lake, Waiia Reservoir (422 acres—171 hectares), Kauai; and the longest rivers, Kaukonahua Stream (33 mi—53 km) in the north on Oahu and Wailuku River (32 mi—51 km) on Hawaii. While much of the Pacific Ocean surrounding the state is up to 20,000 feet (6,100 meters) deep, Oahu, Molokai, Lanai, and Maui stand on a submarine bank at a depth of less than 2,400 feet (730 meters).

³CLIMATE

Hawaii has a tropical climate cooled by trade winds. Normal daily temperatures in Honolulu average 72°F (22°C) in February and 78°F (26°C) in August; the average wind speed is a breezy 11.8 mph (19 km/hr). The record high for the state is 100°F (38°C), set at Pahala on 27 April 1931, and the record low is 12°F (−11°C), set at Mauna Kea Observatory on 17 May 1979.

Rainfall is extremely variable, with far more precipitation on the windward (northeastern) than on the leeward side of the islands. Mt. Waialeale, Kauai, is reputedly the rainiest place on earth, with a mean annual total of 486 in (1,234 cm). In the driest areas—on upper mountain slopes and in island interiors, as in central Maui—the average annual rainfall is under 10 in (25 cm). Snow falls at the summits of Mauna Loa, Mauna Kea, and Halea Kala—the highest mountains. The highest tidal wave (tsunami) in the state's history reached 56 feet (17 meters).

⁴FLORA AND FAUNA

Formed over many centuries by volcanic activity, Hawaii's topography—and therefore its flora and fauna—has been subject to constant and rapid change. Relatively few indigenous trees remain; most of the exotic trees and fruit plants have been introduced since the early 19th century. Of 2,200 species and subspecies of flora, more than half are endangered, threatened, or extinct; the koa is an indigenous tree under state protection.

The only land mammal native to the islands is the Hawaiian hoary bat, now endangered; there are no indigenous snakes. The endangered humpback whale migrates to Hawaiian waters in winter; other marine animals abound. Listed as threatened are Newell's shearwater and the green sea turtle. Among threatened birds are several varieties of honeycreeper and the Hawaiian goose (nene), crow, and hawk. The nene (the state bird), once close to extinction, now numbers in the hundreds and is on the increase. Animals considered endangered by the state but not on the federal list include the Hawaiian storm petrel, Hawaiian owl, Maui 'amakihi *(Loxops virens wilsoni),* and 'i'iwi *(Vestiaria coccinea).*

⁵ENVIRONMENTAL PROTECTION

Environmental protection responsibilities are vested in the Department of Land and Natural Resources and in the Environmental and Health Services Division of the Department of Health. In 1974, the legislature enacted the Hawaii Environmental Policy Act of 1974, establishing environmental guidelines for state agencies, and mandated environmental impact statements for all government and some private projects. Noise pollution requirements in Honolulu are among the strictest in the US, and air and water purity levels are well within federal standards. Between 1973 and 1983, the amount of sulfur dioxide and particulates in the air in downtown Honolulu was reduced by almost one-third.

In 1983, the federal Environmental Protection Agency banned the use of ethylene dibromide (EDB), a pesticide used in the state's pineapple fields, after high levels of the chemical were found in wells on the island of Oahu.

⁶POPULATION

According to the 1980 census, Hawaii had a resident population of 964,691, 25% more than in 1970 and 39th among the 50 states. Estimates for January 1985 give a population of 1,050,270; Census Bureau projections indicate that Hawaii will have a population of 1,138,100 in 1990. Almost four-fifths of the population lives on Oahu, giving the island a density of 1,235 persons per sq mi (477 per sq km) in 1980; the figure for the entire state was 163 per sq mi (63 per sq km) in 1985.

In 1900, 74.5% of Hawaii's inhabitants were rural; by 1980, only 13.5% lived outside urban areas. By far the largest city is Honolulu, with an estimated 1984 population of 805,266, for a rank of 11th in the US. The city of Honolulu is coextensive with Honolulu County.

⁷ETHNIC GROUPS

According to federal census data, Hawaii ranked 2d only to California in numbers of Japanese and Filipino residents in 1980, and placed 3d behind California and New York in Chinese inhabitants. Under the state's own system of ethnic classification, 70% of all residents were of unmixed ethnic stock in 1983: 25% Caucasians, 23% Japanese, 11% Filipinos, 5% Chinese, about 1.5% each for blacks and Koreans, about 1% each for Samoans and Hawaiians, and 1% all others. Of the 30% who were of mixed ethnic stock, 19% were part Hawaiian and 11% were not.

The earliest Asian immigrants, the Chinese, were superseded in number in 1900 by the Japanese, who have since become a significant factor in state politics. The influx of Filipinos and other Pacific island peoples is largely a 20th-century phenomenon. In recent decades, ethnic Hawaiians have been increasingly intent on preserving their cultural identity.

⁸LANGUAGES

Although massive immigration from Asia and the US mainland since the mid-19th century has effectively diluted the native population, the Hawaiian lexical legacy in English is conspicuous. Newcomers soon add to their vocabulary *aloha* (love, good-bye), *haole* (white foreigner), *malihini* (newcomer), *lanai* (porch), *tapa* (bark cloth), *mahimahi* (a kind of fish), *ukulele, muumuu,* and the common directional terms *mauka* (toward the mountains) and *makai* (toward the sea), customarily used instead of "north,"

"east," "west," and "south." Native place-names are numerous: Waikiki, Hawaii, Honolulu, Mauna Kea, Molokai.

Most native-born residents of Hawaiian ancestry speak one of several varieties of Hawaiian pidgin, a lingua franca incorporating elements of Hawaiian, English, and other Asian and Pacific languages. In 1980, 75% of Hawaiians 3 years old or older spoke only English at home. Other languages spoken at home, and the number of speakers, were as follows:

Japanese	80,733	Chinese	20,516
Pilipino or other		Spanish	12,169
Filipino languages	67,695	Korean	9,448
Other Asian or Pacific		German	3,378
island languages	23,666	French	3,009

⁹RELIGIONS

Congregationalist missionaries arrived in 1820, and Roman Catholics in 1827; the constitution of 1840 guaranteed freedom of worship for all religions. Subsequent migration brought Mormons and Methodists, and Anglican representatives were invited by King Kamehameha IV in 1862. Confucianism, Taoism, and Buddhism arrived with the Chinese during the 1850s; by the turn of the century, Shinto and five forms of Mahayana Buddhism were being practiced by Japanese immigrants.

Figures derived from a 1982 survey by the Department of Religion at the University of Hawaii and from a 1980 census of churches and church membership by the National Council of the Churches of Christ in the USA showed that an estimated 210,000 Hawaiians considered themselves Roman Catholics, 110,000 Protestants, and 65,530 Buddhists. Mormons numbered about 31,000 in 1980, and Jews 5,550 in 1984.

¹⁰TRANSPORTATION

Hawaii's only operating railroad is the Lahaina, Kaanapali & Pacific on Maui, with 6 mi (10 km) of track, which carried 172,352 passengers in 1983. Oahu and Hawaii islands have public bus systems. By the end of 1983, Hawaii's 574,533 licensed drivers traversed 4,074 mi (6,556 km) of roads and streets, 92% paved and 73% of that on the two most populous islands. There were 599,845 passenger cars registered in 1984, along with 101,233 trucks and 4,034 buses. Some 71,000 bicycles were registered in 1983.

All scheduled interisland passenger traffic and most transpacific travel is by air. In 1983, the state had 51 aircraft facilities—37 airports and 14 heliports. The busiest air terminal, Honolulu International Airport, accounted for about half of all aircraft operations. In 1981, Hawaii's seven deep-water ports and other harbors handled 11,683,189 tons of overseas cargo and 7,018,382 tons of interisland cargo.

¹¹HISTORY

Hawaii's earliest inhabitants were Polynesians who came to the islands in double-hulled canoes between 1,000 and 1,400 years ago, either from Southeast Asia or from the Marquesas in the South Pacific. The Western world learned of the islands in 1778, when an English navigator, Captain James Cook, sighted Oahu; he named the entire archipelago the Sandwich Islands after his patron, John Montagu, 4th Earl of Sandwich. At that time, each island was ruled by a hereditary chief under a caste system called *kapu.* Subsequent contact with European sailors and traders exposed the Polynesians to smallpox, venereal disease, liquor, firearms, and Western technology—and fatally weakened the *kapu* system. Within 40 years of Cook's arrival, one of the island chiefs, Kamehameha (whose birth date, designated as 11 June, is still celebrated as a state holiday), had consolidated his power on Hawaii, conquered Maui and Oahu, and established a royal dynasty in what became known as the Kingdom of Hawaii.

The death of Kamehameha I in 1819 preceded by a year the arrival of Protestant missionaries. One of the first to come was the Reverend Hiram Bingham, who, as pastor in Honolulu, was instrumental in the christianizing of Hawaii. Even before he came,

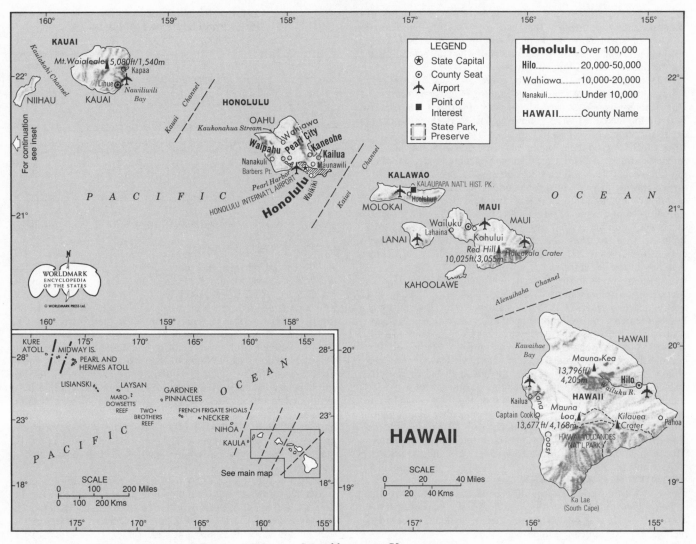

See endsheet maps: C5.
LOCATION: 18°54′ to 28°15′N; 154°40′ to 178°75′W. **BOUNDARIES:** Pacific coastline, 750 mi (1,207 km).

however, Liholiho, successor to the throne under the title of Kamehameha II, had begun to do away with the *kapu* system. After the king's death of measles while on a state trip to England in 1824, another son of Kamehameha I, Kauikeaouli, was proclaimed King Kamehameha III. His reign saw the establishment of public schools, the first newspapers, the first sugar plantation, a bicameral legislature, and Honolulu as the kingdom's capital city. Hawaii's first written constitution was promulgated in 1840, and in 1848 a land reform called the Great Mahele abolished the feudal land system and legitimized private landholdings, in the process fostering the expansion of sugar plantations. The power behind the throne during this period was Dr. Gerrit P. Judd, a medical missionary who served as finance minister and interpreter for Kamehameha III.

Diplomatic maneuverings during the 1840s and 1850s secured recognition of the kingdom from the US, Britain, and France. As the American presence on the islands increased, however, so did pressure for US annexation—a movement opposed by Alexander Liholiho, who ruled as Kamehameha IV after his father's death in 1854. His brief reign and that of his brother Lot (Kamehameha V) witnessed the arrival of Chinese contract laborers and of the first Japanese immigrants, along with the continued growth of Hawaii as an international port of call (especially for whalers) and of the

increasing influence of American sugar planters. Lot's death in 1872 left no direct descendant of Kamehameha, and the legislature elected a new king, whose death only a year later required yet another election. The consequent crowning of Kalakaua, known as the Merry Monarch, inaugurated a stormy decade during which his imperial schemes clashed with the power of the legislature and the interests of the planters. The most significant event of Kalakaua's unstable reign was the signing of a treaty with the US in 1876, guaranteeing Hawaii an American sugar market. The treaty was renewed in 1887 with a clause leasing Pearl Harbor to the US.

Kalakaua died during a visit to San Francisco in 1891 and was succeeded by his sister, Liliuokalani, the last Hawaiian monarch. Two years later, after further political wrangling, she was deposed in an American-led revolution that produced a provisional government under the leadership of Sanford B. Dole. The new regime immediately requested annexation by the US, but the treaty providing for it bogged down in the Senate, and died after the inauguration of President Grover Cleveland, an opponent of expansionism. The provisional government then drafted a new constitution and on 4 July 1894 proclaimed the Republic of Hawaii, with Dole as president. The Spanish-American War, which fanned expansionist feelings in the US and pointed up the nation's strategic interests in the Pacific, gave proponents of annexation the opportunity they had been seeking. The formal transfer of sovereignty took place on 12 August 1898, Dole becoming Hawaii's first territorial governor when the act authorizing the annexation became effective in June 1900.

Notable in the territorial period were a steady US military buildup, the creation of a pineapple canning industry by James D. Dole (the governor's cousin), the growth of tourism (spurred in 1936 by the inauguration of commercial air service), and a rising desire for statehood, especially after passage of the Sugar Act of 1934, which lowered the quota on sugar imports from Hawaii. The Japanese attack on Pearl Harbor on 7 December 1941, crippling the US Pacific fleet and causing some 4,000 casualties, quickly turned Hawaii into an armed camp, under martial law. The record of bravery compiled by Nisei of the 442d Regiment on the European front did much, on the other hand, to allay the mistrust that some mainlanders felt about the loyalties of Hawaiians of Japanese ancestry. Hawaii also bore a disproportionate burden during the Korean conflict, suffering more casualties per capita than any of the 48 states.

Hawaiians pressed for statehood after World War II, but Congress was reluctant, partly because of racial antipathy and partly because of fears that Hawaii's powerful International Longshoremen's and Warehousemen's Union was Communist-controlled. The House of Representatives passed a statehood bill in 1947, but the Senate refused. Not until 1959, after Alaska became the 49th state, did Congress vote to let Hawaii enter the Union. President Eisenhower signed the bill on 18 March, and the question was then put to the Hawaiian electorate, who voted for statehood on 27 June 1959 by a margin of about 17 to 1. Hawaii became the 50th state on 21 August 1959. Since then, defense and tourism have been the mainstays of Hawaii's economy, with the state playing an increasingly important role as an economic, educational, and cultural bridge between the US and the nations of Asia and the Pacific.

¹²STATE GOVERNMENT

The constitution of the state of Hawaii was written by the constitutional convention of 1950 and ratified by the people of the Territory of Hawaii that year and then amended by the 1959 plebiscite on the statehood question.

There is a bicameral legislature of 25 senators, elected from eight senatorial districts for four-year terms, and 51 representatives, elected for two-year terms. The legislature meets annually on the 3d Wednesday in January. To be eligible to serve as a legislator, a person must have attained the age of majority, be an American citizen, have been a resident of the state for at least three years, and be a qualified voter of his district.

The governor and lieutenant governor are elected for concurrent four-year terms and must be of the same political party. They are the only elected officers of the executive branch, except for members of the Board of Education. There are 17 executive departments, each under the supervision of the governor and headed by a single appointed executive.

Voters in Hawaii must be US citizens at least 18 years of age; there is no minimum residency requirement.

¹³POLITICAL PARTIES

Both Republicans and Democrats established party organizations early in the 20th century, when Hawaii was still a territory. Before statehood, the Republican Party dominated the political scene; since the 1960s, however, Hawaii has been solidly Democratic. As of 1985, most local officials were Democrats, as were the governor and majorities of both houses of the state legislature. Hawaii was one of only six states to cast its electoral votes for Jimmy Carter in 1980; in 1984, however, its votes went to Ronald Reagan.

As of 1984, 29 of the 76 state legislators were Japanese, 15 part-Hawaiian, 6 Chinese, and 6 Filipino. Fourteen women served in the legislature in 1985. Statewide, there were nearly 100 Asian

Hawaii Presidential Vote by Major Political Parties, 1960–84

YEAR	ELECTORAL VOTE	HAWAII WINNER	DEMOCRAT	REPUBLICAN
1960	3	*Kennedy (D)	92,410	92,295
1964	4	*Johnson (D)	163,249	44,022
1968	4	Humphrey (D)	141,324	91,425
1972	4	*Nixon (R)	101,433	168,933
1976	4	*Carter (D)	147,375	140,003
1980	4	Carter (D)	135,879	130,112
1984	4	*Reagan (R)	147,154	185,050

*Won US presidential election.

Pacific elected officials, including Governor George R. Ariyoshi and US senators Daniel K. Inouye and Spark M. Matsunaga.

¹⁴LOCAL GOVERNMENT

The state is divided into four principal counties: Hawaii, including the island of Hawaii; Maui, embracing the islands of Maui, Kahoolawe, Lanai, and Molokai; Honolulu, coextensive with the city of Honolulu and covering all of Oahu and the northwestern Hawaiian Islands, from Nihoa to Kure Atoll; and Kauai, including the islands of Kauai and Niihau. Since there are no further subdivisions, the counties provide some services traditionally performed in other states by cities, towns, and villages, notably fire and police protection, refuse collection, and street maintenance and lighting. On the other hand, the state government provides many functions normally performed by counties on the mainland. Each principal county has an elected council and a mayor.

A fifth county of Kalawao forms that part of Molokai more commonly known as the Kalaupapa Settlement, primarily for the care and treatment of persons suffering from leprosy. Kalawao is entirely under the jurisdiction of the Department of Health; the only county officer is an appointed sheriff.

¹⁵STATE SERVICES

Hawaii's first ombudsman, empowered to investigate complaints by the public about any officer or employee of state or county government, took office in 1969. The State Ethics Commission, a legislative agency, implements requirements for financial disclosure by state officials and investigates alleged conflicts of interest and other breaches of ethics.

The Department of Education (headed by an elected Board of Education) operates 226 regular public schools and 7 special schools for the physically and mentally disabled. It administers the statewide library system, regulates private schools, and certifies teachers. The Board of Regents of the University of Hawaii oversees the state's higher educational institutions. Highways, airports, harbors, and other facilities are the concern of the Department of Transportation.

The Department of Health operates 10 public hospitals, the Kalaupapa leper colony, and various programs for the mentally ill, the mentally retarded, and alcoholics. Civil defense and the Air and Army national guards are under the jurisdiction of the Department of Defense.

The Corrections Division of the Department of Social Services and Housing operates the state prison system, along with programs for juvenile offenders. Also within this department are divisions of public welfare and vocational rehabilitation, as well as the Hawaii Housing Authority. The Executive Office on Aging works with state and county departments to coordinate programs for senior citizens. Unemployment insurance, occupational safety and health laws, and workers' compensation programs are run by the Department of Labor and Industrial Relations.

16JUDICIAL SYSTEM

The supreme court, the highest in the state, consists of a chief justice and four associate justices, all of them appointed by the governor with the advice and consent of the senate. All serve 10-year terms, up to the mandatory retirement age of 70.

The state is divided into four judicial circuits with 21 circuit court judges and 3 intermediate appellate court judges, also appointed by the governor to 10-year terms. Circuit courts are the main trial courts, having jurisdiction in most civil and criminal cases. District courts, whose judges are appointed by the chief justice to 6-year terms, function as inferior courts within each judicial circuit; district court judges may also preside over family court proceedings. Hawaii also has a land court and a tax appeal court.

According to the FBI Crime Index, Hawaii's crime rates in 1983 were 252.1 per 100,000 population for violent crimes, well below the US average, and 5,557.5 for property crimes, far above it.

17ARMED FORCES

Hawaii is the nerve center of US defense activities in the Pacific. CINCPAC (Commander-in-Chief Pacific), headquartered at Camp H. M. Smith in Honolulu, directs the US Pacific Command, largest of the six US unified commands and responsible for all US military forces in the Pacific and Indian oceans and southern Asia. Total US Defense Department expenditures in Hawaii in 1983/84 were $2.3 billion; the per capita expenditure of $2,237 was higher than in any other state and only $340 less than in the District of Columbia. Military prime contract awards totaled $537 million.

As of 30 September 1983, Hawaii was home base for 65,959 Department of Defense military and civilian personnel and 67,100 dependents. The US Navy accounted for 34,316 personnel, the Army 22,265, the Air Force 9,273, and other defense agencies 105. Pearl Harbor is home port for 40 ships. The major Army bases, all on Oahu, are Schofield Barracks, Ft. Shafter, and Ft. DeRussy; Air Force bases include Hickam and Wheeler. Military reservations occupy nearly one-fourth of Oahu's land area.

There were 97,000 veterans living in Hawaii as of 30 September 1984. Fewer than 500 were veterans of World War I, 32,000 of World War II, 21,000 of the Korean conflict, and 78,000 of the Viet-Nam era. Veterans' benefits totaled $78 million in 1982/83. Hawaii's National Guard had 4,009 personnel in 1984, and state and local police forces had 3,616 in 1983.

18MIGRATION

The US mainland and Asia have been the main sources of immigrants to Hawaii since the early 19th century. Immigration remains a major source of population growth: between 1950 and 1980, Hawaii's net gain from migration was 91,000, and between 1980 and 1983, 15,000.

Since the early 1970s, about 40,000 mainland Americans have come each year to live in Hawaii. More than half are military personnel and their dependents, on temporary residence during their term of military service. During 1982/83, 7,118 foreign immigrants were admitted, of whom 4,070 were from the Philippines and 883 from Korea. About 29,000 persons were naturalized between 1977 and 1983, according to records compiled by the US Immigration and Naturalization Service.

19INTERGOVERNMENTAL COOPERATION

Among the interstate accords in which Hawaii participates are the Western Corrections Compact and the Compact for Education. Federal aid was estimated at $459 million in fiscal year 1984, of which $21.6 million was general revenue sharing.

20ECONOMY

Hawaii's gross state product was $14.2 billion in 1983, when direct income from the state's three leading export industries—tourism, national defense (treated as an export industry because funds to pay for it come entirely from overseas), and agriculture—totaled $6.5 billion. Visitors' expenditures (excluding transpacific transportation costs) contributed $4 billion; defense programs (mostly for wages, salaries, and supplies), $1.8 billion; raw sugar and molasses, $410 million; and fresh and processed pineapple, $219 million. Manufacturing is a relatively insignificant sector of the Hawaiian economy, though the garment industry grew rapidly during the 1970s and early 1980s; by 1982, value added by manufacture for this industry was $56.5 million, a 26% increase over 1977. Agricultural diversification—including the cultivation of flowers and nursery products, papaya, and macadamia nuts—aquaculture, manganese nodule mining, and film and television production also broadened the state's economic base. In 1983, for example, 63 motion pictures and television specials and series were filmed in Hawaii, bringing in revenues of $35.2 million. Nevertheless, tourism will almost certainly remain Hawaii's leading employer, revenue producer, and growth sector for the foreseeable future.

21INCOME

Average per capita income in Hawaii in 1983 was $12,101 (13th in the US); total personal income was $12.4 billion, representing a real increase of 49% since 1970. The median money income of a four-person family in 1981 was $29,295, for a rank of 6th among the 50 states. Although Hawaii's per capita income was almost 4% above the US average in 1983, the cost of living on the islands was substantially higher than on the mainland. An average budget for a four-person family living on Oahu was $31,893 in 1981, 26% higher than the US urban average.

As of 1979, 9.9% of all Hawaii residents were below the federal poverty level. About 5,700 state residents had gross assets exceeding $500,000 in 1982 and were counted among the nation's leading wealth-holders.

22LABOR

The civilian labor force in 1984 averaged 473,000, of whom 446,000 were employed and 27,000 were unemployed, for an unemployment rate of 5.7%.

A federal survey in March 1983 revealed the following nonfarm employment pattern in Hawaii:

	ESTABLISH-MENTS	EMPLOYEES	ANNUAL PAYROLL ('000)
Agricultural services, forestry, fishing	207	1,850	$ 20,290
Mining	14	282	6,187
Contract construction	1,540	19,416	425,681
Manufacturing, of which:	863	23,215	363,791
Food products	(204)	(10,354)	(158,805)
Transportation, public utilities	944	28,445	624,033
Wholesale trade	1,684	17,695	299,103
Retail trade	6,057	85,817	781,545
Finance, insurance, real estate	2,883	29,526	438,462
Services, of which:	7,091	102,444	1,300,718
Hotels, lodgings	(224)	(25,475)	(282,444)
Personal services	(651)	(4,902)	(39,002)
Other	372	456	5,055
TOTALS	21,655	309,146	$4,264,865

Full-time civilian employees of federal, state, and county governments, not included in this survey, numbered 88,118 in 1983.

Unionization was slow to develop in Hawaii. After World War II, however, the International Longshoremen's and Warehousemen's Union (ILWU) organized workers in the sugar and pineapple industries and then on the docks. The Teamsters Union is also well established. All together, 113,000 Hawaiian workers belonged to labor unions in 1980; there were 241 labor unions in mid-1983.

²³AGRICULTURE

Export crops—especially sugarcane and pineapple—dominate Hawaiian agriculture, which had farm receipts exceeding $479 million in 1983. Hawaii produced only 44% of its locally consumed fresh market vegetables, 20% of the fresh fruits, 30% of the beef and veal, and 28% of the chickens. The following table shows data for crops in 1983:

	ACREAGE ('000)	VOLUME (TONS)	SALES
Sugarcane	194.3	8,926,000	$266,500,000
Pineapples	35.0	722,000	100,376,000
Flowers and nursery products	—	—	36,165,000
Vegetables and melons	5.1	683,000	25,996,000
Macadamia nuts	15.8	36,420	23,928,000
Fruits (excluding pineapples)	6.1	78,900	14,593,000
Other	—	—	11,736,000

The Kona district of the island of Hawaii is the only place in the US where coffee is grown commercially; another tropical product, papaya, has also become a substantial export crop, with an output totaling 44.8 million lb in 1982.

²⁴ANIMAL HUSBANDRY

Livestock products accounted for 15% of Hawaii's farm income in 1983. Sales of cattle, beef, and veal totaled $29 million; of hogs and pigs, $8 million; and of chickens and broilers, $5.5 million. Poultry farms produced 197 million eggs in 1983, for $14 million in sales. Eighty-eight percent of eggs were for domestic consumption, making eggs one of the very few farm commodities in which the state is close to self-sufficient. Most of the state's cattle farms are in Hawaii and Maui counties.

²⁵FISHING

Although expanding, Hawaii's commercial catch remains surprisingly small: 34.8 million lb, worth $29.4 million, in 1984. The most important fish caught are ahi (yellowfin), accounting for 31% of the total value in 1983, followed by aku (skipjack). There were 2,940 licensed commercial fishermen, operating 1,309 fishing vessels and serving 119 fishery establishments. Aquacultural industries produced 268,500 lb of prawns, valued at $1.3 million, in 1983. Sport fishing is extremely popular, with bass, bluegill, tuna, and marlin among the most sought-after varieties.

²⁶FORESTRY

As of 1984, Hawaii had 1,986,000 acres (803,700 hectares) of forestland and water reserves, with 948,000 acres (384,000 hectares) classified as commercial timberland, most of it on the island of Hawaii. Production of lumber and plywood falls far short of local demand. Specialty woods include *Eucalyptus robusta,* used for making pallets for shipping, and blue gum eucalyptus, converted to chips for use by papermakers in Japan.

²⁷MINING

Nonfuel mineral production in Hawaii had a value of $54 million in 1984 (44th in the US). Only cement and stone are of commercial significance: in 1983, an estimated 4,500,000 tons of stone worth $23 million were quarried, and 200,000 tons of cement ($16.3 million) were produced. Local jewelry makers use olivines (green gems formed by lava explosions) and black, pink, red, and golden coral. Among the state's underexploited resources are mineral nodules (usually about 25–35% manganese) on the ocean floor off the coast. Commercial harvesting of nodules began in 1978. In 1984, a federal-state task force was formed to consider the economic possibilities and the environmental impact of mining for deposits of cobalt in the manganese crusts surrounding the Hawaiian archipelago. A grant of $1.8 million was given to Hawaii to conduct a cobalt search.

²⁸ENERGY AND POWER

Devoid of indigenous fossil fuels and nuclear installations, Hawaii depends on imported petroleum for 98% of its energy needs; hydroelectric power, natural gas, windmills, geothermal energy, and sugarcane wastes contribute only 2% but are becoming increasingly more important. On the island of Hawaii, for example, 200 windmills were in operation in 1985, and more were planned for 1986, at which time it is expected that wind power will produce 10% of the island's electricity. Transportation accounts for 47% of energy consumption in Hawaii, industry 35%, residences 10%, and commercial establishments 8%.

Generation of electricity accounts for 31% of the state's fuel consumption (9.4% on the mainland). Installed capacity reached 1.6 million kw in 1983, when production totaled 6.5 billion kwh. All of Hawaii's electric power plants are privately owned.

²⁹INDUSTRIES

Of the total value of shipments by manufacturers—$3.4 billion in 1982—food and food products account for more than 31%. Other major industries are clothing; stone, clay, and glass products; fabricated metals; and shipbuilding. The following table shows the value of shipments for selected industries in 1982:

Preserved and canned fruits and vegetables	$448,500,000
Sugar, confectionery products	428,100,000
Ladies' dresses and outerwear	87,600,000

Hawaii's publicly held corporations include Amfac, involved in food processing, merchandising, and land development, with revenues exceeding $2.2 billion in 1983, and Castle & Cooke, which owns the Dole and Bumble Bee food product lines, $1.6 billion. Other corporations are Dillingham, maritime industries and land development, and Brewer (owned by IU International), which produces 20% of the state's sugar and more than half the world's macadamia nuts.

³⁰COMMERCE

In 1982, Hawaii's wholesale trade amounted to $4.1 billion, of which groceries and related products contributed 29% and petroleum and petroleum products 17%. Honolulu County accounted for 83% of the sales. Retail establishments had sales of $5.2 billion in 1982, 76% of that in Honolulu County. The leading shopping centers, all on Oahu, are the Ala Moana Center, Pearlridge Center, and Kahala Mall. Hawaii's 24 department stores, 21 of which are on Oahu, had sales totaling $483.2 million in 1982.

Hawaii's central position in the Pacific ensures a sizable flow of goods through the Honolulu Customs District. Foreign imports to Hawaii totaled $1.4 billion in 1983, while exports exceeded $203 million. Merchandise imported from the US mainland amounted to nearly $4.9 billion in 1983; exports, $1.2 billion. Hawaii's major trading partners in 1983 were Japan for exports ($72 million) and Indonesia for imports ($629 million).

³¹CONSUMER PROTECTION

Hawaii's Office of Consumer Protection, under the Office of the Governor, coordinates the state's consumer protection activities and deals with landlord-tenant disputes.

³²BANKING

As of 1984, Hawaii had 10 commercial banks with 184 branches, 8 savings and loan associations with 157 branches, 4 trust companies, 72 industrial loan or small loan licensees, and 144 credit unions.

In 1984, the combined assets of all commercial banks totaled almost $8 billion; the largest commercial institution, the Bank of Hawaii, had 73 branches and $3.1 billion in assets in 1983, while the First Hawaiian Bank had 46 branches and $2.2 billion. Honolulu Federal, the largest of the savings and loan associations, had 36 branches and $1.6 billion in assets, with $1.3 billion in mortgage loans outstanding; assets of all savings and loan associations totaled $3.9 billion. Loan companies held assets of $1.6 billion; trust companies, $157 million; and credit unions, more than $1 billion.

[33] INSURANCE

The 699 insurance companies authorized to do business in Hawaii received premiums in 1982 of $795.8 million and paid losses, claims, and benefits of $436.1 million. As of 31 December 1982, insurance firms held Hawaiian investments totaling nearly $2.7 billion.

Hawaii residents held about 1,851,000 life policies with a face value of $26.5 billion in 1983; the average family coverage of $69,300 was 3d highest in the US. Insurance payouts totaled $268 million, of which $66.8 million was in death payments. Property and casualty companies wrote premiums in 1983 for $150.1 million in automotive liability insurance, $79.7 million in automotive physical damage insurance, and $41.4 million for homeowners' coverage. A total of $558.5 million in flood insurance was in effect as of 31 December 1983. By the end of 1983, 535,303 Hawaii residents had prepaid health coverage through the Hawaii Medical Service Association, and 131,304 were covered by the Kaiser Foundation Health Plan.

[34] SECURITIES

Nineteen Hawaiian-based companies were listed on the New York Stock Exchange in 1984. In 1983, 234,000 Hawaiians held shares in public corporations. The Honolulu Stock Exchange, established in 1898, discontinued trading on 30 December 1977.

[35] PUBLIC FINANCE

Development and implementation of Hawaii's biennial budget are the responsibility of the Department of Budget and Finance. The fiscal year runs from 1 July through 30 June. The following table summarizes operating revenues and expenditures for 1982/83 (in thousands):

REVENUES	
Excise taxes	$ 601,486
Individual income taxes	346,951
Federal grants	386,805
Other receipts	770,357
TOTAL	$2,105,599

EXPENDITURES	
Public schools	$ 406,370
Public welfare	328,323
Higher education	261,097
Debt service	199,213
Urban redevelopment and housing	151,535
Other outlays	844,111
TOTAL	$2,190,649

The debt of the Hawaii state government at the end of fiscal 1982 was $2.1 billion, or $2,186 per capita, 4th in the US.

[36] TAXATION

Hawaii's per capita tax burden is one of the highest in the US. The personal income tax ranges from 2.25% on the first $800 to 11% on taxable income over $30,000; there is a capital gains tax of 3.08%. The business income tax is 5.86% on net income up to $25,000 and 6.435% above that amount. There is a broad-based general excise tax of 0.5% on wholesaling and manufacturing activities and 4% on retail sales of goods and services. Taxes on estates, fuel, liquor, and tobacco are also levied, and the property tax is a major source of county income.

Hawaii's total federal tax burden in 1982/83 was $2.5 billion, or $2,465 per capita. In 1982, Hawaiians filed 440,000 federal income tax returns, paying $1.1 billion in tax.

[37] ECONOMIC POLICY

Business activity in Hawaii is limited by physical factors: land for development is scarce, living costs are relatively high, heavy industry is environmentally inappropriate, and there are few land-based mineral operations. On the other hand, Hawaii is well placed as a trading and communications center, and Hawaii's roles as a defense outpost and tourist mecca seem secure for the foreseeable future. The state has actively encouraged tourism and aquaculture. A free trade zone, authorized by the US government in 1966, is managed and promoted by the Foreign Trade Zone, a division of the Department of Planning and Economic Development.

[38] HEALTH

The infant mortality rate in Hawaii in 1983 was 10 per 1,000 live births. In that year, the birthrate (18.8 per 1,000 population) was more than triple the death rate (5.6); death rates from heart disease, stroke, and arteriosclerosis are less than half the national average.

A total of 9,940 people were served by state mental health facilities in 1983, when Hawaii also had 25 civilian hospitals, 34 long-term care facilities, and 15 residential care homes. In 1984 there were 3,273 physicians and surgeons, 966 dentists, 8,156 registered nurses, and 708 pharmacists in Hawaii. The average cost of hospital care in 1984 was $224 per day in a semiprivate room.

[39] SOCIAL WELFARE

Approximately 6% of the resident population of Hawaii was served by major public welfare programs in 1983. Direct public assistance costs reached $304 million in 1982/83, of which federal funds paid 44%. Some 54,800 state residents received $87 million in aid to families with dependent children in 1982. In 1983, 99,000 people participated in the food stamp program at a federal cost of $81 million; the school lunch program served 146,000 students, with a federal subsidy of $13 million. Other programs were Social Security (1983), $536 million; Supplemental Security Income (1983), $15 million; unemployment insurance (1983), $70.4 million; and workers' compensation (1982), $92.6 million.

[40] HOUSING

Although statehood set off a building surge in Hawaii, housing remained in short supply throughout the 1970s and early 1980s. In 1984 there were an estimated 360,000 housing units; at the end of 1982 there were 97,931 condominium units, a type of housing first authorized in 1961. Military and public housing accounted for another 27,000 units in 1984. The mean selling price for a single-family house on Oahu in 1983 was $189,000; for a condominium, it was $114,000. Between 1981 and 1983, 16,800 new privately owned housing units, valued at more than $1 billion, were authorized.

[41] EDUCATION

Education has developed rapidly in Hawaii: 74% of all state residents 25 years of age or older had completed high school by 1980, and 20% had at least 4 years of college.

Hawaii is the only state with a single, unified public school system, founded in 1840. In 1983/84 there were 233 public schools with 7,997 teachers and 162,241 students; in addition, 141 private schools had 2,347 teachers and 37,999 pupils. The University of Hawaii maintains three campuses—Manoa (by far the largest), Hilo, and West Oahu—with an enrollment of 46,468 in fall 1983, when six community colleges enrolled 21,310. Four private colleges—Brigham Young University–Hawaii Campus, Chaminade University of Honolulu, Hawaii Loa College, and Hawaii Pacific College—had a combined enrollment of 7,738 during the same year.

[42] ARTS

The Neal Blaisdell Center in Honolulu has a 2,100-seat theater and concert hall, an 8,400-seat arena, and display rooms. Other performance facilities in Honolulu are the John F. Kennedy Theater at the University of Hawaii, the Waikiki Shell for outdoor concerts, and the Hawaii Opera Theater, which presents three operas each season. The Honolulu Symphony Orchestra performs both on Oahu and on the neighboring islands. Other Oahu cultural institutions are the Honolulu Community Theater, Honolulu Theater for Youth, Windward Theater Guild, and Polynesian Cultural Center.

43 LIBRARIES AND MUSEUMS

The Hawaii state library system had 47 facilities in 1983 (22 on Oahu), with a combined book stock of 2,203,394 and total circulation of 5,321,684. During the same year, the University of Hawaii library system had 2,330,272 volumes, five-sixths of them on the Manoa campus.

Total attendance at Hawaii's 45 major museums and cultural attractions was 12,763,158 (82% on Oahu) in 1983. Among the most popular sites are the National Memorial Cemetery of the Pacific, USS *Arizona* Memorial at Pearl Harbor, Castle Park, Polynesian Cultural Center, Sea Life Park, Bernice P. Bishop Museum (specializing in Polynesian ethnology and natural history), Kahuku Sugar Mill, and Honolulu Academy of Arts. Outside Oahu, the Kilauea Visitor Center (Hawaiian Volcanoes National Park) and Kokee Natural History Museum (Kauai) attract the most visitors.

44 COMMUNICATIONS

Hawaii's 78 post offices and 52 stations handled 305 million pieces of mail and had gross receipts exceeding $80 million in 1983. Commercial interisland wireless service began in 1901, and radiotelephone service to the mainland was established in 1931. In 1980, 95% of Hawaii's 294,052 occupied housing units had telephones.

Hawaii had 26 AM radio stations and 13 FM stations as of 1 January 1984, as well as 12 television stations (10 commercial and 2 educational). The state's 10 cable television companies served 192,739 subscribers in 1984.

45 PRESS

In 1984, Hawaii had nine daily newspapers. Six were published in English: the *Honolulu Advertiser, Honolulu Star-Bulletin* (owned by the Gannett chain), *Hawaii Tribune-Herald, Maui News, West Hawaii Today,* and *The Garden Island. Hawaii Hochi* is bilingual in Japanese and English, the *Korean Times* is published in Korean, and the *United Chinese Press* is published in Chinese. The combined average circulation of all English-language dailies in 1984 was 248,735 weekdays, 219,685 Sundays. That year, 57 magazines and other periodicals were also in circulation. In 1984, the University of Hawaii Press issued 40 new books.

46 ORGANIZATIONS

The 1982 Census of Service Industries counted 422 organizations in Hawaii, including 61 business associations; 293 civic, social, and fraternal associations; and 11 educational, scientific, and research associations. The leading organization headquartered in Honolulu is the East-West Center, a vehicle of scientific and cultural exchange.

47 TOURISM, TRAVEL, AND RECREATION

Jet air service has fueled the Hawaii travel boom in recent decades. Some 243,000 travelers visited Hawaii in 1959, and more than 1,527,000 in 1969; as of 1983, 4,368,105 visitors stayed at least one night in Hawaii, spending $4 billion exclusive of transpacific air fare. Of these visitors, about 2,900,000 came from other states, 268,000 from Canada, and 729,000 from Japan. Some 164,500 jobs were directly or indirectly related to tourism and travel in 1983.

Visitors come for scuba diving, snorkeling, swimming, fishing, and sailing; for the hula, luau, lei, and other distinctive island pleasures; for the tropical climate and magnificent scenic beauty; and for a remarkable variety of recreational facilities, including, as of 1983, 7 national parks and historic sites, 74 state parks, 626 county parks, 57 golf courses, and 1,600 recognized surfing sites.

48 SPORTS

Hawaii's professional baseball team, the Hawaii Islanders, competes in the Pacific Coast League. The Hula Bowl takes place in Hawaii, and the Pro Bowl (the National Football League's all-star game) is also played there. Hawaii is also the site of an annual Professional Golfers' Association tournament, the yearly Duke Kahanamoku and Makaha surfing meets, and the world-famous Ironman Triathlon competition. The Transpacific Yacht Race is held biennially from California to Honolulu. Kona is the site of the International Billfish Tournament, and the Hawaii Big Game Fishing Club holds statewide tournaments each year. Football, baseball, and basketball are the leading collegiate and school sports.

49 FAMOUS HAWAIIANS

Hawaii's best-known federal officeholder is Daniel K. Inouye (b.1924), a US senator since 1962 and the first person of Japanese ancestry ever elected to Congress. Inouye, who lost an arm in World War II, came to national prominence through the Senate Watergate investigation of 1973, when he was a member of the Select Committee on Presidential Campaign Activities. George R. Ariyoshi (b.1926), who was elected governor of Hawaii in 1974, was the first Japanese-American to serve as chief executive of a state.

Commanding figures in Hawaiian history are King Kamehameha I (1758?-1819), who unified the islands through conquest, and Kamehameha III (Kauikeaouli, 1813-54), who transformed Hawaii into a constitutional monarchy. Two missionaries who shaped Hawaiian life and politics were Hiram Bingham (b.Vermont, 1789-1869) and Gerrit Parmele Judd (b.New York, 1803-73). Sanford B. Dole (1844-1926) and Lorrin Andrews Thurston (1858-1931) were leaders of the revolutionary movement that overthrew Queen Liliuokalani (1838-1917), established a republic, and secured annexation by the US. Dole was the republic's first president and the territory's first governor. Another prominent historical figure is Bernice Pauahi Bishop (1831-88), of the Kamehameha line, who married an American banker and left her fortune to endow the Kamehameha Schools in Honolulu; the Bishop Museum was founded by her husband in her memory. Honolulu-born Luther Halsey Gulick (1865-1918), along with his wife, Charlotte Vetter Gulick (b.Ohio, 1865-1928), founded the Camp Fire Girls.

Don Ho (b.1930) is the most prominent Hawaiian-born entertainer; singer-actress Bette Midler (b.1945) was also born on the islands. Duke Kahanamoku (1889-1968), a swimmer and surfer, held the Olympic 100-meter free-style swimming record for almost 20 years.

50 BIBLIOGRAPHY

Daws, Gavan. *Shoal of Time: A History of the Hawaiian Islands.* Honolulu: University of Hawaii Press, 1974.

Fuchs, Lawrence H. *Hawaii Pono: A Social History.* San Diego: Harcourt Brace Jovanovich, 1984.

Hawaii, State of. Department of Planning and Economic Development. *The State of Hawaii Data Book 1984—A Statistical Abstract.* Honolulu, 1985.

Joesting, Edward. *Hawaii: An Uncommon History.* New York: Norton, 1972.

Kuykendall, Ralph S., and A. Grove Day. *Hawaii: A History—From Polynesian Kingdom to American State.* Rev. ed. Englewood Cliffs, N.J.: Prentice-Hall, 1961.

Wenkam, Robert. *Hawaii.* Chicago: Rand McNally, 1972.

IDAHO

State of Idaho

¹LOCATION, SIZE, AND EXTENT

Situated in the northwestern US, Idaho is the smallest of the 8 Rocky Mountain states and 13th in size among the 50 states.

The total area of Idaho is 83,564 sq mi (216,431 sq km), of which land comprises 82,412 sq mi (213,447 sq km) and inland water 1,152 sq mi (2,984 sq km). With a shape described variously as a hatchet, a snub-nosed pistol, and a pork chop, Idaho extends a maximum of 305 mi (491 km) E–W and 479 mi (771 km) N–S.

Idaho is bordered on the N by the Canadian province of British Columbia; on the NE by Montana; on the E by Wyoming; on the S by Utah and Nevada; and on the W by Oregon and Washington (with part of the line formed by the Snake River). The total boundary length of Idaho is 1,787 mi (2,876 km). The state's geographic center is in Custer County, SW of Challis.

²TOPOGRAPHY

Idaho is extremely mountainous. Its northern two-thirds consists of a mountain massif broken only by valleys carved by rivers and streams, and by two prairies: the Big Camas Prairie around Grangeville and the Palouse Country around Moscow. The Snake River Plain extends east–west across Idaho from Yellowstone National Park to the Boise area, curving around the southern end of the mountain mass. A verdant high–mountain area encroaches into the southeastern corner; the rest of Idaho's southern edge consists mostly of low, dry mountains. Among the most important ranges are the Bitterroot (forming the border with Montana), Clearwater (the largest range), Salmon River, Sawtooth, Lost River, and Lemhi mountains. More than 40 peaks rise above 10,000 feet (3,000 meters), of which the highest is Mt. Borah, at 12,662 feet (3,859 meters), in the Lost River Range. Idaho's lowest point is 710 feet (216 meters) near Lewiston, where the Snake River leaves the Idaho border and enters Washington.

The largest lakes are Pend Oreille (180 sq mi—466 sq km), Coeur d'Alene, and Priest in the panhandle, and Bear on the Utah border. The Snake River—one of the longest in the US, extending 1,038 mi (1,671 km) across Wyoming, Idaho, and Washington—dominates the southern part of the state. The Salmon River—the "River of No Return," a salmon-spawning stream that flows through wilderness of extraordinary beauty—separates northern from southern Idaho. The Clearwater, Kootenai, Bear, Boise, and Payette are other major rivers.

There are ice caves near Shoshone and American Falls, and a large scenic cave near Montpelier. Near Arco is an expanse of lava, craters, and caves called the Craters of the Moon, another scenic attraction. At Hell's Canyon in the northernmost part of Adams County, the Snake River cuts the deepest gorge in North America, 7,913 feet (2,412 meters) deep.

³CLIMATE

The four seasons are distinct in all parts of Idaho, but not simultaneous. Spring comes earlier and winter later to Boise and Lewiston, which are protected from severe weather by nearby mountains and call themselves "banana belts." Eastern Idaho has a more continental climate, with more extreme temperatures; climatic conditions there and elsewhere vary with the elevation. Mean temperatures in Boise range from 30°F (−1°C) in January to 75°F (24°C) in July. The record low, −60°F (−51°C), was set at Island Park Dam on 16 January 1943; the record high, 118°F (48°C), at Orofino on 28 July 1934. The corresponding extremes for Boise are −23°F (−31°C) and 111°F (44°C).

Humidity is low throughout the state. Precipitation in southern Idaho averages 13 in (33 cm) per year; in the north, over 30 in (76 cm). Boise gets more than 21 in (53 cm) of snow per year, with much greater accumulations in the mountains.

⁴FLORA AND FAUNA

With 10 life zones extending from prairie to mountaintop, Idaho has some 3,000 native plants. Characteristic evergreens are Douglas fir and western white pine (the state tree); oak/mountain mahogany, juniper/piñon, ponderosa pine, and spruce/fir constitute the other main forest types. Syringa is the state flower. MacFarlane's four-o'clock is listed as endangered.

Classified as game mammals are the elk, moose, mule and white-tailed deer, pronghorn antelope, bighorn sheep, mountain goat, black bear, mountain lion, cottontail, and pigmy rabbit. Several varieties of pheasant, partridge, quail, and grouse are the main game birds, and there are numerous trout, salmon, bass, and whitefish species in Idaho's lakes and streams. Rare animal species include the wolverine, kit fox, and pika. The grizzly bear is listed as threatened, while the woodland caribou, gray (timber) wolf, bald eagle, Arctic and American peregrine falcons, and whooping crane are endangered. There were six national wildlife refuges covering 79,069 acres (31,998 hectares) in the early 1980s.

⁵ENVIRONMENTAL PROTECTION

The environmental protection movement in Idaho dates from 1897, when President Grover Cleveland established the Bitterroot Forest Preserve, encompassing much of the northern region. In the early 1930s, the US Forest Service set aside some 3 million acres (1.2 million hectares) of Idaho's roadless forestland as primitive areas. The Taylor Grazing Act of 1934 regulated grazing on public lands, providing for the first time some relief from the overgrazing that had transformed much Idaho grassland into sagebrush desert. Thirty years later, Idaho Senator Frank Church was floor sponsor for the bill creating the National Wilderness System, which now contains most of the primitive areas earlier set aside. Many miles of Idaho streams are now in the Wild and Scenic Rivers System, another congressional accomplishment in which Senator Church played a leading role. In 1970, Governor Cecil Andrus (later, US secretary of the interior) was elected partly on a platform of environmental protection.

The Department of Health and Welfare's Division of Environment is responsible for enforcing environmental standards. Air quality improved greatly between 1978 and 1983, although the federal standard for maximum acceptable levels of suspended particles in the atmosphere continued to be exceeded at times in 1983 in the Pocatello, Soda Springs, and Lewiston areas. Vehicle emissions were responsible for high carbon monoxide levels in the Boise area. During 1982-83, 46 persons won $32 million in damages because of lead and other contaminants emitted from a mining smelter in Kellogg.

Water quality is generally good. Most of the existing problems stem from runoff from agricultural lands. Water quality is rated as only fair in the Upper Snake River Basin and in the Southwest Basin around Boise, and as poor in the Bear River Basin, partly because of municipal effluents from Soda Springs and Preston.

In the 1984 elections, Idahoans approved a constitutional amendment that would shift control of water resources from the Water Resources Board to the state legislature.

Since 1953, nuclear waste has been buried at the Idaho National Engineering Laboratory west of Idaho Falls or discharged in liquid form into the underground aquifer; some isotopes are migrating toward the boundaries of the site. Tailings from a former uranium-ore milling operation near Lowman are a potential health hazard. A top-priority site for hazardous-waste cleanup is Bunker Hill Mining at Smelterville; two sites in Pocatello are also considered candidates for cleanup.

⁶POPULATION

Idaho's population at the 1980 census was 943,038, 41st among the 50 states. The 32.4% population increase between 1970 and 1980 was the 7th highest among the states. In early 1985, the population was estimated at 1,004,071 (a 6.5% increase from 1980), yielding a density of 12.2 per sq mi (4.7 per sq km). The population projection for 1990 is 1,214,000.

The population in 1980 was nearly 96% white. The median age was 27.6, 6th youngest among the states, and the state was 54% "urban"—although no part of Idaho except Boise is genuinely urban, and even Boise does not have much of a central city. Boise's 1984 census population was estimated at 107,188, a 4.6% increase since the 1980 census; Pocatello was next with 45,334; Idaho Falls had 41,774, Twin Falls 28,168, Lewiston 28,050, and Nampa 27,347. Boise's metropolitan area (Ada County) had an estimated 189,273 inhabitants in mid-1984.

⁷ETHNIC GROUPS

The 1980 census counted 10,418 American Indians. There are five reservations; the most extensive is that of the Nez Percé in northern Idaho.

There is a very small population of black Americans (only 2,716 in 1980) and a somewhat larger number (5,948) of Asian-Pacific peoples, almost half of them (2,585) Japanese. There were 36,615 Hispanic Americans (about three-quarters of Mexican origin) and a very visible Basque community in the Boise area, with an organization devoted to preserving their language and culture.

The foreign-born (23,000) comprised about 2.5% of Idaho's population in 1980. Of persons reporting a single ancestry group, 171,154 named English, 90,268 German, and 33,420 Irish.

⁸LANGUAGES

Only a few place-names, such as Nampa, Pocatello, and Benewah, reflect in the general word stock the presence of Idaho Indians.

In Idaho, English is a merger of Northern and North Midland features, with certain Northern pronunciations marking the panhandle. More than 94% of the people 3 years old or older spoke only English in the home in 1980. The number of persons speaking other languages at home included the following:

| Spanish | 28,205 | French | 2,531 |
| German | 5,385 | Japanese | 1,187 |

⁹RELIGIONS

Roman Catholic and Presbyterian missionaries first came to Idaho between 1820 and 1840. The Church of Jesus Christ of Latter-day Saints (Mormon) has been the leading religion in Idaho since 1860; with about a quarter of the population, the number of Mormons in Idaho is 2d only to that in Utah. Catholicism predominates north of Boise. According to 1980 estimates, Idaho has about 240,843 Mormons, 25,536 members of various Lutheran denominations, and 22,303 United Methodists. In 1984 there were 74,356 Roman Catholics and an estimated 535 Jews.

¹⁰TRANSPORTATION

In 1982, Idaho had 68,395 mi (110,071 km) of roads and streets, 97% of them rural. The major east–west highways are I-90, I-84 (formerly I-80N), and US 12; US 95, Idaho 55, US 93, and I-15 are among the most traveled north–south routes. Idaho had 877,320 registered vehicles—including 529,018 automobiles, 345,348 trucks, and 2,954 buses—in 1983, when there were 648,108 licensed drivers. Only Wyoming had a smaller percentage of households without a motor vehicle than Idaho's 5.4% in 1980. Boise has the only mass transit system—a bus line.

There were 2,381 mi (3,832 km) of operating Class I rail lines in 1983; total rail trackage was 2,894 (4,657 km). Among the two major freight carriers, the Union Pacific Railroad serves southern Idaho, and the Burlington Northern crosses the panhandle; Amtrak provides limited passenger service to Pocatello, Boise, Shoshone, Nampa, and Sandpoint on two Chicago–Seattle trains; 1983/84 Idaho ridership was 47,980. Boise's modern airport—the busiest of the state's 178 airports, 15 heliports, and 3 seaplane bases—enplaned 409,914 passengers during 1983. Other transport facilities are 6,100 mi (9,800 km) of pipeline, carrying virtually all the natural gas and most of the gasoline consumed in Idaho, and a Snake River port at Lewiston that links Idaho, Montana, and the Dakotas with the Pacific via 464 mi (747 km) of navigable waterways in Washington State.

¹¹HISTORY

Human beings came to the land now known as Idaho about 15,000 years ago. Until 1805, only Indians and their ancestors had ever lived in the area, scratching a bare living from seeds and roots, insects, small animals, and what fishing and big-game hunting they could manage. At the time of white penetration, Shoshone and Northern Paiute lived in the south, and two linked tribal families, the Salishan and Shapwailutan (including the Nez Percé), lived in the north. It was the Nez Percé who greeted the Lewis and Clark expedition when it entered Idaho in 1805, and it was their food and canoes that helped these explorers reach the Columbia River and the Pacific.

Fur trappers—notably David Thompson, Andrew Henry, and Donald Mackenzie—followed within a few years. Missionaries came later; Henry Harmon Spalding founded a mission among the Nez Percé in 1836. The Oregon Trail opened in 1842, but for

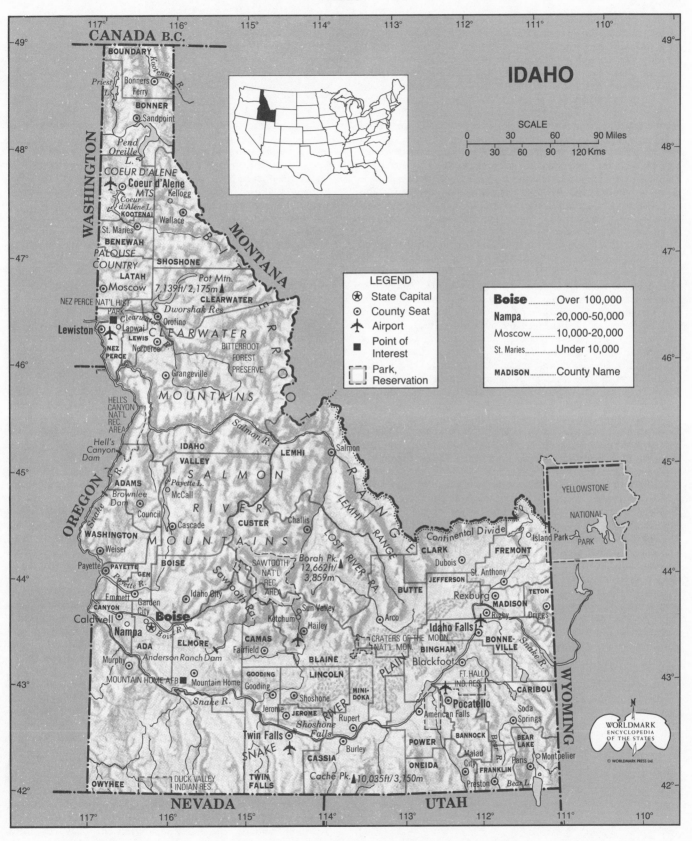

IDAHO

SCALE

0 — 30 — 60 — 90 Miles
0 — 30 — 60 — 90 — 120 Kms

LEGEND

✪ State Capital
⊙ County Seat
✈ Airport
■ Point of Interest
Park, Reservation

Boise .... Over 100,000
Nampa 20,000-50,000
Moscow 10,000-20,000
St. Maries .... Under 10,000
MADISON .... County Name

See endsheet maps: D2.

LOCATION: 42° to 49° N; 111° to 117° W. **BOUNDARIES:** Canadian line, 45 mi (72 km); Montana line, 738 mi (1,188 km); Wyoming line, 170 mi (274 km); Utah line, 153 mi (246 km); Nevada line, 152 mi (245 km); Oregon line, 319 mi (513 km); Washington line, 210 mi (338 km).

two decades, people merely crossed Idaho over it; virtually no one settled. In 1860, 14 years after Idaho had officially become US land through the Oregon Treaty with the United Kingdom, Mormons from Utah established Franklin, Idaho's first permanent settlement, and began farming. Gold was discovered that summer in northern Idaho; a gold rush, lasting several years, led directly to the organizing of Idaho Territory on 10 July 1863.

Boise became the capital of Idaho in 1864, and the following decade saw the inauguration of telegraph service, the linking of Franklin with the transcontinental railway, and the birth of the territory's first daily newspaper. Idaho's population nearly doubled between 1870 and 1880, and the pressure of white settlement impinging on Indian hunting and fishing grounds touched off a series of wars in the late 1870s. The most famous of those was the Nez Percé War, culminating in Chief Joseph's surrender in Montana on 5 October 1877 and in the subsequent confinement of Idaho Indians to reservations.

Lead and silver were discovered in south-central Idaho in 1880 and in the panhandle in 1884, touching off yet another stampede of would-be miners. With a population of 88,548 in 1890, Idaho was eligible to enter the Union, becoming the 43d state on 3 July. Statehood came to Idaho at a time of turmoil, when Mormons and non-Mormons were contending for political influence, the Populist Party was challenging the established political organizations, and violent labor disputes were sweeping the mining districts. In 1907, in a case that grew out of the labor conflict, William "Big Bill" Haywood (defended by Clarence Darrow) was acquitted on charges that he conspired to assassinate former Idaho Governor Frank Steunenberg, murdered on 30 December 1905.

From 1895 onward, federal land and irrigation projects fostered rapid economic growth. The modern timber industry began in 1906 with the completion of one of the nation's largest sawmills at Potlatch. By World War I, agriculture was a leading enterprise; however, a farm depression of the 1920s led to the Great Depression of the 1930s and ended only with the onset of World War II. After the war, an agro-industrial base was established, with fertilizers and potato processing leading the way. Population expansion and the push for economic growth collided with a new interest in the environment, creating controversies over land-use planning, mineral development, and water supply and dam construction. On 5 June 1976, the new, earth-filled Teton Dam in eastern Idaho collapsed, flooding the region and, in the process, claiming 10 lives and causing at least $400 million in damage to property and livestock.

12 STATE GOVERNMENT

Idaho's 1889 constitution, amended 102 times as of 1983, continues to govern the state today. The bicameral legislature, consisting

Idaho Presidential Vote by Major Political Parties, 1948–84

YEAR	ELECTORAL VOTE	IDAHO WINNER	DEMOCRAT	REPUBLICAN
1948	4	*Truman (D)	107,370	101,514
1952	4	*Eisenhower (R)	395,081	180,707
1956	4	*Eisenhower (R)	105,868	166,979
1960	4	Nixon (R)	138,853	161,597
1964	4	*Johnson (D)	148,920	143,557
1968	4	*Nixon (R)	389,273	165,369
1972	4	*Nixon (R)	380,826	199,384
1976	4	Ford (R)	126,549	204,151
1980	4	*Reagan (R)	110,192	290,699
1984	4	*Reagan (R)	108,510	297,523

*Won US presidential election.

of a 42-seat senate and an 84-member house of representatives, regularly meets for 60–90 days a year; special sessions, summoned by the governor, last 20 days. All legislators serve two-year terms. The executive branch is headed by seven elected officials: the governor and lieutenant governor (who run separately), secretary of state, attorney general, auditor, treasurer, and superintendent of public instruction. All serve four-year terms. The governor, who must be at least 30 years of age and must have been a state resident for at least two years prior to election, can sign or veto a bill or let it become law without his signature. Vetoes may be overridden by a two-thirds vote of each house.

The state constitution may be amended with the consent of two-thirds of each house and a majority of the voters at the next general election. Provisions for initiative, referendum, and recall were added by amendment to the state constitution in 1912 but not implemented by the legislature until 1933. The initiative procedure was employed in 1974 to pass the Sunshine Act, mandating registration by lobbyists and campaign financing disclosures by candidates for public office.

An Idaho voter must be at least 18 years of age, a US citizen, and a resident of the state for at least 30 days.

13 POLITICAL PARTIES

Registered voters numbered 541,164 in 1982; there is no party registration. Idahoans usually vote Republican in presidential elections, but sometimes elect Democrats to Congress or the statehouse. While the state has become increasingly conservative politically since the early 1960s, Democrats have been elected governor since 1970. The dominant Republican in the 20th century was US Senator William E. Borah, an isolationist-progressive who opposed US entry into the League of Nations but advocated world disarmament and supported prohibition, the graduated income tax, and some New Deal reforms; as chairman of the Senate Foreign Relations Committee from 1924 to 1940, he was one of the most influential legislators in the nation.

One measure of the conservatism of Idaho voters in the 1960s and 1970s is the showings by George Wallace's American Independent Party in 1968 (12.6% of the total vote) and the American Party in 1972 (9.3%, the highest of any state). In November 1980, Idahoans voted overwhelmingly for conservative Republican Ronald Reagan for president and denied a bid by Frank Church, a leading Democrat, for a fifth term in the US Senate. The Republicans also increased their majorities in both houses of the state legislature.

In 1982, a Democrat, John Evans, was reelected governor, and the party made some gains in the legislature, but 1984 was a banner year for the GOP. President Reagan carried every county and won 72% of the presidential vote (a higher proportion than in any other state but Utah), Republicans increased their heavy majority in the legislature, and Senator James McClure was easily reelected. Democrat Richard Stallings overcame the GOP tide, however, to win a US House seat, shortly after the GOP incumbent George Hansen was convicted of filing false financial disclosure statements. The other Republican incumbent, Larry Craig, was reelected.

In 1985, Majorie Ruth Moon, a Democrat, was state treasurer, and there were 24 female state legislators.

14 LOCAL GOVERNMENT

As of 1982, Idaho had 44 counties, 198 incorporated cities (only 11 of which had more than 10,000 residents), 117 school districts, and 659 special districts or authorities. Most counties elect three commissioners and other officers, usually including an assessor, treasurer, coroner, and sheriff. Nearly all cities have an elected mayor and council of 4 to 6 members. School districts have elected board members.

15 STATE SERVICES

Executive agencies concerned with education are the State Board of Education and the Department of Education. Under the heading of human resources are the departments of health and welfare, employment, correction, and law enforcement, which includes the Idaho State Police. Under the general rubric of natural resources come the departments of lands, water resources, fish and game, and parks and recreation. Self-governing agencies (7 commodity commissions and 15 professional licensing and regulating boards and commissions) and the departments of agriculture, finance, insurance, labor and industrial services, and transportation oversee economic development and regulation. Within the Executive Office of the Governor are a number of funds, divisions, boards, commissions, and other bodies.

[16] JUDICIAL SYSTEM

Idaho's highest court, the supreme court, consists of five justices, each elected at large, on a nonpartisan ballot, to a six-year term; the justice with the shortest remaining term automatically becomes chief justice. There is a 3-member court of appeals. The district court, with 33 judges, is the main trial court in civil and criminal matters, while magistrates' courts handle traffic, misdemeanor, and minor civil cases and preliminary hearings in felony cases. Like supreme court justices, appeals court justices and district court judges are elected by nonpartisan ballot, for six years and four years, respectively. Magistrates are appointed by a commission and run for four-year terms in the first general election succeeding the 18-month period following appointment.

Idaho's crime rates are low in almost every category; the total rate in 1983, 3,866 per 100,000 population, was 11th lowest among the states. The rate for the Boise metropolitan area was 4,755 per 100,000. A few murderers have been hanged, but none since the 1950s. The state permits, but has never used, execution by lethal injection. At the end of 1983 there were 1,206 inmates in state and federal prisons. Local jails held 604 inmates in mid-1983.

[17] ARMED FORCES

In 1983/84, 1,631 Navy personnel were on duty at the Idaho National Engineering Laboratory, where more than 50 experimental nuclear reactors have been built, among them the first practical nuclear-powered electrical generator, a nuclear-fuel breeder reactor, and a prototype engine for the nuclear-powered *Nautilus* submarine. Mountain Home Air Force Base, about 50 mi (80 km) southeast of Boise, had 3,918 officers and enlisted personnel. In all, Department of Defense personnel numbered 6,969 on 30 September 1984. Defense contract awards to Idaho firms in 1983/84 totaled $50.7 million, 2d lowest among the states.

Idaho casualties in recent US wars include 1,419 in World War II, 132 in Korea, and 187 in Viet-Nam. As of 30 September 1983, 122,000 veterans of military service were living in Idaho, including from World War I, 1,000; World War II, 43,000; the Korean conflict, 21,000; and the Viet-Nam era, 40,000. Benefits paid to Idaho veterans totaled $92 million in 1982/83.

The Army National Guard had 2,885 personnel at the end of 1984; the Air National Guard had 1,353 on 28 February 1985. In October 1983 there were 180 Idaho State Police and 2,183 local police and sheriff's officers.

[18] MIGRATION

Idaho's first white immigrants came from Utah, California, and Oregon in the early 1860s. By the end of the Civil War, the chief sources of immigrants were the southern and border states. Homesteaders from the Midwest, Utah, and Scandinavia arrived at the end of the 19th century.

Since 1960, immigrants have come largely from California. Idaho suffered a net loss from migration of 109,000 persons between 1940 and 1970, but had a net gain of 110,000 persons in the 1970s. Net gain from migration from 1981 through 1983 was estimated at 4,000.

[19] INTERGOVERNMENTAL COOPERATION

Idaho participates with Utah and Wyoming in the Bear River Compact; with Oregon, Washington, and Alaska in the Pacific Marine Fisheries Compact; with Wyoming in the Snake River Compact; with Washington, Oregon, Idaho, and Montana in the Northwest Power Planning Council; and in numerous other interstate compacts. Federal aid in 1983/84 was estimated at $413.2 million, of which $19 million was general revenue sharing.

[20] ECONOMY

Fur trapping was Idaho's earliest industry. Agriculture and mining began around 1860, with agriculture dominating since the 1870s. Timber became important after 1900, tourism and manufacturing—especially food processing and forest products—after 1945. Business and health services have grown in importance in recent years, while the forest-products industry was hard hit by the 1981–82 recession.

Total earnings in 1980 were divided as follows: government, 18%; wholesale and retail trade, 17%; manufacturing, 17%; services, 17%; agriculture, 12%; transportation, communications, and public utilities, 7%; contract construction, 7%; finance, insurance, and real estate, 4%; and mining, 1%.

[21] INCOME

Per capita income in Idaho in 1983 was $9,342, 43d in the US. Per capita income increased 25% between 1970 and 1980, compared to 24% for the nation, but fell 2.5% from 1980 through 1983, compared to a 3% gain for the nation. Total personal income reached almost $9.5 billion in 1983, representing an increase in constant dollars of 65% since 1970—one of the highest rates in the US for the period.

Median family money income ranked 36th at $17,492 in 1979, when nearly 13% of all state residents were below the federal poverty level. About 7,000 Idahoans had gross assets greater than $500,000 in 1982.

[22] LABOR

Of Idaho's average civilian labor force of 456,000 in 1983, about 43% were women and nearly 98% were white. By occupation, in 1980, 14.9% were administrative support workers, including clerical; 12.8% were service workers; 12.8% were precision production, craft, and repair workers; 11.4% were professional or specialty workers; 10.4% were executive, administrative, or managerial workers; 10.2% were salespeople; 8.9% were farm, forest, or fishery workers; 6.1% were machine operators, assemblers, or inspectors; 5.3% were transport and material movers; 4.6% were handlers, equipment cleaners, laborers, and helpers; and 2.6% were technicians and related support workers. Unemployment, which reached 10.7% in October 1982, averaged only 7.2% in 1984, slightly below the national average. Average annual pay was $15,242 in 1983, compared to the US average of $17,542.

A federal census in March 1983 revealed the following nonfarm employment pattern in Idaho:

	ESTABLISH-MENTS	EMPLOYEES	ANNUAL PAYROLL ('000)
Agricultural services, forestry, fishing	316	1,979	$ 18,377
Mining, of which:	116	3,432	108,825
Metals	(51)	(2,368)	(76,771)
Contract construction	1,940	14,946	384,480
Manufacturing, of which:	1,324	46,692	922,205
Food products	(145)	(14,850)	(222,391)
Lumber, wood products	(478)	(10,564)	(240,412)
Transportation, public utilities	1,034	14,882	293,171
Wholesale trade	2,056	21,431	324,815
Retail trade	6,241	55,469	511,405
Finance, insurance, real estate	1,920	14,718	229,208
Services, of which:	6,255	55,263	736,690
Business services	(620)	(9,619)	(211,472)
Health services	(1,537)	(17,608)	(252,245)
Other	2,359	2,170	42,952
TOTALS	23,561	230,982	$3,572,128

Government workers, not included in this survey, numbered 11,929 federal, 14,784 state, and 37,851 local in 1983. There were 7,335 year-round agricultural workers.

Idaho was a pioneer in establishing the eight-hour day and in outlawing yellow-dog contracts. By 1980 there were some 61,000 union members in Idaho, accounting for 18.5% of non-agricultural employment, lower than the national average. There were 234 unions in 1983. In 1958, Idaho voters rejected right-to-work legislation; Governor John Evans vetoed similar legislation in 1982.

23 AGRICULTURE

Receipts from farm marketings totaled $2 billion in 1983 (27th in the US); net farm income was about $327 million. As of 1983, Idaho led the US in potato production; was 3d in hops, sugar beets, and barley; 4th in dry edible beans; and 5th in mint.

Development of the russet potato in the 1920s gave Idaho its most famous crop. In 1983, the state produced 83,590,000 hundredweight of potatoes (25.7% of the US total), worth $434 million; some 90% were grown on about 110,000 acres (45,000 hectares) of irrigated land on the Snake River plain. About three-fourths of the crop is processed into frozen french fries, instant mashed potatoes, and other products. Other leading crops were hay, 4,914,000 tons (11th in the US), $348,894,000; wheat, 91,710,000 bushels (9th), $317,375,000; barley, 65,650,000 bushels, $187,103,000; and sugar beets, 3,487,000 tons ($118,370,000 in 1982).

As of 1983, Idaho had 15.1 million acres (6.1 million hectares) in farms, 29% of the state's land area; an estimated 24,400 farms averaged about 619 acres (251 hectares) each and employed about 18,000 persons in mid-1983. Almost 3.5 million acres (1.4 hectares) of land were irrigated.

24 ANIMAL HUSBANDRY

Livestock and livestock products account for 40–45% of Idaho farm marketings. By the close of 1983 there were about 1,890,000 beef cattle and calves, 174,000 dairy cows, 383,000 sheep and lambs, 120,000 hogs and pigs, and 1,265,000 chickens on the state's farms and ranches. Production of meat animals in 1983 included 722.1 million lb of cattle and calves, worth $513.9 million; sheep and lambs, 41.4 million lb, $23 million; and hogs and pigs, 45 million lb, $19.5 million. The wool clip comprised nearly 4.1 million lb (9th among the states).

Leading dairy and poultry products in 1983 were 2.3 billion lb of milk; 135.4 million lb of cheese; and 240 million eggs.

25 FISHING

Sport fishermen catch about 2 million lb of trout each year, along with salmon, steelhead, bass, and 32 other game-fish species. Idaho hatcheries stocked at least 32 million fish in 1980, mostly trout, salmon, and steelhead. The commercial catch amounted to only 420,000 lb, worth $72,000, in 1984.

26 FORESTRY

As of 1979, Idaho forests covered 21,727,000 acres (8,793,000 hectares), or 41% of the land area; of the 13,541,000 acres (5,480,000 hectares) classified as commercial forestland in 1977, the federal government controlled 71%. National forest system lands in Idaho totaled 20,431,000 acres (8,268,160 hectares) in 1984. Idaho forests are used increasingly for ski areas, hunting, and other recreation, as well as for timber and pulp. The total lumber production in 1983 was nearly 2 billion board feet (5th among the states), almost all softwoods. Shipments of lumber and wood products in 1983 were valued at $897.8 million.

27 MINING

Idaho led the US in 1983 in the production of silver (nearly 41% of the US total), antimony, and garnets, and ranked 2d in lead and 3d in vanadium. It was also among the leaders in gold, zinc, and phosphates. In 1984, the state's mineral output, valued at $424 million (20th in the US), included copper, 3,710 tons, and silver, 19,807 troy oz. Lead production was 25,726 tons in 1983.

There are hard-rock lead-silver mines in the Coeur d'Alene Mountains and phosphate strip mines in the southeast. Cement, clays, gypsum, lime, perlite, sand and gravel, and tungsten are also produced, and a new molybdenum mine began operating in 1983. Rising silver prices produced a brief silver boom in early 1980, but the Sunshine Mine in Kellogg, the largest silver mine in the US, closed in 1982. Despite declining prices, silver earned $163.9 million in 1984, or 39% of all earnings from nonfuel mineral production.

28 ENERGY AND POWER

Installed electrical capacity exceeded 2 million kw in 1983, when production (over 99.9% hydroelectric) totaled 12.8 billion kwh. About half of Idaho's irrigation depends on electric pumping, and electrical energy consumption regularly exceeds the state's supply. Large dams used to generate electricity include the Dworshak on the north fork of the Clearwater, the Anderson Ranch on the south fork of the Boise, and the Brownlee on the Snake.

Idaho's large size, widespread and relatively rural population, and lack of public transportation fosters reliance on motor vehicles and imported petroleum products. Natural gas is also imported. Hot water from thermal springs is used to heat buildings in Boise.

29 INDUSTRY

Resource industries—food processing and lumber production—form the backbone of manufacturing in Idaho. Value added by manufacture increased from $1.4 billion in 1977 to $2.1 billion in 1982. The value of shipments by manufacturers in 1982, totaling nearly $5.4 billion, is shown for major sectors in the following table:

Food and kindred products	$2,300,300,000
Lumber, wood products	897,800,000
Chemicals and allied products	623,100,000
Nonelectrical machinery	545,600,000

The growth area was nonelectrical machinery, which increased by over 500% in value added between 1977 and 1983; during this period, many northern California computer companies, including Hewlett Packard, opened or expanded plants in Idaho. By contrast, the value added by the lumber and wood products industry fell almost by half during this period.

Ore-Ida Foods is a leading potato processor, and J. R. Simplot engages in food processing and fertilizer production. Boise Cascade (with headquarters at Boise), Potlatch, and Louisiana-Pacific dominate the wood-products industry; Morrison-Knudsen, a diversified engineering and construction company that also has forest-products interests, has its headquarters in Boise.

30 COMMERCE

In 1982, Idaho's wholesale establishments registered nearly $5.1 billion in trade. Retail sales exceeded $4 billion, of which food stores accounted for 26%, automotive dealers 20%, eating and drinking places 9%, gasoline service stations 9%, and department stores 6%. Boise is the headquarters of the 434-store Albertson's supermarket chain, which in 1984 was the 19th-largest retailing company in the US, with $4.7 billion in sales in 17 western and southern states.

About two-thirds of Idaho's wheat crop and a substantial amount of its fertilizer, peas, lentils, beans, potatoes, and barley are exported abroad; in all, one-quarter of Idaho's farm sales was exports in 1981/82. Natural gas and sulfate are imported from Canada in significant quantities. Foreign exports of agriculture products were valued at $551 million in 1981/82 (27th in the US); manufactured exports totaled $460 million (40th in 1981).

31 CONSUMER PROTECTION

The Attorney General's Office is responsible for investigating consumer complaints and enforcing the state Consumer Credit Code and other consumer laws. The Department of Finance investigates and resolves consumer credit complaints.

32BANKING

The Idaho First National Bank, first in fact as well as name, was chartered in 1867; at the end of 1983, it remained the state's biggest bank, with assets of $2.6 billion. As of 31 December 1983, 25 insured commercial banks (18 state-chartered) had $6.4 billion in assets, $5.2 billion in deposits, and $3.8 billion in outstanding loans. The state's 9 savings and loan associations (2 state-chartered) had nearly $1.4 million in assets at the end of 1983 and $801 million in outstanding mortgage loans. Idaho also had 83 state-chartered credit unions with about $179.5 million in assets at the end of 1983, and 128 credit unions in all.

33INSURANCE

The Department of Insurance regulated 1,366 insurance companies at the end of 1982. Idaho families held, on average, $47,500 in life insurance coverage in 1983. A total of 1,342,000 policies in effect that year had a combined value of $17.5 billion; payouts reached $162.3 million, including $45.8 million in death payments. Property and liability companies wrote $72.5 million for automobile physical damage insurance, $87.3 million for automotive liability insurance, and $38 million for homeowners' coverage in 1983.

34SECURITIES

Although Idaho has no stock exchanges, New York Stock Exchange member firms had 34 sales offices and 173 registered representatives in the state in 1983. About 136,000 Idahoans were shareholders of public corporations in that year.

35PUBLIC FINANCE

Idaho's annual budget, prepared by the Division of Financial Management, is submitted by the governor to the legislature for amendment and approval. The fiscal year runs from 1 July to 30 June.

The following table summarizes proposed revenues and expenditures for 1985/86 in the governor's budget message of December 1984:

REVENUES	
Personal income tax	$ 276,700,000
Corporate income tax	52,600,000
Sales tax	221,800,000
Product taxes	14,200,000
Other general revenues	34,700,000
Federal funds	425,500,000
Dedicated funds	228,600,000
Other funds	193,400,000
TOTAL	$1,447,500,000

EXPENDITURES	
Public schools	$ 463,000,000
Other education	212,500,000
Human resources	397,300,000
Economic development	233,900,000
Natural resources	63,400,000
General government	52,700,000
Constitutional offices	24,600,000
TOTAL	$1,447,400,000

The outstanding debt of the Idaho state and local governments was $932 million, or $988 per capita, as of 30 September 1982—2d and 3d lowest in the US, respectively.

36TAXATION

Idaho's original revenue base of property taxes and a variety of local business license fees has been substantially abandoned. The state instituted an income tax in 1931 and a sales tax in 1965; as of 1984, the personal income tax ranged from 2% of the first $1,000 to 7.5% on income over $5,000, the corporate income tax was 7.7%, and the general sales tax was 4%. The state also levies taxes on inheritances, alcoholic beverages, cigarettes and tobacco products, motor fuels, insurance premiums, hotel/motel rooms and campgrounds, ores mined and extracted, oil and gas produced, and electric utilities. Property taxes, the only major source of local revenue, brought in about $230 million in 1981/82. Per capita state and local taxes came to only $859 in 1981/82, 44th among the states.

In 1982/83, Idaho paid federal taxes totaling $1.8 billion and received federal outlays of $2.4 billion, a highly favorable ratio. Idahoans filed 361,380 federal income tax returns for 1983, paying $745,377,000 in tax.

37ECONOMIC POLICY

The Division of Economic and Community Affairs, within the Office of the Governor, seeks to widen markets for Idaho products and goods and services, encourage film production in the state, attract new business and industry to Idaho, expand and enhance existing enterprises, and promote the state travel industry. Incentives for investment include conservative state fiscal policies and a probusiness regulatory climate.

38HEALTH

Idaho's infant mortality rate of 9.2 per 1,000 births was below the national average in 1981 but increased to 10.6 in 1983. The live birthrate in 1983 was 19.2 per 1,000 population; the 1982 birthrate of 20.1 was 5th highest among the states. The death rate was 7.3 in 1983; the 1982 death rate of 7.2 was 8th lowest. There were 2,456 legal abortions in Idaho in 1983. Death rates for accidents, pneumonia and influenza, suicide, atherosclerosis, bronchitis, emphysema, and asthma were above the respective national averages in 1983. The high birthrate and low death rate reflect Idaho's younger-than-average population, and to some degree its large Mormon population.

Idaho in 1985 had 1 state mental hospital, psychiatric programs at the other state hospital, and 7 community mental health centers. In all, 52 hospitals had 4,088 beds in 1983; with 135,183 admissions; hospital personnel included 2,463 registered nurses in 1981. The average cost of hospital care in 1982 was $266 per day and $1,732 per stay, among the lowest in the US. There were 1,255 licensed physicians and 496 dentists in 1982, and 5,789 registered nurses in 1980.

39SOCIAL SERVICES

In 1983/84, about 2% of Idaho's population (an average of 20,992 people) received public assistance (excluding medical assistance). In 1980, per capita expenditures for public welfare came to $123, 42d in the US. Participants of aid to families with dependent children (AFDC) averaged 18,280, with the average payment $98.63 per person, in September 1984. Total payments to AFDC recipients came to $20 million in 1982. Those receiving medical assistance averaged 17,753 in 1983/84. Medicaid had 39,000 recipients in 1983; payments totaled $67 million. In 1982/83, 28,000 people participated in the food stamp program at a federal cost of $40 million; 114,000 took part in the school lunch program at a federal cost of $9 million. The state's vocational rehabilitation program had a 1983/84 spending level of almost $6 million. Persons receiving unemployment insurance averaged 18,000 a week in 1981/82; benefits came to $126 million during the fiscal year. Workers' compensation payments came to $44.9 million in 1982.

A total of 135,932 Idahoans were beneficiaries of $625 million in Social Security payments during 1983, with an average monthly benefit for retired workers of $409.96 in December 1982. A total of 7,542 individuals received Supplemental Security Income (SSI) payments in December 1983. Federal SSI payments for that year totaled $14.9 million.

40HOUSING

Single-family housing predominates in Idaho. In 1980 there were 356,432 units, of which 72% were single detached units, and 65% were owner-occupied. The median value of owner-occupied units ($45,900) and median gross monthly rent ($219) were below the national averages. From 1981 through 1983, 4,200 private units worth $201 million were authorized.

41 EDUCATION

Idaho's state and local per-pupil expenditure on education, $2,198 in 1983/84, is one of the lowest in the US. Nevertheless, as of 1980, nearly 74% of Idahoans over 25 were high school graduates, well above the national average.

As of fall 1984, public educational institutions enrolled 118,647 elementary school students (including kindergarten) and 92,053 secondary school (7th through 12th grade). The enrollment totals for nonpublic schools were 3,914 and 1,867, respectively. Idaho's 10 institutions of higher learning had 42,975 students in the fall of 1982; an 11th opened in the fall of 1983. The leading public higher educational institutions are the University of Idaho at Moscow, with 9,237 students in 1983/84; Idaho State University (Pocatello), with an enrollment of 6,041; and Boise State University, 11,236. There were 3 other public colleges and community colleges and 5 private institutions. The State Board of Education offers scholarships to graduates of accredited Idaho high schools.

42 ARTS

The Boise Philharmonic is Idaho's leading professional orchestra; other symphony orchestras are in Coeur d'Alene, Moscow, Pocatello, and Twin Falls. Boise and Moscow have seasonal theaters. The Idaho Commission on the Arts and Humanities, founded in 1966, offers grants to support both creative and performing artists. It has also cut records of folk music, mounted a folk art exhibit, and prepared a slide-tape series on Idaho folk life and folk art.

43 LIBRARIES AND MUSEUMS

Idaho's 78 public libraries had a combined book stock of nearly 2,153,746 volumes in 1982/83 and a total circulation of more than 4,889,074. The largest public library system was the Boise Public Library and Information Center, with about 213,558 volumes; the leading academic library, at the University of Idaho (Moscow), had 405,185.

The state also has 31 museums, notably the Boise Gallery of Art, Idaho State Historical Museum (Boise), and the Idaho Museum of Natural History (Pocatello). The University of Idaho Arboretum is at Moscow, and there is a zoo at Boise and an animal park in Idaho Falls. Major historical sites include Cataldo Mission near Kellogg, Spalding Mission near Lapwai, and Nez Percé National Historical Park in north-central Idaho.

44 COMMUNICATIONS

As of 1980, 93% of Idaho's 324,107 occupied housing units had telephones. There were 251 post offices with 1,976 employees on 29 March 1985. Idaho's first radio station, built by a Boise high school teacher and his students, began transmitting in 1921, was licensed in 1922, and six years later was sold and given the initials KIDO—the same call letters as Idaho's first permanent television station, which began broadcasting in 1953 and subsequently became KTVB. As of 1984, the state had 91 operating radio stations (44 AM, 47 FM) and 10 commercial television stations. Another 3 television stations—in Boise, Moscow, and Pocatello—were public. Sixty-three cable systems served 149,514 Idahoans in 129 communities during 1984.

45 PRESS

Idaho, site of the first printing press in the Northwest, had 12 daily newspapers in 1983, with a combined circulation of 183,774, and 7 Sunday papers, with 193,064 circulation. There were 48 weeklies. The most widely read newspaper was the (morning) *Idaho Statesman,* published in Boise, with a circulation of 56,540 daily and 70,678 Sundays in September 1984. Caxton Printers, founded in 1902, is the state's leading publishing house.

46 ORGANIZATIONS

The 1982 US Census of Service Industries counted 288 organizations in Idaho, including 76 business associations; 150 civic, social, and fraternal associations; and 2 educational, scientific, and research associations. Among the few national organizations with headquarters in Idaho are the Food Industries Suppliers Association (Caldwell) and the Appaloosa Horse Club (Moscow).

47 TOURISM, TRAVEL, AND RECREATION

In 1982, travel and tourism generated $1.1 billion in business revenues and employed 27,200 persons on a payroll of $173 million. Tourists come to Idaho primarily for outdoor recreation—river trips, skiing, camping, hunting, fishing, and hiking. There are more than two dozen ski resorts, of which by far the most famous is Sun Valley, which opened in 1936. Licenses were issued to 249,928 hunters and 425,717 fishermen in 1982/83.

Tourist attractions include two US parks, the Craters of the Moon National Monument and the Nez Percé National Historical Park, and the Hell's Canyon and Sawtooth national recreational areas. A sliver of Yellowstone National Park is also in Idaho. There were 25 state parks in the early 1980s, covering 41,713 acres (16,881 hectares).

48 SPORTS

Idaho has no major league professional teams. Most county seats hold pari-mutuel quarterhorse racing a few days a year, and Boise's racing season (including Thoroughbreds) runs three days a week for five months. World chariot racing championships have been held at Pocatello, and polo was one of Boise's leading sports from 1910 through the 1940s. Idaho cowboys have won numerous riding, roping, and steer-wrestling championships. Skiing and golf are among the most popular participant sports.

49 FAMOUS IDAHOANS

Leading federal officeholders born in Idaho include Ezra Taft Benson (b.1899), secretary of agriculture from 1953 to 1961, and Cecil D. Andrus (b.Oregon, 1931), governor of Idaho from 1971 to 1977 and secretary of the interior from 1977 to 1981. Maverick Republican William E. Borah (b.Illinois, 1865–1940) served in the US Senate from 1907 until his death. Frank Church (1924–84) entered the US Senate in 1957 and became chairman of the Senate Foreign Relations Committee in 1979; he was defeated in his bid for a fifth term in 1980. Important state officeholders were the nation's first Jewish governor, Moses Alexander (b.Germany, 1853–1932), and New Deal governor C. Ben Ross (1876–1946).

Author Vardis Fisher (1895–1968) was born and spent most of his life in Idaho, which was also the birthplace of poet Ezra Pound (1885–1972). Nobel Prize–winning novelist Ernest Hemingway (b.Illinois, 1899–1961) is buried at Ketchum. Gutzon Borglum (1871–1941), the sculptor who carved the Mt. Rushmore National Memorial in South Dakota, was an Idaho native. Idaho is the only state in the US with an official seal designed by a woman, Emma Edwards Green (b.California, 1856–1942).

Baseball slugger Harmon Killebrew (b.1936) and football star Jerry Kramer (b.1936) are Idaho's leading sports personalities.

50 BIBLIOGRAPHY

Beal, Merrill D., and Merle W. Wells. *History of Idaho.* 3 vols. New York: Lewis, 1959.

Etulain, Richard W., and Merwin Swanson. *Idaho History: A Bibliography.* Pocatello: Idaho State University Press, 1975.

Federal Writers' Project. *Idaho: A Guide in Word and Picture.* Reprint. New York: Somerset, n.d. (orig. 1937).

Idaho, State of. Secretary of State. *Idaho Blue Book, 1981–82.* Boise, 1981.

Peterson, F. Ross. *Idaho: A Bicentennial History.* New York: Norton, 1976.

Walker, Deward E., Jr. *American Indians of Idaho.* Moscow: University of Idaho Press, 1971.

Young, Virgil. *The Story of Idaho.* Moscow: University of Idaho Press, 1984.

ILLINOIS

State of Illinois

ILLINOIS

ORIGIN OF STATE NAME: French derivative of *Iliniwek,* meaning "tribe of superior men," an Indian group formerly in the region. **NICKNAME:** The Prairie State. **SLOGAN:** Land of Lincoln. **CAPITAL:** Springfield. **ENTERED UNION:** 3 December 1818 (21st). **SONG:** "Illinois." **MOTTO:** State Sovereignty–National Union. **FLAG:** The inner portion of the state seal and the word "Illinois" on a white field. **OFFICIAL SEAL:** An American eagle perched on a boulder holds in its beak a banner bearing the state motto; below the eagle is a shield resting on an olive branch. Also depicted are the prairie, the sun rising over a distant eastern horizon, and, on the boulder, the dates 1818 and 1868, the years of the seal's introduction and revision, respectively. The words "Seal of the State of Illinois Aug. 26th 1818" surround the whole. **ANIMAL:** White-tailed deer. **BIRD:** Cardinal. **FLOWER:** Violet. **TREE:** White oak. **MINERAL:** Fluorite. **INSECT:** Monarch butterfly. **LEGAL HOLIDAYS:** New Year's Day, 1 January; Birthday of Martin Luther King, Jr., 3d Monday in January; Lincoln's Birthday, 12 February; George Washington's Birthday, 3d Monday in February; Memorial Day, last Monday in May; Independence Day, 4 July; Labor Day, 1st Monday in September; Columbus Day, 2d Monday in October; Election Day, 1st Tuesday after the 1st Monday in November in even-numbered years; Veterans Day, 11 November; Thanksgiving Day, 4th Thursday in November; Christmas Day, 25 December. **TIME:** 6 AM CST = noon GMT.

¹LOCATION, SIZE, AND EXTENT

Situated in the eastern north-central US, Illinois ranks 24th in size among the 50 states. Its area totals 56,345 sq mi (145,934 sq km), of which land comprises 55,645 sq mi (144,120 sq km) and inland water 700 sq mi (1,814 sq km). Illinois extends 211 mi (340 km) E–W; its maximum N–S extension is 381 mi (613 km).

Illinois is bounded on the N by Wisconsin; on the E by Lake Michigan and Indiana (with the line in the SE defined by the Wabash River); on the extreme SE and S by Kentucky (with the line passing through the Ohio River); and on the W by Missouri and Iowa (with the entire boundary formed by the Mississippi River).

The state's boundaries total 1,297 mi (2,088 km). The geographic center of Illinois is in Logan County, 28 mi (45 km) NE of Springfield.

²TOPOGRAPHY

Illinois is flat. Lying wholly within the Central Plains, the state exhibits a natural topographic monotony relieved mainly by hills in the northwest (an extension of Wisconsin's Driftless Area) and throughout the southern third of the state, on the fringes of the Ozark Plateau. The highest natural point, Charles Mound, tucked into the far northwest corner, is only 1,235 feet (376 meters) above sea level—far lower than Chicago's towering skyscrapers. The low point, at the extreme southern tip along the Mississippi River, is 279 feet (85 meters) above sea level. The average elevation is about 600 feet (180 meters).

Although some 2,700 rivers and streams totaling 9,000 mi (14,500 km) crisscross the land, pioneers in central Illinois confronted very poor drainage. The installation of elaborate and expensive networks of ditches and tiled drains was necessary before commercial agriculture became feasible. Most of the 2,000 lakes of 6 acres (2.4 hectares) or more were created by dams. The most important rivers are the Wabash and the Ohio, forming the southeastern and southern border; the Mississippi, forming the western border; and the Illinois, flowing northeast-southwest across the central region and meeting the Mississippi at Grafton, just northwest of the junction between the Mississippi and the Missouri rivers. The artificial Lake Carlyle (41 sq mi—106 sq km) is the largest body of inland water. Illinois also has jurisdiction over 1,526 sq mi (3,952 sq km) of Lake Michigan.

³CLIMATE

Illinois has a temperate climate, with cold, snowy winters and hot, wet summers—ideal weather for corn and hogs. The seasons are sharply differentiated: mean winter temperatures are 22°F (−6°C) in the north and 37°F (3°C) in the south; mean summer temperatures are 70°F (21°C) in the north and 77°F (25°C) in the south. The record high, 117°F (47°C), was set at East St. Louis on 14 July 1954; the record low, −35°F (−37°C), was registered at Mt. Carroll on 22 January 1930.

The average farm sees rain one day in three, for a total of 36 in (91 cm) of precipitation a year. An annual snowfall of 37 in (94 cm) is normal for northern Illinois, decreasing to 24 in (61 cm) or less in the central and southern regions. Chicago's record 90 in (229 cm) of snow in the winter of 1978/79 created monumental transportation problems, enormous personal hardship, and even a small political upheaval when incumbent Mayor Michael Bilandic lost a primary election to Jane Byrne in February 1979 partly because of his administration's slowness in snow removal.

Chicago is nicknamed the "Windy City" because in the 1800s New York journalists labeled Chicagoans as "the windy citizenry out west" and called some Chicago leaders "loudmouth and windy"—not because of fierce winds. In fact, the average wind speed, 10.3 mph (16.6 km/hr), is lower than that of Boston, Honolulu, Cleveland, and 16 other major US cities.

⁴FLORA AND FAUNA

Urbanization and commercial development have taken their toll on the plant and animal resources of Illinois. Northern and central Illinois once supported typical prairie flora, but nearly all the land has been given over to crops, roads, and suburban lawns. About 90% of the oak and hickory forests that once were common in the north have been cut down for fuel and lumber. In the forests that do remain, mostly in the south, typical trees are black oak, sugar maple, box elder, slippery elm, beech, shagbark hickory, white ash, sycamore, black walnut, sweet gum, cottonwood, black willow, and jack pine. Characteristic wildflowers are the Chase aster, French's shooting star, lupine, primrose violet, purple trillium, small fringed gentian, and yellow fringed orchid. Tamarack and ginseng are considered threatened, and the small-whorled pogonia is endangered.

Before 1800, wildlife was abundant on the prairies, but the

bison, elk, bear, and wolves that once roamed freely have long since vanished. The white-tailed deer (the state animal) disappeared in 1910 but was successfully reintroduced in 1933 by the Department of Conservation. Among the state's fur-bearing mammals are opossum, raccoon, mink, red and gray foxes, and muskrat. More than 350 birds have been identified, with such game birds as ruffed grouse, wild turkey, and bobwhite quail especially prized. Other indigenous birds are the cardinal (the state bird), horned lark, blue jay, purple martin, black-capped chickadee, tufted titmouse, bluebird, cedar waxwing, great crested flycatcher, and yellow-shafted flicker. Mallard and black ducks are common, and several subspecies of Canada goose are also found. The state claims 17 types of native turtle, 46 kinds of snake, 19 varieties of salamander, and 21 types of frog and toad. Heavy industrial and sewage pollution have eliminated most native fish, except for the durable carp and catfish. Coho salmon were introduced into Lake Michigan in the 1960s, thus reviving sport fishing.

In 1973, the Department of Conservation established an endangered and threatened species protection program. Included among threatened animals are the river otter, bobcat, Swainson's warbler, western hog-nosed snake, and lake sturgeon. Endangered species include the gray and Indiana bats, eastern woodrat, white-tailed jackrabbit, little blue heron, red-shouldered hawk, greater prairie chicken, barn owl, bigeye chub, bluebreast darter, dusky salamander, and Higgins' eye pearly mussel.

5 ENVIRONMENTAL PROTECTION

The history of conservation efforts in Illinois falls into three stages. From 1850 to the 1930s, city and state parks were established and the beauty of Chicago's lakefront was successfully preserved. During the next stage, in the 1930s, federal intervention through the Civilian Conservation Corps and other agencies focused on upgrading park facilities and, most important, on reversing the severe erosion of soils, particularly in the hilly southern areas. Soil conservation laws took effect in 1937, and within a year the first soil conservation district was formed. By 1970, 98 districts, covering 44% of the state's farmland, promoted conservation cropping systems, contour plowing, and drainage.

The third stage of environmentalism began in the late 1960s, when Attorney General William J. Scott assumed the leadership of an antipollution campaign; he won suits against steel mills, sanitary districts, and utility companies, and secured passage of clean air and water legislation. The Illinois Environmental Protection Act of 1970 created the Pollution Control Board to set standards and conduct enforcement proceedings, and the Environmental Protection Agency to establish a comprehensive program for protecting environmental quality. In 1980, the Department of Nuclear Safety was established. The federal Environmental Protection Agency has also helped upgrade water and air quality in Illinois.

The 1970s saw a noticeable improvement in environmental quality. Dirty air became less prevalent—though one by-product was the crippling of the state's high-sulfur coal industry. The Illinois EPA maintains about 250 air-monitoring stations, of which some 100 are in the Chicago area, and conducts about 3,000 facility inspections each year. Since Illinois produces about 600 million gallons of hazardous wastes annually (2d only to New Jersey), the state agency tries to pinpoint and clean up abandoned hazardous waste sites; in 1984, Illinois began a three-year, $20 million program to eliminate the 22 worst such sites and to evaluate nearly 1,000 other potential hazardous waste sites. In that year, the state EPA investigated 1,373 reports of emergency environmental mishaps, an alarming increase of 25% over 1983; most of these involved truck, railroad, and pipeline accidents.

6 POPULATION

At the 1980 census, Illinois ranked 5th among the 50 states, with a population of 11,426,518, having ceded 3d place to California by 1950 and 4th place to Texas during the 1960s. The preliminary estimated 1985 population was 11,502,433, or 207 persons per sq mi (80 per sq km).

The population of Illinois was only 12,282 in 1810. Ten years later, the new state had 55,211 residents. The most rapid period of growth came in the mid-19th century, when heavy immigration made Illinois one of the fastest-growing areas in the world. Between 1820 and 1860, the state's population doubled every 10 years. The rate of increase slowed somewhat after 1900, especially during the 1930s, although the population more than doubled between 1900 and 1960. Population growth was very slow in the 1970s, about 0.3% a year; the rate of growth from 1980 to 1984 was less than 0.2% annually.

The age distribution of the state's population in 1983 closely mirrored the national pattern, with 27% under age 18 and more than 11% aged 65 or older. The number of households was 4,228,000 in 1983, up from 3,502,000 in 1970; nearly all the increase was in families headed by women and in households composed of unmarried persons. The number of husband-wife households fell slightly from 2,405,000 in 1970 to 2,378,000 in 1980. Illinois's population was 48% male and 52% female in 1980.

The rapid rise of Chicago meant that a large proportion of the state's population was concentrated in cities from a relatively early date. Thus, by 1895, 50% of Illinoisans lived in urban areas, whereas the entire country reached that point only in 1920. By 1980, 82% of the population lived in metropolitan areas, compared with 76% nationally; the new lure of rural areas kept these proportions constant as of 1983. With an estimated population of 8,035,000 in mid-1984, Greater Chicago was the 3d-largest metropolitan area in the nation and alone accounted for more than three-fifths of the total state population. The state's other major metropolitan areas, with their 1984 populations, were Peoria, 355,900, and Rockford, 278,800. The largest city that year was Chicago, with 2,992,472 residents, followed by Rockford, 136,531; Peoria, 117,113; Springfield, 101,570; Decatur, 91,851; Joliet, 87,670; Aurora, 85,735; and Evanston, 72,074.

7 ETHNIC GROUPS

The Indian population of Illinois had disappeared by 1832 as a result of warfare and emigration. By 1980, however, Indian migration from Wisconsin, Minnesota, and elsewhere had brought the Native American population to 15,833, concentrated in Chicago.

French settlers brought in black slaves from the Caribbean in the mid-18th century; in 1752, one-third of the small non-Indian population was black. Slavery was slowly abolished in the early 19th century. For decades, however, few blacks entered the state, except to flee slavery in neighboring Kentucky and Missouri. Freed slaves did come to Illinois during the Civil War, concentrating in the state's southern tip and in Chicago. By 1900, 109,000 blacks lived in Illinois. Most held menial jobs in the cities or eked out a precarious existence on small farms in the far south. Large-scale black migration, mainly to Chicago, began during World War I. By 1940, Illinois had a black population of 387,000; extensive wartime and postwar migration brought the total in 1980 to 1,675,229, of whom about three-fourths lived within the city of Chicago, which was about 40% black. Smaller numbers of black Illinoisans lived in Peoria, Rockford, and certain Chicago suburbs.

Illinois's Hispanic population did not become significant until the 1960s. In 1980, the number of Hispanic Americans was 636,000, chiefly in Chicago. There were 167,924 persons of Mexican origin, about 88,930 Puerto Ricans, and 14,539 Cubans;

LOCATION: 36°58′ to 42°30′N; 87°30′ to 91°32′W. BOUNDARIES: Wisconsin line, 185 mi (298 km); Indiana line, 397 mi (639 km); Kentucky line, 134 mi (216 km); Missouri line, 367 mi (591 km); Iowa line, 214 mi (344 km).

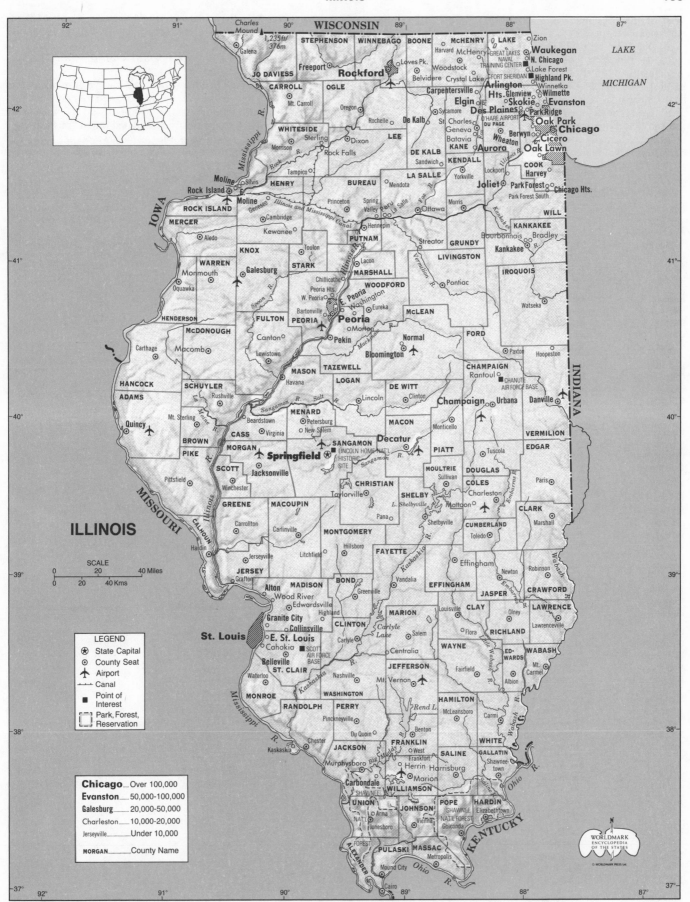

ILLINOIS

SCALE
0   20   40 Miles
0   20   40 Kms

LEGEND
⊛ State Capital
⊙ County Seat
✈ Airport
╂ Canal
■ Point of Interest
⌐⌐ Park, Forest, Reservation

Chicago____Over 100,000
Evanston____50,000–100,000
Galesburg____20,000–50,000
Charleston____10,000–20,000
Jerseyville____Under 10,000

MORGAN____County Name

most of the remainder came from other Caribbean and Latin American countries.

In 1980 there were 26,343 Chinese in Illinois, 17,253 Japanese, 37,973 Filipinos, 22,884 Koreans, and 5,329 Vietnamese refugees.

Members of non-British European ethnic groups are prevalent in all the state's major cities and in many farming areas. In 1980, 823,696 persons were foreign-born, including 340,501 Europeans. The largest groups were Poles (64,293), Germans (54,715), Italians (48,866), and Russians (25,685). There were also significant numbers of Scandinavians, Irish, Lithuanians, Serbs, East European Jews, Ukrainians, Slovaks, Hungarians, Czechs, Greeks, and

Dutch. Except for the widely dispersed Germans, most of these ethnic groups lived in and around Chicago. A 1980 survey of national origins of Illinois's white population showed that 19% were of German ancestry, 11% English, 8% Irish, 7% Polish, 5% Italian, and 3% Scandinavian.

Most ethnic groups in Illinois maintain their own newspapers, clubs, festivals, and houses of worship. These reminders of their cultural heritage are now largely symbolic for the European ethnics, who have become highly assimilated into a "melting pot" society. Such was not always the case, however. In 1889, the legislature attempted to curtail foreign-language schools, causing

Illinois Counties, County Seats, and County Areas and Populations

COUNTY	COUNTY SEAT	LAND AREA (SQ MI)	POPULATION (1984 EST.)	COUNTY	COUNTY SEAT	LAND AREA (SQ MI)	POPULATION (1984 EST.)
Adams	Quincy	852	70,774	Lee	Dixon	725	34,338
Alexander	Cairo	236	11,896	Livingston	Pontiac	1,046	40,844
Bond	Greenville	377	16,086	Logan	Lincoln	619	30,750
Boone	Belvidere	282	29,032	Macon	Decatur	581	128,597
Brown	Mt. Sterling	306	5,470	Macoupin	Carlinville	865	49,395
Bureau	Princeton	869	38,108	Madison	Edwardsville	728	247,508
Calhoun	Hardin	250	5,845	Marion	Salem	573	44,657
Carroll	Mt. Carroll	444	18,118	Marshall	Lacon	388	13,934
Cass	Virginia	374	14,535	Mason	Havana	536	18,264
Champaign	Urbana	998	170,264	Massac	Metropolis	241	14,883
Christian	Taylorville	710	36,319	McDonough	Macomb	590	36,519
Clark	Marshall	505	16,778	McHenry	Woodstock	607	156,489
Clay	Louisville	469	15,542	McLean	Bloomington	1,185	124,094
Clinton	Carlyle	472	33,442	Menard	Petersburg	315	11,686
Coles	Charleston	509	52,371	Mercer	Aledo	559	19,381
Cook	Chicago	958	5,270,411	Monroe	Waterloo	388	20,846
Crawford	Robinson	446	20,933	Montgomery	Hillsboro	705	32,028
Cumberland	Toledo	346	10,953	Morgan	Jacksonville	568	37,228
DeKalb	Sycamore	634	73,360	Moultrie	Sullivan	325	14,594
DeWitt	Clinton	397	17,979	Ogle	Oregon	759	45,835
Douglas	Tuscola	417	19,616	Peoria	Peoria	620	193,107
DuPage	Wheaton	337	701,382	Perry	Pinckneyville	442	22,305
Edgar	Paris	623	21,372	Piatt	Monticello	439	16,446
Edwards	Albion	223	8,185	Pike	Pittsfield	830	18,677
Effingham	Effingham	478	31,508	Pope	Golconda	374	4,364
Fayette	Vandalia	709	22,317	Pulaski	Mound City	203	8,619
Ford	Paxton	486	15,049	Putnam	Hennepin	160	6,180
Franklin	Benton	414	43,140	Randolph	Chester	583	35,684
Fulton	Lewistown	871	41,225	Richland	Olney	360	18,447
Gallatin	Shawneetown	325	7,639	Rock Island	Rock Island	423	165,296
Greene	Carrollton	543	16,092	Saline	Harrisburg	385	28,747
Grundy	Morris	423	31,331	Sangamon	Springfield	866	178,460
Hamilton	McLeansboro	436	9,220	Schuyler	Rushville	436	8,102
Hancock	Carthage	795	23,713	Scott	Winchester	251	6,026
Hardin	Elizabethtown	181	5,460	Shelby	Shelbyville	747	23,757
Henderson	Oquawka	373	9,157	Stark	Toulon	288	6,985
Henry	Cambridge	824	54,441	St. Clair	Belleville	672	267,794
Iroquois	Watseka	1,118	32,641	Stephenson	Freeport	564	49,756
Jackson	Murphysboro	590	61,382	Tazewell	Pekin	650	129,538
Jasper	Newton	496	11,468	Union	Jonesboro	414	18,193
Jefferson	Mt. Vernon	570	38,207	Vermilion	Danville	900	93,172
Jersey	Jerseyville	373	20,320	Wabash	Mt. Carmel	224	14,220
Jo Daviess	Galena	603	23,341	Warren	Monmouth	543	21,206
Johnson	Vienna	346	10,053	Washington	Nashville	563	15,241
Kane	Geneva	524	294,277	Wayne	Fairfield	715	18,662
Kankakee	Kankakee	678	100,146	White	Carmi	497	18,205
Kendall	Yorkville	322	36,769	Whiteside	Morrison	682	64,234
Knox	Galesburg	720	58,703	Will	Joliet	844	332,511
Lake	Waukegan	454	460,798	Williamson	Marion	427	58,551
La Salle	Ottawa	1,139	109,203	Winnebago	Rockford	515	249,770
				Woodford	Eureka	527	33,249
Lawrence	Lawrenceville	374	18,307	TOTALS		55,645	11,512,061

a sharp political reaction among German Lutherans, German Catholics, and some Scandinavians. The upshot was the election of a German-born Democrat, John Peter Altgeld, as governor in 1892. During World War I, anti-German sentiment was intense in the state, despite the manifest American loyalty of the large German element, then about 25% of the state's population. The Germans responded by rapidly abandoning the use of their language and dissolving most of their newspapers and clubs. At about the same time, the US government, educators, social workers, and business firms sponsored extensive "Americanization" programs directed at the large numbers of recent arrivals from Poland, Italy, and elsewhere. The public schools especially played a major role in the assimilation process, as did the Catholic parochial schools, which sought to protect the immigrants' religious but not ethnic identities.

8LANGUAGES

A number of place-names—Illinois itself, Chicago, Peoria, Kankakee, and Ottawa—attest to the early presence of various Algonkian-speaking tribes, such as the Kickapoo, Sauk, and Fox, and particularly those of the Illinois Federation, the remnants of which moved west of the Mississippi River after the Black Hawk War of 1832.

Nineteenth-century western migration patterns determined the rather complex distribution of regional language features. Excepting the Chicago metropolitan area and the extreme northwestern corner of Illinois, the northern quarter of the state is dominated by Northern speech. An even greater frequency of Northern features appears in the northeastern quadrant; in this region, speakers get *sick to the stomach, catch cold* (take cold), use *dove* as the past tense of dive, pronounce *hog, fog, frog, crop,* and *college* with the vowel /ah/, and sound a clear /h/ in *whine, wheel,* and *wheat.*

Settlement from Pennsylvania and Ohio led to a mix of Northern and North Midland speech in central Illinois, with such dominating Northern features as *white bread, pail, greasy* with an /s/ sound, and *creek* rhyming with *stick.* Here appear Midland *fishworm* (earthworm), *firebug* (firefly), *wait on* (wait for), *dived* as the past tense of *dive, quarter till four* (3:45), and *sick at one's stomach* (but *sick on the stomach* in German communities near East St. Louis).

Migration from South Midland areas in Indiana and Kentucky affected basic speech in the southern third of Illinois, known as Egypt. Here especially occur South Midland and Southern *pullybone* (wishbone), *dog irons* (andirons), *light bread* (white bread), and, in extreme southern counties, *loaf bread, snake doctor* (dragonfly), *redworm* (earthworm), *ground squirrel* (chipmunk), *plum peach* (clingstone peach), *to have a crow to pick* (to have a bone to pick) with someone, and the pronunciations of *coop* with the vowel of *put* and of *greasy* with a /z/ sound. Such speech is found also in the northwestern corner around Galena, where Kentucky miners who came to work in the lead mines brought such pronunciations as *bulge* with the vowel of *put, soot* with the vowel of *but,* and /yelk/ for *yolk.*

Metropolitan Chicago has experienced such complex in-migration that, although it still has a basic Northern/Midland mix, elements of almost all varieties of English appear somewhere. The influx since World War II of speakers of black English, a Southern dialect, and of nonstandard Appalachian English has aggravated language problems in the schools. Foreign-language schools were common in the 1880s and 1890s, but by 1920 all instruction was in English. The policy of monolingual education came into question in the 1970s, when the state legislature mandated bilingual classes for immigrant children, especially Spanish speakers.

In Chicago, rough-and-tumble politics has created a new meaning for *clout; prairie* means a vacant lot, *porch* includes the meaning of *stoop,* and *cornbread* has been generalized to include the meanings of *corn pone* and *hush puppies.* A fuel and food stop on the Illinois tollway system is an *oasis.*

In 1980, English was spoken at home by 88% of all state residents 3 years of age and older. Speakers of other languages were as follows:

Spanish	524,280	Greek	46,556
Polish	139,414	French	36,048
German	99,120	Chinese	24,283
Italian	85,898	Serbo-Croatian	22,453

9RELIGIONS

Before 1830, little religion of any sort was practiced on the Illinois frontier. Energetic Protestant missionaries set out to evangelize this un-Christian population, and they largely succeeded. By 1890, 36% of the adults in Illinois were affiliated with evangelical denominations—chiefly Methodist, Disciples of Christ, Baptist, Congregationalist, and Presbyterian—while 35%, mostly immigrants, belonged to liturgical denominations (chiefly Roman Catholic, Lutheran, and Episcopal). The remaining adults acknowledged no particular denomination. Illinois has had episodes of religious bigotry: at Carthage in 1844 the Mormon founder Joseph Smith was killed by a mob, and strong but brief waves of anti-Catholicism developed in the 1850s (the "Know-Nothing" movement) and 1920s (the Ku Klux Klan). Nevertheless, tolerance of religious diversity has been the norm for most of the state's history.

Today, the largest Christian denomination is the Roman Catholic Church, with 3,569,350 members in 1984, 66% of whom were in the archdiocese of Chicago. The largest Protestant denomination in 1980 was the United Methodist Church, with 505,406 adherents, followed by the Lutheran Church–Missouri Synod, 313,425; Southern Baptist Convention, 264,480; United Presbyterian Church, 194,180; Lutheran Church in America, 194,115; and United Church of Christ, 191,865. Protestants are most numerous in the Chicago suburbs and in smaller cities and towns downstate. The Jewish population was estimated at 261,320 in 1984, with 95% concentrated in Chicago and its northern suburbs. Chicago is one of the main theological centers of the country, with 15 seminaries and numerous religious publications.

10TRANSPORTATION

The fact that Illinois is intersected by several long-distance transportation routes has been of central importance in the state's economic development for a century and a half. Easy access by way of the major rivers and the Great Lakes system facilitated extensive migration to Illinois even before the coming of the railroads in the 1850s. Most of the nation's rail lines converge on Illinois, and Chicago and St. Louis (especially East St. Louis) have been the two main US railroad centers since the late 19th century. Interstate highways, notably the main east–west routes, also cross the state, and Chicago's central location has made it a major transfer point for airline connections.

After several false starts in the 1830s and 1840s, the state's railroad system was begun in the 1850s. The Illinois Central (now the Illinois Central Gulf), aided by the first federal land grants, opened up the prairie lands in the years before the Civil War. By 1890, about 10,000 mi (16,000 km) of track crisscrossed the state, placing 90% of all Illinois farms no more than 5 mi (8 km) from a rail line. The railroads stimulated not only farming but also coal mining, and in the process created tens of thousands of jobs in track and bridge construction, maintenance, traffic operations, and the manufacture of cars, rails, and other railroad equipment.

The rise of automobile and truck traffic in the 1920s and 1930s dealt the railroads a serious blow, but their utter ruin was staved off by complex mergers that incorporated bankrupt or threatened lines into ever–larger systems. By 1974, the state had 10,607 mi (17,070 km) of track, 2d only to Texas. Shedding their unprofitable passenger business in the 1970s (except for important commuter lines around Chicago), the railroads concentrated on long-distance freight traffic. The bankruptcy of the Penn Central, Rock Island, and Milwaukee Road systems during the 1970s impelled some companies, notably the Illinois Central Gulf and the Chicago & North Western, to shift their attention to real estate and manufacturing. Chicago is the hub of Amtrak's passenger service, which operated 28 train routes through Illinois in 1984. There were 16 railroad companies operating 8,971 mi (14,437 km) of track within the state at the end of 1983.

Mass transit is of special importance to Chicago, where subways, buses, and commuter railroads are essential to daily movement. The transit systems were built privately but eventually were acquired by the city and regional transportation authorities. Ridership declines every year, as fewer people work in the central city and as more people choose the privacy and convenience of travel by automobile. Federal aid to mass transit, beginning in 1964, and state aid, initiated in 1971, have only partly stemmed the decline. Outside Chicago, bus service is still available in some of the older, larger cities.

The road system of Illinois was inadequate until the 1920s, when an elaborate program to build local and trunk highways first received heavy state aid. In 1983, 134,599 mi (216,616 km) of roadway served 7,513,118 registered vehicles—including 5,929,783 automobiles and 1,561,330 trucks—operated by 6,984,733 licensed drivers. I-90, I-80, I-70, and I-64 are the main east–west routes. I-94 links Chicago with Milwaukee to the north, while I-57 and I-55 connect Chicago with the south and southwest (St. Louis), respectively. The Interstate Highway System totaled 1,691 mi (2,721 km) in the state as of 30 September 1983.

Barge traffic along the Mississippi, Ohio, and Illinois rivers remains important, especially for the shipment of grain. The Port of Chicago no longer harbors the sailing ships that brought lumber, merchandise, and people to a fast-growing city. However, the port is still the largest on the Great Lakes, handling mostly grain and iron ore. Midway Airport in Chicago became the world's busiest after World War II, but was superseded by O'Hare Airport, which opened in the late 1950s. With about 103,000 travelers a day arriving or departing in 1983, O'Hare is not only the largest but also the most hectic airport in the country. Efforts to expand its cramped facilities made headway in the early 1980s. With 735 airports and 167 heliports, Illinois is also an important center for general aviation. There were 29,348 licensed pilots and 7,700 operating aircraft in the state in 1983.

11HISTORY

Different tribes of paleo-Indians lived in Illinois as long ago as 8000 BC. By 2000 BC, the cultivation of plants and use of ceramics were known to village dwellers; the first pottery appeared during the Woodland phase, a millennium later. Between 500 BC and AD 500, skilled Hopewellian craftsmen practiced a limited agriculture, developed an elaborate social structure, and constructed burial mounds. Huge mounds, which still exist, were built along the major rivers by the Middle Mississippian culture, about AD 900.

It is not known why the early native civilizations died out, but by the time white explorers arrived in the 17th century, the state was inhabited by seminomadic Algonkian-speaking tribes. The Kickapoo, Sauk, and Fox lived in the north, while the shores of Lake Michigan were populated by the Potawatomi, Ottawa, and Ojibwa. The Kaskaskia, Illinois (Iliniwek), and Peoria tribes roamed across the central prairies, and the Cahokia and Tamoroa lived in the south. Constant warfare with tribes from neighboring areas, plus disease and alcohol introduced by white fur traders and settlers, combined to decimate the Native American population. Warfare with the whites led to a series of treaties, the last in 1832, that removed all of the Indians to lands across the Mississippi.

French missionaries and fur traders from Quebec explored the rivers of Illinois in the late 17th century. Father Jacques Marquette and trader Louis Jolliet were the first to reach the area now known as the State of Illinois in 1673, when they descended the Mississippi as far as the Arkansas River and then returned by way of the Illinois River. The first permanent settlement was a mission built by French priests at Cahokia, near present-day St. Louis, in 1699. It was followed by more southerly settlements at Kaskaskia in 1703 and Ft. Chartres in 1719. In 1765, pursuant to the Treaty of Paris (1763) that ended the French and Indian War, the British took control of the Illinois country, but they established no settlements of their own. Most of the French settlers were Loyalists during the American Revolution. However, they put up no resistance when Virginia troops, led by George Rogers Clark, captured the small British forts at Cahokia and Kaskaskia in 1778. Virginia governed its new territory in desultory fashion, and most of the French villagers fled to Missouri. In 1784, Virginia relinquished its claim to Illinois, which three years later became part of the newly organized Northwest Territory. In 1800, Illinois was included in the Indiana Territory. Nine years later the Illinois Territory, including the present state of Wisconsin, was created; Kaskaskia became the territorial capital, and Ninian Edwards was appointed territorial governor by President James Madison. A territorial legislature was formed in 1812. During the War of 1812, British and Indian forces combined in a last attempt to push back American expansion into the Illinois country, and much fighting took place in the area. On 3 December 1818, Illinois was formally admitted to the Union as the 21st state. The capital was moved to Vandalia in 1820 and to Springfield in 1839.

Apart from a few thousand nomadic Indians and the remaining French settlers and their slaves, Illinois was largely uninhabited before 1815; two years after statehood, the population barely exceeded 55,000. The withdrawal of British influence after the War of 1812 and the final defeat of the Indian tribes in the Black Hawk War of 1832 opened the fertile prairies to settlers from the south, especially Kentucky. The federal government owned most of the land, and its land offices did a fast business on easy terms. Before the 1830s, most of the pioneers were concerned with acquiring land titles and pursuing subsistence agriculture, supplemented by hunting and fishing. An effort in 1824 to call a constitutional convention to legalize slavery failed because of a widespread fear that rich slaveholders would seize the best land, squeezing out the poor yeoman farmers. Ambitious efforts in the 1830s to promote rapid economic development led to fiscal disaster. Three state banks failed; a lavish program of building roads, canals, and railroads totally collapsed, leaving a heavy state debt that was not paid off until 1880. Despite these setbacks, the steady influx of land-hungry poor people and the arrival after 1840 of energetic Yankee entrepreneurs, all attracted by the rich soil and excellent water routes, guaranteed rapid growth.

Although Illinois gradually eliminated French slavery and even served as a conduit to Canada for slaves escaping from the South, the state was deeply divided over the slavery issue and remained unfriendly territory for blacks and their defenders. The abolitionist leader Elijah P. Lovejoy was killed in Alton in 1837, and as late as 1853 the state passed legislation providing that free blacks entering Illinois could be sold into slavery. In 1856, however, the new Republican Party nominated and Illinois voters elected a governor, William H. Bissell, on a reform program that included support for school construction, commercial and industrial expansion, and abolition of slavery. During the Civil War, Illinois sent half its young men to the battlefield and supplied the Union armies with huge amounts of food, feed, and horses. The strong-handed wartime administration of Republican Governor Richard Yates guaranteed full support for the policies of Abraham Lincoln, who had been prominent in Illinois political life since the 1840s and had been nominated for the presidency in 1860 at a Republican convention held in Chicago. Democratic dissenters were suppressed, sometimes by force, leaving a legacy of bitter feuds that troubled the "Egypt" section (the southern third of the state) for decades thereafter.

Economic and population growth quickened after 1865, as exemplified by the phenomenal rise of Chicago to become the principal city of the Midwest. Responding to opportunities presented by the coming of the railroads, boosters in hundreds of small towns and cities built banks, grain elevators, retail shops, small factories, ornate courthouses, and plain schools, in an abundance of civic pride. The Democrats sought the support of the working class and small farmers, assuming an attitude of hostility toward banks, high railroad freight rates, protective tariffs, and antiunion employers, but they failed to impose any significant restraints on business expansion. They were more successful, however, in opposing prohibition and other "paternalistic" methods of social control demanded by reformers such as Frances Willard, a leader in the Women's Christian Temperance Union, and the Prohibition Party. In Chicago and other cities the Democrats were less concerned with social reform than with building lucrative political machines on the backs of the poor Irish, Polish, and Czech Catholic immigrants, who kept arriving in large numbers. Statewide, Illinois retained a highly competitive two-party system, even as the excitement and high voter turnouts characteristic of 19th-century elections faded rapidly in the early 20th century.

During the second half of the 19th century, Illinois was a center of the American labor movement. Workers joined the Knights of Labor in the 1870s and 1880s and fought for child-labor laws and the eight-hour day. Union organizing led to several spectacular incidents, including the Haymarket riot in 1886 and the violent Pullman strike in 1894, suppressed by federal troops at the behest of President Grover Cleveland. A coalition of Germans, labor, and small farmers elected John Peter Altgeld to the governorship in 1892. After the turn of the century, Illinois became a center of the Progressive movement, led by Jane Addams and Republican Governor Frank Lowden. Lowden reorganized the state government in 1917 by placing experts in powerful positions in state and municipal administrations.

After the great fire of 1871 destroyed Chicago's downtown section (but not its main residential or industrial areas), the city's wealthy elite dedicated itself to rebuilding Chicago and making it one of the great metropolises of the world. Immense steel mills, meat-packing plants, and factories sprang up, and growth was spectacular in the merchandising, banking, and transportation fields. Their fortunes made, Chicago's business leaders began building cultural institutions in the 1890s that were designed to rival the best in the world: the Chicago Symphony, the Art Institute, and the Field Museum of Natural History. The World's Columbian Exposition of 1893 was a significant international exhibition of the nation's technological achievements, and it focused worldwide attention on what was by then the 2d-largest American city. A literary renaissance, stimulated by the new realism that characterized Chicago's newspapers, flourished for a decade or two before World War I, but the city was recognized chiefly for its contributions in science, architecture, and (in the 1920s) jazz.

The first three decades of the 20th century witnessed almost unbroken prosperity in all sections except Egypt, the downstate region where poor soil and the decline of the coal industry produced widespread poverty. The slums of Chicago were poor, too, because most of the hundreds of thousands of new immigrants arrived virtually penniless. After 1920, however, large-scale immigration ended, and the immigrants' steady upward mobility, based on savings and education, became apparent. During the prohibition era, a vast organized crime empire rose to prominence, giving Chicago and Joliet a reputation for gangsterism, violence, and corruption; the most notorious gangster was Al Capone. Money, whether legally or illegally acquired, mesmerized Illinois in the 1920s as never before—and never since.

The Great Depression of the 1930s affected the state unevenly, with agriculture hit first and recovering first. Industries began shutting down in 1930 and did not fully recover until massive military contracts during World War II restored full prosperity. The very fact of massive depression brought discredit to the probusiness Republican regime that had run the state with few exceptions since 1856. Blacks, white ethnics, factory workers, and the undereducated, all of whom suffered heavily during the early years of the Depression, responded enthusiastically to Franklin Roosevelt's New Deal. They elected Henry Horner, a Democrat, to the governorship in 1932, reelected him in 1936, and flocked to the new industrial unions of the Congress of Industrial Organizations, founded in 1938.

World War II and its aftermath brought prosperity, a sense of national unity and purpose, and new anxiety about national security in a nuclear age. Most Illinoisans adopted hard-line anti-Communist attitudes in foreign policy, and many (outside the city of Chicago) transferred their allegiance back to the Republican Party. The goals of personal security and prosperity, encapsulated in the dream of owning a house (preferably in the burgeoning suburbs) and a car, holding a steady job, and providing a good education for one's children, dominated Illinois life in the postwar decades. The chilling events of the 1960s and 1970s—assassinations, the Viet-Nam war, the race riots, and the violence that accompanied the 1968 Democratic National Convention in Chicago—coupled with a new awareness of poverty, alienation, and environmental pollution, helped reshape many people's attitudes in Illinois. The problems attendant on heavy industrialization, particularly air and water pollution and urban decay, began to be addressed for the first time. These changing attitudes contributed toward bringing disadvantaged groups into the mainstream of American life. This transformation was perhaps best exemplified in Chicago, where voters elected Jane Byrne the city's first woman mayor in 1979 and chose Harold Washington as its first black mayor in 1983.

[12] STATE GOVERNMENT

Illinois has had four constitutions. The first, written in 1818, was a short document modeled on those of New York, Kentucky, and Ohio. An attempt to rewrite the charter to allow slavery failed in a bitterly contested referendum in 1824. A new constitution in 1848 democratized government by providing for the popular election of judges. A third constitution, enacted in 1870, lasted a century; its unique feature was a voting system for the lower house of the state legislature that virtually guaranteed minority-party representation in each electoral district. Important amendments in 1884 and 1904, respectively, gave the governor an item veto over appropriation bills and provided a measure of home rule for Chicago. In 1970, a fourth constitution streamlined state offices somewhat, improved accounting procedures, reformed the state tax system, and gave the state, rather than local governments, the major responsibility for financing education. The state bill of rights was expanded to include provisions banning discrimination in housing and employment and recognizing women's rights. An elected judiciary and the state's unique representational system were retained.

Under the 1970 constitution, as amended, the upper house of the general assembly consists of a senate of 59 members, who are elected on a two-year cycle to four-year terms. Until 1980, the lower house, the house of representatives, consisted of 177 members, with 3 representatives elected for two-year terms from each district. Each voter was empowered to cast three ballots for representatives, giving one vote to each of three candidates, one and a half votes to each of two, or all three to one candidate; each party never nominated more than two candidates in any single district. In November 1980, however, Illinois voters chose to reduce the size of house membership to 118 (2 representatives from each district) and to eliminate the proportional system.

The executive officers elected statewide are the governor and lieutenant governor (who run jointly), secretary of state, treasurer, comptroller, and attorney general. Each serves a four-year term and is eligible for reelection. An important revision of appointive offices in 1917 made most agency heads responsible to the governor. In the 1970s, the governor's office expanded its control over the budget and the higher education complex, further augmenting an already strong executive position. The governor must be a US citizen, at least 25 years of age, and must have been a state resident for three years prior to election.

Bills passed by both houses of the legislature become law if signed by the governor, if left unsigned for 60 days while the legislature is in session or 90 days after it adjourns, or if vetoed by the governor but passed again by three-fifths of the elected members of each house. Constitutional amendments require a three-fifths vote by the legislature for placement on the ballot; either a simple majority of those voting in the election or three-fifths of those voting on the amendment is sufficient for ratification.

Qualified voters must be US citizens at least 18 years of age. There is a 30-day district residency requirement.

13POLITICAL PARTIES

The Republican and Democratic parties have been the only major political groups in Illinois since the 1850s. Illinois is a closely balanced state, with a slight Republican predominance from 1860 to 1930 giving way in seesaw fashion to a highly competitive situation statewide. In Chicago and Cook County, an equally balanced division before 1930 gave way to heavy Democratic predominance forged during the New Deal.

The Democrats, organized by patronage-hungry followers of President Andrew Jackson in the 1830s, dominated state politics to the mid-1850s. They appealed to subsistence farmers, former southerners, and poor Catholic immigrants. Though they advocated minimal government intervention, Democratic officials were eager for the patronage and inside deals available in a fast-growing state. Their outstanding leader, Stephen Douglas, became a major national figure in the 1850s, but never lost touch with his base of support. After Douglas died in 1861, many Illinois Democrats began to oppose the conduct of the Civil War and became stigmatized as "Copperheads." The success of the Repub-

lican war policies left the Democrats in confusion in the late 1860s and early 1870s. Negative attitudes toward blacks, banks, railroads, and prohibition kept a large minority of Illinoisans in the Democratic fold, while the influx of Catholic immigrants replenished the party's voter base. However, the administration of Governor Altgeld (1893–97), coinciding with a deep depression and labor unrest, split the party, and only one other Democrat held the governorship between 1852 and 1932. The intraparty balance between Chicago and downstate changed with the rise of the powerful Cook County Democratic organization in the 1930s. Built by Mayor Anton Cermak and continued from 1955 to 1976 by six-term Mayor Richard J. Daley, the Chicago Democratic machine totally controlled the city, dominated the state party, and exerted enormous power at the national level. However, the machine lost its clout with the election in 1979 of independent Democrat Jane Byrne as Chicago's first woman mayor, and again in 1983 when Harold Washington became its first black mayor.

The Republican Party, born amid the political chaos of the 1850s, brought together most former Whigs and some Democrats who favored industrialization and opposed slavery. Abraham Lincoln, aided by many talented lieutenants, forged a coalition of commercial farmers, businessmen, evangelical Protestants, skilled craftsmen, professionals, and later, patronage holders and army veterans. Ridiculing the Democrats' alleged parochialism, the GOP called for vigorous prosecution of the Civil War and Reconstruction and for an active policy of promoting economic growth by encouraging railroads and raising tariffs. However, such moralistic crusades as the fight for prohibition frequently alienated large voting blocs (especially the Germans) from the Republicans.

In the early 20th century, Republican politicians built their own ward machines in Chicago and succumbed to corruption. William "Big Bill" Thompson, Chicago's Republican mayor in the 1910s and 1920s, openly allied himself with the gangster Al Capone. Moralistic Republicans, who were strongest in the smaller towns, struggled to regain control of their party. They succeeded in the 1930s, when the Republican political machines in Chicago collapsed or switched their allegiance to the Democrats. Since then, the Republicans have become uniformly a party of the middle and upper-middle classes, hostile to machine politics, welfare, and high taxes, but favorable to business, education, and environmental protection. Although the GOP has a stronger

Illinois Presidential Vote by Political Parties, 1948–84

YEAR	ELECTORAL VOTE	ILLINOIS WINNER	DEMOCRAT	REPUBLICAN	SOCIALIST LABOR	PROHIBITION	COMMUNIST	SOCIALIST
1948	28	*Truman (D)	1,994,715	1,961,103	3,118	11,959	—	11,522
1952	27	*Eisenhower (R)	2,013,920	2,457,327	9,363	—	—	—
1956	27	*Eisenhower (R)	1,775,682	2,623,327	8,342	—	—	—
1960	27	*Kennedy (D)	2,377,846	2,368,988	10,560	—	—	—
1964	26	*Johnson (D)	2,796,833	1,905,946	—	—	—	—
						AMERICAN IND.		
1968	26	*Nixon (R)	2,039,814	2,174,774	13,878	390,958	—	—
						AMERICAN		
1972	26	*Nixon (R)	1,913,472	2,788,179	12,344	2,471	4,541	—
						LIBERTARIAN		SOC. WORKERS
1976	26	Ford (R)	2,271,295	2,364,269	2,422	8,057	9,250	3,615
						CITIZENS		
1980	26	*Reagan (R)	1,981,413	2,358,094	10,692	38,939	9,711	1,302
1984	24	*Reagan (R)	2,086,499	2,707,103	2,716	10,086	—	—

*Won US presidential election.

formal organization in Illinois than in most other states, its leading candidates have exuded an aura of independence. Republican James R. Thompson, elected to the governorship in 1976 and reelected in 1978 and 1982, has served in that office longer than any other.

The Whigs usually ran a close second to the Democrats from 1832 to 1852. Taken over in the 1840s by a group of professional organizers under Lincoln's leadership, the Whigs simply vanished after their crushing defeat in 1852. Notable among the smaller parties was the Native American ("Know-Nothing") Party, which controlled Chicago briefly in the 1850s. The Prohibitionists, Greenbackers, Union Labor, and Populist parties were weak forces in late 19th-century Illinois. The Socialist Party, strongest among coal miners and central European immigrants, grew to a minor force in the early 20th century and elected the mayor of Rockford for many years.

Illinois provided two important leaders of the national GOP in the 1860s—Abraham Lincoln and Ulysses S. Grant. The only major-party presidential nominee from the state between 1872 and 1976, however, was Governor Adlai Stevenson, the unsuccessful Democratic candidate in 1952 and 1956. In 1980, three native-born Illinoisans pursued the Republican Party nomination. The first, US Representative Philip Crane, was the earliest to declare his candidacy but failed in the primaries. The second, US Representative John Anderson, dropped out of the GOP primaries to pursue an independent candidacy, ultimately winning more than 6% of the popular vote nationally and in Illinois, but no electoral votes. The third, Ronald Reagan, a native of Tampico, won both the Republican nomination and the November election, becoming the 40th president of the US; he was reelected by a heavy majority of Illinois voters in 1984.

There is no party registration. In 1985, Republicans held the governorship, but Democrats held both US Senate seats and 13 of 22 US House seats.

14LOCAL·GOVERNMENT

Illinois has more units of local government (most with property-taxing power) than any other state. In 1984 there were 102 counties, 1,280 municipalities, 1,434 townships, 1,049 school districts, and 2,602 special districts.

County government in Illinois dates from 1778, when Virginia, claiming authority over the territory, established the earliest counties. The major county offices are elective: the county judge (now a judicial officer, but formerly with administrative duties), the county clerk (chief administrative officer), clerk of the circuit court, sheriff, state's attorney, treasurer, coroner, and superintendent of schools. Cook County controls hospital and welfare programs in Chicago, thus spreading the cost over both the city's own tax base and that of the more affluent suburbs. The New England township system was made optional by the 1848 constitution, and eventually 85 counties, including Cook County, adopted the idea. Townships, which elect local judges and administrators, also handle tax collection.

Chicago is governed by an elected mayor, clerk, treasurer, and city council composed of 50 aldermen. The mayor's power depends more on control of the city's Democratic Party organization than on formal authority. Independent candidates get elected to the city council from time to time, but the Democratic machine generally staffs the city with its own members. After the election in 1983 of independent Democrat Harold Washington as mayor, however, differences between him and the machine-dominated city council over patronage and other policies severely hampered the city's administration.

Other cities may choose either the commission or aldermanic system; most are administered by nonpartisan city managers.

15STATE SERVICES

Officials responsible to the governor of Illinois, the mayor of Chicago, and the members of Congress actively provide ombuds-man service, although there is no state office by that name. Illinois has a Board of Ethics, but the US attorney's office in Chicago has far more potent weapons at its disposal: many top political leaders were indicted and convicted in the 1970s, including federal judge and former Governor Otto Kerner and, in 1980, Attorney General William Scott.

Educational services provided by the Illinois Office of Education include teacher certification and placement, curriculum development, educational assessment and evaluation, and programs for the disadvantaged, gifted, handicapped, and ethnic and racial minorities. The Board of Higher Education and the Illinois Community College Board oversee postsecondary education. The Department of Transportation handles highways, traffic safety, and airports.

Among state agencies offering health and welfare services are the Department of Children and Family Services, which focuses on foster care, the deaf, the blind, and the handicapped; and the Department of Public Aid, which supervises Medicaid, food stamps, and general welfare programs. The Mental Health and Developmental Disabilities Department operates homes and outpatient centers for the retarded and the mentally ill; it also offers an alcoholism program. Established in 1973, the Department on Aging provides nutritional and field services. The Board of Vocational Rehabilitation operates programs to retrain the disabled, while the Department of Veterans' Affairs administers bonus and scholarship programs and maintains a veterans' home and a 300-bed nursing facility.

State responsibility for public protection is divided among several agencies: the Office of the Attorney General, Department of Corrections (prisons and parole), Department of Law Enforcement (including the State Police and Bureau of Investigation), Dangerous Drugs Commission, and Military and Naval Department. Resource protection is supervised by the Department of Conservation, which oversees fish hatcheries, state parks, nature reserves, game preserves, and forest fire protection. The Department of Mines and Minerals handles mine safety and land reclamation programs.

The Department of Labor mediates disputes and handles unemployment compensation. The Department of Human Rights, established in 1980, seeks to ensure equal employment, housing, and credit opportunities.

16JUDICIAL SYSTEM

The state's highest court is the supreme court, consisting of seven justices elected by judicial districts for 10-year terms; the justices elect one of their number as chief justice for 3 years. The supreme court has appellate jurisdiction generally, but has original jurisdiction in cases relating to revenue, mandamus, and habeas corpus. The chief justice, assisted by an administrative director, has administrative and supervisory authority over all other courts. The appellate court is divided into five districts; appellate judges, also elected for 10-year terms, hear appeals from the 21 circuit courts, which handle civil and criminal cases. Circuit judges are elected for 6-year terms. Repeated efforts to remove the state's judgeships from partisan politics have failed in the face of strong party opposition.

The penal system, under the general supervision of the Department of Corrections (established in 1970), includes large prisons at Joliet (1860), Pontiac (1871), Menard (1878), and Stateville (1919), near Joliet, plus juvenile facilities and an active parole division. The Cook County House of Corrections is highly active, as are federal facilities in Chicago and Marion. Nevertheless, Illinois has few prisoners at any single time—17,187 in 1984, fewer than five other states. Prisoner unrest, demands for legal rights, gang activity, and low guard morale continue to be serious problems in the state's penal institutions.

Illinois has a reputation for lawlessness, born of the gang warfare in Chicago during the prohibition era. In the 19th century, southern Illinois was ravaged by numerous bands of outlaws, and one county still carries the nickname "Bloody Williamson" because of its history of murders, massacres, and assassinations. However, as of 1983, the crime rate in the state was close to the national average, and Chicago's crime rate was actually lower than the rates of most large cities. The number of violent crimes in Illinois was 63,521 in 1983, for a rate of 553 per 100,000 population. Crime rates statewide were murder, 9.7; forcible rape, 31.5; robbery, 263; aggravated assault, 248; burglary, 1,214; larceny-theft, 2,902; and motor vehicle theft, 538. At the end of 1984, Illinois had an estimated 43,470 practicing attorneys.

[17]ARMED FORCES

The most important military installations in Illinois are Ft. Sheridan, north of Chicago, the headquarters of the Fifth Army; the nearby Great Lakes Naval Training Center; Scott Air Force Base near Belleville; and Chanute Air Force Base in Rantoul. Total military personnel numbered 37,434 in 1983/84, when Illinois firms received defense contract awards amounting to $1,475 million, just 1% of the US total.

About 1,000,000 Illinoisans served in World War II, of whom 30,000 were killed. As of 30 September 1983, 1,338,000 veterans were living in Illinois, of whom 14,000 saw service in World War I, 531,000 in World War II, 236,000 in the Korean conflict, and 357,000 during the Viet-Nam era. Veterans' benefits reached $949 million in 1982/83.

Illinois's National Guard comprised 13,604 men and women at the beginning of 1985. Police forces are relatively large, totaling 37,917 men and women in 1983 and costing $1,132 million (1982); the ratio of 3.3 police employees per 1,000 population was one of the highest in the US.

[18]MIGRATION

Apart from the small French settlements along the Mississippi River that were formed in the 18th century, most early white migration into Illinois came from the South, as poor young farm families trekked overland to southern Illinois from Kentucky, Tennessee, and the Carolinas between 1800 and 1840. After 1830, migration from Indiana, Ohio, and Pennsylvania filled the central portion of the state, while New Englanders and New Yorkers came to the northern portion.

Immigration from Europe became significant in the 1840s and continued in a heavy stream for about 80 years. Before 1890, most of the new arrivals came from Germany, Ireland, Britain, and Scandinavia. These groups continued to arrive after 1890, but they were soon outnumbered by heavy immigration from southern and eastern Europe. The opening of prairie farms, the burgeoning of towns and small cities, and the explosive growth of Chicago created a continuous demand for unskilled and semiskilled labor. Concern for the welfare of these newcomers led to the establishment by Jane Addams in Chicago of Hull House (1889), which served as a social center, shelter, and advocate for immigrants. Hull House launched the settlement movement in America, and its activities helped popularize the concept of cultural pluralism. The University of Chicago was one of the first major universities to concern itself with urban ecology and with the tendency to "ghettoize" culturally and economically disadvantaged populations.

The outbreak of World War I interrupted the flow of European immigrants but also increased the economy's demand for unskilled labor. The migration of blacks from states south of Illinois—especially from Arkansas, Tennessee, Louisiana, Mississippi, and Alabama—played an important role in meeting the demand for labor during both world wars. After World War II, the further collapse of the cotton labor market drove hundreds of thousands more blacks to Chicago and other northern cities.

In contrast to the pattern of foreign and black migration to Illinois was the continued westward search by native-born whites for new farmland, a phenomenon that produced a net outflow by this group from 1870 to 1920. After World War II, native whites again left the state in large numbers, with Southern California a favorite mecca. After 1970, for the first time, more blacks began leaving than entering Illinois.

The major intrastate migration pattern has been from farms to towns. Apart from blacks, who migrated considerable distances from farms in the South, most ex-farmers moved only 10–30 mi (16–48 km) to the nearest town or city.

During the 1970s the state lost 649,000 persons in net migration, for an annual rate of 0.5%. From 1980 to 1983, the net loss from migration totaled 212,000, or 0.6% annually.

[19]INTERGOVERNMENTAL COOPERATION

Illinois participates in 25 interstate compacts, including such regional accords as the Great Lakes Basin Compact and Ohio River Valley Water Sanitation Compact. In 1985, Illinois and seven other states formed the Great Lakes Charter to protect the lakes' water supply.

Federal grants to Illinois totaled $4.3 billion in 1983/84.

[20]ECONOMY

The economic development of Illinois falls into four periods: the frontier economy, up to 1860; the industrial transition, 1860-1900; industrial maturity, 1900-50; and the transition to a service economy, 1950 to the present.

In the first phase, subsistence agriculture was dominant; the cost of transportation was high, cities were small and few, and cash markets for farm products hardly existed. The main activity was settling and clearing the land. A rudimentary market economy developed at the end of the period, with real estate and land speculation the most lucrative activities.

The industrial transition began about 1860, stimulated by the construction of the railroad network, which opened up distant markets for farm products and rural markets for manufactured items. The Civil War stimulated the rapid growth of cash farming, commercial and financial institutions, and the first important factories. The last quarter of the 19th century saw the closing of the agricultural frontier in Illinois and the rapid growth of commercial towns and industrial cities, especially Chicago.

Industrial maturity was reached in the early 20th century. Large factories grew, and small ones proliferated. Chicago's steel industry, actually centered in Gary, Ind., became second in size only to Pittsburgh's, while the state took a commanding lead in food production, agricultural implement manufacture, and agricultural finance. The depression of the 1930s stifled growth in the state and severely damaged the coal industry, but with the heavy industrial and food demands created by World War II, the state recovered its economic health. Since 1950, the importance of manufacturing has declined, but a very strong shift into services—government, medicine, education, law, finance, and business—has underpinned the state's economic vigor.

The gross state product reached $141 billion in 1980, nearly doubling the 1973 total. The following table shows contributions of the same sectors to the GSP for those two years (in billions):

	1973	1980
Manufacturing	$23.3	$ 38.6
Trade	13.2	23.8
Finance, insurance, real estate	10.3	23.6
Services	9.0	19.2
Transportation, communications	6.8	12.8
Government	6.8	11.5
Contract construction	3.7	6.5
Agriculture	2.7	3.5
Mining	0.5	1.5
TOTALS	$76.3	$141.0

The gross state product increased to $162 billion in 1983, representing a real increase of 0.5% from 1982.

Trends in the 1970s showed declines in the relative importance of manufacturing and construction, little relative change in transportation and agriculture, and significant proportional increases in the other sectors. Severe competition from Japan wreaked havoc in the state's steel, television, and automotive industries, while Illinois's high-wage, high-cost business climate encouraged the migration of factories to the South. Meat-packing, once the most famous industry in Illinois, dwindled after the closing of the Chicago stockyards in 1972. On the brighter side, Chicago remained the nation's chief merchandising center during the early 1980s, and an influx of huge international banks boosted the city's financial strength. In 1985, Chrysler and Mitsubishi announced that they would build their Diamond-Star Motors Corporation plant about 100 mi (160 km) southwest of Chicago.

21 INCOME

Illinois is a rich state and has been for the last century. In per capita income, it ranked 7th in 1983, trailing only California and New Jersey among the most populous states. Its per capita income of $12,626 was 8% above the national average in 1983. The median income of four-person families was $29,403 in 1981.

Proportionally fewer Illinois residents live in poverty than do Americans as a whole, but the proportions are somewhat higher than in many other midwestern states. In 1979, 11% of the entire population and 14.9% of all children were below the poverty line. Blacks and households headed by women or old people were the most likely victims of poverty. On the other hand, 241,000 of the nation's leading wealth holders with assets over $300,000—5.5% of the US total—lived in the state in 1982.

Income levels vary by race and geography. The average income for white families is somewhat higher than the mean income for black and Hispanic families. Residents of southern Illinois tend to have lower income than those living in central and northern Illinois.

22 LABOR

In 1984, the Illinois labor force numbered 5,594,000 persons, of whom 5,083,000 were employed. Some 511,000 were unemployed, for an unemployment rate of 9.1%.

A federal census in March 1982 revealed the following nonfarm employment pattern in Illinois:

	ESTABLISH-MENTS	EMPLOYEES	ANNUAL PAYROLL ('000)
Agricultural services, forestry, fishing	1,928	9,800	$ 142,753
Mining, of which:	981	31,544	853,556
Bituminous coal, lignite	(81)	(17,297)	(510,220)
Contract construction	15,721	148,735	3,852,220
Manufacturing, of which:	17,582	1,117,477	23,314,143
Food and food products	(1,011)	(86,484)	(1,817,957)
Printing and publishing	(3,065)	(101,440)	(2,047,171)
Primary metals	(511)	(66,471)	(1,435,410)
Fabricated metals	(2,356)	(119,005)	(2,321,840)
Nonelectrical machinery	(3,297)	(184,421)	(3,817,465)
Electric, electronic equipment	(983)	(136,880)	(2,562,629)
Transportation, public utilities	8,304	227,890	5,571,688
Wholesale trade	21,517	309,299	6,769,766
Retail trade	57,512	761,566	7,462,340
Finance, insurance, real estate	21,673	326,563	6,342,809
Services, of which:	66,917	984,951	14,363,086
Health services	(16,192)	(323,642)	(5,220,331)
Business services	(10,313)	(170,095)	(2,701,281)
Other	3,741	4,179	77,250
TOTALS	215,876	3,922,004	$68,749,611

Some 718,000 Illinoisans, not included in this survey, were government employees in 1982; about half of these were teachers and other educational personnel.

The first labor organizations sprang up among German tailors, teamsters, and carpenters in Chicago in the 1850s, and among British and German coal miners after the Civil War. The period of industrialization after the Civil War saw many strikes, especially in coal mining and construction, many of them spontaneous rather than union-related. The Knights of Labor organized extensively in Chicago, Peoria, and Springfield in the 1870s and 1880s, reaching a membership of 52,000 by 1886. However, in the aftermath of the Haymarket Riot—at which a dynamite blast at a labor rally killed seven policemen and four civilians—the Knights faded rapidly. More durable was the Chicago Federation of Labor, formed in 1877 and eventually absorbed by the American Federation of Labor (AFL). Strongest in the highly skilled construction, transportation, mining, and printing industries, the federation stood aside from the 1894 Pullman strike, led by industrial union organizer Eugene V. Debs, a bitter struggle broken by federal troops over the protest of Governor Altgeld.

Today, labor unions are powerful in Chicago but relatively weak downstate. In 1980, 1,487,000 workers, or 30.6% of nonfarm workers, belonged to unions (versus 25.2% nationally); this represented a sharp decline from 1970, when 37.3% were members. There were 3,230 labor unions in Illinois in 1983. The major unions are the International Brotherhood of Teamsters, the United Steelworkers of America, the International Association of Machinists, the United Automobile Workers, the United Brotherhood of Carpenters, and the American Federation of State, County, and Municipal Employees. The Illinois Education Association, though not strictly a labor union, has become one of the state's most militant employee organizations, often calling strikes and constituting the most active lobby in the state. In 1983, a new law granted all public employees except police and firemen—about 430,000 workers in all—the right to strike.

As of September 1984, average weekly wages in manufacturing were $420; average hourly earnings of $10.17 were 10% above the national average. Weekly earnings in retail trade, services, education, and government also exceeded the national norms.

23 AGRICULTURE

Total agricultural income in 1983 reached $8.1 billion in Illinois, 4th behind California, Iowa, and Texas. Crops accounted for nearly 72% of the value of farm marketings, with soybeans and corn the leading cash commodities.

Prior to 1860, agriculture was the dominant occupation, and food for home consumption was the leading product. Enormous effort was devoted to breaking the thick prairie soil in the northern two-thirds of the state. Fences and barns were erected, and in the 1870s and 1880s the drainage of low-lying areas in central Illinois was a major concern. Commercial agriculture was made possible by the extension of the railroad network in the 1860s and 1870s. Corn, wheat, hogs, cattle, and horses were the state's main products in the 19th century. Since then, wheat and poultry have declined greatly in significance, while soybeans and, to a lesser extent, dairy products and vegetables have played an increasingly important role. The mechanization and electrification of agriculture, beginning about 1910, proceeded faster in Illinois than anywhere else in the world. Strong interest in scientific farming, including the use of hybrid corn, sophisticated animal-breeding techniques, and chemical fertilizers, has also fostered a steady, remarkable growth in agricultural productivity.

The number of farms reached a peak at 264,000 in 1900 and began declining rapidly after World War II, down to 96,000 in 1984. Total acreage in farming, 32.8 million acres (13.3 million hectares) in 1900, or 92% of the state's land area, declined slowly to 29 million acres (11.7 million hectares), or 80%, in 1984. The average farm size has more than doubled from 124 acres (50 hectares) in 1900 to 299 acres (121 hectares) in 1984. The farm population, which averaged 1.2 million persons from 1880 to 1900, declined to 314,000 in 1980; by then, moreover, about half the people who lived on farms commuted to work in stores, shops, and offices.

The major agricultural region is the corn belt, covering all of central and about half of northern Illinois. Among the 50 states, Illinois ranked 2d only to Iowa in production of corn and soybeans in 1983. The following table shows output and value of leading field crops in that year:

	VOLUME (MILLION BUSHELS)	VALUE (MILLIONS)
Corn for grain	624.0	$2,122
Soybeans	251.0	2,119
Wheat	64.4	219
Hay (million tons)	2.7	184
Oats	12.6	24
Sorghum for grain	6.5	20

Agriculture is big business in the state, though very few farms are owned by corporations (except "family corporations," a tax device). The financial investment in agriculture is enormous, largely because of the accelerating cost of land. The value of land quadrupled during the 1970s to an average of $2,013 per acre in 1980, but fell to $1,692 per acre by 1984.

Illinois ranks 1st of the 50 states in agricultural exports; foreign sales in 1983 totaled $2.9 billion, or 8.5% of the US total.

24 ANIMAL HUSBANDRY

Livestock is raised almost everywhere in Illinois, but production is concentrated especially in the west-central region. In 1983, livestock marketings and products were valued at nearly $3 billion. At the close of 1983, Illinois farms had 5,600,000 hogs and pigs (10% of the nation's total) and 2,600,000 head of cattle (2.3%). Production of meat animals included 1,199.1 million lb of cattle, worth $554 million, and 3.5 billion lb of hogs, worth $1.1 billion. The dairy belt covers part of northern Illinois, providing Chicago's milk supply. Milk production in 1983 totaled 2.6 billion lb. Other dairy products included 2.7 million lb of butter (1981) and 103 lb of cheese (1983). During 1983, Illinois poultry farmers produced 9.9 million lb of chickens, worth $867,000; 1 billion eggs sold for $50.2 million.

25 FISHING

Commercial fishing is insignificant in Illinois: only 342,000 lb of fish, valued at $296,000, made up the commercial catch in 1984, down from 6.4 million lb and $1.9 million in 1983. Sport fishing is of modest importance in southern Illinois and in Lake Michigan. Some 450 lakes and ponds and 200 streams and rivers are open to the public.

26 FORESTRY

Except for shade trees and occasional orchards, the northern two-thirds of the state is barren of trees. The south, however, has some 3,810,000 acres (1,542,000 hectares) of forestland, covering about 11% of the state's land area. Of that, 3,692,000 acres (1,494,000 hectares) were classified as commercial forest, 92% of it privately owned. Lumbering, a minor industry, produced 63 million board feet of lumber in 1983. The Shawnee National Forest covered 714,644 acres (289,216 hectares) as of 1984.

27 MINING

Illinois ranked 18th among the 50 states in nonfuel mineral production in 1984, with output valued at $446.5 million. Fossil fuels, especially coal, are the leading mineral commodities, but stone, sand, and gravel are also important. Excluding fossil fuels, mineral output in 1984 comprised 45,100,000 tons of stone, worth $172,000,000; 25,630,000 tons of sand and gravel, $101,271,000; 2,100,000 tons of Portland cement, $85,700,000; and 251,000 tons of clay, $1,018,000. Lead, zinc, silver, fluorspar, and lime are also mined in small commercial quantities.

28 ENERGY AND POWER

Illinois is one of the nation's leading energy producers and consumers. Electric power production totaled 99.2 billion kwh (6th in the US) in 1983, down 3% from 1980; installed capacity was 31 million kw (6th), nearly all of it privately owned. Consumption of electrical energy in 1982 amounted to 3,337 trillion Btu (5th), of which industry accounted for 32%, residences 25%, commercial establishments 19%, and transportation 24%. Commonwealth Edison and Northern Illinois Light and Power are the largest suppliers. Coal-fired plants account for about 60% of the state's power production; nuclear power is also important, particularly for the generation of electricity in the Chicago area. The state's four operating nuclear power plants in 1984 were all owned by Commonwealth Edison.

In 1983, Illinois ranked 2d in natural-gas usage, with sales of 430.6 billion cu ft to 3 million customers, for a total revenue of $2.3 billion. People's Gas, a diversified energy conglomerate based in Chicago, is the largest firm. Petroleum production, though steadily declining, totaled 19 million barrels in 1983; reserves were 135 million barrels.

Illinois ranked 5th in the US in bituminous coal production in 1983, with 56.8 million tons; reserves were estimated at 79 billion tons. Coal is abundant throughout the state, with the largest mines in the south and central regions. Coal mining reached its peak in the 1920s, but suffered thereafter from high pricing policies, the depression of the 1930s, and the environmental restrictions against burning high-sulfur coal in the 1970s. In 1983 there were 55 productive coal mines—23 surface (strip) mines and 32 underground mines.

29 INDUSTRY

Manufacturing in Illinois, concentrated in but not limited to Chicago, has always been diverse. Before 1860s, small gristmills, bakeries, and blacksmith shops handled what little manufacturing was done. Industry tripled in size in the 1860s, doubled in the 1870s, and doubled again in the 1880s, until manufacturing employment leveled off at 10–12% of the population. Value added by manufacture grew at a compound annual rate of 8.1% between 1860 and 1900, and at a rate of 6.3% until 1929. The chief industries in 1929 were iron and steel, printing, food, electrical equipment, and machinery.

In 1982, the value of shipments by manufacturers totaled $113 billion, an increase of 22% over 1977. Of the major sectors, food and food products contributed 17%; nonelectrical machinery, 14%; petroleum and coal products, 10%; electric and electronic equipment, 9%; chemicals and allied products, 9%; fabricated metal products, 8%; printing and publishing, 7%; and primary metal industries, 6%. The following table shows value of shipments by manufacturers for selected industries in 1982:

Petroleum refining	$11,690,900,000
Construction machinery	4,092,600,000
Communications equipment	3,892,100,000
Rubber and miscellaneous plastic products	3,579,400,000
Fats and oils	3,442,900,000
Meat products	3,230,400,000
Blast furnace, basic steel products	3,388,700,000
Paper and allied products	3,350,000,000
Grain mill products	3,206,400,000
Commercial printing	3,147,800,000
Motor vehicles and equipment	3,145,500,000
Soaps, cleansers, toiletries	3,007,500,000

By far the leading industrial center is Chicago, followed by Rockford, the East St. Louis area, Rock Island and Moline in the Quad Cities region, and Peoria. The leading industrial corporations headquartered in Illinois in 1984 were Standard Oil of Indiana (Chicago), the 10th-largest US industrial corporation, with sales of $26.9 billion; Dart & Kraft (Northbrook), ranking 33d with sales of $9.8 billion; Beatrice Foods (Chicago), 36th with $9.3 billion; Consolidated Foods (Chicago), 49th with $7 billion; Caterpillar Tractor (Peoria), 52d with $6.6 billion; and Motorola (Schaumburg), 67th with $5.5 billion.

³⁰COMMERCE

Chicago is the leading wholesaling center of the Midwest. In 1982, the state's 21,722 wholesale establishments employed 287,000 people (3d in the US) and had sales of $131 billion (6.5% of the US total). Chicago is an especially important trade center for furniture, housewares, and apparel.

The state's 59,200 retail stores employed 708,000 people and recorded sales of $52 billion (4.7% of the US total) in 1982. The principal retail store groups and their respective shares were food stores, 20%; automotive dealers, 17%; and general merchandise stores, 12%. The Chicago metropolitan area accounted for retail sales of $34.6 billion, or 66% of the state total. Leading Illinois-based retailing companies in 1984 were Sears, Roebuck, with nationwide sales of $38.8 billion; Household International, $8.3 billion; Montgomery Ward, $6.5 billion; McDonald's, $3.4 billion; and Walgreen, $2.7 billion.

Illinois ranked 3d among the states in exports of manufactured goods, with estimated sales of $14.8 billion in 1983. Of the total value of manufacturing exports in 1981, metals and metal products accounted for 29%; nonelectric machinery, 28%; chemical products, 14%; electronic equipment, 13%; and transport equipment, 11%. In that year, about 84,000 persons, or 6.8% of the state's workers, were employed in producing manufactures for export.

In 1983, 13.7% of the nation's exports and 13.5% of its imports passed through the Chicago customs district, covering most of the Midwest; total trade volume through the district was $63 billion, 2d only to New York.

³¹CONSUMER PROTECTION

Consumer protection became a popular political cause in Illinois during the 1970s. Chicago's consumer commissioner, Jane Byrne, fired for overaggressiveness, used the incident in her successful campaign for the city's mayoralty in 1979.

Statewide, the Office of the Attorney General is the most active protector of consumers with its Consumer Fraud Section and Consumer Protection Division. The governor controls the Consumer Advocate's Office, and the Department of Insurance also has a Consumer Division. The Department of Human Rights was established in 1979 to protect individuals in regard to employment, public accommodations, and other areas.

³²BANKING

Banking was highly controversial in 19th-century Illinois. Modernizers stressed the need for adequate venture capital and money supplies, but traditionalist farmers feared they would be impoverished by an artificial "money monster." Efforts to create a state bank floundered in confusion, while the dubious character of most private banknotes inspired the state to ban private banks altogether. The major breakthrough came during the Civil War, when federal laws encouraged the establishment of strong national banks in all the larger cities, and Chicago quickly became the financial center of the Midwest. Apart from the 1920s and early 1930s, when numerous neighborhood and small-town banks folded, the banking system has flourished ever since.

There were 1,300 commercial banks in Illinois in 1984 (2d only to Texas), an unusually large number attributable to regulations restricting branch banking. Until the 1970s, even the largest banks were allowed only one office. At the end of 1983, commercial banks held $136.9 billion in assets (6.7% of the US total), and deposits of $98.6 billion (6.4%). The largest banks are First National Bank of Chicago, the 10th largest in the US, with assets in 1984 of $39.8 billion and deposits of $28.6 billion; and Continental Illinois, the 12th largest, with assets of $30.4 billion and deposits of $15.1 billion. In 1984, Continental Illinois was threatened with bankruptcy because of bad loans and a "run" on the bank by depositors; the Federal Deposit Insurance Corp. rescued the bank by pledging $4.5 billion to bail it out.

In 1982, Illinois had 296 savings and loan associations, with total assets of $47.6 billion (3d in the US) and outstanding mortgage loans of $30.6 billion. The largest associations are First Federal, Talman, and Home Federal, all of Chicago. There were 1,526 state-chartered credit unions in 1984.

³³INSURANCE

Illinois is a major center of the insurance industry, ranking 2d to the New York–Hartford–Newark complex. In 1983, the state's 89 life insurance companies collected $2.6 billion in premiums and paid out $3 billion in benefits, including $899.2 million in death benefits and $611.1 million in annuities, to Illinois residents. Illinoisans held 20.7 million policies, valued at $271 billion, in 1983; the average family had $61,200 in life insurance coverage.

Illinois fire and casualty companies are among the US leaders. State Farm, based in Bloomington, ranks 1st in the field, and Allstate, a Chicago subsidiary of Sears, Roebuck, is 2d nationally. Within the state, fire and casualty underwriters wrote premiums totaling $5.4 billion in 1983, including $1.2 billion in automotive liability insurance, $891.7 million in automobile physical-damage insurance, and $588.3 million in homeowners' coverage. In the same year, 736 communities were covered by flood insurance totaling $844.8 million.

Blue Cross–Blue Shield, the nation's largest hospital and medical insurance program, is headquartered in Chicago.

³⁴SECURITIES

Chicago ranks 2d only to New York as a center for securities trading. The Midwest Stock Exchange, the largest securities exchange outside New York City, traded 1.8 billion shares worth a total of $59.9 billion in 1984. New York Stock Exchange member firms had 267 sales offices and 3,329 registered representatives in Illinois in 1983. In that year, 2,468,000 residents owned shares in public corporations, and 2,381 brokers and dealers were registered in the state.

The most intensive trading in Chicago takes place on the three major commodity exchanges. The Chicago Board of Trade has set agricultural prices for the world since 1848, especially in soybeans, corn, and wheat. The Chicago Mercantile Exchange specializes in pork bellies (bacon), live cattle, potatoes, and eggs; since 1972, it has also provided a market for world currency futures. The Mid-America Commodity Exchange, the smallest of the three, has a colorful ancestry dating from 1868. It features small-lot futures contracts on soybeans, silver, corn, wheat, and live hogs.

³⁵PUBLIC FINANCE

Among the larger states, Illinois is known for its low taxes and conservative fiscal policy. The Bureau of the Budget, under the governor's control, has major responsibility for the state's overall fiscal program, negotiating annually with key legislators, cabinet officers, and outside pressure groups. The governor then submits the budget to the legislature for amendment and approval. The fiscal year runs from 1 July to 30 June. As in other states, the Illinois budget soared in the 1970s and early 1980s, primarily because of planned educational expansion and unexpected increases in welfare payments. The following table summarizes consolidated revenues and appropriations for 1984/85 (estimated) and 1985/86 (projected) (in millions):

REVENUES	1984/85	1985/86
Sales tax	$ 3,132	$ 3,390
Personal income tax	2,673	2,849
Corporate income tax	505	551
Public utility tax	670	725
Cigarette tax	164	164
Liquor tax	71	71
Inheritance tax	58	55
Other taxes	1,648	1,446
Federal receipts	3,526	3,771
Other receipts and designated funds	2,437	2,795
TOTALS	$14,884	$15,817

APPROPRIATIONS[1]	1984/85	1985/86
Highways and transportation	$ 3,610	$ 3,678
Public welfare	3,232	3,409
Primary and secondary education	2,792	3,071
Higher education	1,424	1,617
Mental health	593	617
Capital development	482	433
Environmental protection	212	292
Public health	156	172
Other outlays	5,328	5,443
TOTALS	$17,829	$18,732

[1] Spending against appropriations is estimated to be $14.9 billion in 1984/85 and $15.8 billion in 1985/86.

The city of Chicago collected general revenues exceeding $1.89 billion in 1982, of which property taxes supplied 18%, sales taxes 23%, the federal government 24%, and the state government 15%. In the same year, Chicago's general expenditures totaled $1.77 billion, of which 30% was spent on police and fire protection. Chicago's budget appears relatively low—less than one-ninth the size of New York City's in 1982—because it does not include expenditures by the Chicago school board and the welfare agencies of Cook County.

The total debt outstanding for all state and local governments in mid-1982 was $16.8 billion, or $1,476 per capita, about 84% of the national average. The state debt in mid-1983 was nearly $12.6 billion.

36 TAXATION

Illinoisans have fiercely resisted the imposition of new and higher taxes. The levying of the first 1% sales tax in 1933 to finance relief programs was bitterly resented, and the inauguration of a state personal income tax in 1970 led to the defeat of Governor Richard Ogilvie in his 1972 reelection campaign. Total state and local revenue in 1981/82 was $21.9 billion, or $1,916 per capita, below the national average (and 27 among the 50 states).

As of 1984, the state personal income tax was a flat 3%. The corporate income tax was 4.8% and the sales tax 5%, with few exemptions. Excise taxes included charges of 12 cents a pack on cigarettes and 11 cents a gallon on gasoline.

Local levies, chiefly property taxes, are relatively light. Indeed, in Chicago the tax burden shrank during the 1970s. Illinois ranked 20th among the states in 1982 in property tax rates, with 1.9% for every $100 of assessed property value.

Low state and local taxes were counterbalanced by high federal payments and a very low return of federal dollars to Illinois. In 1982/83, Illinois ranked 6th among the states in per capita federal tax payments, averaging $2,856. The federal government collected $32.7 billion in Illinois for 1982/83, but sent back only $19 billion in 1983/84; the net deficit of $13.7 billion was the highest for any state. Illinoisans filed 4,726,101 federal income tax returns for 1983, paying $15 billion in tax.

37 ECONOMIC POLICY

The state's policy toward economic development has engendered political controversy since the 1830s. Before the Civil War, the Democrats in power usually tried to slow, though not reverse, the tide of rapid industrial and commercial growth. The Republican ascendancy between the 1850s and the 1930s (with a few brief interruptions) produced a generally favorable business climate, which in turn fostered rapid economic growth. The manufacturing sector eroded slowly in the 1960s and 1970, as incentives and tax credits for new industry were kept at a modest level. Despite the government's attempt to help create employment opportunities in the early 1980s, the economy remained sluggish and unemployment high (about 9% as of mid-1985).

The Department of Commerce and Community Affairs promotes economic development, describing itself as the "sales department for Illinois." It maintains offices in Washington, D.C.,

Brussels, Hong Kong, São Paulo, and Osaka. The promotion of jobs, tourism, minority-owned enterprises, and foreign markets for Illinois products is the department's major responsibility.

38 HEALTH

In pioneer days, Illinois had a rather unhealthful reputation. The state's many swampy areas harbored malaria, then known as the "Illinois shakes"—a menace that declined late in the 19th century as the swamps and low-lying areas were drained. The cities grew faster than did their sanitation and water purification systems, and bad living conditions worsened the effects of improper hygiene. Tuberculosis, widespread in the late 19th and early 20th centuries, was the major cause of death and disability during this period. Public health services, strongly promoted by the medical profession, alleviated the threat of most communicable diseases by the 1940s, when penicillin and sulfa drugs finished the job.

Even now, however, health conditions in Illinois do not meet the national norm. In 1981, infant mortality was 11 per 1,000 live births for whites and 24 for nonwhites. Although the infant mortality rate fell from 21.5 in 1970 to 13.9 in 1981, it did not decline as rapidly as elsewhere. Since the late 1960s, Illinois has had a slightly higher infant mortality rate than the rest of the country; the 1981 rate among nonwhites was the 5th highest among the states. The number of legal abortions climbed rapidly in the 1970s to 72,000 in 1977, then declined to 65,900 in 1982.

Illinois's marriage and divorce rates were both below the US norms, but the birthrate, 16 per 1,000 population in 1982, was marginally higher, as was the death rate, 8.9 per 1,000 residents in 1981. At that time Illinois ranked above the national average in deaths due to heart disease, cancer, lung diseases, diabetes, and chronic liver disease and cirrhosis, but below the average in stroke, accidents, pulmonary diseases, atherosclerosis, and suicide. Major public health problems in the early 1980s included rapidly increasing rates of venereal disease and drug abuse. Alcoholism has always been a major problem in Illinois. The state also has a high proportion of residents receiving psychiatric care. In early 1985, Illinois suffered an outbreak of salmonella poisoning that was the worst in the nation's history; from late March to mid-May, 14,620 cases were treated in the state's hospitals before the outbreak was contained. The epidemic was traced to tainted milk in a dairy, which closed down voluntarily.

Hospitals abound in Illinois, with Chicago serving as a diagnostic and treatment center for patients throughout the Midwest. With 279 facilities (many quite large) and 70,612 beds, Illinois hospitals recorded 1,985,937 admissions in 1983. Hospital personnel in 1981 included 47,929 registered nurses. The average cost of a semiprivate hospital room in January 1984 was $237 per day, well above the US average. In 1982, the state had 574 nursing homes, 23,893 licensed physicians, and 6,404 professionally active dentists.

39 SOCIAL WELFARE

Prior to the 1930s, social welfare programs were the province of county government and private agencies. Asylums, particularly poor farms, were built in most counties following the Civil War; they provided custodial care for orphans, the very old, the helpless sick, and itinerant "tramps." Most people who needed help, however, turned to relatives, neighbors, or church agencies. The local and private agencies were overwhelmed by the severe depression of the 1930s, forcing first the state and then the federal government to intervene. Social welfare programs are implemented by county agencies, but funded by local and state taxes and federal aid. In the early 1980s, the annual outlays for the five largest welfare programs in Illinois totaled more than $2 billion, of which the federal government paid over half.

In 1975, 803,200 children and adults received aid to families with dependent children; by 1983, the total had fallen to 742,000 children and adults. Payments, however, increased from $773 million in 1975 to $805 million in 1982.

The federal food stamp program aided 1,123,000 persons in 1983, at a federal cost of $663 million. The federal school lunch program, which reached 1,129,000 students in 1980, was cut back to 925,000 students in 1983, when federal outlays totaled $104 million.

Social Security monthly payments averaged $471 for the state's 1,171,000 retired workers in 1983; the total paid to all beneficiaries was $8.2 billion. In the same year, Social Security disability benefits were paid to 674,000 persons at an average monthly payment of $477. In addition, total Supplemental Security Income payments of $266 million covered 119,761 disabled, aged, and blind persons.

Outlays for workers' compensation totaled $673 million in 1982. Unemployment insurance is slightly more generous in Illinois than in most other states: the total payout in 1982 was $1,576 million, for an average weekly benefit of $146 to eligible state residents.

40HOUSING

Flimsy cabins and shacks provided rude shelter for many Illinoisans in pioneer days. Later, the balloon-frame house, much cheaper to build than traditional structures, became a trademark of the Prairie State. After a third of Chicago's wooden houses burned in 1871, the city moved to enforce more stringent building codes. The city's predominant dwelling then became the three- or five-story brick apartment house. Great mansions were built in elite areas of Chicago (first Prairie Avenue, later the Gold Coast), and high-rise lakefront luxury apartments first became popular in the 1920s. In the 1970s, Chicago pioneered the conversion of luxury apartment buildings to condominiums, which numbered 158,331 units in 1980.

The 1980 census counted 4,319,672 housing units in Illinois, of which 94% were occupied; of these, more than 62% were owner-occupied (well below the 84% for the US as a whole), and all but 1.7% had full plumbing. In 1983, more than 30,900 new units valued at more than $1.6 billion were authorized. The median value of an owner-occupied unit was $52,800 in 1980. Public housing, serving primarily poor blacks and elderly whites in Chicago, was enmeshed in controversy and court cases during the early 1980s, and few units were built.

41EDUCATION

The pioneers did not see much use in book learning, and arithmetic was not needed for a subsistence economy. Thus, until the Yankee reformers in the Republican Party secured power in the mid-1850s, there was little public effort to support education. Once in office, the reformers helped create an outstanding public school system in Chicago, although until foreign immigration subsided in the 1920s, the city was hard-pressed to construct enough school buildings to serve the growing numbers of students. Rural Illinois clung to its system of one-room schoolhouses until state-mandated consolidation in the 1940s created large modern schools to which students were bused. By the 1940s, the Chicago public schools had begun to deteriorate, a condition that began to be corrected in the early 1980s. The suburban school districts outside the city, however, remain among the finest in the US.

Literacy was virtually universal in 1980, though careful studies showed a fourth of the state's adults were "functionally illiterate"—that is, unable to comprehend simple written forms. In 1980, two-thirds of the adult population held high school diplomas, practically all adults had attended grade school, and nearly one-third had gone to college (16.2% had graduated).

In 1981, Illinois had 3,111 public elementary schools, 863 high schools, and 95 combined elementary/high schools. Enrollment at the elementary level slipped from about 1,419,000 in 1977/78 to 1,286,858 in 1982/83; enrollment at the secondary level declined from about 720,000 in 1977/78 to 593,431 in 1982/83. Nearly half of all public school students belonged to minority groups; of

these, 60% were in schools with 90–100% minority enrollment. Chicago schools remained heavily segregated, despite persistent pressure by state and federal officials. There were 18.3 pupils per teacher in 1983, when the average teacher's salary was $23,345.

Nonpublic schools, dominated by Chicago's extensive Roman Catholic school system, declined sharply in the late 1970s and early 1980s. Total enrollment in private schools fell from 398,000 in 1974/75 to 353,000 in 1982/83. Rising tuition fees, caused in part by higher salaries for lay teachers and a drop in the number of teaching sisters, threatened the parochial schools; the Chicago archdiocese had 129,852 primary pupils and 51,803 high school students as of 1 January 1984. High-tuition private schools continued to flourish in Chicago, however.

Illinois has always been well endowed with colleges. In 1982 there were 160 public and private colleges with 683,969 students. The state appropriated a $1.1 billion budget for colleges and universities in 1983/84. The Board of Higher Education, created in 1961, attempts to coordinate the crazy-quilt pattern of public university systems. The largest system, the University of Illinois, operates three major campuses—Champaign-Urbana, Chicago Circle, and the Chicago Medical Center—plus branch medical schools in Peoria and Rockford. The state also supports Southern Illinois University (with campuses in Carbondale and Edwardsville), Chicago State University, Eastern Illinois University (Charleston), Governors State University (Park Forest South), Northeastern Illinois University (Chicago), Western Illinois University (Macomb), Illinois State University (Normal), Northern Illinois University (De Kalb), and Sangamon State University (Springfield). A flourishing network of 52 community colleges, with 795,000 students in 1983, was built up primarily in the 1960s.

Major private universities, all in the Chicago area, include the University of Chicago, Northwestern University (Evanston), Illinois Institute of Technology, and Loyola University. Each maintains undergraduate and research programs, as well as nationally recognized professional schools. The Illinois State Scholarship Commission administers general grant and guaranteed loan programs, as well as special awards for the children of firemen and policemen killed in the line of duty, of deceased correctional workers, and of prisoners of war and those missing in action.

Illinois spends heavily for its public educational system. In 1981/82, per capita school expenditures reached $622, far above the national average though not disproportionate to the state's high per capita income. The 1970 constitution gives the state government primary responsibility for the public school system, a mandate that was put to the test in order to save the Chicago public schools from financial collapse in 1979/80. The state spent $3,160 per student in 1983/84 (18th among the states).

42ARTS

Chicago emerged in the late 19th century as the leading arts center of the Midwest, and it continues to hold this premier position. The major downstate facilities include the Krannert Center at the University of Illinois (Champaign-Urbana) and the Lakeview Center in Peoria.

Architecture is the outstanding art form in Illinois, and Chicago—where the first skyscrapers were built in the 1880s—has been a mecca for modern commercial and residential architects ever since the fire of 1871. The Art Institute of Chicago, incorporated in 1879, is the leading art museum in the state. Although its holdings, largely donated by wealthy Chicagoans, cover all the major periods, its French Impressionist collection is especially noteworthy. The most recent example of bold architecture is the $172-million State of Illinois Center in Chicago, which opened in 1985.

Theater groups abound—there were 116 theatrical producers in 1982—notably in Chicago, where the Second City comedy troupe and the Steppenwolf Theatre are located; the city's best playwrights and performers, however, often gravitate to Broadway or Hollywood. Film production was an important industry in Illinois before 1920, when operations shifted to the sunnier climate and more opulent production facilities of Southern California. By the early 1980s, however, the Illinois Film Office had staged an impressive comeback, and television films and motion pictures were being routinely shot in the state.

The Chicago Symphony Orchestra, organized by Theodore Thomas in 1891, quickly acquired world stature; its permanent conductors have included Frederick Stock, Fritz Reiner, and Sir George Solti, who has regularly taken the symphony on triumphant European tours. German immigrants founded many musical societies in Chicago in the late 19th century, when the city also became a major center of musical education. Opera flourished in Chicago in the early 20th century, collapsed during the early 1930s, but was reborn through the founding of the Lyric Opera in 1954. Chicago's most original musical contribution was jazz, imported from the South by black musicians in the 1920s. Such jazz greats as King Oliver, Louis Armstrong, Jelly Roll Morton, Benny Goodman, and Gene Krupa all worked or learned their craft in the speakeasies and jazz houses of the city's South Side. More recently, Chicago became the center of an urban blues movement, using electric rather than acoustic guitars and influenced by jazz.

The seamy side of Chicago has fascinated writers throughout the 20th century. Among well-known American novels set in Chicago are two muckraking works, Frank Norris's *The Pit* (1903) and Upton Sinclair's *The Jungle* (1906), as well as James T. Farrell's *Studs Lonigan* (1935) and Saul Bellow's *The Adventures of Augie March* (1953). Famous American plays associated with Chicago are *The Front Page* (1928), by Ben Hecht and Charles MacArthur, and *A Raisin in the Sun* (1959), by Lorraine Hansberry.

43 LIBRARIES AND MUSEUMS

Libraries and library science are particularly strong in Illinois. In 1984 there were 609 public libraries, nearly all of them members of 18 regional systems, which had a combined book stock of 25,078,104. The facilities in Peoria, Oak Park, Evanston, Rockford, and Quincy are noteworthy, but the Chicago Public Library (which operates 77 neighborhood branches) is hampered by an inadequate central library crowded into an old warehouse. The outstanding libraries of the University of Illinois (Champaign-Urbana) and the University of Chicago (with 6,411,948 and 4,688,361 volumes in 1983, respectively) constitute the state's leading research facilities, both universities have famous library schools. Principal historical collections are at the Newberry Library in Chicago, the Illinois State Historical Society in Springfield, and the Chicago Historical Society.

Illinois has 206 museums and 45 historic sites. Chicago's Field Museum of Natural History, founded in 1893, has sponsored numerous worldwide expeditions in the course of acquiring some 13 million anthropological, zoological, botanical, and geological specimens. The Museum of Science and Industry, near the University of Chicago, attracts 5 million visitors a year, mostly children, to see its exhibits of industrial technology. Also noteworthy are the Adler Planetarium, Shedd Aquarium, and the Oriental Institute Museum of the University of Chicago. The Brookfield Zoo, near Chicago, opened in 1934; smaller zoos can be found in Chicago's Lincoln Park and in Peoria, Elgin, and other cities.

Just about every town has one or more historic sites authenticated by the state. The most popular is New Salem, near Springfield, where Abraham Lincoln lived from 1831 to 1837. Its reconstruction, begun by press magnate William Randolph Hearst in 1906, includes one original cabin and numerous replicas. The most important archaeological sites are the Dixon Mounds, 40 mi (64 km) south of Peoria, and the Koster Excavation in Calhoun County, north of St. Louis, Mo.

44 COMMUNICATIONS

Illinois has an extensive communications system. The US Postal Service had 1,301 post offices and 39,323 employees in Illinois in 1985; its largest single facility is located in Chicago, which ranked 2d only to New York City in postal receipts. The state's households with telephones numbered 3,825,251 in 1980, or 95% of all households.

Illinois had 138 AM and 202 FM commercial radio stations in 1984; 30 commercial television stations and 7 educational stations served the metropolitan areas. The state's 207 cable systems, serving 1,288,482 subscribers in 604 communities in 1984, have brought good television reception to the small towns. In 1979, WGN-TV in Chicago became a "superstation," with sports programs, movies, and advertising; by 1984, WGN-TV programs were beamed to 1,034 cable systems and 3,456,056 subscribers across the country. Although the three major networks own stations in Chicago, they originate very little programming from the city. However, as a major advertising center, Chicago produces many commercials and industrial films. Most educational broadcasting in Illinois comes from state universities and the Chicago public and Catholic school systems.

45 PRESS

The state's first newspaper, the *Illinois Herald*, was begun in Kaskaskia in 1814. From the 1830s through the end of the 19th century, small-town weeklies exerted powerful political influence. After 1900, however, publishers discovered that they needed large circulations to appeal to advertisers, and so they toned down their partisanship and began adding a broad range of features to attract a wider audience.

As of 1984, Illinois had 15 morning newspapers (including all-day papers), with a combined 1984 circulation of 1,960,261; 60 evening dailies, with 1,749,931; and 24 Sunday papers, with 2,774,154. The Illinois editions of St. Louis newspapers are also widely read. The following table shows the state's leading dailies with their 1984 circulation:

AREA	NAME	DAILY	SUNDAY
Chicago	Sun-Times (m,S)	649,891	688,793
	Tribune (all day, S)	776,348	1,137,667
Peoria	Journal Star (all day, S)	99,586	117,503
Rockford	Register Star (all day, S)	73,046	86,000
Springfield	State Journal-	62,046	72,122
	Register (m,e,S)	7,941	

The most popular magazines published in Chicago are *Playboy*, with a circulation of 4,209,324 (1983), and *Ebony*, with a circulation of 1,659,243. Many specialized trade and membership magazines, such as the *Lion* and the *Rotarian*, are published in Chicago, which is also the printing and circulation center for many magazines edited in New York.

46 ORGANIZATIONS

Before the Civil War, Yankee-dominated towns and cities in northern Illinois sponsored lyceums, debating circles, women's clubs, temperance groups, and antislavery societies. During the 20th century, Chicago's size and central location attracted the headquarters of numerous national organizations, though far fewer than New York or, more recently, Washington, D.C. The 1982 Census of Service Industries counted 3,159 organizations in Illinois, including 645 business associations; 1,758 civic, social, and fraternal associations; and 73 educational, scientific, and research associations. Major national service and fraternal bodies with headquarters in Chicago or nearby suburbs include the Benevolent and Protective Order of Elks of the USA, Lions Clubs International, Loyal Order of Moose, and Rotary International.

Chicago has long been a center for professional organizations, among them the most powerful single US medical group, the American Medical Association, founded in 1847, and the American Hospital Association, begun in 1898. Other major groups include associations of surgeons, dentists, veterinarians, osteopaths, and dietitians, as well as the Blue Cross and Blue Shield Association and the National Easter Seal Society. The American Bar Association has its headquarters in Chicago, as do several smaller legal groups, including the American Judicature Society and Commercial Law League of America. Librarians also have a base in Chicago: the American Library Association, the Society of American Archivists, and the associations of law and medical librarians. The National Parent-Teacher Association is the only major educational group.

A variety of trade organizations, such as the American Marketing Association, are based in Chicago, though many have moved to Washington, D.C. The American Farm Bureau Federation operates out of Park Ridge. The National Women's Christian Temperance Union, one of the most important of all US pressure groups in the 19th century, has its headquarters in Evanston.

⁴⁷TOURISM, TRAVEL, AND RECREATION

The tourist industry is of special importance to Chicago, the nation's leading convention center. The city's chief tourist attractions are its museums, restaurants, and shops. Chicago also boasts the world's tallest building, the Sears Tower, 110 stories and 1,454 feet (443 meters) high.

For the state as a whole, tourism generated 123,100 jobs and $15.9 million per day in spending in 1982. There are 42 state parks, 4 state forests, 36,659 campsites, and 25 state recreation places; the total number of visitors was 30.5 million in 1984. The Lincoln Home National Historic Site in Springfield is a popular tourist attraction.

Swimming, bicycling, hiking, camping, horseback riding, fishing, and motorboating are the most popular recreational activities. Licenses were issued to 764,377 fishermen and 329,240 hunters in 1982/83. Even more popular than hunting is wildlife observation, an activity that engages nearly 3 million Illinoisans annually.

⁴⁸SPORTS

Illinois is sport-conscious, and some of its teams have recently won championships. In 1984, the Chicago Bears won the National Football League's Central Division championship, and in baseball, the Chicago Cubs won the National League's Eastern Division championship. The Chicago White Sox last won the American League pennant in 1959 and a World Series in 1917. The Chicago Bulls won a National Basketball Association divisional championship in 1975, but the city's other major team, the Black Hawks of the National Hockey League, has not been so fortunate in recent years. The major stadiums in Illinois tend to be old, smallish structures. Wrigley Field, home of the Cubs, still lacks lights for night games.

Horse racing has been profitable (for the owners and a few politicians) since Chicago's first meet was held in 1845. Gambling, both legal and illegal, flourishes, with the pari-mutuel handle exceeding $1 billion.

Colleges and high schools offer full sports programs, with the emphasis on football and basketball. The Fighting Illini of the University of Illinois and the Wildcats of Northwestern compete in the Big Ten conference; the Illini went to the Rose Bowl in 1983, and their basketball team played in the NCAA tourney in the 1984/85 season. The Salukis of Southern Illinois won the NCAA Division II football championship in 1984, and they also won the National Invitation Basketball Tournament in 1967; the DePaul Blue Demons are consistently ranked among the top college basketball teams.

⁴⁹FAMOUS ILLINOISANS

Abraham Lincoln (b. Kentucky, 1809-65), 16th president of the US, is the outstanding figure in Illinois history, having lived and built his political career in the state between 1830 and 1861. The only Illinois native to be elected president is Ronald Reagan (b.1911), who left the state after graduating from Eureka College

to pursue his film and political careers in California. Ulysses S. Grant (b. Ohio, 1822-85), the nation's 18th president, lived in Galena on the eve of the Civil War. Adlai E. Stevenson (b.Kentucky, 1835-1914), founder of a political dynasty, served as US vice president from 1893 to 1897, but was defeated for the same office in 1900. His grandson, also named Adlai E. Stevenson (b.California, 1900-65), served as governor of Illinois from 1949 to 1953, was the Democratic presidential nominee in 1952 and 1956, and ended his career as US ambassador to the United Nations. Charles Gates Dawes (b.Ohio, 1865-1951), a Chicago financier, served as vice president from 1925 to 1929 and shared the 1925 Nobel Peace Prize for the Dawes Plan to reorganize German finances. William Jennings Bryan (1860-1925), a leader of the free-silver and Populist movements, was the Democratic presidential nominee in 1896, 1900, and 1908.

US Supreme Court justices associated with Illinois include David Davis (b.Maryland, 1815-86); John M. Harlan (1899-1971); Chicago-born Arthur Goldberg (b.1908), who also served as secretary of labor and succeeded Stevenson as UN ambassador; Harry A. Blackmun (b.1908); and John Paul Stevens (b.1920). Melville Fuller (b.Maine, 1833-1910) served as chief justice from 1888 to 1910.

Many other politicians who played important roles on the national scene drew their support from the people of Illinois. They include Stephen Douglas (b.Vermont, 1813-61), senator from 1847 to 1861, Democratic Party leader, 1860 presidential candidate, but equally famous as Lincoln's opponent in a series of debates on slavery in 1858; Lyman Trumbull (b.Connecticut, 1813-96), senator from 1855 to 1873, who helped secure passage of the 13th and 14th amendments to the US Constitution; Joseph "Uncle Joe" Cannon (b.North Carolina, 1836-1926), Republican congressman from Danville for half a century and autocratic speaker of the House from 1903 to 1911; Henry Rainey (1860-1934), Democratic speaker of the House during 1933-34; Everett McKinley Dirksen (1896-1969), senator and colorful Republican leader during the 1950s and 1960s; Charles H. Percy (b.Florida, 1919), Republican senator from 1967 to 1985; John B. Anderson (b.1922), Republican congressman for 20 years and an independent presidential candidate in 1980; and Robert H. Michel (b.1923), House Republican leader since 1981.

Among noteworthy governors of the state, in addition to Stevenson, were Richard Yates (b.Kentucky, 1815-73), who maintained Illinois's loyalty to the Union during the Civil War; John Peter Altgeld (b.Germany, 1847-1902), governor from 1893 to 1897; and Republican-Progressive leader Frank Lowden (b.Minnesota, 1861-1943). Richard J. Daley (1902-76) was Democratic boss and mayor of Chicago from 1955 to 1976. Jane Byrne (b.1934), a Daley protégée, became mayor in 1979; she was succeeded in 1983 by Harold Washington (b.1922), the city's first black mayor.

Phyllis Schlafly (b.Missouri, 1924) of Alton became nationally known as an antifeminist conservative crusader during the 1970s. An outstanding Illinoisan was Jane Addams (1860-1935), founder of Hull House (1889), author, reformer, prohibitionist, feminist, and tireless worker for world peace; in 1931, she shared the Nobel Peace Prize. Winners of the Nobel Prize in physics include Albert Michelson (b.Germany, 1852-1931), Robert Millikan (1868-1953), Arthur Holly Compton (b.Ohio, 1892-1962), Enrico Fermi (b.Italy, 1901-54), John Bardeen (b.Wisconsin, 1908), John R. Schrieffer (b. 1931), and James W. Cronin (b.1931). Chemistry prizes went to Robert Mulliken (b.Massachusetts, 1896), Wendell Stanley (b.Indiana, 1904), Willard Libby (b.Colorado, 1908), and Stanford Moore (1913-82). Nobel Prizes in physiology or medicine were won by Charles Huggins (b.Canada, 1901), George Beadle (b.Nebraska, 1903), and Robert W. Holley (b.1922). A Nobel award in literature went to Saul Bellow (b.Canada, 1915), and the economics prize was given to Milton Friedman (b.New

York, 1912), leader of the so-called Chicago school of economists, and to Theodore Schultz (b.South Dakota, 1902) in 1979.

Some of the most influential Illinoisans have been religious leaders; many of them also exercised social and political influence. Notable are Methodist circuit rider Peter Cartwright (b.Virginia, 1785–1872); Dwight Moody (b.Massachusetts, 1837–99), the foremost evangelist of his day; Frances Willard (b.New York, 1839–98), leading force in the National Women's Christian Temperance Union and the feminist cause; Mother Frances Xavier Cabrini (b.Italy, 1850–1917), the first American to be canonized; Bishop Fulton J. Sheen (1895–1979), influential spokesman for the Roman Catholic Church; Elijah Muhammad (Elijah Poole, b.Georgia, 1897–1975), leader of the Black Muslim movement; and Jesse Jackson (b.North Carolina, 1941), civil rights leader and one of the most prominent black spokesmen of the 1980s.

Outstanding business and professional leaders who lived in Illinois include John Deere (b.Vermont, 1804–86), industrialist and inventor of the steel plow; Cyrus Hall McCormick (b.Virginia, 1809–84), inventor of the reaping machine; Nathan Davis (1817–1904), the "father of the American Medical Association"; railroad car inventor George Pullman (b.New York, 1831–97); meat-packer Philip Armour (b.New York, 1832–1901); merchant Marshall Field (b.Massachusetts, 1834–1906); merchant Aaron Montgomery Ward (b.New Jersey, 1843–1913); sporting-goods manufacturer Albert G. Spalding (1850–1915); breakfast-food manufacturer Charles W. Post (1854–1911); William Rainey Harper (b.Ohio, 1856–1906), first president of the University of Chicago; lawyer Clarence Darrow (b.Ohio, 1857–1938); public utilities magnate Samuel Insull (b.England, 1859–1938); Julius Rosenwald (1862–1932), philanthropist and executive of Sears, Roebuck; advertising executive Albert Lasker (b.Texas, 1880–1952); and *Chicago Tribune* publisher Robert R. McCormick (1880–1955).

Artists who worked for significant periods in Illinois (usually in Chicago) include architects William Le Baron Jenney (b.Massachusetts, 1832–1907), Dankmar Adler (b.Germany, 1844–1900), Daniel H. Burnham (b.New York, 1846–1912), John Wellborn Root (b.Georgia, 1850–91), Louis Sullivan (b.Massachusetts, 1856–1924), Frank Lloyd Wright (b.Wisconsin, 1869–1959), and Ludwig Mies van der Rohe (b.Germany, 1886–1969). Important writers include humorist Finley Peter Dunne (1867–1936), creator of the fictional saloonkeeper-philosopher Mr. Dooley; and novelists Hamlin Garland (b.Wisconsin, 1860–1940), Edgar Rice Burroughs (1875–1950), John Dos Passos (1896–1970), Ernest Hemingway (1899–1961), and James Farrell (1904–79). Poets include Harriet Monroe (1860–1936); Edgar Lee Masters (b.Kansas, 1869–1950); biographer-poet Carl Sandburg (1878–1967); Nicholas Vachel Lindsay (1879–1931); Archibald MacLeish (1892–1982), also Librarian of Congress and assistant secretary of state; and Gwendolyn Brooks (b.Kansas, 1917), the first black woman to win a Pulitzer Prize. Performing artists connected with the state include opera stars Mary Garden (b.Scotland, 1877–1967) and Sherrill Milnes (b.1935); clarinetist Benny Goodman (b.1909); pop singers Mel Torme (b.1925) and Grace Slick (b.1939); jazz musician Miles Davis (b.1926); showmen Gower Champion (1921–80) and Robert Louis "Bob" Fosse (b.1927); comedians Jack Benny (Benjamin Kubelsky, 1894–1974), Harvey Korman (b.1927), Bob Newhart (b.1929), and Richard Pryor (b.1940); and a long list of stage and screen stars, including Gloria Swanson (1899–1983), Ralph Bellamy (b.1904), Robert Young

(b.1907), Karl Malden (Malden Sekulovich, b.1913), William Holden (1918–81), Jason Robards, Jr. (b.1922), Charlton Heston (b.1924), Rock Hudson (Roy Fitzgerald, 1925–85), Donald O'Connor (b.1925), Bruce Dern (b.1936), and Raquel Welch (Raquel Tejada, b. 1942).

Dominant figures in the Illinois sports world include Ernest "Ernie" Banks (b.Texas, 1931) of the Chicago Cubs; Robert "Bobby" Hull (b.Canada, 1939) of the Chicago Black Hawks; owner George Halas (1895–1983) and running backs Harold Edward "Red" Grange (b.Pennsylvania, 1903), Gale Sayers (b.Kansas, 1943), and Walter Payton (b.Mississippi, 1954) of the Chicago Bears; and collegiate football coach Amos Alonzo Stagg (b.New Jersey, 1862–1965).

[50]BIBLIOGRAPHY

Allen, John W. *Legends and Lore of Southern Illinois*. Carbondale: Southern Illinois University Press, 1963.

Bluhm, Elaine (ed.). *Illinois Archaeology*. Urbana: University of Illinois Press, 1964.

Bluhm, Elaine (ed.). *Illinois Prehistory*. Urbana: University of Illinois Press, 1963.

Clayton, John. *The Illinois Fact Book and Historical Almanac, 1673-1968*. Carbondale: Southern Illinois University Press, 1970.

Crane, Edgar G. *Illinois: Political Processes and Governmental Performance*. Dubuque: Kendall-Hunt, 1980.

Cutler, Irving. *Chicago: Metropolis of the Mid-Continent*. Dubuque: Kendall-Hunt, 1976.

Dedmon, Emmett. *Fabulous Chicago*. New York: Atheneum, 1983.

Drake, St. Clair, and Horace Cayton. *Black Metropolis*. New York: Harcourt Brace, 1970.

Federal Writers' Project. *Illinois: A Descriptive and Historical Guide*. Reprint. New York: Somerset, n.d. (orig. 1939).

Horrell, C.W., et al. *Land Between the Rivers: The Southern Illinois Country*. Carbondale, Ill.: Southern Illinois University Press, 1982.

Howard, Robert P. Illinois: *A History of the Prairie State*. Grand Rapids, Mich.: Eerdmans, 1972.

Illinois, State of. Department of Commerce and Community Affairs. *Illinois Data Book, 1982*. Springfield, 1982.

Illinois, State of. Secretary of State. *Illinois Blue Book, 1983–1984*. Springfield, 1984.

Jensen, Richard J. *Illinois: A Bicentennial History*. New York: Norton, 1978.

Kenney, David. *Basic Illinois Government*. Carbondale: Southern Illinois University Press, 1974.

Kilian, Michael, Connie Fisher, and F. Richard Ciccone. *Who Runs Chicago?* New York: St. Martin's, 1979.

Kleppner, Paul. *Chicago Divided: The Making of a Black Mayor*. DeKalb: Northern Illinois University Press, 1985.

Koeper, Frederick. *Illinois Architecture from Territorial Times to the Present*. Chicago: University of Chicago Press, 1978.

Nelson, Ronald E. *Illinois: Land and Life in the Prairie State*. Dubuque: Kendall-Hunt, 1978.

Sutton, Robert P. (ed.). *The Prairie State*. 2 vols. Grand Rapids, Mich.: Eerdmans, 1976.

Tingley, Donald F. *The Structuring of a State: The History of Illinois, 1899-1928*. Champaign: University of Illinois Press, 1980.

Wheeler, Adade, and Marlene Wortman. *The Roads They Made: Women in Illinois History*. Chicago: Kerr, 1977.

INDIANA

State of Indiana

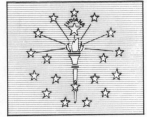

ORIGIN OF STATE NAME: Named "land of Indians" for the many Indian tribes that formerly lived in the state. **NICKNAME:** The Hoosier State. **CAPITAL:** Indianapolis. **ENTERED UNION:** 11 December 1816 (19th). **SONG:** "On the Banks of the Wabash, Far Away." **MOTTO:** The Crossroads of America. **FLAG:** A flaming torch representing liberty is surrounded by 19 gold stars against a blue background. The word "Indiana" is above the flame. **OFFICIAL SEAL:** In a pioneer setting, a farmer fells a tree while a buffalo flees from the forest and across the prairie; in the background, the sun sets over distant hills. The words "Seal of the State of Indiana 1816" surround the scene. **BIRD:** Cardinal. **FLOWER:** Peony. **TREE:** Tulip tree (yellow poplar). **STONE:** Indiana limestone. **POEM:** "Indiana." **LEGAL HOLIDAYS:** New Year's Day, 1 January; Birthday of Martin Luther King, Jr., 3d Monday in January; Lincoln's Birthday, 12 February; Washington's Birthday, 3d Monday in February; Good Friday, March or April; Primary Election Day, 1st Tuesday after 1st Monday in May in even-numbered years; Memorial Day, last Monday in May; Independence Day, 4 July; Labor Day, 1st Monday in September; Columbus Day, 2d Monday in October; Election Day, 1st Tuesday after 1st Monday in November in even-numbered years; Veterans Day, 11 November; Thanksgiving Day, 4th Thursday in November; Christmas Day, 25 December. **TIME:** 7 AM EST = noon GMT; 6 AM CST = noon GMT.

¹LOCATION, SIZE, AND EXTENT

Situated in the eastern north-central US, Indiana is the smallest of the 12 midwestern states and ranks 38th in size among the 50 states.

Indiana's total area is 36,185 sq mi (93,720 sq km), of which land takes up 35,932 (93,064 sq km) and water the remaining 253 sq mi (656 sq km). Shaped somewhat like a vertical quadrangle, with irregular borders on the s and w, the state extends about 160 mi (257 km) E–W and about 280 mi (451 km) N–S.

Indiana is bordered on the N by Michigan (with part of the line passing through Lake Michigan); on the E by Ohio; on the SE and s by Kentucky (the entire line formed by the north bank of the Ohio River); and on the w by Illinois (with the line in the sw demarcated by the Wabash River). The total boundary length of Indiana is 1,696 mi (2,729 km).

Indiana's geographical center is located in Boone County, 14 mi (23 km) NNW of Indianapolis.

²TOPOGRAPHY

Indiana has two principal types of terrain: slightly rolling land in the northern half of the state and rugged hills in the southern, extending to the Ohio River. The highest point in the state, a hill near Lynn (Randolph County) on the eastern boundary, is 1,257 feet (383 meters) above sea level; the lowest point, on the Ohio River, is 320 feet (98 meters). The richest soil is in the north-central region, where the retreating glacier during the last Ice Age enriched the soil, scooped out lakes, and cut passageways for rivers.

Four-fifths of the state's land is drained by the Wabash River, which flows westward across the north-central region and turns southward to empty into the Ohio, and by its tributaries, the White, Eel, Mississinewa, and Tippecanoe rivers. The northern region is drained by the Maumee River, which flows into Lake Erie at Toledo, Ohio, and by the Kankakee River, which joins the Illinois River in Illinois. In the southwest, the two White River forks empty into the Wabash, and in the southeast, the Whitewater River flows into the Ohio.

In addition to Lake Michigan on the northwestern border, there are more than 400 lakes in the northern part of the state. The largest lakes include Wawasee, Maxinkuckee, Freeman, and

Shafer. There are mineral springs at French Lick and West Baden in Orange County, and two large caves at Wyandotte and Marengo in adjoining Crawford County.

The underlying rock strata found in Indiana are sedimentary, atop which are rocks formed during the Paleozoic era, when the land was submerged. About 400 million years ago, the first uplift of land, the Cincinnati arch, divided the Indiana region into two basins, a small one in the north and a large one in the southwest. The land was steadily elevated and at one time formed a lush swamp, which dried up some 200 million years ago when the climate cooled. During the Ice Ages, about five-sixths of the land lay under ice some 2,000 feet (600 meters) thick. The retreat of the glacier more than 10,000 years ago left excellent topsoil and drainage conditions in Indiana.

³CLIMATE

Indiana has a humid continental climate, marked by distinct seasons. Temperatures vary from the extreme north to the extreme south of the state; the annual mean temperature is 53°F (12°C)—49°F (9°C) in the north and 57°F (14°C) in the south. The annual mean for Indianapolis is 52°F (11°C). Although Indiana sometimes has temperatures below 0°F (−18°C) during the winter, the average temperatures in January range between 17°F (−8°C) and 35°F (2°C). Average temperatures during July vary from 63°F (17°C) to 88°F (31°C). The highest recorded temperature was 116°F (47°C) at Collegeville on 14 July 1936; the lowest was −35°F (−37°C) at Greensburg on 2 February 1951.

The growing season averages 155 days in the north and 185 days in the south. Rainfall is distributed fairly evenly throughout the year, although drought sometimes occurs in the southern region. The average annual precipitation in the state is 40 in (102 cm), ranging from about 35 in (89 cm) near Lake Michigan to 45 in (114 cm) along the Ohio River; Indianapolis has an average of 39 in (99 cm) per year. The annual snowfall in Indiana averages less than 22 in (56 cm). Average wind speed in the state is 8 mph (13 km/hr), but gales sometimes occur along the shores of Lake Michigan, and there are occasional tornadoes in the interior.

⁴FLORA AND FAUNA

Because the state has a relatively uniform climate, plant species are distributed fairly generally throughout Indiana. There are 124 native tree species, including 17 varieties of oak, as well as black walnut, sycamore, and tulip tree (yellow poplar), the state tree. Fruit trees—apple, cherry, peach, and pear—are common. Local indigenous species—now reduced because of industrialization and urbanization—are the persimmon, black gum, and southern cypress along the Ohio River; tamarack and bog willow in the northern marsh; and white pine, sassafras, and pawpaw near Lake Michigan. American elderberry and bittersweet are common shrubs, while various jack-in-the-pulpits and spring beauties are among the indigenous wild flowers. The peony is the state flower. Mountain laurel is considered threatened, the prairie white-fringed orchid endangered.

Although the presence of wolves and coyotes has been reported occasionally, the red fox is Indiana's only common carnivorous mammal. Other native mammals are the common cottontail, muskrat, raccoon, opossum, and several types of squirrel. Many waterfowl and marsh birds, including the black duck and great blue heron, inhabit northern Indiana, while the field sparrow, yellow warbler, and red-headed woodpecker nest in central Indiana. Various catfish, pike, bass, and sunfish are native to state waters.

The state provides protection for the following animals, considered to be rare and endangered: bobcat, badger, otter, Indiana bat, gray myotis, southeastern myotis, and big-eared bat. In keeping with federal statutes, Indiana lists as endangered the eastern timber wolf, Arctic peregrine falcon, Kirtland's warbler, bald eagle, longjaw cisco, and eight types of mussel.

⁵ENVIRONMENTAL PROTECTION

During the 19th century, early settlers cut down much of Indiana's forests for farms, leaving the land vulnerable to soil erosion and flood damage, particularly in the southern part of the state. In 1919, the legislature created the State Department of Conservation (which in 1965 became the Department of Natural Resources) to reclaim worn-out soil, prevent further erosion, and control pollution of rivers and streams. In 1934, the state's newly created Natural Resources Planning Board (now the 12-member Natural Resources Commission) made a survey of soil, water, forest, and mineral conditions and outlined conservation practices for their proper future use.

The Department of Natural Resources regulates the use of Indiana's lands, waters, forests, and minerals. Specifically, the department manages land subject to flooding, preserves natural rivers and streams, grants mining permits and regulates strip-mining, plugs and repairs faulty oil or gas wells, administers existing state parks and preserves and buys land for new ones, regulates hunting and fishing, and registers motorboats and snowmobiles. Also, the department is responsible for preventing soil erosion and flood damage, and for conserving and disposing of water in the state's watersheds. In 1984/85, the Department of Natural Resources expended for the above and other purposes an estimated total of $46.1 million, including $5.6 million in federal funds.

Other state agencies assist the department in protecting the environment. The Environmental Management Board makes most of the environmental policy for the state. This 11-member body, established in 1972, reviews regulations adopted by a separate Air Pollution Control Board and a Stream Pollution Control Board. It also has jurisdiction over public water supplies, solid and hazardous waste disposal, certification of water and wastewater treatment plant operators, and construction of nuclear power plants. There is also a 13-member Pesticide Review Board, which classifies pesticides according to their use and regulates their handling.

Hazardous wastes have become an increasing problem in recent years. It was estimated that in 1982 Indiana generated 6 million to 10 million tons of hazardous wastes, more than all but nine other states. Most of the wastes are inorganic, arising from the manu-

facture of durable goods. Effective in 1982, operators of hazardous waste disposal facilities were required to pay a tax of $1.50 per ton of waste disposed. Two-thirds of these revenues are deposited in a state Hazardous Substances Emergency Response Fund; the remainder is paid to the county where the disposal takes place.

⁶POPULATION

In 1985, Indiana had an estimated population of 5,489,287 and ranked 14th in population among the 50 states; the population density was 153 persons per sq mi (59 per sq km).

Although the French founded the first European settlement in Indiana in 1717, the census population was no more than 5,641 in 1800, when the Indiana Territory was established. Settlers flocked to the state during the territorial period, and the population rose to 24,520 by 1810. After Indiana became a state in 1816, its population grew even more rapidly, reaching 147,178 in 1820 and 988,416 in 1850. At the outbreak of the Civil War, Indiana had 1,350,428 inhabitants and ranked 5th in population among the states.

Indiana was relatively untouched by the great waves of European immigration that swept the US from 1860 to 1880. In 1880, when the state's population was 1,978,301, Indiana had fewer foreign-born residents (about 7% of its population) than any other northern state. Indiana doubled its 1900 population to 5,193,669 by the time of the 1970 census. The 1980 census showed a population of 5,490,224, a 10-year growth of 5.7%.

Of the 1980 census population, 64% lived in urban areas and 36% resided in rural areas. Indianapolis, the capital and largest city, expanded its boundaries in 1970 to coincide with those of Marion County, thereby increasing its area to 388 sq mi (1,005 sq km) and its population by some 50% (the city and county limits also include four self-governing communities). The population was 710,280 in 1984, and the Indianapolis metropolitan area had an estimated population of 1,194,600. Other cities with 1984 populations of more than 100,000 were Fort Wayne, 165,416; Gary, 143,096; Evansville, 130,333; and South Bend, 107,117. All of these cities suffered population declines in the 1970s and early 1980s.

⁷ETHNIC GROUPS

Originally an agricultural state, Indiana was settled by native Americans moving west, by a small group of French Creoles, and by European immigrant farmers. Although railroad building and industrialization attracted other immigrant groups—notably the Irish, Hungarians, Italians, Poles, Croats, Slovaks, and Syrians—foreign immigration to Indiana declined sharply in the 20th century. As of 1980, foreign-born Hoosiers numbered 102,000, or less than 2% of the state total. The most common European ancestry groups among Indiana residents were British (including Irish and Scottish), German, French, Dutch, and Polish.

Restrictions on foreign immigration and the availability of jobs spurred the migration of black Americans to Indiana after World War I; by 1980, the state had 414,732 blacks, representing about 7.6% of the total population. Approximately 30% of all Indiana blacks live in the industrial city of Gary, which was 71% black in 1980.

The total Hispanic population in 1980 was 87,000. That year, Indiana's Asian residents included 4,290 Indians, 3,974 Chinese, 3,625 Filipinos, 3,253 Koreans, 2,356 Japanese, and 2,338 Vietnamese.

The Indians of early 19th-century Indiana came from a variety of Algonkian-speaking tribes, including Delaware, Shawnee, and

LOCATION: 37°47′ to 41°46′N; 84°49′ to 88°02′W. **BOUNDARIES:** Lake Michigan shoreline, 45 mi (72 km); State of Michigan line, 99 mi (159 km); Ohio line, 179 mi (288 km); Kentucky line (Ohio River), 848 mi (1,365 km); Illinois line: land, 168 mi (270 km); Wabash River, 357 mi (575 km).

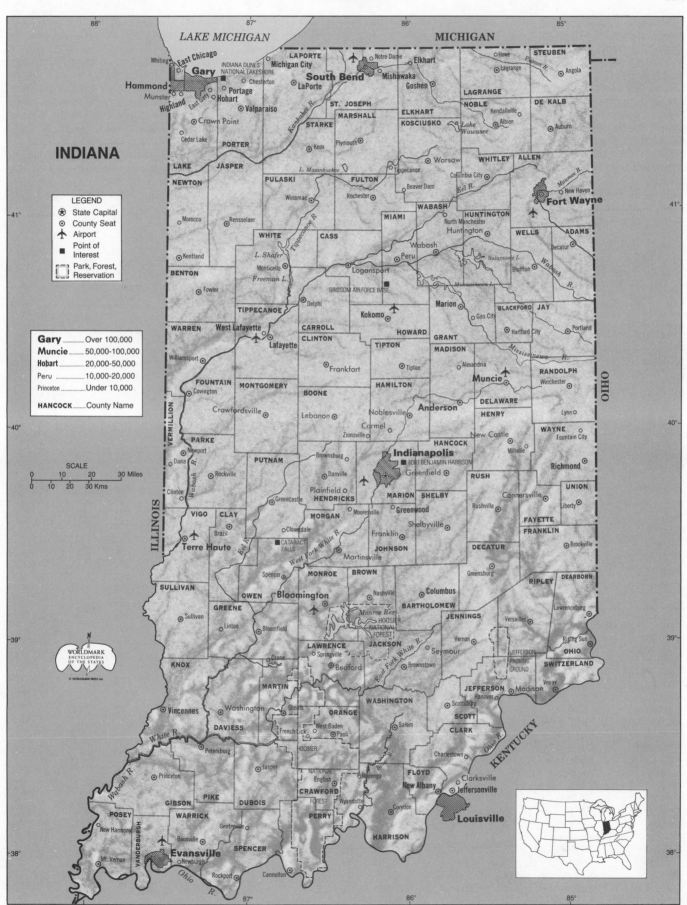

See endsheet maps: J3.

Potawatomi. By 1846, however, all Indian lands in the state had been seized or ceded, and most Indians had been removed. In 1980, only 7,681 Indiana residents identified themselves as American Indian.

⁸LANGUAGES

Several Algonkian Indian tribes, including some from the east, met the white settlers who arrived in Indiana in the early 1800s. The heritage of the Delaware, Potawatomi, Miami, and other groups survives in many place-names, from Kokomo to Nappanee, Muncie, and Shipshewana.

Except for the dialect mixture in the industrial northwest corner and for the Northern-dialect fringe of counties along the Michigan border, Indiana speech is essentially that of the South Midland pioneers from south of the Ohio River, with a transition zone toward North Midland north of Indianapolis. Between the Ohio River and Indianapolis, South Midland speakers use *evening* for late afternoon, eat *clabber cheese* instead of cottage cheese, are wary of *frogstools* rather than toadstools, once held that *toadfrogs* and not plain toads caused warts, eat *goobers* instead of peanuts at a ball game, and may therefore be *sick at the stomach*. In the same region, some Hoosiers use a few Midland words that also occur north of Indianapolis, such as *rock fence* (stone wall), *French harp* (harmonica), *mud dauber* (wasp), *shucks* (leaves on an ear of corn), and perhaps even some expanding North Midland words, such as *run* (a small stream), *teetertotter* (seesaw), and *fishworm*. North of Indianapolis, speakers with a Midland Pennsylvania background wish on the *pullybone* of a chicken, may use a *trestle* (sawhorse), and are likely to get their hands /greezy/ rather than greasy. Such was the Hoosier talk of James Whitcomb Riley.

In 1980, 96% of all Hoosiers 3 years old and older spoke only English at home. Other languages spoken at home were as follows:

Indiana Counties, County Seats, and County Areas and Populations

COUNTY	COUNTY SEAT	LAND AREA (SQ MI)	POPULATION (1984 EST.)	COUNTY	COUNTY SEAT	LAND AREA (SQ MI)	POPULATION (1984 EST.)
Adams	Decatur	340	29,849	Madison	Anderson	453	134,117
Allen	Fort Wayne	659	289,224	Marion	Indianapolis	396	774,774
Bartholomew	Columbus	409	64,406	Marshall	Plymouth	444	40,613
Benton	Fowler	407	10,052	Martin	Shoals	339	10,855
Blackford	Hartford City	166	15,558	Miami	Peru	369	37,186
Boone	Lebanon	424	38,459	Monroe	Bloomington	385	100,856
Brown	Nashville	312	12,473	Montgomery	Crawfordsville	505	35,775
Carroll	Delphi	372	19,488	Morgan	Martinsville	409	53,575
Cass	Logansport	414	39,877	Newton	Kentland	401	14,634
Clark	Jeffersonville	376	89,540	Noble	Albion	413	36,178
Clay	Brazil	360	24,335	Ohio	Rising Sun	87	5,147
Clinton	Frankfort	405	31,902	Orange	Paoli	408	18,672
Crawford	English	307	9,848	Owen	Spencer	386	16,398
Daviess	Washington	432	29,487	Parke	Rockville	444	16,761
Dearborn	Lawrenceburg	307	36,692	Perry	Cannelton	381	19,119
Decatur	Greensburg	373	24,025	Pike	Petersburg	341	13,239
De Kalb	Auburn	364	32,721	Porter	Valparaiso	419	124,977
Delaware	Muncie	392	123,577	Posey	Mt. Vernon	410	26,199
Dubois	Jasper	429	35,487	Pulaski	Winamac	435	13,420
Elkhart	Goshen	466	142,205	Putnam	Greencastle	482	30,370
Fayette	Connersville	215	27,888	Randolph	Winchester	454	28,599
Floyd	New Albany	150	63,019	Ripley	Versailles	447	25,087
Fountain	Covington	398	18,958	Rush	Rushville	408	18,941
Franklin	Brookville	385	20,896	St. Joseph	South Bend	459	239,760
Fulton	Rochester	369	18,959	Scott	Scottsburg	192	20,339
Gibson	Princeton	490	34,335	Shelby	Shelbyville	412	40,250
Grant	Marion	415	78,856	Spencer	Rockport	400	20,884
Greene	Bloomfield	546	30,446	Starke	Knox	309	21,202
Hamilton	Noblesville	398	88,968	Steuben	Angola	308	24,964
Hancock	Greenfield	307	44,358	Sullivan	Sullivan	452	20,872
Harrison	Corydon	486	28,487	Switzerland	Vevay	224	7,534
Hendricks	Danville	409	72,857	Tippecanoe	Lafayette	502	123,872
Henry	New Castle	395	50,232	Tipton	Tipton	261	15,921
Howard	Kokomo	293	85,105	Union	Liberty	163	6,774
Huntington	Huntington	366	35,059	Vanderburgh	Evansville	236	168,073
Jackson	Brownstown	514	37,652	Vermillion	Newport	260	17,976
Jasper	Rensselaer	561	26,664	Vigo	Terre Haute	405	110,156
Jay	Portland	384	21,892	Wabash	Wabash	398	35,189
Jefferson	Madison	363	30,442	Warren	Williamsport	366	8,977
Jennings	Vernon	378	22,768	Warrick	Boonville	391	44,900
Johnson	Franklin	321	81,406	Washington	Salem	516	22,247
Knox	Vincennes	520	42,982	Wayne	Richmond	404	73,612
Kosciusko	Warsaw	540	61,763	Wells	Bluffton	370	24,994
Lagrange	Lagrange	380	27,535	White	Monticello	506	23,664
Lake	Crown Point	501	504,631	Whitley	Columbia City	336	26,462
LaPorte	LaPorte	600	107,093				
Lawrence	Bedford	452	41,362	TOTALS		35,932	5,497,929

Spanish	71,060	Serbo-Croatian	6,421
German	41,444	Greek	5,530
Polish	15,576	Italian	5,338
French	13,885	Dutch	3,925

Hungarian, Chinese, Arabic, and Korean were also reported.

[9]RELIGIONS

The first branch of Christianity to gain a foothold in Indiana was Roman Catholicism, introduced by the French settlers in the early 18th century. The first Protestant church was founded near Charlestown by Baptists from Kentucky in 1798. Three years later, a Methodist church was organized at Springville; in 1806, Presbyterians established a church near Vincennes; and the following year, Quakers built their first meetinghouse at Richmond. The Disciples of Christ, Lutherans, the United Brethren, Mennonites, and Jews were among the later 19th-century arrivals.

A dissident religious sect, the Shakers, established a short-lived community in Sullivan County in 1808. In 1815, some German separatists, led by George Rapp, founded a community called the Harmonie Society, which flourished briefly. Rapp moved his followers to Pennsylvania and sold the town to a Scottish social reformer, Robert Owen, in 1825. Owen renamed the town New Harmony and tried to establish there a nonreligious utopia, but the experiment failed after three years. A group of religious dissidents founded the pentecostal Church of God at Beaver Dam in 1881; the world headquarters of the church, which had 49,194 adherents in 1980, is now at Anderson. The Youth for Christ movement started in Indianapolis in 1943.

In 1980, 44.8% of Indiana's population were estimated to be religious adherents, and there were 6,590 churches in the state. According to the 1980 church census, 31% of the population was Protestant. The largest Protestant denominations were the United Methodists, with 386,029 adherents; Christian Churches and Churches of Christ, 166,973; American Baptist Churches in USA, 146,278; Lutheran Church–Missouri Synod, 110,408; United Presbyterian Church, 108,255; and Churches of Christ–Disciples of Christ, 104,662. There were 724,049 Roman Catholics in 1984, representing about 13% of the population. The estimated Jewish population of the state was 21,360 as of 1984.

[10]TRANSPORTATION

Indiana's central location in the US and its position between Lake Michigan to the north and the Ohio River to the south gave the state its motto, "The Crossroads of America." Historically, the state took advantage of its strategic location by digging canals to connect Indiana rivers and by building roads and railroads to provide farmers access to national markets.

The success of the state's first railroad, completed in 1847 between Madison and Indianapolis, led to a tenfold increase in track mileage during the 1850s, and more railroad expansion took place after the Civil War. In December 1983, Class I railroads operated 5,329 mi (8,576 km) of track in Indiana. Regularly scheduled Amtrak passenger trains served Indianapolis, Fort Wayne, Valparaiso, South Bend, and 16 other stations in the state, with a total ridership in Indiana of 214,142 persons during 1983/84. Indianapolis and other major cities have public transit systems subsidized heavily by the state and federal governments. The South Shore commuter railroad connects South Bend, Gary, and East Chicago with Chicago, Ill.

The east–west National Road (US 40) reached Indiana in 1827, and the north–south Michigan road (US 421) was built in the late 1830s. In December 1983, Indiana had 1,114 mi (1,793 km) of interstate highways—more than other states of comparable size. In that year, there were 11,056 mi (17,793 km) of state highways and 80,680 mi (129,842 km) of county and municipal roads. In 1983, motor vehicle registrations totaled 3,851,752, including 2,852,382 passenger cars and 981,326 trucks. Several of the nation's largest moving companies have their headquarters in Indiana.

Water transportation has been important from the earliest years of European settlement. The Wabash and Erie Canal, constructed in the 1830s from Fort Wayne east to Toledo, Ohio, and southwest to Lafayette, was vital to the state's market economy. In 1836, the state legislature earmarked $10 million for an ambitious network of canals, but excessive construction costs and the financial panic of 1837 caused the state to go virtually bankrupt and default on its bonds. Nevertheless, the Wabash canal was extended to Terre Haute and Evansville by the early 1850s.

The transport of freight via Lake Michigan and the Ohio River helped to spark Indiana's industrial development. A deepwater port on Lake Michigan, which became operational in 1970, provided access to world markets via the St. Lawrence Seaway. Public port facilities on the Ohio River near Mt. Vernon were opened in June 1979, and a second port project was opened in mid-1985.

In 1983, 10 air carriers provided scheduled airline service to Indianapolis; Evansville was served by 5 air carriers, Fort Wayne and South Bend by 3. In the same year there were 432 public and private airports in the state. The number of active general aviation aircraft in Indiana was 4,207.

[11]HISTORY

When the first human beings inhabited Indiana is not known. Hundreds of sites used by primitive hunters, fishermen, and food gatherers before 1000 BC have been found in Indiana. Burial mounds of the Woodland culture (1000 BC to AD 900), when the bow and arrow appeared, have been located across the state. The next culture, called Mississippian and dating about AD 900 to 1500, is marked by gardens, ceramics, tools, weapons, trade, and social organization. It is well illustrated by remains of an extensive village on the north side of the Ohio River near Newburgh. The unidentified inhabitants are believed to have come up from the south about 1300, for reasons not known, and to have migrated back before 1500, again for unknown reasons.

The next Indian invaders, and the first to be seen by white men, were the Miami and Potawatomi tribes that drifted down the west side of Lake Michigan and turned across the northern sector of what is now Indiana after the middle of the 17th century. The Kickapoo and Wea tribes pushed into upper Indiana from northern Illinois. The southern two-thirds of the present state was a vast hunting ground, without villages.

The first European penetration was made in the 1670s by the French explorers Father Jacques Marquette and Robert Cavelier, Sieur de la Salle. After the founding of Detroit in 1701, the Maumee-Wabash river route to the lower Ohio was discovered. At the portage between the two rivers, Jean Baptiste Bissot, Sieur de Vincennes, lived at Kekionga, the principal village of the Miami and the present site of Fort Wayne. The first French fort was built farther down the Wabash among the Wea, near modern Lafayette, in 1717. Three years later, Fort Miami was erected. Vincennes's son constructed another fort on the Wabash in 1732, at the site of the town later named for him.

English traders venturing down the Ohio River disputed the French trade monopoly, and as a result of the French and Indian War, French Canada was given up to the British in 1760. Indians under Chief Pontiac captured the two forts in northern Indiana, and the area was not securely in English hands until 1765. The pre-Revolutionary turbulence in the Atlantic seaboard colonies was hardly felt in Indiana, although the region did not escape the Revolutionary War itself. Colonel George Rogers Clark, acting for Virginia, captured Vincennes from a British garrison early in 1779 after a heroic march. Two years later, a detachment of 108 Pennsylvanians, passing down the Ohio to reinforce Clark, was surprised by a force of French Canadians and Indians under Mohawk Captain Joseph Brant; most of the Pennsylvanians were killed during the battle or after capture.

Following the Revolutionary War, the area northwest of the

Ohio River was granted to the new nation; known as the Northwest Territory, it included present-day Indiana, Ohio, Illinois, Michigan, Wisconsin, and part of Minnesota. The first US settlement in Indiana was made in 1784 on land opposite Louisville, Ky., granted to Clark's veterans by Virginia. (The new town, called Clarksville, still exists.) Americans also moved into Vincennes. Government was established by the Continental Congress under the Northwest Ordinance of 1787. Again, Indian unrest endangered all settlements north of the Ohio, and the small US army, with headquarters at Cincinnati, met defeat at what is now Fort Wayne in 1790 and disaster in neighboring Ohio in 1791. General Anthony Wayne was put in command of an enlarged army and defeated the Indians in 1794 at Fallen Timbers (near Toledo, Ohio). British meddling was ended by Jay's treaty later the same year. Wayne then built a new fort—named for him—among the Miami.

In 1800, as Ohio prepared to enter the Union, the rest of the Northwest Territory was set off and called Indiana Territory, with its capital at Vincennes. There Elihu Stout established a newspaper, the *Indiana Gazette,* in 1804. After Michigan Territory was detached in 1805, and Illinois Territory in 1809, Indiana assumed its present boundaries. The federal census counted 24,520 people in Indiana in 1810, including a new Swiss colony on the Ohio, where settlers planted vineyards and made wine.

William Henry Harrison was appointed first governor and, with a secretary and three appointed judges, constituted the government of Indiana Territory. Under the Northwest Ordinance, when the population reached 5,000 adult males, it was allowed to elect an assembly and nominate candidates for an upper house. When the population totaled 60,000—as it did in 1815—the voters were allowed to write a state constitution and to apply for admission to the Union. A short constitution excluding slavery and recommending public schools was adopted, and Indiana became the 19th state on 11 December 1816.

Meanwhile, Indiana had seen Governor Harrison lead US troops up the Wabash in 1811 and beat off an Indian attack at Tippecanoe. The War of 1812 took Harrison away from Indiana, and battles were fought in other theaters. Hoosiers suffered Indian raids, and two forts were besieged for a few days. After the war, new settlers began pouring into the state from the upper South and in fewer numbers from Ohio, Pennsylvania, New York, and New England. A group of German Pietists under George Rapp settled Harmonie on the lower Wabash in 1815 and stayed 10 years before selling out to Robert Owen, a visionary with utopian dreams that failed at the village he renamed New Harmony. In 1816, Tom Lincoln brought his family from Kentucky, and his son Abe grew up in southern Indiana from age 7 to 21.

Unlike most other states, Indiana was settled from south to north. The inhabitants were called Hoosiers; the origin of the word is obscure, but the term may have come from an Anglo-Saxon word for hill dwellers. Central and northern Indiana was opened up as land was purchased from the Indians. The Potawatomi were forced to go west in 1838, and the Miami left in 1846. Commerce flowed south to the Ohio River in the form of corn, hogs, whiskey, and timber. Indianapolis was laid out as a planned city and centrally located capital in 1820, but 30 years passed before its population caught up to the size of Madison and New Albany on the Ohio.

An overambitious program of internal improvements (canals and roads) in the 1830s plunged the state into debts it could not pay. Railroads, privately financed, began to tie Indiana commercially with the East. The Irish came to dig canals and lay the rails, and Germans, many of them Catholics, came to do woodworking and farming. Levi Coffin, a Quaker who moved to Fountain City in 1826, opened a different kind of road, the Underground Railroad, to help escaping slaves from the South.

A new constitution in 1851 showed Jacksonian preferences for more elective offices, shorter terms, a one-term governorship, limited biennial legislative sessions, county government, obligatory common schools, and severe limits on state debt. But this constitution also prohibited Negroes from entering the state.

Hoosiers showed considerable sympathy with the South in the 1850s, and there was considerable "copperhead" activity in the early 1860s. Nevertheless, Indiana remained staunchly in the Union under Governor Oliver P. Morton, sending some 200,000 soldiers to the Civil War. The state suffered no battles, but General John Hunt Morgan's Confederate cavalry raided the southeastern sector of Indiana in July 1863.

After the Civil War, small local industries expanded rapidly. The first nonfarm enterprises were gristmills, sawmills, meat-packing plants, distilleries and breweries, leatherworking shops, furniture factories, and steamboat and carriage makers. Wagons made by Studebaker in South Bend won fame during the war, as did Van Camp's canned pork and beans from Indianapolis. Discovery of natural gas in several northeastern counties in 1886, and the resultant low fuel prices, spurred the growth of glass factories. Elwood Haynes of Kokomo designed a one-cylinder horseless carriage in 1894 and drove it. As America became infatuated with the new autos, 375 Indiana factories started turning them out. A racetrack for testing cars was built outside Indianapolis in 1908, and the famous 500-mi (805-km) race on Memorial Day weekend began in 1911. Five years earlier, US Steel had constructed a steel plant at the south end of Lake Michigan. The town built by the company to house the workers was called Gary, and it grew rapidly with the help of the company and the onset of World War I. Oil refineries were developed in the same area, known as the Calumet region.

Of the millions of immigrants who flocked to the US from 1870 to 1914, very few settled in Indiana. The percentage of foreign-born residents declined from 9% in 1860 to 7% in 1880, all of them from northern Europe and over half from Germany. By 1920, the percentage was down to 5%, although some workers from southern and eastern Europe had gravitated to the industries of the Calumet.

Although many Hoosiers of German and Irish descent favored neutrality when World War I began, Indiana industries boomed with war orders, and public sympathy swung heavily toward the Allies. Indiana furnished 118,000 men and women to the armed forces and suffered the loss of 3,370—a much smaller participation than in the Civil War, from a population more than twice the size.

After 1920, only about a dozen makes of cars were still being manufactured in Indiana, and those factories steadily lost out to the big three carmakers in Detroit. The exception was Studebaker in South Bend, which grew to have more than 23,000 employees during World War II. The company finally closed its doors in 1965. Auto parts continued to be a big business, however, along with steelmaking and oil refining in the Calumet. Elsewhere there was manufacturing of machinery, farm implements, railway cars, furniture, and pharmaceuticals. Meat-packing, coal mining, and limestone quarrying continued to be important. In 1860, about 21,000 workers were employed in manufacturing; as of 1980, more than 30 times that number were production workers in industrial occupations. With increasing industrialization, cities grew, particularly in the northern half of the state, and the number of farms diminished. The balance of rural and urban population, about even in 1920, tilted in favor of urban dwellers.

World War II had a greater impact on Indiana than did World War I. Most factories converted to production of war materials; 300 held defense orders in 1942. Du Pont built a huge powder plant near Charlestown. The slack in employment was taken up, women went into factories, and more rural families moved to cities. Military training facilities were created; Camp Atterbury covered 100 sq mi (259 sq km) in Bartholomew County, and two air stations trained aviators. Two large ammunition depots loaded and stored shells; and the enormous Jefferson Proving Ground tested ammunition and parachutes.

After the war, many locally owned small industries were taken over by national corporations, and their plants were expanded. By 1984, the largest employer in Indiana was General Motors, with 47,800 employees in six cities. Inland Steel, with 18,500 workers, was second, followed by US Steel with 13,800 workers, Radio Corporation of America with 10,900, and Eli Lilly with 10,600. Other major employers were K-Mart, Cummins Engine, Chrysler, and General Electric. Although the state's population in the mid-1980s was about two-thirds urban and one-third rural, agriculture retained much of its importance.

Nostalgia for an older, simpler, rural way of life pervades much Hoosier thinking. The state stands high in conservation, owing to the vision of Richard Lieber, a state official who from 1933 to 1944 promoted the preservation of land for state parks and recreational areas as well as for state and federal forests.

Hoosiers enjoy politics and participate intensively in conventions and elections. The percentage of registered voters who vote has generally exceeded the national average by a wide margin, reaching a peak of 95% in 1876, when the national average was 82%. The evenness of strength between the two major political parties during much of its history has frequently made Indiana a swing state, eagerly courted by Democrats and Republicans alike.

The state legislature was dominated by rural interests until reapportionment in 1966 gave urban counties more representation. Biennial sessions were then changed to annual, although they are still limited in duration. Indianapolis has extended its boundaries to cover the county, and much of city and county government is unified. The direct primary for nomination of governor, lieutenant governor, and US senator was mandated in 1975. In 1976, for the first time, the governor was allowed to succeed himself, and Republican Otis Bowen was reelected. He was followed in 1980 by Republican Robert D. Orr, who also went on to win reelection to a second term.

12STATE GOVERNMENT

The first state constitution took effect when Indiana became a state in 1816. Reportedly written by convention delegates beneath a huge elm tree in Corydon, the first state capital, the brief document prohibited slavery and recommended a free public school system, including a state university.

This constitution did not allow for amendment, however, and a new constitution that did so was adopted in 1851. The second constitution authorized more elective state officials, gave greater responsibility to county governments, and prohibited the state from going into debt (except under rare circumstances). It also established biennial rather than annual sessions of the state legislature, a provision not repealed until 1971. With amendments, the second constitution is still in effect today.

The Indiana general assembly consists of a 50-member senate elected to four-year terms, with half the senators elected every two years, and a 100-member house of representatives elected to two-year terms. A member of the general assembly must be a US citizen and have been a resident of Indiana for at least one year. A senator must be at least 25 years of age, a representative at least 21 years old. Senators and representatives are paid the same base salary ($9,600 per year) and allowances ($65 per day); legislative leaders receive additional compensation.

The state's chief executive is the governor, elected to a 4-year term and eligible for reelection, although ineligible to serve more than 8 years in a 12-year period. A governor must be at least 30 years old, a US citizen, and a state resident for 5 years prior to election. The governor may call special sessions of the legislature and may veto bills passed by the legislature, but his veto can be overridden by a majority vote in each house. The governor's annual salary as of 1984 was $66,000.

Indiana's other top elected officials are the lieutenant governor, secretary of state, treasurer, auditor, attorney general, and superintendent of public instruction. Each is elected to a four-year term. The lieutenant governor is constitutionally empowered to preside over the state senate and to act as governor if the office should become vacant or the incumbent is unable to discharge his duties. By statute, the lieutenant governor also serves as executive director of the Department of Commerce and as commissioner of agriculture.

Legislation may be introduced in either house of the general assembly, although bills for raising revenue must originate in the house of representatives. A bill approved by both houses goes to the governor for signing into law; if the governor declines to sign it within seven days, the bill becomes law, but if the governor vetoes it, majorities of at least 26 votes in the senate and 51 votes in the house are required to override the veto. Should the governor veto a bill after the end of a legislative session, it must be returned to the legislature when that body reconvenes.

An amendment to the state constitution must be approved by two successive legislatures and be submitted to the voters for approval or rejection at the next general election.

In order to vote in Indiana, a person must be a US citizen, be at least 18 years old, and have been a resident of the voting precinct for 30 days.

13POLITICAL PARTIES

The Democratic Party has been one of the two major political parties since Indiana became a state in 1816, as has the Republican Party since its inception in 1854. In that year, Hoosiers voted for Democrat James Buchanan for president, but in 1860, the voters supported Republican Abraham Lincoln. After voting Republican in four successive presidential elections, Indiana voted Democratic in 1876 and became a swing state. More recently, a Republican trend has been evident: the state voted Republican in 11 out of 12 presidential elections between 1940 and 1984.

Third-party movements have rarely succeeded in Indiana. Native son Eugene Debs, the Socialist Party leader who was personally popular in Indiana, received only 36,931 votes in the state in 1912, while garnering more than 900,000 votes nationally. Even in 1932, during the Great Depression, Socialist candidate Norman Thomas won only 21,388 votes in Indiana. The most successful third-party movement in recent decades was George Wallace's American Independent Party, which took 243,108 votes (11.5% of the Indiana total) in 1968. In each of the four presidential elections of the 1970s and early 1980s, minority party candidates together received only 1.1% or less of the votes cast.

In 1980, 58% of the state's electorate voted in the presidential election—slightly above the US average of 54%. In that year, Indiana voted for Republican presidential candidate Ronald Reagan and elected Republican Governor Robert Orr; Republicans also retained majorities in both houses of the state legislature.

In the November 1984 elections, Hoosiers again chose Republican Ronald Reagan for president and elected a Republican to the governorship for the fifth consecutive time, in addition to retaining sizable majorities in the state legislature. The state's delegation to the US House of Representatives was split, 5–5; one of the Democrats, Representative Frank McCloskey, won in the 8th Congressional District when the Democrat-dominated US House of Representatives voted to seat him on 1 May 1985, after a controversial House-supervised recount gave him victory by a margin of four votes.

After the 1984 elections, the state Senate consisted of 20 Democrats and 30 Republicans; the House had 39 Democrats and 61 Republicans. There were 19 women in the state legislature, or 13% of the membership. Elected officials in the state included only 67 blacks, or less than 1% of the total, and 6 Hispanics.

[14]LOCAL GOVERNMENT

In 1816, when Indians controlled central and northern Indiana, the state had only 15 counties. By 1824, the number of counties had grown to 49. All but one of Indiana's 92 counties were established by 1851. The last county—Newton, in the state's northwest corner—was created in 1859.

Counties in Indiana have traditionally provided law enforcement in rural areas, operated county courts and institutions, maintained county roads, administered public welfare programs, and collected taxes. Under a "home rule" law enacted by the state in 1980, they also have "all the power they need for the effective operation of government as to local affairs," or, in effect, all powers not specifically reserved to the state. In 1984, counties were given the power to impose local income taxes, and during that year six counties, accounting for over 20% of the state's population, elected to do so.

The county's business is conducted by a Board of County Commissioners, consisting of three members elected to four-year terms. Nine officials also elected to four-year terms exercise executive functions: the county auditor, treasurer, recorder, clerk, surveyor, sheriff, prosecuting attorney, coroner, and assessor. The county's appointed officials include the county superintendent of schools, highway supervisor, highway engineer, extension agent, attorney, and physician. An elected seven-member County Council exercises taxing power and acts as a check on the Board of County Commissioners. The major exception to this general pattern is Marion County, which in 1970 was consolidated with the city of Indianapolis and is governed by an elected mayor and council of 29 members.

Townships (1,008 in 1984) provide assistance for the poor and assess taxable property. Each township is administered by a trustee elected to a four-year term. In a few townships, the trustee oversees township schools, but most public schools are now run by community school corporations.

Indiana had 115 cities in 1984. They are governed by elected city councils varying in membership from five to 25 persons. City officials elected for four-year terms are the mayor and (except in Indianapolis) the city clerk or clerk-treasurer. (In Indianapolis, the city clerk is appointed for a one-year term.) In 1984 there were also 450 incorporated towns. Each town is governed by an elected board of trustees (one for each ward) and an elected clerk-treasurer. The board may appoint a town manager and a marshal.

[15]STATE SERVICES

In 1974, Indiana's state legislature created the State Ethics and Conflicts of Interest Commission to formulate and regulate a code of ethics for state officials. The commissioner investigates reported cases of misconduct or violations of the code of ethics by any state official or employee. After holding hearings, the commission reports violations to the governor and makes its findings public. Top-level state officials and heads of state departments must provide statements of their financial interests to the commission.

In 1977, the state established an Interdepartmental Board for the Coordination of Human Service Programs. Members include the chief administrative officers of state agencies for senior citizens and community services, mental health, health, corrections, and public welfare. The board provides assistance to persons and families requiring help from one of these agencies and monitors federal service programs in the state. The Indiana Office of Social Services Fiscal Office administers programs for the board. An executive assistant to the governor serves as chairman of the board, which also includes the director of the State Budget Agency.

Educational services are provided by the Commission for Higher Education, the Indiana Educational Services Foundation, and the Commission for Postsecondary Proprietary Education, which accredits private vocational, technical, and trade schools in the state. A public counselor, appointed by the governor, represents the public at hearings of the Public Service Commission, which regulates public transportation agencies and public utilities. Health services are supplied by the State Board of Health, Department of Mental Health, and Emergency Medical Services Commission. Disabled citizens are assisted by the Indiana Rehabilitation Services Agency. The Civil Rights Commission enforces state antidiscrimination laws.

Indiana Presidential Vote by Political Parties, 1948–84

YEAR	ELECTORAL VOTES	INDIANA WINNER	DEMOCRAT	REPUBLICAN	PROGRESSIVE	PROHIBITION
1948	13	Dewey (R)	807,833	821,079	9,649	14,711
1952	13	*Eisenhower (R)	801,530	1,136,259	1,222	15,335
1956	13	*Eisenhower (R)	783,908	1,182,811	—	6,554
1960	13	*Nixon (R)	952,358	1,175,120	—	6,746
1964	13	*Johnson (D)	1,170,848	911,118	—	8,266
					AMERICAN IND.	
1968	13	*Nixon (R)	806,659	1,067,885	243,108	4,616
					PEOPLE'S	SOC. WORKERS
1972	13	*Nixon (R)	708,568	1,405,154	4,544	5,575
					AMERICAN	
1976	13	Ford (R)	1,014,714	1,185,958	14,048	5,695
					CITIZENS	LIBERTARIAN
1980	13	*Reagan (R)	844,197	1,255,656	4,852	19,627
1984	12	*Reagan (R)	841,481	1,377,230	—	6,741

*Won US presidential election.

¹⁶JUDICIAL SYSTEM

The Indiana supreme court consists of five justices who are appointed by the governor from names submitted by a nonpartisan judicial nominating committee. To qualify for selection, a nominee must have practiced law in the state for at least 10 years or have served as judge of a lower court for at least 5 years. A justice serves for 2 years and then is subject to approval by referendum in the general election; if approved by the voters, the justice serves a 10-year term before being again subject to referendum. The chief justice of the Indiana supreme court is chosen by the nominating commission and serves a 5-year term.

The state court of appeals consists of 12 justices, 3 for each of the four judicial districts in the state; they serve 10-year terms. The court exercises appellate jurisdiction under rules set by the state supreme court. Both the clerk and the reporter for the state's high courts are chosen in statewide elections for 4-year terms.

There are also 90 circuit courts, one or more superior courts in each of the 92 counties, and 58 county courts. When the justice of the peace system in the counties was abolished by the state legislature in 1976, small-claims dockets (civil cases involving up to $1,500) were added to circuit and county courts. Circuit and superior court judges are elected for six-year terms; county judges are elected to four-year terms.

The Department of Correction operated 11 major penal institutions with 10,103 adult and juvenile inmates in July 1983. Indiana State Prison is located at Michigan City, and the Women's Prison is in Indianapolis. For 1983, the FBI Crime Index reported 15,547 instances of violent crime, including murder and manslaughter, 286 (5.2 per 100,000 inhabitants); forcible rape, 1,509; and aggravated assault, 8,278. In every category, Indiana's crime rate was well below the US average.

¹⁷ARMED FORCES

US defense installations in Indiana had 6,271 military personnel in 1983/84. Army installations include Fort Benjamin Harrison and Jefferson Proving Ground; Grissom Air Force Base is the lone Air Force installation. The Navy operates a Weapons Support Center at Crane and an Avionics Center at Indianapolis. The state was awarded $2.5 billion in prime defense contracts in 1983/84.

Indiana supported the Union during the Civil War; about 200,000 Hoosiers served in northern armies, and some 24,400 died while in service. During World War I, a Hoosier reportedly was the first American soldier to fire a shot, and the first American soldier killed was from Indiana; in all, about 118,000 Indiana citizens served and 3,370 lost their lives. In World War II, about 338,000 Hoosiers served in the armed forces and some 10,000 died in line of duty. In 1983, 677,000 veterans were living in Indiana, of whom 7,000 served during World War I, 250,000 in World War II, 121,000 during the Korean conflict, and 193,000 during the Viet-Nam era. After World War II, the state paid a bonus to veterans for the first time; in 1983, veterans' benefits in Indiana totaled $420 million.

Indiana's National Guard units served in World War II, the Korean conflict, and the Viet-Nam war. In early 1985 there were about 15,000 National Guard personnel in the state. In October 1983, Indiana's police forces consisted of 16,668 persons, including 1,610 state police employees (of whom 561 were civilians).

¹⁸MIGRATION

Indiana's early settlers were predominantly northern Europeans who migrated from eastern and southern states. The influx of immigrants to the US in the late 19th and early 20th centuries had little impact on Indiana. In 1860, only 9% of the state's population was foreign-born, mostly Germans and Irish. The percentage was only 5.6% in 1900 and had further declined to 5.2% by 1920. The principal migratory pattern since then has been within the state, from the farms to the cities.

In 1860, more than 91% of the population lived in rural areas;

the percentage fell to 67% in 1900, 50% in 1920, and 40% in 1960. In 1983, 68% of the population was urban, and only 32% was rural.

Since World War II, Indiana has lost population through a growing migratory movement to other states, mostly to Florida and the Southwest. From 1960 to 1970, Indiana suffered a net loss of about 16,000 persons through migration, and from 1970 to 1983, a net total of 340,000 left the state. Demographic studies indicate that a net out-migration will probably continue for the remainder of the 20th century.

¹⁹INTERGOVERNMENTAL COOPERATION

Indiana's Commission on Interstate Cooperation promotes cooperation with other states and with the federal government. It acts largely through the Council of State Governments. Indiana is a member of such interstate regulatory bodies as the Great Lakes Commission, the Ohio River Basin Commission, and the Ohio River Valley Water Sanitation Commission. The Indiana-Kentucky Boundary Compact was signed by Indiana in 1943 and received congressional approval the same year. In 1985, Indiana joined seven other states in signing a Great Lakes Charter, aimed at further protecting the lakes' resources.

Federal aid to Indiana totaled $1.8 billion in 1983/84.

²⁰ECONOMY

Indiana is both a leading agricultural and industrial state. It ranked 10th among the states in farm marketings in 1983, and 9th in value of manufacturing shipments in 1981.

The economy was almost entirely agricultural until after the Civil War, when rapid industrial development tripled the number of factories in the state to 18,000, employing 156,000 workers, by 1900. During that period, the mechanization of agriculture resulted in the doubling of the number of farms to a peak of 220,000 in 1900. Metals and other manufacturing industries surged during and after World War I, lagged during the Great Depression of the 1930s, then surged again during and after World War II. Between 1940 and 1950, the number of wage earners in the state nearly doubled. Job opportunities brought in many workers from other states and encouraged the growth of labor unions.

The state's industrial development in Indianapolis, Gary, and other cities was based on its plentiful natural resources—coal, natural gas, timber, stone, and clay—and on good transportation facilities. In 1982, the largest number of manufacturing jobs were in the primary metals industry; other key sectors included electrical machinery, especially steel, and transportation equipment. The northwestern corner of the state is the site of one of the world's greatest concentrations of heavy industry, especially steel. Other important manufactures are food and livestock products, pharmaceuticals, phonograph records and tapes, musical instruments, burial caskets, and books.

The value of manufacturing shipments in 1981 totaled $70.3 billion, or 3.5% of the US total. During that year, Indiana ranked 10th in the value of its manufacturing exports and 8th in agricultural exports. In 1984, disposable personal income in the state amounted to $46.3 billion; the average annual growth rate in disposable income from 1948 to 1982 was 6.3%, compared with 7.5% in the US as a whole.

The chief economic concern of Hoosiers in the late 1970s and early 1980s was high unemployment in such major industries as steel and automobile and recreational vehicle manufacturing. There were 405 industrial and commercial failures in 1982. The average monthly unemployment rate in the state was 10.1% in 1981, 11.9% in 1982, and 11.1% in 1983. However, the rate dropped to 8.6% for the whole of 1984, as recessionary pressures eased.

²¹INCOME

In 1983, Indiana ranked 33d among the 50 states in individual personal income, with an average of $10,567 per capita, up 0.54%

in constant dollars since 1972. Personal income in 1983 totaled $57.9 billion—14th in the US—representing a decline (in constant dollars) of 3.7% since 1980.

The major sources of earned income in 1980 were manufacturing, 40%; trade, 14%; government, 13%; services, 13%; construction, 6%; transportation, communications, and public utilities, 6%; finance, insurance, and real estate, 4%; agriculture, 3%; and other sectors, 1%.

The median money income of four-person families in 1981 was $24,832, placing Indiana 33d among the 50 states. About 516,000 persons (10% of the population) had incomes below the federal poverty level in 1979. In 1982, 23,100 Hoosiers were among the nation's top wealth-holders, with gross assets of more than $500,000.

22LABOR

In September 1983, the state's civilian labor force totaled 2,574,800 persons, 55.8% male and 44.2% female.

A federal census of workers covered by unemployment insurance in March 1982 revealed the following employment pattern for major industry groups in Indiana:

	ESTABLISH-MENTS	EMPLOYEES	ANNUAL PAYROLL ('000)
Agricultural services, forestry, fishing	943	4,773	$ 56,113
Mining, of which:	490	9,983	263,794
Bituminous coal, lignite	(78)	(5,486)	(167,623)
Contract construction	8,223	71,342	1,526,197
Manufacturing, of which:	7,487	606,197	12,685,671
Primary metals	(277)	(91,809)	(2,302,180)
Electric, electronic equipment	(310)	(86,430)	(1,766,938)
Transport equipment	(344)	(72,040)	(1,716,818)
Transportation, public utilities	3,732	91,530	2,016,590
Wholesale trade	8,772	103,333	1,832,101
Retail trade	29,974	352,692	2,983,767
Finance, insurance, real estate	8,368	100,364	1,506,349
Services	29,592	337,918	4,000,867
Other	1,355	1,263	20,684
TOTALS	98,936	1,679,395	$26,892,133

In September 1984 there were 2,472,800 employed workers and 187,700 unemployed workers, giving Indiana an unemployment rate of 7.1%, slightly below the US average. In that month, the average weekly wage of workers in manufacturing industries was $436.39; average weekly hours worked, 41.8.

Most industrial workers live in Indianapolis and the Calumet area of northwestern Indiana. The AFL first attempted to organize workers at the US Steel Company's plant in Gary in 1919, but a strike to get union recognition failed. Other strikes by Indiana coal miners and railway workers in 1922 had limited success. By 1936, however, the CIO had won bargaining rights and the 40-hour workweek from US Steel, and union organization spread to other industries throughout the state.

In 1980, labor unions had 649,000 members in Indiana, or 30% of the total number of workers, as compared with the US average of 25%. Of that total, the majority belonged to unions affiliated with the AFL-CIO. Overall, Indiana's labor force ranked 8th in the US in extent of organization, with 1,842 reporting unions as of June 1983. In 1981 there were 94 strikes and work stoppages, involving 9,800 workers, which resulted in 193,300 working days lost.

23AGRICULTURE

In 1983, Indiana ranked 10th among the 50 states in cash farm receipts. The state's total farm marketings that year were valued at $4 billion. The leading crops were corn and soybeans.

The golden age of Indiana agriculture was the decade of the 1850s, when Indiana ranked among the four leading states in production of corn, wheat, and hogs. At that time, about 95% of the population lived on farms. The number of farms increased from 132,000 in 1860 to 222,000 in 1900, when about 21.6 million acres (8.7 million hectares) of land were under cultivation. After 1900, as a result of the rapid mechanization of agriculture and the rise of industry, the number of farms declined. Nevertheless, agriculture remains vital to the state's economy, and nearly three-fourths of its total area is farmland. There were 82,000 farms being worked in 1984, when farmland totaled 16.4 million acres (6.6 million hectares), making the size of the average farm 200 acres (81 hectares), about 67% larger than the average 120-acre (49-hectare) farm of 1950. The most productive land is in the broad central region of the state.

In 1983, Indiana ranked 3d among the producing states in output of hogs (by value), 4th in bushels of soybeans produced, and 5th in production of corn. Indiana's principal field crops were as follows:

CROPS	PRODUCTION		VALUE
Corn	340,910,000	bushels	$1,176,140,000
Soybeans	120,475,000	bushels	1,005,966,000
Winter wheat	49,470,000	bushels	163,251,000
Hay	1,891,000	tons	154,117,000
Tobacco	13,041,000	lb	NA
Oats	4,560,000	bushels	8,892,000

24ANIMAL HUSBANDRY

In 1983, Indiana ranked 13th among the 50 states in cash receipts from livestock and livestock products. In that year, about 46% of the state's farm income came from the sale of livestock and livestock products, which amounted to $1.8 billion.

In 1983, Indiana dairy farmers produced 21,779,000 lb of butter, 64,630,000 lb of cheese, and 2,364,000,000 lb of milk. The numbers of livestock on Indiana farms at the start of 1984 were: hogs and pigs, 4,200,000; cattle and calves, 1,640,000; milk cows, 204,000; and sheep and lambs, 105,000. Indiana poultry farmers raised 17,150,000 chickens and 6,710,000 turkeys in 1983. Other animal products are honey, beeswax, and wool.

25FISHING

Fishing is not of commercial importance in Indiana; in 1984, only 591,000 lb of fish valued at $724,000 were landed. Fishing for bass, pike, perch, catfish, and trout is a popular sport with Indiana anglers.

26FORESTRY

About 17% of Indiana land is forested (0.5% of the US total). In 1979, the state had 3,943,000 acres (1,596,000 hectares) of forestland, of which 3,815,000 acres (1,544,000 hectares) were commercial timberland; of the latter total, 239,000 acres (96,800 hectares) were federal land and 169,700 acres (68,700 hectares) were state land. Most forestland is situated in the southern region, where oak, beech, sycamore, poplar, and hickory trees are plentiful. About 89% of the state's commercial timberland is privately owned.

In 1979 there were an estimated 92.3 million cu feet of softwoods and 4.1 billion cu feet of hardwoods on commercial timberlands. The harvest in 1983 was 252 million board feet, of which 97% was softwood. The lumber and wood products industries employed 15,300 persons in 1982 and paid wages totaling $212.7 million; factory shipments amounted to $1 billion. Among these industries were millwood and plywood manufacturing, which employed 6,700 workers and shipped $401 million worth of goods, and mobile homes, employing 3,000 workers and shipping $322 million worth of products. Important related industries include paper and paper products, especially paperboard containers, and wood furniture and fixtures.

There is one national forest in Indiana, Hoosier National

Forest (near Bloomington), with 187,500 acres (76,000 hectares). The largest of the 12 state forests are Harrison-Crawford, Morgan-Monroe, Clark, and Yellowwood.

27 MINING

In 1984, Indiana's nonfuel mineral production was valued at $293 million, an increase of 17% over 1983. Indiana generally produces about two-thirds of all building limestone quarried in the US, as well as significant quantities of other construction materials, including crushed stone, cement, gypsum, sand and gravel, clay, and shale. Coal and petroleum are also produced. Mining and quarrying industries employed 9,200 Indianans in 1983. For the most part, mines are situated in the southern part of the state.

In 1984, the state's principal mineral products (excluding fossil fuels) were stone, 25,660,000 tons valued at $103.8 million; sand and gravel, 14,800,000 tons valued at $40.9 million; and clay, 776,000 tons valued at $2.2 million.

28 ENERGY AND POWER

Indiana is largely dependent on fossil fuels for its energy supplies. The use of wood, an important source of energy in the 19th century, declined with the increased burning of coal. In recent years, petroleum has become an important power source for automobiles, home heating, and electricity. Nevertheless, coal has continued to be the state's major source of power, meeting about half of Indiana's energy needs.

In 1983, Indiana's gross energy consumption totaled 2,156 trillion Btu, of which 47% was provided by coal, 30% by petroleum products, 22% by dry natural gas, and 1% by other sources.

The state has no nuclear power plants. In 1984, construction of the planned Marble Hill nuclear power plant on the Ohio River near Madison was permanently suspended by the Public Service Co. of Indiana because of escalating construction costs; total cost estimates had risen from $1.4 billion during the planning stage in 1973 to more than $7 billion. Per capita energy consumption in the state in 1982 was 393 million Btu, the 7th highest in the US. Indiana also ranks high in energy expenditures per capita—$2,215 in 1981. Residential customers paid 6.49 cents per kwh for electricity in that year.

Electric power produced in Indiana in 1983 totaled 70.9 billion kwh; total installed capacity was 18.6 million kw, 95% of it provided by steam-driven plants. Privately owned power plants account for nearly all of Indiana's production and installed capacity. Of total electricity sales in 1983, roughly 49% was sold to industries, 32% to homes, 17% to commercial users, and 2% to others. The major electric utilities were Northern Indiana Public Service Co., Indiana & Michigan Electric Co., Public Service Co. of Indiana, Indianapolis Power & Light Co., and Southern Indiana Gas & Electric Co.

At the end of 1983, Indiana's estimated proved reserves of petroleum totaled 34 million barrels; production of crude petroleum totaled 5.3 million barrels.

In 1983 there were 3 underground coal mines and 77 strip mines active in the state. Indiana's coal production in that year was estimated at 31.8 million tons of coal, 12th in the US. Recoverable reserves totaled 409 million short tons.

29 INDUSTRY

The industrialization of Indiana that began in the Civil War era was spurred by technological advances in processing agricultural products, manufacturing farm equipment, and improving transportation facilities. Meat-packing plants, textile mills, furniture factories, and wagon works—including Studebaker wagons—were soon followed by metal foundries, machine shops, farm implement plants, and a myriad of other durable-goods plants.

New industries included a pharmaceutical house started in Indianapolis in 1876 by a druggist named Eli Lilly, and several automobile-manufacturing shops established in South Bend and other cities by 1900. In 1906, the US Steel Co. laid out the new town of Gary for steelworkers and their families.

In 1982, the total value of shipments by manufacturers in Indiana was $63.3 billion. The leading industry groups in 1982, and the value of their shipments were as follows:

Primary metal products	$11,768,300,000
Transportation equipment	7,931,700,000
Electrical and electronic equipment	7,141,700,000
Food and food products	6,188,200,000
Chemicals and chemical products	5,153,600,000
Nonelectrical machinery	5,099,400,000
Fabricated metal products	4,484,300,000
Rubber and plastic products	2,342,100,000
Printing and publishing	1,624,500,000
Stone, clay, and glass products	1,465,200,000
Paper and allied products	1,254,900,000
Furniture and fixtures	1,090,300,000
Lumber and wood products	1,008,700,000

According to the 1982 census of manufacturers, Indiana was a leading producer of storage batteries, small motors and generators, mobile homes, household furniture, burial caskets, and musical instruments. Most manufacturing plants are located in and around Indianapolis and in the Calumet region.

30 COMMERCE

The 1982 federal economic census listed 8,953 wholesale establishments in Indiana, with $31.5 billion in sales. Of the total sales, merchant wholesalers accounted for 60%. Principal goods traded included food and related products, $4.3 billion; machinery, equipment, and supplies, $4.1 billion; farm product raw materials, $4 billion; petroleum and petroleum products, $4 billion; and electrical goods, $3.1 billion.

Retail stores in Indiana had total sales of $23.7 billion in 1982. Of that total, automotive dealers and service stations accounted for 32%; food stores, 22%; automobile dealers, 19%; and general merchandise establishments, 11%.

Indiana ranked 8th among the 50 states in agricultural exports during 1981/82, when farm products shipped abroad were valued at $1.6 billion. Major farm exports (in order of value) were soybeans; feed grains; wheat; meat (including poultry) and meat products; fats, oils, and greases; and hides and skins. Indiana was the 4th-leading US exporter of soybeans. The state ranked 10th among US states in exports of manufactured goods, which were valued at $5 billion. Principal nonfarm exports (in order of value) included transportation equipment, electric and electronic equipment, nonelectric machinery, primary metals products, chemicals and allied products, food and kindred products, and fabricated metal products.

31 CONSUMER PROTECTION

The Division of Consumer Protection of the Office of the Attorney General is empowered to investigate consumer complaints, initiate and prosecute civil actions, and warn consumers about deceptive sales practices. Indiana also has a public counselor, who appears on behalf of the public at hearings of the Public Service Commission in regard to rates charged by public utilities and transportation agencies. The public counselor is appointed by the governor to a four-year term and is aided by a 20-member staff and 11-member advisory council.

In early 1980, the first criminal prosecution of an American corporation because of alleged product defects was brought against the Ford Motor Co. at Winamac. A jury found the company not guilty of reckless homicide in a rear-end collision involving a Pinto automobile in which three young women were killed.

32 BANKING

The large-scale mechanization of agriculture in Indiana after 1850 encouraged the growth of banks to lend money to farmers to buy farm machinery, using their land as collateral. The financial panic of 1893 caused most banks in the state to suspend operations, and the depression of the 1930s caused banks to foreclose many farm mortgages and dozens of banks to fail. The nation's subsequent economic recovery, together with the federal reorganization of the banking system, helped Indiana banks to share in the state's prosperity during and after World War II.

In December 1984 there were 384 banks in the state, of which 152 were members of the Federal Reserve System (111 of these had national charters and 41 were state-chartered). The total assets of insured commercial banks amounted to $39.6 billion; their liabilities included time and savings deposits of $26.5 billion as of 31 December 1983.

At the end of 1982, the state's 134 savings and loan associations held $8.2 billion in mortgage loans and $9.5 billion in savings accounts; their total assets amounted to $10.7 billion. In 1984, Indiana had 319 credit unions.

The Department of Financial Institutions regulates the operations of Indiana-chartered banks, savings and loan associations, and credit unions, and monitors observance of a Uniform Consumer Credit Code. The department is headed by a seven-member board; each board member serves a four-year term, and no more than four members may be of the same political party. A full-time director, also appointed by the governor to a four-year term, is the department's chief executive and administrative officer.

33 INSURANCE

As of 1984 there were 113 state-licensed property/casualty insurance companies with home offices in Indiana, and 56 life insurance companies.

In 1983, life insurance companies in the state had in force a total of 9.6 million policies and paid total benefits of $1.3 billion. In that year, life insurance policies were valued at $114 billion; the average amount of life insurance held by a Hoosier family was $53,700.

Some 662,000 people were enrolled in Medicare in 1982 and received benefit payments of $1.1 billion. There were 272,000 Medicaid recipients in 1983, drawing total benefits of $596 million.

Property and liability companies in 1983 wrote premiums for $897 million in automobile liability and physical damage coverage, $301 million in homeowners' insurance, and $49 million in farmowners' insurance.

The Department of Insurance licenses insurance carriers and agents in Indiana, and it enforces regulations governing the issuance of policies. A 1975 law limited medical malpractice insurance claims.

34 SECURITIES

There are no securities exchanges in Indiana. New York Stock Exchange member firms had 99 sales offices and 644 full-time registered representatives in Indiana in December 1983. There were 751,000 shareholders in public corporations during the same year. Laws governing the trading and sale of corporate securities are administered by the secretary of state, who also regulates franchise sales and corporate takeover attempts.

35 PUBLIC FINANCE

The State Budget Agency acts as watchdog over state financial affairs. The agency prepares the budget for the governor and presents it to the general assembly. The budget director, appointed by the governor, serves with four legislators (two from each house) on the State Budget Committee, which helps to prepare the budget. The State Budget Agency receives appropriations requests from the heads of state offices, estimates anticipated revenues for the biennium, and administers the budget.

The fiscal year runs from 1 July to 30 June of the following year; budgets are prepared for the biennium beginning and ending in an odd-numbered year. Federal funds for Indiana were estimated at $1.5 billion for the 1985–87 biennium.

The following is a summary of total estimated revenues and requested expenditures for 1985/86 and 1986/87 (in millions):

REVENUES	1985/86	1986/87
Sales and use taxes	$1,777.2	$1,884.9
Individual income taxes	1,545.2	1,639.4
Corporate income taxes	675.6	687.9
Other taxes	645.1	642.6
Intergovernmental receipts	2,307.3	2,367.3
Other revenues	365.5	365.6
TOTALS	$7,315.9	$7,587.7

EXPENDITURES		
Local school aid	$1,813.2	$1,928.4
Other aid to education	753.0	811.0
Public welfare	1,232.2	1,277.8
Transportation	769.5	804.6
General government	584.2	603.2
Health	493.1	511.5
Corrections	169.8	176.5
Public safety and regulation	119.7	117.6
Natural resources and recreation	90.2	88.8
Other appropriations	1,439.6	1,444.9
TOTALS	$7,464.5	$7,764.3

In 1982/83, local governments spent a total of $6.2 billion. In that fiscal year, local governments received total general revenues of $5.5 billion, of which 58% was raised by property taxes and other local revenue sources, 35% was provided by the state government, and 7% came from the federal government.

The total indebtedness of state and local governments was nearly $5.3 billion in 1982/83. About 94% of this total represented long-term obligation; 18% of this obligation was backed by the full faith and credit of the issuing body, while 82% was not fully guaranteed.

36 TAXATION

The first state property tax in Indiana was levied in 1852 to support public schools. In 1923, a state gasoline tax of 2 cents per gallon was introduced. (It ranged from 11.1 cents to 14 cents in 1984, depending on average prices in a specified month.) In 1933, Indiana instituted the personal income tax, which was the major source of state revenue until 1963, when a 2% retail sales tax was enacted. Also in 1933, with the end of Prohibition, taxes were imposed on the manufacture and sale of alcoholic beverages. In 1973, the state sales tax was doubled to 4% and optional local income taxes of up to 1% were initiated, while local property taxes were reduced by at least 20% to ease the tax burden on property owners. In 1984, the state sales tax was 5%, and the state tax on cigarettes was 10.5 cents per pack. The state's personal income tax was 3% of adjusted gross income after the first $1,000 for single persons and the first $2,000 for married couples, with exemptions of $1,000 for each dependent.

In the 1983/84 fiscal year, Indiana's state taxes totaled $4 billion; the tax per capita was about $729. Of total state taxes collected that year, the sources were income taxes, 47%; sales and use taxes, 38%; motor fuel taxes, 8%; tobacco and alcohol taxes, 4%; and miscellaneous taxes, 3%. Taxes levied by local governments in fiscal 1982/83 totaled $1.8 billion, of which the local property tax provided 96% of the total.

The total federal tax burden in Indiana amounted to $12.5 billion, or $2,291 per capita; the state ranked 27th among the 50 states in total burden and 32d in per capita burden. Hoosiers filed 2,173,559 federal income tax returns for 1983, paying a total tax of $5.8 billion.

37 ECONOMIC POLICY

The state's early economic policy was to provide farmers with access to markets by improving transportation facilities. During the Civil War era, however, the state began to encourage industrial growth. In modern times, the state has financed extensive highway construction, developed deepwater ports on Lake Michigan and the Ohio River, and worked to foster industrial growth and the state tourist industry. Tax incentives to business include a 15-year phaseout, beginning in 1979, of the "intangibles" tax on stocks, bonds, and notes.

The Department of Commerce, which has sole responsibility for economic development, solicits new businesses to locate in Indiana, promotes sales of exports abroad, plans the development of energy resources, continues to foster the expansion of agriculture, and helps minority-group owners of small businesses. The department's Industrial Development Fund makes loans to municipalities for the purchasing or leasing of property for industrial development. In the early 1980s, the state government focused on a series of economic development initiatives—including programs for job training and retraining, promoting new businesses and tourism, developing infrastructure, and stimulating investment capital—as well as programs providing additional tax incentives.

[38] HEALTH

Mortality rates and infant death rates are close to the national average in Indiana. In 1983, the live-birth rate for the state was 15.5 per 1,000 population; the infant mortality rates (1981) were 10.8 per 1,000 live births for whites and 19.6 for blacks. In 1982, the legal abortion rate was 189 per 1,000 live births (far below the US average of 426), and the fertility rate was 65.1 per 1,000 women aged 15–44.

The principal causes of death, with rates of death per 100,000 population, were heart disease, 327.1; cancer, 183.4; cerebrovascular diseases, 80.3; accidents, 42.6; chronic obstructive lung diseases, 27.8; pneumonia, 20.9; atherosclerosis, 19.2; and diabetes, 18.8. The total death rate was 8.6 per 1,000 population, the same as the national rate.

In the 1970s, the state started new health programs for the care and treatment of patients with sickle-cell anemia and hemophilia. Other special programs being conducted by the Board of Health in the 1980s were concerned with such problems as hypertension, sudden infant death syndrome, and venereal diseases.

Indiana had 133 hospitals in 1983, with 32,018 beds; admissions to these hospitals totaled 920,104 in that year, and there were 6,541,851 outpatient visits. As of 1982, the state also had 265 nursing and personal-care facilities. A total of 11,830 registered nurses had full-time positions in Indiana hospitals as of 1981. The state had 7,962 physicians and 2,386 dentists in 1982. The average daily rate for a semiprivate hospital room in January 1984 was about $180, and the average cost per patient day (1982) came to $287.

The State Board of Health is responsible for protecting the health and lives of residents. In the 1983/84 fiscal year, the state spent $305 million for health care; 70% of these funds were for mental-health facilities and programs.

[39] SOCIAL WELFARE

In 1983/84, expenditures for public assistance in Indiana were about $1.1 billion; federal funds provided 66% of the total. Only 3.7% of all Hoosiers were receiving public aid in 1983. Public assistance payments to families with dependent children totaled $140 million in 1982; the average monthly payment per family was $213. Food stamps were issued to 469,000 persons in 1983; the federal cost was $268 million. School lunches were provided to 613,000 pupils, at a cost to the federal government of $41 million.

In 1983, Social Security benefits were paid to 831,140 persons, of whom 485,449 were retired workers, 177,230 were spouses of retired or deceased workers, and 90,168 were disabled workers or their spouses or children. Benefit payments totaled more than $4 billion. The average monthly payment was $441.30 per retired worker, about $22 above the US average. Supplemental Security Income was provided to a monthly average of 40,532 recipients in December 1983 and totaled $80 million for the year, of which about 78% went to the disabled, 19% to the aged, and 3% to the blind.

Workers' compensation payments totaled $121 million in 1982. In the same year, state unemployment insurance covered 1,909,-000 workers in Indiana; weekly payments to unemployed workers averaged $94.24 per worker and lasted an average of 13.9 weeks.

[40] HOUSING

The great majority of Indiana families enjoy adequate housing, particularly in newly built suburbs, but inadequate housing exists in the deteriorating central cores of large cities.

In 1980, the state had 2,091,795 housing units, of which 98% provided year-round housing. Of the latter, 75% were single-family units and 20% were multifamily units; the remainder included 99,133 mobile homes. In 1980, about 69% of all owner-occupied units had mortgages; the median monthly payment for mortgage and other selected costs was $302. There were 6,936 owner-occupied condominiums. Of all rented housing units in 1980, about 39% were occupied by single families, 58% were multifamily dwellings, and 3% were mobile homes. Median monthly rent was $218.

In 1983, 16,000 new housing units were authorized for construction, at an estimated value of $760 million. Between 1981 and 1983, about 500,000 housing starts were recorded in the state.

[41] EDUCATION

Although the 1816 constitution recommended establishment of public schools, the state legislature did not provide funds for education. The constitution of 1851 more specifically outlined the state's responsibility to support a system of free public schools. Development was rapid following passage of this document; more than 2,700 schoolhouses were built in the state from 1852 to 1857, and an adult literacy rate of nearly 90% was achieved by 1860. The illiteracy rate was reduced to 5.2% for the adult population in 1900, to 1.7% in 1950, and to only 0.7% in 1970, when Indiana ranked 14th among the 50 states. In 1980, 66% of those aged 25 years and over were high school graduates, and 13% had completed four years of college.

In the fall of 1983, Indiana had 984,090 pupils enrolled in public elementary and secondary schools, or about 19.9 pupils per teacher, compared with a national average of 18.5. The average public school teacher drew an annual salary of $21,587, about $500 below the national average. Roman Catholic elementary and secondary schools enrolled 65,481 pupils in the same academic year. As of 1980/81, total nonpublic school enrollment came to 100,234.

In the 1982/83 academic year there were 146,955 students attending colleges and universities in the state. Indiana University, the state's largest institution of higher education, was founded in 1820. It is one of the largest state universities in the US, with a total 1983/84 enrollment on eight campuses of 97,757; the Bloomington campus alone had 33,109 students. Other major state universities and their 1983/84 enrollments were Purdue University (Lafayette), 31,856; Ball State University (Muncie), 18,359; and Indiana State University (Terre Haute), 11,587. Well-known private universities in the state include Notre Dame (at Notre Dame), with 9,400 students, and Butler (Indianapolis), with 4,508 students. Small private colleges and universities include DePauw (Greencastle), Earlham (Richmond), Hanover (Hanover), and Wabash (Crawfordsville).

In 1965, the general assembly established a program of state college scholarships, at least two for each county. Recipients must be US citizens under 25 years of age, Indiana residents for at least six months before enrollment, and enrolled full-time in a college or community college.

The state superintendent of public instruction is elected for a four-year term. The superintendent serves on the 19-member Board of Education, which is appointed by the governor. The board, which sets basic policy for the public school system, is divided into three commissions dealing, respectively, with general education, textbook adoption, and teacher training and licensing.

In 1983/84, Indiana spent more than $2.5 billion on public schools, for an average expenditure of $2,566 per student (36th in the US).

⁴²ARTS

The earliest center for artists in Indiana was the Art Association of Indianapolis, founded in 1883. It managed the John Herron Art Institute, consisting of a museum and art school (1906–08). Around 1900, art colonies sprang up in Richmond, Muncie, South Bend, and Nashville. Indianapolis remains the state's cultural center, especially after the opening in the late 1960s of the Lilly Pavilion of the Decorative Arts; the Krannert Pavilion, which houses the paintings originally in the Herron Museum; the Clowes Art Pavilion; and the Grace Showalter Pavilion of the Performing Arts (all collectively known as the Indianapolis Museum of Art). Since 1969, the Indiana Arts Commission has taken art—and artists—into many Indiana communities; the commission also sponsors biennial awards to artists in the state. Some 40 arts councils in Indianapolis and other cities encourage painters and sponsor art exhibitions.

The state's first resident theater company established itself in Indianapolis in 1840, and the first theater building, the Metropolitan, was opened there in 1858. Ten years later, the Academy of Music was founded as the center for dramatic activities in Indianapolis. In 1875, the Grand Opera House opened there, and the following year it was joined by the English Opera House, where touring performers such as Sarah Bernhardt, Edwin Booth, and Ethel Barrymore held the stage. Amateur theater has been popular since the founding in 1915 of the nation's oldest amateur drama group, the Little Theater Society, which later became the Civic Theater of Indianapolis.

Music has flourished in Indiana. Connersville reportedly was the first American city to establish a high school band, while Richmond claims the first high school symphony orchestra. The Indianapolis Symphony Orchestra was founded in 1930. There are 23 other symphony orchestras in the state. The Arthur Jordan College of Music is part of Butler University in Indianapolis.

⁴³LIBRARIES AND MUSEUMS

The state constitution of 1816 provided for the establishment of public libraries. A majority of Indiana counties opened such libraries but neglected to provide adequate financing. Semiprivate libraries did better: workingmen's libraries were set up by a bequest at New Harmony and 14 other towns. After the state legislature provided for township school libraries in 1852, more than two-thirds of the townships established them, and the public library system has thrived ever since. In 1982 there were 51 county libraries, and every county received some form of library service. Federal grants-in-aid to all public libraries totaled $1,493,109; state grants, $934,835. The largest book collections are at public libraries in Indianapolis, Fort Wayne, Gary, Evansville, South Bend, and Hammond; the total book stock of all Indiana public libraries was 14,618,201 volumes in 1982.

The Indiana State Library maintains the state's archives, a complete collection of documents about Indiana's history, and a large genealogical collection. The Indiana University Library has special collections on American literature and history and an extensive collection of rare books; the University of Notre Dame has a noteworthy collection on medieval history; and Purdue University Libraries contain outstanding industrial and agricultural collections, as well as voluminous materials on Indiana history.

Private libraries and museums include those maintained by historical societies in Indianapolis, Fort Wayne, and South Bend. Also of note are the General Lew Wallace Study museum in Crawfordsville and the Elwood Haynes Museum of early technology in Kokomo. In all, Indiana had 149 museums in 1984.

Indiana's historic sites of most interest to visitors are the Lincoln Boyhood National Memorial near Gentryville, the Benjamin Harrison Memorial Home and the James Whitcomb Riley Home in Indianapolis, and the Grouseland Home of William Henry Harrison in Vincennes. Among several archaeological sites are two large mound groups: one at Mounds State Park near Anderson, which dates from about AD 800–900, and a reconstructed village site at Angel Mounds, Newburgh, which dates from 1300–1500.

⁴⁴COMMUNICATIONS

In 1985 there were 750 post offices in Indiana, with 13,717 employees. About 97% of all households had telephone service in 1980. The state's first radio station was licensed in 1922 at Purdue University, Lafayette. Indiana had 90 AM and 155 FM radio stations and 24 commercial and 6 educational television stations as of 1984. Powerful radio and television transmissions from Chicago and Cincinnati also blanket the state. In 1984, 133 cable television systems served 289 communities and 693,921 subscribers throughout much of the state.

⁴⁵PRESS

The first newspaper in Indiana was published at Vincennes in 1804, and a second pioneer weekly appeared at Madison nine years later. By 1830, newspapers were also being published in Terre Haute, Indianapolis, and 11 other towns; the following year, the state's oldest surviving newspaper, the *Richmond Palladium,* began publication. Most pioneer newspapers were highly political and engaged in acrimonious feuds; in 1836, for example, the *Indianapolis Journal* referred to the editors of the rival *Democrat* as "the Lying, Hireling Scoundrels." By the time of the Civil War, Indiana had 154 weeklies and 13 dailies.

The last third of the 19th century brought a sharp increase in both the number and the quality of newspapers. Two newspapers, which later became the state's largest in circulation, the *Indianapolis News* and the *Star,* began publishing in 1869 and 1904, respectively. In 1941 there were 294 weekly and 98 daily newspapers in Indiana; the number declined after World War II because of fierce competition for readers and advertising dollars, rising operating costs, and other financial difficulties.

In 1985, the state had 12 morning dailies with a paid circulation of 521,057, and 62 evening dailies with 1,065,652 paid circulation; Sunday papers numbered 16, with paid circulation totaling 1,186,838. In 1984, the Indianapolis morning *Star* had a daily circulation of 227,829 (Sunday circulation, 381,323); the Indianapolis evening *News* had a daily circulation of 136,447; and the Gary evening *Post-Tribune's* circulation averaged 81,500 daily and 94,000 on Sundays.

A number of magazines are published in Indiana, including *Children's Digest* and *The Saturday Evening Post.*

Indiana is noted for its literary productivity. A survey of authors claimed by Indiana up to 1966 showed a total of 3,600. Examination of the 10 best-selling novels each year from 1900 to 1940 (allowing 10 points to the top best-seller, down to 1 point for the 10th best-selling book) showed Indiana with a score of 213 points, exceeded only by New York's 218.

Many Hoosier authors were first published by Indiana's major book publisher, Bobbs-Merrill. Indiana University Press is an important publisher of scholarly books.

⁴⁶ORGANIZATIONS

The 1982 US Census of Service Industries counted 1,717 organizations in Indiana, including 224 business associations; 1,158 civic, social, and fraternal associations; and 35 educational, scientific, and research associations. National organizations with headquarters in the state include the American Camping Association, located in Martinsville, and the Amateur Athletic Union of the US, the American Legion, the US Gymnastics Federation, and Kiwanis International, all in Indianapolis.

Philanthropic foundations headquartered in Indiana include the Eugene V. Debs Foundation (Terre Haute) and the Irwin-Sweeny-Miller Foundation (Columbus).

⁴⁷TOURISM, TRAVEL, AND RECREATION

Tourism is of moderate economic importance to Indiana. In 1982, 62,200 Hoosiers worked in travel-related industries, and travelers spent over $2.4 billion in the state, generating $279 million in tax revenues.

Summer resorts are located in the north, along Lake Michigan and in Steuben and Kosciusko counties, where there are nearly 200 lakes. Popular tourist sites include the reconstructed village of New Harmony, site of famous communal living experiments in the early 19th century; the Indianapolis Motor Speedway and Museum; and the George Rogers Clark National Historic Park at Vincennes. In addition to the Indiana State Museum there are 15 state memorials, including the Wilbur Wright State Memorial at his birthplace near Millville, the Ernie Pyle birthplace near Dana, and the old state capitol at Corydon. Among the natural attractions are the Indiana Dunes National Lakeshore on Lake Michigan (12,534 acres—5,072 hectares); the state's largest waterfall, Cataract Falls, near Cloverdale; and the largest underground cavern, at Wyandotte.

Indiana has 18 state parks, comprising 53,870 acres (21,800 hectares). The largest state park is Brown County (15,543 acres—6,290 hectares), near Nashville. There are 15 state fish and wildlife preserves, totaling about 75,200 acres (30,400 hectares). The largest are Pigeon River, near Howe, and Willow Slough, at Morocco. Game animals during the hunting season include deer, squirrel, and rabbit; ruffed grouse, quail, ducks, geese, and partridge are the main game birds. In 1982/83, the state issued licenses to 655,801 anglers and 338,994 hunters.

⁴⁸SPORTS

Indiana is represented in professional sports by the Indiana Pacers of the National Basketball Association and by the National Football League's Colts, who moved to Indianapolis from Baltimore in 1984. Indianapolis is also represented in baseball's Class AAA American Association.

The state's biggest annual sports event is the Indianapolis 500, which has been held at the Indianapolis Motor Speedway on Memorial Day every year since 1911 (except for the war years 1917 and 1942–45). The race is now part of a three-day Indiana festival and attracts crowds of over 300,000 spectators; total prize money in 1984 amounted to $2,795,399.

The state's most popular amateur sport is basketball. The high school boys' basketball tournament culminates on the last Saturday in March, when the four finalists play afternoon and evening games to determine the winner. A tournament for girls' basketball teams began in 1976. Basketball is also popular at the college level: Indiana University won the NCAA Division I basketball championship in 1940, 1953, and 1976, and the National Invitational Tournament (NIT) in 1979; Purdue University won the NIT title in 1974; and Evansville College won the NCAA Division II championships in 1959–60, 1964–65, and 1971.

Collegiate football in Indiana has a colorful tradition stretching back at least to 1913, when Knute Rockne of Notre Dame unleashed the forward pass as a potent football weapon. Notre Dame, which competes as an independent, was recognized as the top US college football team in 1946–47, 1949, 1966, 1973 (with Alabama), and 1977. Indiana and Purdue compete in the Big Ten, while Indiana State is part of the Missouri Valley Conference.

The Little 500, a 50-mi (80-km) bicycle race, is held each spring at Indiana University's Bloomington campus. The US Open Clay Court Tennis Championships are held annually in Indianapolis.

⁴⁹FAMOUS INDIANANS

Indiana has contributed one US president and four vice presidents to the nation. Benjamin Harrison (b.Ohio, 1833–1901), the 23d president, was a Republican who served one term (1889–93) and then returned to Indianapolis, where his home is now a national historic landmark. Three vice presidents were Indiana residents: Thomas Hendricks (b.Ohio, 1819–85), who served only eight months under President Cleveland and died in office; Schuyler Colfax (b.New York, 1823–85), who served under President Grant; and Charles Fairbanks (b.Ohio, 1852–1918), who served under Theodore Roosevelt. One vice president was a native son: Thomas Marshall (1854–1925), who served two four-year terms with President Wilson. Marshall, remembered for his wit, originated the remark "What this country needs is a good five-cent cigar."

Other Indiana-born political figures include Eugene V. Debs (1855–1926), Socialist Party candidate for president five times, and Wendell L. Willkie (1892–1944), the Republican candidate in 1940.

A dozen native and adoptive Hoosiers have held cabinet posts. Hugh McCulloch (b.Maine, 1808–95) was twice secretary of the treasury, in 1865–69 and 1884–85. Walter Q. Gresham (b.England, 1832–95) was successively postmaster general, secretary of the treasury, and secretary of state. John W. Foster (1836–1917) was an editor and diplomat before serving as secretary of state under President Benjamin Harrison. Two other postmasters general came from Indiana: Harry S. New (1858–1937), and Will H. Hays (1879–1954). Hays resigned to become president of the Motion Picture Producers and Distributors (1922–45), and enforced its moral code in Hollywood films through what became widely known as the Hays Office. Two Hoosiers served as secretary of the interior: Caleb B. Smith (b.Massachusetts, 1808–64) and John P. Usher (b.New York, 1816–89). Richard W. Thompson (b.Virginia, 1809–1900) was secretary of the Navy. William H. H. Miller (b.New York, 1840–1917) was attorney general. Two native sons and Purdue University alumni have been secretaries of agriculture: Claude R. Wickard (1873–1967) and Earl Butz (b.1909). Paul V. McNutt (1891–1955) was a governor of Indiana, high commissioner to the Philippines, and director of the Federal Security Administration.

Only one Hoosier, Sherman Minton (1890–1965), has served on the US Supreme Court. Ambrose Burnside (1824–81) and Lew Wallace (1827–1905) were Union generals during the Civil War; Wallace later wrote popular historical novels. Oliver P. Morton (1823–77) was a strong and meddlesome governor during the war, and a leader of the radical Republicans during the postwar Reconstruction. Colonel Richard Owen (b.England, 1810–90) commanded Camp Morton (Indianapolis) for Confederate prisoners; after the war, some of his grateful prisoners contributed to place a bust of Owen in the Indiana statehouse. Rear Admiral Norman Scott (1889–1942) distinguished himself at Guadalcanal during World War II. Nearly 70 Hoosiers have won the Medal of Honor.

Dr. Hermann J. Muller (b.New York, 1890–1967), of Indiana University, won the Nobel Prize in physiology or medicine in 1946 for proving that radiation can produce mutation in genes. Harold C. Urey (1893–1981) won the Nobel Prize in chemistry in 1934, and Wendell Stanley (1904–71) won it in 1946. The Nobel Prize in economics was awarded to Paul Samuelson in 1970. The Pulitzer Prize in biography was awarded in 1920 to Albert J. Beveridge (b.Ohio, 1862–1927) for his *Life of John Marshall*. Beveridge also served in the US Senate. Booth Tarkington (1869–1946) won the Pulitzer Prize for fiction in 1918 and 1921. A. B. Guthrie (b.1901) won it for fiction in 1950. The Pulitzer Prize in history went to R. C. Buley (1893–1968) in 1951 for *The Old Northwest*.

Aviation pioneer Wilbur Wright (1867–1912) was born in Millville. Other figures in the public eye were chemist Harvey W. Wiley (1844–1930), who was responsible for the Food and Drug Act of 1906; Emil Schram (b.1893), president of the New York Stock Exchange from 1931 to 1951; and Alfred C. Kinsey (b.New Jersey, 1894–1956), who investigated human sexual behavior and issued the two famous "Kinsey reports" in 1948 and 1953.

Indiana claims such humorists as George Ade (1866–1944),

Frank McKinney "Kin" Hubbard (b.Ohio, 1868–1930), and Don Herold (1889–1966). Historians Charles (1874–1948) and Mary (1876–1958) Beard, Claude Bowers (1878–1958), and Glenn Tucker (1892–1976) were Hoosiers. Maurice Thompson (1844–1901) and George Barr McCutcheon (1866–1928) excelled in historical romances. The best-known poets were James Whitcomb Riley (1849–1916) and William Vaughn Moody (1869–1910). Juvenile writer Annie Fellows Johnston (1863–1931) produced the "Little Colonel" series.

Other Indiana novelists include Edward Eggleston (1837–1902), Meredith Nicholson (1866–1947), David Graham Phillips (1868–1911), Gene Stratton Porter (1868–1924), Theodore Dreiser (1871–1945), Lloyd C. Douglas (1877–1951), Rex Stout (1886–1975), William E. Wilson (b.1906), Jessamyn West (1907–84), and Kurt Vonnegut (b.1922). Well-known journalists were news analyst Elmer Davis (1890–1958), war correspondent Ernie Pyle (1900–45), and columnist Janet Flanner (1892–1978), "Genet" of *The New Yorker*.

Among the few noted painters Indiana has produced are Theodore C. Steele (1847–1928), William M. Chase (1851–1927), J. Ottis Adams (1851–1927), Otto Stark (1859–1926), Wayman Adams (1883–1959), Clifton Wheeler (1883–1953), Marie Goth (1887–1975), C. Curry Bohm (1894–1971), and Floyd Hopper (b.1909).

Composers of Indiana origin have worked mainly in popular music: Paul Dresser (1857–1906), Cole Porter (1893–1964), and Howard Hoagland "Hoagy" Carmichael (1899–1981). Howard Hawks (1896–1977) was a renowned film director. Entertainers from Indiana include actor and dancer Clifton Webb (Webb Hollenbeck, 1896–1966); orchestra leader Phil Harris (b.1906); comedians Ole Olsen (1892–1963), Richard "Red" Skelton (b.1913), and Herb Shriner (b.Ohio, 1918–70); actresses Marjorie Main (1890–1975) and Carole Lombard (Jane Peters, 1908–42); and singer Michael Jackson (b.1958).

Hoosier sports heroes include Knute Rockne (b.Norway, 1888–1931), famed as a football player and coach at Notre Dame. Star professionals who played high school basketball in Indiana include Oscar Robertson (b.Tennessee, 1938) and Larry Bird (b.1956), who at Indiana State University in 1978/79 was honored as college basketball's player of the year.

[50] BIBLIOGRAPHY

Banta, R. E. *Indiana Authors and Their Books, 1816–1916.* Crawfordsville, Ind.: Wabash College, 1949.

Banta, R. E. *The Wabash.* New York: Farrar & Rinehart, 1950.

Barnhart, John D., and Donald F. Carmony. *Indiana from Frontier to Industrial Commonwealth.* 4 vols. New York: Lewis Historical Publishing, 1954.

Barnhart, John D., and Dorothy L. Riker. *Indiana to 1816: The Colonial Period.* Indianapolis: Indiana Historical Society, 1971.

Buley, R. C. *The Old Northwest: Pioneer Period, 1815–1840.* Indianapolis: Indiana Historical Society, 1950.

Carter, Jared, and Darryl Jones. *Indiana.* Portland, Ore.: Graphic Arts Center Publishing Co., 1984.

Dorson, Ron. *The Indy Five Hundred.* New York: Norton, 1974.

Federal Writers' Project. *Indiana: A Guide to the Hoosier State.* Reprint. New York: Somerset, n.d. (orig. 1941).

Indiana State Chamber of Commerce. *Here Is Your Indiana Government.* Indianapolis, 1983.

Latta, William C. *Outline History of Indiana Agriculture.* Lafayette: Purdue University and Indiana County Agricultural Agents Association, 1938.

Leibowitz, Irving. *My Indiana.* Englewood Cliffs, N.J.: Prentice-Hall, 1964.

Lilly, Eli. *Prehistoric Antiquities of Indiana.* Indianapolis: Indiana Historical Society, 1937.

Lindley, Harlow, ed. *Indiana as Seen by Early Travelers.* Indianapolis: Indiana Historical Commission, 1916.

McCord, Shirley S., ed. *Travel Accounts of Indiana, 1679–1961.* Indianapolis: Indiana Historical Bureau, 1970.

Martin, John Bartlow. *Indiana: An Interpretation.* New York: Knopf, 1947.

Nicholson, Meredith. *The Hoosiers.* New York: Macmillan, 1900.

Nolan, Jeanette C. *Hoosier City: The Story of Indianapolis.* New York: Messner, 1943.

Peat, Wilbur D. *Pioneer Painters of Indiana.* Indianapolis. Art Association of Indianapolis, 1954.

Peckham, Howard H. *Indiana: A Bicentennial History.* New York: Norton, 1978.

Phillips, Clifton J. *Indiana in Transition: The Emergence of an Industrial Commonwealth, 1880–1920.* Indianapolis: Indiana Historical Society, 1968.

Shumaker, Arthur W. *A History of Indiana Literature.* Indianapolis: Indiana Historical Society, 1962.

Starr, George W. *Industrial Development of Indiana.* Bloomington: Indiana University Press, 1937.

Thompson, Donald E. *Indiana Authors and Their Books, 1916–66.* Crawfordsville, Ind.: Wabash College, 1974.

Thornbrough, Emma Lou. *Indiana in the Civil War Era, 1850–80.* Indianapolis: Indiana Historical Society, 1965.

Wilson, William E. *The Angel and the Serpent: The Story of New Harmony.* Bloomington: Indiana University Press, 1964.

Wilson, William E. *Indiana: A History.* Bloomington: Indiana University Press, 1966.

IOWA

State of Iowa

IOWA

ORIGIN OF STATE NAME: Named for Iowa Indians of the Siouan family. **NICKNAME:** The Hawkeye State. **CAPITAL:** Des Moines. **ENTERED UNION:** 28 December 1846 (29th). **SONG:** "The Song of Iowa." **MOTTO:** Our Liberties We Prize and Our Rights We Will Maintain. **FLAG:** There are three vertical stripes of blue, white, and red; in the center a spreading eagle holds in its beak a blue ribbon with the state motto. **OFFICIAL SEAL:** A sheaf and field of standing wheat and farm utensils represent agriculture; a lead furnace and a pile of pig lead are to the right. In the center stands a citizen-soldier holding a US flag with a liberty cap atop the staff in one hand and a rifle in the other. Behind him is the Mississippi River with the steamer *Iowa* and mountains; above him an eagle holds the state motto. Surrounding this scene are the words "The Great Seal of the State of Iowa" against a gold background. **BIRD:** Eastern goldfinch. **FLOWER:** Wild rose. **TREE:** Oak. **STONE:** Geode. **LEGAL HOLIDAYS:** New Year's Day, 1 January; Birthday of Martin Luther King, Jr., 3d Monday in January; Lincoln's Birthday, 12 February; Washington's Birthday, 3d Monday in February; Memorial Day, last Monday in May; Independence Day, 4 July; Labor Day, 1st Monday in September; Veterans Day, 11 November; Thanksgiving Day, 4th Thursday in November; Christmas Day, 25 December. **TIME:** 6 AM CST = noon GMT.

¹LOCATION, SIZE, AND EXTENT

Located in the western north-central US, Iowa is the smallest of the midwestern states situated W of the Mississippi River, and ranks 25th in size among the 50 states.

The total area of Iowa is 56,275 sq mi (145,752 sq km), of which land takes up 55,965 sq mi (144,949 sq km) and inland water 310 sq mi (803 sq km). The state extends 324 mi (521 km) E–W; its maximum extension N–S is 210 mi (338 km).

Iowa is bordered on the N by Minnesota; on the E by Wisconsin and Illinois (with the line formed by the Mississippi River); on the S by Missouri (with the extreme southeastern line defined by the Des Moines River); and on the W by Nebraska and South Dakota (with the line demarcated by the Missouri River and a tributary, the Big Sioux).

The total boundary length of Iowa is 1,151 mi (1,853 km). The state's geographic center is in Story County near Ames.

²TOPOGRAPHY

The topography of Iowa consists of a gently rolling plain that slopes from the highest point of 1,670 feet (509 meters) in the northwest to the lowest point of 480 feet (146 meters) in the southeast at the mouth of the Des Moines River. About two-thirds of the state lies between 800 feet (244 meters) and 1,400 feet (427 meters) above sea level; the mean elevation of land is 1,100 feet (335 meters).

Supremely well suited for agriculture, Iowa has the richest and deepest topsoil in the US and an excellent watershed. Approximately two-thirds of the state's area is drained by the Mississippi River, which forms the entire eastern boundary, and its tributaries. The western part of the state is drained by the Missouri River and its tributaries. Iowa has 13 natural lakes. The largest are Spirit Lake (9 mi, or 14 km, long) and West Okoboji Lake (6 mi, or 10 km, long), both near the state's northwest border.

The Iowa glacial plain was formed by five different glaciers. The last glacier, which covered about one-fifth of the state's area, retreated from the north-central region some 10,000 years ago, leaving the topsoil as its legacy. Glacial drift formed the small lakes in the north. The oldest rock outcropping, in the state's northwest corner, is about 1 billion years old.

³CLIMATE

Iowa lies in the humid continental zone and generally has hot summers, cold winters, and wet springs.

Temperatures vary widely during the year, with an annual average of 49°F (9°C). The state averages 166 days of full sunshine and 199 cloudy or partly cloudy days. Des Moines, in the central part of the state, has a normal daily maximum temperature of 86°F (30°C) in July and a normal daily minimum of 10°F (−4°C) in January. The record low temperature for the state is −47°F (−44°C), established at Washta on 12 January 1912; the record high is 118°F (48°C), registered at Keokuk on 20 July 1934. Rainfall averages 32 in (81 cm) annually; snowfall, 30 in (76 cm); and relative humidity, 72%.

⁴FLORA AND FAUNA

Although most of Iowa is under cultivation, such unusual wild specimens as bunchberry and bearberry can be found in the northeast, whose loess soil supports tumblegrass, western beardtongue, and prickly pear cactus. Other notable plants are pink lady's slipper and twinleaf in the eastern woodlands, arrowgrass in the northwest, and erect dryflower and royal and cinnamon ferns in sandy regions. More than 80 native plants can no longer be found, and at least 35 others are confined to a single location. The federal government classified the northern wild monkshood as threatened as of 1984.

Common Iowa mammals include red and gray foxes, raccoon, opossum, woodchuck, muskrat, common cottontail, and gray, fox, and flying squirrels. The bobolink and purple martin have flyways over the state; the cardinal, rose-breasted grosbeak, and eastern goldfinch (the state bird) nest there. Game fish include rainbow trout, smallmouth bass, and walleye; in all, Iowa has 140 native fish species.

Rare animals include the pygmy shrew, ermine, black-billed cuckoo, and crystal darter. The state lists as endangered the red-backed vole, black bear, bobcat, red-shouldered hawk, piping plover, burrowing owl, northern copperhead, and lake sturgeon. The Indiana bat and peregrine falcon, both indigenous to Iowa, are on the federal endangered species list.

⁵ENVIRONMENTAL PROTECTION

Because this traditionally agricultural state's most valuable resource has been its topsoil, Iowa's conservation measures beginning in the 1930s were directed toward preventing soil erosion and preserving watershed runoff. In the 1980s, Iowans were particularly concerned with improving air quality, preventing chemical pollution, and preserving water supplies.

On 1 July 1983, the Department of Water, Air and Waste Management came into operation, with responsibility for environmental functions formerly exercised by separate state agencies. Functions of the new department include regulating operation of the state's 2,900 public water supply systems, overseeing nearly 1,200 municipal and industrial wastewater treatment plants, inspecting dams, and establishing chemical and bacterial standards to protect the quality of lakes. The department also enforces laws prohibiting open dumping of solid wastes, regulates the construction and operation of 140 solid waste disposal projects, and monitors the handling of hazardous wastes. It also establishes standards for air quality and regulates the emission of air pollutants from more than 600 industries and utilities.

On 1 January 1985, the department published a state water plan, analyzing the availability and quality of water in the state, estimating present and future use of water resources, proposing that the department have authority to impose water conservation measures on a daily basis, and recommending a water priority allocation system, to be implemented in specified circumstances during a drought. The department also prepared a hazardous-waste management plan, making recommendations for appropriate hazardous-waste management facilities in the state.

⁶POPULATION

Iowa, the 25th in size of the 50 states, ranked 27th in state population at the 1980 census, with 2,913,808 residents. As of January 1985, Iowa's population was 2,894,273, a decline of 0.7% from 1980, and the population density was 51.7 persons per sq mi (20 per sq km).

Iowa's population growth was rapid during the early years of settlement. When the first pioneers arrived in the early 19th century, an estimated 8,000 Indians were living within the state's present boundaries. From 1832 to 1840, the number of white settlers increased from fewer than 50 to 43,112. The population had almost doubled to more than 80,000 by the time Iowa became a state in 1846. The great influx of European immigrants who came via other states during the 1840s and 1850s caused the new state's population to soar to 674,913 at the 1860 census. By the end of the next decade, the population had reached nearly 1,200,000; by 1900, it had surpassed 2,200,000.

The state's population growth leveled off in the 20th century. In 1980, Iowa's population was 51.4% female and 48.6% male; 59% of all Iowans lived in urban areas. Of the total 1980 population, over 98% was native-born.

In 1980, Iowa had 123 cities and towns of 2,000 or more population. In 1984, eight cities had populations of over 50,000: Des Moines, 190,832; Cedar Rapids, 108,669; Davenport, 102,129; Sioux City, 81,767; Waterloo, 75,661; Dubuque, 60,228; Council Bluffs, 56,943; and Iowa City, 50,984. In 1984, the Des Moines metropolitan area had 377,100 residents, ranking 92d among US metropolitan areas.

⁷ETHNIC GROUPS

In 1980, there were 41,700 black Americans, 5,367 American Indians, and 26,000 people of Spanish origin living in Iowa. Among Iowans of European descent there were 1,331,624 Germans; 806,035 English, Scottish, or Welsh; 630,020 Irish; and 380,367 Scandinavians. The foreign-born population numbered 48,000 in 1980.

⁸LANGUAGES

A few Indian place-names are the legacy of early Siouan Iowa Indians and the westward-moving Algonkian Sauk and Fox tribes who pushed them out: Iowa, Ottumwa, Keokuk, Sioux City, Oskaloosa, Decorah.

Iowa English reflects the three major migration streams: Northern in that half of the state above Des Moines and North Midland in the southern half, with a slight South Midland trace in the extreme southeastern corner. Although some Midland features extend into upper Iowa, rather sharp contrasts exist between the

two halves. In pronunciation, Northern features contrast directly with Midland: /hyumor/ with /yumor/, /ah/ in *on* and *fog* with /aw/, the vowel of *but* in *bulge* with the vowel of *put*, and /too/ with /tyoo/ for *two*. Northern words also contrast with Midland words: *crab* with *crawdad, corn on the cob* with *roasting ears, quarter to* with *quarter till, barnyard* with *barn lot,* and *gopher* with *ground squirrel.*

In 1980, 96.5% of all Iowans aged 3 or more spoke only English at home. The following are other languages reported by Iowans, and the number speaking each at home:

German	26,060	Dutch	4,338
Spanish	21,407	Czech	4,012
French	5,648	Norwegian	3,421

⁹RELIGIONS

The first church building in Iowa was constructed by Methodists in Dubuque in 1834; a Roman Catholic church was built in Dubuque the following year. By 1860, the largest religious sects were the Methodists, Presbyterians, Catholics, Baptists, and Congregationalists. Other religious groups who came to Iowa during the 19th century included Lutherans, Dutch Reformers, Quakers, Mennonites, Jews, and the Community of True Inspiration, or Amana Society, which founded seven communal villages.

An estimated 61% of Iowans were church adherents in 1980. There were more than 1,200,000 Protestants, including 261,613 adherents of the United Methodist Church, 201,668 members of the American Lutheran Church, 96,780 adherents of the Lutheran Church–Missouri Synod, 62,124 members of the Lutheran Church in America, and 54,611 adherents of the Disciples of Christ (Christian Church).

Roman Catholic church membership totaled 554,091 and the Jewish population 7,760 in 1984.

¹⁰TRANSPORTATION

The early settlers came to Iowa by way of the Ohio and Mississippi rivers and the Great Lakes, then traveled overland on trails via wagon and stagecoach. The need of Iowa farmers to haul their products to market over long distances prompted the development of the railroads, particularly during the 1880s. But river traffic continues to play a vital role in the state's transport.

As of 31 December 1983, Iowa ranked 13th among the 50 states in miles of Class I railroads, with 4,699 mi (7,562 km). Amtrak operates the long-distance California Zephyr (Chicago–Oakland, Calif.), serving seven major stations in Iowa and an Iowa ridership of 64,530 in 1983/84.

Iowa ranked 10th among the states in road mileage in 1983, with 112,289 mi (180,712 km) of rural and municipal roads and 735 mi (1,183 km) of interstate highways. There were 1,764,651 registered automobiles and 714,467 trucks and buses in the state, with 1,928,799 licensed drivers as of 1983.

Iowa is bordered by two great navigable rivers, the Mississippi and the Missouri. These provided excellent transport facilities for the early settlers via keelboats and paddle-wheel steamers. Today, freight is still transported by water as far east as Pennsylvania and west to Oklahoma. In 1983, nearly 45 million tons of cargo moved on the Mississippi past Davenport, including over 31 million tons of grain. Important terminal ports on the Mississippi are Dubuque and Davenport; on the Missouri, Sioux City and Council Bluffs.

In 1983, Iowa had 280 aircraft facilities of all types. Air passenger service was supplied by 11 major airlines. The busiest airfield is Des Moines Municipal Airport, which handled 15,256 scheduled departures in 1983.

¹¹HISTORY

The fertile land now known as the State of Iowa was first visited by primitive hunting bands of the Paleo-Indian period some 12,000 years ago. The first permanent settlers of the land were the

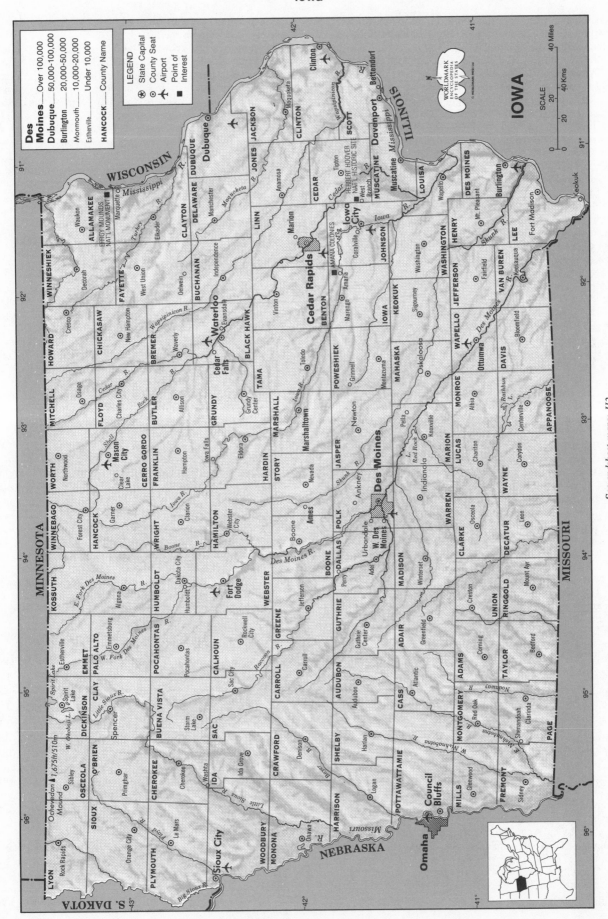

See endsheet maps: H2.

LOCATION: 40°22′32″ to 43°30′3″ N; 90°8′24″ to 96°38′21″ W. **BOUNDARIES:** Minnesota line, 264 mi (425 km); Wisconsin line, 93 mi (150 km); Illinois line, 219 mi (352 km); Missouri line, 244 mi (393 km); Nebraska line, 200 mi (322 km); South Dakota line, 131 mi (211 km).

Woodland Indians, who built villages in the forested areas along the Mississippi River, introduced agriculture, and left behind only their animal-shaped burial mounds.

Not until June 1673 did the first known white men come to the territory. When Louis Jolliet, accompanied by five French voyageurs and a Jesuit priest, Jacques Marquette, stopped briefly in Iowa on his voyage down the Mississippi, the region was uninhabited except for the Sioux in the west and a few outposts of Illinois and Iowa Indians in the east. Iowa was part of the vast, vaguely defined Louisiana Territory that extended from the Gulf of Mexico to the Canadian border and was ruled by the French until the title was transferred to Spain in 1762. Napoleon took the territory back in 1800 and then promptly sold all of Louisiana Territory to the amazed American envoys who had come to Paris seeking only the purchase of New Orleans and the mouth of the Mississippi. After Iowa had thus come under US control in 1803, the Lewis and Clark expedition worked its way up the Missouri River to explore the land that President Thomas Jefferson had purchased so cheaply. Iowa looked as empty as it had to Jolliet 130 years earlier. The only white man who had come to explore its riches before the American annexation was an enterprising former French trapper, Julien Dubuque. Soon after the American Revolution, he had obtained from the Fox Indians the sole right to work the lead mines west of the Mississippi, and for 20 years Dubuque was the only white settler in Iowa.

The first wave of migrants into Iowa were the Winnebago, Sauk, and Fox, driven there by the US Army, which was clearing Wisconsin and Illinois of their Indian populations to make way for white farmers. Although President Andrew Jackson had intended that Louisiana Territory lying north of Missouri should forever be Indian land, the occupation of Iowa by the Indians was to be brief. Following the abortive attempt by an aging Sauk chieftain, Black Hawk, to win his lands in Illinois, the Sauk and Fox were driven westward in 1832 and forced to cede their lands in eastern Iowa to the incoming white settlers.

Placed under the territorial jurisdiction of Michigan in 1834, and then two years later under the newly created Territory of Wisconsin, Iowa became a separate territory in 1838. The first territorial governor, Robert Lucas, extended county boundaries and local government westward, planned for a new capital city to be located on the Iowa River, resisted Missouri's attempt to encroach on Iowa territory, and began planning for statehood by drawing boundary lines that included not only the present State of Iowa but also southern Minnesota up to present-day Minneapolis.

Because a new state seeking admission to the Union at that time could expect favorable action from Congress only if accompanied by a slave state, Iowa was designated to come into the Union with Florida as its slaveholding counterpart. A serious dispute over how large the state would be delayed Iowa's admission into the Union until 28 December 1846, but by the delay the people of Iowa got what they wanted—all the land between the Mississippi and Missouri rivers—even though they had to abandon Lucas's northern claim.

The settlement of Iowa was rapidly accomplished. With one-fourth of the nation's fertile topsoil located within its borders, Iowa was a powerful magnet that drew farmers by the thousands from Indiana, Ohio, and Tennessee, and even from faraway Virginia, the Carolinas, New York, and New England. Except for German and Irish immigrants along the eastern border and later Scandinavian immigrants during the 1870s and 1880s, Iowa was settled primarily by Anglo-American stock. The settlers were overwhelmingly Protestant in religion and remarkably homogeneous in ethnic and cultural backgrounds. Although New Englanders made up only 5% of Iowa's early population, they had a cultural influence that far exceeded their numbers. Many small Iowa towns—with their large frame houses, elm-lined streets, and Congregational churches—looked like New England villages faithfully replicated on the prairie.

Fiercely proud of its claim to be the first free state created out of the Louisiana Purchase, Iowa was an important center of abolitionist sentiment throughout the 1850s. The Underground Railroad for fugitive slaves from the South ran across the southern portion of Iowa to the Mississippi River. Radical abolitionist John Brown spent the winters of 1857 and 1859 in the small Quaker village of Springdale, preparing for his attack on the US arsenal at Harpers Ferry, in western Virginia.

Although the Democrats had a slight edge over their Whig opposition in the early years of statehood, a majority of Iowa voters in 1856 supported the new Republican Party and, for the most part, did so in succeeding years. A Republican legislative majority in 1857 scrapped the state's first constitution, which had been written by Jacksonian Democrats 12 years earlier. The new document moved the state capital from Iowa City westward to Des Moines, but it provided that the state university would remain forever in Iowa City.

When the Civil War came, Iowa overwhelmingly supported the Union cause. Iowans fought not only for their ideals, the abolition of slavery and the preservation of the Union, but also for the very practical objective of keeping open the Mississippi River, the main artery for transport of agricultural products.

In the decades following the Civil War, Iowans on the national scene, most notably US Senators James Harlan and William B. Allison, his successor, belonged to the conservative Republican camp, but they frequently faced liberal Republican and Populist opposition inside the state. Iowan James B. Weaver broke from the Republican Party to become the Greenback-Labor candidate for president in 1880 and the Populist presidential nominee in 1892. Senator Allison and the conservatives had powerful allies in the railroad entrepreneurs, especially Grenville Dodge of the Union Pacific and Charles Eliot Perkins of the Burlington Railroad, throughout the late 19th century. The railroad had been lavishly welcomed by Iowans in the 1850s; by the 1870s, Iowa farmers were desperately trying to free themselves from the stranglehold of the rail lines. The National Grange was powerful enough in Iowa to put through the legislature the so-called Granger laws to regulate the railroads. In 1886, Allison pushed his protégé, William Larrabee, into the governorship, only to discover that Larrabee was a reformer in disguise. During his four years in office, Larrabee carried on a constant battle with the railroad interests for effective regulatory legislation. At the turn of the century, as the aging Allison's hold on the state weakened, Iowa became a center for Republican progressivism.

Following World War I, the conservatives regained control of the Republican Party. They remained in control until, during the 1960s, new liberal leadership was forced on the party because of the debacle of Barry Goldwater's 1964 presidential campaign, controversy over US involvement in Viet-Nam, and effective opposition from a revitalized Democratic Party led by Harold Hughes. After Hughes gave up the governorship in 1969 to become a US senator, he was succeeded in office by Robert Ray, a liberal Republican who dominated the state throughout the 1970s. Republicans kept control of the statehouse with the election of Terry Branstad in 1982, but Democrats gained control of the state legislature and made strong inroads at the top levels of state government. The Democrats kept control of the state legislature in 1984 even though Republican President Ronald Reagan carried the state with more than 53% of the vote.

¹²STATE GOVERNMENT

Iowa has had two state constitutions. The constitution of 1857 replaced the original constitution of 1846 and is still in effect.

The state legislature, or general assembly, consists of a 50-member senate and a 100-member house of representatives. Senators serve four-year terms, with half the members elected every two years. Representatives are elected to two-year terms. The legislature convenes each year on the 2d Monday in January. Special sessions may be called by the governor or initiated by petition of two-thirds of the members of each house. Each house may introduce or amend legislation, with a simple majority vote required for passage. The governor's veto of a bill may be overridden by a two-thirds majority in both houses. Legislators must be US citizens and must have resided in the state for a year; a representative must be at least 21 years of age, and a senator 25.

The state's only elected executives are the governor, lieutenant governor, secretary of state, auditor, treasurer, attorney general, and secretary of agriculture; since 1974, they have been elected to four-year terms. The governor and lieutenant governor each must be a US citizen, at least 30 years old, and a resident of the state for two years.

To vote in Iowa, one must be a US citizen, at least 18 years old, and a state resident. A voter must register at least 10 days before an election, and remains permanently registered if there is no change of residency.

13POLITICAL PARTIES

For 70 years following the Civil War, a majority of Iowa voters supported the Republicans over the Democrats in nearly all state and national elections. During the Great Depression of the 1930s, Iowa briefly turned to the Democrats, supporting Franklin D. Roosevelt in two presidential elections. But from 1940 through 1984, the majority of Iowans voted Republican in 10 of 12 presidential elections. Republicans won 32 of the 40 gubernatorial elections from 1900 through 1982 and controlled both houses of the state legislature for 112 of the 130 years from 1855 to 1984.

In the 1960s, however, Iowa showed signs of a Democratic upsurge. Harold Hughes, a liberal Democrat, revitalized the party in Iowa and was elected governor for three two-year terms before moving on to the US Senate. During the post-Watergate period of the mid-1970s, Democrats captured both US Senate seats, five of the six congressional seats, and both houses of the Iowa legislature. By the early 1980s, the picture was mixed. After the November 1984 elections, Republicans had a 4–2 edge in the US House delegation and one of Iowa's two US Senate seats. On the state level, Republicans controlled the statehouse, but Democrats controlled the legislature.

Iowa's presidential caucuses are held in January of presidential campaign years, earlier than any other state, thus giving Iowans a degree of influence in national politics.

14LOCAL GOVERNMENT

The state's 99 counties are governed by boards of supervisors consisting of three of five members. In general, county officials are elected to four-year terms; they enforce state laws, collect taxes, supervise welfare activities, and manage roads and bridges.

Local government was exercised by 955 municipal units in 1982. The mayor-council system functioned in the great majority of these municipalities. Iowa's towns and cities derive their local powers from the state constitution, but the power to tax is authorized by the general assembly.

15STATE SERVICES

The Department of Public Instruction is responsible for educational services in Iowa. It assists local school boards in supplying special educational programs and administers 15 area education agencies (reduced in 1974 from 79 county school systems).

Transportation services are directed by the Department of Transportation, which is responsible for the safe and efficient operation of highways, motor vehicles, airports, railroads, public transit, and river transportation. The department's motor vehicle division licenses drivers, road vehicles, and car dealers.

Health and welfare services are provided by the Department of Health, the Iowa Mental Health Authority, and the Department of Social Services. Public protection is the responsibility of the Departments of Public Defense and of Public Safety. Housing programs are supported by the Iowa Housing Finance Authority.

Iowa Presidential Vote by Political Parties, 1948–84

YEAR	ELECTORAL VOTE	IOWA WINNER	DEMOCRAT	REPUBLICAN	PROGRESSIVE	PROHIBITION	SOCIALIST LABOR
1948	10	*Truman (D)	522,380	494,018	12,125	3,382	4,274
1952	10	*Eisenhower (R)	451,513	808,906	5,085	2,882	—
							CONSTITUTION
1956	10	*Eisenhower (R)	501,858	729,187	—	—	3,202
1960	10	Nixon (R)	550,565	722,381	—	—	—
1964	9	*Johnson (D)	733,030	449,148	—	1,902	—
					SOC. WORKERS	AMERICAN IND.	
1968	9	*Nixon (R)	476,699	619,106	3,377	66,422	—
						AMERICAN	PEACE AND FREEDOM
1972	8	*Nixon (R)	496,206	706,207	—	22,056	1,332
							LIBERTARIAN
1976	8	Ford (R)	619,931	632,863	—	3,040	1,452
					CITIZENS		
1980	8	*Reagan (R)	508,672	676,026	2,191	NA	12,324
1984	8	*Reagan (R)	605,620	703,088	—	—	—

*Won US presidential election.

¹⁶JUDICIAL SYSTEM

The Iowa supreme court consists of nine justices who are appointed by the governor and confirmed to eight-year terms by judicial elections held after they have served on the bench for at least one year. Judges may stand for reelection before their terms expire. The justices select one of their number as chief justice. The court exercises appellate jurisdiction in civil and criminal cases, supervises the trial courts, and prescribes rules of civil and appellate procedure. The supreme court transfers certain cases to the court of appeals, a five-member appellate court that began reviewing civil and criminal cases in 1977, and may review its decisions. Judges on the court of appeals are appointed and confirmed to six-year terms in the same manner as supreme court justices; they elect one of their members as chief judge.

The state is divided into eight judicial districts, each with a chief judge appointed to a two-year term by the chief justice of the supreme court. District court judges are appointed to six-year terms by the governor from nominations submitted by district nominating commissions. Appointees must stand for election after they have served as judges for at least one year.

As of 31 December 1984 there were 2,836 prisoners in federal and state institutions. Iowa's total crime rate in 1983 was 3,919 per 100,000 population, among the lowest in the US. The state had 3,723 active attorneys as of December 1984.

¹⁷ARMED FORCES

Iowa residents as of 30 September 1983 included an estimated 352,000 veterans, of whom 6,000 saw service during World War I, 132,000 during World War II, 64,000 in the Korean conflict, and 102,000 during the Viet-Nam era. The federal government expended $274 million for veterans' benefits in Iowa during the 1982/83 fiscal year.

The Iowa National Guard provides reserve units for the US Army and Air Force in case of a national emergency or war. As of early 1985, National Guard units stationed in Iowa had a total assigned strength of 8,540. There were 389 military personnel and 1,468 civilians involved in defense activities as of 1983/84.

¹⁸MIGRATION

Iowa was opened, organized, and settled by a generation of native migrants from other states. According to the first federal census of Iowa in 1850, 31% of the total population of 192,214 came from nearby midwestern states (Illinois, Wisconsin, Indiana, Michigan, and Ohio), 14% from the five southern border states, and 13% from the Middle Atlantic states.

The remaining 10% of the state's 1850 population consisted of immigrants from northern Europe. The largest group were Germans who had fled military conscription; the next largest group had sought to escape the hardships of potato famine in Ireland or of agricultural and technological displacement in Scotland, England, and Wales. They were joined in the 1850s by Dutch immigrants seeking religious liberty, and in the 1860s and 1870s by Norwegians and Swedes. During and immediately after the Civil War, some former slaves fled the South for Iowa, and more blacks settled in Iowa cities after 1900.

But many of the migrants who came to Iowa did not stay long. Some Iowans left to join the gold rush, and others settled lands in the West. Migration out of the state has continued to this day, as retired Iowans seeking warmer climates have moved to California and other southwestern states; from 1970 through 1983, Iowa's net loss through migration totaled 189,000.

An important migratory trend within the state has been from the farm to the city. Although Iowa has remained a major agricultural state, the urban population surpassed the rural population by 1960 and increased to over 59% of the total population by 1980.

¹⁹INTERGOVERNMENTAL COOPERATION

Iowa is a signatory to the Midwest Nuclear Compact, the Iowa-Missouri and Iowa-Nebraska boundary compacts, and 11 other major interstate compacts and agreements. Federal aid to the Iowa state government amounted to $1.1 billion in 1983/84.

²⁰ECONOMY

Iowa's economy is based on agriculture. Although the value of the state's manufactures exceeds the value of its farm production, manufacturing is basically farm-centered. The major industries are food processing and the manufacture of agriculture-related products, such as farm machinery. Iowa's gross state product in 1983 totaled $33.6 billion.

Periodic recessions—and especially the Great Depression of the 1930s—have afflicted Iowa farmers and adversely affected the state's entire economy. But technological progress in agriculture and the proliferation of manufacturing industries have enabled Iowans to enjoy general prosperity since World War II. Because the state's population is scattered, the growth of light manufacturing has extended to hundreds of towns and cities.

In the late 1970s, the state's major economic problem was inflation, which boosted the cost of farm equipment and fertilizers. In the early 1980s, high interest rates and falling land prices created serious economic difficulties for farmers and contributed to the continuing decline of the farm population.

²¹INCOME

With a personal income per capita of $11,048 in 1983, slightly below the national average, Iowa ranked 28th in the US.

The state's total personal income was $32.1 billion in 1983. The median income of four-person families in 1981 was $22,821. Some 117,000 persons (13.2% of the population) were below the federal poverty level in 1979, and in 1982, 50,800 Iowans were among the top wealth-holders in the nation, with gross assets greater than $500,000.

²²LABOR

Since 1950, Iowa has consistently ranked above the national average in employment of its work force. Iowa's unemployment rate of 7% for 1984 was slightly below the overall US rate. About 61% of all Iowans were employed that year, as compared with 59.5% of the nation as a whole.

The civilian labor force in 1983 totaled 1,422,000, of whom 1,306,000 were employed and 116,000 unemployed. Of the total work force, 58.9% was male and 41.1% female. A federal survey in March 1983 revealed the following nonfarm employment pattern in Iowa:

	ESTABLISH-MENTS	EMPLOYEES	ANNUAL PAYROLL ('000)
Agricultural services, forestry, fishing	903	3,874	$ 52,877
Mining	193	1,931	42,305
Contract construction	6,313	30,794	618,210
Manufacturing, of which:	3,537	197,621	4,397,229
Food and food products	(468)	(41,038)	(70,128)
Nonelectrical machinery	(552)	(42,908)	(1,141,879)
Transportation, public utilities	3,390	42,662	865,636
Wholesale trade	7,630	68,929	1,208,646
Retail trade	21,468	190,422	1,558,142
Finance, insurance, real estate	6,216	62,370	1,044,558
Services	19,945	203,501	2,293,610
Other	2,842	3,182	52,163
TOTALS	72,437	805,286	$12,133,376

The labor movement generally has not been strong in Iowa, and labor unions have had little success in organizing farm laborers. The Knights of Labor, consisting mostly of miners and railroad workers, was organized in Iowa in 1876 and enrolled 25,000 members by 1885. But the Knights practically disappeared after 1893, when the American Federation of Labor (AFL) established itself in the state among miners and other workers. The Congress of Industrial Organizations (CIO) succeeded in organizing workers in public utilities, meat packing, and light industries in 1937. After 1955, when the AFL and CIO merged, the power and influence of labor unions increased in the state. In 1980, the number of labor union members in Iowa totaled 244,000, and in 1983 there were 953 unions. In 1981 there were 41 work stoppages, involving 9,800 workers and 193,300 working days lost.

As of September 1984, production workers on manufacturing payrolls in the state worked an average of 40.8 hours per week and earned average hourly pay of $10.21.

Iowa did not forbid the employment of women in dangerous occupations or prohibit the employment of children under 14 years of age in factories, shops, or mines until the early 1900s. A right-to-work law was enacted in 1947.

23 AGRICULTURE

Iowa recorded a (realized) gross farm income of $11.1 billion in 1983. More than half of all cash receipts from marketing came from the sale of livestock and meat products; about one-fifth derived from the sale of feed grains. In that year, Iowa ranked 1st in output of corn for grain and soybeans and ranked 4th in production of corn for silage.

The early settlers planted wheat. Iowa ranked 2d in wheat production by 1870, but as the wheat belt moved farther west, the state's farmers turned to raising corn to feed their cattle and hogs. Two important 20th-century developments were the introduction in the 1920s of hybrid corn and the utilization on a massive scale during World War II of soybeans as a feed grain. Significant postwar trends included the rapid mechanization of farming and the decline of the farm population.

In 1984, Iowa had 113,000 farms, with an average size of 297 acres (120 hectares) per farm. This total represents a marked decrease of 27,000 farms since 1970, although the amount of land devoted to farming remained nearly the same during that period, at about 33,600,000 acres (13,600,000 hectares). Of the total number of farms in 1982 for which data are available, 46% were operated by farm owners and 54% were rented to operators. Farm employment in 1984 totaled 149,400 workers.

Nearly all of Iowa's land is tillable, and more than nine-tenths of it is given to farmland. Corn is grown practically everywhere; wheat is raised in the southern half of the state and in counties bordering the Mississippi and Missouri rivers.

In 1983, production of corn for grain totaled 744 million bushels, valued at $2.4 billion; soybeans, 271 million bushels, $1.9 billion; oats, 38 million bushels; and hay, 1.1 million tons.

24 ANIMAL HUSBANDRY

Iowa produced about 26% of the nation's pork and 11% of its grain-fed beef in 1983. In that year, it ranked 1st among the 50 states in hog and pig production, 4th in cattle production, and 4th in production of American cheese.

Pigs, calves, lambs, and chickens are raised throughout the state, particularly in the Mississippi and Missouri river valleys, where good pasture and water are plentiful. Iowa farmers were leaders in applying modern livestock breeding methods to produce lean hogs, tender corn-fed cattle, and larger-breasted chickens and turkeys.

Livestock produced for market in 1983 included 22,569,000 hogs and 2,493,000 grain-fed cattle. Cash receipts to farmers for farm animals and livestock products in 1983 included $2.6 billion for hogs, $2.1 billion for cattle and calves, and $18,595,000 for sheep and lambs. Receipts for dairy products were $557,964,000; gross income for poultry and eggs was $138,361,000. In 1983 there were 1,459,000 beef cattle and 383,000 milk cows.

Iowa dairy farmers produced 4.2 billion lb of whole milk in 1983. In that year, the state's poultry producers raised 7.6 million hens and pullets of laying age.

25 FISHING

Fishing has little commercial importance in Iowa; the catch in 1983, 5,078,000 lb, was worth only $1.5 million. Game fishing in the rivers and lakes is a popular sport.

26 FORESTRY

Lumber and woodworking were important to the early settlers, but the industry has since declined in commercial importance. In 1979, Iowa had 1,561,000 acres (631,700 hectares) of forestland. In 1982, shipments of paper and paper products totaled $576 million, and shipments of lumber and wood products totaled $275 million. The state's lumber industry produced 42 million board feet of lumber in 1983.

27 MINING

In 1984, the value of Iowa's nonfuel mineral output was estimated at $259,687,000, 28th in the US. The principal minerals (in order of value) were stone, cement, sand and gravel, and gypsum.

In 1788, the French explorer Julien Dubuque founded the first white settlement in Iowa at the site of Indian lead mines on the banks of the Mississippi River (later the site of the city named for him), but the supply of lead gave out in the late 19th century. Rich gypsum deposits near Fort Dodge continued to be mined in the mid-1980s. Limestone is still mined in east-central Iowa, and sand and gravel are found throughout the state's northern half. In 1983, shipments of crushed stone were valued at $101 million, of cement at $91 million, and of sand and gravel at $33 million.

28 ENERGY AND POWER

Although Iowa's fossil fuel resources are extremely limited, the state's energy supply has been adequate for consumer needs. In 1983, Iowa consumed 328 million Btu per capita, to rank 20th among the states. According to 1983 estimates, oil supplied about 32% of the state's energy requirements; natural gas, 23%; coal, 34%; hydropower, 3%; and nuclear energy, 8%.

The state's production of electricity totaled 22 billion kwh in 1983; installed capacity was 8.7 million kw. Coal-fired plants supplied 18.7 billion kwh of electricity; nuclear power plants supplied 2.3 billion kwh; and hydroelectric plants less than 1 billion kwh.

Extensive coalfields in southeastern Iowa were first mined in 1840. The boom town of Buxton, in Monroe County, mined sufficient coal in 1901 to support a population of 6,000 people, of whom 5,500 were transplanted southern blacks, but the mines closed in 1918 and Buxton became a ghost town. The state's annual bituminous coal production reached nearly 9 million tons in 1917-18. Coal output in 1983 was only 385,320 tons; demonstrated reserves totaled 2.2 billion tons.

29 INDUSTRY

Because Iowa was primarily a farm state, the first industries were food processing and the manufacture of farm implements. These industries have retained a key role in the economy. In recent years, Iowa has added a variety of other manufactures—including pens, washing machines, even mobile homes.

The total value of shipments by manufacturers was $31.4 billion in 1982. The following table shows value of shipments for selected industries in that year:

Food and kindred products	$13,746,900,000
Nonelectrical machinery	6,145,500,000
Chemicals and allied products	2,325,300,000
Electric and electronic equipment	1,867,100,000
Fabricated metal products	1,341,800,000
Printing and publishing	1,140,100,000
Primary metal products	1,105,200,000
Rubber and plastic products	926,600,000
Paper and allied products	575,600,000
Stone, clay, and glass products	551,400,000
Transportation equipment	498,100,000

More than 120 of *Fortune* magazine's "Top 500" industrial corporations have plants in Iowa, including Caterpillar Tractor, General Motors, Mobil, General Electric, General Foods, Procter & Gamble, and US Steel.

30COMMERCE

In 1982, Iowa's wholesale trade totaled $25.8 billion. The most valuable categories of goods traded were agricultural raw materials, 39%; durable goods, 23%; groceries and related products, 15%; and farm supplies, 10%. Other items accounted for 13%.

Retail sales in 1983 were $14.6 billion. Of that total, food sales accounted for 24%; automotive-related sales, 19%; general merchandise, 12%; and eating and drinking places, 10%.

Iowa ranked 2d among the 50 states (after Illinois) in 1981/82 agricultural exports, which accounted for $3 billion, or 7.7% of the US total. The leading exported commodities were feed grains and products, soybeans and soybean products, and meats and meat products. Iowa's exports of manufactured goods in 1981 had an estimated value of $3.9 billion, 71% more than in 1977.

31CONSUMER PROTECTION

Iowa has laws prohibiting fraud and misrepresentation in sales and advertising and harassment in debt collecting. There is a cooling-off period of three days for door-to-door purchases.

32BANKING

As of 31 December 1984, Iowa had 632 insured commercial banks; total commercial bank assets amounted to $26.9 billion and total deposits to $23.1 billion as of 31 December 1983. The state's 51 savings and loan associations had assets totaling $7.2 billion at the end of 1983. Iowa had 319 credit unions in 1983.

The Department of Banking supervises and controls the state's chartered banks, credit unions, and loan companies.

33INSURANCE

In 1983, Iowa had 38 life insurance companies with combined assets of $17.6 billion. Life insurance in force as of 31 December 1983 totaled $62.9 billion. The 58 fire, casualty, and multiple-line insurance companies operating in the state had assets totaling $3.1 billion, liabilities amounting to $2 billion, and premiums of $1.5 billion. In 1984, $477 million of automobile liability and physical damage insurance premiums were written in the state.

The commissioner of insurance, appointed by the governor, supervises all insurance business transacted in the state.

34SECURITIES

There are no securities exchanges in Iowa. New York Stock Exchange member firms had 105 sales offices and 486 registered full-time representatives in Iowa in December 1983. Some 427,000 Iowans owned shares of public corporations in 1983.

35PUBLIC FINANCE

The public budget is prepared by the state comptroller with the governor's approval and is adopted or revised by the general assembly. Each budget is prepared for the biennium of the upcoming fiscal year and the one following; the fiscal year runs from 1 July to 30 June.

Iowa's estimated 1984/85 budget and projected budget for 1985/86 were as follows (in millions):

REVENUES	1984/85	1985/86
Personal income tax	$1,002.7	$1,057.0
Sales tax	612.5	642.0
Inheritance, use, insurance taxes	202.0	209.5
Corporate income taxes	172.5	171.1
Tobacco and alcohol taxes	75.8	74.1
Other receipts	206.1	218.8
TOTALS	$2,271.6	$2,372.5
EXPENDITURES		
Education	$1,182.8	$1,244.1
Social services	456.9	454.8
Regulatory and finance	291.0	324.0
Government	68.5	79.7
Natural resources	39.2	39.2
Human resources	29.0	29.5
Other expenditures	42.6	58.5
TOTALS	$2,110.0	$2,209.8

The principal sources of revenue in 1984/85 were personal income taxes, the state sales tax, use taxes, and corporate income taxes. The combined state and local government debt in 1982 was $4.1 billion, or $1,412 per capita. State government debt by itself was only $157 per capita; only one state, Arizona, had a lower per capita debt burden.

36TAXATION

In 1984, Iowa's personal income tax ranged from 0.5% on the first $1,023 of income to 13% on amounts over $76,725. The corporate tax rate ranged from 6% on the first $25,000 of net income to 12% on amounts over $250,000. Iowa's retail sales tax was 4% in 1984. The state also taxed gasoline, cigarettes, alcoholic beverages, insurance premiums, inheritances, chain stores, and business franchises.

Iowans filed 1,161,000 federal income tax returns in 1982 and paid more than $3 billion in federal income tax. As of 1982/83, Iowans ranked 27th in federal benefits received per capita, which amounted to $2,265.

37ECONOMIC POLICY

Since World War II, the state government has attracted new manufacturing industries to Iowa by granting tax incentives and by encouraging a favorable business climate. The Iowa Development Commission helps local communities diversify their economies, assists companies already in the state, and helps exporters to sell their products abroad. In the mid-1980s, the Iowa state government stressed such development goals as agricultural diversification, increased small-business support, creation of high-tech jobs, and expansion of tourism.

38HEALTH

Infant mortality in 1981 was 9.8 per 1,000 live births among whites and 18.8 among blacks; both figures were below the national average. In 1982, 8,200 legal abortions were performed, a rate of 183 per 1,000 live births.

In 1983, Iowa's 140 hospitals admitted 517,611 patients, and there were 3,449,729 outpatient visits. The average number of resident hospital patients daily was 14,800 as of 1982, for a bed occupancy rate of 72%. Hospital personnel totaled 52,000. The state's nursing homes had 31,200 patients as of 1980. In 1982, the average hospital stay lasted eight days and cost $260 per day, more than 20% below the average US cost. Semiprivate room charges averaged nearly $180 a day as of early 1984.

As of October 1984 there were an estimated 8,080 licensed medical doctors, 2,460 dentists, 1,123 chiropractors, 575 optometrists, 30,240 active registered nurses, and 11,252 active licensed practical nurses in the state.

39SOCIAL WELFARE

Iowa's per capita expenditures on social welfare are below the national average. In all, 4.4% of the population received public aid in 1983.

In January 1983, aid to families with dependent children in Iowa totaled $11.8 million, or $325 monthly per family. In May 1983, Supplemental Security Income payments amounted to $839,000 million for the elderly ($96 per person per month) and $2.7 million for disabled persons ($184 per person per month). About 375,000 pupils participated in the federal school lunch program in 1983, which cost $23 million, and 208,000 received food stamps costing the federal government $104 million.

Social Security benefits totaling $1.7 billion were paid to 365,000 retired Iowans in 1983. The average monthly payment to retirees was $444. In 1983/84, federal expenditures for hospital insurance payments in Iowa totaled $543 million.

Workers' compensation payments amounted to $110.8 million in 1982. Unemployment insurance benefits went to an average of 43,000 beneficiaries weekly and totaled $349 million.

40HOUSING

Iowa ranks high in the number of housing units that are family owned and occupied. According to the 1980 census, there were a total of 1,131,299 housing units, of which 93% were occupied, 72% of them by the owners. Of all year-round housing units, 77% were one-unit dwellings. In 1983, the state authorized the construction of 4,400 new privately owned housing units, valued at $167 million. The median value of all owner-occupied units was $40,600 as of 1980.

41EDUCATION

The high school graduation rate among those entering ninth grade in 1980 was 88% in 1983/84. Iowa's progressive public school system has been an innovator in school curriculum development, teaching methods, educational administration, and school financing.

In the 1983/84 school year, Iowa had a total of 494,965 pupils enrolled in the public school system; 246,432 students were in grades K-6, 118,357 in grades 7-9, 115,735 in grades 10-12, and 14,441 in special schools. In addition, the state's nonpublic elementary and secondary schools enrolled 50,422 pupils in 1983/84.

In fall 1983, 130,753 students were enrolled in institutions of higher learning. Iowa had 3 state universities, with 67,194 students; 38 private colleges; and 15 vocational schools and area community colleges. Professional and nursing schools enrolled 1,452, and private two-year colleges 1,838. Since the public community college system began offering vocational and technical training in 1960, total enrollment has increased rapidly and the number of different career programs has grown to 162.

Iowa's small liberal arts colleges and universities include Briar Cliff College, Sioux City; Coe College, Cedar Rapids; Cornell College, Mt. Vernon; Drake University, Des Moines; Grinnell College, Grinnell; Iowa Wesleyan College, Mt. Pleasant; Loras College, Dubuque; and Luther College, Decorah.

Public schools are administered by the state superintendent of public instruction, who is appointed by a nine-member state board to a four-year term. Iowa's $3,212 in per-pupil expenditures for public schools in 1983/84 ranked 19th among the states.

42ARTS

Beginning with the public lecture movement in the late 19th century and the Chautauqua shows in the early 20th century, cultural activities have gradually spread throughout the state. Today there is an opera company in Des Moines, and there are art galleries, little theater groups, symphony orchestras, and ballet companies in the major cities and college towns.

The Des Moines Arts Center is a leading exhibition gallery for native painters and sculptors. There are regional theater groups in Des Moines, Davenport, and Sioux City. The Writers' Workshop at the University of Iowa has an international reputation. One problem for the arts in Iowa is the continued migration of native artists to cultural centers in New York, California, and elsewhere.

43LIBRARIES AND MUSEUMS

Beginning with the founding in 1873 of the state's first tax-supported library at Independence, Iowa's public library system has grown to include total book holdings of 9,355,988 volumes in 1983. Among the principal libraries in Iowa are the State Library in Des Moines, the State Historical Society Library in Iowa City, the libraries of the University of Iowa (also in Iowa City), and the Iowa State University Library in Ames.

Iowa had 87 museums and 3 zoological parks in 1984. The Herbert Hoover National Historical Site, in West Branch, houses the birthplace and grave of the 31st US president and a library and museum with papers and memorabilia. Other historic sites include the grave of French explorer Julien Dubuque, near the city named for him; the girlhood home at Charles City of suffragist Carrie Chapman Catt; and the seven communal villages of the Amana colonies.

44COMMUNICATIONS

The first post office in Iowa was established at Augusta in 1836. Mail service developed slowly with the spread of population, and rural free delivery of mail did not begin until 1897. In 1985 there were 947 post offices in the state.

The first telegraph line was built between Burlington and Bloomington (now Muscatine) in 1848. Telegraph service throughout the state is provided by Western Union. In 1980, 1,016,468 housing units, or 97% of all occupied units, had telephones.

Among the first educational radio broadcasting stations in the US were one established in 1919 at the State University in Iowa City and another in 1921 at Iowa State University in Ames. The first commercial radio station west of the Mississippi, WDC at Davenport, began broadcasting in 1921. In 1983 there were 168 commercial radio stations, including 79 AM stations and 89 FM stations. In 1984, Iowa had a total of 14 commercial television stations and 8 educational stations. In that year, 137 cable television systems served 369,992 subscribers in more than 220 communities.

45PRESS

Iowa's first newspaper, the *Dubuque Visitor,* was founded in 1836 but lasted only a year. The following year, the *Fort Madison Patriot* and the *Burlington Territorial Gazette* were established; the latter paper, now the *Hawk Eye,* is the oldest newspaper in the state. In 1860, the *Iowa State Register* was founded; as the *Des Moines Register and Tribune,* it grew to be the state's largest newspaper. The *Tribune* ceased publication in 1982; the *Register* remained preeminent, with a morning circulation of 239,275 and a Sunday circulation of 380,712 as of 1984.

Overall, Iowa had 39 dailies (30 evening, 8 morning, 1 all-day) and 9 Sunday papers in 1984. Also published in Iowa were 92 periodicals, among them *Better Homes and Gardens* (circulation, 8,022,794), *Metropolitan Home* (728,435), and *Successful Farming* (581,892).

46ORGANIZATIONS

The 1982 Census of Service Industries counted 1,077 organizations in Iowa, including 231 business associations; 566 civic, social, and fraternal associations; and 8 educational, scientific, and research associations. Among the organizations headquartered in Iowa are the National Farmers Organization (Corning), the American College Testing Program (Iowa City), and the Antique Airplane Association (Ottumwa).

47TOURISM, TRAVEL, AND RECREATION

In 1982, out-of-state travelers to Iowa spent an estimated $1.6 billion, generating 48,300 jobs and $315 million in payroll, as well as $72 million in state and local tax revenues.

The Mississippi and Missouri rivers offer popular water sports facilities for both out-of-state visitors and resident vacationers. Iowa's "Little Switzerland" region in the northeast, with its high bluffs of woodland overlooking the Mississippi, is popular for hiking and camping. Notable tourist attractions in the area include the Effigy Mounds National Monument (near Marquette), which has hundreds of prehistoric Indian mounds and village sites, and the Buffalo Ranch (at Fayette), with its herd of live buffalo. Tourist sites in the central part of the state include the state capitol and the Herbert Hoover National Historic Site, with its Presidential Library and Museum. The Hoover Site attracted 15,130 visits in 1984.

Iowa has about 85,000 acres (34,400 hectares) of lakes and reservoirs and 19,000 mi (30,600 km) of fishing streams. There are 66 state parks, covering 51,000 acres (20,600 hectares), and 7 state forests, covering 25,000 acres (10,000 hectares); these and other state recreational areas attracted 14,829,000 visits in 1984. The state had 276,216 hunting-license holders and 473,262 holders of fishing licenses in 1982/83.

48SPORTS

Iowa has no major league professional teams. High school and college basketball and football teams draw thousands of spectators, particularly to the state high school basketball tournament at

Des Moines in March. Large crowds also fill stadiums and fieldhouses for the University of Iowa games in Iowa City and Iowa State University games in Ames. In intercollegiate competition, the University of Iowa Hawkeyes belong to the Big Ten Conference, and the Iowa State University Cyclones are in the Big Eight. A popular track-and-field meet for college athletes is the Drake Relays, held every April in Des Moines. Horse racing is popular at state and county fairgrounds, as is stock car racing at small-town tracks. Wagering on sports events is illegal in Iowa.

⁴⁹FAMOUS IOWANS

Iowa was the birthplace of Herbert Clark Hoover (1874-1964), the first US president born west of the Mississippi. Although he was orphaned and left the state for Oregon at the age of 10, he always claimed Iowa as his home. His long and distinguished career included various relief missions in Europe, service as US secretary of commerce (1921-29), and one term in the White House (1929-33). Hoover was buried in West Branch, the town of his birth. Iowa has also produced one US vice president, Henry A. Wallace (1888-1965), who served in that office during Franklin D. Roosevelt's third term (1941-45). Wallace also was secretary of agriculture (1933-41) and of commerce (1945-47); he ran unsuccessfully as the Progressive Party's presidential candidate in 1948.

Two Kentucky-born members of the US Supreme Court were residents of Iowa prior to their appointments: Samuel F. Miller (1816-90) and Wiley B. Rutledge (1894-1949). Iowans who served in presidential cabinets as secretary of the interior were James Harlan (b.Illinois, 1820-99), Samuel J. Kirkwood (b.Maryland, 1813-94), Richard Ballinger (1858-1922), and Ray Lyman Wilbur (1875-1949). Ray Wilbur's brother Curtis (1867-1954) was secretary of the Navy, and James W. Good (1866-1929) was secretary of war. Appropriately enough, Iowans have dominated the post of secretary of agriculture in this century: they included, in addition to Wallace, James "Tama Jim" Wilson (b.Scotland, 1835-1920), who served in that post for 16 years and set a record for longevity in a single cabinet office; Henry C. Wallace (b.Illinois, 1866-1924), the father of the vice president; and Edwin T. Meredith (1876-1928). Harry L. Hopkins (1890-1946) was Franklin D. Roosevelt's closest adviser in all policy matters, foreign and domestic, and served in a variety of key New Deal posts. Prominent US senators from Iowa have included James W. Grimes (b.New Hampshire, 1816-72), whose vote, given from a hospital stretcher, saved President Andrew Johnson from being convicted of impeachment charges in 1868; earlier, Grimes had been governor of the state when its 1857 constitution was adopted. William Boyd Allison (b.Ohio, 1829-1908) was the powerful chairman of the Senate Appropriations Committee for nearly 30 years.

Among Iowa's most influential governors were the first territorial governor, Robert Lucas (b.Virginia, 1781-1853); Cyrus C. Carpenter (b.Pennsylvania, 1829-98); William Larrabee (b.Connecticut, 1832-1912); Horace Boies (b.New York, 1827-1923); and, in recent times, Harold Hughes (b.1922) and Robert D. Ray (b.1928).

Iowa has produced a large number of radical dissenters and social reformers. Abolitionists, strong in Iowa before the Civil War, included Grimes, Josiah B. Grinnell (b.Vermont, 1821-91), and Asa Turner (b.Massachusetts, 1799-1885). George D. Herron (b.Indiana, 1862-1925) made Iowa a center of the Social Gospel movement before helping to found the Socialist Party. William "Billy" Sunday (1862-1935) was an evangelist with a large following among rural Americans. James B. Weaver (b.Ohio,

1833-1912) ran for the presidency on the Greenback-Labor ticket in 1880 and as a Populist in 1892. John L. Lewis (1880-1969), head of the United Mine Workers, founded the Congress of Industrial Organizations (CIO).

Iowa can claim two winners of the Nobel Peace Prize: religious leader John R. Mott (b.New York, 1865-1955) and agronomist and plant geneticist Norman E. Borlaug (b.1914). Three other distinguished scientists who lived in Iowa were George Washington Carver (b.Missouri, 1864-1943), Lee De Forest (1873-1961), and James Van Allen (b.1914). George H. Gallup (1901-84), a public-opinion analyst, originated the Gallup Polls.

Iowa writers of note include Hamlin Garland (b.Wisconsin, 1860-1940), Octave Thanet (Alice French, b.Massachusetts, 1850-1934), Bess Streeter Aldrich (1881-1954), Carl Van Vechten (1880-1964), James Norman Hall (1887-1951), Thomas Beer (1889-1940), Ruth Suckow (1892-1960), Phillip D. Strong (1899-1957), MacKinlay Kantor (1904-77), Wallace Stegner (b.1909), and Richard P. Bissell (b.1913). Iowa's poets include Paul H. Engle (b.1908), who directed the University of Iowa's famed Writers' Workshop, and James S. Hearst (b.1900). Two Iowa playwrights, Susan Glaspell (1882-1948) and her husband, George Cram Cook (1873-1924), were instrumental in founding influential theater groups.

Iowans who have contributed to America's musical heritage include popular composer Meredith Willson (1902-84), jazz musician Leon "Bix" Beiderbecke (1903-31), and bandleader Glenn Miller (1904-44). Iowa's artists of note include Grant Wood (1892-1942), whose *American Gothic* is one of America's best-known paintings, and printmaker Mauricio Lasansky (b.Argentina, 1914).

Iowa's contributions to the field of popular entertainment include William F. "Buffalo Bill" Cody (1846-1917); circus impresario Charles Ringling (1863-1926) and his four brothers; the reigning American beauty of the late 19th century, Lillian Russell (Helen Louise Leonard, 1860-1922); and one of America's best-loved movie actors, John Wayne (Marion Michael Morrison, 1907-79). Iowa sports figures of note are baseball Hall of Famers Adrian C. "Cap" Anson (1851-1922) and Robert "Bob" Feller (b.1918) and football All-American Nile Kinnick (1918-44).

⁵⁰BIBLIOGRAPHY

Bogue, Allen G. *From Prairie to Cornbelt.* Chicago: University of Chicago Press, 1963.

Federal Writers' Project. *Iowa: A Guide to the Hawkeye State.* Reprint. New York: Somerset, n.d. (orig. 1938).

Gue, Benjamin F. *History of Iowa.* 4 vols. New York: Century History Co., 1903.

Hamilton, Carl. *In No time at All.* Ames: Iowa State University Press, 1974.

Iowa Development Commission. *1985 Statistical Profile of Iowa.* Des Moines, 1985.

Ross, Earle D. *Iowa Agriculture.* Iowa City: State Historical Society, 1951.

Sage, Leland. *A History of Iowa.* Ames: Iowa State University Press, 1974.

Schwieder, Dorothy, ed. *Patterns and Perspectives in Iowa History.* Ames: Iowa State University Press, 1973.

Swierenga, Robert P. *Pioneers and Profits.* Ames: Iowa State University Press, 1968.

Wall, Joseph Frazier. *Iowa: A Bicentennial History.* New York: Norton, 1978.

KANSAS

State of Kansas

ORIGIN OF STATE NAME: Named for the Kansa (or Kaw) Indians, the "people of the south wind." **NICKNAME:** The Sunflower State. (Also: the Wheat State; the Jayhawk State.) **CAPITAL:** Topeka. **ENTERED UNION:** 29 January 1861 (34th). **SONG:** "Home on the Range." **MARCH:** "The Kansas March." **MOTTO:** *Ad astra per aspera* (To the stars through difficulties). **FLAG:** The flag consists of a dark blue field with the state seal in the center; a sunflower on a bar of twisted gold and blue is above the seal, the word "Kansas" is below it. **OFFICIAL SEAL:** A sun rising over mountains in the background symbolizes the east; commerce is represented by a river and a steamboat. In the foreground, agriculture, the basis of the state's prosperity, is represented by a settler's cabin and a man plowing a field; beyond this is a wagon train heading west and a herd of buffalo fleeing from two Indians. Around the top is the state motto above a cluster of 34 stars; the circle is surrounded by the words "Great Seal of the State of Kansas, January 29, 1861." **ANIMAL:** American buffalo. **BIRD:** Western meadowlark. **FLOWER:** Wild native sunflower. **TREE:** Cottonwood. **INSECT:** Honeybee. **LEGAL HOLIDAYS:** New Year's Day, 1 January; Birthday of Martin Luther King, Jr., 3d Monday in January; Lincoln's Birthday, 12 February; Washington's Birthday, 3d Monday in February; Memorial Day, last Monday in May; Independence Day, 4 July; Labor Day, 1st Monday in September; Columbus Day, 2d Monday in October; Veterans Day, 11 November; Thanksgiving Day, 4th Thursday in November; Christmas Day, 25 December. **TIME:** 6 AM CST = noon GMT; 5 AM MST = noon GMT.

¹LOCATION, SIZE, AND EXTENT

Located in the western north-central US, Kansas is the 2d-largest midwestern state (following Minnesota) and ranks 14th among the 50 states.

The total area of Kansas is 82,277 sq mi (213,097 sq km), of which 81,778 sq mi (211,805 sq km) are land, and the remaining 499 sq mi (1,292 sq km) inland water. Shaped like a rectangle except for an irregular corner in the NE, the state has a maximum extension E–W of about 411 mi (661 km) and an extreme N–S distance of about 208 mi (335 km).

Kansas is bounded on the N by Nebraska, on the E by Missouri (with the line in the NE following the Missouri River), on the S by Oklahoma, and on the W by Colorado, with a total boundary length of 1,219 mi (1,962 km). The geographic center of Kansas is in Barton County, 15 mi (24 km) NE of Great Bend.

²TOPOGRAPHY

Although the popular image of the state is one of unending flatlands, Kansas has a diverse topography. Three main land regions define the state. The eastern third consists of the Osage Plains, Flint Hills, Dissected Till Plains, and Arkansas River Lowlands. The central third comprises the Smoky Hills (which include the Dakota sandstone formations, Greenhorn limestone formations, and chalk deposits) to the north and several lowland regions to the south. To the west are the Great Plains proper, divided into the Dissected High Plains and the High Plains. Kansas generally slopes eastward from a maximum elevation of 4,039 feet (1,231 meters) at Mt. Sunflower (a mountain in name only) on the Colorado border to 680 feet (207 meters) by the Verdigris River at the Oklahoma border. More than 50,000 streams run through the state, and there are hundreds of artificial lakes. Major rivers include the Missouri, which defines the state's northeastern boundary; the Arkansas, which runs through Wichita; and the Kansas (Kaw), which runs through Topeka and joins the Missouri at Kansas City.

The geographic center of the 48 contiguous states is located in Smith County, in north-central Kansas, at 39°50′ N and 98°35′ W. Forty miles (64 km) south of this point, in Osborne County at 39°13′27″ N and 98°32′31″ W, is the North American geodetic datum, the controlling point for all land surveys in the US, Canada, and Mexico. Extensive beds of prehistoric ocean fossils lie in the chalk beds of two western counties, Logan and Gove.

³CLIMATE

Kansas's continental climate is highly changeable. The average mean temperature is 55°F (13°C). The record high is 121°F (149°C), recorded near Alton on 24 July 1936; the record low, −40°F (−40°C), was registered at Lebanon on 13 February 1905. The normal annual precipitation ranges from slightly more than 40 in (1,016 mm) in the southeast to as little as 16 in (406 mm) in the barren west. The overall annual average is 27 inches (686 mm), although years of drought have not been uncommon. Seventy to 77% of the precipitation falls between 1 April and 30 September. The annual mean snowfall ranges from about 36 inches (914 mm) in the extreme northwest to less than 11 inches (279 mm) in the far southeast. Tornadoes are a regular fact of Kansas life, with an annual average of 33 recorded during 1971–83. Indeed, Kansas is a windy state; Dodge City is said to be the windiest city in the US, with an average wind speed of 14 mph (23 kph).

⁴FLORA AND FAUNA

Native grasses, consisting of 60 different groups subdivided into 194 species, cover one-third of Kansas, which is much overgrazed. Bluestem—both big and little—which grows in most parts of the state, has the greatest forage value. Other grasses include buffalo grass, blue and hairy gramas, and alkali sacaton. One native conifer, eastern red cedar, is found generally throughout the state. Hackberry, black walnut, and sycamore grow in the east, while box elder and cottonwood predominate in western Kansas. There are no native pines. The wild native sunflower, the state flower, is found throughout the state. Other characteristic wildflowers include wild daisy, ivy-leaved morning glory, and smallflower verbena. The prairie white-fringed orchid and Mead's milkweed are protected under federal statutes.

Kansas's indigenous mammals include the common cottontail, black-tailed jackrabbit, black-tailed prairie dog, muskrat, opossum, and raccoon; the white-tailed deer is the state's only big-game animal. There are 12 native species of bat, 2 varieties of shrew and mole, and 3 types of pocket gopher. The western meadowlark is the state bird. Kansas has the largest flock of prairie chickens remaining on the North American continent. The black-footed ferret, gray bat, and bald eagle are on the threatened or endangered list.

[5]ENVIRONMENTAL PROTECTION

No environmental problem is more crucial to Kansas than water quality. The Kansas Water Office and Kansas Water Authority develop and maintain all water resources in the state. A law passed in 1979 mandated each county with a population of more than 30,000 to establish wastewater treatment plants by July 1983. Also of concern are measures to prevent a recurrence of the "Dust Bowl" conditions of the 1930s, when drought caused much of the topsoil to blow away. About a third of the state's cropland soil has been damaged or destroyed. The State Conservation Commission promotes soil and water conservation measures in the state's 105 conservation districts.

Strip mining for coal, in southeast Kansas, is subject to a 1969 law requiring immediate leveling and seeding of the disturbed land. However, the law does not apply to 50,000 acres (20,000 hectares) previously mined.

[6]POPULATION

With a population in 1980 of 2,364,236 (5.1% more than in 1970), Kansas ranked 32d among the states. The estimated population in 1985 was 2,453,581 (a 3.8% increase since 1980), and the population density was 30 per sq mi (12 per sq km).

When it was admitted to the Union in 1861, Kansas's population was 107,206. During the decade that followed, the population grew by 240%, more than 10 times the US growth rate. Steady growth continued through the 1930s, but in the 1940s the population declined by 4%. Since then, the population has risen, though at a slower pace than in the rest of the country.

Of the 1980 population, 49% was male and 51% female. About 67% lived in urban areas and 33% in rural areas, nearly reversing the percentages recorded 60 years earlier. In 1980, the average Kansan was 30.2 years old, close to the national average.

In 1980, Kansas had 34 incorporated areas of 10,000 population or more. Census returns for 1980 showed 279,835 residents of Wichita, 161,148 of Kansas City, and 115,266 of Topeka. In mid-1984, the population of Kansas City was little changed, but the population of Wichita had grown to an estimated 283,496, and the population of Topeka to 118,945. That part of the Kansas City metropolitan area in Kansas had a population of 442,604 in 1980. The Wichita metropolitan area had 428,600 residents in 1984, the Topeka metropolitan area, 159,000.

[7]ETHNIC GROUPS

White settlers began to pour into Kansas in 1854, dispersing the 36 Indian tribes living there and precipitating a struggle over the legal status of slavery. Remnants of six of the original tribes still make their homes in the state. Some Indians live on three reservations covering 30,000 acres (12,140 hectares); others live and work elsewhere, returning to the reservations several times a year for celebrations and observances. There were 15,254 Indians in Kansas as of 1980.

Black Americans in Kansas numbered 126,127—more than 5% of the population—in 1980, when the state also had 63,339 residents of Hispanic origin. The 1980 census recorded 15,078 Asian-Pacific peoples, the largest group being 3,690 Vietnamese.

The foreign-born numbered only 48,000 (2% of the population) in 1980. Among persons who reported descent from a single ancestry group, the leading nationalities were German (356,453), English (248,634), and Irish (93,933).

[8]LANGUAGES

Plains Indians of the Macro-Siouan group originally populated what is now Kansas; their speech echoes in such place-names as Kansas, Wichita, Topeka, Chetopa, and Ogallah.

Regional features of Kansas speech are almost entirely those of the Northern and North Midland dialects, reflecting the migration into Kansas in the 1850s of settlers from the East. Kansans typically use *fish(ing) worms* as bait, play as children on a *teetertotter,* see a *snakefeeder* (dragonfly) over a /krik/ (creek), make *white bread* sandwiches, carry water in a *pail,* and may designate the time 2:45 as a quarter *to,* or *of,* or *till* three.

The migration by southerners in the mid-19th century is evidenced in southeastern Kansas by such South Midland terms as *pullybone* (wishbone) and *light bread* (white bread); the expression *wait on* (wait for) extends farther westward.

In 1980, 2,146,082 Kansans—95% of the residents 3 years old or older—spoke only English at home. The number of Kansans who spoke another language at home included:

Spanish	41,642	Vietnamese	2,714
German	26,592	Korean	2,239
French	5,910	Chinese	1,941

[9]RELIGIONS

Protestant missions played an important role in early Kansas history. Isaac McCoy, a Baptist minister, was instrumental in founding the Shawnee Baptist Mission in Johnson County in 1831. Later Baptist, Methodist, Quaker, Presbyterian, and Jesuit missions became popular stopover points for pioneers traveling along the Oregon and Santa Fe trails. Mennonites were drawn to the state by a law passed in 1874 allowing exemptions from military service on religious grounds. As of 1980, Mennonites comprised 1% of the population.

Religious freedom is specifically granted in the Kansas constitution, and a wide variety of religious groups is represented in the state. The leading Protestant denominations in 1980 were United Methodists, 262,082; Southern Baptist Convention, 73,306; American Baptist USA, 72,789; and Christian Church (Disciples of Christ), 68,858. Roman Catholics constitute the largest single religious group in the state, with 347,843 adherents in 1984. Kansas's estimated Jewish population that year was 11,450.

[10]TRANSPORTATION

In the heartland of the nation, Kansas is at the crossroads of US road and railway systems. The state ranks 3d in Class I railroad miles and 4th in total road miles.

In the late 1800s, the two major railroads, the Kansas Pacific (now the Union Pacific) and the Santa Fe, acquired more than 10 million acres (4 million hectares) of land in the state and then advertised for immigrants to come and buy it. By 1872, the railroads stretched across the state, creating in their path the towns of Ellsworth, Newton, Caldwell, Wichita, and Dodge City. One of the first "cow towns" was Abilene, the terminal point for all cattle shipped to the East. In 1983, the state had 7,688 mi (12,372 km) of Class I railroad track. An Amtrak passenger train crosses Kansas en route from Chicago to Los Angeles.

As of the end of 1983, the state had 133,073 mi (214,161 km) of roads, of which 123,945 mi (199,471 km) were rural, 8,320 mi (13,390 km) were urban, and 808 mi (1,300 km) were part of the interstate highway system. There were 1,395,628 autos, 624,575 trucks, 87,907 trailers, and 3,957 buses registered in Kansas in 1983.

During the same year, the state had 360 airports and 19 heliports. The busiest airport, at Wichita, had 16,782 departing flights carrying 543,030 passengers.

River barges move bulk commodities along the Missouri River. The chief river ports are Atchison, Leavenworth, and Kansas City.

[11]HISTORY

Present-day Kansas was first inhabited by Paleo-Indians approximately 10,000 years ago. They were followed by several prehistoric cultures, forerunners of the Plains tribes—the Wichita, Pawnee, Kansa, and Osage—that were living or hunting in

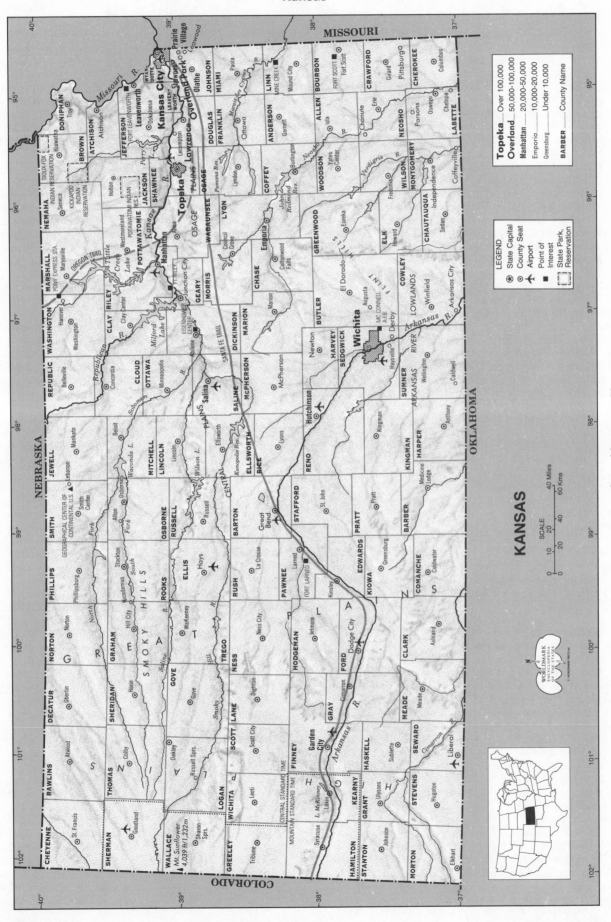

KANSAS

SCALE

0   10   20   40 Miles

0   20   40   60 Kms

See endsheet maps: G3.

LOCATION: 36°59'55'' to 40°N; 94°37'03'' to 102°03'02'' w. **BOUNDARIES:** Nebraska line, 357 mi (575 km); Missouri line, 244 mi (393 km); Oklahoma line, 411 mi (661 km); Colorado line, 207 mi (333 km).

LEGEND

⊛ State Capital
◉ County Seat
✈ Airport
■ Point of Interest
⬚ State Park, Reservation

Topeka	Over 100,000
Overland	50,000–100,000
Manhattan	20,000–50,000
Emporia	10,000–20,000
Greensburg	Under 10,000
BARBER	County Name

Kansas when the earliest Europeans arrived. These tribes were buffalo hunters who also farmed and lived in small permanent communities. Around 1800, they were joined on the Central Plains by the nomadic Cheyenne, Arapaho, Comanche, and Kiowa.

The first European, explorer Francisco Coronado, entered Kansas in 1541, searching for riches in the fabled land of Quivira. He found no gold but was impressed by the land's fertility. A second Spanish expedition to the Plains was led by Juan de Oñate in 1601. Between 1682 and 1739, French explorers established trading contacts with the Indians. France ceded its claims to the area to Spain in 1762, but received it back from Spain in 1800.

Most of Kansas was sold to the US by France as part of the Louisiana Purchase of 1803. (The extreme southwestern corner was gained after the Mexican War.) Lewis and Clark examined the country along the Missouri River in 1804, and expeditions under the command of Zebulon Pike (1806) and Stephen Long (1819) traversed the land from east to west. Pike and Long were not impressed with the territory's dry soil, the latter calling the area "unfit for civilization, and of course uninhabitable by a people depending on agriculture for their subsistence."

Largely because of these negative reports, early settlement of Kansas was sparse, limited to a few thousand eastern Indians who were removed from their lands and relocated in what is now eastern Kansas. Included were such once powerful tribes as the Shawnee, Delaware, Ojibwa, Wyandot, Ottawa, and Potawatomi. They were joined by a number of Christian missionaries seeking to convert the Indians into Christian farmers.

William Becknell opened the Santa Fe Trail to wagon traffic in 1822, and for 50 years that route, two-thirds of which lay in Kansas, was of commercial importance to the West. During the 1840s and 1850s, thousands of migrants crossed northeastern Kansas on the California-Oregon Trail. In 1827, Ft. Leavenworth was established, followed by Ft. Scott (1842) and Ft. Riley (1853). Today, Ft. Leavenworth and Ft. Riley are still the two largest military installations in the state.

Kansas Territory was created by the Kansas-Nebraska Act (30 May 1854), with its western boundary set at the Rocky Mountains. Almost immediately, disputes arose as to whether Kansas would enter the Union as a free or slave state. Both free-staters and proslavery settlers were brought in, and a succession of governors tried to bring order out of the chaos arising from the two groups' differences. Free-staters established an extralegal government at Topeka following the establishment of a territorial capital at Lecompton.

Because of several violent incidents, the territory became known as Bleeding Kansas. One of the most memorable attacks came in May 1856, when the town of Lawrence was sacked by proslavery forces. John Brown, an abolitionist who had recently arrived from upstate New York, retaliated by murdering five proslavery settlers. Guerrilla skirmishes continued for the next few years along the Kansas-Missouri border. The final act of violence was the Marais des Cygnes massacre in 1858, which resulted in the death of several free-staters. In all, about 50 people were killed in the territorial period—not an extraordinary number for a frontier community.

After several attempts to write a constitution acceptable to both anti- and proslavery groups, the final document was drafted in 1859. Kansas entered the Union on 29 January 1861 as a free state. Topeka was named the capital, and the western boundary was reduced to its present location.

Although Kansas lay west of the major Civil War action, more than two-thirds of its adult males served in the Union Army and gave it the highest military death rate among the northern states. Kansas units saw action in the South and West, most notably at Wilson's Creek, Cane Hill, Prairie Grove, and Chickamauga. The only full-scale battle fought in Kansas was at Mine Creek in 1864,

at the end of General Sterling Price's unsuccessful Confederate campaign in the West. The most tragic incident on Kansas soil came on 21 August 1863, when Confederate guerrilla William C. Quantrill raided Lawrence, killing at least 150 persons and burning the town.

Following the Civil War, settlement expanded in Kansas, particularly in the central part of the state. White settlers encroached on the hunting grounds of the Plains tribes, and the Indians retaliated with attacks on white settlements. Treaty councils were held, the largest at Medicine Lodge in 1867, but not until 1878 did conflict cease between Indians and whites. Most of the Indians were eventually removed to the Indian Territory in what is now Oklahoma. Also during this period, buffalo, slaughtered for food and hides, all but disappeared from the state.

By 1872, both the Union Pacific and the Santa Fe railroads had crossed Kansas, and other lines were under construction. Rail expansion brought more settlers, who established new communities. It also led to the great Texas cattle drives that meant prosperity to a number of Kansas towns—including Abilene, Ellsworth, Wichita, Caldwell, and Dodge City—from 1867 to 1885. This was when Bat Masterson, Wyatt Earp, Doc Holliday, and Wild Bill Hickok reigned in Dodge City and Abilene—the now romantic era of the Old West.

A strain of hard winter wheat that proved particularly well suited to the state's soil was brought to Kansas in the 1870s by Russian Mennonites fleeing czarist rule, and Plains agriculture was thereby transformed. There were also political changes: the state adopted limited female suffrage in 1887. Prohibition, made part of the state constitution in 1880, was a source of controversy until its repeal in 1948.

Significant changes in agriculture, industry, transportation, and communications came after 1900. Mechanization became commonplace in farming, and vast areas were opened to wheat production, particularly during World War I. Some automobile manufacturing took place, and the movement for "good roads" began. The so-called agrarian revolt of the late 19th century, characterized politically by populism, evolved into the Progressive movement of the early 1900s, which focused attention on control of monopolies, public health, labor legislation, and more representative politics. Much of the Progressive leadership came from Kansas; Kansan newspaper editor and national Progressive leader William Allen White devoted considerable energy to Theodore Roosevelt's Bull Moose campaign in 1912.

Kansas suffered through the Great Depression of the 1930s. The state's western region, part of the Dust Bowl, was hardest hit. Improved weather conditions and the demands of World War II revived Kansas agriculture in the 1940s. The World War II era also saw the development of industry, especially in transportation. Wichita had been a major center of the aircraft industry in the 1920s and 1930s, and its plants became vital to the US war effort. Other heavy industry grew, and mineral production—oil, natural gas, salt, coal, and gypsum—expanded greatly.

Since World War II, Kansas has become increasingly urban. Agriculture has become highly commercialized, and there are dozens of large industries that process and market farm products and supply materials to crop producers. Livestock production, especially in closely controlled feedlots, is a major enterprise. Recent governors have worked to expand international exports of Kansas products, and by 1981/82, Kansas ranked 7th among the states in agricultural exports, with sales of more than $1.6 billion.

12 STATE GOVERNMENT

The form of Kansas's constitution was a matter of great national concern, for the question of whether Kansas would be a free or slave state was in doubt throughout the 1850s. After three draft constitutions failed to win popular support or congressional approval, a fourth version, banning slavery, was drafted in July 1859 and ratified by Kansas voters that October. Signed by

President James Buchanan on 29 January 1861, this constitution (with 79 subsequent amendments, as of 1983) has governed Kansas to the present day.

The Kansas legislature consists of a 40-member senate and a 125-member house of representatives. Senators serve four-year terms and house members serve for two years; elections are held in even-numbered years. Legislative sessions are limited to 90 days in even-numbered years but are unlimited in odd-numbered years.

Officials elected statewide are the governor, lieutenant governor, secretary of state, attorney general, treasurer, and commissioner of insurance. Members of the state Board of Education are elected by districts. All serve four-year terms. The governor cannot serve more than two consecutive terms. Every office in the executive branch is controlled by either the governor or another elected official. The governor appoints the heads (secretaries) of all state departments.

A bill becomes law when it has been approved by 21 senators and 63 representatives and signed by the governor. A veto can be overridden by one more than two-thirds of the members of both houses.

To vote in the state, a person must be a US citizen, 18 years old at the time of the election, and a resident of Kansas for at least 20 days. Registration closes 20 days before all elections, and mail registration is allowed.

13POLITICAL PARTIES

Kansas was dominated by the Republican Party for the first three decades of statehood. Although the Republicans remain the dominant force in state politics, the Democrats controlled several state offices in the mid-1980s.

The Republican Party of early Kansas espoused the abolitionist ideals of the New England settlers who sought to ban slavery from the state. After the Civil War, the railroads played a major role in Republican politics and won favorable tax advantages from the elected officials. The party's ranks swelled with the arrival of immigrants from Scandinavia and Germany, who tended to side with the party's by then strongly conservative beliefs.

The Republicans' hold over state life was shaken by the Populist revolt toward the end of the 19th century. The high point of Populist Party power came in 1892, when the insurgents won all the statewide elective offices and also took control of the senate. When electoral irregularities denied them control of the house, they temporarily seized the house chambers. The two parties then set up separate houses of representatives, the Populists meeting one day and the Republicans the next. This continued for six weeks, until the Kansas supreme court ruled that the Republicans constituted the rightful legal body. After a Republican sweep in 1894, the Populists returned to office in 1896, but the party declined rapidly thereafter.

The Democrats rose to power in the state as a result of the split between the conservative and progressive wings of the Republican Party in 1912. Nevertheless, the Democrats were very much a minority party until after World War II. Democrats held the governorship for 18 of the 28 years between 1957 and 1985; the most recent Democratic governor is John Carlin, elected in 1978 and reelected in 1982. Republicans have regularly controlled the legislature, however. In 1982, the state had 454,894 registered Republicans (38% of eligible voters) and 345,395 registered Democrats (29%). The remaining 33% included independents and members of minor parties.

In November 1984, Kansans gave two-thirds of their presidential votes to Republican Ronald Reagan. Republican Robert Dole, first elected to the US Senate in 1968 and elected Senate majority leader in 1984, did not face an election in 1984, but Kansas's other Republican senator, Nancy Landon Kassebaum, was reelected, and the Republicans remained in control of the state's US congressional delegation (3 Republicans, 2 Democrats) and both houses of the state legislature. In 1985, 18.2% of state legislators were women, and 29 elected officials were black.

14LOCAL GOVERNMENT

As of 1984, Kansas had 105 counties, 627 incorporated cities, 1,367 townships, and 305 unified school districts. The total number of local government bodies was 3,795 in 1982. By law, no county can be less than 432 sq mi (1,119 sq km).

Kansas Presidential Vote by Political Parties, 1948–84

YEAR	ELECTORAL VOTE	KANSAS WINNER	DEMOCRAT	REPUBLICAN	PROGRESSIVE	SOCIALIST	PROHIBITION
1948	8	Dewey (R)	351,902	423,039	4,603	2,807	6,468
1952	8	*Eisenhower (R)	273,296	616,302	6,038	530	6,038
1956	8	*Eisenhower (R)	296,317	566,878	—	—	3,048
1960	8	Nixon (R)	363,213	561,474	—	—	4,138
						SOC. LABOR	
1964	7	*Johnson (D)	464,028	386,579	—	1,901	5,393
					AMERICAN IND.		
1968	7	*Nixon (R)	302,996	478,674	88,921	—	2,192
					AMERICAN		
1972	7	*Nixon (R)	270,287	619,812	21,808	—	4,188
					LIBERTARIAN		
1976	7	Ford (R)	430,421	502,752	4,724	3,242	1,403
1980	7	*Reagan (R)	326,150	566,812	7,555	14,470	—
1984	7	*Reagan (R)	333,149	677,296	—	3,329	—

*Won US presidential election.

Each county government is headed by three elected county commissioners. Other county officials include the county clerk, treasurer, register of deeds, attorney, sheriff, clerk of district court, and appraiser. Most cities are run by mayor-council systems.

15 STATE SERVICES

All education services except higher education are handled by the Board of Education; higher education lies within the jurisdiction of the Board of Regents. The Department of Human Resources administers employment and worker benefit programs; the Department of Economic Development operates housing and business planning programs. Social, vocational, and children's and youth programs are run by the Department of Social and Rehabilitation Services; the Department of Health and Environment supervises health, environment, and laboratory services.

A "Sunset Law" automatically abolishes specified state agencies at certain times unless they receive renewed statutory authority.

16 JUDICIAL SYSTEM

The supreme court, the highest court in the state, is composed of a chief justice and six other justices. All justices are appointed by the governor but after one year must run for election at the next general election. They then are elected for six-year terms. In case of rejection by the voters, the vacancy is filled by appointment. An intermediate-level court of appeals consists of a chief judge and six other judges appointed by the governor; like supreme court justices, they must be elected to full terms, in this case for four years.

In January 1977, probate, juvenile, and county courts, as well as magistrate courts of countywide jurisdiction, were replaced by district courts. In 1984 there were 31 judicial districts served by 71 district judges, 69 associate district judges, and 72 district magistrate judges.

The Department of Corrections administers the state correctional system. Those convicted of major crimes are sent to the Kansas State Penitentiary at Lansing, which in 1983 housed 1,443 inmates of a total inmate population of 3,160. A federal prison is located at Leavenworth. Kansas's crime rate (4,529.8 per 100,000 inhabitants in 1983) is lower than the national average.

17 ARMED FORCES

The US Army's 1st Infantry Division, known as the Big Red One, is located at Ft. Riley in Junction City and had 15,324 military personnel in 1984. The Army's Command and General Staff College is housed at Fort Leavenworth, which had 3,413 personnel in 1984; McConnell Air Force Base, located in Wichita, had 3,535. A total of 29,206 military personnel were stationed in Kansas in 1984. In 1983/84, nearly $2.4 billion in defense contracts was awarded to state firms.

Kansas had 298,000 veterans in 1983. Of these, 5,000 were veterans of World War I, 116,000 of World War II, 54,000 of the Korean conflict, and 88,000 of the Viet-Nam era. During fiscal year 1982/83, $142,927,000 was paid to veterans for compensation and pensions.

In 1984, Kansas had 2,164 members of the Air National Guard and 6,328 members of the Army National Guard. Of the 8,122 people employed by state and local police in 1983, 558 were full-time employees of the Kansas Highway Patrol.

18 MIGRATION

By the 1770s, Kansas was inhabited by a few thousand Indians, mainly from five tribes: the Kansa (Kaw) and the Osage, both of whom had migrated from the east, the Pawnee from the north, and the Wichita and Comanche, who had come from the southwest. In 1825, the US government signed a treaty with the Kansa and Osage that allowed eastern Indians to settle in the state.

The first wave of white migration came during the 1850s with the arrival of New England abolitionists who settled in Lawrence, Topeka, and Manhattan. They were followed by a much larger wave of emigrants from the eastern Missouri and the upper Mississippi Valley, drawn by the lure of wide-open spaces and abundant economic opportunity.

The population swelled as a result of the Homestead Act of 1862, which offered land to anyone who would improve it and live on it for five years. The railroads promoted the virtues of Kansas overseas and helped sponsor immigrant settlers. By 1870, 11% of the population was European. More than 30,000 blacks, mostly from the South, arrived during 1878–80. Crop failures caused by drought in the late 1890s led to extensive out-migration from the western half of the state. Another period of out-migration occurred in the early 1930s, when massive dust storms drove people off the land. Between 1970 and 1983, Kansas had a net loss from migration of 71,000. Steady migration from farms to cities has been a feature of Kansas life throughout this century, with urban population surpassing farm population after World War II.

19 INTERGOVERNMENTAL COOPERATION

Kansas is a member of the Midwestern Conference, Arkansas River Compact of 1949, Arkansas River Compact of 1965, Big Blue River Compact, Kansas City Area Transportation Compact, Kansas-Missouri Waterworks Compact, Missouri River Toll Bridge Compact, Republican River Compact, Multistate Tax Compact, and other interstate bodies. The Interstate Cooperation Commission assists state officials and employees in maintaining contact with governmental units in other states. In 1983/84, Kansas received $804.8 million in federal assistance.

20 ECONOMY

Although wheat production has long been the mainstay of the Kansas economy, efforts to bring other industries into the state began as early as the 1870s, when the railroads linked Kansas to eastern markets. Today, agricultural products and meat-packing industries are rivaled by the large aircraft industry centered in Wichita. Kansas leads all states and trails only seven countries in wheat production.

Three Kansas companies, all located in Wichita, manufacture two-thirds of the world's general aviation aircraft. The Kansas City metropolitan area is a center of automobile production and printing. Metal fabrication, printing, and mineral products industries predominate in the nine southeastern counties.

21 INCOME

In 1983, Kansas's income per capita was $12,285 (12th in the US). The following table shows personal income in 1981 and 1982 (in thousands):

	1981	% CHANGE	1982
Total earnings, of which:	$18,224,000	3.7	$18,898,000
Farm	1,153,000	2.9	1,186,000
Agricultural services, forestry, fisheries	61,000	−1.7	60,000
Mining	574,000	0.1	578,000
Construction	1,012,000	−1.6	998,000
Manufacturing	4,126,000	−3.5	3,985,000
Transportation and public utilities	1,639,000	6.0	1,737,000
Wholesale and retail trade	3,049,000	4.4	3,183,000
Finance, insurance, and real estate	906,000	9.8	995,000
Services	2,745,000	7.8	2,959,000
Government	2,959,000	8.9	3,217,000

Total personal income was $29.8 billion in 1983.

22 LABOR

Kansas has traditionally had a fairly low unemployment rate. In 1984, the civilian labor force was 1,197,000, of whom 1,135,000 were employed, yielding an unemployment rate of 5.2%. A federal survey in March 1982 revealed the following nonfarm employment pattern in Kansas:

	ESTABLISH-MENTS	EMPLOYEES	ANNUAL PAYROLL ('000)
Agricultural services, forestry, fishing	634	2,670	$27,832
Mining, of which:	1,359	22,109	490,344
Oil, gas extraction	(1,223)	(18,548)	(384,800)
Contract construction	4,637	38,324	701,063
Manufacturing, of which:	3,029	181,574	3,417,912
Food, food products	(298)	(22,331)	(407,014)
Nonelectrical machinery	(568)	(26,162)	(455,500)
Transport equipment	(139)	(44,103)	(1,020,412)
Transportation, public utilities	2,939	48,077	971,079
Wholesale trade	5,497	61,232	1,097,887
Retail trade	15,745	165,421	1,408,180
Finance, insurance, real estate	4,868	50,210	795,984
Services	15,677	169,554	1,979,336
Other	1,091	1,277	14,372
TOTALS	55,476	740,448	10,903,989

In 1980, 146,000 workers, or 15.5% of the nonfarm work force, belonged to labor unions, of which there were 658 in 1983. Kansas has a right-to-work law.

23AGRICULTURE

Known as the Wheat State and the breadbasket of the nation, Kansas produces more wheat than any other state. It ranked 7th in total farm income in 1983, with cash receipts of $5.4 billion.

Because of fluctuating prices, Kansas farmers have always risked economic disaster. During the 1920s, depressed farm prices forced many new farmers out of business. By World War II, Kansas farmers were prospering again, as record prices coincided with record yields. Since then, improved technology has favored corporate farms at the expense of small landholders. Between 1940 and 1984, the number of farms declined from 159,000 to 74,000, while the average size of farms more than doubled (to 647 acres/262 hectares). In 1982, about 28,273,000 acres (11,441,715 hectares) were used for crops, of which 2,675,000 acres (1,100,000 hectares) were irrigated. Income from crops in 1983 totaled $2.2 billion. The following table shows several leading crops in 1983:

	VALUE	PRODUCTION (BU)	US RANK
Wheat	$1,568,700,000	448,200,000	1
Sorghum (grain)	334,368,000	116,100,000	2
Corn (grain)	282,348,000	81,840,000	11
Soybeans	186,960,000	22,800,000	19

Other leading crops are alfalfa, hay, oats, barley, popcorn, rye, dry edible beans, corn and sorghums for silage, wild hay, red clover, and sugar beets. In 1983, Kansas ranked 1st in the US in milled flour production (35,043,000 hundredweight), valued at $355.2 million. The state was also a leading producer of soybean oil and meal: four mills turned out 441,162,000 lb of oil and 962,252 tons of meal, with a combined value of $335.8 million.

24ANIMAL HUSBANDRY

Cash receipts from livestock and animal products totaled $3.2 billion in 1983, when 4,709,000 cattle and 1,417,200 hogs were slaughtered for commercial purposes. The following is a summary of estimated livestock production for 1983:

	VALUE	OUTPUT (LB)	US RANK
Cattle and calves	$1,565,518,000	2,711,525,000	3
Hogs	12,356,000	680,460,000	9
Sheep	8,684,000	16,739,000	13
Chickens	1,233,000	8,218,000	28

Kansas's 206 meat-packing plants processed 3,184,109,000 lb of beef, 62,000 lb of veal, 255,405,000 lb of pork, and 344,000 lb of lamb and mutton. The total value of milk was $193,618,000. Twenty-five plants manufactured 18,826,000 lb of butter, 32,117,-000 lb of American cheese, and 8,217,000 lb of ice cream.

25FISHING

There is little commercial fishing in Kansas: commercial landings in 1983 totaled 555,000 lb and $165,000. Sport fishermen can find bass, crappie, catfish, perch, and pike in the state's reservoirs and artificial lakes.

26FORESTRY

Kansas was at one time so barren of trees that early settlers were offered 160 acres (65 hectares) free if they would plant trees on their land. This program was rarely implemented, however, and today much of Kansas is still treeless.

In 1979, Kansas had about 1,344,000 acres (544,000 hectares) of forestland, 3% of the total state area. There were 1,187,000 acres (480,000 hectares) of commercial timberland, of which 96.9% was privately owned.

27MINING

In 1984, the estimated value of all nonfuel mineral production in Kansas was $297.4 million, 24th in the US. The state produced 13,200,000 tons of crushed stone, 11,330,000 tons of sand and gravel, 1,680,000 tons of salt, and 669,000 tons of clay. Cement production was 1,970,000 tons in 1983.

Most of the helium produced in the US is extracted from Kansas natural gas. In 1984, 1,180,000,000 cu feet (33,414,000 cu meters) of helium was refined, with a value of $38,038,000.

28ENERGY AND POWER

The total value of all fuel production in Kansas in 1982 was $2,855,000,000. The state ranked 6th in energy consumption per capita, with 397 million Btu, and 8th in energy expenditures per capita, with $2,254.

In 1983, Kansas had an installed electrical generating capacity of 9,587,000 kw. Electrical output was 23.8 billion kwh, 82% coal-fired. Sales in 1983 totaled 22.6 billion kwh, of which 37% was for residences, 31% for commercial establishments, 30% for industrial plants, and 2% for other purposes.

In 1982, Kansas was the nation's 8th-leading oil producer. Output totaled 67,010,000 barrels of crude petroleum. There were 45,381 producing wells in 1982 and proved reserves of 344,000,000 barrels at the end of 1983.

Natural gas production was 436 billion cu feet (12.3 billion cu meters) in 1983, when six gas companies served 767,000 customers. Sales totaled $1.1 billion, of which 30% went for industrial purposes, 35% for residential use, 22% for commercial applications, and 12% to utilities.

Seven surface mines produced 1,263,000 tons of bituminous coal in 1983. Recoverable coal reserves were estimated at 39.9 million tons. The value added by coal mining was $27,100,000 in 1982.

29INDUSTRY

Transportation equipment, nonelectrical machinery, food and food products, and petroleum and coal products accounted for 53% of all state manufacturing employment in 1982 and 72% of value of shipments, which totaled 26.8 billion in that year. Industries are concentrated in Sedgwick, Wyandotte, Johnson, Butler, and Shawnee counties.

The following table shows the value of shipments for major industries in 1982:

Food and food products	6,891,000,000
Petroleum and coal products	5,747,200,000
Transportation equipment, of which:	4,564,700,000
Aircraft and parts	2,859,400,000
Nonelectrical machinery	2,022,300,000
Chemicals and allied products	1,637,200,000
Printing and publishing	1,329,300,000
Rubber and miscellaneous plastics products	669,600,000
Apparel and textile products	651,600,000
Fabricated metal products	668,400,000
Stone, clay, glass	621,500,000
Electrical and electronics equipment	599,100,000

Kansas is a world leader in aviation, claiming a large share of both US and world production and sales. Wichita is the home of Beech, Cessna, and Gates Learjet, which among them manufacture two-thirds of the world's general aviation aircraft; Boeing Wichita employs some 12,000 workers. A Kansas City General Motors plant assembled 229,577 1983-model automobiles. Pittsburgh has the nation's largest plant devoted mainly to the production of coal-preparation equipment.

30 COMMERCE

Domestically, Kansas is not a major commercial state. In 1982, wholesale trade totaled $24.5 billion. Retail sales were $10.8 billion, including food stores, 22%; automotive dealers, 21%; and general merchandise stores, 12%.

Kansas does play an important role in US foreign trade. In 1981/82, the state ranked 1st in the export of wheat and flour, and agricultural exports totaled more than $1.6 billion. Manufactured exports totaled $1.5 billion.

31 CONSUMER PROTECTION

The consumer credit commissioner is responsible for administering the state's investment and consumer credit codes.

32 BANKING

Kansas's 620 commercial banks had assets of $19.6 billion and assigned loans and discounts worth $9.4 billion in 1982. In 1983, 65 savings and loan associations reported $9.7 billion in assets and $6.9 billion in first mortgage loans. There were 227 credit unions in 1983, of which 170 were state-chartered.

Records of all banks and trust companies in the state are examined once a year by the bank commissioner, by the Federal Deposit Insurance Corporation, or by a federal reserve bank. A savings and loan commissioner has similar responsibilities for these institutions, as does the Federal Home Loan Bank Board.

33 INSURANCE

As of mid-1983, 20 life insurance companies were headquartered in Kansas. In the same year, 4,192,000 policies were in force, worth $56,329,000,000; benefit payments totaled $158,400,000. The average amount of life insurance per family was $57,800. Of the $1.2 billion in direct premiums written by property and casualty insurers in 1983, $256 million was in automobile liability insurance, $229 million in automotive physical damage insurance, and $157 million in homeowners' coverage.

34 SECURITIES

There are no stock exchanges in Kansas. In 1983, New York Stock Exchange member firm offices had 372 full-time registered representatives in Kansas.

35 PUBLIC FINANCE

The state budget is prepared by the Division of the Budget and is submitted by the governor to the legislature for approval. The fiscal year runs from 1 July to 30 June.

The following is a summary of state general fund net receipts and expenditures for fiscal years 1983/84 (actual) and 1984/85 (estimated):

RECEIPTS	1983/84	1984/85
Taxes	$1,495,492,772	$1,604,668,000
Net transfers	18,130,487	−31,389,604
Other revenues, not including federal grants	33,280,408	83,850,000
TOTALS	$1,546,903,667	$1,657,128,396
EXPENDITURES		
Education and research	$ 956,129,906	$1,041,867,195
Public welfare	243,111,510	243,611,287
General government	139,492,005	153,795,891
Health and hospitals	78,353,535	93,680,342
Public safety	68,621,219	82,439,922
Agriculture and natural resources	11,553,740	14,069,591
Other	6,115,139	17,232,606
TOTALS	$1,503,377,054	$1,646,696,834

According to state law, no Kansas governmental unit may issue revenue bonds to finance current activities; all these must operate on a cash basis. Bonds may be issued for such capital improvements as roads and buildings. The total indebtedness of state, county, and local governments exceeded $5.1 billion as of 30 June 1982.

36 TAXATION

Kansas ranked 29th in state and local taxes per capita in 1981/82, at $1,070.

The state individual income tax rate at the end of 1983 ranged from 2% to 9%; for corporations it was 4.5%, with a 2.25% surtax on taxable income over $25,000. Receipts from individual income taxes in 1983/84 were $568 million. Corporate income tax receipts were $121 million.

A statewide 3% sales tax was adopted in 1965; by law, cities and counties may, by referendum, adopt up to another 1%. As of 1984, 24 cities and 10 counties had a 0.5% tax; 42 counties and 16 cities levied a 1% tax. The state also collects liquor and bingo enforcement taxes, cigarette and tobacco products taxes, inheritance taxes, motor vehicle and motor carrier taxes, motor fuel taxes, and royalty and excise taxes on oil, natural gas, and other minerals. Property taxes are the largest source of income for local governments. Nearly two-thirds of all tax dollars collected in the state go for education.

In 1982, Kansans filed 994,000 federal tax returns on more than $19.1 billion in gross income and paid almost $3 billion in federal income taxes, or about $1,234 per capita.

37 ECONOMIC POLICY

The first state commission to promote industrial development was formed in 1939. In 1963, this commission was reorganized into the Department of Economic Development, whose agencies conduct economic planning, assist local communities, promote high-technology economic development, foster small-business development, encourage minority business enterprises, and provide information and assistance to tourists.

Kansas has a duty-free foreign trade zone and provides tax-exempt bonds to help finance business and industry. Specific tax incentives include job expansion and investment tax credits; tax exemptions or moratoriums on land, capital improvements, and specific machinery; and certain corporate income tax exemptions.

38 HEALTH

The birthrate in Kansas in 1982 was 17 per 1,000 population. Infant mortality for 1981 was 10.6 per 1,000 white live births and 20.9 per 1,000 nonwhite live births; neither figure was far from the US average. In 1983, Kansans had 5,329 legal abortions, a rate of 1 for each 8 live births. Heart disease, the leading cause of death in the state, accounted for 38% of all deaths in 1981. Topeka, a major US center for psychiatric treatment, is home to the world-famous Menninger Clinic. Kansas also has 4 state-run psychiatric hospitals and 31 licensed community mental-health centers. The state sponsors 4 hospitals and treatment centers and 28 community-based programs for the mentally retarded.

In 1983, Kansas had 165 hospitals with 18,280 beds and 435,473 admissions; in 1981 there were 6,110 registered full-time nurses. Nine counties had no hospitals in 1982. State figures for adult-care homes in 1983 showed 354 facilities with about 25,584 beds. Kansas had 4,306 physicians and 1,162 dentists in 1982.

The University of Kansas has the state's only medical and pharmacology schools. The university's Mid-America Cancer Center and Radiation Therapy Center are the major cancer research and treatment facilities in the state. The Menninger Foundation has a research and treatment center for mental health.

39 SOCIAL WELFARE

Public assistance and social programs in Kansas are coordinated through the Department of Human Resources and the Department of Social and Rehabilitation Services.

In 1984/85, out of an estimated $710.5 million in state and federal funds spent for public welfare, the largest sums, excluding unemployment insurance, were $216,138,115 for medical assistance and $117,398,694 for cash assistance, mainly aid for dependent children. AFDC payments were $7,512,600 for January 1983.

An estimated 136,000 Kansans participated in the federal food stamp program in 1982/83, at a cost of $71,000,000. About 273,000 children took part in the school lunch program, at a total cost of $19,000,000.

In December 1982, 380,252 Kansans received Social Security benefits which totaled $1,803,000,000 in 1983. Of that amount, $1,163,000,000 went to 231,959 retirees; 76,406 Kansans received $380,000,000 in survivors' benefits; and 19,134 disabled workers received $110,000,000. A total of $36,064,000 in Supplemental Security Income benefits was paid in 1983: $8,105,000 went to the aged, $27,225,000 to the disabled, and $734,000 to the blind.

A total of $10,415,000 in federal funds and $8,211,290 in state funds was spent on vocational rehabilitation in 1983/84. The Kansas Vocational Rehabilitation Center and Vocational Rehabilitation Unit offer services to the handicapped and mentally retarded, respectively. About $105,100,000 was paid in workers' compensation in 1982.

In order to qualify for unemployment compensation, a worker must have worked at least 20 weeks during the previous year. State payments in 1984/85 were $147,501,142.

⁴⁰HOUSING
Kansas has relatively old housing stock. According to the 1980 census, 34% of all dwellings were built before 1940, and only 23% were built in the 1970s. The overwhelming majority (77%) were one-unit structures, and 70% were owner-occupied. There were 259,804 rented houses or apartment units. The median value of a house was $37,800; the median rent was $219. Only 1.2% of all structures lacked plumbing facilities in 1980. The total value of privately owned housing starts in 1983 was $664,000,000.

⁴¹EDUCATION
Kansans are, by and large, better educated than most other Americans. In 1980, 73% of adult Kansans had completed high school and 7% were college graduates.

Enrollment in the state's 1,498 public schools in 1983/84 was 405,222: 120,298 students were in high school, 65,116 in junior high school, 210,874 in elementary school (including kindergarten), and 8,934 in special-education and nongraded classes. There were about 14,300 elementary-school teachers and 11,500 secondary-school teachers. The dropout rate was 1% of total enrollment. Students' average daily attendance at nonpublic schools in the fall of 1980 was 33,889, of whom 25,610 were in Catholic schools.

In 1983/84 there were 6 state universities, 19 two-year community colleges, 3 private two-year colleges, 17 church-affiliated universities and four-year colleges, and 14 vocational-technical schools. In addition, Kansas has a state technical institute, a municipal university (Washburn University, Topeka), and a federally run junior college for Indians. Kansas State University was the nation's first land-grant university. Washburn University and the University of Kansas have the state's two law schools. In fall 1983, 141,469 students were enrolled in institutions of higher education. Of these, 72% were full-time students. Public institutions enrolled 90%; private, 10%.

During 1983/84, Kansas spent $3,361 per pupil (18th in the US), at a cost of $576 per capita. Total expenditures on education were $1,396,000,000. The Kansas Board of Regents offers scholarships and tuition grants to needy Kansas students.

In 1954, Kansas was the focal point of a US Supreme Court decision that had enormous implications for US public education. The court ruled, in *Brown* v. *Board of Education of Topeka,* that Topeka's "separate but equal" elementary schools for black and white students were inherently unequal, and it ordered the school system to integrate.

⁴²ARTS
The Kansas Arts Commission is a 12-member panel appointed by the governor. In 1984/85, it sponsored 360 touring arts programs. Wichita has a resident symphony orchestra.

⁴³LIBRARIES AND MUSEUMS
The Dwight D. Eisenhower Library in Abilene houses the collection of papers and memorabilia from the 34th president; there is also a museum. The Menninger Foundation Museum and Archives in Topeka maintains various collections pertaining to psychiatry. The Kansas State Historical Society Library (Topeka) contains the state's archives. Volumes of books and documents on the Old West are found in the Cultural Heritage and Arts Center Library in Dodge City. Kansas had 311 public libraries in 1982, with 6,509,826 volumes and a circulation of 12,419,101. Additionally there were 34 county and regional libraries, 10 bookmobiles, 27 college libraries, and 24 junior college libraries. Seven regional library systems serve state residents who have no local library service.

Almost 140 museums, historical societies, and art galleries were scattered across the state in 1984. The Dyche Museum of Natural History at the University of Kansas, Lawrence, draws many vistors. The Kansas State Historical Society maintains an extensive collection of ethnological and archaeological materials in Topeka. A new State Historical Museum at Topeka was scheduled to open in 1985.

Among the art museums are the Mulvane Art Center in Topeka, the Helen Foresman Spencer Museum of Art at the University of Kansas, and the Wichita Art Museum. The Dalton Museum in Coffeyville displays memorabilia from the famed Dalton family of desperadoes. La Crosse is the home of the Barbed Wire Museum, displaying more than 500 varieties of barbed wire. The Emmett Kelly Historical Museum in Sedan honors the world-famous clown born there. The US Cavalry Museum is on the grounds of Ft. Riley. The Sedgwick County Zoo in Wichita and the Topeka Zoo are the largest of seven zoological gardens in Kansas.

The entire town of Nicodemus, where many blacks settled after the Civil War, was made a national historic landmark in 1975. The chalk formations of Monument Rocks in western Kansas constitute the state's only national natural landmark. Fort Scott and Fort Larned are national historic parks.

⁴⁴COMMUNICATIONS
During 1985, Kansas was served by 7,265 postal workers. About 94% of all households had telephone service in 1980, slightly higher than the US average.

The state had 62 AM and 75 FM radio stations, 14 commercial television stations, and 3 public television stations in 1983. As of early 1984 there were 166 cable television systems serving 228 communities with 427,546 subscribers.

⁴⁵PRESS
Starting with the *Shawnee Sun,* a Shawnee-language newspaper founded by missionary Jotham Meeker in 1833, the press has been at the heart of Kansas society. The most famous Kansas newspaperman was William Allen White, whose *Emporia Gazette* was a leading voice of progressive Republicanism around the turn of the century. Earlier, John J. Ingalls launched his political career by editing the *Atchison Freedom's Champion.* Captain Henry King came from Illinois to found the *State Record* and *Daily Capital* in Topeka.

In 1984, Kansas had 46 daily newspapers with a total circulation of 594,570, and 18 Sunday papers with a circulation of 500,685. Leading newspapers and their circulations in 1984 were as follows:

AREA	NAME	DAILY	SUNDAY
Topeka	Capital-Journal (m,S)	65,829	75,880
Wichita	Eagle-Beacon (m,S)	123,074	188,238

The *Kansas City* (Missouri) *Star* is widely read in the Kansas as well as Missouri part of the metropolitan area.

46 ORGANIZATIONS

The 1982 Census of Service Industries counted 984 organizations in Kansas, including 177 business associations; 547 civic, social, and fraternal associations; and 11 educational, scientific, and research associations. Among the organizations headquartered in Kansas are the National Collegiate Athletic Association (NCAA) and the Junior College Athletic Association.

47 TOURISM, TRAVEL, AND RECREATION

During 1982, travel and tourism generated 40,000 jobs and $303 million in payroll. In 1983, Kansas had 22 state parks, 24 federal reservoirs, 48 state fishing lakes, more than 100 privately owned campsites, and more than 304,000 acres (123,000 hectares) of public hunting and game management lands. During 1983, 4,396,000 visitors used the state park system. There are two national historic sites, Fort Larned and Fort Scott, both 19th-century army bases on the Indian frontier.

Topeka features a number of tourist attractions, including the state capitol, state historical museum, and Menninger Foundation. Dodge City offers a reproduction of Old Front Street as it was when the town was the "cowboy capital of the world." Historic Wichita Cowtown is another frontier-town reproduction. In Hanover stands the only remaining original and unaltered Pony Express station. A recreated "Little House on the Prairie," near the childhood home of Laura Ingalls Wilder, is 13 mi (21 km) southwest of Independence. The Eisenhower Center in Abilene contains the 34th president's family home, library, and museum. The state fair is in Hutchinson.

In 1983, the state issued 241,540 fishing licenses, 190,546 hunting licenses, and 55,971 combination fishing-hunting licenses.

48 SPORTS

There are no major professional sports teams in Kansas. During spring, summer, and early fall, horses are raced at Eureka Downs. The National Greyhound Association Meet is held in Abilene.

The University of Kansas and Kansas State are both members of the Big Eight Conference. The National Junior College Basketball Tournament is held in Hutchinson each March. The Kansas Relays take place at Lawrence in April.

A US sporting event unique to Kansas is the International Pancake Race, held in Liberal each Shrove Tuesday. Women wearing housedresses, aprons, and scarves run along an S-shaped course carrying skillets and flipping pancakes as they go.

49 FAMOUS KANSANS

Kansas claims only one US president and one US vice president. Dwight D. Eisenhower (b.Texas, 1890-1969) was elected the 34th president in 1952 and reelected in 1956; he had served as the supreme commander of Allied Forces in World War II. He is buried in Abilene, his boyhood home. Charles Curtis (1860-1936) was vice president during the Hoover administration.

Two Kansans have been associate justices of the US Supreme Court: David J. Brewer (1837-1910) and Charles E. Whittaker (1901-73). Other federal officeholders from Kansas include William Jardine (1879-1955), secretary of agriculture; Harry Woodring (1890-1967), secretary of war; and Georgia Neese Clark Gray (b.1900), treasurer of the US. Prominent US senators include Edmund G. Ross (1826-1907), who cast a crucial acquittal vote at the impeachment trial of Andrew Johnson; John J. Ingalls (1833-1900), who was also a noted literary figure; Joseph L. Bristow (1861-1944), a leader in the Progressive movement; Arthur Capper (1865-1951), a former publisher and governor; Robert Dole (b.1923), who was the Republican candidate for vice president in 1976 and became Senate majority leader in 1984; and Nancy Landon Kassebaum (b.1932), elected to the US Senate in 1978. Among the state's important US representatives were Jeremiah Simpson (1842-1905), a leading Populist, and Clifford R. Hope (1893-1970), important in the farm bloc.

Notable Kansas governors include George W. Glick (1827-1911); Walter R. Stubbs (1858-1929); Alfred M. Landon (b.1887), who ran for US president on the Republican ticket in 1936; and Frank Carlson (b.1893). Other prominent political figures were David L. Payne (1836-84), who helped open Oklahoma to settlement; Carry Nation (1846-1911), the colorful prohibitionist; and Frederick Funston (1865-1917), hero of the Philippine campaign of 1898 and a leader of San Francisco's recovery after the 1906 earthquake and fire.

Earl Sutherland (1915-74) won the Nobel Prize in 1971 for physiology and medicine. Other leaders in medicine and science include Samuel J. Crumbine (1862-1954), a public health pioneer; the doctors Menninger—C. F. (1862-1953), William (1899-1966), and Karl (b.1893)—who established the Menninger Foundation, a leading center for mental health; Arthur Hertzler (1870-1946), a surgeon and author; and Clyde Tombaugh (b.1906), who discovered the planet Pluto.

Kansas also had several pioneers in aviation, including Clyde Cessna (1880-1954), Glenn Martin (1886-1955), Walter Beech (1891-1950), Amelia Earhart (1898-1937), and Lloyd Stearman (1898-1975). Cyrus K. Holliday (1826-1900) founded the Santa Fe railroad; William Coleman (1870-1957) was an innovator in lighting; and Walter Chrysler (1875-1940) was a prominent automotive developer.

Most famous of Kansas writers was William Allen White (1868-1944), whose son William L. White (1900-73) also had a distinguished literary career; Damon Runyon (1884-1946) was a popular journalist and storyteller. Novelists include Edgar Watson Howe (1853-1937), Margaret Hill McCarter (1860-1938), Dorothy Canfield Fisher (1879-1958), Paul Wellman (1898-1966), and Frederic Wakeman (b.1909). Gordon Parks (b.1912) has made his mark in literature, photography, and music. William Inge (1913-73) was a prizewinning playwright who contributed to the Broadway stage. Notable painters are Sven Birger Sandzen (1871-1954), John Noble (1874-1934), and John Steuart Curry (1897-1946). Sculptors include Robert M. Gage (b.1892), Bruce Moore (b.1905), and Bernard Frazier (1906-76). Among composers and conductors are Thurlow Lieurance (b.Iowa 1878-1963), Joseph Maddy (1891-1966), and Kirke L. Mechem (b.1926). Jazz great Charlie "Bird" Parker (Charles Christopher Parker, Jr., 1920-55) was born in Kansas City.

Stage and screen notables include Fred Stone (1873-1959), Joseph "Buster" Keaton (1895-1966), Milburn Stone (b.1904), Charles "Buddy" Rogers (b.1904), Vivian Vance (1912-79), Edward Asner (b.1929), and Shirley Knight (b.1937). The clown Emmett Kelly (1898-1979) was a Kansan. Operatic performers include Marion Talley (b.1906) and Kathleen Kersting (1909-65).

Glenn Cunningham (b.1910) and Jim Ryun (b.1947) both set running records for the mile. Also prominent in sports history were James Naismith (1861-1939), the inventor of basketball; baseball pitcher Walter Johnson (1887-1946); and Gale Sayers (b.1943), a football running back.

50 BIBLIOGRAPHY

Boyer, Richard O. *The Legend of John Brown.* New York: Knopf, 1973.

Davis, Kenneth S. *Kansas: A Bicentennial History.* New York: Norton, 1976.

Davis, Kenneth S. *Soldier of Democracy: A Biography of Dwight Eisenhower.* New York: Doubleday, 1945, 1952.

Howes, Charles C. *This Place Called Kansas.* Norman: University of Oklahoma Press, 1984.

Kansas, University of. Center for Public Affairs. *Kansas Statistical Abstract 1983-84.* 19th ed. Lawrence, 1984.

Richmond, Robert W. *Kansas: A Land of Contrasts.* St. Charles, Mo.: Forum Press, 1977.

Socolofsky, Homer, and Huber Self. *Historical Atlas of Kansas.* Norman: University of Oklahoma Press, 1972.

KENTUCKY

Commonwealth of Kentucky

ORIGIN OF STATE NAME: Derived from the Wyandot Indian word *Kah-ten-tah-teh* (land of tomorrow). **NICKNAME:** The Bluegrass State. **CAPITAL:** Frankfort. **ENTERED UNION:** 1 June 1792 (15th). **SONG:** "My Old Kentucky Home." **MOTTO:** United We Stand, Divided We Fall. **FLAG:** A simplified version of the state seal on a blue field. **OFFICIAL SEAL:** In the center, two men exchange greetings; above and below them is the state motto. On the periphery are two sprigs of goldenrod and the words "Commonwealth of Kentucky." **COLORS:** Blue and gold. **BIRD:** Cardinal. **WILD ANIMAL:** Gray squirrel. **FISH:** Bass. **FLOWER:** Goldenrod. **TREE:** Kentucky coffee tree. **LEGAL HOLIDAYS:** New Year's Day, 1 January, plus one extra day; Birthday of Martin Luther King, Jr., 3d Monday in January; Washington's Birthday, 3d Monday in February; Good Friday, March or April, half-day holiday; Memorial Day, last Monday in May; Independence Day, 4 July; Labor Day, 1st Monday in September; Thanksgiving Day, 4th Thursday in November, plus one extra day; Christmas Day, 25 December, plus one extra day. **TIME:** 7 AM EST = noon GMT; 6 AM CST = noon GMT.

¹LOCATION, SIZE, AND EXTENT

Located in the eastern south-central US, the Commonwealth of Kentucky is the smallest of the eight south-central states and ranks 37th in size among the 50 states.

The total area of Kentucky is 40,409 sq mi (104,659 sq km), of which land comprises 39,669 sq mi (102,743 sq km) and inland water 740 sq mi (1,917 sq km). Kentucky extends about 350 mi (563 km) E–W; its maximum N–S extension is about 175 mi (282 km).

Kentucky is bordered on the N by Illinois, Indiana, and Ohio (with the line roughly following the north bank of the Ohio River); on the NE by West Virginia (with the line formed by the Big Sandy and Tug Fork rivers); on the SE by Virginia; on the S by Tennessee; and on the W by Missouri (separated by the Mississippi River). Because of a double bend in the Mississippi River, about 10 sq mi (26 sq km) of SW Kentucky is separated from the rest of the state by a narrow strip of Missouri.

After 15 years of litigation, Kentucky in 1981 accepted a US Supreme Court decision giving Ohio and Indiana control of at least 100 feet (30 meters) of the Ohio River from the northern shore. This in effect returned Kentucky's border to what it was in 1792, when Kentucky entered the Union.

The total boundary length of Kentucky is 1,290 mi (2,076 km). The state's geographic center is in Marion County, 3 mi (5 km) NNW of Lebanon.

²TOPOGRAPHY

The eastern quarter of the state is dominated by the Cumberland Plateau, on the western border of the Appalachians. At its western edge, the plateau meets the uplands of the Lexington Plain (known as the Bluegrass region) to the north and the hilly Pennyroyal to the south. These two regions, which together comprise nearly half the state's area, are separated by a narrow curving plain known as the Knobs because of the shapes of its eroded hills. The most level area of the state is the western coalfields, bounded by the Pennyroyal to the east and the Ohio River to the north. In the far west are the coastal plains of the Mississippi River; this region is commonly known as the Purchase, having been purchased from the Chickasaw Indians.

The highest point in Kentucky is Black Mountain on the southeastern boundary in Harlan County, at 4,145 feet (1,263 meters). The lowest point is 257 feet (78 meters), along the Mississippi River in Fulton County. The state's mean altitude is 750 feet (229 meters).

The only large lakes in Kentucky are artificial. The biggest is Cumberland Lake (79 sq mi/205 sq km): Kentucky Lake, Lake Barkley, and Dale Hollow Lake straddle the border with Tennessee.

Including the Ohio and Mississippi rivers on its borders and the tributaries of the Ohio, Kentucky claims at least 3,000 mi (4,800 km) of navigable rivers—more than any other state. Among the most important of Kentucky's rivers are the Kentucky, 259 mi (417 km); the Cumberland, partly in Tennessee; the Tennessee, also in Tennessee and Alabama; and the Big Sandy, Green, Licking, and Tradewater rivers. All, except for a portion of the Cumberland, flow northwest into the Ohio and thence to the Mississippi. Completion in 1985 of the Tennessee-Tombigbee Waterway, linking the Tennessee and Tombigbee rivers in Alabama, for the first time gave Kentucky's Appalachian coalfields direct water access to the Gulf of Mexico.

Drainage through porous limestone rock has honeycombed much of the Pennyroyal with underground passages, the best known of which is Mammoth Cave, now a national park. The Cumberland Falls, 92 feet (28 meters) high and 100 feet (30 meters) wide, are located in Whitley County.

³CLIMATE

Kentucky has a moderate, relatively humid climate, with abundant rainfall.

The southern and lowland regions are slightly warmer than the uplands. In Louisville, the normal monthly mean temperature ranges from 33°F (1°C) in January to 76°F (24°C) in July. The record high for the state was 114°F (46°C), registered in Greensburg on 28 July 1930; the record low, −34°F (−37°C), in Cynthiana on 28 January 1963.

Average daily relative humidity in Louisville ranges from 60% to 80%. The normal annual precipitation is 43 in (109 cm); snowfall totals about 18 in (46 cm) a year.

⁴FLORA AND FAUNA

Kentucky's forests are mostly of the oak/hickory variety, with some beech/maple stands. Four species of magnolia are found, and the tulip poplar, eastern hemlock, and eastern white pine are also common; the distinctive "knees" of the cypress may be seen along riverbanks. Kentucky's famed bluegrass is actually blue only in May, when dwarf iris and wild columbine are in bloom. Rare plants include the swamp loosestrife and showy gentian.

Game mammals include the raccoon, muskrat, opossum, mink, gray and red foxes, and beaver; the eastern chipmunk and flying squirrel are common small mammals. At least 300 bird species have been recorded, of which 200 are common. Blackbirds are a serious pest, with some roosts numbering 5-6 million; more desirable avian natives include the cardinal (the state bird), robin, and brown thrasher, while eagles are winter visitors. More than 100 types of fish have been identified.

Rare animal species include the swamp rabbit, black bear, raven (*Corvus corax*), and mud darter. Among Kentucky's threatened species are the river otter, common shrew, and osprey. The Indiana bat, cougar, brown bear, Kirtland's warbler, bald eagle, whooping crane, peregrine falcon, and orange-footed and pink mucket pearly mussels are listed as endangered.

5ENVIRONMENTAL PROTECTION

The Natural Resources and Environmental Protection Cabinet, with broad responsibility, includes the departments of Natural Resources, Environmental Protection, and Surface Mining Reclamation and Enforcement, as well as the Kentucky Nature Preserves Commission. The Environmental Quality Commission, created in 1972 to serve as a watchdog over environmental concerns, is a citizen's group of seven members appointed by the governor. Funding appropriated for environmental programs in 1985/86 totaled $54,880,600.

The most serious environmental concern in Kentucky is repairing and minimizing damage to land and water from strip-mining. Efforts to deal with such damage are relatively recent. The state has had a strip-mining law since 1966, but the first comprehensive attempts at control did not begin until the passage in 1977 of the Federal Surface Mining Control and Reclamation Act. Funding enacted for 1985/86 for the separate Department of Mining Reclamation and Enforcement totaled $18,958,900.

Also active in environmental matters is the Department of Environmental Protection, consisting of four divisions. The Division of Water, administers the state's Safe Drinking Water and Clean Water acts and regulation of sewage disposal. The Division of Waste Management oversees solid waste disposal systems in the state. The Air Pollution Control Division monitors industrial discharges into the air and other forms of air pollution. A special division is concerned with Maxey Flats, a closed nuclear waste disposal facility in Fleming County, where leakage of radioactive materials was discovered.

There are 15 major dams in Kentucky, and more than 900 other dams. Flooding is a chronic problem in southeastern Kentucky, where strip-mining has exacerbated soil erosion.

6POPULATION

Kentucky ranked 23d in population among the states in 1980, with a census population of 3,660,257.

During the early decades of settlement, population grew rapidly, from a few hundred in 1780 to 564,317 in 1820, by which time Kentucky was the 6th most populous state. By 1900, however, when the population was 2,147,174, growth had slowed considerably. For most of the 20th century, Kentucky's growth rate has been significantly slower than the national average.

At the time of the 1980 census, Kentucky's population was 51% urban, far below the national norm of 74%. The provisional estimate of Kentucky's population in 1985 was 3,747,769—a 2.4% increase over the 1980 figure. The projected population for 1990 is 4,073,500. The population density in 1985 was 94 persons per sq mi (36 per sq km).

Louisville, the state's largest city, had a 1984 population of 289,843, down from 298,694 in 1980 and 361,706 in 1970. Lexington-Fayette urban county was next with 210,150 residents in 1984. Owensboro, with 55,723 residents in 1984, was the state's third most populous city. The population of the Louisville (Ky.-Ind.) metropolitan area was 962,600 in 1984, up from 956,486 in 1980. Population in the Lexington-Fayette metropolitan area was 327,200 in 1984, up from 317,548 in 1980. In 1984, the metropolitan population for Owensboro was 88,300.

7ETHNIC GROUPS

Though a slave state, Kentucky never depended on a plantation economy. This may account for its relatively low black population—259,477 (7.1%) in 1980. Kentucky was a center of the American (or Know-Nothing) Party, a pre–Civil War movement opposed to slavery and immigration alike. With relatively little opportunity for industrial employment, Kentucky attracted small numbers of foreign immigrants in the 19th and 20th centuries. The state had only 34,000 foreign-born residents in 1980, accounting for 0.9% of the total population. Among persons reporting a single ancestry in the 1980 census, a total of 931,737 claimed English descent, 245,143 German, 230,900 Irish, 212,127 Afro-American, and 24,000 French.

There were 40,224 reporting American Indian origins in 1980. The 1980 census also found 2,352 Koreans, 1,685 Asian Indians, 1,461 Japanese, 1,380 Vietnamese, and 1,289 Chinese. A total of 3,084 Hispanics reported Mexican ancestry and 1,489 Puerto Rican ancestry.

8LANGUAGES

Kentucky was a fought-over hunting ground for Ohio Shawnee, Carolina Cherokee, and Mississippi Chickasaw Indians. Place-names from this heritage include Etowah (Cherokee) and Paducah (Chickasaw).

Speech patterns in the state generally reflect the first settlers' Virginia and Kentucky backgrounds. South Midland features are best preserved in the mountains, but some common to Midland and Southern are widespread.

Other regional features are typically both South Midland and Southern. After a vowel, the /r/ may be weak or missing. *Coop* has the vowel of *put,* but *root* rhymes with *boot.* In southern Kentucky, earthworms are *redworms,* a burlap bag a *tow sack* or the Southern *grass sack,* and green beans *snap beans.* A young man may *carry,* not escort, his girl friend to a party. Subregional terms appear in abundance. In the east, kindling is *pine,* a seesaw is a *ridyhorse,* and the freestone peach is an *openstone peach.* In central Kentucky, a moth is a *candlefly.*

In 1980, 98% of all residents 3 years old and older spoke only English at home. The number of people who spoke other languages at home included:

Spanish	18,474	Korean	2,074
German	11,294	Italian	1,921
French	9,444	Vietnamese	1,351

9RELIGIONS

Throughout its history, Kentucky has been predominantly Protestant. A group of New Light Baptists who, in conflict with established churches in Virginia immigrated to Kentucky under the leadership of Lewis Craig, built the first church in the state in 1781, near Lancaster. The first Methodist Church was established near Danville in 1783; within a year, Roman Catholics had also built a church and a presbytery of 12 churches had been organized. There were 42 churches in Kentucky by the time of statehood, with a total membership of 3,095.

Beginning in the last few years of the 18th century, the Great Revival sparked a new religious fervor among Kentuckians, a development that brought the Baptists and Methodists many new members. The revival, which had begun among the Presbyterians, led to a schism in that sect. Presbyterian minister Barton W. Stone organized what turned out to be the era's largest frontier revival meeting, at Cane Ridge (near Paris), in August 1801. Differences over doctrine led Stone and his followers to withdraw from the Synod of Kentucky in 1803, and they formed their own church, called simply "Christian." The group later formed an alliance with the sect now known as the Christian Church (Disciples of Christ).

As of 1980 there were 1,982,830 known Protestant adherents in Kentucky, of whom 888,198 belonged to the Southern Baptist Convention, 235,801 to the United Methodist Church, 78,275 to

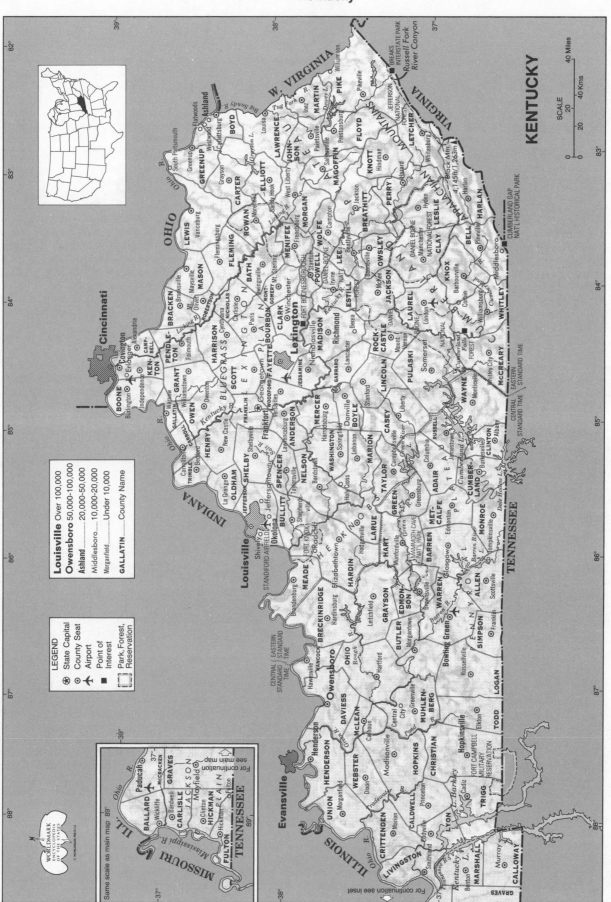

See endsheet maps: J3.

LOCATION: 36°30′ to 39°08′45″N; 81°58′ to 89°34′w. **BOUNDARIES:** Illinois line, 131 mi (211 km); Indiana line, 345 mi (555 km); Ohio line, 167 mi (269 km); West Virginia line, 107 mi (172 km); Virginia line, 127 mi (204 km); Tennessee line, 336 mi (541 km); Missouri line, 77 mi (124 km).

the Christian Church (Disciples of Christ), and 81,787 to the Christian Churches and Churches of Christ. The Roman Catholic Church, with 381,113 members at the beginning of 1984, is the 2d-largest denomination in the state. There were an estimated 12,910 Jews in Kentucky in 1984.

¹⁰TRANSPORTATION

Statewide transportation developed slowly in Kentucky. Although freight and passengers were carried by river and later by rail during the 19th century, mountains and lack of good roads made land travel in eastern Kentucky so arduous that the region was for a long time effectively isolated from the rest of the state.

The first railroad in Kentucky, the Lexington and Ohio, opened on 15 August 1832 with a 6-mi (10-km) route from Lexington to Frankfort. Not until 1851 did the railroad reach the Ohio River. In November 1859, Louisville was connected with Nashville, Tenn., by the Louisville and Nashville Railroad; heavily used by the Union, it was well maintained during the Civil War. Railroad construction increased greatly after the conflict ended. By 1900, Kentucky had three times the track mileage it had had in 1870. As of December 1983, Kentucky had 3,347 mi (5,386 km) of Class I track. Rail service to the state, nearly all of which was freight, was provided by 10 railroads.

The trails of Indians and buffalo became the first roads in Kentucky. Throughout the 19th century, roads were maintained—if at all—by private enterprise, as toll roads. This system came to an end as a result of the "tollgate war" of the late 19th and early 20th centuries—a rebellion in which masked Kentuckians, demanding free roads, raided tollgates and assaulted their keepers. Not until 1909, however, was a constitutional prohibition against the spending of state funds on highways abolished. In 1912, a state highway commission was created, and by 1920, roads had improved considerably. In 1983, Kentucky had 69,150 mi (111,286 km) of roads. Local governments controlled 43,560 mi (70,103 km) of Kentucky's roads; the state government, 25,103 mi (40,399 km); and the federal government, 487 mi (784 km). There were 738 mi (1,188 km) of interstate highway. In 1983, 1,812,875 automobiles, 799,057 trucks, and 58,636 motorcycles were registered in the state.

Until displaced by the railroads in the late 1800s, the Ohio River and its tributaries, along with the Mississippi, were Kentucky's primary commercial routes for trade with the South and the West. The Kentucky Port and River Development Commission was created by the legislature in 1966 to promote river transportation. Louisville, on the Ohio River, is the chief port. In 1982, traffic through the port totaled 5,701,896 tons, down significantly from the 10-year record of 9,537,268 handled in 1979. Paducah is the outlet port for traffic on the Tennessee River.

In 1983 there were 101 airports and 25 heliports in Kentucky. The largest of these is Standiford Field in Louisville, which enplaned 855,970 passengers in 1983.

¹¹HISTORY

Six distinctive Indian cultures inhabited the region now known as Kentucky. The earliest nomadic hunters occupied the land for several thousand years, and were followed by the seminomadic Woodland and Adena cultures (1000 BC–AD 1000). Remains of the Mississippian and Fort Ancient peoples (AD 1000–1650) indicate that they were farmers and hunters who often dwelled in stockaded villages, subsisting on plentiful game and fish supplemented by crops of beans, corn, and squash.

No Indian nations resided in central and eastern Kentucky when these areas were first explored by British-American surveyors Thomas Walker and Christopher Gist in 1750 and 1751. The dominant Shawnee and Cherokee tribes utilized the region as a hunting ground, returning to homes in the neighboring territories of Ohio and Tennessee. Early descriptions of Kentucky generated considerable excitement about the fertile land and abundant wildlife. The elimination of French influence after the French and

Indian War intensified pressures to open the region to American settlement—pressures that were initially thwarted by Britain's Proclamation of 1763, barring such western migration until Native American interests could be protected. This artificial barrier proved impossible to maintain, however, and the first permanent white settlement in Kentucky was finally established at Harrodstown (now Harrodsburg) in 1774 by a group of Virginians.

The most ambitious settlement scheme involved the Transylvania Land Company, a creation of North Carolina speculator Richard Henderson, assisted by the famed woodsman Daniel Boone. Henderson purchased a huge tract of land in central Kentucky from the Cherokee and established Fort Boonesborough. The first political meeting by whites in Kentucky, held at Fort Boonesborough on 23 May 1775, provided for rule by the Transylvania proprietors and a representative assembly. Henderson then sought approval for creation of a 14th colony, but the plan was blocked by Virginians determined to claim Kentucky as a possession of the Old Dominion. On 1 December 1776, the new state of Virginia incorporated its new County of Kentucky.

Kentucky's image soon changed from "western Eden" to "dark and bloody ground," as it became the scene of frequent clashes between Ohio-based Indians and the growing number of white settlements dotting the central Bluegrass region. Nevertheless, immigrants continued to come westward, down the Ohio River and through the Cumberland Gap. Kentucky became the principal conduit for migration into the Mississippi Valley. By the late 1780s, settlements were gaining in population, wealth, and maturity, and it was obvious that Kentucky could not long remain under the proprietorship of distant Virginia. Its entire orientation was westward, and the majority of its citizens were no longer Virginia-born but former residents of Pennsylvania, Maryland, and North Carolina. Virginia yielded permission for the drafting of a Kentucky state constitution, and in June 1792, Kentucky entered the Union as the 15th state.

Over the next several decades, Kentucky prospered because of its diverse agricultural and processing industries. Although there were 225,483 slaves in the state in 1860, Kentucky was spared the evils of one-crop plantation agriculture. Nevertheless, its economy was tightly linked to the lower South's, a tie facilitated by the completion in 1829 of a canal around the Ohio River falls at Louisville. Hemp was one such connection; the plant was the principal source of rope and bagging used to bind cotton bales. Kentucky was a major supplier of hogs, mules, and workhorses; prepared meats, salt, flour, and corn for the plantation markets of the South. The state became a center for breeding and racing fine Thoroughbred horses, an industry that thrives today on Bluegrass horse farms as virtually the state symbol. More important was the growing and processing of tobacco, an enterprise accounting for half the agricultural income of Kentucky farmers by 1860. Finally, whiskey began to be produced in vast quantities by the 1820s, culminating in the standardization of a fine, aged amber-red brew known throughout the world as bourbon, after Bourbon County.

Despite this economic development, several social and cultural problems disturbed the state. Much of the agricultural productivity came from farms employing slave labor, while the less affluent majority of white families often dwelled on less fertile upland farms. Efforts were repeatedly made to consider the slavery question. Leaders such as Henry Clay, Reverend Robert J. Breckinridge, and the fiery antislavery advocate Cassius Marcellus Clay urged an end to the "peculiar institution." Because of racial phobias and hostility to "Yankee meddling," the appeal was rejected. During the Civil War, Kentuckians were forced to choose sides between the Union, led in the North by Kentucky native Abraham Lincoln, and the Confederacy, led in the South by Kentucky native Jefferson Davis.

Although the state legislature finally opted for the Union side, approximately 30,000 men went south to Confederate service, while 70,000—including nearly 24,000 black soldiers—served in the Union army. For four years the state was torn with conflict over the collapse of slavery and wracked with guerrilla warfare and partisan feuds. Vigilantism and abuse of black people continued into the turbulent Reconstruction period, until legislative changes in the early 1870s began to restrain Ku Klux Klan violence and bring increased civil rights to black people.

The decades to 1900 saw other progress. Aided by liberal tax exemptions, railroad construction increased threefold and exploitation of timber and coal reserves began in eastern Kentucky. Industrial employment and productivity increased by more than 200%, drawing rural folk into the growing cities of Louisville and Lexington. In 1900, Kentucky ranked 1st among southern states in per capita income.

An economic and political crisis was developing, however, that would send shock waves across the state. Wealth remained very unevenly distributed. Farmers, especially western Kentucky "dark leaf" tobacco farmers, were feeling the brunt of a prolonged price depression. The major national farm protest movements—the Grange, the Farmers' Alliance, and the Populist Party—all found support here, for by 1900, a third of all Kentucky farmers were landless tenants, and the size of the average family farm had fallen below 100 acres (40 hectares). Calls for currency inflation, reform of corporate monopolies, and improved rights for industrial workers reached a climax in the gubernatorial election of 1899. Republican William S. Taylor narrowly defeated the more reform-minded Democrat William Goebel and was sworn into office. Democrats, claiming electoral fraud, instituted a recount. On 30 January 1900, Goebel, a state senator, was shot while approaching the capitol; as he lingered near death, the legislature, controlled by Democrats, declared him governor. Goebel died immediately thereafter, and his lieutenant governor, John Beckham, was administered the oath of office. Surprisingly, further bloodshed was averted, the courts upheld the Goebel-Beckham election, and "Governor" Taylor fled the state forever.

Goebel's assassination weighed heavily, however. The state was polarized, outside investment plummeted, and Kentucky fell into a prolonged economic and moral depression. By 1940, the state ranked last among the 48 states in per capita income and was burdened by an image of clan feuding and homicide, poverty, and provincial courthouse politics. The Great Depression hit the state hard, though an end to Prohibition revived the dormant whiskey industry.

Kentucky has changed greatly since World War II. Between 1945 and 1980, the farm population decreased by 76% and the number of farms by 53%. In addition, tobacco has come under attack as a menace to public health. Between 1967 and 1982, the number of manufacturing plants increased from 2,994 to 3,504 and the number of industrial workers increased from 224,600 to 246,300. Although Kentucky remains one of the poorest states in the nation, positive change is evident even in relatively isolated rural communities—a result of better roads, education, television, and government programs.

Kentucky state politics drew national attention in 1979 with the election as governor of Democrat John Young Brown, Jr., already famous for his wealth and promotion of Kentucky Fried Chicken franchises (he had bought out the original Colonel Sanders) and as the husband of sportscaster and 1971 Miss America Phyllis George. His attempt to secure a constitutional amendment allowing Kentucky governors to serve a second term was defeated in 1983. Kentucky then elected its first woman governor, former Lieutenant Governor Martha Layne Collins, a Democrat.

12 STATE GOVERNMENT

Kentucky's current and fourth constitution was adopted on 28 September 1891. As of 31 December 1983, it had been amended 25 times. Earlier constitutions were adopted in 1792 and 1849.

The state legislature, called the general assembly, consists of the house of representatives, which has 100 members elected for two-year terms, and the senate, with 38 members elected for staggered four-year terms. A constitutional amendment approved by the voters in November 1979 provided for the election of legislators in even-numbered years, a change scheduled for completion by November 1988. The assembly meets in regular sessions of no more than 60 legislative days, beginning in January of each even-numbered year. The governor may also call special sessions. Except for revenue-raising measures, which must be introduced in the house of representatives, either chamber may introduce or amend a bill. Most bills may be passed by voting majorities equal to at least two-fifths of the membership of each house. Measures requiring an absolute majority in each house include those that appropriate money or create a debt, summon a constitutional convention, or enact emergency measures to take effect immediately. A majority of the members of each house is required to override the governor's veto.

A member of the senate must have been a citizen and resident of Kentucky for six years preceding election, a representative for two. A senator must be at least 30 and a representative at least 24. The constitutional limit of $12,000 for salaries of public officials, which is thought to apply to legislators, has been interpreted by the courts in terms of 1949 dollars and thus may be increased considerably—and has been. However, as of 1983, most legislators in Kentucky probably received less than $12,000 per year, including travel and expense allowances, based on per diem in-session salaries of $100 and per diem expense allowances of $75.

The elected executive officers of Kentucky are the governor, lieutenant governor, secretary of state, attorney general, treasurer, superintendent of public instruction, auditor of public accounts, commissioner of agriculture, and three members of the Railroad Commission. All serve four-year terms and may not succeed themselves. The governor and lieutenant governor, who are independently elected, must each be 30 years old, US citizens, and residents of Kentucky for six years. As of 1984, the governor's salary was $60,000, and the lieutenant governor's $51,010.

A three-fifths majority of each house plus a voting majority of the electorate must approve any proposed constitutional amendment. Before a constitutional convention may be called, two regular sessions of the general assembly must approve it, and the call must be ratified at the polls by a majority voting on the proposal and equal to at least one-fourth the number of voters who cast ballots in the last general election.

To vote in Kentucky, one must be a US citizen, be at least 18 years of age, and have been a resident in the state for 30 days.

13 POLITICAL PARTIES

A rift was created in Kentucky politics by the presidential election of 1824, which had to be determined in the US House of Representatives because neither John Quincy Adams nor Andrew Jackson won a majority of the Electoral College. Representative Henry Clay voted for Adams, despite orders by the Kentucky general assembly to support Jackson, thereby splitting the state into two factions: supporters of Clay, who became Whigs, and supporters of Jackson, who became Democrats. The Whigs dominated Kentucky politics until Clay's death in 1852, after which, as the Whigs divided over slavery, most Kentuckians turned first to the Native American (or Know-Nothing) Party and then to the Democrats. Regional divisions in party affiliation during the Civil War era, according to sympathy with the South and slavery (Democrats) or with the Union and abolition (Republicans), have persisted in the state's voting patterns. In general, the poorer mountain areas tend to vote Republican, while the more affluent lowlanders in the Bluegrass and Pennyroyal tend to vote Democratic.

In 1985, Kentucky had 2,087,509 registered voters—1,418,835 Democrats, 597,055 Republicans, and 71,619 independents or members of minor parties. In 1983, Martha Layne Collins, a Democrat, defeated Republican candidate Jim Bunning to become Kentucky's first woman governor. Ronald Reagan carried the state by a wide margin in the November 1984 presidential voting, helping Mitch McConnell, Jr., narrowly defeat Democratic incumbent Walter D. Huddleston for the US Senate. At the start of 1985, Democrats held 28 seats in the state senate, Republicans 10; the Democrats were equally dominant in the house of representatives, with 74 seats to the Republicans' 26. At the national level, Kentucky was represented by Democratic Senator Wendell H. Ford, elected in 1974 and reelected in 1980; Republican Senator McConnell, elected in 1984; and in the US House of Representatives, by four Democrats and three Republicans.

In addition to having a woman governor in 1985, Kentucky was served by elected women as state auditor, state treasurer, and superintendent of public instruction. Nine women also served in the state legislature. Blacks held four seats in the state legislature.

14 LOCAL GOVERNMENT

The form of Kentucky's county government is of English origin. The state's 120 counties were run in 1985 by about 1,500 county officials. The chief governing body is the fiscal court, consisting of the county judge executive and three to eight magistrates or commissioners. Other elected officials are the sheriff, jailer, attorney, and court clerk. All are elected for four-year terms; a 1984 constitutional amendment allows the sheriff to succeed himself in office. Except for jailers, many county officials earn their living by collecting a share of the fees for the services they render. As of 1984, the maximum income for a county official was set at $30,605.

Cities are assigned by the general assembly to one of six classes on the basis of population. First-class cities have populations of 100,000 or more; second, 20,000 to 99,999; third, 8,000 to 19,999; fourth, 3,000 to 7,999; fifth, 1,000 to 2,999; and sixth, 999 or fewer. In 1984, Kentucky had only one first-class city, Louisville. There were 8 second-class cities, 19 third, 98 fourth, 116 fifth, and 199 sixth. The mayor or other chief executive officer in the top

three classes must be elected; in the bottom classes, the executive may be either elected by the people or appointed by a city council or commission. Mayors serve four-year terms; members of city legislative boards, also provided for in the state constitution, are elected for terms of two years. City officials must be residents of the cities and of the districts in which they are elected.

Other units of local government in Kentucky include urban counties and special-purpose districts, including districts for sewer and flood control and the 15 area-development districts for regional planning. The general assembly may create new local government units, although its power to create new counties is restricted. (Since 1912, it has created only one new county, McCreary.)

15 STATE SERVICES

An ombudsman in the Cabinet for Human Resources receives citizens' complaints concerning services offered by that agency. The Financial Disclosure Review Board was established in 1975 to review the financial status of constitutional officers and government management personnel in order to prevent conflicts of interest.

Educational services are provided through the Education and Humanities Cabinet, which includes the Department of Education, the Office and Advisory Council for Vocational Education, and the Department for the Blind. The Council for Higher Education oversees the state-supported colleges and universities. The Human Rights Commission and the Commission on Women are part of the Education and Humanities Cabinet. Rehabilitation services, including the Eastern Kentucky Comprehensive Rehabilitation Center, are under the jurisdiction of the Department of Education. Transportation services are administered by the Transportation Cabinet. Health, welfare, and other human services are provided primarily by the Cabinet for Human Resources. Among the agencies that provide public protection services are the Department of Military Affairs, the Public Protection and Regulation Cabinet, and the Consumer Protection Division of the Department of Law. Corrections and parole were transferred in 1981 from the Department of Justice to the

Kentucky Presidential Vote by Political Parties, 1948–84

YEAR	ELECTORAL VOTE	KENTUCKY WINNER	DEMOCRAT	REPUBLICAN	STATES' RIGHTS DEMOCRAT	PROHIBITION	PROGRESSIVE	SOCIALIST
1948	11	*Truman (D)	466,756	341,210	10,411	1,245	1,567	1,284
1952	10	Stevenson (D)	495,729	495,029	—	1,161	—	—
1956	10	*Eisenhower (R)	476,453	572,192	—	2,145	—	—
1960	10	Nixon (R)	521,855	602,607	—	—	—	—
					STATES' RIGHTS			
1964	9	*Johnson (D)	669,659	372,977	3,469	—	—	—
					AMERICAN IND.			SOC. WORKERS
1968	9	*Nixon (R)	397,541	462,411	193,098	—	—	2,843
						AMERICAN	PEOPLE'S	
1972	9	*Nixon (R)	371,159	676,446	—	17,627	1,118	—
1976	9	*Carter (D)	615,717	531,852	2,328	8,308	—	—
							LIBERTARIAN	CITIZENS
1980	9	*Reagan (R)	617,417	635,274	—	—	5,531	1,304
1984	9	*Reagan (R)	539,539	821,702	—	—	1,776	599

*Won US presidential election.

Corrections Cabinet. The Department of State Police is part of the Justice Cabinet.

Housing rights for members of minority groups are provided by the Commission on Human Rights. The Department of Economic Development oversees industrial and community development programs within the Commerce Cabinet. Also assisting in community development are programs within the Department of Local Government, which was organized as an independent agency of the office of the governor in 1982.

Natural resource protection services are provided by the separate departments of Natural Resources, Environmental Protection, and Surface Mining Reclamation and Enforcement, all within the Natural Resources and Environmental Protection Cabinet. The Kentucky state park system is administered by the Department of Parks in the Tourism Cabinet, which also includes the Department of Travel Development, Kentucky Horse Park, and the Department of Fish and Wildlife Resources. The Energy Cabinet, created in 1978, is within the Department of Energy Research and Development.

Labor services are administered by the Labor Cabinet; its areas of concern include labor-management relations, occupational safety and health, and occupational injury and disease compensation.

[16] JUDICIAL SYSTEM

In accordance with a constitutional amendment approved in 1975 and fully implemented in 1978, judicial power in Kentucky is vested in a unified court of justice. The highest court is the supreme court, consisting of a chief justice and six associate justices. It has appellate jurisdiction and also bears responsibility for the budget and administration of the entire system. Justices are elected from seven supreme court districts for terms of eight years; they elect one of their number to serve for the remaining term as chief justice.

The court of appeals consists of 14 judges, 2 elected from each supreme court district. The court divides itself into panels of at least 3 judges which may sit anywhere in the state. The judges also serve eight-year terms and elect one of their number to serve a four-year term as chief judge.

Circuit courts, with original and appellate jurisdiction, are held in each county. Circuit court judges are elected for terms of eight years. In circuits with more than one judge, the judges elect one of their number as chief judge for a two-year term. Under the revised judicial system, district courts, which have limited and original jurisdiction, replaced various local and county courts. There is no mandatory retirement age.

In 1984 there were 4,793 prisoners in state and federal prisons in Kentucky. The Corrections Cabinet maintains 10 correctional institutions, including a career development center, a forestry camp, and two farm centers. In 1980, the Corrections Cabinet entered into a consent decree to eliminate overcrowding and provide more humane conditions at the state reformatory and penitentiary. Death by electrocution is the only method of execution. No executions have occurred since 1976, but 19 prisoners were under sentence of death as of 31 December 1983. A total of 7,938 attorneys were actively practicing as of the end of 1984.

In the past, Kentucky had a reputation for lawlessness. In 1890, more homicides were reported in Kentucky than in any other state except New York; blood feuds among Kentucky families were notorious throughout the country. In 1983, however, Kentucky's rate of 3,435 crimes per 100,000 residents was well below the national average of 4,449. The 1983 crime rate included 322 violent crimes per 100,000 population and 3,113 property crimes.

[17] ARMED FORCES

The US Department of Defense had 53,123 personnel in Kentucky in 1983/84. US Army installations in the state include Ft. Knox (site of the US gold depository) near Louisville, with 21,985 personnel, and Ft. Campbell (partly in Tennessee), with 23,859 personnel. Kentucky received $448 million in prime federal defense contracts in 1983/84.

As of 30 September 1983 there were 404,000 veterans of US military service living in Kentucky. Of these, World War I veterans numbered 4,000; World War II, 155,000; Korean conflict, 73,000; and Viet-Nam era, 118,000. Kentucky veterans received more than $412 million in benefits during 1982/83.

Kentucky Army and Air National Guard Units had 8,920 personnel in 1984/85. In 1983, state and local police forces totaled 7,122, or just under 2 police employees per 1,000 population (only three states had a lower ratio).

[18] MIGRATION

During the frontier period, Kentucky first attracted settlers from eastern states, especially Virginia and North Carolina. Prominent among early foreign immigrants were people of English and Scotch-Irish ancestry, who tended to settle in the Kentucky highlands, which resembled their Old World homelands.

Kentucky's black population increased rapidly during the first 40 years of statehood. By the 1830s, however, slavery had become less profitable in the state, and many Kentucky owners either moved to the Deep South or sold their slaves to new owners in that region. During the 1850s, nearly 16% of Kentucky's slave population—more than 43,000 blacks—were sold or moved from the state. A tiny percentage of Kentucky's blacks, probably fewer than 200, emigrated to Liberia under the auspices of the Kentucky Colonization Society.

The waves of European immigration that inundated many states during the late 19th century left Kentucky virtually untouched. In 1890, Kentucky's population was nearly 98% native-born. At that time, there were more than 284,000 blacks in the state—a number that was to fall precipitously until the 1950s because of migration to industrial cities in the Midwest.

Until the early 1970s there was a considerable out-migration of whites, especially from eastern Kentucky to industrial areas of Ohio, Indiana, and other nearby states. The state's net loss to migration from 1960 to 1970 totaled 153,000 persons. This tide of out-migration was temporarily reversed during the 1970s, with Kentucky recording a net migration gain of 131,000 persons. Estimates for the period 1980–83 indicate a net loss to migration of 28,000.

[19] INTERGOVERNMENTAL COOPERATION

Among the many interstate regional commissions in which Kentucky participates are the Appalachian Regional Commission, Interstate Mining Compact, Interstate Oil and Gas Compact, Southern Growth Policies Compact, Ohio River Valley Water Sanitation Compact, and Tennessee-Tombigbee Waterway Compact. Kentucky also participates in the Tennessee Valley Authority. The Council of State Governments, founded in 1925 to foster interstate cooperation, has its headquarters in Lexington.

In 1983/84, Kentucky received just under $1.6 billion in federal aid.

[20] ECONOMY

Between statehood and the Civil War, Kentucky was one of the preeminent agricultural states, partly because of good access to river transportation down the Ohio and the Mississippi to southern markets. Coal mining had become an important part of the economy by the late 19th century. Although agriculture is still important in Kentucky, manufacturing has grown rapidly since World War II and was, by the mid-1980s, the most important sector of the economy as a source of both employment and personal income. Kentucky leads the nation in the production of coal and whiskey, and ranks 2d in tobacco output.

In contrast to the generally prosperous Bluegrass area and the growing industrial cities, eastern Kentucky, highly dependent on coal mining, has long been one of the poorest regions in the US. Beginning in the early 1960s, both the state and federal governments undertook programs to combat poverty in Appalachian Kentucky. Per capita personal income increased faster in this region than in the US as a whole between 1965 and 1976, and unemployment decreased between 1970 and 1978; however, personal income was still much lower, and unemployment higher, than in the rest of the state. In 1979, 29% of the people living in the state's 49 Appalachian counties were below the federal poverty level, compared to 17% of the people in Kentucky's other counties; in 1983, 34 of the 49 Appalachian counties had unemployment rates greater than the state average of 11.6%.

²¹INCOME
Kentucky has long been one of the poorest of the 50 states, and in 1983, per capita income was $9,162, or 79% of the national average, for a rank of 45th.

Sources of personal income in 1982 were manufacturing, 24%; government, 17%; trade, 15%; services, 14%; transportation, communications, and public utilities, 8%; mining, 8%; contract construction, 5%; farms, 5%; and other sectors, 4%. Incomes were highest in the Louisville and Lexington-Fayette metropolitan areas; in Kentucky's share of the Cincinnati, Ohio, and Evansville, Ind., metropolitan regions; and in Franklin County (Frankfort). Among the poorest counties were Owsley, 48% of whose residents had incomes below the poverty level in 1979; Wolfe, 35%; and Wayne, 35%. In the state as a whole, 17% of all Kentuckians were below the federal poverty line in 1979, including 33% of all black people.

²²LABOR
According to federal statistics, Kentucky's civilian labor force in 1984 was about 1,717,000. A total of 1,557,000 persons held jobs and 160,000 were unemployed, for an unemployment rate of 9.3%.

Among those actually employed in 1983, wholesale and retail trade accounted for 25%; manufacturing, 22%; services, 18%; government, 15%; transportation, communications, and public utilities, 5%; contract construction, 4%; mining and quarrying, 4%; and other sectors, 7%.

A federal survey in 1983 revealed the following nonfarm employment pattern for Kentucky:

	ESTABLISH-MENTS	EMPLOYEES	ANNUAL PAYROLL ('000)
Agricultural services, forestry, fishing	660	4,560	$ 48,381
Mining, of which:	1,633	43,567	1,104,077
Bituminous coal, lignite	(1,131)	(37,425)	(985,612)
Contract construction	5,647	39,077	712,897
Manufacturing	3,386	240,310	4,852,538
Transportation, public utilities	3,267	51,636	1,073,120
Wholesale trade	5,497	60,207	1,050,465
Retail trade	21,292	203,190	1,808,236
Finance, insurance, real estate	5,230	51,671	809,869
Services	19,472	207,353	2,526,332
Other	3,458	4,312	78,672
TOTALS	69,542	905,883	$14,064,587

Although a small number of trade unions existed in Kentucky before the 1850s, it was not until after the Civil War that substantial unionization took place. During the 1930s, there were long, violent struggles between the United Mine Workers (UMW) and the mine owners of eastern Kentucky. The UMW won bargaining rights in 1938, but after World War II the displacement of workers because of mechanization, a drastic drop in the demand for coal, and evidence of mismanagement and corruption within the UMW served to undercut the union's position. Following the announcement by the UMW in 1962 that its five hospitals would be sold or closed, unemployed mine workers began protracted picketing of nonunion mines. Episodes of violence accompanied the movement, which succeeded in closing the mines but not in keeping them closed. The protests dissipated when publicworks jobs were provided for unemployed fathers among the miners, beginning in late 1973. Increased demand for coal in the 1970s led to a substantial increase in jobs for miners, and the UMW, under different leaders, began a new drive to organize the Cumberland Plateau.

As of June 1983 there were 931 labor unions in Kentucky. Union membership in 1980 totaled 290,000, or 24% of all nonagricultural workers.

²³AGRICULTURE
With cash receipts totaling $2.8 billion—$1.3 billion from crops and $1.5 billion from livestock—Kentucky ranked 20th among the 50 states in farm marketings in 1983.

Kentucky tobacco, first marketed in New Orleans in 1787, quickly became the state's most important crop. Kentucky ranked 1st among tobacco-producing states until it gave place to North Carolina in 1929. Corn has long been one of the state's most important crops, not only for livestock feed but also as a major ingredient in the distilling of whiskey. Although hemp is no longer an important crop in Kentucky, its early significance to Kentucky farmers, as articulated in Congress by Henry Clay, was partly reponsible for the establishment by the US of a protective tariff system. From 1849 to 1870, the state produced nearly all the hemp grown in the US.

In 1984 there were approximately 101,000 farms in Kentucky (down from 133,000 in 1964), with an average size of 140 acres (57 hectares). Kentucky's farm population was about 245,000 in 1980, one-fourth what it had been in 1950. Farm employment accounted for 66,647 jobs in 1983, or 3.9% of all employment. Preliminary data for 1983 show that Kentucky farms produced some 324,602,000 lb of tobacco (down from 589,350,000 in 1982). Leading field crops in 1983 (in bushels) included corn for grain, 46,080,000; soybeans, 22,935,000; wheat, 16,120,000; sorghum, 1,927,000; and barley, 825,000.

²⁴ANIMAL HUSBANDRY
Since early settlement days, livestock raising has been an important part of Kentucky's economy. The Bluegrass region, which offers excellent pasturage and drinking water, has become renowned as a center for horse breeding and racing.

In 1985 there were 223,000 horses in Kentucky, including Thoroughbreds, quarter horses, American saddle horses, Arabians, and standardbreds.

Cattle production was Kentucky's most profitable source of agricultural income in 1983. In 1984, the production of 825,040,000 lb of cattle and calves yielded $395,931,000 in gross income, and the sale of 327,114,000 lb of hogs and pigs amounted to $160,964,000. In 1983, Kentucky sold 2.4 billion lb of whole milk.

²⁵FISHING
Fishing is of little commercial importance in Kentucky.

²⁶FORESTRY
In 1979 there were 12,161,000 acres (4,921,000 hectares) of forested land in Kentucky—48% of the state's land area and 1.7% of the total forested area in the US.

The most heavily forested areas are in the river valleys of eastern Kentucky, in the Appalachians. In 1983, Kentucky produced 296 million board feet of lumber, nearly all of it in hardwoods. The Division of Forestry of the Department of Natural Resources manages approximately 36,000 acres (15,570 hectares) of state-owned forestland and operates three forest tree nurseries producing 10-12 million seedling trees a year.

There are two national forests—the Daniel Boone and the Jefferson—enclosing two national wilderness areas. National parks in the state include the Mammoth Cave National Park and the Cumberland Gap National Historical Park (also in Tennessee and Virginia).

[27] MINING

In 1982, Kentucky ranked 9th in the US in total mineral production, with output valued at about $5 billion. Fuels accounted for $4.8 billion, nonfuels for $207 million. The state was 1st in the production of coal, the chief mineral produced.

Kentucky's production of nonfuel minerals in 1983 included 36,500,000 tons of crushed stone, valued at $135,800,000; 6,000,000 tons of sand and gravel, worth $13,000,000; and 597,000 tons of clay, worth $2,236,000.

[28] ENERGY AND POWER

At the end of 1983, Kentucky had 32 electric generating plants. Total installed capacity was 16 million kw in 1983, when 57.7 billion kwh of power were produced. Southern Kentucky shares in the power produced by the Tennessee Valley Authority, which supports a coal-fired steam electric plant in Kentucky at Paducah.

Most of Kentucky's coal came from the western fields of the interior coal basin until late in the 19th century, when the lower-sulfur Cumberland Plateau coal reserves of the Appalachian region were discovered. In 1983, eastern Kentucky produced 93,199,000 tons of coal, western Kentucky 35,494,000. Kentucky has more underground mines than any other state. All coal mined is bituminous. Much of the mining in Kentucky is done by out-of-state companies; a number of oil companies have acquired coal companies as a hedge against declining petroleum resources. Bituminous coal reserves as of 1983 were estimated at 21 billion tons in western Kentucky and 19 billion tons in eastern Kentucky, or 16% of the nation's total reserves.

In 1983, Kentucky produced 7,886,000 barrels of crude petroleum and was estimated to have about 35,000,000 barrels of proved oil reserves. The oil industry was centered in Henderson County. In 1983, Kentucky produced 46.7 billion cu feet of natural gas. As of 31 December 1983, the state was estimated to have proved reserves totaling 554 billion cu feet of natural gas.

Oil shale is found in a band stretching from Lawrence County in the northeast through Madison and Washington counties in central Kentucky to Jefferson County in the north-central region.

[29] INDUSTRY

Although primarily an agricultural state during the 19th century, Kentucky was a leading supplier of manufactures to the South before the Civil War. Kentucky ranked 23d among the 50 states with shipments of manufactured goods valued at $30.4 billion in 1981; the total fell to $29.6 billion in 1982. Manufacturing activities are largely concentrated in Louisville and Jefferson County and other cities bordering the Ohio River. In 1983, Kentucky was the leading producer of American whiskey. It also produced 14% of the nation's trucks in assembly plants at Louisville (290,209 units) and Bowling Green (43,298 units) as well as 16,585 automobiles at Bowling Green. Food and food products accounted for $3.7 billion of the value of shipments in 1982; chemicals and allied products, $3.1 billion; nonelectrical machinery, $3 billion; transportation equipment, $3 billion; primary metals, $2.2 billion; electric and electronic equipment, $2.2 billion; and tobacco products, $2.2 billion.

The following table shows value of shipments in 1982 for selected major industries:

Motor vehicles and equipment	$2,926,500,000
Fabricated metals	1,705,900,000
Plastic materials, synthetics	1,217,200,000
Nonferrous metal products	826,300,000
Distilled liquor, except brandy	811,000,000
Refrigeration and service machinery	716,200,000
Blast furnace and steel mill products	681,900,000
Industrial organic chemicals	676,200,000

[30] COMMERCE

In 1982, Kentucky had 5,278 wholesale establishments, with total sales of $19.2 billion. Machinery sales accounted for $2.4 billion, farm product raw materials $2.4 billion, metals and minerals (except petroleum) for $1.8 billion, electrical goods for $582 million, and chemicals and allied products for $419 million. Retail sales totaled $14.6 billion. The KFC Corp., which owns and franchises Kentucky Fried Chicken restaurants, has its headquarters in Louisville. Kentucky's exports to foreign countries in 1981 included $1.9 billion in manufactures, $505 million in coal, $239 million in unmanufactured tobacco, and $202 million in soybeans.

[31] CONSUMER PROTECTION

The Consumer Protection Division of the Department of Law, created in 1972, handled more than 11,000 written complaints from consumers in 1981 and 1982. The Consumer Intervention Branch of the division works to secure savings for consumers in utility cases.

[32] BANKING

Kentucky had 338 commercial banks, including 336 insured commercial banks, as of 31 December 1984. Total assets stood at $25.3 billion, of which only 12.8%, far lower than the national average of 56%, was held by multibank holding companies. Loans amounted to $12.6 billion; insured deposits totaled $20.5 billion, of which demand deposits were $4.7 billion and time deposits $15.8 billion.

As of 31 December 1984 there were 70 insured savings and loan associations in Kentucky, 338 savings banks, and 228 credit unions.

[33] INSURANCE

In 1983, Kentuckians held some 6.1 million life insurance policies, with a total value of $59.8 billion. The average amount of life insurance per family was $43,100. In 1983 Kentuckians paid $407 million in life insurance premiums and received life insurance benefits of $578.6 million.

Premiums written by property and liability insurance companies in 1983 totaled $1.3 billion.

[34] SECURITIES

There are no securities exchanges in Kentucky. However, New York Stock Exchange member firms had 51 sales offices and 438 registered representatives in the state. A total of 392,000 Kentuckians owned shares of public corporations in 1983.

[35] PUBLIC FINANCE

The Kentucky biennial state budget is prepared by the Executive Department for Finance and Administration in the fall of each odd-numbered year and submitted by the governor to the general assembly for approval. The fiscal year runs from 1 July to 30 June. Following is a summary of revenues (available funds) and appropriations for 1984/85 and the governor's recommended budget for 1985/86:

AVAILABLE FUNDS	1984/85	1985/86
General fund	$2,493,090,900	$2,649,857,600
Federal funds	1,729,659,700	1,794,160,700
Agency funds	935,920,500	947,503,300
Road fund	468,240,000	481,000,000
Other funds	219,738,137	53,178,563
TOTALS	$5,846,649,237	$5,925,700,163

APPROPRIATIONS		
Human resources	$1,534,780,600	$1,629,443,500
Education and humanities	1,415,186,300	1,491,441,500
Public higher education	889,039,200	926,222,500
Transportation	834,104,300	754,899,100
General government	234,790,700	244,282,300
Corrections	84,843,500	90,719,800
Other appropriations	753,667,037	690,773,863
TOTALS	$5,746,411,637	$5,827,782,563

For fiscal 1981/82, the total state debt was more than $8.3 billion. Per capita debt was $2,274, 15th in the US.

36 TAXATION

Kentucky collected more than $2.6 billion in state taxes in 1983. As of 1984, the tax rate on personal income ranged from 2% on the first $3,000 to 6% on the amount over $8,000, with collections for 1983 totaling $647 million. Corporate income was taxed at a rate of 3% to 6% for a total 1983 yield of $172 million. Kentucky imposes severance taxes on oil and coal, and also levies a 5% sales and use tax (excluding food and drugs), a gasoline tax, an inheritance tax, taxes on motor vehicles, and excise taxes on alcoholic beverages and cigarettes.

In 1982, 1,308,000 Kentuckians filed federal income tax returns and paid a total tax of $3.2 billion. Kentucky's total share of the federal tax burden in 1982 was $7.2 billion, 27th largest in the nation.

37 ECONOMIC POLICY

The Commerce Cabinet seeks to encourage businesses to locate in Kentucky through its job creation program, which provides assistance in site selection as well as other services to US and Canadian firms. The Commerce Cabinet's Office of International Marketing seeks to attract foreign industrial investment and to increase exports of Kentucky manufactures. Since the early 1960s, Kentucky has used federal and state antipoverty and development funds to help alleviate the economic inequality between the 49 counties of the Appalachian region and the rest of the state. In addition, the Kentucky Enterprise Zone Authority and the Commerce Cabinet's Division of Community Development offer incentives to qualified businesses and communities under federal Enterprise Zone legislation. The communities of Louisville, Hickman, Ashland, and Covington were zone designees during 1983/84.

38 HEALTH

In 1982, Kentucky's birthrate was 15.4 per 1,000 inhabitants, slightly below the national average of 15.9. The state's death rate was 9.1 per 1,000 inhabitants, higher than the national average of 8.6. The infant mortality rate fell from 28.6 per 1,000 live births for nonwhites in 1970 to 17.1 in 1981; the rate for whites declined from 18.8 per 1,000 live births to 11.8. In 1982, 10,800 legal abortions were performed in Kentucky, for a rate of 189 per 1,000 live births, well below the US average of 426. In 1981, Kentucky ranked higher than the national averages in death rates from heart diseases, 349.7 per 100,000 residents; cancer, 189.3; cerebrovascular diseases, 82.3; and pneumonia, 27.4. There were 27,742 nonfatal traffic accidents in 1983 and 778 fatalities. Black lung (pneumoconiosis) has been recognized as a serious work-related illness among coal miners.

In 1983, Kentucky's 120 hospitals had 19,085 beds and recorded 695,391 admissions. At the end of 1982 there were 5,673 active physicians in Kentucky and 1,741 active dentists.

Medicare and Medicaid benefits for 1981 totaled $868 million. The average hospital cost per patient per day was $261.20, the 8th lowest in the nation and well below the national average of $327.40.

39 SOCIAL WELFARE

In 1983, 60,100 families, including 158,700 members, received aid to families with dependent children. Payments averaged $184 a month in 1982, for a total payment to Kentuckians of $122 million. About 593,000 Kentuckians participated in the federal food stamp program in 1983; the federal subsidy was $331 million. An estimated 508,000 Kentucky children participated in the school lunch program in 1983, at a federal cost of $50 million.

In 1983, 369,000 retired workers in Kentucky received $1,489 million in federal Social Security benefits; their average monthly payment was $397. Under the Black Lung Benefit Program, Kentucky ranked 3d in the nation, with payments of $123 million in 1984 to Kentucky miners totally disabled by black lung. Federal Supplemental Security Income payments for 34,300 aged and 55,300 disabled in 1983 totaled $195.1 million.

There is no employee payroll deduction for unemployment insurance. All unemployment benefits, totaling $154.5 million for 1983/84, were derived from the tax on employer payrolls collected by the state for the Federal Unemployment Insurance Fund. In 1982/83, 59,072 persons received weekly insurance benefits for an average unemployment period of 17.8 weeks.

40 HOUSING

According to the 1980 census, Kentucky had 1,353,656 year-round housing units. One-fourth of all year-round housing units were built before 1939, 28.7% between 1970 and 1980. A total of 724,300 new units with a combined value of $940 million were built between 1980 and 1983. Just under 10% of all housing units lacked full bathroom facilities, well above the national average of 3.3%.

41 EDUCATION

Kentucky was relatively slow to establish and support its public education system and has consistently ranked below the national average in per capita spending on education and in the educational attainments of its citizens. The 1980 census reported that only 53.1% of all adults, the lowest proportion in the nation, had completed four years of high school, far below the national average of 66.5%; only 11.1% had completed four or more years of college, placing Kentucky 49th among the states and well below the national average of 16.2%. Per capita spending on education by state and local governments totaled only $519.95 in 1981/82, ranking Kentucky 49th among the states (ahead of only Tennessee) and well below the national average of $682.30.

In 1982/83, 601,000 students attended public schools in Kentucky. The estimated average salary of a public school teacher in 1983/84 was $19,780, ranking Kentucky 31st among the states and below the national average of $22,029. Catholic private schools enrolled 34,826 elementary students and 12,608 high school students.

During 1982/83 there were 57 institutions of higher education in Kentucky, of which 21 were public and 36 private; their total enrollment was 88,452. The University of Kentucky, established in 1865 at Lexington, is the state's largest public institution, with an enrollment on the Lexington campus of 17,137 in 1982/83. The University of Kentucky Community College System enrolled an additional 12,990 in 13 colleges in 1982/83. The state-supported University of Louisville (1798) had an enrollment of 20,000. Loans and grants to Kentucky students are provided by the Kentucky Higher Education Assistance Authority.

The School Foundation Program was established by the general assembly in 1954 for the purpose of upgrading the level of basic education in the state. A total of $906,770,700 was earmarked in the 1985/86 budget to implement the program. In the same year, Kentucky expected to receive $55,868,600 in federal funds for a compensatory education program for neglected, delinquent, handicapped, and migratory youngsters.

42 ARTS

The Actors Theater of Louisville holds a yearly festival of new American plays. The city also has a resident ballet company. The Louisville Orchestra has recorded numerous works by contemporary composers. The Kentucky Arts Council, a division of the Kentucky Department of the Arts within the Commerce Cabinet, is authorized to promote the arts through such programs as Arts in Education and the State Arts Resources Program.

Bluegrass, a form of country music marked by fiddle and banjo and played at a rapid tempo, is named after the style pioneered by Kentuckian Bill Monroe and his Blue Grass Boys.

43 LIBRARIES AND MUSEUMS

In 1984 there were 116 public libraries in Kentucky, with a total of 5,963,888 volumes, including those in bookmobiles. The regional library system of 15 districts included university libraries and the state library at Frankfort, as well as city and county libraries.

The state has more than 70 museums. Art museums include the University of Kentucky Art Museum and the Headley-Whitney Museum in Lexington, the Allen R. Hite Art Institute at the University of Louisville, and the J. B. Speed Art Museum, also in Louisville. Among Kentucky's equine museums are the International Museum of the Horse and the American Horse Museum, both in Lexington, and the Kentucky Derby Museum in Louisville. The John James Audubon Museum is located in Audubon State Park at Henderson.

Leading historical sites include Abraham Lincoln's birthplace at Hodgenville and the Mary Todd Lincoln and Henry Clay homes in Lexington. The Kentucky Historical Society in Frankfort supports a mobile museum system that brings exhibits on Kentucky history to schools, parks, and local gatherings, and aids 353 local historical societies.

[44]COMMUNICATIONS

There were 7,537 postal employees in Kentucky in 1985. Only 87% of all occupied housing units in the state had telephone in 1980.

In 1922, Kentucky's first radio broadcasting station, WHAS, was established. By 1984 there were 248 radio stations, 137 AM and 111 FM. That year there were 11 commercial television broadcasting stations. The Kentucky Educational Television system had 15 transmitters and its programs were viewed in 250,000 households each week in 1984. As of 1984 there were 202 cable television systems serving 579,606 subscribers in 540 communities.

[45]PRESS

In 1985, Kentucky had 25 daily newspapers (5 morning, 20 evening), with a combined 1984 paid circulation of 733,563, and 12 Sunday papers, with a paid circulation of 637,823. The following table shows the leading Kentucky newspapers with their 1984 circulations:

AREA	NAME	DAILY	SUNDAY
Frankfort	State Journal (e,S)	9,962	10,706
Lexington	Herald-Leader (m)	110,194	130,216
Louisville	Courier-Journal (m,S)	174,399	327,875
	Times (e)	134,375	

[46]ORGANIZATIONS

The 1982 US Census of Service Industries counted 710 organizations in Kentucky, including 157 business associations; 363 civic, social, and fraternal associations; and 16 educational, scientific, and research associations. Notable organizations with headquarters in Kentucky include the Thoroughbred Club of America and the Burley Tobacco Growers Cooperative Association (both in Lexington); the Burley Auction Warehouse Association (Mt. Sterling); and the Association of Dark Leaf Tobacco Dealers and Exporters and the American Saddlebred Horse Association (both in Louisville).

[47]TOURISM, TRAVEL, AND RECREATION

Kentucky's income from tourism exceeded $2 billion in 1982. One of the state's top tourist attractions is Mammoth Cave National Park, which contains an estimated 150 mi (241 km) of underground passages. Other units of the national park system in Kentucky include a re-creation of Abraham Lincoln's birthplace in Hodgenville and Cumberland Gap National Historical Park, which extends into Tennessee and Virginia.

As of 1985, the state operated 15 resort parks (13 of them year-round), which were expected to attract 22 million visitors during the 1985/86 budget period. Another 19 state-operated recreational parks and 9 shrines were expected to attract 8 million visitors and 330,000 campers. Breaks Interstate Park, on the Kentucky-Virginia border, is noted for the Russell Fork River Canyon, which is 1,600 feet (488 meters) deep; the park is supported equally by the two states.

In 1979, the Kentucky Horse Park opened in Lexington. The Kentucky State Fair is held every August at Louisville. In 1983, Kentucky issued licenses to 307,964 hunters and 637,522 fishermen.

[48]SPORTS

The first horse race in Kentucky was held in 1783. The annual Kentucky Derby, first run on 17 May 1875, has become probably the single most famous event in US Thoroughbred racing. Held on the 1st Saturday in May at Churchill Downs in Louisville, the Derby is one of the three races constituting the Triple Crown for three-year-olds. Keeneland Race Course in Lexington is the site of the Blue Grass Stakes and other major Thoroughbred races. The Kentucky Futurity, an annual highlight of the harness racing season, is usually held on the 1st Friday in October at the Red Mile in Lexington.

In 1983, 3,136 Thoroughbred races were held in Kentucky, attracting a total attendance of 2,260,011. Total purses of $24,110,592 went to the horsemen; the betting handle amounted to $263,668,250, of which the state's share was $9,536,378. Harness races attracted 713,319 spectators in 1984. Total purses were $8,212,516, and the betting handle was $55,103,492.

Rivaling horse racing as a spectator sport is collegiate basketball. The University of Kentucky Wildcats won NCAA Division I basketball championships in 1948-49, 1951, 1958, and 1978, and the National Invitation Tournament in 1946 and 1976. The University of Louisville Cardinals captured the NCAA crown in 1980, having won an NIT title in 1956. Kentucky Wesleyan, at Owensboro, was the NCAA Division II titleholder in 1966, 1968-69, and 1973.

[49]FAMOUS KENTUCKIANS

Kentucky has been the birthplace of one US president, four US vice presidents, the only president of the Confederacy, and several important jurists, statesmen, writers, artists, and sports figures.

Abraham Lincoln (1809-65), the 16th president of the US, was born in Hodgenville, Hardin (now Larue) County, and spent his developing years in Indiana and Illinois. Elected as the first Republican president in 1860 and reelected in 1864, Lincoln reflected his Kentucky roots in his opposition to secession and the expansion of slavery, and in his conciliatory attitude toward the defeated southern states. His wife, Mary Todd Lincoln (1818-82), was a native of Lexington.

Kentucky-born US vice presidents have all been Democrats. Richard M. Johnson (1780-1850) was elected by the Senate after a deadlock in the Electoral College; John C. Breckinridge (1821-75) in 1857 became the youngest man ever to hold the office; Adlai E. Stevenson (1835-1914) served in Grover Cleveland's second administration. The best known was Alben W. Barkley (1877-1956), who, before his election with President Harry S Truman in 1948, was a US senator and longtime Senate majority leader.

Frederick M. Vinson (1890-1953) was the only Kentuckian to serve as chief justice of the US. Noteworthy associate justices were John Marshall Harlan (1833-1911), famous for his dissent from the segregationist *Plessy* v. *Ferguson* decision (1896), and Louis D. Brandeis (1856-1941), the first Jew to serve on the Supreme Court and a champion of social reform.

Henry Clay (b.Virginia, 1777-1852) came to Lexington in 1797 and went on to serve as speaker of the US House of Representatives, secretary of state, and US senator; he was also a three-time presidential candidate. Other important federal officeholders from Kentucky include Attorneys General John Breckinridge (b.Virginia, 1760-1806) and John J. Crittenden (1787-1863), who also served with distinction as US senator; Treasury Secretaries Benjamin H. Bristow (1830-96) and John G. Carlisle (1835-1910); and US Senator John Sherman Cooper (b.1901).

Among noteworthy state officeholders, Isaac Shelby (b.Maryland, 1750-1826) was a leader in the movement for statehood and the first governor of Kentucky. William Goebel (1856-1900) was the only US governor assassinated in office. Albert B. ("Happy") Chandler (b.1898), twice governor, also served as US senator and as commissioner of baseball.

A figure prominently associated with frontier Kentucky is the explorer and surveyor Daniel Boone (b.Pennsylvania, 1734-1820). Other frontiersmen include Kit Carson (1809-1868) and Roy Bean (1825?-1903). During the Civil War, Lincoln's principal adversary was another native Kentuckian, Jefferson Davis (1808-89); he moved south as a boy to a Mississippi plantation home, subsequently serving as US senator from Mississippi, US secretary of war, and president of the Confederate States of America.

Other personalities of significance include James G. Birney (1792-1857) and Cassius Marcellus Clay (1810-1903), both major antislavery spokesmen. Clay's daughter Laura (1849-1941) and Madeline Breckinridge (1872-1920) were important contributors to the women's suffrage movement. Henry Watterson (1840-1921) founded and edited the *Louisville Courier-Journal* and was a major adviser to the Democratic Party. Carry Nation (1846-1911) was a leader of the temperance movement. During the 1920s, Kentuckian John T. Scopes (1900-70) gained fame as the defendant in the "monkey trial" in Dayton, Tenn.; Scopes was charged with teaching Darwin's theory of evolution. Whitney M. Young (1921-71), a prominent black leader, served as head of the National Urban League.

Thomas Hunt Morgan (1866-1945), honored for his work in heredity and genetics, is Kentucky's lone Nobel Prize winner. Journalists born in Kentucky include Irvin S. Cobb (1876-1944), who was also a humorist and playwright, and Arthur Krock (1887-1974), a winner of four Pulitzer Prizes. Notable businessmen include Harland Sanders (b.Indiana, 1890-1980), founder of Kentucky Fried Chicken restaurants.

Kentucky has produced several distinguished creative artists. These include painters Matthew Jouett (1787-1827), Frank Duveneck (1848-1919), and Paul Sawyier (1865-1917); folk song collector John Jacob Niles (1891-1980); and novelists Harriette Arnow (b.1908) and Wendell Berry (b.1934). Robert Penn Warren (b.1905), a novelist, poet, and critic, won the Pulitzer Prize three times and was the first author to win the award in both the fiction and poetry categories.

Among Kentuckians well recognized in the performing arts are film innovator D. W. Griffith (David Lewelyn Wark Griffith, 1875-1948), Academy Award-winning actress Patricia Neal (b.1926), and country music singer Loretta Lynn (b.1932). Kentucky's sports figures include basketball coach Adolph Rupp (b.Kansas, 1901-77), shortstop Harold ("Pee Wee") Reese (b.1919), football great Paul Hornung (b.1935), and world heavyweight boxing champions Jimmy Ellis (b.1940) and Muhammad Ali (Cassius Clay, b.1942).

[50]BIBLIOGRAPHY

Axton, W. F. *Tobacco and Kentucky.* Lexington: University Press of Kentucky, 1976.

Caudill, Harry M. *Night Comes to the Cumberlands: A Biography of a Depressed Area.* New York: Little, Brown, 1962.

Channing, Steven A. *Kentucky: A Bicentennial History.* New York: Norton, 1977.

Clark, Thomas D. *Kentucky: Land of Contrast.* New York: Harper & Row, 1968.

Coleman, J. Winston. *Slavery Times in Kentucky.* Chapel Hill: University of North Carolina Press, 1940.

Cotterill, Robert S. *History of Pioneer Kentucky.* Cincinnati: Johnson and Hardin, 1917.

Davis, Williams D. *Breckinridge: Statesman, Soldier, Symbol.* Baton Rouge: Louisiana State University Press, 1974.

Federal Writers' Project. Kentucky: *A Guide to the Bluegrass State.* Reprint. New York: Somerset, n.d. (orig. 1939).

Fuller, Paul E. *Laura Clay and the Woman's Rights Movement.* Lexington: University Press of Kentucky, 1975.

Goldstein, Joel. *Kentucky Government and Politics.* Bloomington, Ind.: Tichenor, 1984.

Hollingsworth, Kent. *The Kentucky Thoroughbred.* Lexington: University Press of Kentucky, 1976.

Kentucky Department of Economic Development. Division of Research and Planning. *1985 Kentucky Economic Statistics.* Frankfort, 1985.

Moore, Arthur K. *The Frontier Mind: A Cultural Analysis of the Kentucky Frontiersman.* Lexington: University Press of Kentucky, 1957.

Rennick, Robert M. *Kentucky Place Names.* Lexington: University Press of Kentucky, 1984.

Walls, David S., and John B. Stephenson. *Appalachia in the Sixties: Decade of Reawakening.* Lexington: University Press of Kentucky, 1973.

Wilson, Mary Helen (comp). *A Citizens' Guide to the Kentucky Constitution.* Frankfort: Legislative Research Commission, 1977.

LOUISIANA

State of Louisiana

ORIGIN OF STATE NAME: Named in 1682 for France's King Louis XIV. **NICKNAME:** The Pelican State. **CAPITAL:** Baton Rouge. **ENTERED UNION:** 30 April 1812 (18th). **SONGS:** "Give Me Louisiana"; "You Are My Sunshine." **MOTTO:** Union, Justice, and Confidence. **COLORS:** Gold, white, and blue. **FLAG:** On a blue field, fringed on three sides, a white pelican feeds her three young, symbolizing the state providing for its citizens; the state motto is inscribed on a white ribbon. **OFFICIAL SEAL:** In the center, pelican and young are as depicted on the flag; the state motto encircles the scene, and the words "State of Louisiana" surround the whole. **BIRD:** Eastern brown pelican. **FLOWER:** Magnolia. **TREE:** Bald cypress. **GEM:** Agate. **FOSSIL:** Petrified palmwood. **INSECT:** Honeybee. **LEGAL HOLIDAYS:** New Year's Day, 1 January; Battle of New Orleans Day, 8 January; Birthday of Martin Luther King Jr., 3d Monday in January; Robert E. Lee's Birthday, 19 January; Washington's Birthday, 3d Monday in February; Good Friday, March or April; National Memorial Day, last Monday in May; Confederate Memorial Day and Jefferson Davis's Birthday, 3 June; Independence Day, 4 July; Huey Long's Birthday, 30 August, by proclamation of the governor; Labor Day, 1st Monday in September; Columbus Day, 2d Monday in October; All Saints' Day, 1 November; Veterans Day, 4th Monday in October; Thanksgiving Day, 4th Thursday in November; Christmas Day, 25 December. Legal holidays in New Orleans, Jefferson, St. Bernard, St. Charles, and East Baton Rouge parishes also include Mardi Gras, February or March. **TIME:** 6 AM CST = noon GMT.

¹LOCATION, SIZE, AND EXTENT

Situated in the western south-central US, Louisiana ranks 31st in size among the 50 states.

The total area of Louisiana is 47,751 sq mi (123,675 sq km), including 44,521 sq mi (115,309 sq km) of land and 3,230 sq mi (8,366 sq km) of inland water. The state extends 237 mi (381 km) E–W; its maximum N–S extension is 236 mi (380 km). Louisiana is shaped roughly like a boot, with the heel in the SW corner and the toe at the extreme SE.

Louisiana is bordered on the N by Arkansas; on the E by Mississippi (with part of the line formed by the Mississippi River and part, in the extreme SE, by the Pearl River); on the S by the Gulf of Mexico; and on the W by Texas (with part of the line passing through the Sabine River and Toledo Bend Reservoir). The state's geographic center is in Avoyelles Parish, 3 mi (5 km) SE of Marksville.

The total boundary length of Louisiana is 1,486 mi (2,391 km). Louisiana's total tidal shoreline is 7,721 mi (12,426 km).

²TOPOGRAPHY

Louisiana lies wholly within the Gulf Coastal Plain. Alluvial lands, chiefly of the Red and Mississippi rivers, occupy the north-central third of the state. East and west of this alluvial plain are the upland districts, characterized by rolling hills sloping gently toward the coast. The coastal-delta section, in the southernmost portion of the state, consists of the Mississippi Delta and the coastal lowlands. The highest elevation in the state is Driskill Mountain at 535 feet (163 meters), in Bienville Parish; the lowest, 5 feet (2 meters) below sea level, in New Orleans.

Louisiana has the most wetlands of all the states, about 11,000 sq mi (28,000 sq km) of floodplains and 7,800 sq mi (20,200 sq km) of coastal swamps, marshes, and estuarine waters. The largest lake, actually a coastal lagoon, is Lake Pontchartrain, with an area of more than 620 sq mi (1,600 sq km). Toledo Bend Reservoir, an artificial lake along the Louisiana-Texas border, has an area of 284 sq mi (736 sq km). The most important rivers are the Mississippi, Red, Pearl, Atchafalaya, and Sabine. Most drainage takes place through swamps between the bayous, which serve as outlets for overflowing rivers and streams.

Louisiana has nearly 2,500 coastal islands covering some 2,000 sq mi (5,000 sq km).

³CLIMATE

Louisiana has a relatively constant semitropical climate. Rainfall and humidity decrease, and daily temperature variations increase, with distance from the Gulf of Mexico. The normal daily temperature in New Orleans is 68°F (20°C), ranging from 52°F (11°C) in January to 82°F (28°C) in July. The all-time high temperature is 114°F (46°C), recorded at Plain Dealing on 10 August 1936; the all-time low, −16°F (−27°C), was set at Minden on 13 February 1899. New Orleans has sunshine 59% of the time, and the average annual rainfall is 60 in (152 cm). Snow falls occasionally in the north, but rarely in the south.

Prevailing winds are from the south or southeast. During the summer and fall, tropical storms and hurricanes frequently batter the state, especially along the coast. Among the most severe hurricanes in recent decades were Audrey, which entered Cameron Parish on 28 June 1957, causing 400–500 deaths and property damage of $150 million, and Betsy, which entered the coast near Grand Isle on 9 September 1965, causing 58 deaths and damages of $1.2 billion. The greatest number of tornadoes reported in a single year was 55, in 1974.

⁴FLORA AND FAUNA

Forests in Louisiana consist of four major types: shortleaf pine uplands, slash and longleaf pine flats and hills, hardwood forests in alluvial basins, and cypress and tupelo swamps. Important commercial trees also include beech, eastern red cedar, and black walnut. Among the state's wild flowers are the ground orchid and several hyacinths. Spanish moss (actually a member of the pineapple family) grows profusely in the southern regions but is rare in the north. Two types of orchid are threatened, and the *Schwalbea americana* is endangered.

Louisiana's varied habitats—tidal marshes, swamps, woodlands, and prairies—offer a diversity of fauna. Deer, squirrel, rabbit, and bear are hunted as game, while muskrat, nutria, mink, opossum, bobcat, and skunk are commercially significant furbearers. Prized game birds include quail, turkey, woodcock, and various waterfowl, of which the mottled duck and wood duck are native. Coastal beaches are inhabited by sea turtles, and whales may be seen offshore. Freshwater fish include bass, crappie, and bream; red and white crayfishes are the leading commercial crustaceans. Threatened animal species include both the green and loggerhead sea turtles. The American alligator is on the federal endangered species list. Other endangered animals are the sei and sperm whales, Florida panther, gray wolf, eastern brown pelican (the state bird), whooping crane, and red-cockaded woodpecker.

⁵ENVIRONMENTAL PROTECTION

Louisiana's earliest and most pressing environmental problem was the chronic danger of flooding by the Mississippi River. In April and May 1927, the worst flood in the state's history inundated more than 1,300,000 acres (526,000 hectares) of agricultural land, left 300,000 people homeless, and would have swept away much of New Orleans had levees below the city not been dynamited. The following year, the US Congress funded construction of a system of floodways and spillways to divert water from the Mississippi when necessary. These flood control measures, and dredging for oil and gas exploration, created another environmental problem—the slowing of the natural flow of silt into the wetlands. As a result, salt water from the Gulf of Mexico has seeped into the wetlands.

Legislation enacted in 1979 consolidated much of the state's environmental protection effort in the Office of Environmental Affairs (OEA), within the Department of Natural Resources. Among its responsibilities are maintenance of air and water quality, solid-waste management, hazardous waste disposal, and control of radioactive materials. According to an OEA survey, hazardous waste disposal, past and future, and the protection of the wetlands head the list of state residents' environmental concerns. Louisiana's problem in protecting its wetlands differs from that of most other states in that its wetlands are more than wildlife refuges—they are central to the state's agriculture and fishing industries. Assessment of the environmental impact of various industries on the wetlands has been conducted under the Coastal Zone Management Plan of the Department of Transportation and Development.

The two largest wildlife refuges in the state are the Rockefeller Wildlife Refuge, comprising 84,000 acres (34,000 hectares) in Cameron and Vermilion parishes, and the Marsh Island Refuge, 82,000 acres (33,000 hectares) of marshland in Iberia Parish. Both are managed by the Department of Wildlife and Fisheries.

Because of the sizable chemical industry in Louisiana, hazardous waste disposal is a major problem. In 1982, a train carrying hazardous chemicals derailed and burned at Livingston, east of Baton Rouge, forcing the evacuation of the entire town. Lake Pontchartrain, near New Orleans, is so polluted by chemicals and sewage that people have been told not to swim there.

Among the most active citizens' groups on environmental issues are the League of Women Voters, the Sierra Club (Delta Chapter), and the Ecology Center of New Orleans.

⁶POPULATION

At the time of the 1970 census, Louisiana ranked 20th among the 50 states, with a population of 3,641,306. The population at the time of the 1980 census was 4,203,972, representing an increase of more than 15% since 1970. By early 1985, the population was estimated at 4,553,903, representing an 8% gain. The US Census Bureau estimates that Louisiana's population will be 4,747,000 in 1990.

As of 1980, 51% of Louisianians were female and 49% male. Louisianians are somewhat younger than the national average, but among the least mobile of Americans: in 1980, 78% of all state residents had been born in the state.

Louisiana's estimated population density in 1985 was 102 persons per sq mi (39 per sq km).

About 69% of Louisianians lived in metropolitan areas in 1983.

New Orleans is the largest city, with an estimated 1984 population of 559,101, followed by Baton Rouge, 368,571; Shreveport, 219,996; Lafayette, 82,608; and Lake Charles, 75,080. Spurred by an expansion of state government employment, Baton Rouge, the capital, has grown with exceptional speed since 1940, when its population was 34,719. Among the state's largest metropolitan areas in 1984 were New Orleans, 1,318,800 (27th in the US); Baton Rouge, 538,000 (66th); and Shreveport, 360,900 (95th).

⁷ETHNIC GROUPS

Louisiana, most notably the Delta region, is an enclave of ethnic heterogeneity in the South. At the end of World War II, the established population of the Delta, according to descent, included blacks, French, Spanish (among them Central and South Americans and Islenos, Spanish-speaking migrants from the Canary Islands), Filipinos, Italians, Chinese, American Indians, and numerous other groups.

Blacks made up about 29% of the population in 1980. They include descendants of "free people of color," some of whom were craftsmen and rural property owners before the Civil War (a few were slaveholding plantation owners); many of these, of mixed blood, are referred to locally as "colored Creoles," and have constituted a black elite in both urban and rural Louisiana. The black population of New Orleans constituted 55% of its residents in 1980; the city elected its first black mayor, Ernest N. "Dutch" Morial, in 1977.

Two groups that have been highly identified with the culture of Louisiana are Creoles and Acadians (also called Cajuns). Both descend primarily from early French immigrants to the state, but the Cajuns trace their origins from the mainly rural people exiled from Acadia (Nova Scotia) in the 1740s, while the Creoles tend to be city people from France and, to a lesser extent, from Nova Scotia or Hispaniola (the term "Creole" also applies to the relatively few early Spanish settlers and their descendants). Although Acadians have intermingled with Spaniards and Germans, they still speak a French patois and retain a distinctive culture and cuisine.

At the time of the 1980 census, 85,000 Louisianians (2% of the population) were foreign-born. France, the United Kingdom, Ireland, Germany, and Italy provided Louisiana with the largest ancestry groups. As of 1980 there were 11,950 American Indians in Louisiana, along with 23,771 Pacific Islanders and Asians, including 10,877 Vietnamese.

⁸LANGUAGES

White settlers in Louisiana found several Indian tribes of the Caddoan confederacy, from at least five different language groups. In 1980, about 865 Louisiana residents spoke an American Indian language. Place-names from this heritage include Coushatta, Natchitoches, and Ouachita.

Louisiana English is predominantly Southern. Notable features of the state's speech patterns are *pen* and *pin* as sound-alikes and, in New Orleans, the so-called Brooklyn pronunciation of *bird* as /boyd/. A *pave* is a paved highway, and a pecan patty is well known as *praline*.

In 1980, 3,592,777 Louisiana residents—90% of the population 3 years old and older—spoke only English at home. Other languages spoken at home included:

French	266,033	German	7,852
Spanish	52,152	Italian	7,782
Vietnamese	9,559	Chinese	2,722

Unique to Louisiana is a large enclave, west of New Orleans, where a variety of French called Acadian (Cajun) is the first language. From it, and from early colonial French, English has

LOCATION: 29° to 33°N; 89° to 94°W. **BOUNDARIES:** Arkansas line, 166 mi (267 km); Mississippi line, 596 mi (959 km); Gulf of Mexico coastline, 397 mi (639 km); Texas line, 327 mi (526 km).

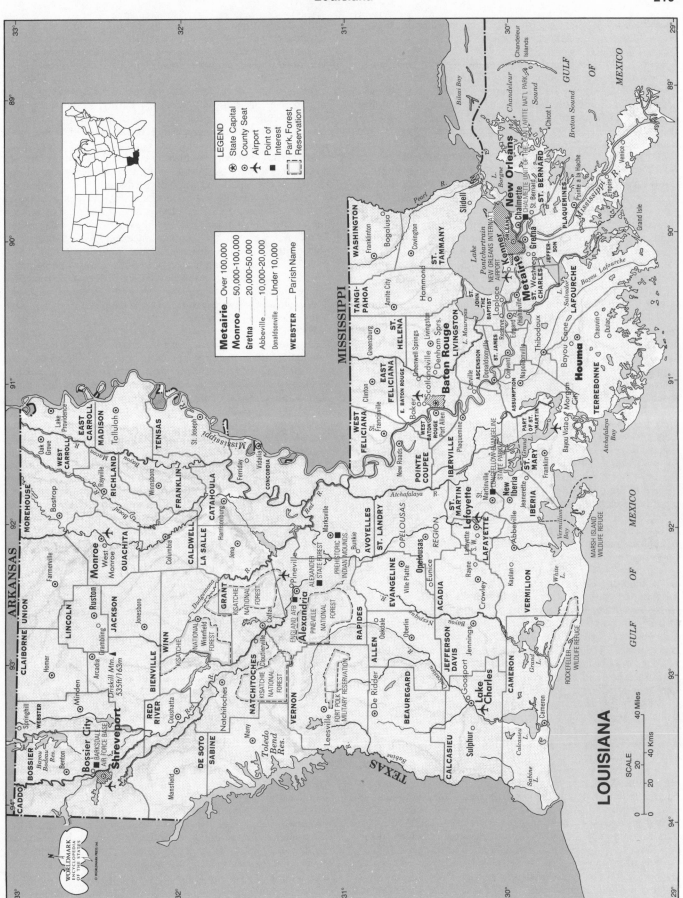

LOUISIANA

LEGEND
✪ State Capital
⊙ County Seat
✈ Airport
■ Point of Interest
▭ Park, Forest, Reservation

Metairie ...... Over 100,000
Monroe 50,000-100,000
Gretna 20,000-50,000
Abbeville 10,000-20,000
Donaldsonville ...... Under 10,000
WEBSTER ...... Parish Name

SCALE
0   20   40 Miles
0   20   40 Kms

See endsheet maps: H4.

WORLDMARK
ENCYCLOPEDIA
OF THE STATES
© WORLDMARK PRESS Ltd.

taken such words as *pirogue* (dugout canoe), *armoire* (wardrobe), *boudin* (blood sausage), and *lagniappe* (extra gift).

9RELIGIONS

Spanish missionaries brought Roman Catholicism to Louisiana in the early 16th century, and many of them were killed in their attempts to convert the Indians. During the early days, the most active religious orders were the Jesuits, Capuchins, and Ursuline nuns. Until the Louisiana Purchase, the public practice of any but the Catholic religion was prohibited, and Jews were entirely banned.

Joseph Willis, a mulatto preacher who conducted prayer meetings at what is now Lafayette in 1804, organized the first Baptist church west of the Mississippi, at Bayou Chicot in 1812. In the Opelousas region, in 1806, the first Methodist church in the state was organized. The first Episcopal church was established in New Orleans in 1805, a Methodist church in 1813, a Presbyterian church in 1817, a synagogue in 1828, and a Baptist church in 1834. After the Civil War, blacks withdrew from white-dominated churches to form their own religious groups, mainly Baptist and Methodist.

As of 1 January 1984, the Roman Catholic Church was the largest Christian denomination, with 1,334,953 church members. The leading Protestant denominations in 1980 were Southern Baptist, 652,246; United Methodist, 169,328; Episcopal, 38,554; and Presbyterian, 31,717. In 1984, 17,340 Jews resided in Louisiana, about 70% of them in New Orleans. Voodoo, in some cases blended with Christian ritual, is more widespread in Louisiana than anywhere else in the US, although the present number of practitioners is impossible to ascertain.

10TRANSPORTATION

New Orleans is a major center of domestic and international freight traffic. In volume of domestic and foreign cargo handled, it was the busiest port on the Gulf of Mexico and 2d-leading port in the US in 1983. Although Louisiana's roads remained poor until the 1930s, the state was one of the nation's major rail centers by the end of the 19th century, and New Orleans was one of the first cities to develop a mass transit system.

Several short-run railroads were built in Louisiana during the 1830s. The first of these, and the first rail line west of the Alleghenies, was the Pontchartrain Railroad, which opened, using horse-drawn vehicles, on 23 April 1831. New Orleans was connected with New York before the Civil War, with Chicago by 1873, and with California in 1883 via a line that subsequently became part of the Southern Pacific. Railroads soon rivaled the Mississippi River in the movement of goods to and from New Orleans, and even today, the long-distance freight service provided by New Orleans' six trunk lines is a major reason for the preeminence of the port. There were eight Class I line-haul railroads in Louisiana as of December 1983, with 3,127 mi (5,032 km) of track. As of 1984, Amtrak provided passenger links with Los Angeles, Chicago, and New York.

The New Orleans and Carrollton Railroads, a horse-drawn trolley system, began service in 1835; 59 years later, electric trolleys came into use. As of 1985, the Regional Transit Authority managed New Orleans' 548 buses and 35 trolleys, which have some 73 million riders per year.

Louisiana's first road-building boom began after Huey Long entered the statehouse. When Long took office in 1928, the state had no more than 300 mi (480 km) of paved roads; by 1931 there were 1,583 (2,548). At the end of 1983, Louisiana had a total of 58,010 mi (93,358 km) of roads, 79% of them rural. In 1983, 2,023,957 automobiles and 833,200 trucks were registered in the state, and 2,767,356 drivers' licenses were in force.

Early in the nation's history, the Mississippi River emerged as the principal route for north–south traffic, and New Orleans soon became the South's main port. The advent of the steamboat in 1812 solved the problem of upstream navigation, which previously

had required three or four months for a distance that could be covered downstream in 15 days. (Barges moved by towboats eventually supplanted steamboats as cargo carriers.) An important breakthrough in international transportation was the deepening of the channel at the mouth of the Mississippi by means of jetties, the first of which were completed in 1879. As of 1984, the port of New Orleans had 15 mi (24 km) of public and private wharves, and its public docks had space for nearly 100 vessels. The port is served by more than 100 steamship lines and by 20 common carrier and about 100 contract carrier barge lines. During 1983, the port handled 37.6 million tons of cargo. The Louisiana Offshore Oil Port (LOOP), the first deepwater oil port in the US, was opened in 1981. Located south of New Orleans in the Gulf of Mexico, the supertanker facility has a designed capacity of 1,400,000 barrels of oil a day, but because of decreasing demand for imported oil, LOOP was handling only 275,000 barrels a day in early 1983. Baton Rouge and Lake Charles are the state's other important deepwater ports.

As of 1984, Louisiana had a total of 290 private and 99 public airfields. The busiest was New Orleans International Airport, which enplaned 2,866,966 passengers and 5,927 tons of freight. Louisiana also claims the world's largest heliport, in Morgan City, with 46 landing pads in 1984.

11HISTORY

The region now known as Louisiana is largely the creation of the Mississippi River; the process of land building still goes on in the Atchafalaya Basin and below New Orleans on the Mississippi Delta. Louisiana was never densely inhabited in prehistoric times, and at no time, probably, did as many as 15,000 Indians live inside the present boundaries of the state. The main relic of prehistoric inhabitants is the great earthwork at Poverty Point, near Marksville, but other Indian mounds are to be found in alluvial and coastal regions.

When white exploration and settlement of North America began, various tribes of Caddo Indians inhabited northwestern Louisiana, and small Tunican-speaking groups lived in the northeast. In the southwest were a number of rather primitive peoples of the Atakapa group; in south-central Louisiana, the Chitimacha ranged through the marshes and lowlands. Various small Muskogean tribes, related to the Choctaw, lived east of the Mississippi in the "Florida parishes," so called because they were once part of Spanish West Florida. The Natchez Indians, whose main villages were in present-day Mississippi near the city that still bears their name, fought with the French settlers in Louisiana's early history but were exterminated in the process.

Several Spanish explorers sailed along the coast of Louisiana, but Hernando de Soto was probably the first to penetrate the state's present boundaries, in 1541. Almost a century and a half passed before Robert Cavelier, Sieur de la Salle, departing from Canada, reached the mouth of the Mississippi on 9 April 1682, named the land there Louisiana in honor of King Louis XIV, and claimed it for France. La Salle's later attempt at a permanent settlement failed, but in 1699 an expedition headed by Pierre le Moyne, Sieur d'Iberville, made a settlement on Biloxi Bay. In 1714, Louis Juchereau de St. Denis established Natchitoches, the first permanent European settlement in Louisiana; Iberville's brother, the Sieur de Bienville, established New Orleans four years later.

Louisiana did not thrive economically under French rule, whether as a royal colony or, from 1712 to 1731, under the proprietorship first of Antoine Crozat and then of John Law's Company of the Indies. On the other hand, French culture was firmly implanted there, and non-French settlers, especially Germans from Switzerland and the Rhineland, were quickly Gallicized. In 1762, on the verge of losing the rest of its North American empire to Great Britain in the French and Indian War, France ceded Louisiana to Spain. Governed by Spaniards, the colony was much more prosperous, although it was a burden on the Spanish treasury. New settlers—Americans, Spaniards, Canary Islanders, and, above all, Acadian refugees from Nova Scotia—added to the population. By 1800 there were about 50,000 inhabitants, a considerable number of them black slaves imported from Africa and the West Indies. The availability of slave labor, Eli Whitney's invention of the cotton gin, and Étienne de Boré's development of a granulation process for making cane sugar set the stage for future prosperity, though not under Spanish auspices. In 1800, by the secret Treaty of San Ildefonso, Napoleon forced the feeble Spanish government to return Louisiana to France. Three years later, having failed to reestablish French rule and slavery in Haiti, Napoleon sold Louisiana to the US to keep it from falling into the hands of Great Britain.

President Thomas Jefferson concluded what was probably the best real estate deal in history, purchasing 800,000 sq mi (2,100,-000 sq km) for $15,000,000 and thus more than doubling the size of the US at a cost of about 3 cents per acre. He made William C. C. Claiborne the governor of the huge new acquisition. The next year, that part of the purchase south of 33°N was separated from the remainder and designated the Territory of Orleans. The people of the territory then began the process of learning self-government, something with which they had had no experience under France and Spain. After the census of 1810 showed that the population had risen to 76,556, the people were authorized by Congress to draw up a state constitution. The constitutional convention met under the presidency of Julian Poydras in a coffeehouse in New Orleans and adopted, with a few changes, the constitution then in effect in Kentucky. In the meantime, in 1810, a revolt against Spain had taken place in West Florida. When the proposed Louisiana constitution reached Washington, Congress added that part of West Florida between the Mississippi and Pearl rivers to the new state, which entered the Union on 30 April 1812.

The key event in the Americanization of Louisiana was the campaign for New Orleans in December 1814 and January 1815, actually fought after the War of 1812 had ended. A force of British veterans under General Sir Edward Pakenham sailed into Lake Borgne and established itself below New Orleans at Chalmette. There they were met by detachments of Creoles, Acadians, blacks, and even Jean Lafitte's pirates, all from Louisiana, as well as Tennesseans, Kentuckians, and Choctaw Indians, with the whole army under the command of Andrew Jackson. After several preliminary battles, the British were bloodily defeated when they launched an all-out assault on Jackson's line.

From 1815 to 1861, Louisiana was one of the most prosperous states in the South, producing sugar and cotton on its rich alluvial lands and grazing hogs and cattle in the wooded hills of the north and on the prairies of the southwest. Yeoman farmers and New Orleans workers far outnumbered the wealthy planters, but the planters, whose slaves made up almost half the population, dominated Louisiana politically and economically. When the secession crisis came in 1861, the planters led Louisiana into the Confederacy and, after four bloody years, to total defeat. The state suffered crippling economic losses during the Civil War, but the greatest loss was the lives of tens of thousands of young white men who died in defense of the South, and of thousands of blacks who died seeking and fighting for freedom. Louisiana did not fully recover from this disaster until the mid-20th century.

After the Civil War, Radical Republican governments elected by black voters ruled the state, but declining support from the North and fierce resistance from Louisiana whites brought the Reconstruction period to an end. Black people and their few white allies lost control of state government, and most of the former slaves became laborers on sugar plantations or sharecroppers in the cotton fields. There, as the years passed, they were joined by more and more landless whites. In 1898, blacks were disfranchised

almost entirely by a new state constitution drawn up primarily for that purpose. This constitution also significantly reduced the number of poorer whites who voted in Louisiana elections.

The vast majority of Louisiana whites—whether hill farmers, Cajuns along the southern rivers and bayous, lumbermen in the yellow pine forests, or workers in New Orleans—were little better off than the black or white sharecroppers. Many economic changes had taken place: rice had become a staple crop on the southwestern prairies, and an oil boom had begun after the turn of the century. But just as before the Civil War, large landowners—combined with New Orleans bankers, businessmen, and politicians—dominated state government, effectively blocking political and social reform. The Populist movement, which succeeded in effecting some change in other southern states, was crushed in Louisiana.

Not until 1928, with the election of Huey P. Long as governor, did the winds of change strike Louisiana; having been so long delayed, they blew with gale force. The years from 1928 through 1960 could well be called the Long Era: three Longs—Huey, who was assassinated in 1935; his brother Earl, who served as governor three times; and his son Russell, who became a powerful US senator—dominated state politics for most of the period. From a backward agricultural state, Louisiana evolved into one of the world's major petrochemical-manufacturing centers. Offshore drilling sent clusters of oil wells 60 mi (97 km) out into the Gulf. The pine lands were reforested, and soybeans provided a new source of income. What had been one of the most parsimonious states became one of the most liberal in welfare spending, care for the aged, highway building, and education. The state could afford these expanding programs because of ever-increasing revenues from oil and gas.

In the mid-1980s, the major problems confronting the state were racial and labor tensions, inadequate disposal sites for industrial wastes, and (despite important new discoveries) the depletion of oil and gas resources. Even so, the state remained prosperous, a fact that little in the near future seemed likely to change.

Meanwhile, the announcement in February 1985 by Russell B. Long, senator since 1948, that he would not seek reelection in 1986, and the indictment of Governor Edwin W. Edwards by a federal grand jury on conspiracy charges during the same month, caused turmoil in Louisiana's political arena.

¹²STATE GOVERNMENT

Louisiana has had 11 constitutions (more than any other state) the most recent of which was enacted in 1974.

The state legislature consists of a 39-member senate and a 105-member house of representatives. All legislators are elected for concurrent four-year terms; they must be US citizens, be at least 18 years old, and have resided in the state for two years and in their districts for at least one year preceding election.

Major elected executive officials include the governor and lieutenant governor (independently elected), secretary of state, attorney general, treasurer, superintendent of education, commissioner of agriculture, commissioner of insurance, and commissioner of elections. All are elected for four-year terms. The governor must be a qualified elector, be at least 25 years old, and have been a US and Louisiana citizen for five years preceding election; after two consecutive terms, a governor may not succeed himself. The same eligibility requirements apply to the lieutenant governor, except that there is no limit on succession to the latter office. Other executive agencies are the State Board of Elementary and Secondary Education, whose eight elected members and three appointed members serve four-year terms, and the Public Service Commission, whose five members serve for six years.

To become law, a bill must receive majority votes in both the senate and the house and be signed by the governor, be left unsigned but not vetoed by the governor, or be passed again by two-thirds votes of both houses over the governor's veto. Appropriation bills must originate in the house but may be amended by the senate. The governor has an item veto on appropriation bills. Constitutional amendments require approval by two-thirds of the elected members of each house and ratification by a majority of the people voting on it at the next general election.

Voters in Louisiana must be US citizens, be 18 years of age, and have registered at least 24 days before a general election (30 days before a primary election).

13POLITICAL PARTIES

The major political organizations are the Democratic Party and the Republican Party, each affiliated with the national party. However, differences in culture and economic interests have made Louisiana's politics extremely complex. Immediately following statehood, the primary political alignment was according to ethnic background, Anglo or Latin. By the 1830s, however, Louisiana politics reflected the national division of Jacksonian Democrats and National Republicans (by mid-decade replaced by the Whigs). By and large, the Whigs were favored by the Anglo-Americans, the Democrats by those of French and Spanish descent. When the Whig Party fell apart over slavery, many former Whigs supported the Native American (Know-Nothing) Party.

Louisiana was one of the three southern states whose disputed electoral votes put Republican Rutherford B. Hayes in the White House in 1877, in return for his agreement to withdraw federal troops from the South, thus putting an end to Reconstruction. The ensuing period of Bourbon Democratic dominance in Louisiana, a time of reaction and racism in politics (though a few blacks continued to hold office), lasted until the early 1890s, when worsening economic conditions inspired Populists and Republicans to challenge Democratic rule. The attempt failed largely because Democratic landowners were able to control the ballots of their black sharecroppers and "vote" them Democrat. The recognition that it was the black vote, however well controlled, that held the balance in Louisiana politics impelled the Democrats to seek its elimination as an electoral factor. The constitution of 1898 imposed a poll tax, a property requirement, a literacy test, and other measures that succeeded in reducing the number of registered black voters from 130,000 at the beginning of 1897 to 5,320 in March 1900 and to 1,342 by 1904. White registration also declined, from 164,000 in 1897 to 92,000 in 1904, since the new constitutional requirements tended to disfranchise poor whites as well as blacks.

Between 1900 and 1920, the New Orleans Ring, or Choctaw Club, was the dominant power in state politics. Growing political discontent led 5,261 Louisianians (6.6% of those voting) to cast their ballots for the Socialist presidential candidate in 1912. A few Socialists won local office that year in Winn Parish, a center of Populist activity in the 1890s and the birthplace of Huey Long in 1893.

During his relatively brief career as a member of the Railroad Commission, governor, and US senator, Long committed government resources to public service to an extent without precedent in the state. He also succeeded in substituting for the traditional Democratic Party organization a state machine geared primarily toward loyalty to himself and, after his assassination in 1935, to the Long family name, which kept its hold on the voters despite a series of scandals that publicized the corruption of his associates. When blacks began voting in increasing numbers during the 1940s, they tended to favor Democratic candidates from the Long camp. The Longs repaid their loyalty: when race became a bitterly divisive issue in the late 1940s and 1950s—Louisiana gave its presidential vote to the States' Rights "Dixiecrat" candidate in 1948—the Longs supported the national Democratic ticket.

The 1960s and 1970s saw a resurgence of the Republican Party and the election in 1979 of David C. Treen, the state's first Republican governor since Reconstruction. As of 1985 there were 1,743,924 registered Democrats and 258,104 registered Republicans. Both US senators—Russell B. Long (who announced that he would not run for reelection in 1986) and J. Bennett Johnston, Jr.—and six US representatives from the state were Democrats in

Louisiana Presidential Vote by Political Parties, 1948–84

YEAR	ELECTORAL VOTE	LOUISIANA WINNER	DEMOCRAT	REPUBLICAN	STATES' RIGHTS DEMOCRAT	PROGRESSIVE	AMERICAN INDEPENDENT
1948	10	Thurmond (SRD)	136,344	72,657	204,290	3,035	—
1952	10	Stevenson (D)	345,027	306,925	—	—	—
					UNPLEDGED		
1956	10	*Eisenhower (R)	243,977	329,047	44,520	—	—
					NAT'L. STATES' RIGHTS		
1960	10	*Kennedy (D)	407,339	230,980	169,572	—	—
1964	10	Goldwater (R)	387,068	509,225	—	—	—
1968	10	Wallace (AI)	309,615	257,535	—	—	530,300
					AMERICAN	SOC. WORKERS	
1972	10	*Nixon (R)	298,142	686,852	44,127	12,169	—
					LIBERTARIAN	COMMUNIST	
1976	10	*Carter (D)	661,365	587,446	3,325	7,417	10,058
						CITIZENS	
1980	10	*Reagan (R)	708,453	792,853	8,240	1,584	10,333
1984	10	*Reagan (R)	651,586	1,037,299	1,876	9,502	—

*Won US presidential election.

1985, and two US representatives were Republicans. In 1985 there were 85 Democrats and 20 Republicans in the house; 38 of the state senators were Democrats, 1 was a Republican.

Louisiana voters, who had been loyal to southerner Jimmy Carter in 1976, switched their allegiance to Ronald Reagan in November 1980, giving him a 51% majority of the popular vote; they gave him almost 61% of their votes in 1984.

Louisiana had 475 black elected officials in 1985, and 3 Hispanic and 1 Asian Pacific American official in 1984. Only 4.8% of the state legislators in 1985 were women (the 2d-lowest percentage in the US), but one woman served in the state's US House delegation and another was state treasurer.

14 LOCAL GOVERNMENT

The ecclesiastical districts, called parishes, into which Louisiana was divided in the late 17th century remain the primary political divisions in the state, serving functions similar to those of counties in other states.

In the mid-1980s there were 64 parishes, many of them governed by police jury. Juries for parishes of over 10,000 population have 5 to 15 members, or the number authorized as of 13 May 1974, whichever is greater; smaller parishes have at least 3 members on their juries. All police jury members are elected for four-year terms. Parishes without police juries include East Baton Rouge, Jefferson, Orleans, Plaquemines, St. Charles, St. James, and St. Tammany, all of which are governed by other means. Other parish officials are the sheriff, clerk of court, assessor, and coroner. Each parish elects a school board whose members generally serve six-year terms; all other officers serve four-year terms.

As of 1982, Louisiana also had 301 municipal governments. Prominent local officials include the mayor, chief of police, and a council or board of aldermen.

15 STATE SERVICES

Louisiana's ethics laws are administered by the Commission on Ethics for Public Employees and the Board of Ethics for Elected Officials, both under the Department of Civil Service.

Educational services are provided through the Department of Public Education, which has jurisdiction over elementary, secondary, higher, and vocational-technical instruction, as well as the state schools for the visually impaired, hearing-impaired, and other handicapped children. Highways, waterways, airports, and mass transit are the province of the Department of Transportation and Development. Environmental affairs, conservation, forestry, and mineral resources are the responsibility of the Department of Natural Resources. The Motor Vehicle Office, Fire Protection Office, Emergency Preparedness Office, and Alcoholic Beverage Control Office are within the Department of Public Safety.

Health and welfare services are administered mainly through the Department of Health and Human Resources, which provides welfare services under the Office of Family Security, services for the blind and vocational rehabilitation through the Office of Human Development, and special training for the mentally handicapped under the Office of Mental Retardation. Antipoverty programs, including Head Start, supplemental food, summer youth recreation, and Indian affairs programs, are administered by the Office of Community Services in the Department of Urban and Community Affairs, which also helps develop and administer housing and urban renewal programs.

16 JUDICIAL SYSTEM

Louisiana's legal system is the only one in the US to be based on civil or Roman law, specifically the Code Napoléon of France. Under Louisiana state law, cases may be decided by judicial interpretation of the statutes, without reference to prior court cases, whereas in other states and in the federal courts the common law prevails, and decisions are generally based on previous judicial interpretations and findings. In actual practice,

Louisiana laws no longer differ radically from US common law, and most Louisiana lawyers and judges now cite previous cases in their arguments and rulings.

The highest court in Louisiana is the supreme court, with appellate jurisdiction. It consists of a chief justice who is chosen by seniority of service, and 6 associate justices, all of them elected from 6 supreme court districts (the first district has 2 judges) for staggered 10-year terms. There are 5 appellate circuits in the state, each divided into 3 districts; as of 1984, the 5 circuits were served by 48 judges, all of them elected for overlapping 10-year terms. Each of the state's district courts serves at least 1 parish and has at least 1 district judge, elected for a 6-year term; in 1984 there were 39 districts and 150 district judges. District courts have original jurisdiction in criminal and civil cases. City courts are the principal courts of limited jurisdiction.

Louisiana may have been the first state to institute a system of leasing convict labor; large numbers of convicts were leased, especially after the Civil War, until the practice was discontinued in the early 1900s. The abuses entailed in this system may be suggested by the fact that of 700 convicts leased in 1882, 149 died in service. At the end of 1984, 13,919 prisoners were in Louisiana's state and federal prisons. In 1983, 17.7% of all state inmates were held in local jails because of overcrowding.

According to the FBI Crime Index in 1983, Louisiana had a violent crime rate of 641 per 100,000 population and a property crime rate of 4,386. The crime index total—5,027—was slightly below the US norm. Louisiana tied with Texas for the highest rate of homicide and nonnegligent manslaughter—14.2 per 100,000 inhabitants. Louisiana's rates of aggravated assault (268), forcible rape (40), burglary (1,668), larceny-theft (3,203), and motor vehicle theft (525) were also above the US norm.

In 1982, a total of $479,000,000 was spent on state and local government police protection and correction employment and payrolls, of which 65% went for police protection and 35% for correction. In 1983 there were 18,708 employees in Louisiana's criminal justice system, 11,462 of them in police protection and 7,247 in correction. Louisiana has a death penalty law; five people were executed between 1977 and 1984. Judges may impose sentences of hard labor.

17 ARMED FORCES

As of 1984, Louisiana's 11 military installations had 34,854 military and civilian personnel. There was one major army installation in the state, Ft. Polk at Leesville, with 17,319 personnel. In addition there were two Air Force bases, Barksdale near Bossier City and England near Alexandria, and a naval air station and support station in the vicinity of New Orleans. During the year ending 30 September 1984, Louisiana firms received defense contracts totaling almost $1.9 billion.

There were 450,000 veterans of US military service in Louisiana on 30 September 1984, of whom 3,000 had served in World War I, 162,000 in World War II, 72,000 during the Korean conflict, and 142,000 in the Viet-Nam era. Veterans' benefits during 1982/83 amounted to $444 million, of which $249 million was in compensation and pensions, $135 million in medical services, $24 million in education and training, and $18 million in insurance and indemnities.

Army and Air National Guard forces in Louisiana totaled 11,064 in early 1985.

18 MIGRATION

Louisiana was settled by an unusually diverse assortment of immigrants. The Company of the Indies, which administered Louisiana from 1717 until 1731, at first began importing French convicts, vagrants, and prostitutes because of the difficulty of finding willing colonists. Next the company turned to struggling farmers in Germany and Switzerland, who proved to be more suitable and productive settlers. The importation of slaves from Africa and the West Indies began early in the 18th century.

Perhaps 10,000 Acadians, or Cajuns—people of French descent who had been exiled from Nova Scotia (Acadia) during the 1740s—migrated to Louisiana after the French and Indian War, attracted by generous land grants; they settled in the area of Lafayette and Breaux Bridge and along Bayou Lafourche and the Mississippi River. Probably the second largest group to migrate in the late 18th century came from the British colonies and, after the Revolution, from the US. Between 1800 and 1870, Americans settled the area north of the Red River. Small groups of Canary Islanders and Spaniards from Malaga also settled in the south, and in 1791, a number of French people fled to Louisiana during the slave insurrection on Hispaniola.

During the 1840s and 1850s, masses of Irish and German immigrants came to New Orleans. In the late 1880s, a large number of midwestern farmers migrated to the prairies of southwestern Louisiana to become rice farmers. Louisiana did not immediately begin losing much of its black population after the Civil War. In fact, the number of blacks who migrated to Louisiana from the poorer southeastern states during the postwar years may have equaled the number of blacks who migrated before the war or were brought into the state as slaves. In 1879, however, "Kansas fever" struck blacks from the cotton country of Louisiana and Mississippi, many of whom migrated to the Wheat State; many later returned to their home states.

Beginning in World War II, large numbers of both black and white farm workers left Louisiana and migrated north and west. During the 1960s, the state had a net out-migration of 15% of its black population, but the trend had slowed somewhat by 1975.

Recent migration within the state has been from north to south, and from rural to urban areas, especially to Shreveport, Baton Rouge, and the suburbs of New Orleans. Overall, Louisiana suffered a net loss from migration of 326,000 from 1940 to 1970, although it gained 178,000 from migration between 1970 and 1983.

[19] INTERGOVERNMENTAL COOPERATION

Among the interstate and regional efforts in which Louisiana participates are the Interstate Oil and Gas Compact, Interstate Compact on Juveniles, Interstate Compact on Placement of Children, Gulf States Marine Fisheries Compact, Red River Compact, Sabine River Compact, South Central Forest Fire Protection Compact, Southern Growth Policies Compact, Southern Interstate Energy Compact, and Southern Regional Education Compact.

Federal aid to Louisiana during 1983/84 was almost $1.8 billion, of which $93 million was general revenue sharing.

[20] ECONOMY

Before the Civil War, when Louisiana was one of the most prosperous of southern states, its economy depended primarily on two then profitable crops—cotton and sugar—and on its position as the anchor of the nation's principal north–south trade route. But the upheaval and destruction wrought by the war, combined with severe flood damage to cotton crops, falling cotton prices, and the removal of the federal bounty on sugar, left the economy stagnant through the end of the 19th century, although New Orleans retained its commercial importance as an exporter of cotton and grain.

With the addition of two major crops, rice and soybeans, the rebirth of the timber industry as a result of reforestation and the demand for pine for paper pulp, and, most dramatically, the rise of the petrochemical industry, Louisiana's economy has regained much of its former vitality. Today, Louisiana ranks 2d only to Texas in the value of its mineral products, and in value per capita it exceeds that much larger and more populous state. As of 1982, the value of mineral production (including fossil fuels) was $31.4 billion, accounting for 18% of the value of mineral production in the US.

Unfortunately, not all of Louisiana's citizens share in this newfound wealth. Wages have been rising—the average manufacturing wage is among the top third in the nation—but the state's unemployment rate has been higher than the national average, and the rate for women is especially high. Per capita income is still well below the national norm.

Louisiana is primarily an industrial state, but its industries are to a large degree based on its natural resources, principally oil, water, and timber. Timber is being replaced as fast as it is being removed. More attention is being paid to the care of Louisiana's inland and coastland waters than ever before, and the port of New Orleans is thriving. The central question for Louisiana's economic planners is, What happens when the oil runs out?

[21] INCOME

Louisiana's per capita personal income in 1970 was $3,023, for a rank of 46th in the US. By 1983, largely because of the oil boom, per capita income had risen to $10,406 (34th). Total personal income in current dollars rose from $11.1 billion in 1970 to $46.2 billion in 1983, while the real growth (measured in constant 1972 dollars) was 85%, well above the national average of 45%. Median money income of a four-person family in 1981 was $26,144, 22d in the US.

Income, although increasing, is unequally distributed. In 1979, almost 19% of all Louisianians were below the federal poverty level, a proportion tied with that in the District of Columbia and exceeded only by Mississippi, Arkansas, and Alabama. In 1982, on the other hand, 30,400 (about 1.5%) of the nation's top wealthholders—individuals with gross assets exceeding $500,000—lived in the state.

[22] LABOR

In 1983, Louisiana had a total civilian labor force of 1,910,000 or about 53.4% of the population aged 16 or over. Only Mississippi, Alabama, Pennsylvania, and West Virginia had lower participation rates that year. Louisiana's rate for females, 47.6%, was ranked among the 4 lowest of the 50 states. About 1,745,000 Louisianians were employed and 194,000 were unemployed in 1984, for an unemployment rate of 10%. The overall unemployment rate during 1983 was 11.8%, 12.1% for males and 11.3% for females.

A federal survey of workers in March 1982 revealed the following nonfarm employment pattern for Louisiana:

	ESTABLISH-MENTS	EMPLOYEES	ANNUAL PAYROLL ('000)
Agricultural services, forestry, fishing	787	5,519	$ 60,270
Mining, of which:	2,153	115,929	2,859,237
Oil, gas extraction	(1,975)	(99,783)	(2,410,564)
Contract construction	6,833	122,253	2,171,498
Manufacturing	3,721	214,061	4,420,426
Transportation, public utilities	4,319	122,563	2,358,160
Wholesale trade	7,476	100,218	1,807,525
Retail trade	22,481	279,441	2,488,509
Finance, insurance, real estate	7,499	80,086	1,285,336
Services	24,114	288,934	3,807,425
Other	2,043	2,620	35,052
TOTALS	81,426	1,331,624	$21,293,438

Government employees, not counted in this survey, totaled about 261,000 in October 1983.

During the antebellum period, Louisiana had both the largest slave market in the US—New Orleans—and the largest slave revolt in the nation's history, in St. Charles and St. John the Baptist parishes in January 1811. New Orleans also had a relatively large free black population, and many of the slaves in the city were skilled workers, some of whom were able to earn their freedom by outside employment. Major efforts to organize Louisiana workers began after the Civil War. There were strikes in the cane fields in the early 1880s, and in the mid-1880s, the Knights of Labor began to organize the cane workers. The strike they called in 1886 was ended by hired strikebreakers, who killed at least 30 blacks. Back in New Orleans, the Knights of Labor led a general strike in 1892. The Brotherhood of Timber Workers began organizing in 1910, but had little to show for their efforts except the scars of violent conflict with the lumber-mill owners.

In 1980, 257,000 Louisianians belonged to labor unions, of which there were 699 in 1983. Only 16.4% of all nonagricultural employees were union members, 35% below the national norm. A right-to-work law was passed in 1976, partly as a result of violent conflict between an AFL-CIO building trades union and an independent union over whose workers would build a petrochemical plant near Lake Charles. In 1979, a police strike began in New Orleans on the eve of Mardi Gras, causing the cancellation of most of the parades, but it collapsed the following month. In 1981, 1,400 workers were involved in a total of seven separate work stoppages.

23AGRICULTURE

With a farm income of almost $1.9 billion in 1983—74% from crops—Louisiana ranked 29th among the 50 states. Nearly every crop grown in North America could be raised somewhere in Louisiana. In the south are strawberries, oranges, sweet potatoes, and truck crops; in the southeast, sugarcane; and in the southwest, rice and soybeans. Soybeans—which were introduced into Louisiana after World War I and are today the state's most valuable crop—are also raised in the cotton-growing area of the northeast and in a diagonal belt running east–northwest along the Red River. Oats, alfalfa, corn, potatoes, and peaches are among the other crops grown in the north.

As of 1984 there were an estimated 36,000 farms covering 10.1 million acres (4.1 million hectares). Louisiana long ranked 1st in the US in sugarcane production, but by the late 1970s, Florida and Hawaii had surpassed it. Cash receipts for the sugar crop in 1982 amounted to $176,453,000 for 7,030,000 tons. Louisiana ranked 3d in the value of its rice production in 1982, $200,139,000 for 24,862,000 hundredweight; 5th for cotton in 1983, $167,516,-000 for 532,000 bales; and 8th for soybeans in 1983, $511,173,000 for 65,535,000 bushels.

24ANIMAL HUSBANDRY

In the mid-19th century, before rice production began there, southwestern Louisiana was a major cattle-raising area. Today, cattle are raised mainly in the southeast (between the Mississippi and Pearl rivers), in the north-central region, and in the west. Livestock production in 1983 accounted for 26% of agricultural income. In January 1984 there were 1,300,000 cattle and calves and 110,000 hogs and pigs. The cattle and calf output in 1983 totaled 285.1 million lb, valued at $140.9 million. Other farm products included 975 million lb of milk valued at $140.9 million, 12 million lb of chickens worth $1.6 million, and 417 million eggs worth $28.5 million.

Fur trapping has some local importance. During the 1981/82 season, 1,616,131 pelts were taken in Louisiana, with a total value of $8,148,407. In 1981, a month-long, statewide alligator season was held for the first time in 18 years; 14,600 alligators taken had hides and meat valued at $2 million.

25FISHING

In 1984, Louisiana led all states in the size of its commercial landings, with over 1.9 billion lb, and ranked 2d by value of catch at $265.4 million. Cameron led all US ports in 1984 with a catch of 679 million lb. The ports of Empire-Venice and Dulac-Chauvin ranked 3d and 4th respectively, together accounting for another 711 million lb.

The most important species caught in Louisiana are shrimp, oysters, menhaden, and blue crabs. Oyster production is 10–15 million lb annually; in 1982, the menhaden harvest ranked 1st in the US. Louisiana led the nation in the number of plants producing industrial fishery products in 1983 with 17.

Louisiana produces most of the US crayfish harvest. With demand far exceeding the natural supply, crayfish farming began about 1959. By 1980, 361 crayfish farms covered some 55,000 acres (22,000 hectares). Catfish are also cultivated in Louisiana, on some 2,600 acres (1,050 hectares), with an annual harvest of 3 million lb valued at more than $1 million. The value of production at all inland farm fisheries in 1982 was $110 million.

26FORESTRY

As of 1979 there were 14,558,000 acres (5,891,000 hectares) of forestland in Louisiana, representing about half the state's land area and 2% of all US forests. The principal forest types are loblolly and shortleaf pine in the northwest, longleaf and slash pine in the south, and hardwood in a wide area along the Mississippi River. More than 99% of Louisiana's forests—some 14,527,000 acres (5,879,000 hectares) in 1979—are commercial timberland, nearly all of it privately owned. In 1981, Louisiana ranked 1st in the nation in production of plywood and pulpwood. The value of shipments from all logging camps and logging contractors during 1982 was estimated at $259.4 million.

As of 30 September 1984, Louisiana had one national forest, Kisatchie, with a gross area of 1,022,703 acres (413,875 hectares) within its boundaries, including 597,933 acres (241,976 hectares) of National Forest System lands. Within the boundaries of Kisatchie's Evangeline Unit is the Alexander State Forest, established in 1923 for reforestation and the protection of seedlings.

The headquarters of the Southern Forest Survey of the US Forest Service and the offices of the largest of the nine US forest experimental stations are located in New Orleans.

27MINING

In 1982, Louisiana ranked 2d among the 50 states in value of minerals produced, with $31.4 billion, most of that from fossil fuels. In order of value, the principal minerals were natural gas, crude petroleum, and sulfur. In 1984, the state was the nation's leading salt producer and 2d-leading sulfur producer. Nonfuel mineral production reached about $568.3 million that year, with cement, salt, sand and gravel, stone, and sulfur the major commodities produced.

The principal mining regions are the south, especially the Gulf Coast, for salt, sulfur, and petroleum, much of which is found below salt domes; offshore for oil and gas; and the northeast for gas. In 1984, Louisiana produced an estimated 17,600,000 tons of construction gravel worth $67,200,000, 12,841,000 tons of salt worth $152,006,000, and 2,058,000 metric tons of Frasch sulfur.

28ENERGY AND POWER

Oil and gas production has expanded greatly since World War II, but production reached its peak in the early 1970s, and proved reserves are declining.

At the end of 1983, power plants in Louisiana had a total installed capacity of 15.6 million kw. These generated a total of 37 billion kwh of power. Energy consumption in 1982 was 3,288 trillion Btu—750 million Btu per capita. Louisiana ranked 1st nationally in per capita energy expenditures with $4,647. About 49% of the state's energy needs were supplied by natural gas, 43% by oil, and 8% by other sources. As of 1985, Louisiana had two nuclear power plants: Waterford 3 in St. Charles Parish (capacity, 1,104,000 kw), which received a low-power operating license on 18 December 1984, and River Bend 1 in West Feliciana Parish (940,000 kw), which was to begin commercial operation by December 1985.

Louisiana produced 417,000,000 barrels of crude oil during 1983: 293,000,000 from offshore wells and 124,000,000 from the rest of the state. Production that year was approximately 14% of the US total. At the end of 1983, remaining proved reserves of oil in Louisiana amounted to 2.7 billion barrels (10% of the US total), down about 21 million barrels from a year earlier, continuing a trend that began in 1971. In 1982, 5,005 wells were drilled in the state, 60% of them in the north.

Louisiana accounts for 29% of US gross withdrawals of natural gas. Marketed production in 1983 was 5.3 trillion cu feet, leaving proved reserves of 42.6 trillion cu feet. There were 16,700 producing gas wells in 1983, up from 16,586 in 1982. Louisiana had 44,233 mi (71,186 km) of gas utility pipeline and main in 1983.

Energy conservation plans in Louisiana call for development of untapped energy sources, such as the state's lignite and geothermal reserves.

[29] INDUSTRY

The Standard Oil Refinery (now owned by Exxon) that is today the largest in North America began operations in Louisiana in 1909, the same year construction started on the state's first long-distance oil pipeline. Since then, a huge and still-growing petrochemical industry has become a dominant force in the state's economy. Other expanding industries are wood products and, especially since World War II, shipbuilding.

In 1982, the total value of shipments of manufactured goods was $57.1 billion. The largest employers among industry groups were chemicals and chemical products, food and food products, transportation equipment, and fabricated metal products, which together utilized 49% of the industrial work force. By value of shipments, the leading groups were petroleum and coal products, 54%; chemicals and allied products, 19%; food and kindred products, 7%; transportation equipment, 4%; and paper and allied products, 3%.

The following table shows value of shipments in 1982 for selected industries:

Petroleum refining	$30,156,600,000
Agricultural chemicals	1,960,600,000
Ship building and repairing	1,262,000,000
Plastics materials, synthetics	1,257,500,000
Sugar, confectionery products	938,400,000
Paperboard mill products	625,800,000

The principal industrial regions extend along the Mississippi River, from north of Baton Rouge to New Orleans, and also include the Monroe, Shreveport, Morgan City, and Lake Charles areas.

[30] COMMERCE

Louisiana ranked 16th among the 50 states in wholesale trade in 1982, with sales amounting to $36.1 billion. Nearly 39% of that took place in metropolitan New Orleans, with metropolitan Baton Rouge accounting for another 15%.

With sales of $19.4 billion, Louisiana ranked 19th in retail trade in 1982. The leading retail categories were food stores, with 25% of total sales; automotive dealers, 19%; and general merchandise stores, 12%. Metropolitan New Orleans accounted for $6.1 billion in retail sales, followed by Baton Rouge, $2.6 billion, and Shreveport, $1.9 billion.

In 1981, Louisiana ranked 14th among the 50 states in value of its manufactures exported abroad, with $3.7 billion. The value of the state's foreign agricultural exports was $626 million in 1981/82 (21st in the US). Exports accounted for 37.4% of Louisiana farm sales, the 5th-highest rate in the nation. A total of 24,800 jobs in Louisiana were dependent on exports of manufactured goods. Louisiana ranked 2d nationally as an exporter of chemicals, 3d in petroleum, 5th in primary metals, and 8th in food manufactures. Louisiana was the 4th-largest US exporter of fish products in 1981, shipping $64 million worth—more than one and a half times the 1977 value. Exports through the port of New Orleans during 1983 totaled $18.3 billion; imports, $8.3 billion.

[31] CONSUMER PROTECTION

The Office of Consumer Protection, in the Department of Urban and Community Affairs, is the principal agency dealing with consumer interests. In addition, an advisory consumer credit council operates under the Department of Commerce.

[32] BANKING

In 1984, Louisiana had 302 insured commercial banks. Total assets in 1983 were $33 billion, outstanding loans $17.2 billion; deposits totaled $27.8 billion, including $20.4 billion in time and savings deposits. There were 119 insured savings and loan associations in the state as of 31 December 1982, with total assets of $10 billion. Outstanding mortgage loans totaled $7.2 billion, and savings capital amounted to $8.7 billion. In 1984 there were 390 credit unions in Louisiana.

Louisiana banks are regulated by the Office of Financial Institutions within the Department of Commerce.

[33] INSURANCE

There were 109 life insurance companies domiciled in the state in mid-1983. That year, Louisianians held 9,890,000 life insurance policies with a total value of $95.4 billion; the average family held $58,600 in coverage. Total benefits paid that year amounted to $803.8 million, of which $299.3 million represented death benefits.

As of 1984, 32 property/casualty insurance companies had headquarters in Louisiana. Premiums written by these and out-of-state insurance firms in 1983 totaled $2.4 billion, of which automobile liability coverage accounted for $582 million; automobile physical damage insurance, $390 million; and homeowners' insurance, $291 million. At the end of 1983, over $14 billion in flood insurance was in force, 3d behind Florida and Texas among the 50 states. The Department of Insurance administers Louisiana's laws governing the industry.

[34] SECURITIES

The New Orleans Commodity Exchange, specializing in cotton futures, is the only exchange in the South.

As of 30 September 1983 there were 59 brokers and dealers registered under the Securities Exchange Act of 1934. New York Stock Exchange member firms maintained 66 sales offices in Louisiana during 1983, with 613 registered representatives. That year, 568,000 Louisianians were shareowners of public corporations.

[35] PUBLIC FINANCE

The budget is prepared by the state executive budget director and submitted annually by the governor to the legislature for amendment and approval. The fiscal year runs from 1 July through 30 June. The following table shows estimated revenues and expenditures for 1982/83 and 1983/84 (in thousands):

REVENUES	1982/83	1983/84
State revenues:		
Taxes, licenses, fees, etc.	$3,835,059	$3,881,270
Revenues collected by agencies	308,690	377,143
Total federal grants	1,460,478	1,558,221
Total interagency transfers	259,236	290,042
Fund surpluses	407,420	70,454
TOTALS	$6,270,883	$6,177,130
EXPENDITURES		
Public education	$2,230,594	$2,217,629
Health and human resources	1,740,123	1,752,079
Aid to local governments	294,378	280,014
Transportation and development	276,472	251,548
Department of Labor	122,454	174,806
Corrections	139,034	165,502
Public safety	137,386	130,652
Other appropriations and requirements	575,512	525,305
Debt service	247,548	287,608
Capital outlays	510,710	381,823
TOTALS	$6,274,211	$6,166,966

In mid-1982, the state debt was $4.1 billion and the total debt of all Louisiana governments was $9.7 billion, or $2,311 per capita (14th in the US).

36TAXATION

For most of the state's history, Louisianians paid little in taxes. Despite increases in taxation and expenditures since the late 1920s, when Huey Long introduced the graduated income tax, Louisiana's state and local tax burden per capita, $1,101 in 1982, is still well below the national average.

Income taxes yielded $600 million in state revenues in 1983/84. As of 1984, when the state's largest tax increase in history went into effect, the individual income tax ranged from 2% on the first $10,000 to 6% on income over $50,000. The corporate income tax ranged from 4% on the first $25,000 up to 8% on net income over $200,000.

The state's sales and use tax, which yielded nearly $933 million to the state in 1983/84, was raised from 3% to 4% in 1984. Parishes and municipalities may impose additional sales taxes. Natural resource severance taxes, whose rates vary according to the resource, brought in more than $869 million in 1982/83. The state also taxes gasoline sales (doubled from 8 cents a gallon to 16 cents in 1984), gifts and inheritances, soft drinks, alcoholic beverages, and tobacco products, among other items. Taxes on beer and chain stores contribute to local revenues, as does the property tax, although Louisiana relies less on this than do most other states. The Louisiana Stadium and Exposition District, the Orleans Parish School Board, and the New Orleans Exhibition Hall Authority impose a tax on hotel and motel room occupancy.

In fiscal 1983, Louisiana had a federal tax burden of $10.3 billion, or $2,341 per capita. Louisianians filed over 1.6 million federal income tax returns in 1982, paying more than $5.1 billion in tax.

37ECONOMIC POLICY

The Office of Commerce and Industry in the Department of Commerce seeks to encourage investment and create jobs in the state and to expand the markets for Louisiana products. Financial assistance services for industrial development include state and local tax incentives and state "Enterprise Zone" legislation; efforts to establish a statewide development credit corporation were pending in 1985. The Louisiana Small Business Equity Corporation and the Louisiana Minority Business Development Authority offer financial assistance. The Music Commission, established in 1979, encourages the making of musical recordings in the state.

38HEALTH

There were 82,514 live births in Louisiana in 1983, for a birthrate of 18.6 per 1,000 population. The infant mortality rate in 1981 was 10.2 per 1,000 live births for whites, 19.9 for nonwhites. The overall death rate from diseases of early infancy, 13.6 per 100,000 population, was among the highest in the US. In 1981, some 22,300 legal abortions were performed in Louisiana.

The state was tied for the 3d-highest death rate in the nation in 1981 from diabetes, 19.1 per 100,000 population. Death rates from heart disease (305), cancer (168), chronic obstructive pulmonary diseases (21), pneumonia and flu (17), suicide (11), and cirrhosis and other liver diseases (10) were below the national averages; from cerebrovascular diseases (73), accidents (58), and arteriosclerosis (17), above the US median. The overall death rate, 8.4 per 100,000 population, was a little below average for 1981.

In 1984, mental health services were provided at 37 clinics and outreach programs, 4 hospitals, and a comprehensive mental health center. Substance-abuse treatment is provided primarily through 42 outpatient clinics and some of the state hospitals. The Office of Mental Retardation administers 8 residential facilities and assists and monitors 14 group homes. The only leprosarium on the US mainland, established in 1894 by the state and taken over by the US Public Health Service in 1921, is at Carville;

however, most leprosy patients can be treated at home. In 1983, the 158 hospitals in Louisiana had a total of 26,430 beds and recorded 825,873 admissions. Hospital personnel included 10,852 registered nurses, 8,323 of them full-time. The average cost of hospital care in 1982 was $337 per day and $2,123 per stay, the latter somewhat below the national average. There were 7,405 licensed physicians and 1,808 active dentists in the state in 1982.

39SOCIAL WELFARE

During the governorships of Huey and Earl Long, Louisiana developed a relatively progressive welfare system. In 1983, 7.3% of the population received public aid, for a rank of 6th in the US. In 1982, Louisiana ranked 44th among the 50 states in average monthly public assistance payments to families with dependent children (AFDC); a total of $127 million went to 191,279 recipients, almost three-fourths of them children. In 1983, 599,000 persons participated in the federal food stamp program, receiving coupons worth $297 million. An estimated 688,000 pupils benefited from the school lunch program in 1983, at a federal cost of $71 million.

During 1983, Social Security benefits exceeding $2.3 billion were paid to 575,488 Louisianians. Supplemental Security Income payments totaled $256.8 million in 1983. In 1984, Louisiana received $14,637,000 for education for the handicapped and $41,148,000 for rehabilitation services and research from the federal Office of Special Education and Rehabilitative Services. That year, Medicare made $587,352,000 in hospital insurance payments and $201,050,000 in supplementary medical insurance payments. In 1982, a total of $482,187,000 was paid in unemployment insurance.

40HOUSING

The Indians of Louisiana built huts with walls made of clay kneaded with Spanish moss and covered with cypress bark or palmetto leaves. The earliest European settlers used split cypress boards filled with clay and moss; a few early 18th-century houses with clay and moss walls remain in the Natchitoches area. Examples of later architectural styles also survive, including buildings constructed of bricks between heavy cypress posts, covered with plaster; houses in the raised cottage style, supported by brick piers and usually including a wide gallery and colonettes; the Creole dwellings of the Vieux Carré in New Orleans, built of brick and characterized by balconies and French windows; and urban and plantation houses from the Greek Revival period of antebellum Louisiana.

According to the 1980 census there were 1,530,949 year-round housing units, of which 1,411,788 were occupied; of these, 66% were owner-occupied. Overcrowding is a problem: Louisiana ranked 9th among the 50 states in density of occupants per room. In 1980, New Orleans had approximately 226,000 housing units, 12.3% of which had been built during the previous decade. Between 1981 and 1983, some 73,700 new units worth nearly $2.5 billion were authorized in the state.

41EDUCATION

Most education in Louisiana was provided through private (often parochial) schools until Reconstruction. Not until Huey Long's administration, when spending for education increased greatly and free textbooks were supplied, did education become a high priority of the state. As of 1980, only 58% of adult Louisianians had completed high school, for a rank of 41st among the states; 14% had completed 4 or more years of college.

In fall 1983, total enrollment in Louisiana's public elementary and secondary schools was 778,634. Enrollment in private elementary and secondary schools was 158,921 as of 1980. Louisiana ranked 5th that year in nonpublic school enrollment as a percentage of the school-age population. Black students made up 41% of the public school enrollment but less than 16% of the private school total.

Integration of New Orleans public schools began in 1960; two years later, the archbishop of New Orleans required that all Catholic schools under his jurisdiction be desegregated. However, it took a federal court order in 1966 to bring about integration in public schools throughout the state. By 1980, 36% of minority students in Louisiana were in schools with under 50% minority enrollment, and 25% were in schools with 99–100% minority enrollment. The civil rights revolution also affected other aspects of Louisiana education. In 1946/47, combined salaries of principals and teachers averaged $1,765 for whites and $936 for blacks. By 1984, the average salary for all teachers was $19,100, with very little, if any, discrepancy between blacks and whites.

As of 1982, in addition to 53 vocational-technical schools, there were 34 institutions of higher education in Louisiana, of which 22 were public and 12 private. The center of the state university system is Louisiana State University (LSU), founded at Baton Rouge in 1855 and having a 1983/84 enrollment of 29,863; LSU also has campuses at Alexandria, Eunice, and Shreveport, and includes the University of New Orleans, with 16,317 students. Tulane, founded in New Orleans in 1834 and with a 1983/84 enrollment of 10,700, is one of the most distinguished private universities in the South. As of the early 1980s, Southern University Agricultural and Mechanical System at Baton Rouge (1881), with 9,500 students, was the largest predominantly black university in the country; other campuses were in New Orleans and Shreveport. Another mainly black institution is Grambling State University (1901), 4,593.

The Governor's Special Commission on Education Services administers state loan, grant, and scholarship programs. In 1983/84, programs of student financial aid administered 3,500 monetary awards averaging $488 apiece. The state Council for the Development of French in Louisiana organizes student exchanges with Quebec, Belgium, and France, and aids Louisianians studying French abroad.

Total state expenditures on public schools in 1983/84 exceeded $2.2 billion.

42 ARTS

New Orleans has long been one of the most important centers of artistic activity in the South. The earliest theaters were French, and the first of these was started by refugees from Hispaniola, who put on the city's first professional theatrical performance in 1791. The American Theater, which opened in 1824, attracted many of the finest actors in America, as did the nationally famous St. Charles. Showboats traveled the Mississippi and other waterways, bringing dramas, musicals, and minstrel shows to river towns and plantations as early as the 1840s, with their heyday being the 1870s and 1880s.

In the mid-1980s, principal theaters included the New Orleans Theater of the Performing Arts, Le Petit Théâtre du Vieux Carré, and the Tulane Theater. The Free Southern Theatre is a black touring company based in New Orleans. LSU at Baton Rouge has theaters for both opera and drama. Baton Rouge, Shreveport, Monroe, Lake Charles, and Hammond are among the cities with little theaters, and Baton Rouge, Lafayette, and Lake Charles have ballet companies. There are symphony orchestras in most of the larger cities, the New Orleans Philharmonic Symphony Orchestra being the best known.

It is probably in music that Louisiana has made its most distinctive contributions to culture. Jazz was born in New Orleans around 1900; among its sources was the music played by brass bands at carnivals and at Negro funerals, and its immediate precursor was the highly syncopated music known as ragtime. Early jazz in the New Orleans style is called Dixieland; the transformation of jazz from the Dixieland ensemble style to a medium for solo improvisation was pioneered by Louis Armstrong. Traditional Dixieland may still be heard in New Orleans at Preservation Hall, Dixieland Hall, and the New Orleans Jazz Club. Equally distinctive is Cajun music, dominated by the sound of the fiddle and accordion. The French Acadian Music Festival, held in Abbeville, takes place in April.

Louisiana has an income tax checkoff for arts programs that allows taxpayers to designate part or all of their income tax refunds to be placed in a fund for the arts. The state provides grants for arts projects and created and maintains the Louisiana State Museum. Each year, the Louisiana State Arts Council's Awards Program honors one or more legislators who have been active in support of the arts.

43 LIBRARIES AND MUSEUMS

Louisiana's 64 parishes were served by 62 public libraries in 1982. That year, the public library system held 6,635,587 volumes and had a total circulation of 14,440,551. The New Orleans Public Library, with 11 branches and 802,934 books, features a special collection on jazz and folk music, and the Tulane University Library (1,105,867 volumes) has special collections on jazz and Louisiana history. Among the libraries with special black-studies collections are those of Grambling State University, Southern University Agricultural and Mechanical System at Baton Rouge, and Xavier University of Louisiana at New Orleans. The library of Northwestern State University at Natchitoches has special collections on Louisiana history, folklore, Indians, botany, and oral history.

As of 1984, Louisiana had 56 museums and historic sites and 15 public gardens. Leading art museums are the New Orleans Museum of Art, the Lampe Gallery in New Orleans, and the R. W. Norton Art Gallery at Shreveport. The art museum of the Louisiana Arts and Science Center at Baton Rouge is located in the renovated Old Illinois Central Railroad Station. The oldest and largest museum in the state is the Louisiana State Museum, an eight-building historic complex in the Vieux Carré. There is a military museum in Beauregard House at Chalmette National Historical Park, on the site of the Battle of New Orleans, and a Confederate Museum in New Orleans. The Bayou Folk Museum at Cloutierville is in the restored home of author Kate Chopin; the Longfellow-Evangeline State Commemorative Area has a historical museum on its site. Among the state's scientific museums are the Lafayette Natural History Museum, Planetarium, and Nature Station, and the Museum of Natural Science in Baton Rouge. Audubon Park and Zoological Gardens are in New Orleans.

44 COMMUNICATIONS

The second rural free delivery route in the US, and the first in Louisiana, was established on 1 November 1896 at Thibodaux. By 1985, the US Postal Service had 528 post offices with 9,600 employees in the state. As of 1980, 90% of Louisiana's 1,411,788 occupied housing units had telephones.

As of 1984, the state had 192 radio broadcasting stations (96 AM and 96 FM) and 21 commercial and 8 educational television stations. That year, 75 cable television systems in 199 communities had 677,506 subscribers.

45 PRESS

At one time, New Orleans had as many as 9 daily newspapers (4 English, 3 French, 1 Italian, and 1 German), but by 1984 there were only 3, the *Times-Picayune* and *States-Item,* both owned by the Newhouse chain, and the *Daily Record.* In 1984, Louisiana had a total of 7 morning dailies with 431,206 circulation, 18 evening dailies with 371,266, 1 all-day paper (circulation is included in morning and evening circulations), and 17 Sunday papers with 859,634. The following table shows the principal dailies with their 1984 circulations:

AREA	NAME	DAILY	SUNDAY
Baton Rouge	Advocate (m, S)	80,504	133,416
	State Times (e)	37,237	
New Orleans	Times-Picayune/		
	States-Item (all day,S)	276,556	339,030
Shreveport	Journal (e)	26,590	
	Times (m, S)	78,727	109,209

Two influential literary magazines originated in the state. The *Southern Review* was founded at LSU in the 1930s by Robert Penn Warren and Cleanth Brooks. The *Tulane Drama Review*, founded in 1955, moved to New York University in 1967 but is still known by its original acronym, *TDR*.

⁴⁶ORGANIZATIONS

The 1982 Census of Service Industries counted 734 organizations in Louisiana, including 182 business associations; 346 civic, social, and fraternal associations; and 18 educational, scientific, and research associations. Among business or professional organizations with headquarters in Louisiana are the Federated Pecan Growers' Associations of the US and the Louisiana Historical Association, in Baton Rouge; the National Rice Growers Association at Jennings; and the Federal Court Clerks Association and the Southern Forest Products Association, both in New Orleans. Blue Key, a national honor society, has its headquarters in Metairie.

Civil rights groups represented in the state include the National Association for the Advancement of Colored People (NAACP) and the Urban League. Especially active during the 1970s were the local branches of the American Civil Liberties Union, the Louisiana Coalition on Jails and Prisons, and its legal arm, the Southern Prisoners Defense Council, and the Fishermen's and Concerned Citizens Association of Plaquemines Parish, which organized a campaign against the continued domination of the parish by the descendants of Leander Perez, a racist judge who ruled there for 50 years until his death in 1969.

The Invisible Empire, Knights of the Ku Klux Klan, is headquartered in Denham Springs.

⁴⁷TOURISM, TRAVEL, AND RECREATION

In 1982, approximately 71,200 persons were employed in the Louisiana tourist industry, which had an annual payroll of $654 million. Domestic visitors spent an estimated $3.2 billion in the state that year.

New Orleans is one of the major tourist attractions in the US. Known for its fine restaurants, serving such distinctive fare as gumbo, jambalaya, crayfish, and beignets, along with an elaborate French-inspired haute cuisine, New Orleans also offers jazz clubs, the graceful buildings of the French Quarter, and a lavish carnival called Mardi Gras ("Fat Tuesday"). Beginning on the Wednesday before Shrove Tuesday, parades and balls, staged by private organizations called krewes, are held almost nightly. In other towns, country folk celebrate Mardi Gras in their own no less uproarious manner.

Among the many other annual events that attract visitors to the state are the blessing of the shrimp fleet at the Louisiana Shrimp and Petroleum Festival in Morgan City on Labor Day weekend and the blessing of the cane fields during the Louisiana Sugar Cane Festival at New Iberia in September. October offers the International Rice Festival (including the Frog Derby) at Crowley, Louisiana Cotton Festival at Ville Platte (with a medieval jousting tournament), the Louisiana Yambilee Festival at Opelousas, and the Louisiana State Fair at Shreveport. Attractions of the Natchitoches Christmas Festival include 170,000 Christmas lights and spectacular fireworks displays. Between May and November 1984, some 7.3 million people visited the Louisiana World Exposition in New Orleans. Although the exposition was not successful financially, it was regarded as an artistic success; one section of the fair will be converted into a mall with more than 200 shops, restaurants, and vendors.

During 1984, 849,004 people visited the Jean Lafitte National Park. Louisiana's 34 state parks and recreation sites totaled 36,624 acres (14,821 hectares) that year. In 1983, licenses were issued to 413,332 hunters and 554,998 anglers.

⁴⁸SPORTS

New Orleans' National Football League team, the Saints, plays in the Louisiana Superdome, a 71,330-seat stadium that hosted the Super Bowl in 1978 and 1981. Super Bowl championships have also been held in Tulane Stadium. In baseball, Shreveport has a class AA minor league team.

During the 1850s, New Orleans was the horse-racing center of the US, and racing is still popular in the state. The principal tracks are the Louisiana Jockey Club at the Fair Grounds in New Orleans, and Evangeline Downs at Lafayette. Gambling has long been widespread in Louisiana, particularly in steamboat days, when races along the Mississippi drew huge wagers.

From the 1880s to World War I, New Orleans was the nation's boxing capital, and in 1893, the city was the site of perhaps the longest bout in boxing history, between Andy Bowen and Jack Burke, lasting 7 hours and 19 minutes, 110 rounds, and ending in a draw. The New Orleans Open Golf Tournament, held in April, has been won by Billy Casper (twice), Gary Player, and Jack Nicklaus, among others.

In 1935, Tulane inaugurated the Sugar Bowl, an annual New Year's Day event. Since then the LSU Fighting Tigers have been frequent participants in the contest.

⁴⁹FAMOUS LOUISIANIANS

Zachary Taylor (b.Virginia, 1784-1850) is the only US president to whom Louisiana can lay claim. Taylor, a professional soldier who made his reputation as an Indian fighter and in the Mexican War, owned a large plantation north of Baton Rouge, which was his residence before his election to the presidency in 1848. Edward Douglass White (1845-1921) served first as associate justice of the US Supreme Court and then as chief justice.

Most other Louisianians who have held national office won more fame as state or Confederate officials. John Slidell (b.New York, 1793-1871), an antebellum political leader, also played an important role in Confederate diplomacy. Judah P. Benjamin (b.West Indies, 1811-84), of Jewish lineage, was a US senator before the Civil War; during the conflict he held three posts in the Confederate cabinet, after which he went to England and became a leading barrister. Henry Watkins Allen (b.Virginia, 1820-66) was elected governor of Confederate Louisiana in 1864, after he had been maimed in battle; perhaps the best administrator in the South, he installed a system of near-socialism in Louisiana as the fortunes of the Confederacy waned. During and after the Civil War, many Louisianians won prominence as military leaders. Leonidas Polk (b.North Carolina, 1806-64), the state's first Episcopal bishop, became a lieutenant general in the Confederate Army, dying in the Atlanta campaign. Zachary Taylor's son Richard (b.Kentucky, 1826-79), a sugar planter who also became a Confederate lieutenant general, is noted for his defeat of Nathaniel P. Banks's Union forces in the Red River campaign of 1864. Pierre Gustave Toutant Beauregard (1818-93) attained the rank of full general in the Confederate Army and later served as director of the Louisiana state lottery, one of the state's major sources of revenue at that time. In the modern era, General Claire Chennault (b.Texas, 1893-1958) commanded the famous "Flying Tigers" and then the US 14th Air Force in China during World War II.

Throughout the 20th century, the Longs have been the first family of Louisiana politics. Without question, the most important state officeholder in Louisiana history was Huey P. Long (1893-1935), a latter-day Populist who was elected to the governorship in 1928 and inaugurated a period of social and economic reform; in the process, he made himself very nearly an absolute dictator within Louisiana. After his election to the US Senate, the "King Fish" became a national figure, challenging Franklin D. Roosevelt's New Deal with his "Share the Wealth" plan and flamboyant oratory. Huey's brother Earl K. Long (1895-1960) served three times as governor. Huey's son, US Senator Russell B. Long (b.1918), was chairman of the Finance Committee—and, consequently, one of the most powerful men in Congress—from 1965 to 1980.

Also prominent in Louisiana history were Robert Cavelier, Sieur de la Salle (b.France, 1643-87), who was the first to claim the region for the French crown; Pierre le Moyne, Sieur d'Iberville (b.Canada, 1661-1706), who commanded the expedition that first established permanent settlements in the lands La Salle had claimed; his brother, Jean Baptiste le Moyne, Sieur de Bienville (b.Canada, 1680-1768), governor of the struggling colony and founder of New Orleans; and Bernardo de Galvez (b.Spain, 1746-86), who, as governor of Spanish Louisiana during the last years of the American Revolution, conquered British-held Florida in a series of brilliant campaigns. William Charles Coles Claiborne (b.Virginia, 1775-1817) was the last territorial and first state governor of Louisiana. The state's first Republican governor, Henry Clay Warmoth (b.Illinois, 1842-1932), came there as a Union officer before the end of the Civil War and was sworn in at age 26. Jean Étienne de Boré (b.France, 1741-1820) laid the foundation of the Louisiana sugar industry by developing a process for granulating sugar from cane; Norbert Rillieux (birthplace unknown, 1806-94), a free black man, developed the much more efficient vacuum pan process of refining sugar.

Andrew Victor Schally (b.Poland, 1926), a biochemist on the faculty of the Tulane University School of Medicine, shared the Nobel Prize for medicine in 1977 for his research on hormones. Among other distinguished Louisiana professionals have been historian T. Harry Williams (1909-79), who won the Pulitzer Prize for his biography of Huey Long; architect Henry Hobson Richardson (1838-86); and four doctors of medicine: public health pioneer Joseph Jones (b.Georgia, 1833-96), surgical innovator Rudolph Matas (1860-1957), surgeon and medical editor Alton V. Ochsner (b.South Dakota, 1896), and heart specialist Michael De Bakey (b.1908).

Louisiana's important writers include George Washington Cable (1844-1925), an early advocate of racial justice; Kate O'Flaherty Chopin (b.Missouri, 1851-1904); playwright and memoirist Lillian Hellman (1905-84); and novelists Walker Percy (b.Alabama 1916), Truman Capote (1924-84), and Shirley Ann Grau (b.1929) and John Kennedy Toole (1937-69), both winners of the Pulitzer Prize.

Louisiana has produced two important composers, Ernest Guiraud (1837-92) and Louis Gottschalk (1829-69). Jelly Roll Morton (Ferdinand Joseph La Menthe, 1885-1941) and Sidney Bechet (1897-1959) were important jazz musicians, and Louis "Satchmo" Armstrong (1900-1971) was one of the most prolific jazz innovators and popular performers in the nation. The distinctive rhythms of pianist and singer Professor Longhair (Henry Byrd, 1918-80) were an important influence on popular music. Other prominent Louisianians in music are gospel singer Mahalia Jackson (1911-72), pianist-singer-songwriter Antoine "Fats" Domino (b.1928), and pop singer Jerry Lee Lewis (b.1935).

Louisiana baseball heroes include Hall of Famer Melvin Thomas "Mel" Ott (1909-58) and pitcher Ron Guidry (b.1950). Terry Bradshaw (b.1948), a native of Shreveport, quarterbacked the Super Bowl champion Pittsburgh Steelers during the 1970s. Player-coach William F. "Bill" Russell (b.1934) led the Boston Celtics to 10 National Basketball Association championships between 1956 and 1969. Chess master Paul Morphy (1837-84) was born in New Orleans.

[50]BIBLIOGRAPHY

Dufour, Charles L. *Ten Flags in the Wind: The Story of Louisiana.* New York: Harper & Row, 1967.

Federal Writers' Project. *Louisiana: A Guide to the State.* Reprint. New York: Somerset, n.d. (orig. 1941).

Hair, William Ivy. *Bourbonism and Agrarian Protest in Louisiana, 1877-1900.* Baton Rouge: Louisiana State University Press, 1965.

Jackson, Joy. *New Orleans in the Gilded Age: Politics and Urban Progress, 1880-96.* Baton Rouge: Louisiana State University Press, 1969.

Louisiana, State of. Secretary of State. *Roster of Officials, 1983.* Baton Rouge, 1983.

Louisiana Almanac, 1984-85. 11th ed. Gretna, La.: Pelican Publishing Co., 1984.

Shugg, Roger W. *Origins of Class Struggle in Louisiana: A Social History of White Farmers and Laborers During Slavery and After, 1840-75.* Baton Rouge: Louisiana State University Press, 1968 (orig. 1939).

Sitterson, J. Carlyle. *Sugar Country: The Cane Sugar Industry in the South, 1753-1950.* Lexington: University of Kentucky Press, 1953.

Taylor, Joe Gray. *Louisiana: A Bicentennial History.* New York: Norton, 1976.

Taylor, Joe Gray. *Louisiana Reconstructed, 1863-77.* Baton Rouge: Louisiana State University Press, 1974.

Taylor, Joe Gray. *Negro Slavery in Louisiana.* Baton Rouge: Louisiana Historical Association, 1963.

Warmoth, Henry Clay. *War, Politics, and Reconstruction: Stormy Days in Louisiana.* New York: Macmillan, 1930.

Williams, T. Harry. *Huey Long.* New York: Random House, 1981.

Winters, John D. *The Civil War in Louisiana.* Baton Rouge: Louisiana State University Press, 1979.

MAINE

State of Maine

ORIGIN OF STATE NAME: Either from the French for a historical district of France or from the early use of "main" to distinguish coast from islands. **NICKNAME:** The Pine Tree State. **CAPITAL:** Augusta. **ENTERED UNION:** 15 March 1820 (23d). **SONG:** "State of Maine Song." **MOTTO:** *Dirigo* (I direct). **COAT OF ARMS:** A farmer and sailor support a shield on which are depicted a pine tree, a moose, and water. Under the shield is the name of the state; above it, the state motto and the North Star. **FLAG:** The coat of arms on a blue field, with a yellow fringed border surrounding on three sides. **OFFICIAL SEAL:** Same as the coat of arms. **ANIMAL:** Moose. **BIRD:** Chickadee. **FISH.** Landlocked salmon. **FLOWER:** White pine cone and tassel. **TREE:** Eastern white pine. **INSECT:** Honeybee. **MINERAL:** Tourmaline. **LEGAL HOLIDAYS:** New Year's Day, 1 January; Birthday of Martin Luther King, Jr., 3d Monday in January; Washington's Birthday, 3d Monday in February; Patriots' Day, 3d Monday in April; Memorial Day, last Monday in May; Independence Day, 4 July; Labor Day, 1st Monday in September; Columbus Day, 2d Monday in October; Veterans Day, 11 November; Thanksgiving Day, 4th Thursday in November and day following; Christmas Day, 25 December. **TIME:** 7 AM EST = noon GMT.

¹LOCATION, SIZE, AND EXTENT

Situated in the extreme northeastern corner of the US, Maine is the nation's most easterly state, the largest in New England, and 39th in size among the 50 states.

The total area of Maine is 33,265 sq mi (86,156 sq km), including 30,995 sq mi (80,277 sq km) of land and 2,270 sq mi (5,879 sq km) of inland water. Maine extends 207 mi (333 km) E–W: the maximum N–S extension is 322 mi (518 km).

Maine is bordered on the N by the Canadian provinces of Quebec (with the line passing through the St. Francis River) and New Brunswick (with the boundary formed by the St. John River); on the E by New Brunswick (with the lower eastern boundary formed by the Chiputneticook Lakes and the St. Croix River); on the SE and S by the Atlantic Ocean; and on the W by New Hampshire (with the line passing through the Piscataqua and Salmon Falls rivers in the SW) and Quebec.

Hundreds of islands dot Maine's coast. The largest is Mt. Desert Island; others include Deer Isle, Vinalhaven, and Isle au Haut. The total boundary length of Maine is 883 mi (1,421 km).

The state's geographic center is in Piscataquis County, 18 mi (29 km) N of Dover-Foxcroft. The easternmost point of the US is West Quoddy Head, at 66°57′W.

²TOPOGRAPHY

Maine is divided into four main regions: coastal lowlands, piedmont, mountains, and uplands.

The narrow coastal lowlands extend, on average, 10–20 mi (16–32 km) inland from the irregular coastline, but occasionally disappear altogether, as at Mt. Desert Island and on the western shore of Penobscot Bay. Mt. Cadillac on Mt. Desert Island rises abruptly to 1,532 feet (467 meters), the highest elevation on the Atlantic coast north of Rio de Janeiro, Brazil. The transitional hilly belt, or piedmont, broadens from about 30 mi (48 km) wide in the southwestern part of the state to about 80 mi (129 km) in the northeast.

Maine's mountain region, the Longfellow range, is at the northeastern end of the Appalachian Mountain system. This zone, extending into Maine from the western border for about 150 mi (240 km) and averaging about 50 mi (80 km) wide, contains nine peaks over 4,000 feet (1,200 meters), including Mt. Katahdin, which at 5,268 feet (1,607 meters) is the highest point in the state. The summit of Katahdin marks the northern terminus of the 2,000-mi (3,200-km) Appalachian Trail. Maine's uplands form a high, relatively flat plateau extending northward beyond the mountains and sloping downward toward the north and east. The eastern part of this zone is the Aroostook potato-farming region; the western part is heavily forested.

Of Maine's more than 2,200 lakes and ponds, the largest are Moosehead Lake, 117 sq mi (303 sq km), and Sebago Lake, 13 mi (21 km) by 10 mi (16 km). Of the more than 5,000 rivers and streams, the Penobscot, Androscoggin, Kennebec, and Saco rivers drain historically and commercially important valleys. The longest river in Maine is the St. John, but it runs for most of its length in the Canadian province of New Brunswick.

³CLIMATE

Maine has three climatic regions: the northern interior zone, comprising roughly the northern half of the state, between Quebec and New Brunswick; the southern interior zone; and the coastal zone. The northern zone is both drier and cooler in all four seasons than either of the other zones, while the coastal zone is more moderate in temperature year-round than the other two.

The annual mean temperature in the northern zone is about 40°F (5°C); in the southern interior zone, 44°F (7°C); and in the coastal zone, 46°F (8°C). Record temperatures for the state are −48°F (−44°C), registered at Van Buren on 19 January 1925, and 105°F (41°C) at North Bridgton on 10 July 1911. The mean annual precipitation increases from 40.2 in (102 cm) in the north to 41.5 in (105 cm) in the southern interior and 45.7 in (116 cm) on the coast. Average annual snowfall is 78 in (198 cm).

⁴FLORA AND FAUNA

Maine's forests are largely softwoods, chiefly red and white spruces, balsam fir *(Abies balsamea),* eastern hemlock, and white and red pine. Important hardwoods include beech, yellow and white birches, sugar and red maples, white oak, black willow, black and white ashes, and American elm, which has fallen victim in recent years to Dutch elm disease. Maine is home to most of the flowers and shrubs common to the north temperate zone, including an important commercial resource, the low-bush blueberry. Maine has 17 rare orchid species, of which 4 are considered threatened; one species, the small whorled pogonia, is on the federal endangered list. The Furbish lousewart is also classified as endangered.

About 30,000 white-tailed deer are killed by hunters in Maine each year, but the herd does not appear to diminish. Moose hunting was banned in Maine in 1935; however, in 1980, 700 moose-hunting permits were issued for a six-day season, and moose hunting has continued despite attempts by some residents to ban the practice. Other common forest animals include the bobcat, beaver, muskrat, river otter, mink, fisher, raccoon, red fox, and snowshoe hare. The woodchuck is a conspicuous inhabitant of pastures, meadows, cornfields, and vegetable gardens. Seals, porpoises, and occasionally finback whales are found in coastal waters, along with virtually every variety of North Atlantic fish and shellfish, including the famous Maine lobster. Coastal water-fowl include the osprey, herring and great black-backed gulls, great and double-crested cormorants, and various duck species. Matinicus Rock, a small uninhabited island about 20 mi (32 km) off the coast near the entrance to Penobscot Bay, is the only known North American nesting site of the common puffin, or sea parrot.

Endangered species include the cougar, bald eagle, peregrine and Arctic falcons, shortnose sturgeon, and Indiana bat.

5ENVIRONMENTAL PROTECTION

The Department of Environmental Protection administers laws regulating the selection of commercial and industrial sites, air and water quality, the prevention and cleanup of oil spills, the control of hazardous wastes, the licensing of oil terminals, the use of coastal wetlands, and mining. The Land Use Regulation Commission, established in 1969, extends the principles of town planning and zoning to Maine's 407 unorganized townships, 55 "plantations," and numerous coastal islands that have no local government and might otherwise be subject to ecologically unsound development.

Several years of spruce budworm epidemics, combined with poor management, have left the state with a population of over-age trees, according to a 1983 US Forest Service report. Between 1974 and 1984, 15 million cords of softwood trees were killed by the spruce budworm.

6POPULATION

Maine's 1980 census population was 1,124,660 (38th among the 50 states), a 13.3% increase from 1970. The population was estimated at 1,156,539 in early 1985, a 2.8% increase from 1980.

The area that now comprises the State of Maine was sparsely settled throughout the colonial period. At statehood, Maine had 298,335 residents. The population doubled by 1860, but then grew slowly until the 1970s, when its growth rate outstripped the nation's.

The population density for 1985 was 37 per sq mi (14 per sq km); more than half the population lives on less than one-seventh of the land within 25 mi (40 km) of the Atlantic coast, and almost half the state is virtually uninhabited. Although almost half of Maine's population is classified as urban (47.5%), much of the urban population lives in towns and small cities. The state's largest cities are Portland, with 61,803 people in 1984; Lewiston, 39,405; Bangor, 30,827; and Auburn, 22,952. The Portland metropolitan area had an estimated population of 210,854 in 1983.

7ETHNIC GROUPS

Maine's population is primarily Yankee, both in its English and Scotch-Irish origins and in its retention of many of the values and folkways of rural New England. The largest minority group consists of French-Canadians. Among those reporting a single ancestry group in 1980, 259,519 claimed English ancestry; 147,058 French (not counting 30,802 who claimed Canadian or French-Canadian); and 56,335 Irish. There were 43,402 foreign-born.

The most notable ethnic issue in Maine during the 1970s was the legal battle of the Penobscot and Passamaquoddy Indians—living on two reservations covering 27,546 acres (11,148 hectares)—to recover 12,500,000 acres (5,059,000 hectares) of treaty lands. A compromise settlement in 1980 awarded them $81.5 million, two-thirds of which went into a fund enabling the Indians to purchase 300,000 acres (121,000 hectares) of timberland.

As of 1980, Maine had 5,005 Hispanics, 4,057 Indians, 3,128 blacks, and 2,947 Asians and Pacific Islanders.

8LANGUAGES

Descendants of the Passamaquoddy and Penobscot Indians of the Algonkian family who inhabited Maine at the coming of the white man still lived there in the mid-1980s. Algonkian place-names abound: Saco, Millinocket, Wiscasset, Kennebec, Skowhegan.

Maine English is celebrated as typical Yankee speech. Final /r/ is absent, a vowel sound between /ah/ and the /a/ in cat appears in car and garden, aunt and calf. Coat and home have a vowel that to outsiders sounds like the vowel in cut. Maple syrup comes from rock or sugar maple trees in a sap or sugar orchard, cottage cheese is curd cheese, and pancakes are fritters.

In 1980, 89% of Maine residents 3 years old or older reported speaking only English in the home; 95,181 residents spoke French.

The decline of parochial schools and the great increase in numbers of young persons attending college have begun to erode the linguistic and cultural separateness that marks the history of the Franco-American experience in Maine.

9RELIGIONS

Maine had 273,292 Roman Catholics in 1984 and an estimated 8,185 Jews in 1983. The leading Protestant denominations were United Methodist, with 37,824; American Baptist USA, with 37,804; United Church of Christ, 33,078; and Episcopal, 22,174.

10TRANSPORTATION

Railroad development in Maine, which reached its peak in 1924, has declined rapidly since World War II, and passenger service has been dropped altogether. The Boston & Maine railroad carried freight on 1,454 mi (2,340 km) of main-line track in 1983; in all, there were 2,249 (3,619 km) of track in use.

About three-quarters of all communities and about half the population depend entirely on highway trucking for the overland transportation of freight. In 1982, Maine had 21,953 mi (35,330 km) of roads. There were 765,781 registered motor vehicles and 770,240 licensed drivers in 1983. The Maine Turnpike and I-95, which coincide between Portland and Kittery, are the major highways.

River traffic has been central to the lumber industry; only since World War II has trucking replaced seasonal log drives downstream from timberlands to the mills, a practice that is now outlawed for environmental reasons. Maine has 10 established seaports, with Portland and Searsport the main depots for overseas shipping; in 1983, 6 Maine ports handled 255 million lb of exports and 11.8 million lb of imports. Waterborne freight traffic had a total value of $2.3 billion in 1983. In 1982, Portland harbor handled 10,455,784 tons; Bucksport, 860,897; and Searsport, 793,717. Crude oil, fuel oil, and gasoline were the chief commodities. There were 146 airfields in 1983; Portland International Jetport was the largest and most active, enplaning 331,078 passengers.

11HISTORY

The first inhabitants of Maine—dating from 3000 to 1000 BC—are known to archaeologists as the Red Paint People because of the red ocher that has been found in their graves. This Paleolithic group had evidently disappeared long before the arrival of the Algonkian-speaking Abnaki (meaning "living at the sunrise"), or Wabanaki. Just at the time of European settlement, an intertribal war and a disastrous epidemic of smallpox swept away many of the Abnaki, some of whom had begun peaceful contacts with the English. After that, most Indian contacts with Europeans were with the French.

The first documented visit by a European to the Maine coast was that of Giovanni da Verrazano during his voyage of 1524, but one may infer from the record that the Abnaki he met there had encountered white men before. Sometime around 1600, English expeditions began fishing the Gulf of Maine regularly. The first

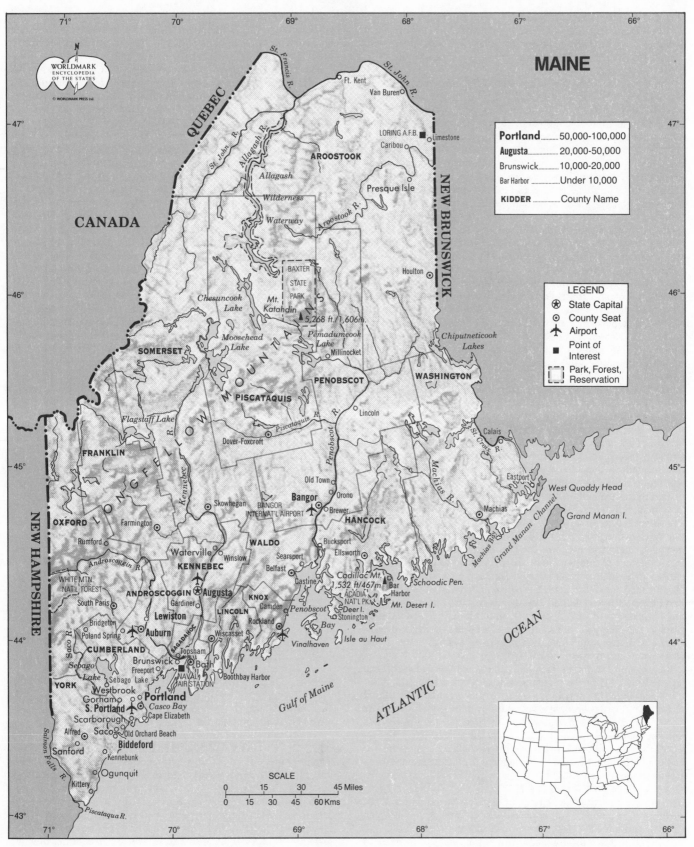

MAINE

Portland............	50,000-100,000
Augusta............	20,000-50,000
Brunswick............	10,000-20,000
Bar Harbor	Under 10,000
KIDDER	County Name

LEGEND
⊗ State Capital
⊙ County Seat
✈ Airport
■ Point of Interest
▢ Park, Forest, Reservation

See endsheet maps: N1.
LOCATION: 43°04' to 47°28'N; 66°57' to 71°07'W. **BOUNDARIES:** Canadian line, 481 mi (774 km); Atlantic Ocean coastline, 228 mi (367 km); New Hampshire line, 174 mi (280 km).

recorded attempts to found permanent colonies, by the French on an island in the St. Croix River in 1604 and by the English at Sagadahoc in 1607, both failed. By 1630, however, there were permanent English settlements on several islands and at nearly a dozen spots along the coast.

The first grant of Maine lands was to Sir Ferdinando Gorges from the Council for New England, a joint-stock company that received and made royal grants of New England territory and which Gorges himself dominated. He and Captain John Mason received the territory between the Merrimack River (in present-day New Hampshire and Massachusetts) and Kennebec River in 1622. Seven years later, the two grantees divided their land at the Piscataqua River, and Gorges became sole proprietor of the "Province of Maine." The source of the name is not quite clear. It seems likely that some connection with the historical French province of the same name was intended, but the name was also used to distinguish the mainland from the islands.

Sir Ferdinando's various schemes for governing the territory and promoting a feudal-style settlement never worked. A few years after his death in 1647, the government of the Massachusetts Bay Colony began absorbing the small Maine settlements. Massachusetts purchased the title to Maine from the Gorges heirs in 1677, and Maine became a district of Massachusetts with the issuance of a new royal charter in 1691. During the first hundred years of settlement, Maine's economy was based entirely on fishing, trading, and exploitation of the forests. The origin of the Maine shipbuilding industry, the early settlement of the interior parts of southern Maine, and the beginning of subsistence farming all date from about the time that New England's supply center of white-pine masts for the Royal Navy moved from Portsmouth, N.H., to Falmouth (now the city of Portland).

The first naval encounter of the Revolutionary War occurred in Machias Bay, when, on 12 June 1775, angry colonials captured the British armed schooner *Margaretta*. On 8 October 1775, a British naval squadron shelled and set fire to Falmouth. Wartime Maine was the scene of two anti-British campaigns, both of which ended in failure: an expedition through the Maine woods in the fall of 1775 intended to drive the British out of Quebec, and a disastrous 1779 expedition in which a Massachusetts amphibious force, failing to dislodge British troops at Castine, scuttled many of its own ships near the mouth of the Penobscot River.

The idea of separation from Massachusetts began surfacing as early as 1785, but popular pressure for such a movement did not mount until the War of 1812. The overwhelming vote for statehood in an 1819 referendum was a victory for William King, who would become the first governor, and his fellow Jeffersonian Democratic-Republicans. Admission of Maine as a free state was joined with the admission of Missouri as a slave state in the Missouri Compromise of 1820.

Textile mills and shoe factories came to Maine between 1830 and 1860 as part of the industrialization of Massachusetts. After the Civil War, the revolution in papermaking that substituted wood pulp for rags brought a vigorous new industry to Maine. By 1900, Maine was one of the leading papermaking states in the US, and the industry continues to dominate the state today. The rise of tourism and the concurrent and often conflicting concerns for economic development and environmental protection have been the main themes since the 1940s.

[12] STATE GOVERNMENT

The Maine constitution, based on that of Massachusetts but incorporating a number of more democratic features, was adopted in 1819 and amended 150 times by the end of 1983.

The bicameral legislature, consisting of a senate of from 31 to 35 members, depending on the number of districts (35 in 1985), and a 151-member house of representatives, convenes bienially in joint session to elect the secretary of state, attorney general, and state treasurer. All legislators serve two-year terms.

The governor, who serves a four-year term and is limited to two consecutive terms, is the only official elected statewide. A gubernatorial veto may be overridden by a two-thirds vote of members present and voting in each legislative chamber.

The state constitution may be amended by a two-thirds vote of the legislature and a majority vote at the next general election. To vote in Maine, one must be a US citizen and at least 18 years of age; there is no minimum residency requirement.

[13] POLITICAL PARTIES

Maine's two major political parties are the Democratic and the Republican, each affiliated with the national party. Minor parties have not figured in Maine elections in this century, although an independent candidate, James B. Longley, beat the candidates of both major parties in the gubernatorial election of 1974.

During the early decades of statehood, Jeffersonian and Jacksonian Democrats remained in power quite consistently. In 1854, however, reformers rallied around the new Republican Party, which dominated Maine politics for the next hundred years. Maine's strong Republican tradition continued into the middle and late 1950s, when Margaret Chase Smith distinguished herself in the US Senate as a leader of national importance. The rise of Democrat Edmund S. Muskie, elected governor in 1954 and 1956 and to the first of four terms in the US Senate in 1960, signaled a change in Maine's political complexion. Muskie appealed personally to many traditionally Republican voters, but his party's resurgence was also the result of demographic changes, especially an increase in the proportion of French-Canadian voters. There were 766,285 registered voters in 1982.

In 1985, Maine had a Democratic governor (Joseph Brennan) and one Democratic and one Republican senator. Both US House of Representative seats were held by Republicans, but both state legislative houses had Democratic majorities. Maine has a high proportion of women state legislators—23.7% in 1985.

Maine Presidential Vote by Major Political Parties, 1948–1984

YEAR	ELECTORAL VOTE	MAINE WINNER	DEMOCRAT	REPUBLICAN
1948	5	Dewey (R)	111,916	150,234
1952	5	*Eisenhower (R)	118,806	232,353
1956	5	*Eisenhower (R)	102,468	249,238
1960	5	Nixon (R)	181,159	240,608
1964	4	*Johnson (D)	262,264	118,701
1968	4	Humphrey (D)	217,312	169,254
1972	4	*Nixon (R)	160,584	256,458
1976	4	Ford (R)	232,279	236,320
1980	4	*Reagan (R)	220,974	238,522
1984	4	*Reagan (R)	214,515	336,500

* Won US presidential election.

[14] LOCAL GOVERNMENT

The principal units of local government in 1982 were the 22 cities and 475 towns; in all, there were 806 local government units. As is customary in New England, the basic instrument of town government is the annual town meeting, with an elective board of selectmen supervising town affairs between meetings; some of the larger towns employ full-time town managers. There is no local government in roughly half the state. Maine's 16 counties function primarily as judicial districts.

[15] STATE SERVICES

There were 20 state executive departments in 1984. The State Board of Education and Department of Educational and Cultural

Services supervise the public education system. The Department of Transportation, established in 1972, includes divisions responsible for aviation and railroads, a bureau to maintain highways and bridges, the Maine Port Authority, the State Ferry Advisory Board, and the Maine Aeronautical Advisory Board.

Various agencies responsible for health and social welfare were combined into the Department of Human Services in 1975. The Maine State Housing Authority, established in 1969, provides construction loans and technical assistance, and conducts surveys of the state's housing needs. The Commission on Governmental Ethics and Election Practices, an advisory and investigative body, was created in 1975 to serve as a watchdog over the legislature.

16JUDICIAL SYSTEM

The highest state court is the supreme judicial court, with a chief justice and six associate justices appointed by the governor, with the consent of the legislature, for seven-year terms, as are all other state judges. The supreme judicial court has statewide appellate jurisdiction in all civil and criminal matters. The 15-member superior court, which has original jurisdiction in cases involving trial by jury and also hears some appeals, holds court sessions in all counties. The 22-member district court consists of a chief judge, 14 judges who sit in 13 districts, and 7 at-large judges. They hear non-felony criminal cases and small claims and juvenile cases, and have concurrent jurisdiction with the superior court in divorce and civil cases involving less than $30,000. A probate court judge is elected in each county. Maine's crime rate of 3,681 per 100,000 persons in 1983 was far below the national average.

There were 1,049 state and federal prisoners at the end of 1983. At the end of 1984 there were 2,064 practicing lawyers.

17ARMED FORCES

Major US military installations in Maine include Loring Air Force Base in Limestone, housing units of the Strategic Air Command; the Portsmouth Naval Shipyard in Kittery, site of an important Atlantic coast nuclear submarine repair facility; the Naval Air Station at Brunswick, home of a wing of antisubmarine patrol squadrons; and a large Coast Guard station in Portland that is home port for oceangoing cutters. Defense Department personnel in Maine totaled 7,510 in 1983/84. In that year, state firms received $449 million in defense contracts, of which the Bath Iron Works, which builds warships and is the state's largest private employer, received $351 million.

There were 153,000 veterans of US military service living in Maine as of 30 September 1983, including 2,000 veterans of World War I, 59,000 of World War II, 29,000 of the Korean conflict, and 46,000 from the Viet-Nam era. A total of $158 million in veterans' benefits were paid in 1982/83.

The Maine Army National Guard consisted of 3,237 personnel at the end of 1984, and the Air National Guard had 1,334 personnel in early 1985. State and local police totaled 2,250 in 1983.

18MIGRATION

Throughout the colonial, Revolutionary, and early national periods, Maine's population grew primarily by immigration from elsewhere in New England. About 1830, after agriculture in the state had passed its peak, Maine farmers and woodsmen began moving west. Europeans and French Canadians came to the state, but not in sufficient numbers to offset this steady emigration.

Net losses from migration have continued through most of this century. Between 1940 and 1970, for example, the net loss was 163,000. However, there was a net gain of 43,100 from 1970 to 1983.

19INTERGOVERNMENTAL COOPERATION

Regional agreements in which Maine participates include the Maine—New Hampshire School District Compact, which authorizes interstate public school districts. Maine also takes part in the New England Interstate Water Pollution Control Compact.

In 1983/84, Maine received $590.3 million in federal grants.

20ECONOMY

Maine's greatest economic strengths, as they have been since the beginning of European settlement, are its forests and waters, yielding wood products, waterpower, fisheries, and ocean commerce. Today, the largest industry by far is paper manufacturing, for which both forests and waterpower are essential.

Maine's greatest current economic weakness is its limited access to the national transportation network that links major production and manufacturing centers with large metropolitan markets. On the other hand, this relative isolation, combined with the state's traditional natural assets, has contributed to Maine's attractiveness as a place for tourism and recreation.

Manufacturing accounted for 27% of the gross state product of $10.8 billion in 1982, of which paper (7%) was the largest portion. Trade accounted for 17%; government, 15%; services, 14%; and finance, insurance, and real estate, 12%.

21INCOME

Personal income in 1983 was $9,619 per capita, 40th in the US and the lowest in New England; total personal income was $11 billion. However, real per capita income grew by 5.2% during 1981–83, compared to only 3.2% for the nation. The median money income of four-person families was $22,201 in 1981, 46th in the US. In 1979, 141,000 persons (13% of the population) were below the federal poverty level.

22LABOR

Maine's civilian labor force totaled 528,100 in November 1984; 56% of the work force was male and 44% female. The unemployment rate in November 1984 was 5.3%, the lowest rate in 11 years. The number of unemployed was 28,000.

A federal census in March 1982 revealed the following nonfarm employment pattern in Maine:

	ESTABLISH-MENTS	EMPLOYEES	ANNUAL PAYROLL ('000)
Agricultural services, forestry, fishing	306	1,487	$ 16,545
Mining	18	97	1,958
Contract construction	2,395	14,164	247,536
Manufacturing	1,830	110,204	1,779,831
Transportation, public utilities	1,064	14,899	276,896
Wholesale trade	1,620	19,036	291,274
Retail trade	7,062	68,429	582,783
Finance, insurance, real estate	1,713	17,996	259,706
Services	7,035	76,645	858,874
Other	442	328	4,467
TOTALS	23,485	323,285	$4,319,870

Among the workers not covered by this survey are government employees, of whom Maine had 83,400 in 1984.

In September 1984, the average Maine production worker on a manufacturing payroll earned $8.20 per hour for 40.3 hours of work, for a weekly total of $330. Blue-collar workers accounted for 39.6% of the labor force in 1980, 3d highest in the nation.

Labor union membership among Maine workers in nonagricultural jobs in 1982 amounted to 84,200; there were 366 labor unions in 1983.

23AGRICULTURE

Maine's gross farm income in 1983 was $454 million (44th in the US); a net deficit of $28.3 million that year left Maine one of only three states (the others were West Virginia and Wyoming) to lose money on farming. There were 8,000 farms in 1984, with an estimated 2,000,000 acres (809,000 hectares) of land.

Potatoes, grown primarily in Aroostook County, are by far the most important crop. Maine was the 3d-largest potato-growing state in the US in 1983, producing 22,090,000 hundredweight.

Other crops included commercial apples, 84,000,000 lb; oats, 2,356,000 bushels; and hay, 425,000 tons. Maine is also a leading producer of blueberries.

24ANIMAL HUSBANDRY

Livestock and livestock products accounted for more than three-fifths of Maine's income from farm marketings in 1983. In 1983, egg production exceeded 1.4 billion, yielding $94.2 million in gross income, and sales of 5.1 million chickens brought more than $2.8 million. South-central Maine is the leading poultry region.

There were 148,000 cattle and calves on Maine farms at the end of 1983, when production totaled 36.7 million lb worth $15.5 million. The milk output, 741 million lb, was valued at $105.8 million.

25FISHING

Fishing has been important to the economy of Maine since its settlement. In 1984, 179.1 million lb of finfish and shellfish worth $107.6 million were landed at Maine ports, ranking the state 8th and 7th in the nation, respectively. Rockland and Portland provided about half the catch. The most valuable Maine fishery product is the lobster. Flounder, halibut, scallops, and shrimp are also caught. The state sought during the late 1970s and early 1980s to conserve and restore Atlantic salmon stocks in Maine's inland waterways.

Shipments of fishery products in 1982 included $68.8 million in canned and cured seafood and $46.6 million in fresh or frozen packaged fish. Average employment in 218 processing and whole-saling plants was 2,501.

26FORESTRY

Maine's 17,749,000 acres (7,183,000 hectares) of forests in 1979 contained an estimated 3.8 billion trees and covered 90% of the state's land area, the largest percentage of any state in the US. About 17,200,000 acres (6,960,000 hectares) are classified as commercial timberland, 98% of it privately owned, and much of that by a few large paper companies. Timber operations in 1983 produced 757 million board feet of softwood lumber, 146 million board feet of hardwood lumber, and 2.5 million cords of pulpwood. Lumber and wood products establishments employed 12,300 persons in 1982; value added by manufacture was $319.7 million, and value of shipments was $883.4 million.

27MINING

The value of Maine's nonfuel mineral output in 1984 was only $34 million (46th in the US). Mining from a garnet deposit in West Paris, in the southwestern part of the state, made Maine the nation's leading garnet producer. A 36-million-ton copper-zinc deposit was discovered in 1977 in northern Maine. The 1984 mineral output included stone, 9,000,000 tons; sand and gravel, 6,400,000 tons; and clays, 49,000 tons.

28ENERGY AND POWER

For more than three centuries, Maine has been exploiting its enormous waterpower potential. In recent decades, however, waterpower has been surpassed in importance by oil-fired steam plants and, most recently, by nuclear power. In 1983, the Maine Yankee Atomic Power Co. station in Wiscasset generated 59% of the state's electric power; in referendums in 1980 and 1982, voters decided that the station should remain open and that future nuclear power development should be allowed. Oil-fired steam units accounted for 21% of electric power generation, and hydroelectric units for 19%.

Installed generating capacity in 1983 totaled 2,392,000 kw, consisting of 370,000 kw in hydroelectric plants, 1,064,000 kw in conventional steam plants, 864,000 kw in the Wiscasset nuclear plant, and 89,000 kw in turbines and internal-combustion plants. Power production in 1983 totaled 9.7 billion kwh.

All fuel oil and coal must be imported; natural gas, piped into the southwest corner of the state, is available in Portland and the Lewiston-Auburn area.

The Office of Energy Resources provides tax incentives and research and development grants to encourage use of solar power and energy conservation. The late 1970s and early 1980s saw some Maine homeowners switch to wood for heating as an alternative to oil.

29INDUSTRY

Manufacturing in Maine has always been related to the forests. From the 17th century through much of the 19th, the staples of Maine industry were shipbuilding and lumber; today they are papermaking and wood products, but footwear, textiles and apparel, shipbuilding, and electronic components and accessories are also important items.

Maine has the largest paper-production capacity of any state in the nation. There are large papermills and pulpmills in more than a dozen towns and cities; major companies include International Paper, Boise-Cascade, Scott Paper, US Gypsum, and Great Northern Nekoosa. Wood-related industries—paper, lumber, wood products, and furniture—accounted for 42% of the value of manufacturers shipments in 1982.

Value of shipments in 1982 exceeded $8.6 billion, of which paper and allied products contributed 31%; leather and leather products, 11%; food and kindred products, 11%; and lumber and wood products, 10%. The following table shows value of shipments for selected industries in 1982:

Paper mill products	$2,116,900,000
Footwear, except rubber	754,200,000
Electric and electronic equipment	422,800,000
Nonelectrical machinery	413,000,000
Textile mill products	410,000,000
Rubber and miscellaneous plastic products	314,000,000
Electronic components and accessories	232,600,000

30COMMERCE

In 1982, Maine's wholesale trade totaled $4.2 billion. Retail sales came to $5.3 billion in 1982, with grocery stores accounting for 24%, automotive dealers 18%, eating and drinking places 8%, department stores 7%, gasoline service stations 7%, and fuel oil dealers 7%.

Maine conducts substantial foreign commerce through Portland, Searsport, and Buckport. In 1982, Portland imported 7,068,651 tons of cargo, mostly crude oil and fuel oil; in the same year, Searsport imported a total of 696,126 tons, mostly crude oil and salt. The value of Maine's exports amounted to $498 million of manufactured goods in 1981 and $28 million in agricultural products in 1982/83.

31CONSUMER PROTECTION

The Bureau of Consumer Credit Protection was established in 1974 to protect state residents from unjust and misleading consumer credit practices, particularly in relation to the federal Truth-in-Lending Act. The bureau also administers state laws regulating home-repair finance, collection agencies, insurance premium finance companies, simplified consumer loan contracts, and fair credit reporting.

32BANKING

In 1984, Maine had 47 banks, of which 26 were commercial and 21 mutual savings banks. Commercial banks had assets of $4.2 billion and deposits of over $3.5 billion at the end of 1983. There were 14 savings and loan associations at the end of 1983 with $637 million in assets.

33INSURANCE

Five life insurance companies were domiciled in Maine in mid-1983, when 1,833,000 policies worth $18.9 billion were in force. The average coverage per family was $41,400 the 3d lowest in the US and the lowest of any northeastern state. Property and liability insurers wrote $523.6 million in premiums in 1983, of which 35% was for automotive coverage.

34 SECURITIES

There are no securities exchanges in Maine. In 1983, New York Stock Exchange firms had 23 sales offices and 166 full-time registered representatives in Maine. Some 165,000 Maine residents owned shares of public corporations that year.

35 PUBLIC FINANCE

Maine's biennial budget is prepared by the Bureau of the Budget, within the Department of Finance and Administration, and submitted by the governor to the legislature for consideration. The fiscal year extends from 1 July to 30 June. The following table shows revenues and expenditures for 1983/84:

REVENUES	
Sales and use tax	$ 314,702,859
Individual income tax	261,889,017
Other taxes	302,831,144
Federal revenues	457,349,083
Other revenues and resources	147,729,955
TOTAL	$1,484,502,058
EXPENDITURES	
Human services	$ 509,565,159
Education and culture	442,365,472
General government	183,914,136
Transportation	170,096,611
Natural resources	46,992,373
Manpower	36,461,429
Other outlays	50,879,278
TOTAL	$1,440,274,458

As of mid-1982, the outstanding state and local government debt was almost $1.5 billion; the debt per capita was $1,328, 25% below the national average.

36 TAXATION

The leading source of tax revenue as of 1984 was a sales and use tax of 5%. The individual income tax ranged from 1% of the first $2,000 of taxable income to 10% of income over $25,000; the corporate income tax rates ranged from 3.5% of the first $25,000 of net income to 8.93% of net income in excess of $25,000. Other state levies include taxes on utilities, inheritance and estate taxes, liquor and cigarette taxes, and a tax on gasoline, fuel, and motor carriers. Counties do not assess taxes, but they do make levies on municipalities and unorganized territories to meet county budgets. In 1983/84, federal expenditures in the state were over $3.3 billion. In 1983, state residents filed 473,670 federal income tax returns, paying $992 million in tax. Federal expenditures traditionally outstrip the state's federal tax burden by a large amount.

37 ECONOMIC POLICY

The Finance Authority of Maine (FAME) encourages industrial and recreational projects by insuring mortgage loans, selling tax-exempt bonds to aid industrial development and natural-resource enterprises, authorizing municipalities to issue such revenue bonds, and guaranteeing loans to small businesses, veterans, and natural-resource enterprises. In 1983/84, FAME approved 65 loans to businesses, with $45,683,434 in total financing. A corporate franchise tax was repealed in December 1974, and the personal property tax on business inventories in April 1977. The State Development Office provides technical, financial, training, and marketing assistance for existing Maine businesses and companies interested in establishing operations in the state.

38 HEALTH

The death rate of 9.2 per 100,000 in 1981 was somewhat higher than the US norm, reflecting a higher than average population in the upper age levels. The birthrate in 1982 was 14.7 per 1,000 population. The infant mortality rate for whites in 1981 was 11.1 per 1,000 live births (the nonwhite population was too small for accurate statistical sampling). During 1982, 5,500 legal abortions were performed. Death rates in 1981 for the leading causes of

death were heart disease, 360 per 1,000 population; and cancer, 205, both above the national average.

In 1983, Maine had 47 hospitals, with 6,359 beds. Hospital personnel included 4,455 registered nurses and 1,320 licensed practical nurses. The average cost of hospital care in 1982 was $296 per day and $2,279 per stay. Licensed medical personnel included 2,103 physicians and 504 dentists in 1982. The $70 million that the state government spent in 1981/82 for health and hospitals was only about $61 per person—lowest in the nation.

39 SOCIAL WELFARE

Despite Maine's relatively low personal income and large proportion of residents below the poverty level, welfare payments per capita generally fall short of the national norms. In 1982, for example, payments of $59 million to Maine's 47,600 recipients of aid to families with dependent children resulted in an average monthly payment of $292 per family. During 1983, 120,000 Maine residents received food stamps at a federal cost of $71 million, and 117,000 students took part in the school lunch program, costing $12 million.

In 1983, Social Security payments totaled $875 million; Supplemental Security Income benefits reached $35 million. Payments for unemployment insurance were $83 million in 1982. Workers' compensation payments were $123.7 million that same year.

40 HOUSING

Housing for Maine families has improved substantially since 1960, when the federal census categorized 57,000 of Maine's 364,650 housing units as deteriorated or dilapidated. Between 1970 and 1980, over 115,000 new units were built. Almost 5% of all occupied units in 1980 lacked full plumbing, however.

Of an estimated 429,341 year-round housing units in 1980, 285,166 were single-family units, 112,361 were in multifamily dwellings, and 31,814 were mobile homes. About one-seventh of all Maine homes are for seasonal rather than year-round use.

41 EDUCATION

Maine has a long and vigorous tradition of education at all levels, both public and private. In 1980, 68.7% of the adult population had completed high school, but only 14.4% college, the lowest rate in New England. In 1982/83, Maine had 747 public schools with a total enrollment of 196,000; the private school enrollment was 17,540 in the fall of 1980.

Since 1968, the state's system of state colleges and universities has been incorporated into a single University of Maine, which in fall 1983 had 28,591 students. The original and chief campus is at Orono; the other major element in the system is the University of Southern Maine at Portland and Gorham. The state also operates the Maine Maritime Academy at Castine and 6 vocational-technical institutes. Of the state's 16 private colleges and professional schools, Bowdoin College in Brunswick, Colby College in Waterville, and Bates College in Lewiston are the best known.

42 ARTS

Maine has long held an attraction for painters and artists, Winslow Homer and Andrew Wyeth among them. The state abounds in summer theaters, the oldest and most famous of which is at Ogunquit. The Portland Symphony is Maine's leading orchestra; another ensemble is in Bangor.

In 1979, Maine became the first state to allow inheritance taxes to be paid with acceptable art. The Department of Educational and Cultural Services has an Arts and Humanities Bureau that provides funds to artists in residence, Maine touring artists, and community arts councils.

43 LIBRARIES AND MUSEUMS

In 1982, Maine public libraries had 4,263,218 volumes and a combined circulation of 5,705,869. Leading libraries and their book holdings in 1983 included the Maine State Library at Augusta (450,000 volumes), the University of Maine at Orono (574,000), Bowdoin College at Brunswick (639,000), and the University of Southern Maine at Portland (338,000).

Maine has at least 110 museums and historic sites. The Maine State Museum in Augusta houses collections in history, natural history, anthropology, marine studies, mineralogy, science, and technology. The privately supported Maine Historical Society in Portland maintains a research library and the Wadsworth-Longfellow House, the boyhood home of Henry Wadsworth Longfellow. The largest of several maritime museums is in Bath.

44COMMUNICATIONS

In 1980, 93% of the 395,184 occupied housing units had telephones. There were 492 post offices in 1985, with 3,391 employees.

Maine had 97 commercial radio stations (39 AM, 58 FM) in 1983, along with 7 commercial television stations. Educational television stations broadcast from Augusta, Biddeford, Calais, Orono, and Presque Isle. In 1984, 43 cable television systems served 181,271 subscribers in 151 communities.

45PRESS

Maine had eight daily newspapers with a total weekday circulation of 275,064 in 1983. The most widely read newspaper was the *Bangor Daily News* (mornings, 78,352, and weekend, 91,494), though its circulation was surpassed by the combined circulations of the *Portland Press Herald* (mornings, 61,149) and *Evening Express* (29,328), both published daily in Portland by Gannett Publishing Co. Gannett also published Maine's only Sunday newspaper, the *Maine Sunday Telegram* (131,626).

46ORGANIZATIONS

The 1982 US Census of Service Industries counted 375 organizations in Maine, including 73 business associations; 226 civic, social, and fraternal associations; and 9 educational, scientific, and research associations. Among the organizations with headquarters in Maine are the Maine Potato Council (Presque Isle); the Maine Lobstermen's Association (Stonington); and the Potato Association of America (Orono).

47TOURISM, TRAVEL, AND RECREATION

Calling itself "Vacationland," the State of Maine is a year-round resort destination. Expenditures by tourists were estimated at over $1.2 billion in 1982. Travel and tourism was the state's largest employer, generating 38,000 jobs.

Most out-of-state visitors continue to come in the summer, when the southern coast offers sandy beaches, icy surf, and several small harbors for sailing and saltwater fishing. Northeastward the scenery becomes more rugged and spectacular, and sailing and hiking are the primary activities. Hundreds of lakes, ponds, rivers, and streams offer opportunities for freshwater bathing, boating, and fishing. Whitewater canoeing lures the adventurous along the Allagash Wilderness Waterway in northern Maine. Maine has always attracted hunters, especially during the fall deer season. Wintertime recreation facilities include nearly 60 ski areas and countless opportunities for cross-country skiing. In 1982/83, Maine issued licenses to 241,733 hunters and 279,648 fishermen.

In 1981 there were 29 state parks and beaches. Baxter State Park, in central Maine, includes Mount Katahdino. In 1983, annual attendance at state recreation areas was 2,858,325; Acadia National Park, a popular attraction, drew 3,734,763 visitors in 1984. There are 4 other federal parks, forests, and wildlife areas, and 7 national wildlife refuges. The state fair is at Bangor.

48SPORTS

The Maine Mariners of the American Hockey League play on their home ice at the Cumberland County Civic Center in Portland. The state's only minor league baseball team, playing out of Old Orchard Beach, is in the class AAA International League. Harness racing is held at Scarborough Downs and other tracks and fairgrounds throughout the state.

49FAMOUS MAINERS

The highest federal officeholders born in Maine were Hannibal Hamlin (1809-91), the nation's first Republican vice president, under Abraham Lincoln, and Nelson A. Rockefeller (1908-79), governor of New York State from 1959 to 1973 and US vice president under Gerald Ford. James G. Blaine (b.Pennsylvania, 1830-93), a lawyer and politician, served 13 years as a US representative from Maine and a term in the Senate; on his third try, he won the Republican presidential nomination in 1884 but lost to Grover Cleveland, later serving as secretary of state (1889-92) under Benjamin Harrison. Edmund S. Muskie (b.1914), leader of the Democratic revival in Maine in the 1950s, followed two successful terms as governor with 21 years in the Senate, until appointed secretary of state by President Jimmy Carter in 1980.

Other conspicuous state and national officeholders have included Rufus King (1755-1827), a member of the Continental Congress and Constitutional Convention and US minister to Great Britain; William King (1768-1852), leader of the movement for Maine statehood and the state's first governor; Thomas Bracket Reed (1839-1902), longtime speaker of the US House of Representatives; and Margaret Chase Smith (b.1897), who served longer in the US Senate—24 years—than any other woman.

Names prominent in Maine's colonial history include those of Sir Ferdinando Gorges (b.England, 1566?-1647), the founder and proprietor of the colony; Sir William Phips (1651-95), who became the first American knight for his recovery of a Spanish treasure, later serving as royal governor of Massachusetts; and Sir William Pepperrell (1696-1759), who led the successful New England expedition against Louisburg in 1745, for which he became the first native American baronet.

Maine claims a large number of well-known reformers and humanitarians: Dorothea Lynde Dix (1802-87), who led the movement for hospitals for the insane; Elijah Parish Lovejoy (1802-37), an abolitionist killed while defending his printing press from a proslavery mob in St. Louis, Mo.; Neal Dow (1804-97), who drafted and secured passage of the Maine prohibition laws of 1846 and 1851, later served as a Civil War general, and ran for president on the Prohibition Party ticket in 1880; and Harriet Beecher Stowe (b.Connecticut, 1811-96), whose *Uncle Tom's Cabin* (1852) was written in Maine.

Other important writers include poet Henry Wadsworth Longfellow (1807-82), born in Portland while Maine was still part of Massachusetts; humorist Artemus Ward (Charles Farrar Browne, 1834-67); Sarah Orne Jewett (1849-1909), novelist and short-story writer; Kate Douglas Wiggin (1856-1923), author of *Rebecca of Sunnybrook Farm;* Kenneth Roberts (1885-1957), historical novelist; and Robert Peter Tristram Coffin (1892-1955), poet, essayist, and novelist. Edwin Arlington Robinson (1869-1935) and Edna St. Vincent Millay (1892-1950) were both Pulitzer Prize-winning poets, and novelist Marguerite Yourcenar (b.Belgium, 1903), a resident of Mt. Desert Island, became in 1980 the first woman ever elected to the Académie Française. Winslow Homer (b.Massachusetts, 1836-1910) had a summer home at Prouts Neck, where he painted many of his seascapes.

50BIBLIOGRAPHY

Bearse, Ray, ed. *Maine: A Guide to the Vacation State.* 2d ed., rev. Boston: Houghton Mifflin, 1969.

Clark, Charles E. *Maine: A Bicentennial History.* New York: Norton, 1977.

Isaacson, Dorris, ed. *Maine: A Guide "Down East."* 2d ed. Rockland: Courier-Gazette, Inc., for the Maine League of Historical Societies and Museums, 1970 (orig. 1937).

Maine, State of. State Development Office. *Maine: A Statistical Summary. Augusta, 1984.*

Morris, Gerald E., ed. *The Maine Bicentennial Atlas.* Portland: Maine Historical Society, 1976.

Osborn, William C. *The Paper Plantation.* New York: Grossman, 1974.

Rich, Louise Dickinson. *State O'Maine.* New York: Harper & Row, 1964.

Rowe, William H. *The Maritime History of Maine.* New York: Norton, 1948.

MARYLAND

State of Maryland

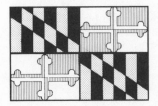

ORIGIN OF STATE NAME: Named for Henrietta Maria, queen consort of King Charles I of England. **NICKNAME:** The Old Line State; Free State. **CAPITAL:** Annapolis. **ENTERED UNION:** 28 April 1788 (7th). **SONG:** "Maryland, My Maryland." **MOTTO:** *Fatti maschii, parole femine* (Manly deeds, womanly words). **FLAG:** Bears the quartered arms of the Calvert and Crossland families (the paternal and maternal families of the founders of Maryland). **OFFICIAL SEAL:** Reverse: A shield bearing the arms of the Calverts and Crosslands is surmounted by an earl's coronet and a helmet and supported by a farmer and fisherman. The state motto (originally that of the Calverts) appears on a scroll below. The circle is surrounded by the Latin legend *Scuto bon voluntatis tu coronasti nos,* meaning "With the shield of thy favor hast thou compassed us," and "1632," the date of Maryland's first charter. Obverse: Lord Baltimore is seen as a knight in armor on a charger. The surrounding inscription, in Latin, means "Cecilius, Absolute Lord of Maryland and Avalon New Foundland, Baron of Baltimore." **BIRD:** Baltimore oriole. **FISH:** Rockfish. **FLOWER:** Black-eyed Susan. **TREE:** White oak. **DOG:** Chesapeake Bay retriever. **INSECT:** Baltimore checkerspot butterfly. **SPORT:** Jousting. **LEGAL HOLIDAYS:** New Year's Day, 1 January; Birthday of Martin Luther King, Jr., 3d Monday in January; Lincoln's Birthday, 12 February; Washington's Birthday, 3d Monday in February; Maryland Day, 25 March; Good Friday, March or April; Memorial Day, 30 May; Independence Day, 4 July; Labor Day, 1st Monday in September; Defenders' Day, 12 September; Columbus Day, 12 October; Election Day, 1st Tuesday after 1st Monday in November, even-numbered years; Veterans Day, 11 November; Thanksgiving Day, 4th Thursday in November; Christmas Day, 25 December. **TIME:** 7 AM EST = noon GMT.

¹LOCATION, SIZE, AND EXTENT

Located on the eastern seaboard of the US in the South Atlantic region, Maryland ranks 42d in size among the 50 states.

Maryland's total area—10,460 sq mi (27,092 sq km)—comprises 9,837 sq mi (25,478 sq km) of land and 623 sq mi (1,614 sq km) of inland water. The state extends 199 mi (320 km) E–W and 126 mi (203 km) N–S.

Maryland is bordered on the N by Pennsylvania; on the E by Delaware and the Atlantic Ocean; on the S and SW by Virginia, the District of Columbia, and West Virginia (with the line passing through Chesapeake Bay and the Potomac River); and on the extreme W by West Virginia. Important islands in Chesapeake Bay, off Maryland's Eastern Shore (the Maryland sector of the Delmarva Peninsula), include Kent, Bloodsworth, South Marsh, and Smith.

The total boundary length of Maryland is 842 mi (1,355 km), including a general coastline of 31 mi (50 km); the total tidal shoreline extends 3,190 mi (5,134 km). The state's geographic center is in Prince Georges County, 4.5 mi (7.2 km) NW of Davidsonville.

²TOPOGRAPHY

Three distinct regions characterize Maryland's topography. The first and major area, falling within the Atlantic Coastal Plain, is nearly bisected by Chesapeake Bay, dividing Maryland into the Eastern Shore and the Western Shore. The Piedmont Plateau, west of the coastal lowlands, is a broad, rolling upland with several deep gorges cut by rivers. Farther west, from the Catoctin Mountains in Frederick County to the West Virginia border, is the Appalachian Mountain region, containing the state's highest hills. Backbone Mountain, in Garrett County in westernmost Maryland, is the state's highest point, at 3,360 feet (1,024 meters).

A few small islands lie in Chesapeake Bay, Maryland's dominant waterway. Extending 195 mi (314 km) inland from the Atlantic and varying in width from 3 to 20 mi (5–32 km), the bay comprises 3,237 sq mi (8,384 sq km), of which 1,726 sq km (4,470 sq km) are under Maryland's jurisdiction. Principal rivers include the Potomac, forming much of the southern and western border; the Patapsco, which runs through Baltimore; the Patuxent, draining the Western Shore; and the Susquehanna, crossing the Pennsylvania border and emptying into Chesapeake Bay in northeastern Maryland. The state has 23 rivers and bays, as well as many lakes and creeks, none of any great size.

³CLIMATE

Despite its small size, Maryland exhibits considerable climatic diversity. Temperatures vary from an annual average of 48°F (9°C) in the extreme western uplands to 59°F (15°C) in the southeast, where the climate is moderated by Chesapeake Bay and the Atlantic Ocean. The daily mean temperature for Baltimore is 55°F (13°C), ranging from 33°F (1°C) in January to 77°F (25°C) in July. The record high temperature for the state is 109°F (43°C), set on 10 July 1936 in Cumberland and Frederick counties; the record low, −40°F (−40°C), occurred on 13 January 1912 at Oakland in Garrett County.

Precipitation averages about 49 in (124 cm) annually in the southeast, but only 36 in (91 cm) in the Cumberland area west of the Appalachians; Baltimore averages 41 in (104 cm) each year. As much as 100 in (254 cm) of snow falls in western Garrett County, while 8–10 in (20–25 cm) is average for the Eastern Shore; Baltimore receives about 22 in (56 cm).

⁴FLORA AND FAUNA

Maryland's three life zones—coastal plain, piedmont, and Appalachian—mingle wildlife characteristic of both North and South. Most of the state lies within a hardwood belt in which red and white oaks, yellow poplar, beech, blackgum, hickory, and white ash are represented; shortleaf and loblolly pines are the leading softwoods. Honeysuckle, Virginia creeper, wild grape, and wild raspberry are also common. Wooded hillsides are rich with such wild flowers as Carolina crane's-bill, trailing arbutus, May apple, early blue violet, wild rose, and goldenrod; *Trillium virginiana* is an endangered plant.

The white-tailed (Virginia) deer, eastern cottontail, raccoon, and red and gray foxes are indigenous to Maryland, although urbanization has sharply reduced their habitat. Common small mammals are the woodchuck, eastern chipmunk, and gray squirrel. The brown-headed nuthatch has been observed in the extreme south, the cardinal and tufted titmouse are common in the piedmont, and the chestnut-sided warbler and the rose-breasted grosbeak are native to the Appalachians. Among saltwater species, shellfish—especially oysters, clams, and crabs—have the greatest economic importance. The Indiana bat, eastern cougar, Maryland darter, southern bald eagle, and Delmarva Peninsula fox squirrel are listed as endangered fauna in the state.

5 ENVIRONMENTAL PROTECTION

Maryland's primary environmental protection agency is the Department of Natural Resources, whose divisions oversee fishery and wildlife management, state parks and forests, land reclamation, scenic and wild rivers, and water allocation. The Office of Environmental Progress, within the Department of Health and Mental Hygiene, is responsible for monitoring emissions and enforcing state air pollution regulations, providing for the safe collection and disposal of industrial and hazardous wastes, and overseeing water supply, management, sewage disposal, consumer product safety, and radiation and noise controls. In addition, the Maryland Environmental Service, a public corporation established in 1970, has the power to design, construct, finance, and operate liquid and solid waste management systems in cooperation with local government and industry.

State programs are designed to ensure the safe utilization of Maryland's Atlantic and Chesapeake Bay shorelines and to reclaim areas in the western counties affected by open-pit mining. In 1984, authorities from Maryland, Virginia, Pennsylvania, and the District of Columbia launched a massive program to combat the effects of sewage pollution, chemical dumping, and pesticide runoffs on the marine life of Chesapeake Bay. The action followed a federal report that the bay area ecosystem was declining and required urgent cleanup measures to arrest the damage.

6 POPULATION

The enormous expansion of the federal government and exodus of people from Washington, D.C., to the surrounding suburbs contributed to the rapid growth that made Maryland the 17th most populous state in 1980, with 4,216,446 residents. The state's population doubled between 1940 and 1970, and increased 7.5% between 1970 and 1980. From 1980 to 1985, the population showed a further increase of 3%, to a total of 4,342,562, and it was estimated that the population would be 4,491,100 by 1990. The population density in 1985 was 441 per sq mi (170 per sq km).

Almost all the growth since World War II has occurred in the four suburban counties around Washington, D.C., and Baltimore; as of 1980, about 54% of the state's population resided there, and more than 80% of the total population was urban. Metropolitan Baltimore, embracing Carroll, Howard, Harford, Anne Arundel, and Baltimore counties, expanded from 2,147,000 to 2,244,700 between 1977 and 1984 (16th in the US); the city of Baltimore, on the other hand, declined from 804,000 to 763,570 (12th in the US) during the same period. Baltimore is the state's only major city; several west-central counties belong to the Washington metropolitan area, and Cecil County, in the northeast, is part of metropolitan Wilmington, Del.

7 ETHNIC GROUPS

Black Americans, numbering 958,050 in 1980, constitute the largest racial minority in Maryland. About 45% of the blacks lived in the city of Baltimore.

Hispanic Americans, mostly from Puerto Rico and Central America, numbered 64,746 in 1980. The Asian population was relatively large: 15,087 Koreans, 14,485 Chinese, 10,963 Filipinos, 4,805 Japanese, and 4,131 Vietnamese.

Foreign-born residents numbered 195,000, or 4.6% of the population in 1980; about 44% of them had immigrated to Maryland in the 1970s. The leading countries of origin were Germany, France, Italy, Poland, and the Soviet Union; a significant proportion of the German, Polish, and Russian immigrants were Jewish refugees arriving just before and after World War II. Maryland's American Indian population is small—only 7,823 in 1980.

8 LANGUAGES

Several Algonkian tribes originally inhabited what is now Maryland. There are some Indian place-names, such as Potomac, Susquehanna, and Allegany.

The state's diverse topography has contributed to unusual diversity in its basic speech. Geographical isolation of the Delmarva Peninsula, proximity to the Virginia piedmont population, and access to southeastern and central Pennsylvania helped to yield a language mixture that now is dominantly Midland and yet reflects earlier ties to Southern English.

Regional features occur as well. In the northeast are found eastern Pennsylvania *pavement* (sidewalk) and *baby coach* (baby carriage). In the north and west are *poke* (bag), *quarter till, sick on the stomach, openseed peach* (freestone peach), and Pennsylvania German *ponhaws* (scrapple). In the southern portion are *light bread* (white bread), *curtain* (shade), *carry* (escort), *crop* as /krap/, and *bulge* with the vowel of *put*. East of Chesapeake Bay are *mosquito hawk* (dragonfly), *paled fence* (picket fence), *poor* rhyming with *mower*, and *Mary* with the vowel of *mate*. In central Maryland, an earthworm is a *baitworm*.

In 1980, 3,799,801 residents, or 94% of the population 3 years old or older, spoke only English at home. Other languages spoken at home, with the number of speakers, included:

Spanish	56,945	Greek	12,252
French	26,593	Chinese	12,200
German	25,057	Korean	12,154
Italian	17,144	Polish	10,470

9 RELIGIONS

Maryland was founded as a haven for Roman Catholics, and they remain the state's leading religious group, although their political supremacy ended in 1692, when Anglicanism became the established religion. Laws against "popery" were enacted by 1704, and Roman Catholic priests were harassed; the state constitution of 1776, however, placed all Christian faiths on an equal footing. The state's first Lutheran church was built in 1729, the first Baptist church in 1742, and the earliest Methodist church in 1760. Jews settled in Baltimore in the early 1800s, with a much larger wave of Jewish immigration in the late 19th century.

As of 1980 there were 736,361 Roman Catholics in Maryland. Adherents of the major Protestant denominations included United Methodist Church, 322,099; Southern Baptist Convention, 128,411; Lutheran Church in America, 95,710; and Episcopal Church, 71,699. In 1984 there were an estimated 199,415 Jews.

10 TRANSPORTATION

Some of the nation's earliest efforts toward the development of a reliable transportation system began in Maryland. In 1695, a public postal road was opened from the Potomac River through Annapolis and the Eastern Shore to Philadelphia. Construction on the National Road (now US 40) began at Cumberland in 1811; within seven years, the road was a conduit for settlers in Ohio. The first commercial steamboat service from Baltimore started in 1813, and steamboats were active all along the Chesapeake during the 1800s. The Delaware and Chesapeake Canal, linking Chesapeake Bay and the Delaware River, opened in 1829.

Maryland's first railroad, the Baltimore and Ohio (B&O), was started in 1828; in 1835, it provided the first passenger train service to Washington, D.C., and Harpers Ferry, Va. (now W.Va.). By 1857, the line was extended to St. Louis, and its freight capacity helped build Baltimore into a major center of commerce. In the 1850s, the Pennsylvania Railroad began to buy up small Maryland lines and provide direct service to the northern cities.

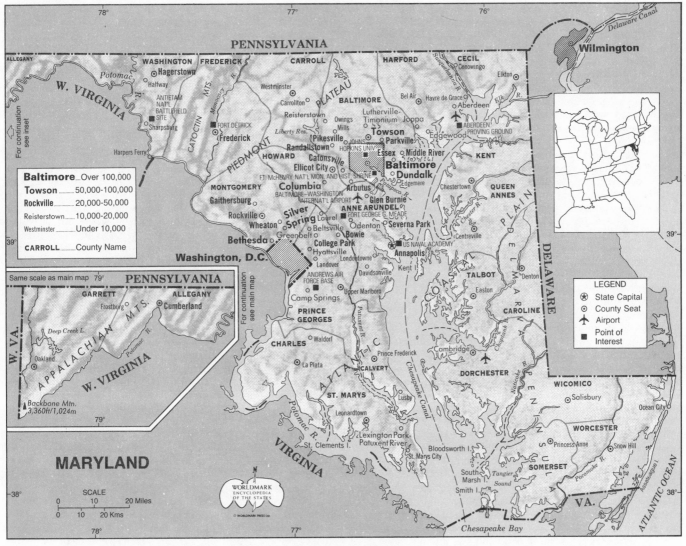

See endsheet maps: L3.

LOCATION: 37°53' to 39°43'N; 75°04' to 79°29'W. **BOUNDARIES**: Pennsylvania line, 196 mi (316 km); Delaware line, 122 mi (196 km); Atlantic Ocean coastline, 31 mi (50 km); Virginia line, 233 mi (375 km); District of Columbia line, 25 mi (40 km); West Virginia line, 235 mi (378 km).

Today, some of the old Pennsylvania lines are operated by Amtrak, while the B&O provides commuter service between Baltimore and Washington. Class I trackage in Maryland totaled 909 mi (1,463 km) in 1983. The Maryland Transportation Department's Railroad Administration subsidizes four commuter lines, as well as freight lines in western Maryland and on the Eastern Shore.

The Maryland Mass Transit Administration inaugurated Baltimore's first subway line on 21 November 1983. The combined underground-elevated line ran for 8 mi (13 km) from downtown Baltimore to Reisterstown Plaza; later, the Baltimore Metro was extended for 6 mi (10 km) to Owings Mills, just outside the city limits. The Metro serves about 50,000 riders daily and cost nearly $1 billion to build. In 1984, the Washington, D.C., mass transit system was extended to the Maryland suburbs, including Bethesda and Rockville.

About half of Maryland's roads serve metropolitan Baltimore and Washington. As of 1982 there were 27,100 mi (43,613 km) of roadway. Interstate highways in Maryland totaled 338 mi (544 km) in 1983; the major toll road is the John F. Kennedy Memorial Highway (I-95), linking Baltimore with Wilmington, Del., and

with the New Jersey Turnpike. There were 2,799,441 licensed drivers and 3,011,109 motor vehicles registered in Maryland in 1983.

The Port of Baltimore, one of the nation's busiest, handled 12,200,000 tons of foreign cargo, valued at $6.9 billion, in 1983.

There are 103 airports in Maryland. The Department of Transportation operates Baltimore-Washington International Airport, the major air terminal in the state. In 1983, it handled 41,371 commercial aircraft departures with 2,296,538 passengers. Another 146 airfields (102 airports, 40 heliports, 3 stolports, and 1 seaplane base) also served the state in 1983.

[11]HISTORY

The Indian tribes living in the region that was to become Maryland were Algonkian-speakers: the Accomac, Nanticoke, and Wicomico on the Eastern Shore, and the Susquehannock, Yacomico, and Piscataway on the Western Shore. The Susquehannock, the most powerful tribe at the time of English colonization, claimed all the land lying between the Susquehanna and Potomac rivers. Although the Algonkian Indians hunted for much of their food, many tribes (including the Susquehannock) also had permanent settlements where they cultivated corn (maize), vegetables, tobacco, and other crops. George Alsop, in his *Character of the Province of Maryland* (1666), noted that Susquehannock women "are the Butchers, Cooks, and Tillers of the ground but the men think it below the honour of a Masculine to stoop to any thing but that which their Gun, or Bow and Arrows can Command." European penetration of the Chesapeake region began early in the 16th century, with the expeditions of Giovanni da Verrazano, a Florentine navigator, and the Spaniard Lucas Vázquez de Ayllón. Captain John Smith, leader of the English settlement at Jamestown, Va., was the first English explorer of Chesapeake Bay (1608) and produced a map of the area that was used for years.

The founding of Maryland is intimately tied to the career of another Englishman, George Calvert. A favorite of King James I, Calvert left the Church of England in 1624 to become a Roman Catholic. He announced his conversion in 1625 and—because Catholics were not allowed to hold public office in England at that time—then resigned his post as secretary of state and, against the king's wishes, retired from the royal court. As a reward for Calvert's service, the king bestowed upon him large Irish estates and a peerage with the title of Baron of Baltimore. Two years later, Calvert sailed for the New World, landing in Newfoundland, to which he had received title in 1621. After a severe winter, however, Calvert decided to seek his fortunes where the weather was warmer—in Virginia. Not well received there because of his religion, Calvert returned to England and asked King Charles I (James's successor) for land south of Virginia; instead he received a grant north of the Potomac. Virginia's agents in England contested Calvert's right to this land strenuously but unsuccessfully, and when he died in 1632, the title passed to his son Cecilius Calvert, 2d Baron Baltimore (usually called Lord Baltimore), who named the region Maryland after the queen consort of Charles I, Henrietta Maria. At this time, the land grant embraced not only present-day Maryland but also the present State of Delaware, a large part of Pennsylvania, and the valley between the north and south branches of the Potomac River. Not until the 1760s was the final boundary between Pennsylvania and Maryland (as surveyed by Charles Mason and Jeremiah Dixon) established by royal decree, and nearly a century passed before Maryland conceded to Virginia the land between the north and south branches of the Potomac.

The government of provincial Maryland was absolute, embodying the most extensive grant of royal powers to a colonial settlement. Lord Baltimore's main source of income as lord proprietary was the quitrents settlers paid for their land; in return for his authority, Calvert had to give the king only two Indian arrows yearly. Lord Baltimore assigned to his half-brother, Leonard Calvert, the task of organizing the settlement of the colony. On 22 November 1633, Calvert and approximately 250 settlers, including many Roman Catholics and two Jesuit priests, set sail for America on two ships, the *Ark* and the *Dove*. They landed at St. Clements Island on 25 March 1634. Two days later, Calvert purchased a site from the Indians, named it St. Marys (the first capital of Maryland), and assumed the governorship of the colony.

The early days of settlement were tumultuous. The refusal by a Virginia colonist, William Claiborne of Kent Island, to acknowledge Lord Baltimore's charter led to a small war that ended in 1638 with a temporary victory for Governor Calvert. The conflict in England during the 1640s found an echo in the struggle between Puritans and Roman Catholics in Maryland, a conflict that saw the two-year exile of Governor Calvert to Virginia, the assumption of power by English representatives (including Claiborne and one of the Puritan leaders) in 1652, a subsequent civil war, and finally the recognition of Lord Baltimore's charter by Oliver Cromwell in 1657.

Cecilius Calvert died in 1675. His successor was Charles Calvert, 3d Baron Baltimore and the next lord proprietary. His tenure, which lasted until 1715, saw a decisive change in the character of the province. In 1689, with Protestants ascendant in both England and Maryland, the British crown assumed direct control over the province, and in 1692, the Church of England became Maryland's established religion. When Charles Calvert died, his successor, Benedict Leonard Calvert, 4th Baron Baltimore, was granted full proprietary rights—but only because he had embraced the Protestant faith. Proprietary rule continued through his legitimate heirs until the eve of the American Revolution.

Throughout this period, the upper and lower houses of the colonial assembly—consisting, respectively, of the governor and his council and of delegates elected from the counties—quarreled over taxation and the extension of English statutes to free Marylanders. Having already secured most rights from the proprietor, the lower house was somewhat reluctant to vote for independence from the British crown, on whose authority the proprietary government now rested. After its initial hesitancy, however, Maryland cast its lot with the Revolution and sent approximately 20,000 soldiers to fight in the war. The Continental Congress met in Baltimore from December 1776 to March 1777 and in Annapolis from November 1783 to June 1784; these cities were thus among the eight that served as US capitals before the designation of a permanent seat of government in Washington, D.C.

Maryland was one of the last states to sign the Articles of Confederation, not ratifying them until other states dropped their claims to what later became the Northwest Territory. On 28 April 1788, Maryland became the 7th state to ratify the federal Constitution. The state constitution, drawn up in 1776, was weighted heavily in favor of propertyholders and the rural counties, at the expense of the propertyless and the city of Baltimore; the legislature removed the property qualifications in 1810.

Maryland's prosperity during the colonial and early federal period waxed and waned according to the world price of tobacco, the staple crop of tidewater and southern Maryland. Planters increasingly employed slave labor on farms and plantations, and the black population grew rapidly in the 18th century. German immigrants began moving into western Maryland, where wheat became the primary crop. The cultivation of wheat also helped make Baltimore's fortune. Founded in 1729 and incorporated in 1796, the city of Baltimore was blessed with a harbor well suited to the export and import trade. As commerce developed, shipbuilding emerged as a major economic activity. By the early 19th century, Baltimore was already the state's major center of commerce and industry.

The city and harbor were the site of extensive naval and military operations during the War of 1812. It was during the bombardment of Ft. McHenry in 1814 that Francis Scott Key, detained on a British frigate, composed "The Star-Spangled Banner," which became the US national anthem in March 1931.

After the War of 1812, Maryland history was marked by the continued growth of Baltimore and increasing division over immigration, slavery, and secession. The chartering in 1827 of the Baltimore and Ohio (B&O) Railroad, which eventually linked Maryland with the markets of the Ohio Valley and the West, added to the city's economic vitality. But distrust of the thousands of newcomers—especially of Irish immigrants and their Roman Catholicism—and fear of the economic threat they supposedly represented spurred the rise of nativist political groups, such as the Know-Nothings, who persecuted the immigrants and dominated Maryland politics in the 1850s.

Although not many Marylanders were in favor of secession, they were hostile to the idea of using force against the secessionist states. On 19 April 1861, as the 6th Massachusetts Regiment passed through Baltimore, it was attacked by a mob of southern sympathizers in a riot that left 4 soldiers and 12 civilians dead.

Ten days later, the Maryland house of delegates, following the lead of Governor Thomas Hicks, rejected a bill of secession. Throughout the Civil War, Maryland was largely occupied by Union troops because of its strategic location and the importance for the northern cause of the B&O Railroad. Marylanders fought on both sides during the war, and one major battle took place on Maryland soil—the Battle of Antietam (1862), during which a Union army thwarted a Confederate thrust toward the north, but at an enormous cost to both sides. Confederate armies invaded the state on two other occasions, when General Robert E. Lee brought his troops through the state on the way to Gettysburg in 1863 and when Lieutenant General Jubal Early ravaged the Hagerstown area and threatened Baltimore in 1864. The Maryland legislature, almost totally pro-Union by 1864, passed a new constitution, which among other things abolished slavery.

The state's economic activity increased during Reconstruction, as Maryland, and especially Baltimore, played a major role in rebuilding the South. Maryland's economic base gradually shifted from agriculture to industry, with shipbuilding, steelmaking, and the manufacture of clothing and shoes leading the way. The decades between the Civil War and World War I were also notable for the philanthropic activities of such wealthy businessmen as Johns Hopkins, George Peabody, and Enoch Pratt, who endowed some of the state's most prestigious cultural and educational institutions. The years after World War I saw the emergence of a political figure without equal in Maryland's more recent history: Albert C. Ritchie, a Democrat who won election to the governorship in 1919 and served in that office until 1935, just one year before his death. Stressing local issues, states' rights, and opposition to prohibition, Ritchie remained in power until Harry W. Nice, a Republican but an advocate of New Deal reforms, defeated him in 1934.

The decades since World War II have been marked by intensive population growth, political scandal, and the passage of open housing and equal opportunity laws to protect Maryland's black citizens. Perhaps the most significant occurrence has been the redevelopment of Baltimore, which, though still the hub of the state's economy, had fallen into decay. Much of Baltimore's downtown area and harbor facilities were revitalized by many urban rejuvenation projects, begun in the late 1970s and continued into the 1980s. These featured the Charles Center development and the waterfront renovation of the Inner Harbor.

12 STATE GOVERNMENT

Maryland's first state constitution was enacted in 1776. Subsequent constitutions were ratified in 1851, 1864, and 1867.

Under the 1867 constitution, as amended, the general assembly, Maryland's legislative body, consists of two branches: a 47-member senate and a 141-member house of delegates. All legislators serve four-year terms and must have been residents of the state for at least a year and of their district for at least six months prior to election. Senators must be at least 25 years of age, delegates 21.

Executives elected statewide are the governor and lieutenant governor (who run jointly), the comptroller of the Treasury, and the attorney general; all serve four-year terms. The state treasurer is elected by joint ballot of the general assembly, while the secretary of state is appointed by the governor. The governor, who may serve no more than two four-year terms in succession, also appoints other members of the executive council (cabinet) and the heads of major boards and commissions. The chief executive must be a US citizen at least 30 years of age and must have been a resident of Maryland for five years before election.

Bills passed by majority vote of both houses of the assembly become law when signed by the governor or if left unsigned for 6 days while the legislature is in session or 30 days if the legislature has adjourned. The only exception is the budget bill, which becomes effective immediately upon legislative passage. Gubernatorial vetoes may be overridden by three-fifths votes in both houses. Proposed constitutional amendments also require approval by three-fifths of both houses of the legislature before submission to the voters at the next general election.

US citizens who are at least 18 years of age and have been residents of the state for 30 days prior to the election are eligible to vote.

13 POLITICAL PARTIES

The Republican and Democratic parties are the dominant political groups in Maryland. Before the Civil War, the Democrats drew much of their strength from the slaveholding Eastern Shore,

Maryland Presidential Vote by Political Parties, 1948–84

YEAR	ELECTORAL VOTE	MARYLAND WINNER	DEMOCRAT	REPUBLICAN	PROGRESSIVE	STATE'S RIGHTS DEMOCRAT	SOCIALIST
1948	8	Dewey (R)	286,521	294,814	9,983	2,467	2,941
1952	9	*Eisenhower (R)	395,337	499,424	7,313	—	—
1956	9	*Eisenhower (R)	372,613	559,738	—	—	—
1960	9	*Kennedy (D)	565,808	489,538	—	—	—
1964	10	*Johnson (D)	730,912	385,495	—	—	—
					AMERICAN IND.		
1968	10	Humphrey (D)	538,310	517,995	178,734	—	—
					AMERICAN		
1972	10	*Nixon (R)	505,781	829,305	18,726	—	—
1976	10	*Carter (D)	759,612	672,661	—	—	—
						LIBERTARIAN	
1980	10	Carter (D)	726,161	680,606	—	14,192	—
1984	10	*Reagan (R)	787,935	879,918	—	5,721	—

*Won US presidential election.

while their opponents, the Whigs, were popular in Baltimore and other centers of antislavery activity. The collapse of the Whigs on both the national and local levels corresponded with the rise in Maryland of the Native American ("Know-Nothing") Party, whose anti-immigrant and anti-Catholic attitudes appealed to Marylanders who saw their livelihood threatened by Roman Catholic immigrants from abroad. The Know-Nothings swept Baltimore in 1855 and won the governorship in 1857; Maryland was the only state to cast its electoral votes for the Know-Nothing presidential candidate, former President Millard Fillmore, in 1856. The Native American Party declined rapidly, however, and by 1860, Maryland was back in the Democratic column, voting for the secessionist John Breckinridge.

The Democrats dominate state politics today, although the Republicans remain strong in some suburbs and on the Eastern Shore. As of 1982 there were 1,968,498 registered voters, of whom 69% were Democrats, 24% Republicans, and 7% independents and members of minor parties. Maryland was one of the few states carried by President Jimmy Carter in the November 1980 presidential election, but four years later the state went for President Ronald Reagan in the national Republican landslide.

Revelations of influence peddling and corruption afflicted both major parties during the 1970s. In 1973, Republican Spiro T. Agnew, then vice president of the US, was accused of taking payments from people who had done business with the state government while he was Baltimore County executive and then governor until 1969. Agnew pleaded *nolo contendere* to a federal charge of income tax evasion and resigned from the vice-presidency on 10 October 1973. His gubernatorial successor, Democrat Marvin Mandel, was convicted of mail fraud and racketeering in 1977 for having used the powers of his office to assist the owners of a now-defunct racetrack in exchange for $350,000 in gifts and favors; he served 20 months of a 36-month prison sentence before receiving a presidential pardon in 1981.

14 LOCAL GOVERNMENT

As of 1984 there were 23 counties, 152 incorporated cities and towns, and 102 unicorporated places of 1,000 or more in Maryland. Eight counties had charter governments, with (in most cases) elected executives and county councils; 12 had elected boards of county commissioners; and 3 home-rule counties also had commission governments, but with additional powers. County government is highly developed in Maryland, and there are numerous appointed county officials with responsibilities ranging from civil defense to liquor licensing.

The city of Baltimore is the only one in Maryland not contained within a county; it provides the same services as a county, and shares in state aid according to the same allocation formulas. The city (not to be confused with Baltimore County, which surrounds the city of Baltimore but has its county seat at Towson) is governed by a mayor and a nine-member city council. Other cities and towns are each governed by a mayor, with or without a council, depending on the local charter.

15 STATE SERVICES

The State Ethics Commission, established in 1979, monitors compliance by state officeholders and employees with the Maryland public ethics law in order to avoid conflicts of interest; the Joint Committee on Legislative Ethics, created in 1972, has similar responsibilities with respect to general assembly members. The Fair Campaign Financing Commission provides for the public financing of elections and sets campaign spending limits.

The State Board of Education is an independent policymaking body whose nine members are appointed by the governor; its responsibilities include selection of a superintendent of schools to run the Education Department. The growth and development of postsecondary institutions are the responsibility of the State Board for Higher Education. The Department of Transportation oversees air, road, rail, bridge, and mass transit. The Department

of Health and Mental Hygiene coordinates public health programs, regulates in-state medical care, and supervises the 24 local health departments. Social services and public assistance programs as well as employment security lie within the jurisdiction of the Department of Human Resources. The Department of Economic and Community Development advances job opportunities and community projects, including public housing programs.

Maryland's state prisons, police, and civil defense agencies are under the Department of Public Safety and Correctional Services, while the Department of Licensing and Regulation sets standards for businesses, professions, and trades.

16 JUDICIAL SYSTEM

The court of appeals, the state's highest court, comprises a chief judge and 6 associate judges. Each is appointed to the court by the governor but must be confirmed by the voters within two years of the appointment. Most criminal appeals are decided by the court of special appeals, consisting of a chief judge and 12 associate judges, selected in the same manner as judges of the high court; each case must be heard by a panel of at least 3 judges. All state judges serve 10-year terms.

In 1970, 12 district courts took the place of all justices of the peace, county trial judges, magistrates, people's courts, and the municipal court of Baltimore. District courts handle all criminal, civil, and traffic cases, with appeals being taken to one of eight circuit courts. Circuit court judges are appointed by the governor and stand for election to 15-year terms; district court judges are appointed by the governor and confirmed by the senate to 10-year terms. The city of Baltimore and all counties except Montgomery and Harford have orphans' courts composed of two judges and one chief judge, all of them elected to four-year terms.

According to the FBI Crime Index for 1983, Maryland had a violent crime rate of 807 per 100,000 population (3d highest in the US, after New York and the District of Columbia) and a property crime rate of 4,550. Rates for specific crimes were murder and nonnegligent homicide, 8.5; forcible rape, 32.8; robbery, 347; aggravated assault, 418; burglary, 1,224; larceny/theft, 2,961; and motor vehicle theft, 364. Baltimore's violent crime rate of 2,003 was one of the highest among US cities.

Maryland has the death penalty, and 11 prisoners were being held under the death sentence as of 31 December 1983; however, no prisoner had been executed since 1977. There were 13,124 prisoners in state and federal prisons at the end of 1984.

17 ARMED FORCES

As of 1984 there were 35,563 US military personnel in Maryland. Nearly 8,500 were stationed at Ft. Meade in Baltimore, and 4,100 at the Aberdeen Proving Ground in Harford County. Perhaps Maryland's best-known defense installation is Andrews Air Force Base in Camp Springs, a military airlift center that had 5,931 military personnel in 1984. Annapolis is the home of the US Naval Academy, which in 1984 had 5,749 military personnel. Total military personnel at all naval facilities, including the National Naval Medical Center at Bethesda, was 12,537 in 1984. Federal defense contract awards to Maryland firms exceeded $4 billion in 1983/84.

Some 541,000 veterans were living in the state as of 30 September 1983. Of these, veterans of wartime service were as follows: World War I, 4,000; World War II, 204,000; Korean conflict, 111,000; Viet-Nam era, 165,000. Veterans' benefits during 1982/83 totaled $387 million, of which $218 million was for compensation and pensions, $102 million for medical services, and $26 million for education and training.

Maryland's National Guard numbered 6,461 members at the end of 1984; the Air National Guard, 1,670 as of 28 February 1985. In 1983 there were 2,092 state police personnel.

18 MIGRATION

Maryland's earliest white settlers were English, many of whom farmed lands on the Eastern Shore. As tobacco crops wore out the soil, these early immigrants moved on to the fertile Western Shore and piedmont. During the 19th century, Baltimore ranked 2d only to New York as a port of entry for European immigrants. First to come were the Germans, followed by the Irish, Poles, East European Jews, and Italians; a significant number of Czechs settled in Cecil County during the 1860s. After the Civil War, many blacks migrated to Baltimore, both from rural Maryland and from southern states.

Since World War II, intrastate migration has followed the familiar urban/suburban pattern: both the metropolitan area of Baltimore and the Maryland part of the metropolitan Washington, D.C., area have experienced rapid growth, while the inner cities have lost population. Overall, Maryland experienced a net loss from migration of about 36,000 between 1970 and 1980, much of it to Pennsylvania, Virginia, and Florida; the out-migration slowed considerably during the early 1980s, however, with a net loss of only 3,000 from 1980 to 1983.

[19] INTERGOVERNMENTAL COOPERATION

Maryland is active in several regional organizations, including the Southern Regional Education Board, Atlantic States Marine Fisheries Commission, Susquehanna River Basin Commission (with Pennsylvania and New York), and the Potomac River Fisheries Commission (with Virginia). Representatives of Maryland, Virginia, and the District of Columbia form the Washington Metropolitan Area Transit Authority, which coordinates regional mass transit. The Delmarva Advisory Council, representing Delaware, Maryland, and Virginia, works with local organizations on the Delmarva Peninsula to develop and implement economic improvement programs.

In 1983/84, federal aid to Maryland totaled nearly $1.7 billion.

[20] ECONOMY

Throughout the colonial period, Maryland's economy was based on one crop—tobacco. Not only slaves but also indentured servants worked the fields, and when they earned their freedom, they too secured plots of land and grew tobacco for the European market. By 1820, however, industry was rivaling agriculture for economic preeminence. Shipbuilding, metalworking, and commerce transformed Baltimore into a major city; within 60 years, it was a leading manufacturer of men's clothing and had the largest steelmaking plant in the US.

Although manufacturing output continues to rise, the biggest growth areas in Maryland's economy are government, construction, trade, and services. With the expansion of federal employment in the Washington metropolitan area by 40% from 1961 to 1980, many US government workers settled in suburban Maryland, primarily Prince Georges and Montgomery counties; construction and services in those areas expanded accordingly. By 1980, the federal government was the state's largest employer, with 118,271 employees.

The growth of state government boosted employment in Anne Arundel and Baltimore counties. Also of local importance are fishing and agriculture (primarily dairy and poultry farming) on the Eastern Shore and coal mining in Garrett and Allegany counties. As of 1982, Maryland's gross state product was $52 billion, more than double the 1973 total in current dollars but no more than a 9% increase in real terms.

[21] INCOME

Per capita disposable income rose by more than 6% annually from 1948 to 1982. As of 1983, Maryland ranked 7th in per capita income with $12,994. Total personal income in 1982 was nearly $52.2 billion, to which services contributed 14%; manufacturing, 10%; state and local government, 8%; federal government, 8%; retail trade, 7%; transportation and public utilities, 5%; construction, 4%; wholesale trade, 4%; and other categories, 40%.

Montgomery County had one of the highest average household disposable incomes of all US counties in 1982 ($41,795) and led the state in total per capita and personal income in 1981. The state's average household income in 1982 was $28,777, 11% above the US average. An estimated 405,000 Marylanders—9.8% of the population—were below the federal poverty level in 1979.

[22] LABOR

Maryland's civilian labor force in 1984 numbered 2,244,000, of whom 2,123,000 were employed and 121,000 unemployed, for an unemployment rate of 5.4%. A federal survey in March 1982 revealed the following nonfarm employment pattern for Maryland:

	ESTABLISH-MENTS	EMPLOYEES	ANNUAL PAYROLL ('000)
Agricultural services, forestry, fisheries	909	6,197	$ 75,971
Mining	117	2,620	63,889
Contract construction	7,762	95,243	1,857,579
Manufacturing	3,621	241,110	4,893,081
Transportation, public utilities	2,777	74,122	1,570,883
Wholesale trade	5,440	78,522	1,551,735
Retail trade	21,386	298,594	2,886,957
Finance, insurance, real estate	7,207	98,408	1,586,526
Services	26,093	374,416	5,373,683
Other	1,520	1,437	26,553
TOTALS	76,832	1,270,669	$19,886,857

Baltimore was a leading trade union center by the early 1830s, although union activity subsided after the Panic of 1837. The Baltimore Federation of Labor was formed in 1889, and by 1900, the coal mines had been organized by the United Mine Workers. In 1902, Maryland passed the first workers' compensation law in the US; it was declared unconstitutional in 1904 but subsequently revived. As of 1980 there were 527,000 members of labor unions in the state, and in 1983 there were 632 unions. In 1981 there were 30 work stoppages involving 13,100 workers.

[23] AGRICULTURE

Maryland ranked 35th among the 50 states in agricultural income in 1983, with estimated receipts of $1,032 million, about one third of that in crops.

Until the Revolutionary War, tobacco was the state's only cash crop; in 1983, an estimated 27,000,000 lb of tobacco, worth $32,000,000, were harvested (7th in the US). Corn and cereal grains are grown mainly in southern Maryland. Production in 1983 included 37,060,000 bushels of corn for grain, worth $148,240,000; 9,250,000 bushels of soybeans, $77,238,000; 5,371,-000 bushels of wheat, $18,530,000; and 4,950,000 bushels of barley, $9,653,000. Fresh vegetables, cultivated primarily on the Eastern Shore, were valued at $9,950,000 in 1983. Fruits are also cultivated.

Maryland had some 17,800 farms covering 2,700,000 acres (1,093,000 hectares) in 1984. The farm population declined from 79,000 in 1970 to 45,000 in 1980, or only 1% of the total population.

[24] ANIMAL HUSBANDRY

About 32% of Maryland's farm income derives from livestock and livestock products. The Eastern Shore is an important dairy and poultry region; cattle are raised in north-central and western Maryland, while the central region is notable for horse breeding.

Maryland ranked 6th among the 50 states in broiler production in 1983, with 260.5 million broilers worth $331.1 million. Also produced during 1983 were 22.5 million lb of chickens, worth $2.8 million, and 836 million eggs, $46.6 million.

An estimated 1.6 billion lb of milk were produced in 1983. Maryland farms and ranches had 125,000 milk cows, 420,000 cattle and calves, and 200,000 hogs and pigs by the end of 1983. Production of meat animals included 112 million lb of cattle ($59.6 million) and 75.5 million lb of hogs ($37.2 million).

25FISHING

A leading source of oysters, clams, and crabs, Maryland had a total commercial catch in 1984 of 89,301,000 lb, valued at $54,979,000 (13th in the US). Shellfish accounted for 85% of the volume and 92% of the value of the catch in 1981; leading items were oyster meats (41% of the total value) and crabs (35%). Striped bass was the most important finfish, accounting for 3% of the total 1981 value, followed by swordfish, flounder, and menhaden. As of the mid-1980s, about 20,000 commercial fishermen were active in Maryland, up from 16,200 in 1975. The state's 143 seafood-processing and wholesale plants employed 2,556 persons during 1983.

The Fisheries Administration of the Department of Natural Resources monitors fish populations and breeds and implants oysters; it also stocked inland waterways with 880,000 finfish in 1983.

26FORESTRY

Maryland's 2,653,200 acres (1,073,700 hectares) of forestland in 1979 covered 42% of the state's land area. More than 95% of that was classified as commercial forest, nine-tenths of it privately owned. Hardwoods predominate, with red and white oaks and yellow poplar among the leading hardwood varieties.

The 1980 harvest included 150,312,000 board feet of timber, 147,200 cords of pulpwood, and 22,194 cords of fuel wood. In 1982, shipments of lumber and wood products were valued at $220.8 million; paper and allied products, $850.6 million.

Forest management and improvement lie within the jurisdiction of the Forest and Park Service, which in 1984 was combined with the Wildlife Administration under one director within the Department of Natural Resources. The Forest and Park Service manages nine state forests.

27MINING

Maryland ranked 32d among the states in nonfuel mineral production in 1984, with a mineral output valued at $232.9 million. Excluding fossil fuels, Maryland mines and quarries in 1984 produced an estimated 21,317,000 tons of crushed stone, valued at $92,864,000, and 12,600,000 tons of sand and gravel, worth $49,100,000. Stone is quarried in Allegany, Baltimore, Cecil, Howard, and Montgomery counties, and sand and gravel are quarried statewide. Small amounts of clays and lime are also mined.

28ENERGY AND POWER

Maryland's installed electrical capacity was 9.7 million kw in 1983, when production of electricity exceeded 32.6 billion kwh. More than 99% of the generating capacity was privately owned, and about 47% of the state's electricity was produced by coal-fired plants. The Calvert Cliffs Nuclear Plant in Lusby, operated by Baltimore Gas and Electric, had a capacity of 845 Mw and produced about 36% of the state's electricity in 1983. The price of electricity in Baltimore increased from 5.6 cents per kwh in 1979 to 7.5 cents per kwh in 1983.

Coal, Maryland's lone fossil fuel resource, is mined in Allegany and Garrett counties, along the Pennsylvania border. Reserves in 1983 were estimated at 803 million tons of bituminous coal; the 1983 output of 29 coal mines totaled 3.1 million tons. About 145 trillion cu feet of natural gas from out of state was sold to 792,400 Maryland consumers in 1983; the amount was down slightly from previous years, as gas companies put restrictions on new hookups. Revenues from sales of natural gas totaled $973 million in 1983.

29INDUSTRY

During the early 1800s, Maryland's first industries centered around the Baltimore shipyards. Small ironworks cast parts for sailing vessels, and many laborers worked as shipbuilders. By the 1850s, Baltimore was also producing weather-measuring instruments and fertilizers, and by the 1930s, it was a major center of metal refining. The city remains an important manufacturer of automobiles and parts, machinery, and steel.

Value of shipments by manufacturers in 1982 was $21.3 billion, up 33% from 1977. Of the 1982 figure, 18% was contributed by food and food products, 14% by electric and electronic equipment, 11% by primary metals, 11% by chemicals and allied products, 7% by nonelectrical machinery, 7% by transportation equipment, and 32% by other industries. The following table shows value of shipments by selected industries in 1982:

Communications equipment	$2,235,100,000
Blast furnace, basic steel products	1,469,700,000
Soaps, cleaners, toiletries	873,200,000
Beverages	837,800,000
Dairy products	594,100,000
Meat products	581,300,000
Paperboard containers and boxes	447,700,000
Plastic products	403,900,000

About a third of all manufacturing activity (by value) takes place in the city of Baltimore, followed by Baltimore County, Montgomery County, and Prince Georges County. Leading Maryland corporations include Martin Marietta of Bethesda (aerospace), the 85th-largest US industrial corporation, with $4.4 billion in sales in 1984; Crown Central Petroleum of Baltimore (221st, $1.7 billion); Black & Decker Manufacturing of Towson (tools; 232d, $1.5 billion); and Fairchild Industries of Germantown (aerospace; 332d, $900 million).

30COMMERCE

Maryland's 5,507 wholesale establishments registered $25.8 billion in trade during 1982, 52% more than in 1977. The city of Baltimore accounted for 28% of all sales, followed by Baltimore County, 16%; Montgomery County, 14%; and Prince Georges County, 13%. Nearly 56% of the trade was in durable goods, with motor vehicles, parts, and supplies accounting for 15% of all sales. The largest nondurable items were groceries and related products (15%) and petroleum and petroleum products (11%).

The state's 22,000 retail establishments in 1982 had $20.7 billion in sales, an increase of 44% since 1977. Baltimore, Prince Georges, and Montgomery counties, along with the city of Baltimore, together accounted for about two-thirds of the total. Food stores yielded 22% of all sales; automobile dealers, 12%; and department stores, 11%.

The Port of Baltimore was the 6th-busiest foreign-trade port in 1983, as well as the fastest-growing foreign-trade port on the east coast. Imports were valued at $5.9 billion, nearly half of that from Europe and almost one-fourth from Japan. Exports worth almost $6.9 billion were shipped primarily to Europe and Asia; Saudi Arabia, the Federal Republic of Germany, Japan, the United Kingdom, Belgium and Luxembourg, and Spain were among the principal destinations. Foreign exports of Maryland's own manufactures totaled $2.3 billion in 1981 (32d in the US); exports of agricultural commodities in 1981/82 amounted to $188 million (35th in the US).

31CONSUMER PROTECTION

The state agency responsible for controlling unfair and deceptive trade practices is the Division of Consumer Protection within the Attorney General's Office. Under the division's jurisdiction is the Maryland Consumer Council, comprising representatives of consumer groups, business groups, and other interests. The consumer credit commissioner, within the Department of Licensing and Regulation, is responsible for enforcing the state's Credit Deregulation Act, Retail Credit Accounts Law, Retail Installment Sales Act, and Equal Credit Opportunity Act (except for those provisions that apply to banks, over which the state banking commissioner has sole jurisdiction). The Consumer Services Division of the Motor Vehicle Administration, under the Department of Transportation, licenses motor vehicle dealers and manufacturers and professional driving schools; the division is also responsible for school bus safety inspections.

³²BANKING

Maryland's 90 insured commercial banks in 1983 reported total assets of $42.7 billion; outstanding loans exceeded $21.7 billion, and deposits $30.9 billion.

The Maryland National Bank (Baltimore), the largest in the state, ranked 51st in the US, with assets of $7.3 billion and deposits of nearly $4.4 billion at the end of 1984. In 1985, the state legislature passed legislation giving the New York-based Citicorp full banking powers in Maryland.

The 158 federally insured savings and loan associations had assets totaling $16.2 billion in 1983. In 1981, Maryland had 28 credit unions with assets of $304 million; 231 consumer-loan and small-loan licensees had combined assets of $680 million in 1982. The volume of consumer loans declined from $266 million in 1981 to $219 million in 1982.

All state-chartered savings and loan associations are regulated by the Board of Savings and Loan Associations, within the Department of Licensing and Regulation.

³³INSURANCE

Life insurance in force as of 31 December 1983 included 7,731,000 policies worth $97.9 billion. Payments of $979 million were made, including $346.7 million in death benefits. The average value of life insurance per family was $59,500 in 1983.

Property and liability insurers wrote premiums in 1983 totaling $2 billion, including $560.6 million in automobile liability insurance, $294.9 million in automobile physical damage insurance, and $211.9 million in homeowners' coverage. Federal flood insurance totaling $878 million was in effect as of 31 December 1983.

The Maryland Automobile Insurance Fund, a quasi-independent agency created in 1972, pays claims against uninsured motorists (i.e., hit-and-run drivers, out-of-state uninsured motorists, and state residents driving in violation of Maryland's compulsory automobile insurance law), and sells policies to Maryland drivers unable to obtain insurance from private companies.

The State Insurance Division of the Department of Licensing and Regulation licenses all state insurance companies, agents, and brokers, and must approve all policies for sale in the state.

³⁴SECURITIES

There are no securities or commodities exchanges in Maryland. New York Stock Exchange member firms had 70 sales offices and 876 registered representatives in the state in 1983. In that year there were 68 stock brokers and dealers, and 879,000 Marylanders owned shares in public corporations. All securities dealers in Maryland are regulated by the Division of Securities, within the Attorney General's Office.

³⁵PUBLIC FINANCE

The state budget, prepared by the Department of Budget and Fiscal Planning, is submitted annually by the governor to the general assembly for amendment and approval. The fiscal year runs from 1 July to 30 June.

The following table shows recommended revenues and expenditures for the fiscal year 1985/86:

REVENUES	
Income taxes	$2,150,795,000
Federal aid	1,650,748,804
Retail sales and use tax	1,181,038,000
Motor vehicle and gasoline taxes and fees	672,445,000
Educational institutions	401,364,373
State lottery	299,562,021
Transportation receipts	230,938,137
Property tax	145,814,116
Corporation tax	124,199,789
Other taxes and fees	387,445,404
Other receipts	238,684,950
TOTAL	$7,483,035,594

EXPENDITURES	
Education	$2,563,624,696
Transportation	1,472,114,455
Health and hospitals	1,381,722,098
Human resources	570,587,876
Public safety and corrections	385,464,796
Natural resources and recreation	120,023,330
Economic and community development	64,522,659
Agriculture	27,167,864
Personnel and retirement	111,256,524
General services	28,292,847
Debt service	241,488,723
Other outlays	581,562,562
TOTAL	$7,547,828,430

During the 1982/83 fiscal year, local governments spent nearly 42% of their total budgets for education, 17% for public safety, 15% for highways, 5% for social services, and 21% for other purposes. More than 55% of their revenues came from local sources, nearly 23% from the state, 15% from the federal government, and 7% from other sources.

In 1981/82, the city of Baltimore collected revenues totaling $1.4 billion, of which state and local revenues provided 48%, the federal government 9%, and its own property tax 16%; expenditures came to nearly $1.3 billion, of which public education took 26%, highways 18%, police and fire protection 11%, and health services 8%.

The outstanding state debt exceeded $4.6 billion as of 30 June 1983; local government debt was nearly $6 billion. The per capita debt of the state and local governments was $2,219 in 1981/82 (16th in the US). Baltimore's gross debt reached $1,123 million as of 30 June 1982.

³⁶TAXATION

Among the taxes levied by the state in 1984 were an individual income tax, ranging from 2% on the first $1,000 of taxable income to 5% on income over $3,000; a corporate income tax of 7%; a 5% sales and use tax; a state property tax of 21 cents per $100 assessed valuation; and motor vehicle use, franchise, pari-mutuel, cigarette, and alcoholic beverage taxes. All county and some local governments levy property taxes. The counties also tax personal income at rates ranging from 20% to 50% of those imposed by the state; the city of Baltimore taxes personal income at rates equal to 50% of the state levy.

Marylanders paid $12 billion in federal taxes in 1982/83, or $2,818 per capita, and received $8.1 billion in federal outlays during 1983/84. State residents filed 1.9 million federal income tax returns for 1983, paying $6.1 billion in tax.

³⁷ECONOMIC POLICY

The Department of Economic and Community Development, created in 1970, encourages new firms to locate in Maryland and established firms to expand their in-state facilities, promotes the tourist industry, and disseminates information about the state's history and attractions. The department helps secure industrial mortgage loans for businesses that create new jobs, and also provides small-business loans, low-interest construction loans, assistance in plant location and expansion, and an Office of Business and Industrial Development to allow companies to maximize their use of state services. In addition, the department assists local governments in attracting federal funds for economic development and maintains programs to encourage minority businesses, the marketing of seafood, and the use of Ocean City Convention Hall. The Division of Economic Development maintains a representative in Brussels to promote European investment in Maryland. The Department of State Planning oversees state and regional development programs and helps local governments develop planning goals.

During the 1930s, Maryland pioneered in urban design with the new town of Greenbelt, in Prince Georges County. A wholly planned community, Columbia, was built in Howard County during the 1960s. More recently, redevelopment of Baltimore's decaying inner city has been aggressively promoted. Harborplace, a waterside pavilion featuring hundreds of shops and restaurants, formally opened in 1980, and an industrial park was developed in a high-unemployment section of northwest Baltimore during the early 1980s. Not far from Harborplace are the 33-story World Trade Center and other key elements of the Inner Harbor renewal project. Urban restoration has also been encouraged by urban homesteading: a Baltimorean willing to make a commitment to live in an old brick building and fix it up can submit a closed bid to buy it. An analogous "shopsteading" program to attract merchants has also been encouraged.

In 1982, Maryland initiated a program of state enterprise zones to encourage economic growth by focusing state and local resources on designated areas requiring economic stimulus. More than $13.6 million was invested initially in these enterprise zones, five of which were located in western Maryland, four in the central part of the state, and one on the Eastern Shore.

[38]HEALTH

One of the nation's most prestigious medical schools and a number of federal health facilities are located in Maryland.

The infant mortality rate in 1981 was 10.4 per 1,000 live births for whites and 18.2 for nonwhites, in each case slightly below the US average. There were 34,800 legal abortions in 1982—a rate of 35 per 1,000 women, or about 2 abortions for every 5 births.

Maryland's birthrate in 1982 (14.9 per 1,000 population) and death rate (7.9) were each below the respective national norms. The death rates for the leading causes of death in 1981 were heart disease, 301 per 100,000 population; cancer, 189; and stroke, 52—rates all below or nearly equal to the US average for these categories. The Alcoholism Control Administration monitors rehabilitation programs for alcoholics, while the Drug Abuse Administration oversees all drug treatment programs.

In 1983, Maryland had 85 hospitals, with 23,759 beds and 636,345 patient admissions; hospital personnel included 16,052 registered nurses. The average daily census of hospital patients was 20,600; the bed occupancy rate was about 83%. The average cost of care in community hospitals in 1982 was $329 per day and $2,731 per stay, among the highest in the US. The state had 46 nursing homes in 1982.

Maryland's two medical schools are at Johns Hopkins University, which operates in connection with the Johns Hopkins Hospital and has superbly equipped research facilities, and at the University of Maryland in Baltimore. Federal health centers located in Bethesda include the National Institutes of Health and the National Naval Medical Center. In 1982, Maryland had 14,481 physicians and 2,701 dentists.

[39]SOCIAL WELFARE

About 193,700 Marylanders received public assistance totaling $224 million ($255 per family) under the Aid to Families with Dependent Children program in 1982; an additional $348 million was expended on Medicaid in 1981. Some 306,000 state residents took part in the federal food stamp program, which cost $178 million in 1982/83. In the same year, the school lunch program served 334,000 students at a federal cost of $32 million. The city of Baltimore accounts for a clear majority of public assistance recipients in the state.

In 1983, Social Security benefits went to approximately 550,000 Marylanders, of whom 381,000 were retirees, 117,000 survivors, and 52,000 disabled workers. Benefits totaled $2.6 billion, with an average monthly payment of $443 for retired workers. About $105 million in Supplemental Security Income benefits was paid in 1983, over three-fourths of that to 31,875 disabled persons. During 1982, workers' compensation programs paid $239.1 million; unemployment insurance benefits exceeded $321 million, averaging $116 a week for each recipient.

[40]HOUSING

Maryland has sought to preserve many of its historic houses. Block upon block of two-story brick row houses, often with white stoops, fill the older parts of Baltimore, and stone cottages built to withstand rough winters are still found in the western counties. Greenbelt and Columbia exemplify changing modern concepts of community planning.

There were 1,571,000 year-round housing units in Maryland at the time of the 1980 census, of which 1,461,000 were occupied. Statewide, 23.6% of all units were built between 1970 and 1980. According to the 1980 census, 62% of Maryland houses were owner-occupied and less than 2% lacked full plumbing facilities. The median value of an owner-occupied unit was $59,200.

During 1983, 35,800 new housing units were started and 39,600 new units were authorized; residential contracts, valued at more than $2.3 billion during that period, represented slightly more than half the value of all construction contracts. The Community Development Administration of the Department of Economic and Community Development provides low-interest loans for the construction of rental housing and for the rehabilitation of housing in certain designated areas. The Maryland Housing Fund insures qualified lending institutions against losses on home mortgage loans.

[41]EDUCATION

Partly because of Maryland's large number of government and professional workers, educational attainments compare favorably with those of the other South Atlantic states. At the 1980 census, 67% of all Marylanders had completed high school, and 20% had at least four years of college (the US average was 16%), while the median number of school years completed was 13. Maryland students must pass state competency exams in order to graduate from high school.

During the 1982/83 school year there were 776 elementary, 311 secondary, and 174 combined schools serving the state. Enrollment in 1981/82 for grades K–12 was 699,201 in public and 134,165 in private schools. Baltimore's total enrollment was 119,570 public school pupils and 24,753 private school pupils in 1982/83. Statewide, there were 40,366 public school teachers, earning average salaries of $24,095, in 1983/84. The pupil/teacher ratio was 18:1 as of fall 1983. In 1983/84, expenditures on public schools averaged $3,400 per pupil (13th among the states).

As of 1982 there were 30 four-year and 22 two-year accredited colleges and universities in the state. The total enrollment for all four-year institutions was 131,181. The Board of Trustees of the State Universities and Colleges, an 11-member panel (including one student) appointed by the governor, manages 4 state colleges, 2 state universities (Towson and Morgan), the University of Baltimore, the University of Maryland, and St. Mary's College. By far the largest is the University of Maryland, a land-grant school with a 1982 enrollment of 61,717 at five campuses, the largest being College Park, with 37,046 students. The leading private institution in Maryland is The Johns Hopkins University in Baltimore, with a 1982 enrollment of 9,957. The State Board for Community Colleges oversees 17 community and 2 regional two-year schools; their 1982 enrollment was 101,562. Three private two-year colleges enrolled 1,242 students that year. In 1981/82, public spending on higher education averaged $187 per capita; the average faculty salary was $27,504 (18th in the US) in 1982/83.

The State Scholarship Board administers a general scholarship program as well as special grants for needy students, including war orphans and children of servicemen missing in action, orphans of firemen and policemen, teachers of persons with impaired hearing, and Viet-Nam veterans.

[42]ARTS

Though close to the arts centers of Washington, D.C., Maryland has its own cultural attractions. Baltimore, a major theatrical center in the 1800s, still contains many legitimate theaters. Center Stage in Baltimore is the designated state theater of Maryland, and the Olney Theatre in Montgomery County is the offical state summer theater. Arts organizations are aided by the 11-member Maryland Arts Council.

The state's leading orchestra is the Baltimore Symphony, under the direction in 1984 of David Zinman. Baltimore is also the home of the Baltimore Opera Company, and its jazz clubs were the launching pads for such musical notables as Eubie Blake, Ella Fitzgerald, and Cab Calloway. The Peabody Institute of Johns Hopkins University in Baltimore is one of the nation's most distinguished music schools. Both the Maryland Ballet Company and Maryland Dance Theater are nationally known.

43 LIBRARIES AND MUSEUMS

Maryland's public libraries held 10,664,267 volumes in 1982/83 and had a combined circulation of 32,164,132. The center of the state library network is the Enoch Pratt Free Library in the city of Baltimore; founded in 1886, it had 33 branches, 1,900,197 volumes, and a circulation of 2,060,225 in 1983. Each county also has its own library system. The largest academic libraries are those of Johns Hopkins (1,832,018 volumes in 1983) and the University of Maryland at College Park (1,537,687). The Maryland Historical Society Library specializes in genealogy, heraldry, and state history. Maryland is also the site of several federal libraries, including the National Agricultural Library at Beltsville, with 1,800,000 volumes; the National Library of Medicine at Bethesda, 973,000; and the National Oceanic and Atmospheric Administration Library at Rockville, 500,000.

Of the approximately 70 museums and historic sites in the state, the major institutions are the US Naval Academy Museum in Annapolis and Baltimore's Museum of Art, National Aquarium Seaport and Maritime Museum, Maryland Academy of Sciences, the Maryland Historical Society Museum, and Peale Museum, the oldest museum building in the US. Important historic sites include Ft. McHenry National Monument and Shrine in Baltimore (inspiration for "The Star-Spangled Banner") and Antietam National Battlefield Site near Sharpsburg.

44 COMMUNICATIONS

In 1980, 96% of Maryland's 1,460,865 occupied housing units had telephones. There were 449 post offices and 13,853 postal employees in 1985.

Of the state's 60 AM radio stations in 1984, 10 originated from Baltimore; there were 63 FM stations, of which 12 were located in Baltimore. Four of the state's 12 television stations are in Baltimore. Maryland Public Broadcasting operates four noncommercial television stations—in Annapolis, Hagerstown, Owings Mills, and Salisbury. There were 44 cable television systems with 380,361 subscribers in 167 Maryland communities as of 1984. Maryland also receives the signals of the Washington, D.C., broadcast stations.

45 PRESS

The *Maryland Gazette,* established at Annapolis in 1727, was the state's first newspaper. Not until 1773 did Baltimore get its first paper, the *Maryland Journal and Baltimore Advertiser,* but by 1820 there were five highly partisan papers in the city. The *Baltimore Sun,* founded in 1837, reached its heyday after 1906, when H. L. Mencken became a staff writer. Mencken, who was also an important editor and critic, helped found the *American Mercury* magazine in 1924.

As of 1984, Maryland had 6 morning and 6 afternoon dailies with a total circulation of 745,788, and 5 Sunday papers with 1,676,515 in circulation, as well as 5 semiweekly newspapers and 61 weeklies. The most influential newspapers, both published in Baltimore, are the *Sun* (morning, 192,067; evening, 155,607; Sunday, 402,593) and the *News-American* (evening, 107,903; Sunday, 144,578). The *Washington Post* is also widely read in Maryland.

46 ORGANIZATIONS

The 1982 US Census of Service Industries counted 1,100 organizations in Maryland, including 186 business associations; 620 civic, social, and fraternal associations; and 43 educational, scientific, and research associations.

National medically oriented organizations with headquarters in Maryland include the National Federation of the Blind and American Urological Association, both in Baltimore; the American Association of Colleges of Pharmacy, American Institute of Nutrition, and National Foundation for Cancer Research, Bethesda; the American Speech-Language-Hearing Association and Cystic Fibrosis Foundation, Rockville; and the National Association of the Deaf, Silver Spring.

Leading commercial, professional, and trade groups include the Aircraft Owners and Pilots Association and American Fisheries Society, Bethesda; International Association of Chiefs of Police, Gaithersburg; and Retail Bakers of America, Hyattsville. Lacrosse, a major sport in the state, is represented by the Lacrosse Foundation in Baltimore and the US Intercollegiate Lacrosse Association in Chestertown.

47 TOURISM, TRAVEL, AND RECREATION

Although not a tourist mecca, Maryland attracted 4,758,894 visitors to its parks, historical sites, and national seashore (Assateague Island) in 1984. Income from tourism totaled about $3.9 billion in 1983, and the tourist industry provided an estimated 74,000 jobs in 1982.

Among the state's attractions is Annapolis, the state capital and site of the US Naval Academy. On Baltimore's waterfront are monuments to Francis Scott Key and Edgar Allan Poe, historic Ft. McHenry, and many restaurants serving the city's famed crab cakes and other seafood specialties. Ocean City is the state's major seaside resort, and there are many resort towns along Chesapeake Bay.

There are 15 state parks with camping facilities and 10 recreation areas. In 1982/83, licenses were issued to 124,973 fishermen and 174,318 hunters.

48 SPORTS

One of major league baseball's most successful teams, the American League's Orioles, makes its home in Baltimore. Under the leadership of manager Earl Weaver (who resumed managing the club in 1985) and such stars as Brooks Robinson, Frank Robinson, Jim Palmer, Eddie Murray, and Cal Ripken, Jr., the Orioles won the World Series in 1966, 1970, and 1983. The Baltimore Colts of the National Football League, who flourished during the 1950s and 1960s and won the Super Bowl in 1971, moved to Indianapolis in 1984. The Bullets of the National Basketball Association moved from Baltimore to Landover in the early 1970s and were renamed the Washington Bullets.

Ever since 1750, when the first Arabian Thoroughbred was imported by a Maryland breeder, horse racing has been a popular state pastime. The major tracks are Pimlico (Baltimore), Bowie, and Laurel; Pimlico is the site of the Preakness, the second leg of racing's Triple Crown. Harness racing is held at Ocean Downs in Ocean City, quarter-horse racing takes place at several tracks throughout the state, and several steeplechase events, including the prestigious Maryland Hunt Cup, are held annually.

In collegiate basketball, the University of Maryland won the National Invitation Tournament in 1972, and Morgan State took the NCAA Division II title in 1974. Another major sport is lacrosse: Johns Hopkins, Navy, the University of Maryland, and Washington College in Chestertown all have performed well in intercollegiate competition.

Every weekend from April to October, Marylanders compete in jousting tournaments held in four classes throughout the state. In modern jousting, designated as the official state sport, horseback riders attempt to pick up small rings with long, lancelike poles. The state championship is held in October.

⁴⁹FAMOUS MARYLANDERS

Maryland's lone US vice president was Spiro Theodore Agnew (b.1918), who served as governor of Maryland before being elected as Richard Nixon's running mate in 1968. Reelected with Nixon in 1972, Agnew resigned the vice-presidency in October 1973 after a federal indictment had been filed against him. Roger Brooke Taney (1777–1864) served as attorney general and secretary of the treasury in Andrew Jackson's cabinet before being confirmed as US chief justice in 1836; his most historically significant case was the Dred Scott decision in 1856, in which the Supreme Court ruled that Congress could not exclude slavery from any territory. Three associate justices of the US Supreme Court were also born in Maryland. Thomas Johnson (1732–1819), a signer of the Declaration of Independence, served as the first governor of the State of Maryland before his appointment to the Court in 1791. Samuel Chase (1741–1811) was a Revolutionary leader, another signer of the Declaration of Independence, and a local judicial and political leader before being appointed to the high court in 1796; impeached in 1804 because of his alleged hostility to the Jeffersonians, he was acquitted by the Senate the following year. As counsel for the National Association for the Advancement of Colored People, Thurgood Marshall (b.1908) argued the landmark *Brown* v. *Board of Education* school desegregation case before the Supreme Court in 1954; President Lyndon Johnson appointed him to the Court 13 years later.

Other major federal officeholders born in Maryland include John Hanson (1721–83), a member of the Continental Congress and first president to serve under the Articles of Confederation (1781–82); Charles Carroll of Carrollton (1737–1832), a signer of the Declaration of Independence and US senator from 1789 to 1792; John Pendleton Kennedy (1795–1870), secretary of the Navy under Millard Fillmore and a popular novelist known by the pseudonym Mark Littleton; Reverdy Johnson (1796–1876), attorney general under Zachary Taylor; Charles Joseph Bonaparte (1851–1921), secretary of the Navy and attorney general in Theodore Roosevelt's cabinet; and Benjamin Civiletti (b.New York, 1935), attorney general under Jimmy Carter. Among the many important state officeholders are William Paca (1740–99), a signer of the Declaration of Independence and later governor; Luther Martin (b.New Jersey, 1748–1826), Maryland's attorney general from 1778 to 1805 and from 1818 to 1822, as well as defense counsel in the impeachment trial of Chase and in the treason trial of Aaron Burr; John Eager Howard (1752–1827), Revolutionary soldier, governor, and US senator; and Albert C. Ritchie (1876–1936), governor from 1919 to 1935. William D. Schaefer (b.1921) has been mayor of Baltimore since 1972.

Lawyer and poet Francis Scott Key (1779–1843) wrote "The Star-Spangled Banner"—now the national anthem—in 1814. The prominent abolitionists Frederick Douglass (Frederick Augustus Washington Bailey, 1817?–95) and Harriet Tubman (1820?–1913) were born in Maryland, as was John Carroll (1735–1815), the first Roman Catholic bishop in the US and founder of Georgetown University. Elizabeth Ann Bayley Seton (b.New York, 1774–1821), canonized by the Roman Catholic Church in 1975, was the first native-born American saint. Stephen Decatur (1779–1820), a prominent naval officer, has been credited with the toast "Our country, right or wrong!"

Prominent Maryland business leaders include Alexander Brown (b.Ireland, 1764–1834), a Scotch-Irish immigrant who built the firm that is now the 2d-oldest private investment banking house in the US; George Peabody (b.Massachusetts, 1795–1869), founder of the world-famous Peabody Conservatory of Music (now the Peabody Institute of Johns Hopkins University); and Enoch Pratt (b.Massachusetts, 1808–96), who endowed the Enoch Pratt Free Library in Baltimore. Benjamin Banneker (1731–1806), a free black, assisted in surveying the new District of Columbia and published almanacs from 1792 to 1797. Ottmar Mergenthaler (b.Germany, 1854–99), who made his home in Baltimore, invented the linotype machine. Financier-philanthropist Johns Hopkins (1795–1873) was a Marylander, and educators Daniel Coit Gilman (b.Connecticut, 1831–1908) and William Osler (b.Canada, 1849–1919), also a famed physician, were prominent in the establishment of the university and medical school named in Hopkins' honor. Peyton Rous (1879–1970) won the 1966 Nobel Prize for physiology or medicine.

Maryland's best-known modern writer was H(enry) L(ouis) Mencken (1880–1956), a Baltimore newspaper reporter who was also a gifted social commentator, political wit, and student of the American language. Edgar Allan Poe (b.Massachusetts, 1809–49), known for his poems and eerie short stories, died in Baltimore, and novelist-reformer Upton Sinclair (1878–1968) was born there. Other writers associated with Maryland include James M. Cain (1892–1976), Leon Uris (b.1924), and John Barth (b.1930). Painters John Hesselius (b.Pennsylvania, 1728–78) and Charles Willson Peale (1741–1827) are also linked with the state.

Most notable among Maryland actors are Edwin Booth (1833–93) and his brother John Wilkes Booth (1838–65), notorious as the assassin of President Abraham Lincoln. Maryland was the birthplace of several jazz musicians, including James Hubert "Eubie" Blake (1883–1983), William Henry "Chick" Webb (1907–39), and Billie Holiday (1915–59).

Probably the greatest baseball player of all time, George Herman "Babe" Ruth (1895–1948) was born in Baltimore. Other prominent ballplayers include Robert Moses "Lefty" Grove (1900–75), James Emory "Jimmy" Foxx (1907–67), and Al Kaline (b.1934). Former lightweight boxing champion Joe Gans (1874–1910) was a Maryland native.

⁵⁰BIBLIOGRAPHY

Bode, Carl. *Maryland: A Bicentennial History.* New York: Norton, 1978.

Cohen, Richard M., and Jules Witcover. *A Heartbeat Away: The Investigation and Resignation of Vice President Spiro T. Agnew.* New York: Viking, 1974.

Dozer, Donald. *Portrait of the Free State: A History of Maryland.* Cambridge, Md.: Tidewater, 1976.

Federal Writers' Project. *Maryland: A Guide to the Old Line State.* Reprint. New York: Somerset, n.d. (orig. 1940).

Fields, Barbara J. *Slavery and Freedom on the Middle Ground: Maryland during the Nineteenth Century.* New Haven, Ct.: Yale University Press, 1985.

Harvey, Katherine. *The Best-Dressed Miners: Life and Labor in the Maryland Coal Region, 1835–1910.* Ithaca, N.Y.: Cornell University Press, 1969.

Maryland, State of. Department of Economic and Community Development. *Maryland Statistical Abstract 1984–85.* Annapolis, 1985.

Maryland, State of. Department of General Services. Hall of Records Commission. Archives Division. *Maryland Manual 1985–1986.* Edited by Gregory A. Stiverson. Annapolis, 1985.

Mencken, H. L. *A Choice of Days: Essays from "Happy Days," "Newspaper Days," and "Heathen Days."* Selected by Edward L. Galligan. New York: Knopf, 1980.

Walsh, Richard, and William Lloyd Fox (eds.). *Maryland: A History.* Baltimore: Maryland Hall of Records, 1983.

Warner, William. *Beautiful Swimmers: Watermen, Crabs, and the Chesapeake Bay.* Boston: Little, Brown, 1976.

MASSACHUSETTS

Commonwealth of Massachusetts

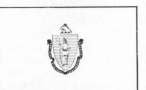

ORIGIN OF STATE NAME: Derived from the name of the Massachuset Indian tribe that lived on Massachusetts Bay; the name is thought to mean "at or about the Great Hill." **NICKNAME:** The Bay State. **CAPITAL:** Boston. **ENTERED UNION:** 6 February 1788 (6th). **SONG:** "All Hail to Massachusetts." **FOLK SONG:** "Massachusetts." **POEM:** "Blue Hills of Massachusetts." **MOTTO:** *Ense petit placidam sub libertate quietem* (By the sword we seek peace, but peace only under liberty). **COAT OF ARMS:** On a blue shield an Indian depicted in gold holds in his right hand a bow, in his left an arrow pointing downward. Above the bow is a five-pointed silver star. The crest shows a bent right arm holding a broadsword. Around the shield beneath the crest is a banner with the state motto in green. **FLAG:** The coat of arms on a white field. **OFFICIAL SEAL:** Same as the coat of arms, with the inscription *Sigillum Reipublicae Massachusettensis* (Seal of the Republic of Massachusetts). **HEROINE:** Deborah Samson. **BIRD:** Chickadee. **HORSE:** Morgan horse. **DOG:** Boston terrier. **MARINE MAMMAL:** Right whale. **FISH:** Cod. **FLOWER:** Mayflower (ground laurel). **TREE:** American elm. **GEM:** Rhodonite. **MINERAL:** Babingtonite. **ROCK:** Roxbury pudding stone. **HISTORICAL ROCK:** Plymouth Rock. **EXPLORER ROCK:** Dighton Rock. **BUILDING AND MONUMENT STONE:** Granite. **FOSSIL:** Theropod dinosaur tracks. **BEVERAGE:** Cranberry juice. **INSECT:** Ladybug. **LEGAL HOLIDAYS:** New Year's Day, 1 January; Birthday of Martin Luther King, Jr., 3d Monday in January; Washington's Birthday, 3d Monday in February; Patriots' Day, 3d Monday in April; Lafayette Day, 20 May; Memorial Day, last Monday in May; Independence Day, 4 July; Labor Day. 1st Monday in September; Columbus Day, 2d Monday in October; Veterans Day, 11 November; Thanksgiving Day, appointed by the governor, customarily the 4th Thursday in November; Christmas Day, 25 December. **TIME:** 7 AM EST = noon GMT.

¹LOCATION, SIZE, AND EXTENT

Located in the northeastern US, Massachusetts is the 4th largest of the 6 New England states and ranks 45th in size among the 50 states.

The total area of Massachusetts is 8,284 sq mi (21,456 sq km), of which land comprises 7,824 sq mi (20,265 sq km) and inland water occupies 460 sq mi (1,191 sq km). Massachusetts extends about 190 mi (306 k) E–W; the maximum N–S extension is about 110 mi (177 km). Massachusetts is bordered on the N by Vermont and New Hampshire; on the E by the Atlantic Ocean; on the S by the Atlantic Ocean and by Rhode Island and Connecticut; and on the W by New York.

Two important islands lie south of the state's fishhook-shaped Cape Cod peninsula: Martha's Vineyard (108 sq mi—280 sq km) and Nantucket (57 sq mi—148 sq km). The Elizabeth Islands, SW of Cape Cod and NW of Martha's Vineyard, consist of 16 small islands separating Buzzards Bay from Vineyard Sound. The total boundary length of Massachusetts is 515 mi (829 km), including a general coastline of 192 mi (309 km); the tidal shoreline, encompassing numerous inlets and islands, is 1,519 mi (2,444 km). The state's geographic center is located in Worcester County, in the northern section of the city of Worcester.

²TOPOGRAPHY

Massachusetts is divided into four topographical regions: coastal lowlands, interior lowlands, dissected uplands, and residuals of ancient mountains. The coastal lowlands, located on the state's eastern edge, extend from the Atlantic Ocean 30–50 mi (48–80 km) inland and include Cape Cod and the offshore islands. The northern shoreline of the state is characterized by rugged high slopes, but at the southern end, along Cape Cod, the ground is flatter and covered with grassy heaths.

The Connecticut River Valley, characterized by red sandstone, curved ridges, meadows, and good soil, is the main feature of west-central Massachusetts. The Berkshire Valley to the west is filled with streams in its northern end, including the two streams that join below Pittsfield to form the Housatonic River.

East of the Connecticut River Valley are the eastern uplands, an extension of the White Mountains of New Hampshire. From elevations of 1,100 feet (335 meters) in midstate, this ridge of heavily forested hills slopes down gradually toward the rocky northern coast.

In western Massachusetts, the Taconic Range and Berkshire Hills (which extend southward from the Green Mountains of Vermont) are characterized by numerous hills and valleys. Mt. Greylock, close to the New York border, is the highest point in the state, at 3,491 feet (1,064 meters). Northeast of the Berkshires is the Hoosac Range, an area of plateau land. Its high point is Spruce Hill, at 1,974 feet (602 meters).

There are more than 4,230 mi (6,808 km) of rivers in the state. The Connecticut River, the longest, runs southward through west-central Massachusetts; the Deerfield, Westfield, Chicopee, and Millers rivers flow into it. Other rivers of note include the Charles and the Mystic, which flow into Boston harbor; the Taunton, which empties into Mount Hope Bay at Fall River; the Blackstone, passing through Worcester on its way to Rhode Island; the Housatonic, winding through the Berkshires; and the Merrimack, flowing from New Hampshire to the Atlantic Ocean via the state's northeast corner. Over 1,100 lakes dot the state; the largest, the artificial Quabbin Reservoir in central Massachusetts, covers 24,704 acres (9,997 hectares). The largest natural lake is Assawompset Pond in southern Massachusetts, occupying 2,656 acres (1,075 hectares).

Hilly Martha's Vineyard is roughly triangular in shape, as is Nantucket Island to the east. The Elizabeth Islands are characterized by broad, grassy plains.

Millions of years ago, three mountainous masses of granite rock extended northeastward across the state. The creation of the Appalachian Mountains transformed limestone into marble, mud and gravel into slate and schist, and sandstone into quartzite. The new surfaces were worn down several times. Then, during the last Ice Age, retreating glaciers left behind the shape of Cape Cod as well as a layer of soil, rock, and boulders.

³CLIMATE

Although Massachusetts is a relatively small state, there are significant climatic differences between its eastern and western sections. The entire state has cold winters and moderately warm summers, but the Berkshires in the west have both the coldest winters and the coolest summers. The normal January temperature in Pittsfield in the Berkshires is 22°F (-6°C), while the normal July temperature is 68°F (20°C). The interior lowlands are several degrees warmer in both winter and summer; the normal January temperature for Worcester is 26°F (-3°C), and the normal July temperature is 71°F (22°C). The coastal sections are the warmest areas of the state; the normal January temperature for Boston is 30°F (-1°C), and the normal July temperature is 74°F (23°C). The record high temperature in the state is 107°F (42°C), established at Chester and New Bedford on 2 August 1975; the record low is -35°F (-37°C), registered at Chester on 12 January 1981.

Precipitation ranges from 39 to 46 in (99–117 cm) annually, with an average for Boston of 43.8 in (111 cm); Worcester, 45.4 in (115 cm); and Pittsfield, 44.4 in (113 cm). The average snowfall for Boston is 42 in (107 cm), with the range in the Berkshires considerably higher. Boston's average wind speed is 13 mph (21 km/hr).

⁴FLORA AND FAUNA

Maple, birch, beech, oak, pine, hemlock, and larch cover the Massachusetts uplands. Common shrubs include rhodora, mountain laurel, and shadbush. Various ferns, maidenhair and osmund among them, grow throughout the state. Typical wild flowers include the Maryland meadow beauty and false loosestrife, as well as several varieties of orchid, lily, goldenrod, and aster. Listed among rare and endangered plants are Eaton's quillwort, climbing fern, burhead, needlegrass, pipewort, mountain alder, white-water crowfoot, Seneca snakeroot, prickly pear, and small whorled pogonia.

Massachusetts had 94 species of mammals in 1985, of which 59 were native land species, 28 were native marine mammals, and 7 had been introduced. Common native mammals include the white-tailed deer, bobcat, river otter, striped skunk, mink, ermine, fisher, raccoon, black bear, gray fox, muskrat, porcupine, beaver, red and gray squirrels, snowshoe hare, little brown bat, and masked shrew. Among the Bay State's 37 resident birds are the mallard, ruffed grouse, bobwhite quail, ring-necked pheasant, herring gull, great horned and screech owls, downy woodpecker, blue jay, mockingbird, cardinal, and song sparrow. Native inland fish include brook trout, chain pickerel, brown bullhead, and yellow perch; brown trout, carp, and smallmouth and largemouth bass have been introduced. Native amphibians include the Jefferson salamander, red-spotted newt, eastern American toad, gray tree frog, and bullfrog. Common reptiles are the snapping turtle, stinkpot, spotted turtle, northern water snake, and northern black racer. The venomous timber rattlesnake and northern copperhead are found mainly in Norfolk, Hampshire, and Hampden counties. The Cape Cod coasts are rich in a variety of shellfish, including clams, mussels, shrimps, and oysters. Among endangered mammals are the sperm, blue, sei, finback, right, and humpback whales; among reptiles, the red-bellied turtle (found only in Plymouth County).

⁵ENVIRONMENTAL PROTECTION

All environmental protection programs are administered by the Department of Environmental Quality Engineering (DEQE) of the Executive Office of Environmental Affairs, which in 1983/84 had a budget of $171.1 million, including $16.3 million in federal funds.

DEQE's Division of Air Quality Control regulates atmospheric and noise pollution. Between 1980 and 1984, air pollution levels in Massachusetts decreased by 10% in urban areas, a result, in part, of the mandatory installation of air-pollution control devices on automobiles.

Carbon monoxide levels did not meet state standards even though they were reduced by 13% between 1982 and 1984. Hydrocarbon emissions were reduced by 18% between 1982 and 1984. During the 1985 and 1986 fiscal years, Massachusetts appropriated $1 million for acid rain research.

The DEQE's Metropolitan District Commission oversees the water supply lines for 35 cities in the Boston metropolitan area. It also operates the two major sewage treatment plants that serve 43 cities and towns in that region. In December 1984, Governor Michael Dukakis signed a bill establishing an independent agency to clean up Boston Harbor and the Charles River.

There are 560 landfills in Massachusetts. In 1984, Massachusetts generated 190,000 tons of hazardous waste. The state made plans to clean up 95 hazardous waste disposal sites in 1985 and 135 sites in 1986. The Division of Forests and Parks maintains four bureaus concerned with public land management. Soil and water conservation districts are managed by the Division of Conservation Services. Approximately $2,250,000 was appropriated by the state for conservation programs in 1985/86.

⁶POPULATION

As New England's most populous state, Massachusetts has seen its population grow steadily since colonial times. However, since the early 1800s, its growth rate has lagged behind the rest of the nation's. Massachusetts' population according to the 1980 federal census was 5,737,037 (11th in the US), an increase of 0.8% over 1970 but well below the US growth rate of 11.4%. The population estimate for the state in 1985 was 5,764,125, a change of only 0.5% since 1980. Reasons behind the population lag include a birthrate (13.2 per 1,000 in 1982) well below the US average, and a net out-migration of 301,000 people between 1970 and 1983, the largest drop of all New England states. In 1980, about 84% of the state was urban, and 16% rural. A density of 737 people per sq mi (285 per sq km) in 1985 made Massachusetts the 3d most densely populated state. A population of 5,703,900 is projected for 1990.

The state's biggest city is Boston, which ranked 20th among the largest US cities with a population of 570,719 in 1984, a decline of 11% since the 1970 census was taken. Other large cities (with their 1984 populations) are Worcester, 159,843; Springfield, 150,454; New Bedford, 97,738; Brockton, 95,892; Lowell, 93,473; Cambridge, 92,535; Fall River, 92,038; Quincy, 83,682; and Newton, 82,297. More than 70% of all state residents live in the Greater Boston area, which in 1984 had a metropolitan population of 4,026,600 (7th largest in the US).

Massachusetts Counties, County Seats, and County Areas and Populations

COUNTY	COUNTY SEAT(S)[1]	LAND AREA (SQ MI)	POPULATION (1984 EST.)
Barnstable	Barnstable	400	163,129
Berkshire	Pittsfield	929	142,426
Bristol	New Bedford, Taunton, Fall River	557	477,900
Dukes	Edgartown	102	10,064
Essex	Lawrence, Salem, Newburyport	495	645,819
Franklin	Greenfield	702	64,868
Hampden	Springfield	618	442,720
Hampshire	Northampton	528	141,216
Middlesex	Cambridge (East), Lowell	822	1,369,400
Nantucket	Nantucket	47	5,876
Norfolk	Dedham	400	603,627
Plymouth	Plymouth, Brockton	655	417,706
Suffolk	[2]	57	658,814
Worcester	Worcester, Fitchburg	1,514	654,019
	TOTALS	7,824	5,797,582

[1] Officially designated "shire town."
[2] No shire town. Suffolk County includes the city of Boston.

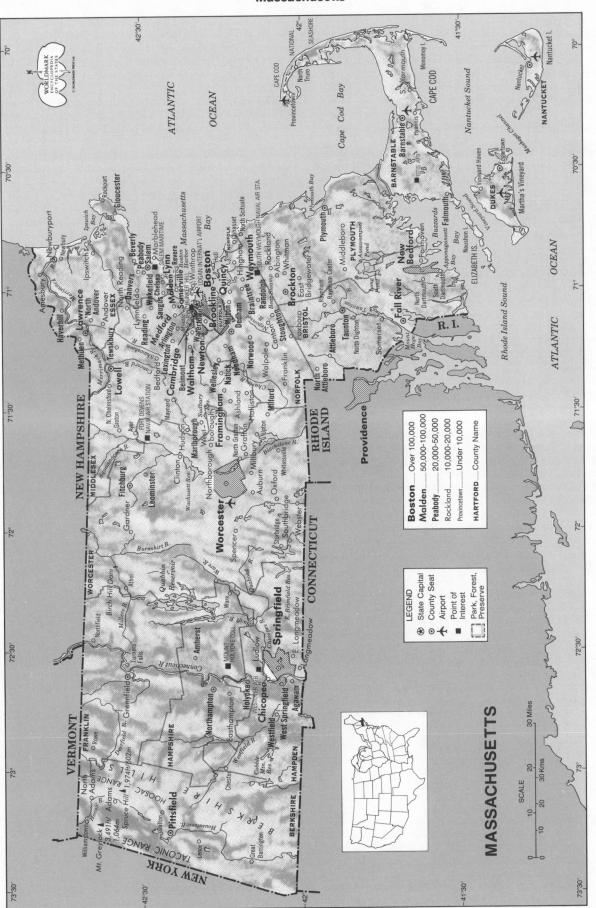

MASSACHUSETTS

LEGEND

⊛ State Capital
⊚ County Seat
✈ Airport
■ Point of Interest
⬚ Park, Forest, Preserve

Boston	Over 100,000
Malden	50,000–100,000
Peabody	20,000–50,000
Rockland	10,000–20,000
Provincetown	Under 10,000
HARTFORD	County Name

SCALE

0   10   20   30 Miles
0   10   20   30 Kms

See endsheet maps: M2.

LOCATION: 41°14′17″ to 42°53′13″N; 69°55′25″ to 73°30′31″W. **BOUNDARIES:** Vermont line, 41 mi (66 km); New Hampshire line, 94 mi (151 km); Atlantic Ocean coastline, 192 mi (309 km); Rhode Island line, 46 mi (74 km); Connecticut line, 92 mi (148 km); New York line, 50 mi (81 km).

[7]ETHNIC GROUPS

Early industrialization helped make Massachusetts a mecca for many European migrants, particularly the Irish. As late as 1980, more than half of the population identified with single ancestry groups, the largest being the Irish (21% of the population), English (15%), Italian (13%), French (10%), Portuguese (6%), and Polish (5%). In that year, 8.7% of the state's population was foreign-born.

Massachusetts has always had some black population, and has contributed such distinguished figures as poet Phillis Wheatley and NAACP founder W. E. B. Du Bois (the first black Ph.D. from Harvard) to US cultural and public life. A sizable class of black professionals has developed, and the 20th century has seen an influx of working-class blacks from southern states. In 1980 there were 221,279 black Americans in Massachusetts, 3.9% of the population; blacks constituted more than 22% of Boston's population. The state also had a Spanish-speaking population of 141,043 as of 1980.

Greater Boston has a small, well-organized Chinatown; in the suburbs reside many business and professional Chinese, as well as those connected with the region's numerous educational institutions. Statewide, there were 25,012 Chinese in 1980, 4,655 Koreans, 4,483 Japanese, and 3,172 Vietnamese.

Although small Indian settlements persist on Cape Cod, Massachusetts had only 7,483 Indians in 1980. Cape Cod also has settlements of Portuguese fishermen, as has New Bedford.

[8]LANGUAGES

Some general Algonkian loanwords and a few place-names—such as Massachusetts itself, Chicopee, Quebbin, and Naukeag—are the language echoes of the Massachuset, Pennacook, and Mahican Indians so historically important in the founding of Massachusetts Bay Colony and Old Colony, now Plymouth.

On the whole, Massachusetts English is classed as Northern, but early migration up the Connecticut River left that waterway a sometimes sharp, sometimes vague boundary, setting off special variations within the eastern half of the state. Two conspicuous but now receding features long held prestige because of the cultural eminence of Boston: the absence of /r/ after a vowel, as in *fear* and *port,* and the use of a vowel halfway between the short /a/ of *cat* and /ah/ in *half* and *past* as well as in *car* and *park.* Eastern Massachusetts speakers are likely to have /ah/ in *orange* and to pronounce *on* and *fog* with the same vowel as in *form.* In the east, a sycamore is a *buttonwood,* a tied and filled quilt is a *comforter,* a *creek* is a saltwater inlet, and pancakes may be called *fritters.*

Around Boston are heard the intrusive /r/ as in "the lawr of the land," the /oo/ vowel in *butcher, tonic* for soft drink, *submarine* for a large sandwich, and *milkshake* for a concoction lacking ice cream. West of the Connecticut River are heard the /aw/ sound in *orange,* /ah/ in *on* and *fog,* and the short /a/ of *cat* in *half* and *bass; buttonball* is a sycamore, and *comfortable* is a tied quilt.

In 1980, 87% of the population 3 years of age or older spoke only English at home. Principal other languages spoken at home were as follows:

French	136,327	Greek	36,555
Portuguese	122,663	Chinese	21,699
Spanish	113,988	German	19,671
Italian	107,866	Armenian	10,827
Polish	49,885	Yiddish	10,774

[9]RELIGIONS

While Protestant sects have contributed greatly to the state's history and development, more than half the state's population is Roman Catholic, a fact that has had a profound effect on Massachusetts politics and policies.

Both the Pilgrims who landed on Plymouth Rock in 1620 and the Puritans who formed the Massachusetts Bay Company in 1629 came to the land to escape harassment by the Church of England. These early communities were based on strict religious principles and forbade the practice of differing religions. Religious tolerance was included in the Charter of 1692, to protect the Baptists, Anglicans, and Catholics who had by then arrived in the colony.

The major influx of Roman Catholics came in the 1840s with the arrival of the Irish in Boston. By the 1850s, they had migrated to other towns and cities and formed the backbone of the state's industrial work force. Later migration by Italian Catholics, German Catholics, and Eastern European Jews turned the state, by 1900, into a melting pot of religions and nationalities, although many of these minorities did not win substantial acceptance from the Protestant elite until the World War II era.

As of 1984 there were 2,955,931 Roman Catholics in Massachusetts, more than half the total population. The Boston archdiocese, one of the largest in the country, had 1,936,146 members. The largest Protestant denominations were United Church of Christ, 149,739 adherents; Episcopal, 141,060; American Baptist Convention, 75,749; United Methodist, 75,653; Unitarian, 29,292; Lutheran Church in America, 25,403; and Congregationalist, 12,158. Most of the state's estimated 248,395 Jewish population in 1984 lived in Boston.

Although small, the Church of Christ, Scientist is significant to Massachusetts history. Its first house of worship was founded in 1879 in Boston by Mary Baker Eddy, who four years earlier had published the Christian Science textbook, *Science and Health with Key to the Scriptures.* In Boston, the church continues to publish an influential newspaper, the *Christian Science Monitor.*

[10]TRANSPORTATION

The first rail line in the US, a 3-mi (5-km) stretch from the Neponset River to the granite quarries in Quincy, was built in 1826. The first steam railroad in New England, connecting Boston and Lowell, was completed seven years later. By the late 1830s, tracks were laid from Boston to Worcester and to Providence, R.I., and during the next two decades, additional railroad lines opened up new cities for industrial expansion.

As of 1984, freight service was supplied by the Boston and Maine, Central Vermont (Canadian National), and Conrail, as well as by the smaller Fore River, Grafton and Upton, and Providence and Worcester lines. Boston is the northern terminus of Amtrak's Northeast Corridor, linking New England with Washington, D.C., via New York City and Philadelphia. At the end of 1983, the state had 1,098 mi (1,767 km) of Class I track.

Commuter service is coordinated by the Massachusetts Bay Transportation Authority (MBTA), formed in 1964 to consolidate bus, commuter rail, high-speed trolley, and subway services to the 79 cities and towns in the Greater Boston area. The Boston subway, which began operation in 1897, is the oldest subway system in the US; the system had 354 subway cars in 1983. Boston is one of the few cities in the US with an operating trolley system, which in 1983 had 225 streetcars and 50 trackless trolleys (electric buses) in operation. There were 1,050 buses in the MBTA system in that year, as well as 176 commuter rail vehicles (138 coaches and 38 locomotives). In December 1984, the MBTA system carried about 858,000 commuters a day, a 2% increase over December 1983. About 40% of all Bostonians commute to work by public transportation, the 2d-highest percentage in the nation, following New York City.

In 1983, 33,796 mi (54,389 km) of paved roadways crisscrossed the state. The major highways, which extend from and through Boston like the spokes of a wheel, include I-95, which runs north–south, the Massachusetts Turnpike (I-90), which runs west to the New York State border, I-93, which leads north to New Hampshire, State Highway 3 to Cape Cod; and State Highway 24 to Fall River. The other major road in the state is I-91, which runs north–south through the Connecticut River Valley. The interstate highway network in Massachusetts totaled 553 mi (890 km) in 1983. A total of $527 million in state highway funds was expended in 1982. In 1983, 3,839,707 motor vehicles were registered in the state, of which 3,374,962 were automobiles, 453,880 were trucks, and 10,865 were buses; there also were 112,597 motorcycles. The state issued 3,678,678 driver's licenses in 1983.

Because it is the major American city closest to Europe, Boston is an important shipping center for both domestic and foreign cargo. In 1981, 20,306,450 tons of cargo, of which 6,778,460 tons were imports, passed through the Port of Boston. All port activity is under the jurisdiction of the Massachusetts Port Authority, which also operates Logan International Airport and Hanscom Field in Bedford. Other important ports and their 1981 cargo totals were Fall River, 3,635,123 tons (893,847 imported), and Salem, 1,228,200 tons (461,912 imported).

There were 75 airports and 47 heliports in the state as of 31 December 1983. In that year, the state's active aircraft numbered 2,737, and pilots, 12,010. Logan International, near Boston, was the 13th-busiest airport in the world in 1983, when it handled 90,012 departing aircraft, enplaning 8,044,651 passengers, and processing 103,968 tons of freight.

11 HISTORY

Some 15,000 years ago, when the last of the glaciers receded from the land we call Massachusetts, what remained was a rocky surface scoured of most of its topsoil. In time, however, forests grew to support a rich variety of wildlife. When the first Indians arrived from the south, game abounded and fish were plentiful in streams and along the coast. These first Indians were hunter-gatherers; their successors not only foraged for food but also cleared fields for planting corn (maize) and squash. Periodically they burned away the woodland underbrush, a technique of forest management that stimulated the vegetation that supported game. When English settlers arrived, they encountered five main Algonkian tribes: the Nauset, a fishing people on Cape Cod; the Wampanoag in the southeast; the Massachuset in the northeast; the Nipmuc in the central hills; and the Pocumtuc in the west.

The earliest European explorers—including the Norsemen, who may have reached Cape Cod—made no apparent impact on these Algonquian groups, but in the wake of John and Sebastian Cabot's voyages (1497 and following), fishermen from England, France, Portugal, and Spain began fishing off the Massachusetts coast. By the mid-16th century, they were regularly going ashore to process and pack their catch. Within 50 years, fur trading with the Indians was established.

Permanent English settlement, which would ultimately destroy the Algonquian peoples, began in 1620 when a small band of Puritans left their haven at Leiden in the Netherlands to start a colony in the northern part of Virginia lands, near the Hudson River. Their ship, the *Mayflower,* was blown off course by an Atlantic storm, and they landed on Cape Cod before settling in an abandoned Wampanoag village they called Plymouth. Ten years later, a much larger Puritan group settled the Massachusetts Bay Colony, some miles to the north in Salem. Between 1630 and 1640, about 20,000 English people, chiefly Puritans, settled in Massachusetts, with offshoots moving to Connecticut and Rhode Island.

The leaders of the Massachusetts settlement, most notably John Winthrop, a country gentleman with some legal training, intended to make their colony an exemplary Christian society. Though church and state were legally separate, they were mutually reinforcing agencies; thus, when Roger Williams and Anne Hutchinson were separately found guilty of heresy in the 1630s, they were banished by the state. All male church members had a voice in both church and state leadership, though both institutions were led by college-educated men. In order to provide for future leaders, Harvard College (now Harvard University) was founded in 1636.

After the beginning of the English revolution in 1640, migration to Massachusetts declined abruptly. Farming soon overtook fishing and fur trading in economic importance; after the trade in beaver skins was exhausted, the remaining Indian tribes were decimated in King Philip's War (1675–76). Shipbuilding and Atlantic commerce also brought prosperity to the Massachusetts Bay Colony, which was granted a new charter by King William and Queen Mary in 1692, merging Massachusetts and the colony of Plymouth. In that year, 19 people were executed for witchcraft on the gallows at Salem before Massachusetts authorities put a stop to the proceedings.

During the 18th century, settlement spread across the entire colony. Boston, the capital, had attained a population of 15,000 by 1730; it was an urbane community of brick as well as wooden buildings, with nearly a dozen church spires distinguishing its skyline by the 1750s. Religious revivals, also occurring elsewhere in America, swept Massachusetts in the 1730s and 1740s, rekindling piety and dividing the inhabitants into competing camps. Although the conflicts had ebbed by the 1750s, Massachusetts did not achieve unity again until the resistance to British imperial actions during the next two decades.

Up to this time, imperial government had rested lightly on Massachusetts, providing more advantages than drawbacks for commerce. The colony had actively supported British expeditions against French Canada, and supply contracts during the French and Indian War had enriched the economy. But the postwar recession after 1763 was accompanied by a new imperial policy that put pressure on Massachusetts as well as other colonies. None of the crown's three objectives—tight regulation of trade, the raising of revenue, and elimination of key areas of colonial political autonomy—were popular among the merchants, tradespeople, and farmers of Massachusetts. From 1765, when Bostonians violently protested the Stamp Act, Massachusetts was in the vanguard of the resistance.

At first, opposition was largely confined to Boston and surrounding towns, although the legislature, representing the entire colony, was active in opposing British measures. By December 1773, when East India Company tea was dumped into Boston harbor to prevent its taxation, most of the colony was committed to resistance. Newspaper polemics composed by Samuel Adams and his cousin John, among others, combined with the persuasive activities of the Boston Committee of Correspondence, helped convince a majority of Massachusetts residents that the slogan "no taxation without representation" stood for the preservation of their communities. When Parliament retaliated for the Tea Party by closing the port of Boston in 1774, rescinding the colony's 1692 charter and remaking the government to put it under London's control, Massachusetts was ready to rebel. Military preparations began immediately on both sides. After almost a year of confrontation, battle began at Lexington and Concord on 19 April 1775. By this time, Massachusetts had the backing of the Continental Congress.

For Massachusetts, the battlefield experience of the Revolution was largely confined to 1775, the climaxes being the Battle of Bunker Hill and the British evacuation of Boston the following year. Thereafter, Massachusetts soldiers were active throughout the colonies, but the theater of action shifted southward. A new republican constitution, adopted in 1780, was the first state constitution to be submitted to the electorate for ratification.

Social and economic conditions in post-Revolutionary Massachusetts were much like those of the colonial era. Although the Shays Rebellion, an uprising of central and western farmers led by Daniel Shays in 1786–87, challenged the political hegemony of commercially oriented eastern leaders, the latter succeeded in maintaining their hold on the state. Massachusetts, which entered the Union on 6 February 1788, was the center of Federalism from 1790 until the mid-1820s. Although Jeffersonian Republicans and

Jacksonian Democrats achieved substantial followings, Federalist policies, embodied in the Whigs in the 1830s and the Republicans from the late 1850s, were dominant. This political continuity was based on the importance of national commercial and industrial development to the state.

Even before 1800, it was clear that Massachusetts could not sustain growth in agriculture. Its soil had never been excellent, and the best lands were tired, having been worked for generations with little regard for conservation. Much of Massachusetts' population departed for New York, Ohio, and beyond during the first decades of the 19th century. Those who stayed maintained a productive agriculture, concentrating more and more on fruits and dairying, but they also developed commerce and industry. At Waltham, Lowell, and Lawrence the first large-scale factories in the US were erected, and smaller textile mills throughout the state helped to make Massachusetts a leader in the cloth industry. At Springfield and Watertown, US armories led the way in metal-working, while shoes and leather goods brought prosperity to Lynn, and whale products and shipbuilding to New Bedford. By the 1850s, steam engines and clipper ships were both Bay State products.

The industrial development of Massachusetts was accompanied by a literary and intellectual flowering that was partly in reaction to the materialism and worldliness associated with urban and industrial growth. Concord, the home of Ralph Waldo Emerson, Henry David Thoreau, and a cluster of others, became the center of the transcendentalist movement in philosophy. Social reform also represented an assertion of moral values, whether in the field of education, health care, temperance, or penology. Abolitionism, the greatest of the moral reform efforts, found some of its chief leaders in Massachusetts, among them William Lloyd Garrison and Wendell Phillips, as well as a host of supporters.

In the years following the Civil War, Massachusetts emerged as an urban industrial state. Its population, fed by immigrants from England, Scotland, Germany, and especially Ireland, grew rapidly in the middle decades of the century. Later, between 1880 and 1920, another wave of immigrants came from French Canada, Italy, Russia, Poland, Scandinavia, Portugal, Greece, and Syria. Still later, between 1950 and 1970, black southerners and Puerto Ricans settled in the cities.

From the election of Lincoln in 1860 through the 1920s, Massachusetts was led by Protestant Yankee Republicans; most Democrats were Catholics. Class, ethnic, and religious tensions were endemic, occasionally erupting into open conflict. Three such episodes gained national attention. In 1912, immigrant textile workers in Lawrence were pitted against Yankee capitalists in a highly publicized strike that dragged on for months. The Boston police strike of 1919 had the largely Irish-American force rebelling against Yankees in city and state government, and brought Governor Calvin Coolidge—who suppressed the strike and refused to reinstate the striking policemen—to national prominence. In 1921, Nicola Sacco and Bartolomeo Vanzetti, Italian immigrant anarchists, were convicted for a payroll robbery and murder, though there was bitter controversy regarding the quality of the evidence against them. Before they were executed in 1927, their case and the issues it raised polarized political opinion throughout the US. Subsequently, electoral competition between Democrats and Republicans emerged as a less divisive outlet for class and ethnic tensions. Since 1959, the Democrats have enjoyed ascendancy statewide, and Republicans have won only when their candidates stood close to the Democrats on the issues. Party loyalties as such have waned, however.

The Massachusetts economy, relatively stagnant between 1920 and 1950, revived in the second half of the 20th century through a combination of university talent, capital resources, a skilled work force, and political clout. As the old industries and the mill cities declined, new high-technology manufacturing developed in Bos-

ton's suburban perimeter. Electronics, computers, and defense-oriented industries led the way, stimulating a general prosperity in which service activities such as banking, insurance, health care, and higher education were especially prominent. As a result, white-collar employment and middle-class suburbs flourished, though run-down mill towns and Yankee dairy farms and orchards still dotted the landscape.

In this respect, as in its politics, Massachusetts resembled many other areas of the Northeast. It was a multiracial state in which the general welfare was defined by shifting coalitions of ethnic groups and special interests. From a national perspective, Massachusetts voters appeared liberal; the Bay State was the only one to choose Democrat George McGovern over President Richard M. Nixon in 1972, and was a secure base for Senator Edward M. Kennedy, a leading liberal Democrat, throughout the 1970s and early 1980s. Yet Boston was also the site of some of the most extreme anti-integration tension during the same era; Massachusetts was simultaneously a center of efforts in favor of the Equal Rights Amendment and against abortion. The old labels "conservative" and "liberal," though still in use, seemed scarcely more descriptive than party labels for understanding Massachusetts public life in the 1980s.

¹²STATE GOVERNMENT

The first state constitution, drawn up soon after the signing of the Declaration of Independence, was rejected by the electorate. A revised draft was not approved by the state voters until 15 June 1780, following two constitutional conventions. This constitution, as amended, governs Massachusetts and is, according to the state, the oldest written constitution in the world still in effect. The legislature of Massachusetts, known as the General Court, is composed of a 40-member senate and 160-member house of representatives, all of whom are elected every two years in even-numbered years. Members of the senate must have resided in their home district for at least five years; representatives, at least one year. A legislator's annual salary was $30,000 in 1983/84.

The governor and lieutenant governor are elected jointly every four years. The governor appoints all state and local judges, as well as the heads of the 10 executive offices. Both the governor and lieutenant governor must have resided in the state for at least seven years. Other elected officials include the attorney general, secretary of the commonwealth, treasurer and receiver-general, and auditor of the commonwealth. All serve four-year terms.

Massachusetts also has an eight-member executive council, elected every two years by district. The council has the power to review the governor's judicial appointments and pardons, and to authorize expenditures from the state treasury. The lieutenant governor also has a vote on the executive council except when the governor is absent, in which case the lieutenant governor presides over it as a nonvoting participant. Members of the council must have resided in the state for five years.

Any Massachusetts citizen may file a bill through a state legislator, or a bill may be filed directly by a legislator or by the governor. To win passage, a bill must gain a majority vote of both houses of the legislature. After a bill is passed, the governor has 10 days in which to sign it, return it for reconsideration (usually with amendments), veto it, or refuse to sign it ("pocket veto"). A veto may be overriden by a two-thirds majority in both houses.

Amendments to the constitution may be introduced by any house or senate member (legislative amendment) or by a petition signed by at least 25,000 qualified voters (initiative amendment) that is presented in a joint session of the General Court. If it is approved by two successive sessions of the legislature, the amendment is then submitted to the voters at the next general election.

To vote in a Massachusetts district, a person must be a US citizen, at least 18 years old, and a resident of that district; there is no time requirement for residency.

[13]POLITICAL PARTIES

The Federalist Party, represented nationally by John Adams, dominated Massachusetts in the late 18th and early 19th centuries. The state turned to the Whig Party in the second quarter of the 19th century. Predominantly Yankee in character, the Whigs supported business growth, promoted protective tariffs, and favored such enterprises as railroads and factories. The new Republican Party, to which most Massachusetts Whigs gravitated when their party split in the 1850s, was a prime mover of abolitionism and played an important role in the election of Abraham Lincoln as president in 1860. Republicans held most of the major state elective offices, as well as most US congressional seats, until the early 1900s.

The Democratic Party's rise starting in the 1870s was tied directly to massive Irish immigration. Other immigrant groups also gravitated toward the Democrats, and in 1876, the state's first Democratic congressman was elected. In 1928, the state voted for Democratic presidential candidate Alfred E. Smith, a Roman Catholic, the first time the Democrats won a majority in a Massachusetts presidential election. Since then Democrats have, for the most part, dominated state politics. In 1960, John F. Kennedy, who had been a popular US senator from Massachusetts, became the first Roman Catholic president in US history. Since then the state has voted for all Democratic presidential candidates except Republican Ronald Reagan in 1980 and 1984; in 1972, it was the only state carried by Democrat George McGovern.

As of 1984, 3,054,146 voters were registered, of whom 48.3% were Democrats, 12.8% were Republicans, and 38.9% were independents. As of 1985, the governorship and both US Senate seats were held by Democrats, the US House delegation consisted of 10 Democrats and 1 Republican. The Massachusetts senate had 32 Democrats and 8 Republicans, while the house of representatives had 126 Democrats and 34 Republicans. As of January 1985, 5 women served in the state senate and 28 in the house; 1 black was a senator, and 4 were representatives. In all, there were 32 black elected officials in Massachusetts in 1985; there were 3 Hispanic elected officials in 1984.

[14]LOCAL GOVERNMENT

As of 1984, Massachusetts had 14 counties, 39 cities, and 312 towns.

In 12 of the 14 counties, executive authority was vested in three county commissioners elected to four-year terms. The exceptions were Suffolk County—where executive powers are exercised by the mayor and city council of Boston, the board of aldermen in Chelsea, the city council of Revere, and the board of selectmen of Winthrop—and Nantucket County, where the five town selectmen have executive powers. Other county officials include the register of probate and family court, sheriff, clerk of courts, county treasurer, and register of deeds.

All Massachusetts cities are governed by mayors and city councils. Towns are governed by selectmen, who are usually elected to either one or two-year terms. Town meetings—a carryover from the colonial period, when every taxpayer was given an equal voice in town government—still take place regularly. By state law, to be designated a city, a place must have at least 12,000 residents. Towns with more than 6,000 inhabitants may hold representative town meetings limited to elected officials.

[15]STATE SERVICES

State services are provided through the 12 executive offices and major departments that constitute the governor's cabinet. The heads of these departments are appointed by the governor.

Educational services are administered by the Executive Office of Education. Included under its jurisdiction are the State Board of Education and Board of Regents of Higher Education, the Massachusetts community college and state college systems, the University of Massachusetts, the Council of the Arts and Humanities, and the State Library.

The Executive Office of Transportation and Construction supervises the Department of Public Works and has responsibility for the planning and development of transportation systems within the state, including the Massachusetts Port Authority, the Massachusetts Turnpike Authority, the Massachusetts Bay Transportation Authority, and the Massachusetts Aeronautics Commission.

All public health, mental health, youth, and veterans' programs

Massachusetts Presidential Vote by Political Party, 1948–84

YEAR	ELECTORAL VOTE	MASSACHUSETTS WINNER	DEMOCRAT	REPUBLICAN	SOCIALIST LABOR	PROGRESSIVE
1948	16	*Truman (D)	1,151,788	909,370	5,535	38,157
1952	16	*Eisenhower (R)	1,083,525	1,292,325	1,957	4,636
1956	16	*Eisenhower (R)	948,190	1,393,197	5,573	—
1960	16	*Kennedy (D)	1,487,174	976,750	3,892	—
1964	14	*Johnson (D)	1,786,422	549,727	4,755	—
						AMERICAN IND.
1968	14	Humphrey (D)	1,469,218	766,844	6,180	87,088
					SOC. WORKERS	AMERICAN
1972	14	McGovern (D)	1,332,540	1,112,078	10,600	2,877
1976	14	*Carter (D)	1,429,475	1,030,276	8,138	7,555
					LIBERTARIAN	
1980	14	*Reagan (R)	1,048,562	1,054,213	21,311	—
1984	13	*Reagan (R)	1,239,600	1,310,936	—	—

*Won US presidential election.

are administered by the Executive Office of Human Services. Also under its jurisdiction are the Department of Public Welfare and the Department of Correction. The Executive Office of Public Safety includes the Division of Civil Defense, Registry of Motor Vehicles, and Highway Safety Bureau.

The Executive Office of Consumer Affairs and Business Regulation regulates state standards and registers professional workers. The Department of Public Utilities and Alcoholic Beverages Control Commission are also part of this office, as are the divisions regulating banks and insurance. Housing services are provided through the Executive Office of Communities and Development. This office administers the Massachusetts Home Mortgage Finance Agency, the Housing Finance Agency, and the Bureaus of Housing Development and of Housing Management and Tenant Services.

The Executive Office of Environmental Affairs protects the state's marine and wildlife, and monitors the qualtiy of its air, water, and food.

Labor and industrial relations are monitored through the Executive Office of Labor, whose Department of Labor and Industries administers the minimum wage law, occupational safety laws, and child labor laws, among others. The Executive Office of Economic Affairs helps to improve the economic climate in the state and promotes exports and tourism. The Department of Elder Affairs plans and implements programs for the elderly, including nutrition, home care, and education programs.

[16] JUDICIAL SYSTEM

All statewide judicial offices are filled by the governor, with the advice and consent of the executive council.

The supreme judicial court, composed of a chief justice and 6 other justices, is the highest court in the state. It has appellate jurisdiction in matters of law and also advises the governor and legislature on legal questions. The superior courts, actually the highest level of trial court, have a chief justice and 55 other justices; these courts hear law, equity, civil, and criminal cases, and make the final determination in matters of fact. The appeals court, consisting of a chief justice and 9 other justices, hears appeals of decisions by district and municipal courts. As of 1984 there were 90 district and municipal courts and 264 trial court judges. Other court systems in the state include the land court, probate and family court, housing court (with divisions in Boston and Hampden counties), and juvenile court (with divisions in Boston, Springfield, Worcester, and Bristol counties).

Except for the murder rate, which was only one-third of the US average, crime rates for both the state and the Boston metropolitan area ranged at or slightly above the US averages. Specific figures for 1983 were murder and nonnegligent manslaughter, 3.5 per 100,000 inhabitants; forcible rape, 25.9; robbery, 208.5; aggravated assault, 338.9; burglary, 1,253.5; larceny-theft, 2,321.5; and motor vehicle theft, 858.9. In 1983, a total of 288,971 crimes were reported in the state, of which 33,264 were violent crimes. Capital punishment was restored to the state in 1979; however, no executions had occurred as of 1985. Under Massachusetts gun control laws, all guns must be registered, and there is a mandatory one-year jail sentence for possession without a permit.

As of 1984 there were 4,890 prisoners in state and federal correctional institutions.

[17] ARMED FORCES

The military installations located in Massachusetts in 1984 had 9,514 active-duty military personnel. Nearly 80% were stationed at Ft. Devens in Ayer and at Laurence G. Hanscom Air Force Base in Bedford. Other installations include the Army research and development center at Natick, the Navy's South Weymouth Naval Air Station, and Westover Air Force Base. The state ranked 4th among the 50 states in the value of defense contracts awarded in 1983/84, with a total of about $7 billion.

Approximately 714,000 military veterans were living in the state in 1983. Veterans of World War I numbered 9,000; World War II, 312,000; the Korean conflict, 132,000; and the Viet-Nam era, 172,000. About $803 million was paid in veterans' benefits in 1982/83, of which 56% went for compensation and pensions, 33% for medical and administrative costs, 4% for education and training, and 7% for other purposes.

The Massachusetts National Guard had 12,474 members at the end of 1984; the strength of the Air National Guard was 2,717 at the beginning of 1985.

In 1983, state and local police forces had 15,932 employees; the rate of 3 police officers per 1,000 population was slightly above the US average. About $385 million was expended on state and local police in 1981/82.

[18] MIGRATION

Massachusetts was founded by the migration of English religious groups to its shores, and for over a century their descendants dominated all activity in the state. The first great wave of non-English to enter Massachusetts were the Irish, who migrated in vast numbers during the 1840s and 1850s. By 1860, one-third of Boston's population was Irish, while nearly one-fourth of Middlesex and Norfolk counties and one-fifth of the inhabitants of Berkshire, Bristol, Essex, and Hampden counties were Irish-born. Other ethnic groups—such as the Scottish, Welsh, Germans, and Poles—were also entering the state at this time, but their numbers were small by comparison. During the late 1880s and 1890s, another wave of immigrants—from Portugal, Spain, Italy, Russia, and Greece—arrived. Irish and Italians continued to enter the state during the 20th century.

A slow but steady migration from Massachusetts farm communities began during the mid-1700s and continued well into the 1800s. The first wave of farmers resettled in northern Connecticut, Vermont, New Hampshire, and Maine. Later farmers moved to New York's Mohawk Valley, Ohio, and points farther west. Out-migration has continued to recent times: from 1970 to 1983, Massachusetts lost 300,000 residents in net migration to other states.

The only significant migration from other areas of the US to Massachusetts has been the influx of southern blacks since World War II. According to census estimates, Massachusetts gained 84,000 blacks between 1940 and 1975; in 1980, it ranked 23d in black population with 221,000 persons, mostly in the Boston area.

[19] INTERGOVERNMENTAL COOPERATION

Massachusetts participates in numerous regional agreements, including the New England Corrections Compact, New England Police Compact, New England Higher Education Compact, New England Radiological Health Protection Compact, and New England Interstate Water Pollution Control Compact. The state is also a party to the Atlantic States Marine Fisheries Compact, the Northeastern Forest Fire Protection Compact, and the Connecticut River Valley Flood Control Compact.

Border agreements include the Connecticut-Massachusetts Boundary Compact (ratified by Massachusetts in 1908), the Massachusetts–New York Compact of 1853, and the Massachusetts–Rhode Island Compact of 1859.

During 1983/84, the state received $2.6 billion in federal aid.

[20] ECONOMY

From its beginnings as a farming and seafaring colony, Massachusetts became one of the most industrialized states in the country in the late 19th century and, more recently, a leader in the manufacture of high-technology products.

During the colonial and early national periods, the towns of Salem, Gloucester, Marblehead, and Boston, among others, gave the state strong fishing and shipbuilding industries. Boston was also an important commercial port and a leading center of foreign commerce. Agriculture was important, but productivity of the rocky soil was limited, and by the mid-1800s, farming could not sustain the expanding population. The opening of the Erie Canal, and subsequent competition with cheaper produce grown in the West, hastened agriculture's decline in the Bay State.

Massachusetts' rise as a center of manufacturing began in the early 1800s, when cottage industries developed in small farming communities. Large factories were then built in towns with water power. The country's first "company town," Lowell, was built in the early 1820s to accommodate the state's growing textile industry. Throughout the rest of the 19th century, the state supplied the nation with most of its shoes and woven goods.

Manufacturing still accounts for a large share of the state's economy. The growth of high-technology industries along Route 128 circling Boston has largely offset the loss to the state of the shoe and textile industries.

21INCOME

In 1983, Massachusetts ranked 6th among the 50 states in per capita income, with $13,089 in current dollars. During the same year, the state ranked 10th in total personal income, with $75.5 billion, representing nearly a 10% increase since 1980 (in constant 1972 dollars). However, the state's share of US income declined from 3% in 1970 to 2.7% in 1983. The median family income in 1981 was $28,839 (9th in the US).

The following table shows the average weekly hours and hourly wages for production workers in the major Massachusetts industries as of July 1984, according to the Massachusetts Department of Commerce:

	HOURS	HOURLY WAGES
Stone, clay, glass	41.4	$10.55
Transportation equipment	41.3	10.47
Instruments	39.4	9.62
Nonelectrical machinery	40.9	9.22
Chemicals and allied products	43.1	9.20
Printing and publishing	36.5	8.80
Paper and allied products	36.5	8.72
Primary metals	42.3	8.61
Electronics and electrical equipment	40.8	8.42
Food and food products	38.4	8.32

As of 1979, 9.6% of all Bay Staters had incomes below the federal poverty line, up from 8.6% in 1969 but below the US average of 12.4%. As of 1982, 32,700 Bay Staters were among the nation's top wealth-holders, with assets of over $500,000 each.

22LABOR

In 1984, the state's labor force numbered 3,051,000 persons, of whom 2,906,000 were employed and 145,000 were unemployed. The unemployment rate for all workers in 1984 was 4.8%, which represented a significant decrease from the previous rates of 6.9% in 1983 and 7.9% in 1982. As recently as 1977, the unemployment rate had been 13%, one of the highest in the US.

A federal census of workers in March 1983 revealed the following nonfarm employment pattern for Massachusetts:

	ESTABLISH-MENTS	EMPLOYEES	ANNUAL PAYROLL ('000)
Agricultural services, forestry, fishing	1,772	8,364	$ 119,541
Mining	113	1,090	29,701
Contract construction	11,407	85,674	1,860,922
Manufacturing	10,737	618,517	13,227,151
Transportation, public utilities	4,554	112,346	2,565,670
Wholesale trade	9,998	136,206	3,044,085
Retail trade	36,572	454,986	4,301,884
Finance, insurance, real estate	10,063	181,499	3,537,667
Services	43,329	729,573	11,082,494
Other	7,264	9,361	223,398
TOTALS	135,809	2,337,616	$39,992,513

As of September 1984, production workers in manufacturing on the average earned $8.58 per hour and a weekly wage of $345.77.

Some of the earliest unionization efforts took place in Massachusetts in the early 1880s, particularly in the shipbuilding and construction trades. However, the most important trade unions to evolve were those in the state's textile and shoe industries. The workers had numerous grievances: shoebinders' salaries of $1.60–2.40 a week during the 1840s, workdays of 14 to 17 hours, wages paid in scrip that could be cashed only at company stores (which charged exorbitantly high prices), children working at dangerous machinery. In 1867, a seven-week-long shoemakers' strike at Lynn, the center of the shoe business, was at that time the longest strike in US history.

After the turn of the century, the state suffered a severe decline in manufacturing, and employers sought to cut wages to make up for lost profits. This resulted in a number of strikes by both the United Textile Workers and the Boot and Shoe Workers Union. The largest strike of the era was at Lawrence in 1912, when textile workers (led by a radical labor group, the Industrial Workers of the World) closed the mills, and the mayor called in troops in an attempt to reopen them. Although the textile and shoe businesses are no longer major employers in the state, the United Shoe Workers of America, the Brotherhood of Shoe and Allied Craftsmen, the United Textile Workers, and the Leather Workers International Union of America have their headquarters in Massachusetts. As of mid-1983 there were 1,347 labor unions in the state; membership totaled about 660,000 workers, or nearly 25% of all employed workers, in 1980. There were 102 strikes or work stoppages involving 28,700 workers in 1981.

Massachusetts was one of the first states to enact child labor laws. In 1842, it established the 10-hour day for children under 12; in 1867, it forbade employment for children under 10. The nation's first Uniform Child Labor Law, establishing an 8-hour day for children aged 14 to 16, was enacted by Massachusetts in 1913. Massachusetts was also the first state to enact minimum wage guidelines (1912).

23AGRICULTURE

As of 1984, there were 6,100 farms in Massachusetts, covering 680,000 acres (275,000 hectares). Farming was mostly limited to the western Massachusetts counties of Hampshire, Franklin, and Berkshire, and southern Bristol County. Total agricultural income for 1983 was estimated at $367,000,000 (44th of the 50 states), of which crops provided 62%.

Although the state is not a major farming area, it is the largest producer of cranberries in the US; production for 1983 was 146 million lb, nearly half of the US total. Output totals for other crops in 1983 were as follows: corn for silage, 663,000 tons; hay, 313,000 tons; fresh vegetables, 44,700 tons; and tobacco, 768,000 lb. While of local economic importance, these figures are tiny fractions of US totals.

24ANIMAL HUSBANDRY

Massachusetts is not a major producer of livestock. As of 1 January 1984, the state had 57,600 cattle and calves, 42,000 hogs and pigs and 8,400 sheep and lambs. Production figures for 1983 were cattle and calves, 18,370,000 lb, worth $7,969,000; hogs and pigs, 14,451,000 lb, worth $6,069,000; sheep and lambs, 413,000 lb, worth $293,000; and turkeys, 3,312,000 lb, worth $2,782,000. The poultry industry also produced 1,378,000 chickens, worth $3,376,000, and 265,000,000 eggs, worth $20,096,000.

In 1983, the state was the 6th-leading producer of ice cream, with 44,510,000 gallons. Other dairy products included milk, 605,000,000 lb, and cheese, 12,643,000 lb.

25FISHING

Massachusetts' fish catch is one of the largest in the US, but the fishing industry is not as important to the state economy as it once was.

The early settlers earned much of their income from the sea.

The first shipyard in Massachusetts opened at Salem Neck in 1637, and during the years before independence the towns of Salem, Newburyport, Plymouth, and Boston were among the colonies' leading ports. By 1807, Massachusetts' fishing fleet made up 88% of the US total; for much of the 19th century, Nantucket and, later, New Bedford were the leading US whaling centers. But with the decline of the whaling industry came a sharp drop in the importance of fishing to the livelihood of the state. By 1978, the fishing industry ranked 13th in importance of the 15 industries monitored by the state. However, the fishing ports of Gloucester and New Bedford were still among the 10 busiest in the US in 1984. As of 1983 there were firms with 3,558 employees and a total annual payroll of $68,568,772.

The value of the commercial catch in 1984—$233,500,000—was the highest among the New England states and the 3d highest in the US. In the total catch of 375,537,000 lb, the major species were cod, haddock, mackerel, and lobster.

The state's long shoreline and many rivers make sport fishing a popular pastime for both deepsea and freshwater fishermen. The fishing season runs from mid-April through late October, with the season extended through February for bass, pickerel, panfish, and trout.

26FORESTRY

Forestry is a minor industry in the state. A federal census in March 1982 found 81 firms with about 700 employees and an annual payroll of $9.6 million involved in forestry.

Forested lands cover about 2,952,000 acres (1,195,000 hectares), 0.4% of the US total. Of more importance is the manufacture of wood and paper products. In 1983 there were 399 lumber and wood products establishments with 4,852 employees and 292 paper and paper products manufacturers with 26,060 workers. The value of lumber and wood products shipped by manufacturers in 1982 was $134.6 million; by paper and paper products manufacturers, $1,367.1 million.

Massachusetts has the 6th-largest state park system in the nation, with 38 state parks and 74 state forests totaling some 260,000 acres (105,200 hectares) in 1985. There are no national forests in Massachusetts.

27MINING

Massachusetts accounts for only a small fraction of the nation's mineral output. In 1984, the value of nonfuel mineral production—$101.1 million—exceeded $100 million for the first time. Estimated production and values for the year were sand and gravel, 12,900,000 tons, $41.2 million; stone, 8,557,000 tons, $50.7 million; lime, 132,000 tons, $7 million; and clays, 240,000 tons, $1.3 million.

28ENERGY AND POWER

Massachusetts is highly dependent on oil for electric generation and home heating, and energy costs in the state are among the highest in the US. During the early 1980s, as much as 81% of the state's electric power output was generated from oil.

In 1983, about 34.7 billion kwh of electric power were produced; installed capacity was nearly 10.1 million kw. Almost all generating capacity in the state is privately owned. Of the 36 billion kwh sold in 1983, 34% was for residential use, 37% commercial, 25% industrial, and 4% for other purposes. Boston Edison supplies electricity to the city of Boston; the rest of the state is served by 13 other companies, although a few municipalities do generate their own power. Power companies are regulated by the Department of Public Utilities, which establishes rates and monitors complaints from customers.

Massachusetts had no proved oil or coal reserves as of 31 December 1984. After a lengthy court battle, oil exploration off the coast of Cape Cod began in 1979. Environmentalists and fishermen had sought to prevent development of an oil industry in the region, which is one of the richest fishing grounds in the country. No wells had been drilled as of the early 1980s, however.

The state consumes but does not produce natural gas. In 1983, about 191 trillion Btu of natural gas were delivered, at a cost of $1.3 billion, to 1,102,800 customers. Almost 45% of the gas sold was for residential use, 32% for industries, and 23% for commercial use.

There are two nuclear power plants: one in Plymouth, with an operating capacity of 670 Mw, and the other in Rowe, in western Massachusetts, with a capacity of 175 Mw. Two pumped-storage hydroelectric plants, one at Bear Swamp and the other at Northfield Mountain, have a combined capacity of 1,600 Mw. These and some 40 run-of-the-river hydroelectric plants produced 3 trillion Btu in 1982.

The cost of electricity to consumers in Boston increased from 6.7 cents per kwh in 1979 to 9.7 cents by 1983. Because of the rising cost of home heating oil, some Massachusetts residents have reportedly converted to wood stoves for home heating.

The state encourages energy conservation and the development of alternative energy systems by granting tax credits to qualifying industries. Private researchers in the state were experimenting with solar energy systems and other alternatives to fossil fuels during the early 1980s.

29INDUSTRY

Massachusetts was the nation's first major industrial state, and during the later part of the 19th century, it was the US leader in shoemaking and textile production. By 1860, the state was a major producer of machinery and milled nearly one-fourth of the country's paper.

Massachusetts remains an important manufacturing center, placing 13th among the 50 states in the value of manufacturers' shipments in 1982. Nearly all the major manufacturing sectors had plants in Massachusetts' eastern counties, with the largest concentration in Bristol, Essex, Middlesex, and Worcester counties, which collectively held 65% of the state's manufacturing jobs in 1983. Much of this industry is located along Route 128, a superhighway that circles Boston from Gloucester in the north to Quincy in the south and is unique in its concentration of high-technology enterprises.

The following table shows estimated value of manufacturers' shipments for selected industries in 1982:

Nonelectrical machinery	$8,729,300,000
Electric and electronic equipment	6,949,500,000
Instruments and related products	3,881,000,000
Food and food products	3,716,300,000
Fabricated metal products	3,420,800,000
Printing and publishing	2,971,100,000
Transportation equipment	2,800,400,000
Paper and paper products	2,446,000,000
Rubber and plastic products	2,229,900,000
Chemicals and chemical products	2,051,000,000
Primary metals	1,944,200,000
Apparel and accessories	1,770,300,000
Textile mill products	1,397,700,000

Massachusetts' future as a manufacturing center depends on its continued preeminence in the production of computers, optical equipment, and other sophisticated instruments. Among the major computer manufacturers in the state are Digital Equipment Corp. in Maynard, Honeywell in Cambridge, Wang Laboratories in Lowell, and Data General in Westboro. Important instrument manufacturers include General Electric in Lynn, Itek in Lexington, and Foxboro in Foxboro. Other major manufacturers are Raytheon (Lexington), with assets of $3.6 billion in 1984; Gillette (Boston), with assets of $2 billion; Cabot (Boston), with assets of $1.8 billion; Polaroid (Cambridge), with assets of $1.3 billion; and Norton (Worcester), with assets of $1 billion.

30COMMERCE

Massachusetts' machinery and electrical goods industries are important components of the state's wholesale trade, along with motor vehicle and automotive equipment, and paper and paper products. Overall, in 1982, 9,709 establishments with 132,000 employees produced about $49.8 billion in sales, 2.5% of the US total and 11th highest among the 50 states. More than one-fourth of the wholesale establishments were located in Middlesex County.

The state ranked 11th in retail trade in 1982, with sales of $28.2 billion, in 47,312 stores of all types. Food stores accounted for 21% of sales; automotive dealers, 16%; department stores, 10%; and eating and drinking places, 10%. Middlesex County led the counties with 24% of all sales; the Boston metropolitan area had sales of $15 billion, 53% of the state total; and the city of Boston had sales of $2.8 billion, almost 10% of the state total.

In 1983, the Boston customs district handled about $27.5 billion in imports (10.6% of the US total) and $14.9 billion in exports (7.4% of the US total). Foreign exports of the state's own manufactures totaled $5.1 billion in 1981 (9th in the US).

³¹CONSUMER PROTECTION

The cabinet-level Executive Office of Consumer Affairs and Business Regulation serves as an information and referral center for consumer complaints and oversees the activities of many regulatory agencies. The Office of the Attorney General also has a Consumer Protection Division that handles consumer complaints. The Massachusetts Consumer Council advises the governor and legislature; there are many local consumer councils.

³²BANKING

By the mid-1800s, Boston had developed into a major banking center whose capital financed the state's burgeoning industries. Today banking remains an important sector of the state's economy; in 1983, 2,084 financial institutions employed 75,532 workers, with an annual payroll of $1.5 billion. As of 31 December 1984 there were 178 insured commercial banks in the state. Assets of the state's insured commercial banks totaled $42.7 billion and deposits $39.9 billion at the end of 1983. The largest banks, all located in Boston, were the Bank of Boston, with deposits of $15.3 billion and assets of $22.1 billion in 1984; the Bank of New England, with deposits of $5.4 billion and assets of $6.8 billion; the Shawmut Bank of Boston, with deposits of $4.8 billion and assets of $6.5 billion; BayBanks, with deposits of $4.4 billion and assets of $5.2 billion; and the State Street Bank and Trust, with deposits of $3.2 billion and assets of $4.7 billion. As of 31 December 1983, the state had 120 savings and loan associations with assets of $7.2 billion.

State-chartered savings banks, trust companies, cooperative banks, credit unions, and consumer credit grantors are examined by the state Division of Banks and Loan Agencies, within the Executive Office of Consumer Affairs and Business Regulation. The division also approves interest rates, administers the state's banking laws, and oversees bank practices and audits.

³³INSURANCE

Insurance is an important business in Massachusetts, and some of the largest life and property/casualty insurance companies in the nation have their headquarters in Boston. In 1983, the state's 2,084 insurance company offices had 66,360 employees and a total payroll of $1.5 billion.

The state's 17 life insurance companies had in force 8,830,000 life policies worth $124 billion in 1983. The average amount of life insurance per family was $53,400, and payments to 74,400 beneficiaries amounted to $403,800,000. New England Mutual Life Insurance Co. of Boston was the first mutual company to be chartered in the US and remains one of the largest firms in the business, with assets of $9.4 billion in 1984. John Hancock Mutual Life, also of Boston, was the 6th-largest life insurance company in the US in 1984, with assets of $24.7 billion.

Of the 51 mutual property/casualty companies in Massachusetts, the largest is Liberty Mutual of Boston. Direct premiums written by these companies in 1983 totaled $3.5 billion, of which

automobile liability premiums accounted for 21%, automobile physical damage premiums 18%, and homeowners' premiums 13%. In 1971, Massachusetts became the first state in the US to implement a no-fault automobile insurance law.

All aspects of the insurance business in Massachusetts, including the licensing of agents and brokers and the examination of all insurance companies doing business in the state, are controlled by the Division of Insurance, under the Executive Office of Consumer Affairs and Business Regulation.

³⁴SECURITIES

The Boston Stock Exchange, founded in 1846, is the only stock exchange in Massachusetts, with 192 members who traded 256 million shares in 1984. As of 31 December 1983 there were 206 registered stock brokers and dealers in the state, and nearly 1.3 million shareowners of public corporations. Mutual funds originated in Boston during the 1920s.

The Securities Division of the Office of the Secretary of the Commonwealth is responsible for licensing and monitoring all brokerage firms in the state.

³⁵PUBLIC FINANCE

The Massachusetts budget is prepared by the Executive Office of Administration and Finance and is presented by the governor to the legislature for revision and approval. The fiscal year runs from 1 July to 30 June.

The following is a summary of estimated expenses and revenues for the 1983/84 and 1984/85 fiscal years (in millions):

REVENUES	1983/84	1984/85
Taxes	$5,467.2	$6,043.3
Federal reimbursements	1,088.0	1,156.7
Departmental revenues (fees, licenses, permits, etc.)	472.4	572.0
Lottery Fund	123.8	137.5
Interest income	58.1	60.5
Other revenues and reversions	374.3	249.4
TOTALS	$7,583.8	$8,219.4

EXPENDITURES		
Agency operations	$2,547.0	$2,795.5
Direct and medical assistance	1,710.7	1,903.9
Other state operations	892.5	1,074.5
Local aid	2,261.9	2,464.3
Other expenditures	169.3	—
TOTALS	$7,581.4	$8,238.2

The total debt of state and local governments as of mid-1982 was more than $11 billion, or $1,929 per capita (22d among the states).

³⁶TAXATION

Massachusetts' tax burden on a per capita basis as of mid-1983 was $2,767 (9th among the states). Total tax revenues received in 1982/83 were the 10th highest in the country at $15.9 billion.

As of 1984, the state levied a 5% tax plus a 7.5% surcharge on earned income and a 10% tax plus a 7.5% surcharge on unearned income. The corporate income tax rate was 8.3%. Commercial banks and other banking and trust companies pay a 12.54% tax on net income; public utilities, 6.5%.

A 5% gross receipts tax on sales was in effect in 1984, but such necessities as food, clothing, and home heating fuel were exempted. A 1985 law gave Massachusetts municipalities the right to levy their own sales taxes. The estate tax ranges from 5% to 16%; estates worth less than $30,000 are not taxed. Other levies include a room occupancy tax of 5.7%, a tax of $3.30 per barrel on alcoholic beverages, a cigarette tax of 26 cents per pack, a gasoline tax of 11 cents per gallon, and a motor vehicle excise tax of $25 for every $1,000 of valuation.

In November 1980, Bay Staters voted to lower their local property taxes by approving the so-called Proposition 2 1/2, which limited the tax to 2.5% of real estate value. The effect was to lower property taxes by $400 million for 1982/83, when total levies of $2.9 billion were actually lower than those for 1977/78. As a result of the reduced revenues, however, many municipalities severely cut the number of public employees, including police, firemen, and teachers.

Massachusetts' share of federal spending in 1983/84 was more than $11.3 billion. Bay Staters filed more than 2.6 million federal income tax returns for 1983, paying a total tax of $8.2 billion, or $3,757 per taxpayer.

37 ECONOMIC POLICY

The Division of Economic Development of the Department of Commerce and Development (within the Executive Office of Economic Affairs) provides business information and aids new businesses in securing venture capital and in exporting their products. Small businesses can also receive help in obtaining loans and mortgage insurance. The state provides training for new employees or provides a subsidy for companies that arrange for on-the-job training. In addition, cities and towns can issue revenue bonds for industrial expansion in their localities.

Among the many tax incentives offered to businesses are a 3% investment tax credit; an export sales tax exemption; property and sales tax exemptions for some machinery, parts, and inventory; credits against the state excise tax; and real-estate tax reductions for building facilities in certain urban areas. Tax reductions are offered for the use of pollution controls and of alternative energy sources (i.e., avoidance of fossil fuels), and five-year tax exemptions are granted for the development of products related to energy conservation. Also, losses for the first five years of business can be carried forward for tax purposes.

Massachusetts is divided into 13 regional planning districts for 347 municipalities. A regional planning commission in each district seeks to direct the economic and social growth of that district and to help the planning efforts of its localities.

38 HEALTH

Massachusetts recorded 13.7 live births per 1,000 population in 1984. Infant mortality was 9.3 per 1,000 live births among whites and 16 among blacks in 1981; both rates were better than the US average. In 1982 there were 41,200 abortions, for a rate of 529 per 1,000 live births, far above the US average.

The death rates from heart disease and cancer in 1981 were above US averages. The major causes of death and their rates per 100,000 population were heart diseases, 365.9; malignant neoplasms, 212.9; cerebrovascular diseases, 70.1; accidents, 35.2; and pneumonia and influenza, 33.5.

Programs for treatment and rehabilitation of alcoholics are administered by the Division of Alcoholism of the Department of Health, under the Executive Office of Human Services. The Division of Communicable Disease Control operates venereal disease clinics throughout of the state and provides educational material to schools and other groups. The Division of Drug Rehabilitation administers drug treatment from a statewide network of hospital agencies and self-help groups. The state also runs a lead-poisoning prevention program.

Massachusetts had 177 hospitals, with 41,208 beds and 943,852 admissions, in 1983. There were 31,975 registered nurses in 1981. Among the best-known institutions are Massachusetts General Hospital, a leading research and treatment center, and the Massachusetts Eye and Ear Infirmary, a Boston clinic. In 1982 there were 17,875 active physicians and 4,013 professionally active dentists. Four prominent medical schools are located in the state: Harvard Medical School, Tufts University School of Medicine, Boston University School of Medicine, and the University of Massachusetts School of Medicine. The average cost per stay of $3,291 in a Massachusetts community hospital in 1982 was the highest in the US except for California and Washington, D.C.; the average cost per day of $370 ranked 8th. There were 746,800

Medicaid patients, who received $1.1 billion in benefits, during 1981.

All health-care facilities are registered by the Department of Public Health. The Division of Health Care Quality inspects and licenses hospitals, clinics, school infirmaries, and blood banks every two years. Licensing of nursing homes is also under its control; the state's nursing homes numbered 91 in 1982.

39 SOCIAL WELFARE

Massachusetts ranked 4th among the 50 states in public welfare spending per capita with an average of $346, for a total of about $2 billion, in 1981. Some two-thirds of the state's budget for human services went to the Medicaid and Aid to Families with Dependent Children (AFDC) programs, although federal payments reimbursed the state for more than half the cost of these programs. AFDC expenditures increased by 21% from 1970 to 1981, but then fell by 25% in 1981–83 as federal cuts in the program were felt. In 1983, 88,600 families, or 6.2% of the state's population, received AFDC; payments totaled $440 million, or $369 monthly per family, in 1982.

An estimated 376,000 state residents participated in the federal food stamp program in 1983; the federal subsidy was $197 million. In that year, about 476,000 students, compared with 703,000 in 1980, took part in the school lunch program, at a cost to the federal government of $42 million ($49 million in 1980).

A total of $4.5 billion was paid to 932,000 recipients of Social Security benefits in 1982: 684,000 persons received retirement benefits totaling $3.2 billion, 169,000 persons received $881 million in survivors' benefits, and 78,000 disabled workers received benefits totaling $368 million. The average monthly retirement benefit (excluding persons with special benefits) of $420 was 6% above the US average.

Of more than 108,400 Massachusetts residents receiving $262.1 million in Supplemental Security Income in 1983, about 53,400 elderly persons received average monthly payments of $161, and 50,100 disabled persons got $270.

Federal grants for the education of handicapped persons totaled $23.9 million in 1983/84, and rehabilitation services for the handicapped amounted to $13.6 million. The state spent about $369 million on workers' compensation in 1982. In that year, about $492 million in unemployment benefits were paid; the average weekly benefit was $115.

40 HOUSING

Massachusetts' housing stock, much older than the US average, reflects the state's colonial heritage and its ties to English architectural traditions.

According to the 1980 census, 57% of the state's housing was built before 1950. Two major styles are common: colonial, typified by a wood frame, two stories, center hall entry, and center chimney; and Cape Cod, one-story houses built by fishermen, typified by shallow basements, shingled roofs, clapboard fronts, and unpainted shingled sides weathered gray by the salt air. Many new houses are also built in these styles.

As of 1980 there were 2,208,146 housing units in the state. Of the 2,136,870 year-round structures, 54.7% were owner-occupied and 40.4% were rented; the rest were vacant. Nearly 98% of all structures had plumbing. In 1983, 23,007 new housing units were built, representing a sizable increase of 49% over the number constructed in 1982. The median value of a house in 1980 was $48,500, slightly above the US average.

The Executive Office of Communities and Development administers federal housing programs for the state. The Massachusetts Housing Finance Agency finances the construction and rehabilitation of housing by private and community groups.

41 EDUCATION

Massachusetts has a long history of support for education. The Boston Latin School opened in 1635 as the first public school in the colonies. Harvard College—the first college in the US—was founded the following year. In 1647, for the first time, towns with more than 50 people were required by law to establish tax-supported school systems. More firsts followed: the country's first board of education, compulsory school attendance law, training school for teachers, state school for the retarded, and school for the blind. The drive for quality public education in the state was intensified through the efforts of educator Horace Mann, who during the 1830s and 1840s was also a leading force for the improvement of school systems throughout the US. Today the state boasts some of the most highly regarded private secondary schools and colleges in the country.

As of the 1980 census, 72.2% of state residents were high school graduates; of the 71,225 high school graduates in 1983, some 60% continued their education. As of 1 January 1983, there were 905,867 students enrolled in public schools. Massachusetts ranked 5th best among the states in pupil-teacher ratio in 1983, with 15.4 pupils per teacher; it was 15th in average annual teacher's salary in 1983/84, with $23,000.

Violence broke out in South Boston schools when they were integrated in the mid-1970s. Between then and the fall of 1984, the percentage of black students increased from 34% to 48%, while the percentage of non-Hispanic white students decreased from 57.2% to 26.5%.

The early years of statehood saw the development of private academies, where the students could learn more than the basic reading and writing skills that were taught in the town schools at the time. Some of these private preparatory schools remain, including such prestigious institutions as Andover, Deerfield, and Groton. Enrollment in private schools in 1982/83 totaled 136,296 of whom some 78% attended church-related schools. There were 72,281 pupils enrolled in Catholic parochial schools as of 1 January 1984.

There are 126 colleges and universities in the state. The major public university system is the University of Massachusetts, with campuses at Amherst and Boston and a medical school at Worcester. The Amherst campus, established in 1863, had an enrollment of 24,500 in the fall of 1983; the Boston campus, established in 1965, had 11,370 students. The other public universities are Southeastern Massachusetts University in North Dartmouth and the University of Lowell. The state's 12 public colleges and universities had 102,023 students in 1982, while the Massachusetts Board of Regional Community Colleges had 69,011 students at its 15 campuses.

Harvard University, which was established in Cambridge originally as a college for clergymen and magistrates, has grown to become one of the country's premier institutions; its 1983 student population was 15,859. Also located in Cambridge are Radcliffe College (whose enrollment is included in Harvard's), founded in 1879, and the Massachusetts Institute of Technology, or MIT (1861), with 8,935 students in 1983. Mount Holyoke College, the first US college for women, was founded in 1837 and had a 1983 enrollment of 1,968. Other prominent private schools, their dates of origin, and their 1983 enrollment are Amherst College (1821), 1,522; Boston College (1863), 14,059; Boston University (1869), 27,724; Brandeis University (1947), 3,245; Clark University (1887), 2,700; Hampshire College (1965), 1,100; the New England Conservatory of Music (1867), 683; Northeastern University (1898), 36,559; Smith College (1871), 2,684; Tufts University (1852), 7,074; Wellesley College (1875), 2,170; and Williams College (1793), 2,023.

Among the tuition assistance programs available to state residents are the Massachusetts General Scholarships, awarded to thousands of college students annually; Massachusetts Honor Scholarships, for outstanding performance on the Scholastic Aptitude Test; special scholarships for war orphans and the children of deceased members of fire, police, and corrections departments; and the Higher Education Loan Plan, offered by the

Massachusetts Higher Education Assistance Corp. In 1983/84, state financial aid awards to college students numbered 32,710 and totaled $27.1 million.

The State Board of Education establishes standards and policies for the public schools throughout the state; its programs are administered by the Department of Education. Higher education planning and programs are under the control of the Board of Regents of Higher Education. State and local expenditures for public elementary and secondary pupils in 1983/84 averaged $3,392 per pupil.

[42]ARTS

Boston is the center of artistic activity in Massachusetts, and Cape Cod and the Berkshires are areas of significant seasonal artistic activity. In 1979, Massachusetts became the first state to establish a lottery solely for funding the arts ($3.3 million in 1984). Boston is the home of several small theaters, some of which offer previews of shows bound for Broadway. Of the regional theaters scattered throughout the state, the Williamstown Theater in the Berkshires and the Provincetown Theater on Cape Cod are especially noteworthy.

The Boston Symphony, one of the major orchestras in the US, was founded in 1881 and has had as its principal conductors Serge Koussevitzky, Charles Munch, Erich Leinsdorf, and Seiji Ozawa, among others. During the summer, the symphony is the main attraction of the Berkshire Music Festival at Tanglewood in Lenox. An offshoot of the Boston Symphony, the Boston Pops Orchestra, gained fame under the conductorship of Arthur Fiedler. Its mixture of popular, jazz, and light symphonic music continued in the early 1980s under the direction of Fiedler's successor, John Williams. Boston is also the headquarters of the Opera Company of Boston, under the artistic direction of Sarah Caldwell. The Boston Ballet Company is the state's major dancing troupe. The Great Woods Performing Arts Center, southeast of Boston in Mansfield, was scheduled to open in 1986.

[43]LIBRARIES AND MUSEUMS

The first public library in the US was established in Boston in 1653. Massachusetts has one of the most important university libraries in the country, and numerous museums and historical sites commemorate the state's rich colonial history.

Three regional library systems served 351 towns and cities with a total library circulation of 33 million in 1982/83. The major city libraries are in Boston, Worcester, and Springfield. Statewide in 1983 there were 25,974,565 volumes in all public libraries. State funding for public libraries was $4.8 million in 1982/83; total library income amounted to $66.9 million.

The Boston Athenaeum, with 650,000 volumes, is the most noteworthy private library in the state. The American Antiquarian Society in Worcester has a 613,000-volume research library of original source material dating from colonial times to 1876.

Harvard University's library system is one of the largest in the world, with 10,567,240 volumes in 1983. Other major academic libraries are those of Boston University, the University of Massachusetts (Amherst), Smith College, and Boston College.

Boston houses a number of important museums, among them the Museum of Fine Arts with vast holdings of artwork including extensive Far East and French impressionists collections and American art and furniture, the Isabella Stewart Gardner Museum, the Museum of Science, the Massachusetts Historical Society, and the Children's Museum. Harvard University's museums include the Fogg Art Museum, the Peabody Museum of Archaeology and Ethnology, the Museum of Comparative Zoology, and the Botanical Museum. Other museums of note are the Whaling Museum in New Bedford, the Essex Institute in Salem, the Worcester Art Museum, the Clark Art Institute in Williamstown, the Bunker Hill Museum near Boston, and the National Basketball Hall of Fame in Springfield. In addition, many towns have their own historical societies and museums, including

Historic Deerfield, Framingham Historical and Natural History Society, Ipswich Historical Society, Lexington Historical Society, and Marblehead Historical Society. Plimoth Plantation in Plymouth is a re-creation of life in the 17th century, and Old Sturbridge Village, a working historical farm, displays 18th- and 19th-century artifacts. The state had over 300 museums in 1984.

44 COMMUNICATIONS

The first American post office was established in Boston in 1639. In 1985 there were 434 post offices in Massachusetts and about 22,000 postal service employees.

Alexander Graham Bell first demonstrated the telephone in 1876 in Boston. As of 1980, 95.8% of the state's 2,032,717 occupied housing units had telephones.

The state had 69 AM stations and 100 FM stations in 1984, when 11 commercial and 3 educational television stations were also in operation. Boston's WGBH is a major producer of programming for the Public Broadcasting Service. In 1984 there were 64 cable television systems, with 766,761 subscribers in 172 communities.

45 PRESS

Milestones in US publishing history that occurred in the state include the first book printed in the English colonies (Cambridge, 1640), the first regularly issued American newspaper, the *Boston News-Letter* (1704), and the first published American novel, William Hill Brown's *The Power of Sympathy* (Worcester, 1789). During the mid-1840s, two noted literary publications made their debut, the *North American Review* and the *Dial*, the latter under the editorial direction of Ralph Waldo Emerson and Margaret Fuller. *The Atlantic*, which began publishing in 1857, *Harvard Law Review, Harvard Business Review,* and *New England Journal of Medicine* are other influential publications.

As of 30 September 1984 there were 45 daily newspapers in the state, 6 morning, 38 evening, and 1 all-day, with a combined circulation of 2,128,518. Ten Sunday papers had a combined circulation of 1,676,515. The *Boston Globe,* the most widely read newspaper in the state, has won numerous awards for journalistic excellence on the local and national levels. The *Christian Science Monitor* is highly respected for its coverage of national and international news. Major newspapers and their average daily circulations in 1984 were:

AREA	NAME	DAILY	SUNDAY
Boston	Christian Science Monitor (m)	141,247	
	Globe (m,S)	520,081	792,786
	Herald (m,S)	343,581	280,767

Massachusetts is also a center of book publishing, with more than 100 publishing houses. Among them are Little, Brown and Co., Houghton Mifflin, Merriam-Webster, and Harvard University Press.

46 ORGANIZATIONS

The 1982 US Census of Service Industries counted 1,769 organizations in Massachusetts, including 273 business associations; 1,044 civic, social, and fraternal associations; and 144 educational, scientific, and research associations.

Headquartered in Massachusetts are the National Association of Independent Schools, the National Commission for Cooperative Education, both in Boston, and the National Bureau of Economic Research in Cambridge. The Union of Concerned Scientists in Cambridge and International Physicians for the Prevention of Nuclear War in Boston—recipient of the 1985 Nobel Peace Prize—are the major public affairs associations based in the state.

Academic and scientific organizations headquartered in Boston include the American Meteorological Society, American Society of Law and Medicine, American Surgical Association, and Optometric Research Institute. The American Academy of Arts and Science is located in Cambridge, the National Association of

Emergency Medical Technicians is in Newton Highlands, and the Protestant Guild for the Blind is in Belmont.

Among the many professional, business, and consumer organizations based in Massachusetts are the American Institute of Management in Quincy, Wood Products Manufacturers Association in Gardner, and the National Consumer Law Center, Northern Textile Association, and Wool Manufacturers Council in Boston. The American Orchid Society and the Nieman Foundation are in Cambridge, the Shoe Suppliers Association of America is in East Bridgewater, and the United Textile Workers of America is in Lawrence.

The headquarters of the John Birch Society, an archconservative political association, is in Belmont. Oxfam-America, the US affiliate of the international humanitarian relief agency, is located in Boston. Major sports associations in the state are the Eastern College Athletic Conference in Centerville and the American Hockey League in Springfield. Action for Children's Television, a group concerned with the quality of broadcast programming, is in Newtonville. The International Friendship League, which matches pen pals in 139 countries, has its headquarters in Boston.

47 TOURISM, TRAVEL, AND RECREATION

Massachusetts beaches are a popular destination for summer travelers, but other areas have their own attractions.

In 1984 there were 11,462,167 visitors to Massachusetts, who spent more than $4.4 billion in the state. It was estimated that about 39% of annual visits were for recreation and entertainment, 37% to see friends or relatives, 11% for business, 9% for other personal reasons, and 4% for conventions. Nearly 43% of the visitor-days were spent during the summer. Tourist-related industries employed 92,300 workers (3.5% of all jobs in the state), with a payroll of $1 billion in 1982. The largest employment—about 40%—was on Nantucket Island, followed by Martha's Vineyard and Cape Cod. Tourism in 1982 contributed $505 million in tax revenues, including $179 million for the state, $42 million for localities, and $284 million for the federal government. The state government spent $10.5 million to promote tourism in 1984/85—an estimated return of $17 in tourism-generated state taxes for each dollar spent.

The largest number of visitor-days are spent in Barnstable County (Cape Cod); in 1984, the Cape Cod National Seashore attracted 4,560,713 visitors. Among its many attractions are beaches, fishing, good dining spots, several artists' colonies with arts and crafts fairs, antique shops, and summer theaters. Beaches, fishing, and quaint villages are also the charms of Nantucket and Martha's Vineyard.

Boston is the second most popular area for tourists. A trip to the city might include visits to such old landmarks as Faneuil Hall, Old North Church, the USS *Constitution,* and Paul Revere's House, and such newer attractions as the John Hancock Observatory, the skywalk above the Prudential Tower, Quincy Market, and Copley Place. Boston Common, one of the oldest public parks in the country, is the most noteworthy municipal park.

The Berkshires are the summer home of the Berkshire Music Festival at Tanglewood and the Jacob's Pillow Dance Festival in Lee, and during the winter also provide recreation for cross-country and downhill skiers. Essex County on the North Shore of Massachusetts Bay offers many seaside towns and the art colony of Rockport. Its main city, Salem, contains the Witch House and Museum as well as Nathaniel Hawthorne's House of Seven Gables. Middlesex County, to the west of Boston, holds the university city of Cambridge as well as the battlegrounds of Lexington and Concord. In Concord are the homes of Henry David Thoreau, Ralph Waldo Emerson, and Louisa May Alcott. Norfolk County, south of Boston, has the homes of three US presidents: John Adams and John Quincy Adams in Quincy and John F. Kennedy in Brookline. The seaport town and former whaling center of New Bedford and the industrial town of Fall

River are in Bristol County. Plymouth County offers Plymouth Rock, Plimoth Plantation, and a steam-train ride through some cranberry bogs.

Massachusetts has about 130 state-owned camping areas, which accommodated 856,392 campers in 1984. In 1982/83, the state issued licenses to 117,800 hunters and 204,489 fishermen.

48 SPORTS

All of the major league professional sports have franchises in Massachusetts. The Boston Red Sox of the American League make their home in Fenway Park, one of the oldest baseball stadiums in the country. The New England Patriots, who play in Foxboro, are members of the American Football Conference of the National Football League. The Boston Celtics, the most successful basketball team in National Basketball Association history, play in Boston Garden, as do the Boston Bruins of the National Hockey League.

Suffolk Downs in East Boston features Thoroughbred horse racing; harness racing takes place at the New England Harness Raceway in Foxboro. Dog racing can be seen at Raynham Park in Raynham, Taunton Dog Track in North Dighton, and Wonderland Park in Revere.

Probably the most famous amateur athletic event in the state is the Boston Marathon, a race of more than 26 mi (42 km) held every Patriots' Day (3d Monday in April). It attracts many of the world's top long-distance runners. During the summer, a number of boat races are held, including the Colorado Cup Races off Martha's Vineyard.

49 FAMOUS BAY STATERS

Massachusetts has produced an extraordinary collection of public figures and leaders of thought. Its three US presidents were John Adams (1735-1826), a signer of the Declaration of Independence; his son John Quincy Adams (1767-1848); and John Fitzgerald Kennedy (1917-63). All three served in Congress. John Adams was also the first US vice president; John Quincy Adams served as secretary of state under James Monroe. Calvin Coolidge (b.Vermont, 1872-1933) was governor of Massachusetts before his election to the vice-presidency in 1920 and his elevation to the presidency in 1923. Two others who held the office of vice president were another signer of the Declaration of Independence, Elbridge Gerry (1744-1814), for whom the political practice of gerrymandering is named, and Henry Wilson (b.New Hampshire, 1812-75), a US senator from Massachusetts before his election with Ulysses S. Grant. George Bush (b.1924) was elected vice president on the Republican ticket in 1980 and reelected in 1984.

Massachusetts' great jurists include US Supreme Court Justices Joseph Story (1779-1845), Oliver Wendell Holmes, Jr. (1841-1935), Louis D. Brandeis (b.Kentucky, 1856-1941), and Felix Frankfurter (b.Austria, 1882-1965). Important federal officeholders at the cabinet level were Henry Knox (1750-1806), the first secretary of war; Timothy Pickering (1745-1820), the first postmaster general and later secretary of war and secretary of state under George Washington and John Adams; Levi Lincoln (1749-1820), attorney general under Jefferson; William Eustis (1753-1825), secretary of war under Madison; Jacob Crowninshield (1770-1808), secretary of the Navy under Jefferson, and his brother Benjamin (1772-1851), who held the same office under Madison; Daniel Webster (b.New Hampshire, 1782-1852), US senator from Massachusetts who served as secretary of state under William Henry Harrison, John Tyler, and Millard Fillmore; Edward Everett (1794-1865), a governor and ambassador who served as secretary of state under Fillmore; George Bancroft (1800-1891), a historian who became secretary of the Navy under James K. Polk; Caleb Cushing (1800-1879), attorney general under Franklin Pierce; Charles Devens (1820-91), attorney general under Rutherford B. Hayes; Christian Herter (1895-1966), secretary of state under Dwight Eisenhower; Elliot L. Richardson (b.1920), secretary of health, education, and

welfare, secretary of defense, and attorney general under Richard Nixon; Henry Kissinger (b.Germany, 1923), secretary of state under Nixon and Gerald Ford and a Nobel Peace Prize winner in 1973; and Robert F. Kennedy (1925-68), attorney general under his brother John and later US senator from New York.

Other federal officeholders include some of the most important figures in American politics. Samuel Adams (1722-1803), the Boston Revolutionary leader, served extensively in the Continental Congress and was later governor of the Bay State. John Hancock (1737-93), a Boston merchant and Revolutionary, was the Continental Congress's first president and later became the first elected governor of the state. In the 19th century, Massachusetts sent abolitionist Charles Sumner (1811-74) to the Senate. As ambassador to England during the Civil War, John Quincy Adams's son Charles Francis Adams (1807-86) played a key role in preserving US-British amity. At the end of the century, Henry Cabot Lodge (1850-1924) emerged as a leading Republican in the US Senate, where he supported regulatory legislation, protectionist tariffs, and restrictive immigration laws, and opposed women's suffrage and the League of Nations; his grandson, also Henry Cabot Lodge (1902-85), was an internationalist who held numerous federal posts and was US senator. Massachusetts has provided two US House speakers: John W. McCormack (1891-1980) and Thomas P. "Tip" O'Neill, Jr. (b.1912). Other well-known legislators include Edward W. Brooke (b.1919), the first black US senator since Reconstruction, and Edward M. Kennedy (b.1932), President Kennedy's youngest brother and a leading Senate liberal.

Among other historic colonial and state leaders were John Winthrop (b.England, 1588-1649), a founder of Massachusetts and longtime governor; William Bradford (b.England, 1590-1657), a founder of Plymouth, its governor, and author of its classic history; Thomas Hutchinson (1711-80), colonial lieutenant governor and governor during the 1760s and 1770s; and Paul Revere (1735-1818), the Patriots' silversmith-courier, who was later an industrial pioneer.

Literary genius has flourished in Massachusetts. In the 17th century, the colony was the home of poets Anne Bradstreet (1612-72) and Edward Taylor (1645-1729) and of the prolific historian, scientist, theologian, and essayist Cotton Mather (1663-1728). Eighteenth-century notables include the theologian Jonathan Edwards (b.Connecticut, 1703-58), poet Phillis Wheatley (b.Senegal, 1753-84), and numerous political essayists and historians. During the 1800s, Massachusetts was the home of novelists Nathaniel Hawthorne (1804-64), Louisa May Alcott (b.Pennsylvania, 1832-88), Horatio Alger (1832-99), and Henry James (b.New York, 1843-1916); essayists Ralph Waldo Emerson (1803-82) and Henry David Thoreau (1817-62); and such poets as Henry Wadsworth Longfellow (b.Maine, 1807-82), John Greenleaf Whittier (1807-92), Oliver Wendell Holmes, Sr. (1809-94), James Russell Lowell (1819-91), and Emily Dickinson (1830-86). Classic historical writings include the works of George Bancroft, William Hickling Prescott (1796-1859), John Lothrop Motley (1814-77), Francis Parkman (1823-93), and Henry B. Adams (1838-1918). Among 20th-century notables are novelists John P. Marquand (b.Delaware, 1893-1960) and John Cheever (1912-82); poets Elizabeth Bishop (1911-79), Robert Lowell (1917-77), Anne Sexton (1928-74), and Sylvia Plath (1932-63); and historian Samuel Eliot Morison (1887-1976). In philosophy, Charles Sanders Peirce (1839-1914) was one of the founders of pragmatism; Henry James's elder brother, William (b.New York, 1842-1910), pioneered psychology; and George Santayana (b.Spain, 1863-1952), philosopher and author, grew up in Boston. Mary Baker Eddy (b.New Hampshire, 1821-1910) founded the Church of Christ, Scientist, during the 1870s.

Reformers have abounded in Massachusetts, especially in the 19th century. William Lloyd Garrison (1805-79), Wendell Phillips

(1811–84), and Lydia Maria Child (1802–80) were outstanding abolitionists. Lucretia Coffin Mott (1793–1880), Lucy Stone (1818–93), Abigail Kelley Foster (1810–87), Margaret Fuller (1810–50), and Susan Brownell Anthony (1820–1906) were leading advocates of women's rights. Horace Mann (1796–1859), the state secretary of education, led the fight for public education; and Mary Lyon (1797–1849) founded Mount Holyoke, the first women's college.

Efforts to improve the care and treatment of the sick, wounded, and handicapped were led by Samuel Gridley Howe (1801–76), Dorothea Lynde Dix (1802–87), and Clara Barton (1821–1912), founder of the American Red Cross. The 20th century reformer and NAACP leader William Edward Burghardt Du Bois (1868–1963) was born in Great Barrington.

Leonard Bernstein (b.1918) is a composer and conductor of worldwide fame. Arthur Fiedler (1894–79) was the celebrated conductor of the Boston Pops Orchestra. Composers include William Billings (1746–1800), Carl Ruggles (1876–1971), and Alan Hovhaness (b.1911). Charles Bulfinch (1763–1844), Henry H. Richardson (b.Louisiana, 1838–86), and Louis Henri Sullivan (1856–1924) have been among the nation's important architects. Painters include John Singleton Copley (1738–1815), James Whistler (1834–1903), Winslow Homer (1836–1910), and Frank Stella (b.1936); Horatio Greenough (1805–52) was a prominent sculptor.

Among the notable scientists associated with Massachusetts are Nathaniel Bowditch (1773–1838), a mathematician and navigator; Samuel F. B. Morse (1791–1872), inventor of the telegraph; and Robert Hutchins Goddard (1882–1945), a physicist and rocketry pioneer.

Two professors at the Massachusetts Institute of Technology, in Cambridge, have won the Nobel Prize in economics—Paul A. Samuelson (b.Indiana, 1915), in 1970, and Franco Modigliani (b.Italy, 1918), in 1985.

Massachusetts' most famous journalist has been Isaiah Thomas (1750–1831). Its great industrialists include textile entrepreneurs Francis Lowell (1775–1817) and Abbott Lawrence (1792–1855). Elias Howe (1819–67) invented the sewing machine.

Massachusetts was the birthplace of television journalists Mike Wallace (b.1918) and Barbara Walters (b.1931). Massachusetts-born show business luminaries include director Cecil B. DeMille (1881–1959); actors Walter Brennan (1894–1974), Jack Haley (1901–79), Ray Bolger (b.1904), Bette Davis (b.1908), and Jack Lemmon (b.1925); and singers Donna Summer (b.1948) and

James Taylor (b.1948). Outstanding among Massachusetts-born athletes was world heavyweight boxing champion Rocky Marciano (Rocco Francis Marchegiano, 1924–69), who retired undefeated in 1956.

[50]BIBLIOGRAPHY

Andrews, Charles McLean. *The Colonial Period in American History.* New Haven, Conn.: Yale University Press, 1934.

Bailyn, Bernard. *The Ordeal of Thomas Hutchinson.* Cambridge, Mass.: Harvard University Press, 1974.

Boyer, Paul, and Stephen Nissenbaum. *Salem Possessed: The Social Origins of Witchcraft.* Cambridge, Mass.: Harvard University Press, 1974.

Brown, Richard D. *Massachusetts: A Bicentennial History.* New York: Norton, 1978.

Butterfield, L. H., et al., eds. *Diary and Autobiography of John Adams.* Cambridge, Mass.: Harvard University Press, 1962.

Federal Writers' Project. *Massachusetts: A Guide to Its Places and People.* Reprint. New York: Somerset, n.d. (orig. 1937).

Gibney, Fred J. *Monograph of the Commonwealth of Massachusetts.* Boston: Massachusetts Office of Economic Affairs, 1980 (rev. 1984).

Gross, Robert. *The Minutemen and Their World.* New York: Hill and Wang, 1976.

Handlin, Oscar. *Boston's Immigrants.* Rev. ed. Cambridge, Mass.: Harvard University Press, 1959.

Hart, Albert Bushnell, ed. *Commonwealth History of Massachusetts: Colony, Province, and State.* 5 vols. New York: Russell, 1967 (orig. 1927–30).

Haskell, John D., Jr., ed. *Massachusetts: A Bibliography of Its History.* Boston: G. K. Hall, 1976.

Massachusetts, Commonwealth of. Secretary of the Commonwealth. Citizen Information Service. *Citizens' Guide to State Services: A Selective Listing of Governmental Agencies and Programs.* Boston, 1984.

Morison, Samuel Eliot. *Builders of the Bay Colony.* New York: AMS Press, n.d. (orig. 1930).

Morison, Samuel Eliot. *The Maritime History of Massachusetts, 1783–1860.* Boston: Northeastern University Press, 1979.

Russell, Francis. *A City in Terror—1919—The Boston Police Strike.* New York: Viking, 1975.

Russell, Francis. *Tragedy in Dedham: The Story of the Sacco-Vanzetti Case.* New York: McGraw-Hill, 1971.

Whitehall, Walter M., and Norman Kotker. *Massachusetts: A Pictorial History.* New York: Scribner, 1981.

MICHIGAN

State of Michigan

ORIGIN OF STATE NAME: Possibly derived from the Fox Indian word *mesikami*, meaning "large lake."
NICKNAME: The Wolverine State. **CAPITAL:** Lansing. **ENTERED UNION:** 26 January 1837 (26th). **SONG:**
"Michigan, My Michigan" (unofficial). **MOTTO:** *Si quæris peninsulam amœnam circumspice* (If you seek a
pleasant peninsula, look about you). **COAT OF ARMS:** In the center, a shield depicts a peninsula on which a
man stands, at sunrise, holding a rifle. At the top of the shield is the word "Tuebor" (I will defend), beneath
it the state motto. Supporting the shield are an elk on the left and a moose on the right. Over the whole, on
a crest, is an American eagle beneath the US motto, *E pluribus unum.* **FLAG:** The coat of arms centered on
a dark blue field, fringed on three sides. **OFFICIAL SEAL:** The coat of arms surrounded by the words "The
Great Seal of the State of Michigan" and the date "A.D. MDCCCXXXV" (1835, the year the state
constitution was adopted). **BIRD:** Robin. **FISH:** Trout. **FLOWER:** Apple blossom. **TREE:** White pine. **GEM:**
Chlorastrolite. **STONE:** Petoskey stone. **LEGAL HOLIDAYS:** New Year's Day, 1 January; Birthday of Martin
Luther King, Jr., 3d Monday in January; Lincoln's Birthday, 12 February; Washington's Birthday, 3d
Monday in February; Memorial Day, last Monday in May; Independence Day, 4 July; Labor Day, 1st
Monday in September; Columbus Day, 2d Monday in October; Veterans Day, 11 November; Thanksgiv-
ing Day, 4th Thursday in November; Christmas Day, 25 December. **TIME:** 7 AM EST = noon GMT; 6 AM
CST = noon GMT.

¹LOCATION, SIZE, AND EXTENT

Located in the eastern north-central US, Michigan is the 3d-
largest state E of the Mississippi River and ranks 23d in size
among the 50 states.

The total area of Michigan (excluding Great Lakes waters) is
58,527 sq mi (151,585 sq km), of which land takes up 56,954 sq mi
(147,511 sq km) and inland water 1,573 sq mi (4,074 sq km). The
state consists of the upper peninsula adjoining three of the Great
Lakes—Superior, Huron, and Michigan—and the lower peninsula, projecting northward between Lakes Michigan, Erie, and
Huron. The upper peninsula extends 334 mi (538 km) E–W and
215 mi (346 km) N–S; the lower peninsula's maximum E–W
extension is 220 mi (354 km), and its greatest N–S length is 286 mi
(460 km).

Michigan's upper peninsula is bordered on the N and E by the
Canadian province of Ontario (with the line passing through Lake
Superior, the St. Marys River, and Lake Huron); on the S by Lake
Huron, the Straits of Mackinac separating the two peninsulas,
and Lake Michigan; and on the SW and W by Wisconsin (with the
line passing through the Menominee, Brule, and Montreal rivers).
The lower peninsula is bordered on the N by Lake Michigan, the
Straits of Mackinac, and Lake Huron; on the E by Ontario (with
the line passing through Lake Huron, the St. Clair River, Lake St.
Clair, and the Detroit River); on the SE by Ontario and Ohio (with
the line passing through Lake Erie); on the S by Ohio and
Indiana; and on the W by Illinois and Wisconsin (with the line
passing through Lake Michigan and Green Bay). The state's
geographic center is in Wexford County, 5 mi (8 km) NNW of
Cadillac.

Among the most important islands are Isle Royale in Lake
Superior; Sugar, Neebish, and Drummond islands in the St.
Marys River; Bois Blanc, Mackinac, and Les Cheneaux islands in
Lake Huron; Beaver Island in Lake Michigan; and Belle Isle and
Grosse Ile in the Detroit River.

The state's total boundary length is 1,673 mi (2,692 km). The
total freshwater shoreline is 3,121 mi (5,023 km).

²TOPOGRAPHY

Michigan's two peninsulas are generally level land masses. Flat
lowlands predominate in the eastern portion of both peninsulas

and in scattered areas elsewhere. The state's lowest point, 572 feet
(174 meters), is found in southeastern Michigan along Lake Erie.
Higher land is found in the western area of the lower peninsula,
where elevations rise to as much as 1,600 feet (500 meters); the
hilly uplands of the upper peninsula attain elevations of 1,800 feet
(550 meters). The state's highest point, at 1,979 feet (603 meters),
is Mt. Arvon, in Baraga County.

Michigan's political boundaries extend into four of the five
Great Lakes, giving Michigan jurisdiction over 16,231 sq mi
(42,038 sq km) of Lake Superior, 13,037 sq mi (33,766 sq km) of
Lake Michigan, 8,975 sq mi (23,245 sq km) of Lake Huron, and
216 sq mi (559 sq km) of Lake Erie, for a total of 38,459 sq mi
(99,608 sq km). In addition, Michigan has about 35,000 inland
lakes and ponds, the largest of which is Houghton Lake, on the
lower peninsula, with an area of 31 sq mi (80 sq km).

The state's leading river is the Grand, about 260 mi (420 km)
long, flowing through the lower peninsula into Lake Michigan.
Other major rivers that flow into Lake Michigan include the St.
Joseph, Kalamazoo, Muskegon, Pere Marquette, and Manistee.
On the eastern side of the peninsula, the Saginaw River and its
tributaries drain an area of some 6,000 sq mi (15,500 sq km),
forming the state's largest watershed. Other important rivers that
flow into Lake Huron include the Au Sable, Thunder Bay, and
Cheboygan. In the southeast, the Huron and Raisin rivers flow
into Lake Erie. Most major rivers in the upper peninsula (including the longest, the Menominee) flow southward into Lake
Michigan and its various bays. Tahquamenon Falls, in the eastern
part of the upper peninsula, is the largest of the state's more than
150 waterfalls.

Most of the many islands belonging to Michigan are located in
northern Lake Michigan and in Lake Huron, although the largest,
Isle Royale, about 44 mi (71 km) long by 8 mi (13 km) wide, is
found in northern Lake Superior. In northern Lake Michigan,
Beaver Island is the largest, while Drummond Island, off the
eastern tip of the upper peninsula, is the largest island in the
northern Lake Huron area.

Michigan's geological development resulted from its location in
what was once a basin south of the Laurentian Shield, a landmass
covering most of eastern and central Canada and extending

southward into the upper peninsula. Successive glaciers that swept down from the north dumped soil from the shield into the basin and eroded the basin's soft sandstone, limestone, and shale. With the retreat of the last glacier from the area about 6000 BC, the two peninsulas, the Great Lakes, and the islands in these lakes began to emerge, assuming their present shapes about 2,500 years ago.

[3] CLIMATE

Michigan has a temperate climate with well-defined seasons. The warmest temperatures and longest frost-free period are found most generally in the southern part of the lower peninsula; Detroit has a normal daily mean temperature of 49°F (9°C), ranging from 23°F (−5°C) in January to 72°F (22°C) in July. Colder temperatures and a shorter growing season prevail in the more northerly regions; Sault Ste. Marie has a normal daily mean of 40°F (4°C), ranging from 13°F (−11°C) in January to 64°F (18°C) in July. The coldest temperature ever recorded in the state is −51°F (−46°C), registered at Vanderbilt on 9 February 1934; the all-time high of 112°F (44°C) was recorded at Mio on 13 July 1936. Both sites are located in the interior of the lower peninsula, away from the moderating influence of the Great Lakes.

Detroit has an average annual precipitation of 31 in (79 cm); rainfall tends to decrease as one moves northward. The greatest snowfall is found in the extreme northern areas, where cloud cover created by cold air blowing over the warmer Lake Superior waters causes frequent heavy snow along the northern coast; Houghton and Calumet, on the Keweenaw Peninsula, average 183 in (465 cm) of snow a year, more than any other area in the state. Similarly, Lake Michigan's water temperatures create a snow belt along the west coast of the lower peninsula.

Cloudy days are more common in Michigan than in most states, in part because of the condensation of water vapor from the Great Lakes. Detroit has sunshine, on average, only 32% of the days in December and January, and only 54% year-round. The annual average relative humidity at Detroit is 77% at 7 AM, dropping to 58% at 1 PM; at Sault Ste. Marie, the comparable percentages are 85% and 67%, respectively. The southern half of the lower peninsula is an area of heavy thunderstorm activity. Late spring and early summer are the height of the tornado season; there were 19 tornadoes in 1983.

[4] FLORA AND FAUNA

Maple, birch, hemlock, aspen, spruce, and fir predominate in the upper peninsula; maple, birch, aspen, pine, and beech in the lower. Once common in the state, elms have largely disappeared because of the ravages of disease, while the white pine (the state tree) and red pine, which dominated northern Michigan forests and were prime objects of logging operations, have been replaced in cutover lands by aspen and birch. The area south of a line from about Muskegon to Saginaw Bay formerly held the only significant patches of open prairie land (found chiefly in southwestern Michigan) and areas of widely scattered trees, called oak openings. Intensive agricultural development, followed by urban industrial growth, leveled much of this region's forests, although significant wooded acreage remains, especially in the less populated western regions.

Strawberries, raspberries, gooseberries, blueberries, and cranberries are among the fruit-bearing plants and shrubs that grow wild in many areas of the state, as do mushrooms and wild asparagus. The state flower, the apple blossom, calls to mind the importance of fruit-bearing trees and shrubs in Michigan, but wild flowers also abound, with as many as 400 varieties found in a single county. Protected plants include all members of the orchid, trillium, and gentian families; trailing arbutus; prince's-pine; bird's-foot violet; climbing bittersweet; flowering dogwood; mountain and Michigan hollies; and American lotus.

Michigan's fauna, like its flora, has been greatly affected by settlement and, in a few cases, by intensive hunting and fishing. Moose are now confined to Isle Royale, as are nearly all the remaining wolves, which once roamed throughout the state. The caribou and passenger pigeon have been extirpated, but the elk and turkey have been successfully reintroduced in the 20th century. There is no evidence that the state's namesake, the wolverine, was ever found in Michigan, at least in historic times. Despite intensive hunting, the deer population remains high. Other game animals include the common cottontail, snowshoe hare, raccoon, and various squirrels. In addition to the raccoon, important native furbearers are the river otter and the beaver, once virtually exterminated but now making a strong comeback.

More than 300 types of birds have been observed. Aside from the robin (the state bird), the most notable bird is Kirtland's warbler, which nests only in a 60-sq-mi (155-sq-km) section of jack-pine forest in north-central Michigan. Ruffed grouse, bobwhite quail, American woodcock, and various ducks and geese are hunted extensively. Populations of ring-necked pheasant, introduced in 1895, have dropped at an alarming rate in recent decades. Reptiles include the massasauga, the state's only poisonous snake.

Whitefish, perch, and lake trout (the state fish) are native to the Great Lakes, while perch, bass, and pike are indigenous to inland waters. In 1877, the carp was introduced, with such success that it has since become a nuisance. Rainbow and brown trout have also been planted, and in the late 1960s, the state enjoyed its most spectacular success with the introduction of several species of salmon.

The first Michigan list of threatened or endangered fauna in 1976 included 64 species, 15 endangered and 49 threatened. Subsequent surveys indicated that Kirtland's warbler, the bald eagle, and the osprey were holding their own, while the greater prairie chicken, barn owl, and common tern were continuing to decline. Other endangered animals include the Indiana bat, gray wolf, Kirtland's water snake, blue pike, and five species of cisco.

[5] ENVIRONMENTAL PROTECTION

Michigan's initial conservation law, enacted in 1859, limited the hunting of deer and certain other wildlife to specific seasons. Later acts extended these provisions to other species, outlawed certain kinds of hunting and fishing equipment, and banned commercial hunting (1881).

The Michigan Forestry Commission was established in 1899, but only after the state's vast timber resources had been virtually wiped out by lumbermen. Four years later, a reforestation program was launched, the first state forest was set aside, and organized forest-fire fighting began. In 1921, the Michigan Department of Conservation was founded; it was renamed the Department of Natural Resources in 1968 and reorganized in 1973 to consolidate state environmental protection activities. The Michigan Environmental Review Board was established in 1974 to advise the governor on environmental policy, hold public hearings, and review environmental impact statements. With the passage in 1970 of the Environmental Protection Act, Michigan became the first state to provide citizens with the statutory right to bring suit to protect the environment for the general good.

In the 1983/84 fiscal year, the Department of Natural Resources received an appropriation of $154,304,418. The largest amounts were budgeted for such traditional purposes as parklands, wildlife, fisheries, and forests. The other principal area of importance was environmental health and safety. In 1984, the Cabinet Council on Environmental Protection was established, consisting of the heads of the departments of Natural Resources, Public Health, Agriculture, and Management and Budget; the attorney general; and heads of the Toxic Substance Control

---

LOCATION: 41°41′ to 47°30′N; 82°26′ to 90°31′W. **BOUNDARIES:** Canadian line, 721 mi (1,160 km); Ohio line, 95 mi (153 km); Indiana line, 126 mi (203 km); Illinois line, 51 mi (82 km); Wisconsin line, 680 mi (1,094 km).

MICHIGAN

SCALE
0   30   60 Miles
0   30   60 Kms.

See endsheet maps: J1.

Commission and the Michigan Environmental Review Board. The council's function is to help devise strategies to deal with basic environmental protection problems. The Environmental Protection Bureau of the Natural Resources Department exercises the major responsibility for protecting and enhancing the environment through the control of air, land, and water pollution. In 1982, it was reorganized into six divisions, dealing with air quality, groundwater quality, hazardous-waste management, surface water quality, assistance to communities, and support services (such as laboratory services) to regulatory programs.

Despite measurable improvements in air quality in the state during the period 1973–83, as of 1984, several areas of southeastern Michigan still failed to meet federal air quality standards for various pollutants, including ozone and carbon monoxide. A mandatory automobile inspection program was being conducted in the Detroit tri-county area, in an effort to decrease air pollution. Groundwater pollution also has been a serious problem in the state. A program was in progress in 1984 to clean up the most seriously affected sites.

The Department of Natural Resources licenses approximately 350 solid-waste processing facilities and 10,000 refuse transport units each year. As of 1984, an estimated 26,000 tons of solid waste were generated in the state each day. All 83 counties were expected to have approved solid-waste management plans by the end of 1985; the state was seeking cost-effective alternatives to land disposal of solid wastes and also sought to establish pilot recycling and composting projects in communities. A "bottle bill" requiring a deposit to be paid on beverage cans and bottles sold in Michigan, so as to promote recycling of these containers, was approved by voters in 1978.

The accidental mixing of the fire-retardant chemical polybrominated biphenyl (PBB) with livestock feed in 1974 and resultant controversy over the disposal of contaminated animals judged unfit for human consumption pointed up the problem of hazardous waste disposal. In 1983, approximately 499,000 tons of hazardous wastes were generated in the state, and another 56,000 tons were brought into the state from outside. By this time, it was estimated that about 7 million tons of hazardous waste needed to be cleaned up and removed from currently contaminated sites.

[6] POPULATION

Michigan ranked 8th among the 50 states in the 1980 census, with a population of 9,262,078. That total represented an increase of

Michigan Counties, County Seats, and County Areas and Populations

COUNTY	COUNTY SEAT	LAND AREA (SQ MI)	POPULATION (1984 EST.)	COUNTY	COUNTY SEAT	LAND AREA (SQ MI)	POPULATION (1984 EST.)
Alcona	Harrisville	679	9,850	Lapeer	Lapeer	658	69,039
Alger	Munising	912	8,826	Leelanau	Leland	341	14,381
Allegan	Allegan	832	84,224	Lenawee	Adrian	753	88,195
Alpena	Alpena	568	31,408	Livingston	Howell	575	100,634
Antrim	Bellaire	480	16,792	Luce	Newberry	905	5,969
Arenac	Standish	368	15,160	Mackinac	St. Ignace	1,025	10,238
Baraga	L'Anse	901	8,326	Macomb	Mt. Clemens	483	686,161
Barry	Hastings	560	46,470	Manistee	Manistee	543	22,338
Bay	Bay City	447	117,178	Marquette	Marquette	1,822	72,440
Benzie	Beulah	322	11,141	Mason	Ludington	495	26,435
Berrien	St. Joseph	576	163,029	Mecosta	Big Rapids	560	37,229
Branch	Coldwater	508	38,710	Menominee	Menominee	1,045	25,975
Calhoun	Marshall	712	137,798	Midland	Midland	525	75,623
Cass	Cassopolis	496	47,814	Missaukee	Lake City	565	10,563
Charlevoix	Charlevoix	421	19,709	Monroe	Monroe	557	130,998
Cheboygan	Cheboygan	720	20,909	Montcalm	Stanton	713	49,757
Chippewa	Sault Ste. Marie	1,590	28,819	Montmorency	Atlanta	550	7,760
Clare	Harrison	570	24,749	Muskegon	Muskegon	507	155,688
Clinton	St. Johns	573	55,284	Newaygo	White Cloud	847	36,238
Crawford	Grayling	559	9,825	Oakland	Pontiac	875	1,004,884
Delta	Escanaba	1,173	39,450	Oceana	Hart	541	21,994
Dickinson	Iron Mt.	770	25,863	Ogemaw	West Branch	569	17,437
Eaton	Charlotte	579	89,292	Ontonagon	Ontonagon	1,311	9,685
Emmet	Petoskey	468	23,610	Osceola	Reed City	569	20,134
Genesee	Flint	642	434,148	Oscoda	Mio	568	6,912
Gladwin	Gladwin	505	21,287	Otsego	Gaylord	516	15,345
Gogebic	Bessemer	1,105	19,319	Ottawa	Grand Haven	567	164,658
Grand Traverse	Traverse City	466	57,039	Presque Isle	Rogers City	656	13,887
Gratiot	Ithaca	570	39,682	Roscommon	Roscommon	528	18,137
Hillsdale	Hillsdale	603	41,678	Saginaw	Saginaw	815	219,059
Houghton	Houghton	1,014	38,109	St. Clair	Port Huron	734	137,954
Huron	Bad Axe	830	36,002	St. Joseph	Centreville	503	57,715
Ingham	Mason	560	271,671	Sanilac	Sandusky	964	40,127
Ionia	Ionia	577	52,380	Schoolcraft	Manistique	1,173	8,453
Iosco	Tawas City	546	30,234	Shiawassee	Corunna	541	68,587
Iron	Crystal Falls	1,163	14,015	Tuscola	Caro	812	55,278
Isabella	Mt. Pleasant	576	54,569	Van Buren	Paw Paw	612	66,534
Jackson	Jackson	705	145,314	Washtenaw	Ann Arbor	710	265,210
Kalamazoo	Kalamazoo	562	215,237	Wayne	Detroit	615	2,186,081
Kalkaska	Kalkaska	563	11,500	Wexford	Cadillac	566	26,154
Kent	Grand Rapids	862	461,718				
Keweenaw	Eagle River	544	2,071				
Lake	Balwin	568	8,425	TOTALS		56,954	9,078,503

4% over the 1970 population; from 1980 to 1985, however, Michigan's population declined 2.9%, to 8,992,766.

During the long prehistoric period, Michigan was inhabited by only a few thousand Indians. As late as 1810, the non-Indian population of Michigan Territory was only 4,762. The late 1820s marked the start of steady, often spectacular growth. The population increased from 31,639 people in 1830 to 212,267 in 1840 and 397,654 in 1850. Subsequently, the state's population grew by about 400,000 each decade until 1910, when its population of 2,810,173 ranked 8th among the 46 states. Industrial development sparked a sharp rise in population to 4,842,325 by 1930, which pushed Michigan ahead of Massachusetts into 7th place.

According to the 1980 census, Michigan's population was slightly younger than the national median age, and in 1983, Michigan had the lowest proportion of elderly citizens of any state in the northeast or north-central regions. State residents were among the least mobile in the US: as of 1980, only 6% of those 5 years of age or older had lived out of state five years before.

With 71% of its population classified as urban in 1980, Michigan's percentage was below the national average. Population density for the entire state in 1985 was 158 persons per sq mi (61 per sq km); half the population was concentrated in the Detroit metropolitan area.

Detroit has always been Michigan's largest city since its founding in 1701, but its growth, like the state's, was slow until well into the 19th century. The city's population grew from 21,019 in 1850 to 285,704 in 1900, when it ranked as the 13th-largest city in the country. Within the next 30 years, the booming automobile industry pushed the city up into 4th place, with a population of 1,568,662 in 1930. Since 1950, when the total reached 1,849,568, Detroit has lost population, dropping to 1,514,063 in 1970, 1,203,369 in 1980, and 1,088,973 in 1984, when it held 6th place among US cities. As Detroit lost population, however, many of its suburban areas grew at an even greater rate, and the Detroit metropolitan area totaled 4,315,800 in 1984 (6th in the US), up from 3,950,000 in 1960.

In 1984, seven cities aside from Detroit had populations of 100,000 or more: Grand Rapids had a population of 183,000, Warren 152,035, Flint 149,007, Lansing (the capital) 127,972, Sterling Heights 109,440, Ann Arbor 103,840, and Livonia 100,363.

7ETHNIC GROUPS

The 1980 census counted 39,702 American Indians, more than double the number in 1970. Most were scattered across the state, with a small number concentrated on the four federal reservations, comprising 16,635 acres (6,732 hectares). The Ottawa, Ojibwa, and Potawatomi were the principal groups with active tribal organizations.

The black population of Michigan in 1980 was 1,199,023, 13% of the state's total population and 3d highest among the northern states. Nearly two-thirds lived in Detroit, where they made up 63% of the population, the highest percentage in any US city of 1 million or more. Detroit, which experienced severe race riots in 1943 and 1967, has had a black mayor since 1974.

The 1980 census found that 417,152 state residents (4.5%) were foreign born. There were 162,440 Hispanics living in the state in 1980, of whom 112,183 were of Mexican descent. The state's Asian population has been increasing: as of 1980 there were 14,680 Asian Indians, 11,162 Filipinos, 10,993 Chinese, 8,700 Koreans, 5,859 Japanese, and 4,208 Vietnamese. Although state residents of first- or second-generation European descent are, almost without exception, decreasing in number and proportion, their influence remains great. Detroit continues to have numerous well-defined ethnic neighborhoods, and Hamtramck, a city surrounded by Detroit, is still dominated by its Polish population. Elsewhere in Michigan, Frankenmuth is the site of an annual German festival, and the city of Holland has an annual tulip

festival that attracts about 400,000 people each spring. In the upper peninsula, the Finnish culture dominates in rural areas; in the iron and copper mining regions, descendants of immigrants from Cornwall in England, the original mining work force, and persons of Scandinavian background predominate.

8LANGUAGES

Before white settlement, Algonkian-language tribes occupied what is now Michigan, with the Menomini and Ojibwa in the upper peninsula and Ottawa on both sides of the Straits of Mackinac. Numerous place-names recall their presence: Michigan itself, Mackinaw City, Petoskey, Kalamazoo, Muskegon, Cheboygan, and Dowagiac.

Except for the huge industrial area in southeastern Michigan, English in the state is remarkably homogeneous in its retention of the major Northern dialect features of upper New York and western New England. Common are such Northern forms as *pail, wishbone, darning needle* (dragonfly), *mouth organ* (harmonica), *sick to the stomach, quarter to four* (3:45), and *dove* as past tense of *dive.* Common also are such pronunciations as the /e/ vowel in *fog, frog,* and *on;* the /aw/ vowel in *horrid, forest,* and *orange; creek* as /krik/; *root* and *roof* with the vowel of *put;* and *greasy* with an /s/ sound. *Swale* (a marsh emptying into a stream) and *clock shelf* (mantel) are dying Northern words not carried west of Michigan. *Pank* (to pack down, as of snow) is confined to the upper peninsula, as is *pasty* (meat-filled pastry), borrowed from Cornish miners. A minister is a *dominie* in the Dutch area around Holland and Zeeland.

Southern blacks have introduced into the southeastern automotive manufacturing areas a regional variety of English that, because it has class connotations in the North, has become a controversial educational concern. Three of its features are perhaps more widely accepted than others: the coalescence of /e/ and /i/ before a nasal consonant, so that *pen* and *pin* sound alike; the loss of /r/ after a vowel, so that *cart* and *cot* also sound alike; and the lengthening of the first part of the diphthong /ai/, so that *time* and *Tom* sound alike, as do *ride* and *rod.*

In 1980, only 6% of the state's population 3 years old or older spoke a language other than English at home. Other languages spoken at home, with the number of speakers, were as follows:

Spanish	106,536	Arabic	31,681
Polish	90,629	Greek	15,381
German	60,507	Finnish	14,832
Italian	48,140	Dutch	14,690
French	34,244	Serbo-Croatian	10,879

9RELIGIONS

The Roman Catholic Church was the only organized religion in Michigan until the 19th century. Detroit's Ste. Anne's parish, established in 1701, is the second-oldest Catholic parish in the country. In 1810, a Methodist society was organized near Detroit, and after the War of 1812, as settlers poured in from the East, Presbyterian, Congregational, Baptist, Episcopal, and Quaker churches were founded. The original French Catholics, reduced to a small minority by the influx of American Protestants, were soon reinforced by the arrival of Catholic immigrants from Germany, Ireland, and later from eastern and southern Europe. The Lutheran religion was introduced by German and Scandinavian immigrants; Dutch settlers were affiliated with the Reformed Church in America. The first Jewish congregations were organized in Detroit by German Jews, with a much greater number of eastern European Jews arriving toward the end of the 1800s. The Orthodox Christian Church and the Islamic religion have been introduced by immigrants from the Near East during the 20th century.

Michigan had 2,297,212 Roman Catholics in 1984, and an estimated 85,275 Jews, 82% of whom lived in Detroit. Among Protestant denominations, a census taken in 1980 showed various

Lutheran groups with a combined total of 445,340 adherents, Methodist groups with 370,813 adherents, and Baptists with 150,538. Among other major denominations, the United Presbyterian Church had 156,233 adherents, the Episcopal Church 100,078, the Christian Reformed Church 96,740, the United Church of Christ 85,917, and the Reformed Church in America 83,904. The Seventh Day Adventists, who had their world headquarters in Battle Creek from 1855 to 1903, numbered 32,975 in 1980, the Salvation Army 22,472, and the Church of Jesus Christ of Latter-day Saints (Mormon) 18,635. Various Church of God denominations had 52,831 adherents in all. Michigan also is the home for a large number of Shiite Muslims; an estimated 20,000 lived in the state as of 1985, about half of their number in Dearborn, where they came to work in the Ford plant.

10 TRANSPORTATION

Because of Michigan's location, its inhabitants have always depended heavily on the Great Lakes for transportation. Not until the 1820s did land transportation systems begin to be developed. Although extensive networks of railroads and highways now reach into all parts of the state, the Great Lakes remain major avenues of commerce.

The first railroad company in the Midwest was chartered in Michigan in 1830, and six years later, the Erie and Kalamazoo, operating between Toledo, Ohio, and Adrian, became the first railroad in service west of the Appalachians. Between 1837 and 1845, the state government sought to build three lines across southern Michigan, before abandoning the project and selling the two lines it had partially completed to private companies. The pace of railroad construction lagged behind that in other midwestern states until after the Civil War, when the combination of federal and state aid and Michigan's booming economy led to an enormous expansion in trackage, from fewer than 800 mi (1,300 km) in 1860 to a peak of 9,021 mi (14,518 km) in 1910. With the economic decline of northern Michigan and the resultant drop in railroad revenues, however, Class I trackage declined to 3,600 mi (5,794 km) by December 1983. Most railroad passenger service is provided by Amtrak, which operates five trains through the state, and much of the freight is carried by Conrail. The Michigan state government, through the Department of Transportation, has helped to revive the railroad system by operating some lines.

Railroads have been used only to a limited degree in the Detroit area as commuter carriers, although efforts have been made to improve this service. In the early 1900s, more than 1,000 mi (1,600 km) of interurban rail lines provided rapid transit service in southern Michigan, but automobiles and buses drove them out of business, and the last line shut down in 1934. Street railway service began in a number of cities in the 1860s, and Detroit took over its street railways in 1922. Use of these public transportation systems declined sharply after World War II. By the 1950s, streetcars had been replaced by buses, but by 1960, many small communities had abandoned city bus service altogether. During the 1970s, with massive government aid, bus service was restored to many cities and was improved in others, and the number of riders generally increased. In the 1980s there were ridership declines, however. Total mass transit ridership in the state was 118,276,540 in 1982, a 7% drop from 1981. In 1985, plans were pending for possible construction of a new light-rail-transit system in metropolitan Detroit.

As of 31 December 1983, the state had 92,274 mi (148,501 km) of rural roads and 25,193 mi (40,544 km) of urban roads under state or local control. There were 1,127 mi (1,814 km) of interstate highway open to traffic. Major toll-free expressways included I-94 (Detroit to Chicago), I-96 (Detroit to Grand Rapids), and I-75 (from the Ohio border to Sault Ste. Marie). In 1983 there were 5,024,521 registered passenger cars, 1,124,193 trucks, 21,135 buses, and 233,066 motorcycles. Licensed drivers numbered 6,344,657 during the same year.

The completion in 1957 of the Mackinac Bridge, the fourth-longest suspension span in the world, eliminated the major barrier to easy movement between the state's two peninsulas. The International Bridge at Sault Ste. Marie, the Blue-Water Bridge at Port Huron, the Ambassador Bridge at Detroit, and the Detroit-Windsor Tunnel link Michigan with Canada.

The opening of the St. Lawrence Seaway in 1959 made it possible for a large number of oceangoing vessels to dock at Michigan ports. In 1981, the Detroit River handled 90,405,900 tons of freight, the St. Clair River 86,824,200 tons, and the St. Marys River 76,377,800. In that year, the port of Detroit handled 17,839,000 tons of cargo, the iron-ore port of Escanaba 10,061,700 tons, and the limestone-shipping port of Calcite 7,974,900.

Michigan was a pioneer in developing air transportation service. The Ford Airport at Dearborn in the 1920s had one of the first air passenger facilities and was the base for some of the first regular airmail service. In 1983, Michigan had 375 airfields, 41 heliports, and 4 seaplane bases. The major airport is Detroit Metropolitan, which in 1983 enplaned 5,995,885 passengers and handled 128,459 scheduled aircraft departures.

11 HISTORY

Indian hunters and fishermen inhabited the region now known as Michigan as early as 9000 BC. By 5000 BC, these peoples were making use of copper found in the upper peninsula—the first known use of a metal by peoples anywhere in the western hemisphere. Around 100 BC, their descendants introduced agriculture into southwestern Michigan. In the latter part of the prehistoric era, the Indians appear to have declined in population.

In the early 17th century, when European penetration began, Michigan's lower peninsula was virtually uninhabited. In the upper peninsula there were small bands of Ojibwa along the St. Marys River and the Lake Michigan shore; in the west, Menomini Indians lived along the present Michigan-Wisconsin border. Both tribes were of Algonkian linguistic stock, as were most Indians who later settled in the area, except for the Winnebago of the Siouan group in the Green Bay region of Lake Michigan, and the Huron of Iroquoian stock in the Georgian Bay area of Canada. In the 1640s, the Huron were nearly wiped out by other Iroquois tribes from New York, and the survivors fled westward with their neighbors to the north, the Ottawa Indians. Eventually, both tribes settled at the Straits of Mackinac before moving to the Detroit area early in the 18th century. During the same period, the Potawatomi and Miami Indians moved from Wisconsin into southern Michigan.

For two centuries after the first Europeans came to Michigan, the Indians remained a vital force in the area's development. They were the source of the furs that the whites traded for, and they also were highly respected as potential allies when war threatened between the rival colonial powers in North America. However, after the War of 1812, when the fur trade declined and the possibility of war receded, the value of the Indians to the white settlers diminished. Between 1795 and 1842, Indian lands in Michigan were ceded to the federal government, and the Huron, Miami, and many Potawatomi were removed from the area. Some Potawatomi were allowed to remain on lands reserved for them, along with most of the Ojibwa and Ottawa Indians in the north.

The first European explorer known to have reached Michigan was a Frenchman, Etienne Brulé, who explored the Sault Ste. Marie area around 1620. Fourteen years later, Jean Nicolet explored the Straits of Mackinac and the southern shore of the upper peninsula en route to Green Bay. Missionary and fur trading posts, to which were later added military forts, were established at Sault Ste. Marie by Father Jacques Marquette in 1668, and then at St. Ignace in 1671. By the 1680s, several temporary posts had been established in the lower peninsula. In 1701, Antoine Laumet de la Mothe Cadillac founded a permanent settlement at the site of present-day Detroit.

Detroit and Michigan grew little, however, because the rulers of the French colony of New France were obsessed with the fur trade, which did not attract large numbers of settlers. After France's defeat in the French and Indian War, fears that the British would turn the area over to English farmers from the coastal colonies, with the consequent destruction of the Indian way of life, led the Indians at Detroit to rebel in May 1763, under the leadership of the Ottawa chief Pontiac. The uprising, which soon spread throughout the west, ended in failure for the Indians; Pontiac gave up his siege of Detroit after six months, and by 1764, the British were in firm control. Nevertheless, the British authorities did not attempt to settle the area. The need to protect the fur trade placed the people of Michigan solidly on the British side during the American Revolution, since a rebel triumph would likely mean the migration of American farmers into the west, converting the wilderness to cropland. The British occupied Michigan and other western areas for 13 years after the Treaty of Paris in 1783 had assigned these territories to the new United States. The US finally got possession of Michigan in the summer of 1796.

Michigan became a center of action in the War of 1812. The capture of Detroit by the British on 16 August 1812 was a crushing defeat for the Americans. Although Detroit was recaptured by the Americans in September 1813, continued British occupation of the fort on Mackinac Island, which they had captured in 1812, enabled them to control most of Michigan. The territory was finally returned to American authority under the terms of the Treaty of Ghent at the end of 1814. With the opening in 1825 of the Erie Canal, which provided a cheap, all-water link between Michigan and New York City, American pioneers turned their attention to these northern areas, and during the 1820s, settlers for the first time pushed into the interior of southern Michigan.

Originally part of the Northwest Territory, Michigan had been set aside in 1805 as a separate territory, but with boundaries considerably different from those of the subsequent state. On the south, the territory's boundary was a line set due east from the southernmost point of Lake Michigan; on the north, only the eastern tip of the upper peninsula was included. In 1818 and 1834, areas as far west as Iowa and the Dakotas were added to the territory for administrative purposes. By 1833, Michigan had attained a population of 60,000, qualifying it for statehood. The territorial government's request in 1834 that Michigan be admitted to the Union was rejected by Congress, however, because of a dispute over Michigan's southern boundary. When Indiana became a state in 1816, it had been given a 10-mi (16-km) strip of land in southwestern Michigan, and Michigan now refused to accede to Ohio's claim that it should be awarded lands in southeastern Michigan, including the present site of Toledo. In 1835, Michigan militia defeated the efforts of Ohio authorities to take over the disputed area during the so-called Toledo War, in which no one was killed. Nevertheless, Ohio's superior political power in Congress ultimately forced Michigan to agree to relinquish the Toledo Strip. In return, Congress approved the state government that the people of Michigan had set up in 1835. As part of the compromise that finally brought Michigan into the Union on 26 January 1837, the new state was given land in the upper peninsula west of St. Ignace as compensation for the loss of Toledo.

Youthful Stevens T. Mason, who had led the drive for statehood, became Michigan's first elected governor, but he and the Democratic Party fell out of grace when the new state was plunged into financial difficulties during the depression of the late 1830s. The party soon returned to power and controlled the state until the mid-1850s. In Michigan, as elsewhere, it was the slavery issue that ended Democratic dominance. In July 1854, antislavery Democrats joined with members of the Whig and Free-Soil

parties at a convention in Jackson to organize the Republican Party. In the elections of 1854, the Republicans swept into office in Michigan, with rare exceptions controlling the state until the 1930s.

Abraham Lincoln was not the first choice of Michigan Republicans for president in 1860, but when he was nominated, they gave him a solid margin of victory that fall and again in 1864. Approximately 90,000 Michigan men served in the Union army, taking part in all major actions of the Civil War. Michigan's Zachariah Chandler was one of the leaders of the Radical Republicans in the US Senate who fought for a harsh policy toward the South during Reconstruction.

Michigan grew rapidly in economic importance. Agriculture sparked the initial growth of the new state and was responsible for its rapid increase in population. By 1850, the southern half of the lower peninsula was filling up, with probably 85% of the state's population dependent in some way on agriculture for a living. Less than two decades later, exploitation of vast pine forests in northern Michigan had made the state the top lumber producer in the US. Settlers were also attracted to the same area by the discovery of rich mineral deposits, which made Michigan for a time the nation's leading source of iron ore, copper, and salt.

Toward the end of the 19th century, as timber resources were being exhausted and as farming and mining reached their peak stages of development, new opportunities in manufacturing opened up. Such well-known Michigan companies as Kellogg, Dow Chemical, and Upjohn had their origins during this period. The furniture industry in Grand Rapids, the paper industry in Kalamazoo, and numerous other industries were in themselves sufficient to ensure the state's increasing industrial importance. But the sudden popularity of Ransom E. Olds's Oldsmobile runabout, manufactured first in Detroit and then in Lansing, inspired a host of Michiganians to produce similar practical, relatively inexpensive automobiles. By 1904, the most successful of the new models, Detroit's Cadillac (initially a cheap car) and the first Fords, together with the Oldsmobile, had made Michigan the leading automobile producer in the country—and, later, in the world. The key developments in Michigan's auto industry were the creation of General Motors by William C. Durant in 1908; Henry Ford's development of the Model T in 1908, followed by his institution of the moving assembly line in 1913–14; and Walter P. Chrysler's formation in 1925 of the automobile corporation named after him.

Industrialization brought with it urbanization; the census of 1920 for the first time showed a majority of Michiganians living in towns and cities. Nearly all industrial development was concentrated in the southern third of the state, particularly the southeastern Detroit area. The northern two-thirds of the state, where nothing took up the slack left by the decline in lumber and mining output, steadily lost population and became increasingly troubled economically. Meanwhile, the Republican Party, under such progressive governors as Fred Warner and Chase Osborn—and, in the 1920s, under a brilliant administrator, Alexander Groesbeck—showed itself far better able than the Democratic opposition to adjust to the complexities of a booming industrial economy.

The onset of the depression of the 1930s had devastating effects in Michigan. The market for automobiles collapsed; by 1932, half of Michigan's industrial workers were unemployed. The ineffectiveness of the Republican state and federal governments during the crisis led to a landslide victory for the Democrats. In traditionally Republican areas of rural Michigan, the defection to the Democratic Party in 1932 was only temporary, but in the urban industrial areas, the faith of the factory workers in the Republican Party was, for the great majority, permanently shaken. These workers, driven by the desire for greater job security, joined the recruiting campaign launched by the new

Congress of Industrial Organizations (CIO). By 1941, with the capitulation of Ford Motor, the United Automobile Workers (UAW) had organized the entire auto industry, and Michigan had been converted to a strongly pro-union state.

Eventually, the liberal leadership of the UAW and of other CIO unions in the state allied itself with the Democratic Party to provide the funds and organization the party needed to mobilize worker support. The coalition elected G. Mennen Williams governor in 1948 and reelected him for five successive two-year terms. By the mid-1950s, the Democrats controlled virtually all statewide elective offices. Because legislative apportionment still reflected an earlier distribution of population, however, the Republicans maintained their control of the legislature and frustrated the efforts of the Williams administration to institute social reforms. In the 1960s, as a result of US Supreme Court rulings, the legislature was reapportioned on a strictly equal-population basis. This shifted a majority of legislative seats into the urban areas, enabling the Democrats generally to control the legislature since that time.

In the meantime, Republican moderates, led by George Romney, gained control of their party's organization. Romney was elected governor in 1962 and served until 1969, when he was succeeded by William G. Milliken, who held the governorship for 14 years. When Milliken chose not to run in the 1982 election, the statehouse was captured by the Democrats, ending 20 years of Republican rule. The new governor faced the immediate tasks of saving Michigan from bankruptcy and reducing the unemployment rate, which had averaged more than 15% in 1982 (60% above the US average).

The nationwide recession of the early 1980s hit Michigan harder than most other states because of its effect on the auto industry, which had already suffered heavy losses primarily as a result of its own inability to foresee the demise of the big luxury cars and because of the increasing share of the American auto market captured by foreign, mostly Japanese, manufacturers. In 1979, Chrysler had been forced to obtain $1.2 billion in federally guaranteed loans to stave off bankruptcy, and during the late 1970s and the first two years of the 1980s, US automakers were forced to lay off hundreds of thousands of workers, tens of thousands of whom left the state. Many smaller businesses, dependent on the auto industry, closed their doors, adding to the unemployment problem and to the state's fiscal problem; as the tax base shrank, state revenues plummeted, creating a budget deficit of nearly $1 billion. Two months after he took office in January 1983, Governor James J. Blanchard was forced to institute budget cuts totaling $225 million and lay off thousands of government workers; and, at his urging, the state legislature increased Michigan's income tax by 38%.

As the recession eased in 1983, Michigan's economy showed some signs of improvement. The automakers became profitable, and Chrysler was even able to repay its $1.2 billion in loans seven years before it was due, rehire 100,000 workers, and make plans to build a $500-million technological center in the city of Rochester. By May 1984, Michigan's unemployment rate dropped to 11.3%, but the state faced the difficult task of restructuring its economy to lessen its dependence on the auto industry.

12 STATE GOVERNMENT

Michigan has had four constitutions. The first, adopted in 1835 when Michigan was applying for statehood, was followed by constitutions adopted in 1850, 1908, and 1963.

The legislature consists of a senate of 38 members, elected for terms of four years, and a house of representatives of 110 members, elected for two-year terms. The legislature meets annually for a session of indeterminate length. Special sessions may be called by the governor. Legislation may be adopted by a majority of each house, but to override a governor's veto, a two-thirds vote of the members of each house is required. A legislator

must be at least 21 years of age, a US citizen, and a qualified voter of the district in which he or she resides.

Elected executive officials include the governor and lieutenant governor (who run jointly), secretary of state, and attorney general, all serving four-year terms. Elections are held in even-numbered years between US presidential elections. The governor and lieutenant governor must be at least 30 years old, US citizens, and must have been registered voters in the state for at least four years prior to election. The governor appoints the members of the governing boards and/or directors of 19 executive departments, with the exception of the Department of Education, whose head is appointed by the elected State Board of Education.

Legislative action is completed when a bill has been passed by both houses of the legislature and signed by the governor. A bill also becomes law if not signed by the governor after a 14-day period when the legislature is in session. The governor may stop passage of a bill by vetoing it or, if the legislature adjourns before the 14-day period expires, by refusing to sign it.

The constitution may be amended by a two-thirds vote of both houses of the legislature and a majority vote at the next general election. An amendment also may be proposed by registered voters through petition and submission to the general electorate. Every 16 years, the question of calling a convention to revise the constitution must be submitted to the voters; the question was last put on the ballot in 1978 and was rejected.

To be eligible to vote in Michigan, one must be a US citizen, 18 years of age, and must have been a resident of the state and precinct for 30 days.

13 POLITICAL PARTIES

From its birth in 1854 through 1932, the Republican Party dominated state politics, rarely losing statewide elections and developing strong support in all parts of the state, both rural and urban. The problems caused by the economic depression of the 1930s revitalized the Democratic Party and made Michigan a strong two-party state. Democratic strength was concentrated in metropolitan Detroit, while Republicans maintained their greatest strength in "outstate" areas, except for the mining regions of the upper peninsula, where the working class, hit hard by the depression, supported the Democrats.

Most labor organizations, led by the powerful United Automobile Workers union, have generally supported the Democratic Party since the 1930s. But in recent years, moderate Republicans have had considerable success in attracting support among previously Democratic voters.

Between 1948 and 1984, the Republican candidate for president carried Michigan in 7 out of 10 elections. Ronald Reagan won 49% of the state's popular vote in 1980 and 59% in 1984. However, as of 1985, the state had a Democratic governor, and both of Michigan's US senators were Democrats, as were 11 of the state's 18-member US House delegation and a majority in the lower house of the state legislature.

Michigan had 1 Asian and 17 Hispanic elected officials in 1984. In 1985 there were 300 black elected officials, including 2 US representatives, 17 state legislators, and 15 mayors. Women in the state legislature numbered 16 in 1985, and Martha W. Griffiths was one of 5 women serving as lieutenant governor of a US state.

Among minor parties, only Theodore Roosevelt's Progressive Party, which captured the state's electoral vote in 1912, has succeeded in winning a statewide contest. The strongest third-party showing since 1912 was that of George Wallace, who in 1968 captured 10% of the total vote cast for president.

14 LOCAL GOVERNMENT

In 1982 there were 2,643 separate units of local government in Michigan, including 83 counties, 532 municipal governments, and 1,245 townships, as well as hundreds of educational and special districts. Each county is administered by a county board of commissioners, whose members, ranging in number from 3 to 35 according to population, are elected for two-year terms. Executive authority is vested in 5 officers elected for four-year terms: the sheriff, prosecuting attorney, treasurer, clerk, and register of deeds. An increasing number of counties are placing overall administrative responsibility in the hands of a county manager or administrator.

Most cities are governed by home-rule legislation, adopted in 1909, enabling them to establish their own form of government under an adopted charter. Some charters provide for the election of a mayor, who usually functions as the chief executive officer of the city. Other cities have chosen the council-manager system, with a council appointing the manager to serve as chief executive and the office of mayor being largely ceremonial. Many villages are incorporated under home-rule legislation in order to provide services such as police and fire protection.

Each county is divided into two types of townships, geographical and political. Each geographical (or congressional) township has an area of 36 sq mi (93 sq km); in sparsely populated areas, parts of two or more geographical townships may be combined into one political township. Township government, its powers strictly limited by state law, consists of a supervisor, clerk, treasurer, and up to four trustees, all elected for four-year terms and together forming the township board.

¹⁵STATE SERVICES

Educational services are handled in part by the Department of Education, which distributes state school-aid funds, certifies teachers, and operates the School for the Deaf at Flint, the School for the Blind at Lansing, the State Technical Institute and Rehabilitation Center at Plainwell, and the state library system. The 13 state-supported colleges and universities are independent of the department's control, each being governed by an elected or appointed board. Although most of the funds administered by the Department of Transportation go for highway construction and maintenance, some allocations support improvements of railroad, bus, ferry, air, and port services.

Health and welfare services are provided by the Department of Public Health, the Department of Mental Health, the Department of Social Services, and the Department of Civil Rights, as well as through programs administered by the Department of Labor, the Commission on Services to the Aging, the Michigan Women's Commission, the Indian Affairs Commission, the Spanish-Speaking Affairs Commission, and the Veterans Trust Fund. The state's Army and Air National Guard units are maintained by the Department of Military Affairs. Civil defense is part of the Department of State Police, and state prisons and other correctional facilities are maintained by the Department of Corrections.

Housing services are provided by the State Housing Development Authority. The Department of Labor establishes and enforces rules and standards relating to safety, wages, licenses, fees, and conditions of employment. The Michigan Employment Security Commission administers unemployment benefits and assists job seekers.

¹⁶JUDICIAL SYSTEM

Michigan's highest court is the state supreme court, consisting of 7 justices elected for eight-year terms; the chief justice is elected by the members of the court. The high court hears cases on appeal from lower state courts and also administers the state's entire court system. The 1963 constitution provided for an 18-member court of appeals to handle most of the cases that previously had clogged the high court's calendar. Unless the supreme court agrees to review a court of appeals ruling, the latter's decision is final. Six appeals court justices are elected from each of three districts for six-year terms. The justices elect a chief justice of the appeals court.

The major trial courts in the state as of 1983 were the 55 circuit courts, encompassing 165 judicial seats, with the justices elected for six-year terms. The circuit courts have original jurisdiction in all felony criminal cases, civil cases involving sums of more than $10,000, and divorces. They also hear appeals from lower courts and state administrative agencies. Probate courts have original jurisdiction in cases involving juveniles and dependents, and also handle wills and estates, adoptions, and commitments of the mentally ill. In 1983 there were 105 probate judges elected for six-year terms and serving in 79 courts.

The 1963 constitution provided for the abolition of justice-of-the-peace courts and nearly all municipal courts. To replace them,

Michigan Presidential Vote by Political Parties, 1948–84

YEAR	ELECTORAL VOTE	MICHIGAN WINNER	DEMOCRAT	REPUBLICAN	PROGRESSIVE	SOCIALIST	PROHIBITION
1948	19	Dewey (R)	1,003,448	1,038,595	46,515	6,063	13,052
						SOC. WORKERS	
1952	20	*Eisenhower (R)	1,230,657	1,551,529	3,922	655	10,331
1956	20	*Eisenhower (R)	1,359,898	1,713,647	—	—	6,923
					SOC. LABOR		
1960	20	*Kennedy (D)	1,687,269	1,620,428	1,718	4,347	2,029
1964	21	*Johnson (D)	2,136,615	1,060,152	1,704	3,817	
							AMERICAN IND.
1968	21	Humphrey (D)	1,593,082	1,370,665	1,762	4,099	331,968
							AMERICAN
1972	21	*Nixon (R)	1,459,435	1,961,721	2,437	1,603	63,321
					PEOPLE'S		LIBERTARIAN
1976	21	Ford (R)	1,696,714	1,893,742	3,504	1,804	5,406
					CITIZENS	COMMUNIST	
1980	21	*Reagan (R)	1,661,532	1,915,225	11,930	3,262	41,597
1984	20	*Reagan (R)	1,529,638	2,251,571	1,191	—	10,055

*Won US presidential election.

96 district courts, some consisting of two or more divisions, have been established. These courts handle civil cases involving sums of less than $10,000, minor criminal violations, and preliminary examinations in all felony cases. As of 1983, only 5 municipal courts remained. Legislation enacted in 1980 had already abolished or merged certain courts unique to the city of Detroit.

The Department of Corrections in 1984 administered 15 prisons and other correctional institutions, as well as state mental institutions and correction camps; the total inmate population was 14,564. The largest prison, the State Prison of Southern Michigan, at Jackson, had 4,770 inmates; it is said to be the largest walled prison in the world.

Detroit became notorious during the 1970s as the "murder capital of the world," with more murders and other violent crimes than any other US city. Violent crimes (murder, nonnegligent manslaughter, forcible rape, and aggravated assault) reached a peak of 2,226 offenses per 100,000 population in 1976; the city's rate subsequently declined somewhat, but by 1983, it was up to 2,169, the highest rate among major US cities. Michigan had an overall 1983 crime rate of 6,478 per 100,000 population, 5th-highest among the 50 states.

In 1846, Michigan became the first state to abolish capital punishment. Despite an increase in the number of murders and other violent crimes, recent efforts to restore capital punishment have failed.

[17] ARMED FORCES

In 1983/84, Department of Defense personnel in Michigan numbered 21,343, of whom more than half were civilians. Many of the military personnel are stationed at the K.I. Sawyer Air Force Base near Marquette; the Detroit Arsenal at Warren is the state's largest center for civilians. In 1983/84, Michigan firms received nearly $2.5 billion in defense contracts (16th in the US).

As of 30 September 1983 there were an estimated 1,112,000 veterans of US military service living in Michigan. Of these, 10,000 saw service during World War I, 413,000 in World War II, 191,000 in the Korean conflict, and 322,000 during the Viet-Nam era. Veterans' benefits exceeded $715 million in 1982/83.

Approximately 14,200 personnel were allocated to Michigan's Army and Air National Guard units in early 1985. In recent decades, the units have frequently been called out for riot duty and to aid in the aftermath of natural diasters such as tornadoes and heavy snowstorms. State and local police forces employed 20,694 persons in 1983. In the preceding year, police expenditures were estimated at $1.15 billion, ranking Michigan 6th among the 50 states.

[18] MIGRATION

The earliest European immigrants were the French and English. The successive opening of interior lands for farming, lumbering, mining, and manufacturing proved an irresistible attraction for hundreds of thousands of immigrants after the War of 1812, principally Germans, Canadians, English, Irish, and Dutch. During the second half of the 19th century, lumbering and mining opportunities in northern Michigan attracted large numbers of Cornishmen, Norwegians, Swedes, and Finns. The growth of manufacturing in southern Michigan at the end of the century brought many Poles, Italians, Russians, Belgians, and Greeks to the state. After World War II, many more Europeans immigrated to Michigan, plus smaller groups of Mexicans, other Spanish-speaking peoples from Latin America, and Arabic-speaking peoples, who by the late 1970s were more numerous in Detroit than in any other US city.

The first large domestic migration into Michigan came in the early 19th century from northeastern states, particularly New York and Pennsylvania, and from Ohio. Beginning in 1916, the demand for labor in Michigan's factories started the second major domestic migration to Michigan, this time by southern blacks, who settled mainly in Detroit, Flint, Pontiac, Grand Rapids, and Saginaw. During World War II, many southern whites migrated to the same industrial areas. Between 1940 and 1970, a net total of 518,000 migrants were drawn to Michigan. The economic problems of the auto industry in the 1970s and 1980s caused a significant reversal of this trend, with the state suffering a net loss of 496,000 by out-migration in the 1970s and 403,000 from 1980 to 1983.

Intrastate migration has been characterized since the late 19th century by a steady movement from rural to urban areas. Most parts of northern Michigan have suffered a loss of population since the early years of this century, although a back-to-the-land movement, together with the growth of rural Michigan as a retirement area, appeared to reverse this trend beginning in the 1970s. Since 1950, the central cities have experienced a steady loss of population to the suburbs, in part caused by the migration of whites from areas that were becoming increasingly black.

[19] INTERGOVERNMENTAL COOPERATION

The Commission on Intergovernmental Cooperation of the Michigan legislature represents the state in dealings with the Council of State Governments and its allied organizations. Since 1935, the state has joined at least 18 interstate compacts, dealing mainly with such subjects as gas and oil problems, law enforcement, pest control, civil defense, tax reciprocity, and water resources. In 1985, Michigan, seven other Great Lakes states, and the Canadian provinces of Quebec and Ontario signed the Great Lakes Charter, designed to protect the lakes' water resources.

The International Bridge Authority, consisting of members from Michigan and Canada, operates a toll bridge connecting Sault Ste. Marie, Mich., and Sault Ste. Marie, Ontario.

In 1983/84, federal aid to Michigan totaled almost $3.8 billion.

[20] ECONOMY

On the whole, Michigan benefited from its position as the center of the auto industry during the first half of the 20th century, when Detroit and other south Michigan cities were the fastest-growing industrial areas in the US. But the state's dependence on automobile production has caused grave and persistent economic problems since the 1950s. Michigan's unemployment rates in times of recession have far exceeded the national average, since auto sales are among the hardest hit in such periods. Even in times of general prosperity, the auto industry's emphasis on labor-saving techniques and its shifting of operations from the state have reduced the number of jobs available to Michigan workers. Although the state was relatively prosperous during the record automotive production years of the 1960s and 1970s, the high cost of gasoline and the encroachment of imports on domestic car sales had disastrous effects by 1980, when it became apparent that the state's future economic health required greater diversification of industry. After manufacturing, agriculture, still dominant in the rural areas of southern Michigan, probably remains the most important element in the state's economy, although tourism, heavily promoted in recent years, now rivals agriculture as a source of income. In northern Michigan, forestry and mining continue but generally at levels far below earlier boom periods.

In the mid-1980s, Michigan's most immediate problem was the high rate of unemployment resulting from the deep decline in domestic car sales. As a result, the state and local governments had to contend with reduced tax revenues and increased social welfare expenditures.

[21] INCOME

In 1983, Michigan had total personal income of $105 billion and a per capita personal income of $11,574, just below the US average. Labor and proprietors' income in 1982 accounted for 71% of personal income; 38% came from manufacturing; 17% from services; 12% from state and local government; 9% from retail trade; 6% from transportation and public utilities; 6% from wholesale trade; 4% from construction; 4% from finance, insurance, and real estate; and 4% from other sources.

In 1981, the median income of four-person families was $27,976. Persons below the federal poverty level in 1979 totaled 945,874, or 10% of the population. Some 48,100 Michiganians were among the nation's top wealth-holders, with assets of $500,000 or more as of 1982.

22LABOR

Michigan ranked 8th in the US in the size of its labor force in 1984, with 4,359,000 workers, of whom an average of 488,000, or 11.2%, were unemployed, the 2d-highest rate among the 50 states. Of the total work force, 45% were in the Detroit area.

A federal survey in March 1982 revealed the following pattern of nonfarm employment in the state:

	ESTABLISH-MENTS	EMPLOYEES	ANNUAL PAYROLL ('000)
Agricultural services, forestry, fishing	1,456	6,553	$ 88,940
Mining	485	12,472	276,209
Contract construction	11,944	91,508	2,179,631
Manufacturing, of which:	14,077	900,413	22,473,997
Fabricated metals	(2,260)	(99,658)	(2,145,180)
Nonelectrical machinery	(3,826)	(137,562)	(3,196,425)
Transportation equipment	(503)	(222,199)	(6,252,142)
Transportation, public utilities	4,692	131,064	3,126,063
Wholesale trade	12,556	161,624	3,460,222
Retail trade	44,972	540,951	5,030,666
Finance, insurance, real estate	12,770	155,729	2,564,347
Services	50,488	622,860	8,866,324
Other	2,251	2,067	38,193
TOTALS	155,691	2,625,241	$48,104,592

Government employees, not covered by this survey, included 514,000 state and local workers in 1983 and 54,000 federal workers in 1982.

The unemployment rate in Michigan exceeded the national average in the late 1970s and early 1980s. It reached a peak of 17.1% in December 1982, when the national rate was at a peak of 10.7%. In 1982, when the average unemployment rate in Michigan was 15.5%, unemployment was 35.6% among black males aged 20 or older and 30.5% among their female counterparts; among whites aged 20 or more, it was 13.6% for men and 13% for women. Migrant workers, chiefly Mexicans and Mexican-Americans, were once widely used in harvesting farm products such as sugar beets and fruit, but their numbers have dropped, as has the number of other foreign workers.

Certain crafts and trades were organized in Michigan in the 19th century, with one national labor union, the Brotherhood of Locomotive Engineers, having been founded at meetings in Michigan in 1863, but efforts to organize workers in the lumber and mining industries were generally unsuccessful. Michigan acquired a reputation as an open-shop state, and factory workers showed little interest in unions at a time when wages were high. But the catastrophic impact of the depression of the 1930s completely changed these attitudes. With the support of sympathetic state and federal government officials, Michigan workers were in the forefront of the greatest labor-organizing drive in American history. The successful sit-down strike by the United Automobile Workers against General Motors in 1936–37 marked the first major victory of the new Congress of Industrial Organizations. Since then, a strong labor movement has provided manufacturing workers in Michigan with some of the most favorable working conditions in the country. As of September 1984, the average hourly earnings of production workers reached $12.28. However, union membership declined from 1,388,000 in 1974 to 1,289,000 in 1980, and the share of union members among nonagricultural workers dropped to 37.4%.

Michigan had 2,179 labor unions in 1983; its most powerful and influential industrial union since the 1930s has been the United Automobile Workers (UAW), with nearly 1.1 million members in 1985; its national headquarters is in Detroit. Under its long-time president Walter Reuther and his successors, Leonard Woodcock, Douglas Fraser, and Owen Bieber, the union has been a dominant force in the state Democratic Party; since 1979, the UAW president has sat on the Chrysler board of directors. In recent years, as government employees and teachers have been organized, unions and associations representing these groups have become increasingly influential. Under the Michigan Public Employment Relations Act of 1965, public employees have the right to organize and to engage in collective bargaining, but are prohibited from striking. However, strikes of teachers, college faculty members, and government employees have been common since the 1960s, and little or no effort has been made to enforce the law. In 1981 there were 129 work stoppages by public and nonpublic employees in Michigan, involving 23,500 workers.

23AGRICULTURE

In 1983, Michigan's agricultural income was estimated at over $3 billion, placing Michigan 15th among the 50 states. About 59% came from crops and the rest from livestock and livestock products; dairy products, cattle, corn, and soybeans were the principal commodities. The state in 1983 ranked 1st in output of tart cherries and dry edible beans, 2d in blueberries, and 3d in apples and in prunes and plums.

The growing of corn and other crops indigenous to North America was introduced in Michigan by the Indians around 100 BC, and a few Frenchmen tried to develop European-style agriculture during the colonial era. But little progress was made until well into the 19th century, when farmers from New York and New England poured into the interior of southern Michigan. By mid-century, 34,000 farms had been established, and the number increased to a peak of about 207,000 in 1910. The major cash crop at first was wheat, until soil exhaustion, insect infestations, bad winters, and competition from huge wheat farms to the west forced a de-emphasis on wheat and a move toward agricultural diversity. Both the number of farms and the amount of farm acreage had declined by 1984 to 63,000 farms and 11,400,000 acres (4,600,000 hectares).

The southern half of the lower peninsula is the principal agricultural region, and the area along Lake Michigan is a leader in fruit growing. Potatoes are profitable in northern Michigan, while eastern Michigan (the "Thumb" area near Lake Huron) is a leading bean producer. The Saginaw Valley leads the state in sugar beets. The south-central and southeastern counties are major centers of soybean production. Leading field crops in 1983 included 165,600,000 bushels of corn for grain, valued at $554,760,000; 33,280,000 bushels of soybeans, worth $274,560,-000; and 35,770,000 bushels of wheat, worth $119,830,000. Output of commercial apples totaled 750,000,000 lb.

24ANIMAL HUSBANDRY

The same areas of southern Michigan that lead in crop production also lead in livestock and livestock products, except that the northern counties are more favorable for dairying than for crop production.

At the beginning of 1984 there were 1,475,000 cattle and calves in the state, 1,250,000 hogs and pigs, 92,000 stock sheep, and 400,000 milk cows and heifers that had calved. During 1983, livestock production included 449,870,000 lb of cattle and calves, valued at $213,460,000, and 369,108,000 lb of hogs and pigs, worth $173,136,000.

In 1983, Michigan ranked 6th in milk production with 5,528,-000 lb, 6th in butter with 58,505,000 lb, and 9th in ice cream with 34,275,000 gallons. Poultry farmers produced nearly 1.5 billion eggs during the same year. In 1982, 4,384,671 lb of honey and 156,952 mink pelts were also marketed.

25FISHING

Commercial fishing, once an important factor in the state's economy, is relatively minor today; the commercial catch in 1984 was 24,982,000 lb, valued at $7,953,000.

Sport fishing continues to flourish and is one of the state's major tourist attractions. A state salmon-planting program, begun in the mid-1960s, has made salmon the most popular game fish for Great Lakes sport fishermen. The state has also sought, through breeding and stocking programs, to bring back the trout, which was devastated by an invasion of lamprey. In 1980, 22.1 million trout, salmon, walleye, and sauger were stocked; in 1983, 3.2 million fish, almost all of them lake trout fingerlings, were stocked by the US Fish and Wildlife Service.

A bitter dispute raged during the 1970s between state officials and Ottawa and Ojibwa commercial fishermen, who claimed that Indian treaties with the federal government exempted them from state fishing regulations. The state contended that without such regulations, Indian commercial fishing would have a devastating impact on the northern Great Lakes' fish population. A federal court in 1979 upheld the Indians' contention; but in 1985, the state secured federal court approval of a compromise settlement intended to satisfy both Indian and non-Indian groups.

26FORESTRY

In 1979, Michigan's forestland totaled 19,270,000 acres (7,798,000 hectares), or more than half the state's total land area. More than 97% of it is classed as commercial timberland, about two-thirds of it privately owned. The major wooded regions are in the northern two-thirds of the state, where great pine forests enabled Michigan to become the leading lumber-producing state in the last four decades of the 19th century. These cutover lands were reforested in the 20th century.

Michigan's lumber production totaled 266 million board feet in 1983, which put the state well down on the list of lumber-producing states. In the sparsely populated western part of the upper peninsula, the lumber industry continues to be the major economic activity, however. Shipments of lumber and wood products were valued at $679 million in 1982.

State and national forests cover 6,419,000 acres (2,598,000 hectares), or more than one-sixth of the state's land area; national forests encompass about 40% of this total. Reforestation programs have been devoted to preventing or curtailing forest fires, which devastated wide areas of forestland in earlier days.

27MINING

In 1984, Michigan's estimated nonfuel mineral output totaled $1.3 billion. The state is a leading producer of iron ore, cement, copper, salt, crushed stone, gypsum, limestone, and sand and gravel.

The major mining area is the western part of the upper peninsula, where copper is mined in the Keweenaw peninsula. Iron ore deposits are found in three areas: west of Marquette, the Iron Mountain–Iron River area, and the extreme western area around Ironwood. A considerable amount of silver has been found associated with the copper deposits, and small amounts of gold have been mined in Marquette County. Limestone deposits are located chiefly in the eastern part of the upper peninsula and the northern part of the lower peninsula. Gypsum is found on the western side of Saginaw Bay and in the Grand Rapids area, while salt deposits appear both as brine and as rock throughout southern Michigan.

The leading minerals with estimated production in 1984 (excluding fossil fuels) included iron ore, 10,713,000 long tons; sand and gravel, 26,545,000 tons; salt, 1,355,000 tons; gypsum, 1,097,-000 tons; lime, 503,000 tons; and crushed stone, 24,763,000 tons. Copper production in 1982 was 32,000 tons.

28ENERGY AND POWER

Michigan's energy supply is provided primarily by private utility companies. In 1982, energy consumption per capita totaled 267 million Btu, which ranked Michigan 36th among the 50 states. Coal is the principal source of fuel used in generating electric power, while natural gas is the major fuel used for other energy needs.

The installed electric generating capacity of electric utilities and industrial plants at the end of 1983 was 22.6 million kw; electric energy production totaled 70.9 billion kwh. Hydroelectric plants, which had produced more than 10% of the state's electric energy in 1947, yielded less than 2% in 1983; coal-fired steam units produced 73%, nuclear-powered units 23%, and other units about 2%.

The two major electric utilities are Detroit Edison, serving the Detroit area and portions of the eastern part of the lower peninsula, and Consumers Power, serving most of the remainder of the lower peninsula. The two companies had combined sales in 1983 of 57.8 billion kwh, out of total electric utility sales of 69.3 billion kwh. Of these sales, 32% went to residential users, 65% to commercial and industrial users, and the remainder for street and highway lighting and other public uses. Rates of the utility companies are set by the Public Service Commission.

Michigan is dependent on outside sources for most of its fuel needs. Petroleum production in 1983 totaled 31.7 million barrels, 1% of total US production; natural gas output was 144.6 billion cu feet, less than one-fourth the natural gas consumed in the state. Proved petroleum reserves were 209 million barrels at the end of 1983, natural gas reserves 1.2 trillion cu feet. Bituminous coal reserves (estimated at 128 million tons in 1983) remain in southern Michigan, but production is negligible.

29INDUSTRY

Manufacturing, a minor element in Michigan's economy in the mid-19th century, grew rapidly in importance until, by 1900, an estimated 25% of the state's jobholders were factory workers. The rise of the auto industry in the early 20th century completed the transformation of Michigan into one of the most important manufacturing areas in the world. In 1981, Michigan ranked 7th among the 50 states in the value of manufacturing shipments. In 1982, value of shipments totaled $99.1 billion, new capital expenditures $3.5 billion. The following table shows the value of shipments for major sectors in 1982:

Transportation equipment	$43,771,800,000
Nonelectrical machinery	10,485,300,000
Fabricated metal products	9,074,300,000
Food and food products	7,909,900,000
Primary metals	5,548,000,000
Chemicals and chemical products	4,800,000,000
Paper and paper products	2,567,400,000
Electric and electronic equipment	2,294,700,000
Printing and publishing	2,227,700,000
Rubber and plastic products	2,163,300,000
Furniture and fixtures	1,871,100,000
Petroleum and coal products	1,723,600,000
Apparel and textile products	1,521,800,000

Motor vehicles and equipment dominate the state's economy, with a payroll of $6.2 billion in 1982, representing more than one-fourth of the state's manufacturing payroll; the value of shipments by automotive manufacturers was $42.2 billion, or 43% of the total. Production of nonelectrical machinery, primary and fabricated metal products, and metal forgings and stampings was directly related to automobile production.

The Detroit metropolitan area is the major industrial region, with a manufacturing payroll of $11.4 billion in 1982 and manufacturing shipments valued at $44.1 billion. This area includes not only the heavy concentration of auto-related plants in Wayne, Oakland, and Macomb counties, but also major steel, chemical, and pharmaceutical industries, among others. Flint, Grand Rapids, Saginaw, Ann Arbor, Lansing, and Kalamazoo are other major industrial centers. Among counties, manufacturing was of the greatest importance in the economy of Genesee County, home of the General Motors plants in the Flint area, with 55% of its earnings coming from manufacturing in 1982. In second place was Midland County (Dow Chemical), where 52% of all labor and proprietors' earnings came from manufacturing.

Because the auto industry's "Big Three"—General Motors (GM), Ford, and Chrysler—have their headquarters in the Detroit area, Michigan has had for many years three of the nation's largest industrial corporations. Until it was replaced in 1979 by Exxon, General Motors was the perennial leader among all manufacturers in the world: GM's assets in 1984 totaled $52.1 billion, and its net income was $4.5 billion. Ford (with $27.5 billion in assets) and Chrysler ($9.1 billion) ranked 4th and 14th, respectively. Dow Chemical, headquartered in Midland, was the 25th-largest corporation in 1984, and American Motors, with its offices in Southfield (although its factories are located outside Michigan), was 91st.

The auto industry's preponderance in Michigan manufacturing has come to be viewed in recent years as more of a liability than an asset. When times are good, as they were in the 1960s and early 1970s, automobile sales soar to record levels and Michigan's economy prospers. But when the national economy slumps, these sales plummet, pushing the state into a far deeper recession than is felt by the nation as a whole. In the 1970s, the escalating cost of gasoline and the slowness of Michigan automakers in providing small, fuel-efficient cars to meet foreign competition caused a severe decline in domestic motor vehicle sales. Michigan's automakers produced 1,894,004 1983-model-year cars (33% of the US total) and 663,459 trucks (27%).

The recession of 1979–80 forced Chrysler to obtain federally guaranteed loans of $1.2 billion and to borrow $150 million from the state in order to stave off bankruptcy. Chrysler reported losses totaling $1.1 billion in 1979—the largest operating loss that had ever been suffered by a US corporation. The other two members of the Big Three were also hard hit; in the 3d quarter of 1980, Ford and General Motors reported record losses of $595 million and $567 million, respectively, with Chrysler losing another $490 million. During part of 1981 and in 1982, Chrysler and the two other top automakers returned to profitability, aided by cost-cutting measures and price increases, and in 1984, they made record profits, despite the fact that record numbers of imported cars were sold in the US.

30 COMMERCE

In 1982, Michigan's wholesale trade establishments had sales exceeding $59.8 billion. Leading categories were motor vehicles and automobile parts and supplies (accounting for nearly one-fifth of all sales by value), groceries, metals and minerals, and machinery.

In 1982, Michigan's retail establishments had at least 540,000 employees and total sales of $39.2 billion. The importance of retail sales to the economy was greatest in northern Michigan, and in the counties of Roscommon and Claire retail earnings accounted for 22% of all labor and proprietors' income.

With its ports open to oceangoing vessels through the St. Lawrence Seaway, Michigan is a major exporting and importing state for foreign as well as domestic markets. In 1981, Michigan was the leading exporter of iron ore and ranked 5th among the states in the export of manufactured goods; it was the 2d-largest exporter of transportation equipment. Exports of Michigan's manufactured goods totaled $10.3 billion in 1980/81; exports of transportation equipment amounted to $5.6 billion. In 1981/82, the state's agricultural exports had a total value of $753 million.

31 CONSUMER PROTECTION

The Michigan Consumer's Council—composed of the attorney general, secretary of state, director of the Department of Commerce, and three members appointed by the governor and three

by the legislature—was established in 1966 to protect consumers from harmful products, false advertising, and deceptive sales practices. Other state agencies, such as the Department of Licensing and Regulation and the Public Service Commission, also are responsible for protecting consumers.

A number of local governments have instituted consumer affairs offices, with Detroit's being especially active.

32 BANKING

Michigan's banks in the territorial and early statehood years were generally wildcat speculative ventures. More restrained banking activities date from the 1840s, when the state's oldest bank, the Detroit Bank and Trust, was founded. A crisis that developed in the early 1930s forced Governor William Comstock to close all banks in February 1933 in order to prevent collapse of the entire banking system. Federal and state authorities supervised a reorganization and reform of the state's banks that has succeeded in preventing any major problems from arising since that time.

In March 1982, Michigan banks and credit agencies employed 70,600 persons and had total payrolls of $1.1 billion. There were 365 commercial banks and nondeposit trust companies in 1984; as of 1983, such banks had assets of $61.8 billion, loans exceeding $31.5 billion, and time and savings deposits of $40.4 billion. The National Bank of Detroit—with assets of $14.2 billion, deposits exceeding $10 billion, and loans of nearly $7 billion—was the state's largest bank in 1984.

There were 50 savings and loan associations with total assets of $27.5 billion in 1983. The Empire of America Savings and Loan in Southfield and First Federal Savings and Loan of Detroit were the largest institutions as of 1982. As of the same year there were 530 state-chartered credit unions, with assets of $3.5 billion, and 284 federally chartered credit unions, with assets of $1.7 billion.

33 INSURANCE

In 1983, 27 life insurance companies were based in Michigan. Life insurance benefit payments totaled $653.9 million. There were 15,429,000 life insurance policies in force, valued at $198.7 billion; the average family had $57,700 worth of life insurance coverage. Property and liability companies wrote premiums of $885.1 million in automobile physical damage coverage, and $499.9 million in homeowners' insurance.

34 SECURITIES

There are no securities or commodity exchanges in Michigan.

In 1983, New York Stock Exchange member firms had 184 sales offices and 1,954 full-time registered representatives in the state. Some 1,672,000 Michiganians were shareowners of public corporations that year.

35 PUBLIC FINANCE

The state constitution requires the governor to submit a budget proposal to the legislature each year. This executive budget, prepared by the Department of Management and Budget, is reviewed, revised, and passed by the legislature. During the fiscal year, which extends from 1 October to 30 September, if actual revenues drop below anticipated levels, the governor, in consultation with the legislative appropriations committees, must reduce expenditures to meet the constitutional requirement that the state budget be kept in balance.

In 1977, the legislature created a budget stabilization fund; a portion of tax revenues collected in good times is held in reserve to be used during periods of recession, when the funding of essential state services is threatened. In 1978, a tax limitation amendment put a lid on government spending by establishing a fixed ratio of state revenues to personal income in the state. Further efforts to limit taxes were rejected by the voters in 1980 and 1984.

State expenditures have expanded from $1.1 billion in 1958/59 to $2.7 billion in 1968/69, $8.9 billion in 1978/79, and $13.5 billion in 1985/86. The following is a summary of recommended revenues and expenditures for 1985/86 (in millions):

REVENUES

Income tax	$ 3,233.8
Sales and use tax	2,598.9
Other taxes	1,818.8
Business tax	1,447.3
Lottery	372.0
Federal aid	3,525.0
Other receipts	781.6
TOTAL	$13,777.4

EXPENDITURES

Department of Social Services	$ 4,262.3
Grant to School Aid Fund	2,387.0
Higher educational institutions	1,125.8
Department of Mental Health	910.0
Department of Corrections	321.6
Other current operations	4,156.4
Capital outlay	157.5
Debt service	139.7
TOTAL	$13,460.3

Detroit's general revenues for the fiscal year ending 30 June 1982 exceeded $1.4 billion, including $406 million received from the federal government and $457 million from city taxes. In 1982, Detroit had the 3d-highest property tax rate among major cities, amounting to $3.94 per $100 of assessed valuation. The following is a summary of estimated general revenues and expenditures for 1981/82:

REVENUES

Local property taxes	$ 182.3
Sales and other local taxes	274.9
Federal funds	406.1
State funds	323.3
Other receipts	239.4
TOTAL	$1,426.0

EXPENDITURES

Police and fire protection	$ 240.4
Housing and community development	154.8
Health and hospitals	65.7
Highways	49.4
Education	11.2
Other outlays	622.5
TOTAL	$1,144.0

The total state debt in mid-1982 was nearly $3.9 billion, 12th-highest among the 50 states, or $417 per capita (35th in the US).

[36] TAXATION

Until the 1930s, Michigan relied mainly on the property tax for revenues to support both local and state governments. A state sales tax, first imposed in 1933, and a state income tax, first levied in 1967, are now the main sources of state revenues. Property taxes are reserved entirely to local governments.

The state income tax in 1985 was 5.35% on all income, with exemptions of $1,500 for a single person, $3,000 for married couples, and $1,500 for each dependent. The state sales tax was 4% on most retail purchases, except food. An inheritance tax ranging from 2% to 17% was levied on inheritances of more than $100, with the first $65,000 to the spouse and the first $10,000 to other close relatives being exempt. Other state taxes and fees are levied on corporate and financial-institution income, cigarettes, alcoholic beverages, pari-mutuel wagering, and gasoline and other fuels.

Local government taxes in 1982/83 totaled $13.5 billion, of which $4.7 billion came from property taxes, $3.5 billion from state income taxes, $3.1 billion from local income taxes and other taxes, and $2.2 billion from state sales and use taxes.

Michigan's share of the federal tax burden in 1982/83 was $21.8 billion, the 9th largest among the US states; the per capita share was $2,409, or 25th highest. Michiganians filed nearly 3.6 million federal income tax returns for 1983, paying $10 billion in tax.

[37] ECONOMIC POLICY

Michigan has sought since territorial days to promote and assist economic development. Since the 1940s, improved coordination of such programs has been achieved through the agency now known as the Office of Economic Development, within the Department of Commerce. Local governmental units develop industrial parks and, together with the state, are authorized to provide a variety of tax incentives to encourage new companies to locate in the area or established companies to expand their operations.

Those in charge of industrial promotion must contend with Michigan's record as a state with one of the highest manufacturing pay scales in the country, entrenched and powerful labor organizations, and what some observers believe is a tax structure unfavorable to business. As a result, for each new business persuaded to locate in the state or old one persuaded to remain, another of equal magnitude often has chosen to locate elsewhere.

[38] HEALTH

Live births in 1982 totaled 137,950, for a rate of 15 per 1,000 people, a decrease from 15.7 per 1,000 in 1980. Infant mortality in 1981 was 10.9 per 1,000 live births for whites and 24.9 per 1,000 for nonwhites. There were 64,200 legal abortions in 1982, up from 37,600 in 1974. The 1982 figure represented 29 abortions per 1,000 women.

Major causes of death in 1981 (with their rates per 100,000 population) included heart disease, 328; cancer, 175; cerebrovascular diseases, 68; accidents, 37; chronic pulmonary diseases, 24; pneumonia and influenza, 19; diabetes, 16; cirrhosis of the liver, 14; atherosclerosis, 14; suicide, 12; and early infancy diseases, 11. The overall death rate of 8.2 per 1,000 population was 5% below the national average.

In 1983, Michigan had 231 hospitals, with 47,812 beds and a total of 1,449,738 admissions. Hospitals employed 21,398 full-time registered nurses in 1981. The average cost of hospital care was $357 per day and $2,853 per stay. Michigan had 16,440 physicians and 5,129 dentists in 1982.

[39] SOCIAL WELFARE

Until the 1930s, Michigan's few limited welfare programs were handled by the counties, but the relief load during the depression shifted the burden to the state and federal levels. In recent decades there have been enormous increases in social welfare programs. In 1983/84, the state Department of Social Welfare appropriated over $4.1 billion, of which the federal government supplied 47%. In 1983, Michigan ranked 2d among the 50 states in the percentage of the population receiving public aid.

In 1982, recipients of aid to families with dependent children numbered 729,500; payments for that year totaled nearly $1.1 billion, 3d highest among the 50 states. The average monthly payment per family was $383. An average of 1,096,000 persons per month participated in the federal food stamp program in 1983, at a federal cost of $570 million. Some 731,000 pupils participated in the school lunch program, at a federal cost of $74 million.

In December 1982, 1,365,173 persons were receiving Social Security benefits, of whom 770,638 were retired workers, 299,240 were surviving dependents of deceased workers, and 103,027 were disabled workers. Total benefits paid out in 1983 were $6.9 billion; the average benefit to retired workers in December 1982 was $451.34.

Under the Supplemental Security Income program, 108,301 aged, blind, and disabled persons received $23.3 million in May 1983. Workers' compensation programs totaled $672.9 million in 1982. Unemployment insurance payments in the same year amounted to $2 billion, and the average duration of benefits was 17.3 weeks.

⁴⁰HOUSING

The 1980 census counted 3,589,912 housing units in Michigan, of which 3,457,022 were year-round units. Of these, 72.7% were owner-occupied, the 2d-highest percentage in the US. The number of year-round dwellings increased 14% from 1975 to 1980.

As of 1980, 42% of all year-round units had been built in 1949 or earlier; 19% dated from the 1950s, and 18% from the 1960s. Some 97% had at least one full bathroom.

In 1983, the number of new privately owned housing units authorized was valued at $1 billion. The median value of owner-occupied units in 1980 was $39,000, or $8,200 below the US average. A limited amount of state aid for low-income housing is available through the State Housing Development Authority.

⁴¹EDUCATION

Historically, Michigan has strongly supported public education, which helps account for the fact that the percentage of students attending public schools is one of the highest in the US. But the cost of maintaining this extensive public educational system has become a major problem in recent years because of the declining school-age population.

In 1980, 68% of persons 25 years and over had completed four years of high school. Of the 5,254,000 residents in this age range, 792,000 had completed eight years or less of schooling; 890,000 had finished up to three years of high school; 1,998,000 had finished high school; 825,000 had completed three years of college; and 749,000 had finished four years or more of higher education.

In 1980/81 there were 3,688 public schools, including 2,748 elementary schools, 895 secondary schools, and 45 combined elementary and secondary schools. Public school enrollment totaled 1,620,954 in 1982/83, a 5.8% decline in two years. In 1980/81 there were 211,871 pupils in nonpublic schools. The largest number of these were enrolled in Catholic schools, which had 128,520 students in 1983/84. Lutherans, Seventh-Day Adventists, and Reformed and Christian Reformed churches also have maintained schools for some time; in the 1970s, a number of new Christian schools, particularly those of fundamentalist Baptist groups, were established.

In 1983/84, Michigan had 42 public institutions of higher learning, with a combined enrollment of 444,229, and 53 private institutions with a total of 73,809 students. The oldest and most prestigious state school is the University of Michigan, founded at its Ann Arbor campus in 1837. In 1983, it had a student enrollment of 46,699, including branches at Flint and Dearborn. Founded in 1855 as an agricultural school but later expanded into a university, Michigan State University, at East Lansing, had an enrollment of 41,765 in 1983. Wayne State University, in Detroit, had 29,639 students. There are 10 other state colleges and universities and 29 two-year community colleges. Among the state's private colleges and universities, the University of Detroit, a Jesuit school, is one of the largest; it had an enrollment of 5,512 in 1983. Kalamazoo College (founded in 1833), Albion College (1835), Hope College (1866), and Alma College (1886) are among the better known of the private liberal arts colleges. Scholarships, merit awards, tuition grants, and guaranteed student loans are provided through the Student Financial Assistance Services office of the Department of Education.

In 1983/84, total expenditures for public schools amounted to $3,208 per pupil in average daily attendance, 21st highest among the 50 states. Teachers' salaries averaged $26,740 per year, 4th highest of the states.

⁴²ARTS

Michigan's major center of arts and cultural activities is the Detroit area. The city's Ford Auditorium is the home of the Detroit Symphony Orchestra; the Music Hall and the Masonic Auditorium present a variety of musical productions; the Fisher Theater is the major home for Broadway productions; and the Detroit Cultural Center supports a number of cultural programs. Nearby Meadow Brook, in Rochester, has a summer music program. At the University of Michigan, in Ann Arbor, the Power Center for the Performing Arts and Hill Auditorium host major musical, theatrical, and dance presentations.

Programs relating to the visual arts tend to be academically centered; the University of Michigan, Michigan State, Wayne State, and Eastern Michigan University have notable art schools. The Cranbrook Academy of Arts, which was created by the architect Eliel Saarinen, is a significant art center, and the Ox-bow School at Saugatuck is also outstanding. The Ann Arbor Art Fair, begun in 1959, is the largest and most prestigious summer outdoor art show in the state.

The Meadow Brook Theater at Rochester is perhaps the largest professional theater company; Detroit has a number of little theater groups. Successful summer theaters include the Cherry County Playhouse at Traverse City and the Star Theater in Flint.

The Detroit Symphony Orchestra, founded in 1914, is nationally known. Grand Rapids and Kalamazoo have regional orchestras that perform on a part-time, seasonal basis. The National Music Camp at Interlochen is a mecca for young musicians in the summer. There are local ballet and opera groups in Detroit and in a few other communities. Michigan's best-known contribution to popular music was that of Berry Gordy, Jr., whose Motown recording company in the 1960s popularized the "Detroit sound" and featured such artists as Diana Ross and the Supremes, Smokey Robinson and the Miracles, the Four Tops, the Temptations, and Stevie Wonder, among many others. In the 1970s, however, Gordy moved his operations to California.

⁴³LIBRARIES AND MUSEUMS

Michigan in 1985 had 371 public libraries, 102 academic libraries, and numerous special libraries. In 1982/83, public libraries in the state had a total of nearly 19 million volumes and a circulation exceeding 36 million. The State Library in Lansing functions as the coordinator of library facilities in the state. The largest public library is the Detroit Public Library, which in 1983 had 2,421,578 bound volumes in its main library and 24 branches. Outstanding among its special collections are the Burton Historical Collection, a major center for genealogical research, and the National Automotive History Collection. Grand Rapids, Kalamazoo, Lansing, Flint, and Ann Arbor are among the larger public libraries.

Among academic libraries, the University of Michigan at Ann Arbor, with 4,684,590 book titles and 2,010,555 microfilm units in 1983, features the William L. Clements collection of books and manuscripts on the colonial period, and the Bentley Library's collection of books and manuscripts on Michigan, the largest such collection. In 1980, the Gerald R. Ford Presidential Library was opened on the university campus. The Michigan State University Library at East Lansing had 2,898,698 books and 1,871,763 microfilm units in 1983. At Wayne State University in Detroit, the Walter P. Reuther Library houses the largest collection of labor history records in the US.

The Detroit Institute of Arts is the largest art museum in the state and has an outstanding collection of African art. It is located in the Detroit Cultural Center, along with the Public Library and the Detroit Historical Museum, one of the largest local history museums in the country. The Kalamazoo Institute of Art, the Flint Institute of Art, the Grand Rapids Art Museum, and the Hackley Art Gallery in Muskegon are important art museums. The University of Michigan and the Cranbrook Academy of Arts in Bloomfield Hills also maintain important collections.

The Detroit Historical Museum heads the more than 140 historical museums in the state, including the State Historical Museum in Lansing and museums in Grand Rapids, Flint,

Kalamazoo, and Dearborn. In the latter city, the privately run Henry Ford Museum and Greenfield Village are leading tourist attractions.

The major historical sites open to the public include the late-18th-century fort on Mackinac Island and the reconstructed early-18th-century fort at Mackinaw City. The latter site has also been the scene of an archaeological program that has accumulated one of the largest collections of 18th-century artifacts in the country. Major investigations of prehistoric Indian sites have also been conducted in recent years.

44 COMMUNICATIONS

Michigan's remote position in the interior of the continent hampered the development of adequate communications services, and the first regular postal service was not instituted until the early 19th century. In 1985, the state had 851 post offices and 22,574 postal employees.

Telephone service began in Detroit in 1877. By 1980, 96% of the 3,195,213 occupied housing units in the state had telephones. As of December 1981, there were 7,253,301 business and residential telephones.

Michigan had 135 AM radio stations and 179 FM stations in 1984. Radio station WWJ, owned by the *Detroit News,* began operating in 1920 as one of the country's first commercial broadcasting stations, and the *News* also started Michigan's first television station in 1947. As of 1984 there were 27 commercial television stations and 8 educational stations in the state. There were 156 cable television systems, serving 983,866 subscribers in 644 communities.

45 PRESS

Continuous newspaper coverage in Michigan dates from the appearance of the weekly *Detroit Gazette* in 1817. The state's oldest paper still being published is the *Detroit Free Press,* founded in 1831 and the state's first daily paper since 1835.

In 1984 there were 52 daily newspapers in Michigan, with a total average daily circulation of 2,524,230. In addition, 15 Sunday editions had a total circulation of 2,533,036. There were also 250 weekly or other nondaily newspapers. The number of daily papers has declined in recent decades; since 1959, Detroit has been the only Michigan city with more than one daily, and the *Detroit Free Press* was for some years the state's only major morning paper. The *Detroit News,* founded in 1873 by James E. Scripps, had the 6th-largest daily circulation of any paper in the US in 1984, and the *Detroit Free Press* ranked 7th.

The following table shows leading daily newspapers in Michigan with average daily and Sunday circulation in 1984:

AREA	NAME	DAILY	SUNDAY
Detroit	Free Press (m,S)	635,114	788,203
	News (e,S)	650,683	858,870
Flint	Journal (e,S)	108,493	112,496
Grand Rapids	Press (e,S)	132,662	167,464
Kalamazoo	Gazette (e,S)	63,872	72,729
Lansing	State Journal (e,S)	69,087	80,234
Pontiac	Oakland Press (e,S)	73,339	80,605
Saginaw	News (e,S)	56,216	59,596

46 ORGANIZATIONS

The 1982 US Census of Service Industries counted 1,984 organizations in Michigan, including 346 business associations; 1,209 civic, social, and fraternal associations; and 50 educational, scientific, and research associations. Few national organizations maintain their headquarters in Michigan, but the first chapters of the Kiwanis and Exchange service clubs were organized in the state.

The most important trade association headquartered in Michigan is the Motor Vehicle Manufacturers Association, with offices in Detroit. Its labor union counterpart, the United Automobile Workers, also has its international headquarters in that city.

Other organizations with headquarters in the state include the American Concrete Institute, Detroit; Society of Manufacturing Engineers, Dearborn; American Society of Agricultural Engineers, St. Joseph; and the National Association of Investment Clubs, Royal Oak.

47 TOURISM, TRAVEL, AND RECREATION

Tourism has been an important source of economic activity in Michigan since the 19th century and now rivals agriculture as the second most important segment of the state's economy.

In 1982, out-of-state visitors were estimated to have spent over $5.1 billion in Michigan. An estimated 120,400 jobs were generated involving recreation-related goods and services.

Michigan's tourist attractions are diverse and readily accessible to much of the country's population. The opportunities offered by Michigan's water resources are the number one attraction; no part of the state is more than 85 mi (137 km) from one of the Great Lakes, and most of the population lives only a few miles away from one of the thousands of inland lakes and streams. Southwestern Michigan's sandy beaches along Lake Michigan offer sunbathing and swimming. Inland lakes in southern Michigan are favored by swimmers, while the Metropolitan Beach on Lake St. Clair, northeast of Detroit, claims to be the largest artificial-lake beach in the world. Camping has enjoyed an enormous increase in popularity; in addition to the extensive public camping facilities, there are many private campgrounds.

Although the tourist and resort business has been primarily a summer activity, the rising popularity of ice fishing, skiing, and other winter sports, autumn scenic tours, hunting, and spring festivals has made tourism a year-round business in many parts of the state. Historic attractions have been heavily promoted in recent years, following the success of Dearborn's Henry Ford Museum and Greenfield Village, which attract about 1.5 million paying visitors each year. Tours of Detroit automobile factories and other industrial sites, such as Battle Creek's breakfast-food plants, are also important tourist attractions.

Camping and recreational facilities are provided by the federal government at three national forests, comprising 2.7 million acres (1.1 million hectares); three facilities operated by the National Park Service (Isle Royale National Park and the Pictured Rocks National Lakeshore and Sleeping Bear Dunes National Lakeshore); and several wildlife sanctuaries. In 1984, Sleeping Bear Dunes had 853,186 visitors.

State-operated facilities include 89 parks and recreational areas, with 248,000 acres (100,400 hectares), and state forests and wildlife areas totaling 4,250,000 acres (1,720,000 hectares). In 1983, total attendance at state parks was 23,029,000. Holland (1,547,513 visitors) and Warren Dunes (1,231,041) state parks, located on Lake Michigan, had the largest overall park attendances for 1982; Ludington State Park, also on Lake Michigan, attracted the largest number of campers. State forest campgrounds were used by 201,179 campers in 1983. In 1982, 203,570 hunting licenses and 1,438,078 fishing licenses were sold.

48 SPORTS

Professional team sports in Michigan are centered in Detroit, home of the American League's Tigers (baseball), the Red Wings of the National Hockey League, the Lions of the National Football League, and the Pistons of the National Basketball Association. In the late 1970s, the Detroit Lions and the Pistons moved their games to the new Silverdome Stadium in suburban Pontiac. To hold its remaining professional teams, the city of Detroit arranged to help refurbish Tiger Stadium, and the new city-owned Joe Louis Arena became the home of the Red Wings in 1979. This arena has also helped to revive professional boxing, which had enjoyed great popularity when Louis was the world heavyweight champion.

Horse racing, Michigan's oldest organized spectator sport, is controlled by the state racing commissioner, who regulates

Thoroughbred and harness-racing seasons at tracks in the Detroit area and at Jackson. Attendance and betting at these races is substantial, although the modest purses rarely attract the nation's leading horses. Auto racing is also popular in Michigan. The state hosts two major races, the Detroit Grand Prix and the Michigan 500 stock car race.

Interest in college sports centers on the football and basketball teams of the University of Michigan and Michigan State University, which usually are among the top-ranked teams in the country. Other colleges also have achieved national ranking in basketball, hockey, baseball, and track. Elaborate facilities have been built for these competitions; the University of Michigan's football stadium, seating 104,001, is the largest college-owned stadium in the country.

⁴⁹FAMOUS MICHIGANIANS

Only one Michiganian has held the offices of US president and vice president. Gerald R. Ford (Leslie King, Jr., b.Nebraska, 1913), the 38th US president, was elected to the US House as a Republican in 1948 and served continuously until 1973, becoming minority leader in 1965. Upon the resignation of Vice President Spiro T. Agnew in 1973, President Richard M. Nixon appointed Ford to the vice-presidency. When Nixon resigned on 9 August 1974, Ford became president, the first to hold that post without having been elected to high national office. Ford succeeded in restoring much of the public's confidence in the presidency, but his pardoning of Nixon for all crimes he may have committed as president helped cost Ford victory in the presidential election of 1976. Ford subsequently moved his legal residence to California.

Lewis Cass (b.New Hampshire, 1782-1866), who served as governor of Michigan Territory, senator from Michigan, secretary of war and secretary of state, is the only other Michigan resident nominated by a major party for president; he lost the 1848 race as the Democratic candidate. Thomas E. Dewey (1902-72), a native of Owosso, was the Republican presidential nominee in 1944 and 1948, but from his adopted state of New York.

Two Michiganians have served as associate justices of the Supreme Court: Henry B. Brown (b.Massachusetts, 1836-1913), author of the 1896 segregationist decision in *Plessy* v. *Ferguson;* and Frank Murphy (1890-1949), who also served as US attorney general and was a notable defender of minority rights during his years on the court. Another justice, Potter Stewart (b.1915), was born in Jackson but appointed to the court from Ohio.

Other Michiganians who have held high federal office include Robert McClelland (b.Pennsylvania, 1807-80), secretary of the interior; Russell A. Alger (b.Ohio, 1836-1907), secretary of war; Edwin Denby (b.Indiana, 1870-1929), secretary of the Navy, who was forced to resign because of the Teapot Dome scandal; Roy D. Chapin (1880-1936), secretary of commerce; Charles E. Wilson (b.Ohio, 1890-1961) and Robert S. McNamara (b.California, 1916), secretaries of defense; George Romney (b.Mexico, 1907), secretary of housing and urban development; Donald M. Dickinson (b.New York, 1846-1917) and Arthur E. Summerfield (1899-1972), postmasters general; and W. Michael Blumenthal (b.Germany, 1926), secretary of the treasury.

Zachariah Chandler (b.New Hampshire, 1813-79) served as secretary of the interior but is best remembered as a leader of the Radical Republicans in the US Senate during the Civil War era. Other prominent US senators have included James M. Couzens (b.Canada, 1872-1936), a former Ford executive who became a maverick Republican liberal during the 1920s; Arthur W. Vandenberg (1884-1951), a leading supporter of a bipartisan internationalist foreign policy after World War II; and Philip A. Hart, Jr. (b.Pennsylvania, 1912-76), one of the most influential senators of the 1960s and 1970s. Recent well-known US representatives include John Conyers, Jr. (b.1929), and Martha W. Griffiths (b.Missouri, 1912), a representative for 20 years who became the state's lieutenant governor in 1983.

In addition to Murphy and Romney, important governors have included Stevens T. Mason (b.Virginia, 1811-43), who guided Michigan to statehood; Austin Blair (b.New York, 1818-94), Civil War governor; Hazen S. Pingree (b.Maine, 1840-1901) and Chase S. Osborn (b.Indiana, 1860-1949), reform-minded governors; Alexander Groesbeck (1873-1953); G. Mennen Williams (b.1911); and William G. Milliken (b.1922), governor from 1969 to January 1983. Since taking office in 1974, Detroit's first black mayor, Coleman A. Young (b.Alabama, 1918), has promoted programs to revive the city's tarnished image.

The most famous figure in the early development of Michigan is Jacques Marquette (b.France, 1637-75). Other famous historical figures include Charles de Langlade (1729-1801), a French-Indian soldier in the French and Indian War and the American Revolution; the Ottawa chieftain Pontiac (1720?-69), leader of an ambitious Indian uprising; and Gabriel Richard (b.France, 1767-1832), an important pioneer in education and the first Catholic priest to serve in Congress. Laura Haviland (b.Canada, 1808-98) was a noted leader in the fight against slavery and for black rights, while Lucinda Hinsdale Stone (b.Vermont, 1814-1900) and Anna Howard Shaw (b.England, 1847-1919) were important in the women's rights movement.

Nobel laureates from Michigan include diplomat Ralph J. Bunche (1904-71), winner of the Nobel Peace Prize in 1950; Glenn T. Seaborg (b.1912), Nobel Prize winner in chemistry in 1951; and Thomas H. Weller (b.1915) and Alfred D. Hershey (b.1908), Nobel Prize winners in physiology or medicine in 1954 and 1969, respectively. Among leading educators, James B. Angell (b.Rhode Island, 1829-1916), president of the University of Michigan, led that school to the forefront among American universities, while John A. Hannah (b.1902), longtime president of Michigan State University, successfully strove to expand and diversify its programs. General Motors executive Charles S. Mott (b.New Jersey, 1875-1973) contributed to the growth of continuing education programs through huge grants of money.

In the business world, William C. Durant (b.Massachusetts, 1861-1947), Henry Ford (1863-1947), and Ransom E. Olds (b.Ohio, 1864-1950) are the three most important figures in making Michigan the center of the American auto industry. Ford's grandson, Henry Ford II (b.1917), was the dominant personality in the auto industry from 1945 through 1979. Two brothers, John Harvey Kellogg (1852-1943) and Will K. Kellogg (1860-1951), helped make Battle Creek the center of the breakfast-food industry. William E. Upjohn (1850-1932) and Herbert H. Dow (b.Canada, 1866-1930) founded major pharmaceutical and chemical companies that bear their names. James E. Scripps (b.England, 1835-1906), founder of the *Detroit News,* was a major innovator in the newspaper business. Pioneer aviator Charles A. Lindbergh (1902-74) was born in Detroit.

Among prominent labor leaders in Michigan were Walter Reuther (b.West Virginia, 1907-70), president of the United Automobile Workers, and his controversial contemporary, James Hoffa (b.Indiana, 1913-1975?), president of the Teamsters Union, whose disappearance and presumed murder remain a mystery.

The best-known literary figures who were either native or adopted Michiganians include Edgar Guest (b.England, 1881-1959), writer of enormously popular sentimental verses; Ring Lardner (1885-1933), master of the short story; Edna Ferber (1885-1968), best-selling novelist; Paul de Kruif (1890-1971), popular writer on scientific topics; Stewart Edward White (1873-1946), writer of adventure tales; Howard Mumford Jones (1892-1980), critic and scholar; and Bruce Catton (1899-1978), Civil War historian.

Other prominent Michiganians past and present include Frederick Stuart Church (1842-1924), painter; Liberty Hyde Bailey (1858-1954), horticulturist and botanist; Albert Kahn (b.Germany, 1869-1942), innovator in factory design; and (Gottlieb)

Eliel Saarinen (b.Finland, 1873–1950), architect and creator of the Cranbrook School of Art, and his son Eero (1910–61), designer of the General Motors Technical Center in Warren and many distinctive structures throughout the US. Malcolm X (Malcolm Little, b.Nebraska, 1925–65) developed his black separatist beliefs while living in Lansing.

Popular entertainers born in Michigan include Danny Thomas (Amos Jacobs, b.1914), David Wayne (b.1914), Betty Hutton (b.1921), Ed McMahon (b.1923), Julie Harris (b.1925), Ellen Burstyn (Edna Rae Gilhooley, b.1932), Della Reese (Dellareese Patricia Early, b.1932), William "Smokey" Robinson (b.1940), Diana Ross (b.1944), Bob Seger (b.1945), and Stevie Wonder (Stevland Morris, b.1950), along with film director Francis Ford Coppola (b.1939).

Among sports figures who had notable careers in the state were Fielding H. Yost (b.West Virginia, 1871–1946), University of Michigan football coach; Joe Louis (Joseph Louis Barrow, b.Alabama, 1914), heavyweight boxing champion from 1937 to 1949; "Sugar Ray" Robinson (b.1920), who held at various times the welterweight and middleweight boxing titles; and baseball Hall of Famer Al Kaline (b.Maryland, 1934), a Detroit Tigers star.

[50] BIBLIOGRAPHY

Bald, F. C. *Michigan in Four Centuries*. Rev. ed. New York: Harper & Row, 1961.

Catton, Bruce. *Michigan: A Bicentennial History*. New York: Norton, 1976.

Dunbar, Willis F., and George S. May. *Michigan: A History of the Wolverine State*. Rev. ed. Grand Rapids: Eerdmans, 1980.

Federal Writers' Project. *Michigan: A Guide to the Wolverine State*. Reprint. New York: Somerset, 1981 (orig. 1941).

Fuller, George N., ed. *Michigan: A Centennial History of the State*. 5 vols. Chicago: Lewis, 1939.

League of Women Voters of Michigan. *The State We're In: A Citizen's Guide to Michigan State Government*. Lansing, 1979.

May, George S. *Pictorial History of Michigan*. 2 vols. Grand Rapids: Eerdmans, 1967, 1969.

Michigan, State of. Department of Management and Budget. *Michigan Manual*, 1983–84. Lansing, 1984.

Michigan State University. Graduate School of Business Administration. Division of Research. *Michigan Statistical Abstract*. 18th ed. Edited by David I. Verway. East Lansing, 1984.

Sommers, Lawrence M., ed. *Atlas of Michigan*. East Lansing: Michigan State University Press, 1977.

MINNESOTA

State of Minnesota

ORIGIN OF STATE NAME: Derived from the Sioux Indian word *minisota*, meaning "sky-tinted waters." **NICKNAME:** The North Star State. **CAPITAL:** St. Paul. **ENTERED UNION:** 11 May 1858 (32d). **SONG:** "Hail! Minnesota." **MOTTO:** *L'Etoile du Nord* (The North Star). **FLAG:** On a blue field bordered on three sides by a gold fringe, a version of the state seal is surrounded by a wreath with the statehood year (1858), the year of the establishment of Ft. Snelling (1819), and the year the flag was adopted (1893); five clusters of gold stars and the word "Minnesota" fill the outer circle. **OFFICIAL SEAL:** A farmer, with a powder horn and musket nearby, plows a field in the foreground, while in the background, before a rising sun, an Indian on horseback crosses the plains; pine trees and a waterfall represent the state's natural resources. The state motto is above, and the whole is surrounded by the words "The Great Seal of the State of Minnesota 1858." Another version of the seal in common use shows a cowboy riding across the plains. **BIRD:** Common loon. **FISH:** Walleye. **FLOWER:** Pink and white lady's-slipper. **TREE:** Red (Norway) pine. **GEM:** Lake Superior agate. **GRAIN:** Wild rice. **MUSHROOM:** Morel or sponge mushroom. **DRINK:** Milk. **LEGAL HOLIDAYS:** New Year's Day, 1 January; Birthday of Martin Luther King, Jr., 3d Monday in January; Washington's and Lincoln's Birthdays, 3d Monday in February; Memorial Day, last Monday in May; Independence Day, 4 July; Labor Day, 1st Monday in September; Columbus Day, 2d Monday in October; Veterans Day, 11 November; Thanksgiving Day, 4th Thursday in November; Christmas Day, 25 December. By statute, schools hold special observances on Susan B. Anthony Day, 15 February; Arbor Day, last Friday in April; Minnesota Day, 11 May; Frances Willard Day, 28 September; Leif Erikson Day, 9 October. **TIME:** 6 AM CST = noon GMT.

¹LOCATION, SIZE, AND EXTENT

Situated in the western north-central US, Minnesota is the largest of the midwestern states and ranks 12th in size among the 50 states.

The total area of Minnesota is 84,402 sq mi (218,601 sq km), of which land accounts for 79,548 sq mi (206,029 sq km) and inland water 4,854 sq mi (12,572 sq km). Minnesota extends 406 mi (653 km) N–S; its extreme E–W extension is 358 mi (576 km).

Minnesota is bordered on the N by the Canadian provinces of Manitoba and Ontario (with the line passing through the Lake of the Woods, Rainy River, Rainy Lake, a succession of smaller lakes, the Pigeon River, and Lake Superior); on the E by Michigan and Wisconsin (with the line passing through Lake Superior and the St. Croix and Mississippi rivers); on the S by Iowa; and on the W by South Dakota and North Dakota (with the line passing through Big Stone Lake, Lake Traverse, the Bois de Sioux River, and the Red River of the North).

The length of Minnesota's boundaries totals 1,783 mi (2,870 km). The state's geographic center is in Crow Wing County, 10 mi (16 km) sw of Brainerd.

²TOPOGRAPHY

Minnesota, lying at the northern rim of the Central Plains region, consists mainly of flat prairie, nowhere flatter than in the Red River Valley of the west. There are rolling hills and deep river valleys in the southeast; the northeast, known as Arrowhead Country, is more rugged and includes the Vermilion Range and the Mesabi Range, with its rich iron deposits. Eagle Mountain, in the extreme northeast, rises to a height of 2,301 feet (701 meters), the highest point in the state; the surface of nearby Lake Superior, 602 feet (183 meters) above sea level, is the state's lowest elevation.

With more than 15,000 lakes and extensive wetlands, rivers, and streams, Minnesota has more inland water than any other state except Alaska. Some of the inland lakes are quite large: Lower and Upper Red Lake, 451 sq mi (1,168 sq km); Mille Lacs, 207 sq mi (536 sq km); and Leech Lake, 176 sq mi (456 sq km). The Lake of the Woods, 1,485 sq mi (3,846 sq km), is shared with Canada, as is Rainy Lake, 345 sq mi (894 sq km). A total of 2,212 sq mi (5,729 sq km) of Lake Superior lies within Minnesota's jurisdiction.

Lake Itasca, in the northwest, is the source of the Mississippi River, which drains about three-fifths of the state and, after meeting with the St. Croix below Minneapolis–St. Paul, forms part of the eastern boundary with Wisconsin. The Minnesota River, which flows across the southern part of the state, joins the Mississippi at the Twin Cities. The Red River of the North, which forms much of the boundary with North Dakota, is part of another large drainage system; it flows north, crosses the Canadian border above St. Vincent, and eventually empties into Lake Winnipeg in Canada.

Most of Minnesota, except for small areas in the southeast, was covered by ice during the glacial ages. When the ice melted, it left behind a body of water known as Lake Agassiz, which extended into what we now call the Dakotas and Canada and was larger than the combined Great Lakes are today; additional melting to the north caused the lake to drain away, leaving flat prairie in its wake. The glaciers also left behind large stretches of pulverized limestone, enriching Minnesota's soil, and the numerous shallow depressions that have developed into its modern-day lakes and streams.

³CLIMATE

Minnesota has a continental climate, with cold, often frigid winters and warm summers. The growing season is 160 days or more in the south-central and southeastern regions, but 100 days or less in the northern counties. Normal daily mean temperatures range from 6°F (−14°C) in January to 65°F (18°C) in July for Duluth, and from 11°F (−12°C) in January to 73°F (23°C) in July for Minneapolis–St. Paul, often called the Twin Cities. The lowest temperature recorded in Minnesota was −59°F (−51°C), at Pokegama Dam on 16 February 1903; the highest, 114°F (46°C), at Moorhead on 6 July 1936.

Precipitation is heaviest in the southeast, where the mean

annual precipitation is 32 in (81 cm), and lightest in the northwest, where it averages 19 in (48 cm) per year. Heavy snowfalls occur from November to April, averaging about 70 in (178 cm) annually in the northeast and 30 in (76 cm) in the southeast. Blizzards hit Minnesota twice each winter on the average. Tornadoes number an average 17 per year, and occur mostly in the south; in 1983 there were 20 tornadoes, 16 of them in June.

⁴FLORA AND FAUNA

Minnesota is divided into three main life zones: the wooded lake regions of the north and east, the prairie lands of the west and southwest, and a transition zone in between. Oak, maple, elm, birch, pine, ash, and poplar still thrive, although much of the state's woodland has been cut down since the 1850s. Common shrubs include thimbleberry, sweetfern, and several varieties of honeysuckle. Familiar among some 1,500 native flowering plants are puccoon, prairie phlox, and blazing star; pink and white lady's slipper is the state flower. White and yellow water lilies cover the pond areas, with bulrushes and cattails on the shore.

Among Minnesota's common mammals are the opossum, eastern and starnose moles, little brown bat, raccoon, mink, river otter, badger, striped and spotted skunks, red fox, bobcat, 13-lined ground squirrel (also known as the Minnesota gopher, symbol of the University of Minnesota), beaver, porcupine, eastern cottontail, moose, and white-tailed deer. The western meadowlark, Brewer's blackbird, Carolina wren, and Louisiana water thrush are among some 240 resident bird species; introduced birds include the English sparrow and ring-necked pheasant. Teeming in Minnesota's many lakes are such game fishes as walleyed pike, muskellunge, northern pike, and steelhead, rainbow, and brown trouts. The only poisonous snake is the rattler.

Classification of rare, threatened, and endangered species is delegated to the Minnesota Department of Natural Resources. Among rare species noted by the department are the white pelican, short-eared owl, rock vole, pine marten, American elk, woodland caribou, lake sturgeon, and paddlefish; threatened species include the bobwhite quail and piping plover. Endangered species are the gray (timber) wolf, trumpeter swan, American peregrine falcon, whooping crane, burrowing owl, and Higgins' eye pearly mussel.

⁵ENVIRONMENTAL PROTECTION

The state's northern forests have been greatly depleted by fires, lumbering, and farming, but efforts to replenish them began as early as 1876, with the formation of the state's first forestry association. In 1911, the legislature authorized a state nursery, established forest reserves and parks, and created the post of chief fire warden to oversee forestry resources and promote reforestation projects. The Conservation Department, created in 1931, evolved into the present Department of Natural Resources, which is responsible for the management of forests, fish and game, public lands, minerals, and state parks and waters. The department's Soil and Water Conservation Board has jurisdiction over the state's 92 soil and water conservation districts. A separate Pollution Control Agency enforces air and water quality standards and oversees solid waste disposal and pollution-related land-use planning. The Environmental Quality Board coordinates conservation efforts among various state agencies. In 1984/85, an estimated $53.1 million was appropriated from the general fund for the Department of Natural Resources, and nearly $7.8 million for the Pollution Control Agency.

Minnesotans dump 4 million tons of trash and 174,000 tons of hazardous waste each year; 34 hazardous waste sites were on the national priority list in 1984. The Reserve Mining Co. complied with a court order in 1980 by ending the dumping of taconite wastes, a possible carcinogen, into Lake Superior. Other pollution problems came to light during the 1970s with the discovery of asbestos in drinking water from Lake Superior, of contaminants from inadequately buried toxic wastes at St. Louis Park, and of

the killing by agricultural pesticides of an estimated 100,000 fish in two southeastern Minnesota brooks. During the early 1980s, the state's Pollution Control Agency approved plans by FMC, a munitions maker, to clean up a hazardous waste site at Fridley (near Minneapolis), which the Environmental Protection Agency claimed was the country's most dangerous hazardous waste area. The Minnesota Mining and Manufacturing Co. in 1983 began to remove chemical wastes from three dumps in Oakdale (a suburb of St. Paul), where the company had disposed of hazardous wastes since the late 1940s. Each cleanup project was to cost the respective companies at least $6 million.

⁶POPULATION

The 1980 census gave Minnesota a population of 4,075,970, ranking it 21st among the 50 states. The 1985 estimate was 4,199,749, representing a population increase of 3% since 1980 and yielding an average density for the state of 53 per sq mi (20 per sq km).

Minnesota was still mostly wilderness until a land boom in 1848 attracted the first substantial wave of settlers, mainly lumbermen from New England, farmers from the Middle Atlantic states, and tradespeople from eastern cities. The 1850 census recorded a population of 6,077 in what was then Minnesota Territory. With the signing of major Indian treaties and widespread use of the steamboat, large areas were opened to settlement, and the population exceeded 150,000 by the end of 1857. Attracted by fertile farmland and enticed by ambitious recruitment programs overseas, large numbers of European immigrants came to settle in the new state from the 1860s onward. In 1880, the state population totaled 780,733; by 1920 (when overseas immigration virtually ceased), the state had 2,387,125 residents. Population growth leveled off during the 1920s and has fallen below the national average since the 1940s. As of 1980, Minnesotans were, on average, somewhat older than the nation as a whole, with a median age of 29 years. Women make up about 51% of the population. Of the 1980 population, 58% were married, 29% had never married, and 13% were formerly married (7% widowed, and 6% divorced or separated).

In 1980, 2 out of 3 Minnesotans lived in metropolitan areas. The Minneapolis–St. Paul metropolitan area was the country's 17th largest in 1984, with an estimated population of 2,230,900, a 4.4% increase since 1980. In Minneapolis itself, the population fell by nearly 15% from 1970 to 1980; in 1984, the city had an estimated 358,355 residents (42d in the US), while St. Paul had a population of 265,903 (58th). The 1984 estimates for other leading cities were as follows: Duluth, 85,162; Bloomington, 84,127; and Rochester, 58,151.

⁷ETHNIC GROUPS

Minnesota was settled during the second half of the 19th century primarily by European immigrants, chiefly Germans, Swedes, Norwegians, Danes, English, and Poles, along with the Irish and some French Canadians. The Swedish newcomers were mainly farmers; Norwegians concentrated on lumbering, while the Swiss worked for the most part in the dairy industry. In 1890, Finns and Slavs were recruited to work in the iron mines; the state's meatpacking plants brought in Balkan nationals, Mexicans, and Poles after the turn of the century. By 1930, 50% of the population was foreign-born. Among first- and second-generation Minnesotans of European origin, Germans and Scandinavians are still the largest groups. The other ethnic groups are concentrated in Minneapolis–St. Paul or in the iron country of the Mesabi Range, where ethnic enclaves still persist. As of 1980, the foreign-born residents of Minnesota numbered 107,000, nearly 3% of the state total.

LOCATION: 43°34′ to 49°23′N; 89°34′ to 97°12′W. **BOUNDARIES:** Canadian line, 596 mi (959 km); Wisconsin line, 426 mi (686 km); Iowa line, 263 mi (423 km); South Dakota line, 182 mi (293 km); North Dakota line, 316 mi (509 km).

MINNESOTA

SCALE
0 10   20   30 Miles
0 10 20 30 Kms

CANADA

MANITOBA

KITTSON
St. Vincent
Hallock

ROSEAU
Roseau
Roseau
Baudette

LAKE OF THE WOODS

Lake of the Woods

MARSHALL
Thief L.
Warren

KOOCHICHING

Rainy L.

Rainy R.
International Falls

VOYAGEURS NATIONAL PARK

ONTARIO

ST. LOUIS

BOUNDARY WATERS CANOE AREA

COOK
GRAND PORTAGE NATIONAL MONUMENT
Eagle Mtn. ▲ 2,301ft/701m
Grand Portage
Pigeon R.

POLK
Thief River Falls
East Grand Forks
Red Lake Falls
Crookston

PENNINGTON

BELTRAMI

Black R.

South Br. Rapid R.

Little Fork R.

Caldwell R.

VERMILION RANGE
Trout L.
Burntside L.
Vermilion Lake

LAKE

LAKE SUPERIOR

RED LAKE
Red Lake

Upper Red Lake

Lower Red Lake

ITASCA

LEECH LAKE INDIAN RESERVATION

MESABI RANGE
Virginia
Aurora
Hoyt Lakes
Eveleth
Babbitt

Arrowhead Country

Whiteface River Res.
Silver Bay

CLEARWATER
Bagley

NORMAN
Ada

MAHNOMEN
Mahnomen

WHITE EARTH INDIAN RESERVATION

Bemidji
Mississippi R.
Cass L.

Winnibigoshish Lake

Hibbing

Grand Rapids

Two Harbors

HUBBARD
L. Itasca

CLAY

BECKER
Detroit Lakes

CASS
Leech Lake
Walker
Pokegama Dam

AITKIN

CARLTON
Cloquet
Carlton

Duluth

Superior

Moorhead

WADENA
Wadena

CROW WING

Aitkin

Moose Lake

WILKIN

OTTER TAIL

Park Rapids

PINE

Breckenridge

Fergus Falls

TODD

MORRISON

Brainerd

Mille Lacs Lake

MILLE LACS

KANABEC

ST. CROIX STATE PARK

DOUGLAS
Elbow Lake
Alexandria

Long Prairie

Little Falls

Mora

Milaca

Pine City

GRANT
Wheaton

STEVENS

POPE
Sauk Centre
Glenwood

BENTON
St. Francis R.
Foley

ISANTI
Cambridge

Lake Traverse

TRAVERSE

BIG STONE
Big Stone Lake

Morris

STEARNS

Sauk R.

St. Cloud

SHERBURNE

Elk River

ANOKA

CHISAGO
Center City

Marine on St. Croix

SWIFT
Benson

KANDIYOHI

Willmar

Litchfield

Buffalo

Coon Rapids

Anoka

Blaine

WASH-INGTON

Ortonville

Madison

CHIPPEWA

Montevideo

MEEKER

WRIGHT

Crow R.

HENNEPIN

RAMSEY

Stillwater

White Bear Lake

Minneapolis
St. Louis Pk.
Richfield

St. Paul

MINNEAPOLIS-ST PAUL INTERNATIONAL AIRPORT

LAC QUI PARLE

RENVILLE

MC LEOD

Glencoe

CARVER
Chaska

Bloomington
Burnsville

Shakopee

Hastings

Granite Falls

Olivia

YELLOW MEDICINE

Yellow Medicine R.

SIBLEY

Gaylord

SCOTT

DAKOTA

Red Wing

Wabasha

LINCOLN

LYON

REDWOOD

Redwood Falls

NICOLLET

St. Peter

LE SUER

Le Center

RICE

Northfield

Faribault

GOODHUE

WABASHA

Ivanhoe

Marshall

Minnesota R.

New Ulm

North Mankato

WASECA

STEELE

DODGE

OLMSTED

PIPESTONE

Pipestone

MURRAY

COTTONWOOD

WATONWAN

St. James

BROWN
Mankato

Waseca

Owatonna

Mantorville

Rochester

Winona

WINONA

NOBLES

Slayton

Windom

BLUE EARTH

MARTIN

FARIBAULT

FREEBORN

MOWER

West Des Moines R.

Worthington

JACKSON
Jackson

Fairmont

Blue Earth

Albert Lea

Austin

Preston

Caledonia

ROCK
Luverne

FILLMORE

HOUSTON

WISCONSIN

Mississippi R.

IOWA

SOUTH DAKOTA

NORTH DAKOTA

Red River

Bois de Sioux R.

Pomme de Terre R.

Chippewa R.

Lake Traverse

Big Stone Lake

Minnesota R.

Thief L.
Mud L.

Clearwater R.

Red Lake R.

St. Paul ___ Over 100,000
Rochester ___ 50,000-100,000
Winona ___ 20,000-50,000
Bemidji ___ 10,000-20,000
Ortonville ___ Under 10,000

STEVENS ___ County Name

LEGEND
⊛ State Capital
⊙ County Seat
✈ Airport
■ Point of Interest
⬚ Park, Forest, Reservation

N

WORLDMARK
ENCYCLOPEDIA
OF THE STATES

© WORLDMARK PRESS Ltd.

See endsheet maps: H1.

As of 1980 there were still 34,841 Indians in Minnesota. Besides those living in seven small reservations and four villages, a cluster of Indian urban dwellers (chiefly Ojibwa) lived in St. Paul. Indian lands totaled 764,000 acres (309,000 hectares) in 1982, of which 93% were tribal lands.

There were only 39 black Americans in Minnesota in 1850; by 1980, blacks numbered 53,334, or 1.3% of the total population. In 1980 there were 26,533 Japanese, Chinese, Filipino, and other Asian and Pacific peoples, including 5,866 refugees from Vietnam. There also were 32,000 Hispanic Americans.

⁸LANGUAGES

Many place-names echo the languages of the Yankton and Santee Sioux Indian tribes and of the incoming Algonkian-language Ojibwa, or Chippewa, from whom most of the Sioux fled to Dakota Territory. Such place-names as Minnesota itself, Minnetonka, and Mankato are Siouan in origin; Kabetogama and Winnibigoshish, both lakes, are Ojibwan.

English in the state is essentially Northern, with minor infiltration of Midland terms because of early movement up the Mississippi River into southern Minnesota and also up the Great Lakes into and beyond Duluth. Among older residents, traces of Scandinavian intonation persist, and on the Iron Range several pronunciation features reflect the mother tongues of mine workers from eastern Europe.

Although some minor variants now compete in frequency, on the whole Minnesota speech features such dominant Northern terms as *andirons, pail, mouth organ* (harmonica), *comforter* (tied and filled bedcover, *wishbone, clingstone peach, sweet corn, angleworm* (earthworm), *darning needle* or *mosquito hawk* (dragonfly), and *sick to the stomach.* Minnesotans call the grass strip between street and sidewalk the *boulevard* and a rubber band a *rubber binder,* and many *cook coffee* when they brew it. Three-fourths of a sample population spoke *root* with the vowel of *put;* one-third, through school influence, pronounced /ah/ in *aunt* instead of the usual Northern short /a/, as in *pants.* Many younger speakers pronounce *caller* and *collar* alike.

In 1980, 3,669,667 Minnesotans 3 years old or older spoke only English at home. Other leading languages spoken at home were:

German	54,970	Finnish	12,907
Norwegian	26,207	French	10,206
Spanish	22,931	Polish	7,644
Swedish	14,305	Czech	6,416

⁹RELIGIONS

Minnesota's first Christian church was organized by Presbyterians in Ft. Snelling in 1835; the first Roman Catholic church, the Chapel of St. Paul, was dedicated in 1841 at a town then called Pig's Eye but now known by the same name as the chapel. Immigrants arriving in subsequent decades brought their religions with them, with Lutherans and Catholics predominating.

As of 1980 there were 1,602,233 known Protestant adherents, including 1,015,845 Lutherans, 146,422 United Methodists, 76,853 United Presbyterians, 54,787 members of the United Church of Christ, and 38,868 Episcopalians. Roman Catholics numbered 1,064,837 in 1984, when the estimated Jewish population was 32,040. Minnesota is the headquarters for three national Lutheran religious groups: the American Lutheran Church, the Church of the Lutheran Brethren, and the Association of Free Lutheran Congregations.

¹⁰TRANSPORTATION

The development of an extensive railroad network after the Civil War was a key factor in the growth of lumbering, iron mining, wheat growing, and other industries. By 1983, Minnesota had 6,088 mi (9,798 km) of Class I track, of which 2,623 mi (4,221 km) were in the Burlington Northern system. Amtrak serves Minneapolis–St. Paul en route from Chicago to Seattle.

As of 1984, 29 municipalities and metropolitan areas had state-aided mass transit systems, which carried 84.7 million passengers. Planning and supervision of mass transportation in the Twin Cities metropolitan area are under the jurisdiction of the Metropolitan Transit Commission, a public corporation. The national Greyhound bus line was founded in Hibbing in 1914.

Minnesota had 131,475 mi (211,588 km) of state and local roads and streets in 1983, of which 118,521 mi (190,741 km) were rural and 12,954 mi (20,847 km) municipal. Interstate highways totaled 873 mi (1,405 km); I-35 links Minneapolis–St. Paul with Duluth, and I-94 connects the Twin Cities with Moorhead and Fargo, N. Dak. In 1983 there were 2,407,373 registered automobiles, 857,870 trucks, and 16,914 buses; there were 2,373,908 licensed drivers in that year.

The first settlements grew up around major river arteries, especially in the southeast; early traders and settlers arrived first by canoe or keelboat, later by steamer. The port of Duluth-Superior, at the western terminus of the Great Lakes–St. Lawrence Seaway (officially opened in 1959), is one of the 10 busiest US ports, handling 1,285 ships and 31.5 million tons of domestic and international cargo in 1984, including bulk grain, coal, metallic ores, and refrigerated commodities. The ports of Minneapolis and St. Paul handle more than 15 million tons of cargo each year, with agricultural products and scrap iron moving downstream and petroleum products, chemicals, and cement moving upstream.

As of 1983, the state had 409 airports and 67 seaplane bases. Minneapolis–St. Paul International Airport handled 86,663 scheduled departing flights, enplaning 5,781,536 passengers.

¹¹HISTORY

People have lived on the land that is now Minnesota for at least 10,000 years. The earliest inhabitants—belonging to what archaeologists classify as the Paleo-Indian (or Big Game) culture—hunted large animals, primarily bison, from which they obtained food, clothing, and materials for shelter. A second identifiable cultural tradition, from around 5000 BC, was the Eastern Archaic (or Old Copper) culture. These people hunted small as well as large game animals and fashioned copper implements through a cold hammering process. The more recent Woodland Tradition (1000 BC–AD 1700) was marked by the introduction of pottery and of mound burials. From the 1870s to the early 1900s, more than 11,000 burial mounds were discovered in Minnesota—the most visible remains of prehistoric life in the area. Finally, overlapping the Woodland culture in time was the Mississippian Tradition, beginning around AD 1000, in which large villages with permanent dwellings were erected near fertile river bottoms; their residents, in addition to hunting and fishing, raised corn, beans, and squash. There are many sites from this culture throughout southern Minnesota.

At the time of European penetration in the 17th and early 18th centuries, the two principal Indian nations were the Dakota, or Minnesota Sioux, and, at least after 1700, the Ojibwa, or Chippewa, who were moving from the east into northern Minnesota and the Dakota homelands. Friendly relations between the two nations were shattered in 1736, when the Dakota slew a party of French missionaries and traders (allies of the Ojibwa) and their Cree Indian guides (distant relatives of the Ojibwa) at the Lake of the Woods, an act the Ojibwa viewed as a declaration of war. There followed more than 100 years of conflict between Dakota and Ojibwa, during which the Dakota were pressed toward the south and west, with the Ojibwa establishing themselves in the north.

Few scholars accept the authenticity of the Kensington Rune Stone, found in 1898, the basis of the claim that Minnesota was visited in 1362 by the Vikings. The first white men whose travels through the region have been documented were Pierre Esprit Radisson and his brother-in-law, Médart Chouart, Sieur de

Groseilliers, who probably reached the interior of northern Minnesota in the 1650s. In 1679, Daniel Greysolon, Sieur Duluth, held council with the Dakota near Mille Lacs and formally claimed the region for King Louis XIV of France. The following year, Duluth negotiated the release of three captives of the Dakota Indians, among them a Belgian explorer and missionary, Father Louis Hennepin, who named the falls of the Mississippi (the site of present-day Minneapolis) after his patron saint, Anthony of Padua, and returned to Europe to write an exaggerated account of his travels in the region.

Duluth was in the vanguard of the French, English, and American explorers, fur traders, and missionaries who came to Minnesota during the two centuries before statehood. Among the best known was Nicolas Perrot, who built Ft. Antoine on the east side of Lake Pepin in 1686. In 1731, Pierre Gaultier de Varennes, Sieur de la Verendrye, journeyed to the Lake of the Woods, along whose shores he erected Ft. St. Charles; subsequently, he or his men ventured farther west than any other known French explorer, reaching the Dakotas and the Saskatchewan Valley. His eldest son was among those slain by Dakota Indians at the Lake of the Woods in 1736.

Competition for control of the upper Mississippi Valley ended with the British victory in the French and Indian War, which placed the portion of Minnesota east of the Mississippi under British control; the land west of the Mississippi was ceded by France to Spain in 1762. Although the Spanish paid little attention to their northern territory, the British immediately sent in fur traders and explorers. One of the best known was Jonathan Carver, who spent the winter of 1766–67 with the Dakota on the Minnesota River. His account of his travels—a mixture of personal observations and borrowings from others—quickly became a popular success.

There was little activity in the region during the Revolutionary War, and for a few decades afterward, the British continued to pursue their interests there. The North West Company built a major fur-trading post at Grand Portage, which quickly became the center of a prosperous inland trade, and other posts dotted the countryside. The company hired David Thompson away from the Hudson's Bay Company to map the area from Lake Superior west to the Red River; his detailed and accurate work, executed in the late 1790s, is still admired today. After the War of 1812, the US Congress passed an act curbing British participation in the fur trade, and the North West Company was eventually replaced by the American Fur Company, which John Jacob Astor had incorporated in 1808.

Under the Northwest Ordinance of 1787, Minnesota east of the Mississippi became part of the Northwest Territory; most of western Minnesota was acquired by the US as part of the Louisiana Purchase of 1803. The Red River Valley became a secure part of the US after an agreement with England on the northern boundary was reached in 1818.

In 1805, the US War Department sent Lieutenant Zebulon Pike and a detachment of troops to explore the Mississippi to its source. Pike failed to locate the source, but he concluded a treaty with a band of Dakota for two parcels of land along the river. Later, additional troops were sent in to establish US control, and in 1819, a military post was established in part of Pike's land, on a bluff overlooking the junction of the Mississippi and Minnesota rivers. First called Ft. St. Anthony, it was renamed in 1825 for Colonel Josiah Snelling, who supervised the construction of the permanent fort. For three decades, Ft. Snelling served as the principal center of civilization in Minnesota and the key frontier outpost in the northwest.

In 1834, Henry H. Sibley was appointed a manager of the American Fur Company on the upper Mississippi. He settled comfortably at Mendota, a trading post across the river from Ft. Snelling, and enjoyed immediate success. The company's fortunes took a downward turn in 1837, however—partly because of a financial panic but, even more important, because the first of a series of treaties with the Dakota and Ojibwa transferred large areas of Indian land to the US government and thus curtailed the profitable relationship between fur traders and Indians. The treaties opened the land for lumbering, farming, and settlement. Lumbering spawned many of the early permanent settlements, such as Marine and Stillwater, on the St. Croix River, and St. Anthony (later Minneapolis) at the falls of the Mississippi. Another important town, St. Paul (originally Pig's Eye), developed as a trading center at the head of navigation on the Mississippi.

In 1849, Minnesota Territory was established. It included all of present-day Minnesota, along with portions of North and South Dakota east of the Missouri River. Alexander Ramsey, a Pennsylvania Whig, was appointed as the first territorial governor, and in 1851, the legislature named St. Paul the capital. Stillwater was chosen for the state prison, while St. Anthony was selected as the site for the university. As of 1850, the new territory had slightly more than 6,000 inhabitants, but as lumbering grew and subsequent Indian treaties opened up more land, the population boomed, reaching a total of more than 150,000 by 1857, with the majority concentrated in the southeast corner, close to the rivers.

On 11 May 1858, Minnesota officially became the 32d state, with its western boundaries pruned from the Missouri to the Red River. Henry Sibley, a Democrat, narrowly defeated Alexander Ramsey, running as a Republican, to become the state's first governor. But under Ramsey's leadership, the fastgrowing Republican Party soon gained control of state politics and held it firmly through the early 20th century. In the first presidential election in which Minnesota participated, Abraham Lincoln, the Republican candidate, easily carried the state, and when the Civil War broke out, Minnesota was the first state to answer Lincoln's call for troops. In all, Minnesota supplied more than 20,000 men to defend the Union.

More challenging to the defense of Minnesota was the Dakota War of 1862. Grieved by the loss of their lands, dissatisfied with reservation life, and ultimately brought to a condition of near starvation, the Dakota appealed to US Indian agencies without success. The murder of five whites by four young Dakota Indians ignited a bloody uprising in which more than 300 whites and an unknown number of Indians were killed. In the aftermath, 38 Dakota captives were hanged for "voluntary participation in murders and massacres," and the Dakota remaining in Minnesota were removed to reservations in Nebraska. (Some later returned to Minnesota.) Meanwhile, the Ojibwa were relegated to reservations on remnants of their former lands.

Also during 1862, Minnesota's first railroad joined St. Anthony (Minneapolis) and St. Paul with 10 mi (16 km) of track. By 1867, the Twin Cities were connected with Chicago by rail; in the early 1870s, tracks crossed the prairie all the way to the Red River Valley. The railroads brought settlers from the eastern states (many of them Scandinavian and German in origin) to every corner of Minnesota; the settlers, in turn, grew produce for the trains to carry back to the cities of the east. The railroads soon ushered in an era of large-scale commercial farming. Wheat provided the biggest cash crop, as exports rose from 2 million bushels in 1860 to 95 million in 1890. Meanwhile, the falls of St. Anthony became the major US flour-milling center; by 1880, 27 Minneapolis mills were producing more than 2 million barrels of flour annually.

Despite these signs of prosperity, discontent grew among Minnesota farmers, who were plagued by high railroad rates, damaging droughts, and a deflationary economy. The first national farmers' movement, the National Grange of the Patrons of Husbandry, was founded in 1867 by a Minnesotan, Oliver H. Kelley, and spread more rapidly in Minnesota than in any other state. The Farmers' Alliance movement, joining forces with the Knights of Labor, exerted a major influence on state politics in the 1880s. In 1898, the Populist Party—in which a Minnesotan, Ignatius Donnelly, played a leading role nationwide—helped elect John Lind to the governorship on a fusion ticket.

Most immigrants during the 1860s and 1870s settled on the rich farmland of the north and west, but after 1880 the cities and industries grew more rapidly. When iron ore was discovered in the 1880s in the sparsely settled northeast, even that part of the state attracted settlers, many of them immigrants from eastern and southern Europe. Before the turn of the century, Duluth had become a major lake port, and by the eve of World War I, Minnesota had become a national ironmining center.

The economic picture changed after the war. As the forests became depleted, the big lumber companies turned to the Pacific Northwest. An agricultural depression hit the region, and flour mills moved to the Kansas City area and to Buffalo, N.Y. Minnesotans adapted to the new realities in various ways. Farmers planted corn, soybeans, and sugar beets along with wheat, and new foodprocessing industries developed. To these were added business machines, electronics, computers, and other high-technology industries. In 1948, for the first time, the dollar value of all manufactured products exceeded total cash farm receipts: Minnesota was becoming an urban commonwealth.

Economic dislocations and the growth of cities and industries encouraged challenges to the Republican leadership from Democrats and third parties. John Johnson, a progressive Democratic governor first elected in 1904, was especially active in securing legislation to regulate the insurance industry; his successor, Republican Adolph Eberhart, promoted numerous progressive measures, including one establishing direct primary elections. The Non-Partisan League, founded in 1915 by Minnesotan Arthur C. Townley, soon grew particularly strong in both North Dakota and Minnesota, where the elder Charles A. Lindbergh was its spokesman.

A political outgrowth of the Non-Partisan League known as the Farmer-Labor Party had many electoral successes in the 1920s and reached its peak with the election of Farmer-Labor candidate Floyd B. Olson to the governorship in 1930. Olson introduced a graduated income tax and other progressivist measures, but his death in office in 1936 was a crippling blow to the party. In 1938, the Republicans recaptured the governorship with the election of Harold E. Stassen. However, a successful merger of the Farmer-Labor and Democratic parties in Minnesota, engineered in 1943-44 by both local and national politicians, revived the progressivist tradition after World War II. Hubert Humphrey (later US vice president) and his colleagues Orville Freeman, Eugene McCarthy, and Eugenie Anderson emerged as leaders of this new coalition. Their political heir, Walter Mondale, was vice president in 1977-81 but, as the Democratic presidential candidate in 1984, lost the election in a Republican landslide, carrying only his native state and the District of Columbia.

[12] STATE GOVERNMENT

The constitutional convention that assembled at St. Paul on 13 July 1857 was marked by such bitter dissension that the Democrats and Republicans had to meet in separate chambers; the final draft was written by a committee of five Democrats and five Republicans and then adopted by a majority of each party, without amendment. Since Democrats and Republicans were also unwilling to sign the same piece of paper, two separate documents were prepared, one on blue-tinted paper, the other on white. The constitution was ratified by the electorate on 13 October and approved by the US Congress on 11 May 1858. An amendment restructuring the constitution for easy reference and simplifying its language was approved in 1974; for purposes of constitutional law, however, the original document (incorporating numerous other amendments) remains authoritative.

As reapportioned by court order after the 1970 census, the Minnesota legislature consists of a 67-member senate and a 134-member house of representatives. Senators serve four years and representatives two, at annual salaries of $21,140 as of 1985. Legislators must be US citizens, must be at least 21 years of age, and must have resided in the state for one year and in the legislative district for six months preceding election.

The governor and lieutenant governor are jointly elected for four-year terms; both must be US citizens at least 25 years old, and must have been residents of Minnesota for a year before election. Other constitutional officers are the secretary of state, auditor, treasurer, and attorney general, all serving for four years. Numerous other officials are appointed by the governor, among them the commissioners of the 20 government departments and many heads and members of independent agencies.

Once a bill is passed by a majority of both houses, the governor may sign it, veto it in whole or in part, or pocket-veto it by failing to act within 14 days of adjournment. (When the legislature is in session, however, a bill becomes law if the governor fails to act on it within 3 days.) A two-thirds vote of both houses is sufficient to override a veto. Constitutional amendments require the approval of a majority of both houses of the legislature and are subject to ratification by the electorate. Those voting in state elections must be at least 18 years old and must have been US citizens for three months and residents of the district for 20 days.

[13] POLITICAL PARTIES

The two major political parties are the Democratic-Farmer-Labor Party (DFL) and the Independent-Republican Party (IR), as Minnesota's Republican Party is now officially called. The Republican Party dominated Minnesota politics from the 1860s through the 1920s, except for a period around the turn of the century. The DFL, formed in 1944 by merger between the Democratic Party and the populist Farmer-Labor Party, rose to prominence in the 1950s under US Senator Hubert Humphrey.

The DFL is the heir to a long populist tradition bred during the panic of 1857 and the early days of statehood, a tradition perpetuated by a succession of strong, though transient, third-party movements. The Grange, a farmers' movement committed to the cause of railroad regulation, took root in Minnesota in 1868; it withered in the panic of 1873, but its successors, the Anti-Monopoly Party and the Greenback Party, attracted large followings for some time afterward. They were followed by a new pro-silver group, the Farmers' Alliance, which spread to Minnesota from Nebraska in 1881 and soon became associated with the Minnesota Knights of Labor. The Populist Party also won a foothold in Minnesota, in alliance with the Democratic Party in the late 1890s.

The Farmer-Labor Party, the most successful of Minnesota's third-party movements, grew out of a socialist and isolationist movement known at first as the Non-Partisan League. Founded in North Dakota with the initial aim of gaining control of the Republican Party in that state, the league moved its headquarters to St. Paul and competed in the 1918 elections under the name Farmer-Labor Party, hastily adopted to attract what party leaders hoped would be its two main constituencies. The party scored a major success in 1922 when its candidate, Henrik Shipstead, a Glenwood dentist, defeated a nationally known incumbent, Republican Senator Frank B. Kellogg; Farmer-Labor candidate Floyd B. Olson won the governorship in 1930. The decline of the party in the late 1930s was hastened by the rise of Republican Harold Stassen, an ardent internationalist, who won the governorship in 1938 and twice won reelection.

The first DFL candidate to become governor was Orville Freeman in 1954. The DFL held the governorship from 1963 to 1967 and from 1971 to 1978, when US Representative Al Quie (IR) defeated his DFL opponent, Rudy Perpich; however, Perpich regained the governorship for the DFL in 1982.

Minnesota is famous as a breeding ground for presidential candidates. Governor Harold Stassen contended seriously for the Republican nomination in 1948 and again in 1952. Vice President Hubert Humphrey was the Democratic presidential nominee in 1968, losing by a narrow margin to Richard Nixon. During the same year, US Senator Eugene McCarthy unsuccessfully sought the Democratic presidential nomination on an antiwar platform; his surprising showings in the early primaries against the incumbent, Lyndon B. Johnson, helped persuade Johnson to withdraw his candidacy. Eight years later, McCarthy ran for the presidency as an independent, drawing 35,490 votes in Minnesota (1.8% of the total votes cast) and 756,631 votes (0.9%) nationwide. Walter Mondale, successor to Hubert Humphrey's seat when Humphrey became Johnson's vice president in 1964, was chosen in 1976 by Jimmy Carter as his vice-presidential running mate; he again ran with Carter in 1980, when the two lost their bid for reelection. In the 1984 election, Minnesota was the only state to favor the Mondale-Ferraro ticket. Minnesotans gave the Republican Party a majority in the state's house of representatives for the first time since 1970, but the Democrats retained control of the state senate. As of 1985, the Republicans held both US Senate seats, while the Democrats controlled the state's congressional delegation.

There were 2 Asian-American and 3 Hispanic elected officials in Minnesota in 1984. In 1985, 9 blacks held elective offices, and 19 women served in the state legislature; women also held the positions of lieutenant governor and secretary of state.

[14]LOCAL GOVERNMENT

Minnesota is divided into 87 counties and 13 regional administrations. As of 1984, the state had 1,792 townships (more than any other state) and 855 cities.

Each of Minnesota's counties is governed by a board of commissioners, ordinarily elected for four-year terms. Other elected officials include the auditor, treasurer, recorder, sheriff, attorney, and coroner; an assessor and engineer are customarily appointed. Besides administering welfare, highway maintenance, and other state programs, the county is responsible for planning and development and, except in large cities, for property assessment. During the 1970s, counties also assumed increased responsibility for solid waste disposal and shoreline management.

Each regional development commission, or RDC, consists of local officials (selected by counties, cities, townships, and boards of education in the region) and of representatives of public interest groups (selected by the elected officials). RDCs prepare and adopt regional development plans and review applications for loans and grants.

As of 1984, 104 cities had home-rule charters; the remaining 751 were statutory cities, restricted to the systems of government prescribed by state law. In either case, the mayor-council system was the most common. Besides providing such traditional functions as street maintenance and police and fire protection, some cities operate utilities, sell liquor, or run hospitals, among other services. Each township is governed by a board of three supervisors and by other officials elected for three-year terms at the town meeting, held annually on the 2d Tuesday in March.

[15]STATE SERVICES

Minnesota's ombudsman for corrections investigates complaints about corrections facilities or the conduct of prison officials. A six-member Ethical Practices Board supervises the registration of some 1,300 lobbyists, monitors the financing of political campaigns, and sees that some 1,000 elected and appointed state officials observe regulations governing conflict of interest and disclosure of personal finances. Minnesota law also provides that legislative meetings of any kind must be open to the public.

The state-aided public school system is under the jurisdiction of the Department of Education, which carries out the policies of a nine-member Board of Education appointed by the governor with

Minnesota Presidential Vote by Political Parties, 1948–84

YEAR	ELECTORAL VOTE	MINNESOTA WINNER	DEMOCRAT[1]	REPUBLICAN[2]	PROGRESSIVE	SOCIALIST	SOCIALIST LABOR[3]
1948	11	*Truman (D)	692,966	483,617	27,866	4,646	2,525
1952	11	*Eisenhower (R)	608,458	763,211	2,666	—	2,383
						SOC. WORKERS	
1956	11	*Eisenhower (R)	617,525	719,302	—	1,098	2,080
1960	11	*Kennedy (D)	779,933	757,915	—	3,077	962
1964	10	*Johnson (D)	991,117	559,624	—	1,177	2,544
							AMERICAN IND.
1968	10	Humphrey (D)	857,738	658,643	—	—	68,931
					PEOPLE'S		AMERICAN
1972	10	*Nixon (R)	802,346	898,269	2,805	4,261	31,407
					LIBERTARIAN		
1976	10	*Carter (D)	1,070,440	819,395	3,529	4,149	13,592
						CITIZENS	
1980	10	Carter (D)	954,173	873,268	31,593	8,406	6,136
1984	10	Mondale (D)	1,036,364	1,032,603	2,996	1,219	—

*Won US presidential election.
[1]Called Democratic-Farmer-Labor Party in Minnesota.
[2]Since 1976, called Independent-Republican in Minnesota.
[3]Appeared as Industrial Government Party on the ballot.

the advice and consent of the senate. Responsible for higher education are the University of Minnesota Board of Regents, elected by the legislature; the State University Board and State Board for Community Colleges, both appointed by the governor; and other agencies. The Department of Transportation maintains roads and bridges, enforces public transportation rates, inspects airports, and has responsibility for railroad safety.

Minnesota's Department of Health investigates health problems, disseminates health information, regulates hospitals and nursing homes, and inspects restaurants and lodgings. Health regulations affecting farm produce are administered by the Department of Agriculture. State facilities for the mentally retarded are operated by the Department of Human Services, which administers state welfare programs and provides social services to the aged, the handicapped, and others in need.

The Department of Public Safety registers motor vehicles, licenses drivers, enforces traffic laws, and regulates the sale of liquor. The Department of Military Affairs has jurisdiction over the Minnesota National Guard, and the Department of Corrections operates prisons, reformatories, and parole programs. The Housing Finance Agency aids the construction and rehabilitation of low- and middle-income housing. Laws governing occupational safety, wages and hours, and child labor are enforced by the Department of Labor and Industry, while the Department of Economic Security supervises public employment programs and administers unemployment insurance.

16JUDICIAL SYSTEM

Minnesota's highest court is the supreme court, consisting of a chief justice and eight associate justices; all are elected without party designation for six-year terms, with vacancies being filled by gubernatorial appointment. The district court, divided into 10 judicial districts, is the principal court of original jurisdiction. Each judicial district has at least three district judges, elected to six-year terms. The governor designates a chief judge for a three-year term.

County courts, operating in all counties of the state except two—Hennepin (Minneapolis) and Ramsey (St. Paul), which have municipal courts—assume functions formerly exercised by probate, family, and local courts. They exercise civil jurisdiction in cases where the amount in contention is $5,000 or less, and criminal jurisdiction in preliminary hearings and misdemeanors. They also hear cases involving family disputes, and have concurrent jurisdiction with the district court in divorces, adoptions, and certain other proceedings. The probate division of the county court system presides over guardianship and incompetency proceedings and all cases relating to the disposition of estates. All county judges are elected for six-year terms.

The Department of Corrections operates a maximum-security state prison at Stillwater for adult males, a maximum-security state reformatory at St. Cloud for males under 21 years of age, and a correctional institution for women at Shakopee. Other facilities include two vocational institutions for adult offenders, a state training school for delinquent boys, and a forestry camp for juveniles. State correctional institutions had a total population of 2,167 at the end of 1984.

Crime rates are generally below the national average, and the state's rate of less than 2 reported cases of murder or nonnegligent manslaughter per 100,000 population was the lowest in the US in 1983. Minnesota has no death penalty statute. The Crime Victims Reparations Board offers compensation to innocent victims of crime or to their dependent survivors.

17ARMED FORCES

There were 3,536 authorized defense personnel stationed in Minnesota in 1983/84, near Minneapolis–St. Paul and Duluth international airports. Firms in the state received $1,826 million in defense contract awards during the same year.

As of 30 September 1983 there were 522,000 veterans living in Minnesota, including 8,000 who saw service in World War I, 184,000 in World War II, 93,000 during the Korean conflict, and 158,000 during the Viet-Nam era. Veterans in Minnesota received a total of $446 million in benefits in 1982/83.

The Minnesota Army National Guard had a total authorized strength of 9,784 at the end of 1984, and the Air National Guard had a strength of 2,247 early in 1985. There were 7,804 state and local police officers in 1983.

18MIGRATION

A succession of migratory waves began in the 17th and 18th centuries with the arrival of the Dakota and Ojibwa, among other Indian groups, followed during the 19th century by New England Yankees, Germans, Scandinavians, and finally southern and eastern Europeans. Especially since 1920, new arrivals from other states and countries have been relatively few, and the state experienced a net loss from migration of 80,000 between 1970 and 1980, and of 46,000 between 1980 and 1983. As of 1980, 7% of all Minnesotans 5 years of age or older had moved to the state within the previous five years, a proportion below the 9% average for the US as a whole.

Within the state, there has been a long-term movement to metropolitan areas and especially to the suburbs of major cities; from 1970 to 1983, the state's metropolitan population grew by nearly 1% annually. From 1980 to 1984, the population of the Minneapolis–St. Paul metropolitan area grew by 4.4%.

19INTERGOVERNMENTAL COOPERATION

Relations with the Council of State Governments are conducted through the Minnesota Commission on Interstate Cooperation, consisting of five members from each house of the state legislature and five administrative officers or other state employees; in addition, the governor, the president of the senate, and the speaker of the house are nonvoting members. Minnesota also participates in the Great Lakes Charter, which it formed with seven other states in 1985 to preserve the lakes' water supply, and in other regional compacts. In 1983/84, Minnesota received almost $1.9 billion in federal aid.

20ECONOMY

Furs, wheat, pine lumber, and high-grade iron ore were once the basis of Minnesota's economy. As these resources diminished, however, the state turned to wood pulp, dairy products, corn and soybeans, taconite, and manufacturing, often in such food-related industries as meat-packing, canning, and the processing of dairy products. The share of GSP generated by government and trade rose significantly between the late 1960s and early 1980s, while the relative contributions of manufacturing and construction declined. The state's growing skilled labor force and its position as a major marketing and distribution center were expected to promote continued economic growth in the decades ahead.

21INCOME

In 1983, Minnesota ranked 18th among the 50 states in personal income per capita, amounting to $11,666; this was three times the figure for 1970. Total personal income in 1983 was $48.3 billion, representing a real growth of 40% since 1970; the figure increased to an estimated $53.8 billion in 1984.

Minnesota ranked 7th among the 50 states in median family income with $29,261 per four-member family in 1981. Wide regional differences in personal income exist within the state; family incomes are highest in the Twin Cities area, where the state's high-income counties are clustered. In 1982, Minnesota had 44,100 of the nation's leading wealth-holders, with gross assets exceeding $500,000 each. On the other hand, 9.5% of all state residents and 7% of total families lived below the poverty level in 1979.

22LABOR

In 1984, the civilian labor force totaled 2,229,000 persons, of whom an average 2,088,000 were employed and 141,000 were unemployed. The unemployment rate that year was 6.3%, representing a significant improvement over the previous year's rate of 8.2%.

As of September 1984, manufacturing workers in the Twin Cities area earned average hourly wages of $10.30 and an average weekly salary of $419. Median monthly salaries varied from $868 for a cannery worker to more than $2,600 for a civil engineer.

A federal survey in March 1983 showed the following nonfarm employment pattern for Minnesota:

	ESTABLISH-MENTS	EMPLOYEES	ANNUAL PAYROLL ('000)
Agricultural services, forestry, fisheries	978	4,266	$ 60,348
Mining, of which:	170	7,883	230,456
Metals	(22)	(6,606)	(189,256)
Contract construction	9,077	60,401	1,559,668
Manufacturing, of which:	6,760	344,137	8,092,492
Food and food products	(545)	(35,804)	(709,772)
Nonelectrical machinery	(1,232)	(67,040)	(1,671,643)
Transportation, public utilities	4,241	80,807	1,866,382
Wholesale trade	9,418	110,043	2,383,278
Retail trade	26,221	303,672	2,754,394
Finance, insurance, real estate	8,682	105,397	2,057,296
Services	27,788	381,163	5,015,151
Other	3,957	5,693	133,259
TOTALS	97,292	1,403,412	$24,152,724

This survey did not include government workers; there were 255,000 state and local government employees in 1983, and 29,000 federal employees in 1982.

The history of unionization in the state includes several long and bitter labor disputes, notably the Iron Range strike of 1916. The earliest known unions—two printers' locals, established in the late 1850s—died out during the Civil War, and several later unions faded in the panic of 1873. The Knights of Labor were the dominant force of the 1880s; the next decade saw the rise of the Minnesota State Federation of Labor, whose increasing political influence bore fruit in the landmark Workmen's Compensation Act of 1913 and the subsequent ascension of the Farmer-Labor Party. The legislature enacted a fair employment practices law in 1955 and passed a measure in 1973 prescribing collective-bargaining procedures for public employees and granting them a limited right to strike.

As of mid-1983, Minnesota had 1,003 labor unions; union membership totaled 103,000 in 1980. There were 81 work stoppages involving 909,000 workers in 1983.

23 AGRICULTURE

Cash receipts from farm marketings totaled nearly $6.3 billion in 1983, placing Minnesota 5th among the 50 states; crops made up about 47% of the total value. For 1984, Minnesota ranked 2d in the production of sugar beets, spring wheat, rye, oats, and hay; and 3d in sunflower seed, flaxseed, and soybeans, which yielded a record crop. The 1984 corn crop was the state's 3d largest on record.

The early farmers settled in the wooded hills and valleys in the southeastern quarter of the state, where they had to cut down trees and dig up stumps to make room for crops. With the coming of the railroads, farmers began planting the prairies with wheat, which by the late 1870s took up 70% of all farm acreage. In succeeding decades, wheat prices fell and railroad rates soared, fanning agrarian discontent. Farmers began to diversify, with dairy farming, oats, and corn becoming increasingly important. Improved corn yields since the 1940s have spurred the production of hogs and beef cattle and the growth of meat-packing as a major industry.

As of 1984, the state had 101,000 farms, covering 30,400,000 acres (12,302,000 hectares), or 55% of the state's total land area; the average farm had 301 acres (122 hectares). The number of people living on farms steadily declined from 624,000 in 1960 to 482,000 in 1970, and then to only 315,000, or 7.7% of the total population, by 1980. The value of farmland also dropped, from $1,197 per acre in 1982 to $990 per acre in 1984. Minnesota's farmers faced acute financial troubles during the early 1980s as a result of heavy debts, high interest rates, and generally low crop prices.

The main farming areas are in the south and southwest, where corn, soybeans, and oats are important, and in the Red River Valley along the western border, where oats, wheat, sugar beets, and potatoes are among the chief crops. The following table shows selected major crops in 1983.

	PRODUCTION		VALUE ('000)
Soybeans	144,900,000	bushels	$1,173,690
Corn for grain	367,080,000	bushels	1,156,302
Hay	8,316,000	tons	565,488
Wheat	78,960,000	bushels	307,410
Oats	76,950,000	bushels	123,120
Barley	43,460,000	bushels	104,304
Sunflowers	264,250	lb	36,023
Sweet corn for processing	498,640	tons	—
Sugar beets	4,738,000	tons	—

Agribusiness is Minnesota's largest basic industry, with about one-third of the state's labor force employed in agriculture or agriculture-related industries, most notably food processing.

24 ANIMAL HUSBANDRY

Excluding the northeast, livestock raising is dispersed throughout the state, with cattle concentrated particularly in west-central Minnesota and in the extreme southeast, and hogs along the southern border. At the start of 1984, Minnesota had 3,690,000 cattle and calves; the state's total of 910,000 milk cows ranked 4th in the US. In addition there were 4,270,000 hogs and pigs (3d in the US) and 255,000 sheep and lambs. Minnesota raised more turkeys in 1983 than any state except North Carolina: 27,000,000, worth $158,263,000.

During 1984, Minnesota farms and ranches produced 1.4 billion lb of cattle and calves, valued at $835 million; 1.6 billion lb of hogs and pigs (3d in the US), $733 million; and 23.8 million lb of sheep and lambs, $15.4 million. Dairy products included 203.8 million lb of nonfat dry milk for human consumption, more than in any other state except California; 177.5 million lb of butter, 3d only to California and Wisconsin; and 587 million lb of cheese, 2d behind Wisconsin. The state's total of 10.3 billion lb of milk outranked all but three states'. Production of chickens and broilers was 112.2 million lb, worth $30 million, and the egg output was 2.5 billion. During 1982, Minnesota produced 10.8 million lb of honey.

25 FISHING

Commercial fishermen in 1983 landed 9,355,626 lb of fish, valued at $1,906,885. The catch included herring and smelts from Lake Superior, whitefish and yellow pike from large inland lakes, and carp and catfish from the Mississippi and Minnesota rivers. Sport fishing attracts some 1.5 million anglers annually to the state's 2.6 million acres (1.1 million hectares) of fishing lakes and 7,000 mi (11,000 km) of fishing streams, which are stocked with trout, bass, pike, muskie, and other fish by the Division of Fish and Wildlife of the Department of Natural Resources.

26 FORESTRY

Forests, which originally occupied two-thirds of Minnesota's land area, have been depleted by lumbering, farming, and forest fires. As of 1980, forestland covered 18,385,000 acres (7,440,000 hectares), or one-third of the state's total area. Most of the forestland is in the north, especially in Arrowhead Country in the northeast. Of the 13,695,000 acres (5,542,000 hectares) of commercial timberland, less than half is privately owned and more than one-third is under state, county, or municipal jurisdiction. In 1980, Minnesota harvested 1,333,000 cords of pulpwood, at a value of $770,897,000; 1,360,000 cords of fuelwood, valued at $51,000,000; and 671,250 cords of specialty wood products, valued at $18,260,500. Total production of softwoods and hardwoods amounted to 170 million board feet in 1983.

The state's two national forests are Superior (2,054,022 acres—831,236 hectares—in 1984) and Chippewa (661,218 acres—267,586 hectares). The Department of Natural Resources, Division of Forestry, promotes effective management of the forest environment and seeks to restrict forest fire occurrence to 1,100 fires annually, burning no more than 30,000 acres (12,000 hectares) in all.

More than 20 million trees are planted each year by the wood fiber industry, other private interests, and federal, state, and county forest services—more than enough to replace those harvested or destroyed by fire, insects, or disease.

27 MINING

In 1984, Minnesota ranked 3d among the 50 states in the value of its nonfuel mineral output, amounting to more than $1.65 billion. The state ranked 1st in the production of iron ore, which accounted for 93% of the total mineral value; about two-thirds of the iron ore produced in the US in 1984 was mined in Minnesota, mainly in the Mesabi Range and other parts of the northeast. Iron ore production amounted to some 775 million tons between 1971 and 1984. A high-quality lode discovered earlier in the Vermilion Range continued to be exploited until the 1960s. Most iron production today is in the form of taconite (20–30% iron), from which a 60% concentrate is extracted in the form of hard pellets. The lack of demand for iron ore during the national recession of the early 1980s led to a severe slowdown of taconite operations in 1983 and resultant high unemployment among miners in the Iron Range.

Production of iron ore declined from 59.6 million tons in 1979 to 35 million tons in 1984 (valued at $1,540 million). Output of other nonfuel minerals in 1984 included sand and gravel, 22 million tons (valued at $61.5 million); and crushed stone, 9.4 million tons ($28 million).

28 ENERGY AND POWER

Minnesota produced 30 billion kwh of electricity in 1983, when installed capacity reached 8.6 million kw. Steam-generating plants accounted for 60% of total installed capacity; most plants were coal-fired. There are three nuclear reactors, all owned by the Northern States Power Co. The retail cost of electricity increased from less than 4 cents per kwh in Minneapolis in 1979 to 5.3 cents in 1983.

Minnesota's 7 million acres (2.8 million hectares) of peat lands, the state's only known fossil fuel resource, constitute nearly half of the US total (excluding Alaska). If burned directly, the accessible fuel-quality peat deposit could supply all of Minnesota's energy needs for 50 years, at current rates of consumption.

29 INDUSTRY

The exploitation of natural resources, especially the state's extensive timberlands and fertile prairie, was the basis of Minnesota's industrial development in the late 19th century, when Minneapolis became the largest sawmill and flour-milling center in the US. Canning and meatpacking became important in the early 20th century, and food processing remains 1st among Minnesota's leading industrial sectors. Computers are among the state's most important products.

The total value of shipments by manufacturers in 1982 exceeded $35.3 billion. Contributions of principal sectors were food and food products, 26%; nonelectrical machinery, 18%; fabricated metals, 7%; electrical and electronic equipment, 6%; paper and allied products, 5%; and printing and publishing, 5%. Value of shipments by selected industries in 1982 was as follows:

Dairy products	$2,782,500,000
Meat products	2,377,100,000
Electronic equipment	2,015,900,000
Commercial printing	960,900,000
Preserved fruits and vegetables	640,700,000
Construction and related machinery	589,500,000
Refrigeration and service machinery	533,800,000

Industry is concentrated in the southeast, especially in the Twin Cities area, which in 1982 accounted for 62% of the value of shipments and 69% of the state's value added by manufacture. Among the well-known national firms with headquarters in Minnesota are Minnesota Mining and Manufacturing (3M) in St. Paul, the 45th largest US industrial corporation in total sales ($7.7 billion in 1984); Honeywell (56th); General Mills (64th); Control Data (71st); and Pillsbury (94th).

30 COMMERCE

Access to the Great Lakes, the St. Lawrence Seaway, and the Atlantic Ocean, as well as to the Mississippi River and the Gulf of Mexico, helps make Minnesota a major marketing and distribution center for the upper Midwest. As of 1982, the state's 9,020 wholesale trade establishments had sales totaling nearly $47.4 billion (13th among the 50 states). The 37,274 retail establishments had sales of almost $19.6 billion (21st in the US), with food stores accounting for 20% of the total; automotive dealers, 18%; and department stores, 11%. Much of this volume was concentrated in the Twin Cities area.

Agricultural exports to foreign countries amounted to $1.9 billion in 1981/82 (5th among the 50 states), an increase of 105% since 1977. Dairy products, feed grains, flaxseed, and hides and skins were the largest export items. Export sales of manufactured goods totaled nearly $2.7 billion in 1981 (20th in the US). The chief manufactured exports included computers and computer software, food-processing and packaging machinery, and other industrial machinery and equipment. Minnesota's exports of manufactured goods provided jobs for approximately 46,400 state residents in 1981.

31 CONSUMER PROTECTION

The Office of Consumer Services, established within the Commerce Department in 1971, administers statutes governing fraudulent or deceptive business practices, attempts to resolve consumer complaints through voluntary arbitration, represents consumers before private organizations and governmental bodies, and provides consumer information and education. The Agriculture Department regulates the manufacture and distribution of food, feed, fertilizer, and other items. State antitrust and consumer laws are enforced by the Office of the Attorney General.

32 BANKING

As of 31 December 1984, Minnesota had 738 insured commercial banks; the total assets of the state's insured commercial banks in 1983 were $41.6 billion, including $23.4 billion in commercial, industrial, and real estate loans. Deposits exceeded $31.2 billion, of which $24.4 billion were time and savings deposits. The two leading commercial banking companies, both in Minneapolis, are First Bank System, Inc., the 14th-largest commercial bank in the US in 1984 with assets of $22.4 billion, deposits of $14.8 billion, and loans of $13.2 billion; and Norwest Corp., the nation's 20th-largest commercial bank, with assets of $21.3 billion, deposits of $14.6 billion, and loans of $13.3 billion.

Minnesota had 40 insured savings and loan associations as of 31 December 1983, with assets totaling nearly $12 billion.

33 INSURANCE

Minnesotans held 6,306,000 life insurance policies valued at $95 billion as of 31 December 1983. Coverage per family averaged $58,800. Payments to beneficiaries in the same year totaled $909.5 million, including $223.3 million in death payments. The two most important Minnesota-based companies are Northwestern

National in Minneapolis, with $60 billion of insurance in force in December 1984, and Minnesota Mutual Life Insurance in St. Paul, with $43.7 billion in force.

Property and liability insurance companies wrote premiums totaling almost $2.2 billion in 1983; automotive liability insurance accounted for $438.9 million, automobile physical damage insurance for $261 million, and homeowners' coverage for $266.8 million. No-fault automobile insurance was enacted in 1974. Minnesotans held nearly $262 million worth of flood insurance at the end of 1983.

34 SECURITIES

The Minneapolis Grain Exchange, founded in 1881, is the state's major commodity exchange. Enforcement of statutes governing securities, franchises, and corporate takeovers (as well as charitable organizations, public cemeteries, collection agencies, and bingo) is the responsibility of the Securities Division of the Department of Commerce.

New York Stock Exchange member firms had 90 sales offices and 1,037 registered representatives in Minnesota in 1983; there were 101 stock brokers and dealers registered in the state. The number of Minnesotans holding shares in public corporations was 681,000 as of 31 December 1983.

35 PUBLIC FINANCE

Minnesota spends a relatively large amount on state government and local assistance, especially on a per capita basis. In 1981/82, Minnesota ranked 4th among the 50 states in total general expenditures per capita ($2,373).

The state budget is prepared by the Department of Finance and submitted biennially by the governor to the legislature for amendment and approval. The fiscal year runs from 1 July to 30 June. The following table summarizes proposed general fund revenues and expenditures for the 1985/87 biennium (in millions):

REVENUES	
Individual income tax	$ 5,119.9
Corporate income tax	678.5
General sales tax	3,087.5
Motor vehicle tax	430.7
Other taxes	914.8
Nontax revenue	370.4
TOTAL	$10,601.8

EXPENDITURES	
School aid	$ 2,603.0
Postsecondary education	1,925.0
Property tax credits and refunds	1,672.0
Health care	1,250.0
Local government aid	589.0
Income support	395.0
Debt service	293.0
Other outlays	2,206.0
TOTAL	$10,933.0

As of 30 June 1983, the outstanding debt of state and local governments totaled $12.6 billion; the per capita debt was $2,555 in mid-1982 (9th among the 50 states).

36 TAXATION

Minnesota ranked 13th among the 50 states in 1981/82 in total receipts from state and local taxes ($5.2 billion) and 9th in taxes collected per capita ($1,290). State taxes and fees collected in 1982/83 totaled $5.3 billion; local revenues amounted to $3.5 billion.

As of 1984 corporate profits were taxed at rates varying from 6% to 12%, according to a weighted formula based on the proportion of payroll, property, and sales within the state; there is a minimum tax of $100. Individual income is taxed at graduated rates in brackets that were indexed to the inflation rate beginning in the 1979 tax year. As of 1984, rates ranged from 1.6% on the first $500 of taxable income to 16% on income over $44,043; that top-bracket tax rate was higher than any other state's. There was also a 6% state sales tax, a gasoline tax of 17 cents per gallon, and a cigarette tax of 18 cents per pack. Gift and inheritance taxes were repealed in 1980.

Minnesota has an estate tax, generally only on amounts above $200,000, with an additional $250,000 exemption for a surviving spouse. Rates range from 7% on the first $100,000 taxable to 12% on amounts over $1 million. The state also levies an employer's excise tax on compensation paid out by a company exceeding $250,000 in a given year, certain other selective business taxes, and severance taxes on mineral production. Commercial, industrial, and residential property is subject to property tax, the principal source of revenue for local governing units ($342 per capita in 1981/82).

Minnesota paid out more than $10.1 billion in federal taxes—$2,456 per capita—in 1982/83 and received $6.8 billion in federal expenditures during 1983/84. State residents filed 1,724,860 federal income tax returns for 1982/83 and paid $4.4 billion in tax ($3,155 per taxpayer).

37 ECONOMIC POLICY

The Department of Energy, Planning, and Development counsels communities on their development potential and acts as a liaison between the business community and the state government. The department also seeks to encourage expansion of existing industries and to attract new industry to the state by providing data to help industrial planners and by disseminating information about the advantages of doing business in Minnesota. The state offers its own supplementary funding through the Area Redevelopment Administration for business development loans under the federal Economic Development Administration program. Minnesota's corporate income tax is structured to favor companies having relatively large payrolls and property (as opposed to sales) within the state.

38 HEALTH

Shortly after the founding of Minnesota Territory, promoters attracted new settlers partly by proclaiming the tonic benefits of Minnesota's soothing landscape and cool, bracing climate; the area was trumpeted as a haven for retirees and for those afflicted with malaria or tuberculosis.

In 1982, Minnesota ranked 20th among the states in birthrate with 68,498 infants born, or 16.6 per 1,000 population, slightly above the national average. Infant mortality was relatively low, with an average of 10.3 per 1,000 live births in 1981. There were 19,000 legal abortions in 1982; the ratio of 28.3 abortions per 1,000 live births was well below the national average of 42.6. In 1981, 32,837 deaths, or 7.9 per 1,000 population (below the 8.6 national average), were recorded. The death rates per 100,000 population for the two leading causes, heart disease and cancer, were 294 and 169, respectively. Rates for these and most other causes of death were below the national norms.

In 1983, Minnesota had 181 hospitals, with 29,128 beds, 679,341 patient admissions, and 3,125,595 outpatient visits. There were 1,231 mentally ill and 2,179 retarded persons in state institutions in mid-1984. Hospital personnel included 31,477 registered nurses as of 1 January 1985. The average cost of hospital care in 1982 was $257 per day. The average hospital stay in 1982 was 9.4 days, and the average cost per stay was $2,415. As of 1980 there were 379 nursing homes, with 41,400 beds and an average of 37,900 resident patients. Minnesota had 8,165 licensed physicians and 2,715 practicing dentists in 1984. In 1981 there were 522,000 Medicare patients, who received a total of $644 million in health care benefits, and 324,000 Medicare recipients, who got $680 million.

The Mayo Clinic, developed by Drs. Charles H. and William J. Mayo in the 1890s and early 1900s, was the first private clinic in the US and became a world-renowned center for surgery; today it is owned and operated by a self-perpetuating charitable foundation. The separate Mayo Foundation for Medical Education and Research, founded and endowed by the Mayo brothers in 1915, was subsequently affiliated with the University of Minnesota, which became the first US institution to offer graduate education in surgery and other branches of clinical medicine.

³⁹SOCIAL WELFARE

As of 1983, Minnesota ranked 33d among the 50 states in proportion of public aid recipients to the total population—4.1%. The value of public assistance given to the state's needy citizens totaled $1.4 billion in 1983/84, or about 15% of all state and local spending. The state budget for income support programs in the 1983/85 biennium totaled $314 million, and was projected at $395 million for 1985/87. In 1983/84, the Department of Human Resources reported that, in an average month, the following numbers of clients were served under various federal and/or state welfare programs: under aid to families with dependent children (AFDC), 146,490; general assistance maintenance, 14,943; Medicaid, 154,542; general assistance medical care, 14,106; supplemental aid to the aged, disabled, and blind, 10,018; emergency assistance, 3,029; and food stamps, 241,795. (Many recipients, of course, drew benefits from more than one program.)

Expenditures under the AFDC program totaled $233 million in 1982, for an average payment of $425 monthly per family. The federal cost of the food stamp program was $103 million in 1982/83. During the same period, an estimated 465,000 schoolchildren—more than half of those enrolled in participating schools—took part in the school lunch program, at a federal cost of $29 million. As of 1983, 623,000 Minnesotans were receiving Social Security retirement benefits totaling $2,071 million, including 118,000 who drew survivors' benefits amounting to $396 million. The average monthly payment to retirees (excluding special benefits) was $428. In 1983, 29,900 Minnesotans received Supplemental Security Income benefits of $52.1 million.

There were 42,000 disabled workers drawing $191 million in Social Security disability benefits in 1983, for an average monthly payment of $455. The cost of vocational training programs in 1982/83 was $197 million for state and local governments, with $11.2 million in federal subsidies. A total of $335 million was paid out in workers' compensation for 1982. During the same year, state and federal unemployment benefits totaled nearly $400 million; the average weekly benefit was $137, or 15% above the US norm.

⁴⁰HOUSING

According to the 1980 census, Minnesota had 1,533,725 year-round housing units, of which 94% were occupied, 72% of those by their owners. Nearly 98% of the occupied housing had full plumbing. Minnesota ranked 18th among the states in number of new housing units started in 1983, with 27,700 units. The median valuation of an owner-occupied house in 1980 was $54,300.

In 1981/82, the state's Housing Finance Agency, founded in 1973, granted low-interest loans for 17,638 family units totaling $543 million, and made loans of $177 million for home improvements on 34,488 housing units.

⁴¹EDUCATION

Minnesota's first public school system was authorized in 1849, but significant growth in enrollment did not occur until after the Civil War. Today, Minnesota has one of the best-supported systems of public education in the US. By 1980, according to state data, 73% of all Minnesotans aged 25 or older were high school graduates, compared with 58% in 1970 and 44% in 1960. About 17% of adult Minnesotans held college degrees in 1980, compared with 8% in 1960.

In 1982/83, Minnesota had an estimated 701,181 public school students, down 7.5% in only three years, and a pupil-teacher ratio of 17.8 to 1. In 1983/84 there were 38,500 public school teachers, earning an average salary of $24,400 (9th place among the states).

Catholic parochial schools had a total enrollment of 61,982 in 1983/84; other private schools enrolled 24,548 pupils in 1980/81. As of 1982/83 there were 214,133 registered students in Minnesota's institutions of higher learning, at least three-fourths of them enrolled in public institutions.

The state has four major systems of public postsecondary education. The state university system—with campuses at Bemidji, Mankato, Minneapolis–St. Paul, Moorhead, St. Cloud, and Winona—had an enrollment of 36,903 in 1982/83. The community college system, consisting of 19 two-year colleges, had 22,984 students. A statewide network of 33 area vocational-technical institutes enrolled 30,363 students as of 1984. Finally, the University of Minnesota (founded as an academy in 1851), with campuses in the Twin Cities, Duluth, Morris, Crookston, and Waseca, had 48,075 students in 1982/83. The state's oldest private college, Hamline University in St. Paul, was founded in 1854 and is affiliated with the United Methodist Church. In 1982/83 there were 18 private colleges, many of them with ties to Lutheran or Roman Catholic religious authorities. Carleton College, at Northfield, is a notable independent institution.

Minnesota has an extensive program of student grants, work-study arrangements, and loan programs, in addition to reciprocal tuition arrangements with Wisconsin and North Dakota. In 1981/82, a total of $28 million was granted to 50,131 students to help pay their expenses at institutions of higher learning inside and outside the state; the average grant amounted to $559. The largest such program was the State Student Loan Program, instituted in 1973.

The 1983/85 budget provided an estimated $3.9 billion for education, of which two-thirds represented state and federal aid to public schools. The proposed budget for the 1985/87 biennium included $4.6 billion (including federal aid) for education at all levels.

⁴²ARTS

The new Ordway Music Theater in St. Paul, which has two concert halls, opened in January 1985. The Ordway is the home of the Minnesota Opera Company and of the St. Paul Chamber Orchestra. The privately owned, nonprofit theater was built for about $45 million and was founded by Minnesota Mining and Manufacturing Corp. and other private sources.

The Minnesota Orchestra, founded in 1903 in Minneapolis, enlisted Neville Marriner as musical director in 1979; violinist Pinchas Zukerman became music director of the St. Paul Chamber Orchestra in 1980. The Minnesota Opera, conducted by Phillip Brunelle, and the St. Olaf College Choir, at Northfield, also have national reputations. The Tyrone Guthrie Theater, founded in Minneapolis in 1963, is one of the nation's most prestigious repertory companies.

The Walker Art Center in Minneapolis is an innovative museum with an outstanding contemporary collection, while the Minneapolis Institute of Arts exhibits more traditional works. The art gallery of the University of Minnesota is in Minneapolis, and the Minnesota Museum of Art is in St. Paul.

State and regional arts groups as well as individual artists are supported by state and federal grants administered through the State Arts Board, an 11-member panel appointed by the governor. For the 1986/87 biennium, the board had an estimated budget of $5.8 million.

⁴³LIBRARIES AND MUSEUMS

Minnesota has 330 public libraries, serving 97% of the state's population; about 90% of them are joined in a network of 12 regional library systems. The total number of books and audiovisual items was 15,702,724 in 1983, when public library circulation reached 28,684,321. The largest single public library system is the Minneapolis Public Library and Information Center (founded in 1885), which had 1,654,717 volumes in 1983. The leading academic library, with 3,591,760 volumes, is that maintained by the

University of Minnesota at Minneapolis. Special libraries include the James Jerome Hill Reference Library (devoted to commerce and transportation) and the library of the Minnesota Historical Society, both located in St. Paul.

There are 133 museums and historic sites. In addition to several noted museums of the visual arts, Minnesota is home to the Mayo Medical Museum at the Mayo Clinic in Rochester. The Minnesota Historical Society Museum offers rotating exhibits on varied aspects of the state's history. Historic sites include the Ft. St. Charles recreation, on an island in the Lake of the Woods; the boyhood home of Charles Lindbergh in Little Falls; and the Sauk Centre home of Sinclair Lewis.

44 COMMUNICATIONS

As of the 1980 census, 97% of Minnesota's 1,445,222 occupied housing units had telephones. The state had 849 post offices and 13,600 postal employees in 1985. Commercial broadcasting began with the opening of the first radio station in 1922; as of 1984 there were 219 radio stations—100 AM and 119 FM—and 35 television stations, including 8 educational stations. As of 1984, 156 cable television systems served 644 communities with 983,866 subscribers.

45 PRESS

The *Minnesota Pioneer,* whose first issue was printed on a small hand press and distributed by the publisher himself on 28 April 1849 in St. Paul, vies with the *Minnesota Register* (its first issue was dated earlier but may have appeared later) for the honor of being Minnesota's first newspaper. Over the next 10 years, in any case, nearly 100 newspapers appeared at locations throughout the territory, including direct ancestors of many present-day publications. In April 1982, Minneapolis's only daily newspapers were merged into the *Minneapolis Star and Tribune.* As of 1984, the state had 10 morning dailies with a combined circulation of 621,981; 16 evening dailies, 311,839; and 12 Sunday papers, 1,022,483. The following table lists the leading dailies, with their paid circulations in 1984:

AREA	NAME	DAILY	SUNDAY
Duluth	News-Tribune and Herald (m,S)	63,107	82,910
Minneapolis	Star and Tribune (m,S)	373,145	590,985
St. Paul	Pioneer Press–Dispatch (m,S)	106,777	246,496

As of 1984, 284 weekly newspapers and 176 periodicals were being published in Minnesota. Among the most widely read magazines published in Minnesota were *Family Handyman,* appearing 10 times a year, with a paid circulation of 1,150,000; *Catholic Digest,* a religious monthly serving 606,828 readers; and *Snow Goer,* published five times a year for snowmobile enthusiasts, with a paid circulation of 75,000.

46 ORGANIZATIONS

The 1982 Census of Service Industries counted 1,625 organizations in Minnesota, including 249 business associations; 1,023 civic, social, and fraternal associations; and 33 educational, scientific, and research associations. The Minnesota Historical Society, founded in 1849, is the oldest educational organization in the state and the official custodian of its history. The society is partly supported by state funds, as are such other semistate organizations as the Academy of Science (which promotes interest in science among high school students), the Minnesota State Horticultural Society, and the Humane Society. The Sons of Norway and American Swedish Institute, both with headquarters in Minneapolis, seek to preserve the state's Scandinavian heritage. Among the various professional, commercial, educational, and hobbyist associations with headquarters in Minnesota are the American Collectors Association and National Scholastic Press Association, Minneapolis; American Board of Radiology, American Ophthalmological Society, and American Board of Physical Medicine and Rehabilitation, Rochester; and World Pen Pals, St. Paul.

47 TOURISM, TRAVEL, AND RECREATION

With its lakes and parks, ski trails and campsites, and historical and cultural attractions, Minnesota provides ample recreational opportunities for residents and visitors alike. According to the state's tourism office, visitors to the state in 1983 spent $4.4 billion and generated the equivalent of 110,000 full-time jobs for Minnesotans. About half the tourists were from out of state.

Besides the museums, sports stadiums, and concert halls in the big cities, Minnesota's attractions include the 220,000-acre (89,000-hectare) Voyageurs National Park, near the Canadian border; Grand Portage National Monument, in Arrowhead Country, a former fur-trading center with a restored trading post; and Pipestone National Monument, in southwestern Minnesota, containing the red pipestone quarry used by Indians to make peace pipes. Lumbertown USA, a restored 1870s lumber community, is in Brainerd, and the US Hockey Hall of Fame is in Eveleth.

The state maintains and operates 65 parks, 9,240 mi (14,870 km) of trails, 10 scenic and natural areas, 5 recreation areas, and 18 canoe and boating routes. Minnesota also has 288 primary wildlife refuges. The parks had an estimated 6.5 million visitors in 1984. As of fiscal 1982/83 there were 1,546,385 licensed fishermen and 484,148 licensed hunters, most of whom hunted deer, muskrat, squirrel, beaver, duck, pheasant, and grouse.

An estimated 625,000 people enjoy boating each year on Minnesota's scenic waterways. Winter sports have gained in popularity, and many parks are now used heavily all year round. Snowmobiling, though it has declined somewhat since the mid-1970s, still attracts an estimated 200,000 enthusiasts annually, and cross-country skiing has rapidly accelerated in popularity.

48 SPORTS

The Twin Cities area has major league professional teams in baseball, football, and hockey. The Minnesota Twins, led by Harmon Killebrew and Tony Oliva, won the American League pennant in 1965; the Minnesota Vikings of the National Football League, featuring such stars as Fran Tarkenton and Chuck Foreman, were National Conference champions four times during the 1970s, although they failed in each of their Super Bowl bids. The North Stars of the National Hockey League also represent the state.

In collegiate sports, the University of Minnesota Golden Gophers are a Big Ten football team. The university is probably best known for its ice hockey team, which won the NCAA title three times during the 1970s and supplied the coach—Herb Brooks—and many of the players for the gold medal–winning US team in the 1980 Winter Olympics.

49 FAMOUS MINNESOTANS

No Minnesotan has been elected to the US presidency, but several have sought the office, including two who served as vice president. Hubert Horatio Humphrey (b.South Dakota, 1911–78) was vice president under Lyndon Johnson and a serious contender for the presidency in 1960, 1968, and 1972. A onetime mayor of Minneapolis, the "Happy Warrior" entered the US Senate in 1949, winning recognition as a vigorous proponent of liberal causes; after he left the vice presidency, Humphrey won reelection to the Senate in 1970. Humphrey's protege, Walter Frederick "Fritz" Mondale (b.1928), a former state attorney general, was appointed to fill Humphrey's Senate seat in 1964, was elected to it twice, and after an unsuccessful try for the presidency, became Jimmy Carter's running mate in 1976; four years later, Mondale and Carter ran unsuccessfully for reelection, losing to Ronald Reagan and George Bush. Mondale won the Democratic presidential nomination in 1984 and chose US Representative Geraldine A. Ferraro of New York as his running mate, making her the first woman to be nominated by a major party for national office; they were overwhelming defeated by Reagan and Bush, winning only 41% of the popular vote and carrying only Minnesota and the

District of Columbia. Warren Earl Burger (b.1907) of St. Paul was named chief justice of the US Supreme Court in 1969. Three other Minnesotans have served on the court: Pierce Butler (1866-1939), William O. Douglas (1898-1980), and Harry A. Blackmun (b.Illinois, 1908).

Senator Frank B. Kellogg (b.New York, 1856-1937), who as secretary of state helped to negotiate the Kellogg-Briand Pact renouncing war as an instrument of national policy (for which he won the 1929 Nobel Peace Prize), also served on the Permanent Court of International Justice. Other political leaders who won national attention include Governors John A. Johnson (1861-1909), Floyd B. Olson (1891-1936), and Harold E. Stassen (b.1907), a perennial presidential candidate since 1948 but a serious contender in his early races. Eugene J. McCarthy (b.1916), who served in the US Senate, was the central figure in a national protest movement against the Vietnam war and, in that role, unsuccessfully sought the 1968 Democratic presidential nomination won by Humphrey. McCarthy also ran for the presidency as an independent in 1976.

Several Minnesotans besides Kellogg have served in cabinet posts. Minnesota's first territorial governor, Alexander Ramsey (1815-1903), later served as a secretary of war, and Senator William Windom (1827-91) was also secretary of the treasury. Others serving in cabinet posts have included William DeWitt Mitchell (1874-1955), attorney general; Maurice H. Stans (b.1908), secretary of commerce; James D. Hodgson (b.1915), secretary of labor; and Orville Freeman (b.1918) and Bob Bergland (b.1928), both secretaries of agriculture. The first woman ambassador in US history was Eugenie M. Anderson (b.Iowa, 1909), like Humphrey an architect of the Democratic-Farmer-Labor Party.

Notable members of Congress include Knute Nelson (b.Norway, 1843-1923), who served in the Senate from 1895 to his death; Henrik Shipstead (1881-1960), who evolved into a leading Republican isolationist during 24 years in the Senate; Representative Andrew J. Volstead (1860-1947), who sponsored the 1919 prohibition act that bears his name; and Representative Walter Judd (b.1898), a prominent leader of the so-called China Lobby.

Daniel Greysolon, Sieur Duluth (b.France, 1636-1710), Father Louis Hennepin (b.Flanders, 1640? -1701), and Jonathan Carver (b.Massachusetts, 1710-80) were among the early explorers and chroniclers of what is now the State of Minnesota. Fur trader Henry H. Sibley (b.Michigan, 1811-91) was a key political leader in the territorial period and became the state's first governor; he also put down the Sioux uprising of 1862. Railroad magnate James J. Hill (b.Canada, 1838-1916) built one of the greatest corporate empires of his time, and Oliver H. Kelley (b.Massachusetts, 1826-1913), a Minnesota farmer, organized the first National Grange. John Ireland (b.Ireland, 1838-1918) was the first Roman Catholic archbishop of St. Paul, while Henry B. Whipple (b.New York, 1822-1901), longtime Episcopal bishop of Minnesota, achieved particular recognition for his work among Indians in the region.

The first US citizen ever to be awarded the Nobel Prize for literature was Sinclair Lewis (1885-1951), whose novel *Main Street* (1920) was modeled on life in his hometown of Sauk Centre. Philip S. Hench (b.Pennsylvania, 1896-1965) and Edward C. Kendall (b.Connecticut, 1886-1972), both of the Mayo Clinic, shared the 1950 Nobel Prize for medicine, and St. Paul native Melvin Calvin (b.1911) won the 1961 Nobel Prize for chemistry.

The Mayo Clinic was founded in Minnesota by Dr. William W. Mayo (b.England, 1819-1911) and developed through the efforts of his sons, Drs. William H. (1861-1939) and Charles H. (1865-1939) Mayo. Oil magnate J. Paul Getty (1892-1976) was a Minnesota native, as was Richard W. Sears (1863-1914), founder of Sears, Roebuck.

Prominent literary figures, besides Sinclair Lewis, include Ignatius Donnelly (b.Pennsylvania, 1831-1901), a writer, editor, and Populist Party crusader; F. Scott Fitzgerald (1896-1940), well known for classic novels including *The Great Gatsby;* and Ole Edvart Rølvaag (b.Norway, 1876-1931), who conveyed the reality of the immigrant experience in his *Giants in the Earth.* The poet and critic Allen Tate (b.Kentucky, 1899-1979) taught for many years at the University of Minnesota.

Journalist Westbrook Pegler (1894-1969) and cartoonist Charles Schulz (b.1922) were both born in Minnesota, and economist Thorstein Veblen (b.Wisconsin, 1857-1929) lived there. Architects LeRoy S. Buffington (1847-1937) and Cass Gilbert (b.Ohio, 1859-1934) influenced their fields well beyond the state's borders, as did Minnesota artists Wanda Gag (1893-1946) and Adolph Dehn (1895-1968).

Minnesota born entertainers include Judy Garland (Frances Gumm, 1922-69) and Bob Dylan (Robert Zimmerman, b.1941). Football star William "Pudge" Heffelfinger (1867-1954) was a Minnesota native, and Bronislaw "Bronco" Nagurski (b.Canada, 1908) played for the University of Minnesota.

50 BIBLIOGRAPHY

Blegen, Theodore C. *Minnesota: A History of the State.* Rev. ed. Minneapolis: University of Minnesota Press, 1975 (orig. 1937).

Brook, Michael. *Reference Guide to Minnesota History: A Subject Bibliography of Books, Pamphlets, and Articles in English.* St. Paul: Minnesota Historical Society, 1974.

Chrislock, Carl H. *The Progressive Era in Minnesota, 1899-1918.* St. Paul: Minnesota Historical Society, 1971.

Coen, Rena N. *Painting and Sculpture in Minnesota, 1820-1914.* Minneapolis: University of Minnesota Press, 1976.

Federal Writers' Project. *Minnesota: A State Guide.* Reprint. New York: Somerset, n.d. (orig. 1938).

Folwell, William W. *A History of Minnesota.* 4 vols. Rev. ed. St. Paul: Minnesota Historical Society, 1956-69 (orig. 1921-30).

Hazard, Evan B. *The Mammals of Minnesota.* Minneapolis: University of Minnesota Press, 1982.

Holmquist, June D., and Jean A. Brookins. *Minnesota's Major Historic Sites: A Guide.* 2d ed. St. Paul: Minnesota Historical Society, 1972.

Lass, William E. *Minnesota: A Bicentennial History.* New York: Norton, 1977.

Meyer, Roy. *History of the Santee Sioux: United States Indian Policy on Trial.* Lincoln: University of Nebraska Press, 1967.

Mitau, G. Theodore. *Politics in Minnesota.* 2d rev. ed. Minneapolis: University of Minnesota Press, 1970.

Schwartz, George M., and G. A. Thiel. *Minnesota's Rocks and Waters: A Geological Story.* Minneapolis: University of Minnesota Press, 1954.

Spadaccini, Victor M. (ed.) *Minnesota Pocket Data Book, 1985-86.* St. Paul: Blue Sky, 1983.

Upham, Warren. *Minnesota Geographic Names: Their Origin and Historic Significance.* St. Paul: Minnesota Historical Society, 1969 (orig. 1920).

MISSISSIPPI

State of Mississippi

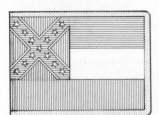

ORIGIN OF STATE NAME: Derived from the Ojibwa Indian words *misi sipi*, meaning great river. **NICKNAME:** The Magnolia State. **CAPITAL:** Jackson. **ENTERED UNION:** 10 December 1817 (20th). **SONG:** "Go, Mississippi." **MOTTO:** *Virtute et armis* (By valor and arms). **COAT OF ARMS:** An American eagle clutches an olive branch and a quiver of arrows in its talons. **FLAG:** Crossed blue bars, on a red field, bordered with white and emblazoned with 13 white stars—the motif of the Confederate battle flag—cover the upper left corner. The field consists of three stripes of equal width, blue, white, and red. **OFFICIAL SEAL:** The seal consists of the coat of arms surrounded by the words "The Great Seal of the State of Mississippi." **MAMMAL:** White-tailed deer. **WATER MAMMAL:** Porpoise. **BIRD:** Mockingbird. **WATERFOWL:** Wood duck. **FISH:** Largemouth or black bass. **INSECT:** Honeybee. **FOSSIL:** Prehistoric whale. **FLOWER:** Magnolia. **TREE:** Magnolia. **STONE:** Petrified wood. **BEVERAGE:** Milk. **LEGAL HOLIDAYS:** New Year's Day, 1 January; Birthdays of Robert E. Lee and Martin Luther King, Jr., 3d Monday in January; Washington's Birthday, 3d Monday in February; Confederate Memorial Day, last Monday in April; Jefferson Davis's Birthday, 1st Monday in June; Independence Day, 4 July; Labor Day, 1st Monday in September; Veterans Day, 11 November; Thanksgiving Day, 4th Thursday in November; Christmas Day, 25 December. **TIME:** 6 AM CST = noon GMT.

¹LOCATION, SIZE, AND EXTENT

Located in the eastern south-central US, Mississippi ranks 32d in size among the 50 states.

The total area of Mississippi is 47,689 sq mi (123,514 sq km), of which land takes up 47,233 sq mi (122,333 sq km) and inland water 456 sq mi (1,181 sq km). Mississippi's maximum E-W extension is 188 mi (303 km); its greatest N-S distance is 352 mi (566 km).

Mississippi is bordered on the N by Tennessee; on the E by Alabama; on the S by the Gulf of Mexico and Louisiana; and on the W by Louisiana (with the line partially formed by the Pearl and Mississippi rivers) and Arkansas (with the line formed by the Mississippi River). Several small islands lie off the coast.

The total boundary length of Mississippi is 1,015 mi (1,634 km). The state's geographic center is in Leake County, 9 mi (14 km) WNW of Carthage.

²TOPOGRAPHY

Mississippi lies entirely within two lowland plains. Extending eastward from the Mississippi River, the Mississippi Alluvial Plain, popularly known as the Delta, is very narrow south of Vicksburg but stretches as much as a third of the way across the state farther north. The Gulf Coastal Plain, covering the rest of the state, includes several subregions, of which the Red Clay Hills of north-central Mississippi and the Piney Woods of the south and southeast are the most extensive. Mississippi's generally hilly landscape ascends from sea level at the Gulf of Mexico to reach its maximum elevation, 806 feet (246 meters), at Woodall Mountain, in the extreme northeastern corner of the state.

The state's largest lakes—Grenada, Sardis, Enid, and Arkabutla—are all man-made. Numerous smaller lakes—called oxbow lakes because of their curved shape—extend along the western edge of the state; once part of the Mississippi River, they were formed when the river changed its course. Mississippi's longest inland river, the Pearl, flows about 490 mi (790 km) from the eastern center of the state to the Gulf of Mexico, its lower reaches forming part of the border with Louisiana. The Big Black River, some 330 mi (530 km) long, begins in the northeast and cuts diagonally across the state, joining the Mississippi about 20 mi (32 km) below Vicksburg. Formed by the confluence of the Tallahatchie and Yalobusha rivers at Greenwood, the Yazoo flows 189 mi (304 km) southwest to the Mississippi just above Vicksburg.

³CLIMATE

Mississippi has short winters and long, humid summers. Summer temperatures vary little from one part of the state to another. Biloxi, on the Gulf coast, averages 82°F (28°C) in July, while Oxford, in the north-central part of the state, averages 80°F (27°C). During the winter, however, because of the temperate influence of the Gulf of Mexico, the southern coast is much warmer than the north; in January, Biloxi averages 52°F (11°C) to Oxford's 41°F (5°C). The lowest temperature ever recorded in Mississippi was −19°F (−28°C) on 30 January 1966 in Corinth; the highest, 115°F (46°C), was set on 29 July 1930 at Holly Springs.

Precipitation in Mississippi increases from north to south. The north-central region averages 53 in (135 cm) of precipitation a year; the coastal region, 62 in (157 cm). Some snow falls in northern and central sections. Mississippi lies in the path of hurricanes moving northward from the Gulf of Mexico during the late summer and fall. On 17–18 August 1969, Hurricane Camille ripped into Biloxi and Gulfport and caused more than 100 deaths throughout the state. Two tornado alleys cross Mississippi from the southwest to northeast, from Vicksburg to Oxford and McComb to Tupelo. In 1983, Mississippi was hit by 22 tornadoes.

⁴FLORA AND FAUNA

Post and white oaks, hickory, maple, and magnolia grow in the forests of the uplands; various willows and sums (including the tupelo) in the Delta; and longleaf pine in the Piney Woods. Characteristic wild flowers include the green Virginia creeper, black-eyed Susan, and Cherokee rose.

Common among the state's mammals are the opossum, eastern mole, armadillo, coyote, mink, white-tailed deer, striped skunk, and diverse bats and mice. Birds include varieties of wren, thrush, warbler, vireo, and hawk, along with numerous waterfowl and seabirds, Franklin's gull, the common loon, and the wood stork among them. Black bass, perch, and mullet are common freshwater fish. Rare species in Mississippi include the hoary bat, American oystercatcher, mole salamander, pigmy killifish, Yazoo darker, and five species of crayfish. The brown or grizzly bear, cliff swallow, and eastern indigo snake are among threatened species, and the Florida panther, gray bat, red wolf, Mississippi sandhill crane, bald eagle, peregrine falcon, brown pelican, red-cockaded and ivory-billed woodpeckers, and Bayou darker are endangered.

[5] ENVIRONMENTAL PROTECTION

Agencies with environmental responsibilities in Mississippi include the Department of Natural Resources and the Department of Wildlife Conservation, both established in 1979, and the Soil and Water Commission. The Gulf Coast Research Laboratory conducts research on marine resources, primarily in relation to the Mississippi Gulf coast. The Department of Health gives licenses for solid waste disposal. The most controversial environmental issues in the early 1980s were the construction of the Grand Gulf and Yellow Creek nuclear reactors and the possible location of nuclear waste disposal sites in the state.

[6] POPULATION

With a 1980 census population of 2,520,638, Mississippi ranked 31st among the 50 states. After remaining virtually level for 30 years, Mississippi's population during the 1970s grew 13.7%. In 1985, the population was estimated at 2,623,089.

According to the 1980 census, 48.2% of Mississippians were male, and 51.8% female. Mississippians were less mobile than residents of most other states: 90% of those living in the state in 1975 were still there in 1980. In 1985, the population density was 56 persons per sq mi (22 per sq km).

In 1980, 47% of the population was urban, and 53% rural. Although this percentage was far below the national average—Mississippi remains one of the most rural states in the US—the urban population has increased fivefold since 1920, when only 13% of state residents lived in cities. Mississippi's largest city, Jackson, had a 1984 census population of 208,810, more than double the 1950 total. Next came Biloxi, with 48,685 residents, followed by Meridian, 47,365; Hattiesburg, 41,927; Gulfport, 41,232; and Greenville, 40,636. The Jackson metropolitan area had an estimated population of 382,400 in 1984.

[7] ETHNIC GROUPS

Since 1860, blacks have constituted a larger proportion of the population of Mississippi than of any other state. By the end of the 1830s, blacks outnumbered whites 52% to 48%, and from the 1860s through the early 20th century, they made up about three-fifths of the population. Because of out-migration, the proportion of black Mississippians declined to less than 36% in 1980, when the state had 1,615,190 whites, 887,206 blacks, and only 18,242 members of other races. A total of 24,731 were of Spanish origin.

Until the 1940s, the Chinese—who numbered 1,835 in 1980—were an intermediate stratum between blacks and whites in the social hierarchy of the Delta Counties. Indochinese refugees began settling in Mississippi in 1975, and in 1980 there were 1,281. Although the number of foreign-born almost tripled in the 1970s, Mississippi (along with Kentucky) had the smallest percentage of foreign-born residents (0.9%) in 1980.

Mississippi has only a small Indian population remaining—6,131 in 1980. Many of them live on the Choctaw reservation in the east-central region.

[8] LANGUAGES

English in the state is largely Southern, with some South Midland speech in northern and eastern Mississippi because of population drift from Tennessee. Typical are the absence of final /r/ and the lengthening and weakening of the diphthongs /ai/ and /oi/ as in *ride* and *oil*. South Midland terms in northern Mississippi include *tow sack* (burlap bag), *dog irons* (andirons), *plum peach* (clingstone peach), *snake doctor* (dragonfly), and *stone wall* (rock fence). In the eastern section are found *jew's harp* (harmonica) and *croker sack* (burlap bag). Southern speech in the southern half features *gallery* for porch, *mosquito hawk* for dragonfly, and *press peach* for clingstone peach. Louisiana French has contributed *armoire* (wardrobe).

In 1980, 98% of Mississippi residents 3 years old and older spoke only English in the home. Other languages spoken at home, and the number of people who spoke them, included Spanish, 14,961, and French, 8,378.

[9] RELIGIONS

Protestants have dominated Mississippi since the late 18th century. The Baptists are the leading denomination, and many adherents are fundamentalists. Partly because of the strong church influence, Mississippi was among the first states to enact prohibition and among the last to repeal it.

During 1980, membership in the two principal Protestant denominations was: Southern Baptist Convention, 760,385 known adherents, and United Methodist Church, 250,240. There were 95,096 Roman Catholics and an estimated 3,080 Jews in 1984.

[10] TRANSPORTATION

At the end of 1983, the Illinois Central Gulf railroad line operated 2,229 mi (3,587 km) of track in Mississippi. Fourteen other lines in the state operated an additional 871 mi (1,402 km) of track. Of this mileage, 2,574 mi (4,142 km) was Class I trackage. Amtrak operated two long-distance trains (Chicago–New Orleans and New York–New Orleans) through Mississippi, with stops at 17 stations in the state.

Mississippi had 71,075 mi (114,384 km) of roads, 64,169 mi (87,365 km) rural and 6,906 mi (11,114 km) urban, at the end of 1983. In addition, there were 686 mi (1,104 km) of interstate highways. Highways I-55, running north–south, and I-20, running east–west, intersect at Jackson. I-59 runs diagonally through the southeastern corner of Mississippi from Meridian to New Orleans. In 1983 there were 1,802,952 licensed drivers in Mississippi and 1,560,277 registered motor vehicles, including 1,189,446 automobiles and 363,168 trucks.

Mississippi has two deepwater seaports, Gulfport and Pascagoula; in 1982, Pascagoula handled 19,157,414 tons of cargo and Gulfport 1,195,212 tons. Other ports include Biloxi, on the Gulf of Mexico, which handled 1,658,973 tons, along with Vicksburg (2,658,080 tons), and Greenville (1,883,605 tons). Much of Pascagoula's heavy volume consists of oil and gas imports; imports by tanker totaled 10,556,500 tons in 1983. The Tennessee-Tombigbee Waterway, linking the Ohio River and the Gulf Coast and passing though Mississippi, opened in 1985. The Yazoo is also open to river traffic.

There were 165 airports and 16 heliports in Mississippi in 1983. The most important airfield, Allen C. Thompson Field, near Jackson, enplaned 324,666 passengers in 1983.

[11] HISTORY

The earliest record of human habitation in the region that is now the State of Mississippi goes back perhaps 2,000 years. The names of Mississippi's pre-Columbian inhabitants are not known. Upon the appearance of the first Spanish explorers in the early 16th century, Mississippi Indians numbered some 30,000 and were divided into 15 tribes. Soon after the French settled in 1699, however, only three large tribes remained: the Choctaw, the Chickasaw, and the Natchez. The French destroyed the Natchez in 1729–30 in retaliation for the massacre of a French settlement on the Natchez bluffs.

Spanish explorers, of whom Hernando de Soto in 1540–41 was the most notable, explored the area that is now Mississippi in the first half of the 16th century. De Soto found little of the mineral wealth he was looking for, and the Spanish quickly lost interest in the region. The French explorer Robert Cavelier, Sieur de la Salle, penetrated the lower Mississippi Valley from New France (Canada) in 1682. La Salle discovered the mouth of the Mississippi and named the entire area Louisiana in honor of the French King, Louis XIV.

An expedition under French-Canadian Pierre Lemoyne, Sieur d'Iberville, planted a settlement at Biloxi Bay in 1699. Soon the

LOCATION: 30°06′32″ to 35°N; 88°5′53″ to 91°38′34″ W. **BOUNDARIES:** Tennessee line, 119 mi (192 km); Alabama line, 337 mi (542 km); Gulf of Mexico coastline, 44 mi (71 km); Louisiana line, 307 mi (494 km); Arkansas line, 208 mi (335 km).

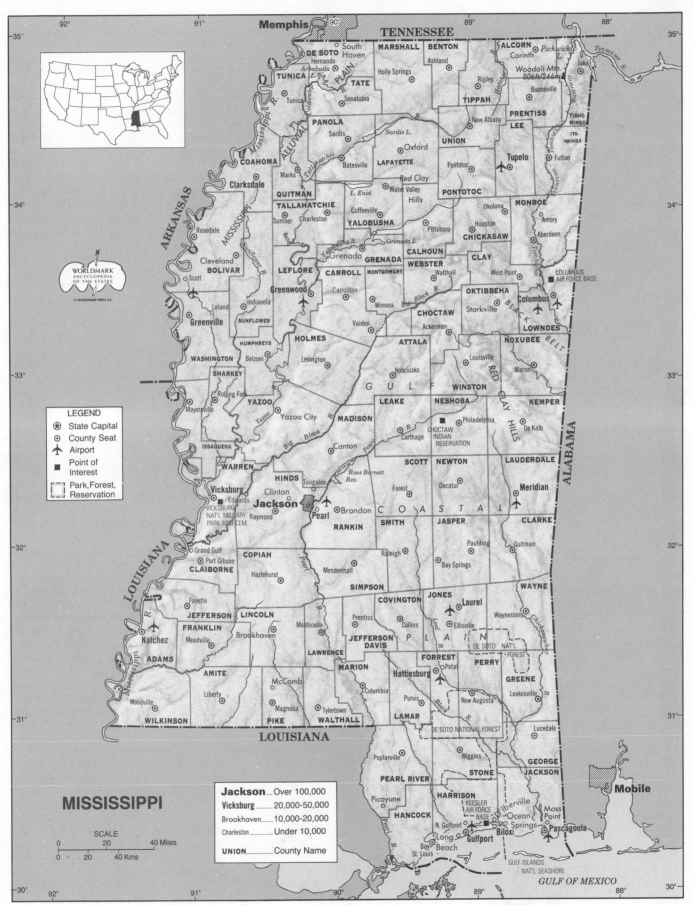

MISSISSIPPI

SCALE

0   20   40 Miles

0   20   40 Kms

LEGEND
- State Capital
- ⊙ County Seat
- ✈ Airport
- ■ Point of Interest
- Park, Forest, Reservation

Jackson......Over 100,000
Vicksburg.........20,000-50,000
Brookhaven........10,000-20,000
Charleston........Under 10,000

UNION.............County Name

See endsheet maps: J4.

French opened settlements at Mobile (1702), Natchez (1716), and finally New Orleans (1718), which quickly eclipsed the others in size and importance. After losing the French and Indian War, France ceded Louisiana to its Spanish ally in 1762. The following year, Spain ceded the portion of the colony that lay east of the Mississippi to England, which governed the new lands as West Florida. During the American Revolution, the Spanish, who still held New Orleans and Louisiana, marched into Natchez, Mobile, and Pensacola (the capital) and took West Florida by conquest.

Although the US claimed the Natchez area after 1783, Spain continued to rule it. However, the Spanish were unable to change the Anglo-American character of the settlement. Spain agreed to relinquish its claim to the Natchez District by signing the Treaty of San Lorenzo on 27 October 1795, but did not evacuate its garrison there for another three years.

The US Congress organized the Mississippi Territory in 1798. Between 1798 and 1817, the territory grew enormously in population, attracting immigrants mainly from the older states of the South but also from the Middle Atlantic states and even from New England. During this period, the territory included all the land area that is today within the borders of Mississippi and Alabama. However, sectionalism and the territory's large size convinced Congress to organize the eastern half as the Alabama Territory in 1817. Congress then offered admission to the western half, which became the nation's 20th state—Mississippi—on 10 December.

Until the Civil War, Mississippi exemplified the American frontier; it was bustling, violent, and aggressive. By and large, Mississippians viewed themselves as westerners, not southerners. Nor was Mississippi, except for a few plantations around Natchez, a land of large planters. Rather, Mississippi's antebellum society and government were dominated by a coalition of prosperous farmers and small landowners. At the time of statehood, the northern two-thirds of Mississippi, though nominally under US rule since 1783, remained in the hands of the Choctaw and Chickasaw and was closed to settlement. Under intense pressure from the state government and from Andrew Jackson's presidential administration, these tribes signed three treaties between 1820 and 1832, ceding their Mississippi lands and agreeing to move to what is now Oklahoma.

The opening of fertile Indian lands for sale and settlement produced a boom of speculation and growth unparalleled in Mississippi history. Cotton agriculture and slavery—introduced by the French and carried on by the British and Spanish, but hitherto limited mostly to the Natchez area—swept over the state. As the profitability and number of slaves increased, so did attempts by white Mississippians to justify slavery morally, socially, and economically. The expansion of slavery also produced a defensive attitude which focused the minds of white Mississippians on two dangers: that the slaves outnumbered the whites and would threaten white society unless kept down by slavery; and that any attack on slavery, whether from the abolitionists or from Free-Soilers like Abraham Lincoln, was a threat to white society. The danger, they believed, was so great that no price was too high to maintain slavery, even secession and civil war.

After Lincoln's election to the US presidency, Mississippi became, on 9 January 1861, the second southern state to secede. When the war began, Mississippi occupied a central place in Union strategy. The state sat squarely astride the major Confederate east–west routes of communication in the lower South, and the Mississippi River twisted along the state's western border. Control of the river was essential to Union division of the Confederacy. The military campaign fell into three phases: the fight for northeastern Mississippi in 1862, the struggle for Vicksburg in 1862–63, and the battle for east Mississippi in 1864–65. The Union advance on Corinth began with the Battle of

Shiloh (Tenn.) in April 1862. The first Union objective was the railroad that ran across the northeastern corner of Mississippi from Corinth to Iuka and linked Memphis, Tenn., to Atlanta, Ga. Losses in the ensuing battle of Shiloh, which eventually led to the occupation of Corinth by Union troops, exceeded 10,000 men on each side.

The campaign that dominated the war in Mississippi—and, indeed, along with Gettysburg provided the turning point of the Civil War—was Vicksburg. Perched atop high bluffs overlooking a bend in the Mississippi and surrounded by hills on all sides, Vicksburg provided a seemingly impregnable fortress. Union forces maneuvered before Vicksburg for more than a year before Grant besieged the city and forced its surrender on 4 July 1863. Along with Vicksburg went the western half of Mississippi. The rest of the military campaign in the state was devoted to the fight for the east, which Union forces still had not secured when the conflict ended in 1865. Of the 78,000 Mississippians who fought in the Civil War, nearly 30,000 died.

Ten years of political, social, and economic turmoil followed. Reconstruction was a tumultuous period during which the Republican Party encouraged blacks to vote and hold political office, while the native white Democrats resisted full freedom for their former slaves. The resulting confrontation lasted until 1875, when, using violence and intimidation, the Democrats recaptured control of the state from the Republicans and began a return to the racial status quo antebellum. Reconstruction left a scar on the minds of Mississippians. To the whites it seemed proof that blacks were incapable of exercising political power; to the blacks it proved that political and social rights could not long be maintained without economic rights.

The era from the end of Reconstruction to World War II was a period of economic, political, and social stagnation for Mississippi. In many respects, white Mississippians pushed Negroes back into slavery in all but name. Segregation laws and customs placed strict social controls on blacks, and a new state constitution in 1890 removed the last vestiges of their political rights. Mississippi's agricultural economy, dominated by cotton and tenant farming, provided the economic equivalent of slavery for black sharecroppers. As a continuing agricultural depression ground down the small white farmers, many of them also were driven into the sharecropper ranks; in 1890, 63% of all Mississippi farmers were tenants. Whether former planter-aristocrats like John Sharp Williams or small-farmer advocates like James K. Vardaman (1908–12) and Theodore Bilbo (1916–20 and 1928–32) held office as governor, political life was dominated by the overriding desire to keep the blacks subservient. From Reconstruction to the 1960s, white political solidarity was of paramount importance. Otherwise, the whites reasoned, another reconstruction would follow. According to the Tuskegee Institute, 538 blacks were lynched in Mississippi between 1883 and 1959, more than in any other state.

The Great Depression of the 1930s pushed Mississippians, predominantly poor and rural, to the point of desperation, and the state's agricultural economy to the brink of disaster. In 1932, cotton sank to 5 cents a pound, and one-fourth of the state's farmland was forfeited for nonpayment of taxes. World War II unleashed the forces that would later revolutionize Mississippi's economic, social, and political order. The war brought the first prosperity to Mississippi in a century. By introducing outsiders to Mississippi and Mississippians to the world, the armed forces and the war began to erode the state's insularity. It also stimulated industrial growth and agricultural mechanization and encouraged an exodus of blacks to better-paying jobs in other states.

The seeds for the civil rights movement were sown in the war years. Within little more than a generation, from 1945 to 1975, legal segregation was destroyed, and black people exercised political rights for the first time since Reconstruction. The

"Mississippi Summer" civil rights campaign—and the violent response to it, including the abduction and murder of three civil rights activists in June 1964—helped turn public opinion toward racial accommodation.

By the early 1980s, according to any standard, Mississippi had become an industrial state. In the agricultural sector, cotton had been dethroned and crop diversification accomplished. In short, after a century, Mississippi had reentered the mainstream of American life.

¹²STATE GOVERNMENT

Mississippi has had four state constitutions. The first (1817) accompanied Mississippi's admission to the Union. A second constitution (1832) was superseded by that of 1868, redrafted under Republican rule to allow Mississippi's readmission to the Union after the Civil War. The state's present constitution, as amended, dates from 1890.

Mississippi's bicameral legislature includes a 52-member senate and a 122-member house of representatives. All state legislators are elected to 4-year terms. State representatives must be at least 21 years of age and senators 25. All legislators must have been Mississippi residents for four years and residents of their district for two years before election.

The governor, lieutenant governor, secretary of state, attorney general, state treasurer, state auditor of public accounts, commissioner of insurance, and commissioner of agriculture and commerce are independently elected to four-year terms. The governor and treasurer may not serve successive terms. The governor (and lieutenant governor) must be at least 30 years of age, a US citizen for 20 years, and a Mississippi resident for 5 years before election.

Constitutional amendments must first receive the approval of two-thirds of the members of each house of the legislature; a majority of voters must approve the amendment on a statewide ballot. The constitution also provides for the calling of a constitutional convention, by majority vote of each house.

Every US citizen over the age of 18 may vote in Mississippi upon producing evidence of 30 days of residence in the state. There were an estimated 1,669,539 persons registered to vote in 1983.

¹³POLITICAL PARTIES

Mississippi's major political parties are the Democratic Party and the Republican Party, each an affiliate of the national party organization, but the Republicans are weak below the national level. Mississippi Democrats have often been at odds with each other and with the national Democratic Party.

In the 1830s, party affiliation in the state began to divide along regional and economic lines: woodsmen and small farmers in eastern Mississippi became staunch Jacksonian Democrats, while the conservative planters in the western river counties tended to be Whigs. An early demonstration of the power of the Democrats was the movement of the state capital from Natchez in 1821 to a new city named after Andrew Jackson. During the pre–Civil War years, the secessionists were largely Democrats; the Unionists, western Whigs.

During Reconstruction, Mississippi had its first Republican governor. After the Democrats returned to power in 1875, they systematically deprived blacks of the right to vote, specifically by inserting in the constitution of 1890 a literacy clause that could be selectively interpreted to include illiterate whites but exclude blacks. A poll tax and convoluted residency requirements also restricted the electorate. Voter registration among blacks fell from 130,607 in 1880 to 16,234 by 1896.

In 1948, Mississippi Democrats seceded from the national party over the platform, which opposed racial discrimination. That November, Mississippi voters backed the States' Rights Democratic (Dixiecrat) presidential ticket. At the national Democratic convention in 1964, the black separatist Freedom Democratic Party asked to be allotted 40% of Mississippi's seats but was turned down. A further division in the party occurred during the

Mississippi Presidential Vote by Political Parties, 1948–84

YEAR	ELECTORAL VOTE	MISSISSIPPI WINNER	DEMOCRAT	REPUBLICAN	STATES' RIGHTS DEMOCRAT	SOCIALIST WORKERS	LIBERTARIAN
1948	9	Thurmond (SRD)	19,384	4,995	167,538	—	—
1952	8	Stevenson (D)	172,553	112,966	—		
					INDEPENDENT		
1956	8	Stevenson (D)	144,453	60,683	42,961	—	—
					UNPLEDGED		
1960	8	Byrd¹	108,362	73,561	116,248		
1964	7	Goldwater (R)	52,616	356,512	—	—	—
					AMERICAN IND.		
1968	7	Wallace (AI)	150,644	88,516	415,349	—	—
					AMERICAN		
1972	7	*Nixon (R)	126,782	505,125	11,598	2,458	—
1976	7	*Carter (D)	381,309	366,846	6,678	2,805	2,788
					WORKERS' WORLD		
1980	7	*Reagan (R)	429,281	441,089	2,402	2,240	4,702
1984	7	*Reagan (R)	352,192	582,377	—	—	2,336

* Won US presidential election.
¹ Unpledged electors won plurality of votes and cast Mississippi's electoral votes for Senator Harry F. Byrd of Virginia.

1960s between the (black) Loyalist Democrats and the (white) Regular Democrats, who were finally reunited in 1976. During the 1950s and early 1960s, the segregationist White Citizens' Councils were so widespread and influential in the state as to rival the major parties in political importance.

Since the passing of the federal Voting Rights Act of 1965, black Mississippians have registered and voted in substantial numbers. According to estimates by the Voter Education Project, only 22,000 (5%) of voting-age blacks were registered in 1960; by 1984, 410,000 (76%) were registered. In 1985 there were 444 elected black officials in Mississippi (2d in the US).

Although blacks have made substantial progress in politics, women have not. In 1985, Mississippi had the lowest percentage of women state legislators, 1.7%. One woman held statewide elective office—Lenore L. Prather, a supreme court justice.

Mississippi was one of the most closely contested states in the South during the 1976 presidential election, and that again proved to be the case in 1980, when Ronald Reagan edged Jimmy Carter by a plurality of fewer than 12,000 votes. In 1984, however, Reagan won the state by a landslide, polling 62% of the vote.

In 1985, Democrats controlled all statewide elective offices, as well as the house and senate, which had a combined total of only nine Republican legislators. The state had one Republican US senator (the first since Reconstruction), Thad Cochran, and one Democratic senator, John C. Stennis. Three of the five US representatives were Democrats.

14 LOCAL GOVERNMENT
Each of Mississippi's 82 counties is divided into 5 districts, each of which elects a member to the county board of supervisors. As of 1980, Mississippi had 30 municipalities with populations of 10,000 or more, 68 with populations of 2,500 to 10,000, and 75 with populations of 1,000 to 2,500. Most cities, including most of the larger ones, have a mayor and city council.

15 STATE SERVICES
The Mississippi Ethics Commission, established by the state legislature in 1979, is composed of eight members who administer a code of ethics requiring all state officials and elected local officials to file statements of sources of income.

The Mississippi Department of Education is primarily a planning and service organization whose role is to assist local schools from kindergarten through junior college and adult education. A separate Board of Trustees of Institutions of Higher Learning administers Mississippi's public college and university system. The Department of Health administers a statewide system of public health services, but other bodies, including the Department of Mental Health, also have important functions in this field. The Department of Public Welfare provides welfare services in the areas of assistance payments, child support, food stamp distribution, and such social services as foster home care.

Public protection is afforded by the Civil Defense Council, Military Department, Bureau of Narcotics, Department of Public Safety (including the Highway Safety Patrol), and Department of Corrections.

16 JUDICIAL SYSTEM
The Mississippi supreme court consists of a chief justice, two presiding justices, and six associate justices, all elected to eight-year terms. The constitution stipulates that the supreme court must hold two sessions a year in the state capital; one session is to commence on the 2d Monday of September, the other on the 1st Monday of March. The principal trial courts are 20 chancery courts, which try civil cases, and 20 circuit courts, which try both civil and criminal cases; their 74 judges are elected to four-year terms. Municipal court judges are appointed. Small-claims courts are presided over by justices of the peace, who need not be lawyers.

There were 6,115 prisoners in state and federal prisons in Mississippi at the end of 1984. In the mid-1970s, federal courts

mandated reform of Mississippi prisons; but in mid-1985, 19% of the state's 91 jails, holding 1,006 prisoners, were under court order to correct overcrowding.

In 1983, Mississippi had a total FBI Crime Index rate of 3,208.2 per 100,000 population, 5th lowest in the nation. The violent crime rate was also well below the national norm, but the murder rate was high, tied for 4th with Florida. The death penalty was reinstated in 1977.

17 ARMED FORCES
Mississippi had 13 defense installations, with 18,215 military personnel, in 1984. There were two major US Air Force bases, Keesler (Biloxi) and Columbus. Among the five US naval installations were an oceanographic command at Bay St. Louis, an air station at Meridian, and a construction battalion center at Gulfport. In 1983/84, Mississippi received $2.2 billion in federal defense contracts.

There were 245,000 veterans of US military service living in Mississippi as of 30 September 1983. Of those who served in wartime, 3,000 were veterans of World War I, 102,000 of World War II, 49,000 of the Korean conflict, and 66,000 of the Viet-Nam era. Benefits totaling some $346 million were paid to Mississippi veterans during 1981/82.

As of the end of 1984, the Mississippi Army National Guard had assigned strength of 12,092, and in early 1985, the Air National Guard had assigned strength of 2,496. State and local police numbered 7,382 in 1983, of whom 921 were employed by the Highway Patrol.

18 MIGRATION
In the late 18th century, most Mississippians were immigrants from the South and predominantly of Scotch-Irish descent. The opening of lands ceded by the Indians beginning in the 1820s brought tens of thousands of settlers into northern and central Mississippi, and a resulting population increase between 1830 and 1840 of 175% (including an increase of 197% in the slave population).

After the Civil War, there was little migration into the state, but much out-migration, mainly of blacks. The exodus from Mississippi was especially heavy during the 1940s and 1950s, when at least 720,000 people, nearly three-quarters of them black, left the state. During the 1960s, a net total of between 267,000 and 279,000 blacks departed, while net white out-migration came to an end. Black out-migration slowed considerably during the 1970s, and more whites settled in the state than left. Also during the 1970s there was considerable intrastate migration to Hinds County (Jackson) and the Gulf Coast. Between 1970 and 1983, Mississippi had a net gain from migration of 22,000.

19 INTERGOVERNMENTAL COOPERATION
The Mississippi Commission on Interstate Cooperation oversees and encourages the state's participation in interstate bodies, especially the Council of State Governments and the National Conference of State Legislatures. Mississippi also participates in the Gulf States Marine Fisheries Compact, Southern Growth Policies Compact, Southern Interstate Energy Compact, Southern Regional Education Compact, and Tennessee-Tombigbee Waterway Development Compact.

Mississippi received $1.2 billion in federal aid in 1983/84.

20 ECONOMY
Between the Civil War and World War II, Mississippi's economy remained poor, stagnant, and highly dependent on the market for cotton—a bitter legacy from which the state is only now beginning to recover. As in the pre–Civil War years, Mississippi exported its raw materials and imported manufactures. In the 1930s, however, state leaders began to realize the necessity of diversifying the economy; and by the mid-1960s, many more Mississippians recognized that political and economic inequality and racial conflict did not provide an environment attractive to the industries the state needed.

Once the turmoil of the 1950s and early 1960s had subsided, the impressive industrial growth of the immediate postwar years resumed. By the mid-1960s, manufacturing—attracted to the state, in part, because of low wage rates and a weak labor movement—surpassed farming as a source of jobs. During the following decade, the balance of industrial growth changed somewhat. The relatively low-paying garment, textile, and wood-products industries, based on cotton and timber, grew less rapidly in both value added and employment than a number of heavy industries, including transportation equipment and electric and electronic goods. Still, Mississippi remains a poor state, in some respects the poorest.

21INCOME

As it has for much of this century, Mississippi ranked last among the 50 states in per capita personal income in 1983, at $8,072. For many years Mississippi steadily closed the gap between its low income level and the national norm. Per capita income grew from 54% of the US average in 1960 to 65% in 1970, 69% in 1975, and 73% in 1978 before leveling off at about 70% in the next few years (69% in 1983). Measured in constant 1972 dollars, per capita income grew 37% between 1970 and 1983, compared to 28% for the nation.

22LABOR

Data for December 1984 showed a civilian labor force of approximately 1,074,000 in Mississippi, of whom 958,000 were employed; the unemployment rate was 10.8%, 4th highest among the states. The labor force participation rate in 1983 was 75% for white males but only 67% for nonwhite males, a decline from the 1960 rate of 72%. In 1960, nonwhite women had participated in the labor force at a higher rate (36%) than white women (33%); in 1983, white women participated at a rate of 45% compared with 49% for black women. The overall participation of women, 58%, ranked 49th among the 50 states. The unemployment rate for blacks (24%) was three times as high as that for whites in 1983.

A federal survey in March 1983 revealed the following nonfarm employment pattern in Mississippi:

	ESTABLISH-MENTS	EMPLOYEES	ANNUAL PAYROLL ('000)
Agricultural services, forestry, fishing	490	3,476	$ 44,235
Mining of which:	520	8,122	176,577
Oil, gas extraction	(451)	(7,140)	(157,863)
Contract construction	3,549	35,117	539,802
Manufacturing	3,056	193,280	3,065,184
Transportation, public utilities	2,114	33,653	681,358
Wholesale trade	3,735	38,512	615,604
Retail trade	15,142	120,536	1,071,422
Finance, insurance, real estate	3,988	33,080	525,601
Services	12,841	109,573	1,256,291
Other	3,110	3,052	49,275
TOTALS	48,545	578,401	$8,025,349

In September 1984, the average weekly earnings of Mississippi production workers in manufacturing were $286 (49th among the 50 states), and average hourly earnings were $7.02 (50th). Mississippi has a right-to-work law. In 1980, only 135,000 Mississippi nonfarm workers (16.3%) were members of unions; there were 553 labor unions in 1983.

23AGRICULTURE

In 1983, Mississippi ranked 25th among the states in income from agriculture, with marketings of over $2.3 billion; crops accounted for $1.4 billion and livestock and livestock products $926 million.

The history of agriculture in the state is dominated by cotton, which from the 1830s through World War II was Mississippi's principal cash crop. During the postwar period, however, as mechanized farming replaced the sharecropper system, agriculture became more diversified. In 1983, Mississippi ranked 3d in cotton production, 5th in rice, and 9th in soybeans. About 1,760,000 bales of cotton worth $505 million were harvested in 1982; in 1983, these figures had dropped to 900,000 bales and $282 million. Soybean output in 1982 totaled 92,300,000 bushels, worth $463.5 million; in 1983, it was 56,425,000 bushels and $459.9 million. Rice production was 10,094,000 and 6,440,000 hundredweight in 1982 and 1983, respectively; 1982 value was $87.4 million.

Federal estimates for 1984 showed some 50,000 farms with a total area of 14 million acres (5.7 million hectares). Mississippi's farm population in 1980 was approximately 85,000, down from 680,000 in 1960 and 277,000 in 1970. The richest soil is in the Delta, where most of the cotton is raised. Livestock has largely taken over the Black Belt, a fertile area in the northwest.

24ANIMAL HUSBANDRY

Cattle are raised throughout the state, though principally in the Black Belt and Delta. The main chicken-raising area is in the eastern hills.

There were 1,764,000 head of cattle and 300,000 hogs and pigs on Mississippi farms at the beginning of 1984. In 1983, the state produced an estimated 462,000,000 lb of cattle, valued at $218,537,000. During the same year, 95,746,000 lb of hogs and pigs were produced, with a value of $45,168,000. Mississippi is a leading producer of broilers, ranking 5th in 1983; 1.3 billion lb of chickens and broilers, worth $343.7 million, were produced in that year.

25FISHING

In 1984, Mississippi ranked 4th among the 50 states in size of commercial fish landings, 476,997,000 lb (a state record), but 13th in value of its catch, $46,762,000. Of this total, 89% of the catch was landed at Pascagoula–Moss Point, the nation's 2d-largest port for commercial landings. Menhaden and shrimp make up the bulk of the commercial landings. The saltwater catch also includes oysters and swordfish; the freshwater catch is dominated by buffalo fish, carp, and catfish. As of 1985, Mississippi ranked 1st among the states in catfish farming, mostly from ponds in the Yazoo river basin.

26FORESTRY

Mississippi had approximately 16,716,000 acres (6,844,000 hectares) of forested land in 1979, 56% of the total land area of the state. Six national forests extended over 2.3 million acres (930,000 hectares) in 1984. The state's most heavily forested region is the Piney Woods in the southeast. Some 16,504,000 acres (6,679,000 hectares) were commercial timberland, 90% of it privately owned; some of this land was also used for agricultural purposes. Mississippi was 9th in forestry products in 1982. Sawtimber volume was 60.6 million board feet, 11th in the nation. Production of softwoods and hardwoods was 1.6 billion board feet in 1983, 6th in the nation.

27MINING

Principal nonfuel mineral products in 1984 were sand and gravel, 10,400,000 tons; crushed stone, 1,900,000 tons; and clays, 1,393,000 tons. The value of nonfuel mineral production was $95.4 million in 1984, 42d among the states.

28ENERGY AND POWER

There were 19 electric generating plants in Mississippi in 1983, with a total installed capacity of 6.1 million kw, almost all coal-fired or gas-fired. During that year they generated a total of 13.3 billion kwh of electricity. Two large nuclear reactors were under construction in the early 1980s, but the board of the Tennessee Valley Authority voted in 1984 to abandon its partially completed Yellow Creek pressurized-water plant at Iuka. The Grand Gulf boiling-water plant, built by Mississippi Power and Light in Claiborne County, was scheduled to begin commercial operations in 1985.

Mississippi is a major petroleum producer, ranking 12th in the

US during 1983, when production totaled 31,450,695 barrels and there were proved reserves of 205,000,000 barrels. Mississippi produced 200.2 billion cu feet of natural gas from 552 wells during 1983, when proved reserves were estimated at 1.6 trillion cu feet. Value added in 1982 for oil and gas combined was $1.3 billion. Most production comes from the south-central part of the state.

29INDUSTRY

In 1982, Mississippi ranked 30th among the states in value of shipments of manufactured goods, which totaled $19.5 billion. Food and kindred products contributed the largest amount of the 1982 total, 15%, followed by transportation equipment, 9%; lumber and wood products, 7%; electric and electronic equipment, 7%; chemical and allied products, 6%; apparel and other textile products, 6%; nonelectrical machinery, 5%; and other sectors, 45%.

The following table shows value of shipments in 1982 for selected major industries:

Food and kindred products	$2,848,000,000
Transportation equipment	$1,842,000,000
Lumber and wood products	$1,374,400,000
Electric and electronic equipment	$1,364,000,000
Chemicals and allied products	$1,231,300,000
Apparel and other textile products	$1,100,700,000
Nonelectrical machinery	$1,049,500,000

The state's biggest manufacturing concern is Litton Industries' Ingalls shipyard at Pascagoula. In addition to merchant vessels, this yard builds US Navy ships, including nuclear-powered submarines.

30COMMERCE

With wholesale trade amounting to $10.8 billion in 1982, Mississippi ranked 32d among the 50 states. Mississippi ranked 33d in 1982 retail sales, which amounted to $9 billion. Food stores accounted for 26% of the retail sales volume, automotive dealers, 20%; department stores, 9%; gasoline service stations, 8%; eating and drinking places, 7%; and other establishments, 30%.

Mississippi was the 15th-leading exporter of farm products in 1981/82, accounting for $787 million, or 2% of the US total. Exports of manufacturers reached nearly $1.2 billion in 1981. Gulfport became the state's first free trade zone in 1983.

31CONSUMER PROTECTION

The Consumer Protection Division of the Office of the Attorney General, established in 1975, may investigate complaints of unfair or deceptive trade practices and, in specific cases, may issue injunctions to halt them.

32BANKING

Mississippi's 165 insured commercial banks had assets of $14.6 billion at the end of 1983. They held loans amounting to nearly $7.4 billion and deposits of nearly $12.5 billion, including $2.7 billion in demand deposits and $9.8 billion in time and savings deposits.

At the end of 1983, Mississippi had 46 insured savings and loan associations, assets totaling nearly $3.3 billion. There were 192 credit unions in 1983.

33INSURANCE

In mid-1983, 30 life insurance companies were domiciled in Mississippi. Some 3,778,000 life insurance policies were in force in 1983, worth a total of $39.7 billion. Life insurance companies paid $113.8 million to 20,800 beneficiaries in 1983. The average Mississippi family held $42,900 in life coverage, 18% below the national average and 46th among the 50 states. In 1983, premiums written for automobile liability and physical damage insurance totaled $361.9 million.

34SECURITIES

There are no securities exchanges in Mississippi. New York Stock Exchange member firms had 32 sales offices and 192 registered representatives in the state in 1983. Some 220,000 Mississippians held shares in public corporations in that year.

35PUBLIC FINANCE

Two state budgets are prepared annually—one by the State Fiscal Management Board, for the executive branch, and one by the Joint Legislative Budget Committee, for the legislative branch—and submitted to the legislature for reconciliation and approval. The fiscal year runs from 1 July through 30 June.

The following table shows general fund revenues and expenditures for 1983/84 (estimated) and 1984/85 (recommended):

REVENUES	1983/84	1984/85
Sales tax	$ 558,000,000	$ 626,000,000
Individual income tax	255,000,000	221,600,000
Corporate income and franchise tax	135,000,000	134,600,000
Other receipts	422,499,100	402,800,000
TOTALS	$1,370,499,100	$1,385,000,000

EXPENDITURES		
Education	$ 839,291,081	$ 868,905,848
Health and human services	210,002,028	212,312,427
Corrections and public safety	83,254,162	89,430,610
Local assistance	55,484,227	67,445,000
Other expenditures	146,521,540	157,292,303
TOTALS	$1,334,553,038	$1,395,386,188

Recommended consolidated expenditures (including federal and designated funds) for 1984/85 totaled nearly $3.3 billion. As of 30 June 1983, the outstanding debt of state and local governments was $2.6 billion, or $1,005 per capita, 48% below the US average in mid-1982 and lower per capita than that of any other state but Indiana.

36TAXATION

In 1983, Mississippi collected slightly more than $1.5 billion in state taxes, 32d in the US. As of the end of 1983, the state income tax for both individuals and corporations ranged from 3% on the first $5,000 of net income to 5% on amounts over $10,000; the corporate income tax rate was one of the lowest in the nation. Mississippi also imposes severance taxes on oil, natural gas, and timber. A 6% retail sales tax is levied, along with taxes on inheritance, gasoline, tobacco, beer, wine, insurance premiums, and numerous other items.

In 1982/83, Mississippians paid $4.2 billion in federal taxes and received nearly $7.7 billion in federal expenditures; the state's ratio of federal taxes to federal spending ($0.53 to $1) was 47th among the 50 states. Mississippians filed 866,960 federal income tax returns in 1982 and paid $1.7 billion in tax.

37ECONOMIC POLICY

In 1936, the state began implementing a program called Balance Agriculture with Industry (BAWI), designed to attract manufacturing to Mississippi. The BAWI laws offered industry substantial tax concessions and permitted local governments to issue bonds to build plants that would be leased to companies for a 20-year period, after which the company would own them. Mississippi continues to offer low tax rates and numerous tax incentives to industry. The Department of Economic Development is charged with encouraging economic growth in the specific fields of industrial development, marketing of state products, and development of tourism. A high-technology asset is the National Space and Technology Laboratory in Hancock County.

38HEALTH

In 1983 there were 43,992 live births in Mississippi, the birthrate was 17 per 1,000 population. In 1981, the infant death rate of 15.4 (20.6 for blacks and 10.5 for whites) was the nation's 2d highest. Mississippi's death rates from prenatal conditions and accidents were 3d highest in the US.

Mississippi had 118 hospitals, with 17,356 beds, in 1983; the

37,700 personnel in 1981 included 5,955 registered nurses. The average cost of a hospital stay in 1982 was $227 per day, 3d lowest among the 50 states. In mid-1983 there were 2,912 active physicians. The 1981 ratio of 108 physicians per 100,000 civilian population was lower than that in every other state but Idaho. At the end of 1983, the state had 908 professionally active dentists.

[39]SOCIAL WELFARE

In 1983, 10.1% of all Mississippians were public aid recipients, a higher percentage than in any other state. Social Security, welfare, and other transfer payments were the principal source of income in 34 counties and the second most important source of income in the state. During 1982/83, aid to families with dependent children (AFDC) was paid to 163,846 Mississippians, 111,137 of whom were children. Average monthly AFDC payments of $92 per family were the lowest among the 50 states. About 523,000 Mississippians participated in the federal food stamp program during 1982/83, at a federal cost of $271 million. In 1982/83, about 407,000 students participated in the state's school lunch program, at a total federal cost of $55 million.

In 1983, 206,600 retired workers in the state received federal Social Security benefits of $854 million; their average monthly payment in December 1982 was $347.54, the lowest in the US. Supplemental Security Income benefits totaling $109 million were paid in 1983. Unemployment insurance expenditure was $180 million. Mississippi spent $70.7 million on workers' compensation in 1982.

[40]HOUSING

The 1980 census counted 904,078 year-round housing units, of which 827,169 were occupied. Only 77% of the occupied units had full plumbing in 1970—the lowest rate in the US—but in 1980 the rate had increased to 94%. Seventy-one percent of the units were owner-occupied in 1980. The median value was $31,400, lower than any other state but Arkansas. The median rental of $180 was the lowest in the nation. About 21,300 new housing units worth $671 million were authorized during 1982/83.

[41]EDUCATION

Only 55% of adult Mississippians 18 and older had completed high school in 1980, the lowest rate in the nation.

Mississippi's reaction to the US Supreme Court decision in 1954 mandating public school desegregation was to repeal the constitutional requirement for public schools and to foster the development of segregated private schools. In 1964, the state's schools did begin to integrate, and compulsory school attendance was restored 13 years later. In 1982, the compulsory school age was raised to 14, and a system of free public kindergartens was established for the first time. As of 1980, 26% of minority (nonwhite) students were in schools in which minorities were less than 50% of the student body, and 19% were in 99–100% minority schools—a considerable degree of de facto segregation, but less so than in some northern states.

As of the fall of 1982, there were 468,295 students enrolled in public schools in Mississippi, 242,187 elementary (including kindergarten) and 226,108 secondary. Private-school enrollment was 50,116 in 1980. At $15,812, the average salary of public classroom teachers in 1983/84 was the lowest of all the states. In 1985, following a teachers' strike, the legislature approved an average $4,400 pay increase for teachers over three years.

During 1982/83 there were 42 institutions of higher education with a total enrollment of 115,504; 8 were public universities, 16 were public junior colleges, and 18 (including 4 Bible colleges and theological seminaries) were private institutions. The University of Mississippi, established in 1844 and located in Lafayette County, had a total enrollment during 1983/84 of 9,326 (excluding its medical school at Jackson but including its law school). One of the most dramatic moments in the history of the civil rights movement came on 30 September 1962, when James Meredith became the university's first black student. A riot

ensued in which two people were killed, the National Guard federalized, and Army troops called in. Mississippi State University had 12,325 students in 1983/84; the University of Southern Mississippi, 11,333. Predominantly black institutions include Tougaloo College, Alcorn State University, Jackson State University, and Mississippi Valley State University.

[42]ARTS

Jackson has two ballet companies, a symphony orchestra, and two opera companies. Opera South, an integrated but predominantly black company, presents free operas during its summer tours and mounts two major productions yearly. The Mississippi Opera instituted a summer festival during its 1980/81 season. There are local symphony orchestras in Meridian, Starkville, Tupelo, and Greenville. The established professional theaters in the state are the Sheffield Ensemble in Biloxi and the New Stage in Jackson. The Greater Gulf Coast Arts Center has been very active in bringing arts programs into the coastal area.

A distinctive contribution to US culture is the music of black sharecroppers from the Delta, known as the blues. The Delta Blues Museum in Clarksdale has an extensive collection documenting blues history.

[43]LIBRARIES AND MUSEUMS

There were 39 county or multicounty (regional) libraries, serving all counties in 1982/83. There were 4.3 million volumes in Mississippi libraries, and total circulation was almost 7 million. The finest collection of Mississippiana is at the Mississippi State Department of Archives and History in Jackson. In the Vicksburg–Warren County Public Library are collections on the Civil War, state history, and oral history. Tougaloo College has special collections of African materials, civil rights papers, and oral history. The Gulf Coast Research Library of Ocean Springs has a marine biology collection.

There are 55 museums, including the distinguished Mississippi State Historical Museum at Jackson. Pascagoula, Laurel, and Jackson all have notable art museums. The Mississippi Museum of Natural Science in Jackson has been designated the state's official natural science museum by the legislature. Also in Jackson is the Mississippi Agriculture and Forestry Museum. In Meridian is a museum devoted to country singer Jimmie Rodgers, and in Jackson one to pitcher Dizzy Dean.

Beauvoir, Jefferson Davis's home at Biloxi, is a state shrine and includes a museum. The Mississippi governor's mansion—completed in 1845, restored in 1975, and purportedly the 2d-oldest executive residence in the US—is a National Historical Landmark.

[44]COMMUNICATIONS

Mississippi had 448 post offices and 4,345 postal employees in 1985. In 1981, 84% of the state's 827,169 occupied housing units had telephones, the lowest rate in the US. In 1983/84, the state had 213 operating commercial radio stations (108 AM, 105 FM) and 13 commercial television stations; 87 cable television systems had 342,948 subscribers in 162 communities.

[45]PRESS

In 1984, Mississippi had 24 daily newspapers: 5 morning dailies with a combined circulation of 135,390 and 19 evening dailies with 273,440 circulation. In addition there were 12 Sunday papers with a circulation of 330,965. The state's leading newspapers were all in Jackson, and all were owned by the Gannett Co.: the *Clarion-Ledger* (m), 68,047; *Daily News* (e), 36,964; and the *Clarion-Ledger–Daily News* (Saturday, 101,657, and Sunday, 112,839). A monthly, *Mississippi Magazine,* is published in Edwards, and a bimonthly, *Mississippi: A View of the Magnolia State,* in Jackson.

[46]ORGANIZATIONS

Among the organizations that played key roles in the civil rights struggles in Mississippi during the 1950s and 1960s were the National Association for the Advancement of Colored People

(NAACP), Congress of Racial Equality, Southern Christian Leadership Conference, and Student Nonviolent Coordinating Committee (SNCC, later the Student National Coordinating Committee). Of the national civil rights organizations still active in Mississippi, the NAACP is the largest, with members in every county. In contrast to the 1960s, most civil rights activities in the state are now organized around local social and economic programs, such as Head Start. The Freedom Information Service is a clearinghouse for information about civil rights activities in the state. The Citizens' Councils of America, headquartered in Jackson, is a states' rights group. The 1982 Census of Service Industries counted 524 organizations in Mississippi, including 135 business associations; 226 civic, social, and fraternal associations; and 5 educational, scientific, and research associations. Among the organizations with headquarters in Mississippi are the American Association of Public Health Physicians (Greenwood); the Sons of Confederate Veterans (Hattiesburg); and the Amateur Field Trial Clubs of America (Hernando).

47TOURISM, TRAVEL, AND RECREATION

During 1984, some 9 million persons traveled to and through Mississippi. Travel and tourism was the 3d-largest employer in the state in 1982, generating 29,300 jobs and over $1 billion in expenditures. Among Mississippi's major tourist attractions are its mansions and plantations, many of them in the Natchez area. McRaven, in Vicksburg, was built in 1797. The Delta and Pine Land Co. plantation near Scott is one of the largest cotton plantations in the US. At Greenwood is the Florewood River Plantation, a museum re-creating 19th-century plantation life. The Mississippi State Fair is held annually in Jackson during the second week in October.

The Natchez Trace Parkway, Gulf Islands National Seashore, and Vicksburg National Military Park—3 of the state's 5 national parks—attracted 13,056,167 visits in 1983. There are also 6 national forests and 27 state parks. In 1982/83, licenses were issued to 303,572 hunters and 445,296 fishermen.

48SPORTS

There are no major league professional teams in Mississippi. Jackson has a minor league baseball team in the Texas League. The University of Mississippi has long been prominent in college football. From 1955 through 1970, "Ole Miss" teams won the Sugar Bowl five times and the Cotton Bowl once.

49FAMOUS MISSISSIPPIANS

Mississippi's most famous political figure, Jefferson Davis (b.Kentucky, 1808-89), came to the state as a very young child, was educated at West Point, and served in the US Army from 1828 to 1835. He resigned a seat in Congress in 1846 to enter the Mexican War, from which he returned home a hero after leading his famous regiment, the 1st Mississippi Rifles, at the Battle of Buena Vista, Mexico. From 1853 to 1857, he served as secretary of war in the cabinet of President Franklin Pierce. Davis was representing Mississippi in the US Senate in 1861 when the state withdrew from the Union. In February 1861, he was chosen president of the Confederacy, an office he held until the defeat of the South in 1865. Imprisoned for two years after the Civil War (though never tried), Davis lived the last years of his life at Beauvoir, an estate on the Mississippi Gulf Coast given to him by an admirer. There he wrote *The Rise and Fall of the Confederate Government,* completed eight years before his death in New Orleans.

Lucius Quintus Cincinnatus Lamar (b.Georgia, 1825-93) settled in Oxford in 1855 and only two years later was elected to the US House of Representatives. A supporter of secession, he served as Confederate minister to Russia in 1862. After the war, Lamar was the first Mississippi Democrat returned to the House; in 1877, he entered the US Senate. President Grover Cleveland made Lamar his secretary of the interior in 1885, later appointing him to the US Supreme Court. Lamar served as associate justice from 1888 until his death.

Some of the foremost authors of 20th-century America owe their origins to Mississippi. Supreme among them is William Faulkner (Falkner, 1897-1962), whose literary career began in 1924 with the publication of *The Marble Faun,* a book of poems. His novels included such classics as *The Sound and the Fury* (1929), *Light in August* (1932), and *Absalom, Absalom!* (1936). Faulkner received two Pulitzer Prizes (one posthumously), and in 1949 was awarded the Nobel Prize for literature.

Richard Wright (1908-60), born near Natchez, spent his childhood years in Jackson. He moved to Memphis as a young man, and from there migrated to Chicago; he lived his last years in Paris. A powerful writer and a leading spokesman for the black Americans of his generation, Wright is best remembered for his novel *Native Son* (1940) and for *Black Boy* (1945), an autobiographical account of his Mississippi childhood.

Other native Mississippians of literary renown (and Pulitzer Prize winners) are Eudora Welty (b.1909), Tennessee Williams (Thomas Lanier Williams, 1911-83), and playwright Beth Henley (b.1952). Like Faulkner's, Welty's work is set in Mississippi; her best-known novels include *Delta Wedding* (1946), *The Ponder Heart* (1954), and *Losing Battles* (1970). Although Tennessee Williams spent most of his life outside Mississippi, some of his most famous plays are set in the state. Other Mississippi authors are Hodding Carter (b.Louisiana, 1907-72), Shelby Foote (b.1916), Walker Percy (b.Alabama, 1916), and Willie Morris (b.1934).

Among the state's numerous musicians are William Grant Still (1895-1978), a composer and conductor, and Leontyne Price (Mary Leontine Price, b.1927), a distinguished opera soprano. Famous blues singers are Charlie Patton (1887-1934), William Lee Conley "Big Bill" Broonzy (1898-1958), Howlin' Wolf (Chester Arthur Burnett, b.1910), Muddy Waters (McKinley Morganfield, 1915-83), John Lee Hooker (b.1917), and Riley "B. B." King (b.1925). Mississippi's contributions to country music include Jimmie Rodgers (1897-1933), Conway Twitty (b.1933), and Charley Pride (b.1939). Elvis Presley (1935-77), born in Tupelo, was one of the most popular singers in US history.

50BIBLIOGRAPHY

Bettersworth, John K. *Your Mississippi.* Austin, Tex.: Steck-Vaughn, 1975.

Brooks, Cleanth. *William Faulkner: The Yoknapatawpha Country.* New Haven: Yale University Press, 1963.

Federal Writers' Project. *Mississippi: A Guide to the Magnolia State.* New York: Somerset, n.d. (orig. 1938).

Ferris, William C. *Blues from the Delta.* Garden City, N.Y.: Doubleday, 1978.

Kirwan, Albert D. *Revolt of the Rednecks: Mississippi Politics (1876-1925).* Magnolia, Mass.: Peter Smith, 1964.

Loewen, James W., and Charles Sallis. *Mississippi: Conflict and Change.* Rev. ed. New York: Pantheon, 1982.

McLemore, Richard A., ed. *A History of Mississippi.* 2 vols. Hattiesburg: University and College Press of Mississippi, 1973.

Miles, Edwin A. *Jacksonian Democracy in Mississippi.* New York: Da Capo, 1970 (orig. 1900).

Mitchell, George. *Blow My Blues Away.* Baton Rouge: Louisiana State University Press, 1971.

Silver, James W. *Mississippi: The Closed Society.* 2d ed. New York: Harcourt, Brace, and World, 1966.

Skates, John Ray. *Mississippi: A Bicentennial History.* New York: Norton, 1979.

Sydnor, Charles S. *Slavery in Mississippi.* New York: Appleton-Century, 1933.

Welty, Eudora. *One Time, One Place: Mississippi in the Depression.* New York: Random House, 1971.

Wharton, Vernon L. *The Negro in Mississippi, 1865-90.* Chapel Hill: University of North Carolina Press, 1947.

MISSOURI

State of Missouri

ORIGIN OF STATE NAME: Probably from the Iliniwek Indian word *missouri,* meaning "owners of big canoes." **NICKNAME:** The Show Me State. **CAPITAL:** Jefferson City. **ENTERED UNION:** 10 August 1821 (24th). **SONG:** "Missouri Waltz." **MOTTO:** *Salus populi suprema lex esto* (The welfare of the people shall be the supreme law). **COAT OF ARMS:** Two grizzly bears standing on a scroll inscribed with the state motto support a shield portraying an American eagle and a constellation of stars, a grizzly bear on all fours, and a crescent moon, all encircled by the words "United We Stand, Divided We Fall." Above are a six-barred helmet and 24 stars; below, the roman numeral MDCCCXX (1820), when Missouri's first constitution was adopted. **FLAG:** Three horizontal stripes of red, white, and blue, with the coat of arms, encircled by 24 white stars on a blue band, in the center. **OFFICIAL SEAL:** The coat of arms surrounded by the words "The Great Seal of the State of Missouri." **BIRD:** Bluebird. **FLOWER:** Hawthorn blossom. **TREE:** Flowering dogwood. **ROCK:** Mozarkite (chert, or flint rock). **MINERAL:** Galena. **LEGAL HOLIDAYS:** New Year's Day, 1 January; Birthday of Martin Luther King, Jr., 3d Monday in January; Lincoln's Birthday, 12 February; Washington's Birthday, 3d Monday in February: Harry S Truman's Birthday, 8 May; Memorial Day, last Monday in May; Independence Day, 4 July; Primary Election Day; 1st Tuesday after 1st Monday in August (every four years); Labor Day, 1st Monday in September; Columbus Day, 2d Monday in October; Election Day, 1st Tuesday after 1st Monday in November (every four years); Veterans Day, 11 November; Thanksgiving Day, 4th Thursday in November; Christmas Day, 25 December. Though not a legal holiday, Missouri Day, the 3d Wednesday in October, is commemorated in schools each year. **TIME:** 6 AM CST = noon GMT.

¹LOCATION, SIZE, AND EXTENT

Located in the western north-central US, Missouri ranks 19th in size among the 50 states.

The total area of Missouri is 69,697 sq mi (180,516 sq km), of which land takes up 68,945 sq mi (178,568 sq km) and inland water 752 sq mi (1,948 sq km). Missouri extends 284 mi (457 km) E–W; its greatest N–S extension is 308 mi (496 km).

Missouri is bounded on the N by Iowa (with the line in the extreme NE defined by the Des Moines River); on the E by Illinois, Kentucky, and Tennessee (with the line passing through the Mississippi River); on the S by Arkansas (with a "boot heel" in the SE bounded by the Mississippi and St. Francis rivers); and on the W by Oklahoma, Kansas, and Nebraska (the line in the NW being formed by the Missouri River).

The total boundary length of Missouri is 1,438 mi (2,314 km). The state's geographic center is in Miller County, 20 mi (32 km) sw of Jefferson City.

²TOPOGRAPHY

Missouri is divided into four major land regions. The Dissected Till Plains, lying north of the Missouri River and forming part of the Central Plains region of the US, comprise rolling hills, open fertile flatlands, and well-watered prairie. The Osage Plains cover the western part of the state, their flat prairie monotony broken by low rounded hills. The Mississippi Alluvial Plain, in the southeastern corner, is made up of fertile black lowlands whose floodplain belts represent both the present and former courses of the Mississippi River. The Ozark Plateau, which comprises most of southern Missouri and extends into northern Arkansas and northeastern Oklahoma, constitutes the state's largest single region. The Ozarks contain Taum Sauk Mountain, at 1,772 feet (540 meters) the highest elevation in the state. Along the St. Francis River, near Cardwell, is the state's lowest point, 230 feet (70 meters).

Including a frontage of at least 500 mi (800 km) along the Mississippi River, Missouri has more than 1,000 mi (1,600 km) of navigable waterways. The Mississippi and Missouri rivers, the two largest in the US, respectively form the state's eastern border and

part of its western border; Kansas City is located at the point where the Missouri bends eastward to cross the state, while St. Louis developed below the junction of the two great waterways. The White, Grand, Chariton, St. Francis, Current, and Osage are among the state's other major rivers. The largest lake is the artificial Lake of the Ozarks, covering a total of 93 sq mi (241 sq km).

Missouri's exceptional number of caves and caverns—23 were commercial tourist attractions in 1985—were formed during the last 50 million years through the erosion of limestone and dolomite by melting snows bearing vegetable acids. Coal, lead, and zinc deposits date from the Pennsylvanian era, beginning some 250 million years ago. The Mississippi Valley area is geologically active: massive earthquakes during 1811 and 1812 devastated the New Madrid area of the southeast.

³CLIMATE

Missouri has a continental climate, but with considerable local and regional variation. The average annual temperature is 50°F (10°C) in the northwest, but about 60°F (16°C) in the southeast. Kansas City has a normal daily mean temperature of 54°F (12°C), ranging from 26°F (−3°C) in January to 79°F (26°C) in July; St. Louis has an annual mean of 55°F (13°C) with 29°F (−2°C) in January and 79° F (26° C) in July.

The coldest temperature ever recorded in Missouri was −40°F (−40°C), set at Warsaw on 13 February 1905; the hottest, 118°F (48°C), at Warsaw and Union on 14 July 1954. A 1980 heat wave caused 311 heat-related deaths in Missouri, the highest toll in the country; most were elderly residents of St. Louis and Kansas City. Fifty-one more heat-related deaths occurred in St. Louis during a 1983 heat wave.

The average annual precipitation for the state is about 40 in (100 cm), with some rain or snow falling about 110 days a year. The heaviest precipitation is in the southeast, averaging 48 in (122 cm); the northwest usually receives 35 in (89 cm) yearly. Snowfall averages 20 in (51 cm) in the north, 10 in (25 cm) in the southeast. During the winter, northwest winds prevail; the air movement is largely from the south and southeast during the rest of the year.

Springtime is the peak tornado season; in 1983, Missouri had 22 tornadoes.

⁴FLORA AND FAUNA

Representative trees of Missouri include the shortleaf pine, scarlet oak, smoke tree, pecan (*Carya illinoensis*), and peachleaf willow, along with species of tupelo, cottonwood, cypress, cedar, and dogwood (the state tree). American holly, which once flourished in the southeastern woodlands, is now considered rare; various types of wild grasses proliferate in the northern plains region. Missouri's state flower is the hawthorn blossom; other wild flowers include Queen Anne's lace, meadow rose, and white snakeroot. Showy and small white lady's slipper, green adder's-mouth, purslane, corn salad, dotted monardo, and prairie white-fringed orchid are rare in Missouri. Among endangered plants are the small-whorled pogonia, water sedge, Loesel's twayblade, and marsh pink; the American elm, common throughout the state, is considered endangered because of Dutch elm disease.

Indigenous mammals are the common cottontail, muskrat, white-tailed deer, and gray and red foxes. The state bird is the bluebird; other common birds are the cardinal, solitary vireo, and the prothonotary warbler. A characteristic amphibian is the plains leopard frog; native snakes include garter, ribbon, and copperhead. Bass, carp, perch, jack salmon (walleye), and crayfish abound in Missouri's waters. The chigger, a minute insect, is a notorious pest.

Listed as endangered in Missouri are the Ozark big-eared, gray, and Indiana bats, two subspecies of peregrine falcon, bald eagle, whooping crane, and eight varieties of mussel.

⁵ENVIRONMENTAL PROTECTION

Missouri's first conservation law, enacted in 1874, provided for a closed hunting season on deer and certain game birds. In 1936, the state established a Conservation Commission to protect the state's wildlife and forest resources. Today, Missouri's principal environmental protection agencies are the Department of Conservation, which manages the state forests and fish hatcheries and maintains wildlife refuges; the Department of Natural Resources, responsible for state parks, energy conservation, and environmental quality programs, including air pollution control, water purification, land reclamation, soil and water conservation, and solid and hazardous waste management; and the State Environmental Improvement and Energy Resources Authority, within the Department of Natural Resources, which is empowered to offer financial aid to any individual, business, institution, or governmental unit seeking to meet pollution control responsibilities.

An important environmental problem is soil erosion; rain washes away an average of 12.2 tons of soil from each cropland acre in Missouri every year—the second-highest erosion rate in the nation. As of 1982, 42 sites in Missouri were found to have unsafe concentrations of dioxin, a highly toxic byproduct of hexachlorophene, manufactured in a Verona chemical plant; in that year, an evacuation was begun (completed in 1985) of the 2,000 residents of Times Beach, a community 30 mi (48 km) west of St. Louis that was declared a federal disaster area. St. Louis ranks high among US cities for the quantities of lead and suspended particles found in the atmosphere, but conditions improved between the mid-1970s and early 1980s.

⁶POPULATION

Missouri ranked 15th among the 50 states at the 1980 census, with a population of 4,917,444, a 5% increase over the 1970 census total of 4,677,983. The state population estimate in early 1985 was 5,004,162, yielding a population density of 73 per sq mi (28 per sq km) and a 1.8% increase since 1980. The population estimate for 1990 is 5,076,800.

In 1830, the first year in which Missouri was enumerated as a state, the population was 140,455. Missouri's population just about doubled each decade until 1860, when the growth rate subsided; the population surpassed the 2 million mark at the 1880

census, 3 million in 1900 (when it ranked 5th in the US), and 4 million during the early 1950s. According to 1980 figures, the population was 51.9% female and slightly older than the national average (13.6% were 65 or older in 1983, 6th among the states, and the median age in 1980 was 30.9, 8th among the states). In addition, the population was slightly less mobile than the national average: almost 70% of Missourians were born in the state, and 54% of those over 5 years of age were living in the same house in 1980 as in 1975.

In 1980, 68% of all Missourians lived in urban areas and 32% in rural areas; in 1830, the population distribution had been 4% urban and 96% rural. The largest cities and their 1984 populations were Kansas City, 443,075, and St. Louis, 429,296, both well below the 1970 figures. St. Louis lost 27.2% of its population during the 1970s, a higher proportion than any of the nation's 74 other largest cities; Kansas City lost 11.2%. The St. Louis metropolitan area, embracing parts of Missouri and Illinois, comprised 2,398,400 people in 1984 (13th in the US), while metropolitan Kansas City, in Missouri and Kansas, had a population of 1,476,700 (25th).

The 1980 census showed that the population center of the United States had moved across the Mississippi River from southern Illinois to DeSoto in Jefferson County, Missouri, about 50 miles southwest of St. Louis.

⁷ETHNIC GROUPS

After the flatboat and French traders and settlers had made possible the earliest development of Missouri and its Mississippi shore, the river steamer, the Civil War, the Homestead Act (1862), and the railroad changed the character of the state ethnically as well as economically. Germans came in large numbers, developing small diversified industries, and they were followed by Czechs and Italians. The foreign-born numbered 86,000 in 1980.

Black Americans have represented a rising proportion of Missouri's population in recent decades: 9% in 1960, 10.3% in 1970, and 10.5% in 1980. Of 514,274 blacks in 1980, 40% lived in St. Louis (which was more than 45% black); Kansas City's black community, 122,336 in 1980, supported a flourishing jazz and urban blues culture between the two world wars, while St. Louis was the home of Scott Joplin and W. C. Handy in the early years of the century. In 1980, Missouri also had 51,853 Hispanics, including 31,803 of Mexican origin. The Asian community was small in 1980: 4,290 Chinese, 4,099 Asian Indians, 4,029 Filipinos, 3,519 Koreans, 3,179 Vietnamese, and 2,651 Japanese.

Only a few American Indians remained in Missouri after 1836. The 1980 census showed an Indian population of 12,127; the state has no Indian reservations.

Of those claiming descent from a single ancestry group in 1980, 633,291 named German, 527,041 English, and 242,610 Irish.

⁸LANGUAGES

White pioneers found Missouri Indians in the northern part of what is now Missouri, Osage in the central portion, and Quapaw in the south. Long after these tribes' removal to Indian Territory, only a few place-names echo their heritage: Missouri itself, Kahoka, Wappapello.

Four westward-flowing language streams met and partly merged in Missouri. Northern and North Midland speakers settled north of the Missouri River and in the western border counties, bringing their Northern *pail* and *sick to the stomach* and their North Midland *fishworm* (earthworm), *gunnysack* (burlap bag), and *sick at the stomach*. But *sick in the stomach* occurs along the Missouri River from St. Louis to Kansas City and along the

LOCATION: 36° to 40° 35′N; 89°06′ to 95°47′W. **BOUNDARIES:** Iowa line, 239 mi (385 km); Illinois line, 360 mi (579 km); Kentucky line, 59 mi (95 km); Tennessee line, 78 mi (125 km); Arkansas line, 331 mi (533 km); Oklahoma line, 35 mi (56 km); Kansas line, 267 mi (430 km); Nebraska line, 69 mi (111 km).

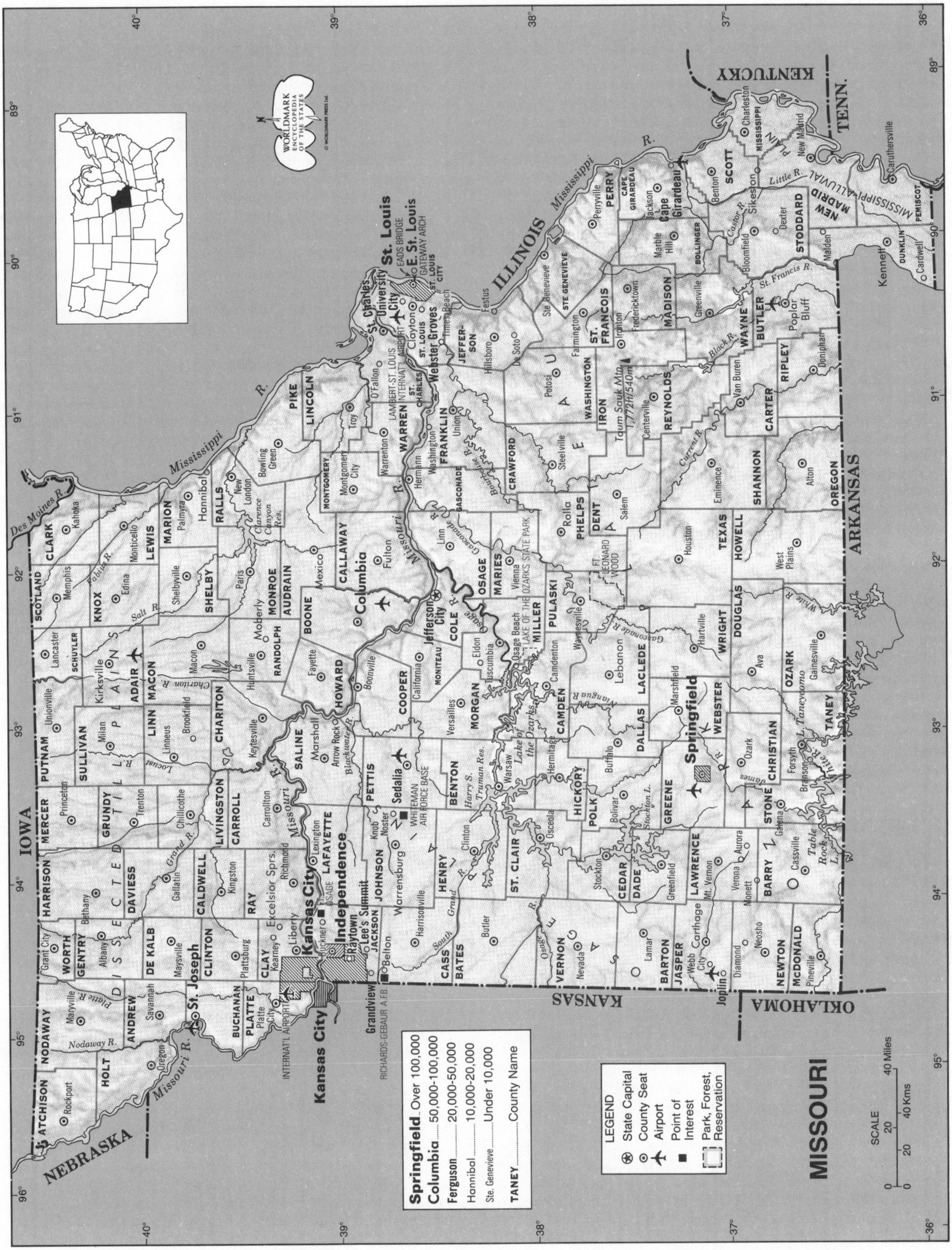

See endsheet maps: H3.

WORLDMARK ENCYCLOPEDIA OF THE STATES
© WORLDMARK PRESS Ltd.

MISSOURI

KENTUCKY
TENN.
ILLINOIS
IOWA
NEBRASKA
KANSAS
OKLAHOMA
ARKANSAS

Legend

Springfield	Over 100,000
Columbia	50,000–100,000
Ferguson	20,000–50,000
Hannibal	10,000–20,000
Ste. Genevieve	Under 10,000
TANEY	County Name

LEGEND
⊛ State Capital
⊙ County Seat
✈ Airport
■ Point of Interest
▢ Park, Forest, Reservation

SCALE
0 20 40 Miles
0 20 40 Kms

Counties and places:

ATCHISON, NODAWAY, WORTH, GENTRY, HARRISON, MERCER, PUTNAM, SCHUYLER, SCOTLAND, CLARK, HOLT, ANDREW, DE KALB, DAVIESS, GRUNDY, SULLIVAN, ADAIR, KNOX, LEWIS, BUCHANAN, CLINTON, CALDWELL, LIVINGSTON, LINN, MACON, SHELBY, MARION, PLATTE, CLAY, RAY, CARROLL, CHARITON, RANDOLPH, MONROE, RALLS, PIKE, JACKSON, LAFAYETTE, SALINE, HOWARD, BOONE, AUDRAIN, LINCOLN, CASS, JOHNSON, PETTIS, COOPER, CALLAWAY, MONTGOMERY, WARREN, ST. CHARLES, BATES, HENRY, BENTON, MORGAN, MONITEAU, COLE, OSAGE, GASCONADE, FRANKLIN, JEFFERSON, ST. LOUIS, VERNON, ST. CLAIR, HICKORY, CAMDEN, MILLER, MARIES, CRAWFORD, WASHINGTON, STE. GENEVIEVE, PERRY, BARTON, CEDAR, POLK, DALLAS, LACLEDE, PULASKI, PHELPS, DENT, IRON, ST. FRANCOIS, MADISON, CAPE GIRARDEAU, JASPER, DADE, GREENE, WEBSTER, WRIGHT, TEXAS, REYNOLDS, BOLLINGER, SCOTT, NEWTON, LAWRENCE, CHRISTIAN, DOUGLAS, HOWELL, SHANNON, CARTER, WAYNE, STODDARD, MISSISSIPPI, NEW MADRID, McDONALD, BARRY, STONE, TANEY, OZARK, OREGON, RIPLEY, BUTLER, DUNKLIN, PEMISCOT

Major cities:
Kansas City, Independence, Lee's Summit, St. Joseph, Springfield, Columbia, Jefferson City, St. Louis, St. Charles, University City, E. St. Louis, Webster Groves, Clayton, Sedalia, Joplin, Hannibal, Ste. Genevieve, Cape Girardeau, Poplar Bluff, West Plains, Lebanon, Rolla, Festus, Hillsboro, Farmington, Sikeston, Kennett, Caruthersville

Water features: Mississippi R., Missouri R., Des Moines R., Nodaway R., Platte R., Grand R., Osage R., Gasconade R., Current R., Black R., St. Francis R., White R., Lake of the Ozarks, Table Rock L., Bull Shoals L., Truman Res., Mark Twain Lake

WHITEMAN AIR FORCE BASE
RICHARDS-GEBAUR A.F.B.
LAMBERT–ST. LOUIS INTERNAT'L AIRPORT
FT. LEONARD WOOD
LAKE OF THE OZARKS STATE PARK
Taum Sauk Mtn. 1772 ft/540 m
GATEWAY ARCH
EADS BRIDGE

Mississippi south of St. Louis. South of the Missouri River, and notably in the Ozark Highlands, South Midland dominates, though with a few Southern forms, especially in the cotton-growing floodplain of the extreme southeast. *Wait on* (wait for), *light bread* (white bread), and *pullybone* (wishbone) are critical dialect markers for this area, as are *redworm* (earthworm), *towsack* (burlap bag), *snap beans* (string beans), *how* and *now* sounding like /haow/ and /naow/, and *Missouri* ending with the vowel of *me* rather than the final vowel of /uh/ heard north of the Missouri. In the extreme southeast are Southern *loaf bread, grass sack* (burlap bag), and *cold drink* as a term for a soft drink. In the eastern half of the state, a soft drink is generally *soda* or *sody;* in the western half, *pop.*

About 97% of state residents three years old or older spoke only English at home in 1980. Of those who claimed to speak another language at home, the leading languages and number of speakers were:

Spanish	37,564	Italian	10,794
German	31,603	Chinese	3,774
French	13,120	Polish	3,565

⁹RELIGIONS

Beginning in the late 17th century, French missionaries brought Roman Catholicism to what is now Missouri; the first permanent Roman Catholic church was built about 1755 at Ste. Genevieve. Immigration from Germany, Ireland, Italy, and eastern Europe swelled the Catholic population during the 19th century, and Roman Catholicism remains the largest Christian denomination. Baptist preachers crossed the Mississippi River into Missouri in the late 1790s, and the state's first Methodist church was organized about 1806. Immigrants from Germany included not only Roman Catholics but also many Lutherans, the most conservative of whom organized the Lutheran Church–Missouri Synod in 1847. In 1983, the Missouri Synod, with its headquarters in St. Louis, had a total US membership of 2,630,823.

In 1984, Missouri had 768,447 Roman Catholics, of whom 542,635 were in the archdiocese of St. Louis. The principal Protestant denominations in 1980 were the Southern Baptist Convention, with 700,053 adherents; United Methodist Church, 270,469; Lutheran Church–Missouri Synod, 144,829; and the Christian Church (Disciples of Christ), 116,875. In 1984, Missouri's estimated Jewish population was 64,770.

¹⁰TRANSPORTATION

Centrally located, Missouri is a leading US transportation center. Both St. Louis and Kansas City are hubs of rail, truck, and airline transportation.

In 1836, delegates from 11 counties met in St. Louis to recommend construction of two railroad lines and to petition Congress for a grant of 800,000 acres (324,000 hectares) of public land on which to build them. More than a dozen companies were incorporated by the legislature, but they all collapsed with the financial panic of 1837. Interest in railroad construction revived during the following decade, and in 1849 a national railroad convention was held in St. Louis at which nearly 1,000 delegates from 13 states recommended the construction of a transcontinental railroad. By 1851, three railroad lines had been chartered, and construction by the Pacific Railroad at St. Louis was under way; the Pacific line reached Kansas City in 1865, and a bridge built over the Missouri River four years later enabled Kansas City to link up with the Hannibal and St. Joseph Railroad, providing a freight route to Chicago that did not pass through St. Louis. In 1983, the state had 5,631 mi (9,062 km) of Class I railway track, the combined total of 13 lines. Class I and II trackage of 6,109 mi (9,832 km) in 1980 was 8th among the states. In 1983/84, Amtrak provided eight passenger trains running directly from Chicago to St. Louis and to Kansas City, en route to New Orleans, San Antonio, and Los Angeles, and two daily Kansas City-St. Louis

round trips. Amtrak trains made 12 other Missouri stops; total 1983/84 Missouri ridership was 425,852.

The first road developed in colonial Missouri was probably a trail between the lead mines and Ste. Genevieve in the early 1700s. A two-level cantilever bridge—the first in the world to have a steel superstructure—spanning the Mississippi at St. Louis was dedicated on 4 July 1874. By 1940, no place in Missouri was more than 10 miles from a highway. In 1983 there were 119,843 mi (192,869 km) of roadway, 7th among the states. The main interstate highways were I-70, linking St. Louis with Kansas City; I-44, connecting St. Louis with Springfield and Joplin; I-55, linking St. Louis with Chicago, Ill., to the north and paralleling the course of the Mississippi between St. Louis and Memphis, Tenn.; I-35, connecting Kansas City with Des Moines, Iowa; and I-29, paralleling the Missouri River north of Kansas City. Motor vehicle registration for the state in 1983 was 3,433,042, including 2,541,403 passenger cars, 881,966 trucks, and 9,673 buses; 3,433,-042 driver's licenses were in force during the same year.

The Mississippi and Missouri rivers have long been important transportation routes. Pirogues, keelboats, and flatboats plied these waterways for more than a century before the first steamboat, the *New Orleans,* traveled down the Mississippi in 1811. The Mississippi still serves considerable barge traffic, making metropolitan St. Louis an active inland port area.

Pioneering aviators in Missouri organized the first international balloon races in 1907 and the first US-sponsored international aviation meet in 1910. Five St. Louis pilots made up the earliest US Army air corps, and a barnstorming pilot named Charles A. Lindbergh, having spent a few years in the St. Louis area, had the backing of businessmen from that city when he flew his *Spirit of St. Louis* across the Atlantic in 1927. Today, Kansas City International Airport and Lambert–St. Louis Municipal Airport are among the busiest airports in the country. These two enplaned 2,399,224 and 7,815,390 passengers, respectively, in 1983, when the state had 354 airports and 58 heliports. At the end of 1983, 3,858 aircraft and 13,095 pilots were based in the state.

¹¹HISTORY

The region we now call Missouri has been inhabited for at least 4,000 years. The prehistoric Woodland peoples left low burial mounds, rudimentary pottery, arrowheads, and grooved axes; remains of the later Mississippian Culture include more sophisticated pottery and finely chipped arrowheads. When the first Europeans arrived in the late 17th century, most of the few thousand Indians living in Missouri were relatively recent immigrants, pushed westward across the Mississippi River because of pressures from eastern tribes and European settlers along the Atlantic coast. Indians then occupying Missouri belonged to two main linguistic groups: Algonkian-speakers, mainly the Sauk, Fox, and Iliniwek (Illinois) in the northeast; and a Siouan group, including the Osage, Missouri, Iowa, Kansas, and other tribes, to the south and west. Of greatest interest to the Europeans were the Osage, among whom were warriors and runners of extraordinary ability. The flood of white settlers into Missouri after 1803 forced the Indians to move into Kansas and into what became known as Indian Territory (present-day Oklahoma). During the 1820s, the US government negotiated treaties with the Osage, Sauk, Fox, and Iowa tribes whereby they surrendered, for the most part peaceably, all their lands in Missouri. By 1836, few Indians remained.

The first white men to pass through land eventually included within Missouri's boundaries apparently were Jacques Marquette and Louis Jolliet, who in 1673 passed the mouth of the Missouri River on their journey down the Mississippi; so did Robert Cavelier, Sieur de la Salle, who claimed the entire Mississippi Valley for France in 1682. Probably the first Frenchman to explore the Missouri River was Louis Armand de Lom d'Arce, Baron de Lahontan, who in 1688 claimed to have reached the junction of the Missouri and Osage rivers. The French did little to develop the Missouri region during the first half of the 18th century, although a few fur traders and priests established posts and missions among the Indians. A false report that silver had been discovered set off a brief mining boom in which no silver but some lead—available in abundance—was extracted. Missouri passed into Spanish hands with the rest of the Louisiana Territory in 1762, but development was still guided by French settlers; in 1764, the French fur trader Pierre Laclède established a trading post on the present site of St. Louis.

Although Spain fortified St. Louis and a few other outposts during the American Revolution and beat back a British-Indian attack on St. Louis in 1780, the Spanish did not attempt to settle Missouri. However, they did allow Americans to migrate freely into the territory. Spanish authorities granted free land to the new settlers, relaxed their restrictions against Protestants, and welcomed slaveholding families from southern states—especially important after 1787, when slavery was banned in the Northwest Territory. Pioneers such as Daniel Boone arrived from Kentucky, and the Chouteau fur-trading family gained a lucrative monopoly among the Osage. Spanish rule ended abruptly in 1800 when Napoleon forced Spain to return Louisiana to France. Included in the Louisiana Purchase, Missouri then became part of the US in 1803. After the Lewis and Clark expedition (1804–6) had successfully explored the Missouri River, Missouri in general—and St. Louis in particular—became the gateway to the West.

Missouri was part of the Louisiana Territory (with headquarters at St. Louis) until 1 October 1812, when Missouri Territory (including present-day Arkansas, organized separately in 1819) was established. A flood of settlers between 1810 and 1820 more than tripled Missouri's population from 19,783 to 66,586, leading Missourians to petition the US Congress for statehood as early as 1818. But Congress, divided over the slavery issue, withheld permission for three years, finally approving statehood for Maine and Missouri under the terms of the Missouri Compromise (1820), which sanctioned slavery in the new state but banned it in the rest of the former Louisiana Territory north of Arkansas. Congress further required that Missouri make no effort to enforce a state constitutional ban on the immigration of free Negroes and mulattoes; once the legislature complied, Missouri became the 24th state on 10 August 1821, Alexander McNair became the state's first governor, and Thomas Hart Benton was one of the state's first two US senators; Benton remained an important political leader for more than three decades.

Aided by the advent of steamboat travel on the Mississippi and Missouri rivers, settlers continued to arrive in the new state, whose population surpassed 1 million by 1860. The site for a new capital, Jefferson City, was selected in 1821, and five years later the legislature met there for the first time. French fur traders settled the present site of Kansas City in 1821 and established a trading post at St. Joseph in 1827. Mormons came to Independence during the early 1830s but were expelled from the state and crossed the Mississippi back into Illinois. For much of the antebellum period, the state's economy flourished, with an emphasis on cotton, cattle, minerals (especially lead and zinc), and commerce—notably the outfitting of wagon trains for the Santa Fe and Oregon trails. On the eve of the Civil War, more than half the population consisted of Missouri natives; 15% of the white population was foreign-born, chiefly German and Irish. Black slaves represented only 9% of the total population—the lowest proportion of any slave state except Delaware—while only about 25,000 Missourians were slaveholders. Nevertheless, there was a great deal of proslavery sentiment in the state, and thousands of Missourians crossed into neighboring Kansas in the mid-1850s to help elect a proslavery government in that territory. State residents were also active in the guerrilla warfare between proslavery forces and Free Staters that erupted along the border

with "bleeding Kansas." The slavery controversy was exacerbated by the US Supreme Court's 1857 decision in the case of Dred Scott, a slave formerly owned by a Missourian who had temporarily brought him to what is now Minnesota, where slavery was prohibited; Scott's suit to obtain his freedom was denied by the Court on the grounds that it was unconstitutional to restrict the property rights of slaveholders, in a decision that voided the Missouri Compromise reached 37 years earlier.

During the Civil War, Missouri remained loyal to the Union, though not without difficulty. When the conflict began, Governor Claiborne Fox Jackson called out the state militia "to repel the invasion" of federal forces, but pro-Union leaders such as Francis P. Blair deposed Jackson on 30 July 1861. Missouri supplied some 110,000 soldiers to the Union and 40,000 to the Confederacy. As devastating as the 1,162 battles or skirmishes fought on Missouri soil—more than in any other state except Virginia and Tennessee—was the general lawlessness that prevailed throughout the state; pro-Confederate guerrilla bands led by William Quantrill and Cole Younger, as well as Unionist freebooters, murdered and looted without hindrance. In October 1864, a Confederate army under Major General Sterling Price was defeated at the Battle of Westport, on the outskirts of Kansas City, ending the main military action. Some 27,000 Missourians were killed during the war. At a constitutional convention held in January 1865, Missouri became the first slave state to free all blacks.

During Reconstruction, the Radical Republicans sought to disfranchise all citizens who failed to swear that they had never aided or sympathized with the Confederacy. But the harshness of this and other measures caused a backlash, and Liberal Republicans such as Benjamin Gratz Brown and Carl Schurz, allied with the Democrats, succeeded in ousting the Radicals by 1872. The subsequent decline of the Liberal Republicans inaugurated a period during which Democrats occupied the governorship uninterruptedly for more than three decades.

The 1870s saw a period of renewed lawlessness, typified by the exploits of Jesse and Frank James, that earned Missouri the epithet of the "robber state." Of more lasting importance were the closing of the frontier in Missouri, the decline of the fur trade and steamboat traffic, and the rise of the railroads, shifting the market economy from St. Louis to Kansas City, whose population tripled during the 1880s, while St. Louis was eclipsed by Chicago as a center of finance, commerce, transportation, and population. Missouri farmers generally supported the movement for free silver coinage, along with other Populist policies such as railroad regulation. Reform Governor Joseph W. Folk (1905–9) fostered the "Missouri idea" that all people really needed were honest leaders to enforce simple laws, and his immediate successors in the statehouse, Herbert S. Hadley (1909–13) and Elliott W. Major (1913–17), introduced progressive policies to Missouri. However, the ideal of honest government was soon subverted by Kansas City's corrupt political machine, under Thomas J. Pendergast, the most powerful Democrat in the state between the two world wars. Machine politics did not prevent capable politicians from rising to prominence—among them Harry S Truman, Missouri's first and thus far only native son to serve in the nation's highest office.

The state's economy increasingly shifted from agriculture to industry, and Missouri's rural population declined from about three-fourths of the total in 1880 to less than one-third by 1970. Although the overall importance of mining declined, Missouri remained the world's top lead producer, and the state has emerged as 2d only to Michigan in US automobile manufacturing. Postwar prosperity was threatened beginning in the 1960s by the deterioration of several cities, notably St. Louis, which lost 47% of its population between 1950 and 1980; both St. Louis and Kansas City subsequently undertook urban renewal programs to cope with the serious problems of air pollution, traffic congestion, crime, and substandard housing. During the early 1980s, millions of dollars in federal, state, and private funds were used to rehabilitate abandoned and dilapidated apartment buildings and houses.

¹²STATE GOVERNMENT

Missouri's first constitutional convention met in St. Louis on 12 May 1820, and on 19 July a constitution was adopted. The constitution was rewritten in 1865 and again in 1875, the latter document remaining in effect until 1945, when another new constitution was enacted and the state government reorganized. A subsequent reorganization, effective 1 July 1974, replaced some 90 independent agencies with 13 cabinet departments and the Office of Administration.

The legislative branch, or general assembly, consists of a 34-member senate and a 163-seat house of representatives. Senators are elected to staggered four-year terms, representatives for two; the minimum age requirement for a senator is 30, for a representative 24. The state's elected executives are the governor and lieutenant governor (who run separately), secretary of state, auditor, treasurer, and attorney general; all serve four-year terms. The governor is limited to two terms in office; he must be at least 30 years of age and must have been a US citizen for 15 years and a Missouri resident for 10 years prior to election.

A bill becomes law when signed by the governor within 15 days of legislative passage. A two-thirds vote by both houses is required to override a gubernatorial veto. Except for appropriations or emergency measures, laws may not take effect until 90 days after the end of the legislative session at which they were enacted. Constitutional amendments require a majority vote of both houses of the legislature and ratification by the voters.

To vote in Missouri, one must be a US citizen and at least 18 years of age; there is no residency requirement.

¹³POLITICAL PARTIES

The major political groups in Missouri are the Democratic Party and the Republican Party, each affiliated with the national party organization. Before 1825, the state had no organized political parties, and candidates ran as independents; however, each of Missouri's first four governors called himself a Jeffersonian Republican, allying himself with the national group from which the modern Democratic Party traces its origins. Except for the Civil War and Reconstruction periods, the Democratic Party held

the governorship from the late 1820s to the early 1900s. Ten Democrats and seven Republicans served in the statehouse from 1908 through 1985. The outstanding figures of 20th-century Missouri politics are both Democrats: Thomas Pendergast, the Kansas City machine boss whose commitment to construction projects bore no small relation to his involvement with a concrete manufacturing firm, and Harry S Truman, who began his political career as a Jackson County judge in the Kansas City area and in 1945 became 33d president of the US.

As of 1984, Missouri had 2,969,300 registered voters; there is no party registration. The state, which had voted for the Republican nominee only three times in the previous 14 presidential elections and had supported Democrat Jimmy Carter in 1976, gave 51% of its popular vote in the 1980 election and 60% in the 1984 election to Republican Ronald Reagan. In 1984, Republican John Ashcroft, the state attorney general, was elected governor. The state's senior senator, Democrat Thomas F. Eagleton, was elected to a 3d term in 1980; Republican Senator John C. Danforth was reelected to a 2d term in 1982. All 9 members of the US House of Representatives (6 Democrats, 3 Republicans) were reelected in 1984, and both houses of the state legislature remained firmly in Democratic control. The voters approved constitutional amendments authorizing a state lottery and pari-mutuel betting.

There were 151 black elected officials in Missouri in early 1984, including 14 state legislators, 14 mayors, and 2 US representatives. Harriet Woods became the first woman to be elected to statewide office in 1984, when she was elected lieutenant governor.

¹⁴LOCAL GOVERNMENT

As of 1982, Missouri had 114 counties, 926 municipalities, 325 townships, 557 school districts, and 1,195 special districts.

Elected county officials generally include a public administrator, prosecuting attorney, sheriff, collector of revenue, assessor, treasurer, and coroner. The city of St. Louis, which is administratively independent of any county, has an elected mayor, a comptroller, and a 29-member board of aldermen (including the president); the circuit attorney, city treasurer, sheriff, and collector of revenue, also elected, perform functions analogous to county officers. Most other cities are governed by an elected mayor and council, but as of 1981, Kansas City and 34 other communities additionally had appointed city managers.

Missouri Presidential Vote by Political Parties, 1948–84

YEAR	ELECTORAL VOTE	MISSOURI WINNER	DEMOCRAT	REPUBLICAN	PROGRESSIVE	SOCIALIST
1948	15	*Truman (D)	917,315	655,039	3,998	2,222
1952	13	*Eisenhower (R)	929,830	959,429	—	—
1956	13	Stevenson (D)	918,273	914,289	—	—
1960	13	*Kennedy (D)	972,201	962,218	—	—
1964	12	*Johnson (D)	1,164,344	653,535		
					AMERICAN IND.	
1968	12	*Nixon (R)	791,444	811,932	206,126	—
1972	12	*Nixon (R)	698,531	1,154,058	—	—
1976	12	*Carter (D)	998,387	927,443	—	—
					LIBERTARIAN	SOC. WORKERS
1980	12	*Reagan (R)	931,182	1,074,181	14,422	1,515
1984	11	*Reagan (R)	848,583	1,274,188		

* Won US presidential election.

¹⁵STATE SERVICES

Under the 1974 reorganization plan, educational services are provided through the Department of Elementary and Secondary Education and the Department of Higher Education. Within the former's jurisdiction are the state schools for the deaf, the blind, and the severely handicapped; adult education programs; teacher certification; and the general supervision of instruction in the state. The department is headed by a board of education whose eight members are appointed by the governor to eight-year terms; the board, in turn, appoints the commissioner of education, the department's chief executive officer. The Department of Higher Education—governed by a nine-member appointive board that selects the commissioner of higher education—sets financial guidelines for state colleges and universities, authorizes the establishment of new senior colleges and residency centers, and establishes academic, admissions, residency, and transfer policies. Transportation services are under the direction of the Department of Highways and Transportation, which is responsible for aviation, railroads, mass transit, water transport, and the state highway system. The Department of Revenue licenses all road vehicles and motor vehicle operators and is responsible for the administration of all state taxes and local-option sales taxes.

Health and welfare services are provided primarily through the Department of Social Services, which oversees all state programs concerning public health (including operating a chest hospital and a cancer hospital), public assistance, youth corrections, probation and parole, veterans' affairs, and the aging. The Department of Mental Health operates 5 state mental hospitals, 3 community mental health centers, and 19 other facilities throughout the state, providing care for the emotionally disturbed, the mentally retarded, alcoholics, and drug abusers. Among the many responsibilities of the Department of Consumer Affairs, Regulation, and Licensing were enforcement of antidiscrimination laws, development of low- and moderate-income housing, and provision of financial aid to private nonprofit hospitals and higher-education facilities. In 1984, however, a constitutional amendment created a new Department of Economic Development, which inherited most of the responsibilities of the former department.

Administered within the Department of Public Safety are the Missouri State Highway Patrol, National Guard, and civil defense, veterans' affairs, highway and water safety, and alcoholic beverage control programs. The Department of Labor and Industrial Relations administers unemployment insurance benefits, workers' compensation, and other programs. The Department of Corrections and Human Resources is responsible for corrections, probation, and parole of adult offenders. The Department of Agriculture enforces state laws regarding agribusiness products. The lieutenant governor is designated as state ombudsman and volunteer coordinator.

¹⁶JUDICIAL SYSTEM

As of 1984, the supreme court, the state's highest court, consisted of seven judges and three commissioners. Judges are selected by the governor from three nominees proposed by a nonpartisan judicial commission; after an interval of at least 12 months, the appointment must be ratified by the voters on a separate nonpartisan ballot. The justices, who serve 12-year terms, select one of their number to act as chief justice. The mandatory retirement age is 70 for all judges in state courts.

The court of appeals, consisting in 1983 of 32 judges in three districts, assumed its present structure by constitutional amendment in 1970. The eastern district, sitting in St. Louis, consists of 14 judges; the western district, in Kansas City, has 11; and the southern district, in either Springfield or Poplar Bluff, has 7. All appellate judges are selected for 12-year terms in the same manner as the supreme court justices.

The circuit court is the only trial court and has original jurisdiction over all cases and matters, civil and municipal. Circuit court judges serve 6-year terms; in 1983 there were 133 judges in 44 circuits. Although many circuit court judges are still popularly elected, judges in St. Louis, Kansas City, and some other areas are selected on a nonpartisan basis. Many circuit courts have established municipal divisions, presided over by judges paid locally.

The Department of Corrections and Human Resources operated the Missouri State Penitentiary (Jefferson City), 10 other correctional facilities, and two urban "honor centers" (St. Louis and Kansas City) designed to help convicts become reintegrated into the community. As of 1984 there were 8,804 inmates in Missouri federal and state prisons. The 1983 crime rate for the state (4,530 per 100,000 population) was below the national norm. The crime rate for the St. Louis metropolitan area was 5,476 per 100,000; for the Kansas City metropolitan area, 6,465 per 100,000. The city of St. Louis had the 2d-highest homicide rate in 1981 of 95 cities surveyed—58.5 per 100,000 population (Miami was 1st).

¹⁷ARMED FORCES

Missouri has played a key role in national defense since World War II, partly because of the influence of Missourian Stuart Symington, first as secretary of the Air Force (1947–50) and later as an influential member of the Senate Armed Services Committee. During 1983/84, authorized Department of Defense personnel at principal bases in Missouri totaled 36,357. Installations include Ft. Leonard Wood, near Rolla; Richards-Gebaur Air Force Base, Belton; and Whiteman AFB, Knob Noster. The Defense Mapping Agency Aerospace Center is in St. Louis. Defense contract awards for the same fiscal year totaled $6.5 billion. The McDonnell Douglas aerospace firm of St. Louis received $5.4 billion of that total.

There were about 645,000 veterans living in the state as of 30 September 1983. Of these, 8,000 saw service in World War I, 249,000 in World War II, 122,000 in the Korean conflict, and 184,000 during the Viet-Nam era. Veterans' benefits amounted to $593 million in 1982/83, of which $298 million consisted of compensation and pensions.

Missouri had 8,638 Army National Guard personnel at the end of 1984 and 2,652 Air National Guard personnel in early 1985. There were 13,935 state and local police employees in 1983.

¹⁸MIGRATION

Missouri's first European immigrants, French fur traders and missionaries, began settling in the state in the early 18th century. Under Spain, Missouri received few Spanish settlers but many immigrants from the eastern US. During the 19th century, newcomers continued to arrive from the South and the East—slave-owning southerners (with their black slaves) as well as New Englanders opposed to slavery. They were joined by a wave of European immigrants, notably Germans and, later, Italians. By 1850, one out of three St. Louis residents was German-born; of all foreign-born Missourians in the late 1800s, more than half came from Germany.

More recently, the state has been losing population through migration—322,000 people were lost to net migration between 1940 and 1970, followed by a net gain of 22,000 during the 1970s and a net loss of 33,000 from 1980 to 1983. The dominant intrastate migration pattern has been the concentration of blacks in the major cities, especially St. Louis and Kansas City, and the exodus of whites from those cities to the suburbs and, more recently, to small towns and rural areas.

¹⁹INTERGOVERNMENTAL COOPERATION

The Commission on Interstate Cooperation, established by the state legislature in 1941, represents Missouri before the Council of State Governments and its allied organizations. Regional agreements in which the state participates include boundary compacts with Arkansas, Iowa, and Kansas and various accords governing bridges across the Mississippi and Missouri rivers. Representatives from both Missouri and Kansas take part in the Kansas City

Area Transportation Authority, which operates public transportation in the metropolitan region, and the Kansas–Missouri Waterworks Compact. Missouri also belongs to the Southern Interstate Energy Compact and many other multistate bodies. Federal aid to state and local governments in 1983/84 was $1.8 billion.

20ECONOMY

Missouri's central location and access to the Mississippi River contributed to its growth as a commercial center. By the mid-1700s, the state's first permanent settlement at Ste. Genevieve was shipping lead, furs, salt, pork, lard, bacon, bear, grease, feathers, flour and grain, and other products to distant markets. The introduction of steamboat traffic on the Mississippi, western migration along the Santa Fe and Oregon trails, and the rise of the railroads spurred the growth of commerce during the 19th century. Flour and grist mills, breweries and whiskey distilleries, and meat-packing establishments were among the state's early industrial enterprises. Lead mining has been profitable since the early 19th century. Grain growing was well established by the mid-18th century, and tobacco was a leading crop 100 years later.

Missouri's economy remains diversified, with manufacturing, farming, trade, tourism, services, government, and mining as prime sources of income. Today, automobile and aerospace manufacturing are by far the state's leading industries, while soybeans and meat and dairy products are the most important agricultural commodities. The state's historic past, varied topography, and modern urban attractions—notably the Gateway Arch in St. Louis—have made tourism a growth industry in recent decades. Mining, employing less than 1% of the state's nonagricultural workers, is no longer as important as it once was, although the state remains the leading US producer of lead and ranks 2d in the US in zinc production. The economic impact of state and local government and of defense-related federal expenditures has increased enormously since World War II.

21INCOME

With an income per capita of $10,790 in 1983, Missouri ranked 31st among the 50 states. Total personal income was $53.6 billion, representing a real increase of 34% since 1970, when Missouri ranked 28th; the state had placed 23d in 1960. Median money income for a family of four was $25,339 in 1981 (26th in the US).

22LABOR

In 1984, Missouri's civilian labor force averaged 2,379,000; 2,207,000 were employed and 172,000 (7.2%) were unemployed, below the US rate of 7.5%. More than half of all unemployed workers in Missouri were in the St. Louis area in November 1984, and almost one-fourth in the Kansas City region; the unemployment rates for these two areas were 6.6% and 4.7%, respectively.

A federal survey in March 1983 revealed the following nonfarm employment pattern for Missouri:

	ESTABLISH-MENTS	EMPLOYEES	ANNUAL PAYROLL ('000)
Agricultural services, forestry, fishing	1,121	5,139	$ 59,978
Mining	361	6,823	178,237
Contract construction	9,079	75,579	1,646,469
Manufacturing, of which:	6,935	390,255	8,488,229
Food and kindred products	(530)	(38,211)	(788,036)
Transportation equipment	(169)	(50,768)	(1,650,067)
Transportation, public utilities	4,635	120,467	2,967,414
Wholesale trade	10,544	126,761	2,574,085
Retail trade	30,955	326,380	3,128,786
Finance, insurance, real estate	9,904	112,837	1,970,464
Services	33,256	420,412	5,768,036
Other	5,868	6,562	138,084
TOTALS	112,658	1,591,215	$26,919,782

The survey did not include government employees, who numbered about 255,000 (state and local) in October 1983 and 67,041 (federal) in March 1982.

As early as the 1830s, journeyman laborers and mechanics in St. Louis, seeking higher wages and shorter hours, banded together to form trade unions and achieved some of their demands. Attempts to establish a workingman's party were unsuccessful, however, and immigration during subsequent decades ensured a plentiful supply of cheap labor. Union activity increased in the 1870s, partly because of the influence of German socialists. The Knights of Labor took a leading role in the labor movement from 1879 to 1887, the year that saw the birth of the St. Louis Trades and Labor Assembly; one year later, the American Federation of Labor came to St. Louis for its third annual convention, with Samuel Gompers presiding. The Missouri State Federation of Labor was formed in 1891, at a convention in Kansas City. By 1916, the state had 915 unions. Missouri remains a strong union state: 544,000 Missourians, representing 28% of the nonagricultural work force, belonged to labor organizations in 1980; and in 1983 there were 1,462 labor unions (11th in the US).

23AGRICULTURE

Missouri had 117,000 farms (5th in the US) covering 31.4 million acres (12.7 million hectares) in 1984. About 12,673,000 acres (5,128,000 hectares) were actually harvested in 1983. Missouri's agricultural income reached $4.8 billion in 1983, 10th among the 50 states. Of this total, about one-sixth came from soybeans.

In 1983, a heat wave and summer drought severely reduced crop yield; nevertheless, Missouri was 4th among the states in grain sorghum production and 6th in soybean and rice production. Soybean production is concentrated mainly in the northern counties and in the extreme southeast, with New Madrid County the leading producer in 1983. Saline County led the state in corn (for grain) production, Barton County in wheat, and Stoddard County in grain sorghum.

Cash receipts from all crops totaled $1.7 billion in 1983, including $812 million from soybeans, $343 million from hay, $259 million from corn, $232 million from winter wheat, $129 million from grain sorghum, $24 million from cotton, and $23 million from rice. Farmers harvested 101.4 million bushels of soybeans, 74.4 million bushels of corn, 70 million bushels of winter wheat, 40.8 million bushels of grain sorghum, 73,000 bales of cotton, 5.7 million tons of hay, and 2.5 million hundredweight of rice. Tobacco, oats, rye, apples, peaches, grapes, watermelons, and various seed crops are also grown in commercial quantities.

24ANIMAL HUSBANDRY

Missouri is a leading livestock-raising state, 2d among the states in 1983 in cows, cattle farms, and hog farms, and 4th in turkeys raised. Cash receipts from sales of livestock and livestock products totaled nearly $2.3 billion in 1983, accounting for 57% of Missouri's agricultural income. Hog raising is concentrated north of the Missouri River, cattle raising in the western counties, and dairy farming in the southwest.

At the end of 1983, Missouri farms and ranches had 5.2 million cattle and calves, 3.55 million hogs, 6.6 million chickens (of which 5.2 million were hens and pullets of laying age), and 13 million turkeys. Some 1.6 billion lb of cattle worth $889 million were slaughtered in 1983, along with 1.4 billion lb of hogs worth $650.2 million. The state's 254,000 milk cows yielded nearly 3.1 billion lb of milk. In 1982, milk production was worth $386.3 million. Other livestock products in 1983 included 250.9 million lb of turkeys worth $95.3 million, 36.3 million lb of chickens worth $4.3 million, and 1.3 billion eggs worth $54.9 million. In 1981, broiler production was 104.4 million lb, worth $29.8 million.

25FISHING

Commercial fishing takes place mainly on the Mississippi, Missouri, and St. Francis rivers. In 1983, 1,316,000 lb of fish valued at $376,000 were harvested. Sport fishing is enjoyed throughout the state, but especially in the Ozarks, whose waters harbor walleye, rainbow trout, bluegill, and largemouth bass.

26 FORESTRY

At one time, Missouri's forests covered 30 million acres (12 million hectares), more than two-thirds of the state. As of 1979, Missouri had 12,876,000 acres (5,211,000 hectares) of forestland (29% of the state), of which more than 95% was commercial forest, more than five-sixths of it privately owned. Most of Missouri's forestland is in the southeastern third of the state. Of the commercial forests, approximately two-thirds are of the oak/hickory type; shortleaf pine and oak/pine forests comprise about 7%, while the remainder consists of cedar and bottomland hardwoods.

According to the Forestry Division of the Department of Conservation, Missouri leads the US in the production of charcoal, cedar novelties, gunstocks, and walnut bowls and nutmeats; railroad ties, veneers, wine and beer casks, and other forest-related items are also produced. Timber production in 1983 totaled 169 million board feet, and shipments of all lumber and wood products in 1982 were valued at $377.7 million.

More than 300,000 acres (120,000 hectares) of state forests, managed by the Forestry Division, are used for timber production, wildlife and watershed protection, hunting, fishing, and other recreational purposes. A state-run nursery sells seedling trees and shrubs to Missouri landowners. Missouri's one national forest, Mark Twain in the southeast, encompassed 1,453,743 acres (588,311 hectares) of National Forest System lands as of 30 September 1984.

27 MINING

The value of Missouri's nonfuel mineral output in 1984 was $750 million, 8th among the 50 states. Lead is far and away the leading mineral product: the state's mines extracted about 87% of the US total in 1984, and if Missouri were a separate country, only the Soviet Union would rival it as a world producer.

The output of principal mineral products in 1984 included lead, 296,236 metric tons; zinc, 48,388 metric tons; copper, 5,722 metric tons; silver, 1,493 troy oz; cement, 4,540,000 tons; crushed stone, 43,800,000 tons; sand and gravel, 8,545,000 tons; and clays, 1,531,000 tons. By value, the biggest items in 1984 were cement, $186 million; lead, $163 million; crushed stone, $138 million; and zinc, $51 million. Iron ore, barite, and lime are also extracted.

28 ENERGY AND POWER

Missouri's electric power plants had an installed generating capacity of 15.7 million kw in 1983, when electrical output totaled 52.7 billion kwh, 12% more than in 1979. Sales of electric power, both private and public, totaled 45.2 billion kwh in 1983, of which 19 billion kwh went to residential users, 11.7 billion kwh to commercial users, 12.9 billion kwh to industrial users, and 1.6 billion kwh for other purposes. Coal-fired plants accounted for more than 96% of all power production in 1983; the state had no operating nuclear facilities, although a plant in Calloway County received a full-power operating license in October 1984.

Fossil fuel resources are limited. Reserves of bituminous coal totaled 6 billion tons as of 1 January 1983, but only 392 million tons were considered recoverable; nearly 5 million tons (21st in the nation) were mined in 1983, all from 15 surface mines. Small quantities of crude petroleum are also produced commercially; in 1983, production was 269,000 barrels. Gas utility industry revenues reached nearly $1.5 billion in 1983.

29 INDUSTRY

About one out of four nonfarm employees in Missouri is engaged in manufacturing. The leading industry groups, by employment, are transportation equipment (mainly automobiles, aircraft, and rockets and missiles), food and food products, electric and electronic equipment, and fabricated metal products, together accounting for about 41% of all industrial jobs in 1982. Of the $41.5 billion value of shipments by Missouri manufacturers during the same year, transportation equipment accounted for

25%, food and food products 20%, chemicals 10%, electric and electronic equipment 8%, fabricated metal products 7%, nonelectrical machinery 5%, and all other sectors 25%. The following table shows value of shipments by manufacturers for selected industries in 1982:

Motor vehicle and car bodies	$5,215,900,000
Dairy products	1,886,700,000
Beverages	1,457,700,000
Fabricated structural metal products	1,143,300,000
Grain mill products	1,230,000,000
Soaps, cleaners, toiletries	1,125,100,000
Apparel and other textile products	836,600,000
Drugs	832,400,000

St. Louis County, the city of St. Louis, and Jackson County (Kansas City) led the state in manufacturing employment, together accounting for 57% of all industrial jobs in March 1983. McDonnell Douglas, with headquarters in St. Louis, was the nation's 34th-ranked industrial corporation in 1984, with sales of $9.7 billion; its aerospace products have included all the Mercury capsules and Gemini space capsules, the DC-9 and DC-10 commercial jet aircraft, and the Tomahawk cruise missile. Other leading corporations headquartered in the St. Louis area are General Dynamics, an aerospace firm with 1984 sales of $7.8 billion (44th in the US); Monsanto, chemicals, $6.4 billion (51st); Anheuser-Busch, beer, $6.5 billion (53d); and Ralston Purina, food and animal feed, $5 billion (72d). During the 1983 model year, automobile plants in metropolitan Kansas City and St. Louis turned out 525,556 cars, 2d only to Michigan among the 50 states and 9% of the US total.

30 COMMERCE

Missouri has been one of the nation's leading trade centers ever since merchants in Independence (now part of the Kansas City metropolitan area) began provisioning wagon trains for the Santa Fe Trail. In 1982, wholesale trade was valued at $49.4 billion—12th in the US. Retail establishments had sales totaling $21.7 billion in 1982 (15th), of which food stores accounted for 23%, automotive dealers 19%, department stores 12%, gasoline service stations 11%, and eating and drinking places 9%. St. Louis County led all counties with 25% of total sales; the city of St. Louis accounted for another 8%. Kansas City led all cities with 12% of sales.

Foreign exports of Missouri agricultural commodities exceeded $1.4 billion in 1981/82 (9th among the states), of which soybeans and soybean products accounted for 47%. In 1983, Missouri was 3d among the states in soybean exports, 6th in meat exports, and 7th in wheat exports. In that year, agricultural exports came to $1 billion. Exports of manufactured goods were valued at more than $3 billion (19th in the US) in 1981. Transportation equipment accounted for 48%.

31 CONSUMER PROTECTION

The Department of Economic Development—created in 1984 from the Department of Consumer Affairs, Regulation, and Licensing, which itself had been created in 1974 through the consolidation of more than 30 state agencies—oversees community and economic development, regulates finance, insurance, and savings and loan institutions, and licenses such professions and occupations as accountancy, cosmetology, dentistry, nursing, pharmacy, and realty. The department's Office of Public Counsel represents consumer interests before the Public Service Commission, another branch of the department. The Consumer Information Center investigates consumer complaints.

32 BANKING

The first banks in Missouri, the Bank of St. Louis (established in 1816) and the Bank of Missouri (1817), had both failed by the time Missouri became a state, and the paper notes they had distributed proved worthless. Not until 1837 did the Missouri state government again permit a bank within its borders, and then only after filling its charter with elaborate restrictions. The Bank of Missouri, chartered for 20 years, kept its reputation for sound banking by issuing notes bearing the portrait of US Senator Thomas Hart Benton, nicknamed "Old Bullion" because of his extreme fiscal conservatism.

At the end of 1984, Missouri ranked 4th in the nation in the number of banks with 719. At the end of 1983, the state's commercial banks had assets totaling $42.8 billion and deposits of nearly $34.5 billion. The state's largest commercial banks are the Mercantile Trust Co. and Centerre Bank (formerly First National Bank), both in St. Louis, with assets at the end of 1983 of $3.43 billion and $3.58 billion, respectively.

During 1982, 42 multibank holding companies controlled 243 state-chartered banks with assets of $13.3 billion, and one-bank holding companies controlled 152 state-chartered banks with assets of nearly $4 billion. Assets of the state's 85 savings and loan associations at the end of 1983 totaled $18.2 billion, and outstanding mortgage loans at the end of 1982 reached $11.8 billion. In addition, there were about 326 state-chartered credit unions at the end of 1982 with nearly $1.8 billion in assets, as well as 293 small-loan offices.

33INSURANCE

In 1983, 47 life insurance firms were domiciled in Missouri. Premiums written in the state in 1981 totaled almost $5 billion for all types of insurance, including $863 million in direct premiums for life insurance (nearly $1.1 billion in 1983). About 8.7 million life insurance policies valued at $107.9 billion were in force in 1983. The average coverage per family was $54,400. Payouts totaled more than $1 billion, of which death payments accounted for $325.7 million, annuities $217 million, and policy and contract dividends $220.9 million.

In 1983, 185 property and casualty insurance companies were domiciled in Missouri. Property and casualty companies wrote premiums totaling $2.1 billion in that year, including $471 million in automotive liability insurance, $414.7 million in automotive physical damage insurance, and $265.8 million in homeowners' coverage.

Flood insurance amounting to $512.7 million was in force as of the end of 1983.

34SECURITIES

The Missouri Uniform Securities Act, also known as the "Blue Sky Law" and administered by the Securities Division of the Office of Secretary of State, requires the registration of stocks, bonds, debentures, notes, investment contracts, and oil, gas, and mining interests intended for sale in the state. In cases of fraud, misrepresentation, or other failure to comply with the act, the Missouri investor has the right to sue to recover the investment, plus interest, costs, and attorney fees. Government securities, stocks listed on the principal national exchanges, and securities sold under specific transactional agreements are exempt from registration.

Missouri had approximately 6,400 broker-dealers, agents, and investment advisers in 1982, all of whom, by law, were required to register with the state annually. Missouri had about 796,000 stockholders of public corporations in 1983. Kansas City has a commodity exchange, the Board of Trade, which deals in grains, including futures and storage. The world's largest winter-wheat market is in Kansas City.

35PUBLIC FINANCE

The Missouri state budget is prepared by the Office of Administration's Division of Budget and Planning and submitted annually by the governor to the general assembly for amendment and approval. The fiscal year runs from 1 July to 30 June.

The following table summarizes actual consolidated revenues and expenditures for 1983/84 and estimated revenues and expenditures for 1984/85 (in millions):

REVENUES	1983/84	1984/85
Individual income tax	$1,090.8	$1,212.9
Sales and use tax	902.2	955.0
Corporate income tax	184.2	186.9
County foreign insurance tax	80.4	88.0
Corporation franchise tax	39.2	42.7
Liquor and beer taxes	26.6	26.6
Inheritance tax	24.7	22.0
All other general revenue sources	116.5	134.5
Federal aid	954.4	1,150.6
Interest on deposits and investments	29.4	45.0
Other funds	1,105.0	1,595.7
TOTALS	$4,553.6	$5,459.9

EXPENDITURES		
Elementary and secondary education	$1,261.7	$1,363.4
Social services	1,034.9	1,097.1
Highways and transportation	499.6	617.6
Higher education	362.0	401.9
Mental health	241.4	273.5
Capital investments	80.0	204.9
Debt service	41.0	55.7
Other outlays	953.1	1,240.8
TOTALS	$4,473.7	$5,254.9

During 1981/82, St. Louis had general revenues of $398 million and general expenditures of $385 million; the gross debt as of 30 June 1982 was $206 million. Kansas City had general revenues of $349 million, general expenditures of $318 million, and an outstanding debt of $267 million for the 1981/82 fiscal year. The combined debt of Missouri state and local governments in mid-1982 was over $5 billion, or $1,020 per capita, well below the US average for the same period. In mid-1983, this debt was nearly $5.8 billion.

36TAXATION

Missouri's total state tax revenues, traditionally low, ranked 24th in the nation in 1981/82. On a per capita basis, state general revenue of $814 ranked 50th in the US in 1981/82. The state and local per capita tax burden of $843 was 45th among the states.

The Missouri personal income tax in mid-1984 ranged from 1.5% on the first $1,000 to 6% on amounts over $9,000; the corporate tax rate was 5% of net income. The basic state sales tax was 4.125%; cities and towns may add an additional tax (St. Louis and Kansas City added 1%). Other taxes levied by the state include charges on motor fuel, cigarettes, and alcoholic beverages, along with motor vehicle and operator's license fees and taxes on credit institutions, insurance companies, and inheritances. Property and sales taxes are the leading sources of local revenue; St. Louis and Kansas City also levied income taxes of 1% within their respective jurisdictions in 1984.

During 1982/83, Missouri contributed $11.3 billion in federal taxes and received $19 billion in federal expenditures, for a spending/tax ratio of 1.68, one of the highest in the US. Missourians filed 1,991,400 federal tax returns for 1983, paying nearly $5.5 billion in taxes.

37ECONOMIC POLICY

Primary responsibility for economic development is vested in the Department of Economic Development, and especially in its Division of Community and Economic Development, which seeks outside investment in the state, promotes the national and international marketing of Missouri products, provides technical assistance to existing businesses, and maintains an office in Dusseldorf in the Federal Republic of Germany and in Tokyo, Japan. Its Enterprise Zone Program provides a variety of tax credits, exemptions, and other incentives to businesses that locate in designated areas. The division also offers grants, information, technical aid, and other public resources to foster local and regional planning. Special programs are provided for the Ozarks region and to rehabilitate urban neighborhoods. Loan and bond guarantees are provided to selected businesses by the Missouri Economic Development Commission and direct loans by the Missouri Industrial Development Authority. A nonprofit Missouri Corporation for Science and Technology was created in 1983.

General incentives for business include the state's reputation for fiscal conservatism, wage rates no better than the national average, and a tax structure toward which the corporate income tax contributes only about 8%.

38HEALTH

The infant mortality rates in Missouri in 1981 were 11.7 per 1,000 live births for whites and 18.7 for nonwhites; the white rate was above the national norm. There were 20,024 legal abortions performed in 1983; the rate of 265 abortions per 1,000 live births was well below the US average.

The overall death rate of 9.9 per 1,000 population in 1981 was the highest in the US except for Florida and Pennsylvania—a phenomenon attributable in part to the relatively high proportion of elderly Missourians in the population as a whole. The rate was even higher in 1983—10.2 per 1,000. Deaths from heart disease, cancer, stroke, accidents, chronic obstructive pulmonary diseases, pneumonia and influenza, and diabetes mellitus—the leading causes of death—were all above the national average.

In 1983, Missouri had 169 hospitals, with 33,370 beds; 977,828 admissions were recorded in 1982, when the average occupancy rate was 75%. Hospital personnel included 19,027 registered nurses. The average cost of hospital care in 1982 was $328 per day and $2,622 per stay, the latter above the US average. The state had 9,127 licensed physicians and 2,413 dentists in 1982. Both figures put Missouri below the national rate per 100,000 civilian population.

39SOCIAL WELFARE

A monthly average of 177,479 Missourians were recipients of aid to families with dependent children in 1982, at an annual cost of $176 million; the average monthly payment of $232 per family was below the national average. Medicaid payments amounted to $249 million in federal funds and $184.8 million in state funds in 1982/83. During the same year, 412,000 state residents benefited from the food stamp program, at a federal cost of $222 million, while 527,000 Missouri schoolchildren took part in the school lunch program, with a federal subsidy of $45 million.

Social Security recipients numbered 858,542 in December 1982, of whom 504,910 were retirees, 174,121 were survivors of deceased workers, and 60,960 were disabled. Payments exceeded $3.9 billion in 1983, with an average monthly payment to retired workers of $407 in December 1982. In 1983, federally administered Supplemental Security Income payments totaled $154.7 million; state payments came to another $7.8 million. Vocational rehabilitation programs served 16,654 applicants in 1983/84, of whom 5,774 were rehabilitated; the total cost was $35.1 million. The state expended $151.6 million on workers' compensation and $338 million on unemployment insurance benefits in 1982.

40HOUSING

In 1980, Missouri had 1,988,915 housing units, of which 1,793,399 were occupied; over 96% of the year-round units had full plumbing. From 1981 through 1983, 40,100 new privately owned housing units were authorized, at a total value of $1.6 billion.

The Missouri Housing Development Commission of the Department of Economic Development is empowered to make and insure loans to encourage the construction of residential housing for persons of low or moderate income; funds for mortgage financing are provided through the sale of tax-exempt notes and bonds issued by the commission. Construction of multiunit public housing stagnated during the 1970s. In 1972, municipal authorities ordered the demolition of two apartment buildings in St. Louis's Pruitt-Igoe public housing complex, built 18 years earlier

and regarded by many commentators as a classic case of the failure of such high-rise projects to offer a livable environment; the site remained vacant in the early 1980s. Indeed, only 3.7% of St. Louis's housing units in 1980 had been built during the 1970s, a smaller percentage than in any other major city except Buffalo and Detroit; during the decade, many units were abandoned.

The median price of a single-family home in the St. Louis metropolitan area was $62,400 in mid-1984; for the Kansas City metropolitan area it was $59,600. Both figures were below the national average.

41EDUCATION

Although the constitution of 1820 provided for the establishment of public schools, it was not until 1839 that the state's public school system became a reality through legislation creating the office of state superintendent of common schools and establishing a permanent school fund. Missouri schools were officially segregated from 1875 to 1954, when the US Supreme Court issued its landmark ruling in *Brown* v. *Board of Education;* the state's school segregation law was not taken off the books until 1976. In that year, nearly 37% of all black students were in schools that were 99–100% black, a condition fostered by the high concentration of black Missourians in the state's two largest cities. In 1983, a desegregation plan was adopted for St. Louis–area public schools that called for 3,000 black students to be transferred from city to county schools.

In 1980, 63.5% of all Missourians 25 years of age or older were high school graduates. As of 1980/81, the state had a total of 2,084 public schools, of which 1,417 were elementary, 657 were secondary, and 10 mixed; public school enrollment during 1984/85 totaled 795,793. Private and parochial school enrollment was about 114,816 in fall 1983; of these enrollees, 90,730, or 79% were in Catholic schools.

Missouri had 28 public and 58 private institutions of higher education in 1980; there were 92 in all in 1983. Total full-time enrollment in the fall of 1983 was 113,976. The University of Missouri, established in 1839, was the first state-supported university west of the Mississippi River. It has four campuses: Columbia (site of the world's oldest and one of the best-known journalism schools), Kansas City, Rolla, and St. Louis. The Rolla campus, originally founded in 1870 as a mining and engineering school, is still one of the nation's leading universities specializing in technology. The four campuses had a combined full-time enrollment of 38,241 in 1983/84, with 24,059 of all students at the Columbia facility. Lincoln University, a public university for blacks until segregation ended in 1954, is located in Jefferson City. There are five regional state universities, at Warrensburg, Maryville, Cape Girardeau, Springfield, and Kirksville, and three state colleges, at St. Louis, St. Joseph, and Joplin. Two leading independent universities, Washington and St. Louis, are located in St. Louis, as is the Concordia Seminary, an affiliate of the Lutheran Church–Missouri Synod and the center of much theological and political controversy during the 1970s. The Department of Higher Education offers grants and guaranteed loans to Missouri students.

42ARTS

Theatrical performances are offered throughout the state, mostly during the summer. In Kansas City, productions of Broadway musicals and light opera are staged at the Starlight Theater, seating 7,600 in an open-air setting. The Missouri Repertory Theater, on the University of Missouri campus in Kansas City, also has a summer season. In St. Louis, the 12,000-seat Municipal Opera puts on outdoor theater, while the *Goldenrod*, built in 1909 and said to be the largest showboat ever constructed, is used today for vaudeville, melodrama, and ragtime shows. Other notable playhouses are the Clemens Outdoor Amphitheater in Hannibal and the Lyceum Theater in Arrow Rock.

Leading orchestras are the St. Louis Symphony and Kansas

City Symphony; Independence, Liberty, Columbia, Kirksville, St. Joseph, and Springfield also have orchestras. The Opera Theatre of St. Louis and the Lyric Opera of Kansas City are distinguished musical organizations. Springfield has a regional opera company.

Between World Wars I and II, Kansas City was the home of a thriving jazz community that included Charlie Parker and Lester Young; leading bandleaders of that time were Benny Moten, Walter Page, and, later, Count Basie. Country music predominates in rural Missouri: the Ozark Opry at Osage Beach and the Baldknobbers Hillbilly Jamboree and Mountain Music Theater in Branson have seasons from May to October.

The Missouri Council on the Arts spent $2.8 million in 1983/84, awarding grants to arts programs, community arts councils, touring programs, artists in residence, and public information services.

43 LIBRARIES AND MUSEUMS

Missouri had 79 county and 16 regional library systems in 1982/83, when the combined book stock of all public libraries in the state was 12,140,655, and their combined circulation 26,169,313. The Missouri State Library, in Jefferson City, is the center of the state's interlibrary loan network. It also serves as the only public library for the 10% of the state's population who live in areas without public libraries; it has more than 200,000 books. The largest public library systems, those of Kansas City and St. Louis County, had 1,275,095 and 1,695,046 volumes, respectively; the public library system of the city of St. Louis had 1,316,613. The University of Missouri–Columbia has the leading academic library, with 1,999,276 volumes in 1982/83. The State Historical Society of Missouri Library in Columbia contains 431,390 volumes. The federally administered Harry S Truman Library and Museum is at Independence.

Missouri has well over 100 museums and historic sites. The William Rockhill Nelson Gallery/Atkins Museum of Fine Arts in Kansas City and the St. Louis Art Museum each house distinguished general collections, while the Springfield Art Museum specializes in American sculpture, paintings, and relics of the westward movement. The Mark Twain Home and Museum in Hannibal has a collection of manuscripts and other memorabilia. Also notable are the Museum of Art and Archaeology, Columbia; the Kansas City Museum of History and Science; the Pony Express Stables Museum, St. Joseph; and the Jefferson National Expansion Memorial, Missouri Botanical Garden, St. Louis Center Museum of Science and Natural History and McDonnell Planetarium, National Museum of Transport, and a zoo, all in St. Louis. Kansas City, Springfield, and Eldon also have zoos.

44 COMMUNICATIONS

In 1858, John Hockaday began weekly mail service by stagecoach between Independence and Salt Lake City, and John Butterfield, with a $600,000 annual appropriation from Congress, established semimonthly mail transportation by coach and rail from St. Louis to San Francisco. On 3 April 1860, the Pony Express was launched, picking up mail arriving by train at St. Joseph and racing it westward on horseback; the system ceased in October 1861, when the Pacific Telegraph Co. began operations. The first experiment in airmail service took place at St. Louis in 1911; Charles Lindbergh was an airmail pilot on the St. Louis–Chicago route in 1926. In 1985, Missouri had 969 post offices and 17,357 postal employees.

As of 1980, Missouri had 1,697,313 residences with telephones. About 95% of all state residences had telephone service.

Radio broadcasting in Missouri dates from 1921, when a station at St. Louis University began experimental programming. On Christmas Eve 1922, the first midnight Mass ever to be put on the air was broadcast from the Old Cathedral in St. Louis. The voice of a US president was heard over the air for the first time on 21 June 1923, when Warren G. Harding gave a speech in St. Louis. FM broadcasting began in Missouri during 1948. As of

1983 there were 117 commercial AM stations and 130 FM stations in service. Missouri's first television station, KSD–TV in St. Louis, began in 1947, with WDAF–TV in Kansas City following in 1949. As of 1984, Missouri had 26 commercial and 4 noncommercial television stations; 6 commercial stations broadcast in the Kansas City area and 6 around St. Louis, with Kansas City, St. Louis, Sedalia, and Springfield each having a noncommercial station. In 1984, the state had 150 cable systems serving 600,282 subscribers in 371 communities.

45 PRESS

The *Missouri Gazette,* published in St. Louis in 1808 by the politically independent and controversial Joseph Charless, was the state's first newspaper; issued to 174 subscribers, the paper was partly in French. In 1815, a group of Charless's enemies raised funds to establish a rival paper, the *Western Journal,* and brought in Joshua Norvell from Nashville to edit it. By 1820 there were five newspapers in Missouri. Since that time, many Missouri newspapermen have achieved national recognition. The best known is Sam Clemens (later Mark Twain), who started out as a "printer's devil" in Hannibal at the age of 13. Hungarian-born Joseph Pulitzer began his journalistic career in 1868 as a reporter for a German-language daily in St. Louis. Pulitzer created the *St. Louis Post-Dispatch* from the merger of two defunct newspapers in 1878, endowed the Columbia University School of Journalism in New York City, and established by bequest the Pulitzer Prizes, which annually honor journalistic and artistic achievement.

As of 1984 there were 10 morning newspapers with a combined daily circulation of 898,554; 38 evening dailies with 553,259 circulation; and 18 Sunday papers with 1,203,765 circulation. The following table shows Missouri's leading dailies with their 1984 circulations:

AREA	NAME	DAILY	SUNDAY
Kansas City	Star (e,S)	233,691	397,655
	Times (m)	281,974	
St. Louis	Globe-Democrat (m)	210,712	
	Post-Dispatch (m,S)	264,721	479,075

Periodicals include the St. Louis–based *Sporting News,* the bimonthly "bible" of baseball fans; *VFW Magazine,* put out monthly in Kansas City by the Veterans of Foreign Wars; and the *Missouri Historical View,* a quarterly with offices in Columbia.

46 ORGANIZATIONS

The 1982 US Census of Service Industries counted 1,264 organizations in Missouri, including 270 business associations; 605 civic, social, and fraternal associations; and 29 educational, scientific, and research associations. Among the organizations with headquarters in Kansas City are the Veterans of Foreign Wars of the USA, Camp Fire Inc., People-to-People International, the American Academy of Family Physicians, the American Business Women's Association, the American Nurses Association, the Fellowship of Christian Athletes, the National Association of Intercollegiate Athletics, and Professional Secretaries International.

Headquartered in St. Louis are the American Association of Orthodontists, the American Optometric Association, the Catholic Health Association of the US, the Danforth Foundation, the International Consumer Credit Association, and the National Hairdressers and Cosmetologists Association. Other organizations are the National Council of State Garden Clubs (Clayton), the Accrediting Council on Education in Journalism and Mass Communications (Columbia), and the American Cat Fanciers Association (Branson).

47 TOURISM, TRAVEL, AND RECREATION

During 1982, travelers spent more than $4.1 billion in Missouri on transportation, accommodations, meals, entertainment, recreation, and other items. About 97,800 jobs, with a payroll of $881 million, were attributable to the travel industry.

The principal attraction in St. Louis is the Gateway Arch, at 630 feet (192 meters) the tallest man-made national monument in the US. Designed by Eero Saarinen in 1948 but not constructed until 1964, three years after his death, the arch and the Museum of Westward Expansion form part of the Jefferson National Expansion Memorial on the western shore of the Mississippi River. There were 2,406,953 visits to this site in 1984. In the Kansas City area are the modern Crown Center hotels and shopping plaza, Country Club Plaza, the Truman Sports Complex, Ft. Osage near Sibley, Jesse James's birthplace near Excelsior Springs, and Harry Truman's hometown of Independence. Memorabilia of Mark Twain are housed in and around Hannibal, in the northeast, and the birthplace and childhood home of George Washington Carver, a national monument, is in Diamond. The Lake of the Ozarks, with 1,375 mi (2,213 km) of shoreline, is one of the most popular vacation spots in mid-America. Other attractions are the Silver Dollar City handicrafts center near Branson; the Pony Express Stables and Museum at St. Joseph; Wilson's Creek National Battlefield at Republic, site of a Confederate victory in the Civil War; and the "Big Springs Country" of the Ozarks, in the southeast. The state fair is held in Sedalia each August.

As of 1983, Missouri had 47 state parks. Operated by the Department of Natural Resources, they offer camping, picnicking, swimming, boating, fishing, and hiking facilities. Lake of the Ozarks State Park is the largest, covering 16,872 acres (6,828 hectares). There were also 24 historic sites in 1983, when state parks and historic sites covered 98,422 acres (39,830 hectares); they attract nearly 10 million visitors annually. Hunting and fishing are popular recreational activities. In 1982/83, licenses were issued to 959,455 fishermen and 482,367 hunters.

⁴⁸SPORTS

Missouri has two major league baseball teams. The St. Louis Cardinals of the National League, now playing at Busch Memorial Stadium, have won the World Series nine times; their glory years came with Dizzy Dean and the "Gas House Gang" in the 1930s, the heyday of Stan Musial in the 1940s, and the era of Bob Gibson and Lou Brock in the 1960s. The Kansas City Royals won the American League Pennant in 1980 and again in 1985, in which year they defeated the Cardinals in the World Series. The Kansas City Kings played in the National Basketball Association for many years before moving to Sacramento in 1985, while the St. Louis Blues compete in the National Hockey League. The St. Louis Cardinals of the National Football Conference and the Kansas City Chiefs of the American Football conference (winners of the 1970 Super Bowl) are the state's two professional football teams.

Horse racing has a long history in Missouri. In 1812, St. Charles County sportsmen held two-day horse races; by the 1820s, racetracks were laid out in nearly every city and in crossroads villages. Today, thoroughbred racing can be seen during a summer and fall season at Cahokia Downs, outside St. Louis.

In collegiate sports, the University of Missouri competes in the Big Eight Conference.

⁴⁹FAMOUS MISSOURIANS

Harry S Truman (1884-1972) has been the only native-born Missourian to serve as US president or vice president. Elected US senator in 1932, Truman became Franklin D. Roosevelt's vice-presidential running mate in 1944 and succeeded to the presidency upon Roosevelt's death on 12 April 1945. The "man from Independence"—whose tenure in office spanned the end of World War II, the inauguration of the Marshall Plan to aid European economic recovery, and the beginning of the Korean conflict—was elected to the presidency in his own right in 1948, defeating Republican Thomas E. Dewey in one of the most surprising upsets in US political history. Charles Evans Whittaker (b.Kansas, 1901-73) was a federal district and appeals court judge in Missouri before his appointment as Supreme Court associate

justice in 1957. Among the state's outstanding US military leaders are Generals John J. Pershing (1860-1948) and Omar Bradley (1893-1981).

Other notable federal officeholders from Missouri include Edward Bates (b.Virginia, 1793-1869), Abraham Lincoln's attorney general and the first cabinet official to be chosen from a state west of the Mississippi River; Montgomery Blair (b.Kentucky, 1813-83), postmaster general in Lincoln's cabinet; and Norman Jay Colman (b.New York, 1827-1911), the first secretary of agriculture. Missouri's best-known senator was Thomas Hart Benton (b.North Carolina, 1782-1858), who championed the interests of Missouri and the West for 30 years. Other well-known federal legislators include Francis P. Blair, Jr. (b.Kentucky, 1821-75), antislavery congressman, pro-Union leader during the Civil War, and Democratic vice-presidential nominee in 1868; Benjamin Gratz Brown (b.Kentucky, 1826-85), senator from 1863 to 1867 and later governor of the state and Republican vice-presidential nominee (1872); Carl Schurz (b.Germany, 1829-1906), senator from 1869 to 1875 and subsequently US secretary of the interior, as well as a journalist and Union military leader; William H. Hatch (b.Kentucky, 1833-96), sponsor of much agricultural legislation as a US representative from 1879 to 1895; Richard P. Bland (b.Kentucky, 1835-99), leader of the free-silver bloc in the US House of Representatives; James Beauchamp "Champ" Clark (b.Kentucky, 1850-1921), speaker of the House from 1911 to 1919; W. Stuart Symington (b.Massachusetts, 1901), senator from 1953 to 1977 and earlier the nation's first secretary of the Air Force; and Thomas F. Eagleton (b.1929), senator since 1969 and, briefly, the Democratic vice-presidential nominee in 1972, until publicity about his having received electroshock treatment for depression forced him off the ticket. (Eagleton announced in 1984 that he would not seek reelection to the Senate in 1986.)

Outstanding figures in Missouri history included two pioneering fur traders: William Henry Ashley (b.Virginia, 1778-1838), who later became a US representative, and Manuel Lisa (b.Louisiana, 1772-1820), who helped establish trade relations with the Indians. Meriwether Lewis (b.Virginia, 1774-1809) and William Clark (b.Virginia, 1770-1838) explored Missouri and the West during 1804-6; Lewis later served as governor of Louisiana Territory, with headquarters at St. Louis, and Clark was governor of Missouri Territory from 1813 to 1821. Dred Scott (b.Virginia, 1795-1858), a slave owned by a Missourian, figured in a Supreme Court decision that set the stage for the Civil War. Missourians with unsavory reputations include such desperadoes as Jesse James (1847-82), his brother Frank (1843-1915), and Cole Younger (1844-1916), also a member of the James gang. Another well-known native was Kansas City's political boss, Thomas Joseph Pendergast (1872-1945), a power among Missouri Democrats until convicted of income tax evasion in 1939 and sent to Leavenworth prison.

Among notable Missouri educators were William Torrey Harris (b.Connecticut, 1835-1909), superintendent of St. Louis public schools, US commissioner of education, and an authority on Hegelian philosophy; James Milton Turner (1840-1915), who helped establish Lincoln University for blacks at Jefferson City; and Susan Elizabeth Blow (1843-1916), cofounder with Harris of the first US public kindergarten at St. Louis in 1873. Distinguished scientists include agricultural chemist George Washington Carver (1864-1943), astronomers Harlow Shapley (1885-1972) and Edwin P. Hubble (1889-1953), Nobel Prize-winning nuclear physicist Arthur Holly Compton (b.Ohio, 1892-1962), and mathematician-cyberneticist Norbert Wiener (1894-1964). Engineer and inventor James Buchanan Eads (b.Indiana, 1820-87) supervised construction during 1867-74 of the St. Louis bridge that bears his name. Charles A. Lindbergh (b.Michigan, 1902-74) was a pilot and aviation instructor in the St. Louis area during the 1920s before winning worldwide acclaim for his solo New York–Paris flight.

Prominent Missouri businessmen include brewer Adolphus Busch (b.Germany, 1839–1913); William Rockhill Nelson (b.Indiana, 1847–1915), who founded the *Kansas City Star* (1880); Joseph Pulitzer (b.Hungary, 1847–1911), who merged two failed newspapers to establish the *St. Louis Post-Dispatch* (1878) and later endowed the journalism and literary prizes that bear his name; and James Cash Penney (1875–1971), founder of the J. C. Penney Co. Noteworthy journalists from Missouri include newspaper and magazine editor William M. Reedy (1862–1920), newspaper reporter Herbert Bayard Swope (1882–1958), and television newscaster Walter Cronkite (b.1916). Other distinguished Missourians include theologian Reinhold Niebuhr (1892–1971), civil rights leader Roy Wilkins (1901–81), and medical missionary Thomas Dooley (1927–61).

Missouri's most popular author is Mark Twain (Samuel Langhorne Clemens, 1835–1910), whose *Adventures of Tom Sawyer* (1876) and *Adventures of Huckleberry Finn* (1884) evoke his boyhood in Hannibal. Novelist Harold Bell Wright (b.New York, 1872–1944) wrote about the people of the Ozarks; Robert Heinlein (b.1907) is a noted writer of science fiction, and William S. Burroughs (b.1914) an experimental novelist. Poet-critic T(homas S(tearns) Eliot (1888–1965), awarded the Nobel Prize for literature in 1948, was born in St. Louis but became a British subject in 1927. Other Missouri-born poets include Sara Teasdale (1884–1933), Marianne Moore (1887–1972), and Langston Hughes (1902–67). Popular novelist and playwright Rupert Hughes (1872–1956) was a Missouri native, as was Zoe Akins (1886–1958), a Pulitzer Prize-winning playwright.

Distinguished painters who lived in Missouri include George Caleb Bingham (b.Virginia, 1811–79), who also served in several state offices; James Carroll Beckwith (1852–1917); and Thomas Hart Benton (1889–1975), the grandnephew and namesake of the state's famous political leader. Among the state's important musicians are ragtime pianist-composers Scott Joplin (b.Texas, 1868–1917) and John William "Blind" Boone (1864–1927); W(illiam) C(hristopher) Handy (b.Alabama, 1873–1958), composer of "St. Louis Blues," "Beale Street Blues," and other classics; composer-critic Virgil Thompson (b.1896), known for his operatic collaborations with Gertrude Stein; jazzman Coleman Hawkins (1907–69); and popular songwriter Burt Bacharach (b.1929). Photographer Walker Evans (1903–75) was a St. Louis native.

Missouri-born entertainers include actors Wallace Beery, (1889–1949), Vincent Price (b.1911), and Edward Asner (b.1929); actresses Jean Harlow (Harlean Carpenter, 1911–37), Jane Wyman (b.1914), Betty Grable (1916–73), and Shelley Winters (b.1922); dancers Sally Rand (1904–79) and Josephine Baker (1906–75); actress-dancer Ginger Rogers (b.1911); film director John Huston (b.1906); and opera stars Helen Traubel (1903–72), Gladys Swarthout (1904–69), and Grace Bumbry (b.1937). In popular music, the state's most widely known singer-songwriter is Charles "Chuck" Berry (b.California, 1926), whose works had a powerful influence on the development of rock 'n' roll.

St. Louis Cardinals stars who became Hall of Famers include Jerome Herman "Dizzy" Dean (b.Arkansas, 1911–74), Stanley Frank "Stan the Man" Musial (b.Pennsylvania, 1920), Robert "Bob" Gibson (b.Nebraska, 1935), and Louis "Lou" Brock

(b.Arkansas, 1939). Among the native Missourians who achieved stardom in the sports world are baseball manager Charles Dillon "Casey" Stengel (1890–1975), catcher Lawrence Peter "Yogi" Berra (b.1925), sportscaster Joe Garagiola (b.1926), and golfer Tom Watson (b.1949).

[50]BIBLIOGRAPHY

Chappell, Philip Edward. *A History of the Missouri River.* Kansas City: Kansas State Historical Society, 1905.

DeVoto, Bernard Augustine. *Mark Twain's America.* Boston: Little, Brown, 1932.

Dorsett, Lyle W. *The Pendergast Machine.* New York: Oxford University Press, 1968.

Federal Writers' Project. *Missouri: A Guide to the "Show Me" State.* Rev. ed. New York: Somerset, 1981 (orig. 1941).

Foley, William E. *A History of Missouri: 1673 to 1820.* Columbia: University of Missouri Press, 1971.

Gerlach, Russel L. *Immigrants in the Ozarks.* Columbia: University of Missouri Press, 1976.

Gibson, Arrell M. *The Encyclopedia of Missouri.* New York: Somerset, 1984.

Glaab, Charles N. *Kansas City and the Railroads.* Madison: State Historical Society of Wisconsin, 1962.

Greene, Lorenzo J., et al. *Missouri's Black Heritage.* St. Louis: Forum, 1980.

Hall, Leonard. *Stars Upstream: Life Along an Ozark River.* Columbia: University of Missouri Press, 1969.

March, David D. *The History of Missouri.* 4 vols. New York and West Palm Beach, Fla.: Lewis, 1967.

McCandless, Perry. *A History of Missouri: 1820 to 1860.* Columbia: University of Missouri Press, 1972.

Missouri, State of. Secretary of State. *Official Manual, 1983–84.* Jefferson City, 1983.

Moore, Glover. *The Missouri Compromise, 1819–21.* Lexington: University of Kentucky Press, 1953.

Nagel, Paul C. *Missouri: A Bicentennial History.* New York: Norton, 1977.

Park, Eleanore G., and Kate S. Morrow. *Women of the Mansion: Missouri, 1821–1936.* Jefferson City: Midland, 1936.

Parrish, William E. *A History of Missouri: 1860 to 1875.* Columbia: University of Missouri Press, 1973.

Parrish, William E., et al. *Missouri: The Heart of the Nation.* St. Louis: Forum, 1980.

Primm, James Neal. *Economic Policy in the Development of a Western State: Missouri, 1820–60.* Cambridge: Harvard University Press, 1954.

Rafferty, Milton D. *Historical Atlas of Missouri.* Norman: University of Oklahoma Press, 1982.

Shoemaker, Floyd C. *Missouri and Missourians: Land of Contrasts and People of Achievement.* 5 vols. Chicago: Lewis, 1943.

Sprague, Marshall. *So Vast a Land: Louisiana and the Purchase.* Boston: Little, Brown, 1974.

State Historical Society of Missouri. *Historic Missouri: A Pictorial Narrative.* Columbia, 1977.

Truman, Harry S. *Memoirs.* 2 vols. Garden City, N.Y.: Doubleday, 1955–56.

Vestal, Stanley. *The Missouri.* New York: Farrar & Rinehart, 1945.

Wecter, Dixon. *Sam Clemens of Hannibal.* Boston: Houghton Mifflin, 1952.

MONTANA

State of Montana

ORIGIN OF STATE NAME: Derived from the Latin word meaning "mountainous." **NICKNAME:** The Treasure State. (Also: Big Sky Country.) **CAPITAL:** Helena. **ENTERED UNION:** 8 November 1889 (41st). **SONG:** "Montana." **MOTTO:** *Oro y Plata* (Gold and Silver). **FLAG:** A blue field, fringed in gold on the top and bottom borders, surround the state coat of arms, with "Montana" in gold letters above the coat of arms. **OFFICIAL SEAL:** In the lower center are a plow and a miner's pick and shovel; mountains appear above them on the left, the Great Falls of the Missouri River on the right, and the state motto on a banner below. The words "The Great Seal of the State of Montana" surround the whole. **BIRD:** Western meadowlark. **FISH:** Black-spotted (cutthroat) trout. **FLOWER:** Bitterroot. **TREE:** Ponderosa pine. **GEMS:** Yogo sapphire; Montana agate. **GRASS:** Bluebunch wheatgrass. **LEGAL HOLIDAYS:** New Year's Day, 1 January; Birthday of Martin Luther King, Jr., 3d Monday in January; Lincoln's Birthday, 12 February; Washington's Birthday, 3d Monday in February; Memorial Day, last Monday in May; Independence Day, 4 July; Labor Day, 1st Monday in September; Columbus Day, 2d Monday in October; State Election Day, 1st Tuesday after the 1st Monday in November in even-numbered years; Veterans Day, 11 November; Thanksgiving Day, 4th Thursday in November; Christmas Day, 25 December. **TIME:** 5 AM MST = noon GMT.

¹LOCATION, SIZE, AND EXTENT

Located in the northwestern US, Montana is the largest of the 8 Rocky Mountain states and ranks 4th in size among the 50 states.

The total area of Montana is 147,046 sq mi (380,849 sq km), of which land takes up 145,388 sq mi (376,555 sq km) and inland water 1,658 sq mi (4,294 sq km). The state's maximum E–W extension is 570 mi (917 km); its extreme N–S distance is 315 mi (507 km).

Montana is bordered on the N by the Canadian provinces of British Columbia, Alberta, and Saskatchewan; on the E by North Dakota and South Dakota; on the S by Wyoming and Idaho; and on the W by Idaho. The total boundary length of Montana is *1,947 mi (3,133 km)*. The state's geographic center is in Fergus County, 12 mi (19 km) W of Lewistown. Nearly 30% of the state's land belonged to the federal government in 1983.

²TOPOGRAPHY

Montana, as mountainous in parts as its name implies, has an approximate mean elevation of 3,400 feet (1,000 meters). The Rocky Mountains cover the western two-fifths of the state, with the Bitterroot Range along the Idaho border; the high, gently rolling Great Plains occupy most of central and eastern Montana. The highest point in the state is Granite Peak, at an elevation of 12,799 feet (3,901 meters), located in south-central Montana, near the Wyoming border. The lowest point, at 1,800 feet (550 meters), is in the northwest, where the Kootenai River leaves the state at the Idaho border. The Continental Divide passes in a jagged pattern through the western part of the state, from the Lewis to the Bitterroot ranges.

Ft. Peck Reservoir is Montana's largest body of inland water, covering 375 sq mi (971 sq km); Flathead Lake is the largest natural lake. The state's most important rivers are the Missouri, rising in southwestern Montana and flowing north and then east across the state, and the Yellowstone, which crosses southeastern Montana to join the Missouri in North Dakota near the Montana border. Located in Glacier National Park is the Triple Divide, from which Montana waters begin their journey to the Arctic and Pacific oceans and the Gulf of Mexico.

³CLIMATE

The Continental Divide separates the state into two distinct climatic regions: the west generally has a milder climate than the east, where winters can be especially harsh. Montana's maximum daytime temperature averages 27°F (−2°C) in January and 85°F

(29°C) in July. Great Falls has a normal daily mean temperature of 45°F (7°C), ranging from 21°F (−6°C) in January to 69°F (21°C) in July. The all-time low temperature in the state, −70°F (−57°C), registered at Rogers Pass on 20 January 1954, is the lowest ever recorded in the conterminous US; the all-time high, 117°F (47°C), was set at Medicine Lake on 5 July 1937. Great Falls receives an average annual precipitation of 15 in (38 cm), but much of north-central Montana is arid. About 57 in (145 cm) of snow descends on Great Falls each year. For some parts of Montana, 1984 witnessed the third consecutive year of drought.

⁴FLORA AND FAUNA

Montana has three major life zones: subalpine, montane, and plains. The subalpine region, in the northern Rocky Mountains, is rich in wild flowers during a short midsummer growing season. The montane flora consists largely of coniferous forests, principally alpine fir, and a variety of shrubs. The plains are characterized by an abundance of grasses, cacti, and sagebrush species.

Game animals of the state include elk, moose, white-tailed and mule deers, pronghorn antelope, bighorn sheep, and mountain goat. Notable among the amphibians is the axolotl; rattlesnakes and other reptiles occur in most of the state. Rare or threatened species include the grizzly bear, spotted bat, prairie falcon, and Arctic grayling. The black-footed ferret, Eskimo curlew, and greenback cutthroat trout are on the endangered list.

⁵ENVIRONMENTAL PROTECTION

Montana's major environmental concerns are management of mineral and water resources and reclamation of strip-mined land. The 1973 Montana Resource Indemnity Trust Act, by 1975 amendment, imposes a coal severance tax of 30% on the contract sales price, with the proceeds placed in a permanent tax trust fund. This tax, in conjunction with the Montana Environmental Policy Act (1971) and the Major Facilities Siting Act (1973), reflects the determination of Montanans to protect the beauty of the Big Sky Country while maintaining economic momentum.

⁶POPULATION

According to the 1980 census, Montana ranked 44th among the 50 states, with a population of 786,690. The 1985 estimated population was 826,933 and the population density was 5.7 per sq mi (2.2 per sq km).

Slightly more than half of all Montanans live in urban areas. Leading cities and their populations in 1984 were Billings, 69,836, and Great Falls, 58,769.

[7]ETHNIC GROUPS

According to the 1980 census, there were 37,153 American Indians in Montana, of whom the Blackfeet and Crow are the most numerous.

The foreign-born made up 2.3% of Montana's 1980 census population, an increase of 6.7% since 1970. Canada, Germany, the United Kingdom, and Norway were the leading places of origin. The black, Hispanic, and Asian populations are very small.

[8]LANGUAGES

English in Montana fuses Northern and Midland features, the Northern proportion declining from east to west. Topography has given new meanings to *basin, hollow, meadow,* and *park* as kinds of clear spaces in the mountains.

In 1980, 708,281 Montanans—95% of the resident population 3 years of age or older—spoke only English at home. Other languages spoken at home included:

German	10,297	Norwegian	2,156
Spanish	5,614	Italian	1,204

[9]RELIGIONS

As of 1980, Protestant groups had 207,776 known adherents in Montana. Leading denominations included American Lutheran, 46,022; United Methodist, 20,287; and Latter-day Saints (Mormons), 26,035. Montana had 131,117 Roman Catholics in 1984 and an estimated 640 Jews in 1984.

[10]TRANSPORTATION

Montana's first railroad, the Utah and Northern, entered the state in 1880. Today, Montana is served by three major railroads, operating on about 3,480 mi (5,600 km) of track.

Because of its large size, small population, and difficult terrain, Montana was slow to develop a highway system. As of 1983, the state had 72,034 mi (115,928 km) of roads, streets, and highways. The Highways Department reoported that 2,660 mi (4,280 km), or about 50%, of primary roads needed repair in 1983. There were 828,737 registered motor vehicles and 488,603 licensed drivers in 1983.

Montana had 184 airports and 2,538 active aircraft in 1983. The leading airports are at Great Falls and Billings.

[11]HISTORY

Much of Montana's prehistory has only recently been unearthed. The abundance of fossils of large and small dinosaurs, marine reptiles, miniature horses, and giant cave bears indicates that, from 100 million to 60 million years ago, the region had a tropical climate. Beginning some 2 million years ago, however, dramatic temperature changes profoundly altered what we now call Montana. At four different times, great sheets of glacial ice moved south through Canada to cover much of the north. The last glacial retreat, about 10,000 years ago, did much to carve the state's present topographic features. Montana's first humans probably came from across the Bering Strait; their fragmentary remains indicate a presence dating between 10,000 and 4000 BC.

The Indians encountered by Montana's first white explorers—probably French traders and trappers from Canada—arrived from the east during the 17th and 18th centuries, pushed westward into Montana by the pressure of European colonization. In January 1743, two traders, Louis-Joseph and François Vérendrye, crossed the Dakota plains and saw before them what they called the "shining mountains," the eastern flank of the northern Rockies. However, it was not until 1803 that the written history of Montana begins. In that year, the Louisiana Purchase gave the United States most of Montana, and the Lewis and Clark Expedition, dispatched by President Thomas Jefferson in 1804 to explore the upper reaches of the Missouri River, added the rest. On 25 April 1805, accompanied by a French trapper named Toussaint Charbonneau and his Shoshoni wife, Sacagawea, Meriwether Lewis and William Clark reached the mouth of the

Yellowstone River near the present-day boundary with North Dakota. Shortly thereafter, the first American trappers, traders, and settlers entered Montana.

The fur trade dominated Montana's economy until 1858, when gold was discovered near the present community of Drummond. By mid-1862, a rush of miners from the gold fields of California, Nevada, Colorado, and Idaho had descended on the state. The temporary gold boom brought not only the state's first substantial white population but also an increased demand for government. In 1863, the eastern and western sectors of Montana were joined as part of Idaho Territory, which, in turn, was divided along the Bitterroot Mountains to form the present boundary between the two states. On 26 May 1864, President Abraham Lincoln signed the Organic Act, which created Montana Territory.

The territorial period was one of rapid and profound change. By the time Montana became a state on 8 November 1889, the remnants of Montana's Indian culture had been largely confined to federal reservations. A key event in this transformation was the Battle of the Little Big Horn River on 25 June 1876, when Lieutenant Colonel George Custer and his 7th US Cavalry regiment of fewer than 700 men were overwhelmed as they attacked an encampment of 15,000 Sioux and Northern Cheyenne led by Crazy Horse and Chief Gall. The following year, after a four-month running battle that traversed most of the state of Montana, Chief Joseph of the Nez Percé tribe surrendered to federal forces, signaling the end of organized Indian resistance.

As the Indian threat subsided, stockmen wasted little time in putting the seemingly limitless open range to use. By 1866, Nelson Story had driven the first longhorns up from Texas, and by the mid-1870s, sheep had also made a significant appearance on the open range. In 1886, at the peak of the open-range boom, approximately 664,000 head of cattle and nearly a million sheep grazed Montana's rangeland. Disaster struck during the "hard winter" of 1886/87, however, when perhaps as many as 362,000 head of cattle starved trying to find the scant forage covered by snow and ice. That winter marked the end of a cattle frontier based on the "free grass" of the open range and taught the stockmen the value of a secure winter feed supply.

Construction of Montana's railroad system between 1880 and 1909 breathed new life into mining as well as the livestock industry. Moreover, the railroads created a new network of market centers at Great Falls, Billings, Bozeman, Missoula, and Havre. By 1890, the Butte copper pits were producing more than 40% of the nation's copper requirements. The struggle to gain financial control of the enormous mineral wealth of Butte Hill led to the "War of the Copper Kings," in which the Amalgamated Copper Co., in conjunction with Standard Oil, bought out the competition and emerged as sole proprietor of "the richest hill on earth." In 1915, as a result of antitrust prosecution, Standard Oil gave up its copper holdings. The new company, Anaconda Copper Mining, virtually controlled the press, politics, and governmental processes of Montana until changes in the structure of the international copper market and the diversification of Montana's economy in the 1940s and 1950s reduced the company's power. Anaconda Copper was absorbed by the Atlantic Richfield Co. in 1976, and the name was changed to Anaconda Minerals in 1982. In 1985, Atlantic Richfield announced that it would sell Anaconda.

The railroads also brought an invasion of agricultural homesteaders. Montana's population surged from 243,329 in 1900 to 376,053 in 1910 and to 548,889 by 1920, while the number of farms and ranches increased from 13,000 to 57,000. Drought and a sharp drop in wheat prices after World War I brought an end to the homestead boom. From 1919 to 1925, nearly 11,000 farms were abandoned, and some 20,000 farm mortgages were foreclosed by banks. By 1926, half of Montana's commercial banks had failed. Conditions worsened with the drought and depression

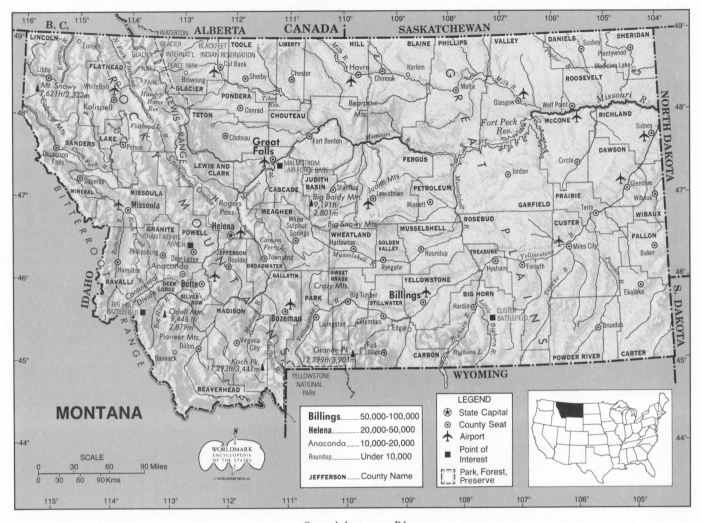

MONTANA

Billings _____ 50,000-100,000
Helena _____ 20,000-50,000
Anaconda _____ 10,000-20,000
Roundup _____ Under 10,000

JEFFERSON _____ County Name

LEGEND
⊛ State Capital
⊙ County Seat
✈ Airport
■ Point of Interest
[ ] Park, Forest, Preserve

SCALE
0   30   60   90 Miles
0   30   60   90 Kms

WORLDMARK
ENCYCLOPEDIA
OF THE STATES

See end-sheet maps: D1

LOCATION: 44°21′26″ to 49°N; 104°02′26″ to 116°02′56″W. **BOUNDARIES:** Canadian line, 546 mi (879 km); North Dakota line, _212 mi (341 km)_; South Dakota line, _67 mi (108 km)_; Wyoming line, 384 mi (618 km); Idaho line, 738 mi (1,187 km).

of the early 1930s, until the New Deal—enormously popular in Montana—helped revive farming and silver mining and financed irrigation and other public works projects.

The decades since the end of World War II have seen moderate growth in Montana's population, economy, and social services. Although manufacturing developed slowly, the state's fossil fuels industry grew rapidly during the national energy crisis of the 1970s. However, production of coal, crude oil, and natural gas leveled off after the crisis and even declined in the early 1980s.

12STATE GOVERNMENT

Montana's original constitution, dating from 1889, was substantially revised by a 1972 constitutional convention, effective 1 July 1973. Under the present document, the state legislature consists of 50 senators, elected to staggered four-year terms, and 100 representatives, who serve for two years. The only elected officers of the executive branch are the governor and lieutenant governor (who run jointly), secretary of state, attorney general, superintendent of public instruction, and auditor; each serves a four-year term.

To become law, a bill must pass both houses by a simple majority and be signed by the governor, remain unsigned for five days, or be passed over the governor's veto by a two-thirds vote of both houses. The state constitution may be amended by constitutional convention, by legislative referendum, or by voter initiative. To be adopted, each proposed amendment must be ratified at the next general election.

To vote in Montana, one must be a US citizen and at least 18 years of age; there is a county residency requirement of 30 days for registration.

13POLITICAL PARTIES

Since statehood, Democrats have generally dominated in contests for the US House and Senate, and Republicans in elections for state and local offices and in national presidential campaigns (except during the New Deal years). It is remarkable that, as of 1985, only one Montana Republican had ever been elected to the US Senate, and then for only one term. Although the erosion of Montana's rural population since the 1920s has diluted the Republicans' agrarian base, the party has gained increasing financial and organizational backing from corporate interests, particularly from the mining and energy-related industries. The strength of the Democratic Party, on the other hand, lies in the strong union movement centered in Butte and its surrounding counties, augmented by smaller family farms throughout the state. Urbanization has also benefited the Democrats. Montanans voted overwhelmingly for Republican President Ronald Reagan in November 1984. But they also gave the Democrats control of the state senate, split the house of representatives evenly between the two parties, and reelected a Democratic governor, Ted Schwinden.

In 1984, Montana had 7 Hispanic elected officials; 21 women served in the state legislature.

Montana Presidential Vote by Major Political Parties, 1948–84

YEAR	ELECTORAL VOTE	MONTANA WINNER	DEMOCRAT	REPUBLICAN
1948	4	*Truman (D)	119,071	96,770
1952	4	*Eisenhower (R)	106,213	157,394
1956	4	*Eisenhower (R)	116,238	154,933
1960	4	Nixon (R)	134,891	141,841
1964	4	*Johnson (D)	164,246	113,032
1968	4	*Nixon (R)	114,117	138,835
1972	4	*Nixon (R)	120,197	183,976
1976	4	*Ford (R)	149,259	173,703
1980	4	*Reagan (R)	118,032	206,814
1984	4	*Reagan (R)	146,742	232,450

* Won US presidential election.

[14]LOCAL GOVERNMENT

As of 1984, Montana had 56 counties, 126 municipalities, 450 special districts, and 399 school districts. Typical elected county officials are three county commissioners, attorney, sheriff, clerk and recorder, school superintendent, treasurer, public administrator, assessor, and coroner. Unified city-county governments include Anaconda–Deer Lodge and Butte–Silver Bow.

[15]STATE SERVICES

The Citizens' Advocate Office, established in 1973, serves as a clearinghouse for problems, complaints, and questions concerning state government. The commissioner of higher education administers the state university system, while the superintendent of public instruction is responsible for the public schools. The Department of Highways is the main transportation agency. Health and welfare programs are the province of the Department of Health and Environmental Sciences and the Department of Social and Rehabilitation Services.

[16]JUDICIAL SYSTEM

Montana's highest court, the supreme court, consists of a chief justice and 6 associate justices. District courts are the courts of general jurisdiction; as of 1980 there were 19 judicial districts. Justice of the peace courts are essentially county courts whose jurisdiction is limited to minor civil cases, misdemeanors, and traffic violations. Montana has seven supreme court justices elected on nonpartisan ballots for eight-year terms, 36 district court judges for six years, and many justices of the peace for four years. Montana's crime rates were below the US averages in every category in 1983. The state rarely enforces the death penalty; only six convicted persons have been executed since 1930, and none since 1977. There were 1,796 practicing lawyers at the end of 1984.

[17]ARMED FORCES

The principal military facility in Montana is Malmstrom Air Force Base (Great Falls), a Strategic Air Command facility employing 3,958 military personnel. An estimated 108,000 veterans of US military service were living in Montana in 1983. In 1982/83, expenditures on veterans totaled $89 million.

[18]MIGRATION

Montana's first great migratory wave brought Indians from the east during the 17th and 18th centuries. The gold rush of the 1860s and a land boom between 1900 and 1920 resulted in surges of white settlement. The economically troubled 1920s and 1930s produced a severe wave of out-migration that continued through the 1960s. The trend reversed between 1970 and 1980, however, when Montana's net gain from migration was 16,000; from 1980 to 1983, the state gained 5,000 residents.

[19]INTERGOVERNMENTAL COOPERATION

Among the interstate agreements in which Montana participates are the Interstate Oil and Gas Compact, Interstate Corrections Compact, Western Interstate Energy Compact, Western Regional Education Compact, and Yellowstone River Compact (with North Dakota and Wyoming).

Federal aid to the state and local governments in 1983/84 totaled $532 million.

[20]ECONOMY

Resource industries—agriculture, mining, lumbering—dominate Montana's economy, although tourism is of increasing importance. In 1983, manufacturing contributed $4 billion to the state economy; mining, $1.6 billion; agriculture, $1.5 billion; and travel, $600 million.

[21]INCOME

With a per capita income of $9,999 in 1983, Montana ranked 37th among the 50 states. Total income of $8.1 billion in 1983 represented a 3.3% increase since 1978 in constant dollars.

Median family income reached $18,413 in 1979, when 19,019 families, accounting for 12.3% of the state's population, were living below the federal poverty level. Some 13,000 Montanans were among the top wealth holders in the US, with gross assets greater than $500,000.

[22]LABOR

Montana's labor force varies sharply with the season, swelling in the summer and shrinking in the winter. As of 1984, the civilian labor force totaled 405,000 persons, of whom 375,000 were employed and 30,000 were unemployed. The unemployment rate was 7.4% in 1984, compared with 8.8% in 1983.

A federal survey in 1983 revealed the following nonfarm employment pattern for Montana:

	ESTABLISHMENTS	EMPLOYEES	ANNUAL PAYROLL ('000)
Agricultural services, forestry, fishing	269	998	$ 9,635
Mining	434	7,408	189,188
Contract construction	2,229	10,091	205,608
Manufacturing	1,110	20,475	429,032
Transportation, public utilities	1,199	15,670	325,310
Wholesale trade	1,848	16,078	275,476
Retail trade	6,625	54,971	486,580
Finance, insurance, real estate	1,919	13,800	207,448
Services	6,598	55,128	625,669
Other	1,874	1,445	24,836
TOTALS	24,105	196,064	$2,778,782

In 1980, union membership was 82,000, or 29% of the nonfarm work force. There were 370 labor unions in the state in 1983.

[23]AGRICULTURE

Montana's farms numbered 24,000 in 1984, with average acreage of 2,546 (1,030 hectares). Farm income totaled $1.5 billion in 1983, ranking 32d in the US. In that year, Montana was the nation's 6th-leading wheat producer, with an output of 136,930,000 bu, valued at $504,000,000. Other major crops were barley (2d), 77,700,000 bu, $150,740,000; sugarbeets (7th), 818,000 bu, $34,683,000; and hay (15th), 4,158,000 tons, $34,948,000. Oats, potatoes, flax, and dry beans are also grown.

[24]ANIMAL HUSBANDRY

Livestock production in 1983 was valued at $656.3 million, or 44% of Montana's farm income. As of early 1984, the state had

3,150,000 cattle and calves. Other livestock included 180,000 hogs and pigs and 509,000 sheep and lambs. Sales of livestock products in 1983 included cattle and calves, worth $543,673,000; dairy products, $44,881,000; hogs and pigs, $22,909,000; and sheep and lambs, $16,703,000. The wool clip was estimated at 5,568,000 lb in 1983.

25 FISHING

Montana's designated fishing streams offer some 10,000 mi (16,000 km) of good to excellent freshwater fishing. The state stocked 8.1 million trout and 2.1 million salmon in 1980.

26 FORESTRY

As of 1979, 22,559,000 acres (9,129,000 hectares) in Montana were classified as forestland. There were 10 national forests, comprising roughly 19,089,083 acres (7,758,660 hectares) in 1984. The lumbering industry produced 875 million board feet in 1982; its product shipments were valued at $203 million.

27 MINING

The discovery of rich deposits of placer gold at Bannack, Virginia City, and Helena led to the settlement of southwestern Montana in the 1860s. Oil and natural gas, discovered in the 1920s, are scattered beneath the eastern two-thirds of the state, but especially near the North Dakota border; lignite is concentrated in the east, and bituminous coal in the southeast. Most metal mining—especially for copper, silver, and gold—takes place in the southwestern region. With the recent decline of the copper industry, Anaconda Minerals closed its copper-mining operations near Butte in 1983, and in 1985, Atlantic Richfield, Anaconda's parent corporation, announced that it would sell off Anaconda.

Montana's nonfuel mineral output was valued at $248 million in 1984, down 15% from 1983. In 1984, mineral production (excluding fossil fuels) included gold, 161,436 troy oz; silver, 5,708 troy oz; copper, 33,337 tons; and lead, 1,163 tons.

28 ENERGY AND POWER

In 1983, Montana produced 15.1 billion kwh of electricity, 74% from hydropower and 26% by coal burning; installed capacity was 3.2 million kwh.

Oil and natural gas supply about 75% of Montana's energy requirements. In 1983, the state produced 29.2 million barrels of crude oil, leaving proved reserves of about 230 million barrels; natural gas production totaled 53.5 billion cu feet with proved reserves amounting to 896 billion. As of 1 January 1983, coal reserves were estimated at 120.3 billion tons—1st in the US and nearly 25% of the US total—of which bituminous coal accounted for 1.3 billion tons; subbituminous, 103.2 billion; and lignite, 15.7 billion. Production of coal in 1983 totaled 28.9 million tons.

29 INDUSTRY

Montana's major manufacturing industries process raw materials from mines, forests, and farms. The total value of shipments by manufacturers in 1982 amounted to $3,677.4 million. Major sectors and their value of shipments included:

Petroleum and coal products	$1,644,400,000
Lumber and wood products	590,400,000
Food and food products	494,400,000
Printing and publishing materials	115,800,000
Stone, clay, and glass products	109,200,000

30 COMMERCE

in 1982, 1,809 wholesale establishments had sales of $4.9 billion, while 9,504 retailers had sales exceeding $3.9 billion. Montana's foreign exports in 1981 included $559 million in agricultural products and $61 million in manufactured goods. Wheat and flour were the leading export items; Montana ranked 6th among all the states in export of these products.

31 CONSUMER PROTECTION

Montana's consumer protection laws are administered by the Legal and Consumer Affairs Division of the Commerce Department.

32 BANKING

At the end of 1984, Montana had 167 insured commercial banks; their assets totaled $6.2 billion in 1982, and deposits were nearly $5.3 billion. In 1984, there were 12 insured savings and loan associations, all federally chartered, with total assets of almost $1.2 billion (1983).

33 INSURANCE

Montanans held 1,118,000 life insurance policies, with a combined value of nearly $15 billion, in 1983. The average per family was $45,000, ranking 43d in the US. Property and casualty insurers wrote premiums worth $369.5 million in 1983.

34 SECURITIES

There are no securities exchanges in Montana. In 1983, New York Stock Exchange member firms had 24 sales offices and 111 representatives in the state. About 107,000 Montanans owned shares in public corporations in that year.

35 PUBLIC FINANCE

The Montana state budget is prepared biennially by the Office of Budget and Program Planning and submitted by the governor to the legislature for amendment and approval. The fiscal year runs from 1 July to 30 June. The following is a summary of estimated revenues and expenditures in the general fund for 1985/86 and 1986/87 (in thousands):

REVENUES	1985/86	1986/87
Personal income taxes	$123,466	$130,626
Corporate income taxes	29,957	32,304
Coal trust fund interest	28,792	33,560
Oil severance tax	27,792	29,269
Other receipts and transfers	162,574	169,192
TOTALS	$372,581	$394,951
EXPENDITURES		
University system	$ 94,563	$ 92,047
Social and rehabilitation services	75,984	84,007
Public schools	52,034	52,324
Department of Revenue	19,940	18,942
Montana State Hospital	17,805	17,826
Government administration	16,088	16,091
Other expenditures	114,630	120,572
TOTALS	$391,044	$401,809

As of mid-1983, the combined debt of Montana state and local governments was $1,476 million, or $1,665 per capita.

36 TAXATION

In 1984, Montana's personal income tax, which has been indexed to inflation since 1981, ranged from 2% on the first $1,200 to 11% on amounts over $42,000. The corporate income tax was 6.75% on net income, with a minimum tax of $50. The state levies a property tax but no sales or use tax.

Montana received $477 million in federal expenditures in 1982/83, or $584 per capita. State residents paid $773 million in federal income taxes in 1982, or $965 per capita.

37 ECONOMIC POLICY

The state's department of commerce encourages economic development, recruits new industries for Montana, and assists small businesses. The state government's small-business advocate provides relevant information for small businesses and tries to cut bureaucratic red tape on their behalf.

38 HEALTH

The infant mortality rate per 1,000 live births in 1982 was 10.7. The birthrate in 1982 was 18.1 per 1,000 population, and the death rate was 6.8 in 1981. Leading causes of death in 1981 (with their rates per 100,000 population) were heart disease, 288; cancer, 169; stroke, 69; and accidents, 70. The first three rates were below the national norms, but Montana's accident rate was the 3d highest in the US; the state's suicide rate of 17.1 per

100,000 population tied with Wyoming's for 4th highest in the US.

There were 67 hospitals in 1983, with 5,345 beds and 141,143 admissions during the year. Hospital personnel included 2,124 registered nurses in 1981. In January 1984, the average cost of a semiprivate hospital room was $185 per day, up 9% from the year before; Montana ranked 23d among the states in this category. There were 1,039 physicians in 1981 and 484 dentists in 1982.

39SOCIAL WELFARE

Montana played an important role in the development of social welfare. It was one of the first states to experiment with workers' compensation, enacting a compulsory compensation law in 1915; eight years later, Montana and Nevada became the first states to provide for old age pensions.

Public assistance payments to an average of 5,773 families with dependent children in 1982 totaled $20 million; average monthly payment per family was $292 as of January 1983. In 1983, Social Security benefits were paid to 122,000 Montanans; their benefits totaled $555 million, with the average monthly payment to retired workers being $431. Other expenditures included Medicaid (1983), $86 million; Supplementary Security Income (1983), $13.5 million; unemployment insurance (1982), $45 million; and workers' compensation (1983), $44 million.

40HOUSING

In 1980, Montana had some 315,015 housing units, of which 283,742 were occupied; 68% of the units were owner-occupied, and 97% had full plumbing. The state authorized about 3,000 new housing units in 1983; units started numbered about 2,700.

41EDUCATION

As of 1980, about three-fourths of Montanans 25 years and older had completed high school, and nearly one-fifth were college graduates.

Estimated public school enrollments in fall 1983 were 100,213 in elementary school, 12,044 in junior high school, and 41,389 in high school. The ratio of pupils to teachers was 16:1.

As of fall 1984, 35,763 students attended institutions of higher education. Of these, the University of Montana (Missoula) enrolled 9,213, and Montana State University (Bozeman) 11,035.

42ARTS

The C. M. Russell Museum in Great Falls honors the work of Charles Russell, whose mural *Lewis and Clark Meeting the Flathead Indians* adorns the capitol in Helena. Other fine-art museums include the Museum of the Rockies in Bozeman, Yellowstone Art Center at Billings, and the Missoula Museum of the Arts. There were six performing-arts groups in 1982.

43LIBRARIES AND MUSEUMS

Montana had six public library federations in 1983, serving 50 counties; the combined book stock of all Montana public libraries was 2,118,124—2.6 volumes per capita—and their combined circulation was 4,213,761. Distinguished collections include those of the University of Montana (Missoula), with 636,193 volumes; Montana State University (Bozeman), 426,725; and the Montana State Library and Montana Historical Society Library, both in Helena.

Among the state's 63 museums are the Montana Historical Society Museum, Helena; World Museum of Mining, Butte; Western Heritage Center, Billings; and Museum of the Plains Indian, Browning. National historic sites include Big Hole and Little Big Horn battlefields and the Grant-Kohrs Ranch at Deer Lodge, west of Helena.

44COMMUNICATIONS

In 1981, 95% of the state's households had telephone service, with the total number of telephones being 630,733—461,943 residential and 168,790 business. There were 95 commercial radio stations (48 AM, 47 FM) in 1984, and 19 television stations, including 6 educational broadcasters. During the same year, 101 cable television systems served 113,333 subscribers in 105 communities.

45PRESS

As of 1985, Montana had 5 morning dailies with a combined circulation of 148,997, 6 evening dailies with 49,874, and 8 Sunday newspapers with 204,554. The leading papers were the *Billings Gazette* (60,294 mornings, 61,623 Sundays) and the *Great Falls Tribune* (38,056 mornings, 43,997 Sundays).

46ORGANIZATIONS

The 1982 Census of Service Industries counted 421 organizations in Montana, including 77 business associations; 257 civic, social, and fraternal associations; and 13 educational, scientific, and research associations. Among the organizations headquartered in Montana are the American Simmental Association (Bozeman) and Bikecentennial: The Bicycle Travel Association (Missoula).

47TOURISM, TRAVEL, AND RECREATION

Travel is Montana's 4th-largest industry and supplied 20,200 travel-related jobs in 1983. During that year, out-of-state visitors numbered 2,208,352 and spent $423 million in Montana.

Many tourists seek out the former gold rush camps, ghost towns, and dude ranches. Scenic wonders include all of Glacier National Park, covering 1,013,595 acres (410,189 hectares), which is the US portion of Waterton-Glacier International Peace Park; part of Yellowstone National Park, which also extends into Idaho and Wyoming; and Bighorn Canyon National Recreation Area.

48SPORTS

There are no major league professional sports teams in Montana. The University of Montana Grizzlies and Montana State University Bobcats both compete in the Big Sky Conference.

49FAMOUS MONTANANS

Prominent national officeholders from Montana include US Senator Thomas Walsh (b.Wisconsin, 1859-1933), who directed the investigation that uncovered the Teapot Dome scandal; Jeannette Rankin (1880-1973), the first woman member of Congress and the only US representative to vote against American participation in both world wars; Burton K. Wheeler (b.Mass., 1882-1975), US senator from 1923 to 1947 and one of the most powerful politicians in Montana history; and Michael Joseph "Mike" Mansfield (b.New York, 1903), who held the office of majority leader of the US Senate longer than anyone else.

Chief Joseph (b.Oregon, 1840?-1904), a Nez Percé Indian, repeatedly outwitted the US Army during the late 1870s; Crazy Horse (1849?-77) led a Sioux-Cheyenne army in battle at Little Big Horn. The town of Bozeman is named for explorer and prospector John M. Bozeman (b.Georgia, 1835-67).

Creative artists from Montana include Alfred Bertram Guthrie, Jr. (b.Indiana, 1901), author of *The Big Sky* and the Pulitzer Prize-winning *The Way West;* Dorothy Johnson (b.Iowa, 1905-84), whose stories have been made into such notable Western movies as *The Hanging Tree, The Man Who Shot Liberty Valance,* and *A Man Called Horse;* and Charles Russell (b.Missouri, 1864-1926), Montana's foremost painter and sculptor. Hollywood stars Gary Cooper (Frank James Cooper, 1901-61) and Myrna Loy (b.1905) were born in Helena. Newscaster Chet Huntley (1911-74) was born in Cardwell.

50BIBLIOGRAPHY

Federal Writers' Project. *Montana: A State Guide Book.* Reprint. New York: Somerset, n.d. (orig. 1939).

Farr, William, and K. Ross Toole. *Montana: Images of the Past.* Boulder, Colo.: Pruett, 1984.

Malone, Michael P., and Richard B. Roeder. *Montana: A History of Two Centuries.* Seattle: University of Washington Press, 1976.

Spence, Clark C. *Montana: A Bicentennial History.* New York: Norton, 1978.

Toole, Kenneth R. *Montana: An Uncommon Land.* Norman: University of Oklahoma Press, 1984.

Toole, Kenneth R. *Twentieth-Century Montana: A State of Extremes.* Norman: University of Oklahoma Press, 1983.

NEBRASKA

State of Nebraska

ORIGIN OF STATE NAME: From the Oto Indian word *nebrathka,* meaning "flat water" (for the Platte River). **NICKNAME:** The Cornhusker State. **CAPITAL:** Lincoln. **ENTERED UNION:** 1 March 1867 (37th). **SONG:** "Beautiful Nebraska." **MOTTO:** Equality Before the Law. **FLAG:** The great seal appears in the center in gold and silver, on a field of blue. **OFFICIAL SEAL:** Agriculture is represented by a farmer's cabin, sheaves of wheat, and growing corn, the mechanic arts by a blacksmith. Above is the state motto; in the background, a steamboat plies the Missouri River and a train heads toward the Rockies. The scene is surrounded by the words "Great Seal of the State of Nebraska, March 1st 1867." **BIRD:** Western meadowlark. **FLOWER:** Goldenrod. **TREE:** Cottonwood. **GEM:** Blue agate. **ROCK:** Prairie agate. **GRASS:** Little bluestem. **INSECT:** Honeybee. **FOSSIL:** Mammoth. **LEGAL HOLIDAYS:** New Year's Day, 1 January; Birthday of Martin Luther King, Jr., 3d Monday in January; Presidents' Day, 3d Monday in February; Arbor Day, 22 April; Memorial Day, last Monday in May; Independence Day, 4 July; Labor Day, 1st Monday in September; Columbus Day, 2d Monday in October; Veterans Day, 11 November; Thanksgiving, 4th Thursday in November and following Friday; Christmas Day, 25 December. **TIME:** 6 AM CST = noon GMT; 5 AM MST = noon GMT.

¹LOCATION, SIZE, AND EXTENT

Located in the western north-central US, Nebraska ranks 15th in size among the 50 states. The total area of the state is 77,355 sq mi (200,349 sq km), of which land takes up 76,644 sq mi (198,508 sq km) and inland water 711 sq mi (1,841 sq km). Nebraska extends about 415 mi (668 km) E–W and 205 mi (330 km) N–S.

Nebraska is bordered on the N by South Dakota (with the line formed in part by the Missouri River); on the E by Iowa and Missouri (the line being defined by the Missouri River); on the S by Kansas and Colorado; and on the W by Colorado and Wyoming. The boundary length of Nebraska totals 1,332 mi (2,143 km). The state's geographic center is in Custer County, 10 mi (16 km) NW of Broken Bow.

²TOPOGRAPHY

Most of Nebraska is prairie; more than two-thirds of the state lies within the Great Plains proper. The elevation slopes upward gradually from east to west, from a low of 840 feet (256 meters) in the southeast to 5,426 feet (1,654 meters) in Kimball County. Rolling alluvial lowlands in the eastern portion of the state give way to the flat, treeless plain of central Nebraska, which in turn rises to a tableland in the west. The Sand Hills of the north-central plain is an unusual region of sand dunes anchored by grasses that cover about 18,000 sq mi (47,000 sq km).

The Sand Hills region is dotted with small natural lakes; in the rest of the state, the main lakes are artificial. The Missouri River—which, with its tributaries, drains the entire state—forms the eastern part of the northern boundary of Nebraska. Three rivers cross the state from west to east: the wide, shallow Platte River flows through the heart of the state for 310 mi (499 km); the Niobrara River traverses the state's northern region; and the Republican River flows through southern Nebraska.

³CLIMATE

Nebraska has a continental climate, with highly variable temperatures from season to season and year to year. The central region has an average annual normal temperature of 50°F (10°C), with a normal monthly maximum of 76°F (24°C) in July and a normal monthly minimum of 22°F (−6°C) in January. The record low for the state is −47°F (−44°C), registered in Morrill County on 12 February 1899; the record high of 118°F (48°C) was recorded at Minden on 24 July 1936.

Normal yearly precipitation in the semiarid panhandle in the west is 17 in (43 cm); in the southeast, 30 in (76 cm). Snowfall in the state varies from about 21 in (53 cm) in the southeast to about 45 in (114 cm) in the northwest corner. Blizzards, droughts, and windstorms have plagued Nebraskans throughout their history.

⁴FLORA AND FAUNA

Nebraska's deciduous forests are generally oak and hickory; conifer forests are dominated by western yellow (ponderosa) pine. The tallgrass prairie may include various slough grasses and needlegrasses, along with big bluestem, and prairie dropseed. Mixed prairie regions abound with western wheatgrass and buffalo grass. The prairie region of the Sand Hills supports a variety of bluestems, gramas, and other grasses. Common Nebraska wild flowers are wild rose, phlox, petunia, columbine, goldenrod, and sunflower. Rare species of Nebraska's flora include the Hayden penstemon, yellow ladyslipper, pawpaw, and snow trillium.

Common mammals native to the state are the pronghorn sheep, white-tailed and mule deer, badger, kit fox, coyote, striped ground squirrel, prairie vole, and several skunk species. There are more than 400 kinds of birds, the mourning dove, barn swallow, and western meadowlark (the state bird) among them. Carp, catfish, trout, and perch are fished for sport. Rare animal species include the least shrew, least weasel, and bobcat. The bald eagle, Arctic peregrine falcon, Higgins' eye pearly mussel, black-footed ferret, and northern swift fox *(Vulpes velox)* are listed as endangered species.

⁵ENVIRONMENTAL PROTECTION

The Department of Environmental Control was established in 1971 to protect and improve the quality of the state's water, air, and land resources.

The Agricultural Pollution Control Division of the department regulates disposal of feedlot wastes and other sources of water pollution by agriculture. The Water and Waste Management Division is responsible for administering the Federal Clean Water Act, the Federal Resources Conservation and Recovery Act, portions of the Federal Safe Drinking Water Act, and the Nebraska Environmental Protection Act as it relates to water, solid waste, and hazardous materials. In 1983, the division regulated a total of 1,008,009 tons of hazardous waste generated by 79 industrial facilities. A program to protect groundwater from such pollutants as nitrates, synthetic organic compounds, hydrocarbons, pesticides, and other sources, was outlined in 1985. The Engineering Division regulates wastewater treatment standards and assists municipalities in securing federal construction grants for wastewater facilities. The Air Quality Division is responsible for monitoring and securing compliance with national ambient air quality standards.

⁶POPULATION

Nebraska ranked 35th in the US in 1980 with a census population of 1,570,000, a 5.7% increase over 1970. A 1985 estimate placed the total population at 1,606,779, for a population density of 21 per sq mi (8 per sq km). A population of 1,639,800 is projected for 1990.

In 1980 there were 765,900 men and 804,100 women in Nebraska. Some 63% of Nebraska's population lived in urban areas in 1980, compared with 54% in 1960. The largest cities were Omaha, which ranked 48th among the nation's cities with a 1984 population of 332,237, and Lincoln, with 180,378.

⁷ETHNIC GROUPS

Among Nebraskans reporting a single ancestry in the 1980 census, 352,873 identified their ancestry as German, 92,699 as English, 61,614 as Irish, 48,442 as Czech, and 34,070 as Swedish. The 1980 population also included 48,389 black Americans (37,852 in Omaha), up from 39,911 in 1970, and 6,996 Asians and Pacific Islanders.

There were 9,147 American Indians in Nebraska in 1980. Three Indian reservations are maintained for the Omaha, Winnebago, and Santee Sioux tribes.

⁸LANGUAGES

Many Plains Indians of the Macro-Siouan family once roamed widely over what is now Nebraska. Place-names derived from the Siouan language include Omaha, Ogallala, Niobrara, and Keya Paha. In 1980, about 3,800 Nebraskans claimed Indian tongues as their first languages.

Nebraska English, except for a slight South Midland influence in the southwest and some Northern influence from Wisconsin and New York settlers in the Platte River Valley, is almost pure North Midland. A few words, mostly food terms like *kolaches* (fruit-filled pastries), are derived from the language of the large Czech population. Usual pronunciation features are *on* and *hog* with the /o/ of *order, cow* and *now* as /kaow/ and /naow/, *because* with the /ah/ vowel, *cot* and *caught* as sound-alikes, and a strong final /r/. *Fire* sounds almost like *far,* and *our* like *are; greasy* is pronounced /greezy/.

In 1980, 1,423,183 Nebraskans—95% of the resident population 3 years old or older—spoke only English at home. The number of residents who spoke other languages at home included:

German	18,150	Polish	3,295
Spanish	18,063	French	3,079
Czech	8,986	Italian	2,151

⁹RELIGION

Nebraska's religious history derives from its patterns of immigration. German and Scandinavian settlers tended to be Lutheran; Irish, Polish, and Czech immigrants were mainly Roman Catholic. Methodism and other Protestant religions were spread by settlers from other midwestern states.

In 1984, the state's Catholic population numbered 334,352. Lutherans constituted the largest Protestant group, with 112,593 adherents of the Missouri Synod, 81,941 of the Lutheran Church of America, and 49,169 of the American Lutheran Church. A total of 156,955 were United Methodists, and 58,702 were Presbyterians. The Jewish population was estimated at 7,850 in 1984, with most living in Omaha.

¹⁰TRANSPORTATION

Nebraska's development was profoundly influenced by two major railroads, the Union Pacific and the Burlington Northern, both of which were major landowners in the state in the late 1800s. These lines still operate in Nebraska, together with three other major railroads and three small companies. All together, the eight lines had 4,775 mi (7,685 km) of track in the state in 1983. Amtrak had a total Nebraska ridership of 60,373 in 1984. Railroad freight traffic increased from 74,770,000 tons in 1970 to 168,463,000 tons in 1981.

The state's road system, 92,442 mi (148,771 km) in 1983, is dominated by Interstate 80, the major east–west route and the largest public investment project in the state's history. Eighty-nine percent of the driving-age population of the state—about 1,094,-680 people—held driver's licenses in 1983. A total of 1,234,591 motor vehicles were registered in 1983, of which 815,232 were automobiles and 415,269 trucks.

There were 211 private and 112 public airports in the state in 1982. Eppley Airfield, Omaha's airport, is by far the busiest in the state, handling 20,930 aircraft departures with 928,601 passengers in 1983.

¹¹HISTORY

Nebraska's first inhabitants, from about 10,000 BC, were nomadic Paleo-Indians. Successive groups were more sedentary, cultivating corn and beans. Archaeological excavations indicate that prolonged drought and dust storms before the 16th century caused these inhabitants to vacate the area. In the 16th and 17th centuries, other Indian tribes came from the east, some pushed by enemy tribes, others seeking new hunting grounds. By 1800, semisedentary Pawnee, Ponca, Omaha, and Oto, along with several nomadic groups, were in the region.

The Indians developed amiable relations with the first white explorers, French and Spanish fur trappers and traders who traveled through Nebraska in the 18th century, using the Missouri River as a route to the West. The area was claimed by both Spain and France and was French territory at the time of the Louisiana Purchase, when it came under US jurisdiction. It was explored during the first half of the 19th century by Lewis and Clark, Zebulon Pike, Stephen H. Long, and John C. Frémont.

The Indian Intercourse Act of 1834 forbade white settlement west of the Mississippi River, reserving the Great Plains as Indian Territory. Nothing prevented whites from traversing Nebraska, however, and from 1840 to 1866, some 350,000 persons crossed the area on the Oregon, California, and Mormon trails, following the Platte River Valley, a natural highway to the West. Military forts were established in the 1840s to protect travelers from Indian attack.

The Kansas-Nebraska Act of 1854 established Nebraska Territory, which stretched from Kansas to Canada and from the Missouri River to the Rockies. The territory assumed its present shape in 1861. Still sparsely populated, Nebraska escaped the violence over the slavery issue that afflicted Kansas. The creation of Nebraska Territory heightened conflict between Indians and white settlers, however, as Indians were forced to cede more and more of their land. From mid-1860 to the late 1870s, western Nebraska was a battleground for Indians and US soldiers. By 1890, the Indians were defeated and moved onto reservations in Nebraska, South Dakota, and Oklahoma.

Settlement of Nebraska Territory was rapid, accelerated by the Homestead Act of 1862, under which the US government provided 160 acres (65 hectares) to a settler for a nominal fee, and the construction of the Union Pacific, the first transcontinental railroad. The Burlington Railroad, which came to Nebraska in the late 1860s, used its vast land grants from Congress to promote immigration, selling the land to potential settlers from the East and from Europe. The end of the Civil War brought an influx of Union veterans, bolstering the Republican administration, which

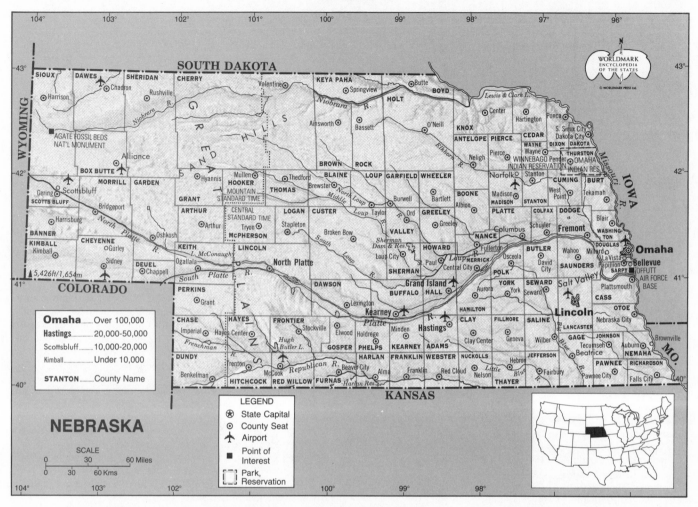

See endsheet maps: F2.

LOCATION: 40° to 43°00′02″N, 95°03′54″ to 104°03′09″W. **BOUNDARIES:** South Dakota line: land, 281 mi (452 km); Missouri River, 140 mi (225 km); Iowa line (Missouri River), 181 mi (291 km); Missouri line (Missouri River), 63 mi (101 km); Kansas line, 355 mi (571 km); Colorado line, 173 mi (279 km); Wyoming line, 139 mi (224 km).

began pushing for statehood. On 1 March 1867, Nebraska became the 37th state to join the Union. Farming and ranching developed as the state's two main enterprises. Facing for the first time the harsh elements of the Great Plains, homesteaders in central and western Nebraska evolved what came to be known as the sod-house culture, using grassy soil to construct sturdy insulated homes. They harnessed the wind with windmills to pump water, constructed fences of barbed wire, and developed dry-land farming techniques.

Ranching existed in Nebraska as early as 1859, and by the 1870s, it was well established in the western part of the state. Some foreign investors controlled hundreds of thousands of acres of the free range. The cruel winter of 1886/87 killed thousands of cattle and bankrupted many of these large ranches.

By 1890, depressed farm prices, high railroad shipping charges, and rising interest rates were hurting the state's farmers, and a drought in the 1890s exacerbated their plight. These problems contributed to the rise of populism, a proagrian movement. Many Nebraska legislators embraced populism, helping to bring about the first initiative and referendum laws in the US, providing for the regulation of stockyards and telephone and telegraph companies, and instituting compulsory education.

World War I created a rift among Nebraskans as excessive patriotic zeal was directed against residents of German descent. German-language newspapers were censored, ministers were

ordered to preach only in English (often to congregations that understood only German), and three university professors of German origin were fired. A Nebraska law (1919) prohibiting the teaching of any foreign language until high school was later declared unconstitutional by the US Supreme Court.

Tilling of marginal land to take advantage of farm prices inflated during World War I caused economic distress during the 1920s. Nebraska's farm economy was already in peril when the dust storms of the 1930s began, and conditions worsened as drought, heat, and grasshopper invasions plagued the state. Thousands of people, particularly from the southwest counties in which dust-bowl conditions were most severe, fled Nebraska for the west coast. Some farmers joined protest movements—dumping milk, for example, rather than selling at depressed prices—while others marched on the state capital to demand a moratorium on farm debts, which they received. In the end, federal aid saved the farmers.

The onset of World War II brought prosperity to other sectors. Military airfields and war industries were placed in the state because of its safe inland location, bringing industrial growth that extended into the postwar years. Much of the new industry developed since that time is agriculture-related, including the manufacture of mechanized implements and irrigation equipment.

Farm output and income increased dramatically into the 1970s through wider use of hybrid seed, pesticides, fungicides, chemical fertilizers, close-row planting, and irrigation, but contaminated runoff adversely affected water quality and greater water use drastically lowered water-table levels. Many farmers took on large debt burdens to finance expanded output, their credit buoyed by strong farm-product prices and exports. When prices began to fall in the early 1980s, many found themselves overextended. By spring 1985, an estimated 10% of all farmers were reportedly close to bankruptcy.

12STATE GOVERNMENT

The first state constitution was adopted in 1866; a second, adopted in 1875, is still in effect. A 1919-20 constitutional convention proposed—and voters passed—41 amendments; by 1982, the document had been revised an additional 138 times.

Nebraska's legislature is unique among the states; since 1934, it has been a unicameral body of 49 members elected on a nonpartisan basis. Members, who go by the title of senator, are chosen in even-numbered years for four-year terms. Elected executives are the governor, lieutenant governor, secretary of state, auditor, treasurer, and attorney general, all of whom serve four-year terms. The governor and lieutenant governor are jointly elected; each must be a US citizen, at least 30 years old, and have been a resident of Nebraska for five years. After serving two consecutive terms, the governor is ineligible for the office for four years.

A bill becomes law when passed by a majority of the legislature and signed by the governor. If the governor does not approve, the bill is returned with objections, and a three-fifths vote of the legislature is required to override the veto. A bill automatically becomes law if the governor does not take action within five days after receiving it.

A three-fifths majority of the legislature is required to propose an amendment to the state constitution. The people may propose an amendment by presenting a petition signed by 10% of the electorate. The amendments are then submitted for approval at the next regular election or at a special election that can be called by a four-fifths vote of the legislature.

Voters in the state must be at least 18 years of age; there is no residency requirement.

13POLITICAL PARTIES

Conservatism and Republican Party domination has typified Nebraska politics, although the voters have not been entirely predictable. For 40 years, for example, voters elected a liberal Republican, George Norris, first to the US House (1902-13) and then to the Senate (1913-43).

Nebraska Presidential Vote by Major Political Parties, 1948–84

YEAR	ELECTORAL VOTE	NEBRASKA WINNER	DEMOCRAT	REPUBLICAN
1948	6	Dewey (R)	224,165	264,774
1952	6	*Eisenhower (R)	188,057	421,603
1956	6	*Eisenhower (R)	199,029	378,108
1960	6	Nixon (R)	232,542	380,553
1964	5	*Johnson (D)	307,307	276,847
1968	5	*Nixon (R)	170,784	321,163
1972	5	*Nixon (R)	169,991	406,298
1976	5	Ford (R)	233,692	359,705
1980	5	*Reagan (R)	166,424	419,214
1984	5	*Reagan (R)	187,866	460,054

*Won US presidential election.

In the 1982 gubernatorial election there were 416,938 registered Republicans, 362,188 Democrats, and 52,995 independents. As of 1984, two Democrats represented Nebraska in the US Senate; three Republicans and no Democrats held seats in the House. In the 1984 elections, 460,054 Nebraskans voted for Republican incumbent Ronald Reagan, and 187,866 voted for Democrat Walter Mondale. Minority elected officials include 8 women in the legislature and a total of 4 blacks and 5 Hispanics statewide.

14LOCAL GOVERNMENT

In 1983, Nebraska had 93 counties, 471 townships, 534 municipalities, and 1,195 school districts. More than 1,000 special districts covered such services as fire protection, housing, irrigation, and sewage treatment.

Twenty-nine counties are run by elected boards of supervisors, and 64 by elected boards of commissioners. Municipalities are governed by mayors and are divided into five classifications by population: metropolitan cities, of which Omaha is the only one; primary cities, of which Lincoln is the only one; and, as of 1983, 29 first-class cities, 111 second-class cities, and 393 villages, some with populations under 100.

15STATE SERVICES

As of 1 June 1971, the Office of Public Counsel (Ombudsman) was empowered to investigate complaints from citizens in relation to the state government. The Accountability and Disclosure Commission, established in 1977, regulates the organization and financing of political campaigns and investigates reports of conflicts of interest involving state officials.

The eight-member state Board of Education, elected on a nonpartisan basis, oversees elementary and secondary public schools and vocational education. The Board of Regents, which also consists of eight elected members, governs the University of Nebraska system. Special examining boards license architects, engineers, psychologists, and land surveyors.

The Department of Roads maintains and builds highways, and the Department of Aeronautics regulates aviation, licenses airports, and registers aviators. The Department of Motor Vehicles provides vehicle and driver services.

Public assistance, child welfare, medical care for the indigent, and a special program of services for crippled children are the responsibility of the Department of Public Welfare. The Department of Health operates community health services, provides nutritional services, and is responsible for disease control.

The state's huge agricultural industry is aided and monitored by the Department of Agriculture, which is empowered to protect livestock, inspect food-processing areas, conduct research into crop development, and encourage product marketing.

16JUDICIAL SYSTEM

The state's highest court is the supreme court, consisting of a chief justice and 6 other justices, all of whom are initially appointed by the governor. They must be elected after serving three years, and every six years thereafter, running unopposed on their own record. Below the supreme court are the district courts; 48 judges serve 21 districts in the state. These are trial courts of general jurisdiction. County courts handle criminal misdemeanors and civil cases involving less than $5,000. In addition, there are a court of industrial relations, a worker's compensation court, two conciliation courts (family courts), two municipal courts (in Omaha and Lincoln), and juvenile courts in three counties. Nebraska's crime rate is well below the national average. In 1984, there were 3,763 practicing attorneys in the state. A total of 10,935 persons were on probation, 844 were in local jails, and 1,677 were in state and federal prisons. Four persons were executed between 1930 and 1983; 10 were under sentence of death in 1984.

17ARMED FORCES

The US military presence in Nebraska is concentrated near Omaha, where Offutt Air Force Base serves as the headquarters of the US Strategic Air Command. Total Air Force personnel at the base in 1984 was 14,380. In 1983, Nebraska firms were awarded $162.8 million in defense contracts.

A total of 190,000 veterans of US military service resided in Nebraska as of 30 September 1983. Of these, about 3,000 served in World War I, 69,000 in World War II, 37,000 in the Korean conflict, and 57,000 in the Viet-Nam era. In 1983, a total of $173 million was spent on veterans benefits.

Thirty communities have Army National Guard units. In 1985 there were 43 troop units, with a strength of 4,482 personnel. Nebraska also had 11 Air National Guard units, with 1,078 personnel, in 1985.

Of the 5,302 police officers in Nebraska in 1983, 2,579 were local police.

[18]MIGRATION

The pioneers who settled Nebraska in the 1860s consisted mainly of Civil War veterans from the North and foreign-born immigrants. Some of the settlers migrated from the East and easterly parts of the Midwest, but many came directly from Europe to farm the land. The Union Pacific and Burlington Northern railroads, which sold land to the settlers, actively recruited immigrants in Europe. Germans were the largest group to settle in Nebraska (in 1900, 65,506 residents were German-born), then Czechs from Bohemia, and Scandinavians from Sweden, Denmark, and Norway. The Irish came to work on the railroads in the 1860s and stayed to help build the cities. Another wave of Irish immigrants in the 1880s went to work in the packinghouses of Omaha. The city's stockyards also attracted Polish workers. The 1900 census showed that over one-half of all Nebraskans were either foreign-born or the children of foreign-born parents. For much of this century, Nebraska has been in a period of outmigration. From 1930 to 1960, the state suffered a net loss of nearly 500,000 people through migration, with more than one-third of the total leaving during the dust-bowl decade, 1930–40. This trend continues, with Nebraska experiencing a net outmigration of 13,000 for the period 1980–83.

[19]INTERGOVERNMENTAL COOPERATION

Nebraska's Commission on Intergovernmental Cooperation represents the state in the Council of State Governments. As an oil-producing state, Nebraska is a member of the Interstate Compact to Conserve Oil and Gas. In addition, the state belongs to several regional commissions. Of particular importance are the Republican River Compact with Colorado and Kansas, the Big Blue River Compact with Kansas, the South Platte River Compact with Colorado, and the Upper Niobrara River Compact with Wyoming. The Nebraska Boundary Commission was authorized in 1982 to enter into negotiations to more precisely demarcate Nebraska's boundaries with Iowa, South Dakota, and Missouri. Nebraska is also a member of the Central Interstate Low-Level Radioactive Waste Compact, under which Nebraska, Kansas, Oklahoma, Louisiana, and Arkansas will ultimately locate a suitable disposal site that all compact members can share.

In 1983/84, the state received $637 million in federal grants.

[20]ECONOMY

Agriculture is the backbone of Nebraska's economy, with cattle, corn, hogs, and soybeans leading the state's list of farm products. However, Nebraska is attempting to diversify its economy and has been successful in attracting new business, in large part because of its location near western coal and oil deposits.

Nearly one-half of the state's labor force is employed in agriculture, either directly or indirectly—as farm workers, as factory workers in the food-processing and farm-equipment industries, or as providers of related services. Unemployment, generally well below the national average, was only 5.7% in 1983 and 4.4% in 1984.

[21]INCOME

Nebraska's per capita income was $10,940 in 1983, giving the state a rank of 29th in the nation. Measured in constant dollars,

per capita income declined by 0.4% in Nebraska from 1982 to 1983. Total personal income was $17.5 billion for the state in 1983 current dollars, or 0.6% of the US total. This reflected an increase of only 2.5% for 1980–83 in constant 1972 dollars, well below the national average increase of 6.3%.

In 1979, 163,000 Nebraskans—10.7% of the state's population—were living below the federal poverty level. There were 29,900 Nebraskans among the nation's top wealth-holders in 1982, with gross assets greater than $500,000.

[22]LABOR

Nebraska's nonfarm labor force has increased rapidly in recent decades. The labor force totaled 782,200 in 1984, of whom 28,400 were unemployed. In 1983, 56.8% of all women were in the state's labor force.

A federal survey in 1983 revealed the following nonfarm employment pattern for Nebraska:

	ESTABLISH-MENTS	EMPLOYEES	ANNUAL PAYROLL ('000)
Agricultural services, forestry, fishing	616	2,301	$ 26,914
Mining	206	2,129	48,587
Contract construction	4,079	23,594	463,120
Manufacturing	1,889	86,178	1,681,188
Transportation, public utilities	1,951	30,054	659,412
Wholesale trade	4,251	42,640	737,754
Retail trade	11,731	109,655	920,500
Finance, insurance, real estate	3,702	44,373	779,020
Services	11,361	123,625	1,580,518
Other	2,103	2,371	41,297
TOTALS	41,889	466,920	$6,938,309

There were 405 unions in 1983; membership in unions totaled 114,000 in 1980, or 18% of the nonagricultural labor force, far below the US average of 25.2%. Most union members work in the food-processing industries. Nebraska has a right-to-work law.

[23]AGRICULTURE

With total farm marketings valued at over $6 billion in 1983, Nebraska ranked 6th among the 50 states—yet it ranked only 13th among the states in net farm income ($721.2 million after expenses). About $4 billion of all farm marketings came from livestock production and $2 billion from cash crops.

Territorial Nebraska was settled by homesteaders. Farmers easily adapted to the land in the relatively rainy eastern region, and corn soon became their major crop. In the drier central and western prairie regions, settlers were forced to learn new farming methods to conserve moisture in the ground. Droughts in the 1890s provided impetus for water conservation. Initially, oats and spring wheat were grown along with corn, but by the end of the 19th century, winter wheat became the main wheat crop. The drought and dust storms of the 1930s, which devastated the state's agricultural economy, once again drove home the need for water and soil conservation. In 1983, a total of 7.7 million acres (3.1 million hectares) was irrigated, a sevenfold increase over 1950.

In 1984 there were 60,000 farms covering 47.2 million acres (19.1 million hectares).The average farm was about 787 acres (318 hectares) in 1984. In 1983, some 92,000 people worked on farms; of these, 19,000 were hired hands.

Crop production in 1983 (in bushels) included corn, 475 million; sorghum grain, 60 million; wheat, 98.9 million; soybeans, 59.5 million; oats, 13.6 million; barley, 3 million; and rye, 1.3 million. Hay production was 8 million tons, and potato production 2.2 million hundredweight. In 1983, Nebraska ranked 3d among the states in production of corn for grain and of sorghum for grain, 6th in winter wheat, and 9th in soybeans.

Farms in Nebraska are major businesses, requiring large landholdings to justify investments. The value of the average farm in 1982 was $533,622, up from $192,544 in 1974. However, Nebraska farms still tend to be owned by single persons or families, rather than by large corporations. The strength of state support for the family farm was reflected in the passage of a 1982 constitutional amendment, initiated by petition, prohibiting the purchase of Nebraska farm and ranchlands by other than a Nebraska family farm corporation.

²⁴ANIMAL HUSBANDRY

In 1983, Nebraska edged out Texas to rank 1st in the US in number of cattle and calves marketed (4,580,000 head) and ranked 2d to that state in commercial cattle slaughter (5,063,200 head). In 1982, 39,555 farms raised cattle and calves, 15,998 raised hogs and pigs, and 2,929 sheep and lambs. In 1984, Nebraska had a total of 6,900,000 cattle on farms, including 121,000 milk cows. Nebraska's hog-raising business is the nation's 5th largest. The state had 3.9 million hogs in 1983 and 205,000 sheep and lambs. Dairy products included 1.4 billion lb of milk in 1983.

²⁵FISHING

Commercial fishing is negligible in Nebraska.

²⁶FORESTRY

Arbor Day, now observed throughout the US, originated in Nebraska in 1872 as a way of encouraging tree planting in the sparsely forested state. Forestland occupied 1,029,000 acres (416,000 hectares), or 2% of all Nebraska, as of 1979. About one-third was located in the state's two national forests—Nebraska and Samuel R. McKelvie national forests at Chadron, which occupied a total of 257,260 acres (104,110 hectares) as of 1984.

²⁷MINING

In 1982, the value of Nebraska's total mineral output was estimated at $295.6 million, 33d in the US. Crude petroleum, valued at $210.5 million, accounted for 71.2% of the value of all mineral output, though Nebraska ranks no higher than 19th in US petroleum production. Other leading minerals in 1982 were sand and gravel (valued at $28.2 million), crushed stone ($14.3 million), and natural gas ($4.5 million).

²⁸ENERGY AND POWER

Total energy consumed in Nebraska in 1981 amounted to 476.4 trillion Btu. Of this amount, 176.3 trillion Btu were derived from petroleum, the leading fuel type; 121.9 trillion Btu came from natural gas, the chief fuel for heating homes and businesses; and 99.6 trillion Btu came from coal, 91.8 trillion Btu of it used by electricity-generating units. Nuclear sources accounted for a total of 65.0 trillion Btu, all of it consumed by electric utilities, and hydroelectricity for 13.6 trillion Btu. All together, slightly more than 36% of all energy was used to generate electricity and 26% was used by the transportation sector.

Nebraska is the only state with an electric power system totally owned by the public, through cooperatives and municipal plants. The state's installed capacity was 5.9 million kw in 1983, when electrical output totaled 17.1 billion kwh. A total of 9.6 billion kwh came from coal-fired steam units, 6.1 billion kwh from two nuclear-powered units (at Brownville and Fort Calhoun), 1.3 million kwh from hydroelectric units, and minor amounts from oil- and gas-fired units. About 52% of all electricity was sold for commercial and industrial use, 42% for residential use, 5% for use by public authorities, and the remaining 1% for other uses.

As of 31 December 1981, crude petroleum production in Nebraska was estimated at 6.4 million barrels, while consumption totaled 37.9 million barrels. Natural gas production totaled 2.1 billion cu feet in 1983. Most of the petroleum and gas is produced in the southwest. Nebraska, which has no commercial coal industry, used 5.4 million tons of coal in 1981, most of it for electric utilities. There were 14,304 mi (23,020 km) of gas pipelines in 1983.

²⁹INDUSTRY

Nebraska has a small but growing industrial sector. Manufacturing employment totaled 91,100 in 1982 with a payroll of $1.6

billion. There were 1,928 manufacturing establishments but only 632 employed more than 20 workers. Processing of food and kindred products was the leading industry, accounting for 29,900 jobs in manufacturing, up from 21,500 in 1977, and a 1982 payroll of $452.5 million. Nonelectrical machinery manufacturing employed 13,400, with a payroll of $257.9 million; manufacture of electric and electronic equipment 7,200, with a payroll of $124.4 million.

The following table shows value of shipments by manufacturers in selected industries in 1982:

Meat products	$6,887,200,000
Grain-mill products	1,118,000,000
Nonelectrical machinery	1,078,000,000
Fabricated metal products	658,700,000

More than one-third (673) of all manufacturing establishments are in the Omaha metropolitan area, including ConAgra, the nation's largest flour miller and a producer of broiler chickens and crop-protection chemicals. Other manufacturing centers are Lincoln (233 establishments) and the Sioux City, Iowa, metropolitan area in Nebraska and Iowa (131 establishments).

³⁰COMMERCE

In 1982, Nebraska had 4,073 wholesale trade establishments, with total sales of $17.4 billion. Trading in all farm-product raw materials provided $5 billion of wholesale sales, of which grain provided $2.9 billion in sales and livestock $2 billion.

In 1982, Nebraska's 16,402 retail establishments had sales of $7 billion. Automotive sales accounted for 19% of the total; grocery stores, 21%.

Nebraska's exports of manufactured goods totaled $945 million in 1981, nearly three times the 1977 figure. Exports of farm commodities were about $1.8 billion in 1981/82, with feed grains the leading export category.

³¹CONSUMER PROTECTION

Nebraska has no separate state agency in charge of consumer protection. The Office of the Attorney General has a Consumers Protection Division.

The Nebraska Public Service Commission regulates railroads, telephone and telegraph companies, motor transport companies, and other common carriers operating in the state.

³²BANKING

As of 31 December 1984 there were 480 insured commercial banks in Nebraska. In 1983, state banks had assets of $14.7 billion and deposits of $12.3 billion, with $2.5 billion in demand deposits and $9.8 billion in time deposits.

There were 27 savings and loan associations in the state as of 31 December 1982. Total assets amounted to $5.9 billion; $4.4 billion in mortgage loans was outstanding. In 1984, 157 credit unions, with $474 million in resources, were operating in Nebraska.

³³INSURANCE

The insurance industry is important in Nebraska's economy. The major company in the state is Mutual of Omaha.

Twenty-nine life insurance companies were based in the state in 1983. In that year, 2,686,000 policies were in force, with a total value of $38.4 billion. The average amount of life insurance per family was valued at $60,000.

In 1984, there were 82 Nebraska-based property and liability insurance companies. The state's property and liability insurance companies earned $744 million in premiums in 1983, $208.5 million of it from private automobile insurance. Flood insurance coverage totaled $294.2 million as of 31 December 1983.

³⁴SECURITIES

The Bureau of Securities within the Department of Banking and Finance regulates the sale of securities in Nebraska. There are no stock exchanges in the state.

New York Stock Exchange member firms had 58 sales offices and 431 registered representatives in Nebraska in 1983. There were 255,000 shareowners of public corporations in 1983.

³⁵PUBLIC FINANCE

The Nebraska state budget is prepared by the Budget Division of the Department of Administrative Services and is submitted annually by the governor to the legislature. The fiscal year runs from 1 July to 30 June.

Following is a summary of revenues and expenditures for 1981/82 and 1982/83 (in millions):

REVENUES	1981/82	1982/83
Taxes	$ 804.0	$ 860.5
Federal government	392.8	383.2
Local government	27.5	24.0
Other receipts	264.2	293.1
TOTALS	$1,488.5	$1,560.8
EXPENDITURES		
Education	$ 496.0	$ 526.2
Highways	255.3	264.6
Public welfare	210.4	238.7
Health and hospitals	131.4	144.0
Natural resources	61.5	67.5
Other	403.9	405.2
TOTALS	$1,558.5	$1,646.2

As of 30 June 1983, the combined state and local debt was $5.3 billion, or $3,318 per capita (6th highest in the US). A total of $387.4 million was owed by the state, $4,882.6 million by local governments.

³⁶TAXATION

A constitutional amendment in 1967 prohibited the use of property tax revenues for state government. This forced the passage of both a sales and use tax and an income tax, which had long been resisted by fiscal conservatives in the state. The sales and use tax became effective in 1967, the income tax in 1968.

State income tax in 1985 was computed as 19% of the federal tax liability. The corporate tax was 4.75% of the first $50,000 income and 6.65% of income over $50,000, based on federal tax liability for Nebraska operations. The state sales tax was 3.5%; Omaha residents pay an additional 1.5%, and 14 other communities an additional 1%.

Nebraska's federal tax burden in 1984 was $3.6 billion, or $2,248 per capita (33d in the US), and the state's share of federal spending was $4.3 billion, or $2,711 per capita. State residents filed 660,000 federal income tax returns in 1982 on adjusted gross incomes totaling $11.3 billion.

³⁷ECONOMIC POLICY

The Department of Economic Development was created in 1967 to promote industrial development and to diversify the state's economy. Field service offices are located in Omaha, Lincoln, Scottsbluff, North Platte, Ainsworth, Kearney, and Norfolk.

³⁸HEALTH

There were 26,954 live births in the state in 1982, or 17 per 1,000 population. Infant mortality in 1981 was 9.5 per 1,000 live births for whites, 19.1 for blacks. About 6,600 legal abortions were performed in 1982, a rate of 247 per 1,000 live births. Nebraska's death rate in 1981 was 9.3 per 1,000 population.

A total of 325,785 patients were admitted to 110 hospitals in 1983. Hospital personnel included 5,663 registered nurses and 1,843 licensed practical nurses. Twenty-two counties in the state did not have hospitals in 1984. University Hospital and University of Nebraska Medical Center are in Omaha. The average cost per day of a semiprivate hospital room in 1984 was $150.74.

There were 2,623 physicians and 1,002 dentists active in the state during 1982.

Leading causes of death per 100,000 residents in 1981 were heart disease (358.0), cancer (188.2), cerebrovascular diseases (85.4), and accidents (49.3). In 1983, there were 255 traffic fatalities and 42 deaths due to murder or nonnegligent homicide.

³⁹SOCIAL WELFARE

In January 1983, 39,900 Nebraskans received $4.3 million worth of payments under the aid to families with dependent children (AFDC) program. The monthly average per family was $315.76. AFDC payments for all of 1984 totaled $32.7 million.

In 1983, an estimated 90,000 people participated in the federal food stamp program, receiving $14.2 million in coupons. A total of 178,000 children enrolled in Nebraska schools took part in the national school lunch program.

A total of 189,000 retired workers in the state received $855 million in Social Security retirement benefits in 1983. Survivors' benefits, paid to 49,000 people, amounted to $248 million. Disability benefits under Social Security, paid to 17,000 people, totaled $78 million for 1980. Under the Supplemental Security Income (SSI) program in 1983, 4,187 eligible aged persons received about $5 million, 229 blind persons $504,000, and 8,585 disabled persons $19 million. In 1984, Nebraska received $277 million in Medicare hospital insurance payments and $102 million in Medicare supplementary payments. Unemployment benefits totaled $65.4 million in 1982, with an average weekly benefit of $96.73.

⁴⁰HOUSING

According to the 1980 census, there were 624,829 housing units in Nebraska, about 68% of which were owner-occupied. In 1980, 77% of all housing units had air conditioning, 76% were single-unit structures, 80% were linked to public sewers, 81% obtained water from public systems or private companies rather than wells, and 98.8% had complete plumbing facilities. Median value for owner-occupied homes was $38,000. In 1983, 5,477 new privately owned housing units authorized by the state had a total value of $217.5 million.

⁴¹EDUCATION

In 1980/81 there were 1,738 public schools in Nebraska; 1,339 were elementary schools and 399 secondary schools. Enrollments for 1982/83 were elementary, 144,425; secondary, 124,955. Nebraska also had 221 nonpublic schools in 1980/81; 174 were elementary schools, 32 were secondary schools, and 15 offered both elementary and secondary education. There were 20,881 pupils enrolled in nonpublic elementary grades in 1982/83; secondary, 15,597. Catholic nonpublic schools enrolled 18,756 elementary students and 9,002 secondary-school students.

The University of Nebraska is the state's largest postsecondary institution, with campuses in Lincoln and Omaha. In 1983/84 there were also 4 state colleges, 15 private colleges and universities, and 16 community-based technical colleges. In fall 1983, 27,372 students were enrolled at the University of Nebraska, Lincoln; 14,531 at the University of Nebraska, Omaha; 13,115 at the four state colleges; 17,553 at private institutions of higher learning; and 22,591 at community colleges.

⁴²ARTS

The Orpheum Theater in Omaha provides performance space for opera, symphony concerts, ballet, plays, and popular music. Opera/Omaha, Inc., presents three operas there a year, drawing an audience of 28,000. In 1983/84, Nebraska received $583,000 in federal support of the arts and appropriated $466,000 from state funds. The 15-member Nebraska Arts Council, appointed by the governor, is empowered to receive federal funds and to plan and administer statewide and special programs in all the arts. Funds are available in six categories: artists in schools/communities, learning through the arts, minigrants (up to $500), projects, touring, and yearlong programs. Affiliation with the Mid-America Arts Alliance allows the council to help sponsor national and regional events.

[43]LIBRARIES AND MUSEUMS

The Nebraska Library Commission coordinates library services. In 1982, the state had 9 county libraries, 16 regional libraries, and 249 public libraries. A total of 3,941,251 volumes were in the public library system in 1982; total circulation was 8,411,970.

The Joslyn Art Museum in Omaha is the state's leading museum. Other important museums include the Nebraska State Museum of History, the University of Nebraska State Museum (natural history), and the Sheldon Memorial Art Gallery, all in Lincoln; the Western Heritage Museum in Omaha; the Stuhr Museum of the Prairie Pioneer in Grand Island; and the Hastings Museum in Hastings. In all, the state had more than 90 museums in 1984. The Agate Fossil Beds National Monument in northwestern Nebraska features mammal fossils from the Miocene era and a library of paleontological and geologic material.

[44]COMMUNICATIONS

In 1985, Nebraska was served by 543 post offices and 4,991 employees of the US Postal Service. Telephone service is regulated by the Public Service Commission. Almost 90% of the state's 957,032 occupied housing units had telephones in 1980.

In 1982, 54 FM stations and 48 AM stations were operating commercially, and another 11 noncommercial FM stations were licensed. There were 14 commercial TV stations and a network of 9 PBS stations. In 1984, 81 cable television systems operated in the state.

[45]PRESS

In 1984, Nebraska had 3 morning dailies with a combined circulation of 173,182, 16 evening dailies with a combined circulation of 307,834, and 7 Sunday newspapers with a total circulation of 431,454. The leading newspaper in 1984 was the *Omaha World-Herald,* with circulations as follows: morning, 121,104; evening, 101,977; and Sunday, 279,431. The *Lincoln Journal-Star* had a daily circulation of 35,177 (morning) and 45,252 (evening) and a Sunday circulation of 75,743.

[46]ORGANIZATIONS

The 1982 Census of Service Industries counted 684 organizations in Nebraska, including 133 business associations; 390 civic, social, and fraternal associations; and 9 educational, scientific, and research organizations. Among the organizations based in Nebraska are the Great Plains Council at the University of Nebraska (Lincoln), the American Shorthorn Society (Omaha), the Morse Telegraph Club (Lincoln), and the National Arbor Day Foundation (Nebraska City).

[47]TOURISM, TRAVEL, AND RECREATION

Although tourism is not regarded as a major industry in Nebraska, travel and tourism was the 2d-largest employer in 1982, generating 32,000 jobs and a $208 million payroll. Expenditures by travelers in the state totaled about $1 billion in 1982.

The 6 state parks, 9 state historical parks, 12 federal areas, and 55 recreational areas are main tourist attractions; fishing, swimming, picnicking, and sightseeing are the principal activities. Pawnee State Recreation Area, with 716,500 visitors, and Fremont State Recreation Area, with 611,250 visitors, were the most popular attractions in 1983.

In 1982/83, licenses were issued to 240,330 fisherman and 173,926 hunters.

[48]SPORTS

The favorite spectator sport of Nebraskans is college football. Equestrian activities, including racing and rodeos, also are popular. Major annual sporting events are the NCAA College Baseball World Series and the World's Championship Rodeo, both held in Omaha. Pari-mutuel racing is licensed by the state.

The University of Nebraska Cornhuskers compete in the Big Eight football conference. The team often places high in national rankings and has been a frequent competitor in postseason play. The Cornhuskers won the Orange Bowl in 1964, 1971, 1972, and 1973, the Cotton Bowl in 1974 (January), the Sugar Bowl in 1974 (December), the Bluebonnet Bowl in 1976, the Liberty Bowl in 1977, and the Sun Bowl in 1980.

[49]FAMOUS NEBRASKANS

Nebraska was the birthplace of only one US president, Gerald R. Ford (Leslie King, Jr., b.1913). When Spiro Agnew resigned the vice-presidency in October 1973, President Richard M. Nixon appointed Ford, then a US representative from Michigan, to the post. Upon Nixon's resignation on 9 August 1974, Ford thus became the first nonelected president in US history.

Four native and adoptive Nebraskans have served in the cabinet. J. Sterling Morton (b.New York, 1832-1902), who originated Arbor Day, was secretary of agriculture under Grover Cleveland. William Jennings Bryan (b.Illinois, 1860-1925), a US representative from Nebraska, served as secretary of state and was three times the unsuccessful Democratic candidate for president. Frederick A. Seaton (b.Washington, 1909-74) was Dwight Eisenhower's secretary of the interior, and Melvin Laird (b.1922) was Richard Nixon's secretary of defense.

George W. Norris (b.Ohio, 1861-1944), the "fighting liberal," served 10 years in the US House of Representatives and 30 years in the Senate. Norris's greatest contributions were in rural electrification (his efforts led to the creation of the Tennessee Valley Authority), farm relief, and labor reform; he also promoted the unicameral form of government in Nebraska. Theodore C. Sorensen (b.1928) was an adviser to President John F. Kennedy.

Indian leaders important in Nebraska history include Oglala Sioux chiefs Red Cloud (1822-1909) and Crazy Horse (1849?-77). Moses Kinkaid (b.West Virginia, 1854-1920) served in the US House and was the author of the Kinkaid Act, which encouraged homesteading in Nebraska. Educator and legal scholar Roscoe Pound (1870-1964) was also a Nebraskan. In agricultural science, Samuel Aughey (b.Pennsylvania, 1831-1912) and Hardy W. Campbell (b.Vermont, 1850-1937) developed dry-land farming techniques. Botanist Charles E. Bessey (b.Ohio, 1845-1915) encouraged forestation. Father Edward Joseph Flanagan (b.Ireland, 1886-1948) was the founder of Boys Town, a home for underprivileged youth. Two native Nebraskans became Nobel laureates in 1980: Lawrence R. Klein (b.1920) in economics and Val L. Fitch (b.1923) in physics.

Writers associated with Nebraska include Willa Cather (b.Virginia, 1873-1947), who used the Nebraska frontier setting of her childhood in many of her writings and won a Pulitzer Prize in 1922; author and poet John G. Neihardt (b.Illinois, 1881-1973), who incorporated Indian mythology and history in his work; Mari Sandoz (1901-66), who wrote of her native Great Plains; writer-photographer Wright Morris (b.1910); and author Tillie Olsen (b.1913). Rollin Kirby (1875-1952) won three Pulitzer Prizes for political cartooning. Composer-conductor Howard Hanson (1896-1982), born in Wahoo, won a Pulitzer Prize in 1944.

Nebraskans important in entertainment include actor-dancer Fred Astaire (Fred Austerlitz, b.1899); actors Harold Lloyd (1894-1971), Henry Fonda (1905-82), Robert Taylor (Spangler Arlington Brugh, 1911-69), Marlon Brando (b.1924), and Sandy Dennis (b.1937); television stars Johnny Carson (b.Iowa, 1925) and Dick Cavett (b.1936); and motion-picture producer Darryl F. Zanuck (b.1902-79).

[50]BIBLIOGRAPHY

Creigh, Dorothy Weyer. *Nebraska: A Bicentennial History.* New York: Norton, 1977.

Federal Writers' Project. *Nebraska: A Guide to the Cornhusker State.* Reprint. New York: Somerset, n.d. (orig. 1939).

Hanna, Robert. *Sketches of Nebraska.* Lincoln: University of Nebraska Press, 1984.

Nebraska, State of. Department of Economic Development. *Nebraska Statistical Handbook, 1984-1985.* Lincoln, 1983.

Olson, James C. *History of Nebraska.* Lincoln: University of Nebraska Press, 1966 (orig. 1955).

NEVADA

State of Nevada

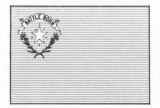

ORIGIN OF STATE NAME: Named for the Sierra Nevada, *nevada* meaning "snow-covered" in Spanish. **NICKNAME:** The Sagebrush State. (Also: The Silver State.) **CAPITAL:** Carson City. **ENTERED UNION:** 31 October 1864 (36th). **SONG:** "Home Means Nevada." **MOTTO:** All for Our Country. **FLAG:** On a blue field, two sprays of sagebrush and a golden scroll in the upper lefthand corner frame a silver star encircled by the word "Nevada"; the scroll, reading "Battle Born," recalls that Nevada was admitted to the Union during the Civil War. **OFFICIAL SEAL:** An ore-crushing mill, ore cart, and mine tunnel symbolize Nevada's mining industry; a plow, sickle, and sheaf of wheat represent its agricultural resources. In the background are a railroad, a telegraph line, and a sun rising over the mountains. Encircling this scene are 36 stars and the state motto. The words "The Great Seal of the State of Nevada" surround the whole. **ANIMAL:** Desert bighorn sheep. **BIRD:** Mountain bluebird. **FLOWER:** Sagebrush. **TREE:** Single-leaf piñon. **METAL:** Silver. **GRASS:** Indian ricegrass. **FOSSIL:** Ichthyosaur. **LEGAL HOLIDAYS:** New Year's Day, 1 January; Birthday of Martin Luther King, Jr., 3d Monday in January; Washington's Birthday, 3d Monday in February; Memorial Day, last Monday in May; Independence Day, 4 July; Labor Day, 1st Monday in September; Nevada Day, 31 October; Veterans Day, 11 November; Thanksgiving Day, 4th Thursday in November; Christmas Day, 25 December. **TIME:** 4 AM PST = noon GMT.

¹LOCATION, SIZE, AND EXTENT

Situated between the Rocky Mountains and the Sierra Nevada in the western US, Nevada ranks 7th in size among the 50 states.

The total area of Nevada is 110,561 sq mi (286,352 sq km), with land comprising 109,894 sq mi (284,624 sq km) and inland water covering 667 sq mi (1,728 sq km). Nevada extends 320 mi (515 km) E–W; the maximum N–S extension is 483 mi (777 km).

Nevada is bordered on the N by Oregon and Idaho; on the E by Utah and Arizona (with the line in the SE formed by the Colorado River); and on the S and W by California (with part of the line passing through Lake Tahoe). The total boundary length of Nevada is *1,480 mi (2,382 km)*. The state's geographic center is in Lander County, 26 mi (42 km) SE of Austin.

²TOPOGRAPHY

Almost all of Nevada belongs physiographically to the Great Basin, a plateau characterized by isolated mountain ranges separated by arid basins. These ranges generally trend north–south; most are short, up to 75 mi (121 km) long and 15 mi (24 km) wide, and rise to altitudes of 7,000–10,000 feet (2,100–3,000 meters). Chief among them are the Schell Creek, Ruby, Toiyabe, and Carson (within the Sierra Nevada). Nevada's highest point is Boundary Peak, 13,143 feet (4,006 meters), in the southwest.

Nevada has a number of large lakes and several large saline marshes known as sinks. The largest lake is Pyramid, with an area of 188 sq mi (487 sq km), in the west. Nevada shares Lake Tahoe with California, and Lake Mead, created by Hoover Dam on the Colorado River, with Arizona. The streams of the Great Basin frequently disappear during dry spells; many of them flow into local lakes or sinks without reaching the sea. The state's longest river, the Humboldt, flows for 290 mi (467 km) through the northern half of the state into the Humboldt Sink. The Walker, Truckee, and Carson rivers drain the western part of Nevada. The canyon carved by the mighty Colorado, the river that forms the extreme southeastern boundary of the state, is the site of Nevada's lowest elevation, 470 feet (143 meters).

³CLIMATE

Nevada's climate is sunny and dry, with wide variation in daily temperatures. The normal daily temperature at Reno is 49°F

(9°C), ranging from 32°F (0°C) in January to 69°F (21°C) in July. The all-time high, 122°F (50°C), was set at Overton on 23 June 1954; the record low, −50°F (−46°C), at San Jacinto on 8 January 1937.

Nevada is the driest state in the US, with an overall average annual precipitation of less than 4 in (10 cm). Snowfall is abundant in the mountains, however, reaching 60 in (152 cm) a year on the highest peaks.

⁴FLORA AND FAUNA

Various species of pine—among them the single-leaf piñon, the state tree—dominate Nevada's woodlands. Creosote bush is common in southern Nevada, as are many kinds of sagebrush throughout the state. Wild flowers include shooting star and white and yellow violets.

Native mammals include the black bear, white-tailed and mule deer, pronghorn antelope, Rocky Mountain elk, cottontail rabbit, and river otter. Grouse, partridge, pheasant, and quail are the leading game birds, and a diversity of trout, char, salmon, and whitefish thrive in Nevada waters. Rare and protected reptiles are the Gila monster and desert tortoise. Listed as endangered are the Colorado squawfish, Moapa dace, Ash Meadows speckled dace, Pahrump killifish, Ash Meadows Armagosa pupfish, Devil's Hole pupfish, Warm Springs pupfish, woundfin, and bonytail chub. The Lahoutan cutthroat trout is threatened.

⁵ENVIRONMENTAL PROTECTION

Preservation of the state's clean air, scarce water resources, and no longer abundant wildlife are the major environmental challenges facing Nevada. The Department of Fish and Game sets quotas on the hunting of deer, antelope, bighorn sheep, and other game animals. The Department of Conservation and Natural Resources has broad responsibility for environmental protection, state lands, forests, and water and mineral resources. The Division of Environmental Protection within the department has primary responsibility for the control of air pollution, water pollution, waste management, and groundwater protection.

⁶POPULATION

Nevada ranked 43d in the US with a 1980 census population of 800,493, up from 488,738 in 1970. This made it the fastest-growing state for the 1970–80 decade, with an increase of 63.8%.

A population of 933,451 was estimated for 1985, accounting for an increase of 16.6% for 1980–85 that outpaced all other states except Alaska. Rapid growth is expected to continue, with a population of 1,275,400 projected for 1990. By contrast, Nevada had only 110,427 residents in 1940, when the current population boom began. As might be expected, Nevadans are among the most mobile Americans; as of 1980, more than 31% of the adults had lived in the state for five years or less.

With a population density of 8.5 per sq mi (3.3 per sq km) in 1985, Nevada remains one of the nation's most sparsely populated states. More than 85% of Nevada's people live in cities, the largest of which, Las Vegas, had 164,674 residents in 1980 and an estimated 183,227 in 1984. Reno had a 1984 population of 105,615; Carson City, the capital, 35,916.

7ETHNIC GROUPS

Some 50,791 black Americans made up about 6% of Nevada's population in 1980. The American Indian population was 14,256 in 1980; tribal landholdings totaled 1,138,462 acres (460,721 hectares). Major tribes are the Washo, Northern Paiute, Southern Paiute, and Shoshoni.

Some 54,000 persons, or 6.7% of all state residents, were foreign-born in 1980. An equal number, including 31,741 persons of Mexican ancestry, identified themselves as Hispanic.

8LANGUAGES

Midland and Northern English dialects are so intermixed in Nevada that no clear regional division appears; an example of this is the scattered use of both Midland *dived* (instead of *dove*) as the past tense of *dive* and the Northern /krik/ for *creek*. In 1980, 691,946 Nevadans—90% of the resident population 3 years old or older—spoke only English at home. Other languages spoken at home included:

Spanish	35,904	Chinese	2,491
German	5,869	Pilipino or other Filipino languages	2,377
Italian	5,046	Korean	1,898
French	3,515	Japanese	1,489

9RELIGIONS

In 1984, Nevada had 144,000 Roman Catholics. Other Christian denominations in 1980 included 55,148 adherents of the Church of Jesus Christ of Latter-day Saints (Mormons), 16,276 Southern Baptists, 7,000 Lutherans, and 7,000 Methodists. In 1984 there were an estimated 18,200 Jews, 93% of them in metropolitan Las Vegas, the remainder in the Reno area.

10TRANSPORTATION

As of 1984, Nevada had 1,492 mi (2,401 km) of railroads. Amtrak provides passenger service en route from Chicago and Salt Lake City to Los Angeles.

In 1983 there were 43,806 mi (70,499 km) of roads and streets, 730,741 registered vehicles, and 676,074 licensed drivers. The major highways, I-80 and I-15, link Salt Lake City with Reno and Las Vegas, respectively. There were 99 civil and joint-use airfields in 1984 and 26 heliports. The leading commercial air terminals are McCarran International Airport at Las Vegas, which handled 10.3 million departing passengers in 1983, and Reno-Cannon Airport, with 2.8 million passengers.

11HISTORY

The first inhabitants of what is now Nevada arrived about 12,000 years ago. They were fishermen, as well as hunters and food gatherers, for the glacial lakes of the ancient Great Basin were then only beginning to recede. Numerous sites of early human habitation have been found, the most famous being Pueblo Grande de Nevada (also known as Lost City). In modern times, four principal Indian groups have inhabited Nevada: Southern Paiute, Northern Paiute, Shoshoni, and Washo.

Probably the first white explorer to enter the state was the Spanish priest Francisco Garćes, who apparently penetrated extreme southern Nevada in 1776. The year 1826 saw Peter Skene Ogden of the British Hudson's Bay Company enter the northeast in a prelude to his later exploration of the Humboldt River; the rival American trapper Jedediah Smith traversed the state in 1826–27. During 1843–44, John C. Frémont led the first of his several expeditions into Nevada.

Nevada's first permanent white settlement, Mormon Station (later Genoa), was founded in 1850 in what is now western Nevada, a region that became part of Utah Territory the same year. (The southeastern tip of Nevada was assigned to the Territory of New Mexico.) Soon other Mormon settlements were started there and in Las Vegas Valley. The Las Vegas mission failed, but the farming communities to the northwest succeeded, even though friction between Mormons and placer miners in that area caused political unrest. Most of the Mormons in western Nevada departed in 1857, when Salt Lake City was threatened by an invasion of federal troops.

A separate Nevada Territory was established in 1861; only three years later, on 31 October 1864, Nevada achieved statehood, although the present boundaries were not established until 18 January 1867. Two factors accelerated the creation of Nevada: the secession of the southern states, whose congressmen had been blocking the creation of new free states, and the discovery, in 1859, of the Comstock Lode, an immense concentration of silver and gold which attracted thousands of fortune seekers and established the region as a thriving mining center.

Nevada's development during the rest of the century was determined by the economic fortunes of the Comstock, whose affairs were dominated, first, by the Bank of California (in alliance with the Central Pacific Railroad) and then by the "Bonanza Firm" of John W. Mackay and his partners. The lode's rich ores were exhausted in the late 1870s, and Nevada slipped into a 20-year depression. A number of efforts were made to revive the economy, one being an attempt to encourage mining by increasing the value of silver. To this end, Nevadans wholeheartedly supported the movement for free silver coinage during the 1890s, and the Silver Party reigned supreme in state politics for most of the decade.

Nevada's economy revived following new discoveries of silver at Tonopah and gold at Goldfield early in the 20th century. A second great mining boom ensued, bolstered and extended by major copper discoveries in eastern Nevada. Progressive political ferment in this pre–World War I period added recall, referendum, and initiative amendments to the state constitution and brought about the adoption of women's suffrage (1914).

The 1920s was a time of subdued economic activity; mining fell off, and not even the celebrated divorce trade, centered in Reno, was able to compensate for its decline. Politically, the decade was conservative and Republican, with millionaire George Wingfield dominating state politics through a so-called bipartisan machine. Nevada went Democratic during the 1930s, when the hard times of the depression were alleviated by federal public-works projects, most notably the construction of the Hoover (Boulder) Dam, and by state laws aiding the divorce business and legalizing gambling.

Gaming grew rapidly after World War II, becoming by the mid-1950s not only the mainstay of Nevada tourism but also the state's leading industry. Revelations during the 1950s and 1960s that organized crime had infiltrated the casino industry and that casino income was being used to finance narcotics and other rackets in major east coast cities led to a state and federal crackdown and the imposition of new state controls.

From 1960 to 1980, Nevada was the fastest-growing of the 50 states, increasing its population by 70% in the 1960s and another 64% in the 1970s. Much of this growth was associated with expansion of the gambling industry—centered in the casinos of Las Vegas and Reno—and of the military. Efforts were made by leaders of the so-called sagebrush rebellion to gain possession of federal lands within Nevada's boundaries.

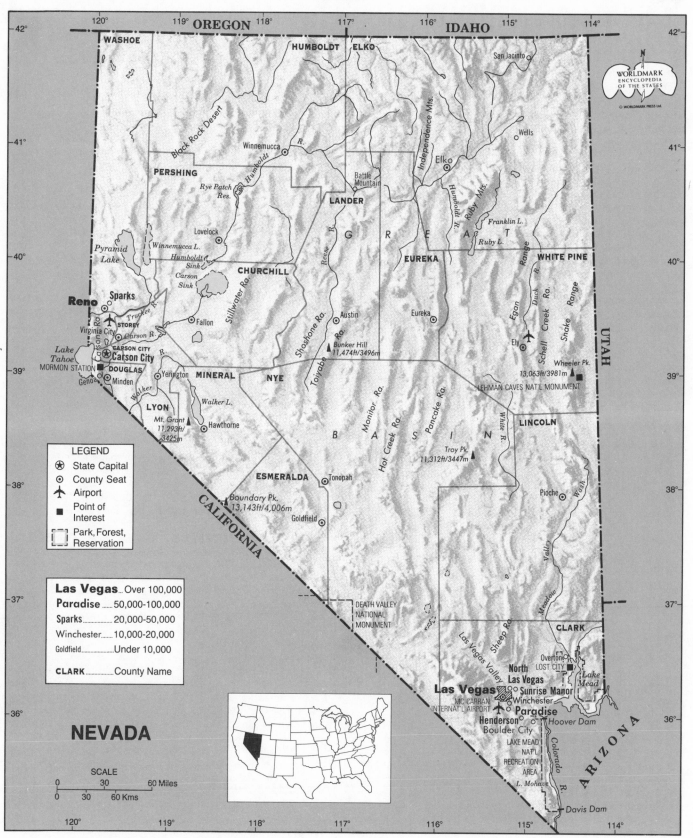

OREGON
IDAHO

42°

120° 119° 118° 117° 116° 115° 114°

WASHOE

HUMBOLDT ELKO

San Jacinto

41°

Wells

PERSHING

Winnemucca

Battle
Mountain

Elko

Independence Mts.

Rye Patch
Res.

LANDER

Humboldt R. Ruby Mts.

Franklin L.

40°

Lovelock

Pyramid
Lake

Winnemucca L.

Humboldt
Sink

Ruby L.

G R E A T

Carson
Sink

CHURCHILL

EUREKA

WHITE PINE

Eureka

Egan Range

Duck Cr.

Schell Creek Ra.

Snake Range

Reno

Sparks

Fallon

Austin

Eureka

Ely

Truckee R.

STOREY

Stillwater Ra.

Shoshone Ra.

Wheeler Pk.
13,063ft/3981m

Virginia City

Carson R.

Bunker Hill
11,474ft/3496m

39°

Lake
Tahoe

CARSON CITY

Carson City

Toiyabe Ra.

LEHMAN CAVES NAT'L MONUMENT

MORMON STATION

DOUGLAS

Yerington

MINERAL

NYE

White R.

Genoa

Minden

Monitor Ra.

Hot Creek Ra.

Pancake Ra.

LINCOLN

LYON

Walker L.

Mt. Grant
11,293ft/
3425m

Hawthorne

B A S I N

Troy Pk.
11,312ft/3447m

38°

ESMERALDA

Tonopah

Pioche

Boundary Pk.
13,143ft/4,006m

Goldfield

Meadow Valley Wash

37°

DEATH VALLEY
NATIONAL
MONUMENT

CLARK

Overton
LOST CITY

Las Vegas Valley

Sheep Ra.

North
Las Vegas

Lake
Mead

Las Vegas

Sunrise Manor

Winchester

MC CARRAN
INTERNAT'L AIRPORT

Paradise

36°

Henderson
Boulder City

Hoover Dam

LAKE MEAD
NAT'L
RECREATION
AREA

Colorado R.

L. Mohave

CALIFORNIA

UTAH

ARIZONA

LEGEND

⊛ State Capital
⊙ County Seat
✈ Airport
■ Point of Interest
▯ Park, Forest, Reservation

Las Vegas ... Over 100,000
Paradise 50,000-100,000
Sparks 20,000-50,000
Winchester 10,000-20,000
Goldfield .... Under 10,000

CLARK .... County Name

NEVADA

SCALE
0 30 60 Miles
0 30 60 Kms

Davis Dam

See endsheet maps: C3.

LOCATION: 35° to 42°N; 114° to 120°W. **BOUNDARIES:** Oregon line, *155 mi (249.5 km)*; Idaho line, *155 mi (249.5 km)*; Utah line, *345 mi (555 km)*; Arizona line, *205 mi (330 km)*; California line, *620 mi (998 km)*.

[12]STATE GOVERNMENT

Nevada's 1864 constitution, as amended, continues to govern the state. The state legislature consists of a senate with 21 members, each elected to a four-year term, and a house of representatives with 42 members, each serving two years. Executive officials elected statewide include the governor and lieutenant governor (who run separately), secretary of state, attorney general, treasurer, and controller, all of whom serve for four years. A two-thirds vote of the elected members of each house is required to override a gubernatorial veto.

Constitutional amendments may be submitted to the voters for ratification if they have received majority votes in each house in two successive sessions or under an initiative procedure calling for petitions signed by 10% of those who voted in the last general election. Legislative amendments need a majority vote; initiative amendments require majorities in two consecutive elections. Voters must be US citizens, be at least 18 years old, and have lived in the state for 30 days.

[13]POLITICAL PARTIES

Registered Democrats outnumbered Republicans by 174,470 to 123,951 in 1982. The Democrats have failed to capitalize on this advantage, however, and since World War II neither party has dominated state politics, which are basically conservative. Ronald Reagan carried Nevada by nearly a two-to-one majority in 1980 and won 70% of the vote in 1984. Minorities in elective office included 5 Hispanics in 1984, and 10 women and 10 blacks in 1985. Nevada was represented in the US Congress by two Republican senators—Paul Laxalt, who did not plan to seek reelection in 1986, and Jacob "Chic" Hecht, who in 1982 defeated four-term Democrat Howard Cannon.

Nevada Presidential Vote by Major Political Parties, 1948–84

YEAR	ELECTORAL VOTE	NEVADA WINNER	DEMOCRAT	REPUBLICAN
1948	3	*Truman (D)	31,290	29,357
1952	3	*Eisenhower (R)	31,688	50,502
1956	3	*Eisenhower (R)	40,640	56,049
1960	3	*Kennedy (D)	54,880	52,387
1964	3	*Johnson (D)	79,339	56,094
1968	3	*Nixon (R)	60,598	73,188
1972	3	*Nixon (R)	66,016	115,750
1976	3	Ford (R)	92,479	101,273
1980	3	*Reagan (R)	66,666	155,017
1984	4	*Reagan (R)	91,655	188,770

* Won US presidential election.

[14]LOCAL GOVERNMENT

As of 1985, Nevada was subdivided into 16 counties, 1 independent municipality (Carson City), and 16 other municipalities constituting the 16 county seats.

[15]STATE SERVICES

The Executive Ethics Commission was created in 1977 to oversee financial disclosure by state officials. The Department of Education and the University of Nevada System are the main state educational agencies. The Department of Human Resources has divisions covering public health, rehabilitation, mental hygiene and mental retardation, welfare, youth services, and programs for the elderly. Regulatory functions are exercised by the Commerce Department (insurance, banking, consumer affairs, real estate),

the Public Service Commission, the Gaming Control Board, and other state agencies.

[16]JUDICIAL SYSTEM

Nevada's supreme court consists of a chief justice and 4 other justices. As of 1984 there were 35 district court judges organized into nine judicial districts. All judges are elected by nonpartisan ballot to six-year terms.

Nevada's overall crime rate—6,701 per 100,000 population—was 2d highest in the US (behind Florida) in 1983. The rate for murder and nonnegligent manslaughter was 12.8 per 100,000 population (behind Texas, Alaska, and Louisiana). Only Alaska and Michigan had higher rates for forcible rape, and no state had a higher burglary rate.

[17]ARMED FORCES

Nevada had four military installations in 1984 and a total of 12,346 military personnel. Of these installations, Nellis Air Force Base near Las Vegas, with 9,524 personnel, was the largest. The state has been the site of both ballistic missile and atomic weapons testing. In 1983/84, Nevada firms received $128.6 million in federal defense contracts.

As of 30 September 1984, 140,000 military veterans were living in the state, including fewer than 500 from World War I, 50,000 of World War II, 30,000 of the Korean conflict, and 47,000 from the Viet-Nam era. Veterans' benefits in 1982/83 totaled $93 million.

Army National Guard personnel numbered 1,174 at the end of 1984, and there were 972 members of the Air National Guard in early 1985. Nevada's state and local police personnel totaled 3,143 in 1983; expenditures for police services in 1982 came to $77 million.

[18]MIGRATION

In 1870, about half of Nevada's population consisted of foreign immigrants, among them Chinese, Italians, Swiss, British, Irish, Germans, and French Canadians. Though their origins were diverse, their numbers were few—no more than 21,000 in all. Not until the 1940s did migrants come in large volume. Between 1940 and 1980, Nevada gained a total of 507,000 residents through migration, equal to 63% of the 1980 population; there was an additional net gain from migration of 65,000 between 1980 and 1983.

[19]INTERGOVERNMENTAL COOPERATION

Nevada takes part in the Colorado River Compact, the Tahoe Regional Planning Compact, and the California-Nevada Interstate Compact, under which the two states administer water rights involving Lake Tahoe and the Carson, Truckee, and Walker rivers. The state also is a signatory to the Interstate Oil and Gas Compact and the Western Interstate Energy Compact.

Federal aid in 1983/84 totaled $340.4 million, of which $14.6 million was general revenue sharing.

[20]ECONOMY

Nevada is disadvantaged by aridity and a shortage of arable land but blessed with a wealth of mineral resources—gold, silver, copper, and other metals. Mining remains important, though overshadowed since World War II by tourism and gambling, which generate more than 50% of the state's income. Legalized gaming alone produces nearly half of Nevada's tax revenues.

[21]INCOME

Personal income in Nevada came to $11.2 billion in 1983; per capita income averaged $12,516. The median family income of $28,829 ranked 10th in the US in 1981. In 1979, 69,000 Nevadans were below the federal poverty level. Some 7,500 Nevadans with gross assets of more than $500,000 each were among the top wealth-holders in the US.

[22]LABOR

Nevada's total civilian labor force in 1983 was 487,000, of whom about 48,000 (10%) were unemployed. A federal survey in March 1982 revealed the following nonfarm employment pattern in Nevada:

	ESTABLISH-MENTS	EMPLOYEES	ANNUAL PAYROLL ('000)
Agricultural services, forestry, fishing	266	1,277	$ 18,654
Mining, of which:	219	4,908	124,238
Metals	(91)	(2,696)	(74,977)
Contract construction	2,174	21,849	544,029
Manufacturing	823	18,602	374,702
Transportation, public utilities	757	22,337	490,627
Wholesale trade	1,308	12,885	262,296
Retail trade	5,782	65,328	726,041
Finance, insurance, real estate	2,182	17,472	288,646
Services, of which:	7,132	161,825	2,489,263
Hotels, lodgings	(588)	(89,547)	(1,260,950)
Amusements, recreation	(415)	(21,834)	(320,294)
Other	1,953	1,918	56,978
TOTALS	22,596	328,401	$5,375,474

There were also about 58,000 government employees not covered by this survey.

There were 139 labor unions in Nevada in 1983. In 1980, 95,000 workers belonged to unions and other employee associations. Nevada has a right-to-work law.

23 AGRICULTURE

Agricultural income in 1983 totaled $224 million (47th in the US), of which $71 million was from crops and $153 million from livestock and animal products. Chief crops in 1983 included 2.7 million bushels of barley, 1.3 million bushels of wheat, 1.3 million tons of hay, and 3.7 million hundredweight of potatoes. Virtually all of the state's cropland requires irrigation.

24 ANIMAL HUSBANDRY

In 1984, Nevada ranches and farms had 660,000 cattle and calves, 17,000 milk cows, and 92,000 sheep and lambs. Livestock products in 1983 included 237 million lb of milk, 224.9 million lb of cattle and calves, 11.8 million lb of sheep and lambs, and 1.8 million eggs. The wool clip was about 1 million lb.

25 FISHING

There is no commercial fishing industry in Nevada.

26 FORESTRY

Nevada in 1979 had 7,683,000 acres (3,109,000 hectares) of forestland, of which 5,150,000 acres (2,084,000 hectares) were in the National Forest System. Less than 2% of all forested land in Nevada was classified as commercial timberland.

27 MINING

Mining of metals and other nonfuel minerals plays an important part in the state's history and economy. Total mineral output in 1982 was valued at $541 million, only $15 million of which was derived from fuels. In 1984, the value of nonfuel mineral production was $622.4 million, with Nevada leading the nation in production of barite (768,000 tons), gold (1,093,647 troy oz), and mercury (20,000 76-lb flasks). The state was also the sole producer of magnesite, a major producer of silver (3,055 troy oz), and a significant producer of copper, diatomite, flourspar, iron ore, lithium, molybdenum, and perlite.

28 ENERGY AND POWER

Nevada had an installed electrical capacity of 4.5 million kw in 1983, when 17.4 billion kwh of power were produced. That year, 10.5 billion kwh of electrical energy were sold in the state, the remainder being exported, principally to California. Hoover Dam, anchored in the bedrock of Black Canyon east of Las Vegas, was the state's largest hydroelectric installation. There were 35 oil-producing wells in Nevada in 1983; total oil production was 811,125 barrels.

29 INDUSTRY

Industry in Nevada is limited but diversified, producing communications equipment, pet food, chemicals, and sprinkler systems, among other products. The total value of shipments by manufac-turers in 1982 was $1.76 billion. Major sectors and their value of shipments in 1982 were as follows:

Food and food products	$259,400,000
Stone, clay, glass products	170,400,000
Chemicals and chemical products	135,600,000
Nonelectrical machinery	130,500,000
Electric and electronic equipment	113,100,000

30 COMMERCE

Wholesale trade in Nevada totaled about $3.9 billion in 1982. Retail sales in 1982 exceeded $5.5 billion, with metropolitan Las Vegas accounting for 55% of the total, metropolitan Reno 27%. Foreign exports included $105 million in manufactured goods in 1981 and $22 million in agricultural products in 1981/82.

31 CONSUMER PROTECTION

The Consumer Affairs Division of the Department of Commerce, with offices in Las Vegas and Carson City, protects consumers from deceptive or fraudulent sales practices and represents consumers' interests in government.

32 BANKING

During 1983 there were 16 insured commercial banks in Nevada, with total assets of $5 billion. Outstanding loans totaled $2.6 billion, deposits $4 billion. In addition, there were 7 savings and loan associations with total assets exceeding $2.7 billion; mortgage loans outstanding reached $1.8 billion.

33 INSURANCE

Nevadans held 1,426,000 life insurance policies in 1983 with a total value of $38.4 billion. Life insurance per family averaged $45,900, and benefit payments totaled $159.5 million. Property and liability insurers wrote $159.5 million in premiums, of which 62% was automobile coverage.

34 SECURITIES

There are no securities exchanges in Nevada. In 1983, 156,000 Nevadans were shareholders of public corporations.

35 PUBLIC FINANCE

Nevada's budget is prepared biennially in odd-numbered years by the Budget Division of the Department of Administration and submitted by the governor to the legislature, which has unlimited power to change it. The fiscal year begins 1 July and ends 30 June.

The following table summarizes general revenues and appropriations (in thousands) for 1984/85 (estimated) and 1985/86 (recommended):

REVENUES	1984/85	1985/86
Gaming tax	$198,279	$206,371
Sales and use tax	157,428	169,235
Casino entertainment tax	15,000	15,200
Other receipts	91,150	82,162
TOTALS	$461,857	$472,968
EXPENDITURES		
Education	$230,023	$274,776
Human resources	97,285	98,659
Public safety	37,827	40,575
Other outlays	55,860	60,873
TOTALS	$420,995	$474,883

As of 1982, the total state and local government debt was $1,593 million, or $1,991 per capita.

36 TAXATION

Taken together, taxes collected on gaming and the casino entertainment tax totaled $194.9 million in 1983/84, or 49% of all tax revenues. As of 1985, Nevada also levied a 2% state sales and use tax (with localities levying additional amounts), along with taxes on liquor, soft drinks, cigarettes, jet fuel, and slot machines. There is no personal or corporate income tax, and real estate transfer taxes ended in 1980. Nevadans filed 415,449 federal income tax returns in 1983, paying $1.2 billion in tax.

37ECONOMIC POLICY

Federal projects have played an especially large role in the development of Nevada. During the depression of the 1930s, Hoover (Boulder) Dam was constructed to provide needed jobs, water, and hydroelectric power for the state. Other public works—Davis Dam (Lake Mohave) and the Southern Nevada Water Project—have served a similar purpose. The fact that some 87% of Nevada land is owned by the US government further increases the federal impact on the economy. Gaming supplies a large proportion of state revenues.

38HEALTH

Infant mortality during 1981 was 9.8 per 1,000 live births. The overall death rate was 8.2 per 1,000 population in 1982. Heart disease, cancer, accidents, and stroke were the leading causes of death. Deaths by accident, suicide, and cirrhosis of the liver were well above the national average. In 1982, Nevada had the nation's highest rates for marriage (119.9 per 1,000 population, compared with the national average of 10.8 per 1,000) and for divorce (14.8 per 1,000 population, compared with the US norm of 5.1 per 1,000). The birthrate in 1983 was 16 per 1,000 population, only slightly above the national average of 15.5. In 1982, abortions totaled 10,000, for a ratio of 702 per 1,000 live births.

In 1983 there were 26 hospitals, with 3,629 beds. The average cost of hospital care in 1984 was $236.10 per day in a semiprivate room. The state had 1,321 physicians and 471 dentists in 1984.

39SOCIAL WELFARE

Aid for families with dependent children totaled $10.2 million in 1984, distributed to 12,687 recipients. In 1984, 34,567 Nevadans took part in the food stamp program at a federal cost of $21.1 million. Some 56,000 children received school lunches, subsidized with $4 million in federal funds in 1984. During 1983, 110,000 Nevadans received $506 million in Social Security benefits.

40HOUSING

The 1980 census counted 339,949 housing units, of which 306,121 were occupied; more than 97% had full plumbing. Between 1980 and 1983, 55,526 new units were constructed.

41EDUCATION

With a high school completion rate of 76% in 1980, adult Nevadans have had more schooling than the US population as a whole. In 1983, 150,422 pupils were enrolled in Nevada's public schools: 10,836 kindergarten, 62,961 elementary, 67,763 secondary, and 8,862 ungraded special education. In 1983/84, 44,130 students were enrolled in institutions of higher learning, nearly all of them in the University of Nevada System.

42ARTS

The Nevada Council on the Arts receives and disburses state and federal funds ($216,270 and $338,700, respectively, in 1985/86) for cultural organizations and individual artists throughout Nevada. Major exhibits are mounted by the Las Vegas Arts League and the Sierra Arts Foundation in Reno. Reno also has a symphony orchestra and an opera association.

43LIBRARIES AND MUSEUMS

Nevada's public library system in 1984 had a combined book stock of 1,411,738 volumes and a circulation of 3,700,957. The University of Nevada had 439,071 books in its Reno campus library system and 339,080 at Las Vegas; the Nevada State Library in Carson City had 46,000.

There are some 17 museums and historic sites. Notable are the Nevada State Museum in Carson City; the museum of the Nevada Historical Society and the Fleischmann Planetarium, University of Nevada, in Reno; and the Museum of Natural History, University of Nevada, at Las Vegas.

44COMMUNICATIONS

In 1980, 91% of Nevada's occupied housing units had telephones; there were 494,058 residential and business telephone access lines in 1985. In that year, commercial broadcasting comprised 45 radio stations (23 AM, 22 FM) and 9 television stations. In 1984, 12 cable television systems served 63,087 subscribers in 24 communities.

45PRESS

In 1984, the state had three morning newspapers, three evening papers, and one all-day paper with a combined circulation of 233,563, and five Sunday papers with 249,113. The leading newspaper was the *Las Vegas Review-Journal*, with an all-day circulation of 97,351 and a Sunday circulation of 110,537. The *Reno Gazette-Journal*, with a daily circulation of 56,822 and Sunday circulation of 62,481, is the most influential newspaper in the northern half of the state.

46ORGANIZATIONS

The 1982 Census of Service Industries counted 244 organizations in Nevada, including 53 business associations and 141 civic, social, and fraternal organizations. Notable organizations with headquarters in Nevada include the National Council of Juvenile and Family Court Judges and the Western History Association.

47TOURISM, TRAVEL, AND RECREATION

Tourism remains Nevada's most important industry. In 1982, tourists spent over $5.5 billion in the state, generating 30% of all jobs and a $1.5 billion payroll. Gross casino revenues alone exceeded $2.8 billion in 1983. Tourists flock to "Vegas" for gambling and for the top-flight entertainers who perform there. Other Nevada attractions are Pyramid Lake, Lake Tahoe, Lake Mead, and Lehman Caves National Monument.

There are 21 state parks and recreation areas. In 1982/83, licenses were issued to 173,051 anglers and 55,063 hunters.

48SPORTS

No major league sports teams represent Nevada. Las Vegas and Reno have hosted many professional boxing title bouts. Golfing and rodeo are also popular.

49FAMOUS NEVADANS

Nevadans who have held important federal offices include Raymond T. Baker (1877-1935) and Eva B. Adams (b.1908), both directors of the US Mint, and Charles B. Henderson (b.California, 1873-1954), head of the Reconstruction Finance Corporation. Prominent US senators have been James W. Nye (b.New York, 1815-76), also the only governor of Nevada Territory; William M. Stewart (b.New York, 1827-1909), author of the final form of the 15th Amendment to the US Constitution, father of federal mining legislation, and a leader of the free-silver-coinage movement in the 1890s; and Francis G. Newlands (b.Mississippi, 1848-1917), author of the federal Reclamation Act of 1902.

Probably the most significant state historical figure is George Wingfield (b.Arkansas, 1876-1959), a mining millionaire who exerted great influence over Nevada's economic and political life in the early 20th century. Among the nationally recognized personalities associated with Nevada is Howard R. Hughes (b.Texas, 1905-76), an aviation entrepreneur who became a casino and hotel owner and wealthy recluse in his later years.

Leading creative or performing artists have included operatic singer Emma Nevada (Emma Wixon, 1862-1940); painter Robert Caples (1908-79); and, among writers, Dan DeQuille (William Wright, b.Ohio, 1829-98); Lucius Beebe (b.Massachusetts, 1902-66); and Walter Van Tilburg Clark (b.Maine, 1909-71).

50BIBLIOGRAPHY

Bushnell, Eleanore, and Don W. Driggs. *The Nevada Constitution: Origin and Growth.* 5th ed. Reno: University of Nevada Press, 1980.

Hulse, James W. *The Nevada Adventure: A History.* 5th ed. Reno: University of Nevada Press, 1981.

Laxalt, Robert. *Nevada: A Bicentennial History.* New York: Norton, 1977.

Nevada, State of. Secretary of State. *Political History of Nevada.* 7th ed. Carson City, 1979.

Toll, David W. *The Complete Nevada Traveler.* Virginia City, Nev.: Gold Hill, 1981.

NEW HAMPSHIRE

State of New Hampshire

ORIGIN OF STATE NAME: Named for the English county of Hampshire. **NICKNAME:** The Granite State. **CAPITAL:** Concord. **ENTERED UNION:** 21 June 1788 (9th). **SONG:** "Old New Hampshire." **MOTTO:** Live Free or Die. **FLAG:** The state seal, surrounded by laurel leaves with nine stars interspersed, is centered on a blue field. **OFFICIAL SEAL:** In the center is a broadside view of the frigate *Raleigh;* in the left foreground is a granite boulder, in the background a rising sun. A laurel wreath and the words "Seal of the State of New Hampshire 1776" surround the whole. **STATE EMBLEM:** Within an elliptical panel appears a replica of the Old Man of the Mountains, with the state name above and motto below. **BIRD:** Purple finch. **FLOWER:** Purple lilac. **TREE:** White birch. **INSECT:** Ladybug. **LEGAL HOLIDAYS:** New Year's Day, 1 January; Birthday of Martin Luther King, Jr., 3d Monday in January; Washington's Birthday, 3d Monday in February; Fast Day, 4th Monday in April; Arbor Day, last Friday in April; Memorial Day, 30 May; Independence Day, 4 July; Labor Day, 1st Monday in September; Columbus Day, 2d Monday in October; Election Day, Tuesday following 1st Monday in November in even-numbered years; Veterans Day, 11 November; Thanksgiving Day, 4th Thursday in November; Christmas Day, 25 December. Time: 7 AM EST = noon GMT.

¹LOCATION, SIZE, AND EXTENT

Situated in New England in the northeastern US, New Hampshire ranks 44th in size among the 50 states.

The total area of New Hampshire is 9,279 sq mi (24,033 sq km), comprising 8,993 sq mi (23,292 sq km) of land and 286 sq mi (741 sq km) of inland water. The state has a maximum extension of 93 mi (150 km) E–W and 180 mi (290 km) N–S. New Hampshire is shaped roughly like a right triangle, with the line from the far N to the extreme SW forming the hypotenuse.

New Hampshire is bordered on the N by the Canadian province of Quebec; on the E by Maine (with part of the line formed by the Piscataqua and Salmon Falls rivers) and the Atlantic Ocean; on the S by Massachusetts; and on the W by Vermont (following the west bank of the Connecticut River) and Quebec (with the line formed by Halls Stream).

The three southernmost Isles of Shoals lying in the Atlantic belong to New Hampshire. The state's total boundary line is 555 mi (893 km). Its geographic center lies in Belknap County, 3 mi (5 km) E of Ashland.

²TOPOGRAPHY

The major regions of New Hampshire are the coastal lowland in the southeast; the New England Uplands, covering most of the south and west; and the White Mountains (part of the Appalachian chain) in the north, including Mt. Washington, at 6,288 feet (1,917 meters) the highest peak in the northeastern US. With a mean elevation of about 1,000 feet (305 meters), New Hampshire is generally hilly, rocky, and in many areas densely wooded.

There are some 1,300 lakes and ponds, of which the largest is Lake Winnipesaukee, covering 70 sq mi (181 sq km). The principal rivers are the Connecticut (forming the border with Vermont), Merrimack, Salmon Falls, Piscataqua, Saco, and Androscoggin. Near the coast are the nine rocky Isles of Shoals, three of which belong to New Hampshire.

³CLIMATE

New Hampshire has a changeable climate, with wide variations in daily and seasonal temperatures. Summers are short and cool, winters long and cold. Concord has a normal daily mean temperature of 46°F (8°C), ranging from 21°F (−6°C) in January to 70°F (21°C) in July. The record low temperature, −46°F (−43°C), was set at Pittsburg on 28 January 1925; the all-time high, 106°F (41°C) at Nashua, 4 July 1911. Annual precipitation at Concord

averages 36 in (91 cm); the average snowfall in Concord is 65 in (165 cm) a year, with more than 100 in (254 cm) yearly in the mountains. The strongest wind ever recorded, other than during a tornado—231 mph (372 kph)—occurred on Mt. Washington on 12 April 1934.

⁴FLORA AND FAUNA

Well forested, New Hampshire supports an abundance of elm, maple, beech, oak, pine, hemlock, and fir trees. Among wild flowers, several orchids are considered rare; three are classified as threatened. Robbins' cinquefoil was added to the federal endangered species list in 1980; and the small whorled pogonia in 1982.

Among native New Hampshire mammals are the white-tailed deer, muskrat, beaver, porcupine, and snowshoe hare. Threatened animals include the pine marten, arctic tern, purple martin, eastern bluebird, whippoorwill, and osprey. The Indiana bat, lynx, bald eagle, shortnose sturgeon, sunapee trout, and Atlantic salmon are on the state's endangered species list.

⁵ENVIRONMENTAL PROTECTION

State agencies concerned with environmental protection include the Water Resources Board, the Water Supply and Pollution Control Commission, the Fish and Game Department, the New Hampshire Air Resources Agency, and the Office of Waste Management. The most controversial environmental issue during the late 1970s was nuclear power, as the Clamshell Alliance organized repeated demonstrations against the Seabrook nuclear power plant, whose first reactor was scheduled to begin operations in the late 1980s; as of early 1985, work on Seabrook's reactor was all but abandoned.

⁶POPULATION

New Hampshire ranked 42d among the 50 states in the 1980 census, with a population of 920,610. The estimated 1985 population figure was 980,841, representing a growth since 1980 of 6.5%, the highest of the New England states. Projections to 1990 foresee a population of 1,138,800, 24% more than 1980. The population density in 1985 was 109 per sq mi (42 per sq km).

In 1980, about 48% of the population lived in rural areas and only 52% in cities and towns, well below the US average. Leading cities with their 1984 census populations are Manchester, 94,937; Nashua, 72,458; and Concord, the capital, 30,902. All three are located in the southeastern region, where more than two-thirds of all state residents live.

⁷ETHNIC GROUPS

In 1980, a total of 305,527 New Hampshirites claimed English ancestry. Those claiming French ancestry numbered 237,137, and Irish 192,718. There is a significant number of French Canadians. About 4,000 black Americans live in New Hampshire.

⁸LANGUAGES

Some place-names, such as Ossipee, Mascoma, and Chocorua, preserve the memory of the Pennacook and Abnaki Algonkian tribes living in the area before white settlement.

New Hampshire speech is essentially Northern, with the special features marking eastern New England: loss of final /r/, *park* and *path* with a vowel between those in *cat* and *father,* and /yu/ in *tube* and *new. Raspberries* sounds like /rawzberries/, a wishbone is a *luckybone,* gutters are *eavespouts,* and cows are summoned by "Loo!" Canadian French is heard in the northern region.

In 1980, 90% of all state residents aged 3 and above—a total of 791,515—spoke only English at home; 62,280 residents spoke French at home.

⁹RELIGIONS

The first settlers of New Hampshire were Separatists, precursors of the modern Congregationalists (United Church of Christ), and their first church was probably built around 1633. The first Episcopal church was built in 1638, and the first Quaker meeting-house in 1701; Presbyterians, Baptists, and Methodists built churches later in the 18th century. The state remained almost entirely Protestant until the second half of the 19th century, when Roman Catholics (French Canadian, Irish, and Italian) began arriving in significant numbers, along with some Greek and Russian Orthodox Christians.

As of 1980, Protestant groups in New Hampshire had 121,109 known adherents. The leading denominations were United Church of Christ, 33,871; United Methodist, 19,837; and American Baptist, 18,184. There were 298,960 Roman Catholics and an estimated 5,980 Jews in 1984.

¹⁰TRANSPORTATION

New Hampshire's first railroad, between Nashua and Lowell, Mass., was chartered in 1835 and opened in 1838. Two years later, Exeter and Boston were linked by rail. The state had more than 1,200 mi (1,900 km) of track in 1920, but by 1983, the total of Class I track in New Hampshire was only 419 mi (674 km). There is limited passenger service between Concord and Boston.

In 1983, the state had a total of 14,545 mi (23,408 km) of roads, of which 12,141 mi (19,539 km) were rural and 2,404 mi (3,869 km) municipal; the main north–south highway is I-93. As of 1983 there were 686,966 automobiles, 52,871 motorcycles, 1,487 buses, and 114,187 trucks registered in the state, as well as 697,012 licensed drivers.

New Hampshire had 42 airports and 9 heliports in 1983. The main airport is Grenier Field in Manchester.

¹¹HISTORY

The land called New Hampshire has supported a human population for at least 10,000 years. Prior to European settlement, Indian tribes of the Algonkian language group lived in the region. During the 17th century, most of New Hampshire's Indians, called Pennacook, were organized in a loose confederation centered along the Merrimack Valley.

The coast of New England was explored by Dutch, English, and French navigators throughout the 16th century. Samuel de Champlain prepared the first accurate map of the New England coast in 1604, and Captain John Smith explored the Isles of Shoals in 1614. By this time, numerous English fishermen were summering on New England's coastal banks, using the Isles of Shoals for temporary shelter and to dry their catch.

The first English settlement was established along the Piscataqua River in 1623. From 1643 to 1680, New Hampshire was a province of Massachusetts, and the boundary between them was not settled until 1740. During the 18th century, as settlers moved up the Merrimack and Connecticut river valleys, they came into conflict with the Indians. By 1760, however, the Pennacook had been expelled from the region.

Throughout the provincial period, people in New Hampshire made their living through fishing, farming, cutting and sawing timber, shipbuilding, and coastal and overseas trade. By the first quarter of the 18th century, Portsmouth, the provincial capital, had become a thriving commercial port. New Hampshire's terrain worked against Portsmouth's commercial interests, however, by dictating that roads (and later railroads) run in a north–south direction—making Boston, and not Portsmouth, New England's primary trading center. During the Revolutionary War, extensive preparations were made to protect the harbor from a British attack that never came. Although nearly 18,500 New Hampshire men enlisted in the war, no battle was fought within its boundaries. New Hampshire was the first of the original 13 colonies to establish an independent government—on 5 January 1776, six months before the Declaration of Independence.

During the 19th century, as overseas trade became less important to the New Hampshire economy, textile mills were built, principally along the Merrimack River. By midcentury, the Merrimack Valley had become the social, political, and economic center of the state. So great was the demand for workers in these mills that immigrant labor was imported during the 1850s; a decade later, French Canadian workers began pouring south from Quebec.

Although industry thrived, agriculture did not; New Hampshire hill farms could not compete against Midwestern farms. The population in farm towns dropped, leaving a maze of stone walls, cellar holes, and new forests on the hillsides. The people who remained began to cluster in small village centers.

World War I, however, marked a turning point for New Hampshire industry. As wartime demand fell off, the state's old textile mills were unable to compete with newer cotton mills in the South, and New Hampshire's mill towns became as depressed as its farm towns; only in the north, the center for logging and paper manufacturing, did state residents continue to enjoy moderate prosperity. Industrial towns in the southern counties responded to the decline in textile manufacture by making other items, particularly shoes, but the collapse of the state's railroad network spelled further trouble for the slumping economy. The growth of tourism aided the rural areas primarily, as old farms became spacious vacation homes for "summer people," who in some cases paid the bulk of local property taxes.

During the 1960s, New Hampshire's economic decline began to reverse, except in agriculture. In the 1970s and early 1980s, growth in the state's northern counties remained modest, but the combination of Boston's urban sprawl, interstate highway construction, and low state taxes encouraged people and industry—notably high-technology businesses—to move into southern New Hampshire. The state's population increased by 243,240 (33%) between 1970 and 1983—a development accompanied by higher local taxes and concern over the state's vanishing open spaces.

¹²STATE GOVERNMENT

New Hampshire's constitution, adopted in 1784 and extensively revised in 1792, is the 2d-oldest state-governing document still in effect. Every 10 years, the people vote on the question of calling a convention to revise it; proposed revisions must then be approved by two-thirds of the voters at a referendum. Amendments may also be placed on the ballot by a three-fifths vote of both houses of the General Court—the state legislature—which consists of a 24-member senate and a 400-seat house of representatives (larger than that of any other state). Legislators serve two-year terms.

The only executive elected statewide is the governor, who serves a two-year term and is assisted by a five-member executive council, elected for two years by district. The council must approve all administrative and judicial appointments. The secretaries of state and treasury are elected by the legislature. The governor must be at least 30 years of age and must have been a state resident for seven years before election.

A bill becomes law if signed by the governor, if passed by the legislature and left unsigned by the governor for five days while the legislature is in session, or if passed over a gubernatorial veto by two-thirds of the legislators present in each house. US citizens at least 18 years of age who have resided in the state for 10 days are eligible to vote in New Hampshire elections.

13POLITICAL PARTIES

New Hampshire has almost always gone with the Republican presidential nominee in recent decades, but the Democratic and Republican parties have been much more evenly balanced in local and state elections. New Hampshire's quadrennial presidential preference primary, traditionally the first state primary of the campaign season, accords to New Hampshirites a degree of national political influence and a claim on media attention far out of proportion to their numbers. In the 1984 presidential election, New Hampshire voters backed President Ronald Reagan. Two other Republican incumbents, US Senator Gordon Humphrey and Governor John Sununu, also defeated Democratic challengers in 1984. In 1985, 33% of New Hampshire's state legislators were women, the highest percentage of any state.

New Hampshire Presidential Vote by Major Political Parties, 1948–84

YEAR	ELECTORAL VOTE	NEW HAMPSHIRE WINNER	DEMOCRAT	REPUBLICAN
1948	4	Dewey (R)	107,995	121,299
1952	4	*Eisenhower (R)	106,663	166,287
1956	4	*Eisenhower (R)	90,364	176,519
1960	4	Nixon (R)	137,772	157,989
1964	4	*Johnson (D)	182,065	104,029
1968	4	*Nixon (R)	130,589	154,903
1972	4	*Nixon (R)	116,435	213,724
1976	4	Ford (R)	147,635	185,935
1980	4	*Reagan (R)	108,864	221,705
1984	4	*Reagan (R)	120,347	267,050

* Won US presidential election.

14LOCAL GOVERNMENT

New Hampshire has 10 counties, each governed by three commissioners. Other elected county officials include the sheriff, attorney, treasurer, register of deeds, and register of probate.

As of 1982, New Hampshire also had 13 municipalities and 221 townships. Municipalities have elected mayors and councils. The basic unit of town government is the traditional town meeting, held once a year, when selectmen and other local officials are chosen.

15STATE SERVICES

The Department of Education, governed by the State Board of Education (which appoints an education commissioner), has primary responsibility for public instruction. The New Hampshire Transportation Authority, Aeronautics Commission, Port Authority, and Department of Public Works and Highways share transport responsibilities, while the Department of Health and Welfare oversees public health and mental health.

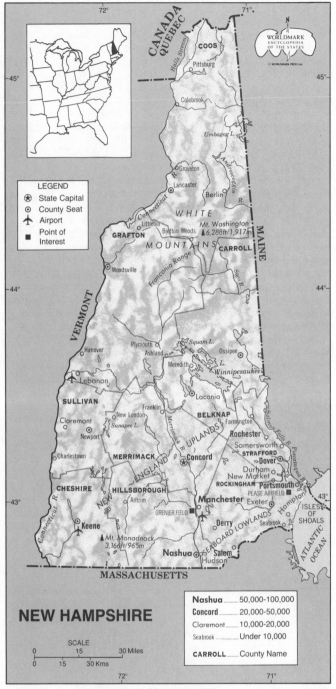

See endsheet maps: M2.

LOCATION: 42°40′ to 45°18′N; 70°37′ to 72°34′W. **BOUNDARIES:** Canadian line, 60 mi (96 km); Maine line, 174 mi (280 km); Atlantic Ocean coastline, 13 mi (21 km); Massachusetts line, 95 mi (153 km); Vermont line, 213 mi (343 km).

16JUDICIAL SYSTEM

All judges in New Hampshire are appointed by the governor, subject to confirmation by the executive council; appointments are to age 70, with retirement compulsory at that time. The state's highest court, the supreme court, consists of a chief justice and 4 associate justices. The main trial court is the superior court, with one judgeship for each 60,000 state residents; there were 14 associate justices and a chief justice in 1982. Crime rates for all categories are among the lowest in the US. In 1984 there were 2,147 practicing lawyers in the state.

[17]ARMED FORCES

The principal military installations in New Hampshire are the Portsmouth Naval Shipyard and Pease Air Force Base at Newington. In 1983/84, 4,157 military personnel were stationed in the state. Firms in the state received $663 million in defense contract awards in 1983/84.

As of 1983, veterans living in New Hampshire numbered 138,000, of whom 1,000 were veterans of World War I, 51,000 of World War II, 21,000 of the Korean conflict, and 40,000 of the Viet-Nam era. Veterans benefits totaled $113 million in 1983.

The Army and Air National Guard had 3,244 personnel in 1985. There were 207 state police and 1,343 local police in 1983.

[18]MIGRATION

From the time of the first European settlement until the middle of the 19th century, the population of New Hampshire was primarily of British origin. Subsequently, immigrants from Quebec and from Ireland, Italy, and other countries began arriving in significant numbers. New Hampshire's population growth since 1960 has been fueled by migrants from other states. The net gain from migration was 135,000 between 1970 and 1983.

[19]INTERGOVERNMENTAL COOPERATION

New Hampshire participates in the American and Canadian French Cultural Exchange Commission, Atlantic States Marine Fisheries Compact, Connecticut River Valley Flood Control Compact, and various New England regional compacts.

Federal grants to New Hampshire totaled $367.6 million in 1983/84.

[20]ECONOMY

New Hampshire is one of the most industrialized states in the US, ranking well above the national median in proportion of labor force employed in manufacturing and in value added by manufacture. Between 1977 and 1982, manufacturing employment rose 13%, to 107,500, as many high-technology firms moved into the southern portion of the state. Since World War II, tourism has been one of the state's fastest-growing sources of income.

[21]INCOME

New Hampshire's per capita income, which ranked 20th in the US in 1970, had slipped to 32d by 1978, largely because of rapid population growth rather than any slowdown in real economic growth. By 1983, New Hampshire had rebounded to 19th, with a per capita income of $11,620, just below the US average of $11,675. In 1981, median income of four-person families was $19,723, ranking the state 27th in the US. In 1979, 8.5% of state residents were below the federal poverty level.

[22]LABOR

New Hampshire's estimated civilian labor force totaled 520,000 in 1984. Of these, 498,000 were employed, yielding an unemployment rate of just over 4%, one of the lowest in the US.

A federal survey in 1982 revealed the following nonfarm employment pattern for New Hampshire:

	ESTABLISH-MENTS	EMPLOYEES	ANNUAL PAYROLL ('000)
Agricultural services, forestry, fishing	241	902	$ 11,090
Mining	26	265	5,445
Contract construction	2,334	19,758	414,428
Manufacturing	1,823	108,576	1,807,738
Transportation, public utilities	749	14,177	286,362
Wholesale trade	1,435	18,342	351,973
Retail trade	6,170	70,538	584,806
Finance, insurance, real estate	1,530	20,777	306,190
Services	6,601	73,486	854,043
Other	414	303	5,567
TOTALS	21,323	327,124	$4,627,642

State and local government employees, who numbered 20,000 in 1983, and federal government employees, who numbered 16,000 in 1982, were not included in this survey. There were 61,000 labor union members in 1980 and 230 unions in 1983.

[23]AGRICULTURE

Only in Rhode Island and Alaska do farmers earn less from farming than in New Hampshire. Farm income in 1983 was an estimated $114 million, less than 32% of which was in crops.

In 1984 there were about 3,500 farms occupying about 550,000 acres (222,600 hectares). Leading crops and their output in 1983 were hay, 201,000 tons, and commercial apples, 57 million lb.

[24]ANIMAL HUSBANDRY

Dairy and poultry products are the mainstays of New Hampshire's agriculture. In 1983, the state had 31,000 milk cows, with a total milk yield of 381 million lb. Poultry items included 2,695,000 lb of chickens, worth $283,000, and 142,000,000 eggs, worth $10,413,000.

[25]FISHING

New Hampshire's commercial catch in 1984 consisted of 11,892,000 lb, much of it cod and lobster, worth $8,442,000.

[26]FORESTRY

New Hampshire had 5,014,000 acres (2,029,000 hectares) of forestland in 1979, of which 4,692,000 acres (1,899,000 hectares) were considered suitable for commercial use. Of that total, 88% was privately owned. Shipments of lumber and wood products were valued at $298.3 million in 1982, while paper and allied products were worth $508 million.

[27]MINING

Although New Hampshire has long been famous for its granite quarries—from which the state nickname derives—mineral resources are a very minor contributor to the state economy. In 1984, nonfuel mineral production amounted to only $21 million, for a rank of 47th among the 50 states. The quarry output included 4,600,000 tons of sand and gravel, accounting for about 85% of the state's mineral production.

[28]ENERGY AND POWER

About 90% of all New Hampshire's electrical power was generated by water in the 1930s. By 1983, however, about 50% of the state's electricity came from coal-fired plants, another 27% from oil-fired plants, and only 23% from hydroelectric facilities. Power production totaled 5.3 billion kwh in 1983, when installed capacity was 1.5 million kw. At the beginning of 1985, work was 80% complete on the controversial nuclear power plant at Seabrook, being built by Public Service Co. of New Hampshire. Originally planned as a two-reactor, 2,300-Mw facility at a cost of less than $1 billion, Seabrook was scaled back to one reactor whose cost, if the project is completed, would approach $5 billion.

[29]INDUSTRY

During the provincial era, shipbuilding was New Hampshire's major industry. By 1870, cotton and woolen mills, concentrated in the southeast, employed about one-third of the labor force and accounted for roughly half the value of all manufactures. In 1982, the most important industry groups, ranked by employment, were nonelectrical machinery, electric and electronic equipment, rubber and miscellaneous plastics products, and leather and leather products (mainly shoes).

The total value of shipments by manufacturers in 1982 was more than $7.6 billion. The following table shows value of shipments of major sectors:

Nonelectrical machinery	$1,968,900,000
Electric and electronic equipment	926,100,000
Paper and allied products	740,600,000
Rubber and miscellaneous plastic products	479,700,000
Instruments and related products	460,900,000
Leather and leather products	275,400,000
Textile mill products	211,700,000

³⁰COMMERCE

New Hampshire wholesalers sold $4.1 billion worth of goods in 1982. Retailers had sales of almost $5.4 billion, the leading sectors being food stores, 23%; automotive dealers, 17%; and general merchandise stores, 10%.

Foreign exports of manufactured goods totaled $637 million in 1981 (37th in the US).

³¹CONSUMER PROTECTION

The Attorney General's Office is responsible for enforcing New Hampshire's consumer protection laws.

³²BANKING

New Hampshire had 85 insured commercial banks in 1984. Total assets in 1983 amounted to $4.8 billion, outstanding loans were $2.9 billion, and deposits were $4.1 billion. In 1982, the state's 13 savings and loan associations held total assets of $1.3 billion, savings capital of $1.1 billion, and mortgage loans of $958 million.

³³INSURANCE

In 1983 there were 1,547,000 life insurance policies in force in New Hampshire, with a total value of $20.1 billion. The average coverage per family was $52,000, slightly below the US average.

Property and liability insurers wrote premiums amounting to $531 million in 1983, of which automobile physical damage insurance accounted for $91 million; automotive liability insurance, $103.8 million; and homeowners' coverage, $72.9 million.

³⁴SECURITIES

New Hampshire has no securities exchanges. New York Stock exchange member firms maintained 23 offices and 151 representatives in the state in 1983. In that year, 167,000 New Hampshirites owned shares of public corporations.

³⁵PUBLIC FINANCE

The New Hampshire state budget is drawn up biennially by the Department of Administrative Services and then submitted by the governor to the legislature for amendment and approval. The fiscal year runs from 1 July to 30 June. The following is a summary of recommended expenditures for 1985/86 and 1986/87 (in millions):

	1985/86	1986/87
Health and welfare	$ 313.0	$ 319.0
Public works and highways	193.1	194.7
Higher Education Fund	171.6	181.6
Board of Education	79.0	74.2
Other outlays	351.2	356.2
TOTALS	$1,107.9	$1,125.7

Estimated leading sources of revenue in 1986/87, in addition to federal aid, are the business profits tax, $93.3 million; gasoline road toll, $70.2 million; meals and rooms tax, $67.9 million; liquor licenses and fees, $46 million; and tobacco taxes, $33.9 million.

As of 1982/83, the combined debt of state and local governments was $1.96 billion, or about $2,105 per capita.

³⁶TAXATION

New Hampshire has no general income or sales tax but does levy 5% taxes on interest and dividends and 9.08% on net corporate income. Levies on property, gasoline, alcoholic beverages, tobacco products, pari-mutuel betting, and many other items are also imposed. General revenues were only $857 per capita in 1982, ranking New Hampshire 48th among the 50 states.

During 1982, New Hampshire paid almost $2.4 billion in federal taxes and received federal benefits amounting to about $26 million less. New Hampshirites filed 424,000 federal income tax returns that year, paying almost $1.9 billion in tax.

³⁷ECONOMIC POLICY

Business incentives in New Hampshire include a generally favorable tax climate, specific tax incentives and exemptions, and relatively low wage rates. The Division of Industrial Develop-ment, within the Department of Resources and Economic Development, is the agency primarily responsible for attracting new investment and industry.

³⁸HEALTH

Infant mortality in New Hampshire is low—9.8 per 1,000 live births for whites (in 1981), statistically not significant for non-whites—and the death rates are below the national average for heart disease and stroke, two of the three leading causes of death. The death rate for cancer, however, is slightly higher than the national average.

In 1983 there were 34 hospitals, with 4,578 beds. Hospital admissions in that year totaled 142,545. Hospital personnel in 1981 included 2,312 full-time and 1,412 part-time registered nurses. The average cost of hospital care in 1982 was $289 per day and $2,048 per stay, both well below the national averages. New Hampshire had 1,899 licensed physicians and 508 dentists in 1982.

³⁹SOCIAL WELFARE

Like its tax revenues, New Hampshire's expenditures on welfare are low for a northeastern state.

A monthly average of 19,700 New Hampshirites received $24.2 million in aid to families with dependent children during 1982. In 1983, 41,000 persons took part in the food stamp program at a federal cost of $25 million, and 70,000 students had school lunches with a federal outlay of $6 million. In 1983, Social Security benefits totaling $693 million were paid to 145,000 persons, the average monthly benefit for retired workers being $444. Federal Supplemental Security Income payments amounted to $11 million in 1983. Unemployment insurance benefits totaled $46 million in 1982; the average weekly payment was $95.37, 20% below the US norm.

⁴⁰HOUSING

In 1980, housing units for year-round use numbered 345,000, a 40% increase over 1970. Of all occupied units, 68% were owner-occupied, the remainder rented; 96% had full plumbing. In 1980 there were 2.75 persons per household, as opposed to 3.14 in 1970. Almost 4,400 new housing units worth $189 million were authorized in 1981.

⁴¹EDUCATION

New Hampshire residents have a long-standing commitment to education. More than 72% of all adult state residents were high school graduates in 1980; 18% had four or more years of college.

In fall 1983, enrollment in public schools and approved public academies totaled 159,037; private elementary and secondary schools had an enrollment of 20,721 in fall 1980. The best-known institution of higher education is Dartmouth College (1983/84 enrollment, 5,000), which originated in Connecticut in 1754 as Moor's Indian Charity School and was established at Hanover in 1769. When the State of New Hampshire attempted to amend Dartmouth's charter to make the institution public in the early 19th century, the US Supreme Court handed down a precedent-setting ruling prohibiting state violation of contract rights. The University of New Hampshire, the leading public institution, was founded at Hanover in 1866 and relocated at Durham in 1891; it had a 1983/84 enrollment of 10,300.

⁴²ARTS

Hopkins Center at Dartmouth College features musical events throughout the year, while the Monadnock Music Concerts are held in several towns during the summer. The New Hampshire Music Festival takes place at Meredith. Theater by the Sea at Portsmouth presents classical and modern plays, and there is a year-round student theater at Dartmouth.

Principal galleries include the Currier Gallery of Art in Manchester, the Arts and Science Center in Nashua, the University Art Galleries at the University of New Hampshire in Durham, the Dartmouth College Museum and Galleries at Hanover, and the Lamont Gallery at Phillips Exeter Academy in Exeter.

43 LIBRARIES AND MUSEUMS
New Hampshire public libraries had a total book stock of 3,459,925 volumes and a combined circulation of 5,486,403 volumes in 1982. Leading academic and historical collections include Dartmouth College's Baker Memorial Library in Hanover (1,515,652 volumes); the New Hampshire State Library (683,015) and New Hampshire Historical Society Library (50,000), both in Concord; and the University of New Hampshire's Ezekiel W. Dimond Library (837,000) in Durham.

Among the more than 50 museums and historic sites are the New Hampshire Historical Society Museum in Concord and the Franklin Pierce Homestead in Hillsborough.

44 COMMUNICATIONS
In 1980, 94% of New Hampshire's occupied housing units—a total of 304,797—had telephones. In 1984, the state had 59 commercial radio stations (29 AM, 30 FM) and 3 commercial television stations. State residents also receive broadcasts from neighboring Massachusetts, Vermont, and Maine. A total of 33 cable systems in 1984 served 125 communities with 166,114 subscribers.

45 PRESS
In 1984, New Hampshire had 9 daily newspapers with a combined circulation of 208,676 and 3 Sunday papers with 105,038. The best-known newspaper in the state is the *Manchester Union-Leader* (69,882), published by the ultraconservative William Loeb until his death in 1981. The *New Hampshire Sunday News*, which was also published by Loeb, has a circulation of 80,278.

46 ORGANIZATIONS
The 1982 Census of Service Industries counted 385 organizations in New Hampshire, including 64 business associations; 254 civic, social, and fraternal associations; and 7 educational, scientific, and research associations. Organizations with headquarters in New Hampshire include the Student Conservation Association (Charlestown), the American Association of Commodity Traders (Concord), the American Society of Environmental Education (Hanover), and the Natural Organic Farmers Association (Antrum).

47 TOURISM, TRAVEL, AND RECREATION
Tourism ranks 2d only to manufacturing in the economy of New Hampshire. An estimated 4.7 million persons traveled to or through the state in 1977.

Skiing, camping, hiking, and boating are the main outdoor attractions. In 1982 there were 35 state parks, 88 trailer parks and camps, 10 public golf courses, and 33 alpine ski areas, some of which also operate as summer resorts. Other attractions include Strawbery Banke, a restored village in Portsmouth; Daniel Webster's birthplace near Franklin; the Mt. Washington Cog Railway; and the natural "Old Man of the Mountains" granite head profile in the Franconia subrange of the White Mountains, on which the state's official emblem is modeled.

48 SPORTS
Major national and international skiing events are frequently held in the state, as are such other winter competitions as snowmobile races and the Sled Dog Derby. Thoroughbred, harness, and greyhound racing are the warm-weather spectator sports. The annual Whaleback Yacht Race is held in early August.

Dartmouth College competes in the Ivy League—the Big Green won or shared conference football championships seven times from 1970 through 1984—while the University of New Hampshire belongs to the Yankee Conference.

49 FAMOUS NEW HAMPSHIRITES
Born in Hillsboro, Franklin Pierce (1804–69), the nation's 14th president, serving from 1853 to 1857, was the only US chief executive to come from New Hampshire. Henry Wilson (Jeremiah Jones Colbath, 1812–75), US vice president from 1873 to 1875, was a native of Farmington.

US Supreme Court chief justices Salmon P. Chase (1808–73) and Harlan Fiske Stone (1872–1946) were New Hampshirites, and Levi Woodbury (1789–1851) was a distinguished associate justice. John Langdon (1741–1819) was the first president pro tempore of the US Senate; two other US senators from New Hampshire, George Higgins Moses (b.Maine, 1869–1944) and Henry Styles Bridges (b.Maine, 1898–1961), also held this position. US cabinet members from New Hampshire included Henry Dearborn (1751–1829), secretary of war; Daniel Webster (1782–1852), secretary of state; and William E. Chandler (1835–1917), secretary of the Navy. Other political leaders of note were Benning Wentworth (1696–1770), royal governor; Meshech Weare (1713–86), the state's leader during the American Revolution; Josiah Bartlett (b.Massachusetts, 1729–95), a physician, governor, and signer of the Declaration of Independence; Isaac Hill (b.Massachusetts, 1789–1851), a publisher, governor, and US senator; and John Parker Hale (1806–73), senator, antislavery agitator, minister to Spain, and presidential candidate of the Free Soil Party.

Military leaders associated with New Hampshire during the colonial and Revolutionary periods include John Stark (1728–1822), Robert Rogers (b.Massachusetts, 1731–95), and John Sullivan (1740–95). Among other figures of note are educator Eleazar Wheelock (b.Connecticut, 1711–79), the founder of Dartmouth College; physicians Lyman Spaulding (1775–1821), Reuben D. Mussey (1780–1866), and Amos Twitchell (1781–1850), as well as Samuel Thomson (1769–1843), a leading advocate of herbal medicine; religious leaders Hosea Ballou (1771–1852), his grandnephew of the same name (1796–1861), and Mary Baker Eddy (1821–1910), founder of Christian Science; George Whipple (1878–1976), winner of the 1934 Nobel Prize for physiology or medicine; and labor organizer and US Communist Party leader Elizabeth Gurley Flynn (1890–1964).

Sarah Josepha Hale (1788–1879), Horace Greeley (1811–72), Charles Dana (1819–97), Thomas Bailey Aldrich (1836–1907), Bradford Torrey (b.Massachusetts, 1843–1912), Alice Brown (1857–1948), and J(erome) D(avid) Salinger (b.New York, 1919) are among the writers and editors who have lived in New Hampshire, along with poets Edna Dean Proctor (1829–1923), Celia Laighton Thaxter (1835–94), Edward Arlington Robinson (b.Maine, 1869–1935), and Robert Frost (b.California, 1874–1963), one of whose poetry volumes is entitled *New Hampshire* (1923). Painter Benjamin Champney (1817–1907) and sculptor Daniel Chester French (1850–1931) were born in New Hampshire, while Augustus Saint-Gaudens (b.Ireland, 1848–1907) created much of his sculpture in the state.

Vaudevillian Will Cressey (1863–1930) was a New Hampshire man. More recent celebrities include newspaper publisher William Loeb (b.New York, 1905–81) and astronaut Alan B. Shepard, Jr. (b.1923).

50 BIBLIOGRAPHY
Brown, William R. *Our Forest Heritage.* Concord: New Hampshire Historical Society, 1958.
Clark, Charles E. *The Eastern Frontier: The Settlement of Northern New England, 1610–1763.* New York: Knopf, 1970.
Federal Writers' Project. *New Hampshire: A Guide to the Granite State.* Reprint. New York: Somerset, n.d. (orig. 1938).
Morison, Elizabeth Forbes and Elting E. *New Hampshire: A Bicentennial History.* New York: Norton, 1976.
Smith, Clyde H. *New Hampshire: A Scenic Discovery.* Dublin, N.H.: Yankee Books, 1985.
Squires, J. Duane. *The Granite State of the United States: A History of New Hampshire from 1623 to the Present.* 4 vols. New York: American Historical Co., 1956.
Stacker, Ann P., and Nancy C. Hefferman. *Short History of New Hampshire.* Grantham, N.H.: Tompson and Rutter, 1985.
Taylor, William L., ed. *Readings in New Hampshire and New England History.* New York: Irvington, 1981.

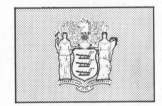

NEW JERSEY

State of New Jersey

ORIGIN OF STATE NAME: Named for the British Channel Island of Jersey. **NICKNAME:** The Garden State. **CAPITAL:** Trenton. **ENTERED UNION:** 18 December 1787 (3d). **MOTTO:** Liberty and Prosperity. **COLORS:** Buff and Jersey blue. **COAT OF ARMS:** In the center is a shield with three plows, symbolic of agriculture; a helmet above indicates sovereignty, and a horse's head atop the helmet signifies speed and prosperity. The state motto and the date "1776" are displayed on a banner below. **FLAG:** The coat of arms on a buff field. **OFFICIAL SEAL:** The coat of arms surrounded by the words "The Great Seal of the State of New Jersey." **ANIMAL:** Horse. **BIRD:** Eastern goldfinch. **FLOWER:** Violet. **TREE:** Red oak. **MEMORIAL TREE:** Dogwood. **INSECT:** Honeybee. **LEGAL HOLIDAYS:** New Year's Day, 1 January; Birthday of Martin Luther King, Jr., 3d Monday in January; Lincoln's Birthday, 12 February; Washington's Birthday, 3d Monday in February; Good Friday, March or April; Memorial Day, last Monday in May; Independence Day, 4 July; Labor Day, 1st Monday in September; Columbus Day, 2d Monday in October; Election Day, 1st Tuesday after 1st Monday in November; Veterans Day, 11 November; Thanksgiving Day, 4th Thursday in November; Christmas Day, 25 December. **TIME:** 7 AM EST = noon GMT.

¹LOCATION, SIZE, AND EXTENT

Situated in the northeastern US, New Jersey is the smallest of the Middle Atlantic states and ranks 46th among the 50 states.

The total area of New Jersey is 7,787 sq mi (20,168 sq km), of which 7,468 sq mi (19,342 sq km) constitute land and 319 sq mi (826 sq km) are inland water. New Jersey extends 166 mi (267 km) N–S; the extreme width E–W is 57 mi (92 km).

New Jersey is bordered on the N and NE by New York State (with the boundary formed partly by the Hudson River, New York Bay, and Arthur Kill, and passing through Raritan Bay); on the E by the Atlantic Ocean; on the S and SW by Delaware (with the line passing through Delaware Bay); and on the W by Pennsylvania (separated by the Delaware River). Numerous barrier islands lie off the Atlantic coast.

New Jersey's total boundary length is 480 mi (773 km), including a general coastline of 130 mi (209 km); the tidal shoreline is 1,792 mi (2,884 km). The state's geographic center is in Mercer County, near Trenton.

²TOPOGRAPHY

Although small, New Jersey has considerable topographic variety. In the extreme northwest corner of the state are the Appalachian Valley and the Kittatinny Ridge and Valley. This area contains High Point, the state's peak elevation, at 1,803 feet (550 meters) above sea level. To the east and south is the highlands region, an area of many natural lakes and steep ridges, including the Ramapo Mountains, part of the Appalachian chain. East of the highlands is a flat area broken by the high ridges of the Watchungs and Sourlands and—most spectacularly—by the Palisades, a column of traprock rising some 500 feet (150 meters) above the Hudson River. The Atlantic Coastal Plain, a flat area with swamps and sandy beaches, claims the remaining two-thirds of the state. Its most notable feature is the Pine Barrens, 760 sq mi (1,968 sq km) of pitch pines and white oaks. Sandy Hook, a peninsula more than 5 mi (8 km) long, extending northward into the Atlantic from Monmouth County, is part of the Gateway National Recreation Area.

Major rivers include the Delaware, forming the border with Pennsylvania, and the Passaic, Hackensack, and Raritan. The largest natural lake is Lake Hopatcong, about 8 mi (13 km) long.

Some 550 to 600 million years ago, New Jersey's topography was the opposite of what it is now, with mountains to the east and a shallow sea to the west. Volcanic eruptions about 225 million years ago caused these eastern mountains to sink and new peaks to rise in the northwest; the lava flow formed the Watchung Mountains and the Palisades. The shoreline settled into its present shape at least 10,000 years ago.

³CLIMATE

Bounded by the Atlantic Ocean and the Delaware River, most of New Jersey has a moderate climate with cold winters and warm, humid summers. Winter temperatures are slightly colder and summer temperatures slightly milder in the northwestern hills than in the rest of the state.

In Atlantic City, the average mean temperature is 53°F (12°C), ranging from 32°F (0°C) in January to 74°F (23°C) in July. Precipitation is plentiful, averaging 46 in (117 cm) annually; snowfall totals about 16 in (41 cm). The annual average humidity is 81% at 7 AM, reaching a normal high of 87% in September.

Statewide, the record high temperature is 110°F (43°C), set in Runyon on 10 July 1936; the record low was −34°F (−37°C), set in River Vale on 5 January 1904. A 29.7-in (75.4-cm) accumulation on Long Beach Island in 1947 was the greatest 24-hour snowfall in the state's recorded history. Occasional hurricanes and violent spring storms have damaged beachfront property over the years, and floods along northern New Jersey rivers, especially in the Passaic River basin, are not uncommon. A serious drought occurs, on average, about once every 15 years.

⁴FLORA AND FAUNA

Although highly urbanized, New Jersey still provides a diversity of natural regions, including a shady coastal zone, the hilly and wooded Allegheny zone, and the Pine Barrens in the south. Birch, beech, hickory, and elm all grow in the state, along with black locust, red maple, and 20 varieties of oak; common shrubs include the spicebush, staggerbush, and mountain laurel. Vast stretches beneath pine trees are covered with pyxie, a small creeping evergreen shrub. Common wild flowers include meadow rue, butterflyweed, black-eyed Susan, and the ubiquitous eastern (common) dandelion. Among rare plants are Candy's lobelia, floating heart, and pennywort; the small whorled pogonia is endangered.

Among mammals indigenous to New Jersey are the white-tailed deer, black bear, gray and red foxes, raccoon, woodchuck, opossum, striped skunk, eastern gray squirrel, eastern chipmunk, and common cottontail. The herring gull, sandpiper, and little green and night herons are common shore birds, while the red-eyed vireo, hermit thrush, English sparrow, robin, cardinal, and Baltimore oriole are frequently sighted inland. Anglers prize the northern pike, chain pickerel, and various species of bass, trout, and perch. Declining or rare animals include the whippoorwill, hooded warbler, eastern hognose snake, northern red salamander, and northern kingfish. The Atlantic green turtle, barred owl, bobolink, great blue heron, pied-billed grebe, corn snake, Atlantic tomcod, Atlantic sturgeon, and native brook trout are on the state's threatened species list. Fauna on the endangered list include six types of whale, four varieties of sea turtle, the Indiana bat, bald eagle, least tern, bog turtle, timber rattlesnake, Pine Barrens treefrog, Tremblay's salamander, and shortnose sturgeon.

⁵ENVIRONMENTAL PROTECTION

Laws and policies regulating the management and protection of New Jersey's environment and natural resources are administered by the Department of Environmental Protection (DEP). The DEP's overall budget for 1985/86 was $70,149,000.

The proximity of the populace to industrial plants and to the state's expansive highway system makes air pollution control a special concern in the state. New Jersey had one of the most comprehensive air pollution control programs in the US in 1985, maintaining a network of 105 air pollution monitoring stations, as well as 60 stations that monitor just for particulates and 10 that monitor for radiation.

The DEP reported that a 1984 review of water quality in the state showed that water quality degradation had been halted and that the quality of streams had been stabilized or improved. The greatest improvements had been made in certain bays and estuaries along the Atlantic coast, where the elimination of discharges from older municipal sewage treatment plants resulted in the reopening of shellfish-harvesting grounds for the first time in 20 years. However, some rivers in highly urbanized areas were still severely polluted.

Approximately 1,500 treatment facilities discharge wastewater into New Jersey's surface and ground waters. Nearly 80% of these facilities comply with the requirements of federal and state clean-water laws. As of 1985, only six sewage agencies still utilized ocean disposal, dumping about 2.7 million tons of sludge a year.

New Jersey Counties, County Seats, and County Areas and Populations

COUNTY	COUNTY SEAT	LAND AREA (SQ MI)	POPULATION (1984 EST.)
Atlantic	Mays Landing	568	201,239
Bergen	Hackensack	237	844,284
Burlington	Mt. Holly	808	378,688
Camden	Camden	223	481,975
Cape May	Cape May	263	89,121
Cumberland	Bridgeton	498	133,459
Essex	Newark	127	831,757
Gloucester	Woodbury	327	207,265
Hudson	Jersey City	46	559,885
Hunterdon	Flemington	427	92,606
Mercer	Trenton	227	314,003
Middlesex	New Brunswick	316	618,365
Monmouth	Freehold	472	525,302
Morris	Morristown	471	417,855
Ocean	Toms River	641	375,248
Passaic	Paterson	187	455,229
Salem	Salem	338	66,232
Somerset	Somerville	305	210,878
Sussex	Newton	525	120,246
Union	Elizabeth	103	505,460
Warren	Belvidere	359	85,361
	TOTALS	7,468	7,514,452

Solid waste disposal in New Jersey became critical as major landfills reached capacity. In 1977, the state had more than 300 operating landfills; in 1985 there were about 125 landfills, with 12 of them accepting more than 90% of the 11.5 million tons of solid waste generated annually. Some counties and municipalities were implementing recycling programs in 1985, and the state legislature was considering a bill to make recycling mandatory.

New Jersey's toxic waste cleanup program is among the most serious in the US. In 1985, 97 hazardous waste sites—more than in any other state—were listed as national priorities for cleanup with federal Superfund financing. Four New Jersey locations, including the worst site in the nation (a landfill near Pitman), were listed among the top 10 on the Superfund list. The state's cleanup plan for hazardous waste sites was funded with more than $300 million in Superfund, state, and responsible-party moneys between fiscal years 1980 and 1985, more than in any other state.

The New Jersey Spill Compensation Fund was established by the state legislature in 1977 and amended in 1980. A tax based on the transfer of hazardous substances and petroleum and petroleum products is paid into the fund and used for the cleanup of spills. In 1985, a Hazardous Site Mitigation Cleanup Fund of $250 million was established to finance toxic waste cleanup.

New Jersey was the first state to begin a statewide search for sites contaminated by dioxin, a toxic by-product in the manufacture of herbicides. In June 1983, a former Agent Orange manufacturing plant in Newark was found to be contaminated with dioxin; since that time, six other sites throughout the state have been discovered. (Agent Orange, a defoliant, was used by US military forces in Viet-Nam during the 1960s.)

New Jersey first acquired land for preservation purposes in 1907. Since 1961, the state has bought more than 200,000 acres (81,000 hectares) under a "Green Acres" program for conservation and recreation. In 1984, an $83-million Green Trust Fund was established to expand land acquisition. The US Congress designated 1.1 million acres (445,000 hectares) in the southern part of the state as the Pinelands National Reserve in 1978. Beginning in the following year, the state purchased more than 36,000 acres (15,000 hectares) in the Pinelands, bringing the state open-space holdings there to more than 250,000 acres (100,000 hectares).

⁶POPULATION

New Jersey ranked 9th among the 50 states in the 1980 census, with a total population of 7,364,823. It remained 9th in 1985, as its population edged upward to 7,509,625. This yielded an average density of 1,006 per sq mi (388 per sq km), about 15 times the national average, making New Jersey the most densely populated state.

Sparsely populated at the time of the Revolutionary War, New Jersey did not pass the 1 million mark until the 1880 census. Most of the state's subsequent growth came through migration, especially from New York during the period after 1950, when the New Jersey population had stood at 4,835,329. The most significant population growth came in older cities in northern New Jersey and in commuter towns near New York and Philadelphia. The average annual population growth declined from 2.3% in the 1950s to 1.7% in the 1960s and to 0.3% in the 1970s. Net migration was no longer a growth factor, as the state actually experienced a net loss from migration of 275,000 during the 1970s. New Jersey ranked 44th in percentage of population increase between 1970 and 1980. Between 1980 and 1985, growth continued at 0.4% a year, but it was projected that the state would record virtually no change in population between 1985 and 1990.

LOCATION: 38°55′40″ to 41°21′23″N; 73°59′50″ to 75°35W. **BOUNDARIES:** New York line, 108 mi (174 km); Atlantic Ocean coastline, 130 mi (209 km); Delaware line, 78 mi (126 km); Pennsylvania line, 164 mi (264 km).

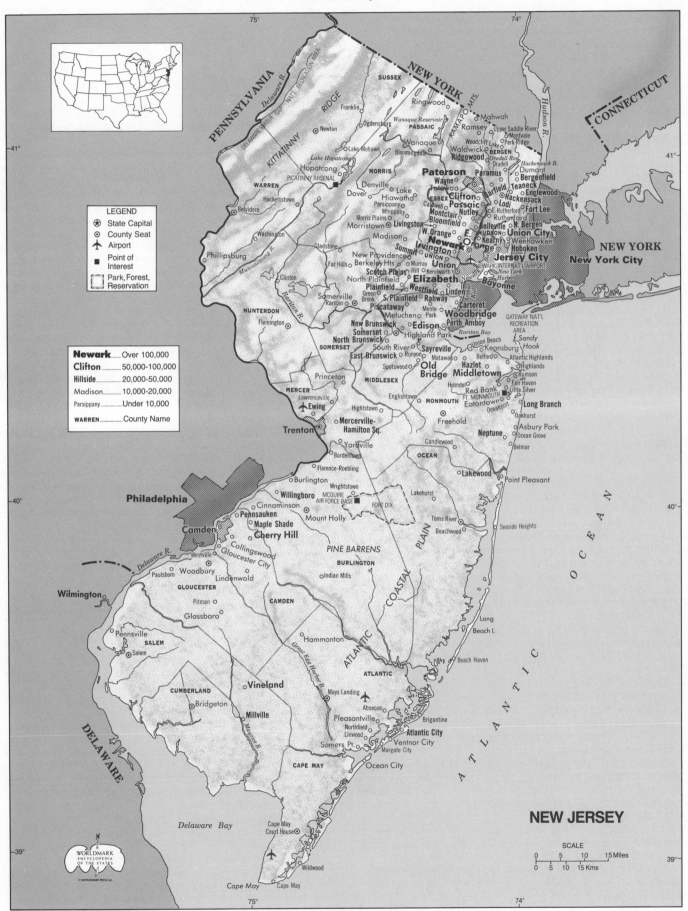

NEW JERSEY

LEGEND

- State Capital
- County Seat
- Airport
- Point of Interest
- Park, Forest, Reservation

Newark............Over 100,000
Clifton............50,000-100,000
Hillside............20,000-50,000
Madison............10,000-20,000
Parsippany............Under 10,000
WARREN............County Name

SCALE

0   5   10   15 Miles

0   5   10   15 Kms

WORLDMARK
ENCYCLOPEDIA
OF THE STATES
© WORLDMARK PRESS Ltd.

See endsheet maps: M2.

The state's entire population is classified as living in metropolitan areas, a distinction claimed by no other state. Ironically, most of the state's major cities have declined in population, with Newark, the state's largest city, changing from 329,248 to 314,387 between 1980 and 1985. Newark's population was 58% black in 1980, the 5th-highest percentage of blacks among major US cities. Populations of other New Jersey cities in 1984 were Jersey City, 223,004; Paterson, 138,818; Elizabeth, 107,455; and Trenton, 92,052.

Parts of New Jersey are included in the New York City and Philadelphia metropolitan areas, the nation's largest and 4th largest, respectively.

7ETHNIC GROUPS

New Jersey is one of the most ethnically heterogeneous states. As of 1980, 757,822 New Jerseyites were of foreign birth. The leading countries of origin were Italy, 11%; Germany, 5%; England, 4%; and Poland, 3%. The foreign-born continue to come, creating new subcommunities of Vietnamese, Soviet Jews, Caribbean blacks, and Hispanic Americans.

Blacks first came to New Jersey as slaves in the 1600s; the state abolished slavery in 1804, one of the last of the northern states to do so. Today black people constitute the state's largest (13%) ethnic minority, 924,786 as of 1980. Newark, 58% black in 1982, elected its first black mayor, Kenneth Gibson, in 1970, three years after the city was torn by racial disorders that killed 26 people and injured some 1,500 others. In 1985, Gibson was in his fourth term as mayor.

There were 388,000 Hispanic Americans in 1980, divided into distinct ethnic communities. The Puerto Rican population, which increased from 55,361 in 1960 to 157,292 in 1980, lived mostly in Newark, Jersey City, Elizabeth, Paterson, and Passaic. There were 56,206 Cubans in 1980, many of them in Union City and Elizabeth; their numbers were augmented by the migration of Cuban refugees in 1980. Smaller Spanish-speaking groups included Colombians and Dominicans.

The number of Asians living in New Jersey quadrupled during the 1970s, to 103,842. The largest group of Asians is from India (29,507 in 1980). There were 24,377 Filipinos, 23,366 Chinese, 12,845 Koreans, and 9,905 Japanese.

American Indians numbered 8,176 in 1980. Among the state's Indians is a group claiming to be descended from Dutch settlers, black slaves, British and German soldiers, and Leni-Lenape and Tuscarora Indians; incorporated as the Ramapough Mountain Indians in 1978, they live in the Ramapo hills near Ringwood and Mahwah.

8LANGUAGES

European settlers found New Jersey inhabited largely by the Leni-Lenape Indians, whose legacy can still be found in such place-names as Passaic, Totowa, Hopatcong, Kittatinny, and Piscataway.

English in New Jersey is rather evenly divided north and south between Northern and Midland dialects. Special characteristics of some New York metropolitan area speech occur in the northeast portion, such as the absence of /r/ after a vowel, a consonant like /d/ or /t/ instead of the /th/ sounds in *this* or *thin*, and pronunciation of *bird* almost as if it were /boyd/ and of *won't* as /woont/. Limited chiefly to the northern half of the state are such pronunciations as *coop* rhyming with *stoop*, *food* with *good*, and *goal* and *fool*; *faucet* has the vowel of *father*. Dominant in the southern half are *run* (small stream), *baby coach* (baby carriage) in the Philadelphia trading area, *winnering owl* (screech owl), and *eel worm* (earthworm). Heard also are *out* as /aot/, *muskmelon* as /muskmillon/, and *keg* rhyming with *bag*, *scarce* with *fierce*, *spook* with *book*, and *haunted* with *panted*.

In 1980, 5,959,871 New Jerseyites—84% of the resident population 3 years old or older—spoke only English at home. Other languages spoken at home were:

Spanish	431,288	Greek	28,036
Italian	195,508	Hungarian	21,940
Polish	78,306	Asian Indian	21,157
German	72,941	Chinese	18,895
Portuguese	38,248	Yiddish	18,673
French	33,338		

9RELIGIONS

With a history of religious tolerance, New Jersey has welcomed many denominations to its shores. Dutch immigrants founded a Reformed Church in 1662, the first in the state. After the English took control, Puritans came from New England and Long Island, Congregationalists from Connecticut, and Baptists from Rhode Island. Quaker settlements in Shrewsbury and western New Jersey during the early 1670s predated the better-known Quaker colony in Pennsylvania. Episcopalians, Presbyterians, German Lutherans, and Methodists arrived during the 18th century. The state's first synagogue was established in 1848, in Newark.

About the only religion not tolerated by New Jerseyites was Catholicism; the first Catholic parish was not organized until 1814, and laws excluding Catholics from holding office were on the books until 1844. The Catholics' numbers swelled as a result of Irish immigration after 1845, and even more with the arrival of Italians after 1880. Today, Roman Catholics constitute the state's single largest religious group, with a population of 3,049,769 in 1984. Passaic is the headquarters of the Byzantine-Ruthenian Rite in the Byzantine Catholic Church.

The Jewish population was estimated at 433,475 in 1984. About 49% of all Jews lived in Bergen and Essex counties. The largest Protestant denomination in 1980 was the United Methodist, with 183,608 adherents, followed by the United Presbyterian Church, with 174,029; Episcopal, 146,990; Lutheran Church in America, 88,739; American Baptist Convention, 73,691; and Reformed Church in America, 54,350.

10TRANSPORTATION

Ever since the first traders sought the fastest way to get from New York to Philadelphia, transportation has been of central importance to New Jersey and has greatly shaped its growth. In the mid-1820s, Hoboken engineer John Stevens built the first steam locomotive operated in the US; over the protests of the dominant stagecoach operators, his son Robert obtained a charter in 1830 for the Camden and Amboy Railroad. The line opened in 1834, and six years later it held a monopoly on the lucrative New York–Philadelphia run. Other lines—such as the Elizabeth and Somerville, the Morris and Essex, the Paterson and Hudson, and the Jersey Central—were limited to shorter runs, largely because the Camden and Amboy's influence with the legislature gave it a huge competitive advantage. Camden and Amboy stock was leased to the Pennsylvania Railroad in 1871, and the ensuing controversy over whether New Jersey transit should be entrusted to an "alien" company led to the passage of a law opening up the state to rail competition. Industry grew around the rail lines, and the railroads became a vital link in the shipment of products from New York and northern New Jersey.

As of 1985, the major freight operations were run by Conrail, representing a consolidation of several bankrupt freight carriers, including the Penn Central, Central of New Jersey, Erie-Lackawanna, Reading, Lehigh Valley, and Lehigh and Hudson River railroads. Amtrak trains linking Newark, Trenton, and a few other New Jersey cities along the main eastern rail corridor served 1,806,857 riders in 1983/84. But the bulk of interstate passenger traffic consists of commuters to New York and Philadelphia on trains operated by the Port Authority of New York and New Jersey (PA) and the Port Authority Transit Corp. (PATCO), a subsidiary of the Delaware River Port Authority.

New Jersey Transit was created in 1979 by the state legislature. The agency took over commuter rail service from Conrail in 1983 and also took over the operations of four private bus companies.

New Jersey Transit provides commuter rail service on 10 lines in 12 counties on 380.4 mi (612.2 km) of track. About 490 cars are in service on an average day, and they were used by 34.5 million passengers in 1984. There are 140 rail stations throughout the system. New Jersey Transit also provides commuter bus service in 20 of the state's 21 counties along 180 different routes. The fleet of buses numbers 1,834, and they were used by 131 million passengers in 1984. Some 450,000 persons use the bus and rail systems each day. In 1984, the American Public Transit Association named New Jersey Transit the best transit agency in the United States in its class.

Although associated more with the West, the first stagecoach service began in New Jersey, as part of a New York–Philadelphia trek that took some five days in 1723. For a time, colonial law required towns along the way to provide taverns for the passengers, and it was not uncommon for coach operators who were also tavern owners to find some way to prolong the journey an extra night. They traveled on roads that were barely more passable than the Leni-Lenape trails from which they originated. Improvement was slow, but by 1828, the legislature had granted 54 turnpike charters.

Road building has continued ever since. In 1983 there were 33,077 mi (53,232 km) of roads in the state, including 381 mi (613 km) of state toll roads, 2,217 mi (3,568 km) of state highways, 6,803 mi (10,948 km) of county highways, 23,173 mi (37,293 km) of municipal roads and streets, 490 mi (789 km) of state park, forest, and institutional roads, and 13 mi (21 km) of national park roads. The major highways are the New Jersey Turnpike, opened in 1952 and extending 133 mi (214 km) between Bergen and Salem counties, and the Garden State Parkway, completed in 1955 and stretching 173 mi (278 km) from the New York State line to Cape May. More than 50 billion vehicle-mi (80 billion vehicle-km) were traveled on state roads in 1981, and each licensed New Jersey driver covered 9,957 mi (16,024 km) on average. New Jersey highways are congested; on average, 4,073 vehicles used each mile of the state's highways every day in 1980, compared to a US average of 1,053. Nonetheless, fatalities per 100 million miles driven in 1981 stood at 2.26, 30% below the US average.

Many bridges and tunnels link New Jersey with New York State, Pennsylvania, and Delaware. The Holland Tunnel, built in 1927 and operated by the PA, carried 13,346,000 eastbound vehicles from Jersey City to Manhattan in 1984; the Lincoln Tunnel, built between 1937 and 1957 at Weehawken, was used by 19,477,000 eastbound vehicles. More than double that number, 46,036,000 entered New York City via the George Washington Bridge from Ft. Lee. The PA also operates three bridges from Perth Amboy, Elizabeth, and Bayonne to Staten Island. Twenty-seven bridges cross the Delaware River, connecting New Jersey with Pennsylvania and Delaware.

At the gateway to New York Harbor, ports at Elizabeth and Newark have overtaken New York City ports in cargo volume, and contribute greatly to the local economy. Operated by the PA, Port Newark has almost 4 mi (6 km) of berthing space along Newark Bay, while nearby Port Elizabeth, with better than 3 mi (5 km) of berths, is a major handler of containerized cargo. These two terminals handled 12,774,532 long tons of cargo in 1984. Privately owned ports are in use in Jersey City and Bayonne. The Ameriport System, centered in Philadelphia, operates facilities between Trenton and Wilmington, Del. Two terminals near Camden are operated by the South Jersey Port Corp.

The state's early aviation centers were Lakehurst and Newark. Lakehurst, whose dirigible operations attracted crowds of spectators during the 1920s and 1930s, was the scene of the crash on 6 May 1937 of the *Hindenburg*, a disaster that killed 36 people and spelled the end of commercial airship flights in the US. The state's first airmail service began in 1924 from New Brunswick's Hadley Field. Newark Airport in the late 1920s billed itself as the busiest air terminal in the world. The PA took over its operation from the city of Newark in 1948. Rebuilt during the 1960s and 1970s, Newark International Airport has become the state's busiest by far; it handled 23,654,000 passengers and 228,000 long tons of cargo in 1983. Statewide in 1983 there were 118 airports and 161 heliports.

[11] HISTORY

The first known inhabitants of what is now New Jersey were the Leni-Lenape (meaning "Original People"), who arrived in the land between the Hudson and Delaware rivers about 6,000 years ago. Members of the Algonkian language group, the Leni-Lenape were an agricultural people supplementing their diet with freshwater fish and shellfish. The peace-loving Leni-Lenape believed in monogamy, educated their children in the simple skills needed for wilderness survival, and clung rigidly to a tradition that a pot of food must always be warm on the fire to welcome all strangers.

The first European explorer to reach New Jersey was Giovanni da Verrazano, who sailed into what is now Newark Bay in 1524. Henry Hudson, an English captain sailing under a Dutch flag, piloted the *Half Moon* along the New Jersey shore and into Sandy Hook Bay in the late summer of 1609, a voyage that established a Dutch claim to the New World. Hollanders came to trade in what is now Hudson County as early as 1618, and in 1660, they founded New Jersey's first town, called Bergen (now part of Jersey City). Meanwhile, across the state, Swedish settlers began moving east of the Delaware River in 1639. Their colony of New Sweden had only one brief spurt of glory, from 1643 to 1653, under Governor Johan Printz.

The Leni-Lenape lost out to the newcomers, whether Dutch, Swedish, or English, despite a series of treaties that the Europeans thought fair. State and local records describe these agreements: huge tracts of land exchanged for trinkets, guns, and alcohol. The guns and alcohol, combined with smallpox (another European import), doomed the "Original People." In 1758, when a treaty established an Indian reservation at Brotherton (now the town of Indian Mills), only a few hundred Indians remained.

England assumed control in March 1664, when King Charles II granted a region from the Connecticut River to the Delaware River to his brother James, the duke of York. The duke, in turn, deeded the land between the Hudson and Delaware rivers, which he named New Jersey, to his court friends John Berkeley, 1st Baron Berkeley of Stratton, and Sir George Carteret, on 23 June 1664. Lord Berkeley and Sir George became proprietors, owning the land and having the right to govern its people. Subsequently, the land passed into the hands of two boards of proprietors in two provinces called East Jersey and West Jersey, with their capitals in Perth Amboy and Burlington, respectively. East Jersey was settled mainly by Puritans from Long Island and New England, West Jersey by Quakers from England. The split cost the colony dearly in 1702, when Queen Anne united East and West Jersey but placed them under New York rule. The colony did not get its own "home rule" until 1738, when Lewis Morris was named the first royal governor.

By this time, New Jersey's divided character was already established. Eastern New Jersey looked toward New York, western New Jersey toward Philadelphia. The level plain connecting those two major colonial towns made it certain that New Jersey would serve as a pathway. Along the makeshift roads that soon crossed the region—more roads than in any other colony—travelers brought conflicting news and ideas. During the American Revolution, the colony was about equally divided between Revolutionists and Loyalists. William Franklin (illegitimate son of Benjamin Franklin), royal governor from 1763 until 1776, strove valiantly to keep New Jersey sympathetic to England, but failed and was arrested. Throughout the Revolutionary period, he remained a leading Loyalist; after the war, he left for England.

Franklin's influence caused New Jersey to dally at first over independence, but in June 1776, the colony sent five new delegates to the Continental Congress—Abraham Clark, John Hart, Frances Hopkinson, Richard Stockton, and the Reverend John Witherspoon—all of whom voted for the Declaration of Independence. Two days before the Declaration was proclaimed, New Jersey adopted its first state constitution. William Livingston, a fiery anti-British propagandist, was the first elected governor of the state.

New Jersey played a pivotal role in the Revolutionary War, for the side that controlled both New York and Philadelphia would almost certainly win. George Washington and his battered troops made their winter headquarters in the state three times during the first four years of the war, twice in Morristown and once in Somerville. Five major battles were fought in New Jersey, the most important being the Battle of Trenton on 26 December 1776 and the Battle of Monmouth on 28 June 1778. At war's end, Princeton became the temporary capital of the US from 26 June 1783 to 4 November 1783.

The state languished after the Revolution, with many of its pathway towns ravaged by the passing of competing armies, its trade dependent on New York City, and its ironworks (first established in 1676) shut down because of decreased demand. The state's leaders vigorously supported a federation of the 13 states, in which all states, regardless of size, would be represented equally in one national legislative body. This so-called New Jersey Plan led to the establishment of the US Senate.

Railroads and canals brought life to the state in the 1830s and set it on a course of urbanization and industrialization. The 90-mi (145-km) Morris Canal linked northern New Jersey with the coalfields of Pennsylvania. Considered one of the engineering marvels of the 19th century, the canal rose to 914 feet (279 meters) from sea level at Newark Bay to Lake Hopatcong, then fell 760 feet (232 meters) to a point on the Delaware River opposite Easton, Pa. Old iron mines beside the canal found markets, the dyeing and weaving mills of Paterson prospered, and Newark, most affected by the emerging industries, became the state's first incorporated city in 1836. Another canal, the Delaware and Raritan, crossed the relatively flat land from Bordentown to New Brunswick. Along this waterway, Bordentown, Trenton, and New Brunswick boomed. Princeton, whose leaders fought to keep the canal away from the town, settled into a long existence as a college community built around the College of New Jersey, founded in Elizabeth in 1746 and transferred to Princeton in 1756.

The canals were doomed by railroad competition almost from the start. The Morris Canal was insolvent long before World War I, and the Delaware Canal, although operative until 1934, went into a long, slow decline after the Civil War. The first railroad, from Bordentown to South Amboy, closely paralleled the Delaware and Raritan Canal and in 1871 became an important part of the Pennsylvania Railroad. The coal brought in on railroad cars freed industry from waterpower; factories sprang up wherever the rails went. The Hudson County waterfront, eastern terminus for most of the nation's railway systems, became the most important railroad area in the US. Rail lines also carried vacationers to the Jersey shore, building an important source of income for the state.

The Civil War split New Jersey bitterly. Leaders in the Democratic Party opposed the war as a "Black Republican" affair. Prosperous industrialists in Newark and Trenton feared that their vigorous trade with the South would be impaired, Cape May hotelkeepers fretted about the loss of tourists from Virginia, and even Princeton students were divided. As late as the summer of 1863, after the Battle of Gettysburg, many state "peace Democrats" were urging the North to make peace with the Confederacy. Draft calls were vigorously opposed in 1863, yet the state sent its full quota of troops into service throughout the conflict. Most important, New Jersey factories poured forth streams of muni-

tions and other equipment for the Union army. At war's end, political leaders stubbornly opposed the 13th, 14th, and 15th Amendments to the US Constitution, and blacks were not permitted to vote in the state until 1870.

During the last decades of the 19th century, New Jersey developed a reputation for factories capable of making the components necessary for thousands of other manufacturing enterprises. Few factories were large, although in 1873, Isaac M. Singer opened a huge sewing machine plant at Elizabeth that employed 3,000 persons. Oil refineries on the Hudson County waterfront had ever-expanding payrolls, pottery firms in Trenton thrived, and Newark gained strength from many diversified manufactures and also saw its insurance companies become nationally powerful.

Twentieth-century wars stimulated New Jersey's industries. During World War I, giant shipyards at Newark, Kearny, and Camden made New Jersey the nation's leading shipbuilding state. The Middlesex County area refined 75% of the nation's copper, and nearly 75% of US shells were loaded in the state. World War II revived the shipbuilding and munitions industries, while chemical and pharmaceutical manufacturing, spawned by the World War I cutoff of German chemicals, showed further growth during the second world conflict. Paterson, preeminent in locomotive building during the 19th century, became the nation's foremost airplane engine manufacturing center. Training and mobilization centers at Ft. Dix and Camp Kilmer moved millions of soldiers into the front lines.

The US Census Bureau termed New Jersey officially "urban" in 1880, when the state population rose above 1 million for the first time. Urbanization intensified throughout the 20th century and especially after World War II, as people left the old cities in New Jersey and other northeastern states to buy homes in developments on former farmlands. Places like Cherry Hill, Woodbridge, Clifton, and Middletown Township have boomed since 1945, increasing their population as much as sixfold. New Jersey has also experienced many of the problems of urbanization. Its cities have declined; traffic congestion is intense in the morning, when commuters stream into urban areas to work, and again in the evening, when they return home to what once was called "the country." That country now knows the problems of urban growth: increased needs for schools, sewers, police and fire protection, and road maintenance, along with rising taxes.

The state has not surrendered to its problems, however. In 1947, voters overwhelmingly approved a new state constitution, a terse, comprehensive document that streamlined state government, reformed the state's chaotic court system, and mandated equal rights for all. Governor Alfred E. Driscoll promptly integrated the New Jersey National Guard, despite strong federal objections; integration of all US armed forces soon followed. Voters since 1950 have passed a wide variety of multimillion-dollar bond issues to establish or rebuild state colleges. Rutgers, the state university, has been rapidly expanded. Funds have been allocated for the purchase and development of new park and forest lands. Large bond issues have financed the construction of highways, reservoirs, and rapid transit systems.

Since the end of World War II, New Jersey has had no predictable political pattern. It gave huge presidential majorities to Republican Dwight D. Eisenhower and Democrat Lyndon B. Johnson, narrowly supported Democrat John F. Kennedy, favored Republican Gerald Ford over Democrat Jimmy Carter by a small margin, and gave two big majorities to Ronald Reagan. For more than 20 years, the state's two US senators, Clifford B. Case (R) and Harrison A. Williams (D), were recognized as likeminded liberals. Democrat Bill Bradley, former Princeton University and New York Knickerbockers basketball star, was elected to Case's seat in 1978, after Case's own party repudiated him in the primary. Williams resigned from the Senate in 1982 after his conviction in the Abscam scandal; in elections held later that year, Democrat Frank R. Lautenberg won that seat with 51% of the vote.

[12] STATE GOVERNMENT

New Jersey's first state constitution took effect in 1776. A second constitution was written in 1844, and a third in 1947. This last document, as amended, continues to govern the state today.

The state legislature consists of a 40-member senate and an 80-member general assembly. Senators, elected to four-year terms, must be at least 30 years of age and have been New Jersey residents for four years and district residents for a year. Assembly members, elected to two-year terms, must be at least 21 years of age and have been New Jersey residents for two years and district residents for a year. Both houses of the legislature meet in unlimited annual sessions.

New Jersey is one of only two states—the other is Maine—in which the governor is the only elective administrative official. Given broad powers by the state constitution, the governor appoints the heads, or commissioners, of the 20 major state departments with the advice and consent of the senate; not subject to senate approval are more than 500 patronage positions. The governor is also commander-in-chief of the state's armed forces, submits the budget to the legislature each January, presents an annual message on the condition of the state, and may grant pardons and, with the aid of the Parole Board, grant executive clemency. Elected to a four-year term in the odd-numbered year following the presidential election, the governor may run for a second term but not for a third until four years have passed. A candidate for governor must be at least 30 years old and must have been a US citizen for 20 years and a New Jersey citizen for 7 years in order to qualify for the ballot.

A bill may be introduced in either house of the legislature. Once passed, it goes to the governor, who may sign it, return it to the legislature with recommendations for change, or veto it in its entirety. A two-thirds majority in each house is needed to override a veto.

Amendments to the state constitution may originate in either house. If, after public hearings, both houses pass the proposal by a three-fifths vote, the amendment is placed on the ballot at the next general election. If approved by a majority, but by less than a three-fifths vote in both houses, the amendment is referred to the next session of the legislature, at which time, if again approved by a majority, it is placed on the ballot. The amendment goes into effect 30 days after ratification by the electorate.

To vote in New Jersey, one must be at least 18 years old, a US citizen, and a New Jersey resident for 30 days before the election.

[13] POLITICAL PARTIES

From the 1830s through the early 1850s, Democrats and Whigs dominated the political life of New Jersey. Exercising considerable, though subtle, influence in the decade before the Civil War was the Native American (Know-Nothing) Party, an anti-immigrant, anti-Catholic group that won several assembly and senate seats. Wary of breaking ties with the South and ambivalent about the slavery issue, New Jerseyites, especially those in Essex and Bergen counties, did not lend much support to the abolitionist cause. Early Republicans thus found it advantageous to call themselves simply "Opposition"; the state's first Opposition governor was elected in 1856. Republicans controlled the state for most of the 1860s; but with heavy support from business leaders, the Democrats regained control in 1869 and held the governorship through 1896. They were succeeded by a series of Progressive Republican governors whose efforts were largely thwarted by a conservative legislature. Sweeping reforms—including a corrupt-practices act, a primary election law, and increased support for public education—were implemented during the two years that Woodrow Wilson, a Democrat, served as governor before being elected to the presidency. Between 1913 and 1985, Democrats held the statehouse almost two-thirds of the time.

As of 1982 there were 3,681,211 registered voters. Balloting in 1983 left the assembly with 44 Democrats and 36 Republicans, and the senate with 23 Democrats and 17 Republicans. In the

New Jersey Presidential Vote by Political Parties, 1948–84

YEAR	ELECTORAL VOTES	NEW JERSEY WINNER	DEMOCRAT	REPUBLICAN	PROGRESSIVE	SOCIALIST	PROHIBITION	SOCIALIST LABOR	SOCIALIST WORKERS
1948	16	Dewey (R)	895,455	981,124	42,683	10,521	10,593	3,354	5,825
1952	16	*Eisenhower (R)	1,015,902	1,373,613	5,589	8,593	—	5,815	3,850
1956	16	*Eisenhower (R)	850,337	1,606,942	CONSTITUTION 5,317	—	9,147	6,736	4,004
1960	16	*Kennedy (D)	1,385,415	1,363,324	CONSERVATIVE 8,708	—	—	4,262	11,402
1964	17	*Johnson (D)	1,867,671	963,843	—	—	—	7,075	8,181
1968	17	*Nixon (R)	1,264,206	1,325,467	AMERICAN IND. 262,187	PEACE & FREEDOM 8,084	—	6,784	8,667
1972	17	*Nixon (R)	1,102,211	1,845,502	—	PEOPLE'S 5,355	AMERICAN 34,378	4,544	2,233
1976	17	Ford (R)	1,444,653	1,509,688	7,716	US LABOR 1,650	LIBERTARIAN 9,449	3,686	COMMUNIST 1,662
1980	17	*Reagan (R)	1,147,364	1,546,557	CITIZENS 8,203	—	20,652	2,198	2,555
1984	16	*Reagan (R)	1,261,323	1,933,630	—	WORKERS WORLD 8,404	6,416	—	1,564

*Won US presidential election.

1984 presidential voting, Ronald Reagan swamped his Democratic opponent, Walter Mondale, in New Jersey by a margin of 60% to 39%. In 1985, the governor, Thomas Kean, was a Republican. The Democrats had an 8–6 majority among New Jersey's 14 US representatives.

In 1984, 196 elected New Jersey officials were black; this was 2.1% of the statewide total of elected officials, the 3d-highest percentage for any state outside the South. Three New Jersey cities with more than 50,000 population had black mayors: Newark (Kenneth Gibson), Camden (Melvin Primas, Jr.), and East Orange (Thomas Cooke, Jr.). Three women sat in the state senate and eight in the assembly. Marge Roukema was the only female US representative from New Jersey.

New Jersey's unenviable reputation for corruption in government dates back at least to 1838, when ballot tampering resulted in the disputed election of five Whigs to the US House of Representatives. (After a House investigation, the Whigs were barred and their Democratic opponents given the seats.) Throughout the rest of the century, corruption was rampant in local elections: Philadelphians, for example, were regularly imported to vote in Atlantic City elections, and vote buying was a standard election-day procedure in Essex and Hudson counties. Wilson's 1911 reform bill eliminated some of these practices, but not the bossism that had come to dominate big-city politics. Frank Hague of Jersey City controlled patronage and political leaders on the local, state, and national level from 1919 to 1947; during the 1960s and 1970s, Hague's successor John V. Kenny, Jersey City mayor Thomas Whelan, and Newark mayor Hugh Addonizio, along with numerous other state and local officials, were convicted of corrupt political dealings. From 1969 to mid-1975, federal prosecutors indicted 148 public officials, securing 72 convictions. Brendan Byrne, who had never before held elective office, won the governorship in 1973, mainly on the strength of a campaign that portrayed him as the "judge who couldn't be bought." On the national level, New Jersey Representative Peter Rodino gained a reputation for honesty and fairness when he chaired the House Judiciary Committee's impeachment hearings against Richard Nixon. However, the state's image suffered a further blow in 1980, when, as a result of the FBI's "Abscam" investigation, charges of influence peddling were brought against several state officials, including members of the Casino Control Commission, whose function was to prevent corruption and crime in Atlantic City's gambling establishments. Later in the year, New Jersey Democrat Harrison Williams became the nation's first US senator to be indicted, on charges of bribery and conspiracy, as a result of the Abscam probe. He was convicted in 1981 and sentenced to prison. As a result of the same investigation, US Representative Frank Thompson, Jr., was convicted in 1980 on bribery and conspiracy charges. A New Jerseyite, Raymond Donovan, was named secretary of labor by President Ronald Reagan in 1981, but he resigned in 1985 after being indicted late in 1984 for allegedly seeking to defraud the New York City Transit Authority while serving as vice president of the Schiavone Construction Company in Secaucus.

14LOCAL GOVERNMENT

As of 1984, New Jersey had 21 counties, 53 cities, 256 boroughs, 233 townships, 21 towns, 3 villages, and 380 special districts.

Counties are classified by population and location. First-class counties (2 in 1984) have populations exceeding 600,000; second-class counties (11), populations of 200,000–600,000; third-class counties (6), populations of 50,000–200,000 but no Atlantic shore; fourth-class counties (0), under 50,000 population, no Atlantic shore; fifth-class counties (1), population more than 100,000, Atlantic shore; and sixth-class counties (1), less than 100,000 population, Atlantic shore. These classes determine the number of members on the main county governing body, the board of freeholders, which may range from three to nine. Elected to

staggered three-year terms, the freeholders administer county and state programs. Under the Optional County Charter Law of 1972, four counties have an elected county executive, and one county has an appointed county manager. Other county officers are the clerk, sheriff, surrogate, prosecutor, boards of election and taxation, county counsel, administrator, medical examiner, chief probation officer, and jury commissioner.

Cities, boroughs, and towns may employ the mayor-council system, council-manager system, commission system, or other forms of their own devising. Most townships and villages are governed by committee or by a council and a mayor with limited powers. Cities, too, are classed by population and location: first-class cities are those over 150,000 in population; second-class, 12,000–150,000; third-class, all others except ocean resorts; and fourth-class, ocean resorts.

The budgets of all local units are supervised by the New Jersey Department of Community Affairs, which also offers municipal aid programs. By state law, all local budgets must be balanced, and budgetary increases ("budget caps") are limited to 5% a year for most items.

15STATE SERVICES

The constitution of 1947 limited the number of state government departments to 20, and that was the total in 1985. New Jersey in 1974 became the first state to establish a Public Advocate Department, empowered to provide legal assistance for indigent criminal defendants, mental patients, and any citizen with a grievance against a government agency or regulated industry. A Code of Ethics, adopted by the legislature in 1976, seeks to prevent state employees from using their positions for personal gain. By executive order, more than 500 state executive officials must file financial disclosure statements.

The Education Department administers state and federal aid to all elementary and secondary schools, oversees pupil transportation, and has jurisdiction over the state library, museum, and historical commission. State-run colleges and universities and higher education policy are the province of the Department of Higher Education. All state-maintained highways and bus and rail transportation are the responsibility of the Department of Transportation, which also operates New Jersey Transit, whose function is to acquire and operate public transportation services.

The Human Services Department administers welfare, Medicaid, mental health, and mental retardation programs, as well as veterans' institutions and programs and other state-supported social services. Alcohol, drug abuse, and many other health-related programs are monitored by the Health Department, which also oversees hospitals and compiles statewide health statistics.

The Office of the Attorney General, officially titled the Department of Law and Public Safety, is the statewide law enforcement agency. Its functions include criminal justice, consumer affairs, civil rights, alcoholic beverage control, and gaming enforcement; also within this department are the State Police, State Racing Commission, Violent Crimes Compensation Board, and a number of regulatory boards. The Defense Department controls the Army and Air National Guard. Correctional institutions, training schools, treatment centers, and parole offices are administered by the Corrections Department.

The Department of Energy monitors the supply and use of fuel and administers the state master plan for energy use and conservation; its Board of Public Utilities has broad regulatory jurisdiction, ranging from garbage collection to public broadcasting. Other agencies are the departments of agriculture, banking, civil service, commerce and economic development, community affairs, environmental protection, insurance, labor and industry, state, and treasury.

16JUDICIAL SYSTEM

All judges in New Jersey, except municipal court judges, are appointed by the governor with the consent of the senate. Initial terms for supreme and superior court judges are seven years; after reappointment, judges may serve indefinitely.

The supreme court, the state's highest, consists of six associate justices and a chief justice, who is also the administrative head of the state court system. As the court of highest authority, the supreme court hears appeals on constitutional questions and of certain cases from the superior court, which comprises three divisions: chancery, law, and appellate. The chancery division has original jurisdiction over general equity cases, most probate cases, and divorce actions. All other original cases are tried within the law division. The appellate division hears appeals from the chancery and law divisions, from lower courts, and from most state administrative agencies. As of 1984 there were 23 appellate judges and 329 superior court judges. A state tax court, empowered to review local property tax assessments, equalization tables, and state tax determinations, has been in operation since 1979; by statute, it may have from 6 to 12 judges. Municipal court judges, appointed by local governing bodies for three-year terms, hear minor criminal matters, motor vehicle cases, and violations of municipal ordinances.

The legislature approved a sweeping reform of the state's criminal law code in 1978. Strict sentencing standards were established, and one result was an overcrowding of the state's prison system. Governor Brendan Byrne signed a law in 1981 imposing a minimum three-year sentence on anyone committing a crime with a gun. In 1982, Governor Thomas Kean signed legislation establishing a death penalty by lethal injection; New Jersey became the ninth state to use that method.

According to the FBI Crime Index, New Jersey's crime rate as of 1983—5,163 per 100,000 population—was virtually the same as the national average of 5,159. Specific rates included murder and nonnegligent homicide, 5; forcible rape, 30; robbery, 269; aggravated assault, 248; burglary, 1,237; larceny-theft, 2,769; and motor vehicle theft, 604. Prisoners under jurisdiction of state and federal correctional authorities in New Jersey numbered 10,363 at the end of 1984, a 13% increase in one year.

[17] ARMED FORCES

Founded as a training base during World War I, Ft. Dix, near Trenton, remains the largest military base in New Jersey, with 12,248 authorized personnel in 1983/84. Other major army facilities are Ft. Monmouth in Eatontown and the Picatinny Arsenal in Dover. The largest naval facility is the Lakehurst Naval Air Center, with 3,134 authorized personnel in 1983/84. McGuire Air Force Base in Wrightstown had 6,981. Authorized personnel at all New Jersey military installations totaled 48,143 in 1983/84. In addition, the US Coast Guard operates a training center in Cape May. New Jersey firms received nearly $3.3 billion in defense contracts awards (10th in the US) in 1983/84.

Of the 908,000 estimated veterans living in New Jersey on 30 September 1984, World War I veterans numbered 8,000; World War II, 387,000; Korean conflict, 170,000; and Viet-Nam era, 215,000. Veterans' benefits in 1982/83 totaled $573 million, of which $350 million was for compensation and pensions, $146 million for medical services, and $22 million for education and training.

In 1985, 12,423 persons served in the New Jersey Army National Guard and 2,637 in the Air National Guard. There were 27,559 state and local police personnel in 1983.

[18] MIGRATION

New Jersey's first white settlers were intercolonial migrants: Dutch from New Amsterdam, Swedes from west of the Delaware River, and Puritans from New England and Long Island. By 1776, New Jersey's population was about 138,000, of whom perhaps 7% were black slaves.

Population growth lagged during the early 19th century, as discouraged farmers left their worn-out plots for more fertile western soil; farmers in Salem County, for example, went off to found new Salems in Ohio, Indiana, Iowa, and Oregon. Not until the rapid industrial growth of the mid-1800s did New Jersey attract great waves of immigrants. Germans and Irish were the first to arrive, the latter comprising 37% of Jersey City's population by 1870. The late 1800s and early 1900s brought newcomers from Eastern Europe, including many Jews, and a much larger number of Italians to the cities. By 1900, 43% of all Hudson County residents were foreign-born. More recently, migration from Puerto Rico and Cuba has been substantial.

From World War I on, there has been a steady migration of blacks from southern states; Newark's black population grew by 130,000 between 1950 and 1970. Black as well as Hispanic newcomers settled in major cities just as whites were departing for the suburbs. New Jersey's suburbs were also attractive to residents of New York City, Philadelphia, and other adjacent areas, who began a massive move to the state just after World War II; nearly all of these suburbanites were white. From 1940 to 1970, New Jersey gained a net total of 1,360,000 residents. Between 1970 and 1983, however, the state lost an estimated 267,000 residents through migration. While the black, Hispanic, and Asian populations were still rising, whites were departing from New Jersey in increasing numbers.

[19] INTERGOVERNMENTAL COOPERATION

New Jersey participates in such regional bodies as the Interstate Sanitation Commission, Atlantic States Marine Fisheries Commission, and Mid-Atlantic Regional Fisheries Management Council. Of primary importance to the state are its relations with neighboring Pennsylvania and New York. With Pennsylvania, New Jersey takes part in the Delaware Valley Regional Planning Commission, Delaware River Joint Toll Bridge Commission, and Delaware River Port Authority; with New York, the Port Authority of New York and New Jersey, the Palisades Interstate Park Commission, and the Waterfront Commission, established to eliminate corruption and stabilize employment at the Hudson River ports. The Delaware River Basin Commission manages the water resources of the 12,750-sq-mi (33,000-sq-km) basin under the jurisdiction of Delaware, New Jersey, New York, and Pennsylvania. The Delaware River and Bay Authority operates a bridge and ferry between New Jersey and Delaware.

In 1983/84, the state received $2.9 billion in federal assistance, including $143.3 million in general revenue sharing.

[20] ECONOMY

New Jersey was predominantly agricultural until the mid-1800s, when the rise of the railroads stimulated manufacturing in northern New Jersey and opened the Jersey shore to resort development. The steady growth of population in the 1900s fostered the growth of service-related industries, construction, and trade, for which the state's proximity to New York and Philadelphia had long been advantageous.

Manufacturing accounted for about one-fourth of nongovernment employment in 1984. Although petroleum refining, chemicals and pharmaceuticals, food processing, apparel, fabricated metals, electric and electronic equipment, and other machinery are all important, the state is more noteworthy for the diversity of its manufactures than for any dominant company or product. The service sector of the economy, led by wholesale and retail trade, continued to grow rapidly during the early 1980s. The heaviest concentrations of jobs are in and near metropolitan New York and Philadelphia, but employment opportunities in the central and north-central counties have been increasing. Fresh market vegetables are the leading source of farm income.

During the 1970s, New Jersey's economy followed national trends, except that the mid-decade recession was especially severe. Conditions in most areas improved in the latter part of the decade, particularly in Atlantic City, with the construction of gambling casinos and other entertainment facilities. Manufacturing in the central cities declined, however, as industries moved to suburban locations. The 1981/82 recession was followed by an economic boom that was especially pronounced in New Jersey. By 1984, state unemployment was below the national average, as the construction and services industries set the pace. In 1984 alone, employment rose by 116,000, or 3.7%. New Jersey's rate of inflation was 4.9% in 1983/84, compared with 4.3% for the nation.

21INCOME

New Jersey's per capita income of $14,122 in 1983 ranked 3d among the 50 states. The 1984 total personal income of $115.7 billion ranked 8th. Median family income, $30,793, ranked 3d in 1981. Real income rose only 18% between 1970 and 1978, well below the US average, but the state then drew even, and between 1982 and 1984 state income ran ahead of the US average; between 1981 and 1984, it increased 35.1%. Disposable income also surpassed the US average in 1983 and 1984.

In the second quarter of 1984, income by employment category was as follows: wholesale and retail trade, 24%; manufacturing, 23%; services, 22%; government, 16%; transportation, communications, and utilities, 6%; finance, insurance, and real estate, 5%; and construction, 4%.

About 689,000 New Jerseyites were below the federal poverty line in 1979 (2.5% of the US total). In 1982, 51,300 state residents, 3.3% of the US total, were among the nation's top wealth-holders, with gross assets exceeding $500,000.

22LABOR

The civilian labor force was estimated at 3,828,000 in 1984, 57% male and 43% female. About 3,592,000 were employed and 236,000 unemployed, for an overall unemployment rate of 6.2%. By year's end, the unemployment rate had fallen to 5.8%, the lowest level since the early 1970s.

The work force was distributed approximately as follows: clerical workers, 22%; professional and technical workers, 18%; service workers, 14%; nonfarm managers and officers, 11%; craftsmen and kindred workers, 10%; equipment operators (except transport), 9%; salespeople, 7%; nonfarm laborers, 5%; transport equipment operators, 4%; and farm workers, less than 1%.

A federal survey of workers in March 1982 revealed the following nonfarm employment pattern in New Jersey:

	ESTABLISH-MENTS	EMPLOYEES	ANNUAL PAYROLL ('000)
Agricultural services, forestry, fishing	2,078	7,452	$ 103,883
Mining	159	3,819	114,729
Contract construction	13,581	103,804	2,279,297
Manufacturing, of which:	14,206	756,736	15,988,233
Chemicals and allied products	(939)	(93,787)	(2,265,270)
Nonelectrical machinery	(1,983)	(64,811)	(1,293,476)
Electric, electronic equipment	(831)	(77,682)	(1,602,289)
Transportation, utilities	6,923	186,538	4,537,534
Wholesale trade	14,315	224,986	4,972,990
Retail trade	40,947	474,334	4,675,856
Finance, insurance, real estate	13,301	167,677	2,858,417
Services	49,818	633,029	9,181,308
Other	3,309	3,170	91,210
TOTALS	158,637	2,561,545	$44,803,457

Government workers, not included in this survey, numbered 318,000 local and 100,000 state employees in 1983. About 72,000 persons were federal employees in 1982. Several migrant work camps are located near south Jersey tomato farms and fruit orchards, but the number of farm workers coming into the state is declining with the increased use of mechanical harvesters.

The state's first child labor law was passed in 1851, and in 1886, workers were given the right to organize. Labor's gains were slow and painful, however. In Paterson, no fewer than 137 strikes were called between 1881 and 1900, every one of them a failure. A 1913 strike of Paterson silkworkers drew nationwide headlines but, again, few results. Other notable strikes were a walkout at a Carteret fertilizer factory in 1915 during which six picketers were killed by guards, a yearlong work stoppage by Passaic textile workers in 1926, and another Paterson silkworkers' strike in 1933, this one finally leading to union recognition and significant wage increases. That year, the state enacted a law setting minimum wages and maximum hours for women. This measure was repealed in 1971, in line with the trend toward nonpreferential labor standards. As of 1980, 784,000 New Jerseyites belonged to labor unions, of which there were 1,583 in 1983.

23AGRICULTURE

Although New Jersey is a leading producer of fresh market fruits and vegetables, its total farm income was only $543 million in 1983, 39th among the 50 states. According to the 1982 Census of Agriculture, New Jersey ranked 2d in the US in the production of cultivated blueberries, 4th in peaches, 5th in tomatoes and in lettuce, 6th in snap beans, 11th in apples, and 12th in fruit, nuts, and berries.

Some 970,000 acres (393,000 hectares) were in 9,400 farms in 1984. There were 28,121 hired farm workers in 1982. The major farm counties are Hunterdon for grain; Gloucester and Cumberland for vegetables, fruits, and berries; Salem for vegetables; Atlantic for fruits and berries; and Monmouth for potatoes and nursery products.

In 1983, New Jersey produced 92,400 tons of fresh market vegetables, worth $28,692,000. Leading crops (in hundredweight units) were sweet corn, 630,000; tomatoes, 542,000; and lettuce, 260,000. New Jersey farmers also produced 69,000 tons of tomatoes for processing and 11,280 tons of snap beans. Fruit crops in 1983 (in lb) included apples, 100,000,000, and peaches, 95,000,000. In 1982, 30,521,114 lb of blueberries, 233,000 hundredweight of cranberries, and 50,000 hundredweight of strawberries were also produced.

The expansion of housing and industry has increased the value of farm acreage throughout the state; since World War II, many farmers have found it more profitable to sell their land than to keep it in crops. The state government sought to slow this trend through tax incentives during the 1970s.

24ANIMAL HUSBANDRY

With cash receipts estimated at $136 million in 1983, New Jersey does not rank as a major livestock-producing state. At the close of 1984 there were 104,000 cattle and calves, 45,000 hogs and pigs, and 11,000 sheep and lambs on New Jersey farms.

Statewide meat production figures for 1983 included cattle, 25,320,000 lb; hogs and pigs, 14,305,000 lb; and sheep and lambs, 664,000 lb. Poultry farmers produced 1,726,000 lb of turkeys, 2,931,000 lb of chickens, and 242,000,000 eggs. Among the leading dairy products were milk, 492,000,000 lb; cheese, 47,803,000 lb; and ice cream, 24,474,000 gallons. In 1982, beekeepers produced 242,850 lb of honey.

25FISHING

New Jersey had a commercial fish catch of 111.6 million lb in 1984, worth $67.6 million (11th in the US). Cape May–Wildwood was the 19th-largest fishing port in the US, bringing in 34.1 million lb of fish, worth $21.4 million. Clams account for about half the state's total catch, with scallops, flounder, tilefish, and lobster accounting for most of the rest.

The Division of Fish, Game, and Wildlife of the Department of Environmental Protection administers 400,000 acres (162,000 hectares) of shellfish grounds, including 37,000 acres (15,000 hectares) of oyster and clam planting grounds, beneath tidal waters from the Raritan to the Delaware Bay. It also operates a fish hatchery at Hackettstown that produces lake trout as well as largemouth bass, channel catfish, sunfish, striped bass, northern pike, blue gill, and a white bass and striped bass hybrid; another hatchery, at Pequest, produces brook, brown, and rainbow trout.

Recreational fishermen catch finfish and shellfish along the Atlantic coast and in the rivers and lakes of northern New Jersey.

26 FORESTRY

About 40% of New Jersey's land area, or 1,928,000 acres (780,000 hectares), was forested in 1979. Of that, 96% was classified as commercial timberland, 83% of it privately owned. Not since the late 17th century, when lumber from forests in what are now Burlington, Camden, Cape May, and Salem counties was used to make ships in southern New Jersey ports, has the timber industry been of significant economic importance to the state. Shipments of lumber and wood products had a value of $274.9 million in 1982, when they comprised less than 1% of all shipments of manufactured goods.

As of 1984, the Bureau of Parks maintained 272,000 acres (110,000 hectares) of state land, nearly 70% of that in 11 state forests. The bureau also operates 40 state parks, 12 natural areas, and a recreation area. These account for 6% of the state's land, far above the US average of 0.4%. In 1979, the legislature approved a bill to protect more than 1,500 sq mi (3,900 sq km) in the Pine Barrens of southern New Jersey.

27 MINING

Once a leading iron producer, New Jersey had a mineral output valued at $156 million in 1984, thereby ranking only 35th among the 50 states.

Throughout the 18th century, New Jersey iron deposits, primarily in the region that is now Morris County, provided most of the colonies' supply. The state's forges were later instrumental in providing weapons and munitions for the Revolutionary War. By the mid-1800s, zinc had replaced iron as the leading state mineral, and mines in Franklin and Ogdensburg led the nation in output. The mines in Franklin are now closed, and zinc is produced only at a mine in Sussex County. Output fell by one-half between 1978 and 1982, to 15,822,000 tons. During the same period, the output of crushed stone and of sand and gravel also declined sharply, to 10,050,000 and 6,650,000 tons, respectively. In 1984, the state ranked 3d in output of industrial sand and 5th in zinc and exfoliated vermiculite.

28 ENERGY AND POWER

Although it contains some of the largest oil refineries in the US, New Jersey produces little of its own energy, importing much of its electric power and virtually all of its fossil fuels.

In 1985 there were 26 electric generating plants in New Jersey; installed capacity totaled 14.2 million kw. Power production amounted to 27.3 billion kwh in 1983. Per capita energy consumption in 1982 was 278 million Btu, and energy expenditure per capita in 1981 was $2,163.

New Jersey had three nuclear reactors in operation in 1985. Two of them, at Salem, are operated by Public Service Electric and Gas (PSE&G), the state's largest utility and the 16th largest in the US in 1984, with assets of $9.7 billion. A smaller unit is at Toms River. Another reactor, at Hope Creek, was scheduled to begin operation in December 1986; with this addition, PSE&G will have the largest nuclear generating complex in the US. Nuclear generating stations accounted for 16.5% of the electric power used in the state in 1983.

In 1985, the following companies had crude-oil refineries in New Jersey: Chevron in Perth Amboy, Exxon in Linden, Mobil in Paulsboro, Seaview Petroleum in Paulsboro, Amerada Hess in Port Reading, and Coastal Corp. in Westville. Transco and Texas Eastern supply over 80% of the state's natural gas. In 1978, six companies began offshore drilling for oil and gas in the Baltimore Canyon area, about 100 mi (161 km) east of Atlantic City. Two years later, tests confirmed the presence of large gas deposits, but it was still uncertain in 1985 whether recovery was commercially feasible.

29 INDUSTRY

New Jersey is one of the nation's major manufacturing states, ranking 1st in pharmaceuticals, 2d in chemicals, 5th in rubber and plastics, 6th in instruments, 7th in petroleum products, and 8th in food and food products and paper and paper products. The total value added by manufacturing was $30.1 billion in 1980. Manufacturers' shipments in 1982 had a total value of $70.4 billion. Although on the rise, the utilization rate for industrial facilities stood at only 82% in December 1984, below the 87% attained before the 1980–82 slump.

New Jersey's earliest industries were glassmaking and iron-working. In 1791, Alexander Hamilton proposed the development of a planned industrial town at the Passaic Falls. The Society for Establishing Useful Manufactures, an agency charged with developing the town, tried but failed to set up a cotton mill at the site, called Paterson, in 1797. By the early 1800s, however, Paterson had become the country's largest silk manufacturing center; by 1850, it was producing locomotives as well. On the eve of the Civil War, industry already had a strong foothold in the state. Newark had breweries, hat factories, and paper plants; Trenton, iron and paper; Jersey City, steel and soap; and Middlesex, clays and ceramics. The late 1800s saw the birth of the electrical industry, the growth of oil refineries on Bayonne's shores, and emerging chemical, drug, paint, and telephone manufacturing centers. All these products retain their places among the state's diverse manufactures.

As of 1982, the chemical industry led all sectors in manufacturing employment, with 12%, followed by electric and electronic equipment, 11%; nonelectrical machinery, 8%; apparel and other textile products, 7%; printing and publishing, 7%; and fabricated metal products, 7%. By value of shipments in 1982, the leading categories were chemical and allied products, 21%; petroleum and coal products, 11%; food and food products, 11%; electric and electronic equipment, 8%; nonelectrical machinery, 6%; and fabricated metal products, 6%.

The following table shows value of shipments by manufacturers for selected industries in 1982:

Petroleum refining	$7,124,100,000
Drugs	4,347,000,000
Transport equipment	3,831,000,000
Soaps, cleansers, toiletries	3,667,000,000
Industrial organic chemicals	2,707,400,000
Communications equipment	2,525,900,000
Plastics products, miscellaneous	2,249,300,000
Beverages	1,514,400,000
Miscellaneous converted paper products	1,331,100,000
Commercial printing	1,230,600,000

Nearly every major US corporation has facilities in the state. Major corporations headquartered in New Jersey as of 1984 included Allied Chemical in Morristown, American Cyanamid in Wayne, Campbell Soup in Camden, CPC International in Englewood Cliffs, Ingersoll-Rand in Woodcliff Lake, Johnson & Johnson in New Brunswick, Merck in Rahway, Nabisco in Parsippany, Schering-Plough in Madison, Walter Kidde in Saddle Brook, Warner-Lambert in Morris Plains, and Engelhard in Edison. Numerous corporations have moved their headquarters from New York to New Jersey since 1960.

30 COMMERCE

With one of the nation's busiest ports and many regional distribution centers, New Jersey is an important commercial state.

In 1982, New Jersey ranked 5th in the US in wholesale trade, with sales of $89.4 billion. Retail sales reached $36.3 billion that year (9th among the 50 states). The major sales outlets were food stores, 24%; automotive dealers, 17%; and general merchandise stores, 10%. The state's wholesale trade is largely concentrated near manufacturing centers and along the New Jersey Turnpike.

Bergen, Union, and Essex counties accounted for more than one-half of wholesale trade. The borough of Paramus—with only 26,474 residents in 1980, but several huge shopping plazas—led all localities in sales and had the highest per capita retail sales in the US; many of the customers are New York City residents lured by New Jersey's lower and less inclusive sales tax. Other large shopping centers are in Burlington, Eatontown, Lawrenceville, Livingston, Menlo Park, and Woodbridge. Three of the nation's largest supermarket chains have their headquarters in New Jersey: Great Atlantic and Pacific Tea Co. (A&P) in Montvale (the 17th-largest retailing company in the US and the largest in New Jersey), Supermarkets General (Pathmark) in Woodbridge, and Grand Union in Elmwood Park. The toy store chain Toys R Us is headquartered in Rochelle Park.

Port Newark and the Elizabeth Marine Terminal, foreign-trade zones operated by the Port Authority of New York and New Jersey, have been modernized and enlarged in recent years, and together account for 90% of the cargo unloaded in New York Harbor. In 1981, New Jersey exported nearly $4.5 billion of its own manufactures to foreign countries (12th in the US). The leading exports were chemicals (28%), nonelectrical machinery (18%), and electrical equipment (11%). New Jersey ranked 3d among the states in chemical exports.

31 CONSUMER PROTECTION

Consumer fraud cases are handled by the Division of Consumer Affairs of the Department of Law and Public Safety, which maintains regional offices in Camden and Newark, and in 1983/84 disposed of 2,497 cases at a savings to consumers of $3,683,532. The Division of Rate Counsel in the Public Advocate Department represents residents' interests before the various regulatory bodies, in some cases helping to postpone or reduce proposed utility rate increases.

32 BANKING

The colonies' first bank of issue opened in Gloucester in 1682. New Jersey's first chartered bank, the Newark Banking and Insurance Co., was the first of many banks to open in that city. By the mid-1800s, Newark was indisputably the financial center of the state, a position it still holds. Newark is the headquarters of the First National State Bank of New Jersey, the state's largest bank, with assets in 1984 of more than $10.7 billion (32d in the US) and deposits of $8.7 billion. For the most part, however, commercial banking in New Jersey is overshadowed by the great financial centers of New York City and Philadelphia.

As of 1983 there were 127 commercial banks, with deposits of $39.2 billion and outstanding loans of $24.7 billion. The state's 169 savings and loan associations had assets of $30.8 billion and outstanding mortgage loans of $18.2 billion in 1982. There were about 4,500 licensed consumer credit associations in 1985.

Regulation of all banks within the state is the responsibility of the Department of Banking.

33 INSURANCE

In 1873, John Dryden (later a US senator) opened the first company to offer low-cost insurance for working people. Called the Widows' and Orphans' Friendly Society, the company was located in the basement of a Newark bank until 1878, one year after Dryden changed the name of the rapidly growing enterprise to the Prudential Insurance Co. Still based in Newark, Prudential is now the largest insurance company in the US, with 1984 assets of $78.9 billion. Another Newark company, Mutual Benefit Life, was the nation's 15th largest in 1984.

New Jersey was headquarters in 1984 for 52 property/casualty insurance companies and in 1983 for 14 life insurance companies. Some 11.6 million life insurance policies, worth $184.3 billion, were in force in 1983. The average family had $65,600 in life coverage, 21% above the national average. More than $2.2 billion in benefits were paid to policyholders and their beneficiaries, including $680.5 million in death payments. Property and liability

premiums totaling $5.1 billion were written in 1983, including $1.6 billion in automobile liability coverage, $663.1 million in automobile physical damage insurance, and $464.6 million in homeowners' insurance. Nearly $5.3 billion in flood insurance was in force in 1983.

No-fault automobile insurance has been compulsory in New Jersey since 1973. All insurance agents, brokers, and companies in the state are licensed and regulated by the Department of Insurance.

34 SECURITIES

There are no stock or commodity exchanges in New Jersey. Regulation of securities trading in the state is under the control of the Bureau of Securities of the Division of Consumer Affairs, within the Department of Law and Public Safety.

New York Stock Exchange member firms had 150 sales offices and 2,319 registered representatives in the state in 1983. In that year, 1,675,000 New Jerseyites held shares in public corporations.

35 PUBLIC FINANCE

The annual budget, prepared by the Treasury Department's Division of Budget and Accounting, is submitted by the governor to the legislature for approval. The fiscal year runs from 1 July to 30 June.

The following is a summary of recommended revenues and expenditures for all state funds in 1985/86 (in thousands):

REVENUES	
Sales tax	$2,430,000
Income tax	2,173,000
Corporation tax	1,170,000
Other taxes	1,321,000
Casino and lottery revenue	572,000
Other receipts and transfers	1,414,000
TOTAL	$9,080,000

APPROPRIATIONS	
Education, of which:	$3,457,451
Higher education	(736,778)
Human services	1,661,032
Transportation	451,148
Debt service	314,956
Law and public safety	272,707
Corrections	257,816
Environmental protection	247,173
Community affairs	141,440
Other net appropriations	2,020,797
TOTAL	$8,824,520

The public debt of state and local government as of 1982 was $15.8 billion, or $2,151 per capita (17th among the 50 states).

36 TAXATION

The repeated refusal of the state legislature to levy a personal income tax ended abruptly in 1976, when the state supreme court, having earlier ruled that local property taxes were an inadequate and unfair way to fund local school systems, ordered that the schools be closed down if the state did not come up with a suitable alternative. The legislature enacted a gross income tax of 2% on the first $20,000 of taxable income and 2.5% on amounts over $20,000, with the proceeds going for school aid, property tax relief to homeowners, and other designated purposes. The tax for income over $50,000 was subsequently increased to 3.5%.

Revenues from the income tax in 1985/86 were expected to exceed $2.1 billion; a 6% sales and use tax was expected to yield more than $2.4 billion. Also levied are other corporation taxes, along with taxes on motor fuels, cigarettes, inheritances, alcoholic beverages, insurance premiums, realty transfers, savings institutions, pari-mutuel income, public utilities, and railroads. Commuters from New York State pay a tax of 2–15% on income earned in New Jersey, in order to defray costs of subsidizing transportation between the two states. The legislature in 1985 approved a law that would sharply reduce inheritance tax revenues.

The major local tax is the property tax, which in 1982/83 raised $4.75 billion in revenues. Under the Homestead Tax Rebate Act of 1976, homeowners are eligible for property tax rebates. Other exemptions and deductions from the property tax apply to senior citizens and to veterans and their widows. State and local taxes in 1982 were 11.1% of personal income—identical to the national average.

In 1983, New Jersey paid $23.6 billion in federal taxes. The $3,168 in total federal taxes per capita was the 4th highest among the 50 states and about 25% above the national average. New Jerseyites filed 3.5 million federal income tax returns in 1982, paying almost $12 billion in tax.

[37] ECONOMIC POLICY

New Jersey's controlled budget and relatively low business tax burden have helped encourage new businesses to enter the state. In addition, the state Department of Commerce and Economic Development administers a number of development programs, including several introduced in the early 1980s. The Local Development Financing Fund stimulates private sector investment that would not otherwise be forthcoming without public assistance. The Enterprise Zone Act seeks to revitalize urban areas by granting tax incentives and relaxing some government regulations. The Office of Industrial Development identifies and assists firms that have expansion needs or are experiencing difficulties. The Office of Business Advocacy performs an ombudsman function, providing technical expertise on taxation, energy, and other problems in an effort to keep companies in the state. Another office assists owners of small businesses. The Division of International Trade seeks to boost the state's exports and bring more foreign companies into the state. Other offices within the department promote tourism and motion-picture production. The New Jersey Economic Development Authority, an independent agency established in 1974 to create jobs by inducing businesses and industry to make additional capital investment in the state, reported in 1984 that its programs had encouraged $5.7 billion in additional spending in New Jersey, resulting in 84,000 new jobs.

More than 100 cities and towns are "qualified" to adopt ordinances that authorize property tax exemptions and abatements for commercial and industrial properties in areas in need of rehabilitation. The state's basic corporate tax rate of 9% compares favorably with rates of 10% in New York and 10.5% in Pennsylvania. By the mid-1980s, New Jersey was phasing out the net-worth portion of the corporation business tax and had repealed the retail gross receipts tax, the unincorporated business tax, and the sales tax on production machinery and equipment.

[38] HEALTH

During 1981, the infant mortality rate in New Jersey was 9 per 1,000 live births for whites, 18.3 for nonwhites. The birthrate in 1983 was 13.3 per 1,000 population, 15% below the national average. There were 61,100 abortions in 1982; the abortion rate of 653 per 1,000 live births was 53% above the national norm and 3d highest in the nation. Another factor contributing to the low birthrate was a marriage rate 25% below the US average in 1982. Partly because the marriage rate was so low, New Jersey's divorce rate, 3.9 per 1,000 residents, was also about 25% below the national average.

The leading causes of death in the state are heart disease and cancer, for both of which New Jersey ranks above the national average. Mortality rates per 100,000 residents in 1981 were as follows: diseases of the heart, 378; cancer, 212; cerebrovascular disease, 67; accidents, 35; chronic obstructive pulmonary diseases, 23; pneumonia and influenza, 22; diabetes, 19; cirrhosis of the liver, 15; atherosclerosis, 8%; suicide, 7; and early infant diseases, 7.

In 1985, New Jersey had 120 private community mental health centers, with about 80,000 admissions per year; 7 state psychiatric hospitals (including 1 geriatric treatment center, 1 child treatment center, and 1 forensic unit), with approximately 5,500 admissions per year; and 5 county psychiatric facilities, with about 3,500 admissions each year. As of 1983, 131 hospitals of all types had 42,581 beds; 1,172,572 patients were admitted that year. Hospital personnel in 1981 included 25,740 registered nurses. Costs of hospital care in 1982, $280 per day and $2,352 per stay, were below the US averages. The state's 324 nursing homes with 25 beds or more had, on average, 32,800 resident patients in 35,400 available beds in 1980. There were 16,258 licensed physicians and 4,824 professionally active dentists in 1982, when the state had 298 osteopathic physicians, 380 chiropractors, and 416 optometrists. The state's only medical school, the University of Medicine and Dentistry of New Jersey, is a public institution that combines three medical schools, one dental school, a school of allied professions, and a graduate school of biomedical sciences.

[39] SOCIAL WELFARE

Through the Department of Human Services, New Jersey administers the major federal welfare programs, as well as several programs specifically designed to meet the needs of New Jersey minority groups. Among the latter in 1985/86 was the Cuban-Haitian Entrant Program, involving expenditures of almost $1.3 million to aid about 1,550 persons. An additional $1 million went to provide assistance to refugees from such areas as Southeast Asia and Eastern Europe. Almost $25 million in block grant funds and $7 million in public and private-donor matching funds were channeled through the Department of Human Services to help support 173 community day-care centers serving 12,500 children.

In 1982, $504 million was spent on aid to families with dependent children, serving 402,916 New Jerseyites, 67% of them children. About 520,000 state residents purchased food stamps in 1983, at a cost to the federal government of $289 million in subsidies. Under the school lunch program, $57 million was spent to feed 570,000 students.

Of the 1,171,000 Social Security beneficiaries who received more than $6 billion in 1983, 842,000 retired workers and their spouses and children received $4.3 billion, and 217,000 survivors, $1.2 billion; 111,000 disabled workers and their spouses and children received $566 million. In 1983, 28,900 elderly persons received $57.2 million and 55,000 disabled persons received $155 million in Supplemental Security Income. Other federal payments in 1983/84 were $2.2 billion for Medicare and $68.4 million for the disabled.

In 1982, the state granted $350 million in workers' compensation benefits. About $25 million was spent on vocational rehabilitation programs in 1984. Unemployment insurance claims for 833,000 workers totaling $963 million were processed in 1982.

[40] HOUSING

Before 1967, New Jersey took a laissez-faire attitude toward housing. With each locality free to fashion its own zoning ordinances, large tracts of rural land succumbed to "suburban sprawl"—single-family housing developments spread out in two huge arcs from New York City and Philadelphia. Meanwhile, the tenement housing of New Jersey's central cities was left to deteriorate. Because poor housing was at least one of the causes of the Newark riot in 1967, the state established the Department of Community Affairs to coordinate existing housing aid programs and establish new ones. The state legislature also created the Mortgage Finance Agency and Housing Finance Agency to stimulate home buying and residential construction. In an effort to halt suburban sprawl, local and county planning boards were encouraged during the 1970s to adopt master plans for controlled growth. Court decisions in the late 1970s and early 1980s challenged the constitutionality of zoning laws that precluded the development of low-income housing in suburban areas.

As of 1980, the state had 2,678,876 year-round housing units, of which 2,548,594 were occupied. Of the latter, 62% were owner-occupied and nearly 98% had full plumbing. Some 77,900 new housing units worth $3.8 billion were authorized from 1981 through 1983; the greatest percentage increases were in lesser-populated counties in the southern and western parts of the state. Housing starts totaled 44,760 in 1983/84, an 18% increase in one year. The median monthly cost to the owner of a mortgaged home in 1980 was $461; the median cost to a renter was $270.

41EDUCATION

Public education in New Jersey dates from 1828, when the legislature first allocated funds to support education; by 1871, a public school system was established statewide. In 1980, the state was close to the US norm in the proportion of persons over age 25 who were high school graduates (66%), and above the US norm in the percentage of persons with four or more years of college (18%).

In 1984, New Jersey had 2,325 public schools, of which 1,900 were elementary and 425 were secondary. There were 77 schools for the disabled. The total public school enrollment in 1982/83, 1,083,814, represented a 5% drop in a single year and a continuation of an 11-year decline. In 1981, the student body was 71% white, 18% black, and 9% Hispanic. There were 41,900 elementary- and 31,400 secondary-school teachers in 1984. The pupil-teacher ratio in the state in 1980 was 16.3, which compared favorably with the national average of 18.8. Enrollment in 155 private schools and academies in 1984 totaled about 200,000.

Rutgers, the state university, began operations as Queen's College in 1766 and was placed under state control in 1956. Encompassing the separate colleges of Rutgers, Douglass, Livingston, and Cook, among others, the university had a total enrollment of 33,927 in 1983/84. Altogether, 256,099 students were enrolled in 1982 in the state's 12 public four-year colleges and 18 two-year community colleges. Enrollment at 26 private colleges totaled 65,288 in 1982. The major private university in the state and one of the nation's leading institutions is Princeton University, founded in 1746, with an enrollment of 6,036 in 1983/84. Other major private universities are Seton Hall (1856), 8,241; Stevens Institute of Technology (1870), 3,062; and Fairleigh Dickinson (1942), 15,526 on three main campuses.

The New Jersey Department of Higher Education offers tuition aid grants and scholarships to state residents who attend colleges and universities in the state. Guaranteed loans for any qualified resident are available through the New Jersey Higher Education Assistance Authority.

Total expenditures for education in 1984 were $5.3 billion. The state ranked 2d nationwide in expenditures per pupil (based on average daily attendance); per capita school expenditures, $711, were the 4th highest of the 50 states.

42ARTS

During the late 1800s and early 1900s, New Jersey towns, especially Atlantic City and Newark, were tryout centers for shows bound for Broadway. The New Jersey Theater Group, a service organization for nonprofit professional theaters, was established in 1978; in 1983, 11 theaters—including the well-known McCarter Theater at Princeton—were members of the Theater Group.

Around the turn of the century, Ft. Lee was the motion-picture capital of the world. Most of the best-known "silents"—including the first, *The Great Train Robbery,* and episodes of *The Perils of Pauline*—were shot there, and in its heyday the state film industry supported 21 companies and 7 studios. New Jersey's early preeminence in cinema, an era that ended with the rise of Hollywood, stemmed partly from the fact that the first motion-picture system was developed by Thomas Edison at Menlo Park in the late 1880s. The state created the New Jersey Motion Picture and Television Commission in 1977; in the next six years,

production companies spent $57 million in the state. Notable productions during this period included two Woody Allen pictures, *Broadway Danny Rose* and *The Purple Rose of Cairo.*

The state's long history of support for classical music dates at least to 1796, when William Dunlap of Perth Amboy wrote the libretto for *The Archers,* the first American opera to be commercially produced. The New Jersey Orchestra Association, established in 1966, supports orchestras throughout the state. The state's leading orchestra is the New Jersey Symphony, which makes its home in Newark's Symphony Hall; there are other symphony orchestras in Plainfield and Trenton. The New Jersey State Opera performs in Newark's Symphony Hall. Noteworthy dance companies include the Garden State Ballet, New Jersey Ballet, and Princeton Regional Ballet. A variety of performers, both classical and contemporary, appear during the summer at the Garden State Arts Center in Holmdel.

The jazz clubs of northern New Jersey and the seaside rock clubs in Asbury Park have helped launch the careers of many local performers. Famous stars perform in the casinos and hotels of Atlantic City.

43LIBRARIES AND MUSEUMS

Statewide, 325 public libraries in 1982 housed more than 23.9 million volumes and recorded a circulation of 35.7 million. The Newark Public Library was the largest municipal system, with 2,231,560 volumes and 11 branches. Distinguished by special collections on Afro-American studies, art and archaeology, economics, and international affairs, among many others, Princeton University's library is the largest in the state, with 3,519,262 volumes in 1982; Rutgers University ranked 2d with 1,659,460. The New Jersey State Library in Trenton contained 454,265 volumes, mostly on the state's history and government. One of the largest business libraries, emphasizing scientific and technical data, is the AT&T Bell Laboratories' library system, based in Murray Hill.

New Jersey has more than 100 museums and historic sites and 23 botanical gardens and arboretums. Among the most noteworthy museums are the New Jersey Historical Society in Newark and New Jersey State Museum in Trenton; the Newark Museum, containing both art and science exhibits; Princeton University's Art Museum and Museum of Natural History; and the Jersey City Museum. Also of interest are the early waterfront homes and vessels of Historic Gardner's Basin in Atlantic City, as well as Grover Cleveland's birthplace in Caldwell; the Campbell Museum in Camden (featuring the soup company's collection of bowls and utensils); Cape May County Historical Museum; Clinton Historical Museum Village; US Army Communications-Electronics Museum at Ft. Monmouth; Batsto Village, near Hammonton; Morristown National Historic Park (where George Washington headquartered during the Revolutionary War); Sandy Hook Museum; and one of the most popular attractions, the Edison National Historic Site, formerly the home and workshop of Thomas Edison, in West Orange. In 1984, the grounds at the Skylands section of Ringwood State Park were designated as the official state botanical garden.

44COMMUNICATIONS

Many communications breakthroughs—including Telstar, the first communications satellite—have been achieved by researchers at Bell Labs in Holmdel, Whippany, and Murray Hill. Three Bell Labs researchers shared the Nobel Prize in physics (1956) for developing the transistor, a device that has revolutionized communications and many other fields. In 1876, at Menlo Park, Thomas Edison invented the carbon telephone transmitter, a device that made the telephone commercially feasible.

The first mail carriers to come to New Jersey were, typically enough, on their way between New York and Philadelphia. Express mail between the two cities began in 1737, and by 1764, carriers could speed through the state in 24 hours. In colonial times, tavern keepers generally served as the local mailmen. By 1985, the US Postal Service employed 29,468 workers in New Jersey, many of them at the nation's largest bulk-mail facility in Jersey City. Altogether, 531 post offices served the state in 1985. In 1980, 95% of the state's 2,548,594 occupied housing units had telephones.

Because the state lacks a major television broadcasting outlet, New Jerseyites receive more news about events in New York City and Philadelphia than in their own towns and cities. In 1984 there were 106 radio stations (40 AM, 66 FM) and 12 television stations, (10 UHF, 2 VHF), none of which commanded anything like the audiences and influence of the stations across the Hudson and Delaware rivers. The state government operates 4 television stations under the Public Broadcasting Authority. In 1978, in cooperation with public television's WNET (licensed in Newark but operated in New York), these stations began producing New Jersey's first nightly newscast. In 1984, cable television service was provided by 60 companies to 1,266,001 subscribers in 502 communities.

45PRESS

If New Jersey is a state without a clear identity, the lack of a powerful press must be at least partly responsible. Queen Anne in 1702 banned printers from the colony; the state's first periodical, founded in 1758, died two years later. New Jersey's first daily paper, the *Newark Daily Advertiser,* did not arrive until 1832.

Although several present-day newspapers, most notably the Newark *Star-Ledger,* have amassed considerable circulation, none have been able to muster statewide influence or match the quality or prestige of the nearby *New York Times* or *Philadelphia Inquirer,* both of which are read widely in the state, along with other New York City and Philadelphia papers. As of September 1984 there were 9 morning dailies with a total circulation of 823,061, 17 evening papers with 876,280 circulation, and 17 Sunday newspapers with 1,793,391 circulation. The following table shows leading New Jersey dailies with their 1984 circulation:

AREA	NAME	DAILY	SUNDAY
Asbury Park	Press (e, S)	125,659	187,644
Atlantic City	Press (m, S)	80,103	87,254
Camden–Cherry Hill	Courier-Post (e, S)	108,566	93,115
Hackensack	Record (e, S)	150,855	219,031
Newark	Star-Ledger (m, S)	434,117	650,243

Numerous scholarly and historical works have been published by the university presses of Princeton and Rutgers. Prentice-Hall's offices are in Englewood Cliffs, and those of Silver Burdett, a textbook publisher, are in Morristown. Several New York City publishing houses maintain their production and warehousing facilities in the state. Periodicals published in New Jersey include *Home, Medical Economics, New Jersey Monthly, Personal Computing,* and *Tiger Beat.*

46ORGANIZATIONS

The 1982 Census of Service Industries counted 1,279 organizations in New Jersey, including 248 business associations; 686 civic, social, and fraternal associations; and 28 educational, scientific, and research associations.

Princeton is the headquarters of several education-related groups, including the Educational Testing Service, Graduate Record Examinations Board, Independent Educational Services, and Woodrow Wilson National Fellowship Foundation. Seeing Eye of Morristown was one of the first organizations to provide seeing-eye dogs for the blind. Other medical and health-related organizations are National Industries for the Blind (Wayne) and the American Association of Veterinary State Boards (Teaneck). Birthright USA, an anti-abortion counseling service, has its headquarters in Woodbury; the National Council on Crime and Delinquency is in Ft. Lee.

Among the many trade and professional organizations are the Hobby Industry Association of America in Elmwood Park, Science Fiction Writers of America in Wharton, and American Littoral Society in Highlands. Hobby and sports groups include the US Golf Association and World Amateur Golf Council in Far Hills, US Equestrian Team in Gladstone, and National Intercollegiate Women's Fencing Association in Upper Montclair.

47TOURISM, TRAVEL, AND RECREATION

Tourism is a leading industry in New Jersey, accounting for more than $8 billion in income in 1983 and providing 300,000 jobs. The state spent $5,551,000 on tourist promotion in 1984/85. The Jersey shore has been a popular attraction since 1801, when Cape May began advertising itself as a summer resort.

Of all the shore resorts, the largest has long been Atlantic City, which by the 1890s was the nation's most popular resort city and by 1905 was the first major city with an economy almost totally dependent on tourism. That proved to be its downfall, as improvements in road and air transportation made more modern resorts in other states easily accessible to easterners. By the early 1970s, the city's only current claims to fame were the Miss America pageant and the game of Monopoly, whose standard version uses its street names. In an effort to restore Atlantic City to its former luster and revive its economy, New Jersey voters approved a constitutional amendment in 1976 to allow casinos in the resort. In 1985, 10 casino-hotels were open and others were under construction. The gambling industry employed 30,000 in 1983, and casino revenue was $1.96 billion in 1984, up from $1.77 billion in 1983. Casino taxes were earmarked to reduce property taxes of senior citizens.

State attractions include 10 ski areas in northwestern New Jersey (on Hamburg Mountain alone, more than 50 slopes are available), canoeing and camping at the Delaware Water Gap National Recreation Area, three national wildlife refuges, 31 public golf courses, and 30 amusement parks, including Great Adventure in central Jersey.

State parks, forests, historic sites, and other areas attracted almost 8.5 million visitors in 1984. New Jersey's inland lakes, 14,000 mi (22,500 km) of trout streams, and 26 towns with saltwater fishing facilities attracted the state's 197,975 fishing-license holders in 1982/83. Licenses were also issued to 131,086 hunters.

48SPORTS

New Jersey did not have a major league professional team until 1976, when the New York Giants of the National Football League (NFL) moved across the Hudson River into the newly completed Giants Stadium in the Meadowlands Sports Complex at East Rutherford. The NFL's New York Jets began playing their home games at the Meadowlands in 1984. The Brendon Byrne Arena, located at the same site, is the home of the New Jersey Nets of the National Basketball Association and the New Jersey Devils of the National Hockey League.

The Meadowlands is also the home of a dual Thorough-bred–harness-racing track. Other racetracks are Garden State Park (Cherry Hill), Monmouth Park (Oceanport), and Atlantic City Race Course for Thoroughbreds, and Freehold Raceway for harness racing. Auto racing is featured at speedways in Trenton, Atlantic City, Bridgeport, Englishtown, and elsewhere. New Jersey has several world-class golf courses, including Baltusrol, the site of the 1980 US Open. Numerous championship boxing matches have been held in Atlantic City.

Princeton and Rutgers played what is claimed to be the first intercollegiate football game on 6 November 1869 at New Brunswick. Several important college games are held at Giants Stadium each fall, including one postseason game, the Garden State Bowl. Princeton also has a strong basketball tradition—Bill Bradley, later a forward for the New York Knickerbockers, starred for the Tigers during the early 1960s— and Rutgers was a collegiate basketball power during the 1970s.

[49] FAMOUS NEW JERSEYITES

While only one native New Jerseyite, (Stephen) Grover Cleveland (1837-1908), has been elected president of the US, the state can also properly claim (Thomas) Woodrow Wilson (b.Virginia, 1856-1924), who spent most of his adult life there. Cleveland left his birthplace in Caldwell as a little boy, winning his fame and two terms in the White House (1885-89, 1893-97) as a resident of New York State. After serving as president, he retired to Princeton, where he died and is buried. Wilson, a member of Princeton's class of 1879, returned to the university in 1908 as a professor and became its president in 1902. Elected governor of New Jersey in 1910, Wilson pushed through a series of sweeping reforms before entering the White House in 1913. Wilson's two presidential terms were marked by his controversial decision to declare war on Germany and his unsuccessful crusade for US membership in the League of Nations after World War I.

Two vice presidents hail from New Jersey: Aaron Burr (1756-1836) and Garret A. Hobart (1844-99). Burr, born in Newark and educated at what is now Princeton University, is best remembered for killing Alexander Hamilton in a duel at Weehawken in 1804. Hobart was born in Long Branch, graduated from Rutgers College, and served as a lawyer in Paterson until elected vice president in 1896; he died in office.

Four New Jerseyites have become associate justices of the US Supreme Court: William Paterson (b.Ireland, 1745-1806), Joseph P. Bradley (1813-92), Mahlon Pitney (1858-1924), and William J. Brennan, Jr. (b.1906). Among the relatively few New Jerseyites to serve in the US cabinet was William E. Simon (b.1927), secretary of the treasury under Gerald Ford.

Few New Jerseyites won important political status in colonial years because the colony was so long under New York's political and social domination. Lewis Morris (b.New York, 1671-1746) was named the first royal governor of New Jersey when severance from New York came in 1738. Governors who made important contributions to the state included William Livingston (b.New York, 1723-90), first governor after New Jersey became a state in 1776; Marcus L. Ward (1812-84), a strong Union supporter; and Alfred E. Driscoll (1902-75), who persevered in getting New Jersey a new state constitution in 1947 despite intense opposition from the Democratic Party leadership. Other important historical figures are Molly Pitcher (Mary Ludwig Hays McCauley, 1754?-1832), a heroine of the American Revolution, and Zebulon Pike (1779-1813), the noted explorer.

Two New Jersey persons have won the Nobel Peace Prize: Woodrow Wilson in 1919, and Nicholas Murray Butler (1862-1947) in 1931. A three-man team at Bell Laboratories in Murray Hill won the 1956 physics award for their invention of the transistor: Walter Brattain (b.China, 1902), John Bardeen (b.Wisconsin, 1908), and William Shockley (b.England, 1910). Dr. Selman Waksman (b.Russia, 1888-1973), a Rutgers University professor, won the 1952 prize in medicine and physiology for the discovery of streptomycin. Dickinson Woodruff (1895-1973) won the medicine and physiology prize in 1956, and Joshua Lederberg (b.1925) was a co-winner in 1958. Theoretical physicist Albert Einstein (b.Germany, 1879-1955), winner of a Nobel Prize in 1921, spent his last decades in Princeton. One of the world's most prolific inventors, Thomas Alva Edison (b.Ohio, 1847-1931) patented over 1,000 devices from workshops at Menlo Park and West Orange.

The state's traditions in the arts began in colonial times. Patience Lovell Wright (1725-86) of Bordentown was America's first recognized sculptor. Jonathan Odell (1737-1818) was an anti-Revolutionary satirist, while Francis Hopkinson (b.Pennsylvania, 1737-91), lawyer, artist, and musician, lampooned the British. Authors of note after the Revolution included William Dunlap (1766-1839), who compiled the first history of the stage in America; James Fenimore Cooper (1789-1851), one of the nation's first novelists; Mary Mapes Dodge (b.New York, 1838-1905), noted author of children's books; Stephen Crane (1871-1900), famed for *The Red Badge of Courage* (1895); and Albert Payson Terhune (1872-1942), beloved for his collie stories.

Quite a number of prominent 20th-century writers were born in or associated with New Jersey. They include poets William Carlos Williams (1883-1963) and Allen Ginsberg (b.1926); satirist Dorothy Parker (1893-1967); journalist-critic Alexander Woollcott (1887-1943); Edmund Wilson (1895-1972), influential critic, editor, and literary historian; Norman Cousins (b.1915); Norman Mailer (b.1923); Thomas Fleming (b.1927); John McPhee (b.1931); Philip Roth (b.1933); Imamu Amiri Baraka (LeRoi Jones, b.1934); and Peter Benchley (b.New York, 1940).

Notable 19th-century artists were Asher B. Durand (1796-1886) and George Inness (b.New York, 1825-94). The best-known 20th-century artist associated with New Jersey was Ben Shahn (1898-1969); cartoonist Charles Addams (b.1912) was born in Westfield. Noted photographers born in New Jersey include Alfred Stieglitz (1864-1946) and Dorothea Lange (1895-1965). Important New Jersey composers were Lowell Mason (b.Massachusetts, 1792-1872), called the "father of American church music," and Milton Babbitt (b.Pennsylvania, 1916), long active at Princeton. The state's many concert singers include Anna Case (b.1889), Paul Robeson (1898-1976), and Richard Crooks (1900-72). Popular singers include Francis Albert "Frank" Sinatra (b.1915), Sarah Vaughan (b.1924), Dionne Warwick (b.1941), Paul Simon (b.1942), and Bruce Springsteen (b.1949). Jazz musician William "Count" Basie (1904-84) was born in Red Bank.

Other celebrities native to New Jersey are actors Jack Nicholson (b.1937), Michael Douglas (b.1944), Meryl Streep (b.1948), and John Travolta (b.1954). Comedians Lou Costello (1906-59), Ernie Kovacs (1919-62), Jerry Lewis (b.1926), and Clerow "Flip" Wilson (b.1933) were also born in the state. New Jersey-born athletes include figure skater Richard "Dick" Button (b.1929), winner of two Olympic gold medals.

[50] BIBLIOGRAPHY

Amick, George. *The American Way of Graft.* Princeton, N.J.: Center for Analysis of Public Issues, 1976.

Burr, Nelson R. *A Narrative and Descriptive Bibliography of New Jersey.* Princeton, N.J.: Van Nostrand, 1964.

Cohen, David. *The Folklore and Folklife of New Jersey.* New Brunswick, N.J.: Rutgers University Press, 1983.

Cunningham, John T. *New Jersey: America's Main Road.* Garden City, N.Y.: Doubleday, 1976.

Cunningham, John T. *This Is New Jersey.* New Brunswick, N.J.: Rutgers University Press, 1983.

Federal Writers' Project. *New Jersey: A Guide to the Present and Past.* Reprint. New York: Somerset, n.d. (orig. 1939).

Fleming, Thomas. *The Forgotten Victory: The Battle for New Jersey.* New York: Reader's Digest, 1973.

Fleming, Thomas. *New Jersey: A Bicentennial History.* New York: Norton, 1977.

League of Women Voters of New Jersey. *New Jersey: Spotlight on Government.* New Brunswick, N.J.: Rutgers University Press, 1983.

Link, Arthur F. *Wilson: The Road to the White House.* Princeton, N.J.: Princeton University Press, 1947.

McPhee, John. *The Pine Barrens.* New York: Farrar, Straus & Giroux, 1981.

New Jersey, State of. Economic Policy Council and Office of Economic Policy. *1985 Economic Outlook for New Jersey.* Trenton, 1984.

Pomfret, John E. *Colonial New Jersey—A History.* New York: Scribners, 1973.

Rosenthal, Alan, and John Blydenburgh (eds.). *Politics in New Jersey.* New Brunswick, N.J.: Rutgers University Press, 1975.

NEW MEXICO

State of New Mexico

ORIGIN OF STATE NAME: Spanish explorers in 1540 called the area "the new Mexico." **NICKNAME:** Land of Enchantment. **CAPITAL:** Santa Fe. **ENTERED UNION:** 6 January 1912 (47th). **SONGS:** "O Fair New Mexico"; "Asíes Nuevo México." **MOTTO:** *Crescit eundo* (It grows as it goes). **FLAG:** The sun symbol of the Zia Indians appears in red on a yellow field. **OFFICIAL SEAL:** An American bald eagle with extended wings grasps three arrows in its talons and shields a smaller eagle grasping a snake in its beak and a cactus in its talons (the emblem of Mexico; and thus symbolic of the change in sovereignty over the state). Below the scene is the state motto; the words "Great Seal of the State of New Mexico 1912" surround the whole. **ANIMAL:** Black bear. **BIRD:** Roadrunner (chaparral bird). **FISH:** Cutthroat trout. **FLOWER:** Yucca. **TREE:** Piñon. **FOSSIL:** *Coelophysis* dinosaur. **GEM:** Turquoise. **VEGETABLES:** Frijol; chili. **LEGAL HOLIDAYS:** New Year's Day, 1 January; Birthday of Martin Luther King, Jr., 3d Monday in January; Lincoln's Birthday, 12 February; Washington's Birthday, 3d Monday in February; Memorial Day, 30 May; Independence Day, 4 July; Labor Day, 1st Monday in September; Columbus Day, 2d Monday in October; Veterans Day, 11 November; Thanksgiving Day, 4th Thursday in November; Christmas Day, 25 December. **TIME:** 5 AM MST = noon GMT.

¹LOCATION, SIZE, AND EXTENT

New Mexico is located in the southwestern US. Smaller only than Montana of the eight Rocky Mountain states, it ranks 5th in size among the 50 states. The area of New Mexico is 121,593 sq mi (314,926 sq km), of which land comprises 121,335 sq mi (314,258 sq km) and inland water 258 sq mi (668 sq km). Almost square in shape except for its jagged southern border, New Mexico extends about 352 mi (566 km) E–W and 391 mi (629 km) N–S.

New Mexico is bordered on the N by Colorado; on the E by Oklahoma and Texas; on the S by Texas and the Mexican state of Chihuahua (with a small portion of the south-central border formed by the Rio Grande); and on the W by Arizona. The total boundary length of New Mexico is 1,434 mi (2,308 km).

The geographic center of the state is in Torrance County, 12 mi (19 km) SSW of Willard.

²TOPOGRAPHY

The Continental Divide extends from north to south through central New Mexico. The north-central part of the state lies within the Southern Rocky Mountains, and the northwest forms part of the Colorado Plateau. The eastern two-fifths of the state fall on the western fringes of the Great Plains.

Major mountain ranges include the Southern Rockies, the Chuska Mountains in the northwest, and the Caballo, San Andres, San Mateo, Sacramento, and Guadalupe ranges in the south and southwest. The highest point in the state is Wheeler Peak, at 13,161 feet (4,011 meters); the lowest point, 2,817 feet (859 meters), is at Red Bluff Reservoir.

The Rio Grande traverses New Mexico from north to south and forms a small part of the state's southern border with Texas. Other major rivers include the Pecos, San Juan, Canadian, and Gila. The largest bodies of inland water are the Elephant Butte Reservoir and Conchas Reservoir, both created by dams.

The Carlsbad Caverns, the largest known subterranean labyrinth in the world, penetrate the foothills of the Guadalupes in the southeast. The caverns embrace more than 37 mi (60 km) of connecting chambers and corridors and are famed for their stalactite and stalagmite formations.

³CLIMATE

New Mexico's climate ranges from arid to semiarid, with a wide range of temperatures. Average January temperatures vary from about 35°F (2°C) in the north to about 55°F (13°C) in the southern and central regions. July temperatures range from about 78°F (26°C) at high elevations to around 92°F (33°C) at lower elevations. The record high temperature for the state is 116°F (47°C), set most recently on 14 July 1934 at Orogrande; the record low, −50°F (−46°C), was set on 1 February 1951 at Gavilan.

Average annual precipitation ranges from under 10 in (25 cm) in the desert to over 20 in (50 cm) at high elevations. Nearly one-half the annual rainfall comes during July and August, and thunderstorms are common in the summer. Snow is much more frequent in the north than in the south; Albuquerque gets about 10 in (25 cm) of snow a year, and the northern mountains receive up to 100 in (254 cm).

⁴FLORA AND FAUNA

New Mexico is divided into the following six life zones: lower Sonoran, upper Sonoran, transition, Canadian, Hudsonian, and arctic-alpine.

Characteristic vegetation in each zone includes, respectively, desert shrubs and grasses; piñon/juniper woodland, sagebrush, and chaparral; ponderosa pine and oak woodlands; mixed conifer and aspen forests; spruce/fir forests and meadows; tundra wild flowers and riparian shrubs. The yucca has three varieties in New Mexico and is the state flower. Six types of aster are considered threatened; several cacti are on the endangered list.

Indigenous animals include pronghorn antelope, javelina, and black-throated sparrow in the lower Sonoran zone; mule and white-tailed deer, ringtail, and brown towhee in the upper Sonoran zone; elk and wild turkey in the transition zone; black bear and hairy woodpecker in the Canadian zone; pine marten and blue grouse in the Hudsonian zone; and bighorn sheep, pika, ermine, and white-tailed ptarmigan in the arctic-alpine zone. Among notable desert insects are the tarantula, centipede, and vinegarroon. The coatimundi, Baird's sparrow, and brook stickle-back are among rare animals. Threatened species include the Arizona shrew and Mexican tetra. The black-footed ferret, river otter, gray wolf, Gila monster, and Socorro isopod are on the endangered list.

⁵ENVIRONMENTAL PROTECTION

Agencies concerned with the environment include the Environmental Improvement Division (EID) of the New Mexico Department of Health and Environment, the Environmental Improvement Board, the Water Quality Control Commission, and the Natural Resources Department.

Because only 0.2% of New Mexico consists of surface water—the lowest percentage of any state—the conservation and protection of scarce water resources is the state's most pressing environmental concern; the agencies with primary responsibility for water conservation and quality are the Surface Water Quality Bureau and the Groundwater Quality and Hazardous Waste Bureau, both departments within the EID. By 1983, the EID had built, or was building, new wastewater treatment facilities in 56 communities throughout the state.

In 1983, work was started on the US Department of Energy's $748 million Waste Isolation Plant near Carlsbad; beginning in 1989, this underground facility will be used to store radioactive wastes produced by federal nuclear-weapons plants in Colorado, Idaho, and Washington.

⁶POPULATION
In 1980, New Mexico had a census population of 1,303,143, 37th in the US. The estimate for 1985 was 1,446,347 (an 11% increase over 1980), yielding a population density of 12 persons per sq mi (5 per sq km). About 72% of the population was urban and 28% rural in 1980. In 1984, 449,400 people, about one-third of New Mexico's population, lived in the Albuquerque metropolitan area in Bernalillo County. Albuquerque itself had 350,575 residents in 1984, and Santa Fe, the 2d-largest city and the state capital, had 49,160 inhabitants in 1982.

⁷ETHIC GROUPS
New Mexico has two large minorities: Indians and Hispanics.

There were 104,634 American Indians in the state in 1980, 8% of the total population. Part of Arizona's great Navaho reservation extends across the border into New Mexico. There are 2 Apache reservations, 19 Pueblo villages (including one for the Zia in Sandoval County), and lands allotted to other tribes. Indian lands cover 7,361,000 acres (2,979,000 hectares), 9% of New Mexico's area.

The Hispanic population is an old one, descending from Spanish-speaking peoples who lived there before the territory was annexed by the US. This group, with a much smaller one of immigrants from modern Mexico, makes up 37% of New Mexicans.

About 6,800 Asians and Pacific Islanders and 24,000 black Americans live in the state. The people of New Mexico are proud of their ethnoracial civic harmony, though there have been local conflicts between Anglos and Hispanics.

⁸LANGUAGES
New Mexico has large Indian and Spanish-speaking populations. But just a few place-names, like Tucumcari and Mescalero, echo in English the presence of the Apache, Zuni, Navaho, and other tribes living there. Numerous Spanish borrowings include *vigas* (rafters) in the northern half, and *canales* (gutters) and *acequia* (irrigation ditch) in the Rio Grande Valley. New Mexico English is a mixture of dominant Midland, with some Northern features (such as *sick to the stomach*) in the northeast, and Southern and South Midland features such as *spoonbread* and *carry* (escort) in the eastern agricultural fringe.

In 1980, 768,397 New Mexicans—62% of the resident population 3 years of age and older—spoke only English at home. Other languages spoken at home included the following:

Spanish	362,244	French	3,259
Other (mostly Native		Italian	1,968
American) languages	79,159	Chinese	1,014
German	5,560		

⁹RELIGIONS
The first religions in New Mexico were practiced by Pueblo and Navaho Indians. Franciscan missionaries arrived at the time of

Coronado's conquest in 1540, and the first Roman Catholic church in the state was built in 1598. Roman Catholicism has long been the dominant religion, though from the mid-1800s there has also been a steady increase in the number of Protestants. The first Baptist missionaries arrived in 1849, the Methodists in 1850, and the Mormons in 1877.

The state's Roman Catholic Churches had 488,261 members in 1984. Major Protestant denominations in 1980 included 131,575 Southern Baptists and 53,185 United Methodists.

The Jewish Population was estimated at 5,155 in 1984. Cults in the state include the Penitentes and Pentecostals.

¹⁰TRANSPORTATION
Important early roads included El Camino Real, extending from Veracruz on the east coast of Mexico up to Albuquerque, and the Santa Fe Trail, leading westward from Independence, Mo. By 1983, New Mexico had 55,127 mi (88,718 km) of roads and streets, including 1,000 mi (1,609 km) of interstate highway.

In 1983, 1,291,318 road vehicles were registered in the state, of which 798,398 were automobiles, 434,369 trucks, 54,776 motorcycles, and 3,775 buses.

Rail service did not begin in New Mexico until 1879. New Mexico had 2,061 mi (3,317 km) of Class I railroad lines in 1983; the main rail lines serving the state are the Southern Pacific and the Atchison, Topeka & Santa Fe.

In 1983 there were 148 airports, 11 heliports, and 1 seaplane base. Albuquerque has the state's main airport.

¹¹HISTORY
The earliest evidence of human occupation in what is now New Mexico, dating from about 20,000 years ago, has been found in Sandia Cave near Albuquerque. This so-called Sandia man was later joined by other nomadic hunters—the Clovis and Folsom people from the northern and eastern portions of the state, and the Cochise culture, which flourished in southwestern New Mexico from about 10,000 to 500 BC. The Mogollon people tilled small farms in the southwest from 300 BC to about 100 years before Columbus came to the New World. Also among the state's early inhabitants were the Basket Makers, a seminomadic people who eventually evolved into the Anasazi, or Cliff Dwellers. The Anasazi, who made their home in the Four Corners region (where present-day New Mexico meets Colorado, Arizona, and Utah), were the predecessors of the modern Pueblo Indians.

The Pueblo people lived along the upper Rio Grande, except for a desert group east of Albuquerque, who lived in the same kind of apartmentlike villages as the river Pueblos. During the 13th century, the Navaho settled in the Four Corners area to become farmers, sheepherders, and occasional enemies of the Pueblos. The Apache, a more nomadic and warlike group who came at about the same time, would later pose a threat to all the non-Indians who arrived in New Mexico during the Spanish, Mexican, and American periods.

Francisco Vasquez de Coronado led the earliest major expedition to New Mexico, beginning in 1540, 80 years before the Pilgrims landed at Plymouth Rock. In 1598, Don Juan de Oñate led an expedition up the Rio Grande, where, one year later, he established the settlement of San Gabriel, near present-day Espanola; in 1610, the Spanish moved their center of activity to Santa Fe. For more than two centuries, the Spaniards, who concentrated their settlements, farms, and ranches in the upper Rio Grande Valley, dominated New Mexico, except for a period from 1680 to about 1683, when the Pueblo Indians temporarily regained control of the region.

In 1821, Mexico gained its independence from Spain, and New Mexico came under the Mexican flag for 25 years. The unpopularity of government officials sent from Mexico City and the inability of the new republic to control the Apache led to the revolt of 1837, which was put down by a force from Albuquerque led by General Manuel Armijo. In 1841, as governor of the Mexican

NEW MEXICO

SCALE

Albuquerque	Over 100,000
Santa Fe	50,000-100,000
Hobbs	20,000-50,000
Portales	10,000-20,000
Cannon	Under 10,000
TRIMBLE	County Name

LEGEND
⊛ State Capital
⊙ County Seat
✈ Airport
■ Point of Interest
⬚ Park, Reservation

See endsheet maps: E4.

LOCATION: 31°20' to 37° N; 103°02' to 109°02' W. **BOUNDARIES:** Colorado line, 333 mi (536 km); Oklahoma line, 34 mi (55 km); Texas line, 498 mi (801 km); Mexico line, 180 mi (290 km); Arizona line, 389 mi (626 km).

territory, Armijo defeated an invading force from the Republic of Texas, but he later made a highly controversial decision not to defend Apache Pass east of Santa Fe during the Mexican-American War, instead retreating and allowing US forces under the command of General Stephen Watts Kearny to enter the capital city unopposed on 18 August 1846.

Kearny, without authorization from Congress, immediately attempted to make New Mexico a US territory. He appointed the respected Indian trader Charles Bent, a founder of Bent's Fort on the Santa Fe Trail, as civil governor, and then led his army on to California. After Kearny's departure, a Mexican and Indian revolt in Taos resulted in Bent's death; the suppression of the Taos uprising by another US Army contingent secured American control over New Mexico, although the area did not officially become a part of the US until the Treaty of Guadalupe-Hidalgo ended the Mexican-American War in 1848.

New Mexico became a US territory as part of the Compromise of 1850, which also brought California into the Union as a free

state. Territorial status did not bring about rapid or dramatic changes in the life of those who were already in New Mexico. However, an increasing number of people traveling on the Santa Fe Trail—which had been used since the early 1820s to carry goods between Independence, Mo., and Santa Fe—were Americans seeking a new home in the Southwest. One issue that divided many of these new settlers from the original Spanish-speaking inhabitants was land. Native New Mexicans resisted, sometimes violently, the efforts of new Anglo residents and outside capital to take over lands that had been allocated during the earlier Spanish and Mexican periods. Anglo lawyers such as Thomas Benton Catron acquired unprecedented amounts of land from native grantees as payment of legal fees in the prolonged litigation that often accompanied these disputes. Eventually, a court of private land claims, established by the federal government, legally processed 33 million acres (13 million hectares) of disputed land from 1891 to 1904.

Land disputes were not the only cause of violence during the territorial period. In 1862, Confederate General Henry Hopkins Sibley led an army of Texans up the Rio Grande and occupied Santa Fe; he was defeated at La Glorieta Pass in northern New Mexico by a hastily assembled army of volunteers from Colorado, in a battle that has been labeled the Gettysburg of the West. The so-called Lincoln County War of 1878–81, a range war pitting cattlemen against merchants and involving, among other partisans, William H. Bonney (Billy the Kid), helped give the territory the image of a lawless region unfit for statehood.

Despite the tumult, New Mexico began to make substantial economic progress. In 1879, the Atchison, Topeka & Santa Fe Railroad entered the territory. General Lew Wallace, who was appointed by President Rutherford B. Hayes to settle the Lincoln County War, was the last territorial governor to enter New Mexico by stagecoach and the first to leave it by train.

By the end of the 19th century, the Indian threat that had plagued the Anglos, like the Spanish-speaking New Mexicans before them, had finally been resolved. New Mexicans won the respect of Theodore Roosevelt by enlisting in his Rough Riders during the Spanish-American War, and when he became president, he returned the favor by working for statehood. New Mexico finally became a state on 6 January 1912, under President William H. Taft.

In March 1916, irregulars of the Mexican revolutionary Pancho Villa crossed the international boundary into New Mexico, killing, robbing, and burning homes in Columbus. US troops under the command of General John J. Pershing were sent into Mexico on a long and unsuccessful expedition to capture Villa, while National Guardsmen remained on the alert in the Columbus area for almost a year.

The decade of the 1920s was characterized by the discovery and development of new resources. Potash salts were found near Carlsbad, and important petroleum reserves in the southeast and northwest were discovered and exploited. Oil development made possible another important industry, tourism, which began to flourish as gasoline became increasingly available. This period of prosperity ended, however, with the onset of the Great Depression.

World War II revived the economy, but at a price. In 1942, hundreds of New Mexicans stationed in the Philippines were among the US troops forced to make the cruel "Bataan march" to Japanese prison camps. Scientists working at Los Alamos ushered in the Atomic Age with the explosion of the first atomic bomb at White Sands Proving Ground in June 1945.

The remarkable growth that characterized the Sunbelt during the postwar era has been most noticeable in New Mexico. Newcomers from many parts of the country moved to the state, a demographic shift with profound social, cultural, and political consequences. Spanish-speaking New Mexicans, once an overwhelming majority, are now a minority.

Nevertheless, the Hispanics have been able to maintain their political influence, and in 1982, former state attorney general Toney Anaya, a Democrat, became the only Hispanic state governor in the US.

¹²STATE GOVERNMENT

The constitution of New Mexico was drafted in 1910, approved by the voters in 1911, and came into effect when statehood was achieved in 1912. A new constitution drawn up by a convention of elected delegates was rejected by the voters in 1969.

The legislature consists of a 42-member senate and a 70-member house of representatives. Senators serve four-year terms, and house members serve two-year terms. The legislature meets every year, for 60 calendar days in odd-numbered years and 30 calendar days in even-numbered years.

The executive branch consists of 10 elected officials, including the governor, lieutenant governor, secretary of state, auditor, treasurer, attorney general, and commissioner of public lands. These 7 are elected for four-year terms; none may serve two

New Mexico Presidential Vote by Political Parties, 1948–84

YEAR	ELECTORAL VOTE	NEW MEXICO WINNER	DEMOCRAT	REPUBLICAN	PROGRESSIVE
1948	4	*Truman (D)	105,240	80,303	1,037
1952	4	*Eisenhower (R)	105,435	132,170	225
					CONSTITUTION
1956	4	*Eisenhower (R)	106,098	146,788	364
					SOC. LABOR
1960	4	*Kennedy (D)	156,027	153,733	570
1964	4	*Johnson (D)	194,015	132,838	1,217
					AMERICAN IND.
1968	4	*Nixon (R)	130,081	169,692	25,737
					AMERICAN
1972	4	*Nixon (R)	141,084	235,606	8,767
					SOC. WORKERS
1976	4	Ford (R)	201,148	211,419	2,462
					LIBERTARIAN
1980	4	*Reagan (R)	167,826	250,779	4,365
1984	4	*Reagan (R)	201,769	307,101	4,459

*Won US presidential election.

successive terms. Three elected members of the Corporation Commission, which has various regulatory and revenue-raising responsibilities, serve six-year terms.

In general, constitutional amendments must be approved by majority vote in each house and by a majority of the electorate. Amendments dealing with voting rights, school lands, and linguistic requirements for education can be proposed only by three-fourths of each house, and subsequently must be approved by three-fourths of the total electorate and two-thirds of the electorate in each county.

In order to vote in state elections, a person must be 18 years of age, a US citizen, and a state resident for at least 42 days.

13 POLITICAL PARTIES

Although Democrats hold a very substantial edge in voter registration—there were 366,828 registered Democrats and only 176,737 registered Republicans in 1982—New Mexico has been a "swing state" in US presidential elections since it entered the Union. Between 1912 and 1984, New Mexicans voted for Democratic presidential candidates 9 times and Republican presidential candidates 10 times, choosing in every election except 1976 the candidate who was also the presidential choice of voters nationwide. In 1980, Ronald Reagan carried the state with 55% of the popular vote to 37% for Jimmy Carter; and in 1984, Reagan defeated Walter Mondale by a margin of 60% to 39%.

14 LOCAL GOVERNMENT

There are 33 counties in New Mexico. Each is governed by three commissioners elected for two-year terms, except for Bernalillo, which is governed by five commissioners, and Los Alamos, which has a seven-member council. Municipalities are incorporated as cities, towns, or villages.

The Indian Reorganization Act of 1934 reaffirmed the right of Indians to govern themselves, adopt constitutions, and form corporations to do business under federal law. Indians also retain the right to vote in state and federal elections. Pueblo Indians elect governors from each of 19 pueblos, and they form a coalition called the All-Indian Pueblo Council. The Apache elect a tribal council headed by a president and vice-president. The Navaho— one-third of whom live in New Mexico—elect a chairman, vice-chairman, and 87 council members from the 107 chapters that make up their reservation in New Mexico and Arizona.

15 STATE SERVICES

Agencies supervising the state transportation system include the Department of Aviation, the Civil Air Patrol, the State Highway Commission, the Department of Motor Transportation, the Department of Motor Vehicles, and the Traffic Safety Commission.

Welfare services are provided through the Human Services Department. Related service agencies include the Office of Indian Affairs and Human Rights Commission. Health services are provided by the Department of Health and Environment. The various public protection agencies include the Office of Civil and Defense Mobilization of the Attorney General's Office, the State Human Rights Commission, the Department of Corrections, and the New Mexico State Police.

The state's natural resources are protected by the Fish and Wildlife Service, the State Forest Conservation Commission, the Environmental Improvement Division of the Department of Health and Environment, the Water Quality Control Commission, and the State Park and Recreation Commission.

16 JUDICIAL SYSTEM

The judicial branch consists of a supreme court, an appeals court, district courts, probate courts, magistrate courts, and other inferior courts as created by law.

The supreme court is composed of a chief justice and four associate justices; the appeals court, created to take over some of the supreme court's caseload, is composed of seven judges. All are elected for eight-year terms.

The state's 33 counties are divided into 13 judicial districts, served by 49 district judges, each elected for a six-year term. District courts have unlimited general jurisdiction and are commonly referred to as trial courts. They also serve as courts of review for decisions of lower courts and administrative agencies. Each county has a probate court, served by a probate judge who is elected from within the county for a two-year term.

In 1983, New Mexico's crime rates—especially the rates for forcible rape and aggravated assault—were far above the US average.

17 ARMED FORCES

In 1984, Army personnel totaled 6,089, Navy and Marine Corps 732, and Air Force 19,271. The major installations are Kirtland Air Force Base in the Albuquerque area, Holloman Air Force Base at Alamogordo, and White Sands Missile Range north of Las Cruces. Defense contract awards totaled $551,690,000 in 1983/84.

There were about 162,000 veterans living in New Mexico in 1983. Of these, 1,000 had served in World War I, 59,000 in World War II, 31,000 in the Korean conflict, and 55,000 during the Viet-Nam era. In 1982/83, total federal expenditures for veterans' programs amounted to $169 million.

In 1983 there were 3,949 state and local police in the state and 2,217 corrections officers.

18 MIGRATION

Prior to statehood, the major influx of migrants came from Texas and Mexico; many of these immigrants spoke Spanish as their primary language.

Wartime prosperity during the 1940s brought a wave of Anglos into the state. New Mexico experienced a net gain through migration of 78,000 people during 1940–60, a net loss of 130,000 during the economic slump of the 1960s, and another net gain of 154,000 between 1970 and 1983.

19 INTERGOVERNMENTAL COOPERATION

New Mexico participates in the Interstate Oil and Gas Compact; Interstate Mining Compact; Western Corrections Compact; Western Interstate Energy Compact; compacts governing use of the Rio Grande and the Canadian, Colorado, La Plata, and Pecos rivers; and other interstate agreements.

In 1983/84, New Mexico received a total of $862,668,000 in government grants.

20 ECONOMY

New Mexico was primarily an agricultural state until the 1940s, when military activities assumed major economic importance. Currently, major industries include primary metals (mining), petroleum, and food. Tourism also continues to flourish. After a lag during the 1960s, New Mexico's economic growth outpaced the nation's during the late 1970s. Today, however, its per capita income remains one of the lowest in the Southwest. The considerable unemployment rate of 10.1% in 1983 dropped to 7.5% in 1984.

21 INCOME

In 1983, total personal income in New Mexico amounted to $13.4 billion; per capita income was $9,560, 41st in the US. Median money income of four-person families was $21,488 in 1981; only Arkansas had a lower figure.

Some 226,000 persons were below the federal poverty level in 1979. In 1982, 5,100 New Mexicans were among the nation's top wealth-holders, with gross assets of $500,000 or more.

22 LABOR

In 1984, the total civilian labor force of New Mexico was estimated at 628,000. Of that total, 582,000 were employed and 47,000 (7.5%) unemployed. About 124,500 workers were government employees, with the federal government employing 63,200, state and local governments 61,300. A federal survey in 1983 revealed the following nonfarm employment pattern in New Mexico:

	ESTABLISH-MENTS	EMPLOYEES	ANNUAL PAYROLL ('000)
Agricultural services, forestry, fishing	309	1,559	$ 15,990
Mining, of which:	790	24,668	599,781
Oil, gas extraction	(662)	(14,472)	(334,386)
Contract construction	3,405	30,439	475,794
Manufacturing	1,156	31,213	544,925
Transportation, public utilities	1,107	24,073	516,371
Wholesale trade	2,160	20,711	350,702
Retail trade	8,694	88,423	775,854
Finance, insurance, real estate	2,601	22,240	335,885
Services	9,253	89,824	1,326,790
Other	2,953	3,231	55,171
TOTALS	32,428	336,381	$4,997,263

Because manufacturing is not a major industry in New Mexico, organized labor is neither large nor strong. In 1983 there were 245 labor unions, and in 1985 there were about 88,000 union members, constituting 18.9% of the work force. From 1969 to 1981, only 10 work stoppages occurred.

23 AGRICULTURE

The first farmers of New Mexico were the Pueblo Indians, who raised maize, beans, and squash. Wheat and barley were introduced from Europe, and indigo and chilies came from Mexico.

In 1983, New Mexico's estimated agricultural income was $962,055,000 (36th in the US). About 34% came from crops and 66% from livestock products. Leading crops included hay and wheat. In 1983, hay production was 1,401,000 tons, valued at $126,090,000, and wheat production was 13,630,000 bushels, valued at $49,750,000. The state also produced 7,000,000 bushels of corn for grain, 1,725,000 bushels of barley, and 1,625,000 hundredweight of potatoes in 1983.

24 ANIMAL HUSBANDRY

Meat animals, especially cattle, represent the bulk of New Mexico's agricultural income. At the start of 1984 there were 1,390,000 head of beef cattle, 525,000 sheep and lambs, and 30,000 hogs and pigs on New Mexico farms. The main stock-raising regions are in the east, northeast, and northwest.

A preliminary estimate of beef production in 1983 was 561,720,000 lb, yielding cash receipts of $451,094,000; sheep, lamb, and mutton, 21,912,000 lb, $11,773,000; and pork, 12,259,000 lb, $6,519,000.

25 FISHING

There is no commercial fishing in New Mexico. The native cutthroat trout is prized by sport fishermen, however, and numerous species have been introduced into state lakes and reservoirs.

26 FORESTRY

Although lumbering ranks low as a source of state income, the forests of New Mexico are of crucial importance because of the role they play in water conservation and recreation.

In 1983, 24% of New Mexico's land area—18,574,000 acres (7,517,000 hectares)—was forestland, constituting 2.4% of the US total. Of the state total, 9,036,000 acres (3,657,000 hectares) were federally owned or managed; 939,000 acres (380,000 hectares) were owned by the state; and 8,599,000 acres (3,480,000 hectares) were privately owned.

27 MINING

Mining is the major economic activity of New Mexico, and the source of much of the state's wealth. In 1983, the state ranked 1st in potash and perlite production, 2d in uranium, and 3d in copper.

In 1983, nonfuel mineral production amounted to more than $517 million, of which nearly $250 million came from copper and $116 million from potash.

Major uranium deposits are located in the northwestern counties of McKinley and Valencia. In 1983, New Mexico mines yielded 2,550 tons of uranium ore, 24% of the US total. Other minerals that year (excluding fossil fuels) were copper, nearly 143,000 tons, and potash, 1,760,000 short tons. Some 54,000 troy oz of gold were mined in 1982.

28 ENERGY AND POWER

New Mexico is a major producer of oil and natural gas, and has significant reserves of low-sulfur bituminous coal.

In 1982, the state consumed, per capita, 34 million Btu of energy. The residential sector used 60 trillion Btu; the commercial, 68 trillion; industrial, 162 trillion; and transportation, 175 trillion. Chief sources were petroleum, 197 trillion Btu; natural gas, 210 trillion Btu; and coal, 226 trillion Btu.

Most of New Mexico's natural gas and oil fields are located in the southeastern counties of Eddy, Lea, and Chaves, and in the northwestern counties of McKinley and San Juan. In 1983, 75 million barrels of crude petroleum were produced, and there were proved reserves of 576 million barrels. Natural gas marketed production in 1983 totaled 890 billion cu feet.

In 1983, 20,415,000 tons of coal were mined.

29 INDUSTRY

The value of shipments by manufacturers in New Mexico totaled $3.9 billion in 1982. Manufacturing's share of total income and nonagricultural employment were far below the national average.

The largest manufacturing sector was petroleum and coal products, accounting for 35% of the value of shipments, followed by food and food products, 12%; electric and electronic equipment, 8%; transportation equipment, 8%; and nonelectrical machinery, 6%. The following table shows the value of shipments for selected industries in 1982:

Petroleum and coal products	$1,366,600,000
Food and food products	457,600,000
Electric and electronic equipment	328,200,000
Transportation equipment	309,000,000
Nonelectrical machinery	246,700,000

More than 50% of the manufacturing jobs in the state are located in and around Albuquerque, in Bernalillo County. Other counties with substantial manufacturing activity include Santa Fe, San Juan, Otero, McKinley, and Dona Ana.

30 COMMERCE

In 1982, wholesale establishments in the state registered sales of $5.1 billion, constituting only 0.2% of the US total. Leading sectors of wholesale trade included petroleum and petroleum products, 28%; food and food products, 15%; machinery, equipment, and supplies, 14%; lumber and construction materials, 5%; and farm products, 4%. In 1982, retail establishments had sales of $6.2 billion. Leading sectors of retail trade included food stores, 24%, and automotive dealers, 20%.

New Mexico's foreign agricultural exports totaled $121 million in 1981/82, an increase of $39 million over the 1977 value. Exports of manufacturers were $64 million in 1981, an increase of $27 million over 1977.

31 CONSUMER PROTECTION

Consumer protection in New Mexico is regulated by the Attorney General's Consumer Protection Division. This office may commence civil and criminal proceedings, represent the state before regulatory agencies, administer consumer protection programs, and handle consumer complaints.

32 BANKING

New Mexico's first bank, the First National Bank of Santa Fe, was organized in 1870. After the turn of the century, banking establishments expanded rapidly in the state, mainly because of growth in the livestock industry.

In 1984 there were 95 insured commercial banks in New Mexico, with total assets of $8.4 billion, time deposits of $5.7

billion, and demand deposits of $1.5 billion. In 1983, 23 savings associations had total assets of $3.3 billion.

33 INSURANCE
There were an estimated 5,400 persons working as carriers, agents, and brokers in the insurance industry in New Mexico in 1982. There were 7 life insurance companies in 1983, and 13 property/casualty insurance companies in 1984.

In 1983, 1,706,000 life insurance policies were in force in the state, and their total value was $25.2 billion; $73.9 million in benefits were paid. The average family had $48,100 in life insurance.

The $597 million of property and liability insurance premiums written in the state in 1983 included $236 million in automobile liability and damage insurance and $68 million in homeowners' insurance.

The insurance industry is regulated by the State Insurance Board.

34 SECURITIES
There are no securities exchanges in New Mexico. In 1983, however, New York Stock Exchange member firms had 30 sales offices and 262 full-time registered representatives in the state. Of a total of 14 registered brokers and dealers, 12 were corporations and 2 were sole proprietorships. In 1983, 183,000 New Mexicans were shareowners of public corporations.

35 PUBLIC FINANCE
The governor of New Mexico submits a budget annually to the legislature for approval. The fiscal year runs 1 July–30 June.

The following is a summary of estimated general revenues and expenditures for 1984/85 and 1985/86, in millions:

REVENUES	1984/85	1985/86
Gross receipts tax	$ 506.6	$ 540.3
Income taxes	154.0	163.0
Mineral severance taxes	149.5	150.2
Other taxes and license fees	14.4	15.8
Other receipts	470.3	500.8
TOTALS	$1,294.8	$1,370.1

EXPENDITURES		
Public schools	$ 667.4	$ 684.2
Higher education	221.1	224.0
Health and hospitals	160.0	177.1
Public safety	89.7	101.2
Other expenses	185.7	188.8
TOTALS	$1,323.9	$1,375.3

Outstanding debt of the state and local governments totaled $3.6 billion in 1983.

36 TAXATION
The state of New Mexico levies a gross receipts tax, various excise taxes, personal and corporate income taxes, property taxes, and mineral severance taxes. Of 1983 tax revenues, sales and gross receipts levies accounted for 56%, motor vehicles and license fees 4%, personal income tax 2%, corporate income tax 5%, and excise taxes 2%. As of 1984, personal income tax rates ranged from 0.7% on income under $1,000 to 7.8% on income over $100,000.

In 1982, 544,000 New Mexicans filed federal income tax returns; their federal tax bill amounted to $1.3 billion.

37 ECONOMIC DEVELOPMENT
The Economic Development Division of the Department of Commerce and Industry promotes industrial and community development, seeks export markets for New Mexico's products, and encourages use of the state by the motion-picture industry. Proposed expenditures in the 1985/86 state budget for economic development and tourism totaled $19 million.

38 HEALTH
In 1982 there were 27,730 live births in New Mexico, or 20.3 per 1,000 population, and about 7,500 legal abortions, or 281 per

1,000 live births. In that same year there were about 9,000 deaths in the state (6.5 per 1,000 population), and the infant death rate was 9.7 per 1,000 live births for whites and 15.9 for other groups. Deaths from heart disease and cancer were far below the national average. In 1983, the rate of fatal motor vehicle accidents was higher in New Mexico than in any other state—4.06 per 100 million vehicle mi (160 million km) of travel.

There were 56 hospitals in 1983, with 6,283 beds and 207,633 admissions; personnel included 2,801 registered nurses employed full-time and 900 who worked part-time. The state had 2,542 active physicians and 473 active dentists in 1982.

39 SOCIAL WELFARE
In January 1983, a total of 48,400 people received aid to families with dependent children; AFDC benefits totaled $35 million in 1983, and the monthly average allotment per family was $203.59 in 1982. There were 161,000 participants in the federal food stamp program in 1983, at a cost to the federal government of $96 million; 163,000 students participated in the school lunch program at a federal cost of $21 million.

In 1983, 24,600 persons received $51.7 million in federally administered Supplemental Security Income, of which $14 million went to the aged and $36.5 million to the disabled. In 1983/84, federal outlays from the Department of Human Services for the state totaled $7.5 million.

The state maintains the Carrie Tingley Crippled Children's Hospital in Truth or Consequences, the Meadows Home for the Aged in Las Vegas, the Miners Hospital of New Mexico in Raton, and the New Mexico School for the Visually Handicapped in Alamogordo.

40 HOUSING
In 1980, New Mexico had 508,000 housing units, with a per-person occupancy rate of 2.8; fewer than 6% of the units lacked full plumbing. The median value of all owner-occupied housing that year was $45,300.

Residential construction accounted for $638 million in 1983; some 10,100 new privately owned housing units were started that year. More than 27,000 new housing units were authorized from 1981 through 1983.

41 EDUCATION
Of 707,000 New Mexicans 25 years old and older in 1980, 241,000 had completed high school, and 124,000 had completed college.

In 1981 there were 612 public schools and 100 private schools in New Mexico. The public school system had 255,655 students in 1982/83. In 1983/84, New Mexico ranked 28th among the states in expenditures for public elementary and secondary schools, with a per pupil amount of $2,775.

New Mexico had 16 institutions of higher education in 1983. The leading public institutions are the University of New Mexico, with its main campus at Albuquerque, and New Mexico State University in Las Cruces. College teachers averaged $27,197 in yearly salary in 1982/83.

42 ARTS
New Mexico is a state rich in Indian, Spanish, Mexican, and contemporary art. Major exhibits can be seen at the University of New Mexico Art Museum in Albuquerque, and the Art Museum of the Harwood Foundation in Taos. Taos itself is an artists' colony of renown. The New Mexico Arts Division's Challenge Grants Program, supplying matching grants, has stimulated utilization of indigenous Native American, Hispanic, and Anglo folk artists.

The Santa Fe Opera, one of the nation's most distinguished regional opera companies, has its season during July and August.

43 LIBRARIES AND MUSEUMS
Public libraries in New Mexico had a combined total of 2,001,420 volumes and a circulation of 4,946,593 volumes in 1982/83. The largest municipal library is the Albuquerque Public Library, with 404,782 books. The largest university library is that of the

University of New Mexico, with 1,126,047 volumes. There is a scientific library at Los Alamos and a law library at Santa Fe.

New Mexico has 77 museums. Especially noteworthy are the Maxwell Museum of Anthropology at Albuquerque; the Museum of New Mexico, Museum of International Folk Art, and Institute of American Indian Arts Museum, all in Santa Fe; and several art galleries and museums in Taos. Historic sites include the Palace of the Governors (1610), the oldest US capitol and probably the nation's oldest public building, in Santa Fe; Aztec Ruins National Monument, near Aztec; and Gila Cliff Dwellings National Monument, 44 mi (71 km) north of Silver City. A state natural history museum, in Albuquerque, opened in 1985.

[44]COMMUNICATIONS

The first monthly mail service in New Mexico began in 1849. There were 325 post offices in 1985 and 2,994 postal employees. In 1980, 86% of the state's 441,466 occupied housing units had telephones. In 1985 there were 62 AM radio stations and 67 FM stations. There were 12 commercial and 3 educational television stations in 1984, as well as 48 cable systems that served 208,010 subscribers in 102 communities.

[45]PRESS

The first newspaper published in New Mexico was *El Crepúsculo de la Libertad* (Dawn of Liberty), a Spanish-language paper established at Santa Fe in 1834. *The Santa Fe Republican,* established in 1847, was the first English-language newspaper.

In 1985 there were 21 daily newspapers (6 morning, 15 evening) in the state. The leading dailies included the *Albuquerque Journal,* with a morning circulation in 1984 of 93,530 (133,872 on Sundays); and the *Santa Fe New Mexican,* with a morning circulation of 16,382 (18,216 on Sundays). The combined net paid circulation of all the dailies was 288,856 in 1984.

[46]ORGANIZATIONS

The 1982 Census of Service Industries counted 373 organizations in New Mexico, including 81 business associations; 203 civic, social, and fraternal associations; and 6 educational, scientific, and research associations. National organizations with headquarters in New Mexico include the National Association of Consumer Credit Administrators (Santa Fe); and the American Indian Law Students Association, the National Indian Youth Council, and Futures for Children, all located in Albuquerque.

[47]TOURISM, TRAVEL, AND RECREATION

The development of New Mexico's recreational resources has made tourism a leading economic activity. In 1982 there were 37,700 employees in travel-related industries, 8% of all jobs in the state. Hunting, fishing, camping, boating, and skiing are among the many outdoor attractions. In 1983 there were 149,466 hunting-license holders and 247,368 fishing-license holders.

The state has a national park—Carlsbad Caverns—and 10 national monuments: Aztec Ruins, Bandelier, Capulin Mountain, Chaco Canyon, El Morro (Inscription Rock), Fort Union, Gila Cliff Dwellings, Gran Quivira, Pecos, and White Sands. In 1984, the US House of Representatives designated 27,840 acres (11,266 hectares) of new wilderness preserves in New Mexico's San Juan basin, including a 2,720-acre (1,100-hectare) "fossil forest." Tourism is actively encouraged by the state. The tourism budget to attract visitors was close to $3 million for 1984/85.

[48]SPORTS

New Mexico has no major professional sports teams, though Albuquerque does have a minor league baseball team in the Class AAA Pacific Coast League. Thoroughbred and quarter-horse racing with pari-mutuel betting is an important spectator sport. Sunland Park, south of Las Cruces, has a winter-long schedule; from May to August there is racing and betting at Ruidoso Downs, La Mesa Park, and the Downs at Santa Fe.

The Lobos of the University of New Mexico compete in the Western Athletic Conference, while the Aggies of New Mexico State belong to the Pacific Coast Conference.

[49]FAMOUS NEW MEXICANS

Among the earliest Europeans to explore New Mexico were Francisco Vasquez de Coronado (b.Spain, 1510–54) and Juan de Oñate (b.Mexico, 1549?-1624?), the founder of New Mexico. Diego de Vargas (b.Spain, 1643-1704) reconquered New Mexico for the Spanish after the Pueblo Revolt of 1680, which was led by Popé (d.1685?), a San Juan Pueblo medicine man. Later Indian leaders include Mangus Coloradas (1795?-1863) and Victorio (1809?-80), both of the Mimbreño Apache. Two prominent native New Mexicans during the brief period of Mexican rule were Manuel Armijo (1792?-1853), governor at the time of the American conquest, and the Taos priest José Antonio Martínez (1793-1867).

Army scout and trapper Christopher Houston "Kit" Carson (b.Kentucky, 1809-68) made his home in Taos, as did Charles Bent (b.Virginia, 1799-1847), one of the builders of Bent's Fort, a famous landmark on the Santa Fe Trail. A pioneer of a different kind was Jean Baptiste Lamy (b.France, 1814-88), the first Roman Catholic bishop in the Southwest, whose life is recalled in Willa Cather's *Death Comes for the Archbishop.* Among the more notorious of the frontier figures in New Mexico was Billy the Kid (William H. Bonney, b.New York, 1859-81); his killer was New Mexico lawman Patrick Floyd "Pat" Garrett (b.Alabama, 1850-1908).

Notable US senators from New Mexico were Thomas Benton Catron (b.Missouri, 1840-1921), a Republican who dominated New Mexico politics during the territorial period; Albert Bacon Fall (b.Kentucky, 1861-1944), who later, as secretary of the interior, gained notoriety for his role in the Teapot Dome scandal; Dennis Chavez (1888-1962), the most prominent and influential native New Mexican to serve in Washington; Carl A. Hatch (b.Kansas, 1889-1963), best known for the Hatch Act of 1939, which limited partisan political activities by federal employees; and Clinton P. Anderson (b.South Dakota, 1895-1975), who was also secretary of agriculture.

New Mexico has attracted many artists and writers. Painters Bert G. Phillips (b.New York, 1868-1956) and Ernest Leonard Blumenschein (b.Ohio, 1874-1960) started the famous Taos art colony in 1898. Mabel Dodge Luhan (b.New York, 1879-1962) did much to lure the creative community to Taos through her writings; the most famous person to take up residence there was English novelist D. H. Lawrence (1885-1930). Peter Hurd (1940-84) was a muralist, portraitist, and book illustrator. New Mexico's best-known living artist is Georgia O'Keeffe (b.Wisconsin, 1887). Maria Povera Martinez (1887?-1980) was known for her black-on-black pottery.

Other prominent persons who have made New Mexico their home include rocketry pioneer Robert H. Goddard (b.Massachusetts, 1882-1945), Pulitzer Prize–winning editorial cartoonist Bill Mauldin (b.1921), novelist and popular historian Paul Horgan (b.New York, 1903), novelist N. Scott Momaday (b.Oklahoma, 1934), and golfer Nancy Lopez-Melton (b.California, 1957).

[50]BIBLIOGRAPHY

Beck, W. A. *New Mexico: A History of Four Centuries.* Reprint. Norman: University of Oklahoma Press, 1979.

Federal Writers' Project. *New Mexico: A Guide to the Colorful State.* Reprint. New York: Somerset, n.d. (orig. 1940).

New Mexico, University of. Bureau of Business and Economic Research. *New Mexico Statistical Abstract, 1984.* Albuquerque, 1984.

Reeve, Frank, and Alice Cleaveland. *New Mexico: Land of Many Cultures.* Rev. ed. Boulder, Colo.: Pruett, 1980.

Samora, Julian, and Patricia Vandel Simon. *A History of the Mexican-American People.* Notre Dame, Ind.: University of Notre Dame Press, 1977.

Simmons, Marc. *New Mexico: A Bicentennial History.* New York: Norton, 1977.

NEW YORK

State of New York

ORIGIN OF STATE NAME: Named for the Duke of York (later King James II) in 1664. **NICKNAME:** The Empire State. **CAPITAL:** Albany. **ENTERED UNION:** 26 July 1788 (11th). **SONG:** "I Love New York." **MOTTO:** *Excelsior* (Ever upward). **COAT OF ARMS:** Liberty and Justice stand on either side of a shield showing a mountain sunrise; surmounted on the shield is an eagle on a globe. In the foreground are a three-masted ship and a Hudson River sloop, both representing commerce. Liberty's left foot has kicked aside a royal crown. Beneath the shield is the state motto. **FLAG:** Dark blue with the coat of arms in the center. **OFFICIAL SEAL:** The coat of arms surrounded by the words "The Great Seal of the State of New York." **ANIMAL:** Beaver. **BIRD:** Bluebird. **FISH:** Brook or speckled Trout. **FLOWER:** Rose. **TREE:** Sugar maple. **FRUIT:** Apple. **FOSSIL:** Prehistoric crab (*Eurypterus remipes*). **GEM:** Garnet. **BEVERAGE:** Milk. **LEGAL HOLIDAYS:** New Year's Day, 1 January; Birthday of Martin Luther King, Jr., 3d Monday in January; Lincoln's Birthday, 12 February; Washington's Birthday, 3d Monday in February; Memorial Day, last Monday in May; Flag Day, 2d Sunday in June; Independence Day, 4 July; Labor Day, 1st Monday in September; Columbus Day, 2d Monday in October; General Election Day, 1st Tuesday after the 1st Monday in November; Veterans Day, 11 November; Thanksgiving Day, 4th Thursday in November; Christmas Day, 25 December. **TIME:** 7 AM EST = noon GMT.

¹LOCATION, SIZE, AND EXTENT

Located in the northeastern US, New York State is the largest of the three Middle Atlantic states and ranks 30th in size among the 50 states.

The total area of New York is 49,108 sq mi (127,190 sq km), of which land takes up 47,377 sq mi (122,707 sq km) and the remaining 1,731 sq mi (4,483 sq km) consist of inland water. New York's width is about 320 mi (515 km) E–W, not including Long Island, which extends an additional 118 mi (190 km) SW–NE; the state's maximum N–S extension is about 310 mi (499 km). New York State is shaped roughly like a right triangle: the line from the extreme NE to the extreme SW forms the hypotenuse, with New York City as the right angle.

Mainland New York is bordered on the NW and N by the Canadian provinces of Ontario (with the boundary line passing through Lake Ontario and the St. Lawrence River) and Quebec; on the E by Vermont (with part of the line passing through Lake Champlain and the Poultney River), Massachusetts, and Connecticut; on the S by the Atlantic Ocean, New Jersey (part of the line passes through the Hudson River), and Pennsylvania (partly through the Delaware River); and on the W by Pennsylvania (with the line extending into Lake Erie) and Ontario (through Lake Erie and the Niagara River).

Two large islands lie off the state's SE corner. Long Island is bounded by Connecticut (through Long Island Sound) to the N, Rhode Island (through the Atlantic Ocean) to the NE, the Atlantic to the S, and the East River and the Narrows to the W. Staten Island (a borough of New York City) is separated from New Jersey by Newark Bay in the N, Raritan Bay in the S, and Arthur Kill channel in the W, and from Long Island by the Narrows to the E. Including these two islands, the total boundary length of New York State is 1,430 mi (2,301 km). Long Island, with an area of 1,396 sq mi (3,616 sq km), is the largest island belonging to one of the 48 coterminous states.

The state's geographic center is in Madison County, 12 mi (19 km) S of Oneida.

²TOPOGRAPHY

Two upland regions—the Adirondack Mountains and the Appalachian Highlands—dominate the topography of New York State.

The Adirondacks cover most of the northeast and occupy about one-fourth of the state's total area. The Appalachian Highlands, including the Catskill Mountains and Kittatinny Mountain Ridge (or Shawangunk Mountains), extend across the southern half of the state, from the Hudson River Valley to the basin of Lake Erie. Between these two upland regions, and also along the state's northern and eastern borders, lies a network of lowlands, including the Great Lakes Plain; the Hudson, Mohawk, Lake Champlain, and St. Lawrence valleys; and the coastal areas of New York City and Long Island.

The state's highest peaks are found in the Adirondacks: Mt. Marcy, 5,344 feet (1,629 meters), and Algonquin Peak, 5,114 feet (1,559 meters). Nestled among the Adirondacks are many scenic lakes, including Lake Placid, Saranac Lake, and Lake George. The region is also the source of the Hudson and Ausable rivers. The Adirondack Forest Preserve covers much of this terrain, and both public and private lakes are mainly for recreational use.

The highest peak in the Catskills is Slide Mountain, at 4,204 feet (1,281 meters). Lesser upland regions of New York include the Hudson Highlands, projecting into the Hudson Valley; the Taconic Range, along the state's eastern border; and Tug Hill Plateau, set amid the lowlands just west of the Adirondacks.

Three lakes—Erie, Ontario, and Champlain—form part of the state's borders. The state has jurisdiction over 594 sq mi (1,538 sq km) of Lake Erie and 3,033 sq mi (7,855 sq km) of Lake Ontario. New York contains some 8,000 lakes; the largest lake wholly within the state is Oneida, about 22 mi (35 km) long, with a maximum width of 6 mi (10 km) and an area of 80 sq mi (207 sq km). Many smaller lakes are found in the Adirondacks and in the Finger Lakes region in west-central New York, renowned for its vineyards and great natural beauty. The 11 Finger Lakes themselves (including Owasco, Cayuga, Seneca, Keuka, Canadaigua, and Skaneateles) are long and narrow, fanning southward from a line that runs roughly from Syracuse westward to Geneseo.

New York's longest river is the Hudson, extending from the Adirondacks to New York Bay for a distance of 306 mi (492 km). The Mohawk River flows into the Hudson north of Albany. The major rivers of central and western New York State—the Black, Genesee, and Oswego—all flow into Lake Ontario. Rivers defining the state's borders are the St. Lawrence in the north, the Poultney in the east, the Delaware in the southeast, and the

Niagara in the west. Along the Niagara River, Niagara Falls forms New York's most spectacular natural feature. The falls, with an estimated mean flow rate of more than 1,585,000 gallons (60,000 hectoliters) per second, are both a leading tourist attraction and a major source of hydroelectric power.

About 2 billion years ago, New York State was entirely covered by a body of water that periodically rose and fell. The Adirondacks and Hudson River Palisades were produced by undersea volcanic action during this Grenville period. At about the same time, the schist and other crystalline rock that lie beneath Manhattan were formed. The Catskills were worn down by erosion from what was once a high, level plain. Glaciers from the last Ice Age carved out the inland lakes and valleys and determined the surface features of Staten Island and Long Island.

³CLIMATE

Although New York lies entirely within the humid continental zone, there is much variation from region to region. The three main climatic regions are the southeastern lowlands, which have the warmest temperatures and the longest season between frosts; the uplands of the Catskills and Adirondacks, where winters are cold and summer cool; and the snow belt along the Great Lakes Plain, one of the snowiest areas of the US. The growing (frost-free) season ranges from 100 to 120 days in the Adirondacks, Catskills, and higher elevations of the hills of southwestern New York to 180–200 days on Long Island.

Among the major population centers, New York City has an annual mean temperature of 55°F (13°C), with a normal maximum of 62°F (17°C) and a normal minimum of 47°F (8°C). Albany has an annual mean of 47°F (8°C), with a normal maximum of 58°F (14°C) and a normal minimum of 37°F (3°C). The mean in Buffalo is 48°F (9°C), the normal maximum 56°F (13°C), and the normal minimum 39°F (4°C). The record low temperature for the state is −52°F (−47°C), recorded at Stillwater Reservoir in the Adirondacks on 9 February 1934 and at Old Forge on 18 February 1979; the record high is 108°F (42°C), registered at Troy on 22 July 1926.

Annual precipitation ranges from over 50 in (127 cm) in the higher elevations to about 30 in (76 cm) in the areas near Lake Ontario and Lake Champlain, and in the lower half of the Genesee River Valley. New York City has an annual mean snowfall of 29 in (74 cm), while Albany gets 65 in (165 cm); in the snow belt, Buffalo receives 92 in (234 cm) of snow, Rochester 86 in (218 cm), and Syracuse 110 in (279 cm). Although New York City has fewer days of precipitation than other major populated areas (120 days annualy, compared with 168 for Buffalo), more of New York City's precipitation comes in the form of rain: 41 in (104 cm), compared with 29 in (74 cm) for Albany and Buffalo. Buffalo is the windiest city in the state, with a mean hourly wind speed of about 12 mph (19 km/hr). Tornadoes are rare, but hurricanes and tropical storms sometimes cause heavy damage to Long Island.

⁴FLORA AND FAUNA

New York has some 150 species of trees. Post and willow oak, laurel magnolia, sweet gum, and hop trees dominate the Atlantic shore areas, while oak, hickory, and chestnut thrive in the Hudson and Mohawk valleys and the Great Lakes Plain. Birch, beech, basswood, white oak, and commercially valuable maple are found on the Appalachian Plateau and in the foothills of the Adirondack Mountains. The bulk of the Adirondacks and Catskills is covered with red and black spruce, balsam fir, and mountain ash, as well as white pine and maple. Spruce, balsam fir, paper birch, and mountain ash rise to the timberline, while only the hardiest plant species grow above it. Larch, mulberry, locust, and several kinds of willow are among the many varieties that have been introduced throughout the state. Apple trees and other fruit-bearing species are important in western New York and the Hudson Valley.

Common meadow flowers include several types of rose (the

state flower), along with dandelion, Queen Anne's lace, goldenrod, and black-eyed Susan. Wild sarsaparilla, Solomon's seal, Indian pipe, bunchberry, and goldthread flourish amid the forests. Cattails grow in profusion along the Hudson, and rushes cover the Finger Lakes shallows. Among protected plants are all species of fern, bayberry, lotus, all native orchids, five species of rhododendron (including azalea), and trillium.

Some 600 species of mammals, birds, amphibians, and reptiles are found in New York, of which more than 450 species are common. Mammals in abundance include many mouse species, the snowshoe hare, common and New England cottontails, woodchuck, squirrel, muskrat, and raccoon. The deer population was estimated in 1983 at 500,000, making them a pest causing millions of dollars annually in crop damage. The wolverine, elk, and moose were all wiped out during the 19th century, and the otter, mink, marten, and fisher populations were drastically reduced; but the beaver, nearly eliminated by fur trappers, had come back strongly by 1940.

More than 260 bird species have been observed. The most common year-round residents are the crow, hawk, and several types of woodpecker. Summer visitors are many, and include the bluebird (the state bird). The wild turkey, which disappeared during the 19th century, was successfully reestablished in the 1970s. The house (or English) sparrow has been in New York since its introduction in the 1800s.

The common toad, newt, and several species of frog and salamander inhabit New York waters. Garter snakes, water snakes, grass snakes, and milk snakes are common; rattlesnakes formerly thrived in the Adirondacks. There are 210 known species of fish; 130 species are found in the Hudson, 120 in the Lake Ontario watershed. Freshwater fish include species of perch, bass, pike, and trout (the state fish). Oysters, clams, and several saltwater fish species are found in Long Island Sound. Of insect varieties, the praying mantis is looked upon as a friend (since it eats insects that prey on crops and trees), while the gypsy moth has been singled out as an enemy in periodic state-run pest-control programs.

In 1984, the following species were classified as endangered: the Indiana bat, eastern cougar, eastern timber wolf, Eskimo curlew, northern and southern bald eagles, American and Arctic peregrine falcons, American osprey, bog turtle, longjaw cisco, blue pike, shortnose sturgeon, Karner blue butterfly, and Chittenango ovate amber snail.

There were nine national wildlife refuges in the mid-1980s, covering a total of 23,395 acres (9,467 hectares).

⁵ENVIRONMENTAL PROTECTION

New York was one of the first states to mount a major conservation effort. In the 1970s, well over $1 billion was spent to reclaim the state from the ravages of pollution. State conservation efforts date back at least to 1885, when a forest preserve was legally established in the Adirondacks and Catskills. Adirondack Park was created in 1892, Catskill Park in 1904. Then, as now, the issue was how much if any state forestland would be put to commercial use. Timber cutting in the forest preserve was legalized in 1893, but the constitution of 1895 forbade the practice. By the late 1930s, the state had spent more than $16 million on land purchases and controlled 2,159,795 acres (874,041 hectares) in the Adirondacks and some 230,000 acres (more than 93,000 hectares) in the Catskills. The constitutional revision of 1938 expressly outlawed the sale, removal, or destruction of timber on forestlands. That requirement was modified by constitutional amendment in 1957 and 1973, however, and the state is now permitted to sell forest products from the preserves in limited amounts.

All state environmental programs are run by the Department of Environmental Conservation (DEC), established in 1970. The department oversees pollution control programs, monitors environmental quality, manages the forest preserves, and administers

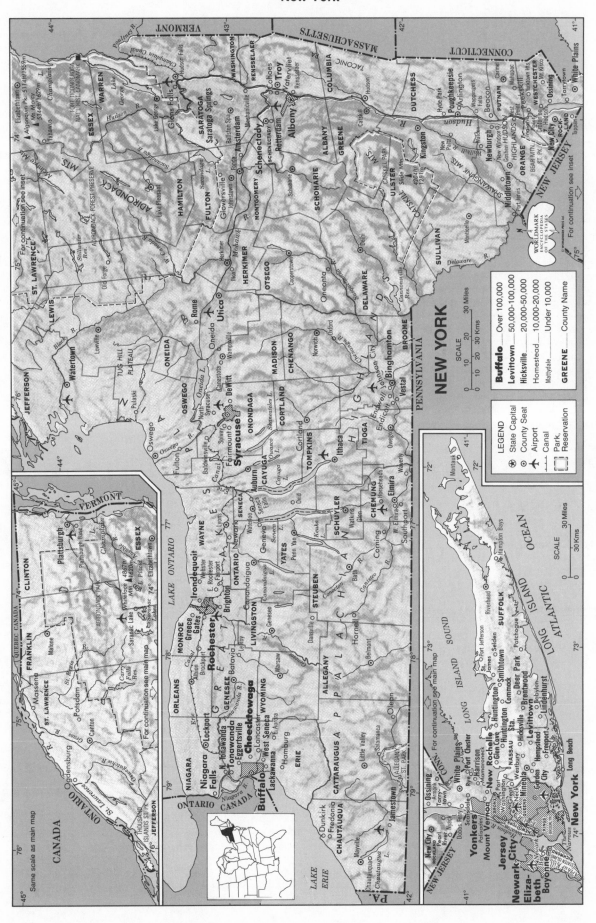

NEW YORK

SCALE

| 0 | 10 | 20 | 30 Miles |
| 0 | 10 | 20 | 30 Kms |

LEGEND

⊛ State Capital
⊙ County Seat
✈ Airport
— Canal
▢ Park, Reservation

Buffalo Over 100,000
Levittown 50,000–100,000
Hicksville 20,000–50,000
Homestead 10,000–20,000
Mattydale Under 10,000

GREENE County Name

See endsheet maps: L2.

LOCATION: 40°29′40″ to 45°0′42″N; 71°47′25″ to 79°45′54″w. **BOUNDARIES:** Canadian line, 445 mi (716 km); Vermont line, 171 mi (275 km); Massachusetts line, 50.5 mi (81 km); Connecticut line to Long Island Sound, 81 mi (130 km); Atlantic Ocean around Long Island to the New Jersey shore, 246 mi (396 km); New Jersey line, 92.5 mi (149 km); Pennsylvania line to beginning of Canadian line in Lake Erie, 344 mi (554 km).

fish and wildlife laws (including the issuance of hunting and fishing licenses). In 1982, 3,679,989 acres (1,489,244 hectares) of state-owned lands were under the department's jurisdiction. Funding for the DEC was $345.7 million in 1984/85.

As of 1983, the state had 165 air pollution monitors in operation. Because of budget cuts by New York City, the state took over the operation of the city's air-monitoring network in 1978. The DEC reported in 1983 that levels of six contaminants had steadily declined until only very small areas of the state remained in violation of air quality standards. The chief problem areas are Buffalo, where levels of particles (especially from the use of coke in steelmaking) are high, and New York City, where little progress has been made in cutting carbon monoxide emissions from motor vehicles.

Despite air-quality efforts, acid rain has been blamed for killing fish and trees in the Adirondacks, Catskills, and other areas. In 1984, the legislature passed the first measure in the nation designed to reduce acid rain, calling for a cut of 12% in sulfur dioxide emissions by 1988 and further reductions after that.

Before the 1960s, the condition of New York's waters was a national scandal. Raw sewage, arsenic, cyanide, and heavy metals were regularly dumped into the state's lakes and rivers, and fish were rapidly dying off. Two Pure Waters Bond Acts during the 1960s, the Enviromental Quality Bond Act of 1972, and a state fishery program have helped reverse the damage. There were 500 municipally owned sewage-treatment plants in operation in 1982. The state has also taken action against corporate polluters, including a $7-million settlement with General Electric over that company's discharge of toxic polychlorinated biphenyls (PCBs) into the Hudson. In addition, the state and federal government spent perhaps $45 million between 1978 and 1982 on the cleanup of the Love Canal area of Niagara Falls, which was contaminated by the improper disposal of toxic wastes, and on the relocation of some 400 families that had lived there. Remaining problems include continued dumping of sewage and industrial wastes into New York Bay and Long Island Sound, sewage overflows into the Lower Hudson, industrial dumping in the Hudson Valley, nuclear wastes in West Valley in Cattaraugus County, and contamination of fish in Lake Erie. In 1983, the DEC reported that 90% of the state's surface waterways met standards for "traditional" pollutants, such as sewage or soil particles. However, toxic pollutants, such as organic chemicals and heavy metals, appear in surface and ground water to an extent not yet fully assessed. More than 350 streams and 100 lakes failed to meet water quality standards. The department had 147 water-monitoring stations in 1983.

In 1983, about 300 old disposal sites were considered significant environmental or health hazards, and only 29 were eligible for cleanup with federal funds; moreover, New York industries were generating about a million tons of hazardous wastes each year. While the DEC reported that industrial waste was coming under state and federal regulatory control, the cost to the state of cleaning up old sites was estimated at about $500 million, far in excess of funds earmarked for this purpose. Federal funding for the state's municipal solid-waste program ended in 1981, cutting by 70% the department's resources to provide oversight.

New York Counties, County Seats, and County Areas and Populations

COUNTY	COUNTY SEAT	LAND AREA (SQ MI)	POPULATION (1984 EST.)	COUNTY	COUNTY SEAT	LAND AREA (SQ MI)	POPULATION (1984 EST.)
NEW YORK CITY		301	7,164,742	Nassau	Mineola	287	1,338,197
Bronx	Bronx	42	1,172,952	Niagara	Lockport	526	219,491
Kings	Brooklyn	70	2,253,858	Oneida	Utica	1,219	253,905
New York (Manhattan)	New York City	22	1,456,102	Onondaga	Syracuse	785	464,384
Queens	Jamaica	108	1,911,235	Ontario	Canandaigua	644	92,115
Richmond (Staten Island)	St. George	59	370,595	Orange	Goshen	826	273,546
				Orleans	Albion	391	39,253
REST OF STATE		47,076	10,570,472	Oswego	Oswego	954	119,115
Albany	Albany	524	285,552	Otsego	Cooperstown	1,004	59,848
Allegany	Belmont	1,032	51,472	Putnam	Carmel	231	80,879
Broome	Binghamton	712	213,516	Rensselaer	Troy	655	152,225
Cattaraugus	Little Valley	1,306	86,081	Rockland	New City	175	264,332
Cayuga	Auburn	695	80,315	St. Lawrence	Canton	2,728	113,736
Chautauqua	Mayville	1,064	146,257	Saratoga	Ballston Spa	810	159,757
Chemung	Elmira	411	93,473	Schenectady	Schenectady	206	150,913
Chenango	Norwich	897	50,260	Schoharie	Schoharie	624	30,284
Clinton	Plattsburgh	1,043	80,919	Schuyler	Watkins Glen	329	17,728
Columbia	Hudson	638	61,248	Seneca	Waterloo	327	32,962
Cortland	Cortland	500	48,180	Steuben	Bath	1,396	99,402
Delaware	Delhi	1,440	46,721	Suffolk	Riverhead	912	1,315,192
Dutchess	Poughkeepsie	804	253,100	Sullivan	Monticello	976	67,248
Erie	Buffalo	1,046	985,306	Tioga	Owego	519	50,602
Essex	Elizabethtown	1,807	36,654	Tompkins	Ithaca	477	87,873
Franklin	Malone	1,642	43,913	Ulster	Kingston	1,131	162,679
Fulton	Johnstown	497	56,177	Warren	Town of Queensbury[1]	882	55,500
Genesee	Batavia	495	59,655	Washington	Hudson Falls[2]	836	56,057
Greene	Catskill	648	41,377	Wayne	Lyons	605	87,989
Hamilton	Lake Pleasant	1,721	4,861	Westchester	White Plains	438	866,912
Herkimer	Herkimer	1,417	67,515	Wyoming	Warsaw	595	40,618
Jefferson	Watertown	1,273	88,936	Yates	Penn Yan	339	21,501
Lewis	Lowville	1,283	24,989				
Livingston	Geneseo	633	58,513	TOTALS		47,377	17,735,214
Madison	Wampsville	656	66,974				
Monroe	Rochester	663	711,151				
Montgomery	Fonda	404	53,107				

[1] Mail Lake George.

[2] Mail Fort Edward.

A 1982 law requires a deposit on beer and soft-drink containers sold in the state, to encourage return and recycling of bottles and cans.

⁶POPULATION

New York is no longer the most populous state, having lost that position to California in the 1970 census. However, New York City remains the most populous US city, as it has been at least since 1790. The 1980 census figures showed a population in New York State of 17,558,073, a decline of 3.7% since the 1970 census. New York and Rhode Island were the only states to lose population during the decade. As of early 1985, however, New York's population was estimated at 17,676,828, a 0.7% increase since 1980.

In 1980, New York's population was 52.5% female and 47.5% male; 85% lived in urban areas. In 1983, 90% lived in metropolitan areas (8th highest in the nation). According to the 1980 census, the densest region of the state was Manhattan, with 64,922 people per sq mi (25,066 per sq km); the least dense county was Hamilton, in the Adirondacks, with only 2.9 people per sq mi (1.1 per sq km). In 1985, the state's population density was 373 per sq mi (144 per sq km), 6th highest among the states.

In 1980, the state had 178 cities and villages of more than 10,000 people. First in the state as well as the nation was New York City, with 7,164,742 residents in 1984. Of the 1980 census total of 7,071,639, Brooklyn had a population of 2,230,936; Queens, 1,891,325; Manhattan, 1,428,285; the Bronx, 1,168,972; and Staten Island, 352,121. Other leading cities, with their 1984 population estimates, were Buffalo, 338,982; Rochester, 242,562; Yonkers, 191,234; and Syracuse, 164,219. Albany, the state capital, had a population of 99,451. All these cities lost population during the 1970s.

With an estimated 17,807,100 people, the tri-state New York City metropolitan area remained the nation's largest in mid-1984; Buffalo, the 31st largest, had an estimated 1,204,800 people; Rochester, the 38th largest, had an estimated 989,000.

The growth of New York City has been remarkable. In 1790, when the first national census was taken, the city had 49,401 residents. By 1850, its population had boomed to 696,115; by 1900, to 3,437,202, double that of Chicago, the city's closest rival. Manhattan alone housed more people in 1900 than any city outside New York. The data for 1980 are equally impressive. If Brooklyn, Queens, Manhattan, and the Bronx had each been a separate city, they would still have ranked 3d, 4th, 7th and 9th in the nation, respectively. Even Staten Island, the smallest of the city's five boroughs in population, would have placed 44th in 1980. The city had an estimated 200,000 to 750,000 illegal aliens in 1984, and city officials claimed that census figures undercounted the real population of New York City.

⁷ETHNIC GROUPS

During the 19th and 20th centuries, New York has been the principal gateway for European immigrants. In the great northern migration that began after World War I, large numbers of blacks also settled there; more recently there has been an influx of Hispanic Americans and, to a lesser extent, of Asians. Today, New York has the largest black and Puerto Rican populations of all the states, the 3d-largest Asian community, and the 2d-largest number of foreign-born.

According to the US Bureau of the Census, New York had 38,117 Indians in 1980, the 10th-highest Indian population in the nation. As of 1980 there were nine state reservations, inhabited by a total of 13,967 Indians.

Blacks have been in New York since 1624. All black slaves were freed by a state law in 1827. Rochester was a major center of the antislavery movement; Frederick Douglass, a former slave, settled and published his newspaper *North Star* there, while helping to run the Underground Railroad. After World War I, blacks moving into New York City displaced the Jews, Italians, Germans, and Irish then living in Harlem, which went on to become the cultural capital of black America. The black population of New York State was 2,402,006 as of 1980—13.7% of the state's population, and 1st among the states in the number of blacks. In 1980, almost three of every four blacks lived in New York City; with 1,784,337 blacks, New York had more black people than any other US city. There are black judges, legislators, and members of Congress, and many blacks are prominent in a variety of professions and arts. A large number, however, are at the bottom of the economic ladder; it is said that New York City could not survive without the black and Hispanic service workers who staff the city's hospitals, offices, hotels, and public institutions.

The Hispanic population as of 1980 was 1,659,300, of whom 1,406,024 (85%) were in New York City. Puerto Ricans numbered 986,389 (59% of Hispanics in the state), of whom 860,552 (87%) were in New York City. Cubans, Dominicans, Colombians, Central Americans, and Mexicans are also present in growing numbers, including a large but undetermined number of illegal immigrants.

The Asian and Pacific Islander population was 310,531 in 1980, 2d only to California. Among state residents were 148,104 Chinese, 60,511 Asian Indians (more than in any other state), 34,157 Koreans, 33,956 Filipinos, 24,524 Japanese, and 6,644 Vietnamese. Three-quarters of all of the Asians in the state lived in New York City, which has the 2d-largest Chinatown in the US.

In 1980 there were 2,388,938 foreign-born New Yorkers, 13.6% of the population and more than any state except California and Hawaii. The leading countries of birth were Italy, 283,940; Germany, 134,991; Dominican Republic, 131,313; Poland, 113,262; Soviet Union, 112,125; and Jamaica, 107,130. Among persons who reported a single ancestry group, 1,937,791 named Italian; 1,489,449 Irish, 898,453 German, 795,136 English, 643,789 Puerto Rican, and 607,871 Polish. These figures do not distinguish the large numbers of European Jewish immigrants who would identify themselves as Jews rather than by their country of origin.

The ethnic diversity of the state is reflected in such Manhattan neighborhoods as Harlem, Chinatown, Little Italy, and "Spanish," or East, Harlem, with its large Puerto Rican concentration. Many of the more successful ethnics have moved to the suburbs; on the other hand, new immigrants still tend to form ethnic communities, often in the outer boroughs, such as Asians and South Americans in certain parts of Queens and Russian Jews in south Brooklyn. Outside New York City there are also important ethnic enclaves in the Buffalo metropolitan area, with its large populations of Polish and Italian origin.

⁸LANGUAGES

Just as New York for three centuries has channeled immigrant speakers of other languages into the English-speaking population, so it has helped to channel some of their words into English, with much more rapid dissemination because of the concentration of publishing and communications industries in New York City.

Little word-borrowing followed contacts by European settlers with the unfriendly Iroquois, who between the 14th and 17th centuries had dispersed the several Algonkian tribes of Montauk, Delaware, and Mahican Indians. In New York State, the effect on English has been almost entirely the adoption of such place-names as Manhattan, Adirondack, Chautauqua, and Skaneateles.

Although the speech of metropolitan New York has its own characteristics, in the state as a whole the Northern dialect predominates. New York State residents generally say /hahg/ and /fahg/ for *hog* and *fog*, /krik/ for *creek, greasy* with an /s/ sound, and *half* and *path* with the vowel of *cat*. They keep the /r/ after a vowel, as in *far* and *cord;* sharply differentiate *horse* and *hoarse* by pronouncing the former with the vowel of *haw* and the latter with the vowel of *hoe;* and call a clump of hard maples a *sugarbush.*

There are many regional variations. In the Hudson Valley, *horse* and *hoarse* tend to be pronounced alike, and a sugarbush is called a *sap bush*. In the eastern sector, New England *piazza* for porch and *buttonball* for sycamore are found, as is the Hudson Valley term *nightwalker* for a large earthworm. In the Niagara peninsula, Midland *eavespout* (gutter) and *bawl* (how a calf sounds) have successfully moved north from Pennsylvania to invade Northern speech. In the North Country, some Canadian influence survives in *stook* (shock), *boodan* (liver sausage), and *shivaree* (wedding celebration). In the New York City area, many speakers pronounce *bird* almost as if it were /boyd/, do not sound the /h/ in *whip* or the /r/ after a vowel—although the trend now is toward the /r/ pronunciation—may pronounce initial /th/ almost like /t/ or /d/, stand *on line* (instead of in a line) while waiting to buy a huge sandwich they call a *hero* and may even pronounce *Long Island* with an inserted /g/ as /long giland/. From the high proportion of New York Yiddish speakers (nearly half of all those in the US in 1980) have come such terms as *schlock, schmaltz,* and *chutzpah.*

Serious communication problems have arisen in New York City, especially in the schools, because of the major influx since World War II of Spanish speakers from the Caribbean region, speakers of so-called black English from the South, and, more recently, Asians, in addition to the ever-present large numbers of speakers of other languages. As a result, schools in some areas have emphasized teaching English as a second language.

According to the 1980 census, 80% of all New Yorkers 3 years of age or older spoke only English at home, well below the national average of 89%. In New York City, the figure was 65%. The following table shows other major languages spoken at home by New York State residents:

Spanish	1,453,417	Greek	98,545
Italian	534,418	Russian	53,591
German	168,966	Hungarian	32,834
French	168,937	Ukrainian	27,069
Yiddish	158,815	Arabic	26,892
Polish	143,186	Korean	26,868
Chinese	129,351	Portuguese	23,168

Numerous other European and Asian languages are also spoken.

⁹RELIGIONS

With less than 8% of the total US population in 1984, New York is the home of nearly 13% of the nation's Roman Catholics and 32% of all US Jews. Both groups have had a considerable effect on state politics during this century.

Before the 1800s, Protestant sects dominated the religious life of New York, although religion did not play as large a role in the public life of New Netherland as it did in New England, with its Puritan population. The first Jews were permitted by the Dutch to settle in New Amsterdam in 1654, but their numbers remained small for the next 200 years. Both the Dutch and later the English forbade the practice of Roman Catholicism. Full religious freedom was not permitted until the constitution of 1777, and there was no Roman Catholic church in upstate New York until 1797. During the early 19th century, Presbyterian, Methodist, Universalist, Baptist, and Quaker pioneers carried their faith westward across the state. Many Protestant churches took part enthusiastically in the abolitionist movement, and the blacks who fled northward out of slavery formed their own Protestant churches and church organizations.

For Roman Catholics and Jews, the history of the 19th century is the story of successive waves of immigration: Roman Catholics first from Ireland and Germany, later from Italy and Poland, Jews first from Germany, Austria, and England, later (in vast numbers) from Russia and other Eastern European nations. The Jews who settled in New York City tended to remain there, the Roman Catholic immigrants were more dispersed throughout the state, with a large German and Eastern European group settling in Buffalo. Irish Catholics were the first group to win great political influence, but since World War II, Jews and Italian Catholics have played a leading role, especially in New York City.

As of 1 January 1984, New York had 6,684,926 Roman Catholic church members. The Jewish population of New York State was estimated at 1,879,955 in 1984, 10.6% of the state population. Approximately 93% of all Jews in the state lived in New York City or its suburbs.

Membership of leading Protestant denominations in 1980 included United Methodist, 508,029; Episcopal, 307,481; United Presbyterian, 249,079; American Baptist, USA, 216,552; American Zion, 174,002; Lutheran Church in America, 172,257; Lutheran Church–Missouri Synod, 102,099; Reformed Church in America, 90,134; and United Church of Christ, 85,424.

Because of diversified immigration, New York City has small percentages but significant numbers of Buddhists, Muslims, Hindus, and Orthodox Christians. There is also a wide variety of religious-nationalist sects and cults, including the World Community of Islam in the West, also called the Nation of Islam (Black Muslims), the Hare Krishna group, and the Unification Church of the Reverend Sun Myung Moon.

¹⁰TRANSPORTATION

New York City is a major transit point for both domestic and international passenger and freight traffic. The Port of New York and New Jersey is the nation's 2d-busiest harbor; in 1983, more than 18.5 million passengers enplaned at New York City's two major airports, Kennedy International and La Guardia, both in Queens. New York City is connected with the rest of the state by an extensive network of good roads, although road and rail transport within the metropolitan region is sagging with age.

The first railroad in New York State was the Mohawk and Hudson, which made its initial trip from Albany to Schenectady on 9 August 1831. A series of short intercity rail lines, built during the 1830s and 1840s, were united into the New York Central in 1853. Cornelius Vanderbilt gained control of the New York Central in 1867, and by 1873 had connected New York with Chicago. Under Vanderbilt and his son William, rail links were also forged between New York and Boston, Buffalo, Montreal, and western Pennsylvania.

The height of the railroads' power and commercial importance came during the last decades of the 19th century. After World War I, road vehicles gradually replaced the railroads as freight carriers. In 1980, New York ranked 15th in Class II railroad mileage, with 4,605 mi (7,411 km). Seven Class I lines accounted for 3,873 mi (6,233 km) in 1983.

The decline in freight business, and the railroads' inability to make up the loss of passenger traffic, led to a series of reorganizations and failures: the best known is the merger of the New York Central with the Pennsylvania Railroad, and the subsequent bankruptcy of the Penn Central. Today, much of New York's rail network is operated by Conrail, a federally assisted private corporation that among its operations provides commuter service up the Hudson and to New Jersey and Connecticut. The National Railroad Passenger Corporation (Amtrak) owns and operates lines along the eastern corridor from Boston through New York City to Washington, D.C. In 1983/84, it regularly operated 56 trains through New York State, stopping at 27 stations; ridership was 7,236,258. The Long Island Railroad, an important commuter carrier, is run by the Metropolitan Transportation Authority (MTA), which also operates the New York City subways. In 1984, metropolitan-area commuter railroads accounted for over 75 million unlinked passenger trips.

Construction of the New York City subway system began in 1900; service started on 27 October 1904. In 1983, the system remained by far the largest in the US and ranked 1st in the world in number of stations and cars, 2d in the length of its total route network, and 4th in number of passengers carried per day. The route network was about 230 mi (370 km) long, of which 137 mi (220 km) were underground; there were 6,173 subway cars. Operating around the clock, the system accounted for an average of 2.7 million unlinked passenger trips per day in 1984. City buses accounted for an average of 743,140 unlinked passenger trips per day.

The only other mass-transit rail line in the state is Buffalo's 6.4 mi (10.3 km) light rail system, of which 5.2 mi (8.4 km) is underground. In 1984, regular trolley service resumed in Buffalo for the first time since 1950 on the other 1.2 mi (1.9 km) of track, running through the downtown shopping district. Among cities served by municipal, county, or metropolitan-area bus systems are Albany, Binghamton, Buffalo, Elmira, and Syracuse.

In 1983, 8,416,503 motor vehicles were registered in New York State, including 7,357,044 automobiles, 34,657 buses, and 1,024,802 trucks. In addition 200,549 motorcycles were registered. As of 31 December 1982, the state had 111,196 mi (178,953 km) of roads and highways, of which 13% were state roads, 19% were county roads, 49% were town roads, 16% were city and village streets, and 3% were interstate highways and roads under various jurisdictions. Two-thirds of the mileage was rural. The major toll road, and the nation's longest toll superhighway, is the New York State Thruway, which extends 559 mi (900 km) from just outside New York City to Buffalo and the Pennsylvania border in southwestern New York. Toll-free expressways include the Adirondack Northway (I-87), from Albany to the Canadian border, and the North-South Expressway (I-81), from the Canadian to the Pennsylvania border.

A number of famous bridges and tunnels connect the five boroughs of New York City with each other and with New Jersey. The Verrazano-Narrows Bridge, opened to traffic in 1964, spans New York Harbor between Brooklyn and Staten Island. Equally famous, and especially renowned for their beauty, are the Brooklyn Bridge (1883), the city's first suspension bridge, and the George Washington Bridge (1931). The Holland (1927) and Lincoln (1937–57) tunnels under the Hudson River link Manhattan with New Jersey. Important links among the five boroughs include the Triborough Bridge, Manhattan Bridge, Williamsburg Bridge, Queensboro Bridge, Bronx-Whitestone Bridge, Throgs Neck Bridge, Brooklyn-Battery Tunnel, and Queens-Midtown Tunnel. The Staten Island Ferry conveys passengers and autos between that borough and lower Manhattan.

Until the early 1800s, almost all the state's trade moved on the Atlantic Ocean, Hudson River, and New York Bay. This waterway transportation system was expanded starting in the 1820s. Off the Hudson, one of the country's major arteries, branched the main elements of the New York Barge Canal System: the Erie Canal, linking the Atlantic with Lake Erie, and New York City with Buffalo; the Oswego Canal, connecting the Erie Canal with Lake Ontario; the Cayuga and Seneca Canal, connecting the Erie Canal with Cayuga and Seneca lakes; and the Champlain Canal, extending the state's navigable waterways from the Hudson to Lake Champlain, and so to Vermont and Quebec Province. By 1872, New York's canal system was carrying over 6 million tons of cargo a year; but an absolute decline in freight tonnage began after 1890 (the relative decline had begun 40 years earlier, with the rise of the railroads). In 1983, the canals carried only 579,777 tons of cargo, less than 10% of the tonnage for 1880 and only about one-half that for 1980; the Champlain Canal actually carried more freight than the Erie, accounting for 60% of the 1983 total.

Buffalo, on Lake Erie, is the most important inland port. In 1982, it handled 3,968,173 tons of cargo. The 70% drop in traffic from 1972 totals is testimony to Buffalo's decline as an industrial center. Albany, the major port on the Hudson, handled 5,154,448 tons of cargo a drop of 54% from 1973. Port Jefferson, on Long Island Sound, handled 3,428,511 tons in 1982.

It would be hard to exaggerate the historic and economic importance of New York Harbor—haven for explorers, point of entry for millions of refugees and immigrants, and the nation's greatest seaport until recent years, when it was passed by Greater New Orleans in terms of cargo tonnage. Harbor facilities, including those of Bayonne, Jersey City, and Newark, N.J., add up to 755 mi (1,215 km) of frontage, with some 700 piers and wharves. The entire port is under the jurisdiction of the Port Authority of New York and New Jersey. In 1982, it handled 149,255,276 tons of cargo, a decline of 31% from 1973.

In 1984, New York State had 479 airfields, including 348 airports, 107 heliports, and 24 seaplane bases; 198 airfields were public and 281 private. By far the busiest airports in the state are John F. Kennedy International and La Guardia, both in New York City. In 1983, Kennedy handled 9,794,648 enplaned passengers, 81,524 aircraft departures, and 253,099 tons of freight, while La Guardia handled 8,786,003 enplaned passengers, 106,175 departures, and 18,189 tons of freight. The Greater Buffalo International Airport, the largest in the state outside New York City, handled 1,706,336 passengers and 5,568 tons of freight in 1983.

¹¹HISTORY

The region now known as New York State has been inhabited for about 10,000 years. The first Indians probably came across the Bering Strait and most likely reached New York via the Niagara Peninsula. Remains have been found in southwestern New York of the Indians called Mound Builders (for their practice of burying their dead in large mounds), who cultivated food crops and tobacco. The Mound Builders were still living in the state well after AD 1000, although by that time most of New York was controlled by later migrants of the Algonkian linguistic group. These Algonkian tribes included the Mahican in the northeast, the Wappinger in the Hudson Valley and on Long Island, and the Leni-Lenape (or Delaware) of the Delaware Valley.

Indians of the Iroquoian language group invaded the state from the north and west during the early 14th century. In 1570, after European explorers had discovered New York but before the establishment of any permanent European settlements, the main Iroquois tribes—the Onondaga, Oneida, Seneca, Cayuga, and Mohawk—established the League of the Five Nations. For the next 200 years, members of the League generally kept peace among themselves but made war on other tribes, using not only traditional weapons but also the guns they were able to get from the French, Dutch, and English. In 1715, a sixth nation joined the League—the Tuscarora, who had fled the British in North Carolina. For much of the 18th century, the Iroquois played a skillful role in balancing competing French and British interests.

The first European known to have entered New York Harbor was the Florentine navigator Giovanni da Verrazano, on 17 April 1524. The Frenchman Samuel de Champlain began exploring the St. Lawrence River in 1603. While Champlain was aiding the Huron Indians in their fight against the League in 1609, the English mariner Henry Hudson, in the service of the Dutch East India Company, entered New York Bay and sailed up the river that would later bear his name, reaching about as far as Albany. To the Dutch the area did not look especially promising, and there was no permanent Dutch settlement until 1624, three years after the Dutch West India Company had been founded. The area near Albany was first to be settled. The Dutch were mainly interested in fur trading, and agriculture in the colony—named New Netherland—was slow to develop. New Amsterdam was founded in 1626, when Director-General Peter Minuit bought Manhattan (from the Indian word *manahatin,* "hill island") from the Indians for goods worth—as tradition has it—about $24.

New Amsterdam grew slowly, and by 1650 had no more than 1,000 people. When the British took over New Netherland in 1664, only 8,000 residents lived in the colony. Already, however, the population was remarkably diverse: there were the Dutch and English, of course, but also French, Germans, Finns, Swedes, and Jews, as well as Negro slaves from Angola. The Swedes lived in what had been New Sweden, a territory along the Delaware River ceded to the Netherlands during the administration of Peter Stuyvesant. Equally famed for his wooden leg and his hot temper, Stuyvesant had become director general of the New Netherland colony in 1647. Three years later, after skirmishes with the English settlers of New England, the colony gave up all claims to the Connecticut Valley in the Treaty of Hartford.

Though small and weak, New Netherland was an annoyance to the English. The presence of Dutch traders in New York Bay made it difficult for England to enforce its monopolies under the Navigation Acts. Moreover, the Dutch colony was a political barrier between New England and two other English colonies, Maryland and Virginia. So, in 1664, King Charles II awarded "all the land from the west side of the Connecticutte River to the East Side of De La Ware Bay" to his brother, the Duke of York and Albany, the future King James II. The British fleet arrived in New York Bay on 18 August 1664. Stuyvesant wanted to fight, but his subjects refused, and the governor had no choice but to surrender. The English agreed to preserve the Dutch rights of property and inheritance, and to guarantee complete liberty of conscience. Thus New Netherland became New York. It remained an English colony for the next 112 years, except for a period in 1673 when Dutch rule was briefly restored.

The first decades under the English were stormy. After repeated demands from the colonists, a general assembly was called in 1683. The assembly adopted a Charter of Liberties and Privileges, but the document, approved by James before his coronation, was revoked after he became king in 1685. The assembly itself was dissolved in 1686, and James II acted to place New York under the Dominion of New England. The plan was aborted by the Glorious Revolution of 1688, when James was forced to abdicate. Power in New York fell to Jacob Leisler, a German merchant with local backing. Leisler ruled until 1691, when a new royal governor arrived and had Leisler hanged for treason.

The succeeding decades were marked by conflict between the English and French and by the rising power of the provincial assembly in relations with the British crown. As early as 1690, a band of 150 Frenchmen and 100 Indians attacked and burned Schenectady. New York contributed men and money to campaigns against the French in Canada in 1709 and 1711 (during Queen Anne's War) and in 1746 (during King George's War). In 1756, the English determined to drive the French out of the region once and for all. After some early reverses, the English defeated the French in 1760. The Treaty of Paris (1763), ending the French and Indian War, ceded all territory east of the Mississippi to England, except for New Orleans and two islands in the mouth of the St. Lawrence River. The Iroquois, their power weakened during the course of the war, signed treaties giving large areas of their land to the New York colony.

The signing of the Treaty of Paris was followed by English attempts to tighten control over the colonies, in New York as elsewhere. New York merchants vehemently protested the Sugar Act and Stamp Act, and the radical Sons of Liberty made their first appearance in the colony in October 1765. Later, in 1774, after Paul Revere brought news of the Boston Tea Party to New York City, British tea was also dumped into that city's harbor. Nevertheless, New York hesitated before committing itself to independence. The colony's delegates to the Continental Congress in Philadelphia were not permitted by the Third Provincial Congress in New York to vote either for or against the Declaration of Independence on 4 July 1776. The Fourth Provincial Congress, meeting at White Plains, did ratify the Declaration five days later. On 6 February 1778, New York became the second state to ratify the Articles of Confederation.

Nearly one-third of all battles during the Revolutionary War took place on New York soil. The action there began when troops under Ethan Allen captured Fort Ticonderoga in May 1775, and Seth Warner and his New England forces took Crown Point. Reverses came in 1776, however, when George Washington's forces were driven from Long Island and Manhattan by the British; New York City was to remain in British hands for the rest of the war. Troops commanded by British General John Burgoyne recaptured Ticonderoga in July 1777, but were defeated in October at Saratoga, in a battle that is often considered the turning point of the war. In 1778, General Washington made his headquarters at West Point, which General Benedict Arnold tried unsuccessfully to betray to the British in 1780. Washington moved his forces to Newburgh in 1782, and marched into New York City on 25 November 1785, the day the British evacuated their forces. On 4 December, he said farewell to his officers at Fraunces Tavern in lower Manhattan, a landmark that still stands.

Even as war raged, New York State adopted its first constitution on 20 April 1777. The constitution provided for an elected governor and house of assembly, but the franchise was limited to property holders. The first state capital was Kingston, but the capital was moved to Albany in January 1797. After much debate, in which the Federalist Alexander Hamilton played a leading role, the state ratified the US Constitution (with amendments) on 26 July 1788. New York City served as the seat of the US government from 11 January 1785 to 12 August 1790, and the first US president, George Washington, was inaugurated in the city on 30 April 1789.

George Clinton was the state's first elected governor, serving from 1777 to 1795 and again from 1801 to 1804. The achievements under his governorship were considerable. Commerce and agriculture expanded, partly because of Clinton's protectionist policies and partly because of the state's extremely favorable geographical situation.

The end of the War of 1812 signaled the opening of an era of unprecedented economic expansion for the state. By this time, the Iroquois were no longer a threat (most had sided with the British during the Revolutionary War, and many later fled to Canada). Migrants from New England were flocking to the state, which the census of 1810 showed was the most populous in the country. Small wonder that New York was the site of the early 19th century's most ambitious engineering project: construction of the Erie Canal. Ground was broken for the canal in 1817, during the first term of Governor De Witt Clinton, the nephew of George Clinton; the first vessels passed through the completed canal in 1825.

Actually, New York had emerged as the nation's leading commercial center before the canal was even started. The textile industry had established itself by the mid-1820s, and the dairy industry was thriving. The effects of the canal were felt most strongly in foreign trade—by 1831, 50% of US imports and 27% of US exports passed through the state—and in the canal towns of Utica, Syracuse, Rochester, and Buffalo, where business boomed.

Commercial progress during this period was matched by social and cultural advancement. New York City became a center of literary activity during the 1820s, and by the 1840s was already the nation's theatrical capital. A new state constitution drafted in 1821 established universal white male suffrage, but retained the property qualifications for blacks. Slavery was abolished as of 4 July 1827 (few slaves actually remained in the state by this time), and New Yorkers soon took the lead in the growing antislavery movement. The first women's rights convention in the US was held in Seneca Falls in 1848—though women would have to wait until 1917 before winning the right to vote in state elections. Also during the 1840s, the state saw the first of several great waves of European immigration. The Irish and Germans were the earliest major arrivals during the 19th century, but before World War I they would be joined—not always amicably—by Italians and European Jews.

New Yorkers voted for Abraham Lincoln in the presidential election of 1860 and were among the readiest recruits to the Union side. Enthusiasm for the conflict diminished during the next two years, however. When the military draft reached New York City on 11 July 1863, the result was three days of rioting in which blacks were lynched and the homes of prominent abolitionists were burned. But New York was not a wartime battleground, and overall the war and Reconstruction were very good for business.

The decades after the Civil War ushered in an era of extraordinary commercial growth and political corruption. This was the Gilded Age, during which entrepreneurs became multimillionaires and New York was transformed from an agricultural state to an industrial giant. In 1860, the leading manufactures in the state were flour and meal, men's clothing, refined sugar, leather goods, liquor, and lumber; 90 years later, apparel, printing and publishing, food, machinery, chemicals, fabricated metal products, electrical machinery, textiles, instruments, and transportation equipment had became the dominant industries.

The key to this transformation was the development of the railroads. The boom period for railroad construction started in the 1850s and reached its high point after 1867, when "Commodore" Cornelius Vanderbilt, who had been a steamboat captain in 1818, took over the New York Central. During the 1860s, native New Yorkers like Jay Gould and Russell Sage made their fortunes through investment and speculation. Especially during the century's last two decades, corporate names that today are household words began to emerge: Westinghouse Electric in 1886, General Electric (as Edison Electric) in 1889, Eastman Kodak in 1892. In 1882, another native New Yorker, John D. Rockefeller, formed the Standard Oil Trust; although the trust would eventually be broken up, the Rockefeller family would help shape New York politics for many decades to come.

The period immediately following the Civil War also marked a new high in political influence for the Tammany Society (or "Tammany Hall"), founded in 1789 as an anti-Federalist organization. From 1857 until his exposure by the press in 1871, Democrat William Marcy "Boss" Tweed ruled Tammany and effectively dominated New York City by dispensing patronage, buying votes, and bribing legislators and judges. Tammany went into temporary eclipse after the Tweed Ring was broken up, and Republicans swept the state in 1872. The first result was a series of constitutional changes, including one abolishing the requirement that blacks hold property in order to vote. A new constitution approved in 1894, and effective in 1895, remains the basic law of New York State today.

During the Union's first 100 years, New York's political life had projected into national prominence such men as Alexander Hamilton, John Jay, George and De Witt Clinton, Martin Van Buren, and Millard Fillmore. The state's vast population—New York held more electoral votes than any other state between 1812 and 1972—coupled with its growing industrial and financial power, enhanced the prestige of state leaders during the nation's second century. Grover Cleveland, though born in New Jersey, became mayor of Buffalo, then governor of New York, and finally the 22d US president in 1885. Theodore Roosevelt was governor of New York, then became vice president and finally president of the US in 1901. In 1910, Charles Evans Hughes resigned the governorship to become an associate justice of the US Supreme Court; he also served as secretary of state, and in 1930 was appointed chief justice of the US. By the 1920s, Tammany had rebounded from the Tweed Ring breakup and from another scandal during the 1890s to reach its peak of prestige: Alfred E. Smith, a longtime member of Tammany, as well as an able and popular official, was four times elected governor and became in 1928 the first Roman Catholic candidate to be nominated by a major party for the presidency of the US. That year saw the election of Franklin D. Roosevelt as governor of New York.

The 1930s, a period of depression, ushered in a new wave of progressive government. From 1933 until 1945, FDR was in the White House. Roosevelt's successor in the statehouse was Herbert H. Lehman, whose Little New Deal established the basic pattern of present state social welfare policies that had begun on a much more modest scale during Smith's administration. The Fusion mayor of New York City at this time—propelled into office by yet another wave of exposure of Tammany corruption—was the colorful and popular Fiorello H. La Guardia.

The decades since World War II have seen extraordinary expansion of New York social services, including construction of the state university system, but also an erosion of the state's industrial base. Fiscal crises are not new to the state—reformers in the 1920s railed against New York City's "spendthrift" policies—but the greatly increased scale of government in the 1970s made the fiscal crisis of 1975 unprecedented in its scope and implications. The city's short-term debt grew from virtually zero to about $6 billion between 1970 and 1975, although its government reported consistently balanced budgets. Eventually a package totaling $4.5 billion in aid was needed to avoid bankruptcy. The decreasing pace of population and industrial growth during the 1950s and 1960s, and the decline during the 1970s, also led to a dimming of New York's political fortunes. No native-born New Yorker has been elected president since FDR, and neither of the two presidents who were New York residents when elected—Dwight Eisenhower in 1952 and Richard Nixon in 1968, both Republicans—ever held elective office in the state. The single most dominant political figure in New York since World War II, Nelson A. Rockefeller (governor, 1959–73), tried and failed three times to win the Republican presidential nomination before his appointment to the vice-presidency in 1974. Unable to overcome the hostility of his party's conservative wing, he was not renominated for the vice-presidency in 1976. In 1984, however, US Representative Geraldine Ferraro of Queens was the Democratic Party's vice-presidential standard-bearer, and Governor Mario M. Cuomo emerged as an influential Democratic spokesman.

[12] STATE GOVERNMENT

New York has had four constitutions, adopted in 1777, 1822, 1846, and 1895. The 1895 constitution was extensively revised in 1938, and the basic structure of state government has not changed since then, although the document had been amended 197 times by the end of 1983.

The legislature consists of a 61-member senate and 150-member assembly. Senators and assembly members serve two-year terms and are elected in even-numbered years. Each house holds regular annual sessions; special sessions may be called by the governor or initiated by petition of two-thirds of the membership of each body. Either senators or assembly members may introduce or amend a bill. To pass, a bill requires a majority vote in both houses; a two-thirds majority is required to override the governor's veto. Members of both the senate and assembly must be US citizens and must have resided in the state for five years and in their district for 12 months. Legislative salaries were $43,000 in 1985, but legislative leaders may receive up to $30,000 in addition.

The state's only elected executives are the governor, lieutenant governor, comptroller, and attorney general. Each serves a four-year term. The governor and lieutenant governor are jointly elected; there is no limit to the number of terms they may serve. The governor must be at least 30 years old, a US citizen, and a resident of the state for five years prior to the date of election. The lieutenant governor is next in line for the governorship (should the governor be unable to complete his term in office) and presides over the senate. Annual salary for the governor was $100,000 in 1984; for the lieutenant governor, $85,000.

The governor appoints the heads of 15 of the 20 major executive departments, 13 of them with the advice and consent of the senate. The exceptions are the comptroller and attorney general, who are elected by the voters; the commissioner of education, who is named by the Regents of the University of the State of New York; the commissioner of social services, elected by the Board of Social Services; and the chief of the Executive Department, which the governor heads ex officio.

A bill becomes law when passed by both houses of the legislature and signed by the governor. While the legislature is in session, a bill may also become law if the governor fails to act on it within 10 days of its receipt. The governor may veto a bill or, if the legislature has adjourned, may kill a bill simply by taking no action on it for 30 days.

A proposed amendment to the state constitution must receive majority votes in both houses of the legislature during two successive sessions. Amendments so approved are put on the ballot in November and adopted or rejected by majority vote. The constitution also provides that the voters must be permitted every 20 years to decide whether a convention should be called to amend the present constitution. To vote in New York State, one must be a US citizen, at least 18 years of age, and a resident of the election district for 30 days. New York State has a permanent personal registration system; a registered voter who fails to vote in any general election during a two-year period must reregister.

¹³POLITICAL PARTIES

In addition to the Democratic and Republican parties, the major political groups, there has always been a profusion of minor political parties in New York, some of which have significantly influenced the outcomes of national and state elections.

Party politics in the state crystallized into their present form around 1855. Up to that time, a welter of parties and factions—including such short-lived groups as the Anti-Masons (later Whigs), Bucktails, Clintonians, Hunkers, and Barnburners (split into Hardshell and Softshell Democrats), Know-Nothings (Native American Party), Woolly Heads and Silver-Grays (factions of the Whigs), and the Liberty Party—jockeyed for power in New York State.

Roughly speaking, the Democratic Party evolved out of the Democratic Republican factions of the old Republican Party and had become a unified party by the 1850s. The Democratic power base was—and has remained—the big cities, especially New York City. The most important big-city political machine from the 1860s through the 1950s, except for a few brief periods, was the Tammany Society ("Tammany Hall"). Tammany controlled the Democratic Party in New York City and, through that party, the city itself.

The Republican Party in New York State emerged in 1855 as the heir of the Whigs, the Liberty Party, and the Softshell Democratic faction. The Republican Party's power base includes the state's rural counties, the smaller cities and towns, and (though not so much in the 1970s and early 1980s as in earlier decades) the New York City suburbs. Although New York Republicans stand to the right of the Democrats on social issues, they have usually been well to the left of the national Republican Party. The liberal "internationalist" strain of Republicanism was

New York Presidential Vote by Political Parties, 1948–84

YEAR	ELECTORAL VOTE	NEW YORK WINNER	DEMOCRAT	LIBERAL¹	REPUBLICAN	PROGRESSIVE²	SOCIALIST	SOCIALIST WORKERS	PEACE AND FREEDOM
1948	47	Dewey (R)	2,557,642	222,562	2,841,163	509,559	40,879	2,675	—
1952	45	*Eisenhower (R)	2,687,890	416,711	3,952,815	64,211	2,664	2,212	—
1956	45	*Eisenhower (R)	2,458,212	292,557	4,340,340	—	—	—	—
1960	45	*Kennedy (D)	3,423,909	406,176	3,446,419	—	—	14,319	—
							SOC. LABOR		
1964	43	*Johnson (D)	4,570,670	342,432	2,243,559		6,118	3,228	—
						AMERICAN IND.³			
1968	43	Humphrey (D)	3,066,848	311,622	3,007,932	358,864	8,432	11,851	24,517
						CONSERVATIVE⁴			COMMUNIST
1972	41	*Nixon (R)	2,767,956	183,128	3,824,642	368,136	4,530	7,797	5,641
							LIBERTARIAN		
1976	41	*Carter (D)	3,244,165	145,393	2,825,913	274,878	12,197	6,996	10,270
								RIGHT TO LIFE	CITIZENS
1980	41	*Reagan (R)	2,728,372	467,801	2,637,700	256,131	52,648	24,159	23,186
									COMMUNIST
1984	36	*Reagan (R)	3,001,285	118,324	3,376,519	288,244	11,949	—	4,226

*Won US presidential election.
¹Supported Democratic candidate except in 1980, when John Anderson ran on the Liberal line.
²Ran in the state as the American Labor Party.
³Appeared on the state ballot as the Courage Party.
⁴Supported Republican candidate.

personified during the 1960s by Governor Nelson Rockefeller, US Senator Jacob Javits, and New York City Mayor John V. Lindsay (who later became a Democrat).

The disaffection of more conservative Republicans and Democrats within the state led to the formation of the Conservative Party in 1963. At first intended as a device to exert pressure on the state Republican establishment, the Conservative Party soon became a power in its own right, electing a US senator, James Buckley, in 1970. Its power decreased in the late 1970s as the Republican Party embraced some of its positions. The Conservative Party has its left-wing counterpart in the Liberal Party, which was formed in 1944 by dissidents in the American Labor Party who claimed the ALP was Communist-influenced. Tied strongly to labor interests, the Liberals have normally supported the national Democratic ticket. Their power, however, has waned considerably in recent years.

Minor parties have sometimes meant the difference between victory and defeat for major party candidates in state and national elections. The Liberal Party line provided the victory margin in the state, and therefore the nation, for Democratic presidential candidate John F. Kennedy in 1960. Other significant, though not victorious, minor-party presidential candidates have included the American Labor Party with Henry Wallace in 1948 (8% of the vote), the Courage Party with George Wallace in 1968 (5%), and the Liberal Party with John Anderson in 1980 (7%). Among radical parties, the Socialists qualified for the presidential ballot continuously between 1900 and 1952, reaching a peak of 203,201 votes (7% of the total) in 1920.

In 1983, the state had 3,498,157 registered Democrats (51% in New York City), 2,469,853 registered Republicans (14% in New York City), 107,236 registered Conservatives (25% in New York City), 66,245 registered Liberals (40% in New York City), and 17,630 registered Right-to-Life Party members (27% in New York City).

In 1982, Lieutenant Governor Mario M. Cuomo was elected to his first term as governor, winning 2,675,213 votes on the Democratic and Liberal lines; the Republican-Conservative-Independent ticket took 2,494,827 votes. As of 1985, the Democrats and Republicans each held one seat in the US Senate; the Democrats held 19 seats in the House, the Republicans 15. In the state legislature, as of January 1985, the Democrats controlled the assembly, with 94 seats to the Republicans' 56, while the Republicans controlled the senate, with 35 seats to the Democrats' 26.

In the November 1980 presidential elections, Republican nominee Ronald Reagan (with Conservative Party backing) won the state's 41 electoral votes, apparently because John Anderson, running in New York State on the Liberal Party line, siphoned enough votes from the Democratic incumbent, Jimmy Carter, to give Reagan a plurality. Reagan carried the state again in 1984, despite the presence on the Democratic ticket of US Representative Geraldine Ferraro of Queens as the running mate of Walter Mondale; Ferraro was the first woman candidate for president or vice president on a major party ticket.

In New York City, in November 1985, Democratic Mayor Edward I. Koch was reelected to a third term.

In 1985, New York had 246 elected black officials, including 3 US Representatives—Charles Rangel of Manhattan, and Major Owens and Ed Towns of Brooklyn—20 state legislators, and 44 judges. Seven of these legislators and 5 of these judges were women. Among the 65 Hispanic elected officials in New York in 1984 was US Representative Robert Garcia of the Bronx, the first congressman of Puerto Rican origin, and 6 legislators, all of Puerto Rican extraction. There were 2 Asian elected officials that year. In 1985, 24 women served in the state legislature.

14 LOCAL GOVERNMENT

The state constitution, endorsing the principle of home rule, recognizes many different levels of local government.

In 1982, New York had 62 counties, 62 cities, 928 towns, 554 villages, and 726 school districts. In addition, there were 923 special districts, of which 841 were for fire prevention. With the exception of some counties within New York City, each county has a county attorney and a district attorney, a sheriff, a fiscal officer usually called a treasurer, a county clerk, and a commissioner of social services.

Cities are contained within counties, with one outstanding exception: New York City is made up of five counties, one for each of its five boroughs. Traditionally, counties are run by an elected board of supervisors or county legislature; however, a growing number of counties (14 in 1980) have vested increased powers in a single elected county executive. Three counties had appointed county managers. Forty cities were governed by an elected mayor and city council in 1980, 19 by a council and city manager, and 3 by elected commissioners who administer individual departments and together form the policymaking board.

Those who do not live in cities live in towns, which ranged in population in 1980 from fewer than 100 to more than 700,000. Towns are run by a town board; the most important board member is the town supervisor, who is the board's presiding officer and acts as town treasurer. A group of people within a town or towns may also incorporate themselves into a village, with their own elected mayor and elected board of trustees. Some villages have administrators or managers. Members of the village remain members of the town, and must pay taxes to both jurisdictions. The constitution grants the state legislature the power to decide which taxes the local governments may levy and how much debt they may incur.

New York City is governed by a mayor and city council, but much practical power resides in the Board of Estimate. On this board sit the city's three top elected officials—the mayor, comptroller, and City Council president. The board also includes the five borough presidents, elected officials who represent (and, to a limited extent, govern) each of the five boroughs. New York City government is further complicated by the fact that certain essential services are provided not by the city itself but by independent "authorities." The Port Authority of New York and New Jersey, for example, operates New York Harbor, sets interstate bridge and tunnel tolls, supervises the city's bus and air terminals, and operates the city's largest office complex, the World Trade Center; it is responsible not to the mayor but to the governors of New York and New Jersey. Similarly, the Metropolitan Transportation Authority, which controls the city's subways and some of its commuter rail lines, is an independent agency responsible to the state rather than the city.

15 STATE SERVICES

Educational services are provided through the Education Department. Under this department's jurisdiction are the State Library, the State Museum, the New York State School for the Blind at Batavia, and the New York State School for the Deaf at Rome. The Education Department also issues licenses for 20 professions, including architecture, engineering and land surveying, massage, pharmacy, public accountancy, social work, and various medical specialties. The state university system is administered by a separate agency headed by a chancellor.

Transportation services are under the direction of the Department of Transportation, which has responsibility for highways, aviation, mass transit, railroads, water transport, transportation safety, and intrastate rate regulation. The Department of Motor Vehicles licenses all road vehicles, motor vehicle dealers, motor vehicle operators, and driving schools.

Human services are provided through several state departments. Among the programs and facilities operated by the Department of Health are three research and treatment facilities: the New York State Veterans' Home at Oxford, Roswell Park Memorial Institute at Buffalo, and Helen Hayes Hospital at West

Haverstraw. The state provides care for the mentally ill, retarded, and alcholics and other substance-dependent persons through the Department of Mental Hygiene. In 1985, it maintained 30 psychiatric centers, and 18 developmental centers for retardation and developmental disabilities. The Department of Social Services supervises and sets standards for locally administered public and private welfare and health programs, including Medicaid and Aid to Families with Dependent Children; it has special responsibilities for the blind and visually handicapped and over Indian affairs. Other human services are provided through the Division of Veterans' Affairs, the Division of Human Rights, the Division for Youth, and the Office for the Aging, all within the Executive Department.

Public protection services include state armed forces, corrections, and consumer protection. Included within the Division of Military and Naval Affairs, in the Executive Department, are the Army National Guard, Air National Guard, Naval Militia, State Civil Defense Commission, and Disaster Preparedness Commission. The Division of State Police operates within the Executive Department, while prisons are administered by the separate Department of Correctional Services, which in 1985 operated 46 correctional facilities. The State Consumer Protection Board (Executive Department) coordinates the consumer protection activities of the various agencies and departments. The major legal role in consumer protection is played by the attorney general.

Housing services are provided through the Division of Housing and Community Renewal of the Executive Department, and through the quasi-independent New York State Housing Finance Agency, State of New York Mortgage Agency, and New York State Urban Development Corporation. The Division of Economic Opportunity (Department of State) acts as the state representative of the economically disadvantaged in dealing with local, state, and federal agencies. The Department of Commerce has an Office of Minority and Women's Business.

Natural resources protection services are centralized in the Department of Environmental Conservation. The administration of the state park and recreation system is carried out by the Office of Parks, Recreation and Historic Preservation, in the Executive Department. The Department of Agriculture and Markets serves the interests of farmers and also administers the state's Pure Food Law. Energy is the province of both the Department of Public Service, which regulates public utilities, and the State Energy Office in the Executive Department, which develops and coordinates energy policy. The quasi-independent Power Authority of the State of New York finances, builds, and operates electricity-generating and transmission facilities.

The Department of Labor provides most labor services for the state. Its responsibilities include occupational health and safety, human resource development and allocation, administration of unemployment insurance and other benefit programs, and maintenance of labor standards, including enforcement of minimum wage and other labor laws. The State Labor Relations Board and State Mediation Board, both within the Department of Labor, try to settle labor disputes and prevent work stoppages.

16 JUDICIAL SYSTEM

New York's highest court is the court of appeals, in Albany, with appellate jurisdiction only. The court of appeals consists of a chief judge and six associate judges, appointed by the governor and approved by the senate for 14-year terms. In 1984, the chief judge received a salary of $95,000; the associate judges, $92,500. Below the court of appeals is the supreme court, which in 1984 consisted of 314 justices in 12 judicial districts. The supreme court of New York State does not sit as one body; instead, most supreme court justices are assigned original jurisdiction in civil and criminal matters, while 24 justices are assigned to one of the court's four appellate divisions. Supreme court justices are elected by district

and serve 14-year terms at annual salaries of $82,000 in 1984, plus an additional $5,000 for appellate division justices and $8,000 for the four presiding justices.

The New York court of claims, which sits in Albany, consisted in 1984 of 17 judges appointed by the governor to nine-year terms, with the advice and consent of the senate. This special trial court hears civil cases involving claims by or against the state. Legislation in 1973, amended in 1982, permitted the one-time appointment of another 15 judges for nine-year terms.

Outside New York City, each county has its own county court to handle criminal cases, although some are delegated to be handled by lower courts. County court judges are elected to 10-year terms. Many counties have a surrogate's court to handle such matters as wills and estates; surrogates are elected to 10-year terms except in New York City counties, where they are elected to 14-year terms. Each county has its own family court. In New York City, judges are appointed by the mayor for 10-year terms; elsewhere they are elected for 10 years. A county's district attorney has authority in criminal matters. Most cities (including New York City) have their own court systems; in New York City, the mayor appoints judges of city criminal and family courts. Village police justices and town justices of the peace handle minor violations and other routine matters.

The Department of Correctional Services maintains correctional facilities throughout the state, as well as regional parole offices. In 1984, 33,155 inmates were in state and federal facilities.

Although 10 states had higher overall crime rates, no other state had a violent crime rate as high as New York's 914.1 per 100,000 in 1983. About 60% of all crimes reported to the police in the state occurred in New York City, which had an FBI Crime Index total of 622,877 crimes. FBI data for the New York metropolitan area showed a violent crime rate of 1,644 per 100,000 population, slightly more than three times the national average and higher than that of any other metropolitan area. The property crime rate of 6,446.8 was well above the national average of 4,629 and was the 5th highest among metropolitan areas. In New York City alone there were 1,622 cases of murder or nonnegligent manslaughter, 3,662 forcible rapes, 84,043 robberies, 43,326 aggravated assaults, 143,698 cases of burglary, 253,801 cases of larceny-theft, and 92,725 motor vehicle thefts. Buffalo ranked 2d to New York City in the state in the number of crimes reported in 1983, but its crime rate was about 39% lower.

17 ARMED FORCES

New York State ranked 2d, behind California, in defense contracts received in 1983/84 ($9.5 billion), and is home to the US Military Academy at West Point. About 10% of all US police employees work in the state. As of 30 September 1984 there were 41,405 Department of Defense personnel in the state; about 46% of them were civilians. The Army had 19,863 personnel, the Navy and Marine Corps, 4,925; and the Air Force, 14,572. About 44% of the Army personnel were at West Point. The academy, which was founded in 1802, enrolled 4,500 cadets in 1983/84. The naval facilities in the state are concentrated in Brooklyn. Griffiss Air Force Base, near Rome, is the largest Air Force facility, the next largest is Plattsburgh Air Force Base, in the extreme northeastern corner of the state. The Coast Guard's Atlantic Area command is in New York City.

In 1983 there were 1,932,000 veterans of US military service in the state. Of these, 1,598,000 had seen service during wartime. Allowing for overlapping (some veterans served in more than one war), the estimates for living veterans of wartime service were as follows: World War I, 22,000; World War II, 834,000; Korea, 340,000; Viet-Nam era, 445,000. Veterans' benefits totaling $1,840 million were paid to New Yorkers during 1982/83.

The manpower strength of the Army National Guard in the state at the end of 1984 was 23,006; the Air National Guard, 5,864 in early 1985. State and local police forces employed 60,653 in

1983; of that total, 56,563 were local employees. Police expenditures in the state during 1982 came to nearly $1.8 billion, 2d behind California. New York City had 26,239 uniformed police as of 28 February 1985; expenditures for 1984/85 were estimated at over $1 billion.

[18]MIGRATION

Since the early 1800s, New York has been the primary port of entry for Europeans coming to the US. The Statue of Liberty—dedicated in 1886 and beckoning "your tired, your poor, /Your huddled masses yearning to breathe free" to the shores of America—was often the immigrants' first glimpse of America. The first stop for some 20 million immigrants in the late 19th and early 20th centuries was Ellis Island, where they were processed, often given Americanized names, and sent onward to an uncertain future.

The first great wave of European immigrants arrived in the 1840s, impelled by the potato famine in Ireland. By 1850, New York City had 133,730 Irish-born inhabitants, by 1890, 409,224. Although smaller in number, German immigration during this period was more widespread; during the 1850s, German-speaking people were the largest foreign-born group in Rochester and Buffalo, and by 1855 about 30,000 of Buffalo's 74,000 residents were German.

The next two great waves of European immigration—Eastern European Jews and Italians—overlapped. Vast numbers of Jews began arriving from Eastern Europe during the 1880s, by which time some 80,000 German-speaking Jews were already living in New York City. By 1910, the Jewish population of the city was about 1,250,000, growing to nearly 2,000,000 by the mid-1920s. The flood of Italians began during the 1890s, when the Italian population of New York City increased from 75,000 to more than 200,000, in 1950, nearly 500,000 Italian-born immigrants were living in the state. Migration from the 1840s onward followed a cyclical pattern; as one group dispersed from New York City throughout the state and the nation, it was replaced by a new wave of immigrants.

Yankees from New England made up the first great wave of domestic migration. Most of the migrants who came to New York between 1790 and 1840 were Yankees; it has been estimated that by 1850, 52,000 natives of Vermont (20% of that state's population) had become residents of New York. There was a slow, steady migration of Negroes from slave states to New York before the Civil War, but massive black migration to New York, and especially to New York City, began during World War I and continued well into the 1960s. The third great wave of domestic migration came after World War II, from Puerto Rico. Nearly 40,000 Puerto Ricans settled in New York City in 1946, and 58,500 in 1952/53. By 1960, the census showed well over 600,000 New Yorkers of Puerto Rican birth or parentage. Many other Caribbean natives—especially Dominicans, Jamaicans, and Haitians—followed.

The fourth and most recent domestic migratory trend is unique in New York history—the net outward migration from New York to other states. During the 1960s, New York suffered a net loss of more than 100,000 residents through migration; between 1970 and 1980, the estimated net loss was probably in excess of 1,500,000, far greater than that in any other state; probably 80% of the migration was from New York City. From 1980 to mid-1983, net loss from migration was estimated at 136,000. These general estimates hide a racial movement of historic proportions: during the 1960s, while an estimated net total of 638,000 whites were moving out of the state, 396,000 blacks were moving in; during 1970-75, according to Census Bureau estimates, 701,000 whites left New York, while 60,000 blacks were arriving. According to a private study, a net total of 700,000 whites and 50,000 blacks left the state during 1975-80. It appears that many of the white emigrants went to suburban areas of New Jersey and Connecticut, but many also went to two Sunbelt states, Florida and California. Overwhelmingly, the black arrivals came from the South.

Intrastate migration has followed the familiar pattern of rural to urban, urban to suburban. In 1790, the state was 88% rural; the rural population grew in absolute terms (though not as a percentage of the total state population) until the 1880s, when the long period of decline began. New York's farm population decreased by 21% during the 1940s, 33% during the 1950s, 38% during the 1960s, and 49% during the 1970s. By 1980, 90% of all New Yorkers lived in metropolitan areas. Meanwhile, the suburban population has grown steadily. In 1950, 3,538,620 New Yorkers (24% of the state total) lived in suburbs; by 1980, this figure had grown to 7,461,161 (42% of all state residents). It should be remembered, of course, that this more than doubling of the suburban population reflects natural increase and direct migration from other states and regions, as well as the intrastate migratory movement from central cities to suburbs.

[19]INTERGOVERNMENTAL COOPERATION

New York State is a member of the Council of State Governments and its allied organizations. The state participates in many interstate regional commissions (and in commissions with the Canadian provinces of Ontario and Quebec). Among the more active interstate commissions are the Atlantic States Marine Fisheries Commission, Delaware River Basin Commission, Great Lakes Basin Commission, Interstate Oil and Gas Compact Commission, Lake Champlain Bridge Commission, New England Interstate Water Pollution Control Commission, Northeastern Forest Fire Protection Commission, and Ohio River Valley Water Sanitation Commission. In 1985, New York joined seven other Great Lakes states and two Canadian provinces in the Great Lakes Charter, for the purpose of protecting the lakes' water reserves.

The three most important interstate bodies for the New York metropolitan area are the Palisades Interstate Park Commission, Interstate Sanitation Commission, and Port Authority of New York and New Jersey. The Palisades Interstates Park Commission was founded in 1900 (with New Jersey) in order to preserve the natural beauty of the Palisades region. The Interstate Sanitation Commission (with New Jersey and Connecticut; established in 1961) monitors and seeks to control pollution within the tri-state Interstate Sanitation District. The Port Authority of New York and New Jersey, created in 1921 and the most powerful of the three, is a public corporation with the power to issue its own bonds. Its vast holdings include 4 bridges, 2 tunnels, 6 airports and heliports, 3 motor vehicle terminals, 7 marine terminals, the trans-Hudson rapid transit system, an industrial park in the Bronx, and the 110-story twin-towered World Trade Center in lower Manhattan.

Federal aid to New York state and local governments totaled $10,268,490,000 in 1983/84, higher than for any other state. Federal aid accounted for 15.6% of New York City revenues in 1983/84.

[20]ECONOMY

From the Civil War through the 1950s, New York State led the nation in just about every category by which an industrial economy can be measured. With the rise of California and the growth of other Sunbelt states, New York can no longer be described in such superlatives.

The state is no stranger to economic transitions. In the colonial and early national periods, New York was a leading wheat-growing state. When the wheat crop declined, dairying and lumbering became the state's mainstays. Then New York emerged as the national leader in wholesaling, retailing, and manufacturing—and remained so well into the 1960s. By 1973, however, the state was running neck and neck with California by most output measures, or had already been surpassed. The total labor force, the number of workers in manufacturing, and the number of factories all declined during the 1960s and 1970s.

In the mid-1980s, the state remained a national power in printing and publishing, fashion and apparel, instruments, machinery, and electronic equipment. No other state—and few countries—can match New York in the value of its banking, securities, and communications industries. Agriculture, forestry, fishing, and mining, though of local importance, are far less vital to the overall state economy. Roughly speaking, in 1982, services employed about 25% of the state's nonfarm labor force; trade 20%; manufacturing, 19%; government, 18%; and other sectors, 18%. New York's labor force is the most organized in the country. Average annual pay and personal income were 3d and 5th highest, respectively, among the states in 1983.

New York City accounted for almost 41% of the state's personal income, in 1980, 72% of its wholesale trade in 1982, virtually the entire securities industry, and much of the banking and communications activity (71% of all bank deposits in the state in mid-1982). It was also the nation's foremost manufacturing city as of 1980 and the country's 2d busiest port in 1982. New York State cannot be economically healthy if the city does not remain so. On that score, most indicators were negative until the early 1980s. The city's manufacturing base and its skilled laborers have been emigrating to the suburbs and to other states since World War II; the labor force now has a large number of poorly educated persons not well adapted to the kind of finance and service positions that increasingly dominate employment in the city. Between 1969 and 1976, the city lost 600,000 jobs. About 1 in 8 New York City residents received some form of public assistance (including Medicaid and Supplemental Security Income benefits) in 1982. With the departure of much of the middle class has come a shrinking of the city's tax base, a factor that contributed to the fiscal crisis of 1975, when a package of short-term aid from Congress, the state government, and the labor union pension funds saved the city from default.

The early 1980s saw New York's fortunes on the rise. A shift in dependence from manufacturing to services, and particularly to finance, helped the state and New York City weather the 1981–82 recession. In 1983, the state's three largest industrial and commercial employers (excluding public utilities) were all banks based in New York City. New York City gained 164,000 jobs from 1980 through 1984, 76,000 in 1984 alone.

²¹INCOME

With an income per capita of $13,146 in 1983 dollars, New York ranked 5th among the 50 states. New York's total personal income for that year was $232.3 billion, 2d behind California, and 8.5% of the US total. During the 1970s, New York registered the worst performance of all the states, as personal income in constant dollars rose only 8.7% over the decade, compared to 38.6% for the nation. During 1980–83, however, personal income rose 8%, higher than the nation's 6.3%.

The following table shows contributions to labor and proprietors' income by major economic sectors in 1982:

Services	$ 35,709,100,000
Manufacturing	33,887,300,000
Government	24,229,700,000
Wholesale, retail trade	24,121,900,000
Finance, insurance, real estate	17,689,500,000
Transportation, communications, public utilities	13,848,900,000
Contract construction	5,665,600,000
Farms	543,400,000
Mining	290,800,000
Other sectors	417,200,000
TOTAL	$156,403,400,000

In 1979, 13.4% of the population of New York State had money incomes below the poverty level, compared to 12.4% nationally.

New York's reputation for "Wall Street millionaires" is not entirely deserved: in 1982, only 5.6% of the nation's wealthiest people (those with gross assets greater than $500,000) lived in the state, compared with 15.3% in California and 10.4% in Texas.

New York City, with 40% of the state's population, accounted for 41% of state income in 1980. On a per capita basis, Manhattan ranked 1st in the state, followed by three of the city's suburban counties: Westchester, Nassau, and Rockland.

²²LABOR

The state ranks 2d to California in the size of its labor force, but 1st in the extent to which it is organized. The civilian labor force totaled 8,166,000 in November 1984, of whom 7,619,000 (93.3%) were employed and 547,000 (6.7%) were unemployed. Of the total labor force, 37% lived in New York City. In May 1985, the state unemployment rate was 7.3%, and the New York City rate was 8.4%.

In 1983, 85% of the civilian work force was white and 15% nonwhite. The participation of women increased by 22% between 1970 and 1983, when 3,524,000 workers were female and 4,539,000 were male. By occupation, administrative support workers made up 19% of the labor force, the highest ratio among the states; service workers, 15%; professional specialty workers, 14%; operators, fabricators, and laborers, 14%; executives, administrators, and managers, 11%; sales workers, 11%; and production, craft, and repair workers, 10%. With only 27% of its workers in blue-collar jobs in 1980 and 58% in white-collar jobs, New York was better prepared than many other states for the continuing decline in the 1980s of manufacturing employment.

A federal survey in March 1982 revealed the following nonfarm employment pattern in New York State:

	ESTABLISH-MENTS	EMPLOYEES	ANNUAL PAYROLL ('000)
Agricultural services, forestry, fisheries	3,056	13,741	$ 194,509
Mining	481	9,399	298,380
Contract construction	23,519	205,304	4,576,178
Manufacturing, of which:	29,942	1,419,336	29,645,907
Food, food products	(1,345)	(64,889)	(1,268,198)
Apparel, other textiles	(5,644)	(164,156)	(1,949,538)
Printing, publishing	(5,388)	(165,553)	(3,514,611)
Fabricated metal products	(2,197)	(76,572)	(1,395,115)
Nonelectrical machinery	(2,626)	(143,981)	(3,205,959)
Electric and electronic equipment	(1,351)	(166,210)	(3,567,410)
Transportation equipment	(354)	(62,280)	(1,614,847)
Instruments and related products	(763)	(103,233)	(2,807,051)
Transportation, public utilities	13,828	416,393	10,340,774
Wholesale trade	37,404	453,400	10,034,613
Retail trade, of which:	93,698	1,010,704	9,900,604
General merchandise stores	(2,252)	(129,198)	(1,113,901)
Food stores	(14,049)	(182,010)	(1,642,999)
Eating and drinking places	(24,018)	(279,860)	(1,933,808)
Finance, insurance, real estate, of which:	41,256	694,438	15,337,390
Banking	(4,738)	(236,849)	(4,635,348)
Insurance	(8,049)	(168,833)	(3,362,249)
Real estate	(21,942)	(118,828)	(1,823,586)
Security, commodity brokers and services	(2,154)	(99,048)	(3,960,462)
Services, of which:	116,555	1,800,048	27,478,534
Business services	(19,392)	(386,709)	(6,430,629)
Health services	(27,321)	(495,127)	(7,963,154)
Educational services	(2,909)	(192,260)	(2,269,661)
Legal services	(10,263)	(71,611)	(1,706,384)
Other	8,909	9,279	184,918
TOTALS	368,648	6,032,042	$107,991,803

There were about 1,319,000 government employees in New York State in November 1984. Of these, about 251,700 were employed by the state and 913,700 by local governments. The federal government employed approximately 153,600.

The average hourly manufacturing wage of $9.24 in September 1984 for production workers was exceeded by 22 states. New York City's average hourly manufacturing wage of $8.31 was below that of any metropolitan area in the state except Binghamton and Newburgh-Middletown. Among minority groups, the unemployment rate in 1983 was 14.5% for blacks and 12.2% for Hispanics; for black youths age 16 to 19, it was 45.8%. Statistics for New York City were quite similar.

At the turn of the century, working conditions in New York were among the worst in the country. The flood of immigrants into the labor market and the absence of labor laws to protect them led to the development in New York City of cramped, ill-lit, poorly ventilated, and unhealthy factories—the sweatshops for which the garment industry became notorious. Since that time, working conditions in the garment factories have improved, primarily through the efforts of the International Ladies' Garment Workers Union and, later, its sister organization, the Amalgamated Clothing and Textiles Workers Union.

According to the US Department of Labor, 2,792,000 New Yorkers belonged to unions in 1980. That figure, represented nearly 39% of the nonagricultural work force, the highest percentage in any state. In 1983 there were 3,267 labor unions in New York State; only Ohio and Pennsylvania had more. In 1984, 16 of 97 large national unions had their headquarters in New York City.

Under the Taylor Law, public employees do not have the right to strike. Penalties for striking may be exacted against both the unions and their leaders. Nevertheless, work stoppages by government employees in 1981 involved a total of 3,000 workers and accounted for 72,000 days idle. In all, there were 266 work stoppages that year affecting 69,000 workers and accounting for 876,000 days idle.

[23] AGRICULTURE

New York ranked 24th in farm income in 1983, with cash receipts from farming estimated at $2,709,937,000. About 71% came from livestock products, mostly dairy goods. The state ranked 2d in the production of apples and corn for silage for that year, 3d in tart cherries, grapes, and snap beans, 4th in pears, cauliflower, celery, and onions, 9th in hay, 10th in potatoes, and 13th in oats, but much lower for other crops.

Corn was the leading crop for the Indians and for the European settlers of the early colonial period. During the early 1800s, however, wheat was the major crop grown in eastern New York. With the opening of the Erie Canal, western New York (especially the Genesee Valley) became a major wheat-growing center as well. By the late 1850s, when the state's wheat crop began to decline, New York still led the nation in barley, flax, hops, and potato production and was a significant grower of corn and oats. The opening of the railroads took away the state's competitive advantage, but as grain production shifted to the Midwest, the state emerged as a leading supplier of meat and dairy products.

New York remains an important dairy state, but urbanization has reduced its overall agricultural potential. In 1982, 15.1% of the state's land area was devoted to crop growing; however, there were only 48,000 farms, and the farm population of about 123,000 was only 0.7% of the state total.

The west-central part of the state is the most intensively farmed. Chautauqua County, in the extreme southwest, leads the state in grape production, while Wayne County, along Lake Ontario, leads in apples and cherries. The dairy industry is concentrated in the St. Lawrence Valley; grain growing dominates the plains between Syracuse and Buffalo. Potatoes are grown mostly in Suffolk County, on eastern Long Island.

The following table shows area, production and value of leading field crops in 1983 (all figures in thousands):

CROP	AREA (ACRES)	PRODUCTION	VALUE
Hay	2,270	5,284 tons	$422,720
Corn, grain	600	54,000 bushels	202,500
Corn, silage	590	7,965 tons	196,736
Potatoes	40	9,710 cwt	71,749
Oats	200	11,400 bushels	21,090
Wheat	160	7,360 bushels	24,288

Farms in 1983 also produced 592,890 tons of vegetables. Leading vegetable crops were cabbage, onions, sweet corn, and snap beans.

State vineyards produced 191,000 tons of grapes for wine and juice in 1983, while the apple crop totaled 1.1 billion lb. Maple syrup production was 235,000 gallons in 1983.

[24] ANIMAL HUSBANDRY

New York is a leading dairy state. In 1983, New York was 3d in the US in milk and cheese production, and 7th in butter production. Of the $1,929,463,000 earned for livestock and livestock products, 83% came from milk alone.

The St. Lawrence Valley is the state's leading cattle-raising region, followed by the Mohawk Valley and Wyoming County, in western New York. The poultry industry is more widely dispersed. At the end of 1983, the state had 2,050,000 head of cattle, of which 943,000 were full-grown milk cows. Other livestock included 110,000 hogs and pigs, 61,000 sheep and lambs, 9,500,-000 chickens (excluding broilers), and 332,000 turkeys.

The following table shows volume and value for selected livestock products in 1983 (all figures in thousands):

PRODUCT	VOLUME (LB)	VALUE
Dairy:		
Milk	11,691,000	$1,608,700
Cheese (excluding cottage cheese)	398,271	NA
Creamed cottage cheese	141,261	NA
Ice cream (gallons)	49,847	NA
Butter	50,038	NA
Nonfat dry milk	83,648	NA
Meat animals:		
Cattle and calves	388,146	159,776
Hogs and pigs	14,932	7,535
Sheep and lambs	5,062	1,871
Poultry:		
Chickens and broilers	31,850	3,119
Turkeys	7,005	2,732
Ducks	31,400	14,130

The state was the nation's 13th largest egg producer during 1983, yielding 1,741,000,000, for an income of $83,423,000. Duck raising is an industry of local importance on Long Island.

[25] FISHING

Fishing, though an attraction for tourists and sportsmen, plays only a marginal role in the economic life of the state.

In 1984, the commercial catch by New York fishers was 38,902,000 lb, less than 1% of the US total and far below the 1880 peak of 335,000,000 lb. The catch was valued at $39,869,000, 16th among the states. Important species for commercial use are menhaden and, among shellfish, clams and oysters. Virtually all of New York's commercial fishing takes place in the Atlantic waters off Long Island, although 1,300,000 lb of shad was taken from the Hudson River in 1984. Montauk, on the eastern end of Long Island, was the state's leading fishing port in 1984 and 47th in the nation, with 10,900,000 lb; Hampton Bays (59th) had 8,200,000 lb.

Pollution and poor wildlife management have seriously endangered the state's commercial and sport fishing in the ocean, rivers, and lakes. Commercial fishing for striped bass in the Hudson River was banned in 1976 because of contamination by polychlorinated biphenyls (PCBs). Commercial fishing in the river for five other species—black crappie, brown bullhead, carp, goldfish, and pumpkinseed—was banned in 1985. Also banned in 1985 was commercial fishing for striped bass in New York Harbor and along both shores of western Long Island.

In recent decades, however, the Department of Environmental Conservation has taken an active role in restocking New York's inland waters. During 1983, the Division of Fish and Wildlife reached an all-time high in fish hatchery production, with more than 1,300,000 lb of fish. The US Fish and Wildlife Service distributes large numbers of lake trout and Atlantic salmon fingerlings and rainbow and brook trout fry throughout the state.

26 FORESTRY

About 60% of New York's surface area is forestland. The most densely forested counties are Hamilton, Essex, and Warren in the Adirondacks, and Delaware, Greene, and Ulster in the Catskills. The total forested area was about 18,000,000 acres (7,300,000 hectares) in 1983, of which 85% was classified as commercial forest, although only a small proportion of the state's forestland was actually being exploited. The state produced 355,000,000 board feet of lumber (mostly in hard woods) in 1983, less than 1% of the national lumber output. In 1983, the Division of Lands and Forests harvested timber from 21,268 acres (8,607 hectares) of state forestland, bringing in revenue of $2,400,000.

The state Department of Environmental Conservation maintains about 3,000,000 acres (1,200,000 hectares) in the Catskills and Adirondacks as forest preserves. The Saratoga Tree Nursery distributed 5,900,000 tree seedlings and 458,500 shrub seedlings in 1983.

27 MINING

New York accounted for virtually all the nation's production of wollastonite, a paper and paint filler, in 1984 and was the only state in which emery was produced. New York also ranked 2d in zinc output; 3d in primary aluminum shipments, salt, and garnet output; and 4th in lead and talc production. The estimated value of the state's mineral output for 1984 (excluding fossil fuels) was $541,381,000, 14th in the US.

Recoverable metals are found almost entirely in St. Lawrence and Essex counties, in the northern Adirondacks. At Tahawus, in Essex County, the state claims to have the largest titanium mine in the US. Wollastonite is also found in Essex County. St. Lawrence County leads the state in zinc, talc, lead, and silver output. Rock salt is mined in Livingston, Onondaga, Tompkins, and Wyoming counties. Emery is found in Westchester, garnets in Warren; stone, sand, and gravel are quarried throughout the state.

Estimated output of leading minerals in 1984 (in tons) included stone, 33,815,000; sand and gravel, 18,700,000; salt, 5,801,000; and clays, 551,000. Production of zinc was 56,748 metric tons; lead, 1,299 metric tons; and silver, 33,000 troy oz in 1983. In terms of value, stone led in 1984, at $143,271,000; salt was $115,496,000; and sand and gravel, $53,400,000. The combined value of cement, ball clay, emery, abrasive garnet, gypsum, lime, talc, ilmenite, wollastonite, silver, lead, and zinc was $225,985,000 in 1984. In 1983, the value of zinc production was $51,783,000; lead, $621,000; and silver, $379,000.

28 ENERGY AND POWER

Although New York State's fossil fuel resources are limited, the state ranked 3d in the US in electric power production in 1983. About one-fourth of the state's annual electric power output came from hydroelectric plants built and operated by the Power Authority of the State of New York. Almost one-third came from oil-fired units, 15% from coal-fired units, 12% from gas-fired units, and 16% from nuclear power plants.

Installed capacity in the state in 1983 was 31.9 million kw. Electrical output totaled nearly 105 billion kwh in 1983; in 1981, 75% was supplied by privately owned electric utilities.

The largest nonfederal hydroelectric plant in the US is the

Niagara Power Project, which had a capacity of 2,400,000 kw at the start of 1985. The New York side of the St. Lawrence River Power Project had a capacity of 800,000 kw in 1985.

Both plants were built and are operated by the Power Authority of the State of New York, which also built and operates a pumped-storage plant in Schoharie County (1,000,000 kw) and a nuclear power plant on Lake Ontario near Oswego (800,000 kw). Other nuclear plants in the state include two reactors at Indian Point (one operated by Consolidated Edison, and one by the State Power Authority), and units operated by the Niagara Mohawk Power Co. and the Rochester Gas & Electric Co. Since 7 July 1985, the $4.2 billion Shoreham Nuclear Power Station in eastern Long Island has been involved in a low-power testing program.

The State Energy Research and Development Authority manages the only commercial nuclear fuel reprocessing plant in the US, at West Valley in Cattaraugus County.

Sales of public and private electric power totaled 109.5 billion kwh in 1983, of which 32% went to commercial users, 29% to industrial purchasers, 29% to residential users, and 10% for other purposes. Electric bills for New York City are the highest in the nation, and customers in Buffalo and Rochester also pay above the national median. While the average usage per residential customer of electric and gas utilities in the state rose from 2,598 kwh in 1960 to 5,312 kwh in 1981 (an increase of 104%), the average annual bill ballooned from $83.84 to $506.70, a 504% increase. The run-up in oil prices during the 1970s is further reflected in the rise of New York's net energy bill from $6,368,-400,000 in 1970 to $26,698,900,000 in 1982—a 319% increase. Nevertheless, New York's per capita energy expenditure of $1,535 in 1981 was well below the national average of $1,991.

Estimated reserves of petroleum in New York State, as of 1981, were 9,070,000 barrels, or much less than 1% of the US total. Oil output in 1983 was 831,000 barrels—again, a tiny fraction of the national total. Because New York has a large number of motor vehicles and because more than half of all occupied housing units in the state are heated by oil, the state is a very large net importer of petroleum products.

The state's estimated natural gas reserves as of December 1983 were 295 billion cu feet; net production in 1983 totaled 21.6 billion cu feet. Although the state's natural gas output has risen steadily since 1966, both production and reserves represent only a tiny fraction of US totals. About 55% of natural gas sold in the state in 1983 was used for residential heating and cooking.

29 INDUSTRY

Until the 1970s, New York was the nation's foremost industrial state, ranking 1st in virtually every general category. However, US Commerce Department data show that by 1975 the state had slipped in manufacturing to 2d in number of employees, payroll, and value added, 4th in value of shipments of manufactured goods, and 6th in new capital spending. By 1982, it had risen back to 3d in value of shipments and new capital spending, but manufacturing jobs declined by 11.3% between March 1979 and March 1985.

Value of shipments by manufacturers totaled $121,451,400,000 in 1982. Leading industries included:

Women's and misses' outerwear	$6,394,400,000
Periodicals publishing and printing	5,499,400,000
Electronic components and accessories	5,201,100,000
Office and computing machines	3,721,100,000
Book publishing and printing	3,622,300,000
Communication equipment	3,375,400,000
Motor vehicles and equipment	3,332,000,000
Pharmaceuticals	2,978,200,000
Commercial printing	2,839,000,000
Aircraft and parts	2,463,300,000

The Buffalo region, with its excellent transport facilities and abundant power supply, is the main center for heavy industry in the state. Plants in the region manufacture iron and steel, aircraft, automobile parts and accessories, and machinery, as well as flour, animal feed, and various chemicals. However, the Buffalo area's biggest private industrial employer, Bethlehem Steel, closed its Lackawanna plant in 1983. Republic Steel also closed its plant, and General Motors cut employment at its Tonawanda facility by two-thirds. Light industry is dispersed throughout the state. Rochester is especially well known for its photographic (Kodak) and optical equipment (Bausch & Lomb) and office machines (Xerox); the city is the world headquarters of the Eastman Kodak Co., a world leader in photography with sales of $10.6 billion and assets of $10.8 billion in 1984. The state's leadership in electronic equipment is in large part attributable to the International Business Machines Corp. (IBM), which was founded in 1911 at Endicott, near Binghamton. In 1984, IBM ranked as the 6th-leading US industrial corporation, with sales of $45.9 billion and assets of $42.8 billion. Its world headquarters is at Armonk, in Westchester County, and it has important facilities at Endicott, Kingston, and Poughkeepsie. The presence of two large General Electric plants has long made Schenectady a leader in the manufacture of electric machinery. Grumman, a major defense contractor, was the largest private employer on Long Island in 1985, with 30,000 workers.

New York City excels not only in the apparel and publishing trades but also in food processing, meat packing, chemicals, leather goods, metal products, and many other manufactures. In addition, the city serves as headquarters for many large industrial corporations whose manufacturing activities often take place entirely outside New York. Two of the three largest US industrial corporations—Exxon and Mobil—had their headquarters in New York City in 1985, as did another 64 of the Fortune 500 largest industrial firms; 14 *Fortune* 500 firms had their headquarters in other cities in New York State.

³⁰COMMERCE

New York led the nation in wholesale-trade volume in 1982; the wholesale trade of New York City alone exceeded that of any state except New York, California, and Texas. The state ranked 3d in retail trade, behind California and Texas.

In 1982, state wholesale sales totaled $261.2 billion, or 13% of the US total; New York City's share was 72%. The most valuable categories of goods traded were petroleum and petroleum products, apparel, piece goods and notions, groceries and related products, jewelry, watches, diamonds and other precious stones, woven goods, grain, machinery, equipment and supplies, minerals and metals, and electrical goods. Except for apparel, woven goods, jewelry, and perhaps electrical goods, this list appears to reflect the importance of New York City as a port and transportation center rather than the makeup of state or city industries.

According to the economic census of March 1982, 93,698 retail establishments in New York employed 1,010,704 workers, 358,200 of them in New York City. The census of retail trade for the same year showed a total of 137,155 retail establishments, of which 28% were "mom and pop" operations with no payroll. Retail sales in the state totaled $72.3 billion, of which 97.5% was attributable to establishments with payroll; of the latter, food stores accounted for 24%; automotive dealers, 14%; general merchandise stores, 11%; eating and drinking places, 10%; apparel and accessory stores, 7%; and others, 34%. In 1983, retail sales were $81.5 billion, and employment was 1,043,200. New York State accounted for almost 7% of all US retail sales in 1982, and New York City for one-third of all retail sales in the state. In 1982, the state had 486 department stores, of which 41 were in New York City; the city had all 9 department stores employing 1,000 persons or more, and Manhattan alone had 7.

The state's long border with Canada, its important ports on Lakes Erie and Ontario, and its vast harbor on New York Bay ensure it a major role in US foreign trade. About 15% of US waterborne imports and exports pass through the New York Customs District (including New York City, Albany, and Newark and Perth Amboy, N.J.), which handled about $31.2 billion in exports and $41.4 billion in imports in 1982. New York's Canadian Border Districts (the Ogdensburg and Buffalo Customs Districts) handled $9.5 billion in exports and $13.1 billion in imports.

Manufacturing exports totaled $10.2 billion in 1981, 6th among the states. New York was 1st in exports of instruments and 2d in electric and electronic equipment. Some 172,000 jobs—about one of every eight manufacturing jobs in the state—were dependent on exports of manufactured goods. New York agricultural exports came to $109 million in 1981/82, 34th in the US.

³¹CONSUMER PROTECTION

Although the state government has a mixed record on consumer issues, one New York organization—the Consumers Union of the United States—has greatly influenced the national consumer movement.

The State Consumer Protection Board in the Executive Department was created in 1970, and is headed by an executive director appointed by the governor with the advice and consent of the senate. The board coordinates the activities of all state agencies performing consumer protection functions, represents consumer interests before federal, state, and local bodies (including the Public Service Commission), and encourages consumer education and research, but it has no enforcement powers. These are vested in the Bureau of Consumer Frauds and Protection within the Department of Law, under the direction of the attorney general. The Department of Public Service has regulatory authority over several areas of key interest to consumers, including gas, electric, and telephone rates. Especially during the 1960s and 1970s, consumer groups complained that the department was more sensitive to the interests of the utility companies than to those of consumers.

State law outlaws unfair or deceptive trade practices and provides for small-claims courts, where consumers can take action at little cost to themselves. New York licenses and regulates automobile repair services, permits advertising of prescription drug prices, and requires unit pricing. A "cooling-off" period for home purchase contracts is mandated, and standards have been established for mobile-home construction. New York has no-fault automobile insurance. In 1974, the legislature outlawed sex discrimination in banking, credit, and insurance policy transactions; the state's fair-trade law, which allowed price fixing on certain items, was repealed in 1975. The Fair Credit Reporting Act passed in 1977, allows consumers access to their credit bureau files. A 1984 "lemon law" entitles purchasers of defective new cars to repairs, a refund, or a replacement under specified circumstances. A similar law for used cars requires a written warranty for most essential mechanical components.

Extremely influential both within the state and throughout the US is the Consumers Union (CU), established as a nonprofit corporation at Mt. Vernon in 1936. CU derives its income solely from sales of its magazine, *Consumer Reports,* and other publications. The magazine embraces many consumer interests, but the bulk of each issue consists of product reports on items as varied as stereos and canned chili. Product tests are conducted by CU's own research staff. Ratings of products may not be cited in advertising or used by product manufacturers or distributors for any commercial purpose.

³²BANKING

New York City, the major US banking center, was the headquarters for five of the six largest US banks in 1984. Banking is one of the state's leading industries, ranking 1st in the US in assets and employing 236,849 people in 4,738 establishments as of March 1982; of these, 1,056 establishments and 152,610 employees were in Manhattan alone.

At the end of 1984 there were 459 banks in New York, with total assets of about $532 billion; among these were 109 banks chartered by the state with assets of $53 billion. As of 31 December 1983, New York's commercial banks held total assets of $340.5 billion; deposits totaled $200.2 billion. New York led the nation in both categories.

There were 83 mutual savings banks at the end of 1984, 72 state-chartered. As of 31 December 1983, savings banks held assets of $102 billion, 41% of the national total. Savings and loan associations with head offices in New York in 1984 numbered 87, of which 17 were state-chartered; their assets totaled $18.1 billion at the close of 1983. The state had 926 credit unions in 1984. New York is the US headquarters for many large foreign banks; in 1984, the state's Banking Department regulated 149 foreign branches and agencies with assets of $207 billion. It also supervised investment companies with assets of about $77 billion, and it licensed and regulated 658 consumer finance companies. A European district office in London helps the department monitor the overseas assets of state banks, which grew from $6.8 billion in 1968 to over $88 billion in 1984.

More than one-fifth of the assets of all insured commercial banks in the US were concentrated in the six largest New York banks at the end of 1983. At the close of 1984, Citibank, the largest New York bank, ranked 1st in the US in assets; Chase Manhattan placed 3d; Manufacturers Hanover, 4th; J. P. Morgan, 5th; Chemical New York, 6th; and Bankers Trust New York, 9th. Only two major banks, Marine Midland in Buffalo (17th) and Norstar Bancorp in Albany (46th), had their headquarters outside New York City.

The following is a list of leading New York commercial bank companies with their assets and deposits at the end of 1984:

NAME	ASSETS ('000)	DEPOSITS ('000)
Citicorp[1]	$150,586,000	$90,349,000
Chase Manhattan Corp.	86,883,018	59,680,011
Manufacturers Hanover Corp.	75,713,707	44,025,759
J. P. Morgan & Co.[2]	64,126,000	38,760,000
Chemical New York Corp.	52,236,326	33,697,684
Bankers Trust New York Corp.	45,208,147	25,559,080
Marine Midland Banks, Inc.	22,055,697	15,690,831

[1] Holding company for Citibank NA.
[2] Holding company for Morgan Guaranty Trust Co.

[33] INSURANCE

Like banking, insurance is big business in New York. Three of the five top US life insurance companies had their headquarters in New York City, and the industry employed 168,833 workers in the state, according to the federal census of March 1982. At the end of 1981, 355 New York companies and 467 out-of-state companies were licensed to sell insurance in the state.

Premiums written in the state at the end of 1981 totaled $23.5 billion for all types of insurance by both state and out-of-state companies. New Yorkers paid $3.7 billion for life insurance, $2.9 billion for health insurance, $3.8 billion for automobile insurance, and $9.3 billion for other property insurance in 1983. Automobile insurance is compulsory for all owners of motor vehicles in the state. A no-fault system is in effect.

In 1983, New Yorkers held 25.1 million life insurance policies with a value of $381.6 billion (2d in the US). The average amount of life insurance per family increased from $8,600 in 1955 to $54,900 in 1983.

At the end of 1981, 69 New York companies and 56 out-of-state companies were licensed to sell life insurance in the state. The leading life insurance companies in the state in 1984 were Metropolitan Life (2d in US), Equitable Life Assurance (3d), New York Life (5th), Teachers Insurance & Annuity (9th), and Mutual of New York (13th).

A total of 594 fire and casualty companies were licensed in the

state at the end of 1981, including 212 New York companies, 345 out-of-state US companies, and 37 foreign companies. The leading fire, marine, and casualty companies with headquarters in New York in 1984 were American Re-Insurance, Firemens Insurance–Newark, Continental Insurance, Motors Insurance, and Royal Insurance of America, all in New York City.

Four companies provided hospital insurance and eight offered medical, surgical, and dental insurance at the end of 1981. In 1980 there were 23,096,724 participants in these plans, including 18,077,554 in various Blue Cross plans, which covered the Greater New York area.

The New York Insurance Exchange, designed primarily as a marketplace for high-risk commercial insurance and modeled after Lloyd's of London, began operations on 31 March 1980. Premium volume in 1983 was estimated at $300 million.

[34] SECURITIES

New York City is the capital of the US securities market. At the end of 1984, 38% of the 344,200 securities personnel in the US were in New York State and 36% were in New York City. The New York Stock Exchange (NYSE) is by far the largest organized securities market in the nation. It began as an agreement among 24 brokers in 1792; the exchange adopted its first constitution in 1817 and took on its present name in 1863. A clear sign of the growth of the NYSE is the development of its communications system. Stock tickers were first introduced in 1867; a faster ticker, installed in 1930, was capable of printing 500 characters a minute. By 1964, this was no longer fast enough, and a 900-character-a-minute ticker was introduced. Annual registered share volume increased from 1.8 billion in 1965 to 7.6 billion in 1978 following the introduction in 1976 of a new data line capable of handling 36,000 characters a minute. In 1984, share volume was over 23 billion.

The value of all shares traded on the NYSE totaled $764.7 billion in 1984, nearly 86% of the US total. Bond volume for the same year was nearly $7 billion (par value). Listings included 2,319 stock issues of 1,543 companies, with a total market value of nearly $1.6 trillion, and 3,751 bond issues by 1,024 issuers, with a market value of over $1 trillion. The NYSE had 628 member organizations in 1984; the membership price ranged between $290,000 and $400,000, well below the all-time high of $625,000 in 1929. The New York Futures Exchange was incorporated in 1979 as a wholly owned subsidiary of the NYSE and began trading in 1980. It also deals in options on futures.

The American Stock Exchange (AMEX) is the 2d-leading US securities market, but the AMEX ranks far below the NYSE in both volume and value of securities. The AMEX traces its origins to the outdoor trading in unlisted securities that began on Wall and Hanover streets in the 1840s, the exchange was organized as the New York Curb Agency in 1908, and adopted its current name in 1953. Constitutional changes in 1976 for the first time permitted qualified issues to be traded on both the AMEX and the NYSE as well as on other exchanges. This Intermarket Trading System (ITS) began in 1978. In 1984, more than 1.5 billion stock shares valued at nearly $21.5 billion were traded on the AMEX. The AMEX had 661 regular memberships at the end of 1984. The price of a seat on the exchange ranged from $160,000 to $255,000; the all-time high, in 1983, was $325,000.

The National Association of Securities Dealers Automated Quotations (NASDAQ) is a highly active exchange for over-the-counter securities. In 1983, total volume was 15.9 billion shares valued at $188.3 billion.

New York City is also a major center for trading in commodity futures. Leading commodity exchanges are the New York Coffee and Sugar Exchange; the New York Cocoa Exchange; the New York Cotton Exchange; the Commodity Exchange, Inc. (COMEX), specializing in gold, silver, and copper futures; and the New York Mercantile Exchange, which trades in futures for potatoes, platinum, palladium, silver coins, beef, and gold, among other items.

Bonds may be issued in New York by cities, counties, towns, villages, school districts, and fire districts, as well as by quasi-independent authorities.

³⁵PUBLIC FINANCE

New York State and New York City have the 2d- and 3d-largest budgets (behind California), respectively, of all states or municipalities in the US. The wide range of services offered, combined with a shrinking tax base, led to serious financial trouble for both the state and the city during the mid-1970s.

The New York State budget is prepared by the Division of the Budget and submitted annually by the governor to the legislature for amendment and approval. The fiscal year runs from 1 April to 31 March.

Under the Rockefeller administration, the state budget expanded rapidly from about $1.8 billion in 1958/59 to $8.5 billion in 1973/74. This trend has continued. The following is a summary of estimated state revenues and expenditures for 1984/85, and recommended revenues and expenditures for 1985/86 (in thousands), on a cash basis:

REVENUES	1984/85	1985/86
Personal income taxes	$10,510,000	$11,340,000
User taxes and fees	5,637,000	6,020,000
Business taxes	3,006,000	3,326,000
Federal grants	8,386,000	9,200,000
Bond funds	314,000	150,000
Other receipts	8,074,000	8,797,000
TOTALS	$35,927,000	$38,833,000

EXPENDITURES		
Grants to local governments	$19,589,000	$21,305,000
State operations	7,856,000	8,527,000
Transfers to other funds	3,478,000	3,563,000
Capital projects	1,767,000	2,010,000
General state charges	1,741,000	1,840,000
Debt service	1,160,000	1,137,000
TOTALS	$35,591,000	$38,382,000

Actual consolidated appropriations for 1984/85 totaled $46.1 billion, but less than the full amount of the appropriation is usually spent within the fiscal year to which it pertains. Of the $32.7 billion in appropriations to major agencies, 32% went to the Department of Social Services, 20% to the Department of Education, 9% to the Department of Transportation, and 7% to the Department of Labor. The State University of New York, City University Senior Colleges, and Higher Education Services Corporation combined received 11%, and the Department of Health and Office of Mental Health together received 6%. Another big appropriation was nearly $4 billion for two funds of the Municipal Assistance Corporation.

The annual budget for New York City is prepared by the Office of Management and Budget and submitted to the Board of Estimate for revision and approval. The fiscal year is 1 July–30 June.

In 1952/53, the city's budgeted appropriations totaled $1.5 billion; 20 years later, that total had risen to $9.9 billion. The following is a summary of actual budgeted revenues and expenditures for 1983/84 and 1984/85 (forecast), in millions:

REVENUES	1983/84	1984/85
Real estate tax	$ 3,957	$ 4,210
Other taxes	5,618	6,424
Federal categorical grants	2,420	2,567
State categorical grants	2,915	3,500
Other receipts	2,230	2,487
TOTALS	$17,140	$19,188

EXPENDITURES	1983/84	1984/85
Department of Social Services	$ 3,900	$ 4,235
Board of Education	3,642	4,021
City University	193	402
Police	962	1,040
Health and Hospitals Corporation	622	645
Fire	493	504
Pensions	1,534	1,595
Debt service	1,237	1,329
Municipal Assistance Corp. Debt service	565	570
Other expenditures	3,987	4,847
TOTALS	$17,135	$19,188

As of 1981/82, the total debt of state and local governments in New York was $49.5 billion, or 12.4% of the total for all 50 states; on a per capita basis, New York's debt ($2,816) ranked 7th among the 50 states. New York City's inability to service its $12.3 billion debt in 1975 brought it to the brink of bankruptcy. For years the city had kept afloat through clever bookkeeping: postponing current expenses, using capital funds to finance current deficits, increasing fringe benefits instead of wages for its municipal employees, counting as current revenues millions of dollars in federal funds that had not been—and would not be—granted. In addition, New York City had been providing services—including a tuition-free university and an extensive network of municipal hospitals—matched by no other US city.

By 1975, with the city increasingly in the red, investors had decided that city notes were too risky; unable to raise cash, the city could not redeem the notes of earlier investors. A moratorium on repayments was declared, and a new bond-issuing agency, the Municipal Assistance Corp. (MAC, or "Big Mac"), was created. Even so, extensive borrowing from union pension funds and the establishment by the US Congress of a short-term federal line of credit were necessary before bankruptcy was averted. New York City's debt as of 31 December 1982 was down to $7.1 billion; however, if MAC's debt of $7.7 billion is included, the city's gross debt was actually $14.8 billion.

³⁶TAXATION

On a per capita basis, New York's taxes are well above the national average, but far from the highest. State tax revenues rank 2d to California's.

Personal income tax is the state's largest source of revenue. Personal income tax rates in 1984 ranged from 2% on the first $1,000 of taxable income to 14% on income over $23,000. The maximum tax rate on earned income was reduced to 9.5% from 10% in 1985. The basic corporate tax was 10% on net income. The state imposes a 4% sales tax, but cities and towns may levy an additional tax. The estate tax ranges from 2% of the first $50,000 to a maximum of 21% on amounts over $10,100,000. Other taxes include charges on motor fuel, cigarettes, alcoholic beverages, petroleum gross receipts, and motor vehicle and highway usage, plus a bank tax, unincorporated business tax, insurance taxes, pari-mutuel taxes, and a real estate transfer and property gains tax. Tax receipts in 1984/85 totaled $20.7 billion, with $22.1 billion estimated for 1985/86.

In 1983/84, 41% of New York City's tax revenues came from general property taxes, which were estimated at nearly $4 billion. In 1984, the city also had a 4% sales tax, an income tax ranging from 0.9% on the first $1,000 of personal income to 4.3% on $25,000 or more, and a general corporation tax of 9% on income attributable to doing business in the city.

New York ranked 2d to California in federal taxes paid and federal expenditures received in 1982/83. The state tax burden was $48.4 billion ($2,744 per capita), while its share of federal spending was $50.1 billion, a slightly favorable ratio. In 1983, New Yorkers filed nearly 7.3 million personal income tax returns and paid taxes totaling $23.7 billion.

37 ECONOMIC POLICY

To attract new businesses and to discourage established enterprises from leaving, the state launched an aggressive program of development incentives during the 1970s.

Among the incentives offered by New York are government-owned industrial park sites, state aid in the creation of county and city master plans, state recruitment and screening of industrial employees, programs for the promotion of research and development, and state help in bidding on federal procurement contracts. The state agency with primary responsibility for development planning is the Division of Policy Development and Research within the Department of Commerce. Representatives of the Division of International Commerce call on firms in Canada, Asia, and Europe; the division maintains offices in London, Tokyo, Montreal, Toronto, and Wiesbaden, West Germany. The Division of Regional Economic Development encourages the retention and expansion of existing facilities and the attraction of new job-creating investments. Other divisions aid small business and minority and women's business.

The state provides specific tax incentives for economic development. Among these are an investment tax credit of 6% on new buildings, an additional credit of 3% for the next three years if a certain number of jobs are added in New York, a job incentive credit, a research and development tax credit of 10% of the value of such facilities, exemption of capital gains on a new business held at least six years, and carryover of any net operating loss. Tax credits are also offered on the purchase and maintenance of pollution-control equipment.

38 HEALTH

Health presents a mixed picture in New York State. The state has some of the finest hospital and medical education facilities in the US, but it also has large numbers of the needy with serious health problems.

Infant deaths per 1,000 live births were 10.6 for whites and 19.9 for blacks in 1981; both rates were close to the national averages. In 1983, the combined infant death rate was 11.5. The birthrate of 14.1 was below the national average. New York State was one of the first states to liberalize its abortion laws, in 1970. A total of 147,348 legal abortions were performed in the state in 1983, of which 94,085 were in New York City. The 1981 state ratio of 731 legal abortions to 1,000 live births was higher than that of any other state.

The death rate of 9.6 per 1,000 residents in 1981, tied for 6th in the US, reflected New York's older-than-average population; the rate was the same in 1983. The state ranks above the national average in deaths due to heart disease, malignant neoplasms (cancer), influenza and pneumonia, and cirrhosis of the liver, but below the national average in deaths due to cerebrovascular diseases, atherosclerosis, suicide, and accidents. Leading causes of death in 1983 (with their rates per 100,000 population) included heart disease, 417; cancer, 216; cerebrovascular diseases, 58; pneumonia and influenza, 30; accidents, 27; diabetes, 18; cirrhosis of the liver, 17; homicides, 12; and suicides, 9.

Major public health problems in the 1980s included drug abuse, alcoholism, venereal disease, and acquired immune deficiency syndrome (AIDS). There are no reliable estimates for the total number of drug addicts, although at the end of 1982, 40,640 persons (75% of them in New York City) were receiving treatment at state-run facilities. The Division of Substance Abuse Services of the Department of Mental Hygiene funds and assists over 400 community-based programs, and is assisted by the Advisory Council on Substance Abuse.

Only 5,387 patients were admitted to the 13 state-run facilities for alcoholism treatment in 1982, but 117,757 were admitted (74,016 nonresidential) to state and local programs during the year. The New York State Research Institute on Alcoholism opened in Buffalo in 1970, and the Advisory Committee on Alcoholism Services operates under the direction of the Department of Mental Hygiene.

New York City was heavily hit by the outbreak, first recognized in 1981, of AIDS, usually fatal in time. Of 12,256 cases reported to the US Centers for Disease Control in Atlanta by 5 August 1985, 33% were in New York City.

During one week in 1982, 127,510 persons were seen at programs operated, certified, or funded by the State Office of Mental Health. There has been a gradual shift away from state treatment of psychiatric patients toward treatment in community mental health facilities. The number of resident patients in state psychiatric centers was 23,942 in 1982, compared to 93,314 in 1955. About 60% of noncriminal adult admissions during the year were involuntary; among resident patients in 1982, about 60% had been resident for five years or more.

In 1983, New York State had 338 hospitals, with 122,150 beds, more than any other state. In 1982, hospitals in New York recorded 2,775,240 admissions, performed 1,204,626 surgical operations, had an occupancy rate of 87.9% (2d highest in the nation), and employed 273,031 persons with a total payroll of $4.5 billion. Medical personnel licensed to practice in the state in 1984 included 47,447 doctors, 14,397 dentists, 165,433 registered nurses, and 70,221 practical nurses.

More than 88% of all New Yorkers are covered by hospital and surgical insurance. About $2.3 billion in private health insurance benefits was paid in 1982. The average daily cost per patient per day in hospitals in 1982 was $312, 21st highest in the US and 15% below the national average; however, the average cost per patient stay was $3,025, 3d highest and 18% above the national average.

39 SOCIAL WELFARE

Social welfare is a major public enterprise in the state; the growth of poverty relief programs has been enormous. In 1982, on the average, 1,296,375 New Yorkers were receiving public assistance, a 28% drop from the high of 1,802,086 in 1972 but a 72% increase over the figure for 1965. New York had the highest per capita state and local spending for welfare in 1980, $358. The number of persons qualifying under the Aid for Dependent Children program was 1,084,121 in 1982, when the average payment of $129 per person per month was 10th highest among the states; in 1983, the state total was 1,115,500. In all, the cost of state-aided public assistance and services programs in 1982 was $11.2 billion, of which $3.4 billion was funded by the state and $2.2 billion by local communities. New York City accounted for 57% of all welfare expenditures in the state and had two-thirds of all state welfare recipients (about 12% of the city's population).

About 1,863,000 New Yorkers (1st in the US) took part in the federal food stamp program on 30 September 1983. The estimated cost of the program in 1982/83 was $957,000,000, of which the federal share was $896,000,000. On 30 September 1983, 1,423,000 pupils participated in the National School Lunch Program; the cost to the federal government for this program was $177,168,899 in 1982/83.

Federal Social Security benefits were paid to 1,971,722 retired workers and their dependents in December 1982. Total Social Security payments were $10.1 billion in 1983, for an average monthly payment for retired workers of $451.37 (3d highest in the US). Survivors' benefits were paid to 542,583 state residents in December 1982; total payments were $2.8 billion in 1983. The aged received an additional $238,764,000 through the Supplemental Security Income program in 1983. Disability benefits in Social Security and supplemental programs totaled $2.1 billion during the same year.

The state had 669 nursing homes with 101,400 beds and 97,300 resident patients in 1980. The poor quality of many nursing homes was revealed in a statewide scandal in the mid-1970s, after which the legislature passed new laws regulating nursing-home practices. Public expenditures for nursing-home care in the state came to $1.4 billion in 1982. In mid-1982 there were 2,347,000 Medicare enrollments in the state; Medicare payments for the year came to $4.6 billion. In 1983, there were 2,161,000 Medicaid recipients; payments came to $6.3 billion in 1982/83, 1st among states and 19% of the US total. Federal aid for this program was $3.3 billion.

In 1982, $760,200,000 was spent on workers' compensation; 237,023 persons in all regular programs received unemployment insurance benefits during the same year. Benefits paid totaled $1.2 billion; the average weekly benefit was $99, 17% below the national average.

⁴⁰HOUSING

Census data show that housing in New York State differs in many important ways from the national housing pattern. In 1980, the state had 6,867,638 housing units, of which about 2,946,000 were in New York City. The first striking feature of the state's housing stock is its age: over 43% of year-round housing units in the state (49% in the city) were in structures built in 1939 or before compared with the US average of 26%. In Buffalo, 73% were built in 1939 or before, the highest percentage among 58 major US cities. Only 12% of New York State's dwelling units were built between 1970 and 1980 (lowest percentage among the states); only 7% of New York City's and 2.7% of Buffalo's were built during those years, while the US average was 26%.

A second striking feature of housing in the state is the dominance of multi-unit dwellings. New York State had the lowest percentage of owner-occupied housing (48.6%) of all states in the US in 1980. That year, 39% of the state's year-round housing units were single-unit detached structures, compared with a national average on 62%; 36% of all units in the state (but only 17.5% of all units in the US) were in buildings with five housing units or more. Even within New York State, New York City is unique. In 1980, 64.5% of the city's housing units were in structures of five units or more, by far the highest on a list of 58 major US cities; for Buffalo, the comparable figure was 14%. There are great differences within New York City: in 1980, 65% of the housing units in the Bronx and 82% of those in Manhattan were in structures of 10 units or more, compared with 31% in Queens and 11% on Staten Island. Other housing differences in New York City offer far greater contrasts than units per structure: the posh penthouses on the East Side of Manhattan and the hovels of the South Bronx both count as multi-unit dwellings. Characteristic of housing in New York is a system of rent controls that began in 1943 and in 1985 covered 61% of the city's rental dwelling units.

In 1980, New York had a lower vacancy rate for rental units (4.1%) than any other state, and tied for 3d for the lowest vacancy rate in owner-occupied units (1.2%). The tight housing market—which may have contributed to the exodus of New Yorkers from the state—was not helped by the slump in housing construction from the mid-1970s to the mid-1980s. In 1972, permits were issued for 111,282 units valued at $2.1 billion, by 1975, however, only 32,623 units worth $756 million were authorized, and in 1982 there were only 25,280 units worth $1.1 billion. In New York City, more units were demolished than built every year from 1974 to 1981, the last year for which data were available. The drop in construction of multi-unit dwelling was even more noticeable: from 64,959 units in 1972 to 11,740 units in 1982. The overall decline in construction was coupled with a drastic drop in new public housing. A state report found that, on an average night in 1983, 20,120 people—85% of them in New York City—spent the night in emergency accommodations. The total number of homeless in the state was estimated at 40,000–50,000. More than 143,000 poor families were 'doubled up' in the homes of other people.

Direct state aid to housing is limited. Governmental and quasi-independent agencies dealing with housing include the Division of Housing and Community Renewal of the Executive Department, which makes loans and grants to municipalities for slum clearance and construction of low-income housing, is responsible for supervising the operation of 427 housing developments (1984), and administers rent-control and rent-stabilization laws; the New York State Housing Finance Agency, empowered to issue notes and bonds for various construction projects, not limited to housing; the State of New York Mortgage Agency, which may purchase existing mortgage loans from banks in order to make funds available for the banks to make new mortgage loans, and which also offers mortgage insurance; and the New York State Urban Development Corporation (UDC), a multibillion-dollar agency designed to raise capital for all types of construction, including low-income housing.

⁴¹EDUCATION

The educational establishment of New York State is larger and better funded than that of most countries. New York has one of the nation's largest public university systems and an extensive system of private schools and universities.

The percentage of high school graduates—66.3% in 1980—among all New Yorkers was slightly below the US average, but the percentage of black high school graduates was much better than the national average: 58% for New York, as compared with 51% for the US. The percentage of New York college graduates—17.9%—was above the US average.

In the fall of 1983, a total of 2,661,041 students were enrolled in public elementary and secondary schools in the state; of these, nearly 6% were in kindergarten, 40% in grades 1–6, 49% in grades 7–12, and 5% in ungraded classes for the handicapped. Public schools in the state employed 193,362 professionals in the fall of 1982, of whom 167,172 were classroom teachers. About 29% of the teachers and 34% of the students were in New York City. The average dropout rate for public high schools in the state was 6.8% in 1982/83. Each of the boroughs of New York City except Staten Island had a dropout rate higher than the statewide average; the rate for Manhattan was 24.7%, up from 10.8% in 1975 and 15.4% in 1980.

New York State had 2,298 nonpublic elementary and secondary schools in the fall of 1982, with enrollment of 570,460. Of these, the 1,071 Roman Catholic schools had 70% of the total enrollment; there were 224,544 pupils in grades K–6 and 175,997 in grades 7–12. In addition, the state had 223 Jewish schools, with 59,240 pupils; 60 Lutheran schools, with 10,540 students; 31 Episcopalian schools, with 5,896 students; 44 Seventh Day Adventist schools, with 3,715 pupils; 266 schools (23,698 pupils) affiliated with other religious groups (including Society of Friends, Mennonites, Greek Orthodox, Russian Orthodox, Methodists, Baptists, Christian fundamentalists, and Muslims); 494 nondenominational schools, enrolling 57,218 students; and 109 schools run by public agencies, with 9,612 students.

There were 298 institutions of higher learning in 1983/84, 86 of them public. Enrollment in all institutions of higher learning totaled 1,012,421 in the fall of 1982. Of that total, 347,003 degree-credit students were enrolled full-time at public colleges and universities, and 289,617 at private institutions. About 23% of all full-time students were in public two-year colleges. A total of 122,501 degrees (excluding two-year degrees) were confirmed in 1981/82, of which 20% were in business and management, 12% in education, 11% in social sciences, 7% in engineering, and 4% in the health professions. The remaining degrees were in a wide variety of other fields.

There are two massive public university systems: the State University of New York (SUNY) and the City University of New York (CUNY). Established in 1948, SUNY in 1984 was the largest university system in the country, with 4 university centers, 4 health sciences centers, 13 university colleges of arts and sciences, 4 specialized colleges, 6 agricultural and technical colleges, 5 statutory colleges (allied with private universities), and

30 locally sponsored community colleges. SUNY's total enrollment in the fall of 1984 was 369,030, of whom 235,729 students were full-time. Of the university centers, Buffalo enrolled 22,952 students; Albany, 15,938; Stony Brook, 14,667; and Binghamton, 11,936. The City University of New York was created in 1961, although many of its 11 component institutions were founded much earlier. CUNY's total degree-credit enrollment in the fall of 1982 was 176,328, of whom 105,387 were full-time students. Under an open-enrollment policy adopted in 1970, every New York City resident with a high school diploma is guaranteed the chance to earn a college degree within the CUNY system (which CUNY campus the student attends is determined by grade point average).

The oldest private university in the state is Columbia University, founded in New York City as King's College in 1754. Columbia had 18,750 students in 1984/85; another 2,416 (all women) were enrolled in Barnard College in 1983/84, and there were approximately 4,000 students (male and female) in Columbia University Teachers College in 1984/85. Other major private institutions are Cornell University in Ithaca (1865), with 17,146 students in 1983/84; Fordham University in Manhattan and the Bronx (1841), 10,020; New York University in Manhattan (1831), 23,000; Rensselaer Polytechnic Institute in Troy (1824), 6,687; St. John's University in Queens (1870), 19,287; Syracuse University (1870), 21,288; the University of Rochester (1850), 6,969. Among the state's many smaller but highly distinguished institutions are Hamilton College, the Juilliard School, the New School for Social Research, Rockefeller University, Sarah Lawrence College, Vassar College, and Yeshiva University.

The educational work of New York State is vested in the Department of Education, under the direction of the Regents of the University of the State of New York. The Board of Regents consists of 16 persons elected, one each year, to 7-year terms by the state legislature. The commissioner of education, who heads the state Department of Education and is appointed by the Regents, also serves as president of the University of the State of New York (which should not be confused with SUNY). Unique features of education in the state are the "Regents exams," uniform subject examinations administered to all high school students, and the Regents Scholarships Tuition Assistance Program (TAP), a higher-education aid program that in 1981/82 distributed $283,445,000 in tuition assistance to 315,164 students. Recipients of these awards must be in full-time attendance at an approved institution in New York State and must have resided in the state at least one year before enrollment. More than 410,000 student loans worth $963,369,888 were guaranteed by the state in 1982/83. The state passed a "truth in testing" law in 1979, giving students the right to see their graded college and graduate school entrance examinations, as well as information on how the test results were validated.

During 1982, the public school system consisted of 735 districts; New York City constituted a single district. Receipts of all public school systems in the state totaled $12.2 billion in 1983/84, of which nearly $4.9 billion came from state sources and $430 million from federal sources. Teachers' salaries averaged $27,330 in 1983/84, 2d only to Alaska's and 24% above the US average.

[42]ARTS

New York City is the cultural capital of the state, and leads the nation in both the creative and the performing arts.

The state's foremost arts center is Lincoln Center for the Performing Arts, in Manhattan. Facilities at Lincoln Center include Avery Fisher Hall (which opened as Philharmonic Hall in 1962), the home of the New York Philharmonic; the Metropolitan Opera House (1966), where the Metropolitan Opera Company performs; and the New York State Theater, which presents both the New York City Opera and the New York City Ballet. Also at Lincoln Center are the Juilliard School and the Library and Museum of the Performing Arts. The best-known arts center outside New York City is the Saratoga Performing Arts Center at Saratoga Springs. During the summer, the Saratoga Center presents performances by the New York City Ballet and the Philadelphia Orchestra. Artpark, a state park at Lewiston, has a 2,324-seat theater for operas and musicals, and offers art exhibits during the summer. Classical music, opera, and plays are performed at the Chautauqua Festival, which has been held every summer since 1874.

In addition to its many museums, New York City has more than 350 galleries devoted to the visual and plastic arts. The city's most famous artists' district is Greenwich Village, which still holds an annual outdoor art show, although after the 1950s many artists moved to SoHo (Manhattan on the West Side between Canal and Houston Streets), NoHo (immediately north of Houston Street), the East Village, and Tribeca (between Canal Street and the World Trade Center). By the early 1980s, artists seeking space at reasonable prices were moving to Long Island City in Queens, to areas of Brooklyn, or out of the city entirely, to places such as Hoboken and Paterson in New Jersey. During the late 1940s and early 1950s, abstract painters—including Jackson Pollock, Mark Rothko, and Willem de Kooning—helped make the city a center of the avant-garde. At the same time, poets such as Frank O'Hara and John Ashbery sought verbal analogues to developments in the visual arts, and an urbane, improvisatory literature was created. New York has enjoyed a vigorous poetic tradition throughout its history, most notably with the works of Walt Whitman (who served as editor of the *Brooklyn Eagle* from 1846 to 1848) and through Hart Crane's mythic vision of the city in his long poem *The Bridge*. The emergence of New York as the center of the US publishing and communications industries fostered the growth of a literary marketplace, attracting writers from across the country and the world. Early New York novelists included Washington Irving, Edgar Allan Poe, and Herman Melville; among the many who made their home in the city in the 20th century were Thomas Wolfe and Norman Mailer. The simultaneous growth of the Broadway stage made New York City a vital forum for playwriting, songwriting, and theatrical production.

There are more than 35 Broadway theaters—large theaters in midtown Manhattan presenting full-scale, sometimes lavish productions with top-rank performers. "Off Broadway" productions are often of high professional quality, though typically in smaller theaters, outside the midtown district, often with smaller casts and less costly settings. "Off-Off Broadway" productions range from small experimental theaters on the fringes of the city to performances in nightclubs and cabarets. The New York metropolitan area has hundreds of motion picture theaters—more than 65 in Manhattan alone, not counting special series at the Museum of Modern Art and other cultural institutions. In the 1970s, New York City made a determined and successful effort to attract motion picture production companies.

New York's leading symphony orchestra is the New York Philharmonic, whose history dates back to the founding of the Philharmonic Society of New York in 1842. Among the principal conductors of the orchestra have been Gustav Mahler, Josef Willem Mengelberg, Wilhelm Furtwangler, Arturo Toscanini, Leonard Bernstein, Pierre Boulez, and Zubin Mehta. Leading US and foreign orchestras and soloists appear at both Avery Fisher Hall and Carnegie Hall, built in 1892 and famed for its acoustics. Important orchestras outside New York City include the Buffalo Philharmonic, which performs at Kleinhans Music Hall, the Rochester Philharmonic, and the Eastman Philharmonia, the orchestra of the Eastman School of Music (University of Rochester).

New York City is one of the world centers of ballet. Of special renown is the New York City Ballet, whose principal choreographer until his death in 1983 was George Balanchine. Many other ballet companies, including the American Ballet Theatre and the Alvin Ailey American Dance Theatre, make regular appearances in New York. Rochester, Syracuse, Cooperstown, Chautauqua, and Binghamton have opera companies, and Lake George has an opera festival.

Jazz and popular artists perform at more than 60 night spots in New York City. The Westbury Music Fair (Long Island) presents a wide-ranging annual program of musical entertainment, and many leading US performers play the Catskill resorts regularly. New York City is a major link in the US songwriting, music-publishing, and recording industries.

The New York State Council on the Arts is the nation's leading state arts funding mechanism. Its 1984/85 appropriation of $40.2 million was about 25% of the total for state arts agencies.

[43]LIBRARIES AND MUSEUMS

New York State has three of the world's largest libraries, and New York City has several of the world's most famous museums.

The state had 725 public libraries, 248 academic and research libraries, and 142 state institutional libraries in 1984. The New York State Library in Albany coordinated 22 public library systems covering every county in the state, with book holdings of 51,442,784 volumes and a combined circulation of 87,147,714 volumes. The public libraries received $293,038,081 during that year, of which $2,141,823 came from the federal government, $39,886,229 from the state government, and the rest from local sources. The New York State Library alone had over 1,900,000 volumes in 1984/85.

The leading public library systems and their operating statistics as of 1982 were the New York Public Library, 10,837,140 volumes and 8,491,584 circulation; Nassau County system, 5,785,200 volumes and 11,060,274 circulation; Brooklyn Public Library, 3,808,398 volumes and 7,157,497 circulation; Queens Borough Public Library, 4,646,327 volumes and 7,850,942 circulation; Suffolk Cooperative system, 4,147,070 volumes and 9,341,493 circulation; and Buffalo and Erie County system, 3,300,374 volumes and 5,562,300 circulation.

Chartered in 1895, the New York Public Library (NYPL) is the most complete municipal library system in the world. The library's main building, at 5th Avenue and 42d St., is one of the city's best-known landmarks; 83 operating branch libraries and three bookmobiles serve the needs of Manhattan, the Bronx, and Staten Island. The NYPL is a repository for every book published in the US, with a book stock of 7,582,385 volumes in 1984. The NYPL also operates the Library and Museum of the Performing Arts at Lincoln Center.

Two private university libraries—at Columbia University (5,270,432 volumes in 1983) and Cornell University (4,640,744)—rank among the world's major libraries. Other major university libraries in the state, with their 1982 book holdings, are Syracuse University, 2,065,873; New York University, 1,791,438; the State University of New York at Buffalo, 1,738,143; and the University of Rochester, 1,305,339.

There are more than 500 museums in New York State; about 150 are major museums, of which perhaps 80% are in New York City. In addition, some 300 sites of historic importance are maintained by local historical societies.

Major art museums in New York City include the Metropolitan Museum of Art, with more than 1 million art objects and paintings from virtually every period and culture; the Cloisters, a branch of the Metropolitan Museum devoted entirely to medieval art and architecture; the Frick collection; the Whitney Museum of American Art; the Brooklyn Museum; and two large modern collections, the Museum of Modern Art and the Solomon R. Guggenheim Museum (the latter designed by Frank Lloyd Wright in a distinctive spiral pattern). The Jewish Museum, the Museum of the American Indian, and the museum and reference library of

the Hispanic Society of America specialize in cultural history. The sciences are represented by the American Museum of Natural History, famed for its dioramas of humans and animals in natural settings and for its massive dinosaur skeletons; the Hayden Planetarium; the New York Botanical Garden and New York Zoological Society Park (Bronx Zoo), both in the Bronx. Also of interest are the Museum of the City of New York, the Museum of the New-York Historical Society, the South Street Seaport Museum, and the New York Aquarium.

The New York State Museum in Albany contains natural history collections and historical artifacts. Buffalo has several museums of note, including the Albright-Knox Art Gallery (for contemporary art), the Buffalo Museum of Science, and the Buffalo and Erie County Historical Society museum. Among the state's many other fine museums, the Everson Museum of Art (Syracuse), the Rochester Museum and Science Center, the National Baseball Hall of Fame and Museum (Cooperstown), and the Corning Museum of Glass deserve special mention. Buffalo, New Rochelle, Rochester, Syracuse, and Utica have zoos.

[44]COMMUNICATIONS

New York City is the hub of the entire US communications network. Postal service was established in New York State in 1692; at the same time, the first General Letter Office was begun in New York City. By the mid-19th century, postal receipts in the state accounted for more than 20% of the US total. "Fast mail" service by train started in the 1870s, with the main routes leading from New York City to either Chicago or St. Louis via Indianapolis and Cincinnati. Mail was carried by air experimentally from Garden City to Mineola, Long Island, in 1911; the first regular airmail service in the US started in 1917, between New York City and Washington, D.C., via Philadelphia. There were 71,806 postal employees and 1,684 post offices in the state in 1985.

Telephone service in New York is provided primarily by the New York Telephone Co., but also by 40 smaller companies throughout the state. In 1985, New York Telephone had 8,240,239 access lines (the customer's link to the telephone network), 50% of which were in New York City; Manhattan alone had 21% of the state total. The smaller companies in the state had 868,918 access lines. As of 1980, 92% of New York's 6,340,429 occupied housing units had telephones; in 1985, New York Telephone had 5,904,-812 residential customers and 612,270 business customers. In addition, there were 150,598 coin and charge-a-call telephones, 64% of them in New York City, from which a daily average of 2,721,000 calls were made during 1984. On an average business day, New York Telephone handles more than 90 million calls, half of them in New York City.

Until 31 December 1983, New York Telephone was part of the Bell System, whose parent organization was the American Telephone and Telegraph Co. (AT&T). Effective 1 January 1984, as the result of a US Justice Department antitrust suit, AT&T divested itself of 22 Bell operating companies, which regrouped into seven independent regional telephone companies to provide local telephone service in the US. One of these companies, NYNEX, is the parent company of New York Telephone. AT&T, which continued to supply long-distance telephone services to New Yorkers (along with competitive carriers such as MCI, ITT, and GTE), is headquartered in New York City and, in 1984, had 365,000 employees, assets of $39.8 billion, and sales of $33.2 billion.

Domestic telegraph service is provided by the Western Union Telegraph Co., ITT World Communications, RCA Global Communications, and Western Union International. All four companies have their headquarters in New York City.

New York State had 167 AM stations and 209 FM stations operating in 1984. New York City operates its own radio stations, WNYC-AM and -FM, devoted largely to classical music and educational programming.

There were 34 commercial television stations in the state in 1984; of these, 6 were operating in New York City. The city is the headquarters for most of the major US television networks, including the American Broadcasting Co. (now part of Capital Cities Communications), Columbia Broadcasting System, National Broadcasting Co., Westinghouse Broadcasting (Group W), Metromedia, and the Public Broadcasting Service (PBS). Twelve educational television stations serve all the state's major populated areas, and the metropolitan area's PBS affiliate, WNET (licensed in Newark, N.J.), is a leading producer of programs for the network. As of 1984, 172 cable television systems in the state served 986 communities and had 2,172,826 subscribers.

45PRESS

A pioneer in the establishment of freedom of the press, New York is the leader of the US newspaper, magazine, and book-publishing industries. The first major test of press freedom in the colonies came in 1734, when a German-American printer, John Peter Zenger, was arrested on charges of sedition and libel. In his newspaper, the *New-York Weekly Journal,* Zenger had published articles criticizing the colonial governor of New York. Zenger's lawyer, Andrew Hamilton, argued that because the charges in the article were true, they could not be libelous. The jury's acceptance of this argument freed Zenger and established the right of the press to criticize those in power. Two later decisions involving a New York newspaper also struck blows for press freedom. In *New York Times* v. *Sullivan* (1964), the US Supreme Court ruled that a public official could not win a libel suit against a newspaper unless he could show that its statements about him were not only false but also malicious or in reckless disregard of the truth. In 1971, the *New York Times* was again involved in a landmark case when the federal government tried—and failed—to prevent the newspaper from publishing the Pentagon Papers, a collection of secret documents concerning the war in Viet-Nam.

All of New York City's major newspapers have claims to fame. The *Times* is the nation's "newspaper of record," excelling in the publication of speeches, press conferences, and government reports. It is widely circulated to US libraries and is often cited in research. The *New York Post,* founded in 1801, is the oldest US newspaper published continuously without change of name. The *New York Daily News* has the largest daily and Sunday circulation of all general-interest newspapers in the US. The *Wall Street Journal,* published Monday through Friday, is a truly national paper, presenting mostly business news in four regional editions. Many historic New York papers first merged and then—bearing compound names like the *Herald-Tribune, Journal-American,* and *World-Telegram & Sun*—died in the 1950s and 1960s.

In 1984, New York had 21 morning newspapers, with a total average daily circulation of 5,144,784; 52 evening papers, 1,628,-570; two all-day papers, 1,245,137; and 33 Sunday editions, 5,760,388. The following table shows leading papers in New York, with their average daily and Sunday circulations in 1984:

AREA	NAME	DAILY	SUNDAY
Albany	Times-Union (m,S)	86,386	168,374
Buffalo	News (all day,S)	315,111	370,202
Long Island	Newsday (e,S)	539,065	602,476
New York City	Daily News (m,S)	1,346,840	1,721,441
	Post (all day)	930,026	628,601[1]
	Times (m,S)	934,616	1,553,720
	Wall Street Journal	777,900[2]	
Rochester	Democrat & Chronicle (m,S)	131,140	} 249,375
	Times-Union (e)	103,136	
Syracuse	Post-Standard (m)	82,915	
	Herald-Journal (e)	104,137	
	Herald-American/Post-Standard (S)		237,360

[1] Weekend edition published Saturdays.
[2] Eastern edition only.

The leading newspaper chain is the Gannett group, which had newspapers in 17 cities in 1984. All the major news agencies have offices in New York City, and the Associated Press has its headquarters there.

Many leading US magazines are published in New York City, including the newsmagazines *Time* and *Newsweek,* business journals like *Fortune, Forbes,* and *Business Week,* and hundreds of consumer and trade publications. *Reader's Digest,* with a paid circulation in 1983 of 18,299,091 (1st in the US), is published in Pleasantville. Two weeklies closely identified with New York are of more than local interest. While the *New Yorker* carries up-to-date listings of cultural events and exhibitions in New York City, the excellence of its journalism, criticism, fiction, and cartoons has long made it a literary standard-bearer for the entire nation. *New York* magazine influenced the writing style and graphic design of the 1960s and set the pattern for a new wave of state and local magazines that avoided boosterism in favor of independent reporting and commentary. Another weekly, the *Village Voice* (actually a tabloid newspaper) became the prototype for a host of alternative or "underground" journals during the 1960s.

New York City is also the center of the nation's book-publishing industry. The New York publishers listed by *Fortune* magazine as among the 500 largest US industrial corporations in 1984 were McGraw-Hill and Macmillan, but such large publishers as Simon & Schuster and Random House are subsidiaries of other companies. As of 1984, there were over 500 book-publishing companies in New York, most in Manhattan.

46ORGANIZATIONS

The 1982 Census of Service Industries counted 4,221 organizations in New York, including 845 business associations; 2,076 civic, social, and fraternal associations; and 285 educational, scientific, and research associations.

The United Nations is the best-known organization to have its headquarters in New York. The UN Secretariat, completed in 1951, remains one of the most familiar landmarks of New York City. Hundreds of US nonprofit organizations also have their national headquarters in New York City. General and service organizations operating out of New York City include the American Field Service, Boys Clubs of America, Girls Clubs of America, Girl Scouts of the USA, Young Women's Christian Associations of the USA (YWCA), and Associated YM-YWHAs of Greater New York (the Jewish equivalent of the YMCA and YWCA). Among the cultural and educational groups are the American Academy of Arts and Letters, Authors League of America, Children's Book Council, Modern Language Association of America, and PEN American Center.

Among the environmental and animal welfare organizations with headquarters in the city are the American Society for the Prevention of Cruelty to Animals (ASPCA), Friends of Animals, Fund for Animals, National Audubon Society, Bide-A-Wee Home Association, Environmental Defense Fund, and American Kennel Club.

Many medical, health, and charitable organizations have their national offices in New York City, including Alcoholics Anonymous, American Foundation for the Blind, National Society to Prevent Blindness, CARE, American Cancer Society, United Cerebral Palsy Associations, Child Welfare League of America, American Diabetes Association, National Multiple Sclerosis Society, Muscular Dystrophy Association, and Planned Parenthood Federation of America.

Leading ethnic and religious organizations based in the city include the American Bible Society, National Conference of Christians and Jews, Hadassah, United Jewish Appeal, American Jewish Committee, American Jewish Congress, National Association for the Advancement of Colored People (NAACP), United Negro College Fund, Congress of Racial Equality, and National Urban League.

There are many commercial, trade, and professional organizations headquartered in New York City. Among the better known are the Actors' Equity Association, American Arbitration Association, American Booksellers Association, American Federation of Musicians, American Institute of Chemical Engineers, American Society of Civil Engineers, American Society of Composers, Authors, and Publishers (ASCAP), American Society of Journalists and Authors, American Insurance Association, Magazine Publishers Association, American Management Associations, American Society of Mechanical Engineers, and American Institute of Physics.

Sports organizations centered in New York City include the National Football League, the American and the National Leagues of Professional Baseball Clubs, National Basketball Association, and the US Tennis Association. There are also several influential political and international-affairs groups: the American Civil Liberties Union, Council on Foreign Relations, Trilateral Commission, United Nations Association of the USA, and US Committee for UNICEF.

Major organizations with their headquarters outside New York City include the Consumers Union of the United States (Mt. Vernon), US Chess Federation (New Windsor), and the Thoroughbred Racing Association (Lake Success). Virtually every other major US organization has one or more chapters within the state.

47 TOURISM, TRAVEL AND RECREATION

New York State is a popular destination for both domestic and foreign travelers. Although New York City is the primary attraction, each of the major state regions has features of interest.

Foreign tourist arrivals increased from 2.6 million in 1977 to an estimated 4.8 million in 1983, about 23.5% of the US total. More than 54% of the foreign visitors arrived by air. With the rise in the value of the dollar, the number of visitors to New York City fell to 2,550,000 in 1982, but the city was still the port of entry for 30% of all travelers from abroad.

More than 63 million Americans took trips to or through New York State in 1981. In 1983, the amount of estimated travel expenditures in New York State by Americans was $12.4 billion; a total of 53,239 travel-related businesses employed 605,871 persons and had a total annual payroll of $7.2 billion.

According to the New York Convention and Visitors Bureau, New York City had about 17.1 million visitors in 1983, who spent about $2.2 billion. A typical visit to New York City might include a boat ride to the Statue of Liberty; a three-hour boat ride around Manhattan; visits to the World Trade Center, the Empire State Building, the UN, Rockefeller Center, and the New York Stock Exchange; walking tours of the Bronx Zoo, Chinatown, and the theater district; and a sampling of the city's many museums, restaurants, shops, and shows. The city had 97,800 hotel rooms in 1982. New York City is also a convention center; 933 conventions in 1982 attracted 4,120,000 delegates, who spent an estimated $828 million.

Second to New York City as a magnet for tourists comes Long Island, with its beaches, racetracks, and other recreational facilities. Attractions of the Hudson Valley include the US Military Academy (West Point), the Franklin D. Roosevelt home at Hyde Park, Bear Mountain State Park, and several wineries. North of Hudson Valley is Albany, with its massive government center, Governor Nelson A. Rockefeller Plaza, often called the Albany Mall; Saratoga Springs, home of an arts center, racetrack, and spa; and the Adirondack region, with its forest preserve, summer and winter resorts, and abundant hunting and fishing. Northwest of the Adirondacks, in the St. Lawrence River, are the Thousand Islands—actually some 1,800 small islands extending over about 50 mi (80 km), and popular among freshwater fishermen and summer vacationers.

Scenic sites in central New York include the resorts of the Catskills and the scenic marvels of the Finger Lakes region, including Taughannock Falls in Trumansburg, the highest waterfall east of the Rockies. Further west lie Buffalo and Niagara Falls. South of the Niagara Frontier is the Southwest Gateway, among whose dominant features are Chautauqua Lake and Allegany State Park, the state's largest.

In 1984, the Office of Parks, Recreation and Historic Preservation operated 147 state parks, 34 historic sites, and 74 boat-launching sites that comprised 255,250 acres (103,296 hectares). In 1982/83, a total of 47,031,000 people visited the state parks and historic sites. In that year also, the state sold 791,568 hunting licenses and 936,977 fishing licenses. An estimated 185,455 deer were taken in 1982, as compared to only 48,290 in 1971. The state registered 321,881 motorboats and 73,244 snowmobiles in 1982.

48 SPORTS

Teams represent the state and its largest cities in almost every major professional sport. Off-track betting is a legal, multi-million-dollar enterprise in New York City, and Lake Placid is a magnet for winter-sports enthusiasts.

Both of the state's major league baseball teams play in New York City. The Mets play at Shea Stadium in Queens, while the Yankees are at Yankee Stadium in the Bronx. Few teams in any sport can match the Yankees' achievements: between 1901 and 1981, they won 33 league championships and 22 world championships, with teams starring such Hall of Famers as Babe Ruth, Lou Gehrig, Joe DiMaggio, and Lawrence Peter "Yogi" Berra. The "miracle" Mets won the World Series in 1969 and the National League championship in 1973.

The state has one major league basketball team, the New York Knickerbockers (Madison Square Garden in Manhattan). The Knicks won playoff championships in 1970 and 1973; their stars during that era included Willis Reed, Walt Frazier, Dave DeBusschere, and Bill Bradley. The Buffalo Bills (War Memorial Stadium) compete in the National Football League. In ice hockey, the New York Islanders (Nassau Coliseum), winners of the Stanley Cup in 1980, 1981, 1982, and 1983, the New York Rangers (Madison Square Garden), and the Buffalo Sabres are the state entries. In addition to owning the Knickerbockers and Rangers, Madison Square Garden is a leading promoter of professional boxing, hosts professional and amateur track-and-field competitions, and presents many other sports and entertainment events.

Three metropolitan area teams moved across the Hudson to New Jersey during the 1970s and early 1980s: the Nets in basketball, the Giants and Jets in football.

Horse racing is important to New York State, both as a sport attraction and because of the tax revenues that betting generates. The main Thoroughbred race tracks are Aqueduct in Queens and Belmont in Nassau County; Belmont is the home of the Belmont Stakes, one of the three jewels in the Triple Crown of US racing. Saratoga (Saratoga Springs) presents Thoroughbred racing and also offers a longer harness-racing season. Thoroughbred racing is also offered at the Finger Lakes track in Canandaigua. The top tracks for harness racing are Roosevelt Raceway (Westbury, Long Island), Yonkers Raceway, and Monticello Raceway (in the Catskills).

The New York City Off-Track Betting Corporation (OTB), which began operations in April 1971, takes bets on races at the state's major tracks, as well as on some out-of-state races. Off-track betting services operate on a smaller scale on Long Island and in upstate New York.

New York City hosts several major professional tennis tournaments every year: the US Open in Flushing, Queens; the WCT Invitational in Forest Hills, Queens; the Colgate Grand Prix Masters (men) and the Avon Championships (women), both at Madison Square Garden.

Among other professional sports facilities, the Watkins Glen automobile racetrack is the site of a Grand Prix race every October. Lake Placid, an important winter-sports region, hosted the 1932 and 1980 Winter Olympics.

Educational institutions in the state offer a wide variety of athletic activities. Among the leading National Collegiate Athletic Association competitors during the early 1980s were Iona, Fordham, St. John's, and Syracuse in basketball, Cornell and Rensselaer Polytechnic Institute in ice hockey, Columbia and Hartwick in soccer, and Cornell and Hobart in lacrosse.

In 1978, New York became the first state to sponsor a statewide amateur athletic event, the Empire State Games. More than 50,000 athletes now compete for a place in the finals, held each summer; the Winter Games, held each March in Lake Placid, host more than 1,000.

The New York marathon, which is held in October, had 19,000 registered participants for its 1985 run.

49FAMOUS NEW YORKERS

New York State has been the home of five US presidents, eight US vice presidents (three of whom also became president), many statesmen of national and international repute, and a large corps of writers and entertainers.

Martin Van Buren (1782-1862), the 8th US president, became governor of New York in 1828. He was elected to the vice-presidency as a Democrat under Andrew Jackson in 1832, and succeeded Jackson in the election of 1836. An unpopular president, Van Buren ran for reelection in 1840 but was defeated, losing even his home state. The 13th US president, Millard Fillmore (1800-74), was elected vice president under Zachary Taylor in 1848. He became president in 1850 when Taylor died. Fillmore's party, the Whigs, did not renominate him in 1852; four years later, he unsuccessfully ran for president as the candidate of the Native American (or Know-Nothing) Party.

New York's other US presidents had more distinguished careers. Although he was born in New Jersey, Grover Cleveland (1837-1908) served as mayor of Buffalo and as governor of New York before his election to his first presidential term in 1884; he was again elected president in 1892. Theodore Roosevelt (1858-1919), a Republican, was elected governor in 1898. He won election as vice president under William McKinley in 1900, and became the nation's 26th president after McKinley was murdered in 1901. Roosevelt pursued an aggressive foreign policy, but also won renown as a conservationist and trustbuster. Reelected in 1904, he was awarded the Nobel Peace Prize in 1906 for helping to settle a war between Russia and Japan. Roosevelt declined to run again in 1908. However, he sought the Republican nomination in 1912 and, when defeated, became the candidate of the Progressive (or Bull Moose) Party, losing the general election to Woodrow Wilson.

Franklin Delano Roosevelt (1882-1945), a fifth cousin of Theodore Roosevelt, first ran for national office in 1920, when he was the Democratic vice-presidential choice. A year after losing that election, FDR was crippled by poliomyelitis. He then made an amazing political comeback: he was elected governor of New York in 1928 and served until 1932, when US voters chose him as their 32d president. Reelected in 1936, 1940, and 1944, FDR is the only president ever to have served more than two full terms in office. Roosevelt guided the US through the Great Depression and World War II, and his New Deal programs greatly enlarged the federal role in promoting social welfare.

In addition to Van Buren, Fillmore, and Theodore Roosevelt, five US vice presidents were born in New York: George Clinton (1739-1812), who was also New York State's first elected governor; Daniel D. Tompkins (1774-1825); William A. Wheeler (1819-87); Schuyler Colfax (1823-85); and James S. Sherman (1855-1912). Two other US vice presidents, though not born in New York, were New Yorkers by the time they became vice president. The first was Aaron Burr (1756-1836), perhaps best known for killing Alexander Hamilton in a duel in 1804; Hamilton (b.Nevis, West Indies, 1757-1804) was a leading Federalist, George Washington's treasury secretary, and the only New York delegate to sign the US Constitution in 1787. The second transplanted New Yorker to become vice president was Nelson Aldrich Rockefeller (1908-79). Born in Maine, Rockefeller served as governor of New York State from 1959 to 1973, was for two decades a major force in national Republican politics, and was appointed vice president by Gerald Ford in 1974, serving in that office through January 1977.

Two native New Yorkers have become chief justices of the US: John Jay (1745-1829) and Charles Evans Hughes (1862-1948). A third chief justice, Harlan Fiske Stone (1872-1946), born in New Hampshire, spent most of his legal career in New York City and served as dean of Columbia University's School of Law. Among New Yorkers who became associate justices of the US Supreme Court, Benjamin Nathan Cardozo (1870-1938) is noteworthy.

Other federal officeholders born in New York include US secretaries of state William Henry Seward (1801-72), Hamilton Fish (1808-93), Elihu Root (1845-1937), Frank B. Kellogg (1856-1937), and Henry L. Stimson (1867-1950). Prominent US senators have included Robert F. Wagner (1877-1953), who sponsored many New Deal laws; Robert F. Kennedy (1925-68), who though born in Massachusetts was elected to represent New York in 1964; Jacob K. Javits (b.1904), who served continuously in the Senate from 1957 through 1980; and Daniel Patrick Moynihan (b.1927), a scholar, author, and former federal bureaucrat who has represented New York since 1977.

The most important—and most colorful—figure in colonial New York was Peter Stuyvesant (b.Netherlands, 1592-1672); as director general of New Netherland, he won the hearty dislike of the Dutch settlers. Signers of the Declaration of Independence in 1776 from New York were Francis Lewis (1713-1803); Philip Livingston (1716-78); Lewis Morris (1726-98), the half-brother of the colonial patriot Gouverneur Morris (1752-1816); and William Floyd (1734-1821).

Other governors who made important contributions to the history of the state include De Witt Clinton (1769-1828); Alfred E. Smith (1873-1944); Herbert H. Lehman (1878-1963); W. Averell Harriman (b.1891), who has also held many US diplomatic posts; and Thomas E. Dewey (1902-71). Mario M. Cuomo (b.1932) was elected governor in 1982. Robert Moses (b.Connecticut, 1888-1981) led in the development of New York's parks and highway transportation system. One of the best-known and best-loved mayors in New York City history was Fiorello H. La Guardia (1882-1947), a reformer who held the office from 1934 to 1945. Edward I. Koch (b.1924) was first elected to the mayorality in 1977.

Native New Yorkers have won Nobel prizes in every category. Winners of the Nobel Peace Prize besides Theodore Roosevelt were Elihu Root in 1912 and Frank B. Kellogg in 1929. The lone winner of the Nobel Prize for literature was Eugene O'Neill (1888-1953) in 1936. The chemistry prize was awarded to Irving Langmuir (1881-1957) in 1932, John H. Northrop (b.1891) in 1946, and William Howard Stein (1911-80) in 1972. Winners in physics include Carl D. Anderson (b.1905) in 1936, Robert Hofstadter (b.1915) in 1961, Richard Phillips Feynman (b.1918) and Julian Seymour Schwinger (b.1918) in 1965, Murray Gell-Mann (b.1929) in 1969, Leon N. Cooper (b.1930) in 1972, Burton Richter (b.1931) in 1976, and Steven Weinberg (b.1933) and Sheldon L. Glashow (b.1923) in 1979.

Eleven New Yorkers have been awarded the Nobel Prize for physiology or medicine: Hermann Joseph Muller (1890-1967) in 1946, Arthur Kornberg (b.1918) in 1959, George Wald (b.1906) in 1967, Marshall Warren Nirenberg (b.1927) in 1968, Julius Axelrod (b.1912) in 1970, Gerald Maurice Edelman (b.1929) in 1972, David Baltimore (b.1938) in 1975, Baruch Samuel Blumberg (b.1925) and Daniel Carlton Gajdusek (b.1923) in 1976, Rosalyn Sussman Yalow (b.1921) in 1977, and Hamilton O. Smith (b.1931) in 1978.

The Nobel Prize for economic science was won by Kenneth J. Arrow (b.1921) in 1972, Milton Friedman (b.1912) in 1976, and Richard Stone (b.1928) in 1984. New York is also the birthplace of national labor leader George Meany (1894–1980) and economist Walter Heller (b.1915). Other distinguished state residents were physicist Joseph Henry (1797–1878), Mormon leader Brigham Young (b.Vermont, 1801–77), botanist Asa Gray (1810–88), inventor-businessman George Westinghouse (1846–1914), and Jonas E. Salk (b.1914), developer of a poliomyelitis vaccine.

Writers born in New York include the storyteller and satirist Washington Irving (1783–1859); poets Walt Whitman (1819–92) and Ogden Nash (1902–71); and playwrights Eugene O'Neill, Arthur Miller (b.1915), Paddy Chayefsky (1923–81), and Neil Simon (b.1927). Two of America's greatest novelists were New Yorkers: Herman Melville (1819–91), who was also an important poet, and Henry James (1843–1916), whose short stories are equally well known. Other novelists include James Fenimore Cooper (b.New Jersey, 1789–1851), Henry Miller (1891–1980), James Michener (b.1907), J(erome) D(avid) Salinger (b.1919), Joseph Heller (b.1923), James Baldwin (b.1924), and Gore Vidal (b.1925). Lionel Trilling (1905–1975) was a well-known literary critic; Barbara Tuchman (b.1912), a historian, has won both scholarly praise and popular favor. New York City has produced two famous journalist-commentators, Walter Lippmann (1889–1974) and William F. Buckley, Jr. (b.1925), and a famous journalist-broadcaster, Walter Winchell (1897–1972).

Broadway is the showcase of American drama and the birthplace of the American musical theater. New Yorkers linked with the growth of the musical include Jerome Kern (1885–1945), Lorenz Hart (1895–1943), Oscar Hammerstein 2d (1895–1960), Richard Rodgers (1902–79), Alan Jay Lerner (b.1918), and Stephen Sondheim (b.1930). George Gershwin (1898–1937), whose *Porgy and Bess* raised the musical to its highest artistic form, also composed piano and orchestral works. Other important US composers who are New Yorkers include Irving Berlin (b.Russia, 1888), Aaron Copland (b.1900), Elliott Carter (b.1908), and William Schuman (b.1910). New York was the adopted home of ballet director and choreographer George Balanchine (b.Russia, 1904–83); his associate Jerome Robbins (b.1918) was born in New York City, as was choreographer Agnes De Mille (b.1905). Leaders in the visual arts include Frederic Remington (1861–1909), the popular illustrator Norman Rockwell (1894–1978), Willem de Kooning (b.Netherlands, 1904), and the photographer Margaret Bourke-White (1906–71).

Many of America's best-loved entertainers come from the state. A small sampling would include comedians Groucho Marx (Julius Marx, 1890–1977), Mae West (1892–1980), Eddie Cantor (Edward Israel Iskowitz, 1892–1964), James "Jimmy" Durante (1893–1980), Bert Lahr (Irving Lahrheim, 1895–1967), George Burns (b.1896), Milton Berle (Berlinger, b.1908), Lucille Ball (b.1911), Danny Kaye (David Daniel Kominsky, b.1913), and Sid Caesar (b.1922); comedian-film directors Mel Brooks (Melvin Kaminsky, b.1926) and Woody Allen (Allen Konigsberg, b.1935); stage and screen stars Humphrey Bogart (1899–1957), James Cagney (b.1904), Zero Mostel (Samuel Joel Mostel, 1915–77), and Lauren Bacall (Betty Joan Perske, b.1924); pop, jazz, and folk singers Cab Calloway (b.1907), Lena Horne (b.1917), Pete Seeger (b.1919), Sammy Davis, Jr. (b.1925), Harry Belafonte (b.1927), Joan Baez (b.1941), Barbra Streisand (b.1942), Carly Simon (b.1945), Arlo Guthrie (b.1947), and Billy Joel (b.1951); and opera stars Robert Merrill (b.1919), Maria Callas (Kalogeropoulos, 1923–77), and Beverly Sills (Belle Silverman, b.1929). Also noteworthy are producers Irving Thalberg (1899–1936), David Susskind (b.1920), Joseph Papp (b.1921), and Harold Prince

(b.1928) and directors George Cukor (1899–1983), Stanley Kubrick (b.1928), John Frankenheimer (b.1930), and Peter Bogdanovich (b.1939).

Among many prominent sports figures born in New York are first-baseman Lou Gehrig (1903–41), football coach Vince Lombardi (1913–70), pitcher Sanford "Sandy" Koufax (b.1935), and basketball stars Kareem Abdul-Jabbar (Lew Alcindor, b.1947) and Julius Erving (b.1950).

[50]BIBLIOGRAPHY

Auletta, Ken. *The Streets Were Paved with Gold.* New York: Random House, 1980.

Barlow, Elizabeth, *Frederick Law Olmsted's New York.* New York: Praeger, 1972.

Bellush, Jewel, and Stephen M. David. *Race and Politics in New York City.* New York: Praeger, 1971.

Berle, Beatrice Bishop. *80 Puerto Rican Families in New York City.* New York: Arno Press, 1975.

Bliven, Bruce. *New York.* New York: Norton, 1981.

Brown, Claude. *Manchild in the Promised Land.* New York: Macmillan, 1965.

Caro, Robert A. *The Power Broker: Robert Moses and the Fall of New York.* New York: Vintage, 1975.

Colby, Peter W. (ed.). *New York State Today: Politics, Government, Public Policy.* Albany: State University of New York Press, 1984.

Cuomo, Mario M. *Diaries of Mario M. Cuomo: The Campaign for Governor.* New York: Random House, 1984.

Edmiston, Susan, and Linda D. Cirino. *Literary New York: A History and Guide.* Boston: Houghton Mifflin, 1976.

Ellis, David M., *New York: State and City.* Ithaca: Cornell University Press, 1979.

Ellis, David M., James A. Frost, Harold C. Syrett, and Harry J. Carman. *A History of New York State.* Rev. ed. Ithaca: Cornell University Press, 1967.

Ellis, David M., James A. Frost, and William B. Fink. *New York: The Empire State.* 4th ed. Englewood Cliffs, N.J.: Prentice-Hall, 1975.

Federal Writers' Project. *New York: A Guide to the Empire State.* New York: Somerset, n.d. (orig. 1940).

Federal Writers' Project. *New York City Guide.* New York: Somerset, n.d. (orig. 1939).

Flick, Alexander C., ed. *History of the State of New York.* 10 vols. in 5. Published under the auspices of the New York State Historical Association. Port Washington, N.Y.: Ira J. Friedman, 1962 (orig. 1933).

French, J. H. *Gazetteer of the State of New York.* Port Washington, N.Y.: Ira J. Friedman, 1969 (orig. 1860).

Furer, Howard B. *New York: A Chronological and Documentary History.* Dobbs Ferry, N.Y.: Oceana Publications, 1974.

Glazer, Nathan, and Daniel Patrick Moynihan. *Beyond the Melting Pot: The Negroes, Puerto Ricans, Jews, Italians, and Irish of New York City.* 2d ed. Cambridge, Mass.: MIT Press, 1970.

Hacker, Andrew. *The New Yorkers.* New York: Mason/Charter, 1975.

Heckscher, August, with Phyllis Robinson. *When La Guardia Was Mayor.* New York: Norton, 1978.

Henderson, Mary C. *The City and the Theatre: New York Playhouses from Bowling Green to Times Square.* Clifton, N.J.: J. T. White, 1973.

Hevesi, Alan G. *Legislative Politics in New York State.* New York: Praeger, 1975.

Howe, Irving. *World of Our Fathers.* New York: Harcourt Brace Jovanovich, 1976.

Irving, Washington. *A History of New York.* Edited by Edwin T. Bowden. New Haven, Conn.: College & University Press, 1964.

Kammen, Michael. *Colonial New York: A History.* Millwood, N.Y.: Kraus, 1975.

Kennedy, William. *O Albany!* New York: Viking, 1983.

Kenney, Alice P. *Stubborn for Liberty: The Dutch in New York.* Syracuse: Syracuse University Press, 1975.

Koch, Edward I., and William Rauch. *Mayor: An Autobiography.* New York: Simon & Schuster, 1984.

Kouwenhoven, John A. *The Columbia Historical Portrait of New York: An Essay in Graphic History.* New York: Harper & Row, 1972 (orig. 1952).

Lopez, Manuel D. *New York: A Guide to Information and Reference Sources.* Metuchen, N.J.: Scarecrow, 1980.

Myers, Gustavus. *The History of Tammany Hall.* New York: Burt Franklin, 1967 (orig. 1901).

Nelson A. Rockefeller Institute of Government, in cooperation with the New York State Division of the Budget. *New York State Statistical Yearbook, 1984–1985.* Albany: Rockefeller Institute, 1985.

Newfield, Jack, and Paul DuBrul. *Abuse of Power: The Permanent Government and the Fall of New York.* New York: Viking, 1977.

Peirce, Neal R. *The Megastates of America: People, Politics and Power in the Ten Great States.* New York: Norton, 1972.

Ravitch, Diane. *The Great School Wars: New York City, 1805–1973.* New York: Basic Books, 1974.

Schneider, David Moses, and Albert Deutsch. *The History of Public Welfare in New York State.* Montclair, N.J.: Patterson, Smith, 1969.

Talese, Gay. *The Kingdom and the Power.* New York: World, 1969.

NORTH CAROLINA

State of North Carolina

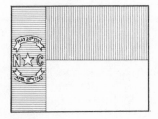

ORIGIN OF STATE NAME: Named in honor of King Charles I of England. **NICKNAME:** The Tarheel State. **CAPITAL:** Raleigh. **ENTERED UNION:** 21 November 1789 (12th). **SONG:** "The Old North State." **MOTTO:** *Esse quam videri* (To be rather than to seem). **FLAG:** Adjacent to the fly of two equally sized bars, red above and white below, is a blue union containing a white star in the center, flanked by the letters N and C in gold. Above and below the star are two gold scrolls, the upper one reading "May 20th 1775," the lower one "April 12th 1776." **OFFICIAL SEAL:** Liberty, clasping a constitution and holding aloft on a pole a liberty cap, stands on the left, while Plenty sits besides a cornucopia on the right; behind them, mountains run to the sea, on which a three-masted ship appears. "May 20, 1775" appears above the figures; the words "The Great Seal of the State of North Carolina" and the state motto surround the whole. **MAMMAL:** Gray squirrel. **BIRD:** Cardinal. **FISH:** Channel bass. **REPTILE:** Eastern box turtle. **INSECT:** Honeybee. **FLOWER:** Dogwood. **TREE:** Pine. **SHELL:** Scotch bonnet. **PRECIOUS STONE:** Emerald. **ROCK:** Granite. **LEGAL HOLIDAYS:** New Year's Day, 1 January; Birthday of Martin Luther King, Jr., 3d Monday in January; Easter Monday, March or April; Memorial Day, last Monday in May; Independence Day, 4 July; Labor Day, 1st Monday in September; Veterans Day, 11 November; Thanksgiving Day, 4th Thursday in November, and the day following; Christmas Eve, 24 December; Christmas Day, 25 December. **TIME:** 7 AM EST = noon GMT.

¹LOCATION, SIZE, AND EXTENT

Located in the southeastern US, North Carolina ranks 28th in size among the 50 states.

The total area of North Carolina is 52,669 sq mi (136,413 sq km), of which land accounts for 48,843 sq mi (126,504 sq km) and inland water 3,826 sq mi (9,909 sq km). North Carolina extends 503 mi (810 km) E–W; the state's maximum N–S extension is 187 mi (301 km).

North Carolina is bordered on the N by Virginia; on the E by the Atlantic Ocean; on the s by South Carolina and Georgia; and on the w by Tennessee. A long chain of islands or sand banks, called the Outer Banks, lies off the state's Atlantic coast. The total boundary line of North Carolina is 1,270 mi (2,044 km), including a general coastline of 301 mi (484 km); the tidal shoreline extends 3,375 mi (5,432 km). The state's geographic center is in Chatham County, 10 mi (16 km) NW of Sanford.

²TOPOGRAPHY

North Carolina's three major topographic regions belong to the Atlantic Coastal Plain, the Piedmont Plateau, and the Appalachian Mountains.

The Outer Banks, narrow islands of shifting sandbars, screen most of the coastal plain from the ocean. Treacherous navigation conditions and numerous shipwrecks have earned the name of "Graveyard of the Atlantic" for the shoal waters off Cape Hatteras, which, like Cape Lookout and Cape Fear, juts out from the banks into the Atlantic. Cape Hatteras Lighthouse is the tallest in the US, rising 208 feet (63 meters). The shallow Pamlico and Albemarle sounds and broad salt marshes lying behind the Outer Banks serve as valuable habitats for marine life but as further hindrances to water transportation.

On the mainland, the coastal plain extends westward from the sounds for 100 to 140 mi (160–225 km) and upward from sea level to nearly 500 feet (150 meters). Near the ocean, the outer coastal plain is very flat, and often swampy; this region contains all the natural lakes in North Carolina, the largest being Lake Mattamuskeet (67 sq mi—174 sq km), followed by Lakes Phelps and Waccamaw. The inner coastal plain is more elevated and better drained. Infertile sand hills mark its southwestern section, but the rest of the region constitutes the state's principal farming country.

The piedmont is a rolling plateau of red clay soil roughly 150 mi (240 km) wide, rising from 300 to 600 feet (90 to 180 meters) in the east to 1,500 feet (460 meters) in the west. The fall line, a sudden change in elevation, separates the piedmont from the coastal plain and produces numerous rapids in the rivers that flow between the regions.

The Blue Ridge, a steep escarpment that parallels the Tennessee border, divides the piedmont from North Carolina's westernmost region, containing the highest and most rugged portion of the Appalachian chain. The two major ranges are the Blue Ridge itself, which averages 3,000–4,000 feet high (900–1,200 meters), and the Great Smoky Mountains, which have 43 peaks higher than 6,000 feet (1,800 meters). Several smaller chains intersect these two ranges; one of them, the Black Mountains, contains Mt. Mitchell, at 6,684 feet (2,037 meters) the tallest peak east of the Mississippi River.

No single river basin dominates North Carolina. The Hiwassee, Little Tennessee, French Broad, Watauga, and New rivers flow from the mountains westward to the Mississippi River system. East of the Blue Ridge, the Chowan, Roanoke, Tar, Neuse, Cape Fear, Yadkin, and Catawba drain the piedmont and coastal plain. The largest artificial lakes are Lake Norman on the Catawba, Lake Gaston on the Roanoke, and High Rock Lake on the Yadkin.

³CLIMATE

North Carolina has a humid, subtropical climate. Winters are short and mild, while summers are usually very sultry; spring and fall are distinct and refreshing periods of transition. In most of North Carolina, temperatures rarely go above 100°F (38°C) or fall below 10°F (−12°C), but differences in altitude and proximity to the ocean create significant local variations. Average January temperatures range from 36°F (21°C) to 48°F (9°C), with an average daily maximum January temperature of 51°F (11°C) and minimum of 29°F (−2°C). Average July temperatures range from 68°F (20°C) to 80°F (27°C), with an average daily high of 87°F (31°C) and a low of 66°F (19°C). The coldest temperature ever recorded in North Carolina was −29°F (−34°C), registered on 30 January 1966 on Mt. Mitchell; the hottest, 110°F (43°C), occurred on 21 August 1983 at Fayetteville.

In the southwestern section of the Blue Ridge, moist southerly winds rising over the mountains drop more than 80 in (203 cm) of

401

precipitation per year, making this region the wettest in the eastern states; the other side of the mountains receives less than half that amount. The piedmont gets between 44 and 48 in (112–122 cm) of precipitation per year, while 44 to 56 in (112–142 cm) annually fall on the coastal plain. Average winter snowfalls vary from 50 in (127 cm) on Mt. Mitchell to only a trace amount at Cape Hatteras. In the summer, North Carolina weather responds to the Bermuda High, a pressure system centered in the mid-Atlantic. Winds from the southwest bring masses of hot, humid air over the state; anticyclones connected with this system frequently lead to upper-level thermal inversions, producing a stagnant air mass that cannot disperse pollutants until cooler, drier air from Canada moves in. During late summer and early autumn, the eastern region is vulnerable to high winds and flooding from hurricanes. Hurricane Diana struck the Carolina coast in September 1984, causing $36 million in damage. A series of tornadoes in March of that year killed 61 people, injured over 1,000, and caused damage exceeding $120 million.

⁴FLORA AND FAUNA

North Carolina has approximately 300 species and subspecies of trees and almost 3,000 varieties of flowering plants. Coastal plant life begins with sea oats predominating on the dunes and saltmeadow and cordgrass in the marshes, then gives way to wax myrtle, yaupon, red cedar, and live oak further inland. Blackwater swamps support dense stands of cypress and gum trees. Pond pine favors the peat soils of the Carolina bays, while longleaf pine and turkey oak cover the sand hills and other well-drained areas. Weeds take root when a field is abandoned in the piedmont, followed soon by loblolly, shortleaf, and Virginia pine; sweet gum and tulip poplars spring up beneath the pines, later giving way to an oak-hickory climax forest. Dogwood decorates the understory, but kudzu—a rank, weedy vine introduced from Japan as an antierosion measure in the 1930s—is a less attractive feature of the landscape. The profusion of plants reaches extraordinary proportions in the mountains. The deciduous forests on the lower slopes contain Carolina hemlock, silver bell, yellow buckeye, white basswood, sugar maple, yellow birch, tulip poplar, and beech, in addition to the common trees of the piedmont. Spruce and fir dominate the high mountain peaks. There is no true treeline in the North Carolina mountains, but unexplained treeless areas called "balds" appear on certain summits. The branched arrowhead is an endangered plant.

The white-tailed deer is the principal big-game animal of North Carolina, and the black bear is a tourist attraction in the Great Smoky Mountains National Park. The wild boar was introduced to the mountains during the 19th century; beavers have been reintroduced and are now the state's principal furbearers. The largest native carnivore is the bobcat.

North Carolina game birds include the bobwhite quail, mourning dove, wild turkey, and many varieties of duck and goose. Trout and smallmouth bass flourish in North Carolina's clear mountain streams, while catfish, pickerel, perch, crappie, and largemouth bass thrive in fresh water elsewhere. The sounds and surf of the coast yield channel bass, striped bass, flounder, and bluefish to anglers. Among insect pests, the pine bark beetle is a threat to the state's forests and forest industries.

The gray wolf, elk, eastern cougar, and bison are extinct in North Carolina; the American alligator, protected by the state, has returned in large numbers to eastern swamps and lakeshores. Endangered species (all on the federal list) include the Florida manatee, Indiana and gray bats, bald eagle, American and Arctic peregrine falcons, eastern brown pelican, ivory woodpecker, and Atlantic ridley and hawksbill turtles. Threatened species include the noonday snail.

⁵ENVIRONMENTAL PROTECTION

State actions to safeguard the environment began in 1915 with the purchase of the summit of Mt. Mitchell as North Carolina's first state park. North Carolina's citizens and officials worked actively (along with those in Tennessee) to establish the Great Smoky Mountains National Park during the 1920s, the same decade that saw the establishment of the first state agency for wildlife conservation. In 1937, a state and local program of soil and water conservation districts began to halt erosion and waste of natural resources.

Interest in environmental protection intensified during the 1970s. In 1971, the state required its own agencies to submit environmental impact statements in connection with all major project proposals; it also empowered local governments to require such statements from major private developers. Voters approved a $150-million bond issue in 1972 to assist in the construction of wastewater treatment facilities by local governments. The Coastal Management Act of 1974 mandated comprehensive land-use planning for estuaries, wetlands, beaches, and adjacent areas of environmental concern. The most controversial environmental action occurred mid-decade, when a coalition of state officials, local residents, and national environmental groups fought the proposed construction of a dam that would have flooded the New River Valley in northwestern North Carolina. Congress quashed the project when it designated the stream as a national scenic river in 1976.

Air quality in most of North Carolina's eight air-quality-control regions is good, although the industrialized areas of the piedmont and mountains experience pollution from engine exhausts and coal-fired electric generating plants. Water quality ranges from extraordinary purity in numerous mountain trout streams to serious pollution in major rivers and coastal waters. Soil erosion and municipal and industrial waste discharges have drastically increased the level of dissolved solids in some piedmont streams, while runoffs from livestock pastures and nitrates leached from fertilized farmland have overstimulated the growth of algae in slow-moving eastern rivers. Pollution also has made large areas of the coast unsafe for commercial shellfishing.

The Department of Natural Resources and Community Development, the state's main environmental agency, issues licenses to industries and municipalities and seeks to enforce clean air and water regulations. The department's budget for the 1985/87 biennium was $111.6 million.

⁶POPULATION

North Carolina had 5,881,776 inhabitants in 1980 (10th in the US), a 15.7% increase over 1970. The estimated population in early 1985 was 6,178,329.

At the time of the first census in 1790, North Carolina ranked 3d among the 13 states, with a population of 393,751, but it slipped to 10th by 1850. In the decades that followed, North Carolina grew slowly by natural increase and suffered from net out-migration, while the rest of the nation expanded rapidly. Out-migration abated after 1890, however, and North Carolina's overall growth rate in the 20th century has been slightly greater than that of the nation as a whole.

As of 1980, the state's population was slightly younger than the national average and very much less mobile: more than 76% of all state residents 14 years of age or older had lived in North Carolina their whole lives.

North Carolina's estimated population density was 126 per sq mi (49 per sq km) in early 1985. About 48% of North Carolinians lived in urban areas in 1980, compared with nearly 74% of all Americans. Most North Carolinians live in and around a relatively large number of small and medium-sized cities and towns, many of which are concentrated in the Piedmont Crescent, between Charlotte, Greensboro, and Raleigh. Leading cities in 1984 were Charlotte, 330,838; Raleigh, 169,331; Greensboro, 159,314; Winston-Salem, 143,366; and Durham, 101,997. The Charlotte metropolitan area had an estimated 1,031,400 people in mid-1984, ranking it 35th in the US.

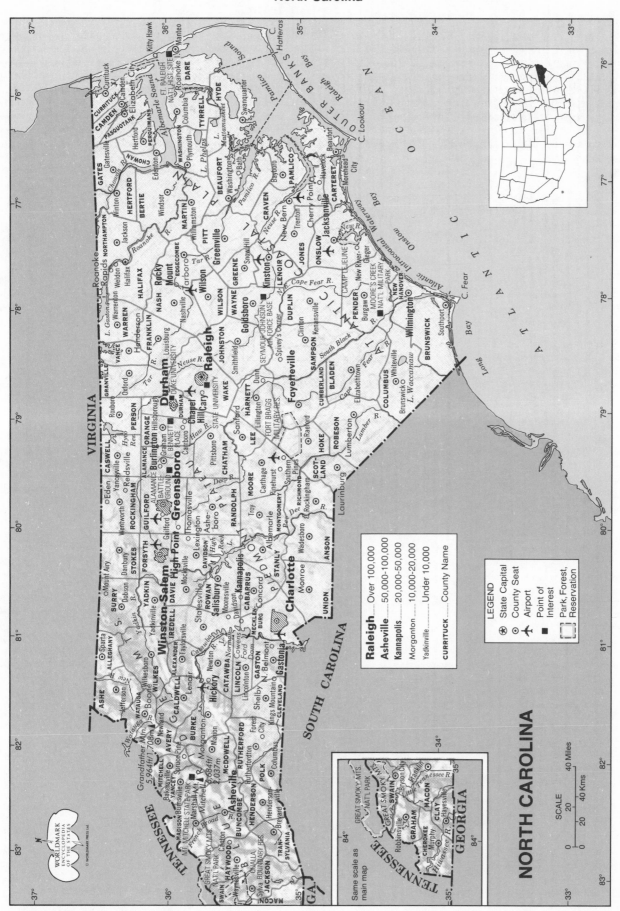

NORTH CAROLINA

LOCATION: 33°51′37″ to 36°34′25″N; 75°27′ to 84°20′W. **BOUNDARIES:** Virginia line, 320 mi (515 km); Atlantic Ocean coastline, 301 mi (484 km); South Carolina line, 328 mi (528 km); Georgia line, 68 mi (109 km); Tennessee line, 253 mi (407 km).

See endsheet maps: L3.

LEGEND
✺ State Capital
⊙ County Seat
✈ Airport
■ Point of Interest
▢ Park, Forest, Reservation

Raleigh Over 100,000
Asheville 50,000-100,000
Kannapolis 20,000-50,000
Morganton 10,000-20,000
Yadkinville Under 10,000
CURRITUCK County Name

SCALE
0   20   40 Miles
0   20   40 Kms

[7] ETHNIC GROUPS

North Carolina's white population is descended mostly from English settlers who arrived in the east in the 17th and early 18th centuries and from Scottish, Scotch-Irish, and German immigrants who poured into the piedmont in the middle of the 18th century. Originally very distinct, these groups assimilated with one another in the first half of the 19th century to form a relatively homogeneous body of native-born white Protestants. By 1860, North Carolina had the lowest proportion of foreign-born whites of any state; more than a century later, in 1980, only 1.1% of North Carolina residents were foreign-born, the 5th-lowest percentage among the states.

According to the 1980 federal census there were 64,519 Indians living in North Carolina, the 5th-largest number in any state, and the largest number in any state east of the Mississippi. The Lumbee of Robeson County and the surrounding area are the major Indian group. Their origins are mysterious, but they probably descend from many small tribes, decimated by war and disease, that banded together in the Lumber River swamps in the 18th century. The Lumbee have no language other than English, have no traditional tribal culture, and are not recognized by the Bureau of Indian Affairs. The Haliwa, Waccamaw Siouan, Coharie, and Person County Indians are smaller groups in eastern North Carolina who share the Lumbee's predicament. The only

North Carolina Counties, County Seats, and County Areas and Populations

COUNTY	COUNTY SEAT	LAND AREA (SQ MI)	POPULATION (1984 EST.)	COUNTY	COUNTY SEAT	LAND AREA (SQ MI)	POPULATION (1984 EST.)
Alamance	Graham	433	102,183	Johnston	Smithfield	795	75,287
Alexander	Taylorsville	259	26,343	Jones	Trenton	470	9,790
Alleghany	Sparta	234	9,874	Lee	Sanford	259	39,382
Anson	Wadesboro	533	26,149	Lenoir	Kinston	402	60,993
Ashe	Jefferson	426	23,086	Lincoln	Lincolnton	298	44,554
Avery	Newland	247	14,922	Macon	Franklin	517	22,817
Beaufort	Washington	826	42,660	Madison	Marshall	451	17,175
Bertie	Windsor	701	21,407	Martin	Williamston	461	27,150
Bladen	Elizabethtown	879	30,746	McDowell	Marion	437	36,200
Brunswick	Southport	861	43,689	Mecklenburg	Charlotte	528	432,879
Buncombe	Asheville	659	165,586	Mitchell	Bakersville	222	14,297
Burke	Morganton	505	74,674	Montgomery	Troy	490	23,435
Cabarrus	Concord	364	92,013	Moore	Carthage	701	54,322
Caldwell	Lenoir	471	68,559	Nash	Nashville	540	70,202
Camden	Camden	241	5,861	New Hanover	Wilmington	185	110,430
Carteret	Beaufort	525	46,778	Northampton	Jackson	538	22,435
Caswell	Yanceyville	427	22,100	Onslow	Jacksonville	763	118,605
Catawba	Newton	396	111,128	Orange	Hillsborough	400	80,929
Chatham	Pittsboro	708	35,246	Pamlico	Bayboro	341	10,826
Cherokee	Murphy	452	19,982	Pasquotank	Elizabeth City	228	29,300
Chowan	Edenton	181	12,911	Pender	Burgaw	875	23,875
Clay	Hayesville	214	7,001	Perquimans	Hertford	246	9,832
Cleveland	Shelby	468	84,520	Person	Roxboro	398	30,090
Columbus	Whiteville	939	51,773	Pitt	Greenville	656	94,868
Craven	New Bern	702	76,479	Polk	Columbus	238	14,408
Cumberland	Fayetteville	657	251,345	Randolph	Asheboro	789	96,410
Currituck	Currituck	256	12,878	Richmond	Rockingham	477	45,623
Dare	Manteo	391	16,371	Robeson	Lumberton	949	105,525
Davidson	Lexington	548	117,593	Rockingham	Wentworth	569	85,350
Davie	Mocksville	267	26,859	Rowan	Salisbury	519	101,998
Duplin	Kenansville	819	41,527	Rutherford	Rutherfordton	568	55,971
Durham	Durham	298	159,246	Sampson	Clinton	947	50,517
Edgecombe	Tarboro	506	58,248	Scotland	Laurinburg	319	33,780
Forsyth	Winston-Salem	412	255,677	Stanly	Albemarle	396	49,884
Franklin	Louisburg	494	31,866	Stokes	Danbury	452	34,884
Gaston	Gastonia	357	169,031	Surry	Dobson	539	60,312
Gates	Gatesville	338	9,264	Swain	Bryson City	526	10,716
Graham	Robbinsville	289	7,133	Transylvania	Brevard	378	24,891
Granville	Oxford	534	36,327	Tyrrell	Columbia	407	4,101
Greene	Snow Hill	266	16,401	Union	Monroe	639	76,348
Guilford	Greensboro	651	325,339	Vance	Henderson	249	38,013
Halifax	Halifax	724	56,033	Wake	Raleigh	854	337,254
Harnett	Lillington	601	61,823	Warren	Warrenton	427	16,241
Haywood	Waynesville	555	47,579	Washington	Plymouth	332	14,543
Henderson	Hendersonville	375	64,933	Watauga	Boone	314	33,764
Hertford	Winton	356	23,940	Wayne	Goldsboro	554	98,116
Hoke	Raeford	391	22,138	Wilkes	Wilkesboro	752	60,510
Hyde	Swanquarter	624	5,918	Wilson	Wilson	374	64,640
Iredell	Statesville	574	86,336	Yadkin	Yadkinville	336	29,304
Jackson	Sylva	490	27,033	Yancey	Burnsville	314	15,336
				TOTALS		48,843	6,164,613

North Carolina Indians with a reservation, a tribal language and culture, and federal recognition are the Cherokee, whose ancestors hid in the Smokies when the majority of their tribe was removed to Indian Territory (now Oklahoma) in 1838. The North Carolina Cherokee have remained in the mountains ever since, living in a community that now centers on the Qualla Boundary Reservation near Great Smoky Mountains National Park.

The 1,316,050 blacks in North Carolina comprised 22% of its total population in 1980. Black slaves came to North Carolina from the 17th century through the early 19th; like most white immigrants, they usually arrived in North Carolina after previous residence in other colonies. Although black slaves performed a wide variety of tasks and lived in every county of the state, they were most often field laborers on the large farms in the eastern region. The distribution of black population today still reflects the patterns of plantation agriculture: the coastal plain contains a much higher than average concentration of black inhabitants, with blacks outnumbering whites in seven counties in 1980. The overall proportion of blacks in North Carolina rose throughout the 19th century but fell steadily in the 20th, until about 1970, as hundreds of thousands migrated to northern and western states. Some of the earliest demonstrations of the civil rights movement, most notably a 1960 lunch counter sit-in at Greensboro, took place in the state.

8LANGUAGES

Although most of the original Cherokee Indians were removed to Indian Territory around 1838, descendants of those who resisted and remained have formed a strong Indian community in the Appalachian foothills. Among Indian place-names are Pamlico, Nantahala, and Cullasaja.

Many regional language features are widespread, but others sharply distinguish two subregions: the western half, including the piedmont and the Appalachian Highlands, and the eastern coastal plain. Terms common to South Midland and Southern speech occur throughout the state: both *dog irons* and *firedogs* (andirons), *bucket*, *spicket* (spigot), *seesaw*, *comfort* (tied and filled bedcover), *pullybone*, *ground squirrel* (chipmunk), *branch* (small stream), *light bread* (white bread), *polecat* (skunk), and *carry* (escort). Also common are *greasy* with the /z/ sound, *new* as /nyoo/ and *due* as /dyoo/, *swallow it* as /swaller it/, *can't* rhyming with *paint*, *poor* with the vowel sound /aw/, and *horse* and *hoarse* with different vowels.

Distinct to the western region are *snake feeder* (dragonfly), *blinds* (roller shades), *poke* (paper bag), *redworm* (earthworm), *a little piece* (a short distance), *plum peach* (clingstone peach), *sick on the stomach* (also found in the Pee Dee River Valley), *boiled* as /bawrld/, *fog* as /fawg/, *Mary* sounding like *merry*, and *bulge* with the vowel of *good*. Setting off eastern North Carolina are *lightwood* (kindling), *mosquito hawk* (dragonfly), *earthworm*, *press peach* (instead of plum peach), *you-all* as plural, and *sick in the stomach*. Distinctive eastern pronunciations include the loss of /r/ after a vowel, *fog* as /fagh/, *scarce* and *Mary* with the vowel of *gate*, *bulge* with the vowel of *bun*, *shrimp* as /srimp/, and *foreign* and *forest* with the vowel sound /ah/. Along the coast, peanuts are *goobers* and a screech owl is a *shivering owl*.

In 1980, 5,504,736 North Carolinians—98% of the population 3 years of age and older—spoke only English at home. Other languages spoken at home, and the number of speakers, included the following:

Spanish	44,347	Korean	3,168
French	24,035	Chinese	3,125
German	15,606	Japanese	2,965
Italian	3,278	Arabic	2,599

9RELIGIONS

The majority of North Carolinians are Protestant. The churches of the Southern Baptist Convention reported 1,341,047 adherents in 1980; the United Methodist Church claimed 490,407; the American Zion Church had 311,006; and the Presbyterian Church US, 191,424. In 1984, the state had 111,084 Roman Catholics and an estimated 15,145 Jews.

The Church of England was the established church of colonial North Carolina but was never a dominant force among the early immigrants. Scottish Presbyterians settled in the upper Cape Fear Valley, and Scotch-Irish Presbyterians occupied the piedmont after 1757. Lutheran Evangelical Reformed Germans later moved into the Yadkin and Catawba valleys of the same region. The Moravians, a German sect, founded the town of Salem (later merging with Winston to become Winston-Salem) in 1766 as the center of their utopian community at Wachovia. Methodist circuit riders and Separate Baptists missionaries won thousands of converts among blacks and whites, strengthening their appeal in the Great Revival of 1801. In the subsequent generation, a powerful evangelical consensus dominated popular culture. After the Civil War, blacks left the white congregations to found their own churches, but the overall strength of Protestantism persisted. When many North Carolinians left their farms at the end of the 19th century, they moved to mill villages that were well supplied with churches, often at the mill owners' expense.

White church organizations have generally kept out of state politics except where matters of personal morality are concerned. In recent years, however, individual Protestant leaders have worked to reject the Equal Rights Amendment and to deny state-funded abortions to the poor. Resurgent political activism among fundamentalist Christians is a growing and prominent feature of contemporary North Carolina politics.

10TRANSPORTATION

Railroad construction was initially hampered in North Carolina by the reluctance of eastern legislators to approve heavy expenditures for the benefit of piedmont farmers. When political reform changed the balance of power in 1835, however, state subsidies for railroad construction followed. The state-owned North Carolina Railroad, completed in 1856, described an arc from Raleigh to Greensboro to Charlotte that became the backbone for the modern industrialized Piedmont Crescent. The builders of this and other early rail lines had hoped to channel trade into North Carolina ports, but development was slow. A generation after the end of the Civil War, financial difficulties forced the ailing local railroads to merge into national lines whose rate structures forced regional commerce northward. The Southern Railway acquired a 99-year lease on the piedmont section of the North Carolina Railroad in 1895, while eastern routes fell to the Atlantic Coast Line and the Seaboard Air Line Railway. Northern interests thus owned all three major lines, and while they ended the dream of locally controlled development, the railroad barons did integrate North Carolina into the interregional trade patterns of the growing national economy. As of 1983, North Carolina had 2,799 mi (4,505 km) of Class I track. Amtrak provides passenger service to most large North Carolina cities; 1983/84 North Carolina ridership was 149,116.

Shortly after consolidation, the railroads began to encourage "good roads" to facilitate wagon traffic between the railheads and the more remote farming districts. The new hard-surface roads soon proved ideal for automobiles and trucks, and in the 1920s, state roads came to rival the railroads as the principal means of transport. Ambitious bond issues paid for a statewide system of paved highways, giving the state more roads by the end of the decade than any other southern state except Texas. The state government took over the county roads in 1931. By 1983 there were 92,404 mi (148,710 km) of roads in the state, of which 81% were rural and 19% municipal; 3,461,818 automobiles were registered in 1983, and there were 3,966,351 licensed drivers. The major interstate highways are I-95, which stretches north–south across the coastal plain, and I-85, which parallels it across the piedmont. I-40 leads from Greensboro westward over the mountains, and I-26 and I-77 handle north–south traffic in the western section. An extension of I-40, scheduled for completion in 1990, will connect the piedmont and the coastal plain. Nearly all North Carolinians who commute to work travel by private automobile; intercity bus connections are poor, and commuter trains are nonexistent.

The Atlantic Intracoastal Waterway follows sounds, rivers, and canals down the entire length of eastern North Carolina, connecting the two state ports at Morehead City and Wilmington. Morehead City freight traffic in 1982 was 3,690,281 tons; Wilmington traffic was 5,661,499 tons. Cargoes that flow through these cities to Europe, South America, and East Asia include coal, phosphates, wood pulp, salt, limestone, residual fuel oil, fertilizer, chemicals, and asphalt.

North Carolina had 244 airports and 39 heliports in 1983. Regularly scheduled commercial airlines serving North Carolina cities in 1983 enplaned almost 5 million passengers; the airports at Charlotte, Raleigh-Durham, Greensboro-High Point, Fayetteville, and Asheville handled 95% of this traffic.

11 HISTORY

Paleo-Indian peoples came to North Carolina about 10,000 years ago. These early inhabitants hunted game with spears and gathered nuts, roots, berries, and freshwater mollusks. Around 500 BC, with the invention of pottery and the development of agriculture, the Woodland Culture began to emerge. The Woodland way of life—growing corn, beans, and squash, and hunting game with bows and arrows—prevailed on the North Carolina coast until the Europeans arrived.

Living in North Carolina by this time were Indians of the Algonkian-, Siouan-, and Iroquoian-language families. The Roanoke, Chowanoc, Hatteras, Meherrin, and other Algonkian-speaking tribes of the coast had probably lived in the area the longest; some of them belonged to the Powhatan Confederacy of Virginia. The Siouan groups were related to larger tribes of the Great Plains. Of the Iroquoian-speakers, the Cherokee probably had lived in the mountains since before the beginning of the Christian era, while the Tuscarora had entered the upper coastal plain somewhat later. After their defeat by the colonists in the Tuscarora War of 1711–13, the tribe fled to what is now upper New York State to become the sixth member of the Iroquois Confederacy.

Contact with whites brought war, disease, and enslavement to the Algonkian and Siouan tribes. Banding together, the survivors probably gave rise to the present-day Lumbee and to the other Indian groups of eastern North Carolina. The Cherokee tried to avoid the fate of the coastal tribes by selectively adopting aspects of white culture. In 1838, however, the federal government responded to the demands of land-hungry whites by expelling most of the Cherokee to Indian Territory along the so-called Trail of Tears.

European penetration began when Giovanni da Verrazano, a Florentine navigator in French service, discovered the North Carolina coast in 1524. Don Lucas Vásquez de Ayllón led an unsuccessful Spanish attempt to settle near the mouth of the Cape Fear River two years later. Hernando de Soto tramped over the North Carolina mountains in 1540 in an unsuccessful search for gold, but the Spanish made no permanent contribution to the colonization of North Carolina.

Sixty years after Verrazano's voyage, North Carolina became the scene of England's first experiment in American empire. Sir Walter Raleigh, a courtier of Queen Elizabeth I, gained the queen's permission to send out explorers to the New World. They landed on the Outer Banks in 1584 and returned with reports so enthusiastic that Raleigh decided to sponsor a colony on Roanoke Island between Albemarle and Pamlico sounds. After a second expedition returned without founding a permanent settlement,

Raleigh sent out a third group in 1587 under John White as governor. The passengers included White's daughter Eleanor and her husband, Ananias Dare. Shortly after landfall, Eleanor gave birth to Virginia Dare, the first child born of English parents in the New World. Several weeks later, White returned to England for supplies, but the threat of the Spanish Armada prevented his prompt return. By the time White got back to Roanoke in 1590, he found no trace of the settlers—only the word "Croatoan" carved on a tree. The fate of this "Lost Colony" has never been satisfactorily explained.

The next English venture focused on the more accessible Jamestown colony in the Chesapeake Bay area of Virginia. England tended to ignore the southern region until 1629, when Charles I laid out the territory between 30° and 36°N, named it Carolana for himself, and granted it to his attorney general, Sir Robert Heath. Heath made no attempt to people his domain, however, and Carolana remained empty of whites until stragglers drifted in from the mid–17th century onward. Events in England transformed Virginia's outpost into a separate colony. After the execution of Charles I in 1649, England had no ruling monarch until a party of noblemen invited Charles II back to England in 1660. Charles thanked eight of his benefactors three years later by making them lords proprietors of the province, now called Carolina. The vast new region eventually stretched from northern Florida to the modern boundary between North Carolina and Virginia, and from the Atlantic to the Pacific Ocean.

The proprietors divided Carolina into three counties and appointed a governor for each one. Albemarle County embraced the existing settlements in northeastern North Carolina near the waters of Albemarle Sound; it was the only one that developed a government within the present state boundaries. From the beginning, relations between the older pioneers and their newly imposed government were stormy. The English philosopher John Locke drew up the Fundamental Constitutions of Carolina, but his political blueprints proved unworkable. The proprietors' arbitrary efforts to collect royal customs touched off factional violence, culminating in Culpepper's Rebellion of 1677, one of the first American uprisings against a corrupt regime.

For a few years afterward, local residents had a more representative government, until the proprietors attempted to strengthen the establishment of the Anglican Church in the colony. In 1711, Cary's Rebellion was touched off by laws passed against the colony's Quakers. During the confusion, Tuscarora Indians launched a war against the white intruders on their lands. The whites won the Tuscarora War in 1713 with assistance from South Carolina, but political weakness in the north persisted. Proprietary officials openly consorted with pirates—including the notorious Edward Teach, alias Blackbeard—and royal inspectors questioned the fitness of proprietary government. South Carolina officially split off in 1719 and received a royal governor in 1721. Ten years later, all but one of the proprietors relinquished their rights for £2,500 each, and North Carolina became a royal colony. The remaining proprietor, Lord Granville, gave up his governing rights but retained ownership of one-eighth of the original grant; the Granville District thus included more than half of the unsettled territory in the North Carolina colony.

In the decades that followed, thousands of new settlers poured into North Carolina; by 1775, the population had swollen to 345,000, making North Carolina the 4th most populous colony. Germans and Scotch-Irish trekked down the Great Wagon Road from Pennsylvania to the piedmont. Scottish Highlanders spread over the upper Cape Fear Valley as more Englishmen filled up the coastal plain. Backcountry settlers practiced self-sufficient farming, but eastern North Carolinians used slave labor to carve out rice and tobacco plantations. The westerners were often exploited by an eastern-dominated colonial assembly that sent corrupt and overbearing officials to govern them. Organizing in 1768 and calling themselves Regulators, unhappy westerners first petitioned for redress and then took up arms. Royal Governor William Tryon used eastern militia to crush the Regulators in a two-hour pitched battle at Alamance Creek in 1771.

The eastern leaders who dominated the assembly opposed all challenges to their authority, whether from the Regulators or from the British ministry. When England tightened its colonial administration, North Carolinians joined their fellow colonists in protests against the Stamp Act and similar impositions by Parliament. Meeting at Halifax in April 1776, the North Carolina provincial congress resolved in favor of American independence, the first colonial representative body to do so. Years later, citizens of Mecklenburg County recalled a gathering in 1775 during which their region declared independence, but subsequent historians have not verified their claim. The two dates on the North Carolina state flag nevertheless commemorate the Halifax Resolves and the "Mecklenburg Declaration of Independence."

Support for Britain appeared among recent Scottish immigrants, who answered the call to aid the royal governor but were ambushed by patriot militia at Moore's Creek Bridge on 27 February 1776. The incident effectively prevented a planned British invasion of the South. There was little further military action in North Carolina until late in the War for Independence, when Gen. Charles Cornwallis invaded the state from South Carolina in the fall of 1780. Guerrilla bands harassed his troops, and North Carolina militia wiped out a Loyalist detachment at King's Mountain. Pursuing the elusive American army under Gen. Nathanael Greene, Cornwallis won a costly victory at Guilford Courthouse in March 1781 but could neither eliminate his rival nor pacify the countryside. For the rest of 1781, Cornwallis wearied his men in marches and countermarches across North Carolina and Virginia before he finally succumbed to a trap set at Yorktown, Va., by an American army and a French fleet.

Numerous problems beset the new state. The government had a dire need of money, but when the victors sought to pay debts by selling land confiscated from the Loyalists, conservative lawyers objected strenuously, and a bitter political controversy ensued. Suspicious of outside control, North Carolina leaders hesitated before joining the Union. The state waited until November 1789 to ratify the US Constitution—a delay that helped stimulate the movement for adoption of a Bill of Rights. North Carolina relinquished its lands beyond the Great Smokies in 1789 (after an unsuccessful attempt by settlers to create a new state called Franklin), and thousands of North Carolinians migrated to the new western territories. The state did not share in the general prosperity of the early federal period. Poor transportation facilities hampered all efforts to expand commercial agriculture, and illiteracy remained widespread. North Carolina society came to appear so backward that some observers nicknamed it the "Rip Van Winkle state."

In 1815, State Senator Archibald D. Murphey of Orange County began to press for public schools and for improved transportation to open up the piedmont. Most eastern planters resisted Murphey's suggestions, partly because they refused to be taxed for the benefit of the westerners and partly because they feared the destabilizing social effects of reform. As long as the east controlled the general assembly, the ideas of Murphey and his sympathizers had little practical impact, but in 1835, as a result of reforms in the state constitution, the west obtained reapportionment and the political climate changed. North Carolina initiated a program of state aid to railroads and other public works, and established the first state-supported system of common schools in the South.

Like other southern whites, North Carolina's white majority feared for the security of slavery under a national Republican administration, but North Carolinians reacted to the election of Abraham Lincoln with caution. When South Carolina and six other states seceded and formed the Confederate States of America in 1861, North Carolina refused to join, instead making a futile attempt to work for a peaceful settlement of the issue. However, after the outbreak of hostilities at Ft. Sumter, S.C., and Lincoln's call for troops in April 1861, neutrality disappeared and public opinion swung to the Confederate side. North Carolina became the last state to withdraw from the Union, joining the Confederacy on 20 May 1861.

North Carolina provided more troops to the Confederacy than any other state, and its losses added up to more than one-fourth of the total for the entire South, but support for the war was mixed. State leaders resisted the centralizing tendencies of the Richmond government, and even Governor Zebulon B. Vance opposed the Confederacy's conscription policies. North Carolina became a haven for deserters from the front lines in Virginia. William W. Holden, a popular Raleigh editor, organized a peace movement when defeat appeared inevitable, and Unionist sentiment flourished in the mountain counties; nevertheless, most white North Carolinians stood by Vance and the dying Confederate cause. At the war's end, Gen. Joseph E. Johnston surrendered the last major Confederate army to Gen. William T. Sherman at Bennett House near Hillsborough on 26 April 1865.

Reconstruction marked a bitter political and social struggle in North Carolina. United in the Conservative Party, most of the prewar slaveholding elite fought to preserve as much as possible of the former system, but a Republican coalition of blacks and nonslaveholding white Unionists defended freedmen's rights and instituted democratic reforms for the benefit of both races. After writing a new constitution in 1868, Republicans elected Holden as governor, but native whites fought back with violence and intimidation under the robes of the Ku Klux Klan. Holden's efforts to restore order were ineffectual, and when the Conservatives recaptured the general assembly in 1870, they impeached him and removed him from office. Election of a Conservative governor in 1876 signaled the end of the Reconstruction era.

Once in power, the Conservatives—or Democrats, as they renamed themselves—slashed public services and enacted legislation to guarantee the power of landlords over tenants and sharecroppers. They cooperated with the consolidation of railroads under northern ownership, and they supported a massive drive to build cotton mills on the swiftly flowing streams of the piedmont. By 1880, industry had surpassed its prewar level. But it was not until the turn of the century that blacks and their white allies were entirely eliminated as contenders for political power.

As the Industrial Revolution gained ground in North Carolina, small farmers protested their steadily worsening condition. The Populist Party expressed their demands for reform, and for a brief period in the 1890s shared power with the Republican Party in the Fusion movement. Under the leadership of Charles Brantley Aycock, conservative Democrats fought back with virulent denunciations of "Negro rule" and a call for white supremacy. In 1900, voters elected Aycock governor and approved a constitutional amendment that barred all illiterates from voting, except for those whose ancestors had voted before 1867. This literacy test and "grandfather clause" effectively disenfranchised blacks, while providing a temporary loophole for uneducated whites. To safeguard white rights after 1908 (the constitutional limit for registration under the grandfather clause), Aycock promised substantial improvements in the school system to put an end to white illiteracy.

In the decades after Aycock's election, an alliance of business interests and moderate-to-conservative Democrats dominated North Carolina politics. The industrial triumvirate of textile, tobacco, and furniture manufacturers, joined by banks and insurance companies, controlled the state's economy. The Republican Party shriveled to a small remnant among mountain whites as blacks were forced out of the electorate. Political leaders emphasized fiscal responsibility, honest government, state assistance to aid economic growth, a tolerable level of social services, and a relative absence of racist extremism. Moderate business influence was so pervasive that political scientist V. O. Key labeled North Carolina a "progressive plutocracy" in the 1930s.

In the years since World War II, North Carolina has taken its place in the booming Sunbelt economy. The development of Research Triangle Park—equidistant from the educational facilities of Duke University, North Carolina State University, and the University of North Carolina at Chapel Hill—has provided a home for dozens of scientific laboratories for government and business. New industries, some of them financed by foreign capital, have appeared in formerly rural areas, and a prolonged population drain has been reversed. The process of development has not been smooth, however. White backlash followed renewed activism among blacks in the 1960s, and the Republican Party (especially its conservative wing) rebounded as the national Democratic Party became closely identified with liberalism. In 1972, North Carolina elected its first Republican US senator (Jesse A. Helms) and governor (James E. Holshouser, Jr.) since Fusion days. Republican strength was evident again in 1984, when Senator Helms won a third term, James G. Martin became the second Republican to capture the governorship in the 20th century, and the Republicans won 5 of 11 seats in the US House of Representatives, the largest number in the 20th century.

¹²STATE GOVERNMENT

North Carolina has operated under three constitutions, adopted in 1776, 1868, and 1971, respectively. The first was drafted hurriedly under wartime pressures and contained several inconsistencies and undemocratic features. The second, a product of Reconstruction, was written by native white Republicans and a sprinkling of blacks and northern-born Republicans. When conservative whites regained power, they left the basic framework of this constitution intact, though they added the literacy test, poll tax, and grandfather clause to it.

A century after the Civil War, the document had become unwieldy and partially obsolete. A constitutional study commission submitted to the general assembly in 1969 a rewritten constitution, which the electorate ratified, as amended, in 1971. As of the end of 1983, 16 other amendments had been added, one of which permits the governor and lieutenant governor to serve two successive four-year terms.

Under the 1971 constitution, the general assembly consists of a 50-member senate and a 120-member house of representatives. Senators must be at least 25 years old, must be qualified voters of the state, and must have been residents of the state for at least two years prior to election. Representatives must be qualified voters of the state and must have lived in their district for at least a year; the constitution establishes 21 as the minimum age for elective office. All members of the general assembly serve two-year terms.

The governor and lieutenant governor (who run separately) must be 30 years old; each must have been a US citizen for five years and a state resident for two. The only state governor without a veto, North Carolina's chief executive has powers of appointment, supervision, and budgetary recommendation. The voters also elect a secretary of state, treasurer, auditor, superintendent of public instruction, attorney general, and commissioners of agriculture, insurance, and labor to four-year terms. These officials preside over their respective departments and sit with the governor and lieutenant governor as the council of state. The governor appoints the heads of the 10 other executive departments.

Bills become law when they have passed three readings in each house of the general assembly, and take effect 30 days after adjournment. Constitutional amendments may be proposed by a convention called by a two-thirds vote of both houses and a majority of the voters, or may be submitted directly to the voters by a three-fifths consent of each house. In either case, the proposed amendments must be ratified by a popular majority before becoming part of the constitution.

Prospective voters in North Carolina must be US citizens who are at least 18 years old and have never been convicted of a felony (unless their rights have been restored by law). They must have lived in North Carolina for one year and in their home precinct for 30 days prior to the election.

¹³POLITICAL PARTIES

Prior to the Civil War, Whigs and Democrats were the two major political groups in North Carolina. The Republican Party

North Carolina Presidential Vote by Political Parties, 1948–84

YEAR	ELECTORAL VOTE	NORTH CAROLINA WINNER	DEMOCRAT	REPUBLICAN	STATES' RIGHTS DEMOCRAT	PROGRESSIVE
1948	14	*Truman (D)	459,070	258,572	69,652	3,915
1952	14	Stevenson (D)	652,803	558,107	—	—
1956	14	Stevenson (D)	590,530	575,069	—	—
1960	14	*Kennedy (D)	713,136	655,420	—	—
1964	13	*Johnson (D)	800,139	624,841	—	—
						AMERICAN IND.
1968	13	*Nixon (R)	464,113	627,192	—	496,188
						AMERICAN
1972	13	*Nixon (R)	438,705	1,054,889	—	25,018
					LIBERTARIAN	
1976	13	*Carter (D)	927,365	741,960	2,219	5,607
1980	13	*Reagan (R)	875,635	915,018	9,677	—
1984	13	*Reagan (R)	824,287	1,346,481	3,794	—

*Won US presidential election.

emerged during Reconstruction as a coalition of newly enfranchised blacks, northern immigrants, and disaffected native whites, especially from nonslaveholding areas in the mountains. The opposing Conservative Party, representing a coalition of antebellum Democrats and former Whigs, became the Democratic Party after winning the governorship in 1876; from that time and for most of the 20th century, North Carolina was practically a one-party state.

Beginning in the 1930s, however, as blacks reentered the electorate as supporters of the New Deal and the liberal measures associated with Democratic presidents, the Republican Party attracted new white members who objected to national Democratic policies. Republican presidential candidates picked up strength in the 1950s and 1960s, and Richard Nixon carried North Carolina in 1968 and 1972, when Republicans also succeeded in electing Governor James E. Holshouser, Jr., and US Senator Jesse A. Helms. The Watergate scandal cut short this movement toward a revitalized two-party system, and in 1976, Jimmy Carter became the first Democratic presidential candidate to carry the state since 1964.

Republican presidential candidate Ronald Reagan narrowly carried North Carolina in 1980, and a second Republican senator, John P. East, was elected that year. In 1984, the Republican Party had its best election year in North Carolina. Reagan won the state by a landslide, Helms won a third term—defeating Governor James B. Hunt in the most expensive race in Senate history (more than $26 million was spent)—and Republican James G. Martin, a US representative, was elected governor, succeeding Hunt. Republicans won 5 of North Carolina's 11 seats in the US House of Representatives, a pickup of 3. Nevertheless, the state remains strongly Democratic at the grassroots level; in 1985, 71% of the state legislators were Democrats. In 1982, 1,924,394 voters (72%) were registered as Democrats and only 640,675 (24%) as Republicans.

Minor parties have had a marked influence on the state. George Wallace's American Independent Party won 496,188 votes in 1968, placing second with more than 31% of the total vote.

In 1985, the state legislature included 20 women and 12 blacks; there were 294 black elected officials in North Carolina in 1984 (8th among the states). Harvey Gantt was elected the first black mayor of Charlotte in 1983, and in 1984, Henry Frye became the first black elected to the state supreme court.

14LOCAL GOVERNMENT

As of 1982, North Carolina had 100 counties, 484 municipalities, and 321 special districts.

Counties have been the basis of local government in North Carolina for more than 300 years, and are still the primary governmental units for most citizens. All counties are led by boards of commissioners; commissioners serve either two- or four-year terms, and most are elected at large rather than by district. Most boards elect their own chairman from among their own members, but voters in some counties choose a chairman separately. More than half the counties employ a county manager to supervise day-to-day operations of county government. Counties are subdivided into townships, but these are for administrative convenience only; they do not exercise any independent government functions.

About 48% of all North Carolinians lived in incorporated cities and towns in 1980, an increase of 3% over 1970. County and municipal governments share many functions, but the precise allocation of authority varies in each case. Although the city of Charlotte and Mecklenburg County share a common school system, most often schools, streets, sewers, garbage collection, police and fire protection, and other services are handled separately. Most cities use the council-manager form of government, with council members elected from the city at large. Proliferation of suburban governments is hampered by a 1972 constitutional

amendment that forbids the incorporation of a new town or city within 1 mi (1.6 km) of a city of 5,000-9,999 people, within 3 mi (4.8 km) of a city of 10,000-24,999, within 4 mi (6.4 km) of a city of 25,000-49,999, and within 5 mi (8 km) of a city of 50,000 or more unless the general assembly acts to do so by a three-fifths vote of all members of each house.

15STATE SERVICES

The Department of Public Instruction administers state aid to local public school systems; a board of governors directs the 16 state-supported institutions of higher education; and the Department of Community Colleges administers the 58 community colleges. The Department of Cultural Resources offers a variety of educational and enrichment services to the public, maintaining historical sites, operating two major state museums, funding the North Carolina Symphony, and providing for the State Library. The Department of Transportation plans, builds, and maintains state highways; registers motor vehicles; develops airport facilities; administers public transportation activities; and operates 15 ferries.

Within the Department of Human Resources, the Division of Mental Health, Mental Retardation, and Substance Abuse Services operates 4 regional psychiatric hospitals, 5 regional mental retardation centers, and 3 alcoholic rehabilitation centers; it also coordinates 41 area mental health programs that include community mental health centers, group homes for the mentally retarded and emotionally disturbed, shelter workshops, halfway houses, a special-care facility, and 2 reeducation programs for emotionally disturbed children and adolescents. The Division of Social Services administers public assistance programs, and other divisions license medical facilities, promote public health, administer programs for juvenile delinquents and the vocationally handicapped, and operate a school for the blind and visually impaired and 3 schools for the deaf.

The Department of Crime Control and Public Safety includes the Highway Patrol and the National Guard, while the Department of Correction manages the prison system. Local law enforcement agencies receive assistance from the Department of Justice's State Bureau of Investigation and the Police Information Network. The Community Assistance Division of the Department of Natural Resources and Community Development offers a variety of planning services to local government in the areas of housing, neighborhood renewal, and fiscal resources. The Department of Labor administers the state Occupational Safety and Health Act; inspects boilers, elevators, amusement rides, mines, and quarries; offers conciliation, mediation, and arbitration services to settle labor disputes; and enforces state laws governing child labor, minimum wages, maximum working hours, and uniform wage payment.

16JUDICIAL SYSTEM

North Carolina's general court of justice is a unified judicial system that includes appellate courts (supreme court and court of appeals) and trial courts (superior court and district court). District court judges are elected to four-year terms; judges above that level are elected for eight years.

The state's highest court, the supreme court, consists of a chief justice and 6 associate justices. It hears cases from the court of appeals as well as certain cases from lower courts. The court of appeals comprises 12 judges who hear cases in 3-judge panels. Superior courts have original jurisdiction in most major civil and criminal cases. There is at least one superior court justice in each of the 34 districts; 8 additional justices are appointed by the governor to four-year terms. All superior court justices rotate between the districts within their divisions. District courts try misdemeanors, civil cases involving less than $5,000, and all domestic cases. They have no juries in criminal cases, but these cases may be appealed to superior court and be given a jury trial de novo; in civil cases, jury trial is provided on demand.

North Carolina had 16,371 prisoners in state and federal institutions in 1984, an increase of 6.3% over 1983. A total of 9,349 persons were employed in the corrections system in North Carolina in 1983. That year, North Carolina's overall crime rate ranked 31st in the US; the 8.1 murders and 21.9 forcible rapes per 100,000 population were below the national averages, and the rate of 300 aggravated assaults per 100,000 population was less than 10% above the US average. North Carolina was well below the US averages in all categories of property crime.

North Carolina punishes crime severely. From 1930 to 1985, the state ranked 5th in number of persons executed (264). The US Supreme Court invalidated North Carolina's death penalty statute in 1976, and the sentences of all inmates then on death row reverted to life imprisonment. The state passed a new capital punishment statute in 1977 that apparently met the Court's objections, and two persons were executed in 1984—the state's first executions since 1961. One of the prisoners executed that year, Velma Barfield, was the first woman executed in the US since 1962 and the first in North Carolina since 1944.

17 ARMED FORCES

North Carolina holds the headquarters of the 3d Army at Ft. Bragg in Fayetteville and a major training facility for the Marine Corps at Camp Lejeune in Jacksonville. The Marine Corps air stations at Cherry Point and New River and Seymour Johnson Air Force Base in Goldsboro are the state's other important military installations. North Carolina firms received $862 million in defense contract awards in 1983/84. Active duty military personnel numbered 100,227.

There were 658,000 veterans living in North Carolina as of 30 September 1983; 5,000 saw service in World War I, 249,000 in World War II, 127,000 in the Korean conflict, and 203,000 during the Viet-Nam era. Veterans' benefits totaled $696 million in 1982/83.

The strength of the Army National Guard was 11,592 in December 1984; the Air National Guard had 1,308 personnel in February 1985. Police personnel numbered 13,594 in 1983, including 1,137 state troopers.

18 MIGRATION

For most of the state's history, more people have moved away every decade than have moved into the state, and population growth has come only from net natural increase. In 1850, one-third of all free, native-born North Carolinians lived outside the state, chiefly in Tennessee, Georgia, Indiana, and Alabama. The state suffered a net loss of population from migration in every decade from 1870 to 1970.

Before 1890, the emigration rate was higher among whites than among blacks; since then, the reverse has been true, but the number of whites moving into North Carolina did not exceed the number of white emigrants until the 1960s. Between 1940 and 1970, 539,000 more blacks left North Carolina than moved into the state; most of these emigrants sought homes in the North and West. After 1970, however, black out-migration abruptly slackened as economic conditions in eastern North Carolina improved. Net migration to North Carolina was estimated at 278,000 (6th among the states) from 1970 to 1980, and at 83,000 (9th among the states) from 1980 to 1983.

19 INTERGOVERNMENTAL COOPERATION

North Carolina adheres to at least 17 interstate compacts, including 4 that promote regional planning and development. The oldest of the 4, establishing the Board of Control for Southern Regional Education, pools the resources of southern states for the support of graduate and professional schools. The Southeastern Forest Fire Protection Compact promotes regional forest conservation, while the Southern Interstate Energy Compact fosters cooperation in nuclear power development. The Southern Growth Policies Board, formed in 1971 at the suggestion of former North Carolina Governor Terry Sanford, collects and publishes data for

planning purposes from its headquarters in Research Triangle Park. The Tennessee Valley Authority operates three dams in western North Carolina to aid in flood control, to generate hydroelectric power, and to assist navigation downstream on the Tennessee River; most of the electricity generated is exported to Tennessee. Total federal aid in 1983/84 was $1.9 billion.

20 ECONOMY

North Carolina's economy was dominated by agriculture until the closing decades of the 19th century, with tobacco the major cash crop; today, tobacco is still the central factor in the economy of the coastal plain. In the piedmont, industrialization accelerated after 1880, when falling crop prices made farming less attractive. During the "cotton mill crusade" of the late 19th and early 20th centuries, local capitalists put spinning or weaving mills on swift streams throughout the region, until nearly every hamlet had its own factory. Under the leadership of James B. Duke, the American Tobacco Co. (now American Brands, with headquarters in New York City) expanded from its Durham headquarters during this same period to control, for a time, virtually the entire US market for smoking products. After native businessmen had established a successful textile boom, New England firms moved south in an effort to cut costs, and the piedmont became a center of southern industrial development.

As more and more Tarheels left agriculture for the factory, their per capita income rose from 47% of the national average in 1930 to almost 83% in 1983. The biggest employers are the textile and furniture industries. Wages in these labor-intensive fields remain comparatively low in all states, but especially low in North Carolina. As of 1984, North Carolina led the nation in the proportion of manufacturing workers in its nonagricultural work force, but had the lowest hourly wage rate for production workers on manufacturing payrolls; the state had the lowest level of unionization in 1980.

Since the 1950s, state government has made a vigorous effort to recruit outside investment and to improve the state's industrial mix. Major new firms now produce electrical equipment, processed foods, technical instruments, fabricated metals, plastics, and chemicals. The greatest industrial growth, however, has come not from wholly new industries but from fields related to industries that were firmly established. Apparel manufacture spread across eastern North Carolina during the 1960s as an obvious extension of the textile industry, and other new firms produce chemicals and machinery for the textile and furniture business.

The major test for North Carolina's continued economic growth will be whether the state can break out of the low-skill, low-wage trap. Despite recent improvements, North Carolina is still poor. Labor-intensive traditional industries cannot be relied on to meet all the needs for sustained growth. The textile industry, moreover, is threatened by foreign competition, and tobacco interests face a federal antismoking campaign.

21 INCOME

In 1983, North Carolina's per capita personal income averaged $9,656, 39th among the 50 states. Measured in constant dollars, per capita income increased by 30% between 1970 and 1983, slightly faster than the national rate of 28% and nearly equal to the 33% increase for the South Atlantic region. North Carolina's total personal income of $58.7 billion in 1983 represented 2.1% of the US total.

Per capita income levels in 1981 showed considerable geographical variation, ranging from $11,460 in urban Mecklenburg County (Charlotte) to $5,356 in mountainous Clay County. Only 25 counties out of 100 exceeded the statewide average of $8,646 in that year. Median family money income was $16,792 (43d in the US) in 1979, when 14.8% of North Carolina residents and 11.6% of all families were below the poverty level. In 1982, 33,300 North Carolinians were among the top wealth-holders in the US, with gross assets of more than $500,000 each.

²²LABOR

North Carolina is unusual for combining a predominantly industrial and service-related work force with a dispersed pattern of rural residence. North Carolina's civilian labor force numbered 3,033,000 in 1984. The proportion of all women who participated in the work force that year was 57.1%. The overall unemployment rate in 1984 was 6.7%, compared to 7.5% nationally. The rates for women and nonwhites were 8.3% and 14.9%, respectively.

A federal survey in March 1982 revealed the following nonfarm employment pattern in North Carolina:

	ESTABLISH-MENTS	EMPLOYEES	ANNUAL PAYROLL ('000)
Agricultural services, forestry, fishing	1,079	7,577	$ 73,309
Mining	164	3,602	57,805
Contract construction	11,115	107,955	1,489,795
Manufacturing, of which:	9,413	803,875	11,751,769
Tobacco manufactures	(24)	(22,683)	(498,324)
Textile mill products	(1,222)	(221,095)	(2,598,398)
Apparel, other textiles	(667)	(76,867)	(692,126)
Furniture and fixtures	(627)	(77,511)	(883,582)
Electric and electronic equipment	(240)	(54,882)	(960,492)
Transportation, public utilities	3,838	117,872	2,404,087
Wholesale trade	10,031	120,100	2,054,219
Retail trade	33,897	354,292	3,082,893
Finance, insurance, real estate	9,307	98,751	1,505,593
Services	31,285	340,216	3,960,490
Other	2,574	1,857	30,044
TOTALS	112,703	1,956,097	$26,410,004

Not included in this survey were government employees, of whom there were 43,000 federal workers in 1982 and 352,000 state and local workers in 1983.

North Carolina working conditions have brought the state considerable notoriety in recent years. As of September 1984, the average North Carolina factory worker received a wage of $7.07 per hour, the lowest industrial wage rate in the US. The overall labor climate is antiunion, and only 9.6% of all nonagricultural employees belonged to a labor organization in 1980, the lowest rate in any state. In 1983 there were 834 unions in the state.

North Carolina has a right-to-work law, and public officials are legally barred from negotiating with any collective bargaining unit. The major symbol of resistance to unionization in recent decades has been J. P. Stevens & Co., a textile firm found guilty of illegal labor practices 21 times between 1966 and 1979, the highest conviction rate in US history. The Amalgamated Clothing and Textile Workers Union (AFL–CIO) won the right to represent employees at seven J. P. Stevens mills in Roanoke Rapids but waged a 17-year battle to use it; not until 1980 was a contract finally approved. In the interim, Stevens was found guilty of refusing to bargain in good faith by the National Labor Relations Board and a US court of appeals, and union supporters had turned to a national boycott of Stevens products as an additional source of pressure on the company. Despite this settlement, Stevens maintained it would continue to block organizational efforts at its other plants. In 1983, a $1.2-million settlement resolved the eight remaining complaints of unfair labor practices brought against Stevens by the union.

²³AGRICULTURE

Gross agricultural income in North Carolina totaled $4.3 billion in 1983, 11th among the 50 states; net income was $802 million. North Carolina led the nation in the production of tobacco and sweet potatoes, ranked 4th in peanuts, and was also a leading producer of corn, grapes, pecans, apples, tomatoes, and soybeans. Farm life plays an important role in the culture of the state.

The number of hired and family workers on North Carolina farms was 163,000 in 1984 (down from 590,000 in 1950), of whom 95,000 were hired workers, 43,000 farm operators, and 25,000 unpaid family workers. According to state government statistics, the number of farms fell from 301,000 in 1950 to 79,000 in 1984, while the number of acres in farms declined from 17,800,000 to 11,000,000 (7,203,000 to 4,452,000 hectares). At 139 acres (56 hectares), the average North Carolina farm was still less than one-third the size of the average US farm—a statistic that in part reflects the smaller acreage requirements of tobacco, the state's principal crop. The relatively large number of family farm owner-operators who depend on a modest tobacco allotment to make their small acreages profitable is the basis for North Carolina's opposition to the US government's antismoking campaign and its fight to preserve tobacco price supports.

Although farm employment continues to decline, a significant share of North Carolina jobs—perhaps more than one-third—are still linked to agriculture either directly or indirectly. North Carolina's most heavily agricultural counties are massed in the coastal plain, the center of tobacco, corn, and soybean production, along with a bank of northern piedmont counties on the Virginia border. Virtually all peanut production is in the eastern part of the state, while tobacco, corn, and soybean production spills over into the piedmont. Cotton is grown in scattered counties along the South Carolina border and in a band leading northward across the coastal plain. Beans, tomatoes, cucumbers, strawberries, and blueberries are commercial crops in selected mountain and coastal plain locations. Apples are important to the economy of the mountains, and the sand hills are a center of peach cultivation.

In 1983, tobacco production was 547,000,000 lb, 38% of US production, and worth nearly $1 billion. Production and value data for North Carolina's other principal crops were as follows: corn, 76,800,000 bushels, $292,000,000; soybeans, 31,000,000 bushels, $248,000,000; peanuts, 318,000,000 lb, $78,600,000; and sweet potatoes, 4,400,000 hundredweight, $64,800,000.

²⁴ANIMAL HUSBANDRY

Livestock and poultry production, increasingly important sectors of North Carolina's farm economy, accounted for 44% of the state's agricultural income in 1983. North Carolina led the nation in turkeys raised and was 3d in receipts from poultry and poultry products. Production figures in 1983 for cattle and calves were 267.8 million lb, worth $119.3 million; and hogs and pigs, 835 million lb, $381.5 million. As of the end of 1983 there were 1,120,000 cattle and calves and 2,300,000 hogs on North Carolina farms and ranches. Dairy cows numbered 133,000 that year, and milk production reached 1.7 billion lb, worth $245 million.

North Carolina was the 3d-leading source of chickens in the US, with 81.2 million lb, worth $14.1 million, produced in 1983. The state ranked 1st in turkeys, producing 519.5 million lb, worth $202.6 million; 4th in broilers, 1.8 billion lb, $487 million; and 8th in eggs, 3.1 billion, $171 million.

²⁵FISHING

North Carolina's fishing industry ranks 2d only to Virginia's among the South Atlantic states, but its overall economic importance has declined. The record landing for the state was in 1981, a total of 432 million lb; the 1984 catch was only 276.2 million lb, valued at $56.5 million. Flounder, striped bass, and menhaden are the most valuable finfish; shrimp, oysters, and crabs are the most sought-after shellfish.

²⁶FORESTRY

As of 1979, forests covered 20,043,000 acres (8,111,000 hectares) in North Carolina, or about 65% of the state's land area. North Carolina's forests constitute 2.8% of all US forestland, and fully 97.5% of the state's wooded areas have commercial value. The largest tracts are found along the coast and beyond the Blue Ridge, where most counties are more than 70% tree-covered.

National forests embrace 6% of North Carolina's timberlands, and state and local forests protect another 2%. The remainder is privately owned.

In the days of wooden sailing vessels, North Carolina pine trees supplied large quantities of "naval stores"—tar, pitch, and turpentine for waterproofing and other nautical purposes. Today, the state produces mainly saw logs, pulpwood, and veneer logs. Shipments of hardwood and softwood logs in 1982 were valued at $95.6 million, pulpwood $31.1 million, hardwood lumber $66 million, softwood lumber $166.6 million, wood chips $109.8 million, and wood for furniture $96 million.

27MINING

At least 300 varieties of rock and minerals occur in North Carolina, and more than 70 have commercial importance. Iron production flourished briefly in the early 19th century but declined after 1830. Surprisingly, North Carolina was the nation's foremost source of gold before 1849. Most North Carolina minerals, however, are too scarce to be mined profitably today. The state's nonfuel mineral production ranked 19th in the nation in 1984, when the value of mineral products was $429 million, a new high.

Sand, gravel, and crushed stone account for the bulk of North Carolina's mining industry, but some other minerals have national importance. The state ranked 1st in the production of feldspar, lithium, and mica in 1984, and was a leader in production of phosphate rock. Other mineral products include kaolin, peat, olivine, and cement. Sites near Grandfather Mountain and Spruce Pine have been identified as potential sources of uranium ore. North Carolina is the only US state where four of the most precious gems—diamonds, rubies, emeralds, and sapphires—are found, and searching for gems and semiprecious stones is popular among amateur rockhounds in western North Carolina.

28ENERGY AND POWER

Except for a modest volume of hydroelectric power, the energy consumed in North Carolina comes from outside sources. The state used nearly 1.56 quadrillion Btu of energy in 1982, of which 41% came from petroleum, 40% from coal, 9% from natural gas, 6% from nuclear power, and 4% from hydropower. Residential users consumed about 23% of North Carolina's energy in that year, commercial users 14%, industrial users 34%, and transportation 29%.

Installed electrical capacity totaled 18.3 million kw in 1983, and production reached 74.7 billion kwh. Four big private utilities produced over 90% of North Carolina's electricity in 1983. Duke Power was the largest supplier, with nearly 1 million customers. The two Brunswick stations operated by Carolina Power & Light (CP&L) in Southport and the two Duke Power McGuire stations at Cowens Ford Dam were the only nuclear power units in operation in 1984, but each company had another reactor under construction.

No petroleum or natural gas has been found in North Carolina, but major companies have expressed interest in offshore drilling. There is no coal mining, and proved coal reserves are minor.

29INDUSTRY

North Carolina has had a predominantly industrial economy for most of the 20th century. It ranked 2d only to Texas among southern states in value of manufacturing shipments and number of manufacturing employees in 1982. Today, the state remains the nation's largest manufacturer of textiles, cigarettes, and furniture. The textile industry was the largest manufacturing employer in 1982, followed by furniture, apparel, electric and electronic equipment, nonelectrical machinery, food processing, chemicals and allied products, and lumber and wood products.

The total value of shipments by manufacturers exceeded $64 billion in 1982, with the textile and apparel industries contributing 25% of this total. Of the other major industries, tobacco contributed 12%, food and kindred products 10%, chemicals and allied products 9%, electric and electronic equipment 8%, nonelectrical machinery 7%, and furniture and fixtures 5%.

The following table shows value of shipments by manufacturers for selected industries in 1982:

Cigarettes	$5,979,700,000
Knitting mill products	4,344,800,000
Yarn and thread	3,470,300,000
Weaving mill products	3,461,600,000
Household furniture	2,847,500,000
Plastics materials, synthetics	1,763,600,000
Transportation equipment	1,292,700,000
Drugs	1,246,300,000
Plastics products	1,099,700,000

The industrial regions of North Carolina spread out from the piedmont cities; roughly speaking, each movement outward represents a step down in the predominant level of skills and wages and a step closer to the primary processing of raw materials. R. J. Reynolds Industries (tobacco and food products), the 23d-largest US industrial company in 1984, has its headquarters at Winston-Salem. Burlington Industries (the world's largest textile company), Blue Bell Inc., Cannon Mills, and Cone Mills are major US textile corporations based in Greensboro; Fieldcrest Mills has its headquarters in Eden. The furniture industry is centered in the High Point–Thomasville and Hickory-Statesville areas. Charlotte's factories produce electrical appliances, textiles, and chemicals and machinery for the textile industry. Broad rural areas of the piedmont also have many industrial installations: Gaston County near Charlotte contains the largest concentration of textile factories in the US.

30COMMERCE

The wholesale trade of North Carolina was valued at $39.5 billion in 1982, 14th in the US. Wholesaling is concentrated in the Piedmont Crescent, with the Charlotte-Gastonia metropolitan area alone accounting for 38% of all 1982 sales. Retail sales, exceeding $25 billion in 1982, are much more widely distributed than wholesale. The leading retail sectors in 1982 were food stores, 25%; automotive dealers, 19%; eating and drinking places, 9%; gasoline service stations, 8%; and department stores, 8%.

The state ports at Wilmington and Morehead City handle a growing volume of international trade. Coal, wood pulp, fertilizer, and phosphate rock made up 68% of all exports by tonnage through these ports in 1982, while chemicals, iron and steel products, asphalt, limestone, and salt constituted 41% of all imports by tonnage, and residual fuel oil 29%. North Carolina exported over $1.2 billion worth of agricultural commodities to foreign markets in 1981/82 (12th in the US), and $1.1 billion in 1982/83. Manufactured exports totaled nearly $4.7 billion in 1981 (11th).

31CONSUMER PROTECTION

The Consumer Protection Section of the Department of Justice has as its function the protection of North Carolina consumers from unfair and deceptive trade practices and dishonest and unethical business competition. Although it assists in the resolution of disputes, investigates cases of consumer fraud, and initiates action to halt proscribed trade practices, it does not represent individual consumers in court. Another section of the department represents the public before the North Carolina Utilities Commission.

32BANKING

North Carolina had 63 insured commercial banks at the end of 1984; such banks held assets of $34.4 billion as of 31 December 1983; their outstanding loans totaled $18 billion, and their deposits exceeded $25 billion. The state's branch banking law has permitted the growth of several large statewide banking firms: the North Carolina National Bank Corp. (NCNB) of Charlotte is the largest bank in the southeastern US. At the end of 1984, NCNB was the 25th-largest bank in the US, with assets of $15.7 billion and deposits of $11.3 billion. NCNB merged with Ellis Banking Corp. of Florida in 1985.

There were 149 savings and loan associations and 300 credit unions in North Carolina in 1984.

33INSURANCE

At the end of 1983, 24 life insurance companies were domiciled in North Carolina; 395 out-of-state companies did business there in 1982. North Carolina families held, on average, $49,900 in life insurance coverage in 1983, not quite 7% below the US average. The 12,469,000 policies in effect that year had a combined value of $117.2 billion. North Carolinians paid life insurance premiums totaling $1.1 billion in 1983, and received $1 billion in benefits, including $340.5 million in death benefits. Also in 1983, policy holders paid property and liability companies $2 billion in premiums, including $592.2 million in automobile liability coverage, $392.2 million in automobile physical damage insurance, and $273.3 million in homeowners' policies.

34SECURITIES

There are no securities exchanges in North Carolina. The Securities Division of the Office of Secretary of State is authorized to protect the public against fraudulent issues and sellers of securities. There were 164 sales offices in the state representing New York Stock Exchange members in 1983, when 822,000 residents were shareowners of public corporations.

35PUBLIC FINANCE

The North Carolina budget is prepared bienially by the governor and reviewed annually by the Office of State Budget and Management, in consultation with the Advisory Budget Commission, an independent agency composed of four gubernatorial appointees and six members from the senate and five from the house of representatives. It is then submitted to the general assembly for amendment and approval. The fiscal year runs from 1 July to 30 June. The following is a summary of projected consolidated revenues and recommended expenditures for 1985/86 and 1986/87, in millions:

REVENUES	1985/86	1986/87
General fund, of which:	$4,779.1	$5,217.0
Individual income tax	(2,189.4)	(2,437.5)
Corporate income tax	(436.4)	(451.2)
Sales and use tax	(1,412.9)	(1,541.0)
Franchise tax	(200.8)	(225.1)
Federal funds	1,836.4	1,875.7
Highway fund	680.4	695.8
Other receipts	630.4	626.9
TOTALS	$7,926.3	$8,415.4
EXPENDITURES		
Education, of which:	$3,855.6	$3,871.0
Public schools	(2,383.8)	(2,373.8)
University system	(1,143.9)	(1,165.0)
Human resources	1,888.3	1,944.9
Transportation	874.6	881.8
Other outlays	1,649.2	1,793.9
TOTALS	$8,267.7	$8,491.6

North Carolina's total outstanding state and local debt was $5.8 billion in 1981/82, or $993 per capita. Only three states had a lower per capita debt burden.

36TAXATION

North Carolina anticipated more than $4.6 billion in tax revenues in 1985/86. The state's total revenues ranked 12th in the US in 1981/82, but the per capita share of North Carolina taxpayers ranked 38th at $1,041.

More than 50% of anticipated state tax revenues in 1985/86 were to come from individual and corporate income tax. In 1984, the personal income tax ranged in five steps from 3% on income

over $2,000 to 7% on income exceeding $10,000. Most corporate incomes faced a 6% levy on net income. The state portion (3%) of the sales and use tax was projected to account for almost 30% of projected 1985/86 state tax revenues; local governments may impose an additional 1% for their own use. Prescription drugs and certain other articles are exempt from sales tax, but food is not. As of 1984, the cigarette tax was 2¢ a pack, the lowest in the nation, and the gasoline tax was 12¢ a gallon. The state also levies inheritance, estate, gift, insurance, beverage, and franchise taxes.

North Carolina state and local governments received general revenue totaling $8.4 billion in 1980/81. Of this sum, 24% was collected from the federal government, 49% came from state government, and 27% came from local governments. North Carolina taxpayers contributed $11.6 billion to the federal treasury in 1982/83 and received $12.9 billion in federal expenditures. North Carolinians filed 2,439,599 federal income tax returns in 1983, paying over $5.4 billion in tax.

37ECONOMIC POLICY

North Carolina's government has actively stimulated economic growth ever since the beginning of the 19th century. During the administration of Governor Luther H. Hodges (1954–61), the state began to recruit outside investment directly, developing such forward-looking facilities as Research Triangle Park. After taking office in 1977, Governor James B. Hunt announced a policy of "balanced growth" to disperse future development into areas that had been neglected in the past. The state economic development structure operates under the Department of Commerce. Technical assistance for small businesses involves counsel and marketing advice.

A constitutional amendment approved in 1976 gives county governments the right to issue revenue bonds to finance pollution controls for industry and public utilities. The bonds must be repaid by the companies themselves, but the income from the interest paid on the bonds is tax-exempt, thus enabling the borrowers to obtain lower rates.

38HEALTH

Health conditions and health care facilities in North Carolina vary widely from region to region. In the larger cities—and especially in proximity to the excellent medical schools at Duke University and the University of North Carolina at Chapel Hill—quality health care is as readily available as anywhere in the US. On the other hand, two counties in the state had no more than a single doctor each in 1982, and one county had none at all.

The birthrate in North Carolina dropped from 24.1 live births per 1,000 population in 1960 to 15.3 per 1,000 in 1977 and to 14.3 in 1982, a rate slightly below the national average. In the latter year, 13.1 infants per 1,000 live births died before their first birthday, compared to the national rate of 11.9. The mortality rate was slightly higher for North Carolina white babies in 1981 than for all US white babies; it was lower for North Carolina black babies than for blacks nationally. There were 33,200 legal abortions in 1982, or 383 per 1,000 live births. The leading causes of death in North Carolina are similar to those in the rest of the US, although, as of 1981, North Carolinians died less frequently from heart disease, cancer, chronic obstructive pulmonary diseases, diabetes, chronic liver disease, and atherosclerosis than other Americans, and more frequently from stroke, accidents, and pneumonia and influenza. The death rate of 8.3 per 1,000 was below the national average.

A particularly serious public health problem in North Carolina is byssinosis, or brown lung disease. Caused by prolonged inhalation of cotton dust, byssinosis cripples the lungs of longtime textile workers, producing grave disability and even death. According to a study in 1980, the Brown Lung Association estimated that some 25,000 present or former North Carolina textile workers showed symptoms of byssinosis, and that between 10,000 and 15,000 North Carolinians were disabled by it; textile industry estimates ran to less than one-tenth of those figures. By 31 August 1985, 3,076 workers' compensation claims for brown lung had been filed with the state, of which 1,589 were found compensable, with total benefits of over $23.4 million.

The 159 hospitals in North Carolina contained 32,605 beds in 1983. Average hospital costs were $257 per day and $1,931 per stay in 1982, with the average stay 7.5 days.

The number of physicians was 7,972 in 1982. There were 32,401 registered nurses and 2,382 dentists that year. The state acted to increase the supply of doctors in eastern North Carolina in the 1970s by the establishment of a new medical school at East Carolina University in Greenville. Medical schools and superior medical research facilities are also located at Duke University Medical Center in Durham, North Carolina Memorial Hospital at the University of North Carolina in Chapel Hill, and the Bowman Gray School of Medicine at Wake Forest University in Winston-Salem.

³⁹SOCIAL WELFARE

North Carolina's social welfare programs are modest by national standards. Public assistance recipients constituted 5.1% of the population, compared to 6.1% nationally. Aid to families with dependent children (AFDC) benefited 170,200 persons in December 1983; the total cost of AFDC in 1982 was $143 million. The average payment was $176 per family per month that year, far below the national average of $303. The school lunch program fed 803,000 pupils in September 1983; the federal cost for the entire year was $83 million. Food stamps costing the federal government $266 million helped pay the grocery bills for 513,000 persons in the same year.

Social security benefits paid nearly $3.9 billion to 923,000 North Carolina residents in December 1983. The Supplemental Security Income program paid an additional $289.8 million to 143,364 aged, disabled, and blind recipients in 1983. Medicare enrolled 705,000 North Carolinians for hospital insurance in 1982, with payments totaling $938 million; 349,000 recipients of Medicaid in 1983 received a total of $567 million.

State vocational rehabilitation centers rehabilitated 9,687 clients in 1982. North Carolinians collected $159.9 million in workers' compensation awards in 1982. In an average week, 107,000 workers received unemployment insurance benefits of $104 each; benefits totaled $449 million in 1982.

⁴⁰HOUSING

The 1980 census counted 2,238,182 units of year-round housing in North Carolina; of these, 76% were single units, 14% were in multiple-unit structures, and 10% were mobile homes. Owners occupied 68% of North Carolina homes, a slightly higher proportion than in the nation at large. The proportion of houses built within the previous 10 years (31%) was slightly higher than the proportion built before 1949 (28%). Most housing units had the customary amenities of the average American home: 94% had one or more bathrooms; 70% had a central heating system; and 59% had air conditioning. Some 121,100 new housing units were authorized between 1981 and 1983; their combined value was almost $4.3 billion.

⁴¹EDUCATION

North Carolina's commitment to public education has deepened in recent years. Although the state ranked 39th in per pupil expenditures for public schools (1983/84), it ranked 4th in state and local government expenditures for all education as a proportion of total expenditures (1981/82) and 15th in per capita state government expenditures for all education (1982). North Carolina established the first state university in the US and the first free system of common schools in the South—but the university preceded the common schools by 44 years. In other words, although North Carolina led the South in education for the masses, its first love has been education for the few.

The common school system began in 1839 after North Carolina received a windfall share of the federal government's surplus revenue. The system suffered neglect between 1865 and 1900, but revived when the imposition of a literacy test for voting made adequate schooling for whites a political necessity. Opportunities for whites improved thereafter (North Carolina led the nation in construction of rural schools in the 1920s), but the system labored under the burden of racial segregation for many decades longer. In 1957, Charlotte, Greensboro, and Winston-Salem were the first cities in the South to admit black students voluntarily to formerly all-white schools, but further progress came slowly. In 1971, the US Supreme Court, in the landmark decision *Swann* v. *Charlotte-Mecklenburg Board of Education,* upheld the use of busing to desegregate the Charlotte school system. Widespread desegregation followed throughout the state.

Median school years attended stood at 12.2 for adults 25 years old and over in 1980, below the US norm. Moreover, in the same year, only 55% of North Carolina adults had completed high school, well below the US average of 67%. The state began a belated experiment in public kindergartens in 1969 and offered the program to all 5-year-olds in 1978. There were 1,972 public schools as of fall 1984, including 1,276 elementary schools, 317 middle and junior high schools, 344 high schools, 13 ungraded schools, and 22 special education schools. Average 1984/85 daily enrollment in all public schools was 1,078,700. Concern over declining academic standards led to the inauguration in 1978 of a statewide competency test that all high school students must pass in order to graduate. Nonpublic school enrollments expanded during most of the 1970s, reaching 58,661 in 450 schools in 1984/85. The popularity of nonpublic education is partly related to desegregation of the public schools, but during the 1970s, thousands of parents placed their children in private Christian academies to protect them from unwanted secular influences.

The University of North Carolina (UNC) was chartered in 1789 and opened at Chapel Hill in 1795. The UNC system now embraces 16 campuses under a common board of governors; total enrollment stood at 120,253 in the fall of 1982. The 3 oldest and largest campuses, all of which offer research and graduate as well as undergraduate programs, are UNC–Chapel Hill, North Carolina State University at Raleigh (the first land-grant college for the study of agriculture and engineering), and UNC–Greensboro, formerly Women's College. North Carolina's 23 community colleges and 38 technical institutes or colleges are intermediate institutions between the public schools and the university system. There were 601,124 students enrolled in these schools in 1981/82.

Duke University at Durham is North Carolina's premier private institution and takes its place with the Chapel Hill and Raleigh public campuses as the third key facility for the Research Triangle. There were 53 other private colleges and universities in 1983/84, of which Wake Forest University in Winston-Salem and Davidson College in Davidson are most noteworthy.

⁴²ARTS

North Carolina has been a pioneer in exploring new channels for state support of the arts. It was the first state to fund its own symphony, to endow its own art museum, to found a state school of the arts, to create a statewide arts council, and to establish a cabinet-level Department of Cultural Resources. The North Carolina Symphony, at Raleigh, received its first state appropriation in 1943; the state funded 47% of the symphony's $2.6-million budget in 1984. The North Carolina Symphony gives free concerts to about 250,000 public school children, and performs 120 evening concerts annually, with an attendance of over 80,000. Twelve North Carolina cities support amateur or semiprofessional symphonies and opera groups, while music festivals in Greensboro and Brevard reach large summer audiences.

The North Carolina Dance Theater is a professional company attached to the North Carolina School of the Arts in Winston-Salem. At least six Tarheel cities support civic ballet companies;

the American Dance Theater, one of the nation's oldest and most respected summer dance festivals, has made its permanent home at Duke University since 1978. Summer stock theater is a long-standing tradition in the mountains: North Carolina's Pulitzer Prize–winning playwright Paul Green created the genre of outdoor historical drama with a 1937 production of *The Lost Colony* at Manteo, and nearly a dozen other historical dramas are now performed on summer evenings throughout the state.

Folk art has survived most completely in the remote coves and hollows of the Appalachians. Traditional mountain string music inspired bluegrass, a progenitor of modern country and western music. Festivals, fiddlers' conventions, gospel concerts, and other public occasions keep this heritage alive and spread it to a new generation and a wider audience. Traditional crafts—pottery, spinning, weaving, quilting, and woodcarving—are also taught, displayed, and marketed at fairs and crafts centers, the most notable of which is Penland School of Handicrafts, near Spruce Pine, in the western part of the state.

43 LIBRARIES AND MUSEUMS

Public libraries, open in nearly every North Carolina community, are linked together through the State Library, ensuring that users in all parts of the state can have access to printed, filmed, and recorded materials. Total volumes in public libraries numbered 9,216,289 in 1983/84, when circulation reached 23,793,455. Major university research libraries are located at Chapel Hill, Raleigh, and Greensboro campuses of the University of North Carolina and at Duke University in Durham. The North Carolina Collection and Southern Historical Collection at the Chapel Hill campus are especially noteworthy.

North Carolina had 158 museums and historical sites in 1984. Established in 1956, the North Carolina Museum of Art, in Raleigh, is one of only two state-supported art museums in the US (the other is in Virginia). In 1981, the museum had an attendance of 36,338, mounted 15 circulating exhibitions, and added 210 artworks to its permanent collection; the new art was valued at over $1.2 million. The North Carolina Museum of History (Raleigh) is housed in the Division of Archives and History of the Department of Cultural Resources, which also administers 20 state historical sites and Tryon Place Restoration in New Bern. The Museum of Natural History in Raleigh is maintained by the state Department of Agriculture; smaller science museums exist in Charlotte, Greensboro, and Durham. The North Carolina Zoological Park opened in Asheboro in 1975; attendance in 1981 was 377,822.

44 COMMUNICATIONS

Government postal service in North Carolina began in 1755 but did not become regular until 1771, with the establishment of a central post office for the southern colonies. Mails were slow and erratic, and many North Carolinians continued to entrust their letters to private travelers until well into the 19th century. Rural free delivery in the state began on 23 October 1896 in Rowan County. In 1985, North Carolina had 784 post offices, with 12,332 employees.

Telephone service began in Wilmington and Raleigh in October 1879, and long distance connections between Wilmington and Petersburg, Va., began later that same year. There were 20 telephone companies in North Carolina in 1982; in 1980, 89% of the state's 2,043,291 occupied housing units had telephones.

There were 233 commercial AM radio stations in North Carolina in 1984, and 97 commercial and 38 noncommercial FM stations. Commercial television stations numbered 25, and there were 10 noncommercial stations, 9 of which belonged to the University of North Carolina Television Network. As of 1984, cable television served 390 communities and 753,255 subscribers.

45 PRESS

As of 1984, North Carolina had 10 morning newspapers with a combined circulation of 634,468, 45 evening dailies with 760,460, and 27 Sunday papers with 1,253,961. The following table shows the circulation of the largest dailies as of September 1984:

AREA	NAME	DAILY	SUNDAY
Charlotte	Observer (m, S)	185,876	259,289
Greensboro	News & Record (m, e, S)	89,212 } 20,695 }	121,161
Raleigh	News & Observer (m, S)	131,760	174,232
Winston-Salem	Journal (m, S)	75,417	97,585

The *Charlotte Observer* won a 1981 Pulitzer Prize for its series on brown lung disease.

North Carolina has been the home of several nationally recognized "little reviews" of literature, poetry, and criticism, including *The Rebel, Crucible, Southern Poetry Review, The Carolina Quarterly, St. Andrews Review, Pembroke Magazine,* and *Miscellany.* The *North Carolina Historical Review* is a quarterly scholarly publication of the Division of Archives and History. The bimonthly *Mother Earth News* is published in Hendersonville.

46 ORGANIZATIONS

The 1982 Census of Service Industries counted 1,221 organizations in North Carolina, including 307 business associations; 554 civic, social, and fraternal associations; and 31 educational, scientific, and research associations.

The North Carolina Citizens Association serves as the voice of the state's business community. A teachers' organization, the North Carolina Association of Educators, is widely acknowledged as one of the most effective political pressure groups in the state, as is the North Carolina State Employees Association. Every major branch of industry has its own trade association, most of which are highly effective lobbying bodies. Carolina Action, the North Carolina Public Interest Research Group, the Kudzu Alliance, and the Brown Lung Association represent related consumer, environmental, antinuclear power, and public health concerns. Among the national organizations headquartered in the state are the Improved Benevolent Protective Order of Elks of the World in Winston, a black service group; the Association of Professors of Medicine, Winston-Salem; the Institute for Southern Studies, Durham; the Tobacco Association of the US, Raleigh; and the US Power Squadrons, Raleigh.

47 TOURISM, TRAVEL, AND RECREATION

Travelers spent $3.1 billion in North Carolina in 1982. The travel industry supported 165,000 jobs in 1984, and in 1982 carried an annual payroll of $811 million.

Tourists are attracted by North Carolina's coastal beaches, by golf and tennis opportunities in the piedmont (including the world-famous golf courses at Pinehurst), and parks and scenery in the North Carolina mountains. Sites of special interest are the Revolutionary War battlegrounds at Guilford Courthouse and Moore's Creek Bridge; Bennett Place, near Hillsborough, where the last major Confederate army surrendered; Ft. Raleigh, the site of the Lost Colony's misadventures; and the Wright Brothers National Memorial at Kitty Hawk. Cape Hatteras and Cape Lookout national seashores, which protect the beauty of the Outer Banks, received 1.5 million visitors in 1984. The Blue Ridge Parkway, a scenic motor route, operated by the National Park Service, that winds over the crest of the Blue Ridge in Virginia, North Carolina, and Georgia, attracted 9.4 million visitors to North Carolina in 1984. Another popular attraction, Great Smoky Mountains National Park, straddling the North Carolina–Tennessee border, attracted 3.7 million visitors to North Carolina. In all, the 9 national parks had 15.7 million visits. In 1982/83, North Carolina's 27 state parks received more than 5.2 million visitors.

Licenses were sold to 485,887 anglers and 351,030 hunters in 1982/83.

48 SPORTS

With no major league teams in North Carolina, the most important professional sports are golf and stock-car racing. The Greater

Greensboro Open, the Kemper Open in Charlotte, and the Hall of Fame Classic at Pinehurst are major tournaments on the Professional Golfers' Association tour. The North Carolina Motor Speedway in Rockingham hosts the Carolina 500 annually, while the Charlotte Motor Speedway is the home of the World 600, the most lucrative race after the Daytona 500 on the National Association for Stock Car Auto Racing (NASCAR) circuit. In minor league baseball, Charlotte has an entry in the class-AA Southern Association; four teams play in the class-A Carolina League; and three teams play in the class-A South Atlantic League.

College basketball is the ruling passion of amateur sports fans in North Carolina. Organized in the Atlantic Coast Conference, the University of North Carolina at Chapel Hill, North Carolina State University, Wake Forest University, and Duke University consistently sponsor nationally competitive basketball teams.

⁴⁹FAMOUS NORTH CAROLINIANS

Three US presidents had North Carolina roots, but all three reached the White House from Tennessee. Andrew Jackson (1767-1845), the 7th president, was born in an unsurveyed border region probably in South Carolina, but studied law and was admitted to the bar in North Carolina before moving to frontier Tennessee in 1788. James K. Polk (1795-1849), the 11th president, was born in Mecklenburg County but grew up in Tennessee. Another native North Carolinian, Andrew Johnson (1808-75), was a tailor's apprentice in Raleigh before moving to Tennessee at the age of 18. Johnson served as Abraham Lincoln's vice president for six weeks in 1865 before becoming the nation's 17th president when Lincoln was assassinated. William Rufus King (1786-1853), the other US vice president from North Carolina, also served for only six weeks, dying before he could exercise his duties.

Three native North Carolinians have served as speaker of the US House of Representatives. The first, Nathaniel Macon (1758-1837), occupied the speaker's chair from 1801 to 1807 and served as president pro tem of the US Senate in 1826-27. The other two were James K. Polk and Joseph G. "Uncle Joe" Cannon (1836-1926), who served as speaker of the House from 1903 to 1911, but as a representative from Illinois.

Sir Walter Raleigh (or Ralegh, b.England, 1552? -1618) never came to North Carolina, but his efforts to found a colony there led state lawmakers to give his name to the new state capital in 1792. Raleigh's "Lost Colony" on Roanoke Island was the home of Virginia Dare (1587-?), the first child of English parents to be born in America. More than a century later, the infamous Edward Teach (or Thatch, b.England, ? -1716) made his headquarters at Bath and terrorized coastal waters as the pirate known as Blackbeard.

Principal leaders of the early national period included Richard Caswell (b.Maryland, 1729-89), Revolutionary War governor; William Richardson Davie (b.England, 1756-1820), governor of the state and founder of the University of North Carolina; and Archibald De Bow Murphey (1777-1832), reform advocate, legislator, and judge. Prominent black Americans of the 19th century who were born or who lived in North Carolina were John Chavis (1763-1838), teacher and minister; David Walker (1785-1830), abolitionist; and Hiram Revels (1827-1901), first black member of the US Senate.

North Carolinians prominent in the era of Civil War and Reconstruction included antislavery author Hinton Rowan Helper (1829-1909), Civil War governor Zebulon B. Vance (1830-94), Reconstruction governor William W. Holden (1818-92), and carpetbagger judge Albion Winegar Tourgee (b.Ohio, 1838-1905). Among major politicians of the 20th century are Furnifold McLendell Simmons (1854-1940), US senator from 1901 to 1931; Charles Brantley Aycock (1859-1912), governor from 1901 to 1905; Frank Porter Graham (1886-1972), University of North Carolina president, New Deal adviser, and US senator,

1949-50; Luther H. Hodges (b.Virginia, 1898-1974), governor from 1954 to 1960, US secretary of commerce from 1961 to 1965, and founder of the Research Triangle Park; Samuel J. Ervin, Jr. (1896-1985), US senator from 1954 to 1974 and chairman of the Senate Watergate investigation; Terry Sanford (b.1917), governor from 1961 to 1965, US presidential aspirant, and president of Duke University; and Jesse Helms (b.1921), Senator since 1973. Civil rights leader Jesse Jackson (b.1941) began his career as a student activist in Greensboro. The most famous North Carolinian living today is probably evangelist Billy Graham (b.1918).

James Buchanan Duke (1856-1925) founded the American Tobacco Co. and provided the endowment that transformed Trinity College into Duke University. The most outstanding North Carolina-born inventor was Richard J. Gatling (1818-1903), creator of the "Gatling gun," the first machine gun. The Wright brothers, Wilbur (b.Indiana, 1867-1912) and Orville (b.Ohio, 1871-1948), achieved the first successful powered airplane flight at Kitty Hawk, on the Outer Banks, on 17 December 1903. Psychologist Joseph Banks Rhine (b.Pennsylvania, 1895-1980) was known for his research on extrasensory perception.

A number of North Carolinians have won fame as literary figures. They include Walter Hines Page (1855-1918), editor and diplomat; William Sydney Porter (1862-1910), a short-story writer who used the pseudonym O. Henry; playwright Paul Green (1894-1984); and novelists Thomas Wolfe (1900-1938) and Reynolds Price (b.1933). Major scholars associated with the state have included sociologist Howard W. Odum (b.Georgia, 1884-1954) and historians W. J. Cash (1901-41) and John Hope Franklin (b.Oklahoma, 1915). Journalists Edward R. Murrow (1908-65), Tom Wicker (b.1926), and Charles Kuralt (b.1934) were all North Carolina natives. Harry Golden (Harry L. Goldhurst, b.New York, 1903-81), a Jewish humorist, founded the *Carolina Israelite*.

Jazz artists Thelonious Monk (1918-82), John Coltrane (1926-67), and Nina Simone (b.1933) were born in the state, as were pop singer Roberta Flack (b.1939), folksinger Arthel "Doc" Watson (b.1923), bluegrass banjo artist Earl Scruggs (b.1924), and actor Andy Griffith (b.1926). North Carolina athletes include former heavyweight champion Floyd Patterson (b.1935), NASCAR driver Richard Petty (b.1937), football quarterbacks Sonny Jurgenson (b.1934) and Roman Gabriel (b.1940), baseball pitchers Gaylord Perry (b.1938) and Jim "Catfish" Hunter (b.1946), and basketball player Meadowlark Lemon (b.1932), long a star with the Harlem Globetrotters.

⁵⁰BIBLIOGRAPHY

Chafe, William H. *Civilities and Civil Rights: Greensboro, North Carolina and the Black Struggle for Freedom.* New York: Oxford University Press, 1980.

Clay, James W., Douglas M. Orr, Jr., and Alfred W. Stuart (eds.). *North Carolina Atlas: Portrait of a Changing Southern State.* Chapel Hill: University of North Carolina Press, 1975.

Lefler, Hugh T., and Albert Ray Newsome. *North Carolina: The History of a Southern State.* 3d ed., rev. Chapel Hill: University of North Carolina Press, 1973.

Nathans, Sydney. *The Quest for Progress: The Way We Lived in North Carolina, 1870-1920.* Chapel Hill: University of North Carolina Press, 1983.

North Carolina, State of. Office of State Budget and Management. *North Carolina State Government Statistical Abstract, 1984.* 5th ed. Raleigh, 1984.

North Carolina, State of. Secretary of State. *North Carolina Manual, 1983-84.* Raleigh, 1984.

Powell, William S. *North Carolina: A Bicentennial History.* New York: Norton, 1977.

Powell, William S. *The North Carolina Gazetteer.* Chapel Hill: University of North Carolina Press, 1968.

NORTH DAKOTA

State of North Dakota

ORIGIN OF STATE NAME: The state was formerly the northern section of Dakota Territory; *dakota* is a Siouan word meaning "allies." **NICKNAME:** The Sioux State. (Also: Flickertail State.) **CAPITAL:** Bismarck. **ENTERED UNION:** 2 November 1889 (39th). **SONG:** "North Dakota Hymn." **MARCH:** "Spirit of the Land." **MOTTO:** Liberty and Union, Now and Forever, One and Inseparable. **FLAG:** The flag consists of a blue field with yellow fringes; on each side is depicted an eagle with outstretched wings, holding in one talon a sheaf of arrows, in the other an olive branch, and in his beak a banner inscribed with the words "*E Pluribus Unum.*" Below the eagle are the words "North Dakota"; above it are 13 stars surmounted by a sunburst. **OFFICIAL SEAL:** In the center is an elm tree; beneath it are a sheaf of wheat, a plow, an anvil, and a bow and three arrows, and in the background an Indian chases a buffalo toward a setting sun. The depiction is surrounded by the state motto, and the words "Great Seal State of North Dakota October 1st 1889" encircle the whole. **BIRD:** Western meadowlark. **FISH:** Northern pike. **FLOWER:** Wild prairie rose. **TREE:** American elm. **GRASS:** Western wheatgrass. **STONE:** Teredo petrified wood. **BEVERAGE:** Milk. **LEGAL HOLIDAYS:** New Year's Day, 1 January; Birthday of Martin Luther King, Jr., 3d Monday in January; George Washington's Birthday, 3d Monday in February; Good Friday, March or April; Memorial Day, last Monday in May; Independence Day, 4 July; Labor Day, 1st Monday in September; Veterans Day, 11 November; Thanksgiving Day, 4th Thursday in November; Christmas Day, 25 December. **TIME:** 6 AM CST = noon GMT; 5 AM MST = noon GMT.

¹LOCATION, SIZE, AND EXTENT

Located in the western north-central US, North Dakota ranks 17th in size among the 50 states.

The total area of North Dakota is 70,703 sq mi (183,121 sq km), comprising 69,300 sq mi (179,487 sq km) of land and 1,403 sq mi (3,634 sq km) of inland water. Shaped roughly like a rectangle, North Dakota has three straight sides and one irregular border on the E. Its maximum length E–W is about 360 mi (580 km), its extreme width N–S about 210 mi (340 km).

North Dakota is bordered on the N by the Canadian provinces of Saskatchewan and Manitoba; on the E by Minnesota (with the line formed by the Red River of the North); on the S by South Dakota; and on the W by Montana. The total boundary length is 1,312 mi (2,111 km). The state's geographic center is in Sheridan County, 5 mi (8 km) SW of McClusky.

²TOPOGRAPHY

North Dakota straddles two major US physiographic regions: the Central Plains in the east and the Great Plains in the west. Along the eastern border is the generally flat Red River Valley, with the state's lowest point, 750 feet (229 meters); this valley was once covered by the waters of a glacial lake. Most of the eastern half of North Dakota consists of the Drift Prairie, at 1,300–1,600 feet (400–500 meters) above sea level. The Missouri Plateau occupies the western half of the state, and has the highest point in North Dakota—White Butte, 3,506 (1,069 meters)—in the Slope Country of the southwest. Separating the Missouri Plateau from the Drift Prairie is the Missouri Escarpment, which rises 400 feet (122 meters) above the prairie and extends diagonally from northwest to southeast.

North Dakota has two major rivers: the Red River of the North, flowing northward into Canada; and the Missouri River, which enters in the northwest and then flows east and, joined by the Yellowstone River, southeast into South Dakota.

³CLIMATE

North Dakota lies in the northwestern continental interior of the US. Characteristically, summers are hot, winters very cold, and rainfall sparse to moderate, with periods of drought. The average annual temperature is 40°F (4°c), ranging from 7°F (−14°C) in January to 69°F (21°C) in July. The record low temperature,

−60°F (−51°C), was set at Parshall on 15 February 1936; the record high, 121°F (49°C), at Steele on 6 July 1936.

The average yearly precipitation is about 18 in (46 cm). The total annual snowfall averages 40 in (102 cm) at Bismarck.

⁴FLORA AND FAUNA

North Dakota is predominantly a region of prairie and plains, although the American elm, green ash, box elder, and cottonwood grow there. Cranberries, juneberries, and wild grapes are also common. Indian, blue, grama, and buffalo grasses grow on the plains; the wild prairie rose is the state flower.

Once on the verge of extinction, the white-tailed and mule deers and pronghorn antelope have been restored. The elk and grizzly bear, both common until about 1880, had disappeared by 1900; bighorn sheep, reintroduced in 1956, are beginning to flourish. North Dakota claims more wild ducks than any other state except Alaska, and it has the largest sharptailed grouse population in the US. The black-footed ferret and northern swift fox are listed by federal authorities as endangered in North Dakota.

⁵ENVIRONMENTAL PROTECTION

North Dakota has little urban or industrial pollution. An environmental issue confronting the state in the mid-1980s was how to use its coal resources without damaging the land through strip mining or polluting the air with coal-fired industrial plants.

To conserve water and provide irrigation, nearly 700 dams have been built, including Garrison Dam, completed in 1960. The Garrison Diversion project, authorized by the US Congress in 1965, was intended to draw water from Lake Sakakawea, the impoundment behind Garrison Dam; however, in 1985 it was still largely incomplete because of environmental disputes.

⁶POPULATION

North Dakota ranked 46th in the US with a 1980 census population of 652,700, representing an increase of 6% since 1970. The estimated population in January 1985 was 692,027 and the population density was 9.9 per sq mi (3.9 per sq km), less than one-sixth the US average. North Dakota is one of the most rural states in the US, with 64% of its population living outside metropolitan areas as of 1983. Leading cities as of 1984 were Fargo, 65,721; Bismarck, the capital, 47,552; Grand Forks, 44,233; and Minot, 37,261.

7 ETHNIC GROUPS

As of 1980, about 96% of the state's population was white. The American Indian population was 20,119, or about 3% of the total, and there were 2,568 blacks, representing less than 1% of the population. Among Americans of European origin, the leading groups were Germans, who made up 47% of the total population, and Norwegians, who made up 28%. Only about 2% of the state's population was foreign-born as of 1980.

8 LANGUAGES

Although a few Indian words are used in the English spoken near the reservations where Ojibwa and Sioux live in North Dakota, the only general impact of Indian speech on English is in such place-names as Pembina, Mandan, Wabek, and Anamoose.

A few Norwegian food terms like *lefse* and *lutefisk* have entered the Northern dialect that is characteristic of North Dakota, and some Midland terms have intruded from the south.

In 1980, 93% of the population 3 years old or older spoke only English at home; German, the next most frequently used language, was spoken by 37,313 residents.

9 RELIGIONS

As of 1980 there were an estimated 308,000 adherents of Protestant groups, representing nearly one-half of the state's total population. Leading denominations were the American Lutheran Church, with 163,956 adherents; Lutheran Church—Missouri Synod, 27,801; United Methodist Church, 26,959; Lutheran Church in America, 14,620; United Presbyterian Church, 14,609; and United Church of Christ, 11,033. As of 1984, the state had 179,882 Roman Catholics, representing more than one-fourth of the total population, and an estimated 1,080 Jews.

10 TRANSPORTATION

In 1983, there were 4,756 mi (7,654 km) of Class I railway trackage in North Dakota. Railroad lines, the largest of which were the Burlington Northern, the Soo Line, and the Chicago and Northwestern, transported (in 1980) 16,110,442 tons of revenue freight (mostly farm products and coal) originating in North Dakota.

Public roads, streets, and highways in North Dakota covered 86,382 mi (139,019 km) in 1983. There were an estimated 665,000 registered motor vehicles and 431,740 licensed drivers.

The state had 448 airports and 3 heliports in 1983, and there were 14,580 scheduled aircraft departures.

11 HISTORY

Human occupation of what is now North Dakota began about 13,000 BC in the southwestern corner of the state, which at that time was covered with lush vegetation. Drought drove away the aboriginal hunter-gatherers, and it was not until about 2,000 years ago that Indians from the more humid regions to the east moved into the easternmost third of the Dakotas. About AD 1300 the Mandan Indians brought an advanced agricultural economy up the Missouri River. They were joined by the Hidatsa and Arikara about three or four centuries later. Moving from the Minnesota forests during the 17th century, the Yanktonai Sioux occupied the southeastern quarter of the state. Their cousins west of the Missouri River, the Teton Sioux, led a nomadic life as hunters and mounted warriors. The Ojibwa, who had driven the Sioux out of Minnesota, settled in the northeast.

European penetration of the Dakotas began in 1738, when Pierre Gaultier de Varennes, Sieur de la Vérendrye, of Trois Rivières in New France, traded for furs in the Red River region. Later the fur trade spread farther into the Red and Missouri river valleys, especially around Pembina, where the North West Company and the Hudson's Bay Company had their posts. After the Lewis and Clark expedition (1804-06) explored the Missouri, the American Fur Company traded there, with buffalo hides the leading commodity.

In 1812, Scottish settlers from Canada moved up the Red River to Pembina. This first white farming settlement in North Dakota also attracted numerous métis, half-breeds of mixed Indian and European ancestry. An extensive trade in furs and buffalo hides, which were transported first by heavy carts and later by steamboats, sprang up between Pembina, Ft. Garry (Winnipeg, Canada), and St. Paul, Minn.

Army movements against the Sioux during and after the Civil War brought white men into central North Dakota, which in 1861 was organized as part of the Dakota Territory, including the present-day Dakotas, Montana, and Wyoming. The signing of treaties confining the agricultural Indians to reservations, the arrival of the Northern Pacific Railroad at Fargo in 1872, and its extension to the Missouri the following year led to the rise of homesteading on giant "bonanza farms." Settlers poured in, especially from Canada. This short-lived "Great Dakota boom" ended in the mid-1880s with drought and depressed farm prices. As many of the original American and Canadian settlers left in disgust, they were replaced by Norwegians, Germans, and other Europeans. By 1910, North Dakota, which had entered the Union in 1889, was among the leading states in percentage of foreign-born residents.

From the time of statehood onward, Republicans dominated politics in North Dakota. Their leader was Alexander McKenzie, a Canadian immigrant who built a reputation as an agent of the railroads, protecting them from regulation. Only in the depths of the 1893 depression did North Dakotans elect a Populist governor, but after two years they returned the Republicans to power. Between 1898 and 1915, the "Second Boom" brought an upsurge in population and railroad construction. In politics, Republican Progressives enacted reforms, but left unsolved the basic problem of how North Dakota farmers could stand up to the powerful grain traders of Minneapolis–St. Paul. Agrarian revolt flared in 1915, when Arthur C. Townley organized the Farmers' Nonpartisan Political League. Operating through Republican Party machinery, Townley succeeded in having his gubernatorial candidate, Lynn J. Frazier, elected in 1916. State-owned enterprises were established, including the Bank of North Dakota, Home Building Association, Hail Insurance Department, and a mill and grain elevator. However, the league was hurt by charges of "socialism" and, after 1917, by allegations of pro-German sympathies in World War I, as well as of mismanagement. In 1921, Frazier and Attorney General William Lemke were removed from office in the nation's first recall election.

The 1920s, a period of bank failures, low farm prices, drought, and political disunity, saw the beginnings of an exodus from the state. Matters grew even worse during the depression of the 1930s. Elected governor by hard-pressed farmers in 1932, William Langer took spectacular steps to save farms from foreclosure and to raise grain prices, until a conflict with the Roosevelt administration led to his removal from office on charges that he had illegally solicited political contributions.

World War II brought a quiet prosperity to North Dakota that lasted into the following decades. The Republican Party generally continued to control the state legislature, although Democrats held the governorship from 1960 until 1981. The Arab oil embargo of 1973 and the rise of oil prices throughout the decade spurred drilling for oil, encouraged the mining of lignite for electrical generation, and led to the construction of the nation's first coal gasification plant, at a cost of $2 billion, in a lignite mining area near Beulah; in 1985, however, this plant was declared uneconomical, and its operations were shut down.

12 STATE GOVERNMENT

North Dakota is governed by the constitution of 1889, as amended. Statewide elected officials include governor and lieutenant governor, secretary of state, auditor, treasurer, attorney general, superintendent of public instruction, commissioners of labor, insurance, taxation, and agriculture, and three public service commissioners. The legislature, which convenes every two

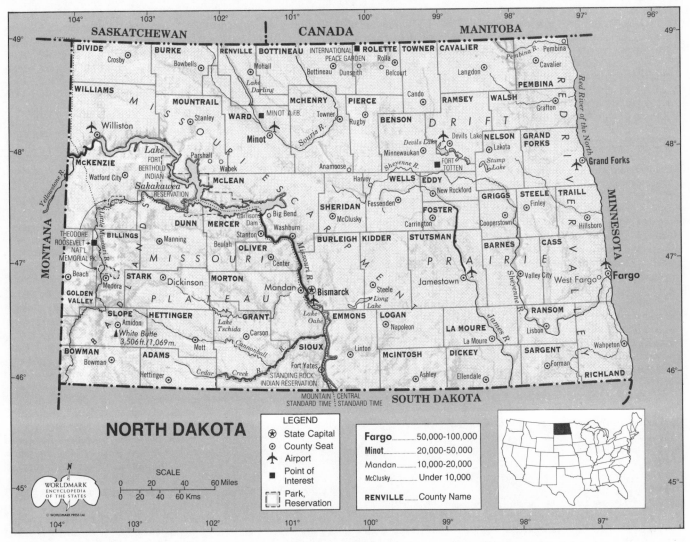

NORTH DAKOTA

LEGEND
- ⊛ State Capital
- ⊙ County Seat
- ✈ Airport
- ■ Point of Interest
- Park, Reservation

Fargo	50,000–100,000
Minot	20,000–50,000
Mandan	10,000–20,000
McClusky	Under 10,000
RENVILLE	County Name

SCALE
0 20 40 60 Miles
0 20 40 60 Kms

WORLDMARK ENCYCLOPEDIA OF THE STATES
© WORLDMARK PRESS Ltd.

See endsheet maps: F1.
LOCATION: 45°56′ to 49°N; 96°35′ to 104°03′W. **BOUNDARIES:** Canadian line, 310 mi (499 km); Minnesota line, 420 mi (676 km); South Dakota line, 370 mi (595 km); Montana line, 212 mi (341 km).

years beginning on the first Tuesday in January, is bicameral, with a 53-member senate and a 106-member house of representatives. Senators are elected to staggered four-year terms, while representatives serve for two years. A two-thirds vote of the elected members of each house is required to override a gubernatorial veto.

Voters in North Dakota must be US citizens, at least 18 years of age, and must have been residents of the state at least 30 days prior to the election. Advance registration is not required.

¹³POLITICAL PARTIES

Between 1889 and 1960, Republicans held the governorship for 58 of the 72 years. North Dakota politics were not monolithic, however, for aside from the Populist and Democratic opposition, the Republican Party was itself torn by factionalism, with Progressive and Nonpartisan League challenges to the conservative, probusiness party establishment. Between 1960 and 1980, the statehouse was in Democratic hands.

In November 1984, North Dakotans cast 65% of the total popular vote for Ronald Reagan. But they turned out Republican Governor Allen I. Olson to elect George Sinner, the Democratic challenger, and elected Democrat Ruth Meiers as the state's first woman lieutenant governor.

North Dakota Presidential Vote by Major Political Parties, 1948–84

YEAR	ELECTORAL VOTE	NORTH DAKOTA WINNER	DEMOCRAT	REPUBLICAN
1948	4	Dewey (R)	95,812	115,139
1952	4	*Eisenhower (R)	76,694	191,712
1956	4	*Eisenhower (R)	96,742	156,766
1960	4	Nixon (R)	123,963	154,310
1964	4	*Johnson (D)	149,784	108,207
1968	4	*Nixon (R)	94,769	138,669
1972	3	*Nixon (R)	100,384	174,109
1976	3	Ford (R)	136,078	153,470
1980	3	*Reagan (R)	79,189	193,695
1984	3	*Reagan (R)	104,429	200,336

*Won US presidential election.

14LOCAL GOVERNMENT

North Dakota in 1982 had 2,795 units of local government, including 53 counties, 365 municipalities designated as cities, 1,360 townships, 325 school districts, and 692 special districts. Typical elected county officials are the sheriff, court clerk, county judge, county justice, and state's attorney.

15STATE SERVICES

Educational services are under the jurisdiction of the Department of Public Instruction and the Board of Higher Education; there are state schools for the deaf, blind, handicapped, and mentally retarded. Health and welfare agencies include the State Health Department, Veterans Affairs Department, Social Service Board, and Indian Affairs Commission. Agricultural services include an extensive program of experiment and extension stations. The state bank, mill, and grain elevator established under Nonpartisan League influence remain to this day.

16JUDICIAL SYSTEM

North Dakota has a supreme court of five justices, seven district courts, and a system of local (county) courts. Supreme court justices are elected for 10-year terms, district court judges for 6-year terms. In 1984 there were 1,462 practicing attorneys in the state.

According to the FBI Crime Index, in 1983 North Dakota had the 3d-lowest crime rate of any US state. Only 2,675 crimes per 100,000 population (54 violent, 2,621 nonviolent) were reported.

17ARMED FORCES

The main national defense installations are the Strategic Air Command bases at Minot and Grand Forks. North Dakota firms received $137 million in defense contract awards in 1982/83. As of 1983, 69,000 veterans were living in North Dakota, including about 1,000 from World War I, 23,000 from World War II, 12,000 from the Korean conflict, and 22,000 from the Viet-Nam era. A total of $65 million was spent on major veterans' benefit programs in the state in 1982/83.

Military personnel totaled 11,921 in 1983; in the same year, North Dakota had 200 state highway patrol officers.

18MIGRATION

During the late 19th century, North Dakota was largely settled by immigrants of German and Scandinavian stock. The state reached a peak population in 1930, but then suffered steady losses until well into the 1970s because of out-migration. This trend has shown some signs of abating, however. From 1980 to 1983, the state's population grew 4.3%, in part because of a net gain in migration of about 5,000 people.

19INTERGOVERNMENTAL COOPERATION

North Dakota participates in such interstate agreements as the Yellowstone River Compact, Western Interstate Energy Compact, and Interstate Oil and Gas Compact. A Minnesota–North Dakota Boundary Compact was ratified in 1961.

Federal assistance in 1983/84 totaled $453.7 million.

20ECONOMY

North Dakota has been and still is an important agricultural state, especially as a producer of wheat, much of which finds its way onto the world market. Many segments of the economy are affected by agriculture; for example, a substantial wholesale trade is involved in moving grain and livestock to market. Like other midwestern farmers, North Dakotans suffered from high interest rates and a federal embargo on grain shipments to the Soviet Union in the early 1980s. Farm numbers have continued to decline, posing a threat to the vitality of the state's rural life-style. Growth industries for the state include petroleum and the mining of coal, chiefly lignite; North Dakota has more coal resources than any other state. Manufacturing is concentrated to a great extent on farm products and machinery.

21INCOME

In 1983, North Dakota ranked 24th among the 50 states in per capita personal income, with $11,350. Total personal income reached $7.7 billion. The median money income of four-person families was $25,364 in 1981. As of 1979, 79,000 North Dakotans (13% of the state population) were below the federal poverty level.

22LABOR

North Dakota's labor force numbered 327,000 in 1984, of whom 17,000 (5.2%) were unemployed. A federal survey in March 1983 revealed the following nonfarm employment pattern in North Dakota:

	ESTABLISH-MENTS	EMPLOYEES	ANNUAL PAYROLL ('000)
Agricultural services, forestry, fisheries	181	548	$ 6,743
Mining	366	6,168	169,455
Contract construction	1,940	9,380	197,615
Manufacturing	580	13,826	251,520
Transportation, public utilities	930	12,451	262,567
Wholesale trade	2,170	17,991	309,499
Retail trade	5,428	46,158	396,806
Finance, insurance, real estate	1,722	12,230	190,708
Services	4,559	52,659	638,029
Other	763	993	9,696
TOTALS	18,639	172,404	$2,432,638

As of 1983, only 9% of all North Dakotans in the nonagricultural work force belonged to the state's labor unions; there were 215 such unions in 1983. A right-to-work law is in force.

23AGRICULTURE

North Dakota's net farm income totaled $858.5 million in 1983 (10th in the US). The state has in some years been the nation's leading wheat producer and is typically the top producer of durum and other spring wheats, as well as of barley, flaxseeds, and sunflowers.

The total number of farms has declined over the years as average farm size has increased. In 1984 there were 36,000 farms in the state, occupying 41 million acres (16.6 million hectares).

Wheat production in 1983 totaled 194,595,000 bushels; barley, 124,200,000 bushels; oats, 63,630,000 bushels; corn for grain, 29,145,000 bushels; soybeans, 14,575,000 bushels; hay, 2,088,000 tons; and potatoes, 20,460,000 hundredweight.

24ANIMAL HUSBANDRY

Livestock on North Dakota farms and ranches in 1984 included 2,140,000 cattle and calves, 224,000 sheep and lambs, and 260,000 hogs and pigs.

Sales of livestock and livestock products accounted for about 20% of North Dakota farm income in 1981. Leading products in 1983 were cattle, 848,810,000 lb; hogs, 81,554,000 lb; sheep, 14,094,000 lb; turkeys, 13,764,000 lb; chickens, 2,666,000 lb; milk, 1,020,000,000 lb; and eggs, 119,000,000.

25FISHING

There is little commercial fishing in North Dakota. In 1983, the commercial catch totaled 987,000 lb, worth only $190,000.

26FORESTRY

North Dakota has only 422,000 acres (171,000 hectares) of forestland, or about 1% of its total area. Although 405,000 acres (164,000 hectares) in all are classed as commercial timberland, forestry products are of very minor importance to the state's economy.

27MINING

Fossil fuels constitute the state's major mineral resource; their production was valued at $1.8 billion in 1982. Nearly all coal production in the state is lignite; total lignite production in 1983 amounted to 19,181,000 short tons (17,400,715 metric tons). Other fuel resources include petroleum and natural gas. Nonfuel mineral production was valued at only $12.7 million in 1984. Sand and gravel were the major nonfuel mineral products.

28ENERGY AND POWER

Power stations in North Dakota generated 19.6 billion kwh of electricity in 1983 and had 3.75 million kw of installed electric generating capacity. Energy consumption per capita in 1982 amounted to 366 million Btu; energy expenditure per capita was $2,389 in 1981.

Recoverable coal reserves totaled 1.6 billion short tons in 1983. Proved petroleum reserves in 1983 totaled 258,000,000 barrels; production was 29,225,000 barrels. In 1983, natural gas reserves totaled 600 billion cu feet; production was 70.4 billion cu feet.

29INDUSTRY

By number of employees, the leading manufacturing industries in North Dakota are food and related products, nonelectrical machinery, printing and publishing, transportation equipment, and stone, clay, and glass products. Value of shipments of manufactures in 1982 totaled $2.4 billion. The following table shows value of shipments by selected industries in 1982:

Food and related products	$870,200,000
Nonelectrical machinery	321,400,000
Printing and publishing	84,900,000
Stone, clay, and glass products	58,400,000

30COMMERCE

In 1982, North Dakota's 2,109 wholesale trade establishments recorded $5.98 billion in sales. The leading wholesale lines by sales volume were farm-product raw materials, 30%; machinery, equipment, and supplies (especially farm machinery), 19%; groceries and related products, 14%; and petroleum and petroleum products, 12%. The state's 7,026 retail establishments recorded $3.3 billion in sales.

Agricultural exports totaled $1.3 billion in 1981/82; sales of wheat and flour accounted for about half of the total. Exports of manufactured goods reached $156 million in 1981.

31CONSUMER PROTECTION

Allegations of consumer fraud and other illegal business practices are handled by the Consumer Fraud Division of the State Attorney General's Office. Other consumer services fall within the jurisdiction of the State Laboratories Consumer Affairs Office.

32BANKING

As of 31 December 1983, North Dakota's 181 commercial banks had assets of $6.4 billion, and the state's 6 savings and loan associations had $2.4 billion in assets. There were 90 credit unions in the state.

33INSURANCE

In 1983, North Dakota had 1,048,000 life insurance policies in force, worth $16 billion. The average life insurance per family was $59,100. The total life insurance benefits paid out were $34.3 million. Direct premiums written by property and casualty insurers in 1983 totaled $314 million.

34SECURITIES

North Dakota has no securities exchanges. As of 1983 there were 93,000 shareowners of public corporations in the state.

35PUBLIC FINANCE

North Dakota's biennial budget is prepared by the director of the Department of Accounts and Purchases. General fund revenues and appropriated expenditures are given below for 1983–85 and 1985–87 (recommended).

REVENUES	1983–85	1985–87
Sales and use tax	$ 398,499,000	$ 404,360,000
Income taxes	253,953,000	249,762,000
Oil extraction tax	161,483,000	133,152,000
Oil/gas production tax	116,609,000	98,891,000
Other receipts	167,816,000	193,693,000
TOTALS	$1,098,360,000	$1,079,858,000

EXPENDITURES	1983–85	1985–87
Education	$ 598,894,020	$ 673,053,341
Health and welfare	197,211,525	269,691,129
Agriculture, industrial development, production	37,084,889	42,735,894
Other expenditures	171,920,612	206,531,905
TOTALS	$1,005,111,046	$1,192,012,269

Consolidated expenditures for 1983–85 (including federal aid and designated funds) totaled more than $2.1 billion, including $357.7 million for highways, a total of $520.8 million for health and welfare, and $809.9 million for education. The recommended total for 1985–87 was $2.5 billion. The total debt of North Dakota state and local governments at the end of 1982/83 amounted to $1.8 billion; the 1982 per capita debt of $1,454 was among the lowest in the US.

36TAXATION

As of 1984, the personal income tax ranged from 2% on the first $3,000 to 9% on taxable income over $50,000. The corporate tax rate ranged up to 10.5%. The state also taxed oil and gas production, gasoline, insurance premiums, alcoholic beverages, tobacco products, mineral leases, and coal severance. There is a general sales and use tax of 4%.

In 1983, North Dakotans filed 278,876 federal (individual) income tax returns and paid $658.2 million in federal income tax.

37ECONOMIC POLICY

North Dakota has long sought to protect its agrarian interests, intervening in the marketplace to help farmers during periods of hardship, especially in the 1930s. Business incentives include a variety of job-training, financing, and tax-abatement programs.

The state's Economic Development Commission, established in 1957, seeks to attract new industry, expand existing industry, promote tourism, and develop markets for state products.

38HEALTH

The birthrate in 1982 was 18.8 per 1,000 population. Death rates in 1981 for all leading causes except accidents and cerebrovascular diseases were below the US average, as was infant mortality.

As of January 1983 the state had 58 general hospitals, with 6,014 beds. There were 74 profit and nonprofit nursing homes and extended care facilities as of 1982. The cost of a semiprivate hospital room (January 1984) averaged about $160 a day, well below the national average. Medical personnel licensed in the state included 1,049 physicians (1982), 5,619 registered nurses (1980), and 320 dentists (1982).

39SOCIAL WELFARE

Benefits totaling $14 million under the aid to families with dependent children program were paid to 4,100 families in 1982. As of September 1983, 31,000 North Dakotans were receiving food stamps at an annual federal cost of $15 million. In 1983, $460 million in Social Security benefits were paid to North Dakotans, and, in May 1983, 5,790 residents received $835,000 in federal Supplemental Security Income payments. In 1982, about 10,000 workers claimed unemployment benefits totaling $45.7 million, and workers' compensation payments amounted to $22.1 million. There were 58 day care facilities.

40HOUSING

North Dakota had some 245,000 households in 1983. The 1980 census counted 259,000 year-round housing units, about 70% of them owner-occupied. Between 1981 and 1983, 4,400 new units were authorized at a value of $167 million.

41EDUCATION

As of 1980, two-thirds of all adults were high school graduates, and 15% had at least four years of college.

In 1982/83, 117,000 students were enrolled in public schools, including 81,000 students in prekindergarten to grade 8 and 36,000 in grades 9-12. Average salary for public school teachers (1983/84) was $19,261.

In fall 1982, 34,000 students were enrolled in North Dakota's higher educational institutions, of which there were 19 as of 1983/84. The chief universities are the University of North Dakota (Grand Forks), with 8,514 students in 1981, and North Dakota State University (Fargo), with 7,477. The Student Financial Assistance Agency offers scholarships for North Dakota college students, and the state Indian Scholarship Board provides aid to needy Indians attending colleges and junior colleges in the state.

[42]ARTS

The Council on the Arts, a branch of the North Dakota state government, provides grants to local artists and groups, encourages visits by out-of-state artists and exhibitions, and provides information and other services to the general public. Two popular musical events are the Old Time Fiddlers Contest (at Dunseith in June) and the Medora Musical (Medora, June through Labor Day); the latter features Western songs and dance.

[43]LIBRARIES AND MUSEUMS

During 1982, North Dakota public libraries had 1,321,878 volumes and a total circulation of 2,637,172. The leading academic library was that of the University of North Dakota (Grand Forks), with 501,700 volumes.

Among the most notable of the state's 34 museums are the Art Galleries and Zoology Museum of the University of North Dakota and the North Dakota Heritage Center at Bismarck, which has an outstanding collection of Indian artifacts. Theodore Roosevelt National Park contains relics from the Elkhorn ranch where Roosevelt lived in the 1880s.

[44]COMMUNICATIONS

In 1980, 94% of South Dakota's 242,523 occupied housing units had telephones. There were 51 commercial radio stations (32 AM, 19 FM) in 1982. As of 1983, 13 commercial television stations and 6 public television stations were in operation, and there were 101 cable television systems serving 113,333 subscribers in 105 North Dakota communities.

[45]PRESS

As of 1984 there were two morning dailies, seven evening dailies, and one all-day paper, with a combined circulation of 193,779. There were five Sunday papers with a combined circulation of 148,771. The leading dailies were the *Fargo Forum,* with an all-day circulation of 56,840 (Sunday, 64,473); the *Grand Forks Herald,* 36,925 morning, 38,754 Sunday; the *Minot Daily News,* 29,510 evening; and the *Bismarck Tribune,* 26,860 evening, 27,626 Sunday. In addition, there were 67 weekly newspapers and 15 periodicals. The leading historical journal is *North Dakota Horizons,* a quarterly founded in 1971.

[46]ORGANIZATIONS

The 1982 census of service industries counted 347 organizations in North Dakota, including 55 business associations; 201 civic, social, and fraternal associations; and 3 educational, scientific, and research organizations. Two of the state's largest organizations are the Friends (Service Club) and the Northwest Farm Managers Association, both headquartered in Fargo.

[47]TOURISM, TRAVEL, AND RECREATION

In 1983, US travelers spent an estimated $651 million in North Dakota, and these expenditures directly generated over 17,400 jobs. State parks and other state recreational areas received 930,836 visitors in 1983. In 1982, including for-profit facilities only, the state had 40 hotels, 188 motels and similar facilities, 11 trailer parks and camps, 50 motion picture theaters, 5 public golf courses, and 31 sport and recreation clubs.

Among the leading tourist attractions is the International Peace Garden, covering 2,200 acres (890 hectares) in North Dakota and Manitoba and commemorating friendly relations between the US and Canada. Ft. Abraham Lincoln State Park, south of Mandan, has been restored to suggest the 1870s, when General Custer left the area for his "last stand" against the Sioux. The most spectac-

ular scenery in North Dakota is part of the Theodore Roosevelt National Park, which had 381,920 recreational visitors within North Dakota in 1984. So-called badlands, an integral part of the park, consist of strangely colored and intricately eroded buttes and other rock formations. Hunting and fishing are major recreational activities in North Dakota. In 1982/83 there were 96,232 licensed hunters and 183,261 licensed fishermen.

[48]SPORTS

There are no major professional sports teams in North Dakota. In collegiate football, the University of North Dakota Sioux and the North Dakota State University Bison compete in the North Central Conference. The University of North Dakota also competes in collegiate ice hockey.

[49]FAMOUS NORTH DAKOTANS

Preeminent among North Dakota politicians known to the nation was Gerald P. Nye (b.Wisconsin, 1892–1971), a US senator and a leading isolationist opponent of President Franklin D. Roosevelt's foreign policy, as was Senator William Langer (1886–1959). Another prominent senator, Porter J. McCumber (1858–1933), supported President Woodrow Wilson in the League of Nations battle. US Representative William Lemke (1878–1950) sponsored farm-relief legislation and in 1936 ran for US president on the Union Party ticket. Usher L. Burdick (1879–1960), a maverick isolationist and champion of the American Indian, served 18 years in the US House of Representatives.

Vilhjalmur Stefansson (b.Canada, 1879–1962) recorded in numerous books his explorations and experiments in the high Arctic. Orin G. Libby (1864–1952) made a significant contribution to the study of American history. Other North Dakota-nurtured writers and commentators include Maxwell Anderson (b.Pennsylvania, 1888–1959), a Pulitzer Prize–winning playwright; Edward K. Thompson (b.Minnesota, 1907), editor of *Life* magazine and founder-editor of *Smithsonian;* radio and television commentator Eric Severeid (b.1912); and novelist Larry Woiwode (b.1941).

To the entertainment world North Dakota has contributed band leaders Harold Bachman (1892–1972); Lawrence Welk (b.1903), and Tommy Tucker (Gerald Duppler, b.1908); jazz vocalist Peggy Lee (Norma Delores Egstrom, b.1920) and country singer Lynn Anderson (b.1947); and actresses Dorothy Stickney (b.1903) and Angie Dickinson (Angeline Brown, b.1931).

Sports personalities associated with the state include outfielder Roger Maris (b.1934), who in 1961 broke Babe Ruth's record for home runs in one season.

[50]BIBLIOGRAPHY

Crawford, Lewis F. *History of North Dakota.* 3 vols. Chicago: American Historical Society, 1931.

Federal Writers' Project. *North Dakota: A Guide to the Northern Prairie State.* Reprint. New York: Somerset, 1980 (orig. 1938).

Goodman, L. R., and R. J. Eidem. *The Atlas of North Dakota.* Fargo: North Dakota Studies, Inc., 1976.

North Dakota Economic Development Commission. *North Dakota Growth Indicators.* Bismarck, 1984.

North Dakota, University of. Bureau of Business and Economic Research. *Statistical Abstract of North Dakota 1983.* Grand Forks: University of North Dakota Press, 1983.

Robinson, Elwyn B. *History of North Dakota.* Lincoln: University of Nebraska Press, 1966.

Tweton, D. Jerome, and Theodore Jelliff. *North Dakota—The Heritage of a People.* Fargo: North Dakota Institute for Regional Studies, 1976.

Tweton, D. Jerome, and Daniel F. Rylance. *The Years of Despair: North Dakota in the Depression.* Grand Forks: Oxcart Press, 1973.

Wilkins, Robert P. and Wynona H. *North Dakota.* New York: Norton, 1977.

OHIO

State of Ohio

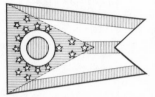

ORIGIN OF STATE NAME: From the Iroquois Indian word *oheo,* meaning "beautiful." **NICKNAME:** The Buckeye State. **CAPITAL:** Columbus. **ENTERED UNION:** 1 March 1803 (17th). **SONG:** "Beautiful Ohio." **MOTTO:** With God All Things Are Possible. **FLAG:** The flag is swallow-tailed, with three red and two white lateral stripes; at the staff is a blue triangular field covered with 17 stars (signifying Ohio's order of entry into the Union) grouped around a red disk superimposed on a white circular O. **OFFICIAL SEAL:** In the foreground are a sheaf of wheat and a sheaf of 17 arrows; behind, a sun rises over a mountain range, indicating that Ohio is the 1st state west of the Alleghenies. Surrounding the scene are the words "The Great Seal of the State of Ohio." **BIRD:** Cardinal. **INSECT:** Ladybug. **FLOWER:** Scarlet carnation. **TREE:** Buckeye. **GEM:** Ohio flint. **BEVERAGE:** Tomato juice. **LEGAL HOLIDAYS:** New Year's Day, 1 January; Birthday of Martin Luther King, Jr., 3d Monday in January; Presidents' Day, 3d Monday in February; Memorial Day, last Monday in May; Independence Day, 4 July; Labor Day, 1st Monday in September; Columbus Day, 2d Monday in October; Veterans Day, 11 November; Thanksgiving Day, 4th Thursday in November; Christmas Day, 25 December. **TIME:** 7 AM EST = noon GMT.

¹LOCATION, SIZE, AND EXTENT

Located in the eastern north-central US, Ohio is the 11th largest of the 12 midwestern states and ranks 35th in size among the 50 states.

The state's total area is 41,330 sq mi (107,044 sq km), of which land comprises 41,004 sq mi (106,201 sq km) and inland water 326 sq mi (823 sq km). Ohio extends about 210 mi (338 km) E–W; its maximum N–S extension is 230 mi (370 km).

Ohio is bordered on the N by Michigan and the Canadian province of Ontario (with the line passing through Lake Erie); on the E by Pennsylvania and West Virginia (with the Ohio River forming part of the boundary); on the S by West Virginia and Kentucky (with the entire line defined by the Ohio River); and on the W by Indiana.

Five important islands lie off the state's northern shore, in Lake Erie: the three Bass Islands, Kelleys Island, and Catawba Island. Ohio's total boundary length is 997 mi (1,605 km).

The state's geographic center is in Delaware County, 25 mi (40 km) NNE of Columbus.

²TOPOGRAPHY

Ohio has three distinct topographical regions: the foothills of the Allegheny Mountains in the eastern half of the state; the Erie lakeshore, extending for nearly three-fourths of the northern boundary; and the central plains in the western half of the state.

The Allegheny Plateau in eastern Ohio consists of rugged hills and steep valleys that recede gradually as the terrain sweeps westward toward the central plains. The highest point in the state is Campbell Hill (1,550 feet—470 meters), located in Logan County about 50 mi (80 km) northwest of Columbus. The Erie lakeshore, a band of level lowland that runs across the state to the northwestern corner on the Michigan boundary, is distinguished by sandy beaches. The central plains extend to the western boundary with Indiana. In the south, undulating hills decline in altitude as they reach the serpentine Ohio River, which forms the state's southern boundary with Kentucky and West Virginia. The state's lowest point is on the banks of the Ohio River in the southwest, where the altitude drops to 433 feet (132 meters) above sea level.

Most of Ohio's 2,500 lakes are situated in the east, and nearly all are reservoirs backed up by river dams. The largest, Pymatuning Reservoir, on the Pennsylvania border, has an area of 14,650 acres (5,929 hectares). Grand Lake (St. Marys), located near the western border, covering 12,500 acres (5,059 hectares), is the largest lake wholly within Ohio.

Ohio has two drainage basins separated by a low ridge extending from the northeast to about the middle of the western border with Indiana. North of the ridge, more than one-third of Ohio's area is drained by the Maumee, Portage, Sandusky, Cuyahoga, and Grand rivers into Lake Erie. South of the ridge, the remaining two-thirds of the state is drained mainly by the Muskingum, Hocking, Raccoon, Scioto, Little Miami, and Miami rivers into the Ohio River, which winds for about 450 mi (725 km) along the eastern and southern borders.

Ohio's bedrock of sandstone, shale, and limestone was formed during the Paleozoic era some 300–600 million years ago. The oldest limestone rocks are found in the Cincinnati anticline, a ridge of sedimentary rock layers about 3,000 feet (900 meters) thick that extends from north to south in west-central Ohio. Inland seas filled and receded periodically to form salt and gypsum, also creating peat bogs that later were pressurized into the coal beds of southeastern Ohio. At the end of the Paleozoic era, the land in the eastern region uplifted to form a plateau that was later eroded by wind and water into hills and gorges.

About 2 million years ago, glaciers covering two-thirds of the state leveled the western region into plains and deposited fertile limestone topsoil. As the glaciers retreated, the melting ice formed a vast lake, which overflowed southward into the channels that became the Ohio River. Perhaps 15,000 years ago, during the last Ice Age, the glacial waters ran off and reduced Lake Erie to its present size. Limestone rocks in Glacier Grooves State Park on Kelleys Island bear the marks of the glaciers' movements.

³CLIMATE

Lying in the humid continental zone, Ohio has a generally temperate climate. Winters are cold and summers mild in the eastern highlands. The southern region has the warmest temperatures and longest growing season—198 days on the average, compared with 150 to 178 days in the remainder of the state. More than half of the annual rainfall occurs during the growing season, from May to October.

Among the major cities, Columbus, in the central region, has an annual mean temperature of 52°F (11°C), with a normal maximum of 62°F (17°C) and a normal minimum of 41°F (5°C). Cleveland, in the north, has an annual mean of 50°F (10°C), with a normal maximum of 59°F (15°C) and minimum of 41°F (5°C).

The mean temperature in Cincinnati, in the south, is 55°F (13°C), the normal maximum 65°F (18°C), and the normal minimum 45°F (7°C). Cleveland has an average of 127 days per year in which the temperature drops to 32°F (0°C) or lower, Columbus 124 days, and Cincinnati 99 days. The record low temperature for the state is −39°F (−39°C), set at Milligan on 10 February 1899. The record high is 113°F (45°C), registered near Gallipolis on 21 July 1934.

Cleveland has an average annual snowfall of 52 in (132 cm), while Columbus receives 28 in (71 cm), and Cincinnati 24 in (61 cm). Cincinnati has the most total precipitation, with 40 in (102 cm), compared with 37 in (94 cm) for Columbus and 35 in (89 cm) for Cleveland. Because of its proximity to Lake Erie, Cleveland is the windiest city, with winds that average 11 mph (18 km/hr).

⁴FLORA AND FAUNA

More than 2,500 plant species have been found in Ohio. The southeastern hill and valley region supports pitch pine, bigleaf magnolia, and sourwood, with undergrowths of sassafras, witch-hazel, pawpaw, hornbeam and various dogwoods. At least 14 species of oak, 10 of maple, 9 of poplar, 9 of pine, 7 of ash, 7 of elm, 6 of hickory, 5 of birch, and 2 of beech grow in the state, along with butternut, eastern black walnut, wild black cherry, black locust, and sycamore. A relative of the horse chestnut (introduced to Ohio from Asia), the distinctive buckeye—first called the Ohio buckeye and now the official state tree—is characterized by its clusters of cream-colored flowers that bloom in spring and later form large, brown, thick-hulled nuts. The yellow-fringed orchid, northern wild monkshood, and wild kidney bean are threatened species, while painted trillium, white lady's-slipper, and rock elm are among more than 200 endangered plants.

The Buckeye State is rich in mammals. White-tailed deer, badger, mink, raccoon, red and gray foxes, coyote, beaver, eastern cottontail, woodchuck, least shrew, and opossum are found throughout the state's five wildlife districts; the bobcat, woodland jumping mouse, and red-backed mole are among many species with more restricted habitats. Common birds include the eastern great blue heron, green-winged teal, mourning dove, eastern belted kingfisher, eastern horned lark, blue-gray gnatcatcher, eastern cowbird, and a great variety of ducks, woodpeckers, and warblers; the cardinal is the state bird, and the ruffed grouse, mostly confined to the Allegheny Plateau, is a favorite game species. Bass, pickerel, perch, carp, pike, trout, catfish, sucker, and darter thrive in Ohio's lakes and streams. The snapping, midland painted, and spiny soft-shelled turtles, five-lined skink, northern water snake, midland brown snake, eastern hognose, and eastern milk snake appear throughout Ohio. The northern copperhead, eastern massasauga (swamp rattler), and timber rattlesnake are Ohio's only poisonous reptiles. Fowler's toad, bullfrog, green pickerel frog, and marbled and red-backed salamanders are common native amphibians.

Acting on the premise that the largest problem facing wildlife is the destruction of their habitat, the Division of Wildlife of the Department of Natural Resources has instituted an ambitious endangered species program. Among the numerous animals categorized by the state as endangered are the river otter, bobcat, Allegheny woodrat, sharp-shinned hawk, king rail, upland sand-piper, common tern, spotted turtle, five species of salamander, Ohio lamprey, shortnose gar, Great Lakes muskellunge, northern madtom, Tippecanoe darter, and Allegheny crayfish.

⁵ENVIRONMENTAL PROTECTION

Early conservation efforts in Ohio were aimed at controlling the ravages of spring floods and preventing soil erosion. After the Miami River floods of March 1913, which took 361 lives and resulted in property losses of more than $100 million in Dayton alone, the Miami Conservancy District was formed; five earth dams and 60 mi (97 km) of river levees were completed by 1922,

at a cost of $40 million, to hold back cresting waters. In the Muskingum Conservancy District in eastern Ohio, construction of flood-control dams has prevented spring flooding and the washing away of valuable topsoil into the Ohio River. In recent years, the state's major environmental concerns have been to reverse the pollution of Lake Erie, to control the air pollution attributable to coal burning and automobiles, and to clean up dumps for solid and hazardous wastes.

The state's regulatory agency for environmental matters is the Ohio Environmental Protection Agency, established in 1972. The agency draws up both emergency plans and long-range programs to deal with pollution of air, water, and land resources. The EPA also coordinates state, local, and federal funding of environmental programs.

Since 1972, antipollution efforts in Lake Erie have focused on reducing the discharge of phosphorus into the lake from sewage and agricultural wastes; sewage treatment facilities have been upgraded with the aid of more than $750 million in federal grants, and efforts have been made to promote reduced-tillage farming to control runoff. By the early 1980s, numerous beaches had been reopened, and sport fishing was once again on the increase. During the first 10 years of the state EPA's existence, Ohio industries spent about $1.5 billion on efforts to control air pollution, and air quality reportedly improved by 68% over this period. From 1967 to 1983, through the efforts of local health departments and with the eventual help of the EPA, over 1,300 open garbage dumps were closed down and more than 200 sanitary landfills constructed to replace them.

In 1980, Ohio passed its first legislation aimed at controlling hazardous wastes, and by early 1985, with the aid of more than $11 million in federal Superfund grants, cleanup had been completed or begun at 16 major sites. In June 1985, 158 companies promised, in a consent decree, to pay $7 million in penalties and to fund cleanup of one of the largest toxic dumps in the state, the Chem-Dyne waste dump at Hamilton.

For all of its own programs, the EPA expended an estimated $37.6 million in the 1984/85 fiscal year. Another agency, the Ohio Department of Natural Resources, is responsible for the development and use of the state's natural resources. In 1983/84, the department managed 71 state parks, 20 forests, and 61 natural areas and preserves. The department also assists in soil conservation, issues permits for dams, promotes conservation of oil and gas, and allocates strip-mining licenses. The department's funding was estimated at $168.5 million in the 1984/85 fiscal year.

⁶POPULATION

Ohio ranked 6th in population among the 50 states at the 1980 census, with a population of 10,797,630. From 1980 to 1985, however, Ohio's population declined 0.3%, to 10,763,309, and the state fell to 7th place, behind Florida.

Ohio's population grew slowly during the colonial period and totaled 45,365 persons in 1800. Once the territory became a state in 1803, settlers flocked to Ohio and the population quintupled to 230,760 by 1810. The state's population doubled again by 1820, approached 2,000,000 in 1850, and totaled 3,198,062 by 1880. Ohio's annual rate of population increase slowed considerably after 1900, when its population was 4,157,545; nevertheless, in the period between 1900 and 1960, the total population more than doubled to 9,706,397.

The 1980 census figure of 10,797,630 represented an increase of only 1.3% over the 1970 census population—a much lower rate than the overall US growth rate (11.4%) during the same period. The slow rate of population increase during the 1970s, and the

LOCATION: 38°23′ to 41°59′N; 80°32′ to 84°49′ W. BOUNDARIES: Michigan line, 70 mi (113 km); Canadian line, 205 mi (330 km); Pennsylvania line, 92 mi (148 km); West Virginia line, 277 mi (446 km); Kentucky line, 174 mi (280 km); Indiana line, 179 mi (288 km).

OHIO

SCALE

| 0 | 10 | 20 | 30 Miles |
| 0 | 10 | 20 | 30 Kms |

Akron ——— Over 100,000
Euclid ——— 50,000-100,000
Whitehall ——— 20,000-50,000
Sylvania ——— 10,000-20,000
Streetsboro ——— Under 10,000

ALLEN ——— County Name

LEGEND
⊛ State Capital
⊙ County Seat
✈ Airport
■ Point of Interest

population decline during 1980–85, resulted from a net migration loss and a declining birthrate.

In 1980, Ohio's population was 51.7% female and almost 80% urban. The population density was 262 persons per sq mi (101 per sq km) in 1985, almost four times the US average.

According to the 1980 census, Cleveland was Ohio's largest city, with a population of 573,822; however, from April 1980 to July 1984, Cleveland lost 27,279 persons, and its 1984 population of 546,543 was second to that of Columbus, which had 566,114 persons. Cincinnati and other large cities also lost population during this period, largely because of the shift of the middle class from the inner cities to the suburbs or to other states. In 1984, according to Census Bureau estimates, Cincinnati's population was 370,481, followed by Toledo, 343,939; Akron, 226,877; Dayton, 181,859; and Youngstown, 108,042.

Ohio's three most populated cities and their suburbs ranked among the 30 largest metropolitan areas in the US in 1984. In that year, metropolitan Cleveland (including Akron and Lorain) had an estimated population of 2,788,400, ranking 12th; metropolitan Cincinnati (including some portions of Kentucky and Indiana) 1,673,500, ranking 23d; and the metropolitan area of Columbus 1,279,000, ranking 28th.

7ETHNIC GROUPS

Ohio was first settled by migrants from the eastern states and from the British Isles and northern Europe, especially Germany. Cincinnati had such a large German population that its public schools were bilingual until World War I. With the coming of the railroads and the development of industry, Slavic and south Europeans were recruited in large numbers. By 1980, however, only about 3% of Ohioans were foreign-born, the major places of origin being Germany, Italy, and the United Kingdom. Ethnic clusters persist in the large cities, and some small communities retain a specific ethnic flavor, such as Fairport Harbor on Lake Erie, with its large Finnish population.

As of 1980 there were 1,077,000 blacks, representing 10% of the population. Most lived in the larger cities, especially Cleveland,

Ohio Counties, County Seats, and County Areas and Populations

COUNTY	COUNTY SEAT	LAND AREA (SQ MI)	POPULATION (1984 EST.)	COUNTY	COUNTY SEAT	LAND AREA (SQ MI)	POPULATION (1984 EST.)
Adams	West Union	586	24,734	Logan	Bellefontaine	458	39,522
Allen	Lima	405	109,070	Lorain	Elyria	495	271,290
Ashland	Ashland	424	46,381	Lucas	Toledo	341	463,989
Ashtabula	Jefferson	703	101,650	Madison	London	467	34,569
Athens	Athens	508	58,000	Mahoning	Youngstown	417	281,229
Auglaize	Wapakoneta	398	43,121	Marion	Marion	403	66,447
Belmont	St. Clairsville	537	81,759	Medina	Medina	422	115,973
Brown	Georgetown	493	33,474	Meigs	Pomeroy	432	23,692
Butler	Hamilton	469	265,458	Mercer	Celina	457	38,596
Carroll	Carrollton	393	26,910	Miami	Troy	410	89,274
Champaign	Urbana	429	33,450	Monroe	Woodsfield	458	16,358
Clark	Springfield	398	147,552	Montgomery	Dayton	458	563,766
Clermont	Batavia	456	136,526	Morgan	McConnelsville	420	14,100
Clinton	Wilmington	410	34,844	Morrow	Mt. Gilead	406	26,556
Columbiana	Lisbon	534	112,017	Muskingum	Zanesville	654	84,190
Coshocton	Coshocton	566	36,409	Noble	Caldwell	399	11,256
Crawford	Bucyrus	403	49,004	Ottawa	Port Clinton	253	39,738
Cuyahoga	Cleveland	459	1,460,561	Paulding	Paulding	419	20,676
Darke	Greenville	600	53,859	Perry	New Lexington	412	31,701
Defiance	Defiance	414	38,413	Pickaway	Circleville	503	42,778
Delaware	Delaware	443	57,298	Pike	Waverly	443	24,006
Erie	Sandusky	264	77,909	Portage	Ravenna	493	137,522
Fairfield	Lancaster	506	95,828	Preble	Eaton	426	38,470
Fayette	Washington Ct. House	405	27,421	Putnam	Ottawa	484	33,232
Franklin	Columbus	542	893,774	Richland	Mansfield	497	129,109
Fulton	Wauseon	407	38,421	Ross	Chillicothe	692	67,528
Gallia	Gallipolis	471	30,104	Sandusky	Fremont	409	62,201
Geauga	Chardon	408	74,922	Scioto	Portsmouth	614	84,463
Greene	Xenia	415	129,504	Seneca	Tiffin	553	61,379
Guernsey	Cambridge	522	41,463	Shelby	Sidney	409	43,319
Hamilton	Cincinnati	412	863,989	Stark	Canton	574	377,305
Hancock	Findlay	532	64,867	Summit	Akron	412	512,610
Hardin	Kenton	471	31,663	Trumbull	Warren	612	236,834
Harrison	Cadiz	400	16,567	Tuscarawas	New Philadelphia	569	85,404
Henry	Napoleon	415	28,234	Union	Marysville	437	30,734
Highland	Hillsboro	553	34,208	Van Wert	Van Wert	410	29,968
Hocking	Logan	423	24,664	Vinton	McArthur	414	11,501
Holmes	Millersburg	424	30,122	Warren	Lebanon	403	102,525
Huron	Norwalk	495	54,994	Washington	Marietta	640	64,810
Jackson	Jackson	420	29,927	Wayne	Wooster	557	99,726
Jefferson	Steubenville	410	87,759	Williams	Bryan	422	36,211
Knox	Mt. Vernon	529	47,363	Wood	Bowling Green	619	108,414
Lake	Painesville	231	215,518	Wyandot	Upper Sandusky	406	22,423
Lawrence	Ironton	456	62,847				
Licking	Newark	686	124,035	TOTALS		41,004	10,751,999

which in 1980 was 44% black. Historically, Ohio was very active in the antislavery movement. Oberlin College, established in 1833 by dissident theological students, admitted blacks from its founding and maintained a "station" on the Underground Railroad. Cleveland elected its first black mayor in 1967.

Some 120,000 Hispanic Americans lived in Ohio in 1980. The largest number were of Mexican descent, but there were also many Puerto Ricans. American Indians numbered only 12,200. In 1980, 9,900 Chinese, 7,400 Filipinos, 5,500 Japanese, and 3,500 Vietnamese were living in Ohio.

⁸LANGUAGES

Except for small Iroquoian groups like the Erie and Seneca, most of the Indian population before white settlement comprised four Algonkian tribes: Delaware, Miami, Wyandot, and Shawnee. Indian place-names include Ohio, Coshocton, Cuyahoga, and Wapakoneta.

Ohio English reflects three post-Revolutionary migration paths. Into the Western Reserve south of Lake Erie came Northern speech from New York and Connecticut. Still common there are the Northern pronunciation of the *ow* diphthong, as in *cow,* with a beginning like the /ah/ vowel in *father,* and the use of that /ah/ in *fog* and *college;* /krik/ is more common than /kreek/ for *creek.* A dragonfly is a *devil's darning needle;* doughnuts may be *fried cakes;* a boy throws himself face down on a sled in a *bellyflop-* (*per*); and a tied and filled bedcover is a *comforter.*

Most of nonurban Ohio has North Midland speech from Pennsylvania. Generally, except in the northern strip, *cot* and *caught* are sound-alikes, and *now* is /naow/. South of Columbus, because of the influence of South Midland patterns from Kentucky and extreme southern Pennsylvania, corn bread may be *corn pone,* lima beans are *butter beans,* and a tied quilt is a *comfort.* *Spouting,* yielding to *gutters,* barely reaches across to Indiana; and *sick at the stomach, dived,* and *wait on me* are competing with expanding Northern *to the stomach, dove,* and *wait for me.* A new Midland term, *bellybuster,* originated around Wheeling and has spread north to compete with *bellyflop.* Northern and Midland merge in the mixed dialect west of Toledo.

From Kentucky, South Midland speakers took *you-all* into Ohio River towns, and in the southwestern tip of the state can be heard their *evening* for *afternoon, terrapin* for *tortoise,* and *frogstool* for *toadstool.* Recent northward migration has introduced South Midland speech and black English, a southern dialect, into such industrial centers as Cleveland, Toledo, and Akron.

Localisms have developed. For the grass strip between sidewalk and street, Akron has *devil-strip* and Cleveland has *treelawn.* Foreign-language influence appears in such Pennsylvania Germanisms as *clook* (hatching hen), *snits* (dried apples), *smearcase* (cottage cheese), and *got awake.*

Of Ohioans aged 3 years or older, 95% spoke only English at home in 1980. Other languages spoken at home, and the number of speakers, were as follows:

Spanish	100,515	French	33,384
German	85,230	Hungarian	26,345
Italian	53,093	Greek	18,575
Polish	37,930	Slovak	17,523

⁹RELIGIONS

The first religious settlement in Ohio territory was founded among Huron Indians in 1751 by a Roman Catholic priest near what is now Sandusky. Shortly afterward, Moravian missionaries converted some Delaware Indians to Christianity; the first Protestant church was founded by Congregationalist ministers at Marietta in 1788. Dissident religious sects such as the Shakers, Amish, and Quakers moved into Ohio from the early 18th century onward, but the majority of settlers in the early 19th century were Presbyterians, Methodists, Baptists, Disciples of Christ, and Episcopalians.

The first Roman Catholic priest to be stationed permanently in Ohio was Father Edward Fenwick, who settled in Cincinnati in 1817. When the Protestant settlers there did not allow him to build a Catholic church in the town, he founded Christ Church (now St. Francis Church) just outside Cincinnati. In 1821, Father Fenwick became the first Catholic bishop in Ohio. The large influx of Irish and German immigrants after 1830 greatly increased the Catholic constituency in Cleveland, Cincinnati, Columbus, and Toledo. Among the German immigrants were many Lutherans and a large number of Jews, who made Cincinnati a center of Reform Judaism. In the mid-19th century, Cincinnati had the nation's third-largest Jewish community; the Union of American Hebrew Congregations, the most important Reform body, was founded there in 1873, and Hebrew Union College, a rabbinical training school and center of Jewish learning, was founded two years later.

In 1984, Ohio had a Roman Catholic population of 2,259,097, of whom about 880,000 were in the Cleveland diocese and 490,000 in the archdiocese of Cincinnati. During the same year, the state's Jewish population was estimated at 140,435. Leading Jewish communities were in Cleveland, Cincinnati, and Columbus.

There were 10,060 Protestant churches in 1980, when adherents of all Protestant groups totaled 5,346,227. The largest Protestant denominations and their adherents in 1980 were United Methodist, 741,753; United Presbyterian, 258,476; United Church of Christ, 210,961; American Lutheran Church, 177,807; Lutheran Church of America, 176,623; Southern Baptist Convention, 152,277; Christian Church and Churches of Christ, 133,183; American Baptist USA, 129,650; and Episcopal, 93,552.

¹⁰TRANSPORTATION

Sandwiched between two of the country's largest inland water systems, Lake Erie and the Ohio River, Ohio has long been a leader in water transport. With its numerous terminals on the Ohio River and deepwater ports on Lake Erie, Ohio ranks as one of the major US states for shipping. In 1983, Ohio also ranked 4th among the 50 states in miles of completed interstate highway and in miles of Class I railroad track.

The building of railroads in the mid-19th century greatly improved transportation within the state by connecting inland counties with Lake Erie and the Ohio River. The Mad River and Lake Erie Railroad, between Dayton and Sandusky, was completed in 1844, and two years later, it was joined with the Little Miami Railroad, to provide through service to Cincinnati. Also in 1846, Cleveland was connected by rail with Columbus and Pittsburgh. Railroad building in the state reached a peak in the 1850s; at the outbreak of the Civil War, Ohio had more miles of track than any other state. By 1900, railroads were by far the most important system of transport.

In 1983, Class I railroads operated more than 6,850 mi (11,000 km) of track in the state. Freight service on branch lines to counties has been maintained, despite declining freight volume, through a state subsidy program.

Mass transit in Ohio's cities began in 1859 with horse-drawn carriages carrying paying passengers in Cleveland and Cincinnati, which added a cable car on rails about 1880. The electric trolley car, introduced to Cleveland in 1884, soon became the most popular mass transit system for the large cities. Interurban electric railways carried passengers to and from rural towns that had been bypassed by the railroads; there were 2,809 mi (4,521 km) of interurban track in the state by 1907. The use of electric railways declined with the development of the motor car in the 1920s, and by 1939, for example, the seven interurban lines serving Columbus had been abandoned. Today, suburbanites commute to their workplaces in Columbus and other cities by automobile, commuter railroads, and bus lines. In 1984/85, Amtrak operated four regularly scheduled trains through Ohio, with a total ridership in the state of 168,635 passengers.

Rough roads were used by settlers in the early 19th century. The National Road was built from Wheeling, W.Va., to Zanesville in 1826, and was extended to Columbus by 1833. The increasing use of the automobile in the 1930s led to massive state and federal road-building programs in Ohio as elsewhere. The major interstate highways across Ohio connect Cleveland and the Toledo area in the north (I-80, I-90), link Columbus with Zanesville and Wheeling (I-70) and with Cincinnati (I-73), and extend north–south from Cleveland and Akron to Marietta (I-77).

In 1983, Ohio had 1,529 mi (2,461 km) of interstate highways, 29,425 mi (47,355 km) of municipal roads, and 82,827 mi (133,297 km) of rural roads. In that year, 6,429,344 automobiles, 1,307,834 trucks, and 286,290 motorcycles were registered in the state, and there were 7,397,289 licensed drivers.

Inland waterways have long been important for transport and commerce in Ohio. The first settlers traveled into Ohio by flatboat down the Ohio River to establish such towns as Marietta and Cincinnati. Lake Erie schooners brought the founders of Cleveland and Sandusky. Steamboat service began on the Ohio River in 1811, and at Lake Erie ports in 1818. The public demand for water transportation in the interior of the state, where few rivers were navigable, led to construction of the Ohio and Erie Canal from Cleveland to Portsmouth, and of the Miami and Erie Canal from Cincinnati to Dayton; both were opened to traffic in 1827 but not completed for another 14 years. The canals gave Ohio's farmers better access to eastern and southern markets. Water transportation is still a principal means of shipping Ohio's products through the St. Lawrence Seaway to foreign countries, and the method by which millions of tons of cargo, particularly coal, are moved via the Ohio River to domestic markets.

Ohio's ports rank among the busiest of the 50 states in volume and in value of foreign exports. In 1983, the state's two largest port districts, Toledo and Ashtabula-Conneaut, accounted for exports (dry cargo) valued at $862 million and total imports of $218 million.

Ohioans consider Dayton to be the birthplace of aviation because it was there that Wilbur and Orville Wright built the first motor-powered airplane in 1903. In 1983 there were 516 airports in the state. The major air terminal is Cleveland's Hopkins International airport, which enplaned 2,626,602 passengers on 48,518 departing flights in 1983.

¹¹HISTORY

The first people in Ohio, some 11,000 years ago, were hunters. Their stone tools have been found with skeletal remains of long-extinct mammoths and mastodons. Centuries later, Ohio was inhabited by the Adena people, the earliest mound builders. Their descendants, the Hopewell Indians, built burial mounds, fortifications, and ceremonial earthworks, some of which are now preserved in state parks.

The first European travelers in Ohio, during the 17th century, found four Indian tribes: Wyandot and Delaware in northern Ohio, Miami and Shawnee in the south. All were hunters who followed game trails that threaded the dense Ohio forest. All together, these four tribes numbered about 15,000 people. European exploration was begun by a French nobleman, Robert Cavelier, Sieur de la Salle, who, with Indian guides and paddlers, voyaged from the St. Lawrence River to the Ohio, which he explored in 1669–70. In the early 1700s, French and English traders brought knives, hatchets, guns, blankets, tobacco, rum, and brandy to exchange for the Indians' deer and beaver skins.

Both the French and the English claimed possession of Ohio, the French claim resting on La Salle's exploration, while the British claimed all territory extending westward from their coastal colonies. To reinforce the French claim, Celeron de Bienville led an expedition from Canada to Ohio in 1749 to warn off English traders, win over the Indians, and assert French possession of the land. Traveling by canoe, with marches overland, he found the

Indians better disposed at that time to the English than to the French. The following year, a company of Virginia merchants sent Christopher Gist to map Ohio trade routes and to make friendship and trade agreements with the tribes. The clash of ambitions brought on the French and Indian War—during which the Indians fought on both sides—ending in 1763 with French defeat and the ceding of the vast western territory to the British. During the Revolutionary War, the American militiaman George Rogers Clark, with a small company of woodsmen-soldiers, seized British posts and trading stations in Ohio, and in the Battle of Piqua, defeated Indian warriors allied with the British. It was largely Clark's campaigns that won the Northwest Territory for the US.

The new nation had a huge public domain, extending from the Allegheny Mountains to the Mississippi River. To provide future government and development of the territory northwest of the Ohio River, the US Congress enacted the Land Ordinance of 1785 and the Northwest Ordinance of 1787. The Land Ordinance created a survey system of rectangular sections and townships, a system begun in Ohio and extended to all new areas in the expanding nation. The farsighted Northwest Ordinance provided a system of government under which territories could achieve statehood on a basis equal with that of the original colonies. When a specified area had a population of 60,000 free adult males, it could seek admission to the Union as a state.

The first permanent settlement in Ohio was made in 1788 by an organization of Revolutionary War veterans who had received land warrants as a reward for their military service. They trekked by ox-drawn wagons over the mountains and by flatboat down the Ohio River to the mouth of the Muskingum, where they built the historic town of Marietta. John Cleves Symmes, a New Jersey official, brought pioneer settlers to his Miami Purchase in southwestern Ohio; their first settlement, in 1789, eventually became the city of Cincinnati. Access to the fertile Ohio Valley was provided by the westward-flowing Ohio River, which carried pioneer settlers and frontier commerce. Flatboats made a one-way journey, as families floated toward what they hoped would be new settlements. Keelboats traveled both downstream and upstream— an easy journey followed by a hard one. The keelboat trade, carrying military supplies and frontier produce, created an enduring river lore. Its legendary hero is burly, blustering Mike Fink, "half horse and half alligator," always ready for a fight or a frolic, for riot or rampage.

Increasing settlement of the Ohio Valley aroused Indian resistance. War parties raided outlying villages, burned houses, and drove families away. Two military expeditions against the Indians were shattered by Chief Little Turtle and his Miami warriors. Then, in 1793, Maj. Gen. "Mad Anthony" Wayne took command in the West. He built roads and forts in the Miami Valley, and trained a force of riflemen. On a summer morning in 1794, Wayne routed allied tribesmen, mostly Miami and Shawnee, in the decisive Battle of Fallen Timbers. In the ensuing Treaty of Greenville, Indian leaders surrendered claim to the southern half of Ohio, opening that large domain to uncontested American occupation.

When, in 1800, Connecticut ceded to the US a strip of land along Lake Erie claimed by its colonial charter and called the Western Reserve, that region became a part of the Northwest Territory. Now the future seemed unclouded, and from the older colonies came a great migration to the promised land. By 1802, Ohio had enough population to seek statehood, and in November, a constitutional convention assembled at Chillicothe. In 25 days and at a total cost of $5,000, the 35 delegates framed a constitution that vested most authority in the state legislature and gave the vote to all white male taxpayers. On 1 March 1803, Ohio joined the Union as the 17th state.

Beyond Ohio's western border, Indians still roamed free. In 1811, the powerful Shawnee chief Tecumseh led a tribal resistance movement (supported by the British) seeking to halt the white man's advance into the new territory and to regain lands already lost to the Americans. Ohio militia regiments led by Gen. William Henry Harrison repulsed an Indian invasion near Toledo in the battle of Tippecanoe on 7 November 1811. Control of Lake Erie and of Great Lakes commerce was at stake when Commodore Oliver Hazard Perry won a decisive naval victory over a British fleet in western Lake Erie during the War of 1812. Tecumseh was slain in the Battle of Thames in Canada on 5 October 1813.

With peace restored in 1815, "Ohio fever" spread through New England. In a great migration, people streamed over the mountains and the lakes to a land of rich soil, mild climate, and beckoning opportunities. Across the Atlantic, especially in England, Ireland, and Germany, thousands of immigrants boarded ship for America. At newly opened land offices, public land was sold at $1.25 an acre. Forest became fields, fields became villages and towns, towns became cities. By 1850, Ohio was the 3d most populous state in the Union.

Having cleared millions of acres of forest, Ohioans turned to economic development. Producing more than its people consumed, the state needed transportation routes to eastern markets. The National Road extended across the central counties in the 1830s, carrying stagecoach passengers and wagon commerce from Pennsylvania and Maryland. The Ohio canal system, created between 1825 and 1841, linked the Ohio River and Lake Erie, providing a waterway to the Atlantic via New York's Erie Canal. In 1826, state lands were valued at $16 million; 15 years later, their value exceeded $100 million. The chief products were wheat, corn, pork, beef, salt, wool, and leather. By 1850, when farm and factory production outstripped the capacity of mule teams and canal barges, railroad building had begun. In the next decade, railroads crisscrossed the state.

In 1861, Ohio, like the rest of the nation, was divided. The northern counties, teeming with former New Englanders, were imbued with abolitionist zeal. But Ohio's southern counties had close ties with Virginia and Kentucky across the river. From southeastern Ohio came Clement L. Vallandigham, leader of the Peace Democrats—called Copperheads by their opponents—who defended states' rights, opposed all of President Lincoln's policies, and urged compromise with the Confederacy. While Ohio surpassed its quota by providing a total of 320,000 Union Army volunteers, the Copperhead movement grew strong enough to nominate Vallandigham for state governor in 1863. Responding to the news of Vallandigham's defeat by the rugged Unionist John Brough, Lincoln telegraphed: "Ohio has saved the nation." Ohio became directly involved in the war for two weeks in 1863, when Confederate Gen. John Hunt Morgan led a Kentucky cavalry force on a daring but ineffectual raid through the southern counties.

Ohio gave the Union its greatest generals—Ulysses S. Grant, William Tecumseh Sherman, and Philip H. Sheridan—each of whom won decisive victories at crucial times. Also essential to the Union cause was the service of Ohio men in Lincoln's cabinet, including Treasury Secretary Salmon P. Chase and War Secretary Edwin M. Stanton.

Mid-19th-century Ohio was primarily an agricultural state, but war demands stimulated Ohio manufactures, and in the decade following the war, the state's industrial products surpassed the value of its rich farm production. The greatest commercial development came in northern Ohio, where heavy industry grew dramatically. To Toledo, Cleveland, and Youngstown via Lake Superior came iron ore that was converted into iron and steel with coal from the Ohio Valley. In the 1870s, John D. Rockefeller of Cleveland organized the Standard Oil Co., which soon controlled oil refining and distribution throughout the nation. At the same time, B. F. Goodrich of Akron began making fire hose, the first rubber product in an industry whose prodigious growth would make Akron the "rubber capital of the world." In the middle of the state, the capital city, Columbus, became a center of the brewing, railroad equipment, and farm implement industries. Cincinnati factories made steamboat boilers, machine tools, meat products, railroad cars, and soap. Dayton became known for its paper products, refrigerators, and cash registers.

With industrial growth came political power. In the next half-century, Ohio virtually took possession of the White House. Presidents Grant, Rutherford B. Hayes, James A. Garfield, Benjamin Harrison, William McKinley, William Howard Taft, and Warren G. Harding were all Ohioans. The first four had been Civil War generals, but even these men were politicians more than commanders; moderation was their rule. Ohio, the heartland of the expanding nation, grew national in mind and character. The four great business pursuits—agriculture, commerce, mining, and manufacturing—were remarkably balanced in Ohio. With one in seven Ohioans being of foreign birth, the state had neither the greatest nor the least admixture of immigrants. Its ethnic strains were various. Following the earlier English, Irish, and German influx came Italian, Czech, Dutch, Finnish, Greek, Hungarian, Polish, Russian, Serbian, and Ukrainian immigrants, along with a growing number of blacks from the rural South. Thus Ohio was an advantageous background for a president; to any other part of the nation, an Ohio candidate did not seem alien. In the 1920 campaign, both the Republican and Democratic nominees—Harding and James M. Cox—were Ohio men. Norman Thomas, a perennial Socialist candidate, was likewise an Ohioan.

During World War I, Ohio's heavy industry expanded and its cities grew. Progressivism developed in Toledo and Cleveland, where the respective mayors, Samuel M. "Golden Rule" Jones and Tom L. Johnson, both wealthy businessmen, strove for civic virtue and social justice. Their reforms resulted in the city-manager form of government that spread to other Ohio cities. In the postwar 1920s, Ohio's oil, rubber, and glass industries kept pace with accelerating automobile production. Yet none of these industries was immune to the prolonged depression of the 1930s. Widespread unemployment and a stagnant economy were not relieved until the outbreak of World War II. The war swept 641,000 Ohioans into military service and gave Ohio industry military contracts totaling $18 billion.

The state's economy prospered after World War II, with highway building, truck and tractor production, aircraft manufacture, and airport construction leading the field. The completion of the St. Lawrence Seaway in 1959 made active ocean ports of Toledo and Cleveland. Major problems during this period involved pollution created by the dumping of industrial wastes (especially in Lake Erie) and urban decay resulting from the departure of middle-class families to the suburbs, an exodus that left the central cities to growing numbers of the poor and underprivileged. Related to these problems were troubles in the Ohio school system. Deteriorating neighborhoods produced inadequate revenues for schools and public services, and attempts at racial integration brought controversy and disturbance. When political office went to minority leaders—in 1967, Carl Stokes of Cleveland became the first black mayor of any major US city—friction and tension continued. Urban renewal barely kept pace with urban blight. A further shock to Ohioans was the shooting of 13 students, 4 of whom died, at Kent State University on 4 May 1970 by National Guardsmen who had been sent to the campus to preserve order during a series of demonstrations against US involvement in Indochina.

During the early 1980s, Ohio was still beset by serious social and economic problems. The state's population was static. Shrinking industrial employment was only partly offset by new enterprises that offered services rather than material goods. Ohio's huge coal reserves were of limited usefulness because of their content of sulfur, an atmospheric pollutant.

On the positive side, Ohio had strengthened its state universities and developed a system of community colleges that brought vocational training within the reach of most of its citizens. Major conservancy programs embracing the watersheds of the Muskingum and the Miami rivers had become models for such undertakings in other states. The state also participated in programs that, to a significant extent, reversed pollution of Lake Erie.

¹²STATE GOVERNMENT

The Ohio constitution of 1803 was replaced by a second constitution in 1851. Amendments proposed by a constitutional convention in 1912 and subsequently approved by the voters so heavily revised the 1851 constitution as to make it virtually a new document. This modified constitution, with subsequent amendments, provides for county and municipal home rule, direct primary elections, recall of elected officials, and constitutional amendment by initiative and referendum.

Ohio's general assembly consists of a 99-member house of representatives, elected for two years, and a senate of 33 members serving four-year terms (half the members are chosen every two years). Regular sessions of the legislature convene the first Monday in January of odd-numbered years, and a second session is called on the same date of the following year. Each house may introduce legislation, and both houses must approve a bill before it can be signed into law by the governor. The governor's veto of a bill can be overridden by three-fifths majority votes of both houses. Legislators' salaries were $22,500 annually in 1983.

Officials elected statewide are the governor and lieutenant governor (elected jointly), secretary of state, attorney general, auditor, and treasurer, all of whom serve four-year terms. Effective in 1959, a constitutional amendment changed the governor's term from two to four years and forbade a governor from serving more than two successive terms. The governor appoints the heads of 22 executive departments, as well as the adjutant general and members of most statutory boards. In 1983, the governor's annual salary was $65,000, and the lieutenant governor's $35,000.

The constitution may be amended legislatively by a three-fifths vote of each house; the proposed amendment must then receive majority approval by the voters at the next general election. Amendments may also be proposed by petition of 10% of the electors who voted for governor in the last general election; a majority vote in a subsequent referendum is required for passage.

The constitution provides that every 20 years (from 1932 onward), the voters must be given the chance to choose whether a constitutional convention should be held. Voters rejected this option in 1932, 1952, and again in 1972.

To vote in Ohio, one must be a US citizen, 18 years of age or older, and have been a resident of the county and voting precinct for at least 30 days.

¹³POLITICAL PARTIES

Ohio has sent seven native sons and one other state resident to the White House—equaling Virginia as the "mother of presidents." The state's two major political parties, Democratic and Republican, have dominated the political scene since 1856.

Ohioans scattered their votes among various political factions until 1836, when they rallied behind state resident William Henry Harrison and the Whig Party; they again supported Harrison in 1840, helping him win his second bid for the presidency. Whigs and Democrats divided the votes in 1844, 1848, and 1852; in 1856, however, Ohio supported the newly formed Republican Party, and after the Civil War, 7 of the country's next 12 presidents were Ohio-born Republicans, beginning with Grant and ending with Harding. From 1856 to 1984, Ohioans voted for the Republican candidate in all presidential elections except those in which the following 5 Democrats were elected: Woodrow Wilson (twice), Franklin D. Roosevelt (three times), Harry S Truman, Lyndon B. Johnson, and Jimmy Carter. In 1920, when the presidential candidates of both major parties were Ohioans, the Republican, Warren G. Harding, carried Ohio as well as the nation.

Political bossism flourished in Ohio during the last quarter of the 19th century, when the state government was controlled by Republicans Mark Hanna in Cleveland and George B. Cox in Cincinnati. Hanna played an influential role in Republican national politics; in 1896, his handpicked candidate, William McKinley, was elected to the presidency. But the despotism of the bosses and the widespread corruption in city governments led to public demands for reform. In Toledo, a reform mayor, Samuel "Golden Rule" Jones, began to clean house in 1897. Four years later, another group of reformers, led by Mayor Tom L. Johnson, ousted the Hanna machine and instituted honest government in Cleveland. At the time, journalist Lincoln Steffens called Cleveland "the best-governed city in the US" and Cincinnati "the worst." The era of bossism ended for Cincinnati in 1905, when the

Ohio Presidential Vote by Political Parties, 1948–84

YEAR	ELECTORAL VOTE	OHIO WINNER	DEMOCRAT	REPUBLICAN	PROGRESSIVE	SOC. LABOR	COMMUNIST	LIBERTARIAN
1948	25	*Truman (D)	1,452,791	1,445,684	37,487	—	—	—
1952	25	*Eisenhower (R)	1,600,367	2,100,391	—	—	—	—
1956	25	*Eisenhower (R)	1,439,655	2,262,610	—	—	—	—
1960	25	Nixon (R)	1,944,248	2,217,611	—	—	—	—
1964	26	*Johnson (D)	2,498,331	1,470,865	—	—	—	—
					AMERICAN IND.			
1968	26	*Nixon (R)	1,700,586	1,791,014	467,495	—	—	—
					AMERICAN			
1972	25	*Nixon (R)	1,558,889	2,441,827	80,067	7,107	6,437	—
					SOC. WORKERS			
1976	25	*Carter (D)	2,011,621	2,000,505	15,529	4,717	7,817	8,961
					CITIZENS			
1980	25	*Reagan (R)	1,745,103	2,203,139	8,979	4,436	5,030	49,604
1984	25	*Reagan (R)	1,825,440	2,678,560	—	—	—	5,886

* Won US presidential election.

voters overthrew the Cox machine, elected a reform mayor on a fusion ticket, and instituted reforms that in 1925 made Cincinnati the first major US city with a nonpartisan city-manager form of government.

With the decline of big-city political machines, ticket splitting has become a regular practice among Ohio voters in state and local contests. Overall, between 1900 and 1985, 12 Republicans and 12 Democrats held the governorship. Democrat Frank J. Lausche was elected to an unprecedented five two-year terms (1945-47, 1949-57), and Republican James A. Rhodes served four four-year terms (1963-71, 1975-83). In 1982, Ohioans elected a Democratic governor, Richard F. Celeste, and Democrats swept all state offices and won control of both houses of the state legislature. Two years later, Republicans lost the majority they had enjoyed in the state's congressional delegation but regained control of the state senate.

In general, third parties have fared poorly in Ohio since 1856. An exception was the 1968 presidential election, in which American Independent Party candidate George Wallace garnered nearly 12% of Ohio's popular vote. Wallace's capture of many dissident Democrats probably provided the margin of victory for Richard M. Nixon, who carried the state by only 90,428 votes over the Democratic candidate, Hubert H. Humphrey. A more typical voting pattern was displayed in the 1976 presidential election, when the two major parties together received 97.7% of the total votes cast, and only 2.3% of the votes were split among minor parties and independents. Of the state's 5,222,041 registered voters, nearly 80% voted, and Democrat Jimmy Carter won an extremely close election by only 11,116 votes. The result was not nearly so close in 1980, when Ronald Reagan, the Republican presidential nominee, won 51% of the popular vote to 41% for Jimmy Carter (with 6% going to John Anderson and 2% to minor party candidates), or in 1984, when Reagan won 59% of the popular vote to defeat Walter Mondale in the state.

John Glenn easily retained his US Senate seat in 1980, and Howard Metzenbaum, also a Democrat, was reelected to the Senate by a wide margin in 1982. Although the Democrats lost their majority in the state senate in 1984, they gained a majority in the state's US House delegation. Two of the US representatives were women and one was a black. In addition, the state treasurer was a woman, 12 women served in the state legislature, and there were 195 state and local black elected officials and 5 Hispanic elected officials.

¹⁴LOCAL GOVERNMENT
Local government in Ohio is exercised by the 88 counties, 939 cities and villages, and more than 1,300 townships.

Each county is administered by a board of three commissioners, elected to four-year terms, whose authority is limited by state law. The county government is run by eight officials elected to four-year terms: the auditor or financial officer, whose duties include levying taxes; the clerk of courts, who is elected as clerk of the court of common pleas and also serves as clerk of the county court of appeals; the coroner, who must be a licensed physician; an engineer; a prosecuting attorney; the recorder, who keeps records of deeds, mortgages, and other legal documents; a sheriff; and the treasurer, who collects and disburses public funds.

Within each county are incorporated areas with limited authority to govern their own affairs. Thirty voters in an area may request incorporation of the community as a village. A village reaching the population of 5,000 automatically becomes a city, which by law must establish executive and legislative bodies. There are three types of city government: the mayor-council plan, which is the form adopted by a majority of the state's approximately 200 cities; the city-manager form, under which the city council appoints a professional manager to conduct nonpartisan government operations; and the commission type, in which a board of elected commissioners administers the city government.

In practice, most large cities have adopted a home-rule charter, which permits them to select the form of government best suited to their requirements.

Cleveland experimented with the city-manager form of government from 1924 to 1932, at which time public disclosures of municipal corruption led the city's voters to return to the mayor-council plan. In 1967, Cleveland became the first major US city to elect a black mayor; Carl Stokes served two two-year terms but retired from politics in 1971. Cleveland again attracted national attention in 1978 when its 31-year-old mayor, Dennis J. Kucinich, publicly disputed the city's financial policies with members of the city council, and the city defaulted on $15 million in bank loans. Mayor Kucinich narrowly survived a recall election; in 1979, he was defeated for reelection by the state's lieutenant governor, George Voinovich.

Cincinnati has retained the city-manager form of government since 1925. The mayor, elected by the city council from among its nine members, has no administrative duties. Instead, the council appoints a city manager to an indefinite term as chief executive. In 1979, for the first time, the council appointed a black man, Sylvester Murray, as city manager. Columbus, the state capital since 1816, has a mayor-council form of government.

Townships are governed by three trustees and a clerk, all elected to staggered four-year terms. These elected officials oversee zoning ordinances, parks, road maintenance, fire protection, and other matters within their jurisdiction.

¹⁵STATE SERVICES
The State Department of Education administers every phase of public school operations, including counseling and testing services, the federal school lunch program, and teacher education and certification. The department also oversees special schools for the blind and deaf. The department's chief administrator is the superintendent of public instruction.

Health and welfare services are provided by several departments. The Department of Health issues and enforces health and sanitary regulations. Violations of health rules are reviewed by a Public Health Council of seven members, including three physicians and a pharmacist. The Department of Mental Health administers 18 mental health institutions; develops diagnostic, prevention, and rehabilitation programs; and trains mental health professionals. The Department of Public Welfare helps the needy through aid to families with dependent children, public assistance payments, food stamps, and Medicaid. The Bureau of Employment Services and the Bureau of Workers' Compensation administer labor benefit programs.

Public protection services include those of the State Highway Patrol and the Bureau of Motor Vehicles, both within the Department of Highway Safety; the Department of Rehabilitation and Correction, which operates penal institutions; the Department of Youth Services, which administers juvenile correction centers; and the Environmental Protection Agency.

¹⁶JUDICIAL SYSTEM
The supreme court of Ohio, the highest court in the state, reviews proceedings of the lower courts and of state agencies. The high court has a chief justice and six associate justices elected to six-year terms. Below the supreme court are 12 courts of appeals, which exercise jurisdiction over their respective judicial districts. Each court has at least three judges elected to six-year terms; the district including Cleveland has nine appeals court judges, and the Cincinnati district has six.

Trial courts include 88 courts of common pleas, one in each county; judges are elected to six-year terms. Probate courts, domestic relations courts, and juvenile courts often function as divisions of the common pleas courts. In 1957, a system of county courts was established by the legislature to replace justices of the peace and mayors' courts at the local level. Large cities have their own municipal, juvenile, and police courts.

Ohio operates 12 correctional facilities, including reformatories for young offenders and adult prisons for older repeat felons. In 1984, state and federal prisons in Ohio had 18,694 inmates. According to the FBI Crime Index, Ohio's crime rates rank well below the national averages. In the FBI data for 1983, the state's crime rates per 100,000 population were murder and nonnegligent manslaughter, 6; forcible rape, 32; robbery, 159; aggravated assault, 201; burglary, 1,156; larceny, 2,607; and motor vehicle theft, 345.

[17] ARMED FORCES

The US Department of Defense had 47,459 military and civilian personnel in Ohio in 1983/84. The principal military installation was Wright-Patterson Air Force Base near Dayton, with 28,258 personnel. In 1983/84, the Defense Department awarded $2.8 billion in defense contracts to Ohio companies.

In 1983, Ohio had 1,377,000 living veterans, of whom an estimated 13,000 had served in World War I, 541,000 in World War II, 237,000 during the Korean conflict, and 380,000 during the Viet-Nam era. In fiscal year 1982/83, the Veterans Administration expended $973 million in pensions, medical assistance, and other major veterans' benefits.

Army and Air National Guardsmen in Ohio numbered 19,600 at the beginning of 1985. State and local police forces in 1983 employed 24,073 persons (full-time equivalent figure).

[18] MIGRATION

After the Ohio country became a US territory in 1785, Virginians, Connecticut Yankees, and New Jerseyites began arriving in significant numbers; tens of thousands of settlers from New England, Pennsylvania, and some southern states thronged into Ohio in subsequent decades. The great migration from the eastern states continued throughout most of the 19th century, and was bolstered by new arrivals from Europe. The Irish came in the 1830s, and many Germans began arriving in the 1840s. Another wave of European immigration brought about 500,000 people a year to Ohio during the 1880s, many of them from southern and eastern Europe. Former slaves left the South for Ohio following the Civil War, and a larger migratory wave brought blacks to Ohio after World War II to work in the industrial cities. In the 1910s, many emigrants from Greece, Albania, and Latvia settled in Akron to work in the rubber industry.

The industrialization of Ohio in the late 19th and the 20th centuries encouraged the migration of Ohioans from the farms to the cities. The large number of Ohioans who lived in rural areas and worked on farms declined steadily after 1900, with the farm population decreasing to under 1,000,000 during World War II and then to fewer than 400,000 by 1979. A more recent development has been the exodus of urbanites from Ohio's largest cities. From 1970 to 1984, Cleveland lost 204,336 residents, Cincinnati 83,033, Dayton 61,864, Akron 48,548, and Toledo 39,123. Columbus was the only major city to gain residents—26,089—during this period. Ohio lost more than 1 million people through migration during the period 1970–83.

[19] INTERGOVERNMENTAL COOPERATION

The Ohio Commission on Interstate Cooperation represents the state in dealings with the Council of State Governments and its allied organizations. Ohio is a signatory to interstate compacts covering the Ohio River Valley, Pymatuning Reservoir, and the Great Lakes Basin, including the Great Lakes Charter signed in February 1985. The state also participates in the Interstate Mining Compact, the Interstate Oil and Gas Compact, and other compacts.

Federal aid to Ohio for all purposes exceeded $4 billion in the 1983/84 fiscal year.

[20] ECONOMY

Ohio's economy has shown remarkable balance over the years. In the mid-19th century, Ohio became a leader in agriculture, ranking 1st among the states in wheat production in 1840, and 1st in corn and wool by 1850. With industrialization, Ohio ranked 4th in value added by manufacturing in 1900.

Coal mining in the southeastern part of the state and easy access to Minnesota's iron ore via the Great Lakes contributed to the growth of the iron and steel industry in the Cleveland-Youngstown area; Ohio led the nation in the manufacture of machine tools and placed 2d among the states in steel production in the early 1900s. Automobile manufacturing and other new industries were developed after World War I. Hit hard by the depression of the 1930s, the state diversified its industrial foundation and enjoyed prosperity during and after World War II, as its population increased and its income grew.

In the 1970s, however, growth began to lag. By 1980, per capita income in Ohio had fallen well behind the national average. While the gross national product in constant dollars grew 99% from 1960 to 1980, the gross state product expanded only 66%. Manufacturing, which traditionally accounts for more than one-third of the gross state product, was performing less well, as demand for durable goods declined. Manufacturing employment peaked at 1.4 million in 1969; by 1982, the total was down to 1.1 million, and it was believed that many of these jobs would be permanently lost because of a reorientation of Ohio's economy from manufacturing toward services. With unemployment reaching peak levels, the state was forced to borrow from the federal government to fund the soaring cost of unemployment benefits.

In 1982, manufacturing accounted for 27% of the state's total employment, down from 39% in 1970. Steel was produced primarily in Youngstown, automotive and aircraft parts in Cleveland, automobile tires and other rubber products in Akron, and office equipment in Dayton. Recessionary trends in 1980 led to the closing of a US Steel plant in Youngstown and of two Firestone tire and rubber factories in the Akron area, and to widespread layoffs in the auto parts industry. This bad economic news was partially offset when in 1983 the Honda Motor Co. opened Japan's first US automobile assembly plant, at Marysville near Columbus, where Honda had already been manufacturing motorcycles. Honda suppliers also began establishing plants in the state.

[21] INCOME

Ohio ranked 7th among the 50 states in personal income in 1983, with a total of $120.9 billion, but placed 25th in per capita income, with an average of $11,254. The US Department of Commerce forecast a 40% increase in per capita income for Ohioans between 1980 and 1990—slightly below the projected US average for the same period.

The sources of personal earned income in 1980 were as follows: manufacturing, 37%; service industries, 16%; wholesale and retail trade, 15%; government, 13%; transportation and utilities, 6%; construction, 6%; finance, insurance, and real estate, 5%; and other sectors, 2%.

In 1981, Ohio ranked 18th among the 50 states in median money income of four-person families, with $26,851. As of 1982, 52,500 Ohioans were among the nation's top wealth-holders, with gross assets of more than $500,000. An estimated 12.3% of Ohioans had incomes that were below the national poverty level in 1981.

[22] LABOR

In 1984, Ohio ranked 6th among the 50 states in the size of its civilian labor force, which averaged 5,099,000. In that year there were an average of 4,618,000 wage and salary earners, for an unemployment rate of 9.4%, 8th highest among the 50 states. In September 1984, the unemployment rate was 8.7%, down from 11.1% a year earlier.

A federal survey in March 1982 revealed the following nonfarm employment pattern for Ohio:

	ESTABLISH-MENTS	EMPLOYEES	ANNUAL PAYROLL ('000)
Agricultural services, forestry, fishing	2,070	11,057	$ 150,987
Mining, of which:	1,268	30,782	740,007
Bituminous coal, lignite	(201)	(13,914)	(372,211)
Contract construction	15,666	128,220	2,923,865
Manufacturing, of which:	15,976	1,133,998	24,867,867
Primary metals	(632)	(107,322)	(2,530,339)
Fabricated metals	(2,289)	(138,790)	(2,911,132)
Nonelectrical machinery	(3,559)	(175,347)	(3,622,002)
Transportation equipment	(389)	(121,842)	(3,243,132)
Transportation, public utilities	6,619	183,251	4,157,145
Wholesale trade	16,610	229,275	4,478,805
Retail trade	55,777	710,854	6,424,069
Finance, insurance, real estate	17,854	210,373	3,256,931
Services, of which:	62,482	831,826	10,556,332
Health services	(16,114)	(311,500)	(4,921,519)
Other	2,577	3,075	39,313
TOTALS	196,899	3,472,711	$57,595,321

This survey excluded farm laborers, self-employed workers, government workers, and certain other employees. Ohio had 576,000 full- and part-time state and local government employees in 1983, and 87,000 federal employees in 1982.

Of all Ohioans employed in 1984, about 72% worked in the state's major metropolitan areas. As of September, there were 841,600 employees in the Cleveland metropolitan area, 553,700 in the Columbus area, and 642,900 in the Cincinnati area. In August 1984, the unemployment rate in metropolitan areas ranged from 7.2% in Columbus and 7.6% in Dayton to more than 12% in Youngstown-Warren.

The first workers' organization in Ohio was formed by Dayton mechanics in 1811. The Ohio Federation of Labor was founded in 1884; the American Federation of Labor (AFL) was founded in Columbus in 1886, and Ohio native William Green became president of the AFL in 1924. But it was not until the 1930s that labor unions in Ohio were formed on a large scale. In 1934, the United Rubber Workers began to organize workers in Akron; through a successful series of sit-down strikes at the city's rubber plants, the union grew to about 70,000 members by 1937. In that year, the United Steelworkers struck seven steel plants in the Youngstown area and won the right to bargain collectively for 50,000 steelworkers. The number of union members increased from about 25% of the state's nonfarm employees in 1939 to 32% in 1980, when about 1.4 million workers belonged to labor organizations (5th in the US). In 1981 there were 226 strikes or other labor stoppages, involving 60,100 workers. There were 3,581 unions in Ohio in mid-1983, more than in any other state except Pennsylvania.

Progressive labor legislation in the state began in 1852 with laws regulating working hours for women and children and limiting men to a 10-hour workday. In 1890, Ohio became the first state to establish a public employment service. Subsequent labor legislation included a workers' compensation act in 1911, and child labor and minimum wage measures in the 1930s. In 1983, a law was passed giving public employees, other than police officers and fire fighters, a limited right to strike.

23AGRICULTURE

Despite increasing urbanization and industrialization, agriculture retains its economic importance. In 1983, Ohio ranked 11th in farm value (including land and buildings) and 13th in agricultural income among the 50 states. In that year, the state's production of crops, dairy products, and livestock was valued at nearly $3.7 billion.

Mechanization of agriculture contributed to the decline of Ohio's farm population from 1,089,000 in 1940 to 272,000 by 1980, and to a decrease in the number of farms from 234,000 in 1940 to 90,000 by 1984. The average size of farms increased from 94 acres (38 hectares) in 1940 to 176 acres (71 hectares) in 1984. As of July 1983, Ohio had about 124,000 unpaid and 45,000 paid farm workers.

Grain is grown and cattle and hogs are raised on large farms in the north-central and western parts of the state, while smaller farms predominate in the hilly southeastern region. Truck farming has developed and expanded near the large cities.

Ohio was the 2d-leading producer of tomatoes for processing in 1983, with 408,960 tons. Output of grapes, apples, and peaches totaled 130,000,000 lb. Field crops in 1983 (in bushels) included corn for grain, 224,000,000; soybeans, 101,680,000; wheat, 58,800,000; and oats, 15,360,000. The most valuable crops included soybeans for beans, with sales of $838,860,000, and corn, $761,600,000. Ohio farmers also produced 3,244,000 tons of hay and an estimated 222,000 tons of sugar beets in 1983.

24ANIMAL HUSBANDRY

In 1983, marketing of livestock and livestock products accounted for more than two-fifths of Ohio's farm income and was valued at more than $1.5 billion. Ohio ranked 7th among the states in production of milk, 8th in butter production, 8th in sales of hogs and pigs, and 9th in sales of eggs.

Cattle and hogs are raised in the central and western regions. At the end of 1983, Ohio had 2,100,000 hogs and pigs, 1,900,000 head of cattle, and 260,000 stock sheep and sheep on feed. A total of 859,398,000 lb of hogs and pigs, 679,465,000 lb of cattle, and 16,520,000 lb of sheep and lambs were slaughtered in 1983.

Dairying is common in most regions of the state, but especially in the east and southeast. In 1983, 4.6 billion lb of milk was produced. The poultry industry is dispersed throughout the state. Ohio poultry farmers raised 12,050,000 chickens and 2,400,000 turkeys in 1983. Egg production totaled nearly 3 billion.

25FISHING

Commercial fishing, which once flourished in Lake Erie, has declined during the 20th century. Only 3,980,000 lb of fish, worth $917,000, were landed in 1984.

26FORESTRY

In 1980, Ohio had 7,120,100 acres (2,881,400 hectares) of forestland, representing 27% of the state's total land area but less than 1% of all US forests. Although scattered throughout the state, hardwood forests are concentrated in the hilly region of the southeast. Commercial timberlands in 1980 totaled 6,917,100 acres (2,799,300 hectares), of which 94% was privately owned.

The state's lumber and wood products industry supplies building materials, household furniture, and paper products. In 1982, shipments of paper and allied products were valued at $3.3 billion, lumber and wood products at $776 million, and furniture and fixtures at $841 million.

In 1980 there were about 237,500 acres (96,100 hectares) of state, county, and municipal timberland. State parks and recreation areas totaled about 112,000 acres (45,000 hectares).

Reforestation programs involving the reseeding of more than 10,000 acres (4,000 hectares) per year ensure that timber growth exceeds the state's timber harvest. In 1983, 248 million board feet of hardwoods and 21 million board feet of softwoods were produced.

27MINING

In 1984, Ohio's nonfuel mineral production was valued at $554.9 million, a 16% gain over 1983. The most valuable mineral, aside from fossil fuels, was stone. Mining of limestone, clay, and sand and gravel predominates in the southeast. Ohio is the leading US producer of dimension sandstone.

Estimated output of major minerals in 1984 (in tons, excluding fossil fuels) were stone, 38,537,000; sand and gravel, 32,380,000; salt, 3,626,000; lime, 1,925,000; clays, 1,863,000; and cement, 1,600,000.

28ENERGY AND POWER

Ohio is blessed with energy resources. The state government estimates that Ohio's coal reserves are sufficient to meet demand for 500 years and that oil and natural gas reserves are also ample. On the other hand, nearly 98% of the state's electricity is produced by coal with an ash content ranging from 5% to 20%—a major atmospheric pollutant.

In 1983, Ohio ranked 4th among all states in electric power production. In that year, installed electric power capacity was 27.8 million kw, of which 97% was privately owned, and electrical output totaled 105.9 billion kwh.

With an energy consumption of 329 million Btu per capita, Ohio ranked 18th among the 50 states in 1982. Total energy consumption amounted to 3,549 trillion Btu, of which industries consumed 42%, residential users 23%, commercial establishments 14%, and transportation 21%. Coal was the source of about 39% of all energy consumed, petroleum 34%, dry natural gas 24%, nuclear power 1%, and other sources about 2%.

In the 1880s, petroleum was discovered near Lima and natural gas near Toledo, both in the northwest; these fossil fuels have since been found and exploited in the central and eastern regions. In 1983, the state produced 14,971,000 barrels of crude petroleum; proved reserves were estimated at approximately 130,000,000 barrels.

About 151.3 billion cu feet of natural gas were extracted in 1983, with reserves estimated at 2.03 trillion cu feet. At the end of 1983 there were 25,762 producing gas wells. In the same year, Ohio's gas utilities served 2,767,000 customers, of whom 93% were residential users.

Coalfields lie beneath southeastern Ohio, particularly in Hocking, Athens, and Perry counties. In 1983, Ohio ranked 9th in the US in coal production, with a total output of 33,582,000 tons. During that year, 173 surface mines produced two-thirds of the coal; 12 underground mines supplied the remainder.

A potential energy source is a rich bed of shale rock, underlying more than half of Ohio, which was estimated to contain more than 200 trillion cu feet of natural gas; but much research was needed before the gas could be extracted economically.

In January 1985, three Ohio utilities halted construction of the Zimmerman nuclear power plant, after spending $1.7 billion on the project; plans were announced to convert the plant to coal use, at an estimated cost of another $1.7 billion. In June 1985, a nuclear reactor at Oak Park was closed down after an accident involving the failure of 14 pieces of equipment; no radiation release or major damage was reported.

29INDUSTRY

Ohio has been a leading manufacturing state since the mid-1800s, and in 1981, despite a recessionary slump, it ranked 4th among the states in value of manufacturing shipments.

During the last two decades of the 19th century, Ohio became the nation's leader in machine-tool manufacturing, the 2d-leading steel producer, and a pioneer in oil refining and in the production of automobiles and automotive parts, such as rubber tires.

In recent decades, Ohio has become important as a manufacturer of glassware, soap, matches, paint, business machines, refrigerators—and even comic books and Chinese food products.

In 1982, the value of manufacturing shipments totaled $112.6 billion. Of this total value, transportation equipment accounted for 16%; nonelectrical machinery 11%; fabricated metal products 11%; primary metals 11%; food products 10%; chemicals 8%; electrical and electronic equipment 7%; petroleum and coal products 7%; rubber and plastics 5%; printing and publishing 3%; stone, clay, and glass products 3%; paper 3%; and other sectors 5%.

The following table shows value of shipments for selected industries in 1982:

Motor vehicles and equipment	$13,449,700,000
Blast furnace and steel mill products	7,636,000,000
Petroleum refining	6,620,800,000
Metal forgings and stampings	3,734,300,000
Miscellaneous plastics products	3,147,800,000
Aircraft and parts	3,143,100,000
Metalworking machinery	2,805,600,000
General industrial machinery	2,645,300,000
Soap cleaners and toilet goods	2,407,100,000
Fabricated structural metal products	2,384,500,000

Of the leading US industrial corporations listed by *Fortune* magazine for 1984, seven had their headquarters in Ohio. The sales leader was Procter and Gamble (Cincinnati), with sales totaling $12.9 billion, 22d in the US. Others were Standard Oil of Ohio (Cleveland), $11.7 billion; Goodyear Tire and Rubber (Akron), $10.2 billion; TRW (Cleveland), $6.1 billion; Armco (Middletown), $4.5 billion; Firestone Tire and Rubber (Akron), $4.2 billion; and NCR (Dayton), $4.1 billion.

30COMMERCE

Ohio is a major commercial state. In 1982, wholesale trade amounted to $79.6 billion, 6th highest among the 50 states. In that year, the chief categories of wholesale goods traded (in order of largest sales) were groceries and related products, 15%; machinery, equipment, and supplies, 14%; motor vehicles and automotive parts, 12%; metals and minerals (except petroleum), 10%; petroleum and oil products, 7%; chemicals, 7%; electrical goods, 6%; grains and other agricultural raw materials, 5%; hardware, plumbing, and heating equipment, 3%; lumber and construction materials, 2%; and other items, 19%.

Ohio's retail sales amounted to $46.3 billion in 1982. The principal retail store groups and their sales percentages, among establishments with payrolls, were food, 24%; automotive, 17%; and general merchandise, 13%. From 1977 to 1982, total retail sales increased by 32%.

In 1981, Ohio ranked 4th in the US as an exporter of manufactured goods, with foreign sales of $10.4 billion, or 6% of the US total. In that year, transportation equipment, nonelectric machinery, and chemicals accounted for nearly two-thirds of the state's export value, followed by electric and electronic equipment; primary metals; fabricated metal products; stone, clay, and glass products; and rubber and plastic products. In 1981/82, Ohio ranked 13th among the states in agricultural exports, with an estimated value of $1.1 billion. Soybeans and feed grains accounted for over two-thirds of that total; the third most important agricultural export category was wheat and flour.

31CONSUMER PROTECTION

Agencies involved in consumer protection include the Agriculture Department's Division of Foods, Dairies and Drugs, which operates inspection programs to protect consumers, and the Commerce Department's Division of Banking and its Division of Savings and Loans, which regulate commercial banks and savings and loan associations, respectively. The Office of the Consumer Counsel acts to protect residential customers of public utilities. The Attorney General's Office resolves consumer complaints and enforces consumer protection laws.

32BANKING

Ohio's first banks, in Marietta and Chillicothe, were incorporated in 1808, and a state bank was authorized in 1845. There were 320 insured commercial banks in Ohio in 1984. As of 1983, the state's insured commercial banks had total assets of $61.8 billion, including $31.5 billion in outstanding loans. Their deposits totaled $50.4 billion, including time and savings deposits of $40.4 billion. In 1983 there were 405 savings associations in the state, with total assets of more than $44 billion. In March 1985, the Home State Savings Bank of Cincinnati folded after a severe depletion of funds caused by news of its involvement with a failed securities firm.

As a result of the collapse, Governor Richard Celeste shut down 69 privately insured savings and loan associations to prevent a run on them by depositors. The associations were not allowed to reopen until they obtained federal insurance; some, however, reopened as banks, and others merged with stronger financial institutions. The state senate began an investigation of the bank's collapse.

33INSURANCE

In 1983, 41 life insurance companies had headquarters in Ohio. That year, Ohio and out-of-state life insurance companies wrote premiums worth $31.7 billion and paid benefits of more than $2.8 billion, including $780 million in death benefits; premium receipts for life insurance policies were $2.1 billion. The average Ohio family had $56,900 in life insurance in 1983. There were 18,474,-000 life insurance policies in force in the state in 1983, worth $232 billion.

In 1983, 152 property and casualty insurance companies were domiciled in Ohio. That year, Ohio and out-of-state companies wrote direct premiums of nearly $4.2 billion, which included $1.8 billion in automobile liability and physical damage insurance.

34SECURITIES

The Cincinnati Stock Exchange (CSE) was organized on 11 March 1885 by 12 stockbrokers who agreed to meet regularly to buy and sell securities. The exchange was incorporated as a nonprofit organization two years later. As of 1985, three Cincinnati companies that had been listed on the CSE in 1885—including Dayton and Michigan Railroad—still had their stock traded on the exchange.

The CSE is governed by an elected 8-member board of trustees, the majority of whom must approve new members of the exchange. As of 1985, the exchange had 37 members. The cost of a seat on the CSE, originally $50, was $5,000 in 1985. As of 1985, the CSE was trading approximately 2,000 different stocks. About 69.2 million shares were traded during 1984.

As of 31 December 1983, firms belonging to the New York Stock Exchange had 249 sales offices and 2,266 full-time registered representatives in Ohio. In 1984, 1,884,000 Ohioans were shareowners of public corporations.

35PUBLIC FINANCE

The state budget is prepared on a biennial basis by the Office of Budget and Management. It is submitted by the governor to the state legislature, which must act on it by the close of the current fiscal year (30 June).

The general assembly has nearly total discretion in allocating general revenues, which are used primarily to support education, welfare, mental health facilities, law enforcement, property tax relief, and government operations. The assembly also allocates money from special revenue funds by means of specific legislative acts. More than one-half of all state expenditures come from the general fund.

The general fund for the 1983–85 biennium was estimated at $17.2 billion, compared with the 1981–83 total of $13.1 billion. The following table summarizes state general revenues and expenditures for the fiscal year 1984/85 (in millions):

REVENUES

Sales and use taxes	$2,539.0
Personal income tax	2,840.0
Corporation franchise tax	735.0
Public utility excise tax	674.0
Cigarette tax	168.0
Alcoholic beverages taxes	70.5
Miscellaneous taxes	208.2
Federal aid	1,265.0
Other income	15.3
TOTAL	$8,515.0

EXPENDITURES

Primary and secondary education	$2,467.3
Human services	2,191.7
Higher education	1,129.0
Tax relief	473.9
Justice and corrections	393.4
Local government	294.1
General state government	210.8
Development, transportation, and environment	183.7
Other state expenses	240.3
Federal welfare match	1,361.2
TOTAL	$8,945.4

In 1981/82, the city of Cleveland collected general revenues totaling $406 million, of which the state and local governments provided $38.6 million and the federal government $107.4 million. Tax revenues totaled $187.8 million, of which property taxes supplied 26%. In the same year, Cleveland's general expenditures totaled $381 million. Columbus had revenues of $299 million and expenditures of $319 million, while Cincinnati tallied $333 million in revenues and $321 million in expenditures.

The total debt owed by state and local governments in Ohio increased from $8.9 billion in 1977 to $12.1 billion in 1982. In the latter year, the per capita state and local government debt in Ohio amounted to $1,125, 44th among the 50 states.

36TAXATION

Ohio ranked 7th among the 50 states in total state tax revenues collected during the 1982/83 fiscal year, but was 35th in per capita state and local taxation as of 1981/82, with an average tax burden of $973; the tax burden was the lowest of any northern industrial state except Indiana, and was nearly 20% below the national average.

In 1984, the state personal income tax ranged from 0.95% on the first $5,000 of taxable income to 9.5% on amounts over $100,000. The state sales and use tax was 5% on retail sales (excepting groceries and prescription drugs), rental of personal property, and selected services. The corporate franchise tax rate was either 5.1% on the first $25,000 of net income plus 9.2% on amounts above that figure, or 5.82 mills on the net worth of the corporation attributable to Ohio, whichever was greater. The public utility excise tax, on intrastate business receipts of public utilities, ranged from 4.5% to 5% for most utilities. Other taxes include those on cigarettes, alcoholic beverages, and gasoline. The state also imposes taxes on estates, banks and insurance companies, dealers in intangibles (stockbrokers), coal and other minerals, and pari-mutuels. In the fiscal year 1982/83, Ohio's state tax receipts from its own sources totaled $6.7 billion. Local tax receipts totaled $5.1 billion.

Most local tax revenues come from property taxes, which exceeded $3.7 billion in 1982/83. In the same year, personal and corporate income taxes levied by local governments generated $1 billion; local general sales taxes generated $212.9 million.

Ohio residents filed 4,328,283 federal income tax returns in 1982 and paid more than $11.6 billion in federal income tax.

37ECONOMIC POLICY

Although Ohio seeks to attract new industries, about four-fifths of the state's annual economic growth stems from the expansion of existing businesses.

Ohio offers numerous business incentives to spur industrial development. The state's impacted-cities program encourages capital investment by offering private developers property tax abatements for commercial redevelopment. A 1976 state law permits municipal corporations to exempt certain property improvements from real property taxes for periods of up to 30 years. The state's guaranteed-loan program for industrial developers provides repayment guarantees on 90% of loans up to $1 million. The state also offers revenue bonds to finance a developer's land, buildings, and equipment at interest rates below the going mortgage interest rates.

The Ohio Department of Development administers plans for economic growth in cooperation with city and county governments. It informs companies about opportunities and advantages in the state and promotes the sale of Ohio's exports abroad. In the 1980s, the department instituted research and development programs at state universities in such fields as robotics, polymers, genetic engineering, and artificial intelligence.

[38] HEALTH

Ohio's birthrate fell from 23.8 live births per 1,000 population in 1960 to 15.1 in 1977, rising to 15.3 in 1982. In the latter year there were about 165,000 live births (140,000 for whites and 25,000 for other races). The general decrease in the birthrate was due in part to the increasing availability of legal abortions in Ohio. Between 1974/75 and 1982/83, the ratio of abortions to births more than doubled, from 188 to 378 legal abortions for every 1,000 live births; about 61,400 legal abortions were performed during the latter year. The infant death rate has decreased significantly since 1960, when there were 22.2 infant deaths per 1,000 live births among whites, 39.4 among blacks; by 1981, the rates had fallen to 10.8 for whites and 22.1 for blacks.

Ohio ranks above the national average in deaths due to heart disease, cancer, and diabetes, but below the US average for deaths caused by accidents, pneumonia, and cirrhosis of the liver. The leading causes of death in 1981 (with rates per 100,000 population) included heart disease, 355; malignant neoplasms, 194; cerebrovascular diseases, 72; accidents, 35; chronic obstructive pulmonary diseases, 27; pneumonia, 21; and diabetes, 19.

Ohio had 236 hospitals with 62,405 beds in 1983, when 1,876,661 patients were admitted. The average daily census in 1982 was 50,100, for an occupancy rate of about 80%. As of 1981, hospital personnel included 27,983 full-time registered nurses. In that year there were 781 nursing homes and personal care facilities in the state.

At the end of 1982, the state had 20,197 licensed physicians and 5,394 practicing dentists. Medicare and Medicaid benefits paid out in 1981 totaled $2.8 billion; private health insurance benefits paid to hospitals in 1982 exceeded $1.3 billion. The average cost per day to hospital patients was $325 as of 1982, and the average cost per stay was more than $2,600; both figures were close to the national average. The semiprivate room rate in January 1984 averaged about $222.

[39] SOCIAL WELFARE

In 1983, Ohio ranked 9th among the states in percentage of the population receiving public aid. The growth of welfare programs in the state was remarkably rapid during the 1970s and early 1980s. From 1970 to 1978, for example, aid to families with dependent children (AFDC) nearly tripled, to $446 million; by 1982, the figure had reached $606 million. There were 626,200 AFDC recipients in January 1983; the average payment per family was $263.38.

In May 1983, Supplemental Security Income payments were made to 113,606 needy persons and totaled about $20,958,000, of which 82% went to the disabled, 16% to the aged, and 2% to the blind. That year, a total of $253,931,000 in federal Supplemental Security Income payments was received in Ohio, for an average monthly payment of $203.13 per recipient.

In 1983, federal expenditures for the food stamp program in Ohio totaled $667 million; food stamps were purchased by an average of 1,183,000 persons each month. In 1983, the school lunch program benefited 939,000 schoolchildren, at an estimated cost to the federal government of about $92 million.

Social Security old-age, survivors, and disability insurance benefits amounting to almost $8 billion were paid to Ohioans in 1983. Of that total, 59% went to retired workers, 7% went to the spouses and children of retired workers, and 24% went to the survivors of deceased workers; an additional 9% went to disabled workers and 1% to their spouses and children. In 1983, the Medicaid program benefited an average of 911,000 persons at a cost for the year of $1.47 billion.

An average of 616,000 persons claimed unemployment insurance benefits in 1982; total benefits amounted to $1,495,839,000, 5th highest among US states, and the average weekly payment of $143.59 also ranked 5th nationally.

In December 1983, payments totaling $5,230,000 under the Black Lung Benefit Program were paid to about 5,206 disabled miners, 9,435 miners' widows, and 5,104 dependents.

[40] HOUSING

In 1980, Ohio had 3,833,828 occupied housing units, of which 68% were occupied by their owners and 32% were rented. About 86% of the housing units were built in 1940 or later; 20% were built in the 1970s. About 73% had three or more bedrooms. The median monthly mortgage and selected carrying costs for owner-occupied housing units with mortgages was $348. The median monthly rent was $225. In 1983, the state authorized construction of 26,600 new privately owned housing units, valued at $1.3 billion.

[41] EDUCATION

Ohio claims a number of "firsts" in US education: the first kindergarten, established by German settlers in Columbus in 1838; the first junior high school, also at Columbus, in 1909; the first municipal university, the University of Cincinnati, founded in 1870; and the first college to grant degrees to women, Oberlin, in 1837. The state's earliest school system was organized in Akron in 1847.

Of the total adult population in 1980, 67% were high school graduates, and 14% had completed at least four years of postsecondary study. In 1984/85, the state's 3,759 public elementary and secondary schools enrolled an estimated 1,805,732 pupils, 36,946 of whom were in vocational educational courses; in the same year there were 98,062 public school teachers. In 1983/84, Ohio's public school teachers received an average annual salary of $21,290, 23d highest among the 50 states. Roman Catholic parochial schools had 162,509 elementary school pupils and 55,856 high school pupils in 1984.

In 1984/85, public schools enrolled 250,543 blacks, comprising almost 14% of all public school pupils. De facto school segregation in Cleveland and other cities has been reduced in recent years, largely under court order. In March 1980, Cleveland began to desegregate its junior high schools by busing about 16,000 of 90,000 junior high school pupils, two-thirds of whom were blacks. In 1984, Cincinnati announced plans to desegregate its 78 public schools by means of a voluntary busing system. Fifty of the city's schools were more than 50% black, and 13 schools were at least 90% black.

In 1983/84, the state had 81 private and 59 public institutions of higher education; the enrollment, as of 1982/83, was 532,361. There are 12 state universities, including Ohio State University (Columbus), Ohio University (Athens), Miami University (Oxford), and other state universities at Akron, Bowling Green, Cincinnati, Cleveland, Dayton, Kent, Toledo, Wilberforce, and Youngstown. The largest, Ohio State, was chartered in 1870 and also has campuses at Lima, Mansfield, Marion, Newark, and Wooster; the Columbus campus alone had a total of 53,757 students in 1983/84. Ohio has 45 public two-year colleges. Well-known private colleges and universities include Antioch (Yellow Springs), Case Western Reserve (Cleveland), Kenyon (Gambier), Muskingum (New Concord), Oberlin, and Wooster. Eleven Roman Catholic colleges enrolled 29,526 students in 1984.

Ohio residents enrolled as full-time students at an eligible institution within the state may apply for instructional grants from the Student Assistance Office of the Ohio Board of Regents. Guaranteed loans are provided through the Ohio Student Loan Commission.

⁴²ARTS

The earliest center of artistic activities in Ohio was Cincinnati, where a group of young painters did landscapes and portraits as early as 1840. The state's first art gallery was established there in 1854; the Cincinnati Art Academy was founded in 1869, and the Art Museum in 1886. Famous American artists who worked in Cincinnati during part of their careers include Thomas Cole, a founder of the "Hudson River School" of landscape painting, who was raised in Steubenville; Frank Duveneck, dean of the Cincinnati Academy; and Columbus-born George Bellows, whose realistic "Stag at Sharkey's" is displayed at the Cleveland Museum of Art (founded in 1913). Other notable centers for the visual arts include the Akron Art Institute, Columbus Museum of Art, Dayton Art Institute, Toledo Museum of Art, and museums or galleries in Marion, Oberlin, Springfield, Youngstown, and Zanesville.

Cincinnati also was an early center for the theater; the Eagle Theater opened there in 1839, and shortly afterward, the first showboat on the Ohio River began making regular stops at the city. The first US minstrel show appeared in Ohio in 1842; Al Field's famous minstrels, formed in Columbus in 1886, successfully toured the US for 41 years. As of 1984, Ohio had three professional theatrical companies: the Cincinnati Playhouse, the Cleveland Playhouse, and the Great Lakes Shakespeare Festival. The Ohio Community Theater Association included groups in Akron, Canton, Columbus, Mansfield, Toledo, and Youngstown.

Musical activities revolve around the Cincinnati Symphony, which was founded in 1895 and reorganized in 1909 with Leopold Stokowski as conductor, and the Cleveland Symphony, founded in 1918; especially since 1946, when George Szell began his 24-year tenure as conductor and music director, the Cleveland Orchestra has been considered one of the finest in the world. The nation's first college music department was organized at Oberlin College in 1865; the Cincinnati Conservatory of Music was established in 1867, and the Cleveland Institute of Music in 1920. There are civic symphony orchestras in Columbus, Dayton, Toledo, and Youngstown. Operas are performed by resident companies in Cleveland, Columbus, Cincinnati, Toledo, and Dayton. The Cincinnati Pops Orchestra acquired a new summer home in 1984 at the newly opened Riverbend Music Center.

⁴³LIBRARIES AND MUSEUMS

Ever since early settlers traded coonskins for books and established, in 1804, the Coonskin Library (now on display at the Ohio Historical Center in Columbus), Ohioans have stressed the importance of the public library system. In 1982, the state public library system had 30,025,812 volumes, a circulation of 70,806,-173, and a total income of $136,108,262.

Major public library systems include those of Cincinnati, with 3,332,926 volumes in 1983; Cleveland, 2,484,947; Dayton, 1,368,-923; and Columbus, 1,209,858. Columbus also has the library of the Ohio Historical Society, with 116,000 volumes. Leading academic libraries include those of Ohio State University, with 3,700,000 books and bound periodicals; Case Western Reserve University, 1,410,669 books; and the University of Cincinnati, 1,100,008 books and bound periodicals.

Among the state's more than 240 museums are the Museum of Art, Natural History Museum, and Western Reserve Historical Society Museum in Cleveland; the Museum of Natural History, Art Museum, and Taft Museum in Cincinnati; the Dayton Art Institute; and the Center of Science and Industry and Ohio Historical Center in Columbus. The Zanesville Art Center has collections of ceramics and glass made in the Zanesville area. Also noteworthy are the US Air Force Museum near Dayton, the Neil Armstrong Air and Space Museum at Wapakoneta, and the Ohio River Museum in Marietta. Cincinnati has a conservatory of rare plants, while Cleveland has botanical gardens and an aquarium; both cities have zoos.

Historical sites in Ohio include the Schoenbrunn Village State Memorial, a reconstruction of the state's first settlement by Moravian missionaries, near New Philadelphia; the early 19th-century Piqua Historical Area, with exhibits of Indian culture; and the Fort Meigs reconstruction at Perrysburg. Archaeological sites include the "great circle" mounds, built by the Hopewell Indians at present-day Newark, and Inscription Rock, marked by prehistoric Indians, on Kelleys Island.

⁴⁴COMMUNICATIONS

In 1980, 94% of Ohio's 3,833,828 occupied housing units had telephones.

Many of the state's radio stations were established in the early 1920s, when the growth of radio broadcasting was fostered by the availability of low-priced sets manufactured by Crosley Radio of Cincinnati. In 1984 there were 128 AM stations, 197 FM stations, and 33 commercial and 12 noncommercial television stations. There were also 239 cable television systems, serving 1,588,956 Ohio subscribers in 835 communities.

⁴⁵PRESS

The first newspaper published in the region north and west of the Ohio River was the *Centinel of the North-Western Territory*, which was written, typeset, and printed in Cincinnati by William Maxwell in 1793. The oldest newspaper in the state still published under its original name is the *Scioto Gazette*, which appeared in 1800. The oldest extant weekly, the *Lebanon Western Star*, began publication in 1807, and the first daily, the *Cincinnati Commercial Register*, appeared in 1826. By 1840 there were 145 newspapers in Ohio.

Two of the state's most influential newspapers, the Cleveland *Plain Dealer* and the Cincinnati *Enquirer*, were founded in 1841. In 1878, Edward W. Scripps established the Cleveland *Penny Press* (now the *Press*), the first newspaper in what would become the extensive Scripps-Howard chain; he later added to his newspaper empire the Cincinnati *Post* (1881) and the Columbus *Citizen* (1899; now the *Citizen-Journal*), as well as papers in Akron, Toledo, and Youngstown.

In 1984 there were 10 morning daily newspapers with net paid circulation of 1,000,087 copies; 83 daily evening papers with total circulation of 2,009,935; and 30 Sunday papers with 2,709,411 circulation. The following table lists leading Ohio newspapers with their daily circulation in 1984:

AREA	NAME	DAILY	SUNDAY
Akron	Beacon Journal (e,S)	163,597	231,846
Cincinnati	Enquirer (m,S)	189,840	300,176
	Post (e)	129,697	
Cleveland	Plain Dealer (m,S)	482,564	497,774
Columbus	Citizen-Journal (m)	119,850	
	Dispatch (e,S)	206,207	360,590
Dayton	Daily News (e,S)	114,100	231,152
	Journal-Herald (m)	101,641	
Toledo	Blade (e,S)	163,881	216,817
Youngstown	Vindicator (e,S)	101,716	154,290

⁴⁶ORGANIZATIONS

The 1982 Census of Service Industries counted 2,950 organizations in Ohio, including 401 business associations; 1,889 civic, social, and fraternal associations; and 59 educational, scientific, and research associations.

Service organizations with headquarters in Ohio include the Army and Navy Union, USA, at Lakemore, and the National Exchange Club, in Toledo. Among the state's cultural associations are the American Classical League, at Oxford, and the Guild of Carilloneurs in North America and the Music Teachers National Association, both in Cincinnati.

Commercial and professional organizations include the American Ceramic Society and Order of United Commercial Travelers of America, both in Columbus; American Society for Metals, in

Metals Park; and Association for Systems Management and Brotherhood of Locomotive Engineers, both in Cleveland.

Sports associations operating out of Ohio are the Lighter-Than-Air Society and Professional Bowlers Association, Akron; American Motorcyclist Association, Westerville; Amateur Trapshooting Association, Vandalia; and Indoor Sports Club, Napoleon. The International Brotherhood of Magicians has its headquarters in Kenton.

[47] TOURISM, TRAVEL, AND RECREATION

In 1982, travelers spent $5.4 billion in Ohio. Leading tourist attractions include Ohio's presidential memorials and homes: the William Henry Harrison Memorial at North Bend, Ulysses S. Grant's birthplace at Point Pleasant, the James A. Garfield home at Mentor, the Rutherford B. Hayes home at Fremont, the William McKinley Memorial at Canton, the Taft National Historic Site in Cincinnati, and the Warren G. Harding home in Marion. Also of interest are the Thomas A. Edison birthplace at Milan, and Malabar Farm, in Richland County, home of author and conservationist Louis Bromfield.

Beaches and parks in the Lake Erie region are especially popular with tourists during the summer. Among the many attractions in the northern region are the Crosby Gardens in Toledo, reconstructions of Auglaize Village near Defiance and of Harbour Town in Vermilion, the Marblehead Lighthouse, and the Lake Erie Nature and Science Center in Cleveland. In 1985, the Portside Festival Marketplace, with 97 shops and many restaurants, opened in downtown Toledo along the banks of the Maumee River. The popular Cuyahoga Valley Natural Recreation Area, linking the urban centers of Cleveland and Akron, had a total of 1,018,828 visitors in 1984.

The eastern Allegheny region has several ski resorts for winter sports enthusiasts. Popular tourist attractions include the Amish settlement around Millersburg, the National Road–Zane Grey Museum near Zanesville, and the restored Roscoe Village on the Ohio-Erie Canal. The southern region offers scenic hill country, the Kings Island entertainment complex at Kings Mills, and the showboat *Majestic,* the last of the original floating theaters, in Cincinnati.

In the western region, tourist sites include the Wright brothers' early flying machines in Dayton's Carillon Park, the Ohio Caverns at West Liberty, and the Zane Caverns near Bellefontaine. The central region is "Johnny Appleseed" country; the folk hero (a frontiersman whose real name was John Chapman) is commemorated in Mansfield by the blockhouse to which he directed settlers in order to save them from an Indian raid. At Columbus are reconstructed Ohio and German villages, as well as the Exposition Center, site of the annual Ohio State Fair, held for 13 days in mid-August.

Ohio has 71 state parks, comprising 111,797 acres (45,243 hectares). Among the most visited state parks are Alum Creek, East Harbor and Kelleys Island (both on Lake Erie), Grand Lake St. Marys, Hocking Hills, Hueston Woods, Mohican, Pymatuning (on the Pennsylvania border), Rocky Fork, Salt Fork, Scioto Trail, and West Branch.

The most popular sport fish are bass, catfish, bullhead, carp, perch, and rainbow trout. The deer-shooting season is held in late November; hunters are limited to one deer per season. Licenses were sold to 491,935 hunters and 1,049,752 anglers in 1982/83.

[48] SPORTS

Ohio is well represented in professional sports, with two major league teams in baseball and football, one in basketball, and major golf and bowling tournaments.

In baseball, as of 1985, the Cincinnati Reds had played 110 consecutive years in the National League, and the Cleveland Indians had competed for 85 years in the American League. During the 1970s, Cincinnati's Big Red Machine won four league titles and two World Series (1975-76). Cleveland last won a pennant in 1954 and a World Series in 1948. The Columbus Clippers and Toledo Mudhens are the state's minor-league entries.

The Cleveland Browns and Cincinnati Bengals belong to the National Football League (NFL). The Browns share Cleveland's Municipal Stadium with the baseball Indians; the Bengals and baseball Reds play in Cincinnati's Riverfront Stadium. The Browns—who entered the NFL in 1950 after winning four consecutive championships in the All-America Conference—have won three NFL titles, the last in 1964. The Bengals, who joined the NFL when the American Football League merged with it in 1970, won the American Football Conference Championship in 1981 but were defeated by the San Francisco 49ers in the January 1982 Super Bowl. Paul Brown was the founder and original coach of both the Browns and the Bengals. The Pro Football Hall of Fame is located in Canton, where the pro sport originated in 1920.

The Cleveland Cavaliers have represented the state in the National Basketball Association since 1970; the Cincinnati Royals formerly played in the league from 1945 to 1972, when the club moved to Kansas City.

Akron has been headquarters for the Professional Bowlers Association (PBA) since its founding in 1958. The PBA's richest tournament is played there each year, and the PBA Hall of Fame is also located in Akron. The World Series of Golf is played annually in Akron, and the Memorial Golf Tournament in Columbus.

Major horse-racing tracks include Cleveland's Thistledown, Cincinnati's River Downs, Columbus's Scioto Downs, and other tracks at Toledo, Lebanon, Grove City, and Northfield. The Cleveland Gold Cup race is held annually at Thistledown, as is the Ohio Derby. The Little Brown Jug classic for three-year-old pacers takes place every year at the Delaware Fairgrounds, and the Ohio State race for two-year-old trotters is held during the state fair at Columbus.

In collegiate sports, Ohio State University has long been a football power, winning 24 Big Ten titles (through 1984). Under coach Woody Hayes from 1951 to 1978, the Buckeyes won 205 games (including 4 Rose Bowls), lost 61, and tied 10. Ohio State also has won NCAA championships in baseball, basketball, fencing, golf, gymnastics, and swimming, while Cincinnati and Dayton universities have had highly successful basketball teams. The College Football Hall of Fame is located at Kings Island Amusement Park, near Cincinnati.

The Amateur Trapshooting Association usually holds its grand tournament in August at Vandalia. The All-American Soap Box Derby, attracting participants between the ages of 9 and 15 and crowds of about 75,000, was held at Dayton annually from 1934-72; since then, it has been held in Akron.

[49] FAMOUS OHIOANS

Ohio has been the native state of seven US presidents and the residence of another. Inventions by Ohioans include the incandescent light, the arc light, and the flying machine.

William Henry Harrison (b.Virginia, 1773-1841), the 9th US president, came to Ohio as an Army ensign in territorial times. After serving in the Indian wars under Gen. Anthony Wayne, he became secretary of the Northwest Territory. As the territorial delegate to Congress, he fostered the Harrison Land Act, which stimulated settlement of the public domain. Named territorial governor in 1800, Harrison conducted both warfare and peace negotiations with the Indians. After the defeat of British and Indian forces in 1813, he became known as the "Washington of the West." After settling at North Bend on the Ohio River, he began a political career that carried him to the White House in 1841. Harrison caught a chill from a raw March wind and died of pneumonia, exactly one month after his inauguration.

From 1869 to 1881, the White House was occupied by three

Ohio men. All were Republicans who had served with distinction as Union Army generals. The first, Ulysses Simpson Grant (1822–85), the 18th US president, was an Ohio farm boy educated at West Point. After service in the Mexican War, he left the Army, having been charged with intemperance. He emerged from obscurity in 1861, when he was assigned to an Illinois regiment. Grant rose quickly in command; after victories at Shiloh and Vicksburg, he was commissioned major general. In 1864, he directed the Virginia campaign that ended with Confederate surrender, and this rumpled, slouching, laconic man became the nation's hero. In 1868, he was elected president, and he was reelected in 1872. His second term was rocked with financial scandals, though none were directly connected to Grant. After leaving the presidency in 1877, he went bankrupt, and to discharge his debts, he wrote his memoirs. That extraordinary book was completed four days before his death from throat cancer in 1885. Grant is buried in a monumental tomb in New York City.

Rutherford B. Hayes (1822–93), the 19th US president, was born in Delaware, Ohio, and educated at Kenyon College and Harvard Law School. Following Army service, he was elected to Congress, and in 1876 became the Republican presidential nominee. In a close and disputed election, he defeated New York's Governor Samuel J. Tilden. Hayes chose not to run for reelection, returning instead to Ohio to work on behalf of humanitarian causes. In 1893, Hayes died in Fremont, where the Hayes Memorial was created—the first presidential museum and library in the nation.

James A. Garfield (1831–81), 20th US president, was born in a log cabin in northern Ohio. Between school terms, he worked as a farmhand and a mule driver on the Ohio Canal. After holding several Civil War commands, he served in Congress for 18 years. Elected president in 1880, he held office but a few months; he was shot by a disappointed office seeker in the Washington, D.C., railroad station on 2 July and died 11 weeks later.

Benjamin Harrison (1833–1901), 23d US president and grandson of William Henry Harrison, was born in North Bend. After graduation from Miami University, he studied law and began to practice in Indianapolis. Military command in the Civil War was followed by service in the US Senate and the Republic presidential nomination in 1888. As president, Harrison gave impetus to westward expansion, moved toward annexation of Hawaii, and enlarged the civil-service system.

US presidents in the 20th century include three more native Ohioans. William McKinley (1843–1901) was born in Niles. Elected in 1896 as the 25th president, he established the gold standard and maintained tariff protection for US manufactures. Early in his second term, while greeting a throng of people, he was shot to death by a young anarchist. William Howard Taft (1857–1930), of Cincinnati, was the 27th US president. He gained a national reputation in 1904 as President Theodore Roosevelt's secretary of war; five years later, he succeeded Roosevelt in the White House. Defeated in 1912, Taft then left Washington for a law professorship at Yale. In 1921, under President Warren G. Harding (1865–1923), he became US chief justice, serving in that office until a month before his death. Harding, the last Ohioan to win the White House, was born in Blooming Grove. He went into politics from journalism, after serving as editor of the *Marion Star*. After eight years in the US Senate, he was a dark-horse candidate for the Republican presidential nomination in 1920. He won the election from James M. Cox (1870–1957), another Ohio journalist-politician, and became the 29th US president. Harding, who died in office, was surrounded by graft and corruption in his own cabinet.

Three US vice presidents were natives of Ohio. Thomas A. Hendricks (1819–85) was elected on the Democratic ticket with Grover Cleveland in 1884. Charles W. Fairbanks (1852–1918) served from 1905 to 1909 under Theodore Roosevelt. Charles

Gates Dawes (1865–1951) became vice president under Calvin Coolidge in 1925, the same year the Dawes Plan for reorganizing German finances brought him the Nobel Peace Prize; from 1929 to 1932, he served as US ambassador to Great Britain.

Three Ohioans served as chief justice on the Supreme Court: Salmon P. Chase (b.New Hampshire, 1808–73), Morrison R. Waite (b.Connecticut, 1816–88), and Taft. Most notable among nearly 40 cabinet officers from Ohio were Secretary of State Lewis Cass (b.New Hampshire, 1782–1866), Treasury Secretaries Chase and John Sherman (1823–1900), and Secretary of War Edwin M. Stanton (1814–69). William Tecumseh Sherman (1820–91) was a Union general in the Civil War whose Georgia campaign in 1864 helped effect the surrender of the Confederacy. Although disappointed in his quest for the presidency, US Senator Robert A. Taft (1889–1953) was an enduring figure, best remembered for his authorship of the Taft-Hartley Labor Management Relations Act of 1947.

Nobel Prize winners from Ohio include Dawes and physicists Arthur Compton (1892–1962) and Donald Glaser (b.1926). Notable Pulitzer Prize winners include novelist Louis Bromfield (1896–1956), dramatist Russell Crouse (1893–1966), historian Paul Herman Buck (1899–1979), and historian and biographer Arthur Schlesinger, Jr. (b.1917). Ohio writers of enduring fame are novelists William Dean Howells (1837–1920), Zane Grey (1875–1939), and Sherwood Anderson (1876–1941); poets Paul Laurence Dunbar (1872–1906) and Hart Crane (1899–1932); and humorist James Thurber (1894–1961). Among Ohio's eminent journalists are Whitelaw Reid (1837–1912), satirists David R. Locke (1833–88) and Ambrose Bierce (1842–1914), columnist O. O. McIntyre (1884–1938), newsletter publisher W. M. Kiplinger (1891–1967), and James Reston (b.Scotland, 1909), an editor and columnist for the *New York Times*, along with author-commentator Lowell Thomas (1892–1981). Important in the art world were painters Thomas Cole (b.England, 1801–48), Frank Duveneck (b.Kentucky, 1848–1919), and George Bellows (1882–1925), as well as architects Cass Gilbert (1859–1934) and Philip Johnson (b.1906). Defense lawyer Clarence Darrow (1857–1938) was also an Ohioan.

Ohio educators whose books taught reading, writing, and arithmetic to the nation's schoolchildren were William Holmes McGuffey (b.Pennsylvania, 1800–73), Platt R. Spencer (1800–64), and Joseph Ray (1807–65). In higher education, Horace Mann (b.Massachusetts, 1796–1859) was the first president of innovative Antioch College, and William Rainey Harper (1856–1906) founded the University of Chicago.

Allied with industry are Ohio-born inventor-scientists. Thomas A. Edison (1847–1931) produced the incandescent lamp, the phonograph, and the movie camera. Charles Brush (1849–1929) invented the arc light. John H. Patterson (1844–1922) helped develop the cash register. The Wright brothers, Orville (1871–1948) and Wilbur (b.Indiana, 1867–1912), made the first flight in a powered aircraft. Charles F. Kettering (1876–1958) invented the automobile self-starter. Ohio's leading industrialist was John D. Rockefeller (b.New York, 1839–1937), founder of Standard Oil of Ohio. Harvey S. Firestone (1868–1938) started the tire company that bears his name. Edward "Eddie" Rickenbacker (1890–1973), an ace pilot in World War I, was president of Eastern Airlines.

The most notable Ohioans in the entertainment field are markswoman Annie Oakley (Phoebe Anne Oakley Mozee, 1860–1926); movie actors Clark Gable (1901–60) and Roy Rogers (Leonard Slye, b.1912); movie director Stephen Spielberg (b.1947); comedian Bob Hope (Leslie Townes Hope, b.England, 1903); actors Paul Newman (b.1925), Hal Holbrook (b.1925), and Joel Grey (b.1932); jazz pianist Art Tatum (1910–56); and composer Henry Mancini (b.1924).

Leading sports figures from Ohio are boxing champion Jim

Jeffries (1875–1953), racing driver Barney Oldfield (1878–1946), baseball pitcher Cy Young (1867–1955), baseball executive Branch Rickey (1881–1965), baseball star Peter "Pete" Rose (b.1942), track star Jesse Owens (b.Alabama, 1912–80), jockey George Edward "Eddie" Arcaro (b.1916), and golfer Jack Nicklaus (b.1940).

Astronauts from Ohio include John Glenn (b.1921), the first American to orbit the earth, who was elected US senator from Ohio in 1974; and Neil Armstrong (b.1930), the first man to walk on the moon.

50BIBLIOGRAPHY

Bromfield, Louis. *The Farm.* New York: Harper, 1933.

Condon, George E. *Cleveland: The Best Kept Secret.* Garden City, N.Y.: Doubleday, 1967.

Downes, Randolph C. *Frontier Ohio: 1788–1803.* Columbus: Ohio State Archaeological and Historical Society, 1935.

Ellis, William D. *The Cuyahoga.* New York: Holt, 1966.

Federal Writers' Project. *The Ohio Guide.* Reprint. New York: Somerset, n.d. (orig. 1940).

Galbreath, Charles B. *History of Ohio.* 5 vols. Chicago and New York: American Historical Society, 1925.

Havighurst, Walter. *Ohio: A Bicentennial History.* New York: Norton, 1976.

Hopkins, Charles E. *Ohio the Beautiful and Historic.* Boston: Page, 1931.

Howe, Henry. *Historical Collections of Ohio.* 2 vols. Columbus: Henry Howe & Son, 1889.

Howells, William Dean. *A Boy's Town.* New York: Harper, 1890.

Hulbert, Archer B. *The Ohio River: A Course of Empire.* New York and London: Putnam, 1906.

Jones, Robert L. *The History of Agriculture in Ohio to Eighteen Eighty.* Kent: Kent State University Press, 1983.

Jordan, Philip D. *The National Road.* Indianapolis: Bobbs-Merrill, 1948.

Leech, Margaret. *In the Days of McKinley.* New York: Harper & Row, 1959.

Longworth de Chambrun, Clara. *Cincinnati: The Story of the Queen City.* New York: Scribner, 1939.

McCormick, Richard P. *The Second American Party System.* Chapel Hill: University of North Carolina Press, 1966.

Morgan, H. Wayne. *From Hayes to McKinley.* Syracuse, N.Y.: Syracuse University Press, 1969.

Notestein, Lucy. *Wooster of the Middle West.* New Haven, Conn.: Yale University Press, 1937.

Ohio State. Department of State. *Constitution of the State of Ohio.* Columbus: Anderson, 1979.

Ohio State. Secretary of State. *Official Roster, 1983–1984.* Columbus, 1984.

Patterson, James T. *Mr. Republican: A Biography of Robert A. Taft.* Boston: Houghton Mifflin, 1972.

Reid, Whitelaw. *Ohio in the War.* 2 vols. Cincinnati: Moore, Wilstach & Baldwin, 1868.

Roseboom, Eugene Holloway, and Francis P. Weisenburger, *A History of Ohio.* New York: Prentice-Hall, 1934.

Thurber, James. *My Life and Hard Times.* New York: Harper, 1933.

Warner, Hoyt Landon. *Progressivism in Ohio.* Columbus: Ohio State University Press, 1964.

Wittke, Carl, ed. *History of the State of Ohio.* 6 vols. Columbus: Ohio State Archaeological and Historical Society, 1941–44.

OKLAHOMA

State of Oklahoma

ORIGIN OF STATE NAME: Derived from the Choctaw Indian words *okla humma*, meaning "land of the red people." **NICKNAME:** The Sooner State. **CAPITAL:** Oklahoma City. **ENTERED UNION:** 16 November 1907 (46th). **SONG:** "Oklahoma!" **POEM:** "Howdy Folks." **MOTTO:** *Labor omnia vincit* (Labor conquers all things). **FLAG:** On a blue field, a peace pipe and an olive branch cross an Osage warrior's shield, which is decorated with small crosses and from which seven eagle feathers descend; the word "Oklahoma" appears below. **OFFICIAL SEAL:** Each point of a five-pointed star incorporates the emblem of an Indian nation: (clockwise from top) Chickasaw, Choctaw, Seminole, Creek, and Cherokee. In the center, a frontiersman and Indian shake hands before the goddess of justice; behind them are symbols of progress, including a farm, train, and mill. Surrounding the large star are 45 small ones and the words "Great Seal of the State of Oklahoma 1907." **ANIMAL:** American buffalo (bison). **BIRD:** Scissor-tailed flycatcher. **FISH:** White bass (sand bass). **FLORAL EMBLEM:** Mistletoe. **TREE:** Redbud. **STONE:** Barite rose (rose rock). **REPTILE:** Collared lizard (mountain boomer). **GRASS:** Indian grass. **LEGAL HOLIDAYS:** New Year's Day, 1 January; Birthday of Martin Luther King, Jr., 3d Monday in January; Washington's Birthday, 3d Monday in February; Memorial Day, last Monday in May; Independence Day, 4 July; Labor Day, 1st Monday in September; Columbus Day, 2d Monday in October; Veterans Day, 11 November; Thanksgiving Day, 4th Thursday in November; Christmas Day, 25 December. **TIME:** 6 AM CST = noon GMT.

¹LOCATION, SIZE, AND EXTENT

Situated in the western south-central US, Oklahoma ranks 18th in size among the 50 states.

The total area of Oklahoma is 69,956 sq mi (181,186 sq km), of which land takes up 68,655 sq mi (177,817 sq km) and inland water 1,301 sq mi (3,369 sq km). Oklahoma extends 464 mi (747 km) E–W including the panhandle in the NW, which is about 165 mi (266 km) long. The maximum N–S extension is 230 mi (370 km).

Oklahoma is bordered on the N by Colorado and Kansas; on the E by Missouri and Arkansas; on the S and SW by Texas (with part of the line formed by the Red River); and on the extreme W by New Mexico. The total estimated boundary length of Oklahoma is *1,581 mi (2,544 km)*. The state's geographic center is in Oklahoma County, 8 mi (13 km) N of Oklahoma City.

²TOPOGRAPHY

The land of Oklahoma rises gently to the west from an altitude of 287 feet (87 meters) at Little River in the southeastern corner to a height of 4,973 feet (1,516 meters) at Black Mesa, on the tip of the panhandle. Four mountain ranges cross this Great Plains state: the Boston Mountains (part of the Ozark Plateau) in the northeast, the Ouachitas in the southeast, the Arbuckles in the south-central region, and the Wichitas in the southwest. Much of the northwest belongs to the High Plains, while northeastern Oklahoma is mainly a region of buttes and valleys.

Not quite two-thirds of the state is drained by the Arkansas River, and the remainder by the Red River. Within Oklahoma, the Arkansas is joined by the Verdigris, Grand (Neosho), and Illinois rivers from the north and northeast, and by the Cimarron and Canadian rivers from the northwest and west. The Red River, which marks most of the state's southern boundary, is joined by the Washita, Salt Fork, Blue, Kiamichi, and many smaller rivers. There are few natural lakes but many artificial ones, of which the largest is Lake Eufaula, covering 102,500 acres (41,500 hectares).

³CLIMATE

Oklahoma has a continental climate with cold winters and hot summers. Normal daily mean temperatures in Oklahoma City range from 37°F (3°C) in January to 82°F (28°C) in July. The record low temperature of −27°F (−33°C) was set at Watts on 18 January 1930; the record high, 120°F (49°C), occurred most recently at Tishomingo on 26 July 1943.

Dry, sunny weather generally prevails throughout the state. Precipitation varies from an average of 15 in (38 cm) annually in the panhandle to over 50 in (127 cm) in the southeast. Snowfall averages 9 in (23 cm) a year in Oklahoma City, which is also one of the windiest cities in the US, with an average annual wind speed of 12.8 mph (20.6 km/hr).

Oklahoma is tornado prone. One of the most destructive windstorms was the tornado that tore through Ellis, Woods, and Woodward countries on 9 April 1947, killing 101 people and injuring 782 others. There were 92 tornadoes in 1983.

⁴FLORA AND FAUNA

Grasses grow in abundance in Oklahoma. Bluestem, buffalo, sand lovegrass, and grama grasses are native, with the bluestem found mostly in the eastern and central regions, and buffalo grass most common in the western counties, known as the "short grass country." Deciduous hardwoods stand in eastern Oklahoma, and red and yellow cactus blossoms brighten the Black Mesa area in the northwest.

The white-tailed deer is found in all counties, and Rio Grande wild turkeys are hunted across much of the state. Pronghorn antelope inhabit the panhandle area, and elk survive in the Wichita Mountains Wildlife Refuge, where a few herds of American buffalo (bison) are also preserved. The bobwhite quail, ring-necked pheasant, and prairie chicken are common game birds. Native sport fish include largemouth, smallmouth, white, and spotted bass; catfish; crappie; and sunfish.

Among the state's endangered or threatened species of wildlife are the leopard darter, Ozark big-eared bat, red wolf, black-footed ferret, Indiana bat, southern bald eagle, whooping crane, ivory-billed and red-cockaded woodpeckers, Bachman's warbler, American peregrine falcon, Eskimo curlew, and American alligator.

⁵ENVIRONMENTAL PROTECTION

The Department of Pollution Control, created in 1968, is the executive arm of the Oklahoma Pollution Control Coordinating

Board. The department has overall responsibility for coordinating all pollution control activities by other state agencies and for developing a comprehensive water quality management program for Oklahoma.

The Department of Health is responsible for the monitoring of air quality standards; the enforcement of regulations covering control of industrial and solid waste; the enforcement of regulations covering radioactive materials at the Kerr-McGee processing facility at Gore and elsewhere; and the maintenance of standards at all public waterworks and sewer systems. The Water Resources Board has broad statutory authority to protect the state's waters.

Toxic industrial wastes remain an environmental concern, and old mines in the Tar Creek area of northeastern Oklahoma still exude groundwater contaminated by zinc, iron, and cadmium. Three additional hazardous waste sites placed on the EPA's National Priorities List for Superfund cleanup were Compass Industries, Sand Springs Petro Chemical Complex, and Criner.

Lands devastated by erosion during the droughts of the 1930s were purchased by the federal government and turned over to the Soil Conservation Service for restoration. When grasses were firmly established in the mid-1950s, the land was turned over to the US Forest Service and is now leased for grazing.

6POPULATION

Oklahoma ranked 26th among the states in the 1980 census with a total population of 3,025,300, an increase of 19% over the census total of 2,559,463 in 1970, when Oklahoma ranked 27th among the 50 states. A population of 3,503,400 is projected for 1990. The estimated 1985 population was 3,427,371, yielding an increase of 13.3% for 1980–85 and ranking Oklahoma 5th among the states in rate of growth for the period. In 1985, Oklahoma had a population density of 50 per sq mi (19 per sq km).

In 1980, 67% of all Oklahomans lived in urban areas. The largest city is Oklahoma City, which in 1984 had 443,172 inhabitants in the inner city (up from 403,484 in 1980) and 962,600 in the metropolitan statistical area (up from 861,000 in 1980). Tulsa, the 2d-largest city, had 1984 populations of 374,535 in the inner city (up from 360,919 in 1980) and 725,600 in the metropolitan area (up from 657,173 in 1980). Lawton ranked 3d with a 1984 population of 85,629 in the city (80,054 in 1980). Other cities with more than 50,000 residents in 1984 were Norman (75,350), Midwest City (53,385), and Enid (52,502).

7ETHNIC GROUPS

According to the 1980 Census, Oklahoma has more American Indians—169,297—than any other state except California. Tulsa (19,059) and Oklahoma (14,285) counties have the largest Indian populations; Adair (33%), Cherokee (26%), and Delaware (21%), the highest percentages. On the whole Indians have the lowest income level and highest unemployment rate of any ethnic group in the state.

Black slaves came to Oklahoma (then known as Indian Territory) with their Indian masters after Congress forced the resettlement of Indians from the southeast to lands west of the Mississippi River in 1830. By the time of the Civil War, there were 7,000 free Negroes in Oklahoma. After the depression of the 1930s, blacks left the farms and small towns and concentrated in Oklahoma City and Tulsa. More than 55% of the state's black population, which totaled 204,658 in 1980, lives in Oklahoma City and Tulsa.

Mexicans came to Oklahoma during the 19th century as laborers and were supplied to railroads, ranches, and coal mines. Later they worked in the cotton fields until the depression of the 1930s and subsequent mechanization reduced the need for seasonal labor. Today, most 1st- and 2d-generation Mexicans live in Oklahoma City, Tulsa, and Lawton. Most of the 57,419 persons classified as of Spanish or Hispanic origin in the 1980 census said they were of Mexican descent.

Italians, Czechs, Germans, Poles, Britons, Irish, and others of European stock also came to Oklahoma during the 19th century. Foreign immigration has been small since that time, however, and in 1980, less than 3% of the population consisted of the foreign-born. Persons claiming a single ancestry group in 1980 included English, 400,283; German, 178,615; and Irish, 158,897.

8LANGUAGES

Once the open hunting ground of the Osage, Comanche, and Apache Indians, what is now Oklahoma perforce welcomed the deported Cherokee and other transferred eastern tribes. The diversity of tribal and linguistic backgrounds is reflected in numerous place-names such as Oklahoma itself, Kiamichi, and Muskogee. Almost equally diverse is Oklahoma English, with its uneven blending of features of North Midland, South Midland, and Southern dialects.

In 1980, 2,767,867 Oklahomans—96% of the resident population 3 years or older—spoke only English at home. Other languages spoken at home, and the number of people who spoke them included Spanish, 42,266, German, 12,446, and French, 6,778.

9RELIGIONS

Protestant groups predominate in Oklahoma, and Protestant fundamentalists constitute a third of the population. This group was influential in keeping the state "dry"—that is, banning the sale of all alcoholic beverages—until 1959 and resisting legalization of public drinking until 29 counties voted to permit the sale of liquor by the drink in 1985.

The leading Protestant groups in 1980 included Southern Baptist, 799,357; United Methodist, 313,466; Church of Christ, 91,606; Assembly of God, 69,689; Christian Church (Disciples of Christ), 66,424; and Episcopal, 22,230. There were 139,906 Roman Catholics in 1984 and an estimated 7,160 Jews in 1983.

Oral Roberts, a popular minister, has established a college and faith-healing hospital in Tulsa, and his "Tower of Faith" broadcasts by radio and television have made him a well-known preacher throughout the US.

10TRANSPORTATION

In 1930, the high point for railroad transportation in Oklahoma, there were 6,678 mi (10,747 km) of railroad track in the state. As of 31 December 1983 there were 3,853 mi (6,200 km) of Class I track; Burlington Northern had the most track, followed by the Atchison, Topeka & Santa Fe. In 1979, Amtrak terminated the state's last passenger train. Inter-urban transit needs, formerly served by streetcars (one of the most popular routes operated between Oklahoma City and Norman), are now supplied by buses.

The Department of Transportation is responsible for construction and maintenance of the state road system, which in 1983 included 12,060 mi (19,409 km) of state roads and highways and 926 mi (1,490 km) of interstate highways. The main east-west highways are I-44, connecting Tulsa and Oklahoma City, and I-40; the major north-south route is I-35, which links Oklahoma City with Topeka, Kansas and Dallas–Ft. Worth, Texas. Overall, in 1983, Oklahoma had 111,924 mi (180,125 km) of roadway. A total of 2,769,058 motor vehicles were registered in 1983, including 1,757,739 automobiles and 999,771 commercial trucks. There were an estimated 2,174,350 licensed drivers.

The opening of the McClellan-Kerr Arkansas River Navigation System in 1971 linked Oklahoma with the Mississippi River and thus to Gulf coast ports. Catoosa (Tulsa), chief port on the system, handled 1,379,138 tons of cargo in 1983.

Oklahoma had 289 airports, 42 heliports, and 1 seaplane base in

LOCATION: 33°38'17" to 37°N; 94 °25'51" to 103° W. **BOUNDARIES:** Colorado line, *58 mi (93 km)*; Kansas line, *411 mi (661 km)*; Missouri line, *34 mi (55 km)*; Arkansas line, *198 mi (319 km)*; Texas line *846 mi (1,361 km)*; New Mexico line, *34 mi (55 km)*.

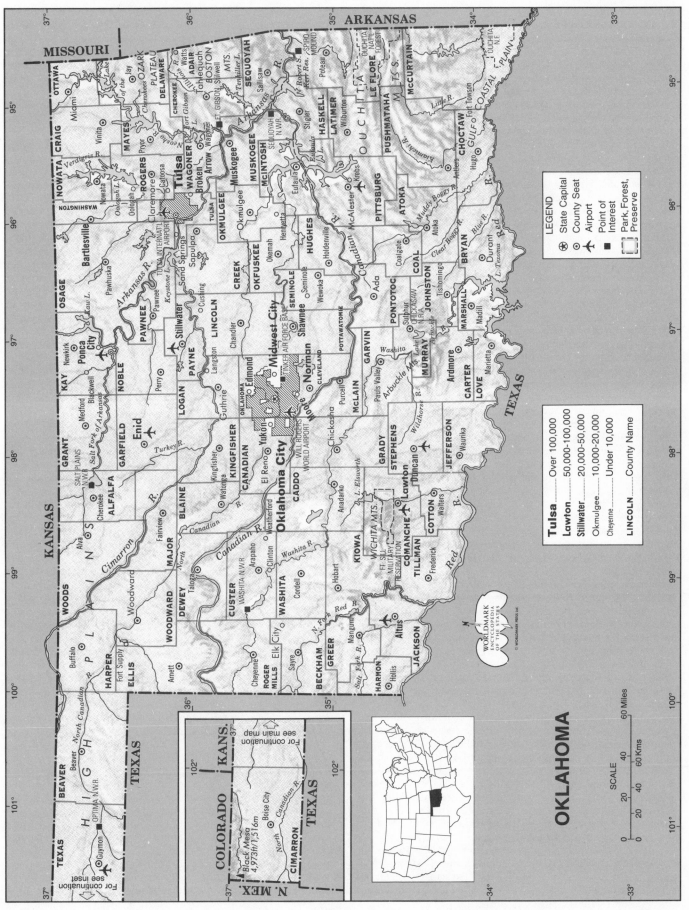

OKLAHOMA

See endsheet maps: G3.

LEGEND
⊗ State Capital
⊙ County Seat
✈ Airport
■ Point of Interest
▫ Park, Forest, Preserve

Tulsa	Over 100,000
Lawton	50,000-100,000
Stillwater	20,000-50,000
Okmulgee	10,000-20,000
Cheyenne	Under 10,000
LINCOLN	County Name

SCALE

0   20   40   60 Miles

0   20   40   60 Kms

WORLDMARK ENCYCLOPEDIA OF THE STATES
© WORLDMARK PRESS Ltd.

1983. Tulsa International Airport, which enplaned 1,272,961 passengers in 1983, and Will Rogers World Airport in Oklahoma City, which enplaned 1,288,170 passengers, are the state's largest airports.

11 HISTORY

There is evidence—chiefly from the Spiro Mound in eastern Oklahoma, excavated in 1930—that an advanced Indian civilization inhabited the region around AD 900–1100. By the time the Spanish conquistadores, led by Hernando de Soto and Francisco Vasquez de Coronado, arrived there in the 16th century, however, only a few scattered tribes remained. Two centuries later, French trappers moved up the rivers of Oklahoma.

Except for the panhandle, which remained a no-man's-land until 1890, all of present-day Oklahoma became part of US territory with the Louisiana Purchase in 1803. Under the Indian Removal Act of 1830, Indian tribes from the southeastern US were resettled in what was then known as Indian Country. Although 4,000 Indians died along the "Trail of Tears" (from Georgia to Oklahoma) between the time of removal and the Civil War, the Five Civilized Tribes—Cherokees, Chickasaw, Choctaw, Creek, and Seminole—prospered in the new land. The eastern region that they settled, comprising not quite half of modern Oklahoma and known as Indian Territory since the early 19th century (although not formally organized under that name until 1890), offered rich soil and luxurious vegetation. White settlers also came to farm the land, but their methods depleted the soil, preparing the way for the dust bowl of the 1930s. Meanwhile, the increasing movement of people and goods between Santa Fe and New Orleans spurred further growth in the region. Military posts such as Ft. Gibson, Ft. Supply, and Ft. Towson were established between 1824 and the 1880s, with settlements growing up around them.

During the early Civil War period, the Five Civilized Tribes—some of whose members were slaveholders—allied with the Confederacy. After Union troops captured Ft. Gibson in 1863, the Union Army controlled one-half of Indian Territory. From the end of the Civil War to the 1880s, the federal government removed the eastern tribes from certain lands that were especially attractive to the railroads and to interested white settlers. Skirmishes between the Indians and the federal troops occurred, culminating in a massacre of Cheyenne Indians on 27 November 1868 by Colonel George Custer and his 7th Cavalry at the Battle of the Washita.

Amid a clamor for Indian lands, Congress opened western Oklahoma—formerly reserved for the Cherokee, Cheyenne, Fox, and other tribes—to homesteaders in 1889. Present-day Oklahoma City, Norman, Guthrie, Edmond, and Stillwater represent the eastern boundary for the 1889 "run" on Oklahoma lands; eight more runs were to follow. The greatest was in 1893, when about 100,000 people stormed onto the newly opened Cherokee outlet. The drive to get a land claim was fierce, and thousands of "Sooners" staked their claims before the land was officially opened. The western region became Oklahoma Territory, governed by a territorial legislature and a federally appointed governor in 1890; Guthrie was named the capital. Most of eastern Oklahoma continued to be governed by the Five Civilized Tribes.

Although an Oklahoma statehood bill was introduced in Congress as early as 1892, the Five Civilized Tribes resisted all efforts to unite Indian Territory until their attempt to form their own state was defeated in 1905. Congress passed an enabling act in June 1906, and Oklahoma became the 46th state on 16 November 1907 after a vote of the residents of both territories. Oklahoma City was named the state capital in 1910.

When President Theodore Roosevelt signed the statehood proclamation, Oklahoma's population was about 1,500,000—75% rural, 25% urban—most of them drawn by the state's agricultural and mineral resources. The McAlester coal mines had opened in

1871, and lead and zinc were being mined in Ottawa County. But it was oil that made the state prosperous. Prospecting began in 1882, and the first commercial well was drilled at Bartlesville in 1897. The famous Glenn Pool gusher, near Tulsa, was struck in 1905. Oil wells were producing more than 40 million barrels annually when Oklahoma entered the Union, and the state led all others in oil production until 1928.

Generally, the decade of the 1920s was a tumultuous period for Oklahoma. A race riot in Tulsa in 1921 was put down by the National Guard; the Ku Klux Klan claimed close to 100,000 Oklahomans that same year. The Klan was outlawed when Governor John C. Walton declared martial law in 1923, during a period of turmoil and violence that culminated in Walton's impeachment and conviction on charges of incompetence, corruption, and abuse of power. The 1930s brought a destructive drought, dust storms, and an exodus of "Okies," many of them to California. Colorful Governor William "Alfalfa Bill" Murray led the call for federal relief for the distressed dust bowl region—though he insisted on his right to administer the funds. When Oklahoma oil fields were glutting the market at 15 cents a barrel, Murray placed 3,106 producing wells under martial law from August 1931 to April 1933. Kansas, New Mexico, and Texas also agreed to control their oil production, and under the leadership of Governor E. W. Marland, the Interstate Oil Compact was created in 1936 to conserve petroleum and stabilize prices.

Oklahoma's first native-born governor, Robert Kerr, later to be senator for 14 years, held the statehouse during World War II and brought the state national recognition by promoting Oklahoma as a site for military, industrial, and conservation projects. Under early postwar governors Roy Turner, Johnston Murray, and Raymond Gary, tax reductions attracted industry, major highways were built, a loyalty oath for state employees was declared unconstitutional, and Oklahoma's higher educational facilities were integrated. The term of Governor Howard Edmondson saw the repeal of prohibition in 1959, the establishment of merit and central purchasing systems, and the introduction of a state income tax withholding plan.

Oil and gas again brought increased wealth to the state in the 1960s, 1970s, and early 1980s, as state revenues from oil and gas increased from $72 million in 1972 to $745 million in 1982. State budget expenditures, financed by the swelling revenues, doubled in five years, from $800 million in 1978 to some $1.6 billion in 1983. Nearly $1 billion was spent for new highways, schools, and state offices; new police were hired; and teacher salaries were raised to nationally competitive levels. Unemployment fell to low levels while an influx of job seekers from other states made Oklahoma one of the fastest growing states in the nation in the early 1980s.

In 1983, as oil prices fell in the face of a growing worldwide oil glut, the oil boom suddenly ended. Bank failures, increased bankruptcies, and mounting distress among the state's farmers added to Oklahoma's financial woes. In 1983, with state revenues falling, a balanced budget requirement in the state constitution compelled Governor George Nigh to cut appropriations and preside over a series of tax increases that lost for Oklahoma its claim to one of the lowest tax burdens in the nation.

12 STATE GOVERNMENT

Oklahoma's first and only constitutional convention began its deliberations in Guthrie on 20 November 1906. The constitution was approved by the electorate on 11 September 1907.

The Oklahoma legislature consists of two chambers, a 48-member senate and a 101-member house of representatives. To serve in the legislature, senators must be 25 years of age, and representatives 21, senators hold office for four years, representatives for two. Elected executive officials include the governor, lieutenant governor, attorney general, state treasurer, superintendent of public instruction, and commissioner of insurance, all of whom serve four-year terms, and three corporation commissioners, who serve staggered six-year terms. The governor must be a US citizen, at least 31 years of age, and must have been a qualified voter in Oklahoma for at least 10 years preceding election.

Any member of either house may introduce legislation. A bill passed by the legislature becomes law if signed by the governor, if left unsigned by the governor for five days while the legislature is in session, or if passed over the governor's veto by two-thirds of the elected members of each house or three-fourths in the case of emergency bills. Constitutional amendments may be placed on the ballot by majority vote in both houses, by initiative petition of 15% of the electorate, or by constitutional convention.

To vote in Oklahoma, one must be a US citizen, at least 18 years of age, and registered. There is no minimum state residence requirement.

13POLITICAL PARTIES

The history of the two major political groups in Oklahoma, the Democratic and Republican parties, dates back to 1890, when Indian Territory and Oklahoma Territory were separately organized. Indian Territory was dominated by Democrats, reflecting the influence of southern immigrants, while Oklahoma Territory was primarily Republican because of immigration from the northern states. When the two territories joined for admission to the Union in 1907, Democrats outnumbered Republicans, as they have ever since. Democrats have continued to dominate the lesser state offices, but the Republicans won the governorship three times between 1962 and 1970, and the Republican presidential nominee outpolled his Democratic counterpart in eight of ten presidential elections between 1948 and 1984. The best showing by a minor party in a recent presidential race was the vote total of 191,731 garnered by American Independent Party candidate George Wallace in 1968.

As of 1982, the Democratic Party had 1,146,261 registered voters, while registered Republicans numbered 430,327; there were 37,261 registered independents.

Oklahomans gave the Republican Party a resounding victory in November 1984 by casting more than 68% of the popular vote for Ronald Reagan; but they reelected Don Nickles, a Democrat, to the US Senate over Republican challenger Bill Crozier. Democrat George Nigh was reelected to a third term as governor in 1982, and Democrats retained control in both legislative houses in 1984.

In 1985, women held 3 seats in the state senate and 10 in the house of representatives. Blacks were represented by 2 state senators, 3 representatives, and 117 other elected officials.

Oklahoma Presidential Vote By Political Parties, 1948–84

YEAR	ELECTORAL VOTE	OKLAHOMA WINNER	DEMOCRAT	REPUBLICAN
1948	10	*Truman (D)	452,782	268,817
1952	8	*Eisenhower (R)	430,939	518,045
1956	8	*Eisenhower (R)	385,581	473,769
1960	8	Nixon (R)	370,111	533,039
1964	8	*Johnson (D)	519,834	412,665
1968	8	*Nixon (R)	301,658	449,697
1972	8	*Nixon (R)	247,147	759,025
1976	8	Ford (R)	532,442	545,708
1980	8	*Reagan (R)	402,026	695,570
1984	8	*Reagan (R)	385,080	861,530

* Won US presidential election.

14LOCAL GOVERNMENT

As of 1983, local governmental units in Oklahoma included 77 counties, 581 incorporated cities and towns, and several hundred unincorporated areas.

County government consists of three commissioners elected by districts, a county clerk, assessor, treasurer, sheriff, surveyor, and (in most counties) superintendent of schools. Towns of 1,000 population or more may incorporate as cities. Any city of 2,000 or more people may vote to become a home-rule city, determining its own form of government, by adopting a home-rule charter. Cities electing not to adopt a home-rule charter operate under aldermanic, mayor-council, or council-manager systems. A large majority of home-rule cities have council-manager forms.

15STATE SERVICES

The Oklahoma Department of Education, functioning under a six-member appointed Board of Education and an elected superintendent of public instruction, has responsibility for all phases of education through the first 12 grades. Postsecondary study is under the general authority of the Oklahoma State Regents for Higher Education and 16 separate boards of regents associated with one or more institutions. Vocational and technical education, a federal-state cooperative program, is administered in Oklahoma under the Department of Vocational and Technical Education. The Department of Transportation has authority over the planning, construction, and maintenance of the state highway system. The Oklahoma Corporation Commission regulates transportation and transmission companies, public utilities, motor carriers, and the oil and gas industry, while the Oklahoma Aeronautics Commission participates in financing airports.

The Department of Health has as a major function the control and prevention of communicable diseases; it administers community health program funds and licenses most health-related facilities. The Department of Human Services oversees the care of neglected children, delinquent youths, and the mentally retarded and operates various facilities and programs for the handicapped, the elderly, and the infirm.

Protective services are supplied through the Oklahoma Military Department, which administers the Army and Air National Guard; the Department of Corrections, overseeing the state penitentiary and reformatory, nine adult correctional centers, and eight community treatment centers; and the Department of Public Safety, with general safety and law enforcement responsibilities, among which are licensing drivers and patrolling the highways. Natural resource protection services are centered principally in the Oklahoma Conservation Commission; the Department of Wildlife Conservation and the Wildlife Conservation Commission administer the game and fish laws.

16JUDICIAL SYSTEM

In 1967, following some of the worst judicial scandals in the history of the state, in which one supreme court justice was imprisoned for income tax evasion and another impeached on charges of bribery and corruption, Oklahoma approved a constitutional amendment to reform the state's judicial system. Under the new provisions, the supreme court, the state's highest court, consists of nine justices initially elected to six-year terms, but with additional terms pursuant to nonpartisan, noncompetitive elections; if a justice is rejected by the voters, the vacancy is filled by gubernatorial appointment, subject to confirmation by the electorate. The court's appellate jurisdiction includes all civil cases (except those which it assigns to the courts of appeals), while its original jurisdiction extends to general supervisory control over all inferior courts and agencies created by law.

The highest appellate court for criminal cases is the court of criminal appeals, a three-member body filled in the same manner as the supreme court. Courts of appeals, created by the legislature in 1968, are located in Tulsa and Oklahoma City; each has three elective judges with powers to hear civil cases assigned to them by the supreme court. When final, their decisions are not appealable to any other state court, a system unique to Oklahoma.

District courts have original jurisdiction over all justiciable matters and some review powers over administrative actions. There are 26 districts with 70 district judges and 76 associate judges, who are elected to four-year terms. Municipal courts hear cases arising from local ordinances. A total of 7,970 attorneys lived and practiced in the state in 1984. As of 1983, 7,487 prisoners were under the jurisdiction of state and federal authorities. The FBI reported a crime index total of 4,929 crimes per 100,000 population, including rates of 423 for violent crime and 4,506 for property crime. Oklahoma law permits capital punishment by lethal injection for several felony crimes. As of July 1985, it had never been utilized, but there were 39 prisoners under sentence of death as of 31 December 1983.

[17]ARMED FORCES

Oklahoma has two Army and four Air Force facilities. Of major importance are Ft. Sill, near Lawton, the training facility for the Artillery Branch, with 17,997 military personnel in 1984; and the Air Logistics Center at Tinker Air Force Base, with 7,181 military personnel. A total of $533 million in prime military contracts was received by local businesses in 1983/84.

As of 30 September 1983, about 397,000 veterans were living in Oklahoma, of whom 5,000 saw service in World War I, 150,000 in World War II, 75,000 in the Korean conflict, and 127,000 during the Viet-Nam era. Expenditures on veterans totaled $451 million in 1982/83.

The Oklahoma Army National Guard had 9,566 assigned personnel as of December 1984, and the Air National Guard had 2,047 personnel as of February 1985. State and local police forces employed some 7,970 sworn peace officers in 1983.

[18]MIGRATION

Early immigrants to what is now Oklahoma included explorers, adventurers, and traders who made the country conscious of the new territory, and Indian tribes forcibly removed from the East and Midwest. The interior plains of Oklahoma remained basically unchanged until white settlers came in the late 1880s.

Coal mining brought miners from Italy in the 1870s to the McAlester and Krebs area, and Poles migrated to Bartlesville to work in the lead and zinc smelters. British and Irish coal miners came to Indian Territory because they could earn higher wages there than in their native countries, and Czechs and Slovaks arrived from Nebraska, Kansas, Iowa, and Texas when railroad construction began. Mexicans also worked as railroad laborers, ranch hands, and coal miners before statehood. The oil boom of the early 20th century brought an influx of workers from the eastern and midwestern industrial regions.

In 1907, the population of Oklahoma was 75% rural and 25% urban; by 1980, however, 67% of all inhabitants resided in urban areas. Oklahoma lost population during the 1930s because of dust bowl and drought conditions, and the trend toward out-migration continued after World War II; from 1940 through 1960, the net loss from migration was 653,000. Migration patterns were reversed, however, after 1960. From 1960 to 1970 nearly 21,000 more people moved into the state than out of it. In the period 1970–80 a total of 293,500 more people came than left, the migration accounting for nearly two-thirds of Oklahoma's total increase of 466,000 persons in that decade. From 1980 to 1983, Oklahoma ranked 4th among the states with a total net gain from migration of 186,000 people.

[19]INTERGOVERNMENTAL COOPERATION

Oklahoma participates in a number of regional intergovernmental agreements, among them the Arkansas River Compact, Arkansas River Basin Compact, Canadian River Compact, Interstate Oil and Gas Compact, Red River Compact, South Central Interstate Forest Fire Protection Compact, Southern Growth Policies Compact, and Southern Interstate Energy Compact. Oklahoma also takes part in the Ozarks Regional Commission along with Arkansas, Kansas, Louisiana, and Missouri.

Federal grants in 1983/84 totaled nearly $1.2 billion.

[20]ECONOMY

Primarily an agricultural state through the first half of the 20th century, Oklahoma has assumed a broader economic structure since the 1950s. Manufacturing heads the list of growth sectors, followed by wholesale and retail trade, services, finance, insurance, and real estate. Oil and gas extraction continues to play a major role, with Oklahoma ranking 5th among the states in value of mineral fuel production in 1982. All but two of the state's 77 counties produced some oil or gas in 1981.

In 1984, the Office of Business and Economic Research at Oklahoma State University's College of Business Administration forecast that Oklahoma would have a gross state product of $52.2 billion in 1985, an increase of $3.5 billion in current dollars, or 7.1%, over 1984. In 1972 dollars, however, the increase was expected to be only 3.1%. Of this increase, manufacturing was forecast to account for 6.4% (durable goods, 8.2%; nondurable goods, 3.3%). The growth forecast for other sectors of the economy included wholesale and retail trade, 3.5%; services, 3.5%; finance, insurance, and real estate, 3.2%; state and local government, 2.3%; agriculture, 1.1%; mining, less than 1%; and transportation, communications, and public utilities, less than 1%. Contract construction was expected to decrease by 0.6%.

[21]INCOME

In 1983 the average per capita personal income in Oklahoma was $11,187, below the US average of $11,675; median income of four-person families was $24,701 in 1981, below the national average of $26,274. Total personal income for 1985 was estimated at $42.4 billion. Major components of personal income estimated for 1985 were $23.7 billion from wages and salaries; $8.6 billion from dividends, interest, and rents; $5.6 billion from federal transfer payments; and $539 million from farm proprietors.

As of 1980, 13% of all state residents were below the federal poverty level. There were 34,700 top wealth-holders in Oklahoma in 1982, with gross assets of $500,000 or more.

[22]LABOR

The civilian labor force in September 1984 was estimated at 1,590,200 persons, of whom 104,500, or 6.6%, were unemployed. This was an increase from the 1,544,000 persons in the labor force in 1983. An estimated 53.2% of all adult women were in the labor force in 1983, compared with the national average of 52.9%. In 1984, 667,000 workers were female and 877,000 were male. There were an estimated 74,000 state employees in 1983.

A federal survey in 1982 reported the following nonfarm employment pattern in Oklahoma:

	ESTABLISH-MENTS	EMPLOYEES	ANNUAL PAYROLL ('000)
Agricultural services, forestry, fishing	609	4,217	$ 42,807
Mining	3,557	98,105	2,215,557
Contract construction	5,953	60,521	1,052,513
Manufacturing	3,719	202,019	4,014,556
Transportation, public utilities	2,587	72,427	1,529,653
Wholesale trade	5,987	76,753	1,973,969
Retail trade	18,497	216,488	2,002,443
Finance, insurance, real estate	5,728	60,955	1,040,566
Services	18,603	204,698	2,668,387
Other	2,383	2,635	39,937
TOTALS	67,623	998,818	$16,580,388

Government workers were not covered by this survey; state and local government workers numbered about 196,000 in 1983, and federal workers numbered 46,000 in 1982.

There were 174,000 labor union members in 1980 and 533 labor unions in 1983. Union membership, at 15.3% of all nonagricultural workers in 1980, was well below the national average of 25.2%.

23AGRICULTURE

Agriculture remains an important economic activity in Oklahoma, even though its relative share of personal income and employment has declined since 1950. Total farm income, estimated at $2.7 billion, ranked 21st in the US in 1983. Crop marketings contributed $1 billion, livestock $1.7 billion.

As of 1984, Oklahoma had 74,000 farms and ranches covering 33,000,000 acres (13,355,000 hectares). The state ranked 5th in the US for wheat production in 1983, with 150,500,000 bushels (down from 227,700,000 bushels in 1982) worth $534 million (down from $831 million in 1982). Peanut production ranked 6th in 1982, with 176,540,000 lb., valued at $40,852,000. Estimated crop figures for 1983 include hay, 3,716,000 tons, $239,682,000; sorghum for grain, 11,880,000 bushels, $37,303,000; soybeans, 3,910,000 bushels, $31,867,000; corn for grain, 4,144,000 bushels, $15,747,-000; oats, 3,920,000 bushels, $7,056,000; and barley, 1,496,000 bushels, $3,890,000.

Virtually all of Oklahoma's wheat production is located in the western half of the state; cotton (145,000 bales in 1983) is grown in the southwest corner. Sorghum-producing regions include the panhandle, central to southwestern Oklahoma, and the northeast corner of the state.

24ANIMAL HUSBANDRY

On 1 January 1984, there were 5,500,000 cattle and calves (5th in the US), 110,000 milk cows, 290,000 hogs and pigs, and 115,000 sheep and lambs on Oklahoma farms and ranches.

Preliminary production figures for 1983 were as follows: cattle and calves, 2.1 billion lb, $1.1 billion (5th in the US); hogs and pigs, 81.2 million lb, $39.6 million; sheep and lambs, 4.5 million lb, $2 million; chickens and broilers, 220.6 million lb, $62.3 million; turkeys, 30.4 million lb, $10.6 million; eggs, 841 million, $51.9 million; and milk, 1.2 billion lb, $166.1 million.

25FISHING

Commercial fishing is of minor importance in Oklahoma. The prolific white bass (sand bass), Oklahoma's state fish, is abundant in most large reservoirs. Smallmouth and spotted bass, bluegill, and channel catfish have won favor with fishermen. Rainbow trout are stocked year-round in the Illinois River, and walleye and sauger are stocked in most reservoirs. A total of 31,967,534 fish were distributed in Oklahoma hatcheries in 1982, up from 18,376,802 in 1981.

26FORESTRY

Forests covered 8,513,000 acres (3,445,000 hectares) in 1979, or nearly one-fifth of the state's land area. Just about half of that was commercial forestland, mostly located in eastern Oklahoma and 87% privately owned. In 1984, there were 248,965 acres (100,753 hectares) of national forest system lands, all of them part of Ouachita National Forest.

Lumber production totaled 211 million board feet in 1983, of which 196 million board feet was derived from hardwood timber. Thirty-five sawmills provided a payroll of $8.7 million in 1982.

27MINING

Production of all minerals had an estimated value of nearly $10.8 billion in 1982, 5th highest in the US. More than $10.5 billion was derived from fossil fuels and only $225 million from nonfuel minerals. The value of nonfuel mineral production rose to $253.2 million in 1984, ranking Oklahoma 30th among the states. Iodine is a specialty, and Oklahoma ranks 1st among the states in its production.

Large deposits of limestone are found throughout northeastern Oklahoma, while gypsum is extracted in the northwest, the west-central region, and the four southwesternmost counties. Oklahoma had been a leading producer of lead and zinc until the

1970s. The state also has varying amounts of copper, asphalt, bentonite, clays, germanium, glass sand, granite, iron, manganese, sandstone, shale, silver, titanium, and uranium.

Output of selected nonfuel minerals in 1984 was as follows: crushed stone, 28,500,000 tons; sand and gravel, 9,640,000 tons; and clays (including shale), 991,000 tons. The combined value of small amounts of feldspar, iodine, lime, salt, tripoli (used in polishing), and pumice produced was $16.8 million in 1984.

28ENERGY AND POWER

Electric power production in Oklahoma in 1983 was 45.7 billion kwh, based on an installed capacity of 12.4 million kw. That year, electric energy sales in the state totaled 37 billion kwh, of which 10.6 billion kwh went for industrial purposes, 9 billion kwh for commercial purposes, and 14.6 billion kwh for residential uses. Coal-fired steam units accounted for 19.6 billion kwh of electricity produced in 1983, natural gas-fired units 22.6 billion kwh, and hydroelectricity 2.5 billion kwh. There were no nuclear power plants as of 1 January 1985.

Oklahoma is rich in fossil fuel resources, producing oil, natural gas, and coal. Crude oil production declined from 223.6 million barrels in 1968 to 150.5 million barrels in 1978 and 158.6 million barrels in 1983. Proved reserves of crude oil were estimated at 931 million barrels at the close of 1983. In 1983, Oklahoma's natural gas output exceeded 1.7 trillion cu feet (3d in the US), leaving reserves of 16.2 trillion cu feet. Consumption of natural gas in the state in 1983 was 1.1 trillion cu feet.

Production of bituminous coal fell from a record high of 6.1 million tons in 1978 to 3.7 million tons in 1983. All of it was strip-mined. Reserves totaling 1.6 billion tons as of 1983 are concentrated in eastern Oklahoma.

29INDUSTRY

Oklahoma's earliest manufactures were based on agricultural and petroleum production. As late as 1939, the food-processing and petroleum-refining industries together accounted for one-third of the total value added by manufacture. Although resource-related industries continue to predominate, manufacturing was much more diversified in 1982. The leading sectors in terms of value of shipments by manufacturers were petroleum and coal products (28.9%), nonelectrical machinery (13.9%), food and kindred products (9.3%), fabricated metal products (8.2%), and rubber and plastic products (5.8%). The total value of shipments of manufactured goods in 1982 was more than $23.1 billion. The following table shows the value of shipments by manufacturers for selected industries:

Petroleum and coal products	$6,666,100,000
Nonelectrical machinery	3,226,100,000
Food and kindred products	2,163,800,000
Fabricated metal products	1,905,300,000
Transportation equipment	1,678,700,000
Rubber and miscellaneous plastics products	1,332,500,000
Electric and electronic equipment	1,228,900,000
Chemical and allied products	1,043,900,000

Among the nation's largest industrial corporations with headquarters in Oklahoma are Phillips Petroleum, located in Bartlesville (1984 sales: $15.5 billion); Kerr-McGee, Oklahoma City ($3.5 billion); and Wilson Foods, Oklahoma City ($1.9 billion). A large General Motors plant, opened in Oklahoma City in 1979, produced 148,316 automobiles, (2.6% of the US total) during the 1983 model year.

30COMMERCE

Wholesale establishments had sales totaling $28.1 billion in 1982, with sales of machinery, equipment, and supplies accounting for $5.6 billion and farm-product raw materials for $2.5 million. Oklahoma and Tulsa counties account for well over half of all sales by wholesalers in the state.

Retail establishments with payrolls had sales totaling $15.53 billion in 1982, ranking 23d in the US. Automotive dealers accounted for 24% of the total, followed by food stores, 17%; department stores, 9%; eating and drinking places, 9%; and gasoline service stations, 7%.

The value of foreign agricultural exports in 1981/82 was $781 million (16th in the US). In manufactured goods exported in 1981, Oklahoma ranked 30th among the states with a value of $1.5 billion.

31 CONSUMER PROTECTION

A Uniform Consumer Credit Code, passed in 1969, prohibits discrimination because of sex or marital status when credit is involved. It is administered by the Commission on Consumer Credit, which also maintains a program of consumer education and has the power to require lawful and businesslike procedures by lending agencies. The attorney general is responsible for enforcing the state's Consumer Protection Act.

32 BANKING

At the end of 1984, Oklahoma had 539 insured commercial banks. As of December 1982, the state's insured commercial banks had assets totaling $29.6 billion and deposit liabilities of $25.2 billion. Fifty-five savings and loan associations had combined assets of $7.2 billion as of 1984. There were 148 credit unions in that year. The failure of the Penn Square Bank in 1982 led to a series of mergers between strong institutions and distressed smaller ones and a 1983 revision of the state's banking laws which allowed for the restructuring of the industry. Concurrently, Oklahoma's banks were adversely affected by proposals to ease restrictions on reciprocal interstate banking in Missouri and Texas.

The State Banking Department has the responsibility for supervising all state-chartered banks, savings and loan associations, credit unions, and trust companies.

33 INSURANCE

In 1983, 75 life insurance companies were headquartered in Oklahoma. During that same year, Oklahoma purchased $11.5 billion of ordinary life insurance and received benefit payments of $596 million, including $196 million in death payments. Life insurance policies in force in 1983 numbered 4.8 million, with a total value of $67.1 billion and an average value per family of $50,900. Property and liability companies wrote premiums totaling $1.6 billion in 1983.

The State Insurance Commission has the responsibility for supervising and licensing all domestic and foreign insurance companies doing business in Oklahoma. The commission must approve certain life, accident, and health insurance policy forms before such contracts can be offered for sale to the public.

34 SECURITIES

There are no stock or commodity exchanges in Oklahoma. New York Stock Exchange member firms maintained 81 sales offices and 559 full-time registered representatives as of 30 December 1983. A total of 452,000 Oklahomans were shareowners of public corporations in 1983.

35 PUBLIC FINANCE

The Oklahoma budget is prepared by the director of state finance and submitted by the governor to the legislature each January. Article 10, section 23 of the Oklahoma Constitution requires a balanced budget, a provision that necessitated drastic cutting of appropriations as state revenues fell in 1983 and 1984 after a long period of expansion. In addition, state law requires a reserve fund equal to 10% of the approved budget. The appropriated budget represented only about 42% of all the revenue collections used for state functions in 1982/83, since major revenue sources are dedicated to specific uses by the constitution or by statute and are outside the annual legislative budgetary process. The fiscal year is 1 July–30 June.

The following table summarizes consolidated state revenues and expenditures for 1983/84 (in millions):

REVENUES	
Receipts from federal government	$1,055.6
Income tax	737.4
Severance tax	714.3
Sales and use taxes	457.1
Motor fuels tax	143.0
Tobacco tax	77.7
Motor vehicle excise tax	58.8
Other receipts	1,361.3
TOTAL	$4,605.2

EXPENDITURES	
Public education	$1,211.4
Human services	1,030.9
Higher education	628.5
Highways	480.9
Other social services	341.0
Land and structures	84.2
Other outlays	752.8
TOTAL	$4,529.7

The total indebtedness of state and local governments in Oklahoma surpassed $3.7 billion in mid-1982; the per capital figure of $1,251 was below the US average of $1,763.

36 TAXATION

In 1982, Oklahoma ranked 18th in the US (up from 35th in 1977) in state and local taxes per capita. Taxes and tax rates were raised substantially in 1982 and 1983.

As of 1984, the state income tax ranged from 0.5% on the first $1,000 of taxable income to 6% on amounts over $7,500. The corporate tax on net income was 4%. The state levies a sales and use tax of 3%, and 431 municipalities levy their own sales taxes of 1–4%. Property taxes remain the principal source of revenue for local governments. The motor fuels tax was increased from 6.58 cents to 9 cents per gallon of gasoline in 1984, the first such increase in 36 years. The severance tax on oil and gas is 7% of gross value; on uranium, 5%; and on asphalt and minerals, up to 1%. The gas conservation excise tax is 7 cents per million cu feet, the petroleum tax .085 of 1% of gross value.

Oklahoma's federal tax burden in 1982 was $7.7 billion, while the state received a total of $1.1 billion from the federal government in 1982/83 for public welfare ($474.8 million), education (207.9 million), general revenue sharing ($57.1 million), and other purposes. In 1982, Oklahomans filed 1,279,000 federal income tax returns, paying more than $4 billion in tax.

37 ECONOMIC POLICY

Probusiness measures in Oklahoma include a comparatively low property tax assessment level, a state oil depletion allowance of 30% to encourage oil exploration, a strong vocational education program to provide industrial manpower, and the creation in 1955 of a state Department of Economic Development, which encourages major US and foreign corporations to establish new manufacturing facilities in Oklahoma. Diversification through development of a larger high-tech sector was actively sought in the 1980s to lessen the impact of recession in the oil industry.

38 HEALTH

In 1982 there were 58,799 births, for a birthrate of 18.2 per 1,000. The infant mortality rate was 12.3 per 1,000 live births. The death rate for 1982 was 9.2. A total of 47,660 marriages were performed in 1982, at a rate of 15 per 1,000 inhabitants; 23,668 divorces were granted, for a rate of 7.4 per 1,000 residents. Approximately 12,700 legal abortions were performed in Oklahoma in 1982, a ratio of 229 per 1,000 live births. The leading causes of death and their rates in 1982 were heart disease, 3.4 per 1,000 population; cancer, 1.8; cerebrovascular disease, 0.8; and accidents, 0.6.

Oklahoma had 143 hospitals in 1983, providing 17,833 beds and recording 573,505 admissions. Hospital personnel in 1981 included 8,846 registered nurses. The average cost of a semiprivate hospital room in January 1984 was $165 per day, well below the US average. Oklahoma had 4,634 licensed physicians and 1,397 active dentists in 1982. The Health Sciences Center of the University of Oklahoma is located in Oklahoma City.

39SOCIAL WELFARE

Total expenditures in 1983/84 of the major state social welfare agency, the Department of Human Services, exceeded $1 billion. Benefits in aid to families with dependent children totaled $81 million in 1981/82; Medicaid expenditures reached $372 million in 1983. During 1983, 253,000 Oklahomans received food stamps at a federal cost of $109 million, and some 363,000 children received school lunches requiring a federal expenditure of $31 million.

Social Security benefits of $2.1 billion were paid to 484,000 retired persons, survivors, and disabled workers in 1983; monthly payments to retired workers averaged $414. Federal Supplemental Security Income payments to 59,100 aged, disabled, and blind persons were $112.3 million. During 1982, 244,000 Oklahomans received more than $195 million in unemployment insurance, for an average weekly benefit of $137.

40HOUSING

Indian tepees and settlers' sod houses dotted the Oklahoma plains when the "eighty-niners" swarmed into the territory; old neighborhoods in cities and towns of Oklahoma still retain some of the modest frame houses they built. More than 70% of all occupied housing units were owner-occupied in 1980, and most Oklahomans continue to prefer single-family dwellings, despite a recent trend toward condominiums. Modern underground homes and solar-heated dwellings can be seen in the university towns of Norman and Stillwater.

The 1980 census counted 1,234,698 year-round housing units, of which 20% dated from 1939 or earlier and 30% from the 1970s; only 2.6% were without one or more complete bathrooms. As of 1980 only 10.1% of all structures contained five or more dwelling units.

41EDUCATION

Oklahoma's educational enterprise is the largest expenditure item in the state budget. Teachers' salaries ranked 39th nationally in 1983/84, up from 42d in 1979, and capital expenditures on all education placed Oklahoma 20th among the 50 states in 1981/82, up from 34th in 1977/78.

About 66% of all Oklahomans 25 years of age or older were high school graduates in 1980; during the same year, 31% of adult state residents had at least one year of college, almost the same as the US average. There were 1,176 public elementary schools in 1981 and 679 secondary schools; three schools combined elementary and secondary instruction. In fall 1983, Oklahoma's public school enrollment totaled 592,195. A total of 30 private Catholic elementary schools enrolled 5,554 students in 1984; five Catholic high schools had 2,470 students.

Public higher educational institutions include 2 comprehensive institutions, 6 regional campuses, 18 senior and junior colleges, and a professional college. The comprehensive institutions, the University of Oklahoma (Norman) and Oklahoma State University (Stillwater), have more than 20,000 students each and offer the major graduate-level programs. The 16 private colleges and universities in Oklahoma increased their enrollment to 22,623 in 1983. Well-known institutions include Oral Roberts University and the University of Tulsa.

42ARTS

Major arts centers are located in Tulsa and Oklahoma City, but there are many art and crafts museums throughout the state. The Oklahoma Arts and Humanities Council, now known as the State Arts Council of Oklahoma, was created by the Oklahoma legislature in 1965.

Oklahoma City's leading cultural institution is the Oklahoma Symphony. The Tulsa Philharmonic, Tulsa Ballet Theater, and Tulsa Opera all appear at the Tulsa Performing Arts Center, a municipally owned and operated facility; this six-level center consists of a 2,500-seat concert hall, 450-seat theater, and two multilevel experimental theaters.

The Myriad Center in Oklahoma City and the Lloyd Noble Center in Norman host rock, jazz, and country music concerts.

43LIBRARIES AND MUSEUMS

Public libraries serve 29 counties in Oklahoma, while 12 bookmobiles aid in serving counties without libraries of their own. In 1983, a total of 4,418,442 volumes occupied public library shelves; total circulation was 10,405,709. The Five Civilized Tribes Museum Library in Muskogee has a large collection of Indian documents and art, while the Cherokee archives are held at the Cherokee National Historical Society in Tahlequah. The Morris Swett Library at Ft. Sill has a special collection on military history, particularly field artillery. The Oklahoma Department of Libraries in Oklahoma City has holdings covering law, library science, Oklahoma history, and other fields. Large academic libraries include those of the University of Oklahoma (Norman), with 2,000,000 volumes in 1984, and Oklahoma State University Library (Stillwater), with 1,346,878.

Oklahoma has 111 museums and historic sites. The Philbrook Art Center in Tulsa houses important collections of Indian, Renaissance, and Oriental art. Also in Tulsa are the Thomas Gilcrease Institute of American History and Art. Major museums in Norman are the University of Oklahoma's Museum of Art and the Stovall Museum of Science and Industry. The Oklahoma Art Center, National Cowboy Hall of Fame and Western Heritage Center, Oklahoma Heritage Association, Oklahoma Historical Society Museum, Oklahoma Museum of Art, State Museum of Oklahoma, and the Omniplex Science Museum are major attractions in Oklahoma City. Other museums of special interest include the Museum of the Great Plains in Lawton, the Will Rogers Memorial in Claremore, Cherokee National Museum in Tahlequah, and the Woolaroc Museum in Bartlesville.

44COMMUNICATIONS

The Butterfield Stage and Overland Mail delivered the mail to Millerton on 18 September 1858 as part of the first US transcontinental postal route. After the Civil War, the early railroads delivered mail and parcels to the Oklahoma and Indian territories. As of 1985 there were 8,242 postal employees and 631 post offices in the state.

In 1980, 92% of Oklahoma's occupied housing units had telephones. In 1984, Oklahoma had 67 AM and 87 FM radio stations, 16 commercial and 4 educational television channels, and 206 cable television systems serving 283 communities and 515,483 subscribers.

45PRESS

In 1984, Oklahoma had 7 morning dailies with a combined 1984 circulation of 444,453, 45 evening dailies with 373,227, and 44 Sunday newspapers with 905,300. Leading dailies and their circulation in 1984 were as follows:

AREA	NAME	DAILY	SUNDAY
Oklahoma City	Daily Oklahoman (m)	236,470	
	Sunday Oklahoman (S)		313,508
Tulsa	Tulsa World (m,S)	132,789	226,249
	Tulsa Tribune (e)	77,067	

As of 1985 there were 156 newspapers that appeared weekly or up to three times a week; most had circulations of less than 10,000 copies.

Tulsa and Oklahoma City each have monthly city-interest publications, and the University of Oklahoma has a highly active university press.

46ORGANIZATIONS

The 1982 census of service industries counted 848 organizations in Oklahoma, including 195 business associations; 420 civic, social, and fraternal associations; and 23 educational, scientific, and research associations. Among the organizations headquartered in Oklahoma are the Football Writers Association of America (Edmond); the International Professional Rodeo Association (Pauls Valley); the Amateur Softball Association of America and the International Softball Federation, both in Oklahoma City; and the American Association of Petroleum Geologists, the Gas Processors Association, and the US Jaycees, all located in Tulsa.

47 TOURISM, TRAVEL, AND RECREATION

Tourism has become a growing sector of Oklahoma's economy. It was the second largest employer in the state in 1982, generating 5.4% of all jobs and a payroll of $581 million. Tourists spent over $2.6 million in the state in 1982. In 1983, Oklahoma's 35 state parks drew 13.7 million visitors; 22 state recreation areas drew 4 million visitors. The national park service maintains one facility in Oklahoma—Chickasaw National Recreation Area, centering on artificial Lake Arbuckle.

The state also maintains and operates the American Indian Hall of Fame, in Anadarko; Black Kettle Museum, in Cheyenne; the T. B. Ferguson Home in Watonga; the Murrell Home, south of Tahlequah; the Pawnee Bill Museum, in Pawnee; the Pioneer Woman Statue and Museum, in Ponca City; the Chisholm Trail Museum, in Kingfisher; and the Western Trails Museum, in Clinton.

Licenses were issued to 312,721 hunters and 663,058 fishermen in 1982/83. National wildlife refuges include Optima, Salt Plains, Sequoyah, Tishomingo, Washita, and Wichita Mountains; they have a combined area of 140,696 acres (56,938 hectares).

48 SPORTS

Oklahoma has no major league professional teams. The class-AAA baseball 89ers play in Oklahoma City. Sports on the college level are still the primary source of pride for Oklahomans. As of mid-1985, the University of Oklahoma Sooners had won 5 national football titles in season-ending polls, 11 Big Eight Conference crowns, Big Eight basketball crowns in 1979 and 1984, and NCAA championships in wrestling, baseball, and gymnastics. The Oklahoma State University Cowboys have captured NCAA and Big Eight titles in basketball, baseball, and golf, and as of mid-1980 had won 27 national championships in wrestling.

Oklahoma City hosts the National Finals of Rodeo every December. In golf, Tulsa has been the site of several US Open tournaments. The Softball Hall of Fame is in Oklahoma City.

49 FAMOUS OKLAHOMANS

Carl Albert (b.1908), a McAlester native, has held the highest public position of any Oklahoman. Elected to the US House of Representatives in 1947, he became majority leader in 1962 and served as speaker of the House from 1971 until his retirement in 1976. Patrick Jay Hurley (1883-1963), the first Oklahoman appointed to a cabinet post, was secretary of war under Herbert Hoover and later ambassador to China.

William "Alfalfa Bill" Murray (b.Texas, 1869-1956) was president of the state constitutional convention and served as governor from 1931 to 1935. Robert S. Kerr (1896-1963), founder of Kerr-McGee Oil, was the state's first native-born governor, serving from 1943 to 1947; elected to the US Senate in 1948, he became an influential Democratic leader. A(lmer) S(tillwell) Mike Monroney (1902-80) served as US representative from 1939 to 1951 and senator from 1951 to 1969.

Oklahomans have been prominent in literature and the arts. Journalist and historian Marquis James (b.Missouri, 1891-1955) won a Pulitzer Prize in 1930 for his biography of Sam Houston and another in 1938 for Andrew Jackson; John Berryman (1914-72) won the 1965 Pulitzer Prize in poetry for *77 Dream Songs, 1964*; and Ralph Ellison (b.1914) won the 1953 National

Book Award for his novel *Invisible Man*. The popular musical *Oklahoma!*, by Richard Rodgers and Oscar Hammerstein 2d, is based on *Green Grow the Lilacs* by Oklahoman Lynn Riggs (1899-1954). N(avarre) Scott Momaday (b. 1934), born in Lawton, received a Pulitzer Prize in 1969 for *House Made of Dawn*. Woodrow Crumbo (b.1912) and Allen Houser (b.1914) are prominent Indian artists born in the state.

Just about the best-known Oklahoman was William Penn Adair "Will" Rogers (1879-1935), the beloved humorist and writer who spread cheer in the dreary days of the depression. Part Cherokee, Rogers was a horse rider, trick roper, and stage and movie star until he was killed in a plane crash in Alaska. Among his gifts to the American language are the oft-quoted expressions "I never met a man I didn't like" and "All I know is what I read in the newspapers." Other prominent performing artists include singer-songwriter Woody Guthrie (1912-67), composer of "This Land Is Your Land," among other classics; ballerina Maria Tallchief (b.1925); popular singer Patti Page (b.1927); and operatic soprano Roberta Knie (b.1938). Famous Oklahoma actors include (Francis) Van Heflin (1910–71), Ben Johnson (b.1918), Jennifer Jones (b.1919), Tony Randall (b.1920), James Garner (James Baumgardner, b.1928), and Cleavon Little (b.1939). Paul Harvey (b.1918) is a widely syndicated radio commentator.

James Francis "Jim" Thorpe (1888-1953) became known as the "world's greatest athlete" after his pentathlon and decathlon performances at the 1912 Olympic Games; of Indian ancestry, Thorpe also starred in baseball, football, and other sports. Bud Wilkinson (b.Minnesota, 1916) coached the University of Oklahoma football team to a record 47-game unbeaten streak in the 1950s. Baseball stars Paul Waner (1903-65) and his brother Lloyd (b.1906), Mickey Mantle (b.1931), Wilver Dornel "Willie" Stargell (b.1941), and Johnny Bench (b.1947) are native Oklahomans.

50 BIBLIOGRAPHY

Bicha, Karel D. *The Czechs in Oklahoma*. Norman: University of Oklahoma Press, 1980.

Blessing, Patrick. *The British and Irish in Oklahoma*. Norman: University of Oklahoma Press, 1980

Brown, Kenny L. *Italians in Oklahoma*. Norman: University of Oklahoma Press, 1978.

Fischer, John. *From the High Plains*. New York: Harper & Row, 1978.

Franklin, Jimmie Lewis. *Blacks in Oklahoma*. Norman: University of Oklahoma Press, 1980.

Gibson, Arrell M. *The Oklahoma Story*. Norman: University of Oklahoma Press, 1978.

Goble, Danney. *Progressive Oklahoma: The Making of a New Kind of State*. Norman: University of Oklahoma Press, 1980.

Morgan, H. Wayne and Anne Hodges. *Oklahoma: A Bicentennial History*. New York: Norton, 1977.

Morgan, Anne H. and H. Wayne. *Oklahoma: New Views of the Forty-sixth State*. Norman: University of Oklahoma Press, 1982.

Oklahoma, University of. College of Business Administration. Center for Economic and Management Research. *Statistical Abstract of Oklahoma, 1984*. Norman, 1984.

Rohrs, Richard. *Germans in Oklahoma*. Norman: University of Oklahoma Press, 1980.

Smith, Michael. *Mexicans in Oklahoma*. Norman: University of Oklahoma Press, 1980.

Strain, Jack M. *An Outline of Oklahoma Government*. Edmond, Okla.: Central State University, 1978.

Strickland, Rennard. *Indians in Oklahoma*. Norman: University of Oklahoma Press, 1980.

Tobias, Henry J. *Jews in Oklahoma*. Norman: University of Oklahoma Press, 1980.

OREGON

State of Oregon

ORIGIN OF STATE NAME: Unknown; name first applied to the river now known as the Columbia. **NICKNAME:** The Beaver State. **CAPITAL:** Salem. **ENTERED UNION:** 14 February 1859 (33d). **SONG:** "Oregon, My Oregon." **DANCE:** Square dance. **POET LAUREATE:** William E. Stafford. **MOTTO:** The Union. **COLORS:** Navy blue and gold. **FLAG:** The flag consists of a navy-blue field with gold lettering and illustrations. Obverse: the shield from the state seal, supported by 33 stars, with the words "State of Oregon" above and the year of admission below. Reverse: a beaver. **OFFICIAL SEAL:** A shield, supported by 33 stars and crested by an American eagle, depicts mountains and forests, an elk, a covered wagon and ox team, wheat, a plow, a pickax, and the state motto; in the background, as the sun sets over the Pacific, an American merchant ship arrives as a British man-o'-war departs. The words "State of Oregon 1859" surround the whole. **ANIMAL:** American beaver. **BIRD:** Western meadowlark. **FISH:** Chinook salmon. **FLOWER:** Oregon grape. **TREE:** Douglas fir. **ROCK:** Thunderegg (geode). **INSECT:** Oregon swallowtail butterfly. **LEGAL HOLIDAYS:** New Year's Day, 1 January; Birthday of Martin Luther King, Jr., 3d Monday in January; Lincoln's Birthday, 1st Monday in February; Washington's Birthday, 3d Monday in February; Memorial Day, last Monday in May; Independence Day, 4 July; Labor Day, 1st Monday in September; Veterans Day, 11 November; Thanksgiving Day, 4th Thursday in November; Christmas Day, 25 December. Designated as commemoration days are Oregon's Admission into the Union, 14 February, and Columbus Day, 12 October. **TIME:** 5 AM MST = noon GMT; 4 AM PST = noon GMT.

¹LOCATION, SIZE, AND EXTENT

Located on the Pacific coast of the northwestern US, Oregon ranks 10th in size among the 50 states.

The total area of Oregon is 97,073 sq mi (251,419 sq km), with land comprising 96,184 sq mi (249,117 sq km) and inland water 889 sq mi (2,302 sq km). Oregon extends 395 mi (636 km) E–W; the state's maximum N–S extension is 295 mi (475 km).

Oregon is bordered on the N by Washington (with most of the line formed by the Columbia River); on the E by Idaho (with part of the line defined by the Snake River); on the S by Nevada and California; and on the W by the Pacific Ocean. The total boundary length of Oregon is *1,444 mi (2,324 km),* including a general coastline of 296 mi (476 km); the tidal shoreline extends 1,410 mi (2,269 km). The state's geographic center is in Crook County, 25 mi (40 km) SSE of Prineville.

²TOPOGRAPHY

The Cascade Range, extending north–south, divides Oregon into distinct eastern and western regions, each of which contains a great variety of landforms.

At the state's western edge, the Coast Range, a relatively low mountain system, rises from the beaches, bays, and rugged headlands of the Pacific coast. Between the Coast and Cascade ranges lie fertile valleys, the largest being the Willamette Valley, Oregon's heartland. The two-thirds of the state lying east of the Cascade Range consists generally of arid plateau cut by river canyons, with rolling hills in the north-central portion giving way to the Blue Mountains in the northeast. The Great Basin in the southeast is characterized by fault-block ridges, weathered buttes, and remnants of large prehistoric lakes.

The Cascades, Oregon's highest mountains, contain nine snow-capped volcanic peaks more than 9,000 feet (2,700 meters) high, of which the highest is Mt. Hood, at 11,235 feet (3,424 meters); a dormant volcano, Mt. Hood last erupted in 1865 (Mt. St. Helens, which erupted in 1980, is only 60 mi—97 km—to the northwest, in Washington.) The Blue Mountains include several rugged subranges interspersed with plateaus, alluvial basins, and deep river canyons. The Klamath Mountains in the southwest form a jumble of ridges where the Coast and Cascade ranges join.

Oregon is drained by many rivers, but the Columbia, demarcating most of the northern border with Washington, is by far the biggest and most important. Originating in Canada, it flows more than 1,200 mi (1,900 km) to the Pacific Ocean. With a mean flow rate of 250,134 cu feet per second, the Columbia is the 3d-largest river in the US. It drains some 58% of Oregon's surface by way of a series of northward-flowing rivers, including the Deschutes, John Day, and Umatilla. The largest of the Columbia's tributaries in Oregon, and longest river entirely within the state, is the Willamette, which drains a fertile valley more than 100 mi (160 km) long. Better than half of Oregon's eastern boundary with Idaho is formed by the Snake River, which flows through Hells Canyon, one of the deepest canyons in North America.

Oregon has 19 natural lakes with a surface area of more than 3,000 acres (1,200 hectares), and many smaller ones. The largest is Upper Klamath Lake, which covers 58,922 acres (23,845 hectares) and is quite shallow. The most famous, however, is Crater Lake, which formed in the crater created by the violent eruption of Mt. Mazama several thousand years ago and is now a national park. Its depth of 1,932 feet (589 meters)—greater than any other lake in the US—and its nearly circular expanse of bright-blue water, edged by the crater's rim, make it a natural wonder.

³CLIMATE

Oregon has a generally temperate climate, but there are marked regional variations. The Cascade Range separates the state into two broad climatic zones: the western third, with relatively heavy precipitation and moderate temperatures, and the eastern two-thirds, with relatively little precipitation and more extreme temperatures. Within these general regions, climate depends largely on elevation and land configuration.

In January, normal daily mean temperatures range from more than 45°F (7°C) in the coastal sections to between 25°F (−4°C) and 28°F (−2°C) in the southeast. In July, the normal daily means range between 65°F (18°C) and 70°F (21°C) in the plateau regions and central valleys and between 70°F (21°C) and 78°F (26°C) along the eastern border. The record low temperature, −54°F (−48°C), was registered at Seneca on 10 February 1933; the all-time high, 119°F (48°C), at Pendleton on 10 August 1938.

451

The Cascades serve as a barrier to the warm, moist winds blowing in from the Pacific, confining most precipitation to western Oregon. The average annual rainfall varies from less than 8 in (20 cm) in the drier plateau regions to as much as 200 in (508 cm) at locations on the upper west slopes of the Coast Range. In the Blue Mountains and the Columbia River Basin, totals are about 15 in (38 cm) to 20 in (51 cm). In 1982, Portland averaged 43 in (109 cm) of precipitation but only 2 in (5 cm) of sleet or snow; fog is common, and the sun shines, on average, during only 49% of the daylight hours—one of the lowest such percentages for any major US city. From 300 in (760 cm) to 550 in (1,400 cm) of snow falls each year in the highest reaches of the Cascades.

⁴FLORA AND FAUNA

With its variety of climatic conditions and surface features, Oregon has a diverse assortment of vegetation and wildlife, including 78 native tree species. The coastal region is covered by a rain forest of spruce, hemlock, and cedar rising above dense underbrush. A short distance inland, the stands of Douglas fir—Oregon's state tree and dominant timber resource—begin, extending across the western slopes to the summit of the Cascade Range. Where the Douglas fir has been destroyed by fire or logging, alder and various types of berries grow. In the high elevations of the Cascades, Douglas fir gives way to pines and true firs. Ponderosa pine predominates on the eastern slopes, while in areas too dry for pine the forests give way to open range, which, in its natural state, is characterized by sagebrush, occasional juniper trees, and sparse grasses. The state's many species of smaller indigenous plants include Oregon grape—the state flower—as well as salmonberry, huckleberry, blackberry, and many other berries. The Malheur wire-lettuce and MacFarlane's four-o'clock are endangered.

More than 130 species of mammal are native to Oregon, of which 28 are found throughout the state. Many species, such as the cougar and bear, are protected, either entirely or through hunting restrictions. The bighorn sheep, once extirpated in Oregon, has been reintroduced in limited numbers; the Columbian white-tailed deer, with an extremely limited habitat along the Columbia River, is still classified as endangered. Deer and elk are popular game mammals, with herds managed by the state: mule deer predominate in eastern Oregon, black-tailed deer in the west. Among introduced mammals, the nutria and opossum are now present in large numbers. At least 60 species of fish are found in Oregon, including five different salmon species, of which the Chinook is the largest and the coho most common. Salmon form the basis of Oregon's sport and commercial fishing, although dams and development have blocked many spawning areas, causing a decline in numbers and heavy reliance on hatcheries to continue the runs. Hundreds of species of birds inhabit Oregon, either year-round or during particular seasons. The state lies in the path of the Pacific Flyway, a major route for migratory waterfowl, and large numbers of geese and ducks may be found in western Oregon and marshy areas east of the Cascades. Extensive bird refuges have been established in various parts of the state. The bald eagle, southern sea otter, and Oregon silver-spot butterfly are considered threatened, while the brown pelican, short-tailed albatross, California condor, Aleutian Canada goose, American and Arctic peregrine falcons, and Borax Lake chub are classified as endangered.

⁵ENVIRONMENTAL PROTECTION

Oregon has been among the most active states in environmental protection. In 1938, the polluted condition of the Willamette River led to the enactment, by initiative, of one of the nation's first comprehensive water pollution control laws, which helped restore the river's quality for swimming and fishing. An air pollution control law was enacted in 1951, and air and water quality programs were placed under the new Department of Environmental Quality (DEQ) in 1969. This department is Oregon's major environmental protection agency, enforcing standards for air and water quality and solid and hazardous waste disposal. A vehicle inspection program has been instituted to reduce exhaust emissions in the Portland area. The DEQ monitors 18 river basins for water quality and issues permits to the more than 700 businesses, industries, and government bodies that discharge waste water into public waters.

Oregon had 100 municipal solid-waste landfills in 1983. About 50 recycling companies served more than 200 community and commercial recycling programs in 65 cities and 28 counties. The state had one hazardous waste disposal and collection facility, near Arlington. As of 1983, three sites in Oregon were on the federal Environmental Protection Agency's priority list of hazardous waste sites in need of cleanup: United Chrome in Corvallis; Gould, Inc., in northwest Portland; and Teledyne Wah Chang in Albany.

In 1973, the legislature enacted what has become known as the Oregon Bottle Bill, the first state law prohibiting the sale of nonreturnable beer or soft-drink containers. The DEQ estimates that more than 90% of beverage containers are returned for recycling. The success of the Bottle Bill was partly responsible for the passage, in 1983, of the Recycling Opportunity Act, which is intended to reduce the amount of solid waste generated, to foster the reuse and recycling of materials, and to aid the recovery of energy from materials that cannot be reused or recycled. By 1 July 1986, collection sites for recyclable materials will be located at landfills and other places, and collections will be made at homes, businesses, and industries in cities with more than 4,000 people.

⁶POPULATION

Oregon ranked 30th among the 50 states at the 1980 census, with a population of 2,633,105. Like other western states, Oregon has experienced a more rapid population growth than that of the US as a whole in recent decades. The 1980 census figure represented a 26% increase over the 1970 census population; the 1985 population estimate, 2,680,067, was 1.8% more than in 1980. The Census Bureau population estimate for 1990 is 3,318,600. Oregon's estimated population density in 1985 was 28 per sq mi (11 per sq km), less than half the national average.

As of 1984, about 42% of all Oregonians lived in the Portland region, while another 33% lived in the remainder of the Willamette Valley, particularly in and around Salem and Eugene. The city of Portland had an estimated 365,861 residents in 1984; the Portland Consolidated Metropolitan Statistical Area (which includes Vancouver, Wash.) had an estimated population of 1,340,900, ranking 26th in the US. The population of Eugene was 101,602; Salem, 90,323.

⁷ETHNIC GROUPS

Oregon's Indians numbered 26,587 in 1980, with about 90% of the population living in urban areas. The state has four reservations, and important salmon fishing rights in the north are reserved under treaty. About 37,000 black Americans lived in Oregon in 1980, most of them in the Portland area. People of Hispanic descent numbered about 66,000 in the same year. In 1980 there were 8,429 Japanese, 8,033 Chinese, 5,564 Vietnamese, 4,427 Koreans, 4,257 Filipinos, 2,310 Laotians, and 1,938 Asian Indians. French Canadians have lived in Oregon since the opening of the territory, and they have continued to come in a small but steady migration. In all, the 1980 census counted some 108,000 Oregonians of foreign birth, accounting for 4.1% of the population.

⁸LANGUAGES

Place-names such as Umatilla, Coos Bay, Klamath Falls, and Tillamook reflect the variety of Indian tribes that white settlers found in Oregon territory.

The Midland dialect dominates Oregon English, except for an apparent Northern dialect influence in the Willamette Valley. Throughout the state, *foreign* and *orange* have the /aw/ vowel, and *tomorrow* has the /ah/ of *father*.

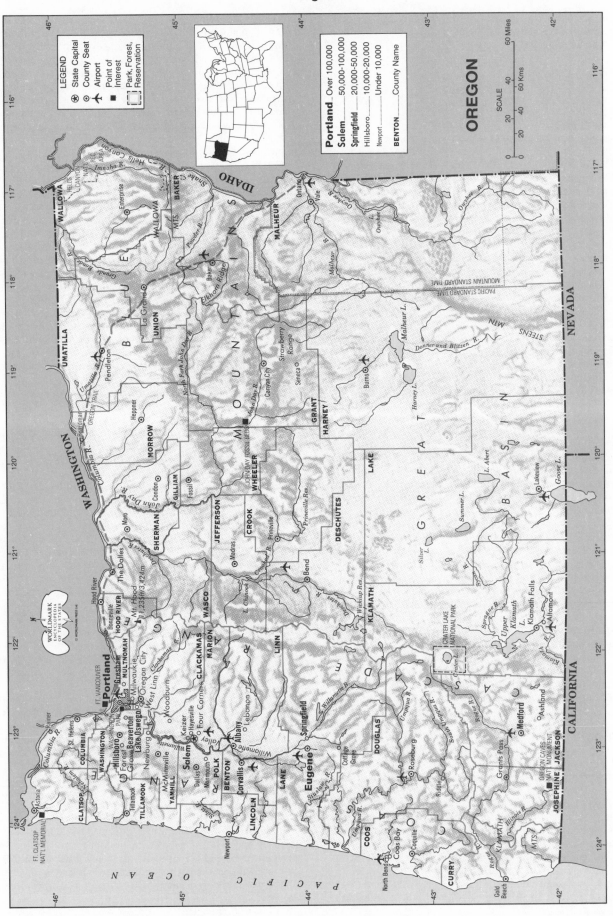

OREGON

LEGEND
⊕ State Capital
⊙ County Seat
✈ Airport
■ Point of Interest
▢ Park, Forest, Reservation

Portland ..... Over 100,000
Salem 50,000-100,000
Springfield 20,000-50,000
Hillsboro 10,000-20,000
Newport ..... Under 10,000
BENTON ..... County Name

SCALE
0   20   40   60 Miles
0   20   40   60 Kms

See endsheet maps: B2.

LOCATION: 42° to 46° 15′N; 116°33′ to 124°32′W. **BOUNDARIES:** Washington line, 443 mi (713 km); Idaho line, 332 mi (534 km); Nevada line, 153 mi (246 km); California line, 220 mi (354 km); Pacific Ocean coastline, 296 mi (477 km).

In 1980, 2,376,608 Oregonians—91% of the population 3 years old or older—spoke only English at home. Other languages spoken at home included:

Spanish	43,401	Vietnamese	4,758
German	20,002	Japanese	4,026
French	8,635	Italian	3,278
Chinese	6,139	Korean	3,234

9RELIGIONS

Just over one-third of Oregon's population is affiliated with an organized religion, well below the national average. The leading Christian denominations were the Roman Catholic Church, with 323,281 members in 1984, and the Church of Jesus Christ of Latter-day Saints (Mormon), with 74,099 adherents in 1980. Other major Protestant groups, with their 1980 adherents, were United Methodist, 49,578; Assemblies of God, 46,902; United Presbyterian, 42,053; and Conservative Baptist, 39,412. Jewish Oregonians were estimated to number 10,940 in 1984.

In 1985 there were some 2,000 followers of the Indian guru Bhagwan Shree Rajneesh living in the new town of Rajneesh Purahm. The guru departed the US voluntarily in November 1985 after pleading guilty to counts of arranging sham marriages and lying on his own visa.

10TRANSPORTATION

With the state's major deepwater port and international airport, Portland is the transportation hub of Oregon. The state has 2,940 mi (4,731 km) of Class I track and is served by several major rail systems, including the Southern Pacific, Union Pacific, and Burlington Northern. Lumber and other wood products are the major commodities carried; significantly more rail freight is shipped out of Oregon than is shipped in. Amtrak provides passenger service north–south through Portland, Salem, and Eugene, and east–west through Pendleton and Portland. Thirty cities in Oregon operated local transit systems in 1985.

Starting with pioneer trails and toll roads, Oregon's roads and highways had become a network extending 148,938 mi (239,693 km) by 1983. The main interstate highways are I-5, connecting the major Willamette Valley cities, and I-84 (formerly I-80N), running northwest from Ontario in eastern Oregon and then along the northern border. At the beginning of 1984 there were 2,120,-523 motor vehicles, including 1,476,548 passenger cars, registered in Oregon, with 1,900,674 licensed drivers.

The Columbia River forms the major inland waterway for the Pacific Northwest, with barge navigation possible for 464 mi (747 km) upstream to Lewiston, Idaho, via the Snake River. Wheat from eastern Oregon and Washington is shipped downstream to Portland for reloading onto oceangoing vessels. The Port of Portland operates five major cargo terminals. Oregon also has several important coastal harbors, including Astoria, Newport, and Coos Bay, the largest lumber export harbor in the US.

In 1983, Oregon had 341 airfields (104 public, 237 private), of which the largest and busiest, Portland International Airport, handled 35,593 scheduled departures and enplaned 2,074,741 passengers.

11HISTORY

The land now known as Oregon has been inhabited for at least 10,000 years, the age assigned to woven brush sandals found in caves along what was once the shore of a large inland lake. Later, a variety of Indian cultures evolved. Along the coast and lower Columbia River lived peoples of the Northern Coast Culture, who ate salmon and other marine life, built large dugout canoes and cedar plank houses, and possessed a complex social structure, including slavery, that emphasized status and wealth. East of the Cascade Range were hunter-gatherers who migrated from place to place as the food supply dictated.

The first European to see Oregon was probably Sir Francis Drake. In 1578, while on a raiding expedition against the Spanish,

Drake reported sighting what is believed to be the Oregon coast before being forced to return southward by "vile, thicke and stinking fogges." For most of the next 200 years, European contact was limited to occasional sightings by mariners, who considered the coast too dangerous for landing. In 1778, however, British Captain James Cook, on his third voyage of discovery, visited the Northwest and named several Oregon capes. Soon afterward, American ships arrived in search of sea otter and other furs. A Yankee merchant captain, Robert Gray, discovered the Columbia River (which he named for his ship) in 1792, contributing to the US claim to the Northwest.

The first overland trek to Oregon was the Lewis and Clark Expedition, which traveled from St. Louis to the mouth of the Columbia, where it spent the winter of 1805/6. In 1811, a party of fur traders employed by New York merchant John Jacob Astor arrived by ship at the mouth of the Columbia and built a trading post named Astoria. The venture was not a success and was sold three years later to British interests, but some of the Astor party stayed, becoming Oregon's first permanent white residents. For the next 20 years, European and US interest in Oregon focused on the quest for beaver pelts. Agents of the British North West Company (which merged in 1821 with the Hudson's Bay Company) and some rival American parties explored the region, mapped trails, and established trading posts. Although Britain and the US had agreed to a treaty of joint occupation in 1818, the de facto governor from 1824 to the early 1840s was Dr. John McLoughlin, the Hudson's Bay Company chief factor at Ft. Vancouver in Washington.

Another major influence on the region was Protestant missionary activity, which began with the arrival of Jason Lee, a Methodist missionary, in 1834. Lee started his mission in the Willamette Valley, near present-day Salem. After a lecture tour of the East, he returned to Oregon in 1840 with 50 settlers and assistants. While Lee's mission was of little help to the local Indians, most of whom had been killed off by white men's diseases, it served as a base for subsequent American settlement and as a counterbalance to the Hudson's Bay Company.

The first major wagon trains arrived by way of the Oregon Trail in the early 1840s. On 2 May 1843, as a "great migration" of 875 men, women, and children was crossing the plains, about 100 settlers met at the Willamette Valley community of Champoeg and voted to form a provisional government. That government remained in power until 1849, when Oregon became a territory, three years after the Oregon Treaty between Great Britain and the US established the present US-Canadian boundary. As originally constituted, Oregon Territory included present-day Washington and much of Montana, Idaho, and Wyoming. A constitution prepared by an elected convention was approved in November 1857, and after a delay caused by North-South rivalries, on 14 February 1859, Congress voted to make Oregon, reduced to its present borders, the 33d state.

Oregon remained relatively isolated until the completion of the first transcontinental railroad link in 1883. State politics, which had followed a pattern of venality and influence buying, underwent an upheaval in the early 1900s. Reformers led by William S. U'Ren instituted what became known as the "Oregon System" of initiative, referendum, and recall, by which voters could legislate directly and remove corrupt elected officials.

Oregon's population grew steadily in the 20th century as migration into the state continued. Improved transportation helped make the state the nation's leading lumber producer and a major exporter of agricultural products. Development was also aided by hydroelectric projects, many undertaken by the federal government. The principal economic changes since World War II have been the growth of the aluminum industry, a rapid expansion of the tourist trade, and the creation of a growing electronics industry.

[12]STATE GOVERNMENT

The Oregon constitution—drafted and approved in 1857, effective in 1859, and amended 172 times as of 1984—governs the state today. The first decade of the 20th century saw the passage of numerous progressive amendments, including provisions for the direct election of senators, the rights of initiative, referendum, and recall, and a direct primary system.

The constitution establishes a 60-member house of representatives, elected for 2 years, and a senate of 30 members, serving 4-year terms. Major executive officials include the governor, secretary of state, attorney general, state treasurer, superintendent of public instruction, and commissioner of labor and industries, all elected for 4 years. The governor, who may serve no more than 8 years in any 12-year period, must be a US citizen, must be at least 30 years of age, and must have been a resident of the state for 3 years before assuming office. Much policy in Oregon is set by boards and commissions whose members are appointed by the governor subject to confirmation by the senate.

Bills become law when approved by a majority of house and senate and either signed by the governor or left unsigned for 5 days when the legislature is in session or for 20 days after it has adjourned. Measures presented to the voters by the legislature or by petition become law when approved by a majority of the electorate. The governor may veto a legislative bill, but the legislature may override a veto by a two-thirds vote of those present in each house. Proposed constitutional amendments require voter approval to take effect, and they may be placed on the ballot either by the legislature or by initiative petition.

US citizens over age 18 are entitled to vote, subject to residency requirements of 20 days.

[13]POLITICAL PARTIES

Oregon has two major political parties, Democratic and Republican. Partly because of the role the direct primary system plays in choosing nominees, party organization is relatively weak. There is a strong tradition of political independence, evidenced in 1976 when Oregon gave independent presidential candidate Eugene McCarthy 3.9% of the vote—his highest percentage in any state—a total that probably cost Jimmy Carter Oregon's 6 electoral votes. Another independent, John Anderson, won 112,389 votes (9.5%) in the 1980 presidential election.

As of November 1984 there were 792,208 registered Democrats and 594,307 registered Republicans, along with 222,098 indepen-

dents and minor-party members. The Republicans held the governorship and both US Senate seats, while Democrats controlled both houses of the legislature. Oregon voters gave Republican presidential nominee Ronald Reagan a large plurality of the popular vote in November 1980, as well as in 1984, when US Senator Mark Hatfield, also a Republican, won reelection to a fourth term. In 1984, Oregon had four Hispanic and two Asian Pacific American elected officials. In 1985 there were 11 black elected officials, and 20% of Oregon's state legislators were women—one of the highest such percentages in the US.

[14]LOCAL GOVERNMENT

As of 1984, Oregon had 36 counties and 242 incorporated cities. Towns and cities enjoy home rule, the right to choose their own form of government and enact legislation on matters of local concern. In 1958, home rule was extended to counties, but by 1984, only seven counties had adopted home-rule charters. Most of Oregon's larger cities have council-manager forms of government. Typical elected county officials are three to five commissioners, assessor, district attorney, sheriff, and treasurer.

The state constitution gives voters strong control over local government revenue by requiring voter approval of property tax levies.

[15]STATE SERVICES

Special offices within the governor's office include the Office of the Citizens' Representative, the state Affirmative Action Office, and the Long-Term Care Ombudsman. The Oregon Government Ethics Commission is a citizens' panel, established in 1974, to investigate conflicts of interest involving public officials and to levy civil penalties for infractions. Responsibility for educational matters is divided among the Board of Education, which oversees primary and secondary schools and community colleges; the Board of Higher Education, which controls the state college and university system; and the Educational Coordinating Commission, which monitors programs and advises the governor and legislature on policy.

State highways, airfields, and public transit systems are under the jurisdiction of the Department of Transportation, which is headed by an appointed commission. The largest state agency is the Department of Human Resources; nearly one-fourth of the state's budget and work force is committed to this department's 250 programs, which include corrections, children's services, adult and family services, health, mental health, and vocational reha-

Oregon Presidential Vote by Political Parties, 1948–84

YEAR	ELECTORAL VOTE	OREGON WINNER	DEMOCRAT	REPUBLICAN	PROGRESSIVE	SOCIALIST
1948	6	Dewey (R)	243,147	260,904	14,978	5,051
1952	6	*Eisenhower (R)	270,579	420,815	3,665	—
1956	6	*Eisenhower (R)	329,204	406,393	—	—
1960	6	Nixon (R)	367,402	408,065	—	—
1964	6	*Johnson (D)	501,017	282,779	—	—
						AMERICAN IND.
1968	6	*Nixon (R)	358,866	408,433	—	49,683
						AMERICAN
1972	6	*Nixon (R)	392,760	486,686	—	46,211
1976	6	Ford (R)	490,407	492,120	—	—
					LIBERTARIAN	CITIZENS
1980	6	*Reagan (R)	456,890	571,044	25,838	13,642
1984	7	*Reagan (R)	536,479	685,700	—	—

*Won US presidential election

bilitation. State agencies involved in environmental matters include the Department of Environmental Quality, the Department of Land Conservation and Development, and the departments of Energy, Forestry, and Water Resources. State-owned lands are administered through the Land Board.

¹⁶JUDICIAL SYSTEM

Oregon's highest court is the supreme court, consisting of 7 justices who elect one of their number to serve as chief justice. It accepts cases on review from the 10-judge court of appeals, which has exclusive jurisdiction over all criminal and civil appeals from lower courts and over certain actions of state agencies. Circuit courts (with 84 circuit judges as of 1984) are the trial courts of original jurisdiction for civil and criminal matters. The more populous counties also have district courts, which hear minor civil, criminal, and traffic matters; 39 localities retain justices of the peace, also with jurisdiction over minor cases. State judges are elected by nonpartisan ballot for six-year terms.

Oregon's penal system is operated by the Corrections Division of the Department of Human Resources. As measured by the FBI Crime Index, Oregon's crime rate was above the national average in 1983, with especially high rates for forcible rape (41 per 100,000 population), burglary (1,746), and larceny-theft (3,715).

¹⁷ARMED FORCES

Oregon has no major military facilities, and had fewer than 4,000 military and civilian defense personnel in 1983/84. The US Coast Guard does maintain search-and-rescue facilities, and the Army Corps of Engineers operates a number of hydroelectric projects in the state. Military contract awards in 1983/84 totaled $229 million.

As of 30 September 1984, some 402,000 military veterans were living in Oregon, of whom 4,000 served in World War I, 142,000 in World War II, 65,000 during the Korean conflict, and 137,000 during the Viet-Nam era. Federal veterans' benefits totaled $349 million in 1982/83.

Oregon had 6,114 law enforcement officers in 1983, of whom 981 were state police. Army National Guard personnel numbered 7,574 in December 1984, Air National Guard 2,027 in February 1985.

¹⁸MIGRATION

The Oregon Trail was the route along which thousands of settlers traveled to Oregon by covered wagon in the 1840s and 1850s. This early immigration was predominantly from midwestern states. After the completion of the transcontinental railroad, northeastern states supplied an increasing proportion of the newcomers.

Foreign immigration began in the 1860s with the importation of Chinese contract laborers, and reached its peak about 1900. Germans and Scandinavians (particularly after 1900) were the most numerous foreign immigrants; Japanese, who began arriving in the 1890s, met a hostile reception in some areas. Canadians have also come to Oregon in significant numbers. Nevertheless, immigration from other states has predominated. Between 1970 and 1980, the state's net gain from migration was about 341,000; from 1980 to 1983, however, the state suffered a net loss of about 37,000.

¹⁹INTERGOVERNMENTAL COOPERATION

Oregon participates in such regional accords as the Columbia River Compact, Klamath River Compact (with California), Pacific Marine Fisheries Compact, and several western groups concerned with corrections, education, and energy matters.

While Oregon receives federal assistance for a variety of programs, federal involvement is particularly heavy in the areas of energy and natural resources, through federal development, operation, and marketing of hydroelectric power and federal ownership of forest and grazing lands. Approximately 49% of Oregon's land area is owned by the federal government. Federal aid to Oregon was more than $1.2 billion in 1983/84, of which $53.7 million was general revenue sharing.

²⁰ECONOMY

Since early settlement, Oregon's natural resources have formed the basis of its economy. Vast forests have made lumber and wood products the leading industry in the state. Since World War II, however, the state has striven to diversify its job base. The aluminum industry has been attracted to Oregon, along with computer and electronics firms, which now constitute the fastest-growing manufacturing sector. Development, principally in the "Silicon Forest" west of Portland, is expected to bring as many as 3,000 jobs a year during the mid- and late 1980s. Meanwhile, the trend in employment has been toward white-collar and service jobs, with agriculture and manufacturing holding a declining share of the civilian labor force. Tourism and research-related businesses growing out of partnerships between government and higher education are on the rise.

Despite diversification efforts, 60% of manufacturing jobs outside the Portland area were in the lumber and wood products field in 1982. As a result, the state's economy remains dependent on the health of the US construction industry: jobs are plentiful when US housing starts rise, but unemployment increases when nationwide construction drops off. The cyclical changes in demand for forest products are a chronic problem, with rural areas and small towns particularly hard hit by the periodic closing of local lumber and plywood mills.

²¹INCOME

Per capita income in Oregon in 1983 was $10,920, 30th among the 50 states. Total personal income was $29.1 billion, representing a real growth of 62% since 1970. Of a total labor and proprietary income of $23.3 billion, the major components were manufacturing, 24%; services, 19%; government, 18%; trade, 17%; construction, 6%; and other sectors, 16%.

Median family income in 1979 was $20,027, for a rank of 21st in the US. About 11% of all Oregonians were living below the federal poverty level during that year. In 1982, some 25,700 Oregonians were among the top wealth-holders in the US, with gross assets greater than $500,000.

²²LABOR

Oregon's civilian labor force numbered 1,336,000 in 1984, out of which 1,210,000 were employed, giving the state an unemployment rate of 9.4%. A federal survey of workers in March 1982 revealed the following nonfarm employment pattern for Oregon:

	ESTABLISH-MENTS	EMPLOYEES	ANNUAL PAYROLL ('000)
Agricultural services, forestry, fishing	866	6,064	$ 60,528
Mining	158	1,521	36,390
Contract construction	4,703	33,382	661,508
Manufacturing, of which:	4,979	182,487	3,793,431
Lumber, wood products	(1,632)	(50,321)	(1,067,434)
Transportation, public utilities	2,539	51,815	1,154,160
Wholesale trade	5,000	60,710	1,181,242
Retail trade	16,273	182,741	1,700,775
Finance, insurance, real estate	5,605	57,425	872,266
Services	18,432	186,581	2,318,895
Other	1,211	1,246	14,598
TOTALS	59,766	763,972	$11,793,793

Government employees, not included in this survey, numbered at least 194,000 in 1983.

In 1980, 272,000 Oregon workers—26% of all nonagricultural employees (24th in the US)—were members of a labor union; there were 578 unions in 1983. Average weekly earnings of production workers rose to $418 in 1984, 12% above the national average.

23AGRICULTURE

Oregon ranked 30th in the US in agricultural output in 1983, with cash receipts of $1.7 billion. Crops accounted for more than two-thirds of the total. Wheat has been Oregon's leading crop since the state was first settled, but more than 170 farm and ranch commodities are now commercially produced. Oregon leads the nation in the production of winter pears, filberts, peppermint oil, blackberries, black raspberries, boysenberries, loganberries, and several grass and seed crops. Oregon produces 20% of the nation's soft white wheat; wheat and flour are the state's leading farm exports. More than 25% of Oregon's agricultural output is exported.

Farmland covers about 18 million acres (7.3 million hectares), or nearly one-third of Oregon's total area. Oregon's average farm is 486 acres (197 hectares). In 1984, the state had some 37,000 farms and an agricultural work force of 30,500. Quantity and value of selected crops in 1983 were as follows:

	VOLUME	VALUE
Wheat	65,670,000 bushels	$256,716,000
Hay	3,121,000 tons	124,718,000
Potatoes	20,710,000 hundredweight	81,872,000
Ryegrass seed	183,580,000 lb	42,707,000
Pears	188,000 tons	41,792,000
Strawberries	794,000 hundredweight	30,988,000

In recent years, the growth of Oregon's wine industry has become noteworthy; in 1984 there were nearly 50 wineries.

24ANIMAL HUSBANDRY

Most beef cattle are raised on the rangeland of eastern Oregon, while dairy operations are concentrated in the western portion of the state. Sheep and poultry are also raised largely in the west.

Cattle and calf production is Oregon's leading agricultural activity in terms of value, although income varies greatly with market conditions. Ranchers lease large tracts of federally owned grazing land under a permit system. In 1983, Oregon's cattle and calf production was estimated at 564.3 million lb, valued at $273.5 million. There were 100,000 dairy cows, providing the basis for a dairy industry that includes a major cheese industry on the northern Oregon coast; the 1983 milk output was estimated at 1.3 billion lb, and cheese production exceeded 31 million lb. Sheep and lambs were estimated at 350,000 in 1984; meat production totaled 25.1 million lb, worth $11.8 million, in 1983, and the wool clip was 3.3 million lb. Oregon's poultry farms produced an estimated 64.7 million lb of chickens and broilers in 1983, along with 16.1 million lb of turkeys and 630 million eggs.

25FISHING

Oregon's fish resources have long been of great importance to its inhabitants. For centuries, salmon provided much of the food for Indians, who gathered at traditional fishing grounds when the salmon were returning upstream from the ocean to spawn.

In 1984, Oregon ranked 17th among the states in the total value of its commercial catch. Commercial landings totaled 82.5 million lb in 1984, with an estimated value of $33.6 million. The catch included salmon, especially chinook and coho; groundfish such as flounder, perch, rockfish, and lingcod; shellfish such as crabs, clams, and oysters; and albacore tuna.

Sport fishing, primarily for salmon and trout, is a major recreational attraction. The sport catch in 1984 was estimated at 15 million fish; trout made up 53% of the catch, salmon 20%, warm-water game fish 10%, steelhead 10%, and other varieties 7%.

Hatchery production of salmon and trout has taken on increased importance, as development has destroyed natural fish-spawning areas. Some 59 million salmon, 9 million trout, and 5 million steelhead were released from hatcheries in 1980.

26FORESTRY

Oregon is the nation's leading timber producer, and the forest products industry is the most important component of the state's economy. As of 1984, 29,400,000 acres (11,900,000 hectares) were in forestland, occupying not quite half the state's total area; 24,600,000 acres (9,955,000 hectares) were classified as commercial timberland.

The US government owns or manages 56% of the commercial timberland, with 4% under state or local control and 40% privately owned, mostly by large timber companies. The state produced 7.1 billion board feet of lumber in 1983, of which 7 billion board feet were in soft woods.

Logging of the Douglas fir forests of western Oregon is characterized by clear-cutting; selective logging of mature trees is practiced in much of the ponderosa pine forest of eastern Oregon. Sawmills and plywood mills are the main employers in many communities, and log trucks are a common sight on state highways. Since World War II, an increasing percentage of the timber harvest has gone to plywood production.

In 1983, Oregon produced about 20% of the nation's lumber and 40% of its plywood. Shipments of forest products were worth $5.5 billion in 1982. The most important were plywood, $1.9 billion; sawtimber, $1.7 billion; and logs, $1.5 billion.

Public timberlands are managed on a sustained-yield basis, under which the amount cut each year is theoretically matched by new growth, so that the timber harvest may be indefinitely sustained. Major private timber owners operate tree farms, but not all of them practice sustained-yield management. Forest research activities are conducted by the US Forest Service Experiment Station in Portland and by the Oregon State University School of Forestry at Corvallis.

Oregon has 15 national forests, with a total area of 15,491,000 acres (6,269,000 hectares), and 5 state forests with a total area of 683,000 acres (276,400 hectares). The largest national forest, Willamette, occupies 1,675,500 acres (678,000 hectares). The largest state forest is Tillamook, with an area of 364,000 acres (147,000 hectares).

27MINING

The value of Oregon's nonfuel mineral production was $129 million in 1984 (36th in the US), of which 68% derived from sand, gravel, and crushed stone for construction materials. Oregon also contains the nation's only producing nickel mine, near Riddle, in Douglas County. While millions of dollars' worth of gold were mined in eastern and, to a lesser degree, southern Oregon in the late 19th century, production amounted to only 322 troy oz in 1983.

28ENERGY AND POWER

Oregon ranks 3d in the US in hydroelectric power development, and hydropower supplies nearly half of the state's energy needs. Multipurpose federal projects, including four dams on the Columbia River and eight in the Willamette Basin, account for over two-thirds of the state's hydroelectric capacity, with the rest coming from projects owned by private or public utilities. In recent decades, low-cost power from dams has proved inadequate to meet the state's energy needs, and nuclear and coal-fired steam plants have been built to supply additional electric power. The Trojan Nuclear Plant, near Rainier, generated 4.1 billion kwh of electricity in 1983, or 8% of the state's energy needs; a coal-fired steam plant at Boardman produced 1.2%. Oregon's total electric power production in 1983 was 49.2 billion kwh; installed capacity was 10.6 million kw. The Bonneville Power Administration, the federal power-marketing agency, operates a power distribution grid interconnecting Oregon, Washington, and parts of Idaho and Montana.

About 8% of Oregon's total energy is provided by natural gas, the majority of which comes from Canada and the southwestern US.

Energy consumption per capita was 316 million Btu in 1982 (23d in the US), and energy expenditures per capita amounted to $1,676 (41st in the US).

[29] INDUSTRY

Manufacturing in Oregon is dominated by the lumber and wood products industry, which accounted for more than 31% of the state's value of shipments by manufacturers—$17.9 billion—in 1982. The other leading industries, with their share of the 1982 total value of shipments, were food, 16%; paper and allied products, 10%; nonelectrical machinery, 7%; instruments, 7%; primary metals, 5%; and fabricated metals, 5%. In 1982, 55% of Oregon's industrial workers were employed in the Portland area. The Willamette Valley is the site of one of the nation's largest canning and freezing industries. In 1982, 349 food-processing establishments were licensed by the state.

The following table shows value of shipments by manufacturers for selected industry groups in 1982:

Lumber and wood products	$5,510,400,000
Preserved fruits and vegetables	879,900,000
Primary metal products	873,300,000
Papermill products	752,700,000
Frozen fruits and vegetables	659,400,000
Office and computing machines	557,600,000
Paperboard mill products	485,300,000

[30] COMMERCE

Wholesale trade in Oregon exceeded $26.9 billion in 1982. Retail establishments had sales of $12.6 billion, of which food stores accounted for 22%, automotive dealers 19%, restaurants and taverns 11%, and gasoline service stations 9%.

Exports moving through Oregon customs districts were valued at almost $3.7 billion in 1983, with imports valued at more than $2.8 billion. Wheat was the top export by value, followed by logs, lumber, yellow corn, and barley. The leading imports were automobiles and trucks, followed by alumina for aluminum production. Oregon ranked 26th in the US in foreign exports of its own manufactures in 1981, shipping $1.7 billion worth of goods abroad. Agricultural exports amounted to $423 million in 1981/82 (29th in the US).

[31] CONSUMER PROTECTION

The Consumer Advisory Council of the Department of Justice coordinates consumer services carried on by other government agencies, conducts studies and research in consumer services, and advises executive and legislative branches in matters affecting consumer interests. Also responsible for consumer protection are the Executive Department, through its Economic Development and Consumer Service program; the Department of Agriculture (weights and measures); the Real Estate, Corporation, and Insurance divisions of the Department of Commerce; and the public utility commissioner.

[32] BANKING

Consolidations and acquisitions have transformed Oregon's banking system from one characterized by a large number of local banks into one dominated by two large chains—the US National Bank of Oregon and the First Interstate Bank of Oregon—each with over 165 branches throughout the state and each with assets of more than $5 billion in 1985. The Oregon Bank has assets of over $1 billion.

In all, the state had 72 insured commercial banks in 1984, with total assets of $15.3 billion and savings deposits of $12 billion. There were 21 insured savings and loan associations in 1984, with combined assets of nearly $8 billion.

[33] INSURANCE

The number of insurance companies domiciled in Oregon in 1983 totaled 24; direct premiums written for all types of insurance amounted to nearly $2 billion, and losses paid totaled $529 million. At the end of 1983, Oregonians held 3,170,000 life insurance policies valued at $49.7 billion. The average life insurance per family was $43,700, 19% below the national average. Flood insurance in force in 1983 totaled $277 million.

[34] SECURITIES

There are no securities or commodities exchanges in Oregon. New York Stock Exchange member firms had 63 sales offices and 636 registered representatives in Oregon in 1983. Some 389,000 Oregonians owned shares of public corporations that year.

[35] PUBLIC FINANCE

Oregon's biennial budget, covering a period from 1 July of each odd-numbered year to 30 June of the next odd-numbered year, is prepared by the Executive Department and submitted by the governor to the legislature for amendment and approval. Unlike some state budgets, Oregon's is not contained in a single omnibus appropriations bill; instead, each agency appropriation is considered as a separate measure. When the legislature is not in session, fiscal problems are considered by an emergency board of 17 legislators; this board may adjust budgets, allocate money from a special emergency fund, and establish new expenditure limitations, but it cannot enact new general fund appropriations. The Oregon constitution prohibits a state budget deficit and requires that all general obligation bond issues be submitted to the voters.

The following table summarizes revenues and expenditures for the 1979–81 budget period and the adopted budget for the 1983–85 period (in millions):

REVENUES	1979–81	1983–85
Individual taxes	$ 2,498	$ 2,518
Federal funds	1,906	2,087
Business taxes	1,690	1,838
Interest	834	1,501
Loan repayments	904	962
Bond sales	1,686	818
Charges for services	568	745
Liquor and other sales income	251	329
Licenses and fees	181	276
Other receipts	644	1,843
TOTALS	**$11,162**	**$12,917**
EXPENDITURES		
Economic development and consumer services	$ 3,393	$ 2,958
Education	2,135	2,660
Human resources	2,143	2,377
Administration and support services	939	984
Transportation	948	847
Natural resources	258	414
Public safety	116	138
Judicial branch	21	120
Legislative branch	22	27
Other outlays	211	1
TOTALS	**$10,186**	**$10,526**

The city of Portland, Oregon's largest municipality, had total revenues of $229 million and expenditures of $210 million in 1981/82. Local governments are also prohibited from incurring deficits.

As of mid-1982, the total state and local government debt was $9.4 billion. The per capita state debt was $2,363 (2d among the 50 states).

[36] TAXATION

Oregon's chief source of general revenue is the personal income tax, adopted in 1929; as of 1985, the tax ranged from 4% on the first $500 of taxable income to 10.8% on amounts over $5,000. A corporate income tax of 7.5% is also levied. Local governments rely on the property tax. Oregon does not have a general sales tax, although it does tax sales of gasoline and cigarettes.

The state constitution gives voters the right to vote on any substantial tax increase, either by the state or by local governments. State tax measures may be placed on the ballot by the legislature or by petition; local levies must be voted on yearly unless voter approval has been secured for a tax base that may increase by 6% a year without an additional vote.

In 1983, Oregon had a total federal tax burden of $5,965 million and a per capita burden of $2,242. State residents filed 1,082,000 federal income tax returns in 1982, paying more than $2.5 billion in tax.

37ECONOMIC POLICY

During the 1970s and early 1980s, Oregon actively sought balanced economic growth in order to diversify its industrial base, reduce its dependence on the wood products industry, and provide jobs for a steadily growing labor force. The major agency promoting Oregon as a potential industrial location—particularly for relatively nonpolluting industries—is the Economic Development Department, which, in 1984, opened up an office in Tokyo. The department had a 1983–85 biennial budget of $13.4 million, $5.2 million of which was allocated for direct loans to Oregon ports and businesses.

38HEALTH

In 1983 there were 39,977 live births in Oregon, a rate of 15 per 1,000 population; this represented about a 50% decline from the 1960 birthrate. The 1981 infant mortality rate was 10.8 infant deaths per 1,000 live births among whites and 19.5 among nonwhites. In 1982, some 16,4000 legal abortions were performed in Oregon, a rate of 395 abortions per 1,000 births. The leading causes of death, with rates per 100,000 population, were heart disease, 288; cancer, 179; cerebrovascular diseases, 74; and accidents, 49.

In 1983, Oregon had 83 hospitals, with 11,747 beds; hospital personnel included 9,945 registered nurses and 2,660 licensed practical nurses. The average cost of hospital care in 1982 was $382 per day and $2,256 per stay. There were 5,762 licensed physicians and 2,010 active dentists in 1984. The only medical and dental schools in the state are at the University of Oregon Health Sciences University in Portland.

39SOCIAL WELFARE

The Department of Human Resources, created in 1971 to coordinate social service activities, was operating more than 250 programs by 1984, when nearly one-fourth of the state budget was devoted to social welfare programs.

Public assistance payments to Oregonians in 1982 totaled $891 million, of which $99 million consisted of aid to families with dependent children and $792 million was medical assistance. A total of $2.1 billion in Social Security benefits was paid to 431,000 Oregonians in 1983. Federal Supplemental Security Income payments totaled $47.4 million, of which the state paid $8 million. Some 240,000 Oregonians took part in the food stamp program in 1983, at a federal cost of $162 million, and 229,000 students were enrolled in the school lunch program, subsidized by $20 million in federal funds.

In 1982, workers' compensation payments amounted to $307.3 million. That year, the state's unemployment insurance program processed 475,000 initial claims and paid $417 million.

40HOUSING

In general, owner-occupied homes predominate in Oregon, and there are few urban slums. During the 1970s and early 1980s, however, a growing percentage of new construction went for rental units. Between 1970 and 1980, the proportion of the housing stock in single-family units fell from 77% to 68%. In 1980 there were 1,083,285 housing units in Oregon, of which 991,593 were occupied—645,941 by owners and 345,652 by renters. A growing share of the state's housing stock was also in mobile homes (from 5.1% in 1970 to 8.3% in 1980).

As of 1980, about 32,000 housing units were receiving some type of subsidy, mostly from the Department of Housing and Urban Development. The Housing Division of the Department of Commerce offers housing purchase assistance (through interest rates below the prevailing market) and construction subsidies to build units for disabled and for low- and moderate-income renters.

41EDUCATION

On the whole, Oregonians are among the best educated of Americans. In 1980, 76% of adult state residents were high school graduates, and 18% were college graduates. As of 31 December 1984 there were 940 public elementary schools and 288 public secondary schools, with a combined enrollment of 431,184. Private (mainly parochial) elementary and secondary schools numbered 166 in 1981; enrollment was 39,250 in fall 1984.

Higher education in Oregon comprises 15 community colleges, a state college and university system made up of 8 institutions, and 27 independent institutions. The state college and university system had a fall enrollment of 58,360 in 1984. The largest institution was the University of Oregon in Eugene (15,840), followed by Oregon State University in Corvallis (15,636) and Portland State University (14,390). Major private institutions include Lewis and Clark College, Reed College, and the University of Portland, all in Portland, along with Williamette University in Salem, Pacific University in Forest Grove, and Linfield College in McMinnville. A financial aid program for Oregon state college students is administered by the State Scholarship Commission, with more than $30 million earmarked for assistance in the 1983–85 budget period.

42ARTS

The Portland Art Museum, with an associated art school, is the city's center for the visual arts. The University of Oregon in Eugene has an art museum specializing in Oriental art.

The state's most noted theatrical enterprise is the annual Shakespeare Festival in Ashland, with a complex of theaters drawing actors and audiences from around the nation. The Oregon Symphony is situated in Portland, and Salem and Eugene have small symphony orchestras of their own.

The Oregon Arts Commission operates a program of direct mail marketing of fine arts prints created by artists from the Northwest. The commission and the Department of Education jointly administer a program of Young Writers Fellowships.

43LIBRARIES AND MUSEUMS

In 1984, Oregon had 603 academic, public, and special libraries, including branches; the total book stock of all public libraries was 5,199,451, and their combined circulation in 1981/82 was 15,246,400. Most cities and counties in Oregon have public library systems, the largest being the Multnomah County library system in Portland, with 14 branches and 1,161,710 volumes in 1981/82. The State Library in Salem, 1,371,000 volumes in 1984, serves as a reference agency for state government.

Oregon has almost 60 museums and historic sites, as well as 9 botanical gardens and arboretums. Historical museums emphasizing Oregon's pioneer heritage appear throughout the state, with Ft. Clatsop National Memorial—featuring a replica of Lewis and Clark's winter headquarters—among the notable attractions. The Oregon Historical Society operates a major historical museum in Portland, publishes books of historical interest, and issues the *Oregon Historical Quarterly*. In Portland's Washington Park area are the Oregon Museum of Science and Industry, Washington Park Zoo, Western Forestry Center, and an arboretum and other gardens.

44COMMUNICATIONS

As of 1980, 92% of Oregon's 991,593 occupied housing units had telephones. Oregon had 86 AM and 57 FM commercial radio stations in 1984; 5 of the state's 16 commercial television stations were in Portland. A state-owned broadcasting system, which includes 4 television stations, provides educational radio and television programming. As of 1984, 131 cable television systems served 397,570 subscribers in 259 communities.

45PRESS

Oregon's first newspaper was the weekly *Oregon Spectator,* which began publication in 1846. Early newspapers engaged in what became known as the "Oregon style" of journalism, characterized by intemperate, vituperative, and fiercely partisan comments. As of 1984, 20 daily and 90 weekly newspapers were published in Oregon. The state's largest newspaper, the *Oregonian,* published in Portland, is owned by the Newhouse group. The total circulation of Oregon's morning dailies in 1984 was 286,655; evening dailies, 339,245; and Sunday papers, 653,344. The following table lists leading Oregon newspapers with their 1984 circulations:

AREA	NAME	DAILY	SUNDAY
Eugene	Register-Guard (e,S)	69,434	72,671
Portland	Oregonian (m,S)	289,166	409,132
Salem	Statesman-Journal (m)	57,073	

46ORGANIZATIONS

The 1982 Census of Service Industries counted 842 organizations in Oregon, including 202 business associations; 446 civic, social, and fraternal associations; and 18 educational, scientific, and research associations.

Among the many forestry-related organizations in Oregon are the International Woodworkers of America (AFL-CIO), Association of Western Pulp and Paper Workers, Pacific Lumber Exporters Association, Western Forest Industries Association, and Western Wood Products Association, all with their headquarters in Portland.

47TOURISM, TRAVEL, AND RECREATION

Oregon's abundance and variety of natural features and recreational opportunities make the state a major tourist attraction. Travel and tourism was the state's 3d-largest employer in 1982, generating 54,100 jobs. In 1983, more than 15 million out-of-state visitors spent more than $2 billion. The Travel Information Council of the Department of Transportation maintains an active tourist advertising program, and Portland hotels busily seek major conventions.

Among the leading attractions are the rugged Oregon coast, with its offshore salmon fishing; Crater Lake National Park; the Rogue River, for river running and fishing; the Columbia Gorge, east of Portland; the Cascades wilderness; and Portland's annual Rose Festival. Oregon has one national park, Crater Lake, and three other areas—John Day Fossil Beds National Monument, Oregon Caves National Monument, and Ft. Clatsop National Memorial—managed by the National Park Service. The US Forest Service administers the Oregon Dunes National Recreation Area, on the Oregon coast; the Lava Lands Visitor Complex near Bend; and the Hells Canyon National Recreation Area, east of Enterprise. Oregon has one of the nation's most extensive state park systems: 225 parks and recreation areas cover 88,493 acres (35,812 hectares). Licenses were issued to 845,736 hunters and 1,661,003 anglers in 1983.

48SPORTS

Oregon's lone major professional team, based in Portland, is the Trail Blazers, winners of the National Basketball Association championship in 1977. The Portland Beavers compete in baseball's class-AAA Pacific Coast League.

Horse racing takes place at Portland Meadows in Portland and, in late August and early September, at the Oregon State Fair in Salem; there is greyhound racing at the Multnomah Kennel Club near Portland. Pari-mutuel betting is permitted at the tracks, but off-track betting is prohibited.

The University of Oregon and Oregon State University belong to the Pacific 10 Conference.

49FAMOUS OREGONIANS

Prominent federal officeholders from Oregon include Senator Charles McNary (1874-1944), a leading advocate of federal reclamation and development projects and the Republican vice-presidential nominee in 1940; Senator Wayne Morse (b.Wisconsin, 1900-1974), who was an early opponent of US involvement in Viet-Nam; Representative Edith Green (b.1910), a leader in federal education assistance; and Representative Al Ullman (b.Montana, 1914), chairman of the House Ways and Means Committee until his defeat in 1980. Recent cabinet members from Oregon have been Douglas McKay (1893-1959), secretary of the interior; and Neil Goldschmidt (b.1940), secretary of transportation.

A major figure in early Oregon history was sea captain Robert Gray (b.Rhode Island, 1755-1806), discoverer of the Columbia River. Although never holding a government position, fur trader Dr. John McLoughlin (b.Canada, 1784-1857) in effect ruled Oregon from 1824 to 1845; he was officially designated the "father of Oregon" by the 1957 state legislature. Also of importance in the early settlement was Methodist missionary Jason Lee (b.Canada, 1803-45). Oregon's most famous Indian was Chief Joseph (1840?-1904), leader of the Nez Percé in northeastern Oregon; when tension between the Nez Percé and white settlers erupted into open hostilities in 1877, Chief Joseph led his band of about 650 men, women, and children from the Oregon-Idaho border across the Bitterroot Range, evading three army detachments before being captured in northern Montana.

Other important figures in the early days of statehood were Harvey W. Scott (b.Illinois, 1838-1910), longtime editor of the Portland *Oregonian,* and his sister, Abigail Scott Duniway (b.Illinois, 1834-1915), the Northwest's foremost advocate of women's suffrage, a cause her brother strongly opposed. William Simon U'Ren (b.Wisconsin, 1859-1949) was a lawyer and reformer whose influence on Oregon politics and government endures to this day. Journalist and Communist John Reed (1887-1920), author of *Ten Days That Shook the World,* an eyewitness account of the Bolshevik Revolution, was born in Portland, and award-winning science-fiction writer Ursula K. LeGuin (b.California, 1929) is a Portland resident.

Linus Pauling (b.1901), two-time winner of the Nobel Prize (for chemistry in 1954, for peace in 1962), is another Portland native. Other scientists prominent in the state's history include botanist David Douglas (b.Scotland, 1798-1834), who made two trips to Oregon and after whom the Douglas fir is named; and geologist and paleontologist Thomas Condon (b.Ireland, 1822-1907), discoverer of major fossil beds in eastern Oregon.

50BIBLIOGRAPHY

Cogswell, Philip. *Capitol Names: Individuals Woven into Oregon's History.* Portland: Oregon Historical Society, 1977.

Corning, Howard. *Dictionary of Oregon History.* Portland: Binfords & Mort, 1956.

Dodds, Gordon B. *Oregon: A Bicentennial History.* New York: Norton, 1977.

Federal Writers' Project. *Oregon: End of the Trail.* Reprint. New York: Somerset, n.d. (orig. 1941).

Johansen, Dorothy, and Charles Gates. *Empire of the Columbia: A History of the Pacific Northwest.* 2d ed. New York: Harper & Row, 1967.

Loy, William. *Atlas of Oregon.* Eugene: University of Oregon Books, 1976.

Parkman, Francis, Jr. *The Oregon Trail.* New York: Penguin, 1982.

Throckmorton, Arthur L. *Oregon Argonauts: Merchant Adventurers on the Western Frontier.* Portland: Oregon Historical Society, 1961.

Vaughan, Thomas, and Terrence O'Donnell. *Portland: A Historical Sketch and Guide.* Portland: Oregon Historical Society, 1976.

PENNSYLVANIA

Commonwealth of Pennsylvania

ORIGIN OF STATE NAME: Named for Admiral William Penn, father of the founder of Pennsylvania. **NICKNAME:** The Keystone State. **CAPITAL:** Harrisburg. **ENTERED UNION:** 12 December 1787 (2d). **MOTTO:** Virtue, Liberty, and Independence. **COAT OF ARMS:** A shield supported by two horses displays a sailing ship, a plow, and three sheaves of wheat; an eagle forms the crest. Beneath the shield an olive branch and a cornstalk are crossed, and below them is the state motto. **FLAG:** The coat of arms appears in the center of a blue field. **OFFICIAL SEAL:** Obverse: A shield displays a sailing ship, a plow, and three sheaves of wheat, with a cornstalk to the left, an olive branch to the right, and an eagle above, surrounded by the inscription "Seal of the State of Pennsylvania." Reverse: A woman representing Liberty holds a wand topped by a liberty cap in her left hand and a drawn sword in her right, as she tramples a lion representing Tyranny; the legend "Both Can't Survive" encircles the design. **ANIMAL:** White-tailed deer. **BIRD:** Ruffed grouse. **DOG:** Great Dane. **FISH:** Brook trout. **FLOWER:** Mountain laurel. **INSECT:** Firefly. **TREE:** Hemlock. **BEAUTIFICATION AND CONSERVATION PLANT:** Penngift crownvetch. **BEVERAGE:** Milk. **LEGAL HOLIDAYS:** New Year's Day, 1 January; Birthday of Martin Luther King, Jr., 3d Monday in January; Lincoln's Birthday, 12 February; Washington's Birthday, 3d Monday in February; Memorial Day, last Monday in May; Flag Day, 2d Sunday in June; Independence Day, 4 July; Labor Day, 1st Monday in September; Columbus Day, 2d Monday in October; Veterans Day, 11 November; Thanksgiving Day, 4th Thursday in November; Christmas Day, 25 December. **TIME:** 7 AM EST = noon GMT.

¹LOCATION, SIZE, AND EXTENT

Located in the northeastern US, the Commonwealth of Pennsylvania is the 2d largest of the three Middle Atlantic states and ranks 33d in size among the 50 states.

The total area of Pennsylvania is 45,308 sq mi (117,348 sq km), of which land occupies 44,888 sq mi (116,260 sq km) and inland water 420 sq mi (1,088 sq km). The state extends 307 mi (494 km) E–W and 169 mi (272 km) N–S. Pennsylvania is rectangular in shape, except for an irregular side on the E and a break in the even boundary in the NW where the line extends N–E for about 50 mi (80 km) along the shore of Lake Erie.

Pennsylvania is bordered on the N by New York; on the E by New York and New Jersey (with the Delaware River forming the entire boundary); on the SE by Delaware; on the S by Maryland and West Virginia (demarcated by the Mason-Dixon line); on the W by West Virginia and Ohio; and on the NW by Lake Erie. The total boundary length of Pennsylvania is 880 mi (1,416 km). The state's geographical center lies in Centre County, 2.5 mi (4 km) SW of Bellefonte.

²TOPOGRAPHY

Pennsylvania may be divided into more than a dozen distinct physiographic regions, most of which extend in curved bands from east to south. Beginning in the southeast, the first region (including Philadelphia) is a narrow belt of coastal plain along the lower Delaware River; this area, at sea level, is the state's lowest region. The next belt, dominating the southeastern corner, is the Piedmont Plateau, a wide area of rolling hills and lowlands. The Great Valley, approximately 10–15 mi (16–24 km) in width, runs from the middle of the state's eastern border to the middle of its southern border. The eastern, central, and western parts of the Great Valley are known as the Lehigh, Lebanon, and Cumberland valleys, respectively. West and north of the Great Valley, the Pocono Plateau rises to about 2,200 feet (700 meters). Next, in a band 50–60 mi (80–100 km) wide most of the way from the north-central part of the eastern border to the west-central part of the southern border are the Appalachian Mountains, a distinctive region of parallel ridges and valleys.

The Allegheny High Plateau, part of the Appalachian Plateaus, makes up the western and northern parts of the state. The Allegheny Front, the escarpment along the eastern edge of the plateau, is the most striking topographical feature in Pennsylvania, being dissected by many winding streams to form narrow, steep-sided valleys; the southwestern extension of the Allegheny High Plateau contains the state's highest peak, Mt. Davis, at 3,213 feet (979 meters). A narrow lowland region, the Erie Plain, borders Lake Erie in the extreme northwestern part of the state.

According to federal sources, Pennsylvania has jurisdiction over 735 sq mi (1,904 sq km) of Lake Erie; the state government gives a figure of 891 sq mi (2,308 sq km). Pennsylvania contains about 250 natural lakes larger than 20 acres (8 hectares), most of them in the glaciated regions of the northeast and northwest. The largest natural lake within the state's borders is Conneaut Lake, about 30 mi (48 km) south of the city of Erie, with an area of less than 1.5 sq mi (3.9 sq km); the largest man-made lake is Lake Wallenpaupack, in the Poconos, occupying about 9 sq mi (23 sq km). Pennsylvania claims more than 21 sq mi (54 sq km) of the Pymatuning Reservoir, on the Ohio border.

The Susquehanna River and its tributaries drain more than 46% of the area of Pennsylvania, much in the Appalachian Mountains. The Delaware River forms Pennsylvania's eastern border and, like the Susquehanna, flows southeastward to the Atlantic Ocean. Most of the western part of the state is drained by the Allegheny and Monongahela rivers, which join at Pittsburgh to form the Ohio. The Beaver, Clarion, and Youghiogheny rivers are also important parts of this system.

During early geological history, the topography of Pennsylvania had the reverse of its present configurations, with mountains in the southeast and a large inland sea covering the rest of the state. This sea, which alternately expanded and contracted, interwove layers of vegetation (which later became coal) with layers of sandstone and shale.

³CLIMATE

Although Pennsylvania lies entirely within the humid continental zone, its climate varies according to region and elevation. The region with the warmest temperatures and the longest growing season is the low-lying southeast, in the Ohio and Monongahela valleys. The region bordering Lake Erie also has a long growing season, as the moderating effect of the lake prevents early spring and late autumn frosts. The first two areas have hot summers, while the Erie area is more moderate. The rest of the state, at higher elevations, has cold winters and cool summers.

Among the major population centers, Philadelphia has an annual mean temperature of 54°F (12°C), with a normal minimum of 46°F (8°C) and a normal maximum of 64°F (18°C). Pittsburgh has an annual mean of 50°F (10°C), with a minimum of 41°F (5°C) and a maximum of 60°F(16°C). In the cooler northern areas, Scranton has a normal annual mean ranging from 41°F (5°C) to 59°F(15°C); Erie, from 42°F (6°C) to 58°F(14°C). The record low temperature for the state is −42°F (−41°C), set at Smethport on 5 January 1904; the record high, 111°F (44°C), was reached at Phoenixville on 10 July 1936.

Philadelphia has about 41 in (104 cm) of precipitation annually, and Pittsburgh has 36 in (91 cm). Pittsburgh, however, has much more snow—45 in (114 cm), compared with 22 in (56 cm) for Philadelphia. The snowfall in Erie, in the snow belt, exceeds 54 in (137 cm) per year, with heavy snows sometimes experienced late in April. In Philadelphia, the sun shines an average of 57% of the time; in Pittsburgh, 49%.

The state has experienced several destructive floods. On 31 May 1889, the South Fork Dam near Johnstown broke after a heavy rainfall, and its rampaging waters killed 2,200 people and devasted the entire city in less than 10 minutes. On 19–20 July 1977, Johnstown experienced another flood, resulting in 68 deaths. Three tornadoes raked the southwestern part of the commonwealth on 23 June 1944, killing 45 persons and injuring another 362. Rains from Hurricane Agnes in June 1972 resulted in floods that caused 48 deaths and more than $1.2 billion worth of property damage in the Susquehanna Valley.

4FLORA AND FAUNA

Maple, walnut, poplar, oak, pine, ash, beech, and linden trees fill Pennsylvania's extensive forests, along with sassafras, sycamore, weeping willow, and balsam fir (*Abies fraseri*). Red pine and paper birch are found in the north, while the sweet gum is dominant in the extreme southwest. Mountain laurel (the state flower), Juneberry, dotted hawthorn, New Jersey tea, and various dogwoods are among the shrubs and small trees found in most parts of the state, and dewberry, wintergreen, wild columbine, and wild ginger are also common. The wild red rose is classified as threatened, the small whorled pogonia as endangered.

Numerous mammals persist in Pennsylvania, among them the white-tailed deer (the state animal), black bear, red and gray foxes, opossum, raccoon, muskrat, mink, snowshoe hare, common cottontail, and red, gray, fox, and flying squirrels. Native amphibians include the hellbender, Fowler's toad, and the tree, cricket, and true frogs; among reptilian species are the five-lined and black skinks and five varieties of lizard. The ruffed grouse, a common game species, is the official state bird; other game birds are the wood dove, ring-necked pheasant, bobwhite quail, and mallard and black ducks. The robin, cardinal, English sparrow, red-eyed vireo, cedar waxwing, tufted titmouse, yellow-shafted flicker, barn swallow, blue jay, and killdeer are common nongame birds. More than 170 types of fish have been identified in Pennsylvania, with brown and brook trout, grass pickerel, bigeye chub, pirate perch, and white bass among the common native varieties.

In 1978, the Pennsylvania Game Commission and the US Fish and Wildlife Service signed a cooperative agreement under which the federal government provides two dollars for each dollar spent by the state to determine the status of and improve conditions for threatened or endangered species. On the endangered list are the brown (grizzly) bear, Indiana bat, Delmarva Peninsula fox squirrel, eastern cougar, bald eagle, American and Arctic peregrine falcons, Kirkland's warbler, orange-footed pearly mussel, and pink mucket pearly mussel.

5ENVIRONMENTAL PROTECTION

Pennsylvania's environment was ravaged by uncontrolled timber cutting in the 19th century, and by extensive coal mining and industrial development until recent times. Pittsburgh's most famous landmarks were its smokestacks, and it was said that silverware on ships entering the port of Philadelphia would tarnish immediately from the fumes of the Delaware River. The anthracite-mining regions were filled with huge, hideous culm piles, and the bituminous and anthracite fields were torn up by strip-mining. In 1979, a different kind of threat to Pennsylvania's environment received worldwide attention when the nuclear power plant at Three Mile Island seriously malfunctioned.

In 1895, Pennsylvania appointed its first commissioner of forestry, in an attempt to repair some of the earlier damage. Gifford Pinchot, who twice served as governor of Pennsylvania, was the first professionally trained forester in the US (he studied at the École National Forestière in Paris), developed the US Forest Service, and served as Pennsylvania forest commissioner from 1920 to 1922. In 1955, the state forests were put under scientific management.

In 1972, Pennsylvania voters ratified a state constitutional amendment adopted 18 May 1971, acknowledging the people's "right to clean air, pure water, and to the preservation of the natural, scenic, historic and esthetic values of the environment" and naming the state as trustee of these resources. Passage of the amendment came only two years after establishment of the Pennsylvania Department of Environmental Resources, which regulates mining operations, administers land and water management programs, and oversees all aspects of environmental control. Agencies in this department include bureaus responsible for land protection (solid waste management, mine subsidence regulation), surface mine regulation, water quality management, air quality control, community environmental control (municipal sewage facilities, water supply), radiation protection, forestry, state parks, soil and water conservation, dam safety, and storm water management. In 1985/86, the department had a budget of $308.9 million.

The Bureau of Air Quality Control, which enforces the Federal Clean Air Act of 1970 as well as state laws and regulations, increased its expenditures from $307,805, in 1965/66 to more than $6.5 million in 1982/83. During the same period, the number of air pollution problems corrected as a result of actions by the bureau increased from 210 to 571. In 1965, the bureau handled 406 new cases; by 1980, that number had grown to 1,889.

Expenditures for water pollution abatement decreased from $99.7 million in 1978 to $67.1 million in 1982. As of 1985, sewage and industrial wastes were the major pollutants in areas with high industrial and population concentrations. In western and parts of central Pennsylvania, drainage from abandoned bituminous coal mines created serious water quality problems; active mines in this region were also potentially polluting. A similar situation prevailed in the anthracite areas of northeastern Pennsylvania. Oil and gas well operations, located primarily in the northwestern portion of the commonwealth, were additional pollution sources. Pennsylvania's Water Quality Management program has resulted in the investment of some $10 billion by industries, municipalities, and others. As of 1985, 79% of the state's 13,000 mi (21,000 km) of major streams complied with state and federal water quality standards.

LOCATION: 39°43′16″ to 42°30′35″N; 74°41′23″ to 80°31′08″W. **BOUNDARIES:** New York line, 327 mi (526 km); New Jersey line, 166 mi (267 km); Delaware line, 25 mi (40 km); Maryland line, 101 mi (163 km); West Virginia line, 81 mi (130 km); Ohio line, 129 mi (208 km); Lake Erie shoreline, 51 mi (82 km).

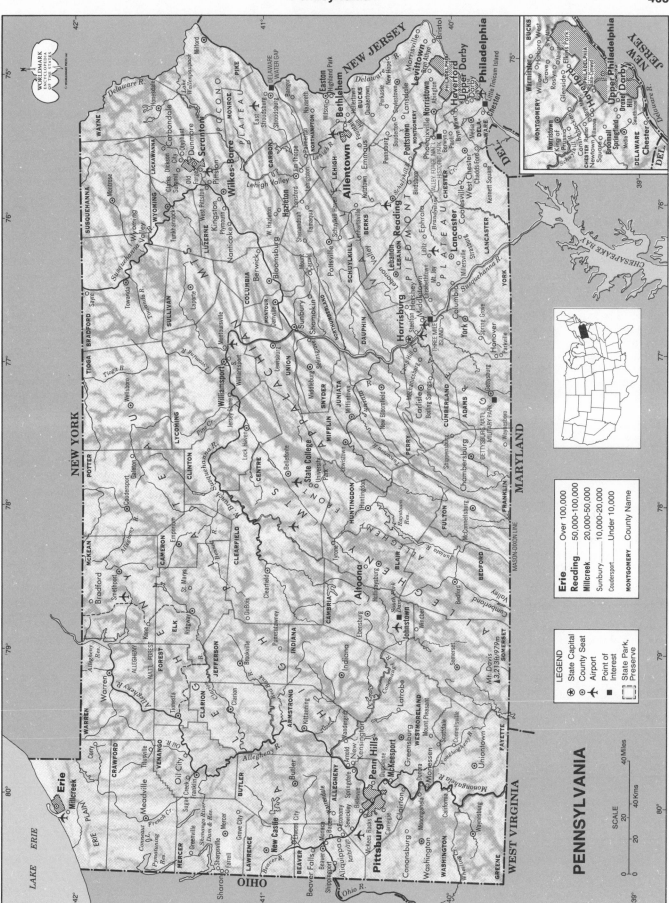

PENNSYLVANIA

See endsheet maps: L2.

LEGEND
⊕ State Capital
⊙ County Seat
✈ Airport
■ Point of Interest
State Park, Preserve

Erie — Over 100,000
Reading — 50,000-100,000
Millcreek — 20,000-50,000
Sunbury — 10,000-20,000
Coudersport — Under 10,000
MONTGOMERY — County Name

SCALE
0   20   40 Miles
0   20   40 Kms

In March 1979, Pennsylvania suffered the worst nuclear-power accident in US history when a nuclear reactor on Three Mile Island malfunctioned and radioactive gases escaped. A second reactor was shut down immediately even though it was not damaged. The cleanup of radioactive waste cost about $1 billion, and it was not until late 1985 that the undamaged unit was placed back in operation.

An oil spill at Marcus Hook, near the Delaware border, released 435,000 gallons of crude oil into the Delaware River in September 1985; damage to birds and wetlands was more extensive in Delaware than in Pennsylvania. Also in September 1985, Pennsylvania—along with Maryland, Virginia, the District of Columbia, and the federal government—published plans for the cleanup of Chesapeake Bay, the primary objective being to reduce the flow of toxic substances into that body of water.

6 POPULATION

As recently as 1940, Pennsylvania was the second most populous state in the US. By the 1970 census, however, Pennsylvania had slipped to fourth place, with a population of 11,793,909; it remained in fourth place in 1980 with a population of 11,863,895, and in 1985 with an estimated population of 11,895,301. The population projection for 1990 is 11,720,400.

As of 1982, 48% of the population was male, 52% female; 69.3% of all Pennsylvanians lived in urban areas. The population density in 1985 was 265 per sq mi (102 per sq km), making Pennsylvania the 8th most densely populated state in the US. There are fewer young people and more persons aged 65 or over represented in the state's population than in the US as a whole: as of 1983, 19.7% of the state's population was under age 18, while 13.8% was over 65. Pennsylvania's birthrate in 1981 was 13.5 per 1,000 population,

15% below the US average. At the same time, the death rate of 10.1 was nearly 15% above the national norm. A net out-migration of 649,000 between 1970 and 1983 was a further drag on population growth. Population projections by the US Census Bureau suggested that Pennsylvania will experience a 1.2% population decrease between 1980 and 1990.

The largest city in the state, Philadelphia, was the 5th-largest US city as of 1984, with a population of 1,646,713. Philadelphia's population has declined since 1970, when 1,949,996 people lived there. The population of its metropolitan area also declined during the 1970s, but then increased from 4,716,559 in 1980 to an estimated 4,768,400 in 1984. Further, the larger Philadelphia–Wilmington, Del.–Trenton, N.J. consolidated metropolitan area (4th largest in the US) increased from 5,680,509 in 1980 to an estimated 5,755,300 in 1984. Pittsburgh's population declined from 616,806 in 1950 to 402,583 in 1984 (33d) in the city proper; the Pittsburgh metropolitan area population (15th in the US) went from 2,348,000 in 1970 to 2,372,000 in 1984.

The 1984 populations of Pennsylvania's other major cities were Erie, 117,761; Allentown, 103,899; Scranton, 83,695; Reading, 78,364; Bethlehem, 69,967; Lancaster, 56,261; Altoona, 54,595; Harrisburg, 52,056; and Wilkes-Barre, 49,316.

7 ETHNIC GROUPS

During the colonial period, under a religiously tolerant Quaker government, Pennsylvania was a haven for dissident sectarians from continental Europe and the British Isles. Some German sectarians, including the Amish, have kept up their traditions to this day. An initially friendly policy toward the Indians waned in the late 18th century under the pressures of population growth and the anxieties of the French and Indian War. The famous

Pennsylvania Counties, County Seats, and County Areas and Populations

COUNTY	COUNTY SEAT	LAND AREA (SQ MI)	POPULATION (1984 EST.)	COUNTY	COUNTY SEAT	LAND AREA (SQ MI)	POPULATION (1984 EST.)
Adams	Gettysburg	521	68,441	Lancaster	Lancaster	952	381,348
Allegheny	Pittsburgh	727	1,409,829	Lawrence	New Castle	363	105,772
Armstrong	Kittanning	646	82,238	Lebanon	Lebanon	363	111,272
Beaver	Beaver	436	199,161	Lehigh	Allentown	348	276,507
Bedford	Bedford	1,017	48,060	Luzerne	Wilkes-Barre	891	336,380
Berks	Reading	861	317,194	Lycoming	Williamsport	1,237	117,131
Blair	Hollidaysburg	527	134,396	McKean	Smethport	979	49,248
Bradford	Towanda	1,152	63,970	Mercer	Mercer	672	126,943
Bucks	Doylestown	610	508,576	Mifflin	Lewistown	413	46,649
Butler	Butler	789	151,253	Monroe	Stroudsburg	609	77,222
Cambria	Ebensburg	691	177,106	Montgomery	Norristown	486	658,999
Cameron	Emporium	398	6,615	Montour	Danville	131	16,993
Carbon	Jim Thorpe	385	54,011	Northhampton	Easton	376	230,813
Centre	Bellefonte	1,106	114,461	Northumberland	Sunbury	461	100,211
Chester	West Chester	758	334,589	Perry	New Bloomfield	557	37,621
Clarion	Clarion	607	43,085	Philadelphia	Philadelphia	136	1,646,713
Clearfield	Clearfield	1,149	84,534	Pike	Milford	551	20,584
Clinton	Lock Haven	892	39,132	Potter	Coudersport	1,081	18,350
Columbia	Bloomsburg	486	61,832	Schuylkill	Pottsville	781	157,681
Crawford	Meadville	1,011	89,052	Snyder	Middleburg	329	35,384
Cumberland	Carlisle	547	185,541	Somerset	Somerset	1,073	81,882
Dauphin	Harrisburg	528	235,811	Sullivan	Laporte	451	6,202
Delaware	Media	184	551,359	Susquehanna	Montrose	826	38,979
Elk	Ridgway	830	337,617	Tioga	Wellsboro	1,131	40,286
Erie	Erie	804	282,124	Union	Lewisburg	317	33,620
Fayette	Uniontown	794	159,178	Venango	Franklin	679	64,467
Forest	Tionesta	428	5,040	Warren	Warren	885	47,672
Franklin	Chambersburg	774	116,248	Washington	Washington	858	217,184
Fulton	McConnellsburg	437	115,549	Wayne	Honesdale	731	37,797
Greene	Waynesburg	577	41,200	Westmoreland	Greensburg	1,033	386,603
Huntingdon	Huntingdon	877	42,716	Wyoming	Tunkhannock	399	27,344
Indiana	Indiana	829	93,321	York	York	906	322,391
Jefferson	Brookville	656	48,922				
Juniata	Mifflintown	392	19,830		TOTALS	44,888	11,900,222
Lackawanna	Scranton	461	223,977				

Carlisle Indian School (1879-1918) educated many leaders from various tribes throughout the US. In Pennsylvania itself, however, there were only 9,173 American Indians as of 1980.

Modest numbers of black slaves were utilized as domestics, field workers, and iron miners in colonial Pennsylvania. Antislavery sentiment was stirred in the 18th century through the efforts of a Quaker, John Woolman, and other Pennsylvanians. An Act for the Gradual Abolition of Slavery was passed in 1780, and the important antislavery newspaper *The Liberator* appeared in Philadelphia in 1831. Today, black Americans are concentrated in the large cities. Pennsylvania's black population in 1980 was 1,047,609, amounting to nearly 9% of the state's population as a whole. Philadelphia was almost 38% black in 1980, Pittsburgh 24%.

The late 19th and early 20th centuries brought waves of immigrants from Ireland, Wales, various Slavic nations, and the eastern Mediterranean and the Balkans. Many of the new immigrants settled in the east-central anthracite coal-mining region. In 1980, 3.4% of all Pennsylvania residents were foreign-born. Italy, Germany, the Soviet Union, Poland, England, Austria, and Ireland were leading European countries of origin in 1980. In the valleys surrounding Pittsburgh there are still self-contained ethnic enclaves, and there has been increased interest in preserving distinctive ethnic traditions.

Hispanic Americans in Pennsylvania numbered 154,004 in 1980. Most were Puerto Ricans, with smaller numbers of Cubans and Central Americans. In 1980, the Asian population included 15,212 Asian Indians, 13,291 Chinese, 12,503 Koreans, 9,257 Vietnamese, 8,267 Filipinos, and 4,669 Japanese.

[8] LANGUAGES

Once home to several Algonkian tribes, Pennsylvania still has such Algonkian place-names as Punxsutawney, Aliquippa, Pocono, Towanda, Susquehanna, and Shamokin. An Iroquoian tribe gave its name to the region Conestoga. The word came to identify first the pioneers' covered wagons manufactured in the area and then, in shortened form, a cheap cigar called a *stogie*.

Although not quite homogeneous, Pennsylvania's North Midland dialect is significant as the source of much midwestern and western speech. The only non-Midland sector is the northern tier of counties, settled from southern New York State, where features of the northern dialect predominate.

On the whole, Pennsylvania North Midland is distinguished by the presence of *want off* a tram or bus, *snake feeder* (dragonfly), *run* (small stream), *waterspouts* and *spouts* (gutters), and *creek* as /krik/. With these features are found others that commonly occur in Southern, such as *corn pone, roasting ears,* and *spicket* (spigot). Western Pennsylvania, however, contrasts with the eastern half by the dominance of /nawthing/ for *nothing,* /greezy/ for *greasy,* /kao/ for *cow, sugar tree* (sugar maple), *hap* (quilt), and *clothes press* (closet), as well as by the influential merging of the /ah/ vowel and the /aw/ vowel so that *cot* and *caught* sound alike. Southeastern Pennsylvania has *flannel cakes* for pancakes and *ground hackie* for chipmunk; within this region, Philadelphia and its suburbs have distinctive *baby coach* for baby carriage, *pavement* for sidewalk, *hoagie* for a large sandwich, the vowel of *put* in *broom* and *Cooper,* and the vowel of *father* in *on* and *fog.* In the east and northeast, a doughnut is a *cruller,* one is *sick in the stomach,* and *syrup* has the vowel of *sit.*

In much of central Pennsylvania, descendants of the colonial Palatinate German population retain their speech as Deutsch, often misnamed Pennsylvania Dutch, which has influenced English in the state through such loanwords as *toot* (bag), *rainworm* (earthworm), *snits* (dried apples), and *smearcase* (cottage cheese).

In 1980, 10,630,077 Pennsylvanians—93% of the population 3 years old or older—spoke only English at home. Other languages spoken at home included:

Italian	142,030	Ukrainian	21,118
Spanish	139,651	Greek	20,149
German	84,879	Yiddish	19,329
Polish	77,095	Hungarian	14,604
French	34,664	Serbo-Croatian	13,973
Slovak	29,168	Russian	12,595

[9] RELIGIONS

With a long history of toleration, Pennsylvania has been a haven for numerous religious groups.

The first European settlers were Swedish Lutherans; German Lutherans began arriving in 1703. William Penn brought the Quakers to Pennsylvania during the 1680s, and the climate of religious liberty soon attracted other dissident groups, including German Mennonites, Dunkards, Moravians, and Schwenkfelders; French Huguenots; Scotch-Irish Presbyterians; and English Baptists. Descendants of the 16th-century Anabaptists, the Mennonites for the most part settled as farmers; they and the Quakers were the first religious groups openly to advocate abolition of slavery and to help runaway slaves to freedom via the Underground Railroad. The Amish—Mennonite followers of Jacob Amman—continue to dress in black clothing, shun the use of mechanized tools, automobiles, and electrical appliances, and observe Sundays by singing 16th-century hymns.

The Presbyterians, who built their first church in the state in 1704, played a major role both in the establishment of schools in the colony and in the later development of Pittsburgh and other cities in the western part of the state. Methodists held their first services in Philadelphia in 1768; for many years thereafter, Methodist circuit riders proselytized throughout the state.

Immigration during the 19th century brought a major change in patterns of worship. The Quakers gradually diminished in number and influence, while Roman Catholic and Greek Orthodox churches and Jewish synagogues opened in many of the mining and manufacturing centers. The bulk of the Jewish migration came, after 1848, from Germany and, after 1882, from East Europe and Russia. The Gilded Age saw the founding of a new group in Pittsburgh by clergyman Charles Taze Russell; first called the Russellites, members of this group are known today as Jehovah's Witnesses.

As of 1984, Roman Catholics constituted the largest religious group in the state, with a total population of 3,682,012; the Philadelphia archdiocese accounted for 1,353,691. The largest Protestant denomination in 1980 was the United Methodists, with 786,153 adherents. Other Protestant groups were the Lutheran Church in America, 742,085; United Presbyterians, 478,881; Congregationalists (United Church of Christ), 302,562; Episcopalians, 168,001; and American Baptists, 128,906. The historically important Mennonites had 38,228 adherents in 1980; Friends USA (Quakers), 13,174; and Moravians, 12,501. About 7% of the nation's Jewish population lived in Pennsylvania as of 1984. Of the state's estimated 412,210 Jews, 295,000 lived in metropolitan Philadelphia.

[10] TRANSPORTATION

Like so many of its industrial assets, Pennsylvania's well-developed road and rail networks are showing signs of old age. Nevertheless, the state remains an important center of transportation, and its ports are among the busiest in the US.

The early years of railroad building left Pennsylvania with more miles of track than any other state. The first railroad charter, issued in 1819, provided for a horse-drawn railroad from the Delaware Valley to the headwaters of the Lehigh River. The state authorized construction of a line between Columbia and Philadelphia in 1828, and partial service began four years later as part of the State Works. The roadbed was state-owned, and private rail car companies paid a toll to use the rails. During this time, Pennsylvanians John Jervis and Joseph Harrison were developing steam-powered locomotives. Taking advantage of the new technology were separate rail lines connecting Philadelphia with Germantown (1834), Trenton, N.J. (1838), and Reading (1839), with the Lehigh Valley (1846), and with New York City (1855). In December 1852, the Pennsylvania Central completed lines connecting Philadelphia and Pittsburgh. Five years later, the Pennsylvania Railroad purchased the State Works, eliminating state competition and tolls. By 1880, the company (which had added many smaller coal-hauling lines to its holdings) was the world's largest corporation, with more than 30,000 employees and $400 million in capital. Although railroad revenues declined with the rise of the automobile, the Pennsylvania Railroad remained profitable until the 1960s, when the line merged with the New York Central to form the Penn Central. In 1970, the Penn Central separated its real estate holdings from its transportation operation, on which it declared bankruptcy.

As of 1983, the major (Class I) lines using the state's 6,453 mi (10,385 km) of track were the Consolidated Rail Corp., or Conrail (which took over much of Penn Central's business), the Baltimore and Ohio, Delaware and Hudson, Pittsburgh and Lake Erie, Norfolk and Western, and Bessemer and Lake Erie. Amtrak operates a total of 51 regularly scheduled trains through Pennsylvania, offering passenger service to Philadelphia, Pittsburgh, and other cities along the east–west route, and from Philadelphia to New York and Washington, D.C., along the northeast corridor. Amtrak's total Pennsylvania ridership in 1984 was 4,504,135.

Mass transit systems in metropolitan Philadelphia and Pittsburgh, in Bucks, Chester, Delaware, Montgomery, and Philadelphia counties, and in Altoona, Allentown, Erie, Harrisburg, Johnstown, Lancaster, Reading, Scranton, State College, Wilkes-Barre, Williamsport, and York served 469,312,000 passengers in 1982/83. The Philadelphia Rapid Transit System, the state's first subway, was established in 1902 and is operated by the Southeastern Pennsylvania Transportation Authority, (SEPTA), which also runs buses, trolleys, trackless trolleys, and commuter trains in Bucks, Chester, Delaware, Montgomery, and Philadelphia counties. In 1985, a 1.1-mile (1.8-km) subway was opened in Pittsburgh as part of a 10.5-mile (16.9-km) light-rail (trolley) transit system linking downtown Pittsburgh with the South Hills section of the city.

Throughout its history, Pennsylvania has been a pioneer in road transportation. One of the earliest roads in the colonies was a "king's highway," connecting Philadelphia to Delaware in 1677; a "queen's road" from Philadelphia to Chester opened in 1706. A flurry of road building connected Philadelphia with other eastern Pennsylvania communities between 1705 and 1735. The first interior artery, the Great Conestoga Road, was opened in 1741 and linked Philadelphia with Lancaster. Indian trails in western Pennsylvania were developed into roadways, and a thoroughfare to Pittsburgh was completed in 1758. During the mid-1700s, a Lancaster County artisan developed an improved wagon for transporting goods across the Alleghenies; called a Conestoga wagon after the region from which it came, this vehicle later became the prime means of transport for westward pioneers. Another major improvement in land transportation came with the opening in 1792 of the Philadelphia and Lancaster Turnpike, one of the first stone-surfaced roads in the US. The steel-cable suspension bridge built by John Roebling over the Monongahela River at Pittsburgh in 1846 revolutionized bridge building, leading to the construction of spans longer and wider than had previously been thought possible. During the 1920s, Pennsylvania farmers were aided by the building of inexpensive rural roads connecting them with their markets.

A major development in automotive transport, the limited-access highway, came to fruition with the Pennsylvania Turnpike, which opened in 1940 and was the first high-speed, multilane highway in the US. In 1984, this 470-mi (756-km) roadway served 68,200,067 passenger and 10,766,360 commercial vehicles. In 1983, Pennsylvania had 115,601 mi (186,042 km) of roads, including 87,512 mi (140,837 km) of rural roads and 28,089 mi (45,205 km) of urban streets. In addition, 1,496 mi (2,408 km) of interstate highways crossed the state. Besides the Turnpike, the major highways are I-80 (Keystone Shortway), crossing the state from East Stroudsburg to the Ohio Turnpike; I-81, from New York to the Maryland border via Scranton, Wilkes-Barre, and Harrisburg; and I-79, from Erie to the West Virginia border via Pittsburgh. As of 1983 there were 6,844,049 motor vehicles registered, including 5,738,155 automobiles, 1,078,324 trucks, and 27,570 buses. The total of 7,442,842 licensed drivers as of 1983 was the 5th highest in the US.

Blessed with access to the Atlantic Ocean and the Great Lakes and with such navigable waterways as the Delaware, Monongahela, Allegheny, and Ohio rivers, Pennsylvania was an early leader in water transportation, and Philadelphia, Pittsburgh, and Erie all developed as major ports. The peak period of canal building came during the 1820s and 1830s, which saw the completion of the Main Line of Public Works, used to transport goods between Philadelphia and Pittsburgh from 1834 to 1854; this system used waterways and a spectacular portage railroad that climbed over and cut through, via a tunnel, the Allegheny Mountains. Monumental as it was, the undertaking was largely a failure. Built too late to challenge the Erie Canal's domination of east–west trade, the Main Line was soon made obsolete by the railroads, as was the rest of the state's 800-mi (1,300-km) canal system.

Philadelphia, Pittsburgh, and Erie are the state's major shipping ports. The Philadelphia Harbor (including ports in the Philadelphia metropolitan area) handled 53.9 million tons of cargo in 1983. Although no longer the dominant gateway to the Mississippi, Pittsburgh is still a major inland port. Erie is the state's outlet to the Great Lakes.

As of 31 December 1983, Pennsylvania had 720 airfields, including 491 airports, 213 heliports, 13 seaplane bases, and 3 stolports. The busiest air terminal in the state, Greater Pittsburgh Airport, enplaned 5,544,359 passengers in 1983; Philadelphia International Airport was next with 3,980,574 enplaned passengers.

¹¹HISTORY

Soon after the glacier receded from what is now Pennsylvania, about 20,000 years ago, nomadic hunters from the west moved up the Ohio River, penetrated the passes through the Allegheny Mountains, and moved down the Susquehanna and Delaware rivers. By about AD 500, the earliest Indians, already accustomed to fishing and gathering nuts, seeds, fruit, and roots, were beginning to cultivate the soil, make pottery, and build burial mounds. Over the next thousand years, the Indians became semisedentary or only seasonal nomads.

Woodland Indians living in Pennsylvania, mostly of the Algonkian language family, were less inclined toward agriculture than other Indian tribes. The first Europeans to sail up the Delaware River found the Leni-Lenape ("original people"), who, as their name signified, had long occupied that valley, and whom the English later called the Delaware. Other Algonkian tribes related to the Leni-Lenape were the Nanticoke, who ranged along the Susquehanna River, and the Shawnee, who were scattered throughout central Pennsylvania. The other major Indian language group in Pennsylvania was Iroquoian. This group included the Susquehanna (Conestoga), living east of the Susquehanna and south to the shores of Chesapeake Bay; the Wyandot, along the Allegheny River; and the Erie, south of Lake Erie. Proving that tribes related by language could be deadly enemies, the Iroquoian Confederacy of the Five Nations, located in what is now New York, destroyed the Iroquoian-speaking Erie in the 1640s and the Susquehanna by 1680. The confederacy conquered the Leni-Lenape by 1720 but failed to destroy them.

The first European to reach Pennsylvania was probably Cornelis Jacobssen, who in 1614 entered Delaware Bay for Dutch merchants interested in the fur trade. In 1638, the Swedes began planting farms along the Delaware River; they lived in peace with the Leni-Lenape and Susquehanna, with whom they traded for furs. Under Governor Johan Printz, the Swedes expanded into present-day Pennsylvania with a post at Tinicum Island (1643) and several forts along the Schuylkill River. The Dutch conquered New Sweden in 1655, but surrendered the land in 1664 to the English, led by James, Duke of York, the brother of King Charles II and the future King James II.

The English conquest was financed partly by Admiral William Penn, whose son, also named William, subsequently joined the Society of Friends (Quakers), a radical Protestant sect persecuted for its ideas of equality and pacifism. Dreaming of an ideal commonwealth that would be a refuge for all persecuted peoples, Penn asked Charles II, who had not paid the debt owed to Penn's father, to grant him land west of the Delaware. The Duke of York willingly gave up his claim to that land, and Charles II granted it in 1681 as a proprietary colony to the younger Penn and named it Pennsylvania in honor of Penn's father.

As proprietor of Pennsylvania, Penn was given enormous power to make laws and wars (subject to approval by the king and the freemen of Pennsylvania), levy taxes, coin money, regulate commerce, sell land, appoint officials, administer justice, and construct a government. From the beginning, Penn virtually gave up his lawmaking power and granted suffrage to property holders of 50 acres or £50. Even before coming to Pennsylvania, he forged his first Frame of Government, a document that went into effect 25 April 1682 but lasted less than a year. Under it, a 72-member council, presided over by a governor, monopolized executive, legislative, and judicial power, although a 200-member assembly could veto or amend the council's legislation. Arriving in the colony in October 1682, Penn approved the location and layout of Philadelphia, met with the Leni-Lenape to acquire land and exchange vows of peace, called for elections to select an assembly, and proposed a Great Law that ranged from prescribing weights and measures to guaranteeing fundamental liberties.

When the First Frame proved unwieldy, Penn on 2 April 1683 approved a Second Frame, which created an 18-member council and a 36-member assembly. A conspicuous friend of the deposed James II, Penn lost control of Pennsylvania from 1692 to 1694, and it was during this period that the legislature began to assert its rights. Penn returned to the colony in 1699, and on 28 October 1701 approved yet another constitution, called the Charter of Privileges. This document lodged legislative power in an annually elected unicameral assembly, executive power in a governor and council, which he now appointed, and judicial power in appointed provincial judges and an elected county judiciary. The Charter of Privileges remained in force until 1776.

As Pennsylvania's government evolved, its population grew steadily. Most of the first immigrants were from the British Isles and Germany. From 1681 to 1710, numerous English and Welsh Quaker migrants populated a 25-mi (40-km) zone surrounding Philadelphia. By 1750, most German immigrants were settled in a semicircular zone some 25–75 mi (40–120 km) from Philadelphia. A third and outermost ring, extending roughly 75 mi (120 km) west and north of the Germans, was populated beginning in 1717 by the Scotch-Irish; these lowland Scots, who had been resettled in Ulster in the 17th century, penetrated the interior of Pennsylvania west of the Susquehanna River until turned in a southwesterly direction by the Alleghenies. Poor lands and distant markets made the Scotch-Irish indifferent farmers, but they were known as aggressive pioneers. By 1776, each major group—which remained quite distinct—constituted roughly a third of the 300,000 Pennsylvanians. Minorities included about 10,000 Scots, 10,000 Irish Catholics, 8,000 French Huguenots, 8,000 black slaves (despite Quaker hostility to slavery), and 1,000 Jews.

A key issue during the pre-Revolutionary period was the size and extent of the colony. Conflicting colonial charters, reflecting vague English ideas of American geography, brought all of Pennsylvania's boundaries except the Delaware River into dispute. After a protracted struggle, Pennsylvania and Maryland agreed upon a basis for Charles Mason and Jeremiah Dixon to run the famous line (1763–67) that divided North and South. Although Virginia and Pennsylvania both claimed the area around Pittsburgh, a joint commission agreed in 1779 to extend the Mason-Dixon line west the full five degrees prescribed in Penn's original charter. Five years earlier, the Penn family had abandoned to New York land north of the 42d parallel. This was confirmed as Pennsylvania's northern border in 1782, when the US Congress rejected Connecticut's claim to the Wyoming Valley area, where skirmishes (called the Yankee-Pennamite wars) had been going on since the 1760s.

Pennsylvania moved rapidly toward independence after the British victory in the French and Indian War. The Proclamation of 1763, preventing settlement west of the Alleghenies, outraged western Pennsylvania, while the Stamp Act (1765), Townshend Acts (1767), and Tea Act (1773) incensed Philadelphians. Although the Continental Congress began meeting in Philadelphia in September 1774, Pennsylvania revolted reluctantly. In July 1776, only three Pennsylvania delegates to the Second Continental Congress voted for independence, while two were opposed and two absented themselves from the vote. Nevertheless, the Declaration of Independence was proclaimed from Independence Hall on 4 July 1776. As the headquarters of the Congress, Philadelphia was an important British target. The American defeat at the Battle of Brandywine Creek on 11 September 1777 led to the British occupation of the city. The provisional capital was moved first to Lancaster and then to York, where the Articles of Confederation were drafted. Following battles at Germantown and Whitemarsh, General George Washington set up winter headquarters at Valley Forge, remaining there from December 1777 to June 1778. Faced with the threat of French naval power intervening on behalf of the Americans, the British evacuated Philadelphia during the spring of 1778, and Congress reconvened there on 2 July. Philadelphia would serve as the US capital until 1783, and again from 1790 to 1800.

With independence, Pennsylvania adopted the state constitution of 1776, which established a powerful unicameral assembly elected annually by all freemen supporting the Revolution, a weak administrative supreme executive council (with a figurehead president), an appointed judiciary, and a council of censors meeting every seven years in order to take a census, reapportion the assembly, and review the constitutionality of state actions. In 1780, Pennsylvania passed the first state law abolishing slavery. Seven years later, Pennsylvania became the second state to ratify the US Constitution and join the Union. In 1790, Pennsylvania adopted a new constitution, modeled on the federal one, allowing all taxpaying males to vote. This document provided for a powerful governor, elected for a three-year term and eligible to succeed himself twice, a bicameral legislature (with senators elected every four years and a house elected annually), and an appointed judiciary.

Opposition to national taxes was evidenced by two disturbances in the 1790s. In 1794, western Pennsylvania settlers, opposed to a federal excise tax on distilled spirits, waged the Whiskey Rebellion. The insurrection was soon quashed by state troops under federal command. The levying of a federal property tax inspired the unsuccessful Fries Rebellion (1799) among Pennsylvania Germans.

By 1800, the first stages of industrialization were at hand. Pittsburgh's first iron furnace was built in 1792, and the increasing use of coal as fuel made its mining commercially feasible. The completion of the Main Line of Public Works, a canal and rail system connecting Philadelphia with Pittsburgh, was a major development of the early 19th century, which was otherwise a period of political turmoil and shifting party alliances.

By 1838, Pennsylvania adopted a new constitution curtailing the governor's power (he could serve only two three-year terms in a nine-year period), making many judgeships elective for specific terms, restricting the charter of banks, and disfranchising black people. The 1840s saw not only an influx of Irish immigrants but also the rise of the Native American (Know-Nothing) Party, an anti-Catholic movement. The antislavery crusade, which gave birth to the Republican Party, influenced state politics during the following decade.

Although a Pennsylvania Democrat, James Buchanan, carried the state and won the presidency in 1856, the Republicans captured Pennsylvania for Abraham Lincoln in 1860, partly by their strong support for a protective tariff. Protectionism attracted Pennsylvania because, in addition to its enormously productive farms, it was heavily industrialized, leading the nation in the production of iron, lumber, textiles, and leather.

Pennsylvania rallied to the Union cause, supplying some 338,000 men, a figure exceeded only by New York. The state was the scene of the Battle of Gettysburg (1–3 July 1863), a turning point in the war for the Union cause. Under General George Gordon Meade, the Union troops (one-third of whom were Pennsylvanians) defeated Confederate forces under General Robert E. Lee, who was then forced to lead a retreat to Virginia.

The Civil War left the Republican Party dominant in Pennsylvania, but, in the post–Civil War years, the Republicans were themselves dominated by industry, particularly the Pennsylvania Railroad. Between 1890 and 1900, the state was the nation's chief producer of coal, iron, and steel, and for much of that period the main source of petroleum and lumber. Farmers' sons and daughters joined immigrants from abroad in flocking to the anthracite and bituminous coal regions and to Philadelphia, Pittsburgh, and other urban centers to work in mines, mills, and factories. As the state's industrial wealth increased, education, journalism, literature, art, and architecture flourished in Philadelphia and Pittsburgh. The 1876 Centennial Exhibition at Philadelphia illustrated America's advancement in the arts and industry.

Pennsylvania adopted a reform constitution in 1873, increasing the size of the senate and house to reduce the threat of bribery, prescribing rules to prevent treachery in legislation and fraud at the polls, equalizing taxation, limiting state indebtedness, restricting the governor to one four-year term in eight years, and creating the office of lieutenant governor. None of this, however, seriously hampered the Republican political machine, led by Simon Cameron, Matthew Quay, and Boies Penrose, which dominated the state from the 1860s to the 1920s. Though Progressive reforms were enacted in subsequent years, the Penrose machine grew ever more efficient, while industrial leaders—supported both by the Pennsylvania state government and by society at large—smashed labor's efforts to unite, particularly in the great steel strike of 1919.

During the nationwide boom years of the 1920s, Pennsylvania did little more than hold its own economically, and its industrial growth rate was low. The state's share of the nation's iron and steel output no longer exceeded that of the rest of the country combined. Coal, textiles, and agriculture—all basic to the state's economy—were depressed. When Penrose died in 1921, at least five factions sought to control the powerful Pennyslvania Republican Party. In this confusion, Gifford Pinchot, a Progressive disciple of Theodore Roosevelt, won the governorship for 1923–27 and reorganized the state's administration, but failed in his attempt to enforce prohibition and to regulate power utilities.

The disastrous depression of the 1930s brought major changes to Pennsylvania. Serving again as governor (1931–35), Pinchot fought for state and federal relief for the unemployed. The

Republican organization's lack of enthusiasm for Pinchot and Progressivism helped revive the state Democratic Party long enough to secure the election in 1934—for the first time since 1890—of its gubernatorial nominee, George H. Earle. As governor, Earle successfully introduced a Little New Deal, supporting labor, regulating utilities, aiding farmers, and building public works. With government support, coal miners, steelworkers, and other organized labor groups emerged from the depression strong enough to challenge industry. Full employment and prosperity returned to Pennsylvania with the unprecedented demands on it for steel, ships, munitions, and uniforms during World War II. But the fact that Pennsylvania was not producing road vehicles and aircraft did not augur well for its industrial future.

Despite their professed opposition to government control, the Republican administrations (1939–55) that succeeded the Earle regime actually espoused and even enlarged Earle's program. They regulated industry, improved education, and augmented social services, at the same time increasing state bureaucracy, budgets, and taxes. The relative economic decline of Pennsylvania after World War II formed an ominous background for these Republican years. Although Pennsylvania remained the 2d-leading manufacturing state as late as 1947, it slipped to 3d place in 1954 and to 5th by 1958. Markets, transportation, banks, factories, machinery, and skilled labor remained abundant, however, and two Democratic governors were able to attract new industries to the state during the 1950s and early 1960s. Governor George Leader (1955–59) established the Pennsylvania Industrial Development Authority, and both he and his successor, David Lawrence, improved educational and mental health facilities, reformed government administration, and raised taxes. The economy was still not healthy in 1963, when Republican William W. Scranton entered the state house (1963–67). Warmly sympathetic to the poor, Scranton continued both to enlarge state responsibilities (through increased taxes) and to beg for federal aid for economic and social programs. He was rewarded with four years of steady economic growth. Pennsylvania's unemployment level, 2d highest in the nation from 1950 to 1962, had dropped below the national average by 1966. Raymond P. Shafer (1967–71), Scranton's Republican successor, was as forward-looking as his predecessors in his efforts to rehabilitate the economy. The 1873 constitution was extensively revised at a constitutional convention held in 1967–68, during his administration.

Because the legislature refused to approve a state income tax to pay for Shafer's new social programs, Pennsylvania faced an unresolved financial crisis in 1971 when Democrat Milton J. Shapp became governor. During his first term (1971–75), Shapp weathered the storm by securing passage of the tax. He virtually eliminated state patronage by signing union contracts covering state employees. Not only did he continue to attract business to Pennsylvania, but he also championed the consumer with no-fault auto insurance, adopted in 1974. Shapp's second term, however, was wrecked by his pursuit of the 1976 presidential nomination and by rampant corruption among Pennsylvania Democrats. Shapp's successor, Republican Richard L. Thornburgh, had scarcely sat in the governor's chair before the release of radioactive gases resulting from the malfunction of one of the two nuclear reactors at Three Mile Island in March 1979 confronted him—and others—with vexing questions concerning the safety and wisdom of atomic power. Nevertheless, in September 1985, during Thornburgh's second term, and following six years of cleanup of radioactive waste, the undamaged reactor at Three Mile Island was restarted.

¹²STATE GOVERNMENT

The 1873 constitution, substantially reshaped by a constitutional convention in 1967–68, is the foundation of state government in Pennsylvania.

The general assembly consists of a 50-member senate, elected to staggered four-year terms, and a 203-member house of representatives, elected every two years. Each house meets annually, and there are no limits to the length of each session. To qualify for the general assembly, a person must be a US citizen and have been a Pennsylvania resident for at least four years and a district resident for at least one; senators must be at least 25 years of age, representatives at least 21.

As head of the executive branch and chief executive officer of the state, the governor of Pennsylvania has the power to appoint heads of administrative departments, boards, and commissions, to approve or veto legislation, to grant pardons, and to command the state's military forces. The governor, who may serve no more than two four-year terms in succession, must be a US citizen, be at least 30 years of age, and have been a Pennsylvania resident for at least seven years before election. Elected with the governor is the lieutenant governor, who serves as president of the senate and chairman of the board of pardons, and assumes the powers of the governor if the governor becomes unable to continue in that office.

Other Pennsylvania officials also elected for four years are the auditor general, who oversees all state financial transactions; the state treasurer, who receives and keeps records of all state funds; and the attorney general, who heads the Department of Justice. All other department heads, or secretaries, are appointed by the governor and confirmed by a majority of the senate.

A bill may be introduced in either house of the general assembly. After the measure is passed by majority vote in each house, the governor has 10 days in which to sign it, refuse to sign it (in which case it becomes law), or veto it. Vetoes may be overridden by a two-thirds vote of the elected members of each house. A bill becomes effective 60 days after enactment.

A proposed constitutional amendment must be approved by a majority of both house and senate members in two successive legislatures before it can be placed on the ballot. If approved by a majority of the voters in a general election, the amendment then becomes part of the constitution.

To vote in state elections, a person must be a US citizen, be at least 18 years old, and have been a resident of Pennsylvania and of the district for at least 30 days preceding the election. Registration may take place at any time up to 30 days before an election; mail registration is allowed.

[13]POLITICAL PARTIES

The Republican Party totally dominated Pennsylvania politics from 1860, when the first Republican governor was elected, to the early 1930s. During this period, there were 16 Republican and only 2 Democratic administrations. Most of the Republicans were staunchly probusiness, though one Republican Progressive, Gifford Pinchot, was elected governor in 1922 and again in 1930. A Democrat, George Earle, won the governorship in 1934, in the depths of the depression, but from 1939 through 1955, Republicans again held the office without interruption. Only since the mid-1950s has Pennsylvania emerged as a two-party state, with Democrats electing governors in 1954, 1958, 1970, and 1974, and Republicans winning the governorships in 1962, 1966, 1978, and 1982. Both US Senate seats have been held by Republicans since 1968, with H. John Heinz III and Arlen Specter holding seats in 1985.

As of 1982 there were 5,702,557 registered voters in the state, of whom 3,035,523 were Democrats, 2,357,448 were Republicans, and 309,586 were unaffiliated or members of other parties. Democratic voters were heavily concentrated in metropolitan Philadelphia and Pittsburgh. Pennsylvania, a pivotal state for Jimmy Carter in 1976, was swept by the Republican tide in the 1980 presidential election; Ronald Reagan, the Republican nominee, won nearly 50% of the popular vote. In 1984, President Reagan received 53% of the popular vote, while Democrat Walter Mondale received 46%.

There were 138 black elected officials in Pennsylvania in 1985, including W. Wilson Goode, Philadelphia's first black mayor, and

Pennsylvania Presidential Vote by Political Parties, 1948–84

YEAR	ELECTORAL VOTE	PENNSYLVANIA WINNER	DEMOCRAT	REPUBLICAN	PROGRESSIVE	SOCIALIST	PROHIBITION	SOC. LABOR
1948	35	Dewey (R)	1,752,426	1,902,197	55,161	11,325	10,538	1,461
						SOC. WORKERS		
1952	32	*Eisenhower (R)	2,146,269	2,415,789	4,222	1,508	8,951	1,377
1956	32	*Eisenhower (R)	1,981,769	2,585,252	—	2,035	—	7,447
1960	32	*Kennedy (D)	2,556,282	2,439,956	—	2,678	—	7,158
1964	29	*Johnson (D)	3,130,954	1,673,657	—	10,456	—	5,092
						PEACE & FREEDOM	AMERICAN IND.	
1968	29	Humphrey (D)	2,259,403	2,090,017	7,821	4,862	378,582	4,977
								AMERICAN
1972	27	*Nixon (R)	1,796,951	2,714,521	—	4,639	—	70,593
					COMMUNIST			US LABOR
1976	27	*Carter (D)	2,328,677	2,205,604	1,891	3,009	25,344	2,744
						LIBERTARIAN	SOC. WORKERS	
1980	27	*Reagan (R)	1,937,540	2,261,872	5,184	33,263	20,291	—
					CITIZENS			
1984	25	*Reagan (R)	2,228,131	2,584,323	21,628	6,982	—	—

*Won US presidential election.

US Representative William H. Gray III, chairman of the House Budget Committee. One Asian and five Hispanic elected officials held office in 1984; thirteen women were state legislators in 1985.

¹⁴LOCAL GOVERNMENT

As of 1982, Pennsylvania had 67 counties, 52 cities, 968 boroughs, 1 incorporated town, 1,551 townships, 501 school districts, and 2,376 authorities (special districts). Under home-rule laws, municipalities may choose to draft and amend their own charter. As of January 1983, 57 municipalities—5 counties, 10 cities, 16 boroughs, and 26 townships—had adopted home-rule charters.

Pennsylvania counties are responsible for state law enforcement, judicial administration, and the conduct of state elections; counties also are involved in public health, regional planning, and solid waste disposal. Counties can also maintain hospitals, homes for the aged, community colleges, libraries, and other community facilities. The chief governing body in each county is a three-member board of commissioners, each elected to a four-year term. Other elected officials generally include the sheriff, district attorney, prothonotary (notary), clerk of courts, register of wills, recorder of deeds, two jury commissioners, three auditors or a controller, and treasurer. Among the appointed officials is a public defender. Counties are divided by law into nine classes, depending on population. The only first-class county, Philadelphia, is also the only city-county in the state; its county offices were merged with the city government in 1952, pursuant to the home-rule charter of 1951.

There are four classes of cities. The only first-class city, Philadelphia, is governed by a mayor and 17-member city council. Other elected officials are the controller, district attorney, sheriff, register of wills, and three city commissioners. Major appointed officials include managing director, director of finance, city representative, and city solicitor. Both Pittsburgh, the only second-class city, and Scranton, the only second-class-A city, are governed under mayor-council systems that give the mayors strong discretionary powers. Of the remaining (third-class) cities in 1982, 24 were operating under commissions consisting of a mayor, four council members, a controller, and a treasurer, and 5 had council-manager systems; the others, including Allentown, Erie, Harrisburg, Lancaster, and York, had mayor-council forms of government.

Boroughs are governed under mayor-council systems giving the council strong powers. Other elected officials are the tax assessor, tax collector, and auditor or controller. The state's first-class townships (92 in 1982), located mostly in metropolitan areas, are governed by elected commissioners who serve four-year overlapping terms. Second-class townships (1,459), most of them located in rural areas, have three supervisors who are elected at large to six-year terms. Other elected officials may include a tax assessor, tax collector, three auditors or a controller, and treasurer.

¹⁵STATE SERVICES

Operating out of the governor's office is the Governor's Action Center, a state ombudsman program. Through a toll-free telephone network, the center guides individuals to appropriate state agencies and notifies these agencies of problems that demand their attention. Other executive agencies under the governor's jurisdiction are the Pennsylvania Council on Aging, Human Relations Commission, Governor's Council on the Hispanic Community, Commission for Women, and State Ethics Commission. The Liquor Control Board operates over 700 state liquor stores and claims to be the world's largest single purchaser of liquors and wines.

The Department of Education administers the school laws of Pennsylvania, oversees community colleges, licenses and regulates private schools, and administers the state public library program. Educational policy is the province of the State Board of Education, a panel with 17 members appointed by the governor to six-year terms. Also within the department are various boards that

make policies for and review developments within the state's higher educational system.

The Department of Transportation maintains state-operated highways, mass transit, rail service, and aviation facilities. The State Highway and Bridge Authority and the Pennsylvania Turnpike Commission also have transport-related responsibilities. Agencies and departments providing health and welfare services include the Department of Aging, Department of Community Affairs, and Department of Health. All public assistance, social service, mental health, and mental retardation programs are administered by the Department of Public Welfare.

The Office of Attorney General has divisions on criminal law, legal services, and public protection. The National Guard, Bureau for Veterans' Affairs, and state veterans' homes are under the Department of Military Affairs; the Pennsylvania State Police is a separate state agency. The Pennsylvania Commission on Crime and Delinquency, created in 1978, allocates federal funds for crime control, juvenile justice, and delinquency prevention. The Pennsylvania Emergency Management Agency (formerly the State Council of Civil Defense) provides assistance in emergency situations resulting from natural or man-made disasters.

All environmental programs, mining operations, and state land and water management programs are under the supervision of the Department of Environmental Resources. The Governor's Energy Council seeks to augment the state's energy security through the planned development and conservation of energy resources. The Department of Labor and Industry administers safety, employment, and industrial standards; operates vocational rehabilitation and workers' compensation programs; and mediates labor disputes.

¹⁶JUDICIAL SYSTEM

Since 1968, all Pennsylvania courts have been organized under the Unified Judicial System. The highest court in the state is the supreme court, which, having been established in 1722, is the oldest appellate court in the US. The supreme court consists of 7 justices, elected to 10-year terms; the justice with the longest continuous service on the court automatically becomes chief justice. In general, the supreme court hears appeals from the commonwealth court. A separate appellate court, called the superior court, hears appeals from the courts of common pleas. There are 15 superior court judges, also elected to 10-year terms, as are the commonwealth and common pleas court judges.

The commonwealth court, created in 1968 and composed of 9 judges, has original jurisdiction over civil actions brought by or against the state, and appellate jurisdiction over civil actions involving the state and its related agencies, nonprofit corporations, and eminent domain proceedings. The state's principal trial courts are the courts of common pleas, which have original jurisdiction over all civil and criminal cases not otherwise specified. As of 1980 there were 310 judges in 59 courts of common pleas, one for each judicial district.

In counties other than Philadelphia, misdemeanors and other minor offenses are tried by district justices, formerly known as justices of the peace. The Philadelphia municipal court consists of 22 judges, all of whom must be lawyers; the 6 judges who constitute the Philadelphia traffic court need not be lawyers. Pittsburgh's magistrates court, appointed by the mayor, comprises 5 to 8 judges who need not be lawyers. All of Pennsylvania's judges, except traffic court judges and Pittsburgh's magistrates, are initially elected on a partisan ballot and thereafter on a nonpartisan retention ballot.

Pennsylvania's overall crime rate in 1983 was 3,196 per 100,000 people; the violent crime rate was 343, the property crime rate 2,853. Rates for specific crimes were as follows: murder and nonnegligent manslaughter, 5; forcible rape, 21; robbery, 172; aggravated assault, 145; burglary, 811; larceny-theft, 1,727; and motor vehicle theft, 315. All these figures were well below the national norms. In 1982, Philadelphia County's crime index rate—5,630—was significantly above the state average of 3,470, and the crime rate of 3,857 in Allegheny County (including Pittsburgh) was slightly above the state average.

[17] ARMED FORCES

Several important US Army and Navy facilities are located in Pennsylvania. The US Army War College is in Carlisle, and there are Army depots in Chambersburg, Harrisburg, and Scranton. The largest naval facility in the state is the historically important Philadelphia Naval Shipyard. As of 1984 there were 60,874 Department of Defense personnel in the state, 30,064 of them in the Navy and 20,773 in the Army. Defense contracts worth nearly $3.3 billion were awarded to Pennsylvania firms in 1983/84.

As of 30 September 1984 there were 1,571,000 veterans living in the state (4th in the US), of whom 13,000 served in World War I, 665,000 in World War II, 275,000 in the Korean conflict, and 392,000 during the Viet-Nam era. Veterans' benefits exceeded $1.2 billion in 1982/83.

At the end of 1984, 17,459 Pennsylvanians served in the Army National Guard; the Air National Guard had 4,529 personnel in February 1985. Of the 43,577 state and local police personnel in 1983, 38,748 were members of local forces.

[18] MIGRATION

When William Penn's followers arrived in Pennsylvania, they joined small groups of Dutch, Swedish, and Finnish immigrants who were already settled along the Delaware River. By 1685, 50% of Pennsylvania's European population was British. In 1683, the Frankfort Land Co. founded the Mennonite community of Germantown on 6,000 acres (2,400 hectares) east of the Schuylkill River. One hundred years later there were 120,000 Germans, about one-fourth of the state's census population: the Moravians, from Saxony, settled primarily in Bethlehem and Nazareth, and the Amish in Lancaster and Reading.

During the 19th century, more immigrants settled in Pennsylvania than in any other state except New York. Between 1840 and 1890, the anthracite mines in east-central Pennsylvania attracted the Irish, Welsh, and Slavs; Scotch-Irish, Italian, Austrian, Hungarian, and Polish (and, after 1880, Russian) immigrants worked the western coalfields. The cities attracted Italian, French, and Slavic workers. East European and Russian Jews settled in Philadelphia and Pittsburgh between 1882 and 1900. By the turn of the century, the urban population surpassed the rural population.

During the 20th century, these patterns have been reversed. The trend among whites, particularly since World War II, has been to move out—from the cities to the suburbs, and from Pennsylvania to other states. Blacks, who began entering the state first as slaves and then as freemen, continued to migrate to the larger cities until the early 1970s, when a small out-migration began. Overall, between 1940 and 1980, Pennsylvania lost a net total of 1,759,000 residents through migration; it lost an additional 98,000 residents between 1980 and 1983.

[19] INTERGOVERNMENTAL COOPERATION

Pennsylvania participates in such regional bodies as the Atlantic States Marine Fisheries Commission, Susquehanna River Basin Commission, Ohio River Valley Sanitation Commission, Wheeling Creek Watershed Protection and Flood Prevention Commission, and Great Lakes Basin Commission. In 1985, Pennsylvania, seven other Great Lakes states, and the Canadian provinces of Quebec and Ontario signed the Great Lakes Compact to protect the lakes' water reserves.

Some of the most important interstate agreements concern commerce and development along the Delaware River. The Delaware River Basin Commission involves the governors of Delaware, New Jersey, New York, and Pennsylvania in the utilization and conservation of the Delaware and its surrounding areas. Through the Delaware River Port Authority, New Jersey and Pennsylvania control an interstate mass transit system. The two states also are signatories to the Delaware River Joint Toll Bridge Compact and Delaware Valley Urban Area Compact.

During 1983/84, Pennsylvania received almost $4.7 billion in federal grants (3d among the 50 states), including $222 million in general revenue-sharing funds.

[20] ECONOMY

Dominated by coal and steel, Pennsylvania is an important contributor to the national economy, but its role has diminished considerably in this century. The state reached the height of its economic development by 1920, when its western oil wells and coalfields made it the nation's leading energy producer. By that time, however, Pennsylvania's oil production was already on the decline, and demand for coal had slackened. No longer did the state dominate US steel production: Pennsylvania produced 60% of the US total in 1900, but only 30% in 1940 and 24% in 1960. Philadelphia, a diversified manufacturing center, was beginning to lose many of its textile and apparel factories. The depression hastened the decline. Industrial production in 1932 was less than half the 1929 level, and mineral production, already in a slump throughout the 1920s, dropped more than 50% in value between 1929 and 1933. By 1933, 37% of the work force was unemployed.

Massive federal aid programs and the production of munitions stimulated employment during the 1940s, but some sections of the state have never fully recovered from the damage of the depression years. Declines in coal and steel production and the loss of other industries to the Sunbelt have not been entirely counterbalanced by gains in other sectors, despite a steady expansion of machinery production, increased tourism, and the growth of service-related industries and trade. The outlook for the steel industry remained uncertain in the 1980s, as Pennsylvania's aging factories faced severe competition from foreign producers.

The gross state product in 1983 was estimated at almost $149.9 billion in current dollars, or 4.5% of the GNP. Measured in constant 1972 dollars, the gross state product increased only 6% between 1972 and 1983. Contributions to the 1983 total included manufacturing, 26%; trade, 16%; finance, insurance, and real estate, 16%; services, 16%; government, 10%; transportation, communications, and public utilities, 9%; construction, 4%; mining, 2%; and other sectors, 1%.

[21] INCOME

Although Pennsylvania is one of the nation's most industrialized states, its wage earners tend to receive less than their counterparts in other states.

Per capita income in 1983 was $11,510—22d among the 50 states and slightly below the US average. Total personal income for 1983—$136.9 billion—was only 19% higher than in 1980, below the overall US growth rate of 21%. The state's share of the nation's income dropped from 6.4% in 1960 to 5% in 1983. Median money income of a four-person family in 1981 was $26,519, 20th in the US.

According to state figures, sources of personal income in 1980 were as follows: manufacturing, 33%; services, 18%; trade, 15%; government, 14%; transportation and public utilities, 7%; construction, 6%; finance, insurance, and real estate, 5%; mining, 1%; and farming, 1%. In 1982, more than 45% of all income was earned in the Philadelphia metropolitan area (including parts of New Jersey). Other major income areas were metropolitan Pittsburgh, 20%; Allentown-Bethlehem-Easton, 6%; Scranton and Wilkes-Barre, 5%; and Harrisburg-Lebanon-Carlisle, 5%.

About 1,210,000 Pennsylvanians (10.5%) were below the federal poverty level in 1979. In 1982, 86,800 Pennsylvanians, each with gross assets greater than $500,000, were among the top wealthholders in the US.

[22] LABOR

In 1984, Pennsylvania's civilian labor force averaged 5,487,000 (5th in the US), of whom 4,988,000 were employed. In 1982, employment in private industry was distributed among the following occupational categories: operatives, 19%; office and clerical workers, 15%; craft workers, 13%; officials and managers, 12%; professionals, 10%; service workers, 9%; sales workers; 8%; laborers, 8%; and technicians, 6%.

A federal survey of workers in March 1982 revealed the following nonfarm employment pattern for Pennsylvania:

	ESTABLISH-MENTS	EMPLOYEES	ANNUAL PAYROLL ('000)
Agricultural services, forestry, fishing	1,914	10,686	$ 145,079
Mining, of which:	1,442	51,252	1,195,518
Bituminous coal, lignite	(552)	(34,058)	(811,754)
Contract construction	18,061	174,461	3,722,344
Manufacturing, of which:	16,544	1,228,795	23,754,704
Food products	(1,151)	(86,883)	(1,518,552)
Apparel, other textiles	(1,587)	(117,724)	(1,107,331)
Primary metals	(555)	(139,207)	(3,190,136)
Fabricated metal products	(1,845)	(98,416)	(1,908,218)
Nonelectrical machinery	(2,392)	(130,683)	(2,621,088)
Transportation, public utilities	8,400	219,170	4,862,817
Wholesale trade	17,546	241,076	4,599,846
Retail trade	62,108	731,847	6,284,040
Finance, insurance, real estate	17,756	246,433	4,135,868
Services	69,173	1,012,223	13,346,883
Other	3,532	3,258	70,962
TOTALS	216,476	3,919,201	$62,118,061

Among the categories of employees not included in this survey were government workers, of whom Pennsylvania had 529,549 in March 1983.

The average adjusted unemployment for the state during the recession in 1982 was 10.9%, ranging from 8.1% for adult white females to 26.5% for adult black males. In Philadelphia, the rates were a bit lower: 9.5% overall, ranging from 6.9% for adult white women to 24.5% for adult black men. Pittsburgh fared worse, with an overall unemployment rate of 12.5% during the same year. By 1984, the statewide unemployment rate had fallen to 9.1%.

The history of unionism in Pennsylvania dates back to 1724, when Philadelphia workers organized the Carpenters' Company, the first crafts association in the colonies. Its Carpenters' Hall gained fame as the site of the First Continental Congress in 1774; the carpenters were also responsible for the first strike in the US, in 1791. The nation's first labor union was organized by Philadelphia shoemakers in 1794. By 1827, the Mechanics' Union of Trade Associations, the country's first central labor body, was striking for a 10-hour workday and was the impetus behind the formation of the Organized Workingman's Party. Nine years later there were no fewer than 58 labor organizations in Philadelphia and 13 in Pittsburgh, but the Panic of 1837 resulted in a sharp decline of union strength and membership for many years. Union ranks were further depleted by the Civil War, despite the efforts of Pennsylvania labor leader William Sylvis, who later became an important figure in the national labor reform movement. After the Civil War ended, the Noble Order of the Knights of Labor was established in Philadelphia in 1869.

The coalfields were the sites of violent organizing struggles. In 1835, low wages and long hours sparked the first general mine strikes, which, like a walkout by anthracite miners in 1849, proved unsuccessful. During the 1850s and 1870s, a secret society known as the Molly Maguires led uprisings in the anthracite fields, but its influence ended after the conviction of its leaders for terrorist activities. The demise of the Molly Maguires did not stop the violence, however. Eleven persons were killed during a mine strike at Connellsville in 1891, and a strike by Luzerne County miners in

1897 resulted in 20 deaths. Finally, a five-month walkout by anthracite miners in 1902 led to increased pay, reduced hours, and an agreement to employ arbitration to settle disputes.

Steelworkers, burdened for many years by 12-hour workdays and 7-day workweeks, called several major strikes during this period. An 1892 lockout at Andrew Carnegie's Homestead steel mill led to a clash between workers and Pinkerton guards hired by the company; after several months, the strikers went back to work, their resources exhausted. A major strike in 1919, involving half of the nation's steelworkers, shut down the industry for more than three months, but it too produced no immediate gains. The Steel Workers Organizing Committee, later the United Steelworkers, finally won a contract and improved benefits from US Steel in 1937, although other steel companies held out until the early 1940s, when the Supreme Court forced recognition of the union.

As of 1980, nearly 35% of all nonagricultural employees belonged to unions, the 3d highest such percentage in the US. Pennsylvania leads the US in number of unions, with 4,108 as of 30 June 1983. The most important union in the state is the United Steelworkers of America, headquartered in Pittsburgh.

²³AGRICULTURE

Pennsylvania ranked 29th among the 50 states in agricultural income in 1983, with receipts of nearly $3 billion.

During the colonial period, German immigrants farmed the fertile land in southeastern Pennsylvania, making the state a leader in agricultural production. Unlike farmers in other states who worked the soil until it was depleted and then moved on, these farmers carefully cultivated the same plots year after year, using crop rotation techniques that kept the land productive. As late as 1840, the state led the nation in wheat production, thanks in part to planting techniques developed and largely confined to southeastern Pennsylvania. However, westward expansion and the subsequent fall in agricultural prices hurt farming in the state, and many left the land for industrial jobs in the cities. Today, most farms in the state produce crops and dairy items for Philadelphia and other major eastern markets.

As of 1984 there were about 58,000 farms averaging 150 acres (61 hectares) in size. The leading farm areas were all in southeastern Pennsylvania. Lancaster County was by far the most productive, followed by the counties of Franklin, Adams, York, Berks, and Chester.

The following table shows leading field crops in 1983:

	OUTPUT (BUSHELS)	VALUE
Hay (tons)	4,620,000	$455,070,000
Corn for grain	72,450,000	286,178,000
Oats	16,200,000	31,590,000
Soybeans	3,480,000	27,318,000
Wheat	7,600,000	24,360,000
Barley	3,575,000	7,508,000

In 1982, Pennsylvania led the nation in the production of mushrooms with 273,048,000 lb, worth $179,557,000; it was also a major producer of roses, carnations, and chrysanthemums. Other crops were fresh vegetables, potatoes, strawberries, apples, pears, peaches, grapes, and cherries (sweet and tart). The value of fresh market vegetables exceeded $20 million in 1982; of vegetables for processing, $11.5 million.

²⁴ANIMAL HUSBANDRY

Most of Pennsylvania's farm income stems from livestock production, primarily in Lancaster County.

There were 2,000,000 cattle and calves, 950,000 hogs and pigs, 738,000 milk cows, and 98,000 sheep and lambs on Pennsylvania farms in 1983. Sales of 2,000,000 head of cattle and calves brought state farmers almost $1.3 billion, while sales of 98,000 head of sheep and lambs brought farmers $6,223,000. In 1982, sales of 830,000 hogs and pigs brought in $63,910,000.

Pennsylvania was a leading producer of chickens in the US during 1983, yielding 14.9 million chickens, 102.6 million broilers, and 153.2 million commercially hatched chicks. Some 6.8 million turkeys were raised in the state that year. Egg production—4.7 billion—was 2d highest in the US.

Over 9 billion lb of milk (5th among the 50 states), worth almost $1.3 billion, was produced by Pennsylvania dairy farms in 1983. In 1984, the state ranked 5th in butter production, with 63.2 million lb; 2d in ice cream, with 80.3 million gallons; 6th in ice milk, with 14.9 million gallons; and among the top 6 in cheeses, with 147.4 million lb (not including cottage cheese).

25 FISHING

Although there is little commercial fishing in Pennsylvania—the 1984 catch of 326,000 lb was worth only $162,000—the state's many lakes and streams make it a popular area for sport fishing. All recreational fishing in the state is supervised by the Fish Commission, established in 1866 and one of the oldest conservation agencies in the US. Almost 37 million fish were stocked in 1980/81. Walleye, trout, and salmon were the leading species.

26 FORESTRY

The hills of central Pennsylvania made up the bulk of the state's 16,826,000 acres (6,809,000 hectares) of forestland in 1979, when forests covered 59% of the total land area. As of 30 January 1985, 1,960,959 acres (793,575 hectares) of forestland managed by the state were spread across 42 counties. Nearly all of Pennsylvania's forests are usable as commercial timberland, about four-fifths of it privately owned. In 1984, the National Forest Service administered the 510,406-acre (206,555-hectare) Allegheny National Forest in western Pennsylvania, as well as 227 acres (92 hectares) of forest elsewhere in the state.

During the 1860s, Pennsylvania led the nation in lumber production, but overcutting and mismanagement nearly decimated the forests by 1900. Although more than 715.5 million seedlings from state nurseries were planted on public and private land between 1899 and 1985, the industry has never regained its former importance. In 1982, shipments of lumber and wood products exceeded $1.1 billion. Paper and paper products played a much larger role, accounting for over $5.2 billion in shipments in 1982.

27 MINING

The value of Pennsylvania's nonfuel mineral production in 1984 was $657.6 million, for a ranking of 9th among the states. The most valuable nonfuel mineral was portland cement, with a 1984 output of 5,600,000 tons, followed by crushed stone, 56,200,000 tons. Other important minerals were pig iron, 7,100,000 tons (1983); sand and gravel, 13,200,000 tons; lime, 1,620,000 tons; clays, 934,000 tons; masonry cement, 295,000 tons; zinc, 16,792 metric tons (1983—production of zinc has since been halted); and dimension stone, 44,000 tons.

28 ENERGY AND POWER

Installed capacity of Pennsylvania's electric power plants in 1983 was 34.8 million kw, all of it privately owned. Total energy consumption in 1982 reached 3,256 trillion Btu; coal supplied 38% of the state's energy needs, petroleum 35%, natural gas 21%, nuclear power 5%, and hydropower and other sources 1%. That year, energy consumption was 274 million Btu per capita. In 1981, energy expenditures per capita averaged $1,839.

Pennsylvania's nuclear power production dropped abruptly on 28 March 1979, when a malfunction at the 906,000-kw Unit 2 plant operated by Metropolitan Edison (a subsidiary of General Public Utilities) at Three Mile Island near Harrisburg caused the reactor's containment building to fill up with radioactive water. Some radioactive steam was vented into the atmosphere, and thousands of residents of nearby areas were temporarily evacuated. A 12-member panel appointed by President Jimmy Carter to investigate the accident found serious flaws in the design of the plant's safety systems and in federal regulation of the nuclear

power industry. Metropolitan Edison's 819,000-kw Unit 1 plant was also shut down after the accident but was reopened in fall 1985. Remaining nuclear plants in Pennsylvania are the Peach Bottom Units 2 and 3 (combined capacity 2,130,000 kw), 85% of which is owned jointly by Philadelphia Electric and Public Service Electric and Gas; Beaver Valley Unit 1 (833,000 kw), at Shippingport, owned by Duquesne Light, Ohio Edison, and Pennsylvania Power; Susquehanna Units 1 and 2 (2,100,000 kw), 90% owned by Pennsylvania Power & Light; and Limerick 1 (1,055,000 kw), owned by Philadelphia Electric.

Electric energy sales in the state in 1983 exceeded 96.6 billion kwh, of which 42% was industrial, 33% residential, 22% commercial, and 3% for other purposes. The largest utility in the state is Philadelphia Electric, the 15th largest in the US, with 1984 assets of $9.6 billion.

The nation's first oil well was struck in Titusville in 1859, and for the next five decades Pennsylvania led the nation in oil production. Reserves totaled 41 million barrels in 1983, when output dropped to an estimated 3 million barrels. The state's natural gas production in 1983 was 118.3 billion cu feet; estimated reserves as of 1983 were 1,882 billion cu feet. Virtually all the state's commercial oil and gas reserves lie beneath the Allegheny High Plateau, in western Pennsylvania.

Coal is the state's most valuable mineral commodity, accounting for more than two-thirds of all mine income; the state's output in 1983 represented 9% of US production. In 1982, Pennsylvania's 733 mining companies produced 78,749,000 tons of coal from 1,642 mines. Pennsylvania is the only major US producer of anthracite coal, with an output of 3,900,000 tons in 1982; bituminous coal production totaled 74,700,000 tons. In 1984, 7,953,000 tons of coke were produced. Bituminous coal is mined in Washington, Clearfield, Greene, Cambria, Armstrong, Somerset, Clarion, Allegheny, and 19 other counties in the western part of the state; anthracite mining is concentrated in Schuylkill, Luzerne, Lackawanna, Northumberland, Carbon, Columbia, Sullivan, and Dauphin counties in the east. In 1983 there were 484 active coal mines, 94 underground and 390 surface. Demonstrated reserves as of 1983 were almost 23 billion tons of bituminous (3d in the US) and 7.1 billion tons of anthracite, 96% of the US total.

29 INDUSTRY

At different times throughout its history, Pennsylvania has been the nation's principal producer of ships, iron, chemicals, lumber, oil, textiles, glass, coal, and steel. Although it is still a major manufacturing center, Pennsylvania's industrial leadership has diminished steadily during this century.

The first major industry in colonial Pennsylvania was shipbuilding, centered in Philadelphia. Iron works, brick kilns, candle factories, and other small crafts industries also grew up around the city. By 1850, Philadelphia alone accounted for nearly half of Pennsylvania's manufacturing output, with an array of products including flour, preserved meats, sugar, textiles, shoes, furniture, iron, locomotives, pharmaceuticals, and books. The exploitation of the state's coal and oil resources and the discovery of new steelmaking processes helped build Pittsburgh into a major industrial center.

Manufacturing accounted for 26% of Pennsylvania's estimated gross state product in 1983. According to the federal Census of Manufacturers, manufacturing employment declined by almost 12% between 1977 and 1982, with jobs lost in nearly every major category except printing and publishing, which showed a slight increase. During the same period, value added by manufacture increased from $36 billion to $44.8 billion, and the value of shipments of manufactured goods grew from $79.8 billion to $102.9 billion. Leading industries by value of shipments were primary metals, 13%; food and kindred products, 13%; petroleum and coal products, 10%; nonelectrical machinery, 9%; fabricated metal products, 7%; electric and electronic equipment, 7%;

chemicals and chemical products, 7%; and transportation equipment, 6%.

The following table shows the value of shipments for selected industries in 1982:

Basic steel and blast furnace products	$10,440,800,000
Petroleum refining	9,860,200,000
Stone, clay, and glass products	3,452,500,000
Drugs	3,140,900,000
Motor vehicles and equipment	2,381,800,000
Electronic components and accessories	2,245,800,000
Textile mill products	2,234,000,000
Preserved fruits and vegetables	1,853,700,000
Construction and related machinery	1,736,800,000
Commercial printing	1,478,100,000
Ladies' outerwear	1,374,600,000
Measuring and controlling devices	1,205,800,000
Railroad equipment	1,188,800,000
Metalworking machinery	1,115,000,000

Pennsylvania is a leading industrial state, with 17,669 industrial establishments in 1982; 36 of the *Fortune* 500 leading industrial firms had their headquarters in Pennsylvania as of 31 December 1984. Pittsburgh is a popular site in the US for corporate headquarters. Among the leading companies are US Steel, with 1984 assets of $19 billion; Westinghouse Electric, $9.2 billion; Aluminum Co. of America, $6.3 billion; Rockwell International, $5.9 billion; PPG Industries, $3.8 billion; H. J. Heinz, $2.3 billion; and Allegheny International, $1.7 billion. Other major corporations headquartered in the state are Sun, a petroleum refiner in Radnor, $14.5 billion, and Bethlehem Steel in Bethlehem, $4.4 billion. Scott Paper, Rohm & Haas, and SmithKline Beckman (pharmaceuticals) have their headquarters in Philadelphia, Armstrong World Industries in Lancaster, Mack Trucks in Allentown, Hershey Foods in Hershey, and Hammermill Paper in Erie.

30 COMMERCE

A large ingredient in Philadelphia's early economy, trade remains important to the state, accounting for about 16% of the gross state product in 1983.

According to federal data, wholesale trade in 1982 totaled $78.4 billion, 7th highest in the US. The main items sold were groceries and related products; machinery, equipment, and supplies; motor vehicles and automotive parts and supplies; metals and minerals (excluding petroleum); electrical goods; and petroleum and petroleum products.

Pennsylvania ranked 5th in the US in retail trade in 1982, with almost $50.8 billion in sales. The top sales categories were grocery stores, 24%; new-car dealers, 17%; and general merchandise stores, 12%. Philadelphian John Wanamaker opened the world's first department store in 1876; by 1982, Pennsylvania had 572 department stores, 172 of them in the Philadelphia metropolitan area and 102 in the Pittsburgh metropolitan area.

During the colonial era, Philadelphia was one of the busiest Atlantic ports and the leading port for the lucrative Caribbean trade. Philadelphia remains one of the country's leading foreign trade centers; the main import suppliers in 1983 were the United Kingdom, Indonesia, Saudi Arabia, Venezuela, and Algeria. During 1983, the Philadelphia Customs District—which includes ports in Delaware and New Jersey, as well as Pennsylvania—processed over 48.3 million tons of import cargo worth almost $11.3 billion, of which fully 85% consisted of coal, coke, and other mineral fuels. Also passing through the Philadelphia Customs District were 5.8 million tons of exports worth over $2 billion. In 1981, total exports of Pennsylvania manufactures had a value of $8.1 billion (8th in the US); exports of the state's agricultural products totaled $246 million (33d) in 1981/82.

31 CONSUMER PROTECTION

The Bureau of Consumer Protection is part of the Public Protection Division of the Office of the Attorney General. Also within the Office of the Attorney General is the Office of the Consumer Advocate, established in 1976, which represents citizens' interests before the Public Utilities Commission. The Bureau of Advocacy in the Department of Aging is designed to ensure that senior citizens' needs are being met by the programs of other state agencies. Pennsylvanians are encouraged to report instances of fraud, waste, or mismanagement of state funds through a toll-free telephone service maintained by the Taxpayer Information Program under the Department of the Auditor General. Other consumer services protect state residents against insurance fraud and illegal marketing of milk products.

32 BANKING

Philadelphia is the nation's oldest banking center, and Third St. between Chestnut and Walnut has been called the cradle of American finance. The first chartered commercial bank in the US was the Bank of North America, granted its charter in Philadelphia by the federal government in December 1781 and by Pennsylvania in April 1782. The First Bank of the US was headquartered in Philadelphia from its inception in 1791 to 1811, when its charter was allowed to expire. Its building was bought by Stephen Girard, a private banker whose new institution quickly became one of the nation's largest banks. Girard's bank was closed after he died in 1831, but a new Girard Bank was opened in 1832; it merged with Philadelphia National Bank in 1926.

By the early 1800s, Philadelphia had reached its zenith as the nation's financial center. It was the home of the Bank of Pennsylvania, founded in 1793; the Bank of Philadelphia (1804); the Farmers and Mechanics Bank (1809); the Philadelphia Savings Fund Society (1816), the first mutual savings bank; and the most powerful of all, the Second Bank of the US (1816). After 1823, under the directorship of Nicholas Biddle, this bank became an international leader and the only rival to New York City's growing banking industry. When President Jackson vetoed the bank's recharter in 1831, Philadelphia lost its preeminence as a banking center.

Pittsburgh also rose to prominence during the Gilded Age, in great part because of the efforts of its most successful financier, Andrew Mellon. As of 1984, the Mellon Bank of Pittsburgh was the nation's 11th largest, with assets of $30.6 billion. In March 1982, the state legalized multibank holding companies; subsequently, the Mellon Bank acquired Central County Bank of State College, Girard Bank, and Northwest Bank. Other major institutions are Pittsburgh National Bank, part of PNC Financial (27th), and Philadelphia National Bank, part of CoreStates Financial (34th). First Pennsylvania, in financial difficulty for several years, was saved from possible failure early in 1980 through a loan package engineered by the Federal Deposit Insurance Corporation.

As of 31 December 1984, Pennsylvania had 333 insured commercial banks; such banks had $141.2 billion in deposits at the end of 1983, and $104.8 billion in assets (5th in the US). The 293 savings and loan associations held $25.6 billion in assets and $18.4 billion in outstanding mortgage loans as of 31 December 1982. There were also 1,355 federally chartered credit unions with 1,546,143 members and $2.7 billion in assets, and 192 state-chartered credit unions with 320,556 members and $446,029,000 in assets.

33 INSURANCE

As in banking, Philadelphia has been a leading center for the insurance business. Most of the first insurance companies underwrote only maritime insurance for goods in trade; fire and casualty insurance and life insurance did not become popular until the early 1800s. One of the oldest stock property and casualty firms, the Insurance Co. of North America (INA), was formed in Philadelphia in 1792. In 1982, INA merged with Connecticut General to form CIGNA, which was the 4th-largest diversified financial company in the US in 1984, with assets of $39 billion.

In 1984, 201 property/casualty insurance companies had their home offices in Pennsylvania. Direct written premiums totaled $5.5 billion in 1983.

As of mid-1983, 58 life insurance companies were domiciled in Pennsylvania. That year, Pennsylvanians held 22 million life insurance policies worth $243.4 billion, averaging $53,000 per family; 212,400 beneficiaries received a total of $892.7 million in payments. Total premium receipts for 1983 reached $2.4 billion for life insurance and $1.4 billion for health insurance. The largest in-state company, Penn Mutual (20th largest in the US), had $3.9 billion in assets in 1984. As of 31 December 1983, Pennsylvania led the nation in the number of communities that had qualified for federal flood disaster assistance, with 2,442.

All insurance companies operating in the state are regulated by the Insurance Department, as are all insurance brokers and agents. The Bureau of Policyholders Service and Enforcement, which receives approximately 150,000 inquiries and complaints annually, investigates alleged violations of insurance laws and refers appropriate cases for legal action. The department prepares shoppers' guides and was a leader in seeking to make the language of insurance policies more comprehensible to ordinary policyholders.

34 SECURITIES

Formally established in 1790, the Philadelphia Stock Exchange (PHLX) is the oldest stock exchange in the US. It was also the nation's most important exchange until the 1820s, when the New York Stock Exchange eclipsed it. Since World War II, the Philadelphia exchange has merged with stock exchanges in Baltimore (1949), Washington, D.C. (1953), and Pittsburgh (1969). As the primary odd-lot market for Government National Mortgage Association securities and as a leading market for odd-lot government securities and stock options, PHLX ranks after only the New York and American exchanges in trading volume. PHLX was the first exchange in the US to trade foreign currency options (1982) and the National Over-the-Counter Index (1985). In 1984, 452,566,096 shares of stock and 16,053,206 options contracts on primary contracts were traded by PHLX member firms.

As of 1983, New York Stock Exchange member firms had 228 sales offices and 2,942 registered representatives in the state. The state had 376 brokers and dealers registered under the Securities Exchange Act of 1934. There were 2,071,000 shareowners of public corporations in Pennsylvania in 1983.

Sales of securities are regulated by the Pennsylvania Securities Commission, which also licenses all securities dealers, agents, and investment advisers in the state.

35 PUBLIC FINANCE

Pennsylvania's budget is prepared annually by the Office of Budget and submitted by the governor to the general assembly for amendment and approval. By law, annual operating expenditures may not exceed available revenues and surpluses from prior years. The fiscal year runs from 1 July to 30 June.

The following table shows general revenues and expenditures for the state government in 1984/85 (available) and 1985/86 (recommended), in thousands:

REVENUES	1984/85	1985/86
Consumption taxes	$3,357,749	$3,517,847
Personal income taxes	2,584,157	2,732,909
Corporation taxes	2,211,466	2,320,367
Other taxes	390,500	379,800
Nontax revenues	196,198	192,447
TOTALS	$8,740,070	$9,143,370

EXPENDITURES	1984/85	1985/86
Education	$4,298,753	$4,551,072
Health	1,609,791	1,669,926
Economic development, income maintenance	1,066,457	1,101,871
Social development	514,533	524,448
Public protection	472,919	501,332
Transportation and communications	186,402	194,691
Recreation and culture	100,526	114,451
Other expenses	367,649	381,939
TOTALS	$8,617,030	$9,039,730

The recommended consolidated state budget—including revenues from the motor license fund, federal aid, and other fees and special fund revenues—totaled $16.5 billion for 1985/86. The state debt as of the end of fiscal 1982 exceeded $6.2 billion. At that time, the total state and local government debt was $20.2 billion, or $1,706 per capita.

During the 1970s, Philadelphia showed many of the fiscal ills typical of aging eastern cities: a stagnant economy, high tax base, deficit spending, deteriorating public works, and out-migration of upper-income whites to the suburbs. By the mid-1980s, however, the city showed signs of a marked improvement. The following table summarizes Philadelphia's operating budget for 1983/84 (actual) and 1984/85 (estimated), in thousands:

REVENUES	1983/84	1984/85
Taxes	$1,078,821	$1,127,639
Locally generated nontax revenues	339,446	369,786
Revenues from other governments	439,706	496,373
Receipts from other city funds	24,399	23,450
Adjustments	17,920	17,920
TOTALS	$1,900,292	$2,035,168

EXPENDITURES	1983/84	1984/85
Personal services	$ 737,682	$ 757,896
Purchase of services	511,424	611,105
Employee benefits	266,422	288,631
Debt service	214,079	239,419
Other expenditures	145,304	168,067
TOTALS	$1,874,911	$2,065,118

Pittsburgh's estimated operating revenues and expenditures in 1985 were $277,067,356.

36 TAXATION

Pennsylvania's personal income tax, adopted in 1971, is levied at a rate of 2.35%. Business taxes include a corporation net income tax of 9.5%, capital stock and franchise tax, and taxes on public utilities, insurance premiums, and financial institutions. Pennsylvania's 6% sales and use tax exempts essential items like clothing, groceries, and medicines, giving Pennsylvania the 39th-lowest effective sales tax rate of the 45 states with sales taxes. Other taxable items are employee wages, at a basic rate of 3.5% for all new employers except construction contractors, and real estate transfers, at a rate of 1% on the value of the real property. Programs such as the Target Jobs Tax Credit and the Employment Incentive Payment program make tax credits, worth up to $8,100 per employee, available to employers who hire welfare recipients or persons from specified groups that have special employment needs or high unemployment rates.

The Local Economic Revitalization Tax Assistance Act allows local taxing authorities to exempt improvements to business property if such property is located in a deteriorating area or is subject to a government order requiring the property to be vacated, condemned, or demolished. Any county, city, borough, incorporated town, township, institutional district, or school district may participate in this program.

In 1983, Pennsylvanians filed 4.9 million federal income tax returns, paying $13.1 billion in tax.

[37] ECONOMIC POLICY

The Bureau of Economic Assistance directs and controls the Commerce Department's economic assistance activities, including the administration and management of the Revenue Bond and Mortgage Program, the Pennsylvania Industrial Development Authority, the Job Training Partnership Act, and the Pennsylvania Capital Loan Fund. Among the activities of the Bureau of Domestic and International Commerce are supervising projects encouraging industrial development, attracting both domestic and foreign investment to the state, and providing export assistance to Pennsylvania companies.

The Pennsylvania Industrial Development Authority Board and the Pennsylvania Minority Business Development Authority provide loans to businesses that want to build new facilities or renovate and expand older ones. The Office of Minority Business Enterprise seeks to strengthen minority businesses by helping them obtain contracts with the state. The Small Business Action Center aids small businesses by providing a network of informational sources. The Bureau of Technological Development directs all phases of the Department of Commerce's scientific and technological activities and monitors advanced technology initiatives and growth throughout the state. Additional services are provided by the Bureau of Statistics, Research and Planning, the Bureau of Appalachian Development, the Bureau of Travel Development, the Bureau of Motion Picture and Television Development, the Bureau of Management and Administration, the Financial Analysis Office, and the Nursing Home Loan Agency.

The Department of Community Affairs—the first such department in the US—coordinates local economic improvement programs. Nearly $20 million in state grants is approved by the department each year. In addition, the department administers $100 million in federal grants to Pennsylvania communities.

[38] HEALTH

Pennsylvania's infant mortality rate in 1981 was 10.7 per 1,000 live births for whites, 21.1 for blacks; both figures were above the US averages. An estimated 64,100 legal abortions were performed in 1982. Death rates for the leading causes of death—heart disease, cancer, and stroke—were well above the US average. Rates in 1981 per 100,000 population were as follows: heart disease, 412; cancer, 220; stroke, 75; accidents, 37; chronic obstructive pulmonary diseases, 27; pneumonia and flu, 23; diabetes, 21; cirrhosis of the liver, 13; atherosclerosis, 13; suicide, 11; and early infant diseases, 9.

As of 30 June 1983, there were 6,073 residents in the 12 state mental retardation centers; another 537 persons were residents of the state's 7 mental retardation units. State mental hospitals had 11,782 patients. Statewide in 1983 there were 307 hospitals of all types, with 79,946 beds. Hospital personnel included 37,538 full-time registered nurses and 14,251 part-time registered nurses in 1981. The average cost per stay in Pennsylvania hospitals in 1982, $2,720, was the 12th highest in the US, but the average cost per day, $320, ranked only 21st. There were 25,822 licensed physicians and 6,743 active dentists in the state in 1982.

The University of Pennsylvania School of Medicine, which originated as the medical school of the College of Philadelphia in 1765, is the nation's oldest medical school. One of the nation's newest is the Hershey Medical Center of Pennsylvania State University. Other medical schools in Pennsylvania are the University of Pittsburgh School of Medicine, Temple University's School of Medicine, the Medical College of Pennsylvania, and Hahnemann Medical College, the last three in Philadelphia. The state also aids colleges of osteopathic medicine, podiatric medicine, and optometry—all in Philadelphia. Among the many medical certification boards located in Philadelphia are the American boards of allergy and immunology, internal medicine, ophthalmology, and surgery.

[39] SOCIAL WELFARE

During 1983, 587,700 Pennsylvanians—members of 197,000 families—received $739 million in aid to families with dependent children. About 1,099,000 Pennsylvanians participated in the food stamp program at a cost to the federal government of $557 million. Some $88 million in federal funds was spent on school lunches, providing food for 1,021,000 students.

In 1983, about 2,105,000 state residents received $10.4 billion in Social Security benefits; 1,475,000 retired workers received $7.1 billion, 429,000 survivors received $2.3 billion, and 198,000 disabled persons received $1 billion. Pennsylvania ranked 5th nationally in its number of Social Security beneficiaries per 10,000 population with 1,770. Supplemental Security Income totaling $375.3 million was paid to 154,026 Pennsylvanians in 1983.

About 29% of the beneficiaries of the federal Black Lung Benefit Program lived in Pennsylvania in 1983. Of the 95,656 Pennsylvanians who received a total of $303 million in benefits, 26,550 were miners, 43,977 were miners' widows, and 25,129 were miners' dependents. In 1984, Medicare paid $2.7 billion in hospital insurance payments and $1.2 billion in supplementary medical insurance payments. About $46.4 million was spent on vocational rehabilitation and $33.3 million on education for the handicapped in fiscal 1984. The state spent $1.8 billion on unemployment insurance benefits in 1982.

[40] HOUSING

Most of Pennsylvania's housing is owner-occupied. Of the 4,492,-156 year-round housing units in the state in 1980, 2,950,649 were owned by their occupants, 1,268,957 were rented, and 272,550 were vacant. About 38% of the state's housing units were in Philadelphia, while 19% were in Pittsburgh. The median monthly cost for an owner-occupied housing unit with a mortgage was $334; for a nonmortgaged unit the cost was $142. The median monthly rent for a renter-occupied unit was $226.

Some 79,700 new privately owned housing units worth nearly $3.3 billion were authorized from 1981 to 1983; 91,700 new privately owned housing units were started during those years. Faced with a decaying housing stock, Philadelphia during the 1970s and early 1980s encouraged renovation of existing units along with the construction of new ones, effectively revitalizing several neighborhoods.

[41] EDUCATION

Pennsylvania lagged behind many of its neighbors in establishing a free public school system. From colonial times until the 1830s, almost all instruction in reading and writing took place in private schools. Called "dame schools" in the cities and "neighborhood schools" in rural areas, these primary courses were usually taught by women in their own homes. In addition, the Quakers, Moravians, and Scotch-Irish Presbyterians all formed their own private schools, emphasizing religious study. Many communities also set up secondary schools, called academies, on land granted by the state; by 1850, there were 524 academies, some of which later developed into colleges. A public school law passed in 1834 was not mandatory in the school districts but was still unpopular. Thaddeus Stevens, then a state legislator, is credited with saving the law from repeal in 1835. Two years later, more than 40% of the state's children were in public schools.

As of 1980, 64.7% of the population 25 years old and older had completed four years of high school; 24.3% of the population had completed at least one year of college, with 13.6% of the residents having finished four or more years. In 1982, 143,400 students graduated from public high schools.

During the 1983/84 school year, 846,145 students were enrolled in public elementary schools, and 891,807 in secondary schools. Enrollment in private elementary schools was 285,372; in private secondary schools, 110,861. In 1984, Pennsylvania had 854 Roman Catholic elementary and high schools, which accounted for about 75% of all nonpublic school enrollment.

Enrollment in the 14 state-owned colleges in fall 1983 totaled 81,524. Indiana University of Pennsylvania, established in 1872, accounted for about 15% of this enrollment, with 12,503 students as of fall 1982. Four universities have nonprofit corporate charters but are classified as state-related: Pennsylvania State University, Temple University, the University of Pittsburgh, and Lincoln University. Of these, Penn State is by far the largest, with a fall 1982 enrollment of 65,091. Founded in 1855 as the Farmers' High School of Pennsylvania, Penn State now has its main campus at University Park and 21 smaller campuses statewide. In 1981 there were 14 community colleges.

The 12 state-aided private institutions (receiving designated grants from the legislature) had a combined enrollment of 47,920 in 1981. The largest of these schools is the University of Pennsylvania, founded in 1740 by Benjamin Franklin as the Philadelphia Academy and Charitable School; among its noteworthy professional schools is the Wharton School of Business. The state's many private colleges and universities, which may also receive state aid through a per-pupil funding formula, include Bryn Mawr College (founded in 1880), Bucknell University (1846) in Lewisburg, Carnegie-Mellon University (1900) in Pittsburgh, Dickinson College (1733) in Carlisle, Duquesne University (1878) in Pittsburgh, Haverford College (1833), Lafayette College (1826), Lehigh University (1865), Swarthmore College (1864), and Villanova University (1842). Enrollment at all private colleges and universities in the state totaled 157,865 in fall 1981. The Pennsylvania Higher Education Assistance Agency offers higher education grants, guarantees private loans, and administers work-study programs for Pennsylvania students.

During the 1983/84 school year, Pennsylvania's total public school revenue receipts exceeded $6.5 billion. That year, per capita school expenditures of $557 were only the 22d highest among the 50 states, but per pupil expenditures of $3,725 ranked 11th.

⁴²ARTS

Philadelphia was the cultural capital of the colonies, and rivaled New York as a theatrical center during the 1800s. In 1984, Philadelphia had five fully developed resident theaters, ranking 3d in the nation after New York and California. A number of regional and summer-stock theaters are scattered throughout the state, the most noteworthy being in Bucks County, Lancaster, and Pittsburgh. The National Choreographic Center was established in the mid-1980s in Carlisle in conjunction with the Central Pennsylvania Youth Ballet School.

Pennsylvania's most significant contribution to the performing arts has come through music. One of America's first important songwriters, Stephen Foster, grew up in Pittsburgh. The Pittsburgh Symphony, which began performing in 1896, first achieved prominence under Victor Herbert. Temporarily disbanded in 1910, the symphony was revived under Fritz Reiner in 1927; subsequent music directors have included William Steinberg and André Previn. Even more illustrious has been the career of the Philadelphia Orchestra, founded in 1900. Among this orchestra's best-known permanent conductors have been Leopold Stokowski and Eugene Ormandy, both of whom recorded extensively. Ormandy was succeeded in 1980 by Riccardo Muti. An important dance company, the Pennsylvania Ballet, is based in Philadelphia, which also has the Curtis Institute of Music, founded in 1924. Opera companies include the Pennsylvania Opera Theater, Pittsburgh Opera, and Opera Company of Philadelphia.

In 1984, Pennsylvania received $520,000 from the National Endowment for the Arts. Expressions '80, a minorities arts festival in Philadelphia that attracted artists from a six-state area, was the first regional festival of its kind in the Northeast. The Pennsylvania Writers Collection, an initiative program of the Pennsylvania Council on the Arts, supports the work of the state's creative writers.

⁴³LIBRARIES AND MUSEUMS

Pennsylvania's public libraries stocked 21,506,476 volumes during 1982/83, with a total circulation of 38,632,329. The largest public library in the state, and one of the oldest in the US, is the Free Library of Philadelphia, with 4,126,214 volumes in 48 branches. The Carnegie Library in Pittsburgh has 1,907,373 volumes and 20 branches. Harrisburg offers the State Library of Pennsylvania, which had 957,871 volumes in 1983. The Alverthorpe Gallery Library in Jenkintown contains the Rosenwald collection of illustrated books dating from the 15th century.

Philadelphia is the site of the state's largest academic collection, the University of Pennsylvania Libraries, with 3,120,976 volumes. Other major academic libraries are at the University of Pittsburgh, 2,472,489 volumes; Penn State, 1,808,827; Temple, 1,742,595; Carnegie-Mellon, 633,777; and Swarthmore, 619,272.

Pennsylvania has 259 museums and 29 public gardens, with many of the museums located in Philadelphia. The Franklin Institute, established in 1824 as an exhibition hall and training center for inventors and mechanics, is a leading showcase for science and technology. Other important museums are the Philadelphia Museum of Art, Academy of Natural Sciences, Pennsylvania Academy of the Fine Arts, Afro-American Historical and Cultural Museum, American Catholic Historical Society, American Swedish Historical Foundation Museum, and Museum of American Jewish History.

The Carnegie Institute in Pittsburgh is home to several major museums, including the Carnegie Museum of Natural History and the Museum of Art. Also in Pittsburgh are the Buhl Planetarium and Institute of Popular Science and the Frick Art Museum. Other institutions scattered throughout the state include the Moravian Museum, Bethlehem; US Army Military History Institute, Carlisle; Erie Art Center, Museum, and Old Custom House; Pennsylvania Lumber Museum, Galeton; Pennsylvania Historical and Museum Commission and William Penn Memorial Museum, Harrisburg; Pennsylvania Dutch Folk Culture Society, Lenhartsville; Schwenkfelder Museum, Pennsburg; and Railroad Museum of Pennsylvania, Strasburg.

Several old forts commemorate the French and Indian War, and George Washington's Revolutionary headquarters at Valley Forge is now a national historical park. Brandywine Battlefield (Chadds Ford) is another Revolutionary War site. Gettysburg National Military Park commemorates the Civil War. Other historic sites are Independence National Historical Park, Philadelphia; the Daniel Boone Homestead, Birdsboro; John Brown's House, Chambersburg; James Buchanan's home, Lancaster; and Ft. Augusta, Sunbury, a frontier outpost.

⁴⁴COMMUNICATIONS

Philadelphia already had mail links to surrounding towns and to Maryland and Virginia by 1737, when Benjamin Franklin was named deputy postmaster of the city, but service was slow and not always reliable. During the remainder of the century, significant improvements in delivery were made, but some townspeople devised ingenious ways of transmitting information even faster than the mails. Philadelphia stock exchange brokers, for instance, communicated with agents in New York by flashing coded signals with mirrors and lights from a series of high points across New Jersey, thereby receiving stock prices on the same day they were transacted. By 1846, the first telegraph service in the state linked Harrisburg and Lancaster. There were 36,449 postal service workers and 1,839 post offices in the state in 1985.

In 1980, 96% of Pennsylvania's 4,219,606 occupied housing units had telephones.

Pittsburgh's KDKA became the world's first commercial radio station in 1920. By 1983, it was one of 189 AM stations in the state; in addition, there were 192 FM radio stations and 40 television stations. Philadelphia alone had 11 AM, 18 FM, and 7 television stations; Pittsburgh, 13 AM, 16 FM, and 7 television stations. WQED in Pittsburgh pioneered community-sponsored educational television when it began broadcasting in 1954. In 1984 there were 358 cable television systems serving 1,757 communities and 2,020,945 subscribers (4th in the US).

45 PRESS

Benjamin Franklin may have been colonial Pennsylvania's most renowned publisher, but its first was Andrew Bradford, whose *American Weekly Mercury,* established in 1719, was the third newspaper to appear in the colonies. Founded nine years later, the *Pennsylvania Gazette* was purchased by Franklin in 1730 and served as the springboard for *Poor Richard's Almanack.*

During the 1800s, newspapers sprang up in all the major cities and many small communities. By 1880, Pittsburgh had 10 daily newspapers, more than any other city its size. After a series of mergers and closings, however, it was left with only two in 1985—the *Press* and *Post-Gazette.* Philadelphia also had two newspapers, the *Inquirer* and *Daily News.* The *Inquirer,* founded in 1829, has won numerous awards for its investigative reporting.

In 1984, Pennsylvania had 33 morning newspapers with a combined circulation of 1,670,741, 63 evening newspapers with 1,753,487 circulation, and 21 Sunday papers with 2,898,521 circulation (all-day papers are included in both morning and evening figures). The following table shows the circulation of some of the leading dailies in 1984:

AREA	NAME	DAILY	SUNDAY
Allentown	Morning Call (m)	129,469	
	Sunday Call-Chronicle (S)		166,220
Harrisburg	Patriot (m)	47,144	
	Evening News (e)	56,993	
	Sunday Patriot News (S)		162,492
Philadelphia	Inquirer (m, S)	525,569	1,000,427
	Daily News (e)	295,747	
Pittsburgh	Post-Gazette (m)	173,838	
	Press (e, S)	243,865	594,398
Wilkes-Barre	Citizens' Voice (m)	47,172	
	Times Leader (m)	44,953	

The most widely read magazine in the US, *TV Guide,* with a 1984 circulation of 17,115,233, is produced by Walter Annenberg's Triangle Publications in Radnor. *Farm Journal* and *Current History,* both monthlies, are published in Philadelphia, and there are monthlies named for both Philadelphia and Pittsburgh. Of more specialized interest are the gardening, nutrition, and health magazines and books from Rodale Press in Emmaus and automotive guides from the Chilton Co. in Radnor.

46 ORGANIZATIONS

The 1982 US Census of Service Industries counted 4,371 organizations in Pennsylvania, including 476 business associations; 3,186 civic, social, and fraternal associations; and 85 educational, scientific, and research associations.

Philadelphia is the home for two major service organizations: Big Brothers/Big Sisters of America and the Grand United Order of Odd Fellows. Cultural and educational organizations in that city include the American Academy of Political and Social Science, American Philosophical Society, and Middle States Association of Colleges and Schools. The Association for Children with Learning Disabilities is located in Pittsburgh, the College Placement Council in Bethlehem, and the American Philatelic Society in State College. Also in State College is the Environmental Coalition on Nuclear Power. The Society for Animal Rights, a humane organization, is in Clarks Summit.

Commercial and trade groups in the state include the American Mushroom Institute, Kennett Square; Insurance Institute of America, Malvern; and Society of Automotive Engineers, Warrendale. The Gray Panthers, a senior citizens' activist group, and Women's Strike for Peace are in Philadelphia. Valley Forge is the home of the Patriotic Order of the Sons of America.

Among the many sports organizations headquartered in Pennsylvania are the US Squash Racquets Association, in Bala-Cynwyd; National Trotting and Pacing Association, Hanover; Pop Warner Football and US Rowing Association, Philadelphia; and Little League Baseball, Williamsport.

47 TOURISM, TRAVEL, AND RECREATION

Travelers and tourists spent more than $7 billion in Pennsylvania in 1982. Travel and tourism was the state's 2d-largest employer.

Philadelphia—whose Independence National Historical Park has been called the most historic square mile in America—offers the Liberty Bell, Independence Hall, Carpenter's Hall, and many other sites. North of Philadelphia, in Bucks County, is the town of New Hope, with its numerous crafts and antique shops. The Lancaster area is "Pennsylvania Dutch" country, featuring tours and exhibits of Amish farm life. Gettysburg contains not only the famous Civil War battlefield but also the home of Dwight D. Eisenhower, opened to the public in 1980. Among the most popular sites are Chocolate World and Hersheypark in the town of Hershey and Valley Forge National Historic Park. Annual parades and festivals include the Mummers Parade on 1 January in Philadelphia and the Kutztown Folk Festival, commemorating Pennsylvania Dutch life, held the first week of July.

No less an attraction are the state's outdoor recreation areas. By far the most popular for both skiing and camping are the Delaware Water Gap and the Poconos, also a favorite resort region. During 1984, more than 33 million people used the state park system, which included 98 parks, 11 state forests, and 3 environmental education centers. Pennsylvania has far more licensed hunters than any other state, 1,313,191 in 1983, when the state also had 1,122,279 licensed anglers.

48 SPORTS

Pennsylvanians at both ends of the state have access to major professional sports.

Philadelphia's Veterans Stadium is the home of baseball's Phillies (winners of the 1980 World Series) and football's Eagles, while the 76ers of the National Basketball Association and the Flyers of the National Hockey League both play at the Spectrum. Wilt Chamberlain starred for the 76ers championship team in 1967; Julius Erving and Moses Malone took the team to another title in 1983. Bobby Clarke and Bernie Parent led the Flyers to the Stanley Cup in 1974 and 1975.

After several dismal decades, the Pittsburgh Steelers emerged as a National Football League powerhouse during the 1970s. Led by such stars as Franco Harris, Terry Bradshaw, and "Mean" Joe Greene, and coached by Chuck Noll, the Steelers won the Super Bowl in 1975, 1976, 1979, and 1980. The Pirates—who, like the Steelers, play at Three Rivers Stadium—had won five world series as of 1985, most recently in 1979, behind Dave Parker and Willie Stargell. The Pittsburgh Penguins compete in the National Hockey League.

Horse racing is conducted at Keystone Race Track in Bucks County, Penn National Race Course in Dauphin County, and Commodore Downs in Erie County. Harness-racing tracks include Liberty Bell Park in northeast Philadelphia, the Meadows in Washington County, and Pocono Downs in Luzerne County. Each June, Pennsylvania hosts a major auto race, the Pocono 500.

In collegiate football, the University of Pittsburgh Panthers were voted national champions in the postseason wire-service polls for 1976/77. Penn State, under coach Joe Paterno, has been ranked at or near the top of the polls since the 1960s; its Nittany Lions were national champions in 1982/83, and are frequent winners of the Lambert Cup as the best independent college team in the East. Villanova won the NCAA basketball championship in 1984/85. The Penn Relays, an important amateur track meet, are held in Philadelphia every April.

Each summer, Williamsport hosts baseball's Little League world series.

⁴⁹FAMOUS PENNSYLVANIANS

Johan Printz (b.Sweden, 1592–1663), the 400-lb, hard-drinking, hard-swearing, and hard-ruling governor of New Sweden, was Pennsylvania's first European resident of note. The founder of Pennsylvania was William Penn (b.England, 1644–1718), a Quaker of sober habits and deep religious beliefs. Most extraordinary of all Pennsylvanians, Benjamin Franklin (b.Massachusetts, 1706–90), a printer, author, inventor, scientist, legislator, diplomat, and statesman, served the Philadelphia, Pennsylvania, and US governments in a variety of posts.

Only one native Pennsylvanian, James Buchanan (1791–1868), has ever become US president. Buchanan was a state assemblyman, five-term US representative, two-term US senator, secretary of state, and minister to Russia and then to Great Britain before entering the White House as a 65-year-old bachelor in 1857. As president, he tried to maintain the Union by avoiding extremes and preaching compromise, but his toleration of slavery was abhorrent to abolitionists and his desire to preserve the Union was obnoxious to secessionists. Dwight D. Eisenhower (b.Texas, 1890–1969) retired to a farm in Gettysburg after his presidency was over. George M. Dallas (1792–1864), Pennsylvania's only US vice-president, was James K. Polk's running mate.

The six Pennsylvanians who have served on the US Supreme Court have all been associate justices: James Wilson (1742–98), Henry Baldwin (1780–1844), Robert C. Grier (1794–1870), William Strong (1808–95), George Shiras, Jr. (1832–1924), and Owen J. Roberts (1875–1955).

Many other Pennsylvanians have held prominent federal positions. Albert Gallatin (b.Switzerland, 1761–1849), brilliant secretary of the treasury under Thomas Jefferson and James Madison, later served as minister to France and then to Great Britain. Richard Rush (1780–1859) was Madison's attorney general and John Quincy Adams's secretary of the treasury. A distinguished jurist, Jeremiah Sullivan Black (1810–83) was Buchanan's attorney general and later his secretary of state. John Wanamaker (1838–1922), an innovative department store merchandiser, served as postmaster general under Benjamin Harrison. Philander C. Knox (1853–1921) was Theodore Roosevelt's attorney general and William Howard Taft's secretary of state. Financier Andrew C. Mellon (1855–1937) was secretary of the treasury under Warren G. Harding, Calvin Coolidge, and Herbert Hoover. Recent Pennsylvanians in high office include Richard Helms (b.1913), director of the US Central Intelligence Agency from 1966 to 1973, and Alexander Haig (b.1924), former commander of NATO forces in Europe, chief of staff under Richard Nixon, and Ronald Reagan's choice for secretary of state.

Three US senators, Simon Cameron (1799–1889), Matthew Quay (1833–1904), and Boies Penrose (1860–1921), are best known as leaders of the powerful Pennsylvania Republican machine. Senator Joseph F. Guffey (1870–1959) sponsored legislation to stabilize the bituminous coal industry. After serving as reform mayor of Philadelphia, Joseph S. Clark (b.1901) also distinguished himself in the Senate, and Hugh Scott (b.1900) was Republican minority leader from 1969 to 1977. Outstanding representatives from Pennsylvania include Thaddeus Stevens (1792–1868), leader of radical Republicans during the Civil War era; David Wilmot (1814–68), author of the proviso attempting to prohibit slavery in territory acquired from Mexico; and Samuel J. Randall (1828–90), speaker of the House of Representatives from 1876 to 1881.

Other notable historical figures were Joseph Galloway (b.Maryland, 1729?–1803), a loyalist; Robert Morris (b.England, 1734–1806), a Revolutionary financier; and Betsy Ross (Elizabeth Griscom, 1752–1836), the seamstress who allegedly stitched the first American flag. Pamphleteer Thomas Paine (b.England, 1737–1809), pioneer Daniel Boone (1734–1820), and General Anthony Wayne (1745–96) also distinguished themselves during

this period. In the Civil War, General George B. McClellan (1826–85) led the Union army on the Peninsula and at the Battle of Antietam, while at the Battle of Gettysburg, Generals George Gordon Meade (b.Spain, 1815–72) and Winfield Scott Hancock (1824–86) both showed their military prowess.

Important state governors include John W. Geary (1819–73), Samuel W. Pennypacker (1843–1916), Robert E. Pattison (b.Maryland, 1850–1904), Gifford Pinchot (b.Connecticut, 1865–1946), James H. Duff (1883–1969), George H. Earle (1890–1974), Milton J. Shapp (b.Ohio, 1912), William W. Scranton (b.Connecticut, 1917), George M. Leader (b.1918), and Richard L. Thornburgh (b.1932).

Pennsylvanians have won Nobel Prizes in every category except literature. General George C. Marshall (1880–1959), chief of staff of the US Army in World War II and secretary of state when the European Recovery Program (Marshall Plan) was adopted, won the 1953 Nobel Peace Prize. Simon Kuznets (b.Russia, 1901–85) received the 1971 Nobel Prize in economic science for work on economic growth, and Herbert A. Simon (b.Wisconsin, 1916) received the 1978 award for work on decision making in economic organizations; in 1980, Lawrence R. Klein (b.Nebraska, 1920) was honored for his design and application of econometric models. In physics, Otto Stern (b.Germany, 1888–1969) won the 1943 prize for work on the magnetic momentum of protons. In chemistry, Theodore W. Richards (1868–1928) won the 1914 Nobel Prize for determining the atomic weight of many elements, and Christian Boehmer Anfinsen (b.1916) won the 1972 award for pioneering studies in enzymes. In physiology or medicine, Philip S. Hench (1896–1965) won in 1950 for his discoveries about hormones of the adrenal cortex, Haldane K. Hartline (1903–83) won in 1967 for work on the human eye, and Howard M. Temin (b.1934) was honored in 1975 for the study of tumor viruses.

Many other Pennsylvanians were distinguished scientists. Ebenezer Kinnersly (1711–78) studied electricity, and Benjamin Franklin's grandson Alexander Dallas Bache (1806–67) was an expert on magnetism. Caspar Wistar (b.Germany, 1761–1818) and Thomas Woodhouse (1770–1809) pioneered the study of chemistry, while William Maclure (b.Scotland, 1763–1840) and James Mease (1771–1846) were early geologists. David Rittenhouse (1732–96) was a distinguished astronomer. John Bartram (1699–1777) and his son William (1739–1823) won international repute as botanists. Benjamin Rush (1745–1813) was Pennsylvania's most distinguished physician. Philip Syng Physick (1768–1837) was a leading surgeon, and Nathaniel Chapman (b.Virginia, 1780–1853) was the first president of the American Medical Association. Rachel Carson (1907–64), a marine biologist and writer, became widely known for her crusade against the use of chemical pesticides. Noted inventors born in Pennsylvania include steamboat builder Robert Fulton (1765–1815) and David Thomas (1794–1882), the father of the American anthracite iron industry.

Pennsylvania played a large role in the economic development of the US. In addition to Mellon, outstanding bankers include Stephen Girard (b.France, 1750–1831), Nicholas Biddle (1786–1844), Anthony J. Drexel (1826–93), and John J. McCloy (b.1895). Andrew Carnegie (b.Scotland, 1835–1919) and his lieutenants, including Henry Clay Frick (1849–1919) and Charles M. Schwab (1862–1939), created the most efficient steel-manufacturing company in the 19th century. Wanamaker, Frank W. Woolworth (b.New York, 1852–1919), and Sebastian S. Kresge (1867–1966) were pioneer merchandisers.

Other prominent businessmen born in Pennsylvania are automobile pioneer Clement Studebaker (1831–1901), chocolate manufacturer Milton S. Hershey (1857–1945), and Chrysler Chairman of the Board Lee A. Iacocca (b.1924).

Pennsylvania labor leaders include Uriah S. Stephens (1821–82) and Terence V. Powderly (1849–1924), leaders of the Knights of

Labor; Philip Murray (b.Scotland, 1886-1952), president of the CIO; and David J. MacDonald (1902-79), leader of the steel-workers. Among economic theorists, Henry George (1839-97) was the unorthodox advocate of the single tax. Florence Kelley (1859-1932) was an important social reformer, as is Bayard Rustin (b.1910).

Important early religious leaders, all born in Germany, include Henry Melchior Muhlenberg (1711-87), organizer of Pennsylvania's Lutherans; Count Nikolaus Ludwig von Zinzendorf (1700-1760), a Moravian leader; and Johann Conrad Beissel (1690-1768), founder of the Ephrata Cloister. Charles Taze Russell (1852-1916), born a Congregationalist, founded the group that later became Jehovah's Witnesses. Among the state's outstanding scholars are historians Henry C. Lee (1825-1909), John Bach McMaster (1852-1932), Ellis Paxson Oberholtzer (1868-1936), and Henry Steele Commager (b.1902); anthropologist Margaret Mead (1901-78); behavioral psychologist B(urrhus) F(rederic) Skinner (b.1904); urbanologist Jane Jacobs (b.1916); and language theorist Noam Chomsky (b.1928). Thomas Gallaudet (b.1787-1851) was a pioneer in education of the deaf.

Pennsylvania has produced a large number of distinguished journalists and writers. In addition to Franklin, newspapermen include John Dunlap (b.Ireland, 1747-1812), Benjamin Franklin Bache (1769-98), William L. McLean (1852-1931), and Moses L. Annenberg (1878-1942). Magazine editors were Sarah Josepha Buell Hale (b.New Hampshire, 1788-1879), Cyrus H. K. Curtis (b.Maine, 1850-1933), Edward W. Bok (b.Netherlands, (1863-1930), and I(sidor) F(einstein) Stone (b.1907). Ida M. Tarbell (1857-1944) was perhaps Pennsylvania's most famous muckraker. Among the many noteworthy Pennsylvania-born writers are Charles Brockden Brown (1771-1810), Bayard Taylor (1825-78), novelist and physician Silas Weir Mitchell (1829-1914), Charles Godfrey Leland (1824-1903), Owen Wister (1860-1938), Richard Harding Davis (1864-1916), Gertrude Stein (1874-1946), Mary Roberts Rinehart (1876-1958), Hervey Allen (1889-1949), Christopher Morley (1890-1957), Conrad Richter (1890-1968), John O'Hara (1905-70), Donald Barthelme (b.1931), and John Updike (b.1932); James Michener (b.New York, 1907) was raised in the state. Pennsylvania playwrights include James Nelson Barker (1784-1858), Maxwell Anderson (1888-1959), George S. Kaufman (1889-1961), Marc Connelly (1890-1980), Clifford Odets (1906-63), and Ed Bullins (b.1935). Among Pennsylvania poets are Francis Hopkinson (1737-91), Philip Freneau (b.New York, 1752-1832), Thomas Dunn English (1819-1902), Thomas Buchanan Read (1822-72), and Wallace Stevens (1879-1955).

Composers include Stephen Collins Foster (1826-64), Ethelbert Woodbridge Nevin (1862-1901), Charles Wakefield Cadman (1881-1946), and Samuel Barber (1910-81). Among Pennsylvania painters prominent in the history of American art are Benjamin West (1738-1820), renowned as the father of American painting; Charles Willson Peale (1741-1827), who was also a naturalist; Thomas Sully (b.England, 1783-1872); George Catlin (1796-1872); Thomas Eakins (1844-1916); Mary Cassatt (1845-1926); Man Ray (1890-1976); Andrew Wyeth (b.1917); and Andy Warhol (b.1927). Outstanding sculptors include William Rush (1756-1833), George Grey Barnard (1863-1938), and Alexander Calder (1898-1976).

Pennsylvania produced and patronized a host of actors, including Edwin Forrest (1806-72); Lionel (1878-1954), Ethel (1879-1959), and John (1882-1942) Barrymore; W. C. Fields (William Claude Dukenfield, 1880-1946); Ed Wynn (Isaiah Edwin Leopold, 1886-1966); William Powell (1892-1984); Ethel Waters (1896-1977); Janet Gaynor (1906-84); James Stewart (b.1908); Broderick Crawford (b.1911); Gene Kelly (b.1912); Charles Bronson (Charles Buchinsky, b.1922); Mario Lanza (1925-59); Shirley Jones (b.1934); and comedian Bill Cosby

(b.1937). Film directors Joseph L. Mankiewicz (b.1909), Arthur Penn (b.1922), and Sidney Lumet (b.1924) and film producer David O. Selznick (1902-65) also came from Pennsylvania.

Pennsylvania has produced outstanding musicians. Four important Pennsylvania-born vocalists are Marian Anderson (b.1902), Blanche Thebom (b.1919), Marilyn Horne (b.1934), and Anna Moffo (b.1934). Pianists include the versatile Oscar Levant (1906-72) and jazz interpreters Earl "Fatha" Hines (1905-83) and Erroll Garner (1921-77). Popular band leaders include Fred Waring (1900-84), Jimmy Dorsey (1904-57) and his brother Tommy (1905-56), and Les Brown (b.1912). Perry Como (b.1913) and Daryl Hall (b.1949) and John Oates (b.New York, 1948) have achieved renown as popular singers. Dancers and choreographers from Pennsylvania include Martha Graham (b.1894), Paul Taylor (b.1930), and Gelsey Kirkland (b.1952).

Of the many outstanding athletes associated with Pennsylvania, Jim Thorpe (b.Oklahoma, 1888-1953) was most versatile, having starred in Olympic pentathlon and decathlon events and football. Baseball Hall of Famers include Honus Wagner (1874-1955), Stan Musial (b.1920), and Roy Campanella (b.1921). Outstanding Pennsylvania football players include Harold "Red" Grange (b.1903), George Blanda (b.1927), John Unitas (b.1933), Joe Namath (b.1943), and Tony Dorsett (b.1954). Other stars include basketball's Wilt Chamberlain (b.1936); golf's Arnold Palmer (b.1929); tennis's Bill Tilden (1893-1953); horse racing's Bill Hartack (b.1932); billiards' Willie Mosconi (b.1913); swimming's Johnny Weissmuller (1904-84); and track and field's Bill Toomey (b.1939).

Pennsylvania has also been the birthplace of a duchess—Bessie Wallis Warfield, the Duchess of Windsor (b.1896)—and of a princess—Grace Kelly, Princess Grace of Monaco (1929-82).

⁵⁰BIBLIOGRAPHY

Billinger, Robert D. *Pennsylvania's Coal Industry.* Gettysburg: Pennsylvania Historical Association, 1954.

Binder, Frederick Moor. *Coal Age Empire: Pennsylvania Coal and Its Utilization to 1860.* Harrisburg: Pennsylvania Historical and Museum Commission, 1974.

Bridenbaugh, Carl. *Cities in Revolt: Urban Life in America, 1743-76.* New York: Capricorn, 1964.

Cochran, Thomas C. *Pennsylvania: A Bicentennial History.* New York: Norton, 1978.

Federal Writers' Project. *Pennsylvania: A Guide to the Keystone State.* Reprint. New York: Somerset, 1980 (orig. 1940).

Klein, Philip S., and Ari Hoogenboom. *A History of Pennsylvania.* Rev. ed. University Park: Pennsylvania State University Press, 1980.

Murphy, Raymond E. and Marion. *Pennsylvania: A Regional Geography.* Harrisburg: Pennsylvania Book Service, 1937.

Pennsylvania, Commonwealth of. Bureau of Statistics, Research, and Planning. *1984 Pennsylvania Statistical Abstract.* 26th ed. Harrisburg, 1984.

Pennsylvania, Commonwealth of. Department of General Services. *The Pennsylvania Manual, 1982-1983.* Vol. 106. Harrisburg, 1983.

Pennsylvania Chamber of Commerce. *Pennsylvania Government Today.* State College, Pa.: Pennsylvania Valley Publishers, 1973.

Soderland, Jean R., and Richard S. Dunn (eds.). *William Penn and the Founding of Pennsylvania, 1680-1684: A Documentary History.* Philadelphia: University of Pennsylvania Press, 1983.

Stevens, Sylvester K. *Pennsylvania: Birthplace of a Nation.* New York: Random House, 1964.

Stradley, Leighton P. *Early Financial and Economic History of Pennsylvania.* New York: Commerce Clearing House, 1942.

Wallace, Paul A. W. *Pennsylvania: Seed of a Nation.* New York: Harper & Row, 1962.

RHODE ISLAND

State of Rhode Island and Providence Plantations

ORIGIN OF STATE NAME: Named for Rhode Island in Narragansett Bay, which was likened to the isle of Rhodes in the Mediterranean Sea. **NICKNAME:** The Ocean State. (Also: Little Rhody.) **CAPITAL:** Providence. **ENTERED UNION:** 29 May 1790 (13th). **SONG:** "Rhode Island." **MOTTO:** Hope. **COAT OF ARMS:** A golden anchor on a blue field. **FLAG:** In the center of a white field is a golden anchor and, beneath it, a blue ribbon with the state motto in gold letters, all surrounded by a circle of 13 gold stars. **OFFICIAL SEAL:** The anchor of the arms is surrounded by four scrolls, the topmost bearing the state motto; the words "Seal of the State of Rhode Island and Providence Plantations 1636" encircle the whole. **BIRD:** Rhode Island Red. **FLOWER:** Violet. **TREE:** Red maple. **MINERAL:** Bowenite. **ROCK:** Cumberlandite. **LEGAL HOLIDAYS:** New Year's Day, 1 January; Birthday of Martin Luther King, Jr., 3d Monday in January; Washington's Birthday, 3d Monday in February; Rhode Island Independence Day, 4 May; Memorial Day, last Monday in May; Independence Day, 4 July; Victory Day, 2d Monday in August; Labor Day, 1st Monday in September; Columbus Day, 2d Monday in October; Election Day, 1st Tuesday after 1st Monday in November, in even-numbered years; Veterans Day, 11 November; Thanksgiving Day, 4th Thursday in November; Christmas Day, 25 December. **TIME:** 7 AM EST = noon GMT.

¹LOCATION, SIZE AND EXTENT

One of the six New England states in the northeastern US, Rhode Island is the smallest of all the 50 states. Rhode Island occupies only 0.03% of the total US area, and could fit inside Alaska, the largest state, nearly 486 times.

The total area of Rhode Island is 1,212 sq mi (3,139 sq km), of which land comprises 1,055 sq mi (2,732 sq km), and inland water 157 sq mi (407 sq km). The state extends 37 mi (60 km) E-W and 48 mi (77 km) N-S.

Rhode Island is bordered on the N and E by Massachusetts; on the S by the Atlantic Ocean (enclosing the ocean inlet, Narragansett Bay); and on the W by Connecticut (with part of the line formed by the Pawcatuck River). Three large islands—Prudence, Aquidneck (officially known as Rhode Island), and Conanicut—are situated within Narrangansett Bay. Block Island, with an area of about 11 sq mi (28 sq km), lies some 9 mi (14 km) SW of Pt. Judith, on the mainland. There are 38 islands in all.

The total boundary length of Rhode Island is 160 mi (257 km). The state's geographic center is in Kent County, 1 mi (1.6 km) SSW of Crompton.

²TOPOGRAPHY

Rhode Island comprises two main regions. The New England Upland Region, which is rough and hilly and marked by forests and lakes, occupies the western two-thirds of the state, while the Seaboard Lowland, with its sandy beaches and salt marshes, occupies the eastern third. The highest point in the state is Jerimoth Hill, at 812 feet (247 meters), in the northwest.

Rhode Island's principal river, the Blackstone, flows from Woonsocket past Pawtucket and thence into the Providence River, which, like the Sakonnet, is an estuary of Narragansett Bay; the Pawcatuck River flows into Block Island Sound. The state has 38 islands, the largest being Aquidneck (Rhode Island), with an area of about 45 sq mi (117 sq km).

³CLIMATE

Rhode Island has a humid climate, with cold winters and short summers. The average annual temperature is 50°F (10°C). At Providence the temperature ranges from an average of 28°F (−2°C) in January to 72°F (22°C) in July. The record high temperature, 104°F (40°C), was registered in Providence on 2 August 1975; the record low, −23°F (−31°C), at Kingston on 11

January 1942. In Providence, the average annual precipitation is 45 in (114 cm); snowfall averages 37 in (94 cm) a year. Rhode Island's weather is highly changeable, with storms and hurricanes an occasional threat. On 21 September 1938, a hurricane and tidal wave took a toll of 262 lives; Hurricane Carol, on 31 August 1954, left 19 dead, and property damage was estimated at $90 million. A blizzard on 6–7 February 1978 dropped a record 28.6 in (73 cm) of snow on the state, as measured at Warwick, and caused 21 storm-attributed deaths.

⁴FLORA AND FAUNA

Though small, Rhode Island has three distinct life zones: sandplain lowlands, rising hills, and highlands. Common trees are the tuliptree, pin and post oaks, and red cedar. Cattails are abundant in marsh areas, and 40 types of fern and 30 species of orchid are indigenous to the state. The small whorled pogonia is endangered.

Urbanization and industrialization have taken their toll of native mammals. Swordfish, bluefish, lobsters, and clams populate coastal waters; brook trout and pickerel are among the common freshwater fish. The Indiana bat, peregrine and Arctic falcons, bald eagle, and shortnose sturgeon are on the federal endangered list.

⁵ENVIRONMENTAL PROTECTION

The Department of Environmental Management coordinates all of the state's environmental protection programs. The Division of Air and Hazardous Materials enforces controls on solid waste disposal and hazardous waste management facilities, and on industrial air pollution; the Division of Water Resources regulates waste-treatment facilities and the discharge of industrial and oil waste into state waters and public sewer facilities; the Division of Land Resources oversees dam maintenance, freshwater wetlands, solid-waste disposal, and home sewage disposal systems.

⁶POPULATION

Rhode Island ranked 40th in population among the 50 states with a 1980 census total of 947,154 (a decrease of 0.3% from 1970), making it and New York the only states to lose population during the decade. In early 1985, however, it had an estimated population of 958,151, a 1.2% increase since 1980. In 1985, Rhode Island was, at 908 persons per sq mi (351 per sq km), the nation's 2d most densely populated state, after New Jersey. According to the 1980 census, 87% of all Rhode Islanders lived in urban areas, 3d only to California and New Jersey; only Nevada had fewer rural residents than Rhode Island's 123,000. Providence, the capital, is the leading city, with a population in 1984 of 154,148 (as opposed to the 1940 peak of 253,504). Other cities include Warwick, 87,198; Pawtucket, 72,803; and Cranston, 72,720. In 1983, the Providence metropolitan area had an estimated population of 632,029.

7ETHNIC GROUPS

Rhode Island's black population numbered 27,584 in 1980, or less than 3% of the state total. Among other minority groups, the 1980 census counted 19,707 persons of Spanish origin, 2,872 American Indians, 1,718 Chinese, 1,218 Filipinos, and 851 Asian Indians. The foreign born made up 8.9% of the population in 1980.

8LANGUAGES

Many place-names in Rhode Island attest to the early presence of Mahican Indians: for instance, Sakonnet Point, Pawtucket, Matunuck, Narragansett.

English in Rhode Island is of the Northern dialect, with the distinctive features of eastern New England: absence of final /r/, and a vowel in *part* and *bath* intermediate between that in *father* and that in *bat*.

Rhode Island's immigrant past is reflected in the fact that in 1980, 16% of states residents reported speaking a language other than English in the home. The leading languages, and the number of people speaking them, were French, 40,841; Portuguese, 39,555; Italian, 28,336; and Spanish, 12,475.

9RELIGIONS

The first European settlement in Rhode Island was founded by an English clergyman, Roger Williams, who left Massachusetts to find freedom of worship. The Rhode Island Charter of 1663 proclaimed "that a most flourishing civil state may stand and best be maintained with full liberty in religious concernments." Rhode Island has maintained this viewpoint throughout its history, and has long been a model of religious pluralism. The first Baptist congregation in the US was established in 1638 in Providence. In Newport stands the oldest synagogue (1763) and the oldest Quaker meetinghouse (1699) in the US.

Contemporary Rhode Island is the most Catholic state in the US, reflecting heavy immigration from Italy, Ireland, Portugal, and French Canada. Roman Catholics make up 66% of the population. There was 624,141 Roman Catholics and an estimated 22,000 Jews in 1984. The only large Protestant denominations were Episcopalians, 40,877, and American Baptist USA, 24,183.

10TRANSPORTATION

Conrail, Amtrak, and several private railroads serve the state, operating on 146 mi (235 km) of Class I and II track as of 1981. In 1984, Amtrak operated 9 trains through Rhode Island, which is in the New York–Boston corridor; Rhode Island ridership, at four stops, was 416,287 in 1983/84. In 1984 there were 6,396 mi (10,291 km) of highways and roads; 598,338 motor vehicles were registered in 1983, and 603,176 drivers' licenses were in force. The major route through New England, I–95, crosses Rhode Island. The Rhode Island Public Transit Authority provides commuter bus service connecting urbanized areas; in 1984 there were 17.7 million passengers. Some of the best deepwater ocean ports on the east coast are in Narragansett Bay.

There were 18 airfields in 1983. Theodore Francis Green Airport is the major air terminal.

11HISTORY

Before the arrival of the first white settlers, the Narragansett Indians inhabited the area from what is now Providence south along Narragansett Bay. Their principal rivals, the Wampanoag, dominated the eastern shore region.

In 1524, Florentine navigator Giovanni da Verrazano, sailing in the employ of France, became the first European to explore Rhode Island. The earliest permanent settlement was established at Providence in 1636 by English clergyman Roger Williams and a small band of followers who left the repressive atmosphere of the Massachusetts Bay Colony to seek freedom of worship. Other nonconformists followed, settling Portsmouth (1638), Newport (1639), and Warwick (1642). In 1644, Williams journeyed to England, where he secured a parliamentary patent uniting the four original towns into a single colony, the Providence Plantations. This legislative grant remained in effect until the Stuart Restoration made it prudent to seek a royal charter. The charter, secured for Rhode Island and the Providence Plantations from Charles II in 1663, guaranteed religious liberty, permitted significant local autonomy, and strengthened the colony's territorial claims. Encroachments by white settlers on Indian lands led to the Indian uprising known as King Philip's War (1675–76), during which the Indians were soundly defeated.

The early 18th century was marked by significant growth in agriculture and commerce, including the rise of the slave trade. Having the greatest degree of self-rule, Rhode Island had the most to lose from British efforts after 1763 to increase the mother country's supervision and control over the colonies. On 4 May 1776, Rhode Island became the first colony formally to renounce all allegiance to King George III. Favoring the weak central government established by the Articles of Confederation, the state quickly ratified them in 1778, but subsequently resisted the centralizing tendencies of the federal constitution. Rhode Island withheld ratification until 29 May 1790, making it the last of the original 13 states to join the Union.

The principal trends in 19th-century Rhode Island were industrialization, immigration, and urbanization. The state's royal charter (then still in effect) contained no procedure for its amendment, gave disproportionate influence to the declining rural towns and conferred almost unlimited power on the legislature. In addition, suffrage was restricted by the general assembly to owners of real estate and their eldest sons. Because earlier, moderate efforts at change had been virtually ignored by the assembly, political reformers decided to bypass the legislature and convene a People's Convention. Thomas Wilson Dorr, who led this movement, became the principal draftsman of a progressive "People's Constitution," ratified in a popular referendum in December 1841. A coalition of Whigs and rural Democrats used force to suppress the movement now known as Dorr's Rebellion, but they bowed to popular pressure and made limited changes via a new constitution, effective May 1843.

The latter half of the 19th century was marked by continued industrialization and urbanization. Immigration became both more voluminous and more diverse. Politically the state was dominated by the Republican Party until the 1930s. The Democrats, having seized the opportunity during the New Deal, consolidated their power during the 1940s, and from that time onward have captured most state and congressional elections. Present-day Rhode Island, though predominantly urban, industrial, Catholic, and Democratic, retains an ethnic and cultural diversity surprising in view of its size but consistent with its pluralist traditions.

12STATE GOVERNMENT

Rhode Island is governed under the constitution of 1842. Through 1983 there had been 43 amendments to this document.

Legislative authority is vested in the general assembly, a bicameral body composed of 50 senators and 100 representatives. All legislators are elected for two-year terms from districts that are apportioned equally according to population after every federal decennial census. Legislators must be qualified voters in the state and must thus have been residents of the state and their district for 30 days prior to election. Among the more important checks enjoyed by the assembly is the power to override the governor's veto by a three-fifths vote of its members, the authority in joint session (Grand Committee) to name justices to the supreme court, and the power to establish all courts below the supreme court.

The chief officers of the executive branch are the governor, lieutenant governor, attorney general, secretary of state, and general treasurer. All are elected for two-year terms in even-numbered years. The governor and lieutenant governor must be qualified voters in Rhode Island and must have been residents of the US and the state for 30 days prior to election.

Constitutional amendments are enacted by majority vote of the whole membership of each house of the legislature, and by a simple majority at the next general election. In 1984, voters authorized a constitutional convention to revise the state constitution. Voters must be US citizens, 18 years old or over, and have been residents of the state at least 30 days prior to an election.

¹³POLITICAL PARTIES

For nearly five decades, Rhode Island has been one of the nation's most solidly Democratic states. It has voted for the Republican presidential candidate only four times since 1928, elected only one Republican (former governor John Chafee) to the US Senate since 1934, and sent no Republicans to the US House from 1940 until 1980, when one Republican and one Democrat were elected. (They were reelected in 1982 and 1984.) Also in 1980, Rhode Island was one of only six states to favor Jimmy Carter. However, in 1984, Republican Edward DiPrete was elected governor, and Ronald Reagan narrowly carried the state in the presidential election. The state legislature remained Democratic.

Women have become a force in state and Republican politics. In 1985, the attorney general and secretary of state were both Republican women, as was one of the two US representatives, Claudine Schneider. Arlene Violet was Rhode Island's first woman state attorney general. There were 23 women in the state legislature. In 1984 there were 3 elected Hispanic officials, and in 1985 8 blacks held elective positions.

Rhode Island Presidential Vote by Major Political Parties, 1948–84

YEAR	ELECTORAL VOTE	RHODE ISLAND WINNER	DEMOCRAT	REPUBLICAN
1948	4	*Truman (D)	188,736	135,787
1952	4	*Eisenhower (R)	203,293	210,935
1956	4	*Eisenhower (R)	161,790	225,819
1960	4	*Kennedy (D)	258,032	147,502
1964	4	*Johnson (D)	315,463	74,615
1968	4	Humphrey (D)	246,518	122,359
1972	4	*Nixon (R)	194,645	220,383
1976	4	*Carter (D)	227,636	181,249
1980	4	Carter (D)	198,342	154,793
1984	4	*Reagan (R)	197,106	212,080

* Won US presidential election.

¹⁴LOCAL GOVERNMENT

As of 1984, Rhode Island was subdivided into 8 cities and 31 towns, the main units of local government. The state's 5 counties are merely units of judicial administration.

Many smaller communities retain the New England town meeting form of government, under which the town's eligible voters assemble to enact the local budget, set the tax levy, and approve other local measures. Larger cities and towns are governed by a mayor and/or city manager and a council.

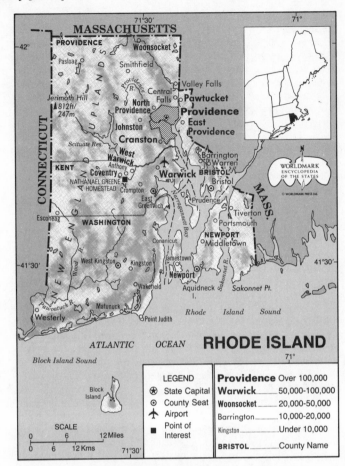

See endsheet maps: M2.

LOCATION: 41°18′ to 42°01′N; 71°08′ to 71°53′W. **BOUNDARIES:** Massachusetts line, 64 mi (103 km); Atlantic Ocean coastline, 40 mi (64 km); Connecticut line, 56 mi (90 km).

¹⁵STATE SERVICES

The Board of Regents for Elementary and Secondary Education and the Board of Governors for Higher Education oversee all state educational services. Airports, railroads, motor vehicle administration, and highway and bridge management come under the jurisdiction of the Department of Transportation. Health and welfare services are provided through the Department of Children and Their Families, the Department of Community Affairs, the Department of Elderly Affairs, the Department of Health, the Department of Mental Health, Retardation, and Hospitals, and the Department of Social and Rehabilitative Services.

¹⁶JUDICIAL SYSTEM

The five-member supreme court is the state's highest appellate tribunal; it may also issue, upon request, advisory opinions on the constitutionality of a questioned act to the governor or either house of the legislature. Supreme court justices are chosen by the legislature and, like other state judges, hold office for life ("during good behavior"), but in actuality they can be removed by a mere resolution of the general assembly. In 1935, all five justices were ousted in this manner when a Democratic legislature replaced a court previously appointed by Republicans.

The second judicial level consists of the 19-member superior court and the 11-member family court. The former, the state's trial court, hears all jury trials in criminal cases and in civil matters involving more than $5,000, but can also hear nonjury cases. The family court deals with divorce, custody, juvenile crime, adoption, and related cases. Superior, family, and district court judges are appointed by the governor with the consent of the senate.

District courts do not hold jury trials. Civil matters that involve $5,000 or less, small claims procedures, and nonjury criminal cases, including felony arraignments and misdemeanors, are handled at the district level. All cities and towns appoint judges to operate probate courts for wills and estates. Providence and a few other communities each have a municipal or police court.

According to the FBI Crime Index for 1983, the crime rate of 5,005 per 100,000 persons was slightly below the national average. There were 1,220 prisoners in state and federal prisons at the end of 1984. There is no death penalty. Practicing attorneys numbered 1,985 in December 1984.

¹⁷ARMED FORCES

As of 1983/84, Department of Defense personnel in the state totaled 8,273, most of them at naval installations, including the US Naval Education and Training Center and Naval War College in Newport and the Naval Construction Battalion Center in Davisville. Rhode Island firms received $381 million in defense contracts during 1982/83. The Electric Boat Division of General Dynamics, at Quonset Point in Narragansett Bay, is a large contractor, employing more than 5,000 people. A total of 126,000 US veterans were living in the state in 1983, of whom 2,000 saw military service during World War I, 56,000 during World War II, 24,000 in the Korean conflict, and 33,000 during the Viet-Nam era; veterans' benefits totaled $130 million in 1982/83.

In early 1985, the state had 4,461 Army and Air National Guard personnel. There were 2,517 state and local police in 1983.

¹⁸MIGRATION

During the 19th and early 20th centuries, the major immigrant groups who came to work in the state's growing industries were Irish, Italian, and French-Canadian. Significant numbers of British, Portuguese, Swedish, Polish, and German immigrants also moved to Rhode Island. Between 1940 and 1970, however, 2,000 more people left the state than moved to it, and between 1970 and 1983 there was a net loss of about 42,000.

¹⁹INTERGOVERNMENTAL COOPERATION

Rhode Island participates in many interstate regional bodies, including the Atlantic States Marine Fisheries Commission, New England Corrections Commission, New England Interstate Water Pollution Control Commission, and Northeastern Forest Fire Protection Commission.

Federal aid to Rhode Island state and local governments totaled $547.6 million in 1983/84. General revenue-sharing funds came to $20 million in 1982/83.

²⁰ECONOMY

Rhode Island's economy is based overwhelmingly on industry; agriculture, mining, forestry, and fishing make only small contributions. The state's leading manufactured products are jewelry, silverware, machinery, primary metals, textiles, and rubber products. Unemployment rates in Rhode Island exceeded those of the US throughout the 1970s, and the state's economic growth lagged behind that of the nation as a whole. While unemployment fell dramatically in 1983 and 1984, Rhode Island's chief economic problem, its concentration of manufacturing industries paying low wages, remained.

²¹INCOME

With an income per capita of $11,504 in 1983, Rhode Island ranked 23d among the 50 states. Total personal income reached $11.1 billion. Measured in constant 1972 dollars, per capita income increased 27% between 1970 and 1983.

²²LABOR

In November 1984, the civilian labor force for the state totaled 504,700. Unemployment was 22,400. The unemployment rate, as high as 11.2% in January 1983, averaged 6% in 1984.

A federal census in March 1983 revealed the following nonfarm employment pattern for Rhode Island:

	ESTABLISH-MENTS	EMPLOYEES	ANNUAL PAYROLL ('000)
Agricultural services, forestry, fishing	299	909	$ 13,488
Mining	17	150	2,766
Contract construction	2,237	12,640	263,694
Manufacturing	2,727	110,385	1,913,831
Transportation, public utilities	689	12,413	258,071
Wholesale trade	1,578	18,965	357,797
Retail trade	6,156	65,184	608,994
Finance, insurance, real estate	1,673	21,659	396,635
Services	6,984	90,737	1,134,757
Other	1,219	1,233	27,993
TOTALS	23,579	334,275	$4,978,026

In 1980, 113,000 Rhode Islanders belonged to labor unions, which numbered 278 in 1983. There were about 54,000 state and local government employees in 1983, and about 9,000 federal employees in 1982.

²³AGRICULTURE

The state's gross agricultural income in 1983 was $39 million, 49th in the US; Alaska was 50th. Rhode Island had only about 1,000 farms in 1984, and the least area devoted to farming (about 97,000 acres, or 39,250 hectares) of any state. Nursery and greenhouse products were the main agricultural commodity. Total receipts from farm marketings came to $31.3 million in 1983.

²⁴ANIMAL HUSBANDRY

Cash income from livestock and livestock products totaled $12,280,000 in 1983, of which milk accounted for $6,090,000; eggs, $4,030,000; cattle and calves, $806,000; and hogs and pigs, $547,000. In 1983, the state produced 46 million lb of milk and 62 million eggs.

²⁵FISHING

The commercial catch in 1984 was 120 million lb, valued at $70.4 million. Point Judith is the main fishing port. The principal edible fish and shellfish caught were butterfish, sea bass, squid, tuna, scup, yellowtail and blackback flounders, mackerel, cod, whiting, herring, lobster, clams, and scallops.

²⁶FORESTRY

In 1979, forest covered 404,000 acres (163,000 hectares), three-fifths of the state's land area. Some 395,000 acres (160,000 hectares) were usable as commercial timberland.

²⁷MINING

Rhode Island's nonfuel mineral production in 1984 totaled about $9.1 million, 49th in the US. The chief mineral resources are construction materials such as sand and gravel, of which 1,200,000 tons were produced in 1984, and crushed stone, of which 900,000 tons were produced. These goods were used for construction.

²⁸ENERGY AND POWER

Rhode Island is part of the New England regional power grid and imports most of its electric power. The state's installed capacity was 270,000 kw in 1983, and power production totaled 584 million kwh; both figures were the lowest in the US. Electric energy sales in Rhode Island in 1983 were 5.2 billion kwh, of which 1.9 billion kwh went for residential users, 1.8 billion kwh for commercial purchasers, and nearly all the rest for industry. The total number of gas utility customers for 1983 was 178,000.

²⁹INDUSTRY

The Industrial Revolution began early in Rhode Island. The first spinning jenny in the US was built at Providence in 1787; three years later, in Pawtucket, Samuel Slater opened a cotton mill, one of the first modern factories in America. By the end of the 18th century, textile, jewelry, and metal products were being manufactured in the state.

Manufacturing remains Rhode Island's chief source of income. Shipments of manufactured goods totaled nearly $7.7 billion in

1982. Of this total, jewelry and silverware accounted for 17%; primary metal products, 12%; fabricated metal products, 11%; and textile mill products, 8%.

The following table shows value of shipments for selected industries in 1982:

Textile mill products	$601,300,000
Costume jewelry and notions	590,500,000
Nonferrous wire drawing and insulating materials	514,200,000
Precious-metal jewelry	480,900,000

30 COMMERCE

Wholesale trade totaled nearly $5 billion in 1982. In that year, retail establishments recorded sales of $4.1 billion. Of this total, 22% came from food store sales, 15% from automotive dealers, 11% from general merchandise stores, 10% from eating and drinking places, 9% from gasoline service stations, and 33% from other establishments. Foreign exports of manufactured goods were $586 million in 1981.

31 CONSUMER PROTECTION

The consumer unit of the public protection division of the Department of the Attorney General, the Consumer's Council, and the consumer affairs division within the Department of Business Regulation bear primary responsibility for enforcing consumer laws and regulations.

32 BANKING

At the end of 1983, Rhode Island had 19 commercial banks with combined assets of $9.2 billion; deposits totaled $6.2 billion, and outstanding loans were over $5 billion. At the end of 1983 there were four savings and loan associations; their combined assets reached $906 million.

33 INSURANCE

In mid-1983, the state had eight life insurance companies; the 1,883,000 policies held by state residents had an aggregate value of over $21 billion. The average amount of life insurance per family was $55,000.

34 SECURITIES

Rhode Island has no securities exchanges. New York Stock Exchange member firms had 19 sales offices in the state in 1982.

35 PUBLIC FINANCE

The annual budget is prepared by the Division of the Budget in conjunction with the governor, and submitted to the legislature for approval. The fiscal year runs from 1 July to 30 June.

The following is a summary of general revenues and expenditures in 1982/83 and 1983/84:

REVENUES	1982/83	1983/84
Sales and use taxes	$ 311,672,237	$ 346,918,536
Personal income taxes	263,699,420	284,148,650
General business taxes	123,625,121	142,955,237
Other taxes	15,948,627	17,498,760
Department revenues	135,180,493	143,428,515
Federal grants	291,707,162	310,582,550
Other revenues	18,973,845	26,823,103
TOTALS	$1,160,806,905	$1,272,355,351
EXPENDITURES		
Department of Social and Rehabilitative Services	$ 361,118,865	$ 378,080,639
Department of Elementary and Secondary Education	227,788,951	245,437,314
Department of Mental Health, Retardation, and Hospitals	132,567,769	145,656,843
Office of Higher Education	96,812,348	102,274,960
Department of Transportation	62,456,658	65,826,615
Other expenditures	290,169,341	304,554,795
TOTALS	$1,170,913,932	$1,241,831,166

36 TAXATION

As of 1984, Rhode Island levied a state income tax equal to 25.5% of the taxpayer's federal income tax liability. The basic corporate tax rate was 8% of net taxable income, or 40 cents per $100 of net worth, or $2.50 per $10,000 of approved capital stock , whichever was greater. The sales and use tax was 6% on most items.

Rhode Islanders filed 415,748 federal income tax returns for 1983, paying $1,027,095,000 in tax.

37 ECONOMIC POLICY

The Department of Economic Development seeks to promote the preservation and expansion of industry, commerce, and tourism. Business tax incentives offered by the state include elimination of local property taxes on manufacturers' machinery and equipment purchased after 1974, exemption from sales tax on all manufacturers' machinery and equipment, and a 2% investment tax credit for purchases of depreciable tangible property, including buildings. In June 1984, voters defeated a $250-million plan promoted by Governor J. Joseph Garrahy and the business community to spur industrial growth in the state.

38 HEALTH

Rhode Island's birthrate of 13.1 per 1,000 persons in 1982 was the 2d lowest in the nation. Despite the state's heavy Catholic population, the ratio of 590 abortions to 1,000 live births in 1982 was 6th highest. As of 1981, the infant mortality rate was 11.5 per 1,000 live births for whites, 17.2 for blacks. Death rates from heart disease (3d highest among the states) and cancer (2d highest)—the leading causes of death in 1981—were well above the national averages. The death rate of 9.6 per 1,000 ranked the state 5th.

The state had 2,246 active physicians and 533 active dentists in 1982. There were 21 hospitals, with 5,829 beds in 1983; hospital personnel included 4,387 registered nurses.

39 SOCIAL WELFARE

Aid to families with dependent children, paid to 44,900 Rhode Islanders in 1982, totaled $66 million. In 1983, $827 million in Social Security benefits were paid to 173,000 retired workers and their dependents. Outlays for other social programs in 1983 included $40 million for federal food stamps and $12 million for the school lunch program. In 1982, $71 million was paid for unemployment insurance. State and local government is a generous provider of public welfare; the $380.29 spent per capita for this purpose in 1983 ranked the state 3d in the nation.

40 HOUSING

In 1980, the census counted 372,672 housing units, of which 9,864 were seasonal (principally summer cottages). More than 98% of year-round units had full plumbing. The state authorized 9,500 privately owned new housing units worth $222 million from 1981 through 1983. Much of the new residential construction has taken place in the suburbs south and west of Providence.

41 EDUCATION

Only 61.1% of adult Rhode Islanders were high school graduates in 1980, the lowest such percentage for any northern state.

As of October 1982, 136,180 students were enrolled in 309 public schools; 30,116 were in private schools in the fall of 1981, most of them in Catholic schools. More than 68,000 students were enrolled in the state's 13 institutions of higher education in fall 1982. Leading institutions included Brown University (1764) in Providence, with 6,869 students in fall 1983; the University of Rhode Island (1892), in Kingston, with 10,239; and Providence College (1917), with 4,735. The Rhode Island School of Design (1877), with 1,775 students in 1983, is located in Providence.

42 ARTS

Newport and Providence have notable art galleries and museums. Theatrical groups include the Trinity Square Repertory Company, the Sock and Buskin Players of Brown University, and the Players, all in Providence. The Rhode Island Philharmonic performs throughout the state. Newport is the site of the internationally famous Newport Jazz Festival.

[43] LIBRARIES AND MUSEUMS

In 1983, Rhode Island had 120 public, academic, and special libraries. In 1982/83, public libraries had a book stock of 2,671,881, and a combined circulation of 4,424,630. The Providence Public Library maintains several special historical collections. The Brown University Libraries, containing more than 2.6 million books and periodicals, include the Annmary Brown Memorial Library, with its collection of rare manuscripts, and the John Carter Brown Library, with an excellent collection of early Americana.

Among the state's more than 40 museums and historic sites are the Haffenreffer Museum of Anthropology in Bristol, the Museum of Art of the Rhode Island School of Design in Providence, the Roger Williams Park Museum, also in Providence, the Nathanael Greene Homestead in Coventry, and the Slater Mill Historic Site in Pawtucket. Providence has the Roger Williams Park Zoo.

[44] COMMUNICATIONS

Rhode Island had some 155 post offices in 1985 with 2,910 employees; the first automated post office in the US postal system was opened in Providence in 1960. As of 1980, there were 95% of the state's 338,590 occupied housing units had telephones. In 1983, the state had 15 AM and 13 FM radio stations. Providence had 5 television stations, including an affiliate for each of the three major networks, a pay-TV station, and one public broadcasting affiliate operated by the state's Public Telecommunications Authority. The state had 8 cable television systems in 1984 serving 30 communities and 137,494 subscribers.

[45] PRESS

The *Rhode Island Gazette,* the state's first newspaper, appeared in 1732. In 1850, Paulina Wright Davis established *Una,* one of the first women's rights newspapers in the country.

In 1984, Rhode Island had seven daily newspapers, with a combined circulation of 947,154. The two largest newspapers were the *Providence Bulletin* and *Providence Journal.*

[46] ORGANIZATIONS

The 1982 US Census of Service Industries counted 319 organizations in Rhode Island, including 46 business associations; 197 civic, social, and fraternal associations; and 8 educational, scientific, and research associations. Among the organizations with headquarters in Rhode Island are the US Surfing Federation (Barrington); the Rooster Class Yacht Racing Association (Wakefield); the Foundation for Gifted and Creative Children and the Foster Parents Plan USA (both in Warwick); the American Mathematical Society and the Manufacturing Jewelers and Silversmiths of America (both in Providence); and the US International Sailing Association and US Yacht Racing Union (both in Newport).

[47] TOURISM, TRAVEL, AND RECREATION

Travel and tourism generated 10,100 jobs in 1982. Historic sites—especially the mansions of Newport and Providence—and water sports (particularly the America's Cup yacht races) are the main tourist attractions. Block Island is a popular resort. During 1982/83, licenses were issued to 12,810 hunters and 29,767 fishermen. Rhode Island has 65 state recreation areas.

[48] SPORTS

Rhode Island's most famous sports competition, the America's Cup yacht races, was held at Newport from 1851 to 1983, when an Australian yacht won the race, thereby ensuring that at least the next competition (probably in 1987) will be held in Australia. Pawtucket has a minor league baseball team. Providence College has competed successfully in intercollegiate basketball, winning National Invitation Tournament titles in 1961 and 1963 and the NCAA Eastern Division crown in 1973. Swimming, boating, golf, tennis, softball, skiing, and hiking are popular participant sports. The International Tennis Hall of Fame and the Yachting Hall of Fame are located in Newport. Dog racing (Lincoln) and jai alai (Newport) are spectator sports with pari-mutuel betting.

[49] FAMOUS RHODE ISLANDERS

Important federal officeholders from Rhode Island have included US Senators Nelson W. Aldrich (1841-1915), Henry Bowen Anthony (1815-84), Theodore Francis Green (1867-1966), and John O. Pastore (b.1907), and US Representative John E. Fogarty (1913-67). J. Howard McGrath (1903-66) held the posts of US senator, solicitor general, and attorney general.

Foremost among Rhode Island's historical figures is Roger Williams (b.England, 1603?-83), apostle of religious liberty and founder of Providence. Other significant pioneers, also born in England, include Anne Hutchinson (1591-1643), religious leader and cofounder of Portsmouth, and William Coddington (1601-78), founder of Newport. Other 17th-century Rhode Islanders of note were Dr. John Clarke (b.England, 1609-76), who secured the colony's royal charter, and Indian leader King Philip, known also as Metacomet (1639?-76). Important participants in the War for Independence were Commodore Esek Hopkins (1718-1802) and General Nathanael Greene (1742-86). The 19th century brought to prominence Thomas Wilson Dorr (1805-54), courageous leader of Dorr's Rebellion; social reformer Elizabeth Buffum Chace (1806-99); and naval officers Oliver Hazard Perry (1785-1819), who secured important US victories in the War of 1812; and his brother, Matthew C. Perry (1794-1858), who led the expedition that opened Japan to foreign intercourse in 1854. Among the state's many prominent industrialists and inventors are Samuel Slater (b.England, 1768-1835), pioneer in textile manufacturing, and silversmith Jabez Gorham (1792-1869). Other significant public figures include Unitarian theologian William Ellery Channing (1780-1842); political boss Charles R. Brayton (1840-1910); Roman Catholic bishop and social reformer Matthew Harkins (b.Massachusetts, 1845-1921); and Dr. Charles V. Chapin (1856-1941), pioneer in public health.

Rhode Island's best-known creative writers are Gothic novelists H. P. Lovecraft (1890-1937) and Oliver LaFarge (1901-63), and its most famous artist is portrait painter Gilbert Stuart (1755-1828). Popular performing artists include George M. Cohan (1878-1942), Nelson Eddy (1901-67), Bobby Hackett (1915-76), and Van Johnson (b. 1916).

Important sports personalities include Baseball Hall of Famers Hugh Duffy (1866-1954), Napoleon Lajoie (1875-1959), and Charles "Gabby" Hartnett (1900-1972).

[50] BIBLIOGRAPHY

Carroll, Charles. *Rhode Island: Three Centuries of Democracy.* 4 vols. New York: Lewis, 1932.

Conley, Patrick T. *Democracy in Decline: Rhode Island's Constitutional Development, 1775-1841.* Providence: Rhode Island Historical Society, 1977.

Conley, Patrick T., and Matthew J. Smith. *Catholicism in Rhode Island.* Providence: Diocese of Providence, 1976.

Conley, Patrick T. *Rhode Island Profile.* Providence: Rhode Island Publications Society, 1983.

Federal Writers' Project. *Rhode Island: A Guide to the Smallest State.* Reprint. New York: Somerset, n.d. (orig. 1937).

James, Sydney V. *Colonial Rhode Island.* White Plains, N.Y.: Kraus International, 1975.

McLoughlin, William G. *Rhode Island: A Bicentennial History.* New York: Norton, 1978.

Providence Journal-Bulletin. *1985. Journal-Bulletin Rhode Island Almanac.* 99th ed. Providence, n.d.

Rhode Island, Department of Economic Development, Economic Research Division. *Rhode Island: Basic Economic Statistics 1982-1983.* Providence, 1983.

Rhode Island. Secretary of State. *Rhode Island Manual 1983-84.* Edited by Edward F. Walsh. Providence, 1983.

Steinberg, Sheila, and Cathleen McGuigan. *Rhode Island: An Historical Guide.* Providence: Rhode Island Bicentennial Commission, 1976.

SOUTH CAROLINA

State of South Carolina

ORIGIN OF STATE NAME: Named in honor of King Charles I of England. **NICKNAME:** The Palmetto State. **CAPITAL:** Columbia. **ENTERED UNION:** 23 May 1788 (8th). **SONG:** "Carolina." **POET LAUREATE:** Helen von Kolnitz Hyer. **MOTTO:** *Animis opibusque parati* (Prepared in mind and resources); *Dum spiro spero* (While I breathe, I hope). **COAT OF ARMS:** A palmetto stands erect, with a ravaged oak (representing the British fleet) at its base; 12 spears, symbolizing the first 12 states, are bound crosswise to the palmetto's trunk by a band bearing the inscription "Quis separabit" (Who shall separate?). Two shields bearing the inscriptions "March 26" (the date in 1776 when South Carolina established its first independent government) and "July 4," respectively, hang from the tree; under the oak are the words "Meliorem lapsa locavit" (Having fallen, it has set up a better one) and the year "1776." The words "South Carolina" and the motto *Animis opibusque parati* surround the whole. **FLAG:** Blue field with a white palmetto in the center and a white crescent at the union. **OFFICIAL SEAL:** The official seal consists of two ovals showing the original designs for the obverse and the reverse of South Carolina's great seal of 1777. Left (obverse): same as the coat of arms. Right (reverse): as the sun rises over the seashore, Hope, holding a laurel branch, walks over swords and daggers; the motto *Dum spiro spero* is above her, the word "Spes" (Hope) below. **ANIMAL:** White-tailed deer. **BIRD:** Carolina wren. **WILD GAME BIRD:** Wild turkey. **FISH:** Striped bass. **FLOWER:** Yellow jessamine. **TREE:** Palmetto. **GEM:** Amethyst. **STONE:** Blue granite. **LEGAL HOLIDAYS:** New Year's Day, 1 January; Birthday of Martin Luther King, Jr., 3d Monday in January; Lee's Birthday, 19 January; Washington's Birthday, 3d Monday in February; Jefferson Davis's Birthday, 3 June; Independence Day, 4 July; Labor Day, 1st Monday in September; Election Day, 1st Tuesday after 1st Monday in November (even-numbered years); Veterans Day, 11 November; Thanksgiving Day, 4th Thursday in November; Christmas Eve, 24 December, when declared by the governor; Christmas Day, 25 December; day after Christmas. **TIME:** 7 AM EST = noon GMT.

¹LOCATION, SIZE, AND EXTENT

Situated in the southeastern US, South Carolina ranks 40th in size among the 50 states.

The state's total area is 31,113 sq mi (80,583 sq km), of which land takes up 30,203 sq mi (78,226 sq km) and inland water 910 sq mi (2,357 sq km). South Carolina extends 273 mi (439 km) E–W; its maximum N–S extension is 210 mi (338 km).

South Carolina is bounded on the N and NE by North Carolina; on the SE by the Atlantic Ocean; and on the SW and W by Georgia (with the line passing through the Savannah and Chattooga rivers).

Among the 13 major Sea Islands in the Atlantic off South Carolina are Bull, Sullivans, Kiawah, Edisto, Hunting, and Hilton Head, the largest island (42 sq mi—109 sq km) on the Atlantic seaboard between New Jersey and Florida. The total boundary length of South Carolina is 824 mi (1,326 km), including a general coastline of 187 mi (301 km); the tidal shoreline extends 2,876 mi (4,628 km). The state's geographic center is located in Richland County, 13 mi (21 km) SE of Columbia.

²TOPOGRAPHY

South Carolina is divided into two major regions by the fall line that runs through the center of the state from Augusta, Ga., to Columbia and thence to Cheraw, near the North Carolina border. The area northwest of the line, known as the upcountry, lies within the Piedmont Plateau; the region to the southeast, called the low country, forms part of the Atlantic Coastal Plain. The rise of the land from ocean to the fall line is very gradual: Columbia, 120 mi (193 km) inland, is only 135 feet (41 meters) above sea level. In the extreme northwest, the Blue Ridge Mountains cover about 500 sq mi (1,300 sq km); the highest elevation, at 3,560 feet (1,085 meters), is Sassafras Mountain.

Among the many artificial lakes, mostly associated with electric power plants, is Lake Marion, the state's largest, covering 173 sq mi (448 sq km). Three river systems—the Pee Dee, Santee, and Savannah—drain most of the state. No rivers are navigable above the fall line.

³CLIMATE

South Carolina has a humid, subtropical climate. Average temperatures range from 68°F (20°C) on the coast to 58°F (14°C) in the northwest, with colder temperatures in the mountains. Summers are hot: in the central part of the state, temperatures often exceed 90°F (32°C), with a record of 111°F (44°C) set at Camden on 28 June 1954. In the northwest, temperatures of 32°F (0°C) or less occur from 50 to 70 days a year; the record low for the state is −20°F (−29°C), set at Caesars Head Mountain on 18 January 1977. The daily mean temperature at Columbia is 44°F (7°C) in January and 81°F (27°C) in July.

Rainfall is ample throughout the state, averaging 49 in (124 cm) annually at Columbia and ranging from 38 in (97 cm) in the central region to 52 in (132 cm) in the upper piedmont. Snow and sleet (averaging 2 in—5 cm—a year at Columbia) occur about three times annually, but more frequently and heavily in the mountains.

⁴FLORA AND FAUNA

Principal trees of South Carolina include palmetto (the state tree), balsam fir, beech, yellow birch, pitch pine, cypress, and several types of maple, ash, hickory, and oak; longleaf pine grows mainly south of the fall line. Rocky areas of the piedmont contain a wide mixture of moss and lichens. The coastal plain has a diversity of land formations—swamp, prairie, savannah, marsh, dunes—and, accordingly, a great number of different grasses, shrubs, and vines. Azaleas and camellias, not native to the state, have been planted profusely in private and public gardens. Bunched arrowhead and persistent trillium are endangered plants.

South Carolina mammals include white-tailed deer (the state animal), black bear, opossum, gray and red foxes, cottontail and marsh rabbits, mink, and woodchuck. Three varieties of raccoon are indigenous, one of them unique to Hilton Head Island. The state is also home to Bachman's shrew, originally identified in South Carolina by John Bachman, one of John J. Audubon's collaborators. Common birds include the mockingbird and Carolina wren (the state bird). Among endangered animals—all of which appear on the federal list—are the brown pelican, bald eagle, Bachman's warbler, eastern cougar, Florida manatee, shortnose sturgeon, American alligator, Atlantic leatherback and ridley turtles, and several whale species.

⁵ENVIRONMENTAL PROTECTION

The Department of Health and Environmental Control, established in 1971, has primary responsibility for such environmental matters as water purity, solid-waste disposal, air quality, nuclear energy, and food inspection. Nuclear energy and radioactive waste materials became controversial issues in South Carolina during the early 1980s. Because of the high concentration of nuclear facilities in the state, officials responded to the public's concern over their effects on the environment and on public health and safety. In 1983, state regulations regarding the storage and burial of low-level nuclear wastes began to be more strictly enforced after plans to build two new nuclear power plants had been canceled in 1982.

Stringent measures also have been taken by the state to regulate air and water pollution controls, sewage treatment, and proper disposal of solid waste. Because South Carolina has a large amount of hazardous solid wastes (estimated at 94,600 tons in 1984), methods were sought to limit their environmental impact and to improve disposal facilities.

Water improvement programs have brought most of South Carolina's major bodies of water within federal "fishable, swimmable" standards. The Water Resources Commission, created in 1967, formulates programs for the development and enlargement of the state's water supply.

⁶POPULATION

South Carolina in 1980 ranked 24th in population among the 50 states, with 3,121,820 residents, representing a 21% increase over the 1970 census. From 1980 to 1985, the population increased by 6% to an estimated total of 3,321,520. The population density in 1985 was 110 per sq mi (42 per sq km).

In 1980, South Carolina had 1,518,013 men and 1,603,807 women. South Carolinians are much less mobile than the US population as a whole: as of 1980, about 72% of state residents had lived in South Carolina their whole lives. The state's population remains somewhat younger than the national average, and in 1984, the median age was 28. The number of residents between 12 and 19 years of age increased by nearly 13% during the 1970s, while those aged 65 or over increased by 50% to comprise nearly 10% of the total population.

The urban share of the population was 54.1% in 1980. In 1982, 29 incorporated municipalities had more than 10,000 people, but only Columbia had a population in excess of 100,000. The largest cities, with 1984 estimated populations, were Columbia, 98,634; Charleston, 67,108; North Charleston, 66,735; and Greenville, 57,351. In 1984, the Charleston metropolitan area had an estimated 472,500 residents, ranking 73d in the US.

⁷ETHNIC GROUPS

The white population of South Carolina is mainly of Northern European stock; the great migratory wave from Southern and Eastern Europe during the late 19th century left South Carolina virtually untouched. As of 1980, only 1.5% of South Carolinians were foreign-born.

Black Americans made up about 30% of the state's population in 1980. In the coastal regions and offshore islands there still can be found some vestiges of African heritage, notably the Gullah dialect. South Carolina has always had an urban black elite, much of it of mixed racial heritage. After 1954, racial integration proceeded relatively peacefully, with careful planning by both black and white leaders.

The 1980 census counted 6,449 American Indians, of whom 43% lived in urban areas. At that time, about 1,350 Indians were living on the Catawba reservation in York County; in 1983, a federal appeals court upheld the Indians' claim that 144,000 acres (58,275 hectares) of disputed land still belonged to the Catawba tribe. There were also 4,232 Mexicans, 3,850 Filipinos, 2,367 Puerto Ricans, 2,231 Japanese, and 1,072 Vietnamese in South Carolina.

⁸LANGUAGES

English settlers in the 17th century encountered first the Yamasee Indians and then the Catawba, both having languages of the Hokan-Siouan family. Few Indians remain today, and a bare handful of their place-names persist: Cherokee Falls, Santee, Saluda.

South Carolina English is marked by a division between the South Midland of the upcountry and the plantation Southern of the coastal plain, where dominant Charleston speech has extensive cultural influence even in rural areas. Many upcountry speakers of Scotch-Irish background retain /r/ after a vowel, as in *hard*, a feature now gaining acceptance among younger speakers in Charleston. At the same time, a longtime distinctive Charleston feature, a centering glide after a long vowel, so that *date* and *eight* sound like /day-uht/ and /ay-uht/, is losing ground among younger speakers.

Along the coast and on the Sea Islands, some blacks still use the Gullah dialect, based on a Creole mixture of pre-Revolutionary English and African speech. The dialect is rapidly dying in South Carolina, though its influence on local pronunciations persists.

In 1980, 97% of all state residents 3 years of age and older reported speaking English at home. Other languages spoken at home included:

Spanish	22,624	Filipino	2,435
French	15,872	Greek	2,293
German	9,028	Italian	1,635

⁹RELIGIONS

South Carolina is predominantly Protestant, and has been since colonial days. The Protestant denominations with more than 100,000 members in 1980 were Southern Baptist, 659,392; Baptist Educational and Missionary Convention (black), 400,000; United Methodist, 240,992; and African Methodist Episcopal, 133,000. The Episcopal Church had great influence during colonial times, but in 1980 it had only 50,551 members. As of 1984, the state also had 70,565 Roman Catholics and an estimated 8,615 Jews.

¹⁰TRANSPORTATION

Since the Revolutionary War, South Carolina has been concerned with expanding the transport of goods between the upcountry and the port of Charleston and the midwestern US. Several canals were constructed north of the fall line, and the 136-mi (219-km) railroad completed from Charleston to Hamburg (across the Savannah River from Augusta, Ga.) in 1833 was the longest in the world at that time. Three years earlier, the *Best Friend of Charleston* had become the first American steam locomotive built for public railway passenger service; by the time the Charleston–Hamburg railway was completed, however, the *Best Friend* had blown up, and a new engine, the *Phoenix,* had replaced it. Many other efforts were made to connect Charleston to the interior by railway, but tunnels through the mountains were never completed. Today, most freight service is furnished by the Southern Railway and the Seaboard System Railroad. Amtrak passenger trains pass north–south through the state, providing limited service to Charleston, Columbia, and other cities. In 1983, Class I railway trackage totaled 2,524 mi (4,062 km).

The state road network in 1984 was made up of 54,263 mi (87,328 km) of rural roads, 9001 mi (14,486 km) of urban roads,

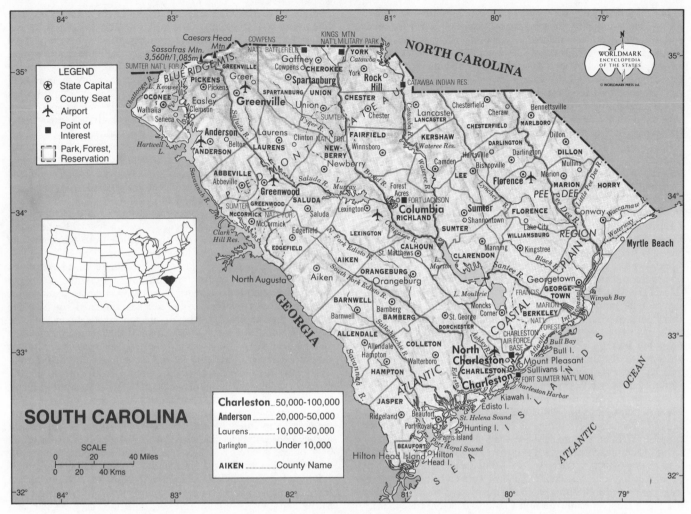

LEGEND
- ⊛ State Capital
- ⊙ County Seat
- ✈ Airport
- ■ Point of Interest
- ⌐ ⌐ Park, Forest, Reservation

SOUTH CAROLINA

SCALE
0   20   40 Miles
0 20 40 Kms

Charleston	50,000-100,000
Anderson	20,000-50,000
Laurens	10,000-20,000
Darlington	Under 10,000
AIKEN	County Name

See endsheet maps: K4.
LOCATION: 32°02′ to 35°12′56″N; 78°32′36″ to 83°21′14″W. **BOUNDARIES:** North Carolina line, 327 mi (526 km); Atlantic Ocean coastline, 187 mi (301 km); Georgia line, 310 mi (499 km).

and 678 mi (1,091 km) of interstate highways. Highway I-26, running northwest–southeast from the upcountry to the Atlantic, intersects I-85 at Spartanburg, I-20 at Columbia, and I-95 on its way toward Charleston. There were 2,054,000 passenger vehicles and 451,000 trucks and buses registered in 1983, when the number of licensed drivers totaled 2,007,683.

The state has three deepwater seaports. Charleston is one of the major ports on the Atlantic, and the harbors of Georgetown and Port Royal also handle significant waterborne trade. The Atlantic Intracoastal Waterway, crossing the state slightly inward from the Atlantic Ocean, is a major thoroughfare.

South Carolina's 137 airports handled 1,195,635 passengers in 1983. Charleston, Columbia, and Greenville-Spartanburg are the major airports within the state; many travelers also enter South Carolina via the air terminals at Savannah, Augusta, and Atlanta, Ga., and at Charlotte, N.C. In 1983, the state registered 1,686 active aircraft and 6,174 pilots.

[11] HISTORY

Prior to European settlement, the region now called South Carolina was populated by several Indian groups. Indians of Iroquoian stock, including the Cherokee, inhabited the northwestern section, while those of the Siouan stock—of whom the Catawba were the most numerous—occupied the northern and eastern regions. Indians of Muskogean stock lived in the south.

In the early 1500s, long before the English claimed the Carolinas, Spanish sea captains explored the coast. The Spaniards made an unsuccessful attempt to establish a settlement in 1526 at Winyah Bay, near the present city of Georgetown. Thirty-six years later, a group of French Huguenots under Jean Ribault landed at a site near Parris Island, but the colony failed after Ribault returned to France. The English established the first permanent settlement in 1670 under the supervision of the eight lords proprietors who had been granted "Carolana" by King Charles II. At first the colonists settled at Albemarle Point on the Ashley River; 10 years later, they moved across the river to the present site of Charleston.

Rice cultivation began in the coastal swamps, and black slaves were imported as field hands. The colony flourished, and by the mid-1700s, new areas were developing inland. Germans, Scotch-Irish, and Welsh, who differed markedly from the original aristocratic settlers of the Charleston area, migrated to the southern part of the new province. Although the upcountry was developing and was taxed, it was not until 1770 that the settlers there were represented in the government. For the most part, the colonists had friendly relations with the Indians. In 1715, however, the Yamasee were incited by Spanish colonists at St. Augustine, Fla., to attack the South Carolina settlements. The settlers successfully resisted, with no help from the proprietors.

The original royal grant had made South Carolina a very large colony, but eventually the separate provinces of North Carolina and Georgia were established, two moves that destined South Carolina to be a small state. The colonists were successful in having the proprietors overthrown in 1719 and the government transferred to royal rule by 1721.

Skirmishes with the French, Spanish, Indians, and pirates, as well as a slave uprising in 1739, marked the pre-Revolutionary period. South Carolina opposed the Stamp Act of 1765 and took an active part in the American Revolution. The first British property seized by American Revolutionary forces was Ft. Charlotte in McCormick County in 1775. Among the many battles fought in South Carolina were major Patriot victories at Ft. Moultrie in Charleston (1776), Kings Mountain (1780), and Cowpens (1781), the last two among the war's most important engagements. Delegates from South Carolina, notably Charles Cotesworth Pinckney, were leaders at the federal constitutional convention of 1787. On 23 May 1788, South Carolina became the 8th state to ratify the Constitution.

Between the Revolutionary War and the Civil War, two issues dominated South Carolinians' political thinking: tariffs and slavery. Senator John C. Calhoun took an active part in developing the nullification theory by which a state claimed the right to abrogate unpopular federal laws. Open conflict over tariffs during the early 1830s was narrowly averted by a compromise on the rates, but in 1860, on the issue of slavery, no compromise was possible. At the time of secession, on 20 December 1860, more than half the state's population consisted of black slaves. The first battle of the Civil War took place at Ft. Sumter in Charleston Harbor on 12 April 1861. Federal forces soon captured the Sea Islands, but Charleston withstood a long siege until February 1865. In the closing months of the war, Union troops under General William Tecumseh Sherman burned Columbia and caused widespread destruction elsewhere. South Carolina contributed about 63,000 soldiers to the Confederacy out of a white population of some 291,000. Casualties were high: nearly 14,000 men were killed in battle or died after capture.

Federal troops occupied South Carolina after the war. During Reconstruction, as white South Carolinians saw it, illiterates, carpetbaggers, and scalawags raided the treasury, plunging the state into debt. The constitution was revised in 1868 by a convention in which blacks outnumbered whites by 76 to 48; given the franchise, blacks attained the offices of lieutenant governor and US representative. In 1876, bands of white militants called Red Shirts, supporting the gubernatorial candidacy of former Confederate General Wade Hampton, rode through the countryside urging whites to vote and intimidating potential black voters. Hampton, a Democrat, won the election, but was not permitted by the Republican incumbent to take office until President Rutherford B. Hayes declared an end to Reconstruction and withdrew federal troops from the state in April 1877.

For the next 100 years, South Carolina suffered through political turmoil, crop failures, and recessions. A major political change came in the 1880s with a large population increase upcountry and the migration of poor whites to cities. These trends gave farmers and industrial workers a majority of votes, and they found their leader in Benjamin Ryan "Pitchfork Ben" Tillman, a populist who stirred up class and racial hatreds by attacking the "Charleston ring." Tillman was influential in wresting control of the state Democratic Party from the coastal aristocrats; he served as governor from 1890 to 1894 and then as US senator until his death in 1918. However, his success inaugurated a period of political and racial demagoguery that saw the gradual (though not total) disfranchisement of black voters.

The main economic transformation since 1890 has been the replacement of rice and cotton growing by tobacco and soybean cultivation and truck farming, along with the movement of tenant farmers, or sharecroppers, from the land to the cities. There they found jobs in textile mills, and textiles became the state's leading industry after 1900. With the devastation of the cotton crop by the boll weevil in the 1920s, farmers were compelled to diversify their crops, and some turned to cattle raising. Labor shortages in the North during and after World War II drew many thousands of blacks to Philadelphia, Washington, D.C., New York, and other cities. After 1954, public school desegregation proceeded very slowly but peaceably, and blacks were gradually accepted alongside whites in the textile mills and other industries. In 1983, for the first time in 95 years, a black state senator was elected; the following year, four blacks were elected to the reapportioned senate. Despite these changes, most white South Carolinians remain staunchly conservative in political and social matters.

[12] STATE GOVERNMENT

South Carolina has had seven constitutions, dating from 1776, 1778, 1790, 1861, 1865, 1868, and 1895, respectively. Beginning in 1970, most articles of the 1895 constitution were rewritten. The present document had been amended 444 times by 1983.

The general assembly consists of a senate of 46 members, elected for four-year terms, and a house of representatives of 124 members, elected for two years. Senators must be 25 years of age, representatives 21; all legislators must be qualified voters in the districts they represent. Officials elected statewide include the governor and lieutenant governor (who run separately), attorney general, secretary of state, comptroller general, treasurer, adjutant general, secretary of agriculture, and superintendent of education, all elected to four-year terms. Eligibility requirements for the governor include a minimum age of 30, US citizenship for at least five years, and a five-year state residency.

Most state agencies are governed by boards of at least five members appointed by the governor, but some of the largest agency boards are elected by the general assembly. Outside of his own office, the governor appoints very few department heads. South Carolina has nearly 130 agencies, boards, and commissions.

Legislative sessions are held annually. Bills may be introduced in either house, except for revenue measures, which are reserved to the house of representatives. The governor has a regular veto and an item veto on appropriation matters, either of which may be overridden by a two-thirds vote of those present in each house of the legislature; bills automatically become law after five days if the governor takes no action. The constitution may be amended by a two-thirds vote of each house of the general assembly and by a majority of those casting their ballots at the next general election. To take effect, however, the amendment must then be ratified by a majority vote of the next general assembly.

Citizens 18 years of age and older who have been state residents for 30 days are eligible to vote.

[13] POLITICAL PARTIES

South Carolina's major political organizations are the Democratic and Republican parties. From the end of Reconstruction, the Democratic Party dominated state politics. Dissatisfaction with the national party's position on civil rights in 1948 led to the formation of the States' Rights Democrat faction, whose candidate, South Carolina Governor J. Strom Thurmond, carried the state in 1948. Thurmond's subsequent switch to the Republicans while in the US Senate was a big boost for the state's Republican Party, which since 1964 has captured South Carolina's eight electoral votes in five of the six presidential elections. However, about four of every five state legislators were Democrats as of January 1985, as were most municipal and county officials. Although a Republican was elected governor in 1974, the Democrats have won the governorship and nearly all major statewide offices in elections ever since. In 1980, Ronald Reagan carried South Carolina by less than 1%, but four years later he won the state by a 28% margin. Both parties have shared equally in the state's delegation to the US Congress since 1980.

Voters do not register according to political party in South Carolina. Instead, at primary elections, they simply take an oath that they have not participated in another primary. In September 1984 there were 1,309,964 registered voters, of whom 28% were black. Black voter registration has more than doubled since 1964, and South Carolina had 310 elected black public officials as of January 1985.

14 LOCAL GOVERNMENT

As of 1984, South Carolina had 46 counties, 265 incorporated municipalities, 92 school districts, and 242 special districts of various types. Ten regional councils provide a broad range of technical and advisory services to county and municipal governments.

Under legislation enacted in 1975, all counties and municipalities have the same powers, regardless of size. Most municipalities operate under the mayor-council or city manager system; more than half the counties have a county administrator or manager. Customarily, each county has a council or commission, attorney, auditor, clerk of court, coroner, tax collector, treasurer, and sheriff. Many of these county officials are elected, but the only municipal officers elected are the mayor and the members of the council.

While the state shares revenues from many different sources with the counties and municipalities, these local units derive virtually all their direct revenue from the property tax. In recent years, the state's school districts have rapidly increased their own property tax levies, squeezing the counties' and municipalities' revenue base.

15 STATE SERVICES

The State Ethics Commission establishes rules covering possible conflicts of interest, oversees election campaign practices, and provides for officeholders' financial disclosure.

The Department of Education administers state and federal aid to the public schools, while the State Commission on Higher Education oversees the public colleges and universities, and the State Board for Technical and Comprehensive Education is responsible for postsecondary technical training schools. The state also runs special schools for the deaf and blind. Complementing both public and higher education is a state educational television network, under the jurisdiction of the South Carolina Educational Television Commission. Transportation services are provided by the Department of Highways and Public Transportation, which maintains most major roads, issues drivers' licenses, and has jurisdiction over the Highway Patrol. The Aeronautics Commission licenses airplanes and pilots and oversees airport operations.

Through a variety of agencies, South Carolina offers a broad array of human services in the fields of mental health, mental retardation, vocational rehabilitation, veterans' affairs, care of the blind, and adoptions. An ombudsman for the aging handles complaints about nursing homes, which are licensed by the state. The South Carolina Law Enforcement Division provides technical aid to county sheriffs and municipal police departments. Emergency situations are handled by the Office of Disaster Assistance and the National Guard.

The State Housing Authority is authorized to subsidize interest rates on mortgages for middle- and low-income families. The Employment Security Commission oversees unemployment compensation and job placement, while the Department of Labor offers arbitration and mediation services and enforces health and safety standards. The Human Affairs Commission looks into unfair labor practices based on sex, race, or age.

16 JUDICIAL SYSTEM

South Carolina's unified judicial system is headed by the chief justice of the supreme court, who, along with four associate justices, is elected by the general assembly to a 10-year term. The supreme court is the final court of appeal. A five-member intermediate court of appeals for criminal cases was established in 1979, but legal questions (specifically, about the election of general assembly members to four of the five seats) prevented the court from convening until 1981; the court became a permanent constitutional court in 1984.

Sixteen circuit courts hear major criminal and civil cases. As of 1984 there were 31 circuit court judges, all of them elected by the general assembly to six-year terms. The state also has a system of

South Carolina Presidential Vote by Political Parties, 1948–84

YEAR	ELECTORAL VOTE	SOUTH CAROLINA WINNER	DEMOCRAT	REPUBLICAN	STATES' RIGHTS DEMOCRAT
1948	8	Thurmond (SRD)	34,423	5,386	102,607
1952	8	Stevenson (D)	172,957	168,043	—
					UNPLEDGED
1956	8	Stevenson (D)	136,278	75,634	88,509
1960	8	*Kennedy (D)	198,121	188,558	
1964	8	Goldwater (R)	215,723	309,048	—
					AMERICAN IND.
1968	8	*Nixon (R)	197,486	254,062	215,430
					AMERICAN
1972	8	*Nixon (R)	186,824	477,044	10,075
1976	8	*Carter (D)	450,807	346,149	2,996
					LIBERTARIAN
1980	8	*Reagan (R)	430,385	441,841	4,975
1984	8	*Reagan (R)	344,459	615,539	4,359

*Won US presidential election.

family courts for domestic and juvenile cases. In addition, there are magistrates courts (justices of the peace) in all counties, municipal courts, and county probate judges.

The state penal system is rapidly becoming centralized under the state Department of Corrections; there is a separate state system for juvenile offenders. In 1984 there were 10,035 inmates in state and federal correctional institutions. Crime rates for 1983 exceeded the national averages for murder, forcible rape, and aggravated assault. South Carolina has a death penalty statute, but the death penalty has not been employed since the late 1950s.

¹⁷ARMED FORCES
Ft. Jackson, in Columbia, is one of the Army's major training centers, with 12,965 military personnel in 1984. Air Force bases at Charleston, Sumter, and Myrtle Beach are all major installations, and the Charleston Harbor area has many naval facilities, including a nuclear submarine base. Parris Island has long been one of the country's chief Marine Corps training bases. Overall, the state had 65,770 military and civilian defense personnel in 1984. South Carolina firms received $478 million in defense contract awards during 1983/84.

Veterans in South Carolina as of 30 September 1983 totaled 350,000, including some 2,000 from World War I, 128,000 from World War II, 70,000 from the Korean conflict, and 116,000 who served during the Viet-Nam era. Veterans' benefits during 1982/83 amounted to $338 million.

National Guard units had 12,760 personnel at the end of 1984. State and local police departments had 6,613 employees in 1983, when expenditures on police protection totaled $139 million.

¹⁸MIGRATION
The original European migration into South Carolina consisted mostly of German, Welsh, and Scotch-Irish settlers. During the 19th century, many of the original settlers emigrated westward to Alabama, Mississippi, and Texas. In the 20th century, many blacks left the state for cities in the North. Between 1940 and 1970, South Carolina's net loss from migration was 601,000. During 1970–80, however, the state enjoyed a net gain of 210,000; in 1981–83, the net gain from migration was 58,000.

¹⁹INTERGOVERNMENTAL COOPERATION
The South Carolina Interstate Cooperation Commission represents the state before the Council of State Governments. South Carolina also participates in the Atlantic States Marine Fisheries Compact, Southeastern Forest Fire Protection Compact, Southern Growth Policies Compact, Southern Interstate Energy Compact, and Southern Regional Education Compact.

In 1983/84, the state received $1,168 million in federal funds.

²⁰ECONOMY
During its early days, South Carolina was one of the country's richest areas. Its economy depended on foreign commerce and agriculture, especially indigo, rice, and later cotton. After the Civil War, the state suffered severe economic depression. Not until the 1880s did the textile industry—today the state's major employer—begin to develop. Textiles and farming completely dominated the economy until after World War II, when efforts toward economic diversification attracted paper, chemical, and other industries to the state. During the postwar period, the state spent sizable amounts to improve its three ports, especially the harbor facilities of Charleston. South Carolina is blessed with an ample supply of electric power, an abundance of water and forests, adequate public highways, and good harbors. Largely because of its legacy of low-wage industries and an unorganized and poorly educated work force, the state continues to fall below national norms by most economic measures. Nevertheless, during the 1970s and early 1980s, real per capita income increased faster than in the nation as a whole. Rising foreign and domestic investment, coupled with an abundance of first-class tourist facilities along the coast, contributed to the continuing growth of South Carolina's economy in the early 1980s.

²¹INCOME
South Carolina ranked 48th among the 50 states in 1983 with a per capita income of $8,954—30% below the national norm and the 3d lowest of the 12 southeastern states. In 1970, the state's per capita income in constant dollars was $3,216, 33% below the US average.

Total income for 1983 exceeded $29.2 billion, representing real increases of 52% during the 1970s and 7.9% in 1980–83. The major sources of 1983 personal income were manufacturing, $7 billion; government, $4.8 billion; and services, $3 billion. State income tax data indicate that 36% of individual incomes were below $15,000 and 17% were below $10,000 in 1983. Median family income averaged $22,060 (47th in the US) in 1981, when an estimated 11% of all whites and 34% of all blacks were living below the federal poverty level. Counties with sizable military or other federal facilities generally ranked highest in per capita income, while counties having majority black populations ranked among the lowest.

²²LABOR
In 1984, South Carolina had a nonagricultural civilian labor force of 1,479,000, of whom 1,374,000 were employed and 105,000 were unemployed. The unemployment rate of 7.1% was below the 1984 national average of 7.5%; the rate for blacks and other minorities was significantly higher. A federal census in March 1982 revealed the following nonfarm employment pattern for South Carolina:

	ESTABLISH-MENTS	EMPLOYEES	ANNUAL PAYROLL ('000)
Agricultural services, forestry, fishing	564	4,582	$ 52,033
Mining	76	1,550	24,243
Contract construction	5,127	101,575	1,906,397
Manufacturing, of which:	3,864	377,065	5,598,888
Textile mill products	(382)	(120,938)	(1,519,500)
Apparel, other textiles	(318)	(44,095)	(402,972)
Chemicals, allied products	(155)	(33,829)	(768,006)
Transportation, public utilities	1,847	40,796	784,633
Wholesale trade	4,398	47,437	752,519
Retail trade	17,287	178,075	1,450,566
Finance, insurance, real estate	4,708	47,371	699, 439
Services	15,904	161,850	1,791,860
Other	1,248	1,072	16,002
TOTALS	55,023	961,373	$13,076,580

South Carolinians not covered by this survey included federal employees (33,000 in 1982) and state and local government employees (179,000 in 1983).

In 1983, the average hourly payment for workers in durable goods was $7.08; in nondurable goods, $6.97. Average weekly earnings for production workers were $285, placing South Carolina 44th among the states in average annual wages with $14,647, which was 17% below the average wage level for the US as a whole.

South Carolina has one of the lowest work stoppage rates in the nation—only five work stoppages involving 200 workers in 1981—and only a small percentage of the total labor force is organized. Membership in South Carolina's 431 labor unions in 1983 was only 51,816 (about 4% of the employed labor force). Textile, clothing, and ladies' garment workers' unions make up the bulk of the membership, followed by transportation and electrical workers. Several large textile companies have made major efforts to prevent their workers from organizing unions; conflicts between management and workers have continued for years, but without serious violence. A right-to-work law was enacted in the 1950s.

[23] AGRICULTURE

In 1983, South Carolina ranked 34th among the 50 states in farm income with nearly $1.1 billion, of which $653 million was from crop production. For most of the 19th century, cotton was king, but today soybeans and tobacco outrank cotton in annual cash value. In 1983 there were 31,000 farms comprising 5,900,000 acres (2,388,000 hectares); the average size of a farm was 190 acres (77 hectares).

The principal farming area is a 50-mi (80-km) band across the upper coastal plain. The Pee Dee region in the east is the center for tobacco production. Cotton is grown mostly south of the fall line, and truck crops abound in the coastal and sand hill counties. South Carolina leads the nation in the sale of fresh peaches, which are produced mainly in the sand hills and piedmont. The following table shows acreage, production, and value of leading crops in 1983:

	ACRES HARVESTED	PRODUCTION	VALUE
Tobacco	54,000	112,860,000 lb	$204,164,000
Soybeans	1,430,000	22,880,000 bu	184,184,000
Corn for grain	275,000	17,050,000 bu	64,710,000
Wheat	375,000	10,500,000 bu	34,125,000
Cotton	69,000	53,000 bales	18,444,000
Peaches	—	210,000,000 lb	16,115,000

[24] ANIMAL HUSBANDRY

In 1983, livestock and livestock products were valued at $405 million, or 38% of the state's farm income. Although production is well distributed throughout the state, Newberry and Orangeburg counties, in central South Carolina, lead in cash receipts from livestock products.

At the close of 1984 there were 590,000 cattle and calves, 248,000 milk cows, and 510,000 hogs and pigs. Production of cattle totaled 159 million lb, valued at $73.4 million, while hog production amounted to 155 million lb, worth $69.5 million. Dairy farms produced 573 million lb of milk, worth $90.7 million. Poultry farmers sold 1.6 billion eggs, worth $76.6 million, and 193 million lb of chickens and broilers, worth $53 million.

[25] FISHING

The state's oceanfront saltwater inlets and freshwater rivers and lakes provide ample fishing opportunities. Major commercial fishing is restricted to saltwater species of fish and shellfish, mainly shrimp, crabs, clams, and oysters. In 1984, the commercial catch totaled 15,104,000 lb, valued at $14,609,000.

[26] FORESTRY

South Carolina had 12,502,906 acres (5,059,764 hectares) of forestland in 1979—about 65% of the state's area and 1.7% of all US forests. The state's two national forests, Francis Marion and Sumter, comprised 11% of the forested area. Nearly all of South Carolina's forests are classified as commercial timberland, 91% of it privately owned. Several varieties of pine, both long- and shortleaf, are the major source of timber and of pulp for the paper industry. Shipments of lumber and wood products were valued at $1 billion in 1982; shipments of paper and allied products, $1.7 billion.

[27] MINING

Mining is not a major industry. In 1984, South Carolina ranked 29th among the 50 states in nonfuel mineral production, with a total value of $255 million. The only commercial minerals are kaolin and other types of clay, sand and gravel, stone, cement, phosphates, and gem stones. The 1984 mineral output included 2,200,000 tons of cement (valued at $93 million); stone, 16,200,-000 tons ($63 million); clays, 1,818,000 tons ($39 million); and sand and gravel, 6,675,000 tons ($30 million).

[28] ENERGY AND POWER

Although it lacks fossil fuel resources, South Carolina produces more electricity than it consumes. Installed electric capacity totaled 13.2 million kw in 1983, when power production reached 46 billion kwh.

About 56% of electric output came from nuclear reactors, 37% from coal-fired plants, and 7% from hydropower. Sales of electric energy in 1983 amounted to 41.7 billion kwh. Major power suppliers are six private companies and the state-owned Public Service Authority, known as Santee-Cooper. Gas utilities sold 96.9 trillion Btu of natural gas—all of it imported—to some 312,100 customers in 1983.

South Carolina is heavily engaged in nuclear energy. As of the early 1980s, four nuclear plants were producing electricity, but plans for two additional plants were canceled in 1982 when disposal of low-level nuclear waste became a controversial issue. The vast Savannah River plant in Aiken County produces most of the plutonium for the nation's nuclear weapons; Chem-Nuclear Systems in Barnwell County stores about half of the country's low-level nuclear wastes; and a Westinghouse plant in Richland County makes fuel assemblies for nuclear reactors.

[29] INDUSTRY

South Carolina's principal industry beginning in the 1880s was textiles, but many textile mills were closed during the 1970s and early 1980s because of the importation of cheaper textiles from abroad. The economic slack was made up, however, by the establishment of new industries, especially paper and chemical manufactures, and by increasing foreign investment in the state. Overseas investments totaled $130 million in South Carolina in 1983, or more than 10% of total industrial investment in the state; 39% of the investment in manufacturing plants was in chemicals, and 26% in fabricated metals. Principal overseas investment in 1983 came from Switzerland, the Federal Republic of Germany, the United Kingdom, and Japan. South Carolina's major manufacturing centers are concentrated north of the fall line and in the piedmont.

The total value of shipments by manufacturers was $27.8 billion in 1982, representing a threefold increase since 1977. The following table shows the value of shipments by major manufacturing sectors in 1982:

Textile mill products	$7,249,700,000
Chemicals and chemical products	5,106,400,000
Nonelectrical machinery	1,989,000,000
Paper and allied products	1,714,400,000
Food and food products	1,666,400,000
Apparel, other textiles	1,503,900,000
Electric, electronic equipment	1,385,800,000
Fabricated metal products	1,163,500,000
Lumber and wood products	1,060,800,000
Primary metals	1,009,900,000

[30] COMMERCE

Personal income from wholesale and retail trade exceeded $2 billion in 1982. Wholesalers reported a trade volume of $12.6 billion in 1982. Tobacco wholesale markets and warehouses are centered in the Pee Dee region, while soybean sales and storage facilities cluster around the port of Charleston; truck crops, fruits, and melons are sold in large quantities at the state farmers' market in Columbia. Retail trade in 1982 totaled $12.1 billion, of which food stores accounted for 25%, automotive dealers 17%, department stores 10%, and other outlets 48%.

In 1984, Charleston, Georgetown, Port Royal, and other ports handled total tonnage of 4,647,030: 1,961,864 tons of imports and 2,685,166 tons of exports. Foreign imports were valued at $2.3 billion and overseas exports at $2.5 billion in 1983. Foreign exports of South Carolina's own manufactures were valued at $2.2 billion in 1981 (23d in the US); agricultural exports totaled $407 million in 1981/82. In the latter year, South Carolina's exports of tobacco were valued at $137 million, placing it 3d among the states as a tobacco exporter.

31 CONSUMER PROTECTION

The Department of Consumer Affairs, established in 1974, has the authority to investigate consumer complaints and represent the public at regulatory proceedings.

32 BANKING

In 1983, the state's 73 commercial banks had total assets of $11.5 billion and deposits of more than $9.3 billion; outstanding loans amounted to $5.3 billion. There were 44 savings and loan associations with assets totaling $6.9 billion and mortgage loans of $5.1 billion. In addition, South Carolina had 157 credit unions with 470,548 members; assets amounted to $881 million, loans to $572 million.

33 INSURANCE

The South Carolina Insurance Department licenses and supervises the insurance companies doing business in the state; most of these represent national insurance organizations. In 1983, life insurance in force in the state totaled nearly $63.3 billion, or $52,400 per insured family; premiums paid amounted to $607 million, and payments to beneficiaries $188 million. Property and casualty companies wrote premiums totaling $1.2 billion in 1983. Accident and health insurance companies had $721 million in premiums in 1982. A modified no-fault system of automobile insurance is in effect.

34 SECURITIES

There are no securities exchanges in South Carolina, but in 1983 there were 29 stockbrokers and dealers registered there, and New York Stock Exchange member firms had 72 offices and 478 representatives. Some 361,000 South Carolinians owned shares of public corporations that year. Enforcement of the state Securities Act is vested in the securities commissioner within the Office of the Secretary of State.

35 PUBLIC FINANCE

Contrary to general practice, South Carolina's governor is not the sole budget officer, but instead chairs the State Budget and Control Board, which also includes the state treasurer, comptroller general, and two legislators representing the senate and house, respectively.

The budget board submits the annual budget to the general assembly in January as the basis for enactment of an appropriations bill, effective for the fiscal year beginning 1 July.

The state constitution requires that budget appropriations not exceed expected revenues (including reserve funds). If there should be a deficit, the next general assembly must provide for the deficit as a first order of business. Many tax revenues are earmarked for specific purposes: all gasoline taxes and related charges are designated for highways, and the sales tax finances public education, which accounts for more than half of all general fund expenditures. The state shares tax collections with its subdivisions—counties and municipalities—which determine how their share of the money will be spent.

The following table summarizes general fund revenues and expenditures for 1982/83 and 1983/84:

REVENUES	1982/83	1983/84
Retail sales tax	$ 684,390,837	$ 798,897,692
Individual income tax	718,862,771	795,480,492
Corporate income tax	124,132,398	154,483,469
Other receipts	442,544,034	479,348,442
TOTALS	$1,969,930,040	$2,228,210,095
EXPENDITURES		
Education	$1,111,093,000	$1,211,835,000
Health	190,409,000	205,721,000
Social rehabilitation	132,002,000	137,980,000
Aid to subdivisions	108,298,000	123,255,000
Debt service	105,091,000	101,436,000
Other outlays	289,445,000	330,564,000
TOTALS	$1,936,338,000	$2,110,791,000

In mid-1982, South Carolina's public debt totaled nearly $4.8 billion, or $1,474 per capita, as compared with the US average of $1,537 per capita.

36 TAXATION

South Carolinians bear a tax burden that is lighter than most other states'. As of 1984, the chief levies were a personal income tax ranging from 2% on the first $1,000 to 7% on taxable income over $10,093; a corporate income tax of 6% (4.5% for banks, 8% for associations); a broad-based 5% sales tax; and taxes on gasoline, alcoholic beverages, business licenses, insurance, and gifts and estates. In addition to the property tax, municipalities and counties may impose business license fees—although no county charges these fees—plus charges for such services as garbage collection and water supply.

Taxes remitted to the federal government totaled $5.8 billion in 1982/83; South Carolina received more than $5.2 billion in federal expenditures during 1983/84. South Carolinians filed 1,220,503 federal income tax returns for 1983, paying more than $2.6 billion in tax, or an average of $2,686 per taxpayer (24% less than the national average).

37 ECONOMIC POLICY

Created in 1945, the State Development Board seeks to encourage economic growth and to attract new industries; it has been successful in attracting foreign companies, especially to the piedmont. The state's Small Business Development Center and Governor's Office of Minority Business help small companies to get started.

The state exempts for five years all new industrial construction from local property taxes (except the school tax). Moreover, industrial properties are assessed very leniently for tax purposes. State and local governments have cooperated in building necessary roads to industrial sites, providing water and sewer services, and helping industries to meet environmental standards. Counties are authorized to issue industrial bonds at low interest rates. Generally conservative state fiscal policies, relatively low wage rates, and an anti-union climate also serve as magnets for industry.

38 HEALTH

South Carolina's infant mortality rates have improved significantly in recent years but were still among the nation's highest in 1983: 11.3 per 1,000 live births for whites and 20.6 for nonwhites. Impeding efforts to further improve health standards are a shortage of doctors (especially in rural areas and small towns), inadequate public education, poor housing, and improper sanitation.

As of 1983, the leading causes of death were heart disease, which caused 37% of all deaths; cancer, 20%; stroke, 9%; and accidents, 6%. The rates for heart disease and cancer were higher among whites than nonwhites, while the reverse was true of strokes and accidents. The state has mounted major programs to detect heart disease and high blood pressure, reduce infant mortality, and expand medical education. In 1980, the birthrates for whites of 14.2 per 1,000 population and for blacks of 22 per 1,000 were below the US average; the state's death rate of 8 per 1,000 in 1981 was also below the norm.

In 1983, South Carolina's health facilities included 92 general hospitals, 177 long-term care facilities, 12 psychiatric hospitals, 3 rehabilitation facilities, and 50 mental retardation centers, along with 2 major veterans' hospitals. The average cost of care in a hospital semiprivate room was $251 per day and $1,867 per stay in 1982, both rates well below the national norms. Nonfederal medical personnel in 1984 were physicians, 4,431; dentists, 1,212; registered nurses, 13,933; licensed practical nurses, 7,253; and pharmacists, 2,145.

39 SOCIAL WELFARE

Aid to 47,700 families with dependent children amounted to $73 million in 1983/84. Federal Supplemental Security Income (SSI) payments to 81,000 recipients in 1983 came to $156.9 million, to which state-administered SSI added $2.8 million; the average monthly payment was $129 for the aged and $211 for the disabled in December 1983. Food stamps were issued to 411,524 persons and amounted to $208 million in 1983/84. In that year, virtually every school in the state participated in the school lunch program, at a federal cost of $53 million.

In 1983, 462,000 people received Social Security benefits amounting to $1.9 billion; the average monthly payment to retirees was $403. In 1982, 887,000 claims were filed for unemployment insurance; benefits totaled $286 million, with an average weekly payment of $110. Workers' compensation totaled $98.9 million in 1982. Vocational rehabilitation services aided 8,518 persons in 1984.

40HOUSING

In 1980 there were 1,153,709 year-round housing units, of which 722,547 were owner-occupied. About 96% of all occupied units had full plumbing. The median value of an owner-occupied house was $35,100 in 1980.

South Carolina has made a determined effort to upgrade housing. The State Housing Authority, created in 1971, is empowered to issue bonds to provide mortgage subsidies for low- and middle-income families. In 1983, construction began on 25,800 housing units; 29,800 new units, valued at $1.1 billion, were authorized.

41EDUCATION

For decades, South Carolina ranked below the national averages in most phases of education, including expenditures per pupil, median years of school completed, teachers' salaries, and literacy levels. During the 1970s, however, significant improvements were made through the adoption of five-year achievement goals, enactment of a statewide educational funding plan, provision of special programs for exceptional children and of kindergartens for all children, measurement of students' achievements at various stages, and expansion of adult education programs. As a result, South Carolina high school graduates now score only slightly lower than the national averages on standardized examinations. Although per pupil expenditures were only $2,133 in 1983/84 (41st in the US), South Carolina's educational funding is higher in relation to per capita income than that of most other states. As of 1980, more than 50% of all residents 25 years or older had completed high school and 13.4% had attended college.

In 1982/83 there were 1,133 public schools, enrolling 582,804 pupils, and 253 private schools, with 43,246 pupils. During the same year there were 124,376 vocational pupils, 84,795 handicapped students, and 77,499 persons enrolled in adult education programs. Enrollment of white students in private schools has expanded since the 1960s, partly in reaction to public school desegregation. South Carolina ranked 15th in pupil/teacher ratio in public schools with 18.8 pupils per teacher in 1983.

Higher education institutions in 1982/83 enrolled 96,484 students, of whom 75% were attending public institutions. The state has three major universities: the University of South Carolina, with 24,296 students (1983/84) at its main campus in Columbia; Clemson University (12,459 students), at Clemson; and the Medical University of South Carolina (2,254 students), in Charleston. In addition, there are 6 four-year state colleges, as well as 9 four-year and two-year branches of the University of South Carolina. The state also has 21 four-year private colleges and universities, of which only Bob Jones University in Greenville enrolled more than 4,000 students in 1983/84; most are church-affiliated. The Lutheran Theological Southern Seminary in Columbia is the only major private graduate institution. There were 8 private junior colleges, with a total enrollment of 4,702, in 1982/83. South Carolina has an extensive technical education system, supported by both state and local funds. There were 16 such institutions and 2 special schools, with a total enrollment of 160,458, in 1983/84. Tuition grants are offered for needy South Carolina students enrolled in private colleges in the state.

42ARTS

South Carolina's three major centers for the visual arts are the Gibbes Art Gallery in Charleston, the Columbia Museum of Art and Science, and the Greenville County Museum of Art. Local theater groups in the larger municipalities produce five or six plays a year; Columbia's Town Theater claims to be the nation's oldest continuous community playhouse. Perhaps South Carolina's best-known musical event is the Spoleto Festival—held annually in Charleston during May and June and modeled on the Spoleto Festival in Italy—at which artists of international repute perform in original productions of operas and dramas. The South Carolina Arts Commission, created in 1967, has developed apprenticeship programs in which students learn from master artists.

43LIBRARIES AND MUSEUMS

Public libraries in South Carolina had a combined book stock of 4,435,283 volumes and a total circulation of 9,636,352 in 1982/83. The State Library in Columbia supervises the 39 county and regional libraries and also provides reference and research services for the state government. The University of South Carolina and Clemson University libraries, with 1,707,877 and 1,110,738 volumes respectively, have the most outstanding academic collections. Special libraries are maintained by the South Carolina Historical Society in Charleston, the Department of Archives and History in Columbia, and the South Carolina Society at the University of South Carolina.

There are 97 museums and historic sites, notably the Charleston Museum (specializing in history, natural history, and anthropology); the Citadel Archives-Museum, also in Charleston; and the University of South Carolina McKissick Museums (with silver, lapidary, and military collections) in Columbia. Charleston is also famous for its many old homes, streets, churches, and public facilities; at the entrance to Charleston Harbor stands Ft. Sumter, where the Civil War began. Throughout the state, numerous battle sites of the American Revolution have been preserved; many antebellum plantation homes have been restored, especially in the low country. Restoration projects have proceeded in Columbia and Charleston, where the restored Exchange Building, dating to the Revolutionary War, was opened to the public in 1981.

Among the state's best-known botanical gardens are the Cypress, Magnolia, and Middleton gardens in the Charleston area. Edisto Garden in Orangeburg is renowned for its azaleas and roses, and Brookgreen Gardens near Georgetown displays a wide variety of plants, animals, and sculpture.

44COMMUNICATIONS

In 1980, 87% of South Carolina's 1,029,981 occupied housing units had telephones. The state had 187 commercial radio stations (111 AM, 76 FM) and 14 commercial television stations in 1984. South Carolina has one of the most highly regarded educational television systems in the nation, with 10 stations serving the public schools, higher education institutions, state agencies, and the general public through a multichannel closed-circuit network and seven open channels. In 1984, 68 cable systems broadcast television programming to 398,278 South Carolinians in 210 communities.

45PRESS

Of the leading morning newspapers still published in South Carolina, the *Charleston News and Courier* (with a paid circulation of 70,914 in 1984) was founded in 1803, the *Spartanburg Herald* (47,184) in 1872, the *Greenville News* (86,582) in 1874, and the *State of Columbia* (108,288) in 1891. Overall, as of 1984, South Carolina had 8 morning newspapers with a combined circulation of 427,365, 9 evening dailies with 183,173, and 9 Sunday newspapers with 567,654.

46ORGANIZATIONS

The 1982 US Census of Service Industries counted 620 organizations in South Carolina, including 139 business associations; 319 civic, social, and fraternal associations; and 3 educational, scientific, and research associations. National organizations with headquarters in the state include the Association of Social and Behavioral Scientists, in Orangeburg, and the International Studies Association and the US Collegiate Sports Council, in Columbia.

47TOURISM, TRAVEL, AND RECREATION

Tourists spent $2.6 billion in South Carolina during 1983. The increasingly important tourism industry provided an estimated 69,300 jobs. About 75% of tourist revenue is spent by vacationers in Charleston and at the Myrtle Beach and Hilton Head Island resorts. The Cowpens National Battlefield and the Ft. Sumter and Kings Mountain national military sites drew 565,435 visitors in 1984; the 40 state parks attracted a total of 9,186,276 visitors in 1983/84. There were 422,403 licensed fishermen and 207,648 licensed hunters in 1982/83.

48SPORTS

There are no major professional sports teams in South Carolina, and wagering on sports events is illegal. Several steeplechase horse races are held annually in Camden, and important professional golf and tennis tournaments are held at Hilton Head Island.

In collegiate football, the Clemson Tigers won the AP and UPI polls in 1981; the University of South Carolina and South Carolina State also have football programs. Fishing, waterskiing, and sailing are popular participant sports. Annually, on Labor Day, the Southern 500 stock-car race is held in Darlington.

49FAMOUS SOUTH CAROLINIANS

Many distinguished South Carolinians made their reputations outside the state. Andrew Jackson (1767–1845), the 7th US president, was born in a border settlement probably inside present-day South Carolina, but studied law in North Carolina before establishing a legal practice in Tennessee. Identified more closely with South Carolina is John C. Calhoun (1782–1850), vice president from 1825 to 1832; Calhoun also served as US senator and was a leader of the South before the Civil War.

John Rutledge (1739–1800), the first governor of the state and a leader during the American Revolution, served a term as US chief justice but was never confirmed by the Senate. Another Revolutionary leader, Charles Cotesworth Pinckney (1746–1825), was also a delegate to the US constitutional convention. A strong Unionist, Joel R. Poinsett (1779–1851) served as secretary of war and as the first US ambassador to Mexico; he developed the poinsettia, named after him, from a Mexican plant. Benjamin R. Tillman (1847–1918) was governor, US senator, and leader of the populist movement in South Carolina. Bernard M. Baruch (1870–1965), an outstanding financier, statesman, and adviser to presidents, was born in South Carolina. Another presidential adviser, James F. Byrnes (1879–1972), also served as US senator, associate justice of the Supreme Court, and secretary of state. The state's best-known recent political leader is J(ames) Strom Thurmond (b.1902), who ran for the presidency as a States' Rights Democrat ("Dixiecrat") in 1948, winning 1,169,134 popular votes and 39 electoral votes, and has served in the Senate since 1955.

Famous military leaders native to the state are the Revolutionary War General Francis Marion (1732?–95), known as the Swamp Fox, and James Longstreet (1821–1904), a Confederate lieutenant general during the Civil War. Mark W. Clark (b.New York, 1896–1984), US Army general and former president of the Citadel, lived in South Carolina after 1954. General William C. Westmoreland (b.1914) was commander of US forces in Viet-Nam.

Notable in the academic world are Francis Lieber (b.Germany, 1800–1872), a political scientist who taught at the University of South Carolina and, later, Columbia University in New York City, and wrote for the US the world's first comprehensive code of military laws and procedures; Mary McLeod Bethune (1875–1955), founder of Bethune-Cookman College in Florida and of the National Council of Negro Women; John B. Watson (1878–1958), a pioneer in behavioral psychology; and Charles H. Townes (b.1915), awarded the Nobel Prize in physics in 1964. South Carolinians prominent in business and the professions include architect Robert Mills (1781–1855), who designed the Washington Monument and many other buildings; William Gregg (b.Virginia, 1800–1867), a leader in establishing the textile industry in the South; David R. Coker (1870–1938), who developed many varieties of pedigreed seed; and industrial builder Charles E. Daniel (1895–1964), who helped bring many new industries to the state.

South Carolinians who made significant contributions to literature include William Gilmore Simms (1806–70), author of nearly 100 books; Julia Peterkin (1880–1961), who won the Pulitzer Prize for *Scarlet Sister Mary;* DuBose Heyward (1885–1940), whose novel *Porgy* was the basis of the folk opera *Porgy and Bess;* and James M. Dabbs (1896–1970), a writer who was also a leader in the racial integration movement.

Entertainers born in the state include singer Eartha Kitt (b.1928) and jazz trumpeter John Birks "Dizzy" Gillespie (b.1917). Tennis champion Althea Gibson (b.1927) is another South Carolina native.

50BIBLIOGRAPHY

Barry, John M. *Natural Vegetation of South Carolina.* Columbia: University of South Carolina Press, 1979.

Cauthen, Charles E. *South Carolina Goes to War, 1860–65.* Chapel Hill: University of North Carolina Press, 1950.

Jones, Lewis P. *South Carolina: A Synoptic History for Laymen.* Orangeburg, S.C.: Sandlapper Publishing, 1979.

Lander, Ernest M. *A History of South Carolina, 1856–1960.* Columbia: University of South Carolina Press, 1970.

Sirmans, M. Eugene. *Colonial South Carolina: A Political History, 1663–1763.* Chapel Hill: University of North Carolina Press, 1966.

South Carolina, State of. Division of Research and Statistical Services. *South Carolina Statistical Abstract, 1984.* Columbia, 1984.

Taylor, Rosser H. *Ante-Bellum South Carolina: A Social and Cultural History.* New York: Da Capo Press, 1970.

Wallace, David Duncan. *History of South Carolina.* 4 vols. New York: American Historical Society, 1934.

Wood, Peter H. *Black Majority: Negroes in Colonial South Carolina from 1670 through the Stono Rebellion.* New York: Knopf, 1974.

Wright, Louis B. *South Carolina: A Bicentennial History.* New York: Norton, 1976.

SOUTH DAKOTA

State of South Dakota

ORIGIN OF STATE NAME: The state was formerly the southern part of Dakota Territory; *dakota* is a Sioux word meaning "friend." **NICKNAME:** The Coyote State. (Also: The Sunshine State.) **CAPITAL:** Pierre. **ENTERED UNION:** 2 November 1889 (40th). **SONG:** "Hail, South Dakota." **MOTTO:** Rule. **COAT OF ARMS:** Beneath the state motto, the Missouri River winds between hills and plains; symbols representing mining (a smelting furnace and hills), commerce (a steamboat), and agriculture (a man plowing, cattle, and a field of corn) complete the scene. **FLAG:** The state seal, centered on a white or light-blue field and encircled by a serrated sun, is surrounded by the words "South Dakota" above and "The Sunshine State" below. **OFFICIAL SEAL:** The words "State of South Dakota Great Seal 1889" encircle the arms. **ANIMAL:** Coyote. **BIRD:** Chinese ring-necked pheasant. **FLOWER:** Pasqueflower. **TREE:** Black Hills spruce. **GEM:** Fairburn agate. **MINERAL:** Rose quartz. **GRASS:** Western wheatgrass. **LEGAL HOLIDAYS:** New Year's Day, 1 January; Birthday of Martin Luther King, Jr., 3d Monday in January; Washington's Birthday, 3d Monday in February; Memorial Day, last Monday in May; Independence Day, 4 July; Labor Day, 1st Monday in September; Columbus Day, 2d Monday in October; Veterans Day, 11 November; Thanksgiving Day, 4th Thursday in November; Christmas Day, 25 December. **TIME:** 6 AM CST = noon GMT; 5 AM MST = noon GMT.

¹LOCATION, SIZE, AND EXTENT

Situated in the western north-central US, South Dakota ranks 16th in size among the 50 states.

The state has a total area of 77,116 sq mi (199,730 sq km), comprising 75,952 sq mi (196,715 sq km) of land and 1,164 sq mi (3,015 sq km) of inland water. Shaped roughly like a rectangle with irregular borders on the E and SE, South Dakota extends about 380 mi (610 km) E–W and has a maximum N–S extension of 245 mi (394 km).

South Dakota is bordered on the N by North Dakota; on the E by Minnesota and Iowa (with the line in the NE passing through the Bois de Sioux River, Lake Traverse, and Big Stone Lake, and in the SE through the Big Sioux River); on the S by Nebraska (with part of the line formed by the Missouri River and Lewis and Clark Lake); and on the W by Wyoming and Montana.

The total boundary length of South Dakota is 1,316 mi (2,118 km). The state's geographic center is in Hughes County, 8 mi (13 km) NE of Pierre. The geographic center of the US, including Alaska and Hawaii, is at 44°58′N, 103°46′W, in Butte County, 17 mi (27 km) W of Castle Rock.

²TOPOGRAPHY

The eastern two-fifths of South Dakota is prairie, belonging to the Central Lowlands. The western three-fifths falls within the Missouri Plateau, part of the Great Plains region; the High Plains extend into the southern fringes of the state. The Black Hills, an extension of the Rocky Mountains, occupy the southern half of the state's western border; the mountains, which tower about 4,000 feet (1,200 meters) over the neighboring plains, include Harney Peak, at 7,242 feet (2,207 meters) the highest point in the state. East of the southern Black Hills are the Badlands, a barren, eroded region with extensive fossil deposits. South Dakota's lowest elevation, 962 feet (293 meters), is at Big Stone Lake, in the northeastern corner.

Flowing south and southeast, the Missouri River cuts a huge swath through the heart of South Dakota before forming part of the southeastern boundary. Tributaries of the Missouri include the Grand, Cheyenne, Bad, Moreau, and White rivers in the west and the James, Vermillion, and Big Sioux in the east. The Missouri River itself is controlled by four massive dams—Gavins

Point, Ft. Randall, Big Bend, and Oahe—which provide water for irrigation, flood control, and hydroelectric power. Major lakes in the state include Traverse, Big Stone, Lewis and Clark, Francis Case, and Oahe.

³CLIMATE

South Dakota has an interior continental climate, with hot summers, extremely cold winters, high winds, and periodic droughts. The normal January temperature is 12°F (−11°C); the normal July temperature, 74°F (23°C). The record low temperature is −58°F (−50°C), set at McIntosh on 17 February 1936; the record high, 120°F (49°C), at Gannvalley on 5 July 1936.

Normal annual precipitation averages 24 in (61 cm) in Sioux Falls in the southeast, decreasing to less than 13 in (33 cm) in the northwest. Sioux Falls receives an average of 39 in (99 cm) of snow per year.

⁴FLORA AND FAUNA

Oak, maple, beech, birch, hickory, and willow all are represented in South Dakota's forests, while thickets of chokecherry, wild plum, gooseberry, and currant are found in the eastern part of the state. Pasqueflower (*Anemone ludoviciana*) is the state flower; other wild flowers are beardtongue, bluebell, and monkshood.

Familiar native mammals are the coyote (the state animal), porcupine, raccoon, bobcat, white-tailed and mule deer, white-tailed jackrabbit, and black-tailed prairie dog. Nearly 300 species of birds have been identified; the sage grouse, bobwhite quail, and ring-necked pheasant are leading game birds. Trout, catfish, pike, bass, and perch are fished for sport. South Dakota's list of threatened animals includes the river otter, mountain lion, northern swift fox, black bear, peregrine falcon, longnose sucker, brown snake, and Blandings turtle. The black-footed ferret, interior least tern, and pearl dace are endangered.

⁵ENVIRONMENTAL PROTECTION

Agencies concerned with natural resources and environmental protection include the Department of Game, Fish, and Parks and the Department of Water and Natural Resources. The Health Department deals with solid waste disposal and air pollution.

⁶POPULATION

South Dakota ranked 45th in the US with a 1980 census population of 690,768; the 1985 population was estimated at 705,027, representing a 2% increase over 1980. The Census Bureau estimates that the state will have 698,500 residents in 1990. The average population density in 1985 was 9.3 per sq mi (3.6 per sq km).

Only 16% of all South Dakotans lived in metropolitan areas in 1983, the lowest percentage in the US except for Wyoming. The leading cities as of 1984 were Sioux Falls, with 87,776 residents; Rapid City, 49,146; and Aberdeen, 25,764.

[7] ETHNIC GROUPS
According to the 1980 census, South Dakota's American Indian population totaled 45,081. Many lived on the 5,099,000 acres (2,063,500 hectares) of Indian lands in 1982, but Rapid City also had a large Indian population. As of 1980, the state had 2,144 black Americans and 1,728 Asian-Pacific peoples. Of the South Dakotans who reported a single ancestry in the 1980 census, 26% listed German, 7% Norwegian, 4% English, 3% Irish, and 2% Dutch. In the same year, 10,000 South Dakotans—1.4% of the population—were foreign-born.

[8] LANGUAGES
Despite hints given by such place-names as Dakota, Oahe, and Akaska, English has borrowed little from the language of the Sioux still living in South Dakota. *Tepee* is such a loanword, and *tado* (jerky) is heard near Pine Ridge. South Dakota English is transitional between the Northern and Midland dialects. Diffusion throughout the state is apparent, but many terms contrast along a curving line from the southeast to the northwest corner.

In 1980, 604,198 South Dakotans—92% of the resident population 3 years of age or older—spoke only English at home. Other languages spoken at home included:

German	22,305	Czech	1,841
Spanish	3,057	French	1,060
Norwegian	2,742	Dutch	722

[9] RELIGIONS
Leading Protestant denominations in 1980 were the American Lutheran Church, with 110,152 adherents; United Methodist, 46,234; and Lutheran Church–Missouri Synod, 33,773. The state had 134,019 Roman Catholics and an estimated 610 Jews in 1984.

[10] TRANSPORTATION
The Burlington Northern, Chicago and Northwestern, Soo Line, and Chicago, Milwaukee, St. Paul and Pacific railroads operated 2,022 mi (3,254 km) of Class I track in South Dakota in 1983. Highways, streets, and roads covered 74,063 mi (119,193 km) in 1983, when the state had 629,458 registered motor vehicles and 482,439 licensed drivers. There were a total of 165 airfields, of which Joe Foss Field at Sioux Falls was the most active, with 9,586 aircraft departures in 1983.

[11] HISTORY
People have lived in what is now South Dakota for at least 25,000 years. The original inhabitants, who hunted in the northern Great Plains until about 5000 BC, were the first of a succession of nomadic groups, followed by a society of semisedentary mound builders. After them came the prehistoric forebears of the modern riverine groups—Mandan, Hidatsa, and Arikara—who were found gathering, hunting, farming, and fishing along the upper Missouri River by the first European immigrants. These groups faced no challenge until the Sioux, driven from the Minnesota woodlands, began to move westward during the second quarter of the 18th century, expelling all other Native American groups from South Dakota by the mid-1830s.

Significant European penetration of South Dakota followed the Lewis and Clark expedition of 1804-6. White men came to assert US sovereignty, to negotiate Indian treaties, to "save Indian souls," and to traffic in hides and furs. Among the most important early merchants were Manuel Lisa, who pressed up the Missouri from St. Louis, and Pierre Chouteau, Jr., whose offices in St. Louis dominated trade on both the upper Mississippi and upper

Missouri rivers from 1825 until his death in 1865, by which time all major sources of hides and furs were exhausted, negotiations for Indian land titles were in progress, and surveyors were preparing ceded territories for non-Indian settlers.

The Dakota Territory, which included much of present-day Wyoming and Montana as well as North and South Dakota, was established in 1861, with headquarters first at Yankton (1861-83) and later at Bismarck (1883-89). The territory was reduced to just the Dakotas in 1868; six years later, a gold rush brought thousands of prospectors and settlers to the Black Hills. South Dakota emerged as a state in 1889, with the capital in Pierre. Included within the state were nine Indian reservations, established, after protracted negotiations and three wars with the Sioux, by Indian Office personnel. Five reservations were established west of the Missouri for the Teton and Yanktonai Sioux, and four reserves east of the Missouri for the Yankton and several Isanti Sioux tribes. Sovereignty was thus divided among Indian agents, state officials, and tribal leaders, a division that did not always make for efficient government.

Through the late 19th and early 20th centuries, South Dakotans had limited economic opportunities, for they depended mainly on agriculture. Some 30,000 Sioux barely survived on farming and livestock production, supplemented by irregular government jobs and off-reservation employment. The 500,000 non-Indians lived mainly off cattle-breeding operations west of the Missouri, cattle-feeding enterprises and small grain sales east of the Missouri, mineral production (especially gold) in the Black Hills, and various service industries at urban centers throughout South Dakota.

The period after World War I saw extensive road building, the establishment of a tourist industry, and efforts to subdue and harness the waters of the Missouri. Like other Americans, South Dakotans were helped through the drought and depression of the 1930s by federal aid. Non-Indians were assisted by food relief, various work-relief programs, and crop-marketing plans, while Indians enjoyed an array of federal programs often called the "Indian New Deal." The economic revival brought about by World War II persisted into the postwar era. Rural whites benefited from the mechanization of agriculture, dam construction along the Missouri, rural electrification, and arid-land reclamation. Federal programs were organized for reservation Indians, relocating them in urban centers where industrial jobs were available, establishing light industries in areas already heavily populated by Indians, and improving education and occupational opportunities on reservations.

Meanwhile, the Sioux continued to bring their historic grievances to public attention. For 70 days in 1973, some 200 armed Indians occupied Wounded Knee, on the Pine Ridge Reservation, where hundreds of Sioux had been killed by US cavalry 83 years earlier. In 1980, reviewing one of several land claims brought by the Sioux, the US Supreme Court upheld compensation of $105 million for land in the Black Hills taken from the Indians by the federal government in 1877. But members of the American Indian Movement (AIM) opposed this settlement and demanded the return of the Black Hills to the Sioux. The economic plight of South Dakota's Indians worsened during the early 1980s after the federal government reduced job training programs; unemployment on Indian reservations was above 80% in 1984.

In sharp contrast, the state economy showed strength under the direction of Republican Governor William Janklow, elected in 1978 and reelected in 1982 with 71% of the vote. Janklow, noted for his strong opposition to Indian claims, developed the state's water resources, revived railroad transportation, and attracted new industry to South Dakota, including Citicorp, the largest bank-holding company in the US, which set up a credit-card operation in Sioux Falls and bought controlling interest in the American State Bank of Rapid City.

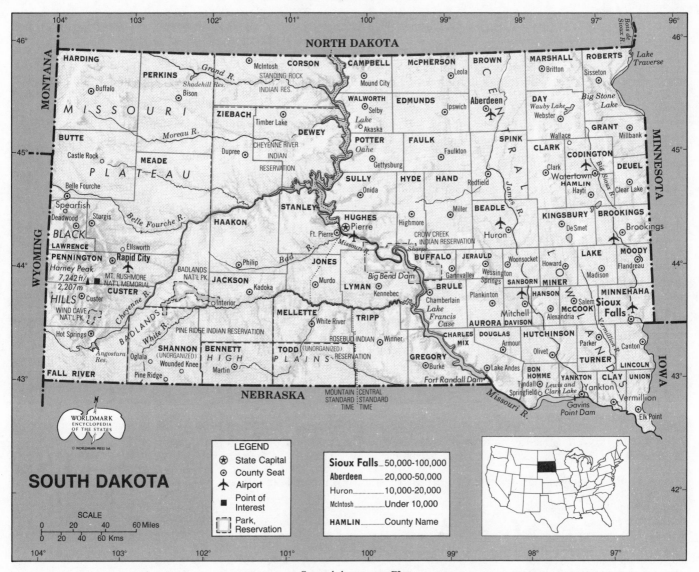

SOUTH DAKOTA

LEGEND
⊛ State Capital
⊙ County Seat
✈ Airport
■ Point of Interest
⬚ Park, Reservation

Sioux Falls ___ 50,000-100,000
Aberdeen ___ 20,000-50,000
Huron ___ 10,000-20,000
McIntosh ___ Under 10,000
HAMLIN ___ County Name

SCALE
0 20 40 60 Miles
0 20 40 60 Kms

See end sheet maps: F2.

LOCATION: 42°28′47″ to 45°56′41″N; 96°26′11″ to 104°03′10″W. **BOUNDARIES:** North Dakota line, 361 mi (581 km); Minnesota line, 184 mi (296 km); Iowa line, 143 mi (230 km); Nebraska line, 423 mi (681 km); Wyoming line, 138 mi (222 km); Montana line, 67 mi (108 km).

South Dakota Presidential Vote by Major Political Parties, 1948–84

YEAR	ELECTORAL VOTE	SOUTH DAKOTA WINNER	DEMOCRAT	REPUBLICAN
1948	4	Dewey (R)	117,653	129,651
1952	4	*Eisenhower (R)	90,426	203,857
1956	4	*Eisenhower (R)	122,288	171,569
1960	4	Nixon (R)	128,070	178,417
1964	4	*Johnson (D)	163,010	130,108
1968	4	*Nixon (R)	118,023	149,841
1972	4	*Nixon (R)	139,945	166,476
1976	4	Ford (R)	147,068	151,505
1980	4	*Reagan (R)	103,855	198,343
1984	3	*Reagan (R)	116,113	200,267

*Won US presidential election.

12 STATE GOVERNMENT

South Dakota is governed by the constitution of 1889, which had been amended 92 times as of December 1983. The legislature consists of a 35-seat senate and 70-seat house of representatives, all of whose members serve two-year terms. Chief executive officials are the governor, lieutenant governor, secretary of state, attorney general, treasurer, auditor, and commissioner of school and public lands, all of them elected for four-year terms. Voters must be US citizens and at least 18 years of age.

13 POLITICAL PARTIES

For the most part, South Dakota has voted Republican in presidential elections, even when native-son George McGovern was the Democratic candidate in 1972. Conservatism runs strong at the local level, although between the two world wars, populist groups gained a broad agrarian following. South Dakotans chose Ronald Reagan in the 1980 presidential election. At the same time, Republican James Abdnor denied George McGovern's bid for a fourth term in the US Senate. In 1984, President Reagan again carried the state, with 63% of the vote; Republican Senator Larry Pressler was reelected, with 74% of the vote.

As of 1985, South Dakota had a female secretary of state, Republican Alice Kundert. That year, 15 women served in the state legislature.

14 LOCAL GOVERNMENT
As of 1982, South Dakota had 1,767 units of local government, including 64 counties, 312 municipalities, 996 townships, 196 school districts, and 199 special districts. Typical county officials include a treasurer, auditor, state's attorney, sheriff, register of deeds, and clerk of courts.

15 STATE SERVICES
The Department of Education and Cultural Affairs oversees all elementary, secondary, higher, vocational, and cultural education programs.

The Department of Social Services administers a variety of welfare programs, the Department of Labor aids the unemployed and underemployed, and the Department of Vocational Rehabilitation serves disabled South Dakotans. Special agencies within the executive branch include the Office of Indian Affairs and Office of Energy Policy.

16 JUDICIAL SYSTEM
South Dakota has a supreme court with 5 justices, and eight circuit courts with 36 judges; all are elected on a nonpartisan ballot to staggered eight-year terms.

In 1983, South Dakota's crime index total—2,548 crimes per 100,000 inhabitants—was lower than that of every other state except West Virginia.

17 ARMED FORCES
Ellsworth Air Force Base, near Rapid City, with 6,223 military personnel in 1984, is the state's only defense installation. South Dakota firms received $46.6 million in federal defense contracts in 1983/84. As of 30 September 1983, about 80,000 veterans were living in the state, including 2,000 from World War I, 29,000 from World War II, 16,000 from the Korean conflict, and 22,000 from the Viet-Nam era. Veterans' benefits totaled $122 million during 1983.

South Dakota's National Guard numbered 4,878 as of early 1985. In 1983, state and local police and corrections officers numbered 1,907.

18 MIGRATION
Since the 1930s, more people have left South Dakota than have settled in the state. Between 1940 and 1983, the net loss from migration totaled 321,000.

19 INTERGOVERNMENTAL COOPERATION
South Dakota participates in the Belle Fourche River Compact (with Wyoming), the Interstate Oil and Gas Compact, and the Western Interstate Energy Compact, among other organizations; there are, in addition, boundary compacts with Minnesota and Nebraska. In 1983/84, South Dakota received almost $436 million in federal aid, of which $15 million was general revenue sharing.

20 ECONOMY
Agriculture dominates South Dakota's economy. Grains and livestock are the main farm products, and processed foods and farm equipment are leading manufactured items. Mining and tourism are also important.

21 INCOME
Per capita income in 1983 was $9,704, ranking South Dakota 38th among the 50 states. Total personal income reached $6.8 billion, representing a growth of 3% since 1980. Median family income was $15,993 (48th in the US) in 1983, when 112,739 South Dakotans (16.9% of the population), many of them Indians, were below the federal poverty level.

22 LABOR
The state's civilian labor force numbered 346,000 in 1984, of whom 331,000 were employed and 15,000, or 4.3%, were unemployed. A federal survey in March 1983 revealed the following nonfarm employment pattern in South Dakota:

	ESTABLISH-MENTS	EMPLOYEES	ANNUAL PAYROLL ('000)
Agricultural services, forestry, fishing	236	824	$ 9,474
Mining	59	2,381	59,015
Contract construction	1,712	6,878	123,891
Manufacturing	742	23,531	416,719
Transportation, public utilities	1,005	11,644	227,918
Wholesale trade	1,826	15,489	239,862
Retail trade	5,535	43,601	351,443
Finance, insurance, real estate	1,636	11,765	185,281
Services	5,058	48,119	512,284
Other	747	949	10,813
TOTALS	18,556	165,181	$2,136,700

Among the workers not covered by this survey were government employees (55,000 in 1984) and agricultural workers (43,000).

In 1983 there were 207 labor unions in South Dakota; 35,000 South Dakota workers belonged to labor unions in 1980. The state has a right-to-work law.

23 AGRICULTURE
South Dakota ranked 24th among the 50 states in 1983 in agricultural income, with receipts of $2.6 billion. In 1984 there were an estimated 37,000 farms and ranches in the state, covering about 44 million acres (18 million hectares).

Leading crops and their values during 1983 were hay, 7.8 million tons, $328 million; wheat, 89.7 million bushels, $327.9 million; corn for grain, 103.4 million bushels, $320.5 million; soybeans, 24.7 million bushels, $196.9 million; oats, 79.2 million bushels, $126.7 million; and barley, 23.1 million bushels, $51.9 million.

24 ANIMAL HUSBANDRY
The livestock industry is of great importance in South Dakota, particularly in the High Plains. In late 1984 there were 4,160,000 cattle and calves, 1,600,000 hogs and pigs, and 639,000 sheep and lambs. Livestock products in 1984 included 1.8 billion lb of cattle and calves, 608.9 million lb of hogs, and 54.8 million lb of sheep and lambs. That year, 1.7 billion lb of milk were produced, along with 378 million eggs, 5 million lb of chickens, and 38 million lb of turkeys.

25 FISHING
Virtually all fishing is recreational. Commercial landings were worth only $579,000 in 1983.

26 FORESTRY
In 1980, South Dakota's forestlands covered 1,702,000 acres (689,000 hectares), including 1,467,000 acres (594,000 hectares) of commercial timberland. Over three-fourths of the state's woodlands were publicly owned or managed. Shipments of lumber and wood products totaled $92.2 million in 1982.

27 MINING
The overall value of the state's nonfuel mineral output in 1984 was only $227.5 million (33d in the US). South Dakota is the nation's 2d-leading producer of gold, after Nevada. Production of gold, found in the Black Hills, totaled 310,527 troy oz in 1984. That year, South Dakota quarries produced 5.1 million tons of sand and gravel and 3.8 million tons of stone. Some silver and clay is also extracted.

28 ENERGY AND POWER
In 1983, South Dakota had an installed electrical capacity of 2.3 million kw and produced 7.8 billion kwh of electricity, more than half of it sold to customers in other states. Over 70% of the power output came from hydroelectric sources and almost all the remainder from coal-fired plants. Utilities in South Dakota sold 25 trillion Btu of gas to 107,000 customers during 1982, with revenues of $108 million.

South Dakota has very modest fossil-fuel resources. Proved

petroleum reserves totaled 1,172,000 barrels in 1983, and lignite reserves were 336,100,000 tons; production of both resources was negligible in 1984, however.

29 INDUSTRY

Food and related products, nonelectrical machinery, lumber and wood products, and printing and publishing together account for more than 63% of all South Dakota manufacturing employment. The total value of shipments for all industries in 1982 was $3.1 billion, with food contributing nearly 40%. The following table shows value of shipments by selected industries in 1982:

Meat products	$1,174,900,000
Dairy products	292,300,000
Grain mill products	115,300,000
Construction and related machinery	73,000,000

30 COMMERCE

Wholesale trade in South Dakota totaled $4.9 billion in 1982. Of the state's $2.9 billion in retail trade that year, grocery stores contributed 19%; automobile dealers, 19%; gasoline service stations, 12%; eating and drinking places, 9%; department stores, 7%; and other establishments, 34%. The state's foreign agricultural exports were valued at $655 million in 1982 (20th in the US); manufactured exports were $185 million in 1981.

31 CONSUMER PROTECTION

The Division of Consumer Affairs of the Office of the Attorney General enforces South Dakota's Deceptive Trade Practices Act, prosecutes cases of fraud and other illegal activities, and, in cooperation with the Department of Commerce, mediates disputes between consumers and businesses.

32 BANKING

South Dakota in 1983 had 141 insured commercial banks with total assets of $10.4 billion and total deposits of $7.9 billion. In 1982 there were 18 savings and loan associations with total assets of $1.2 billion, outstanding mortgage loans of $828 million, and savings deposits of over $1 billion.

33 INSURANCE

During 1983, nine life insurance companies were licensed to do business in South Dakota. The 979,000 policies in force had a combined value of $14.2 billion. Benefit payments totaled $134.3 million in 1983, when the average family had $51,400 in coverage, 9% below the national norm. Automobile insurance accounted for 38% of the $279 million in premiums written by property and casualty insurance companies in 1983.

34 SECURITIES

Although South Dakota has no securities exchanges, New York Stock Exchange member firms had 20 sales offices and 103 registered representatives in the state during 1983. That year, some 91,000 South Dakotans owned shares of public corporations.

35 PUBLIC FINANCE

The governor must submit the annual budget to the state legislature by 1 December; the fiscal year begins the following 1 July. The legislature may amend the budget at will, but the governor has an item veto. The following table summarizes general revenues and expenditures for 1984/85 (estimated) and 1985/86 (recommended):

REVENUES	1984/85	1985/86
Sales and use tax	$186,900,000	$195,300,000
Tuition and fees	17,400,000	17,740,000
Insurance company tax	15,500,000	16,000,000
Cigarette tax	10,300,000	15,700,000
Investment income and interest	11,158,000	12,800,000
Bank franchise tax	11,000,000	11,500,000
Inheritance tax	10,300,000	10,000,000
Contractors' excise tax	12,400,000	9,900,000
Other receipts	48,720,000	56,571,000
TOTALS	$323,678,000	$345,511,000

EXPENDITURES	1984/85	1985/86
General budget bill	$202,929,000	$218,344,000
Other appropriations	120,749,000	127,167,000
TOTALS	$323,678,000	$345,511,000

Consolidated expenditures for 1981/82 (including federal aid) totaled $1.2 billion, of which 36% went for public education, 16% for highways, 9% for public welfare, 5% for health and hospitals, and 34% for other purposes. The total state and local government debt was $1.2 billion in 1982, or $1,692 per capita (28th among the 50 states).

36 TAXATION

South Dakota's per capita taxation is among the lowest in the US: state and local taxes per capita were $916 in 1981/82, for a rank of 40th; federal taxes per capita in 1982/83 were $1,893 for a rank of 44th. There is no personal or corporate income tax, and personal property taxes have been reduced since 1978. A state sales and use tax of 4% supplies more than two-thirds of South Dakota's general-fund receipts. Taxes are also levied on gasoline sales, alcoholic beverages, tobacco products, mineral severance, inheritances, insurance premiums, and other items.

Federal taxes in South Dakota totaled $1.3 billion in 1983. In 1982, South Dakotans filed 277,000 federal income tax returns and paid $522,000,000 in tax.

37 ECONOMIC POLICY

Efforts to attract industry to South Dakota and to broaden the state's economic base are under the jurisdiction of the Department of State Development, which had a recommended budget of $3.2 million in 1984/85. Among the advantages noted by the department are low taxes, the availability of community development corporations to finance construction of new facilities, various property tax relief measures, inventory tax exemptions, and a favorable labor climate in which work stoppages are few and union activity is limited by a right-to-work law. South Dakota is one of the few states to have enacted a statute of limitations on product liability—in this case, six years—a measure cited as further proof of the state's attempt to create an atmosphere conducive to manufacturing.

38 HEALTH

The birthrate and death rate in South Dakota were both above the national average in 1981; infant mortality was below the national norm in 1982. Death rates for heart disease, suicide, accidents, and early childhood diseases were above the national average.

In 1983, the state had 68 hospitals, with 5,730 beds; hospital personnel included 2,532 registered nurses and 891 licensed practical nurses. The average cost of hospital care in 1982 was $217 per day and $1,885 per stay, among the lowest of all the states. In 1978 there were 117 nursing and related care facilities, with 8,500 beds and 7,900 residents. South Dakota had 960 active physicians and 331 active dentists in 1982.

39 SOCIAL WELFARE

In 1983, 3.5% of South Dakotans received public aid, for a rank of 39th among the 50 states. Payments amounting to nearly $17 million in aid for dependent children went to 16,798 South Dakotans in 1982. In 1983, 120,387 state residents received Social Security benefits worth $516 million. Federally administered Supplemental Security Income paid $13.7 million to 7,663 South Dakotans. Other benefit programs in 1983 included $27 million for food stamps and $8 million for school lunches. In 1982, $20 million was paid in unemployment insurance and $16.7 million in workers' compensation.

40 HOUSING

The 1980 census counted 269,322 year-round housing units, of which 242,523 were occupied; 97% of the occupied units had full plumbing. Between 1981 and 1983, some 5,200 new housing units valued at $186 million were authorized.

41EDUCATION
As of 1980, 68% of South Dakotans 25 years of age or older were high school graduates and 14% had four or more years of college.

Fall 1984 enrollment in public schools totaled 122,818, 86,228 in grades K–8 and 36,590 in grades 9–12. Private school enrollment included 9,591 students in grades K–8 and 4,006 in grades 9–12. More than 35,000 students were enrolled in 20 higher educational institutions in fall 1983. There were 8 state-supported colleges and universities, of which the largest was the University of South Dakota, with 9,057 students; South Dakota State University had 7,981. In addition, the state had 12 private institutions of higher education.

42ARTS
The South Dakota State Fine Arts Council, located at Sioux Falls, and the South Dakota Committee on the Humanities, at Brookings, aid and coordinate arts and humanities activities throughout the state. Artworks and handicrafts are displayed at the Dacotah Prairie Museum (Aberdeen), South Dakota Memorial Art Center (Brookings), Sioux Indian Museum and Crafts Center (Rapid City), Civic Fine Arts Association (Sioux Falls), Sioux Falls Coliseum, and W. H. Over Museum (Vermillion). Symphony orchestras include the South Dakota Symphony in Sioux Falls and the Rapid City Symphony Orchestra.

43LIBRARIES AND MUSEUMS
In 1982, South Dakota had 105 public libraries with a combined total of 1,766,549 volumes and 3,358,349 circulation. Leading collections, each with more than 100,000 volumes, were those of South Dakota State University (Brookings), Northern State College and Alexander Mitchell Library (Aberdeen), Augustana College (Sioux Falls), the University of South Dakota (Vermillion), the South Dakota State Library (Pierre), and the Sioux Falls and Rapid City public libraries.

South Dakota has 214 museums and historic sites, including the Robinson Museum (Pierre), Siouxland Heritage Museums and Delbridge Museum of Natural History (Sioux Falls), and the Shrine to Music Museum (Vermillion). Badlands National Park and Wind Cave National Park also display interesting exhibits.

44COMMUNICATIONS
South Dakota had 398 post offices and 1,974 Postal Service employees in 1985. In 1980, 94% of South Dakota's 242,523 occupied housing units had telephones. Commercial broadcasters included 79 radio stations (36 AM, 43 FM) and 19 television stations. Cable television systems served 196,970 subscribers in 158 communities.

45PRESS
In 1984, South Dakota had 2 morning newspapers, with a daily circulation of 46,230; 10 evening papers, with a total of 119,335; and 4 Sunday papers, with 127,182. Leading newspapers included the *Rapid City Journal,* evenings 31,547, Sundays 34,430; and the Sioux Falls *Argus Leader,* mornings 42,733, Sundays 57,420.

46ORGANIZATIONS
The 1982 Census of Service Industries counted 620 organizations in South Dakota, including 139 business associations; 319 civic, social, and fraternal associations; and 3 educational, scientific, and research associations. Among the organizations headquartered in South Dakota are the National Buffalo Association (Custer) and the National Trail Council (Brookings).

47TOURISM, TRAVEL, AND RECREATION
Visitors spent an estimated $460 million in South Dakota in 1982. Most of the state's tourist attractions lie west of the Missouri River, especially in the Black Hills region. Mt. Rushmore National Memorial consists of the heads of four US presidents— George Washington, Thomas Jefferson, Abraham Lincoln, and Theodore Roosevelt—carved in granite in the mountainside. Wind Cave National Park and Jewel Cave National Monument are also in the Black Hills region. Just to the east is Badlands National Monument, consisting of fossil beds and eroded cliffs almost bare of vegetation. In 1982/83, South Dakota had 167,703 licensed hunters and 200,077 licensed fishermen.

48SPORTS
There are no major league sports teams in South Dakota. The University of South Dakota Coyotes and the Jackrabbits of South Dakota State both compete in the North Central Intercollegiate Athletic Conference. Skiing and hiking are popular in the Black Hills.

49FAMOUS SOUTH DAKOTANS
The only South Dakotan to win high elective office was Hubert H. Humphrey (1911–78), a native of Wallace who, after rising to power in Minnesota Democratic politics, served as US senator for 16 years before becoming vice president under Lyndon Johnson (1965–69).

Other outstanding federal officeholders from South Dakota were Newton Edmunds (1819–1908), second governor of the Dakota Territory; Charles Henry Burke (b.New York, 1861–1944), who as commissioner of Indian affairs improved education and health care for Native Americans; and Vermillion-born Peter Norbeck (1870–1936), a Progressive Republican leader, first while governor (1917–21) and then as US senator until his death. The son of a German-American father and a Brulé Indian mother, Benjamin Reifel (b.1906) was the first American Indian elected to Congress from South Dakota; he later served as the last US commissioner of Indian affairs. George McGovern (b.1922) served in the US Senate from 1963 through 1980; an early opponent of the war in Viet-Nam, he ran unsuccessfully as the Democratic presidential nominee in 1972.

Associated with South Dakota are several distinguished Indian leaders. Among them were Red Cloud (b.Nebraska, 1822–1909), an Oglala warrior; Spotted Tail (b.Wyoming, 1833?–1881), the Brulé chief who was a commanding figure on the Rosebud Reservation; Sitting Bull (1834–90), a Hunkpapa Sioux most famous as the main leader of the Indian army that crushed George Custer's Seventh US Cavalry at the Battle of the Little Big Horn (1876) in Montana; and Crazy Horse (1849?–1877), an Oglala chief who also fought at Little Big Horn.

Ernest Orlando Lawrence (1901–58), the state's only Nobel Prize winner, received the physics award in 1939 for the invention of the cyclotron. The business leader with the greatest personal influence on South Dakota's history was Pierre Chouteau, Jr. (b.Missouri, 1789–1865), a fur trader after whom the state capital is named.

South Dakota artists include George Catlin (b.Pennsylvania, 1796–1872), Karl Bodmer (1809–93), Harvey Dunn (1884–1952), and Oscar Howe (1915–83). Gutzon Borglum (b.Idaho, 1871–1941) carved the faces on Mt. Rushmore. The state's two leading writers are Ole Edvart Rölvaag (b.Norway, 1876–1931), author of *Giants in the Earth* and other novels, and Frederick Manfred (b.Iowa, 1912), a Minnesota resident who served as writer-in-residence at the University of South Dakota and has used the state as a setting for many of his novels.

50BIBLIOGRAPHY
Federal Writers' Project. *South Dakota: A Guide to the State.* New York: Somerset, n.d. (orig. 1938).

Kingsbury, George W. *History of Dakota Territory.* 2 vols. Chicago: Clarke, 1915.

Milton, John R. (ed.). *The Literature of South Dakota.* Vermillion: Dakota Press, 1976.

Milton, John R. *South Dakota: A Bicentennial History.* New York: Norton, 1977.

Parker, Watson. *Gold in the Black Hills.* Norman: University of Oklahoma Press, 1966.

Schell, Herbert. *History of South Dakota.* 3d ed. Lincoln: University of Nebraska Press, 1975.

Shaff, Howard and Audrey. *Six Wars at a Time.* Sioux Falls, S.D.: Center for Western Studies at Augustana College, 1985.

TENNESSEE

State of Tennessee

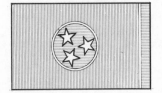

ORIGIN OF STATE NAME: Probably from Indian name *Tenase,* which was the principal village of the Cherokee. **NICKNAME:** The Volunteer State. **CAPITAL:** Nashville. **ENTERED UNION:** 1 June 1796 (16th). **SONGS:** "When It's Iris Time in Tennessee"; "The Tennessee Waltz"; "My Homeland, Tennessee"; "Rocky Top." **PUBLIC SCHOOL SONG:** "My Tennessee." **POET LAUREATE:** Richard M. "Pek" Gunn. **POEM:** "Oh Tennessee, My Tennessee." **FOLK DANCE:** Square dance. **MOTTO:** Agriculture and Commerce. **SLOGAN:** Tennessee— America at Its Best. **FLAG:** On a crimson field separated by a white border from a blue bar at the fly, three white stars on a blue circle edged in white represent the state's three main general divisions—East, Middle, and West Tennessee. **OFFICIAL SEAL:** The upper half consists of the word "Agriculture," a plow, a sheaf of wheat, a cotton plant, and the roman numeral XVI, signifying the order of entry into the Union; the lower half comprises the word "Commerce" and a boat. The words "The Great Seal of the State of Tennessee 1796" surround the whole. The date commemorates the passage of the state constitution. **WILD ANIMAL:** Raccoon. **BIRD:** Mockingbird. **CULTIVATED FLOWER:** Iris. **WILD FLOWER:** Passion flower. **TREE:** Tulip poplar. **GEM:** Tennessee pearl. **ROCKS:** Limestone, agate. **INSECTS:** Ladybug, firefly. **LEGAL HOLIDAYS:** New Year's Day, 1 January; Birthday of Martin Luther King, Jr., 3d Monday in January; Washington's Birthday, 3d Monday in February; Good Friday, March or April; Decoration Day, last Monday in May; Independence Day, 4 July; primary and county elections, 1st Thursday in August in even-numbered years; Labor Day, 1st Monday in September; Columbus Day, 2d Monday in October; General Election Day, 1st Tuesday after 1st Monday in November in even-numbered years; Veterans Day, 11 November; Thanksgiving Day, 4th Thursday in November; Christmas Day, 25 December. **TIME:** 7 AM EST = noon GMT; 6 AM CST = noon GMT.

¹LOCATION, SIZE, AND EXTENT

Situated in the eastern south-central US, Tennessee ranks 34th in size among the 50 states.

The total area of the state is 42,144 sq mi (109,152 sq km), of which land occupies 41,155 sq mi (106,591 sq km) and inland water 989 sq mi (2,561 sq km). Tennessee extends about 430 mi (690 km) E–W and 110 mi (180 km) N–S.

Tennessee is bordered on the N by Kentucky and Virginia; on the E by North Carolina; on the S by Georgia, Alabama, and Mississippi; and on the W by Arkansas and Missouri (with the line formed by the Mississippi River). The boundary length of Tennessee totals 1,306 mi (2,102 km). The state's geographic center lies in Rutherford County, 5 mi (8 km) NE of Murfreesboro.

²TOPOGRAPHY

Long, narrow, and rhomboidal, Tennessee is divided topographically into six major physical regions: the Unaka Mountains, the Great Valley of East Tennessee, the Cumberland Plateau, the Highland Rim, the Central Basin, and the Gulf Coastal Plain. In addition, there are two minor physical regions: the Western Valley of the Tennessee River and the Mississippi Flood Plains.

The easternmost region is the Unaka Mountains, part of the Appalachian chain. The Unakas actually include several ranges, the most notable of which is the Great Smoky Mountains. The region constitutes the highest and most rugged surface in the state and covers an area of about 2,600 sq mi (6,700 sq km). Several peaks reach a height of 6,000 feet (1,800 meters) or more; the tallest is Clingmans Dome in the Great Smokies, which rises to 6,643 feet (2,025 meters) and is the highest point in the state.

Lying due west of the Unakas is the Great Valley of East Tennessee. Extending from southwestern Virginia into northern Georgia, the Great Valley is a segment of the Ridge and Valley province of the Appalachian Highlands, which reach from New York into Alabama. This region, consisting of long, narrow ridges with broad valleys between them, covers more than 9,000 sq mi

(23,000 sq km) of Tennessee. Since the coming of the Tennessee Valley Authority (TVA) in 1933, the area has been dotted with artificial lakes and dams, which supply electric power and aid in flood control.

The Cumberland Plateau, which extends in its entirety from southern Kentucky into central Alabama, has an area of about 5,400 sq mi (14,000 sq km) in Middle Tennessee. The plateau is a region of contrasts, including both the Cumberland Mountains, which rise to a height of 3,500 feet (1,100 meters), and the Sequatchie Valley, the floor of which lies about 1,000 feet (300 meters) below the surface of the adjoining plateau.

The Highland Rim, also in Middle Tennessee, is the state's largest natural region, consisting of more than 12,500 sq mi (32,400 sq km) and encircling the Central Basin. The eastern section is a gently rolling plain some 1,000 feet (300 meters) lower than the Cumberland Plateau. The western part has an even lower elevation and sinks gently toward the Tennessee River.

The Central Basin, an oval depression with a gently rolling surface, has been compared to the bottom of an oval dish, of which the Highland Rim forms the broad, flat brim. With its rich soil, the region has attracted people from the earliest days of European settlement and is more densely populated than any other area in the state.

The westernmost of the major regions is the Gulf Coastal Plain. It embraces practically all of West Tennessee and covers an area of 9,000 sq mi (23,000 sq km). It is a broad plain, sloping gradually westward until it ends abruptly at the bluffs overlooking the Mississippi Flood Plains. In the northwest corner is Reelfoot Lake, the only natural lake of significance in the state, formed by a series of earthquakes in 1811 and 1812. The state's lowest point, 182 feet (55 meters) above sea level, is on the banks of the Mississippi in the southwest.

Most of the state is drained by the Mississippi River system. Waters from the two longest rivers—the Tennessee, with a total length of 652 mi (1,049 km), and the Cumberland, which is 687 mi (1,106 km) long—flow into the Ohio River in Kentucky and join the Mississippi at Cairo, Ill. Formed a few miles north of Knoxville by the confluence of the Holston and French Broad rivers, the Tennessee flows southwestward through the Great Valley into northern Alabama, then curves back into the state and flows northward into Kentucky. Other tributaries of the Tennessee are the Clinch, Duck, Elk, Hiwassee, and Sequatchie rivers. The Cumberland River rises in southeastern Kentucky, flows across central Tennessee, and then turns northward back into Kentucky; its principal tributaries are the Harpeth, Red, Obey, Caney Fork, and Stones rivers and Yellow Creek. In the western part of the state, the Forked Deer and Wolf rivers are among those flowing into the Mississippi, which forms the western border with Missouri and Arkansas.

³CLIMATE

Generally, Tennessee has a temperate climate, with warm summers and mild winters. However, the state's varied topography leads to a wide range of climatic conditions.

The warmest parts of the state, with the longest growing season, are the Gulf Coastal Plain, the Central Basin, and the Sequatchie Valley. In the Memphis area in the southwest, the average date of the last killing frost is 20 March, and the growing season is about 235 days. Memphis has an annual mean temperature of 62°F (17°C), 41°F (5°C) in January and 82°F (28°C) in July. In the Nashville area, the growing season lasts about 225 days. Nashville has an annual mean of 59°F (15°C), ranging from 38°F (3°C) in January to 80°F (27°C) in July. The Knoxville area has a growing season of 220 days. The city's annual mean temperature is 60°F (16°C), with averages of 41°F (5°C) in January and 78°F (26°C) in July. In some parts of the mountainous east, where the temperatures are considerably lower, the growing season is as short as 130 days. The record high temperature for the state is 113°F (45°C), set at Perryville on 9 August 1930; the record low, −32°F (−36°C), was registered at Mountain City on 30 December 1917.

Severe storms occur infrequently. The greatest rainfall occurs in the winter and early spring, especially March; the early fall months, particularly September and October, are the driest. Average annual precipitation is 49 in (124 cm) in Memphis and 46 in (117 cm) in Nashville. Snowfall varies and is more prevalent in East Tennessee than in the western section; Nashville gets about 11 in (28 cm) a year, Memphis only 6 in (15 cm).

⁴FLORA AND FAUNA

With its varied terrain and soils, Tennessee has an abundance of flora, including at least 150 kinds of native trees. Tulip poplar (the state tree), shortleaf pine, and chestnut, black, and red oaks are commonly found in the eastern part of the state, while the Highland Rim abounds in several varieties of oak, hickory, ash, and pine. Gum maple, black walnut, sycamore, and cottonwood grow in the west, and cypress is plentiful in the Reelfoot Lake area. In East Tennessee, rhododendron, mountain laurel, and wild azalea blossoms create a blaze of color in the mountains. More than 300 native Tennessee plants, including digitalis and ginseng, have been utilized for medicinal purposes. Glade cress, Duck River bladderpod, and American yellowwood are listed as threatened plants; Cumberland rosemary is classified as endangered.

Tennessee mammals, of which there were 81 species in 1984, include the raccoon (the state animal), white-tailed deer, black bear, bobcat, muskrat, woodchuck, opossum, and red and gray foxes; the European wild boar was introduced by sportsmen in 1912. As of 1984, 259 bird species resided in Tennessee. Bobwhite quail, ruffed grouse, mourning dove, and mallard duck are the most common game birds. The state's 56 amphibian species include numerous frogs, salamanders, newts, and lizards; 58 reptile species include three types of rattlesnake. Of the 286 fish species in Tennessee's lakes and streams, catfish, bream, bass, crappie, pike, and trout are the leading game fish.

Tennessee's Wildlife Resources Agency conducts an endangered and threatened species protection program. Among threatened species are the river otter, Cooper's hawk, Bewick's wren, northern pine snake, Tennessee cave salamander, blue sucker, amber darter, and silverjaw minnow. The snail darter, cited by opponents of the Tellico Dam, is probably Tennessee's most famous threatened species. The eastern cougar, gray and Indiana bats, Bachman's sparrow, Mississippi kite, lake sturgeon, Ohio River muskellunge, and Cumberland monkey face and Appalachian monkey face pearly mussels are on the endangered list.

⁵ENVIRONMENTAL PROTECTION

The first conservationists were agricultural reformers who, even before the Civil War, recommended terracing to conserve the soil and curtail erosion. Such conservation techniques as crop rotation and contour plowing were discussed at county fairs and other places where farmers gathered. In 1854, the legislature established the State Agricultural Bureau, which sought primarily to protect farmlands from floods. Nevertheless, soil erosion, flooding, and other conservation problems remained severe until relatively recent times.

Counties were authorized to construct levees as early as 1871, but effective flood control was not achieved until establishment of the Tennessee Valley Authority in 1933. Eroded areas were reforested with seedling trees, and cover crops were planted to hold the soil.

Today, the state Soil Conservation Committee, with the assistance of the state Department of Agriculture, is responsible for soil protection. The state Department of Conservation is responsible for development of state parks and forest preserves. The state Department of Health and Environment monitors water resources, solid waste management, air pollution, the effects of radiation on health, and the effects of surface mining.

The Strip Mine Law of 1967, as modified in 1972, requires permits for strip-mining and makes compulsory the restoration of the soil with a vegetative cover. The law extends to coal mining and to the mining of barite, clay, phosphate, and sand and gravel.

In the mid-1970s, a major controversy erupted over the then incomplete Tellico Dam, near Knoxville, and the proposed flooding of the Little Tennessee River. Opponents of the project filed suit under the Endangered Species Act of 1973 to protect the snail darter, a tiny perch that lived in the last free-flowing stretch of the river. Their suit, upheld in the US Supreme Court in 1978, stopped construction for a time, but in 1979, the dam's supporters in Congress succeeded in exempting it from the terms of all environmental legislation, and the project was completed. During the mid-1980s, the status of the snail darter had apparently improved enough for its classification to be upgraded from "endangered" to "threatened."

⁶POPULATION

Tennessee, with a population of 4,590,750, ranked 17th among the 50 states at the 1980 census. Estimates for 1985 showed Tennessee with a population of 4,723,332, yielding an average density of 115 persons per sq mi (44 per sq km). Projections to 1990 foresee a population of 5,072,600.

The first permanent white settlements in the state were made in the 1760s, when people from North Carolina and Virginia crossed the Unaka Mountains and settled in the fertile valleys. Between 1790 and 1800, the population increased threefold, from 35,690 to 105,600, and it doubled during each of the next two decades. After the Civil War, the population continued to increase, though at a slower rate, tripling between 1870 and 1970.

A pronounced urban trend became apparent after World War II. In 1960, for the first time in the state's history, census figures showed slightly more people living in urban than in rural areas. By 1980, 60% of all Tennesseans lived in towns and cities of 2,500 or more. Memphis is the state's largest city; in 1984, it had a population of 648,399. Nashville (Davidson County) had 462,450,

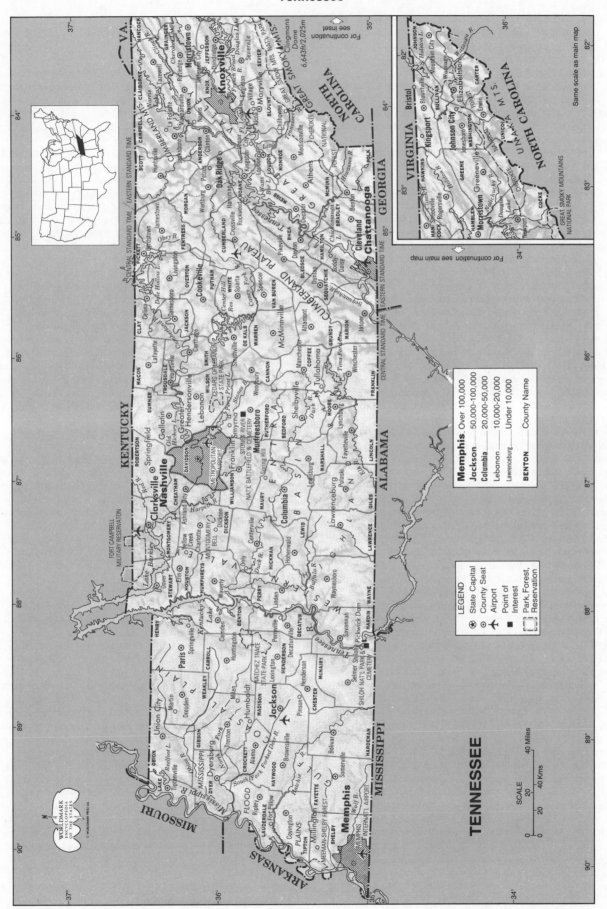

TENNESSEE

LEGEND

⊛ State Capital
⊙ County Seat
✈ Airport
■ Point of Interest
▢ Park, Forest, Reservation

Memphis Over 100,000
Jackson 50,000–100,000
Columbia 20,000–50,000
Lebanon 10,000–20,000
Lawrenceburg Under 10,000

BENTON County Name

SCALE

0   20   40 Miles

0   20   40 Kms

See endsheet maps: J3.

LOCATION: 34°58'59" to 36°40'41"N; 81°38'49" to 90°18'35" w. **BOUNDARIES:** Kentucky line, 346 mi (557 km); Virginia line, 127 mi (204 km); North Carolina line, 255 mi (410 km); Georgia line, 75 mi (121 km); Alabama line, 150 mi (242 km); Mississippi line, 123 mi (198 km); Arkansas line, 163 mi (262 km); Missouri line, 67 mi (108 km).

followed by Knoxville, 173,972, and Chattanooga, 164,400. The Memphis metropolitan area, including parts of Arkansas and Mississippi, had an estimated 934,600 residents in 1984, while metropolitan Nashville had 890,300.

7ETHNIC GROUPS

For nearly a century after the earliest white settlements, Tennessee was inhabited by three ethnoracial populations: whites of English and Scotch-Irish descent, Cherokee Indians, and black Americans. Settlers crossing the Appalachians met Indian resistance as early as the late 1700s. Eventually, however, nearly all the Cherokee were forced to leave; in 1980 there were only 5,012 Indians in the state.

Blacks, originally brought into the state as slaves to work in the cotton fields of West Tennessee, made up about 10% of the population in 1790. White Tennesseans were divided on the issue of slavery. The small farmers of the eastern region were against it, and in the late 1820s and 1830s there were more antislavery societies in Tennessee than in any other southern state except North Carolina. The planters and merchants of southwest Tennessee, however, linked their sentiments and interests with those of the proslavery planters of the Mississippi Valley. The introduction of the cotton gin gave impetus to the acquisition of more slaves; by 1840, blacks accounted for 26% of the population, and Memphis had become a major market for the shipment of black slaves to large plantations farther south.

Immediately after the Civil War, many blacks, now free, migrated from Virginia and North Carolina to East Tennessee to become farmers, artisans, and owners of small businesses. After 1880, however, the black proportion of the population declined steadily. In 1980 there were about 726,000 blacks in Tennessee, less than 16% of the total.

Descendants of European immigrants make up about half the population of Tennessee, the largest groups being of English and German descent. In 1980, only 48,000 residents—1.1% of the population—were foreign-born.

8LANGUAGES

White settlers found Tennessee inhabited by Cherokee Indians in the eastern mountains, Shawnee in most of the eastern and central region, and Chickasaw in the west—all of them speakers of Hokan-Siouan languages. Subsequently removed to Indian Territory, they left behind such place names as Chickamauga, Chattanooga, and Chilhowee, as well as Tennessee itself.

Tennessee English represents a mixture of North Midland and South Midland features brought into the northeastern and north-central areas, of South Midland and Southern features introduced by settlers from Virginia and the Carolinas, and of a few additional Southern terms in the extreme western fringe, to which they were carried from Mississippi and Louisiana. Certain pronunciations exhibit a declining frequency from the Appalachians to the Mississippi River, such as /r/ after a vowel in the same syllable, as in *form* and *short*, and a rounded /aw/ before /r/ in *arm* and *barbed*. Others occur statewide, such as the /ah/vowel in *forest* and *foreign*, *coop* and *Cooper* with the vowel of *book*, and simplification of the long /i/ vowel, so that *lice* sounds like *lass*. Common are such non-Northern terms as *wait on* (wait for), *pullybone* (along with Northern *wishbone*), *nicker* (neigh), *light bread* (white bread), and *snake feeder* (dragonfly), as well as *Jew's harp, juice harp,* and *French harp* (all for harmonica). In eastern Tennessee are found *goobers* (peanuts), *tote* (carry), *plum peach* (clingstone peach), *ash cake* (a kind of cornbread), *fireboard* (mantel), *redworm* (earthworm), *branch* (stream), and *peckerwood* (woodpecker). Appearing in western Tennessee are *loaf bread, cold drink* (soft drink), and *burlap bag*. In Memphis, a large, long sandwich is a *poorboy*.

In 1980, 4,306,831 Tennesseans 3 years old and over—98% of the population in that age group—spoke only English at home. Speakers of other languages were as follows:

Spanish	27,261	Chinese	2,666
German	12,435	Italian	2,440
French	12,379	Korean	2,306

9RELIGIONS

Baptist and Presbyterian churches were organized on the frontier soon after permanent settlements were made. Many divisions have occurred in both groups. The Cumberland Presbyterian Church, which spread into other states, was organized near Nashville in 1810 because of differences within the parent church. Both the Baptists and the Presbyterians divided over slavery. Methodist circuit riders arrived with the early settlers, and they quickly succeeded in attracting many followers. Controversies over slavery and other sectional issues also developed within the Methodist Church, and, as with the Baptists and Presbyterians, divisions emerged during the 1840s. The Methodists, however, were able to heal their differences and regroup. The United Presbyterian Church and the Presbyterian Church in the US finally ended their 122-year separation in 1983, reuniting to form the Presbyterian Church (USA).

Two other Protestant groups with large followings in the state had their origin on the Tennessee frontier in the first half of the 19th century: the Disciples of Christ and the Church of Christ. Both groups began with the followers of Thomas and Alexander Campbell and Barton W. Stone, among others, who deplored formal creeds and denominations and sought to return to the purity of early Christianity. As their numbers grew, these followers divided into Progressives, who supported missionary societies and instrumental music in church, and Conservatives, who did not. In 1906, a federal census of religions listed the Conservatives for the first time as the Church of Christ and the Progressives as the Disciples of Christ. The latter, now the Christian Church (Disciples of Christ), had 19,397 known adherents in 1980.

Protestant groups had a total of 2,369,724 known adherents in 1980. By far the largest group was the Southern Baptist Church, with 1,203,537 adherents. The United Methodists were second with 413,149 adherents, followed by the Churches of Christ, 226,320, and the Presbyterian Church in the US, 65,867. In 1984 there were 121,475 Roman Catholics and an estimated 18,465 Jews in the state.

Tennessee has long been considered part of the Bible Belt because of the influence of fundamentalist Protestant groups that insist upon the literal accuracy of the Bible.

10TRANSPORTATION

Memphis, Nashville, Knoxville, and Chattanooga are the focal points for rail, highway, water, and air transportation. All are located on important rivers and interstate highways, and all have airports served by the major airlines.

Railroad building began in Tennessee as early as the 1820s. During the 1850s, the basis for 20th-century rail transportation was laid: the Louisville and Nashville Railroad linked Tennessee to the northern states, and the Memphis and Charleston line established ties with the East Coast. In 1983, Tennessee had 2,785 mi (4,482 km) of Class I railroad track; railroad employees numbered 7,900. No east–west passenger trains operate in the state today (Amtrak serves Memphis on the Chicago–New Orleans route), but considerable freight is hauled.

The first roads—such as the Natchez Trace, which connected Nashville with the southwestern part of the state—often followed Indian trails. Many roads in the early 1800s were constructed by private individuals or chartered turnpike companies. The introduction of the automobile shortly after the beginning of the 20th century brought the development of modern roads and highways. After 1916, the federal government began to share the high cost of highway construction, and the 1920s were a decade of extensive road building.

In 1983, Tennessee had 71,851 mi (115,633 km) of rural roads and 11,938 mi (19,212 km) of municipal roads. The major interstate highway is I-40, crossing east–west from Knoxville to Nashville and Memphis. In 1983, 3,537,012 motor vehicles were registered in the state, and 2,933,197 Tennesseans held drivers' licenses.

The principal means of transportation during Tennessee's early history was water, and all the early settlements were built on or near streams. The introduction of steamboats on the Cumberland River in the early 19th century helped make Nashville the state's largest city and its foremost trading center. By midcentury, however, Memphis, on the Mississippi River, had surpassed Nashville in population and trade, largely because of cotton. Tennessee has about 1,000 mi (1,600 km) of navigable waterways. The completion in 1985 of the 234-mi (377-km) Tennessee-Tombigbee Waterway gave Tennessee shippers a direct north–south route for all vessels between the Tennessee River and the Gulf of Mexico via the Black Warrior River in Alabama. Although none of the waterway runs through Tennessee, the northern terminus is on the Tennessee River near the common borders of Tennessee, Alabama, and Mississippi.

In December 1983 there were 129 public and private airports, 31 heliports, 9 stolports, and 1 seaplane base in the state. Memphis International Airport is the state's major air terminal, with 56,248 aircraft departures and 1,352,044 enplaned passengers in 1983.

[11] HISTORY

The lower Tennessee Valley was heavily populated with hunter-gatherers some 10,000 years ago. Their descendants, called Paleo-Indians, were succeeded by other native cultures, including the Archaic Indians, Woodland Indians, and Early Mississippians. When the first Spanish arrived in the early 16th century, Creek Indians were living in what is now East Tennessee, along with the Yuchi. About 200 years later, the powerful Cherokee—the largest single tribe south of the Ohio River, occupying parts of North Carolina, South Carolina, Georgia, and East Tennessee—drove the Creek and Yuchi out of the area and established themselves as the dominant tribe. Their settlements, varying in size from a dozen families to more than 200, were known as the Upper or Overhill Towns. The Cherokee retained their tribal dominance until they were forced out by the federal government in the 1830s. In West Tennessee, the Chickasaw were the major group. They lived principally in northern Mississippi but used Tennessee lands as a hunting ground. Shawnee occupied the Cumberland Valley in Middle Tennessee until driven north of the Ohio River by the Cherokee and Chickasaw.

Explorers and traders from continental Europe and the British Isles were in Tennessee for well over 200 years before permanent settlements were established in the 1760s. Hernando de Soto, a Spaniard, came from Florida to explore the area as early as 1540. He was followed during the 17th century by the French explorers Jacques Marquette, Louis Jolliet, and Robert Cavelier, Sieur de la Salle. Englishmen were not far behind: by the mid-1700s, hundreds—perhaps thousands—had crossed the Appalachian barrier and explored the transmontane country beyond, which was claimed first by the colony of Virginia and later assigned to North Carolina. They came in search of pelts, furs, or whatever else of value they might find. A fiercely independent breed, they were accustomed to hardship and unwilling to settle in a civilized community. Perhaps the best known was Daniel Boone, who by 1760 had found his way into present-day Washington County.

With the conclusion of the French and Indian War in 1763, many people from North Carolina and Virginia began to cross the Alleghenies. Elisha Walden was among those who first led groups of "long hunters" into the wilderness. By 1770, small pockets of white settlement were developing in the valley between the Unaka and Cumberland mountains. In the two decades that followed, more than 35,000 people settled on soil soon to become the State of Tennessee.

Two major areas of settlement developed. The larger one—in the northeast along the Holston, Nolichucky, and Watauga rivers—was organized as the Watauga Association in 1791. The second major area was in the Cumberland Basin, where James Robertson, under the sponsorship of the Transylvania Company (formed by eastern land speculators), established a settlement he called Nashborough (now Nashville) in 1779. There more than 250 adult males signed the Cumberland Compact, which established a government. They pledged to abide by the will of the majority and expressed their allegiance to North Carolina.

The Revolutionary War did not reach as far west as Tennessee, but many of the frontiersmen fought in the Carolinas and Virginia. The most famous battle involving these early Tennesseans was that of Kings Mountain, in South Carolina, where Colonel John Sevier and others defeated a superior force of British soldiers and captured more than 1,000 prisoners. Hardly was the Revolution over when Tennesseans began to think about statehood for themselves. As early as 1784, leaders in three mountain countries—Greene, Sullivan, and Washington—established the Free State of Franklin. John Sevier was chosen as governor, and an assembly was formed. Only after border warfare developed and factionalism weakened their cause did Franklin's leaders abandon their plans and return their allegiance to North Carolina. But the spirit of independence—indeed, defiance—persisted.

In 1790, less than two years after Franklin collapsed, North Carolina ceded its western lands to the US. Tennessee became known as the Southwest Territory, with William Blount, a prominent North Carolina speculator and politician, as its governor. During his six-year tenure, a government was organized and a capital established at Knoxville. The population doubled to more than 70,000 in 1795, and steps were taken to convert the territory into a state. When the territorial legislature presented Congress with a petition for statehood, a lively debate ensued in the US Senate between Jeffersonian Democratic-Republicans, who urged immediate admission, and Federalists, who opposed it. The Jeffersonians triumphed, and on 1 June 1796, President George Washington signed a bill admitting Tennessee as the 16th state. Sevier became governor of the new state, Blount was elected to the US Senate, and Andrew Jackson became the state's first US representative.

Sevier dominated state politics for the first two decades of statehood, and he had little difficulty in thwarting the ambitions of Andrew Jackson and others who sought to challenge his leadership. Tennessee's population, about 85,000 when Sevier became governor, was more than 250,000 when he left the statehouse in 1809. Under Sevier's governorship, Nashville, Knoxville, and other early settlements became thriving frontier towns. Churches and schools were established, industry and agriculture developed, and Tennessee became a leading iron producer.

Andrew Jackson's rise to prominence came as a result of the Battle of New Orleans, fought at the conclusion of the War of 1812. Jackson, who had little difficulty raising troops in a state where volunteers for military service have always been abundant, lost only about a half dozen of his men, while British casualties exceeded 2,000. He returned to Nashville a hero, built a fine house that he named The Hermitage, received thousands of congratulatory messages, and conferred with friends about his political and military future. In 1823, Jackson was elected to the US Senate. Defeated the following year in a four-man race for the presidency, he ran again, this time successfully, in 1828, serving in that office for eight years.

Jackson alienated himself from many people in the state after 1835, when he announced his support of Martin Van Buren for president instead of Knoxvillian Hugh Lawson White, an avowed candidate. A majority of Tennesseans joined the new Whig Party, which arose in opposition to Jackson's Democratic Party, and voted in the 1836 presidential election for White instead of for Van Buren. The Whigs won every presidential election in Tennessee from 1836 to 1852, including the election of 1844, which sent Tennessean James Knox Polk, a Democrat, to the White House. Polk's term (1845–49) brought another war, this one with Mexico. Although Tennessee's quota was only 2,800, more than 25,000 men volunteered for service. Among the heroes of that war were William Trousdale and William B. Campbell, both of whom later were elected governor.

Social reform and cultural growth characterized the first half of the 19th century. A penitentiary was built, and the penal code made somewhat more humane. Temperance newspapers were published, temperance societies formed, and laws passed to curtail the consumption of alcoholic beverages. In 1834, a few women, embracing the feminist cause, were influential in giving the courts, rather than the legislature, the right to grant divorces. Many important schools were established, including the Nashville Female Academy, the University of Nashville, and more than two dozen colleges.

More than most other southern states, antebellum Tennessee was divided over the issue of slavery. Slaves had accompanied their owners into Tennessee in the 18th century, and by 1850, they constituted about one-fourth of the state's population. Although slaveholders lived in all sections of the state, they predominated in the west, where cotton was grown profitably, as well as in Middle Tennessee. In East Tennessee, where blacks made up less than 10% of the population, antislavery sentiment thrived. Most of those who supported emancipation urged that it be accomplished peacefully, gradually, and with compensation to the slave owners. Frances Wright, the Scottish reformer, founded the colony of Nashoba near Memphis in the 1820s as a place where freed blacks could learn self-reliance. After a few years the colony failed, however, and Wright took her colonists to Haiti. At the constitutional convention of 1834, hundreds of petitions were presented asking that the legislature be empowered to free the slaves. But while the convention endorsed several measures to democratize the constitution of 1796—abolishing property qualifications as a condition for holding office, for example—it decided against emancipation.

Considerable economic growth took place during this period. West Tennessee became a major cotton-growing area immediately after it was purchased from the Chickasaw in 1818, and Memphis, established in 1821, became the principal cotton-marketing center. The Volunteer State's annual cotton crop grew from less than 3,000 bales in 1810 to nearly 200,000 bales by midcentury. The counties of the Highland Rim produced tobacco in such abundance that, by 1840, Tennessee ranked just behind Kentucky and Virginia in total production. East Tennessee farmers practiced greater crop diversification, growing a variety of fruits and vegetables for market. Silk cultivation flourished briefly in the 1830s and 1840s.

Tennessee became a major battleground during the Civil War, as armies from both North and South crossed the state several times. Most Tennesseans favored secession. But the eastern counties remained staunchly Unionist, and many East Tennesseans crossed over into Kentucky to enlist in the Union Army. General Albert Sidney Johnston, the Confederate commander of the western theater, set up lines of defense across the northern border of the state and built forts on both the Cumberland and Tennessee rivers. In February 1862, Ft. Donelson and Ft. Henry were taken by General Ulysses S. Grant and naval Captain Andrew H. Foote, thereby opening the state to Union armies. Within two weeks Nashville was in the hands of the enemy. Northern troops pushed farther south and west, taking key positions on the Mississippi River. Less than two months later, on 6 April, Union forces near the Mississippi state line engaged Johnston's army in the Battle of Shiloh. Both sides suffered tremendous losses, including Johnston himself, who bled to death after sustaining a thigh wound. In the meantime, President Abraham Lincoln had established a military government for the conquered state and appointed Andrew Johnson to head it. Johnson, who had served two terms as governor a decade earlier, had been elected to the US Senate in 1858; he remained there in 1861, the only southern senator to do so, refusing to follow his state into the Confederacy. In 1864, he was elected vice president under Lincoln.

Johnson's governorship did not mean the end of Confederate activities in Tennessee. Late in December 1862, Confederate forces made the first of two vigorous attempts to rid the state of the invader. General Braxton Bragg, who replaced Johnston as Confederate commander, established himself at Murfreesboro, 30 mi (48 km) southwest of Nashville, and threatened to retake the capital city. But at the Battle of Stones River, Union troops under General William S. Rosecrans forced Bragg to retreat to the southeast. Fighting did not resume until 19–20 September 1863, when the Confederates drove Union troops back to Chattanooga in the Battle of Chickamauga, one of the bloodiest engagements of the war. The second major Confederate drive occurred in November and December 1864, when General John B. Hood, commanding the Confederate Army of Tennessee, came out of Georgia and attacked the Union forces at Franklin and Nashville. Hood's army was destroyed, and these battles were the last major engagements in the state.

Returning to the Union in 1866, Tennessee was the only former Confederate state not to have a military government during Reconstruction. Economic readjustment was not as difficult as elsewhere in the South, and within a few years agricultural production exceeded antebellum levels. Extensive coal and iron deposits in East Tennessee attracted northern capital, and by the early 1880s, flour, woolen, and paper mills were established in all the urban areas. By the late 1890s, Memphis was a leading cotton market and the nation's foremost producer of cottonseed oil. Politically, the Democratic Party became firmly entrenched, and would remain so until the 1950s.

As the 20th century dawned, the major issue in Tennessee was the crusade against alcohol, a movement with deep roots in the 19th century. Though the major cities still were "wet," earlier legislation had dried up the rural areas and small towns, and the Tennessee Anti-Saloon League and Women's Christian Temperance Union (WCTU) kept the matter in the public eye. In 1908, with "wet" forces controlling the state government, Edward Ward Carmack—a rabid prohibitionist, powerful politician, newspaper editor, and former US senator—was shot and killed in the street in Nashville. His assailants were convicted but pardoned immediately by the governor. In the following year, with Carmack as a martyr to their cause, "dry" forces enacted legislation that, in effect, imposed prohibition on the entire state. The dominant Democratic Party was divided and demoralized to such an extent that the Republicans elected a governor—only the second Republican since Reconstruction. The prohibition movement helped promote the cause of women's suffrage. A proposed state constitutional amendment giving women the right to vote failed in 1915, but in 1919, they were granted the franchise in municipal elections. One year later, Tennessee became the 36th state to ratify the 19th Amendment to the US Constitution, thereby granting women the right to vote nationwide.

The 1920s brought a resurgence of religious fundamentalism. When, in 1925, the legislature enacted a measure that prohibited the teaching of the theory of evolution in the public schools, a high school teacher named John T. Scopes decided to challenge the law. Three-time presidential candidate and fundamentalist spokesman William Jennings Bryan arrived in the tiny town of Dayton to aid in Scopes's prosecution, while the great civil liberties lawyer Clarence Darrow came from Chicago to lead the defense. The Scopes trial gave the Volunteer State unwanted notoriety throughout the civilized world. Scopes was convicted, and it was not until 1967 that the law was repealed.

The 1930s brought depression, but they also brought the Tennessee Valley Authority. Before TVA, residents of the Tennessee River Valley could boast of the beauty of the landscape, but of little else. The soil was so thin that little other than subsistence agriculture was possible, and many people lived on cash incomes of less than $100 a year. There were some senators, such as George Norris of Nebraska and Tennessee's own Kenneth D. McKellar, who saw great possibilities in valley development. Harnessing the Tennessee River with dams could not only generate electricity inexpensively but also greatly improve navigation; aid flood control, soil conservation, and reforestation; and produce nitrate fertilizer. Efforts to establish such a program failed, however, until Franklin D. Roosevelt included it in his New Deal. The law establishing TVA was passed a few weeks after Roosevelt's inauguration in 1933, and dam construction began almost immediately. Before TVA, people in the valley consumed only 1.5 billion kwh of electricity annually; but consumption increased to 11.5 billion kwh by 1945 and to 57.5 billion kwh by 1960. Fewer than 2% of rural families in Tennessee had electricity in 1933, but by the late 1930s, power lines were being strung into remote areas, bringing to practically everyone the advantages that hitherto only urban residents had enjoyed. Inexpensive power became a magnet for industry, and industrial employment in the region nearly doubled in two decades. The building of a plant for the production of atomic weapons at Oak Ridge in 1942 was due in large measure to the availability of TVA power.

The TVA notwithstanding, the depression caused many manufacturers to close or curtail operations, and farm prices declined drastically. Cotton, which had earlier brought farmers more than 30 cents a pound, declined to 5.7 cents, and the prices of corn, tobacco, and other crops fell proportionately. The state still was in the grip of financial depression when World War II began. Thousands of men volunteered for service before conscription was introduced; when the US entered the war in 1941, several training posts were established in Tennessee. Tennessee firms manufacturing war matériel received contracts amounting to $1.25 billion and employed more than 200,000 people during the war. Industrial growth continued during the postwar period, while agriculture recovered and diversified. The chemical industry, spurred by high demand during and after World War II, became a leading sector, along with textiles, apparel, and food processing. Cotton and tobacco continued to be major crops, but by the early 1970s, soybeans had taken the lead, accounting for 22% of estimated farm income in 1980. Beef and dairy production also flourished.

Democratic boss Edward H. Crump, who ran an efficient political machine in Memphis, dominated state politics for most of the period between 1910 and the early 1950s, an era that saw the elevation of many Tennessee Democrats to national prominence. Considerable progress was made toward ending racial discrimination during the postwar years, although the desegregation of public schools was accomplished only after outbursts of violence at Clinton, Nashville, and Memphis. The killing of civil rights leader Martin Luther King, Jr., in Memphis in 1968 resulted in rioting by blacks in that city. The most notable political development during the 1970s was the resurgence of the Republican Party, making Tennessee one of the few true two-party states in the South.

The early 1980s saw the exposure of corruption in high places: former governor Ray Blanton and several aides were convicted for conspiracy to sell liquor licenses, and banker and former gubernatorial candidate Jacob F. "Jake" Butcher was convicted for fraud in the aftermath of the collapse of his banking empire. On the brighter side, there was a successful world's fair in 1982, the Knoxville International Energy Exposition, and a somewhat resilient state economy, attended by the much-heralded arrivals of Nissan and General Motors. By the end of 1984, capital investment in the state was averaging $1.2 billion a year.

¹²STATE GOVERNMENT

Tennessee's first constitution was adopted in 1796, just before the state was admitted to the Union. It vested executive authority in a governor, elected for two years, who had to be at least 25 years of age and own at least 500 acres (202 hectares) of land. The governor could approve or veto bills adopted by the legislature, was commander-in-chief of the militia, and could grant pardons and reprieves, among other powers. Legislative power was placed in a general assembly, consisting of a house and senate, whose members served terms of two years. Candidates for the legislature were required to fulfill residence and age requirements and to own at least 200 acres (81 hectares). Property qualifications were not required for voting, and all freemen—including free Negroes—could vote.

The basic governmental structure established in 1796 remains the fundamental law today. The constitution has been revised several times, however. The spirit of Jacksonian democracy prompted delegates at the constitutional convention of 1834 to remove property qualifications as a requirement for public office, reapportion representation, transfer the right to select county officials from justices of the peace to the voters, and reorganize the court system. At the same time, though, free blacks were disfranchised. In 1870, another constitutional convention confirmed the abolition of slavery and the enfranchisement of black men but imposed a poll tax as a requirement for voting. Membership of the house was fixed at 99, and of the senate at 33—numbers retained today.

Yet another constitutional convention was held in 1953. Delegates increased the gubernatorial term from two to four years, gave the governor the power of item veto, eliminated the poll tax, authorized home rule for cities, and provided for the consolidation of county and city functions. Later conventions extended the term of state senators from two to four years, sought to improve and streamline county government, and placed a constitutional limit on state spending. A limited convention in 1965 required the apportionment of the legislature according to population. This change greatly increased the weight of urban, and particularly black, votes.

The governor appoints a cabinet of 21 members. The speaker of the state senate automatically becomes lieutenant governor; the secretary of state, treasurer, and comptroller of the treasury are chosen by the legislature.

Legislation is enacted after bills are read and approved three times in each house and signed by the governor. If the governor vetoes a measure, the legislature may override the veto by majority vote of both houses. Not more often than once every six years, the legislature may submit to the voters the question of calling a convention to amend the constitution. If the vote is favorable, delegates are chosen. Changes proposed by the convention must be approved by a majority vote in a subsequent election. Individual amendments also may be considered by the legislature from time to time, but the process is cumbersome.

People may vote in state and national elections if they are US citizens, are at least 18 years of age, and have registered at least 30 days before the election.

¹³POLITICAL PARTIES

The major political groups are the Democratic and Republican parties. Minor parties have seldom affected the outcome of an election in Tennessee.

When Tennessee entered the Union in 1796, it was strongly loyal to the Democratic-Republican Party. The Jacksonian era brought a change in political affiliations, and for more than 20 years, Tennessee had a vibrant two-party system. Jackson's followers formed the Democratic Party, which prevailed for a decade over the National Republican Party led by John Quincy Adams and Henry Clay. But by 1835, Tennesseans had become disillusioned with Jackson, and they joined the new Whig Party in large numbers. A Whig governor was elected in that year, and Whig presidential nominees consistently garnered Tennessee's electoral votes until the party foundered over the slavery issue in the 1850s.

After the Civil War and Reconstruction, Tennessee was part of the solid Democratic South for nearly a century. Only three Republican governors were elected during that period, and only then because bitter factionalism had divided the dominant party. East Tennessee remained a Republican stronghold, however; the 2d Congressional District, which includes Knoxville, was the only district in the country to elect a Republican continuously from 1860 on. Republicans Warren G. Harding and Herbert Hoover carried the state in the presidential elections of 1920 and 1928. But whereas the 1920s saw a tendency away from one-party domination, Franklin D. Roosevelt and the New Deal brought the Volunteer State decisively back into the Democratic fold. Tennesseans voted overwhelmingly Democratic in the four elections that Roosevelt won (1932–44).

After World War II, the one-party system in Tennessee was shaken anew. Dwight D. Eisenhower narrowly won the state in 1952 and 1956, although Tennessee Senator Estes Kefauver was the Democratic vice-presidential nominee in the latter year. Tennesseans chose Richard Nixon all three times he ran for president. In fact, between 1948 and 1976, the only Democratic nominees to carry the state came from the South (Lyndon Johnson and Jimmy Carter) or from a border state (Harry Truman).

In state elections, the Republicans made deep inroads into Democratic power during the 1960s and 1970s. In 1966, Howard Baker became the first popularly elected Republican US senator in state history. In 1970, voters elected Winfield Dunn as the first Republican governor in more than 50 years, and in the same year, they sent Republican Bill Brock to join Baker in the Senate. The Democrats regained the governorship in 1974 and Brock's seat in

1976, but Republicans again won the governorship in 1978 when Lamar Alexander defeated Jacob F. "Jake" Butcher. In 1982, Alexander became the first Tennessee governor to be elected to two successive four-year terms. Except for 1969–71, when the state house of representatives had 49 members from each party, the Democrats maintained their hold over the state legislature. Tennessee voters, who had given Ronald Reagan only the slimmest of winning margins in 1980, gave Reagan a 16.2% margin in 1984.

There were 138 black elected officials in Tennessee as of 1985, including 1 US representative and 13 state legislators; 1 US representative and 10 state legislators were women.

¹⁴LOCAL GOVERNMENT

Local government in Tennessee is exercised by 95 counties and more than 300 municipalities. The county, a direct descendant of the Anglo-Saxon shire, has remained remarkably unaltered in Tennessee since it was brought from Virginia and North Carolina in frontier days. The constitution specifies that county officials must include at least a register, trustee (the custodian of county funds), sheriff, and county clerk, all of whom hold office for four years and may succeed themselves. Other officials have been added by legislative enactment: county executives (known for many years as county judges or county chairmen), tax assessors, county court clerks, and superintendents of public schools.

City government is of more recent origin than county government and is, in fact, a creature of the state. There are three forms of municipal government: mayor-council (or mayor-alderman), council-manager, and commission. The mayor-council system is the oldest and by far the most widely employed. There were 335 municipalities in 1982, as well as 469 special districts.

¹⁵STATE SERVICES

The commissioner of education oversees the public schools as well as special, higher, and vocational-technical education. Highways, aeronautics, mass transit, and waterways are the responsibility of the Department of Transportation. The Department of Safety, including the State Highway Patrol, is charged with enforcing the safety laws on all state roads and interstate highways. Railroad regulation and the setting of railroad rates are the duties of the Public Service Commission. Public protection services are pro-

Tennessee Presidential Vote by Political Parties, 1948–84

YEAR	ELECTORAL VOTE	TENNESSEE WINNER	DEMOCRAT	REPUBLICAN	STATES' RIGHTS DEMOCRAT	SOCIALIST	PROGRESSIVE
1948	11	*Truman (D)	270,402	202,914	73,815	1,288	1,864
					CONSTITUTION	PROHIBITION	
1952	11	*Eisenhower (R)	443,710	446,147	379	1,432	887
1956	11	*Eisenhower (R)	456,507	462,288	19,820	789	—
					NATL. STATES' RIGHTS		
1960	11	Nixon (R)	481,453	556,577	11,298	2,450	—
1964	11	*Johnson (D)	635,047	508,965	—	—	—
					AMERICAN IND.		
1968	11	*Nixon (R)	351,233	472,592	424,792	—	—
							AMERICAN
1972	10	*Nixon (R)	357,293	813,147	—		30,373
						LIBERTARIAN	
1976	10	*Carter (D)	825,897	633,969	2,303	1,375	5,769
					NATL. STATESMAN		CITIZENS
1980	10	*Reagan (R)	783,051	787,761	5,021[1]	7,116[1]	1,112[1]
1984	11	*Reagan (R)	711,714	990,212	—	3,072	978

*Won US presidential election.   [1]National party candidate appeared on Tennessee ballot as independent.

vided by the Military Department, which includes the Army and Air National Guard. The Department of Correction maintains prisons for adult offenders, a work-release program, and correctional and rehabilitation centers for juveniles.

The Department of Health and Environment licenses medical facilities, provides medical care for the indigent, operates tuberculosis treatment centers, and administers pollution control programs. The Department of Mental Health and Mental Retardation supervises mental hospitals, mental health clinics, and homes for retarded children. The Department of Human Services administers aid to the blind, aged, disabled, and families with dependent children, and determines eligibility for families receiving food stamps. The Department of Employment Security administers unemployment insurance and provides job training and placement services. State laws governing workers' compensation, occupational and mine safety, child labor, and wage standards are enforced by the Department of Labor.

16 JUDICIAL SYSTEM

The supreme court is the highest court in the state. It consists of five justices, not more than two of whom may reside in any one grand division of the state—East, Middle, or West Tennessee. The justices are elected by popular vote for terms of 8 years and must be at least 35 years of age. The court has appellate jurisdiction only, holding sessions in Nashville, Knoxville, and Jackson. The position of chief justice rotates every 19 months.

Immediately below the supreme court are two appellate courts (each sitting in three divisions), established by the legislature to relieve the crowded high court docket. The court of appeals, consisting of 12 judges in 1984, has appellate jurisdiction in most civil cases. The court of criminal appeals, consisting of 9 judges, hears cases from the lower courts involving criminal matters. Judges on both appellate courts are elected for eight-year terms.

Circuit courts have original jurisdiction in both civil and criminal cases. As of 1984, the state was divided into 31 circuits, with 58 judges. Tennessee still has chancery courts, vestiges of the English courts designed to hear cases where there was no adequate remedy at law. As of 1984 there were 19 chancery districts in the state, with a total of 27 chancellors. They administer cases involving receiverships of corporations, settle disputes regarding property ownership, hear divorce cases, and adjudicate on a variety of other matters. In some districts, judges of the circuit and chancery courts, all of whom are elected for eight-year terms, have concurrent jurisdiction.

At the bottom of the judicial structure are general sessions courts. A comprehensive juvenile court system was set up in 1911. Other courts created for specific services include domestic relations courts and probate courts.

As of 31 December 1984, federal and state prisons in Tennessee had 7,302 inmates.

According to the FBI Crime Index, Tennessee's crime rates rank below the national averages for all crimes except murder. FBI data for 1983 show that the state's crime rates per 100,000 population were murder and nonnegligent manslaughter, 8.8; forcible rape, 35.9; robbery, 173; aggravated assault, 185; burglary, 1,206; larceny, 2,087; and auto theft, 317.

17 ARMED FORCES

Authorized personnel at US military installations in Tennessee totaled 19,605 in 1983/84, most of them at Millington Naval Air Station near Memphis.

Tennessee supplied so many soldiers for the War of 1812 and the Mexican War that it became known as the Volunteer State. During the Civil War, more than 100,000 Tennesseans fought for the Confederacy and about half that number for the Union. In World War I, some 91,000 men served in the armed forces, and in World War II, 316,000 Tennesseans saw active duty. As of 30 September 1983, 543,000 veterans were living in Tennessee, of whom 5,000 served in World War I, 201,000 in World War II,

99,000 in the Korean conflict, and 168,000 during the Viet-Nam era. Veterans' benefits totaled $639 million in 1982/83.

The Army National Guard had 13,037 personnel in 1984, organized into more than 135 units and activities in 88 cities and towns. Tennessee's Air National Guard had more than 3,707 members in early 1985. State and local police personnel numbered 10,855 in 1983.

18 MIGRATION

The first white settlers in Tennessee, who came across the mountains from North Carolina and Virginia, were almost entirely of English extraction. They were followed by an influx of Scotch-Irish, mainly from Pennsylvania. About 3,800 German and Irish migrants arrived during the 1830s and 1840s. In the next century, Tennessee's population remained relatively stable, except for an influx of blacks immediately following the Civil War. There was a steady out-migration of blacks to industrial centers in the North during the 20th century. The state suffered a net loss through migration of 462,000 between 1940 and 1970, but gained 305,000 between 1970 and 1983.

The major in-state migration has been away from rural areas and into towns and cities. Blacks, especially, have tended to cluster in large urban centers. The population of Memphis, for example, is more than 50% black.

19 INTERGOVERNMENTAL COOPERATION

Tennessee participates in such interstate agreements as the Interstate Mining Compact, Southeastern Forest Fire Protection Compact, Southern Growth Policies Compact, and Southern Interstate Energy Compact. There are boundary accords with Arkansas, Kentucky, and Virginia, and an agreement with Alabama, Florida, Kentucky, and Mississippi governing development of the Tennessee-Tombigbee waterway.

Federal aid to Tennessee was about $1.9 billion in 1983/84.

20 ECONOMY

Tennessee's economy is based primarily on industry. Since the 1930s, the number of people employed in industry has grown at a rapid rate, while the number of farmers has declined proportionately. The total industrial payroll in 1982 was more than $21 billion. Wage rates and average weekly earnings are well below the national average. The principal manufacturing areas are Memphis, Nashville, Chattanooga, Knoxville, and Kingsport-Bristol. Apparel production employed more workers in 1983 than any other industry; chemical and allied products, food products, electrical and electronic equipment, fabricated metal products, nonelectrical machinery, and transportation equipment followed in that order. With the construction of a Nissan automobile and truck plant in 1980 and the planned construction of a General Motors automobile plant, both in the area southeast of Nashville, Tennessee is becoming an important producer of transportation equipment. The new GM plant will employ 6,000 persons.

Income from agricultural products now comes more from dairy and beef cattle and soybeans than from tobacco, cotton, and corn, which were the leading money crops for many years. Tourism is the third major contributor to the state's economy.

21 INCOME

With a per capita income of $9,362 in 1983, Tennessee ranked 42d in the US. Between 1975 and 1982, Tennessee's total disposable personal income doubled, from $17.9 billion to $35.8 billion.

In 1979, median family income was $16,564; only six states ranked lower. In that year, about 13% of Tennessee families and 16.5% of all Tennesseans were below the federal poverty level. Some 21,000 Tennesseans were among the top wealth-holders in the US in 1982, with gross assets of more than $500,000.

22 LABOR

In 1984, Tennessee had a total civilian labor force of 2,223,000. The overall unemployment rate for that year was 8.6%, down from 11.5% in 1983. In 1983, the labor force was 56% male and 44% female.

A federal survey in March 1983 revealed the following nonfarm employment pattern in Tennessee:

	ESTABLISH-MENTS	EMPLOYEES	ANNUAL PAYROLL ('000)
Agricultural services, forestry, fishing	805	4,365	$ 44,928
Mining	435	8,051	171,028
Contract construction	7,123	67,597	1,165,620
Manufacturing, of which:	6,230	446,677	7,891,026
Food products	(410)	(34,534)	(653,579)
Apparel, other textiles	(443)	(58,431)	(564,012)
Chemicals, allied products	(229)	(48,101)	(1,210,815)
Transportation, public utilities	3,296	70,418	1,532,791
Wholesale trade	8,057	102,092	1,909,195
Retail trade	27,491	276,517	2,591,926
Finance, insurance, real estate	7,733	83,279	1,426,212
Services	27,107	318,553	4,474,813
Other	6,800	6,646	141,242
TOTALS	95,077	1,384,195	$21,348,781

This survey excluded government workers; in 1983, Tennessee had about 244,000 state and government employees, and in 1982 there were 58,000 federal government employees in the state.

There were 1,170 labor unions in Tennessee in 1983, and in 1980, 334,000 Tennesseans—19% of the state's nonagricultural workers—were union members. Tennessee has a right-to-work law. Weakness of the labor movement is one reason why average hourly earnings of production workers, $7.88 in September 1984, were below the US norm.

23 AGRICULTURE

Tennessee ranked 28th among the 50 states in 1983 with farm receipts of almost $2 billion. There were 94,000 farms in 1984.

From the antebellum period to the 1950s, cotton was the leading crop, followed by corn and tobacco. But during the early 1960s, soybeans surpassed cotton as the principal source of income. In 1983, 30 million bushels of soybeans, valued at $249 million, were harvested. Tobacco production in 1983 was 118 million lb, at a value of $212 million. The main types of tobacco are burley, a fine leaf used primarily for cigarettes, and eastern and western dark-fired, which are used primarily for cigars, pipe tobacco, and snuff. Tennessee ranked 3d among the tobacco-producing states in total crop value in 1983. The corn harvest in 1983 was about 23 million bushels, valued at $88 million. In 1983, cotton production was 151,000 bales, valued at $50 million, down from 339,000 bales, valued at almost $94 million, in 1982.

24 ANIMAL HUSBANDRY

Livestock and livestock products account for more than half of Tennessee's agricultural income, and meat animals are the state's most important commodity. Cattle are raised throughout the state, but principally in Middle and East Tennessee. In 1930, fewer than a million cattle and calves were raised on Tennessee farms; by early 1984, however, there were 2.5 million head of beef cattle. Production in 1983 was valued at more than $364 million.

Hogs and pigs accounted for about 8.5% of farm income in 1982. The number of hogs and pigs produced has declined slightly during the past 50 years, but their value has increased considerably. In late 1983 there were 950,000 hogs and pigs; production was 298 million lb, worth $141.6 million. Sheep and lamb production has declined sharply in the past 50 years. In 1925, 368,000 head were raised, but by 1984, the number had declined to 6,000.

Poultry and eggs accounted for $113 million in income in 1982. In 1981, poultry farmers produced 232.3 million lb of broilers. In 1983, 822 million eggs were produced, and Tennessee dairy farms yielded 2.3 billion lb of milk.

Horses are raised for market primarily in Middle and West Tennessee. The Tennessee Walking Horse is bred throughout the state but especially around Tullahoma.

25 FISHING

Fishing is a major attraction for sport but plays a relatively small role in the economic life of Tennessee. There are 17 TVA lakes and 7 other lakes, all maintained by the Army Corps of Engineers; 10 of these lakes span an area of 10,000 acres (4,000 hectares) or more, and there are thousands of miles of creeks and mountain streams, all of which attract anglers. Tennessee has no closed season, except on trout.

In the 1970s, pollution from industrial waste dumping killed millions of fish and seriously endangered sport fishing. By the early 1980s, however, industrial establishments in the state were complying more fully with the 1974 Water Pollution Act, and in 1984, only 63,000 fish were killed by industrial waste dumping.

26 FORESTRY

Forests covered 13,161,000 acres (5,326,000 hectares) in 1979, or about half the state's total land area. Commercial timberlands in 1980 totaled 12,820,000 acres (5,188,000 hectares), of which 90% was privately owned, 40% by farmers and the remainder by forest industries, other firms, and private individuals. The counties of the Cumberland Plateau and Highland Rim are the major sources of timber products, and in Lewis, Perry, Polk, Scott, Sequatchie, Unicoi, and Wayne counties, more than 75% of the total area is commercial forest.

About 80% of Tennessee's timber is in hardwoods, and more than one-half of that is in white and red oak. Of the softwoods, pine—shortleaf, loblolly, Virginia, pitch, and white—accounts for 75%. Red cedar, once in great abundance, now accounts for only 5% of the softwood supply.

Shipments of lumber and wood products totaled $699.7 million in 1982. There were more than 300 sawmills and planing mills in the state, employing 2,600 production workers. Most of the lumber produced by the mills is sold to the building trades, but some goes into furniture manufacture, cooperage (barrel making), poles, mine timbers, and fuel wood. Shipments of wood household furniture totaled $234.7 million in 1982. The leading forest-related industry is the manufacture of paper and paper products, shipments of which were worth nearly $2.3 billion in 1982.

27 MINING

Tennessee possesses a great variety of mineral resources. In 1982, the state ranked 29th among the 50 states in mineral production, with a total value of $639 million. Copper and zinc, the principal metals, are produced almost entirely in the Ducktown Basin near Chattanooga; iron, lead, manganese, and gold are mined in the same area. The sulfide copper ores are smelted and shipped to electrolytic refineries in New Jersey. Tennessee is the South's largest producer of sulfuric acid, a by-product of copper smelting. Coal, clays, portland cement, sand and gravel, and stone are the principal nonmetals.

The estimated 1984 mineral output includes 121,600 metric tons of zinc, 35,700,000 tons of stone, 6,520,000 tons of sand and gravel, and 1,290,000 tons of clays.

28 ENERGY AND POWER

The Tennessee Valley Authority (TVA) is the principal supplier of power in the state, providing electricity to more than 100 cities and 50 rural cooperatives. In 1983, Tennessee's installed electrical generating capacity was 18.3 million kw, virtually all of it publicly owned; electrical output totaled 69.4 billion kwh (99% public). Since electric energy sales amounted to 70 billion kwh that year, Tennessee imported about 1% of its electricity from neighboring states. There were no nuclear power facilities in operation as of August 1985. The Sequoyah plant was temporarily shut down for an evaluation of the equipment; the Watts Bar plant was awaiting official sanctioning before being put into operation; and construction on two other plants had been halted. In 1985, the US government announced that it was closing its uranium enrichment plant at Oak Ridge.

Bituminous coal is Tennessee's most valuable mineral commodity; $215 million worth of it was mined in 1982. Between 1978 and 1983, however, declining demand for coal, conservationist opposition to surface mining, and other factors led to a drop in coal production from 10 million tons to 6.6 million tons. Reserves in 1983 totaled 940 million tons. Surface mining, which has marred thousands of acres of land and which accounted for 59% of coal production in 1978, accounted for only 32% by the end of 1982. Most of the coal mined in the state is used for producing electricity, although some is utilized for home heating.

Tennessee produced 1.1 million barrels of petroleum in 1982; natural gas reserves were negligible. In 1983, Tennessee's gas utilities served 527,000 customers, of whom 87% were residential users.

Energy consumption per capita for 1983 was 329 million Btu; energy expenditures per capita were $1,935.

29 INDUSTRY

On the eve of the Civil War, only 1% of Tennessee's population was employed in manufacturing, mostly in the iron, cotton, lumber, and flour-milling industries. Rapid industrial growth took place during the 20th century, however, and by 1981, Tennessee ranked 3d among the southeastern states and 15th in the US in value of shipments, with $41.3 billion. In 1982, Tennessee's four major metropolitan areas—Memphis, Nashville, Knoxville, and Chattanooga—employed 49% of all the state's industrial workers.

To the $40.8 billion of value of shipments in 1982, the principal contributors, with their percentages of the total, were chemicals, 16%; food and food products, 16%; electric and electronic equipment, 9%; nonelectrical machinery, 7%; and apparel, 5%. The following table shows value of shipments by selected industries in 1982:

Plastics materials and synthetics	$2,100,300,000
Industrial inorganic chemicals	2,041,000,000
Men's and boys' clothing	1,501,300,000
Refrigeration and heating equipment	1,017,700,000
Motor vehicle parts and accessories	1,008,700,000
Household furniture	733,200,000

30 COMMERCE

Tennessee has been an important inland commercial center for some 60 years. In 1982, the state's wholesale trade amounted to $37.8 billion (15th in the US). Sales in the Memphis metropolitan area, which includes communities in Arkansas and Mississippi, were $16.9 billion. In 1982, Tennessee had retail sales of $19.6 billion (20th in the US). The principal retail groups and their sales percentages were grocery stores, 22%; automotive dealers, 19%; department stores, 10%; gasoline service stations, 10%; and eating and drinking places, 8%. Retail sales in metropolitan Memphis, Nashville, Knoxville, and Chattanooga together totaled nearly $13 billion.

Tennessee's foreign exports included nearly $3.3 billion in manufactured goods in 1981 and $590 million in agricultural commodities in 1981/82.

31 CONSUMER PROTECTION

The Consumer Affairs Division of the Tennessee Department of Commerce and Insurance enforces the state's Consumer Protection Act. The division works in conjunction with the state Attorney General's office.

32 BANKING

The first bank in Tennessee was the Bank of Nashville, chartered in 1807. Four years later, the Bank of the State of Tennessee was chartered at Knoxville; branches were established at Nashville, Jonesboro, Clarksville, and Columbia. In 1817, nearly a dozen more banks were chartered in various frontier towns. The Civil War curtailed banking operations, but the industry began again immediately after cessation of hostilities.

As of the end of 1984 there were 296 insured commercial banks. The previous year, total assets were $29.9 billion; outstanding loans totaled $14.9 billion in 1982.

In 1983, the United American Bank of Knoxville failed. Financier Jake Butcher, head of the banking empire that included United American, was convicted of having embezzled at least $20 million from banks under his control. Numerous bank failures followed the collapse of United American.

As of the end of 1984, there were 296 insured commercial banks. The previous year, total assets were $29.9 billion; outstanding loans totaled $14.9 billion in 1982.

In 1982 there were 81 savings and loan associations in the state, with assets of $7.8 billion.

33 INSURANCE

In 1983, 1,193 insurance companies were licensed to operate in the state, including 47 Tennessee companies.

Some 9,423,000 life insurance policies worth $96.5 billion were in force in 1983, when the average Tennessee family held $54,800 in coverage. Some $888 million in benefits was paid to Tennesseans during the same year, including $291.6 million in death payments. Property and liability insurers wrote premiums totaling nearly $1.7 billion, of which $392.9 million was automobile liability insurance, $319.2 million was automobile physical damage insurance, and $252.8 million was homeowners' coverage.

34 SECURITIES

There are no securities exchanges in Tennessee. An estimated 614,000 Tennesseans held shares of public corporations in 1983, when New York Stock Exchange member firms had 79 sales offices and 851 registered representatives in the state.

35 PUBLIC FINANCE

The state budget is prepared annually by the Budget Division of the Tennessee Department of Finance and Administration and submitted by the governor to the legislature every January. The fiscal year lasts from 1 July to 30 June.

The following table summarizes estimated general fund revenues and proposed state-funded appropriations for 1985/86 (in millions):

REVENUES	
Sales and use tax	$1,757.5
Excise tax	210.3
Gross receipts tax	119.0
Franchise tax	92.0
Tobacco tax	80.1
Income tax	37.5
Alcoholic beverage tax	25.1
Other taxes	106.8
Other receipts	206.4
TOTAL	$2,634.7

APPROPRIATIONS	
Primary and secondary education	$1,044.1
Higher education	554.2
Health and environment	318.8
Corrections	154.6
Mental health and retardation	119.2
Other outlays	406.3
TOTAL	$2,597.2

The total debt owed by the state and local governments in Tennessee increased from $4.8 billion in 1977 to $6.7 billion in 1981/82, when the per capita debt amounted to $1,452. For 1982/83, the total debt was $7.5 billion.

36 TAXATION

The Tennessee state government ranked 25th in the US in general revenues in 1982 with $3.95 billion, including intergovernmental receipts. It was 48th in per capita state and local taxation, however, with an average tax burden in 1982 of $772, 34% below the national average.

The major source of general state revenue is a sales and use tax, first levied in 1947; in 1984, the maximum rate was 6%, of which the state collected 4.5% and municipalities 1.5%. Other taxes include a 6% levy on dividend and interest income, a 6% corporate income tax, and levies on inheritances, alcoholic beverages, tobacco, gross receipts, motor vehicle registration, and other items. Tennessee is one of a few states that do not impose a tax on salaries and wages. Counties and municipalities depend on real property taxes as their major source of income.

In 1982, Tennessee paid about $9.7 billion in federal taxes. Tennesseans filed 1.8 million federal income tax returns for 1982, paying $4.3 billion in tax.

37 ECONOMIC POLICY

Since World War II, Tennessee has aggressively sought new business and industry. The Department of Economic and Community Development helps prospective firms locate industrial sites in communities throughout the state, and its representatives work with firms in Canada, Europe, and the Far East, as well as with domestic businesses. The department also administers special Appalachian regional programs in 50 counties and directs the state Office of Minority Business Enterprise.

Tennessee's right-to-work law and relatively weak labor movement constitute important industrial incentives. The counties and municipalities, moreover, offer tax exemptions on land, capital improvements, equipment, and machinery. In the early 1980s, numerous Japanese companies began operations in Tennessee, employing more than 6,000 Tennesseans by 1985. Most notable was Nissan, which opened a 2,600-employee (as of 1985) plant in Smyrna. Its first truck rolled off the assembly line in June 1983. In 1985, General Motors announced plans to build a plant in Spring Hill, south of Nashville, to produce its new Saturn car.

38 HEALTH

Tennessee's birthrate fell from 23 live births per 1,000 population in 1960 to 14.4 in 1982, when there were 67,078 live births. Between 1977 and 1982, the ratio of legal abortions to live births increased from 246 to 372 for every 1,000 births; about 26,100 abortions were performed in 1982. The infant mortality rate in 1981 was 10.2 per 1,000 live births for whites and 21.2 for blacks.

The leading causes of death in Tennessee in 1982 (with rates per 100,000 population) were heart disease, 317; malignant neoplasms, 185; cerebrovascular diseases, 85; accidents, 45; chronic pulmonary disease, 27; pneumonia and influenza, 21; suicide, 13; and diabetes, 12. The death rate overall in 1981, 8.8 per 1,000 population, was slightly above the US average.

There were 162 hospitals, with 32,080 beds, in 1983, when 1,031,108 patients were admitted. The average cost of hospital care in 1982 was $275 per day and $1,981 per stay, both far below the US average. The state's 251 nursing homes admitted 18,664 patients in 1982.

Tennessee has four medical schools: two in Nashville (Vanderbilt University and Meharry Medical School), one at Johnson City (East Tennessee State University), and one at Memphis (University of Tennessee). The state had 8,246 licensed physicians and 2,226 professionally active dentists in 1982, when hospital personnel, with trainees, included 13,534 registered nurses and 7,614 licensed practical nurses.

39 SOCIAL WELFARE

Public aid payments were made to 5.9% of Tennessee's population in 1983, for a rank of 20th among the 50 states.

Aid to families with dependent children amounted to $76 million in 1982; there were 147,584 AFDC recipients, 101,116 of them children. About 580,000 Tennesseans took part in the food stamp program in 1983, at a federal cost of $324 million; the school lunch program served 604,000 students and cost the federal government $58 million.

In 1983, about $3.1 billion in Social Security benefits was paid to some 744,000 Tennesseans; 63% went to 481,000 retired workers, 23% to 162,000 survivors of deceased workers, and 14% to 100,000 disabled workers. Supplemental Security Income payments were made to 124,150 needy state residents, for a total of $247 million.

Tennessee spent $149.1 million on workers' compensation in 1982. An average of 80,000 Tennesseans received unemployment insurance benefits under state and federal programs in 1982, for a total of $422 million.

40 HOUSING

The 1980 census counted 1,618,505 occupied housing units in the state, 68.6% of which were owner-occupied. About 7% of the occupied units were mobile homes or trailers. The median mortgage payment in 1980 was $296 per month; the median price of an existing home was $43,000 in 1981, while that of a new home was $52,500. The average monthly apartment rent in 1982/83 was $270. From 1981 through 1983, a total of 53,100 new privately owned units worth nearly $1.9 billion were authorized.

41 EDUCATION

The state assumed very little responsibility for education until 1873, when the legislature established a permanent school fund and made schools free to all persons between the ages of 6 and 21. In 1917, an eight-year elementary and four-year secondary school system was set up. Thirty years later, enactment of the state sales and use tax enabled state authorities to increase teachers' salaries by about 100% and to provide capital funds for a variety of expanded educational programs. In the early 1980s, Tennessee further improved its educational system by offering incentive pay to its teachers. Today, nearly half of the consolidated state budget is spent on education.

Of the total adult population in 1980, only 56% were high school graduates, compared with a national average of 67%. The state also lagged slightly in median school years completed, with 12.2, below the nationwide average of 12.5. Public school expenditures per pupil in average daily attendance ranked 48th in the US in 1984.

In 1980/81, the Department of Education administered 1,691 public schools for grades K–12, 27 vocational-technical schools, 4 special schools, 4 regional technical institutes, a network of educational television stations, and a variety of public educational services. In 1982 there were 600,254 students enrolled in the public elementary schools (K–8), 245,465 in public secondary schools, and 201,806 in public and private colleges and universities.

The University of Tennessee system, with principal campuses at Nashville, Knoxville, Memphis, Martin, and Chattanooga, enrolled some 42,135 students in 1982/83. Components of the State University and Community College System of Tennessee included Memphis State University (the largest, with 20,624 students), Tennessee Technological University at Cookeville, East Tennessee State University at Johnson City, Austin Peay State University at Clarksville, Tennessee State University at Nashville, and Middle Tennessee State University at Murfreesboro, along with 10 two-year community colleges enrolling 34,265 students on campuses throughout the state. Well-known private colleges are Vanderbilt University at Nashville (with 8,500 students in 1983/84), the University of the South at Sewanee (1,163), and Rhodes College at Memphis (985). Vanderbilt has schools of medicine, law, divinity, nursing, business, and education, as well as an undergraduate program. Loan and grant programs are administered by the Tennessee Student Assistance Corporation.

42 ARTS

Each of Tennessee's major cities has a symphony orchestra. The best known are the Memphis Symphony and the Nashville Symphony, the latter of which makes its home in the James K. Polk Cultural Center. Included in this complex are three performing arts centers and the State Museum. The major operatic troupe is Opera Memphis.

Nashville is a center for country music. The Grand Ole Opry, Country Music Hall of Fame, and numerous recording studios are located there.

Among the leading art galleries are the Dixon Gallery and the Brooks Memorial Art Gallery in Memphis, the Tennessee Botanical Gardens and Fine Arts Center in Nashville, and the Dulin Gallery of Art in Knoxville.

⁴³LIBRARIES AND MUSEUMS

Libraries and library associations were formed soon after Tennessee became a state. The Dickson Library at Charlotte was founded in 1811, and the Nashville Library Company in 1813. Not until 1854, however, was the first state-maintained library established. Andrew Johnson, the governor, requested a library appropriation of $5,000, telling legislators that he wanted other Tennesseans to have the opportunities that had been denied him. Today, the institution he founded, the State Library at Nashville, with more than 250,000 volumes, has a renowned collection of state materials and is the repository for state records. In all, there were more than 150 public libraries and nearly 50 academic libraries in Tennessee in 1982/83. Their combined book stock exceeded 6.4 million, and their total circulation surpassed 14.4 million volumes. The largest libraries are the Vanderbilt University Library at Nashville (1,517,038 volumes in 1984), Memphis-Shelby County Library (1,354,000), Memphis State University Libraries (844,524), University of Tennessee at Knoxville Library (715,389), Knoxville-Knox County Library (623,803), Chattanooga-Hamilton County Library (380,766), and East Tennessee State University Library at Johnson City (366,835).

Tennessee has more than 90 museums and historic sites. The Tennessee State Museum in Nashville displays exhibits on pioneer life, military traditions, evangelical religion, and presidential lore. The Museum of Appalachia, near Norris, attempts an authentic replica of early Appalachian life, with more than 20,000 pioneer relics on display in several log cabins. Displays of solar, nuclear, and other energy technologies are featured at the American Museum of Science and Energy, at Oak Ridge. There are floral collections at the Goldsmith Civic Garden Center in Memphis and the Tennessee Botanical Gardens and Fine Arts Center in Nashville.

⁴⁴COMMUNICATIONS

The first postal service across the state, by stagecoach, began operations in the early 1790s. As of 1985 there were 620 post offices, branches, and stations.

As of 1980, 89% of Tennessee's 1,618,505 occupied housing units had telephones.

Tennessee had 183 commercial AM stations and 116 FM stations (including 21 noncommercial) in 1983. There were 29 television stations in operation in 1984; in 1983 there were 6 television stations in Memphis, 5 in Nashville, 5 in Chattanooga, and 3 in Knoxville. In 1984, 141 cable systems served 646,983 subscribers in 283 communities.

⁴⁵PRESS

In 1984 there were 10 morning newspapers with net paid circulation of about 577,000, 22 evening dailies with a combined circulation of about 532,000, and 15 Sunday papers with 1,046,-000 circulation. The following table lists leading Tennessee newspapers with their daily circulation in 1984:

AREA	NAME	DAILY	SUNDAY
Chattanooga	News–Free Press (e,S)	55,916	105,188
Knoxville	Journal (m)	60,219	
	News-Sentinel (e,S)	94,703	163,579
Memphis	Commercial Appeal (m,S)	230,666	289,005
Nashville	Banner (e)	71,963	
	Tennessean (m,S)	123,454	244,231

Several dozen trade publications, such as *Southern Lumberman,* appear in Nashville, the state's major publishing center.

⁴⁶ORGANIZATIONS

The 1982 US Census of Service Industries counted 1,112 organizations in Tennessee, including 233 business associations; 577 civic, social, and fraternal associations; and 16 educational, scientific, and research associations.

Nashville is a center for Tennessee cultural and educational organizations. Among them are the American Association for State and Local History, Country Music Association, and Gospel Music Association.

Several national and regional trade associations are based in Tennessee, including the Walking Horse Breeders' and Exhibitors' Association (Lewisburg) and the Walking Horse Trainers' Association (Shelbyville). Knoxville is the headquarters of the Burley (Tobacco) Stabilization Corporation, and Springfield is the home of the Eastern Dark-Fired Tobacco Growers Association. The offices of the Southern Cotton Association, National Cotton Council of America, and Southern Hardwood Lumber Manufacturing Association are in Memphis, as is the headquarters of the American Contract Bridge League.

⁴⁷TOURISM, TRAVEL, AND RECREATION

The natural beauty of Tennessee, combined with the activity of the Department of Tourist Development, has made tourism the state's 3d-largest industry. Tennessee was the first state to create a government department devoted solely to the promotion of tourism. In 1984/85, the state spent $7.2 million to attract tourists. In 1982, more than 27 million persons traveled to and through Tennessee, spending some $2.2 billion in the state.

Leading tourist attractions include Fort Loudoun, built by the British in 1757; the American Museum of Science and Energy at Oak Ridge; the William Blount Mansion at Knoxville; the Beale Street Historic District in Memphis, home of W. C. Handy, the "father of the blues"; Graceland, the Memphis estate of Elvis Presley; and Opryland USA and the Grand Ole Opry at Nashville. There are three presidential homes—Andrew Johnson's at Greeneville, Andrew Jackson's Hermitage near Nashville, and James K. Polk's at Columbia. Pinson Mounds, near Jackson, offers outstanding archaeological treasures and the remains of an Indian city. Reservoirs and lakes attract thousands of anglers and water sports enthusiasts.

There are 18 state parks, almost all of which have camping facilities. All together, they cover 53,000 acres (21,000 hectares). Among the most visited state parks are the Meeman-Shelby Forest in Shelby County, Montgomery Bell in Dickson County, Cedars of Lebanon in Wilson County, and Natchez Trace in Henderson and Carroll counties. Extending into North Carolina, the Great Smoky Mountains National Park covers 241,207 acres (97,613 hectares) in Tennessee and drew nearly 8.5 million visitors in 1983.

Licenses were issued to 541,617 hunters and 699,086 fishermen in 1982/83.

⁴⁸SPORTS

Tennessee has been a baseball state for many years. Minor league teams in the class-AA Southern League are the Knoxville Vols, Chattanooga Lookouts, and Memphis Chicks.

Tennessee's colleges and universities provide the major fall and winter sports. The University of Tennessee Volunteers and Vanderbilt University Commodores, in the Southeastern Conference, compete in football, basketball, and baseball. Austin Peay and Tennessee Technological universities belong to the Ohio Valley Conference.

⁴⁹FAMOUS TENNESSEANS

Andrew Jackson (b.South Carolina, 1767-1845), the 7th president, moved to Tennessee as a young man. He won renown in the War of 1812 and became the first Democratic president in 1828. Jackson's close friend and associate, James Knox Polk (b.North Carolina, 1795-1849), came to Tennessee at the age of 10. He was elected the nation's 11th president in 1844 and served one term.

Andrew Johnson (b.North Carolina, 1808–75), also a Democrat, remained loyal to the Union during the Civil War and was elected vice president with Abraham Lincoln in 1864. He became president upon Lincoln's assassination in 1865 and served out his predecessor's second term. Impeached because of a dispute over Reconstruction policies and presidential power, Johnson escaped conviction by one vote in 1868.

Supreme Court justices from Tennessee include John Catron (b.Pennsylvania, 1786–1865), Howell Jackson (1832–95), James C. McReynolds (b.Kentucky, 1862–1946), and Edward T. Sanford (1865–1930). Tennesseans who became cabinet officials include Secretary of State Cordell Hull (1871–1955), secretaries of war John Eaton (1790–1856) and John Bell (1797–1869), Secretary of the Treasury George Campbell (b.Scotland, 1769–1848), and attorneys general Felix Grundy (b.Virginia, 1777–1840) and James C. McReynolds.

Other nationally prominent political figures from Tennessee are Cary Estes Kefauver (1903–63), two-term US senator who ran unsuccessfully for vice president in 1956 on the Democratic ticket; Albert Gore (b.1907), three-term member of the US Senate; and Howard Baker (b.1925), who in 1966 became the first popularly elected Republican senator in Tennessee history. Three Tennesseans have been speaker of the US House of Representatives: James K. Polk, John Bell, and Joseph W. Byrns (1869–1936). Nancy Ward (1738–1822) was an outstanding Cherokee leader, and Sue Shelton White (1887–1943) played a major role in the campaign for women's suffrage.

Tennessee history features several military leaders and combat heroes. John Sevier (b.Virginia, 1745–1815), the first governor of the state, defeated British troops at Kings Mountain in the Revolution. David "Davy" Crockett (1786–1836) was a frontiersman who fought the British with Jackson in the War of 1812. Sam Houston (b.Virginia, 1793–1863) also fought in the War of 1812 and was governor of Tennessee before migrating to Texas. Nathan Bedford Forrest (1821–77) and Sam Davis (1842–63) were heroes of the Civil War. Sergeant Alvin C. York (1887–1964) won the Medal of Honor for his bravery in World War I.

Cordell Hull was awarded the Nobel Peace Prize in 1945 for his work on behalf of the United Nations. In 1971, Earl W. Sutherland, Jr. (b.Kansas, 1915–75), a biomedical scientist at Vanderbilt University, won a Nobel award for his discoveries concerning the mechanisms of hormones. Outstanding educators include Philip Lindsey (1786–1855), a Presbyterian minister and first president of the University of Nashville, and Alexander Heard (b.Georgia, 1917), nationally known political scientist and chancellor of Vanderbilt University.

Famous Tennessee writers are Mary Noailles Murfree (1850–1922), who used the pseudonym Charles Egbert Craddock; influential poet and critic John Crowe Ransom (1888–1974); author and critic James Agee (1909–55), posthumously awarded a Pulitzer Prize for his novel *A Death in the Family*; poet Randall Jarrell (1914–65), winner of two National Book Awards; and Wilma Dykeman (b.1920), novelist and historian. Sportswriter Grantland Rice (1880–1954) was born in Murfreesboro.

Tennessee has long been a center of popular music. Musician and songwriter William C. Handy (1873–1958) wrote "St. Louis Blues" and "Memphis Blues," among other classics. Bessie Smith (1898?–1937) was a leading blues singer. Elvis Presley (b.Mississippi, 1935–77) fused rhythm-and-blues with country-and-western styles to become one of the most popular entertainers who ever lived. Other Tennessee-born singers are Dinah Shore (b.1917), Aretha Franklin (b.1942), and Dolly Parton (b.1946).

⁵⁰BIBLIOGRAPHY

Abernethy, Thomas P. *From Frontier to Plantation in Tennessee*. Reprint. Westport, Ct.: Greenwood Press, 1979 (orig. 1932).

Connelly, T. L. *Civil War Tennessee: Battles and Leaders*. Knoxville: University of Tennessee Press, 1979.

Corlew, Robert E. *Statehood for Tennessee*. Nashville: Tennessee Bicentennial Commission, 1976.

Corlew, Robert E. *Tennessee: A Short History*. 2d ed. Knoxville: University of Tennessee Press, 1981.

Dykeman, Wilma. *Tennessee: A Bicentennial History*. New York: Norton, 1975.

Federal Writers' Project: *Tennessee: A Guide to the State*. New York: Somerset, n.d. (orig. 1939).

Folmsbee, Stanley J., Robert E. Corlew, and Enoch Mitchell. *History of Tennessee*. 4 vols. New York: Lewis, 1960.

Goehring, Eleanor E. *Tennessee Folk Culture: An Annotated Bibliography*. Knoxville: University of Tennessee Press, 1982.

Greene, Lee S., et al. *Government in Tennessee*. 4th ed. Knoxville: University of Tennessee Press, 1982.

Hubbard, Preston. *Origins of the TVA*. New York: Norton, 1968.

Lewis, Thomas M. N., and Madeline Kneberg. *Tribes That Slumber: Indians of the Tennessee Region*. Knoxville: University of Tennessee Press, 1958.

Mooney, Chase. *Slavery in Tennessee*. Bloomington: Indiana University Press, 1957.

Smith, Samuel B. (ed.). *Tennessee History: A Bibliography*. Knoxville: University of Tennessee Press, 1974.

Tennessee, State of. Secretary of State. *Tennessee Blue Book 1983/84*. Nashville, 1983.

Tennessee, University of. College of Business Administration. Center for Business and Economic Research. *Tennessee Statistical Abstract 1984/85*. Knoxville, 1984.

TEXAS

State of Texas

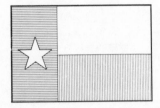

ORIGIN OF STATE NAME: Derived from the Caddo word *tavshas*, meaning "allies" or "friends." **NICKNAME:** The Lone Star State. **CAPITAL:** Austin. **ENTERED UNION:** 29 December 1845 (28th). **SONG:** "Texas, Our Texas." Also: "The Eyes of Texas." **MOTTO:** Friendship. **FLAG:** At the hoist is a vertical bar of blue with a single white five-pointed star; two horizontal bars of white and red cover the remainder of the flag. **OFFICIAL SEAL:** A five-pointed star encircled by olive and live oak branches, with the words "The State of Texas" surrounding. **BIRD:** Mockingbird. **FLOWER:** Bluebonnet. **TREE:** Pecan. **GEM:** Topaz. **STONE:** Palmwood. **GRASS:** Sideoats grama. **DISH:** Chili. **LEGAL HOLIDAYS:** New Year's Day, 1 January; Confederate Heroes Day, 19 January; Birthday of Martin Luther King, Jr., 3d Monday in January; Washington's Birthday, 3d Monday in February; Texas Independence Day, 2 March; San Jacinto Day, 21 April; Memorial Day, last Monday in May; Emancipation Day, 19 June; Independence Day, 4 July; Lyndon B. Johnson's Birthday, 27 August; Labor Day, 1st Monday in September; Columbus Day, 2d Monday in October; General Election Day, 1st Tuesday after 1st Monday in November; Veterans Day, 11 November; Thanksgiving Day, 4th Thursday in November; Christmas Day, 25 December. **TIME:** 6 AM CST=noon GMT.

¹LOCATION, SIZE, AND EXTENT

Located in the west south-central US, Texas is the largest of the 48 conterminous states. Texas's US rank slipped to 2d when Alaska entered the Union in 1959.

The total area of Texas is 266,807 sq mi (691,030 sq km), of which land comprises 262,017 sq mi (678,624 sq km) and inland water 4,790 sq mi (12,406 sq km). The state's land area represents 8.8% of the US mainland and 7.4% of the nation as a whole. The state's maximum E–W extension is 801 mi (1,289 km); its extreme N–S distance is 773 mi (1,244 km).

Texas is bordered on the N by Oklahoma and Arkansas (with part of the line formed by the Red River); on the E by Arkansas and Louisiana (with part of the Louisiana line defined by the Sabine River); on the SE by the Gulf of Mexico; on the SW by the Mexican states of Tamaulipas, Nuevo León, Coahuila, and Chihuahua (with the line formed by the Rio Grande); and on the W by New Mexico. The state's geographic center is in McCulloch County, 15 mi (24 km) NE of Brady.

Large islands in the Gulf of Mexico belonging to Texas are Galveston, Matagorda, and Padre. The boundary length of the state totals 3,029 mi (4,875 km), including a general Gulf of Mexico coastline of 367 mi (591 km); the tidal shoreline is 3,359 mi (5,406 km).

²TOPOGRAPHY

Texas's major physiographic divisions are the Gulf Coastal Plain in the east and southeast; the North Central Plains, covering most of central Texas; the Great Plains, extending from west-central Texas up into the panhandle; and the mountainous trans-Pecos area in the extreme west.

Within the Gulf Coastal Plain are the Piney Woods, an extension of western Louisiana that intrudes into East Texas for about 125 mi (200 km), and the Post Oak Belt, a flat region of mixed soil that gives way to the rolling prairie of the Blackland Belt, the state's most densely populated region. The Balcones Escarpment (so-called by the Spanish because its sharp profile suggests a balcony), a geological fault line running from the Rio Grande near Del Rio across central Texas, separates the Gulf Coastal Plain and Rio Grande Plain from the North Central Plains and south-central Hill Country, and in so doing, divides East Texas from West Texas, watered Texas from dry Texas, and (culturally speaking) the Old South from the burgeoning West.

The North Central Plains extend from the Blackland Belt to the Cap Rock Escarpment, a natural boundary carved by erosion to heights of nearly 1,000 feet (300 meters) in some places. Much of this plains region is rolling prairie, but the dude ranches of the Hill Country and the mineral-rich Burnet-Llano Basin are also found here. West of the Cap Rock Escarpment are the Great Plains, stretching north–south from the Panhandle Plains to the Edwards Plateau, just north of the Balcones Escarpment. Along the western edge of the panhandle and extending into New Mexico is the Llano Estacado (Staked Plains), an extension of the High Plains lying east of the base of the Rocky Mountains.

The trans-Pecos region, between the Pecos River and the Rio Grande, contains the highest point in the state: Guadalupe Peak, with an altitude of 8,751 feet (2,667 meters), part of the Guadalupe Range extending southward from New Mexico into western Texas for about 20 mi (32 km). Also in the trans-Pecos region is the Diablo Plateau, which has no runoff to the sea and holds its scant water in lakes that often evaporate entirely. Farther south are the Davis Mountains, with a number of peaks rising above 7,000 feet (2,100 meters), and Big Bend country (surrounded on three sides by the Rio Grande), whose canyons sometimes reach depths of nearly 2,000 feet (600 meters). The Chisos Mountains, also exceeding 7,000 feet (2,100 meters) at some points, stand just north and west of the Rio Grande.

For all its vast expanse, Texas boasts few natural lakes. Caddo Lake, which lies in Texas and Louisiana, is the state's largest natural lake, though its present length of 20 mi (32 km) includes waters added by dam construction in Louisiana. Two artificial reservoirs—Amistad (shared with Mexico), near Del Rio, and Toledo Bend (shared with Louisiana) on the Sabine River—have respective storage capacities exceeding 3 million and 4 million acre-feet, and the Sam Rayburn Reservoir (covering 179 sq mi—464 sq km) has a capacity of 2.9 million acre-feet. All together, the state contains close to 200 major reservoirs, eight of which can store more than 1 million acre-feet of water. From the air, Texas looks as well-watered as Minnesota, but the lakes are artificial and much of the soil is dry.

One reason Texas has so many reservoirs is that it is blessed with a number of major river systems, although none is navigable for more than 50 mi (80 km) inland. Starting from the west, the Rio Grande, a majestic stream in some places but a trickling

517

trough in others, imparts life to the Texas desert and serves as the international boundary with Mexico. Its total length of 1,896 mi (3,051 km), including segments in Colorado and New Mexico, makes the Rio Grande the nation's 2d-longest river, exceeded only by the Missouri-Mississippi river system. The Colorado River is the longest river wholly within the state, extending about 600 mi (970 km) on its journey across central and southeastern Texas to the Gulf of Mexico. Other important rivers include the Nueces, in whose brushy valley the range cattle industry began; the San Antonio, which stems from springs within the present city limits and flows, like most Texas rivers, to the Gulf of Mexico; the Brazos, which rises in New Mexico and stretches diagonally for about 840 mi (1,350 km) across Texas; the Trinity, which serves Ft. Worth and Dallas; the San Jacinto, a short river but one of the most heavily trafficked in North America, overlapping the Houston Ship Channel, which connects the Port of Houston with the Gulf; the Neches, which makes an ocean port out of Beaumont; the Sabine, which has the largest water discharge (6,800,000 acre-feet) at its mouth of any Texas river; the Red, forming part of the northern boundary; and the Canadian, which crosses the Texas panhandle from New Mexico to Oklahoma, bringing moisture to the cattle raisers and wheat growers of that region. In all, Texas has about 3,700 identifiable streams, many of which dry up in the summer and flood during periods of rainfall.

Because of its extensive outcroppings of limestone, extending westward from the Balcones Escarpment, Texas contains a maze of caverns. Among the better-known caves are Longhorn Cavern in Burnet County; Wonder Cave, near San Marcos; the Caverns of Sonora, at Sonora; and Jack Pit Cave, in Menard County, which, with 19,000 feet (5,800 meters) of passages, is the most extensive cave yet mapped in the state.

About 1 billion years ago, shallow seas covered much of Texas. After the seas receded, the land dropped gradually over millions of years, leaving a thick sediment that was then compressed into a long mountain range called the Ouachita Fold Belt. The sea was eventually restricted to a zone in West Texas called the Permian Basin, a giant evaporation pan holding gypsum and salt deposits hundreds of feet deep. As the mountain chain across central Texas eroded and the land continued to subside, the Rocky Mountains were uplifted, leaving deep cuts in Big Bend country and creating the Llano Estacado. The Gulf of Mexico area subsided rapidly, depositing sediment accumulations several thousand feet deep, while salt domes formed over vast petroleum and sulfur deposits. All this geologic activity also deposited quicksilver in the Terlingua section of the Big Bend, built up the Horseshoe Atoll (a buried reef in west-central Texas that is the largest limestone reservoir in the nation), created uranium deposits in southern Texas, and preserved the oil-bearing Jurassic rocks of the northeast.

³CLIMATE

Texas's great size and topographic variety make climatic description difficult. Brownsville, at the mouth of the Rio Grande, has had no measurable snowfall during all the years that records have been kept, but Vega, in the panhandle, averages 23 in (58 cm) of snowfall a year. Near the Louisiana border, rainfall exceeds 56 in (142 cm) annually, while in parts of extreme West Texas, rainfall averages less than 8 in (20 cm).

Generally, a maritime climate prevails along the Gulf coast, with continental conditions inland; the Balcones Escarpment is the main dividing line between the two zones, but they are not completely isolated from each other's influence. Texas has two basic seasons—a hot summer that may last from April through October, and a winter that starts in November and usually lasts until March. When summer ends, the state is too dry for autumn foliage, except in East Texas. Temperatures in El Paso, in the southwest, range from a mean January minimum of 30°F (−1°C) to a mean July maximum of 95°F (35°C); at Amarillo, in the

panhandle, from 23°F (−5°C) in January to 91°F (33°C) in July; and at Galveston, on the Gulf, from 48°F (9°C) in January to 88°F (31°C) in August. Perhaps the most startling contrast is in relative humidity, averaging 34% at noon in El Paso, 44% in Amarillo, and 72% in Galveston. In the Texas panhandle, the average date of the first freeze is 1 November; in the lower Rio Grande Valley, 16 December. The last freeze arrives in the panhandle on 15 April, and in the lower Rio Grande Valley on 30 January. The valley thus falls only six weeks short of having a 12-month growing season, while the panhandle approximates the growing season of the upper Midwest.

Record temperatures range from −23°F (−31°C) at Tulia, on 12 February 1899, and at Seminole, on 8 February 1933, to 120°F (49°C) at Seymour in north-central Texas on 12 August 1936. The greatest annual rainfall was 109 in (277 cm), measured in 1873 at Clarksville, just below the Red River in northeast Texas; the least annual rainfall, 1.76 in (4.47 cm), was recorded at Wink, near the New Mexico line, in 1956. Thrall, in central Texas, received 38.2 in (97 cm) of rain in 24 hours on 9–10 September 1921. Romero, on the New Mexico border, received a record 65 in (165 cm) of snow in the winter of 1923/24, and Hale Center, near Lubbock, measured 33 in (84 cm) during one storm in February 1956. The highest sustained wind velocity in Texas history, 145 mph (233 km/hr), occurred when Hurricane Carla hit Matagorda and Port Lavaca along the Gulf coast on 11 September 1961.

Hurricanes strike the Gulf coast about once every decade, usually in September or October. A hurricane on 19–20 August 1886 leveled the port of Indianola; the town (near present-day Port Lavaca) was never rebuilt. Galveston was the site of the most destructive storm is US history: on 8–9 September 1900, a hurricane blew across the island of 38,000 residents, leaving at least 6,000 dead (the exact total has never been ascertained) and leveling most of the city. A storm of equal intensity hit Galveston in mid-August 1915, but this time, the city was prepared; its new seawall held the toll to 275 deaths and $50 million worth of property damage. Because of well-planned damage-prevention and evacuation procedures, Hurricane Carla—at least as powerful as any previous hurricane—claimed no more than 34 lives. More recent hurricanes have frequently passed over the coastal area with no loss of life at all. Texas also lies in the path of "Tornado Alley," stretching across the Great Plains to Canada. The worst tornado in recent decades struck downtown Waco on 11 May 1953, killing 114 persons, injuring another 597, and destroying or damaging some 1,050 homes and 685 buildings. At least 115 tornadoes—the greatest concentration on record—occurred with Hurricane Beulah during 19–23 September 1967; the 67 tornadoes on 20 September set a record for the largest number of tornadoes on one day in the state.

Floods and droughts have also taken their toll in Texas. The worst flood occurred on 26–28 June 1954, when Hurricane Alice moved inland up the Rio Grande for several hundred miles, dropping 27 in (69 cm) of rain on Pandale above Del Rio. The Rio Grande rose 50 to 60 feet (15–18 meters) within 48 hours, as a wall of water 86 feet (26 meters) high in the Pecos River canyon fed it from the north. A Pecos River bridge built with a 50-foot (15-meter) clearance was washed out, as was the international bridge linking Laredo with Mexico. Periodic droughts afflicted Texas in the 1930s and 1950s.

⁴FLORA AND FAUNA

More than 500 species of grasses covered Texas when the Spanish and Anglo-Americans arrived. Although plowing and lack of soil

---

LOCATION: 25°50' to 36°30'N; 93°31' to 106°38'W. **BOUNDARIES:** Oklahoma line, *846 mi (1,362 km);* Arkansas line, *102 mi (164 km);* Louisiana line, *327 mi (526 km);* Gulf of Mexico coastline, 367 mi (591 km); Mexico line, *889 mi (1,431 km);* New Mexico line, *498 mi (801 km).*

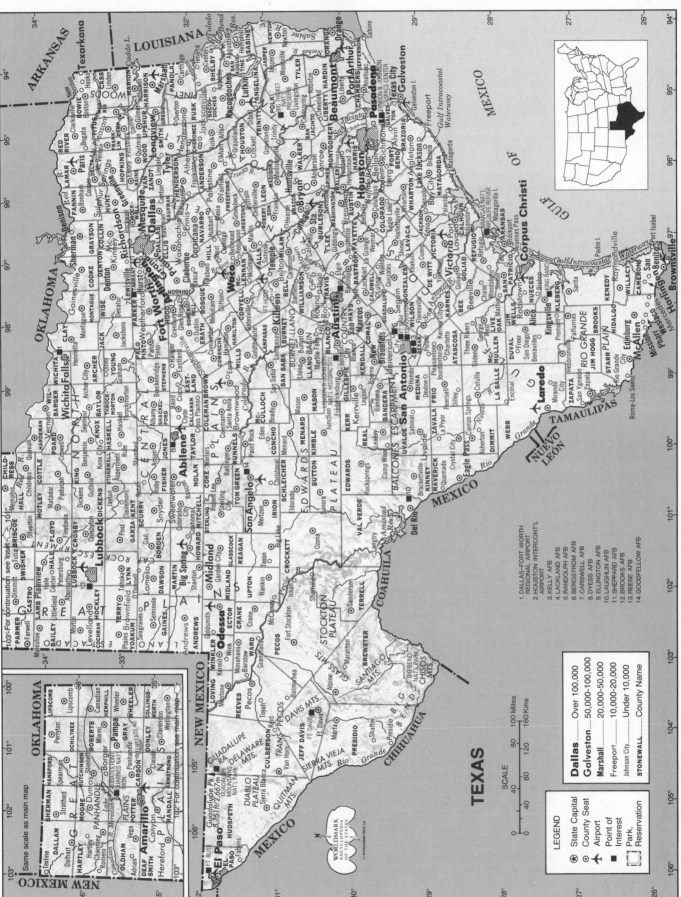

See endsheet maps: G4.

TEXAS

LEGEND

⊛ State Capital
⊙ County Seat
✈ Airport
■ Point of Interest
▢ Park, Reservation

Dallas .... Over 100,000
Galveston 50,000–100,000
Marshall 20,000–50,000
Freeport 10,000–20,000
Johnson City .... Under 10,000
STONEWALL .... County Name

SCALE
0   40   80   120   100 Miles
0   40   80   120   160 Kms

1. DALLAS–FORT WORTH REGIONAL AIRPORT
2. HOUSTON INTERCONTL. AIRPORT
3. KELLY AFB
4. LACKLAND AFB
5. RANDOLPH AFB
6. BERGSTROM AFB
7. CARSWELL AFB
8. DYESS AFB
9. ELLINGTON AFB
10. LAUGHLIN AFB
11. SHEPPARD AFB
12. BROOKS AFB
13. REESE AFB
14. GOODFELLOW AFB

conservation destroyed a considerable portion of this rich heritage, grassy pastureland still covers about two-thirds of the state. Bermuda grass is a favorite ground cover, especially an improved type called Coastal Bermuda, introduced after World War II. The prickly pear cactus is a mixed blessing: like the cedar and mesquite, it saps moisture and inhibits grass growth, but it does retain moisture in periods of drought and will survive the worst dry spells, so that (with the spines burned off) it can be of great value to ranchers as cattle feed in difficult times. The bean of the mesquite also provides food for horses and cattle when they have little else to eat, and its wood is a favorite in barbecues and fireplaces.

Texas has more than 200 native trees, of which the catclaw, flowering mimosa, huisache, black persimmon, huajillo, and weeping juniper (unique to the Big Bend) are common only in Texas. Cottonwood grows along streams in almost every part of the state, while cypress inhabits the swamps. The flowering dogwood in East Texas draws tourists to that region every spring, and the largest bois d'arc trees in the US are grown in the Red River Valley. Probably the most popular shade tree is the American (white) elm, which, like the gum tree, has considerable commercial importance. The magnolia is treasured for its grace and beauty; no home of substance in southeastern Texas would have a lawn without one. Of the principal hardwoods, the white oak is the most commercially valuable, the post oak the most common, and the live oak the most desirable for shade; the pecan is the state tree. Pines grow in two areas about 600 mi (970 km) apart—deep East Texas and the trans-Pecos region. In southeast Texas stands the Big Thicket, a unique area originally covering more than 3 million acres (1.2 million hectares) but now reduced to about one-tenth that by lumbering. Gonzales County, in south-central Texas, is the home of palmettos, orchids, and other semitropical plants not found anywhere else in the state. Texas wild rice and several cactus species are classified as endangered throughout the state.

Possibly the rarest mammal in Texas is the red wolf, which inhabits the marshland between Houston and Beaumont, one of the most thickly settled areas of the state; owing to human encroachment and possible hybridization with coyotes, the red wolf is steadily disappearing despite efforts by naturalists throughout the US to save it. On the other hand, Texans claim to have the largest number of white-tailed deer of any state in the Union, an estimated 3 million. Although the Hill Country is the white-tailed deer's natural habitat, the species has been transplanted successfully throughout the state.

Perhaps the most unusual mammal in Texas is the nine-banded armadillo. Originally confined to the Rio Grande border, the armadillo has gradually spread northward and eastward, crossing the Red River into Oklahoma and the Mississippi River into the Deep South. It accomplishes these feats of transport by sucking in air until it becomes buoyant and then swimming across the water. The armadillo is likewise notable for always having its young in litters of identical quadruplets. The chief mammalian predators are the coyote, bobcat, and mountain lion.

Texas attracts more than 825 different kinds of birds, with bird life most abundant in the lower Rio Grande Valley and coastal plains. Argument continues as to whether Texas is the last home of the ivory-billed woodpecker, which lives in inaccessible swamps, preferably in cutover timber. Somewhat less rare is the pileated woodpecker, which also inhabits the forested lowlands. Other characteristic birds include the yellow-trimmed hooded warbler, which frequents the canebrakes and produces one of the most melodious songs of any Texas bird; the scissor-tailed flycatcher, known popularly as the scissor-tail; Attwater's greater prairie chicken, now declining because of inadequate protection from hunters and urbanization; the mockingbird, the state bird; and the roadrunner, also known as paisano and chaparral. Rare

birds include the Mexican jacana, with a fleshy comb and bright yellow-green wings; the white-throated swift, one of the world's fastest flyers; the Texas canyon wren, with a musical range of more than an octave; and the Colima warbler, which breeds only in the Chisos Mountains. In the Aransas National Wildlife Refuge, along the central Gulf coast, lives the whooping crane, long on the endangered list and numbering 69 adults and 14 young in 1984. Controversy surrounds the golden eagle, protected by federal law but despised by ranchers for allegedly preying on lambs and other young livestock.

Texas has its fair share of reptiles, including more than 100 species of snake, 16 of them poisonous, notably the deadly Texas coral snake. There are 10 kinds of rattlesnake, and some parts of West Texas hold annual rattlesnake roundups. Disappearing with the onset of urbanization are the horned toad, a small iguanalike lizard; the vinegarroon, a stinging scorpion; and the tarantula, a large, black, hairy spider that is scary to behold but basically harmless.

In addition to providing protection for the animals on federal lists of threatened and endangered species, the state has its own wildlife protection programs. Among the animals classified as nongame (not hunted) and therefore given special consideration are the lesser yellow bat, spotted dolphin, reddish egret, white-tailed hawk, wood stork, Big Bend gecko, rock rattlesnake, Louisiana pine snake, white-lipped frog, giant toad, toothless blindcat, and blue sucker. Along with the red wolf, Attwater's greater prairie chicken, and ivory-billed woodpecker, endangered species include the American alligator, jaguar, ocelot, Eskimo curlew, Houston toad, Big Bend gambusia, Comanche Springs pupfish, and fountain darter.

Texas has 15 National Wildlife Refuges, with a total of 302,731 acres (122,511 hectares). The Texas Parks and Wildlife Department administers an additional 19 wildlife management areas.

⁵ENVIRONMENTAL PROTECTION

Conservation in Texas officially began with the creation of a State Department of Forestry in 1915; 11 years later, this body was reorganized as the Texas Forest Service, the name it retains today. The state's Soil Conservation Service was created in 1935.

The scarcity of water is the one crisis every Texan must live with. Much of the state has absorbent soils, a high evaporation rate, vast areas without trees to hold moisture, and a rolling terrain susceptible to rapid runoff. The Texas Water Commission and Water Development Board direct the state's water supply and conservation programs. Various county and regional water authorities have been constituted, as have several water commissions for river systems. Probably the most complete system is that of the three Colorado River authorities—lower, central, and upper. The oldest of these is the Lower Colorado River Authority, created in 1934 by the Texas legislature to "control, store, preserve, and distribute" the waters of the Colorado River and its feeder streams. The authority exercises control over a 10-county area stretching from above Austin to the Gulf coast, overseeing flood control, municipal and industrial water supplies, irrigation, hydroelectric power generation, soil conservation, and recreation.

The most powerful conservation agency in Texas is the Railroad Commission. Originally established to regulate railroads, the commission extended its power to regulate oil and natural gas by virtue of its jurisdiction over the transportation of those products by rail and pipeline. In 1917, the state legislature empowered the commission to prevent the waste of oil and gas. The key step in conservation arrived with the discovery of oil in East Texas in 1930. With a national depression in full swing and the price of oil dropping to $1 a barrel, the commission agreed to halt ruinous overproduction, issuing the first proration order in April 1931. In a field composed of hundreds of small ownerships, however, control was difficult to establish: oil was bootlegged, the commission's authority broke down, Governor Ross S. Sterling declared martial law, and the state's conservation edicts were not heeded until the federal government stepped in to enforce them. Today, the Railroad Commission acts to eliminate wasteful drilling procedures and decides which equipment and techniques are permissible. In addition, the commission requires careful accounting of all production and sales.

As in other states, hazardous wastes have become an environmental concern in Texas. In 1984, for example, a suit was brought against eight oil and chemical companies, including both Exxon and Shell Oil, alleging that they had dumped hazardous wastes at four sites in Harris County. The agency that oversees compliance with hazardous-waste statutes is the Hazardous and Solid Waste Division of the Texas Water Commission.

⁶POPULATION

According to 1980 census figures, Texas ranked 3d behind California and New York, with a population of 14,229,191. The state had placed 4th at the 1970 census, with a population of 11,196,730, but had surpassed Pennsylvania in 1974. The estimated population in 1985 was 16,384,800, yielding a population density of 63 per sq mi (24 per sq km). The US Census Bureau projects a population of 17,498,200 in 1990.

At the first decennial census of 1850, less than five years after Texas had become a state, the population totaled 212,592. It reached 1,600,000 by the early 1880s (when the state ranked 11th), passed 4,000,000 during World War I, and jumped to 7,700,000 in 1950. The slowest period of growth occurred during the depression decade (1930–40), when the population rose only 10%, and the state was surpassed by California. The growth rate ranged between 17% and 27% for each decade from the 1940s through the 1970s, and was 15% between 1980 and 1985.

The ratio between the sexes has changed during the 20th century. In 1920, the state had 155,000 more men than women; 40 years later, women outnumbered men by 90,000, and in 1980, the female lead was 231,000. At the same time, the Texas population has grown steadily older, a phenomenon linked to declining birthrates and increased life expectancies. In 1870, only one out of 68 Texans was 65 years of age or older; by 1980, the proportion was one out of 10. Surprisingly for a state whose population has grown so fast, fully 72% of all state residents were born in Texas.

In 1980, about 80% of all Texans lived in metropolitan areas. The largest, Houston-Galveston-Brazoria, had an estimated 3,565,700 people (8th in the US) in mid-1984. Close behind was the Dallas–Ft. Worth area, with 3,348,000 residents (10th). San Antonio ranked 33d nationally at 1,188,500. Houston, the largest city in Texas and 4th largest in the US, had an estimated 1984 population of 1,088,973. Next was Dallas with 974,234, followed by San Antonio 842,779, El Paso 463,809, Ft. Worth 414,562, Austin 397,001, Corpus Christi 258,067, Arlington 213,832, and Lubbock 178,529. With the exception of El Paso, in the far western corner of the trans-Pecos region, most of the larger cities are situated along the Gulf coast or on or near an axis that extends north–south from Wichita Falls to Corpus Christi, in the heart of the Blackland Belt.

⁷ETHNIC GROUPS

Hispanic Americans constitute the largest ethnic minority in Texas, about 21% of the population in 1980. Black Americans accounted for 12%, and most of the rest were "Anglos"—a term denoting all whites except Spanish-surnamed or Spanish-speaking individuals. Although many Indian groups have lived in Texas, few Indians remain.

As white settlers pushed toward Texas during the 19th century, many Indian groups moved west and south into the region. The most notable tribes were the Comanche, Wichita, Kiowa, Apache, Choctaw, and Cherokee. Also entering in significant numbers were the Kickapoo and Potawatomi from Illinois, the Delaware and Shawnee from Missouri, the Quapaw from Arkansas, and the Creek from Alabama and Georgia. One of the few Texas tribes

that has survived to the present time as an identifiable group is the Alabama-Coushatta, who inhabit a 4,351-acre (1,761-hectare) reservation in Polk County, 90 mi (145 km) northeast of Houston. The Tigua, living in Texas since the 1680s, were recognized by a federal law in 1968 that transferred all responsibility for them to the State of Texas. The two Indian reservations number about 500 persons each; overall, at the 1980 census, there were 39,374 American Indians, 395 Eskimos, and 305 Aleuts living in Texas.

Blacks have been integral to the history of Texas ever since a black Moor named Estevanico was shipwrecked near present-day Galveston in 1528. By 1860, Texas had 182,921 blacks, or 30% of the total population, of whom only 355 were free. Once emancipated, blacks made effective use of the franchise, electing two of their number to the state senate and nine to the house in 1868. After the return of the Democratic Party to political dominance, however, the power of blacks steadily diminished. Since then, their numbers have grown, but their proportion of the total population has dwindled, although Houston and Dallas were, respectively, about 28% and 29% black at the 1980 census. In 1980, 1,710,250 blacks lived in the state.

Hispanic Americans, the largest and fastest-growing minority in Texas, numbered 2,985,643 in 1980 (2d only to California). Mostly of Mexican ancestry, they are nevertheless a heterogeneous group, divided by history, geography, and economic circumstances. These differences have prevented them from becoming the political force their numbers would suggest, although Hispanics have been elected to the state legislature and to the US Congress. In 1980, the Houston independent school district, the state's largest, reported more Hispanic students than Anglos for the first time in its history.

All together, Texas has nearly 30 identifiable ethnic groups. Certain areas of central Texas are heavily Germanic and Czech. The first permanent Polish colony in the US was established at Panna Maria, near San Antonio, in 1854. Texas has one of the largest colonies of Wends in the world, principally at Serbin in central Texas. Significant numbers of Danes, Swedes, and Norwegians have also settled in Texas. As of 1980, foreign-born Texans numbered 856,213, with Mexico, Germany, and the United Kingdom the leading countries of origin. The same census counted 25,549 Chinese, 22,226 Asian Indians, 15,096 Filipinos, 13,997 Koreans, and 10,502 Japanese. Of the 29,112 Vietnamese, most were refugees who resettled in Texas beginning in 1975.

⁸LANGUAGES

The Indians of Texas are mostly descendants of the Alabama-Coushatta who came to Texas in the 19th century. The few Indian place-names include Texas itself, Pecos, Waco, and Toyah.

Most of the regional features in Texas English derive from the influx of South Midland and Southern speakers, with a noticeable Spanish flavor from older as well as more recent loans. Settlers from the Gulf Coast states brought such terms as *snap beans* (green beans), the widespread *pail* (here probably of Southern rather than Northern origin), and *carry* (escort), with a 47% frequency in north Texas and 22% in the south. Louisiana *praline* (pecan patty) is now widespread, but *banquette* (sidewalk) appears only in the extreme southeast corner.

Southern and South Midland terms were largely introduced by settlers from Arkansas, Missouri, and Tennessee; their use ranges from northeast to west, but with declining frequency in the trans-Pecos area. Examples are *clabber cheese* (cottage cheese), *mosquito hawk* (dragonfly), *croker sack* (burlap bag), *mouth harp* (harmonica), *branch* (stream), and *dog irons* (andirons). A dialect survey showed *pallet* (bed on the floor) with a 90% overall frequency; *light bread* (white bread) and *pullybone* (wishbone), each 78%; and *you-all*, more than 80%. General Midland terms also widespread in the state are *sook!* (call to calves), *blinds* (roller shades), *piece* (a certain distance), and *quarter till five* (4:45).

Some terms exhibit uneven distribution. Examples include *mott*

Texas Counties, County Seats, and County Areas and Populations

COUNTY	COUNTY SEAT	LAND AREA (SQ MI)	POPULATION (1984 EST.)	COUNTY	COUNTY SEAT	LAND AREA (SQ MI)	POPULATION (1984 EST.)
Anderson	Palestine	1,077	45,807	Duval	San Diego	1,795	13,346
Andrews	Andrews	1,501	15,956	Eastland	Eastland	924	21,075
Angelina	Lufkin	807	68,337	Ector	Odessa	903	144,477
Aransas	Rockport	280	17,097	Edwards	Rocksprings	2,120	2,143
Archer	Archer City	907	8,000	Ellis	Waxahachie	939	67,317
Armstrong	Claude	910	1,951	El Paso	El Paso	1,014	526,465
Atascosa	Jourdanton	1,218	27,989	Erath	Stephenville	1,080	24,441
Austin	Bellville	656	20,472	Falls	Marlin	770	18,213
Bailey	Muleshoe	827	8,219	Fannin	Bonham	895	24,612
Bandera	Bandera	793	8,408	Fayette	La Grange	950	20,099
Bastrop	Bastrop	895	31,119	Fisher	Roby	897	5,774
Baylor	Seymour	862	5,087	Floyd	Floydada	992	9,013
Bee	Beeville	880	28,712	Foard	Crowell	703	1,913
Bell	Belton	1,055	167,167	Ft. Bend	Richmond	876	179,334
Bexar	San Antonio	1,248	1,092,142	Franklin	Mt. Vernon	294	7,313
Blanco	Johnson City	714	5,345	Freestone	Fairfield	888	16,745
Borden	Gail	900	976	Frio	Pearsall	1,133	14,476
Bosque	Meridian	989	14,120	Gaines	Seminole	1,504	14,300
Bowie	Boston	891	79,199	Galveston	Galveston	399	215,366
Brazoria	Angleton	1,407	186,113	Garza	Post	895	5,645
Brazos	Bryan	588	117,355	Gillespie	Fredericksburg	1,061	15,207
Brewster	Alpine	6,169	8,096	Glasscock	Garden City	900	1,263
Briscoe	Silverton	887	2,331	Goliad	Goliad	859	5,759
Brooks	Falfurrias	942	9,110	Gonzales	Gonzales	1,068	18,659
Brown	Brownwood	936	35,688	Gray	Pampa	921	27,285
Burleson	Caldwell	668	15,137	Grayson	Sherman	934	94,699
Burnet	Burnet	994	21,636	Gregg	Longview	273	111,164
Caldwell	Lockhart	546	26,437	Grimes	Anderson	799	17,095
Calhoun	Port Lavaca	540	22,319	Guadalupe	Seguin	713	53,243
Callahan	Baird	899	12,610	Hale	Plainview	1,005	37,320
Cameron	Brownsville	905	241,090	Hall	Memphis	876	5,092
Camp	Pittsburg	203	10,229	Hamilton	Hamilton	836	8,171
Carson	Panhandle	924	7,088	Hansford	Spearman	921	6,516
Cass	Linden	937	30,676	Hardeman	Quanah	688	6,548
Castro	Dimmitt	899	10,257	Hardin	Kountze	898	43,414
Chambers	Anahuac	616	19,823	Harris	Houston	1,734	2,747,341
Cherokee	Rusk	1,052	39,452	Harrison	Marshall	908	57,243
Childress	Childress	707	6,554	Hartley	Channing	1,462	3,706
Clay	Henrietta	1,085	9,822	Haskell	Haskell	901	7,563
Cochran	Morton	775	4,753	Hays	San Marcos	678	49,540
Coke	Robert Lee	908	3,608	Hemphill	Canadian	903	6,198
Coleman	Coleman	1,277	10,468	Henderson	Athens	888	50,205
Collin	McKinney	851	178,636	Hidalgo	Edinburg	1,569	337,118
Collingsworth	Wellington	909	4,190	Hill	Hillsboro	968	26,884
Colorado	Columbus	964	20,015	Hockley	Levelland	908	24,501
Comal	New Braunfels	555	43,159	Hood	Granbury	425	24,201
Comanche	Comanche	930	13,103	Hopkins	Sulphur Springs	789	27,753
Concho	Paint Rock	992	2,966	Houston	Crockett	1,234	23,404
Cooke	Gainesville	893	28,818	Howard	Big Spring	901	37,283
Coryell	Gatesville	1,057	58,512	Hudspeth	Sierra Blanca	4,566	2,649
Cottle	Paducah	895	2,697	Hunt	Greenville	840	63,284
Crane	Crane	782	5,202	Hutchinson	Stinnett	871	29,497
Crockett	Ozona	2,806	4,954	Irion	Mertzon	1,052	1,719
Crosby	Crosbyton	898	8,523	Jack	Jacksboro	920	7,597
Culberson	Van Horn	3,815	3,476	Jackson	Edna	844	13,833
Dallam	Dalhart	1,505	6,746	Jasper	Jasper	921	32,140
Dallas	Dallas	880	1,723,423	Jeff Davis	Ft. Davis	2,258	1,749
Dawson	Lamesa	903	16,888	Jefferson	Beaumont	937	258,532
Deaf Smith	Hereford	1,497	20,401	Jim Hogg	Hebbronville	1,136	5,407
Delta	Cooper	278	4,925	Jim Wells	Alice	867	39,872
Denton	Denton	911	167,861	Johnson	Cleburne	731	81,848
DeWitt	Cuero	910	20,310	Jones	Anson	931	18,630
Dickens	Dickens	907	3,182	Karnes	Karnes City	753	13,520
Dimmet	Carrizo Springs	1,307	11,928	Kaufman	Kaufman	788	47,409
Donley	Clarendon	929	4,184	Kendall	Boerne	663	13,161

Texas Counties, County Seats, and County Areas and Populations

COUNTY	COUNTY SEAT	LAND AREA (SQ MI)	POPULATION (1984 EST.)	COUNTY	COUNTY SEAT	LAND AREA (SQ MI)	POPULATION (1984 EST.)
Kenedy	Sarita	1,389	498	Red River	Clarksville	1,054	17,783
Kent	Jayton	878	1,110	Reeves	Pecos	2,626	15,905
Kerr	Kerrville	1,107	32,673	Refugio	Refugio	771	9,243
Kimble	Junction	1,250	4,252	Roberts	Miami	915	1,088
King	Guthrie	914	430	Robertson	Franklin	864	15,949
Kinney	Brackettville	1,359	2,383	Rockwall	Rockwall	128	19,056
Kleberg	Kingsville	853	34,661	Runnels	Ballinger	1,056	12,382
Knox	Benjamin	845	5,535	Rusk	Henderson	932	43,992
Lamar	Paris	919	44,402	Sabine	Hemphill	486	9,725
Lamb	Littlefield	1,013	17,400	San Augustine	San Augustine	524	8,869
Lampasas	Lampasas	714	13,408	San Jacinto	Coldspring	572	13,551
La Salle	Cotulla	1,517	5,874	San Patricio	Sinton	693	62,953
Lavaca	Hallettsville	971	18,399	San Saba	San Saba	1,136	5,972
Lee	Giddings	631	13,570	Schleicher	Eldorado	1,309	3,218
Leon	Centerville	1,078	11,293	Scurry	Snyder	900	20,113
Liberty	Liberty	1,174	54,120	Shackelford	Albany	915	4,064
Limestone	Groesbeck	931	21,411	Shelby	Center	791	23,859
Lipscomb	Lipscomb	933	4,204	Sherman	Stratford	923	3,225
Live Oak	George West	1,057	9,822	Smith	Tyler	932	144,611
Llano	Llano	939	11,693	Somervell	Glen Rose	188	4,454
Loving	Mentone	671	79	Starr	Rio Grande City	1,226	32,496
Lubbock	Lubbock	900	218,861	Stephens	Breckenridge	894	10,652
Lynn	Tahoka	888	7,992	Sterling	Sterling City	923	1,556
McCulloch	Brady	1,071	8,830	Stonewall	Aspermont	925	2,456
McLennan	Waco	1,031	182,115	Sutton	Sonora	1,455	5,789
McMullen	Tilden	1,163	907	Swisher	Tulia	902	9,035
Madison	Madisonville	473	11,922	Tarrant	Ft. Worth	868	1,010,272
Marion	Jefferson	385	10,662	Taylor	Abilene	917	123,107
Martin	Stanton	914	5,299	Terrell	Sanderson	2,357	1,494
Mason	Mason	934	3,560	Terry	Brownfield	886	15,300
Matagorda	Bay City	1,127	37,669	Throckmorton	Throckmorton	912	2,320
Maverick	Eagle Pass	1,287	35,419	Titus	Mt. Pleasant	412	23,007
Medina	Hondo	1,331	24,679	Tom Green	San Angelo	1,515	95,998
Menard	Menard	902	2,398	Travis	Austin	989	499,127
Midland	Midland	902	113,585	Trinity	Groveton	692	11,199
Milam	Cameron	1,019	23,397	Tyler	Woodville	922	18,333
Mills	Goldthwaite	748	4,571	Upshur	Gilmer	587	32,316
Mitchell	Colorado City	912	9,250	Upton	Rankin	1,243	5,554
Montague	Montague	928	18,495	Uvalde	Uvalde	1,564	23,705
Montgomery	Conroe	1,047	160,008	Val Verde	Del Rio	3,150	40,190
Moore	Dumas	905	17,502	Van Zandt	Canton	855	35,920
Morris	Daingerfield	256	15,349	Victoria	Victoria	887	74,495
Motley	Matador	959	1,851	Walker	Huntsville	786	50,142
Nacogdoches	Nacogdoches	939	50,634	Waller	Hempstead	514	23,374
Navarro	Corsicana	1,068	38,543	Ward	Monahans	836	15,995
Newton	Newton	935	13,406	Washington	Brenham	610	24,747
Nolan	Sweetwater	915	17,884	Webb	Laredo	3,363	118,171
Nueces	Corpus Christi	847	298,359	Wharton	Wharton	1,086	41,185
Ochiltree	Perryton	919	11,108	Wheeler	Wheeler	905	8,232
Oldham	Vega	1,485	2,410	Wichita	Wichita Falls	606	126,175
Orange	Orange	362	89,931	Wilbarger	Vernon	947	16,712
Palo Pinto	Palo Pinto	949	26,232	Willacy	Raymondville	589	18,588
Panola	Carthage	812	22,672	Williamson	Georgetown	1,137	96,775
Parker	Weatherford	902	52,246	Wilson	Floresville	807	18,323
Parmer	Farwell	885	10,832	Winkler	Kermit	840	10,981
Pecos	Ft. Stockton	4,776	17,054	Wise	Decatur	902	30,672
Polk	Livingston	1,061	29,437	Wood	Quitman	689	27,338
Potter	Amarillo	902	106,672	Yoakum	Plains	800	9,385
Presidio	Marfa	3,857	5,369	Young	Graham	919	19,872
Rains	Emory	243	5,753	Zapata	Zapata	999	8,004
Randall	Canyon	917	83,579	Zavala	Crystal City	1,298	12,133
Reagan	Big Lake	1,173	4,882				
Real	Leakey	697	2,723	TOTALS		262,017	15,988,538

(clump of trees) in the south and southwest, *sugan* (a wool-filled comforter for a cowboy's bedroll) in the west, Midland *draw* (dry streambed) in the west and southwest, South Midland *peckerwood* (woodpecker) in most of the state except west of the Pecos, *poke* (paper bag) in the central and northern areas, and *surly* (euphemism for bull) in the west. A curious result of dialect mixture is the appearance of a number of hybrids combining two different dialects, such as *freeseed* peach from *freestone* and *clearseed, fire mantel* and *mantel board* from *fireboard* and *mantel, flapcakes* from *flapjacks* and *pancakes,* and *horse doctor* from *horsefly* and *snake doctor.* The large sandwich is known as a *torpedo* in San Antonio and a *poorboy* in Houston.

Texas pronunciation is largely South Midland, with such characteristic forms as /caow/, and /naow/ for *cow* and *now* and /dyoo/ for *due,* although /doo/ is now more common in urban areas. In the German settlement around New Braunfels are heard a few loanwords such as *smearcase* (cottage cheese), *krebbel* (doughnut), *clook* (setting hen), and *oma* and *opa* for grandmother and grandfather.

Spanish has been the major foreign-language influence. In areas like Laredo and Brownsville, along the Rio Grande, as many as 90% of the people may be bilingual; in northeast Texas, however, Spanish is as foreign as French. In the days of the early Spanish ranchers, standard English adopted *hacienda, ranch, burro, canyon,* and *lariat;* in the southwestern cattle country are heard *la reata* (lasso), *remuda* (group of horses), and *resaca* (pond), along with the *acequia* (irrigation ditch), *pilon* (something extra, as a trip), and *olla* (water jar). The presence of the large Spanish-speaking population was a major factor in the passage of the state's bilingual education law, as a result of which numerous school programs in both English and Spanish are now offered; in a ruling issued in January 1981, US District Judge William Wayne Justice ruled that by 1987, the state must expand such programs to cover all Spanish-speaking students. About one-sixth of all Texas counties—and a great many cities—are named for Mexicans or Spaniards or after place-names in Spain or Mexico.

In 1980, 10,552,893 Texans—78% of the population 3 years old or older—spoke only English at home. Other languages most commonly spoken at home included:

Spanish	2,595,242	Chinese	22,455
German	85,595	Hindi or other	
French	48,525	Asian Indian	
Czech	26,396	language	16,976
Vietnamese	24,329	Korean	11,668

[9] RELIGIONS

Because of its Spanish heritage, Texas originally was entirely Roman Catholic except for unconverted Indians. Consequently, the early history of Texas is almost identical with that of the Roman Catholic Chuch in the area. Under the Mexican Republic, the Catholic Church continued as the sole recognized religious body. In order to receive the generous land grants given by the Mexicans, Anglo-American immigrants had to sign a paper saying that they followed the Catholic religion. With an average grant of 4,605 acres (1,864 hectares) as bait, many early Protestants and atheists must have felt little hesitancy about becoming instant Catholics.

The Mexican government was careless about enforcing adherence to the Catholic faith in Texas, however, and many Baptists, Methodists, and Presbyterians drifted in from the east. The Methodist practice of having itinerant ministers range over frontier areas was particularly well suited to the Texas scene, and in 1837, the church hierarchy sent three preachers to the new republic. The first presbytery had been formed by that date, and Baptists had organized in Houston by 1840. Swedish and German immigrants brought their Lutheranism with them; the first German Lutheran synod was organized in Houston in 1851.

Geographically, Texas tends to be heavily Protestant in the north and east, and Catholic in the south and southwest. Leading Protestant denominations and their known adherents in 1980 were Southern Baptist Convention, 2,659,894; United Methodist Church, 932,488; Churches of Christ, 355,396; Episcopal Church, 174,581; Assembly of God, 162,232; Presbyterian Church in the US, 130,895; Christian Church (Disciples of Christ), 120,296; Baptist Missionary Association of America, 118,643; Lutheran Church–Missouri Synod, 117,074; and American Lutheran Church, 106,657. Roman Catholics numbered 2,800,470 in 1984, when there were an estimated 78,470 Jews.

[10] TRANSPORTATION

Texas ranks 1st among the 50 states in total railroad mileage, highway mileage, and number of airports, and 2d only to California in motor vehicle registrations and in number of general aviation aircraft.

Transportation has been a severe problem for Texas because of the state's extraordinary size and sometimes difficult terrain; one of the more unusual experiments in US transport history was the use of camels in southwestern Texas during the mid-1800s. The Republic of Texas authorized railroad construction as early as 1836, but the financial panic of 1837 helped kill that attempt. Not until 1853 did the state's first railroad—from Harrisburg (now incorporated into Houston) to Stafford's Point, 20 mi (32 km) to the west—come into service. At the outbreak of the Civil War, 10 railroads were operating, all but 2 connected with seaports. Texas lacked sufficient capital to satisfy its railroad-building needs until the war was over, although the state legislature in 1852 had offered railroad companies eight sections (5,120 acres—2,072 hectares) of land per mile of road construction and doubled that offer two years later; the state generally held to the 10,240-acre (4,144-hectare) figure until all grants ceased in 1882. In all, Texas granted more than 50,000 sq mi (130,000 sq km) to railroad companies.

In 1870, Texas had fewer than 600 mi (970 km) of track; 10 years later, it had 3,026 mi (4,870 km); in 1890, 6,045 mi (9,729 km); and by 1920, 16,049 mi (25,828 km). A peak was reached in 1932, when there were 17,078 mi (27,484 km) of track; by the end of 1983, Class I trackage had dwindled to 12,917 mi (20,788 km). Three carriers—the Atchison, Topeka & Santa Fe, Missouri-Pacific, and Southern Pacific—control about three-fourths of the mileage. The only rail passenger service in Texas is provided by Amtrak, which runs two routes—the Sunset Limited (New Orleans–Los Angeles) from Beaumont through Houston and San Antonio to El Paso, and the Eagle (Chicago–San Antonio). In 1983/84, Amtrak carried 150,367 passengers within Texas.

In mid-1983, Dallas area voters approved the creation of the Dallas Area Rapid Transit system (DART), to serve the city and 13 suburbs; 160 mi (257 km) of surface rail routes were to be constructed and bus service doubled at an expense of some $8.9 billion over a 26-year period. Ft. Worth has the state's only true subway—a 1-mi (1.6-km) line from a parking lot to a downtown shopping and office center—although Dallas–Ft. Worth Regional Airport has its own rail shuttle system.

In 1983, Texas had 275,784 mi (443,832 km) of public roadway, 77% of it rural. The leading interstate highways are I-10 and I-20, respectively linking Houston and the Dallas–Ft. Worth areas with El Paso in the west, and I-35 and I-45, connecting Dallas–Ft. Worth with, respectively, San Antonio (via Austin) and Galveston (via Houston). There were 11,406,433 licensed drivers in 1983, as well as 12,021,771 registered vehicles, including 8,159,008 automobiles, 326,293 motorcycles, 3,482,625 trucks, and 53,845 buses. Trucking is vital to the Texas economy because more than 1,800 communities are served by no other freight carrier.

River transport did not become commercially successful until the end of the 19th century, when the Houston Ship Channel was dredged along the San Jacinto River and Buffalo Bayou for more than 50 mi (80 km), and another channel was dredged down the Neches River to make a seaport out of Beaumont. With 13 major seaports and many shallow-water ports, Texas has been a major factor in waterborne commerce since the early 1950s. In 1981, Texas ports handled 291,168,791 tons of cargo, 35% of that by the Port of Houston, among the nation's three most active harbors. The Gulf Intracoastal Waterway begins in Brownsville, at the mouth of the Rio Grande, and extends across Texas for 423 mi (681 km) on its way to Florida and its connections with a similar waterway on the Atlantic. In 1981, the waterway transported 68,016,979 tons of cargo.

After American entry into World War I, Texas began to build airfields for training grounds; when the war ended, many US fliers returned to Texas and became civilian commercial pilots, carrying air mail (from 1926), dusting crops, and mapping potential oil fields. In 1983, the state had 1,543 aircraft facilities, including 1,212 airports and 325 heliports. The Dallas–Ft. Worth Regional Airport, the nation's largest air terminal and its 7th busiest in 1983, serviced more than one-third of the aircraft departures in Texas, with another fifth handled by Houston Intercontinental Airport.

11HISTORY

Although a site near Lewisville, in Denton County, contains artifacts that might be more than 37,000 years old, the generally accepted date for the earliest human presence in the region now known as Texas is the Llano civilization, dating from 12,000 years ago. Prehistoric Indians in Texas failed to develop as complex technologies as their neighbors to the west and east. When the first Europeans arrived in the 16th century, the Indians had developed little in the way of pottery or basketry, and had shown little interest in agriculture except in the extreme east and northeast, and possibly west of the Pecos. They were still largely hunter-gatherers on whom the more technologically complex cultures of Mexico and the southeastern US had little effect.

Along the Gulf coast and overlapping into northeastern Mexico were the Coahuiltecan and Karankawa peoples. They lived in a hostile environment, consuming berries in season, animal dung, spiders, and an occasional deer, bison, or jabalina that had surrendered to time or been encircled by fire. In central Texas lived the Tonkawa, who hunted buffalo, slept in tepees, used dogs for hauling, and had a communal sensitivity akin to that of the Plains Indians. Unlike the Karankawa, who were tall, the Tonkawa were of average height, tattooed, and dressed in breechclouts—long for men, short for women. They proved extremely susceptible to European diseases and evidently died out, whereas the Karankawa migrated to northern Mexico.

About two dozen tribes of Caddo in eastern and northeastern Texas were, at the time of European penetration, the most technologically complex Indians living within the state's present borders. Having developed agriculture, the Caddo were relatively sedentary and village oriented. Those belonging to the Hasinai Confederation called each other *tayshas,* a term that translates as "allies" or "friends." When the Hasinai told Spanish explorers that they were *tayshas,* the Spaniards wrote the word as *Tejas,* which in time became *Texas.* The Caddo lived in the gentle portion of Texas, where woods, wild fruits, and berries abound, and where game was plentiful until the advent of European civilization. Life was so good, in fact, that several members of an expedition under Robert Cavelier, Sieur de la Salle, reaching Matagorda Bay on 15 February 1685, chose to desert to the Caddo rather than remain with their fellow Frenchmen. Henri de Tonti, who entered the region somewhat later, reported that one Caddo tribe had a woman as chief. The Caddo were also unusual in their belief that three women had created the world.

In trans-Pecos Texas, to the west, lived a fourth Indian group, the Jumano, probably descendants of the Pueblo cultures. Some of the Jumano were nomadic hunters in the Davis and Chisos mountains. Others became farmers along the Rio Grande and the lower Rio Conchos, making and using some pottery and raising good crops of corn, beans, squash, and possibly cotton. Probably the successive droughts so common to the region began to thin out their ranks, and the coming of the Spanish removed them from the historical picture altogether.

The first white man to enter Texas was Spanish explorer Alonso Alvarez de Pineda, who sailed into the mouth of the Rio Grande in 1519. Basically, the Spanish left the Texas Indians alone for more than 150 years. Sometimes an accident placed Spaniards in Texas, or sometimes they entered by design, but generally, the Spanish looked on Texas as too remote from Florida and the Mexico highlands—where most of their colonizing occurred—for successful settlement. A remarkable episode of this period involves the survivors of the Pánfilo de Narváez expedition, which had been commissioned to occupy the Gulf of Mexico coast from Mexico to Florida. Four shipwrecked men, led by Álvar Núñez Cabeza de Vaca, were washed ashore on a Texas sandbar on 6 November 1528; three were Spaniards, and one was the Moor Estevanico. For eight years, they wandered virtually naked among the Texas Indians, sometimes as slaves and sometimes as free men, alternately blistered by the summer sun and freezing under winter ice storms. Using a deer bone as a needle, Cabeza removed an arrowhead from deep in an Indian's chest—a bit of surgical magic that earned him treatment as a demigod, for a time. Finally, the four Europeans reached the west coast of Mexico, from where Cabeza de Vaca returned home a hero. The other two Spaniards remained in Mexico, but Estevanico joined the Fray Marcos de Niza expedition as a guide, dying at the hands of Pueblo Indians in New Mexico in 1539. The trail he helped blaze through the High Plains of West Texas served as the route for the expedition a year later by Francisco Vásquez de Coronado. The first Texas towns and missions were begun by Spaniards in West Texas, outside present-day El Paso. Ysleta del Sur was founded in 1682, almost a decade before the earliest East Texas missions. But Ysleta was 500 mi (800 km) from anything else resembling a settlement in Texas, and the Spanish considered it a part of New Mexico.

What changed the Spaniards' attitude toward the colonization of Texas was the establishment of Ft. St. Louis by La Salle on the Gulf coast in 1685. Four years later, Capt. Alonso de León, governor of Coahuila, sent out an expedition to expel the French. Father Damien Massanet, a Coahuilan priest, accompanied the León expedition and was charged with establishing a mission near wherever the captain built a fort. During the next several decades, these two men and their successors established a string of mission-forts across Texas. After fear of the French presence eased, Spain tended to neglect these establishments. But when the French entered Louisiana in force during the early 18th century, Spanish fears of French expansion were re-ignited. In 1718, the Spanish began to build a mission, San Antonio de Valero, and a fort, San Antonio de Bexar, at the site of the present city of San Antonio. As a halfway post between Mexico and the Louisiana border, San Antonio grew to be Texas's most important city during the Spanish period.

Until the 19th century, the US showed little interest in Texas. But the purchase of Louisiana Territory from the French by the US government in 1803 made Texas a next-door neighbor, and "filibusters" (military adventurers) began to filter across the border into Spanish territory. The best known is Philip Nolan, an Irish-born intriguer who started spending time in Texas as early as 1790. Ostensibly, he was trading horses with the Indians, but the Spanish associated him with Aaron Burr's schemes to excise the Spanish southwest from its owners. In the summer of 1800, the Spanish governor of Texas, Juan Bautista Elguezábal, ordered that Nolan should be arrested if he returned. In December of that year, Nolan returned with a small force of 20 men and built a fort near Nacogdoches; he was killed fighting the Spanish on 4 March 1801. Nolan is remembered for having drafted the first Anglo-American map of Texas.

In 1810–11, the Mexicans launched their revolution against Spain, and though only an outpost, Texas as a Spanish-Mexican colony was naturally involved. In 1813, Texas formally declared its independence of Spain and its intention of becoming a Mexican state, with its capital at San Antonio. Various Anglo-Americans entered the new state to serve on behalf of Mexico. Pirates also aided the Mexican cause: on Galveston Island, Luis Aury preyed on Spanish shipping, and after 1816, his place was taken by Jean Laffite, who privateered against both Spanish and US shipping until the US Navy drove him out.

The Spanish finally gave up on Mexico in 1821, leaving Texas as a Mexican province with a non-Indian population of about 7,000. The only towns of significant size were Goliad, San Antonio (commonly called Bexar), and Nacogdoches. A year earlier, Moses Austin of Missouri had received permission from Spanish authorities to introduce Anglo-American colonists into Texas, presumably as a barrier against aggression by the US. When Spanish rule ended, his son, Stephen F. Austin, succeeded his late father as head of the colonization movement, securing permission from the new Mexican government to settle 300 families in the area between the lower Colorado and Brazos rivers. After Austin had settled his "Old Three Hundred" in 1821, he received permission to settle more, and within a decade, his colonists numbered more than 5,000. The Mexicans invested Austin with the responsibilities and privileges of an *empresario*: authority to run commerce, maintain militia, administer justice, and hand out land titles. Other *empresarios* made similar arrangements. Green DeWitt, also of Missouri, settled several hundred families farther west and founded the town of Gonzales in 1825. Hayden Edwards received a grant to settle 800 families near Nacogdoches. Mexicans were also permitted to organize colonies. Texas thus began a pattern of growth from the outside that has continued to the present day.

Between 1821 and 1835, the population of non-Indian Texas expanded to between 35,000 and 50,000. Most new settlers were Anglo-Americans who brought their prejudices against Mexico with them, whether they were from the South or the North. They could not stomach Mexican culture, Mexican folkways, Mexican justice—and the Protestants among them disliked the omnipresence of the Roman Catholic Church. All of these Anglo-American settlers had ties to the US, and many undoubtedly longed for the time when they would live under the American flag again. The ineptitude of the Mexican government made the situation even worse. In 1826, Hayden Edwards organized the Republic of Fredonia and tried to drive the Mexicans from East Texas, but in the end, he had to flee the province himself. Troubled by the rising spirit of rebellion, the Mexican Congress enacted the Law of 1830, which forbade most immigration and imposed duties on all imports. Anglo-Americans in Texas responded with the same anger that New Englanders had once shown when Britain imposed tax restrictions on the original American colonies.

At first, the Anglo-Texans insisted they were opposing Mexican political excesses, not the Mexican nation. Their hope lay with Gen. Antonio López de Santa Anna, who was leading a liberal revolution against President Anastasio Bustamante. Skirmishes between the Anglo-Texans and Mexican officials remained sporadic and localized until 1833, when Santa Anna became president of Mexico and almost immediately dropped his liberal stance. Texans sent Austin to Mexico City to petition Santa Anna to rescind the Law of 1830, to allow the use of English in public business, and to make Texas (then an appendage of Coahuila) a separate state. After several months in Mexico City, Austin was arrested on his way back to Texas and was imprisoned for a year. When Santa Anna tried to enforce customs collections, colonists

at Anahuac, led by William Barret Travis, drove the Mexican officials out of town. Santa Anna's answer was to place Texas under military jurisdiction. When the Mexican military commander, Col. Domingo de Ugartechea, sent his soldiers to Gonzales to take a cannon there from the colonists, the Anglo-Texan civilians drove them off on 2 October 1835, in a battle that is generally considered to mark the start of the Texas Revolution.

On 3 November, a provisional government was formed. It called not for independence but for a return to the liberal Mexican constitution of 1824. Three commissioners, one of them Austin, were sent to Washington, D.C., to request aid from the US. Sam Houston, who only six years earlier had resigned the governorship of Tennessee (when his wife left him) and had come to Texas after stays in Oklahoma and Arkansas, was named commander in chief of the upstart Texas army. Hostilities remained at a standstill until February 1836, when Santa Anna led an army across the Rio Grande. The Mexicans concentrated outside San Antonio at a mission-fort called the Alamo, where 187 or so Texans, commanded by Col. William Barret Travis, had holed up in defense. The Mexicans besieged the Alamo until 6 March, when Santa Anna's forces, now numbering more than 4,000, stormed the fortress. When the battle ended, all the Alamo's defenders, including several native Mexicans, were dead. Among those killed were Travis and two Americans who became legends—James Bowie and Davy Crockett.

Four days before the battle of the Alamo, other Texans gathered at Washington-on-the-Brazos and issued a declaration of independence. As so often happens, a fight that had started on principle—in this case, a constitutional issue—grew into a fight for independence. The men who died at the Alamo believed they were fighting for restoration of the constitution of 1824. But three weeks after the Alamo fell, on 27 March 1836, the Mexicans killed 342 Texans who had surrendered at Goliad, thinking they would be treated as prisoners of war. Coming on the heels of the Alamo tragedy, the "Goliad massacre" persuaded Texans that only total victory or total defeat would solve their problems with Santa Anna. The Texas army under Sam Houston retreated before Santa Anna's oncoming forces, which held a numerical advantage over Houston's of about 1,600 to 800. On 21 April 1836, however, the Texans surprised the Mexicans during their siesta period at San Jacinto (east of present-day Houston). Mexican losses were 630 killed, 280 wounded, and 730 taken prisoner, while the Texans had only 9 killed and 30 wounded. This decisive battle—fought to the cry of "Remember the Alamo, remember Goliad!"—freed Texas from Mexico once and for all.

For 10 years, Texas existed as an independent republic, recognized by the US, Belgium, France, the United Kingdom, the Netherlands, and several German states. Sam Houston, the victorious commander at San Jacinto, became the republic's first nationally elected president. Although Texans are proud of their once-independent status, the fact is that the republic limped along like any new nation, strife-torn and short of cash. It was unable to reach agreement with Mexico on a treaty to clarify the border. Moreover, its original $1-million public debt increased eightfold in a decade, and its paper money depreciated alarmingly. Consequently, when Texas joined the Union on 29 December 1845, the date of the US congressional resolution recognizing the new state (the Lone Star flag, the republic's official banner, was not actually lowered and a governor inaugurated until 19 February 1846), its citizens looked on the action as a rescue. The annexation in great measure provoked the Mexican War, which in turn led to the conclusion of the Treaty of Guadalupe Hidalgo on 2 February 1848. Under the treaty, Mexico dropped its claim to the territory between the Rio Grande and the Nueces River. Later, in accordance with the Compromise of 1850, Texas relinquished, for $10 million, its claim on lands stretching into New Mexico, Colorado, Wyoming, Oklahoma, and Kansas.

With the coming of the Civil War, Texas followed its proslavery southern neighbors out of the Union into the Confederacy; Governor Houston, who opposed secession, was ousted from office. The state saw little fighting, and Texas thus suffered from the war far less than most of the South. The last battle of the war was fought on Texas soil at Palmito Ranch, near Brownsville, on 13 May 1865—more than a month after Gen. Robert E. Lee's surrender at Appomattox Court House in Virginia.

During Reconstruction, Texas was governed briefly by a military occupation force and then by a Republican regime; the so-called carpetbag constitution of 1869, passed during this period, gave the franchise to blacks, a right that the Ku Klux Klan actively sought to deny them. Texas was allowed to rejoin the Union on 30 March 1870. Three years later, Republican Governor Edmund J. Davis was defeated at the polls by Richard Coke, and a Democratic legislature wrote a new constitution, which was approved by the voters in 1876.

While most southern states were economically prostrate, the Texas economy flourished because of the rapid development of the cattle industry. Millions of Texas cattle walked the trails to northern markets, where they were sold for hard cash, providing a bonanza for the state. The widespread use of barbed wire to fence cattle ranches in the 1880s ended the open range and encouraged scientific cattle breeding. By 1900, Texas began to transform its predominantly agricultural economy into an industrial one. This process was accelerated by the discovery of the Spindletop oil field—the state's first gusher—near Beaumont in 1901, and by the subsequent development of the petroleum and petrochemical industries. World War I saw the emergence of Texas as a military training center. The rapid growth of the aircraft industry and other high-technology fields contributed to the continuing industrialization of Texas during and after World War II.

Texas politics remained solidly Democratic during most of the modern era, and the significant political conflict in the state was between the liberal and conservative wings of the Democratic Party. Populist-style reforms were enacted slowly during the governorships of James E. Ferguson—impeached and removed from office during his second term in 1917—and of his wife, Miriam A. "Ma" Ferguson (1925-27, 1933-35), and more rapidly during the two administrations of James V. Allred (1935-39). During the 1960s and 1970s, the Republican Party gathered strength in the state, electing John G. Tower as US senator in 1961 and William P. Clements, Jr., as governor in 1978—the first Republicans to hold those offices since Reconstruction. In general, the state's recent political leaders, Democrats as well as Republicans, have represented property interests and taken a conservative line.

On the national level, Texans have been influential since the 1930s, notably through such congressional leaders as US House Speaker Sam Rayburn and Senate Majority Leader Lyndon B. Johnson. Johnson, elected vice president under John F. Kennedy, was riding in the motorcade with the president when Kennedy was assassinated in Dallas on 22 November 1963. The city attained further national notoriety when Kennedy's alleged killer, Lee Harvey Oswald, was shot to death by Jack Ruby, a Dallas nightclub operator, two days later. Johnson served out the remainder of Kennedy's term, was elected to the presidency by a landslide in 1964, and presided over one of the stormiest periods in US history before retiring to his LBJ ranch in 1969. Memorials to him include the Lyndon B. Johnson Library at Austin and the renamed Lyndon B. Johnson Space Center, headquarters for the US manned spaceflight program, near Houston.

The most prominent Texan on the national scene since Johnson is Republican George Bush, who, in 1980, after failing in his bid for the Republican presidential nomination, became Ronald Reagan's running mate; Reagan and Bush won in 1980 and were reelected in 1984. Prior to 1980, Bush had been a member of the Texas delegation to the US House of Representatives, chairman of the Republican National Committee, envoy to the People's Republic of China, and director of the Central Intelligence Agency.

¹²STATE GOVERNMENT

Texas has been governed directly under eight constitutions: the Mexican national constitution of 1824, the Coahuila-Texas state constitution of 1827, the independent Republic of Texas constitution of 1836, and the five US state constitutions of 1845, 1861, 1866, 1869, and 1876. This last document, with more than 250 amendments, is the foundation of the state government today. An attempt to replace it with eight propositions that in effect would have given Texas a new constitution was defeated at the polls in November 1975.

The state legislature consists of a senate of 31 members elected to four-year terms, and a house of representatives of 150 members elected to two-year terms. The legislature meets on the 2d Tuesday in January of odd-numbered years for sessions of as many as 140 calendar days; the governor may also call special sessions, each limited to 30 calendar days. Senators and representatives receive the same pay, pursuant to a constitutional amendment of 1975: $7,200 per year and $30 per diem while the legislature is in session. The constitution requires that senators be 26 years of age and residents of the state for five years prior to election; for representatives, the requirements are 21 years of age and two years of residency.

The state's chief executives are the governor and lieutenant governor, separately elected to four-year terms. Other elected executives, also serving for four years, are the attorney general, comptroller, treasurer, commissioner of agriculture, and commissioner of the general land office. The remaining cabinet members are appointed by the governor, who also appoints members of the many executive boards and commissions. The governor, whose salary was $88,900 as of 1983, must be a US citizen, at least 30 years of age, and must have resided in the state for at least five years prior to election. A uniquely important executive agency is the Texas Railroad Commission, established in 1891 and consisting of three members, elected for six-year terms, who regulate the state's oil and gas production, coal and uranium mining, and trucking industry, in addition to the railroads. The commission thus wields extraordinary economic power, and the alleged influence by the regulated industries over the commission was a major source of political controversy as the 1980s began.

To become law, a bill must be approved by a majority of members present and voting in each house, with a quorum of two-thirds of the membership present, and either signed by the governor or left unsigned for 10 days while the legislature is in session or 20 days after it has adjourned. A gubernatorial veto may be overridden by a two-thirds vote of members present in the house of origin, followed by either a vote of two-thirds of members present in the lower house or two-thirds of the entire membership of the senate (the difference is the result of differing interpretations of an ambiguity in the state constitution). Overrides are rare: the vote in April 1979 by state legislators to override the new Republican governor's veto of a minor wildlife regulation measure affecting only one county was the first successful attempt in 38 years—and the last one as of July 1985. A constitutional amendment requires a two-thirds vote of the membership of each house and ratification by the voters at the next election.

In order to vote in Texas, one must be a US citizen and 18 years of age or over; there is a 30-day residency requirement.

¹³POLITICAL PARTIES

The Democratic Party has dominated politics in Texas. William P. Clements, Jr., elected governor in 1978, was the first Republican since Reconstruction to hold that office. No Republican carried Texas in a presidential election until 1928, when Herbert Hoover defeated Democrat Al Smith, a Roman Catholic at a severe disadvantage in a Protestant fundamentalist state. Another Roman Catholic, Democratic presidential candidate John Kennedy, carried the state in 1960 largely because he had a Texan, Lyndon Johnson, on his ticket.

Prior to the Civil War, many candidates for statewide office ran as independents. After a period of Republican rule during Reconstruction, Democrats won control of the statehouse and state legislature in 1873. The major challenge to Democratic rule during the late 19th century came not from Republicans but from the People's Party, whose candidates placed second in the gubernatorial races of 1894, 1896, and 1898, aided by the collapse of the cotton market; imposition of a poll tax in 1902 helped disfranchise the poor white farmers and laborers who were the base of Populist support. The Populists and the Farmers' Alliance probably exercised their greatest influence through a Democratic reformer, Governor James S. Hogg (1891–95), who fought the railroad magnates, secured lower freight rates for farmers and shippers, and curbed the power of large landholding companies. Another Democratic governor, James E. "Farmer Jim" Ferguson, was elected on an agrarian reform platform in 1914 and reelected in 1916, but was impeached and convicted the following year for irregular financial dealings. Barred from holding state office, he promoted the candidacy of his wife, Miriam "Ma" Ferguson, whose first term as governor (1925–27) marked her as a formidable opponent of the Ku Klux Klan. During her second term (1933–35), the state's first New Deal reforms were enacted, and prohibition was repealed. The Fergusons came to represent the more liberal wing of the Democratic Party in a state where liberals have long been in the minority. After the progressive administration of Governor James V. Allred, during which the state's first old-age assistance program was enacted, conservative Democrats, sometimes called "Texas Tories," controlled the state until the late 1970s.

There is no voter registration by party in Texas; as of November 1982, the state had 6,414,988 registered voters. In 1982, Democrat Mark White defeated Clements for the governorship by a margin of 231,933 out of 3,191,091 votes cast. Democratic Senator Lloyd Bentsen was reelected that year. Two years later, Texans gave the Republican presidential ticket of Ronald Reagan and George Bush (a longtime state resident and a US representa-tive from Texas for two terms, beginning in 1967) a 64% majority of the popular vote, and Republican Phil Gramm was elected senator. Democrats retained control of the state legislature, however, and won 17 of 27 US House seats.

Aside from the Populists, third parties have played a minor role in Texas politics. The Native American (Know-Nothing) Party helped elect Sam Houston governor in 1859. The best showing by a third-party candidate in a presidential election was in 1968, when George Wallace of the American Independent Party won 19% of the Texas popular vote.

Following passage of the federal Voting Rights Act of 1965, registration of black voters increased from about 375,000 in 1964 to 720,000 in 1984, or about 59% of the black voting-age population. Between 1895 and 1967, no black person served as a state legislator; in 1985, however, there were 14 black legislators, including one state senator, and one black member of the Texas delegation to the US House of Representatives. In 1984, 3 Hispanic Americans served as US representatives and 25 as state legislators; Henry Cisneros, an Hispanic, was mayor of San Antonio. In 1985, one woman was in the state senate, 16 in the house, and Kathy Whitmire was mayor of Houston.

¹⁴LOCAL GOVERNMENT

The Texas constitution grants considerable autonomy to local governments. As of 1985, Texas had 254 counties, a number that has remained constant since 1931. In 1983, there were 1,100 cities and towns, about 1,100 school districts (down from 8,600 in 1910), and at least 1,600 special districts.

Each county is governed by a commissioners court of five members, consisting of four commissioners elected by precinct and a county judge or administrator elected at large. Other elected officials generally include a county clerk, attorney, treasurer, assessor-collector, and sheriff.

The 1980 census counted 237 cities having more than 5,000 population and thus entitled by law to adopt their own home-rule charters. As of 1 January 1983, 165 cities had adopted the council-manager form of government, 41 had the commissioner-manager type, 24 had mayors and/or councils, and most of the remainder were governed by some form of commission. In 1983, Texas had 24 regional councils consisting of representatives from two or more counties. The councils develop regional plans and seek to promote economy and efficiency in local governments.

Texas Presidential Vote by Political Parties, 1948–84

YEAR	ELECTORAL VOTE	TEXAS WINNER	DEMOCRAT	REPUBLICAN	STATES' RIGHTS DEMOCRAT	PROGRESSIVE	PROHIBITION
1948	23	*Truman (D)	750,700	282,240	106,909	3,764	2,758
					CONSTITUTION		
1952	24	*Eisenhower (R)	969,227	1,102,818	1,563	—	1,983
1956	24	*Eisenhower (R)	859,958	1,080,619	14,591	—	—
1960	24	*Kennedy (D)	1,167,935	1,121,693	18,170	—	3,868
1964	25	*Johnson (D)	1,663,185	958,566	5,060		
					AMERICAN IND.		
1968	25	Humphrey (D)	1,266,804	1,227,844	584,269	—	—
					AMERICAN	SOC. WORKERS	
1972	26	*Nixon (R)	1,154,289	2,298,896	6,039	8,664	—
1976	26	*Carter (D)	2,082,319	1,953,300	11,442	1,723	
					LIBERTARIAN		
1980	26	*Reagan (R)	1,881,147	2,510,705	37,643	—	—
1984	29	*Reagan (R)	1,949,276	3,433,428	—	—	

*Won US presidential election.

¹⁵STATE SERVICES

For a state of its size and population, Texas provides rather limited statewide services to its citizens. Texas has no ombudsman, ethics commission, consumer protection division, department of housing, or unified environmental protection agency.

Educational services in the public schools are administered by the Texas Education Agency, which is run by a commissioner of education appointed by a 27-member elected State Board of Education; the State Textbook Committee, appointed by the board, oversees textbook purchases statewide. The Coordinating Board for the State College and University System, consisting of 18 appointed members, oversees public higher education. Transportation facilities are regulated by the State Highway and Public Transportation Commission, the Texas Railroad Commission, and the Texas Aeronautics Commission.

Health and welfare services are offered by the Department of Health and the Department of Human Resources, while public protection is the responsibility of the National Guard, Texas Department of Corrections, and Texas Youth Council, which maintains institutions for juvenile offenders. Labor services are provided by the Texas Employment Commission and the Department of Labor and Standards.

¹⁶JUDICIAL SYSTEM

The Texas judiciary comprises the supreme court, the state court of criminal appeals, 14 courts of appeals, and more than 330 district courts.

The highest court is the supreme court, consisting of a chief justice and eight associate justices, who are popularly elected to staggered six-year terms. As of September 1983, each justice received a salary of $74,300 annually, and the chief justice earned $500 more. The court of criminal appeals, which has final jurisdiction in most criminal cases, consists of a presiding judge and eight associate judges, also elected to staggered six-year terms. Appeals court judges' salaries were the same as those of supreme court justices in 1983.

Justices of the courts of appeals are elected to six-year terms and sit in 14 judicial districts; each court has a chief justice and at least two associate justices. Chief justices received annual salaries from the state of $74,300 in 1983. There were 330 district court judges in 1983, each elected to a four-year term, and receiving state salaries of $52,900 each, plus supplemental payments by local subdivisions. County, justice of the peace, and police courts handle local matters.

The state maintains 24 correctional units, housing more than 36,000 adult offenders. Since 1982, the system has been under federal court order to remedy overcrowding and otherwise improve conditions.

Texas has a high crime rate—substantially above the national average for murder, forcible rape, burglary, larceny-theft, and motor vehicle theft, though below average for robbery and assault. Overall, according to the FBI Crime Index, Texas had a 1983 crime rate of 5,907.3 per 100,000 population, 512 for violent crimes and 5,395 for property crimes. Rates for specific crimes were murder, 14.2 (with Louisiana, highest in the nation);forcible rape, 40.3; robbery, 189.3; aggravated assault, 268.4; burglary, 1,667.5; larceny-theft, 3,202.6; and motor vehicle theft, 524.9. In 1982, Odessa had the highest murder rate of any city in the nation—29.8—and 3 other Texas cities had rates among the 10 highest in the nation.

Texas criminal law provides for capital punishment by lethal intravenous injection for certain violent crimes. Between 1982 and mid-March 1985, six persons were executed in Texas.

¹⁷ARMED FORCES

In few states do US military forces and defense-related industries play such a large role as in Texas, which as of 30 September 1984 had 131,214 military personnel and 64,457 civilians employed at major US military bases. In 1983/84, Texas ranked behind only California and New York in the value of defense prime contract awards, which were worth more than $8.7 billion, or 6.3% of the US total.

Ft. Sam Houston, at San Antonio, is headquarters of the US 5th Army, while Ft. Bliss, at El Paso, is the home of the US Army Air Defense Center. Ft. Hood, near Killeen, is headquarters of the 3d Army Corps and other military units; with more than 37,000 military personnel in 1983/84, it is the state's largest single defense installation. Ft. Sam Houston is also the headquarters of the US Army Health Services Command and the site of the Academy of Health Sciences, the largest US military medical school, enrolling more than 25,000 officers and enlisted personnel and providing correspondence courses for another 30,000 students. Brooke Army Medical Center, the 2d-largest Army hospital in the US, is located at the same installation. William Beaumont Army Medical Center, at El Paso, is one of the nation's largest Army hospitals and most modern medical treatment centers.

Four principal Air Force bases are located near San Antonio: Brooks, Kelly, Lackland, and Randolph. Other major air bases are Bergstrom, near Austin; Carswell, Ft. Worth; Dyess, Abilene; Ellington, southwest of Houston; Goodfellow, San Angelo; Laughlin, Del Rio; Reese, Lubbock; and Sheppard, Wichita Falls. All US manned spaceflights are controlled from the Lyndon B. Johnson Space Center, operated by the National Aeronautics and Space Administration. Naval air training stations are located at Corpus Christi, Beeville, Dallas, and Kingsville. The Inactive Ships Maintenance Facility, at Orange, is home port for some of the US Navy's "mothball fleet."

Texas was a major military training center during World War II, when about one out of every 10 soldiers was trained there. Some 750,000 Texans served in the US armed forces during that war; the state's war dead numbered 23,022. Military veterans living in the state as of 30 September 1983 totaled 1,732,000, including 15,000 who served in World War I, 627,000 in World War II, 329,000 during the Korean conflict, and 580,000 during the Viet-Nam era. Expenditures on Texas veterans exceeded $1.6 billion in 1982/83.

The Texas Army National Guard has dual status as a federal and state military force; its assigned strength was 19,386 as of 31 December 1984. The Air Force National Guard had 3,637 personnel in February 1985. The Texas State Guard—an all-volunteer force available either to back up National Guard units or to respond to local emergencies—had some 2,500 members.

The famous Texas Rangers, a state police force first employed in 1823 (though not formally organized until 1835) to protect the early settlers, served as scouts for the US Army during the Mexican War. Many individual Rangers fought with the Confederacy in the Civil War; during Reconstruction, however, the Rangers were used to enforce unpopular carpetbagger laws. Later, the Rangers put down banditry on the Rio Grande. The force was reorganized in 1935 as a unit of the Department of Public Safety and is now called on in major criminal cases, helps control mob violence in emergencies, and sometimes assists local police officers. The Rangers have been romanticized in fiction and films, but one of their less glamorous tasks has been to intervene in labor disputes on the side of management. As of October 1983, Texas had a total of 52,447 state and local police employees; in 1982, expenditures on police protection totaled $1.3 billion, or $90.14 per capita.

¹⁸MIGRATION

Estimates of the number of Indians living in Texas when the first Europeans arrived range from 30,000 to 130,000. Eventually, they all were killed, fled southward or westward, or were removed to reservations. The first great wave of white settlers, beginning in 1821, came from nearby southern states, particularly Tennessee, Alabama, Arkansas, and Mississippi; some of these newcomers brought their black slaves to work in the cotton fields. During the

1840s, a second wave of immigrants arrived directly from Germany, France, and eastern Europe.

Interstate migration during the second half of the 19th century was accelerated by the Homestead Act of 1862 and the westward march of the railroads. Particularly notable since 1900 has been the intrastate movement from rural areas to the cities; this trend was especially pronounced from the end of World War II, when about half the state's population was rural, to the late 1970s, when nearly four out of every five Texans made their homes in metropolitan areas.

Texas's net gain from migration between 1940 and 1980 was 1,821,000, 81% of that during the 1970–80 period. A significant proportion of postwar immigrants were seasonal laborers from Mexico, remaining in the US either legally or illegally. During 1980–83, Texas had the highest net migration gain—922,000—in the nation.

[19]INTERGOVERNMENTAL COOPERATION

The Texas Advisory Commission on Intergovernmental Relations—including three state senators, three representatives, and (ex officio) the lieutenant governor and the speaker of the state house of representatives—works to improve coordination between state, local, and federal governments. The Texas Commission on Interstate Cooperation, similarly constituted but with greater legislative participation, represents Texas before the Council of State Governments. Texas is a member of the Interstate Mining Compact and Interstate Oil and Gas Compact. The state is also a signatory to the Gulf States Marine Fisheries Compact, Southern Interstate Energy Compact, and Southern Regional Education Compact, and to accords apportioning the waters of the Canadian, Pecos, and Sabine rivers and the Rio Grande.

During 1983/84, Texas received an estimated $4.1 billion in federal aid, including a total of $250.7 million in general revenue sharing.

[20]ECONOMY

Traditionally, the Texas economy has been dependent on the production of cotton, cattle, timber, and petroleum. In recent years, cotton has declined in importance, cattle ranchers have suffered financial difficulties because of increased production costs, and lumber production has remained relatively stable. But in the 1970s, as a result of rising world petroleum prices, oil and natural gas emerged as by far the state's most important resource. The decades since World War II have also witnessed a boom in the electronics, computer, transport equipment, aerospace, and communications industries, which has placed Texas 2d only to California in manufacturing among all the states of the Sunbelt region.

The Lone Star State's robust economy as the 1980s began was due chiefly to a plentiful labor market, high worker productivity, diversification of new industries, and less restrictive regulation of business activities than in most other states. The result was a steady increase in industrial production, construction values, retail sales, and personal income, coupled with a relatively low rate of unemployment. In 1982, however, Texas began to be affected by worldwide recession. Although still well below the national average, unemployment in Texas increased from 6.9% in 1982 to 8% in 1983, a period during which the national rate actually fell 0.1%; much of this unemployment was among persons who came to Texas seeking jobs, particularly from northern industrial states. Depressed energy demand and the resulting fall in prices, beginning in 1982, caused a slump in the state's petroleum industry, which continued into 1984, when other sectors of the economy had begun to recover. Since oil and gas production and prices are unlikely to continue to grow at past rates, the Texas economy will need increasing diversification in order to maintain its vigor.

In 1983, Texas ranked 3d among the 50 states in total personal income, but only 17th in income per capita. Sharp inequalities of income persist, and wages in certain sectors of manufacturing remain low, partly because of the relative weakness of the labor movement.

Among the most valuable manufactures are petroleum refinery products, machinery, chemicals and chemical products (including plastics and synthetics), processed foods, transportation equipment (including aircraft), and electric and electronic equipment. Leading industrial sites include the Port of Houston and the aerospace plants in the Dallas–Ft. Worth area. The Lyndon B. Johnson Space Center, near Houston, is a center for the state's high-technology industries.

[21]INCOME

In 1983, personal income in Texas amounted to $184 billion, or nearly 6.7% of the US total; per capita personal income during the same year reached $11,702, slightly above the national norm. The state's personal income enjoyed a real growth of 77.5% from 1970 to 1980, compared with the national growth average of 38.6%, while Texas's per capita income rank rose from 31st to 17th among the 50 states.

Major sources of personal income and their total contributions in the years 1980 and 1981 are represented in the following table (in millions):

	1980	1981
Total, of which:	$136,263	$158,431
Farm	1,722	2,153
Agricultural services, forestry, fisheries, and other	477	533
Mining	6,786	9,394
Construction	8,776	9,629
Manufacturing	21,493	25,148
Transportation and public utilities	8,594	9,808
Wholesale trade	8,839	10,114
Retail trade	10,912	12,369
Finance, insurance, and real estate	5,794	6,629
Services	16,613	19,476
Government	15,322	17,237
Dividends, interest, rent	22,453	26,822

Median family income in Texas averaged $26,492 in 1981, 21st among the 50 states and slightly above the US average. During 1979, 2,036,000 Texans, many of them blacks and Mexican-Americans, were living below the federal poverty level; the total represented nearly 15% of all state residents. Famed for its cattle barons and oil millionaires, Texas was the home of 204,800 of the nation's leading wealth-holders in 1982, each with gross assets exceeding $500,000.

Among metropolitan areas in 1981, the highest average per capita personal incomes, in the nation as well as the state, went to residents of Midland—$16,467. For Houston residents, the average was $13,303; Dallas–Ft. Worth, $12,144; Austin, $10,442; San Antonio, $9,427; and El Paso, $7,360.

[22]LABOR

With a civilian labor force of 7,853,000 in 1984, Texas ranked 3d among the 50 states; of that total, 7,387,000 were employed, and the unemployment rate was 5.9%. The labor force in 1983 was 7,629,000, of whom 58% were men and 42% women. Nearly 81% of males 16 years of age or older were in the labor force, and the female participation rate was more than 55%; both rates exceeded the national average. In 1983, the number of unemployed workers averaged 609,000, or 8% of the labor force. The unemployment rate for males was 7.9%, for females 8.1%. The largest number of unemployed persons lived in south Texas and the Rio Grande Valley; many of the unemployed and underemployed workers were Mexican aliens, some of them legal residents of the US, others not. In 1981, when unemployment statewide was 5.3%, it was 16% in south Texas, 15.3% in the middle Rio Grande, and 12.3% in the lower Rio Grande Valley.

A federal survey in March 1982 revealed the following nonfarm employment pattern for Texas:

	ESTABLISH-MENTS	EMPLOYEES	ANNUAL PAYROLL ('000)
Agricultural services, forestry, fishing	3,180	25,193	$ 278,619
Mining, of which:	8,576	299,136	7,520,855
Oil, gas extraction	(7,884)	(239,004)	(5,468,891)
Contract construction	29,308	469,658	8,699,715
Manufacturing, of which:	18,337	1,122,118	21,952,263
Food and food products	(1,157)	(93,233)	(1,536,452)
Chemicals, allied products	(830)	(74,909)	(2,117,540)
Fabricated metals	(2,148)	(103,690)	(1,838,037)
Nonelectrical machinery	(3,338)	(192,084)	(3,747,973)
Electric, electronic equipment	(776)	(100,678)	(2,063,292)
Transport equipment	(548)	(77,467)	(1,761,075)
Transportation, public utilities	12,338	363,978	7,788,052
Wholesale trade	31,159	445,177	8,938,280
Retail trade	89,408	1,140,133	10,999,918
Finance, insurance, real estate	27,622	362,496	6,768,955
Services, of which:	93,787	1,166,230	16,503,689
Business services	(14,753)	(250,788)	(3,830,441)
Health services	(21,828)	(327,678)	(4,984,215)
Other	11,554	13,604	224,989
TOTALS	325,269	5,407,723	$89,675,335

Government workers were not covered by this survey; at the end of 1982 there were 156,000 federal civilian employees in Texas, 40.1% of them working in defense. In 1983, there were 217,000 state and 626,000 local government employees.

Organized labor has never been able to establish a strong base in Texas, and a state right-to-work law continues to make unionization difficult. The earliest national union, the Knights of Labor, declined in Texas after failing to win a strike against the railroads in 1886, when the Texas Rangers served as strike breakers. That same year, the American Federation of Labor (AFL) began to organize workers along craft lines. One of the more protracted and violent disputes in Texas labor history occurred in 1935, when longshoremen struck Gulf Coast ports for 62 days. The Congress of Industrial Organizations (CIO) succeeded in organizing oil-field and maritime workers during the 1930s. As of 1980, labor union membership in the state totaled 669,000, representing 11.4% of the nonfarm work force—a very low percentage for a large industrial state. As of 30 June 1983 there were 2,067 unions in Texas. In 1981, 52 work stoppages involving 27,600 workers resulted in some 688,100 days idle.

As of January 1985, average weekly earnings were $378.84 and average weekly hours worked, 41. Manufacturing workers earned an average hourly wage of $8.88 in 1983, slightly above the US norm. Wage rates ranged from $15.53 an hour in brewing and $14.87 in petroleum refining to $5.09 in apparel and textiles and $5.54 in leather and leather products.

[23]AGRICULTURE

Texas ranked 3d among the 50 states in agricultural production in 1983, with farm marketings totaling nearly $9 billion; crops accounted for 38% of the total. Texas leads the nation in output of cotton, grain sorghum, watermelons, cabbages, and spinach.

Since 1880, Texas has been the leading producer of cotton, which accounted for 31% of total US production in 1983 and 31.5% of the state's crop sales in 1982. After 1900, Texas farmers developed bumper crops of wheat, corn, and other grains by irrigating dry land and transformed the "great Sahara" of West Texas into one of the nation's foremost grain-growing regions. Texans also grow practically every vegetable suited to a temperate or semitropical climate. Since World War II, farms have become fewer and larger, more specialized in raising certain crops and meat animals, more expensive to operate, and far more productive.

About 136.8 million acres (55.4 million hectares) are devoted to farms and ranches, representing more than three-fourths of the state's total area. The number of farms declined from 418,000 in 1940 to fewer than 160,000 in 1978, but rose to 187,000 in 1984. The average farm was valued at $520,000 in 1982. During the 1940–70 period, the farm population decreased from 2,160,000 persons to 471,000; as of mid-July 1983 there were 120,000 unpaid workers and 85,000 hired laborers on Texas farms and ranches.

Productive farmland is located throughout the state. Grains are grown mainly in the temperate north and west, and vegetables and citrus fruits in the subtropical south. Cotton has been grown in all sections, but in recent years, it has been extensively cultivated in the High Plains of the west and the upper Rio Grande Valley. Grain sorghum, wheat, corn, hay, and other forage crops are raised in the north-central and western plains regions. Rice is cultivated along the Gulf Coast, and soybeans are raised mainly in the High Plains and Red River Valley.

Major crops in 1983 (except where indicated) are shown in the following table:

	HARVESTED ACREAGE ('000)	PRODUCTION	VALUE ('000)
Cotton	3,572	2,412,000 bales	$695,000
Wheat	4,600	161,000,000 bushels	579,600
Hay	3,070	7,486,000 tons	527,763
Sorghum, grain	3,150	157,500,000 bushels	472,500
Corn	1,080	104,760,000 bushels	356,184
Rice (1982)	474	22,214,000 cwt	198,593
Vegetables, fresh market	632	566,100 tons	148,424
Soybeans for beans	420	9,450,000 bushels	73,238

The leading vegetables and fruits, in terms of 1982 value, were onions, cabbages, watermelons, carrots, potatoes, cantaloupes, green peppers, honeydew melons, spinach, cucumbers, and lettuce. Cottonseed, barley, oats, peanuts, pecans, sugar beets, sugarcane, and sunflowers are also produced in commercial quantities.

Agribusiness contributed more than $35 billion to the economy in 1982. In that year, fewer than 3% of all Texans whose livelihood was linked to agriculture worked on farms, while nearly 25% were engaged in marketing agricultural supplies and services. The total value of farmland, buildings, machinery, crop and livestock inventories, and other assets was estimated at $96 billion in 1982; farmland and buildings alone were valued at $82 billion in 1984, higher than any other state.

Irrigated farmland in 1983 totaled 8 million acres (3.2 million hectares), of which about 65% was in the High Plains; other areas dependent on irrigation included the lower Rio Grande Valley and the trans-Pecos region. Approximately 80% of the irrigated land is supplied with water pumped from wells. Because more than half of the state's irrigation pumps are fueled by natural gas, the cost of irrigation increased significantly as gas prices rose during the 1970s.

[24]ANIMAL HUSBANDRY

In 1983, Texas ranked 1st in livestock production, which contributed more than 60% of the state's total agricultural income. The state leads the US in output of cattle, goats, and sheep and lambs, and in the production of wool and mohair.

At the close of 1983, Texas had 14,350,000 head of cattle and calves (12.6% of the US total), 333,000 milk cows, 1,800,000 sheep and lambs, and 590,000 hogs. At the beginning of 1983 there were about 1,420,000 goats, 17,200,000 chickens, and 310,000 turkey hens. The combined value of all livestock in 1983 was $5.6 billion.

Cattle, calves, beef, and veal account for about 76% of cash receipts from the sale of meat animals and livestock products. About 60% of cattle fattened for market are kept in feedlots located in the Texas panhandle and northwestern plains. The state's feedlots marketed 4,075,000 head of grain-fed cattle in 1982. Meat animals slaughtered at Texas plants in 1983 included 6,109,800 cattle, 258,100 calves, 1,280,500 hogs, and 751,000 sheep and lambs. In 1982, 64,000 goats were slaughtered. Production of meat animals in 1983 included 5.5 billion lb of cattle and calves, worth nearly $3.3 billion; 209.6 million lb of hogs and pigs, $94.8 million; and 113.7 million lb of sheep and lambs, $50.1 million.

About 90% of the dairy industry is located in eastern Texas. In 1983, milk production was nearly 3.8 million lb. Other dairy products included 57.3 million gallons of ice cream. Poultry production included 872 million lb of broilers, valued at more than $261 million, and 5.4 million turkeys, valued at $45.1 million. In 1982, mohair production was 9.8 million lb, valued at $25 million; wool, 18.7 million lb, $15.7 million. In 1981, 11.4 million lb of honey was produced, valued at $6.7 million.

Breeding of Palominos, Arabians, Appaloosas, Thoroughbreds, and quarter horses is a major industry in Texas. The animals are most abundant in the most heavily populated areas, and it is not unusual for residential subdivisions of metropolitan areas to include facilities for keeping and riding horses.

[25]FISHING

Texas in 1984 recorded a commercial catch of 104.1 million lb (13th among the 50 states), valued at $190.3 million (4th). The leading commercial fishing ports are Brownsville–Port Isabel, Aransas Pass–Rockport, and Freeport.

Shrimp accounted for about 99% of the total value of the catch in 1982; other commercial shellfish include blue crabs, oysters, and squid. Species of saltwater fish with the greatest commercial value are sea trout, black drum, redfish, red snapper, and flounder. Early in 1980, the US government banned shrimp fishing for 45 days, effective in the summer of 1981, in order to conserve shrimp supplies. Texas has since continued to close the Gulf to shrimping from about 1 June to 15 July.

Sport fishing involves some 3 million Texans annually and pumps nearly $1 billion into local economies. Among the most sought-after native freshwater fish are large-mouth and white bass, crappie, sunfish, and catfish.

[26]FORESTRY

Texas forestland in 1979 covered 23,279,000 acres (9,421,000 hectares), representing more than 3% of the US total and about 14% of the state's land area. Commercial timberland comprised 12,513,000 acres (5,064,000 hectares), of which 94% was privately owned. Lands owned or managed by the federal government covered 740,000 acres (299,000 hectares), and state and local forestlands occupied 56,000 acres (23,000 hectares).

Most forested land, including practically all commercial timberland, is located in the Piney Woods region. In 1981, Texas timberlands yielded a harvest of 418,000,000 cu feet of softwoods (down 6% from 1980) and 100,000,000 cu feet of hardwoods (down 11%). Primary forest products included 4,015,000 cords of pulpwood (86% softwood), close to 1 billion board feet of lumber (72% softwood), and more than 1.5 billion sq feet of plywood. Although annual tree growth exceeds the yearly harvest by about one-third, acreage in commercial timberland continues to shrink as the result of real estate development, water impoundment, and the expansion of public parks and wilderness areas.

Production of timber alone was valued at more than $352 million in 1981, but that represents only a very small fraction of the state's forest-related industries. Shipments of lumber and wood products were valued at more than $2.5 billion in 1982, paper and allied products $2.6 billion. Wood-preserving plants in Texas treated some 5.9 million crossties, 1.4 million fence posts, and 296,458 utility poles during 1980/81. Insured employment in the lumber and wood products industry totaled 35,720 in 1982, with a payroll of $473.7 million.

The Texas Forest Service manages state reforestation programs, coordinates pest control activities, and protects against forest fires. For reforestation purposes there were 2,198 privately owned tree farms as of October 1982, aggregating 4,114,441 acres (1,665,061 hectares). The worst forest pest is the southern pine bark beetle, which kills more trees each year than are lost in forest fires. A pine beetle infestation in 1976 caused so much damage that the governor declared 34 counties in eastern Texas disaster areas. An infestation in eastern Texas in 1984 caused even greater damage. In 1982, 1,941 forest fires burned a total of 22,121 acres (8,952 hectares). About 54% of the fires resulted from the burning of debris.

As of 1984 there were four national forests in Texas—Angelina, Davy Crockett, Sabine, and Sam Houston—with a total area of 1,730,936 acres (700,487 hectares). Texas has four state forests: the E. O. Siecke, W. Goodrich Jones, I. D. Fairchild, and John Henry Kirby.

[27]MINING

Texas leads the US in mineral production, valued at more than $45 billion in 1982. It is also the leading producer of mineral fuels, which accounted for more than 96% of the state's mineral value and nearly 25% of the value of total US mineral output in 1982. Texas ranks 2d among the states in nonfuel mineral production, valued at $1.7 billion in 1984. All together, the state mines 24 different minerals, and in 1982 ranked 1st among the 50 states in production of cement, gypsum, sulfur, and magnesium chloride; 2d in clay, salt, and sand and gravel; and 3d in uranium. Production of petroleum, which exceeded 900,000,000 barrels in 1983, is concentrated in the panhandle and eastern part of the state.

Construction materials are mined throughout the central and the northern regions, sulfur and magnesium compounds on the Gulf Coast and in the west, and uranium in the Gulf Coastal Plain.

Mining accounted for about 6% of the state's personal income in 1981; in 1982, all mineral-related industries employed about 324,000 workers, or some 5% of the civilian labor force. Apart from fuels, leading minerals in 1984 included portland cement, 10,900,000 tons, $603,200,000; stone, 79,847,000 tons, $266,236,000; sand and gravel, 55,400,000 tons, $227,600,000; lime, 1,210,000 tons, $66,130,000; and salt, 8,142,000 tons, $64,896,000. The combined values of fluorspar, helium, magnesium, iron ore, and sulfur amounted to at least $415,631,000.

[28]ENERGY AND POWER

Texas is an energy-rich state. Its vast deposits of petroleum and natural gas liquids account for nearly 30% of US proved liquid hydrocarbon reserves. Texas is also the largest producer and exporter of oil and natural gas to other states, and it leads the US in electric power production.

As of 31 December 1983, Texas power plants had a combined installed capacity of 57.4 million kw; their power output was 206.2 billion kwh, 8.9% of the US total. Gas-fired steam plants accounted for 54% of the production, coal 42%, oil less than 1%, and others 3%. In 1983, domestic sales of electricity totaled 191 million kwh, of which industrial plants used 41%, homes 31%, businesses 24%, and other consumers 4%. Major suppliers of electricity were Central Power & Light, Community Public Service, Dallas Power & Light, El Paso Electric, Gulf States Utilities, Houston Lighting & Power, Southwestern Electric Power, Southwestern Electric Service, Southwestern Public Service, Texas Electric Service, Texas Power & Light, Texas–New Mexico Power, and West Texas Utilities.

The state's first oil well was drilled in 1866 at Melrose in East Texas, and the first major oil discovery was made in 1894 at Corsicana, northwest of Melrose, in Navarro County. The famous Spindletop gusher, near Beaumont, was tapped on 10 January 1901. Another great oil deposit was discovered in the panhandle in 1921, and the largest of all, the East Texas field, in Rusk County, was opened in 1930. Subsequent major oil discoveries were made in West Texas, starting in Scurry County in 1948. Thirty years later, the state's crude-oil production exceeded 1 billion barrels. In 1983, production was 908.2 million barrels, averaging 2.5 million barrels per day, and represented the 11th straight year of decline in output. Production in 1982 was 923.9 million barrels, valued at $29.1 billion. In 1983, only the USSR, Saudi Arabia, and Mexico produced more barrels per day than Texas. Proved petroleum reserves at the end of 1983 were estimated at more than 7.5 billion barrels, representing over 27% of total US reserves. In addition, Texas had reserves of more than 3 billion barrels of natural gas liquids. As of 1983, Texas had 40 oil refineries, with a capacity of 4,532,005 barrels of crude oil per day, or more than 28% of the US total. There were 13,851 oil wells and 4,345 gas wells in operation in 1982.

In 1983, Texas produced more than 6.4 trillion cu feet of natural gas, representing one-third of total US production; the 1982 production of 6.5 trillion cu feet was valued at $13.6 billion. Production of natural gas liquids in 1983 amounted to 322 million barrels (44% of the national total). As of 31 December 1983, proved natural gas reserves were estimated at 50.1 trillion cu feet, 25% of the US total.

Coal production totaled 38.9 million tons in 1983, all from 12 surface mines. Almost 99% of the coal was lignite, nearly all of it used as fuel for electric generating plants close to the mines. Lignite reserves were estimated at 681.7 million tons in 1983.

In the mid-1980s, four nuclear reactors—South Texas Project 1 and 2, in Matagorda County, and Comanche Peak 1 and 2, in Somervell County—were under construction. Comanche Peak 1 was scheduled to begin commercial operation late in 1985.

In 1982, energy consumption per capita in Texas was 520 million Btu, 4th highest in the nation. Energy expenditures per capita—$3,575—were 3d highest in 1981.

[29]INDUSTRY

Before 1900, Texas had an agricultural economy based, in the common phrase, on "cotton, cows, and corn." When the first US Census of Manufactures was taken in Texas in 1849, there were only 309 industrial establishments, with 1,066 wage earners; payrolls totaled $322,368, and the value added by manufacture was a mere $773,896. The number of establishments increased tenfold by 1899, when the state had 38,604 wage earners and a total value added of $38,506,130. During World War II, the value added passed the $1-billion mark, and by 1976, the total was $27.6 billion, 7th among the 50 states.

In 1982, the leading industries, ranked by numbers of employees, were nonelectrical machinery, chemical products, electric and electronic equipment, fabricated metal products, transportation equipment, food processing, and petroleum and coal products. The total value added by manufacture in 1982 amounted to $53.4 billion. The value of all shipments by manufacturers was nearly $171.7 billion. The following table shows value of shipments for the state's principal industrial sectors during 1982:

Petroleum and coal products	$60,341,400,000
Chemicals and allied products	25,154,100,000
Food and food products	17,303,200,000
Nonelectrical machinery	16,743,900,000
Fabricated metal products	8,871,800,000
Transportation equipment	7,751,600,000
Electric and electronic equipment	7,250,600,000
Primary metals	5,602,600,000
Stone, clay, and glass products	4,010,800,000
Printing and publishing	3,852,400,000
Apparel and other textile products	2,800,500,000
Paper and allied products	2,631,600,000
Lumber and wood products	2,572,900,000
Rubber and plastics products	2,572,700,000
Instruments and related products	1,404,900,000

Three of the state's leading industrial products—refined petroleum, industrial organic chemicals, and oil-field machinery—all stem directly from the petrochemical sector. Major oil refineries are located at Houston and other Gulf ports. Aircraft plants include those of North American Aviation and Chance-Vought at Grand Prairie, General Dynamics near Ft. Worth, and Bell Aircraft's helicopter division at Hurst. In 1984, Texas was home to 10 US industrial corporations with annual sales of at least $2 billion, all with headquarters in Houston or Dallas. Houston boasted Shell Oil, Tenneco, Coastal, Pennzoil, and Cooper, while Dallas was the headquarters of LTV, Texas Instruments, Diamond Shamrock, Dresser Industries, and American Petrofina.

[30]COMMERCE

Texas ranked 3d among the 50 states in wholesale trade, with total sales of $204.95 billion in 1982. Wholesaling is an important source of personal income, accounting for 6.4% of the state total in 1981. The leading wholesaling centers are the Houston, Dallas–Ft. Worth, San Antonio, El Paso, Lubbock, Midland, Amarillo, Austin, and Corpus Christi metropolitan areas.

Texas ranked 2d behind California in retail sales, amounting to $88.3 billion in 1983. Of the 1982 retail sales total, food stores accounted for 23%, automotive dealers 22%, department stores 9%, restaurants and taverns 9%, gasoline stations 8%, clothing stores 6%, and furniture and home furnishing stores 4%. Retail trade contributed 7.8% of the state's total personal income in 1981 and in 1980 accounted for more than 10% of personal income in the Laredo, Lubbock, Tyler, Brownsville–Harlingen–San Benito, Amarillo, and McAllen-Pharr-Edinburg metropolitan areas.

Many businesses in Texas's border cities are highly dependent on trade with Mexico. Devaluations of the Mexican peso in 1982 and 1983 caused considerable reduction in retail sales in border areas and a resultant rise in unemployment; Laredo's unemployment, for example, increased from 11.9% to 23.7% over the course of 1982. Among the largest metropolitan areas, Houston provided 24% of the retail sales total in 1982, Dallas–Ft. Worth 24%, and San Antonio 7%.

An amendment to the state constitution permits counties to vote on whether to allow sales of liquor by the drink. As of 31 August 1982, distilled spirits were legal in all or part of 167 counties; 2 counties permitted beverages with an alcohol content of 14% or less; 11 counties permitted only 4% beer; and 74 counties were wholly dry.

Foreign exports through Texas customs districts in 1982 totaled $25.4 billion, imports $28.2 billion. The leading items shipped through Texas ports to foreign countries were grains, chemicals, fertilizers, and petroleum refinery products; principal imports included crude petroleum, minerals and metals (especially aluminum ores), liquefied gases, motor vehicles, bananas, sugar, and molasses. Exports of the state's own agricultural products were valued at $2.6 billion in 1981/82, the main export items being cotton, worth $782.7 million, 36% of the US total; feed grains, $402.5 million, 5%; wheat and flour, $544.1 million, 7%; and rice, $171.2 million, 15%. Texas ranked 2d among the 50 states in 1981 as a producer of manufactured goods for export, with nearly $11.7 billion. Texas is the nation's leading exporter of sulfur.

[31]CONSUMER PROTECTION

Texas has no consumer protection agency. The consumer credit commissioner, appointed by the State Finance Commission, maintains a main office in Austin and branch offices in Houston and Dallas.

[32]BANKING

Texas has more banks than any other state and ranked 3d among the 50 states in commercial bank assets in 1983.

Banking was illegal in the Texas Republic and under the first state constitution, reflecting the widespread fear of financial speculation like that which had caused the panic of 1837. Because both the independent republic and the new state government found it difficult to raise funds or obtain credit without a banking system, they were forced to borrow money from merchants, thus permitting banking functions and privileges despite the constitutional ban. A formal banking system was legalized during the latter part of the 19th century.

At the end of 1984, Texas had 1,864 commercial banks; commercial bank assets totaled $172.1 billion at the end of 1983. As of 31 December 1982, 840 state banks held deposits totaling $41.9 billion, and 758 national banks had $78.4 billion in deposits. In 1983, bank loans outstanding by insured commercial banks amounted to $98.4 billion, of which commercial and industrial loans accounted for $43.7 billion and real estate loans $29.2 billion. As of 31 December 1982, the state's leading commercial banks, each having more than $1.9 billion in domestic deposits, were First City National Bank of Houston (the largest bank in Texas, with assets of $5.8 billion), Texas Commerce, and Bank of the Southwest, all in Houston; and InterFirst Bank Dallas, Republic Bank Dallas, First National, and Mercantile National Bank, all in Dallas.

At the end of 1982, Texas had 288 insured savings-and-loan and building-and-loan associations (227 state-chartered), with combined assets of $42.5 billion; these associations held $28.5 billion in outstanding mortgage loans and $34.5 billion in savings capital, and had a combined net worth (including surplus reserves and undivided profits) of more than $1.6 billion. In 1982 there were 1,300 credit unions, with about 3.5 million members and assets totaling more than $6 billion—2d in the nation.

Many out-of-state institutions have extended substantial loans to Texas businesses in recent years because the state's own banks have been unable to supply sufficient capital to meet the investment needs of the state's booming economy. In addition, at least 70 foreign banks have offices in the state, mostly in Houston, where they participate in the international financing of the petrochemical industry.

[33] INSURANCE

As of 31 August 1982, 2,022 companies were licensed to handle insurance, including 705 Texas firms and 1,317 out-of-state companies. Texas ranked 2d among the 50 states in 1983 in number of life insurance companies and 3d in amount of life insurance in force; the state's 230 life companies had 25,357,000 policies in force, valued at $351.1 billion, and paid $1 billion in benefits. The average Texas family held $58,500 in life coverage, 8% above the US norm. Some 40 fraternal benefit societies issued 33,786 policies valued at $601.4 million during 1981; insurance in force totaled $3.1 billion at the end of the year.

In 1984, Texas ranked 2d in the nation in number of domestic property and casualty insurers—255. During 1983, property and casualty insurers wrote premiums totaling $8.5 billion, including $1.7 billion in automotive liability insurance, $1.7 million in automobile physical damage insurance, and $912.7 million in homeowners' coverage.

The insurance industry is regulated by the State Board of Insurance, consisting of three members appointed by the governor. The board appoints a state insurance commissioner who is responsible for directing the Texas Insurance Department.

[34] SECURITIES

Although there are no securities exchanges in Texas, New York Stock Exchange member firms had 388 sales offices and 4,091 full-time registered representatives in the state in 1983. Some 2,606,000 Texans were shareholders of US corporations that year.

The State Securities Board, established in 1957, oversees the issuance and sale of stocks and bonds in Texas.

[35] PUBLIC FINANCE

In general, public officials in Texas have sought to minimize state government spending, and budget surpluses have normally been recorded. At the end of the 1984/85 fiscal year, state revenues exceeded expenditures by some $650 million. Under the "pay-as-

you-go" amendment to the state constitution, in effect since 1945, the legislature may not appropriate more funds than the state comptroller certifies are available.

The governor's Office of Budget and Planning prepares a budget for the biennium that is reviewed by the Legislative Budget Board, consisting of the lieutenant governor, the speaker of the house, and eight other legislators. The proposed budget is then submitted to the legislature by the governor, approved or revised by the legislature subject to the governor's veto, and signed by the governor after final approval by majority vote of both houses. The Texas fiscal year extends from 1 September to 31 August.

The following table summarizes state revenues and expenditures for fiscal year 1982/83 (in millions):

REVENUES	
Taxes	$ 9,018
Federal grants	3,147
Current charges	1,079
Interest earnings	791
Other revenues	903
TOTAL	$14,938

EXPENDITURES	
Social services and income maintenance, of which:	$ 3,264
Public welfare	(1,924)
Health and hospitals	(1,199)
Education, of which:	2,990
Higher education	(2,713)
Highways	1,555
Public safety	542
Environment	281
General administration	298
Debt interest	168
Transfers to local governments	4,840
Other expenditures	419
TOTAL	$14,357

The following table shows general revenues and expenditures for leading Texas cities in 1981/82 (in millions):

REVENUES	HOUSTON	DALLAS	SAN ANTONIO
Property tax	$245.9	$135.5	$ 29.8
Sales and gross receipts tax	202.6	114.6	41.9
Federal transfers	113.9	50.8	78.5
Other receipts	281.6	138.1	127.8
TOTALS	$844.0	$439.0	$278.0

EXPENDITURES	HOUSTON	DALLAS	SAN ANTONIO
Police protection	$128.9	$70.0	$37.2
Fire protection	96.7	41.0	20.1
Highways	65.3	41.5	24.7
Health and hospitals	26.7	6.6	16.3
Housing and community development	6.7	5.9	21.1
Other outlays	556.7	284.0	186.6
TOTALS	$881.0	$449.0	$306.0

Houston's gross debt as of 30 June 1984 was more than $1.9 billion; in 1982, San Antonio's debt was more than $1.7 billion, and Dallas's was $722 million. In 1982/83, state government debt was $3 billion and that for local governments $27.1 billion, 85% higher than in 1977, but still, at $1,710.80 per capita, slightly below the national average.

[36] TAXATION

Texas, whose total tax collections amounted to almost $9.9 billion in 1983/84, is one of only five states that imposed neither a personal nor a corporate income tax in 1985. The principal source of state tax revenue is the 4% sales and use tax (4.125% as of 1 October 1985), which accounted for more than 38% of total 1983/84 tax collections. Other major sources of revenue are oil and natural-gas production taxes, providing more than 22% of the 1983/84 total; motor fuel taxes, 5.5%; motor vehicle sales and rental tax, 7%; and cigarette and tobacco taxes, 3%. Other state levies include the corporation franchise tax, alcoholic beverage tax, public utility taxes, inheritance tax, and a tax on telephones.

Local property tax rates per $100 of assessed valuation varied widely throughout the state; in 1982, the rate in Houston was $1.76. The city sales tax of 1% is a major source of revenue for the municipalities, amounting to about 10% of total local tax revenues.

Although the state and local tax burden is well below the US average, tax collections per capita nearly tripled between 1972 and 1982. In 1978, Texans approved a constitutional amendment limiting further tax increases. The legislature responded the following year by passing a tax-cut package totaling about $400 million, including a comprehensive property tax reform bill and a public education bill to provide greater equity between rich and poor school districts. In 1984, the legislature significantly increased taxes for the first time in 13 years, raising the general sales tax from 4 to 4.125%, the taxes on gasoline and diesel fuel to 10 cents a gallon from 5 cents and 6.5 cents, respectively, and the hotel and motel occupancy tax from 3% to 4%. Taxes on cigarettes and alcoholic beverages were increased, and various exemptions to the sales tax, including those on amusement admissions, cable television, and laundry and dry cleaning, were eliminated. Most of the funds raised by the increases will be used for public school finance.

Federal tax collections in Texas during 1982/83 exceeded $42.5 billion; the per capita federal tax burden—$2,728—was 15th highest in the nation. In 1984, Texans paid an average of $703 per capita more in federal taxes than they received in federal expenditures—the 3d-largest negative difference among the 50 states.

37 ECONOMIC POLICY

The major factors responsible for the state's rapid economic expansion during the 1970s and early 1980s were the availability of reasonably priced real estate, plentiful fuel supplies, and a sizable, largely unorganized labor force. Texas state government has been notably probusiness: regulation is less restrictive than in many states, and there is no corporate income tax. The state government actively encourages outside capital investment in Texas industries, and the state's remarkable industrial productivity has produced a generally high return on investment.

The Texas Economic Development Commission (formerly the Texas Industrial Commission) helps businesses locate or expand their operations in the state. A private organization, the Texas Industrial Development Council, in Bryan, also assists new and developing industries.

38 HEALTH

Medical care ranges from adequate to excellent in the state's largest cities, but many small communities are without doctors and hospitals. Texas suffers from a general shortage of health care personnel: although the number of students enrolled in medical education programs more than doubled during the 1970s, many hospitals were functioning without adequate numbers of registered nurses, laboratory technicians, and therapists as the 1980s began. These shortages were expected to continue, given the state's rapidly increasing population.

The infant mortality rate as of 1981 was 10.4 per 1,000 live births for whites and 17.8 for nonwhites. In 1982/83, there were 105,800 legal abortions—29.2 per 1,000 women of childbearing age and 343 for each 1,000 live births. Rates for live births (19.4 per 1,000 population), marriages (13.4), and divorces (6.4) all exceeded the national norms in 1982.

The overall death rate in 1981 was 7.5 per 1,000 population.

Texas ranked below the national average in deaths due to heart disease, cancer, stroke, pneumonia and influenza, diabetes, cirrhosis of the liver, and arteriosclerosis, but above average in deaths from accidents, suicide, and early childhood diseases. The leading causes of death in 1981, with their rates per 100,000 population, were heart disease, 261.4; cancer, 147.4; stroke, 64.4; accidents, 53.2; chronic lung disease, 21.5; pneumonia, 18.3; diabetes, 12.6; suicide, 12.5; early childhood diseases, 11; atherosclerosis, 10.3; and liver disease, 9.3.

As of 1983, the Texas Department of Mental Health and Mental Retardation operated 8 state hospitals and 7 major mental health centers, plus 13 state training schools for the mentally retarded.

In 1983, Texas's 562 hospitals, with a total of 84,935 beds, admitted 2,738,568 patients and had 14,501,411 outpatient visits. In 1981, hospital personnel numbered 215,196, including 41,340 registered nurses and 3,894 medical and health care trainees. The average cost of a semiprivate hospital room as of January 1984 was $151.73 per day, well below the US average. In 1982, the average hospital stay in Texas was 6.6 days, at an average cost of $2,027.50—again, below the US average. Texas had 24,927 licensed physicians and 6,712 active dentists in 1982.

There are 8 medical schools, 2 dental colleges, and 64 schools of nursing in the state. The University of Texas has medical colleges at Dallas, Houston, Galveston, San Antonio, and Tyler. The University of Texas Cancer Center at Houston is one of the nation's major facilities for cancer research. Houston is also noted as a center for cardiovascular surgery. On 3 May 1968, Houston surgeon Denton Cooley performed the first human heart transplant in the US.

39 SOCIAL WELFARE

In 1983, Texas ranked 40th among the states in the proportion of its population receiving public aid—3.5%. Although the state has never enjoyed a reputation for lavishness in its public assistance efforts, the 1985 legislature considerably expanded its social welfare programs, increasing welfare benefits $4 per month (to $57), extending unemployment benefits to farm workers, and providing $150 million per year to pay for private health care for the poor in the 132 countries lacking public hospitals.

In 1982, payments to families with dependent children, made to 301,700 Texans, totaled $117 million; the monthly payment per family averaged $105, just over one-third of the national norm. Medicaid payments, on the other hand, more than quadrupled during the 1970s, reaching more than $1.1 billion in 1981, when there were 705,800 recipients. In 1983, an estimated 1,251,000 Texans took part in the food stamp program, at a federal cost of $688 million. About 1,705,000 schoolchildren—130,000 fewer than in 1980—enjoyed school-provided lunches, at a federal cost of $180 million—up from $166 million in 1980.

Social Security benefits were paid in 1983 to 1,870,000 persons, of whom 67% were retired workers, 24% were survivors of deceased workers, and 9% were disabled. These pensions, amounting to nearly $8.1 billion, averaged $418 monthly for retired workers, about 5% below the national average. Federally administered Supplemental Security Income payments, made to 244,278 Texans, totaled $445.8 million in 1983. The average monthly benefit in December was $131.54 for a retiree and $203.34 for a disabled person—in each case, well below the US average.

Workers' compensation payments were $981.7 million in 1982. A weekly average of 101,000 beneficiaries received $668 million in unemployment benefits in 1982; the average weekly payment was $127, slightly above the national norm.

40 HOUSING

The variety of Texas architectural styles reflects the diversity of the state's topography and climate. In the early settlement period, Spanish-style adobe houses were built in southern Texas. During the 1840s, Anglo-American settlers in the east erected primitive log cabins. These were later replaced by "dog-run" houses, consisting of two rooms linked by an open passageway covered by a gabled roof, so-called because pet dogs slept in the open, roofed shelter, as did occasional overnight guests. During the late 19th century, southern-style mansions were built in East Texas, and the familiar ranch house, constructed of stone and usually stuccoed or whitewashed, with a shingle roof and a long porch, proliferated throughout the state; the modern ranch house in southwestern Texas shows a distinct Mexican-Spanish influence. Climate affects such modern amenities as air conditioning: a new house in the humid eastern region is likely to have a refrigeration-style cooler, while in the dry west and south, an evaporating "swamp cooler" is the more common means of making hot weather bearable.

As of the 1980 census, Texas had 5,505,016 year-round housing units, of which 4,929,267 were occupied. Of the latter number, about 65% were lived in by the owner. Between 1981 and 1983, the state authorized construction of 613,200 housing units, valued at $22.6 billion.

41 EDUCATION

Texas ranks 2d only to California in number of public schools and in public school enrollment. Only California has a more extensive public college and university system. School finance has been a chronic problem, however; per pupil expenditures for Texas public schools in 1983/84 were $2,913, 30th among the 50 states, but a considerable improvement over 1977/78, when Texas ranked 40th.

Although public instruction began in Texas as early as 1746, education was slow to develop during the period of Spanish and Mexican rule. The legislative foundation for a public school system was laid by the government of the Republic of Texas during the late 1830s, but funding was slow to develop. After annexation, in 1846, Galveston began to support free public schools, and San Antonio had at least four free schools by the time a statewide system of public education was established in 1854. Free segregated schooling was provided for black children beginning in the 1870s; their schools were ill-maintained and underfinanced. School integration was accomplished during the 1960s, nonviolently for the most part.

In 1980, more than 62% of the population 25 years old and over had completed at least four years of high school, and nearly 17% had four or more years of college. More than 20% had fewer than eight years of school, however, and the state's median of 12.4 school years lagged behind the US average.

Enrollment in Texas's 5,730 public schools in 1984/85 was 1,178,286 in grades K–8 and 868,761 in grades 9–12. Teachers numbered 169,942 in 1983/84, and the average salary was $18,682. Enrollment in nonpublic schools was estimated at 148,534, of whom 88% were in church-affiliated schools.

Institutions of higher education in 1982/83 included 24 public senior colleges and universities, with 355,706 students; 47 public community college campuses, with 289,363 students; and more than 3 dozen private institutions with a combined enrollment of 79,196. The leading public universities are Texas A&M (College Station), which opened in 1876 and enrolled 36,108 students in 1982/83, and the University of Texas (Austin), founded in 1883, with 48,039 students enrolled in 1982/83; each institution is now the center of its own university system, including campuses in several other cities. Oil was discovered on lands owned by the University of Texas in 1923, and beginning in 1924, the university and Texas A&M shared more than $1 billion in oil-related rentals and royalties. Another state-supported institution, the University of Houston, enrolled 30,845 on its main campus in 1982/83, and Texas Tech (Lubbock) enrolled 22,849.

The first private college in Texas was Rutersville, established by a Methodist minister in Fayette County in 1840. The oldest private institution still active in the state is Baylor University (1845), at Waco, enrolling 10,473 students in 1982/83. Other major private universities and their enrollment in 1982/83 include Hardin-Simmons (Abilene), 1,949; Lamar (Beaumont), 14,600; Rice (Houston), 3,716; Southern Methodist, or SMU (Dallas), 9,150; and Texas Christian, or TCU (Ft. Worth), 6,881. Well-known black-oriented institutions of higher learning include Texas Southern University in Houston, 8,333 in 1982/83, and Bishop College in Dallas, 1,186.

Tuition charges to Texas colleges are among the lowest in the nation. In 1985, the legislature tripled tuition charges of state colleges—from $4 to $12 per semester hour. The Texas Student Assistance Corp. administers a guaranteed-loan program and tuition equalization grants for needy students.

42 ARTS

Although Texas has never been regarded as a leading cultural center, the arts have a long history in the state. The cities of Houston and Matagorda each had a theater before they had a church, and the state's first theater was active in Houston as early as 1838. Stark Young founded the Curtain Club acting group at the University of Texas in Austin in 1909, and the little-theater movement began in that city in 1921. The performing arts now flourish at Houston's Theater Center, Jones Hall of Performing Arts, and Alley Theater, as well as at Dallas's Theater Center, National Children's Theater, and Theater Three. The Margo Jones repertory company in Dallas has a national reputation, and there are major repertory groups in Houston and San Antonio. During the late 1970s, Texas also emerged as a center for motion picture production.

Texas has 5 major symphony orchestras—the Dallas Symphony (with Eduardo Mata as music director), Houston Symphony, San Antonio Symphony, Austin Symphony, and Ft. Worth Symphony—and 25 orchestras in other cities. The Houston Grand Opera performs at Jones Hall; other opera companies perform regularly in Beaumont, Dallas, El Paso, Ft. Worth, and San Antonio. All these cities also have resident dance companies, as do Abilene, Amarillo, Austin, Corpus Christi, Denton, Galveston, Garland, Longview, Lubbock, Midland-Odessa, and Pampa.

Popular music in Texas stems from early Spanish and Mexican folk songs, Negro spirituals, cowboy ballads, and German-language songfests. Texans pioneered a kind of country and western music that is more outspoken and direct than Nashville's commercial product, and a colony of country-rock songwriters and musicians were active in the Austin area during the 1970s. Texans of Mexican ancestry have also fashioned a Latin-flavored music that is as distinctly "Tex-Mex" as the state's famous chili.

43 LIBRARIES AND MUSEUMS

Public libraries in Texas, serving 235 counties in 1981/82, had a combined book stock of 21,944,412 volumes and a circulation of 48,248,925 volumes. Funding for basic library service is provided by cities and counties, although state and federal library aid increased considerably during the 1970s. More libraries are supported by cities than by counties, partly because of fewer state restrictions on municipal library funding. The largest municipal library systems are those of Houston (2,802,949 volumes), Dallas (1,945,172), and San Antonio (1,207,133). The University of Texas at Austin, noted for outstanding collections in the humanities and in Latin American studies, had 5,057,649 volumes in 1982. The Lyndon B. Johnson Presidential Library is also located in Austin, as are the state archives. Other notable academic libraries include those of Rice and Southern Methodist universities.

Among the state's 280 museums are Austin's Texas Memorial Museum; the Dallas Museum of Fine Arts and the Dallas Museum of Art; and the Amon Carter Museum of Western Art, the Ft. Worth Art Museums, and Kimbell Art Museum, all in Ft. Worth. Houston has the Museum of Fine Arts, Contemporary Arts Museum, and at least 30 galleries. Both Dallas–Ft. Worth and Houston have become major centers of art sales.

National historic sites in Texas are Ft. Davis (Jeff Davis

County), President Johnson's boyhood home and Texas White House (Blanco and Gillespie counties), and the San Jose Mission (San Antonio). Other historic places include the Alamo, Dwight D. Eisenhower's birthplace at Denison, the Sam Rayburn home in Bonham, and the John F. Kennedy memorials in Dallas. A noteworthy prehistoric Indian site is the Alibates Flint Quarries National Monument, located in Potter County and accessible by guided tour.

44COMMUNICATIONS
The US Postal Service had about 42,000 employees and 1,495 post offices in Texas in March 1985. In 1980, 90% of the 4,929,267 occupied housing units in Texas had telephones. There were 6,888,646 residential and commercial telephone access lines in service at the end of 1982. Dallas was one of Western Union's first US communications satellite stations, and it leads the state as a center for data communications.

The state has not always been in the communications vanguard, however. Texas passed up a chance to make a handsome profit from the invention of the telegraph when, in 1838, inventor Samuel F. B. Morse offered his newfangled device to the republic as a gift. When the Texas government neglected to respond, Morse withdrew the offer.

Texas had 550 radio stations in 1982 and 67 commercial and 10 educational television stations in 1984. The state's first radio station, WRR, was established by the city of Dallas in 1920—an interesting fact in view of Dallas's worldwide reputation as a bastion of free enterprise. The first television station, WBAP, began broadcasting in Ft. Worth in 1948. Cable television systems served 2,048,233 Texans in 869 communities in 1984.

45PRESS
The first newspaper in Texas was a revolutionary Spanish-language sheet published in May 1813 at Nacogdoches. Six years later, the *Texas Republican* was published by Dr. James Long in the same city. In 1835, the *Telegraph and Texas Register* became the official newspaper of the Texas Republic, and it continued to publish until 1877. The first modern newspaper was the *Galveston News* (1842), a forerunner of the *Dallas Morning News* (1885). Other pioneering papers still in continuous publication include the *San Antonio Express* and the *Houston Post*.

By 1984, Texas had 113 morning, evening, and all-day dailies, with a combined circulation of 3,668,183, and 98 Sunday papers, with 4,298,755. The newspapers with the largest daily circulations were as follows:

AREA	NAME	DAILY	SUNDAY
Austin	American-Statesman (all day,S)	160,526	191,206
Dallas	Morning News (m,S)	360,347	454,828
	Times Herald (all day,S)	270,622	373,225
Ft. Worth	Star-Telegram (m,e,S)	111,893 } 128,184	271,563
Houston	Chronicle (all day,S)	441,557	538,232
	Post (m,S)	326,556	376,015
San Antonio	Express (all day,S)	169,884	204,918
	Light (all day,S)	140,141	202,223

In 1984 there were 422 weekly newspapers. The *Texas Almanac,* a comprehensive guide to the state, has been issued at regular intervals since 1857 by the A. H. Belo Corp., publishers of the *Dallas Morning News.* Leading magazines include the *Texas Monthly* and *Texas Observer,* both published in Austin. Including university presses, there were 35 book publishers in Texas in 1984.

46ORGANIZATIONS
The 1982 Census of Service Industries counted 3,433 organizations in Texas, including 769 business associations; 1,829 civic, social, and fraternal associations; and 88 educational, scientific, and research associations.

Irving is the home of one of the nation's largest organizations, the Boy Scouts of America, with more than 4.5 million members

in 1984. Important medical groups are the American Heart Association, in Dallas, and the National Association for Retarded Citizens, Arlington. Professional societies include the American Association of Petroleum Landmen, Ft. Worth. The Noncommissioned Officers Association has its home office in San Antonio; the Airline Passengers Association is in Dallas. Among the many organizations devoted to horse breeding are the American Quarter Horse Association, Amarillo; Palomino Horse Breeders of America, Mineral Wells; and the National Cutting Horse Association and American Paint Association, both in Ft. Worth. Ft. Worth is also the home of the Texas Longhorn Breeders Association of America.

47TOURISM, TRAVEL, AND RECREATION
According to a study prepared for the Texas Tourist Development Agency, 35,400,000 out-of-staters visited Texas in 1982, of whom 91% were from the US—mostly from California, Louisiana, Oklahoma, New York, and Illinois, in that order—and 9% were foreign— 1,300,000 of them from Mexico, 396,000 from Canada; and 293,000 from England. The most-visited cities in 1982 were Dallas, San Antonio, Houston, Ft. Worth, El Paso, Austin, Brownsville, Corpus Christi, Harlingen, and Galveston. In 1982, travelers and tourists generated more than $13.7 billion in expenditures in Texas—the 3d-largest figure in the US. The travel industry was the 3d-largest employer in Texas that year, accounting for 4.5% of all jobs in the state.

Each of the state's seven major tourist regions offers outstanding attractions. East Texas has one of the state's oldest cities, Nacogdoches, with the nation's oldest public thoroughfare and a reconstruction of the Old Stone Fort, a Spanish trading post dating from 1779. Jefferson, an important 19th-century inland port, has many old homes, including Excelsior House. Tyler, which bills itself as the "rose capital of the world," features a 28-acre (11-hectare) municipal rose garden and puts on a Rose Festival each October. The Gulf Coast region of southeastern Texas offers the Lyndon B. Johnson Space Center, the Astrodome sports stadium and adjacent Astroworld amusement park, and a profusion of museums, galleries, and shops, all in metropolitan Houston; Spindletop Park, in Beaumont, commemorating the state's first great oil gusher; Galveston's sandy beaches, deep-sea fishing, and Sea-Arama Marineworld; and the Padre Island National Seashore.

To the north, the Dallas–Ft. Worth metropolitan area (including Arlington) has numerous cultural and entertainment attractions, including the Six Flags Over Texas amusement park and the state fair held in Dallas each October. Old Abilene Town amusement park, with its strong western flavor, is also popular with visitors. The Hill Country of south-central Texas encompasses many tourist sites, including the state capitol in Austin, Waco's Texas Ranger Museum (Ft. Fisher), the Lyndon B. Johnson National Historic Site, and frontier relics in Bastrop and Bandera.

South Texas has the state's most famous historic site—the Alamo, in San Antonio, which also contains HemisFair Plaza and Brackenridge Park. The Rio Grande Valley Museum, at Harlingen, is popular with visitors, as is the King Ranch headquarters in Kleberg County. The Great Plains region of the Texas panhandle offers Palo Duro Canyon—Texas's largest state park, covering 16,402 acres (6,638 hectares) in Armstrong and Randall counties—the Prairie Dog Town at Lubbock, Old West exhibits at Matador, and the cultural and entertainment resources of Amarillo. In the extreme northwestern corner of the panhandle is the XIT Museum, recalling the famous XIT Ranch, at one time the world's largest fenced ranch, which formerly covered more than 3 million acres (1.2 million hectares). Outstanding tourist sites in the far west are the Big Bend and Guadalupe Mountains national parks, the Jersey Lilly Saloon and Judge Roy Bean visitor center in Langtry, and metropolitan El Paso.

Texas's state park system—including 30 state parks, 35 recreation areas, 16 state historical parks, 18 state historical sites, 4 state historic structures, 3 state natural areas, and 3 state fishing piers—attracted some 17 million visitors in 1981/82. In addition to Palo Duro Canyon, notable state parks include Big Creek (Ft. Bend County), Brazos Island (Cameron County), Caddo Lake (Harrison County), Dinosaur Valley (Somervell County), Eisenhower (Grayson County), Galveston Island, and Longhorn Cavern (Burnet County). State historical parks include San Jacinto Battleground (east Harris County), Texas State Railroad (Anderson and Cherokee counties), and Washington-on-the-Brazos (Washington County).

Hunting and fishing are extremely popular in Texas. Some 337,600 white-tailed deer were harvested in the 1982/83 season; in 1981/82, hunters killed about 5,200 male deer, 18,900 jabalina, and 53,500 wild turkeys. During 1982/83, licenses were issued to 1,050,496 hunters and 1,789,298 anglers.

⁴⁸SPORTS

Texas is a great state for sports enthusiasts. Perhaps the nation's most-publicized football team—known both for its prowess and its cheerleaders—is the Dallas Cowboys of the National Football League (NFL); coached by Tom Landry and led by such stars as Roger Staubach and Tony Dorsett, the Cowboys played in five Super Bowls between 1971 and 1979, winning two and losing three. The Houston Oilers captured American Football League championships in 1960 and 1961, subsequently joining the NFL's American Conference.

The Astros, playing in the Astrodome, represent Houston in baseball's National League West; their best season was in 1980, when they won a divisional title but lost in the playoffs to the Philadelphia Phillies. The Texas Rangers of the American League play at Arlington, midway between Dallas and Ft. Worth. The class-AA Texas League, which has been in continous existence since 1918 and had its heyday in the 1920s and 1930s, now includes teams in Beaumont, El Paso, Midland, and San Antonio, as well as four out-of-state clubs.

Texas is represented in the National Basketball Association by the Houston Rockets, San Antonio Spurs, and Dallas Mavericks.

College football is extremely popular in the state. Eight Texas college teams—the University of Texas, Texas A&M, Rice, Baylor, SMU, TCU, Texas Tech, and Houston—participate in the Southwest Conference, whose annual champion is the host team at the Cotton Bowl game held in Dallas on New Year's Day. Through January 1985, the University of Texas had won 9 Cotton Bowl games; Rice had won three; and Houston, SMU, TCU, and Texas A&M two each. Other college bowl games held in the state each year are the Astro-Bluebonnet Bowl in Houston and the Sun Bowl in El Paso.

Pari-mutuel betting on horse races is illegal in Texas, but quarter horse racing is popular, and rodeo is a leading spectator sport. Participant sports popular with Texans include hunting, fishing, horseback riding, boating, swimming, tennis, and golf. State professional and amateur golf tournaments are held annually. The Texas Sports Hall of Fame was organized in 1951; new members are selected each year by a special committee of the Texas Sports Writers Association.

⁴⁹FAMOUS TEXANS

Two native sons of Texas have served as president of the US. Dwight D. Eisenhower (1890-1969), the 34th president, was born in Denison, but his family moved to Kansas when he was two years old. Lyndon Baines Johnson (1908-73), the 36th president, was the only lifelong resident of the state to serve in that office. Born near Stonewall, he occupied center stage in state and national politics for a third of a century as US representative, Democratic majority leader of the US Senate, and vice president under John F. Kennedy, before succeeding to the presidency after Kennedy's assassination. Reelected by a landslide, Johnson accomplished much of his Great Society program of social reform but saw his power and popularity wane because of the war in Viet-Nam. His wife, Claudia Alta Taylor "Lady Bird" Johnson (b.1912), was influential in environmental causes as First Lady.

Texas's other native vice president was John Nance Garner (1868-1967), former speaker of the US House of Representatives. George Bush (b.Massachusetts, 1924), who founded his own oil development company and has served in numerous federal posts, was elected vice president in 1980 on the Republican ticket. Tom C. Clark (1899-1977) served as an associate justice on the US Supreme Court from 1949 to 1967; he stepped down when his son Ramsey (b.1927) was appointed US attorney general, a post the elder Clark had also held.

Another prominent federal officeholder from Texas was Jesse H. Jones (1874-1956), who served as chairman of the Reconstruction Finance Corporation and secretary of commerce under Franklin D. Roosevelt. Oveta Culp Hobby (b.1905), publisher of the *Houston Post,* became the first director of the Women's Army Corps (WAC) during World War II and the first secretary of the Department of Health, Education, and Welfare under President Eisenhower. John Connally (b.1917), a protégé of Lyndon Johnson's, served as secretary of the Navy under Kennedy and, as governor of Texas, was wounded in the same attack that killed the president; subsequently, he switched political allegiance, was secretary of the treasury under Richard Nixon, and has been active in Republican Party politics. Other federal officials from Texas include "Colonel" Edward M. House (1858-1938), principal adviser to President Wilson, and Leon Jaworski (1905-82), the Watergate special prosecutor whose investigations led to President Nixon's resignation.

The state's most famous legislative leader was Sam Rayburn (1882-1961), who served the longest tenure in the nation's history as speaker of the US House of Representatives—17 years in three periods between 1940 and 1961. James Wright (b.1922) was Democratic majority leader of the House in the 1970s and early 1980s, and Barbara C. Jordan (b.1936) won national attention as a forceful member of the House Judiciary Committee during its impeachment deliberations in 1974.

Famous figures in early Texas history include Moses Austin (b.Connecticut, 1761-1821) and his son, Stephen F. Austin (b.Virginia, 1793-1836), often called the "father of Texas." Samuel "Sam" Houston (b.Virginia, 1793-1863), adopted as a youth by the Cherokee, won enduring fame as commander in chief of the Texas revolutionary army, as president of the Texas Republic, and as the new state's first US senator; earlier in his career, he had been governor of Tennessee. Mirabeau Bonaparte Lamar (b.Georgia, 1798-1859), the second president of the republic, founded the present state capital (now called Austin) in 1839. Anson Jones (b.Massachusetts, 1798-1858) was the last president of the republic.

Noteworthy state leaders include John H. Reagan (b. Tennessee, 1818-1905), postmaster general for the Confederacy; he dominated Texas politics from the Civil War to the 1890s, helping to write the state constitutions of 1866 and 1875, and eventually becoming chairman of the newly created Texas Railroad Commission. The most able Texas governor was probably James Stephen Hogg (1851-1906), the first native-born Texan to hold that office. Another administration with a progressive record was that of Governor James V. Allred (1899-1959), who served during the 1930s. Miriam A. "Ma" Ferguson (1875-1961) became in 1924 the first woman to be elected governor of a state, and she was elected again in 1932. With her husband, Governor James E. Ferguson (1871-1944), she was a factor in Texas politics for nearly 30 years. Texas military heroes include Audie Murphy (1924-71), the most decorated soldier of World War II (and later a film actor), and Admiral of the Fleet Chester W. Nimitz (1885-1966).

Figures of history and legend include James Bowie (b.Kentucky, 1796?-1836), who had a reputation as a brawling fighter and wheeler-dealer until he died at the Alamo; he is popularly credited with the invention of the bowie knife. David "Davy" Crockett (b.Tennessee, 1786-1836) served three terms as a US representative from Tennessee before departing for Texas; he, too, lost his life at the Alamo. Among the more notorious Texans was Roy Bean (b.Kentucky, 1825-1903), a judge who proclaimed himself "the law west of the Pecos." Gambler, gunman, and desperado John Wesley Hardin (1853-95) boasted that he "never killed a man who didn't deserve it." Bonnie Parker (1910-34) and Clyde Barrow (1909-34), second-rate bank robbers and murderers who were shot to death by Texas lawmen, achieved posthumous notoriety through the movie *Bonnie and Clyde* (1967).

Many Texas businessmen have profoundly influenced the state's politics and life-style. Clint Murchison (1895-1969) and Sid Richardson (1891-1959) made great fortunes as independent oil operators and spread their wealth into other enterprises: Murchison became owner-operator of the successful Dallas Cowboys professional football franchise, and Richardson, through the Sid Richardson Foundation, aided educational institutions throughout the Southwest. Oilman H(aroldson) L(afayette) Hunt (b.Illinois, 1889-1974), reputedly the wealthiest man in the US, was an avid supporter of right-wing causes. Howard Hughes (1905-79), an industrialist, aviation pioneer, film producer, and casino owner, became a fabulously wealthy recluse in his later years. Stanley Marcus (b.1905), head of the famous specialty store Neiman-Marcus, became an arbiter of taste for the world's wealthy and fashionable men and women. Rancher Richard King (b.New York, 1825-85) put together the famed King Ranch, the largest in the US at his death. Charles Goodnight (b.Illinois, 1836-1929) was an outstanding cattleman.

Influential Texan historians include folklorist John A. Lomax (b.Mississippi, 1867-1948); Walter Prescott Webb (1888-1963), whose books *The Great Plains* and *The Great Frontier* helped shape American thought; and J. Frank Dobie (1888-1964), well-known University of Texas educator and compiler of Texas folklore. Dan Rather (b.1931) has earned a nationwide reputation as a television reporter and anchorman. Frank Buck (1884-1950), a successful film producer, narrated and appeared in documentaries showing his exploits among animals.

William Sydney Porter (b.North Carolina, 1862-1910) apparently embezzled funds from an Austin bank, escaped to Honduras, but returned to serve a three-year jail term—during which time he began writing short stories, later published under the pen name O. Henry. Katherine Anne Porter (1890-1980) also won fame as a short-story writer. Fred Gipson (b.1908) wrote *Hound Dog Man* and *Old Yeller,* praised by critics as a remarkable evocation of a frontier boy's viewpoint. Two novels by Larry McMurtry (b.1936), *Horseman, Pass By* (film title, *Hud*) and *The Last Picture Show,* became significant motion pictures. Robert Rauschenberg (b.1925) is a leading contemporary painter. Elisabet Ney (b.Germany, 1833-1907), a sculptor, came to Texas with a European reputation and became the state's first determined feminist; she wore pants in public, and seldom passed up an opportunity to outrage Texans' Victorian mores.

Prominent Texans in the entertainment field include Mary Martin (b.1913), who reigned over the New York musical comedy world for two decades; her son, Larry Hagman (b.1931), star of the "Dallas" television series; actress Debbie Reynolds (b.1931); movie director King Vidor (1894-1982); and Joshua Logan (b.1908), director of Broadway plays and Hollywood movies. Texans who achieved national reputations with local repertory companies were Margo Jones (1912-55) and Nina Vance (1914-80), who founded and directed theater groups in Dallas and Houston, respectively; and Preston Jones (1936-79), author of *A Texas Trilogy* and other plays.

Among Texas-born musicians, Tina Turner (b.1941) is a leading rock singer, as was Janis Joplin (1943-70). Willie Nelson (b.1933) wedded progressive rock with country music to start a new school of progressive "outlaw" music. Bob Wills (b.Oklahoma, 1905-75) was the acknowledged king of western swing. Musicians Trini Lopez (b.1937), Freddy Fender (Baldemar Huerta, b.1937), and Johnny Rodriguez (b.1951) have earned popular followings based on their Mexican-American backgrounds. Charlie Pride (b.Mississippi, 1938) became the first black country-western star. Other country-western stars born in Texas are Waylon Jennings (b.1937) and Kenny Rogers (b.1938). In the jazz field, pianist Teddy Wilson (b.1912) was a member of the famed Benny Goodman trio in the 1930s. Trombonist Jack Teagarden (1905-64) and trumpeter Harry James (1916-83) have also been influential.

The imposing list of Texas athletes is headed by Mildred "Babe" Didrikson Zaharias (1913-56), who gained fame as an All-American basketball player in 1930, won two gold medals in track and field in the 1932 Olympics, and was the leading woman golfer during the 1940s and early 1950s. Another Texan, John Arthur "Jack" Johnson (1878-1946), was boxing's first black heavyweight champion. Texans who won fame in football include quarterbacks Sammy Baugh (b.1914), Don Meredith (b.1938), and Roger Staubach (b.Ohio, 1942); running back Earl Campbell (b.1955); and coaches Dana X. Bible (1892-1980), Darrell Royal (b.Oklahoma, 1924), and Thomas Wade "Tom" Landry (b.1924). Among other Texas sports greats are baseball Hall of Famers Tris Speaker (1888-1958) and Rogers Hornsby (1896-1963); golfers Ben Hogan (b.1912), Byron Nelson (b.1912), and Lee Trevino (b.1939); auto racing driver A(nthony) J(oseph) Foyt (b.1935); and jockey William Lee "Willie" Shoemaker (b.1931).

[50]BIBLIOGRAPHY

Arbingast, Stanley A., et al. *Atlas of Texas.* Austin: University of Texas Press, 1979.

Bainbridge, John. *The Super-Americans.* New York: Doubleday, 1961.

Binkley, William C. *The Texas Revolution.* Austin: Texas State Historical Association, 1979.

Buenger, Walter L. (ed.) *Texas History.* Boston: American Press, 1983.

Caro, Robert A. *The Years of Lyndon Johnson: The Path to Power.* New York: Knopf, 1982.

Conaway, James. *The Texans.* New York: Knopf, 1976.

Connor, Seymour V. (ed.). *The Saga of Texas.* 6 vols. Austin: Steck-Vaughn, 1965.

Dobie, J. Frank. *Coronado's Children.* New York: Grosset & Dunlap, 1963.

Dobie, J. Frank. *The Longhorns.* Boston: Little, Brown, 1941.

Duke, Cordia Sloan, and Joe B. Frantz. *6000 Miles of Fence: Life on the XIT Ranch of Texas.* Austin: University of Texas Press, 1981.

Federal Writers' Project. *Texas: A Guide to the Lone Star State.* Reprint. New York: Somerset, n.d. (orig. 1940).

Fehrenbach, T. R. *Lone Star: A History of Texas and the Texans.* New York: Crown, 1983.

Frantz, Joe B. *Texas: A Bicentennial History.* New York: Norton, 1976.

Frantz, Joe B., and Julian E. Choate, Jr. *The American Cowboy.* Norman: University of Oklahoma Press, 1955.

Friend, Llerena. *Sam Houston: The Great Designer.* Austin: University of Texas Press, 1954.

Gambrell, Herbert. *Anson Jones: The Last President of Texas.* Austin: University of Texas Press, 1964.

Handbook of Texas. 3 vols. Edited by Walter Prescott Webb and H. Bailey Carroll, with supplement edited by Eldon Stephen Brandon. Austin: Texas State Historical Association, 1952-76.

Horgan, Paul. *Great River: The Rio Grande in North American History.* New York: Holt, Rinehart & Winston, 1954.

Jordan, Terry G. *Immigration to Texas.* Boston: American Press, 1981.

Newcomb, W. W., Jr. *The Indians of Texas: From Prehistoric to Modern Times.* Austin: University of Texas Press, 1969.

Nunn, W. C. *Texas under the Carpetbaggers.* Austin: University of Texas Press, 1962.

Richardson, Rupert N., et al. *Texas: The Lone Star State.* 4th ed. Englewood Cliffs, N.J.: Prentice-Hall, 1981.

Sibley, Marilyn McAdams. *The Port of Houston: A History.* Austin: University of Texas Press, 1968.

Spratt, John S. *The Road to Spindletop: Economic Change in Texas, 1875–1901.* Austin: University of Texas Press, 1983.

Texas Almanac and State Industrial Guide, 1984–85. 50th ed. Dallas: A. H. Belo Corp., 1983.

Texas Fact Book, 1984. Edited by Joseph E. Pluta, Rita J. Wright, and Midred C. Anderson. Austin: University of Texas, 1983.

Tinkle, Lon. *Thirteen Days to Glory.* New York: McGraw-Hill, 1958.

Webb, Walter Prescott. *The Texas Rangers: A Century of Frontier Defense.* Austin: University of Texas Press, 1965.

Wright, Rita J. *Texas Sources: A Bibliography.* Austin: University of Texas, 1976.

UTAH

State of Utah

ORIGIN OF STATE NAME: Named for the Ute Indians. **NICKNAME:** The Beehive State. **CAPITAL:** Salt Lake City. **ENTERED UNION:** 4 January 1896 (45th). **SONG:** "Utah, We Love Thee." **MOTTO:** Industry. **COAT OF ARMS:** In the center, a shield, flanked by American flags, shows a beehive with the state motto and six arrows above, sego lilies on either side, and the numerals "1847" (the year the Mormons settled in Utah) below. Perched atop the shield is an American eagle. **FLAG:** Inside a thin gold circle, the coat of arms and the year of statehood are centered on a blue field, fringed with gold. **STATE SEAL:** The coat of arms with the words "The Great Seal of the State of Utah 1896" surrounding. **ANIMAL:** Elk. **BIRD:** Sea gull. **FISH:** Rainbow trout. **FLOWER:** Sego lily. **TREE:** Blue spruce. **GEM:** Topaz. **EMBLEM:** Beehive. **LEGAL HOLIDAYS:** New Year's Day, 1 January; Birthday of Martin Luther King, Jr., 3d Monday in January; Lincoln's Birthday, 12 February; Washington's Birthday, 3d Monday in February; Memorial Day, last Monday in May; Independence Day, 4 July; Pioneer Day, 24 July; Labor Day, 1st Monday in September; Columbus Day, 2d Monday in October; Veterans Day, 11 November; Thanksgiving Day, 4th Thursday in November; Christmas Day, 25 December. **TIME:** 5 AM MST = noon GMT.

¹LOCATION, SIZE, AND EXTENT

Located in the Rocky Mountain region of the western US, Utah ranks 11th in size among the 50 states.

The area of Utah totals 84,899 sq mi (219,899 sq km), of which land comprises 82,073 sq mi (212,569 sq km) and inland water 2,826 sq mi (7,320 sq km). Utah extends 275 mi (443 km) E–W and 345 mi (555 km) N–S.

Utah is bordered on the N by Idaho; on the NE by Wyoming; on the E by Colorado and on the S by Arizona (with the two borders joined at Four Corners); and on the W by Nevada. The total boundary length of Utah is 1,226 mi (1,973 km). The state's geographic center is in Sanpete County, 3 mi (5 km) N of Manti.

²TOPOGRAPHY

The eastern and southern two-thirds of Utah belong to the Colorado Plateau, a region characterized by deep river canyons; erosion has carved much of the plateau into buttes and mesas. The Rocky Mountains are represented by the Bear River, Wasatch, and Uinta ranges in the north and northeast. These ranges, rising well above 10,000 feet (3,000 meters), hold the highest point in Utah—Kings Peak in the Uintas—at an altitude of 13,528 feet (4,123 meters).

The arid, sparsely populated Great Basin dominates the western third of the state. Drainage in this region does not reach the sea, and streams often disappear in the dry season. To the north are the Great Salt Lake, a body of hypersaline water, and the Great Salt Lake Desert (containing the Bonneville Salt Flats), both remnants of a vast prehistoric lake that covered the region during the last Ice Age. The lowest point in Utah—2,000 feet (610 meters) above sea level—occurs at Beaverdam Creek in Washington County, in the southwest corner of the state.

The western edge of the Wasatch Range, or Wasatch Front, holds most of Utah's major cities. It also attracts the greatest rainfall and snowfall, particularly in the north. Two regions rich in fossil fuels are the Kaiparowits Plateau, in southern Utah, and the Overthrust Belt, a geologic structural zone underlying the north-central part of the state.

The largest lake is the Great Salt Lake, which at the end of 1984 covered 2,250 sq mi (5,827 sq km) and was 34% larger than in 1976. In 1984, as the result of increased precipitation, the lake rose to 4,209.25 feet (1,283 meters) above sea level, its highest level since 1877; the lake has been rising steadily since 1963, causing severe flooding, and its waters, diluted by runoff, have lost

some salinity. Other major bodies of water are Utah Lake, Bear Lake (shared with Idaho), and Lake Powell, formed by the Glen Canyon Dam on the Colorado River. Other important rivers include the Green, flowing into the Colorado; the Sevier, which drains central and southern Utah; and the Bear, which flows into the Great Salt Lake.

³CLIMATE

The climate of Utah is generally semiarid to arid. Temperatures are favorable along the Wasatch Front, where there are relatively mild winters. At Salt Lake City, the normal daily mean temperature is 51°F (11°C), ranging from 28°F (−2°C) in January to 77°F (25°C) in July. The record high temperature, 116°F (47°C), was set at St. George on 28 June 1892; the record low temperature, −69°F (−56°C), in Peter Sink, on 1 February 1985. The average precipitation varies from less than 5 in (12.7 cm) in the west to over 40 in (102 cm) in the mountains, with Salt Lake City receiving 15 in (38 cm) per year. The annual snowfall is about 59 in (150 cm) and remains on the higher mountains until late summer.

⁴FLORA AND FAUNA

Botanists have recognized more than 4,000 floral species in Utah's six major life zones. Common trees and shrubs include four species of pine and three of juniper; aspen, cottonwood, maple, hawthorn, and chokecherry also flourish, along with the Utah oak, Joshua tree, and blue spruce (the state tree). Among Utah's wild flowers are sweet William and Indian paintbrush; the sego lily is the state flower. Endangered plants include four varieties of cactus (purple-spined hedgehog, spineless hedgehog, Siler pincushion, and Wright fishhook), Rydberg milk-vetch, clay phacelia, and the dwarf bear-poppy.

Mule deer are the most common of Utah's large mammals; other mammals include pronghorn antelope, Rocky Mountain bighorn sheep, lynx, grizzly and black bears, and white- and black-tailed jackrabbits. Among native bird species are the great horned owl, plain titmouse, and water ouzel; the golden eagle and great white pelican are rare species; and the sea gull (the state bird) is a spring and summer visitor from the California coast. The pygmy rattler is found in southwest Utah, and the Mormon cricket is unique to the state. Among Utah's endangered fauna (as listed by the US government) are the grizzly bear, bald eagle, Utah prairie dog, bonytail and humpback chubs, Colorado River squawfish, and woundfin. Many birds and fish have been killed or imperiled by the inundation of freshwater marshes with salt water from the flooding Great Salt Lake.

⁵ENVIRONMENTAL PROTECTION

Divisions of the Department of Natural Resources oversee water and mineral resources, parks and recreation, state lands and forests, and wildlife; spending on natural resources totaled $53.5 million in 1984/85, of which $18.1 million came from state general revenues. The Department of Agriculture is concerned with soil conservation and pesticide control. The Department of Health's Division of Environmental Health has separate bureaus dealing with air quality, public drinking water quality, and regulation of water pollution and radioactive and hazardous wastes.

Air pollution is a serious problem along the Wasatch Front, where automobiles abound. Also of considerable concern is the quality of drinking water. In 1984, 58% of the community water systems, serving 96% of the population, were rated "approved." However, among 250 smaller community systems, only 26 were rated "approved." The Division of Environmental Health has identified 180 sites in Utah where uncontrolled dumping and abandoned materials pose a potential threat. Another environmental problem is the pollution of Great Salt Lake by industrial waste.

⁶POPULATION

At the 1980 census, Utah had a population of 1,461,037, 36th in the US. The estimated population in 1985 was 1,684,942, representing a growth of 15.3% since 1980. A population of 2,043,000 is projected for 1990. Utah's estimated population density was 20.5 per sq mi (7.9 per sq km) in 1985. Because of the consistently high birthrate, Utahns tend to be much younger than the US population as a whole: more than 13% of state residents were under 5 years of age in 1980, and 41%—compared with 32% for the nation—were younger than 19 years of age.

More than four-fifths of all Utahns live in cities and towns, mostly along the Wasatch Front. Salt Lake City is Utah's most populous urban center, with a 1980 population of 163,033 in the city proper and 910,222—62% of state residents—in its metropolitan region. Salt Lake City's estimated 1984 population was 164,844. West Valley City, incorporated in 1980 in Salt Lake County, had an estimated 89,147 residents in 1984. Other major cities included Provo, 74,138; Ogden, 68,183; Sandy City, 63,498; and Orem, 60,884.

⁷ETHNIC GROUPS

Hispanic Americans constitute the largest ethnic minority in Utah, with a 1980 population estimated by the state at more than 50,000.

American Indians are the 2d-largest minority group in Utah, numbering 19,158 in 1980. Indian lands covered 2,331,000 acres (943,000 hectares) in 1982, all but 35 acres (14 hectares) of which were tribal landholdings. The Uintah and Ouray Indian reservation, in the northeast, and the Navaho Indian reservation, in the southeast, are the largest. Far smaller are the Skull Valley and Goshute reservations, in the west.

Over 15,000 Asian-Pacific peoples lived in Utah in 1980, the largest group (5,474) being Japanese-Americans. There were 2,730 Chinese and 2,108 Vietnamese. Utah also had 9,225 black Americans in 1980. Until 1978, blacks were denied full church membership as Mormons.

Utah had 50,000 residents who were foreign-born in 1980. Among persons reporting a single ancestry in 1980, 404,717 persons claimed English descent, 55,251 German, 24,636 Mexican, 22,440 Irish, and 19,685 Swedish.

⁸LANGUAGES

Forebears of the Ute, Goshute, and Paiute contributed to English only a few place-names, such as Utah itself, Uinta (and Uintah), Wasatch, and Tavaputs.

Utah English is primarily that merger of Northern and Midland carried west by the Mormons, whose original New York dialect later incorporated features from southern Ohio and central Illinois. Conspicuous in Mormon speech in the central valley, although less frequent now in Salt Lake City, is a reversal of vowels before, so that *farm* and *barn* sound like *form* and *born* and, conversely, *form* and *born* sound like *farm* and *barn*.

In 1980, 92% of all state residents 3 years of age or older spoke only English at home. Other languages spoken at home, and the number of people who spoke them, included Spanish, 37,131, and German, 11,593.

⁹RELIGIONS

The dominant religious group in Utah is the Church of Jesus Christ of Latter-day Saints, popularly known as the Mormons. The church was founded by Joseph Smith, Jr., in 1830, the same year he published the *Book of Mormon,* the group's sacred text. The Mormons' arrival in Utah climaxed a long pilgrimage that began in New York State and led westward to Missouri, then back to Illinois (where Smith was lynched), and finally across Iowa, Nebraska, and Wyoming to Salt Lake City in 1847. The Latter-day Saints had 985,070 members in Utah in 1980, according to state figures—roughly 67% of the state population. The Mormon Church and its leadership continue to play a central role in the state's political, economic, and cultural institutions. Among other assets, the church owns Zion Cooperative Mercantile Institute (the largest department store in Salt Lake City), two leading newspapers, one television station, and holdings in banks, insurance companies, and real estate.

Other leading Christian denominations and their 1980 memberships include various Baptist groups, 14,165; Presbyterian, 6,742; and Episcopal, 4,490. In 1984 there were 65,729 Roman Catholics and an estimated 2,600 Jews.

¹⁰TRANSPORTATION

Utah, where the golden spike was driven in 1869 to mark the completion of the first transcontinental railroad, had 1,651 mi (2,657 km) of track in 1984. Major railroads are the Union Pacific and the Denver & Rio Grande Western. Amtrak provides passenger service to Salt Lake City, Ogden, and Milford and carried 160,645 riders in 1983/84.

The Utah Transit Authority, created in 1970, provides bus service for Salt Lake City, Provo, and Ogden.

Utah in 1983 had 46,078 mi (74,155 km) of roads and streets; there were 1,048,000 registered motor vehicles and 926,434 licensed drivers. The state has 806 mi (1,297 km) of interstate highways; another 133 mi (214 km) were under construction as of 31 December 1983. The main east–west and north–south routes—I-80 and I-15, respectively—intersect at Salt Lake City.

Utah had 78 airports in 1983. By far the busiest was Salt Lake City International Airport, handling 54,616 scheduled departures in 1983 and enplaning 3,237,442 passengers.

¹¹HISTORY

Utah's historic Indian groups are primarily Shoshonean: the Ute in the eastern two-thirds of the state, the Goshute of the western desert, and the Southern Paiute of southwestern Utah. The Athapaskan-speaking Navaho of southeastern Utah migrated from western Canada, arriving not long before the Spaniards. The differing life-styles of each group remained essentially unchanged until the introduction of the horse by the Spanish sometime after 1600. White settlement from 1847 led to two wars between whites and Indians—the Walker War of 1853–54 and the even more costly Black Hawk War of 1865–68—resulting finally in the removal of many Indians to reservations.

Mexicans and Spaniards are the first non-Indians known to have entered Utah, with Juan María Antonio Rivera reportedly near present-day Moab as early as 1765. In July 1776, a party led by two Franciscan priests, Francisco Atanasio Domínguez and Silvestre Vélez de Escalante, entered Utah from the east, traversed the Uinta basin, crossed the Wasatch Mountains, and visited the

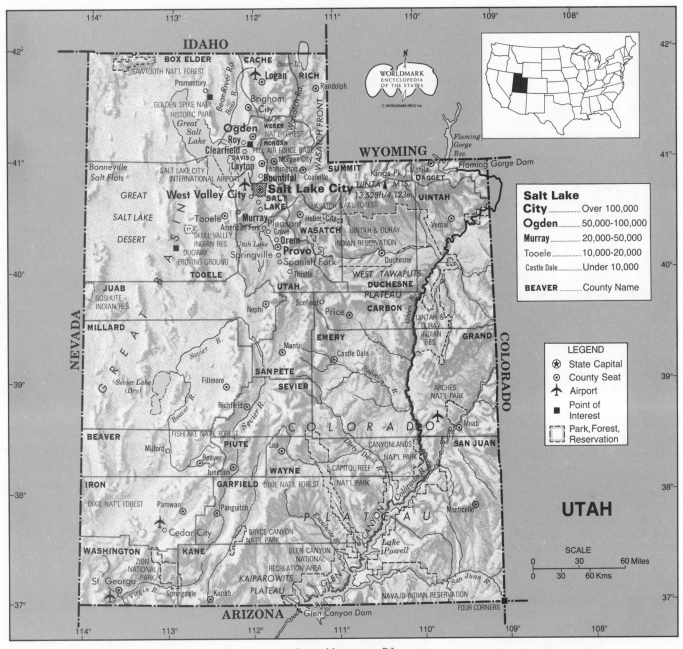

SCALE
0   30   60 Miles
0   30   60 Kms

UTAH

See endsheet maps: D3.
LOCATION: 36°59′57″ to 41°59′39″N; 109°02′40″ to 114°02′26″W. **BOUNDARIES:** Idaho line, 154 mi (248 km); Wyoming line,
174 mi (280 km); Colorado line, 276 mi (444 km); Arizona line, 277 mi (446 km); Nevada line, 345 mi (555 km).

Ute encampment at Utah Lake. Trade between Santa Fe, the capital of the Spanish province of New Mexico, and the Indians of Utah was fairly well established by the early 1800s.

Until 1848, the 1,200-mi (1,900-km) Spanish Trail, the longest segment of which lies in Utah, was the main route through the Southwest. Following this trail, mountain men competing for fur explored vast areas of the American West, including most of Utah's rivers and valleys. In the 1840s, Utah was traversed by California-bound settlers and explorers, the most notable being John C. Frémont.

When Joseph Smith, Jr., founder of the Church of Jesus Christ of Latter-day Saints (Mormons), was lynched at Carthage, Ill., in June 1844, Brigham Young and other Mormon leaders decided to move west. By April 1847, the pioneer company of Mormons, including three blacks, was on its way to Utah, the reports of

Frémont having influenced their choice of the Great Basin as a refuge. Advance scouts entered the Salt Lake Valley on 22 July, and the rest of the company two days later. Planting and irrigating were begun immediately. Natural resources were regarded as community property, and the church organization served as the first government.

After the Treaty of Guadalupe-Hidalgo (1848) gave the US title to much of the Southwest, the Mormons established the provisional state of Deseret. Congress refused to admit Deseret to the Union, choosing instead to create Utah Territory "with or without slavery." The territory encompassed, in addition to present-day Utah, most of Nevada and parts of Wyoming and Colorado; land cessions during the 1860s left Utah with its present boundaries.

The territorial period lasted for 46 difficult years, marked by immigration, growth, and conflict. Reports that Utahns were in rebellion against federal authority led President James Buchanan to send an expeditionary force under Albert Sidney Johnston to Utah in 1857. On 11 September, Mormon militiamen and their Indian allies, caught up in an atmosphere of war hysteria, massacred some 120 California-bound migrants at Mountain Meadows—the darkest event in Utah history and the only major disaster of the so-called Utah War. Peace was attained in June 1858, and Alfred Cumming assumed civil authority, replacing Brigham Young as territorial governor. Cumming's appointment signaled the beginning of prolonged hostility between Mormon leaders and federal authorities.

Almost 98% of Utah's total population was Mormon until after 1870, and the Mormon way of life dominated politics, economics, and social and cultural activities. As church president, Brigham Young remained the principal figure in the territory until his death in 1877. He contracted in 1868 with the Union Pacific to lay part of the track for the transcontinental railroad in Utah, and on 10 May 1869, the Central (now Southern) Pacific and Union Pacific were joined at Promontory. During the 1870s, new rail lines connected many settlements with the capital, Salt Lake City, spurring commerce and mining. Young had discouraged mining until agriculture and manufacturing were firmly established. Not until 1863, with the rediscovery of silver-bearing ore in Bingham Canyon, did the boom in precious metals begin. Those connected with mining, mostly non-Mormons, began to exert influence in the territory's business, politics, and social life.

Several factors made the non-Mormon minority fearful of Mormon domination: communitarian economic practices, lack of free public schools, encouragement of immigration of Mormon converts, church authoritarianism, and the mingling of church and state. But the most sensational reason was the Mormon practice of polygamy. Congress passed the Anti-Bigamy Act in 1862, but it was generally not enforced. After the Edmunds Act of 1882 was upheld by the US Supreme Court, arrests for polygamy greatly increased. Finally, in 1887, the Edmunds-Tucker Act dissolved the Mormon Church as a corporate entity, thereby threatening the survival of all Mormon institutions.

In fall 1890, Mormon President Wilford Woodruff issued a manifesto renouncing the practice of polygamy. The following year, the Republican and Democratic parties were organized in Utah, effectively ending political division along religious lines. A constitutional convention was held in 1895, and statehood became a reality on 4 January 1896. The new state constitution provided for an elected governor and a bicameral legislature, and restored the franchise to women, a privilege they had enjoyed from 1870 until 1887, when the Edmunds-Tucker Act had disfranchised Utah women and polygamous men.

The early 20th century saw further growth of the mineral industry. Many of those who came to mine copper and coal were foreign immigrants. Militant union activity had begun slowly during the 1890s, until an explosion that killed 200 miners at Scofield on 1 May 1900 dramatized the plight of the miners and galvanized radical organizers in the state. It was in Utah in 1915 that a Swedish miner and songwriter named Joe Hill, associated with the Industrial Workers of the World ("Wobblies"), was executed for the murder of a Salt Lake City grocer and his son, a case that continues to generate controversy because of the circumstantial quality of the evidence against him.

Gradually, modern cities emerged, along with power plants, interurban railroads, and highways. By 1920, nearly half the population lived along the Wasatch Front. The influx of various ethnic groups diversified the state's social and cultural life, and the proportion of Mormons in the total population declined to about 68% in 1920.

Utah businesses enjoyed the postwar prosperity of the 1920s. On the other hand, mining and agriculture were depressed throughout the 1920s and 1930s, decades marked by increased union activity, particularly in the coal and copper industries. The depression of the 1930s hit Utah especially hard. Severe droughts hurt farmers in 1931 and 1934, and high freight rates limited the expansion of manufacturing. With the coming of World War II, increased demand for food revived Utah's agriculture, and important military installations and war-related industries brought new jobs to the state.

In the years since World War II, the state's population has more than doubled, while per capita income has declined relative to the national average—both trends indicative of a very high birthrate. Politics generally reflect prevailing Mormon attitudes and tend to be conservative. Democrats held the state governorship for 20 years after 1964, with Scott M. Matheson at the helm from 1977 to 1985. Matheson successfully opposed plans for storing nerve gas bombs in Utah and for the location in the western desert of an MX missile racetrack system. When Matheson declined to run for reelection in 1984, Norman Bangerter, a conservative Republican, won the governorship.

¹²STATE GOVERNMENT

The state legislature, as established in the constitution of 1896, consists of a 29-member senate and a 75-seat house of representatives; senators serve for four years, representatives for two.

The chief executive officers, all elected for four-year terms, include the governor, lieutenant governor (who also serves as secretary of state), the attorney general, treasurer, and auditor. The governor must be at least 30 years of age and must have been a US citizen and state resident for at least five years. Gubernatorial vetoes may be overridden by two-thirds of the elected members of each house of the legislature.

Amending the constitution requires a two-thirds vote of the legislature and ratification by majority vote at the next general election. Voters must be US citizens, at least 18 years of age, and have been residents of the state 30 days prior to voting day.

¹³POLITICAL PARTIES

The Republican and Democratic parties, each affiliated with the national party organization, are the state's leading political groups. In 1984, Republicans regained control of the governorship after 20 years of control by the Democrats. They also retained control of the state legislature, holding 23 seats in the senate and 61 in the house of representatives as of January 1985. As of that same date, all 3 seats of the US House delegation and both seats in the US Senate were held by Republicans. Generally, the state has voted Republican in presidential elections. Even more than

Utah Presidential Vote by Major Political Parties, 1948–84

YEAR	ELECTORAL VOTE	UTAH WINNER	DEMOCRAT	REPUBLICAN
1948	4	*Truman (D)	149,151	124,402
1952	4	*Eisenhower (R)	135,364	194,190
1956	4	*Eisenhower (R)	118,364	215,631
1960	4	Nixon (R)	169,248	205,361
1964	4	*Johnson (D)	219,628	181,785
1968	4	*Nixon (R)	156,665	238,728
1972	4	*Nixon (R)	126,284	323,643
1976	4	Ford (R)	182,110	337,908
1980	4	*Reagan (R)	124,266	439,687
1984	5	*Reagan (R)	155,369	469,105

* Won US presidential election.

party rivalry, the central facts of political life in Utah are conservatism and Mormon Church influence.

In November 1984, Utahns cast 74.5% of their presidential votes for Ronald Reagan. There were 748,730 registered voters in 1982. Minorities hold few elected positions. Only 7 out of 104 (6.7%) of all state legislators were women in 1985, ranking Utah among the 10 states with the lowest percentages of women legislators. In 1984, 6 of the state's elected officials were Hispanic; there was one black elected official, a state senator, in 1985.

14 LOCAL GOVERNMENT

Utah has 29 counties, governed by a total of 94 elected commissioners. Other elected county officials include clerk-auditor, sheriff, assessor, recorder, treasurer, county attorney, and surveyor. There were 245 cities, towns, and census-designated (unincorporated) places in 1980.

15 STATE SERVICES

The Department of Public Education is responsible for public instruction, and the Utah State Board of Regents oversees the state college and university system. Highways and airports are the responsibility of the Department of Transportation.

The Department of Community and Economic Development supports economic and technological development programs in the state. Agencies dealing with the elderly, disabled, family services, mental health, assistance payments, and youth corrections are under the Department of Social Services. The Department of Health oversees public health and health care for the indigent. Other state departments deal with natural resources, business, labor, agriculture, corrections, and public safety.

16 JUDICIAL SYSTEM

Utah's highest court is the supreme court, consisting of a chief justice and 4 other justices, each serving a 10-year term. As of 1984 there were 29 district court judges, each one serving a 6-year term. Supreme court justices and district court judges are appointed by the governor with the consent of the state senate. Appointments must be ratified by the voters at the next general election. In 1984, to ease the supreme court's caseload, Utahns approved a constitutional amendment allowing the legislature to create an intermediate court. An estimated 3,073 attorneys practiced in the state in 1983.

In 1983, the FBI reported a crime-index total of 5,118 crimes per 100,000 inhabitants, about average for the states. This included rates of 4,826 per 100,000 inhabitants for property crimes and a relatively low 256 per 100,000 for violent crimes. Utah has a death penalty statute providing for execution by lethal injection or firing squad. In 1977 it executed (by firing squad) a prisoner, Gary Gilmore, thus becoming the first state in a decade to carry out a sentence of capital punishment. No prisoners have been executed since then, but as of 31 December 1983, four prisoners were under sentence of death. Prisoners under jurisdiction of state and federal correctional facilities numbered 1,419 at the end of 1984.

17 ARMED FORCES

Federal facilities in the state include Hill Air Force Base near Ogden and, in the Great Salt Lake Desert, Tooele Army Depot, Dugway Proving Ground—where nerve gas tests have been conducted—and the USAF Utah Test and Training Range near the Nevada line. There were 6,121 active-duty military personnel in Utah in 1983/84. State firms were awarded $878 million in federal defense contracts during the same year.

As of 30 September 1983, 155,000 veterans were living in Utah, of whom 2,000 were veterans of World War I, 55,000 of World War II, 28,000 of the Korean conflict, and 51,000 of the Viet-Nam era. Veterans' benefits in 1982/83 totaled $137 million.

Utah's Army National Guard had 5,722 personnel as of 31 December 1984; the Air National Guard had 1,449 members as of 28 February 1985. There were 3,190 state and local police and 1,274 corrections personnel as of October 1983.

18 MIGRATION

After the initial exodus of Latter-day Saints from the eastern US to Utah, Mormon missionaries attracted other immigrants to the state, and some 90,000 foreign converts arrived between 1850 and 1905. Many non-Mormons were recruited from overseas to work in the mines, especially during the early 20th century. Utah had a net gain from migration of 176,000 between 1940 and 1985.

19 INTERGOVERNMENTAL COOPERATION

Utah participates in several regional agreements, including the Bear River Compact (with Idaho and Wyoming), Colorado River Compact, and the Upper Colorado River Basin Compact. The state is also a signatory to the Interstate Oil and Gas Compact, Western Corrections Compact, and Western Interstate Energy Compact. Federal aid in 1983/84 amounted to $708 million.

20 ECONOMY

Trade replaced government as the leading employer in Utah in 1980, employing 143,693 workers in September 1984 against 129,772 employed by government. Nonetheless, more than 28% of personal income in the state was derived from government sources in 1983. With more than 70% of Utah lands under US control and some 37,750 civilian workers on federal payrolls—and others employed by defense industries or the military—the federal presence in Utah is both a major economic force and a controversial political issue. On the one hand, elected officials have sought federal funds for mammoth reclamation and power projects. On the other hand, they resent many federal programs concerned with social welfare, land use, or environmental protection.

Since 1965, employment has shifted away from agriculture, transportation, and communications, toward government, trade, and service occupations, and to a lesser extent, manufacturing. Mining employment has also declined, but the value of Utah's mineral industries has steadily increased. Development of new oil, gas, and coal discoveries promises more growth in that sector.

21 INCOME

In 1983, Utahns' total personal income reached $14.6 billion, or $9,031 per capita. Real income (as measured in constant 1972 dollars) more than doubled between 1970 and 1983, but in current dollars, Utah's national rank in per capita income has steadily slipped from 31st in the 1950s to 47th in 1983. Total average annual pay in 1983 topped $16,500, ranking Utah 26th among the states.

Median family income was $20,024 (22d in the US) in 1979, when 8% of Utah families and more than 10% of all Utahns were below the federal poverty level. Utah had 14,000 persons with over $500,000 in gross assets.

22 LABOR

In September 1984, Utah's civilian labor force amounted to 738,317, of whom 695,493 were employed and 42,824 unemployed, with a resulting unemployment rate of 5.8%.

A federal survey in 1983 revealed the following nonfarm employment pattern for Utah:

	ESTABLISH-MENTS	EMPLOYEES	ANNUAL PAYROLL ('000)
Agricultural services, forestry, fishing	261	1,087	$ 12,951
Mining	434	13,214	400,203
Contract construction	3,575	23,883	495,258
Manufacturing	1,865	82,397	1,658,566
Transportation, public utilities	1,055	31,699	744,469
Wholesale trade	2,668	31,510	599,594
Retail trade	7,970	93,097	842,400
Finance, insurance, real estate	3,172	29,111	469,600
Services	9,541	112,907	1,468,631
Other	2,778	3,003	57,074
TOTALS	33,319	421,908	$6,748,746

Average weekly earnings of production workers in 1984 totaled $363.09. Although Utah has a radical labor tradition, the union movement has weakened since World War II. In 1980, 98,000, or 17.8%, of Utah's workers belonged to labor unions, and in 1983 there were 290 labor unions in the state. Utah has a right-to-work law, enacted in 1955.

23 AGRICULTURE

Despite a dry climate and unpromising terrain, Utah ranked 39th in the US in value of farm marketings in 1983, with $579 million. Crops accounted for $146 million, livestock and livestock products for $433 million. The first pioneers in Utah settled in fertile valleys near streams, which were diverted for irrigation. Since 1900, however, farmers have relied on dryland farming techniques and, more recently, on massive irrigation projects. In 1984 there were some 14,000 farms and ranches, covering 11,800,000 acres (4,775,000 hectares).

The chief crops in 1983 were barley, 11.1 million bushels; wheat, 8 million bushels; hay, 2 million tons; and commercial apples, 58 million lb.

24 ANIMAL HUSBANDRY

Livestock and livestock products account for about three-fourths of Utah's agricultural income. In 1983 there were 860,000 cattle and calves on Utah farms, 540,000 sheep and lambs, and 33,000 hogs and pigs. The 1983 cattle output was 298.1 million lb, valued at $149.9 million; sheep and lambs, 39.7 million lb, $18 million; and hogs and pigs, 9.5 million lb, $4.4 million. Dairy farms had 86,000 milk cows, producing 1.2 billion lb of milk. The 1983 wool clip was 5.7 million lb.

25 FISHING

Fishing in Utah is for recreation only.

26 FORESTRY

In 1979, Utah had 15,557,000 acres (6,296,000 hectares) of forestland. Of that, 7,989,733 acres (3,233,341 hectares) were in the state's nine national forests—Ashley, Cache, Caribou, Dixie, Fishlake, Manti–La Sal, Sawtooth, Uinta, and Wasatch. Only 3,405,000 acres (1,378,000,000 hectares) were classed as commercial timberland. The value of all lumber and wood products shipped in 1982 was $123.7 million.

27 MINING

Utah is an important producer of nonfuel minerals, including copper, gold, silver, iron ore, lead, zinc, vanadium, and tungsten, as well as salt, magnesium, and potash evaporated from the Great Salt Lake and other water bodies.

In 1984, the value of Utah's crude nonfuel mineral output was nearly $525 million (16th in the US). The following table shows volume and value for leading nonfuel minerals in 1983:

	VOLUME		VALUE
Copper	169,751	tons	$286,403,000
Gold	238,459	troy oz	101,107,000
Silver	4,567,000	troy oz	52,242,000
Salt	936,000	tons	23,184,000
Sand and gravel	9,800,000	tons	19,800,000

Output of copper, traditionally Utah's most important nonfuel mineral, was severely depressed in 1984 because of a labor dispute. Iron ore mines at Cedar City were also permanently closed that year, while temporary layoffs, resulting from a national controversy over nuclear power, shut down three uranium-vanadium mines and the Moab uranium-vanadium mill.

28 ENERGY AND POWER

During 1983, electric utilities in the state had an installed capacity of 3 million kw and produced 12.4 billion kwh of power. A total of 11 billion kwh was derived from coal-fired steam units, 1.4 billion kwh from hydroelectric units.

Fuels accounted for 70% of Utah's total mineral output in 1982. Proved oil reserves totaled 187 million barrels in 1983, when production was 29.5 million barrels. Reserves of natural gas

amounted to 2,333 billion cu feet, production 63.2 billion cu feet. Early in 1980 there were large new natural gas finds in northeastern Utah. The state's reserves of bituminous coal were estimated at nearly 6.4 billion tons in 1983; production reached 11.8 million tons in 1983. Huge coal resources of the Kaiparowits Plateau remained mostly undeveloped, largely because of environmental considerations.

29 INDUSTRY

Utah's manufacturing, diversified in products, is concentrated geographically in Salt Lake City, Weber, Utah, and Cache counties. Total employment in manufacturing was 83,200 in 1982, when the leading sectors and their employment were nonelectrical machinery, 10,800; transportation equipment, 9,000; electric and electronic equipment, 8,200; primary metals, 7,900; food and food products, 7,600; fabricated metals, 6,200; and apparel and textiles, 6,100. Utah has one truck-assembly plant, in Ogden, which produced 750 units in 1983.

The total value of shipments by manufacturers in 1982 was almost $9 billion. The main industry groups by value of shipments were petroleum and coal products, 19%; food, 15%; and primary metals, 13%. The following table shows value of shipments for selected industries in 1982:

Petroleum refining	$1,645,500,000
Primary metals	1,145,000,000
Nonelectrical machinery	973,800,000
Transportation equipment	839,700,000

30 COMMERCE

Utah's wholesale business had sales exceeding $8.7 billion in 1982, when retail sales surpassed $6.3 billion. Retail and wholesale trading establishments are heavily concentrated in the Salt Lake City–Ogden metropolitan area. The leading wholesale trade categories were machinery, 20%; petroleum and petroleum products, 13%; automotive dealers, 9%; electrical goods, 8%; and groceries, 8%.

Foreign exports of Utah's manufactured goods totaled $449 million in 1981, 2.5 times the 1977 level. Exports of coal were valued at $106 million in 1981, exports of agricultural products at $80 million in 1982.

31 CONSUMER PROTECTION

The Division of Consumer Protection in the Department of Business Regulation is charged with protecting Utah's consumers.

32 BANKING

Utah in 1984 had 60 insured commercial banks. In 1983, the state's insured commercial banks had total assets of $8.8 billion, outstanding loans of $5 billion, and deposits of nearly $7.1 billion. There were 15 insured savings and loan associations in 1984, with total assets of $5.4 billion.

33 INSURANCE

Utahns held some 2,195,000 life insurance policies in 1983; their total value was $30.4 billion, and the average coverage per family was $54,300. Total benefit payments of $274.4 million included $77.8 million in death payments, $56.3 million in dividends, and $55.8 million in annuities. Premiums written by property and liability companies in 1983 totaled $483.6 million.

34 SECURITIES

The Intermountain Stock Exchange does a limited business in Salt Lake City. In addition, New York Stock Exchange member firms had 24 sales offices and 345 registered representatives in Utah in 1983. A total of 246,000 Utahns owned shares of public corporations in 1983.

35 PUBLIC FINANCE

The annual budget is prepared by the State Budget Office and submitted by the governor to the legislature for amendment and approval. The fiscal year runs from 1 July through 30 June.

The following table summarizes state revenues and expenditures for fiscal years 1982/83 and 1983/84 (in millions):

REVENUES	1982/83	1983/84
General sales tax	$ 388.6	$ 525.1
Individual income tax	345.6	387.3
Motor fuel tax	85.1	92.2
Corporate income tax	31.6	45.0
Other taxes	241.0	304.3
Federal aid	497.2	557.8
Other receipts	136.6	157.1
TOTALS	$1,725.7	$2,068.8
EXPENDITURES		
Education	$ 767.8	$ 772.7
Social services	555.8	493.1
Transportation	195.7	280.0
General government	129.0	137.4
Capital projects	71.1	58.5
Natural resources	38.3	37.3
Other outlays	83.6	139.4
TOTALS	$1,841.3	$1,918.4

General revenues of all local government units in 1982/83 totaled $1,777.6 million, general expenditures $1,577.2 million. As of 1982, the total state and local government debt was $3,547.7 million.

36 TAXATION

The main source of state revenue is a 4.625% general sales and gross receipts tax. As of 1985, personal income tax rates ranged from 2.75% to 7.75%; the corporate income tax rate was 5%. Taxes are also levied on motor fuels, alcoholic beverages, tobacco products, and other items. Property taxes, the main source of local revenue, yielded $535.3 million in 1984.

Utah's total federal tax burden was $2,993.6 million in 1982, while federal expenditures in the state totaled $3,710 million.

37 ECONOMIC POLICY

The economic development of Utah has been dominated by two major forces: the relatively closed system of the original Mormon settlers and the more wide-open, speculative ventures of the state's later immigrants. The Mormons developed agriculture, industry, and a cooperative exchange system that excluded non-Mormons. The church actively opposed mining, and it was mostly with non-Mormon capital, by non-Mormon foreign immigrants, that the state's mineral industry was developed.

In recent years, these conflicts have been supplanted by a widespread fiscal conservatism that supports business activities and opposes expansion of government social programs at all levels. One Utah politician, Governor J. Bracken Lee, became nationally famous for his call to repeal the federal income tax.

The Department of Community and Economic Development is the state agency responsible for the expansion of tourism and industry. The 1985/86 state budget included $2 million to support the Technology Finance Corporation, which with private venture capital provides loans for new technological investments.

38 HEALTH

Health conditions in Utah are exceptionally good. The infant mortality rates—9.8 per 1,000 live births in 1981—is among the lowest in the US, and the overall death rate (5.5 per 1,000 persons in 1981) is well below the national average of 8.6. The marriage rate of 12 per 1,000 inhabitants in 1982 exceeds the national norm of 10.8; the divorce rate of 5.4 per 1,000 inhabitants is above the national average of 5. The birthrate—26.4 per 1,000 persons in 1982—was 1st among the 50 states. A total of 11,400 legal abortions were performed in 1982/83, when the abortion rate was 100 per 1,000 live births.

In 1983 there were 44 hospitals in the state, with 5,390 beds and admissions totaling 211,692. The average cost of a semiprivate hospital room was $178.15 per day in 1984. The average cost to a hospital per patient day was $375.60 in 1982. Hospital personnel in 1983 included 4,804 registered nurses and 1,029 licensed practical nurses. There were 2,832 licensed physicians and 979 dentists in Utah in 1982.

39 SOCIAL WELFARE

During fiscal 1984, Utah received federal direct grant payments of $29 million for aid to families with dependent children, $126.2 million for Medicare insurance, and $39.5 million in food stamp payments; 38,300 persons in 13,100 families received AFDC payments in 1983, and 115,000 received Medicaid, the fastest-growing of all social welfare programs in the state. The food stamp program benefited 84,000 Utahns in 1983, for a federal value of $43 million. The national school lunch program enrolled 218,000 students in 1983 at a cost of $14 million.

In 1983, 158,000 Utahns received a total of $730 million in Social Security benefits; average monthly payments were $445. In 1984, federal Supplemental Security Income payments totaled $15.3 million. During 1983, $148.1 million was paid under state and federal programs in unemployment benefits.

40 HOUSING

The 1980 census counted 490,006 housing units in Utah, of which 481,701 were year-round housing units and 448,603 occupied. An additional 30,900 units were authorized between 1980 and 1983, with a total value of $1.4 billion. All but 7,004 housing units in 1980 had complete bathroom facilities. There were 317,172 owner-occupied housing units in 1980; only 131,431 were renter-occupied.

41 EDUCATION

Utahns are among the nation's leaders in educational attainments. In 1980, Utah had the 2d-highest proportion of adult high school graduates, 80% (Alaska was 1st); nearly 20% had four years or more of college, and fewer than 1% had four years or less of grade school.

In fall 1984, Utah public schools had an enrollment of 390,141. There were 15,826 teachers in 1983/84, when per pupil expenditures reached $2,429.

Enrollment at Utah's higher educational institutions totaled 96,612 in 1984/85, 68,678 in public and 27,934 in private colleges and universities. Major public institutions include the University of Utah, with 24,490 students in 1984/85; Utah State University, 12,087; and Weber State College, 10,717. Brigham Young University (Provo), founded in 1875 and affiliated with the Latter-day Saints, is the main private institution, with 26,700 students.

42 ARTS

Music has a central role in Utah's cultural life. Under the baton of former director Maurice Abravanel and current director Joseph Silverstein, the Utah Symphony (Salt Lake City) has become one of the nation's leading orchestras. The Mormon Tabernacle Choir has won world renown, and Ballet West is ranked among the nation's leading dance companies. Opera buffs enjoy the Utah Opera Company, founded in 1976.

The Utah Arts Council sponsors exhibitions, artists in the schools, rural arts and folk arts programs, and statewide arts competitions in cooperation with arts organizations throughout the state. In addition, the partially state-funded Utah Arts Festival is held each year in Salt Lake City. Legislative appropriations for the arts have increased steadily from $83,000 in 1970 to $1,399,900 in 1985. In 1985, Utah had 24 art museums and galleries, including Utah State University's Nora Eccles Harrison Museum in Logan and the LDS Church Museum of Art and History in Salt Lake City. Other major facilities are the Brigham Young University Art Museum Collection, Provo; Museum of Fine Arts of the University of Utah, Salt Lake City; and the Springville Art Museum.

43 LIBRARIES AND MUSUEMS

In 1985, Utah had 83 public libraries with a circulation of 10,446,200 volumes. In 1982, the combined book stock of public libraries was 3,333,953. The Salt Lake County library system had 882,171 volumes, the Weber County system (including Ogden)

239,500. The leading academic libraries are the University of Utah (Salt Lake City), 2,045,797; and Brigham Young University (Provo), 1,628,483. Other collections are the Latter-day Saints' Library-Archives and the Utah State Historical Society Library, both in Salt Lake City. During 1984, Utah had at least 48 museums, notably the Utah Museum of Natural History, Salt Lake City; Edge of the Cedars Indian Cultural Museum, Blanding; and Museum of Peoples and Cultures, Provo. Utah had 829 historic sites as of 1985. Some are maintained as museums, including Beehive House and Wheeler Historic Farm, Salt Lake City, and Brigham Young's Winter Home, St. George.

44 COMMUNICATIONS

There were 195 post offices in 1985, with 3,378 paid employees. In 1980, 95% of Utah's 448,603 occupied houses had telephones.

A total of 69 radio stations broadcast in Utah in 1985; 38 were AM stations, 31 FM (15 of them noncommercial). There were 6 television stations in 1984; 5 in Salt Lake City and 1 in Provo. Cable television served 94,804 subscribers in 70 communities in 1984.

45 PRESS

Utah in 1985 had five daily newspapers and five Sunday papers. The following table shows leading daily newspapers in 1984:

AREA	NAME	DAILY	SUNDAY
Ogden	*Standard-Examiner* (e,S)	51,684	52,937
Provo	*Daily Herald* (e,S)	31,331	31,771
Salt Lake City	*Deseret News* (e,S)	64,426	68,635
	Tribune (m,S)	109,717	132,531

46 ORGANIZATIONS

The 1982 Census of Service Industries counted 327 organizations in Utah, including 83 business associations; 186 civic, social, and fraternal associations; and 8 educational, scientific, and research associations. Salt Lake City is the world headquarters of the Church of Jesus Christ of Latter-day Saints (Mormon). The city is also home to the Mental Retardation Association of America and to Executive Women International.

47 TOURISM, TRAVEL, AND RECREATION

Temple Square, Pioneer Trail State Park, and Hogle Zoological Gardens are leading attractions of Salt Lake City, about 11 mi (18 km) east of the Great Salt Lake. At the Bonneville Salt Flats, experimental automobiles have set world land-speed records.

Under federal jurisdiction are 5 national parks—the Arches, Bryce Canyon, Canyonlands, Capitol Reef, and Zion—Glen Canyon National Recreation Area, 6 national monuments, and 1 national historical site, Golden Spike. Under state control are 6 state parks, 7 state natural areas, 13 state recreation areas, 8 state historic areas, and 9 water-use areas. Mountain and rock climbing, skiing, fishing, and hunting are major recreations. Licenses were issued to 251,808 hunters and 372,772 anglers in 1983.

48 SPORTS

Utah's only major league professional team is basketball's Utah Jazz, which moved from New Orleans at the close of the 1978/79 season. Basketball is also popular at the college level. The University of Utah's Running Utes won the NCAA championship in 1944 and the National Invitation Tournament in 1947, while the Cougars of Brigham Young won or tied for the Western Athletic Conference championship in 1979, 1980, and 1983 and won NIT titles in 1951 and 1966.

49 FAMOUS UTAHNS

George Sutherland (b. England, 1862-1942) capped a long career in Utah Republican politics by serving as an associate justice of the US Supreme Court (1922-38). Other important federal officeholders from Utah include George Dern (b. Nebraska, 1872-1936), a mining man who was President Franklin D. Roosevelt's secretary of war from 1933 to 1936; Ezra Taft Benson (b. Idaho, 1899), a high official of the Mormon Church and President Dwight Eisenhower's secretary of agriculture; and Ivy Baker Priest (1905-75), US treasurer during 1953-61. Prominent

in the US Senate for 30 years was Republican tariff expert Reed Smoot (1862-1941), also a Mormon Church official. The most colorful politician in state history, J(oseph) Bracken Lee (b. 1899), was mayor of Price for 12 years before serving as governor during 1949-57 and mayor of Salt Lake City during 1960-72. Jacob "Jake" Garn (b. 1932), first elected to the US Senate in 1974, was launched into space aboard the space shuttle in 1985.

The dominant figure in Utah history is undoubtedly Brigham Young (b. Vermont, 1801-77), the great western colonizer. As leader of the Mormons for more than 30 years, he initiated white settlement of Utah in 1847 and, until his death, exerted almost complete control over life in the territory. Other major historical figures include Eliza R. Snow (b. Massachusetts, 1804-87), Mormon women's leader; Wakara, anglicized Walker (c. 1808-55), the foremost Ute leader of the early settlement period; Colonel Patrick Edward Conner (b. Ireland, 1820-91), founder of Camp Douglas and father of Utah mining; George Q. Cannon (b. England, 1827-1901), editor, businessman, political leader, and a power in the Mormon Church for more than 40 years; and Lawrence Scanlan (b. Ireland, 1843-1915), first Roman Catholic bishop of Salt Lake City, founder of schools and a hospital.

Utah's most important scientist is John A. Widtsoe (b. Norway, 1872-1952), whose pioneering research in dryland farming revolutionized agricultural practices. Noted inventors are gunsmith John M. Browning (1855-1926) and television innovator Philo T. Farnsworth (1906-71). Of note in business are mining entrepreneurs David Keith (b. Canada, 1847-1918), Samuel Newhouse (b. New York, 1853-1930), Susanna Emery-Holmes (b. Missouri, 1859-1942), Thomas Kearns (b. Canada, 1862-1918), and Daniel C. Jackling (b. Missouri, 1869-1956). Labor leaders include William Dudley "Big Bill" Haywood (1869-1928), radical Industrial Workers of the World organizer, and Frank Bonacci (b. Italy, 1884-1954), United Mine Workers of America organizer.

Utah's artists and writers include sculptors Cyrus E. Dallin (1861-1944) and Mahonri M. Young (1877-1957), painter Henry L. A. Culmer (b. England, 1854-1914), author-critic Bernard A. DeVoto (1897-1955), poet-critic Brewster Ghiselin (b. Missouri, 1903), folklorist Austin E. Fife (b. Idaho, 1909), and novelists Maurine Whipple (b. 1904), Virginia Sorensen (b. 1912), and Edward Abbey (b. 1927).

Actresses from Utah are Maude Adams (1872-1953), Loretta Young (b. 1913), and Laraine Day (b. 1920). Donald "Donny" Osmond (b. 1957) and his sister Marie (b. 1959) are Utah's best-known popular singers. Emma Lucy Gates Bowen (1880-1951), an opera singer, founded her own traveling opera company, and William F. Christensen (b. 1902) founded Ballet West. Maurice Abravanel (b. Greece, 1903) conducted the Utah Symphony for many years.

Sports figures of note are former world middleweight boxing champion Gene Fullmer (b. 1931) and former Los Angeles Rams tackle Merlin Olsen (b. 1940).

50 BIBLIOGRAPHY

Alexander, Thomas. *Mormons and Gentiles: A History of Salt Lake City.* Boulder, Colo.: Pruett, 1984.

Alter, J. Cecil. *Utah, the Storied Domain: A Documentary History.* 3 vols. Chicago: American Historical Society, 1932.

Arrington, Leonard J., and Davis Bitton. *The Mormon Experience: A History of the Latter-day Saints.* New York: Knopf, 1979.

Ellsworth, Samuel G. *Utah's Heritage.* Salt Lake City: Peregrine Smith, 1984.

Papanikolas, Helen Z. (ed.) *The Peoples of Utah.* Salt Lake City: Utah State Historical Society, 1981 (orig. 1976).

Peterson, Charles S. *Utah: A Bicentennial History.* New York: Norton, 1977.

Poll, Richard D., et al. *Utah's History.* Provo, Utah: Brigham Young University Press, 1978.

VERMONT

State of Vermont

ORIGIN OF STATE NAME: Derived from the French words *vert* (green) and *mont* (mountain). **NICKNAME:** The Green Mountain State. **CAPITAL:** Montpelier. **ENTERED UNION:** 4 March 1791 (14th). **SONG:** "Hail, Vermont!" **MOTTO:** Freedom and Unity. **COAT OF ARMS:** Rural Vermont is represented by a pine tree in the center, three sheaves of grain on the left, and a cow on the right, with a background of fields and mountains; a deer crests the shield. Below are crossed pine branches and the state name and motto. **FLAG:** The coat of arms on a field of dark blue. **OFFICIAL SEAL:** Bisecting Vermont's golden seal is a row of wooded hills above the state name; the upper half has a spearhead, pine tree, cow, and two sheaves of wheat, while two more sheaves and the state motto fill the lower half. **ANIMAL:** Morgan horse. **BIRD:** Hermit thrush. **FISH:** Brook trout (cold water); walleye pike (warm water). **FLOWER:** Red clover. **TREE:** Sugar maple. **INSECT:** Honeybee. **BEVERAGE:** Milk. **POET LAUREATE:** Robert Frost. **LEGAL HOLIDAYS:** New Year's Day, 1 January; Birthday of Martin Luther King, Jr., 3d Monday in January; Lincoln's Birthday, 12 February; Washington's Birthday, 3d Monday in February; Town Meeting Day, 1st Tuesday in March; Memorial Day, last Monday in May; Independence Day, 4 July; Bennington Battle Day, 16 August; Labor Day, 1st Monday in September; Columbus Day, 2d Monday in October; Veterans Day, 11 November; Thanksgiving Day, 4th Thursday in November; Christmas Day, 25 December. **TIME:** 7 AM EST = noon GMT.

¹LOCATION, SIZE, AND EXTENT

Situated in the northeastern US, Vermont is the 2d largest of the 6 New England states, and ranks 43d in size among the 50 states.

Vermont's total area of 9,614 sq mi (24,900 sq km) consists of 9,273 sq mi (24,017 sq km) of land and 341 sq mi (883 sq km) of inland water. Vermont's maximum E–W extension is 90 mi (145 km); its maximum N–S extension is 158 mi (254 km). The state resembles a wedge, wide and flat at the top and narrower at the bottom.

Vermont is bordered on the N by the Canadian province of Quebec; on the E by New Hampshire (separated by the Connecticut River); on the s by Massachusetts; and on the w by New York (with part of the line passing through Lake Champlain and the Poultney River).

The state's territory includes several islands and the lower part of a peninsula jutting south into Lake Champlain from the Canadian border, collectively called Grand Isle County. Vermont's total boundary length is 561 mi (903 km). Its geographic center is in Washington County, 3 mi (5 km) E of Roxbury.

²TOPOGRAPHY

The Green Mountains are the most prominent topographic region in Vermont. Extending north–south from the Canadian border to the Massachusetts state line, the Green Mountains contain the state's highest peaks, including Mansfield, 4,393 feet (1,339 meters), the highest point in Vermont; Killington, 4,241 feet (1,293 meters); and Ellen, 4,135 feet (1,260 meters). A much lower range, the Taconic Mountains, straddles the New York–Vermont border for about 80 mi (129 km). To their north is the narrow Valley of Vermont; farther north is the Champlain Valley, a lowland about 20 mi (32 km) wide between Lake Champlain—site of the state's lowest point, 95 feet (29 meters) above sea level—and the Green Mountains. The Vermont piedmont is a narrow corridor of hills and valleys stretching about 100 mi (161 km) to the east of the Green Mountains. The Northeast Highlands consist of an isolated series of peaks near the New Hampshire border.

Vermont's major inland rivers are the Missisquoi, Lamoille, and Winooski. The state includes about 75% of Lake Champlain on its western border and about 25% of Lake Memphremagog on the northern border.

³CLIMATE

Burlington's normal daily mean temperature is 44°F (7°C), ranging from 17°F (−8°C) in January to 70°F (21°C) in July. Winters are generally colder and summer nights cooler in the higher elevations of the Green Mountains. The record high temperature for the state is 105°F (41°C), registered at Vernon on 4 July 1911; the record low, −50°F (−46°C), at Bloomfield, 30 December 1933. Burlington's average annual precipitation of 34 in (86 cm) is less than the statewide average of 40 in (102 cm). Annual snowfall ranges from 55 to 65 in (140–165 cm) in the lower regions, and from 100 to 125 in (254–318 cm) in the mountain areas.

⁴FLORA AND FAUNA

Common trees of Vermont are the commercially important sugar maple (the state tree), the butternut, and various birches and ashes. Other recognized flora include 15 types of conifer, 130 grasses, and 192 sedges. Among endangered plants in Vermont are the alpine woodsia, adder's-mouth, and small whorled pogonia.

Native mammalian species include white-tailed deer, coyote, red fox, and snowshoe hare. Several species of trout are prolific. Characteristic birds include the raven (*Corvus corax*), gray or Canada jay, and saw-whet owl. Among endangered animals in Vermont are the Canada lynx, pine marten, and lake sturgeon.

⁵ENVIRONMENTAL PROTECTION

All natural resource regulation, planning, and operation are coordinated by the Agency of Environmental Conservation. The state is divided into 14 soil and water conservation districts operated by local landowners with the assistance of the state Natural Resources Conservation Council. Several dams on the Winooski and Connecticut river drainage basins help control flooding. Legislation enacted in 1973 bans the use of throwaway beverage containers in Vermont, in an effort to reduce roadside litter. Billboards were also banned. In the 1980s, the effects of acid rain became a source of concern in Vermont, as in the rest of the northeast.

⁶POPULATION

Vermont ranked 48th among the 50 states in population, with a 1980 census total of 511,456. Population estimates for 1985 put the total at 529,396, an increase of 3.5%.

In 1980, Vermont's population was 66% rural, the highest percentage of all states; the rural population increased 12% between 1970 and 1980, and the urban population increased 21%. The population density was 57 per sq mi (22 per sq km) in 1985.

According to final census data, Burlington had 37,817 residents in 1984; Rutland, 17,809; and Montpelier, 8,167.

7ETHNIC GROUPS

The largest ethnic minority, first- and second-generation French Canadians, makes up about 10% of the population. These Vermonters are congregated chiefly in the northern counties and in such urban centers as Burlington, St. Albans, and Montpelier. First- and second-generation Italians make up a little over 1% of the population. The foreign-born numbered 21,000—4.1% of the population—in 1980.

The 1980 census counted few non-Caucasians. There were 1,355 Asians and Pacific islanders, 1,135 blacks, and 968 American Indians.

8LANGUAGES

A few place-names and very few Indian-language speakers remain as evidence of the early Vermont presence of the Algonkian Mohawk tribe and of some Iroquois in the north. Vermont English, although typical of the Northern dialect, differs from that of New Hampshire in several respects, including retention of the final /r/ and use of *eavestrough* in place of eavespout.

In 1980, 457,742 Vermonters—93% of the population aged 3 and over—spoke only English at home. Other languages spoken at home included:

French	20,134	Spanish	2,183
German	2,282	Italian	1,750

9RELIGIONS

From the early days of settlement to the present, Congregationalists (whose church is now called the United Church of Christ) have played a dominant role in the state. They are the second leading Protestant denomination in the state, with 25,066 known adherents in 1980. Other major Protestant groups include the United Methodists, 26,443; American Baptists, 10,902; and Episcopalians, 10,455. The largest single religious organization in Vermont is the Roman Catholic Church, with 150,800 members in 1984. There is a small Jewish population (estimated at 2,465 in 1984), most of which lives in Burlington.

Vermont was the birthplace of both Joseph Smith and Brigham Young, founders of the Church of Jesus Christ of Latter-day Saints. The state had 1,724 Mormons in 1980.

10TRANSPORTATION

Vermont's first railroad, completed in 1849, served more as a link to Boston than as an intrastate line; it soon went into receivership, as did many other early state lines. From a high of nearly 1,100 mi (1,770 km) of track in 1910, trackage shrank to 813 mi (1,308 km) in 1983, of which only 102 mi (164 km) were class I lines.

Of the 15,663 mi (25,207 km) of streets, roads, and highways in 1983, towns had jurisdiction over 13,020 mi (20,954 km), the state over 2,323 mi (3,739 km), and the federal government over 320 mi (515 km). A total of 367,355 motor vehicles were registered in 1983, when there were 360,642 licensed drivers.

In 1983, Vermont had 45 airports and 13 heliports. Burlington International Airport, the state's major air terminal, had 5,091 aircraft departures in that year.

11HISTORY

Vermont has been inhabited continuously since about 10,000 BC. Archaeological finds suggest the presence of a pre-Algonkian group along the Otter River. Algonkian-speaking Abnaki settled along Lake Champlain and in the Connecticut Valley, and Mahican settled in the southern counties between 1200 and 1790. In 1609, Samuel de Champlain crossed the lake that now bears his name, becoming the first European explorer of Vermont. From the 1650s to the 1760s, French, Iroquois Indians from New York, Dutch, and English passed through the state over trails connecting Montreal with Massachusetts and New York. Few settled there. In 1666, the French built and briefly occupied Ft. Ste. Anne on Isle La Motte, and in 1690 there was a short-lived settlement at Chimney Point. Ft. Dummer, built in 1724 near present-day Brattleboro, was the first permanent settlement.

Governor Benning Wentworth of New Hampshire, claiming that his colony extended as far west as did Massachusetts and Connecticut, had granted 131 town charters in the territory by 1764. In that year, the crown declared that New York's northeastern boundary was the Connecticut River. Owners of New Hampshire titles, fearful of losing their land, prevented New York from enforcing its jurisdiction. The Green Mountain Boys, organized by Ethan Allen in 1770-71, scared off the defenseless settlers under New York title and flouted New York courts.

Shortly after the outbreak of the Revolutionary War, Ethan Allen's men helped capture Ft. Ticonderoga, and for two years frontiersmen fought in the northern theater. On 16 August 1777, after a skirmish at Hubbardton, a Vermont contingent routed German detachments sent by British General Burgoyne toward Bennington—a battle that contributed to the general's surrender at Saratoga, N.Y. There were several British raids on Vermont towns during the war.

Vermont declared itself an independent republic with the name "New Connecticut" in 1777, promulgated a constitution abolishing slavery and providing universal manhood suffrage, adopted the laws of Connecticut, and confiscated Tory lands. Most Vermonters preferred to join the US, but the dominant Allen faction, with large holdings in the northwest, needed free trade with Canada, even at the price of returning to the British Empire. Political defeat of the Allen faction in 1789 led to negotiations that settled New York's claims and secured Vermont's admission to the Union on 4 March 1791.

With 30,000 people in 1781 and nearly 220,000 in 1810, Vermont was a state of newcomers spread evenly over the hills in self-sufficient homesteads. Second-generation Vermonters developed towns and villages with water-powered mills, charcoal-fired furnaces, general stores, newspapers, craft shops, churches, and schools. Those who ran these local institutions tended to be Congregationalist in religion and successively Federalist, Whig, and Republican in party politics. Dissidents in the early 1800s included minority Protestants suffering legal and social discrimination, hardscrabble farmers, and Jacksonian Democrats.

Northwestern Vermonters smuggled to avoid the US foreign trade embargo of 1808, and widespread trade continued with Canada during the War of 1812. In September 1814, however, Vermont soldiers fought in the Battle of Plattsburgh, N.Y., won by Thomas Macdonough's fleet built at Vergennes the previous winter. The Mexican War (1846-48) was unpopular in the state, but Vermont, which had strongly opposed slavery, was an enthusiastic supporter of the Union during the Civil War.

The opening of the Champlain-Hudson Canal in 1823, and the building of the early railroad lines in 1846-53, made Vermont more vulnerable to western competition, killed many small farms and businesses, and stimulated emigration. The remaining farmers' purchasing power steadily increased, as they held temporary advantages in wool, then in butter and cheesemaking, and finally in milk production. The immigration of the Irish and French Canadians stabilized the population, and the expansion of light industry bolstered the economy.

During the 20th century and especially after World War II, autos, buses, trucks, and planes took over most passengers and much freight from the railroads. Manufacturing, especially light industry, prospered in valley villages. Vermont's picturesque landscape began to attract city buyers of second homes. Still rural in population distribution, Vermont became increasingly suburban in outlook, as new highways made the cities and hills mutually accessible. Longtime Vermonters, accustomed to their state's pristine beauty, were confronted in the 1980s with the question of how much development is necessary for the state's economic health.

[12]STATE GOVERNMENT

A constitution establishing Vermont as an independent republic was adopted in 1777. This document, as revised and amended, still governs the state.

The general assembly consists of a 150-member house of representatives and a 30-member senate. All legislators are elected to two-year terms. State elective officials include the governor, lieutenant governor (elected separately), treasurer, secretary of state, auditor of accounts, and attorney general, all of whom serve two-year terms.

The legislature meets biennially in odd-numbered years. All bills require a majority vote in each house for passage. Bills can be vetoed by the governor, and vetoes can be overridden by a two-thirds vote of each legislative house. A constitutional amendment must first be passed by a two-thirds vote in the senate, followed by a majority in the house during the same legislative session. It must then receive majority votes in both houses before it can be submitted to the voters for approval.

Voters must be US citizens and 18 years of age; there is no minimum residency requirement.

[13]POLITICAL PARTIES

The Republican Party, which originally drew strength from powerful abolitionist sentiment, gained control of Vermont state offices in 1856 and for more than 100 years dominated state politics. No Democrat was elected governor from 1853 until 1962.

In 1982 there were 318,832 registered voters. In 1985, Madeleine M. Kunin took office as Vermont's first woman governor and only the third Democratic governor in the state's history. Women also held 48 seats (26.7%) in the state legislature. For the first time in history, Democrats controlled the state senate; Republicans had a majority in the house of representatives. Vermont had one Republican and one Democratic US senator.

Vermont has often shown its independence in national political elections. In 1832, it was the only state to cast a plurality vote for the Anti-Masonic presidential candidate, William Wirt; in 1912, the only state besides Utah to vote for William Howard Taft; and in 1936, the only state besides Maine to prefer Alf Landon to Franklin D. Roosevelt. In 1984, though, Vermont helped reelect President Ronald Reagan with a margin of victory very similar to the national margin. Reagan won Vermont with 57.9% of the vote.

Vermont Presidential Vote by Major Political Parties, 1948–84

YEAR	ELECTORAL VOTE	VERMONT WINNER	DEMOCRAT	REPUBLICAN
1948	3	Dewey (R)	45,557	75,926
1952	3	*Eisenhower (R)	43,299	109,717
1956	3	*Eisenhower (R)	42,540	110,390
1960	3	Nixon (R)	69,186	98,131
1964	3	*Johnson (D)	108,127	54,942
1968	3	*Nixon (R)	70,255	85,142
1972	3	*Nixon (R)	68,174	117,149
1976	3	Ford (R)	77,798	100,387
1980	3	*Reagan (R)	81,952	94,628
1984	3	*Reagan (R)	95,730	135,865

*Won US presidential election.

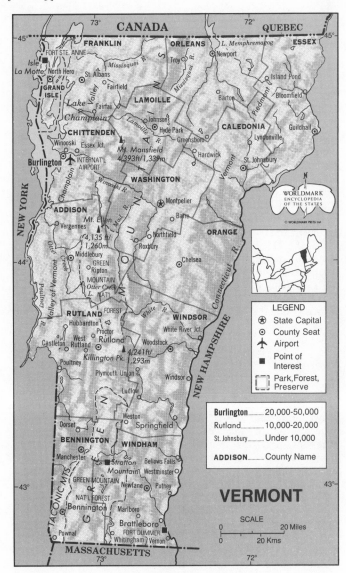

See endsheet maps: M2

LOCATION: 42°43′25″ to 45°00′50″N. 71°27′57″ to 73°26′03″w.
BOUNDARIES: Canadian line, 89 mi (143 km); New Hampshire line, 242 mi (390 km); Massachusetts line, 41 mi (66 km); New York line, 189 mi (304 km).

[14]LOCAL GOVERNMENT

As of 1984 there were 14 counties, 9 cities, 237 organized towns, 5 unorganized towns, 52 incorporated villages, and 5 gores and grants (nongoverned areas) in Vermont. County officers, operating out of shire towns (county seats), include the probate court judge, assistant judges of the county court, county clerk, state's attorney, high bailiff, treasurer, and sheriff. All cities have mayor-council systems. Towns are governed by three selectmen, who serve staggered three-year terms; one is elected at each annual town meeting, held on the 1st Tuesday in March. Larger towns also have town managers.

[15]STATE SERVICES

Vermont's Department of Education oversees public elementary, secondary, higher education, and adult education programs. The Agency of Transportation includes the Department of Motor Vehicles, Transportation Board, and Hazardous Materials Committee. The Agency of Human Services coordinates programs for nursing homes, veterans' affairs, social welfare, employment and training, health, corrections, and parole. The Agency of Development and Community Affairs administers federal housing programs and offers aid to localities.

¹⁶JUDICIAL SYSTEM

Vermont's highest court is the supreme court, which consists of a chief justice and 4 associate justices. Other courts include the superior court, with 10 superior judges, and a district court divided into 3 territorial units embracing 14 circuits. All judges are appointed by the governor to six-year terms, subject to senate confirmation, from a list of qualified candidates prepared by the Judicial Nominating Board, which includes representatives of the governor, the legislature, and the Vermont bar association. There are also 28 assistant judges and 19 probate court judges, all of them elected to four-year terms. There were 1,418 practicing attorneys in December 1984.

Crime rates in virtually every category except forcible rape are far below the national average.

¹⁷ARMED FORCES

An estimated $170 million in federal defense contracts was awarded to Vermont firms in 1983/84. Of the estimated 64,000 veterans living in Vermont in 1983, about 1,000 served in World War I, 23,000 in World War II, 12,000 in the Korean conflict, and 19,000 during the Viet-Nam era. Veterans' benefits in 1983 totaled $73 million.

¹⁸MIGRATION

The earliest Vermont settlers were farmers from southern New England and New York; most were of English descent, although some Dutch settlers moved to Vermont from New York. French Canadians came beginning in the 1830s; by 1850, several thousand had moved into Vermont. As milling, quarrying, and mining grew during the 19th century, other Europeans arrived—small groups of Italians and Scots in Barre, and Poles, Swedes, Czechs, Russians, and Austrians in the Rutland quarry areas. Irish immigrants built the railroads in the mid-19th century. Steady out-migrations during the 19th and early 20th centuries kept population increases down, and in the decades 1890–1900 and 1910–20, the population dropped. During the 1960s, the population of blacks more than doubled, though they still accounted for only 0.2% of the population in 1980. Between 1970 and 1983, 45,000 migrants settled in Vermont.

¹⁹INTERGOVERNMENTAL COOPERATION

Vermont participates in New England compacts on corrections, higher education, water pollution control, police, and radiological health protection. The state also takes part in the Connecticut River Valley Flood Control Compact, Lake Champlain Bridge Compact, and Interstate Pest Control Compact.

Federal aid to Vermont totaled $331 million in 1983/84.

²⁰ECONOMY

During its early years of statehood, Vermont was overwhelmingly agricultural, with beef cattle, sheep, and dairying contributing greatly to the state's income. After World War II, agriculture was replaced by manufacturing and tourism as the backbone of the economy. Durable goods manufacturing (primarily electronics and machine parts), construction, wholesale and retail trade, and other service industries have shown the largest growth in employment during the early 1980s.

²¹INCOME

Vermont ranked 36th in per capita income with $10,036 in 1983. Total personal income that year was $5.3 billion, 0.2% of the US total and the smallest amount of all states. The median money income for a family of four persons was $23,551 (38th in the US) in 1981. More than 12% of all Vermonters were below the federal poverty level in 1979.

²²LABOR

According to the US Department of Labor, 255,000 Vermonters were employed during 1984, about 44% of them women. The unemployment rate that year was 5.2%.

A federal survey in 1983 revealed the following nonfarm employment pattern for Vermont:

	ESTABLISH-MENTS	EMPLOYEES	ANNUAL PAYROLL ('000)
Agricultural services, forestry, fishing	171	861	$ 10,708
Mining	41	671	15,231
Contract construction	1,781	7,999	146,557
Manufacturing, of which:	1,071	44,681	893,153
Electric, electronic equipment	(38)	(10,590)	(285,827)
Transportation, public utilities	576	8,013	164,887
Wholesale trade	812	7,866	132,549
Retail trade	4,509	35,096	320,003
Finance, insurance, real estate	1,004	8,517	131,611
Services	4,396	46,978	483,444
Other	996	982	18,237
TOTALS	15,357	161,664	$2,316,380

There were 36,000 members of labor unions in Vermont in 1980 and 168 unions in 1983.

²³AGRICULTURE

Although Vermont is one of the nation's most rural states, its agricultural income was only $427 million in 1983, 42d highest in the US. More than 90% of that came from livestock and livestock products, especially dairy products. The leading crops were corn for silage, 1,363,000 tons; hay, 926,000 tons; and apples, 46,000,-000 lb.

²⁴ANIMAL HUSBANDRY

The merino sheep and the Morgan horse (a breed developed in Vermont) were common sights on pastures more than a century ago, but today they have been for the most part replaced by dairy cattle. Vermont led the New England states in milk production in 1983 with 2.4 billion lb; it ranked 13th in the US in 1982 in value of dairy products sold ($304 million).

²⁵FISHING

Sport fishermen can find ample species of trout, perch, walleye pike, bass, and pickerel in Vermont's waters, many of which are stocked by the Department of Fish and Game. There is little commercial fishing.

²⁶FORESTRY

The Green Mountain State is covered by 4,512,000 acres (1,826,-000 hectares) of forestland—76% of the state's total land area—much of it owned or leased by lumber companies. More than 25% of all manufacturing establishments in the state depend on the lumber industry. Shipments of forest products in 1982 included paper and paper products, $309.8 million, and lumber and wood products, $172.6 million. Vermont is the nation's leading producer of maple syrup; the 1984 output of 530,000 gallons was valued at $10 million wholesale.

The largest forest reserve in Vermont is the Green Mountain National Forest, with 294,610 acres (119,225 hectares) in 1984.

²⁷MINING

Although Vermont is a leading producer of granite, marble, talc, and asbestos, the state's total estimated mining income for 1984 was only $42.9 million, 45th in the US.

The granite hills near Barre contain most of the state's actively worked granite deposits. Marble is mined in the West Rutland–Proctor area; slate is found in the southeast. Sand and gravel are quarried throughout the state. The estimated 1984 output included 3.4 million short tons of sand and gravel and 1.4 million short tons of crushed stone.

²⁸ENERGY AND POWER

Because of the state's lack of fossil fuel resources, utility bills are higher in Vermont than in most states. During 1983, 57 plants with a capacity of 945,000 kw generated 3.9 billion kwh of power, more than 73% of which was produced by the state's lone nuclear plant at Vernon, operated by Vermont Yankee Nuclear Power Corp.

29 INDUSTRY

Value of shipments of manufactured goods exceeded $3.7 billion in 1982—a figure important to the state's economy though very small by national standards. Leading industry groups (and their percentage of value of shipments) were electrical and electronic equipment, 25.8%; food and food products, 13.5%; nonelectrical machinery, 12.2%; fabricated metal products, 9.7%; paper and paper products, 8.3%; printing and publishing, 6.2%; and lumber and wood products, 4.6%.

Scales, machine tools, and electronic components are important manufactured items. The following table shows value of shipments of manufacturers for selected industries in 1982:

Electric and electronic equipment	$961,600,000
Nonelectrical machinery	455,700,000
Fabricated metal products	363,600,000
Dairy products	321,900,000
Paper and allied products	309,800,000

30 COMMERCE

Wholesale trade in 1982 was nearly $1.9 billion. Retail trade totaled $2.6 billion, of which the leading sectors were food stores, 23%; automotive dealers, 18%; gasoline service stations, 8%; and eating and drinking places, 8%. Foreign exports of Vermont manufactures were estimated at $229 million for 1981.

31 CONSUMER PROTECTION

The Consumer Protection Division of the Attorney General's Office handles most consumer complaints. The Vermont Public Service Board's Consumer Affairs Division monitors utility rates, and the Agency of Human Services' Office on Aging protects the rights of the state's senior citizens.

32 BANKING

In 1984 there were 33 commercial banks. In 1981, the state's commercial banks had assets of $3.2 billion and deposits exceeding $2.9 billion. There were 4 savings and loan associations and 66 credit unions in 1984.

33 INSURANCE

Thirty-five insurance companies have their home offices in Vermont. As of 1983, about 801,000 life insurance policies worth nearly $9.7 billion were held by Vermonters. The average Vermont family had $44,500 in life insurance coverage. Automobile insurance premiums written in the state in 1983 totaled $48.2 million; homeowners insurance premiums, $12.8 million.

34 SECURITIES

There are no stock or commodity exchanges in Vermont. New York Stock Exchange member firms had 17 sales offices and 133 registered representatives in the state in 1983. About 86,000 Vermonters were shareowners of public corporations in 1984.

35 PUBLIC FINANCE

The budgets for two fiscal years are submitted by the governor to the general assembly for approval during its biennial session. The fiscal year runs from 1 July to 30 June.

The following table shows estimated general revenues and expenditures for the years 1984/85 and 1985/86:

REVENUES	1984/85	1985/86
Personal income taxes	$149,000,000	$164,226,531
Sales and use taxes	91,230,000	99,805,620
Corporate income taxes	31,680,000	32,472,000
Other taxes	86,673,694	94,829,495
Other receipts	18,316,306	18,616,354
TOTALS	$376,900,000	$409,950,000

EXPENDITURES	1984/85	1985/86
Education	$152,326,901	$166,298,883
Human services	105,972,335	114,177,874
Other expenditures	107,430,824	111,404,094
TOTALS	$365,730,060	$391,820,851

These figures do not include activities supported by federal funds or by self-supporting revolving funds. Total estimated expenditures for 1984/85 are $792,203,811; for 1985/86, $826,120,739.

The total outstanding debt of state and local governments in Vermont was $1.05 billion, or $2,025 per capita, as of 1982/83.

36 TAXATION

Vermont ranked 50th in the US in state and local tax receipts in 1982, with $1.04 billion; the tax burden was $2,039 per capita. The state imposes personal and corporate income taxes, sales and use taxes, a franchise tax, and an inheritance tax. Taxes are also levied on beverages, electrical energy, insurance, meals and rooms, old age assistance, real estate transfers, and tobacco products, among other items.

In 1983, Vermonters filed 217,373 federal income tax returns and paid $475 million in tax.

37 ECONOMIC POLICY

Incentives for industrial expansion include state and municipally financed industrial sites, state employment development and training funds, revenue bond financing, loans and loan guarantees for construction and equipment, and financial incentives for locating plants in areas of high unemployment. There are also exemptions from inventory taxes and sales tax on new equipment and raw materials.

38 HEALTH

In 1982 there were about 8,000 live births (15.4 per 1,000 population) in Vermont. In 1981, the infant mortality rate for whites was 7.6 per 1,000 live births. There were 3,700 legal abortions in 1982.

Heart disease was the leading cause of death in 1981, though the rate of 322 per 100,000 population was slightly below the US average. In 1983, there were 93 motor vehicle traffic fatalities in the state.

Of the 19 hospitals operating in 1983, 17 were general and 2 were psychiatric. Hospital personnel in 1981 included 2,007 registered nurses, of whom 1,237 worked full time. In 1982, the average costs per stay ($2,099) and per day ($256) were below the US average. The state had 1,338 licensed physicians and 275 professionally active dentists in 1982.

39 SOCIAL WELFARE

Participants in and benefits from leading social welfare programs in 1982 included aid to families with dependent children, 21,500 participants, $36.9 million in benefits. For 1983: food stamps, 50,000 participants, $27 million in benefits; school lunch program, 47,000, $4 million; and Supplemental Security Income, 8,700, $19.9 million. Unemployment insurance (1982): 7,000, $42.7 million. In 1983, 81,000 Vermonters received $375 million in Social Security payments.

40 HOUSING

As rustic farmhouses gradually disappear, modern units (many of them vacation homes for Vermonters and out-of-staters) are being built to replace them. About 2,600 new units worth $92 million were authorized in 1981.

41 EDUCATION

As of 1980, more than 70% of Vermonters above the age of 25 were high school graduates, and 19% had completed at least four years of postsecondary study.

During the 1983/84 school year, 91,038 students were enrolled in Vermont's 428 public and private schools. The total 1982/83 enrollment in the state's 22 two- and four-year colleges was 30,648. The state operates the University of Vermont (Burlington), founded in 1791 and the oldest higher educational institution in the state, as well as the three state colleges, at Castleton, Johnson, and Lyndonville. Notable private institutions include Bennington College, Middlebury College, St. Michael's College (Winooski), and Norwich University (Northfield), the oldest private military college in the US. The School for International Training at Brattleboro is the academic branch of the Experiment in International Living, a student exchange program. The Vermont Student Assistance Corp., a state agency, offers scholarships, incentive grants, and guaranteed loans for eligible Vermont students.

⁴²ARTS

The Vermont State Crafts Centers at Frog Hollow (Middlebury) and Windsor display the works of Vermont artisans. The Vermont Symphony Orchestra, in Burlington, makes extensive statewide tours. Marlboro College is the home of the summer Marlboro Music Festival, directed by pianist Rudolf Serkin. Among the summer theaters in the state are those at Dorset and Weston, and the University of Vermont Shakespeare Festival. The Middlebury College Bread Loaf Writers' Conference, founded in 1926, meets each August in Ripton.

⁴³LIBRARIES AND MUSEUMS

During 1983/84, the state's public libraries held 2,051,601 volumes and had a combined circulation of 2,730,350. The largest academic library was at the University of Vermont, with a book stock of 779,543.

Vermont has 74 museums and more than 65 historic sites. Among them are the Bennington Museum, with its collection of Early American glass, pottery, furniture, and Grandma Moses paintings, and the Art Gallery-St. Johnsbury Athenaeum, featuring 19th-century American artists. The Shelburne Museum, housed in restored Early American buildings, contains collections of American primitives and Indian artifacts. The Vermont Museum, in Montpelier, features historical exhibits concerning Indians, the Revolutionary War, rural life, and railroads and industry. Old Constitution House in Windsor offers exhibits on Vermont history.

⁴⁴COMMUNICATIONS

As of 1985 there were about 1,700 postal service workers in Vermont. In 1980, 165,914 homes (93%) had telephones. There were 20 AM and 32 FM radio stations and 3 commercial television stations in operation in 1984. In the same year, 67 cable television systems served 81,260 subscribers in 138 communities.

⁴⁵PRESS

In 1985 there were 8 daily papers, with a combined 1984 circulation of 118,518, and 3 Sunday papers, with 81,439 circulation. The leading daily in 1984 was the *Burlington Free Press* (50,857 mornings, 48,300 Sundays).

Vermont Life magazine is published quarterly under the aegis of the Agency of Development and Community Affairs.

⁴⁶ORGANIZATIONS

The 1982 Census of Service Industries counted 268 organizations in Vermont, including 51 business associations; 160 civic, social, and fraternal associations; and 11 educational, scientific, and research organizations. Associations headquartered in Vermont largely reflect the state's agricultural interests. Among these are Gardens for All (the National Association for Gardening) in Burlington, the Ayrshire Breeders' Association of America (Brattleboro), and the International Maple Syrup Institute (Fairfax).

⁴⁷TOURISM, TRAVEL, AND RECREATION

With the building of the first ski slopes in the 1930s (Woodstock claims the first ski area in the US) and the development of modern highways, tourism became a major industry in Vermont. As of 1982, it was the state's 2d largest, accounting for almost 20% of the gross state product. Travel-related industries earned about $1 billion in 1982 and employed 33,500 full- and part-time workers. Perhaps 60–65% of all tourist expenditures come during the summer, when the state attracts campers to nearly 40 state-owned

campgrounds, more than 70 private campgrounds, and numerous private resorts. In the winter, the state's ski areas (56 in 1983) offer some of the finest skiing in the East. In 1982/83, the state issued licenses to 117,261 hunters and 163,565 fishermen.

⁴⁸SPORTS

Vermont ski areas have hosted national and international ski competitions in both Alpine and Nordic events. World Cup races have been run at Stratton Mountain, and the national cross-country championships have been held near Putney.

⁴⁹FAMOUS VERMONTERS

Two US presidents, both of whom assumed office on the death of their predecessor, were born in Vermont. Chester Alan Arthur (1829-86) became the 21st president after James A. Garfield's assassination in 1881 and finished his term. A machine politician, Arthur became a civil-service reformer in the White House. Calvin Coolidge (1872-1933), 28th president, was born in Plymouth Notch but pursued a political career in Massachusetts. Elected vice president in 1920, he became president on the death of Warren G. Harding in 1923 and was elected to a full term in 1924.

Other federal officeholders have included Matthew Lyon (1750-1822), a US representative imprisoned under the Sedition Act and reelected from a Vergennes jail; Jacob Collamer (1791-1865), who after serving three terms in the US House, was US postmaster general and then a US senator; Justin Smith Morrill (1810-98), US representative and senator who sponsored the Morrill tariff in 1861 and the Land Grant College Act in 1862; Levi Parsons Morton (1824-1920), Benjamin Harrison's vice president from 1889 to 1893; George Franklin Edmunds (1828-1919), a US senator who helped draft the Sherman Antitrust Act; Redfield Proctor (1831-1908), secretary of war, US senator, state governor, and the founder of a marble company; John Garibaldi Sargent (1860-1939), Coolidge's attorney general; Warren Robinson Austin (1877-1963), US senator and head of the US delegation to the UN; and George David Aiken (1892-1984), US senator from 1941 to 1977.

Important state leaders were Thomas Chittenden (1730-97), leader of the Vermont republic and the state's first governor; Ethan Allen (1738-89), a frontier folk hero, leader of the Green Mountain Boys, and presenter of Vermont's claim to independence to the US Congress in 1778; Ira Allen (1751-1814), the brother of Ethan, who led the fight for statehood; Cornelius Peter Van Ness (b.New York, 1782-1852), who served first as Vermont chief justice and then as governor; and Erastus Fairbanks (1792-1864), a governor and railroad promoter.

Vermont's many businessmen and inventors include Thaddeus Fairbanks (1796-1886), inventor of the platform scale; Thomas Davenport (1802-51), inventor of the electric motor; plow and tractor manufacturer John Deere (1804-86); Elisha G. Otis (1811-61), inventor of a steam elevator and elevator safety devices; and Horace Wells (1815-48), inventor of laughing gas. Educator John Dewey (1859-1952) was born in Burlington.

Robert Frost (b.California, 1874-1963) maintained a summer home near Ripton, where he helped found Middlebury College's Bread Loaf Writers' Conference. He was named poet laureate of Vermont in 1961. A famous Vermont performer is crooner and orchestra leader Rudy Vallee (Hubert Prior Rudy Vallee, b.1901).

⁵⁰BIBLIOGRAPHY

Bassett, T. D. S. *Vermont.* Hanover, N.H.: University Press of New England, 1983.

Bearse, Ray, ed. *Vermont: A Guide to the Green Mountain State.* Boston: Houghton Mifflin, 1968.

Crockett, Walter H. *History of Vermont.* 5 vols. New York: Century, 1921.

Hill, Ralph Nading. *Vermont: A Special World.* Montpelier: Vermont Life, 1969.

Morrissey, Charles T. *Vermont: A Bicentennial History.* New York: Norton, 1981.

VIRGINIA

Commonwealth of Virginia

ORIGIN OF STATE NAME: Named for Queen Elizabeth I of England, the "Virgin Queen." **NICKNAME:** The Old Dominion. **CAPITAL:** Richmond. **ENTERED UNION:** 25 June 1788 (10th). **SONG:** "Carry Me Back to Old Virginia." **MOTTO:** *Sic semper tyrannis* (Thus ever to tyrants). **FLAG:** On a blue field with a white border at the fly, the state seal is centered on a white circle. **OFFICIAL SEAL:** Obverse: the Roman goddess Virtus, dressed as an Amazon and holding a sheathed sword in one hand and a spear in the other, stands over the body of Tyranny, who is pictured with a broken chain in his hand and a fallen crown nearby. The state motto appears below, the word "Virginia" above, and a border of Virginia creeper encircles the whole. Reverse: the Roman goddesses of Liberty, Eternity, and Fruitfulness, with the word "Perseverando" (By persevering) above. **BIRD:** Cardinal. **FLOWER:** Dogwood. **TREE:** Dogwood. **DOG:** Foxhound. **SHELL:** Oyster. **BEVERAGE:** Milk. **LEGAL HOLIDAYS:** New Year's Day, 1 January; Lee-Jackson Day and Birthday of Martin Luther King, Jr., 3d Monday in January; Washington's Birthday, 3d Monday in February; Memorial Day, last Monday in May; Independence Day, 4 July; Labor Day, 1st Monday in September; Columbus Day and Yorktown Victory Day, 2d Monday in October; Election Day, 1st Tuesday after 1st Monday in November; Veterans Day, 11 November; Thanksgiving Day, 4th Thursday in November; Christmas Day, 25 December. **TIME:** 7 AM EST = noon GMT.

[1] LOCATION, SIZE, AND EXTENT

Situated on the eastern seaboard of the US, Virginia is the 4th largest of the South Atlantic states and ranks 36th in size among the 50 states.

The total area of Virginia is 40,767 sq mi (105,586 sq km), of which land occupies 39,704 sq mi (102,833 sq km) and inland water 1,063 sq mi (2,753 sq km). Virginia extends approximately 440 mi (710 km) E–W, but the maximum point-to-point distance from the state's noncontiguous Eastern Shore to the western extremity is 470 mi (756 km). The maximum N–S extension is about 200 mi (320 km).

Virginia is bordered on the NW by West Virginia; on the NE by Maryland and the District of Columbia (with the line passing through the Potomac River and Chesapeake Bay); on the E by the Atlantic Ocean; on the S by North Carolina and Tennessee; and on the W by Kentucky. The state's geographic center is in Buckingham County, 5 mi (8 km) SW of the town of Buckingham.

Virginia's offshore islands in the Atlantic include Chincoteague, Wallops, Cedar, Parramore, Hog, Cobb, and Smith. The boundaries of Virginia, including the Eastern Shore at the tip of the Delmarva Peninsula, total 1,356 mi (2,182 km), of which 112 mi (180 km) is general coastline; the tidal shoreline extends 3,315 mi (5,335 km).

[2] TOPOGRAPHY

Virginia consists of three principal physiographic areas: the Atlantic Coastal Plain, or Tidewater; the Piedmont Plateau, in the central section; and the Blue Ridge and Allegheny Mountains of the Appalachian chain, in the west and northwest.

The long, narrow Blue Ridge rises sharply from the piedmont, reaching a maximum elevation of 5,729 feet (1,746 meters) at Mt. Rogers, the state's highest point. Between the Blue Ridge and the Allegheny Mountains of the Appalachian chain in the northwest lies the Valley of Virginia, consisting of transverse ridges and six separate valleys. The floors of these valleys ascend in altitude from about 300 feet (90 meters) in the northern Shenandoah Valley to 2,400 feet (730 meters) in the Powell Valley. The Alleghenies average 3,000 feet (900 meters) in height.

The piedmont, shaped roughly like a triangle, varies in width from 40 mi (64 km) in the far north to 180 mi (290 km) in the extreme south. Altitudes in this region range from about 300 feet

(90 meters) at the fall line in the east to a maximum of about 1,000 feet (300 meters) at the base of the Blue Ridge in the southwest. The Tidewater, which declines gently from the fall line to sea level, is divided by four long peninsulas cut by the state's four principal rivers—the Potomac, Rappahannock, York, and James—and Chesapeake Bay. On the opposite side of the bay is Virginia's low-lying Eastern Shore, the southern tip of the Delmarva Peninsula. The Tidewater has many excellent harbors, notably the deep Hampton Roads estuary. Also in the southeast lies the Dismal Swamp, a drainage basin that includes Lake Drummond, about 7 mi (11 km) long and 5 mi (8 km) wide near the North Carolina border. Other major lakes in Virginia are Smith Mountain—at 31 sq mi (80 sq km) the largest lake wholly within the state—Claytor, and South Holston. The John H. Kerr Reservoir, covering 76 sq mi (197 sq km), straddles the Virginia–North Carolina line.

[3] CLIMATE

A mild, humid coastal climate is characteristic of Virginia. Temperatures, most equable in the Tidewater, become increasingly cooler with the rising altitudes as one moves westward. The normal daily mean temperature at Richmond is about 58°F (14°C), ranging from 37°F (3°C) in January to 78°F (26°C) in July. The record high, 110°F (43°C), was registered at Balcony Falls (near Glasgow) on 15 July 1954; the record low, −29°F (−34°C), was set at Monterey on 10 February 1899. The frost-free growing season ranges from about 140 days in the mountains of the extreme west to over 250 in the Norfolk area.

Precipitation at Richmond averages 44 in (112 cm) a year; the average snowfall amounts to nearly 15 in (38 cm) at Richmond but only 8 in (20 cm) at Norfolk.

[4] FLORA AND FAUNA

Native to Virginia are 12 varieties of oak, 5 of pine, and 2 each of walnut, locust, gum, and popular. Pines predominate in the coastal areas, with numerous hardwoods on slopes and ridges inland; isolated stands of persimmon, ash, cedar, and basswood can also be found. Characteristic wild flowers include trailing arbutus, mountain laurel, and diverse azaleas and rhododendrons. The Virginia round-leaf birch and small whorled pogonia are endangered.

Among indigenous mammalian species are white-tailed (Virginia) deer, elk, black bear, bobcat, woodchuck, raccoon, opossum, nutria, red and gray foxes, and spotted and striped skunks, along with several species each of moles, shrews, bats, squirrels, deermice, rats, and rabbits; the beaver, mink, and river otter, once thought to be endangered, have returned in recent decades. Principal game birds include the ruffed grouse (commonly called pheasant in Virginia), wild turkey, bobwhite quail, mourning dove, woodcock, and Wilson's snipe. Tidal waters abound with croaker, hogfish, gray and spotted trout, and flounder; bass, bream, bluegill, sunfish, perch, carp, catfish, and crappie live in freshwater ponds and streams. Native reptiles include such poisonous snakes as the northern copperhead, eastern cottonmouth, and timber rattler.

Endangered species in Virginia include the eastern cougar; Delmarva fox squirrel; Indiana, gray, and Virginia big-eared bats; southern bald eagle; red-cockaded and ivory-billed woodpeckers; Bachman's warbler; shortnose sturgeon; Virginia fringed mountain snail; Madison cave isopod; five species of pearly mussels; three species of pigtoe; and the tan riffle shell. Threatened species are the slender and spotfin chubs and yellowfin madtom. Of more than 60 rare or endangered species in the state, at least one-fourth are found in the Dismal Swamp.

[5] ENVIRONMENTAL PROTECTION

The Department of Conservation and Economic Development, established in the late 1920s and now under the jurisdiction of the secretary of commerce and resources, is charged with the protection and development of the state's forest and mineral resources, and with the management of state parks and other recreational areas. The Council on the Environment, within the same branch of the cabinet, is responsible for coordinating the state's environmental protection programs and for implementing the Virginia Environmental Quality Act of 1972. The State Air Pollution Control Board, organized in 1966, monitors air quality throughout the state and enforces the emissions standards promulgated by the US Environmental Protection Agency. Under the Virginia Groundwater Act of 1973, the State Water Control Board has developed programs and regulations to conserve water resources and has instituted water pollution controls, and groundwater management plans. The Commission of Game and Inland Fisheries manages land wildlife and freshwater fish resources, while the Marine Resources Commission manages the wetlands, commercial fishery resources, and the use of the marine environment in the Tidewater area.

Among Virginia's persistent problems are air and water pollution in the Richmond-Petersburg-Hopewell industrial triangle, where chemical and other industrial plants are concentrated. The pollution of Chesapeake Bay is another problem, and in 1984, the state legislature appropriated $10 million to clean up the waterway; previously, the federal government had given $10 million to Virginia, Maryland, Pennsylvania, and the District of Columbia for the same purpose.

[6] POPULATION

Virginia ranked 14th among the 50 states at the 1980 census with a population of 5,346,818, a 15% increase over 1970. In early 1985, the population was estimated at 5,642,183, yielding a population density of 142 persons per sq mi (55 per sq km).

From the outset, Virginia was the most populous of the English colonies, with a population that doubled every 25 years and totaled more than 100,000 by 1727. By 1790, the time of the first US census, Virginia's population of 821,287 was about 21% of the US total and almost twice that of 2d-ranked Pennsylvania. Although surpassed by New York State at the 1820 census, Virginia continued to enjoy slow but steady growth until the Civil War, when the loss of its western counties (which became the new state of West Virginia) and the wartime devastation caused a decline of 23% for the decade of the 1860s. The population passed the 2-million mark in 1910, and the number of Virginians doubled

between 1920 and 1970. The population growth rates for the four decades since 1940 were 23.9%, 19.5%, 17.2%, and 15%, in each case above the US average.

In mid-1983, 70.7% of all Virginians lived in metropolitan areas, the largest of which in 1984 was the Norfolk–Virginia Beach–Newport News area, with an estimated 1,261,200 people. Virginia's share of metropolitan Washington, D.C., was 1,105,714 people in 1980, and the Richmond-Petersburg metropolitan area had an estimated 796,100 people in 1984. Virginia's most populous cities that year were Virginia Beach, 308,664; Norfolk, 279,683; Richmond, 219,056; Newport News, 154,560; Chesapeake, 126,031; Hampton, 125,992; Portsmouth, 107,961; Alexandria, 107,026; and Roanoke, 100,688.

[7] ETHNIC GROUPS

When the first federal census was taken in 1790, more than 306,000 blacks—of whom only 12,000 were free—made up more than one-third of Virginia's total population. After emancipation, blacks continued to be heavily represented, accounting in 1870 for 512,841 (42%) of the 1,225,163 Virginians. Blacks numbered 1,008,311 in 1980—but their proportion of the total estimated population was less than 19%. Richmond was 51% black in 1980, and Norfolk and Newport News each had a black population of 35%.

In 1980, Virginia had 79,722 Hispanic residents, chiefly Puerto Ricans and Cubans. The 1980 census counted 70,569 Asians and Pacific Islanders, including 19,111 Filipinos, 12,797 Koreans, 9,495 Chinese, 9,451 Vietnamese, 9,046 Asian Indians, and 5,173 Japanese. Some 177,000 Virginians—3.3% of all state residents—were of foreign birth in 1980, compared with 72,281 in 1970. The American Indian population was only 9,336 during the same year.

[8] LANGUAGES

English settlers encountered members of the Powhatan Indian confederacy, speakers of an Algonkian language, whose legacy consists of such place-names as Roanoke and Rappahannock.

Although the expanding suburban area south of the District of Columbia has become dialectally heterogeneous, the rest of the state has retained its essentially Southern speech features. Many dialect markers occur statewide, but subregional contrasts distinguish the South Midland of the Appalachians from the Southern of the piedmont and Tidewater. General are *batter bread* (a soft corn cake), *batter cake* (pancake), *comfort* (tied and filled bedcover), and *polecat* (skunk). Widespread pronunciation features include *greasy* with a /z/ sound; *yeast* and *east* as sound-alikes; *creek* rhyming with *peek,* and *can't* with *paint; coop* and *bulge* with the vowel of *book;* and *forest* with an /ah/ sound.

The Tidewater is set off by *creek* meaning a saltwater inlet, *fishing worm* for earthworm, and *fog* as /fahg/. Appalachian South Midland has *redworm* for earthworm, *fog* as /fawg/, *wash* as /wawsh/, *Mary* and *merry* as sound-alikes, and *poor* with the vowel of *book.* The Richmond area is noted also for having two variants of the long /i/ and /ow/ diphthongs as they occur before voiceless and voiced consonants, so that the vowel in the noun *house* is quite different from the vowel in the verb *house,* and the vowel in *advice* differs from that in *advise.* The Tidewater exhibits similar features.

In 1980, Virginia residents 3 years of age and over who spoke only English at home numbered 4,898,539, or 96% of the total. Other languages spoken at home, and the number of people who spoke them, included:

Spanish	66,423	Chinese	8,141
French	28,329	Vietnamese	8,022
German	24,297	Italian	8,004
Korean	10,437		

[9] RELIGIONS

The Anglican Church (later, the Episcopal Church), whose members founded and populated Virginia Colony in the early

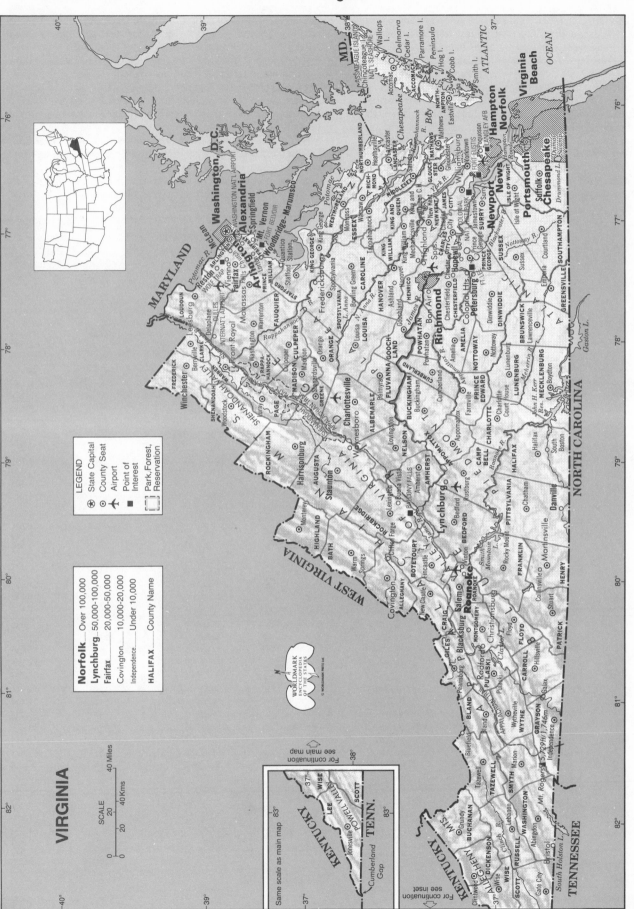

VIRGINIA

LEGEND
- ⊛ State Capital
- ⊙ County Seat
- ✈ Airport
- ■ Point of Interest
- ⬚ Park, Forest, Reservation

Norfolk Over 100,000
Lynchburg 50,000-100,000
Fairfax 20,000-50,000
Covington 10,000-20,000
Independence Under 10,000
HALIFAX County Name

SCALE
0   20   40 Miles
0   20   40 Kms

See endsheet maps: L3.

LOCATION: 36°31′ to 39°27′ N; 75°13′ to 83°37′ w. **BOUNDARIES:** West Virginia line, 438 mi (705 km); Maryland line, 233 mi (375 km); District of Columbia line, 12 mi (19 km); Atlantic Ocean coastline, 112 mi (180 km); North Carolina line, 320 mi (515 km); Tennessee line, 114 mi (184 km); Kentucky line, 127 mi (204 km).

days, was the established church during the colonial period. The first dissenters to arrive were Scotch-Irish Presbyterians in the late 17th century; they were followed by large numbers of German Lutherans, Welsh Baptists, and English Quakers, who settled in the Valley of Virginia in the early 18th century. The general assembly's adoption in 1785 of the Virginia Statute for Religious Freedom, drafted by Thomas Jefferson, disestablished the Episcopal Church and made religious toleration the norm in Virginia. Although the Episcopal and Presbyterian churches retained the allegiance of the landed gentry during the 19th century, the Methodists and Baptists became the largest church groups in the state.

Protestant denominations combined had the greatest number of known adherents in 1980, when the leading groups were the Southern Baptist Convention, with 690,276 adherents; United Methodist Church, 501,217; Episcopal Church, 134,258; and Presbyterian Church in the US, 125,715. As of 1984 there were 316,772 Roman Catholics in Virginia, and the Jewish population was estimated at 60,820.

¹⁰TRANSPORTATION

Virginia has one of the nation's most extensive highway systems, one of the leading ports—Hampton Roads—and two of the nation's busiest air terminals.

Virginia was a leader in early railroad development. Rail lines were completed between Richmond and Fredericksburg in 1836, from Portsmouth to Roanoke in 1837, between Petersburg and Lynchburg in 1854, and from Richmond to Washington, D.C., in 1872. Virginia's 1,290 mi (2,076 km) of track, two-thirds of it

Virginia Counties, County Seats, and County Areas and Populations

COUNTY	COUNTY SEAT	LAND AREA (SQ MI)	POPULATION (1984 EST.)	COUNTY	COUNTY SEAT	LAND AREA (SQ MI)	POPULATION (1984 EST.)
Accomack	Accomac	476	30,970	King George	King George	180	11,077
Albemarle	Charlottesville	725	58,857	King William	King William	278	9,782
Alleghany	Covington	446	13,709	Lancaster	Lancaster	133	10,640
Amelia	Amelia	357	8,282	Lee	Jonesville	437	26,810
Amherst	Amherst	478	29,203	Loudoun	Leesburg	521	62,689
Appomattox	Appomattox	336	12,306	Louisa	Louisa	497	18,707
Arlington	Arlington	26	154,200	Lunenburg	Lunenburg	432	12,155
Augusta	Staunton	989	54,663	Madison	Madison	322	10,539
Bath	Warm Springs	537	5,418	Mathews	Mathews	87	8,500
Bedford	Bedford	747	36,906	Mecklenburg	Boydton	616	29,745
Bland	Bland	359	6,542	Middlesex	Saluda	134	8,180
Botetourt	Fincastle	545	23,858	Montgomery	Christiansburg	390	64,970
Brunswick	Lawrenceville	563	16,231	Nelson	Lovingston	475	12,315
Buchanan	Grundy	504	37,912	New Kent	New Kent	213	9,849
Buckingham	Buckingham	583	11,934	Northampton	Eastville	226	14,207
Campbell	Rustburg	505	46,053	Northumberland	Heathsville	185	9,903
Caroline	Bowling Green	536	18,690	Nottoway	Nottoway	317	14,288
Carroll	Hillsville	478	28,279	Orange	Orange	342	18,870
Charles City	Charles City	181	6,742	Page	Luray	313	19,563
Charlotte	Charlotte	476	11,937	Patrick	Stuart	481	17,730
Chesterfield	Chesterfield	434	159,897	Pittsylvania	Chatham	995	66,530
Clarke	Berryville	178	10,168	Powhatan	Powhatan	261	13,619
Craig	New Castle	330	4,015	Prince Edward	Farmville	354	16,842
Culpeper	Culpeper	382	23,221	Prince George	Prince George	266	25,958
Cumberland	Cumberland	300	7,887	Prince William	Manassas	339	164,275
Dickenson	Clintwood	331	20,149	Pulaski	Pulaski	318	35,039
Dinwiddie	Dinwiddie	507	21,401	Rappahannock	Washington	267	6,118
Essex	Tappahannock	263	8,778	Richmond	Warsaw	193	6,846
Fairfax	Fairfax	393	672,937	Roanoke	Salem	251	73,878
Fauquier	Warrenton	651	39,275	Rockbridge	Lexington	603	18,055
Floyd	Floyd	381	11,772	Rockingham	Harrisonburg	865	52,980
Fluvanna	Palmyra	290	10,470	Russell	Lebanon	479	32,873
Franklin	Rocky Mount	683	36,843	Scott	Gate City	536	25,564
Frederick	Winchester	415	35,344	Shenandoah	Woodstock	512	28,244
Giles	Pearisburg	362	17,810	Smyth	Marion	452	33,432
Gloucester	Gloucester	225	24,700	Southampton	Courtland	603	19,025
Goochland	Goochland	281	12,498	Spotsylvania	Spotsylvania	404	35,774
Grayson	Independence	446	16,791	Stafford	Stafford	271	47,141
Greene	Standardsville	157	8,345	Surry	Surry	281	6,244
Greensville	Emporia	300	10,249	Sussex	Sussex	492	10,362
Halifax	Halifax	816	29,863	Tazewell	Tazewell	520	51,544
Hanover	Hanover	468	52,249	Warren	Front Royal	217	22,137
Henrico	Richmond	238	192,346	Washington	Abingdon	563	47,424
Henry	Martinsville	382	56,597	Westmoreland	Montross	227	14,234
Highland	Monterey	416	2,827	Wise	Wise	405	45,384
Isle of Wight	Isle of Wight	319	23,171	Wythe	Wytheville	464	25,612
James City	Williamsburg	153	24,766	York	Yorktown	113	39,237
King and Queen	King and Queen	317	5,914	Independent Cities	—	1,605	2,127,731
				TOTALS		39,704	5,635,544

owned by the Baltimore and Ohio line, formed an important strategic supply link for Confederate and Union armies during the Civil War. Railroads remained the primary system of transportation until the rise of the automobile in the 1920s. As of 1983 there were 20 rail companies operating in the state, six of them Class I railways with combined trackage of 3,596 mi (5,787 km). Principal north–south railroads are the Richmond, Fredericksburg, and Potomac, the Seaboard System, and the Southern Railway System; major east–west lines include the Chessie System and the Norfolk and Western. Amtrak passenger trains served 16 communities, with total Virginia ridership of 350,618 in 1983/84.

Virginia's road network, at first built mainly for hauling tobacco to market, had expanded across the Blue Ridge by 1782, to the Cumberland Gap by 1795, and into the Shenandoah Valley by means of the Valley Turnpike in 1840. As of 1983, Virginia had 66,127 mi (106,421 km) of roads, 3,894,060 registered vehicles, and 3,704,171 licensed drivers. Major interstate highways are I-95, extending north–south from Washington, D.C., via Richmond to the North Carolina border and, eventually, to Florida; I-81, connecting northern Virginia with the southwest; and I-64, linking the Hampton Roads area with Staunton (and I-81) in the west. The 18-mi (29-km) Chesapeake Bay Bridge-Tunnel, completed in 1964, connects the Eastern Shore with the southeastern mainland. Popular scenic highways include the Blue Ridge Parkway, Colonial National Historical Parkway, and George Washington Memorial Parkway.

Virginia's District of Columbia suburbs are linked to the nation's capital by the Washington Metropolitan Area Transit Authority's bus and rail systems. Norfolk, Newport News–Hampton, and Richmond have extensive bus systems.

Coastal and ocean shipping are vital to Virginia's commerce. The Port of Hampton Roads—consisting of marine terminals in Chesapeake, Newport News, Norfolk, and Portsmouth—was linked as of 1982 by 68 steamship lines with 250 ports in 97 foreign countries.

Virginia's 270 aircraft facilities included 208 airports, 56 heliports, 3 stolports, and 3 seaplane bases in 1983. That year, 1,325,933 passengers enplaned at Dulles International Airport; 6,559,868 enplaned at Washington National Airport, at Arlington, a major center for domestic flights.

¹¹HISTORY

Distinctively fluted stone points found at Flint Run in Front Royal and at the Williamson Site in Dinwiddie County testify to the presence in what is now the Commonwealth of Virginia of nomadic Paleo-Indians after 8000 BC. Climatic changes and the arrival of other Indian groups about 3500 BC produced the Archaic Culture, which lasted until about AD 500. These Indians apparently were great eaters of oysters, and shell accumulations along riverbanks mark their settlement sites. The Woodland Period (AD 500-1600) marked the Indians' development of the bow and arrow and sophisticated pottery. At the time of English contact, early in the 17th century, Tidewater Virginia was occupied principally by Algonkian-speakers, planters as well as hunters and fishers, who lived in pole-framed dwellings forming small, palisaded towns. The piedmont area was the home of the Manahoac, Monacan, and Tutelo, all of Siouan stock. Cherokee lived in Virginia's far southwestern triangle.

The first permanent English settlement in America was established at Jamestown on 13 May 1607 in the new land named Virginia in honor of Elizabeth I, the "Virgin Queen." The successful settlement was sponsored by the London Company (also known as the Virginia Company), a joint-stock venture chartered by King James I in 1606. The charter defined Virginia as all of the North American coast between 30° and 45°N and extending inland for 50 mi (80 km). A new royal charter in 1609 placed Virginia's northern and southern boundaries at points 200 mi (320 km) north and south of Point Comfort, at the mouth of

the James River, and extended its territory westward to the Pacific; a third charter, issued in 1612, pushed Virginia eastward to embrace the Bermuda Islands. Thus, Virginia at one time stretched from southern Maine to California and encompassed all or part of 42 of the present 50 states, as well as Bermuda and part of the Canadian province of Ontario.

Upon landing at Jamestown, the 100 or more male colonists—there were no women—elected from among 12 royally approved councillors a governor and captain general, Edward Maria Wingfield. Much internal strife, conflict with the Indians, and a "starving time" that reduced the settlers to eating their horses caused them to vote to leave the colony in 1610, but just as they were leaving, three supply ships arrived; with them came Thomas West, Baron De La Warr (Lord Delaware), who stayed to govern the Virginia Colony until 1611. Finally, however, it was the energy, resourcefulness, and military skill of Captain John Smith that saved the colony from both starvation and destruction by the Indians. He also charted the coast and wrote the first American book, *A True Relation,* which effectively publicized English colonization of the New World.

Smith's chief Algonkian adversary was Powhatan, emperor of a confederacy in eastern Virginia that bore his name. Although Smith was taken prisoner by Powhatan, he was able to work out a tenuous peace later cemented by the marriage in 1614 of the emperor's favorite daughter, Pocahontas, to John Rolfe, a Jamestown settler who founded the colonial tobacco industry.

Three events marked 1619 as a red-letter year in Virginia history. First, women were sent to the colony in large numbers. Any man marrying one of a shipment of 90 "young maids" had to pay 120 lb of tobacco for the cost of her transportation. The women were carefully screened for respectability, and none had to marry if she did not find a man to her liking. The second key event was the arrival in Jamestown of the first blacks, probably as indentured servants, a condition from which slavery in the colony evolved; the first legally recognized slaveholder, in the 1630s, was Anthony Johnson, himself a black. The third and most celebrated event of 1619 was the convening in Jamestown of the first representative assembly in the New World, consisting of a council chosen by the London Company and a house of burgesses elected by the colonists. Thus, self-government through locally elected representatives became a reality in America and an important precedent for the English colonies.

King James I, for whom the colonial capital was named, was at first content with colonization under the London Company's direction. But in 1624, he charged the company with mismanagement and revoked its charter. Virginia remained a royal colony until 1776, although royal governors such as Sir Francis Wyatt and Sir George Yeardley continued to convoke the general assembly without the Crown's assent. A serious challenge to self-government came in 1629-35 with Governor John Harvey's "executive offenses"—including the knocking out of a councillor's teeth and the detaining of a petition of protest to the king—which sparked a rebellion led by Dr. John Pott. Harvey was bloodlessly deposed by the council—which, significantly, turned to the house of burgesses for confirmation of the action the council had taken.

Despite serious setbacks because of Indian massacres in 1622 and 1644, the colony's population expanded rapidly along the James, York, Rappahannock, and Potomac rivers, and along the Eastern Shore. In 1653, the general assembly attempted to collect taxes from the Eastern Shore, although that area had no legislative representation. At a mass meeting, Colonel Thomas Johnson urged resistance to taxation without representation. The resulting Northampton Declaration embodied this principle, which would provide the rallying cry for the American Revolution; the immediate result was the granting of representation to the Eastern Shore.

Virginia earned the designation Old Dominion through its loyalty to the Stuarts during England's Civil War, but the superior military and naval forces of Oliver Cromwell compelled submission to parliamentary commissioners in 1652. In the eight years that followed, the house of burgesses played an increasingly prominent role. Colonial governors, while at least nominally Puritan, usually conducted affairs with an easy tolerance that did not mar Virginia's general hospitality to refugee Cavaliers from the mother country.

With the restoration of the royal family in 1660, Sir William Berkeley, an ardent royalist who had served as governor before the colony's surrender to the Commonwealth, was returned to that office. In his first administration, his benign policies and appealing personality had earned him great popularity, but during his second term, his dictatorial and vindictive support of royal prerogatives made him the most hated man in the colony. When he seemed unable to defend the people against Indian incursions in 1676, they sought a general of their own. They found him in young Nathaniel Bacon, a charismatic planter of great daring and eloquence, whose leadership attracted many small planters impatient by this time with the privileged oligarchy directing the colony. Bacon's war against the Indians became a populist-style revolt against the governor, who fled to the Eastern Shore, and reform legislation was pushed by the burgesses. Berkeley regained control of the capital briefly, only to be defeated by Bacon's forces; but Jamestown was burned by the retreating Bacon, who died of fever shortly afterward. Berkeley's subsequent return to power was marked by so many hangings of offenders that the governor was summoned to the court of Charles II to answer for his actions. Bacon's Rebellion was cited as a precedent when the colonies waged war against George III a century later.

The 17th century closed on a note of material and cultural progress with the gubernatorial administration of Francis Nicholson. The College of William and Mary, the second institution of higher learning in America, was chartered in 1693, and Middle Plantation (renamed Williamsburg in 1722), the site of the college, became the seat of government when the capital was moved from Jamestown in 1699. The new capital remained small, although it was crowded when the legislature was in session. A new era of cultural and economic progress dawned with the administration of Alexander Spotswood (1710–22), sometimes considered the greatest of Virginia's colonial governors. He discouraged the colony's excessively heavy dependence on a single crop, tobacco; promoted industry, especially ironworks; took a humane interest in blacks and Indians; strengthened fortifications; ended the depredations of the notorious pirate Edward Teach, better known as Blackbeard; and, by leading his "Knights of the Golden Horseshoe" across the Blue Ridge, dramatized the opening of the transmontane region.

In the decades that followed, eastern Virginians moving into the Valley of Virginia were joined by Scotch-Irish and Germans moving southward from Maryland and Pennsylvania. Virginians caught up in western settlement lost much of their awe of the mother country during the French and Indian War (1756–63). A young Virginia militiaman, Colonel George Washington, gave wise but unheeded advice to Britain's Major General Edward Braddock before the Battle of Monongahela, and afterward emerged as the hero of that action.

Virginia, acting independently and with other colonies, repeatedly challenged agents of the Crown. In 1765, the house of burgesses, swept by the eloquence of Patrick Henry, adopted five resolutions opposing the Stamp Act, through which the English Parliament had sought to tax the colonists for their own defense. In 1768, Virginia joined Massachusetts in issuing an appeal to all the colonies for concerted action. The following year, Virginia initiated a boycott of British goods in answer to the taxation provisions of the hated Townshend Acts. In 1773, the Old Dominion became the first colony to establish an intercolonial committee of correspondence. And it joined the other colonies at the First Continental Congress, which met in Philadelphia in 1774 and elected Virginia's Peyton Randolph president.

Virginia was the first colony to instruct its delegates to move for independence at the Continental Congress of 1776. The congressional resolution was introduced by one native son, Richard Henry Lee, and the Declaration of Independence was written by another, Thomas Jefferson. In the same year, Virginians proclaimed their government a commonwealth and adopted a constitution and declaration of rights, prepared by George Mason. The declaration became the basis for the Bill of Rights in the US Constitution. Virginians were equally active in the Revolutionary War. George Washington was commander in chief of the Continental Army, and other outstanding Virginia officers were George Rogers Clark, Hugh Mercer, Henry "Light Horse Harry" Lee, William Campbell, Isaac Shelby, and an adopted son, Daniel Morgan. In addition, the greatest American naval hero was a Scottish-born Virginian, John Paul Jones. Virginia itself was a major battlefield, and it was on Virginia soil, at Yorktown on 19 October 1781, that British General Charles Cornwallis surrendered to Washington, effectively ending the war.

During the early federal period, Virginia's leadership was as notable as it had been during the American Revolution. James Madison is honored as the "father of the Constitution," and Washington, who was president of the constitutional convention, became the first US president in 1789. Indeed, Virginians occupied the presidency for all but 4 of the nation's first 28 years. Far more influential than most presidents was another Virginian, John Marshall, who served as US chief justice for 34 years, beginning in 1801.

During the first half of the 19th century, Virginians became increasingly concerned with the problem of slavery. From the early 1700s, the general assembly had repeatedly prohibited the importation of slaves, only to be overruled by the Crown, protecting the interests of British slave traders. In 1778, no longer subject to royal veto, the legislature provided that any slave brought into the state would automatically be freed upon arrival. (There was no immediate legal termination of the bondage of those already enslaved, or of their offspring.) The number of free blacks grew tenfold by 1810, and though some became self-supporting farmers and artisans, many could find no employment. Fearing that unhappy free blacks might incite those who were still slaves to rebellion, the general assembly in 1806 decreed that each slave emancipated in due course must then leave Virginia within a year or after reaching the age of 21. Nat Turner's slave revolt—which took the lives of at least 55 white men, women, and children in Southampton County in 1831—increased white fears of black emancipation. Nevertheless, legislation to end slavery in Virginia failed of adoption by only seven votes the following year.

The slavery controversy did not consume all Virginians' energies in the first half of the 19th century, an era that saw the state become a leading center of scientific, artistic, and educational advancement. But this era ended with the coming of the Civil War, a conflict about which many Virginians had grave misgivings. Governor John Letcher was a Union man, and most of the state's top political leaders hoped to retain the federal tie. Even after the formation at Montgomery, Ala., of the Confederate States of America, Virginia initiated a national peace convention in Washington, D.C., headed by a native son and former US president, John Tyler. A statewide convention, assembled in Richmond in April 1861, adopted an ordinance of secession only after President Abraham Lincoln sought to send troops across Virginia to punish the states that had already seceded and called upon the commonwealth to furnish soldiers for that task. Virginia adopted secession with some regret and apprehension but with no agonizing over constitutional principles, for in ratifying the Constitution the state had reserved the right to secede. Shortly afterward, Richmond, the capital of Virginia since 1780, became the capital of the Confederacy. It was also the home of the Tredegar Ironworks, the South's most important manufacturer of heavy weaponry.

Robert E. Lee, offered field command of the Union armies, instead resigned his US commission in order to serve his native state as commander of the Army of Northern Virginia and eventually as chief of the Confederate armies. Other outstanding Virginian generals included Thomas Jonathan "Stonewall" Jackson, J. E. B. "Jeb" Stuart, Joseph E. Johnston, and A. P. Hill. Besides furnishing a greater number of outstanding Confederate generals than any other state, the Old Dominion supplied some of the Union's military leaders, George H. Thomas, the "Rock of Chickamauga," among them. More than 30 Virginians held the rank of brigadier general or major general in the federal forces.

Virginia became the principal battlefield of the Civil War, the scene of brilliant victories won by General Lee's army at Bull Run (about 30 mi—48 km—southwest of Washington, D.C.), Fredericksburg, and Chancellorsville (Spotsylvania County). But the overwhelming numbers and industrial and naval might of the Union compelled Lee's surrender at Appomattox on 9 April 1865. Virginia waters were the scene of one of the most celebrated naval engagements in world history, the first battle of the ironclads, when the USS *Monitor* and CSS *Virginia* (*Merrimac*), rebuilt in the Portsmouth Shipyard, met at Hampton Roads. The war cost Virginia one-third of its territory when West Virginia was admitted to the Union as a separate state on 20 June 1863. Richmond was left in ruins, and agriculture and industry throughout the commonwealth were destroyed. Union General Philip H. Sheridan's systematic campaign of demolition in the Shenandoah Valley almost made good his boast that a crow flying over the valley would have to carry its own rations.

In 1867, Virginia was placed under US military rule. A constitutional convention held in Richmond under the leadership of carpetbaggers and scalawags drafted a constitution that disqualified the overwhelming majority of white Virginians from holding office and deprived about 95% of them of the right to vote. In this crisis, a compromise was negotiated under which white Virginians would accept Negro suffrage if they themselves were permitted to vote and hold office. The amended constitution, providing for universal manhood suffrage, was adopted in 1869, and Virginia was readmitted to the Union on 26 January 1870.

Although the bankrupt state was saddled with a debt of more than $45 million, the Conservative Democrats undertook repayment of the entire debt, including approximately one-third estimated to be West Virginia's share. Other Democrats, who came to be known as Readjusters, argued that the commonwealth could not provide education and other essential services to its citizens unless it disclaimed one-third of the debt and reached a compromise with creditors concerning the remainder. William Mahone, a railroad president and former Confederate major general, engineered victory for the Readjusters in 1880 with the aid of the Republicans. His election to the US Senate that year represented another success for the Readjuster-Republican coalition, which was attentive to the needs of both blacks and underprivileged whites.

Throughout the 1880s and 1890s, life in public places in Virginia continued in an unsegregated fashion that sometimes amazed visitors from northern cities. As the 19th century neared an end, however, Virginia moved toward legal separation of the races. In 1900, the general assembly by a one-vote majority enacted segregation on railroad cars. The rule became applicable the following year to streetcars and steamboats. In 1902, the Virginia constitutional convention enacted a literacy test and poll tax that effectively reduced the black vote to negligible size.

Two decades later, just when the Old Dominion seemed permanently set in the grooves of conservatism, two liberals, each with impeccable old-line backgrounds, found themselves battling for the governorship in a Democratic primary campaign that changed the course of Virginia's political history. Harry F. Byrd defeated G. Walter Mapp in the election of 1925 and immediately after taking office launched the state on an era of reform. In a whirlwind 60 days, the general assembly revised the tax system, revised balloting procedures, and adopted measures to lure industry to Virginia. The Anti-Lynch Act of 1927 made anyone present at the scene of a lynching who did not intervene guilty of murder; there has not been a lynching in Virginia since its passage. Byrd also reorganized the state government, consolidating nearly 100 agencies into 14 departments. Later, as US senator, Byrd became so renowned as a conservative that many people forgot his earlier career as a fighting liberal.

Following the depression of the 1930s, Virginia became one of the most prosperous states of the Southeast. It profited partly from national defense contracts and military and naval expansion, but also from increased manufacturing and from what became one of the nation's leading tourist industries. Few states made so great a contribution as Virginia to the US effort in World War II. More than 300,000 Virginians served in the armed forces; 9,000 lost their lives and 10 were awarded the Medal of Honor. Virginians were proud of the fact that General George C. Marshall was a Virginia resident and a graduate of Virginia Military Institute, and even delighted in the knowledge that both General Dwight D. Eisenhower, commander in the European theater, and General Douglas MacArthur, commander in the Pacific, were sons of Virginia mothers.

The postwar period brought many changes in the commonwealth's public life. During the first administration of Governor Mills E. Godwin, Jr. (1966–70), the state abandoned its strict pay-as-you-go fiscal policy, secured an $81-million bond issue, and enacted a sales tax. Much of the increased revenue benefited the public school system; funding for the four-year colleges was greatly expanded, and a system of low-tuition community colleges was instituted.

In 1970, A. Linwood Holton, Jr., became the first Republican governor of Virginia since 1874. Pledging to "make today's Virginia a model in race relations," Holton increased black representation on state boards and in the higher echelons of government. (For many Americans, the new era in the South was symbolized by a nationally distributed news photo of Governor Holton escorting his 13-year-old daughter to a predominantly black public school.) He reversed the policies of his immediate predecessors, who had generally met the US Supreme Court's desegregation ruling in 1954 with a program of "massive resistance," eschewing violence but adopting every legal expedient to frustrate integration. By the mid-1970s, public school integration in Virginia had been achieved to a degree not yet accomplished in many northern states.

Meanwhile, the expansion of federal jobs and services spurred Virginia's economic development. During the early 1980s, a national military buildup brought renewed prosperity to Virginia's shipyards.

[12] STATE GOVERNMENT

Since 1776, Virginia has had six constitutions, all of which have expanded the power of the executive branch. The last constitution, framed in 1970 and effective 1 July 1971, governs the state today. As of the end of 1983, this document had been amended 14 times.

The general assembly consists of a 40-member senate, elected to four-year terms, and a 100-member house of delegates, serving for two years. Senators and delegates must be at least 21 years of age and residents of their district. The assembly convenes annually on the 2d Wednesday in January for 60-day sessions in even-numbered years and 30-day sessions in odd-numbered years, with an option to extend the annual session for a maximum of 30 days or declare a special session by two-thirds vote of both houses.

The governor, lieutenant governor, and attorney general, all serving four-year terms, are the only officials elected statewide; the governor, who must be at least 30 years of age and a state resident for five years, may not serve two successive terms. Most state officials—including the secretaries of administration and finance, commerce and resources, education, human resources, public safety, and transportation—are appointed by the governor but must be confirmed by both houses of the legislature. Bills become law when signed by the governor or left unsigned for 7 days while the legislature is in session; a bill dies if left unsigned for 30 days after the legislature has adjourned. A two-thirds majority in each house is needed to override a gubernatorial veto. The constitution may be amended by constitutional convention or by a two-thirds vote of two successive sessions of the general assembly; ratification by the electorate is required.

A qualified voter must be a US citizen, be at least 18 years old, and have registered in the precinct of residence at least 31 days before the election. Elections for state offices are held in odd-numbered years.

[13]POLITICAL PARTIES

Virginia has exercised a unique role in US politics as the birthplace not only of representative government but also of one of America's two major parties. The modern Democratic Party traces its origins to the original Republican Party (usually referred to as the Democratic-Republican Party, or the Jeffersonian Democrats), led by two native sons of Virginia, Thomas Jefferson and James Madison. Virginians have also been remarkably influential in the political life of other states: a survey published in 1949 showed that 319 Virginia natives had represented 31 other states in the US Senate and House of Representatives.

From the end of Reconstruction through the 1960s, conservative Democrats dominated state politics, with few exceptions. Harry F. Byrd was the state's Democratic political leader for 40

years, first as a reform governor (1926–30) and then as a conservative senator (1933–65). During the 1970s, Virginians, still staunchly conservative, turned increasingly to the Republican Party, whose presidential nominees carried the state in every election from 1952 through 1984 except for 1964. Linwood Holton, the first Republican governor since Reconstruction, was elected in 1969. His Republican successor, Mills E. Godwin, Jr., the first governor since the Civil War to serve more than one term, had earlier won election as a Democrat. The election in 1977 of another Republican, John N. Dalton, finally proved that Virginia had become a two-party state. In 1981, however, the governorship was won by Democrat Charles S. Robb, who appointed a record number of blacks and women to state offices. Robb, prohibited by law from seeking a consecutive second term, was succeeded by Democrat Gerald L. Baliles in 1985, when Virginians also elected L. Douglas Wilder as lieutenant governor and Mary Sue Terry as attorney general; Wilder became the highest-ranking black state official in the US, and Terry was the first woman to win a statewide office in Virginia. As of 1985, 11 women served in the general assembly, and 107 blacks held elective office. Republicans held both US Senate seats and 6 of the 10 seats in the US House of Representatives in 1985.

[14]LOCAL GOVERNMENT

As of 1984, Virginia had 95 counties, 41 independent cities, and 189 incorporated towns. In all, there were 407 local government units.

During the colonial period, most Virginians lived on plantations and were reluctant to form towns. In 1705, the general assembly approved the formation of 16 "free boroughs"; although only Jamestown, Williamsburg, and Norfolk chose at that time to avail themselves of the option and become independent municipalities, their decision laid the foundation for the independence of Virginia's present-day cities from county government. In 1842,

Virginia Presidential Vote by Political Parties, 1948–84

YEAR	ELECTORAL VOTE	VIRGINIA WINNER	DEMOCRAT	REPUBLICAN	STATES' RIGHTS DEMOCRAT	PROGRESSIVE	SOCIALIST	SOCIALIST LABOR
1948	11	*Truman (D)	200,786	172,070	43,393	2,047	726	234
1952	12	*Eisenhower (R)	268,677	349,037	—	—	504	1,160
					CONSTITUTION			
1956	12	*Eisenhower (R)	267,760	386,459	42,964	—	444	351
					VA. CONSERVATIVE			
1960	12	Nixon (R)	362,327	404,521	4,204	—	—	397
1964	12	*Johnson (D)	558,038	481,334	—	—	—	2,895
					AMERICAN IND.	PEACE & FREEDOM		
1968	12	*Nixon (R)	442,387	590,319	320,272	—	1,680	4,671
					AMERICAN			
1972	12	*Nixon (R)	438,887	988,493	19,721	—	—	9,918
						LIBERTARIAN	US LABOR	SOC. WORKERS
1976	12	Ford (R)	813,896	836,554	16,686	4,648	7,508	17,802
							CITIZENS	
1980	12	*Reagan (R)	752,174	989,609	—	12,821	14,024[1]	1,986[1]
1984	12	*Reagan (R)	796,250	1,337,078	—	—	—	—

*Won US presidential election.

[1]Candidates of the nationwide Citizens and Socialist Workers parties were listed as independents on the Virginia ballot; another independent, John Anderson, won 95,418 votes.

Richmond became the commonwealth's first charter city. Today, 41 cities elect their own officials, levy their own taxes, and are unencumbered by any county obligations. The 189 incorporated towns remain part of the counties.

In general, counties are governed by elected boards of supervisors, with a county administrator or executive handling day-to-day affairs; other typical county officials are the clerk of the circuit court (chief administrator of the court), county treasurer, commissioner of the revenue, commonwealth's attorney, and sheriff. Incorporated towns have elected mayors and councils.

[15]STATE SERVICES

Under the jurisdiction of the secretary of education are the Department of Education, which administers the public school system, and the State Council of Higher Education, which coordinates the programs of the state-controlled colleges and universities. The secretary of transportation oversees the Department of Highways and Transportation, Department of Transportation Safety, Department of Aviation, Virginia Port Authority, Department of Military Affairs (National Guard), Division of Motor Vehicles, and State Office of Emergency Services.

Within the purview of the secretary of human resources are the Department of Health, Department of Mental Health and Mental Retardation, Department of Health Regulatory Boards, Department of Social Services, and Department of Rehabilitative Services, as well as special offices dealing with problems that affect women, children, the elderly, and the disabled. The departments of State Police, Corrections, Criminal Justice Services, Fire Programs, and Alcoholic Beverage Control are under the aegis of the secretary of public safety.

The secretary of commerce and resources oversees the departments of Housing and Community Development, Labor and Industry, Commerce, Agriculture and Consumer Services, and Conservation and Economic Development, as well as a profusion of boards, councils, offices, divisions, and commissions. The secretary of administration and finance exercises jurisdiction over budgeting, telecommunications, accounting, computer services, taxation, the state treasury, records, and personnel, as well as over the State Board of Elections.

Regulatory functions are concentrated in the quasi-independent State Corporation Commission, consisting of three commissioners elected by the legislature to staggered six-year terms. The commission regulates all public utilities; licenses banks, savings and loan associations, credit unions, and small loan companies; enforces motor carrier and certain aviation laws and sets railroad rates; supervises the activities of insurance companies; and enforces laws governing securities and retail franchising.

[16]JUDICIAL SYSTEM

The highest judicial body in the commonwealth is the supreme court, consisting of a chief justice and six other justices elected to 12-year terms. The court of appeals has nine judges serving 8-year terms. Judges in the state's 31 circuit courts (as of 1982), the main trial courts, are also elected to 8-year terms. Each circuit coincides with a judicial district in which sits at least one general district court judge, also serving an 8-year term, and at least four magistrates. District courts hear all misdemeanors, including civil cases involving $1,000 or less, and have concurrent jurisdiction with the circuit courts in claims involving $1,000 to $7,000. District courts also hold preliminary hearings concerning felony cases. Each of the 31 judicial districts has a juvenile and a domestic relations court, with judges elected by the general assembly to 6-year terms. Each city or county has at least one local magistrate.

Virginia's state and federal prisons had 10,667 inmates at the end of 1984. According to the FBI Crime Index, the state's crime rates per 100,000 population, below the national average in every category, were as follows in 1983: murder and nonnegligent manslaughter, 7; forcible rape, 24.7; robbery, 111; aggravated assault, 150; burglary, 920; larceny-theft, 2,571; and motor vehicle theft, 178. A capital punishment statute providing for death by electrocution is in effect; two prisoners were executed in the early 1980s, the first executions since 1962.

[17]ARMED FORCES

The Hampton Roads area, one of the nation's major concentrations of military facilities, includes Langley Air Force Base in Hampton, the Norfolk naval air station and shipyard, the naval air station at Virginia Beach, the Marine Corps air facility and command and staff college at Quantico, and Fts. Eustis, Belvoir, and Lee. Norfolk is the home base of the Atlantic Fleet, and several major army and air commands are in Virginia. Virginia's major defense establishments also include an army base at Arlington. In 1983/84, Virginia ranked 4th among the 50 states in value of federal defense contracts, receiving awards worth $7.1 billion. Department of Defense personnel numbered 199,589, 103,764 civilian and 95,825 military.

As of 30 September 1983, 661,000 veterans of US military service lived in Virginia. Of these, 5,000 saw service during World War I, 245,000 in World War II, 146,000 during the Korean conflict, and 228,000 during the Viet-Nam era. Veterans' benefits allocated to Virginia totaled $692 million in 1982/83.

Army and Air National Guard units comprised some 8,905 personnel by early 1985. There were 12,569 state and local police in 1983.

[18]MIGRATION

Virginia's earliest European immigrants were English—only a few hundred at first, but 4,000 between 1619 and 1624, of whom fewer than 1,200 survived epidemics and Indian attacks. Despite such setbacks, Virginia's population increased, mostly by means of immigration, from about 5,000 in 1634 to more than 15,000 in 1642, including 300 blacks. Within 30 years, the population had risen to more than 40,000, including 2,000 blacks. In the late 17th and early 18th centuries, immigrants came not only from England but also from Scotland, Wales, Ireland, Germany, France, the Netherlands, and Poland. In 1701, about 500 French Huguenots fled Catholic France to settle near the present site of Richmond, and beginning in 1714, many Germans and Scotch-Irish moved from Pennsylvania into the Valley of Virginia.

By the early 19th century, Virginians were moving westward into Kentucky, Ohio, and other states; the 1850 census showed that 388,000 former Virginians (not including the many thousands of slaves sold to other states) were living elsewhere. Some of those who left—Henry Clay, Sam Houston, Stephen Austin—were among the most able men of their time. The Civil War era saw the movement of thousands of blacks to northern states, a trend that accelerated after Reconstruction and again after World War I. Since 1900, the dominant migratory trend has been intrastate, from farm to city. Urbanization has been most noticeable since World War II in the Richmond and Hampton Roads areas. At the same time, the movement of middle-income Virginians to the suburbs and increasing concentrations of blacks in the central cities have been evident in Virginia as in other states.

Between 1940 and 1970, Virginia enjoyed a net gain from migration of 325,000. In the 1970s, the net gain was 239,000, and during 1980–83, 81,000.

[19]INTERGOVERNMENTAL COOPERATION

Regional bodies in which Virginia participates include the Atlantic States Marine Fisheries Commission, Ohio River Valley Water Sanitation Commission, Southern Growth Policies Board, Southern Interstate Energy Board, and Washington Metropolitan Area Transit Authority.

In 1983/84, Virginia received federal aid totaling $1.6 billion.

[20]ECONOMY

Early settlements in Virginia depended on subsistence farming of native crops, such as corn and potatoes. Tobacco, the leading export crop during the colonial era, was joined by cotton during the early statehood period. Although cotton was never "king" in Virginia, as it was in many southern states, the sale of slaves to Deep South plantations was an important source of income for Virginians, especially during the 1830s, when some 118,000 slaves were exported for profit. Eventually, a diversified agriculture developed in the piedmont and the Shenandoah Valley. Manufacturing became significant during the 19th century, with a proliferation of cotton mills, tobacco-processing plants, ironworks, paper mills, and shipyards.

Manufacturing, services, and trade are important economic sectors today, but government outranks them as sources of personal income and employment. Because of Virginia's extensive military installations and the large numbers of Virginia residents working for the federal government in the Washington, D.C., metropolitan area, the federal government plays a larger role in the Virginia economy than in any other state except Hawaii. Of the state's 1982 total personal income, federal sources, civilian as well as military, accounted for 12.7%; state and local government contributed another 7.6%, for a combined total of 20.3%. Manufacturing followed with 13%, services 12%, and trade 10%. The federal government in 1983 employed 95,825 military personnel and 161,000 civilians in the state, while an estimated 60,000 federal employees working in the non-Virginia part of the Washington, D.C., metropolitan area were Virginia residents. Coal and timber are the leading resource industries, and services, including tourism, constitute another growth sector.

²¹INCOME
Virginia's per capita personal income in 1983 averaged $11,835, 16th among the 50 states and highest in the South. Total personal income that year was $65.7 billion. Total personal income rose 38% in real terms between 1970—when Virginia placed 26th in income per capita—and 1983, compared to 28% for the US.

Virginia ranked 8th in median money income of four-person families, with $28,850 in 1981. In 1979, 9.2% of Virginia families and 11.8% of all state residents were below the federal poverty level. Some 26,600 of the leading US wealth-holders—those with gross assets greater than $500,000—lived in Virginia in 1982.

²²LABOR
In 1984, Virginia's civilian labor force averaged 2,841,000 persons, of whom 55% were male and 45% female.

A federal survey in March 1982 revealed the following nonfarm employment pattern in Virginia:

	ESTABLISH-MENTS	EMPLOYEES	ANNUAL PAYROLL ('000)
Agricultural services, forestry, fishing	993	6,623	$ 73,898
Mining, of which:	767	26,104	526,563
Bituminous coal, lignite	(549)	(21,513)	(436,401)
Contract construction	10,096	96,211	1,539,264
Manufacturing, of which:	5,176	398,163	6,791,789
Electric and electronic equipment	(180)	(33,854)	(684,779)
Textile mill products	(99)	(40,274)	(499,954)
Chemicals, chemical products	(154)	(31,369)	(746,567)
Transportation equipment	(123)	(36,597)	(755,672)
Transportation, public utilities	3,767	100,167	2,085,104
Wholesale trade	6,869	98,639	1,781,589
Retail trade	28,201	349,433	3,167,541
Finance, insurance, real estate	9,374	103,893	1,601,699
Services, of which:	32,153	408,502	5,576,167
Health services	(6,998)	(114,227)	(1,748,534)
Business services	(4,643)	(86,947)	(1,403,465)
Other	1,774	1,733	34,079
TOTALS	99,170	1,589,468	$23,177,693

Government workers, not covered by this survey, averaged 163,600 federal civilian employees and 343,000 state and local government personnel in 1984.

Virginia had an unemployment rate of 5% in 1984, well below the national average of 7.5%. The rate for all men was 4.5%; for all women, 5.7%; and for blacks alone, 11.1%. A right-to-work law is in effect. Although the state has no equal-employment statute, an equal-pay law does prohibit employers from wage discrimination on the basis of sex, and the Virginia Employment Contracting Act established as state policy the elimination of racial, religious, ethnic, and sexual bias in the employment practices of government agencies and contractors. The labor movement has grown slowly, partly because of past practices of racial segregation that prevented workers from acting in concert. There were 999 labor unions in Virginia in 1983; union membership totaled 318,000 in 1980, or only 15% of all nonagricultural employment—far below the US average of 25.2%. In September 1984, weekly earnings of production workers on manufacturing payrolls averaged $328, 37th among the states.

²³AGRICULTURE
Virginia, ranking 33d among the 50 states in 1983 with farm marketings of more than $1.4 billion, is an important producer of tobacco, peanuts, sweet potatoes, apples, and peaches. Net income was only $26.5 million in 1983, however. There were an estimated 57,000 farms in 1984. Farm employees in July 1983 (including those self-employed) numbered 73,000, of whom 22,000 were hired laborers.

The Tidewater is still a major farming region, as it has been since the early 17th century. Corn, wheat, tobacco, peanuts, and truck crops are all grown there, and potatoes are cultivated on the Eastern Shore. The piedmont is known for its apples and other fruits, while the Shenandoah Valley is one of the nation's main apple-growing regions. In 1983, Virginia ranked 5th among states in peanuts and tobacco, and 6th in apples. The following table shows data for leading crops in 1983:

	ACRES ('000)	PRODUCTION		VALUE
Tobacco	54	99,052,000	lb	$172,350,000
Hay	1,040	1,546,000	tons	142,232,000
Soybeans	650	9,760,000	bushels	78,080,000
Corn for silage	255	2,600,000	tons	68,900,000
Corn for grain	340	16,320,000	bushels	62,932,000
Peanuts	96	198,550,000	lb	53,000,000
Wheat	410	14,280,000	bushels	48,552,000
Apples	—	455,000,000	lb	47,775,000
Tomatoes	5	2,001,600	hundredweight	13,277,000
Barley	124	5,900,000	bushels	11,800,000
Potatoes	14	1,001,000	hundredweight	5,605,000
Peaches	—	24,000,000	lb	1,920,000
Sweet potatoes	1.2	138,000	hundredweight	1,209,000

²⁴ANIMAL HUSBANDRY
Cattle raising, poultry farming, and dairying, which together account for more than half of all cash receipts from farm marketings, play an increasingly important role in Virginia agriculture, especially in the Valley of Virginia and some central counties. At the close of 1983 there were 2,020,000 cattle and calves, 170,000 milk cows, 550,000 hogs and pigs, and 130,000 sheep and lambs. Production of meat animals that year included 560 million lb of cattle worth $164.4 million; 228.3 million lb of hogs, $106 million; and 9.2 million lb of sheep, $4.1 million. In 1983, 132,000 shorn sheep yielded 832,000 lb of wool valued at $399,000.

Milk production totaled 2.1 billion lb in 1983, worth $285 million. The output of creamed cottage cheese totaled 12.7 million lb; other cheeses, 7.2 million lb; and ice cream, 12.7 million gallons. Cash receipts from poultry and eggs in the 1983 marketing year amounted to $303.5 million, of which 57% came from chickens and broilers, 27% from turkeys, and 16% from eggs. Poultry farmers produced 619.4 million lb of broilers and 206.1 lb of turkeys; they also sold 828 million eggs.

25 FISHING
The relative importance of Chesapeake Bay and Atlantic fisheries to Virginia's economy has lessened considerably in recent decades, although the state continues to place high in national rankings. In 1984, Virginia's commercial fish landings totaled 574 million lb (3d in the US), worth $83 million (8th). The bulk of the catch consists of shellfish such as oysters, clams, and crabs, and finfish such as alewives and menhaden. Both saltwater and freshwater fish are avidly sought by sport fishermen. A threat to Virginia fisheries is chemical and oil pollution of Chesapeake Bay and its tributaries.

26 FORESTRY
As of 1979, Virginia had 16,417,000 acres (6,644,000 hectares) of forestland, representing more than 64% of the state's land area and 2.2% of all US forests. Virtually every county has some commercial forestland and supports a wood products industry.

In 1983, 864 million board feet of lumber were produced (15th among states). Shipments of lumber and wood products were valued at $1.2 billion in 1982; wood household furniture, $626 million; and paper and allied products, over $1.8 billion.

A 1980 survey showed that pine volume had increased 4% since the 1977 survey. Reforestation programs initiated by the Division of Forestry in 1971 have paid landowners to plant pine seedlings; state-funded tree nurseries produce 60–70 million seedlings annually. In 1984, the Division of Forestry was responsible for the reforesting of 100,000 acres (40,500 hectares). The division's tree seed orchards have developed improved strains of loblolly, shortleaf, white, and Virginia pine for planting in cutover timberland.

For recreational purposes there were over 2 million acres (800,000 hectares) of forested public lands in 1982, including Shenandoah National Park, Washington and Jefferson national forests, 24 state parks, and 8 state forests.

27 MINING
Virginia ranked 23d among the 50 states with a record nonfuel mineral output valued at $317.7 million in 1984. The state ranks 1st in production of kyanite, a heat-resistant aluminum silicate used in ceramics and refractory products.

The piedmont is the source of kyanite, quartzite, and soapstone. Minerals found in the Valley of Virginia, as well as in the Blue Ridge, include limestone, calcined lime, sandstone, and dolomite; the coastal plain has large deposits of sand, gravel, and clay.

The total value of stone quarried was over $172 million in 1984, when 42,392 tons were extracted. Construction sand and gravel production came to 8,300 tons and $31.8 million; lime production to 8.3 million tons and $31.4 million. Clay production was 774,000 tons, worth $6.2 million. The combined value of aplite (fine-grained granite), cement, gem stones, gypsum, crude iron oxide pigments, kyanite, industrial sand and gravel, talc (soapstone), and vermiculite was $76.2 million.

28 ENERGY AND POWER
Virginia's installed electric generating capacity was 11.6 million kw in 1983, when production of electricity totaled 37.7 billion kwh, almost all of it provided by private utilities. Electric power is supplied to the eastern and central parts of the state chiefly by the Virginia Electric and Power Co. (Vepco), to the central and southwestern regions by Appalachian Power, in the far southwest by Old Dominion Power, in the northwest by Potomac Edison, in the northeast by Potomac Electric Power and Vepco, and on the Eastern Shore by Delmarva Power and Light. Sales of electric power in 1983 amounted to 53.5 billion kwh: 40% residential, 24% industrial, 25% commercial, and 11% other.

Although Virginia has no petroleum deposits, it does have a major oil refinery at Yorktown that uses imported petroleum. The state is supplied with natural gas by three major interstate pipeline companies. Liquefied natural gas plants operate in Chesapeake, Roanoke, and Lynchburg, and a synthetic gas plant is in service at Chesapeake.

Coal-fired steam units accounted for 43% of electric power production in 1983; oil-fired plants produced 4%, nuclear 49%, hydroelectric 3%, and gas 1%. As a result of rising oil prices during the 1970s, the state's utilities have converted many oil-fired electric plants to coal, a trend in which Vepco was a national leader. Vepco is also committed to nuclear power: the state has two nuclear power reactors, both owned by Vepco.

Virginia's 438 coal mines (333 underground), most of them in the Appalachian Mountains area, produced 34.5 million tons in 1983, 4.4% of the US total and 6th among states. All the coal was bituminous.

29 INDUSTRY
Beginning with the establishment of a glass factory at Jamestown in 1608, manufacturing grew slowly during the colonial era to include flour mills and, by 1715, an iron foundry. During the 19th century, the shipbuilding industry flourished, and many cotton mills, tanneries, and ironworks were built; light industries producing a wide variety of consumer goods developed mainly after 1900.

Reflecting a national trend, manufacturing employment in Virginia fell by 1% between 1978 and 1982; it rose about 5% between 1982 and 1984. Textile, clothing, transportation equipment, food processing, chemical products, and electric and electronic equipment plants together employed 53% of all manufacturing workers in the state in 1982, when value added by manufacture reached nearly $17.3 billion. During that year, the value of shipments by manufacturers was $36.8 billion. The following table shows value of shipments by leading industry groups in 1982:

Food and food products	$5,903,400,000
Tobacco products	4,470,800,000
Chemical and allied products	4,342,900,000
Electric and electronic equipment	2,772,700,000
Transportation equipment	2,769,500,000
Textile mill products	2,516,300,000
Paper and allied products	1,829,000,000
Fabricated metal products	1,817,800,000
Printing and publishing	1,701,400,000
Nonelectrical machinery	1,423,800,000
Rubber and miscellaneous plastic products	1,334,100,000
Lumber and wood products	1,230,400,000
Apparel and other textile products	925,700,000
Furniture and fixtures	911,100,000

Richmond is a principal industrial area for tobacco processing, paper and printing, clothing, and food products; nearby Hopewell is a locus of the chemical industry. Newport News, Hampton, and Norfolk are centers for shipbuilding and the manufacture of other transportation equipment; 87,409 1983-model-year trucks were assembled in Virginia, 92% of them in Norfolk. In the western part of the state, Lynchburg is a center for electrical machinery, metals, clothing, and printing, and Roanoke for food, clothing, and textiles. In the south, Martinsville has a concentration of furniture and textile-manufacturing plants, and textiles are also dominant in Danville. Newport News Shipbuilding and Dry Dock, a subsidiary of Tenneco, which constructs naval warships, is the state's single largest private employer.

30 COMMERCE
Virginia ranked 20th among the 50 states in wholesale trade in 1982, with sales totaling $28.7 billion. In 1982, the state placed 11th in retail sales, with $24.2 billion. The leading retail categories among establishments with payroll were food stores, 24%; automotive dealers, 19%; department stores, 10%; gasoline service

stations, 10%; and eating and drinking places, 9%. Fairfax County, in the Washington, D.C., metropolitan area, led all counties, with nearly 14% of total sales; Richmond, the leading city, had 5%.

Virginia, a major container shipping center, handled nearly 54 million tons of import and export cargo worth nearly $10.9 billion in 1983, almost all through the Hampton Roads estuary. As of 1981, coal was the leading exported commodity and residual fuel oil the principal import. Foreign exports of Virginia's own manufactured goods totaled $3.3 billion in 1981 (16th in the US); agricultural exports amounted to $384 million (32d) in 1981/82.

31 CONSUMER PROTECTION

The Department of Law, headed by the attorney general, is responsible for enforcement of consumer protection legislation and for representing consumer interests at hearings of the State Corporation Commission. The Department of Agriculture and Consumer Services regulates food processors and handlers, product labeling, the use of pesticides, and product safety.

32 BANKING

At the end of 1983, Virginia's 178 commercial banks reported combined assets of $33.3 billion and total deposits of $27.1 billion. Insured commercial banks had outstanding loans amounting to $19.9 billion, of which 26% went for commercial and industrial purposes and 34% for real estate. Leading commercial banks include Sovran Bank in Norfolk, with nearly $8.2 billion in assets at the end of 1984; United Virginia Bank, Richmond, $5.9 billion; Dominion Bank, Roanoke, $4.3 billion; and Bank of Virginia, Richmond, $4.1 billion. There were 66 savings and loan associations and 368 credit unions in 1984.

33 INSURANCE

Virginians held 11,358,000 life insurance policies worth $126 billion in 1983; the average coverage per family was $60,300, 11% above the national norm. Benefit payments totaled nearly $1.1 billion, of which death payments accounted for $398.9 million, annuities $177.9 million, policy and contract dividends $211.7 million, and other disbursements $307.7 million.

In 1983, property and casualty companies wrote premiums of $2.1 billion, including $653.5 million in automotive liability insurance, $386.7 million in automobile physical damage insurance, and $231.1 million in homeowners' coverage.

34 SECURITIES

There are no securities exchanges in Virginia. New York Stock Exchange member firms had 124 sales offices and 1,293 registered representatives in the state during 1983. In that year, 987,000 Virginians owned shares of public corporations.

35 PUBLIC FINANCE

The biennial budget is prepared by the Virginia Department of Planning and Budget and submitted by the governor to the general assembly for amendment and approval. A 1984 amendment to the constitution requires that the budget be balanced; in practice, there is generally a small surplus of revenues over expenditures. The fiscal year runs from 1 July through 30 June.

Consolidated revenues and expenditures budgeted for the 1984–86 biennium were as follows (in millions):

REVENUES
General fund:

Personal and fiduciaries income tax	$ 4,127.0
Corporate income tax	543.0
Sales and use tax	1,922.1
Other revenues	1,423.1
Nongeneral fund:	
Federal grants and donations	3,524.1
Institutional revenue	1,877.5
Taxes	1,648.3
Sales of property and commodities	493.8
Other receipts	981.4
TOTAL	$16,540.3

EXPENDITURES

Education	$ 6,723.7
Transportation and public safety	3,390.6
Human resources	3,298.3
Commerce and resources	954.5
Central accounts	464.9
Administration	410.9
Capital outlays	282.4
Other outlays	463.6
TOTAL	$15,988.9

As of 30 June 1983, debts incurred by state and local governments totaled $7.35 billion. Per capita debt in mid-1982 was $1,305, well below the US average.

36 TAXATION

Virginians bear a lighter tax burden than residents of most states. In 1981/82, for example, the state and local tax burden was $1,030 per capita, 32d among the 50 states and 12% below the national average. For the 1984–86 biennium, state taxes accounted for 51% of total state revenues.

As of 1984, state income tax rates ranged from 2% on adjusted federal gross income below $3,000 to 5.75% on amounts over $12,000. The basic corporate income tax rate was 6%. The state sales and use tax rate was 3%, with 1% added by counties and independent cities; gasoline tax, 11%; cigarette tax, per pack, 2.5%. Also taxed by the state are motor vehicles, watercraft, alcoholic beverages, railroad property, timber, motor fuel, estates and wills, and insurance transactions. The real estate tax, tangible personal property taxes, and utility taxes are levied by counties and independent cities.

Virginia remitted $13.9 billion in taxes to the federal government in 1982. State residents filed 2,323,758 personal income tax returns for 1983, paying $6.8 billion in tax.

37 ECONOMIC POLICY

The state government actively promotes a probusiness climate. Conservative traditions, low tax rates, low wage rates, a weak labor movement, and excellent access to eastern and overseas markets are the general incentives for companies to relocate into Virginia. A duty-free foreign trade zone has been established at Suffolk.

The Virginia Industrial Development Corp., a privately financed and privately capitalized lending facility operating under special charter from the general assembly, extends low-interest loans to creditworthy companies to purchase land, buildings, and machinery if conventional financing is not available. The state also may issue revenue bonds to finance industrial projects—a popular method of financing because the return to investors is tax-free. The bonds may also be used to finance the installation of pollution control equipment. At their discretion, localities may allow total or partial tax exemptions for such equipment and for certified solar energy devices.

Counties, cities, and incorporated towns may form local industrial development authorities to finance industrial projects and various other facilities, and may issue their own revenue bonds to cover the cost of land, buildings, machinery, and equipment. The authority's lease of the property normally includes an option to buy at a nominal price on the expiration of the lease. In addition, some 110 local development corporations have been organized. The Department of Housing and Urban Development offers special state and local financial incentives. For minority-owned entrepreneurships, Virginia maintains the Office of Minority Business Enterprises, to give advice on special problems.

38 HEALTH

Virginia's live-birth rate was 14.8 per 1,000 population in 1982, below the national rate of 15.9. The infant mortality rate in 1983 was 9.6 deaths per 1,000 live births for whites and 19.8 for nonwhites—in each case, close to the US average. A total of

31,339 legal abortions were performed in 1983. Virginia's death rate of 7.8 per 1,000 population was below the US average of 8.6 during 1981. The death rate was 7.9 in 1983. Virginia's death rates in 1981 for the leading causes of death—heart disease, cancer, and stroke—were likewise below the national norms.

In 1983, Virginia's 135 general hospitals had 31,289 beds and recorded 863,816 admissions, with 6,177,823 outpatient visits. Medical personnel in 1980 included 34,966 registered nurses. The average cost of hospital care in 1982 was $282 per day and $2,288 per stay, below the US norm. Virginia had 11,353 licensed physicians and 2,876 active dentists at the end of 1982.

39 SOCIAL WELFARE

Public aid recipients constituted 4.3% of the population in 1983, compared to 6.1% for the US as a whole. Payments totaling $16.6 million in aid to families with dependent children were made in 1982. In 1982/83, the food stamp program enrolled 408,000 Virginians at a federal cost of $221 million. During the same fiscal year, the school lunch program involved a federal outlay of $52 million, with 598,000 pupils enrolled on 30 September 1983.

In 1983, Social Security benefits were paid to 737,000 state residents; disbursements amounted to nearly $3.2 billion, with an average monthly benefit of $410 per retired worker, 7% below the US average. Federal Supplemental Security Income payments were made to 83,390 aged, blind, or disabled persons in December 1983; total payments in 1983 amounted to $166.3 million. Funds distributed under the Black Lung Benefit Program to coal miners afflicted with pneumoconiosis (and to their dependents or survivors) totaled $5 million for December 1983. Workers' compensation amounted to $249.4 million in 1982. Unemployment insurance payments, issued to an average of 45,000 beneficiaries weekly, totaled $258 million in 1982, when the average benefit was $108.

40 HOUSING

In 1980, Virginia had 2,020,941 housing units, 1,863,073 of them occupied year-round. Of the latter, 66% were owner-occupied. From 1981 through 1983, the state authorized construction of 113,400 privately owned units, valued at $4.6 billion.

41 EDUCATION

Although Virginia was the first English colony to found a free school (1634), the state's public school system developed very slowly. Thomas Jefferson proposed a system of free public schools as early as 1779, but it was not until 1851 that such a system was established—for whites only. Free schools for blacks were founded after the Civil War, but they were poorly funded until recent years. Opposition by white Virginians to the US Supreme Court's desegregation order in 1954 was marked in certain communities by public school closings and the establishment of all-white private schools; in Prince Edward County, the most extreme case, the school board abandoned public education and left black children without schools from 1959 to 1963. By the 1970s, however, school integration was an accomplished fact throughout the commonwealth.

About 63% of all state residents 25 years of age or older were high school graduates in 1980, and more than 19% had at least 4 years of college. Under a Standards of Quality Program adopted in 1972, students must pass minimum competency tests in reading and math in order to qualify for a high school diploma. During the 1984/85 school year, Virginia had 1,750 public schools, with 965,222 pupils; there were 317 private schools in 1980/81, with an estimated 75,069 pupils.

Virginia has had a distinguished record in higher education since the College of William and Mary was founded at Williamsburg (then called Middle Plantation) in 1693, especially after Thomas Jefferson established the University of Virginia at Charlottesville in 1819. In 1983/84, 74 colleges and universities in the state enrolled more than 294,000 students, almost three-fifths of them full-time; 23 community colleges on 33 campuses had

112,336 students in the fall of 1983. In addition to the University of Virginia and the College of William and Mary—with enrollments of 16,379 and 6,607, respectively, in 1983/84—public state-supported institutions include Virginia Polytechnic Institute and State University, at Blacksburg; Virginia Commonwealth University, Richmond; Virginia Military Institute, Lexington; Old Dominion University, Norfolk; and George Mason University, Fairfax. Well-known private institutions include the Hampton Institute, at Hampton; Randolph-Macon College, Ashland; University of Richmond; Sweet Briar College, Sweet Briar; and Washington and Lee University, Lexington. Tuition assistance grants and scholarships are provided through the State Council of Higher Education, while the State Education Assistance Authority provides guaranteed student loans.

42 ARTS

Richmond and Norfolk are the principal centers for both the creative and the performing arts in Virginia. In Richmond, the Mosque has been the scene of concerts by internationally famous orchestras and soloists for generations, and a modernistic coliseum now shelters many musical events. The intimate 500-seat Virginia Museum Theater presents new plays and classics with professional casts. Just outside the Richmond metropolitan area, the Barksdale Theater and its repertory company present serious plays and occasionally give first performances of new works. In Norfolk, the performing arts are strikingly housed in Scope, a large auditorium designed by Pier Luigi Nervi; Chrysler Hall, an elegant structure with gleaming crystal; and the Wells Theater, an ornate building where John Philip Sousa, Will Rogers, and Fred Astaire once performed and which now houses the Virginia Stage Company, a repertory theater. The Virginia Opera Association, centered in Norfolk, is internationally recognized.

Wolf Trap Farm Park for the Performing Arts, in northern Virginia, provides theatrical, operatic, and symphonic performances featuring internationally celebrated performers. William and Mary's Phi Beta Kappa Hall in Williamsburg is the site of the Virginia Shakespeare Festival, an annual summer event inaugurated in 1979. Abingdon is the home of the Barter Theater, the first state-supported theater in the United States, whose alumni include Ernest Borgnine and Gregory Peck. This repertory company has performed widely in the United States and at the Elsinore Shakespeare Festival in Denmark. The John F. Kennedy Center for the Performing Arts in nearby Washington, D.C., is heavily patronized by Virginians.

43 LIBRARIES AND MUSEUMS

A total of 93 county, city, town, and regional library systems served 98% of the population in 1982/83; their combined book stock reached 10,229,657 volumes, and their combined circulation was 29,334,653. The Virginia State Library in Richmond and the libraries of the University of Virginia (Charlottesville) and the College of William and Mary (Williamsburg) have the personal papers of such notables as Washington, Jefferson, Madison, Robert E. Lee, William H. McGuffey, and William Faulkner. The University of Virginia also has an impressive collection of medieval illuminated manuscripts, and the library of colonial Williamsburg has extensive microfilms of British records.

There were 178 museums in 1983. In Richmond, the Virginia Museum of Fine Arts, the first state museum of art in the US, has a collection that ranges from ancient Egyptian artifacts to mobile jewelry by Salvador Dali. The Science Museum of Virginia has a 280-seat planetarium that features a simulated excursion to outer space. Other museums in Richmond are Wilton, the Randolphs' handsome 18th-century mansion, and the Maymont and Wickham-Valentine houses, elaborate 19th-century residences; Agecroft Hall and Virginia House, Tudor manor houses that were moved from England, are also open to the public. Norfolk has the Chrysler Museum, with its famous glassware collection; Myers House, an early Federal period home with handsome art and furnishings; and the Hermitage Foundation Museum, noted for its Oriental art. The Mariners Museum in Newport News has a superb maritime collection, and the much smaller but quite select exhibits of the Portsmouth Naval Shipyard Museum are also notable. Perhaps the most extensive "museum" in the US is Williamsburg's mile-long Duke of Gloucester Street, with such remarkable restorations as the Christopher Wren Building of the College of William and Mary, Bruton Parish Church, the Governor's Palace, and the colonial capital.

More historic sites are maintained as museums in Virginia than in any other state. These include Washington's home at Mt. Vernon (Fairfax County), Jefferson's residence at Monticello (Charlottesville), and James River plantation houses such as Berkeley, Shirley, Westover, Sherwood Forest, and Carter's Grove. The National Park Service operates a visitors' center at Jamestown.

44 COMMUNICATIONS

The state's communications network has expanded steadily since the first postal routes were established in 1738. Airmail service from Richmond to New York and to Atlanta began in 1928. As of 1985, the US Postal Service had 14,202 employees in 891 post offices in Virginia.

According to the 1980 census, nearly 93% of Virginia's 1,863,073 occupied housing units had telephones. In 1984, commercial broadcasters operated 144 AM radio stations and 111 FM stations. In the same year, Virginia had 16 commercial and 9 educational television stations, and 100 cable television systems serving 646,033 subscribers in 288 communities.

45 PRESS

Although the Crown forbade the establishment of a printing press in Virginia Colony, William Parks was publishing the *Virginia Gazette* at Williamsburg in 1736. Three newspapers were published regularly during the Revolutionary period, and in 1780 the general assembly declared that the press was "indispensable for the right information of the people and for the public service." The oldest continuously published Virginia daily, tracing its origins to 1784, is the *Alexandria Gazette*. The first Negro newspaper, *The True Southerner,* was started by a white man in 1865; several weeklies published and edited by blacks began soon after. By 1900 there were 180 newspapers in the state, but the number has declined drastically since then because of fierce competition, mergers, and rising costs.

In 1984, Virginia had 14 morning dailies with a combined circulation of 622,850, 24 evening dailies with 564,526 circulation, and 15 Sunday papers with 915,132 circulation. Leading Virginia newspapers with their 1984 circulations are shown in the following table:

AREA	NAME	DAILY	SUNDAY
Norfolk	Ledger-Star (e, S)	86,895 ⎱	
	Virginian-Pilot (m, S)	141,714 ⎰	221,992
Richmond	News Leader (e)	113,460	
	Times-Dispatch (m, S)	136,598	230,878
Roanoke	Times & World News	75,541 (m) ⎱	
	(m, e, S)	47,294 (e) ⎰	124,825

46 ORGANIZATIONS

The 1982 US Census of Service Industries counted 1,541 organizations in Virginia, including 385 business associations; 699 civic, social, and fraternal associations; and 58 educational, scientific, and research associations. Service and educational groups headquartered in the state include the United Way of America, American Astronautical Society, American Society for Horticultural Science, American Geological Institute, and American Physical Therapy Association, all located in Alexandria; and the National Honor Society, Music Educators National Conference, and National Art Education Association, located in Reston.

Veterans' organizations include the Veterans of World War I of the USA and the Retired Officers Association, Alexandria; and the Military Order of the Purple Heart, Springfield. The United Daughters of the Confederacy has national offices in Richmond. Among the business and professional groups based in Virginia are the American Gas Association, Arlington; and National Automobile Dealers Association, McLean.

Sports societies headquartered in the state include the American Canoe Association, Lorton; Boat Owners Association of the US, Alexandria; and National Rowing Foundation and the Walking Association, both located in Arlington.

Other groups operating out of Virginia include the Future Farmers of America and the National Sojourners, Alexandria; American Automobile Association, Falls Church; Federation of Homemakers and National Alliance of Senior Citizens, Arlington; Association of Former Intelligence Officers, McLean; and the Moral Majority, Lynchburg.

47 TOURISM, TRAVEL, AND RECREATION

In 1984, tourists spent an estimated $4.1 billion in Virginia. Attractions in the coastal region alone include the Jamestown and Yorktown historic sites, the Williamsburg restoration, and the homes of George Washington and Robert E. Lee. Also featured are the National Aeronautics and Space Administration's Langley Research Center, Assateague Island National Seashore, and the resort pleasures of Virginia Beach.

The interior offers numerous Civil War Sites, including Appomattox; Thomas Jefferson's Monticello; Booker T. Washington's birthplace near Smith Mountain Lake; and the historic cities of Richmond, Petersburg, and Fredericksburg. In the west, the Blue Ridge Parkway and Shenandoah National Park, traversed by the breathtaking Skyline Drive, are favorite tourist destinations, as are Cumberland Gap and, in the Lexington area, the Natural Bridge, the home of Confederate General Thomas "Stonewall" Jackson, the George C. Marshall Library and Museum, and the Virginia Military Institute. A number of historic sites in Arlington and Alexandria attract many visitors to the Washington, D.C., area.

The state's many recreation areas include 25 state parks, 2 national forests, a major national park, 4 scenic parkways, and thousands of miles of hiking trails and shoreline. State parks attracted 3,255,618 visits in 1982/83, and 16 national park sites attracted 23,131,627 visitors in 1984. Some of the most-visited sites are Mt. Rogers National Recreational Area, Prince William Forest Park, Chincoteague National Wildlife Refuge, and the Kerr Reservoir. Part of the famous Appalachian Trail winds through Virginia's Blue Ridge and Appalachian mountains. The commonwealth, with more than 1,500 mi (2,400 km) of well-stocked trout streams, issued licenses to 543,362 anglers and 462,009 hunters in 1982/83.

48 SPORTS

Although Virginia has no major league sports team, it does support two entries in baseball's class-AAA International League: the Richmond Grays, affiliated with the Atlanta Braves, and Norfolk's Tidewater Tides, a farm team of the New York Mets.

In collegiate sports, the University of Virginia belongs to the Atlantic Coast Conference, while VMI competes in the Southern Conference. The 1979/80 season brought three national basketball crowns to the commonwealth: men's teams at the University of Virginia and Virginia Union University won the National Invitation Tournament and the NCAA Division II championships, respectively, and for the second consecutive year, Old Dominion University's Lady Monarchs won the national tournament sponsored by the Association for Intercollegiate Athletics for Women. The men's basketball team at Old Dominion won the NCAA Division II title in 1975.

Participant sports popular with Virginians include tennis, golf, swimming, skiing, boating, and water skiing. The state has at least 180 public and private golf courses.

⁴⁹FAMOUS VIRGINIANS

Virginia is the birthplace of eight US presidents and many famous statesmen, noted scientists, influential educators, distinguished writers, and popular entertainers.

The 1st president of the US, George Washington (1732–99), also led his country's armies in the Revolutionary War and presided over the convention that framed its Constitution. Washington—who was unanimously elected president in 1789 and served two four-year terms, declining a third—was not, as has sometimes been assumed, a newcomer to politics: his political career began at the age of 27 with his election to the house of burgesses.

Thomas Jefferson (1743–1826), the nation's 3d president, offered this as his epitaph: "author of the Declaration of Independence and the Virginia Statute for Religious Freedom, and father of the University of Virginia." After serving as secretary of state under Washington and vice president under John Adams, he was elected president of the US in 1800 and reelected in 1804. Honored now as a statesman and political thinker, Jefferson was also a musician and one of the foremost architects of his time, and he has been called the first American archaeologist.

Jefferson's successor, James Madison (1751–1836), actually made his most important contributions before becoming chief executive. As a skillful and persistent negotiator throughout the Constitutional Convention of 1787, he earned the designation "father of the Constitution"; then, as coauthor of the Federalist papers, he helped produce a classic of American political philosophy. He was more responsible than any other statesman for Virginia's crucial ratification vote. Secretary of state during Jefferson's two terms, Madison occupied the presidency from 1809 to 1817.

Madison was succeeded as president in 1817 by James Monroe (1758–1831), who was reelected to a second term starting in 1821. Monroe—who had served as governor, US senator, minister to France, and secretary of state—is best known for the Monroe Doctrine, which has been US policy since his administration. William Henry Harrison (1773–1841) became the 9th president in 1841 but died of pneumonia one month after his inauguration; he had been a governor of Indiana Territory, a major general in the War of 1812, and a US representative and senator from Indiana. Harrison was succeeded by Vice President John Tyler (1790–1862), a native and resident of Virginia, who established the precedent that, upon the death of the president, the vice president inherits the title as well as the duties of the office.

Another native Virginian, Zachary Taylor (1784–1850), renowned chiefly as a military leader, became the 12th US president in 1849 but died midway through his term. The eighth Virginia-born president, (Thomas) Woodrow Wilson (1856–1924), became the 28th president of the US in 1913 after serving as governor of New Jersey.

John Marshall (1755–1835) was the third confirmed chief justice of the US and is generally regarded by historians as the first great American jurist, partly because of his establishment of the principle of judicial review. Five other Virginians—John Blair (1732–1800), Bushrod Washington (1762–1829), Philip P. Barbour (1783–1841), Peter V. Daniel (1784–1860), and Lewis F. Powell, Jr. (b.1907)—have served as associate justices.

George Washington's cabinet included two Virginians, Secretary of State Jefferson and Attorney General Edmund Randolph (1753–1813), who, as governor of Virginia, had introduced the Virginia Plan—drafted by Madison and calling for a House of Representatives elected by the people and a Senate elected by the House—at the Constitutional Convention of 1787. Among other distinguished Virginians who have served in the cabinet are James Barbour (1775–1842), secretary of war; John Y. Mason (1799–1859), secretary of the Navy and attorney general; Carter Glass (1858–1946), secretary of the treasury, author of the Federal Reserve System, and US senator for 26 years; and Claude Augustus Swanson (1862–1939), secretary of the Navy and earlier, state governor and US senator.

Other prominent US senators from Virginia include Richard Henry Lee (1732–94), former president of the Continental Congress; James M. Mason (b.District of Columbia, 1798–1871), who later was commissioner of the Confederacy to the United Kingdom and France; John W. Daniel (1842–1910), a legal scholar and powerful Democratic Party leader; Thomas S. Martin (1847–1919), US Senate majority leader; Harry F. Byrd (1887–1966), governor of Virginia from 1926 to 1930 and US senator from 1933 to 1965; and Harry F. Byrd, Jr. (b.1914), senator from 1965 to 1982. In 1985, Virginia was represented in the Senate by Republican John W. Warner (b.District of Columbia, 1927), former secretary of the Navy, and Republican Paul S. Trible, Jr. (b.Maryland, 1946), a US representative from 1976 to 1982.

Some native-born Virginians have become famous as leaders in other nations. Joseph Jenkins Roberts (1809–76) was the first president of the Republic of Liberia, and Nancy Langhorne Astor (1879–1964) was the first woman to serve in the British House of Commons.

Virginia's important colonial governors included Captain John Smith (b.England, 1580?–1631), Sir George Yeardley (b.England, 1587?–1627), Sir William Berkeley (b.England, 1606–77), Alexander Spotswood (b.Tangier, 1676–1740), Sir William Gooch (b.England, 1681–1751), and Robert Dinwiddie (b.Scotland, 1693–1770).

Virginia signers of the Declaration of Independence, besides Jefferson and Richard Henry Lee, were Carter Braxton (1736–97); Benjamin Harrison (1726?–1791), father of President William Henry Harrison; Francis Lightfoot Lee (1734–97); Thomas Nelson, Jr. (1738–89); and George Wythe (1726–1806). Wythe is also famous as the first US law professor and the teacher, in their student days, of Presidents Jefferson, Monroe, and Tyler, and Chief Justice Marshall. Virginia furnished both the first president of the Continental Congress, Peyton Randolph (1721–75), and the last, Cyrus Griffin (1748–1810).

Other notable Virginia governors include Patrick Henry (1736–99), the first governor of the commonwealth, though best remembered as a Revolutionary orator; Westmoreland Davis (1859–1942); Andrew Jackson Montague (1862–1937); and Mills E. Godwin, Jr. (b.1914). A major historical figure who defies classification is Robert "King" Carter (1663–1732), greatest of the Virginia land barons, who also served as acting governor of Virginia and rector of the College of William and Mary.

Chief among Virginia's great military and naval leaders besides Washington and Taylor are John Paul Jones (b.Scotland, 1747–92); George Rogers Clark (1752–1818); Winfield Scott (1786–1866); Robert E. Lee (1807–70), the Confederate commander who earlier served in the Mexican War and as superintendent of West Point; Joseph E. Johnston (1807–91); George H. Thomas (1816–70); Thomas Jonathan "Stonewall" Jackson (1824–63); James Ewell Brown "Jeb" Stuart (1833–64); and George C. Marshall (b.Pennsylvania, 1880–1959). Virginians' names are also written high in the history of exploration. Daniel Boone (b.Pennsylvania, 1734–1820), who pioneered in Kentucky and Missouri, was once a member of the Virginia general assembly. Meriwether Lewis (1774–1809) and William Clark (1770–1838), both native Virginians, led the most famous expedition in US history, from St. Louis to the Pacific coast (1804–6). Richard E. Byrd (1888–1957) was both an explorer of Antarctica and a pioneer aviator.

Woodrow Wilson and George C. Marshall both received the Nobel Peace Prize, in 1919 and 1953, respectively. Distinguished Virginia-born scientists and inventors include Matthew Fontaine Maury (1806–73), founder of the science of oceanography; Cyrus

H. McCormick (1809–84), who perfected the mechanical reaper; and Dr. Walter Reed (1851–1902), who proved that yellow fever was transmitted by a mosquito. Among educators associated with the state are William H. McGuffey (b.Pennsylvania, 1800–1873), a University of Virginia professor who designed and edited the most famous series of school readers in American history; and Booker T. Washington (1856–1915), the nation's foremost black educator.

William Byrd II (1674–1744) is widely acknowledged to have been the most graceful writer in English America in his day, and Jefferson was a leading prose stylist of the Revolutionary period. Edgar Allan Poe (b.Massachusetts, 1809–49), who was taken to Richmond at the age of 3 and later educated at the University of Virginia, was the father of the detective story and one of America's great poets and short-story writers. Virginia is the setting of historical romances by three natives, John Esten Cooke (1830–86), Thomas Nelson Page (1853–1922), and Mary Johnston (1870–1936). Notable 20th-century novelists include Willa Cather (1873–1947), Ellen Glasgow (1874–1945), and James Branch Cabell (1879–1958). Willard Huntington Wright (1888–1939), better known as S. S. Van Dine, wrote many detective thrillers. Twice winner of the Pulitzer Prize for biography and often regarded as the greatest American master of that genre was Douglas Southall Freeman (1886–1953). Other important historians were Lyon Gardiner Tyler (1853–1935), son of President Tyler and also an eminent educator; Philip A. Bruce (1856–1933); William Cabell Bruce (1860–1946); Virginius Dabney (b.1901); and Alf J. Mapp, Jr. (b.1925). Some contemporary Virginia authors are poet Guy Carlton Drewry (b.1901); television writer-producer Earl Hamner (b.1923); novelist William Styron (b.1925); and journalists Virginia Moore (b.1903) and Tom Wolfe (Thomas Kennerly Wolfe, Jr., b.1931).

Celebrated Virginia artists include sculptors Edward V. Valentine (1838–1930) and Moses Ezekiel (1844–1917), and painters George Caleb Bingham (1811–79) and Jerome Myers (1867–1940). A protégé of Jefferson's, Robert Mills (b.South Carolina, 1781–1855), designed the Washington Monument.

The roster of Virginians prominent in the entertainment world includes Bill "Bojangles" Robinson (1878–1949), Francis X. Bushman (1883–1966), Freeman Gosden (1899–1982), Randolph Scott (b.1903), Joseph Cotten (b.1905), Margaret Sullavan (1911–60), John Payne (b.1912), George C. Scott (b.1927), Shirley MacLaine (b.1934), and Warren Beatty (b.1938).

Outstanding musical performers include John Powell (1882–1963), whose fame as a pianist once equaled his prominence as a composer; Virginia's most eminent contemporary composer is Thea Musgrave (b.Scotland, 1928). Popular musical stars include Kathryn Elizabeth "Kate" Smith (b.1909), Pearl Bailey (b.1918), Ella Fitzgerald (b.1918), June Carter (b.1929), Roy Clark (b.1933), and Wayne Newton (b.1942).

The Old Dominion's sports champions include golfers Bobby Cruickshank (b.1896), Sam Snead (b.1912), and Chandler Harper (b.1914); tennis star Arthur Ashe (b.1943); football players Clarence "Ace" Parker (b.1912), Bill Dudley (b.1921), and Francis "Fran" Tarkenton (b.1940); and baseball pitcher Eppa Rixey (1891–1963). At age 15, Olympic swimming champion

Melissa Belote (b.1957) won three gold medals. Helen Chenery "Penny" Tweedy (b.1922) is a famous breeder and racer of horses from whose stables have come Secretariat and other champions. Equestrienne Jean McLean Davis (b.1929) won 65 world championships.

[50]BIBLIOGRAPHY

Ashe, Dora J. (comp.). *Four Hundred Years of Virginia, 1584–1984: An Anthology.* Lanham, Md.: University Press of America, 1985.

Bruce, Philip Alexander. *Economic History of Virginia in the Seventeenth Century.* 2 vols. New York: Johnson Reprints, n.d. (orig. 1896).

Bruce, Philip Alexander. *Social Life of Virginia in the Seventeenth Century.* Lynchburg, Va.: J. P. Bell, 1927.

Buni, Andrew. *The Negro in Virginia Politics, 1902–65.* Charlottesville: University Press of Virginia, 1967.

Dabney, Virginius. *Richmond: The Story of a City.* Garden City, N.Y.: Doubleday, 1976.

Dabney, Virginius. *Virginia: The New Dominion.* Charlottesville: University Press of Virginia, 1983 (orig. 1971).

Davis, Richard Beale. *Intellectual Life in Jefferson's Virginia, 1790–1830.* Knoxville: University of Tennessee Press, 1972 (orig. 1964).

Federal Writers' Project. *Virginia: A Guide to the Old Dominion.* New York: Somerset, 1980 (orig. 1940).

Freeman, Douglas Southall. *George Washington.* 7 vols. New York: Scribner, 1948–57.

Friddell, Guy. *What Is It About Virginia?* Richmond: Dietz, 1983 (orig. 1966).

Gottmann, Jean. *Virginia in Our Century.* Charlottesville: University Press of Virginia, 1969.

Malone, Dumas. *Jefferson and His Time.* Vols. 1 and 2. Boston: Little, Brown, 1948, 1951.

Mapp, Alf J., Jr. *Frock Coats and Epaulets: Confederate Political and Military Leaders.* New York: A. S. Barnes, 1963.

Mapp, Alf J., Jr. *The Virginia Experiment: The Old Dominion's Role in the Making of America, 1607–1781.* 2d ed. La Salle, Ill.: Open Court, 1975.

Moger, Allen W. *Virginia: Bourbonism to Byrd, 1870–1925.* Charlottesville: University Press of Virginia, 1968.

Morgan, Edmund S. *American Slavery, American Freedom: The Ordeal of Colonial Virginia.* New York: Norton, 1975.

Morton, Richard L. *Colonial Virginia.* 2 vols. Chapel Hill: University of North Carolina Press, 1960.

Rubin, Louis D., Jr. *Virginia: A Bicentennial History.* New York: Norton, 1977.

Stanard, Mary Newton. *The Story of Virginia's First Century.* Philadelphia: Lippincott, 1938.

Virginia, Commonwealth of. Division of Industrial Development. *Virginia Facts and Figures 1984.* Richmond, 1984.

Wertenbaker, Thomas J. *Norfolk, Historic Southern Port.* 2d ed. Durham, N.C.: Duke University Press, 1962.

Wertenbaker, Thomas J. *Torchbearer of the Revolution.* Princeton, N.J.: Princeton University Press, 1940.

Wright, Louis B. *The First Gentlemen of Virginia.* Charlottesville: University Press of Virginia, 1940.

WASHINGTON

State of Washington

ORIGIN OF STATE NAME: Named for George Washington. **NICKNAME:** The Evergreen State. **CAPITAL:** Olympia. **ENTERED UNION:** 11 November 1889 (42d). **SONG:** "Washington, My Home." **DANCE:** Square dance. **MOTTO:** Alki (By and by). **FLAG:** The state seal centered on a dark green field. **OFFICIAL SEAL:** Portrait of George Washington surrounded by the words "The Seal of the State of Washington 1889." **BIRD:** Willow goldfinch. **FISH:** Steelhead trout. **FLOWER:** Western rhododendron. **TREE:** Western hemlock. **GEM:** Petrified wood. **LEGAL HOLIDAYS:** New Year's Day, 1 January; Birthday of Martin Luther King, Jr., 3d Monday in January; Lincoln's Birthday, 12 February; Washington's Birthday, 3d Monday in February; Memorial Day, last Monday in May; Independence Day, 4 July; Labor Day, 1st Monday in September; Veterans Day and State Admission Day, 11 November; Thanksgiving Day, 4th Thursday in November; Christmas Day, 25 December. **TIME:** 4 AM PST = noon GMT.

¹LOCATION, SIZE, AND EXTENT

Located on the Pacific coast of the northwestern US, Washington ranks 20th in size among the 50 states.

The total area of Washington is 68,138 sq mi (176,477 sq km), of which land takes up 66,511 sq mi (172,263 sq km) and inland water 1,627 sq mi (4,214 sq km). The state extends about 360 mi (580 km) E–W and 240 mi (390 km) N–S.

Washington is bounded on the N by the Canadian province of British Columbia (with the northwestern line passing through the Juan de Fuca Strait and the Haro and Georgia straits); on the E by Idaho (with the line in the southwest passing through the Snake River); on the S by Oregon (with most of the line defined by the Columbia River); and on the W by the Pacific Ocean.

Islands of the San Juan group, lying between the Haro and Rosario straits, include Orcas, San Juan, and Lopez; Whidbey is a large island in the upper Puget Sound. The state's boundary length totals 1,099 mi (1,769 km), including 157 mi (253 km) of general coastline; the tidal shoreline extends 3,026 mi (4,870 km). Washington's geographic center is in Chelan County, 10 mi (16 km) WSW of Wenatchee.

²TOPOGRAPHY

Much of Washington is mountainous. Along the Pacific coast are the Coast Ranges extending northward from Oregon and California. This chain forms two groups: the Olympic Mountains in the northwest, mainly on the Olympic Peninsula between the Pacific Ocean and Puget Sound, and the Willapa Hills in the southwest. The highest of the Olympic group is Mt. Olympus, at 7,965 feet (2,428 meters). About 100 mi (160 km) inward from the Pacific coast is the Cascade Range, extending northward from the Sierra Nevada in California. This chain, 50-100 mi (80-160 km) wide, has peaks generally ranging up to 10,000 feet (3,000 meters), except for such volcanic cones as Mt. Adams, Mt. Baker, Glacier Peak, Mt. St. Helens, and Mt. Rainier, which at 14,410 feet (4,392 meters) is the highest peak in the state.

Between the Coast and Cascade ranges lies a long, troughlike depression—the Western Corridor—where most of Washington's major cities are concentrated. The northern section of this lowland is carved by Puget Sound, a complex, narrow arm of the Pacific wending southward for about 80 mi (130 km) and covering an area of 561 sq mi (1,453 sq km). Of all the state's other major regions, only south-central Washington, forming part of the Columbia Plateau, is generally flat.

The Cascade volcanoes were dormant, for the most part, during the second half of the 19th century and most of the 20th. Early in 1980, however, Mt. St. Helens began to show ominous signs of activity. On 18 May, the volcano exploded, blasting more than 1,300 feet (400 meters) off a mountain crest that had been 9,677 feet (2,950 meters) high. Tremendous plumes of steam and ash were thrust into the stratosphere, where prevailing winds carried volcanic dust thousands of miles eastward. The areas immediately surrounding Mt. St. Helens were deluged with ash and mudflows, choking local streams and lakes, particularly Spirit Lake. About 150 sq mi (388 sq km) of trees and brush were destroyed; the ash fall also damaged crops in neighboring agricultural areas and made highway travel extremely hazardous. The eruption left 57 people dead or missing. Eruptions of lesser severity followed the main outburst; the mountain continued to pose a serious danger to life in the area as the estimated cost of the damage to property, crops, and livestock approached $3 billion. Another minor eruption, on 14 May 1984, shot ash 4 miles (6 km) high and caused a small mudflow down the mountain's flanks, but no injuries or other damage occurred. East of the Cascade Range, much of Washington is a plateau underlain by ancient basalt lava flows. In the northeast are the Okanogan Highlands; in the southeast, the Blue Mountains and the Palouse Hills. All these uplands form extensions of the Rocky Mountain system.

Among Washington's numerous rivers, the longest and most powerful is the Columbia, entering Washington from Canada in the northeast corner and flowing for more than 1,200 mi (1,900 km) across the heart of the state and then along the Oregon border to the Pacific. In average discharge, the Columbia ranks 2d only to the Mississippi, with 262,000 cu feet (7,400 cu meters) per second. Washington's other major river, the Snake, enters the state from Idaho in the southeast and flows generally westward, meeting the Columbia River near Pasco.

Washington has numerous lakes, of which the largest is the artificial Franklin D. Roosevelt Lake, covering 123 sq mi (319 sq km). Washington has some 90 dams, providing water storage, flood control, and hydroelectric power. One of the largest and most famous dams in the US is Grand Coulee on the upper Columbia River, measuring 550 feet (168 meters) high and 4,173 feet (1,272 meters) long, with a storage capacity of more than 9.7 million acre-feet.

³CLIMATE

The Cascade Mountains divide Washington not only topographically but also climatically. Despite its northerly location, western Washington is as mild as the middle and southeastern Atlantic coast; it is also one of the rainiest regions in the world. Eastern Washington, on the other hand, has a much more continental climate, characterized by cold winters, hot summers, and sparse rainfall. Since the prevailing winds are from the west, the windward (western) slopes of the state's major mountains intercept most of the atmospheric moisture and precipitate it as rain or snow. Certain coastal areas, receiving more than 200 in (500 cm) of rain a year, support dense stands of timber in a temperate rain forest. But in the dry southeastern quadrant, there are sagebrush deserts.

Average January temperatures in western Washington range from a minimum of 20°F (−7°C) on the western slope of the Cascades to a maximum of 48°F (9°C) along the Pacific coast; July temperatures range from a minimum of 44°F (7°C) on the western slope of the Cascades to a maximum of 80°F (27°C) in the foothills. In the east, the temperature ranges are much more extreme: in January, from 8°F (−13°C) in the northeastern Cascades to 40°F (4°C) on the southeastern plateau; in July, from 48°F (9°C) on the eastern slope of the Cascades to 92°F (33°C) in the south-central portion of the state. The normal daily mean temperature in Seattle is 51°F (11°C), ranging from 38°F (3°C) in January to 65°F (18°C) in July; Spokane averages 47°F (8°C), ranging from 25°F (−4°C) in January to 69°F (21°C) in July. The lowest temperature ever recorded in the state is −48°F (−44°C), set at Mazama and Winthrop on 30 December 1968; the highest, at Ice Harbor Dam on 5 August 1961, was 118°F (48°C).

The average annual precipitation in Seattle is 43 in (109 cm), falling most heavily from October through March; Spokane receives only 17 in (43 cm), more than half of that from November through February. Snowfall in Seattle averages 13 in (33 cm) annually; in Spokane, 51 in (130 cm). High mountain peaks, such as Mt. Adams, Mt. Baker, and Mt. Rainier, have permanent snowcaps or snowfields of up to 100 feet (30 meters) deep.

⁴FLORA AND FAUNA

More than 1,300 plant species have been identified in Washington. Sand strawberries and beach peas are found among the dunes, while fennel and spurry grow in salt marshes; greasewood and sagebrush predominate in the desert regions of the Columbia Plateau. Conifers include Sitka spruce, Douglas fir, western hemlock, and Alaska cedar; big-leaf maple, red alder, black cottonwood, and western yew are among the characteristic deciduous trees. Wild flowers include the deerhead orchid and wake-robin; the western rhododendron is the state flower.

Forest and mountain regions support Columbia black-tailed and mule deer, elk, and black bear; the Roosevelt elk, named after President Theodore Roosevelt, is indigenous to the Olympic Mountains. Other native mammals are the Canadian lynx, red fox, and red western bobcat. Smaller native mammals—western fisher, raccoon, muskrat, porcupine, marten, and mink—are plentiful. The whistler (hoary) marmot is the largest rodent. Game birds include the ruffed grouse, bobwhite quail, and ring-necked pheasant. Sixteen varieties of owl have been identified; other birds of prey include the prairie falcon, sparrow hawk, and golden eagle. The bald eagle is more numerous in Washington than in any other state except Alaska. Washington is also a haven for marsh, shore, and water birds. Various salmon species thrive in coastal waters and along the Columbia River, and the hair seal and sea lion inhabit Puget Sound.

Animals driven away from the slopes of Mt. St. Helens by the volcanic eruption in 1980 have largely returned, and as of July 1984, 26 species of mammals and 101 species of birds were observed inhabiting the mountain again. The number of elk and deer in the vicinity was roughly the same as prior to the eruption, although the mountain goat population reportedly had been killed off. Earlier, on 17 August 1982, the Mt. St. Helens National Volcanic Monument was created by an act of Congress; it includes about 110,000 acres (44,500 hectares) of the area that had been devastated by the original eruption.

Endangered wildlife in the state includes the northern Rocky Mountain wolf, Columbian white-tailed deer, woodland caribou, southern sea otter, American peregrine falcon, Aleutian Canada goose, and short-tailed albatross.

⁵ENVIRONMENTAL PROTECTION

The Department of Ecology, established in 1970, is responsible for water, air, and noise pollution control, solid waste disposal, and shoreline management. Among other state agencies with environmental responsibilities are the State Conservation Commission, Environmental Hearings Office, State Parks and Recreation Commission, State Energy Office, Department of Game, and Department of Natural Resources.

Principal air pollutants in the state are particulate emissions, sulfur dioxides, carbon monoxide, hydrocarbons, lead, and dioxides of nitrogen. Fuel combustion and industrial processes are responsible for most of the first two pollutants, transportation (especially the automobile) for most of the last four. The Department of Ecology has estimated that 16,000 lb (7,250 kg) of solid waste per person per year are generated in Washington, including agricultural, residential, commercial, and industrial and hazardous wastes. The number of penalties assessed by the state for air and water pollution increased from 1,248 in 1979 to 3,069 in 1982. Federal and state construction grants for municipal sewage-treatment facilities in 1982 totaled $581.7 million, with 90% of that amount provided by the state and local communities. In 1981/82, Washington ranked 10th among the states in per capita expenditure for natural resources and parks, with a rate of $91.

Nuclear energy has been the source of considerable controversy in Washington, especially since the publication during the 1970s of reports of leakage from the Hanford Reservation, the nation's largest nuclear waste disposal site, managed by the US Department of Energy. A measure tightening controls on nuclear waste storage was approved by the voters in November 1980.

⁶POPULATION

Washington was the nation's 20th most populous state at the 1980 census, with 4,132,204 residents. The estimated population for January 1985 was 4,366,248, an increase of 5.7% over 1980. Washington's estimated population density in 1985 was 66 per sq mi (24 per sq km). From 1940 to 1980 alone, the state's population more than doubled; the population projection for 1990 was 5,011,800, representing a projected increase of 20% during the decade, about the same as the 21% increase during 1970–80. About 74% of Washington's population was urban at the 1980 census, and 26% rural.

In 1980, more than 2 out of 3 Washingtonians were concentrated in the Western Corridor, a broad strip in western Washington running north–south between the Coast and Cascade ranges. The leading city in the Western Corridor is Seattle, with a 1984 estimated population of 488,474. Other leading cities and their 1984 populations are Spokane, 173,349; Tacoma, 159,435; Bellevue, 78,597; Everett, 56,766; Yakima, 48,945; Vancouver, 43,398; and Bellingham, 45,123. Olympia, the state capital, had 29,176 residents. The Seattle-Tacoma metropolitan area ranked 18th in the US in 1984, with an estimated population of 2,207,800.

⁷ETHNIC GROUPS

Washington is ethnically and racially heterogeneous. As of 1980, foreign-born Washingtonians made up only 5.8% of the state's population, with Canada, Germany, Norway, Sweden, and the United Kingdom being the leading countries of origin. The largest minority group consists of Hispanic Americans, numbering 120,000 according to the 1980 census. Most of the state's Spanish-speaking residents have come since World War II. Black Americans numbered 105,544, according to 1980 census data. Black immigration dates largely from World War II and postwar recruitment for defense-related industries.

Japanese-Americans have been farmers and small merchants in Washington throughout the 20th century. During World War II,

---

LOCATION: 45°32′ 40″ to 49° N; 116°54′ 45″ to 124°44′ 40″ W. **BOUNDARIES:** Canadian line, 286 mi (460 km); Idaho line, 213 mi (343 km); Oregon line, 443 mi (713 km); Pacific Ocean coastline, 157 mi (253 km).

WASHINGTON

SCALE

60 Miles

80 Kms

See endsheet maps: B1.

LEGEND

Tacoma	Over 100,000
Everett	50,000–100,000
Edmonds	20,000–50,000
Aberdeen	10,000–20,000
South Bend	Under 10,000
CLALLAM	County Name

LEGEND

⊛ State Capital
◉ County Seat
✈ Airport
■ Point of Interest
▢ Park, Forest, Reservation

CANADA BRITISH COLUMBIA

IDAHO

OREGON

PACIFIC OCEAN

Strait of Georgia

Strait of Juan de Fuca

Puget Sound

COLUMBIA PLATEAU

R A N G E S

C A S C A D E

WORLDMARK ENCYCLOPEDIA OF THE STATES
© WORLDMARK PRESS LTD.

the Nisei of Washington were interned. Chinese-Americans, imported as laborers in the mid-1800s, endured a wave of mob violence during the 1880s. According to the 1980 census, there were 25,781 Japanese, 20,473 Filipinos, 17,461 Chinese, 13,439 Koreans, and 7,431 Vietnamese in Washington. Immigration from Southeast Asia was an important factor during the late 1970s and early 1980s.

There were 60,771 American Indians, Eskimos, and Aleuts living in Washington in 1980. Indian lands in the state cover some 2.5 million acres (1 million hectares). A dispute developed in the 1970s over Indian fishing rights in the Puget Sound area; a decision in 1974 by US District Judge George Boldt that two 120-year-old treaties guaranteed the Indians 50% of the salmon catch in certain rivers was essentially upheld by the US Supreme Court in 1979.

⁸LANGUAGES

Early settlers took from Chinook some words like *potlatch* (gift-dispensing feast), *skookum* (strong), and *tillicum* (friend), but little other language influence came from the many Indian tribes inhabiting Washington except for such place-names as Chehalis, Walla Walla, Puyallup, and Spokane. Northern and Midland dialects dominate, with Midland strongest in eastern Washington and the Bellingham area, Northern elsewhere. In the urban areas, minor eastern variants have been lost; in rural sections, however, older people have preserved such terms as *johnnycake* (corn bread) and *mouth organ* (harmonica). One survey showed Northern *quarter to* dominant in the state with 81%, with Midland *quarter till* having only a 5% response; Northern *angleworm* (earthworm) had 63%, but Midland *fishworm* and *fishing worm* only 17%. The north coast of the Olympic Peninsula, settled by New Englanders who sailed around Cape Horn, retains New England /ah/ in *glass* and *aunt*. In Seattle, *fog* and *frog* are Midland /fawg/ and /frawg/, but *on* is Northern /ahn/; *cot* and *caught* sound alike, as in Midland; but the final /y/, as in *city* and *pretty*, has the Northern /ee/ sound rather than the Midland short /i/ as in *pit*.

In 1980, English was the language spoken at home by 93% of Washington residents 3 years old and older. Other languages spoken at home and the number of people 3 years old and older speaking them were:

Spanish	81,598	Japanese	13,095
German	37,544	Korean	10,497
French	15,456	Norwegian	9,888
Filipino	14,987	Vietnamese	7,354
Chinese	14,434	Italian	6,786

⁹RELIGIONS

First settled by Protestant missionaries, Washington remains a predominantly Protestant state. As of 1980 there were 895,066 known adherents of Protestant groups. The leading denominations were the Church of Jesus Christ of Latter-day Saints (Mormon), 105,073; United Methodist, 98,114; American Lutheran, 87,554; and United Presbyterian, 71,838. In 1984 there were 428,443 Roman Catholics and an estimated 22,060 Jews.

¹⁰TRANSPORTATION

As of 1983, Washington had 4,225 mi (6,799 km) of Class I railroad lines. Amtrak provides service from Seattle down the coast to Los Angeles, and eastward via Spokane to St. Paul, Minn., and Chicago. There were 358,731 Amtrak passengers in 1983/84; more than half of them boarded trains in Seattle.

Washington had 86,445 mi (139,120 km) of highways, roads, and streets in 1983, of which 714 mi (1,149 km) formed part of the interstate highway system. Principal interstate highways include I-90, connecting Spokane and Seattle, and I-5, proceeding north–south from Vancouver in British Columbia through Seattle and Tacoma to Vancouver, Wash., and Portland, Ore. In 1983, the

state had 2,867,032 licensed drivers and 3,338,333 registered motor vehicles, including 2,360,117 automobiles.

Washington's principal ports include Seattle, Tacoma, and Bellingham, all part of the Puget Sound area and belonging to the Seattle Customs District; and Longview, Kalama, and Vancouver, along the Columbia River and considered part of the Portland (Ore.) Customs District. State-operated ferry systems transported 10,678,800 passengers and 8,072,222 vehicles across Puget Sound in 1983.

In 1983, Washington had 294 airports and 76 heliports. Seattle-Tacoma (SEATAC) International Airport, by far the busiest, handled 9,278,737 arriving and departing passengers, 211,394 aircraft, and 147,986 tons of air freight in 1982.

¹¹HISTORY

The region now known as the State of Washington has been inhabited for at least 9,000 years, the first Americans having crossed the Bering Strait from Asia and entered North America via the Pacific Northwest. Their earliest known remains in Washington—some burned bison bones and a human skeleton—date from approximately 7000 BC.

The Cascades impeded communications between coastal Indians and those of the eastern plateau, and their material cultures evolved somewhat differently. Coastal Indians—belonging mainly to the Nootkin and Salishan language families—lived in a land of plenty, with ample fish, shellfish, roots, and berries. Timber was abundant for the construction of dugout canoes, villages with wooden dwellings, and some stationary wooden furniture. Warfare between villages was fairly common, with the acquisition of slaves the primary objective. The coastal Indians also emphasized rank based on wealth, through such institutions as the potlatch, a gigantic feast with extravagant exchanges of gifts. The plateau (or "horse") Indians, on the other hand, paid little attention to class distinctions. Social organization was simpler and intertribal warfare less frequent here than on the coast. After the horse reached Washington around 1730, the plateau tribes (mainly of the Shahaptian language group) became largely nomadic, traveling long distances in search of food. Housing was portable, often taking the form of skin or mat tepees. In winter, circular pit houses were dug for protection from the wind and snow.

The first Europeans known to have sailed along the Washington coast were 18th-century Spaniards; stories of earlier voyages to the area by Sir Francis Drake in 1579 and Juan de Fuca in 1592 are largely undocumented. In 1774, Juan Pérez explored the northwestern coastline to the southern tip of Alaska; an expedition led by Bruno Heceta and his assistant, Juan Francisco de la Bodega y Quadra, arrived a year later. Men from this expedition made the first known landing on Washington soil, at the mouth of the Hoh River, but the venture ended in tragedy when the Indians seized the landing boat and killed the Spaniards.

English captain James Cook, on his third voyage of exploration, arrived in the Pacific Northwest in 1778 while searching for a northwest passage across America. He was the first of numerous British explorers and traders to be attracted by the luxuriant fur of the sea otter. Cook was followed in 1792 by another Englishman, George Vancouver, who mapped the Pacific coast and the Puget Sound area. In the same year, an American fur trader and explorer, Captain Robert Gray, discovered the mouth of the Columbia River. As the maritime fur trade began to prosper, overland traders moved toward the Northwest, the most active organizations being the British Hudson's Bay Company and the Canadian North West Company.

American interest in the area also increased. Several US maritime explorers had already visited the Northwest when President Thomas Jefferson commissioned an overland expedition to inspect the territory acquired from France through the Louisiana Purchase (1803). That expedition, led by Meriwether Lewis and William Clark, first sighted the Pacific Ocean in early

November 1805 from the north bank of the Columbia River in what is now Pacific County. In time, as reports of the trip became known, a host of British and American fur traders followed portions of their route to the Pacific coast, and the interest of missionaries was excited. In 1831, a delegation visited Clark in St. Louis, Mo., where he was then superintendent of Indian affairs, to persuade him to send teachers who could instruct the Indians in the Christian religion. When news of the visit became known, there was an immediate response from the churches.

The first missionaries to settle in Washington were Marcus and Narcissa Whitman, representing the Protestant American Board of Missions; their settlement, at Waiilatpu in southeastern Washington (near present-day Walla Walla), was established in 1836. Although the early Protestant missions had scant success in converting the Indians, the publicity surrounding their activities encouraged other Americans to journey to the Pacific Northwest, and the first immigrant wagons arrived at Waiilatpu in 1840. The Indian population became increasingly hostile to the missionaries, however, and on 29 November 1847, Marcus and Narcissa Whitman and 12 other Americans were massacred.

As early as 1843, an American provisional government had been established, embracing the entire Oregon country and extending far into the area that is now British Columbia, Canada. Three years later, after considerable military and diplomatic maneuvering, a US-Canada boundary along the 49th parallel was established by agreement with the British. Oregon Territory, including the present state of Washington, was organized in 1848. In the early 1850s, residents north of the Columbia River petitioned Congress to create a separate "Columbia Territory." The new territorial status was granted in 1853, but at the last minute the name of the territory (which embraced part of present-day Idaho) was changed to Washington.

President Franklin Pierce appointed Isaac I. Stevens as the first territorial governor. Stevens, who served at the same time as a US superintendent of Indian affairs, negotiated a series of treaties with the Northwest Indian tribes, establishing a system of reservations. Although the Indian situation had long been tense, it worsened after the treaties were concluded, and bloody uprisings by the Yakima, Nisqualli, and Cayuse were not suppressed until the late 1850s. Court battles over fishing rights spelled out in those treaties were not substantially resolved until 1980.

On the economic front, discoveries of gold in the Walla Walla area, in British Columbia, and in Idaho brought prosperity to the entire region. The completion in 1883 of the Northern Pacific Railroad line from the eastern US to Puget Sound encouraged immigration, and Washington's population, only 23,955 in 1870, swelled to 357,232 by 1890. In the political sphere, Washington was an early champion of women's suffrage. The territorial legislature granted women the vote in 1883; however, the suffrage acts were pronounced unconstitutional in 1887.

Cattle and sheep raising, farming, and lumbering were all established by the time Washington became the 42d state in 1889. The Populist movement of the 1890s found fertile soil in Washington, and the financial panic of 1893 further stimulated radical labor and Granger activity. In 1896, the Fusionists—a coalition of Populists, Democrats, and Silver Republicans—swept the state. The discovery of gold in the Klondike, for which Seattle was the primary departure point, helped dim the Fusionists' prospects, and for the next three decades the Republican Party dominated state politics.

In 1909, Seattle staged the Alaska-Yukon-Pacific Exposition, celebrating the Alaska gold rush and Seattle's new position as a major seaport. World War I brought the state several major new military installations, and the Puget Sound area thrived as a shipbuilding center. The war years also saw the emergence of radical labor activities, especially in the shipbuilding and logging industries. Seattle was the national headquarters of the Industrial

Workers of the World (IWW) and became, in 1919, the scene of the first general strike in the US, involving about 60,000 workers. The towns of Centralia and Everett were the sites of violent conflict between the IWW and conservative groups.

Washington's economy was in dire straits during the depression of the 1930s, when the market for forest products and field crops tumbled. The New Deal era brought numerous federally funded public works projects, notably the Bonneville and Grand Coulee dams on the Columbia River, providing hydroelectric power for industry and water for the irrigation of desert lands. Eventually, more than 1 million acres (400,000 hectares) were reclaimed for agricultural production. During World War II, Boeing led the way in establishing the aerospace industry as Washington's primary employer. Also during the war, the federal government built the Hanford Reservation nuclear research center; the Hanford plant was one of the major contractors in the construction of the first atomic bomb and later became a pioneer producer of atomic-powered electricity.

In 1962, "Century 21," the Seattle World's Fair, again promoted the area, as the Alaska-Yukon-Pacific Exposition had a half-century earlier. The exhibition left Seattle a number of buildings—including the Space Needle and Coliseum—that have since been converted into a civic and performing arts center. The 1960s and 1970s, a period of rapid population growth (with Seattle and the Puget Sound area leading the way), also witnessed an effort by government and industry to reconcile the needs of an expanding economy with an increasing public concern for protection of the state's unique natural heritage. An unforeseen environmental hazard emerged in May 1980 with the eruption of Mt. St. Helens and the resultant widespread destruction.

12 STATE GOVERNMENT

Washington's constitution of 1889, as amended, continues to govern the state today. The legislative branch consists of a senate of 49 members elected to four-year terms, and a house of representatives with 98 members serving for two years. Executives elected statewide include the governor and lieutenant governor (who run separately), secretary of state, treasurer, attorney general, auditor, and officers for education, insurance, and land. The governor and lieutenant governor, who serve four-year terms, must be qualified voters in the state.

A bill becomes law if passed by a majority of the elected members of each house and then signed by the governor or left unsigned for 5 days while the legislature is in session or 20 days after it has adjourned; a two-thirds vote of members present in each house is sufficient to override a gubernatorial veto. Constitutional amendments require a two-thirds vote of the legislature and ratification by the voters at the next general election.

Voters in Washington must be US citizens, at least 18 years of age; the residency requirement is 30 days. There were 2,105,563 registered voters in 1982, and 66.7% of them went to the polls.

13 POLITICAL PARTIES

Washington never went for a full-fledged Democrat in a presidential election until 1932, when Franklin D. Roosevelt won the first of four successive victories in the state. Until then, Washington had generally voted Republican, the lone exceptions being 1896, when the state's Populist voters carried Washington for William Jennings Bryan, and 1912, when a plurality of the voters chose Theodore Roosevelt on the Progressive ticket.

In recent decades, the state has tended to favor Republicans in presidential elections, but Democrats have more than held their own in other contests. Washingtonians elected a Democratic governor, Dixy Lee Ray, in 1976, but in 1980 they chose a Republican, John Spellman; in 1984, they returned to the Democratic column, electing Booth Gardner. As of 1985, the Republicans held both US Senate seats, while the Democrats had a majority of the state's congressional delegation and majorities in both houses of the state legislature. The rise of the Democratic

Party after World War II was linked to the careers of two US senators—Henry Jackson, who held his seat from 1953 until his death in 1983, and Warren Magnuson, defeated in 1980 after serving since 1945. Republicans scored a victory in November 1984, reelecting President Ronald Reagan, who captured 55.8% of the vote to Democrat Walter Mondale's 42.9%.

Washington ranks 5th among the states in percentage of women state legislators—23.8% in 1985. In 1984, the state had 15 black elected officials, including 2 state legislators; 13 Hispanics, including 1 state legislator; and 13 Asians.

14LOCAL GOVERNMENT

As of 1982, Washington had 39 counties, 265 cities and towns, 300 school districts, and 1,130 special districts, including public utility, library, port, water, hospital, cemetery, and sewer districts.

Counties may establish their own institutions of government by charter; otherwise, the chief governing body is an elected board of three commissioners. Other elected officials generally include the sheriff, prosecuting attorney, coroner, auditor, treasurer, and clerk. Cities and towns are governed under the mayor-council or council-manager systems. Larger cities, Seattle among them, generally have their own charters and elected mayors.

15STATE SERVICES

The Public Disclosure Commission, consisting of five members appointed by the governor and confirmed by the senate, provides disclosure of financial data in connection with political campaigns, lobbyists' activities, and the holdings of elected officials and candidates for public office. Each house of the legislature has its own board of ethics.

Public education in Washington is governed by a Board of Education and superintendent of public instruction; the Council for Postsecondary Education coordinates the state's higher educational institutions. The Department of Transportation oversees the construction and maintenance of highways, bridges, and ferries and assists locally owned airports.

The Department of Social and Health Services, the main human resources agency, oversees programs for adult corrections, juvenile rehabilitation, public and mental health, Medicaid, nursing homes, income maintenance, and vocational rehabilita-

tion. Also involved in human resources activities are the Human Rights Commission, Department of Labor and Industries, Employment Security Department, Department of Veterans Affairs, and Council on Child Abuse and Neglect. Public protection services are provided by the Washington State Patrol, the Department of Emergency Services (civil defense), and the Military Department (Army and Air National Guard).

16JUDICIAL SYSTEM

The state's highest court, the supreme court, consists of 9 justices serving six-year terms; 3 justices are elected by nonpartisan ballot in each even-numbered year. Every two years, the senior justice who has not previously been chief justice is named to that office. Appeals of lower court decisions are normally heard in the court of appeals, whose 16 judges are elected to six-year terms. The superior courts, consisting in 1982 of 127 judges elected to four-year terms in 28 judicial districts, are the state trial courts.

Crime rates in 1983 were below the national averages for murder, robbery, aggravated assault, and motor vehicle theft, but above average for forcible rape, burglary, and larceny-theft. The state expended $513 million for police protection and prisons in 1982, or $124 per capita—slightly above the national average.

17ARMED FORCES

The chief US military facilities in Washington in mid-1984 were a Trident nuclear submarine base at Bangor and the Puget Sound Naval Shipyard (Bremerton), Whidbey Island Naval Air Station, McChord Air Force Base (Tacoma), Fairchild Air Force Base (Airway Heights), and Ft. Lewis (Tacoma); authorized military personnel numbered 72,412. In 1983/84, federal defense contract awards exceeded $5.1 billion, 11th among the 50 states.

Veterans living in Washington as of 30 September 1983 numbered 629,000, of whom 6,000 saw service during World War I, 215,000 during World War II, 122,000 in the Korean conflict, and 227,000 during the Viet-Nam era. During 1982/83, veterans' benefits totaled $496 million, including $246 million for compensation and pensions and $47 million for education and training. In mid-1983, Army and Air National Guard units had 7,696 personnel. State and local police forces in Washington during 1983 had 8,536 members, 80% of them local.

Washington Presidential Vote by Political Parties, 1948–84

YEAR	ELECTORAL VOTE	WASHINGTON WINNER	DEMOCRAT	REPUBLICAN	PROGRESSIVE	SOCIALIST	PROHIBITION	SOCIALIST LABOR	CONSTITUTION
1948	8	*Truman (D)	476,165	386,315	31,692	3,534	6,117	1,133	—
1952	9	*Eisenhower (R)	492,845	599,107	2,460	—	—	633	7,290
1956	9	*Eisenhower (R)	523,002	620,430	—	—	—	7,457	—
1960	9	Nixon (R)	599,298	629,273	—	—	—	10,895	1,401
1964	9	*Johnson (D)	779,699	470,366	—	—	—	7,772	—
					PEACE & FREEDOM		AMERICAN IND.		
1968	9	Humphrey (D)	616,037	588,510	1,669	—	96,900	491	—
					PEOPLE'S	LIBERTARIAN			AMERICAN
1972	9	*Nixon (R)	568,334	837,135	2,644	1,537	—	1,102	58,906
1976	9	Ford (R)	717,323	777,732	1,124	5,042	8,585	—	5,046
					CITIZENS			SOC. WORKERS	
1980	9	*Reagan (R)	650,193	865,244	9,403	29,213	—	1,137	
1984	9	*Reagan (R)	807,352	1,051,670	1,891	8,844	—	—	—

*Won US presidential election.

[18] MIGRATION

The first overseas immigrants to reach Washington were Chinese laborers, imported during the 1860s; Chinese continued to arrive into the 1880s, when mob attacks on Chinese homes forced the territorial government to put Seattle under martial law and call in federal troops to restore order. The 1870s and 1880s brought an influx of immigrants from western Europe—especially Germany, Scandinavia, and the Netherlands—and from Russia and Japan.

In recent decades, Washington has benefited from a second migratory wave even more massive than the first. From 1970 to 1980, the state ranked 7th among the states in net migration with a gain of 719,000. Many of those new residents were drawn from other states by Washington's thriving defense- and trade-related industries. In addition, many immigrants from Southeast Asia arrived during the late 1970s. In 1980–83, the state netted 49,000 additional immigrants.

[19] INTERGOVERNMENTAL COOPERATION

Washington participates in the Columbia River Compact (with Oregon), Pacific Marine Fisheries Compact, Western Corrections Compact, Western Interstate Energy Compact, Western Regional Education Compact, Interstate Compact for the Supervision of Parolees and Probationers, Agreement on Qualification of Educational Personnel, Interstate Compact on Placement of Children, Multistate Tax Compact, and Driver License Compact, among other interstate bodies.

Federal aid in 1983/84 totaled nearly $1.7 billion, or about $357 per capita.

[20] ECONOMY

The mainstays of Washington's economy are wholesale and retail trade, manufacturing (especially aerospace equipment, shipbuilding, food processing, and wood products), agriculture, lumbering, and tourism. Between 1971 and 1984, employment increased in such sectors as lumber and wood products, metals and machinery, food processing, trade, services, and government, while decreasing in aerospace—which remains, nevertheless, the state's single leading industry. Foreign trade, especially with Canada and Japan, was an important growth sector during the 1970s and early 1980s. The eruption of Mt. St. Helens in 1980 had an immediate negative impact on the forestry industry—already clouded by a slowdown in housing construction—crop growing, and the tourist trade.

[21] INCOME

With an income per capita of $12,051 in 1983, Washington ranked 14th among the 50 states. Total personal income was $51.8 billion, representing a real increase of 58% in 1970–80 and of 3% in 1980–83. Disposable personal income during 1982 was $42 billion, or $9,903 per capita, 15th among the states, and the average annual pay of Washington workers was $18,039, 11th among the states. Median family income of four-person families in 1981 was $28,254, 12th in the US, and money income per capita in 1983 was $9,426, 10th in the US.

During 1979, 9.8% of all state residents and 7.2% of all Washington families were listed below the federal poverty level, proportions considerably below those for the West and the US as a whole. In 1982, Washington had 30,100 top wealth-holders (those with more than $500,000 in gross assets) with combined assets of $44.2 billion and net worth of $38.8 billion.

[22] LABOR

In 1983, Washington's civilian labor force numbered 2,063,000, of whom 1,833,000 were employed and 230,000 unemployed, yielding an overall unemployment rate of 11.2%. Of those actually employed, 952,000 were males and 881,000 females. The Department of Labor and Industries administers laws pertaining to conditions of work; its concerns include industrial safety, health, and insurance, as well as employment standards.

A federal survey in 1982 revealed the following nonfarm employment pattern in Washington:

	ESTABLISH-MENTS	EMPLOYEES	ANNUAL PAYROLL ('000)
Agricultural services, forestry, fishing	1,478	10,557	$ 135,311
Mining	183	3,061	76,169
Contract construction	8,485	72,957	1,691,458
Manufacturing, of which:	6,000	291,256	6,745,565
Food and food products	(490)	(28,156)	(545,682)
Lumber, wood products	(1,354)	(34,506)	(712,353)
Transport equipment	(291)	(86,743)	(2,248,463)
Transportation, public utilities	3,529	79,236	1,893,346
Wholesale trade	7,622	92,970	1,903,150
Retail trade	24,014	281,732	2,832,295
Finance, insurance, real estate	8,498	92,568	1,596,632
Services	28,746	313,450	4,239,356
Other	1,985	1,906	26,710
TOTALS	90,540	1,239,693	$21,139,992

Government employees, not included in this survey, numbered 301,000 in 1982/83.

Although state and federal authorities suppressed radical labor activities in the mines around the turn of the century, in the logging camps during World War I, and in Seattle in 1919, the impulse to unionize remained strong in Washington. The state's labor force is still one of the most organized in the US, although (in line with national trends) the unions' share of the nonfarm work force declined from 45% in 1970 to 34% in 1980, when there were 553,000 union members; the latter percentage was exceeded only in New York State, Michigan, and Pennsylvania. In 1983, Washington production workers earned an average of $11.41 hourly and $18,039 yearly, figures that ranked among the highest in the US at a time when, in Seattle at least, consumer prices were rising more slowly than in the US as a whole.

[23] AGRICULTURE

Orchard and field crops dominate Washington's agricultural economy, which yielded $3 billion in farm receipts in 1983, 15th among the 50 states. Fruits and vegetables are raised in the humid and in the irrigated areas of the state, while wheat and other grains grow in the drier central and eastern regions.

Washington is the nation's leading producer of apples. The estimated 1983 crop, representing 36% of the US total, totaled 3 billion lb. Among leading varieties, delicious apples ranked first, followed by golden delicious and winesap. The state also ranked 1st in production of hops and cherries, 2nd in grapes, apricots, prunes, and plums, and 3rd in pears. Other preliminary crop figures for 1983 included wheat, 173 million bushels, valued at $647 million; potatoes, 54 million hundredweight, $230 million; barley, 54.4 million bushels, $144.2 million; and corn, 17.6 million bushels, $66 million. Sugar beets, peaches, and various seed crops are also grown.

[24] ANIMAL HUSBANDRY

Livestock and livestock products accounted for an estimated 31% of Washington's agricultural income in 1983. By the end of 1984, farms and ranches had an estimated 1.65 million cattle and calves, 53,000 hogs and pigs, and 62,000 sheep and lambs. Production of meat animals in 1983 included 589.9 million lb of cattle, worth $356.2 million; 19.6 million lb of hogs, $9.6 million; and 5.2 million lb of sheep and lambs, $2.1 million.

In 1983, Washington dairy farmers had 213,000 milk cows that produced 3.4 billion lb of milk, valued at $442 million. Poultry farmers sold $30.1 million worth of chickens and broilers in 1983; egg production accounted for cash receipts of $58.1 million.

[25] FISHING

In 1984, Washington's production of food fish reached 156.3 million lb, valued at $75.7 million. The leading fishing ports, by value of landings in 1984, were Seattle, $16.5 million; Bellingham,

$14.9 million; Blaine, $6.9 million; and Westport, $6.6 million. In 1982, about 3,353 full-time and 5,263 seasonal workers were employed in the state's 264 fish processing plants, and 1,213 were engaged in the wholesale business.

Sport fishermen catch more than 1 million salmon and 100,000 steelhead trout each year. In 1982/83, nearly 67 million game fish and fish eggs were planted in Washington's streams and lakes; planted species included mainly salmon and trout.

²⁶FORESTRY

Washington's forests, covering 23,181,000 acres (9,381,000 hectares), are an important commercial and recreational resource. Some 17,922,000 acres (7,253,000 hectares) are classified as commercial forestland, of which 49% are privately owned, 39% federally owned or managed in nine national forests, and 12% controlled by the state. The largest forests are Wenatchee, Snoqualmie, and Okanogan.

About 324,000 acres (131,000 hectares) encompassing 4.6 billion board feet of lumber were harvested in 1983. In 1982 there were 223 sawmills with about 11,000 employees. Shipments of lumber and wood products amounted to nearly $3.7 billion in 1982. Lumber and plywood, logs for export, various chip products, pulp logs, and shakes and shingles are the leading forest commodities. The giant of Washington's forest industry is Weyerhaeuser, with headquarters in Tacoma; in 1984 the firm ranked 66th among the largest US industrial corporations with sales of $5.5 billion, assets of nearly $6 billion, and net income of $226 million.

Federal, state, and private nurseries in Washington produced more than 136 million seedlings in 1982. Since 1975, more acres have been planted or seeded than have been cut down. Washington's forest-fire control program covers some 12.5 million acres (5.1 million hectares). In 1982 there were 1,287 fires that burned 6,101 acres (2,469 hectares). Leading causes of forest fires in lands under the jurisdiction of the Department of Natural Resources are (in order of frequency) lightning, burning debris, recreation, smoking, and railroad operations.

²⁷MINING

Washington's nonfuel mineral output in 1984 had a total value of $195 million, for a rank of 34th among the 50 states. Principal mineral products in 1984 (excluding fossil fuels) were sand and gravel, 18,840,000 tons, and stone, 11,500,000 tons. Cement, clays, gold, silver, copper, uranium, peat, gypsum, lime, talc, and tungsten are also mined. Washington ranked 1st among the 50 states in aluminum production, accounting for about 30% of the US total in 1984; the aluminum is refined electrolytically from imported ores.

²⁸ENERGY AND POWER

With an installed capacity of 21.8 million kw in 1983, Washington power plants generated 93 billion kwh of electricity, about 90% of that from publicly owned hydroelectric facilities. Energy consumption in the state totaled 1,564 trillion Btu in 1982, of which 34% went to industrial users, 25% to residential customers, 17% to commercial users, and 24% for transportation.

The Hanford Reservation was the site of the first US nuclear energy plant. The Washington Public Power Supply System's ambitious plans to build five nuclear power plants collasped in the early 1980s when it defaulted on bonds to construct two of the plants. As a result, work on two other plants was delayed, and as of 1985, only one of the nuclear plants was in operation. The suspension resulted in the doubling of electric rates for utility customers between 1981 and 1983.

The state's lone major fossil fuel resource is coal. Reserves were estimated at nearly 1.5 billion tons in 1983, of which 78% was subbituminous, 21% bituminous, and 1% lignite. Production of coal from two surface mines totaled 3.8 million tons in 1983.

²⁹INDUSTRY

Transportation equipment, lumber and wood products, food and

food products, nonelectrical machinery, printing and publishing, and paper and allied products are Washington's leading industries, together employing 68% of the state's manufacturing work force. The total value of shipments by manufacturers in 1982 was $34.7 billion. The following table shows the leading industrial sectors and their value of shipments in 1982:

Transportation equipment	$9,021,900,000
Food and food products	5,000,100,000
Lumber and wood products	3,689,600,000
Paper and allied products	2,660,700,000
Primary metal industries	2,519,100,000
Nonelectrical machinery	1,209,000,000

The Seattle-Everett metropolitan area accounts for about half of all industrial employment and value added by manufacture. By far the leading firm in Washington is Boeing, whose Everett manufacturing plant occupied 6.4 million sq feet (595.000 sq meters) in 1984. Other major plants were located at Seattle, Renton, Auburn, and Kent. The aerospace giant ranked 29th among the largest US industrial corporations in 1984, with assets of $8.5 billion, sales of $10.4 billion, and a net income of $787 million. Boeing manufactures commercial jetliners, such as the wide-body 747 and 767 and the smaller but technically advanced 737-300, which began operations in 1984. Boeing's military production includes cruise missiles and E-3 Airborne Warning and Control System (AWACS) aircraft.

Aluminum refining is a major industry, producing about 30% of the nation's supply and employing some 12,000 Washington workers.

³⁰COMMERCE

Wholesalers in 1982 had a trade volume of $31.3 billion. Total retail sales in 1982 reached $20 billion, of which food stores accounted for 24%, automotive dealers 16%, and department stores 14%. The Seattle-Everett metropolitan area accounted for 45% of all retail sales, metropolitan Tacoma 10.5%, and metropolitan Spokane 8%.

In 1982, exports from the state had a value of $11.3 billion; imports, $16 billion. The leading exports were aircraft and aircraft parts (accounting for more than one-half of the total), machinery, lumber and logs, fish and fish products, grains, motor vehicles and parts, fruits and vegetables, wood pulp, and paper products. Principal imports included crude petroleum, lumber, natural gas, passenger cars, truck chassis and bodies, newsprint, aluminum oxide, motorcycles, radios, and television sets. Japan purchased about 21% of the exports and supplied 39% of the imports in 1982; Canada's approximate shares were 16% and 15%, respectively; and Taiwan's, 2% and 10%. The Port of Seattle handled 68% of Washington's waterborne trade.

³¹CONSUMER PROTECTION

The Office of the Attorney General, which enforces the state's 1961 Consumer Protection Act, investigates consumer complaints and, when necessary, seeks court action in connection with retail sales abuses, unfair automobile sales techniques, false advertising, and other fraudulent or deceptive practices. In 1981–83, the consumer protection office handled 161,000 complaints and recovered some $3.5 million for consumers. Consumer protection responsibilities of the Department of Agriculture include food inspection and labeling, sanitary food handling and storage, and accurate weights and measures.

³²BANKING

As of 31 December 1984, Washington had 110 insured commercial banks. At the end of 1983, the state's insured commercial banks had $25 billion in assets; outstanding loans exceeded $15.8 billion, and deposits reached $20.5 billion. The largest commercial bank as of 1984 was the Rainier Bancorporation (Seattle), with assets of $7.7 billion. In 1984 there were 45 savings and loan associations; their assets totaled $10.9 billion at the end of 1982, including mortgage loans worth nearly $8 billion. Washington also had 13 savings banks with assets totaling $9.6 billion as of 31 December 1983.

33 INSURANCE

Washingtonians held 5.2 million life insurance policies with a total face value of $83.7 billion as of 31 December 1983, when the average life insurance coverage per family was $46,300. Benefits paid by life insurance companies in 1983 totaled $867 million, including $223.3 million in death payments, $249.4 million in annuities, and $165.5 million in policy dividends. During the same year, property and liability companies wrote premiums totaling more than $1.6 billion, of which $269.3 million was automobile physical damage insurance, $418.8 million was automotive passenger liability insurance, and $233.7 million was homeowners' coverage. Flood insurance valued at $1 billion was in force as of 31 December 1983.

The Office of the Insurance Commissioner and State Fire Marshal regulates insurance company operations, reviews insurance policies and rates, and examines and licenses agents and brokers. It also conducts fire safety inspections in hospitals, nursing homes, and other facilities, investigates fires of suspicious origin, and regulates the manufacture, sale, and public display of fireworks.

34 SECURITIES

The Spokane Stock Exchange (founded 1897), specializing in mining stocks, traded more than 29 million shares with a total value of $41.3 million in 1983. In that year, New York Stock Exchange member firms had 105 sales offices and 1,113 registered representatives throughout Washington, and there were 102 registered stock dealers and brokers in the state. The number of Washingtonians owning shares in public corporations totaled about 805,000 in 1983.

35 PUBLIC FINANCE

Washington's biennial budget is prepared by the Office of Financial Management and submitted by the governor to the legislature for amendment and approval. The fiscal year runs from 1 July through 30 June.

The following table shows estimated revenues and expenditures for 1983/85 and 1985/87 (in millions):

REVENUES	1983/85	1985/87
Retail sales and use taxes	$ 4,137.1	$ 4,939.4
Business and occupational taxes	1,269.4	1,598.2
Property tax	978.5	1,139.3
Motor vehicle fuel tax	734.3	849.9
Other taxes	1,596.7	1,805.1
Federal grants	3,051.9	3,443.1
Non-tax revenue and other receipts	3,461.4	4,293.4
TOTALS	$15,229.3	$18,068.4

EXPENDITURES		
Education, of which:	$ 6,237.2	$ 7,413.0
Public schools	(3,876.3)	(4,641.3)
Higher education	(1,748.9)	(2,062.0)
Community colleges	(520.1)	(609.4)
Human resources	4,180.5	4,718.9
Transportation	1,604.4	1,881.5
General government	772.7	963.8
Revenue distribution to political subdivisions	572.3	636.5
Natural resources and recreation	634.5	1,108.5
Debt service	531.9	677.5
Other outlays	659.5	556.2
TOTALS	$15,192.9	$17,955.9

During 1981/82, the city of Seattle had general revenues of $391 million and expenditures of $366 million. All state and local government units in Washington had a combined public debt in mid-1983 of $16.9 billion; the per capita debt was $3,917, ranking 2d among the 50 states.

36 TAXATION

Taxes, licenses, permits, and fees account for about three-fifths of Washington state government revenues. In 1982, the state and local tax burden per capita was $1,172, 21st among the 50 states.

Washington has no individual or corporate income tax. As of 1 January 1984, the state levied a general sales tax of 6.5%; a gasoline tax of 16 cents per gallon; a cigarette tax of 23 cents a pack; a liquor excise tax of 11% on sales to restaurants and 17% on sales to individuals; an inheritance tax ranging from 1% to 25%, depending on inheritance size and beneficiary; a gift tax rated at 90% of the inheritance tax; an insurance premium tax of 1% for in-state insurers and 2% for out-of-state insurers; a business and occupation tax averaging 0.47%; a 1% tax on most service activities; a motor vehicle excise tax of 2.3%; a 6.5% excise tax on the value of the timber harvest from private land; and taxes on public utilities and pari-mutuel income. The state property tax, dedicated to the public schools, was levied in 1982/83 at a rate of $2.81 per $1,000 of assessed value.

During the 1982/83 fiscal year, Washington remitted more than $11.7 billion in taxes to the federal government, or $2,734 per capita (14th among the states), and received nearly $7.7 billion in federal expenditures. Washingtonians filed 1.8 million federal tax returns for 1983, paying $6.8 billion in tax.

37 ECONOMIC POLICY

Divisions within the Department of Commerce and Economic Development seek to promote tourism and the motion picture industry, expand markets for Washington's products, and aid existing businesses and attract new ones. The department provides special services for small and minority-owned enterprises, and attempts to create jobs through grants and loans to municipalities and Indian reservations. In 1981–83, the International Trade and Investment Division helped arrange 28 foreign investments totaling an estimated $247 million, which was expected to create 4,800 jobs for Washingtonians. The most important investment was for a manufacturing plant in Puyallup to produce integrated circuits.

38 HEALTH

The birthrate in 1982 was 16.3 per 1,000 population, slightly above the US average; the infant mortality rate in 1981 was 10.5 per 1,000 live births, substantially below the national norm. The state ranked 10th of all the states in legal abortions, with 517 for every 1,000 live births in 1982. The death rate, 7.6 per 1,000 population, was below the national average in 1981, and the rates for the leading causes of death—heart disease, cancer, and stroke—were all lower than average.

The following table shows cases of diseases reported to the Department of Social and Health Services for 1978 and 1982:

	1978	1982
Gonorrhea	13,487	11,381
Gastroenteric disorders	2,082	2,074
Food poisoning	829	589
Meningitis	300	390
Tuberculosis	305	301
Syphilis	265	172
Mumps	223	102
Rubella	149	58
Measles	442	42
Whooping cough	59	36

As of 1983 there were 405 health-care facilities in Washington, including 109 general hospitals, with 13,145 beds; 2 psychiatric institutions, with 156 beds; and 295 nursing homes, with 26,771 beds. The average length of stay in community hospitals during 1983 was 5.4 days, costing an average of $2,681 per stay. General hospitals admitted 584,322 patients in 1983; the average daily population was 8,665. The average daily population of psychiatric institutions in 1983 was 116 patients; the average stay was 23.8 days, costing an average of $3,633 per stay. Of the total number of beds occupied in nursing homes in 1983, 58% were occupied by Medicaid patients; the average daily cost of a Medicaid bed was $36.68, of which the patient paid $9.74 and the state $26.94. In 1983, state expenditures on public health totaled $57.5 million.

Licensed health professionals as of 1981 included 35,071 registered nurses, of whom 70% were employed in nursing. In 1982 there were 9,123 physicians and 2,890 dentists.

[39]SOCIAL WELFARE

Most of the state and federal funds for income maintenance go to families with dependent children (AFDC) and supplemental security income (SSI) programs. In 1983, such aid went to 4.3% of state residents—to 57,600 families receiving AFDC payments, and to 43,700 recipients of SSI aid. In 1982, AFDC benefits totaled $242 million, for an average monthly payment of $389 per family. In 1983, SSI aid amounted to $103 million, with average monthly payments in December being $152 for aged persons and $254 for the disabled.

Social Security benefits in December 1983 were paid to 451,000 retired workers, to 107,000 spouses or other dependents of the beneficiaries, and to 53,000 disabled workers. In that year, Social Security payments totaled $2.9 billion; average monthly benefits were $458 for retired workers, $417 for widows and widowers, and $479 for disabled workers—all above the US average.

During 1982/83, 272,000 state residents took part in the food stamp program, at a federal cost of $137 million. Workers' compensation in 1982 covered 1,475,000 employees, with more than $419 million in benefits paid. State unemployment benefit payments in 1982 were made to 250,000 persons and totaled $526 million; the average weekly benefit was $131, slightly above the national average.

[40]HOUSING

The 1980 census counted 1,689,450 occupied housing units in Washington, or 38% more than at the 1970 census. Some 65% of the units were owner-occupied, and more than 98% had full plumbing. In 1983, Washington ranked 19th among the states with 27,500 new housing units started. That year, 27,200 new privately owned housing units were authorized, worth over $1.5 billion. In 1980, the median monthly cost for an owner-occupied housing unit was $362; median rent was $256. According to the 1980 census, Seattle had 230,000 units, of which 10% were built since 1970; the median value of a one-family house was $65,100.

[41]EDUCATION

Washingtonians rank exceptionally high by most educational standards. As of 1980, more than 77% of all Washingtonians 25 years of age or older were high school graduates; 19% had four or more years of college; and only 4% had less than eight years of grade school. Washington ranked 3d among the 50 states in state aid for education with an average expenditure of $718 per capita in 1982, and was 10th in state and local expenditures for education with $803 per capita. It placed 4th among the states in the number of public school pupils (21.6) per teacher as of the fall 1983 enrollment. The state was 7th in average yearly salaries for public school teachers with $24,780 in 1983/84. In that year, Washington spent a total of $2,468 million on public schools, or $595 per capita; about 75% of the funds were provided by the state, 19% by localities, and 6% by the federal government.

As of 1981/82, there were 800,213 pupils enrolled in all kindergartens, elementary schools, and high schools, of which 92% attended public schools and 8% went to private schools. As of 1 January 1984, 48,368 students were enrolled in parochial schools. The state had 71 vocational and correspondence schools in 1982.

The following table shows enrollments of public and private schools from the fall semester of 1975 to fall 1982:

	PUBLIC (K-12)	PRIVATE (K-12)
1982/83	736,094	NA
1981/82	738,618	61,595
1980/81	749,120	59,538
1979/80	756,572	57,803
1978/79	763,748	54,940
1977/78	769,248	52,940
1976/77	775,894	48,750
1975/76	780,080	47,924

As of 1983, Washington had 22 accredited colleges and universities (6 public, 16 private) and 27 community colleges. The largest institution is the University of Washington (Seattle), founded in 1861 and enrolling 33,884 students in 1983. Other public institutions are Washington State University (Pullman), with 16,261 students; Eastern Washington University (Cheney), 8,492; Central Washington University (Ellensburg), 7,121; Western Washington University (Bellingham), 9,617; and Evergreen State College (Olympia), 2,717. Private institutions include Gonzaga University (Spokane), with 3,412 students in 1983; Pacific Lutheran University (Tacoma), 3,533 students; Seattle University, 4,685; Seattle Pacific College, 3,548; University of Puget Sound (Tacoma), 4,197; Walla Walla College, 1,625; Witman College (Walla Walla), 1,270; and Whitworth College (Spokane), 1,884. Community colleges enrolled 152,976 students in 1982/83. In that year the average salary for a college teacher was $26,874 (23d among the states).

[42]ARTS

The focus of professional performance activities in Washington is Seattle Center, home of the Seattle Symphony, Pacific Northwest Dance Ballet Company, and Seattle Repertory Theater. The Seattle Opera Association (founded 1964), which also performs there throughout the year, is one of the nation's leading opera companies, offering five operas each season and presenting Richard Wagner's "Ring" cycle at the Pacific Northwest Festival in July. Tacoma and Spokane have notable local orchestras.

Among Washington's many museums, universities, and other organizations exhibiting works of art on a permanent or periodic basis are the Seattle Art Museum, with its Modern Art Pavilion, and the Henry Art Gallery of the University of Washington at Seattle. Others include the Washington State University Museum of Art at Pullman; the Whatcom Museum of History and Art (Bellingham); the Tacoma Art Museum; the State Capitol Museum (Olympia); and the Cheney Cowles Memorial Museum of the Eastern Washington State Historical Society (Spokane). The State Arts Commission, established in 1961, supports nonprofit arts groups as well as professional arts organizations and an artists-in-schools program. The commission's proposed operating budget for 1984/85 was $1,717,133, 17.7% more than in the preceding fiscal year.

[43]LIBRARIES AND MUSEUMS

In 1982, Washington's system of public libraries held more than 8.9 million volumes and had a combined circulation of nearly 28 million. Of Washington's 39 counties, 29 were served by the state's 19 county and multicounty libraries. Total library income in 1982 came to $66 million, most of that total derived from public funds.

The leading public library system is the Seattle Public Library, with 17 branches and 1,566,424 volumes in 1982. The principal academic libraries are at the University of Washington (Seattle) and Washington State University (Pullman), with 4,168,079 and 1,346,609 volumes, respectively. Olympia is the home of the Washington State Library, with a collection of 395,000 books and more than 1 million documents.

Washington has at least 100 museums and historic sites. The Washington State Historical Society Museum (Tacoma) features Indian and other pioneer artifacts; the State Capitol Museum

(Olympia) and Cheney Cowles Memorial Museum (Spokane) also have important historical exhibits, as do the Thomas Burke Memorial Washington State Museum (Seattle) and the Pacific Northwest Indian Center (Spokane). Mt. Rainer National Park displays zoological, botanical, geological, and historical collections. The Pacific Science Center (Seattle) concentrates on aerospace technology; the Seattle Aquarium is a leading attraction of Waterfront Park. Also in Seattle is Woodland Park Zoological Gardens, while Tacoma has the Point Defiance Zoo and Aquarium.

44COMMUNICATIONS

The US Postal Service operated 464 post offices in Washington and employed 11,514 persons in 1985. As of the 1980 census, 94% of Washington's 1,540,510 occupied housing units had telephones. During 1984, Washington had 197 commercial radio stations—104 AM, 93 FM—and 20 commercial and 6 educational television stations, while 2 cable systems served 744,839 subscribers in 271 communities.

45PRESS

In 1984, Washington had 9 morning newspapers (including all-day papers), with a combined 1984 circulation of 393,664; 20 evening dailies, with 737,043; and 17 Sunday papers, with 1,094,928. The following table shows the leading newspapers with their 1984 circulations:

AREA	NAME	DAILY	SUNDAY
Seattle	Post-Intelligencer (m,S)	191,825	464,995
	Times (e,S)	226,916	464,995
Spokane	Chronicle (e)	48,828	
	Spokesman-Review (m,S)	82,711	139,687
Tacoma	News Tribune and Sunday Ledger (e, S)	104,077	114,493

46ORGANIZATIONS

The 1982 Census of Service Industries counted 1,384 organizations in Washington, including 259 business associations; 843 civic, social, and fraternal associations; and 31 educational, scientific, and research associations.

Among the associations with headquarters in Washington are the American Plywood Association and the American Academy on Mental Retardation, in Tacoma; the Center for the Defense of Free Enterprise and the Citizens Committee for the Right to Keep and Bear Arms, in Bellevue; the Pacific International Trapshooting Association (Puyallup); the Northwest Mining Association (Spokane); and the Northwest Fisheries Association, the International Association for the Study of Pain, the International Conference of Symphony and Opera Musicians, and the Mountaineers, all located in Seattle.

47TOURISM, TRAVEL, AND RECREATION

Seattle Center—featuring the 605-foot (184-meter) Space Needle tower, Opera House, and Pacific Science Center—helps make Washington's largest city one of the most exciting on the West Coast. Nevertheless, scenic beauty and opportunities for outdoor recreation are Washington's principal attractions for tourists from out of the state.

Mt. Rainier National Park, covering 235,404 acres (95,265 hectares), encompasses not only the state's highest peak but also the most extensive glacial system in the conterminous US. Glaciers, lakes, and mountain peaks are also featured at North Cascades National Park (504,780 acres—204,278 hectares), while Olympic National Park (908,720 acres—367,747 hectares) is famous as the site of Mt. Olympus and for its dense rain forest and rare elk herds. Washington also offers two national historic parks (San Juan Island and part of Klondike Gold Rush), two national historic sites (Fort Vancouver and the Whitman Mis-

sion), and three national recreation areas (Coulee Dam, Lake Chelan, and Ross Lake). The most popular state park is Deception Pass, which received more than 2 million visitors in 1982.

Hunting is a highly popular pastime. In 1982, 41,640 deer were killed out of an estimated population of 457,000; the elk harvest was 12,570 out of a population estimated at 59,000. Washington hunters also bagged at least 1,000,000 ducks, 360,000 pheasants, and about 425,000 grouse and quail. Licenses were issued to 303,967 deer and elk hunters and to 607,855 fishermen in 1982.

The tourism industry was the 4th largest employer in Washington, providing an estimated 57,900 jobs in 1982. In that year, domestic tourists spent an average of $7 million a day in the state for an annual total of more than $2.5 billion.

48SPORTS

Seattle, home of the Kingdome, an enclosed stadium seating 59,438 for baseball and 64,752 for football, is well represented in professional sports. In baseball, the Mariners compete in the American League; the Seahawks play in the National Football League. Seattle's winningest franchise has been the SuperSonics of the National Basketball Association, who captured the league championship in 1979. In collegiate sports, the Huskies of the University of Washington, who play at Husky Stadium in Seattle (with a seating capacity of nearly 60,000), won the Rose Bowl games in 1960, 1961, 1978, and 1982.

49FAMOUS WASHINGTONIANS

Washington's most distinguished public figure was US Supreme Court Justice William O. Douglas (b. Minnesota, 1898-1980), who grew up in Yakima and attended Whitman College in Walla Walla. In addition to his 37-year tenure on the Court, an all-time high, Douglas was the author of numerous legal casebooks as well as 27 other volumes on various subjects. Other federal officeholders from Washington include Lewis B. Schwellenbach (b. Wisconsin, 1894-1948), secretary of labor under Harry Truman, and Brockman Adams (b. Georgia, 1927), secretary of transportation under Jimmy Carter. Serving in the US Senate from 1945 to 1981, Warren G. Magnuson (b. Minnesota, 1905) held the chairmanship of the powerful Appropriations Committee. A fellow Democrat, Henry M. "Scoop" Jackson (1912-83) was first elected to the House in 1940 and to the Senate in 1952. Influential on the Armed Services Committee, Jackson ran unsuccessfully for his party's presidential nomination in 1976. William E. Boeing (b.Michigan, 1881-1956) pioneered Washington's largest single industry, aerospace technology.

Notable governors include Isaac I. Stevens (b.Massachusetts, 1818-62), Washington's first territorial governor; after serving as Washington's territorial representative to Congress, he died in the Civil War. Elisha P. Ferry (b.Michigan, 1825-95), territorial governor from 1872 to 1880, was elected as Washington's first state governor in 1889. John R. Rogers (b.Maine, 1838-1901), Washington's only Populist governor, was also the first to be elected for a second term. Clarence D. Martin (1886-1955) was governor during the critical New Deal period. Daniel J. Evans (b.1925) is the youngest man ever elected governor of Washington and also is the only one to have served three consecutive terms (1965-77).

Dixy Lee Ray (b.1914), governor from 1977 to 1981 and the only woman governor in the state's history, is a former head of the federal Atomic Energy Commission and a staunch advocate of nuclear power. Other notable women were Emma Smith De Voe (b.New Jersey, 1848-1927), a leading proponent of equal suffrage, and Bertha Knight Landes (b.Massachusetts, 1868-1943), elected mayor of Seattle in 1926; Landes, the first woman to be elected mayor of a large US city, was also an outspoken advocate of moral reform in municipal government.

Several Washington Indians attained national prominence. Seattle (1786-1866) was the first signer of the Treaty of Point Elliott, which established two Indian reservations; the city of

Seattle is named for him. Kamiakin (b.Idaho, c. 1800–80) was the leader of the Yakima tribe during the Indian Wars of 1855, and Leschi (d.1858) was chief of the Nisqualli Indians and commanded the forces west of the Cascades during the 1855 uprising; Leschi was executed by the territorial government after the uprising was suppressed.

Washington authors have made substantial contributions to American literature. Mary McCarthy (b.1912) was born in Seattle, and one of her books, *Memories of a Catholic Girlhood* (1957), describes her early life there. University of Washington professor Vernon Louis Parrington (b. Illinois, 1871–1929) was the first Washingtonian to win a Pulitzer Prize (1928), for his monumental *Main Currents in American Thought.* Another University of Washington faculty member, Theodore Roethke (b.Michigan, 1908–63), won the Pulitzer Prize for poetry in 1953. Seattle-born Audrey May Wurdemann (1911–60) was awarded a Pulitzer Prize for poetry in 1934 for *Bright Ambush.* Max Brand (Frederick Schiller Faust, 1892–1944) wrote hundreds of Western novels.

Singer-actor Harry Lillis "Bing" Crosby (1904–77), born in Tacoma, remained a loyal alumnus of Spokane's Gonzaga University. Modern dance choreographers Merce Cunningham (b.1919) and Robert Joffrey (b.1930) are both Washington natives. Photographer Edward S. Curtis (b.Wisconsin, 1868–1952) did most of the work on the North American Indian series while residing in Seattle. Modern artist Mark Tobey (b.Wisconsin, 1890–1976) spent much of his productive life in Seattle, and Robert Motherwell (b.1915) was born in Aberdeen. Washington's major contribution to popular music is rock guitarist Jimi Hendrix (1943–70).

⁵⁰BIBLIOGRAPHY

Bancroft, H. H. *History of Washington, Idaho, and Montana.* San Francisco: History Co., 1890.

Clark, Norman H. *Washington: A Bicentennial History.* New York: Norton, 1976.

Douglas, William O. *Of Men and Mountains.* New York: Harper & Row, 1950.

Drury, Clifford M. *Marcus and Narcissa Whitman and the Opening of Old Oregon.* 2 vols. Glendale, Calif.: Clark, 1973.

Ficken, Robert E. *Lumber and Politics: The Career of Mark E. Reed.* Seattle: University of Washington Press, 1980.

Johansen, Dorothy O., and Charles M. Gates. *Empire of the Columbia.* 2d ed. New York: Harper & Row, 1967.

Kirk, Ruth. *Washington State: National Parks, Historic Sites, Recreation Areas, and Natural Landmarks.* Seattle: University of Washington Press, 1974.

Lee, W. Storrs (ed.). *Washington State: A Literary Chronicle.* New York: Funk & Wagnalls, 1969.

Meany, Edmond S. *History of the State of Washington.* New York: Macmillan, 1909.

Meinig, D. W. *The Great Columbia Plain: A Historical Geography. 1805–1910.* Seattle: University of Washington Press, 1968.

Snowden, Clinton A. *History of Washington.* 6 vols. New York: Century History, 1911.

Stewart, Edgar I. *Washington, Northwest Frontier.* 4 vols. New York: Lewis, 1957.

Tyler, Robert. *Rebels of the Woods: The I.W.W. and the Pacific Northwest.* Eugene, Ore.: University of Oregon Books, 1967.

Washington, State of. Office of Financial Management. *Washington State Data Book 1985.* Olympia, 1984.

Washington State Research Council. *The Fact Book: Economic and Fiscal Data 1984.* Olympia, 1984.

Yates, Richard and Charity (eds.). *Washington State Yearbook, 1984: A Guide to Government in the Evergreen State.* Sisters, Ore.: Information Press, 1983.

WEST VIRGINIA

State of West Virginia

ORIGIN OF STATE NAME: The state was originally the western part of Virginia. **NICKNAME:** The Mountain State. **CAPITAL:** Charleston. **ENTERED UNION:** 20 June 1863 (35th). **SONGS:** "The West Virginia Hills"; "West Virginia, My Home Sweet Home"; "This Is My West Virginia." **MOTTO:** *Montani semper liberi* (Mountaineers are always free). **COAT OF ARMS:** A farmer stands to the right and a miner to the left of a large ivy-draped rock bearing the date of the state's admission to the Union. In front of the rock are two hunters' rifles upon which rests a Cap of Liberty. The state motto is beneath and the words "State of West Virginia" above. **FLAG:** The flag has a white field bordered by a strip of blue, with the coat of arms in the center, wreathed by rhododendron leaves; across the top of the coat of arms are the words "State of West Virginia." **OFFICIAL SEAL:** The obverse is the same as the coat of arms; the reverse is no longer in common use. **ANIMAL:** Black bear. **BIRD:** Cardinal. **FISH:** Brook trout. **FLOWER:** *Rhododendron maximum* ("big laurel"). **TREE:** Sugar maple. **FRUIT:** Apple. **COLORS:** Old gold and blue. **LEGAL HOLIDAYS:** New Year's Day, 1 January; Birthday of Martin Luther King, Jr., 3d Monday in January; Lincoln's Birthday, 12 February; Washington's Birthday, 3d Monday in February; Memorial Day, last Monday in May; West Virginia Day, 20 June; Independence Day, 4 July; Labor Day, 1st Monday in September; Columbus Day, 2d Monday in October; Veterans Day, 11 November; Thanksgiving Day, 4th Thursday in November; Christmas Day, 25 December. **TIME:** 7 AM EST = noon GMT.

¹LOCATION, SIZE, AND EXTENT

Located in the eastern US in the South Atlantic region, West Virginia ranks 41st in size among the 50 states.

The area of West Virginia totals 24,231 sq mi (62,758 sq km), including 24,119 sq mi (62,468 sq km) of land and 112 sq mi (290 sq km) of inland water. The state extends 265 mi (426 km) E–W; its maximum N–S extension is 237 mi (381 km). West Virginia is one of the most irregularly shaped states in the US, with two panhandles of land—the northern, narrower one separating parts of Ohio and Pennsylvania, and the eastern panhandle separating parts of Maryland and Virginia.

West Virginia is bordered on the N by Ohio (with the line formed by the Ohio River), Pennsylvania, and Maryland (with most of the line defined by the Potomac River); on the E and S by Virginia; and on the W by Kentucky and Ohio (with the line following the Ohio, Big Sandy, and Tug Fork rivers).

The total boundary length of West Virginia is 1,180 mi (1,899 km). The geographical center of the state is in the Elk River Public Hunting Area in Braxton County, 4 mi (6 km) E of Sutton.

²TOPOGRAPHY

West Virginia lies within two divisions of the Appalachian Highlands. Most of the eastern panhandle, which is crossed by the Allegheny Mountains, is in the Ridge and Valley region. The remainder, or more than two-thirds of the state, is part of the Allegheny Plateau, to the west of a bold escarpment known as the Allegheny Front, and tilts toward the Ohio River.

The mean elevation of West Virginia is 1,500 feet (457 meters), higher than any other state east of the Mississippi River. Its highest point, Spruce Knob, towers 4,863 feet (1,482 meters) above sea level. Major lowlands lie along the rivers, especially the Potomac, Ohio, and Kanawha. A point on the Potomac River near Harpers Ferry has the lowest elevation, only 240 feet (73 meters) above sea level. West Virginia has no natural lakes.

Most of the eastern panhandle drains into the Potomac River. The Ohio and its tributaries—the Monongahela, Little Kanawha, Kanawha, Guyandotte, and Big Sandy—drain most of the Allegheny Plateau section. Subterranean streams have carved out numerous caverns—including Seneca Caverns, Smoke Hole Caverns, and Organ Cave—from limestone beds.

During the Paleozoic era, when West Virginia was under water, a 30,000-foot (9,000-meter) layer of rock streaked with rich coal deposits was laid down over much of the state. Alternately worn down and uplifted during succeeding eras, most of West Virginia is thus a plateau where rivers have carved deep valleys and gorges and given the land a rugged character.

³CLIMATE

West Virginia has a humid continental climate, with hot summers and cool to cold winters. The climate of the eastern panhandle is influenced by its proximity to the Atlantic slope and is similar to that of nearby coastal areas. Mean annual temperatures vary from 56°F (13°C) in the southwest to 48°F (9°C) in higher elevations. The yearly average is 53°F (12°C). The highest recorded temperature, 112°F (44°C), was at Martinsburg on 10 July 1936; the lowest, −37°F (−38°C), at Lewisburg on 30 December 1917.

Prevailing winds are from the south and west, and seldom reach hurricane or tornado force. Precipitation averages 45 in (114 cm) annually and is slightly heavier on the western slopes of the Alleghenies. Accumulations of snow may vary from about 20 in (51 cm) in the western sections to more than 50 in (127 cm) in the higher mountains.

⁴FLORA AND FAUNA

With its varied topography and climate, West Virginia provides a natural habitat for more than 3,200 species of plants in three life zones: Canadian, Alleghenian, and Carolinian. Oak, maple, poplar, walnut, hickory, birch, and such softwoods as hemlock, pine, and spruce are the common forest trees. Rhododendron, laurel, dogwood, redbud, and pussy willow are among the more than 200 flowering trees and shrubs. Rare plant species include the box huckleberry, Guyandotte beauty, and Kate's mountain clover. The Cranberry Glades, an ancient lakebed similar to a glacial bog, contains the bog rosemary and other plant species common in more northern climates.

West Virginia fauna includes at least 56 species and subspecies of mammals and more than 300 types of birds. The gray wolf, puma, elk, and bison of early times have disappeared. The white-tailed (Virginia) deer and the black bear (both protected by the state) as well as the wildcat are still found in the deep timber of the Allegheny ridges; raccoons, skunks, woodchucks, opossums, gray and red foxes, squirrels, and cottontail rabbits remain numerous. Common birds include the cardinal, tufted titmouse, brown thrasher, scarlet tanager, catbird, and a diversity of sparrows, woodpeckers, swallows, and warblers. Major game birds are the wild turkey, bobwhite quail, and ruffed grouse; hawks and owls are the most common birds of prey. Notable among more than 100 species of fish are smallmouth bass, rainbow trout, and brook trout (the state fish). The copperhead and rattlesnake are both numerous and poisonous. The southern bald eagle, American and Arctic peregrine falcons, Indiana and Virginia big-eared bats, flat-spired three-toothed snail, and tuberculed-blossom and pink mucket pearly mussels are on the endangered list.

5ENVIRONMENTAL PROTECTION

Major responsibility for environmental protection in West Virginia rests with the Department of Natural Resources. The Youth Conservation Program, with 1,000 groups and over 60,000 young people, the Junior Conservation Camp, nature tours, and other programs are part of a statewide conservation education effort.

"Chemical Valley," a 30-mi (48-km) stretch on the Kanawha River west of Charleston, contains more than a dozen plants handling potentially hazardous materials. Higher rates for respiratory cancer than in the nation at large have been found here. The US Environmental Protection Agency reported that at the Union Carbide plant in Institute there had been 28 leaks of methyl isocyanate—the chemical that killed over 2,000 people in 1984 in Bhopal, India—between 1980 and 1984. In August 1985, a leak of another toxic gas from the same plant hospitalized 135 persons.

Abandoned coal mines are another environmental hazard. The state has estimated that $500 million would be needed to rectify the problem of mine subsidence, which threatens existing structures with cave-ins. Waste accumulations from coal mines were serving as fill when a makeshift dam collapsed at Buffalo Creek, near Logan, in 1972, taking 118 lives.

6POPULATION

With a 1980 census total of 1,949,644, West Virginia ranked 34th among the 50 states in population. In early 1985, the population was estimated at 1,968,969, yielding a density of 82 persons per sq mi (32 per sq km). The projection for 1990 is 2,037,400.

The state's population grew rapidly in the 1880s and 1890s, as coal mining, lumbering, and railroads expanded to meet the needs of nearby industrial centers, but the pace of expansion slowed in the early 20th century. The population peaked at 2,005,552 in 1950; then mass unemployment, particularly in the coal industry, caused thousands of families to migrate to midwestern cities. An upswing began in the 1970s.

In 1980, when nearly 75% of the US population lived in urban areas, only 36% of West Virginia's population was urban—a smaller proportion than any other state but Vermont. In 1984, Charleston, the largest city, had 59,371 residents; Huntington, 61,086; Wheeling, 42,080; and Parkersburg, 39,379. Huntington belongs to a metropolitan region that includes parts of eastern Kentucky and southern Ohio and had an estimated population of 334,000 in 1984; the Charleston region had 267,000.

7ETHNIC GROUPS

Nearly all Indian inhabitants had left the state before the arrival of European settlers. In the 1980 census, 1,555 Indians were counted.

The 61,051 blacks in the state in 1980 constituted about 3.1% of the population. The majority lived in industrial centers and coal-mining areas. Only 21,980 West Virginians, or 1.1% of the population, were foreign-born in 1980. Most of these residents came from Italy, the UK, Germany, and the Soviet Union. There were 12,707 Hispanic Americans and 5,194 persons of Asian and Pacific origin. Persons reporting a single ancestry group in 1980 included 388,511 English, 138,664 Germans, 106,452 Irish, and 33,191 Italians.

8LANGUAGES

With little foreign immigration and with no effect from the original Iroquois and Cherokee Indians, West Virginia maintains Midland speech. There is a secondary contrast between the northern half and the southern half, with the former influenced by Pennsylvania and the latter by western Virginia.

The basic Midland speech sounds the /r/ after a vowel, as in *far* and *short*, and has /kag/ for *keg*, /greezy/ for *greasy*, *sofy* instead of sofa, and *nicker* in place of neigh. The northern part has /yelk/ for *yolk*, /loom/ for *loam*, an /ai/ diphthong so stretched that *sat* and *sight* sound very much alike, *run* for creek, and *teeter (totter)* for seesaw. The southern half pronounces *here* and *hear* as /hyeer/, *aunt* and *can't* as /aint/ and /kaint/, and uses *branch* for creek, and *tinter* for teeter.

In 1980, 1,820,954 West Virginians—98% of the population 3 years of age or over—spoke only English at home. Other languages spoken at home included Spanish, 8,144; Italian, 5,652; French, 4,959; and German, 4,591.

9RELIGIONS

Throughout its history, West Virginia has been overwhelmingly Protestant. Most settlers before the American Revolution were Anglicans, Presbyterians, Quakers, or members of German sects, such as Lutherans, German Reformed, Dunkers, and Mennonites. The Great Awakening had a profound effect on these settlers, and they avidly embraced its evangelism, emotionalism, and emphasis on personal religious experience. Catholics were mostly immigrants from Ireland and southern and eastern Europe.

In 1980, the major Protestant denominations and the number of their adherents were United Methodist, 219,560; American Baptist USA, 156,126; and Presbyterian Church US, 30,538. The Catholic population was 107,318 in 1984; the Jewish population was estimated at 4,265. Leading fundamentalist denominations in 1980 were the Church of God (33,162) and the Church of the Nazarene (26,538).

10TRANSPORTATION

West Virginia has long been plagued by inadequate transportation. The first major pre–Civil War railroad line was the Baltimore and Ohio (B&O), completed to Wheeling in 1852. Later railroads, mostly built between 1880 and 1917 to tap rich coal and timber resources, also helped open up interior regions to settlement. Today, the railroads still play an important part in coal transportation. In 1983, the B&O and three other Class I railroads operated 3,387 mi (5,451 km) of track. Amtrak provides passenger service for parts of the state; in 1983/84, it had a total West Virginian ridership of 55,534.

At the end of 1983 there were 33,814 mi (54,418 km) of roads under the state system and 4,015 mi (6,462 km) of other municipal and rural roads. The West Virginia Turnpike, completed from Charleston to Princeton in 1955 and regarded at that time as a marvel of engineering, has been far outclassed by the state's interstate highways, including the Appalachian Corridor highway system. There were 1,446,057 registered motor vehicles in the state in 1983/84, and 1,417,000 licensed drivers in 1983.

Major navigable inland rivers are the Ohio, Kanawha, and Monongahela; each has locks and dams. In 1983, West Virginia had 61 airports, 23 heliports, and 10 seaplane bases. Kanawha County Airport is the state's main air terminal, with 187,448 enplaned passengers in 1983.

11HISTORY

Paleo-Indian cultures in what is now West Virginia existed some 15,000 years ago, when hunters pursued buffalo and other large game. About 7000 BC, they were supplanted by Archaic cultures, marked by pursuit of smaller game. Woodland (Adena) cultures, characterized by mound-building and agriculture, prevailed after about 1000 BC.

By the 1640s, the principal Indian claimants, the Iroquois and Cherokee, had driven out older inhabitants and made the region a

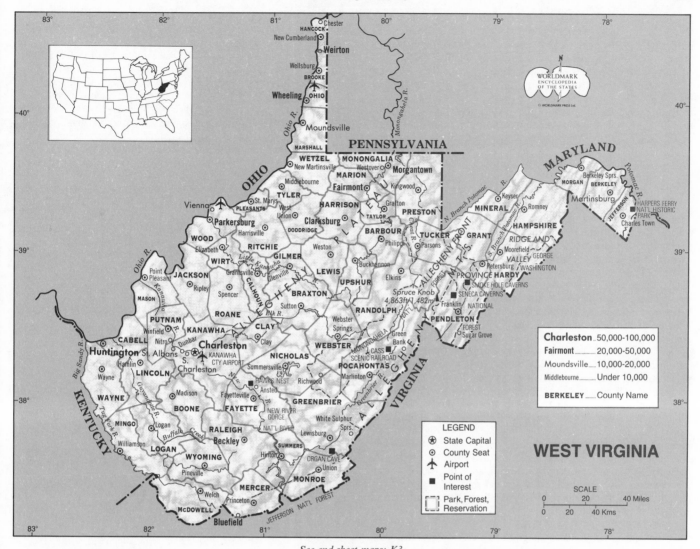

WEST VIRGINIA

Charleston 50,000–100,000
Fairmont 20,000–50,000
Moundsville 10,000–20,000
Middlebourne .... Under 10,000

BERKELEY .... County Name

LEGEND
⊛ State Capital
⊙ County Seat
✈ Airport
■ Point of Interest
⬚ Park, Forest, Reservation

SCALE
0   20   40 Miles
0 20   40 Kms

See end sheet maps: K3.

LOCATION: 37°12′8″ to 40°38′17″N; 77°43′11″ to 82°38′48″ W. **BOUNDARIES:** Ohio line, 277 mi (446 km); Pennsylvania line, 119 mi (191 km); Maryland line, 235 mi (378 km); Virginia line, 438 mi (705 km); Kentucky line, 111 mi (179 km).

vast buffer land. When European settlers arrived, only a few Shawnee, Tuscarora, and Delaware Indian villages remained, but the area was still actively used as hunting and warring grounds, and European possession was hotly contested.

The fur trade stimulated early exploration. In 1671, Thomas Batts and Robert Fallam explored New River and gave England a claim to the Ohio Valley, to which most of West Virginia belongs. France also claimed the Ohio Valley by virtue of an alleged visit by Robert Cavelier, Sieur de la Salle, in 1669. England eventually prevailed as a result of the French and Indian War.

Unsubstantiated tradition credits Morgan Morgan, who moved to Bunker Hill in 1731, with the first settlement in the state. By 1750, several thousand settlers were living in the eastern panhandle. In 1769, following treaties with the Iroquois and Cherokee, settlers began to occupy the Greenbrier, Monongahela, and upper Ohio valleys, and movement into other interior sections continued into the Revolutionary War, although wars with Indians occurred sporadically until the 1790s. The area that is now West Virginia was part of Virginia at the time of that state's entry into the Union, 25 June 1788.

Serious differences between eastern and western Virginia developed after the War of 1812. Eastern Virginia was dominated by a slaveholding aristocracy, while small diversified farms and infant industries predominated in western Virginia. Westerners rankled under property qualifications for voting, inadequate representation in the Virginia legislature, and undemocratic county governments, as well as poor transportation, inadequate schools, inequitable taxes, and economic retardation. A constitutional convention in 1829–30 failed to effect changes, leaving the westerners embittered. Another convention in 1850–51 met the west's political demands but exacerbated economic differences.

When Virginia seceded from the Union in 1861, western counties remaining loyal to the Union set up the Reorganized Government at Wheeling. Two years later, the Reorganized Government consented to the separation of present West Virginia from Virginia. After approval by Congress and President Lincoln, West Virginia entered the Union on 20 June 1863 as the 35th state. West Virginia won control over Jefferson and Berkeley counties in the eastern panhandle in 1871, giving it a greater share of the Baltimore and Ohio Railroad lines in the state.

Both Bourbon Democratic and Republican governors after the Civil War sought to improve transportation, foster immigration, and provide tax structures attractive to business. Industrialists such as Democrats Henry Gassaway Davis and Johnson N. Camden, who amassed fortunes in coal, oil, railroads, and timber, sat in the US Senate and dominated party affairs in West Virginia.

Similarly, industrialists Nathan Goff, Jr., and Stephen B. Elkins—Davis's son-in-law—wielded preponderant influence in the Republican Party from the 1870s until 1911. Native industrialists often collaborated with eastern interests to give the state a colonial economy dominated by absentee owners. Although Republican governors of the early 20th century were dominated by Elkins, they were attuned to Progressive ideas and were instrumental in the adoption of the direct primary, safety legislation for the coal mines, revision of corporate tax laws, and improvements in highways and education.

The Great Depression of the 1930s, from which West Virginia suffered acutely, ushered in a Democratic era. West Virginians embraced the New Deal and Fair Deal philosophies of Presidents Franklin D. Roosevelt and Harry S Truman.

World Wars I and II produced significant changes in West Virginia, particularly through stimulation of chemical, steel, and textile industries in the Kanawha and Ohio valleys and the eastern panhandle. These industries lessened the state's dependence on extractive industries, historically the backbone of its economy, and gave cities and towns a more cosmopolitan character.

Overshadowing the economic diversification was the plight of the coal-mining areas, where, after World War II, mechanization and strip-mining displaced thousands of miners and resulted in a large exodus to other states. By 1960, West Virginia was considered one of the most economically depressed areas of the country, primarily because of conditions in the mining regions. Antipoverty programs of the Kennedy and Johnson administrations provided some relief, but much of it was temporary, as was a brief upsurge in coal mining during the late 1970s.

In general, over the last several decades, West Virginia's economy has gradually become less dependent on mining and heavy industry, and tourism has become a growing source of income. Moreover, the homogenizing effects of the automobile and of radio, television, and national marketing of goods and ideas have changed West Virginia considerably. Its people, who in the 1920s and 1930s were as culturally isolated as their grandparents had been, were by the 1980s generally in the mainstream of American life.

¹²STATE GOVERNMENT

Since becoming a state, West Virginia has had two constitutions. The first, adopted in 1863, served until 1872, when the present constitution was adopted. As of the end of 1983, 55 amendments to this constitution had become law.

The legislature consists of a senate with 34 members and a house of delegates with 100 members. Senators and delegates must be at least 25 and 18 years old, respectively. Both must have been residents of the state and of their districts for at least one year before taking their seats. Senators are elected to staggered four-year terms, and delegates serve for two years. The legislature meets annually in 60-day sessions.

Elected officials of the executive branch of government are the governor, secretary of state, auditor, attorney general, commissioner of agriculture, and treasurer, all elected for four-year terms. The governor, who may serve no more than two terms in succession, must be at least 30 years old and have been a resident of the state for at least five years.

Bills passed by the legislature become law when signed by the governor. Those he vetoes may become law if repassed by majorities of both house memberships—except for revenue and appropriations bills, which require a two-thirds majority of both houses. Either house may propose an amendment to the state constitution. If both houses approve it by a two-thirds majority, it is submitted to the voters at the next regular election or at a special election for adoption by majority vote.

The right to vote in an election is extended to all citizens over 18 years old who have registered and who have resided in their respective counties for 30 days.

¹³POLITICAL PARTIES

The Republican Party presided over the birth of West Virginia, but the Democrats have generally been in power for the past five decades. In 1940, a strong New Deal faction, headed by Matthew M. Neely and supported by organized labor, formed the "state-house machine," which became a dominant factor in state politics. Only two Republicans, Cecil H. Underwood (1957-61) and Arch Moore, Jr. (1969-77, 1985-   ), have been governor since 1933.

In 1984, West Virginia had 996,689 registered voters, of whom 675,106 were Democrats, 300,147 were Republicans, and 21,436 belonged to minor parties or were unaffiliated. Since the New Deal, Republican presidential candidates have carried West Virginia only in 1956, 1972, and 1984. The state is even more firmly Democratic in elections for other offices. Although Moore was elected to a third term as governor in 1984, all other statewide offices were won by Democrats. After two terms as governor (1977-85), John D. "Jay" Rockefeller IV, a Democrat, was elected to the Senate in 1984; the senior senator, Robert Byrd, also a Democrat, was Senate majority leader, 1977-80. The four US representatives, all Democrats, were reelected in 1984, and the legislature is overwhelmingly Democratic.

There were 21 female members of the West Virginia legislature in 1985. In 1984 there were 19 black elected officials, including one state legislator.

West Virginia Presidential Vote by Major Political Parties, 1948–84

YEAR	ELECTORAL VOTE	WEST VIRGINIA WINNER	DEMOCRAT	REPUBLICAN
1948	8	*Truman (D)	429,188	316,251
1952	8	Stevenson (D)	453,578	419,970
1956	8	*Eisenhower (R)	381,534	449,297
1960	8	*Kennedy (D)	441,786	395,995
1964	7	*Johnson (D)	538,087	253,953
1968	7	Humphrey (D)	374,091	307,555
1972	6	*Nixon (R)	277,435	484,964
1976	6	*Carter (D)	435,914	314,760
1980	6	Carter (D)	367,462	334,206
1984	6	*Reagan (R)	328,125	405,483

*Won US presidential election.

¹⁴LOCAL GOVERNMENT

West Virginia has 55 counties. The chief county officials are the three commissioners, elected for six-year terms, who comprise the county court; the sheriff, assessor, county clerk, and prosecuting attorney, elected for four-year terms; and the five-member board of education, elected for six-year terms. The sheriff is the principal peace officer but also collects taxes and disburses funds of the county court and board of education.

The 231 cities, towns, and villages, as of mid-1984, are divided into classes according to population. The minimum requirement for incorporation is a population of 100; the minimum for incorporation as a city is 2,000.

¹⁵STATE SERVICES

The Board of Education determines policy for public elementary and secondary schools, and the Board of Regents governs the state's colleges and universities. The Department of Highways is responsible for construction and operation of state roads. Services of the Department of Health center around treatment of alcoholism and drug abuse, mental health, and environmental health services, maternal and child care, family planning, and control of communicable diseases. The Department of Human Services administers a variety of economic, medical, and social services.

In the area of public protection, the Department of Public Safety enforces criminal and traffic laws, the Office of Emergency Services oversees civil defense and other emergency activities, and the Department of Corrections oversees prisons and other such facilities. The Public Service Commission regulates utilities. The Housing Development Fund concentrates on housing for low- and middle-income families and the elderly. The Department of Natural Resources has the major responsibility for protection of forests, wildlife, water, and other resources, for reclamation projects, and for operation of state parks and recreational facilities.

Responsibility in labor matters is shared by the Department of Labor, Department of Employment Security, Department of Mines, Workers' Compensation Fund, and Labor-Management Relations Board.

[16]JUDICIAL SYSTEM

The highest court in West Virginia, the supreme court of appeals, has five justices, including the chief justice, elected for 12-year terms. The court has broad appellate jurisdiction in both civil and criminal cases, and original jurisdiction in certain other cases.

As of 1985, West Virginia was divided into 31 judicial circuits, each with from 1 to 7 judges, for a total of 60, elected for eight-year terms. Each circuit served from one to four counties and had jurisdiction over civil and criminal cases in amounts that exceeded $300. Circuit courts also had jurisdiction over juvenile, domestic relations, and administrative proceedings.

Local courts include the county magistrate courts and municipal courts. Magistrate courts have original jurisdiction in criminal matters but may not convict or sentence in felony cases. All judges down to the magistrate level are popularly elected by partisan ballot. Municipal, police, or mayor's courts have authority to enforce municipal ordinances.

As of 1985, the Department of Corrections had charge of eight state institutions, ranging from the maximum security prison at Moundsville to youth correctional and rehabilitation centers.

In total crimes per 100,000 population in 1983, West Virginia ranked lowest among the states, with a rate of 2,419; the violent crime rate, 172, was about one-third the US average. The state abolished the death penalty in 1965.

[17]ARMED FORCES

West Virginia has no military bases, academies, or training facilities. The Naval Telecommunications Station (Sugar Grove) is the main receiving facility for the Navy's global high-frequency radio communications and for point-to-point circuits destined for Washington, D.C. In 1983/84, defense contracts awarded West Virginia firms totaled $72 million.

In 1984, West Virginia had about 242,000 living war veterans, of whom 2,000 served in World War I, 93,000 in World War II, 42,000 in the Korean conflict, and 72,000 in the Viet-Nam era. In 1983, veterans received $299 million in benefits.

There were 5,594 National Guard personnel at the beginning of 1985, and 3,449 state and local police in 1983.

[18]MIGRATION

West Virginia has considerable national and ethnic diversity. Settlers before the Civil War consisted principally of English, German, Scotch-Irish, and Welsh immigrants, many of whom came by way of Pennsylvania. A second wave of immigration from the 1880s to the 1920s brought thousands of Italians, Poles, Austrians, and Hungarians to the coal mines and industrial towns, which also attracted many blacks from the South. In 1980, 79% of the residents of the state were born in West Virginia (4th highest among states).

Between 1950 and 1970, West Virginia suffered a 13% loss in population, chiefly from the coal-mining areas; but between 1970 and 1980, population rose by almost 12%. According to federal estimates, the state had a net migration gain of 71,000 in the 1970s and a net migration loss of 13,000 between 1981 and 1983.

[19]INTERGOVERNMENTAL COOPERATION

The West Virginia Commission on Interstate Cooperation participates in the Council of State Governments. West Virginia is a member of 23 regional compacts, including the Ohio River Valley Water Sanitation and Potomac River Basin compacts, Southern Regional Education Board, Southern States Energy Board, and Southern States Governors' Association.

In 1983/84, federal grants to West Virginia totaled $819 million. Federal allocations to West Virginia under the Appalachian development program in 1984/85 totaled $19.5 million.

[20]ECONOMY

Agriculture was the backbone of West Virginia's economy until the 1890s, when extractive industries (including coal, oil, natural gas, and timber) began to play a major role. World War I stimulated important secondary industries, such as chemicals, steel, glass, and textiles. The beauty of West Virginia's mountains and forests attracted an increasing number of tourists in the 1980s, but the state's rugged topography and relative isolation from major markets continued to hamper its economic development.

[21]INCOME

West Virginia's total personal income—$17.6 billion—ranked the state 34th in the US in 1983. The state's 1983 per capita income of $8,937 ranked 49th in the US, down from 43d in 1980. In 1979, 62,000 West Virginia families had incomes below the federal poverty level. Some 5,900 West Virginians were among the nation's top wealth-holders in 1982, with gross assets of more than $500,000. By place of work, manufacturing accounted for 20% of total personal income in 1983, followed by services and government, each 16%; mining and trade, each 15%; and transportation and public utilities, 10%.

[22]LABOR

West Virginia's labor force had the nation's lowest rate of participation by women (39%) in 1984. Unemployment averaged 15% in 1984, highest among states; the number of unemployed persons in 1984 averaged 116,000. In February 1983, during the depths of the 1981–83 recession, unemployment reached 21%—the highest rate for any state since accurate records were first collected in 1940. A federal survey of workers in March 1982 revealed the following nonfarm employment pattern for West Virginia:

	ESTABLISH-MENTS	EMPLOYEES	ANNUAL PAYROLL ('000)
Agricultural services, forestry, fishing	194	1,303	$ 14,319
Mining, of which:	1,273	67,904	1,655,192
Bituminous coal, lignite	(716)	(57,425)	(1,447,359)
Contract construction	2,481	20,948	410,721
Manufacturing, of which:	1,528	101,207	2,031,666
Primary metals	(40)	(18,030)	(497,156)
Chemicals and allied products	(67)	(17,430)	(472,004)
Transportation, public utilities	1,511	29,078	596,309
Wholesale trade	2,357	27,665	478,147
Retail trade	9,397	98,370	851,087
Finance, insurance, real estate	2,428	21,483	297,302
Services, of which:	9,314	97,787	1,175,619
Health services	(2,380)	(40,577)	(633,663)
Other	504	518	7,243
TOTALS	30,987	466,263	$7,517,605

Government workers were not included in this survey; in 1983 there were 107,000 state and local workers, and in 1982 there were 15,000 civilian federal employees. The state's total civilian labor force averaged 769,000 in 1984. In 1980, a higher proportion of West Virginia's work force—44.5%—consisted of blue-collar workers than that of any other state.

Important milestones in the growth of unionism were the organization of the state as District 17 of the United Mine Workers of America (UMWA) in 1890 and the formation of the State Federation of Labor in 1903. The coal miners fought to gain union recognition by coal companies, and instances of violence were not uncommon in the early 1900s. Wages, working conditions, and benefits for miners improved rapidly after World War II. Membership in unions in 1980 was 222,000, or 34% of the work force, compared to 47% in 1970, an indication of the UMWA's waning strength. In 1983 there were 983 unions. Strikes, both legal and wildcat, contributed to decreases in production, particularly coal production, during the 1970s and early 1980s. In 1981, for example, 95 work stoppages idled 96,400 workers for a total of 3,405,000 days. The rate of over 2% working time lost was by far the highest among states. In 1984, however, a contract agreement between the UMWA and the coal industry was reached without an industry-wide work stoppage. But no agreement was reached with the nonunion A. T. Massey Coal Co., the nation's 6th-largest coal operator, and a bitter strike ensued.

23 AGRICULTURE

With estimated farm marketings of $228 million (only $55 million from crops), West Virginia ranked 46th among the 50 states in 1983; it was also one of only three states that had a negative net income from farming that year. Until about 1890, small diversified farms dominated the economy, but, as in other states, farms have grown larger and the farm population has declined since that time. The farm population dropped from 533,000 in 1940 to 78,000 in 1970 and 29,000 in 1980.

In 1984, the state had 3,800,000 acres (1,538,000 hectares), or 25% of its land, devoted to farming. Its 22,000 farms averaged 173 acres (70 hectares) in size. Major farm sections are the eastern panhandle, a tier of counties along the Virginia border, the upper Monongahela Valley, and the Ohio Valley. Leading crops produced in 1983 were hay, 594,000 tons; corn for grain, 4,680,000 bushels; corn for silage, 540,000 tons; commercial apples, 210,000,000 lb; and tobacco, 3,762,000 lb.

24 ANIMAL HUSBANDRY

Animal husbandry in 1983 accounted for 76% of West Virginia's agricultural cash receipts. The cattle industry, the largest component, produced 142,435,000 lb of beef in 1983, for a gross income of $62,192,000. Other major livestock products in 1983 were hogs and pigs, 24,693,000 lb; broilers, 105,798,000 lb; and turkeys, 32,727,000 lb. In 1983, the dairy industry yielded 380,000,000 lb of milk and 136,000,000 eggs.

25 FISHING

West Virginia fishing has little commercial importance; the 25 licensed Ohio River commercial fishermen reported harvesting 41,812 lb of fish in 1983, of which 78% was catfish.

26 FORESTRY

In 1984, West Virginia had 11.5 million acres (4.7 million hectares) of commercial forestland. The state was once covered by a forest believed to have had the greatest variety of hardwoods in the US, but because of excessive lumbering, the 10 million acres (4 million hectares) of virgin timber left in 1870 had all but disappeared by 1920. Reforestation is now a state priority, and more wood was grown than cut in 1984. Hardwoods—chiefly oaks, maples, poplars, and beeches—make up about 90% of the timber volume. A total of 260,000 board feet of lumber, almost all hardwoods, was produced in 1983.

All of the Monongahela National Forest and parts of the George Washington and Jefferson national forests are in West Virginia. These three included a total of 968,270 acres (391,846 hectares) of forestland as of 1984. Nine state forests, totaling 79,285 acres (32,086 hectares), include recreational facilities.

27 MINING

In 1984, West Virginia ranked 38th among the states in production of nonfuel minerals, which were valued at more than $102.6 million. Crushed stone accounted for 37% of the state total. Other mining products were cement, clays, salt, lime, and sand and gravel.

28 ENERGY AND POWER

West Virginia has long been an important supplier of energy in the form of electric power and fossil fuels. In 1983, installed capacity was 15.1 million kw. Net generation of electric energy was 72.5 billion kwh, of which 99% was produced by coal-fired steam units. Out of 20.7 billion kwh of electricity sold in the state in 1983, 33% went to residential customers, 20% to commercial, and 47% to industrial. The state's power facilities are all privately owned. The John Amos Plant, on the Kanawha River, is one of the world's largest investor-owned generating plants.

Major coal-mining regions lie within a north–south belt some 60 mi (97 km) wide through the central part of the state and include the Fairmount, New River–Kanawha, Pocahontas, and Logan-Mingo fields. In 1983, West Virginia was 2d to Kentucky in coal production, with 642 mines producing 114 million tons (15% of the national total), all of it bituminous and 80% of it mined underground. In 1982, the value of coal shipments was $5.9 billion; value added, $3.9 billion. Demonstrated reserves came to 39 billion tons. In 1983, West Virginia produced 3.6 million barrels of oil (24th in the US) and 130 billion cu feet of natural gas (13th). Proved petroleum reserves in 1983 totaled 49 million barrels; natural gas, 2.2 trillion cu feet.

29 INDUSTRY

Although an industrial state throughout much of its history, West Virginia enjoyed only a 4% increase in value added by manufacture between 1977 and 1982, and manufacturing employment declined by 18%. The total value added by manufacture in 1982 exceeded $4 billion, the chief components being chemicals, 36%; primary metals, 14%; and stone, clay, and glass products, 10%. The declining fortunes of the steel industry were mostly responsible for a 26% drop in value added for primary metals.

The value of shipments by manufacturers in 1982 totaled $9.9 billion. The following table shows value of shipments for selected industries:

Industrial organic chemicals	$2,030,400,000
Blast furnace and basic steel products	1,234,200,000
Pressed or blown glass products	398,700,000
Construction machinery	190,200,000

Major industrial areas are the Kanawha, Ohio, and Monongahela valleys and the eastern panhandle. The largest industrial corporation with headquarters in West Virginia is worker-owned Weirton Steel, which in 1984 had sales of $1.1 billion. Among large out-of-state companies operating in West Virginia are Union Carbide, Wheeling-Pittsburgh Steel, and E. I. du Pont de Nemours.

30 COMMERCE

In 1982, West Virginia's wholesale trade establishments had sales of $6.1 billion, ranking the state 37th in the US. Retail sales of $7.5 billion in 1982 represented 0.7% of US sales and placed the state 34th among the 50 states. Sales in food stores made up 27% of the total; automotive dealers, 18%; general merchandise group stores, 14%; gasoline service stations, 10%; eating and drinking places, 7%; and other retailers, 24%.

In 1981, West Virginia exported $1.8 billion worth of coal and led all other states in the export of bituminous coal. Exports of manufactured goods, chiefly chemicals and primary metals, were valued at $1.2 million.

31 CONSUMER PROTECTION

The state attorney general is empowered to investigate, arbitrate, and prosecute complaints involving unfair and deceptive trade practices. Assisting the attorney general is a nine-member Consumer Affairs Advisory Council, with five members representing the general public and four representing the consumer financing and retail businesses.

The Public Service Commission, consisting of three members, regulates rates, charges, and services of utilities and common carriers. Since 1977, it has included one member who is supposed to represent the "average" wage earner.

32 BANKING

West Virginia had 228 commercial banks at the end of 1984; assets were $12.6 billion at the end of 1983. Twenty-one savings and loan associations had assets of $1.5 billion in 1983. West Virginia had one building and loan association and 191 credit unions in 1984.

33 INSURANCE

In 1983, 7 life insurance companies and 17 property and casualty insurance companies were domiciled in West Virginia. There were 3.1 million life insurance policies in force in 1983, with a total value of $30.1 billion; the average family had $40,800 in coverage (49th among the 50 states). Life insurance companies collected premiums of $553 million and paid out $307.3 million, including $107.8 million in death benefits.

Property and casualty firms collected $593.4 million in premiums in 1983. The state insurance commissioner has responsibility for regulating the insurance business.

34 SECURITIES

There are no securities exchanges in West Virginia. New York Stock Exchange member firms had 27 sales offices and 227 registered representatives in the state in 1983. A total of 251,000 West Virginians were shareowners of public corporations that year.

35 PUBLIC FINANCE

The state constitution requires the governor to submit to the legislature within 10 days after the opening of a regular legislative session a budget for the ensuing fiscal year (1 July–30 June). The budget for fiscal 1983 was as follows (in millions):

REVENUES	
General revenue fund	$1,255.2
Federal funds	607.5
Special revenue funds	287.4
State road fund	276.9
Other revenues	1.7
TOTAL	$2,428.7

EXPENDITURES	
Education	$1,062.1
Health and welfare	473.5
Highways	444.8
Other governmental costs	483.2
TOTAL	$2,463.6

Total public debts of the state and local governments in 1982/83 were nearly $3.9 billion. Per capita debt in 1981/82 was $1,939 (21st in the US).

36 TAXATION

West Virginia's diversified tax base yielded receipts of more than $1.4 billion in 1983/84. Personal income taxes, ranging from 2.1% to 13%, accounted for $400.4 million; the corporate income tax of 6–7%, $73.6 million; and consumers' sales taxes of 5% on goods and services excluding food, $252 million. The business and occupations tax on gross income, yielding $511 million, made up 36% of the general revenue fund. Counties and localities mainly tax real and personal property. The federal tax burden was almost $4 billion in 1982, 34th largest among the states. West Virginians filed 654,614 federal income tax returns in 1983, paying a total of nearly $1.6 billion.

37 ECONOMIC POLICY

The Industrial Development Division of the Office of Economic Development has major responsibility for development planning. It supports business and industry in the state and assists new companies with site location and employee training programs, as well as with construction of plants and access roads and provision of essential services.

The West Virginia Economic Development Authority may make loans of up to 50% of the costs of land, buildings, and equipment at low interest rates for a normal term of 15 years.

Tax incentives include a credit of 10% on industrial expansion and revitalization, applicable to the business and occupations tax over a 10-year period.

38 HEALTH

West Virginia's birthrate of 13.9 per 1,000 in 1982 was 44th among states; the abortion rate of 122 per 1,000 live births ranked 47th. The death rate in West Virginia, 9.8 per 1,000 population in 1981, was exceeded by only two states, Florida and Pennsylvania. Only nine states had a higher infant mortality than West Virginia's rate of 13 in 1981, but by 1983 this rate had dropped to the national average, 10.9. The state had the 3d-highest death rate from heart disease, 409 per 100,000, in 1981. Other leading causes of death were cancer, stroke, accidents, chronic obstructive pulmonary diseases (highest rate among states), pneumonia and flu, and diabetes—in that order. Pneumoconiosis (black lung) is an occupational hazard among coal miners.

In 1983, the state's 73 hospitals—including 8 state-run facilities—had 12,983 beds and 427,864 admissions. There were 3,111 physicians and 797 dentists in 1982.

Medical education is provided by medical schools at West Virginia University and Marshall University and at the West Virginia School of Osteopathic Medicine.

39 SOCIAL WELFARE

Although rich in resources, West Virginia has more than its share of poverty. In 1982, 77,500 persons received $58 million in aid to families with dependent children; the average monthly payment to families was $183. An estimated 278,000 persons received food stamps in 1983, with a federal subsidy amounting to $146 million. About 230,000 pupils participated in the school lunch program, at a federal cost of $25 million.

Social Security benefits in 1983 included $920 million for 210,000 retired workers and their dependents, $410 million for 86,000 survivors, and $253 million for 57,000 disabled persons. Federal Supplemental Security Income payments to 39,571 aged, blind, and disabled totaled $90.7 million in 1983. Under the Black Lung Benefit Program, 60,052 miners, widows, and dependents received $15 million in December 1983. Workers' compensation payments were $221.1 million in 1982/83, when unemployment benefits reached $312.5 million.

40 HOUSING

In 1980, West Virginia had 686,311 occupied housing units, of which 504,921, or 74%, were owner-occupied (highest among states) and 181,390, or 26%, were renter-occupied. Of these units, 93% had full plumbing. Between 1981 and 1983, about 6,000 new housing units valued at $196 million were authorized by the state.

41 EDUCATION

West Virginia has generally ranked below national standards in education. In 1980, 56% of adult West Virginians were high school graduates, a criterion by which the state outranked only five other states.

In 1984/85, the state's public schools enrolled 363,042 students, with 214,214—including 25,721 in preschools and kindergarten—in elementary grades and 148,828 in secondary grades. Nonpublic schools enrolled 14,543 students, including 13,243 in church-related schools.

The state supports West Virginia University, Marshall University, and the West Virginia College of Graduate Studies (all offering graduate work), as well as 3 medical schools, 8 four-year colleges, and 4 two-year institutions. Public higher educational institutions enrolled 72,054 students in the fall of 1983. There were 10 private colleges with 10,617 students in fall 1983.

42 ARTS

Known for the quilts, pottery, and woodwork of its mountain artisans, West Virginia has shown considerable artistic enterprise. Huntington Galleries, the Sunrise Foundation at Charleston, and Oglebay Park, Wheeling, are major art centers. The Science and Culture Center at Charleston features West Virginia and Appalachian artists at work. The Mountain State Art and Craft Fair is held each summer at Ripley. Musical attractions include the Charleston Symphony Orchestra, the Wheeling Symphony, and a country music program at Wheeling.

43 LIBRARIES AND MUSEUMS

In 1982/83, West Virginia's public libraries had 3,252,629 volumes and a circulation of 14,932,922. The largest was the Kanawha County Public Library system at Charleston, with 519,736 volumes. Of college and university libraries, the largest collection was West Virginia University's 1,030,484 volumes.

There were 28 museums in 1984, including the State Museum and the Sunrise Museums in Charleston, and Oglebay Institute–Mansion Museum in Wheeling. Point Pleasant marks the site of a battle between colonists and Indians, and Harpers Ferry is the site of John Brown's raid. Wheeling is the location of the Oglebay's Good Children's Zoo.

44 COMMUNICATIONS

In 1985, West Virginia had 1,002 post offices, with 4,242 employees. Of the 686,311 occupied housing units in the state in 1980, 615,130, or fewer than 90%, had telephones. In 1984, broadcasting included 67 AM and 66 FM radio stations, 11 television stations, and 3 public television stations; 189 cable television systems served 402,962 subscribers.

45 PRESS

In 1984, West Virginia had 24 daily newspapers and 10 Sunday newspapers. The following table shows leading West Virginia newspapers with their 1984 circulations:

AREA	NAME	DAILY	SUNDAY
Charleston	Gazette (m, S)	54,554	107,903
	Daily Mail (e, S)	52,993	
Huntington	Herald-Dispatch (m, S)	43,255	47,485

46 ORGANIZATIONS

The 1982 US Census of Service Industries counted 556 organizations in West Virginia, including 85 business associations; 389 civic, social, and fraternal associations; and 8 educational, scientific, and research associations. The Black Lung Association, based in Beckley, promotes safe working conditions in coal mines and benefits for disabled miners. The headquarters of the Appalachian Trail Conference is in Harpers Ferry, and the American Association of Zoological Parks and Aquariums is in Oglebay.

47 TOURISM, TRAVEL, AND RECREATION

The tourist industry in West Virginia earned $1.4 billion and employed 40,165 persons in 1982. Major attractions are Harpers Ferry National Historical Park, New River Gorge National River, the Naval Telecommunications Station at Sugar Grove, and White Sulphur Springs, a popular mountain resort. Among 34 state parks and 9 state forests are Cass Scenic Railroad, which includes a restoration of an old logging line, and Prickett's Fort, with re-creations of pioneer life. State parks drew 7,120,191 visitors in 1983/84; state forests, 974,947. In 1982/83, licenses were issued to 336,676 hunters and 295,681 anglers.

48 SPORTS

No major league professional teams are based in West Virginia. West Virginia University's basketball team won a National Invitation Tournament championship in 1942 and was NCAA Division I runner-up in 1959.

Horse-racing tracks operate in Chester and Charles Town. Greyhound races are run in Wheeling and Charleston. Other popular sports are skiing and white-water rafting.

49 FAMOUS WEST VIRGINIANS

Among West Virginians who have served in presidential cabinets are Nathan Goff, Jr. (1843-1920), Navy secretary; William L. Wilson (1843-1900), postmaster general; John Barton Payne (1855-1935), interior secretary; and Newton D. Baker (1871-1937), secretary of war during World War I. Lewis L. Strauss (1896-1974) was commerce secretary and chairman of the Atomic Energy Commission, and Cyrus R. Vance (b.1917) served as secretary of state. John W. Davis (1873-1955), an ambassador to Great Britain, ran as the Democratic presidential nominee in 1924. Prominent members of the US Senate have included Matthew M. Neely (1874-1958), who was also governor, Harley M. Kilgore (1893-1956), and Robert C. Byrd (b.1917).

Thomas J. "Stonewall" Jackson (1824-63) was a leading Confederate general during the Civil War. Brigadier General Charles E. "Chuck" Yeager (b.1923), a World War II ace, became the first person to fly faster than the speed of sound.

Major state political leaders, all governors (though some have held federal offices), have been E. Willis Wilson (1844-1905), Henry D. Hatfield (1875-1962), Arch A. Moore, Jr. (b.1923), and John D. "Jay" Rockefeller IV (b.New York, 1937).

The state's only Nobel Prize winner has been Pearl S. Buck (Pearl Sydenstricker, 1893-1973), who won the prize for literature for her novels concerning China. Alexander Campbell (b.Ireland, 1788-1866), with his father, founded the Disciples of Christ Church and was president of Bethany College in West Virginia. Major labor leaders have included Walter Reuther (1907-70), president of the United Automobile Workers, and Arnold Miller (b.1923), president of the United Mine Workers.

Musicians include George Crumb (b.1929), a Pulitzer Prize-winning composer, and opera singers Eleanor Steber (b.1916) and Phyllis Curtin (b.1922). Melville Davisson Post (1871-1930) was a leading writer of mystery stories. Important writers of the modern period include Mary Lee Settle (b.1918) and John Knowles (b.1926). Jerry West (b.1938) was a collegiate and professional basketball star, and a pro coach after his playing days ended; Rod Hundley (b.1934) and Hal Greer (b.1936) also starred in the National Basketball Association. Mary Lou Retton (b.1968) won a gold medal in gymnastics at the 1984 Olympics. Another West Virginian of note is Anna Jarvis (1864-1948), founder of Mother's Day.

50 BIBLIOGRAPHY

Conley, Phil, and William Thomas Doherty. *West Virginia History.* Charleston: Education Foundation, 1974.

Federal Writers' Project. *West Virginia: A Guide to the Mountain State.* Reprint. New York: Somerset, 1980 (orig. 1941).

Forbes, Harold M. *West Virginia History: Bibliography and Guide to Studies.* Morgantown: West Virginia University Press, 1981.

Rice, Otis K. *West Virginia: The State and Its People.* Parsons, W.Va.: McClain, 1972.

West Virginia Research League, Inc. *1984 Statistical Handbook.* Charleston, 1984.

Williams, John Alexander. *West Virginia: A Bicentennial History.* New York: Norton, 1976.

Willis, Todd C. (ed.). *West Virginia Blue Book, 1984.* Charleston: Jarrett, 1984.

WISCONSIN

State of Wisconsin

ORIGIN OF STATE NAME: Probably from the Ojibwa word *wishkonsing*, meaning "place of the beaver." **NICKNAME:** The Badger State. **CAPITAL:** Madison. **ENTERED UNION:** 29 May 1848 (30th). **SONG:** "On, Wisconsin!" **MOTTO:** Forward. **COAT OF ARMS:** Surrounding the US shield is the shield of Wisconsin, which is divided into four parts symbolizing agriculture, mining, navigation, and manufacturing. Flanking the shield are a sailor, representing labor on water, and a yeoman or miner, labor on land. Above is a badger and the state motto; below, a horn of plenty and a pyramid of pig lead. **FLAG:** A dark-blue field, fringed in yellow on three sides, surrounds the state coat of arms on each side, with 'Wisconsin' in white letters above the coat of arms and '1848' below. **OFFICIAL SEAL:** arms surrounded by the words "Great Seal of the State of Wisconsin" and 13 stars below. **ANIMAL:** Badger. **WILDLIFE ANIMAL:** White-tailed deer. **DOMESTIC ANIMAL:** Dairy cow. **BIRD:** Robin. **FISH:** Muskellunge. **FLOWER:** Wood violet. **TREE:** Sugar maple. **SYMBOL OF PEACE:** Mourning dove. **ROCK:** Red granite. **MINERAL:** Galena. **INSECT:** Honeybee. **SOIL:** Antigo silt loam. **LEGAL HOLIDAYS:** New Year's Day, 1 January; Birthday of Martin Luther King, Jr., 3d Monday in January; Lincoln and Washington Day, 3d Monday in February; Good Friday, March or April; Memorial Day, last Monday in May; Independence Day, 4 July; Labor Day, 1st Monday in September; Primary Day, 2d Tuesday in September in even-numbered years; Columbus Day, 2d Monday in October; Election Day, 1st Tuesday after 1st Monday in November in even-numbered years; Veterans Day, 11 November; Thanksgiving Day, 4th Thursday in November; Christmas Day, 25 December. **TIME:** 6 AM CST = noon GMT.

¹LOCATION, SIZE, AND EXTENT

Located in the eastern north-central US, Wisconsin ranks 26th in size among the 50 states.

The total area of Wisconsin is 56,153 sq mi (145,436 sq km), of which 54,426 sq mi (140,963 sq km) is land and 1,727 sq mi (4,473 sq km) inland water. The state extends 295 mi (475 km) E–W and 320 mi (515 km) N–S.

Wisconsin is bordered on the N by Lake Superior and the State of Michigan (with the northeastern boundary formed by the Menominee River); on the E by Lake Michigan; on the S by Illinois; and on the W by Iowa and Minnesota (with the line defined mainly by the Mississippi and St. Croix rivers).

Important islands belonging to Wisconsin are the Apostle Islands in Lake Superior and Washington Island in Lake Michigan. The state's boundaries have a total length of 1,379 mi (2,219 km). Wisconsin's geographic center is in Wood County, 9 mi (14 km) SE of Marshfield.

²TOPOGRAPHY

Wisconsin can be divided into four main geographical regions, each covering roughly one-quarter of the state's land area. The most highly elevated of these is the Superior Upland, below Lake Superior and the border with Michigan. It has heavily forested rolling hills but no high mountains. Elevations range from about 700 feet (200 meters) to slightly under 2,000 feet (600 meters). A second upland region, called the Driftless Area, has a more rugged terrain, having been largely untouched by the glacial drifts that smoothed out topographical features in other parts of the state. Elevations here reach more than 1,200 feet (400 meters). The third region is a large, crescent-shaped plain in central Wisconsin; its unglaciated portion is a sandstone plain, broken by rock formations that from a distance appear similar to the buttes and mesas of Colorado. Finally, in the east and southeast along Lake Michigan lies a large, glaciated lowland plain, fairly smooth in the Green Bay–Winnebago area but more irregular on the Door Peninsula and in the south.

Wisconsin's mean altitude is 1,050 feet (320 meters), with elevations generally higher in the north. The Gogebic Range, extending westward from Michigan's Upper Peninsula into northern Wisconsin, was an important center of iron mining in the early days of statehood. Timms Hill, in north-central Wisconsin, is the state's highest point, at 1,952 feet (595 meters). The lowest elevation is 581 feet (177 meters), along the Lake Michigan shoreline.

There are well over 8,000 lakes in Wisconsin. Lakes Michigan and Superior form part of the northern and eastern borders; the Wisconsin mainland has at least 575 mi (925 km) of lakeshore and holds jurisdiction over 10,062 sq mi (26,061 sq km) of lake waters. By far the largest inland lake is Lake Winnebago, in eastern Wisconsin, covering an area of 215 sq mi (557 sq km).

The Mississippi River, which forms part of the border with Minnesota and the entire border with Iowa, is the main navigable river. The major river flowing through the state is the Wisconsin, which follows a south-southwest course for 430 mi (692 km) before meeting the Mississippi at the Iowa border. Other tributaries of the Mississippi are the St. Croix River, also part of the Minnesota border, and the Chippewa and Black rivers. Located on the Black River are Big Manitou Falls, at 165 feet (50 meters) the highest of the state's many waterfalls. Waters from the Fox River and its major tributary, the Wolf, flow into Green Bay and thence into Lake Michigan, as does the Menominee, which is part of the Michigan state line.

Except in the Driftless Area, glaciation smoothed out many surface features, gouged out new ones, and left deposits of rock and soil creating distinctively shaped hills and ridges. Oval mounds, called drumlins, are still scattered over the southeast, and moraines, formed by deposits left at the edges of glaciers, are a prominent feature of eastern, central, and northwestern Wisconsin. In one section, called the Dells, the Wisconsin River has cut a gorge through 8 mi (13 km) of sandstone, creating caves and interesting rock formations.

³CLIMATE

Wisconsin has a continental climate. Summers are warm and winters very cold, especially in the upper northeast and north-central lowlands, where the freeze-free (growing) season is around 80 days. The average annual temperature ranges from 39°F (4°C) in the north to about 50°F (10°C) in the south. At Danbury, in the northwest, the average January daily temperature over a 34-year period was 8°F (−13°C), and the average July daily temperature 68°F (20°C); at Racine, in the southeast, these figures were 21°F (−6°C) and 72°F (22°C), respectively. Over a 30-year period ending in 1980, the state's largest city, Milwaukee, had average daily temperatures ranging from 11°F (−12°C) to 26°F (−3°C) in January and from 61°F (16°C) to 80°F (27°C) in July. Among major US metropolitan areas, only Minneapolis–St. Paul is colder than Milwaukee. The lowest temperature ever recorded in Wisconsin was −54°F (−48°C), at Danbury on 24 January 1922; the highest, 114°F (46°C), at Wisconsin Dells on 13 July 1936.

Annual precipitation in the state ranges from about 34 in (86 cm) for parts of the northwest to about 28 in (71 cm) in the south-central region and the areas bordering Lake Superior and Lake Michigan. Normal annual precipitation in Milwaukee is 31 in (79 cm); April, June, and July are the rainiest months in Milwaukee. Milwaukee's annual snowfall averages 47 in (119 cm); the average wind speed is 12 mph (19 km/hr).

⁴FLORA AND FAUNA

Common trees of Wisconsin include four oaks—bur, black, white, and red—along with black cherry and hickory. Jack, red, and white pine, yellow birch, eastern hemlock, mountain maple, moosewood, and leatherwood grow in the north, with black spruce, black ash, balsam fir, and tamarack concentrated in the northern lowlands. Characteristic of southern Wisconsin's climax forests are sugar maple (the state tree), white elm, basswood, and ironwood, with silver maple, black willow, silver birch, and cottonwood on low, moist land. Prairies are thick with grasses; bogs and marshes are home to white and jack pines and jack oak. Forty-five varieties of orchid have been identified, as well as 20 types of violet, including the wood violet (the state flower). Threatened plants include lenticular sedge, ram's-head lady's-slipper, blue ash, prairie white-fringed orchid, prairie bush-clover, and snow trillium. Lake cress, harbinger-of-spring, pink milkwort wild petunia, Lake Huron tansy, mountain cranberry, northern wild monkshood, and dwarf bilberry are listed as endangered.

White-tailed deer, black bear, woodchuck, snowshoe hare, chipmunk, and porcupine are mammals typical of forestlands. The striped skunk, red and gray foxes, and various mice are characteristic of upland fields, while wetlands harbor such mammals as the muskrat, mink, river otter, and water shrew. The badger, dwelling in grasslands and semiopen areas, is rarely seen today. Game birds include the ring-necked pheasant, bobwhite quail, Hungarian partridge, and ruffed grouse; among 336 bird species native to Wisconsin are 42 kinds of waterfowl and 6 types of shorebird that are also hunted. Reptiles include 23 varieties of snake, 13 types of turtle, and 4 kinds of lizard. Muskellunge (the state fish), northern pike, walleye, and brook trout are native to Wisconsin waterways.

Among threatened animals are the red-shouldered hawk, greater prairie chicken, glass lizard, Blanding's turtle, pickerel frog, and longear sunfish. The pine marten, Canada lynx, timber wolf, bald eagle, barn owl, osprey, queen snake, massasauga, slender madtom, and Higgins' eye pearly mussel are on the endangered list. The Bureau of Endangered Resources in the Department of Natural Resources develops programs designed to aid the recovery of threatened or endangered flora and fauna.

⁵ENVIRONMENTAL PROTECTION

Conservation has been a concern in Wisconsin for more than a century. In 1867, a legislative commission reported that depletion of the northern forests by wasteful timber industry practices and frequent forest fires had become an urgent problem, partly because it increased the hazards of flooding. In 1897, a forestry warden was appointed and a system of fire detection and control was set up. A reforestation program was instituted in 1911; at about the same time, the state university began planting rows of trees in plains areas to protect soil from wind erosion, a method that was widely copied in other states. Fish and game wardens were appointed in the 1880s. In 1927, the state began a program to clean its waters of industrial wastes, caused especially by pulp and paper mills and canneries. The legislature enacted a comprehensive antipollution program in 1966.

The present Department of Natural Resources, organized in 1967, brings together conservation and environmental protection responsibilities. The department supervises air, water, and solid waste pollution control programs and deals with the protection of forest, fish, and wildlife resources. A separate Solid Waste Recycling Authority was created in 1973 to develop solid waste disposal and recycling facilities. Nearly 3,000 abandoned waste disposal sites were known to exist in 1985.

Funding for the Department of Natural Resources amounted to $160.5 million for 1983/84. A substantial amount—$48 million in 1983/84—is allocated annually to municipalities for such problems as septic system replacement and sewer overflow abatement.

Air pollution is heaviest in the industrial southeast, but by 1981, when Milwaukee was 14th highest in air quality on a list of 51 US cities, the situation was much improved. A state program to limit auto exhaust emissions began in 1984. Since water pollution became a serious problem in the 1920s, pulp and paper mills have spent over $50 million to develop methods of recycling their industrial waste. A cleanup campaign during the 1970s led to significant declines in the level of suspended solids in rivers. In the late 1970s and early 1980s, over 100 municipal wastewater treatment plants were built or upgraded. Nevertheless, more than 100 different chemicals were still being discharged into Wisconsin's rivers and streams as of 1984, and a few community water supply systems as well as some of the state's more than 500,000 wells contained volatile organic compounds or pesticides exceeding health standards. Damage from acid threatens about 2,200 lakes in northern in Wisconsin; testing has found state rain and snow to be 10 times more acidic than normal.

Wisconsin produced an estimated 17.2 million tons of solid waste in 1982. To assist local governments and landfill operators, the Bureau of Solid Waste Management offers aid in locating acceptable landfill sites and planning for the collection, transportation, and disposal of solid wastes. In 1982, 6.3 million tons of solid waste were disposed of in landfills.

⁶POPULATION

Wisconsin ranked 16th in population among the 50 states in 1980, with a census population of 4,705,642. The population was estimated at 4,792,115 in early 1985, a 1.8% increase over the 1980 census.

During the 18th and early 19th centuries, the area that is now Wisconsin was very sparsely settled by perhaps 20,000 Indians and a few hundred white settlers, most of them engaged in the fur trade. With the development of lead mining, the population began to expand, reaching a total of 30,945 (excluding Indians) by 1840. During the next two decades, the population increased rapidly to 775,881, as large numbers of settlers from the East and German, British, and Scandinavian immigrants arrived. Subsequent growth has been steady, if slower. In the late 19th century, industry expanded, and by 1930, the population became predominantly urban.

LOCATION: 42°29′34″ to 47°04′90″N; 86°14′55″ to 92°53′31″W. **BOUNDARIES:** Michigan line, 680 mi (1,094 km); Illinois line, 182 mi (293 km); Iowa line, 91 mi (146 km); Minnesota line, 426 mi (686 km).

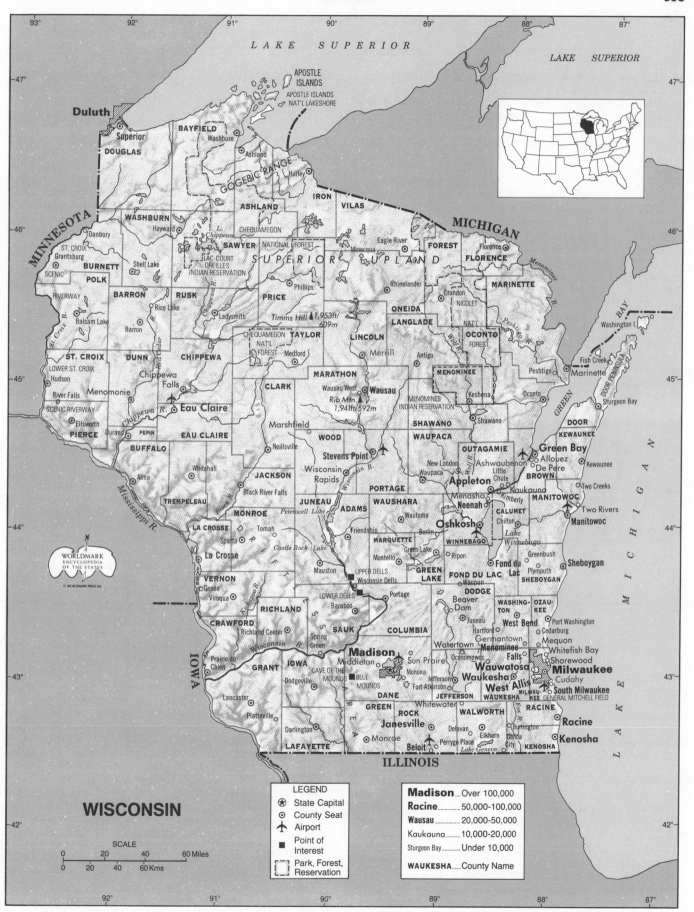

WISCONSIN

SCALE
0   20   40   60 Miles
0   20   40   60 Kms

LEGEND
⊛ State Capital
⊙ County Seat
✈ Airport
■ Point of Interest
⌐ ¬ Park, Forest,
    Reservation

Madison —— Over 100,000
Racine ——— 50,000-100,000
Wausau ——— 20,000-50,000
Kaukauna ——— 10,000-20,000
Sturgeon Bay ——— Under 10,000
WAUKESHA ——— County Name

See endsheet maps: J2.

The 1980 census showed a population increase of 6.5% since 1970, below the national average. Average population density in 1985 was 88 per sq mi (34 per sq km). Of the 1980 population, an estimated 51.1% were females and 48.9% males. About 12.6% of the population was 65 years of age or older in 1983, somewhat above the US average.

Almost two-thirds (64%) of all Wisconsinites lived in urban areas in 1980, most of them in the heavily urbanized southeastern region. Milwaukee, the largest city in Wisconsin and, as of 1984, the 18th-largest in the US, had a population of 620,811. Other large cities, with their 1984 populations, were Madison, 170,745; Green Bay, 90,003; Racine, 83,436; and Kenosha, 75,657. Of these, Green Bay was the only city to gain population (only 90 people) between 1970 and 1980. The state's largest metropolitan area, Milwaukee-Racine, had 1,567,600 residents as of 1 July 1984 (24th in the US); the Milwaukee area alone had 1,393,800.

7ETHNIC GROUPS

As early as 1839, Wisconsin attracted immigrants from Norway, Sweden, Denmark, and Finland, soon to be followed by large numbers of Germans and Irish. In 1850, the greatest number of foreign-born were English-speaking, but within a decade, the Germans had eclipsed them. Industrial development brought Belgians, Greeks, Hungarians, Lithuanians, Italians, and especially Poles, who continued to come steadily until the restriction of immigration in the early 1920s; in the 1930 census, Poles were the largest foreign-born group. In 1980, foreign-born residents numbered 125,297 (2.7% of the total).

Black Americans were in the region as early as 1822. Before World War I, however, there were no more than 3,000 blacks. Migration during and after that war brought the number to 10,739 by 1930; by 1980, blacks were the largest racial minority in the state, numbering 182,592 (4% of Wisconsin's population). Most black Wisconsinites live in Milwaukee, which was 23% black in 1980.

Asians are few in number: in 1980 there were 4,087 Chinese, 3,657 Asian Indians, 2,698 Filipinos, 2,249 Vietnamese, and 2,237 Japanese. As of 1980 there were 62,972 state residents of Hispanic origin, of whom 40,679 were of Mexican ancestry and 10,693 of Puerto Rican descent.

Wisconsin had 29,320 American Indians in 1980. The total living on reservations was 23,987, 56% of them Oneida; other tribes included the Menominee, Ojibwa (Chippewa), and Winnebago. There were 11 reservations, the largest being that of the Menominee, which comprised Menominee County (346 sq mi 896 sq km); Indian reservations covered 634 sq mi (1,642 sq km).

8LANGUAGES

Early French and English fur traders found in what is now Wisconsin several Indian tribes of the Algonkian family: Ojibwa along Lake Superior, Sauk in the northeast, Winnebago and Fox south of them, and Kickapoo in the southwest. Numerous Indian place-names include Antigo, Kaukauna, Kewaunee, Menomonie, Oshkosh, Wausau, and Winnebago.

Wisconsin English is almost entirely Northern, like that of the areas that provided Wisconsin's first settlers—Michigan, northern Ohio, New York State, and western New England. Common are the Northern *pail, comforter* (tied and filled bedcover), *sick to the stomach, angleworm* (earthworm), *skip school* (play truant), and *dove* as the past of *dive*. Pronunciation features are *fog, frog,* and *on* with the vowel /ah/; and *orange, forest,* and *foreign* with the /aw/ vowel. Northern *fried cakes* is now yielding to *doughnuts,* and *johnnycake* is giving way to *corn bread.* Milwaukee has *sick in the stomach* and is known for the localism *bubbler* (drinking fountain). A small exception to Northern homogeneity is the cluster of South Midland terms brought by Kentucky miners to the southwestern lead-mining district, such as *dressing* (sweet sauce for a pudding), *eaves spout* as a blend of *eavestrough* and Midland *spouting, branch* for stream, and *fishworm* for earthworm.

In 1980, 94% of the state population 3 years old and older spoke only English in the home. Other languages spoken at home, and the number of speakers, included:

German	78,904	French	10,648
Spanish	49,358	Norwegian	10,638
Polish	29,936	Italian	10,524

9RELIGIONS

The largest religious groups in Wisconsin are Roman Catholics and Lutherans.

The first Catholics to arrive were Jesuit missionaries seeking to convert the Huron Indians in the 17th century. Protestant settlers and missionaries of different sects, including large numbers of German Lutherans, came during the 19th century, along with Protestants from the East. Jews settled primarily in the cities.

These groups often had conflicting aims. Evangelical sects favored strict blue laws and temperance legislation, which was enacted in many communities. The use of Protestant prayers and the King James Bible in public schools was another source of public discord until these practices were declared unconstitutional by the state supreme court in 1890. A constitutional amendment allowing parochial school students to ride in public school buses was defeated in 1946 amid great controversy; 19 years later, however, it was enacted with little opposition. By that time, religious conflicts appeared to be on the decline.

As of the end of 1983, there were 1,586,804 Roman Catholics (one-third of the population) in Wisconsin. Of Protestant denominations in 1980, the American Lutheran Church had the most members (347,591), followed by the Lutheran Church–Missouri Synod (254,683), Wisconsin Evangelical Lutheran Synod (222,420), United Methodists (164,199), and Lutheran Church in America (116,102). There were an estimated 31,495 Jews in 1984, 76% of them in Milwaukee.

10TRANSPORTATION

Wisconsin's first rail line was built across the state, from Milwaukee to Prairie du Chien, in the 1850s. Communities soon began vying with one another to be included on proposed railroad routes. Several thousand farmers mortgaged property to buy railroad stock; the state had to rescue them from ruin when companies went bankrupt. By the late 1860s, two railroads—the Chicago and North Western and the Chicago, Milwaukee, and St. Paul—had become dominant in the state. However, Chicago emerged as the major rail center of the Midwest because of its proximity to eastern markets. In 1920 there were 35 railroads operating on 11,615 mi (18,693 km) of track; by 1981 there were only 21 railroads, of which 5 were Class I lines. There were 7,509 mi (12,085 km) of track, of which 3,850 mi (6,196 km) were for Class I lines. Passenger traffic declined from 20,188,000 to 122,000 during the 1920-81 period. Freight revenues increased from $92,826,000 to $476,265,000, though tonnage declined by about 4% to 97,288,000 tons. Amtrak provides passenger service to Milwaukee, La Crosse, and several other cities on Chicago–Milwaukee and Chicago–Seattle runs; 1983/84 Wisconsin ridership was 290,093.

As of 1 January 1982, Wisconsin had 107,864 mi (173,591 km) of roadway; about 57% of this mileage consisted of town roads, 29% state or county highways, 10% city streets, 3% village streets, and 1% park and forest roads. Registered vehicles included 2,547,140 automobiles and 654,067 trucks in 1983, when the state had 3,085,549 licensed drivers. There were 21 publicly owned bus systems in Wisconsin in 1983, including large systems in Milwaukee and Madison.

The opening of the St. Lawrence Seaway in 1959 allowed oceangoing vessels access to Wisconsin via the Great Lakes but failed to stimulate traffic to the extent anticipated. Overall, the state has 15 cargo-handling ports. The port of Superior (shared with Duluth, Minn.) on Lake Superior handled 27,436,085 tons of cargo in 1982, making it the busiest of all US Great Lakes ports. Iron ore was the chief commodity. Other important Wisconsin ports, all on Lake Michigan, were Milwaukee, Green Bay, Port Washington, Oak Creek, Kewaunee, Manitowoc, Sturgeon Bay, and Sheboygan; coal and lignite was the chief commodity. Manitowoc and Kewaunee offer ferry service across Lake Michigan to Michigan. Ashland on Lake Superior and La Crosse on the Mississippi River are other important ports.

As of 1 November 1982, Wisconsin had 459 airports. In 1983, 1,352,044 passengers enplaned at General Mitchell Field, in Milwaukee.

11 HISTORY

The region that is now Wisconsin has probably been inhabited since the end of the glacial period, 10,000 years ago. Some of the earliest inhabitants were ancestors of the Menominee; these early immigrants from the north built burial mounds, conical ones at first, then large effigy mounds shaped like different animals. Other peoples arrived from the south and east, including ancestors of the Winnebago Indians (about AD 1400) and a tribe that built flat-top earthen pyramids. During the 17th century, the Ojibwa, Sauk, Fox, Potawatomi, Kickapoo, and other tribes came to Wisconsin. These tribes engaged in agriculture, hunting, and fishing, but with the arrival of Europeans, they became increasingly dependent on the fur trade—a dependence that had serious economic consequences when the fur trade declined in the early 19th century.

The first European believed to have reached Wisconsin was Jean Nicolet, who in 1634 landed on the shores of Green Bay while in the service of Samuel de Champlain. Two decades later, Médard Chouart des Groseilliers and Pierre Esprit Radisson, both fur traders, explored northern Wisconsin; in 1673, the Jesuit priest Jacques Marquette and the explorer Louis Jolliet crossed the whole area that is now Wisconsin, via the Fox and Wisconsin rivers, on their way to the Mississippi. Other Jesuits established missions, and French fur traders opened up posts. The French were succeeded by the British after the French and Indian War (the British ruled Wisconsin as part of Quebec Province from 1774 to 1783). Although ceded to the US in 1783, it remained British in all but name until 1816, when the US built forts at Prairie du Chien and Green Bay.

Under the Ordinance of 1787, Wisconsin became part of the Northwest Territory; it was subsequently included in the Indiana Territory, the Territory of Illinois, and then the Michigan Territory. In the early 1820s, lead mining brought an influx of white settlers called "Badgers." Indian resistance to white expansion collapsed after the 1832 Black Hawk War, in which Sauk and Fox Indians fleeing from Illinois were defeated and massacred by white militia near the site of present-day La Crosse, at the Battle of Bad Axe. Subsequently, the Winnebago and other tribes were removed to reservations outside the state, while the Ojibwa, Menominee, and some eastern tribes were among those resettled in reservations inside Wisconsin.

The Wisconsin Territory was formed in 1836. Initially it included all of Iowa and Minnesota, along with a portion of the Dakotas, but in 1838, these areas became part of a newly organized Iowa Territory. The 1830s also saw the beginning of a land boom, fueled by migration of Yankees from New England and southerners who moved to the lead-mining region of southwestern Wisconsin. The population and economy began to expand rapidly. Wisconsin voters endorsed statehood in 1846, and Congress passed enabling legislation that year. After a first constitution was rejected by the voters, a revised document was adopted on 13 March 1848, and on 29 May, President James K. Polk signed the bill that made Wisconsin the 30th state.

Transportation and industry did not develop as rapidly as proponents of statehood had expected. A canal was opened at the portage between the Fox and Wisconsin rivers in 1851, but the waterway was not heavily used. Railroads encountered difficulties in gaining financing, then suffered setbacks in the panic of 1857.

Wisconsinites took a generally abolitionist stand, and it was in Wisconsin—at Ripon, on 28 February 1854—that the Republican Party was formally established in the state. The new party developed an efficient political machine and later used much of its influence to benefit the railroads and lumber industry, both of which grew in importance in the decade following the Civil War. In that war, 96,000 Wisconsin men fought on the Union side, and 12,216 died. During the late 19th century, Wisconsin was generally prosperous; dairying, food processing, and lumbering emerged as major industries, and Milwaukee grew into an important industrial center.

Wisconsin took a new political turn in the early 20th century with the inauguration of Republican Robert "Fighting Bob" La Follette as governor and the dawning of the Progressive Era. An ardent reformer, La Follette fought against conservatives within his own party. In 1903, the legislature under his prodding passed a law providing for the nation's first direct statewide primary; other measures that he championed during his tenure as governor (1901–6) provided for increased taxation of railroads, regulation of lobbyists, creation of a civil service, and establishment of a railroad commission to regulate intrastate rates.

La Follette was also a conspicuous exponent of what came to be called the "Wisconsin idea": governmental reform guided by academic experts and supported by an enlightened electorate. Around the time he was governor, the philosophy of reform was energetically promoted at the University of Wisconsin (which had opened at Madison, the state capital, in 1849), and many professors were drafted to serve on government commissions and boards. In 1901, Wisconsin became the first state to establish a legislative reference bureau, intended to help lawmakers shape effective, forward-looking measures.

After La Follette left the governor's office to become a US senator, his progressivism was carried on by Republican governors James O. Davidson (1906–11) and especially by Francis E. McGovern (1911–15). During one session in 1911, legislators enacted the first state income tax in the US and one of the first workers' compensation programs. Other legislation passed during the same year sought to regulate the insurance business and the use of water power, create forest reserves, encourage farmer cooperatives, limit and require disclosure of political campaign expenditures, and establish a board of public affairs to recommend efficiency measures for state and local governments. This outburst of activity attracted national attention, and many states followed in Wisconsin's footsteps.

While US senator (1906–25), La Follette opposed involvement in World War I and was one of only six senators to vote against US entry into the war; as a result, he was censured by the state legislature and the faculty of the University of Wisconsin, and there was a move to expel him from the Senate. His renomination and reelection in 1922 served to vindicate him, however, and he carried Wisconsin when he ran in 1924 for president on the national League for Progressive Political Action ticket.

After his death in 1925, the reform tradition continued in Wisconsin. A pioneering old-age pension act was passed in 1925; seven years later, Wisconsin enacted the nation's first unemployment compensation act, with the encouragement of La Follette's son Philip, then serving his first term as governor. When Wisconsin went Democratic in November 1932, turning Philip out of office, he and his brother, Robert, Jr., a US senator, temporarily left the state Republican organization and in 1934 formed a separate Progressive Party; that party, with the support of President Franklin Roosevelt and the Socialists, swept the 1934 elections and returned both brothers to office. During his second and third terms as governor, Philip La Follette successfully pressed for the creation of state agencies to develop electric power, arbitrate labor disputes, and set rules for fair business competition; his so-called Little New Deal corresponded to the New Deal policies of the Roosevelt administration.

After World War II, the state continued a trend toward increased urbanization, and its industries prospered. The major figure on the national scene in the postwar era was Senator Joseph R. McCarthy, who defeated Robert La Follette, Jr., in the 1946 Wisconsin Republican primary and went on to serve 10 years in the Senate. McCarthy launched an unsubstantiated attack in 1950 on alleged Communists and other subversives in the federal government. Especially after McCarthy's censure by the US Senate in 1954 and death in 1957, the Progressive tradition began to recover strength, and the liberal Democratic Party grew increasingly influential in state politics. There was student unrest at the University of Wisconsin during the 1960s and early 1970s, and growing discontent among Milwaukee's black population. A major controversy in the 1970s concerned a court-ordered busing plan, implemented in 1979, aimed at decreasing racial imbalances in Milwaukee's public schools. In 1984, the Milwaukee school board filed suit in federal court against the state as well as 24 school districts in Milwaukee's suburbs, charging that the policies of the state and suburban schools had resulted in an unconstitutionally segregated school system that restricted blacks to city schools. The suit asked the court to enforce programs that would end segregation and remedy the effects of past discrimination.

[12] STATE GOVERNMENT

Wisconsin's first constitutional convention, meeting in Madison in October 1846, was marked by controversy between conservative Whigs and allied Democrats, on the one hand, and progressive Democrats with a constituency made up of miners, farmers, and immigrants, on the other. The latter, who favored the popular election of judges and exemption of homesteads from seizure for debt, among other provisions, carried the day, but this version of the constitution failed to win ratification. A second constitutional convention, convened in December 1847, agreed on a new draft, which made few major changes. This document, ratified by the electorate in 1848 and amended 118 times as of the end of 1983, remains in effect today.

The Wisconsin legislature consists of a senate with 33 members elected for four-year terms, and an assembly of 99 representatives elected for two-year terms. Legislators must be state residents for one year prior to election and residents of their districts at least 10 days before the election. Voters elect an assembly and half the senate membership in even-numbered years. Regular legislative sessions begin in January. Each house elects its own presiding officer and other officers from among its members.

There are six elected state officers: governor and lieutenant governor (elected jointly), secretary of state, state treasurer, attorney general, and superintendent of public instruction. Since 1970, all have been elected for four-year terms. The governor and lieutenant governor must be US citizens and qualified voters; there are no additional age or residency requirements. As the chief executive officer, the governor exercises authority by the power of appointment, by presenting a budget bill and major addresses to the legislature, and by the power to veto bills and call special legislative sessions. Of 15 administrative departments in the executive branch, two—the Department of Justice and the Department of Public Instruction—are headed by the attorney general and the superintendent of public instruction, respectively. Eight departments are headed by secretaries appointed by the governor with the advice and consent of the senate, while the Department of Military Affairs is headed by the adjutant general, who is appointed by the governor alone; part-time boards appoint the heads of the four remaining departments. There were also 18 independent agencies in 1983, of which 5 were headed by individual commissioners and 13 by commissions or boards.

A bill may be introduced in either house of the legislature, but must be passed by both houses to become law. The governor has six days (Sundays excluded) to sign or veto a measure. If the governor fails to act and the legislature is still in session, the bill automatically becomes law; if the legislature has adjourned sine die, however, the governor's failure to act has the effect of a veto. (However, the legislature usually does not adjourn until the next legislature convenes.) Vetoes can be overriden by a two-thirds majority of both houses. Constitutional amendments may be introduced in either house. They must be approved by a simple majority of both houses in two successive legislatures and then ratified by a majority of the electorate at a subsequent election.

Voters must be US citizens 18 years of age or older and must have resided in the state for at least 10 days before the election. The residency requirement is waived in voting for US president and vice president. Wisconsin has a statewide primary open to all without respect to party affiliation, a procedure that was challenged by the national Democratic Party. As a result, in 1984, for the first time, the Democratic party presidential primary results were not used to select delegates to the national convention.

[13] POLITICAL PARTIES

The Democratic Party dominated politics until the late 1850s; then the newly founded Republican Party held sway for almost 100 years. More recently, the Democrats have held a substantial edge at the state level.

Jacksonian democracy was strong in Wisconsin in the early days, and until 1856, all territorial and state governors were Democrats, except for one Whig. In 1854, however, a coalition of Whigs, antislavery Democrats, and Free Soilers, formed a Republican Party in the state—a key event in the establishment of the national Republican Party. Republicans quickly gained control of most elective offices; from 1856 to 1959 there were only three Democratic governors. The Republican Party was dominated in the late 19th century by conservatives, who were sympathetic to the railroads and the lumbering industry but whose stands on pensions and jobs for Union veterans and ability to win federal funds for the state attracted support from farmers and small business. Then, in the 1890s, Progressives within the party, led by Robert La Follette, began a successful battle for control that culminated in La Follette's election as governor in 1900.

The La Follette brand of progressivism remained strong in the state, although not always under the umbrella of Republicanism. In 1924, La Follette ran for president on the Progressive ticket; 10 years later, his sons, Robert and Philip, also broke away from the GOP, to head a Progressive Party slate. However, their newly organized national third party faded and folded when Philip La Follette failed to be reelected governor and World War II made isolationism unpopular. The Progressives rejoined the GOP in 1946.

Socialist parties have won some success in Wisconsin's political history. Socialists worked with progressive Republicans at the state level to pass important legislation in the early 20th century. In 1910, the Socialists scored two major political victories in Wisconsin: Emil Seidel was elected mayor of Milwaukee, becoming the first Socialist mayor of a major US city, and Victor Berger became the first Socialist ever elected to Congress.

There is no statewide system of registration by parties; registered voters numbered 1,787,799 in 1982. As of 1985, Democrats held all the statewide offices except the nonpartisan position of superintendent of public instruction. They also had control of both legislative houses.

Nearly 48% of Wisconsin voters cast their ballots for Ronald Reagan in the 1980 presidential election. Gaylord Nelson, a liberal Democrat, lost his US Senate seat to a Republican, Robert W. Kasten, Jr., but the Democrats retained majorities in the legislature and in Wisconsin's congressional delegation. In 1982, Democrats continued to dominate at both the state and federal levels; Anthony Earl was elected governor and William Proxmire was reelected senator. Reagan carried Wisconsin with 54% of the vote in 1984, and all 9 House members (5 Democrats, 4 Republicans) were reelected.

There were 22 black elected officials in Wisconsin in 1985 (including 4 state legislators and 3 judges) and 1 Hispanic. Twenty-five state legislators were women in that year.

¹⁴LOCAL GOVERNMENT

As of early 1983, Wisconsin had 72 counties, 187 incorporated cities, 394 incorporated villages, 1,269 towns, and 433 school districts. There were also 265 special districts, each providing a certain local service, such as sewerage or fire fighting, usually across municipal lines.

Each county is governed by a board of supervisors (which in the most populous counties has more than 40 members), elected for two years, except in Milwaukee County, where the term is four years. Eight counties, including Milwaukee County, have elected county executives, serving four-year terms; 10 had an appointed administrator or similar official. Other county officials include district attorneys, sheriffs, clerks, treasurers, coroners, registers of deeds, and surveyors.

Towns are civil subdivisions of counties equivalent to townships in other states. Each town is a unit of 6 sq mi (16 sq km) marked off for governmental purposes. Cities and villages have home-rule powers limited by legislative review. Each incorporated village must have at least 150 residents; each incorporated city, 1,000.

Most cities are governed by a mayor-council system; about 6% of all cities have a council-manager system, which was first authorized in Wisconsin in 1923. Executive power in a village is vested in an elected president, who presides over an elected board but has no veto power.

Wisconsin towns are generally small units; as of 1980, 92% had populations under 2,500. Each town is governed by a board of supervisors elected every two years; a town supervisor carries out policies set at an annual April town meeting.

¹⁵STATE SERVICES

In its first year of statehood, Wisconsin had only 14 paid officials and employees, including the 6 constitutional officers. In October 1982, the state had 61,992, of whom 35,236 were permanent civil service employees. A six-member Ethics Board, appointed by the governor, administers an ethics code for public officials and employees and investigates complaints against them. The board may refer cases for criminal prosecution.

The Department of Public Instruction administers public elementary and secondary education in the state, and the Board of Regents of the University of Wisconsin System has jurisdiction over all public higher education. A Board of Vocational, Technical, and Adult Education supervises programs in these areas.

The Transportation Department plans, constructs, and maintains highways and licenses motor vehicles and drivers. Physical and mental health, corrections, public and medical assistance, service to the aged, children's services, and vocational rehabilitation fall within the purview of the Department of Health and Social Services. The Department of Industry, Labor, and Human Relations enforces antidiscrimination laws in employment as well as minimum standards for wages and working conditions, provides training for the unemployed and disadvantaged, and sets safety standards for buildings.

Public protection in general is provided by the Department of Justice, which is responsible for investigating crimes of statewide magnitude and offering technical assistance to local law enforcement agencies. Regulations to protect consumers are administered and enforced by the Trade and Consumer Protection Division of the Department of Agriculture, Trade, and Consumer Protection, in cooperation with the Justice Department. The Army and Air National Guard are under the Department of Military Affairs.

The Department of Development has responsibilities in the areas of community, economic, and housing development, promotion of trade and tourism, and small and minority business assistance.

Wisconsin Presidential Vote by Political Party, 1948–84

YEAR	ELECTORAL VOTE	WISCONSIN WINNER	DEMOCRAT	REPUBLICAN	PROGRESSIVE	SOCIALIST	SOCIALIST WORKERS	SOCIALIST LABOR
1948	12	*Truman (D)	647,310	590,959	25,282	12,547	—	399
1952	12	*Eisenhower (R)	622,175	979,744	2,174	1,157	1,350	770
					CONSTITUTION			
1956	12	*Eisenhower (R)	586,768	954,844	6,918	754	564	710
1960	12	Nixon (R)	830,805	895,175	—	—	1,792	1,310
1964	12	*Johnson (D)	1,050,424	638,495	—	—	1,692	1,204
1968	12	*Nixon (R)	748,804	809,997	—	—	1,222	1,338
					AMERICAN IND.	AMERICAN		
1972	11	*Nixon (R)	810,174	989,430	127,835	47,525	—	998
						SOCIALIST		LIBERTARIAN
1976	11	*Carter (D)	1,040,232	1,004,967	8,552	4,298	1,691	3,814
							CITIZENS	
1980	11	*Reagan (R)	981,584	1,088,845	1,519¹	—	7,767	29,135
1984	11	*Reagan (R)	995,740	1,198,584	—	—	—	4,883

*Won US presidential election.
¹Listed as Constitution Party on Wisconsin ballot.

[16] JUDICIAL SYSTEM

The judicial branch is headed by a supreme court, consisting of seven justices, elected statewide on a nonpartisan basis for terms of 10 years. Vacancies are filled by gubernatorial appointment until an open election day becomes available. The justice with the greatest seniority serves as chief justice. The supreme court, which is the final authority on state constitutional questions, hears appeals at its own discretion and has original jurisdiction in limited areas.

The state's next-highest court is the court of appeals, established by constitutional amendment in 1977. Its 12 judges (3 for each of 4 judicial districts) are elected by district on a nonpartisan basis and serve staggered six-year terms. Vacancies are filled by the governor until a successor is elected. Judges sit in panels of 3 for most cases, although some cases can be heard by a single judge. Decisions by the court of appeals may be reviewed by the supreme court.

The circuit court, the trial court of general jurisdiction, also hears appeals from municipal courts. Circuit court boundaries coincide with county boundaries, except that 3 judicial circuits comprise 2 counties each; thus, there are 69 judicial circuits. As of 1983 there were 190 trial judges, elected by district on a nonpartisan basis for six-year terms. All justices at the circuit court level or higher must have at least five years' experience as practicing attorneys and be less than 70 years old in order to qualify for office. Vacancies are filled by the governor until a successor is elected.

Wisconsin's 200 municipal courts have jurisdiction over local matters. Municipal judges are elected for terms of two or four years, generally serve on a part-time basis, and need not be attorneys. In December 1984 there were 9,787 practicing attorneys in the state.

A total of 202,879 criminal offenses were reported in Wisconsin during 1983, of which 193,760 were property crimes. The crime rate was 4,246.1 per 100,000 population, below the national average. The violent crime rate was 191 per 100,000 population in the state, far below the national average.

A total of 31,787 persons were under Division of Corrections supervision on 30 June 1984; 3,030 were in local jails on 30 June 1983.

[17] ARMED FORCES

There is little military activity in Wisconsin; personnel in the five military facilities, mostly civilians, totaled only 3,928 in 1983/84. Prime military contracts amounted to $951,439,000 in the same fiscal year.

A total of 3,932 Wisconsinites were killed in World War I, 7,980 in World War II, 800 in Korea, and 1,142 in Viet-Nam. An estimated 572,000 veterans were living in Wisconsin as of 30 September 1983. Of these, 8,000 saw service in World War I, 208,000 in World War II, 100,000 in the Korean conflict, and 162,000 in the Viet-Nam era. In the year ending 30 September 1983, Wisconsin veterans received benefits exceeding $466 million.

Wisconsin's Army National Guard units had 8,646 officers and enlisted men as of December 1984; the Air National Guard had 1,942 in February 1985. In 1983 there were 11,757 full-time police personnel, of whom state troopers numbered 618.

[18] MIGRATION

Until the early 19th century, Wisconsin was inhabited mainly by Indians; the French and British brought few permanent settlers. In the 1820s, southerners began to arrive from the lower Mississippi, and in the 1830s, easterners poured in from New York, Ohio, Pennsylvania, and New England.

Foreign immigrants began arriving in the 1820s, either directly from Europe or after temporary settlement in eastern states. Most of the early immigrants were from Ireland and England. Germans also came in large numbers, especially after the Revolution of 1848, and by 1860, they were predominant in the immigrant population, which was proportionately larger than in any other state except California. The state soon became a patchwork of ethnic communities—Germans in the counties near Lake Michigan, Norwegians in southern and western Wisconsin, Dutch in the lower Fox Valley and near Sheboygan, and other groups in other regions.

After the Civil War, and especially in the 1880s, immigration reached new heights, with Wisconsin receiving a large share of Germans and Scandinavians; the proportion of Germans declined, however, as new immigrants arrived from Finland and Russia and from southern and eastern Europe, especially Poland, before World War I. Despite this overseas immigration, Wisconsin suffered a net population loss from migration beginning in 1900, as Wisconsinites moved to other states. Between 1970 and 1983 alone, this loss totaled 154,000. Wisconsin no longer tends to attract outsiders. In 1980, 77% of state residents were born in the state; only six other states had a higher percentage.

A significant trend since 1970 has been the decline in population in Milwaukee and other large cities; at the same time, suburbs have continued to grow, as have many other areas, especially in parts of northern Wisconsin.

[19] INTERGOVERNMENTAL COOPERATION

The Commission on Interstate Cooperation represents the state in its dealings with the Council of State Governments. Wisconsin also participates in the Education Commission of the States, Great Lakes Basin Commission, Minnesota-Wisconsin Boundary Area Commission, and Mississippi River Parkway Commission. In 1985, Wisconsin, seven other Great Lakes states, and the Canadian provinces of Quebec and Ontario signed the Great Lakes Compact to protect the lakes' water reserves.

In 1983/84, Wisconsin received nearly $2.1 billion in federal aid.

[20] ECONOMY

With the coming of the first Europeans, fur trading became a major economic activity; as more settlers arrived, agriculture prospered. Although farming—preeminently dairying—remains important, manufacturing is the mainstay of today's economy. Wisconsin's industries are diversified, with nonelectrical machinery and food products the leading items. Other important industries are paper and pulp products, transportation equipment, electrical and electronic equipment, and fabricated metals. Economic growth has been concentrated in the southeast. There, soils and climate are favorable for agriculture; a skilled labor force is available to industry; and capital, transportation, and markets are most readily accessible.

As happened to the country at large, Wisconsin in 1981-82 experienced the worst economic slump since the Great Depression, with the unemployment rate rising to 11.7% in late 1982. Manufacturing was hard hit, and the loss of jobs in this sector was considered permanent. The growth areas in the early 1980s were restaurants (especially fast-food establishments), retail outlets, office buildings, and service firms.

[21] INCOME

In 1983, Wisconsin ranked 27th among the 50 states in per capita income, which was $11,132. During the 1970s, per capita income, in constant dollars, grew by 28%, compared to 24% for the nation, but there was virtually no growth from 1981 through 1983. Earned income (labor and proprietors' income) in 1981 totaled $34.3 billion, of which manufacturing accounted for 35%; services, 16%; retail and wholesale trade, 15%; government, 13%; transportation and public utilities, 6%; construction 5%; farming, 4%; and other sectors, 6%; Median income of four-person families in 1981 was $27,232, 16th in the US.

On an adjusted basis (excluding transfer payments, retirement benefits, and other income not taxed by the state), per capita income in 1981 was $7,431. An estimated 398,000 Wisconsinites, or 8.7% of the population, were below the federal poverty level in 1979; only two states, Connecticut and Wyoming, had lower percentages that year. In 1982, 25,100 Wisconsinites were among the nation's top wealth-holders, with gross assets greater than $500,000.

²²LABOR

As of April 1985, the civilian labor force (seasonally adjusted) numbered 2,401,600 persons, of whom 2,224,200, or 92.6% were employed, and 177,400, or 7.4%, were unemployed, slightly above the national rate. The unemployment rate in the Milwaukee area was 5.7%. Women represented 43% of the labor force as of 1983. Of all nonfarm workers, 507,600 were employed in manufacturing as of April 1985, one-third of them in the Milwaukee area.

A federal census in March 1982 revealed the following nonfarm employment pattern in Wisconsin:

	ESTABLISH-MENTS	EMPLOYEES	ANNUAL PAYROLL ('000)
Agricultural services, forestry, fishing	944	5,261	$ 79,354
Mining	186	1,814	40,851
Contract construction	8,908	56,884	1,215,113
Manufacturing, of which:	8,192	504,002	10,195,044
Food products	(941)	(53,727)	(1,053,147)
Fabricated metal products	(846)	(48,219)	(957,966)
Nonelectrical machinery	(1,524)	(105,055)	(2,214,696)
Electric and electronic equipment	(298)	(48,098)	(949,177)
Transportation, public utilities	4,104	79,213	1,581,779
Wholesale trade	8,009	94,421	1,729,496
Retail trade	29,022	329,745	2,618,918
Finance, insurance, real estate	8,231	97,445	1,532,672
Services, of which:	27,975	372,761	4,421,258
Health services	(6,468)	(148,743)	(2,191,256)
Other	1,328	1,046	14,389
TOTALS	96,899	1,542,592	$23,428,874

Among the categories of workers excluded from this survey were government employees, of whom Wisconsin had about 314,000 in 1983. Farm employment was about 210,000 in 1983.

Labor began to organize in the state after the Civil War. The Knights of St. Crispin, a shoemakers' union, grew into what was, at that time, the nation's largest union, before it collapsed during the Panic of 1873. In 1887, unions of printers, cigarmakers, and iron molders organized the Milwaukee Federated Trades Council, and in 1893, the Wisconsin State Federation of Labor was formed. A statewide union for public employees was established in 1932. In 1977, the state's legislature granted public employees (except public safety personnel) the right to strike, subject to certain limitations. As of 1980, labor union membership totaled 554,000, or 29% of all nonfarm employment; in 1983 there were 1,495 labor unions.

²³AGRICULTURE

Gross agricultural income in 1983 amounted to nearly $6 billion, 7th among the 50 states; over $4.1 billion in farm marketings came from dairy products and livestock. Net income was $1.1 billion, 7th in the US. The nation's leading dairy state, Wisconsin also led the US in 1983 in the production of hay, corn for silage, sweet corn, cabbage for sauerkraut, beets for canning, and peas and snap beans for processing. It also ranked 2d in the production of cranberries, 4th in oats and carrots, 6th in tart cherries, and 7th in potatoes and corn for grain.

In the early years, Wisconsin developed an agricultural economy based on wheat, some of which was exported to eastern states and overseas via the port of Milwaukee. Farmers also grew barley and hops, finding a market for these products among early Milwaukee brewers. After the Civil War, soil exhaustion and the

depredations of the chinch bug forced farmers to turn to other crops, including corn, oats, and hay, which could be used to feed hogs, sheep, cows, and other livestock.

Although agricultural income has continued to rise in recent years and the average size of farms has increased, farm acreage and the number of farms have declined. In 1983 there were 18.2 million acres (7.4 million hectares) of land in farms, nearly 52% of the total land area, distributed among 90,000 farms, a decline of 2,000 from 1982. Farmland is concentrated in the southern two-thirds of the state, especially in the southeast. Potatoes are grown mainly in central Wisconsin, cranberries in the Wisconsin River Valley, and cherries in the Door Peninsula.

Leading field crops (in bushels) in 1983 were corn for grain, 223,100,000, smallest harvest since 1976; oats, 45,040,000; soybeans, 13,430,000; wheat, 5,812,000; and barley, 1,680,000. About 12,200,000 tons of hay and 10,375,000 tons of corn for silage were harvested in 1983. Potato production was 18,910,000 hundredweight. In 1983, Wisconsin farmers produced for processing 568,200 tons of sweet corn, 210,680 tons of snap beans, 110,160 tons of green peas, 1,132,000 barrels of cranberries, and 2,250 tons of tart cherries, along with 55,000,000 lb of commercial apples and 489,000 lb of mint for oil. Some 60,120 tons of beets for canning and 47,850 tons of cucumber pickles were produced in 1981.

²⁴ANIMAL HUSBANDRY

Aided by the skills of immigrant cheesemakers and by the encouragement of dairymen who emigrated from New York—especially by the promotional efforts of the agriculturalist and publisher William D. Hoard—Wisconsin turned to dairying in the late 19th century. Today, Wisconsin has more milk cows than any other state and ranks 1st in the production of cheese, butter, and milk. More than one-third of the nation's cheese—and at least three-fourths of all Muenster cheese—is produced in the state. Dairy farms are prominent in nearly all regions, but especially in the Central Plains and Western Uplands. Wisconsin ranchers also raise livestock for meat production.

In 1983 there were 4,350,000 head of cattle (8th in the US), including 1,842,000 milk cows and heifers (1st), along with 1,280,000 hogs and pigs; 4,450,000 chickens and 7,115,000 turkeys were on poultry farms during the same year. Cash receipts from marketings of 1.1 billion lb of cattle and calves amounted to $628.4 million in 1983; from 570,785,000 lb of hogs, $262.4 million.

Dairy products account for about 60% of all cash receipts from agriculture. In 1983, the state produced over 1.7 billion lb of cheese, of which American cheese accounted for nearly 1.2 billion lb (1st in the US), Italian-style cheeses, 388.7 million lb (1st), brick and Muenster cheeses, 61.6 million lb (1st); and Swiss cheese, 36.1 million lb (3d). Wisconsin farms also yielded 23.8 billion lb of milk and 304.7 million lb of butter. Cash receipts from milk alone were almost $3.1 billion.

Cash receipts from marketings of poultry and eggs amounted to $92.2 million in 1983, when 904 million eggs were produced. During the same year, mink ranches produced 1,158,300 pelts of mink, more than any other state, and Wisconsin apiaries generated 4,125,000 lb of honey (14th in the US).

²⁵FISHING

In 1984, Wisconsin ranked 18th among the 50 states in the value of its commercial fishing; 29,768,000 lb of fish were landed (not including Mississippi River landings), at a total value of $3,387,000. During the 1981 season, sport fishermen caught an estimated 64 million fish, 87% of them bass, perch, and other panfish; trout, walleye, and northern pike made up most of the remaining 13%. The muskellunge is the premier game fish of Wisconsin's inland waters; Coho and chinook salmon, introduced to Lake Michigan, now thrive there. The largest concentration of lake sturgeon in the US is in Lake Winnebago.

[26]FORESTRY

Wisconsin was once about 85% forested. Although much of the forest was depleted by forest fires and wasteful lumber industry practices, vast areas reseeded naturally, and more than 820,000 acres (332,000 hectares) have been replanted. In 1979, Wisconsin had 14,478,000 acres (5,859,000 hectares) of forest, covering 42% of the state's land area; two-thirds of all forestlands are privately owned. Hardwoods make up about two-thirds of the sawtimber. The most heavily forested region is in the north. The timber industry reached its peak in the late 19th century; in 1983, lumber production was estimated at 315,000,000 board feet.

Wisconsin's woods have recreational as well as commercial value. Two national forests—Chequamegon and Nicolet—located in northern Wisconsin, covered 1,504,202 acres (608,731 hectares) in 1984. The 10 state forests covered 471,329 acres (190,741 hectares) in 1982.

Forest management and fire control programs are directed by the Department of Natural Resources. The US Forest Service operates a Forest Products Laboratory at Madison, in cooperation with the University of Wisconsin.

[27]MINING

When Wisconsin became a state, lead mining was an important activity in the southeast; later, for a time, there was substantial iron mining in the Gogebic Range. Today, however, mining plays only a small role in the state's economy. In 1984, Wisconsin ranked 37th among the states in the value of its nonfuel mineral output, which amounted to $121.6 million. Chief minerals in 1984 were sand and gravel (16.1 million tons) and stone (17.2 million tons). Lime (377,000 tons) and peat (9,000 tons) were also mined.

[28]ENERGY AND POWER

The state's first hydroelectric plant was built at Appleton in 1882; many others were built later, especially along the Wisconsin River. Today, however, 94% of the state's electrical energy comes from other sources, mainly steam-generating plants. Wisconsin itself has no significant coal, oil, or gas resources.

In 1983, electrical energy production totaled 39.3 billion kwh, and installed capacity was 10.8 million kw. As of 31 December 1983 there were 25 steam-generating plants, accounting for 68% of the state's total installed generating capacity, and 46 combination turbine/internal combustion plants, accounting for 13%. The 78 hydroelectric plants accounted for less than 4%. The remaining 15% was attributable to three nuclear installations: Point Beach Units 1 and 2, at Two Creeks, operated by the Wisconsin Electric Power Co., and Kewaunee Unit 1, at Carlton Township, operated by the Wisconsin Public Service Corp.

[29]INDUSTRY

As of 1981, Wisconsin ranked 12th in the nation in value by shipment. Nonelectrical machinery, food products, fabricated metal products, and electrical and electronic equipment accounted for half of all industrial employment.

The total value of shipments by manufacturers was $52.2 billion in 1982. Of that total, food products (especially beer, cheese, meat, and canned fruits and vegetables) accounted for 25%; nonelectrical machinery, for 17%; paper and paper products, 13%; transportation equipment, 9%; fabricated metal products, 7%; electrical and electronic equipment, 7%; printing and publishing, 4%; primary metal products, 3%; chemicals and related products, 3%; and other items, 12%.

The following table shows value of shipments for selected industries in 1982:

Motor vehicles and equipment	$4,254,700,000
Cheese	4,115,900,000
Meat products	2,523,800,000
Engines and turbines	2,026,000,000
Sanitary paper products	1,861,100,000
Farm and garden machinery	1,642,600,000
Construction machinery	1,111,100,000

Industrial activity is concentrated in the southeast, especially the Milwaukee metropolitan area, which by itself accounted for 31% of the total value of shipments in 1982. Major corporations based in Milwaukee include Allis-Chalmers, manufacturers of industrial and agricultural machinery, and Johnson Controls and Allen-Bradley, makers of electric and electronic components. Briggs & Stratton, a manufacturer of engines, is in suburban Wauwatosa. Milwaukee has lost some of its luster as a brewery center; Miller and Pabst still have breweries there, but Schlitz beer is no longer made in Wisconsin, and Blatz and Old Style are brewed by G. Heileman in La Crosse.

Of the state's three biggest paper and lumber products firms, Kimberley-Clark has its headquarters in Neenah, Consolidated Paper in Wisconsin Rapids, and Fort Howard Paper in Green Bay. Johnson & Son (wax products) and J. I. Case (agricultural equipment), the latter a subsidiary of Tenneco, are in Racine. Oscar Mayer (meat-packing and food products), located in Madison, is now a subsidiary of General Foods. Parker Pen Co. is in Janesville.

A General Motors assembly plant in Janesville turned out 127,381 cars and 122,595 trucks during the 1983 model year. American Motors assembled 168,726 cars in Kenosha.

[30]COMMERCE

Wholesale trade in 1982 totaled $27.4 billion. Milwaukee County accounted for 20% of all wholesale establishments and 31% of the total trade volume.

Retail sales for 1982 amounted to $20.6 billion, compared with $14.9 billion in 1977. Figures for 1982 show that nearly 22% of the retail sales volume was in Milwaukee County. Statewide, food stores accounted for 21% of retail sales; automotive dealers, 17%; department stores and general merchandise stores, 12%; eating and drinking places, 10%; gasoline service stations, 10%; and other establishments, 30%.

The state engages in foreign as well as domestic trade through the Great Lakes ports of Superior-Duluth, Milwaukee, Green Bay, and Kenosha. Iron ore and grain are shipped primarily from Superior-Duluth, while Milwaukee handles the heaviest volume of general merchandise. Wisconsin exported about $4 billion in manufactured goods (13th in the US) in 1981 and $592 million in agricultural products (23d) in 1981/82. Greater Milwaukee is a foreign-trade zone, where goods can enter duty-free under certain conditions.

[31]CONSUMER PROTECTION

The Department of Agriculture, Trade, and Consumer Protection monitors food production, inspects meat, and administers grading programs; its Trade and Consumer Protection Division administers laws governing product safety and trade practices, in cooperation with the state Department of Justice. The Office of the Commissioner of Banking administers laws governing consumer credit, and the Department of Transportation's Motor Vehicles Division investigates complaints from buyers of new and used automobiles. The Department of Justice's Legal Services Division provides protection against deceptive and fraudulent business practices through its consumer protection unit.

[32]BANKING

As of 31 December 1984 there were 593 commercial banks in the state. Employment in banking was about 28,400 as of April 1985.

Commercial banks operating at the end of 1983 has assets totaling nearly $32.8 billion and deposits exceeding $27 billion. The largest commercial bank, First Wisconsin National Bank of Milwaukee, had assets of $3.5 billion at the end of 1983. There were 86 savings and loan associations as of 31 December 1983, with assets of over $14 billion. There were 585 state-chartered credit unions with total assets of $1.8 billion at the end of 1982.

The Office of the Commissioner of Banking licenses and charters banks, loan and collection companies, and currency exchanges. The Office of the Commissioner of Savings and Loan supervises state-chartered savings and loan associations. The Office of the Commissioner of Credit Unions enforces laws relating to credit unions.

33 INSURANCE

As of April 1985, an estimated 33,400 Wisconsinites were employed in the insurance industry. There were 7,528,000 life insurance policies worth $92.8 billion in force in 1983; benefits paid totaled $282 million. The average family had $50,900 in life insurance. In 1983, automobile insurance premiums written in Wisconsin totaled $751.5 million; homeowners premiums that year reached $195.5 million.

The Office of the Commissioner of Insurance licenses insurance agents, enforces state and federal regulations, responds to consumer complaints, and develops consumer education programs and literature. The office also operates the State Life Insurance Fund, which sells basic life insurance (maximum $10,000) to state residents; and the Local Government Property Insurance Fund, which insures properties of local government units on an optional basis.

34 SECURITIES

Wisconsin has no securities exchanges. However, as of 31 December 1983, member firms of the New York Stock Exchange had 127 sales offices and 957 full-time registered representatives in the state. The sale of securities is regulated by the Office of the Commissioner of Securities. Wisconsinites holding shares of public corporations numbered 779,000 in 1983.

35 PUBLIC FINANCE

Budget estimates are prepared by departments and sent to the governor or governor-elect in the fall of each even-numbered year; the following January, the governor presents a biennial budget to the legislature, which passes a budget bill, often after many amendments. Most appropriations are made separately for each year of the biennium. The fiscal year begins 1 July.

Revenues and expenditures for 1983/84 and 1984/85 (estimated) were as follows (in millions):

REVENUES	1983/84	1984/85
General purpose revenue	$3,977.5	$4,558.5
Federal revenue	1,796.4	2,007.3
Program revenue	834.6	936.2
Segregated revenue	771.7	867.8
TOTALS	$7,380.2	$8,369.8
EXPENDITURES		
State operations	$2,833.9	$3,152.8
Local assistance	2,611.6	3,034.6
Aid to individuals and organizations	1,934.7	2,182.4
TOTALS	$7,380.2	$8,369.8

Of the appropriations for 1984/85, 35% was allocated to human relations, 34% to education, 12% to environmental resources, and 19% to other purposes. In 1981/82, the city of Milwaukee had revenues of $520 million; expenditures were $409 million.

Expenditures by state and local governments alike have risen dramatically since 1960 and now claim about one-fifth of all personal income. At one time, the state was constitutionally prohibited from borrowing money; this provision was at first circumvented by the use of private corporations and then, in 1969, eliminated by constitutional amendment. As of mid-1982, state indebtedness exceeded $2.6 billion. The total indebtedness of state and local governments surpassed $6 billion in mid-1982.

36 TAXATION

In 1981/82, state and local taxes amounted to $1,260 per capita, 13th in the nation. The largest single source of state revenue is the income tax on individuals; revenue from this tax tripled in the 1970s. Most local tax revenue comes from property taxes, and most of that goes for education.

In 1984, personal income tax rates on net taxable income ranged from 3.4% for the first $3,900 to 11% on amounts over $51,600. The corporate tax rate was 7.9% of net income. Beginning in 1980/81, tax brackets were altered annually to correspond with changes in the consumer price index. The general sales tax in 1984 was 5%, and an inheritance and gift tax was levied at rates ranging from 2.5% to 30%. Other state taxes are those on gasoline, cigarettes, liquor, wine, beer, motor vehicles, insurance companies, real estate transfers, and public utilities.

In 1982/83, total federal tax receipts from Wisconsin were nearly $11.1 billion, or $2,336 per capita. During the same year, federal expenditures in Wisconsin were nearly $10.6 billion. Wisconsinites filed over 1.9 million federal income tax returns for 1983 and paid nearly $4.7 billion in tax.

37 ECONOMIC POLICY

The state seeks to promote relocation of new industries to Wisconsin, as well as expansion of existing ones, by providing advice and assistance through the Department of Development and some 280 local development corporations. Communities are authorized to issue tax-exempt bonds to enable industries to finance new equipment. In addition, all machinery and equipment used in goods production is tax-exempt under state law. Taxes on raw materials, work in progress, waste-treatment equipment, and finished goods inventories were eliminated entirely as of 1981.

38 HEALTH

There were 15.7 live births per 1,000 population in 1982, a rate slightly below the national average. The infant mortality rate was 9.4 per 1,000 live births, well below the national average. The death rate for nonwhite infants was much higher—17.5 per 1,000 in 1981—but still under the comparable national norm.

About 20,400 legal abortions were performed in Wisconsin during 1982, a ratio of 276 abortions for every 1,000 live births. State law prohibits the use of public funds for abortions, except in cases of incest or rape or for grave health reasons. The state does provide funds for family-planning counseling.

The death rate in 1982 was 8.5 per 1,000 population. As of 1981, Wisconsin ranked above the nation as a whole in death rates for heart disease, cerebrovascular disease, atherosclerosis, cancer, and diabetes, but below the national averages for pneumonia, cirrhosis of the liver, accidents, and infant deaths arising from prenatal conditions. Leading causes of death in 1982 were heart disease (346.6 deaths per 100,000 population), cancer (189.2), and cerebrovascular disease (70.3).

As of 1983, Wisconsin had 163 hospitals, with 29,357 beds and average daily admissions of 2,140 persons and outpatient visits of 14,782 persons. Hospital personnel included 18,190 registered nurses in 1981. The average cost of a hospital stay in 1982 came to $283 per day (13% below the national average). Wisconsin had 397 nursing homes with 25 beds or more in 1980.

At the end of 1982 there were 8,593 practicing physicians and 2,879 dentists; there were 34,669 registered nurses in 1980. Medical degrees are granted by the University of Wisconsin at Madison and by the Medical College of Wisconsin (formerly part of Marquette University).

The Division of Health, a branch of the State Department of Health and Social Services, has responsibility for planning and supervising health services and facilities, enforcing state and federal regulations, administering medical assistance programs, and providing information to the public. State laws provide for generic drug substitution and require continuing physician education.

39 SOCIAL WELFARE

Social welfare costs increased greatly during the 1970s and early 1980s. State and local expenditures on public welfare exceeded $1.3 billion in 1981, or $281 per capita, 10th highest in the nation. Public aid recipients were 7% of the population in 1983, above the national average.

Recipients of aid to families with dependent children totaled 281,400 in 1983; payments reached $413 million in 1982 and averaged $149 per person in February 1983. Medicaid enrollment was 480,000 in 1983, with payments of $901 million in 1982/83. Federal payments for public assistance came to $347 million in 1983; for Medicaid, $505 million.

Approximately 351,000 Wisconsin residents purchased food stamps in 1983, at a total federal cost of $140 million. Also in 1983, 430,000 pupils participated in the national school lunch program, at a federal cost of $32 million. Nearly $3.8 billion was paid out in 1983 to 773,908 Social Security beneficiaries, including 474,983 retirees, 147,929 survivors, and 45,164 disabled workers. The average monthly grant for retired persons in December 1982 was $432. In addition, in 1983, Supplemental Security Income funds totaling $144.6 million were disbursed to 23,988 aged, 37,638 disabled, and 984 blind persons.

A total of $205.8 million in workers' compensation was paid during 1982. An average of 107,000 persons each week received state unemployment benefits in 1982; the average weekly amount was $137.

40HOUSING

The 1980 census counted 1,863,897 housing units, of which 94% were occupied year-round. Nearly 97% of the occupied units had full plumbing facilities. Rural areas had a higher proportion of deficient housing than urban areas, and substandard conditions were three times as common in units built before 1940. Overcrowding was not a serious problem; only about 2.4% of the occupied units averaged more than one resident per room. From 1981 through 1983, 43,500 new housing units were authorized. The total value of construction was nearly $1.8 billion.

The Department of Veterans Affairs makes home loans to veterans. The Housing Finance Authority, created by the legislature in 1971, raises money through the sale of tax-exempt bonds and makes loans directly or indirectly to low- and moderate-income home buyers. Wisconsin's state building code, developed in 1913 to cover construction of all dwellings with three or more units, was revised in the late 1970s to cover new one- and two-family dwellings. Local housing codes prescribing standards for structural upkeep and maintenance in existing buildings are in force in all large cities and in many smaller cities and villages.

41EDUCATION

Wisconsin has a tradition of leadership in education. The University of Wisconsin, one of the world's largest and most respected institutions of higher learning, was chartered by the state legislature in 1848. The constitution of that year also provided for free public education; however, there was no state tax for schools until 1885, and no effective compulsory attendance law until 1903. In 1911, the legislature enacted the first system of vocational, technical, and adult education in the nation.

As of 1980, 70% of all Wisconsinites 25 years or older had completed high school, well above the US average. In 1983/84, Wisconsin's elementary and secondary schools had a total enrollment of 929,605 students, of whom 774,646, or 83%, attended the state's 2,035 public schools. There were 53,833 faculty members and administrators in the public school system. Wisconsin's 998 nonpublic schools had a total enrollment of 154,959.

The University of Wisconsin System, which embraces all public institutions of higher education in the state, had 159,868 students (excluding extension students) as of 1982/83. The University of Wisconsin at Madison, the system's largest campus, had 42,230 students as of 1982/83, of whom 31,509 were undergraduates; 47% of the undergraduates but only 39% of the graduate and professional students were women. The University of Wisconsin at Milwaukee, the system's 2d-largest center, had 26,468 graduate and undergraduate students as of 1983/84. Other campuses were Eau Claire, Green Bay, La Crosse, Oshkosh, Parkside (at Kenosha), Platteville, River Falls, Stevens Point, Stout (at Menom-

onie), Superior, and Whitewater. Also part of the University of Wisconsin System were 13 two-year centers offering courses in occupational and vocational fields (total 1982/83 enrollment, 10,379) and an Extension Division, offering agricultural and other courses (1981/82 enrollment, 200,637, mostly noncredit).

Marquette University in Milwaukee, with a graduate and undergraduate enrollment of 11,722 as of 1983/84, is the largest private institution. Other leading private colleges include Lawrence University in Appleton, Ripon College, and Beloit College. There are also four seminaries and six technical and professional schools, the largest of which, the Milwaukee School of Engineering, had a 1983/84 enrollment of 1,925 students.

Wisconsin also has a system of public vocational, technical, and adult education (VTAE), with an enrollment of 461,080 in 1981/82. This system is operated by 16 separate VTAE districts, under the overall guidance of the state Board of Vocational, Technical, and Adult Education.

General public elementary and secondary education is administered by local school districts, under the overall supervision of the Department of Public Instruction, which is headed by a state superintendent elected on a nonpartisan basis. The University of Wisconsin System was created under a 1971 law that merged the University of Wisconsin with the State University System, under a single Board of Regents. The Board of Regents appoints the president of the system and the chancellors of each of its 13 universities. The Wisconsin Higher Educational Aids Board offers grants and guaranteed loans to qualified residents attending institutions within the state.

In 1983/84, total expenditures for public education in Wisconsin amounted to more than $2.5 billion. Of this amount, over $1.1 billion, or 44%, was spent on public schools and their administration; another $1.3 billion was allocated to the University of Wisconsin System; the rest went for vocational education, communications, and other purposes. In 1981/82, Wisconsin state and local governments spent $840 per capita on public education, placing the state 6th among the 50 states. Expenditures per pupil were $3,553 in 1984, 14th among the states. Per capita expenditures for public higher education amounted to $258 in 1981/82, 8th among the 50 states.

42ARTS

Wisconsin offers numerous facilities for drama, music, and other performing arts, including a four-theater Performing Arts Center in Milwaukee and the Dane County Exposition Center in Madison. Milwaukee has a repertory theater, and there are many other theater groups around the state. Summer plays are performed at an unusual garden theater at Fish Creek in the Door Peninsula; there is also an annual music festival at that site.

The Pro Arte String Quartet in Madison and the Fine Arts Quartet in Milwaukee have been sponsored by the University of Wisconsin, which has also supported many other musical activities. Milwaukee is the home of the Florentine Opera Company, the Milwaukee Ballet Company, and the Milwaukee Symphony.

The Wisconsin Arts Board, consisting of 12 members appointed by the governor for three-year terms, aids artists and performing groups and assists communities in developing arts programs.

43LIBRARIES AND MUSEUMS

In 1983, the state had about 354 public libraries, with a total of 13.2 million volumes and circulation of 30.8 million. The Milwaukee Public Library, founded in 1878, maintained 12 branches and had 2,088,046 bound volumes as of 1983; the Madison Public Library had 7 branches and 572,909 volumes. The largest academic library is that of the University of Wisconsin at Madison, with 4,281,749 bound volumes and 2,068,146 microfiche units as of the same year. The best-known special library is that of the State Historical Society of Wisconsin at Madison, with 255,430 books and 60,000 cu feet of government publications and documents.

Wisconsin had 183 museums and historical sites in 1983. The State Historical Society maintains a historical museum in Madison and other historical sites and museums around the state. The Milwaukee Public Museum contains collections on history, natural history, and art. The Milwaukee Art Center, founded in 1888, a major museum of the visual arts, emphasizes European works of the 17th to 19th centuries. The Madison Art Center, founded in 1901, has European, Japanese, Mexican, and American paintings and sculpture, as well as 17th-century Flemish tapestries. The Charles Allis Art Library in Milwaukee, founded in 1947, houses collections of Chinese porcelains, French antiques, and 19th-century American landscape paintings. Other leading art museums include the Elvehjem Museum of Art in Madison and the Theodore Lyman Wright Art Center at Beloit College.

The Circus World Museum at Baraboo occupies the site of the original Ringling Brothers Circus. Other museums of special interest include the Dard Hunter Paper Museum (Appleton), the National Railroad Museum (Green Bay), and the Green Bay Packer Hall of Fame. More than 500 species of animals are on exhibit at the Milwaukee County Zoological Park; Madison and Racine also have zoos. Historical sites in Wisconsin include Villa Louis, a fur trader's mansion at Prairie du Chien; the Old Wade House in Greenbush; Old World Wisconsin, an outdoor ethnic museum near Eagle; Pendarvis, focusing on lead mining at Mineral Point, and the Taliesin estate of architect Frank Lloyd Wright, in Spring Green.

44COMMUNICATIONS

A total of 778 post offices and 11,447 postal employees provided mail service to Wisconsin in early 1985. About 97% of the state's 1,652,261 occupied housing units had telephones in 1980. In 1984 there were 104 AM and 144 FM radio stations and 35 educational stations, including 8 operated by the Educational Communications Board (ECB) and 9 by the state Board of Regents. The state also had 23 commercial television stations, 6 of them based in Milwaukee, and 8 educational television stations, including 5 operated by the ECB and 1 by the University of Wisconsin Board of Regents. In 1984, 134 cable television systems served 533,891 subscribers in 373 communities.

45PRESS

The state's first newspaper was the *Green Bay Intelligencer,* founded in 1833. Some early papers were put out by rival land speculators, who used them to promote their interests; among these was the *Milwaukee Sentinel,* launched in 1837 and a major daily newspaper today. As immigrants poured in from Europe in succeeding decades, German, Norwegian, Polish, Yiddish, and Finnish papers sprang up. Wisconsin journalism has a tradition of political involvement. The *Milwaukee Leader,* founded as a Socialist daily by Victor Berger in 1911, was denied the use of the US mails because it printed antiwar articles; the *Madison Capital Times,* still important today, also started as an antiwar paper. Founded in 1882 by Lucius Nieman, the *Milwaukee Journal* won a Pulitzer Prize in 1919 for distinguished public service and remains the state's largest-selling and most influential newspaper.

In 1984, Wisconsin had 5 morning papers with a combined circulation of 287,579, 30 evening papers with 903,044, and 11 Sunday papers with 975,930; in addition to these English-language newspapers, the *Milwaukee Deutsche Zeitung,* in German and English, appeared daily except Saturdays and Mondays. The following table shows leading dailies with their late 1984 circulations:

AREA	NAME	DAILY	SUNDAY
Green Bay	Press-Gazette (e,S)	54,418	72,512
Madison	Wisconsin State Journal (m,S)	76,554	135,941
Milwaukee	Capital Times (e)	29,907	
	Journal (e,S)	303,127	529,636
	Sentinel (m)	183,621	

As of 1983 there were also 22 semiweekly newspapers and 58 weeklies, as well as 156 periodicals directed to a wide variety of special interests. Among the largest are *Hoard's Dairyman,* founded by William D. Hoard in 1885, with a paid semimonthly circulation of 205,680; *Model Railroader,* monthly, 183,871; *The Woman Bowler,* monthly, 155,000; *Bowling Magazine,* monthly, 140,000; *Coin Prices,* bimonthly, 133,000; *Coins,* monthly, 103,000; and *Old Cars Weekly,* 85,000. Other notable periodicals are the *Wisconsin Magazine of History,* published quarterly in Madison by the state historical society, and *Wisconsin Trails,* another quarterly, also published in Madison.

46ORGANIZATIONS

The 1982 Census of Service Industries counted 1,225 organizations in Wisconsin, including 276 business associations; 654 civic, social, and fraternal associations; and 13 educational, scientific, and research associations. The State Historical Society of Wisconsin, founded in 1846, is one of the largest organizations of its kind; it has a museum, library, and research collections in Madison and is a prominent publisher of historical articles and books. The Forest Products Research Society, in Madison, has an international membership of about 5,000.

Other national organizations based in Wisconsin include the American Bowling Congress, American Society of Agronomy, Conservation Education Association, Crop Science Society of America, Experimental Aircraft Association, Master Brewers Association of the Americas, Model Railroad Industry Association, National Funeral Directors Association, Wilderness Watch, and World Council of Credit Unions.

47TOURISM, TRAVEL, AND RECREATION

Nearly $3.5 billion was spent in Wisconsin on travel and tourism (excluding foreign travelers) in 1982, and the industry was responsible for 105,500 jobs. The state has ample scenic attractions and outdoor recreational opportunities. In addition to the famous Wisconsin Dells gorge, visitors are attracted to the Cave of the Mounds at Blue Mounds, the sandstone cliffs along the Mississippi River, the rocky Lake Michigan shoreline of the Door Peninsula, the lakes and forests of the Rhinelander and Minocqua areas in the north, and Lake Geneva, a resort, in the south. Several areas in southern and northwestern Wisconsin, preserved by the state as the Ice Age National Scientific Reserve, still exhibit drumlins, moraines, and unusual geological formations.

There are three national parks in Wisconsin: Apostle Islands National Lakeshore, on Lake Superior, and the St. Croix and Lower St. Croix scenic riverways; visits totaled 358,762 in 1984. In 1983 there were 55 state parks, covering 60,570 acres (24,512 hectares); 265 mi (426 km) of state park trails; 10 state forests; and 5,216 campsites. More than 8.2 million people visited the state parks in 1981, and 4 million visited the state forests. During the same year, the state issued 372,419 deer-hunting licenses, 183,783 licenses for small game, and 644,336 fishing licenses.

48SPORTS

Wisconsin has professional teams in football, basketball, and baseball. The National Football League's Green Bay Packers, who play their home games at Lambeau Field, won five league championships and the first two Super Bowls during the 1960s under coach Vince Lombardi. Basketball fans follow the Milwaukee Bucks, who, led by Kareem Abdul-Jabbar (then known as Lew Alcindor), were National Basketball Association champions in 1970/71. In baseball, the Milwaukee Brewers, who play at Milwaukee County Stadium, won the American League pennant in 1982. Milwaukee is also the site of the annual Miller High Life Open in professional bowling and of the Greater Milwaukee Open in professional golf.

The University of Wisconsin Badgers compete in the Big Ten Conference. Badger ice hockey teams won the NCAA championship in 1973, 1977, 1981, and 1983. Basketball teams from Marquette University, in Milwaukee, won the NCAA Division I title in 1977 and the National Invitation Tournament championship in 1970.

⁴⁹FAMOUS WISCONSINITES

Wisconsinites who have won prominence as federal judicial or executive officers include Jeremiah Rusk (b.Ohio, 1830–93), a Wisconsin governor selected as the first head of the Agriculture Department in 1889; William F. Vilas (b.Vermont, 1840–1908), who served as postmaster general under Grover Cleveland; Melvin Laird (b.Nebraska, 1922), a congressman who served as secretary of defense from 1969–73; and William Rehnquist (b.1924), named to the Supreme Court in 1971.

The state's best-known political figures achieved nationwide reputations as members of the US Senate. John C. Spooner (b.Indiana, 1843–1919) won distinction as one of the inner circle of Senate conservatives before he retired in 1907 amid an upsurge of Progressivism within his party. Robert La Follette (1855–1925) embodied the new wave of Republican Progressivism—and, later, isolationism—as governor and in the Senate. His sons, Robert, Jr. (1895–1953), and Philip (1897–1965), carried on the Progressive tradition as US senator and governor, respectively. Joseph R. McCarthy (1908–57) won attention in the Senate and throughout the nation for his anti-Communist crusade. William Proxmire (b.Illinois, 1915), a Democrat, succeeded McCarthy in the Senate and eventually became chairman of the powerful Senate Banking Committee. Representative Henry S. Reuss (b.1912), also a Democrat, served in the House for 28 years and was chairman of the Banking Committee. Democrat Clement Zablocki (1912–83), elected to the House in 1948, was chairman of the Foreign Affairs Committee. Victor L. Berger (b.Transylvania, 1860–1929), a founder of the Social-Democratic Party, was first elected to the House in 1910; during World War I, he was denied his seat and prosecuted because of his antiwar views.

Besides the La Follettes, other governors who made notable contributions to the state include James D. Doty (b.New York, 1799–1865), who fought to make Wisconsin a separate territory and became the territory's second governor; William D. Hoard (b.New York, 1836–1918), a tireless promoter of dairy farming, as both private citizen and chief executive; James O. Davidson (b.Norway, 1854–1922), who attempted to improve relations between conservatives and progressives; Francis E. McGovern (1866–1946), who pushed through the legislature significant social and economic reform legislation; and Walter J. Kohler (1875–1940), an industrialist who, as governor, greatly expanded the power of the office.

Prominent figures in the state's early history include the Jesuit Jacques Marquette (b.France, 1637–75) and the explorer Louis Jolliet (b.Canada, 1645–1700) and the Sauk Indian leader Black Hawk (b.Illinois, 1767–1838), who was defeated in the Battle of Bad Axe. John Bascom (b.New York, 1827–1911) was an early president of the University of Wisconsin. Charles Van Hise (1857–1918), a later president, promoted the use of academic experts as government advisers; John R. Commons (b. Ohio, 1862–1945), an economist at the university, drafted major state legislation. Philetus Sawyer (b.Vermont, 1816–1900), a prosperous lumberman and US senator, led the state Republican Party for 15 years, before Progressives won control. Carl Schurz (b.Germany, 1829–1906) was a prominent Republican Party figure in the years immediately before the Civil War. Lucius W. Nieman (1857–1935) founded the *Milwaukee Journal,* and Edward P. Allis (b.New York, 1824–89) was an important iron industrialist.

Wisconsin was the birthplace of several Nobel Prize winners, including Herbert S. Gasser (1888–1963), who shared a 1944 Nobel Prize for research into nerve impulses; William P. Murphy (b.1892), who shared a 1934 prize for research relating to anemia; John Bardeen (b.1908), who shared the physics award in 1956 for his contribution to the development of the transistor; and Herbert

A. Simon (b.1916), who won the 1978 prize in economics. Stephen Babcock (b.New York, 1843–1931) was an agricultural chemist who did research important to the dairy industry. In addition, Wisconsin was the birthplace of the child psychologist Arnold Gesell (1880–1961) and of naturalist and explorer Chapman Andrews (1884–1960). John Muir (b.Scotland, 1838–1914), another noted naturalist and explorer, lived in Wisconsin in his youth. Conservationist Aldo Leopold (1887–1948) taught at the University of Wisconsin and wrote *A Sand County Almanac.*

Frederick Jackson Turner (1861–1932), historian of the American frontier, was born in Wisconsin, as were the economist and social theorist Thorstein Veblen (1857–1929) and the diplomat and historian George F. Kennan (b.1904). Famous journalists include news commentator H. V. Kaltenborn (1878–1965), award-winning sports columnist Red Smith (Walter Wellesley Smith, 1905–82), and television newsman Tom Snyder (b.1936).

Thornton Wilder (1897–1975), a novelist and playwright best known for *The Bridge of San Luis Rey* (1927), *Our Town* (1938), and *The Skin of Our Teeth* (1942), each of which won a Pulitzer Prize, heads the list of literary figures born in the state. Hamlin Garland (1860–1940), a novelist and essayist, was also a native, as were the poet Ella Wheeler Wilcox (1850–1919) and the novelist and playwright Zona Gale (1874–1938). The novelist Edna Ferber (b.Michigan, 1887–1968) spent her early life in the state.

Wisconsin is the birthplace of architect Frank Lloyd Wright (1869–1959) and the site of his famous Taliesin estate (Spring Green), Johnson Wax Co. headquarters (Racine), and First Unitarian Church (Madison). The artist Georgia O'Keeffe (b.1887) was born in Sun Prairie. Wisconsin natives who have distinguished themselves in the performing arts include Alfred Lunt (1893–1977), Frederic March (Frederick Bickel, 1897–1975), Spencer Tracy (1900–1967), Agnes Moorehead (1906–74), and Orson Welles (1915–85). Magician and escape artist Harry Houdini (Ehrich Weiss, b.Hungary, 1874–1926) was raised in the state, and piano stylist Liberace (Wlad Ziu Valentino Liberace, b.1919) was born there. Speed skater Eric Heiden (b.1958), a five-time Olympic gold medalist in 1980, is another Wisconsin native.

⁵⁰BIBLIOGRAPHY

Abrams, Lawrence and Kathleen. *Exploring Wisconsin.* Skokie, Ill.: Rand McNally, 1983.

Current, Richard N. *The History of Wisconsin. Vol. 2: The Civil War Era, 1848–73.* Madison: State Historical Society of Wisconsin, 1976.

Current, Richard N. *Wisconsin: A Bicentennial History.* New York: Norton, 1977.

Dictionary of Wisconsin Biography. Madison: State Historical Society of Wisconsin, 1960.

Epstein, Leon D. *Politics in Wisconsin.* Madison: University of Wisconsin Press, 1958.

Federal Writers' Project. *Wisconsin: A Guide to the Badger State.* Reprint. New York: Somerset, n.d. (orig. 1941).

Gara, Larry. *A Short History of Wisconsin.* Madison: State Historical Society of Wisconsin, 1962.

Nesbit, Robert C. *Wisconsin: A History.* Madison: University of Wisconsin Press, 1973.

Ritzenthaler, Robert E. *Prehistoric Indians of Wisconsin.* Milwaukee: Milwaukee Public Museum, 1953.

Smith, Alice E. *The History of Wisconsin. Vol. 1: From Exploration to Statehood.* Madison: State Historical Society of Wisconsin, 1973.

Still, Bayrd. *Milwaukee: The History of a City.* Madison: State Historical Society of Wisconsin, 1948.

Thelen, David P. *Robert M. La Follette and the Insurgent Spirit.* Boston: Little, Brown, 1976.

Wisconsin, State of. Legislative Reference Bureau. *1983–84 Blue Book.* Madison, 1983.

WYOMING

State of Wyoming

ORIGIN OF STATE NAME: Derived from the Delaware Indian words *maugh-wau-wa-ma*, meaning "large plains." **NICKNAME:** The Equality State. **CAPITAL:** Cheyenne. **ENTERED UNION:** 10 July 1890 (44th). **SONG:** "Wyoming." **MOTTO:** Equal Rights. **FLAG:** A blue field with a white inner border and a red outer border (symbolizing, respectively, the sky, purity, and the Indians) surrounds a bison with the state seal branded on its side. **OFFICIAL SEAL:** A female figure holding the banner "Equal Rights" stands on a pedestal between pillars topped by lamps symbolizing the light of knowledge; two male figures flank the pillars, on which are draped banners that proclaim "Livestock," "Grain," "Mines," and "Oil." At the bottom is a shield with an eagle, star, and Roman numerals XLIV, flanked by the dates 1869 and 1890. The whole is surrounded by the words "Great Seal of the State of Wyoming." **BIRD:** Meadowlark. **FLOWER:** Indian paintbrush. **TREE:** Cottonwood. **GEM:** Jade. **LEGAL HOLIDAYS:** New Year's Day, 1 January; Birthday of Martin Luther King, Jr., 3d Monday in January; Presidents' Day, 3d Monday in February; Memorial Day, last Monday in May; Independence Day, 4 July; Labor Day, 1st Monday in September; Columbus Day, 2d Monday in October; Election Day, 1st Tuesday after 1st Monday in November in even-numbered years; Veterans Day, 11 November; Thanksgiving Day, 4th Thursday in November; Christmas Day, 25 December. **TIME:** 5 AM MST = noon GMT.

¹LOCATION, SIZE, AND EXTENT

Located in the Rocky Mountain region of the northwestern US, Wyoming ranks 9th in size among the 50 states.

The total area of Wyoming is 97,809 sq mi (253,325 sq km), of which land comprises 96,989 sq mi (251,201 sq km) and inland water 820 sq mi (2,124 sq km). Shaped like a rectangle, Wyoming has a maximum E–W extension of 365 mi (587 km); its extreme distance N–S is 265 mi (426 km).

Wyoming is bordered on the N by Montana; on the E by South Dakota and Nebraska; on the S by Colorado and Utah; and on the W by Utah, Idaho, and Montana. The boundary length of Wyoming totals 1,269 mi (2,042 km). The state's geographic center lies in Fremont County, 58 mi (93 km) ENE of Lander.

²TOPOGRAPHY

The eastern third of Wyoming forms part of the Great Plains; the remainder belongs to the Rocky Mountains. Much of western Wyoming constitutes a special geomorphic province known as the Wyoming Basin. It represents a westward extension of the Great Plains into the Rocky Mountains, separating the Middle and Southern Rockies. Extending diagonally across the state from northwest to south is the Continental Divide, which separates the generally eastward-flowing drainage system of North America from the westward-flowing drainage of the Pacific states.

Wyoming's mean elevation is 6,700 feet (2,042 meters), 2d only to Colorado's among the 50 states. Gannett Peak, in western Wyoming, at 13,804 feet (4,207 meters), is the highest point in the state. With the notable exception of the Black Hills in the northeast, the eastern portion of Wyoming is generally much lower. The lowest point in the state—3,100 feet (945 meters)—occurs in the northeast, on the Belle Fourche River.

Wyoming's largest lake—Yellowstone—lies in the heart of Yellowstone National Park. In Grand Teton National Park to the south are two smaller lakes, Jackson and Jenny. All but one of Wyoming's major rivers originate within its boundaries and flow into neighboring states. The Green River flows into Utah; the Yellowstone, Big Horn, and Powder rivers into Montana; the Snake River into Idaho; the Belle Fourche and Cheyenne rivers into South Dakota; and the Niobrara and Bear rivers into Nebraska. The lone exception, the North Platte River, enters Wyoming from Colorado and eventually exits into Nebraska.

³CLIMATE

Wyoming is generally semiarid, with local desert conditions. Normal daily temperatures in Cheyenne range from 14°F (−10°C) to 37°F (3°C) in January and 44°F (7°C) to 80°F (27°C) in July. The record low temperature, −63°F (−53°C), was set 9 February 1933 at Moran; the record high, 114°F (46°C), 12 July 1900 at Basin. Normal precipitation in Cheyenne is 13 in (33 cm) a year, most of that falling between March and September; the snowfall in Cheyenne averages 52 in (132 cm) annually.

⁴FLORA AND FAUNA

Wyoming has more than 2,000 species of ferns, conifers, and flowering plants. Prairie grasses dominate the eastern third of the state; desert shrubs, primarily sagebrush, cover the Great Basin in the west. Rocky Mountain forests consist largely of pine, spruce, and fir.

The mule deer is the most abundant game mammal; others include the white-tailed deer, pronghorn antelope, elk, and moose. The jackrabbit, antelope, and raccoon are plentiful. Wild turkey, bobwhite quail, and several grouse species are leading game birds; more than 50 species of nongame birds also inhabit Wyoming all year long. There are 78 species of fish, of which rainbow trout is the favorite game fish. Endangered mammals include the black-footed ferret and the otter. The grizzly bear population in Yellowstone National Park was on the decline in the mid-1980s.

⁵ENVIRONMENTAL PROTECTION

The state's principal environmental concerns are conservation of scarce water resources and preservation of air quality. The Environmental Quality Council, a 7-member board appointed by the governor, hears and decides all cases arising under the regulations of the Department of Environmental Quality. The department enforces measures to prevent pollution of Wyoming's surface water and groundwater, and it administers 21 air-monitoring sites to maintain air quality. The state spent nearly $6 million for environmental quality controls in 1982/83. Programs to dispose of hazardous waste and assure safe drinking water are administered by the federal Environmental Protection Agency.

⁶POPULATION

Wyoming ranks 49th in the US in both population and population density; only Alaska is more sparsely populated. However, during the 1970s Wyoming was the third-fastest-growing state; its population grew by 41%, from 332,416 at the 1970 census to 469,557 according to the 1980 census, largely from migration. The growth rate slowed to 13.9% from mid-1980 to January 1985, when the population was estimated as 534,744. The population density in 1985 was 5.5 per sq mi (2 per sq km). Leading cities in 1984 were Cheyenne, 50,935; Casper, 49,588; and Laramie, 25,261.

7ETHNIC GROUPS

There were 7,088 Indians in Wyoming at the 1980 census. The largest tribe is the Arapaho, numbering more than 2,000. Wind River is the state's only reservation; tribal lands covered 1,793,000 acres (726,000 hectares) in 1982.

There were 3,364 black Americans living in Wyoming in 1980, and 1,804 Asian-Pacific peoples, about half of whom were of Japanese or Chinese descent. In 1980, 95% of the population was white and mostly of European descent, the largest groups being German, English, and Irish.

8LANGUAGES

Some place-names—Oshoto, Shoshoni, Cheyenne, Uinta—reflect early contacts with regional Indians.

Some terms common in Wyoming, like *comforter* (tied quilt) and *angleworm* (earthworm), evidence the Northern dialect of early settlers from New York State and New England, but generally Wyoming English is North Midland with some South Midland mixture, especially along the Nebraska border. Geography has changed the meaning of *hole, basin, meadow,* and *park* to signify mountain openings.

In 1980, 413,431 Wyomingites—94% of the residents 3 years old or older—spoke only English at home. The number of residents who spoke other languages at home included:

| Spanish | 14,875 | French | 1,571 |
| German | 3,809 | Italian | 785 |

9RELIGIONS

Wyoming's churchgoing population is preponderantly Protestant. There were 207,484 known adherents of Christian denominations other than Roman Catholic in 1980. The largest denominations were Catholic, with 62,315 members; Mormon, 40,368; United Methodist, 14,851; Episcopal, 13,057; Southern Baptist Convention, 10,871; Lutheran, 10,640; and United Presbyterian, 10,265. Wyoming also had an estimated 310 Jews in 1984.

10TRANSPORTATION

Wyoming is served chiefly by the Burlington Northern, Chicago and Northwestern, and Union Pacific railroads. The total trackage of these railroads in 1983 was 2,081 mi (3,349 km). Highways and rural and urban roads, totaling 39,074 mi (62,884 km) in 1983, cross the state. There were an estimated 522,000 registered motor vehicles in 1983.

At the end of 1983, Wyoming had 1,192 active aircraft, 2,951 licensed pilots, and 104 airports and heliports.

11HISTORY

The first human inhabitants of what is now Wyoming probably arrived about 11,500 BC. The forebears of these early Americans had most likely come by way of the Bering Strait and then worked their way south. Sites of mammoth kills south of Rawlins and near Powell suggest that the area was well populated. Artifacts from the period beginning in 500 BC include, high in the Big Horn Mountains of northern Wyoming, the Medicine Wheel monument, a circle of stones some 75 feet (23 meters) in diameter with 28 "spokes" that were apparently used to mark the seasons.

The first Europeans to visit Wyoming were French Canadian traders. The Vérendrye brothers, François and Louis-Joseph, probably reached the Big Horn Mountains in 1743; nothing came of their travels, however. The first effective discovery of Wyoming was made by an American fur trader, John Colter, earlier a member of the Lewis and Clark expedition. In 1806-7, Colter traversed much of the northwestern part of the state, probably crossing what is now Yellowstone Park, and came back to report on the natural wonders of the area. After Colter, trappers and fur traders crisscrossed Wyoming. By 1840, the major rivers and mountains were named, and the general topography of the region was well documented.

Between 1840 and 1867, thousands of Americans crossed Wyoming on the Oregon Trail, bound for Oregon or California. Migration began as a trickle, but with the discovery of gold in California in 1848, the trickle became a flood. In 1849 alone it is estimated that more than 22,000 forty-niners passed through the state via the Oregon Trail. Fort Laramie in the east and Fort Bridger in the west were the best-known supply points; between the two forts, immigrants encountered Independence Rock, Devil's Gate, Split Rock, and South Pass, all landmarks on the Oregon Trail. Although thousands of Americans crossed Wyoming during this period, very few stayed in this harsh region.

The event that brought population as well as territorial status to Wyoming was the coming of the Union Pacific Railroad. Railroad towns such as Cheyenne, Laramie, Rawlins, Rock Springs, and Evanston sprang up as the transcontinental railroad leapfrogged across the region in 1867 and 1868; in the latter year, Wyoming was organized as a territory. The first territorial legislature distinguished itself in 1869 by passing a women's suffrage act, the first state or territory to do so. Wyoming quickly acquired the nickname the Equality State.

After hostile Indians had been subdued by the late 1870s, Wyoming became a center for cattlemen and foreign investors who hoped to make a fortune from free grass and the high price of cattle. Thousands of Texas longhorn cattle were driven to the southeastern quarter of the territory. In time, blooded cattle, particularly Hereford, were introduced. As cattle "barons" dominated both the rangeland and state politics, the small rancher and cowboy found it difficult to go into the ranching business. However, overgrazing, low cattle prices, and the dry summer of 1886 and harsh winter of 1886/87 all proved disastrous to the speculators. The struggle between the large landowners and small ranchers culminated in the so-called Johnson County War of 1891–92, in which the large landowners were arrested by federal troops after attempting to take the law into their own hands.

Wyoming became a state in 1890, but growth remained slow. Attempts at farming proved unsuccessful in this high, arid region, and Wyoming to this day remains a sparsely settled ranching state. What growth has occurred has been primarily through the minerals industry, especially the development of coal, oil, and natural-gas resources during the 1970s because of the national energy crisis. However, the world's oil glut in the early 1980s slowed the growth of the state's energy industries; in 1984, the growth of the state's nonfuel mineral industry slowed as well. As a result, royalty payments from the federal government for minerals extracted from federal land in Wyoming fell by 16% in 1984, adding to the state's budgetary problems.

12STATE GOVERNMENT

Wyoming's state constitution was approved by the voters in November 1889 and accepted by Congress in 1890.

The legislature consists of a 30-member senate and a 64-member house of representatives. Senators are elected to staggered four-year terms. The entire house of representatives is elected every two years for a two-year term.

Heading the executive branch are five elected officials: the governor, secretary of state, auditor, treasurer, and superintendent of public instruction. Each serves a four-year term and, under Wyoming's cabinet form of government, each is also a member of seven state boards and commissions.

Voters must be US citizens, at least 18 years of age, and have been registered for at least 30 days before the election.

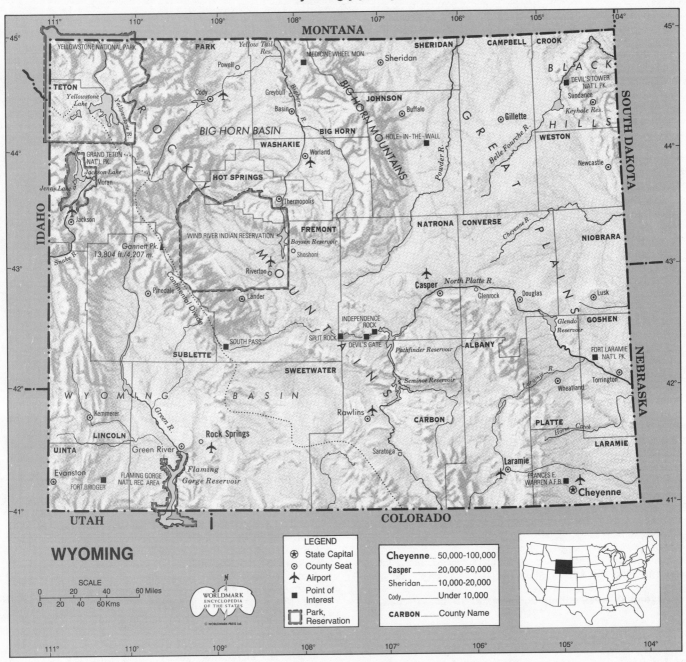

WYOMING

SCALE
0 20 40 60 Miles
0 20 40 60 Kms

WORLDMARK
ENCYCLOPEDIA
OF THE STATES
© Worldmark Press Ltd.

LEGEND
✪ State Capital
⊙ County Seat
✈ Airport
■ Point of Interest
▭ Park, Reservation

Cheyenne......... 50,000-100,000
Casper............ 20,000-50,000
Sheridan............ 10,000-20,000
Cody................. Under 10,000

CARBON...........County Name

See endsheet maps: E2

LOCATION: 41° to 45°N; 104°03' to 111°03'W. BOUNDARIES: Montana line, 384 mi (618 km); South Dakota line, 137 mi (221 km); Nebraska line, 140 mi (225 km); Colorado line, 260 mi (418 km); Utah line 174 mi (280 km); Idaho line 174 mi (280 km).

13 POLITICAL PARTIES

Except for the New Deal period during the 1930s and the national Democratic landslide of 1964, state government in Wyoming has been solidly in Republican hands. In the 94 years from 1890 to 1984, Republicans controlled the state senate for 88 years and the house of representatives 78 years. In 1985, the secretary of state and 24 of Wyoming's state legislators (25.6%) were women; this was the 3d-highest percentage in the U.S.

14 LOCAL GOVERNMENT

Wyoming is subdivided into 23 counties, 91 municipalities, 56 school districts, and 225 special districts and authorities.

Each county has a clerk, treasurer, assessor, sheriff, attorney, coroner, a district court clerk, three commissioners, and from one to five county judges or justices of the peace.

15 STATE SERVICES

The Board of Education has primary responsibility for educational services in Wyoming. Transportation services are provided by the Highway Commission; health and welfare matters fall under the jurisdiction of the Department of Health and Social Services. Among the many state agencies concerned with natural resources are the Department of Environmental Quality, Conservation Commission, Energy Conservation Advisory Committee, Board of Land Commissioners, Oil and Gas Conservation Commission, and Water Development Commission. The Department of Labor and Statistics is responsible for labor services.

16 JUDICIAL SYSTEM

Wyoming's judicial branch consists of a supreme court with a chief justice and 4 other justices, 9 district courts with 17 judges, and county judges and justices of the peace. Supreme court justices are appointed by the governor but must stand for retention at the next general election; once elected, they serve eight-year terms.

17ARMED FORCES
Wyoming has only one US military installation—the Francis E. Warren Air Force Base at Cheyenne, which had 4,815 personnel in 1984.

There were about 55,000 military veterans living in Wyoming in 1983. Of these, 1,000 were veterans of World War I, 19,000 of World War II, 11,000 of the Korean conflict, and 26,000 of the Viet-Nam era. Veterans' benefits totaled $63 million in 1982/83. At the beginning of 1985, 2,612 persons were members of the National Guard.

Wyoming had 1,810 state and local police and 567 corrections officials in 1983.

18MIGRATION
Many people have passed through Wyoming, but relatively few have come to stay. Not until the 1970s, a time of rapid economic development, did the picture change. Between 1970 and 1983, Wyoming gained a net total of 45,500 residents through migration.

19INTERGOVERNMENTAL COOPERATION
Emblematic of Wyoming's concern for water resources is the fact that it belongs to seven compacts with neighboring states concerning the Bear, Belle Fourche, Colorado, Snake, Upper Niobrara, and Yellowstone rivers.

Wyoming has also joined the Interstate Oil and Gas Compact, the Western Interstate Energy Compact, and numerous other multistate bodies, including the Council of State Governments.

Federal aid in 1983/84 totaled $556,326,000.

20ECONOMY
The economic life of Wyoming is largely sustained by agriculture—chiefly feed grains and livestock—and mining, including petroleum and gas production. Mining and petroleum production mushroomed during the 1970s, leading to a powerful upsurge in population. In the early 1980s, unemployment remained low, per capita income was high, and the inflation rate declined. The absence of personal and corporate income taxes has helped foster a favorable business climate.

21INCOME
In personal income per capita, Wyoming ranked 3d among the 50 states in 1980, but fell to 15th place by 1983. In 1983 personal

Wyoming Presidential Vote by Major Political Parties, 1948–84

YEAR	ELECTORAL VOTE	WYOMING WINNER	DEMOCRAT	REPUBLICAN
1948	3	*Truman (D)	52,354	47,947
1952	3	*Eisenhower (R)	47,934	81,049
1956	3	*Eisenhower (R)	49,554	74,573
1960	3	Nixon (R)	63,331	77,451
1964	3	*Johnson (D)	80,718	61,998
1968	3	*Nixon (R)	45,173	70,927
1972	3	*Nixon (R)	44,358	100,464
1976	3	Ford (R)	62,239	92,717
1980	3	*Reagan (R)	49,427	110,700
1984	3	*Reagan (R)	53,370	133,241

* Won US presidential election.

income totaled $6.2 billion, or $11,969 per capita, more than triple the 1970 figure. In 1979, 7.7% of Wyoming families had incomes below the federal poverty level. Some 8,500 persons were among the nation's top wealth-holders in 1982, with gross assets of more than $500,000.

22LABOR
In 1984, Wyoming's civilian labor force exceeded 254,000, compared with 230,000 in 1979. Of the 1984 total, about 238,000 were employed and 16,000 were unemployed; the unemployment rate was 6.3%. About one-third of the labor force were women.

A federal survey in March 1983 revealed the following nonfarm employment pattern for Wyoming:

	ESTABLISH- MENTS	EMPLOYEES	ANNUAL PAYROLL ('000)
Agricultural services, forestry, fishing	149	398	$ 4,637
Mining	858	26,166	763,822
Contract construction	1,741	10,616	208,028
Manufacturing	484	8,286	170,707
Transportation, public utilities	772	10,137	224,082
Wholesale trade	1,117	8,159	156,562
Retail trade	3,944	34,771	332,857
Finance, insurance, real estate	1,103	8,032	130,785
Services	4,283	29,164	354,591
Other	1,022	917	20,835
TOTALS	15,473	136,646	$2,366,905

In 1980, 39,000 employees belonged to labor unions, representing 18% of the working force in that year.

23AGRICULTURE
Agriculture—especially livestock and grain—is one of Wyoming's most important industries. In 1984, Wyoming had about 9,100 farms and ranches covering almost 35 million acres (14 million hectares). The state's average of 3,824 acres (1,548 hectares) per farm ranked 2d in the US after Arizona. The value of the lands and buildings of Wyoming's farms and ranches in 1984 was nearly $5.8 billion. Total farm marketings in 1983 amounted to $593 million (37th in the US). Of this, livestock and animal products accounted for $478 million; crops, $115 million.

Field crops in 1983 included barley, 10,032,000 bu; wheat, 8,964,000 bu; oats, 3,381,000 bu; potatoes, 954,000 cwt; sugar beets, 616,000 tons; and dry beans, 324,000 cwt.

24ANIMAL HUSBANDRY
For most of Wyoming's territorial and state history, cattle ranchers have dominated the economy even though the livestock industry is not large by national standards (Wyoming's income from livestock marketings ranked 37th in 1983). At the end of 1983, Wyoming had 1,395,000 head of cattle, 1,090,000 sheep and lambs, and 27,000 hogs and pigs. In 1983, Wyoming farmers had 57,000 chickens and produced 8,600,000 eggs. The state's 12,000,-000 milk cows produced an estimated 136,000,000 lb of milk in 1983. Estimated wool production in 1983 was 9,667,000 lb.

25FISHING
There is no important commercial fishing in Wyoming. Fishing is largely recreational, and fish hatcheries and fish-planting programs keep the streams well stocked. Wyoming's streams and lakes were stocked with 12.2 million trout for sportfishing in 1980.

26FORESTRY
Wyoming has 10,028,000 acres (4,058,000 hectares) of forested land, equal to 16% of the state's land area. Of this, 4,334,000 acres (1,754,000 hectares) are usable as commercial timberland. The state's 10 national forests, of which Shoshone, Bridger, and Teton were the largest, covered 9,120,616 acres (3,690,995 hectares) in 1984. Shipments by the lumber industry were valued at $14.2 million in 1982.

²⁷MINING

Wyoming's total mineral production (excluding fossil fuels) declined in value to $525.4 million in 1984, or 32% below the peak production year of 1981. The decline was due mainly to the lower value of soda ash, which accounts for 75% of total mineral value, and to the discontinuance of iron-ore mining in the state. The state was 1st in production of bentonite (a moisture-absorbing mineral) and soda ash in 1983. It ranked 2d in uranium output and had about 35% of known US uranium deposits.

Nonfuel mineral production (in tons) in 1983 included soda ash (trona), 10,542,417; uranium ore, 3,049,068; bentonite, 2,183,865; sand and gravel, 2,400,000; and clays, 2,140,000.

²⁸ENERGY AND POWER

Wyoming is comparatively energy rich, ranking 3d among the states in coal production and 6th in output of both crude oil and natural gas. The state's production of fossil fuels declined in the early 1980s as the result of the sudden glut of crude oil on the world market. Wyoming's 1983 oil output of 120 million barrels was 20% below that of the peak year 1970.

The state's proved reserves of petroleum were estimated at 808 million barrels in 1983. In that year, reserves of natural gas were estimated at 3.7 trillion cu ft, and natural gas production totaled 540 billion cu ft.

Wyoming has the three largest producing coal mines in the US and total coal deposits estimated at 69 billion tons. In 1983, the state's 28 active mines produced 112 million tons of coal.

Electric power production in 1983 totaled 263 billion kwh; installed capacity was 5.8 million kw. In the same year, consumption of electricity totaled 8.2 billion kwh.

²⁹INDUSTRY

Although manufacturing increased markedly in Wyoming from 1977 to 1982—value of shipments by manufacturers nearly doubled from $1,287 million to $2,558 million—it remains insignificant by national standards. The following table shows value of shipments for major sectors in 1982:

Petroleum, coal products	$1,775,300,000
Food and kindred products	184,200,000
Nonelectrical machinery	126,900,000
Stone, clay, glass products	98,600,000

³⁰COMMERCE

In 1982, sales in wholesale trade were about $3.3 billion, and retailers sold goods worth $2.8 billion, up 72% from 1977 sales. Of the 1982 total, food stores accounted for 23%; automotive dealers, 18%; general merchandise stores, 7%; and other retailers, 52%. Wyoming's exports of manufactures to other countries were valued at $18 million in 1981, nearly double the 1977 value.

³¹CONSUMER PROTECTION

The Consumer Protection Division of the Attorney General's Office handles consumer complaints.

³²BANKING

In 1983, Wyoming had 112 insured commercial banks with total assets of $4.3 billion, including $3 billion in time and savings deposits and $2.3 billion in outstanding loans. There were also 11 savings and loan associations with assets of $1,062 million.

³³INSURANCE

There were about 665,000 life insurance policies in force in Wyoming during 1983; their total value was $10.5 billion. The average amount of life insurance per family was $52,100.

In 1983, $211.2 million in liability insurance premiums were written in the state, including $99 million in automobile insurance.

³⁴SECURITIES

New York Stock Exchange member firms had 15 sales offices and 70 full-time registered representatives in the state in 1983. About 92,000 Wyoming residents held shares in public corporations. Wyoming has no securities exchanges.

³⁵PUBLIC FINANCE

Wyoming's biennial budget is prepared by the governor and submitted to the legislature at the beginning of each even-numbered calendar year. The fiscal year is 1 July–30 June.

The following is a summary of general estimated revenues and recommended expenditures for 1985/86 (in millions):

REVENUES	
Sales and use taxes	$266.5
Mineral severance tax	221.5
Other current receipts	265.4
TOTAL	$753.4

EXPENDITURES	
University of Wyoming	$138.4
Other education	68.7
Hospitals, health, social services	120.4
Government administration	21.8
Other current expenses	371.0
TOTAL	$720.3

State and local government debt in 1982 totaled $2,774,000,000.

³⁶TAXATION

In 1982/83, the state government collected a total of $735 million in taxes. Sales and gross receipts tax revenues alone came to $190 million during the same period; property taxes amounted to $46 million. Wyoming has no personal or corporate income tax. The state retail sales tax is 3%.

In 1982, Wyoming residents filed 213,000 federal income tax returns and paid $724 million in taxes.

³⁷ECONOMIC POLICY

State policy in Wyoming has traditionally favored fiscal and social conservatism. A pro-business climate has generally prevailed: not until 1969, for example, was the minerals industry compelled to pay a severance tax on the wealth it was extracting from Wyoming soils. During the 1970s, the state government seemed increasingly aware of environmental problems (especially the lack of water) posed by economic growth. The Department of Economic Planning and Development sets long-range economic policies.

³⁸HEALTH

Wyoming's birthrate—21.7 per 1,000 population in 1982—was the 3d highest among the states (after Utah and Alaska). In 1981, however, the state's death rate—6.4 per 1,000 population— was well below the national average. The death rates for heart disease, cancer, and cerebrovascular diseases were well below the national norm, but accidental deaths were 88% above it. Wyoming has the lowest number of abortions of all the states—about 1,000 in 1982.

In 1983, Wyoming's 31 hospitals, with 2,762 beds, admitted 78,249 patients and had 503,669 outpatient visits. Hospital personnel included 1,445 registered nurses in 1981. In 1982, the state had 694 licensed physicians and 255 active dentists.

³⁹SOCIAL WELFARE

In 1982, 6,177 Wyoming residents received about $9 million in aid to families with dependent children, or an average monthly payment of $315 per family. Medicaid benefits exceeded $24 million in 1983; federal Supplementary Security Income payments were $3.2 million. In 1983, 52,000 Social Security recipients were paid $239 million; the average monthly payment to retired workers was $439.

About $61 million in unemployment insurance benefits were awarded to Wyoming residents in 1982. Workers' compensation payments totaled $37.2 million in the same year.

⁴⁰HOUSING

The 1980 census counted 182,368 housing units in Wyoming, of which 69% were owner-occupied and 98% had full plumbing; both percentages exceeded the US average. In 1983, 3,502 new housing units were built.

41EDUCATION

Nearly 78% of all adults in the state were high school graduates according to the 1980 census; and more than 17% were college graduates.

In 1981, Wyoming had 375 public schools: 262 elementary, 103 secondary, and 10 combined. In 1982 there were 4 vocational and correspondence schools. In 1982/83, student enrollment was 95,184, up 4.2% from 1980/81. There were 7,100 public-school teachers in 1984. In that year, Wyoming ranked 2d among the states in school expenditures per capita—$1,088—and was 6th in public-school teachers' salaries, with an average salary of $25,197 per teacher.

In the fall of 1982, 22,713 full- and part-time students were enrolled in Wyoming's higher educational institutions. The state controls and funds the University of Wyoming and seven community colleges. There are no private colleges or universities, although the National Outdoor Leadership School, based in Lander, offers courses in mountaineering and ecology.

42ARTS

The Wyoming Council on the Arts, consisting of 10 members appointed by the governor to three-year terms, funds local activities and organizations in the visual and performing arts, including painting, music, theater, and dance.

43LIBRARIES AND MUSEUMS

Wyoming was served by 23 county public library systems, with nearly 1.5 million volumes, in 1982/83. Public library circulation exceeded 3 million during the same period. The University of Wyoming, in Laramie, had 747,980 volumes in 1984.

There are at least 40 museums and historical sites, including the Wyoming State Art Gallery and Wyoming State Museum in Cheyenne; the Buffalo Bill Historical Center (Cody), which exhibits paintings by Frederic Remington; and the anthropological, geological, and art museums of the University of Wyoming at Laramie.

44COMMUNICATIONS

According to the 1980 census, 151,615 Wyoming homes—83% of the total—had telephones. There were 164 post offices in the state in 1985, with 1,177 employees.

In 1984, Wyoming had 55 radio stations, 30 AM and 25 FM, plus 5 commercial television stations. During the same year, 38 cable television systems served 119,086 subscribers in 59 communities.

45PRESS

There were 10 daily newspapers and 4 Sunday newspapers in Wyoming in 1984, with circulations of 101,419 and 76,180, respectively. The major daily and its 1984 circulation was the *Casper Star-Tribune*, 36,485.

46ORGANIZATIONS

The 1982 Census of Service Industries counted 265 organizations in Wyoming, including 45 business associations; 162 civic, social, and fraternal associations; and 8 educational, scientific, and research organizations. The National Association for Outlaw and Lawman History, headquartered in Laramie, is one of the few national organizations with headquarters in the state.

47TOURISM, TRAVEL, AND RECREATION

There are 2 national parks in Wyoming—Yellowstone and Grand Teton—and 10 state parks. Devils Tower and Fossil Butte are national monuments, and Fort Laramie is a national historic site. The national parks drew nearly 5 million visitors in 1983. Tourists spent more than $1.8 million per day in Wyoming during 1982.

Yellowstone National Park, covering 2,219,823 acres (898,334 hectares), mostly in the northwestern corner of the state, is the oldest (1872) and largest national park in the US. The park features some 3,000 geysers and hot springs, including the celebrated Old Faithful. Just to the south of Yellowstone is Grand Teton National Park, 310,516 acres (125,662 hectares). Adjacent to Grand Teton is the National Elk Refuge, the feeding range of the continent's largest known herd of elk. Devils Tower, a rock formation in the northeast, looming 5,117 feet (1,560 meters) high, is the country's oldest national monument (1906).

Hunting and fishing are important recreational industries in Wyoming. In 1982/83, licenses were sold to 364,612 hunters and 263,013 fishermen.

48SPORTS

Sports in Wyoming are typically western. Skills developed by ranch hands in herding cattle are featured at rodeos held throughout the state. Cheyenne Frontier Days is the largest of these rodeos. Skiing is also a major sport.

Wyoming has no major professional sports teams. Male athletes at the University of Wyoming compete in the Western Athletic Conference; women compete in the High Country Athletic Conference. The University of Wyoming Cowboys have won conference titles in football four times and basketball three times since 1966. Also during that period, the University captured two NCAA national ski titles.

49FAMOUS WYOMINGITES

The most important federal officeholder from Wyoming was Willis Van Devanter (b. Indiana, 1859-1941), who served on the US Supreme Court from 1910 to 1937.

Many of Wyoming's better-known individuals are associated with the frontier. John Colter (b.Virginia, 1775?-1813), a fur trader, was the first white man to explore northwestern Wyoming. Jim Bridger (b.Virginia, 1804-81), perhaps the most famous fur trapper in the West, centered his activities in Wyoming. Late in life, William F. "Buffalo Bill" Cody (b.Iowa, 1846-1917) settled in the Big Horn Basin and established the town of Cody. A number of outlaws made their headquarters in Wyoming. The most famous were "Butch Cassidy" (George Leroy Parker, b.Utah, 1866-1908) and the "Sundance Kid" (Harry Longabaugh, birthplace in dispute, 1863?-1908), who, as members of the Wild Bunch, could often be found there.

Two Wyoming women, Esther Morris (b.New York, 1814-1902) and Nellie Tayloe Ross (b.Missouri, 1880-1979), are recognized as the first woman judge and the first woman governor, respectively, in the US; Ross also was the first woman to serve as director of the US Mint. Few Wyoming politicians have received national recognition, but Francis E. Warren (b.Massachusetts, 1844-1929), the state's first governor, served 37 years in the US Senate and came to wield considerable influence and power.

Without question, Wyoming's most famous businessman was James Cash Penney (b.Missouri, 1875-1971). Penney established his first "Golden Rule" store in Kemmerer and eventually built a chain of department stores nationwide. The water-reclamation accomplishments of Elwood Mead (b.Indiana, 1858-1936) and the botanical work in the Rocky Mountains of Aven Nelson (b.Iowa, 1859-1952) were highly significant. Jackson Pollock (1912-56), born in Cody, was a leading painter in the abstract expressionist movement.

50BIBLIOGRAPHY

Athearn, Robert G. *Union Pacific Country*. New York: Rand McNally, 1971.

Gressley, Gene M. *Bankers and Cattlemen*. New York: Knopf, 1966.

Larson, T. A. *History of Wyoming*. 2d ed., rev. Lincoln: University of Nebraska Press, 1978.

Larson, T. A. *Wyoming: A History*. New York: Norton, 1984.

Lavender, David. *Westward Vision: The Story of the Oregon Trail*. New York: McGraw-Hill, 1963.

Mead, Jean. *Wyoming in Profile*. Boulder, Colo.: Pruett, 1982.

Woods, L. Milton. *The Wyoming Country Before Statehood*. Worland, Wyo.: Worland Press, 1971.

Wyoming. Department of Administration and Fiscal Control, Division of Research and Statistics. *Wyoming Data Handbook 1983*. 6th ed. Cheyenne, 1983.

DISTRICT OF COLUMBIA

ORIGIN OF NAME: From "Columbia," a name commonly applied to the US in the late 18th century, ultimately deriving from Christopher Columbus. **BECAME US CAPITAL:** 1 December 1800. **MOTTO:** *Justitia omnibus* (Justice for all). **FLAG:** The flag, based on George Washington's coat of arms, consists of three red stars above two horizontal red stripes on a white field. **OFFICIAL SEAL:** In the background, the Potomac River separates the District of Columbia from the Virginia shore, over which the sun is rising. In the foreground, Justice, holding a wreath and a tablet with the word "Constitution," stands beside a statue of George Washington. To her left is the Capitol; to her right, an eagle and various agricultural products. Below is the District motto and the date 1871; above are the words "District of Columbia." **BIRD:** Wood thrush. **FLOWER:** American beauty rose. **TREE:** Scarlet oak. **LEGAL HOLIDAYS:** New Year's Day, 1 January; Birthday of Martin Luther King, Jr., 3d Monday in January; Washington's Birthday, 3d Monday in February; Memorial Day, last Monday in May; Independence Day, 4 July; Labor Day, 1st Monday in September; Columbus Day, 2d Monday in October; Veterans Day, 11 November; Thanksgiving Day, 4th Thursday in November; Christmas Day, 25 December. **TIME:** 7 AM EST = noon GMT.

¹LOCATION, SIZE, AND EXTENT
Located in the South Atlantic region of the US, the District of Columbia has a total area of 69 sq mi (179 sq km), of which land takes up 63 sq mi (163 sq km) and inland water 6 sq mi (16 sq km). The District is bounded on the N, E, and S by Maryland, and on the W by the Virginia shore of the Potomac River. The total boundary length is 37 mi (60 km).

For statistical purposes, the District of Columbia (coextensive since 1890 with the city of Washington, D.C.) is considered part of the Washington, D.C., metropolitan area, which in 1985 embraced Calvert, Charles, Frederick, Montgomery, and Prince George's counties in Maryland, and Arlington, Fairfax, Loudoun, Prince William, and Stafford counties in Virginia, along with a number of other Virginia jurisdictions, most notably the city of Alexandria.

²TOPOGRAPHY
The District of Columbia, an enclave of western Maryland, lies wholly within the Atlantic Coastal Plain. The major topographical features are the Potomac River and its adjacent marshlands; the Anacostia River, edged by reclaimed flatlands to the south and east; Rock Creek, wending its way from the northwestern plateau to the Potomac; and the gentle hills of the north. The District's average elevation is about 150 feet (46 meters). The highest point—410 feet (125 meters)—is in the northwest, at Tenleytown; the low point is the Potomac, only 1 foot (30 cm) above sea level.

³CLIMATE
The climate of the nation's capital is characterized by chilly, damp winters and hot, humid summers. The normal daily mean temperature is 58°F (14°C), ranging from 35°F (2°C) in January to 79°F (26°C) in July. The record low, −15°F (−26°C), was set on 11 February 1899; the all-time high, 106°F (41°C), on 20 July 1930. Precipitation averages 39 in (99 cm) yearly; snowfall, 17 in (43 cm). The average annual relative humidity is 73% at 7 AM and 52% at 1 PM.

⁴FLORA AND FAUNA
Although most of its original flora has been obliterated by urbanization, the District has long been known for its beautiful parks, and about 1,800 varieties of flowering plants and 250 shrubs and trees grow there. Boulevards are shaded by stately sycamores, pin and red oaks, American lindens, and black walnut trees. Famous among the introduced species are the Japanese cherry trees around the Tidal Basin. Magnolia, dogwood, and

gingko are also characteristic. The District's fauna is less exotic, with squirrels, cottontails, English sparrows, and starlings predominating.

⁵ENVIRONMENTAL PROTECTION
The Department of Public Works oversees air- and noise-pollution control, radiological health, food and milk sanitation, and beautification, among other programs. Overall air quality was rated 23d on a list of 51 large US cities in 1981.

⁶POPULATION
The District of Columbia outranked four states in population in 1980, with a census total of 638,333, a decline of 16% from 1970. In early 1985, the population was estimated at 621,251, a further 2.7% decrease. Following the 1980 census, the Census Bureau projected a population of 501,500 in 1990; but with a decline of only 17,082 during 1980-85, it seems unlikely that the population would decline by an additional 119,751 during 1985-90. The population density was 9,861 per sq mi (3,835 per sq km). Considered as a city, the District ranked 15th in the US in 1980; by 1982 it had dropped to 17th. Even as the capital's population has declined, the number of Washington, D.C., metropolitan area residents has been increasing, from 2,109,000 in 1960 to 2,910,000 in 1970, and an estimated 3,429,400 in mid-1984 (9th in the US). The District's population is 100% urban and extremely mobile; only 38.5% of all residents in 1980 were born in the District.

⁷ETHNIC GROUPS
Black Americans have long been the largest ethnic group in the District of Columbia, accounting for 70.3% of the population in 1980, the highest percentage in any major US city except for Gary, Indiana (70.8%). Blacks comprised about 28% of the metropolitan area population in 1980, a proportion that has been relatively constant for more than 200 years. District ethnic minorities in 1980 included 17,679 Hispanics, 2,475 Chinese, 1,297 Filipinos, 996 American Indians, 950 Asian Indians, and 752 Japanese. Only 6.4% of the population was foreign-born in 1980, but contributing to Washington's ethnic diversity are the many foreign-born residents attached to foreign embassies and missions.

Between 1970 and 1980, the population of groups other than white and black almost quadrupled within the Washington metropolitan area, reaching 134,209 in 1980. Southeast Asians made up a significant proportion of the immigrants, as did Mexicans and Central and South Americans.

[8] LANGUAGES

Dialectically, the Washington, D.C., area is extremely heterogeneous. In 1980, 92% of all District of Columbia residents 3 years of age or older spoke only English at home. Other languages spoken at home included the following:

Spanish	19,263	Chinese	2,078
French	8,884	Italian	1,716
German	2,934	Greek	1,145

[9] RELIGIONS

As of 1 January 1984, the Washington diocese (including five Maryland counties) had 393,601 Roman Catholics. The leading Protestant denominations in 1980 were Episcopal, 49,861; American Baptist Convention, 28,126; United Methodist, 20,888; and Southern Baptist Convention, 18,935. Data on some predominantly black Protestant groups were unavailable. The Jewish population in 1984 was estimated at 24,285 in the District and 157,335 in the Washington metropolitan area. Washington's resident foreign population includes followers of numerous other religions.

[10] TRANSPORTATION

Union Station, located north of the Capitol, is the District's one rail terminal, from which Amtrak provides passenger service to the northeast corridor and southern points. In all, five Class I lines operated 47 mi (76 km) of rail in 1983. The Washington Metropolitan Area Transit Authority, or Metro, operates bus and subway transportation within the city and its Maryland and Virginia suburbs; 54.4 mi (87.5 km) of subway lines were in use by January 1985. Scheduled for completion in 1990, the system would comprise 101 mi (163 km) of track and 86 stations in the Washington metropolitan area. The subway system accounted for 112.5 million unlinked passenger trips in 1984. The Metrobus fleet of 1,510 vehicles had 171 million unlinked passenger trips in 1984 on a system about 42% within the District. Forty percent of working District residents commute by public transportation.

Within the District as of 1983 were 1,517 mi (2,441 km) of streets and roads; 232,725 motor vehicles were registered, and 370,196 driver's licenses were in force. Three major airports handle the District's commercial air traffic: Washington National Airport, just south of the city in Virginia; Dulles International Airport in Virginia; and Baltimore-Washington International Airport in Maryland. Of these, Washington National was the busiest airfield in 1983, handling 6,559,868 enplaned passengers. Baltimore-Washington enplaned 2,296,538, and Dulles International 1,325,933.

[11] HISTORY

Algonkian-speakers were living in what is now the District of Columbia when Englishmen founded the Jamestown, Va., settlement in 1607. The first white person known to have set foot in the Washington area was the English fur trader Henry Fleete, who in 1622 was captured by the Indians and held there for several years. Originally part of Maryland Colony, the region had been carved up into plantations by the latter half of the 17th century.

After the US Constitution (1787) provided that a tract of land be reserved for the seat of the federal government, both Maryland and Virginia offered parcels for that purpose; on 16 July 1790, Congress authorized George Washington to choose a site not more than 10 mi (16 km) square along the Potomac River. President Washington made his selection in January 1791 and then appointed Andrew Ellicott to survey the area and employed Pierre Charles L'Enfant, a French military engineer who had served in the Continental Army, to draw up plans for the federal city. L'Enfant's masterful design called for a wide roadway (now called Pennsylvania Avenue) to connect the Capitol with the President's House (Executive Mansion, now commonly called the White House) a mile away, and for other widely separated public buildings with spacious vistas. However, L'Enfant was late in completing the engraved plan of his design, and he also had difficulty in working with the three commissioners who had been appointed to direct a territorial survey; for these and other reasons, L'Enfant was dismissed and Ellicott carried out the plans. Construction was delayed by lack of adequate financing, and only one wing of the Capitol was completed and the President's House was still under construction when President John Adams and some 125 government officials moved into the District in 1800. Congress met there for the first time on 17 November, and the District officially became the nation's capital on 1 December. On 3 May 1802, the city of Washington was incorporated (the District also included other local entities), with an elected council and a mayor appointed by the president.

Construction proceeded slowly, while the city's population grew to about 24,000 by 1810. In August 1814, during the War of 1812, British forces invaded and burned the Capitol, the President's House, and other public buildings. These were rebuilt within five years, but for a long time, Washington remained a rude, rough city. In 1842, English author Charles Dickens described it as a "monument raised to a deceased project," consisting of "spacious avenues that begin in nothing and lead nowhere." At the request of its residents, the Virginia portion was retroceded in 1846, thus confining the federal district to the eastern shore of the Potomac. The Civil War brought a large influx of Union soldiers, workers, and escaped slaves, and the District's population rose sharply from 75,080 in 1860 to 131,700 by the end of the decade, spurring the development of modern Washington.

In 1871, Congress created a territorial form of government; this territorial government was abolished three years later because of alleged local extravagances, and in 1878, a new form of government was established, headed by three commissioners appointed by the president. During the same decade, Congress barred District residents from voting in national elections or even for their own local officials. In the 1890s, Rock Creek Park and Potomac Park were established, and during the early 1900s, city planners began to rebuild the monumental core of Washington in harmony with L'Enfant's original design. The New Deal period brought a rise in public employment, substantial growth of federal facilities, and the beginnings of large-scale public housing construction and slum clearance. After World War II, redevelopment efforts concentrated on demolishing slums in the city's southwest section. The White House was completely renovated in the late 1940s, and a huge building program coincided with the expansion of the federal bureaucracy during the 1960s.

Because it is the residence of the president, Washington, D.C., has always been noted for its public events, in particular the Presidential Inauguration and Inaugural Ball. The District has also been the site of many historic demonstrations: the appearance in 1894 of Coxey's Army—some 300 unemployed workers; the demonstrations in 1932 of the Bonus Marchers—17,000 Army veterans demanding that the government cash their bonus certificates; and the massive March on Washington by civil rights demonstrators in 1963. A riot following the assassination of Martin Luther King, Jr., in 1968 took nine lives and resulted in $24 million in property damage paid by insurance companies.

In recent years, the District's form of government has undergone significant changes. The 23d Amendment to the US Constitution, ratified on 3 April 1961, permits residents to vote in presidential elections, and beginning in 1971, the District was allowed to send a nonvoting delegate to the US House of Representatives. Local self-rule began in 1975, when an elected mayor and council took office.

[12] DISTRICT GOVERNMENT

The District of Columbia is the seat of the federal government and houses the principal organs of the legislative, executive, and

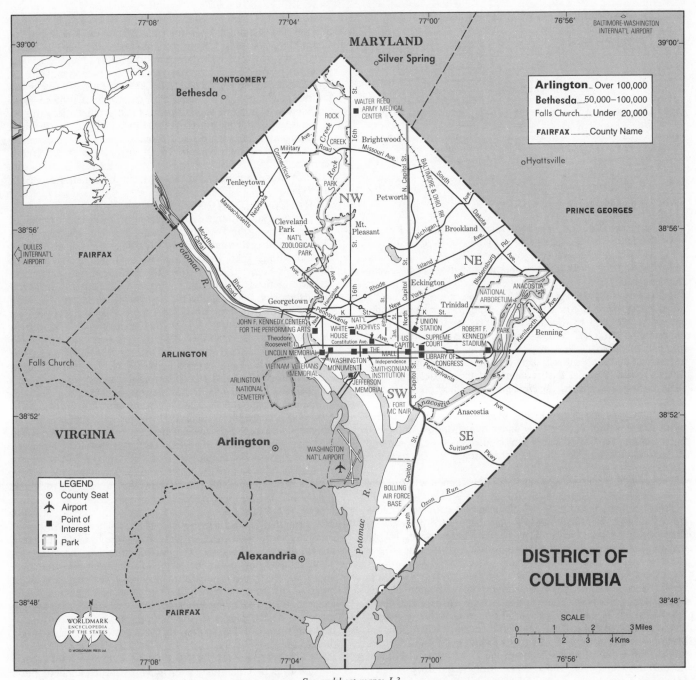

See endsheet maps: L3.
LOCATION: 38°47′ to 39°N; 76°55′ to 77°07′ W. **BOUNDARIES:** Maryland line, 25 mi (40 km); Virginia line, 12 mi (19 km).

judicial branches. The District of Columbia committees of the US Senate and House of Representatives oversee affairs within Washington, D.C. The District elects a delegate to the US House who participates in discussions and votes on bills within the District of Columbia Committee but may not vote on measures on the floor of the House. In 1978, Congress approved a constitutional amendment granting the District two US senators and at least one representative; however, the amendment failed to become law because it was not ratified by the necessary 38 state legislatures by August 1985.

Elected delegates drafted a constitution for the proposed State of New Columbia in 1982. After it was approved by voters within the District, a petition for statehood was forwarded to Congress in 1983, where it faced an uncertain fate.

[13]POLITICAL PARTIES

Washington, D.C., is the headquarters of the Democratic and Republican parties, the nation's major political organizations. The District itself is overwhelmingly Democratic: in November 1984, out of 274,810 registered voters, 222,539 were Democrats (81%), 21,747 were Republicans (8%), and 30,524 (11%) were independents or members of other groups. The District has a 30-day residency requirement for voters. Residents of the District, permitted to vote for president since 1964, have unfailingly cast their ballots for the Democratic nominee.

The first mayor, Walter Washington, was defeated for reelection in 1978 by Marion Barry, Jr., who was reelected in 1982. Walter Fauntroy has been the District's delegate to the House of Representatives since 1971.

D.C. Presidential Vote by Major Parties, 1964–84

YEAR	ELECTORAL VOTE	D.C. WINNER	DEMOCRAT	REPUBLICAN
1964	3	*Johnson (D)	169,796	28,801
1968	3	Humphrey (D)	139,566	31,012
1972	3	McGovern (D)	127,627	35,226
1976	3	*Carter (D)	137,818	27,873
1980	3	Carter (D)	124,376	21,765
1984	3	Mondale (D)	180,408	29,009

* Won US presidential election.

Blacks have played a dominant role in politics. In January 1985, the mayor, the delegate to the US House of Representatives, 9 of the 13 council members, 9 of the 12 Board of Education members, and 232 of the 321 neighborhood advisory commissioners were black.

¹⁴LOCAL GOVERNMENT

Local government in the District of Columbia operates under authority delegated by Congress. In 1973, for the first time in more than a century, Congress provided the District with a home-rule charter, allowing Washington, D.C., residents to elect their own mayor and a city council of 13 members, all serving four-year terms. Residents of the District approved the charter on 7 May 1974, and a new elected government took office on 1 January 1975.

The mayor is the District's chief executive, and the council is the legislative branch; however, under constitutional authority, Congress can enact laws on any subject affecting the District, and all legislation enacted by the District is subject to congressional veto.

The council consists of 13 members, the council chairman, 4 members elected at-large, 8 elected by wards. The 12-member Board of Education consists of 8 officials elected by ward and 4 elected at-large, including 1 at-large member elected by students. They serve for four years. The charter also provides for 36 neighborhood advisory commissions, whose seats (321 in all) are filled through nonpartisan elections.

¹⁵DISTRICT SERVICES

Public education in the District is the responsibility of an elected Board of Education and the University of the District of Columbia Board of Trustees. Transportation services are provided through the Department of Transportation and the Washington Metropolitan Area Transit Authority, while health and welfare services fall within the jurisdiction of the Department of Human Resources. The Office of Consumer and Regulatory Affairs, Department of Corrections, District of Columbia National Guard, and Metropolitan Police Department provide public protection services, and the Department of Housing and Community Development is the main housing agency. Employment and job training programs are offered through the Department of Employment Services.

¹⁶JUDICIAL SYSTEM

All judges in Washington, D.C., are nominated by the president of the US from a list of persons recommended by the District of Columbia Nomination Commission, and appointed upon the advice and consent of the Senate. The US Court of Appeals for the District of Columbia functions in a manner similar to that of a state supreme court; it also has original jurisdiction over federal crimes. The court consists of a chief judge and 8 associate judges, all serving 15-year terms. The Superior Court of the District of Columbia, the trial court, consisted in late 1983 of five divisions and a chief judge, 42 associate judges, and 10 senior judges, also

serving for 15 years. Washington, D.C., is the site of the US Supreme Court and the US Department of Justice. The District of Columbia is the only US jurisdiction where the US Attorney's Office, an arm of the Justice Department, and not the local government, prosecutes criminal offenders for non-federal crimes.

According to the FBI Crime Index, the violent crime rate in the District of Columbia was 1,985 per 100,000 population in 1983, and the total crime rate was 9,453 per 100,000 population. Crimes in the District during 1983 included 186 cases of murder and nonnegligent manslaughter, 406 forcible rapes, 7,698 robberies, and 3,646 aggravated assaults. In all, there were 78,420 crimes reported.

D.C. prisoners in state and federal institutions numbered 4,834 in December 1984.

¹⁷ARMED FORCES

As of mid-1984, authorized Department of Defense personnel on military bases in the District of Columbia totaled 28,988. Of the District's principal installations in 1982/83, 5 were naval, and 2—including Bolling Air Force Base—were Air Force installations; the Army's Fort McNair and Walter Reed Medical Center were also within the District. The Pentagon, headquarters of the US Department of Defense, covers 34 acres (14 hectares) of Arlington, Va., across the Potomac. Firms in the District received $815 million in federal defense contract awards over $25,000 in 1983/84. As of 30 September 1983, about 64,000 veterans of US military service were living in the District, of whom about 1,000 served in World War I, 27,000 in World War II, 13,000 during the Korean conflict, and 17,000 during the Viet-Nam era. Veterans' benefits totaled $654 million during 1982/83.

Because Washington is often the scene of political demonstrations and because high federal officials and the District's foreign embassy personnel pose special police-protection problems, the ratio of police personnel to residents is higher than in any state: in 1983, its 4,409 police employees represented a rate of 7 per 1,000 population, almost three times the national average. There were 2,297 Army National Guard personnel at the end of 1984 and 1,349 Air National Guard personnel as of February 1985.

¹⁸MIGRATION

The principal migratory movements have been an influx of southern blacks after the Civil War and, more recently, the rapid growth of the Washington, D.C., metropolitan area, coupled with a shrinkage in the population of the District itself. Between 1950 and 1970, the District suffered a net loss from migration of as much as 260,000, much of it to Maryland and Virginia; white emigration was even greater, since there was an estimated net inflow of 87,000 blacks in this period. Net emigration totaled between 150,000 to 190,000 during the 1970s, and perhaps 23,000 more during 1981–1983.

¹⁹INTERGOVERNMENTAL COOPERATION

The District of Columbia, a member of the Council of State Governments and its allied organizations, also participates in such interstate regional bodies as the Commission on Mental Health, Interstate Commission on Juveniles, Vehicle Equipment Safety Commission, Washington Metropolitan Area Transit Authority Commission, and Potomac Valley Commission. Counties and incorporated cities in the Washington area are represented on the Metropolitan Washington Council of Governments, established in 1957.

The District relies heavily on federal assistance, which came to nearly $1.4 billion in 1983/84. Federal aid per capita that year was more than $2,200, highest in the US. and over five times the national average.

²⁰ECONOMY

The federal government is the District's largest employer, and printing is the leading industry. More than one-third of employed persons in 1983 were employed in managerial and professional occupations. Government activity attracts an extraordinary number of professionals: in 1983, for example, the D.C. Bar Association had an estimated 26,000 lawyers, more than most states. Banking and especially health and legal services were the growth sectors in the late 1970s.

²¹INCOME
With an income per capita of $16,409 in 1983, the District of Columbia outranked every state except Alaska. Total personal income was $10.2 billion, representing a real increase of 23% since 1970; in terms of per-capita personal income, the real increase was 49% (compared to 28% for the nation). However, the proportion of District of Columbia residents below the federal poverty level increased from 17% to 18.6% between 1969 and 1979, and significant inequalities remain. In 1979, for example, the median household income for black residents was only 63% that of whites.

²²LABOR
The civilian labor force in the District of Columbia averaged 320,000 in 1984, of whom 291,000 were employed and 29,000 (9.1%) were unemployed, higher than all but 12 states. In March 1985, the District's unemployment rate was 8.1%; the metropolitan area's unemployment rate was only 3.7%. There were 163,000 District women in the labor force in 1983, representing 50.2% of the total, higher than in any state or any of 10 other big cities. The unemployment rate was 14.3% for men and 9.1% for women in that year; the white rate was 3.4%, and the black rate was 15.8%. The teenage unemployment rate was 43.7% in 1983 (52.4% for blacks).

A federal census of workers in March 1982 revealed the following nonfarm employment pattern for the District of Columbia:

	ESTABLISH-MENTS	EMPLOYEES	ANNUAL PAYROLL ('000)
Agricultural services, forestry, fishing	27	130	$ 1,455
Mining	19	98	4,646
Contract construction	350	7,113	149,831
Manufacturing, of which:	456	17,854	408,365
Printing, publishing	(294)	(13,186)	(292,238)
Transportation, public utilities	543	26,046	689,817
Wholesale trade	521	9,714	210,615
Retail trade	3,173	49,103	505,509
Finance, insurance, real estate	2,378	37,257	687,892
Services, of which:	8,536	180,826	3,389,153
Business services	(1,450)	(31,148)	(486,703)
Health services	(1,404)	(27,775)	(539,399)
Legal services	(1,403)	(19,279)	(581,797)
Membership organizations	(1,457)	(31,402)	(681,157)
Other	436	557	15,020
TOTALS	16,439	328,698	$6,062,303

Local government employees in the District of Columbia numbered 38,721 on 1 October 1983, including 1,220 part-time workers; the ratio of full-time city workers to population was the highest of any major US city. Federal employment in the District was 209,600; District residents received nearly $6.8 billion in federal salaries and wages in 1982/83, more than the residents of any state except California and Virginia.

The District of Columbia is the headquarters of the American Federation of Labor and Congress of Industrial Organizations (AFL–CIO), the dominant US labor organization. The AFL–CIO is a federation of affiliated unions. The Assembly of Governmental Employees and other independent unions also have headquarters there.

²³AGRICULTURE
There is no commercial farming in the District of Columbia.

²⁴ANIMAL HUSBANDRY
The District of Columbia has no livestock industry.

²⁵FISHING
There is no commercial fishing in the District of Columbia.

²⁶FORESTRY
There is no forestland or forest products industry in the District of Columbia.

²⁷MINING
There is no mining in the District of Columbia, although a few mining firms have offices there.

²⁸ENERGY AND POWER
The District of Columbia had an installed electric energy capacity of 868,000 kw in 1983 from two oil-fired plants, both privately owned. Electrical output during the same year totaled 221 million kwh.

In 1983, the District had 146,100 gas utility customers, of whom 131,100 were residential and 15,000 were commercial. Revenues totaled $224 million in 1983. All fossil fuels must be imported from domestic and foreign suppliers outside the District.

²⁹INDUSTRY
Value of shipments by manufacturers in 1982 reached $1,537,100,000, of which printing and publishing (most of it by or about the federal government) accounted for 83%. Within the District is the Government Printing Office (established by Congress in 1860), which operates one of the largest printing plants in the US. Also in the District is the Washington Post Co., publisher of the newspaper of that name and of *Newsweek* magazine; the company also owns television stations. Food and kindred products, fabricated metal products, and nonelectrical machinery were the District's only other significant industries.

³⁰COMMERCE
Wholesaling in the District of Columbia totaled $2.6 billion in 1982, when the District's retail sales came to nearly $2.7 billion. Among retail establishments with payroll, eating and drinking places accounted for 21% of sales, food stores 19%, apparel and accessory stores 9%, department stores 8%, gasoline service stations 7%, and automotive dealers 5%. Retailers in the Washington, D.C., metropolitan area had sales of $17.1 billion during the same year, with food stores accounting for 20%, department stores and eating and drinking places 20% each, automotive dealers 18%, gasoline service stations 9%, and others 13%.

Alexandria, Va., which lies within the Washington, D.C., Customs District, was the port of unloading for 135,500 tons of cargo in 1983, worth $54 million.

³¹CONSUMER PROTECTION
The Department of Consumer and Regulatory Affairs has primary responsibility for consumer protection in the District. This department also licenses all business and all professionals except lawyers.

³²BANKING
Banking in the District of Columbia began with the chartering of the Bank of Alexandria in 1792 and the Bank of Columbia in 1793; both banks terminated in the early 19th century. The oldest surviving bank in the District is the National Bank of Washington, founded as the Bank of Washington in 1809. As of the end of 1983, the District's two largest commercial banks were the Riggs National Bank and the American Security Bank, with assets of $4.8 billion and $3.7 billion, respectively.

Overall, there were 19 insured commercial banks at the end of 1983, with assets totaling $11 billion and outstanding loans of nearly $5.5 billion. As of the end of 1983, 7 savings and loan institutions had combined assets of $3.3 billion. At the end of 1982, outstanding mortgage loans came to $2.5 billion for 8 savings and loan institutions.

³³INSURANCE
In 1983, District of Columbia policyholders held life insurance policies worth $34.8 billion. The average coverage per family was $113,000, far higher than in any state. Benefits totaled $236.6 million, of which death payments made up $96.1 million. Property

and casualty insurers wrote premiums worth $357.9 million in the District in 1983, of which $59 million was automotive liability insurance, $35.7 million was automobile physical damage insurance, and $26.4 million was homeowners' coverage.

34 SECURITIES

There are no securities exchanges in the District of Columbia. New York Stock Exchange member firms had 36 sales offices with 827 registered representatives in 1983. The District of Columbia had 193,000 shareowners of public corporations during the same year.

35 PUBLIC FINANCE

The budget for the District of Columbia is prepared by the mayor's office and reviewed by the city council, but is subject to review and approval by Congress. The fiscal year runs from 1 October through 30 September.

The following table summarizes general and capital revenues and expenditures for 1981/82 and 1982/83 (in thousands).

REVENUES	1981/82	1982/83
Taxes, of which:	$1,173,725	$1,295,927
Sales and use tax	(382,300)	(416,600)
Income and franchise tax	(427,400)	(461,100)
Property tax	(325,100)	(368,200)
Intergovernmental transfers	726,267	796,457
Nontax revenues	138,309	158,321
TOTALS	$2,038,301	$2,250,705
EXPENDITURES		
Human support services	$ 605,205	$ 718,348
Public safety and justice	485,325	558,464
Public education system	417,778	448,763
Transportation services and assistance	127,806	147,236
Economic development and regulation	107,496	118,315
Government direction and support	50,057	55,196
Environmental services and supply	44,822	42,980
Capital outlays	159,640	169,814
Debt service	137,787	153,217
TOTALS	$2,135,916	$2,412,333

The local tax base is limited by a shortage of taxable real estate, much of the District being occupied by government buildings and federal reservations. Moreover, Congress has not allowed the District to tax the incomes of people who work in Washington but live in the suburbs, an objective the District government has urgently sought. The total public debt was over $2.8 billion, or about $4,440 per capita, as of 30 September 1982; the latter was higher than for any state but Alaska.

36 TAXATION

In 1984, the District of Columbia's personal income tax ranged from 2% on the first $1,000 to 11% on amounts over $25,000. The basic corporate tax was 10%, plus a 5% surtax. The District levies a 6% general sales and use tax, plus real and personal property taxes, an inheritance tax, and various excise taxes. The average per-capita tax burden of $1,924 in 1981/82 was higher than that of the residents of any state but Alaska and Wyoming. As of 1982/83, federal receipts from the District totaled nearly $2.1 billion, and federal expenditures exceeded $12.1 billion, more than half in the form of salaries and wages to the District's large number of federal employees. District residents filed 309,902 income tax returns for 1983, paying over $1 billion in tax.

37 ECONOMIC POLICY

The District's Office of Business and Economic Development (OBED) administers a revolving loan fund that helps small businesses in need of investment capital. The Local Development Corporation administers the federal Small Business Administration's loan guarantee program for plant and equipment. OBED also assists business in applying for federal urban development action grant funds. Federal community development block grant funds are available as well. By District law, 35% of all construction contracts and procurement contracts by District government agencies must go to minority-owned business enterprises.

The $98.7-million Washington Convention Center, opened in December 1982, has served as a catalyst for significant development in downtown Washington. Techworld (offices and exhibit space for high technology industries) and two new hotels with a total of 1,807 rooms were scheduled for completion in 1987.

38 HEALTH

Health conditions in the nation's capital are no source of national pride. Infant mortality rates—17.5 per 1,000 live births for whites and 26.7 for blacks in 1981—exceed those of virtually every state. Legal abortions outnumber live births in the District, a distinction that no states and few cities share; there were 152 abortions for every 100 live births in 1982/83, almost four times the national average. The District also suffered the nation's highest death rate from early infancy diseases—25.1 deaths per 1,000 population in 1981, more than twice the US average, and from prenatal conditions, 23.7 per 1,000, also more than twice the US average—and an overall death rate of 11 per 1,000 population, higher than any state. In addition, the District has the nation's highest rate of death from cirrhosis and other chronic liver diseases—three times the national rate—and its estimated alcoholism rate was also among the nation's highest. A 1983 survey found that the District had the highest per-capita alcohol consumption rate in the US— 5.22 gallons per year, compared to the national average of 2.69 gallons.

In 1983 there were 15 hospitals with 7,344 beds. Hospital personnel included 6,008 registered nurses in 1981. The average cost of hospital care in that year was $459 per day and $4,008 per stay—among the highest in the US. Medical personnel licensed to practice in the District in 1983 included 8,211 physicians, 1,831 dentists, 13,654 registered nurses, and 2,774 practical nurses.

39 SOCIAL WELFARE

Public aid recipients in 1983 comprised 12.1% of the District's population, a higher proportion than for any state. Aid to families with dependent children in 1982 totaled $86 million, with the average monthly payment per person in June 1984 being $104 for the caseload of 61,102. During June 1984, 79,927 District residents took part in the food stamp program; the federal cost in 1983/84 was $42.2 million. About 50,000 children were served school lunches in 1982/83, with a federal subsidy of $8 million. Medicaid recipients in 1983 totaled 117,250, and the 1982/83 cost of the program was $241.2 million, of which an estimated 48.5% was paid by federal funds.

Social Security benefits in December 1982 were paid to 82,731 residents; benefit payments totaled $342 million in 1983, with an average monthly stipend for retired workers of $361.92 in December 1982, 14% below the national average. Other social welfare expenditures included Supplemental Security Income (1983), $38.5 million; unemployment insurance (1982), $107 million; and workers' compensation (1982), $80.5 million.

40 HOUSING

The District of Columbia in 1980 had 276,792 housing units, of which 253,143 were occupied (65% rented); 39% of the units dated from 1939 or earlier. Nearly 98% of all occupied units had full plumbing. Of Washington's housing units, 52.7% were in structures of five or more units; among major US cities, only New York had a higher percentage. The preponderance of rental units in a densely populated urban setting has led to overcrowding: 8.1% of all units in 1980 averaged more than one person per room. From 1980 through 1983, 4,279 new units, valued at $144.4 million, were authorized. Housing prices were high; the average sales price of a single-family home in 1982 was $120,846.

A total of 1,639 public housing units were completed from fiscal 1980/81 through fiscal 1982/83.

⁴¹EDUCATION

The District of Columbia's first public schools were opened in 1805. In the fall of 1983 there were 186 public schools with a total enrollment of 89,491 students (94% black), of whom 48,180 were in elementary schools, 20,246 in junior high schools, 16,261 in senior high schools, and 4,804 in special or ungraded classes. Per-pupil expenditure was $3,726 in 1983/84. Until 1954, public schools for whites and blacks were operated separately. The school system remains virtually segregated; in 1980, only 700 minority-group students were attending schools with less than 50% minority enrollment, while 95% attended schools of 90% to 100% minority-group enrollment. Most white and many black students attend private schools; attendance in 1983/84 was 16,506, including 7,662 in Catholic schools; of nonpublic students, 62% were black. In 1980 over 67% of all residents 25 years of age or older were high school graduates, and 27.5% were college graduates.

The District had 19 institutions of higher education in 1983, 17 private and 2 public. Some of the best-known private universities, with 1983/84 enrollment, are American, 11,322; Georgetown, 11,237; George Washington, 14,482; and Howard, 11,594. The University of the District of Columbia, created in 1976 from the merger of 3 institutions, has an open admissions policy for District freshman undergraduate students. Its 5 academic colleges had 12,263 undergraduate and 786 graduate students in the fall of 1983; 65% were attending part-time.

⁴²ARTS

The John F. Kennedy Center for the Performing Arts, officially opened on 8 September 1971, is the District's principal performing arts center. Its five main halls—the Opera House, Concert Hall, Eisenhower Theater, Terrace Theatre, and American Film Institute Theater—display gifts from at least 30 foreign governments, ranging from stage curtains and tapestries to sculptures and crystal chandeliers. Major theatrical productions are also presented at the Arena Stage-Kreeger Theater, National Theatre, Folger Theatre, and Ford's Theatre. Rep, Inc. is one of the few professional black theatres in the US; the New Playwrights' Theatre of Washington is a nonprofit group presenting new plays by American dramatists.

The District's leading symphony is the National Symphony Orchestra, which performs from October through April at the Concert Hall of the Kennedy Center; its principal conductor in 1985 was Mstislav Rostropovich. On a smaller scale, the Phillips Collection, National Gallery of Art, and Library of Congress offer concerts and recitals. The Washington Opera performs at the Kennedy Center's Opera House.

During the summer months, the Carter Barron Amphitheater presents popular music and jazz. Concerts featuring the Army, Navy, and Marine bands and the Air Force Symphony Orchestra are held throughout the District.

⁴³LIBRARIES AND MUSEUMS

Washington, D.C., is the site of the world's largest library, the Library of Congress, with a 1983 collection of more than 80 million items, including 19.7 million books and pamphlets. The Library, which is also the cataloging and bibliographic center for libraries throughout the US, has on permanent display a 1455 Gutenberg Bible, Thomas Jefferson's first draft of the Declaration of Independence, and Abraham Lincoln's first two drafts of the Gettysburg Address. Also in its permanent collection are the oldest known existing film (Thomas Edison's *The Sneeze,* lasting all of three seconds), maps believed to date from the Lewis and Clark expedition, original musical scores by Charles Ives, and huge libraries of Russian and Chinese texts. The Folger Shakespeare Library contains not only rare Renaissance manuscripts but also a fullsize re-creation of an Elizabethan theater. The District's own public library system had a main library and 24 branches with 1,360,000 volumes in 1982/83.

The Smithsonian Institution—endowed in 1826 by an Englishman, James Smithson, who had never visited the US—operates a vast museum and research complex that includes the National Air and Space Museum, National Museum of Natural History, National Museum of History and Technology, many of the District's art museums, and the National Zoological Park. Among the art museums operated by the Smithsonian are the National Gallery of Art, housing one of the world's outstanding collections of Western art from the 13th century to the present; the Freer Gallery of Art, housing a renowned collection of Near and Far Eastern treasures, along with one of the largest collections of the works of James McNeill Whistler, whose Peacock Room is one of the museum's highlights; the National Collection of Fine Arts; the National Portrait Gallery; and the Hirshhorn Museum and Sculpture Garden. Among the capital's other distinguished art collections are the Phillips Collection, the oldest museum of modern art in the US; the Museum of African Art, located in the Frederick Douglass Memorial Home; and the Corcoran Gallery of Art, devoted primarily to American paintings, sculpture, and drawings of the last 300 years. Washington is also the site of such historic house-museums as Octagon House, Decatur House, Dumbarton Oaks, and the Woodrow Wilson House. Many national associations maintain exhibitions relevant to their areas of interest. The US National Arboretum, US Botanic Garden, and National Aquarium are in the city.

⁴⁴COMMUNICATIONS

Washington, D.C., is the headquarters of the US Postal Service; the District had 377 post offices and 8,896 employees in early 1985. As of 1980, there were 1,242,910 residences with telephones. In 1984, the District had 8 AM and 15 FM radio stations and 6 commercial and 2 public television stations.

⁴⁵PRESS

Because the District of Columbia is the center of US government activity, hundreds of US and foreign newspapers maintain permanent news bureaus there. The District has two major newspapers, the *Washington Post* and *USA Today,* a national newspaper. In 1984, the *Post,* a morning paper, had an average daily circulation of 728,857 and a Sunday circulation of 1,033,027. *USA Today,* published weekday mornings only, had a circulation of 1,247,324. The *Washington Times,* also published on weekday mornings, had a circulation of 75,576. Press clubs active within the District include the National Press Club, Gridiron Club, American Newspaper Women's Club, Washington Press Club, and White House Correspondents Association.

There are more than 30 major Washington-based periodicals. Among the best known are the *National Geographic* (with a circulation of 10,800,000 as of 1984), *U.S. News & World Report, Smithsonian,* and *New Republic.* Important periodicals covering the workings of the federal government and Congress are the *Congressional Quarterly* and its companion, *CQ Weekly Report.*

⁴⁶ORGANIZATIONS

The 1982 US Census of Service Industries counted 1,188 organizations in the District of Columbia, including 396 business associations; 252 civic, social, and fraternal associations; and 161 educational, scientific, and research associations. Service and patriotic organizations with headquarters in the District include the Air Force Association, Daughters of the American Revolution, and the 4-H Program. Among the cultural, scientific, and educational groups are the American Film Institute, American Theatre Association, Federation of American Scientists, American Association for the Advancement of Science, National Academy of Sciences, National Geographic Society, Association of American Colleges, American Council on Education, National Education Association, American Association of University Professors, American Association of University Women, and US Student Association.

Among the environmental and animal protection organizations

in the District are the Animal Welfare Institute, Humane Society of the US, and National Wildlife Federation. Medical, health, and charitable organizations include the American Red Cross. Groups dealing with the elderly include the National Association of Retired Federal Employees and the American Association of Retired Persons. Among ethnic and religious bodies with headquarters in the District are the National Association of Arab Americans, B'nai B'rith International, and the US Catholic Conference.

Among the trade, professional, and commercial organizations are the American Advertising Federation, American Federation of Police, American Youth Hostels, National Aeronautic Association, Air Line Pilots Association, American Bankers Association, National Cable Television Association, Chamber of Commerce of the US, American Chemical Society, International Association of Firefighters, Health Insurance Association of America, American Council of Life Insurance, National Association of Manufacturers, American Petroleum Institute, and National Press Club.

Virtually every major public interest group maintains an office in Washington, D.C. Notable examples are the Consumer Federation of America, National Consumers League, National Abortion Rights Action League, National League of Cities, Common Cause, US Conference of Mayors, National Organization for Women, and National Rifle Association of America.

Among the important world organizations with headquarters in the District are the Organization of American States, International Monetary Fund, and International Bank for Reconstruction and Development.

⁴⁷TOURISM, TRAVEL, AND RECREATION
The District of Columbia is one of the world's leading tourist centers, with 17.2 million visitors in 1980, 2 million of them foreigners. Visitors spend about $1 billion in the District each year, supporting nearly 45,000 jobs. The Washington Monument, Lincoln Memorial, Jefferson Memorial, Vietnam Veterans Memorial, White House, Capitol, US Supreme Court Building, Smithsonian Institution, Library of Congress, Ford's Theater, National Archives, National Gallery of Art, and Kennedy Center for the Performing Arts are only a few of the capital's extraordinary attractions. The most popular site is the Smithsonian's Air and Space Museum (14.4 million visits in 1983), followed by its Museum of Natural History (6.1 million visits) and Museum of American History (5.4 million visits). Nearly 28.8 million visits were paid to National Park Service sites in the District in 1984. Across the Potomac, in Virginia, are Arlington National Cemetery, site of the Tomb of the Unknown Soldier and the grave of John F. Kennedy, and George Washington's home at Mt. Vernon.

⁴⁸SPORTS
Major professional sports teams representing the Washington, D.C., metropolitan area include the Redskins of the National Football League, the Bullets of the National Basketball Association (NBA), and the Capitals of the National Hockey League. The Redskins have reached football's Super Bowl three times, winning the contest in 1983. The Bullets won the NBA championship in 1978. The Georgetown Hoyas were a dominant force in collegiate basketball during the first half of the 1980s, reaching the NCAA championship game in 1982, 1984, and 1985, and winning the title in 1984.

⁴⁹FAMOUS WASHINGTONIANS
Although no US president has been born in the District of Columbia, all but George Washington (b. Virginia, 1732–99) lived there while serving as chief executive. Seven presidents died in Washington, D.C., including three during their term of office: William Henry Harrison (b.Virginia, 1773-1841), Zachary Taylor (b.Virginia, 1784-1850), and Abraham Lincoln (b.Kentucky, 1809-65). In addition, John Quincy Adams (b.Massachusetts, 1767-1848), who served as a congressman for 17 years after he left the White House, died at his desk in the House of Representa-

tives; and William Howard Taft (b.Ohio, 1857-1930) passed away while serving as US chief justice. Retired presidents Woodrow Wilson (b.Virginia, 1856-1924) and Dwight D. Eisenhower (b.Texas, 1890-1969) also died in the capital; Wilson is the only president buried there. Federal officials born in Washington, D.C., include John Foster Dulles (1888-1959), secretary of state; J(ohn) Edgar Hoover (1895-1972), director of the FBI; and Robert C. Weaver (b.1907), who as secretary of housing and urban development during the administration of President Lyndon B. Johnson was the first black American to hold cabinet rank. Walter E. Fauntroy (b. 1933) has been the District's delegate to Congress since that office was established in 1971.

Among the outstanding scientists and other professionals associated with the District were Cleveland Abbe (b.New York, 1838-1916), a meteorologist who helped develop the US Weather Service; inventor Alexander Graham Bell (b.Scotland, 1842-1922), president of the National Geographic Society (NGS) in his later years; Henry Gannett (b.Maryland, 1846-1914), chief geographer with the US Geological Survey, president of the NGS and a pioneer in American cartography; Charles D. Walcott (b.New York, 1850-1927), director of the Geological Survey and secretary of the Smithsonian Institution; Emile Berliner (b.Germany, 1851-1929), a pioneer in the development of the phonograph; Gilbert H. Grosvenor (b.Turkey, 1875-1966), editor in chief of *National Geographic* magazine; and Charles R. Drew (1904-50), developer of the blood bank concept. Leading business executives who lived or worked in the District include William W. Corcoran (1798-1888), banker and philanthropist, and Katharine Graham (b.New York, 1917), publisher of the *Washington Post;* the two *Post* reporters who received much of the credit for uncovering the Watergate scandal are Carl Bernstein (b.1944), a native Washingtonian, and Robert "Bob" Woodward (b.Illinois, 1943). Washingtonians who achieved military fame include Benjamin O. Davis (1877-1970), the first black to become an Army general, and his son, Benjamin O. Davis, Jr. (b.1912), who was the first black to become a general in the Air Force.

The designer of the nation's capital was Pierre Charles L'Enfant (b.France, 1754-1825), whose grave is in Arlington National Cemetery; also involved in laying out the capital were surveyor Andrew Ellicott (b.Pennsylvania, 1754-1820) and mathematician-astronomer Benjamin Banneker (b.Maryland, 1731-1806), a black who was an early champion of equal rights. Among Washingtonians to achieve distinction in the creative arts were John Philip Sousa (1854-1932), bandmaster and composer; Herblock (Herbert L. Block, b.Illinois, 1909), political cartoonist; and playwright Edward Albee (b.1928), winner of the Pulitzer Prize for drama in 1967 and 1975. Famous performers born in the District of Columbia include composer-pianist-bandleader Edward Kennedy "Duke" Ellington (1899-1974) and actress Helen Hayes (Helen Hayes Brown, b.1900). Alice Roosevelt Longworth (b.New York, 1884-1980) dominated the Washington social scene for much of this century.

⁵⁰BIBLIOGRAPHY
Federal Writers' Project. *Washington, D.C.: A City and Capital.* Reprint. New York: Somerset, n.d. (orig. 1942).

Green, Constance M. *Washington: A History of the Capital.* Princeton, N.J.: Princeton University Press, 1976.

Gurney, Gene, and Harold Wise. *The Official Washington, D.C., Directory.* New York: Crown, 1977.

Gutheim, Frederick. *Worthy of the Nation: The Planning and Development of the National Capital City.* Washington, D.C.: Smithsonian, 1977.

Lewis, David L. *District of Columbia: A Bicentennial History.* New York: Norton, 1976.

Shidler, Atlee E. *Trends and Issues in the Greater Washington Region: A Preliminary Report.* Washington, D.C.: Center for Municipal and Metropolitan Research, 1979.

PUERTO RICO*

Commonwealth of Puerto Rico
Estado Libre Asociado de Puerto Rico

ORIGIN OF NAME: Spanish for "rich port." **NICKNAME:** Island of Enchantment. **CAPITAL:** San Juan. **BECAME A COMMONWEALTH:** 25 July 1952. **SONG:** *La Borinqueña.* **MOTTO:** *Joannes est nomen ejus.* (John is his name.) **FLAG:** From the hoist extends a blue triangle, with one white star; five horizontal stripes—three red, two white—make up the balance. **OFFICIAL SEAL:** In the center of a green circular shield, a lamb holding a white banner reclines on the book of the Apocalypse. Above are a yoke, a cluster of arrows, and the letters "F" and "I," signifying King Ferdinand and Queen Isabella, rulers of Spain at the time of discovery; below is the commonwealth motto. Surrounding the shield, on a white border, are the towers of Castile and lions symbolizing Spain, crosses representing the conquest of Jerusalem, and Spanish banners. **ANIMAL:** Coquí. **LEGAL HOLIDAYS:** New Year's Day, 1 January; Three Kings Day (Epiphany), 6 January; Birthday of Eugenio María de Hostos, 11 January; Birthday of Martin Luther King, Jr., 3d Monday in January; Washington's Birthday, 3d Monday in February; Abolition Day, 22 March; Good Friday, March or April; Birthday of José de Diego, 16 April; Memorial Day, last Monday in May; Independence Day, 4 July; Birthday of Luis Muñoz Rivera, 17 July; Constitution Day, 25 July; Birthday of José Celso Barbosa, 27 July; Labor Day, 1st Monday in September; Anniversary of the "Grito de Lares," 23 September; Discovery of America, 12 October; Veterans Day, 11 November; Discovery of Puerto Rico Day, 19 November; Thanksgiving Day, 4th Thursday in November; Christmas Day, 25 December. **TIME:** 8 AM Atlantic Standard Time = noon GMT.

¹LOCATION, SIZE, AND EXTENT

Situated on the NE periphery of the Caribbean Sea, about 1,000 mi (1,600 km) SE of Miami, Puerto Rico is the easternmost and smallest island of the Greater Antilles group. Its total area is 3,515 sq mi (9,104 sq km), including 3,459 sq mi (8,959 sq km) of land and 56 sq mi (145 sq km) of inland water.

Shaped roughly like a rectangle, the main island measures 111 mi (179 km) E–W and 36 mi (58 km) N–S. Offshore and to the E are two major islands, Vieques and Culebra.

Puerto Rico is bounded by the Atlantic Ocean to the N, the Virgin Passage and Vieques Sound to the E, the Caribbean Sea to the S, and the Mona Passage to the W. Puerto Rico's total boundary length is 378 mi (608 km).

²TOPOGRAPHY

About 75% of Puerto Rico's land area consists of hills or mountains too steep for intensive commercial cultivation. The Cordillera Central range, separating the northern coast from the semiarid south, has the island's highest peak, Cerro de Punta (4,389 feet—1,338 meters). Puerto Rico's best-known peak, El Yunque (3,496 feet—1,066 meters), stands to the east, in the Luquillo Mountains (Sierra de Luquillo). The north coast consists of a level strip about 100 mi (160 km) long and 5 mi (8 km) wide. Principal valleys are located along the east coast, from Fajardo to Cape Mala Pascua, and around Caguas, in the east-central region. Off the eastern shore are two small islands: Vieques, with an area of 51 sq mi (132 sq km), and Culebra, covering 24 sq mi (62 sq km). Uninhabited Mona Island (19 sq mi—49 sq km), off the southwest coast, is a breeding ground for wildlife.

Puerto Rico has 50 waterways large enough to be classified as rivers, but none is navigable by large vessels. The longest river is the Río de la Plata, extending 46 mi (74 km) from Cayey to Dorado, where it empties into the Atlantic. There are few natural lakes but numerous artificial ones, of which Dos Bocas, south of Arecibo, is one of the most beautiful. Phosphorescent Bay, whose luminescent organisms glow in the night, is a tourist attraction on the south coast.

Like many other Caribbean islands, Puerto Rico is the crest of an extinct submarine volcano. About 45 mi (72 km) north of the island lies the Puerto Rico Trench, at over 28,000 feet (8,500 meters) one of the world's deepest chasms.

³CLIMATE

Tradewinds from the northeast keep Puerto Rico's climate equable, although tropical. San Juan has a normal daily mean temperature of 80°F (27°C), ranging from 77°F (25°C) in January to 82°F (28°C) in July; the normal daily minimum is 73°F (23°C), the maximum 86°F (30°C). The lowest temperature ever recorded on the island is 39°F (4°C), the highest 103°F (39°C); the recorded temperature in San Juan has never been lower than 60°F (16°C) or higher than 98°F (37°C).

Rainfall varies by region. Ponce, on the south coast, averages only 32 in (81 cm) a year, while the highlands average 108 in (274 cm); the rain forest on El Yunque receives an annual average of 183 in (465 cm). San Juan's average annual rainfall is 54 in (137 cm), the rainiest months being May through November.

The word "hurricane" derives from *hurakán*, a term the Spanish learned from Puerto Rico's Taíno Indians. Eight hurricanes have struck Puerto Rico in this century, most recently in 1979. On 7 October 1985, torrential rains created a mud slide that devastated the hillside barrio of Mameyes, killing hundreds of people; not only was this Puerto Rico's worst disaster of the century, but it was the single most destructive landslide in US history.

⁴FLORA AND FAUNA

During the 19th century, forests covered about three-fourths of Puerto Rico. Today, however, only one-fourth of the island is forested. Flowering trees still abound, and the butterfly tree, African tulip, and flamboyán (royal poinciana) add bright reds and pinks to Puerto Rico's lush green landscape. Among hardwoods, now rare, are nutmeg, satinwood, Spanish elm, and Spanish cedar. Pre-Columbian peoples cultivated yucca, yams, peanuts, hot peppers, tobacco, and cotton. Pineapple, guava, tamarind, and cashews are indigenous, and such fruits as mamey, jobo, guanábana, and quenepa are new to most visitors. Coconuts,

*All comparative data for the US exclude Puerto Rico unless otherwise noted.

coffee, sugarcane, plantains, mangoes, and most citrus fruits were introduced by the Spanish.

The only mammal found by the conquistadores on the island was a kind of barkless dog, now extinct. Virtually all present-day mammals have been introduced, including horses, cattle, cats, and dogs. The only troublesome mammal is the mongoose, brought in from India to control reptiles in the cane fields and now wild in remote rural areas. Mosquitoes and sand flies are common pests, but the only dangerous insect is the giant centipede, whose sting is painful but rarely fatal. Perhaps the island's best-known inhabitant is the golden coquí, a tiny tree frog whose call of "ko-kee, ko-kee" is heard all through the night; it is a threatened species. Marine life is extraordinarily abundant, including many tropical fish, crabs, and corals. Puerto Rico has some 200 bird species, many of which live in the rain forest. Thrushes, orioles, grosbeaks, and hummingbirds are common, and the reinita and pitirre are distinctive to the island. Several parrot species are rare, and the Puerto Rican parrot is endangered. Also on the endangered list are the yellow-shouldered blackbird and the Puerto Rican plain pigeon, Puerto Rican whippoorwill, Culebra giant anole, Puerto Rican boa, and Monita gecko. The Mona boa and Mona ground iguana are threatened. There were three national wildlife refuges, covering a total of 2,425 acres (981 hectares), in the early 1980s.

⁵ENVIRONMENTAL PROTECTION

US environmental laws and regulations are applicable in Puerto Rico. Land-use planning, overseen by the Puerto Rico Planning Board, is an especially difficult problem, since residential, industrial, and recreational developers are all competing for about 30% of the total land area on an island that is already more densely populated than any state of the US except New Jersey. Pollution from highland latrines and septic systems and from agricultural and industrial wastes is a potential hazard; the rum industry, for example, has traditionally dumped its wastes into the ocean. Moreover, the US requirement that sewage receive secondary treatment before being discharged into deep seas may be unrealistic in view of the commonwealth government's claim, in the late 1970s, that it could not afford to build secondary sewage treatment facilities when 45% of its population lacks primary sewage treatment systems.

In 1982, the US Geological Survey found contamination in most of the 57 wells it sampled in highly populated industrial areas. Eight hazardous waste sites are on the Environment Protection Agency's list for priority cleanup.

⁶POPULATION

Puerto Rico, with a 1980 census population of 3,196,520, had more people than 27 of the 50 states. The 1980 census total represented a 17.9% increase over 1970. The population was estimated at 3,319,500 in November 1984, yielding a population density of 960 per sq mi (371 per sq km).

In 1980, Puerto Rico was 49% male and 51% female. The median age was 24, 6 years younger than that of any state except Utah. Fully 29% of the population was under 14 years of age; half were under 25, compared with 41% for the US. The birthrate declined steadily from 38.9 live births per 10,000 population in 1950 to 19.9 in 1983/84, but the 1981 rate of 21.8 was 39% above the US average. The death rate, on the other hand, was nearly one-third lower than the US norm in 1981, and a continuation of rapid population growth appeared likely.

According to the 1980 census, the population was two-thirds urban and one-third rural. San Juan is Puerto Rico's capital and largest city, with a 1980 census population of 434,849, followed by Bayamón, 196,206; Ponce, 189,046; Carolina, 165,954; and Caguas, 117,959. The San Juan metropolitan area had a population of 1,086,376.

⁷ETHNIC GROUPS

Three main ethnic strands are the heritage of Puerto Rico: the Taíno Indians, most of whom fled or perished after the Spanish conquest; black Africans, imported as slaves under Spanish rule; and the Spanish themselves. With an admixture of Dutch, English, Corsicans, and other Europeans, Puerto Ricans today enjoy a distinct Hispanic-Afro-Antillean heritage.

Of the 199,524 residents in 1980 who were born on the US mainland, most were of Puerto Rican extraction. Of the 63,351 persons born on neither the mainland nor the island, most were West Indians from Cuba or the Dominican Republic. About half were naturalized US citizens.

Less than two-thirds of all ethnic Puerto Ricans live on the island. Virtually all the remainder reside on the US mainland; in 1980 there were 2,013,945 people of Puerto Rican origin in the 50 states, where they made up less than 14% of the US Hispanic population. The State of New York had almost half the US mainland's Puerto Rican population; 860,552 lived in New York City alone.

⁸LANGUAGES

Spanish is the official language of Puerto Rico; English is required in schools as a second language. From 1898 through the 1920s, US authorities unsuccessfully sought to make English the island's primary language. Of 2,855,868 persons 5 years of age or over in 1980, 2,805,444 were able to speak Spanish, while 541,160 could speak English easily and 643,873 could speak English with difficulty.

Taíno Indian terms that survive in Puerto Rican Spanish include such place-names as Arecibo, Guayama, and Mayagüez, as well as *hamaca* (hammock) and *canoa* (canoe). Among many African borrowings are food terms like *quimbombó* (okra), *guineo* (banana), and *mondongo* (a spicy stew).

⁹RELIGIONS

During the first three centuries of Spanish rule, Roman Catholicism was the only religion permitted in Puerto Rico. More than 80% of the population was still Roman Catholic at the end of 1982, and the Church maintains numerous hospitals and schools on the island. Most of the remaining Puerto Ricans belong to other Christian denominations, which have been allowed on the island since the 1850s. Pentecostal churches have attracted a significant following, particularly among the urban poor of the barrios.

¹⁰TRANSPORTATION

Puerto Rico's inland transportation network consists primarily of roads and road vehicles. Rivers are not navigable, and the only function of narrow-gauge rural railroads is to haul sugarcane to the mills during the harvesting season; other goods are transported by truck. A few public bus systems provide intercity passenger transport, the largest being the Metropolitan Bus Authority (MBA), a government-owned company serving San Juan and nearby cities. Each working weekday in December 1984, MBA buses transported an average of 138,500 passengers. The predominant form of public transportation outside the San Juan metropolitan area is the *público*, or privately owned jitney, a small bus that carries passengers between fixed destinations; in many rural areas, it is the only form of public transit.

In 1983/84, Puerto Rico had 8,561 mi (13,778 km) of streets and roads. Motor vehicle registrations at the end of 1982 totaled 1,164,111, including 976,241 cars and 187,870 trucks and buses. The island's ratio of better than one automobile for each four residents is one of the highest in the world, though far below the US average.

San Juan, the island's principal port and a leading containerized cargo-handling facility, handled 10,049,195 tons of cargo in 1982. Crude oil and gasoline were the leading items. Ponce handled 726,182 tons, and Mayagüez 317,046 tons. Ferries link the main island with Vieques and Culebra.

Puerto Rico receives flights from the US mainland and from the Virgin Islands, the British West Indies, Jamaica, and the Dominican Republic, as well as from Great Britain, France, Spain, and

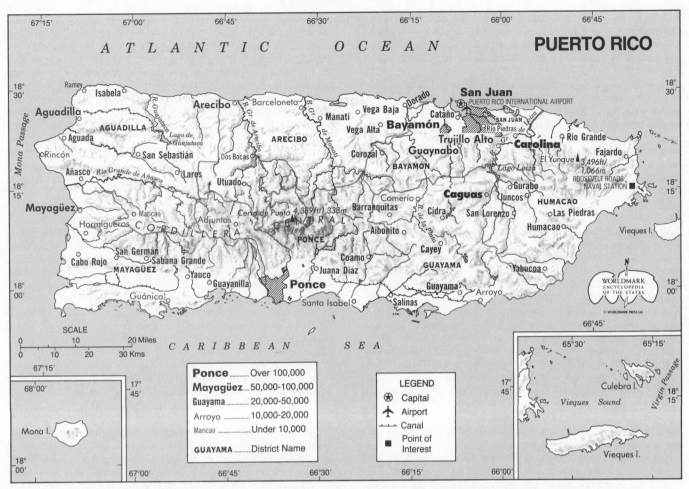

LOCATION: (main island only): 18°04′15″ to 18°31′N; 65°35′30″ to 67°15′W. **BOUNDARIES:** Total coastline, 378 mi (608 km).

West Germany; airlines connect Puerto Rico with 38 mainland cities. Puerto Rico International Airport (San Juan) enplaned 5,126,000 passengers in 1983/84 and 63,359 tons of cargo in 1983. Other leading air terminals are located at Ponce, Mayagüez, and Vieques. In all, there were 18 airports, 12 heliports, and a seaplane base in Puerto Rico at the end of 1983.

11 HISTORY

Archaeological finds indicate that at least three Indian cultures settled on the island now known as Puerto Rico long before its discovery by Christopher Columbus on 19 November 1493. The first group, belonging to the Archaic Culture, is believed to have come from Florida. Having no knowledge of agriculture or pottery, it relied on the products of the sea; their remains have been found mostly in caves. The second group, the Igneri, came from northern South America. Descended from South American Arawak stock, the Igneri brought agriculture and pottery to the island; their remains are found mostly in the coastal areas. The third culture, the Taíno, also of Arawak origin, combined fishing with agriculture. A peaceful, sedentary tribe, the Taíno were adept at stonework and lived in many parts of the island; Taíno relics have been discovered not only along the coastal perimeter but also high in the mountains, where the Taíno performed ritual games in ball parks that have been restored in recent times. To the Indians, the island was known as Boriquén.

Columbus, accompanied by a young nobleman named Juan Ponce de León, landed at the western end of the island—which he called San Juan Bautista (St. John the Baptist)—and claimed it for Spain. Not until colonization was well under way would the island acquire the name Puerto Rico (literally, "rich port"), with the name San Juan Bautista applied to the capital city. The first settlers arrived on 12 August 1508, under the able leadership of Ponce de León, who sought to transplant and adapt Spanish civilization to Puerto Rico's tropical habitat. The small contingent of Spaniards compelled the Taíno, numbering perhaps 30,000, to mine for gold; the rigors of forced labor and the losses from rebellion reduced the Taíno population to about 4,000 by 1514, by which time the mines were nearly depleted. With the introduction of slaves from Africa, sugarcane growing became the leading economic activity. Since neither mining nor sugarcane was able to provide sufficient revenue to support the struggling colony, the treasury of New Spain began a subsidy, known as the *situado,* which until the early 19th century defrayed the cost of the island's government and defense.

From the early 16th century onward, an intense power struggle for the control of the Caribbean marked Puerto Rico as a strategic base of the first magnitude. After a French attack in 1528, construction of La Fortaleza (still in use today as the governor's palace) was begun in 1533, and work on El Morro fortress in San Juan commenced six years later. The new fortifications helped repel a British attack led by Sir Francis Drake in 1595; a second force, arriving in 1598 under George Clifford, Earl of Cumberland, succeeded in capturing San Juan, but the British were forced to withdraw by tropical heat and disease. In 1625, a Dutch attack under the command of Boudewijn Hendrikszoon was repulsed, although much of San Juan was sacked and burned by the attackers. By the 18th century, Puerto Rico had become a haven for pirates, and smuggling was the major economic activity. A Spanish envoy who came to the island in 1765 was appalled, and his report to the crown inaugurated a period of economic, administrative, and military reform. The creation of a native militia helped Puerto Rico withstand a fierce British assault on San Juan in 1797, by which time the island had more than 100,000 inhabitants.

Long after most of the Spanish colonies in the New World had obtained independence, Puerto Rico and Cuba remained under Spanish tutelage. Despite several insurrection attempts, most of them inspired by the Liberator, Simón Bolívar, Spain's military might concentrated on these islands precluded any revolution. Puerto Rico became a shelter for refugees from Santo Domingo, Haiti, and Venezuela who were faithful to Spain, fearful of disturbances in their own countries, or both. As in Cuba, the sugar industry developed in Puerto Rico during this period under policies that favored foreign settlers. As a result, a new landowner class emerged—the *hacendados*—who were instrumental in strengthening the institution of slavery in the island. By 1830, the population was 300,000. Sugar, tobacco, and coffee were the leading export crops, although subsistence farming still covered much of the interior. Sugar found a ready market in the US, and trade steadily developed, particularly with the Northeast.

The 19th century also gave birth, however, to a new Puerto Rican civic and political consciousness. Puerto Rican participation in the short-lived constitutional experiments in Spain (1812–14 and 1820–23) fostered the rise of a spirit of liberalism, expressed most notably by Ramón Power y Giralt, at one time vice president of the Spanish Cortes (parliament). During these early decades, Spain's hold on the island was never seriously threatened. Although the Spanish constitution of 1812 declared that the people of Puerto Rico were no longer colonial subjects but were full-fledged citizens of Spain, the crown maintained an alert, centralized, absolutist government with all basic powers concentrated in the captain general.

Toward the middle of the 19th century, a *criollo* generation with strong liberal roots began a new era in Puerto Rican history. This group, which called for the abolition of slavery and the introduction of far-reaching economic and political reforms, at the same time developed and strengthened Puerto Rican literary tradition. The more radical reformers espoused the cause of separation from Spain and joined in a propaganda campaign in New York on behalf of Cuban independence. An aborted revolution, beginning in the town of Lares in September 1868 (and coinciding with an insurrection in Spain that deposed Queen Isabella II), though soon quelled, awakened among Puerto Ricans a dormant sense of national identity. "El Grito de Lares" (the Cry of Lares) helped inspire a strong anti-Spanish separatist current that was unable to challenge Spanish power effectively but produced such influential leaders as Ramón Emeterio Betances and Eugenio María de Hostos.

The major reform efforts after 1868 revolved around abolitionism and *autonomía,* or self-government. Slavery was abolished in 1873 by the First Spanish Republic, which also granted new political rights to the islanders. The restoration of the Spanish monarchy two years later, however, was a check to Puerto Rican aspirations. During the last quarter of the century, leaders such as Luis Muñoz Rivera sought unsuccessfully to secure vast new powers of self-government. By this time, Puerto Rico was an island with a distinct Antillean profile, strong Hispanic roots, and a mixed population that, borrowing from its Indian-Spanish-African background and an influx of Dutch, English, Corsicans, and other Europeans, had developed its own folkways and mores.

The imminence of war with the US over Cuba, coupled with autonomist agitation within Puerto Rico, led Spain in November 1897 to grant to the island a charter with broad powers of self-rule. Led by Luis Muñoz Rivera, Puerto Ricans began to establish new organs of self-government; but no sooner had an elected government begun to function in July 1898 than US forces,

overcoming Spanish resistance, took over the island. A cease-fire was proclaimed on 13 August, and sovereignty was formally transferred to the US with the signing in December of the Treaty of Paris, ending the Spanish-American War. The US government swept aside the self-governing charter granted by Spain and established military rule from 1898 to 1900. Civilian government was restored in 1900 under a colonial law, the Foraker Act, that gave the federal government full control of the executive and legislative branches, leaving some local representation in the lower chamber, or house of delegates. Under the Jones Act, signed into law by President Woodrow Wilson on 2 March 1917, Congress extended US citizenship to the islanders and granted an elective senate, but still reserved vast powers over Puerto Rico to the federal bureaucracy.

The early period of US rule saw an effort to Americanize all insular institutions, and even aimed at superseding the Spanish language as the vernacular. In the meantime, American corporate capital took over the sugar industry, developing a plantation economy so pervasive that, by 1920, 75% of the population relied on the cane crop for its livelihood. Glaring irregularities of wealth resulted, sharpening social and political divisions. This period also saw the development of three main trends in Puerto Rican political thinking. One group favored the incorporation of Puerto Rico into the US as a state; a second group, fearful of cultural assimilation, favored self-government; while a third group spoke for independence.

The depression hit Puerto Rico especially hard. With a population approaching 2 million by the late 1930s and with few occupational opportunities outside the sugar industry, the island's economy deteriorated, and mass unemployment and near-starvation were the results. Controlling the Puerto Rican legislature from 1932 to 1940 was a coalition of the Socialist Party, led by Santiago Iglesias, a Spanish labor leader who became a protégé of the American Federation of Labor, and the Republican Party, which had traditionally espoused statehood and had been founded in Puerto Rico by José Celso Barbosa, a black physician who had studied in the US. The coalition was unable to produce any significant improvement, although under the New Deal a US government effort was made to supply emergency relief for the "stricken island."

Agitation for full political and economic reform or independence gained ground during this period. A violent challenge to US authority in Puerto Rico was posed by the small Nationalist Party, led by Harvard-educated Pedro Albizu Campos. A broader attack on the island's political and economic ills was led by Luis Muñoz Marín and the Popular Democratic Party (PDP), founded in 1938; within two years, the PDP won control of the senate. Under Muñoz Marín, a new era began in Puerto Rico. Great pressure was put on Washington for a change in the island's political status, while social and economic reform was carried to the fullest extent possible within the limitations of the Jones Act. Intensive efforts were made to centralize economic planning, attract new industries through local tax exemptions (Puerto Rico was already exempt from federal taxation), reduce inequalities of income, and improve housing, schools, and health conditions. Meanwhile, a land distribution program helped the destitute peasants who were the backbone of the new party. All these measures—widely publicized as Operation Bootstrap—coupled with the general US economic expansion after World War II, so transformed Puerto Rico's economy that income from manufacturing surpassed that from agriculture by 1955 and was five times as great by 1970. Annual income per capita rose steadily from $296 in 1950 to $1,384 in 1970.

The PDP, the dominant force in Puerto Rican politics from 1940 to 1968, favored a new self-governing relationship with the US, distinct from statehood or independence. The party succeeded not only in bringing about significant social and economic change but also in obtaining from Congress in 1950 a law allowing Puerto Ricans to draft their own constitution with full local self-government. This new constitution, approved in a general referendum on 3 March 1952, led to the establishment on 25 July of the Commonwealth of Puerto Rico (Estado Libre Asociado de Puerto Rico), which, according to a resolution approved in 1953 by the United Nations Committee on Information from Non-Self-Governing Territories, was constituted as an autonomous political entity in voluntary association with the United States.

An island-wide plebiscite in 1967 showed that 60% of those voting favored continuation and improvement of the commonwealth relationship, 39% preferred statehood, and less than 1% supported independence; the turnout among eligible voters was 65%. The result of the plebiscite, held to support a movement for additional home-rule powers, met with indifference from the US executive branch and outright opposition from the pro-statehood minority in Puerto Rico. Consequently, efforts to obtain passage by Congress of a "Compact of Permanent Union between Puerto Rico and the United States," although approved at the subcommittee level by the House of Representatives, failed to produce any change in the commonwealth arrangement.

The result was renewed agitation for either statehood or independence, with growing internal political polarization. The island's Republican Party rearranged itself after the plebiscite as the New Progressive Party (NPP), and came to power in 1968 as a result of a split in PDP ranks that led to the creation of the splinter People's Party. The two major blocs have been evenly balanced since that time, with the PDP returning to power in 1972 but losing to the NPP in 1976 and again, by a very narrow margin, in 1980, before regaining the governorship in 1984. The independence movement, in turn, divided into two wings: the moderates favored social democracy, while the radicals pursued close ties with the Fidel Castro regime in Cuba. Capitalizing on the increased power of Third World countries in the United Nations and with Soviet support, the radicals challenged US policies and demanded a full transfer of sovereign rights to the people of Puerto Rico. Their position won the support of the UN Special Committee on the Situation with Regard to the Implementation of the Declaration on the Granting of Independence to Colonial Countries and Peoples (more generally known as the Committee of 24), which on 15 August 1979 reaffirmed "the inalienable right of the people of Puerto Rico to self-determination and independence. . . ." The US government replied that the people of Puerto Rico had already exercised their right of self-determination in the 1967 plebiscite, and noted that Congress in 1979 had restated its "commitment to respect and support the right of the people of Puerto Rico to determine their own political future through peaceful, open and democratic processes." More advanced than most Caribbean countries in education, health, and social development, Puerto Rico was, nevertheless, a land of growing political tensions in the early 1980s, with occasional terrorist attacks on US military installations and personnel. The island was hard hit by the 1981–82 recession, and federal budget cuts ended a jobs program and reduced access to food stamps.

[12]COMMONWEALTH GOVERNMENT

Since 1952, Puerto Rico has been a commonwealth of the US, governed under the Puerto Rican Federal Relations Act and under a constitution based on the US model. The Puerto Rican constitution specifically prohibits discrimination "on account of race, color, sex, birth, social origin or condition, or political ideas." The constitution had been amended six times as of the end of 1983.

The commonwealth legislature comprises a senate (Senado) of 27 members, 2 from each of 8 senatorial districts and 11 elected at large, and a house of representatives (Cámara de Representantes) of 51 members, 1 from each of 40 districts and 11 at large. Each senate district consists of five house districts. If a single party wins

Puerto Rico Gubernatorial Vote by Political Parties, 1948–84[1]

YEAR	WINNER	POPULAR DEMOCRAT (PDP)	NEW PROGRESSIVE (NPP)	REPUBLICAN	PUERTO RICAN INDEPENDENCE	SOCIALIST	LIBERAL REFORMIST
1948	Luis Muñoz Marín (PDP)	392,033	—	88,819	66,141	64,121	28,203
1952	Luis Muñoz Marín (PDP)	429,064	—	85,172	125,734	21,655	—
1956	Luis Muñoz Marín (PDP)	433,010	—	172,838	86,386	—	—
1960	Luis Muñoz Marín (PDP)	457,880	—	252,364	24,103	—	—
						CHRISTIAN ACTION	
1964	Roberto Sanchez Vitella (PDP)	487,280	—	284,627	22,201	26,867	—
						PEOPLE'S	
1968	Luis A. Ferré (NPP)	367,903	390,623	4,057	24,713	87,844	—
							PR UNION
1972	Rafael Hernández Colón (PDP)	609,670	524,039	—	52,070	2,910	1,608
						PR SOCIALIST	
1976	Carlos Romero Barceló (NPP)	634,941	682,607	—	58,556	9,761	—
1980	Carlos Romero Barceló (NPP)	756,434	759,868	—	87,275	5,225	—
							RENEWAL
1984	Rafael Hernández Colón (PDP)	822,040	767,710	—	61,101	—	68,536

[1]Residents of Puerto Rico are barred from voting in US presidential elections.

two-thirds or more of the seats in either house, the number of seats can be expanded (up to a limit of 9 in the senate and 17 in the house) to assure representation for minority parties. Senators must be at least 30 years of age, representatives 25. Legislators must have been commonwealth residents for two years and district or municipal residents for one year. All legislators serve four-year terms.

The governor, who may serve an unlimited number of four-year terms, is the only elected executive. Candidates for the governorship must be US citizens for at least five years, must be at least 35 years of age, and must have resided in Puerto Rico for at least five years.

A bill becomes law if approved by both houses and either signed by the governor or left unsigned for 10 days while the legislature is in session. A two-thirds vote of the elected members of each house is sufficient to override a gubernatorial veto. The governor can employ the item veto or reduce amounts in appropriations bills. The governor also has the power to declare martial law in cases of rebellion, invasion, or immediate danger of rebellion or invasion. The constitution may be amended by a two-thirds vote of the legislature and ratification by popular majority vote.

Residents of Puerto Rico may not vote in US presidential elections. A Puerto Rican who settles in one of the 50 states automatically becomes eligible to vote for president; conversely, a state resident who migrates to Puerto Rico forfeits such eligibility. Puerto Rico has no vote in the US Senate or House of Representatives, but a nonvoting resident delegate, elected every four years, may speak on the floor of the House, introduce legislation, and vote in House committees.

Qualified voters must be US citizens, be at least 18 years of age, and have registered 50 days before a general election; absentee registration is not allowed.

13POLITICAL PARTIES

Taking part in Puerto Rican elections during the mid-1980s were two major and three smaller political parties. The Popular Democratic Party (PDP), founded in 1938, favors the strengthening and development of commonwealth status. The New Progressive Party (NPP), created in 1968 as the successor to the Puerto Rican Republican Party, is pro-statehood. Two smaller parties, each favoring independence for the island, are the Puerto Rican Independence Party, founded in the mid-1940s and committed to democratic socialism, and the more radical Puerto Rican Socialist Party, which has close ties with Cuba. A breakaway group, the Renewal Party, led by the mayor of San Juan, Hernán Padilla, left the NPP and took part in the 1984 elections.

In 1980, Governor Carlos Romero Barceló of the NPP, who had pledged to seek actively Puerto Rico's admission to the Union if elected by a large margin, retained the governorship by a plurality of fewer than 3,500 votes, in the closest election in the island's history, while the PDP won control of the legislature and 52 out of 78 mayoralty contests. Former governor Rafael Hernández Colón defeated Romero Barceló's bid for reelection in 1984 by more than 54,000 votes. Hernández Colón's PDP greatly extended its legislative majority, winning 39 of the 51 house seats and 18 of the 27 senate seats. In the 1981–84 legislature, three senate members and two house members were women.

Although Puerto Ricans have no vote in US presidential elections, the island does send voting delegates to the national conventions of the Democratic and Republican parties. In 1980, for the first time, those delegates were chosen by presidential preference primary.

Puerto Rico's political parties have generally committed themselves to peaceful change through democratic methods. One exception was the pro-independence Nationalist Party, whose followers were involved in an attempt to assassinate US President Harry S Truman in 1950 and in an outbreak of shooting in the House of Representatives that wounded five congressman in 1954. A US-based terrorist group, the Armed Forces of Puerto Rican National Liberation (FALN), claimed credit during the late 1970s for bombings in New York and other major cities. FALN members briefly took over the Statue of Liberty in New York Harbor on 25 October 1977. Another group, the Macheteros, apparently based on the island, claimed responsibility for an attack on a US Navy bus in 1980 and for blowing up eight US Air Force planes at a Puerto Rico Air National Guard installation early in 1981.

14LOCAL GOVERNMENT

The Commonwealth of Puerto Rico had 78 municipalities in 1982, each governed by a mayor and municipal assembly elected every four years. In fact, these governments resemble US county governments in that they perform services for both urban and rural areas. Many of the functions normally performed by municipal governments in the US—for instance, fire protection, education, water supply, and law enforcement—are performed by the commonwealth government directly.

15COMMONWEALTH SERVICES

As of 1983, the executive branch of Puerto Rico's highly centralized government was organized into 15 departments, 54 agencies, and 46 public corporations. The departments are as follows: addiction services, agriculture, commerce, consumer affairs, finance, health, housing, justice, labor and human resources, natural resources, public instruction, recreation and sports, social services, state, and transportation and public works. Lodged within the Office of the Governor are the Bureau of Budget and Management, Planning Board, Commission to Combat Crime, Commission on Women's Affairs, and Environmental Quality Board, as well as offices of economic opportunity, energy, youth affairs, cultural affairs, labor affairs, child development, and development of the disabled, and commissions for the protection and strengthening of the family and of agricultural planning and action.

Puerto Rico is more heavily socialized than any US state. Almost one-fourth of all those employed work for the commonwealth government, which operates hotels, marine transports, the telephone company, and all sugar mills, among other enterprises.

16JUDICIAL SYSTEM

Puerto Rico's highest court, the Supreme Court, consists of a chief justice and six associate justices, appointed, like all other judges, by the governor with the consent of the senate and serving until compulsory retirement at age 70. The court may sit in separate panels for some purposes, but not in cases dealing with the constitutionality of commonwealth law, for which the entire body convenes. Decisions of the Supreme Court of Puerto Rico regarding US constitutional questions may be appealed to the US Supreme Court.

The nine superior courts are the main trial courts; superior court judges are appointed to 12-year terms. In 1983 there were 92 superior courts justices in 12 districts. Superior courts heard appeals from the 39 district courts, which had 99 judges. These courts have original jurisdiction in civil cases not exceeding $10,000 and in minor criminal cases. District courts also hear preliminary motions in more serious criminal cases. Municipal judges, serving for five years, and justices of the peace, in rural areas, decide cases involving local ordinances.

San Juan is the seat of the US District Court for Puerto Rico, which has the same jurisdiction as federal district courts on the US mainland.

Puerto Rico's crime rate, 2,878 per 100,000 population in 1983, was far below the US average, despite a murder rate that ranked with the highest among the states. Estimated crime rates that year included murder and nonnegligent manslaughter, 13.1 per 100,000 population; forcible rape, 14.1; aggravated assault, 211.5; robbery, 246.6; burglary, 1,014.8; larceny-theft, 958.7; and automobile theft, 419.2. The death penalty is constitutionally forbidden. The San Juan metropolitan area had a crime rate of 4,181 per 100,000 in 1983, and surpassed the island rate in every category except aggravated assault.

17 ARMED FORCES

Principal of the six US military installations in Puerto Rico are the Roosevelt Roads Naval Reservation, near Ceiba, and the Naval Security Station at Sabana Seca. Ft. Buchanan, an army base, is near San Juan. Use of Vieques for training maneuvers, including shelling and bombing, forced many of that island's residents to move; aerial and naval target practice on Culebra by the US Navy was halted by protests and legal action. Department of Defense personnel numbered 9,032 in 1983/84. Defense contracts totaled $309.5 million that fiscal year.

As of 1984, an estimated 127,000 veterans of US military service were living on the island, including 1,000 who served in World War I, 31,000 in World War II, 37,000 during the Korean conflict, and 37,000 during the Viet-Nam era. Veterans' benefits totaled $338.3 million in 1982/83. Puerto Ricans suffered 731 combat deaths in Korea and 270 in Viet-Nam.

Army National Guard personnel in Puerto Rico totaled 9,746 at the end of 1984; Air National Guard strength was 1,308 in February 1985.

18 MIGRATION

Although migration from Puerto Rico to the US mainland is not an entirely new phenomenon—several Puerto Rican merchants were living in New York City as early as 1830—there were no more than 70,000 islanders in the US in 1940. Mass migration, spurred by the booming postwar job market in the US, began in 1947. The out-migration was particularly large from 1951 through 1959, when the net outflow of migrants from the island averaged more than 47,000 a year. According to the 1980 census, 2,013,945 ethnic Puerto Ricans were living in the 50 states; at least 32 cities had Puerto Rican communities of 5,000 or more. Puerto Ricans are found in significant numbers not only in New York State but also in New Jersey, Illinois, Pennsylvania, California, Florida, Connecticut, and Massachusetts.

During the 1970s, in part because of the economic decline of many US urban centers, the migration trend slowed; official estimates show that the net flow of migrants from the island totaled only 65,900. But with the Puerto Rican economy worsening in the early 1980s, the net migration from early 1980 to mid-1983 was about 90,000.

One striking aspect of the US–Puerto Rico migration pattern is its fluidity. As US citizens, Puerto Ricans can move freely between the island and the mainland. Even in 1953, when the heaviest net outflow was recorded—74,603—fully 230,307 persons emigrated from the US mainland to Puerto Rico, as 304,910 Puerto Ricans were migrating the other way. In 1980, 191,723 people living on the US mainland said that they had lived in Puerto Rico in 1975, while 137,474 people living in the commonwealth in 1980 said that they had lived on the mainland in 1975. This extreme mobility, though sensitive to the job market, would not be possible were it not for the increased income available to Puerto Ricans on both the island and the US mainland, and the fact that Puerto Ricans who come to the continental US generally preserve their ties of family and friendship with those in the commonwealth, thus finding it easy to return, whether for a short stay at Christmastime or for a new job on the island.

19 INTERGOVERNMENTAL COOPERATION

A member of the US Council of State Governments, Puerto Rico subscribes to the Compact for Education, the Interstate Compact for the Supervision of Parolees and Probationers, the Southern Interstate Energy Compact, and the Southern Growth Policies Compact. In its relations with the US government, the commonwealth is in most respects like a state, except in the key areas of taxation and representation. US laws are in effect, federal agencies regulate aviation and broadcasting, and Puerto Ricans participate in such federally funded programs as Social Security and food stamps. US aid to Puerto Rico totaled $2.2 billion in 1983/84.

20 ECONOMY

Puerto Rico made enormous strides economically in only four decades, changing from a backward agricultural society into a highly industrialized one. In 1940, annual income per capita was $118, and agricultural workers made as little as 6 cents an hour. By 1978, income per capita was $2,600, and the average hourly farm wage at least $1.65—in each case, far below the US average, but also in each case, a vast improvement over former times. The gross product of Puerto Rico was nearly $14 billion in 1983/84; in constant dollars, gross product had just returned to the 1970/80 level following a decline of 3.6% in 1981/82 and 2.2% in 1982/83. The increase in 1983/84 was 4.7%.

The following table shows how major sectors contributed to Puerto Rico's income in fiscal 1984:

Manufacturing	40%
Government	13%
Finance, insurance, real estate	13%
Trade	12%
Services	9%
Transportation, public utilities	8%
Agriculture	3%
Contract construction, mining	2%
TOTAL	100%

The island's most important industrial products are apparel, textiles, pharmaceuticals, petroleum products, rum, refined sugar, computers, instruments, and office machines. Tourism is the backbone of a large service industry, and the government sector has also grown. Tourist revenues and remittances from workers on the US mainland largely counterbalance Puerto Rico's chronic trade deficit. Federal funds to the government and directly to the people are vital to the Puerto Rican economy; in the mid-1980s, they constituted 30.7% of the commonwealth's gross product, compared to 11.6% for the average US state.

Puerto Rico's major problem is lack of jobs for an expanding population, a problem exacerbated when rising unemployment in the US persuades Puerto Ricans to return to the island. From its former dependence on subsistence agriculture, Puerto Rico became a center for low-wage textile manufacturing, then a home for refining cheap crude oil from abroad—mainly Venezuela. The sharp rise of overseas oil prices that began in 1973 devastated this economic sector. Since then, high-technology industries have become a major presence on the island.

21 INCOME

Per capita income in Puerto Rico, $3,918 in 1982, was far lower than in any of the 50 states during that year, but far exceeded that of its Caribbean neighbors. Total income increased from $1.3 billion in 1960 to $5 billion in 1972, $7.3 billion in 1978, and $13.6 billion in 1983/84. Average family income on the island in 1983 was $14,430. The increase in family income between 1970 and 1980 in constant dollars was 14%; from 1980 through 1983, it fell by almost 5%.

Inequalities of income continue to plague Puerto Rico. Between 1959 and 1969, for example, the share of the total income received by the top 10% income group declined from 45% to 36%, but the share earned by the bottom 10% also declined, from 0.44% to 0.31%—a statistic suggesting that although economic expansion has stimulated the growth of the middle class, the poorest segment of Puerto Rican society did not benefit from industrialization. In the mid-1980s, an estimated 62% of Puerto Ricans were below the federal poverty line, compared to 15% of US residents.

22 LABOR

Puerto Rico's civilian labor force in March 1985 numbered

966,200, of whom 765,000 were employed, yielding an unemployment rate of 20.8%. There were 620,000 males and 322,000 females in the labor force in 1983, about 58% of eligible men but only 27% of the women.

A federal survey in March 1982 found the following nonfarm employment pattern in Puerto Rico:

	ESTABLISH-MENTS	EMPLOYEES	ANNUAL PAYROLL ('000)
Agricultural services, forestry, fisheries	80	883	$ 7,087
Mining	44	574	6,322
Contract construction	960	25,720	197,715
Manufacturing	1,909	142,545	1,517,286
Transportation, public utilities	774	19,399	267,137
Wholesale trade	1,909	28,078	361,753
Retail trade	9,683	76,679	564,087
Finance, insurance, real estate	1,777	27,742	327,510
Service industries	6,772	80,046	682,256
Other establishments	3,715	12,056	103,126
TOTALS	27,623	413,722	$4,034,279

Average agricultural employment was 38,000 in 1983/84. Government employment was 222,500 in October 1982, of whom 185,000 were commonwealth government employees and 37,500 were municipal government employees.

Unemployment dropped as low as 10.3% in 1970 but rose sharply in the 1970s. In 1983 it was 23.4%, in 1984 20.7%—almost triple the US rate.

In 1980, 20% of employed persons were in managerial and professional specialty occupations; 15% were in administrative support occupations, including clerical; 15% were in precision production, craft, and repair occupations; 10% were in sales; 3% were in farming, forestry, and fishing occupations; and 2% were technologists and technicians.

Approximately 13% of the labor force belonged to trade unions in 1982; there were 273 labor unions in 1983. Wages tend to adhere closely to the US statutory minimum, which applies to Puerto Rico. In September 1984, the average hourly wage of production workers was $5.06 and the average weekly earnings $197.85—not much over half of what the average US production worker could expect. The following table shows average hourly earnings for selected industries as of March 1985:

Oil refining	$9.14
Chemicals, including pharmaceuticals	7.22
Electrical and electronic equipment	5.43
Instruments	5.36
Fabricated metal products	5.33
Food processing	5.18
Plastics and rubber	4.92
Nonelectrical machinery	4.86
Apparel	4.29
Leather, including shoes	4.05

[23]AGRICULTURE

In 1940, agriculture employed 43% of the work force; by 1982, fewer than 5% of Puerto Rican workers had agricultural jobs. Nowhere is this decline more evident than in the sugar industry. Production peaked at 1,300,000 tons in 1952, when 150,000 cane cutters were employed; by 1978, however, production was 300,000 tons, fewer than 20,000 cutters were in the fields, and the industry was heavily subsidized. In 1982, production fell below 100,000 tons. The hilly terrain makes mechanization difficult, and manual cutting contributes to production costs that are much higher than

those of Hawaii and Louisiana. Despite incentives and subsidies, tobacco is no longer profitable, and coffee production—well adapted to the highlands—falls far short of domestic consumption, although about half of the crop is exported. Pineapple growing, managed by the Puerto Rican Land Authority, was also unprofitable during much of the 1970s. One of the few promising long-term agricultural developments has been the attempt to convert sugarcane lands to the cultivation of rice, a staple food in the diet of most Puerto Rico residents.

In 1982 there were 21,820 farms in Puerto Rico; the average size was 44 acres (18 hectares). Agricultural sales that year totaled $366.4 million, including $129.8 million for crops. The following table shows acreage, production data, and market value for leading crops in 1982:

	ACRES	OUTPUT		VALUE
Coffee	79,362	243,942	hundredweight	$39,301,680
Sugarcane	39,890	1,462,043	tons	26,662,612
Fruits and nuts	53,148	—		14,059,974
Grains and farinaceous crops, of which:	2,623	140,997	hundredweight	9,713,476
Rice	(2,196)	(137,564)	hundredweight	
Vegetables	6,477	N/A		9,546,660
Pineapples	3,099	44,022	tons	9,533,789
Ornamental and flowering plants and lawn grass	825	—		8,835,237
Tobacco	243	368,116	lb	262,789

[24]ANIMAL HUSBANDRY

Livestock and livestock products produced a gross income of $385.5 million in 1982/83, or 65% of total agricultural income. By the close of 1983 there were 585,088 cattle and 205,597 hogs and pigs on Puerto Rico farms and ranches. Production of meat animals in 1982/83 included 46.8 million lb of beef worth $48.9 million, and 36.6 million lb of pork worth $31.6 million.

Leading dairy and poultry products in 1982/83 were 393.5 million quarts of milk, valued at $153.7 million; 232.5 million eggs, valued at $17.7 million; and 51.2 million lb of broilers, valued at $34.5 million. In 1983 there were about 6 million chickens on Puerto Rican poultry farms.

[25]FISHING

Although sport fishing, especially for blue marlin, is an important tourist attraction, the waters surrounding Puerto Rico are too deep to lend themselves to commercial fishery. Tuna brought in from African and South American waters is processed at five large plants on the western shore that together provide much of the canned tuna sold in eastern US markets. In all, seven canned and cured seafood plants employed 5,735 persons in March 1982. Commercial fishing receipts came to $8.1 million in 1982. Total catch in 1981 was about 2,700 metric tons.

[26]FORESTRY

Puerto Rico lost its self-sufficiency in timber production by the mid-19th century, as population expansion and increasing demand for food led to massive deforestation. Today, commercial timberland is scarce, and the island must import at least 90% of its wood and paper products. The Caribbean National Forest covered 55,665 acres (22,527 hectares) in 1984, of which 27,846 (11,269 hectares) constituted National Forest System lands.

[27]MINING

The search for gold first brought the Spaniards to Puerto Rico, but they soon exhausted the known supply. The island was thought barren of mineral resources until recently, when deposits of copper, silver, and gold were discovered in the mountains of the northwest. Nickel deposits in the southwest are also being explored, and offshore drilling for oil and gas is under consideration. For the present, construction materials are Puerto Rico's most abundant resource. Sand, gravel, and crushed stone are used for concrete. Limestone and clay deposits are also exploited. In

1984, the value of nonfuel mineral production was estimated at $120.3 million (excluding sand and gravel), of which $90 million came from cement and $28.1 million from stone.

²⁸ENERGY AND POWER

Puerto Rico is almost totally dependent on imported crude oil for its energy needs. The island has not yet developed any fossil fuel resources of its own, and its one experimental nuclear reactor, built on the south coast at Rincón in 1964, was shut down after a few years. Solar-powered hot-water heaters have been installed in a few private homes and at La Fortaleza.

The Puerto Rico Electric Power Authority, a public agency, is virtually the sole producer of electricity on the island. Of the 4.2 million kw of installed capacity in 1981, 98% was publicly owned. Consumption was less than 10.2 billion kwh in 1983/84.

Inefficiency in the public transport system has encouraged commonwealth residents to rely on private vehicles, thereby increasing the demands for imported petroleum. Domestic consumption of petroleum products in 1981 included 7,033,000 metric tons of crude oil, 3,808,000 tons of residual fuel oil, and 1,832,000 tons of motor-fuel gasoline. Production by island refineries included 3,307,000 tons of residual fuel oil, 2,865,000 tons of motor-fuel gasoline, and 1,400,000 tons of gas-diesel oils. Some of these petroleum products, especially gasoline, were exported. In all, Puerto Rico produced 8,269,000 metric tons of energy petroleum products in 1981, consumed 7,342,000 tons, exported 2,371,000 tons, and imported 1,484,000 tons.

²⁹INDUSTRY

Value added by manufacture surpassed $8.6 billion in 1982, more than double the total for 1977. In 1949, about 55,200 Puerto Rican workers were employed in industrial jobs, 26% of them in sugar refining. By 1982, despite the loss of many jobs in the sugar industry, the number was 142,545, with a payroll of nearly $1.6 billion. The major employment categories were apparel and other textile products, 31,764; electric and electronic equipment, 21,682; food and kindred products, 18,621; chemicals and allied products, 17,225; and instruments and related products, 10,865. The growth areas were electric and electronic equipment, up 47% from 1977, and instruments and related products, up 60%.

In terms of value added by manufacture, the leading sectors in 1977 and 1982 were (in thousands):

Chemicals and allied products	$1,476,470	$3,713,928
Electric and electronic equipment	403,055	1,139,568
Food and kindred products	485,569	965,463
Instruments and related products	257,314	616,440
Machinery, except electrical	161,154	542,244
Apparel and other textile products	278,904	510,075
Other industries	1,034,690	1,117,883
TOTALS	$4,097,156	$8,605,601

Among major US pharmaceutical companies, Johnson & Johnson had 15 plants in Puerto Rico in 1985, Baxter Travenol Laboratories had 14, and American Hospital Supply had 8. Major manufacturers of electrical and electronic equipment in Puerto Rico are Westinghouse (24 plants in 1985), General Electric (17), GTE (9), and Motorola (6).

³⁰COMMERCE

Wholesale trade in Puerto Rico in 1982 involved 2,282 establishments and total sales of $7.1 billion. Of that total, $2.3 billion came from groceries and related products and $1.1 billion from petroleum and petroleum products. Retail trade during the same year involved 34,461 establishments and total receipts of $6.5 billion, the major sectors being food stores, with $1.8 billion, and gasoline service stations, with $734 million. Two large shopping centers, Plaza las Americas and Plaza Carolina, are in the San Juan area.

Foreign trade is a significant factor in Puerto Rico's economy. Trade between the US and Puerto Rico is unrestricted. Imports have always exceeded exports: in 1984, the island's imports were $9.5 billion and exports $9.1 billion. During 1984, the US received 83% of Puerto Rico's exports (91% in 1965/66) and supplied about 57% of its imports (82% in 1966). The principal reason for the relative decline in imports from the continental US has been the heavy volume and cost of oil imports, especially from Venezuela, auto imports and other goods from Japan, and, in general, the stimulus to foreign imports created by the strong US dollar in the early 1980s.

The following table shows how major commodity groups shared in Puerto Rico's imports and exports during two recent years, 1982 and 1984:

	IMPORTS		EXPORTS	
	1982	1984	1982	1984
Animal and vegetable products	22%	19%	14%	17%
Wood and paper, printed matter	5	5	1	1
Textiles, fibers, and textile products	6	6	8	8
Chemicals and related products	38	34	40	33
Nonmetallic minerals and products	2	2	3	2
Metals and metal products	18	23	19	23
Other items	9	11	15	16
TOTALS	100%	100%	100%	100%

³¹CONSUMER PROTECTION

Consumer protection is the responsibility of Puerto Rico's cabinet-level Department of Consumer Affairs.

³²BANKING

Puerto Rico's first bank began operations in 1850. The commonwealth's largest commercial bank, the Banco Popular de Puerto Rico, with assets of over $3.5 billion at the end of 1984, was founded in 1893, near the end of the Spanish colonial era. The next biggest, the Banco de Ponce, with assets of over $1.8 billion, was established in 1917. In 1984, Puerto Rico had 12 insured commercial banks, whose total assets exceeded $16 billion; loans totaled $7.7 billion in 1983/84, and deposits amounted to $12.8 billion.

During 1982, 11 savings and loan associations, all federally chartered, had total assets of $2.9 billion, including $1.6 billion in outstanding mortgage loans. Personal loan companies had loan portfolios of $261 million in 1982/83.

US corporations operating in Puerto Rico are virtually exempted from paying federal corporate taxes on nearly all income earned in the commonwealth; in 1985, 631 companies had about $5.5 billion in unrepatriated profits in banks in Puerto Rico.

The Government Development Bank, founded in 1948, serves as a fiscal agent for the commonwealth government, municipalities, and public authorities and corporations, while also extending credit to private industry. In 1983/84, the bank had assets of $3.8 billion and provided loans of over $1.1 billion to public and private borrowers.

³³INSURANCE

Puerto Ricans paid $796 million in insurance premiums in 1981/82; life insurance in force came to nearly $13.8 billion that year. Flood insurance worth $289 million was in force at the end of 1983.

³⁴SECURITIES

There are no securities exchanges in Puerto Rico. Bonds issued by the Government Development Bank, exempt from federal income taxes and from the income taxes of all US states and cities, are offered for sale on the world securities market.

³⁵PUBLIC FINANCE

Puerto Rico's annual budget is prepared by the Bureau of Budget and Management and submitted by the governor to the legislature, which has unlimited power to amend it. The fiscal year extends from 1 July to 30 June. The following table shows Puerto Rico's proposed revenues and expenditures for 1985/86 (in millions):

REVENUES

General fund revenues, of which:	$2,733.6
Income tax	(1,437.6)
Excise taxes	(647.3)
Special local funds and US grants	764.5
Capital investment fund revenues	145.0
Proprietary fund revenues	
of the public corporations	4,336.3
Other receipts	30.0
TOTAL	$8,009.4

EXPENDITURES

Central government, of which:	$3,643.2
Operating expenses	(3,136.0)
Debt service	(306.6)
Capital improvements	(200.6)
Public corporations, of which:	4,366.2
Operating expenses	(2,665.2)
Debt service	(830.3)
Capital improvements	(870.7)
TOTAL	$8,009.4

Of expenditures, 43.3% was assigned to economic development and 41% to social development; of central government expenditures, the largest item was education, 28.8%.

In 1959/60, transfers from the US government amounted to $44 million, or less than 13% of all revenues. By 1972/73, receipts from the US government represented 23% of all revenues; by 1977/78, more than 29%. In 1983/84, federal grants came to nearly $1.4 billion (34% of commonwealth government receipts), while federal excise and customs refunds to Puerto Rico came to $443 million. The total public debt was $8.8 billion in March 1985, of which almost 74% belonged to public corporations, rather than to the commonwealth government or municipal governments.

36 TAXATION

The Puerto Rican Federal Relations Act stipulates that the commonwealth is exempt from US internal revenue laws. The federal income tax is not levied on permanent residents of Puerto Rico, but federal Social Security and unemployment taxes are deducted from payrolls, and the commonwealth government collects an income tax that ranged in 1982 from 9.72% on the first $2,000 of taxable income to 63.99% on income exceeding $200,000. That same year, the normal corporate income tax was 22%, with a surcharge ranging from 9% on the first $75,000 to 23% on income over $275,000. The estate tax ranged from 3% to 70%, and the gift tax from 2.25% to 52.5%. Property, franchise, and excise taxes are also levied, with the excise tax on new and used cars being an especially important source of revenue.

Vital to Puerto Rico's economy is Section 936 of the federal income tax code of 1976, which allows subsidiaries of US corporations virtual exemption from US corporate income taxes. At the time of repatriation of profits to the US stockholder, the Puerto Rican government imposes a "tollgate" tax of 5–10%.

Transfers from the US federal government to the commonwealth government and municipal governments during 1983/84 totaled $2.2 billion. In that year, the total inflow of US federal funds to Puerto Rico exceeded $5.4 billion. These sums were far in excess of payments to the US Treasury by Puerto Rican governments and individuals.

37 ECONOMIC POLICY

Inaugurated during the 1940s, Operation Bootstrap had succeeded by 1982 in attracting investments from more than 500 US corporations. The key Puerto Rican agencies responsible for this transformation is the Administración de Fomento Económico, known as Fomento (Development), and its subsidiary, the Puerto Rico Industrial Development Co., which help select plant sites, build factories, hire and train workers, and arrange financing. Fomento reorganized certain industries, taking a direct role, for example, in promoting export sales of Puerto Rican rum. At first, Fomento brought in apparel and textile manufacturers, who needed relatively unskilled workers. More recently, with the improvement in Puerto Rico's educational system, Fomento has emphasized such technologically advanced industries as pharmaceuticals and electronics. Industrialization has also required heavy investment in roads, power, water facilities, and communications systems.

The key incentives to investment in Puerto Rico have been lower wage scales than in the continental US and the exemption of up to 90% of corporate profits from island corporate and property taxes for five years, with a descending rate of exemption that can last as long as 25 years in some regions. The commonwealth government is planning a 218-acre (88-hectare) free-trade zone in the San Juan area that will allow companies to assemble imports duty-free in government-built warehouses for export from the island. Under the US Tax Reform Act of 1976, US companies may repatriate earnings from their Puerto Rican subsidiaries free of federal taxation.

38 HEALTH

Health conditions in Puerto Rico have improved remarkably since 1940, when the average life expectancy was only 46 years. A resident of Puerto Rico born in 1985 could expect to live 73 years, actually longer than a US mainland resident. Similarly, infant mortality declined from 113 per 1,000 live births in 1940 to 18.5 in 1981—a rate that was still about 55% above the US norm. The leading causes of death in 1940 were diseases brought on by malnutrition or infection: diarrhea, enteritis, tuberculosis, and pneumonia. By 1983/84, when Puerto Rico enjoyed one of the lowest death rates in the world—only 6.7 per 1,000 population—the leading causes of death were similar to those in most industrialized countries. Alcoholism and drug addiction are among the major public health problems.

In 1983, Puerto Rico had 62 hospitals, with 10,896 beds; average daily occupancy was 7,622, or 70% of beds filled. Outpatient visits came to 2,665,510. Medical personnel included 4,057 physicians, 741 dentists, and 14,592 nursing personnel in 1980.

39 SOCIAL WELFARE

Since the mid-1960s, residents of Puerto Rico have been eligible for most of the programs that apply throughout the 50 states. The average monthly payment in aid to families with dependent children in 1982 was only $97, less than one-third the US average; payments came to $64 million. At the end of the 1970s, more than half of the population received food stamps. In fact, 1,855,000 commonwealth residents, a total exceeded only in New York State, actually took part in the program during 1979/80, receiving a US federal bonus worth $827 million, more than was allocated to any state and about 10.5% of the total US government subsidy. Because of federal cutbacks, the cost dropped to $670 million in 1981/82. In 1982, cash benefits partially replaced the food stamp program in Puerto Rico. In 1983, 529,000 children received school lunches at a cost to the US government of $87 million.

Because unemployment is high and wages are low, Social Security benefits are well below the US average. In 1983, $1.45 billion was paid to 543,000 Social Security recipients; the average monthly payment for a retiree was $276, well below the US norm. Unemployment insurance, paid to an average of 30,000 claimants a week in 1982, totaled $190 million; the average weekly benefit was $66, slightly more than half the US average. Island residents are not eligible for Supplemental Security Income.

40 HOUSING

In 1985, about 10% of all Puerto Ricans lived in US-funded housing projects. The value of housing construction was $331 million in 1983/84.

Of Puerto Rico's 968,474 year-round housing units in 1980, 85.1% were single units, and 13.4% lacked complete plumbing for the exclusive use of the unit's inhabitants. Of the 867,697 year-round occupied units, 73.3% were owner-occupied. The median value of owner-occupied units was $19,800, and the median rent $85 a month. Eighteen percent averaged more than one person a room.

[41] EDUCATION

Puerto Rico has made enormous strides in public education. In 1900, only 14% of the island's school-age children were actually in school; the proportion had increased to 50% by 1940 and 85% by the late 1970s. Of the population 25 years of age or older in 1980, more than 40% had completed high school and 18% had completed college. The government encouraged school attendance among the poor in the 1940s and 1950s by providing inexpensive shoes, free lunches, school uniforms, and small scholarships. Today, education is compulsory for children between 6 and 16 years of age, and nearly one out of three commonwealth budget dollars goes to education.

There are 82 public school districts, but these are merely administrative areas of the commonwealth Department of Education, which is responsible for public education. As of 1983 there were 708,700 students in public elementary and secondary day schools, and 92,300 in other public schools; in 1982, another 102,400 were in private schools, mostly Roman Catholic. During 1983/84, 160,000 students were in higher educational institutions, of whom 54,000 were in the University of Puerto Rico system, with its main campus at Río Piedras. The system also includes doctorate-level campuses at Mayagüez and San Juan (for medical sciences); four-year colleges at Arecibo, Cayey, Humacao, and Ponce; and two-year colleges at Aguadilla and Carolina. The 30 private institutions in 1983/84 included Inter-American University, with campuses at Hato Rey, San Germán, and other locations, and the Catholic University of Puerto Rico, at Ponce. Numerous student aid and loan programs are available.

[42] ARTS

The Tapia Theater in Old San Juan is the island's major showcase for local and visiting performers, including the Taller de Histriones group and *zarzuela* (comic opera) troupes from Spain. The Institute of Puerto Rican Culture produces an annual theatrical festival. The Fine Arts Center features entertainment ranging from ballet, opera, and symphonies to drama, jazz, and popular music.

Puerto Rico has its own symphony orchestra and conservatory of music. Both were formerly directed by Pablo Casals, and the annual Music Festival Casals, which he founded, still attracts world-renowned musicians to the island each May. The Opera de Cámara tours several houses. Puerto Rico supports both a classical ballet company (the Ballets de San Juan), and the Areyto Folkloric Group, which performs traditional folk dances. Salsa, a popular style pioneered by Puerto Rican musicians like Tito Puente, influenced the development of pop music on the US mainland during the 1970s.

[43] LIBRARIES AND MUSEUMS

In 1983/84, Puerto Rico's public libraries contained about 1,752,898 volumes and had a combined circulation of 598,138. The University of Puerto Rico Library at Río Piedras held 587,270 books; the library of the Puerto Rico Conservatory of Music, in San Juan, has a collection of music written by Puerto Rican and Latin American composers. La Casa del Libro, also in San Juan, is a library-museum of typographic and graphic arts. Among the 24 museums in 1983, the Museo de Arte de Ponce (Luis A. Ferré Foundation) has paintings, sculptures, and archaeological artifacts, as well as a library. The Marine Station Museum in Mayagüez exhibits Caribbean marine specimens and sponsors research and field trips.

[44] COMMUNICATIONS

The Puerto Rico Telephone Co. was founded in 1914 by two German sugar brokers, Sosthenes and Hernand Behn, best known today as the creators of International Telephone and Telegraph (ITT). In 1974, the Puerto Rican government bought the phone company from ITT.

In 1983/84 there were more than 789,000 telephones on the island. Direct dialing to the continental US was inaugurated in 1968. The US Postal Service, which handles Puerto Rico's mail traffic, had 96 post offices and 2,740 employees in Puerto Rico in March 1985.

WKAQ, the island's first radio station, came on the air in 1923. As of 1984 there were 60 AM and 40 FM radio stations and 16 television stations. The first television station, WKAQ-TV, began broadcasting in 1954. There are two operating public stations, affiliated with the US Public Broadcasting System. Cable television, with 79,000 subscribers in 1983, was reaching the English-speaking public in San Juan and Ponce.

[45] PRESS

Puerto Rico has three major Spanish-language dailies; September 1984 circulation for *El Nuevo Día* was 194,679 mornings, 203,097 Sundays; for *El Mundo,* 99,943 mornings, 119,960 Sundays; and for *El Vocero,* 206,247 mornings. The English-language *San Juan Star,* with a circulation of 38,250 mornings and 45,932 Sundays, won a Pulitzer Prize in 1961. *El Reportero* is an evening Spanish-language newspaper with a circulation in 1984 of about 43,000. Puerto Rico had three book publishers in 1984, two of them university presses.

[46] ORGANIZATIONS

Important organizations on the island include the Puerto Rico Medical Association, Puerto Rico Manufacturers' Association, and Puerto Rico Bar Association. Also maintaining headquarters in Puerto Rico are the Association of Island Marine Laboratories of the Caribbean, Puerto Rico Rum Producers Association, Caribbean Hotel Association, and Caribbean Studies Association.

US-based agencies such as the National Puerto Rican Forum and the Puerto Rican Community Development Project assist Puerto Ricans living on the mainland. "Hometown clubs" consisting of "absent sons" *(hijos ausentes)* of various Puerto Rican towns are a typical feature of the barrios in New York and other cities in the continental US.

[47] TOURISM, TRAVEL, AND RECREATION

Only government and manufacturing exceed tourism in importance to the Puerto Rican economy. The industry has grown rapidly, from 65,000 tourists in 1950 to 1,088,000 in 1970 and 1,496,000 in 1983/84, with estimated expenditures of $659.4 million, and an average in 1983 of $452 spent per visit. Tourism accounts for about 5% of the island's gross product. Net income from tourism was $217.6 million in 1983.

As of 30 June 1984, tourist hotels had a total of 5,888 rooms, about 73% of them in San Juan. In all, Puerto Rico had 7,421 rooms in hotels and guest houses. Of all those who registered in hotels in 1983/84, 54% were from the 50 states (mostly from New York, New Jersey, and Florida, where many Puerto Ricans live), 31% were Puerto Rico residents, 6% were from the West Indies, and virtually all the rest were from Europe, Canada, South America, Mexico, and Central America, in that order.

Most tourists come for sunning, swimming, deep-sea fishing, and the fashionable shops, nightclubs, and casinos of San Juan's Condado Strip. Attractions of Old San Juan include two fortresses, El Morro and San Cristóbal, San José Church (one of the oldest in the New World), and La Fortaleza, the governor's palace. The government has been encouraging tourists to journey outside San Juan to such destinations as the Arecibo Observatory (with its radio telescope used for research astronomy, ionospheric studies, and radar mapping), the rain forest of El Yunque, Phosphorescent Bay, colonial-style San Germán, and the bird sanctuary and mangrove forest on the shores of Torrecilla Lagoon. The 53-acre (21-hectare) San Juan harbor fortifications are a national historic site; there were 2,005,185 visits to this site in 1984.

48 SPORTS

Baseball is very popular in Puerto Rico. There is a six-team professional winter league, in which many ball players from American and National league teams participate. Horse racing, cockfighting, boxing, and basketball are also popular. Puerto Rico, which has its own Olympic Committee, sent a delegation to the 1980 Olympics in Moscow despite the US boycott.

49 FAMOUS PUERTO RICANS

Elected to represent Puerto Rico before the Spanish Cortes in 1812, Ramón Power y Giralt (1775-1813), a liberal reformer, was the leading Puerto Rican political figure of the early 19th century. Power, appointed vice president of the Cortes, participated in the drafting of the new Spanish constitution of 1812. Ramón Emeterio Betances (1827-98) became well known not only for his efforts to alleviate a cholera epidemic in 1855 but also for his crusade to abolish slavery in Puerto Rico and as a leader of a separatist movement that culminated in 1868 in the "Grito de Lares." Eugenio María de Hostos (1839-1903), a writer, abolitionist, and educator, spent much of his adult life in Latin America, seeking to establish a free federation of the West Indies to replace colonial rule in the Caribbean. Luis Muñoz Rivera (1859-1916), a liberal journalist, led the movement that obtained for Puerto Rico the Autonomic Charter of 1897, and he headed the cabinet that took office in 1898. With the island under US rule, Muñoz Rivera served between 1911 and 1916 as Puerto Rico's resident commissioner to the US Congress. Other important Puerto Rican historical figures include Juan Alejo Arizmendi (1760?-1814), the first Puerto Rican–born bishop, appointed to the See of San Juan; José Celso Barbosa (1857-1921), a US-trained physician who founded the Republican Party of Puerto Rico in 1899; and José de Diego (1866-1918), a noted poet and gifted orator who, under the Foraker Act, became the first speaker of the island house of delegates and was a champion of independence for Puerto Rico.

The dominant political figure in 20th-century Puerto Rico was Luis Muñoz Marín (1898-1980), founder of the Popular Democratic Party in 1938 and president of the Puerto Rico senate from 1940 to 1948. Muñoz, the first native-born elected governor of the island (1948-64), devised the commonwealth relationship that has governed the island since 1952. Another prominent 20th-century figure, Antonio R. Barceló (1869-1939), led the Unionista Party after Muñoz Rivera's death, was the first president of the senate under the Jones Act, and was later the leader of the Liberal Party. In 1946, Jesús T. Piñero (1897-1952) became the first Puerto Rican appointed governor of the island by a US president; he had been elected as resident commissioner of Puerto Rico to the US Congress two years before. Pedro Albizu Campos (1891-1965), a Harvard Law School graduate, presided over the militant Nationalist Party and was until his death the leader of forces that advocated independence for Puerto Rico by revolution. In 1945, Gilberto Concepción de Gracia (1909-68), also a lawyer, helped found the more moderate Puerto Rican Independence Party. Herman Badillo (b.1929) was the first person of Puerto Rican birth to be a voting member of the US House of Representatives, as congressman from New York, and Maurice Ferré (b.1935), elected mayor of Miami in 1973, was the first native-born Puerto Rican to run a large US mainland city. Hernán Padilla (b.1938), mayor of San Juan, became the first Hispanic American elected to head the US Conference of Mayors (1984).

Women have participated actively in Puerto Rican politics. Ana Roqué de Duprey (1853-1933) led the Asociación Puertorriqueña de Mujeres Sufragistas, organized in late 1926, while Milagros Benet de Mewton (1868-1945) presided over the Liga Social Sufragista, founded in 1917. Both groups actively lobbied for the extension of the right to vote to Puerto Rican women, not only in Puerto Rico but in the US and other countries as well. Felisa Rincón de Gautier (b.1897), mayor of San Juan from 1946 to 1968, was named Woman of the Americas in 1954, the year she presided over the Inter-American Organization for Municipalities. Carmen Delgado Votaw (b.1935) was the first person of Puerto Rican birth to be elected president of the Inter-American Commission of Women, the oldest international organization in the field of women's rights.

Manuel A. Alonso (1822-89) blazed the trail for a distinctly Puerto Rican literature with the publication, in 1849, of *El Gíbaro*, the first major effort to depict the traditions and mores of the island's rural society. Following him in the development of a rich Puerto Rican literary tradition were, among many others, that most prolific of 19th-century Puerto Rican writers, Alejandro Tapia y Rivera (1826-82), adept in history, drama, poetry, and other forms of literary expression; essayist and critic Manuel Elzaburu (1852-92); novelist Manuel Zeno Gandía (1855-1930); and poets Lola Rodríguez de Tió (1843-1924) and José Gautier Benítez (1848-80). The former's patriotic lyrics, popularly acclaimed, were adapted to become Puerto Rico's national anthem. Among 20th-century Puerto Rican literary figures are poets Luis Lloréns Torres (1878-1944), Luis Palés Matos (1898-1959), and Julia de Burgos (1916-1953) and essayists and critics Antonio S. Pedreira (1898-1939), Tomás Blanco (b.1900), José A. Balseiro (b.1900), Margot Arce (b.1904), Concha Meléndez (b.1904), Nilita Vientós Gastón (b.1908), and María T. Babín (b.1910). In the field of fiction, René Marqués (1919-79), Abelardo Díaz Alfaro (b.1919), José Luis González (b.1926), and Pedro Juan Soto (b.1928) are among the best known outside Puerto Rico.

In the world of entertainment, Academy Award winners José Ferrer (b.1912) and Rita Moreno (b.1931) are among the most famous. Notable in classical music are cellist-conductor Pablo Casals (b.Spain, 1876-1973), a long-time resident of Puerto Rico; pianist Jesús María Sanromá (1902-84); and opera star Justino Díaz (b.1940). Well-known popular musicians include Tito Puente (b.New York, 1923) and José Feliciano (b.1945).

Roberto Clemente (1934-72), one of baseball's most admired performers and a member of the Hall of Fame, played on 12 National League All-Star teams and was named Most Valuable Player in 1966.

50 BIBLIOGRAPHY

Babín, María Teresa. *The Puerto Ricans' Spirit: Their History, Life and Culture.* New York: Collier, 1971.

Carr, Raymond. *Puerto Rico: A Colonial Experiment.* New York: Vintage, 1984.

Hostos, Adolfo de. *Diccionario Histórico Bibliográfico Comentado de Puerto Rico.* San Juan: Academia Puertorriqueña de la Historia, 1976.

Morales Carrión, Arturo. *Puerto Rico: A Political and Cultural History.* New York: Norton, 1984.

Morales Carrión, Arturo. *Puerto Rico and the Non-Hispanic Caribbean.* Río Piedras: University of Puerto Rico, 1971.

Puerto Rico Federal Affairs Administration. *Puerto Rico, U.S.A.* Washington, D.C., 1979.

US Commission on Civil Rights. *Puerto Ricans in the Continental United States: An Uncertain Future.* Washington, D.C., 1976.

US Department of Commerce. *Economic Study of Puerto Rico.* 2 vols. Washington, D.C., 1979.

Vivó, Paquita. *The Puerto Ricans: An Annotated Bibliography.* New York: Bowker, 1973.

Votaw, Carmen Delgado. *Puerto Rican Women: Some Biographical Profiles.* Washington, D.C.: National Conference of Puerto Rican Women, 1978.

Wagenheim, Kal. *Puerto Rico: A Profile.* New York: Praeger, 1970.

Wagenheim, Kal, and Olga Jiménez de Wagenheim (eds.). *The Puerto Ricans: A Documentary History.* New York: Praeger, 1973.

Wells, Henry. *The Modernization of Puerto Rico.* Cambridge, Mass.: Harvard University Press, 1969.

UNITED STATES
CARIBBEAN DEPENDENCIES

NAVASSA

Navassa, a 2-sq-mi (5-sq-km) island between Jamaica and Haiti, was claimed by the US under the Guano Act of 1856. The island, located at 18°24′N and 75°1′W, is uninhabited except for a lighthouse station under the administration of the Coast Guard.

VIRGIN ISLANDS OF THE UNITED STATES

The Virgin Islands of the United States lie about 40 mi (64 km) E of Puerto Rico and 1,075 mi (1,730 km) ESE of Miami, between 17°40′ and 18°25′N and 63°34′ and 65°3′W. The island group extends 51 mi (82 km) N–S and 50 mi (80 km) E–W, with a total area of at least 136 sq mi (353 sq km). Only 3 of the more than 50 islands and cays are of significant size: St. Croix, 84 sq mi (218 sq km) in area; St. Thomas, 32 sq mi (83 sq km); and St. John, 20 sq mi (52 sq km). The territorial capital, Charlotte Amalie, on St. Thomas, has one of the finest harbors in the Caribbean.

St. Croix is relatively flat, with a terrain suitable for sugarcane cultivation. St. Thomas is mountainous and little cultivated, but it has many snug harbors. St. John, also mountainous, has fine beaches and lush vegetation; about two-thirds of St. John's area has been declared a national park. The subtropical climate, with temperatures ranging from 70° to 90°F (21–32°C) and a mean temperature of 78°F (26°C), is moderated by northeast trade winds. Rainfall, the main source of fresh water, varies widely, and severe droughts are frequent. The average yearly rainfall is 45 in (114 cm), with most occurring during the summer months. As of 1984, both the Anegada ground iguana and the Virgin Islands tree boa were endangered species.

The population of the US Virgin Islands was estimated at 104,000 in 1983, a 7.7% increase over the 1980 census total of 96,569 and a 66.5% increase over the 1970 census population of 62,468. In 1980, St. Croix had a population of 49,725; St. Thomas, 44,372; and St. John, 2,472. St. Croix has two principal towns: Christiansted and Frederiksted. Economic development has brought an influx of new residents, mainly from Puerto Rico, other Caribbean islands, and the US mainland. In 1980, only 44.8% of the population was native-born. Most permanent inhabitants are descendants of slaves who were brought from Africa in the early days of Danish rule, and about 80% of the population is black. A small settlement of French fishermen on St. Thomas maintains its own language and traditions. English is the official and most widely spoken language.

Some of the oldest religious congregations in the western hemisphere are located in the Virgin Islands. A Jewish synagogue there is the 2d oldest in the New World, and the Lutheran Congregation of St. Thomas, founded in 1666, is one of the three oldest congregations in the US. The Catholic population in the Virgin Islands was about 25,000 in 1984, the Jewish population 510. Data on the Protestant population were not available.

In 1983, 39,661 motor vehicles were registered in the US Virgin Islands. Cargo-shipping services operate from Baltimore, Port Elizabeth, Houston, Charleston, Jacksonville, and New Orleans via Puerto Rico to St. Thomas and St. Croix. In addition, shipping service is available twice a week from West Palm Beach. In 1983, 715 cruise ships bearing 632,760 passengers arrived at Virgin Islands ports. Both St. Croix and St. Thomas have airports, with St. Croix's facility handling the larger number of jet flights from the continental US and Europe. In 1983, about 478,000 airline passengers landed in the US Virgin Islands; nearly 72% of these were overnight tourists, and 28% were one-day excursionists.

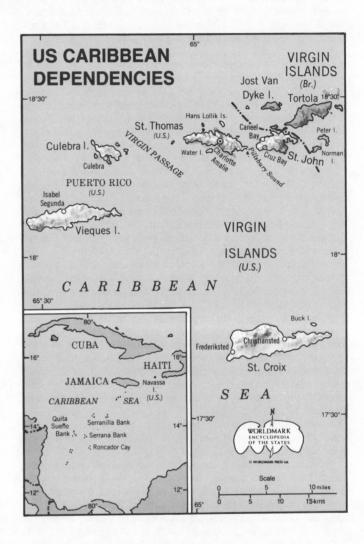

In 1983, the islands had 9 radio stations (4 AM and 5 FM) and 5 television stations, including 1 public broadcasting channel and 2 cable television stations. There were 49,043 telephones in 1983, and an estimated 52,000 television sets and 88,000 radios in 1981.

Excavations at St. Croix in the 1970s uncovered evidence of a civilization perhaps as ancient as AD 100. Christopher Columbus, who reached the islands in 1493, named them for the martyred virgin St. Ursula. At that time, St. Croix was inhabited by Carib Indians, who were driven from the island by Spanish soldiers in 1555. In the 17th century, the archipelago was divided into two territorial units, one controlled by the British, the other (now the US Virgin Islands) controlled by Denmark. The separate history of the latter unit began with the settlement of St. Thomas by the Danish West India Co. in 1672. St. John was claimed by the company in 1683, and St. Croix was purchased from France in 1733. The holdings of the company were taken over as a Danish crown colony in 1754. Sugarcane, cultivated by slave labor, was the backbone of the islands' prosperity in the 18th and early 19th centuries. After brutally suppressing several slave revolts, Denmark abolished slavery in the colony in 1848. A long period of economic decline followed, until Denmark sold the islands to the US in 1917 for $25 million. Congress granted US citizenship to the Virgin Islanders in 1927. In 1931, administration of the islands was transferred from the Department of the Navy to the Department of the Interior, and the first civilian governor was appointed. In the late 1970s, the Virgin Islands government began to consider ways to expand self-rule. A UN delegation in 1977 found little interest in independence, however, and a locally drafted constitution was voted down by the electorate in 1979.

The chief executive of the Virgin Islands is the territorial governor. On 7 November 1970, Melvin H. Evans became the first governor of the islands elected by direct popular vote (governors had previously been appointed by the US president). Constitutionally, the US Congress has plenary authority to legislate for the territory. Enactment of the Revised Organic Act of the Virgin Islands on 22 July 1954 vested local legislative power—subject to veto by the governor—in a unicameral legislature. In 1983, the legislature was composed of 15 members elected for two-year terms by popular vote. Seven legislators were elected from St. Croix and 7 from St. Thomas/St. John; 1 was elected at-large. A two-thirds vote of the legislature is needed to override a veto by the governor. Since 1972, the islands have sent one nonvoting representative to the US House of Representatives.

Courts are under the US federal judiciary; the two federal district court judges are appointed by the US president. The eight territorial court judges, who preside over misdemeanor and traffic cases, are appointed by the governor and confirmed by the legislature. The district court has appellate jurisdiction over the territorial court.

Tourism has supplanted agriculture as the islands' principal economic activity. The number of tourists rose dramatically throughout the late 1960s and early 1970s, from 448,165 in 1964 to 1,116,127 in 1972/73. A series of murders on St. Croix in 1972/73 had a devastating effect on the tourist trade, and the occupancy rate for hotels dropped below 50% until 1977. Tourism experienced a brief recovery, peaking in 1979 with 75% occupancy; however, in 1981/82, occupancy dropped below 58%, resulting in the closing of 11 hotels (4 permanently) in 1982 with a loss of 710 jobs. A total of 1,262,850 tourists visited the islands and spent over $377 million in 1984.

An estimated 30% of the labor force is employed in areas directly related to tourism. In 1983, of a total labor force of 43,260, 32% was employed in government, 17% in retail and wholesale trade, 8% in services, 6% in hotels, 6% in manufacturing, and 5% in construction and mining. The overall unemployment rate was 8.3%. Per capita income in 1983 was $7,120.

Much of the land formerly devoted to agriculture has been developed for tourism and industry. Land devoted to agriculture declined from 69,892 acres (28,284 hectares) in 1917 to 20,824 acres (8,427 hectares) in 1982, a decrease of 70%. About 90% of the farmland is on St. Croix, 8% on St. Thomas, and 2% on St. John. Fruit, vegetables, and livestock are raised on the farmland that remains. The value of all agricultural products sold in 1982 was $2,391,916, with livestock, milk, poultry, and eggs accounting for 82% of the total.

Electrical energy production in 1983 was 466.1 million kwh, all of it produced by the Virgin Islands Water and Power Authority.

Rum is an important manufacture and export. In 1982/83, the Virgin Islands received from the US government $34,699,638 in excise taxes on rum, based on shipments to the US of 3,037,610 gallons; rum excise taxes made up 15% of the islands' net government revenues in 1982/83. Amerada Hess, which employed some 800 persons at its oil refinery on St. Croix, threatened to close that facility in 1984 because of declining profits.

Most basic goods must be imported to the islands, making for an unfavorable balance of trade. Petroleum is the major import category ($3.8 billion in 1983), for local use and for the Hess refinery. Imports in 1983 totaled $4.7 billion, of which 43.5% came from the US (including Puerto Rico); exports were $3.6 billion, of which 95% went to the US.

Economic development is promoted by the US government-owned Virgin Islands Corp. To encourage industrial development, an extensive program of business incentives is offered: a company can receive, for a period of 10 years (15 years in economically depressed areas), a nontaxable subsidy equal to 90% of income tax liability, exemptions from property and excise taxes, and a nontaxable subsidy equal to 90% of customs duties on raw materials.

The total operating budget for the US Virgin Islands in 1982/83 was $252.4 million. Allocation of operating expenditures was as follows: education, 24%; health and medical programs, 17%; public works, 7%; commerce, 3%; other purposes, 49%. Major sources of revenues included $226.5 million in taxes and $1.7 million in US customs collections.

The Department of Health provides hospital, medical, and public health services. A schedule of graduated fees has been established, based on ability to pay. There are three general hospitals, one on St. Thomas, one on St. John, and one on St. Croix. Both the neonatal and infant death rates in the Virgin Islands are much higher than the US averages.

As of 1984, monthly public assistance payments averaged $65.20 per recipient. Aid to families with dependent children, old age assistance, and aid to the disabled amounted to $2.4 million in 1984, with one-half coming from the federal government. About 23,000 children participated in the federal school lunch program at a cost to the US government of $3.7 million. In 1983, $66.8 million in federal direct payments to individuals included $30 million in Social Security benefits and $23.4 million for food stamps.

Education is compulsory in the US Virgin Islands, but the dropout rate is high. In 1982/83, a total of 26,136 children were enrolled in public schools and 7,001 in private schools. Overcrowding and lack of supplies are major problems in the schools, and many schools operate on double session. The Division of Vocational-Technical Education is responsible for retraining the unemployed, underemployed, and disadvantaged, as well as providing vocational training to secondary school students. In 1984, 2,842 secondary students in grades 9–12 participated in vocational programs. An additional 3,336 students were enrolled in prevocational programs at the junior high school level. A special adult education program had 223 students in 1984. The College of the Virgin Islands had 2,756 students, with 765 full-time undergraduates.

In 1984, libraries in the Virgin Islands had a total circulation of 61,303 volumes. There are five public libraries and three museums in the Virgin Islands.

UNITED STATES
PACIFIC DEPENDENCIES

AMERICAN SAMOA

American Samoa, an unincorporated insular US territory in the South Pacific Ocean, comprises that portion of the Samoan archipelago lying E of longitude 171°W. (The rest of the Samoan islands make up the independent state of Western Samoa.) American Samoa consists of seven small islands (between 14° and 15°s and 168° and 171°w) with a total area (land and inland water) of 77 sq mi (199 sq km). Five of the islands are volcanic, with rugged peaks rising sharply, and two are coral atolls. The climate is hot and rainy; normal temperatures range from 75°F (24°C) in August to 90°F (32°C) from December through February. The average annual rainfall is 130 in (330 cm); the rainy season lasts from November through March. Hurricanes are common. The native flora includes tree ferns, coconut, hardwoods, and rubber trees. There are few wild animals.

As of the 1980 census, the population was 32,297, an increase of almost 19% over the 1970 census figure of 27,159. The 1985 population was estimated at 35,600. The inhabitants are almost pure Polynesian. Samoan and English are the principal languages. Most Samoans are Christians.

The capital and international port of the territory, Pago Pago, on the island of Tutuila, has one of the finest natural harbors in the South Pacific. American Samoa is a duty-free port. Passenger liners call there on South Pacific tours, and cargo ships arrive regularly from New Zealand, Australia, and the US west coast. There are regular air and sea services between American Samoa and Western Samoa, and regular flights connect Pago Pago with Honolulu and the US mainland. During 1983, Pago Pago International Airport processed 11,616 flights. In 1985, telephone service was available to every village in American Samoa. Radiotelegraph circuits connect the territory with Hawaii, Fiji, and Western Samoa.

American Samoa was settled by Melanesian migrants in the 1st millennium BC. The Samoan islands were visited in 1768 by the French explorer Louis Antoine de Bougainville, who named them the Îles des Navigateurs as a tribute to the skill of their native boatmen. In 1889, the US, the United Kingdom, and Germany agreed to share control of the islands. The United Kingdom later withdrew its claim, and under the 1899 Treaty of Berlin, the US was internationally acknowledged to have rights extending over all the islands of the Samoan group lying east of 171°w, while Germany was acknowledged to have similar rights to the islands west of that meridian. The islands of American Samoa were officially ceded to the US by the various ruling chiefs in 1900 and in 1904, and on 20 February 1929 the US Congress formally accepted sovereignty over the entire group. From 1900 to 1951, the territory was administered by the US Department of the Navy, thereafter by the Department of the Interior. Since 1981, American Samoa has sent a nonvoting delegate to the US House of Representatives.

The executive branch of the government is headed by the governor, who, along with the lieutenant governor, is elected by popular vote for a four-year term. (Before 1977, the two posts were appointed by the US president.) Village, county, and district councils have full authority to regulate local affairs. The legislature (Fono) is composed of the house of representatives and the senate. The 15 counties select, according to Samoan custom, 18 *matais* (chiefs) to four-year terms in the senate, while the 21 house members are elected for two-year terms by popular vote within the counties. The secretary for Samoan affairs, who heads the Department of Local Government, is appointed by the governor. Under his administration are district governors, county chiefs, village mayors, and police officials. The judiciary functions through the high court and five district and village courts; high court justices are appointed by the US secretary of the interior, others by the governor with US Senate consent. Samoans living in the islands as of 17 April 1900 or born there since that date are nationals of the US; they may migrate freely to the US proper, and may become US citizens after fulfilling the requirements of the Immigration and Nationality Act.

The economy is primarily agricultural. The median income for persons 15 years and older in 1979 was $4,219. Farms occupied about one-eighth of the land area in 1980; all but 8% of the land is communally owned, and most farms are only 1 to 7 acres (0.4–2.8 hectares). The principal crops are bananas, breadfruit, taro, and coconuts. Other crops include papayas, pineapples, oranges, mangoes, cucumbers, sugarcane, yams, cacao, avocados, and lemons. Hogs, goats, and chickens are the principal livestock raised; dairy cattle are few. The fish catch was 403 metric tons in 1981.

American Samoa's labor force was 10,752 at the end of 1982; the unemployment rate was 12%. Of the 9,514 employed, 3,705 were government employees. A division of H. J. Heinz—Starkist Tuna Samoa—operates a modern tuna cannery, supplied with fish caught by Japanese and Taiwanese fishing fleets; Van Camp Seafood, a division of Ralston Purina, operates a second tuna cannery. The two canneries provided employment for 3,300 men and women in 1985. The Pago Pago Intercontinental Hotel, opened in 1965 by the American Samoan Development Corp., and an air terminal that can accommodate Boeing 747 jets have contributed to the further development of the tourist trade in the territory.

Owing largely to the cannery operations, American Samoa's balance of international trade has been highly favorable. In 1982, exports were valued at $186,782,000. American Samoa's trade is sensitive to fluctuations in the value of canned tuna shipments, since they make up more than 90% of the territory's total exports (97% in 1982). The US took 83% of exports in 1982. The islands are highly dependent on imports, which more than quintupled between 1970 and 1982. That year, total imports were valued at $119,417,000; fuel oil made up 39% of imports, followed by jewelry and food (16% each). The US mainland accounted for 83% of imports.

Local revenues are supplemented by grants-in-aid and direct US appropriations. In 1983/84, federal grants to American Samoa came to $47.4 million. Total expenditures by the US government came to $61 million. US currency is legal tender in the territory. Banking and credit are handled by the government-owned Development Bank of American Samoa.

American Samoans are entitled to free medical treatment, including hospital care. Besides district dispensaries, the government maintains a central hospital (with 157 beds in 1981), a tuberculosis unit, and a leprosarium. In 1981 there were 25 physicians, 7 dentists, and 150 nurses.

Education is a joint undertaking between the territorial government and the villages. School attendance is compulsory for all children from 6 through 18. The villages furnish the elementary school buildings and living quarters for the teachers; the territorial government pays teachers' salaries and provides buildings and supplies for all but primary schools. Since 1964, educational television has served as a basic teaching tool in the school system. In 1984/85 there were 8,197 pupils in elementary schools (grades 1–8) and 3,187 at the secondary level. American Samoa Community College, a two-year school, had a 1984/85 enrollment of 1,673. American Samoa's Office of Library Services had 12 branches and 95,000 books in 1982; the community college had 17,000 volumes. In 1985 there were three radio stations (two AM, one FM) and three television stations, all owned by the government.

The Office of Tourism actively promotes development of the tourist industry. American Samoa attracted 52,087 tourists in 1982. The Jean P. Haydon Museum in Pago Pago has collections on Samoan arts and culture. The two daily newspapers, the *News Bulletin* and *Samoa News,* had circulation of about 4,000 each in the mid-1980s.

GUAM

The largest and most populous of the Mariana Islands in the Western Pacific, Guam (13°28′29″N and 144°44′45″E) has a total area of 209 sq mi (541 km) and is about 30 mi (48 km) long and from 4 to 10 mi (6 to 16 km) wide. The island is of volcanic origin; in the south the terrain is mountainous, while the northern part is a plateau with shallow, fertile soil. Cliffs on the northern end rise 500 feet (152 meters) above sea level. The central part of the island (where the capital, Agaña, is located) consists of undulating country. The rugged southern end has several peaks over 1,000 feet (300 meters), with Mt. Lamlam reaching 1,329 feet (405 meters), the highest point on the island.

Guam lies in the typhoon belt of the Western Pacific. In May 1976, a typhoon caused an estimated $300 million in damage and left 80% of the island's buildings in ruins. In general, Guam has a tropical climate with little seasonal variation. The average temperature is 81°F (27°C); annual rainfall is substantial, ranging from 85 in (216 cm) at Apra Harbor to 110 in (279 cm) in the mountains. Endangered species in 1983 included the giant Micronesian kingfisher, Marianas crow, Mariana mallard, and Guam rail.

The 1980 census showed a population, including US military and civilian personnel and their families, of 105,979, a growth of 25% over the 1970 total. The population was estimated at 108,987 in 1982, of whom 19,489 were uniformed military personnel and their dependents and 2,798 were aliens and other temporary residents; local residents numbered 86,700. The present-day Chamorro, who comprise about 48% of the population, descend from the intermingling of the few surviving original Chamorro with Spanish, Filipino, and Mexican settlers, as well as later arrivals from the US, United Kingdom, Korea, China, and Japan. Filipinos (21%) were the largest ethnic minority in 1980. Chamorro is the primary language of many Guamanians, but English is the official language. Roman Catholicism is the dominant religion.

There were 461 mi (742 km) of public roads in 1983. Licensed motor vehicles totaled 58,207 in 1982, the same year that Guam's first public mass transit network—a bus system—went into full operation. Apra, the only good harbor, ships goods to Japan, Taiwan, Hawaii, and the Trust Territory of the Pacific Islands. In 1983, 937 vessels arrived on Guam. Four international airlines served Guam in 1984, with flights from the US mainland, Hawaii, Taiwan, Japan, Australia, and New Zealand, as well as other Pacific islands.

The earliest known settlers on Guam were the original Chamorro, who migrated from the Malay Peninsula to the Pacific possibly as early as 1500 BC. It is believed that when Ferdinand Magellan landed on Guam in 1521, as many as 100,000 Chamorro

lived on the island. By 1741, their numbers had been reduced to 5,000; most of the population had either fled the island or been killed through disease or war with the Spanish. A Spanish fort was established in 1565, and from 1696 until 1898, Guam was under Spanish rule. Under the Treaty of Paris that ended the Spanish-American War in 1898, the island was ceded to the US and placed under the jurisdiction of the Department of the Navy. During World War II, Guam was occupied by Japanese forces; the US recaptured the island in 1944 after 54 days of fighting. In 1950, the island's administration was transferred from the Navy to the US Department of the Interior. Under the 1950 Organic Act of Guam, passed by the US Congress, the island was established as an unincorporated territory of the US; Guamanians were granted US citizenship, and internal self-government was introduced.

The executive branch of government is headed by the governor, who with the lieutenant governor serves a four-year term. In 1970, the office was filled for the first time through direct election rather than presidential appointment. Ricardo J. Bordallo, a Democrat, was elected governor in 1982.

A unicameral legislature of 21 senators (14 Democrats and 7 Republicans in 1983) is empowered to legislate on all local matters, including taxation and appropriations. There are no local governments. The US Congress reserves the right to annul any law passed by the Guam legislature, but must do so within a year of the date it receives the text of any such law. A representative from Guam to the US House of Representatives has no vote on the floor, although he can vote in committee. A territorial constitution was drafted in 1977 and submitted to Guamanian voters in 1979. They rejected the document after critics contended that its provisions for self-government were inadequate. In 1982, the electorate approved a referendum urging the US Congress to make Guam a US commonwealth, like Puerto Rico.

Judicial authority is vested in the district court of Guam, and appeals may be taken to the regular US courts of appeal and ultimately to the US Supreme Court. A superior court, a police court, and a juvenile court have jurisdiction over certain cases arising under Guamanian law. The judge of the district court is appointed by the US president; the judges of the other courts are appointed by the governor and confirmed by the legislature; retention is by popular vote. Guam's laws were codified in 1953.

Guam is one of the most important US military bases in the Pacific, and the island's economy has been profoundly affected by the large sums of money spent by the US defense establishment. During the late 1960s and early 1970s, when the US was a major combatant in the Viet-Nam conflict, Guam served as a base for long-range US bombers on sorties over Indochina. In 1975, Guam was a way station for more than 100,000 Indochinese refugees. Military personnel on active duty on 30 September 1984 totaled 9,175, of whom 4,715 were in the Navy and 4,046 were in the Air Force. Department of Defense civilian employees numbered 7,908 in 1983/84. Military expenditures on Guam reached $677,300,000 in 1983/84, including $278,201,000 in military pay, $191,789,000 in civilian pay, and $193,664,000 in military construction. Contending that military land ownership had impeded the island's economic development, some of Guam's landholders have challenged the federal government's right to condemn their property for military use; in the early 1980s, federal holdings came to about one-third of the island.

Before World War II, agriculture and animal husbandry were the primary economic activities. By 1947, however, most adults were wage earners employed by the US armed forces, although many continued to cultivate small plots to supplement their earnings. Median income of persons 15 years and older was $8,392 in 1979. In 1982, private employment made up half of the total civilian employment; the federal government and territorial government employed the other half. Of 34,960 persons in the civilian labor force in July 1984, 31,740 were employed.

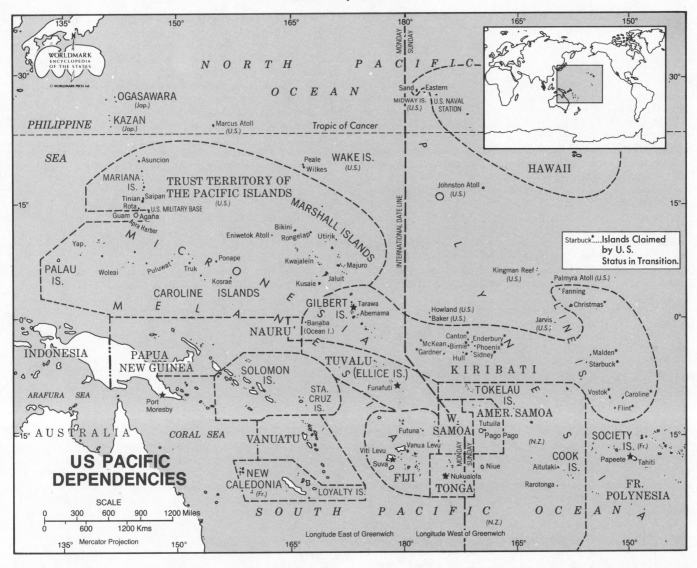

US PACIFIC DEPENDENCIES

SCALE

0   300   600   900   1200 Miles

0   600   1200 Kms

Mercator Projection

Starbuck........**Islands Claimed by U. S. Status in Transition.**

Melons, bananas, coconuts, avocados, cucumbers, and pineapples grow in the fertile valleys, but agriculture has not returned to pre–World War II levels, partly because a considerable amount of arable land is taken up by military installations. In 1981/82, the agricultural sector produced 2,416 tons of fruits and vegetables worth $3,583,096; 1,134,000 dozen eggs valued at $1,622,186; and 532 tons of pork worth $957,041. The lack of available land in the port area is the major obstacle to development of a fisheries industry, but the estimated offshore catch of 319,300 lb in 1981/82 was almost double that of the previous year. Electric power is derived from fossil fuels; the Guam Power Authority, an autonomous agency, controls power generation and distribution. Power consumption was 436 million kwh in 1982.

Guam has a vigorous and growing business community, as well as a rapidly growing tourist industry. Gross business income reached $1.1 billion in 1982. Manufacturing accounted for about one-third of all business revenues in 1981; 90% of that share came from petroleum refinery operations. Retail trade made up 40% of annual payroll in 1982 and employed 5,400 people in 1982. Services accounted for 27% and employed 3,469.

Tourism expanded rapidly in the early 1970s and, after a temporary decline later in the decade, picked up again in the early 1980s. In 1983, 350,540 visitors spent about $196 million on the island. In 1984, visitors numbered 368,655, of whom 82% were from Japan.

Banking is important to the island economy. A 1979 act established international banking facilities to make loans to and accept deposits from entities outside Guam. Total deposits came to $705 million at the end of 1984. The five banks active in offshore lending operations had over $2.2 billion in assets at the end of 1982.

Guam shows a large trade deficit. In 1983, exports came to $39,244,728 (compared with $5,832,316 in 1970), while imports cost $610,743,985 ($96,402,314 in 1970). The bulk of Guam's trade is with the US and Japan. Refined petroleum and petroleum products are the major exports; petroleum, food, and manufactured goods are the largest imports.

General revenues were $153.5 million in 1981/82; current expenditures were $158.4 million. US income tax laws are applicable in Guam; all internal revenue taxes derived by the US from Guam are paid into the territorial treasury. US customs duties, however, are not levied; Guam is a duty-free port. In its trade with the US mainland, Guam is required to use US shipping. Total federal expenditures in Guam were $768 million in 1983/84, of which grants to the territorial government came to $89 million.

By local custom, the aged, the indigent, and orphans are cared for by their families. However, the federal government has begun to play a larger role in public welfare. In 1983, $5,312,695 in public assistance benefits was paid to Guamanians, excluding

Medicaid ($2,276,030 in 1982). Of the former amount, 80% went for aid to families with dependent children. Additionally, a monthly average of 24,779 persons received food stamps worth $18,467,935.

Typical tropical diseases are practically unknown today in Guam. Tuberculosis, long the principal killer, was brought under control by the mid-1950s. Hospital facilities include the Guam Memorial Hospital, which has a special tuberculosis wing, and a medical center, opened in 1979. The US Naval Hospital serves military personnel. In all, there were four hospitals in 1979 with 223 beds. Village dispensaries function both as public health units and as first-aid stations. There were 95 civilian physicians in 1984, and 33 dentists and 419 nurses in 1980.

School attendance is compulsory from the age of 5 through 16. In 1981/82, enrollment in public schools totaled 26,314; in private schools, 5,000. The University of Guam in Agaña, which became a four-year college in 1961, had a 1981/82 enrollment of 2,395, of whom 1,588 were full-time students; in 1983/84, total enrollment was 2,774. The Guam Community College, established in 1978, includes participants in high school and adult education programs. There are also two business colleges, one of which is accredited.

Public libraries had 187,864 volumes and circulation of 178,483 in 1981/82. The *Pacific Daily News* had a circulation of 18,050 in the early 1980s. In 1985 there were six operating radio stations (three AM and three FM) and two television stations. The War in the Pacific National Historic Park, at Asan, includes a museum.

HOWLAND, BAKER, AND JARVIS ISLANDS

Howland Island (0°48′N and 176°38′W), Baker Island (0°14′N and 176°28′W), and Jarvis Island (0°23′S and 160°2′W) are three small coral islands, each about 1 sq mi (2.6 sq km), of the Line Islands group located in the Central Pacific Ocean. All three are administered directly from Washington as US unincorporated territories. Howland was discovered in 1842 by US sailors, claimed by the US in 1857, and formally proclaimed a US territory in 1935/36. It was worked for guano by US and British companies until about 1890.

Baker, 40 mi (64 km) s of Howland, and Jarvis, 1,100 mi (1,770 km) E of Howland, also were claimed by the US in 1857, and their guano deposits were similarly worked by US and British enterprises. Britain annexed Jarvis in 1889. In 1935, the US sent colonists from Hawaii to all three islands, which were placed under the US Department of the Interior in 1936. Baker was captured by the Japanese in 1942 and recaptured by the US in 1944. The three islands have no permanent residents but are visited annually by the US Coast Guard.

JOHNSTON ATOLL

Johnston Atoll, located in the North Pacific 715 mi (1,151 km) sw of Honolulu, consists of two islands, Johnston (16°45′N and 169°32′W) and Sand (16°45′N and 169°30′W), with a total land and water area of less than 0.5 sq mi (1.3 sq km). It was discovered by English sailors in 1807 and claimed by the US in 1858. For many years, it was a bird reservation. Commissioned as a naval station in 1941, it remains an unincorporated US territory under the control of the US Department of the Air Force. In recent years, it has been used primarily for the testing of nuclear weapons. The 1980 census reported a population of 327, down from 1,007 in 1970. A total of 140 military personnel were on the atoll on 30 September 1984.

MIDWAY

Midway (28°12′–17′N and 177°19′–26′W) consists of an atoll and two small islets, Eastern Island and Sand Island, about 1,300 mi (2,100 km) WNW of Honolulu. Their total land and water area is 2 sq mi (5 sq km). Their population was 468 as of the 1980 census, down from 2,200 in 1970 because of the end of US military involvement in Indochina.

Discovered and claimed by the US in 1859 and formally annexed in 1867, Midway became a submarine cable station early in the 20th century and an airlines station in 1935. Made a US naval base in 1941, Midway was attacked by the Japanese in December 1941 and January 1942. In one of the decisive battles of World War II, a Japanese naval attack on 3–6 June 1942 was repelled by US airplanes. There is a naval station at Midway, and the island is a US unincorporated territory under the administrative control of the US Department of the Navy.

PALMYRA ATOLL

Palmyra, an atoll in the Central Pacific Ocean containing some 50 islets (with a total area of about 4 sq mi—10 sq km), is situated about 1,000 mi (1,600 km) ssw of Honolulu at 5°52′N and 162°5′W. It was discovered in 1802 by the USS *Palmyra*, was formally annexed by the US in 1912, and was under the jurisdiction of the city of Honolulu until 1959, when Hawaii became the 50th state of the US. It is now the responsibility of the US Department of the Interior. The atoll is privately owned by the Fullard-Leo family of Hawaii.

Kingman Reef, NW of Palmyra Atoll at 6°25′N and 162°23′W, and less than 0.5 sq mi (1.3 sq km) in area, was discovered by the US in 1874, was annexed in 1922, and became a naval reservation in 1934. Now abandoned, it remains under the control of the US Department of the Navy.

TRUST TERRITORY OF THE PACIFIC ISLANDS

The US-administered Trust Territory of the Pacific Islands consists of 2,141 islands and atolls with a total land area of 713 sq mi (1,847 sq km) scattered over some 3,000,000 sq mi (7,800,000 sq km) of the Western Pacific Ocean, an expanse almost equal to the area of the continental US. The islands form part of Micronesia. Only 96 of the islands were inhabited in 1983/84. The territory extends about 2,700 mi (4,350 km) from 130° to 172°E, and 1,500 mi (2,400 km) from 1° to 21°N; its approximate geographical center is the island of Truk, lying about 5,000 mi (8,000 km) sw of San Francisco and 2,000 mi (3,200 km) E of the Philippines. Three groups of islands are included in the territory: The Caroline Islands, to the s and w, include the Palau Islands, 191 sq mi (495 sq km), with its capital, Koror, at 7°19′N and 134°30′E; Truk, 49 sq mi (127 sq km), at 7°28′N and 151°51′E; Pohnpei (formerly Ponape), 133 sq mi (344 sq km), at 6°58′N and 158°31′E; and Kosrae, 42 sq mi (109 sq km), at 5°19′N and 162°59′E. The Marshall Islands, of which Kwajalein (8°43′ to 9°24′N and 166°50′ to 167°44′E) is the largest atoll, lie to the E. The Northern Mariana Islands (all the Marianas except Guam, a separate political entity) include Rota, 33 sq mi (85 sq km), at 14°10′N and 145°15′E; Saipan, 47 sq mi (122 sq km), at 15°5′–17′N and 145°41′–50′E; and Tinian, 39 sq mi (101 sq km), at 15°1′N and 145°38′E.

The Marianas are a volcanic archipelago. Other volcanic islands are found in the western Carolines, and there are volcanic outcroppings on Truk, Pohnpei, and Kosrae in the eastern Carolines. Other islands, mostly atolls, are of coral formation. The climate is tropical, with relatively little seasonal change; the temperature averages 75–85°F (24–29°C), and relative humidity is generally high, averaging about 78% throughout the islands. Average rainfall varies from 85 in (216 cm) per year in the Northern Marianas to 182 in (462 cm) in the eastern Carolines. The Marshalls average 110 in (279 cm) of rainfall per year, except in the north, where average rainfall is only about 20 in (51 cm). In most of the territory, typhoons threaten from July through November, but the eastern islands are relatively free of these disturbances. Serious damage to the Northern Marianas was caused by Typhoon Dinah in 1980.

The islands generally are covered with moderately heavy tropical vegetation. Trees, including excellent hardwoods, grow on the slopes of the higher volcanic islands. Coconut palms flourish on the coral atolls. Insects are numerous (over 7,000 species), and ocean birds—including the tern, albatross, frigate, and heron—are common. The only native land mammals are four species of bats; water buffalo, deer, goats, and rats have been introduced by man. Ocean fauna, which is abundant, includes tuna, barracuda, sharks, sea bass, eels, flying fish, octopus, many kinds of crustaceans, and porpoises.

As of 1984, endangered species on Palau Island in the Carolines were Palau varieties of La Pérouse's megapode, ground dove, owl, and fantail flycatcher; and on Pohnpei, Pohnpei varieties of mountain starling and great white-eye. In the Marianas, endangered species included La Pérouse's megapode, Marianas mallard, reed warbler, and Tinian monarch flycatcher. The dugong, or sea cow, was endangered throughout the trust territory.

Total population in 1980 was 132,929; the 1984 estimated population was 155,933. The populations of the political units composing the trust territory in 1980 were Federated States of Micronesia, 73,160 (including Truk, 37,488, and Pohnpei, 22,081); Marshall Islands, 30,873; Palau, 12,116; and Northern Marianas, 16,780. Saipan (population 14,549 in 1980) is the administrative capital of the territory.

Almost all the local island people are classified broadly as Micronesians (literally, "peoples of the tiny islands"), and physically resemble the Malaysians. However, there are about 1,500 Polynesian inhabitants of Kapingamarangi and Nukuoro atolls, located in the extreme south of the Federated States of Micronesia (Pohnpei State). No native Micronesian culture encompasses the entire territory. Nine major Micronesian languages, each with dialect variations, are spoken. Common cultural features are close kinship ties, a cult of ancestors, complex class distinctions, and local chieftainship. The Christian religion has been widely accepted, but earlier beliefs persist in certain forms. Japanese is widely spoken, but English is the official language.

Continental Air Micronesia provides air service in the territory. International airports are on Saipan, Pohnpei, and Majuro. A new airport was built on Truk in 1984. Services between the islands of each district are supplied by government-owned vessels operated by Micronesian companies. Most passenger movement among the territory's islands is by sea. Some of the larger centers have local telephone service; communication between most points is by radio. The number of telephones in use was 5,773 in 1983/84. There were 362 mi (583 km) of primary roads. The number of registered motor vehicles (excluding the Federated States of Micronesia) was 8,610.

It is believed that Yap, Palau, and the Marianas were the islands first settled in Micronesia, probably by migrants from the Philippines and Indonesia. Excavations on Saipan have yielded evidence of settlement around 1500 BC. The Marshalls and the eastern and central Carolines were settled later by Melanesian migrants. The first European to reach the Marianas, in 1521, was Ferdinand Magellan; but as a whole, the Micronesian islands were almost entirely unknown until the 19th century. By the late 19th century, Spain had extended its administrative control to include all three major island groups. Germany established a protectorate in the Marshall Islands in 1885, and, following Spain's defeat by the US in the Spanish-American War (1898), the Carolines and Marianas (with the exception of Guam, which was ceded to the US) were sold to Germany. With the outbreak of World War I, Japan took over the German-held islands, and on 17 December 1920 they were entrusted to Japan under a League of Nations mandate. Upon its withdrawal from the League in 1935, Japan began to fortify the islands, and in World War II they served as important military bases. Several of the islands were the scene of heavy fighting during the war. In the battle for control of Saipan in June 1944, some 23,000 Japanese and 3,500 US troops lost their lives in one day's fighting. As each island was occupied by US troops, it became subject to US authority in accordance

with the international law of belligerent occupation. On 18 July 1947, the islands formally became a UN trust territory under US administration, in accordance with a special strategic areas trusteeship agreement. The territory was administered by the US Department of the Navy until 1 July 1951, when administration was transferred to the Department of the Interior. In 1953, the Northern Marianas, with the exception of Rota, were transferred to the Department of the Navy's administrative control; the Department of the Interior resumed jurisdiction over these islands in 1962.

The atolls of Bikini and Enewetak (formerly Eniwetok) have become world famous since 1946 as the sites of US nuclear and thermonuclear tests. The 167 inhabitants of Bikini and the 137 inhabitants of Enewetak were resettled on other islands, and the people of two other islands, Utirik and Rongelap, had to leave their homes temporarily in 1954 because of unforeseen radioactive fallout. The US government committed itself to clearing the nuclear debris on the islands, and in 1980 some former residents of Enewetak began to return home. However, a 1978 program to resettle Bikini was canceled when radiation tests showed that the island was still unsafe. The cost of a complete cleanup was estimated in 1983 at $100 million.

The trusteeship agreement under which the US has controlled the territory was drawing to a close in the mid-1980s. Accordingly, territorial government was changing rapidly as the island groups began to decide their political futures.

The 14 Mariana Islands were separated from the Caroline and Marshall groups on 24 March 1976 by the Northern Marianas Covenant, which also provided for the Marianas' transition to US commonwealth status similar to that of Puerto Rico. The covenant followed a plebiscite, held in June 1975, in which 78.8% of Marianas voters opted for US citizenship and constitutional integration with the US. The remaining island groups adopted self-government in "free association" with the US, which meant that they would continue to receive US military protection and economic and technical assistance. Constitutional governments were installed in the Federated States of Micronesia (the Truk, Pohnpei, Kosrae, and Yap districts) and the Marshall Islands in May 1979, and in the Republic of Palau in January 1981.

Each of the three island groups signed a compact of free association in 1982, and each compact was approved by voters in separate plebiscites held during 1983. However, complications arose when the residents of Palau, while approving their free-association compact on 10 February 1983, did not give the required 75% majority approval to the waiver of a section of their constitution that banned hazardous (including nuclear) materials from their territory; under a US-Palauan agreement, the compact could not take effect without the nuclear waiver, permitting US vessels powered by nuclear fuel or armed with nuclear weapons to pass through that part of the Pacific. The dispute was referred to the courts for solution, delaying the termination of the trust territory. The Palau plebiscite was nullified by the republic's supreme court, and a second plebiscite, on 4 September 1984, saw the compact approved by 67% of the electorate. In December 1985, legislation approving the compacts with the Federated States of Micronesia and with the Republic of the Marshall Islands was approved by the US Congress and signed by President Reagan. It was expected that the compact with the Republic of Palau would be approved at a later date.

As of 1985, authority over the trust territory (excluding the Northern Marianas) was vested in a high commissioner, appointed by the US president and under the immediate authority of the US secretary of the interior. Formerly, the high commissioner appointed a deputy commissioner for each district. However, by 1980 the districts of the Northern Marianas, Truk, Kosrae, Yap, and Pohnpei had elected their own governors, and the Federated States of Micronesia, the Republic of the Marshall

Islands, and the Republic of Palau each had an elected president. All these island groups have their own legislatures. There are more than 100 municipalities headed by magistrates or mayors who are elected through universal adult suffrage. A municipality may consist of a group of villages on the larger islands; in the case of smaller islands, the municipality may consist of an entire island or group of islands.

The territorial judiciary in 1984 consisted of the High Court of the Trust Territory and a District Court for the State of Kosrae, with judges appointed by the US secretary of the interior. These courts were to be phased out, and other trust territory courts had already been dissolved, in favor of courts established by the constitutions of the Federated States of Micronesia, the Republic of the Marshall Islands, and the Republic of Palau. In the Northern Marianas, the judicial branch is embodied in the Commonwealth Court.

All persons born in the trust territory are citizens of the territory; they are not US citizens and, if they desire US citizenship, must acquire it in the same way as other immigrants. (Residents of the Northern Marianas were granted US citizenship on 9 January 1978, when the district's new constitution took effect, but this special status was revoked in March 1980, pending official termination of the trusteeship agreement.) Except for aliens whose permanent residence was in the territory before the present administration, only indigenous inhabitants may own land. Until 1962, persons not citizens or residents were required to obtain the specific authorization of the high commissioner to enter the territory. In that year, however, in response to recommendations by the UN Trusteeship Council to accelerate preparations for eventual self-government and independence, US President John Kennedy opened the territory to US citizens, shipping, and investment without prior security clearance.

The economy of the Trust Territory of the Pacific Islands is less well developed than that of other US territories. In 1983/84, the territory governments (excluding the Northern Marianas) received $161.5 million from the US government; total federal expenditures came to $182 million. Exports came to $5.5 million and imports to $75.7 million. The Northern Marianas governments received $43.6 million in federal grants in 1983/84; total federal expenditures came to $103 million.

The trust territory labor force numbered 22,343 in 1983/84. Median income for persons 15 years and older was only $1,389 in 1979 (excluding the Northern Marianas). In July 1984, the Northern Marianas labor force was 5,388, and the unemployment rate was 11.4%. The 1980 labor force in the Federated States of Micronesia was 25,080, and the unemployment rate was 22%.

The traditional economic activities in the territory are subsistence agriculture, livestock raising, and fishing. Numerous tropical and semitropical food plants are cultivated, including taro, arrowroot, yams, tapioca, bananas, and coconuts. Hogs and chickens are widely raised; water buffalo, cattle, goats, and ducks are also commercially significant.

Seafood is an important part of the diet, and development of the islands' rich maritime resources began in the late 1970s. In 1983/84, 1,178 tons of fish were caught in the trust territory, excluding the Republic of Palau. For the whole territory, the catch was about 5,500 metric tons in 1981.

Tourism has become important, particularly in the Northern Marianas, which had 131,827 entries for 1983/84, 79% from Japan. For the other trust entities, the total was about 18,620.

Small processing and service industries have been developed, among them soap factories, a starch-making factory, hollow-tile factories, an oil press, sawmills, and boatbuilding establishments. The principal commercial products are copra and trochus shell (used to make pearl buttons). Trochus shell has been the second-largest cash export, but the shell beds have been dangerously depleted in recent years.

The principal commercial products are coconut oil and tuna. In 1978, the value of exported coconut products was $1.1 million. Two copra-crushing mills operate in the territory, one in Palau and one in the Marshall Islands; they have a combined capacity of 70,000 tons of coconut oil a year.

Commerce in each major island group is conducted by one to three relatively large enterprises operating as general importers, wholesalers, and retailers, and by many small retailers who buy their stock from the large importing enterprises. Imports, totaling about $175.7 million in 1983/84, vastly exceed exports, which were about $5.5 million during the same year. Trade is mainly with Japan and the US.

US currency is the official medium of exchange. Banking services are provided locally and by institutions in Guam, Hawaii, and the continental US. The Trust Territory Economic Development Loan Fund has made loans to trading companies from a revolving fund appropriated by the US Congress.

The major communicable disease problems on the islands are influenza and digestive disorders. Heart disease and cancer are on the increase and were the leading causes of death in 1983. Tuberculosis, once the most serious health problem in the territory, has now been controlled. As of 1981 there were seven district hospitals and two smaller field hospitals in the territory, with 629 beds. The breakdown of health personnel in the Trust Territory of the Pacific Islands as of 1981 was as follows: doctors, 55; dentists, 20; nurses, 426; and pharmacists, 2.

The educational goal of the islands is to provide universal free public education through primary and secondary levels, with advanced training in the trades and professions for those who can profit by further study. Adult and remedial education is also stressed. Education is free and compulsory for children from 8 through 14 years of age, and more than 90% of the children in this age group attend school. Public and private school enrollment for the 1983/84 school year for the trust territory (including the Northern Marianas) was: elementary, 38,559; secondary, 8,750; and postsecondary (excluding the Marshall Islands), 2,126. The College of Micronesia, established in 1977, had 930 students on three island campuses in 1983/84. Northern Marianas College, on Saipan, had 880 students in 1983/84. The Federated States of Micronesia relies heavily on federal grants to enroll students in colleges and universities abroad; about 1,200 students were enrolled in Guam, Hawaii, and the US mainland in 1983/84, while the Republic of the Marshall Islands subsidized about 90 students abroad.

The Saipan Museum has Marianas artifacts and shells. The Federated States of Micronesia maintained four government-owned radio stations in 1985, while the Northern Marianas had two commercial AM stations and a television station.

WAKE ISLAND

Wake Island, actually a coral atoll and three islets (Wake, Peale, and Wilkes) about 5 mi (8 km) long by 2 mi (3 km) wide, lies in the North Pacific 2,100 mi (3,380 km) W of Honolulu at 19°18'N and 166°35'E. The total land and water area is about 3 sq mi (8 sq km). Discovered by the British in 1796, it was long uninhabited. In 1898, a US expeditionary force en route to Manila landed on the island. The US formally claimed Wake in 1899. It was made a US naval reservation in 1934, and became a civil aviation station in 1935. Captured by the Japanese on 23 December 1941, it was subsequently the target of several US air raids. It was surrendered by the Japanese in September 1945 and has thereafter remained a US unincorporated territory under the jurisdiction, since 1972, of the Department of the Air Force.

As of the 1970 census, 1,647 persons lived on the island; but by 1980, the resident population had declined to 302, virtually all of them employees and dependents connected with military installations. Wake is a stopover and fueling station for civilian and military aircraft flying between Honolulu, Guam, and Japan.

UNITED STATES OF AMERICA

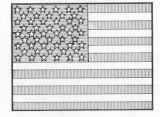

CAPITAL: Washington, D.C. **ANTHEM:** "The Star-Spangled Banner." **MOTTO:** In God We Trust. **FLAG:** The flag consists of 13 alternate stripes, 7 red and 6 white; these represent the 13 original colonies. Fifty five-pointed white stars, representing the present number of states in the Union, are placed in 9 horizontal rows alternately of 6 and 5 against a blue field in the upper left corner of the flag. **OFFICIAL SEAL:** Obverse: An American eagle with outstretched wings bears a shield consisting of 13 alternating white and red stripes with a broad blue band across the top. The right talon clutches an olive branch, representing peace; in the left are 13 arrows, symbolizing military strength. The eagle's beak holds a banner with the motto "*E pluribus unum*" (From many, one); overhead is a constellation of 13 five-pointed stars in a glory. Reverse: Above a truncated pyramid is an all-seeing eye within a triangle; at the bottom of this triangle appear the roman numerals MDCCLXXVI (1776). The pyramid stands on a grassy ground, against a backdrop of mountains. The words "*Annuit Cœptis*" (He has favored our undertakings) and, on a banner, "*Novus Ordo Seclorum*" (A new order of the ages) surround the whole. **MONETARY UNIT:** The dollar ($) of 100 cents is a nonconvertible paper currency with a floating rate. There are coins of 1, 5, 10, 25, and 50 cents and 1 dollar, and notes of 1, 2, 5, 10, 20, 50, and 100 dollars (issuance of higher notes ceased in 1969). **FEDERAL HOLIDAYS:** New Year's Day, 1 January; Birthday of Martin Luther King, Jr., 3d Monday in January; Washington's Birthday, 3d Monday in February; Memorial Day, last Monday in May; Independence Day, 4 July; Labor Day, 1st Monday in September; Columbus Day, 2d Monday in October; Veterans Day, 11 November; Thanksgiving Day, 4th Thursday in November; Christmas Day, 25 December. **TIME:** Noon GMT = 7 AM EST, 6 AM CST, 5 AM MST, 4 AM PST, 3 AM Alaska Standard Time, 2 AM Hawaii-Aleutian Standard Time.

¹LOCATION, SIZE, AND EXTENT

Located in the Western Hemisphere on the continent of North America, the US is the 4th-largest country in the world. The total area, including the 50 states and the District of Columbia, is 3,618,770 sq mi (9,372,607 sq km), of which land comprises 3,539,289 sq mi (9,166,751 sq km); the remaining 79,481 sq mi (205,856 sq km) are classified as inland water. The conterminous US extends 2,897 mi (4,662 km) ENE-WSW and 2,848 mi (4,583 km) SSE-NNW. It is bordered on the N by Canada, on the E by the Atlantic Ocean, on the S by the Gulf of Mexico and Mexico, and on the W by the Pacific Ocean, with a total boundary length of 10,913 mi (17,563 km). Alaska, the 49th state, extends 2,261 mi (3,639 km) E-W and 1,420 mi (2,285 km) N-S. It is bounded on the N by the Arctic Ocean and Beaufort Sea; on the E by Canada; on the S by the Gulf of Alaska, Pacific Ocean, and Bering Sea; and on the W by the Bering Sea, Bering Strait, Chukchi Sea, and Arctic Ocean, with a total boundary length of 8,187 mi (13,176 km). The 50th state, Hawaii, consists of islands in the Pacific Ocean, extending 1,576 mi (2,536 km) N-S and 1,425 mi (2,293 km) E-W, with a general coastline of 750 mi (1,207 km).

The geographic center of the US, including Alaska and Hawaii, is in Butte County, S.D., at 44°58′N, 103°46′W; the geographic center of the 48 conterminous states is in Smith County, Kan., at 39°50′N, 98°35′W.

Land boundaries between states have long since been determined; river boundaries, however, are subject to occasional variation.

²TOPOGRAPHY

The great topographical divisions of the US are largely determined by the various mountain ranges that traverse the country from north to south.

Although the northern New England coast is rocky, along the rest of the eastern seaboard the Atlantic Coastal Plain rises gradually from the shorelines. Narrow in the north, the plain widens to about 200 mi (320 km) in the south, and in Georgia merges with the Gulf Coastal Plain that borders the Gulf of Mexico and ultimately extends as far as Yucatán. West of the Atlantic Coastal Plain is the Piedmont Plateau, bounded by the Appalachian Mountains. The Appalachians, which extend from Maine southwest into central Alabama—with special names in some areas—are old mountains, largely eroded away, with rounded contours, and forested as a rule to the top. Few of their summits rise much above 3,500 feet (1,100 meters), although the highest, Mt. Mitchell in North Carolina, reaches 6,684 feet (2,037 meters).

Between the Appalachians and the Rocky Mountains, more than 1,000 mi (1,600 km) to the west, lies the vast interior plain of the U.S. Running south through the center of this plain and draining almost two-thirds of the area of the continental US is the Mississippi River. Waters starting from the source of the Missouri, the longest of its tributaries, travel almost 4,000 mi (6,400 km) to the Gulf of Mexico. The eastern reaches of the great interior plain are bounded on the north by the Great Lakes, which are thought to contain about half of the world's total supply of fresh water. Under US jurisdiction are 22,178 sq mi (57,441 sq km) of Lake Michigan, 21,118 sq mi (54,696 sq km) of Lake Superior, 8,975 sq mi (23,245 sq km) of Lake Huron, 5,002 sq mi (12,955 sq km) of Lake Erie, and 3,033 sq mi (7,855 sq km) of Lake Ontario. The five lakes are now accessible to oceangoing vessels from the Atlantic via the St. Lawrence Seaway. The basins of the Great Lakes were formed by the glacial ice cap that moved

LOCATIONS: Conterminous US: 66°57′ to 124°44′W; 24°33′ to 49°23′N. Alaska: 130°W to 172°28′E; 51° to 71°23′N. Hawaii: 154°48′ to 178°22′W; 18°55′ to 28°25′N. **BOUNDARY LENGTHS:** Conterminous US: Canada, 3,987 mi (6,416 km); Atlantic Ocean, 2,069 mi (3,330 km); Gulf of Mexico coastline, 1,631 mi (2,625 km); Mexico, 1,933 mi (3,111 km); Pacific coastline, 1,293 mi (2,081 km). Alaska: Arctic Ocean coastline, 1,060 mi (1,706 km); Canada, 1,538 mi (2,475 km); Pacific coastline, including the Bering Sea and Strait and Chukchi coastlines, 5,580 mi (8,980 km). Hawaii: coastline, 750 mi (1,207 km).

down over large parts of North America some 25,000 years ago. The glaciers also determined the direction of flow of the Missouri River and, it is believed, were responsible for carrying soil from what is now Canada down into the central agricultural basin of the U.S. The great interior plain consists of two major subregions: the fertile Central Plains, extending from the Appalachian Highlands to a line drawn approximately 300 mi (480 km) west of the Mississippi, broken only by the Ozark Plateau; and the more arid Great Plains, extending from that line to the foothills of the Rocky Mountains. Although they appear flat, the Great Plains rise gradually from about 1,500 feet (460 meters) to more than 5,000 feet (1,500 meters) at their western extremity.

The Continental Divide, the Atlantic-Pacific watershed, runs along the crest of the Rocky Mountains. The Rockies and the ranges to the west are parts of the great system of young, rugged mountains shaped like a gigantic spinal column along western North, Central, and South America from Alaska to Tierra del Fuego. In the continental US, the series of western ranges, most of them paralleling the Pacific Coast, are the Sierra Nevada, the Coast Ranges, the Cascade Range, and the Tehachapi and San Bernardino mountains. Between the Rockies and the Sierra Nevada–Cascade mountain barrier to the west lies the Great Basin, a group of vast arid plateaus containing most of the desert areas of the US, in the south eroded by deep canyons. The coastal plains along the Pacific are narrow, and in many places, the mountains plunge directly into the sea. The most extensive lowland near the west coast is the Great Valley of California, lying between the Sierra Nevada and the Coast Ranges. There are 71 peaks in these western ranges of the continental US that rise to an altitude of 14,000 feet (4,300 meters) or more, Mt. Whitney in California, at 14,494 feet (4,418 meters), being the highest. The greatest rivers of the Far West are the Colorado in the south, flowing into the Gulf of California, and the Columbia in the northwest, flowing to the Pacific. Each is more than 1,200 mi (1,900 km) long; both have been intensively developed to generate electric power, and both are important sources of irrigation.

Separated from the continental US by Canadian territory, the state of Alaska occupies the extreme northwest portion of the North American continent. A series of precipitous mountain ranges separates the heavily indented Pacific coast on the south from Alaska's broad central basin, through which the Yukon River flows from Canada in the east to the Bering Sea in the west. The central basin is bounded on the north by the Brooks Range, which slopes down to the Arctic Ocean. The Alaskan Peninsula and the Aleutian Islands consist of a chain of volcanoes.

The state of Hawaii consists of a group of Pacific islands formed by volcanoes rising sharply from the ocean floor. The highest of these volcanoes, Mauna Loa, 13,675 feet (4,168 meters), is located on the largest of the islands, Hawaii, and is still active.

The lowest point in the US is Death Valley in California, 282 feet (86 meters) below sea level. At 20,320 feet (6,194 meters), Mt. McKinley in Alaska is the highest peak in North America. These topographic extremes suggest the geological instability of the Pacific Coast region. Major earthquakes destroyed San Francisco in 1906 and Anchorage, Alaska, in 1964, and the San Andreas Fault in California still causes frequent earth tremors. Washington State's Mt. St. Helens erupted violently in May 1980, spewing volcanic ash over much of the Northwest; smaller eruptions occurred frequently in subsequent years.

[3]CLIMATE

The eastern continental region is well watered, with annual rainfall generally in excess of 40 in (102 cm). It includes all the Atlantic seaboard and southeastern states and extends west to cover Indiana, southern Illinois, most of Missouri, Arkansas, Louisiana, and easternmost Texas. The eastern seaboard is affected primarily by the masses of air moving from west to east across the continent rather than by air moving in from the Atlantic. Hence its climate is basically continental rather than maritime. The midwestern and Atlantic seaboard states experience hot summers and cold winters; spring and autumn are clearly defined periods of climatic transition. Only Florida, with the Gulf of Mexico lying to its west, experiences moderate differences between summer and winter temperatures. Mean annual temperatures vary considerably between north and south: Boston, 52°F (11°C); New York City, 55°F (13°C); Charlotte, N.C., 60°F (16°C); Miami, 76°F (24°C). The Gulf and southern Atlantic states are often hit by severe tropical storms originating in the Caribbean Sea in late summer and early autumn.

The prairie lands lying to the west constitute subhumid region. Precipitation usually exceeds evaporation by only a small amount; hence the region is more often familiar with drought than with excessive rainfall. Dryness generally increases from east to west. The average midwinter temperature in the extreme north—Minnesota and North Dakota—is about 9°F (−13°C) or lower, while the average July temperature is 65°F (18°C). In the Texas prairie region to the south, January temperatures average 50–55°F (10–13°C) and July temperatures 80–85°F (27–29°C). Rainfall along the western border of the prairie region is as low as 18 in (46 cm) per year in the north and 25 in (64 cm) in the south. Precipitation is greatest in the early summer—a matter of great

Outlying Areas of the US[1]

NAME	AREA SQ MI	AREA SQ KM	CAPITAL	YEAR OF ACQUISITION	POPULATION 1980	POPULATION 1984
Puerto Rico	3,515	9,104	San Juan	1898	3,196,520	3,267,000[4]
Virgin Islands of the US	136	352	Charlotte Amalie	1917	96,569	104,000[4]
Trust Territory of the Pacific Islands, of which:	713	1,847	Saipan	1947	132,929	155,933
Federated States of Micronesia[2]	271	702	Ponape[3]	—	73,160	88,375
Marshall Islands[2]	69	179	Majuro[3]	—	30,873	34,923
Northern Marianas[2]	182	471	Saipan[3]	—	16,780	19,635
Republic of Palau[2]	191	495	Koror[3]	—	12,116	13,000
Other Pacific territories:						
American Samoa	77	199	Pago Pago	1899	32,297	34,000[4]
Guam	209	541	Agaña	1898	105,979	113,230[4]
Midway Islands	2	5	—	1867	468	NA
Wake Island	3	8	—	1899	302	NA

[1] Excludes minor and uninhabited islands.
[2] Although governed under separate constitutional arrangements by the mid–1980s, these territories formally remained part of the Trust Territory of the Pacific Islands pending action by the US Congress, the US president, and the UN Security Council.
[3] Centers of constitutional government. The entire Trust Territory of the Pacific Islands is administered from Saipan.
[4] Estimate for 1983.

importance to agriculture, particularly in the growing of grain crops. In dry years, the prevailing winds may carry the topsoil eastward for hundreds of miles in clouds that obscure the sun.

The Great Plains constitute a semiarid climatic region. Rainfall in the southern plains averages about 20 in (51 cm) per year and in the northern plains about 10 in (25 cm), but extreme year-to-year variations are common. The tropical air masses that move northward across the plains originate on the fairly high plateaus of Mexico and contain little water vapor. Periods as long as 120 days without rain have been experienced in this region. The rains that do occur are often violent, and a third of the total annual rainfall may be recorded in a single day at certain weather stations. The contrast between summer and winter temperatures is extreme throughout the Great Plains. Maximum summer temperatures of over 110°F (43°C) have been recorded in the north as well as in the south. From the Texas panhandle north, blizzards are common in the winter, and tornadoes in other seasons. The normal daily minimum temperature for January in Duluth, Minn., is −3°F (−19°C).

The higher reaches of the Rockies and the mountains paralleling the Pacific coast to the west are characterized by a typical alpine climate. Precipitation as a rule is heavier on the western slopes of the ranges. The great intermontane arid region of the West shows considerable climatic variation between its northern and southern portions. In New Mexico, Arizona, and southeastern California, the greatest precipitation occurs in July, August, and September, mean annual rainfall ranging from 3 in (8 cm) in Yuma to 30 in (76 cm) in the mountains of northern Arizona and New Mexico. Phoenix, Ariz., has a mean annual temperature of 71°F (22°C), rising to 92°F (33°C) in July and falling to 52°F (11°C) in January. North of the Utah-Arizona line, the summer months usually are very dry; maximum precipitation occurs in the winter and early spring. In the desert valleys west of the Great Salt Lake, mean annual precipitation adds up to only 4 in (10 cm). Although the northern plateaus are generally arid, some of the mountainous areas of central Washington and Idaho receive at least 60 in (152 cm) of rain per year. Throughout the intermontane region, the uneven availability of water is the principal factor shaping the habitat.

The Pacific coast, separated by tall mountain barriers from the severe continental climate to the east, is a region of mild winters and moderately warm, dry summers. Its climate is basically maritime, the westerly winds from the Pacific Ocean moderating the extremes of both winter and summer temperatures. Los Angeles in the south has an average temperature of 56°F (13°C) in January and 69°F (21°C) in July; Seattle in the north has an average temperature of 39°F (4°C) in January and 65°F (18°C) in July. Precipitation in general increases along the coast from south to north, extremes ranging from an annual average of less than 2 in (5 cm) at Death Valley, Calif. (the lowest in the US), to more than 140 in (356 cm) in Washington's Olympic Mountains.

Climatic conditions vary considerably in the vastness of Alaska. In the fogbound Aleutians and in the coastal panhandle strip that extends south along the Gulf of Alaska and includes the capital, Juneau, a relatively moderate maritime climate prevails. The interior is characterized by short, hot summers and long, bitterly cold winters; and in the region bordering the Arctic Ocean, a polar climate prevails, the soil hundreds of feet below the surface remaining frozen the year round. Although snowy in winter, continental Alaska is relatively dry.

Hawaii has a remarkably mild and stable climate, with only slight seasonal variations in temperature, as a result of northeast ocean winds. The mean January temperature in Honolulu is 73°F (23°C), the mean July temperature 80°F (27°C). Rainfall is moderate—about 28 in (71 cm) per year—but much greater in the mountains. Mt. Waialeala on Kauai has a mean annual rainfall of 460 in (1,168 cm), highest in the world.

The lowest temperature recorded in the US was −79.8°F (−62°C) at Prospect Creek Camp, Alaska, on 23 January 1971; the highest, 134°F (57°C), at Greenland Ranch, in Death Valley, Calif., on 10 July 1913. The record annual rainfall is 578 in (1,468 cm) at Fuu Kukui, Maui, in 1950; for a 24-hour period, 38.7 in (98.3 cm) at Yankeetown, Fla., on 5–6 September 1950; in 1 hour, 12 in (30 cm), at Holt, Mo., on 22 June 1947, and on Kauai, Hawaii, on 24–25 January 1956.

[4]FLORA AND FAUNA

As of 1982, including Alaska and Hawaii, about 29% of the US was forestland, 26% was grassland pasture, 17% was active cropland, 4% was idle or pastured cropland, and the remaining 24% encompassed military, urban, and designated recreational and wilderness areas, among other lands.

At least 7,000 species and subspecies of indigenous US flora have been categorized. The eastern forests contain a mixture of softwoods and hardwoods that includes pine, oak, maple, spruce, beech, birch, hemlock, walnut, gum, and hickory. The central hardwood forest, which orginally stretched unbroken from Cape Cod to Texas and northwest to Minnesota—still an important timber source—supports oak, hickory, ash, maple, and walnut. Pine, hickory, tupelo, pecan, gum, birch, and sycamore are found in the southern forest that stretches along the Gulf coast into the eastern half of Texas. The Pacific forest is the most spectacular of all because of the enormous size of its giant redwoods and Douglas firs. In the Southwest are saguaro (giant cactus), yucca, candlewood, and the Joshua tree.

The central grasslands lie in the interior of the continent, where moisture is not sufficient to support the growth of large forests. The tall grassland or prairie (now almost entirely under cultivation) lies to the east of the 100th meridian. To the west of this line, where rainfall is frequently less than 20 in (51 cm) per year, is the short grassland. Mesquite grass covers parts of west Texas, southern New Mexico, and Arizona. Short grass may be found in the highlands of the latter two states, while tall grass covers large portions of the coastal regions of Texas and Louisiana and occurs in some parts of Mississippi, Alabama, and Florida. The Pacific grassland region includes northern Idaho, the higher plateaus of eastern Washington and Oregon, and the mountain valleys of California.

The intermontane region of the Western Cordillera is for the most part covered with desert shrubs. Sagebrush predominates in the northern part of this area, creosote in the southern, and saltbush near the Great Salt Lake and in Death Valley.

The lower slopes of the mountains running up to the coastline of Alaska are covered with coniferous forests as far north as the Seward Peninsula. The central part of the Yukon Basin is also a region of softwood forests. The rest of Alaska is heath or tundra. Hawaii has extensive forests of bamboo and ferns. Sugarcane and pineapple, although not native to the islands, now cover a large portion of the cultivated land.

Small trees and shrubs common to most of the US include hackberry, hawthorn, serviceberry, blackberry, wild cherry, dogwood, and snowberry. Wild flowers bloom in all areas, from the seldom-seen blossoms of rare desert cacti to the hardiest alpine species. Wild flowers include forget-me-not, fringed and closed gentians, jack-in-the-pulpit, black-eyed Susan, columbine, and common dandelion, along with numerous varieties of aster, orchid, lady's slipper, and wild rose.

An estimated 1,500 species and subspecies of mammals characterize the animal life of the continental US. Among the larger game animals are the white-tailed deer, moose, pronghorn antelope, bighorn sheep, mountain goat, black bear, and grizzly bear. The Alaskan brown bear often reaches a weight of 1,200–1,400 lb (545–635 kg). Some 25 important furbearers are common, including the muskrat, red and gray foxes, mink, raccoon, beaver, opossum, striped skunk, woodchuck, common cottontail, snowshoe hare, and various squirrels. Human encroachment has transformed the mammalian habitat since the early 19th century. The American buffalo (bison), millions of which once roamed the plains, is now found only on select reserves. Other mammals, such as the elk and gray wolf, have been restricted to much smaller ranges.

Year-round and migratory birds abound. Loons, wild ducks, and wild geese are found in lake country; terns, gulls, sandpipers, herons, and other seabirds live along the coasts. Wrens, thrushes, owls, hummingbirds, sparrows, woodpeckers, swallows, chickadees, vireos, warblers, and finches appear in profusion, along with the robin, common crow, cardinal, Baltimore oriole, eastern and western meadowlarks, and various blackbirds. Wild turkey, ruffed grouse, and ring-necked pheasant (introduced from Europe) are popular game birds.

Lakes, rivers, and streams teem with trout, bass, perch, muskellunge, carp, catfish, and pike; sea bass, cod, snapper, and flounder are abundant along the coasts, as are such shellfish as lobster, shrimp, clams, oysters, and mussels. Garter, pine, and milk snakes are found in most regions. Four poisonous snakes survive, of which the rattlesnake is the most common. Alligators appear in southern waterways, and the Gila monster makes its home in the Southwest.

During the 1960s and 1970s, numerous laws and lists designed to protect threatened and endangered flora and fauna were adopted throughout the US. Generally, each species listed as protected by the federal government is also protected by the states, but some states may list species not included on federal lists or on the lists of neighboring states. (Conversely, a species threatened throughout most of the US may be abundant in one or two states.) As of February 1984, the US Fish and Wildlife Service listed 248 endangered US species, including 61 plants, and 57 threatened species, including 11 plants. The agency listed another 459 endangered and 37 threatened foreign species by international agreement.

Threatened species, likely to become endangered if recent trends continue, include such plants as Rydberg milk-vetch, northern wild monkshood, Lee pincushion cactus, and Lloyd's Mariposa cactus. Among the endangered floral species (in imminent danger of extinction in the wild) are the Virginia round-leaf birch, San Clemente Island broom, Texas wild-rice, Furbish lousewort, Truckee barberry, Sneed pincushion cactus, spineless hedgehog cactus, Knowlton cactus, persistent trillium, dwarf bear-poppy, and small whorled pogonia.

Threatened fauna include the grizzly bear, southern sea otter, Newell's shearwater, American alligator, eastern indigo snake, bayou darter, several southwestern trout species, and Bahama and Schaus swallowtail butterflies. Among endangered fauna are the Indiana bat, key deer, black-footed ferret, northern swift fox, San Joaquin kit fox, jaguar, jaguarundi, Florida manatee, ocelot, Florida panther, Utah prairie dog, Sonoran pronghorn, Delmarva Peninsula fox squirrel, gray wolf (except in Minnesota, where it is threatened), red wolf, numerous whale species, bald eagle (endangered in most states, but only threatened in the Northwest and the Great Lakes region), Hawaii creeper, Everglade kite, brown pelican, California clapper rail, red-cockaded woodpecker, blunt-nosed leopard lizard, American crocodile, desert slender salamander, Houston toad, humpback chub, several species of pupfish, 17 US species of pearly mussel, Socorro isopod, Kentucky cave shrimp, and mission blue butterfly. Several species on the federal list of endangered and threatened wildlife and plants are found only in Hawaii.

5ENVIRONMENTAL PROTECTION

The Council on Environmental Quality, an advisory body contained within the Executive Office of the President, was established by the National Environmental Policy Act of 1969, which mandated an assessment of environmental impact for every federally funded project. The Environmental Protection Agency (EPA), created in 1970, is an independent body with primary regulatory responsibility in the fields of air and noise pollution, water and waste management, and control of toxic substances. Other federal agencies with environmental responsibilities include the Forest Service and Soil Conservation Service within the Department of Agriculture, the Fish and Wildlife Service and National Park Service within the Department of the Interior, the Department of Energy, and the Nuclear Regulatory Commission. In addition to the 1969 legislation, landmark federal laws protecting the environment include the Clean Air Act Amendments of 1970, controlling automobile emissions; the Water Pollution Act of 1972, setting clean water criteria for fishing and swimming; and the Endangered Species Act of 1973, protecting wildlife near extinction. A measure enacted in December 1980 established a $1.6-billion "Superfund," financed largely by excise taxes on chemical companies, to clean up toxic waste dumps such as the one in the Love Canal district of Niagara Falls, N.Y. Provisions of this act were invoked when the federal government in 1983 began to buy up homes and businesses in Times Beach, Mo., which had been contaminated by dioxin, a highly toxic byproduct of hexachlorophene manufacture. Alleged irregularities in the administration of the Superfund led to congressional investigations and replacement of most top EPA officials during the same period.

Among the most influential environmental lobbies are the Sierra Club (founded in 1892; 350,000 members in 1984) and its legal arm, the Sierra Club Legal Defense Fund. Large conservation groups include the National Wildlife Federation (1936; 4,200,000) and the National Audubon Society (1905; 510,000). Smaller but highly active in the environmental protection movement are the Environmental Defense Fund (1967; 50,000), Friends of the Earth (1969; 36,000), Environmental Action (1970; 25,000), and the League of Conservation Voters (1970; 55,000), all of which undertake research and litigation and monitor federal enforcement of environmental standards. Greenpeace USA (1979; 265,000) has gained international attention by seeking to disrupt the hunts for whales and seals. Among the environmental movement's most notable successes have been the inauguration (and mandating in some states) of recycling programs, the banning in the US of the insecticide dichlorodiphenyltrichloroethane (DDT), the successful fight against construction of a supersonic transport (SST), the mandating of various energy conservation measures, and the protection of more than 100 million acres (40 million hectares) of Alaska lands, after a fruitless fight to halt construction of the Trans-Alaska Pipeline. The movement was much less successful in opposing certain large-scale water projects, and in the early 1980s, environmentalists faced pressure from business, labor, and civic interests seeking a relaxation of air quality standards where strict enforcement threatened local industries.

A 1983 assessment by the Council on Environmental Quality found that between 1975 and 1982, quantities of airborne particulate emissions, sulfur dioxide, carbon monoxide, and lead had all declined; the report noted, however, that increased use of wood-burning stoves as an alternative to oil heat posed a significant threat to air quality in some parts of the US. Since 1970, some 7,000 air-quality monitoring stations have been set up, and $25.8 billion was spent on cleaner air in 1982 alone. Under the Reagan administration, however, enforcement of environmental laws lagged, as EPA's enforcement budget was cut almost in half. Total EPA spending declined from $5.6 billion in 1979/80 to $4 billion in 1983/84.

Outstanding problems include "acid rain," precipitation contaminated by fossil fuel wastes; the identification and decontamination of dump sites for toxic wastes; runoffs of agricultural pesticides, a pollutant deadly to fishing streams and difficult to regulate; the continued dumping of raw or partially treated sewage from major cities into US waterways; the falling water tables in many western states; the decrease in arable land because of depletion, erosion, and urbanization; the need for reclamation of strip-mined lands; and the expansion of the US nuclear industry in the absence of any fully satisfactory technique for the handling and permanent disposal of radioactive wastes.

[6]POPULATION

The total population of the US rose from 203,302,031 in 1970 to 226,545,805 in 1980, according to census figures. In 1983, the population was estimated at 233,981,000, of whom 113,714,000 were male and 120,267,000 were female. By age, the population

State Areas, Entry Dates, and Populations

| | TOTAL AREA | | | | ORDER OF | | POPULATION | | |
	SQ MI	SQ KM	RANK	CAPITAL	ENTRY	DATE OF ENTRY	AT ENTRY	1980 CENSUS	1985 EST.
Alabama	51,705	133,916	29	Montgomery	22	14 December 1819	127,901	3,890,061	4,004,435
Alaska	591,004	1,530,699	1	Juneau	49	3 January 1959	226,167	400,481	514,819
Arizona	114,000	295,260	6	Phoenix	48	14 February 1912	204,354	2,717,866	3,086,827
Arkansas	53,187	137,754	27	Little Rock	25	15 June 1836	57,574	2,285,513	2,345,431
California	158,706	411,048	3	Sacramento	31	9 September 1850	92,597	23,668,562	25,816,590
Colorado	104,091	269,595	8	Denver	38	1 August 1876	39,864	2,888,834	3,253,425
Connecticut*	5,018	12,997	48	Hartford	5	9 January 1788	237,946	3,107,576	3,160,280
Delaware*	2,044	5,294	49	Dover	1	7 December 1787	59,096	595,225	605,711
Florida	58,664	151,940	22	Tallahassee	27	3 March 1845	87,445	9,739,992	11,071,358
Georgia*	58,910	152,577	21	Atlanta	4	2 January 1788	82,548	5,464,265	5,878,225
Hawaii	6,471	16,760	47	Honolulu	50	21 August 1959	632,772	965,000	1,050,270
Idaho	83,564	216,431	13	Boise	43	3 July 1890	88,548	943,935	1,004,071
Illinois	56,345	145,933	24	Springfield	21	3 December 1818	55,211	11,418,461	11,502,433
Indiana	36,185	93,719	38	Indianapolis	19	11 December 1816	147,178	5,490,179	5,489,287
Iowa	56,275	145,752	25	Des Moines	29	28 December 1846	192,214	2,913,387	2,894,273
Kansas	82,277	213,097	14	Topeka	34	29 January 1861	107,206	2,363,208	2,453,581
Kentucky	40,409	104,659	37	Frankfort	15	1 June 1792	73,677	3,661,433	3,747,769
Louisiana	47,751	123,675	31	Baton Rouge	18	30 April 1812	76,556	4,203,972	4,553,903
Maine	33,265	86,156	39	Augusta	23	15 March 1820	298,335	1,124,660	1,156,539
Maryland*	10,460	27,091	42	Annapolis	7	28 April 1788	319,728	4,216,446	4,342,562
Massachusetts*	8,284	21,456	45	Boston	6	6 February 1788	378,787	5,737,037	5,764,125
Michigan	58,527	151,585	23	Lansing	26	26 January 1837	212,267	9,258,344	8,992,766
Minnesota	84,402	218,601	12	St. Paul	32	11 May 1858	172,023	4,077,148	4,199,749
Mississippi	47,689	123,514	32	Jackson	20	10 December 1817	75,448	2,520,638	2,623,069
Missouri	69,697	180,515	19	Jefferson City	24	10 August 1821	66,586	4,917,444	5,004,162
Montana	147,046	380,849	4	Helena	41	8 November 1889	142,924	786,690	826,933
Nebraska	77,355	200,349	15	Lincoln	37	1 March 1867	122,993	1,570,006	1,606,779
Nevada	110,561	286,353	7	Carson City	36	31 October 1864	42,491	799,184	933,451
New Hampshire*	9,279	24,033	44	Concord	9	21 June 1788	141,885	920,610	980,841
New Jersey*	7,787	20,168	46	Trenton	3	18 December 1787	184,139	7,364,158	7,509,625
New Mexico	121,593	314,926	5	Sante Fe	47	6 January 1912	327,301	1,299,968	1,446,347
New York*	49,108	127,190	30	Albany	11	26 July 1788	340,120	17,557,288	17,676,828
North Carolina*	52,669	136,413	28	Raleigh	12	21 November 1789	393,751	5,874,429	6,178,329
North Dakota	70,703	183,121	17	Bismarck	39	2 November 1889	190,983	652,695	692,027
Ohio	41,330	107,045	35	Columbus	17	1 March 1803[2]	43,365	10,797,419	10,763,309
Oklahoma	69,956	181,186	18	Oklahoma City	46	16 November 1907	657,155	3,025,266	3,427,371
Oregon	97,073	251,419	10	Salem	33	14 February 1859	52,465	2,632,663	2,680,087
Pennsylvania*	45,308	117,348	33	Harrisburg	2	12 December 1787	434,373	11,866,728	11,895,301
Rhode Island*	1,212	3,139	50	Providence	13	29 May 1790	68,825	947,154	958,151
South Carolina*	31,113	80,583	40	Columbia	8	23 May 1788	393,751	3,119,208	3,321,520
South Dakota	77,116	199,730	16	Pierre	40	2 November 1889	348,600	690,178	705,027
Tennessee	42,144	109,153	34	Nashville	16	1 June 1796	35,691	4,590,750	4,723,332
Texas	266,807	691,030	2	Austin	28	29 December 1845	212,592	14,228,383	16,384,800
Utah	84,899	219,888	11	Salt Lake City	45	4 January 1896	276,749	1,461,037	1,684,942
Vermont	9,614	24,900	43	Montpelier	14	4 March 1791	85,425	511,456	529,396
Virginia*	40,767	105,586	36	Richmond	10	25 June 1788	747,610	5,346,279	5,642,183
Washington	68,138	176,477	20	Olympia	42	11 November 1889	357,232	4,130,163	4,366,248
West Virginia	24,231	62,758	41	Charleston	35	20 June 1863	442,014	1,949,644	1,968,969
Wisconsin	56,153	145,436	26	Madison	30	29 May 1848	305,391	4,705,335	4,792,115
Wyoming	97,809	253,325	9	Cheyenne	44	10 July 1890	62,555	470,816	534,744

[1] Census closest to entry date.
[2] Date fixed in 1953 by congressional resolution.
*One of original 13 colonies.

was distributed as follows: under 5 years, 17,826,000; 5–13, 30,116,000; 14–17, 14,633,000; 18–24, 29,897,000; 25–44, 69,570,-000; 45–64, 44,555,000; 65 and over, 27,384,000. The US population is extremely mobile: 46.4% of the population over 5 years of age moved during the years 1975–1980, and 16.6% of all Americans at least 1 year old changed their residence between 1981 and 1982.

At the time of the first federal census, in 1790, the population of the country was 3,929,214. Between 1800 and 1850, the population almost quadrupled; between 1850 and 1900, it tripled; and between 1900 and 1950, it almost doubled. During the 1960s and 1970s, however, the growth rate slowed steadily, declining from an annual average of 1.7% during the 1950s to 1% during 1975–83. Over the same period, the median age of the population rose from 16.7 in 1820 to 20.2 in 1870, 25.3 in 1920, and 30.9 in 1983.

The estimated population density in early 1985 was 67 per sq mi of land area (26 per sq km). As of 1980, 73.7% of the population was urban and 26.3% rural. By 1983, metropolitan areas had a total population of 177.9 million, representing a 14.4% increase over 1970. Suburbs have absorbed most of the shift in population distribution since 1950. Estimates for 1984 listed 176 cities of 100,000 residents or more, encompassing about one-fourth the total US population. Leading cities that year were New York, with 7,164,742 residents; Los Angeles, 3,096,721; Chicago, 2,992,472; Houston, 1,705,697; Philadelphia, 1,646,713; and Detroit, 1,088,973.

7ETHNIC GROUPS
The majority of the population of the United States is of European origin, with the largest groups having ancestry traceable in 1980 to England (49,598,035), Germany (49,224,146), and Ireland (40,165,702); many Americans reported multiple ancestries. Major racial and national minority groups include blacks (of either US or Caribbean parentage), Chinese, Filipinos, Japanese, as well as Mexicans and other Spanish-speaking peoples of the Americas. Whites constituted 83.2% of the US population at the time of the 1980 census, blacks 11.7%, Asians and Pacific Islanders 1.5%, Native Americans (Indians, Eskimos, and Aleuts) 0.6%, and other races 3%. Responding to a census question that cut across racial lines, 6.4% of Americans in 1980 described themselves as of Hispanic origin. Inequality in social and economic opportunities for ethnic minorities has become a key public issue in the post–World War II period.

Some American Indian societies survived warfare with land-hungry white settlers and retained their original cultures. Their survival, however, has been on the fringes of North American society, especially as a result of the implementation of a national policy of resettling Indian tribes on reservations. In 1890, according to the official census count, there were 248,253 Native Americans; in 1940, 333,909; and in 1980, 1,418,195. The ease with which Indians have moved into both white and black groups has made it difficult to measure their population growth, and state estimates often exceed the federal figures by a substantial margin. Groups of Indians are found most numerously in the southwestern states of Oklahoma, Arizona, New Mexico, and California. South Dakota has a large Sioux population, as does Oklahoma. North Carolina has a large population of Cherokee, and groups of the Onondaga, Seneca, and Mohawk tribes live in New York. Since 1871, the Indians have been official wards of the federal government. The 1960s and 1970s saw successful court fights by Native Americans in Alaska, Maine, South Dakota, and other states to regain tribal lands or to receive cash settlements for lands taken from them in violation of treaties during the 1800s. In December 1982, with more than 50 lawsuits pending in the courts over claims to water flowing through Indian lands, the federal government offered to help negotiate such disputes.

The black population in 1980 was 26,488,218. Some 53% of blacks still reside in the South, the region that absorbed the majority of slaves brought from Africa in the 18th and 19th centuries. Two important regional migrations of blacks have taken place: (1) the "Great Migration" to the North, commencing in 1915, and (2) the small but hitherto unprecedented westward movement beginning about 1940. Both migrations were fostered by wartime demand for labor and by postwar job opportunities in northern and western urban centers. More than three out of four black Americans live in metropolitan areas, constituting, as of 1980, 71% of the population of Gary, Ind., 70% of Washington, D.C., 67% of Atlanta, 63% of Detroit, 58% of Newark, N.J., and 55% of Baltimore; in New York City, which had the largest number of black residents (1,788,377), 25% of the population was black. Large-scale federal programs to ensure equality for blacks in voting rights, public education, employment, and housing were initiated after the historic 1954 Supreme Court ruling that barred racial segregation in public schools. By 1966, however, in the midst of growing and increasingly violent expressions of dissatisfaction by black residents of northern cities and southern rural areas, the federal Civil Rights Commission reported that integration programs were lagging. Throughout the 1960s and 1970s, the unemployment rate among nonwhites in the US was double that for whites, and school integration proceeded slowly, especially outside the South.

Included in the population of the US in 1980 were 3,500,636 persons whose lineage can be traced to Asian and Pacific nationalities, chiefly Chinese, 806,027; Filipino, 774,640; Japanese, 700,747; Indian, 361,544; Korean, 354,529; and Vietnamese, 261,714. The Chinese population is highly urbanized and concentrated particularly in cities of over 100,000 population, mostly on the West Coast and in New York City. The Japanese population has risen steadily from a level of 72,157 in 1910. Hawaii and California are the leading magnets of Japanese immigration; the Japanese population of Hawaii accounted in 1980 for 25% of the state's residents and 34% of the nation's total number of Japanese, with California contributing another 37%. Most Japanese in California were farmers until the outbreak of World War II, when they were interned and deprived of their landholdings; after the war, most entered the professions and other urban occupations. Of the Filipino immigrants, 46% lived in California in 1980, where they made up 1.5% of the state's population; and 17% in Hawaii, where they accounted for nearly 14% of state residents.

Mexican settlements are largely in the Southwest. Spanish speaking Puerto Ricans, who often represent an amalgam of racial strains, have largely settled in the New York metropolitan area, where they partake in considerable measure of the hardships and problems experienced by other immigrant groups in the process of settling in the US. Since 1959, many Cubans have settled in Florida and other eastern states. As of 1 July 1985 there were an estimated 17,907,100 Hispanic Americans in the US, of whom about 60% were of Mexican ancestry, 14% Puerto Rican, and 6% Cuban.

8LANGUAGES
The primary language of the US is English, enriched by words borrowed from the languages of Native Americans and immigrants, predominantly European.

When European settlement began, Indians living north of Mexico spoke about 300 different languages now held to belong to 58 different language families. Only 2 such families have contributed noticeably to the American vocabulary: Algonkian in the Northeast and Uto-Aztecan in the Southwest. From Algonkian languages, directly or sometimes through Canadian French, English has taken such words as *moose, skunk, caribou, opossum, woodchuck, muskellunge,* and *raccoon* for New World animals; *hickory, kinnikinnick, squash,* and *tamarack* for New World flora; and *mugwump, succotash, hominy, mackinaw, moccasin, tomahawk, toboggan,* and *totem* for various cultural items. From Nahuatl, the language of the Aztecs, terms such as *tomato, mesquite, ocotillo,*

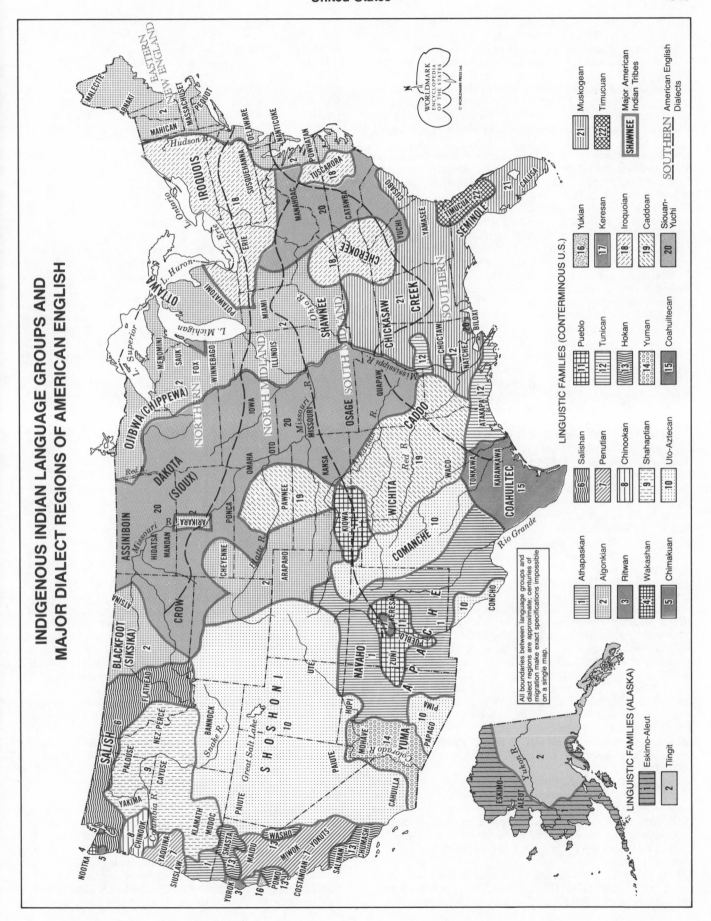

INDIGENOUS INDIAN LANGUAGE GROUPS AND MAJOR DIALECT REGIONS OF AMERICAN ENGLISH

coyote, chili, tamale, chocolate, and *ocelot* have entered English, largely by way of Spanish. A bare handful of words come from other language groups, such as *tepee* from Dakota Siouan, *catalpa* from Creek, *sequoia* from Cherokee, *hogan* from Navaho, and *sockeye* from Salish, as well as *cayuse* from Chinook.

Professional dialect research, initiated in Germany in 1878 and in France in 1902, did not begin in the US until 1931, in connection with the *Linguistic Atlas of New England* (1939–43). This kind of research, requiring trained field-workers to interview representative informants in their homes, subsequently was extended to the entire Atlantic Coast, the north-central states, the upper Midwest, the Pacific Coast, the Gulf states, and Oklahoma. As of 1985, only the New England atlas, the *Linguistic Atlas of the Upper Midwest* (1973–76), and the first two fascicles of the *Linguistic Atlas of the Middle and South Atlantic States* (1980) had been published, along with three volumes based on Atlantic Coast field materials; nearing publication were atlases of the north-central states, the Gulf states, and Oklahoma. In other areas, individual dialect researchers have produced more specialized studies. The definitive work on dialect speech, the American Dialect Society's monumental *Dictionary of American Regional English,* began publication in 1985.

Dialect studies confirm that standard English is not uniform throughout the country. Major regional variations reflect patterns of colonial settlement, dialect features from England having dominated particular areas along the Atlantic Coast and then spread westward along the three main migration routes through the Appalachian system. Dialectologists recognize three main dialects—Northern, Midland, and Southern—each with subdivisions related to the effect of mountain ranges and rivers and railroads on population movement.

The Northern dialect is that of New England and its derivative settlements in New York; the northern parts of Ohio, Indiana, Illinois, and Iowa; and Michigan, Wisconsin, northeastern South Dakota, and North Dakota. A major subdivision is that of New England east of the Connecticut River, an area noted typically by the loss of /r/ after a vowel, and by the pronunciation of *can't, dance, half,* and *bath* with a vowel more like that in *father* than that in *fat.* Generally, however, Northern speech has a strong /r/ after a vowel, the same vowel in *can't* and *cat,* a conspicuous contrast between *cot* and *caught,* the /s/ sound in *greasy, creek* rhyming with *pick,* and *with* ending with the same consonant sound as at the end of *breathe.*

Midland speech extends in a wide band across the US: there are two main subdivisions, North Midland and South Midland. North Midland speech extends westward from New Jersey, Delaware, and Pennsylvania into Ohio, Illinois, southern Iowa, and northern Missouri. Its speakers generally end *with* with the consonant sound that begins the word *thin,* pronounce *cot* and *caught* alike, and say *cow* and *down* as /caow/ and /daown/. South Midland speech was carried by the Scotch-Irish from Pennsylvania down the Shenandoah Valley into the southern Appalachians, where it acquired many Southern speech features before it spread westward into Kentucky, Tennessee, southern Missouri, Arkansas, and northeast Texas. Its speakers are likely to say *plum peach* rather than *clingstone peach* and *snake doctor* rather than *dragonfly.*

Southern speech typically, though not always, lacks the consonant /r/ after a vowel, lengthens the first part of the diphthong in *write* so that to Northern ears it sounds almost like *rat,* and diphthongizes the vowels in *bed* and *hit* so that they sould like /beuhd/ and /hiuht/. *Horse* and *hoarse* do not sound alike, and *creek* rhymes with *meek. Corn bread* is *corn pone,* and *you-all* is standard for the plural.

In the western part of the US, migration routes so crossed and intermingled that no neat dialect boundaries can be drawn, although there are a few rather clear population pockets.

The 1980 census recorded that of 216,560,952 Americans 3 years of age or over, 192,777,930 spoke only English at home; the remaining 23,783,022 spoke a language other than English. The principal foreign languages and their speakers were as follows:

Spanish	11,559,330	Arabic	225,597
Italian	1,634,858	Vietnamese	203,268
German	1,610,269	Hungarian	180,083
French	1,574,454	Russian	175,965
Polish	826,150	Serbo-Croatian	153,022
Chinese	648,844	Dutch	150,721
Pilipino	482,277	Ukrainian	123,548
Greek	441,376	Czech	123,228
Portuguese	361,430	Norwegian	113,415
Japanese	342,205	Farsi (Persian)	109,720
Yiddish	320,743	Armenian	102,387
Korean	275,712	Swedish	101,129

The majority of Spanish speakers live in the Southwest, Florida, and eastern urban centers. Refugee immigration since the 1950s has greatly increased the number of foreign-language speakers from Latin America and Asia.

Very early, English borrowed from neighboring French speakers such words as *shivaree, butte, levee,* and *prairie;* from German, *sauerkraut, smearcase,* and *cranberry;* from Dutch, *stoop, spook,* and *cookie;* and from Spanish, *tornado, corral, ranch,* and *canyon.* From various West African languages, blacks have given English *jazz, voodoo,* and *okra.*

Educational problems raised by the presence of large blocs of non-English speakers led to the passage in 1976 of the Bilingual Education Act, enabling children to study basic courses in their first language while they learn English. A related school problem is that of black English, a Southern dialect variant that is the vernacular of many black students now in northern schools.

9 RELIGIONS

US religious traditions are predominantly Judeo-Christian, and most Americans identify themselves as Protestants (of various denominations), Roman Catholics, or Jews. As of 1982, the US religious bodies counted 341,111 places of worship and 139,603,-000 members, or about three-fifths of the total population. The largest Christian denomination is the Roman Catholic Church, with 52,392,394 members in 19,118 parishes in 1984. Immigration from Ireland, Italy, Eastern Europe, French Canada, and the Caribbean accounts for the predominance of Roman Catholicism in the Northeast, the Northwest, and some parts of the Great Lakes region, while Hispanic traditions and more recent immigration from Mexico and other Latin American countries account for the historical importance of Roman Catholicism in California and throughout much of the Sunbelt. More than any other US religious body, the Roman Catholic Church maintains an extensive network of parochial schools. Jewish immigrants settled first in the Northeast, where the largest Jewish populations remain; in 1984, 1,879,955 Jews lived in New York State, out of an estimated US total of 5,817,235. Eastern Orthodox churches have some 5,000,000 members.

As of 1982, US Protestant groups had at least 76,754,000 adherents. Baptists predominate below the Mason-Dixon line and west to Texas. By far the nation's largest Protestant group, the Southern Baptist Convention had 13,992,000 adherents in 1982; the American Baptist Churches in the USA claimed some 1,622,-000 adherents in 1981. A concentration of Methodist groups extends westward in a band from Delaware to eastern Colorado; the largest of these groups, the United Methodist Church, had 9,457,000 adherents in 1981, and the African Methodist Episcopal Church had 2,210,000. Lutheran denominations, reflecting in part the patterns of German and Scandinavian settlement, are most highly concentrated in the north-central states, especially Minnesota and the Dakotas. The three major groups in 1982—the

Lutheran Church in America, with 2,926,000 members; Lutheran Church–Missouri Synod, with 2,631,000; and American Lutheran Church, with 2,347,000—voted that year to form a single church by 1987. In June 1983, the two major Presbyterian churches, the northern-based United Presbyterian Church in the USA (2,342,-000 adherents in 1982) and the southern-based Presbyterian Church in the US (815,000), formally merged as the Presbyterian Church (USA), ending a division that began with the Civil War. Other Protestant denominations and their estimated adherents in 1982 were the Episcopal Church, 2,794,000; Churches of Christ, 1,605,000; and the United Church of Christ (Congregationalist), 1,717,000. One Christian group, the Church of Jesus Christ of Latter-day Saints (Morman), which claimed 3,521,000 members in 1982, was organized in New York State in 1830 and, since migrating westward, has played a leading role in Utah's political, economic, and religious life. Notable during the 1970s and early 1980s was a rise in the fundamentalist, evangelical, and Pentecostal movements.

Some 2,000,000 Muslims, followers of various Asian religions, a multiplicity of small Protestant groups, and a sizable number of cults also participate in US religious life. Controversies in the 1980s surrounded the proper role of organized religion in politics; the attitudes of religious bodies toward women, homosexuals, abortion, and racial discrimination; attempts to aid sectarian schools through governmental funding or tuition tax credits; and the effort to allow prayer in public schools.

¹⁰TRANSPORTATION

The extent of the transportation industry in the US is indicated by the fact that railroads, motor vehicles, inland waterways, oil pipelines, and domestic airways carried a total of more than 2.3 trillion ton-miles of domestic intercity freight in 1983. Of this total, railroads accounted for 35.93% (compared with 68.4% in 1945), motor vehicles 23.62% (6.2% in 1945), oil pipelines 24.95%, inland waterways 15.26%, and domestic airlines 0.24%. Outlays for all types of transportation totaled an estimated $651.4 billion in 1983; by far the largest outlay, $343.4 billion, went for the purchase, maintenance, and operation of private automobiles.

Railroads lost not only the largest share of intercity freight traffic, their chief source of revenue, but passenger traffic as well. Despite an attempt to revive passenger transport through the development of a national network (Amtrak) in the 1970s, the rail sector continued to experience heavy losses and declining revenues. In 1983 there were 31 Class I rail companies in the US operating 167,424 mi (269,443 km) of track. Railroads carried 300 million passengers in 1980, an increase of 31 million since 1975 but a decrease of 188 million since 1950.

Railroads have lost the bulk of their passenger trade to the private automobile, which, it is estimated, accounted for 83.2% of intercity passenger traffic in 1983. Airways accounted for 14.6% of the intercity traffic during the same year (4.4% in 1960); intercity bus lines carried 365 million passengers. A 1980 survey of workers at least 16 years of age confirmed that the single-passenger private car was by far the most popular means of commuting to work; only in the New York metropolitan area did the proportion of those using mass transit and car pools exceed 50%.

The most conspicious form of transportation is the automobile, and the extent and quality of the US road-transport system are without parallel in the world. In 1983, 72.2 million US households owned a total of 129.3 million motor vehicles. Over 161.9 million vehicles—a record number—were registered in 1983, including more than 125.3 million passenger cars and some 36.6 million trucks and buses. Motor vehicle travel in the US set a new record of 1.59 trillion vehicle miles in 1982. During the 1970s, as average gasoline prices tripled and motor fuel shortages developed, fuel economy became an important consideration for car buyers and manufacturers; average annual fuel comsumption per passenger car fell from 851 gallons in 1973 to 788 gallons in 1974, 711

gallons in 1980, and 686 gallons in 1982. Speed limits were reduced to 55 mph (88.5 km/hr) as a fuel economy measure, and the lower speeds had the beneficial by-product of reducing death rates from motor vehicle accidents. The combined effect of decreased speeds, stricter design standards, and publicity campaigns designed to persuade motorists to make use of seat belts and other safety devices helped reduce traffic fatalities from 41,927 in 1980 to 35,646 in 1982. A crash-protection order issued by the US Department of Transportation in 1984 required manufacturers to install air bags or automatic seat belts as standard equipment unless states representing two-thirds of the US population passed their own mandatory seat-belt laws or automotive planners were able to design more crash-resistant interiors. By mid-1985, 14 states had adopted compulsory seat-belt legislation.

The US has a vast network of roads, whose total length as of 31 December 1982 was 3,866,000 mi (6,222,000 km). Of that total, 88% was surfaced. During the 1970s, about $10 billion was spent annually on highway construction. By the late 1970s, new highway construction had slowed, and an increasing share of highway funds was allocated to the improvement of existing roads.

Major ocean ports or port areas are New York, the Delaware River areas (Philadelphia), the Chesapeake Bay area (Baltimore, Norfolk, Newport News), New Orleans, Houston, and the San Francisco Bay area. The inland port of Duluth on Lake Superior handles more freight than all but the top-ranking ocean ports. The importance of this port, along with those of Chicago and Detroit, was enhanced with the opening in 1959 of the St. Lawrence Seaway. US overseas trade and domestic waterborne commerce in 1982 totaled 1,777 million tons, 820 million tons (46%) of which involved foreign trade; the remainder was in domestic shipping—coastal, inland, and Great Lakes commerce. Waterborne freight consists primarily of bulk commodities such as petroleum and its products, coal and coke, iron ore and steel, sand, gravel and stone, grains, and lumber. The US merchant marine industry has been decreasing gradually since the 1950s. In 1982, the US had the 6th-largest registered merchant shipping fleet in the world, with 832 vessels of more than 1,000 gross registered tons. The average age of US merchant vessels was 23 years, compared with 7 years for Japanese vessels and 13 years for those of the world as a whole.

In 1983, the US had 96 certified air carriers, more than double the number in 1978, when the Airline Deregulation Act was passed. Revenue passengers carried by the airlines in 1940 totaled 2.7 million; by 1983, the figure was 318 million. The US in 1983 had 16,029 airports, of which 4,812 were public. US international carriers flew 326 million mi (525 million km) in scheduled international service, serving some 22 million passengers. By the end of the year, the US had 699,546 active pilots, of whom 153,820 were on commercial routes. An estimated 213,293 general aviation aircraft flew a total of 35,249,000 hours in 1983.

The federal government regulates and subsidizes almost all forms of transportation. It regulates—through the Interstate Commerce Commission—railroads, pipelines, and motor carriers of passengers and freight. The Federal Railroad Administration coordinates federal railroad support programs. Air transportation is regulated through the Federal Aviation Administration (FAA); under a policy of deregulation, the Civil Aeronautics Board was abolished at the end of 1984. The FAA operates the National Airspace System, which develops and maintains guidance and control devices along designated air lanes. The government subsidizes shipping through the Maritime Subsidy Board of the Maritime Administration, sharing heavily in the costs of ship construction by private concerns. During 1983, the Highway Trust Fund disbursed $8.4 billion to state and local governments for highway maintenance and construction, while the Urban Mass Transit Administration distributed $3.7 billion for rapid-transit systems.

11HISTORY

The first Americans—distant ancestors of the American Indians—probably crossed the Bering Strait from Asia at least 12,000 years ago. By the time Christopher Columbus came to the New World in 1492 there were probably no more than 2 million Native Americans living in the land that was to become the US.

Following exploration of the American coasts by English, Portuguese, Spanish, Dutch, and French sea captains from the late 15th century onward, European settlements sprang up in the latter part of the 16th century. The Spanish established the first permanent settlement at St. Augustine in the future state of Florida in 1565, and another in New Mexico in 1599. During the early 17th century, the English founded Jamestown in Virginia Colony (1607) and Plymouth Colony in present-day Massachusetts (1620). The Dutch established settlements at Ft. Orange (now Albany, N.Y.) in 1624, New Amsterdam (now New York City) in 1626, and at Bergen (now part of Jersey City, N.J.) in 1660; they conquered New Sweden—the Swedish colony in Delaware and New Jersey—in 1655. Nine years later, however, the English seized this New Netherland Colony and subsequently monopolized settlement of the East Coast except for Florida, where Spanish rule prevailed until 1821. In the Southwest, California, Arizona, New Mexico, and Texas also were part of the Spanish empire until the 19th century. Meanwhile, in the Great Lakes area south of present-day Canada, France set up a few trading posts and settlements but never established effective control; New Orleans was one of the few areas of the US where France pursued an active colonial policy.

From the founding of Jamestown to the outbreak of the American Revolution more than 150 years later, the British government administered its American colonies within the context of mercantilism: the colonies existed primarily for the economic benefit of the empire. Great Britain valued its American colonies especially for their tobacco, lumber, indigo, rice, furs, fish, grain, and naval stores, relying particularly in the southern colonies on black slave labor.

The colonies enjoyed a large measure of internal self-government until the end of the French and Indian War (1745–63), which resulted in the loss of French Canada to the British. To prevent further troubles with the Indians, the British government in 1763 prohibited the American colonists from settling beyond the Appalachian Mountains. Heavy debts forced London to decree that the colonists should assume the costs of their own defense, and the British government enacted a series of revenue measures to provide funds for that purpose. But soon, the colonists began to insist that they could be taxed "only with their consent," and the struggle grew to become one of local versus imperial authority.

Widening cultural and intellectual differences also served to divide the colonies and the mother country. Life on the edge of the civilized world had brought about changes in the colonists' attitudes and outlook, emphasizing their remoteness from English life. In view of the long tradition of virtual self-government in the colonies, strict enforcement of imperial regulations and British efforts to curtail the power of colonial legislatures presaged inevitable conflict between the colonies and the mother country. When citizens of Massachusetts, protesting the tax on tea, dumped a shipload of tea belonging to the East India Company into Boston harbor in 1773, the British felt compelled to act in defense of their authority as well as in defense of private property. Punitive measures—referred to as the Intolerable Acts by the colonists—struck at the foundations of self-government.

In response, the First Continental Congress, composed of delegates from 12 of the 13 colonies—Georgia was not represented—met in Philadelphia in September 1774, and proposed a general boycott of English goods, together with the organizing of a militia. British troops marched to Concord, Mass., on 19 April 1775 and destroyed the supplies that the colonists had assembled there. American "minutemen" assembled on the nearby Lexington green and fired "the shot heard round the world," although no one knows who actually fired the first shot that morning. The British soldiers withdrew and fought their way back to Boston.

Voices in favor of conciliation were raised in the Second Continental Congress that assembled in Philadelphia on 10 May 1775, this time including Georgia; but with news of the Restraining Act (30 March 1775), which denied the colonies the right to trade with countries outside the British Empire, all hopes for peace vanished. George Washington was appointed commander in chief of the new American army, and on 4 July 1776, the 13 American colonies adopted the Declaration of Independence, justifying the right of revolution by the theory of natural rights.

British and American forces met in their first organized encounter near Boston on 17 June 1775. Numerous battles up and down the coast followed. The British seized and held the principal cities but were unable to inflict a decisive defeat on Washington's troops. The entry of France into the war on the American side eventually tipped the balance. On 19 October 1781, the British commander, Cornwallis, cut off from reinforcements by the French fleet on one side and besieged by French and American forces on the other, surrendered his army at Yorktown, Va. American independence was acknowledged by the British in a treaty of peace signed in Paris on 3 September 1783.

The first constitution uniting the 13 original states—the Articles of Confederation—reflected all the suspicions that Americans entertained about a strong central government. Congress was denied power to raise taxes or regulate commerce, and many of the powers it was authorized to exercise required the approval of a minimum of nine states. Dissatisfaction with the Articles of Confederation was aggravated by the hardships of a postwar depression, and in 1787—the same year that Congress passed the Northwest Ordinance, providing for the organization of new territories and states on the frontier—a convention assembled in Philadelphia to revise the articles. The convention adopted an altogether new constitution, the present Constitution of the United States, which greatly increased the powers of the central government at the expense of the states. This document was ratified by the states with the understanding that it would be amended to include a bill of rights guaranteeing certain fundamental freedoms. These freedoms—including the rights of free speech, press, and assembly, freedom from unreasonable search and seizure, and the right to a speedy and public trial by an impartial jury—are assured by the first 10 amendments to the constitution, adopted on 15 December 1791; the constitution did however, recognize slavery, and did not provide for universal suffrage. On 30 April 1789, George Washington was inaugurated as the first president of the US.

During Washington's administration, the credit of the new nation was bolstered by acts providing for a revenue tariff and an excise tax; opposition to the excise on whiskey sparked the Whiskey Rebellion, suppressed on Washington's orders in 1794. Alexander Hamilton's proposals for funding the domestic and foreign debt and permitting the national government to assume the debts of the states were also implemented. Hamilton, the secretary of the treasury, also created the first national bank, and was the founder of the Federalist Party. Opposition to the bank as well as to the rest of the Hamiltonian program, which tended to favor northeastern commercial and business interests, led to the formation of an anti-Federalist party, the Democratic-Republicans, led by Thomas Jefferson. The Federalist Party, to which Washington belonged, regarded the French Revolution as a threat to security and property; the Democratic-Republicans, while condemning the violence of the revolutionists, hailed the overthrow of the French monarchy as a blow to tyranny. The split of the nation's leadership into rival camps was the first manifestation

WESTWARD EXPANSION, MAIN EXPLORERS' ROUTES

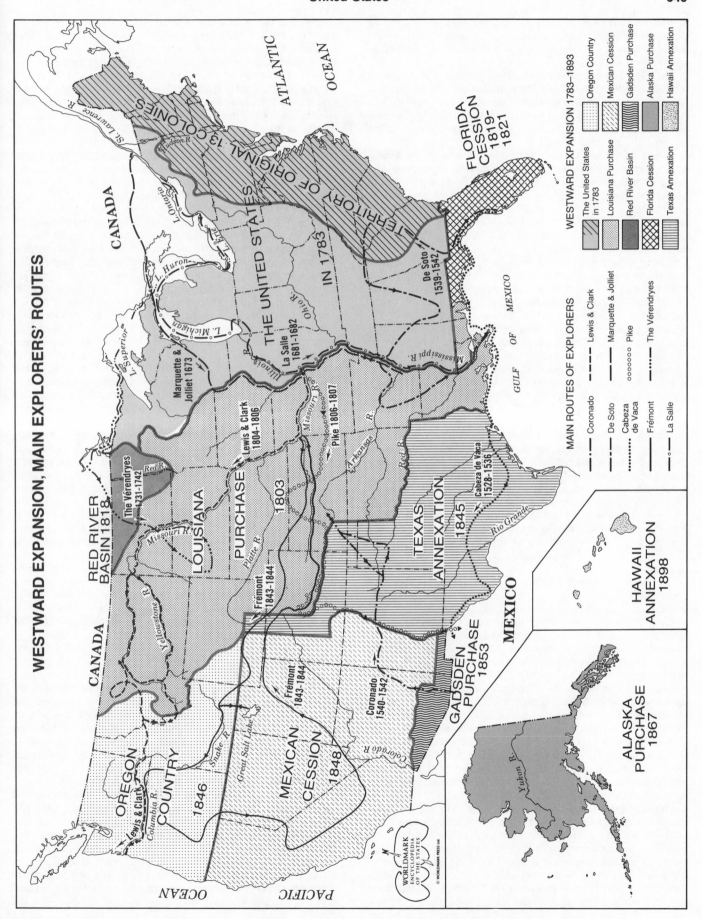

WESTWARD EXPANSION 1783–1893

Oregon Country
Mexican Cession
Gadsden Purchase
Alaska Purchase
Hawaii Annexation

The United States in 1783
Louisiana Purchase
Red River Basin
Florida Cession
Texas Annexation

MAIN ROUTES OF EXPLORERS

Coronado
De Soto
Cabeza de Vaca
Frémont
La Salle

Lewis & Clark
Marquette & Jolliet
Pike
The Vérendryes

ATLANTIC OCEAN

CANADA

TERRITORY OF ORIGINAL 13 COLONIES

FLORIDA CESSION 1819–1821

De Soto 1539–1542

THE UNITED STATES IN 1783

Marquette & Jolliet 1673

La Salle 1681–1682

GULF OF MEXICO

RED RIVER BASIN 1818

The Vérendryes 1731–1742

Lewis & Clark 1804–1806

Pike 1806–1807

LOUISIANA PURCHASE 1803

Cabeza de Vaca 1528–1536

TEXAS ANNEXATION 1845

Frémont 1843–1844

GADSDEN PURCHASE 1853

MEXICO

Frémont 1843–1844

Coronado 1540–1542

OREGON COUNTRY 1846

MEXICAN CESSION 1848

Lewis & Clark

Great Salt Lake

PACIFIC OCEAN

CANADA

HAWAII ANNEXATION 1898

ALASKA PURCHASE 1867

Yukon R.

WORLDMARK ENCYCLOPEDIA OF THE STATES
© WORLDMARK PRESS Ltd.

St. Lawrence R.
Hudson R.
L. Ontario
L. Erie
L. Huron
L. Michigan
L. Superior
Ohio R.
Illinois R.
Missouri R.
Mississippi R.
Arkansas R.
Red R.
Rio Grande
Platte R.
Missouri R.
Yellowstone R.
Snake R.
Columbia R.
Colorado R.
Red R.

of the two-party system, which has since been the dominant characteristic of the US political scene. (Jefferson's party should not be confused with the modern Republican Party, formed in 1854.)

The 1800 election brought the defeat of Federalist President John Adams, Washington's successor, by Jefferson; a key factor in Adam's loss was the unpopularity of the Alien and Sedition Acts (1798), Federalist-sponsored measures that had abridged certain freedoms guaranteed in the Bill of Rights. In 1803, Jefferson achieved the purchase from France of the Louisiana Territory, including all the present territory of the US west of the Mississippi drained by that river and its tributaries; exploration and mapping of the new territory, notably through the expeditions of Meriwether Lewis and William Clark, began almost immediately. Under Chief Justice John Marshall, the US Supreme Court, in the landmark case of *Marbury* v. *Madison,* established the principle of federal supremacy in conflicts with the states and enunciated the doctrine of judicial review.

During Jefferson's second term in office, the US became involved in a protracted struggle between Britain and Napoleonic France. Seizures of US ships and the impressment of US seamen by the British navy led the administration to pass the Embargo Act of 1807, under which no US ships were to put out to sea. After the act was repealed in 1809, ship seizures and impressment of seamen by the British continued, and were the ostensible reasons for the declaration of war on Britain in 1812 during the administration of James Madison. An underlying cause of the War of 1812, however, was land-hungry westerners' coveting of southern Canada as potential US territory.

The war was largely a standoff. A few surprising US naval victories countered British successes on land. The Treaty of Ghent (24 December 1814), which ended the war, made no mention of impressment and provided for no territorial changes. The occasion for further maritime conflict with Britain, however, disappeared with the defeat of Napoleon in 1815.

Now the nation became occupied primarily with domestic problems and westward expansion. Because the US had been cut off from its normal sources of manufactured goods in Great Britain during the war, textiles and other industries developed and prospered in New England. To protect these infant industries, Congress adopted a high-tariff policy in 1816.

Three events of the late 1810s and the 1820s were of considerable importance for the future of the country. The federal government in 1817 began a policy of forcibly resettling the Indians, already decimated by war and disease, in what later became known as Indian Territory (now Oklahoma); those Indians not forced to move were restricted to reservations. The Missouri Compromise (1820) was an attempt to find a nationally acceptable solution to the volatile dispute over the extension of black slavery to new territories. It provided for admission of Missouri into the Union as a slave state but banned slavery in territories to the west that lay north of 36°30′. As a result of the establishment of independent Latin American republics and threats by France and Spain to reestablish colonial rule, President James Monroe in 1923 asserted that the Western Hemisphere was closed to further colonization by European powers. The Monroe Doctrine declared that any effort by such powers to recover territories whose independence the US had recognized would be regarded as an unfriendly act.

From the 1820s to the outbreak of the Civil War, the growth of manufacturing continued, mainly in the North, and was accelerated by inventions and technological advances. Farming expanded with westward migration. The South discovered that its future lay in the cultivation of cotton. The cotton gin, invented by Eli Whitney in 1793, greatly simplified the problems of production; the growth of the textile industry in New England and Great Britain assured a firm market for cotton. Hence, during the first half of the 19th century, the South remained a fundamentally agrarian society based increasingly on a one-crop economy. Large numbers of field hands were required for cotton cultivation, and black slavery became solidly entrenched in the southern economy.

The construction of roads and canals paralleled the country's growth and economic expansion. The successful completion of the Erie Canal (1825), linking the Great Lakes with the Atlantic, ushered in a canal-building boom. Railroad building began in earnest in the 1830s, and by 1840, about 3,300 mi (5,300 km) of track had been laid. The development of the telegraph a few years later gave the nation the beginnings of a modern telecommunications network. As a result of the establishment of the factory system, a laboring class appeared in the North by the 1830s, bringing with it the earliest unionization efforts.

Western states admitted into the Union following the War of 1812 provided for free white male suffrage without property qualifications and helped spark a democratic revolution. As eastern states began to broaden the franchise, mass appeal became an important requisite for political candidates. The election to the presidency in 1928 of Andrew Jackson, a military hero and Indian fighter from Tennessee, was no doubt a result of this widening of the democratic process. By this time, the US consisted of 24 states and had a population of nearly 13 million.

The relentless westward thrust of the US population ultimately involved the US in foreign conflict. In 1836, US settlers in Texas revolted against Mexican rule and established an independent republic. Texas was admitted to the Union as a state in 1845, and relations between Mexico and the US steadily worsened. A dispute arose over the southern boundary of Texas, and a Mexican attack on a US patrol in May 1846 gave President James K. Polk a pretext to declare war. After a rapid advance, US forces captured Mexico City, and on 2 February 1848, Mexico formally gave up the unequal fight by signing the Treaty of Guadalupe Hidalgo, providing for the cession of California and the territory of New Mexico to the US. With the Gadsden Purchase of 1853, the US acquired from Mexico for $10 million large strips of land forming the balance of southern Arizona and New Mexico. A dispute with Britain over the Oregon Territory was settled in 1846 by a treaty that established the 49th parallel as the boundary with Canada. Thenceforth, the US was to be a Pacific as well as an Atlantic power.

Westward expansion exacerbated the issue of slavery in the territories. By 1840, abolition of slavery constituted a fundamental aspect of a movement for moral reform, which also encompassed woman's rights, universal education, alleviation of working class hardships, and temperance. In 1849, a year after the discovery of gold had precipitated a rush of new settlers to California, that territory (whose constitution prohibited slavery) demanded admission to the Union. A compromise engineered in Congress by Senator Henry Clay in 1850 provided for California's admission as a free state in return for various concessions to the South. But enmities dividing North and South could not be silenced. The issue of slavery in the territories came to a head with the Kansas-Nebraska Act of 1854, which repealed the Missouri Compromise and left the question of slavery in those territories to be decided by the settlers themselves. The ensuing conflicts in Kansas between northern and southern settlers earned the territory the name "bleeding Kansas." In 1860, the Democratic Party, split along northern and southern lines, offered two presidential candidates. The new Republican Party, organized in 1854 and opposed to the expansion of slavery, nominated Abraham Lincoln. Owing to the defection in Democratic ranks, Lincoln was able to carry the election in the electoral college, although he did not obtain a majority of the popular vote. To ardent supporters of slavery, Lincoln's election provided a reason for immediate secession. Between December 1860 and February 1861, the seven states of the Deep South—South Carolina, Mississippi, Florida,

Alabama, Georgia, Louisiana, and Texas—withdrew from the Union and formed a separate government, known as the Confederate States of America, under the presidency of Jefferson Davis. The secessionists soon began to confiscate federal property in the South. On 12 April 1861, the Confederates opened fire on Ft. Sumter in the harbor of Charleston, S.C., and thus precipitated the US Civil War. Following the outbreak of hostilities, Arkansas, North Carolina, Virginia, and Tennessee joined the Confederacy.

For the next four years, war raged between the Confederate and Union forces, largely in southern territories. An estimated 360,000 men in the Union forces died of various causes, including 110,000 killed in battle. Confederate dead were estimated at 250,000, including 94,000 killed in battle. The North, with great superiority in manpower and resources, finally prevailed. A Confederate invasion of the North was repulsed at the battle of Gettysburg, Pa., in July 1863; a Union army took Atlanta in September 1864; and Confederate forces evacuated Richmond, the Confederate capital, in early April 1865. With much of the South in Union hands, Confederate Gen. Robert E. Lee surrendered to Gen. Ulysses S. Grant at Appomattox Courthouse in Virginia on 9 April.

The outcome of the war brought great changes in US life. Lincoln's Emancipation Proclamation of 1863 was the initial step in freeing some 4 million black slaves; their liberation was completed soon after the war's end by amendments to the Constitution. Lincoln's plan for the reconstruction of the rebellious states was compassionate, but only five days after Lee's surrender, Lincoln was assassinated by John Wilkes Booth as part of a conspiracy in which US Secretary of State William H. Seward was seriously wounded.

During the Reconstruction era (1865–77), the defeated South was governed by Union Army commanders, and the resultant bitterness of southerners toward northern Republican rule, which enfranchised blacks, persisted for years afterward. Vice President Andrew Johnson, who succeeded Lincoln as president, tried to carry out Lincoln's conciliatory policies but was opposed by Radical Republican leaders in Congress, who demanded harsher treatment of the South. On the pretext that he had failed to carry out an act of Congress, the House of Representatives voted to impeach Johnson in 1868, but the Senate failed by one vote to convict him and remove him from office. It was during Johnson's presidency that Secretary of State Seward negotiated the purchase of Alaska (which attained statehood in 1959) from Russia for $7.2 million.

The efforts of southern whites to regain political control of their states led to the formation of terrorist organizations like the Ku Klux Klan, which employed violence to prevent blacks from voting. By the end of the Reconstruction era, whites had reestablished their political domination over blacks in the southern states and had begun to enforce patterns of segregation in education and social organization that were to last for nearly a century.

In many southern states, the decades following the Civil War were ones of economic devastation, in which rural whites as well as blacks were reduced to sharecropper status. Outside the South, however, a great period of economic expansion began. Transcontinental railroads were constructed, corporate enterprise spurted ahead, and the remaining western frontier lands were rapidly occupied and settled. The age of big business tycoons dawned. As heavy manufacturing developed, Pittsburgh, Chicago, and New York emerged as the nation's great industrial centers. The Knights of Labor, founded in 1869, engaged in numerous strikes, and violent conflicts between strikers and strikebreakers were common. The American Federation of Labor, founded in 1886, established a nationwide system of craft unionism that remained dominant for many decades. During this period, too, the woman's rights movement organized actively to secure the vote (although woman's suffrage was not enacted nationally until 1920), and

groups outraged by the depletion of forests and wildlife in the West pressed for the conservation of natural resources.

During the latter half of the 19th century, the acceleration of westward expansion made room for millions of immigrants from Europe. The country's population grew to more than 76 million by 1900. As homesteaders, prospectors, and other settlers tamed the frontier, the federal government forced Indians west of the Mississippi to cede vast tracts of land to the whites, precipitating a series of wars with various tribes. By 1890, only 250,000 Indians remained in the US, virtually all of them residing on reservations.

The 1890s marked the closing of the US frontier for settlement and the beginning of US overseas expansion. By 1892, Hawaiian sugar planters of US origin had become strong enough to bring about the downfall of the native queen and to establish a republic, which in 1898, at its own request, was annexed as a territory by the US. The sympathies of the US with the Cuban nationalists who were battling for independence from Spain were aroused by a lurid press and by expansionist elements. A series of events climaxed by the sinking of the USS *Maine* in Havana harbor finally forced a reluctant President William McKinley to declare war on Spain on 25 April 1898. US forces overwhelmed those of Spain in Cuba, and as a result of the Spanish-American War, the US added to its territories the Philippines, Guam, and Puerto Rico. A newly independent Cuba was drawn into the US orbit as a virtual protectorate through the 1950s. Many eminent citizens saw these new departures into imperialism as a betrayal of the time-honored US doctrine of government by the consent of the governed.

With the marked expansion of big business came increasing protests against the oppressive policies of large corporations and their dominant role in the public life of the nation. A demand emerged for strict control of monopolistic business practice through the enforcement of antitrust laws. Two US presidents, Theodore Roosevelt (1901–9), a Republican, and Woodrow Wilson (1913–21), a Democrat, approved of the general movement for reform, which came to be called progressivism. Roosevelt developed a considerable reputation as a trustbuster, while Wilson's program, known as the New Freedom, called for reform of tariffs, business procedures, and banking. During Roosevelt's first term, the US leased the Panama Canal Zone and started construction of a 42-mi (68-km) canal, completed in 1914.

US involvement in World War I marked the country's active emergence as one of the great powers of the world. When war broke out in 1914 between Germany, Austria-Hungary, and Turkey on one side and Britain, France, and Russia on the other, sentiment in the US was strongly opposed to participation in the conflict, although a large segment of the American people sympathized with the British and the French. While both sides violated US maritime rights on the high seas, the Germans, enmeshed in a British blockade, resorted to unrestricted submarine warfare. On 6 April 1917, Congress declared war on Germany. Through a national draft of all able-bodied men between the ages of 18 and 45, some 4 million US soldiers were trained, of whom more than 2 million were sent overseas to France. By late 1917, when US troops began to take part in the fighting on the western front, the European armies were approaching exhaustion, and US intervention may well have been decisive in ensuring the eventual victory of the Allies. In a series of great battles in which US soldiers took an increasingly major part, the German forces were rolled back in the west, and in the autumn of 1918 were compelled to sue for peace. Fighting ended with the armistice of 11 November 1918. President Wilson played an active role in drawing up the 1919 Versailles peace treaty, which embodied his dream of establishing a League of Nations to preserve the peace, but the isolationist bloc in the Senate was able to prevent US ratification of the treaty.

In the 1920s, the US had little enthusiasm left for crusades, either for democracy abroad or for reform at home; a rare instance of idealism in action was the Kellogg-Briand Pact (1928), an antiwar accord negotiated on behalf of the US by Secretary of State Frank B. Kellogg. In general, however, the philosophy of the Republican administrations from 1921 to 1933 was expressed in the aphorism "The business of America is business," and the 1920s saw a great business boom. The years 1923–24 also witnessed the unraveling of the Teapot Dome scandal: the revelation that President Warren G. Harding's secretary of the interior, Albert B. Fall, had secretly leased federal oil reserves in California and Wyoming to private oil companies in return for gifts and loans.

The great stock market crash of October 1929 ushered in the most serious and most prolonged economic depression the country had ever known. By 1933, an estimated 12 million men and women were out of work; personal savings were wiped out on a vast scale through a disastrous series of corporate bankruptcies and bank failures. Relief for the unemployed was left to private charities and local governments, which were incapable of handling the enormous task.

The inauguration of the successful Democratic presidential candidate, Franklin D. Roosevelt, in March 1933 ushered in a new era of US history, in which the federal government was to assume a much more prominent role in the nation's economic affairs. Proposing to give the country a "New Deal," Roosevelt accepted national responsibility for alleviating the hardships of unemployment; relief measures were instituted, work projects were established, and deficit spending was accepted in preference to ignoring public distress. The federal Social Security program was inaugurated, as were various measures designed to stimulate and develop the economy through federal intervention. Unions were strengthened through the National Labor Relations Act, which established the right of employees' organizations to bargain collectively with employers. Union membership increased rapidly, and the dominance of the American Federation of Labor was challenged by the newly formed Congress of Industrial Organizations, which organized workers along industrial lines.

The depression of the 1930s was worldwide, and certain nations attempted to counter economic stagnation by building large military establishments and embarking on foreign adventures. Following German, Italian, and Japanese aggression, World War II broke out in Europe during September 1939. In 1940, Roosevelt, disregarding a tradition dating back to Washington that no president should serve more than two terms, ran again for reelection. He easily defeated his Republican opponent, Wendell Willkie, who, along with Roosevelt, advocated increased rearmament and all possible aid to victims of aggression. The US was brought actively into the war by the Japanese attack on the Pearl Harbor naval base in Hawaii on 7 December 1941. The forces of Germany, Italy, and Japan were now arrayed over a vast theater of war against those of the US and the British Commonwealth; in Europe, Germany was locked in a bloody struggle with the Soviet Union. US forces waged war across the vast expanses of the Pacific, in Africa, in Asia, and in Europe. Italy surrendered in 1943; Germany was successfully invaded in 1944 and conquered in May 1945; and after the US dropped the world's first atomic bombs on Hiroshima and Nagasaki, the Japanese capitulated in August. The Philippines became an independent republic soon after the war, but the US retained most of its other Pacific possessions, with Hawaii becoming the 50th state in 1959.

Roosevelt, who had been elected to a fourth term in 1944, died in April 1945 and was succeeded by Harry S Truman, his vice president. Under the Truman administration, the US became an active member of the new world organization, the United Nations. The Truman administration embarked on large-scale programs of military aid and economic support to check the expansion of communism. Aid to Greece and Turkey in 1948 and

the Marshall Plan, a program designed to accelerate the economic recovery of Western Europe, were outstanding features of US postwar foreign policy. The North Atlantic Treaty (1949) established a defensive alliance among a number of West European nations and the US. Truman's Point Four program gave technical and scientific aid to developing nations. When, following the North Korean attack on South Korea on 25 June 1950, the UN Security Council resolved that members of the UN should proceed to the aid of South Korea, US naval, air, and ground forces were immediately dispatched by President Truman. An undeclared war ensued, which eventually was brought to a halt by an armistice signed on 27 June 1953.

In 1952, Dwight D. Eisenhower, supreme commander of Allied forces in Europe during World War II, was elected president on the Republican ticket, thereby bringing to an end 20 years of Democratic presidential leadership. In foreign affairs, the Eisenhower administration continued the Truman policy of containing the USSR and threatened "massive retaliation" in the event of Soviet aggression, thus heightening the Cold War between the world's two great nuclear powers. Although Republican domestic policies were more conservative than those of the Democrats, the Eisenhower administration extended certain major social and economic programs of the Roosevelt and Truman administrations, notably Social Security and public housing. The early years of the Eisenhower administration were marked by agitation (arising in 1950) over charges of Communist and other allegedly subversive activities in the US—a phenomenon known as McCarthyism, after Republican Senator Joseph R. McCarthy of Wisconsin, who aroused much controversy with unsubstantiated allegations that Communists had penetrated the US government, especially the Army and the Department of State. Even those who personally opposed McCarthy lent their support to the imposition of loyalty oaths and the blacklisting of persons with left-wing backgrounds.

A major event of the Eisenhower years was the US Supreme Court's decision in *Brown* v. *Board of Education of Topeka* (1954) outlawing segregation of whites and blacks in public schools. In the aftermath of this ruling, desegregation proceeded slowly and painfully. In the early 1960s, sit-ins, "freedom rides," and similar expressions of nonviolent resistance by blacks and their sympathizers led to a lessening of segregation practices in public facilities. Under Chief Justice Earl Warren, the high court in 1962 mandated the reapportionment of state and federal legislative districts according to a "one person, one vote" formula. It also broadly extended the rights of defendants in criminal trials to include the provision of a defense lawyer at public expense for an accused person unable to afford one, and established the duty of police to advise an accused person of his or her legal rights immediately upon arrest.

In the early 1960s, during the administration of Eisenhower's Democratic successor, John F. Kennedy, the Cold War heated up as Cuba, under the regime of Fidel Castro, aligned itself with the Soviet Union. Attempts by anti-Communist Cuban exiles to invade their homeland in the spring of 1961 failed despite US aid. In October 1962, President Kennedy successfully forced a showdown with the Soviet Union over Cuba in demanding the withdrawal of Soviet-supplied "offensive weapons"—missiles—from the nearby island. On 22 November 1963, President Kennedy was assassinated while riding in a motorcade through Dallas, Texas; hours later, Vice President Lyndon B. Johnson was inaugurated president. In the November 1964 elections, Johnson overwhelmingly defeated his Republican opponent, Barry M. Goldwater, and embarked on a vigorous program of social legislation unprecedented since Roosevelt's New Deal. His "Great Society" program sought to ensure black Americans' rights in voting and public housing, to give the underprivileged job training, and to provide persons 65 and over with hospitalization and other medical benefits (Medicare). Measures ensuring equal opportunity for minority groups may have contributed to the growth of the woman's rights movement in the late 1960s. This same period also saw the growth of a powerful environmental protection movement.

US military and economic aid to anti-Communist forces in Viet-Nam, which had its beginnings during the Truman administration (while Viet-Nam was still part of French Indochina) and was increased gradually by presidents Eisenhower and Kennedy, escalated in 1965. In that year, President Johnson sent US combat troops to South Viet-Nam and ordered US bombing raids on North Viet-Nam, after Congress (in the Gulf of Tonkin Resolution of 1964) had given him practically carte blanche authority to wage war in that region. By the end of 1968, American forces in Viet-Nam numbered 536,100 men, but US military might was unable to defeat the Vietnamese guerrillas, and the American people were badly split over continuing the undeclared (and, some thought, ill-advised or even immoral) war, with its high price in casualties and matériel. Reacting to widespread dissatisfaction with his Viet-Nam policies, Johnson withdrew in March 1968 from the upcoming presidential race, and in November, Republican Richard M. Nixon, who had been the vice president under Eisenhower, was elected president. Thus, the Johnson years—which had begun with the new hopes of a Great Society but had soured with a rising tide of racial violence in US cities and the assassinations of civil rights leader Martin Luther King, Jr., and US Senator Robert F. Kennedy, among others—drew to a close.

President Nixon gradually withdrew US ground troops from Viet-Nam but expanded aerial bombardment throughout Indochina, and the increasingly unpopular and costly war continued for four more years before a cease-fire—negotiated by Nixon's national security adviser, Henry Kissinger—was finally signed on 27 January 1973 and the last US soldiers were withdrawn. The most protracted conflict in American history had resulted in 46,163 US combat deaths and 303,654 wounded soldiers, and had cost the US government $112 billion in military allocations. Two years later, the South Vietnamese army collapsed, and the North Vietnamese Communist regime united the country.

In 1972, during the last year of his first administration, Nixon initiated the normalization of relations—ruptured in 1949—with the People's Republic of China and signed a strategic arms limitation agreement with the Soviet Union as part of a Nixon-Kissinger policy of pursuing détente with both major Communist powers. (Earlier, in July 1969, American technology had achieved a national triumph by landing the first astronaut on the moon.) The Nixon administration sought to muster a "silent majority" in support of its Indochina policies and its conservative social outlook in domestic affairs. The most momentous domestic development, however, was the Watergate scandal, which began on 17 June 1972 with the arrest of five men associated with Nixon's reelection campaign, during a break-in at Democratic Party headquarters in the Watergate office building in Washington, D.C. Although Nixon was reelected in 1972, subsequent disclosures by the press and by a Senate investigating committee revealed a complex pattern of political "dirty tricks" and illegal domestic surveillance throughout his first term. The president's apparent attempts to obstruct justice by helping his aides cover up the scandal were confirmed by tape recordings (made by Nixon himself) of his private conversations, which the Supreme Court ordered him to release for use as evidence in criminal proceedings. The House voted to begin impeachment proceedings, and in late July 1974, its Judiciary Committee approved three articles of impeachment. On 9 August, Nixon became the first president to resign the office. The following year, Nixon's top aides and former attorney general, John N. Mitchell, were convicted of obstruction and were subsequently sentenced to prison.

Nixon's successor was Gerald R. Ford, who in October 1973

had been appointed to succeed Vice President Spiro T. Agnew when Agnew resigned following his plea of *nolo contendere* to charges that he had evaded paying income tax on moneys he had received from contractors while governor of Maryland. Less than a month after taking office, President Ford granted a full pardon to Nixon for any crimes he may have committed as president. In August 1974, Ford nominated Nelson A. Rockefeller as vice president (he was not confirmed until December), thus giving the country the first instance of a nonelected president and an appointed vice president serving simultaneously. Ford's pardon of Nixon, as well as continued inflation and unemployment, probably contributed to his narrow defeat by a Georgia Democrat, Jimmy Carter, in 1976.

President Carter's forthright championing of human rights—though consistent with the Helsinki accords, the "final act" of the Conference on Security and Cooperation in Europe, signed by the US and 34 other nations in July 1974—contributed to strained relations with the USSR and with some US allies. During 1978–79, the president concluded and secured Senate passage of treaties ending US sovereignty over the Panama Canal Zone. His major accomplishment in foreign affairs, however, was his role in mediating a peace agreement between Israel and Egypt, signed at the Camp David, Md., retreat in September 1978. Domestically, the Carter administration initiated a national energy program to reduce US dependence on foreign oil by cutting gasoline and oil consumption and by encouraging the development of alternative energy resources. But the continuing decline of the economy because of double-digit inflation and high unemployment caused his popularity to wane, and confusing shifts in economic policy (coupled with a lack of clear goals in foreign affairs) characterized his administration during 1979 and 1980; a prolonged quarrel with Iran over more than 50 US hostages seized in Tehran on 4 November 1979 contributed to public doubts about his presidency. Exactly a year after the hostages were taken, former California Governor Ronald Reagan defeated Carter in an election that saw the Republican Party score major gains throughout the US. The hostages were released on 20 January 1981, the day of Reagan's inauguration.

Reagan, who survived a chest wound from an assassination attempt in Washington, D.C., on 30 March 1981, rode a crest of sympathy and popularity to legislative success. By the end of the year, he had succeeded in obtaining congressional passage of much of his economic program: tax cuts, substantial reductions in domestic spending and the tightening of eligibility requirements for public assistance, and the elimination or consolidation of some federal programs.

Reagan's appointment of Sandra Day O'Connor as the first woman justice of the Supreme Court was widely praised and won unanimous confirmation from the Senate. However, some of his other high-level choices were extremely controversial—none more so than that of his secretary of the interior, James G. Watt, who finally resigned on October 1983. To direct foreign affairs, Reagan named Alexander M. Haig, Jr., former NATO supreme commander for Europe, to the post of secretary of state; Haig, who clashed frequently with other administration officials, resigned in June 1982 and was replaced by George P. Shultz. In framing his foreign and defense policy, Reagan generally supported a military buildup as a precondition for arms-control talks with the USSR. His administration sent money and advisers to help the government of El Salvador in its war against leftist rebels, and US advisers were also sent to Honduras, reportedly to aid groups of Nicaraguans trying to overthrow the Sandinista government in their country. Troops were also dispatched to Lebanon in September 1982, as part of a multinational peacekeeping force in Beirut, and to Grenada in October 1983 to oust a leftist government there.

Reelected in 1984, President Reagan embarked on his second

term with a legislative agenda that included reduction of federal budget deficits (which had mounted rapidly during his first term in office), further cuts in domestic spending, and reform of the federal tax code. In military affairs, Reagan sought to persuade the public, the Congress, and the nation's European allies to support his Strategic Defense Initiative, commonly known as Star Wars, a highly complex and extremely costly space-based antimissile system. US–Soviet disarmament talks, suspended in late 1983 after deployment of US Tomahawk cruise and Pershing II missiles began in Western Europe, resumed in Geneva in January 1985.

In November 1985, Reagan met with Soviet leader Mikhail S. Gorbachev in Geneva, where they discussed arms control, human rights, US–Soviet relations, and other issues. The two leaders agreed to renew cultural and academic exchanges; to avoid incidents on air routes in the North Pacific; to meet again, in the US in 1986 and in the Soviet Union in 1987; and to work toward a pact to halt the spread of chemical weapons. However, arms-control and human-rights issues remained unresolved.

¹²FEDERAL GOVERNMENT

The Constitution of the United States, signed in 1787, is the nation's governing document. In the first 10 amendments to the Constitution, ratified in 1791 and known as the Bill of Rights, the federal government is denied the power to infringe on rights generally regarded as fundamental to the civil liberties of the people. These amendments prohibit the establishment of a state religion and the abridgment of freedom of speech, press, and the right to assemble. They protect all persons against unreasonable searches and seizures, guarantee trial by jury, and prohibit excessive bail and cruel and unusual punishments. No person may be required to testify against himself, nor may he be deprived of life, liberty, or property without due process of law. The 13th Amendment (1865) banned slavery; the 15th (1870) protected the freed slaves' right to vote; and the 19th (1920) guaranteed the franchise to women. In all, there have been 26 amendments, the last of which, in 1971, reduced the voting age to 18. The Equal Rights Amendment (ERA), approved by Congress in 1972, would have mandated equality between the sexes; only 35 of the required 38 states had ratified the ERA by the time the ratification deadline expired on 30 June 1982.

The US has a federal form of goverment, with the distribution of powers between the federal government and the states constitutionally defined. The legislative powers of the federal government are vested in Congress, which consists of the House of Representatives and the Senate. There are 435 members of the House of Representatives. Each state is allotted a number of representatives in proportion to its population as determined by the decennial census. Representatives are elected for two-year terms in every even-numbered year. A representative must be at least 25 years old, must be a resident of the state represented, and must have been a citizen of the US for at least seven years. The Senate consists of two senators from each state, elected for six-year terms. Senators must be at least 30 years old, must be residents of the states from which they are elected, and must have been citizens of the US for at least nine years. One-third of the Senate is elected in every even-numbered year.

Congress legislates on matters of taxation, borrowing, regulation of international and interstate commerce, formulation of rules of naturalization, bankruptcy, coinage, weights and measures, post offices and post roads, courts inferior to the Supreme Court, provision for the armed forces, among many other matters. A broad interpretation of the "necessary and proper" clause of the Constitution has widened considerably the scope of congressional legislation based on the enumerated powers.

A bill that is passed by both houses of Congress in the same form is submitted to the president, who may sign it or veto it. If the president chooses to veto the bill, it is returned to the house in which it originated with the reasons for the veto. The bill may become law despite the president's veto if it is passed again by a two-thirds vote in both houses. A bill becomes law without the president's signature if retained for 10 days while Congress is in session. After Congress adjourns, if the president does not sign a bill within 10 days, an automatic veto ensues.

The president must be "a natural born citizen " at least 35 years old, and must have been a resident of the US for 14 years. Under the 22nd Amendment to the Constitution, adopted in 1951, a president may not be elected more than twice. Each state is allotted a number of electors based on its combined total of US senators and representatives, and, technically, it is these electors who, constituted as the electoral college, cast their vote for president, with all of the state's electoral votes customarily going to the candidate who won the largest share of the popular vote of the state (the District of Columbia also has three electors, making a total of 538 votes). Thus, the candidate who wins the greatest share of the popular vote throughout the US may, in rare cases, fail to win a majority of the electoral vote. If no candidate gains a majority in the electoral college, the choice passes to the House of Representatives.

The vice president, elected at the same time and on the same ballot as the president, serves as ex officio president of the Senate. The vice president assumes the power and duties of the presidency on the president's removal from office or as a result of the president's death, resignation, or inability to perform his duties. In the case of a vacancy in the vice-presidency, the president nominates a successor, who must be approved by a majority in both houses of Congress. The Congress has the power to determine the line of presidential succession in case of the death or disability of both the president and vice president.

Under the Constitution, the president is enjoined to "take care that the laws be faithfully executed." In reality, the president has a considerable amount of leeway in determining to what extent a law is or is not enforced. Congress's only recourse is impeachment, to which it has resorted only twice, in proceedings against presidents Andrew Johnson and Richard Nixon. Both the president and the vice president are removable from office after impeachment by the House and conviction at a Senate trial for "treason, bribery, or other high crimes and misdemeanors." The president has the power to grant reprieves and pardons for offenses against the US except in cases of impeachment.

The President nominates and "by and with the advice and consent of the Senate" appoints ambassadors, public ministers, consuls, and all federal judges, including the justices of the Supreme Court. As commander in chief, the president is ultimately responsible for the disposition of the land, naval, and air forces, but the power to declare war belongs to Congress. The president conducts foreign relations and makes treaties with the advice and consent of the Senate. No treaty is binding unless it wins the approval of two-thirds of the Senate. The president's independence is also limited by the House of Representatives, where all money bills originate.

The president also appoints as his cabinet, subject to Senate confirmation, the secretaries who head the departments of the executive branch. As of 1985, the executive branch included the following cabinet departments: Agriculture (created in 1862), Commerce (1913), Defense (1947), Education (1980), Energy (1977), Health and Human Services (1980), Housing and Urban Development (1965), Interior (1849), Justice (1870), Labor (1913), State (1789), Transportation (1966), and Treasury (1789). The Department of Defense—headquartered in the Pentagon, the world's largest office building—also administers the various branches of the military: Air Force, Army, Navy, defense agencies, and joint-service schools. The Department of Justice administers the Federal Bureau of Investigation, which originated in 1908; the Central Intelligence Agency (1947) is under the aegis of the Executive office. Among the several hundred quasi-independent agencies are the Federal Reserve System (1913), serving as the nation's central bank, and the major regulatory bodies, notably the Environmental Protection Agency (1970), Federal Communications Commission (1934), Federal Power Commission (1920), Federal Trade Commission (1914), and Interstate Commerce Commission (1887).

Regulations for voting are determined by the individual states for federal as well as for local offices, and requirements vary from state to state. In the past, various southern states used literacy tests, poll taxes, "grandfather" clauses, and other methods to disfranchise black voters, but Supreme Court decisions and congressional measures, including the Voting Rights Act of 1965, more than tripled the number of black registrants in Deep South states between 1964 and 1984. In 1960, only 29.1% of the black voting-age population was registered, compared with 61.1% of the white population; by May 1984, the respective proportions were 58.5% and 66.5%.

[13]POLITICAL PARTIES

Two major parties, Democratic and Republican, have dominated national, state, and local politics since 1860. These parties are made up of clusters of small autonomous local groups primarily concerned with local politics and the election of local candidates to office. Within each party, such groups frequently differ drastically in policies and beliefs on many issues, but once every four years, they successfully bury their differences and rally around a candidate for the presidency. Minority parties have been formed at various periods in US political history, but most have generally allied with one of the two major parties, and none has achieved sustained national prominence. The most successful minority party in recent decades—the American Independent Party in

1968—was little more than a vehicle for former Alabama Governor George C. Wallace, who won 13.5% of the total vote cast. Various extreme groups on the right and left, including a small US Communist Party, have had little political significance on a national scale; in 1980, the Libertarian Party became the first minor party since 1916 to appear on the ballot in all 50 states. Independent candidates have won state and local office, but no candidate has won the presidency without major party backing. Running as an independent in 1980, former US Representative John Anderson won 6.6% of the vote. The Republican and Democratic presidential candidates in 1984 together captured 99.4% of the total vote; the leading minority party, the Libertarians, accounted for only 0.2% of the total.

Traditionally, the Republican Party is more solicitous of business interests and gets greater support from business than does the Democratic Party. A majority of blue-collar workers, by contrast, have generally supported the Democratic Party, which favors more lenient labor laws, particularly as they affect labor unions; the Republican Party often (though not always) supports legislation that restricts the powers of labor unions. Republicans favor the enhancement of the private sector of the economy, while Democrats generally urge the cause of greater government participation and regulatory authority, especially at the federal level.

Within both parties there are sharp differences on a great many issues; for example, northeastern Democrats in the past almost uniformly favored strong federal civil rights legislation, which was anathema to the Deep South; eastern Republicans in foreign policy are internationalist-minded, while midwesterners of the same party constituted from 1910 through 1940 the hard core of isolationist sentiment in the country. More recently, "conserva-

US Popular Vote for President by National Political Parties, 1948–1984

YEAR	WINNER	TOTAL VOTES CAST	% OF ELIGIBLE VOTERS	DEMOCRAT	REPUBLICAN	PROHIBITION	SOCIALIST LABOR	SOCIALIST WORKERS	SOCIALIST	PROGRESSIVE	STATES' RIGHTS DEMOCRAT	CONSTITUTION	OTHER[1]
1948	Truman (D)	48,692,442	51	24,105,587	21,970,017	103,489	29,038	13,614	138,973	1,157,057	1,169,134	—	5,533
1952	Eisenhower (R)	61,551,118	62	27,314,649	33,936,137	73,413	30,250	10,312	20,065	140,416	—	17,200	8,676
1956	Eisenhower (R)	62,025,372	59	26,030,172	35,585,245	41,937	44,300	7,797	2,044	—	2,657	108,055	203,165
											NATL. STATES' RIGHTS		
1960	Kennedy (D)	68,828,960	63	34,221,344	34,106,671	44,087	47,522	40,166	—	—	209,314	—	159,856
											UNPLEDGED DEM.		
1964	Johnson (D)	70,641,104	62	43,126,584	27,177,838	23,266	45,187	32,701	—	—	6,953	210,732	17,843
									COMMUNIST	PEACE & FREEDOM	AMERICAN IND.		
1968	Nixon (R)	73,203,370	61	31,274,503	31,785,148	14,915	52,591	41,390	1,076	83,720[2]	9,901,151	—	48,876
										LIBERTARIAN		AMERICAN	
1972	Nixon (R)	77,727,590	55	29,171,791	47,170,179	12,818	53,811	94,415[2]	25,343	3,671	—	1,090,673	104,889
								US LABOR					
1976	Carter (D)	81,552,331	54	40,829,046	39,146,006	15,958	40,041	91,310	58,992	173,019	170,531	160,773	866,655[3]
						POPULIST	RESPECT FOR LIFE	CITIZENS			IND. ALLIANCE		
1980	Reagan (R)	86,495,678	54	35,481,435	43,899,248	230,377	32,319	40,105	43,871	920,859	41,172	6,539	5,799,753[4]
1984	Reagan (R)	92,652,793	53	37,577,137	54,455,074	72,200	66,336	24,706	36,386	228,314	46,852	13,161	132,627[5]

[1] Includes votes for state parties, independent candidates and unpledged electors.
[2] Total includes votes for several candidates in different states under the same party label.
[3] Includes 756,631 votes for Eugene McCarthy, an independent.
[4] Includes 5,719,437 votes for John Anderson, an independent.
[5] Includes 78,807 votes for Lyndon H. LaRouche, an independent.

tive" headings have been adopted by members of both parties who emphasize decentralized government power, strengthened private enterprise, and a strong US military posture overseas, while the designation "liberal" has been applied to those favoring an increased federal government role in economic and social affairs, disengagement from foreign military commitments, and the intensive pursuit of nuclear-arms reduction.

President Nixon's resignation and the accompanying scandal surrounding the Republican Party hierarchy had a telling, if predictable, effect on party morale, as indicated by Republican losses in the 1974 and 1976 elections. The latent consequences of the Viet-Nam and Watergate years appeared to take their toll on both parties, however, in growing apathy toward politics and mistrust of politicans among the electorate. By 1979, Democrats enjoyed a large advantage over Republicans in voter registration, held both houses of Congress, had a majority of state governorships, and controlled most state legislative bodies. The centers of Democratic strength were the South and the major cities; most of the solidly Republican states were west of the Mississippi. Ronald Reagan's successful 1980 presidential bid cut into traditional Democratic strongholds throughout the US, as Republicans won control of the US Senate and eroded state and local Democratic majorities. In 1982, a recession year, the Democrats scored a net gain of 26 seats in the House of Representatives and won 27 out of 36 governorships being contested. On the strength of an economic recovery, President Reagan won reelection in November 1984, carrying 49 of 50 states (with a combined total of 525 electoral votes) and 58.8% of the popular vote; the Republicans retained control of the Senate, but the Democrats held on to the House. In the state legislatures, as of January 1985, Democrats occupied 1,188 of 1,995 senate memberships and 3,150 of 5,466 house seats.

The 1984 election marked a turning point for women in national politics, as Geraldine A. Ferraro, a Democrat, became the first female vice-presidential nominee of a major US political party; no woman has ever captured a major-party presidential nomination. As of January 1985, women held 2 US Senate seats, 22 seats in the US House of Representatives, 2 state governorships, and mayoralties in several major cities. The candidacy of another Democrat, Jesse L. Jackson, the first black ever to win a plurality in a statewide presidential preference primary, likewise marked the emergence of black Americans as a political force, especially within the Democratic Party. As of early 1985, the US had 6,056 black elected officials, including the mayors of some of the nation's largest cities; there were 20 blacks in the House. The House also has 10 Hispanics, and one state governor was of Hispanic ancestry.

[14]LOCAL GOVERNMENT

Governmental units within each state comprise counties, municipalities, and such special districts as those for water, sanitation,

Leading US Metropolitan Areas and Major Cities, 1980–84[1]

	METROPOLITAN AREAS			MAJOR CITIES					
RANK			POPULATION ('000)			POPULATION ('000)		AREA	% BLACK	% HISPANIC
1984	1980	NAME	1984 (EST.)	1980 (CENSUS)	NAME	1984 (EST.)	1980 (CENSUS)	(SQ MI)[2]	(1980)	(1980)
1	1	New York—Northern New Jersey—Long Island, N.Y., N.J., Conn.	17,807	17,540	New York	7,165	7,072	301.5	25.2	19.9
2	2	Los Angeles-Anaheim-Riverside, Calif.	12,373	11,498	Los Angeles	3,097	2,969	465.6	17.0	27.5
3	3	Chicago-Gary-Lake County, Ill.-Ind.-Wis.	8,035	7,937	Chicago	2,992	3,005	228.1	39.8	14.0
4	4	Philadelphia-Wilmington-Trenton, Pa.-N.J.-Del.-Md.	5,755	5,681	Philadelphia	1,647	1,688	136.0	37.8	3.8
5	5	San Francisco–Oakland–San Jose, Calif.	5,685	5,368	San Francisco	713	679	46.4	12.7	12.3
6	6	Detroit–Ann Arbor, Mich.	4,577	4,753	Detroit	1,089	1,203	135.6	63.1	2.4
7	7	Boston-Lawrence-Salem, Mass.	4,027	3,972	Boston	571	563	47.2	22.4	6.4
8	9	Houston-Galveston-Brazoria, Tex.	3,566	3,101	Houston	1,706	1,595	565.2	27.6	17.6
9	8	Washington, D.C.–Md.-Va.	3,429	3,251	Washington, D.C.	623	638	62.7	70.3	2.8
10	10	Dallas–Ft. Worth, Tex.	3,348	2,931	Dallas	974	905	332.6	29.4	12.3
11	12	Miami–Ft. Lauderdale, Fla.	2,799	2,644	Miami	373	347	34.3	25.1	55.9
12	11	Cleveland-Akron-Lorain, Ohio	2,788	2,834	Cleveland	547	574	79.0	43.8	3.1
13	14	St. Louis, Mo.-Ill.	2,398	2,377	St. Louis	429	453	61.4	45.6	1.2
14	16	Atlanta, Ga.	2,380	2,138	Atlanta	426	425	131.2	66.6	1.4
15	13	Pittsburgh–Beaver Valley, Pa.	2,372	2,423	Pittsburgh	403	424	55.4	24.0	0.8
16	15	Baltimore, Md.	2,245	2,199	Baltimore	764	787	80.3	54.8	1.0
17	17	Minneapolis–St. Paul, Minn.-Wis.	2,231	2,137	Minneapolis	358	371	55.1	7.7	1.3
18	18	Seattle-Tacoma, Wash.	2,208	2,093	Seattle	488	494	144.6	9.5	2.6
19	19	San Diego, Calif.	2,067	1,862	San Diego	960	876	321.4	8.9	14.9
20	22	Tampa–St. Petersburg–Clearwater, Fla.	1,811	1,614	Tampa	275	272	84.4	23.5	13.3
21	21	Denver-Boulder, Colo.	1,791	1,618	Denver	505	493	106.3	12.0	18.8
22	24	Phoenix, Ariz.	1,715	1,509	Phoenix	853	790	343.2	4.8	14.8
23	20	Cincinnati-Hamilton, Ohio-Ky.-Ind.	1,674	1,660	Cincinnati	370	385	78.0	33.8	0.8
24	23	Milwaukee-Racine, Wis.	1,568	1,570	Milwaukee	621	636	95.8	23.1	4.1
25	25	Kansas City, Mo.-Kans.	1,477	1,433	Kansas City, Mo.	443	448	316.3	27.4	3.3

[1] Includes both consolidated metropolitan statistical areas (CMSAs) and metropolitan statistical areas (MSAs) as classified by the US Bureau of the Census.
[2] As of 31 December 1980.

highways, parks, and recreation. There are more than 3,000 counties in the US; more than 19,000 municipalities, including cities, villages, towns, and boroughs; nearly 15,000 school districts; and at least 28,000 special districts. Additional townships, authorities, commissions, and boards make up the rest of the more than 82,000 local governmental units.

The states are autonomous within their own spheres of government, and their autonomy is defined in broad terms by the 10th Amendment to the US Constitution, which reserves to the states such powers as are not granted to the federal government and not denied to the states. The states may not, among other restrictions, issue paper money, conduct foreign relations, impair the obligations of contracts, or establish a government that is not republican in form. Subsequent amendments to the Constitution and many Supreme Court decisions added to the restrictions placed on the states. The 13th Amendment prohibited the states from legalizing the ownership of one person by another (slavery); the 14th Amendment deprived the states of their power to determine qualifications for citizenship; the 15th Amendment prohibited the states from denying the right to vote because of race, color, or previous condition of servitude; and the 19th, from denying the vote to women.

Since the Civil War, the functions of the state have expanded. Local business—that is, business not involved in foreign or interstate commerce—is regulated by the state. The states create subordinate governmental bodies such as counties, cities, towns, villages, and boroughs, whose charters they either issue or, where home rule is permitted, approve. States regulate employment of children and women in industry, and enact safety laws to prevent industrial accidents. Unemployment insurance is a state function, as are education, public health, highway construction and safety, operation of a state highway patrol, and various kinds of personal relief. The state and local governments still are primarily responsible for providing public assistance, despite the large part the federal government plays in financing welfare.

Each state is headed by an elected governor. State legislatures are bicameral except Nebraska's, which has been unicameral since 1934. Generally, the upper house is called the senate, and the lower house the house of representatives or the assembly. Bills must be passed by both houses, and the governor has a suspensive veto, which usually may be overridden by a two-thirds vote.

The number, population, and geographic extent of the more than 3,000 counties in the US—including the analogous units called boroughs in Alaska and parishes in Louisiana—show no uniformity from state to state. The county is the most conspicuous unit of rural local government and has a variety of powers, including location and repair of highways, county poor relief, determination of voting precincts and of polling places, and organization of school and road districts. City governments, usually headed by a mayor or city manager, have the power to levy taxes; to borrow; to pass, amend, and repeal local ordinances; and to grant franchises for public service corporations. Township government through an annual town meeting is an important New England tradition.

During the late 1960s and 1970s, several large cities began to suffer severe fiscal crises brought on by a combination of factors. Loss of tax revenues stemmed from the migration of middle-class residents to the suburbs and the flight of many small and large firms seeking to avoid the usually higher costs of doing business in urban areas. Low-income groups, many of them unskilled black and hispanic migrants, came to constitute large segments of city populations, placing added burdens on locally funded welfare, medical, housing, and other services without providing the commensurate tax base for additional revenues.

15 STATE SERVICES

All state governments provide services in the fields of education, transportation, health and social welfare, public protection (including state police and prison personnel), housing, and labor. The 1970s saw an expansion of state services in four key areas: energy, environment, consumer protection, and governmental ethics. By 1983, 47 states had offices or departments of energy; 48 states had agencies of departments of environment, while the others provided environmental protection through departments of health or natural resources. Each state provided some form of consumer advocacy, either through a separate department or agency or through the office of the attorney general. State government in the 1970s and early 1980s also showed the effects of the so-called post-Watergate morality. Laws mandating financial disclosure by public officials, once rare, had become common by 1983, when 37 states had offices designed specifically to oversee ethical compliance in government. Also notable were "sunshine laws," opening legislative committee meetings and administrative hearings to the public, and the use of an ombudsman either with general jurisdiction or with special powers relating, for example, to the problems of businesses, prisoners, the elderly, or racial minorities. Other trends in state administration, reflected on the federal level, include the separation of education from other services and the consolidation of social welfare programs in departments of human resources.

Federal aid to state, local, and territorial governments was estimated at more than $97 billion in 1983/84. The largest outlays were for income security, $24.6 billion; Medicaid, $20.2 billion; highways, $11.2 billion; general revenue sharing, $4.6 billion; community development block grants, $3.9 billion; urban mass transit, $3.9 billion; natural resources and environment, $3.6 billion; and compensatory education for the disadvantaged, $3.4 billion. New York State received more aid than any other state, $10.3 billion, followed by California, $9.8 billion. Illinois, Ohio, Michigan, Pennsylvania, and Texas each received more than $3.75 billion in federal assistance.

16 JUDICIAL SYSTEM

The Supreme Court, established by the US Constitution, is the nation's highest judicial body, consisting of the chief justice of the US and eight associate justices. All justices are appointed by the president with the advice and consent of the Senate. Appointments are for life "during good behavior," otherwise terminating only by resignation or impeachment and conviction.

The original jurisdiction of the Supreme Court is relatively narrow; as an appellate court, it is open to appeal from decisions of federal district courts, circuit courts of appeals, and the highest court in each state, although it may dismiss an appeal if it sees fit to do so. The Supreme Court, by means of a writ of certiorari, may call up a case from a district court for review. Regardless of how cases reach it, the Court enforces a kind of unity on the decisions of the lower courts. It also exercises the power of judicial review, determining the constitutionality of state laws, state constitutions, congressional statutes, and federal regulations, but only when these are specifically challenged.

The Constitution empowers Congress to establish all federal courts inferior to the Supreme Court. On the lowest level and handling the greatest proportion of federal cases are the district courts—numbering 91 in 1984, including one each in Puerto Rico and the District of Columbia—where all offenses against the laws of the US are tried. Certain civil actions that involve cases arising under treaties and laws of the US and under the Constitution also fall within the jurisdiction of the district courts. District courts have no appellate jurisdiction; their decisions may be carried to the courts of appeals, organized into 12 circuits (including one for the District of Columbia). These courts also hear appeals from decisions made by administrative commissions. For most cases, this is usually the last stage of appeal, except where the court rules that a state statute conflicts with the Constitution of the US, with federal law, or with a treaty. Special federal courts include the Claims Court, Court of International Trade, and Tax Court.

State courts operate independently of the federal judiciary. Most states adhere to a court system that begins on the lowest level with a justice of the peace, and includes courts of general trial jurisdiction, appellate courts, and, at the apex of the system, a state supreme court. The court of trial jurisdiction, sometimes called the county or superior court, has both original and appellate jurisdiction; all criminal cases (except those of a petty kind) and some civil cases are tried in this court. The state's highest court, like the Supreme Court of the US, interprets the constitution and the laws of the state.

The grand jury is a body of from 13 to 23 persons that brings indictments against individuals suspected of having violated the law. Initially, evidence is presented to it by either a justice of the peace or a prosecuting county or district attorney. The trial or petit jury is used in trials of common law, both criminal and civil, except where the right to a jury trial is waived by consent of all parties at law. It judges the facts of the case, while the court is concerned exclusively with questions of law.

The judicial system is only one facet of a US legal establishment that comprised an estimated 647,575 lawyers in 1984 and the full-time equivalent of 606,223 police and 296,645 corrections personnel in 1983; expenditures on police and corrections were nearly $25 billion during 1981/82. A total of 438,830 prisoners were in local, state, and federal institutions at the end of 1983 (42.8% more than in 1978); at the close of 1982, an estimated 243,900 convicts were on parole.

The Federal Bureau of Investigation Crime Index of offenses known to the police stood at 5,159 per 100,000 population in 1983. The rate for violent crimes was 529, for property crimes 4,630. Rates for specific crimes in 1983 per 100,000 population were as follows: murder and nonnegligent manslaughter, 8; forcible rape, 34; robbery, 214; aggravated assault, 273; burglary, 1,334; larceny-theft, 2,867; and motor vehicle theft, 429. Guns—widely available in the US—are the weapons used in more than half of all murders. Crime in the US is highly concentrated in metropolitan areas: the crime rate for all metropolitan statistical areas was 5,852 per 100,000 population in 1983, 26% higher than the rate for small towns and cities, and over three times the rate for rural areas.

Prior to 1972, almost all of the states authorized capital punishment for the most serious crimes—notably first-degree murder—but the death penalty was often applied in an arbitary and unequal manner. Between 1930 and 1967, 3,859 prisoners were executed under civil authority, of whom 2,066 were black. Of 3,335 prisoners executed for murder, 49% were black; of the 455 executed for rape, nearly 89% were black, an execution rate thought to reflect the historical prejudice of predominantly white juries against black people, especially in the South. It was this inequality of application, rather than a ruling that the death penalty was "cruel and unusual punishment," that led the Supreme Court to invalidate all death penalty statutes on the books in 1972, ordering that any subsequent capital punishment laws must provide safeguards against arbitrary and discriminatory treatment. By the end of 1983, 38 states had reinstituted the death penalty, electrocution and lethal injection being the most common methods; 31 prisoners (21 white, 10 black; 30 men, 1 woman) were executed from 1977 through 1984.

[17] ARMED FORCES

Compulsory induction into the armed services was terminated as of 1 July 1973; since that date, service in the military has been voluntary, although all men must register at age 18. As of 28 February 1985 there were an estimated 2,147,311 full-time personnel on active duty in the armed forces. The Army had 784,219, the Navy 563,569, the Marine Corps 198,733, and the Air Force 600,790. As of March 1984 there were 406,250 black personnel in the armed forces, or 19.6% of the total, although blacks made up only about 12% of the US population; blacks represented less than 6% of the total officer corps in 1983. Women numbered 226,439 in the armed forces as of March 1984; since 1970, their share of total enlisted and commissioned personnel has climbed from 1.1% to more than 10%. Personnel in the ready reserve but not on active duty in 1984 numbered 1,429,545: Army (including Army National Guard), 951,455; Navy, 178,474;

Net Migration by States and Other Selected Areas 1950–1983

	1950–1960	1960–1970	1970–1980	1980–1983
Alabama	−369,000	−233,000	97,000	−19,000
Alaska	41,000	16,000	28,000	50,000
Arizona	329,000	228,000	656,000	147,000
Arkansas	−433,000	−71,000	184,000	−2,000
California	3,142,000	2,113,000	1,573,000	750,000
Colorado	164,000	215,000	385,000	141,000
Connecticut	234,000	214,000	−121,000	−11,000
Delaware	63,000	38,000	−6,000	−2,000
Florida	1,616,000	1,326,000	2,519,000	831,000
Georgia	−212,000	51,000	329,000	120,000
Hawaii	3,000	11,000	55,000	15,000
Idaho	−40,000	−42,000	110,000	4,000
Illinois	124,000	−43,000	−649,000	−212,000
Indiana	61,000	−16,000	−206,000	−134,000
Iowa	−234,000	−183,000	−122,000	−70,000
Kansas	−44,000	−130,000	−71,000	—
Kentucky	−390,000	−153,000	131,000	−24,000
Louisiana	−49,000	−130,000	100,000	78,000
Maine	−67,000	−69,000	52,000	1,000
Maryland	321,000	385,000	−36,000	−3,000
Massachusetts	−96,000	74,000	−263,000	−38,000
Michigan	155,000	27,000	−496,000	−403,000
Minnesota	−98,000	−25,000	−80,000	−46,000
Mississippi	−433,000	−267,000	31,000	−8,000
Missouri	−134,000	2,000	−92,000	−38,000
Montana	−25,000	−58,000	16,000	5,000
Nebraska	−117,000	−73,000	−47,000	−13,000
Nevada	86,000	144,000	243,000	65,000
New Hampshire	12,000	69,000	117,000	18,000
New Jersey	578,000	488,000	−275,000	8,000
New Mexico	52,000	−130,000	116,000	38,000
New York	210,000	−101,000	−1,820,000	−136,000
North Carolina	−328,000	−94,000	278,000	83,000
North Dakota	−105,000	−94,000	−31,000	5,000
Ohio	407,000	−126,000	−779,000	−278,000
Oklahoma	−219,000	13,000	230,000	186,000
Oregon	16,000	159,000	341,000	−37,000
Pennsylvania	−475,000	−378,000	−551,000	−98,000
Rhode Island	−26,000	13,000	−53,000	−2,000
South Carolina	−222,000	−149,000	210,000	58,000
South Dakota	−95,000	−94,000	−41,000	−12,000
Tennessee	−274,000	−45,000	297,000	8,000
Texas	121,000	146,000	1,481,000	922,000
Utah	9,000	−11,000	119,000	50,000
Vermont	−38,000	15,000	27,000	3,000
Virginia	15,000	141,000	239,000	81,000
Washington	87,000	249,000	388,000	49,000
West Virginia	−446,000	−265,000	71,000	−13,000
Wisconsin	−53,000	4,000	−90,000	−64,000
Wyoming	−20,000	−39,000	85,000	20,000
Other areas:				
D.C.	−160,000	−100,000	−164,000	−24,000
Puerto Rico	−493,000	−200,000	−65,900	−90,000
Virgin Islands	−900	15,700	12,500	800
American Samoa	−5,400	−1,700	−4,100	−1,100
Guam	−10,800	−4,200	−5,700	2,200

Marine Corps, 88,047; and Air Force (including Air National Guard), 211,569.

Since World War II, national defense has been one of the largest items of expenditure in the federal budget. Although defense spending has increased enormously, its proportion of the total federal budget has decreased since the late 1950s (26.5% of the total estimated for 1984/85, as compared with nearly 51% in 1958/59). Estimated federal defense outlays in 1982 reached $196.3 billion, constituting about 24% of total world military expenditures. In accordance with Reagan administration policies, US defense spending increased from $157.5 billion in fiscal 1981 to $227.4 billion in fiscal 1984; federal budget proposals released in February 1985 called for further increases to $355.8 billion by fiscal 1988. Prime defense contract awards in 1983/84 totaled $136.9 billion, of which firms in California accounted for 21%. Defense-oriented industries employed 855,000 Americans in 1982 and shipped goods valued at $72.1 billion. The US is one of the world's leading arms exporters, registering military sales agreements worth $60.3 billion during 1978–82.

From the end of World War II to mid-1985, the US built about 60,000 nuclear warheads for use in more than 100 weapon systems, at an estimated cost of $750 billion. Strategic nuclear forces in 1985 included 1,030 intercontinental ballistic missiles (24 fewer than the maximum permissible under the 1972 treaty with the Soviet Union), 616 submarine-launched ballistic missiles, and 324 bombers. In late 1983, the US began deploying intermediate-range Pershing II and Tomahawk cruise missiles in Europe. Other US military hardware in 1983 included 12,300 Army tanks; 90 Navy attack submarines (85 of them nuclear-powered), 14 aircraft carriers (4 nuclear-powered), and 190 other major surface combat vessels; and some 3,650 Air Force combat aircraft. Overseas deployment of US forces, exceeding 500,000 at the end of 1984, is concentrated in Western Europe (especially the Federal Republic of Germany), the Philippines, Japan, the Republic of Korea, and Guam, a US possession.

As of 30 September 1983, an estimated 28,202,000 veterans of US military service were living in the 50 states, the outlying areas, and abroad. Of these, 23,109,000 served during wartime, some in more than one war: 297,000 in World War I, 10,978,000 in World War II, 5,294,000 during the Korean conflict, and 8,238,000 during the Viet-Nam era. Benefits disbursed by the Veterans Administration in 1982/83 totaled $28.2 billion, of which $13.9 billion went for compensation and pensions, $8.7 billion for medical assistance and administrative expenses, and $5.6 billion for other programs.

Army and Air National Guard units, distributed throughout the 50 states, may be summoned by a governor for military emergencies, riot control, disaster relief, and other purposes. As of 1983 there were 3,429 Army National Guard units, with 418,000 personnel, and 1,065 Air National Guard units, with 102,000 personnel.

The Federal Bureau of Investigation (FBI) investigates violations of federal law in more than 180 categories; state and local police matters and US intelligence activities overseas are among those categories excluded from its jurisdiction. Although lodged within the Department of Justice, the FBI enjoyed virtual autonomy until the death of longtime director J. Edgar Hoover in 1972. During the 1960s and 1970s, the FBI's political surveillance activities included illegal break-ins and attempts to harass, entrap, or discredit members of civil rights and radical groups.

State and local police forces had the full-time equivalent of 606,223 personnel in 1983, when expenditures on police protection exceeded $19 billion.

[18]MIGRATION

Between 1840 and 1930, some 37 million immigrants, the overwhelming majority of them Europeans, arrived in the US. Immigration reached its peak in the first decade of the 20th century,

when nearly 9 million came. Following the end of World War I, the tradition of almost unlimited immigration was abandoned, and through the National Origins Act of 1924, a quota system was established as the basis of a carefully restricted policy of immigration. Under the McCarran Act of 1952, one-sixth of 1% of the number of inhabitants from each European nation who resided in the continental US as of 1920 could be admitted annually. In practice, this system favored nations of Northern and Western Europe, with the United Kingdom, Germany, and Ireland being the chief beneficiaries. The quota system was radically reformed as of 1 July 1968, under a new law that established an annual ceiling of 170,000 for Eastern Hemisphere immigrants and 120,000 for entrants from the Western Hemisphere; in October 1978, these limits were replaced by a worldwide quota of 290,000, and in 1980, the quota was reduced to 270,000. Preferential exemptions from numerical limitations have been granted to outstanding scientists and their families, to parents of US citizens, to children of resident aliens, and to siblings and children of naturalized US citizens. The McCarran Act prohibits the immigration of Nazis, Communists, and Communist sympathizers to the US.

In the 12 months ending 30 September 1983, 559,763 legal immigrants entered the US, including 204,574 from countries in the Western Hemisphere, 58,867 from Europe, and 15,084 from Africa. A direct result of the revised quota system has been a sharp rise in the influx of Asians (primarily Filipinos, Koreans, Chinese, Indians, and Vietnamese), of whom 2,443,513 entered the country during 1971–83, as compared with 580,891 during 1951–70. In 1983, the countries that supplied the largest numbers of immigrants were Mexico, 59,079; China (People's Republic of China and Taiwan), 42,475; the Philippines, 41,456; Viet-Nam, 37,560; the Republic of Korea, 33,339; India, 25,451; Laos, 23,662; the Dominican Republic, 22,058; Jamaica, 19,535; Kampuchea, 18,120; and the United Kingdom, 14,830.

Since 1961, the federal government has supported and financed the Cuban Refugee Program. More than 500,000 Cubans were living in southern Florida by 1980, when another 125,000 Cuban refugees arrived. Following the defeat of the US-backed Saigon government, more than 200,000 Vietnamese refugees settled in the US between 1975 and 1981.

Under the Refugee Act of 1980, a ceiling for the number of admissible refugees is set annually. For 1984/85, the ceiling was set at 70,000: East Asia, 50,000; Eastern Europe and the Soviet Union, 10,000; Near East, 6,000; Africa, 3,000; and Latin America and the Caribbean, 1,000. Large numbers of aliens, mainly from the Caribbean, Latin America, and Asia, illegally establish residence in the US, having initially entered the country as tourists, students, or temporary visitors engaged in work or business. In 1980, 14,079,900 foreign-born persons were officially in residence; of these, slightly more than half were naturalized citizens.

The major migratory trends within the US have been the general westward movement during the 19th century; the long-term movement from farms and other rural settlements to metropolitan areas, a trend that showed signs of reversing in some states during the 1970s and early 1980s; the exodus of southern blacks to the cities of the North and Midwest, especially after World War I; a shift of whites from central cities to surrounding suburbs since World War II; and, also during the post–World War II period, a massive shift to the Sunbelt.

During 1980–83 alone, the states of the Northeast and Midwest collectively lost an estimated 1,520,000 residents through net migration, while the states of the South and West gained 3,567,-000—more than the populations of Alaska, Idaho, Montana, Nevada, and Wyoming combined. This population shift both reflected and stimulated a nationwide redistribution of political and economic power.

¹⁹INTERGOVERNMENTAL COOPERATION

The US government interacts with the governments of other nations and with the governments of the several states, the outlying areas, and thousands of municipalities. Interstate cooperation is reflected through numerous compacts and organizations.

The US, whose failure to join the League of Nations was a major cause of the failure of that body, is a charter member of the United Nations (UN) and of its specialized agencies. It contributes about 25% of the total funds required for the upkeep of the UN, far more than any other nation does. At the end of 1984, however, the US withdrew from the UN Educational, Scientific, and Cultural Organization (UNESCO) because of that body's alleged politicization and wasteful spending; moreover, there were policy differences between the US and several other UN agencies.

In the mid-1980s, the US participated in more than 70 intergovernmental organizations. Among these are international councils and commissions on rice, rubber, tea, wheat, wool, and other commodities, along with the Asian Development Bank and the Organization for Economic Cooperation and Development. The US also participates actively in the Permanent Court of Arbitration and in such hemispheric bodies as the Inter-American Development Bank, the Organization of American States, and Pan-American Health Organization. The North Atlantic Treaty Organization (NATO) is the principal military alliance to which the US belongs.

The US spent $265.8 billion in foreign aid from the end of World War II through September 1983, of which $187.3 billion was extended in outright grants and $78.5 billion in net credits. Of the total 1946–83 outlay, 62% was in economic assistance and 38% in military aid. The Marshall Plan, or European Recovery Program, had as its chief purpose the reconstruction of Europe; during the period 1948–52, $14.5 billion in economic aid was disbursed. The Mutual Security Act, in effect from 1953 to 1961, provided for the distribution of $16.9 billion in military and economic aid, with 38.3% going to East and Southeast Asia, the largest regional recipient. Since 1961, funds for economic aid have been administered by the Agency for International Development (AID). The Department of Defense now administers military aid. Through AID, the US, in addition to providing funds, goods, and equipment, makes available US experts in such fields as agriculture, industrial development, health, and housing. Also functioning chiefly in developing areas is the Peace Corps, created in 1961 and established as an independent agency in 1981. As of 1984, this organization had more than 5,200 volunteers providing assistance in health, education, agriculture, and community development throughout Latin America, Africa, Asia, and Oceania.

Foreign-aid disbursement has been used increasingly as a vehicle for the pursuit of US foreign policy objectives. In the early 1960s, more than $10 billion was committed to Latin America in support of President Kennedy's Alliance for Progress program. In the latter half of that decade, the bulk of aid went to East Asia, with South Viet-Nam, the Republic of Korea, and Thailand as primary recipients. In the 1970's, the active role taken by the US in seeking to negotiate a Middle East peace settlement was reflected in sizable outlays to that region, chiefly to Israel and Egypt. Of all nations, Israel is the largest economic aid beneficiary, having received nearly $6.9 billion in grants and credits between 1962 and 1983; Egypt follows closely with $6.7 billion. During the 1960s and 1970s, large amounts of development assistance also went to Brazil, India, Indonesia, Jordan, Pakistan, the Republic of Korea, and Turkey. In 1983, the US provided $5.2 billion in economic aid and $5.6 billion in military assistance; Israel was the chief beneficiary, with a combined aid total of $2.5 billion, followed by Egypt, with $2.1 billion. The chief aid recipient in Latin America was El Salvador, with $280 million

($199 million economic, $81 million military), as the US supported the Salvadoran central government against a left-wing insurgency.

Interstate compacts and agencies generally reflect shared geographic concerns. Established in 1933 and headquartered in Lexington, Ky., the Council of State Governments is a coordinating body supported by contributions from all the states; its chief functions are those of research and publication. Numerous interstate agreements govern such areas as boundaries, corrections, economic development, education, energy, environment, health, taxation, transportation, and water resources, among other concerns.

Revenues collected by the federal government are a major source of funding for state and local government operations. In 1959/60, federal grants-in-aid represented 16.8% of all state and local government receipts; the proportion rose to a high of 31.8% in 1977/78 but declined from 31.5% to 23.9% between 1979/80 and 1982/83, as a result of the Reagan administration's domestic spending cuts.

Federal grants to all territories, states, and localities in 1983/84 amounted to $97.2 billion; the leading beneficiary was New York State, which received $10.3 billion, followed by California, which received $9.8 billion.

²⁰ECONOMY

In variety and quantity, the natural resources of the US probably exceed those of any other nation, with the possible exception of the Soviet Union. The US is among the world's leading exporters of coal, wheat, corn, and soybeans. However, because of its vast economic growth, the US depends increasingly on foreign sources for a long list of raw materials. The extent of US dependence on oil imports was dramatically demonstrated during the 1973 Arab oil embargo, when serious fuel shortages developed in many sections of the country.

By the middle of the 20th century, the US was a leading consumer of nearly every important industrial raw material. The industry of the US produced about 40% of the world's total output of goods, despite the fact that the country's population comprised about 6% of the world total and its land area about 7% of the earth's surface.

In recent decades US production has continued to expand, though at a slower rate than that of most other industrialized nations. While the value of US exports of manufactured goods increased from $12.7 billion to $132.4 billion between 1960 and 1983, the US share of all world industrial exports decreased from 25.3% to 19.4%.

The US gross national product (GNP) more than tripled between 1979 and 1983 to $3.3 trillion in current dollars; measured in constant 1972 dollars, the increase was 41.3%. Although in absolute terms the US far exceeds every other nation in the size of its GNP, the US GNP per capita—$14,093 in 1983— was surpassed by Switzerland among the industrialized nations and by several oil-producing countries of the Middle East. According to preliminary data, the GNP rose by 6.8% in 1984, the best growth performance since 1951.

Inflation is an ever-present factor in the US economy, although the US inflation rate tends to be lower than that of the majority of industrialized countries: for the period 1970–78, for example, consumer prices increased by an annual average of 6.7%, less than in every other Western country except Austria, Luxembourg, Switzerland, and West Germany, and well below the price increase in Japan. Thus, the double-digit inflation of 1979–81 came as a rude shock to most Americans, and economists and politicians vied with each other in blaming international oil price rises, federal monetary policies, and US government spending for the problem.

The following table shows the erosion of the purchasing power of the dollar between 1940 and mid-1984 (1967 = $1.00):

	PRODUCER PRICES	CONSUMER PRICES
1945	$2.47	$2.38
1950	1.83	1.86
1955	1.17	1.25
1960	1.07	1.13
1965	1.05	1.06
1970	.91	.86
1975	.61	.62
1980	.41	.41
1984	.34	.32

Consumer price rises of 6.2% in 1982, 3.2% in 1983, and 3.7% in 1984 compared very favorably with inflation rates in most other nations.

The US usually exceeds most industrialized nations in unemployment—the US rate was 9.6% in 1983, as compared with 3.5% in Sweden and 2.7% in Japan—and pockets of concentrated unemployment in the central cities, especially among young nonwhites, constitute one of the nation's most serious social and economic problems.

The following table shows major components of the GNP for 1970, 1980, and 1983 (in billions):

	1970	1980	1983
Agriculture, forestry, fisheries	$ 28.6	$ 76.8	$ 72.7
Mining	17.6	96.0	112.4
Construction	48.9	119.8	130.7
Manufacturing	252.2	581.5	685.2
Transportation	38.7	98.5	114.9
Communications	23.8	67.0	92.4
Electric, gas, and sanitary services	23.1	66.4	99.4
Wholesale and retail trade	166.5	428.8	536.2
Finance, insurance, real estate	142.4	398.7	542.5
Services	114.4	342.6	477.5
Government and government enterprises	130.5	308.1	392.1
Other	6.0	47.5	48.8
TOTALS	$992.7	$2,631.7	$3,304.8

Industrial activity within the US has been expanding southward and westward for much of the 20th century, most rapidly since World War II. Louisiana, Oklahoma, and especially Texas are centers of industrial expansion based on petroleum refining; aerospace and other high technology industries are the basis of the new wealth of Texas and California, the nation's leading manufacturing state. The industrial heartland of the US is the east–north–central region, comprising Ohio, Indiana, Illinois, Michigan, and Wisconsin, with steelmaking and automobile manufacturing among the leading industries. The Middle Atlantic states (New Jersey, New York, and Pennsylvania) and the Northeast are also highly industrialized; but of the major industrial states in these two regions, Massachusetts has taken the lead in reorienting itself toward such high technology industries as electronics and information processing.

The mid-1980s found the US at an economic crossroads. During the first four years of the Reagan administration, the nation endured two years of severe recession followed by two years of robust recovery. The inflation rate was brought down, and millions of new jobs were created; but federal budget deficits averaged more than $150 billion annually, and the nation's trade position deteriorated. In 1984, the US posted a deficit on current accounts of $101.6 billion, more than double the old record, set in 1983. How to meet increasingly vigorous trade competition from Japan and other industrialized nations, how to control federal spending (without weakening the social "safety net" that keeps millions of Americans from sliding into poverty), how to promote investment in growth industries without wholly abandoning ailing but essential enterprises—these were the challenges the US faced on the way to the 1990s.

21 INCOME

Personal income in 1983 was estimated at $2.7 trillion; estimated disposable personal income (gross personal income less taxes and certain nontax payments) amounted to $2.3 trillion, as compared with $1.1 trillion in 1975. Estimated personal savings, after deductions for personal consumption expenditures, were $118.1 billion in 1983.

Average personal income per capita in 1983 was $11,675 ($3,945 in 1970), but it varied considerably from state to state. The highest per capita incomes (more than $13,000) were recorded in Alaska, the District of Columbia, Connecticut, New Jersey, California, New York, and Massachusetts, in that order. Among the lowest (under $9,100) were those of Mississippi, West Virginia, South Carolina, Utah, and Arkansas. Median family income rose from $6,957 in 1965 to $24,580 in 1983. The median income for white families (as defined by the head of the household) was $25,757, for Hispanic families $16,956, and for black families $14,506. Percentage distribution of money income by family groups in 1960, 1975, and 1983 was as follows:

INCOME LEVEL	1960	1975	1983[1]
Under $5,000	8.7	3.9	5.7
$5,000–9,999	12.8	9.8	10.2
$10,000–14,999	14.9	11.4	11.6
$15,000–19,999	21.9	11.8	11.8
$20,000–24,999	11.6	12.2	11.5
$25,000–34,999	11.7	26.8	19.5
$35,000–49,999	7.5	13.2	17.0
$50,000 and over	10.9	10.9	12.6
TOTALS	100.0	100.0	100.0

[1] Column does not add to total because of rounding.

Among white families, 4.4% had money incomes below $5,000, while 13.6% had incomes of $50,000 or more; the respective proportions for Hispanic families were 10.2% and 5%, for black families 16.2% and 4%.

Between 1959 and 1979, the proportion of the US population living below the poverty level, as defined by the federal government, declined from 22.4% to 12.4%; between 1980 and 1983, however, the proportion rose from 13% to 15.2%, as a result of inflation, recession, and federal cutbacks in domestic programs. Among the black population in 1983, an estimated 35.7% were impoverished, a higher proportion than at any time since the late 1960s. Nationwide economic recovery brought the overall poverty rate down to 14.4% in 1984, as the numbers of poor whites and poor blacks declined; among Hispanics, however, the poverty rate reached 28.4%, the highest since the federal government began keeping record. Families headed by women, representing 16% of all families in 1984, accounted for 48% of all families below the poverty level.

22 LABOR

About 115,299,000 persons constituted the nation's civilian labor force in August 1985. Of this number, 8,127,000 (7%) were unemployed; during 1982, a recession year, the unemployment rate reached 9.7%. A federal survey in March 1982 revealed the following nonfarm employment patterns:

	ESTABLISH-MENTS	EMPLOYEES	ANNUAL PAYROLL ('000)
Agricultural services, forestry, fishing, of which:	49,703	320,411	$ 3,874,359
Agricultural services	(46,664)	(289,681)	(3,453,247)
Mining, of which:	35,184	1,187,807	29,219,133
Bituminous coal, lignite	(3,959)	(249,477)	(6,295,676)
Oil, gas extraction	(23,577)	(634,267)	(14,778,114)
Contract construction	386,091	3,940,770	78,757,067
Manufacturing, of which:	328,932	19,572,113	385,474,515
Food and food products	(20,808)	(1,462,798)	(26,413,810)
Apparel and textiles	(21,367)	(1,188,608)	(12,069,965)

	ESTABLISH- MENTS	EMPLOYEES	ANNUAL PAYROLL ('000)
Paper and paper products	(6,160)	(607,985)	(12,907,622)
Printing and publishing	(48,264)	(1,298,298)	(22,867,193)
Chemicals and chemical products	(11,363)	(894,219)	(21,122,761)
Rubber and plastics products	(12,348)	(685,865)	(11,645,640)
Stone, clay, glass products	(15,591)	(545,966)	(10,448,774)
Primary metals	(7,048)	(955,986)	(21,100,377)
Fabricated metal products	(32,793)	(1,497,989)	(28,472,457)
Nonelectrical machinery	(48,947)	(2,341,417)	(48,035,661)
Electric and electronic equipment	(15,116)	(1,971,348)	(39,024,313)
Transportation equipment	(8,466)	(1,651,968)	(41,047,674)
Transportation, public utilities, of which:	176,589	4,626,875	103,817,037
Trucking and warehousing	(78,927)	(1,193,397)	(23,175,000)
Air transportation	(6,880)	(429,071)	(11,959,313)
Communications	(25,250)	(1,350,309)	(32,834,279)
Electric, gas, sanitary services	(16,641)	(795,015)	(20,206,413)
Wholesale trade	404,250	5,234,731	104,131,442
Retail trade	1,284,965	15,280,312	141,061,607
Finance, insurance, real estate, of which:	425,739	5,447,030	96,138,157
Banking	(49,014)	(1,581,979)	(24,999,222)
Insurance	(115,606)	(1,692,030)	(31,641,864)
Real estate	(165,971)	(973,111)	(12,931,133)
Services, of which:	1,441,316	18,581,939	255,156,602
Business services	(194,663)	(3,240,648)	(46,316,735)
Health services	(351,091)	(5,804,501)	(91,850,138)
Legal services	(111,555)	(581,263)	(12,862,151)
Educational services	(24,630)	(1,333,467)	(14,761,599)
Other	101,191	105,264	97,867,441
TOTALS	4,633,960	74,297,252	$1,199,359,203

As of October 1984, federal, state, and local governments employed 16,436,000 persons. Agriculture engaged 3,384,000 Americans during 1983.

Earnings of workers vary considerably with type of work and section of country. In 1983, the average hourly wage for industrial workers ranged from $6.68 in North Carolina to $12.34 in Alaska; the national average was $8.83. The average workweek for nonfarm employees in 1983 was 35 hours (41.2 hours in 1965), ranging from 40.1 hours in manufacturing to 29.8 hours in retail trade.

There were 55,097 labor unions in the United States, Puerto Rico, and US dependencies in 1983. In 1984, 18.8% of the nonagricultural work force belonged to labor unions, down from 23% in 1980. The most important federation of organized workers in the US is the American Federation of Labor–Congress of Industrial Organizations (AFL-CIO), whose affiliated unions had 13,758,000 members in 1983. The major independent unions and their estimated 1984 memberships are the International Brotherhood of Teamsters, 2,000,000, and the United Automobile Workers, 1,158,000. Most of the other unaffiliated unions are confined to a single establishment or locality. US labor unions exercise economic and political influence not only through the power of strikes and slowdowns but also through the human and financial resources they allocate to political campaigns (usually on behalf of Democratic candidates) and through the selective investment of multibillion-dollar pension funds.

The National Labor Relations Act of 1935 (The Wagner Act), the basic labor law of the US, was considerably modified by the Labor-Management Relations Act of 1947 (the Taft-Hartley Act) and the Labor-Management Reporting and Disclosure Act of 1959 (the Landrum-Griffin Act). Closed-shop agreements, which require employers to hire only union members, are banned. The union shop agreement, however, is permitted; it allows the hiring of nonunion members on the condition that they join the union within a given period of time.

In the mid-1980s, 19 states had right-to-work laws, forbidding the imposition of union membership as a condition of employment. Under the Taft-Hartley Act, the president of the US may postpone a strike for 90 days in the national interest. The act of 1959 requires all labor organizations to file constitutions, bylaws, and detailed financial reports with the secretary of labor, and stipulates methods of union elections. The National Labor Relations Board seeks to remedy or prevent unfair labor practices and supervises union elections, while the Equal Employment Opportunity Commission seeks to prevent discrimination in hiring, firing, and apprenticeship programs.

The number of work stoppages and of workers involved reached a peak in the late 1960s and early 1970s, declining steadily thereafter. In 1984 there were 62 major stoppages involving 376,000 workers, the lowest number at least since World War II; a major stoppage was defined as one involving 1,000 workers or more for a minimum of one day or shift.

[23]AGRICULTURE

The US produces a huge share of the world's soybeans, tallow and greases, corn for grain (maize), edible vegetable oils, cotton, oats, tobacco, and wheat. In 1981, US agricultural exports reached an all-time high of $43.3 billion and accounted for 18.5% of total US exports; in 1983, agricultural products worth $36.1 billion were exported (18% of total exports).

Gross farm income of $151.4 billion in 1983 included 138.7 billion in cash receipts (crops, $69.5 billion; livestock and livestock products, $69.2 billion); $9.3 billion in government payments; $1 billion in the value of products consumed on the farm; and $2.4 billion in other income. The net income of farm operators from farming in 1983 was $16.1 billion. Average net income per farm declined from $12,700 to $6,800 between 1981 and 1983, and the value of farm holdings dwindled from $816.3 billion in 1982 to $762.3 billion in 1984. By 1985, a combination of shrinking exports, weak domestic farm prices, rising costs, high interest rates on farm loans, and a debt burden that increased from $132 billion to $215 billion during 1979–84 had driven many farm families to the brink of bankruptcy. About 17% of the total US land area was in crops in 1982; another 26% was grassland pasture. Between 1930 and 1984, the number of farms in the US declined from 6,546,000 to 2,333,000; meanwhile, the size of the average farm nearly tripled from 151 to 437 acres (61 to 177 hectares), a result of the consolidation affected by large-scale mechanized production. The farm population, which comprised 35% of the total US population in 1910, declined to 25% during the depression of the 1930s and had dwindled to 2.5% by 1983.

A remarkable increase in the application of machinery to farms took place during and after World War II. Tractors, trucks, milking machines, grain combines, corn pickers, and pickup bailers became virtual necessities in farming. In 1920 there was fewer than one tractor in use for every 1,000 acres (405 hectares) of cropland harvested; by 1983 there were 14. Two other elements essential to US farm productivity are chemical fertilizers and irrigation. Fertilizer use increased from 24.9 million tons in 1960 to 42.3 million tons in 1983, when the $7.4 billion allocated to fertilizers and lime represented 5.5% of farm operating expenses. Farmers also spend well over $3 billion each year on chemical pesticides. Although only 5% of all farmland is irrigated, some of the irrigated lands—in particular, the Imperial and Central valleys in California—are among the most productive in the US.

Substantial quantities of corn, the most valuable crop produced in the US, are grown in almost every state; in 1983, 17.6% of the corn crop was exported, as was 16.4% of the soybean harvest. The following table reports estimated production and value of principal US field crops in 1983:

	PRODUCTION ('000)	ACREAGE HARVESTED ('000)	FARM VALUE ('000)
Corn for grain	4,166,000 bu	51,443	$14,064,000
Soybeans for beans	1,567,000 bu	61,815	12,838,000
Hay	141,000 tons	59,697	9,780,000
Wheat	2,420,000 bu	61,390	8,601,000
Tobacco	1,428,000 cwt	789	2,496,000
Cotton	7,800 bales	7,368	2,486,000
Potatoes	334,000 cwt	1,243	1,936,000
Sorghum for grain	479,000 bu	9,836	1,420,000
Rice, rough	100,000 cwt	2,169	888,000
Oats	477,000 bu	9,076	804,000

In 1983, the US also harvested an estimated 9,929,750 tons of vegetables for fresh market at a production value of $2.8 billion, and 10,246,280 tons of vegetables for processing, valued at $791.8 million. Arizona, Florida, California, and Texas are the major citrus-producing states. US citrus yields in 1982/83 included 9,734,000 tons of oranges and tangerines, valued at $1.4 billion; 2,447,000 tons of grapefruit, $190 million; and 947,000 tons of lemons, $109.7 million. Also produced in 1983 were an estimated 8.3 billion lb of commercial apples, worth $876.2 million.

24 ANIMAL HUSBANDRY

The US produced an estimated 15% of the world's meat supply in 1982. Major cattle-producing states are, in order of rank, Texas, Nebraska, Iowa, Kansas, Oklahoma, Missouri, and California; leading producers of sheep and lambs are Texas, California, Wyoming, South Dakota, Colorado, New Mexico, and Utah; and ranking producers of hogs and pigs include Iowa, Illinois, Minnesota, Indiana, Nebraska, and Missouri.

The livestock population at the end of 1983 included an estimated 114 million head of cattle (approximately 10% were milk cows), 55.8 million hogs and pigs, and 11.4 million sheep and lambs. In addition to about 365 million chickens on the nation's farms in 1983, 4.2 billion broilers were marketed by commerical producers, yielding $4.9 billion in gross income. Some 68.1 billion eggs were produced that year.

Improved techniques in breeding, feeding, and prevention of disease have accounted for great advances in animal husbandry in recent decades, and especially for the greater yield of high-quality meat per animal and the increased production of milk per cow. Milk production totaled 140 billion lb in 1983, with Wisconsin, California, New York, and Minnesota together accounting for 44% of the milk sold by farms. Wisconsin, California, and Minnesota account for more than half of all US butter production. Butter output totaled 1.3 billion lb in 1983. That year, the US was the world's largest producer of cheese, with more than 4.8 billion lb. Total production of meat (commercial and farm) in 1983 was estimated at more than 39.2 billion lb, comprising 23.2 billion lb of beef, 454 million lb of veal, 376 million lb of lamb and mutton, and 15.2 billion lb of pork.

25 FISHING

The US, which ranked 4th in the world in commercial fish landings in 1983, nevertheless imports far more fish and fishery products than it exports. In 1984, US fishery exports were valued at $948.8 million; imports, $5.9 billion.

The 1984 commercial catch was 6.4 billion lb and had a value of $2.35 billion. Food and nonfood fish are caught in almost equal quantities, with nonfood fish being processed for fertilizer and oil. Of the total volume of fish caught in 1984, 1.1 billion lb were utilized by canneries, 2.3 billion lb were sold either fresh or frozen, 82 million lb were cured, and 2.9 billion lb were reduced to meal, scrap, and oil. Industrial products included 375,100 tons of fish meal and scrap and 372.7 million lb of fish oils. A total of 3,891 firms in the US processed fishery products in 1983, when an estimated 223,000 fishermen were commercially employed on more than 127,000 vessels.

Menhaden—important for oil, meal, and fertilizer—was the leading species caught in 1984, accounting for 45% of all US commercial landings. Other leading varieties are salmon, shrimp, crabs, tuna, flounder, herring, mackerel, cod, and clams.

Pollution is a problem of increasing concern to the US fishing industry; dumping of raw sewage, industrial wastes, spillage from oil tankers, and blowouts of offshore wells are the main threats to the fishing grounds. Overfishing is also a threat to the viability of the industry in some areas, especially Alaska.

Leading fishing ports include Cameron, Empire-Venice, and Dulac-Chauvin in Louisiana; Pascagoula–Moss Point in Mississippi; the Los Angeles area in California; Beaufort–Morehead City in North Carolina; Gloucester and New Bedford in Massachusetts; and Kodiak in Alaska.

26 FORESTRY

US forestlands covered about 654 million acres (265 million hectares) in 1982. Major forest regions include the eastern, central hardwood, southern, Rocky Mountain, and Pacific coast areas. The National Forest System accounts for approximately 29% of the nation's forestland. Large private lumber companies control extensive tracts of land in Maine, Oregon, and several other states.

Domestic production of lumber during 1982 amounted to 25.6 billion board feet, of which softwoods accounted for roughly 80%. Consumption that year required imports of 7.7 billion board feet. The US, the world's 2d-leading producer of newsprint, attained an output of 4.7 million tons in 1983. To satisfy its needs for about 10.6 million tons, it also imported heavily from Canada, the major world producer. Other forest products in 1983 included 54.6 million tons of wood pulp, 67 million tons of paper and paperboard (excluding newsprint), 315 million board feet of hardwood flooring, 3.1 billion sq feet of particleboard (¾-in basis), and 21 billion sq feet of plywood (⅜-in basis). Rising petroleum prices in the late 1970s sparked a revival in the use of wood as home heating fuel, especially in the Northeast.

Throughout the 19th century, the federal government distributed forestlands lavishly as a means of subsidizing railroads and education. By the turn of the century, the realization that the forests were not inexhaustible led to the growth of a vigorous conservation movement, which was given increased impetus during the 1930s and again in the late 1960s. Federal timberlands are no longer open for private acquisition, although the lands can be leased for timber cutting and for grazing. In recent decades, the states also have moved in the direction of retaining forestlands and adding to their holdings when possible. As of 30 September 1984, the US Forest Service managed 156 national forests and 19 national grasslands.

The federal government assists owners of private forestland to plant, grow, protect, and market trees. The Soil Bank Plan, operated through the federal government, encourages the planting of trees through federal payments to farmers. The Cooperative Forestry Act of 1950 (amended in 1978) provides federal subsidies for reforestation, with sums granted by the government matched by similar sums from the state.

27 MINING

Rich in a variety of mineral resources, the US ranks among the world leaders in the production of many important mineral commodities, such as aluminum, cement, copper, pig iron, lead, magnesium, mercury, molybdenum, phosphates, potash, salt, sulfur, uranium, and zinc. The value of mineral production in 1982 amounted to $179.1 billion, three times that of 1975; fuels accounted for more than 89% of the total mineral output.

Leading mineral-producing states are Texas, Louisiana, California, Oklahoma, Wyoming, and New Mexico, important for petroleum and natural gas; and Kentucky, West Virginia, and Pennsylvania, important for coal. Iron ore supports the nation's most basic nonagricultural industry, iron and steel manufacture. Although large, the domestic output of iron ore does not fulfill the nation's total requirements, and in 1983, about 13.2 million tons

MAJOR LAND-USE REGIONS AND MINERAL RESOURCES

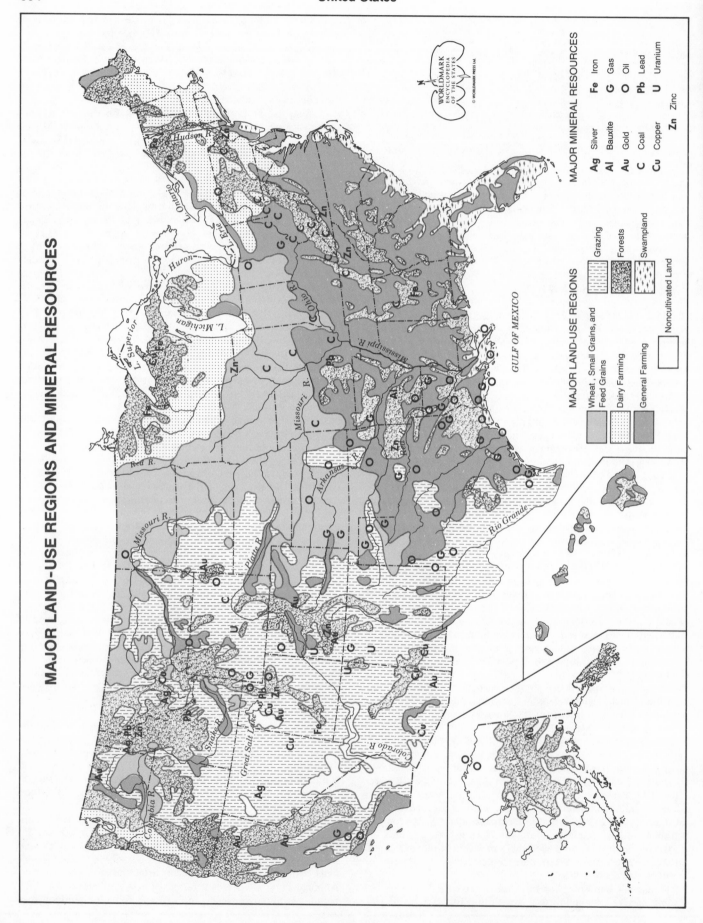

MAJOR MINERAL RESOURCES

Ag Silver   **Fe** Iron
Al Bauxite   **G** Gas
Au Gold   **O** Oil
C Coal   **Pb** Lead
Cu Copper   **U** Uranium
     Zn Zinc

MAJOR LAND-USE REGIONS

Wheat, Small Grains, and Feed Grains

Dairy Farming

General Farming

Grazing

Forests

Swampland

Noncultivated Land

WORLDMARK
ENCYCLOPEDIA
OF THE STATES
© WORLDMARK PRESS Ltd.

GULF OF MEXICO

Hudson R.
L. Ontario
L. Erie
L. Huron
L. Michigan
L. Superior
Ohio R.
Mississippi R.
Missouri R.
Red R.
Arkansas R.
Red R.
Rio Grande
Platte R.
Missouri R.
Snake R.
Great Salt Lake
Columbia R.
Colorado R.
Yukon R.

were imported. The major domestic sources are in the Lake Superior area: Minnesota and Michigan lead all other states in iron-ore yields.

The following table shows volume and value for selected mineral industries in 1982 (excluding fossil fuels):

	VOLUME	VALUE
Portland cement	61,100,000 tons	$3,084,400,000
Stone	791,000,000 tons	3,063,400,000
Sand and gravel	626,000,000 tons	2,022,900,000
Copper	1,256,000 tons	1,866,900,000
Iron ore	35,800,000 long tons	1,491,700,000
Uranium	26,900,000 lb	955,000,000
Phosphate rock	41,200,000 tons	950,300,000
Clays	35,300,000 tons	825,100,000
Salt	37,900,000 tons	671,100,000
Gold	1,447,000 fine oz	543,900,000
Molybdenum	78,000,000 lb	514,800,000
Sulfur	3,541,000 long tons	434,700,000
Boron minerals	1,234,000 tons	384,600,000
Silver	40,200,000 fine oz	319,900,000
Lead	565,000 tons	288,500,000
Potash	1,967,000 tons	265,600,000
Zinc	334,000 tons	254,700,000

[28]ENERGY AND POWER

The US, with 5% of the world's population, consumed 26% of the world's energy in 1982. The US produced 27% of the world's electricity, 25% of its coal, and 33% of its natural gas, but only 16% of its crude oil.

Refined petroleum products supplied 42.6% of the energy consumed in the US in 1983, natural gas 24.7%, coal 22.6%, waterpower 5.5%, nuclear power 4.5%, and other sources 0.1%. About 35% of the primary energy consumed in the US during 1983 went for the generation of electricity, 27% for transportation, 14% for residential and commercial use, and 24% for industrial and other purposes.

Approximately 55% of all residences were heated by utility gas in 1982, and nearly 5% by bottled gas. Some 15%, chiefly in the Northeast, used fuel oil; 19% of housing units were electrically heated; and 5% relied primarily on wood-burning stoves. The proportion of homes employing coal or coke declined from 35% to 0.5% between 1950 and 1982; a tiny minority of homes were using solar power devices, and some homes had no heat at all. For cooking purposes, electric stoves were the most popular, serving 52% of all residences in 1980; utility gas followed with 40%; and bottled and other gas fuels accounted for nearly all the remainder. In 1982, gas utility companies sold 14.3 quadrillion Btu of gas, of which 33% went to residential customers.

Proved recoverable reserves of crude oil totaled 27.7 billion barrels at the end of 1983, and reserves of natural gas were about 202 trillion cu feet. Proved coal reserves amounted to 489.5 billion tons in 1983, of which about 52% was bituminous, 37% subbituminous, 9% lignite, and less than 2% anthracite.

Mineral fuel production in 1983 included 777 million tons of coal, 16.8 trillion cu feet of natural gas, and 3.1 million barrels of crude petroleum. Oil imports that year exceeded 1.2 billion barrels (down 50% since 1977), of which 23% came from Mexico, 11% from the United Kingdom, 8% from Canada, and most of the remainder from members of the Organization of Petroleum Exporting Countries (OPEC). As of 1982, natural gas pipelines measuring 201,900 mi (324,900 km) carried a total of 16.2 trillion cu feet of gas; petroleum pipelines measured 172,500 mi (277,600 km) and conveyed 10.2 billion barrels of crude oil and petroleum products.

The 1973 Arab oil embargo and subsequent fuel shortages and price increases prompted a host of governmental measures aimed at increasing development of oil and gas resources; these included deployment of the trans-Alaska pipeline, to facilitate exploitation of that state's considerable petroleum reserves, and an easing of restrictions for oil drilling on the continental shelf. By the end of the 1970s, intensive exploration for oil and gas was under way along the Overthrust Belt, a region spanning 200,000 sq mi (518,000 sq km) of the Rocky Mountain states.

In 1983, utilities generated about 2.3 trillion kwh of electricity. Coal was the source of 54.8% of the electrical output, hydropower 14.4%, nuclear energy 12.7% (1.4% in 1970), natural gas 11.9%, and oil 6.2% (16.5% in 1978). Privately owned utility companies supplied 76.4% of the total output. Installed generating capacity in all facilities amounted to 658 million kw, an increase of nearly 30% since 1975.

The 1970s brought rapid development of plants and equipment designed to use nuclear energy for commercial purposes. In 1983, a total of 80 operable nuclear reactors had a total capacity of 62.8 million kw and generated 292 billion kwh of electricity (as compared with 6.5 million kw and 22 billion kwh in 1970). The future of the nuclear power industry was cast in doubt, however, not only by safety concerns stemming from a nuclear accident at Pennsylvania's Three Mile Island facility in March 1979 but also by the rapidly rising costs of nuclear power plant construction and operation.

By the end of 1983, power companies had canceled plans for an estimated 100 additional reactors, leaving stockholders and consumers to cover $11 billion in lost development outlays. Several large utilities, faced with cost overruns of up to 1,000 percent on nuclear power plant construction, were in severe financial difficulty at mid-decade.

As the 1980s began, growing attention was focused on the development of solar power, synthetic fuels, geothermal resources, and other energy technologies. Such energy conservation measures as mandatory automobile fuel-efficiency standards, enforcement of a 55-mph (89-km/hr) speed limit on the nation's highways, and tax incentives for home insulation have been promoted by the federal government, which also decontrolled oil and gas prices in the expectation that a rise in domestic costs to world market levels would provide a powerful economic incentive for consumers to conserve fuel.

Between 1979 and 1983, through the combined effects of recession and conservation, US energy consumption per capita dropped from 351 million to 301 million Btu, and consumption per dollar of GNP declined from 53,300 to 46,000 Btu. Such consumption cutbacks helped lead to disarray within OPEC and to a drop in oil prices—a development that had a beneficial impact on consumer prices and the US trade balance, but which also had the effect of undercutting efforts to make alternatives to fossil fuels economically viable.

[29]INDUSTRY

Although the US remains one of the world's preeminent industrial powers, manufacturing no longer plays as dominant a role in the economy as it once did.

Between 1970 and 1983, industry's share of the GNP declined from 25.4% to 20.7%, and its share of total civilian employment decreased from 26.3% to 19.8%, with a further drop to 18.4% projected by 1995. Throughout the 1960s, manufacturing accounted for about 29% of total national income; by 1983, the proportion was down to about 22%.

Leading industrial centers are the metropolitan areas of Chicago, Los Angeles, New York, Detroit, and Philadelphia. The Midwest leads all other regions by virtue of its huge concentration of heavy industry, including the manufacturing of automobiles, trucks, and other vehicles.

In 1983, US manufacturers had net sales of $2.1 trillion and a net profit (after state and federal taxes) of $86 billion. Value of shipments exceeded $1.9 trillion in 1982, when new capital expenditures totaled $75 billion and value added by manufacturing exceeded $824 billion. The following table shows selected data for major industrial sectors in 1982 (in millions of dollars):

	COST OF MATERIALS	CAPITAL OUTLAYS, NEW	VALUE OF SHIPMENTS
Food and food products	$192,117	$6,807	$280,961
Petroleum and coal products	185,432	6,666	209,691
Transportation equipment	120,250	7,306	202,769
Nonelectrical machinery	83,822	8,481	187,601
Chemicals and chemical products	91,966	9,080	170,085
Electric and electronic equipment	63,593	7,587	148,877
Fabricated metals	58,856	3,844	118,484
Primary metals	68,768	4,640	104,855
Apparel and textiles	56,025	2,243	101,197
Printing and publishing	31,375	3,247	85,762
Paper and paper products	46,512	5,087	79,675
Rubber and plastic products	27,870	2,215	55,311
Instruments and related products	17,739	2,109	51,696

Giant firms dominate industry in the US. During 1983, the 100 largest manufacturing firms accounted for 48% of the total value of corporate assets. Large corporations are dominant especially in sectors such as steel, automobiles, pharmaceuticals, aircraft, petroleum refining, computers, soaps and detergents, tires, and communications equipment. The 500 largest firms as ranked by *Fortune* magazine in 1984 had nearly $1.8 trillion in sales, employed 14.2 million persons, and registered $86 billion in net profits. Fourteen companies earned more than $1 billion each in 1984. The largest earners, with 1984 net income, were International Business Machines, $6.6 billion; Exxon (formerly Standard Oil of New Jersey), $5.5 billion; General Motors, $4.5 billion; Ford, $2.9 billion; Chrysler, $2.4 billion, and General Electric, $2.3 billion. The growth of multinational activities of US corporations has been rapid in recent decades, with capital expenditures by US-owned foreign affiliates peaking at $42.4 billion in 1980 before declining to $37.7 billion in 1983.

The history of US industry has been marked by the introduction of increasingly sophisticated technology in the manufacturing process. Advances in chemistry and electronics have revolutionized many industries through new products and methods: examples include the impact of plastics on petrochemicals, the use of lasers and electronic sensors as measuring and controlling devices, and the application of microprocessors to computing machines, home entertainment products, and a variety of other industries. Science has vastly expanded the number of metals available for industrial purposes, notably such light metals as aluminum, magnesium, and titanium. Integrated machines now perform a complex number of successive operations that formerly were done on the assembly line at separate stations. Those industries have prospered that have been best able to make use of the new technology, and the economies of some states—in particular, California and Massachusetts—are largely based on it. On the other hand, certain industries—especially clothing and steelmaking—have suffered from outmoded facilities that (coupled with high US labor costs) force the price of their products above the world market level. Automobile manufacturing was another ailing industry in the early 1980s: although domestic producers of passenger cars rebounded from a prolonged slump by selling 7,914,738 units during the 1984 model year, sales of imported cars that year reached a record 2,407,679 units. Moreover, included among the cars assembled domestically were an increasing number of vehicles produced by US subsidiaries of foreign firms and containing a substantial proportion of foreign-made parts.

30 COMMERCE

Domestic trade is a vast enterprise in the US, employing more than 20 million Americans and accounting for payrolls of over $245 billion and sales of well over 3 trillion. The value of wholesale trade reached $1,997.9 billion in 1982, up 59% from 1977; retail sales were $1,039 billion, representing a 49% increase.

Major competitors of the small shopkeeper are the chain retailers; discount houses; manufacturers of millinery, clothing, shoes, ties, candy, and other products who operate their own outlets; and the great department stores. Corporate chains usually buy directly from producers and processors, thus avoiding intermediaries. Multiunit chain stores account for only about 35% of the total retail trade, but in certain kinds of retail business, the chain is the dominant mode of business organization. Chains handled about 79% of the variety store trade in 1983, 96% of department store trade, 52% of the shoe business, and 57% of grocery store volume. With the great suburban expansion of the 1960s emerged the planned shopping center, usually designed by a single developer and embracing many different kinds of stores in a single setting. Characteristic of the 1970s and early 1980s was the growth of franchising, especially in fast foods: the number of restaurant franchise outlets grew from 32,600 to 74,800 between 1970 and 1984, while their collective sales volume expanded from $4.6 billion to an estimated $44.1 billion.

Installment credit is a major support for consumer purchases in the US. The total amount of consumer installment credit outstanding by May 1984 was $418.1 billion, of which automobile loans accounted for $152.2 billion. Commercial banks held 44.7% of the debt outstanding; finance companies 24.6%, credit unions 14%, and retailers and other lenders 16.7%. Noninstallment credit totaled $102 billion. Credit cards are used by a majority of US families.

The US advertising industry is the world's most highly developed. With the expansion of television audiences particularly, spending for advertising has increased almost annually to successive record levels. Advertising expenditures in 1983 reached an estimated $75.9 billion, as compared with $19.6 billion in 1970 and $5.7 billion in 1950. Of the 1983 total, newspaper advertising accounted for about $20.6 billion, television $16.1 billion, radio $5.2 billion, and magazines $4.2 billion. Direct mail advertising (chiefly letters, booklets, catalogs, and handbills) amounted to $11.8 billion. New York City is the center of the nation's advertising industry.

In the realm of foreign commerce, the US leads the world in value of both exports and imports. According to preliminary figures, exports of domestic merchandise, raw materials, agricultural and industrial products, and military goods amounted in 1984 to $217.9 billion; general imports for the same calendar year were valued at an all-time record $341.2 billion. During the late 1970s, the US ran annual trade deficits of more than $20 billion, largely because of the rising cost of foreign oil. By 1983, however, the deficit had soared to $57.5 billion, and the estimated 1984 shortfall of $123.3 billion was by far the worst in US history. The poor 1984 trade performance was attributed to overvaluation of the US dollar, which made US imports cheaper and exports dearer; to the fact that the US had recovered from recession earlier than its major export customers; and to protectionist practices by foreign competitors, especially Japan.

The following table shows exports, including reexports, for 1983 (in millions):

Machinery, of which:	$ 54,695
Electronic computers, parts, and accessories	(10,599)
Power generating machinery	(8,718)
Telecommunications apparatus	(3,804)
Transport equipment, of which:	27,298
Road motor vehicles and parts	(13,492)
Aircraft, parts, and accessories	(12,189)
Food and live animals, of which:	24,166
Wheat and wheat flour	(6,509)
Corn	(6,480)
Animal feed	(2,802)

Chemicals and chemical products, of which:	19,751
Plastic materials and resins	(3,732)
Medicines and pharmaceuticals	(2,494)
Chemical fertilizers	(1,267)
Crude materials, of which:	18,596
Soybeans	(5,925)
Wood and wood pulp	(3,712)
Ores and metal scrap	(2,276)
Raw cotton	(1,817)
Mineral fuels, of which:	9,500
Petroleum and products	(4,557)
Coal	(4,051)
Professional, scientific, and controlling instruments	5,856
Metals and metal products	5,477
Clothing and textiles	3,164
Armaments and ammunition	3,092
Beverages and tobacco	2,813
Paper and paper products	2,553
Animal and vegetable oils and fats	1,459
Other exports and reexports	22,118
TOTAL	$200,538

Imports for 1983 (in millions) were as follows:

Mineral fuels, of which:	$ 57,952
Petroleum and products	(52,325)
Machinery, of which:	46,975
Telecommunications apparatus	(11,278)
Office machines	(6,759)
Electronic components	(5,388)
Engines and parts	(3,570)
Transport equipment, of which:	39,156
Automobiles, new	(22,934)
Aircraft and parts	(2,051)
Metals and metal products, of which:	18,717
Iron and steel	(6,338)
Silver, refined bullion	(1,926)
Aluminum	(1,629)
Copper	(1,341)
Food and live animals, of which:	15,412
Fish	(3,594)
Coffee, green	(2,590)
Vegetables, fruits, and nuts	(2,237)
Meat and meat products	(2,034)
Sugar	(1,046)
Clothing and textiles	12,808
Chemicals and chemical products	10,779
Crude materials, of which:	9,590
Wood and wood pulp	(4,187)
Ores and metal scrap	(2,498)
Footwear	4,010
Professional, scientific, photographic, and controlling instruments	3,528
Beverages and tobacco	3,408
Toys, games, and sporting goods	2,412
Artworks and antiques	2,017
Clocks and watches	1,058
Animal and vegetable oils and fats	495
Other imports	29,731
TOTAL	$258,048

By value of combined exports and imports, the largest proportion of US foreign trade is with nations of the Western Hemisphere. These countries accounted for 31.9% of US exports and 36.4% of US imports in 1983, closely followed by Asia (exports 31.8%, imports 35.4%) and Western Europe (exports 27.9%, imports 20.9%). Canada is the nation's single best customer and supplier, accounting for about 20% of total trade in 1983. Intensive negotiations were held between Washington, D.C., and Tokyo during the late 1970s and early 1980s in an attempt to lessen the growing US trade imbalance with Japan, but in 1984, the deficit reached a record $36.8 billion.

Principal trading partners in 1983 (ranked by total trade, in millions of dollars) were as follows:

	EXPORTS	IMPORTS	BALANCE
Canada	$ 38,244	$ 52,130	$−13,886
Japan	21,894	41,183	−19,289
Mexico	9,082	16,776	−7,694
United Kingdom	10,621	12,470	−1,849
Germany, Federal Republic of	8,737	12,695	−3,958
Taiwan	4,667	11,204	−6,537
Korea, Republic of	5,925	7,148	−1,223
France	5,961	6,025	−64
Sa'udi Arabia	7,903	3,627	4,276
Netherlands	7,767	2,970	4,797
Italy	3,908	5,455	−1,547
Hong Kong	2,564	6,394	−3,830
Venezuela	2,811	4,938	−2,127
Brazil	2,557	4,946	−2,389
Belgium-Luxembourg	5,049	2,412	2,637
Indonesia	1,466	5,285	−3,819
Singapore	3,759	2,868	891
Australia	3,954	2,222	1,732
Switzerland	2,960	2,494	466
Nigeria	864	3,736	−2,872
Spain	2,763	1,533	1,230
South Africa	2,129	2,027	102
Algeria	594	3,551	−2,957
India	1,828	2,191	−363
Sweden	1,581	2,429	−848
Philippines	1,807	2,001	−194
Malaysia	1,684	2,124	−440
Israel	2,017	1,255	762
Egypt	2,813	303	2,510
Other countries	32,629	33,656	−1,027
TOTALS	$200,538	$258,048	$−57,510

Since 1950, the US has consistently had deficits in its overall payments with the rest of the world, despite the fact that it had an unbroken record of annual surpluses up to 1970 on current-account goods, services, and remittances transactions. The nation's stock of gold declined from a value of $22.9 billion at the start of 1958 to $10.5 billion as of 31 July 1971, only two weeks before President Richard Nixon announced that the US would no longer exchange dollars for gold. On 12 February 1973, pressures on the US dollar compelled the government to announce a 10% devaluation against nearly all of the world's major currencies. International gold reserves thereupon rose from $14.4 billion in 1973 to $15.9 billion at the end of 1974; as of 30 June 1984, gold and other reserve assets stood at $34.5 billion. In March 1985, the Commerce Department reported that in 1984, the US had posted a record current-accounts deficit of $101.6 billion. Estimates were that the US net investment position, in surplus by $32 billion at the close of 1984, had passed into deficit during the first quarter of 1985—the first time the US had been classed as a net debtor nation since 1914.

[31]CONSUMER PROTECTION

Consumer protection has become a major government enterprise during the 20th century. The Federal Trade Commission (FTC), established in 1914, administers laws governing the granting and use of credit and the activities of credit bureaus; it also investigates unfair or deceptive trade practices, including price fixing and false advertising. The Securities and Exchange Commission, created in 1934, seeks to protect investors, while the Consumer Product Safety Commission, created in 1972, has the authority to establish product safety standards and to ban hazardous products. Overseeing the safety of air and highway transport is the National Transportation Safety Board, established in 1975. The Consumer Information Center Program of the General Services Administration (Pueblo, Colo.) and the Food Safety and Inspection Service and Food and Nutrition Service of the Department of Agriculture also serve consumer interests. Legislation that would have established a Department of Consumer Affairs failed to win congressional approval several times during the 1970s, however.

Public interest groups have been exceptionally effective in promoting consumer issues. The Consumer Federation of America (CFA; founded in 1967), with 220 member organizations, is the largest US consumer advocacy body; its concerns include product pricing, credit, and the cost and quality of health care, education, and housing. The CFA also serves as a clearinghouse for consumer information. Consumers Union of the US, founded in 1936, publishes the widely read monthly *Consumer Reports*, which tests, grades, and comments on a variety of retail products. The National Consumers League, founded in 1899, was a pioneer in the consumer movement, focusing especially on labor laws and working conditions. Much of the growth of consumerism in the 1970s resulted from the public relations efforts of one man— Ralph Nader. Already a well-known consumer advocate concerned particularly with automobile safety, Nader founded Public Citizen in 1971 and an affiliated litigation group the following year. As of 1984, Public Citizen claimed 200,000 supporters; its activities include research committees on tax reform, health care, work safety, and energy.

Other avenues open to consumers in most states include small claims courts, generally open to claims between $100 and $1,500 at modest legal cost. Complaints involving professional malpractice may be brought to state licensing or regulatory boards. Supported by the business community, the nearly 150 US Better Business Bureaus provide general consumer information and arbitrate some customer-company disputes.

[32]BANKING

The Federal Reserve Act of 1913 provided the US with a central banking system. The Federal Reserve System dominates US banking, is a strong influence in the affairs of commercial banks, and exercises virtually unlimited control over the money supply.

Each of the 12 federal reserve districts contains a federal reserve bank. A board of nine directors presides over each reserve bank. Six are elected by the member banks in the district: of this group, three may be bankers; the other three represent business, industry, or agriculture. The Board of Governors of the Federal Reserve System (usually known as the Federal Reserve Board) appoints the remaining three, who may not be officers, directors, stockholders, or employees of any bank and who are presumed therefore to represent the public.

The Federal Reserve Board regulates the money supply and the amount of credit available to the public by asserting its power to alter the rediscount rate, by buying and selling securities in the open market, by setting margin requirements for securities purchases, by altering reserve requirements of member banks in the system, and by resorting to a specific number of selective controls at its disposal. The Federal Reserve Board's role in regulating the money supply is held by economists of the monetarist school to be the single most important factor in determining the nation's inflation rate.

Member banks increase their reserves or cash holdings by rediscounting commercial notes at the federal reserve bank at a rate of interest ultimately determined by the Board of Governors. A change in the rediscount rate, therefore, directly affects the capacity of the member banks to accommodate their customers with loans. Similarly, the purchase or sale of securities in the open market, as determined by the Federal Open Market Committee, is another device whereby the amount of credit available to the public is expanded or contracted. The same effect is achieved in some measure by the power of the Board of Governors to raise or lower the reserves that member banks must keep against demand deposits. Credit tightening by federal authorities in early 1980

pushed the prime rate—the rate that commercial banks charge their most creditworthy customers—above 20% for the first time since the financial panics of 1837 and 1839, when rates reached 36%. As federal monetary policies eased, the prime rate dropped below 12% in late 1984.

Combined assets of all commercial banks as of April 1985 exceeded $2.5 trillion. Net loans and leases amounted to nearly $1.5 trillion; aggregate domestic deposits were $1,645 billion, foreign deposits $318 billion.

The nation's leading commercial bank holding-companies on the basis of total assets as of 31 December 1984 were Citicorp— the largest commercial bank in the world, with $150.6 billion—in New York, followed by BankAmerica, $117.7 billion in San Francisco.

Under the provisions of the Banking Act of 1935, all members of the Federal Reserve System (and other banks that wish to do so) participate in a plan of deposit insurance (up to $100,000 for each individual account as of 1985) administered by the Federal Deposit Insurance Corporation (FDIC). As of April 1985, the deposits of 14,494 commercial banks were insured by the FDIC. All national banks, of which there were 4,904 in 1985, are regulated members of the Federal Reserve System; most of the 9,590 insured state banks were subject to FDIC regulation. Mutual savings banks, also regulated by the FDIC, numbered 393 in 1983 and had combined assets of $179.5 billion.

As of 31 December 1983, 3,040 savings and loan associations were insured by the Federal Savings and Loan Insurance Corporation (FSLIC). Individual accounts were insured up to a limit of $100,000 in 1985. Assets of all FSLIC-insured institutions totaled $814.6 billion in 1983, of which $516.4 billion consisted of mortgage loans and contracts. The 10,962 federally chartered credit unions had 26.8 million members and combined assets of $54.5 billion in 1983; during the same year there were 8,200 state-chartered credit unions with 21.8 million members and $43.8 billion in assets.

As of December 1983, total currency in circulation was $171.9 billion.

[33]INSURANCE

In 1983, 2,125 companies dispensed ordinary life, group, industrial, and other kinds of life insurance policies. The overwhelming majority of US families have some life insurance with a legal reserve company, the Veterans Administration, or fraternal, assessment, burial, or savings bank organizations. As of 31 December 1983, 387 million life insurance policies were in force, with a total value of nearly $5 trillion. Payments to policyholders, annuitants, and beneficiaries totaled $80.9 billion. The average value of life insurance per family was $54,200, ranging from $39,900 in Arkansas to $113,000 in the District of Columbia.

Hundreds of varieties of insurance may be purchased. Besides life, the more important types of insurance include accident, fire, hospital and medical expense, group accident and health, automobile liability, automobile damage, workers' compensation, ocean marine, and inland marine. In the early 1980s, 23 states had a "no-fault" form of automobile insurance, under which damages may be awarded automatically, without recourse to legal suit.

As of 1983, insurance companies accounted for 11.7% of new investment funds and for 4.8% of all public debt securities held by private investors. Life insurance company assets totaled $654.9 billion on 31 December 1983; the leading life insurance companies, ranked by assets at the end of 1984, were Prudential, $78.9 billion; Metropolitan, $67.4 billion; Equitable, $44.5 billion; and Aetna, $34 billion. The property and liability insurance industry, consisting of 3,474 companies in 1984, had about $231.7 billion in assets in 1982. Premiums written that year totaled $104 billion: automotive liability coverage, $26.2 billion; automobile physical damage insurance, $18 billion; workers' compensation coverage, $13.9 billion; homeowners' multiperil insurance, $11.7 billion;

commercial multiperil policies, $7 billion; and other lines, $27.2 billion. Flood insurance worth $113.1 billion was in force in 1983.

³⁴SECURITIES

Stocks, stock options, commodities, commodity features, bonds of various types, and other investment instruments are traded on exchanges in major US cities. Of the nearly two dozen registered stock exchanges in the US, those of greatest importance as judged by the total value of shares and options bought and sold are the New York and American (both in New York), Midwest and Chicago Option (both in Chicago), Pacific (San Francisco and Los Angeles), and Philadelphia exchanges. In 1983, 24.3 billion shares with a market value of $815 billion were traded on the New York Stock Exchange; these record totals represented 80% by volume and 85% by value of all shares traded in the US that year. The American Stock Exchange ranked second in both volume and value, accounting for 7% and 3%, respectively. The total market value of all new securities offered in 1983 was $100.5 billion, of which bonds accounted for $48.1 billion, common stock $44.8 billion, and preferred stock $7.6 billion. Chicago and Minneapolis are centers for farm product trading.

Following World War II there was a steady growth in the number of US stockholders, until the early 1970s, when a decline became evident; as of 1975 there were an estimated 25,270,000 shareowners, or one out of every six adult Americans. That proportion remained relatively constant into the 1980s, although the absolute number of shareowners rose to 42,360,000 in 1983. The average shareowner that year had attended college, lived in a major metropolitan area, and received a total annual income of $25,000 or more. About 51% of all stockholders were women; some 6% of all shareowners were below 21 years of age, 45% between 21 and 44, and 49% at least 45. Americans reported $52.1 billion in taxable domestic and foreign dividends on their 1982 federal income tax returns. Increasingly, such institutional investors as corporate pension funds and investment companies have entered the market for stock. The Securities and Exchange Commission regulates securities trading.

³⁵PUBLIC FINANCE

Under the Budget and Accounting Act of 1921, the president is responsible for preparing the federal government budget. In fact, the budget is prepared by the Office of Management and Budget (established in 1970), based on requests from the heads of all federal departments and agencies and on advice from the Board of Governors of the Federal Reserve System, the Council of Economic Advisers, and the Treasury Department. The president submits a budget message to Congress every January. Under the Congressional Budget Act of 1974, the Congress establishes, by concurrent resolution, targets for overall expenditures and broad functional categories, as well as targets for revenues, the budget deficit, and the public debt. The Congressional Budget Office, established by the same act, monitors the actions of Congress on individual appropriations bills with reference to those targets. The president exercises fiscal control over executive agencies, which issue periodic reports subject to his perusal. Congress exercises control through the comptroller general, head of the General Accounting Office, who sees to it that all funds have been spent and accounted for according to legislative intent. The fiscal year runs from 1 October to 30 September.

The following table summarizes federal budget receipts, outlays, net deficits, and cumulative national debt outstanding for fiscal years 1980–85 (in billions of dollars):

	1980	1981	1982	1983	1984	1985 (EST.)
Receipts	517.1	599.3	617.8	600.6	666.5	736.9
Outlays	590.9	678.2	745.7	808.3	851.8	959.1
Net deficit	73.8	78.9	127.9	207.7	185.3	222.2
Debt outstanding (as of 30 September)	914.3	1,003.9	1,147.0	1,381.9	1,576.7	1,841.1

Estimated federal receipts (by source) and expenditures (by general function) for the 1985 fiscal year (in billions) are indicated in the accompanying table, along with distribution percentages for the 1977, 1980, and 1984 fiscal years:

	1984/85	1983/84	1980/81	1976/77
RECEIPTS				
Individual income taxes	$329.7	44.4%	47.7%	44.3%
Social insurance taxes and contributions	268.4	36.3	30.5	29.9
Corporate income taxes	66.4	8.5	10.2	15.4
Excise taxes	37.0	5.6	6.8	4.9
Customs, estate, and gift taxes	17.4	2.6	2.5	3.5
Other receipts	18.0	2.6	2.3	2.0
TOTALS	$736.9	100.0%	100.0%	100.0%
EXPENDITURES				
Social Security and Medicare	$257.4	27.7%	26.3%	25.5%
National defense	253.8	26.7	23.2	23.8
Income security	127.2	13.2	14.7	14.9
Health	33.9	3.4	4.0	4.2
Education, training, employment, and social services	30.4	3.2	5.0	5.2
Transportation	27.0	2.8	3.5	3.6
Veterans benefits and services	26.8	3.0	3.4	4.4
Agriculture	20.2	1.6	1.7	1.7
Natural resources and environment	13.0	1.5	2.0	2.3
Net interest	130.4	13.0	10.1	7.3
Other outlays	39.0	3.9	6.1	7.1
TOTALS	$959.1	100.0%	100.0%	100.0%

The public debt of the US subject to the statutory debt limit rose from $43 billion in 1940 to $258.7 billion in 1945, $397.3 billion in 1971, $771.5 billion in 1978, and to $1,381.9 billion in 1983; the per capita federal debt increased from $325 in 1940 to nearly $6,000 in 1983. Annual interest on the federal debt reached $128.8 billion in fiscal 1983; between 1972 and 1983, the proportion of federal outlays represented by gross interest paid on the public debt increased from 9.4% to 16.2%.

Deficit financing, a fixture of US budgets throughout the 1970s, accelerated during the first four years of the Reagan administration. Contributing to those deficits, which averaged well over $150 billion annually during fiscal years 1982–1985, were the continued growth of so-called entitlements (social insurance programs whose funding and benefit schedules are established by law), substantial increases in defense expenditures, a tax cut enacted in 1981, and the revenue shortfalls resulting from the 1981–1982 recession. Some critics of the Reagan administration alleged that deficit creation during Reagan's first term was part of an ideologically motivated campaign to restrict Congress's ability to pass new social programs and fund existing ones. According to the Congressional Budget Office, funding for programs targeted mainly toward the poor was cut by $57 billion during fiscal years 1982–1985.

At the end of World War II, combined expenditures, revenues, and debts of all state and local governments were almost in balance at $18 billion. Largely because of new borrowings for school and highway construction, however, the net outstanding debt of all state and local governments jumped from $69.9 billion in 1960 to $399.3 billion in 1982. The average debt per capita was $1,763 in 1982, ranging from $748 in Indiana to $16,830 in Alaska. The following table summarizes the finances of state and city governments in 1982 (in billions):

REVENUES	STATES	CITIES
Taxes, of which:	$162.7	$ 37.1
Sales and gross receipts	(78.8)	(10.2)
Individual income	(45.7)	
Corporate income	(14.0)	(7.4)[1]
Licenses	(10.1)	
Property	(3.1)	(19.5)
Transfers, of which:	69.2	31.6
From federal government	(66.0)	(12.7)
From local governments	(3.2)	(—)
From state governments	(—)	(18.9)
Borrowing	20.3	(NA)
Other revenue sources	99.0	46.7
TOTALS	$351.2	$115.4
EXPENDITURES		
Education	$103.0	$ 10.2
Public welfare	55.3	4.5
Social insurance (including retirement and unemployment)	34.7	3.7
Highways	25.1	7.0
Health and hospitals	22.3	5.2
Police and corrections	8.6	9.9
Sewerage and other sanitation	NA	9.6
Debt interest	9.4	5.8
Public utilities	3.7	23.5
Other outlays	55.4	33.6
TOTALS	$317.5	$113.0

[1] Includes income taxes, license fees, and all other taxes.

36 TAXATION

Measured as a proportion of the gross domestic product, the total US tax burden amounted to 30.5% in 1982, less than that in most other industrialized countries. Federal, state, and local taxes are levied in a variety of forms, and together totaled $671.4 billion in 1982 (as compared with $51.1 billion in 1950 and $272.5 billion in 1970). The greatest source of revenue for the federal government is the personal income tax, which in 1985 was paid by citizens and noncitizens under the age of 65 with gross incomes of $3,300 or more and by citizens over 65 years of age who earned $4,300 or more.

The personal income tax is a progressive tax: the rate increases with increased income. In 1984, for example, the rate ranged, for a single person without dependents, from 11% on taxable income of between $2,300 and $3,400 to 50% on taxable income of more than $55,300; before the 1981 tax cut, the rate had ranged from 14% to 70%. During the early 1980s, several plans were introduced in Congress to simplify the tax system by drastically reducing the number and value of tax shelters and by substituting a much lower "flat" tax rate for the prevailing system of 14 graduated brackets. The tax simplification plan submitted to Congress by President Reagan in 1985 would create 3 rate brackets of 15%, 25%, and 35%. In 1983, according to preliminary figures, Americans filed 96.3 million federal income tax returns and paid $276.1 billion in tax.

Also reduced in 1981, corporation taxes in 1984 consisted of a 15% levy on the first $25,000 of net income, 18% on the next $25,000, and 46% on income in excess of $100,000. The share of federal revenues supplied by corporate income taxes decreased from 15.4% in 1976/77 to 9% in 1984/85.

Excise taxes include a 10% levy on certain motor vehicles, an 8% charge on personal air transportation, and a tax of 9 cents a gallon on some motor fuels (excluding gasohol); also taxed are alcoholic beverages, tobacco products, tires and tubes, telephone charges, capital gains, and gifts and estates. Effective 1 March 1980, a "windfall profits" tax was levied on producers of domestic crude oil; the tax, ranging up to 70%, was based on the difference between the selling price of the oil and its basic price according to Department of Energy regulations. Provisions of this levy, which accompanied the decontrol of oil prices, were eased as part of the 1981 tax reduction package, with a consequent decline in revenues under the tax from $23.3 billion in 1980/81 to a projected $5 billion in 1985/86.

Customs revenues were estimated at $11.8 billion for 1984/85. From the Civil War until the eve of World War I, when the lower Underwood tariff was enacted, a high-tariff policy prevailed. Following the war, the Underwood tariff was abandoned and a high-tariff policy was resumed; the Hawley-Smoot tariff of 1930 was the highest tariff in the history of the nation. Beginning with the Trade Agreements Act of 1934, the policy of protection was modified but not reversed. Under this act, the principle of reciprocity became part of the tariff-making process. The US was authorized to make concessions in its own tariff rates for nations that countered with tariff concessions of their own. As a result, tariff rates have declined considerably since the 1930s.

Under the Trade Agreements Extension Act of 1951, the president is required to inform the US International Trade Commission (known until 1974 as the US Tariff Commission) of contemplated concessions in the tariff schedules. The commission then determines what the "peril point" is; that is, it informs the president how far the tariff may be lowered without injuring a domestic producer, or it indicates the amount of increase necessary to enable a domestic producer to avoid injury by foreign competition. Similarly, the same act provides an "escape clause," which, in effect, constitutes a method for rescinding a tariff concession granted on a specific commodity if the effect of the concession, once granted, has caused or threatens to cause "serious injury" to a domestic producer. The Trade Expansion Act of 1962 grants the president the power to negotiate tariff reductions of up to 50% under the terms of the General Agreement on Tariffs and Trade (GATT). In 1974, Congress authorized the president to reduce tariffs still further, especially on goods from developing countries. As the cost of imported oil rose in the mid-1970s, however, Congress became increasingly concerned with reducing the trade imbalance by discouraging the "dumping" of foreign goods on the US market. The International Trade Commission is required to impose a special duty on foreign goods offered for sale at what the commission determines is less than fair market value. The Office of the US Trade Representative, within the Executive Office of the President, is responsible for setting overall trade policy and conducting international trade and tariff negotiations.

State tax collections in 1982 exceeded $162 billion, with California, New York, Texas, and Pennsylvania each collecting more than $8 billion. As of mid-1985, only six states—Alaska, Nevada, South Dakota, Texas, Washington, and Wyoming—had no income tax in any form. At the same time, 45 states—all but Alaska, Delaware, Montana, New Hampshire, and Oregon—levied a sales tax. Increasingly in the 1980s, state governments were running lotteries as fund-raising devices; in 1982, state lotteries had a gross income of more than $3.5 billion, with about 52% going for prizes. Property taxes are the main source of local revenue; a movement to limit property tax increases had its major success in California, with the passage of Proposition 13 in June 1978.

37 ECONOMIC POLICY

By the end of the 19th century, regulation rather than subsidy had become the characteristic form of government intervention in US economic life. The abuses of the railroads with respect to rates and services gave rise to the Interstate Commerce Commission in 1887, which was subsequently strengthened by numerous acts that now stringently regulate all aspects of US railroad operations.

The growth of large-scale corporate enterprises, capable of exercising monopolistic or near-monopolistic control of given segments of the economy, resulted in federal legislation designed to control trusts. The Sherman Antitrust Act of 1890, reinforced by the Clayton Act of 1914 and subsequent acts, established the federal government as regulator of large-scale business. This tradition of government intervention in the economy was reinforced during the Great Depression of the 1930s, when the Securities and Exchange Commission and the National Labor Relations Board were established. The expansion of regulatory programs accelerated during the 1960s and early 1970s with the creation of the federal Environmental Protection Agency, Equal Employment Opportunity Commission, Occupational Safety and Health Administration, and Consumer Product Safety Commission, among other bodies. Subsidy programs were not entirely abandoned, however. Federal price supports and production subsidies have made the government a major force in stabilizing US agriculture. Moreover, the federal government has stepped in to arrange for guaranteed loans for two large private firms— Lockheed in 1971 and Chrysler in 1980—where thousands of jobs would have been lost in the event of bankruptcy.

As the 1980s began, there was a general consensus that, at least in some areas, government regulation was contributing to inefficiency and higher prices. Thus, the Carter administration moved to deregulate the airline, trucking, and communications industries; subsequently, the Reagan administration relaxed government regulation of bank savings accounts and automobile manufacture as it decontrolled oil and gas prices. The Reagan administration also sought to slow the growth of social-welfare spending and attempted, with only partial success, to transfer control over certain federal social programs to the states and to reduce or eliminate some programs entirely.

Some areas of federal involvement, however, seem safely entrenched. Old age and survivors' insurance, unemployment insurance, and other aspects of the Social Security program have been accepted areas of governmental responsibility since the 1930s. Federal responsibility has also been extended to insurance of bank deposits, to mortgage insurance, and to regulation of stock transactions. The government fulfills a supervisory and regulatory role in labor-management relations. Labor and management customarily disagree on what that role should be, but neither side advocates total removal of government from this field.

From the end of World War II until the end of 1952, US government transfers of capital abroad represented an annual average of about $5,470 million, or 88.3% of the overall national average, while private investments averaged roughly $730 million a year, or about 11.7%. Portfolio investment represented less than $150 million a year, or only 2.5% of the annual aggregate.

After 1952, however, direct private investment began to increase, and portfolio investment rose markedly. In the late 1950s, new private direct investment was increasing yearly by $2 billion or more, while private portfolio investment and official US government loans were climbing by a minimum annual amount of $1 billion each. During the 1960–73 period, the value of US-held assets abroad increased by nearly 12% annually; from the mid-1970s, they rose most years by at least 15%, reaching $887.4 billion in 1983. Through 1980, direct private investments abroad represented the largest share of US overseas assets; since 1981, however, foreign lending by US banks has come to dominate US investment holdings. In 1983, US claims reported by US banks totaled $430 billion (111% more than in 1980), of which $208.8 billion was in Latin America, $116.9 billion in Western Europe, and $20.4 billion in Canada. Direct private investments, which declined in the early 1980s, amounted to $226.1 billion in 1983, of which Western Europe accounted for $102.5 billion, Canada $47.5 billion, and Latin America $27.4 billion. The rapid rise of US bank lending abroad gave the US a vital economic as well as political stake in the financial stability of developing nations, especially in Latin America, since a serious default by a principal debtor nation would threaten all creditor institutions.

Direct foreign investments in the US have risen rapidly, from $6.9 billion in 1960 to $27.7 billion in 1975 and $135.5 billion in 1983. Of that total, about 35% was in manufacturing, 18% in wholesale and retail trade, 14% in finance and insurance, 14% in petroleum, and 19% in other sectors. As of 1981, US affiliates of foreign companies had assets of $395 billion, sales of $503.7 billion, and 2,343,000 employees. All foreign assets in the US amounted to $781.5 billion in 1983, with Western Europe accounting for about 47%, Latin America 21%, and Canada 6%. Buoyed by high US interest rates, foreign holdings in US banks rose from $78.2 billion in 1977 to $305.7 billion in 1983.

The US net investment position—US assets abroad less foreign assets in the US—rose steadily throughout the 1970s but dropped precipitously from $149.5 billion in 1982 to an estimated $32 billion at the end of 1984. According to preliminary indications, during the first half of 1985, the US became a net debtor for the first time since 1914. So rapid was the decline in the net investment position that some observers forecast a foreign debt of $100 billion by early in 1986—by which time, the US would be the world's leading debtor nation.

[38] HEALTH

The US health care system is among the most advanced in the world, and overall death rate and longevity figures compare favorably with those of most nations. However, the system serves different regions and ethnic groups unequally, and the rising cost of the system is also a continuing cause for concern. Between 1970 and 1984, total health care costs increased by over 400%, and although numerous proposals for a national health insurance program were advanced during the 1970s and early 1980s, none was adopted.

Life expectancy in the US averaged 74.7 years in 1983. The life expectancy of women (78.3 years) was greater than that of men (71); and of whites (75.2), higher than nonwhites (71.3). The birthrate fell slowly from a postwar peak of 25 per 1,000 population in 1955 to a low of 14.6 in 1975–76, then began rising again; in 1983, the rate was 15.5, with a total of 3,614,000 live births occurring during that year. Infant mortality declined from 47 per 1,000 live births in 1940 to 10.6 in 1984; however, the rate among nonwhites was 17.8, and the maternal death rate of 20.4 per 100,000 live births among blacks was more than three times the rate for whites. The rate of legal abortions increased from 19.3 per 1,000 women in 1974 to 28.8 in 1982, when the total of 1,573,900 legal abortions represented a rate of at least 2 abortions for each 5 live births. The marriage rate declined from 11.1 per 1,000 population in 1950 to 8.5 in 1960, before rising steadily to 10.8 in 1982; the divorce rate edged upward from 2.3 per 1,000 population in 1955 to 5.1 in 1982. In 1981, 686,600 children were born out of wedlock, representing 18.9% of all live births that year; among whites, 11.6% of all children were born to unmarried mothers, and among blacks, the proportion was a record high 56%. A survey of married women in 1982 indicated that 68.8% of white women of childbearing age were using contraception, compared with 60.8% of black women. The single most widely practiced contraceptive technique was sterilization, usually of the woman; use of "the pill," the most popular method of contraception during the early 1970s, declined because of widespread concern over long-term health risks.

There has been a marked decline in the death rate from childhood diseases (including poliomyelitis, scarlet fever, diphtheria, measles, and whooping cough), from at least 100 per 1,000 children under 15 in 1900 to less than 0.05 by 1970. Tuberculosis has been brought under control with the use of new drugs, as have many other infectious diseases; less spectacular progress has been made against the diseases that strike later in life. By the early 1980s, about 70% of all deaths were attributable to cardiovascular diseases and cancer. The death rates from cancer and various pulmonary disorders increased between 1970 and 1981, as did the rate for homicide. First identified in June 1981, acquired immune deficiency syndrome (AIDS) was declared by the US Public Health Service in May 1983 to be the nation's top-priority public health problem. By July 1985, more than 12,000 cases of AIDS had been positively identified, with the overwhelming majority of sufferers being male homosexuals (or bisexuals) or intravenous drug users. Another 500,000 or more Americans were believed to be symptomless carriers of the AIDS virus.

Although deaths from venereal diseases have been almost eradicated and syphilis has been curbed, the number of reported cases of gonorrhea has increased. In 1982, about 961,000 cases were reported, compared with 325,000 in 1965. An estimated 13% of adult Americans in 1979 consumed more than 60 alcoholic drinks a month. Concern about alcoholism was focused not only on the long-term health consequences of alcohol abuse but also on the dangers posed by the drunken driver; reversing a long-term trend, many states in the late 1970s and early 1980s began raising the minimum drinking age, so that by 1984, more than half of all

State Social and Economic Indicators

	% URBAN (1980)	BIRTHS PER 1,000 POP. (1981)	ABORTIONS PER 1,000 LIVE BIRTHS (1982)	CRIMES PER 100,000 POP. (1983)	% BELOW POVERTY LEVEL (1979)	INCOME PER CAPITA (1983)	PHYSICIANS PER 100,000 POP. (1981)[1]
Alabama	60.0	15.7	340	4,101	18.9	$ 9,235	127
Alaska	64.3	24.3	172	6,019	10.7	16,820	126
Arizona	83.8	18.4	298	6,392	13.2	10,719	177
Arkansas	51.6	15.6	191	3,501	19.0	9,040	123
California	91.3	17.4	617	6,677	11.4	13,239	230
Colorado	80.6	17.5	454	6,627	10.1	12,580	194
Connecticut	78.8	12.7	540	4,978	8.0	14,826	251
Delaware	70.6	15.4	403	5,466	11.9	12,442	168
Florida	84.3	13.6	524	6,781	13.5	11,592	175
Georgia	62.4	16.2	415	4,505	16.6	10,283	145
Hawaii	86.5	18.6	484	5,810	9.9	12,101	208
Idaho	54.0	20.5	158	3,866	12.6	9,342	107
Illinois	83.3	16.2	366	5,207	11.0	12,626	186
Indiana	64.2	15.4	189	4,130	9.7	10,567	129
Iowa	58.6	15.8	183	3,919	10.1	11,048	126
Kansas	66.7	17.3	361	4,530	10.1	12,285	156
Kentucky	50.9	15.6	189	3,435	17.6	9,162	136
Louisiana	68.6	19.1	265	5,027	18.6	10,406	157
Maine	47.5	14.6	327	3,681	13.0	9,619	150
Maryland	80.3	14.4	603	5,357	9.8	12,994	270
Massachusetts	83.8	12.8	529	5,011	9.6	13,089	277
Michigan	70.7	15.3	482	6,478	10.4	11,574	161
Minnesota	66.9	16.7	283	4,034	9.5	11,666	189
Mississippi	47.3	18.2	122	3,208	23.9	8,072	108
Missouri	68.1	15.8	253	4,530	12.2	10,790	165
Montana	52.9	18.0	303	4,644	12.3	9,999	131
Nebraska	62.9	17.2	247	3,788	10.7	10,940	148
Nevada	85.3	16.7	702	6,701	8.7	12,516	143
New Hampshire	52.2	14.4	399	3,356	8.5	11,620	165
New Jersey	89.0	13.0	653	5,163	9.5	14,057	193
New Mexico	72.1	20.0	281	6,346	17.6	9,560	153
New York	84.6	13.8	731	5,903	13.4	13,146	271
North Carolina	48.0	14.1	383	4,184	14.8	9,656	153
North Dakota	48.8	18.8	246	2,675	12.6	11,350	137
Ohio	73.3	15.5	378	4,505	10.3	11,254	165
Oklahoma	67.3	17.3	229	4,929	13.4	11,187	128
Oregon	67.9	16.2	395	6,251	10.7	10,920	181
Pennsylvania	69.3	13.5	392	3,196	10.5	11,510	191
Rhode Island	87.0	13.0	590	5,005	10.3	11,504	210
South Carolina	54.1	16.4	328	4,771	16.6	8,954	134
South Dakota	46.4	18.5	143	2,548	16.9	9,704	116
Tennessee	60.4	14.5	372	4,012	16.5	9,362	158
Texas	79.6	19.1	343	5,907	14.7	11,702	152
Utah	84.4	27.3	100	5,118	10.3	9,031	163
Vermont	33.8	15.4	479	4,133	12.1	10,036	212
Virginia	66.0	14.6	447	3,962	11.8	11,835	177
Washington	73.5	16.5	517	6,078	9.8	12,051	179
West Virginia	36.2	14.3	122	2,419	15.0	8,937	137
Wisconsin	64.2	15.7	276	4,256	8.7	11,132	160
Wyoming	62.7	22.1	101	4,014	7.9	11,132	115
US AVERAGE	73.7	15.8	426	5,159	12.4	11,675	185

[1] Excluding inactive and federal physicians.

[2] Persons 25 years old and over with four years of high school.

the states had minimums of 20 or 21. As of 1982, an estimated 27.4% of young adults (18–25 years of age) were current users of marijuana, and 6.8% used cocaine. Cigarette smoking, linked to heart and lung disease, is one of the nation's most widespread public health problems; despite a federal antismoking campaign, 29.1% of adult women and 36.5% of adult men—46 million Americans in all—were regular smokers in 1980. The movement to restrict or ban cigarette smoking in public places gathered strength during the 1980s.

HOSPITAL BEDS PER 100,000 POP. (1982)	PUBLIC SCHOOL SPENDING PER PUPIL (1983/84)	% HIGH SCHOOL GRADUATES (1980)[2]
657	$2,102	56.5
383	7,026	82.5
418	2,738	72.4
590	2,198	55.5
451	2,912	73.5
492	3,261	78.6
582	4,036	70.3
667	3,735	68.6
569	3,201	66.7
584	2,322	56.4
411	3,982	73.8
409	2,198	73.7
621	3,397	66.5
374	2,730	66.4
705	3,212	71.5
768	3,361	73.3
509	2,550	53.1
593	2,802	57.7
581	2,813	68.7
583	3,720	67.4
720	3,739	72.2
530	3,208	68.0
709	3,378	73.1
681	1,962	54.8
692	2,600	63.5
646	3,631	74.4
743	2,927	73.4
411	2,861	75.5
496	2,796	72.3
571	4,943	67.4
461	2,921	68.9
717	4,783	66.3
540	2,447	54.8
893	2,969	66.4
582	3,042	67.0
547	2,891	66.0
446	3,771	75.6
698	3,725	64.7
619	3,720	61.1
530	2,255	53.7
821	2,640	67.9
683	2,173	56.2
552	2,913	62.6
337	2,119	80.0
558	3,148	71.0
576	2,968	62.4
367	3,106	77.6
656	2,587	56.0
611	3,553	69.6
530	4,488	77.9
587	3,173	66.5

The following table shows death rates per 100,000 population for the leading causes of death (excluding fetal deaths) in 1981:

	US AVERAGE	MEN WHITE	MEN BLACK	WOMEN WHITE	WOMEN BLACK
Heart disease	195.0	268.8	316.7	129.8	191.2
Cancer	131.6	158.3	232.0	107.2	127.1
Accidents	39.8	59.1	74.7	20.2	21.6
Stroke	38.1	38.9	72.7	33.1	58.1
Pneumonia, influenza	12.3	15.6	26.4	9.0	11.3
Suicide	11.5	18.9	11.0	6.0	2.5
Chronic liver diseases	11.4	14.8	27.3	6.7	12.7
Homicide	10.4	10.4	64.8	3.1	12.7
Diabetes mellitus	9.8	9.3	16.8	8.4	21.3
Atherosclerosis	5.2	6.0	6.4	4.5	5.1
Others	103.1	124.3	218.9	73.4	135.5
TOTALS	568.2	724.4	1,067.7	401.4	599.1

In 1979, 6,405,000 persons were treated in US mental health facilities; of this total, 4% were residents and 96% were outpatients (in 1955, 70% of all persons treated in mental health facilities were seen on an inpatient basis). Medical facilities in 1982 included 1,359,800 beds in 6,915 hospitals, of which 346 were under federal control; of 5,863 nonfederal institutions providing short-term care, 1,761 were state or locally controlled, 3,354 were private nonprofit institutions, and 748 were investor-owned (for profit). A total of 39.1 million admissions were recorded; there were 313.7 million outpatient visits.

The US had 149 physicians for each 100,000 persons in 1950. By 1981, when 505,000 physicians practiced in the US, the ratio was 217 per 100,000 population (with inactive and federal physicians excluded, the ratio was 185 per 100,000). The percentage of medical doctors serving as general practitioners dropped from 75% in 1930 to 12.5% in 1981, while the percentage of all physicians engaged in office-based private practice decreased from 86% to 59%. Most of the 41% not in private practice were engaged in public health, industrial or military medicine, teaching, research, and hospital practice. During the period 1970–80, the number of US physicians increased at a rate equal to three times the population growth rate. However, some regions of the US did not benefit from this increase. Physicians tended to locate in urban or suburban settings, and rural areas actually showed a decline in physician/population ratio. Other health care personnel in 1983 included 1,372,000 registered nurses, 1,269,000 nursing aides, 1,111,000 technologists and technicians, 318,000 therapists and dietitians, and 126,000 dentists.

Per capita expenditures for health and medical care rose from about $78 in 1950 to $1,580 in 1984. Out of a total of $387.4 billion spent for health care in 1984, expenditures for hospital care absorbed $157.9 billion, physician services $75.4 billion, nursing home services $32 billion, pharmaceuticals $25.8 billion, and dental services $25.1 billion. The average cost of care in community hospitals in 1982 was $327 per day and $2,501 per stay. In 1948, 8% of all medical expenses were defrayed through health insurance arrangements; in 1984, private health insurers covered 31.3% of health care expenses.

[39]SOCIAL WELFARE

Social welfare programs in the US depend on both the federal government and the state governments for resources and administration. Old age, survivors, disability, and Medicare (health) programs are administered by the federal government; unemployment insurance, dependent child care, and a variety of other public assistance programs are state-administered, although the federal government contributes to all of them through grants to the states. Total public expenditures for all social welfare programs—including income maintenance, health, education, and welfare services—reached $592.6 billion in 1982, representing 55.5% of all government outlays and 19.3% of the GNP (10.3% in

1960). Federal expenditures on social welfare rose during the first four years of the Reagan presidency, but his administration did succeed in slowing the growth rate from 15% in 1980 to 13.7% in 1981 and 6.3% in 1982. Eligibility requirements for many social programs were tightened, operating budgets were slashed, and some programs—notably public-sector employment under the Comprehensive Employment and Training Act (CETA)—were eliminated entirely.

In 1983, 10,868,000 Americans received $12.9 billion under the aid to families with dependent children (AFDC) program. Medical assistance (Medicaid) amounted to $32.3 billion in 1983, more than six times the figure for 1970. Attempts to get clients "off the welfare rolls and onto the payrolls" have foundered on the reality that many of them are not easily employable, especially in a tight job market. Moreover, the job training and day care facilities necessary to allow some poor people to find and hold jobs are often no less expensive than direct public assistance.

The Food and Nutrition Service of the US Department of Agriculture oversees several food assistance programs. In 1983, 21,600,00 Americans took part in the food stamp program, at a cost to the federal government of $11.2 billion. An estimated 23,200,000 pupils participated in the school lunch program, at a federal cost of $2.4 billion. During the same year, the federal government also expended $336 million for school breakfasts, $108 million on nutrition programs for the elderly, and more than $2 billion in commodity aid for the needy.

The present Social Security program differs greatly from that created by the Social Security Act of 1935, which provided that retirement benefits be paid to retired workers aged 65 or older. Since 1939, Congress has attached a series of amendments to the program, including provisions for workers who retire at age 62, for widows, for dependent children under 18 years of age, and for children who are disabled prior to age 18. Disabled workers between 50 and 65 years of age are also entitled to monthly benefits. Other measures increased the number of years a person may work; among these reforms was a 1977 law banning mandatory retirement in private industry before age 70. The actuarial basis for the Social Security system has also changed. In 1935 there were about nine US wage earners for each American aged 65 or more; by the mid-1980s, however, the ratio was closer to three to one.

In 1940, the first year benefits were payable, $35 million was paid out. By 1983, Social Security benefits totaled $167 billion, paid to more than 36 million beneficiaries. The average monthly benefit for a retired worker with no dependents in 1960 was $74; by 1983, the average benefit was $441 for a single retired worker. Under legislation enacted in the early 1970s, increases in monthly benefits are pegged to the inflation rate, as expressed through the Consumer Price Index. Employers, employees, and the self-employed are legally required to make contributions to the Social Security fund. In 1982, some 80,500,000 wage and salary earners were paying Social Security taxes under the Federal Insurance Contributions Act (FICA). As the amount of benefits and the number of beneficiaries have increased, so has the maximum FICA payment, which for 1984 was $2,533, based on a rate of 6.7% on earnings up to $37,800, plus another 0.3% supplied from general revenues; the 1960 maximum was $144. Among workers with many dependents, the Social Security tax deduction can now exceed the federal income tax deduction.

In January 1974, the Social Security Administration assumed responsibility for assisting the aged, blind, and disabled under the Supplemental Security Income program. In 1983, some 3,901,000 Americans—1,515,000 aged, 2,307,000 disabled, and 79,000 blind—received over $9.1 billion. Medicare, another program administered under the Social Security Act, provides hospital insurance and voluntary medical insurance for persons 65 and over, with reduced benefits available at age 62. Medicare hospital insurance covered some 29,069,000 Americans in 1982, when benefit payments exceeded $51 billion. Medicaid, a program that helps the needy meet the costs of medical, hospital, and nursing-home care, paid $32.3 billion to 21,471,000 recipients in 1983.

The laws governing unemployment compensation originate in the states. Therefore, the benefits provided vary from state to state in duration (generally from 26 to 39 weeks) and amount (ranging from $80 to $154 weekly in 1982); the national average was about $119 per week, with total benefits exceeding $25 billion. Workers' compensation payments in 1982 amounted to more than $16.1 billion; federal and state outlays for vocational rehabilitation exceeded $1.2 billion in 1983.

Private philanthropy plays a major role in the support of relief and health services. In 1984, Americans contributed a record $74.25 billion to philanthropic causes, with religious organizations receiving nearly half the total. The private sector plays an especially important role in pension management; private pension funds controlled $679.9 billion in assets in 1983, while public funds amounted to $425.7 billion.

The federal agency ACTION, established in 1971, coordinates several US social service agencies. Chief among these is Volunteers in Service to America—VISTA—which was created in 1964 to marshal human resources against economic and environmental problems. ACTION also administers activities for the young and the aged, including the Retired Senior Volunteer Program and the Foster Grandparent Program, enlisting elderly persons to work with children who have special physical, mental, or emotional needs.

⁴⁰HOUSING

The US had 93,519,000 housing units in 1983, 84,638,000 of which were occupied all year round. Some 65% of year-round units in 1983 were occupied by their owners, as compared with 55% in 1950; these owner-occupied homes had an average of 2.5 persons per unit, down from 3.1 in 1960. The majority of rental tenants are found in the large metropolitan areas.

In design and physical elements, houses in the US reflect variations in climate, materials, income, traditions, taste, and age. By 1983, 97.6% of US homes were equipped with running water, private toilet facilities, and private bathing facilities. Of an estimated 83.8 million households that owned electrical appliances in 1982, 31% had room air conditioners, 27% had central air conditioning, 36% had electric dishwashers, 71% had washing machines, 60% had electric or gas clothes dryers, 85% had color television sets, and virtually every electrified home had a refrigerator and radio. As of 1983, about 30% of all year-round housing units in the US had been built prior to 1940, 9% during 1940-49, 15% during 1950-59, 20% during 1960-70, and 26% since April 1970. The housing stock in the Northeast and Midwest and in all central cities is much older than that of the US as a whole.

Construction of housing following World War II set a record-breaking pace, so that 1983 was the 35th successive year during which construction of more than 1 million housing units was begun. Between 1970 and 1973, 8 million housing units were constructed; by 1975, however, construction had dropped to 1,171,000 units, the lowest production in more than a decade. In 1978, new housing starts climbed back to 2,036,000, only to be squeezed again by high interest rates and a sluggish economy that did not begin to recover until 1983. That year, housing construction contracts totaled $93.2 billion, representing 48% of all contract construction.

One-family houses make up the great bulk of the housing stock as well as the new homes produced. Private firms or individuals are the main producers of houses, with most single-family, non-farm houses being built by professional builders and developers. The government participates in a variety of programs designed to stimulate housing construction and slum clearance, including issuance of housing loans through the Department of Housing and Urban Development (HUD), the guarantee of mortgages by the Veterans Administration, and the effort to expand the secondary market for HUD mortgages through the Government National Mortgage Association. Perhaps the most significant change in the housing scene since World War II has been the shift to the suburbs, made possible by the widespread ownership of automobiles. Much land, once rural, has been put to use for single-family and multi-unit dwellings, shopping centers, office complexes, and factories.

⁴¹EDUCATION

Between 1900 and 1980, illiteracy among persons 14 years and older declined from 11.3% to 0.5%; however, many educators believe the percentage of "functional illiterates"—those unable to read and write on more than the most basic level—is significantly higher. By 1983, 72.1% of all persons aged 25 years or older had completed high school, as compared with 41.1% in 1960 and 24.5% in 1940; 56.8% of all black adults had completed high school in 1983, as compared with only 7.3% in 1940. Some 18.8% of all adults had completed 4 or more years of college in 1983, as compared with 4.6% in 1940; the rate for blacks in 1983 was 9.5%. The rates of high school completion for men and women were roughly equal, but 24% of white men had 4 or more years of college in 1983, compared with 15.4% of white women. As of that year, Americans 25 years of age or older had completed a median of 12.6 years of school, as compared with 10.6 years in 1960; for whites in 1983, the figure was also 12.6, blacks 12.2, and Hispanics 11.1.

Public schools are controlled and supported primarily by the states, each of which has its own system of public education and laws regulating private schools. Actual school administration, however, is usually in the hands of a state-authorized local school district, which is empowered to levy taxes. Supporting schools in this manner led to inequality of educational funding because of variations in the extent to which residential suburbs could draw on commercial and industrial properties for their tax base. To an increasing extent, therefore, state governments have been called upon—sometimes by court order—to contribute to and equalize the costs of education. In 1983, local districts provided 38% of the revenues needed for elementary and secondary education; the states contributed 43%, the federal government 7%, and other sources 12%. Current expenditures per pupil in public elementary and secondary schools averaged $3,173 in 1983/84, compared with $773 in 1969/70.

The individual states establish compulsory attendance requirements, generally for children between the ages of 7 and 16. Some 99% of all children 7–15 years old actually attended school in 1983. In October of that year, some 57.7 million persons, or 48.4% of US civilians 3–34 years old, were enrolled in schools or colleges. As of fall 1982, 30 million were enrolled in public school classes from kindergarten through the 8th grade, and 12.3 million were in public secondary schools. Nonpublic schools enrolled 5.3 million students (grades K–12) during the same year. A total of 11.8 million attended colleges and universities (9.3 million public, 2.5 million private).

Public tax-supported schools predominate in elementary and secondary education. Schools affiliated with the Roman Catholic Church have traditionally enrolled the large majority of private school students, about 3 million in 1985. As a result of reorganization and consolidation, the number of public schools declined from 248,279 in 1900 to 83,688 (59,326 elementary, 22,619 secondary, and 1,743 combined) in 1981.

The number of public-school elementary teachers increased from 402,700 in 1900 to 1,171,000 in 1984; secondary-school teachers, from 20,400 in 1900 to 946,000 in 1984. The total number of teachers in all public and private educational institutions was about 2,401,000 in 1982. The public-school pupil/teacher ratio has improved gradually in recent decades, decreasing from 30.2 to 1 in elementary schools and 20.9 to 1 in secondary schools in 1955 to 20.5 to 1 and 16.7 to 1, respectively, in 1982.

Since 1957, the federal government has carried out a vigorous program of school desegregation. Of a total of 9.1 million minority pupils attending school in 1980, 82.3% were enrolled in integrated schools; however, some 32.6% of all minority pupils were enrolled in schools where minorities made up at least 90% of the student population. While desegregation has largely been accomplished in the public schools of the South, de facto segregation—resulting from residential patterns rather than from the deliberate establishment of separate school systems—remains widespread, especially in urban areas. In order to compel conformity to federal desegregation guidelines, the courts in several localities have attempted to establish racial balance by ordering interdistrict busing of pupils.

Of 3,274 institutions of higher learning in the US in 1983, 1,803 were private and 1,471 public. College and university enrollment soared uninterruptedly from World War II through the 1960s; in the late 1970s, however, the rate began to slow considerably. By the fall of 1982, the nation's higher institutions enrolled 12,426,-000 students, about 51% of them women. California, with 273 institutions, led all the states with 1,843,000 students, nearly 15% of the national total.

Of the $72.2 billion in income received by colleges and universities in 1981/82, 44.5% was derived from governmental sources and 21.9% from student tuitions and fees, leaving the balance to be supplied from private sources and auxiliary enterprises. During the 1970s and early 1980s, rising costs for higher education, coupled with lack of growth among various sources of support, placed considerable pressure on US colleges and universities, many of which were forced to cut back on staff and range of offerings. Tuition and fees, which in 1965 had averaged $243 annually at public colleges and universities and $1,088 at private institutions, rose by 1984 to $870 and $4,880, respectively. Among the most prestigious private universities, annual charges to students, including tuition, fees, and room and board, exceeded $15,000 in the mid-1980s. Meanwhile, funding of federal student aid and loan programs, which had risen rapidly throughout the 1970s, slowed during the early 1980s, and would have been cut back drastically if Congress had followed the Reagan administration's budget requests.

Most states provide guaranteed loans to state residents attending postsecondary schools, and the largest states generally offer multimillion-dollar tuition assistance programs for needy and deserving students. Grant and loan programs are offered on a national scale through the US Department of Education, the Veterans Administration, and other federal agencies, as well as through numerous private organizations.

The federal government allocates funds for vocational education to the states and some of the territories, which must match federal funds with state and local expenditures. Adult education agencies, both public and private, operate at all levels. In 1982, $937 million in federal funds and $7.1 billion in state and local resources were expended on vocational and adult education.

⁴²ARTS

The nation's arts centers are emblems of the importance of the performing arts in US life. New York City's Lincoln Center for the Performing Arts, whose first concert hall opened in 1962, is now the site of the Metropolitan Opera House, three halls for concerts and other musical performances, two theaters, the New York Public Library's Library and Museum of the Performing Arts, and The Juilliard School. The John F. Kennedy Center for the Performing Arts in Washington, D.C., opened in 1971; it comprises two main theaters, two smaller theaters, an opera house, and a concert hall.

In 1983 there were 1,572 symphony orchestras in the US, including 30 major orchestras; 19,167 symphony concerts drew a total attendance of 22 million. The New York Philharmonic, founded in 1842, and conducted by Zubin Mehta in 1985, is the nation's oldest professional musical ensemble. Other leading orchestras include those of Boston (conducted by Seiji Ozawa), Chicago (Sir Georg Solti), Cleveland (Christoph von Dohnanyi), Los Angeles (Carlo Maria Giulini), Philadephia (Ricardo Muti), Pittsburgh (André Previn), St. Louis (Leonard Slatkin), and Washington, D.C. (the National Symphony, led by Mstislav Rostropovich). In 1983, the US had 1,031 opera companies, 144 of which had budgets above $100,000. Some 12.7 million people attended 10,693 operatic performances, both figures more than double those of 1970. Expenses more than quintupled during the same period, reaching $212.4 million in 1983. Particularly renowned for artistic excellence are the Lyric Opera of Chicago, San Francisco Opera, Opera Company of Boston, Santa Fe Opera, New York City Opera, and Metropolitan Opera.

The recording industry is an integral part of the music world. In 1983, 334 million US-made records (singles and albums) and 244 million prerecorded cartridges, cassettes, and compact discs had a value of, respectively, $2 billion and $1.9 billion. Popular music (mostly rock), performed in halls and arenas in every major city and on college campuses throughout the US, dominates record sales.

Though still financially insecure, dance is winning an increasingly wide following. The American Ballet Theater, founded in 1940, is the nation's oldest dance company still active today; the New York City Ballet is equally acclaimed. Other important companies include those of Martha Graham, Merce Cunningham, Alvin Ailey, Paul Taylor, and Twyla Tharp, as well as the Feld Ballet, Joffrey Ballet, and Pilobolus.

Drama remains a principal performing art, not only in New York City's renowned theater district but also in regional, university, summer, and dinner theaters throughout the US. Television and the motion picture industry have made film the dominant modern medium. The US had about 19,000 public movie screens in 1983, when films registered record box-office receipts of $3.8 billion, based on a paying audience of nearly 1.2 billion.

The National Endowment for the Arts, a federal agency established in 1965, provides funding for nonprofit arts organizations and individual artists. Recipients of its matching grants programs are determined by panels of artists on the basis of aesthetic merit. Musical organizations, including opera, directly received $12.9 million in 1983, the highest subsidy provided any of the performing arts; funding for dance companies increased from $177,000 in 1967 to $9.1 million in 1983, when theatrical companies received $9.5 million, and public media $9.3 million. State legislative appropriations to state arts agencies in 1984 exceeded $136 million; New York led in total amount ($35.3 million), and Alaska held a huge lead in annual per capita spending ($10.25). Support for the arts by private enterprise amounted to $506.5 million in 1982, with $47.3 million going to theaters and theater groups, $45.6 million to cultural centers, $39.5 million to symphony orchestras, and $29.8 million to opera. The privately endowed National Institute of Arts and Letters, founded in 1898, is a society of artists, writers, and composers limited to 250 American citizens at any one time.

43 LIBRARIES AND MUSEUMS

Of the 32,297 libraries in the US in 1983, 8,822 were public, with 6,146 branches; 4,900 were academic; 1,591 were government; 1,552 were medical; and 9,286 were religious, military, legal, and specialized independent collections.

The foremost library in the US is the Library of Congress with holdings of more than 80 million items (including over 20 million books and pamphlets) in the mid-1980s. Other great libraries are the public libraries in New York, Philadelphia, Boston, and Baltimore, and the John Crerar and Newberry libraries in Chi-

cago. Noted special collections are those of the Pierpont Morgan Library in New York, the Huntington Library in San Marino, Calif., the Folger Shakespeare Library in Washington, D.C., the Hoover Library at Stanford University, and the rare book divisions of Harvard, Yale, Indiana, Texas, and Virginia universities. Among the leading university libraries, as judged by the extent of their holdings in 1982, are those of Harvard, Yale, Illinois (Urbana-Champaign), Michigan (Ann Arbor), California (Berkeley), Columbia, Stanford, Cornell, Texas (Austin), California (Los Angeles), Chicago, Wisconsin (Madison), and Washington (Seattle)—each having more than 4 million bound volumes.

There are about 4,400 nonprofit museums in the US. The most numerous type is the historic building, followed in descending order by college and university museums, museums of science, public museums of history, and public museums of art. In 1983, the National Endowment for the Arts provided $10 million in grants to museums; funds from private enterprise in 1982 totaled $97 million. Eminent US museums include the American Museum of Natural History, the Metropolitan Museum of Art, and the Museum of Modern Art, all in New York City; the National Gallery of Art and the Smithsonian Institution in Washington, D.C.; the Boston Museum of Fine Arts; the Art Institute of Chicago and the Chicago Museum of Natural History; the Franklin Institute and Philadelphia Museum of Art, both in Philadelphia; and the M. H. de Young Memorial Museum in San Francisco.

44 COMMUNICATIONS

The US uses wire and radio services for communications more extensively than any other country in the world. In 1980, it had about 35% of the world's telephones, in a system that connects with every continent. In 1980, 93% of the 80,389,673 occupied housing units in the US had telephones; the total number of telephones reached 151 million in 1982, when the domestic telephone industry's total investment in plant and equipment reached $174 billion and operating revenues were $70.7 billion. Western Union provides domestic telegraph service; there are seven international carriers.

Formerly, the US telephone system was dominated by American Telephone and Telegraph (AT&T), also known as the Bell System, under the overall regulation of the Federal Communications Commission (FCC). In 1982, however, a settlement of an antitrust suit originally brought by the US Justice Department in 1974 provided for the breakup of what had been, by most standards of measurement, the world's largest private corporation. In return for spinning off its local telephone companies, AT&T was allowed to retain its long-lines, manufacturing, and research divisions and to enter the unregulated and potentially lucrative fields of computer manufacture and computer-related data transmission. The result, as of 1 January 1984, was the creation of seven regional telephone companies, providing local service, and the inauguration of a new era of competition for long-distance customers, pitting AT&T against several other large private firms. The Bell breakup was only one facet of a phased deregulation of the communications industry. Although the FCC retained regulatory responsibility for airborne radio and television transmissions, it relaxed operating restrictions on broadcasters; the FCC exercises no statutory authority over cable broadcasting.

Radio serves a variety of purposes other than broadcasting. It is widely used by ships and aircraft for safety; it has become an important tool in the movement of buses, trucks, and taxicabs. Forest conservation, fire protection, and the police operate with radio as a necessary aid; it is used in logging operations, surveying, construction work, and dispatching of repair crews. Citizen's band radio became popular in the mid-1970s, especially among truck and automobile drivers seeking traffic information or emergency assistance from other drivers. In 1984, commercial broadcasting stations on the air comprised 4,754 AM radio stations, 3,716 FM radio stations, and 904 television stations. In addition, 1,172 educational FM stations and 290 educational television channels were in operation. As of 1981 there were 631 television receivers and 2,110 radio sets per 1,000 population—ratios no other nation came close to matching. The average number of radios per household was 5.5 in 1984, the average number of television receivers, 1.78. The expanding cable television industry, with 5,855 systems in the US as of December 1984, served 31,880,764 subscribers in 15,603 communities.

The Post Office Department of the US was replaced on 1 July 1971 by the US Postal Service, a financially autonomous federal agency. In 1983, 119.4 billion pieces of mail passed through 29,990 post offices with approximately 547,000 full-time employees. The Postal Service's operating revenues that year were $24.7 billion, but its operations required an additional $789 million in direct government subsidies. In addition to mail delivery, the Postal Service provides registered, certified, insured, and COD mail service; issues money orders ($7.4 billion worth in 1983); and operates a postal savings system. Numerous private firms provide package delivery, courier, and electronic mail services.

⁴⁵PRESS

As of September 1984 there were 1,688 daily newspapers in the US, with a combined circulation of 63,081,740, and 783 Sunday papers, with a circulation of 57,573,979; the US had 6,798 weekly newspapers during the same year. The following newspapers, all English-language, reported daily circulation of more than 500,000 as of 30 September 1984:

	DAILY	SUNDAY
Wall Street Journal (m)	1,959,873	
New York Daily News (all day, S)	1,346,840	1,721,441
USA Today (m)	1,247,324	
Los Angeles Times (m,S)	1,046,965	1,298,487
New York Times (m,S)	934,616	1,553,720
New York Post (all day)	930,026	
Chicago Tribune (all day, S)	776,348	1,137,667
Washington Post (m,S)	728,857	1,033,207
Detroit News (all day, S)	656,367	865,717
Chicago Sun-Times (m,S)	649,891	453,604
Detroit Free Press (m,S)	647,130	803,714
Newsday (e,S)	539,065	602,476
San Francisco Chronicle (m,S)	535,796	699,256
Philadelphia Inquirer (m,S)	525,569	1,000,427
Boston Globe (m,S)	520,081	792,786

The Thomson, Hearst, Scripps-Howard, Gannett, Newhouse, and other chains are an important part of the newspaper business, but they do not control it. Newspapers rely both on their own reporters and, to a marked degree, on the information furnished to them by the large press services, such as the Associated Press and United Press International. The United Features Syndicate and other syndicates supply by-lined columns, photographs, comic strips, and cartoons. Income of newspapers is derived mainly from advertising.

The 10,809 periodicals published during 1984 included 1,376 weeklies, 4,096 monthlies, and 1,711 quarterlies. As of 31 December 1983, *Reader's Digest,* with a monthly circulation of 17,937,-045, and *TV Guide,* with a weekly circulation of 17,066,126, were the best-selling consumer magazines by a wide margin. *Time* and *Newsweek* were the leading newsmagazines, with weekly circulations of 4,615, 594 and 3,038,832, respectively.

The US book publishing industry consists of the major book companies (mainly in the New York metropolitan area), nonprofit university presses distributed throughout the US, and numerous small publishing firms. Of the more than 1.7 billion books sold in the US in 1982, 33% were hardbound and 67% softbound. The average price per hardcover edition was $31.19 in 1983. Receipts from sales of all books in 1982 totaled $9.7 billion, with general adult and juvenile books accounting for 42%, textbooks 36%, and professional books 13%.

⁴⁶ORGANIZATIONS

The 1982 Census of Service Industries counted 61,336 membership organizations, embracing 35,457 civic, social, and fraternal associations, 12,108 business associations, 5,194 professional associations, and 8,577 other bodies.

A number of industrial and commercial organizations exercise considerable influence on economic policy. The National Association of Manufacturers and the US Chamber of Commerce, with numerous local branches, are the two central bodies of business and commerce. Various industries have their own associations, concerned with cooperative research and questions of policy alike.

Practically every profession in the US is represented by one or more professional organizations. Among the most powerful of these are the American Medical Association, comprising regional, state, and local medical societies; the American Bar Association, also comprising state and local associations; the American Hospital Association; and the National Education Association.

Many private organizations are dedicated to programs of political and social action. Prominent in this realm are the National Association for the Advancement of Colored People (NAACP), the Urban League, the American Civil Liberties Union (ACLU), Common Cause, and the Anti-Defamation League of B'nai B'rith. The League of Women Voters, which provides the public with nonpartisan information about candidates and election issues, sponsored televised debates between the major presidential candidates in 1976, 1980, and 1984. The National Organization for Women and the National Rifle Association have each mounted nationwide lobbying campaigns on issues affecting their members. During 1981–82, political action committees (PACs) affiliated with corporations, labor unions, and other groups gave $83 million to candidates for the House and Senate—an increase of 50% over 1979–80. Probably the best-financed and most influential PAC by the mid-1980s was the National Conservative Political Action Committee (NCPAC), which provided $9.8 million to President Reagan's reelection campaign in 1984.

The great privately endowed philanthropic foundations and trusts play an important part in encouraging the development of education, art, science, and social progress in the US. In the early 1980s there were nearly 22,000 foundations, with combined assets exceeding $47 billion. Foundations, donations, and charitable bequests in 1983 provided a total of $11.1 billion in philanthropic gifts. Prominent foundations include the Carnegie Corporation and the Carnegie Endowment for International Peace, the Ford Foundation, the Guggenheim Foundation, the Mayo Association for the Advancement of Medical Research and Education, and the Rockefeller Foundation. Private philanthropy was responsible for the establishment of many of the nation's most eminent libraries, concert halls, museums, and university and medical facilities; private bequests were responsible for the establishment of the Pulitzer Prizes. Merit awards offered by industry and professional groups include the "Oscars" from the Academy of Motion Picture Arts and Sciences, the "Emmys" of the National Academy of Television Arts and Sciences, and the "Grammys" of the National Academy of Recording Arts and Sciences.

Funds for a variety of community health and welfare services are funneled through United Way Campaigns, which raised nearly $2 billion in 1983. That year, the American Red Cross had 2,963 chapters; about 30% of its $653.4 million in expenditures, which paid for services and activities ranging from disaster relief to blood donor programs, were raised through fund campaign contributions. Private organizations supported by contributions from the general public lead the fight against specific diseases.

The Boy Scouts of America, the Girl Scouts of the USA, rural 4-H Clubs, and the Young Men's and the Young Women's Christian Associations are among the organizations devoted to recreation, sports, camping, and education.

The largest religious organization in the US is the National

Council of the Churches of Christ in the USA, which embraces 31 Protestant and Orthodox denominations, whose adherents total more than 40 million. Many organizations, such as the American Philosophical Society, the American Association for the Advancement of Science, and the National Geographic Society, are dedicated to the enlargement of various branches of human knowledge. National, state, and local historical societies abound, and there are numerous educational, sports, and hobbyist groups.

The larger veterans' organizations are the American Legion, the Veterans of Foreign Wars of the US, the Catholic War Veterans, and the Jewish War Veterans. Fraternal organizations, in addition to such international organizations as the Masons, include indigenous groups, such as the Benevolent and Protective Order of Elks, the Loyal Order of Moose, and the Woodmen of the World. Many, such as the Ancient Order of Hibernians in America, commemorate the national origin of their members. One of the largest fraternal organizations is the Roman Catholic Knights of Columbus.

⁴⁷TOURISM, TRAVEL, AND RECREATION

Foreign visitors to the US (excluding Canadian and Mexican citizens) numbered at least 7,873,000 in 1983, more than triple the total for 1970. Of the 1983 visitors, 3,020,000 came from Europe and 2,273,000 from Central and South America and the Caribbean. Travelers to the US from all foreign countries spent $11.4 billion, including $2.5 billion for fares on US carriers to and from their home countries. Canadians accounted for about 28% of foreign tourist spending, Mexicans 17%. With a few exceptions, such as Canadians entering from the Western Hemisphere, all visitors to the US are required to have passports and visas.

Nearly $14 billion was spent by US tourists abroad in 1983, of which Mexico collected $3.6 billion and Canada $2.2 billion. Europe and the Mediterranean attracted $4.4 billion in US tourist spending; the average per tourist was $878 per trip (transatlantic fares excluded). The Caribbean and Central America absorbed $1.5 billion, South America $422 million, Japan $302 million. In all, 10,154,000 Americans traveled overseas in 1983, and millions more US tourists crossed the border to Canada and Mexico.

Both for US domestic travelers and for visitors from abroad, state and national parks are leading destinations. In 1982/83, the nation's state parks and recreation areas, covering over 9.9 million acres (4 million hectares), registered 645 million vistors. Expenditures for state park maintenance and services exceeded $838 million. By 1982, the 333 areas of the National Park System—including 78 national monuments, 62 national historical sites, 48 national parks, the White House, and numerous memorials, battlefields, scenic rivers, and other sites—embraced more than 79 million acres (32 million hectares). The system registered 335.6 million visits (including 16.1 million overnight stays) in 1983. The following table shows some of the most popular sites and the number of visitors they attracted in 1983:

	ACREAGE ('000)	RECREATIONAL VISITS ('000)
Blue Ridge Parkway, N.C.-Va.	82	15,200
Colonial National Historical Park, Va.	9	2,100
Gateway National Recreation Area, N.J.-N.Y.	26	10,300
Glacier National Park, Mont.	1,014	2,200
Golden Gate National Recreation Area, Calif.	73	17,600
Grand Canyon National Park, Ariz.	1,218	2,200
Great Smoky Mountains National Park, N.C.-Tenn.	520	8,400
Lake Mead National Recreation Area, Ariz.-Nev.	1,497	5,900
Natchez Trace Parkway, Miss.-Tenn.-Ala.	50	12,800
Olympic National Park, Wash.	915	2,400
Rocky Mountain National Park, Colo.	265	2,600

	ACREAGE ('000)	RECREATIONAL VISITS ('000)
Valley Forge National Historical Park, Pa.	3	3,400
Yellowstone National Park, Idaho-Mont.-Wyo.	2,220	2,300
Yosemite National Park, Calif.	761	2,500

Americans spent at least $141 billion on recreational activities in 1983, ranging from spectator sports to home gardening. Participant sports are a favorite form of recreation: as of 1983, an estimated 53 million persons had gone swimming within the previous 12 months, 32 million had gone bicycling, 17 million had played tennis, and 13 million had played golf. Skiing is a popular recreation in New England and the western mountain ranges, while sailing, power boating, rafting, and canoeing are popular water sports. Wilderness areas provide year-round hiking, climbing, hunting, and fishing. In 1982/83, state authorities issued licenses to 29.1 million anglers and 16.4 million hunters.

⁴⁸SPORTS

Baseball, long honored as the national pastime, is the nation's leading professional team sport, with two major leagues having 26 teams (2 in Canada); in summer 1983, 46.3 million attended major league games. In addition, there is an extensive network of minor league baseball teams, each of them related to a major league franchise. The National Basketball Association, created in 1946, included 23 teams, which drew 10.3 million fans during the 1982/83 season. In Autumn 1983, 14 million Americans attended regular season games of the National Football League's 28 teams. Attendance at National Hockey League (NHL) games exceeded 11 million in 1982/83; of 21 NHL teams, 7 were Canadian, as were 80% of the players. The North American Soccer League (NASL), which appeared to be growing popular in the late 1970s, discontinued outdoor play in 1985. The NASL did continue an indoor soccer schedule, with competition for fans from the Major Indoor Soccer League. Radio and television contracts are integral to the popular and financial success of all professional team sports.

Several other professional sports are popular nationwide. Horse racing, which lured some 76 million Americans to the track in 1983, is the nation's most popular spectator sport. Annual highlights of Thoroughbred racing are the three jewels of the Triple Crown—Kentucky Derby, Preakness, and Belmont Stakes—most recently won by Seattle Slew in 1977 and by Affirmed in 1978. In 1983, $11.7 billion was legally wagered on horse racing, including licensed off-track betting in New York. Attendance at greyhound racetracks totaled 22.1 million in 1983. The prize money that Henry Ford won on a 1901 auto race helped him start his now-famous car company two years later; since then, automobile manufacturers have backed sports car, stock car, and motorcycle racing at tracks throughout the US. From John L. Sullivan to Muhammad Ali, the personality and power of the great boxing champions have drawn millions of spectators to ringside. Glamour and top prizes also draw national followings for tennis and golf, the two professional sports in which women are nationally prominent.

Football has been part of US college life since the game was born on 6 November 1869 with a New Jersey match between Rutgers and Princeton. In 1983, the 651 teams of the National Collegiate Athletic Association (NCAA) and National Association of Intercollegiate Athletics (NAIA) drew crowds totaling 36.3 million. Collegiate basketball is almost as important; 1,266 teams drew 31.5 million spectators. Colleges recruit top athletes with sports scholarships in order to win media attention, and to keep the loyalty of the alumni, thereby boosting fund-raising. Baseball, hockey, swimming, gymnastics, crew, lacrosse, track and field, and a variety of other sports also fill the intercollegiate competitive program.

LANDS UNDER FEDERAL JURISDICTION

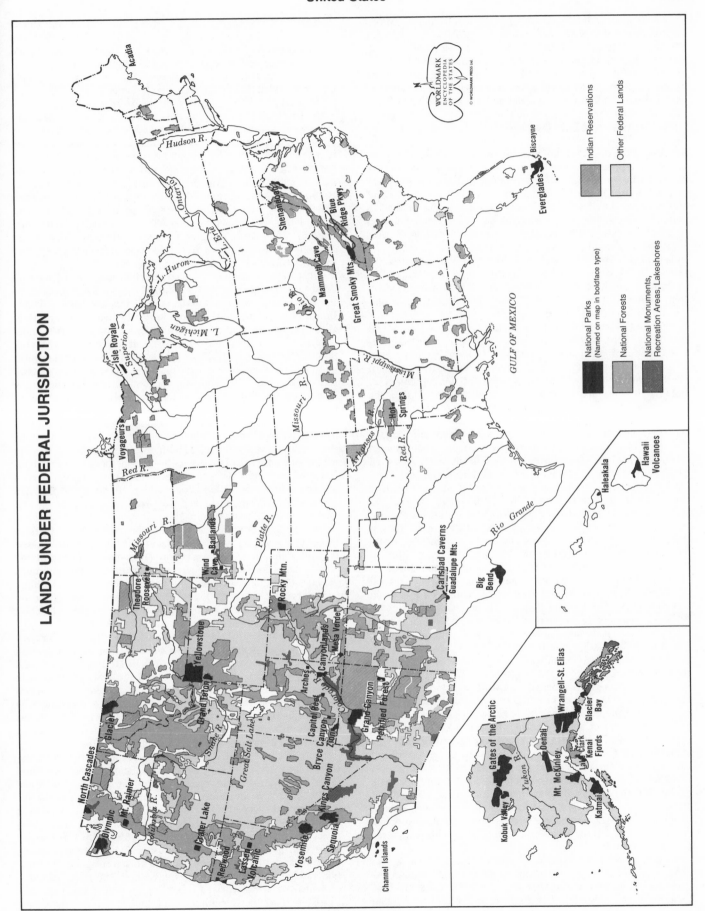

Acadia

Hudson R.

L. Ontario

L. Erie

L. Huron

Shenandoah

Blue Ridge Pkwy.

Mammoth Cave

Great Smoky Mts.

Biscayne

Everglades

GULF OF MEXICO

WORLDMARK ENCYCLOPEDIA OF THE STATES © WORLDMARK PRESS Ltd.

Indian Reservations

Other Federal Lands

National Parks (Named on map in boldface type)

National Forests

National Monuments, Recreation Areas, Lakeshores

Isle Royale

L. Superior

L. Michigan

Voyageurs

Red R.

Missouri R.

Mississippi R.

Ohio R.

Arkansas R.

Red R.

Hot Springs

Missouri R.

Platte R.

Theodore Roosevelt

Badlands

Wind Cave

Rocky Mtn.

Carlsbad Caverns

Guadalupe Mts.

Big Bend

Rio Grande

Haleakala

Hawaii Volcanoes

North Cascades

Glacier

Yellowstone

Grand Teton

Snake R.

Columbia R.

Mt. Rainier

Olympic

Redwood

Lassen Volcanic

Crater Lake

Great Salt Lake

Arches

Canyonlands

Mesa Verde

Capitol Reef

Bryce Canyon

Zion

Grand Canyon

Petrified Forest

Kings Canyon

Yosemite

Sequoia

Channel Islands

Gates of the Arctic

Kobuk Valley

Yukon R.

Mt. McKinley

Wrangell-St. Elias

Glacier Bay

Denali

Lake Clark

Kenai Fjords

Katmai

The Amateur Athletic Union (AAU), a national nonprofit organization founded in 1888, conducts the AAU/USA Junior Olympics, offering competition in 22 sports in order to help identify candidates for international Olympic competition. St. Louis hosted the 1904 summer Olympics; Los Angeles was home to the games in 1932 and 1984. The winter Olympic games were held in Squaw Valley, Calif., in 1960, and at Lake Placid, N.Y., in 1932 and 1980.

⁴⁹FAMOUS AMERICANS

Printer, publisher, inventor, scientist, statesman, and diplomat, Benjamin Franklin (1706–90) was America's outstanding figure of the colonial period. George Washington (1732–99), leader of the colonial army in the American Revolution, became first president of the US and is known as the "father of his country." Chief author of the Declaration of Independence, founder of the US political party system, and third president was Thomas Jefferson (1743–1826). His leading political opponents were John Adams (1735–1826), second president, and Alexander Hamilton (b.West Indies, 1755–1804), first secretary of the treasury, who ensured the new nation's credit. James Madison (1751–1836), a leading figure in drawing up the US Constitution, served as fourth president. John Quincy Adams (1767–1848), sixth president, was an outstanding diplomat and secretary of state.

Andrew Jackson (1767–1845), seventh president, was an ardent champion of the common people and opponent of vested interests. Outstanding senators during the Jackson era were John Caldwell Calhoun (1782–1850), spokesman of the southern planter aristocracy and leading exponent of the supremacy of state's rights over federal powers; Henry Clay (1777–1852), the great compromiser, who sought to reconcile the conflicting views of the North and the South; and Daniel Webster (1782–1852), statesman and orator, who championed the preservation of the Union against sectional interests and division. Abraham Lincoln (1809–1865) led the US through its most difficult period, the Civil War, in the course of which he issued the Emancipation Proclamation. Jefferson Davis (1808–89) served as the only president of the short-lived Confederacy. Stephen Grover Cleveland (1837–1908), a conservative reformer, was the strongest president in the latter part of the 19th century. The foremost presidents of the 20th century have been Nobel Peace Prize winner Theodore Roosevelt (1858–1919); Woodrow Wilson (1856–1924), who led the nation during World War I; and Franklin Delano Roosevelt (1882–1945), elected to four terms spanning the Great Depression and World War II. Presidents since 1960 include John Fitzgerald Kennedy (1917–63), Lyndon Baines Johnson (1908–73), Richard Milhous Nixon (b.1913), Gerald Rudolph Ford (Leslie Lynch King, Jr., b.1913), Jimmy Carter (James Earl Carter, Jr., b.1924), and Ronald Wilson Reagan (b.1911).

Of the outstanding US military leaders, four were produced by the Civil War: Union generals Ulysses Simpson Grant (1822–85), who later served as the eighteenth president, and William Tecumseh Sherman (1820–91); and Confederate generals Robert Edward Lee (1807–70) and Thomas Jonathan "Stonewall" Jackson (1824–63). George Catlett Marshall (1880–1959), army chief of staff during World War II, in his later capacity as secretary of state under President Harry S Truman (1884–1972) formulated the Marshall Plan, which did much to revitalize Western Europe. Douglas MacArthur (1880–1964) commanded US forces in Asia during World War II, oversaw the postwar reorganization of Japan, and directed UN forces in the first year of the Korean conflict. Dwight D. Eisenhower (1890–1969) served as supreme Allied commander during World War II, later becoming the thirty-fourth president.

John Marshall (1755–1835), chief justice of the US from 1801 to 1835, established the power of the Supreme Court through the principle of judicial review. Other important chief justices include Edward Douglass White (1845–1921), former president William

Howard Taft (1857–1930), and Earl Warren (1891–1974), whose tenure as chief justice from 1953 to 1969 saw important decisions on desegregation and civil liberties. The justice who enjoyed the longest tenure was William O. Douglas (1898–1980), who served from 1939 to 1975; other prominent associate justices include Oliver Wendell Holmes (1841–1935), Louis Dembitz Brandeis (1856–1941), and Hugo La Fayette Black (1886–1971).

Indian chiefs renowned for their resistance to white encroachment were Pontiac (d.1769), Black Hawk (1767–1838), Tecumseh (1768–1813), Osceola (c.1804–38), Cochise (1812?–74), Geronimo (c.1829–1909), Sitting Bull (c.1831–90), Chief Joseph (1840?–1904), and Crazy Horse (1849?–77). Historical figures who have become part of American folklore include pioneer Daniel Boone (1734–1820); Paul Revere (1735–1818); frontiersman David "Davy" Crockett (1786–1836); scout and Indian agent Christopher "Kit" Carson (1809–68); James Butler "Wild Bill" Hickok (1837–76); William Frederick "Buffalo Bill" Cody (1846–1917); and outlaws Jesse Woodson James (1847–82) and Billy the Kid (William H. Bonney, 1859–81).

Outstanding inventors were Robert Fulton (1765–1815), who developed the steamboat; Eli Whitney (1765–1825), inventor of the cotton gin and mass production techniques; Samuel Finley Breese Morse (1791–1872), who invented the telegraph; and Elias Howe (1819–67), who produced the sewing machine. Alexander Graham Bell (b.Scotland, 1847–1922) gave the world the telephone. Thomas Alva Edison (1847–1931) was responsible for hundreds of inventions, among them the long-burning incandescent electric lamp, the phonograph, automatic telegraph devices, a motion picture camera and projector, the microphone, and the mimeograph. Lee De Forest (1873–1961), the "father of the radio," developed the audion vacuum tube and many other inventions. Two brothers, Wilbur Wright (1867–1912) and Orville Wright (1871–1948), designed, built, and flew the first successful motor-powered airplane. Amelia Earhart (1898–1937) and Charles Lindbergh (1902–74) were aviation pioneers. Pioneers in the space program include John Glenn (b.1921), the first US astronaut to orbit the earth, and Neil Armstrong (b.1930), the first man to set foot on the moon.

Outstanding botanists and naturalists were John Bartram (1699–1777); his son William Bartram (1739–1832); Louis Agassiz (b.Switzerland, 1807–73); Asa Gray (1810–88); Luther Burbank (1849–1926), developer of a vast number of new and improved varieties of fruits, vegetables, and flowers; and George Washington Carver (1864–1943), known especially for his work on industrial applications for peanuts. John James Audubon (1785–1851) won fame as an ornithologist and artist.

Distinguished physical scientists include Samuel Pierpont Langley (1834–1906), astronomer and aviation pioneer; Josiah Willard Gibbs (1839–1903), whose work laid the basis for physical chemistry; Henry Augustus Rowland (1848–1901), who did important research in magnetism and optics; and Albert Abraham Michelson (b.Germany, 1852–1931), who measured the speed of light and became the first US Nobel Prize winner. The chemists Gilbert Newton Lewis (1875–1946) and Irving Langmuir (1881–1957) developed a theory of atomic structure.

The theory of relativity was conceived by Albert Einstein (b.Germany, 1879–1955), generally considered the greatest mind in the physical sciences since Newton. Percy Williams Bridgman (1882–1961) was the father of operationalism and studied the effect of high pressures on materials. Arthur Holly Compton (1892–1962) made discoveries in the field of X rays and cosmic rays. The physical chemist Harold Clayton Urey (1893–1981) discovered heavy hydrogen. Isidor Isaac Rabi (b.Austria, 1898), nuclear physicist, has done important work in magnetism, quantum mechanics, and radiation. Enrico Fermi (b.Italy, 1901–54) created the first nuclear chain reaction, in Chicago in 1942, and contributed to the development of the atomic and hydrogen bombs. Also prominent in the splitting of the atom were Leo Szilard (b.Hungary, 1898–1964), J. Robert Oppenheimer (1904–67), and Edward Teller (b.Hungary, 1908). Ernest Orlando Lawrence (1901–58) developed the cyclotron. Carl David Anderson (b.1905) discovered the positron. Mathematician Norbert Wiener (1894–1964) developed the science of cybernetics.

Outstanding figures in the biological sciences include Theobald Smith (1859–1934), who developed immunization theory and practical immunization techniques for animals; the geneticist Thomas Hunt Morgan (1866–1945), who discovered the heredity functions of chromosomes; and neurosurgeon Harvey William Cushing (1869–1939). Selman Abraham Waksman (b.Russia, 1888–1973), a microbiologist specializing in antibiotics, was co-discoverer of streptomycin. Edwin Joseph Cohn (1892–1953) is noted for his work in the protein fractionalization of blood, particularly the isolation of serum albumin. Philip Showalter Hench (1896–1965) isolated and synthesized cortisone. Wendell Meredith Stanley (1904–71) was the first to isolate and crystallize a virus. Jonas Edward Salk (b.1914) developed an effective killed-virus poliomyelitis vaccine, and Albert Bruce Sabin (b.1906) contributed oral, attenuated live-virus polio vaccines. Joseph Leonard Goldstein (b.1940) and Michael Stuart Brown (b.1941), both molecular geneticists, were awarded the 1985 Nobel Prize in medicine for their research into cholesterol metabolism.

Adolf Meyer (b.Switzerland, 1866–1950) developed the concepts of mental hygiene and dementia praecox and the theory of psychobiology; Harry Stack Sullivan (1892–1949) created the interpersonal theory of psychiatry. Social psychologist George Herbert Mead (1863–1931), behaviorist B(urrhus) F(rederic) Skinner (b.1904), and learning theorist Jerome S. Bruner (b.1915) have been influential in the 20th century.

A pioneer in psychology who was also an influential philosopher was William James (1842–1910). Other leading US philosophers include Charles Sanders Peirce (1839–1914); Josiah Royce (1855–1916); John Dewey (1859–1952), also famous for his theories of education; George Santayana (b.Spain, 1863–1952); Rudolf Carnap (b.Germany, 1891–1970); and Willard Van Orman Quine (b.1908). Educators of note include Horace Mann (1796–1859), Henry Barnard (1811–1900), and Charles William Eliot (1834–1926). Noah Webster (1758–1843) was the outstanding US lexicographer, and Melvil Dewey (1851–1931) was a leader in the development of library science. Thorstein Veblen (1857–1929) wrote books that have strongly influenced economic and social thinking. Also important in the social sciences have been sociologist Talcott Parsons (1902–79) and anthropologist Margaret Mead (1901–78).

Social reformers of note include Dorothea Lynde Dix (1802–87), who led movements for the reform of prisons and insane asylums; Elizabeth Cady Stanton (1815–1902) and Susan Brownell Anthony (1820–1906), leaders in the woman suffrage movement; Clara Barton (1821–1912), founder of the American Red Cross; economist Henry George (1839–97), advocate of the single-tax theory; Eugene Victor Debs (1855–1926), labor leader and an outstanding organizer of the Socialist movement in the US; Jane Addams (1860–1935), who pioneered in settlement house work; Robert Marion La Follette (1855–1925), a leader of progressive political reform in Wisconsin and in the US Senate; Margaret Higgins Sanger (1883–1966), pioneer in birth control; Norman Thomas (1884–1968), Socialist Party leader; and Martin Luther King, Jr. (1929–68), a central figure in the black civil rights movement and winner of the Nobel Peace Prize in 1964.

Religious leaders include Roger Williams (c.1603–83), an early advocate of religious tolerance in the US; Jonathan Edwards (1703–58), New England preacher and theologian; Elizabeth Ann Seton (1774–1821), the first American canonized in the Roman Catholic Church; William Ellery Channing (1780–1842), a founder of American Unitarianism; Joseph Smith (1805–44),

founder of the Church of Jesus Christ of Latter-day Saints (Mormon) and his chief associate, Brigham Young (1801–77); and Mary Baker Eddy (1821–1910), founder of the Christian Science Church. Paul Tillich (b.Germany, 1886–1965) and Reinhold Niebuhr (1892–1971) were outstanding Protestant theologians of international influence.

Famous US businessmen include Pierre Samuel du Pont (1739–1817), John Jacob Astor (Johann Jakob Ashdour, b.Germany, 1763–1848), Cornelius Vanderbilt (1794–1877), Andrew Carnegie (b.Scotland, 1835–1919), John Pierpont Morgan (1837–1913), John Davison Rockefeller (1839–1937), Andrew William Mellon (1855–1937), Henry Ford (1863–1947), and Thomas John Watson (1874–1956).

The first US author to be widely read outside the US was Washington Irving (1783–1859). James Fenimore Cooper (1789–1851) was the first popular US novelist. Three noted historians were William Hickling Prescott (1796–1859), John Lothrop Motley (1814–77), and Francis Parkman (1823–93). The writings of two men of Concord, Mass.—Ralph Waldo Emerson (1803–82) and Henry David Thoreau (1817–62)—influenced philosophers, political leaders, and ordinary men and women in many parts of the world. The novels and short stories of Nathaniel Hawthorne (1804–64) explore New England's Puritan heritage. Herman Melville (1819–91) wrote the powerful novel *Moby Dick*, a symbolic work about a whale hunt that has become an American classic. Mark Twain (Samuel Langhorne Clemens, 1835–1910) is the best-known US humorist. Other leading novelists of the later 19th and the early 20th centuries include William Dean Howells (1837–1920), Henry James (1843–1916), Edith Wharton (1862–1937), Stephen Crane (1871–1900), Theodore Dreiser (1871–1945), Willa Cather (1873–1947), and Sinclair Lewis (1885–1951), first US winner of the Nobel Prize for literature (1930). Later Nobel Prize–winning US novelists include Pearl S. Buck (1892–1973), in 1938; William Faulkner (1897–1962), in 1949; Ernest Hemingway (1899–1961), in 1954; John Steinbeck (1902–68), in 1962; Saul Bellow (b.Canada, 1915), in 1976; and Isaac Bashevis Singer (b.Poland, 1904), in 1978. Anong other noteworthy writers are Thomas Wolfe (1900–1938), Eudora Welty (b.1909), John Cheever (1912–82), Norman Mailer (b.1923), James Baldwin (b.1924), and John Updike (b.1932).

Noted US poets include Henry Wadsworth Longfellow (1807–82), Edgar Allan Poe (1809–49), Walt Whitman (1819–92), Emily Dickinson (1830–86), Edwin Arlington Robinson (1869–1935), Robert Frost (1874–1963), Wallace Stevens (1879–1955), William Carlos Williams (1883–1963), Marianne Moore (1887–1972), and Hart Crane (1899–1932). Ezra Pound (1885–1972) and Nobel laureate T(homas) S(tearns) Eliot (1888–1965) lived and worked abroad for most of their careers. W(ystan) H(ugh) Auden (b.England, 1907–73), who became an American citizen in 1946, wrote poetry and criticism. Elizabeth Bishop (1911–79), Robert Lowell (1917–77), Allen Ginsberg (b.1926), and Sylvia Plath (1932–63) are among the best-known poets since World War II. Robert Penn Warren (b.1905) won the Pulitzer Prize for both fiction and poetry. Carl Sandburg (1878–1967) was a noted poet, historian, novelist, and folklorist. The foremost US dramatists are Eugene (Gladstone) O'Neill (1888–1953), who won the Nobel Prize for literature in 1936; Tennessee Williams (Thomas Lanier Williams, 1911–83); Arthur Miller (b.1915); and Edward Albee (b.1928). Neil Simon (b.1927) is among the nation's most popular playwrights and screenwriters.

Two renowned painters of the early period are John Singleton Copley (1738–1815) and Gilbert Stuart (1755–1828). Outstanding 19th-century painters are James Abbott McNeill Whistler (1834–1903), Winslow Homer (1836–1910), Thomas Eakins (1844–1916), Mary Cassatt (1845–1926), Albert Pinkham Ryder (1847–1917), and John Singer Sargent (1856–1925). More recently, Edward Hopper (1882–1967), Georgia O'Keeffe (b.1887),

Presidents of the US, 1789–1881

	NAME	BORN	DIED	OTHER MAJOR OFFICES HELD	RESIDENCE AT ELECTION
1	George Washington	Westmoreland County, Va., 22 February 1732	Mt. Vernon, Va., 14 December 1799	Commander in Chief, Continental Army (1775–83)	Mt. Vernon, Va.
2	John Adams	Braintree (later Quincy). Mass., 30 October 1735	Quincy, Mass. 4 July 1826	Representative, Continental Congress (1774–77); US vice president (1789–97)	Quincy, Mass.
3	Thomas Jefferson	Goochland (now Albemarle) County, Va., 13 April 1743	Monticello, Va., 4 July 1826	Representative, Continental Congress (1775–76); governor of Virginia (1779–81); secretary of state (1790–93); US vice president (1797–1801)	Monticello, Va.
4	James Madison	Port Conway, Va., 16 March 1751	Montpelier, Va., 28 June 1836	Representative, Continental Congress (1780–83; 1786–88); US representative (1789–97); secretary of state (1801–9)	Montpelier, Va.
5	James Monroe	Westmoreland County, Va. 28 April 1758	New York, N.Y., 4 July 1831	US senator (1790–94); governor of Virginia (1799–1802); secretary of state (1811–17); secretary of war (1814–15)	Leesburg, Va.
6	John Quincy Adams	Braintree (later Quincy), Mass., 11 July 1767	Washington, D.C., 23 February 1848	US senator (1803–8); secretary of state (1817–25); US representative (1831–48)	Quincy, Mass.
7	Andrew Jackson	Waxhaw, Carolina frontier, 15 March 1767	The Hermitage, Tenn., 8 June 1845	US representative (1796–97); US senator (1797–98)	The Hermitage, Tenn.
8	Martin Van Buren	Kinderhook, N.Y., 5 December 1782	Kinderhook, N.Y., 24 July 1862	US senator (1821–28); governor of New York (1829); secretary of state (1829–31); US vice president (1833–37)	New York
9	William Henry Harrison	Charles City County, Va., 9 February 1773	Washington, D.C., 4 April 1841	Governor of Indiana Territory (1801–13); US representative (1816–19); US senator (1825–28)	North Bend, Ohio
10	John Tyler	Charles City County, Va., 29 March 1790	Richmond, Va., 18 January 1862	US representative (1816–21); governor of Virginia (1825–27); US senator (1827–36); US vice president (1841)	Richmond, Va.
11	James K. Polk	Mecklenburg County, N.C., 2 November 1795	Nashville, Tenn., 15 June 1849	US representative (1825–39); governor of Tennessee (1839–41)	Nashville, Tenn.
12	Zachary Taylor	Orange County, Va., 24 November 1784	Washington, D.C., 9 July 1850	—	Louisiana
13	Millard Fillmore	Cayuga County, N.Y., 7 January 1800	Buffalo, N.Y., 8 March 1874	US representative (1833–35; 1837–43); US vice president (1849–50)	Buffalo, N.Y.
14	Franklin Pierce	Hillsboro, N.H., 23 November 1804	Concord, N.H., 8 October 1869	US representative, (1833–37); US senator (1837–42)	Concord, N.H.
15	James Buchanan	Mercersburg, Pa., 23 April 1791	Lancaster, Pa., 1 June 1868	US representative (1821–31); US senator (1834–45); secretary of state (1845–49)	Lancaster, Pa.
16	Abraham Lincoln	Hodgenville, Ky., 12 February 1809	Washington, D.C., 15 April 1865	US representative (1847–49)	Springfield, Ill.
17	Andrew Johnson	Raleigh, N.C., 29 December 1808	Carter Station, Tenn., 31 July 1875	US representative (1843–53); governor of Tennessee (1853–57; 1862–65); US senator (1857–62); US vice president (1865)	Greeneville, Tenn.
18	Ulysses S. Grant	Point Pleasant, Ohio, 27 April 1822	Mount McGregor, N.Y., 23 July 1885	Commander, Union Army (1864–65); secretary of war (1867–68)	Galena, Ill.
19	Rutherford B. Hayes	Delaware, Ohio, 4 October 1822	Fremont, Ohio, 17 January 1893	US representative (1865–67); governor of Ohio (1868–72; 1876–77)	Fremont, Ohio
20	James A. Garfield	Orange, Ohio, 19 November 1831	Elberon, N.J., 19 September 1881	US representative (1863–80)	Mentor, Ohio

PARTY	% OF POPULAR VOTE	% OF ELECTORAL VOTE[1,2]	TERMS IN OFFICE[5]	VICE PRESIDENTS	NOTABLE EVENTS	
Federalist	—	50.0 50.0	30 April 1789–4 March 1793 4 March 1793–4 March 1797	John Adams John Adams	Federal government organized; Bill of Rights enacted (1791); Whiskey Rebellion suppressed (1794); North Carolina, Rhode Island, Vermont, Kentucky, Tennessee enter Union.	1
Federalist	—	25.7	4 March 1797–4 March 1801	Thomas Jefferson	Alien and Sedition Acts passed (1798); Washington, D.C., becomes US capital (1800)	2
Dem.-Rep.	—	26.4[3] 92.0	4 March 1801–4 March 1805 4 March 1805–4 March 1809	Aaron Burr George Clinton	Louisiana Purchase (1803); Lewis and Clark Expedition (1803–6); Ohio enters Union.	3
Dem.-Rep.	—	69.7 58.9	4 March 1809–4 March 1813 4 March 1813–4 March 1817	George Clinton Elbridge Gerry	War of 1812 (1812–14); protective tariffs passed (1816); Louisiana, Indiana enter Union.	4
Dem.-Rep.	—	84.3 99.5	4 March 1817–4 March 1821 4 March 1821–4 March 1825	Daniel D. Tompkins Daniel D. Tompkins	Florida purchased from Spain (1819–21); Missouri Compromise (1820); Monroe Doctrine (1823); Mississippi, Illinois, Alabama, Maine, Missouri enter Union.	5
National Republican	30.9	38.0[4]	4 March 1825–4 March 1829	John C. Calhoun	Period of political antagonisms, producing little legislation; road and canal construction supported; Erie Canal opens (1825).	6
Democrat	56.0 54.2	68.2 76.6	4 March 1829–4 March 1833 4 March 1833–4 March 1837	John C. Calhoun Martin Van Buren	Introduction of spoils system; Texas Republic established (1836); Arkansas, Michigan enter Union.	7
Democrat	50.8	57.8	4 March 1837–4 March 1841	Richard M. Johnson	Financial panic (1837) and subsequent depression.	8
Whig	52.9	79.6	4 March 1841–4 April 1841	John Tyler	Died of pneumonia one month after taking office.	9
Whig	—	—	4 April 1841–4 March 1845	—	Monroe Doctrine extended to Hawaiian Islands (1842); Second Seminole War in Florida ends (1842).	10
Democrat	49.5	61.8	4 March 1845–4 March 1849	George M. Dallas	Boundary between US and Canada set at 49th parallel (1846); Mexican War (1846-48), ending with Treaty of Guadalupe Hidalgo (1848); California gold rush begins (1848); Florida, Texas, Iowa, Wisconsin enter Union.	11
Whig	47.3	56.2	4 March 1849–9 July 1850	Millard Fillmore	Died after 16 months in office.	12
Whig	—	—	9 July 1850–4 March 1853	—	Fugitive Slave Law (1850); California enters Union.	13
Democrat	50.8	85.8	4 March 1853–4 March 1857	William R. King	Gadsden Purchase (1853); Kansas-Nebraska Act (1854); trade opened with Japan (1854).	14
Democrat	45.3	58.8	4 March 1857–4 March 1861	John C. Breckinridge	John Brown's raid at Harpers Ferry, Va. (now W.Va.; 1859); South Carolina secedes (1860); Minnesota, Oregon, Kansas enter Union.	15
Republican	39.8 55.0	59.4 91.0	4 March 1861–4 March 1865 4 March 1865–15 April 1865	Hannibal Hamlin Andrew Johnson	Confederacy established, Civil War begins (1851); Emancipation Proclamation (1863); Confederacy defeated (1865); Lincoln assassinated (1865); West Virginia, Nevada attain statehood.	16
Republican	—	—	15 April 1865–4 March 1869	—	Reconstruction Acts (1867); Alaska purchased from Russia (1867); Johnson impeached but acquitted (1868); Nebraska enters Union.	17
Republican	52.7 55.6	72.8 78.1	4 March 1869–4 March 1873 4 March 1873–4 March 1877	Schuyler Colfax Henry Wilson	Numerous government scandals; financial panic (1873); Colorado enters Union.	18
Republican	48.0	50.1	4 March 1877–4 March 1881	William A. Wheeler	Federal troops withdrawn from South (1877); civil service reform begun.	19
Republican	48.3	58.0	4 March 1881–19 Sept. 1881	Chester A. Arthur	Shot after 4 months in office, dead 2½ months later.	20

Presidents of the US, 1881–1985

	NAME	BORN	DIED	OTHER MAJOR OFFICES HELD	RESIDENCE AT ELECTION
21	Chester A. Arthur	Fairfield, Vt., 5 October 1829	New York, N.Y., 18 November 1886	US vice president (1881)	New York, N.Y.
22	Grover Cleveland	Caldwell, N.J., 18 March 1837	Princeton, N.J., 24 June 1908	Governor of New York (1882–84)	Albany, N.Y.
23	Benjamin Harrison	North Bend, Ohio, 20 August 1833	Indianapolis, Ind., 13 March 1901	US senator (1881–87)	Indianapolis, Ind.
24	Grover Cleveland	(see above)	(see above)	(see above)	New York, N.Y.
25	William McKinley	Niles, Ohio, 29 January 1843	Buffalo, N.Y., 14 September 1901	US representative (1877–83; 1885–91); governor of Ohio (1892–96)	Canton, Ohio
26	Theodore Roosevelt	New York, N.Y., 27 October 1858	Oyster Bay, N.Y., 6 January 1919	Governor of New York (1899–1900); US vice president (1901)	Oyster Bay, N.Y.
27	William H. Taft	Cincinnati, Ohio, 15 September 1857	Washington, D.C., 8 March 1930	Governor of Philippines (1901–4); secretary of war (1904–8); chief justice of the US (1921–30)	Washington, D.C.
28	Woodrow Wilson	Staunton, Va., 28 December 1856	Washington, D.C., 3 February 1924	Governor of New Jersey (1911–13)	Trenton, N.J.
29	Warren G. Harding	Blooming Grove, Ohio, 2 November 1865	San Francisco, Calif., 2 August 1923	US senator (1915–21)	Marion, Ohio
30	Calvin Coolidge	Plymouth Notch, Vt., 4 July 1872	Northampton, Mass., 5 January 1933	Governor of Massachusetts (1919–20); US vice president (1921–23)	Boston, Mass.
31	Herbert Hoover	West Branch, Iowa, 10 August 1874	New York, N.Y., 20 October 1964	Secretary of commerce (1921–29)	Stanford, Calif.
32	Franklin D. Roosevelt	Hyde Park, N.Y., 30 January 1882	Warm Springs, Ga., 12 April 1945	Governor of New York (1929–1933)	Hyde Park, N.Y.
33	Harry S Truman	Lamar, Mo., 8 May 1884	Kansas City, Mo., 26 December 1972	US senator (1935–45); US vice president (1945)	Independence, Mo.
34	Dwight D. Eisenhower	Denison, Tex., 14 October 1890	Washington, D.C., 28 March 1969	Supreme allied commander in Europe (1943–44); Army chief of staff (1945–48)	New York
35	John F. Kennedy	Brookline, Mass., 29 May 1917	Dallas, Tex., 22 November 1963	US representative (1947–52); US senator (1953–60)	Massachusetts
36	Lyndon B. Johnson	Stonewall, Tex., 27 August 1908	Johnson City, Tex., 22 January 1973	US representative (1937–48); US senator (1949–60); US vice president (1961–63)	Johnson City, Tex.
37	Richard M. Nixon	Yorba Linda, Calif., 9 January 1913	—	US representative (1947–51); US senator (1951–53); US vice president (1953–61)	New York, N.Y.
38	Gerald R. Ford	Omaha, Neb., 14 July 1913	—	US representative (1949–73); US vice president (1973–74)	Grand Rapids, Mich.
39	Jimmy Carter	Plains, Ga., 1 October 1924	—	Governor of Georgia (1951–75)	Plains, Ga.
40	Ronald Reagan	Tampico, Ill., 6 February 1911	—	Governor of California (1967–75)	Los Angeles, Calif.

[1]Percentage of electors actually voting.

[2]In elections of 1789, 1792, 1796, and 1800, each elector voted for two candidates for president. The candidate receiving the highest number of votes was elected president; the next highest, vice president. Percentages in table are of total vote cast. From 1804 onward, electors were required to designate which vote was for president and which for vice president, and an electoral majority was required.

PARTY	% OF POPULAR VOTE	% OF ELECTORAL VOTE[1,2]	TERMS IN OFFICE[5]	VICE PRESIDENTS	NOTABLE EVENTS	
Republican	—	—	19 Sept. 1881–4 March 1885	—	Chinese immigration banned despite presidential veto (1882); Civil Service Commission established by Pendleton Act (1883).	21
Democrat	48.5	54.6	4 March 1885–4 March 1889	Thomas A. Hendricks	Interstate Commerce Act (1887)	22
Republican	47.8	58.1	4 March 1889–4 March 1893	Levi P. Morton	Sherman Silver Purchase Act (1890); North Dakota, South Dakota, Montana, Washington, Idaho, Wyoming enter Union.	23
Democrat	46.1	62.4	4 March 1893–4 March 1897	Adlai E. Stevenson	Financial panic (1893); Sherman Silver Purchase Act repealed (1893); Utah enters Union.	24
Republican	51.0 51.7	60.6 65.3	4 March 1897–4 March 1901 4 March 1901–14 Sept. 1901	Garret A. Hobart Theodore Roosevelt	Spanish-American War (1898); Puerto Rico, Guam, Philippines ceded by Spain; independent Republic of Hawaii annexed; US troops sent to China to suppress Boxer Rebellion (1900); McKinley assassinated.	25
Republican	— 56.4	— 70.6	14 Sept. 1901–4 March 1905 4 March 1905–4 March 1909	— Charles W. Fairbanks	Antitrust and conservation policies emphasized; Roosevelt awarded Nobel Peace Prize (1906) for mediating settlement of Russo-Japanese War; Panama Canal construction begun (1907); Oklahoma enters Union.	26
Republican	51.6	66.5	4 March 1909–4 March 1913	James S. Sherman	Federal income tax ratified (1913); New Mexico, Arizona enter Union.	27
Democrat	41.8 49.2	81.9 52.2	4 March 1913–4 March 1917 4 March 1917–4 March 1921	Thomas R. Marshall Thomas R. Marshall	Clayton Antitrust Act (1914); US Virgin Islands purchased from Denmark (1917); US enters World War I (1917); Treaty of Versailles signed (1919) but not ratified by US; constitutional amendments enforce prohibition (1919), enfranchise women (1920).	28
Republican	60.3	76.1	4 March 1921–2 Aug. 1923	Calvin Coolidge	Teapot Dome scandal (1923–24).	29
Republican	— 54.1	— 71.9	3 Aug. 1923–4 March 1925 4 March 1925–4 March 1929	— Charles G. Dawes	Kellogg-Briand Pact (1928).	30
Republican	58.2	83.6	4 March 1929–4 March 1933	Charles Curtis	Stock market crash (1929) inaugurates Great Depression.	31
Democrat	57.4 60.8 54.7 53.4	88.9 98.5 84.6 81.4	4 March 1933–20 Jan. 1937 20 Jan. 1937–20 Jan. 1941 20 Jan. 1941–20 Jan. 1945 20 Jan. 1945–12 April 1945	John N. Garner John N. Garner Henry A. Wallace Harry S Truman	New Deal social reforms; prohibition repealed (1933); US enters World War II (1941).	32
Democrat	— 49.5	— 57.1	12 April 1945–20 Jan. 1949 20 Jan. 1949–20 Jan. 1953	— Alben W. Barkley	United Nations founded (1945); US nuclear bombs dropped on Japan (1945); World War II ends (1945); Philippines granted independence (1946); Marshall Plan (1947); Korean conflict begins (1950); era of McCarthyism.	33
Republican	55.1 57.4	83.2 86.1	20 Jan. 1953–20 Jan. 1957 20 Jan. 1957–20 Jan. 1961	Richard M. Nixon Richard M. Nixon	Korean conflict ended (1953); Supreme Court orders school desegregation (1954); Alaska, Hawaii enter Union.	34
Democrat	49.7	56.4	20 Jan. 1961–22 Nov. 1963	Lyndon B. Johnson	Conflicts with Cuba (1961–62); aboveground nuclear test ban treaty (1963); Kennedy assassinated.	35
Democrat	— 61.1	— 90.3	22 Nov. 1963–20 Jan. 1965 20 Jan. 1965–20 Jan. 1969	— Hubert H. Humphrey	Great Society programs; Voting Rights Act (1965); escalation of US military role in Indochina; race riots, political assassinations.	36
Republican	43.4 60.7	55.9 96.7	20 Jan. 1969–20 Jan. 1973 20 Jan. 1973–9 Aug. 1974	Spiro T. Agnew Spiro T. Agnew Gerald R. Ford	First lunar landing (1969); arms limitation treaty with Soviet Union (1972); US withdraws from Viet-Nam (1973); Agnew resigns in tax scandal (1973); Nixon resigns at height of Watergate scandal (1974).	37
Republican	—	—	9 Aug. 1974–20 Jan. 1977	Nelson A. Rockefeller	First combination of unelected president and vice president; Nixon pardoned (1974).	38
Democrat	50.1	55.2	20 Jan. 1977–20 Jan. 1981	Walter F. Mondale	Carter mediates Israel-Egypt peace accord (1978); Panama Canal treaties ratified (1979); tensions with Iran (1979–81).	39
Republican	50.8 58.8	90.9 97.6	20 Jan. 1981–20 Jan. 1985 20 Jan. 1985–	George H. Bush George H. Bush	Defense buildup; social spending cuts; rising trade and budget deficits; tensions with Nicaragua.	40

[3]Electoral vote tied between Jefferson and Aaron Burr; election decided in House of Representatives.

[4]No candidate received a majority; election decided in House.

[5]In the event of a president's death or removal from office, his duties are assumed to devolve immediately upon his successor, even if he does not immediately take the oath of office.

Chief Justices of the US, 1789–1984

	NAME	BORN	DIED	APPOINTED BY	SUPREME COURT TERM	MAJOR COURT DEVELOPMENTS
1	John Jay	New York City 12 December 1745	Bedford, N.Y., 17 May 1829	Washington	October 1789– June 1795	Organized court, established procedures.
2	John Rutledge	Charleston, S.C. September 1739	Charleston, S.C., 18 July 1800	Washington	—	Presided for one term in 1795, but Senate refused to confirm his appointment.
3	Oliver Ellsworth	Windsor, Conn., 29 April 1745	Windsor, Conn. 26 November 1807	Washington	March 1796– December 1800	—
4	John Marshall	Fauquier County, Va., 24 September 1755	Philadelphia, Pa. 6 July 1835	Adams	February 1801– July 1835	Established principle of judicial review (*Marbury* v. *Madison,* 1803); formulated concept of implied powers (*McCulloch* v. *Maryland,* 1819).
5	Roger Brooke Taney	Calvert County, Md., 17 March 1777	Washington, D.C., 12 October 1864	Jackson	March 1836– October 1864	Held that slaves could not become citizens, ruled Missouri Compromise illegal (*Dred Scott* v. *Sanford,* 1857).
6	Salmon Portland Chase	Cornish, N.H., 13 January 1808	New York, N.Y. 7 May 1873	Lincoln	December 1864– May 1873	Ruled military trials of civilians illegal (*Ex parte Milligan* 1866); Chase presided at A. Johnson's impeachment trial.
7	Morrison Remick Waite	Old Lynne, Conn., 29 November 1816	Washington, D.C., 23 March 1888	Grant	March 1874– March 1888	Held that businesses affecting the "public interest" are subject to state regulation (*Munn* v. *Illinois,* 1877).
8	Melville Weston Fuller	Augusta, Me., 11 February 1833	Sorvento, Me., 4 July 1910	Cleveland	October 1888– July 1910	Issued first opinions on cases under the Sherman Antitrust Act, (*US* v. *E.C. Knight Co.,* 1895; *Northern Securities Co.,* v. *US,* 1904); held the income tax unconstitutional (*Pollock* v. *Farmers' Loan,* 1895).
9	Edward Douglass White	Lafourche Parish, La., 3 November 1845	Washington, D.C., 19 May 1921	Taft	December 1910– May 1921	Further qualified the Sherman Antitrust Act (*Standard Oil Co.* v. *US,* 1911) by applying the "rule of reason."
10	William Howard Taft	Cincinnati, Ohio, 15 September 1857	Washington, D.C., 8 March 1930	Harding	July 1921– February 1930	Held against congressional use of taxes for social reform (*Bailey* v. *Drexel Furniture,* 1922).
11	Charles Evans Hughes	Glens Falls, N.Y., 11 April 1862	Osterville, Mass., 27 August 1948	Hoover	February 1930– June 1941	Upheld constitutionality of National Labor Relations Act, Social Security Act; invalidated National Industrial Recovery Act (*Schechter* v. *US,* 1935); F. Roosevelt's attempt to pack Court opposed.
12	Harlan Fiske Stone	Chesterfield, N.H., 11 October 1872	Washington, D.C., 22 April 1946	F. Roosevelt	July 1941– April 1946	Upheld Court's power to invalidate state laws (*Southern Pacific Co.* v. *Arizona,* 1945).
13	Frederick Moore Vinson	Louisa, Ky., 22 January 1890	Washington, D.C., 8 September 1953	Truman	June 1946– September 1953	Overturned federal seizure of steel mills (*Youngstown Sheet and Tube Co.* v. *Sawyer,* 1952), Vinson dissenting.
14	Earl Warren	Los Angeles, Calif., 19 March 1891	Washington, D.C., 9 July 1974	Eisenhower	October 1953– June 1969	Mandated public school desegregation (*Brown* v. *Topeka, Kans., Board of Education,* 1954) and reapportionment of state legislatures (*Baker* v. *Carr,* 1962); upheld rights of suspects in police custody (*Miranda* v. *Arizona,* 1966).
15	Warren Earl Burger	St. Paul, Minn., 17 September 1907	—	Nixon	June 1969–	Legalized abortion (*Roe* v. *Wade,* 1973); rejected claim of executive privilege in a criminal case (*US* v. *Nixon,* 1974); first female justice (1981).

Thomas Hart Benton (1889–1975), Charles Burchfield (1893–1967). Ben Shahn (1898–1969), Mark Rothko (b.Russia, 1903–70), Jackson Pollock (1912–56), Robert Rauschenberg (b.1925), and Jasper Johns (b.1930) have achieved international recognition.

Sculptors of note include Augustus Saint-Gaudens (1848–1907), Gaston Lachaise (1882–1935), Jo Davidson (1883–1952), Alexander Calder (1898–1976), Louise Nevelson (b.Russia, 1900), and Isamu Noguchi (b.1904). Henry Hobson Richardson (1838–86), Louis Henry Sullivan (1856–1924), Frank Lloyd Wright (1869–1959), Louis I. Kahn (b.Estonia, 1901–74), and Eero Saarinen (1910–61) were outstanding architects. Contemporary architects of note include R(ichard) Buckminster Fuller (1895–1983) Edward Durell Stone (1902–78), Philip Cortelyou Johnson (b.1906), and I(eoh) M(ing) Pei (b.China, 1917). The US has produced many fine photographers, notably Mathew B. Brady (c.1823–96), Alfred Stieglitz (1864–1946), Edward Steichen (1879–1973), Edward Weston (1886–1958), Ansel Adams (1902–84), and Margaret Bourke-White (1904–71).

Outstanding figures in the motion picture industry are D. W. (David Lewelyn Wark) Griffith (1875–1948), Charles Spencer "Charlie" Chaplin (b.England, 1889–1978), Walter E. "Walt" Disney (1906–66), and (George) Orson Welles (1915–85). John Ford (1895–1973), Howard Hawks (1896–1977), and Alfred Hitchcock (b.England, 1899–1980) were influential motion picture directors, and George Lucas (b.1944) and Steven Spielberg (b.1947) have achieved remarkable popular success. Woody Allen (Allen Konigsberg, b.1935) has written, directed, and starred in comedies on stage and screen. Classic American actors and actresses include the Barrymores—Ethel (1879–1959), and her brothers Lionel (1878–1954) and John (1882–1942); Humphrey Bogart (1899–1957); Spencer Tracy (1900–1967); Clark Gable (1901–60); Greta Garbo (Greta Gustafsson, b.Sweden, 1905); Henry Fonda (1905–82) and his daughter, Jane (b.1937); Bette Davis (b.1908); Katharine Hepburn (b.1909); Judy Garland (France Gumm, 1922–69); and Marlon Brando (b.1924). Among great entertainers are W. C. Fields (William Claude Dukenfield, 1880–1946), Jack Benny (Benjamin Kubelsky, 1894–1974), Fred Astaire (Fred Austerlitz. b.1899), Bob Hope (Leslie Townes Hope, b.England, 1903), Bing Crosby (Harry Lillis Crosby, 1904–78), and Frank Sinatra (Francis Albert Sinatra, b.1915). The first great US "showman" was Phineas Taylor Barnum (1810–91).

Foremost composers are Edward MacDowell (1861–1908), Charles Ives (1874–1954), Ernest Bloch (b.Switzerland, 1880–1959), Roger Sessions (1896–1985), Roy Harris (1898–1979), Aaron Copland (b.1900), Elliott Carter (b.1908), Samuel Barber (1910–81), John Cage (b.1912), and Leonard Bernstein (b.1918). George Rochberg (b.1918), George Crumb (b.1929), Steve Reich (b.1936), and Philip Glass (b.1937) have won more recent followings. The songs of Stephen Collins Foster (1826–64) have achieved folksong status. Leading composers of popular music are John Philip Sousa (1854–1932), Jerome Kern (1885–1945), Irving Berlin (Israel Baline, b.Russia, 1888), Cole Porter (1893–1964), George Gershwin (1898–1937), Richard Rodgers (1902–1979), Woody Guthrie (1912–67), and Bob Dylan (Robert Zimmerman, b.1941). Preeminent in the blues and folk traditions are Leadbelly (Huddie Ledbetter, 1888–1949), Bessie Smith (1898?–1937), and Muddy Waters (McKinley Morganfield, 1915–83). Leading jazz figures include the composers Scott Joplin (1868–1917), James Hubert "Eubie" Blake (1883–1983), and Edward Kennedy "Duke" Ellington (1899–1974); and performers Louis Armstrong (1900–1971), Billie Holiday (Eleanora Fagan, 1915–59), John Birks "Dizzy" Gillespie (b.1917), Charlie "Bird" Parker (1920–55), John Coltrane (1926–67), and Miles Davis (b.1926).

Many foreign-born musicians have enjoyed personal and professional freedom in the US; principal among them are

pianists Artur Schnabel (b.Austria, 1882–1951), Arthur Rubinstein (b.Poland, 1887–1982), Rudolf Serkin (b.Bohemia, 1903), Vladimir Horowitz (b.Russia, 1904), and violinists Jascha Heifetz (b.Russia, 1901) and Isaac Stern (b.Soviet Union, 1920). Among distinguished instrumentalists born in the US are Benny Goodman (b.1909), a classical as well as jazz clarinetist, and concert pianist Van Cliburn (Harvey Lavan, Jr., b.1934). Singers Paul Robeson (1898–1976), Marian Anderson (b.1902), Maria Callas (Maria Kalogeropoulos, 1923–77), Leontyne Price (b.1927), and Beverly Sills (Belle Silverman, b.1929) achieved international acclaim. Isadora Duncan (1878–1927) was one of the first US dancers to win fame abroad. George Balanchine (b.Russia, 1904–83), Agnes De Mille (b.1908), and Jerome Robbins (b.1918) are leading choreographers; Martha Graham (b.1893) pioneered in modern dance.

Among the many noteworthy sports stars are baseball's Tyrus Raymond "Ty" Cobb (1886–1961) and George Herman "Babe" Ruth (1895–1948); football's Samuel Adrian "Sammy" Baugh (b.1914), Jim Brown (b.1936), Francis A. "Fran" Tarkenton (b.1940), and O(renthal) J(ames) Simpson (b.1947); golf's Robert Tyre "Bobby" Jones and Mildred "Babe" Didrikson Zaharias (1914–56); William "Bill" Tilden (1893–1953) and Billie Jean (Moffitt) King (b.1943) in tennis; Joe Louis (Joseph Louis Barrow, 1914–81) and Muhammad Ali (Cassius Marcellus Clay, b.1942) in boxing; William Felton "Bill" Russell (b.1934) and Wilton Norman "Wilt" Chamberlain (b.1936) in basketball; Mark Spitz (b.1950) in swimming; Eric Heiden (b.1958) in speed skating; and Jesse Owens (1913–80) in track and field.

[50]BIBLIOGRAPHY

Adams Family Correspondence. 4 vols. Edited by L. H. Butterfield et al. Cambridge, Mass.: Belknap Press, 1963–73.

Adams Family Diaries. 6 vols. Edited by L. H. Butterfield et al. Cambridge, Mass.: Belknap Press, 1964–74.

Ahlstrom, Sydney E. *A Religious History of the American People.* New Haven: Yale University Press, 1972.

Allen, Harold B. (ed.). *Readings in American Dialectology.* New York: Appleton-Century-Crofts, 1971.

Bailey, Thomas Andrew. *A Diplomatic History of the American People.* New York: Appleton-Century-Crofts, 1955.

Barnouw, Erik. *A History of Broadcasting in the US.* 3 vols. New York: Oxford University Press, 1966–70.

Barone, Michael, and Grant Ujifusa. *The Almanac of American Politics, 1986.* Washington, D.C.: National Journal, 1985.

Becker, Carl Lotus. *The Declaration of Independence: A Study in the History of Political Ideas.* New York: Random House, 1958.

Berlin, Ira. *Slaves without Masters: The Free Negro in the Antebellum South.* New York: Pantheon, 1976.

The Book of the States. Lexington, Ky.: Council of State Governments, 1935–date.

Boorstin, Daniel J. *The Americans: The Colonial Experience.* New York: Random House, 1958.

Boorstin, Daniel J. *The Americans: The Democratic Experience.* New York: Random House, 1973.

Boorstin, Daniel J. *The Americans: The National Experience.* New York: Random House, 1965.

Broder, David S. *Changing of the Guard: Power and Leadership in America.* New York: Simon & Schuster, 1980.

Brown, Dee. *Bury My Heart at Wounded Knee: An Indian History of the American West.* New York: Holt, Rinehart and Winston, 1971.

Caro, Robert A. *The Years of Lyndon Johnson: The Path to Power.* New York: Knopf, 1982.

Carter, Jimmy. *Keeping Faith: Memoirs of a President.* New York: Bantam, 1982.

Catton, Bruce. *The Centennial History of the Civil War.* 3 vols. Garden City, N.Y.: Doubleday, 1961–65.

Chase, Gilbert. *Our American Music from the Pilgrims to the Present.* New York: McGraw-Hill, 1966.

Coles, Robert. *Children of Crisis.* 5 vols. Boston: Little, Brown, 1967-77.

Commager, Henry Steele (ed.). *Documents of American History.* 9th ed. 2 vols. Englewood-Cliffs, N.J.: Prentice-Hall, 1974.

Congressional Directory . . . Washington, D.C.: Government Printing Office, 1809-date.

Dickstein, Morris. *Gates of Eden: American Culture in the 1960s.* New York: Basic Books, 1977.

Dictionary of American Biography. 16 vols. New York: Scribners, 1927-72. Supplements, 1973-81.

Encyclopedia of Associations: National Organizations of the U.S. 19th ed. Detroit: Gale Research, 1984.

Fallows, James. *National Defense.* New York: Random House, 1981.

Freidel, Frank, with Richard K. Showman (ed.). *Harvard Guide to American History.* Rev. ed. Cambridge, Mass.: Belknap Press, 1974.

Goodwyn, Lawrence. *Democratic Promise: The Populist Movement in America.* New York: Oxford University Press, 1977.

Greeley, Andrew M. *The American Catholic: A Social Portrait.* New York: Basic Books, 1978.

Halberstam, David. *The Best and the Brightest.* New York: Penguin, 1983.

Haley, Alex. *Roots: The Saga of an American Family.* New York: Doubleday, 1976.

Hart, James David (ed.). *Oxford Companion to American Literature.* 5th ed. New York: Oxford University Press, 1983.

Harvard Encyclopedia of American Ethnic Groups. Cambridge, Mass.: Harvard University Press, 1980.

Howe, Irving. *The World of Our Fathers.* New York: Harcourt Brace Jovanovich, 1976.

Jordan, Winthrop P. *White over Black: American Attitudes toward the Negro, 1550-1812.* New York: Norton, 1977.

Joseph, Alvin M., Jr. *Now That the Buffalo's Gone: A Study of Today's American Indians.* New York: Knopf, 1982.

The Justices of the United States Supreme Court, 1789-1969: Their Lives and Major Opinions. 4 vols. Edited by Leon Friedman and Fred L. Israel. New York: Chelsea House and Bowker, 1970.

Kammen, Michael. *People of Paradox: An Inquiry Concerning the Origins of American Civilization.* New York: Knopf, 1972.

Kelly, Alfred Hinsey, and Winfred Audif Harbison. *The American Constitution: Its Origins and Development.* 6th ed. New York: Norton, 1982.

Key, V. O. *Southern Politics in State and Nation.* New York: Vintage, 1949.

Kissinger, Henry. *The White House Years.* Boston: Little, Brown, 1979.

Kissinger, Henry. *Years of Upheaval.* Boston: Little, Brown, 1982.

Lerner, Max. *America As a Civilization: Life and Thought in the United States Today.* New York: Simon and Schuster, 1957.

Mardan, Charles F., and Gladys Meyer. *Minorities in American Society.* 5th ed. New York: Van Nostrand, 1978.

Mencken, Henry Louis. *The American Language.* New York: Knopf, 1936. Supplements, 1945, 1948.

Morison, Samuel Eliot. *The European Discovery of America: The Northern and Southern Voyages.* 2 vols. New York: Oxford University Press, 1971-74.

Morison, Samuel Eliot. *The Oxford History of the American People.* New York: Oxford University Press, 1965.

Morris, Richard Brandon, et al. (eds.). *Encyclopedia of American History.* 6th ed. New York: Harper & Row, 1982.

Nevins, Allan. *Ordeal of the Union.* 8 vols. New York: Scribners, 1947-71.

Notable American Women, 1607-1950: A Biographical Dictionary. 3 vols. Edited by Edward T. James et al. Cambridge, Mass.: Belknap Press, 1973.

Peirce, Neal R., and Jerry Hagstrom. *The Book of America: Inside 50 States Today.* New York: Norton, 1983.

Reed, Carroll E. *Dialects of American English.* Amherst: University of Massachusetts Press, 1973.

Reeves, Thomas C. *The Life and Times of Joe McCarthy.* New York: Stein & Day, 1981.

Richardson, E. P. *Painting in America.* New York: Crowell, 1965.

Rockwell, John. *All American Music: Composition in the Late Twentieth Century.* New York: Knopf, 1983.

Rose, Barbara. *American Art since 1900.* New York: Praeger, 1975.

Sherwin, Martin J. *A World Destroyed: The Atomic Bomb and the Grand Alliance.* New York: Knopf, 1977.

Silberman, Charles E. *Criminal Violence, Criminal Justice.* New York: Norton, 1979.

Slotkin, Richard. *Regeneration through Violence: The Mythology of the American Frontier, 1600-1860.* Middletown, Conn.: Wesleyan University Press, 1975.

Spiller, Robert, et al. (eds.). *Literary History of the United States.* 4th ed., rev. New York: Macmillan, 1974.

Stampp, Kenneth Milton. *The Peculiar Institution: Slavery in the Ante-Bellum South.* New York: Knopf, 1956.

Starr, Paul. *The Social Transformation of American Medicine.* New York: Basic Books, 1982.

Terkel, Studs. *Hard Times: An Oral History of the Great Depression in America.* New York: Pantheon, 1970.

Tocqueville, Alexis Charles Henri Maurice Cherel de. *Democracy in America.* New York: Knopf, 1944.

US Bureau of the Census. *Historical Statistics of the United States, Colonial Times to 1970.* Washington, D.C.: Government Printing Office, 1975.

US Bureau of the Census. *Statistical Abstract of the United States.* Washington, D.C.: Government Printing Office, 1879-date.

US Geological Survey. *National Atlas of the United States.* Washington, D.C., 1970.

United States Government Manual. Washington, D.C.: Federal Register Division, National Archives and Records Service, General Service Administration, 1935-date.

Urdang, Laurence (ed.). *The Timetables of American History.* New York: Simon & Schuster, 1983.

White, Theodore. *Breach of Faith: The Fall of Richard Nixon.* New York: Atheneum, 1975.

Who's Who in America: A Biographical Dictionary of Notable Living Men and Women. Chicago: Marquis, 1899-date.

Wills, Garry. *Explaining America: The Federalist.* New York: Penguin, 1982.

Wills, Garry. *Inventing America: Jefferson's Declaration of Independence.* New York: Doubleday, 1978.

Wood, Michael. *America in the Movies.* New York: Basic Books, 1976.

The World Almanac and Book of Facts. New York: Newspaper Enterprise Association, 1868-date.

Zaretsky, Irving I., and Mark P. Leone (eds.). *Religious Movements in Contemporary America.* Princeton, N.J.: Princeton University Press, 1976.

ANTEBELLUM: before the US Civil War.

BLUE LAWS: laws forbidding certain practices (e.g., conducting business, gaming, drinking liquor), especially on Sundays.

CAPITAL BUDGET: a financial plan for acquiring and improving buildings or land, paid for by the sale of bonds.

CAPITAL PUNISHMENT: punishment by death.

CIVILIAN LABOR FORCE: all persons 16 years of age or older who are not in the armed forces and who are now holding a job, have been temporarily laid off, are waiting to be reassigned to a new position, or are unemployed but actively looking for work.

CLASS I RAILROAD: a railroad having gross annual revenues of $83.5 million or more in 1983.

COMMERCIAL BANK: a bank that offers to businesses and individuals a variety of banking services, including the right of withdrawal by check.

COMPACT: a formal agreement, covenant, or understanding between two or more parties.

CONTINENTAL CLIMATE: the climate typical of the US interior, having distinct seasons, a wide range of daily and annual temperatures, and dry, sunny summers.

CONSOLIDATED BUDGET: a financial plan that includes the general budget, federal funds, and all special funds.

CONSTANT DOLLARS: money values calculated so as to eliminate the effect of inflation on prices and income.

COUNCIL-MANAGER SYSTEM: a system of local government under which a professional administrator is hired by an elected council to carry out its laws and policies.

CREDIT UNION: a cooperative body that raises funds from its members by the sale of shares and makes loans to its members at relatively low interest rates.

CURRENT DOLLARS: money values that reflect prevailing prices, without excluding the effects of inflation.

DEMAND DEPOSIT: a bank deposit that can be withdrawn by the depositor with no advance notice to the bank.

ELECTORAL VOTES: the votes that a state may cast for president, equal to the combined total of its US senators and representatives and nearly always cast entirely on behalf of the candidate who won the most votes in that state on Election Day.

ENDANGERED SPECIES: a type of plant or animal threatened with extinction in all or part of its natural range.

FEDERAL POVERTY LEVEL: a level of money income below which a person or family qualifies for US government aid.

FISCAL YEAR: a 12-month period for accounting purposes.

FOOD STAMPS: coupons issued by the government to low-income persons for food purchases at local stores.

GENERAL BUDGET: a financial plan based on a government's normal revenues and operating expenses, excluding special funds.

GENERAL COASTLINE: a measurement of the general outline of the US seacoast. See also TIDAL SHORELINE.

GREAT AWAKENING: during the mid-18th century, a Protestant religious revival in North America, especially New England.

GROSS STATE PRODUCT: the total value of goods and services produced in the state.

GROWING SEASON: the period between the last 32°F (0°C) temperature in spring and the first 32°F (0°C) temperature in autumn.

HOME-RULE CHARTER: a document stating how and in what respects a city, town, or county may govern itself.

INSTALLED CAPACITY: the maximum possible output of electric power at any given time.

MAYOR-COUNCIL SYSTEM: a system of local government under which an elected council serves as a legislature and an elected mayor is the chief administrator.

MEDICAID: a federal-state program that helps defray the hospital and medical costs of needy persons.

MEDICARE: a program of hospital and medical insurance for the elderly, administered by the federal government.

METROPOLITAN AREA: in most cases, a city and its surrounding suburbs.

NO-FAULT INSURANCE: an automobile insurance plan that allows an accident victim to receive payment from an insurance company without having to prove who was responsible for the accident.

NORTHERN, NORTH MIDLAND: major US dialect regions; see pp. 644–46.

OMBUDSMAN: a public official empowered to hear and investigate complaints by private citizens about government agencies.

POCKET VETO: a method by which a state governor (or the US president) may kill a bill by taking no action on it before the legislature adjourns.

PER CAPITA: per person.

PROVED RESERVES: the quantity of a recoverable mineral resource (such as oil or natural gas) that is still in the ground.

PUBLIC DEBT: the amount owed by a government.

RELIGIOUS ADHERENTS: the followers of a religious group, including (but not confined to) the full, confirmed, or communicant members of that group.

RETAIL TRADE: the sale of goods directly to the consumer.

REVENUE SHARING: the distribution of federal tax receipts to state and local governments.

RIGHT-TO-WORK LAW: a measure outlawing any attempt to require union membership as a condition of employment.

SAVINGS AND LOAN ASSOCIATION: a bank that invests the savings of depositors primarily in home mortgage loans.

SERVICE INDUSTRIES: industries that provide services (e.g., health, legal, automotive repair) for individuals, businesses, and others.

SOCIAL SECURITY: as commonly understood, the federal system of old age, survivors, and disability insurance.

SOUTHERN, SOUTH MIDLAND: major US dialect regions; see pp. 644–46.

STOLPORT: an airfield for short-takeoff-and-landing (STOL) aircraft, which require runways shorter than those used by conventional aircraft.

SUNBELT: the southernmost states of the US, extending from Florida to California.

SUPPLEMENTAL SECURITY INCOME: a federally administered program of aid to the aged, blind, and disabled.

TIDAL SHORELINE: a detailed measurement of the US seacoast that includes sounds, bays, other outlets, and offshore islands.

TIME DEPOSIT: a bank deposit that may be withdrawn only at the end of a specified time period or upon advance notice to the bank.

VALUE ADDED BY MANUFACTURE: the difference, measured in dollars, between the value of finished goods and the cost of the materials needed to produce them.

WHOLESALE TRADE: the sale of goods, usually in large quantities, for ultimate resale to consumers.

Abbreviations and Acronyms

AD—Anno Domini
AFDC—aid to families with dependent children
AFL-CIO—American Federation of Labor–Congress of Industrial
  Organizations
AM—before noon
AM—amplitude modulation
American Ind.—American Independent Party
Amtrak—National Railroad Passenger Corp.
b.—born
BC—Before Christ
Btu—British thermal unit(s)
bu—bushel(s)
c.—circa, about
c—Centigrade
CIA—Central Intelligence Agency
cm—centimeter
Co.—company
comp.—compiler
Conrail—Consolidated Rail Corp.
Corp.—corporation
CST—Central Standard Time
cu—cubic
cwt—hundredweight(s)
d.—died
D—Democrat
e—evening
E—east
ed.—edition, editor
e.g.—exempli gratia (for example)
EPA—Environmental Protection Agency
est.—estimated
EST—Eastern Standard Time
et al.—et alii (and others)
etc.—et cetera (and so on)
F—Fahrenheit
FBI—Federal Bureau of Investigation
FCC—Federal Communications Commission
FM—frequency modulation
Ft.—fort
GMT—Greenwich Mean Time
GNP—gross national product
Hist.—Historic

I—interstate (highway)
in—inch(es)
Inc.—incorporated
Jct.—junction
K—kindergarten
km—kilometer(s)
km/hr—kilometers per hour
kw—kilowatt(s)
kwh—kilowatt-hour(s)
lb—pound(s)
m—morning
mi—mile(s)
Mon.—monument
mph—miles per hour
mw—megawatt(s)
MST—Mountain Standard Time
Mt.—mount
Mtn.—mountain
N—north
NA—not available
Natl.—National
NATO—North Atlantic Treaty Organization
NCAA—National Collegiate Athletic Association
n.d.—no date
N.F.—National Forest
N.W.R.—National Wildlife Refuge
oz—ounce(s)
PM—after noon
PST—Pacific Standard Time
R—Republican
Ra.—range
Res.—reservoir, reservation
rev. ed.—revised edition
s—south
S—Sunday
Soc.—Socialist
sq—square
St.—saint
UN—United Nations
US—United States
USIA—United States Information Agency
w—west

NAMES OF STATES AND OTHER SELECTED AREAS

	STANDARD ABBREVIATION(S)	POSTAL ABBREVIATION		STANDARD ABBREVIATION(S)	POSTAL ABBREVIATION
Alabama	Ala.	AL	Nebraska	Nebr. (Neb.)	NE
Alaska	*	AK	Nevada	Nev.	NV
Arizona	Ariz.	AZ	New Hampshire	N.H.	NH
Arkansas	Ark.	AR	New Jersey	N.J.	NJ
California	Calif.	CA	New Mexico	N.Mex. (N.M.)	NM
Colorado	Colo.	CO	New York	N.Y.	NY
Connecticut	Conn.	CT	North Carolina	N.C.	NC
Delaware	Del.	DE	North Dakota	N.Dak. (N.D.)	ND
District of Columbia	D.C.	DC	Ohio	*	OH
Florida	Fla.	FL	Oklahoma	Okla.	OK
Georgia	Ga.	GA	Oregon	Oreg. (Ore.)	OR
Hawaii	*	HI	Pennsylvania	Pa.	PA
Idaho	*	ID	Puerto Rico	P.R.	PR
Illinois	Ill.	IL	Rhode Island	R.I.	RI
Indiana	Ind.	IN	South Carolina	S.C.	SC
Iowa	*	IA	South Dakota	S.Dak. (S.D.)	SD
Kansas	Kans. (Kan.)	KS	Tennessee	Tenn.	TN
Kentucky	Ky.	KY	Texas	Tex.	TX
Louisiana	La.	LA	Utah	*	UT
Maine	Me.	ME	Vermont	Vt.	VT
Maryland	Md.	MD	Virginia	Va.	VA
Massachusetts	Mass.	MA	Virgin Islands	V.I.	VI
Michigan	Mich.	MI	Washington	Wash.	WA
Minnesota	Minn.	MN	West Virginia	W.Va.	WV
Mississippi	Miss.	MS	Wisconsin	Wis.	WI
Missouri	Mo.	MO	Wyoming	Wyo.	WY
Montana	Mont.	MT	*No standard abbreviation